Reagents for Organic Synthesis

LOUIS F. FIESER
Sheldon Emery Professor of Organic Chemistry
Harvard University

MARY FIESER
Research Fellow in Chemistry
Harvard University

JOHN WILEY AND SONS, INC.
New York • Chichester • Brisbane • Toronto

20 19 18 17 16 15 14 13

LIBRARY OF CONGRESS CATALOG CARD NUMBER: 66-27894
PRINTED IN THE UNITED STATES OF AMERICA
ISBN 0 471 25875 X

PREFACE

A precursor of this book which appeared as a 49-page section in *Experiments in Organic Chemistry*, 3rd edition, 1955 was received with sufficient enthusiasm to encourage us to undertake this more extensive book describing 1,120 reagents of use to organic chemists.

The reagents are listed alphabetically, and for each we give the structural formula, molecular weight, physical constants, preferred methods of preparation or purification, suppliers, and examples of significant uses, each documented with references and with flow sheets showing at a glance the molar ratio of the reactants, the reaction conditions, and the yield. Where a choice exists, we have given preference to examples that compare the reagent in question with alternative reagents, that explain details of the experimental procedure, or that seem otherwise particularly informative. Where the reaction cited is part of a significant synthetic sequence, the formulation often includes enough preceding or succeeding steps to bring out points of interest. In deciding whether to regard a given compound as a reagent or merely a starting material, we probably have been swayed somewhat by the consideration of interest.

In citing examples of both the preparation and use of reagents, we have made liberal use of material that has appeared over a period of 46 years in *Organic Syntheses*; indeed some 17% of the references are to this source. These examples are of special value not only because each has been checked but also because the compounds and the reactions clearly are ones that have been of active interest to modern workers. Procedures are cited also from the manual by one of us entitled *Organic Experiments* (1964); some of them originated as *Organic Syntheses* procedures and were modified for reduction in operating time or otherwise simplified.

The preparation of this book has involved repeated study of the question of excess reagent. Is it a good plan to routinely employ a reagent in 1.2 times the theoretical amount? In our opinion the answer is in the affirmative in only two situations. In a study of a particular Diels-Alder reaction presented in this book it is shown that with 1.5 equivalents of the dienophile the reaction in tetrachloroethylene at 78.8° was complete in 8.0 hrs., but that the time can be reduced to 2.4 hrs. by use of 3.0 equivalents of the dienophile. In determining the time required for reaction in 50 other solvents, use of a threefold excess of the dienophile seemed indicated. In many permanganate oxidations, use of excess oxidant is justified because considerable permanganate decomposes in the process with liberation of oxygen. In all situations except those of the two types exemplified, we recommend use of a reagent in the exact theoretical amount. We regard as faulty a procedure that specifies use of excess reagent without presenting evidence that the excess is beneficial.

The precursor of this book included a section on the properties of solvents and methods for their purification. After some study of the problem, we decided against an extensive coverage of solvent purification. Abundant references are available to procedures for the purification and drying of solvents, but the papers present no evidence of the degree of purity achieved and hence provide little basis for their evaluation. We place more confidence in the 44 purified solvents now available from five competing U.S. firms and have listed them in a section on **Solvents**. It stands to reason that a firm specializing in solvent purification is in a position to make a long-range study of each problem undertaken and to install efficient techniques of distillation and drying and effective measures for product control. At least some of the prominent industrial research groups find it expedient to purchase purified solvents rather than utilize their own manpower for the purpose. The relatively few detailed procedures for solvent purification included in this book are supplemented by descriptions of drying with lithium aluminum hydride, calcium hydride, and molecular sieves. Properties of the pure solvents available commercially are recorded in tables in which solvents are arranged in order of boiling point in each of five solubility types (aprotic and water-immiscible, etc.).

The impact of *Organic Syntheses* on the development and use of reagents is evident from the chart. Not shown at the left is an almost straight-line section extending back to 1900 and beginning to rise slowly from 5–10 references per year only in about 1925. Then come four collective-volume peaks, each surging higher than before.

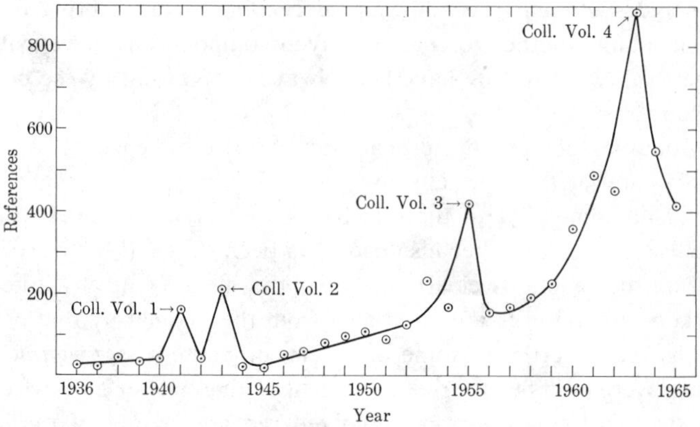
Coll. Vol. 1→
← Coll. Vol. 2
Coll. Vol. 3→
Coll. Vol. 4
References
800
600
400
200
1936
1940
1945
1950
1955
1960
1965
Year

The prognosis for the next few years is clear, and it is evident that if this book meets with favor it will need to be revised or supplemented in the near future. We expect to keep abreast of the new literature in preparation for a possible paperback supplement to be published whenever enough new and revised material accumulates. We cordially invite fellow chemists to submit corrections, comments, suggestions, and contributions. If you think of reagents or references we have missed, do send in either notes and references or else a finished contribution which we can use with minor editing and with acknowledgment of the source. Scattered throughout the present volume are contributions submitted originally for a book planned by Melvin S. Newman and Robert E. Ireland and kindly turned over to us when the project was abandoned.

It is a pleasure to acknowledge with sincere thanks the generous help of several people in the preparation of this book. Outside readers included Drs. Elmore L. Martin of du Pont, William P. Schneider of Upjohn, John J. Baldwin, John M. Chemerda, Stanton A. Harris, James H. Jones, Larry J. Loeffler, Arthur A. Patchett, Peter I. Pollak, David C. Remy, Walfred S. Saari, and David Taub of Merck Sharp and Dohme Research Laboratories. Co-workers who proofread manuscript and checked references are Drs. Makhluf J. Haddadin, Musa Z. Nazer, E. Paul Papadopoulos, Joseph P. Schirmer, Jr., and Mr. Edwin Arima. We are indebted also to investigators who provided explanations or supplied additional information. Suppliers of chemicals and apparatus were also highly cooperative.

We alone are to blame for some duplication of references, discovered only in the course of preparing the indexes. A reaction involving A and B and cited as an attribute of A is easily forgotten and later cited as an attribute of B with repetition of the reference, rather than by cross reference. We apologize for such blemishes.

Our closing date was August 23, 1966; journals received after that date could not be reviewed. Even so, there are 182 references to papers of 1966.

Cambridge, Massachusetts

Louis F. Fieser
Mary Fieser

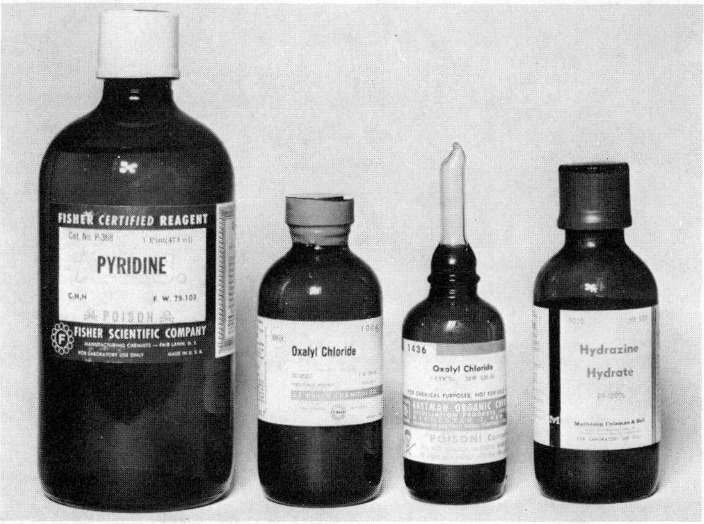
FISHER CERTIFIED REAGENT
PYRIDINE
POISON
FISHER SCIENTIFIC COMPANY
Oxalyl Chloride
Oxalyl Chloride
EASTMAN ORGANIC
POISON
Hydrazine
Hydrate

CONTENTS

Reagents for Organic Synthesis

Introduction

Arrangement. Suppliers mentioned in the text are listed in a section placed before the indexes and easily located by an indenture.

For enhanced usefulness the book is provided not only with a subject and an author index but also with an index of types, that is, types of reactions or types of compounds, for example: acetylation, bromination, characterization (of a, b, c, etc.), decarboxylation, or: π-acids, benzyne precursors, carbene precursors, diimide precursors. Listed alphabetically under each such entry are all the reagents which figure in the operation or group cited, whether as prime reactant, catalyst, solvent, scavenger, etc. A given reagent may fit appropriately in two or more categories. When a reagent does not fit easily into a reasonable category, we leave it unclassified rather than make a forced assignment. With no less than 92 reagents available as oxidants and 101 for use as reducing reagents, it seemed out of the question to attempt to indicate in the index of types further details about these general reactions. In a few instances a procedure cited for the preparation of one reagent provides a good example of the use of another one. For example, a preferred route to allene is by reaction of 2,3-dichloropropene with zinc dust and ethanol; in the index of types the entries under "Dechlorination" include "Zinc dust–ethanol, *see* Allene, preparation."

Names and spelling. One guideline we have followed is the rule recently adopted by *Organic Syntheses* that when an ester, ether, or peroxide contains two or more alkyl, aryl, or acyl groups the name must indicate the number of such groups:

Formula	*Correct*	*Incorrect*
$(CH_3)_2O$	Dimethyl ether	Methyl ether
$(C_2H_5O)_2SO_2$	Diethyl sulfate	Ethyl sulfate
$(C_6H_5)_2O$	Diphenyl ether	Phenyl ether
$(CO_2CH_3)_2$	Dimethyl oxalate	Methyl oxalate
$CH_2(CO_2C_2H_5)_2$	Diethyl malonate	Ethyl malonate
$(C_6H_5COO)_2$	Dibenzoyl peroxide	Benzoyl peroxide
$HC(OC_2H_5)_3$	Triethyl orthoformate	Ethyl orthoformate
$(C_2H_5O)_4C$	Tetraethyl orthocarbonate	Ethyl orthocarbonate

That the situation previously was highly confused is evident from the following entries in the index of *Org. Syn., Coll. Vol.*, **4**: "Diethyl oxalate" and "Diethyl malonate" (both correct), but "Ethyl orthoformate" and "Ethyl orthocarbonate" (both incorrect). The following entry is describable as a double error: "Triethyl orthoformate, *see* Ethyl orthoformate." To locate all references to a given ester, it is thus necessary to search under two names. We urge suppliers to revise their catalogs in accordance with the rule cited. In this book we do not even list, with cross references, names which we consider to be incorrect.

Similar reform in the nomenclature of polyhalogen compounds may come some day, but for the present we consider it imprudent to do more than make a start. Thus the correct names for BF_3 and for $ClCH_2CH_2Cl$ surely are boron trifluoride and ethylene dichloride, and we feel no restraint from using them. However, although the names methylene chloride for CH_2Cl_2 and aluminum chloride for $AlCl_3$ seem incorrect, we cannot bring ourselves to break with tradition and employ other names.

As explained in our *Style Guide for Chemists* (p. 77), we disapprove of the weak-sounding dioxăn, furăn, tryptophăn, and urethăn and add the letter *e* to these words to produce the strong pronunciations dioxāne, furāne, tryptophāne, and urethāne. For the same reason we favor desoxo - and desoxy- over deoxo- and deoxy-.

Abbreviations. Short forms of abbreviations of journal titles are as follows:

J**o**urnal of the **Am**erican **Chem**ical **Soc**iety
Angewandte **Chemie**
Annalen der Chemie
Annales de **chim**ie (Paris)
Chemische **Ber**ichte (formerly **Ber**ichte der deutschen chemischen Gesellschaft)
Bulletin de la **soc**iété chimique de France
Chemical **Comm**unications
Acta **Chemica Scand**inavica
Collection of **Czech**oslovak Chemical Communications
Comptes rendus hebdomadaires des séances de l'académie des sciences
Gazzetta chimica italiana
Helvetica Chimica Acta
Journal of the **Chem**ical **Soc**iety (London)
Journal of **Org**anic Chemistry
Monatshefte für Chemie
Organic **Syn**theses
Organic **Syn**theses, **Coll**ective **V**olume
Recueil des **trav**aux chimique des Pays-Bas (The Netherlands)

The book by one of us, *Organic Experiments*, D. C. Heath and Co., Boston (1964), is referred to as *Org. Expts*.

Abbreviations

Ac	Acetyl	MeOH	Methanol
AcOH	Acetic acid	Ms	Mesyl, CH_3SO_2
BuOH	Butanol	NBS	N-Bromosuccinimide
Bz	Benzoyl	Ph	Phenyl
Cathyl	Carboethoxy	Phth	Phthaloyl
Cb	Carbobenzoxy	PrOH	Propanol
Diglyme	Diethylene glycol dimethyl ether	Py	Pyridine
DMF	Dimethylformamide	THF	Tetrahydrofurane
DMSO	Dimethyl sulfoxide	Triglyme	Triethylene glycol dimethyl ether
DNF	2,4-Dinitrofluoro-benzene	Trityl	$(C_6H_5)_3C-$
DNP	2,4-Dinitrophenyl-hydrazone	Ts	Tosyl, p-$CH_3C_6H_4SO_2-$
EtOH	Ethanol	TsCl	Tosyl chloride
Glyme	1,2-Dimethoxyethane	TsOH	Tosic acid, p-$CH_3C_6H_4SO_3H$

A

Acetaldehyde, CH_3CHO. Mol. wt. 44.05, b.p. 20.8°, sp. gr. 0.78. Suppliers: B, E, F, MCB.

Preparation. (a) Measure 20 ml. (20 g.) of paraldehyde into a 50-ml. round-bottomed flask, add a cooled mixture of 0.5 ml. each of concd. sulfuric acid and water, attach a fractionating column, condenser, and ice-cooled receiver, and heat gently with a microburner at such a rate that acetaldehyde distils at a temperature not higher than 35°. To avoid charring of the mixture, continue only until about half of the material has been depolymerized.

(b) *p*-Toluenesulfonic acid has been recommended as catalyst for depolymerization of paraldehyde without specification of details.[1]

[1] N. L. Drake and G. B. Cooke, *Org. Syn., Coll. Vol.*, **2**, 407 (1943)

Acetaldoxime, $CH_3CH{=}NOH$. Mol. wt. 59.07, m. p. 47°. Suppliers: A, B, KK.

The reagent reacts with a diazonium salt to form an oxime which on acid hydrolysis affords an aryl methyl ketone.[1]

$$ArN_2Cl + CH_3CH{=}NOH \longrightarrow Ar\overset{CH_3}{\overset{|}{C}}{=}NOH \xrightarrow{H^+} Ar\overset{CH_3}{\overset{|}{C}}{=}O$$

[1] W. F. Beech, *J. Chem. Soc.*, 1297 (1954)

Acetamide, CH_3CONH_2. Mol. wt. 59.07, m.p. 82°. Suppliers: A, B, E, F, KK, MCB.

Acetamide is useful in the bromination of acid-sensitive compounds since it forms a stable complex with hydrogen bromide, $CH_3CONH_2 \cdot HBr$, which is insoluble in common bromination solvents.[1]

[1] K. Zeile and H. Meyer, *Ber.*, **82**, 275 (1949)

***p*-Acetamidobenzenesulfonyl chloride,** 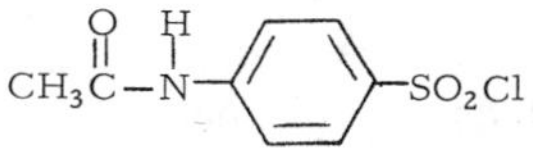

Mol. wt. 233.68, m.p. 149°. Suppliers: E, F, MCB.

Preparation from acetanilide and chlorosulfonic acid.[1] Use as catalyst for the Beckmann rearrangement of oximes in pyridine.[2]

[1] S. Smiles and J. Stewart, *Org. Syn., Coll. Vol.*, **1**, 8 (1941)

[2] St. Kaufmann, *Am. Soc.*, **73**, 1779 (1951); H. Heusser *et al.*, *Helv.*, **38**, 1399 (1955); G. Rosenkranz, O. Mancera, F. Sondheimer, and C. Djerassi, *J. Org.*, **21**, 520 (1956)

Acetic anhydride, $(CH_3CO)_2O$. Mol. wt. 102.09, b.p. 139.6°, sp. gr. 1.08.

Reagent that has stood for a time after a bottle has been opened should be tested either in a preliminary run or by shaking a sample with ice water and rapidly titrating the free acetic acid. Fractionation affords pure anhydride; material of practical grade should first be distilled from anhydrous sodium acetate to eliminate halogen compounds and metals.

Procedures for acetylation described in *Organic Experiments*: DL-alanine (Chapt. 28); of amines in aqueous solution (Chapts. 34.3, 53.4); reductive acetylation (Chapt. 42.4); salicylic acid, order catalytic activity: $H_2SO_4 > BF_3 > Py > NaOAc$ (Chapt. 48).

Acetic–formic anhydride, $CH_3\overset{O}{\overset{\|}{C}}-O-\overset{O}{\overset{\|}{C}}H$. The reagent is prepared by cooling 2 volumes of acetic anhydride to 0°, slowly adding 1 volume of 100% formic acid, heating at 50° for 15 min., and cooling immediately to 0°.[1] It formylates alcohols, including tertiary alcohols that are dehydrated on attempted acetylation, and so is useful for the analysis of oils containing such alcohols.[1]

It has been used for the preparation in high yield of O^{12a}-formyltetracycline (1)[2] and of 5′-O-formyl derivatives of nucleoside 2′,3′-acetonides (2),[3] in each case in piperidine at −20 to 0°. The O-formyl group is suggested for the protection of the

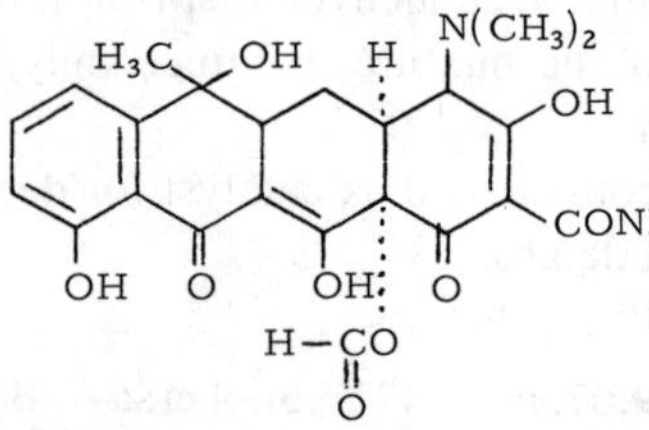

(1)

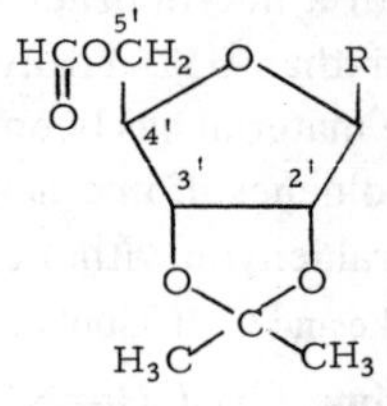

(2)

5′-hydroxyl group since it is cleaved more readily than an acetyl group; for example, it is removed by boiling methanol.

Japanese chemists[4] prepare the reagent either by the reaction of acetyl chloride with sodium formate or of formic acid with ketene. The reagent formylates amino acids in formic acid as solvent. The N-formyl group is useful as a blocking group in peptide synthesis. It is surprisingly resistant to basic hydrolysis but readily solvolyzed in dilute acid.[5]

[1]V. C. Mehlenbacher, *Org. Analysis*, **1**, 37 (1953); W. Stevens and A. Van Es, *Rec. trav.*, **83**, 1287, 1294 (1964)
[2]R. K. Blackwood, H. H. Rennhard, and C. R. Stephens, *Am. Soc.*, **82**, 5194 (1960)
[3]J. Žemlička, J. Beránek, and J. Smrt, *Coll. Czech.*, **27**, 2784 (1962)
[4]L. Muramatsu, M. Murakami, T. Yoneda, and A. Hagitani, *Bull. Chem. Soc. Japan*, **38**, 244 (1965)
[5]J. C. Sheehan and D.-D. H. Yang, *Am. Soc.*, **80**, 1154 (1958)

Acetoacetyl fluoride, CH_3COCH_2COF. Mol. wt. 104.08, b.p. 132–134°.

Prepared[1] by reaction of diketene with anhydrous hydrogen fluoride, the reagent can be stored for weeks at 0°, but at room temperature it slowly decomposes to form

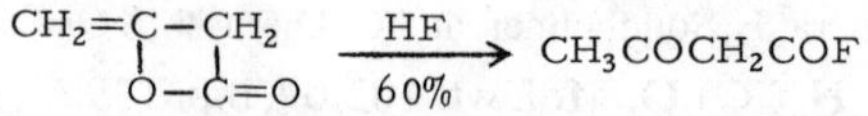
CH₂=C–CH₂
O–C=O
HF
60%
CH₃COCH₂COF

dehydroacetic acid. It is unsatisfactory for Friedel-Crafts acylation, but serves as an acetoacetylating agent for alcohols and amines:

$$CH_3COCH_2COF \xrightarrow[68-86\%]{ROH} CH_3\overset{O}{\overset{\|}{C}}CH_2\overset{O}{\overset{\|}{C}}OR + HF$$

$$CH_3COCH_2COF \xrightarrow[56-87\%]{2\ RNH_2} CH_3\overset{O}{\overset{\|}{C}}CH_2\overset{O}{\overset{\|}{C}}\overset{H}{\overset{|}{N}}R + RNH_3F$$

[1]G. A. Olah and S. J. Kuhn, *J. Org.*, **26**, 225 (1961)

Acetone cyanohydrin, $(CH_3)_2C(OH)CN$. Mol. wt. 85.10, b.p. 80°/15 mm. Suppliers: Rohm and Haas, Union Carbide, A, B, MCB.

Preparation. (a) By addition of 40% sulfuric acid to an aqueous solution of acetone and sodium cyanide at 10–20°.[1] (b) Reaction of potassium cyanide with the sodium bisulfite addition compound of acetone gives material which is less pure but satisfactory for immediate use.[2]

Use for ***transcyanohydrination***: preparation in high yield of the 17-monocyanohydrin of a 3,17-diketo-Δ⁴-steroid by hydrogen cyanide exchange with the reagent.[3]

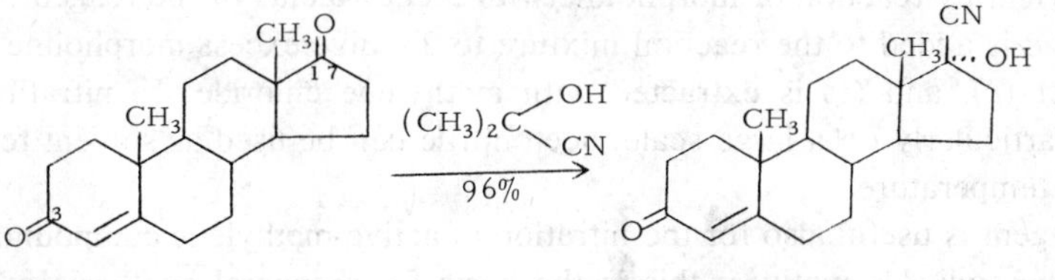
CH3
O
17
CH3
3
O
OH
(CH3)2C
CN
96%
CN
CH3
OH
CH3
O

Δ⁴-Androstene-3, 17-dione

Use for the ***addition of hydrogen cyanide*** to benzalacetophenone and other α,β-unsaturated ketones; 5–10% aqueous sodium carbonate is the most satisfactory

$$C_6H_5CH{=}CHCOC_6H_5 \xrightarrow[95\%]{(CH_3)_2C(OH)CN} C_6H_5\underset{CN}{CH}CH_2COC_6H_5$$

catalyst.[4] The addition of hydrogen cyanide to conjugated steroid ketones usually gives mixtures and poor yields, but Julia *et al.*[5] achieved smooth addition to Δ¹-5α-cholestene-3-one as follows. A solution of the ketone and acetone cyanohydrin in tetrahydrofurane–methanol was treated with a little aqueous sodium carbonate

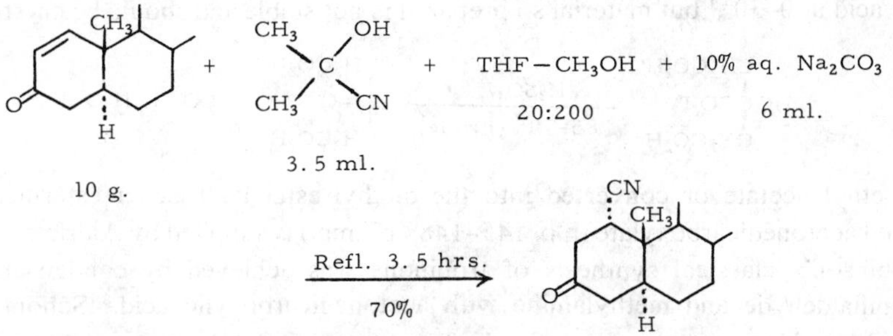
CH3
O
H
10 g.
+
CH3
OH
C
CH3
CN
3. 5 ml.
+
THF – CH3OH
20:200
+ 10% aq. Na2CO3
6 ml.
Refl. 3. 5 hrs.
70%
CN
CH3
O
H

and refluxed for 3.5 hrs. After evaporation in vacuum, chromatography separated 1.2 g. of starting material and afforded 1α-cyano-5α-cholestane-3-one in good yield.

[1] R. F. B. Cox and R. T. Stormont, *Org. Syn., Coll. Vol.*, **2**, 7 (1943)
[2] E. C. Wagner and M. Baizer, *ibid.*, **3**, 324 (1955)
[3] A. Ercoli and P. de Ruggieri, *Am. Soc.*, **75**, 650 (1953)
[4] B. E. Betts and W. Davey, *J. Chem. Soc.*, 4193 (1958)
[5] S. Julia, H. Linarès, and P. Simon, *Bull. soc.*, 2471 (1963)

Acetone cyanohydrin nitrate, $(CH_3)_2\overset{ONO_2}{\overset{|}{C}}CN$. Mol. wt. 130.11, b.p. 65–66°/10 mm. Supplier: Aldrich. Preparation by nitration of acetone cyanohydrin with fuming nitric acid and acetic anhydride.[1] *Caution*: moderately explosive.

Acetone cyanohydrin nitrate is useful for conversion of primary and secondary amines into nitramines.[2] The reaction is unique in that nitration is carried out under

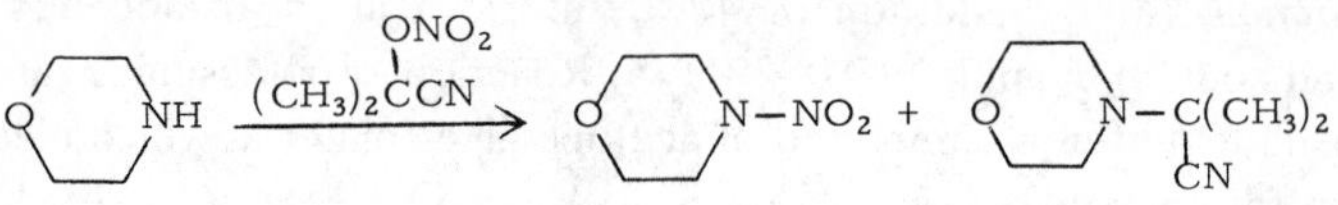

(1) (2)

neutral or alkaline conditions. Thus N-nitromorpholine (1, m.p. 54°) is obtained in 57–64% yield by reaction of morpholine with 2 equivalents of the reagent;[1] hydrochloric acid is added to the reaction mixture to dissolve excess morpholine and the by-product (2), and (1) is extracted with methylene chloride. In nitrating other amines, particularly on a large scale, acetonitrile can be used as solvent for better control of temperature.

The reagent is useful also for the nitration of active-methylene compounds in the form of the sodio derivatives; this is the basis for a general synthesis of α-nitro esters.[3]

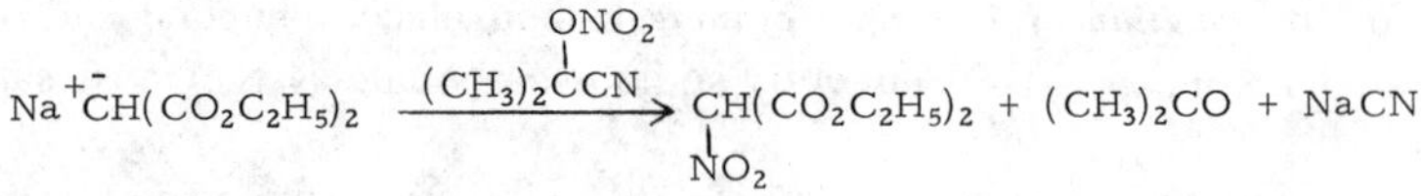

[1]J. P. Freeman and I. G. Shepard, *Org. Syn.*, **43**, 83 (1963)
[2]W. D. Emmons and J. P. Freeman, *Am. Soc.*, **77**, 4387 (1955)
[3]*Idem, ibid.*, **77**, 4391 (1955)

Acetonedicarboxylic acid, $HO_2CCH_2COCH_2CO_2H$. Mol. wt. 146.10, m.p. 138° dec. Suppliers: Baker, Eastern, Pfizer.

The diacid can be obtained in high yield by the action of fuming sulfuric acid on citric acid at 0–30°,[1] but material so prepared is not stable and should be crystallized

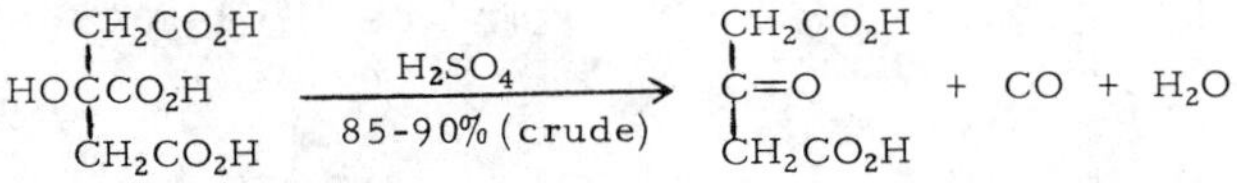

from ethyl acetate or converted into the diethyl ester by Fischer esterification.[2] Diethyl acetonedicarboxylate (b.p. 145–148°/17 mm.) is supplied by Aldrich.

Robinson's classical synthesis of tropinone was achieved by condensation of succindialdehyde and methylamine with acetonedicarboxylic acid.[3] Schöpf[4] later carried out the reaction in a solution buffered to pH 5 and at room temperature and

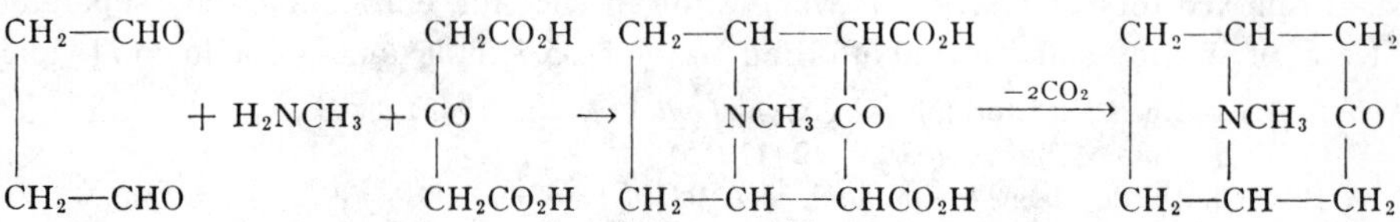

Tropinone

under these simulated physiological conditions the intermediate diacid loses carbon dioxide spontaneously and tropinone was obtained in yield as high as 90%. *See also* Glutaraldehyde.

[1]R. Adams, H. M. Chiles, and C. F. Rassweiler, *Org. Syn., Coll. Vol.*, **1**, 10 (1941)
[2]R. Adams and H. M. Chiles, *ibid.*, **1**, 237 (1941)

[3]R. Robinson, *J. Chem. Soc.*, **111**, 762 (1917)
[4]C. Schöpf and G. Lehmann, *Ann.*, **518**, 1 (1935)

Acetone dimethyl ketal, *see* 2,2-Dimethoxypropane.

α-Acetoxyacrylonitrile, $CH_2{=}C(OCOCH_3){-}CN$. Mol. wt. 111.10., b.p. 173°/772 mm. Preparation by addition of hydrogen cyanide to ketene; best yields are obtained with a mildly basic catalyst such as potassium acetate.[1]

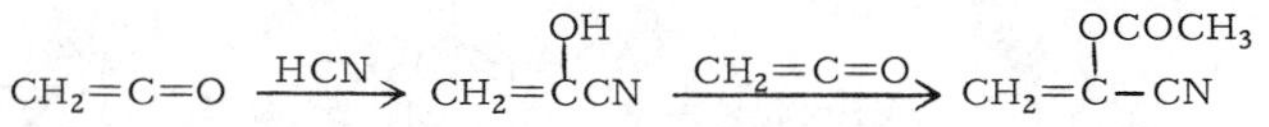

The reagent is useful as a dienophile because the product on hydrolysis affords a ketone, as in the synthesis of dehydronorcamphor:[2]

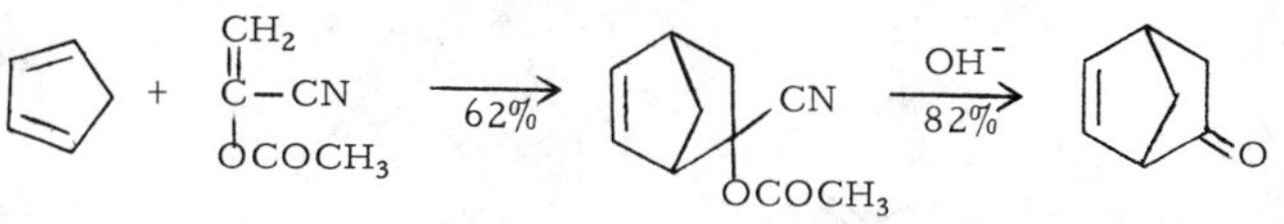

A further example is in the synthesis of 7-isopropylidenebicyclo[2.2.1]-5-heptene-2-one:[3]

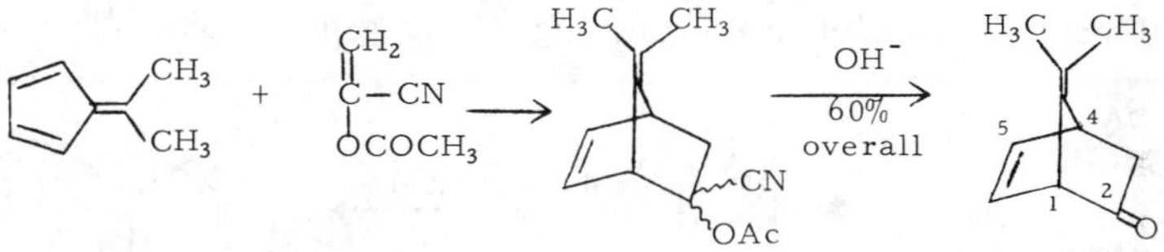

[1]S. Deakin and N. T. M. Wilsmore, *J. Chem. Soc.*, **97**, 1971 (1910); F. Johnston and L. W. Newton, U. S. patent 2,395,930 (1946) [*C.A.*, **40**, 4078 (1946)]; H. J. Hagemeyer, Jr., *Ind. Eng. Chem.*, **41**, 765 (1949)
[2]P. D. Bartlett and B. E. Tate, *Am. Soc.*, **78**, 2473 (1956)
[3]C. H. De Puy and P. R. Story, *Am. Soc.*, **82**, 627 (1960)

1-Acetoxybutadiene, Mol. wt. 112.16, b.p. 42–43°/16 mm., 51–52°/30 mm.

Preparation. This diene, the enol acetate of crotonaldehyde, can be prepared

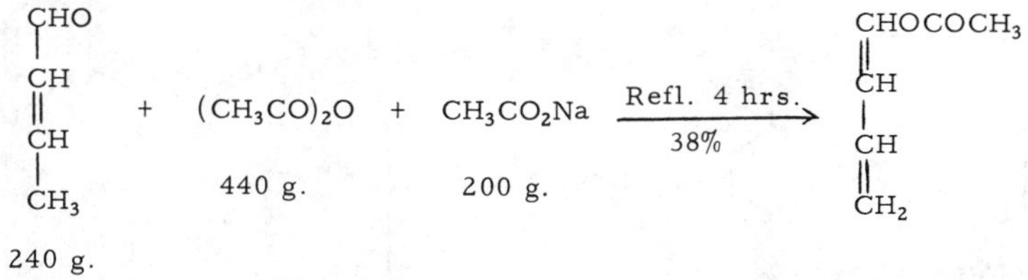

by refluxing the aldehyde with acetic anhydride and sodium acetate.[1] Workup includes removal of considerable crotonaldehyde (powerful lachrymator) with bisulfite; this fact together with the low yield[2] suggests that the reaction reaches equilibrium. A much better method described by Hagemeyer and Hull[3] consists in reaction of crotonaldehyde with isopropenyl acetate, a catalytic amount of *p*-toluenesulfonic acid, and a little copper acetate (function not stated). The aldehyde is added to the other components in 2–3 hrs. with provision for continuous removal of acetone to displace the equilibrium. The yield of twice-distilled l-acetoxybutadiene is 90%.

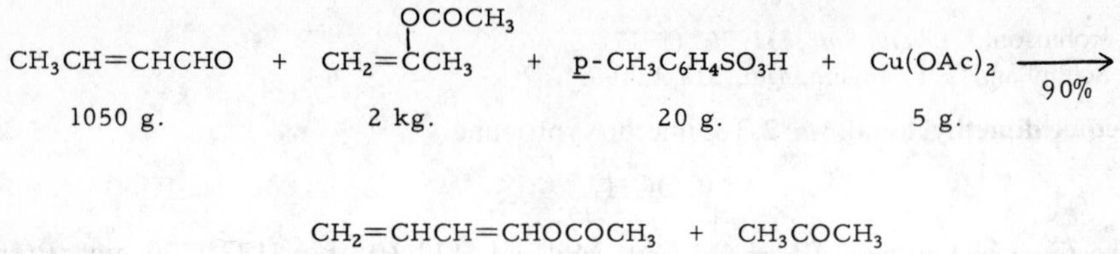
OCOCH3
CH3CH=CHCHO + CH2=CCH3 + p-CH3C6H4SO3H + Cu(OAc)2
90%
1050 g.
2 kg.
20 g.
5 g.
CH2=CHCH=CHOCOCH3 + CH3COCH3

Diels-Alder reactions. Hill *et al.*[4] used the reagent in one step of a synthesis of the parent structure of lycorine.

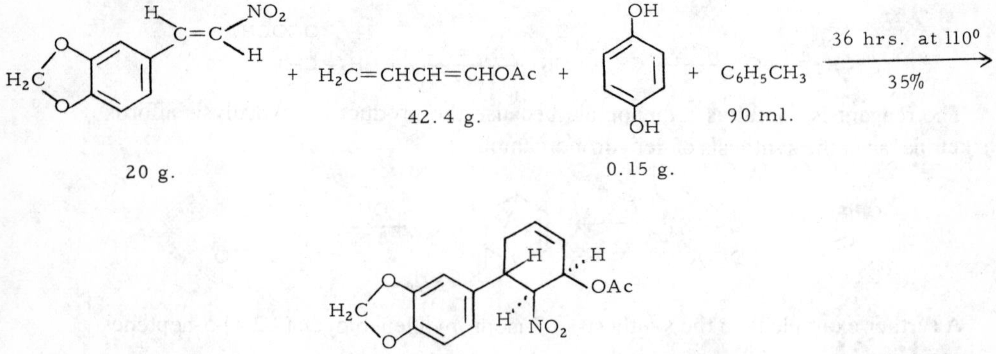
NO2
H2C=CHCH=CHOAc
OH
C6H5CH3
36 hrs. at 110°
35%
42.4 g.
90 ml.
20 g.
0.15 g.
OAc
NO2

Hill and Carlson[5] demonstrated use of the reagent in combination with an acetylene for the direct production of an aromatic ring. The reagent adds normally to quinones,

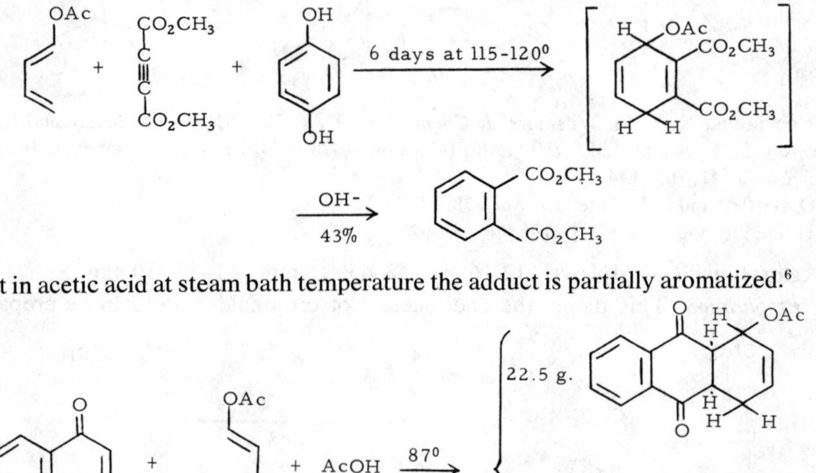
OAc
CO2CH3
OH
6 days at 115-120°
OH-
43%

but in acetic acid at steam bath temperature the adduct is partially aromatized.[6]

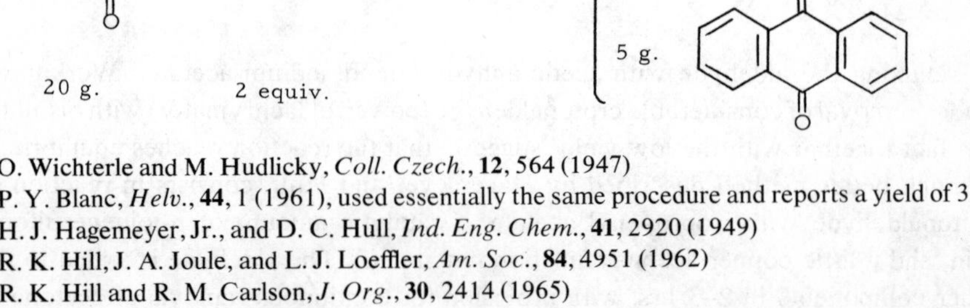
OAc
AcOH
87°
22.5 g.
5 g.
20 g.
2 equiv.

[1]O. Wichterle and M. Hudlicky, *Coll. Czech.*, **12**, 564 (1947)
[2]P. Y. Blanc, *Helv.*, **44**, 1 (1961), used essentially the same procedure and reports a yield of 35%.
[3]H. J. Hagemeyer, Jr., and D. C. Hull, *Ind. Eng. Chem.*, **41**, 2920 (1949)
[4]R. K. Hill, J. A. Joule, and L. J. Loeffler, *Am. Soc.*, **84**, 4951 (1962)
[5]R. K. Hill and R. M. Carlson, *J. Org.*, **30**, 2414 (1965)
[6]W. Flaig, *Ann.*, **568**, 1 (1950)

3β-Acetoxy-Δ⁵-etienic acid,

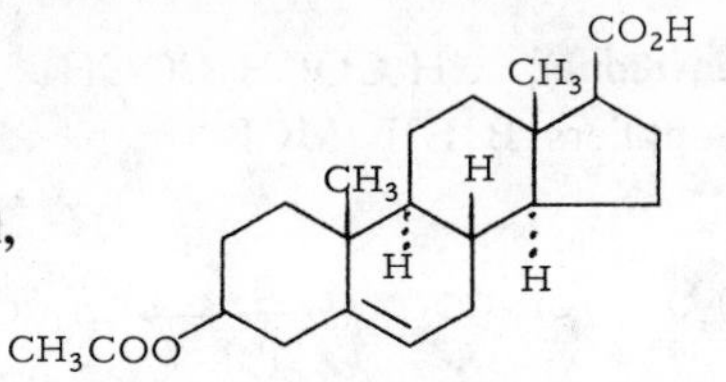
CO2H
CH3
CH3
H
H
H
CH3COO

Mol. wt. 360.48, m.p. 238°. Preparation by hypobromite oxidation of pregnenolone acetate (available from Syntex S. A.).[1] The corresponding acid chloride has been used for the resolution of 1α-hydroxydicyclopentadiene,[2] of *cis,cis*-l-decalol,[3] and of *trans*-3-*t*-butylcyclohexanol[4] (in this case conventional resolution of acid phthalate salts with alkaloids failed).

[1]J. Staunton and E. J. Eisenbraun, *Org. Syn.*, **42**, 4 (1962)
[2]R. B. Woodward and T. J. Katz, *Tetrahedron*, **5**, 70 (1959)
[3]C. Djerassi and J. Staunton, *Am. Soc.*, **83**, 736 (1961)
[4]C. Djerassi, E. J. Warawa, R. E. Wolff, and E. J. Eisenbraun, *J. Org.*, **25**, 917 (1960)

N-Acetoxyphthalimide (2). The reagent is prepared by reaction of sodium N-hydroxyphthalimide (*which see*) with acetyl chloride.[1] It is recommended specifically[1,2] for N-acetylation of muramic acid (1), since acetylation with acetic anhydride and pyridine gives products of intramolecular cyclization (lactams). Osawa and Jeanloz[2] treated a solution of (1) in methanol at 0° with 2 equivalents of the reagent and 1 equivalent of triethylamine and let the mixture stand at room temperature for 20 hrs.

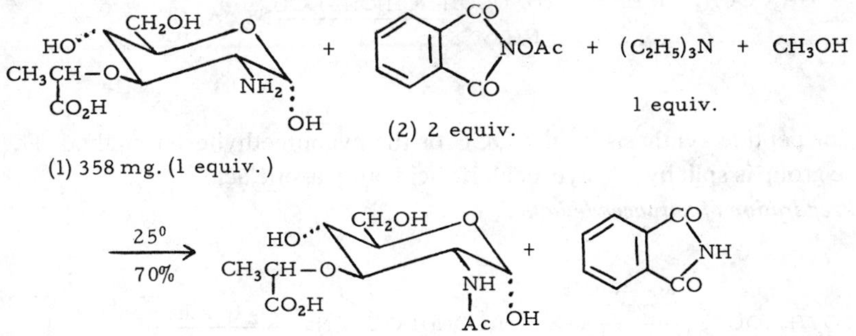
CH2OH
HO
CH3CH–O
CO2H
NH2
OH
(1) 358 mg. (1 equiv.)
CO
NOAc
CO
(2) 2 equiv.
(C2H5)3N
CH3OH
1 equiv.
25°
70%
CH2OH
HO
CH3CH–O
CO2H
NH
Ac
OH
CO
NH
CO

(3)

After evaporation, the residue was extracted with water and the filtered solution adjusted to pH 3.5 with Amberlite IR 120 and extracted with ethyl acetate. The residual sirup crystallized spontaneously, and recrystallization from ethyl acetate–methanol afforded pure (3) in 70% yield. When the reaction was carried out with only 1 equivalent of N-acetoxyphthalimide,[1] the product was a mixture of (1) and (3).

[1]P. M. Carroll, *Nature*, **197**, 694 (1963)
[2]T. Osawa and R. W. Jeanloz, *J. Org.*, **30**, 448 (1965)

2- and 3-Acetoxypyridine, $C_5H_4NOCOCH_3$.[1] Mol. wt. 137.14.

Preparation. The 2-isomer (b.p. 110–112°/10 mm.) by the action of acetyl chloride on the sodium salt of 2-hydroxypyridine; the 3-isomer (b.p. 92°/9 mm.) from 3-hydroxypyridine and acetic anhydride.

Acetylation. The reagents can be used for the acetylation of alcohols, phenols, and amines and for Friedel-Crafts acetylation of reactive aromatics. In general 2-acetoxypyridine is more reactive than the 3-isomer.

[1]Y. Ueno, T. Takaya, and E. Imoto, *Bull. Chem. Soc. Japan*, **37**, 864 (1964)

Acetylacetone (2,4-Pentanedione), $CH_3COCH_2COCH_3$. Mol. wt. 100.11, b.p. 134–136°; sp. gr. 0.975. Suppliers: B, E, F, MCB.

Preparation.[1]

$$(CH_3CO)_2O + CH_3COCH_3 \xrightarrow[80-85\%]{BF_3} CH_3COCH_2COCH_3$$

Cleavage of—N= derivatives. This reagent is more effective than pyruvic acid for the cleavage of phenylhydrazones and semicarbazones.[2] Reaction with a phenylhydrazone produces 3,5-dimethyl-N-phenylpyrazole and liberates the free carbonyl compound.

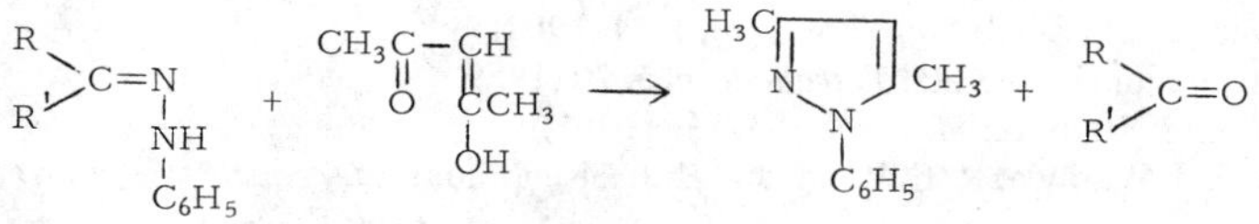

Peptide synthesis.[3] This β-diketone reacts with an amino acid in the presence of methanolic potassium hydroxide to give the potassium salt of an azomethine, formulated as an enamine stabilized by hydrogen bonding. These derivatives can be

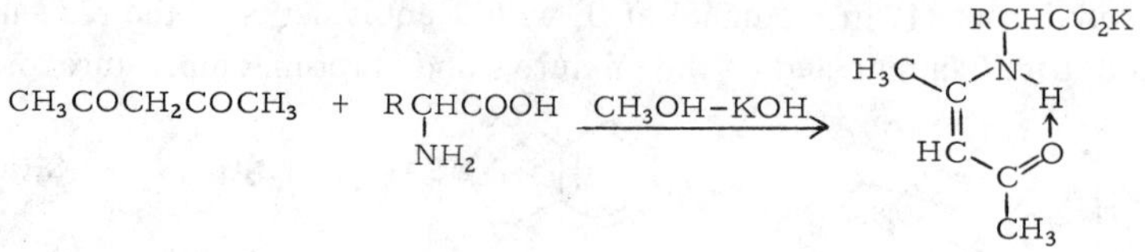

used for peptide synthesis by the DCC or the cyanomethyl ester method. The protective group is split by 2 *N* hydrochloric acid or by acetic acid.

Preparation of tetraacetylethane.[4]

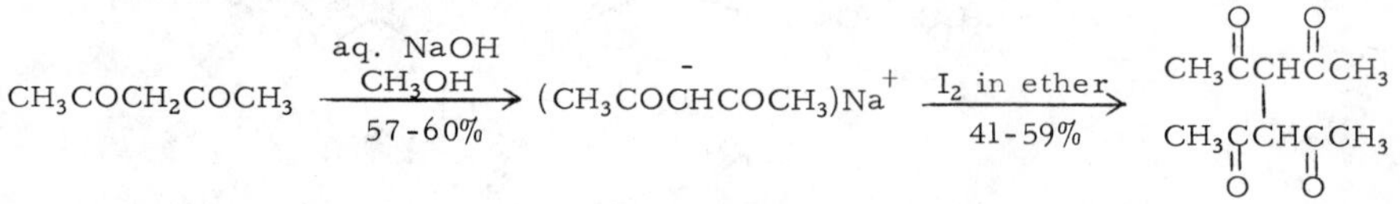

Heterocycles. Syntheses are formulated for the preparation of 2,4-dimethyl-3-acetyl-5-carboethoxypyrrole (1)[5] and 3,5-dimethylpyrazole (2).[6]

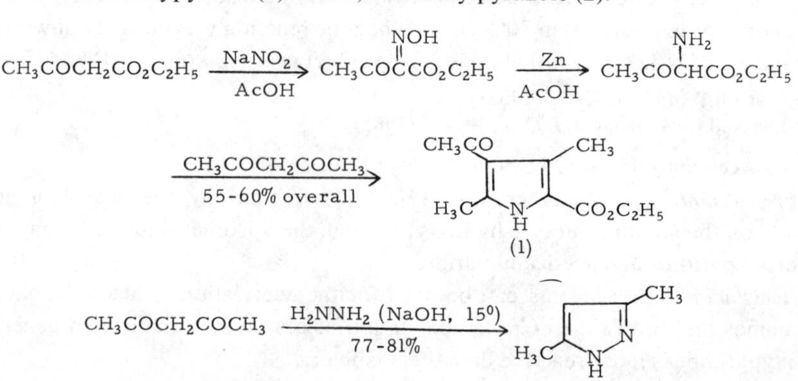

[1]C. E. Denoon, Jr., *Org. Syn., Coll. Vol.*, **3**, 16 (1955)
[2]W. Ried and G. Mühle, *Ann.*, **656**, 119 (1962)
[3]E. Dane, F. Drees, P. Konrad, and T. Dockner, *Angew. Chem., Internat. Ed.*, **1**, 658 (1962)
[4]R. G. Charles, *Org. Syn., Coll. Vol.*, **4**, 869 (1963)
[5]H. Fischer, *ibid.*, **3**, 513 (1955)
[6]R. H. Wiley and P. E. Hexner, *ibid.*, **4**, 351 (1963)

Acetyl chloride, CH_3COCl. Mol. wt. 78.50, b.p. 52°, sp. gr. 1.10. Preparation from acetic anhydride and calcium chloride.[1] Suppliers: B, E, F, MCB.

An approximately 3% solution of hydrogen chloride in methanol (Fischer esterification) can be prepared by addition of 1 ml. of acetyl chloride to 20 ml. of methanol.

[1]J. Gmünder, *Helv.*, **36**, 2021 (1953)

Acetylene, $HC{\equiv}CH$. Mol. wt. 26.04, b.p. −83°. Supplier: MCB.

An early technique[1] for purifying acetylene from a cylinder containing acetone consisted in passing the gas through water to absorb acetone and then through concd. sulfuric acid. Later workers for the most part pass the gas through a trap at −80°, a mercury safety valve, an empty bottle, concd. sulfuric acid, and finally a soda-lime tower.[2,3] Another scheme is to pass the gas through a tower of 10-mesh alumina and then into sulfuric acid.[3]

Typical syntheses utilizing acetylene are as follows:

1.[1]

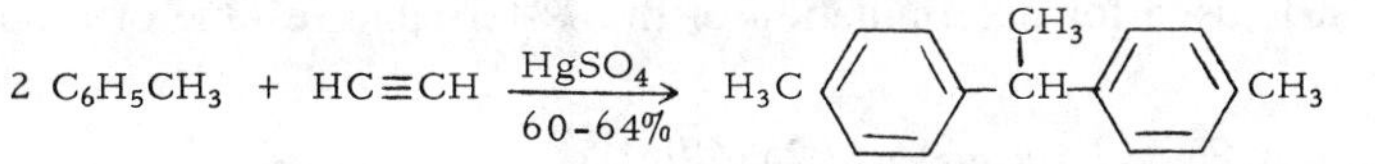

2.[4] $CH_3COCH_3 \xrightarrow{NaNH_2} CH_3C(ONa){=}CH_2 \xrightarrow{HC\equiv CH} CH_3C(ONa)(CH_3)\text{-}C{\equiv}CH \xrightarrow{H^+} CH_3C(OH)(CH_3)\text{-}C{\equiv}CH$ (40-46% overall)

3.[5]

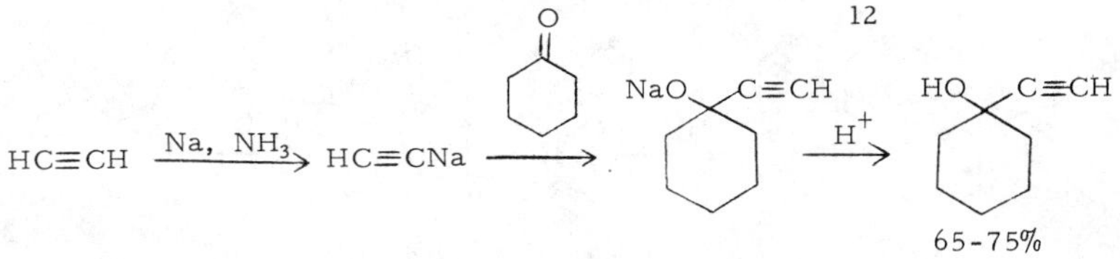

4.[2] $ClCH_2CO_2H + HC{\equiv}CH \xrightarrow[42\text{-}49\%]{Ag_2O} ClCH_2CO_2CH{=}CH_2$

5.[6] $HC{\equiv}CH \xrightarrow[-1/2\ H_2]{Na(NH_3)} HC{\equiv}CNa \xrightarrow[70\text{-}77\%]{n\text{-}C_4H_9Br} n\text{-}C_4H_9C{\equiv}CH$

6.[7] $(CH_3)_2CHCH_2CH_2COCl + HC{\equiv}CH \xrightarrow[54\text{-}64\%]{AlCl_3} (CH_3)_2CHCH_2CH_2COCH{=}CHCl$

β-Chlorovinyl isoamyl ketone

7.[3] $HC{\equiv}CH \xrightarrow{C_2H_5MgBr} HC{\equiv}CMgBr \xrightarrow{C_6H_5CH=CHCHO} C_6H_5CH{=}CHCH(OH)C{\equiv}CH$

[1]J. S. Reichert and J. A. Nieuwland, *Org. Syn., Coll. Vol.*, **1**, 229 (1941)
[2]R. H. Wiley, *ibid.*, **3**, 853 (1955)
[3]L. Skattebøl, E. R. H. Jones, and M. C. Whiting, *ibid.*, **4**, 793 (1963)
[4]D. D. Coffman, *ibid.*, **3**, 320 (1955)
[5]J. H. Saunders, *ibid.*, **3**, 416 (1955)
[6]K. N. Campbell and B. K. Campbell, *ibid.*, **4**, 117 (1963)
[7]C. C. Price and J. A. Pappalardo, *ibid.*, **4**, 186 (1963)

Acetylenedicarbonitrile, *see* Dicyanoacetylene.

Acetylenedicarboxylic acid, $HO_2CC{\equiv}CCO_2H$, m.p. 176°. Suppliers: Farchan, A, B, F. Preparation by reaction of α,β-dibromosuccinic acid with methanolic potassium hydroxide, isolation of the potassium acid salt, acidification, and extraction with ether.[1] The potassium acid salt is available from National Aniline and from Eastman.

The diacid is a useful dienophile (*see also* Dimethyl acetylenedicarboxylate).

[1]T. W. Abbott, R. T. Arnold, and R. B. Thompson, *Org. Syn., Coll. Vol.*, **2**, 10 (1943)

α-Acetyl-β-ethoxy-N-carboethoxyacrylamide. Mol. wt. 229.23, m.p. 90°.

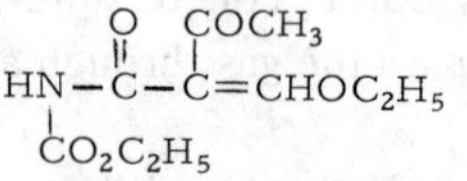
O COCH₃
HN—C—C=CHOC₂H₅
CO₂C₂H₅

The reagent is used for determination of the N-terminal residue of a protein or peptide.[1]

[1]J. H. Dewar and G. Shaw, *J. Chem. Soc.*, 3254 (1961)

Acetyl hypobromite, CH_3COOBr. Mol. wt. 138.96.

The reagent (or its equivalent) can be generated by addition of N-bromoacetamide to an acetic acid solution of lithium acetate and an olefinic substrate such as (1); *trans*-diaxial addition gives the bromohydrin acetate (2) in 74% yield.[1] Reaction conditions are essentially neutral, acid-sensitive groups are not attacked, and the

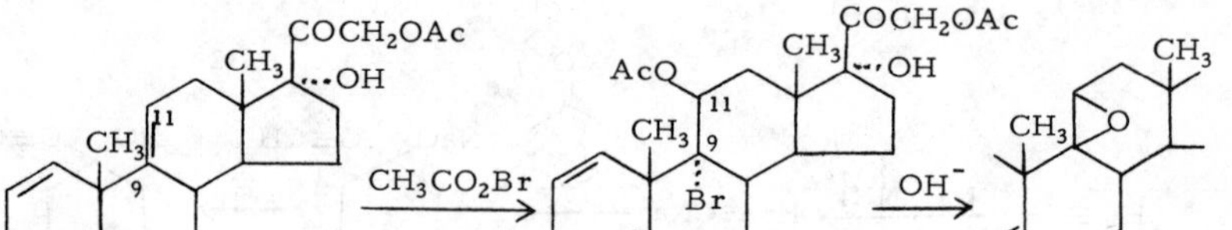
COCH₂OAc
CH₃
OH
CH₃
11
9
CH₃CO₂Br
AcO
COCH₂OAc
CH₃
OH
11
CH₃
9
Br
OH⁻
CH₃
CH₃
O

(1) (2) (3)

acetoxy group introduced is resistant to oxidation. Treatment of (2) with base gives the β-epoxide (3), and treatment with zinc–copper in ethanol regenerates the olefin (1).

In a second method [2] a solution of bromine in carbon tetrachloride is added to a stirred suspension of silver acetate in the same solvent, and the clear supernatant

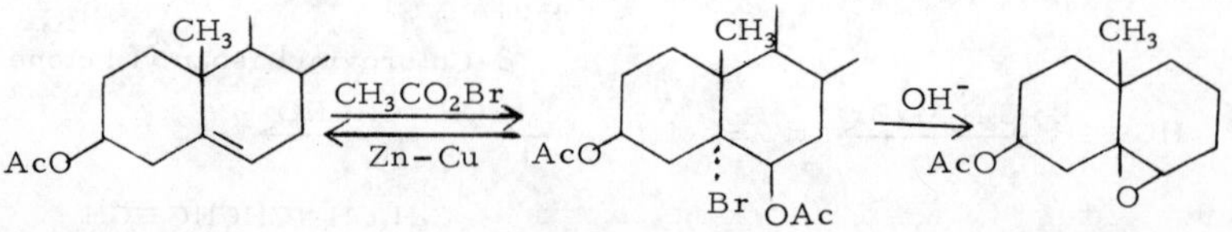
CH₃
AcO
CH₃CO₂Br
Zn–Cu
CH₃
AcO
Br
OAc
OH⁻
CH₃
AcO
O

(4) (5) (6)

solution of acetyl hypobromite is added at 0° to a carbon tetrachloride solution of an olefinic substrate such as cholesteryl acetate (4). The reaction is less stereospecific than that above, but the major product is the *trans*-diaxial bromohydrin acetate (5), convertible into the β-oxide (6). In contrast to the reaction of the $\Delta^{9(11)}$-compound (1) with acetyl hypobromite generated by the first method, methyl 3α-acetoxy-$\Delta^{9(11)}$-cholenate failed to react with a solution of reagent prepared by the second method.

[1]C. H. Robinson, L. Finckenor, M. Kirtley, D. Gould, and E. P. Oliveto, *Am. Soc.*, **81**, 2195 (1959)
[2]S. G. Levine and M. E. Wall, *ibid.*, **81**, 2826 (1959)

N-Acetylimidazole (1). Mol. wt. 110.12, m.p. 101.5–102.5°.

The reagent is prepared by dissolving imidazole in a slight excess of acetic anhydride and removing the generated acetic acid and remaining anhydride under vacuum (yield quant.).[1] When the reagent (1) is refluxed for 90 min. with an equimolar amount of pyrrole (2), the acetyl group is transferred to pyrrole to give N-acetylpyrrole (4) is about 90% yield.[2] N-Acetylpyrrole is difficult to prepare by usual

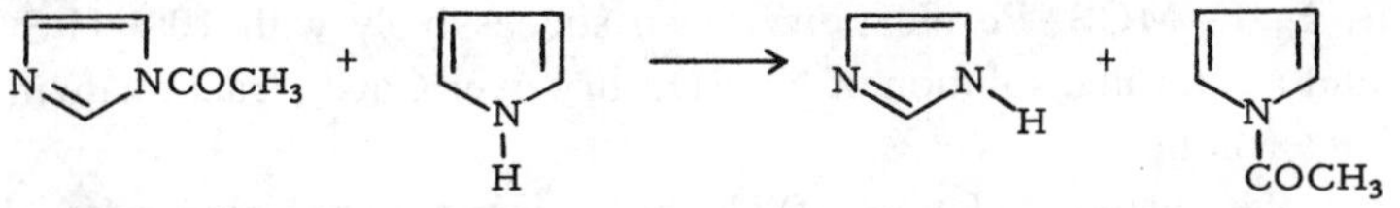

(1) (2) (3) (4)

methods since direct acetylation of pyrrole gives 2- and 3-acetylpyrrole in about equal amounts.

[1]J. S. Reddy, L. Mandell, and J. H. Goldstein, *J. Chem. Soc.*, 1414 (1963)
[2]J. S. Reddy, *Chem. Ind.*, 1426 (1965)

S-Acetylmercaptosuccinic anhydride (1). Mol. wt. 174.18, m.p. 77°. Supplier: Columbia. Preparation in 83% yield by addition of thiolacetic acid to maleic anhydride.[1]

A method for the introduction of thiol groups into proteins (and polyhydroxylic molecules such as dextran or polyvinyl alcohol) involves adding solid anhydride to a protein solution at pH 7 under nitrogen.[2] Hydrolyzed anhydride is removed with an anion exchanger, salts with a mixed-bed exchanger or by dialysis, and the

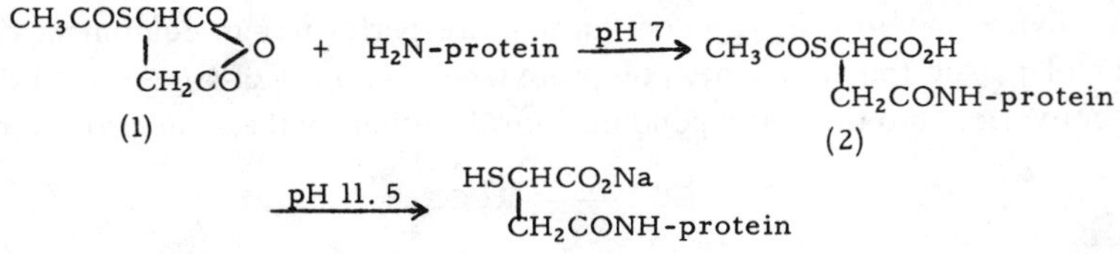

(3)

mercaptosuccinoylated protein (2) is isolated by lyophilization. The acetyl-S linkage is split rapidly in dilute sodium hydroxide at pH 11.5 (3).

[1]R. Brown, W. E. Jones, and A. R. Pinder, *J. Chem. Soc.*, 2123 (1951)
[2]I. M. Klotz and R. E. Heiney, *Am. Soc.*, **81**, 3802 (1959)

Acetyl nitrate, CH_3COONO_2. Mol. wt. 105.05. Potentially explosive.

Reagent prepared by adding 4.5 g. of 70% nitric acid to 35 ml. of acetic anhydride at 25–30° reacts with simple alkenes to give mixtures of β-nitro acetates, nitroolefins,

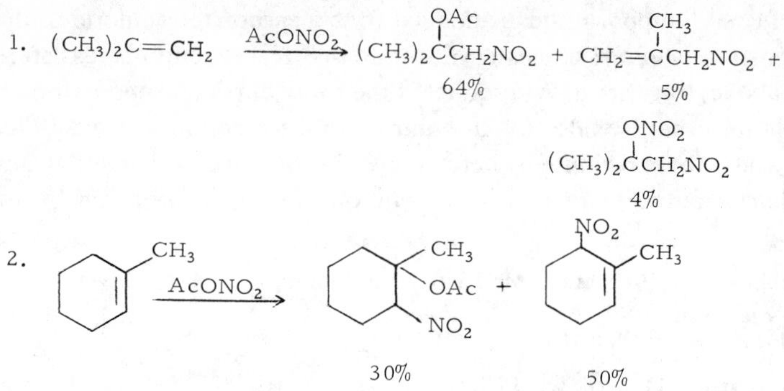
1.
$(CH_3)_2C{=}CH_2$
$AcONO_2$
OAc
$(CH_3)_2CCH_2NO_2$
64%
CH_3
$CH_2{=}CCH_2NO_2$
5%
ONO_2
$(CH_3)_2CCH_2NO_2$
4%
2.
CH_3
$AcONO_2$
CH_3
OAc
NO_2
30%
NO_2
CH_3
50%

and β-nitronitrates,[1] as illustrated in examples (1) and (2). Styrene, stilbenes, and 1,1-diarylalkenes give 40–70% yields of β-nitro acetates.

[1]F. G. Bordwell and E. W. Garbisch, Jr., *Am. Soc.*, **82**, 3588 (1960); *J. Org.*, **27**, 2322, 3049 (1962); *ibid.*, **28**, 1765 (1963)

Acrylonitrile, $CH_2{=}CHCN$. Mol. wt. 53.06, b.p. 78.5°, sp. gr. 0.81. Suppliers (pract.): B, E, F, MCB. Purification:[1] wash successively with 10% H_2SO_4, 10% Na_2CO_3, and a saturated solution of Na_2SO_4; dry over $CaCl_2$, fractionate in air, and then distil in vacuum.

Used for the cyanoethylation of alcohols, mercaptans, aldehydes, ketones, malonic ester, nitromethane, bromoform, etc.[2]

[1]J. C. Bevington and D. E. Eaves, *Trans. Farad. Soc.*, **55**, 1777 (1957)
[2]H. A. Bruson, *Org. Reactions*, **5**, 79 (1949)

Acyl fluorides.[1] An acyl fluoride can be prepared by stirring 2 moles of the corresponding acid anhydride at −10° with a Teflon-covered magnetic stirring bar and adding 45 g. (2.25 moles) of liquid anhydrous hydrogen fluoride.[2] After 1 hr. at −10° the mixture is let stand for 2 hrs. without cooling, treated with 15 g. of anhydrous

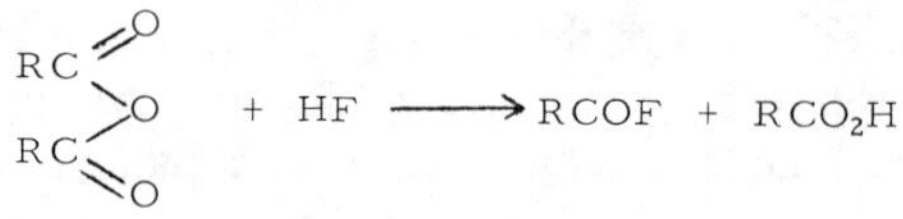
RC
O
RC
+ HF
RCOF + RCO_2H

sodium fluoride to remove excess hydrogen fluoride, filtered, and distilled. Operations involving anhydrous hydrogen fluoride are performed in equipment of fused silica or of plastic (polyethylene, polypropylene, Teflon, Kel-F). An acid chloride can be converted into the corresponding fluoride either by the same procedure or by

$$RCOCl + HF \longrightarrow RCOF + HCl$$

stirring 2 moles of the acid chloride at −5 to 0° and passing in anhydrous gaseous hydrogen fluoride at a rate of about 1 g./min. for 1 hr. Twenty-one acyl fluorides were obtained by these procedures in yields of 78–93%.

Formyl fluoride is prepared by reaction of acetic-formic anhydride with hydrogen fluoride at atmospheric pressure with continuous removal of formyl fluoride (b.p.

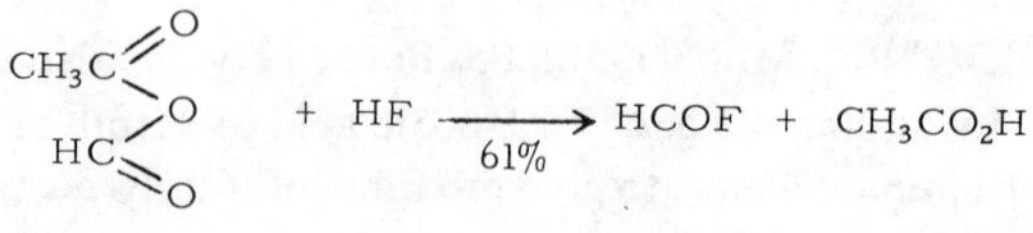
CH_3C
O
HC
+ HF
61%
HCOF + CH_3CO_2H

−29°).[2] Formyl fluoride is useful as a reagent for the introduction of the aldehyde group in Friedel-Crafts and Schotten-Baumann type formylations.

The use of anhydrous hydrogen fluoride for the preparation of acyl fluorides, first proposed by Colson[3] and by Fredenhagen,[4] was developed into a general preparative method by Olah and Kuhn.[2] A procedure for the preparation of benzoyl fluoride has appeared in *Organic Syntheses*.[5]

[1]Contributed by G. A. Olah and S. J. Kuhn (formerly of the Dow Chemical Co., Framingham, Mass.)
[2]G. A. Olah and S. J. Kuhn, *Am. Soc.*, **82**, 2380 (1960); *J. Org.*, **26**, 237 (1961)
[3]A. Colson, *Bull. soc.*, [3], **17**, 55 (1897); *Ann. chim.*, [7], **12**, 255 (1897)
[4]K. Fredenhagen, *Z. physik. Chem.*, **A164**, 189 (1933); K. Fredenhagen and G. Cadenbach, *ibid.*, **A164**, 201 (1933)
[5]G. A. Olah and S. J. Kuhn, *Org. Syn.*, **45**, 3 (1965)

Acyl peroxides, *see* Diacyl peroxides.

1-Adamantylchloroformate (2). Mol. wt. 214.69, m.p. 46–47°.

Preparation from 1-hydroxyadamantane and phosgene in pyridine.[1]

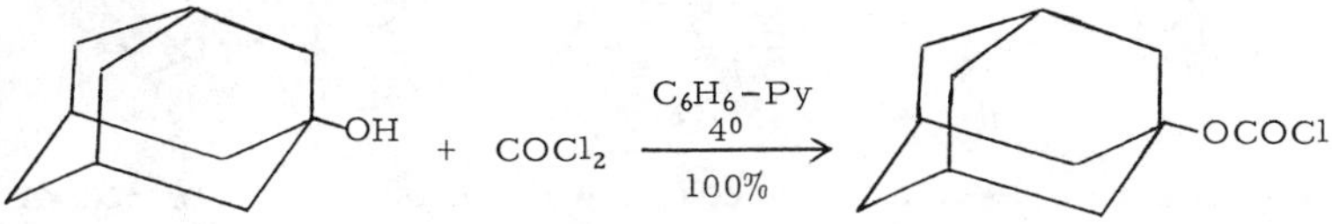
OH
+ COCl2
C6H6−Py
4°
100%
OCOCl

(1) (2)

Peptide synthesis. The reagent has been used successfully for the N-protection of amino acids; the bulky 1-adamantyloxycarbonyl group is removed by solvolysis with trifluoroacetic acid. The reagent was used also in the mixed-anhydride method of peptide synthesis, but considerable racemization was noted.

[1]W. L. Haas, E. V. Krumkalns, and K. Gerzon, *Am. Soc.*, **88**, 1988 (1966)

Adams catalyst, *see* Platinum catalysts.

Adipic acid, $HO_2C(CH_2)_4CO_2H$. Mol. wt. 146.14, m.p. 151–153°, pKa_1 4.43, pKa_2 5.52, solubility in 100 g. of water 1.4 g. at 15°. Suppliers: B, E, F, MCB.

Catalyst for ketalization. In the usual procedure of ketalization a Δ^4-3-ketosteroid (1) is refluxed with ethylene glycol in benzene in the presence of a catalytic amount of *p*-toluenesulfonic acid under a water separator until the theoretical amount of

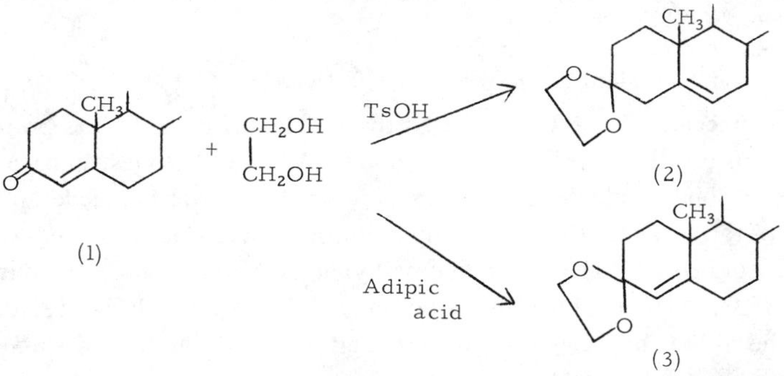
CH3
CH2OH
+
CH2OH
TsOH
CH3
O
O
CH3
Adipic
acid
O
O

(1) (2) (3)

water has been collected. Under these conditions the double bond usually migrates to the 5,6-position (2). Bernstein's group[1] found that, if *p*-toluenesulfonic acid is replaced by adipic acid, a weaker acid, the reaction proceeds without bond migration to give the Δ^4-ketal (3). Catalysis with oxalic acid, which is intermediate in acidic strength between *p*-toluenesulfonic acid and adipic acid, leads to mixtures of the Δ^4- and Δ^5-ketals. It is now known that (3) can be obtained with *p*-toluenesulfonic acid if this is taken in limited amount (*see* *p*-Toluenesulfonic acid).

[1]J. J. Brown, R. H. Lenhard, and S. Bernstein, *Am. Soc.*, **86**, 2183 (1964)

Adkins catalyst, *see* Copper–chromium oxide.

β-Alanine, $H_2NCH_2CH_2CO_2H$. Mol. wt. 89.10. Suppliers: B, E, F, MCB.

This and other amino acids (*see* Glycine) are effective catalysts for the Knoevenagel condensation.[1]

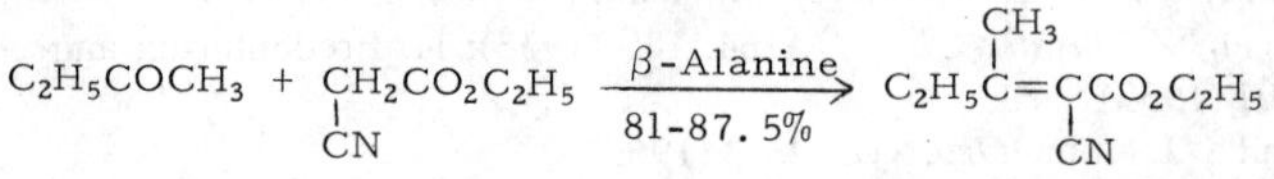
C₂H₅COCH₃ + CH₂CO₂C₂H₅ (CN) → C₂H₅C(CH₃)=C(CN)CO₂C₂H₅
β-Alanine
81-87.5%

A catalytic amount of β-alanine significantly improves the yields in the condensation of diphenylcyclopropenone with malononitrile and ethyl cyanoacetate.[2]

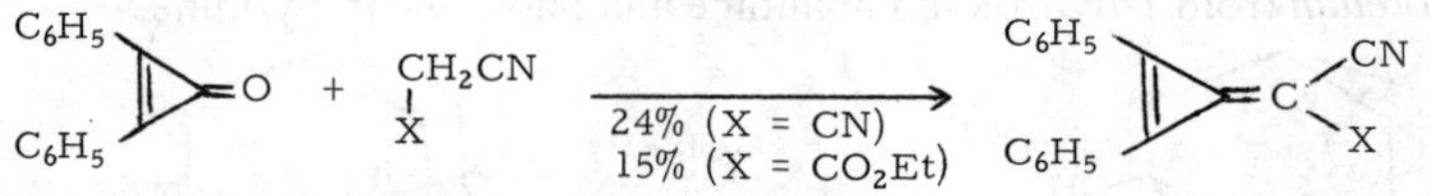
C₆H₅
CH₂CN
X
24% (X = CN)
15% (X = CO₂Et)
CN

The condensation of 1,4-cyclohexanedione with malononitrile proceeds almost quantitatively under catalysis by β-alanine.[3] The reactants are melted together on the steam bath, an aqueous solution of β-alanine is added, and the mixture is heated

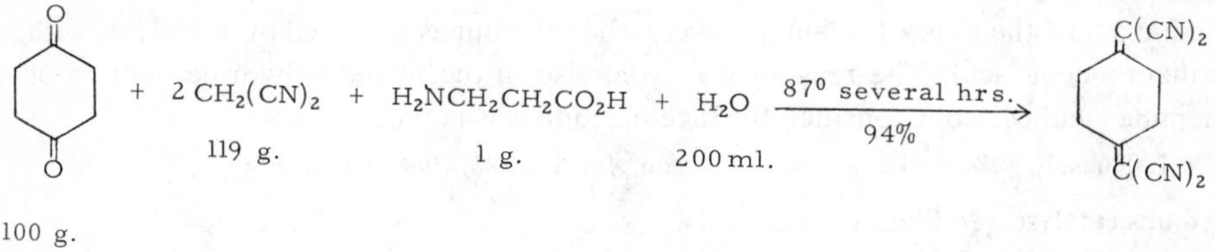
+ 2 CH₂(CN)₂ + H₂NCH₂CH₂CO₂H + H₂O
119 g.
1 g.
200 ml.
87° several hrs.
94%
C(CN)₂
100 g.

on the steam bath. The product separates in crystalline form, and washing with water and with ether gives a pure product, m.p. 216–217°. When the reaction was conducted in benzene with acetic acid–ammonium acetate as catalyst the yield dropped to 76.5%.

[1]F. S. Prout *et al.*, *Org. Syn., Coll. Vol.*, **4**, 93 (1963); Y. Egawa, M. Suzuki, and T. Okuda, *Chem. Pharm. Bull.*, **11**, 589 (1963)
[2]S. Andreades, *Am. Soc.*, **87**, 3941 (1965)
[3]D. S. Acker and W. R. Hertler, *Am. Soc.*, **84**, 3370 (1962)

Alkylamine-boranes, $R_{3-n}NH_n \cdot BH_3$. Nöth and Beyer[1] describe the preparation of these reagents and their use in the reduction of aldehydes, ketones, and acid chlorides to the corresponding alcohols. The reducing properties vary considerably with the structure of the reagent. Monoethylamine-borane, $C_2H_5NH_2 \cdot BH_3$, reduces 3 moles of 1,4-benzoquinone, whereas monomethylamine-borane, dimethylamine-borane, $(CH_3)_2NH \cdot BH_3$, and *t*-butylamine-borane, $(CH_3)_3CNH_2 \cdot BH_3$, reduce only 2 moles. The following reagents are supplied by Callery: trimethylamine-borane (solid), pyridine-borane (liq.).

[1]H. Nöth and H. Beyer, *Ber.*, **93**, 928, 939, 1078 (1960)

Alkyl cyanides.[1] *n*-Propyl and *n*-butyl cyanide can be prepared conveniently by efficient stirring under reflux of a mixture of 0.5 mole of the corresponding primary alkyl bromide, 0.6 mole of sodium cyanide, and 150 ml. of ethylene glycol.[2] The temperature recorded by a thermometer in the vapor is initially close to the boiling point of the halide and rises sharply to a constant value as the halide is exhausted. For conversion of chlorides from *n*-butyl to *n*-octyl to the cyanides, the preferred solvent is polyethylene glycol-300 (Dow).[3] The reaction time varies from 15 min. to several hours. The nitrile is isolated by distillation in yields of 90–95%.

For conversion of secondary alkyl chlorides to the corresponding cyanide the preferred solvent is dimethyl sulfoxide.[4] The chloride is added slowly to a heated mixture of sodium cyanide and dimethyl sulfoxide, and heating is continued for 1–3 hrs. The reaction mixture is poured into water and the product extracted with chloroform or ether. After drying and removal of solvent, distillation gives the alkyl cyanide in yield of about 70%.

[1]Contributed by Peter V. Susi, American Cyanamid Co.
[2]R. N. Lewis and P. V. Susi, *Am. Soc.*, **74**, 840 (1952)
[3]A. Brändström, *Chem. Scand.*, **10**, 1197 (1956)
[4]R. A. Smiley and C. Arnold, *J. Org.*, **25**, 257 (1960)

Alkyl hydroperoxides, ROOH. *See also t*-Butyl hydroperoxide.

Preparation. (a) Baeyer and Villiger's method[1] of treating 30% hydrogen peroxide with a dialkyl sulfate gives in low yield impure and dangerously unstable preparations of methyl, ethyl, *n*-propyl, and *n*-butyl hydroperoxide. An improved procedure[2] utilizing methanol as solvent affords pure *n*- and *sec*-butyl hydroperoxide in yields of 20 and 40%. (b) A synthesis by the reaction of an *n*-alkyl methanesulfonate (CH_3SO_3R) with hydrogen peroxide in methanolic potassium hydroxide solution affords pure *n*-alkyl[3] hydroperoxides (propyl through decyl) in 70–80% yield (secondary, 20–25%). These hydroperoxides are of the same order of stability as the *t*-butyl derivative. (c) Pure alkyl hydroperoxides are obtainable by slow addition of alkyl Grignard reagents to oxygen-saturated ether at −75°.[4]

[1]A. Baeyer and V. Villiger, *Ber.*, **33**, 3387 (1900); **34**, 738 (1901)
[2]E. G. Lindstrom, *Am. Soc.*, **75**, 5123 (1953)
[3]H. R. Williams and H. S. Mosher, *ibid.*, **76**, 2984, 2987 (1954)
[4]C. Walling and S. A. Buckler, *ibid.*, **75**, 4372 (1953)

Alkyl iodides. Methyl and ethyl iodide deteriorate rapidly with liberation of iodine if exposed to light, and commercial preparations which have been kept long in storage are often dark and unsuitable for use. Such material can be purified by shaking it with successive portions of a dilute solution of either sodium thiosulfate or sodium bisulfite until the color is bleached, washing with water, drying over calcium chloride, and distilling the product. The colorless distillate should be stored in a brown bottle and kept out of sunlight. As a further protective measure a few drops of clean mercury can be added, for the alkyl iodide will then keep almost indefinitely without becoming discolored if not exposed to light for long periods. Prolonged exposure presents a certain danger, for appreciable quantities of the poisonous methylmercuric iodide may result from photochemical reaction. A little mercury can be used also to remove a slight purple or pink color from a sample of an iodide which has begun to decompose.

Alkyl methanesulfonates, CH_3SO_3R. Preparation by reaction of mesyl chloride with alcohols in pyridine solution.[1]

[1]V. C. Sekera and C. S. Marvel, *Am. Soc.*, **55**, 345 (1933)

Alkyl sulfates. Discolored samples of dimethyl or diethyl sulfate can be purified by slow distillation at the water pump ($ROSO_2OH$ if present decomposes to R_2SO_4 and H_2SO_4). Higher alkyl sulfates are conveniently made through the sulfites.[1]

[1] C. M. Suter and H. L. Gerhart, *Org. Syn., Coll. Vol.*, **2**, 111, 112 (1943)

Allene, $CH_2{=}C{=}CH_2$. Mol. wt. 40.06, b.p. −32°. Suppliers: Columbia Org. Chem. (225-g. cylinder), KK, MCB, Peninsular ChemResearch.

Preparation. Most routes to allene give difficultly separable mixtures of allene and methylacetylene, but material free from methylacetylene can be prepared by dechlorination of 2,3-dichloropropene with zinc dust.[1]

$$\underset{\text{2.34 m.}}{CH_2{=}C(Cl){-}CH_2Cl} + \underset{\text{4.6 g. at.}}{Zn} + \underset{\text{400 ml.-80 ml.}}{95\%\ EtOH{-}H_2O} \xrightarrow[80\%]{\text{Refl.}} CH_2{=}C{=}CH_2$$

Synthesis of a cyclobutane derivative. A du Pont group[2] synthesized 3-methylene-cyclobutane-1,2-dicarboxylic anhydride (3) by charging a cooled (−70°) autoclave with maleic anhydride, 645 ml. of benzene, 0.25 g. of hydroquinone, and allene. The

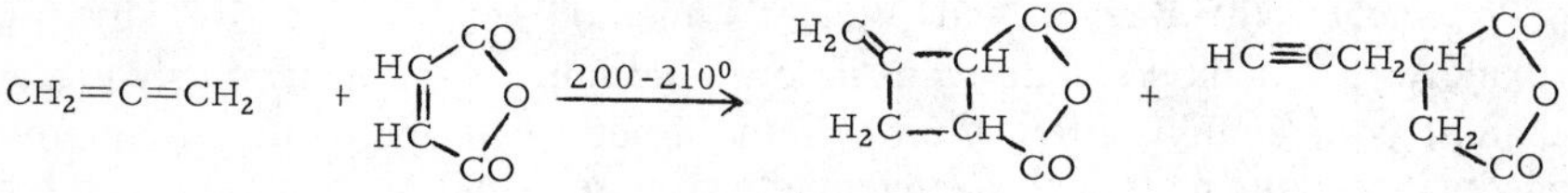

(1) 2.5 m. (2) 5.1 m. (3) 22-26% (4) 7-9%

mixture was heated with stirring for 8–10 hrs. at 200–210°. Workup and careful fractionation afforded products (3) and (4) in the yields indicated.

Conversion to a cyclooctane derivative. Under catalysis by bis-(triphenylphosphite)-nickel dicarbonyl, allene is converted into a tetramer shown to have the structure 1,3,5,7-tetramethylenecyclooctane.[3]

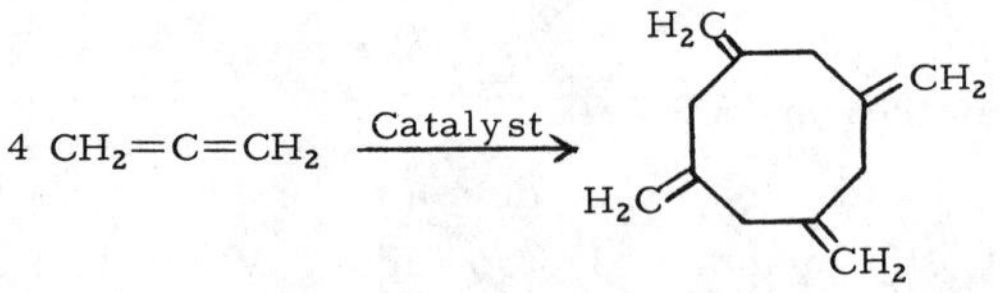

[1] H. N. Cripps and E. F. Kiefer, *Org. Syn.*, **42**, 12 (1962)
[2] H. B. Stevenson, H. N. Cripps, and J. K. Williams, *ibid.*, **43**, 27 (1963)
[3] R. E. Benson and R. V. Lindsey, Jr., *Am. Soc.*, **81**, 4247 (1959)

Allyllithium. This previously unknown reagent was initially prepared[1] by combining the preparation of allylmagnesium chloride and its reaction with triphenyltin chloride (1) to give a solution of allyltriphenyltin (2) in ether, tetrahydrofurane, or benzene. Addition of a solution of phenyllithium then gives a suspension of tetraphenyltin in a solution of allyllithium, which can be used directly, for example for reaction with a

$$CH_2{=}CHCH_2Cl + \underset{(1)}{(C_6H_5)_3SnCl} + Mg \longrightarrow \underset{\text{(2) m.p. }74^0}{(C_6H_5)_3SnCH_2CH{=}CH_2}$$

$$\downarrow C_6H_5Li$$

$$CH_2{=}CHCH_2Li + (C_6H_5)_4Sn$$

carbonyl compound. Allyllithium is obtainable also by transmetalation between phenyllithium and tetraallyltin (available from Metal and Thermit Corp.).[2] The reagent reacts with a variety of organic and organometallic substances to give allyl derivatives.

$$4\ C_6H_5Li + (CH_2{=}CHCH_2)_4Sn \longrightarrow 4\ CH_2{=}CHCH_2Li + (C_6H_5)_4Sn$$

A convenient new method of preparing the reagent is by cleavage of allyl phenyl ether with lithium in tetrahydrofurane.[3]

$$CH_2{=}CHCH_2OC_6H_5 + 2\ Li \xrightarrow[62\text{-}66\%]{THF\ -15^0} CH_2{=}CHCH_2Li + C_6H_5OLi$$

[1]D. Seyferth and M. A. Weiner, *J. Org.*, **24**, 1395 (1959)
[2]*Idem, ibid.*, **26**, 4797 (1961)
[3]J. J. Eisch and A. M. Jacobs, *J. Org.*, **28**, 2145 (1963)

Alumina for chromatography. The equivalent of *acid-washed* material can be prepared by stirring ordinary alumina with ethyl acetate, letting the mixture stand for 1–2 days, filtering, and drying at 80°. Alumina washed with 5–10% acetic acid in water is suitable for adsorption of base-sensitive substances (enol acetates, α,β-unsaturated acetates).[1] For chromatography of β,γ-unsaturated ketones without isomerization, the alumina can be washed with warm aqueous alkali, then warm acetic acid followed by water (to neutrality); it is then reactivated at 200° for 30 hrs.[2]

[1]K. R. Farrar, J. C. Hamlet, H. B. Henbest and E. R. H. Jones, *J. Chem. Soc.*, 2657 (1952); R. B. Clayton, A. Crawshaw, H. B. Henbest, E. R. H. Jones, B. J. Lovell, and G. W. Wood, *ibid.*, 2009 (1953)
[2]C. W. Shoppee and G. H. R. Summers, *J. Chem. Soc.*, 689 (1950)

Alumina for dehydration of alcohols. Alumina prepared by hydrolysis of aluminum isopropoxide is relatively strongly acidic and causes extensive isomerization of olefins initially formed.[1] This isomerization can be suppressed by exposing the catalyst to ammonia or by using alumina prepared from sodium aluminate.

von Rudloff[2] found that dehydration of terpene and sesquiterpene alcohols is advantageously carried out with neutral alumina (Woelm activity grade 1) to which 1–2% of pyridine has been added; rearrangements encountered with acidic reagents are thus avoided. Corey and Hortmann[3] applied this procedure to the allylic alcohol-lactone (1) prepared from santonin and obtained the dienic lactone (2) in amounts sufficient for a synthesis (photochemical) of dihydrocostunolide (3).

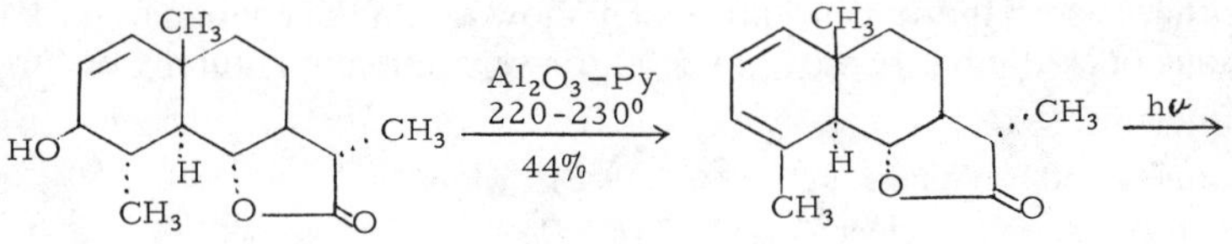
CH3
HO
H
CH3
O
O
CH3
Al2O3–Py
220-230⁰
44%
CH3
H
CH3
O
O
CH3
hν

(1) (2)

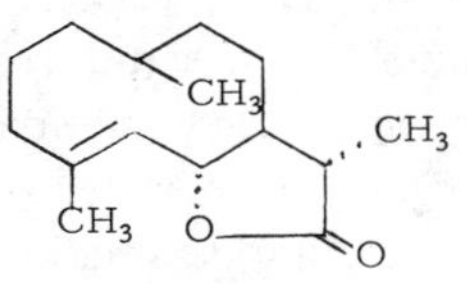
CH3
CH3
CH3
O
O

(3)

Stoll *et al.*[4] prepared a dehydration catalyst by moistening a mixture of 90 g. of alumina, 60 g. of diatomaceous earth, and 20 g. of cork powder and heating the mixture to red heat. After cooling, the catalyst was packed into a combustion tube which could be heated to 380–400°, and 50 g. of cyclopentanol was heated to boiling and the vapor passed with a stream of nitrogen into the tube. The mixture of water and cyclopentene was separated and the hydrocarbon dried and distilled; yield 81%.

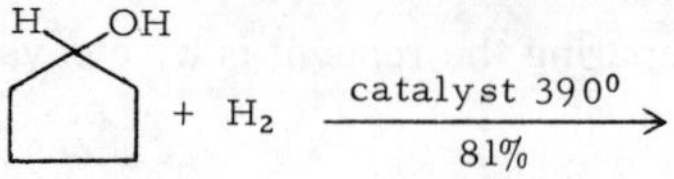
H
OH
+ H₂
catalyst 390⁰
81%

[1]H. Pines and C. N. Pillai, *Am. Soc.*, **83**, 3270 (1961); F. G. Schappell and H. Pines, *J. Org.*, **31**, 1735 (1966)
[2]E. von Rudloff, *Can. J. Chem.*, **39**, 1860 (1961)
[3]E. J. Corey and A. G. Hortmann, *Am. Soc.*, **87**, 5736 (1965)
[4]A. Stoll, A. Lindenmann, and E. Jucker, *Helv.*, **36**, 268 (1953)

Alumina for esterification.[1] A simple and rapid method of esterification which is particularly convenient for working on a scale of 20 mg. to 20 g. is as follows. An alcohol, for example ergosterol, is treated in benzene solution with a slight excess of benzoyl chloride or *p*-phenylazobenzoyl chloride and 2 equivalents of pyridine. When the reaction is complete, filtration through a small column of alumina (activity II, Woelm; E. Merck) gives a solution of the desired ester in pyridine. After evaporation of the benzene, pyridine is removed by addition of toluene and evacuation on the steam bath.

[1]Contributed by G. E. Molau, Dow Chemical Co.

Alumina, morin-dyed.[1]

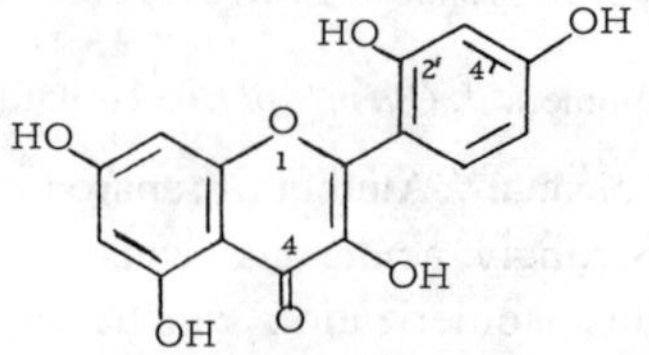
HO
OH
HO
O
1
2'
4'
4
OH
OH
O

Morin

Technical morin (3,5,7,2′,4′-pentahydroxyflavone) is supplied by Eastman, Baker, and Matheson, Coleman and Bell. A solution of 300 mg. of morin in 500 ml. of methanol is added with stirring to a slurry of 500 g. of alumina in 500 ml. of methanol. When the supernatant liquor is decolorized, the alumina is collected and dried at 150° for 2 hrs., when the citron-yellow preparation has grade II activity.[2] A column of this adsorbent fluoresces in ultraviolet light (*ca.* 360 mμ), and colorless substances that absorb in the near ultraviolet show up on the column as dark bands on a luminous background. It proved useful for separation of naturally occurring polyunsaturated diols.[3]

[1]H. Brockmann and F. Volpers, *Ber.*, **80**, 77 (1947)
[2]H. Brockmann and H. Schodder, *ibid.*, **74**, 73 (1941)
[3]E. F. L. J. Anet, *et al.*, *J. Chem. Soc.*, 309 (1953)

Aluminum amalgam, from Al (26.98) and $HgCl_2$ (271.52).

Preparation. (a)[1] Aluminum turnings (oil-free) are etched with dilute sodium hydroxide to a point of strong hydrogen evolution, the solution is decanted, and the metal is washed once superficially with water so that it retains some alkali. It is then treated with 0.5% mercuric chloride solution for 1–2 min., and the entire process is repeated. The shiny amalgamated metal is washed rapidly in turn with water, ethanol, and ether and used at once. This material reacts vigorously with water with liberation of hydrogen equivalent to the amount of aluminum present and can be used to dry ether or ethanol.

(b) Newman[2] used sandpapered aluminum foil in combination with a small amount of mercuric chloride for amalgamation.

Reduction. Since the reagent is neutral, it is useful for the reduction of alkali-sensitive compounds. Thus diethyl oxaloacetate is reduced to diethyl malate by aluminum amalgam in ether solution in 70–80% yield, whereas with sodium amalgam the yield is only 50%.[1] Nitrobenzene can be reduced to aniline in hot ethanol or,

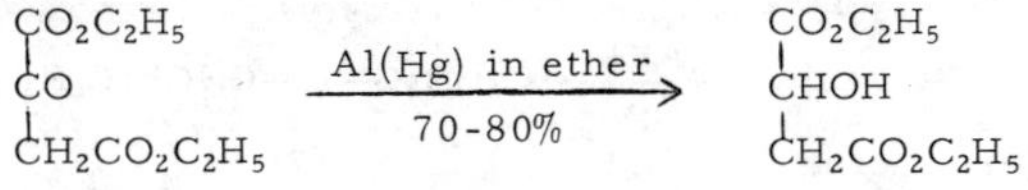

$CO_2C_2H_5$
CO
$CH_2CO_2C_2H_5$
Al(Hg) in ether
70-80%
$CO_2C_2H_5$
CHOH
$CH_2CO_2C_2H_5$

in ether at a controlled temperature, to phenylhydroxylamine.[1]

Newman[2] used the reagent for the reduction of aryl alkyl ketones to pinacols (yields 30–60%); with aliphatic ketones yields were negligible. Thus a solution

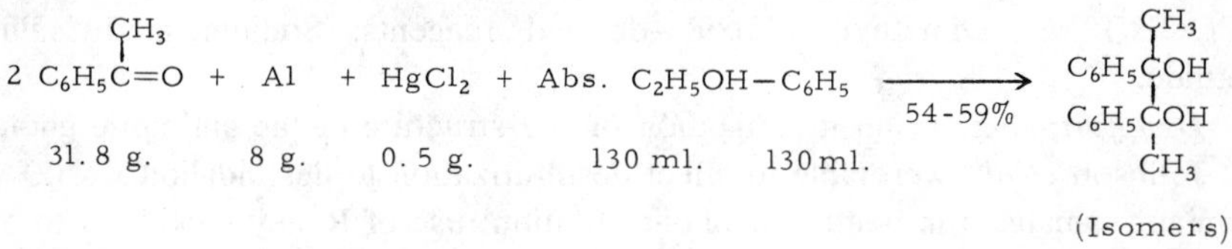

CH_3
2 $C_6H_5C=O$ + Al + $HgCl_2$ + Abs. $C_2H_5OH-C_6H_5$
31. 8 g. 8 g. 0. 5 g. 130 ml. 130 ml.
54-59%
C_6H_5COH
C_6H_5COH
CH_3
CH_3
(Isomers)

of acetophenone in ethanol–benzene was treated with mercuric chloride and $\frac{1}{64}$-inch thick aluminum foil which had been sandpapered and cut into 1-inch squares. On heating the mixture a vigorous reaction started and was allowed to proceed spontaneously until it moderated. External heat was then applied to maintain reflux until all the aluminum had disappeared (2 hrs.). After cooling, the mixture was treated with dilute hydrochloric acid, and the product was extracted with benzene, distilled, and crystallized.

Thiele and Henle[3] reduced yellow cinnamylidenefluorene (1) with aluminum

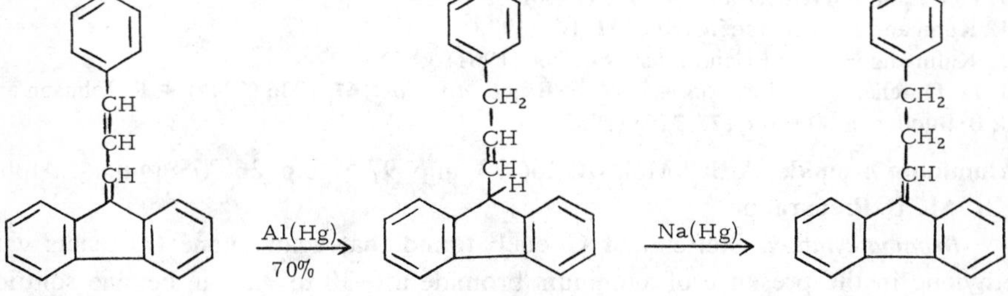

CH
CH
CH
Al(Hg)
70%
CH_2
CH
CH
H
Na(Hg)
CH_2
CH_2
CH

(1) (2) (3)

amalgam in moist ether and obtained a colorless dihydride characterized as the product of 1,4-addition. Kuhn and Winterstein[4] confirmed this result and obtained (2) in 70% yield. On reduction of (1) with sodium amalgam they obtained the isomeric dihydride (3), but by operating at 0° for a limited reaction period they were able to isolate (2) as the primary product and to show that under the usual conditions it undergoes base-catalyzed isomerization to (3). In the case of 1,6-diphenyl-$\Delta^{1,3,5}$-

hexatriene (4), reduction with either aluminum amalgam or sodium amalgam proceeds by 1,6-addition to give the α,ω-dibenzylpolyene (5); the best results are obtained

$$\underset{(4)}{C_6H_5CH{=}CHCH{=}CHCH{=}CHC_6H_5} \xrightarrow[\text{or NaHg}]{\text{AlHg}} \underset{(5)}{C_6H_5CH_2CH{=}CHCH{=}CHCH_2C_6H_5}$$

with aluminum amalgam. Kuhn and Fischer[5] found that the tetraphenylcumulenes (6) and (8) are reduced by aluminum amalgam in tetrahydrofurane containing 4% of water to the dihydrides (7) and (9).

$$\underset{(6)}{(C_6H_5)_2C{=}C{=}C{=}C(C_6H_5)_2} \xrightarrow[70\%]{\text{AlHg}} \underset{(7)}{(C_6H_5)_2C{=}C{=}CHCH(C_6H_5)_2}$$

$$\underset{(8)}{(C_6H_5)_2C{=}C{=}C{=}C{=}C{=}C(C_6H_5)_2} \xrightarrow[75\%]{\text{AlHg}} \underset{(9)}{(C_6H_5)_2C{=}C{=}CHCH{=}C{=}C(C_6H_5)_2}$$

For the reduction of β-ketosulfoxides to methyl ketones ($RCOCH_2SOCH_3 \longrightarrow RCOCH_3$) *see* Dimethyl sulfoxide–derived reagents: Sodium methylsulfinylmethide.

Desulfurization. In an investigation of the structure of the antibiotic gliotoxin (1), Johnson *et al.*[6] were able to effect desulfurization to desthiogliotoxin (2) with aluminum amalgam in neutral alcoholic solution; use of Raney nickel led to overreduction.

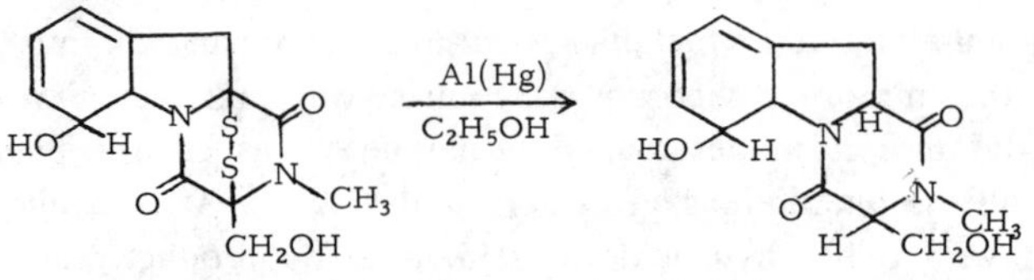
HO
H
N
S
S
O
N
CH3
O
CH2OH
Al(Hg)
C2H5OH
HO
H
N
H
O
N
CH3
O
H
CH2OH

(1) (2)

[1]H. Wislicenus and L. Kaufmann, *Ber.*, **28**, 1323 (1895)
[2]M. S. Newman, *J. Org.*, **26**, 582 (1961)
[3]J. Thiele and F. Henle, *Ann.*, **347**, 307 (1906)
[4]R. Kuhn and A. Winterstein, *Helv.*, **11**, 123 (1928)
[5]R. Kuhn and Herbert Fischer, *Ber.*, **94**, 3060 (1961)
[6]J. D. Dutcher, J. R. Johnson, and W. F. Bruce, *Am. Soc.*, **67**, 1736 (1945); J. R. Johnson and J. B. Buchanan, *Am. Soc.*, **75**, 2103 (1953)

Aluminum bromide, $AlBr_3$. Mol. wt. 266.73, m.p. 97.5°, b.p. 268°. Suppliers: Alpha, KK, MCB. Preparation.[1]

Bromoalkylation. Stetter and Goebel[2] found that adamantane (1) reacts with ethylene in the presence of aluminum bromide at −30 to −20° in hexane solution

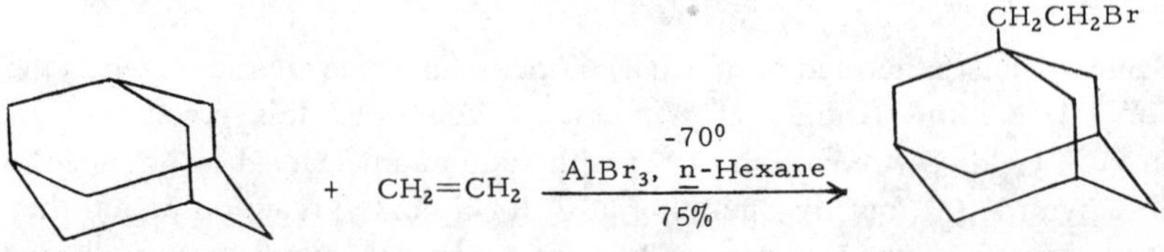
CH2CH2Br
+ CH2=CH2
-70°
AlBr3, n-Hexane
75%

(1) (2)

to give 1-β-bromoethyladamantane (2) in yield of 57%. Experiments in this laboratory[3] have shown that the yield (of crystallized product) can be raised to 75% by carrying out the reaction at dry ice–acetone bath temperature and introducing just the calculated amount of ethylene. However, the results depended upon the quality of the aluminum bromide, which varied from supplier to supplier and from lot to lot. The best sample of reagent was one received in a sealed ampoule.

Cleavage of ethers. Of many reagents tried for demethylation of 2,6-dimethoxy-4-methylbenzaldehyde, aluminum bromide proved to be the most satisfactory.[4] A solution of the reagent in carbon disulfide was added to a stirred solution of the

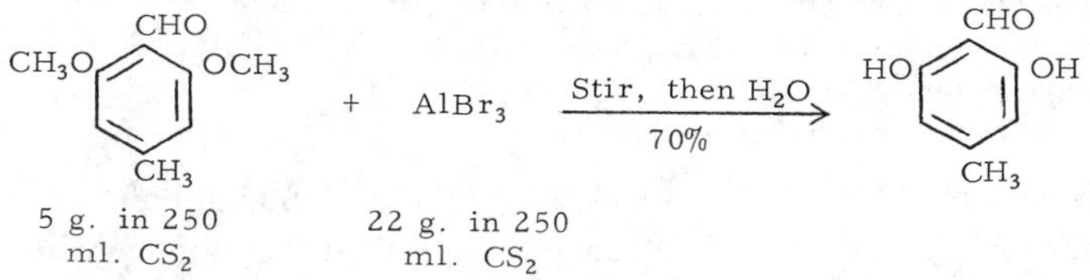
CHO
CH3O
OCH3
CH3
5 g. in 250 ml. CS2
+ AlBr3
22 g. in 250 ml. CS2
Stir, then H2O
70%
CHO
HO
OH
CH3

ether in the same solvent and stirring was continued for 1 hr. An addition complex separated as a red gum. Workup included crystallization from water and afforded phenolic aldehyde of satisfactory purity.

Prey[5] found that the pyridinium salt of aluminum bromide (2 Py: 1 $AlBr_3$) is an excellent reagent for cleavage of phenol ethers; diphenyl ether alone resisted cleavage by the reagent. A solution of 33 g. of pyridine in 50 ml. of benzene is stirred and a solution of 50 g. of $AlBr_3$ in 300 ml. of benzene is added; the precipitated salt is purified by solution in pyridine and precipitation with ether. For demethylation, a mixture of the ether with 3 parts of salt is heated at 220°.

Bromination.[6] Bromodichloromethane (b.p. 89–90°) is prepared in good yield by the bromination of chloroform with hydrogen bromide and aluminum bromide.

[1]D. G. Nicholson, P. K. Winter, and H. Fineberg, *Inorg. Syn.*, **3**, 30 (1950)
[2]H. Stetter and P. Goebel, *Ber.*, **95**, 1040 (1962)
[3]M. Z. Nazer, unpublished
[4]R. Adams and J. Mathieu, *Am. Soc.*, **70**, 2120 (1948)
[5]V. Prey, *Ber.*, **75**, 537 (1942)
[6]D. E. Lake and A. A. Asadorian, U. S. patent 2,553,518 (1951)

Aluminum *t*-butoxide, $Al[OC(CH_3)_3]_3$. Mol. wt. 246.32. Suppliers: Alfa, B, MCB. Preparation.[1]

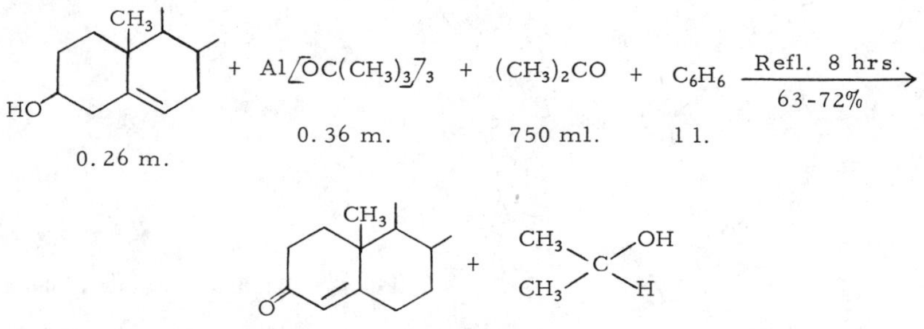
CH3
HO
0.26 m.
+ Al[OC(CH3)3]3
0.36 m.
+ (CH3)2CO
750 ml.
+ C6H6
1 l.
Refl. 8 hrs.
63-72%
CH3
O
+
CH3
CH3
C
OH
H

Oppenauer oxidation.[2] The oxidation of cholesterol to cholestenone as described by Oppenauer[3] is carried out by gentle refluxing for 8 hrs. of a mixture of commercial cholesterol, aluminum *t*-butoxide, acetone (hydrogen acceptor), and benzene. The cooled mixture is shaken with dilute sulfuric acid and the product recovered from the benzene layer and crystallized twice.

Condensation catalyst. Aluminum *t*-butoxide is one of many reagents that catalyze the self-condensation of acetophenone to dypnone[4] in xylene. The mixture is stirred and kept at a temperature such that *t*-butanol slowly distils.

$$\underbrace{C_6H_5\overset{CH_3}{\overset{|}{C}}{=}O + CH_3\overset{O}{\overset{\|}{C}}C_6H_5}_{1\ m.} + \underset{0.55\ m.}{Al[OC(CH_3)_3]_3} + \underset{400\ ml.}{C_6H_4(CH_3)_2} \xrightarrow[77\text{-}82\%]{150\text{-}155^{\circ}} C_6H_5\overset{CH_3}{\overset{|}{C}}{=}CH\overset{O}{\overset{\|}{C}}C_6H_5$$

[1]W. Wayne and H. Adkins, *Org. Syn., Coll. Vol.* **3**, 48 (1955)
[2]C. Djerassi, "The Oppenauer Oxidation," *Org. Reactions*, **6**, 207 (1951)
[3]R. V. Oppenauer, *Org. Syn., Coll. Vol.*, **3**, 207 (1955)
[4]W. Wayne and H. Adkins, *ibid.*, **3**, 367 (1955)

Aluminum chloride, $AlCl_3$. Mol. wt. 133.35. *See also* Sodium aluminum chloride. Some procedures of the not-too-distant past specified that commercial material·be freshly sublimed or finely ground, or that an excess be taken to allow for the presence of inert material. Finely divided material of high purity is now available in small lots (B, E, F, Mallinckrodt, MCB).

Reaction techniques. In many reactions involving aluminum chloride, this reagent is added in small portions to a solution of the other reactants, with minimal exposure to moist air. A convenient technique is to weigh the aluminum chloride into an Erlenmeyer and to connect this with a section of wide tubing to an opening in the reaction flask as shown in Fig. A-1.[1] To add a portion, the flask is raised and tapped with the fingers.

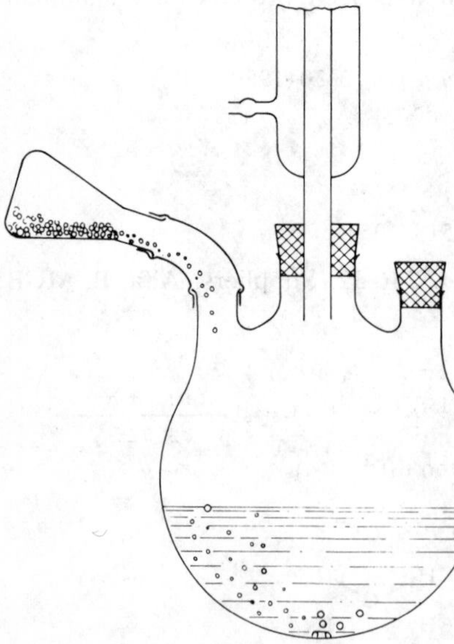

Fig. A-1 Method of adding a solid reagent

Most $AlCl_3$-catalyzed reactions proceed with liberation of gaseous hydrogen chloride. In any but very small-scale experiments it is advisable to connect the top of the condenser to either gas trap shown in Fig. A-2.[2,3]

Since the reaction mixture often becomes very thick, the stirrer shaft should be rigidly attached to an efficient motor; a magnetic stirrer is unsatisfactory.[4] Pearson[5] recommends a stirrer terminating in a stiff, crescent-shaped Teflon (polytetrafluoroethylene) paddle. Satisfactory motors: Sargent cone drive and Waco.

Nitrobenzene is a preferred solvent for some Friedel-Crafts reactions both because it has high solvent power for organic compounds and because it forms a complex with, and dissolves, aluminum chloride. Thus 532 g. of aluminum chloride, added to 2.1 l. of nitrobenzene in 50-g. portions with stirring and with cooling as required, dissolves and remains in solution at room temperature.[6] Removal of nitrobenzene solvent at the end of a reaction can be done efficiently in the apparatus shown in Fig. A-3.[7] The small round-bottomed flask serves as a vapor trap since it invariably retains some liquid through which vapor must bubble. With this apparatus nitrobenzene can be steam distilled at the rate of 400 g./hr.

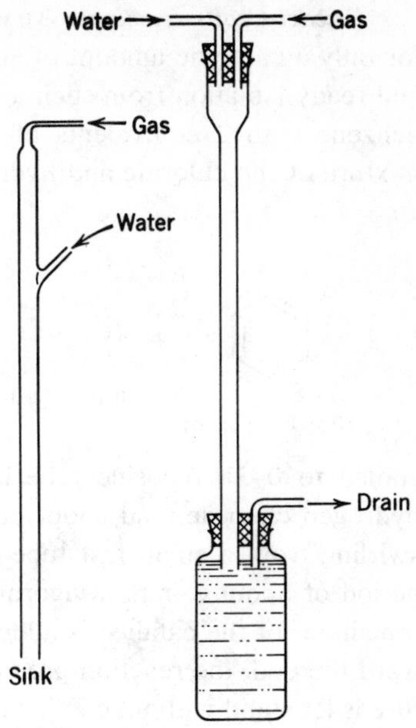
Water
Gas
Gas
Water
Drain
Sink

Fig. A-2 Gas absorption tubes

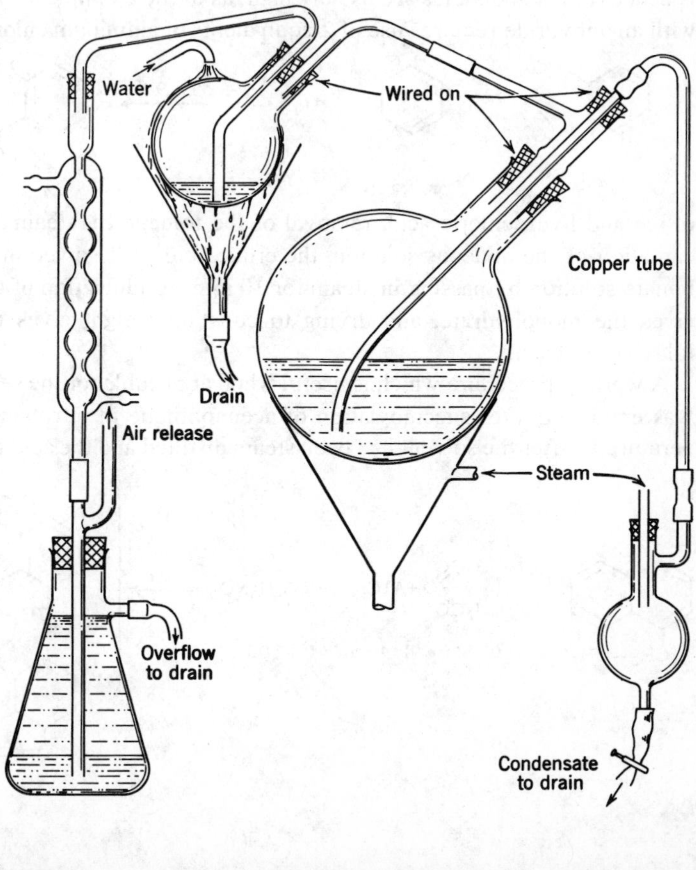
Water
Wired on
Copper tube
Drain
Air release
Steam
Overflow to drain
Condensate to drain

Fig. A-3 Apparatus for steam distillation

Friedel-Crafts reactions. An example of alkylation which demonstrates the need for only a catalytic amount of aluminum chloride, formation of a complex mixture, and ready isolation from such a mixture of a symmetrical product is the reaction of benzene with 2 equivalents of *t*-butyl chloride to produce di-*t*-butylbenzene.[8] A mixture of the chloride and hydrocarbon in a 125-ml. Erlenmeyer with side tube is

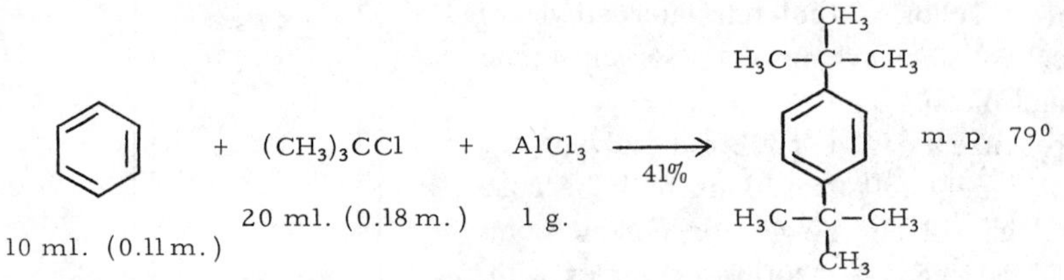

+ $(CH_3)_3CCl$ + $AlCl_3$
41%
m. p. 79°
10 ml. (0.11 m.)
20 ml. (0.18 m.)
1 g.

cooled to 0–3°, the side tube is connected to a plastic water pump to carry off hydrogen chloride, and about one quarter of the aluminum chloride is added with swirling from a small test tube in which it has been weighed. After an induction period of about 2 min. a vigorous reaction sets in at ice-bath temperature, and the remainder of the catalyst is added in three portions at intervals of about 2 min. Toward the end, the reaction product begins to separate as a white solid and the mixture is let stand without cooling for 5 min. and then treated with ice and water. Ether extraction and crystallization from 20 ml. of methanol gives beautiful needles or plates of pure product (8.4 g., 41%). A second small crop (0.8 g.) can be isolated from the mother liquor as the thiourea inclusion complex (3.2 g.).

Friedel-Crafts acylations stop sharply with the introduction of one group, since it is deactivating, and yields are usually high, as in the example formulated.[9] Acylation with an anhydride requires use of 2 equivalents of aluminum chloride. After addition

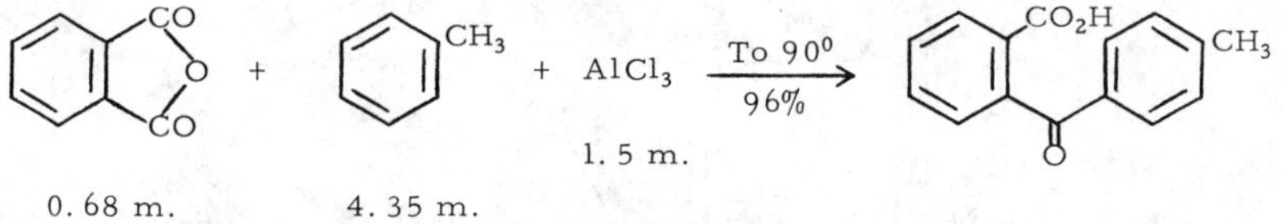

+ $AlCl_3$
To 90°
96%
0.68 m.
4.35 m.
1.5 m.

of ice and hydrochloric acid, removal of the toluene by steam distillation, and decantation of the aqueous solution, the crude acid is dissolved in dilute sodium carbonate solution by passing in steam for 10 min. Acidification of the filtered solution gives the monohydrate, and drying to constant weight gives pure anhydrous *o*-toluylbenzoic acid.

A workup procedure which is useful when applicable for the separation of isomers was employed after succinoylation of acenaphthene in nitrobenzene at a low temperature.[10] After the solvent had been steam distilled and the acidic product dissolved

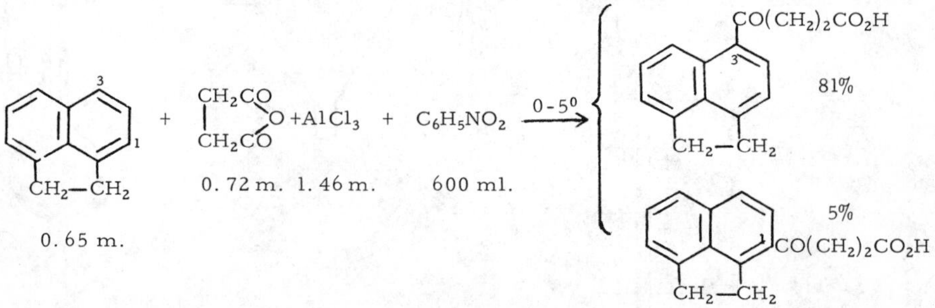

+ $AlCl_3$ + $C_6H_5NO_2$
0-5°
0.65 m.
0.72 m. 1.46 m.
600 ml.
81%
5%

in hot soda solution, sodium chloride was dissolved in the filtered solution at the boiling point. The sodium salt of 3(3′-acenaphthoyl)-propionic acid crystallized selectively and afforded very pure 3-acid. The solubility relationship is reversed with the methyl esters, and esterification of the mother-liquor acid provided a simple route to the pure 1-acid.

In some instances one isomer can be isolated from a mixture as an aluminum chloride complex. In the Perrier procedure for the preparation of 1-benzoylnaphthalene,[11] 14 g. of aluminum chloride is added in portions with swirling to 14 g. of benzoyl chloride and, after an initial exothermal reaction, the solid is all dissolved by gentle heating over a free flame. The resulting yellow melt of addition complex on cooling sets to a yellow or orange crystalline solid. This is dissolved in 80 ml. of carbon disulfide by swirling and gentle heating, and the solution is let cool slightly and 12.8 g. of naphthalene is added in portions. Rapid reaction occurs, with copious evolution of HCl but with little heat effect. After the addition the mixture is heated briefly to complete the reaction and then cooled thoroughly in ice. The aluminum chloride complex of 1-benzoylnaphthalene separates as a dark oil which solidifies on being rubbed with a rod and which is then collected, washed, and added in portions to dilute hydrochloric acid to effect liberation of 1-benzoylnaphthalene, m.p. 76°, yield 18.0 g. (78%). This product is directly pure; 2-benzoylnaphthalene is also formed, but its more soluble aluminum chloride complex is retained in the CS_2 mother liquor (along with other impurities). Although the Friedel-Crafts reaction of benzoyl chloride with phenanthrene affords a mixture of isomers, 1-benzoylphenanthrene can be separated easily from the mixture as the Perrier complex and obtained pure in 8% yield.[12]

On low-temperature acetylation of anthracene in benzene a red complex separates and can be collected on a sintered-glass funnel.[13] This is decomposed by stirring

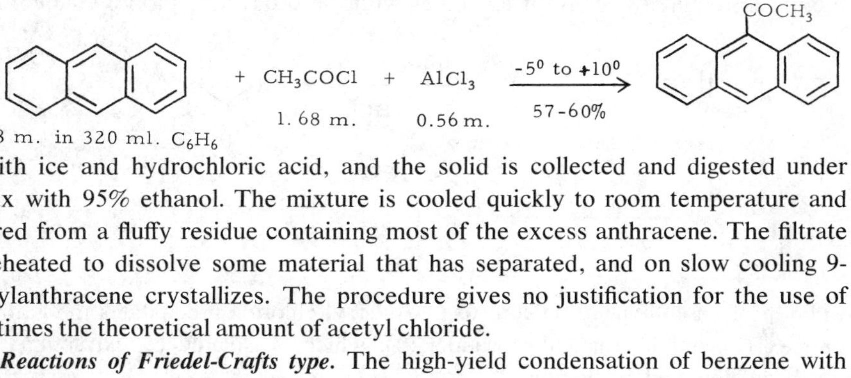
COCH3
+ CH3COCl + AlCl3
-5° to +10°
57-60%
1.68 m.
0.56 m.
0.28 m. in 320 ml. C6H6

it with ice and hydrochloric acid, and the solid is collected and digested under reflux with 95% ethanol. The mixture is cooled quickly to room temperature and filtered from a fluffy residue containing most of the excess anthracene. The filtrate is reheated to dissolve some material that has separated, and on slow cooling 9-acetylanthracene crystallizes. The procedure gives no justification for the use of five times the theoretical amount of acetyl chloride.

Reactions of Friedel-Crafts type. The high-yield condensation of benzene with γ-butyrolactone to give α-tetralone[14] accomplishes in one step what can be done by

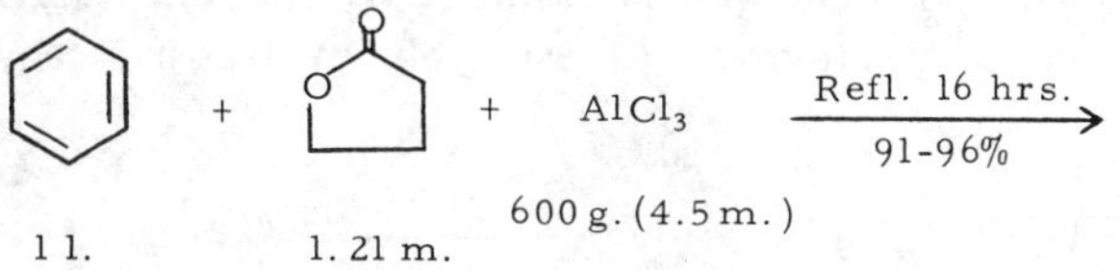
+ + AlCl3
Refl. 16 hrs.
91-96%
600 g. (4.5 m.)
1 l.
1.21 m.

succinoylation, reduction, and cyclization (*see also* γ-Butyrolactone). The amount of aluminum chloride specified is based upon the following results:

$AlCl_3$, g.	400	500	600	700
Yield, %	68	86	94	95

The Fries reaction is illustrated by the conversion of hydroquinone diacetate into acetylhydroquinone. In an improved procedure[15] the reaction is done in two steps.

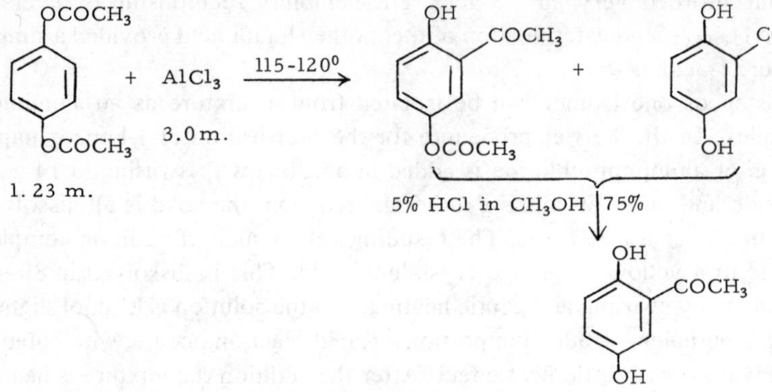
OCOCH3
OCOCH3
1.23 m.
+ AlCl3
3.0 m.
115-120°
OH
COCH3
OCOCH3
+
OH
COCH3
OH
5% HCl in CH3OH
75%
OH
COCH3
OH

Hydroquinone diacetate is ground to 20 mesh and mixed thoroughly with aluminum chloride, and the mixture is added with stirring in three portions to a beaker suspended in an oil bath. After 20 min. the mixture is cooled, ground in a mortar, and added with stirring to ice and water. The product, mainly 2-acetylhydroquinone but containing some acetylhydroquinone monoacetate and a trace of starting material, is washed and dried and stirred with methanol containing 5% hydrogen chloride. On stirring the mixture into ice and water, acetylhydroquinone separates as a bright yellow solid melting at 203–204° (pure: 205–206°), and hydroquinone is retained in solution.

A reaction analogous to a Friedel-Crafts alkylation is the reaction of PCl_3 with benzene to produce phenyldichlorophosphine.[16] At least a full mole of aluminum chloride is required because it combines with the product to form a complex (1).

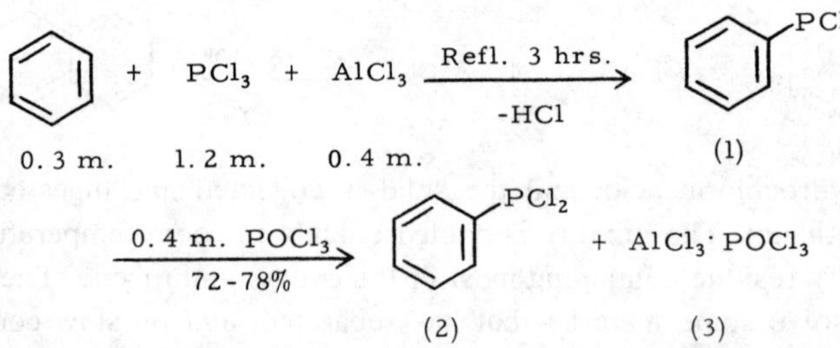
+ PCl3 + AlCl3
Refl. 3 hrs.
-HCl
PCl2· AlCl3
0.3 m.
1.2 m.
0.4 m.
(1)
0.4 m. POCl3
72-78%
PCl2
+ AlCl3· POCl3
(2)
(3)

Addition of an equivalent amount of phosphoryl chloride precipitates the granular complex (3) and liberates the product (2), which is isolated by extraction with petroleum ether and distillation.

Carboxylation. The reaction of mesitylene with oxalyl chloride catalyzed by aluminum chloride to produce mesitoic acid[17] presumably involves monoacylation,

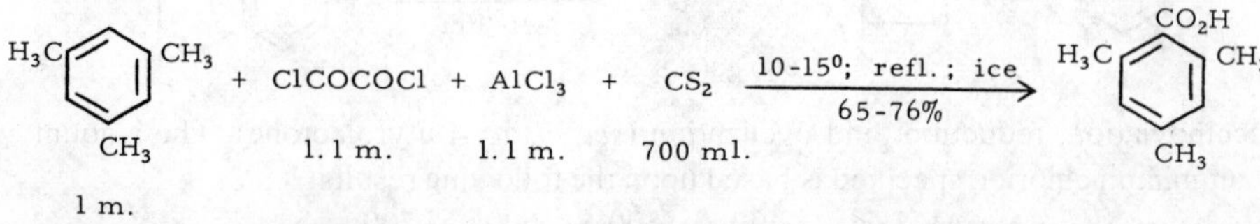
H3C
CH3
CH3
1 m.
+ ClCOCOCl + AlCl3 + CS2
1.1 m.
1.1 m.
700 ml.
10-15°; refl.; ice
65-76%
CO2H
H3C
CH3
CH3

hydrolysis, and decarbonylation, but the order in which the last two reactions occur is not clear. Diacylation with oxalyl chloride is accomplished by reaction with N,N-dimethylaniline in large excess.[4]

$$2 \ C_6H_5N(CH_3)_2 \ + \ ClCOCOCl \ + \ AlCl_3 \ + \ CS_2 \xrightarrow[38-42\%]{5°, \ then \ refl.}$$

1. 5 m.   0. 25 m.   1 m.   200 ml.

$$(CH_3)_2N-\underset{}{\bigcirc}-\overset{O}{\underset{}{C}}-\overset{O}{\underset{}{C}}-\underset{}{\bigcirc}-N(CH_3)_2$$

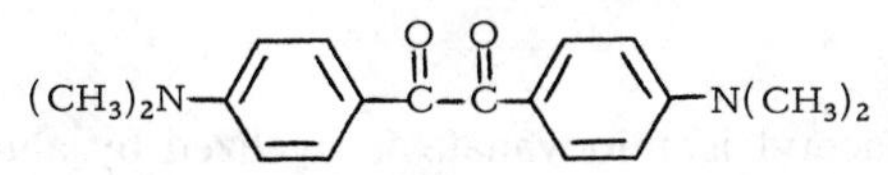

Addition reactions. Refluxing with aluminum chloride effects addition of chloroform to tetrachloroethylene to give *unsym*-heptachloropropane, isolated by cooling,

$$CCl_2{=}CCl_2 \ + \ CHCl_3 \ + \ AlCl_3 \xrightarrow[88-93\%]{Refl. \ 15 \ hrs.} CCl_3CCl_2CHCl_2$$

1 m.   2. 5 m.   0. 2 m.

adding ice, washing the organic layer, drying, and distilling (b.p. 110–113°/10 mm., m.p. 29–30°).[18] Note that tetrachloroethylene is very resistant to addition of dichlorocarbene generated from sodium trichloroacetate (*see* Sodium trichloroacetate).

Aluminum chloride catalyzes addition of benzene to benzalacetophenone to give β,β-diphenylpropiophenone[19] and of aniline to benzonitrile to give N-phenylbenzamidine.[20]

$$C_6H_5CH{=}CHCOC_6H_5 \ + \ C_6H_6 \ + \ AlCl_3 \xrightarrow[76-85\%]{10° \ then \ 25°} (C_6H_5)_2CHCH_2COC_6H_5$$

0. 58 m. in 300 ml. C$_6$H$_6$ 1. 7 l.   1. 2 m.

$$C_6H_5C{\equiv}N \ + \ C_6H_5NH_2 \ + \ AlCl_3 \xrightarrow[69-74\%]{30 \ min. \ at \ 200°} C_6H_5\overset{NH}{\overset{\|}{C}}NHC_6H_5$$

0. 66 m.   0. 67 m.   0. 67 m.

Addition of isocaproyl chloride to acetylene to produce β-chlorovinyl isoamyl ketone is accomplished by saturating carbon tetrachloride with acetylene, adding

$$(CH_3)_2CHCH_2CH_2COCl \ + \ HC{\equiv}CH \ + \ AlCl_3 \ + \ CCl_4 \xrightarrow[54-64\%]{} (CH_3)_2CHCH_2CH_2\overset{O}{\overset{\|}{C}}CH{=}CHCl$$

0. 63 m.   Excess   0. 74 m.   260 g.

aluminum chloride, stirring, and dropping in the acid chloride while passing in acetylene.[21]

Cyclodehydration. In a procedure[6] for the synthesis of 4-methylcoumarin based on one by Sethna, Shah, and Shah[22] a mixture of phenol, ethyl acetoacetate, and

$$\underset{2 \ m.}{\bigcirc_{OH}} \ + \ \underset{2 \ m.}{\overset{CH_3}{\underset{\underset{OC_2H_5}{C=O}}{\overset{CO}{\overset{|}{CH_2}}}}} \ + \ \underset{300 \ ml.}{C_6H_5NO_2} \ + \ \underset{\substack{4 \ m. \ in \ 2.1 \ l. \\ of \ C_6H_5NO_2}}{AlCl_3} \xrightarrow[33-48\%]{130°}$$

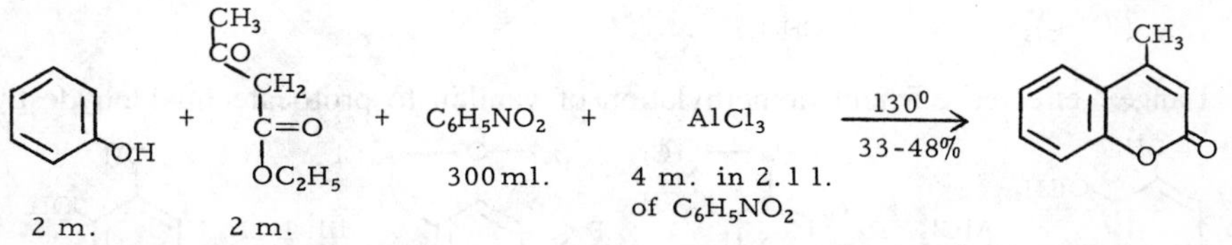

nitrobenzene is stirred, heated to 100°, treated with a solution of aluminum chloride in nitrobenzene, added in 30–45 min., and heated for 3 hrs. at 130°.

The reaction of benzilic acid with aluminum chloride in benzene to give fluorene-9-carboxylic acid[23] involves cyclodehydration with rearrangement.

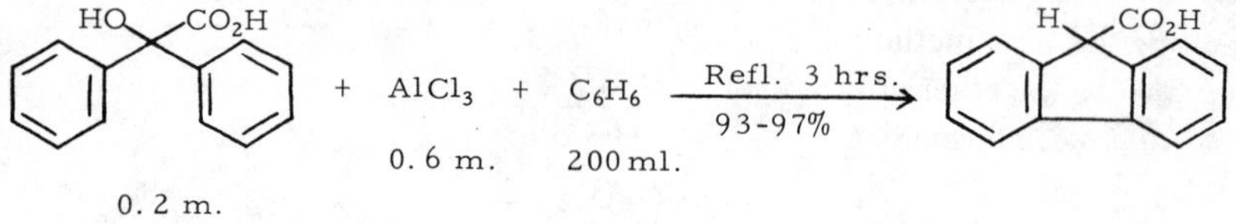
HO
CO2H
+ AlCl3 + C6H6
Refl. 3 hrs.
93-97%
0. 6 m.
200 ml.
0. 2 m.
H
CO2H

Cyclization. Phenylacetyl isothiocyanate is cyclized by aluminum chloride to 2 α-thiohomophthalimide.[24]

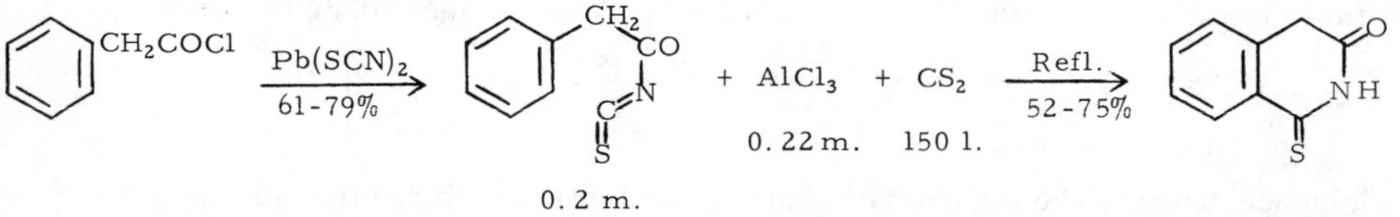
CH2COCl
Pb(SCN)2
61-79%
CH2
CO
N
C
S
+ AlCl3 + CS2
0. 22 m.
150 l.
0. 2 m.
Refl.
52-75%
NH
S

Ether cleavage. Aluminum chloride is a classical reagent for demethylation: $ArOCH_3 + AlCl_3 \longrightarrow ArOAlCl_2 + CH_3Cl$. The procedure formulated for the demethylation of 3-bromocatechol dimethyl ether[25] is applicable also to 2,3-dimethoxydiphenyl.[26]

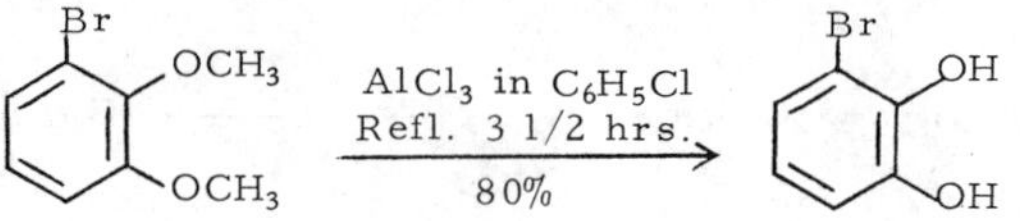
Br
OCH3
OCH3
AlCl3 in C6H5Cl
Refl. 3 1/2 hrs.
80%
Br
OH
OH

Benzene is also used as solvent but requires a longer reflux period:[27]

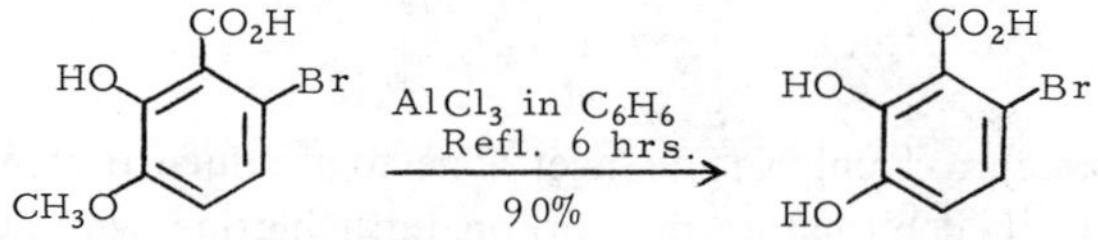
CO2H
HO
Br
CH3O
AlCl3 in C6H6
Refl. 6 hrs.
90%
CO2H
HO
Br
HO

Kulka[28] developed a synthesis of *o*-hydroxybenzophenones in which a *t*-butyl group is used to block the *para* position of a phenol methyl ether to permit *o*-benzoylation and is then removed, along with the ether group, with aluminum chloride.

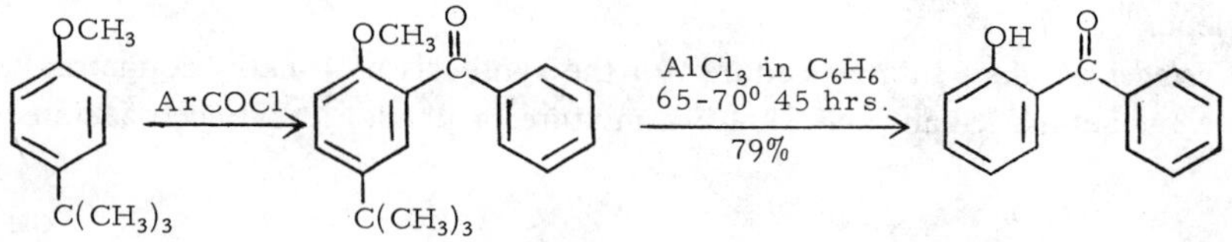
OCH3
ArCOCl
C(CH3)3
OCH3
O
C
C(CH3)3
AlCl3 in C6H6
65-70° 45 hrs.
79%
OH
O
C

Lange[29] effected efficient demethylation of vanillin to protocatechualdehyde by

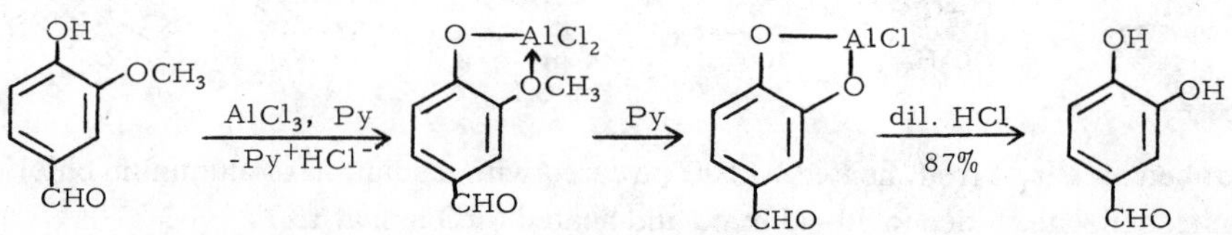
OH
OCH3
CHO
AlCl3, Py
-Py+HCl-
O—AlCl2
OCH3
CHO
Py
O—AlCl
O
CHO
dil. HCl
87%
OH
OH
CHO

adding 4.4 equivalents of pyridine to a suspension of aluminum chloride (1.1 equiv.) in a solution of vanillin (1 equiv.) in methylene chloride and refluxing the mixture for 24 hrs. Since demethylation of vanillin by conventional methods proceeds poorly, and since the new method is applicable only to compounds having a free hydroxyl group next to the ether group, the reaction is regarded as involving a solvated cyclic five-membered intermediate. The failure of *o*-vanillin to undergo cleavage under the

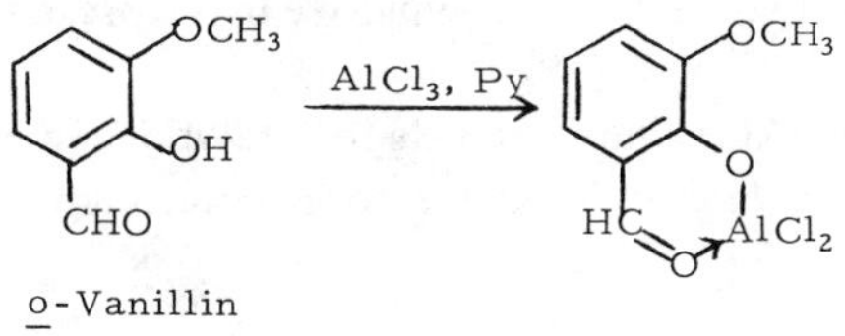
OCH3
OH
CHO
o-Vanillin
AlCl3, Py
OCH3
O
HC
AlCl2
O

same conditions is attributed to complexing of aluminum with the adjacent carbonyl oxygen rather than with the less basic methoxyl oxygen.

Hardegger *et al.*[30] found that a phenolic ether group *ortho* or *peri* to a carbonyl group is selectively cleaved by treatment with aluminum chloride in nitrobenzene.

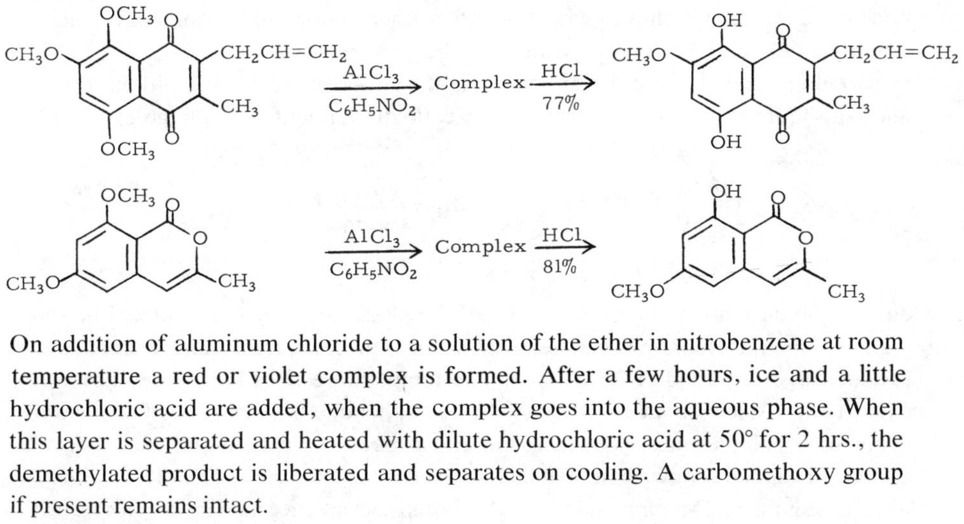
OCH3
CH3O
CH2CH=CH2
CH3
OCH3
AlCl3
C6H5NO2
Complex
HCl
77%
OH
CH3O
CH2CH=CH2
CH3
OH
OCH3
CH3O
CH3
AlCl3
C6H5NO2
Complex
HCl
81%
OH
CH3O
CH3

On addition of aluminum chloride to a solution of the ether in nitrobenzene at room temperature a red or violet complex is formed. After a few hours, ice and a little hydrochloric acid are added, when the complex goes into the aqueous phase. When this layer is separated and heated with dilute hydrochloric acid at 50° for 2 hrs., the demethylated product is liberated and separates on cooling. A carbomethoxy group if present remains intact.

Gattermann-Koch reaction. For the conversion of toluene into *p*-tolylaldehyde, gaseous carbon monoxide and hydrogen chloride are passed into a stirred suspension of aluminum chloride and cuprous chloride.[31]

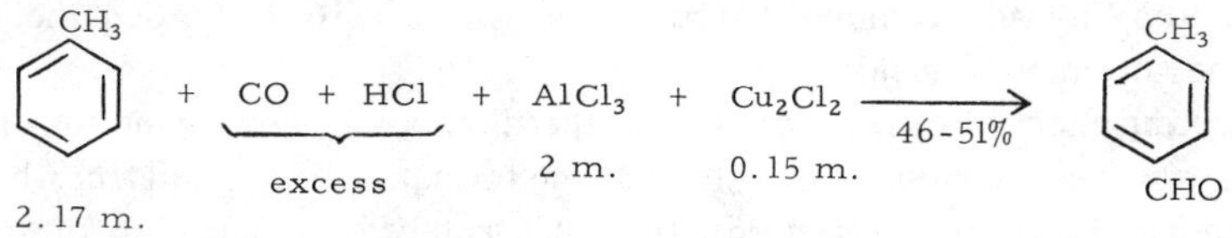
CH3
2.17 m.
+ CO + HCl + AlCl3 + Cu2Cl2
excess
2 m.
0.15 m.
46-51%
CH3
CHO

Diels-Alder catalyst. Certain Diels-Alder reactions are markedly accelerated by aluminum chloride[32,33] and by other Friedel-Crafts type catalysts[33] ($SnCl_4$, BF_3, $FeCl_3$, $TiCl_4$). Thus the reaction of equimolecular amounts of anthracene, dimethyl fumarate, and aluminum chloride in methylene chloride is complete at room temperature in less than 2 hrs.; in refluxing dioxane in the absence of catalyst the reaction requires 2–3 days.[32] Without catalyst, the reaction of butadiene with methyl vinyl ketone affords 4-acetylcyclohexene in yield of 75–80% after heating the reactants in a sealed tube at 140° for 8–10 hrs.; in a benzene solution containing 0.18 equivalent of stannic chloride the adduct was obtained in 73% yield after 1 hr. at room temperature.[33]

However, catalysis of Diels-Alder reactions is satisfactory only with specific dienophiles, for example, the two cited in the examples, acrolein, and acrylic acid. Another limitation is that some dienes polymerize in the presence of a Friedel-Crafts catalyst.

Bromination catalyst. Aluminum chloride in catalytic amount promotes bromination of the methyl group of acetophenone to give phenacyl bromide.[34] Ether and

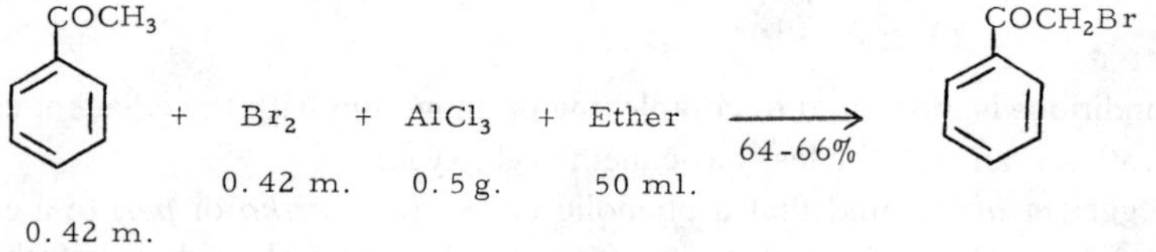
COCH3
+ Br2 + AlCl3 + Ether
0.42 m. 0.5 g. 50 ml.
0.42 m.
64-66%
COCH2Br

dissolved HBr are removed at reduced pressure, and the residual brownish-yellow crystalline material is shaken with 1:1 water–ether to remove the color. Crystallization from methanol then gives satisfactory phenacyl bromide.

Pearson[5,35] found that if the bromination of acetophenone is done without solvent and with aluminum chloride taken not in catalytic amount but in large excess (2.5

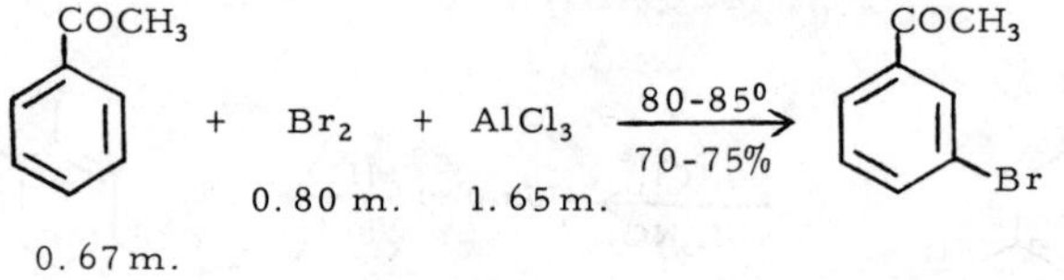
COCH3
+ Br2 + AlCl3
0.80 m. 1.65 m.
0.67 m.
80-85°
70-75%
COCH3
Br

equiv.) the reaction product is *m*-bromoacetophenone, which is obtained in good yield. Aluminum chloride is agitated with a stout stirrer assembly, and acetophenone is added over a period of 20–30 min. The temperature rises but is not allowed to exceed 180°. When about one third of the acetophenone has been added the mixture becomes a ball-like mass, but near the end it becomes molten and easily stirred. Bromine is added dropwise in 40 min. and the mixture is stirred at 80–85° for 1 hr. or until it solidifies. Decomposition of the complex with ice and hydrochloric acid liberates the product. In further exploitation of his "swamping catalyst effect," Pearson[36] applied the method with success to the nuclear halogenation of aromatic acid esters, chlorides, and nitriles. Methyl benzoate, for example, on bromination as the aluminum chloride complex (2 equiv. of $AlCl_3$) at 65° afforded methyl *m*-bromobenzoate in 86% yield.

The striking effect of excess catalyst on the direction of bromination is clarified by results obtained by House[37] in a study of the formation of *cis*- and *trans*-hexahydrofluorenone, formulas (2) and (6), by $AlCl_3$-catalyzed cyclization of *trans*-2-phenylcyclohexanecarboxylic acid chloride (1). The *trans*-ketone (2, m.p. 92°) is less stable than the *cis*-isomer (6, m.p. 41°), and when aluminum chloride is added to a CS_2 solution of the acid chloride the reaction product is the stable *cis*-ketone (6). The reaction is interpreted as involving initial formation of the *trans*-ketone (2), its conversion to the conjugate acid (3), and the conversion of (3) to the enolate (4) by abstraction of a proton by one of the basic species present, such as the uncomplexed

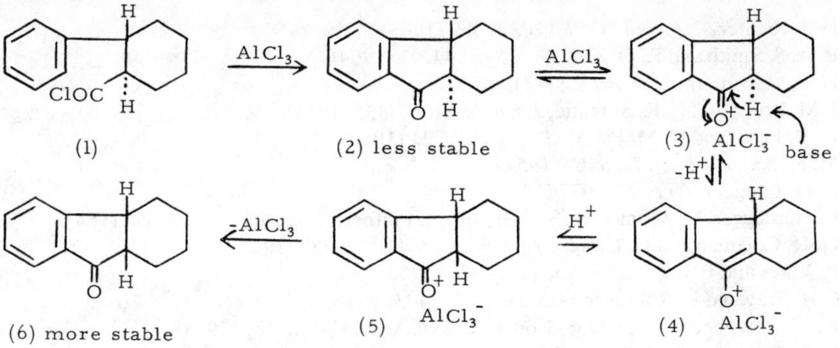
AlCl$_3$
AlCl$_3$
ClOC
(1)
(2) less stable
(3) AlCl$_3^-$
base
-H$^+$
-AlCl$_3$
H$^+$
(6) more stable
(5) AlCl$_3^-$
(4) AlCl$_3^-$

ketone (2) or the uncomplexed acid chloride (1). Once the enolate (4) is present in equilibrium it will revert less to (3) than to the *cis*-conjugate acid (5), which by loss of aluminum chloride affords the stable *cis*-ketone (6). When the acid chloride (1) is added to aluminum chloride, which is therefore present in excess, the product obtained on decomposition of the complex with water is the less stable *trans*-ketone (2). In this case the ketone can form the conjugate acid (3), in which the original stereochemistry is preserved, but basic species are present in amounts insufficient for abstraction of a proton for conversion to the enolate (4) and hence isomerization is not realized. The swamping catalyst effect in the bromination of acetophenone is thus interpreted as follows. Excess catalyst converts the ketone into the conjugate acid (7) and scavenges all basic species for conversion to the enolate (8) required for

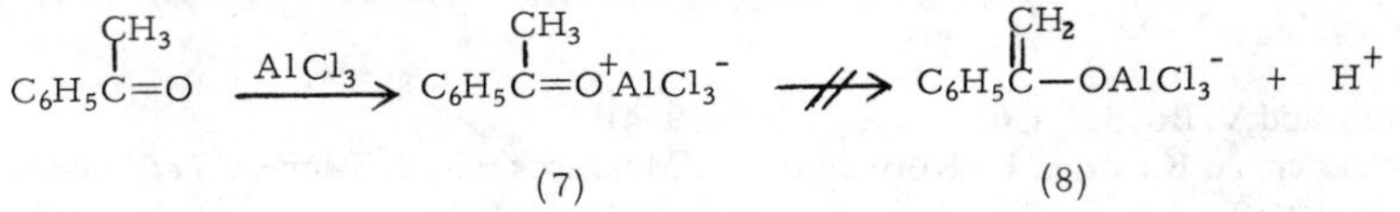
CH$_3$
C$_6$H$_5$C=O
AlCl$_3$
C$_6$H$_5$C=O$^+$AlCl$_3^-$
CH$_2$
C$_6$H$_5$C—OAlCl$_3^-$ + H$^+$
(7)
(8)

side-chain bromination. The excess catalyst also enhances the electrophilic activity of bromine through complex formation, $Br^+[AlCl_3Br]^-$.

[1] L. F. Fieser, "Experiments in Organic Chemistry," 3rd Ed , 265, D. C. Heath (1955)
[2] C. F. H. Allen, *Org. Syn., Coll. Vol.*, **2**, 4 (1943)
[3] J. R. Johnson, *ibid.*, **1**, 97 (1941)
[4] C. Tüzün, M. Ogliaruso, and E. I. Becker, *Org. Syn.*, **41**, 1 (1961)
[5] D. E. Pearson, H. W. Pope, and W. W. Hargrove, *ibid.*, **40**, 7 (1960)
[6] E. H. Woodruff, *Org. Syn., Coll. Vol.*, **3**, 581 (1955)
[7] Ref. 1, p. 256
[8] *Org. Expts.*, 184
[9] L. F. Fieser, *Org. Syn., Coll. Vol.*, **1**, 517 (1941)
[10] *Idem, ibid.*, **3**, 6 (1955)
[11] G. Perrier, *Bull. soc.*, [3], **31**, 859 (1904); the example cited is from L. F. Fieser, "Experiments in Org. Chem," 2nd Ed., 192 (1941)
[12] W. E. Bachmann, *Am. Soc.*, **57**, 555 (1935)
[13] C. Merritt, Jr., and C. E. Braun, *Org. Syn., Coll. Vol.*, **4**, 8 (1963)
[14] C. E. Olson and A. R. Bader, *ibid.*, **4**, 898 (1963)
[15] D. D. Reynolds, J. A. Cathcart, and J. L. R. Williams, *J. Org.*, **18**, 1709 (1953)
[16] B. Buchner and L. B. Lockhart, Jr., *Org. Syn., Coll. Vol.*, **4**, 784 (1963)
[17] P. E. Sokol, *Org. Syn.*, **44**, 69 (1964)
[18] M. W. Farlow, *Org. Syn., Coll. Vol.*, **2**, 312 (1943)
[19] P. R. Shildneck, *ibid.*, **2**, 236 (1943)
[20] F. C. Cooper and M. W. Partridge, *ibid.*, **4**, 769 (1963)
[21] C. C. Price and J. A. Pappalardo, *ibid.*, **4**, 186 (1963)
[22] S. M. Sethna, N. M. Shah, and R. C. Shah, *J. Chem. Soc.*, 228 (1938)

[23]H. J. Richter, *Org. Syn., Coll. Vol.*, **4**, 482 (1963)
[24]P. A. S. Smith and R. O. Kan, *Org. Syn.*, **44**, 91 (1964)
[25]H. S. Mason, *Am. Soc.*, **69**, 2241 (1947)
[26]J. M. Bruce and F. K. Sutcliffe, *J. Chem. Soc.*, 4435 (1955)
[27]G. R. Pettit and D. M. Piatak, *J. Org.*, **25**, 721 (1960)
[28]M. Kulka, *Am. Soc.*, **76**, 5469 (1954)
[29]R. G. Lange, *J. Org.*, **27**, 2037 (1962)
[30]E. Hardegger, E. Widmer, K. Steiner, and A. Pfiffner, *Helv.*, **47**, 2027, 2031 (1964)
[31]G. H. Coleman and D. Craig, *Org. Syn., Coll. Vol.*, **2**, 583 (1943)
[32]P. Yates and P. Eaton, *Am. Soc.*, **82**, 4436 (1960)
[33]G. I. Fray and R. Robinson, *Am. Soc.*, **83**, 249 (1961)
[34]R. M. Cowper and L. H. Davidson, *Org. Syn., Coll. Vol.*, **2**, 480 (1943)
[35]D. E. Pearson, H. W. Pope, W. W. Hargrove, and W. E. Stamper, *J. Org.*, **23**, 1412 (1958)
[36]D. E. Pearson, W. E. Stamper, and B. R. Suthers, *J. Org.*, **28**, 3147 (1963)
[37]H. O. House, V. Paragamian, R. S. Ro, and D. J. Wluka, *Am. Soc.*, **82**, 1457 (1960)

Aluminum cyclohexoxide, $Al(OC_6H_{11})_3$. Mol. wt. 324.42.

When refluxed for 48 hrs. with a solution of aluminum in cyclohexanol, anthraquinone affords anthracene in about 80% yield.[1] This method provides a simple route to pure pentacene[2] from pentacene-6,13-quinone, readily prepared from phthalaldehyde and 1,4-cyclohexanedione.

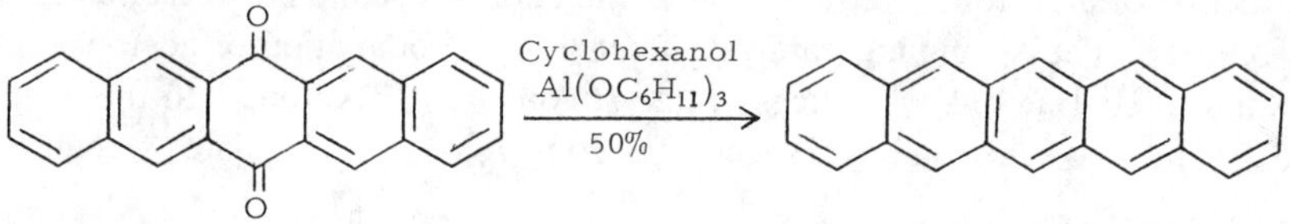
O
O
Cyclohexanol
Al(OC6H11)3
50%

[1]S. Coffey and V. Boyd, *J. Chem. Soc.*, 2468 (1954)
[2]V. Bruckner, A. Karczag, K. Körmendy, M. Meszaros, and J. Tomasz, *Tetrahedron Letters*, No. 1, 5 (1960)

Aluminum hydride, AlH_3. Mol. wt. 30.01. *See also* $LiAlH_4$–$AlCl_3$.

The reagent is prepared in ether solution from lithium aluminum hydride and aluminum chloride.[1] Initial studies[1] indicated that the reagent is similar in reducing

$$3\ LiAlH_4 + AlCl_3 \longrightarrow 4\ AlH_3 + 3\ LiCl$$

action to lithium aluminum hydride, but Jorgenson[2] showed that it is superior to $LiAlH_4$ for the selective reduction of α,β-unsaturated aldehydes to allylic alcohols,

$$C_6H_5CH{=}CHCHO \xrightarrow{AlH_3} C_6H_5CH{=}CHCH_2OH$$

for example: lithium aluminum hydride tends to reduce the double or triple bonds of allylic or propargylic alcohols, but with aluminum hydride no control is required to prevent such reduction.

A cooled solution of lithium aluminum hydride in ether is treated with sufficient aluminum chloride to give more than the stoichiometrically required amount of aluminum hydride, and the mixture is stirred at room temperature until the aluminum chloride has dissolved and precipitation of lithium chloride is complete. An ethereal solution of the substrate is added, and the reaction is allowed to proceed at room temperature for 30 min. (or for 5 hrs.). The presence of a large amount of $LiAlH_4$ in the solution does not change the course of the AlH_3 reduction. Methyl *o*-coumarate could not be reduced to *o*-hydroxycinnamyl alcohol by $LiAlH_4$ under carefully chosen conditions but was reduced by AlH_3 to the pure unsaturated alcohol in 75% yield.

Brown and Yoon[3] have prepared relatively stable solutions of the reagent in tetrahydrofurane by addition of 0.5 mole of 100% sulfuric acid to 1 mole of $LiAlH_4$:

$$2\ LiAlH_4 + H_2SO_4 \xrightarrow{THF} Li_2SO_4 + 2\ AlH_3 + 2\ H_2$$

Hydrogen is evolved and lithium sulfate is precipitated quantitatively. The reagent is comparable to $LiAlH_4$ in reducing power and in general offers no advantages except for the selective reduction of the carbonyl group of dienones.

[1]E. Wiberg, *Angew. Chem.*, **63**, 485 (1951)
[2]M. J. Jorgenson, *Tetrahedron Letters*, 559 (1962)
[3]H. C. Brown and N. M. Yoon, *Am. Soc.*, **88**, 1464 (1966)

Aluminum iodide, AlI_3. Mol. wt. 407.72. Suppliers: Alfa, Fisher.

Reagent generated in solution by reaction of amalgamated aluminum with iodine under ether converts cholesterol into cholesteryl iodide in good yield.[1] Aluminum powder is shaken briefly with a 1% solution of mercuric chloride in ether, and the resulting grey powder is washed by decantation and treated gradually with iodine in ether. The mixture is refluxed until solution is complete, cholesterol is added, and the solution is refluxed for 8 hrs. Under similar conditions cholestane-3β-ol was recovered unchanged.

A later paper[2] which does not mention this work states that aluminum iodide is so sensitive to air and moisture that a special apparatus is required for its preparation. As a Friedel-Crafts catalyst, the reagent was found generally less effective than the chloride or bromide.

[1]J. Broome, B. R. Brown, and G. H. R. Summers, *J. Chem. Soc.*, 2071 (1957)
[2]E. R. Kline, B. N. Campbell, Jr., and E. C. Spaeth, *J. Org.*, **24**, 1781 (1959)

Aluminum isopropoxide, $Al[OCH(CH_3)_2]_3$. Mol. wt. 204.24, m.p. 118°, b.p. 130–140°/7 mm. Preparation and uses.[1] Suppliers: Alfa, B, E, F, MCB, Peninsular ChemResearch, Chatten Chemicals.

Meerwein-Ponndorf reduction. One use of the reagent is for the reduction of carbonyl compounds, particularly of unsaturated aldehydes and ketones, for the reagent attacks only carbonyl compounds. An example is the reduction of crotonaldehyde to crotyl alcohol.[1] A mixture of 27 g. of cleaned aluminum foil, 300 ml. of isopropanol, and 0.5 g. of mercuric chloride is heated to boiling, 2 ml. of carbon tetrachloride is added as catalyst, and heating is continued. The mixture turns grey, and vigorous evolution of hydrogen begins. Refluxing is continued until gas evolution has largely subsided (6–12 hrs.). The solution, which is black from the presence of suspended solid, can be concentrated and the aluminum isopropoxide distilled in vacuum (colorless liquid) or used as such. Thus the undistilled solution

$$CH_3CH{=}CHCHO + (CH_3)_2CHOH \xrightarrow[60\%]{Al[OCH(CH_3)_2]_3} CH_3CH{=}CHCH_2OH + (CH_3)_2CO$$

prepared as described from 1.74 moles of aluminum and 500 ml. of isopropanol is treated with 3 moles of crotonaldehyde and 1 l. of isopropanol. On being refluxed at a bath temperature of 110°, acetone slowly distils at 60–70°. After 8–9 hrs., when the distillate no longer gives a test for acetone, most of the remaining isopropanol is distilled at reduced pressure and the residue cooled and hydrolyzed with 6 *N* sulfuric acid to liberate crotyl alcohol from its aluminum derivative.

Reduction: $Ar_2CO \longrightarrow Ar_2CH_2$. Wendler *et al.*[2] suggested that the Meerwein-Ponndorf reduction of the ketone (1) involves formation of a cyclic coordination complex (3) which, by hydrogen transfer, affords the mixed alkoxide (4), hydrolyzed to the alcohol (5). Further reflection[3] suggested that under forcing conditions it might

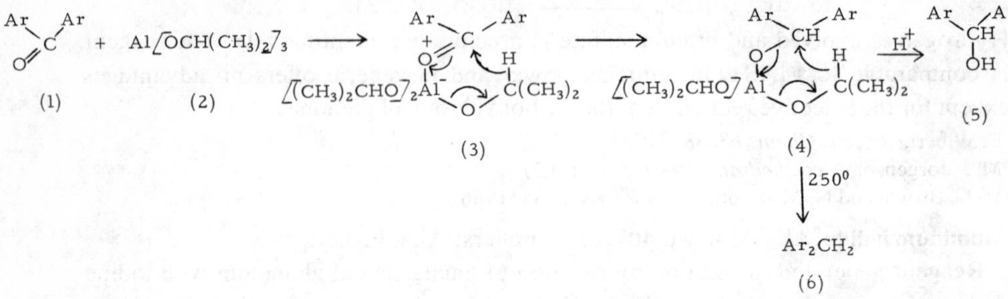

be possible to effect repetition of the hydrogen transfer and produce the hydrocarbon (6). Trial indeed showed that reduction of diaryl ketones can be effected efficiently by heating with excess reagent at 250°. Examples:

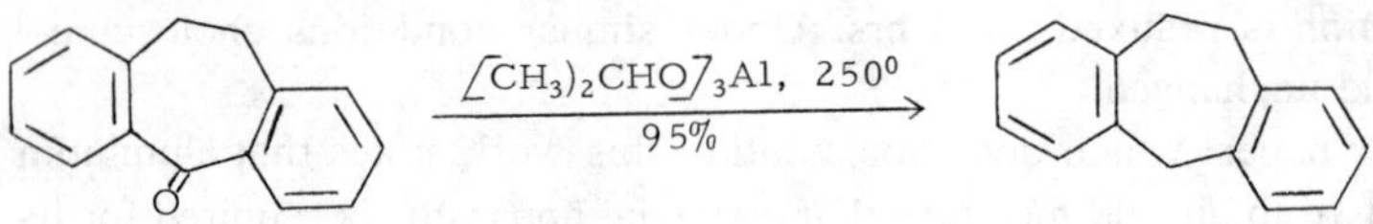

Anthraquinone ⟶ Anthracene (75%)

Anthrone ⟶ Anthracene (92%)

Oppenauer oxidation.[4] Eastham and Teranishi[5] prepared cholestenone by oxidation of cholesterol in toluene solution with aluminum isopropoxide as catalyst

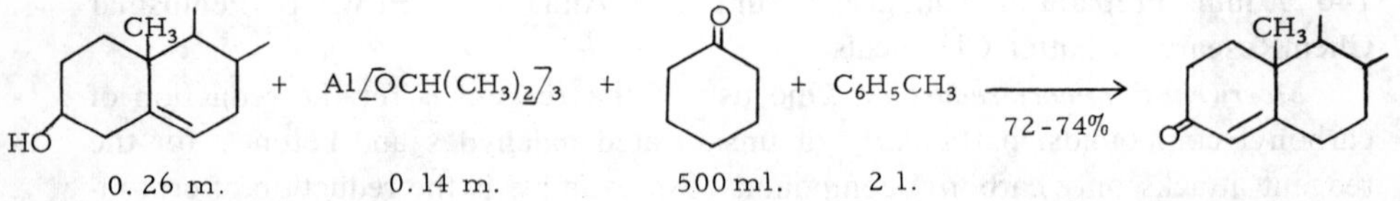

and cyclohexanone as hydrogen acceptor. The mixture was heated with stirring, and a total of 900 ml. of toluene was distilled. Aqueous potassium–sodium tartrate was added to keep aluminum ion in solution, and the mixture was steam distilled until about 6 l. of distillate had been collected. The product was then collected by extraction with chloroform and crystallized twice to give cholestenone of satisfactory purity.

Eastham and Teranishi compare their procedure with that described earlier by Oppenauer using aluminum *t*-butoxide (*which see*), acetone, and benzene with the comment "The present modification is less laborious than the earlier method," but we question this conclusion. Eight hours of unattended refluxing (Oppenauer) seems significantly less laborious than the lengthy distillation and steam distillation cited above.

A Syntex group[6] found that a formate, unlike an acetate, is easily oxidized and gives the same product as the free alcohol. For oxidation of (1) to (2) they used the combination of cyclohexanone and aluminum isopropoxide and a hydrocarbon

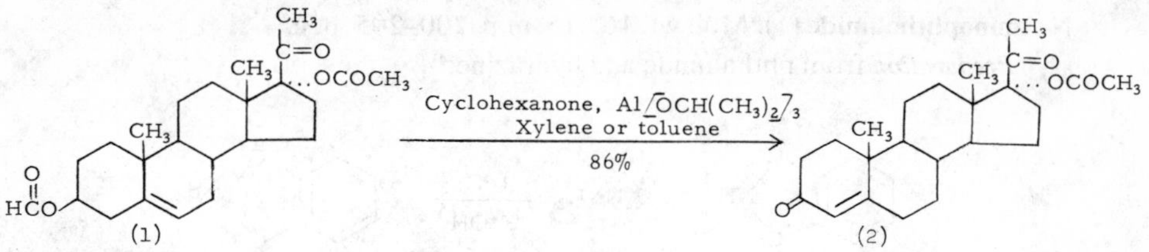

solvent: xylene (b.p. 140° at 760 mm.) in Mexico City (570 mm.) or toluene (b.p. 111° at 760 mm.).

[1]A. L. Wilds, "Reduction with Aluminum Alkoxides," *Org. Reactions*, **2**, 178 (1944)
[2]R. B. Woodward, N. L. Wendler, and F. J. Brutschy, *Am. Soc.*, **67**, 1425 (1945)
[3]R. D. Hoffsommer, D. Taub, and N. L. Wendler, *Chem. Ind.*, 482 (1964)
[4]C. Djerassi, "The Oppenauer Oxidation" *Org. Reactions*, **6**, 207 (1951)
[5]J. F. Eastham and R. Teranishi, *Org. Syn., Coll. Vol.*, **4**, 192 (1963)
[6]H. J. Ringold, B. Löken, G. Rosenkranz, and F. Sondheimer, *Am. Soc.*, **78**, 816 (1956)

***p*-Aminoazobenzene,** $C_6H_5N{=}NC_6H_4NH_2$-*p*. Mol. wt. 197.23, m.p. 128°. Suppliers: B, E, F. Diacids of the succinic and glutaric type produced on oxidation of natural products are often difficult to isolate or purify, either because of unfavorable solubility relationships or because a contaminant prevents crystallization. A useful expedient is to convert the crude diacid into the anhydride (Ac_2O or AcCl), form the anilic acid (2) by reaction in chloroform with *p*-aminoazobenzene, and cyclize the

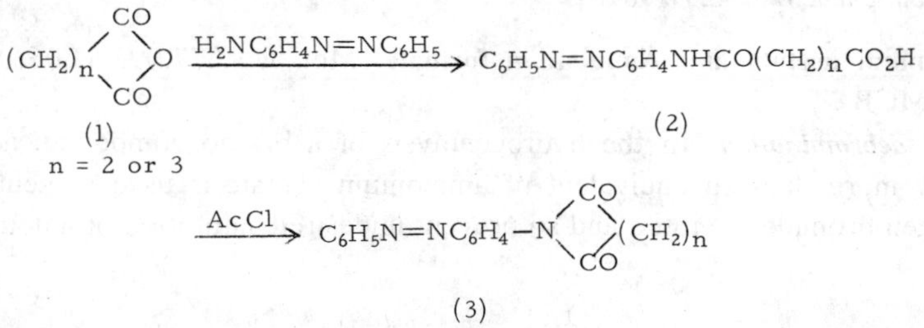

product with acetyl chloride to the *p*-phenylazoanil (3).[1] Chromatographed on alumina, the anils form deep yellow bands and are easily separated; their light-absorption properties (UV and IR) facilitate estimation and identification.

[1]H. B. Henbest and T. C. Owen, *J. Chem. Soc.*, 2968 (1955)

1-Aminobenzotriazole, 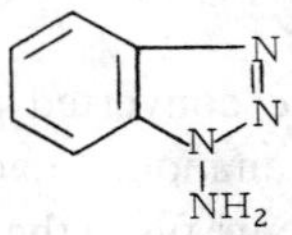

Mol. wt. 134.14, m.p. 82°. For preparation, *see* Hydroxylamine-O-sulfonic acid; for generation of benzyne, *see* Lead tetraacetate.

2-Amino-2-methyl-1-propanol, $(CH_3)_2C(NH_2)CH_2OH$. Mol. wt. 89.14, m.p. 25–29°, neut. equiv. 88.5–99.0. Suppliers: Commercial Solvents Corp., A, B, E, F, MCB.

Harris and Sanderson[1] found this amine to be highly specific for precipitation of levopimeric acid from oleoresins. The amine salt can be decomposed with either boric acid or phosphoric acid. A detailed procedure is described by Lloyd and Hedrick.[2]

[1]G. C. Harris and T. F. Sanderson, *Am. Soc.*, **70**, 334 (1948); *see also* V. M. Roeblich, D. E. Baldwin, R. T. O'Connor, and R. V. Lawrence, *ibid.*, **77**, 6311 (1955)
[2]W. D. Lloyd and G. W. Hedrick, *Org. Syn.*, **45**, 65 (1965)

N-Aminophthalimide (1). Mol. wt. 162.15, m.p. 200–205° dec.

Preparation from phthalimide and hydrazine:[1]

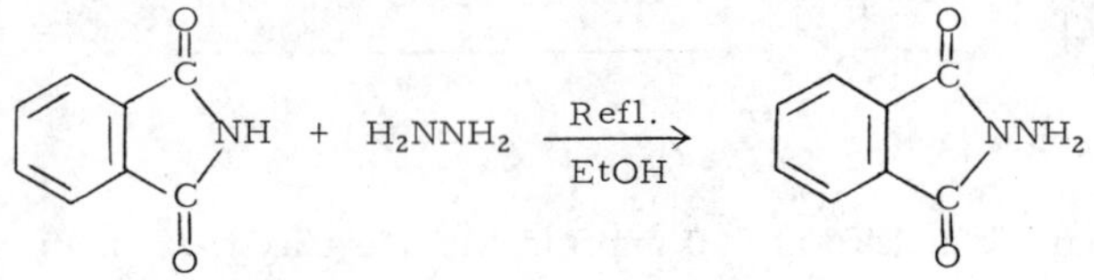
NH
H_2NNH_2
Refl.
EtOH
NNH_2

(1)

N-Aminopyrroles.[2] The reagent condenses with acetonylacetone in hot acetic acid to give N-phthalimido-2,5-dimethylpyrrole (2), which on reaction with hydrazine is split to phthalhydrazide (3) and N-amino-2,5-dimethylpyrrole (4).

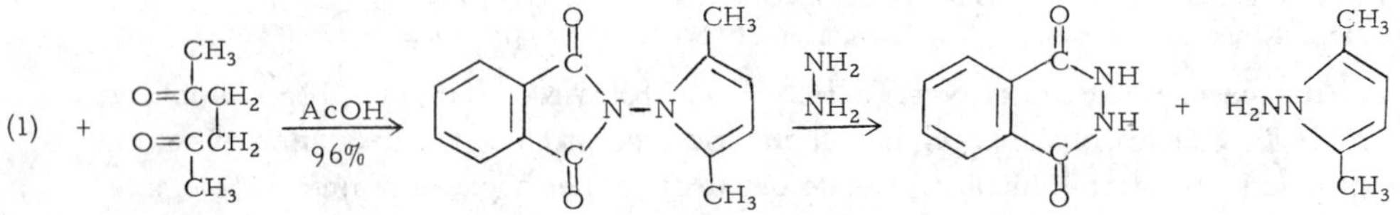
(1) +
AcOH
96%
NH_2
NH_2
+
H_2NN

(2) (3) (4) 65%

[1]H. D. K. Drew and H. H. Hatt, *J. Chem. Soc.*, 16 (1937)
[2]R. Epton, *Chem. Ind.*, 425 (1965)

Ammonium acetate and Ammonium formate. Mol. wts. 77.09, 63.06. Suppliers: B, F, MCB

In debromination. In the hydrogenolysis of a bromo compound, addition of slightly more than an equivalent of ammonium acetate is used to neutralize the hydrogen bromide liberated and so prevent undesired saturation of a double bond.[1]

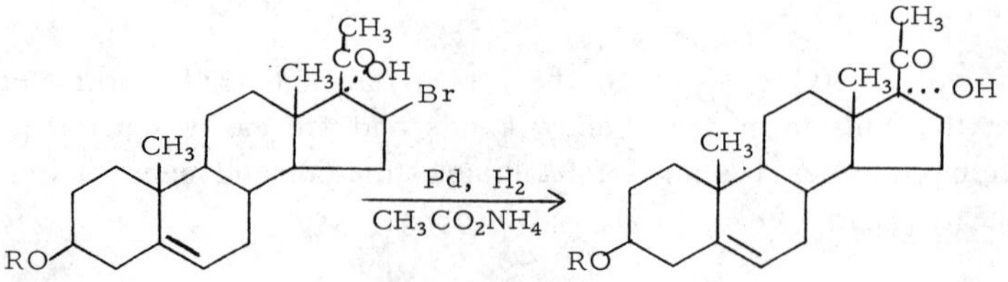
Br
Pd, H_2
$CH_3CO_2NH_4$
RO

Formation of amides. Acid chlorides are converted into amides in yields of 40–90% by stirring an acetone solution with ammonium acetate for 30–60 min.[2] In a total synthesis of tetracyclines, a Lederle group[3] fused the ester (1) with ammonium formate and, after dealkylation, isolated the amide (2) in unspecified yield.

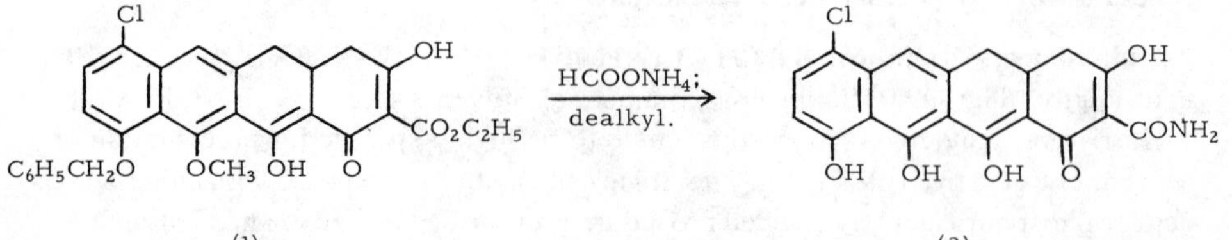
Cl
OH
$CO_2C_2H_5$
$C_6H_5CH_2O$
OCH_3 OH O
$HCOONH_4$;
dealkyl.
$CONH_2$
OH OH OH O

(1) (2)

Amination. Taking a clue from work on a related compound[4] Curtin *et al.*[5] converted 2-phenyl-1,3-indanedione into 3-amino-2-phenylindenone by fusing it with ammonium acetate in a nitrogen atmosphere. When the red slurry was cooled to 80–90° and diluted with water, the amine was obtained in crystalline form of high purity.

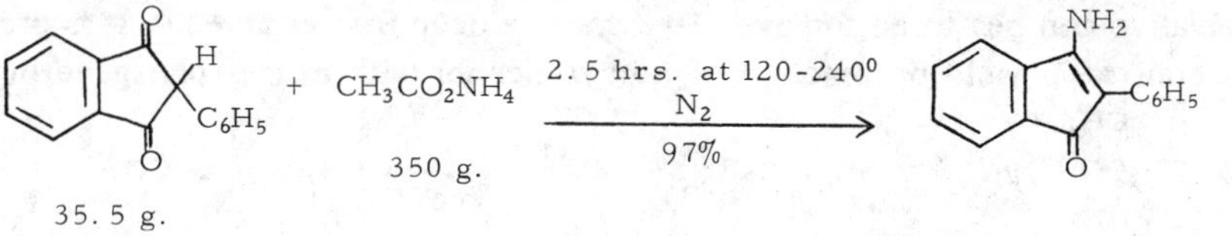

Condensation catalyst. Ammonium acetate is recommended as catalyst for the condensation of ketones with cyanoacetic acid or ethyl cyanoacetate.[6]

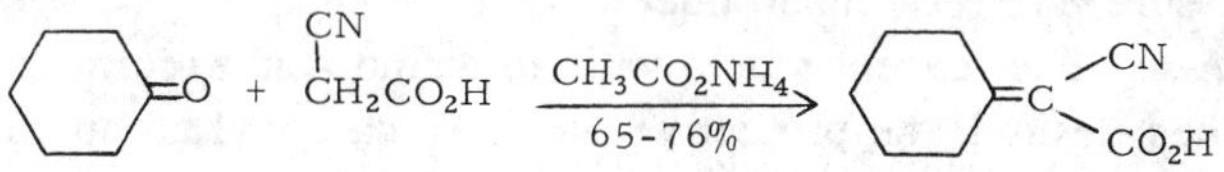

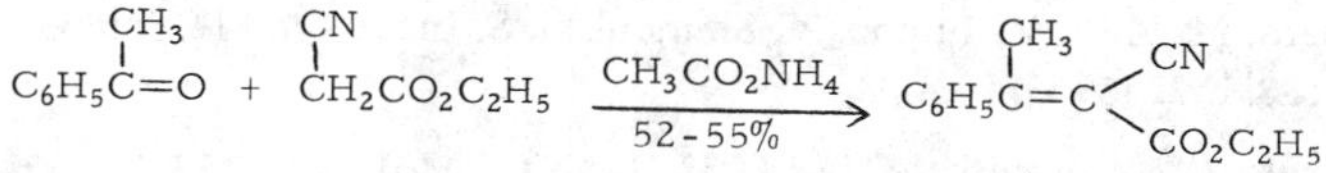

[1]H. J. Ringold, B. Löken, G. Rosenkranz, and F. Sondheimer, *Am. Soc.*, **78**, 816 (1956); B. Löken, S. Kaufmann, G. Rosenkranz, and F. Sondheimer, *ibid.*, **78**, 1738 (1956)
[2]P. A. Finan and G. A. Fothergill, *J. Chem. Soc.*, 2824 (1962)
[3]J. H. Boothe, A. S. Kende, T. L. Fields, and R. G. Wilkinson, *Am. Soc.*, **81**, 1006 (1959)
[4]R. L. Horton and K. C. Murdock, *J. Org.*, **25**, 938 (1960)
[5]D. Y. Curtin, J. A. Kampmeier, and M. L. Farmer, *Am. Soc.*, **87**, 874 (1965)
[6]A. C. Cope *et al.*, *Am. Soc.*, **63**, 3452 (1941); *Org. Syn., Coll. Vol.*, **4**, 234 (1963)

Ammonium metavanadate, NH_4VO_3, Mol. wt. 116.99. Supplier: Fisher.

Billman *et al.*[1] found that the reagent is a better catalyst for the perchlorate-oxidation of hydroquinone than the conventionally used vanadium pentoxide.[2]

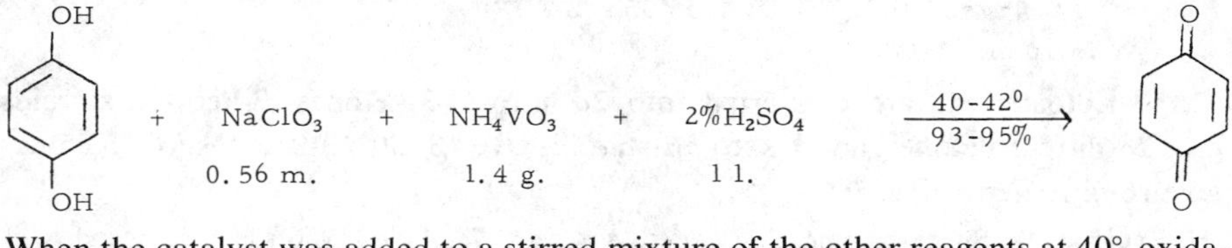

When the catalyst was added to a stirred mixture of the other reagents at 40°, oxidation started at once and cooling was required to prevent a temperature rise, and the oxidation was over in less than half an hour, about one eighth the time normally required.

[1]J. H. Billman, B. Wolnak, and D. K. Barnes, *Am. Soc.*, **66**, 652 (1944)
[2]H. W. Underwood, Jr., and W. L. Walsh, *Org. Syn., Coll. Vol.*, **2**, 553 (1943)

Ammonium nitrate. Mol. wt. 80.05. Suppliers: B, F, MCB.

This salt has been used as catalyst for the condensation of acrolein with triethyl orthoformate to produce acrolein diethyl acetal.[1]

$$\underset{0.79\text{ m.}}{CH_2{=}CHCHO} + \underset{0.97\text{ m.}}{HC(OC_2H_5)_3} + \underset{\substack{3\text{ g. in 50 ml.}\\ \text{anhyd. } C_2H_5OH}}{NH_4NO_3} \xrightarrow[72-80\%]{6-8\text{ hrs.}} CH_2{=}CHCH(OC_2H_5)_2 + HCO_2C_2H_5$$

[1]H. O. L. Fischer and E. Baer, *Helv.*, **18**, 514 (1935); J. A. VanAllan, *Org. Syn., Coll. Vol.*, **4**, 21 (1963)

Ammonium persulfate. Mol. wt. 228.21. *See* Potassium persulfate.

***t*-Amyl chloroformate.**[1] Mol. wt. 150.61 (crude).

In contrast to *t*-butyl chloroformate, which is too unstable to be used, the *t*-amyl derivative can be stored for over 10 days in a deep freezer at −20°. It is prepared in about 60% yield by reaction of *t*-amyl alcohol with excess phosgene in ether

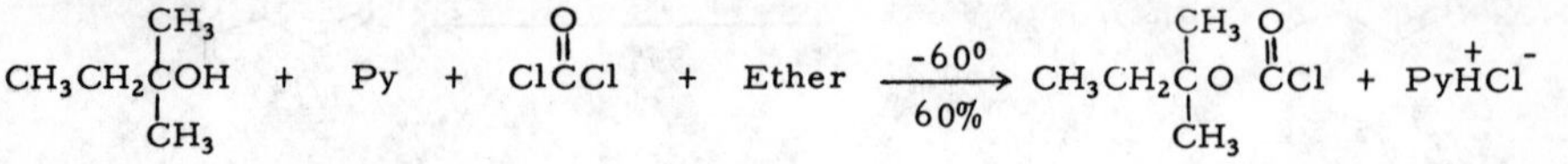

0.52 m. 0.52 m. 1.06 m. 1 l.

at − 60°. The pyridine hydrochloride formed is filtered and the solvent is removed at reduced pressure. The residual product was used as such.

Peptide synthesis. The reagent reacts with an amino acid to form an N-protected derivative. When desired, the protective *t*-amyloxycarbonyl group can be cleaved with trifluoroacetic acid or with anhydrous hydrogen chloride in an organic solvent.

[1]S. Sakakibara, M. Shin, M. Fujino, Y. Shimonishi, S. Inoue, and N. Inukai, *Bull. Chem. Soc. Japan*, **38**, 1522 (1965)

***i*-Amyl nitrate** ("Amyl nitrate"), *i*-$C_5H_{11}ONO_2$. Mol. wt. 133.15, sp. gr. 0.996. Suppliers, KK, MCB.

Studies by Feuer[1] of the nitration of cyclic ketones with amyl nitrate have shown potassium *t*-butoxide to be the best base and tetrahydrofurane the best solvent. Under these conditions dipotassium 2,5-dinitrocyclopentanone is obtained in good yield.

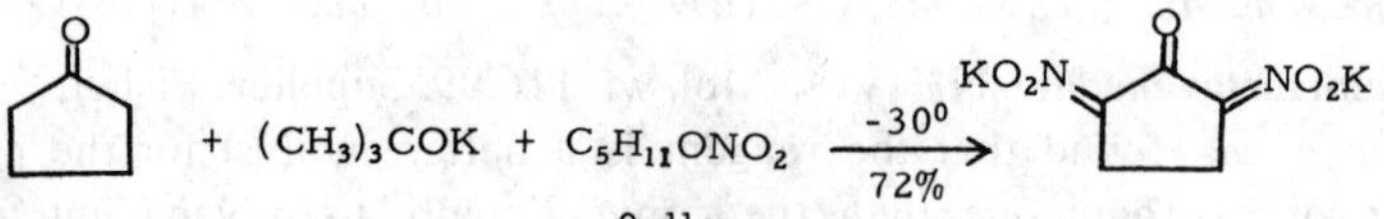

0.05 m. in 70 ml. THF; 0.165 m.; 0.11 m. in 35 ml. THF

Δ^4-3-Ketosteroids are converted into 2α-nitro-Δ^4-3-ketones, 3-keto-5α-steroids give 2α-nitro-3-ketones, and 3-keto-5β-steroids give 4β-nitro-3-ketones.[2] 17-Ketones give 16-nitro derivatives.[2,3]

[1]H. Feuer, J. W. Shepherd, and C. Savides, *Am. Soc.*, **78**, 4364 (1956); H. Feuer and R. S. Anderson, *ibid.*, **83**, 2960 (1961)
[2]R. E. Schaub, W. Fulmor, and M. J. Weiss, *Tetrahedron*, **20**, 373 (1964)
[3]A. Hassner and J. M. Larkin, *Am. Soc.*, **85**, 2181 (1963)

***i*-Amyl nitrite,** $C_5H_{11}ONO$. Mol. wt. 117.15, b.p. 99°, sp. gr. 0.872. Suppliers: E, F, MCB. Preparation.[1]

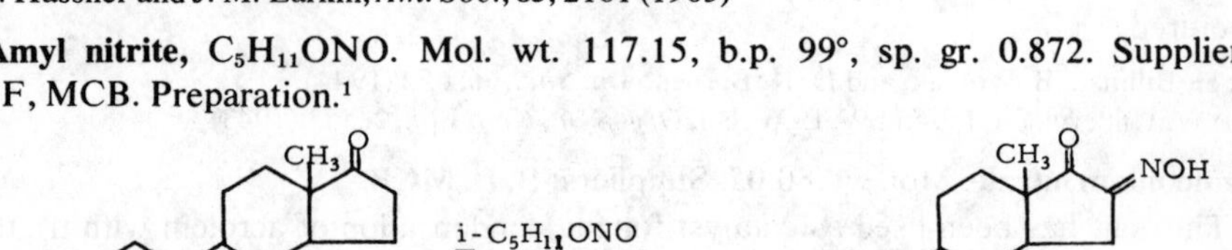

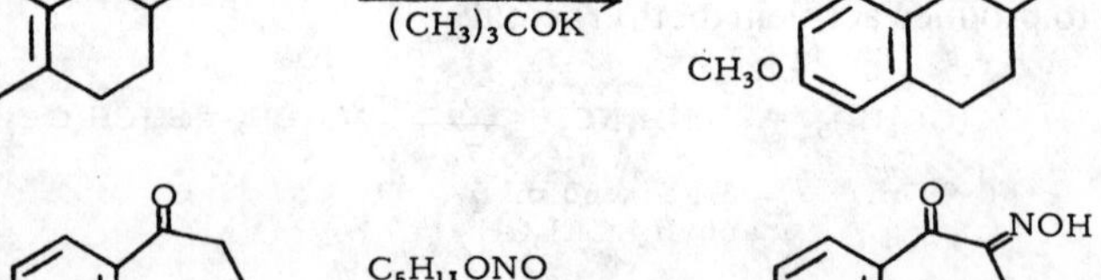

α-Oximino ketones. In the presence of potassium *t*-butoxide[2] or hydrochloric acid,[3] *i*-amyl nitrite converts a ketone having an adjacent methylene group into the α-oximino derivative.

Aromatic arylation. Decomposition of an aqueous solution of a diazonium salt at pH 8 in the presence of an aromatic compound proceeds poorly because of the heterogeneity of the mixture and because of side reactions. The process is improved by aprotic diazotization with amyl nitrite in the presence of the substrate.[4]

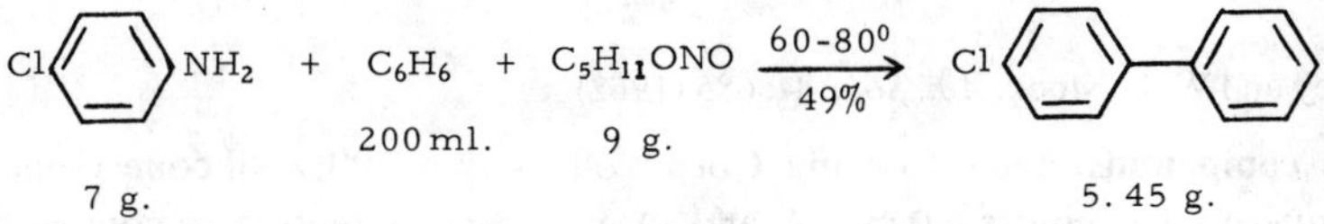

Cl
NH2
+ C6H6 + C5H11ONO
60-80°
49%
Cl
200 ml.
9 g.
7 g.
5.45 g.

Generation of benzyne. *See* Benzenediazonium 2-carboxylate.

Reaction with vinylamines. In an attempt to obtain a stable diazonium group attached to a nonaromatic carbon atom, Curtin *et al.*[5] treated diphenylvinylamine (1) with *i*-amyl nitrite and surprisingly obtained diphenylacetylene (2). 9-(Aminomethylene)fluorene (3) gave the divinylamine (4) in lower yield. The mechanism is unknown.

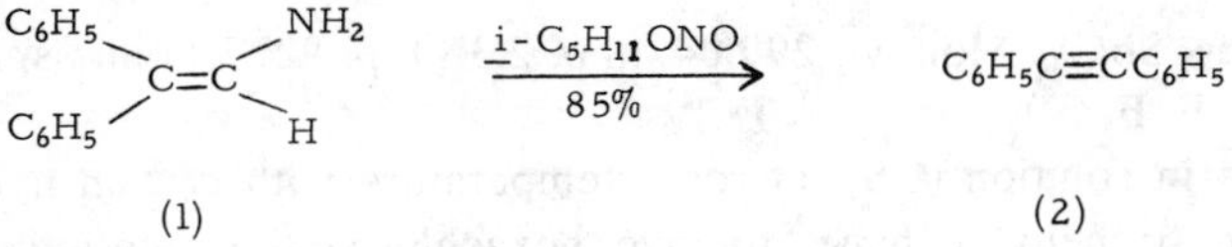

C6H5
NH2
C=C
C6H5
H
i-C5H11ONO
85%
C6H5C≡CC6H5
(1)
(2)

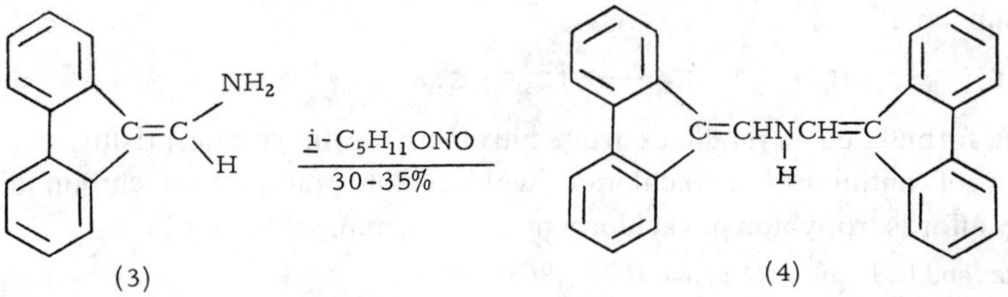

NH2
C=C
H
i-C5H11ONO
30-35%
C=CHNCH=C
H
(3)
(4)

[1]W. A. Noyes, *Org. Syn., Coll. Vol.*, **2**, 108 (1943)
[2]F. Litvan and R. Robinson, *J. Chem. Soc.*, 1997 (1938)
[3]D. Caunt, W. D. Crow, R. D. Haworth, and C. A. Vodoz, *J. Chem. Soc.*, 1631 (1950)
[4]J. I. G. Cadogan, *J. Chem. Soc.*, 4257 (1962)
[5]D. Y. Curtin, J. A. Kampmeier, and B. R. O'Connor, *Am. Soc.*, **87**, 863 (1965)

***n*-Amyl nitrite,** $C_5H_{11}ONO$. Mol. wt. 117.15, b.p. 104°, sp. gr. 0.853. References in the literature to "amyl nitrite" or "pentyl nitrite" probably refer not to the normal ester but to *i*-amyl nitrite, which alone is available commercially.

Aniline. Mol. wt. 93.12, b.p. 184°, sp. gr. 1.02, pKb 9.30. Suppliers: B, E, F, MCB. The color can be removed from old samples by distillation from a small amount of zinc dust.

Anisole, $C_6H_5OCH_3$. Mol. wt. 108.13, b.p. 154°, sp. gr. 0.990. Suppliers: E, F, MCB.

Japanese investigators[1] report that anisole is a satisfactory substitute for diglyme (b.p. 161°) in hydroboration reactions. Note, however, that unlike the diether, anisole is not water soluble.

[1]H. Kugita and M. Takeda, *Chem. Pharm. Bull.*, **12**, 1166 (1964)

Anthracene-9,10-biimine, 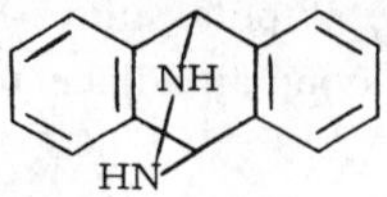

Mol. wt. 208.26, m.p. *ca.* 100°.

Prepared by hydrolysis of the anthracene–diethyl azodicarboxylate and decarboxylation, the reagent decomposes in refluxing ethanol to anthracene and diimide.[1] When taken in 2–8 fold excess, it reduces azobenzene and reactive alkenes and alkynes.

[1]E. J. Corey and W. L. Mock, *Am. Soc.*, **84**, 685 (1962)

Antifoam compounds. Dow-Corning Corp.[1] offers two 100% silicone compounds for nonaqueous foaming systems, A and Q, which are effective in concentrations of 1–50 ppm. and 0.1–150 ppm. Thus one small drop of Q stops foam in a whole gallon of liquid. Antifoam B, C, and H–10 emulsions (pH 6.5, 3.9, and 4.0) contain 10–30% active defoamer and are useful for control of aqueous foaming systems.

FMC Corporation offers the following antifoam agents: tributoxyethyl phosphate, tributyl phosphate, tri (2-ethylhexyl) phosphate.[2] *See also* Capryl alcohol.

[1]Chemical Products Division, Midland, Mich., 48641

[2]See also Hodag Chem. Corp., 7247 North Central Park Ave., Skokie, Ill.

Antimony pentachloride, $SbCl_5$. Mol. wt. 299.04, m.p. 2.3°, b.p. 92°/30 mm., sp. gr. 2.346. Suppliers: Alfa, B, F.

This reagent reacts in solution (CS_2) at room temperature with certain hydrocarbons, for example triphenylmethane, to give hexachloroantimonate salts of carbonium ions:[1]

$$(C_6H_5)_3CH + 2\ SbCl_5 \longrightarrow (C_6H_5)_3\overset{+}{C}SbCl_6^- + SbCl_3 + HCl$$

The salt formulated crystallizes from nitromethane in yellow prisms, m.p. 218°. Reaction of antimony pentachloride with cycloheptatriene in carbon disulfide similarly affords tropylium hexachloroantimonate, m.p. 190°.

[1]J. Holmes and R. Pettit, *J. Org.*, **28**, 1695 (1963)

Aryldiazonium hexafluorophosphates, $Ar\overset{+}{N}{\equiv}NPF_6^-$.

The Schiemann procedure for conversion of an amine to an aryl fluoride is improved considerably by use as intermediate of the aryldiazonium hexafluorophosphate instead of the aryldiazonium tetrafluoroborate.[1] Since these salts are less soluble than the tetrafluoroborates they are obtained in better yield and can be washed thoroughly for removal of impurities. In the preparation[2] of 2-bromofluorobenzene from *o*-bromoaniline the amine hydrochloride is diazotized as usual,

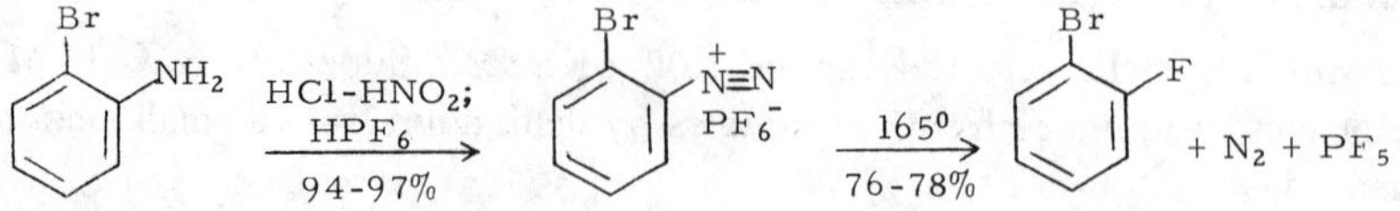

Br
NH2
HCl-HNO2;
HPF6
94-97%
Br
N≡N
PF6
165°
76-78%
Br
F
+ N2 + PF5

65% hexafluorophosphoric acid (Ozark-Mahoning Co.) is added, and the aryldiazonium hexafluorophosphate is collected and dried. For decomposition, the salt is placed in an Erlenmeyer connected by a short section of Gooch tubing to one opening of a three-necked flask fitted with a condenser and a thermometer and containing mineral oil. The oil is heated at 165–170° during addition of the salt in portions. After cooling, 10% aqueous sodium carbonate is added and the product steam distilled. The overall yield is 73–75%, whereas the yield by the Schiemann procedure is 37%. In the preparation of *o*-fluorobenzoic acid, overall yields by the modified and by the Schiemann procedure are 61% and 9%, respectively.

A new deamination procedure involves spontaneous and exothermal decomposition of an aryldiazonium hexafluorophosphate in tetramethylurea.[3] This organic but

$$Ar\overset{+}{N}\equiv N\,PF_6^- \xrightarrow{(CH_3)_2NC(=O)N(CH_3)_2} ArH + N_2 + PF_5 + (CH_3)_2NC(=O)N(CH_2F)(CH_3)$$

water-miscible solvent functioned satisfactorily where several other nonaqueous solvents failed. Yields are best (75–85%) where the aromatic ring carries an electron-withdrawing substituent (NO_2, CO_2H, Br). In the case of aniline, *o*- or *p*-toluidine, and *o*-anisidine yields are in the range 25–30%. The dry salt is added in portions at room temperature and the temperature is controlled to 65°.

[1]K. G. Rutherford, W. Redmond, and J. Rigamonti, *J. Org.*, **26**, 5149 (1961)
[2]K. G. Rutherford and W. Redmond, *Org. Syn.*, **43**, 12 (1963)
[3]*Idem*, *J. Org.*, **28**, 568 (1963)

Aryldiazonium tetrahaloborates, $Ar\overset{+}{N}\equiv N\overset{-}{B}X_4$. In the Schiemann reaction,[1] an arylamine is diazotized in aqueous hydrochloric acid solution, and the solution or suspension of aryldiazonium chloride is treated with a solution of fluoroboric acid prepared[2–4] by dissolving boric acid with cooling to 20–25° in commercial 50–60% hydrofluoric acid contained in a reaction flask coated with wax or lined with copper, lead, or silver. The aryldiazonium tetrafluoroborate is collected, dried, and decomposed by pyrolysis. Examples are:

$$1.^2\quad C_6H_5\overset{+}{N}\equiv NCl^- \xrightarrow{HBF_4} C_6H_5\overset{+}{N}\equiv N\overset{-}{B}F_4 \xrightarrow[51-57\%]{\Delta} C_6H_5F + N_2 + BF_3$$

$$2.^3\quad p\text{-}C_2H_5O_2CC_6H_4\overset{+}{N}\equiv NCl^- \xrightarrow[75-78\%]{HBF_4} p\text{-}C_2H_5O_2CC_6H_4\overset{+}{N}\equiv NBF_4^-$$

$$\xrightarrow[84-87\%]{\Delta} p\text{-}C_2H_5O_2CC_6H_4F$$

In another procedure[5] the diazonium tetrafluoroborate is prepared from the amine and nitrosonium tetrafluoroborate (4), obtainable by reaction (3).

$$3.\quad N_2O_3 + 2\,H^+BF_4^- \longrightarrow 2\,NO^+BF_4^- + H_2O$$

$$4.\quad ArNH_2 + NO^+BF_4^- \longrightarrow ArN_2^+BF_4^- + H_2O$$

Aryldiazonium tetrachloroborates and tetrabromoborates have been prepared in the following way.[6] A solution of the amine in chloroform is added to a solution of nitrosyl chloride (5, or bromide) in chloroform–petroleum ether at −18° and the white precipitate of aryldiazonium salt is collected. Boron trichloride is added to a stirred suspension of the salt in chloroform–petroleum ether at −18° (6) and the

$$5.\quad ArNH_2 + NOCl \longrightarrow ArN_2^+Cl^- + H_2O$$

$$6.\quad ArN_2^+Cl^- + BCl_3 \longrightarrow ArN_2^+BCl_4^-$$

off-white precipitate of aryldiazonium tetrahaloborate is collected. These salts when heated often decompose explosively, but the reaction can be controlled by use of an inert diluent (ligroin, b.p. 110–115°).

A reaction related to the Schiemann reaction but conducted in aqueous solution rather than by pyrolysis is employed in a synthesis of *p*-dinitrobenzene from *p*-nitroaniline.[4] Aryldiazonium fluoroborates containing nitro groups also decompose smoothly in chlorobenzene.[7]

$$7.\quad \text{p-}O_2NC_6H_4\overset{+}{N}{\equiv}NCl^- \xrightarrow[95\text{-}99\%]{HBF_4} \text{p-}O_2NC_6H_4\overset{+}{N}{\equiv}NBF_4^- \xrightarrow[67\text{-}82\%]{aq.\ NaNO_2(Cu)} \text{p-}O_2NC_6H_4NO_2$$

Burgstahler *et al.*[8] diazotized the sparingly soluble hydrochloride of 3,4-di-*t*-butylaniline at 0–5°, treated the filtered solution with sodium fluoroborate, and obtained the diazonium fluoroborate. This salt underwent slow solvolysis in a mixture of acetic acid – acetic anhydride with formation of the corresponding acetate.

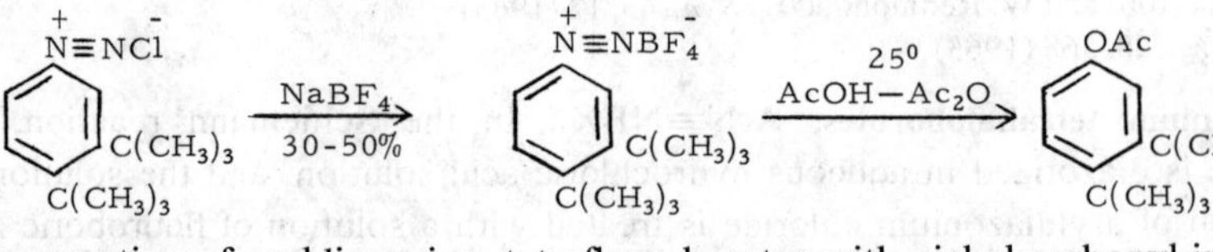

N≡NCl
NaBF4
30-50%
C(CH3)3
N≡NBF4
25°
AcOH—Ac2O
OAc

The reaction of aryldiazonium tetrafluoroborates with nickel carbonyl in ethanol or acetic acid gives aryl carboxylic acids in yields of 2–76%.[9]

Deamination of an amine can be accomplished by reduction of the aryldiazonium tetrafluoroborate with zinc and ethanol[10] or, more effectively, with sodium borohydride (8).[11] The simplest procedure is to add the solid borohydride in portions to

$$8.\quad ArN_2^+BF_4^- \xrightarrow{NaBH_4} ArH$$

a chilled solution or suspension of the tetrafluoroborate in methanol and to pour the resulting solution onto ice and hydrochloric acid. Alternatively, a chilled solution of sodium borohydride in dimethylformamide is added to a chilled solution of the diazonium salt in the same solvent. Yields are in the range 50–75%.

[1]A. Roe, "The Schiemann Reaction," *Org. Reactions*, **5**, 193 (1949)
[2]D. T. Flood, *Org. Syn., Coll. Vol.*, **2**, 295 (1943)
[3]G. Schiemann and W. Winkelmüller, *ibid.*, **2**, 299 (1943)
[4]E. B. Starkey, *ibid.*, **2**, 225 (1943)
[5]U. Wannagat and G. Hohlstein, *Ber.*, **88**, 1839 (1955)
[6]G. A. Olah and W. S. Tolgyesi, *J. Org.*, **26**, 2053 (1961)
[7]G. Valkanas and H. Hopff, *J. Chem. Soc.*, 1925 (1963)
[8]A. W. Burgstahler, M. O. Abdel-Rahman, and P.-L. Chien, *Tetrahedron Letters*, 61 (1964)
[9]J. C. Clark and R. C. Cookson, *J. Chem. Soc.*, 686 (1962)
[10]A. Roe and J. R. Graham, *Am. Soc.*, **74**, 6297 (1952)
[11]J. B. Hendrickson, *ibid.*, **83**, 1251 (1961)

Azobenzene-4-carboxylic acid hydrazide. Mol. wt. 240.26, m.p. 209°. Prepared from the corresponding ethyl ester and hydrazine, the reagent reacts with carbonyl

4'
3'
—N=N—
4
CNHNH2

compounds to form high-melting orange hydrazones.[1] The 3′-nitro and 4′-nitro derivatives give intensely colored hydrazones of even higher melting point.[2]

[1]U. Westphal, H. Feier, O. Lüderitz, and I. Fromme, *Biochem. Z.*, **326**, 139 (1954/55)
[2]S. M. A. D. Zayed and I. M. I. Fakhr, *Ann.*, **662**, 165 (1963)

Azobisisobutyronitrile, $(CH_3)_2C(CN)—N{=}N—C(CN)(CH_3)_2$. Mol. wt. 136.20, m.p. 102°. Suppliers: Westville Chem. Corp., A, B, E, KK, MCB.

Purification is effected by recrystallization from methanol and drying in vacuum over phosphorus pentoxide.[1] Under mild conditions (40°) the reagent decomposes into cyanopropyl radicals and hence is useful for initiation of radical reactions.[2]

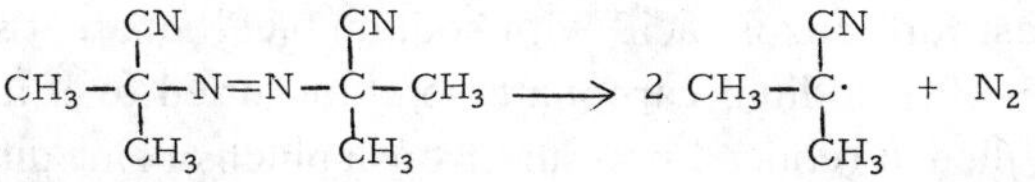

With use of this initiator, Stockmann developed a simple, stereospecific synthesis of *exo*-norbornyl methyl ketone by radical addition of acetaldehyde to norbornene.[3] Use of peroxidic radical sources drastically reduced the yield.

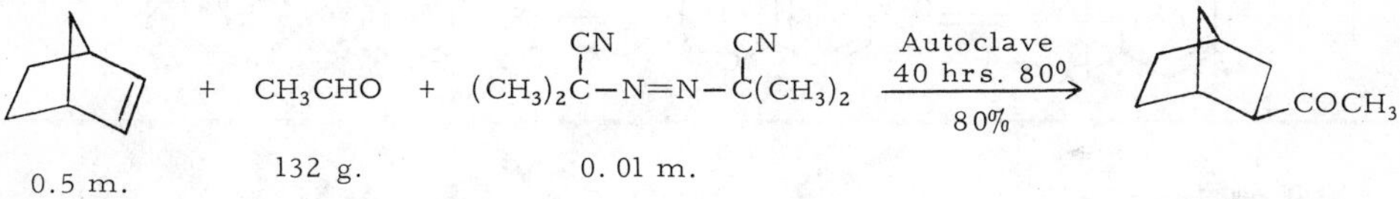
+ CH3CHO + (CH3)2C(CN)–N=N–C(CN)(CH3)2
Autoclave
40 hrs. 80⁰
80%
COCH3
0.5 m.
132 g.
0. 01 m.

[1]G. S. Hammond, L. R. Mahoney, and U. S. Nandi, *Am. Soc.*, **85**, 740 (1963)
[2]C. Walling, "Free Radicals in Solution," p. 70, Wiley (1957); C. Walling and E. S. Huyser, *Org. Reactions*, **13**, 115–116 (1963)
[3]H. Stockmann, *J. Org.*, **29**, 245 (1964)

Azodicarboxamate, $H_2NC(=O)—N{=}N—C(=O)NH_2$. Mol. wt. 116.08. Supplier: Aldrich. Preparation.[1] *See* Diimide.

[1]J. Thiele, *Ann.*, **271**, 129 (1892)

B

Barium permanganate. Mol. wt. 375.22, 205.8 g./100 g. H_2O at 0°. Suppliers: Alfa, KK. Preparation.[1] Oxidation: —SH → —SO_3H.[2]

[1]L. Vanino, "Handbuch der Präparativen Chemie," 3rd ed., 479 (1925)
[2]K. Hofmann, A. Bridgwater, and A. E. Axelrod, *Am. Soc.*, **71**, 1253 (1949)

Benzaldehyde, C_6H_5CHO. Mol. wt. 106.12, b.p. 179° (57–59°/8 mm.), sp. gr. 1.046. Suppliers: A, B, E, F, MCB.

Reagent from a freshly opened bottle should be satisfactory, but if the material available gives a test for benzoic acid with sodium bicarbonate it should be washed free of acid with 5–10% sodium carbonate solution, dried over anhydrous sodium carbonate, and distilled at reduced pressure from a pinch of zinc dust with avoidance of exposure to air.

Benzenediazonium 2-carboxylate (Benzyne precursor).

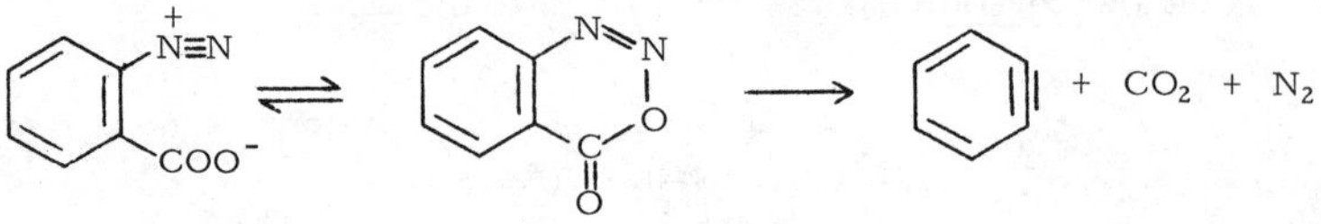
N≡N
COO⁻
N
N
O
C
O
+ CO2 + N2

Mol. wt. 148.12

The preformed reagent can be obtained by diazotization of anthranilic acid,[1] but it has to be stored at 0° and is sometimes explosive. It is more convenient to generate the reagent *in situ* by aprotic diazotization of anthranilic acid in the presence of a trapping agent.[2] By dropwise addition of a solution of anthranilic acid in tetrahydrofurane in the course of 4 hrs. to a refluxing solution of amyl nitrite and anthracene in methylene chloride, triptycene can be obtained in yield of 51%. In a simplified procedure for students,[3] a solution of anthracene and isoamyl nitrite in the higher boiling 1,2-dimethoxyethane is refluxed during addition of anthranilic acid in 40 min. and pure triptycene is isolated with ease.

[1]M. Stiles, R. G. Miller, and U. Burckhardt, *Am. Soc.*, **85**, 1792 (1963); R. G. Miller and M. Stiles, *ibid.*, **85**, 1798 (1963)
[2]L. Friedman and F. M. Logullo, *ibid.*, **85**, 1549 (1963); F. M. Logullo and L. Friedman, procedure submitted to *Org. Syn.*
[3]*Org. Expts.*, Chapt. 62

Benzenesulfonic anhydride, $(CH_6H_5SO_2)_2O$. Mol. wt. 298.33, m.p. 90°.

Preparation and use in Friedel-Crafts synthesis of sulfones and for preparation of benzenesulfonates of phenols.[1]

[1]L. Field, *Am. Soc.*, **74**, 394 (1952)

Benzenesulfonyl chloride, $C_6H_5SO_2Cl$. Mol. wt. 176.63, m.p. 15°, b.p. 251.5° (113–115°/10 mm.), sp. gr. 1.384. Suppliers: B, E, F, MCB. Prepared by reaction of sodium benzenesulfonate with either PCl_5 or $POCl_3$,[1] and by chlorosulfonation of benzene.[2]

Benzenesulfonamides. In the Hinsberg test a primary amine is recognized by the formation of an alkali-soluble benzenesulfonamide, and a secondary amine by formation of a sulfonamide insoluble in alkali.[3]

In the preparation of 1-*n*-butylpyrrolidine (3) from di-*n*-butylamine (1) by the Hofmann-Löffler reaction,[4] benzenesulfonyl chloride is used for Hinsberg separation of the product (3) from starting material (1). The *t*-amine (3) is separated by steam

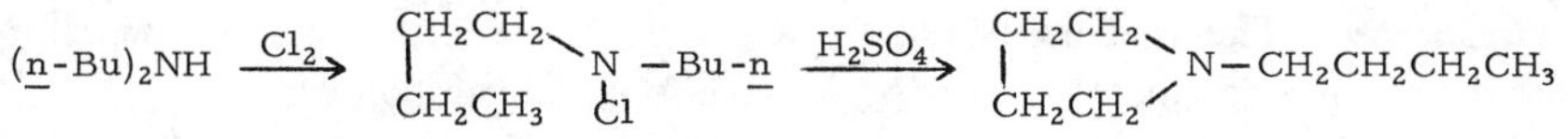

(1) (2) (3)

distillation from the nonvolatile Hinsberg derivative of (1).

Benzenesulfonyl esters. That $OSO_2C_6H_5$ is an effective leaving group is illustrated by a procedure for the conversion of pentaerythritol to pentaerythrityl tetrabromide.[5]

$$\underset{0.96\text{ m.}}{C(CH_2OH)_4} + \underset{4.24\text{ m.}}{4\ C_6H_5SO_2Cl} + \underset{650\text{ ml.}}{\text{Pyridine}} \xrightarrow[\text{not dried}]{30\text{-}35^0;\ 40^0} C(CH_2OSO_2C_6H_5)_4$$

$$\xrightarrow[68\text{-}78\%]{\text{NaBr (5.8 m.) in triethylene glycol } 140\text{-}150^0} C(CH_2Br)_4$$

In this case esterification to the tetra ester proceeds well in pyridine solution. Kampouris[6] states that esterification of phloroglucinol in pyridine gives a mixture of the mono-, the di-, and the tri-ester. He esterified this and other polyprotonic phenols in aqueous solution in the presence of calcium hydroxide, added in portions to the reaction mixture until the solution was permanently alkaline to litmus. The yield of catechol dibenzenesulfonate was 96%; the yield of the triester from phloroglucinol is not reported.

A method for the cyclization of 1,4- and 1,5-glycols consists in refluxing the glycol with a tertiary amine with gradual addition of benzenesulfonyl chloride.[7]

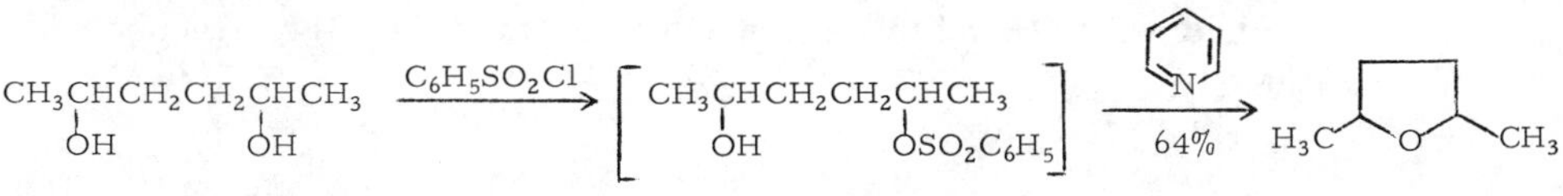

Other reactions. Benzenesulfonyl chloride serves also for the preparation of methyl phenyl sulfone[8] and of thiophenol.[9]

$$C_6H_5SO_2Cl \xrightarrow{Na_2SO_3,\ NaHCO_3} C_6H_5SO_2Na \xrightarrow[66\text{-}69\%\text{ overall}]{(CH_3O)_2SO_2} C_6H_5SO_2CH_3$$

$$\underset{600\text{ g.}}{C_6H_5SO_2Cl} + \underset{2.4\text{ kg.}}{\text{concd. } H_2SO_4} + \underset{1.2\text{ kg.}}{\text{Zn dust}} + \underset{7.2\text{ kg.}}{\text{Ice}} \xrightarrow[91\%]{0^0\text{ to }100^0} C_6H_5SH$$

[1]R. Adams and C. S. Marvel, *Org. Syn.*, *Coll. Vol.*, **1**, 84 (1941)
[2]H. T. Clarke, G. S. Babcock, and T. F. Murray, *ibid.*, **1**, 85 (1941)
[3]*Org. Expts.*, 126
[4]G. H. Coleman, G. Nichols, and T. F. Martens, *Org. Syn.*, *Coll. Vol.*, **3**, 159 (1955)
[5]H. L. Herzog, *ibid.*, **4**, 753 (1963)

[6]E. M. Kampouris, *J. Chem. Soc.*, 2651 (1965)
[7]D. D. Reynolds and W. O. Kenyon, *Am. Soc.*, **72**, 1593 (1950)
[8]L. Field and R. D. Clark, *Org. Syn., Coll. Vol.*, **4**, 674 (1963)
[9]R. Adams and C. S. Marvel, *ibid.*, **1**, 504 (1941)

Benzenesulfonyl isocyanate, $C_6H_5SO_2N{=}C{=}O$. Mol. wt. 183.19, b.p. 69–72°/ 0.15 mm.

Preparation. The reagent is readily available by reaction of benzenesulfonyl chloride with silver cyanate[1] or from benzenesulfonamide and phosgene.[2]

Characterization of hindered alcohols and phenols.[3] The reagent readily converts hindered hydroxy compounds into urethanes and hence is useful for the preparation of characterizing derivatives. Thus in the reaction with *t*-butanol in toluene (0.01 M solution) at 0°, titration after 1 min. indicated that only a negligible amount of

$$CH_3\underset{\underset{\displaystyle CH_3}{|}}{\overset{\overset{\displaystyle CH_3}{|}}{C}}-OH \ + \ C_6H_5SO_2N{=}C{=}O \xrightarrow[\ 0^0\ ]{C_6H_5CH_3} CH_3\underset{\underset{\displaystyle CH_3}{|}}{\overset{\overset{\displaystyle CH_3}{|}}{C}}-O\overset{\overset{\displaystyle O}{\|}}{C}\ \overset{\overset{\displaystyle H}{|}}{N}SO_2C_6H_5$$

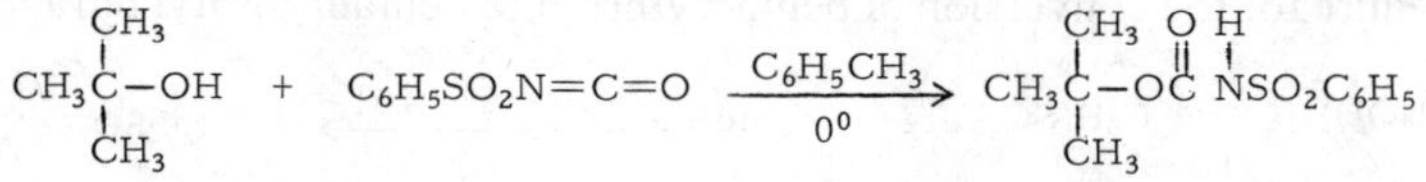

M. p. 128°

isocyanate remained. In contrast the reaction of *t*-butanol with phenyl isocyanate at the same concentration in toluene was only 59% complete after 19 days at 100°. Rapid reactions occurred with 2,6-dimethylphenol, 2,6-dimethoxyphenol, pentachlorophenol, and even 2,6-di-*t*-butylphenol.

[1]O. C. Billeter, *Ber.*, **37**, 690 (1904)
[2]H. Krzikalla, German patent 817,602 [*C.A.*, **47**, 2206 (1953)]
[3]J. W. McFarland and J. B. Howard, *J. Org.*, **30**, 957 (1965)

Benzhydryl azidoformate,[1] $(C_6H_5)_2CHO\overset{\overset{\displaystyle O}{\|}}{C}N{=}\overset{+}{N}{=}\overset{-}{N}$. Mol. wt. 253.25, m.p. 37–39°.

Preparation. This stable, crystalline white solid is readily prepared as shown, starting with the condensation of benzhydrol and methyl chlorothiolcarbonate to produce the ester (1). The crude ester is treated in ether with hydrazine in methanol to form the hydrazide (2) and this is converted into the azide (3) by reaction with nitrous acid.

$$(C_6H_5)_2CHOH \ + \ Cl\overset{\overset{\displaystyle O}{\|}}{C}SCH_3 \xrightarrow{Py} (C_6H_5)_2CHO\overset{\overset{\displaystyle O}{\|}}{C}SCH_3 \xrightarrow[82\%]{H_2NNH_2}$$

$$(1)$$

$$(C_6H_5)_2CHO\overset{\overset{\displaystyle O}{\|}}{C}NHNH_2 \xrightarrow[100\%]{NaNO_2-AcOH} (C_6H_5)_2CHO\overset{\overset{\displaystyle O}{\|}}{C}N{=}\overset{+}{N}{=}\overset{-}{N}$$

$$(2) \qquad\qquad\qquad (3)$$

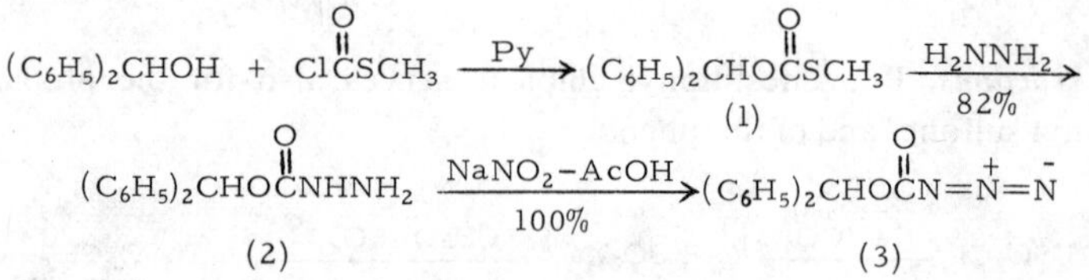

N-Protection in peptide synthesis. The reagent reacts with an amino acid to give the N-benzhydryloxycarbonyl derivative:

$$(C_6H_5)_2CHO\overset{\overset{\displaystyle O}{\|}}{C}N{=}\overset{+}{N}{=}\overset{-}{N} \ + \ H_2N\underset{\underset{\displaystyle R}{|}}{C}HCO_2H \xrightarrow{MgO,\ aq.\ dioxane} (C_6H_5)_2CHO\overset{\overset{\displaystyle O}{\|}}{C}\ \overset{\overset{\displaystyle H}{|}}{N}\underset{\underset{\displaystyle R}{|}}{C}HCO_2H \ + \ HN_3$$

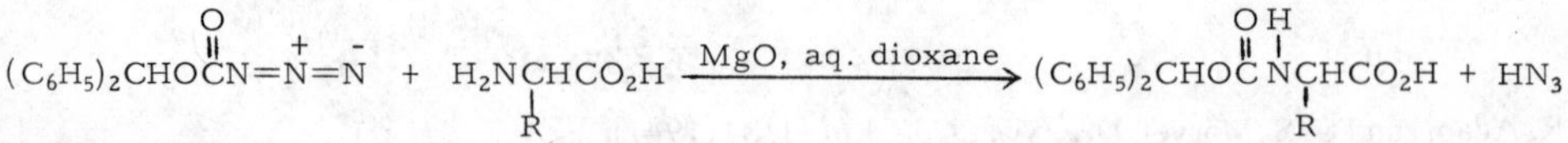

The blocking group is smoothly cleaved under mild acidic conditions, for example with 1.7 N hydrochloric acid in THF. This protective group is recommended particularly for the synthesis of unsymmetrical cystine peptides where use of stronger acid conditions leads to formation of a symmetrical cystine peptide. The reagent is comparable to *t*-butyl azidoformate but more easily accessible.

[1]R. G. Hiskey and J. B. Adams, Jr., *Am. Soc.*, **87**, 3969 (1965)

Benzoic acid, $C_6H_5CO_2H$. Mol. wt. 122.12, m.p. 122°, b.p. 249°, solubility in 100 g. of water 0.21 g. at 17.5°, 2.2 g. at 75°, pKa 4.17. Suppliers: B, E, F, MCB.

Rüchardt *et al.*[1] found that acids, particularly benzoic, catalyze the Wittig reaction of carboethoxymethylenetriphenylphosphorane with ketones in benzene solution. The effect is less marked at 40° than at 23° and also less marked in chloroform

$$(C_6H_5)_3P{=}CHCO_2C_2H_5 + R_2C{=}O \xrightarrow[C_6H_6]{C_6H_5CO_2H} R_2C{=}CHCO_2C_2H_5 + (C_6H_5)_3PO$$

solution than in benzene.[2] The effect is ascribed to specific hydrogen bonding in the transition state. The acid catalyst has no effect on the relative proportion of *cis* and *trans* esters formed.[3]

[1]C. Rüchardt, S. Eichler, and P. Panse, *Angew. Chem., Internat. Ed.*, **2**, 619 (1963)
[2]S. Fliszár, R. F. Hudson, and G. Salvadori, *Helv.*, **47**, 159 (1964)
[3]H. O. House, V. K. Jones, and G. A. Frank, *J. Org.*, **29**, 3327 (1964)

Benzoic anhydride, $(C_6H_5CO)_2O$. Mol. wt. 226.23, m.p. 42–43°. Suppliers: A, B, E, F.

Decarboxylation of α-keto acids. Phenylglyoxylic acid is decarboxylated smoothly by refluxing in benzene with benzoic anhydride and pyridine.[1] Other anhydrides can be used but benzoic anhydride is preferred because Perkin-type condensations are avoided. A mechanism is postulated.

$$\underset{\text{13 mmoles}}{C_6H_5\overset{O}{\overset{\|}{C}}CO_2H} + \underset{\text{27 mmoles}}{(C_6H_5CO)_2O} + \underset{\text{140 mmoles}}{\text{Pyridine}} + \underset{\text{50 ml.}}{C_6H_6} \xrightarrow{\text{Refl. 13 hrs.}} \underset{75\%}{C_6H_5CHO} + \underset{88\%}{CO_2}$$

[1]T. Cohen and I. H. Song, *Am. Soc.*, **87**, 3780 (1965)

1,4-Benzoquinone. Mol. wt. 108.09, m.p. 113°. Suppliers (prem. grade): B, F.

Dark, practical grade material (Aldrich, Eastman) is conveniently purified (bright yellow) by vacuum sublimation and crystallization from tetrachloroethylene.

One use is for dehydrogenation of α-hydroxyaminonitriles to α-oximidonitriles.[1] The quinone is added in small portions with heating and shaking to a solution of the starting material in benzene and the mixture is then refluxed for 3–5 hrs. Yields are in the range 41–64%.

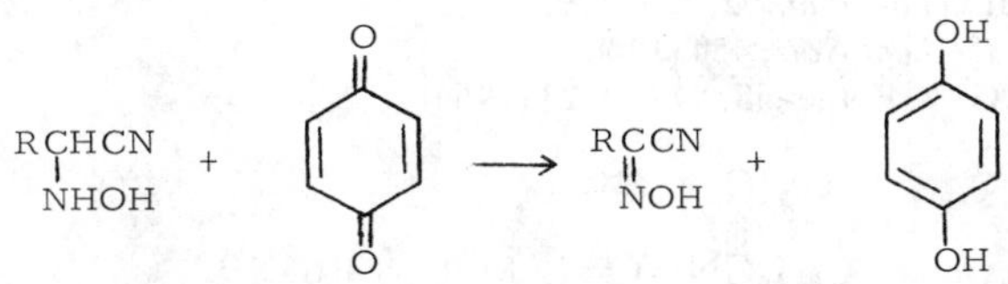
RCHCN
NHOH
+
O
O
RCCN
NOH
+
OH
OH

Another is use as hydrogen acceptor in the Oppenauer oxidation of a Δ^5-3β-hydroxysteroid to produce a $\Delta^{4,6}$-diene-3-one. In reporting discovery of this reaction,

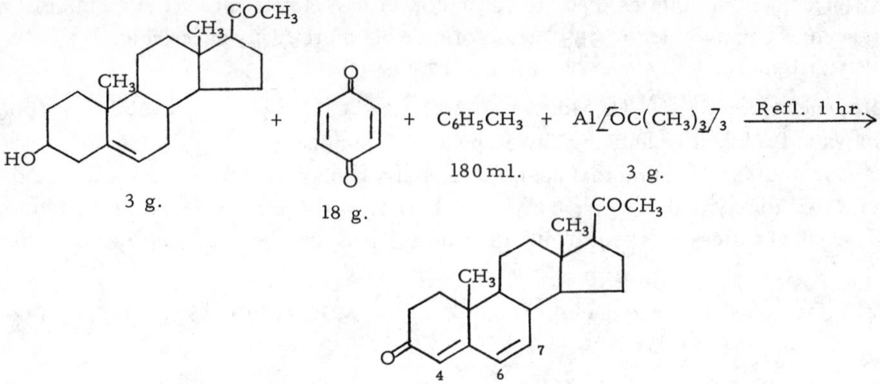
COCH3
CH3
CH3
HO
3 g.
18 g.
+ C6H5CH3 + Al/OC(CH3)3/3
180 ml.
3 g.
Refl. 1 hr.
COCH3
CH3
CH3
4
6
7

Wettstein[2] described the oxidation of pregnenolone to 6-dehydroprogesterone in detail but did not record the yield. In discussing the mechanism of the Wettstein-Oppenauer oxidation, Mandell[3] surveyed the literature and stated that yields are "as high as 50%."

For use as a dienophile, *see* 1,4-Naphthoquinone.

[1]L. W. Kissinger and H. E. Ungnade, *J. Org.*, **25**, 1471 (1960)
[2]A. Wettstein, *Helv.*, **23**, 388 (1940)
[3]L. Mandell, *Am. Soc.*, **78**, 3199 (1956)

Benzo-1,2,3-thiadiazole-1,1-dioxide, 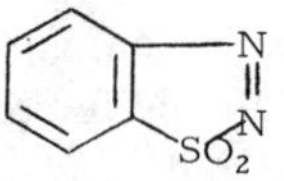
N
N
SO2

Mol. wt. 168.18.

The preparation[1] of this benzyne precursor is lengthy and the compound explodes at about 60°. It decomposes slowly at about 10° to give benzyne, nitrogen, and sulfur dioxide.

[1]G. Wittig and R. W. Hoffmann, *Ber.*, **95**, 2718 (1962)

Benzoyl chloride, C_6H_5COCl. Mol. wt. 140.57, b.p. 197°, sp. gr. 1.21. Suppliers: B, E, F, MCB.

Practical grade benzoyl chloride is purified by washing a benzene solution with aqueous sodium bicarbonate, drying, and distilling.[1]

Aliphatic acid chlorides (C_2 to C_6) can be prepared in 80–90% yield by mixing the acid with excess benzoyl chloride and distilling the volatile RCOCl.[2] Other examples: acrylyl chloride (70%),[3] propionyl chloride (80%),[4] isovaleryl chloride.[5]

[1]T. S. Oakwood and C. A. Weisgerber, *Org. Syn., Coll. Vol.*, **3**, 113 (1955)
[2]H. C. Brown, *Am. Soc.*, **60**, 1325 (1938)
[3]G. H. Stempel, Jr., *et al.*, *ibid.*, **72**, 2299 (1950)
[4]J. Forrest *et al.*, *J. Chem. Soc.*, 454 (1946)
[5]P. A. Finan and G. A. Fothergill, *ibid.*, 2723 (1963)

Benzoylisothiocyanate, $C_6H_5\overset{\displaystyle O}{\overset{\|}{C}}N{=}C{=}S$. Mol. wt. 163.19.

Prepared *in situ* by reaction of benzoyl chloride with ammonium thiocyanate (mol. wt. 76.11).[1] Reacts with a primary or secondary amine to form a benzoylthiourea, which is readily hydrolyzed by base to the free thiourea.

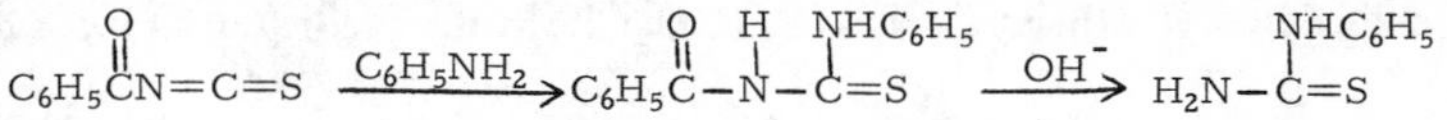

$$C_6H_5\overset{\overset{\displaystyle O}{\|}}{C}N=C=S \xrightarrow{C_6H_5NH_2} C_6H_5\overset{\overset{\displaystyle O}{\|}}{C}-\overset{\overset{\displaystyle H}{|}}{N}-\overset{\overset{\displaystyle NHC_6H_5}{|}}{C}=S \xrightarrow{OH^-} H_2N-\overset{\overset{\displaystyle NHC_6H_5}{|}}{C}=S$$

[1] I. B. Douglass and F. B. Dains, *Am. Soc.*, **56**, 1408 (1934)

Benzoyl peroxide, *see* Dibenzoyl peroxide.

Benzylamine, $C_6H_5CH_2NH_2$. Mol. wt. 107.15, b.p. 184.5° (70–71°/10 mm.), sp. gr. 0.982, water-miscible. Suppliers: B, E, F, KK, MCB.

α,β-Epoxy acids. α,β-Epoxy acids, prepared from α,β-unsaturated acids by epoxidation with hydrogen peroxide–sodium tungstate, react with benzylamine to give exclusively α-benzylamino-β-hydroxy acids; the free amino acids are obtained by hydrogenolysis with 30% Pd-charcoal.[1] The reaction is stereospecific: *trans*-acids

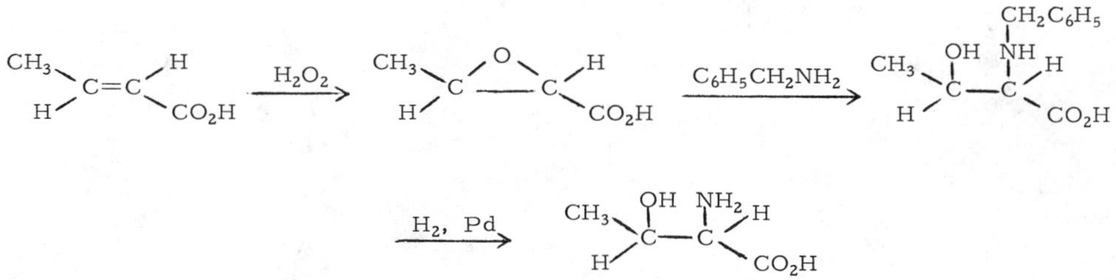

give *erythro*-isomers whereas *cis*-acids give *threo*-isomers. Thus the product from crotonic acid (*trans*) is DL-threonine.

Reductive amination. Reductive amination of 3-ketosteroids with ammonia, hydrogen, and Pd-C gives 3β-aminosteroids, but in poor yield. It is preferable to use benzylamine; the intermediate secondary amine undergoes hydrogenolysis.[2] The reaction is selective for 3-ketosteroids; carbonyl functions at C_{17} and C_{20} are not affected. Thus hydrogenation of progesterone in the presence of benzylamine and Pd-C afforded 3β-amino-5β-pregnane-20-one in 71% yield.

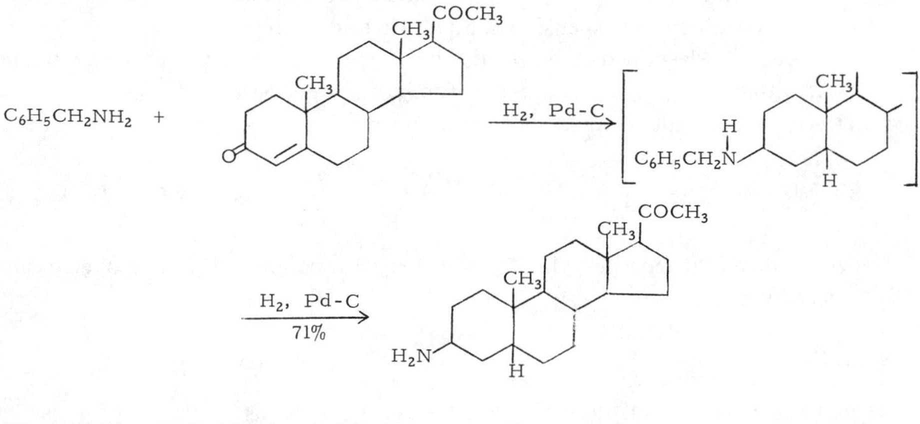

[1] Y. Liwschitz, Y. Rabinsohn, and D. Perera, *J. Chem. Soc.*, 1116 (1962); Y. Liwschitz, Y. Rabinsohn, and A. Haber, *ibid.*, 3589 (1962)
[2] J. Schmitt *et al.*, *Bull. Soc.*, 1846, 1855 (1962)

Benzyl chloroformate, *see* Carbobenzoxy chloride.

Benzyl chloromethyl ether, $C_6H_5CH_2OCH_2Cl$. Mol. wt. 156.61, b.p. 105–109°/ 11 mm.

Preparation.[1] A mixture of benzyl alcohol and aqueous formaldehyde is stirred in an ice bath and saturated with hydrogen chloride.

Hydroxymethylation of ketones.[2] A ketone is treated with sodium hydride for conversion into the enolate and this is alkylated with benzyl chloromethyl ether. Dioxane is the preferred solvent since O-alkylation is minimized. The examples formulated illustrate a high degree of stereoselectivity. The benzyl group should

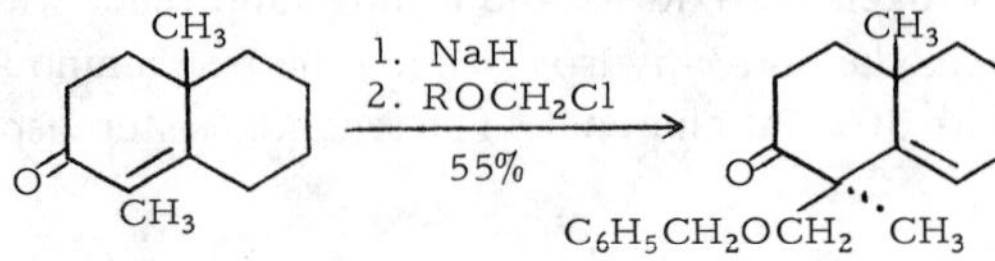
CH3
CH3
1. NaH
2. ROCH2Cl
55%
CH3
C6H5CH2OCH2
CH3

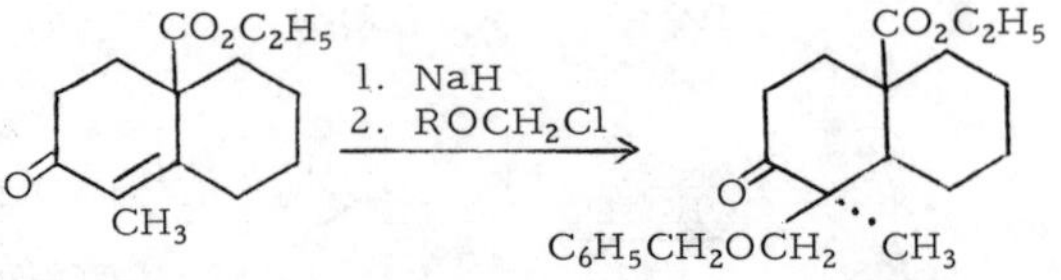
CO2C2H5
CH3
1. NaH
2. ROCH2Cl
CO2C2H5
C6H5CH2OCH2
CH3

be removable by hydrogenolysis at a suitable stage of synthesis.

[1]A. J. Hill and DeW. T. Keach, *Am. Soc.*, **48**, 257 (1926)

[2]C. L. Graham and F. J. McQuillin, *J. Chem. Soc.*, 4634 (1963); F. J. McQuillin and P. L. Simpson, *ibid.*, 4726 (1963); C. L. Graham and F. J. McQuillin, *ibid.*, 4521 (1964)

Benzyllithium, $C_6H_5CH_2Li$. Mol. wt. 98.07.

Benzyllithium, one of the more useful metalating agents,[1] can be obtained in high yield by cleavage of benzyl methyl ether in tetrahydrofurane at −5° to −15°.[2] Since benzyllithium is formed readily in tetrahydrofurane but is more stable in diethyl ether solution, dark brown solutions of concentrations as high as 1.2 *N* can be obtained in yields up to 83% by gradual addition of a solution of benzyl methyl ether in diethyl ether to a suspension of lithium in tetrahydrofurane.[3]

In a procedure described by Seyferth[4] the reagent is prepared by transmetalation of methyllithium with tribenzyltin chloride, readily accessible by direct reaction of benzyl chloride with metallic tin in anhydrous medium.[5] Benzyllithium is produced

$$(C_6H_5CH_2)_3SnCl \; + \; 4\,CH_3Li \xrightarrow[89\%]{(C_2H_5)_2O} 3\,C_6H_5CH_2Li \; + \; (CH_3)_4Sn \; + \; LiCl$$

in high yield and the tetramethyltin formed, being low boiling (78°), is removed easily with the solvent.

[1]H. Gilman and H. A. McNinch, *J. Org.*, **27**, 1889 (1962)

[2]*Idem, ibid.*, **26**, 3723 (1961)

[3]H. Gilman and G. L. Schwebke, *ibid.*, **27**, 4259 (1962)

[4]D. Seyferth, R. Suzuki, C. J. Murphy, and C. R. Sabet, *J. Organometallic Chem.*, 431 (1964)

[5]K. Sisido, Y. Takeda, and Z. Kinugawa, *Am. Soc.*, **83**, 538 (1961)

O-Benzyl phosphorus O,O-diphenylphosphoric anhydride, $(C_6H_5O)_2\overset{O}{\overset{\|}{P}}-O-\underset{\underset{O}{\|}}{\overset{H}{\overset{|}{P}}}-OCH_2C_6H_5$

Mol. wt. 404.28.

Preparation.[1] Under the catalytic action of 2,6-lutidine, the reagent reacts with a free hydroxyl group of a sugar (1) to form a benzyl phosphate (2), which can be chlorinated with N-chlorosuccinimide (3). Hydrolysis then gives the benzyl hydrogen phosphate (4), and the benzyl group is removed by hydrogenation.[2]

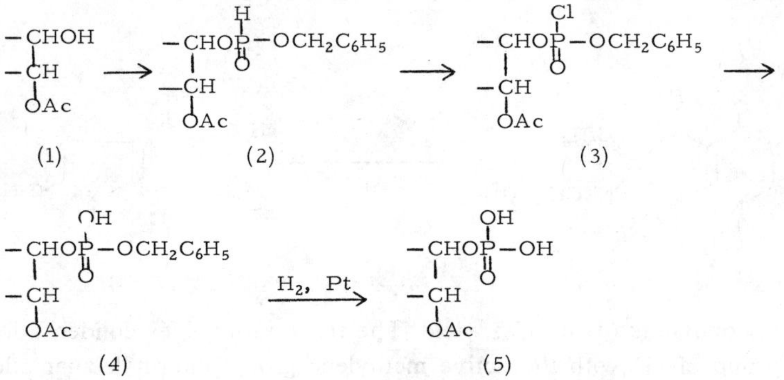

[1]N. S. Corby, G. W. Kenner, and A. R. Todd, *J. Chem. Soc.*, 3669 (1952)
[2]D. M. Brown, G. D. Fasman, D. I. Magrath, and A. R. Todd, *ibid.*, 1448 (1954)

Benzyltrimethylammonium chloride, $C_6H_5CH_2(CH_3)_3\overset{+}{N}\overset{-}{Cl}$. Mol. wt. 185.70. Suppliers of 60% solution in water: Eastern Chem. Co., Commercial Solvents, B, MCB. Supplier of crystalline salt: KK.

The reagent catalyzes the reaction of sodium salts of fatty acids with epichlorohydrin to form glycidyl esters.[1] Other quaternary ammonium halides are less effective, probably owing to decreased solubility in epichlorohydrin.

$$\underline{n}\text{-}C_{11}H_{23}COONa \ + \ ClCH_2CH\overset{\diagdown \diagup}{\underset{O}{\quad}}CH_2 \xrightarrow[84\% \text{ crude}]{} \underline{n}\text{-}C_{11}H_{23}COOCH\overset{\diagdown \diagup}{\underset{O}{\quad}}CH_2$$

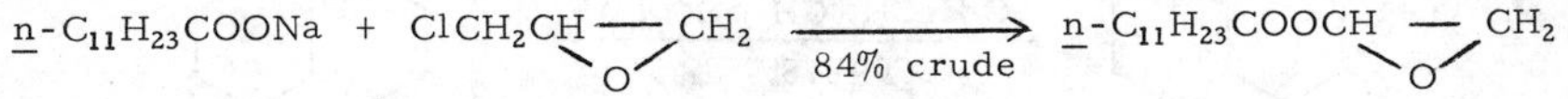

[1]G. Maerker, J. F. Carmichael, and W. Port, *J. Org.*, **26**, 2681 (1961)

Benzyltrimethylammonium cyanide, $C_6H_5CH_2\overset{+}{N}(CH_3)_3\overset{-}{C}N$. Mol. wt. 176.26.

An aqueous solution of the reagent is prepared by dissolving 0.12 mole each of benzyltrimethylammonium chloride and sodium cyanide in 40–50 ml. of water.[1] An alkyl halide added with stirring and cooling is soon converted in part into the nitrile and a benzyltrimethylammonium halide which acts as an emulsifying agent. After the emulsion has been heated on the steam bath for 1–6 hrs., the nitrile is obtained, usually in good yield.

$$RX \ + \ C_6H_5CH_2\overset{+}{N}(CH_3)_3\overset{-}{C}N \longrightarrow RCN \ + \ C_6H_5CH_2\overset{+}{N}(CH_3)_3\overset{-}{X}$$

[1]N. Sugimoto, T. Fujita, N. Shigematsu, and A. Ayada, *Chem. Pharm. Bull.*, **10**, 427 (1962)

Benzyltrimethylammonium hydroxide, *see* Triton B.

Benzyltrimethylammonium iodide, $C_6H_5CH_2\overset{+}{N}(CH_3)_3\overset{-}{I}$. Mol. wt. 277.15, m.p. 181–182°. Supplier: Fisher.

A new purine synthesis[1] is exemplified by the condensation of 4-amino-1,3-dimethyl-5-nitrosouracil (1) with the reagent, a quaternized Mannich base, to give

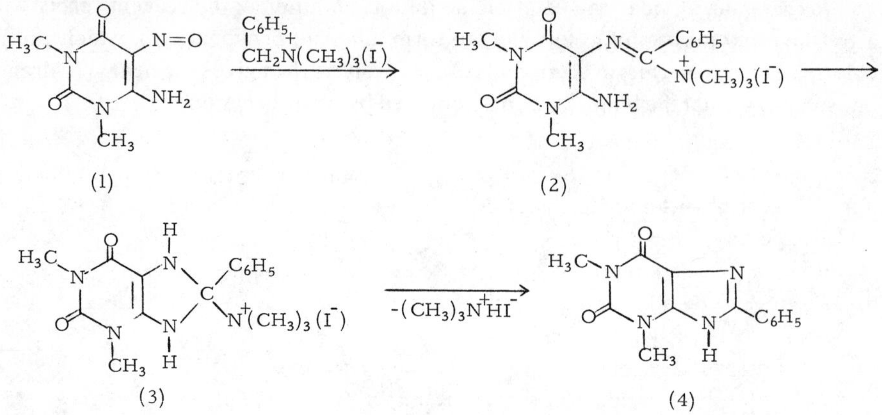

8-phenyltheophylline (4) in 35% yield. The reaction involves condensation of the nitroso group of (1) with the active methylene group, intermolecular addition of the *o*-amino group to the anil group, followed by Hofmann elimination.

[1]E. C. Taylor and E. E. Garcia, *Am. Soc.*, **86**, 4720 (1964)

Benzyltrimethylammonium mesitoate, $C_6H_5CH_2(CH_3)_3\overset{+-}{NO_2}CC_6H_2(CH_3)_3$-2,4,6. Mol. wt. 281.43.

Dehydrohalogenation. A preliminary report[1] states that the reagent is useful for the dehydrohalogenation of α-halo ketones under mild conditions. On treatment with the salt in acetone, 4β-bromocoprostanone afforded cholestenone in good yield.

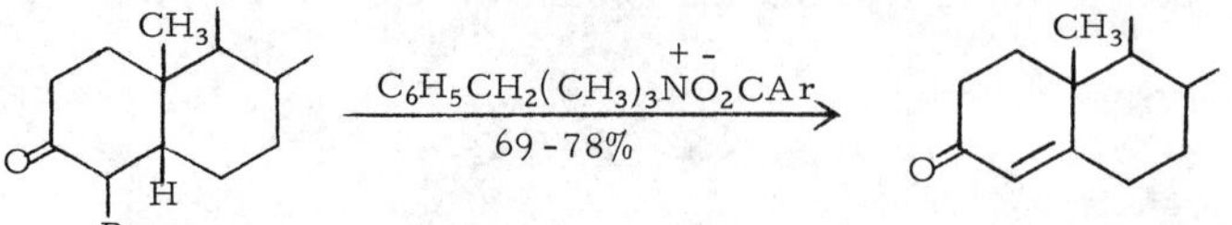

[1]W. S. Johnson and J. F. W. Keana, *Tetrahedron Letters*, 195, note 3 (1963)

Birch reduction, Li(Na), NH_3, ROH. Reviews.[1]

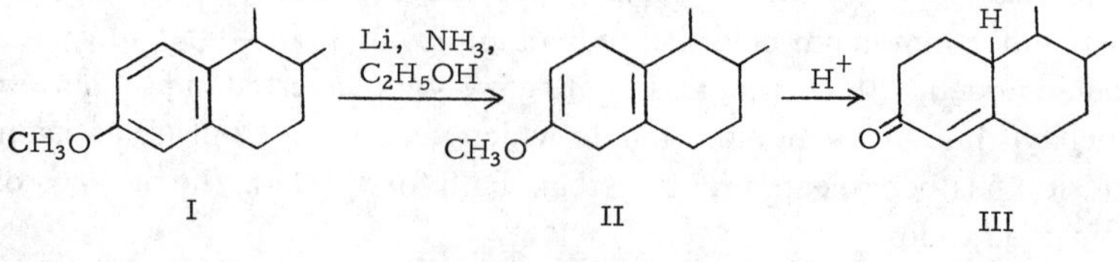

Reduction of derivatives of estradiol 3-methyl ether (I) and hydrolysis of the initially formed enol ether II provides an efficient route to 19-norsteroids (III) of considerable importance in hormone therapy. A. J. Birch, who introduced the method (1949), used sodium in liquid ammonia with ethanol as proton donor. A. L. Wilds and N.A. Nelson[2] (1953) found that yields are improved by use of lithium in place of sodium and that lithium is effective in some cases where sodium is not. The Wilds-Nelson procedure, which became the standard one, employs ether as co-solvent and involves adding the ethanol last; terminal decomposition is done with water after evaporation of ammonia. Since this reaction is the key step in processes developed by G. D. Searle and Co. for the production of two 19-norsteroids of clinical promise, H. L. Dryden, Jr., *et al.*[3] made an extensive investigation aimed at efficient Birch reduction on a large scale. Initial results seemed to confirm the general superiority of the Wilds-Nelson procedure (Li) over that of Birch (Na). Careful study showed, however, that the Wilds-Nelson procedure did not give reproducible results; the extent of reduction varied from run to run without apparent cause. Study of different proton sources and co-solvents, as applied to the reduction of a standard substrate with lithium in distilled ammonia, led to selection of *t*-butanol to fulfill the dual role of proton donor and co-solvent. Yields by the new procedure were high (87%) and reproducible. However, when the ammonia used was taken directly from a cylinder and not distilled, the lithium was consumed in 30 min. as compared to the usual 3–5 hrs., the solution at no time became blue, the crude product contained 23% of starting material, and the aqueous mother liquor remaining after isolation of the product deposited ferric hydroxide. The run had been spoiled by colloidal iron compounds present in commercial ammonia!

Study of the effect of small amounts of iron salts on the reaction of metals with alcohols in liquid ammonia showed that 0.5 ppm. of iron increases the rate of reaction of lithium by a factor of 2 and increases the rate of reaction of sodium by a factor of 50. High yields were then achieved in Birch reductions conducted with sodium and *t*-butanol in iron-free ammonia. Evidently lithium had seemed superior to sodium merely because the lithium–alcohol reaction is catalyzed by iron much less strongly than the sodium–alcohol reaction is catalyzed. Actually sodium usually is as effective as lithium and in some cases definitely superior. Only in the case of 5-methoxytetralin did lithium give a higher yield (62%) than sodium (45%). The reduction of this compound is particularly slow, and hence the greater rate of reduction by lithium becomes significant. Bothner-By[4] found that benzene is reduced by lithium 60 times faster than by sodium.

It appears established (Bothner-By, 1959) that Birch reduction of the benzene ring proceeds by the addition of one electron to give a radical ion (2), which must be protonated (3) before addition of a second electron and proton can occur (4).

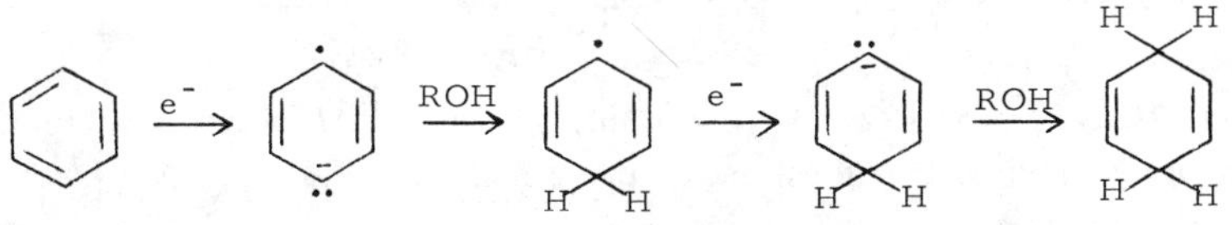
e⁻
ROH
e⁻
ROH
H
H
H
H
H
H
H
H

(1) (2) (3) (4) (5)

Since ammonia is not acidic enough to effect this protonation it is necessary to use an alcohol.

Reduction of α-naphthol by the Wilds-Nelson technique to 5,8-dihydro-1-naphthol is the first step in a procedure for the preparation of *ar*-tetrahydro-1-naphthol (overall yield 84–88%).[5]

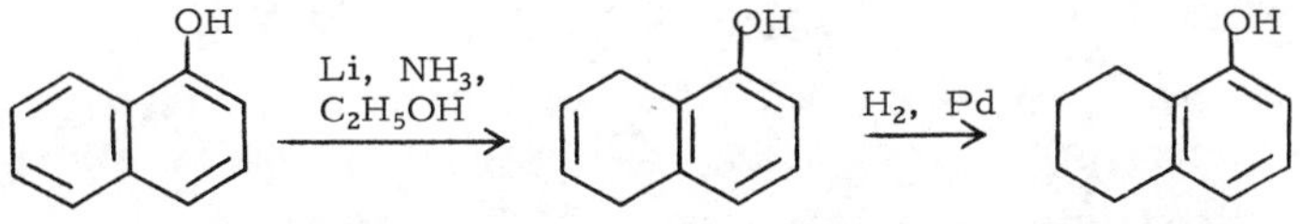
OH
Li, NH3,
C2H5OH
OH
H2, Pd
OH

Birch reduction of benzoic acid (Na, NH_3, C_2H_5OH) affords 1,4-dihydrobenzoic acid in yield of 89–95%.[6,7] Other examples of related reductions have been summarized.[7]

Fried and Abraham[8] found that the nature of the co-solvent used to improve solubility in Li–NH_3–proton donor reduction of steroid ketones plays an important role. Use of ether or 1,2-dimethoxyethane as co-solvent leads to pinacolic reduction to the extent of as much as 30%. The side reaction is suppressed by use of tetrahydrofurane, dioxane, or no co-solvent. Diglyme is intermediate.

[1]A. J. Birch, *Quart. Rev.*, **4**, 69 (1950); A. J. Birch and H. Smith, *ibid.*, **12**, 17 (1958); G. W. Watt, *Chem. Rev.*, **46**, 317 (1950)
[2]A. L. Wilds and N. A. Nelson, *Am. Soc.*, **75**, 5366 (1953)
[3]H. L. Dryden, Jr., G. M. Webber, R. R. Burtner, and J. A. Cella, *J. Org.*, **26**, 3237 (1961)
[4]A. P. Krapcho and A. A. Bothner-By, *Am. Soc.*, **81**, 3658 (1959); **82**, 751 (1960)
[5]C. D. Gutsche and H. H. Peter, *Org. Syn., Coll. Vol.*, **4**, 887 (1963)
[6]M. E. Kuehne and B. F. Lambert, *Am. Soc.*, **81**, 4278 (1959)
[7]*Idem, Org. Syn.*, **43**, 22 (1963)
[8]J. Fried and N. A. Abraham, *Tetrahedron Letters*, 1879 (1964)

Bis-(benzonitrile)-palladium (II) chloride, $(C_6H_5CN)_2PdCl_2$. Mol. wt. 383.85.

This stable, orange-yellow compound is prepared in good yield by heating palladium (II) chloride in benzonitrile on the steam bath until solution is effected; the product separates on cooling.[1] The benzonitrile ligands can be displaced by olefins

$$PdCl_2 + 2\ C_6H_5CN \xrightarrow[88\text{-}95\%]{87^0} (C_6H_5CN)_2PdCl_2$$

to give very unstable coordination compounds of the formula $(C_nH_{2n}PdCl_2)_2$. The reagent is soluble in benzene, chloroform, acetone.

Blomquist and Maitlis[2] found that this organometallic compound in catalytic amount in an aprotic solvent (benzene, chloroform, acetone) effects trimerization of diphenylacetylene to hexaphenylbenzene in 80–85% yield.

[1]M. S. Kharasch, R. C. Seyler, and F. R. Mayo, *Am. Soc.*, **60**, 882 (1938)
[2]A. T. Blomquist and P. M. Maitlis, *Am. Soc.*, **84**, 2329 (1962)

Bis(2,6-diethylphenyl)carbodiimide, 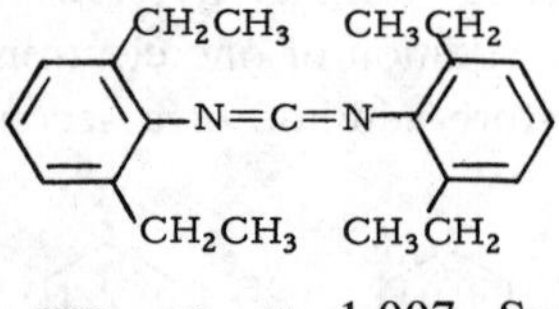
CH2CH3 CH3CH2
N=C=N
CH2CH3 CH3CH2

Mol. wt. 306.20, b.p. 192–194°/0.4 mm., sp. gr. 1.007. Supplier: Upjohn Co., Carwin Org. Chemicals; smallest package 8 oz. (225 g.).

Bis-(2,4-dinitrophenyl)carbonate (1). Mol. wt. 394.25, m.p. 148°. Preparation.[1]

The reagent (1) reacts with an amino acid ester (2) to give an N-activated derivative (3) suitable for condensation with an N-protected amino acid (4) to give a dipeptide ester (5).

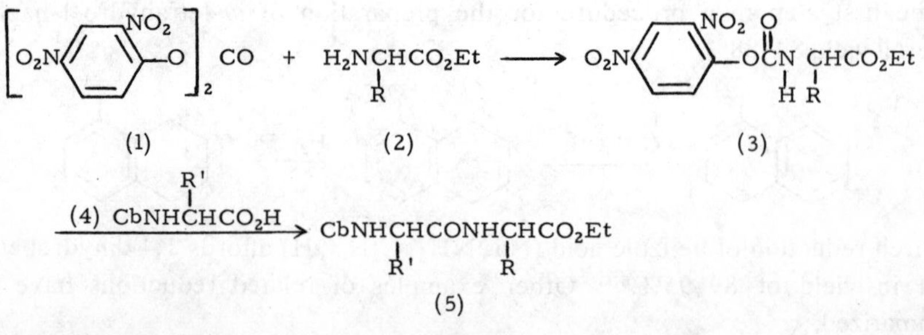
NO2
O2N
O
CO
H2NCHCO2Et
R
NO2 O
O2N
OCN CHCO2Et
H R
(1)
(2)
(3)
R'
(4) CbNHCHCO2H
CbNHCHCONHCHCO2Et
R'
R
(5)

[1]R. Glatthard and M. Matter, *Helv.*, **46**, 795 (1963)

α,γ-Bis(diphenylene)-β-phenylallyl (BDPA). Mol. wt. 417.50, m.p. 188–191°.

This stable radical was prepared by Koelsch[1] by reaction of the corresponding chloride with mercury, and by Kuhn and Neugebauer[2] by the process formulated.

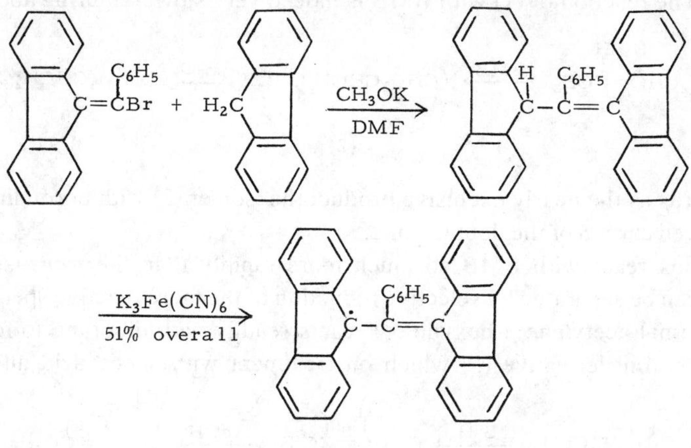
C6H5
C=CBr + H2C
CH3OK
DMF
H C6H5
C–C=C
K3Fe(CN)6
51% overall
C6H5
C–C=C

BDPA

The radical crystallizes from benzene as a green 1:1 complex, m.p. 231–233°, or without solvent from acetic acid as brown needles, m.p. 188–191°. It is highly dissociated but remarkably stable to oxygen. In fact the original paper describing the substance and submitted by Koelsch in 1932 was initially rejected because a referee held that the properties were not those of a radical. Of the available radicals, BDPA was found to be the most suitable as a radical scavenger in the decomposition of cyclohexylformyl and isobutyryl peroxides.[3]

[1]C. F. Koelsch, *Am. Soc.*, **79**, 4439 (1957)

[2]R. Kuhn and F. A. Neugebauer, *Monatsh.*, **95**, 3 (1964)

[3]R. C. Lamb, J. G. Pacifici, and P. W. Ayers, *Am. Soc.*, **87**, 3928 (1965); see also R. C. Lamb, L. P. Spadafino, R. G. Webb, E. B. Smith, and J. G. Pacifici, *J. Org.*, **31**, 147 (1966)

Bis-iodomethylzinc-Zinc iodide, *see* Simmons-Smith reagent.

Bis-3-methyl-2-butylborane (BMB, "Disiamylborane"), $[(CH_3)_2CHCH(CH_3)]_2BH$. Mol. wt. 154.10, m.p. 35–40°.

The reagent is prepared by reaction of diborane with 3-methyl-2-butene (Aldrich); even with excess olefin, the reaction proceeds only to the dialkylborane stage; the semisolid reaction product is used as such.[1] In another procedure a solution of

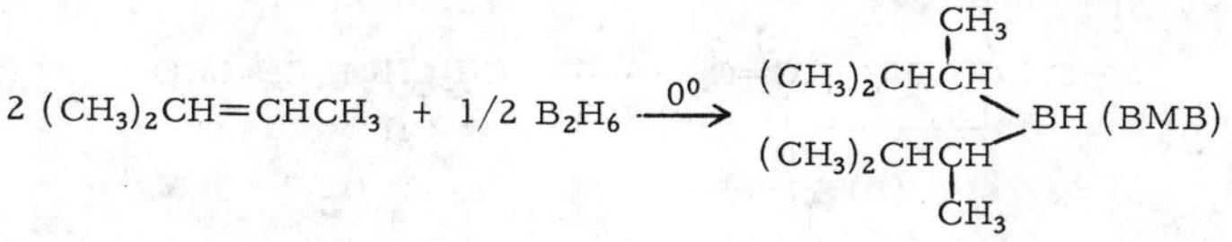
2 (CH3)2CH=CHCH3 + 1/2 B2H6
0°
CH3
(CH3)2CHCH
BH (BMB)
(CH3)2CHCH
CH3

diborane in tetrahydrofurane is treated with the theoretical quantity of 3-methyl-2-butene in tetrahydrofurane; removal of the solvent gives white crystals unstable to air.

The reagent has reducing properties similar to diborane, but the larger steric requirement offers certain advantages. It has an enhanced tendency to add to the terminal position of a 1-alkene. Thus 1-hexene is converted by hydroboration with BMB and hydrogen peroxide oxidation into 1-hexanol in 99% yield, whereas when diborane is used the yield of primary alcohol is 94%, with 6% of the secondary alcohol also formed. Greater steric control is possible also in the hydroboration of an unsymmetrical internal olefin such as *cis*-4-methyl-2-pentene (1) which on reaction with diborane and H_2O_2- oxidation affords nearly equal amounts of the two possible alcohols. The reaction of (1) with BMB is indeed very slow, requiring about 12 hrs.,

$$(CH_3)_2CH\overset{\overset{\text{H}}{|}}{C}=\overset{\overset{\text{H}}{|}}{C}CH_3 \xrightarrow{\text{BMB}} (CH_3)_2CHCH_2\underset{\underset{\text{BR}_2}{|}}{C}HCH_3 \xrightarrow{\text{H}_2\text{O}_2} (CH_3)_3CHCH_2\underset{\underset{\text{OH}}{|}}{C}HCH_3$$

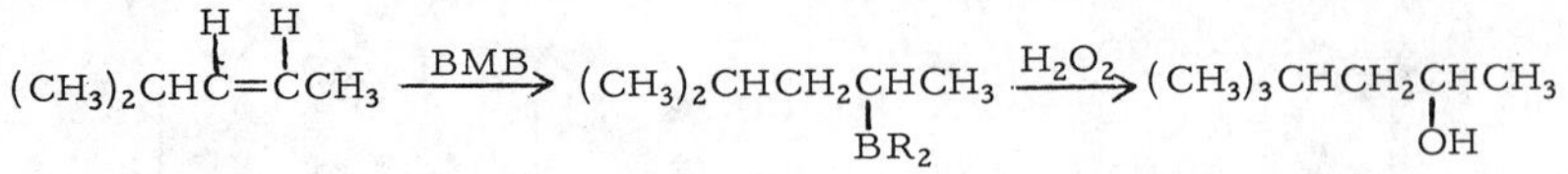

$$\text{(1)} \qquad\qquad\qquad \text{(2)} \qquad\qquad\qquad \text{(3) 97\%}$$

but it affords as the nearly exclusive product the isomer (2) with boron linked to the less hindered carbon of the double bond.

cis-Olefins react with BMB so much more rapidly than the *trans*-isomers that mixtures can be separated by selective reduction of the more reactive species.

The internal acetylene 3-hexyne (4) reacts readily with diborane to form an unsaturated boron derivative (5) which on treatment with acetic acid affords *cis*-3-

$$C_2H_5C\equiv CC_2H_5 \xrightarrow{\text{B}_2\text{H}_6} C_2H_5\overset{\overset{\text{H}}{|}}{C}=\overset{\overset{\text{BH}_3}{|}}{C}C_2H_5 \xrightarrow{\text{AcOH}} C_2H_5\overset{\overset{\text{H}}{|}}{C}=\overset{\overset{\text{H}}{|}}{C}C_2H_5$$

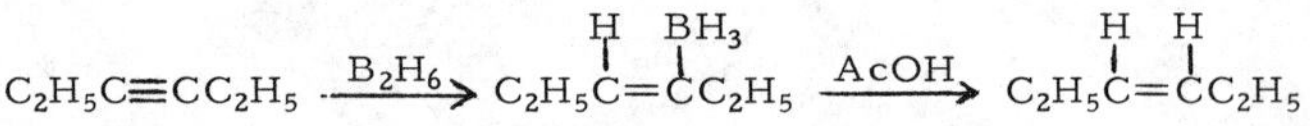

$$\text{(4)} \qquad\qquad\qquad \text{(5)} \qquad\qquad\qquad \text{(6)}$$

hexene. Attempted hydroboration with diborane of 1-hexyne (7), a terminal acetylene, showed that half of the hydrocarbon had not reacted whereas the remaining half had added two H—B bonds. With the bulkier reagent BMB, however, the reaction stopped with formation of a monoadduct, which on treatment with acetic acid afforded 1-hexene in yield of 90%. Boron was shown to be on the terminal carbon,

$$\underline{n}\text{-}C_4H_9C\equiv CH \xrightarrow{\text{BMB}} \underline{n}\text{-}C_4H_9CH=\underset{\underset{\text{BR}_2}{|}}{C}H \overset{\xrightarrow{\text{AcOH}}}{\underset{\xrightarrow{\text{H}_2\text{O}_2}}{}} \begin{array}{l} \underline{n}\text{-}C_4H_9CH=CH_2 \\ \text{(9)} \\ \\ \underline{n}\text{-}C_4H_9CH_2CHO \\ \text{(10)} \end{array}$$

$$\text{(7)} \qquad\qquad\qquad \text{(8)}$$

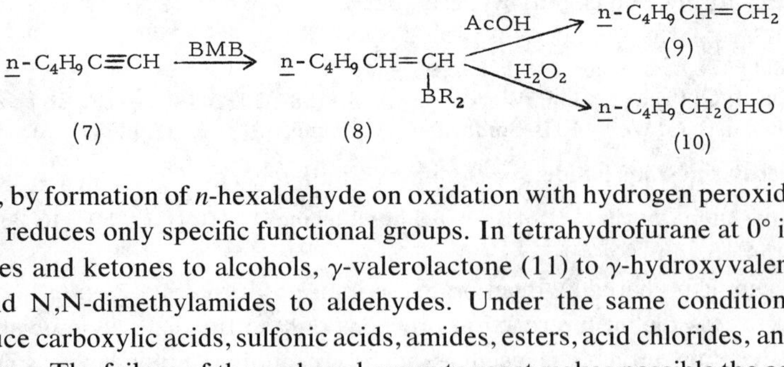

as in (8), by formation of *n*-hexaldehyde on oxidation with hydrogen peroxide.

BMB reduces only specific functional groups. In tetrahydrofurane at 0° it reduces aldehydes and ketones to alcohols, γ-valerolactone (11) to γ-hydroxyvaleraldehyde (12), and N,N-dimethylamides to aldehydes. Under the same conditions it does not reduce carboxylic acids, sulfonic acids, amides, esters, acid chlorides, anhydrides, or sulfones. The failure of the carboxyl group to react makes possible the conversion

$$CH_3CHCH_2CH_2C=O \xrightarrow[76\%]{\text{BMB}} CH_3\underset{\underset{\text{OH}}{|}}{C}HCH_2CH_2CHO$$

$$\text{(11)} \qquad\qquad\qquad \text{(12)}$$

of Δ^{10}-undecylenic acid (13) into 11-hydroxyundecanoic acid (14).[2] Oxides and nitriles are reduced very slowly by BMB.

$$CH_2=CH(CH_2)_8CO_2H \xrightarrow[82\%]{\text{BMB; alk. H}_2\text{O}_2} HO(CH_3)_{10}CO_2H$$

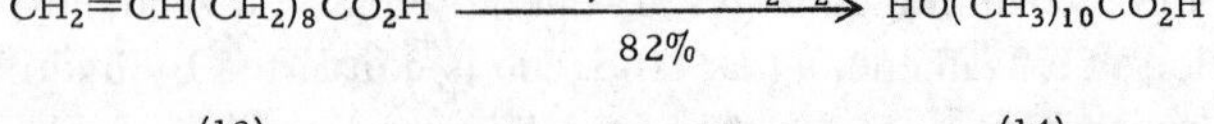

$$\text{(13)} \qquad\qquad\qquad\qquad \text{(14)}$$

On hydroboration with BMB and peroxide oxidation, steroid 1,2-disubstituted ethylenes give about equal amounts of the two possible isomers.[3] Thus Δ^1-cholestene gives cholestane-1α-ol (35%) and cholestane-2α-ol (40%).

BMB is superior to diborane for dihydroboration of terminal dienes to give, after oxidation, α,ω-diols.[4] Thus 1,5-hexadiene is converted into 1,6-hexanediol in 93% yield, whereas with diborane the yield is 69% and the remainder is a mixture of isomeric diols. With disiamylborane, Zweifel, Nagase, and Brown were able to effect monohydroboration of cyclic dienes in reasonable yield. This finding was used to advantage by a Merck group[5] for selective conversion of the diene (1) to the primary alcohol (2).

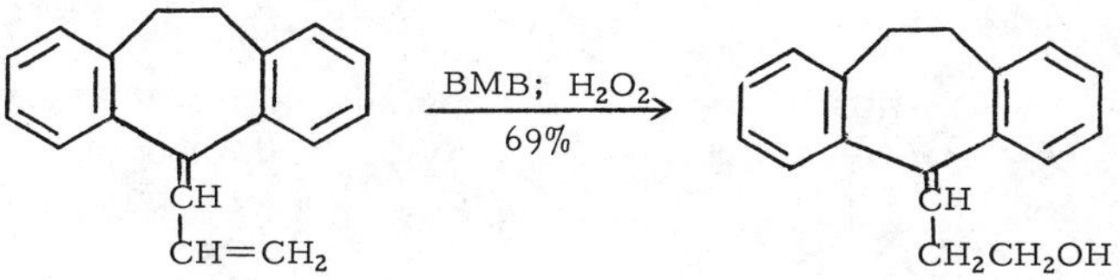
CH
CH=CH₂
BMB; H₂O₂
69%
CH
CH₂CH₂OH

(1) (2)

BMB has been used to reduce 2,3,5,6-tetra-O-acyl-D-hexono-γ-lactones to the furanoses in high yields.[6] Furanose derivatives are difficult to obtain by other methods, but the new method takes advantage of the preferential conversion of sugar acids into γ-lactones.

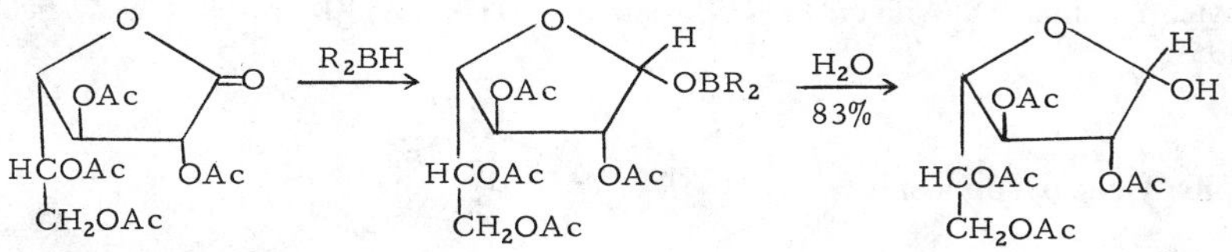
O
OAc
HCOAc
OAc
CH₂OAc
R₂BH
H
OBR₂
H₂O
83%
H
OH

Reduction with sodium amalgam or sodium borohydride affords furanoses in only very low yield.

[1]H. C. Brown and G. Zweifel, *Am. Soc.*, **81**, 1512 (1959); **82**, 3222, 3223 (1960); **83**, 1241 (1961)
[2]H. C. Brown and D. B. Bigley, *ibid.*, **83**, 486 (1961)
[3]F. Sondheimer and M. Nussim, *J. Org.*, **26**, 630 (1961)
[4]G. Zweifel, K. Nagase, and H. C. Brown, *Am. Soc.*, **84**, 190 (1962)
[5]R. D. Hoffsommer, D. Taub, and N. L. Wendler, *J. Org.*, **28**, 1751 (1963)
[6]P. Kohn, R. H. Samaritano, and L. M. Lerner, *Am. Soc.*, **86**, 1457 (1964); **87**, 5475 (1965)

Bis-(N-methyl-N-nitroso)terephthalamide. Precursor of Diazomethane, *which see.*

Bismuth trioxide, Bi_2O_3. Mol. wt. 466.00. Suppliers: Alfa, B, F, MCB.

The oxide is a specific reagent for oxidation of acyloins to α-diketones, the oxide being reduced to the metal.[1] A solution of the acyloin in acetic acid is treated with 1.2 times the theoretical amount of Bi_2O_3 and the mixture is heated on the steam bath with stirring for about 15 min. Benzoin → benzil (95% yield).[1] 12-Hydroxy-11-ketosteroids → 11,12-diketosteroids (70% yield).[2] Oxidation of cevine and isomers.[3] Other examples are the oxidation of 2-hydroxypulegone (1) to diosphenolone (2) in 94% yield (crude),[4] of the 2α-hydroxy-Δ^4-3-ketosteroid (3) to (4),[5] and of (5) to (6).[6]

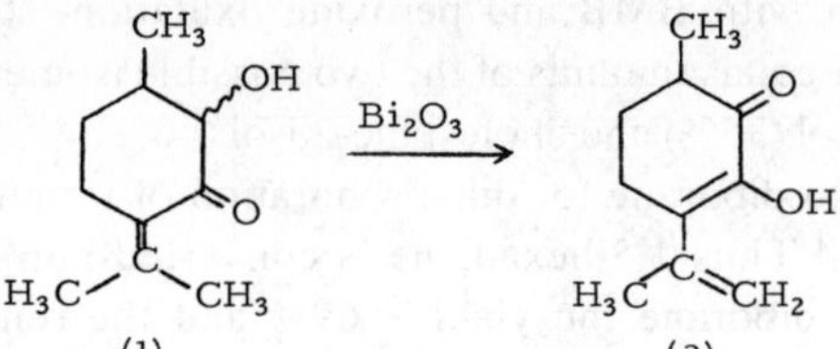

(1) (2)

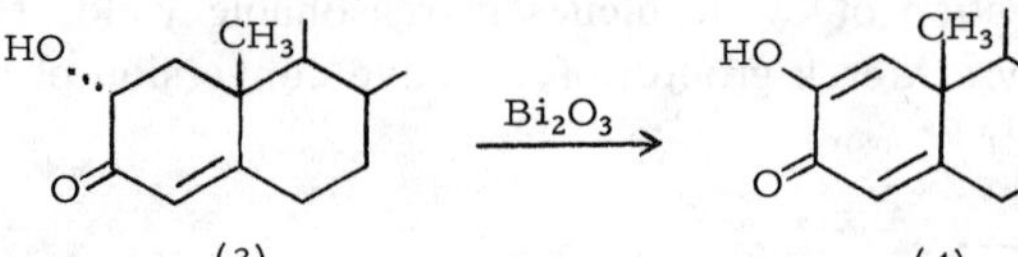

(3) (4)

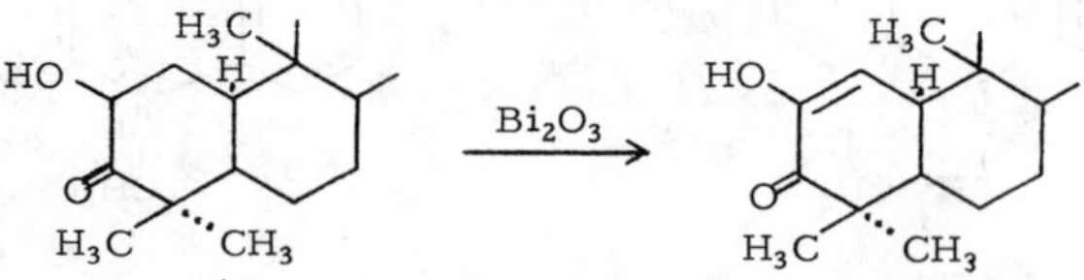

(5) (6)

[1]W. Rigby, *J. Chem. Soc.*, 793 (1951)
[2]C. Djerassi, H. J. Ringold, and G. Rosenkranz, *Am. Soc.*, **76**, 5533 (1954)
[3]S. M. Kupchan and D. Lavie, *Am. Soc.*, **77**, 683 (1955)
[4]R. H. Reitsema, *Am. Soc.*, **79**, 4465 (1957)
[5]J. S. Baran, *Am. Soc.*, **80**, 1687 (1958)
[6]D. Lavie, Y. Shvo, D. Willner, P. R. Enslin, J. M. Hugo, and K. B. Norton, *Chem. Ind.*, 951 (1959)

Bis-*o*-phenylene pyrophosphite, 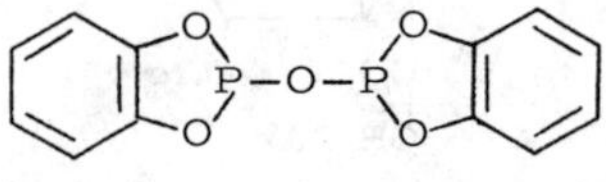

Mol. wt. 294.13, m.p. 72°, b.p. 156°/0.25 mm.

Prepared from catechol and phosphorus trichloride via *o*-phenylenephosphorochloridite.[1] Used in peptide synthesis for the direct interaction of the two components[2] (*compare* tetraethyl pyrophosphite). Synthesis of dipeptides is satisfactory, but extensive racemization occurs when N-acyl peptides are employed.

[1]P. C. Crofts, J. H. H. Markes, and H. N. Rydon, *J. Chem. Soc.*, 4250 (1958)
[2]*Idem, ibid.*, 3610 (1959)

O,N-Bis-(trifluoroacetyl)-hydroxylamine, $HN(COCF_3)—OCOCF_3$. Mol. wt. 225.06, m.p. 62°.

Prepared from trifluoroacetic anhydride and hydroxylamine.[1] Reacts with aldehydes in pyridine–benzene to give the corresponding nitriles in 70–90% yield.[1]

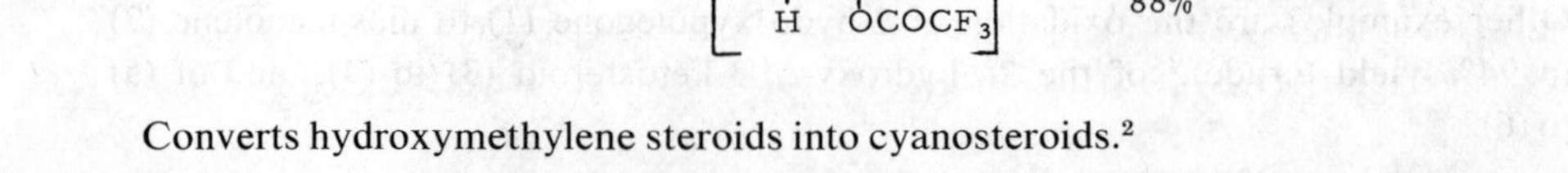
$C_6H_5CH{=}CHCHO$ + $HN(COCF_3)—OCOCF_3$
OH COCF3
H OCOCF3
-2 CF3CO2H
88%
$C_6H_5CH{=}CHCN$

Converts hydroxymethylene steroids into cyanosteroids.[2]

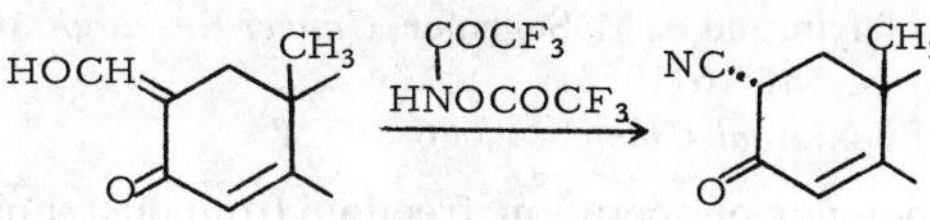
HOCH
CH3
COCF3
HNOCOCF3
NC
CH3
O
O

[1]J. H. Pomeroy and C. A. Craig, *Am. Soc.*, **81**, 6340 (1959)
[2]H. M. Kissman, A. S. Hoffman, and M. J. Weiss, *J. Org.*, **27**, 3168 (1962)

Bis(trimethylsilyl)acetamide (1). Mol. wt. 203.44, b.p. 71–73°/35 mm. Prepared in 80% yield by reaction of acetamide with trimethylchlorosilane with triethylamine as catalyst, the reagent effects ***trimethylsilylation*** of amides, ureas, amino acids, phenols, carboxylic acids, enols.[1]

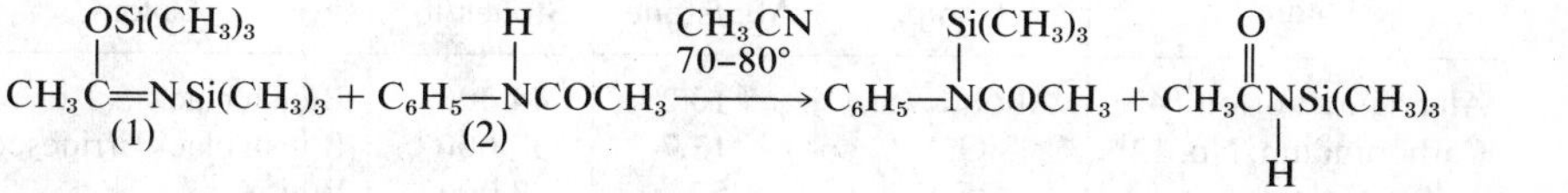
OSi(CH3)3
CH3C=NSi(CH3)3 + C6H5—NCOCH3
H
(1)
(2)
CH3CN
70–80°
Si(CH3)3
C6H5—NCOCH3 + CH3CNSi(CH3)3
O
H

[1]J. F. Klebe, H. Finkbeiner, and D. M. White, *Am. Soc.* **88**, 3390 (1966)

Bistriphenylphosphinedicarbonylnickel, $[(C_6H_5)_3P]_2Ni(CO)_2$. This complex catalyzes trimerization of acetylenes to aromatic compounds,[1] and in the presence of acetylene promotes dimerization of butadiene to cyclooctadiene-1,5.[2]

[1]W. Reppe and W. J. Schweckendiek, *Ann.*, **560**, 104 (1948)
[2]H. W. B. Reed, *J. Chem. Soc.*, 1931 (1954)

Blue tetrazolium, Mol. wt. 658.74, m.p. 244°. Suppliers: A, B, F, MCB. Preparation.[1] Purified by crystallization from pyridine.[2]

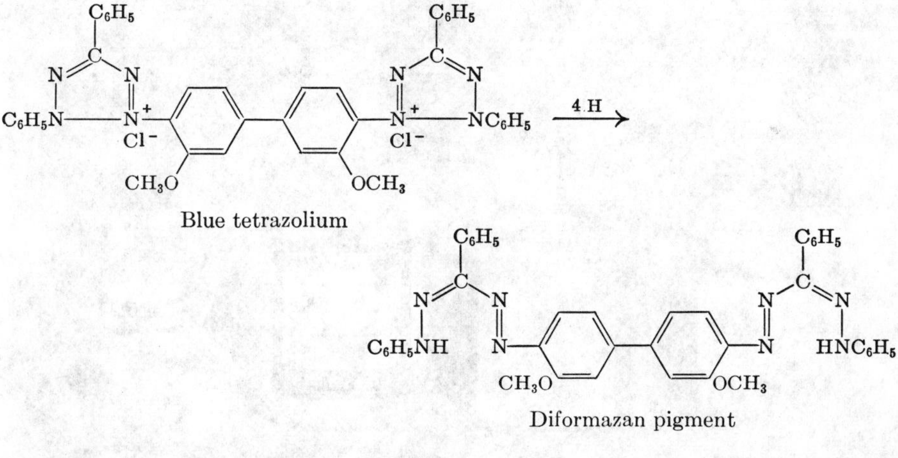
C6H5
C6H5N
N
Cl−
CH3O
OCH3
NC6H5
4 H
Blue tetrazolium
C6H5NH
HNC6H5

Diformazan pigment

The reagent is a pale yellow, water-soluble substance which, in the presence of a trace of alkali, oxidizes aldoses, ketoses, and other α-ketols and is thereby reduced to the water-insoluble, intense blue diformazan. This sensitive test for reducing sugars distinguishes between α-ketols and simple aldehydes. Thus the order of reactivity is as follows: fructose > glucose > lactose > maltose > *n*-butyraldehyde.

The reagent is useful for detection of dehydrogenase in tissues, cells, and bacteria,[1] for research in seed germination, and for determination of cortical steroids.[3]

[1]A. M. Rutenburg, R. Gofstein, and A. M. Seligman, *Cancer Research*, **10**, 113 (1950)
[2]W. Ried and H. Gick, *Ann.*, **581**, 16 (1953)
[3]W. J. Mader and R. R. Buck, *Anal. Chem.*, **24**, 666 (1952)

Boiling stones. For efficiency of operation, freedom from dust, uniformity, cheapness, and ready visibility in a reaction product, carborundum stones are recommended above all others. One grain is sufficient for a given operation. In our opinion the next best of those listed in the table are the alundum stones. These are likewise bluish black, if not irridescent; in comparison with the coarser carborundum No. 12, the No. 14 stones seem a little too small and too light. The other two stones listed doubt-

Boiling stones

Name	Comp.	Mg./stone	Stones/lb.	Color
Alundum, Norton 14[a]	Al_2O_3(cryst.)	10.2	44,300	Bluish black
Carborundum, No. 12[b]	SiC	13.7	37,000	Bluish black, irridescent
Teflon-Halon[c]	$—(CF_2)_2—$	57	8,000	White
Hengar granules[d]	Al_2O_3	190	2,406	White

[a]Suppliers: A. H. Thomas Co.; Henry Troemner, Inc.
[b]Carborundum Co.; supplier in 1-lb. lots: Henry Troemner, Inc.
[c]Supplier: Chemical Rubber Co.
[d]Hengar Co., Philadelphia, Pa.; suppliers: Thomas; Troemner

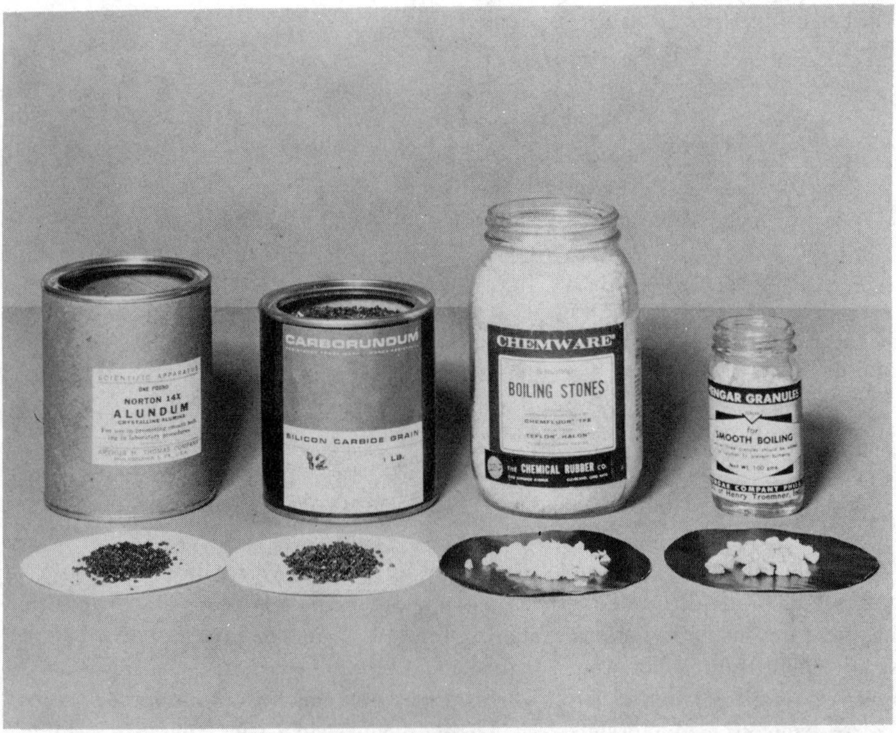
NORTON 14X
ALUNDUM
CARBORUNDUM
SILICON CARBIDE GRAIN
CHEMWARE
BOILING STONES
SMOOTH BOILING

Fig. B-1

less have application where nothing but even boiling matters, although it should be noted that the directions call for use of "two or three granules" (Hengar) or state that "a dozen (Teflon/Halon) stones should do the job." However, for boiling down a filtered solution before it is left for crystallization, a white stone has the serious disadvantage of not being easily visible in the usual crystallizate (white, yellow, etc.). Furthermore these white stones both seem to us unduly large and clumsy.

Boric acid, H_3BO_3. Mol. wt. 61.84. Suppliers: B, E, F, MCB.

Dehydration. Alcohols, primary as well as secondary and tertiary, are dehydrated in good yield when heated to 350° with an equimolecular quantity of boric acid.[1] The reagent effected dehydration of the alcohol (1) to the triene (2); in this case conventional methods failed or gave (2) in very low yield.[2]

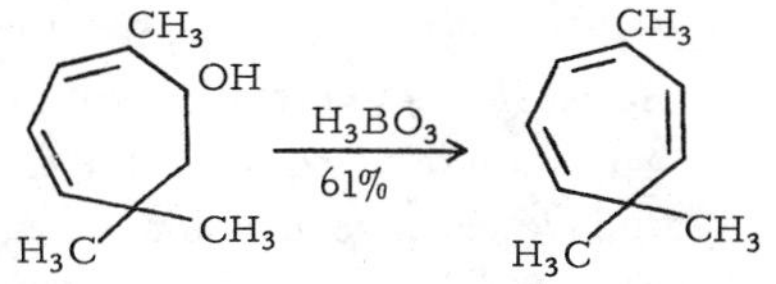
CH3
OH
H3BO3
61%
CH3
H3C
CH3
H3C
CH3

(1) (2)

However, in a system prone to carbonium ion rearrangements, dehydration with boric acid leads largely to rearranged products, as illustrated for 3,3-dimethyl-2-

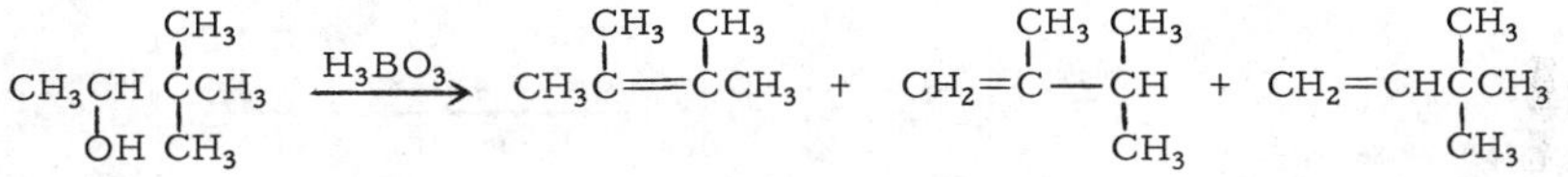
H3BO3

(1) (2) 73.7% (3) 25.9% (4) 0.3%

butanol.[3] Pyrolysis of the acetate of (1) gave exclusively the normal product (4).

Color reactions. O. Dimroth[4] developed a diagnostic color reaction utilizing a solution of boroacetic anhydride prepared by simply dissolving a little boric acid in acetic anhydride. 2-Hydroxyanthraquinone, and polyhydroxyanthraquinones having β- but no α-hydroxyl groups, react with this reagent, as with acetic anhydride alone, to form a normal acetate of unexceptional color. 1-Hydroxyanthraquinone, however, gives an intensely colored boroacetate complex (1) recognizable

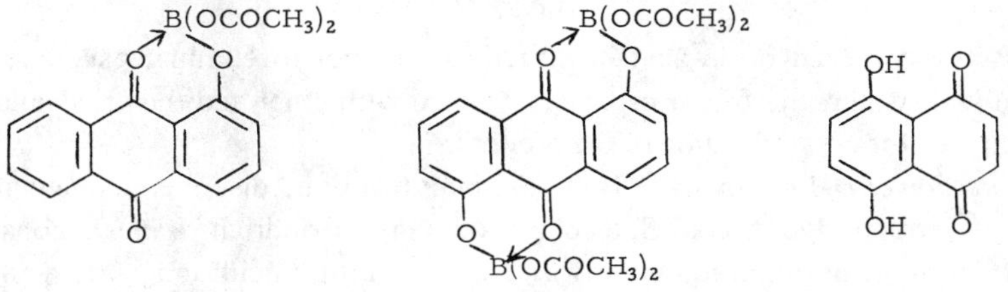
B(OCOCH3)2
B(OCOCH3)2
B(OCOCH3)2
OH
OH

(1) (2) (3)

from the brilliant color and verifiable by isolation and analysis. 1,5-Dihydroxyanthraquinone affords a di-derivative (2). On applying the test to naphthazarin, Dimroth found that the substance has the structure (3) and is not, as previously supposed, the lower benzolog of alizarin. One of us[5] established that the color test is valid in the phenanthrenequinone series and used it to advantage in deducing the structures of six products of disulfonation of phenanthrene.

In the laboratory of the California Fruit Growers Exchange it was observed that lemon juice dried in the presence of boric acid acquires a brilliant yellow color.[6] The color-reactive substance was identified as quercetin or its 3-L-rhamnoside quercitrin, which can be detected in amounts of 0.002 mg. and 0.004 mg. on treatment with a 1:1 mixture of acetone saturated with boric acid and a 10% solution

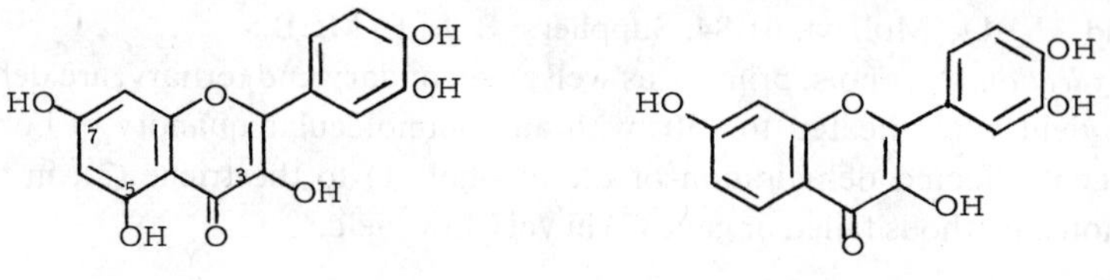

Quercetin (positive)   Fisetin (negative)

of citric acid in acetone. The test is positive with flavones which contain an auxochrome (OH, OCH_3) in the 5-position. For example, the test is negative with fisetin. Flavanones, that is 2,3-dihydroflavones, give no color.

Acetonide cleavage. A general process for the synthesis of glycerol α-esters[7] involves refluxing a mixture of glycerol, acetone, chloroform, and a trace of *p*-toluenesulfonic acid under a water separator until the theoretical amount of water has collected, adding an acid, for example stearic acid, and refluxing again to an end point. The reaction mixture is worked up and the ester-acetonide is refluxed

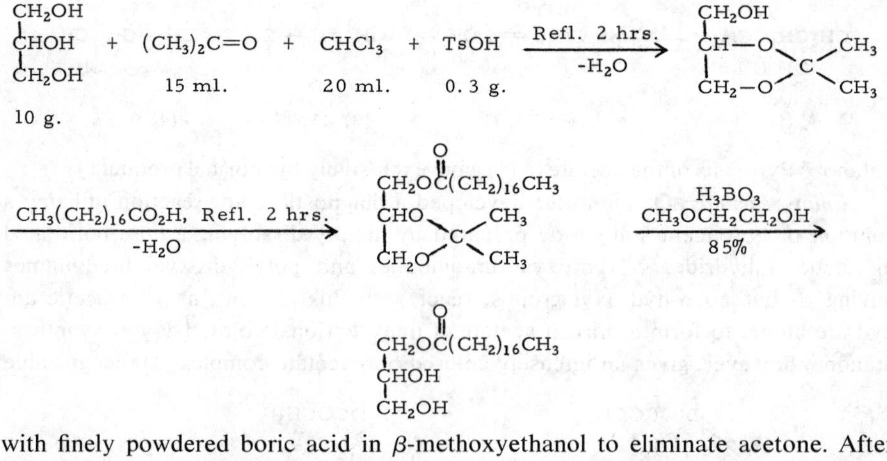

with finely powdered boric acid in β-methoxyethanol to eliminate acetone. After dilution with water, the free α-ester is extracted with ether and the dried solution is concentrated for crystallization of the product.

Protection of OH groups. The Bohn-Schmidt reaction,[8] discovered independently in rival German dye firms (Badische Company, Friedrich Bayer), consists in hydroxylation of an anthraquinone with fuming sulfuric acid and boric acid, often in the presence of a catalytic amount of mercury. Boric acid appears to be beneficial

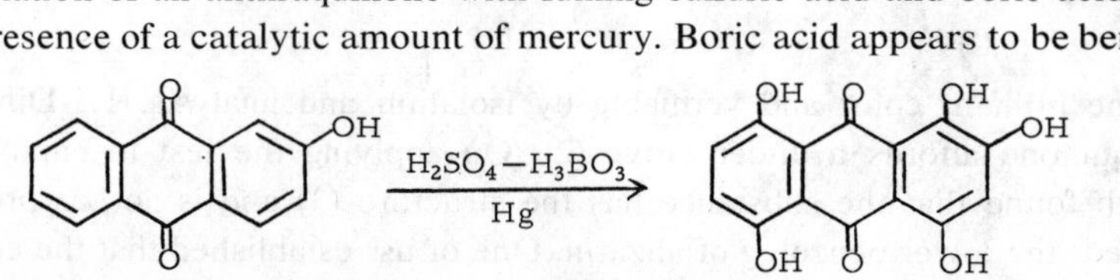

because hydroxyl groups are protected during the reaction as boric acid ester groups. The acid exerts the same function in the synthesis of quinizarin formulated.[9]

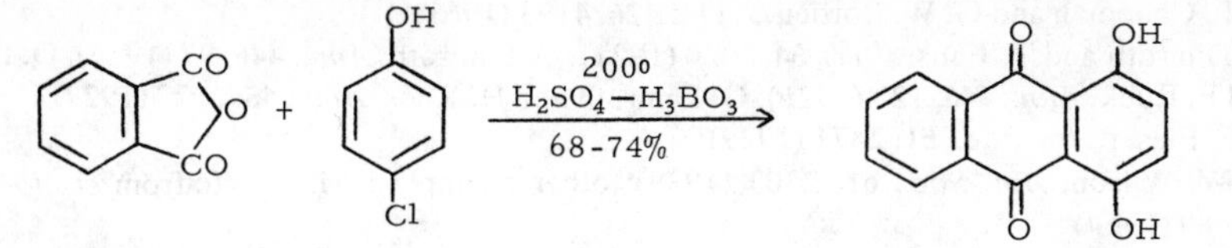

Condensation catalyst. Deichler and Weizmann[10] found that phthalic anhydride can be condensed with α-naphthol in a mixture of sulfuric and boric acid to give either 2-phthaloyl-1-naphthol or 11-hydroxynaphthacene-9,10-quinone, depending upon the conditions. The yield in the first step for acid purified by crystallization of

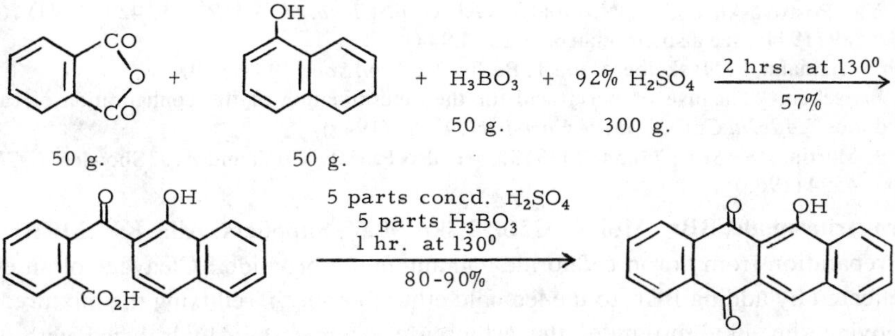

the sodium salt is that obtained by one of us as a doctoral candidate. The yield for the ring closure is that reported in the Russian literature.[11] In the initial condensation, boric acid plays a unique role in a reaction which is noteworthy also because of the exclusive and unusual *ortho* substitution. An interpretation suggested later by Weizmann[12] is that an initially formed acid phthalate undergoes Fries rearrangement to an *o*-hydroxycarbonyl compound stabilized by formation of a chelated boric acid complex.

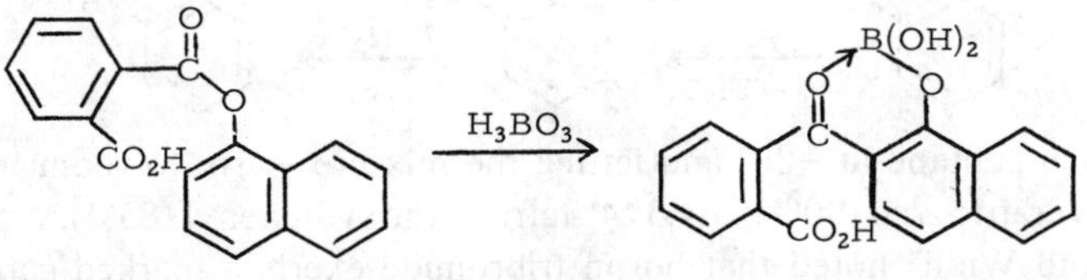

Carbohydrate complexes. Böeseken[13] deduced configurations of sugar anomers by studying their effect on the electrical conductivity of boric acid solutions.

Synthesis of 2-monoglycerides.[14] Boric acid is recommended for removal of the 1,3-benzylidene group from 1,3-benzylidene-2-acyl glycerols. The derivative is heated in triethyl borate with finely powdered boric acid until conversion to the diborate ester is complete and this is then hydrolyzed with water, for example to 2-monoolein.

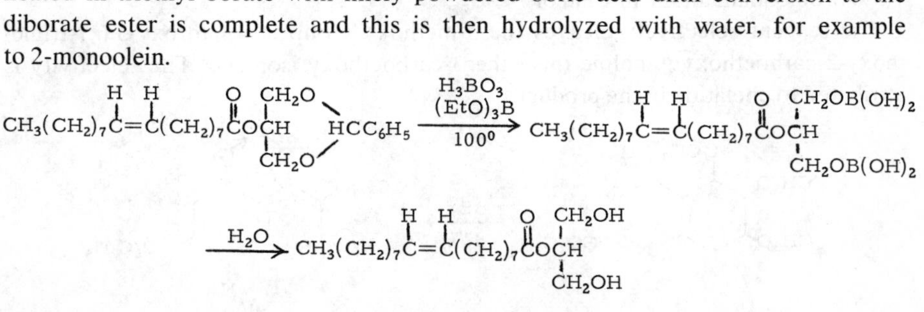

[1]W. Brandenberg and A. Galat, *Am. Soc.*, **72**, 3275 (1950)
[2]J. R. B. Campbell, A. M. Islam, and R. A. Raphael, *J. Chem. Soc.*, 4096 (1956)
[3]O. L. Chapman and G. W. Borden, *J. Org.*, **26**, 4193 (1961)
[4]O. Dimroth and T. Faust, *Ber.*, **54**, 3020 (1921); O. Dimroth, *Ann.*, **446**, 97 (1926); O. Dimroth and F. Ruck, *Ann.*, **446**, 123 (1926); O. Dimroth and H. Roos, *Ann.*, **456**, 177 (1927)
[5]L. F. Fieser, *Am. Soc.*, **51**, 2471 (1929)
[6]C. W. Wilson, *Am. Soc.*, **61**, 2303 (1939); other examples: M. L. Wolfrom *et al.*, *J. Org.*, **29**, 689 (1964)
[7]L. Hartman, *Chem. Ind.*, 711 (1960)
[8]R. Bohn, German patent 46,654 (1889); R. E. Schmidt, *ibid.*, 60,855 (1891); M. Phillips, *Chem. Rev.*, **6**, 168 (1929)
[9]L. A. Bigelow and H. H. Reynolds, *Org. Syn.*, *Coll. Vol.*, **1**, 476 (1941)
[10]Chr. Deichler and Ch. Weizmann, *Ber.*, **36**, 547 (1903)
[11]I. Ya. Postovaskiĭ and L. N. Goldyrev, *J. Gen. Chem.*, *U.S.S.R.*, **11**, 429 (1941) [*C.A.*, **35**, 6589 (1941); see also German patent 134,985]
[12]Ch. Weizmann, L. Haskelberg, and T. Berlin, *J. Chem. Soc.*, 398 (1939)
[13]J. Böeseken, "The use of boric acid for the determination of the configuration of carbohydrates," *Adv. in Carbohydrate Chemistry*, **4**, 189 (1949)
[14]J. B. Martin, *Am. Soc.*, **75**, 5482 (1953); see also P. F. E. Cook and A. J. Showler, *J. Chem. Soc.*, 4594 (1965)

Boron tribromide, BBr_3. Mol. wt. 250.57, b.p. 90.6°. Suppliers: Alfa, KK, MCB.

Preparation from boron trifluoride and aluminum bromide.[1] Cleavage of an ether is effected by adding BBr_3 to the ice-cold ether (3 equiv.), refluxing the mixture, and removing the alkyl bromide; the orthoboric ester is then hydrolyzed with 10%

$$3\ \underline{n}\text{-Bu—O—Bu-}\underline{n} + BBr_3 \xrightarrow[77\%]{} (\underline{n}\text{-BuO})_3B + 3\underline{n}\text{-BuBr}$$

$$(\underline{n}\text{-BuO})_3B + H_2O \xrightarrow[62\%]{} 3\ \underline{n}\text{-BuOH} + H_3BO_3$$

alkali. Phenyl ethers consistently yield the phenol and the alkyl bromide.[2] Phenolic methyl ethers are demethylated in high yield by mixing the reactants in methylene

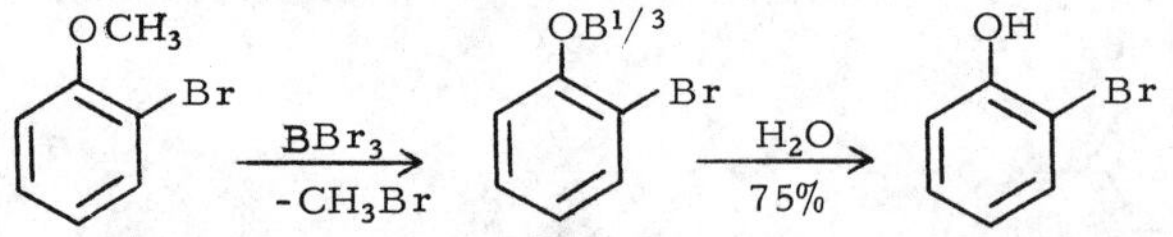
OCH3
Br
BBr3
-CH3Br
OB1/3
Br
H2O
75%
OH
Br

chloride or *n*-pentane at −80° and letting the mixture warm to room temperature; examples: veratric acid (90%); 3,4,3′,4′-tetramethoxydiphenyl (85%).[3]

Stetter and Wulff[4] noted that boron tribromide exerts a marked catalytic effect on the bromination of adamantane and reported that under these conditions the product formed is 1,3-dibromoadamantane. However, Baughman[5] established that the product actually is 1-bromoadamantane. He then found that addition of a trace of aluminum bromide (tribromide) to the boron tribromide leads to the formation of 1,3-dibromoadamantane in yield of 74%.

The reagent selectively cleaves the 5-methoxy group of 4-hydroxy-5,6,7-trimethoxy-2-carboethoxyquinoline (also the 3-carboethoxy isomer).[6] The selectivity is attributed to chelation in the product.

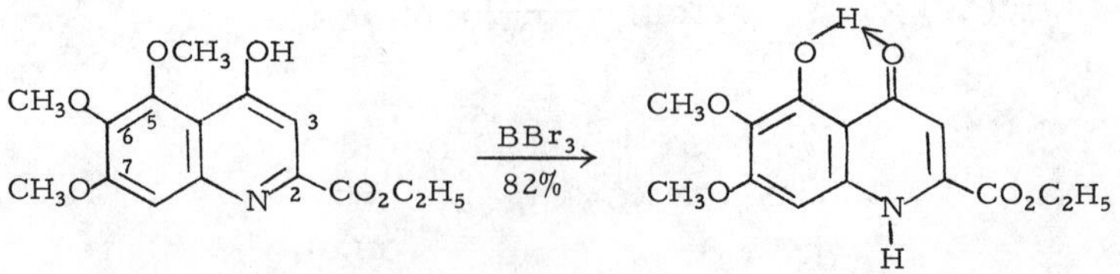
OCH3 OH
CH3O
CH3O
CO2C2H5
BBr3
82%
H
O
O
CH3O
CH3O
N
CO2C2H5
H

[1]E. L. Gamble, P. Gilmont, and J. F. Stiff, *Am. Soc.*, **62**, 1257 (1940)
[2]F. L. Benton and T. E. Dillon, *ibid.*, **64**, 1128 (1942)
[3]J. F. W. McOmie and M. L. Watts, *Chem. Ind.*, 1658 (1963)
[4]H. Stetter and C. Wulff, *Ber.*, **93**, 1366 (1960)
[5]G. L. Baughman, *J. Org.*, **29**, 238 (1964)
[6]W. Schäfer and B. Franck, *Ber.*, **99**, 160 (1966)

Boron trichloride, BCl_3. Mol. wt. 117.19, b.p. 13°. Supplier: Matheson. Review.[1] Purification through the nitrobenzene addition compound.[2]

The reagent is used to cleave carbohydrate methyl ethers.[3] A suspension of the methyl ether in methylene chloride is cooled to −70°, boron trichloride is distilled in, and the mixture is kept at −70° for 1 hr. with occasional stirring.

Whereas the addition compounds of boron trichloride with dimethyl ether and with diethyl ether are stable at room temperature, BCl_3 effects cleavage of tetrahydrofurane,[4] and it reacts with ethyleneglycol dimethyl ether even at 0° with evolution of methyl chloride, presumably with formation of the mono- and dichloroboronates:[5]

$$3\ CH_3(OCH_2CH_2)_2OCH_3 \xrightarrow{BCl_3} \begin{cases} CH_3(OCH_2CH_2)_2OBCl_2 + CH_3Cl \\ [CH_3(OCH_2CH_2)_2O]_2BCl + 2\ CH_3Cl \end{cases}$$

Boron trichloride reacts with an acetal, or with a mixture of an aldehyde and an alcohol, to give a 1-chloroalkyl ether (yields 40–85%).[6] The 1-chloroalkyl ethers are of interest because they are dehydrochlorinated by trimethylamine to give vinyl ethers.

$$RCH_2CHO + BCl_3 + R'OH \longrightarrow RCH_2\overset{Cl}{\overset{|}{C}}HOR' + HOBCl_2$$

$$\downarrow (CH_3)_3N$$

$$RCH{=}CHOR' + (CH_3)_3\overset{+}{N}H\overset{-}{Cl}$$

Boron trichloride was used to effect demethylation (and deketalization) of a methyl ether in the inositol series without cleavage of a tosyl ester group.[7] Quebrachitol (1), or 2-O-methyl-(−)-inositol, was converted into the 3,4; 5,6-di-O-cyclohexylidene

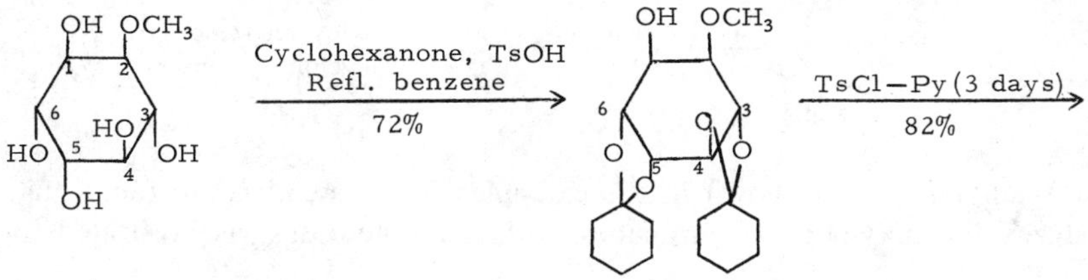
OH OCH3
HO
HO
OH
OH
Cyclohexanone, TsOH
Refl. benzene
72%
OH OCH3
TsCl—Py (3 days)
82%

(1) (2)

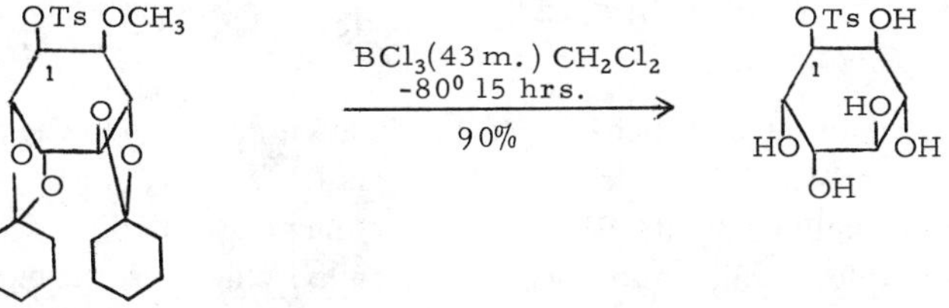
OTs OCH3
BCl3(43 m.) CH2Cl2
-80° 15 hrs.
90%
OTs OH
HO
HO
OH
OH

(3) (4)

derivative (2) and the 1-hydroxyl group was tosylated (slow, the hydroxyl group is axial). Treatment of (3) with BCl_3—CH_2Cl_2 at −80° effected demethylation and deketalization to give 1-O-tosyl-(−)-inositol (4).

Use of the reagent for the building of heterocyclic compounds of a new class is illustrated by the synthesis of 10-methyl-10,9-borazarophenanthrene.[8]

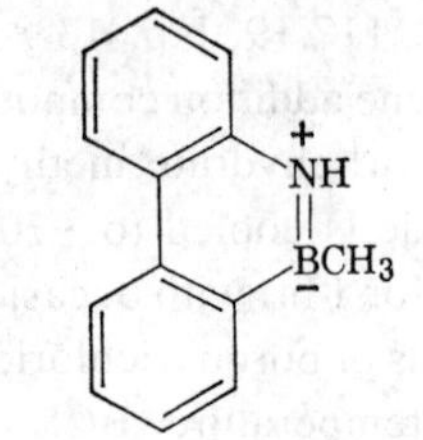
NH
BCH3

[1]W. Gerrard and M. F. Lappert, *Chem. Rev.*, **58**, 1081 (1958)
[2]H. C. Brown and R. R. Holmes, *Am. Soc.*, **78**, 2173 (1956)
[3]S. Allen, T. G. Bonner, E. J. Bourne, and N. M. Saville, *Chem. Ind.*, 630 (1958); T. G. Bonner, E. J. Bourne, and S. McNally, *J. Chem. Soc.*, 2929 (1960); A. B. Foster, D. Horton, N. Salim, M. Stacey, and J. M. Webber, *ibid.*, 2587 (1960)
[4]J. D. Edwards, W. Gerrard, M. F. Lappert, *J. Chem. Soc.*, 1470 (1955)
[5]H. C. Brown and P. A. Tierney, *Am. Soc.*, **80**, 1552 (1958)
[6]D. K. Black and S. R. Landor, *J. Chem. Soc.*, 5225 (1965)
[7]S. D. Géro, *Tetrahedron Letters*, 591 (1966)
[8]M. J. S. Dewar, R. B. K. Dewar, and Z. L. F. Gaibel, *Org. Syn.*, **46**, 65 (1966)

Boron trifluoride, BF_3. Mol. wt. 67.82, b.p. −101°. Supplier: Harshaw. The reagent is available in cylinders and can be bubbled into a reaction mixture.

Friedel-Crafts reactions. Anhydrous BF_3 is a mildly active catalyst for Friedel-

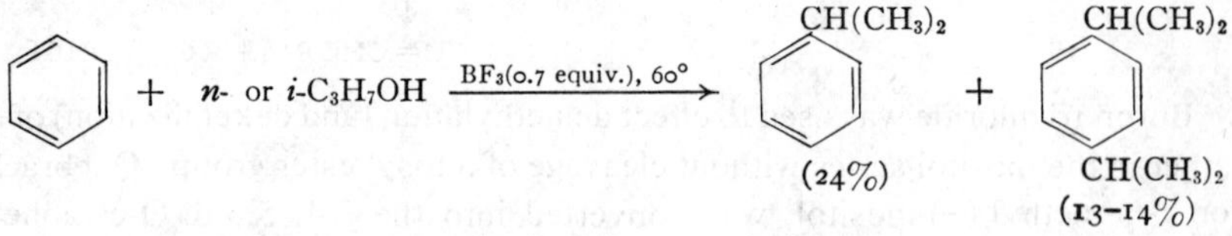
CH(CH3)2
CH(CH3)2
+ n- or i-C3H7OH
BF3(0.7 equiv.), 60°
+
(24%)
CH(CH3)2
(13–14%)

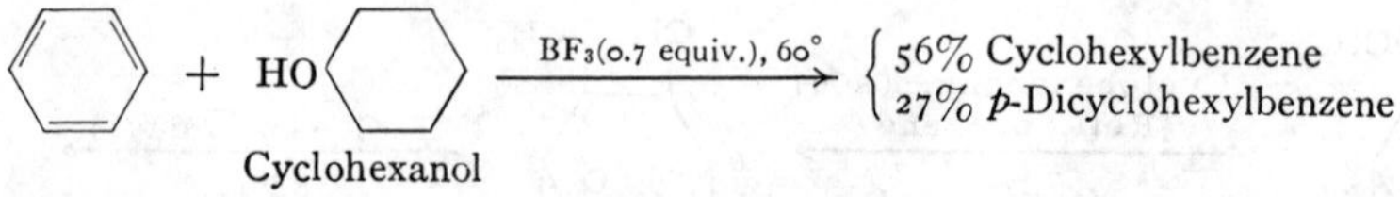
+ HO
BF3(0.7 equiv.), 60°
56% Cyclohexylbenzene
27% p-Dicyclohexylbenzene
Cyclohexanol

Crafts alkylations, as shown in the examples formulated.[1] Olah[2] found that BF_3 catalyzes the alkylation of aromatics with alkyl fluorides, as illustrated for the

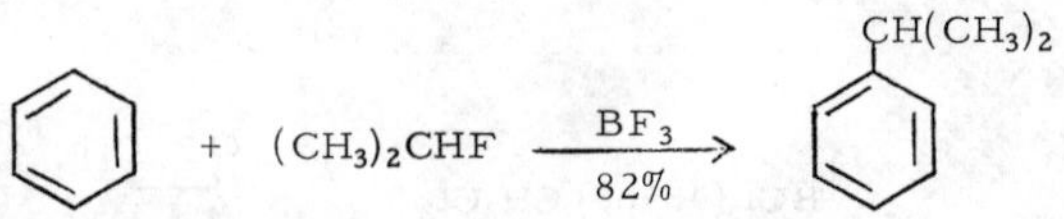
CH(CH3)2
+ (CH3)2CHF
BF3
82%

reaction of isopropyl fluoride with benzene to produce cumene. The other boron trihalides function similarly and the order of reactivity is $BI_3 > BBr_3 > BCl_3 > BF_3$. However, yields are usually best with BF_3; yields of cumene are: BCl_3, 63%; BBr_3, 72%; BI_3, 60%. As a rule, no alkylation occurs with alkyl chlorides, bromides, or iodides. Mixed halides containing fluorine react with elimination of HF. Examples:

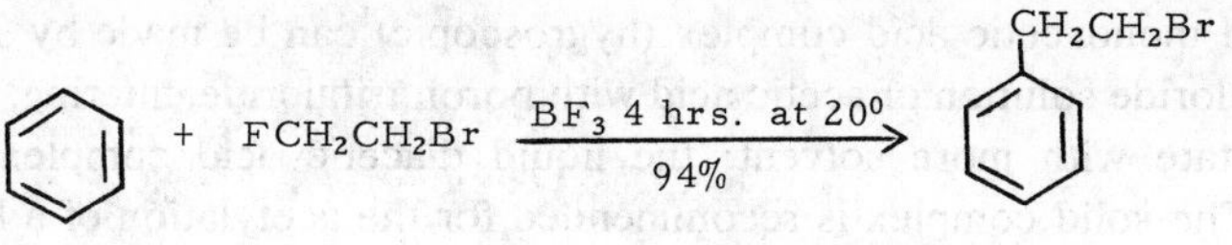

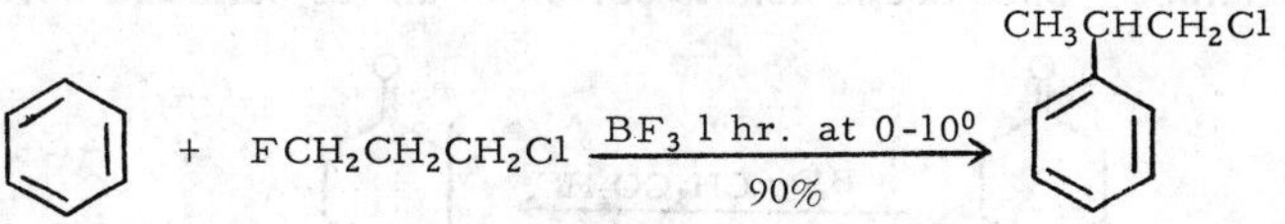

In the second example the alkylation is attended with isomerization of the group introduced.

Rearrangement of epoxides. An interesting example reported by House[3] is as follows. A benzene solution of the glycidic ester (1) is saturated with BF_3 and allowed to stand at room temperature for 30 min.; workup affords the α-keto ester (2) in high yield.

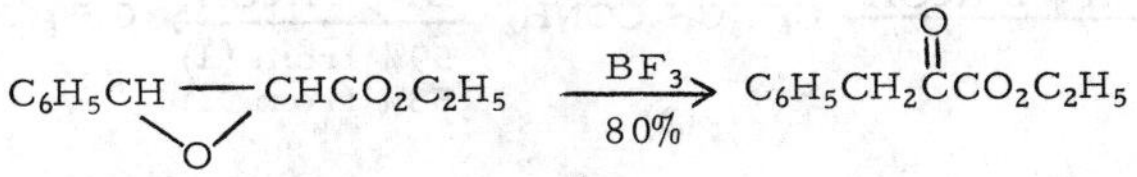

(1) (2)

Alkylation and acylation. A procedure for alkylation of ethyl acetoacetate with isopropanol specified that the gas be passed through a saturated solution of boric acid in concd. sulfuric acid and bubbled into a mixture of the reactants at 0–7°.[4]

$$CH_3COCH_2CO_2C_2H_5 + (CH_3)_2CHOH \xrightarrow[60\text{-}70\%]{BF_3} CH_3COCH(CH(CH_3)_2)CO_2C_2H_5$$

Unpurified gas is used in the synthesis of acetylacetone; the yield cited is for material purified through the copper derivative and then distilled.[5]

$$CH_3COCH_3 + (CH_3CO)_2O \xrightarrow[80\text{-}85\%]{BF_3} CH_3COCH_2COCH_3$$

Beckmann rearrangement of aldoximes and ketoximes.[6]

Nitriles → amides.[7] A solution of the nitrile in aqueous acetic acid is saturated with gaseous BF_3. The temperature rises to 115–135° and workup affords the amide in excellent yield.

[1] J. F. McKenna and F. J. Sowa, *Am. Soc.*, **59**, 470 (1937)
[2] G. A. Olah, S. J. Kuhn, and J. A. Olah, *J. Chem. Soc.*, 2174 (1957); G. A. Olah and S. J. Kuhn, *J. Org.*, **29**, 2317 (1964)
[3] H. O. House, J. W. Blaker, and D. A. Madden, *Am. Soc.*, **80**, 6386 (1958)
[4] J. T. Adams, R. Levine, and C. R. Hauser, *Org. Syn.*, *Coll. Vol.*, **3**, 405 (1955)
[5] C. E. Denoon, Jr., *ibid.*, **3**, 16 (1955)
[6] C. R. Hauser and D. S. Hoffenberg, *J. Org.*, **20**, 1482, 1496 (1955)
[7] *Idem, ibid.*, **20**, 1448 (1955)

Boron trifluoride–Acetic acid complexes: $BF_3 \cdot CH_3CO_2H$ (solid), $BF_3 \cdot 2CH_3CO_2H$ (liquid). The liquid complex containing 36% acetic acid by weight is supplied by Harshaw Chem. Co.

The solid monoacetic acid complex (hygroscopic) can be made by saturating an ethylene chloride solution of acetic acid with boron trifluoride, filtering, and washing the precipitate with more solvent; the liquid diacetic acid complex remains in solution.[1] The solid complex is recommended for the acetylation of a ketone, such as cyclohexanone.[1] Thus acetic acid is stirred in an ice bath and boron trifluoride

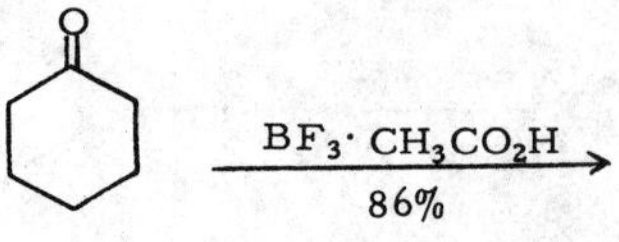
COCH$_3$
BF$_3$· CH$_3$CO$_2$H
86%

gas is passed in until the contents of the flask become a powdery solid (about 80 mole percent absorption). Cyclohexanone is then added dropwise with cooling and stirring; the mixture is let stand at room temperature for 4 hrs. and then worked up.

The liquid boron trifluoride–diacetic acid complex, used with excess acetic

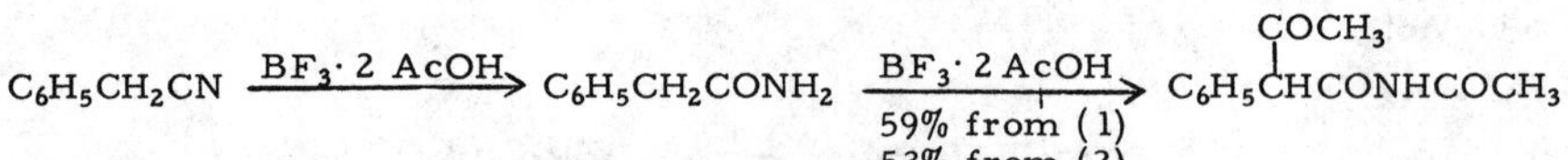
C$_6$H$_5$CH$_2$CN
BF$_3$· 2 AcOH
C$_6$H$_5$CH$_2$CONH$_2$
BF$_3$· 2 AcOH
59% from (1)
53% from (2)
COCH$_3$
C$_6$H$_5$CHCONHCOCH$_3$

(1) (2) (3)

anhydride, effects both C- and N-acetylation of phenylacetamide to give (3).[2] Since the reagent converts phenylacetonitrile (1) into phenylacetamide (2), the diacetyl derivative (3) can be obtained directly from (1).

[1]R. M. Manyik, F. C. Frostick, Jr., J. J. Sanderson, and C. R. Hauser, *Am. Soc.*, **75**, 5030 (1953)
[2]J. F. Wolfe, C. J. Eby, and C. R. Hauser, *J. Org.*, **30**, 55 (1965)

Boron trifluoride etherate, $BF_3\cdot(C_2H_5)_2O$. Mol. wt. 141.94, b.p. 126°, sp. gr. 1.154. Supplies (pract.): B, E, F, KK, MCB.

Purification. Commercial material darkens rapidly owing to air oxidation, but redistillation gives water-white liquid. Zweifel and Brown[1] recommend addition of 10 ml. of ether to 500 ml. of etherate (to ensure an excess) and distillation from 2 g. of calcium hydride in an all-glass apparatus at 46°/10 mm. The hydride removes volatile acids and reduces bumping.

Dehydration. 11β-Hydroxysteroids are dehydrated smoothly to $\Delta^{9(11)}$-enes on

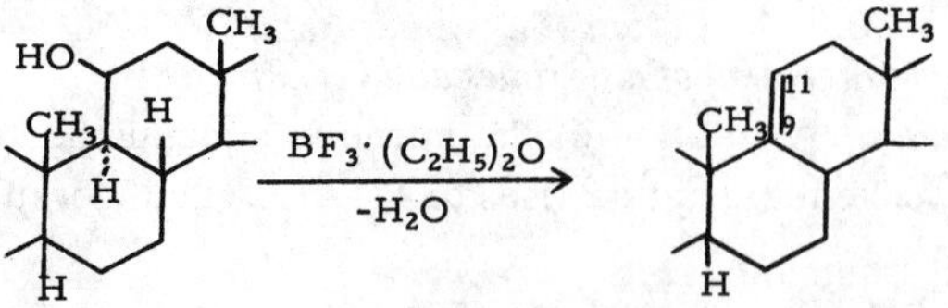
CH$_3$
HO
CH$_3$
H
H
H
BF$_3$· (C$_2$H$_5$)$_2$O
-H$_2$O
CH$_3$
11
CH$_3$
9
H

treatment with boron fluoride etherate in acetic acid solution (4 hrs. at 25°).[2]

Examples of intermolecular dehydration are as follows. The reagent is an excellent catalyst for the condensation of ketones with ethanedithiol[3] and with thiols.[4] Addition of 1 ml. of boron trifluoride etherate to a hot solution of 0.6 g. of cholestane-3-one

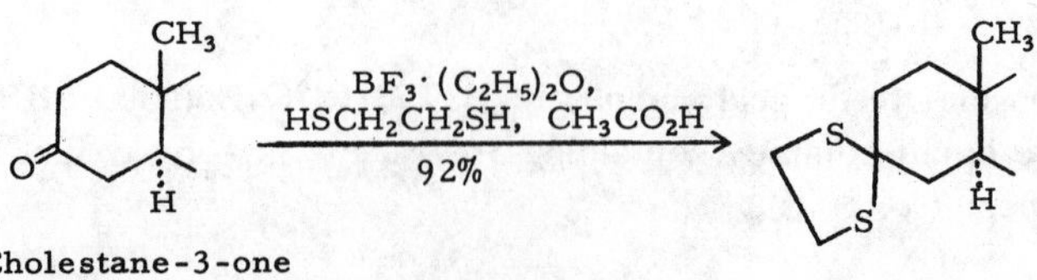
CH$_3$
O
H
BF$_3$· (C$_2$H$_5$)$_2$O,
HSCH$_2$CH$_2$SH, CH$_3$CO$_2$H
92%
CH$_3$
S
S
H

Cholestane-3-one

in 15 ml. of acetic acid resulted in prompt separation of needles of the ethylenethioketal. Still more effective catalysis is provided by adding BF_3-etherate to a suspension or solution of the ketone in ethanedithiol, for the latter reagent has remarkable solvent power. Thus cholestane-3,6-dione was converted by the second method into the 3,6-bis-ethylenethioketal in 94% yield, whereas the reaction in acetic acid at room temperature gave a mixture of the 3-monoethylenethioketal and the bis-derivative.

The reagent is an effective catalyst for the condensation of allylic alcohols with 2-methyl-1,4-naphthohydroquinone (in dioxane at 50°)[5] and for the condensation of ketones with amines to form azomethines.[6]

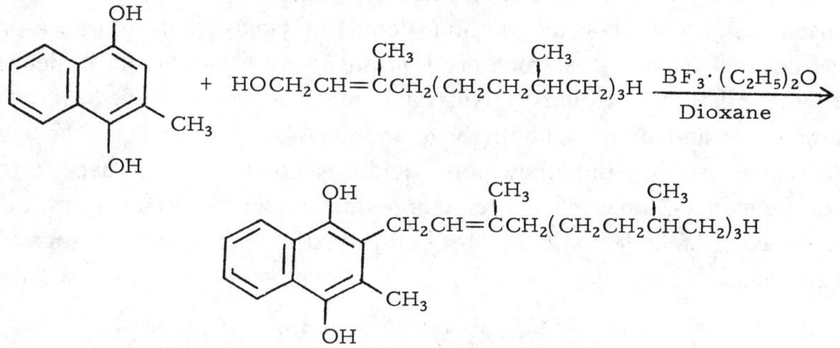
OH
CH3
OH
+ HOCH2CH=CCH2(CH2CH2CHCH2)3H
CH3 CH3
BF3·(C2H5)2O
Dioxane
CH2CH=CCH2(CH2CH2CHCH2)3H

Vitamin K_1 hydroquinone

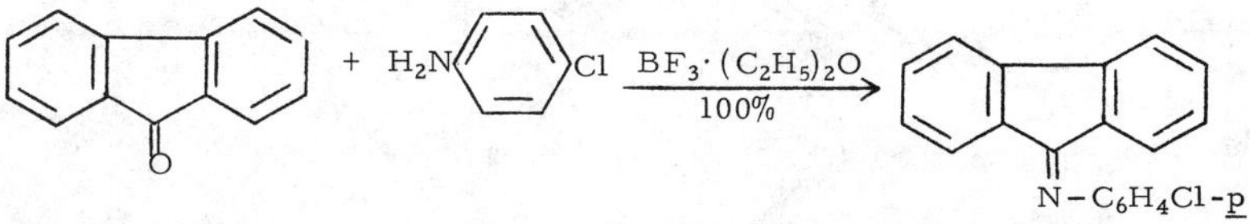
+ H2N
Cl
BF3·(C2H5)2O
100%
N-C6H4Cl-p

Thiele reaction. BF_3-etherate is more satisfactory than sulfuric acid as catalyst for the Thiele reaction of α-naphthoquinone, for with sulfuric acid the yield is only

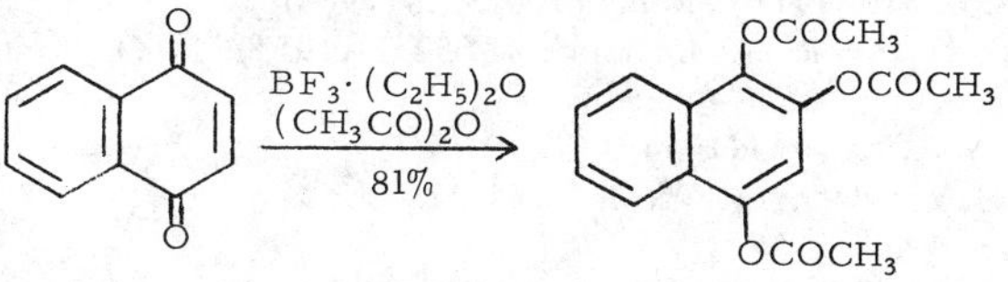
BF3·(C2H5)2O
(CH3CO)2O
81%
OCOCH3
OCOCH3
OCOCH3

74.5% as compared to 81%.[7] However, experiments with hindered ketones show that BF_3-etherate is a less potent catalyst than H_2SO_4. Thus 2,5-dimethyl-1,4-benzoquinone adds acetic anhydride smoothly in the presence of BF_3-etherate, but 2,6-dimethyl-1,4-benzoquinone does not;[8] on the other hand, 2,6-dimethyl-1,4-benzoquinone undergoes the H_2SO_4-catalyzed Thiele reaction.[9]

Fission-rearrangement of epoxides. For isomerization of *trans*-stilbene oxide to diphenylacetaldehyde,[10] a solution in benzene is treated at room temperature with

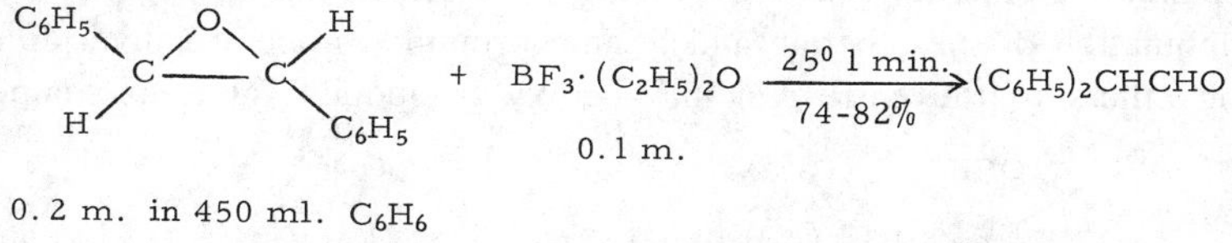
C6H5
H
C6H5
+ BF3·(C2H5)2O
0.1 m.
25° 1 min.
74-82%
(C6H5)2CHCHO
0.2 m. in 450 ml. C6H6

BF_3-etherate, swirled, let stand for 1 min., and treated with water. The water-washed organic layer is dried and evaporated, and the product distilled at 0.6 mm.

Isomerization of isophorone oxide[11] to a rearranged aldehyde is accomplished similarly except that the reaction mixture is let stand for 30 min. before quenching.

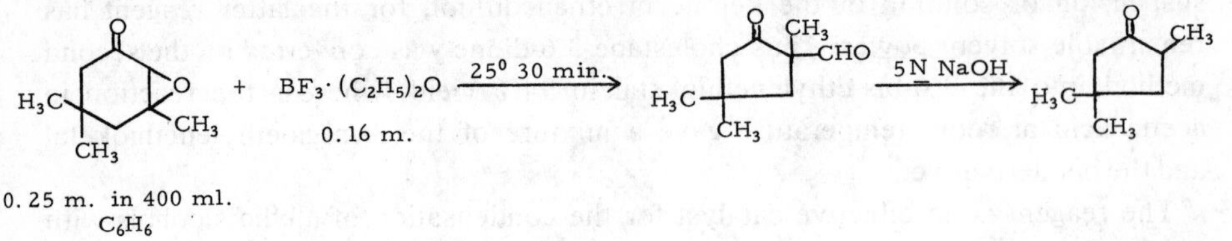

H_3C
CH_3
CH_3
O
+ $BF_3 \cdot (C_2H_5)_2O$
0.16 m.
25° 30 min.
CH_3
CHO
H_3C
CH_3
5N NaOH
CH_3
H_3C
CH_3
0.25 m. in 400 ml. C_6H_6

The water-washed organic layer is shaken for 1–2 min. with alkali to eliminate formaldehyde and give 2,4,4-trimethylcyclopentanone.

A Farmitalia group[12] reported, without details or yields, that both the α- and β-epoxides derived from testosterone are rearranged by BF_3-etherate in benzene to 4-hydroxy-Δ^4-3-ketones. Collins[13] (Australia) made a more detailed study in the cholestane series and found that both the α- and the β-oxide indeed give the hydroxy-Δ^4-cholestene-3-one but that they both yield a second product characterized as 5β-A-norcholestane-3-one. The latter compound evidently comes from an intermediate β-keto aldehyde, probably by deformylation during chromatography on neutral alumina.

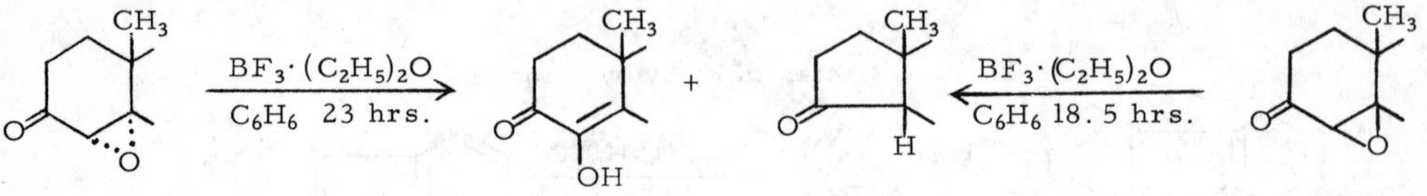

CH_3
$BF_3 \cdot (C_2H_5)_2O$
C_6H_6 23 hrs.
CH_3
OH
+
CH_3
H
$BF_3 \cdot (C_2H_5)_2O$
C_6H_6 18.5 hrs.
CH_3

(1) (2) (3)

	(2)	(3)	
α-Oxide ⟶	30%	25%	
	35%	3% ⟵	β-Oxide

[1]G. Zweifel and H. C. Brown, *Org. Reactions*, **13**, 28 (1963)
[2]H. Heymann and L. F. Fieser, *Am. Soc.*, **73**, 5252 (1951); R. O. Clinton *et al.*, *ibid.*, **79**, 6475 (1957)
[3]L. F. Fieser, *Am. Soc.*, **76**, 1945 (1954)
[4]A. Schöberl and G. Wiehler, *Angew. Chem.*, **66**, 273 (1954)
[5]*Org. Expts.*, Chapt. 49.4
[6]M. E. Taylor and T. L. Fletcher, *J. Org.*, **26**, 940 (1961)
[7]L. F. Fieser, *Am. Soc.*, **70**, 3165 (1948)
[8]L. F. Fieser and M. I. Ardao, *Am. Soc.*, **78**, 774 (1956)
[9]H. Erdtman, *Svensk Kem. Tids.*, **44**, 135 (1932); [*C.A.*, **26**, 4803 (1932)]
[10]D. J. Reif and H. O. House, *Org. Syn., Coll. Vol.*, **4**, 375 (1963)
[11]G. D. Ryerson, R. L. Wasson, and H. O. House, *ibid.*, **4**, 957 (1963)
[12]B. Camerino, B. Patelli, and A. Vercellone, *Am. Soc.*, **78**, 3540 (1956)
[13]D. J. Collins, *J. Chem. Soc.*, 3919 (1959)

Boron trifluoride etherate–Acetic anhydride. Youssefyeh and Mazur[1] reported that the combination of the etherate and lithium bromide in acetic anhydride cleaves aliphatic ethers to the corresponding acetoxy compounds at room temperature.

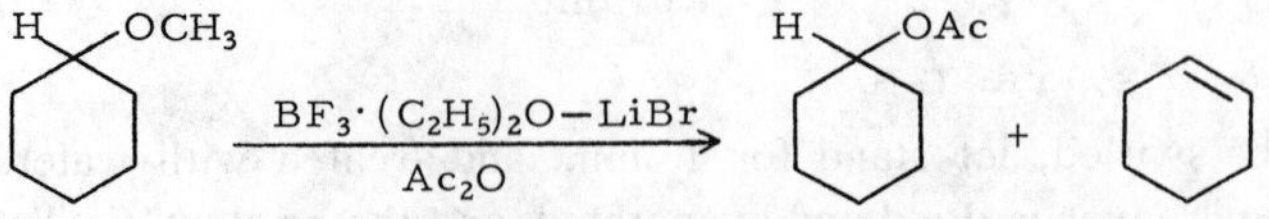

H
OCH_3
$BF_3 \cdot (C_2H_5)_2O—LiBr$
Ac_2O
H
OAc
+

7 parts 1 part

However, Narayanan and Iyer[2] found that cleavage can be effected with boron trifluoride and acetic anhydride alone and that a lithium halide slows down the reaction and does not decrease elimination to the olefin. They obtained cholesteryl acetate from cholesteryl methyl ether (homoallylic) in 93% yield (15 hrs. at 0°). Cleavage of an allylic methyl ether (Δ^4-cholestene-3β-ol methyl ether) required only 3 min. at −18° (90% yield); under less mild conditions the main product was the $\Delta^{3,5}$-diene. Cleavage of saturated ethers was attended with considerable elimination.

[1]R. D. Youssefyeh and Y. Mazur, *Tetrahedron Letters*, 1287 (1962)
[2]C. R. Narayanan and K. N. Iyer, *J. Org.*, **30**, 1734 (1965)

Boron trifluoride–Methanol (51% BF_3; 2 equiv., 4.4 ml.). Suppliers: B, D, H.

Aromatic acids are esterified in good yield by the reagent in the presence of excess methanol.[1]

[1]G. Hallas, *J. Chem. Soc.*, 5770 (1965)

Bromine chloride, BrCl. Mol. wt. 115.37.

Generated *in situ* from sodium bromide and chlorine, the reagent is useful for bromination of phenols: *p*-nitrophenol → 2,6-dibromo-4-nitrophenol (96% yield).[1]

[1]C. O. Obenland, *J. Chem. Ed.*, **41**, 566 (1964); procedure submitted to *Org. Syn.*

Bromine–Silver acetate. Lemieux and Fraser-Reid[1] found a route to 2,5-anhydro-sugars in the reaction of methyl 2-desoxy-2-iodo-β-D-glucopyranoside triacetate with a large excess of bromine and silver acetate to give in high yield an anomeric mixture of 1,3,4,6-tetra-O-acetyl-2,5-anhydro-1-methoxy-D-mannoses.

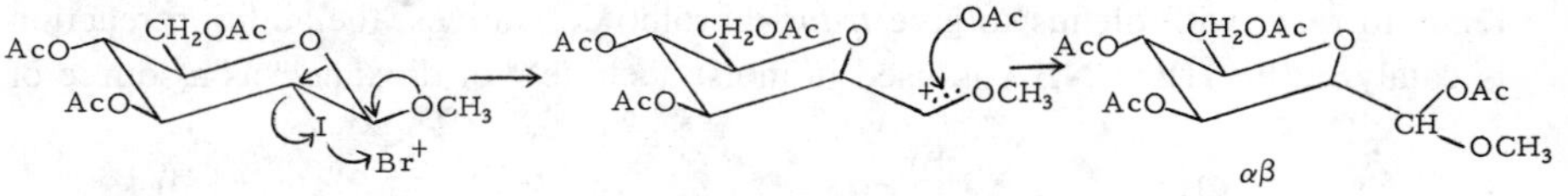

[1]R. U. Lemieux and B. Fraser-Reid, *Can. J. Chem.*, **42**, 547 (1964)

Bromine–Silver oxide. The reaction of bromine in the presence of a silver salt, usually the oxide, with a tertiary alcohol gives in good yield a product of the insertion of oxygen. Thus 1,3,3-trimethylcyclohexanol (1) is transformed into the bicyclic ether (4) in 75% yield.[1] It is suggested that the reaction proceeds through the hypobromite (2), the decomposition of which is catalyzed by silver ions.

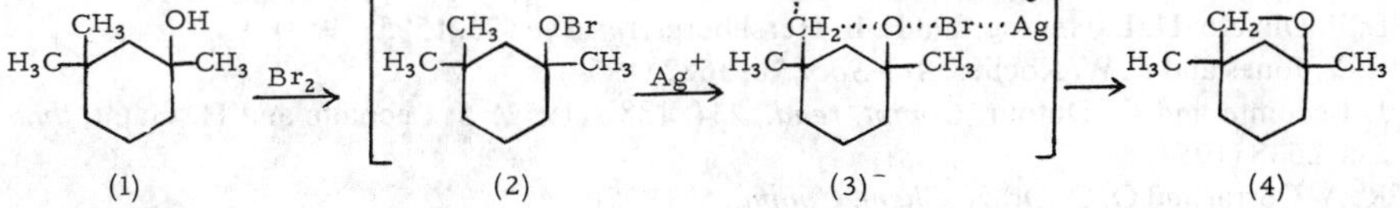

6α-Methylcholestane-6β-ol similarly gives the 6,19-oxido derivative.

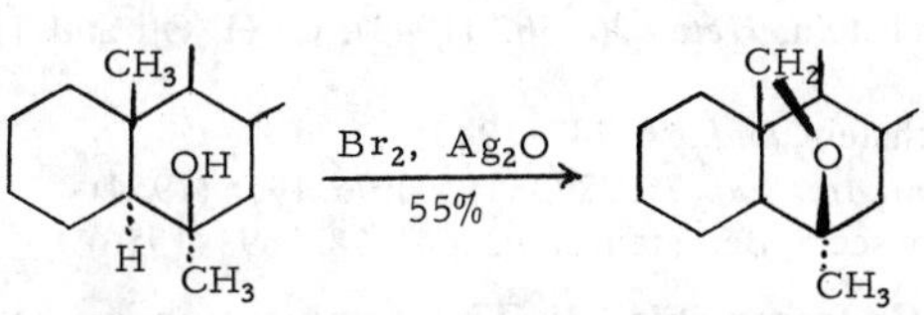

[1]R. A. Sneen and N. P. Matheny, *Am. Soc.*, **86**, 3905 (1964)

N-Bromoacetamide (NBA), $CH_3CONHBr$. Mol. wt. 137.98, m.p. 102–105°. Suppliers: Arapahoe, A, F. Preparation.[1]

Oxidation of secondary alcohols. Gallagher and co-workers[2] found this reagent very satisfactory for oxidation of 3α-hydroxy-d_2-11,12-pregnane-20-one to the corresponding 3,20-dione. A solution of 600 mg. of the steroid in 3 ml. of *t*-butanol and 0.9 ml. of pyridine was treated with 4 drops of water and 600 mg. of NBA (4 equiv.) and stirred overnight; yield 90%. On oxidation of the same compound with chromic acid in acetic acid the yield was only 84%. Hershberg and co-workers[3] used a similar procedure for selective oxidation of 3α,11α-dihydroxypregnane-20-one to 11α-hydroxypregnane-3,20-dione (yield 80%). Jones and Kocher[4] used the reagent for selective oxidation of steroid 3α,20-diols to 3-keto-20-ols (95% yield). A statement to the effect that secondary alcohols containing an aromatic ring are oxidized by NBA only if they are benzylic[5] has been shown to be in error.[6] One of five oxidations cited is formulated.

$$\underset{0.003\ m.}{C_6H_5CH_2CH_2CH(CH_3)OH} + \underset{0.0025\ m.}{CH_3CONHBr} + \underset{0.003\ m.}{Py} + \underset{25\ ml.}{Acetone} + \underset{8\ ml.}{H_2O}$$

$$\xrightarrow[74\%]{25^0\ (dark)} C_6H_5CH_2CH_2C(CH_3){=}O$$

Reaction with olefins. In contrast to N-bromosuccinimide, N-bromoacetamide tends to react with olefins to give *trans*-dibromo derivatives; the addition reaction is catalyzed by HBr.[7] NBA is used in moist acetone[8,9] or dioxane[10] as a source of

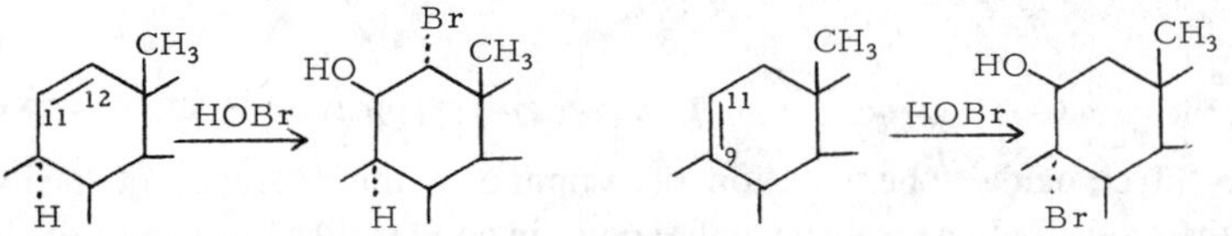
CH3
12
11
H
HOBr
Br
HO
CH3
H
CH3
11
9
HOBr
HO
CH3
Br

HOBr. The Δ^{11}-steroid formulated was treated with NBA in aqueous acetone;[9] the $\Delta^{9(11)}$-steroid was suspended in dioxane and 0.46 *N* perchloric acid and treated with solid NBA.[10,11] In each case the adduct is *trans*-diaxial.

[1]E. P. Oliveto and C. Gerold, *Org. Syn., Coll. Vol.*, **4**, 104 (1963)
[2]B. A. Koechlin, T. H. Kritchevsky, and T. F. Gallagher, *J. Biol. Chem.*, **184**, 393 (1950)
[3]E. P. Oliveto, H. L. Herzog, and E. B. Hershberg, *Am. Soc.*, **75**, 1505 (1953)
[4]R. E. Jones and F. W. Kocher, *Am. Soc.*, **76**, 3682 (1954)
[5]J. Lecomte and C. Dufour, *Compt. rend.*, **234**, 1887 (1952); J. Lecomte and H. Gault, *ibid.*, **238**, 2538 (1954)
[6]R. A. Corral and O. O. Orazi, *Chem. Comm.*, 5 (1965)
[7]R. E. Buckles, *Am. Soc.*, **71**, 1157 (1949); R. E. Buckles, R. C. Johnson, and W. J. Probst, *J. Org.*, **22**, 55 (1957)
[8]H. Reich and T. Reichstein, *Helv.*, **26**, 562 (1943); G. H. Ott and T. Reichstein, *ibid.*, **26**, 1799 (1943)
[9]A. Lardon and T. Reichstein, *ibid.*, **26**, 747 (1943)
[10]J. Fried and E. F. Sabo, *Am. Soc.*, **75**, 2273 (1953); **76**, 1455 (1954)
[11]For another procedure see S. Bernstein *et al.*, *ibid.*, **78**, 5693 (1956)

N-Bromoacetamide–Hydrogen chloride. This combination is a source of the elements of bromine chloride. For example, styrene and NBA are added slowly and separately to a cooled aqueous solution of hydrochloric acid; the product (of *trans*-addition) is 2-bromo-1-chloro-1-phenylethane (40% yield).[1] Bromochlorination of cholesterol, effected by saturating a chloroform solution with hydrogen chloride and adding NBA in chloroform, afforded 5α-bromo-6β-chlorocholestane-3β-ol in good yield.[2] Other examples in the steroid field.[3]

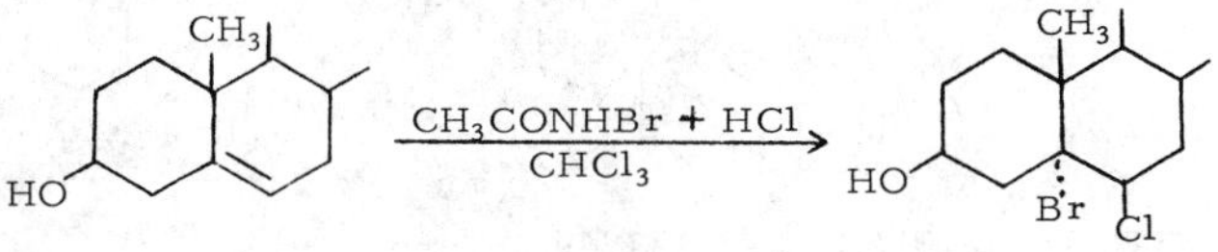

CH3
HO
CH3CONHBr + HCl
CHCl3
CH3
HO
Br
Cl

[1]R. E. Buckles and J. W. Long. *Am. Soc.*, **73**, 998 (1951)
[2]J. B. Ziegler and A. C. Shabica, *ibid.*, **74**, 4891 (1952)
[3]C. H. Robinson, L. Finckenor, E. P. Oliveto, and D. Gould, *ibid.*, **81**, 2191 (1959)

N-Bromoacetamide–Hydrogen fluoride. In the presence of a proton acceptor such as ether or tetrahydrofurane (to increase the ionization of HF), this combination serves as a source of bromine fluoride.[1,2] Thus cyclohexene is converted in good yield

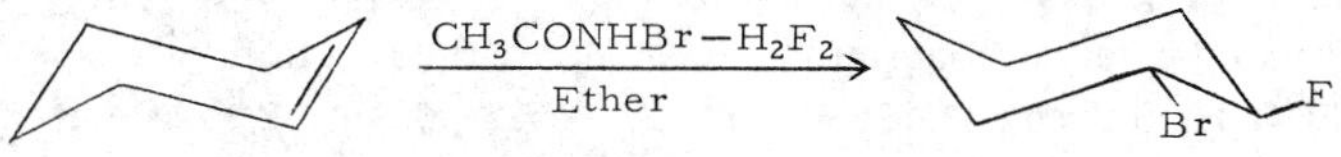

CH3CONHBr—H2F2
Ether
Br
F

into *trans*-1-fluoro-2-bromocyclohexane.[2] The reaction has proved useful for the synthesis of fluorosteroids.[3]

[1]C. H. Robinson, L. Finckenor, E. P. Oliveto, and D. Gould, *Am. Soc.*, **81**, 2191 (1959)
[2]A. Bowers, L. C. Ibáñez, E. Denot, and R. Becerra, *ibid.*, **82**, 4001 (1960)
[3]A. Bowers, *ibid.*, **81**, 4107 (1959)

N-Bromoacetamide–Pyridine–SO_2. This method of dehydration, described in a patent,[1] was applied as follows to 11β-hydroxy-11α-methyl-5β-pregnane-3,20-dione.[2] A solution of the alcohol in dry pyridine was treated with 1 equivalent of

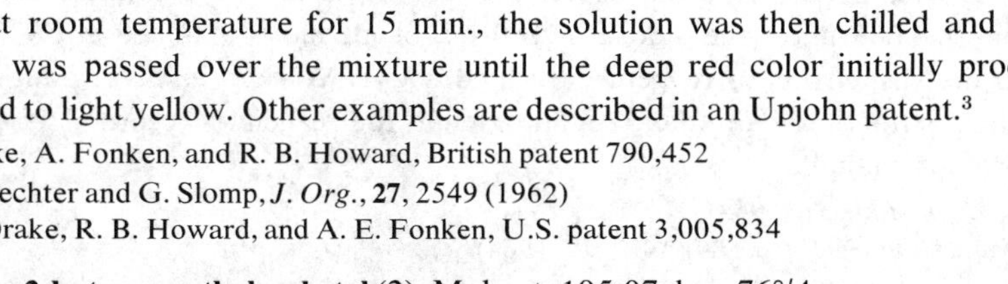

COCH3
CH3
CH3
HO
CH3
H
H
O
H
NBA-Py-SO2
H3C
CH3
85%
+
H3C
CH3
15%

NBA at room temperature for 15 min., the solution was then chilled and sulfur dioxide was passed over the mixture until the deep red color initially produced bleached to light yellow. Other examples are described in an Upjohn patent.[3]

[1]A. Drake, A. Fonken, and R. B. Howard, British patent 790,452
[2]W. J. Wechter and G. Slomp, *J. Org.*, **27**, 2549 (1962)
[3]H. A. Drake, R. B. Howard, and A. E. Fonken, U.S. patent 3,005,834

4-Bromo-2-butanone ethyleneketal (2). Mol. wt. 195.07, b.p. 76°/4 mm.

Preparation by reaction of the corresponding alcohol with PBr_3[1] or by addition of HBr to methyl vinyl ketone and exchange ketalization.[2]

For Robinson annelation. Stork[3] used the reagent for the snythesis of (4) from

(1) but states that the low reactivity of the reagent and its tendency to lose HBr preclude its general use.

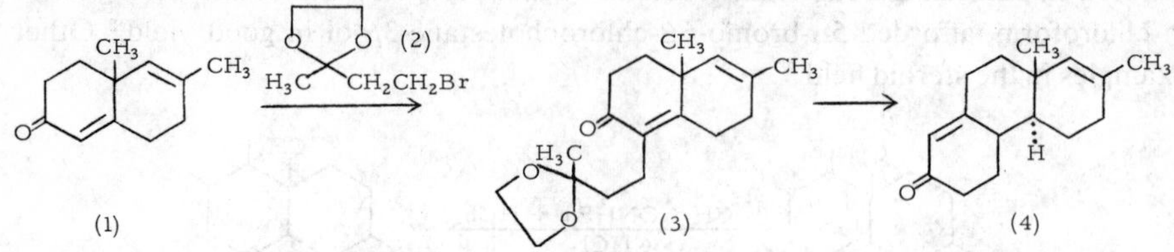
CH3
CH3
(2)
H3C
CH2CH2Br
CH3
CH3
H3C
CH3
CH3
H
(1)
(3)
(4)

[1]L. Willimann and H. Schinz, *Helv.*, **32**, 2151 (1949)
[2]G. Stork and R. Borch, *Am. Soc.*, **86**, 935 (1964)
[3]G. Stork, "Third Internat. Symp. on the Chem. of Nat. Prod.," Kyoto, 1964, p. 134

N-Bromocaprolactam, $\overset{\displaystyle CH_2CH_2CO}{\underset{\displaystyle CH_2CH_2CH_2}{|}}\!\!>\!NBr$. Mol. wt. 192.07, m.p. 64–66°. Supplier: KK.

Prepared in 75% yield by the bromination of caprolactam, this reagent functions as a brominating agent in the same way as N-bromosuccinimide but it does not require a peroxide catalyst or actinic light to initiate the reaction.[1]

[1]B. Taub and J. B. Hino, *J. Org.*, **25**, 263 (1960)

Bromocarbamide (Bromourea), $H_2NCONHBr$. Mol. wt. 122.97.

Preparation.[1] A mixture of urea, calcium carbonate, and water is stirred in an

$$\underset{\substack{0.221\text{ m.}\\ \text{In }13.3\text{ ml. }H_2O}}{H_2NCONH_2} + \underset{0.126\text{ m.}}{CaCO_3} + \underset{0.188\text{ m.}}{Br_2} \xrightarrow{0-5^0} H_2NCONHBr$$

ice bath, bromine is added dropwise, and stirring is continued for 2 hrs. as carbon dioxide is evolved and the reagent separates as pale yellow crystals. The suspension is used as prepared for oxidation of sugars.

Oxidation of O-protected aldoses.[2] 3,5,6-Tri-O-benzyl-2-O-methyl-D-gluco-

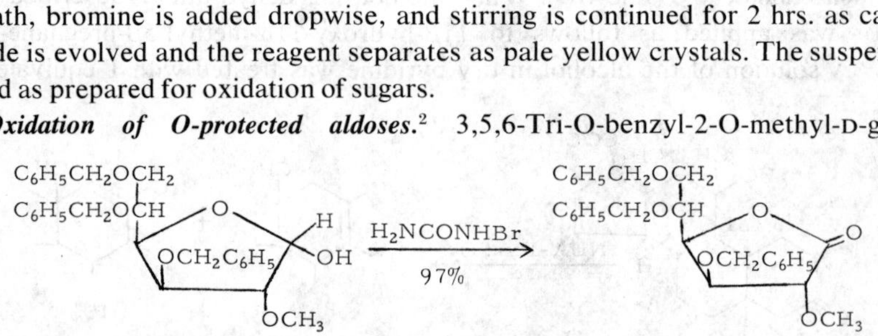
C6H5CH2OCH2
C6H5CH2OCH
O
H
OCH2C6H5
OH
OCH3
H2NCONHBr
97%
C6H5CH2OCH2
C6H5CH2OCH
O
O
OCH2C6H5
OCH3

furanose is oxidized to the corresponding lactone in almost quantitative yield. A solution of 0.043 m. of the aldose in 180 ml. of methanol is added at 5–20° in 30 min. to the suspension of reagent described above. Workup involves dilution with water and extraction with ether. As an oxidant, the reagent is regarded as superior to N-bromoacetamide.

[1]J. Kiss, *Chem. Ind.*, **32**, (1964)
[2]J. Kiss and H. Spiegelberg, *Helv.*, **47**, 398 (1964)

***p*-Bromocarbobenzoxy chloride**, *p*-$BrC_6H_4CH_2OCOCl$. Mol. wt. 249.50.

A stable solid, this reagent gives derivatives of amino acids and peptides that are higher melting and better crystalline than the carbobenzoxy derivatives.[1]

[1]D. M. Channing, P. B. Turner and G. T. Young, *Nature*, **167**, 487 (1951)

o-Bromofluorobenzene, C_6H_4BrF. Mol. wt. 175.01, b.p. 156–157°, $n^{25}D$ 1.5320–1.5325, sp. gr. 1.60. Suppliers: A, B, E, KK, Peninsular ChemResearch. Preparation: *see* Aryldiazonium hexafluorophosphates.

The reagent is useful for the generation of benzyne,[1,2] for example:[2]

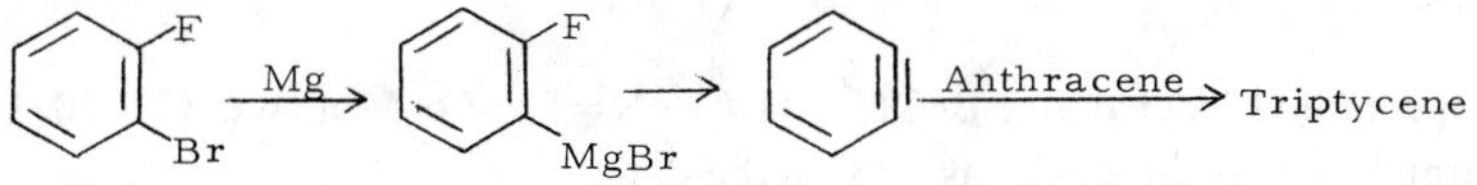
F
Mg
Br
F
MgBr
Anthracene
Triptycene

[1]G. Wittig and L. Pohmer, *Ber.*, **89**, 1334 (1956)
[2]G. Wittig, *Org. Syn., Coll. Vol.*, **4**, 964 (1963)

o-Bromoiodobenzene, BrC_6H_4I-*o*. Mol. wt. 282.92, m.p. 7–9°, b.p. 257° (120–121°/15 mm.). Supplier: Eastman. Preparation from *o*-bromoaniline by the Sandmeyer reaction and use for generation of benzyne, for example for trimerization to

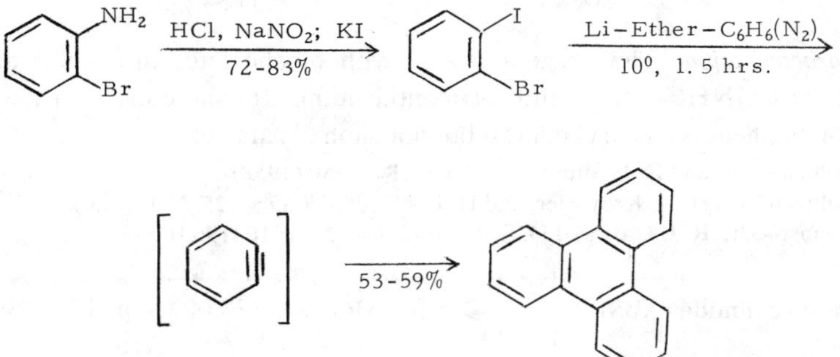
NH₂
Br
HCl, NaNO₂; KI
72-83%
I
Br
Li-Ether-C₆H₆(N₂)
10°, 1.5 hrs.
53-59%

triphenylene.[1] In the example cited, slugs of lithium coated with paraffin oil are hammered into thin foil and the pieces are washed free of oil with dry ether and cut by scissors into chips which are allowed to fall directly into ether in the reaction flask. A solution of *o*-bromoiodobenzene in ether is then added dropwise with stirring and with cooling as required.

[1]H. Heaney and I. T. Millar, *Org. Syn.*, **40**, 105 (1960)

p-Bromophenacyl bromide, *p*-$BrC_6H_4COCH_2Br$. Mol. wt. 277.96, m.p. 110°. Suppliers: B, E, F, MCB. Preparation.[1]

Carboxylic acids can be characterized by conversion into the *p*-bromophenacyl derivatives.[2] Alcohols are characterized by conversion into the potassium alkyl xanthates and reaction of these with *p*-bromophenacyl bromide in acetone to form

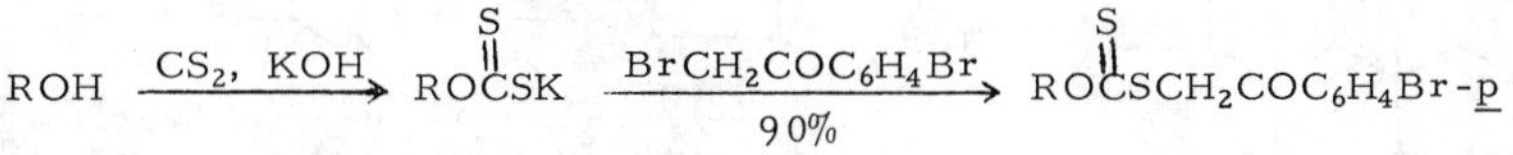
ROH
CS₂, KOH
ROCSK
S
BrCH₂COC₆H₄Br
90%
ROCSCH₂COC₆H₄Br-p
S

the nicely crystalline *p*-bromophenacyl alkyl xanthates.[3] These derivatives are more suitable than the xanthates themselves for identification of alcohols.

[1]W. D. Langley, *Org. Syn., Coll. Vol.*, **1**, 127 (1941)
[2]F. Wild, "Characterization of Organic Compounds," 146, 147, Cambridge (1947)
[3]J. Berger and I. Uldall, *Chem. Scand.*, **18**, 1353 (1964)

β-Bromopropionitrile, $BrCH_2CH_2CN$. Mol. wt. 133.98, b.p. 64–66°/5 mm. Suppliers: B, E, F, MCB.

Butskus[1] reports that this is an excellent reagent for use in combination with a base for the cyanoethylation of amines and alcohols but gives no reason for preferring it to the much less costly acrylonitrile.

[1]R. F. Butskus, *J. Gen. Chem., U.S.S.R.*, **30**, 1799 (1960)

β-Bromopropionyl isocyanate, $BrCH_2CH_2\overset{O}{\overset{\|}{C}}—N=C=O$. Mol. wt. 178.00, b.p. 68–70°/10 mm. Stable for 4 weeks at −78° in the dark.

Preparation by rearrangement of N-bromosuccinimide in refluxing chloroform containing a little allyl chloride and a trace of dibenzoyl peroxide.[1]

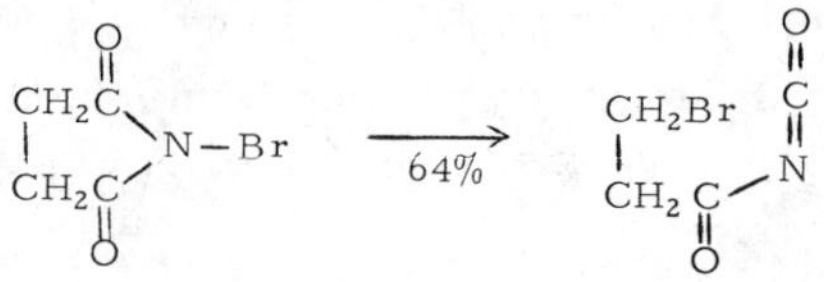
64%

Characterization. The reagent reacts with alcohols to form solid urethanes ($BrCH_2CH_2CONHCO_2R$) useful for identification.[2] It can be used for the characterization of phenols and arylamines (but not aliphatic amines).[3]

[1]H. W. Johnson, Jr., and D. E. Bublitz, *Am. Soc.*, **80**, 3150 (1958)
[2]H. W. Johnson, Jr., H. A. Kreyssler, and H. L. Needles, *J. Org.*, **25**, 279 (1960)
[3]H. W. Johnson, Jr., R. J. Day, and D. S. Tinti, *J. Org.*, **28**, 1416 (1963)

N-Bromosuccinimide (NBS), $\begin{matrix} CH_2CO \\ | \\ CH_2CO \end{matrix}\!\!>NBr$. Mol. wt. 178.00, m.p. 173°. Suppliers: Arapahoe, A, B, E, F, KK, MCB. See reviews.[1-3] Preparation.[4] A commercial sample colored by occlusion of bromine can be purified by crystallization from 10 times its weight of water;[5,6] this purification eliminates inorganic salts which tend to increase the amount of addition of bromine rather than of allylic bromination.[6]

Allylic bromination. Procedures for allylic bromination as used in the commercial production of cortisone are reported by Velluz.[7] A solution of 100 g. of the diphenylethylene (1) in 1 l. of carbon tetrachloride is treated with 30 g. of N-bromosuccinimide and refluxed under irradiation for 15 min., when hydrogen bromide begins to be

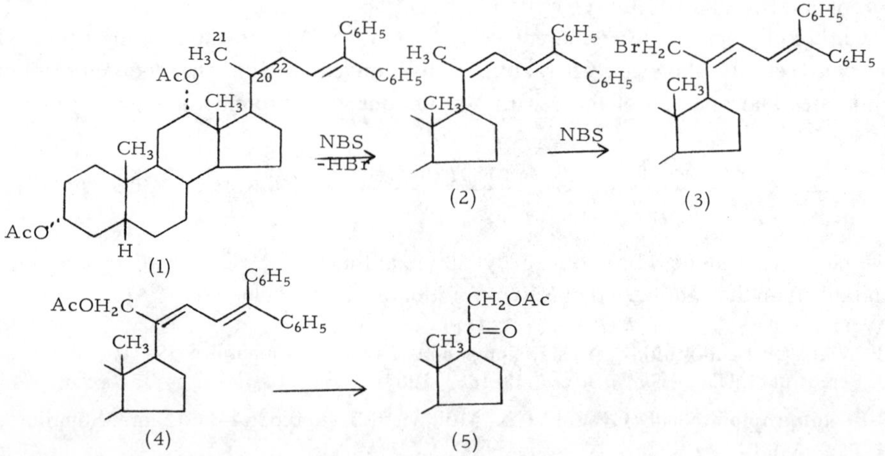
NBS
−HBr
NBS
(1)
(2)
(3)
(4)
(5)

evolved. The mixture is cooled, the succinimide formed is removed by filtration, and the solution of the 22-bromo derivative is refluxed for about 4 hrs. to complete the elimination of hydrogen bromide and produce the diene (2). After removal of solvent, crystallization from acetone gives 75.4 g. of (2). Material recovered from the mother liquor is refluxed briefly with dimethylaniline, reacetylated, and crystallized from acetone to give 6.4 g. of (2); total yield 82%. A second allylic bromination affords (3) in 72% yield, and this on reaction with anhydrous potassium acetate in refluxing acetic acid gives (4) in 77.5% yield. Finally oxidation with chromic acid in acetic acid–ethylene chloride eliminates the side chain and gives the 21-acetoxy-20-ketone (5) in 72% yield.

Allylic bromination of cholesteryl benzoate at C_7 is the key step in the commercial production of vitamin D_3. A carefully developed procedure[8] employs NBS in 20% excess and involves refluxing in petroleum ether under illumination for 4 min.

Investigations in four laboratories[9] indicate that NBS and Br_2 reactions show such similarity that a bromine atom chain is the most plausible mechanism for the NBS reaction.

Oxidation of secondary alcohols. In the presence of water, N-bromosuccinimide is a highly selective oxidizing agent. Thus oxidation of cholestane-3β,5α,6β-triol with

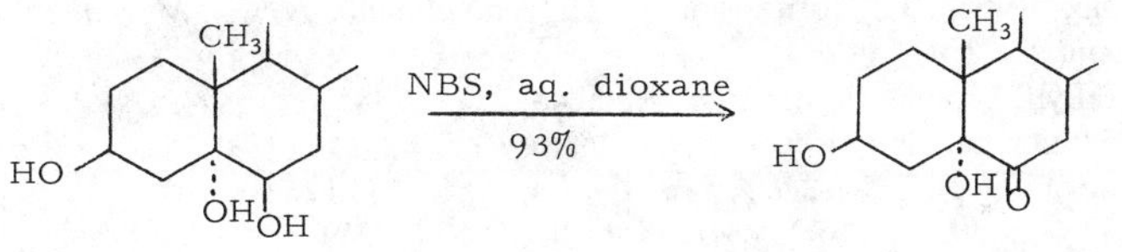
CH3
HO
OH
OH
NBS, aq. dioxane
93%
CH3
HO
OH
O

NBS in aqueous dioxane affords 6-ketocholestane-3β,5α-diol in 93% yield.[10] The selective attack of one of the three secondary alcoholic groups of cholic acid by

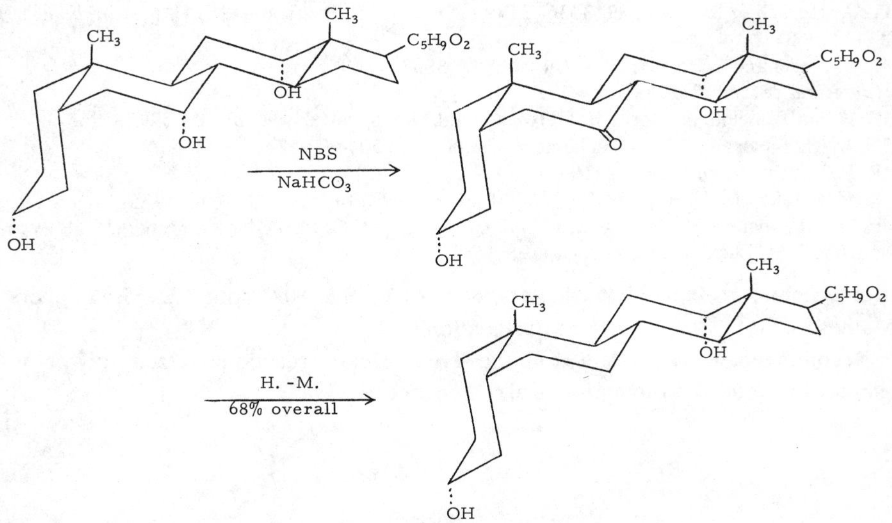
CH3
CH3
C5H9O2
OH
OH
OH
NBS
NaHCO3
CH3
CH3
C5H9O2
OH
O
OH
H. -M.
68% overall
CH3
CH3
C5H9O2
OH
OH

oxidation with excess NBS in aqueous sodium bicarbonate solution provides an efficient route to desoxycholic acid, required as starting material for the synthesis of cortisone.[11]

Other reactions. N-Bromosuccinimide has been used in aqueous 1,2-dimethoxyethane for the generation of HOBr,[12] but N-bromoacetamide is preferred for this purpose.

Whereas pure NBS effects side-chain bromination of aromatic hydrocarbons, aged material effects bromination of the aromatic ring.[13] The responsible impurity can be removed by pumping out at 0.5 mm. over phosphorus pentoxide for 4 hrs.

Caution: Explosions have been reported in reactions of NBS.[14]

Barakat *et al.*[15] observed that on treatment with NBS in aqueous solution an α-amino acid (alanine) and a dicarboxylic acid (oxalic) undergo decarboxylation with liberation of bromine. Luck[16] showed that α-amino acids, peptides, and proteins on treatment with NBS (2 equiv.) in aqueous solution undergo quantitative decarboxylation. The reaction is complete in 30 min. with quantitative evolution of CO_2, as determined in a Warburg apparatus. The first stable product formed is the aldehyde.

$$CH_3\underset{\displaystyle NH_2}{\underset{|}{C}}HCO_2H \xrightarrow{NBS} CH_3CHO + NH_3 + CO_2$$

[1]T. D. Waugh, "N-Bromosuccinimide, Its Reactions and Uses," Arapahoe Chemicals, Inc., Boulder, Colo. (1951); L. Horner and E. H. Winkelmann, *Newer Methods of Preparative Organic Chemistry*, **3**, 151 (1964)
[2]C. Djerassi (allylic bromination), *Chem. Rev.*, **43**, 271 (1948)
[3]R. Filler (oxidation and dehydrogenation), *Chem. Rev.*, **63**, 21 (1963)
[4]E. Campaigne and B. F. Tullar, *Org. Syn., Coll. Vol.*, **4**, 921 (1963)
[5]H. J. Dauben, Jr. and L. L. McCoy, *Am. Soc.*, **81**, 4863 (1959)
[6]W. J. Bailey and J. Bello, *J. Org.*, **20**, 693 (1955)
[7]L. Velluz, *Substances naturelles de synthèse*, **7**, 31 (1953)
[8]S. Bernstein, L. J. Binovi, L. Dorfman, K. J. Sax, and Y. Subbarow, *J. Org.*, **14**, 433 (1949)
[9]P. S. Skell *et al.*, *Am. Soc.*, **85**, 2850 (1963); C. Walling *et al.*, *ibid.*, **85**, 3129, 3134 (1963); G. A. Russell *et al.*, *ibid.*, **85**, 3136, 3139 (1963); R. E. Pearson and J. C. Martin, *ibid.*, **85**, 3142 (1963)
[10]L. F. Fieser and S. Rajagopalan, *Am. Soc.*, **71**, 3938 (1949)
[11]*Idem, ibid.*, **71**, 3935 (1949)
[12]E. E. van Tamelen, A. Storni, E. J. Hessler, and M. Schwartz, *Am. Soc.*, **85**, 3295 (1963)
[13]N. B. Chapman and J. F. A. Williams, *J. Chem. Soc.*, 5044 (1952)
[14]R. H. Martin, *Nature*, **168**, 32 (1951)
[15]M. Z. Barakat, M. F. A. El-Wahab, and M. M. El-Sadr, *Am. Soc.*, **77**, 1670 (1955)
[16]E. W. Chappelle and J. M. Luck, *J. Biol. Chem.*, **229**, 171 (1957); N. Konigsberg, G. Stevenson, and J. M. Luck, *ibid.*, **235**, 1341 (1960)

Bromotrichloromethane, $BrCCl_3$. Mol. wt. 198.30, b.p. 104°, sp. gr. 2.055. Suppliers: Matheson, Dow, A, E, F, MCB. Purification.[1]

Recommended as solvent for the Hunsdiecker reaction.[1] Used for photochemically induced bromination of alkylbenzenes:[2]

$$BrCCl_3 \xrightarrow{h\nu} Br\cdot + \cdot CCl_3$$

$$\cdot CCl_3 + ArCH_3 \longrightarrow HCCl_3 + Ar\dot{C}H_2$$

$$Ar\dot{C}H_2 + BrCCl_3 \longrightarrow ArCH_2Br + \cdot CCl_3$$

Use for the generation of dichlorocarbene.[3]

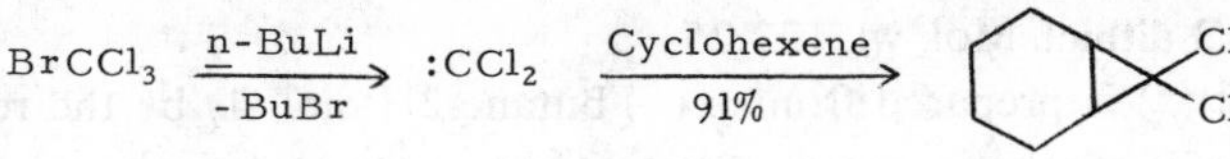
BrCCl3
n-BuLi
-BuBr
:CCl2
Cyclohexene
91%
Cl
Cl

[1]F. W. Baker, H. D. Holtz, and L. M. Stock, *J. Org.*, **28**, 514 (1963)
[2]E. S. Huyser, *Am. Soc.*, **82**, 391, 394 (1960)
[3]W. T. Miller, Jr., and C. S. Y. Kim. *ibid.*, **81**, 5008 (1959); G. L. Closs and L. E. Closs, *ibid.*, **82**, 5723 (1960)

Bromotrifluoromethane, CF_3Br. Mol. wt. 148.93, a gas. Supplier (6-lb. cylinder): du Pont Co., Organic Chemicals Division.

The reagent reacts with an alkyllithium or an alkylmagnesium bromide in the same way that difluorodibromomethane reacts and in about the same yield:[1]

$$3\ C_4H_9Li + CF_3Br \xrightarrow[62\%]{} C_4H_9CH{=}CHC_3H_7 + C_4H_9Br + 3\ LiF$$

[1]V. Franzen and L. Fikentscher, *Ber.*, **95**, 1958 (1962)

4-Bromo-2,5,7-trinitrofluorenone, 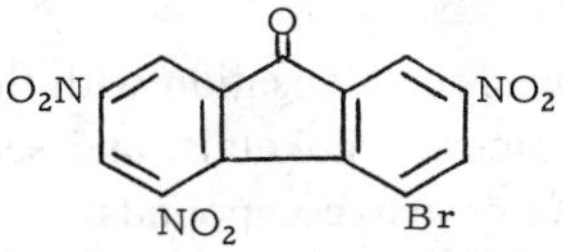
O2N
NO2
NO2
Br

Mol. wt. 283.22, m.p. 191–192°.

2,4,7-Trinitrofluorenone forms both a black 1:1 complex with 1-methoxybenzo-[c]phenanthrene melting at 212° and a brown-red 1:1 complex melting at 157°. In the hope of obtaining a similar pair of complexes with which X-ray analysis of structure would be greatly simplified, Newman and Blum[1] prepared the bromotrinitro ketone and the corresponding iodotrinitro ketone (m.p. 193–194°). Both compounds are excellent complexing agents, but only one complex was obtained with 1-methoxybenzo[c]phenanthrene. All the compounds tested formed 1:1 complexes except prehnitene, which formed a complex containing 2 equivalents of hydrocarbon to one of the π-acid component. Of particular interest is the fact that benzene forms yellow and light orange complexes with the bromo and the iodo compounds. These complexes do not lose benzene when held at 1–2 mm. pressure at room temperature for a day, but when heated they lose benzene at temperatures below 100°.

[1]M. S. Newman and J. Blum, *Am. Soc.*, **86**, 5600 (1964)

Butadiene, $CH_2{=}CHCH{=}CH_2$. Mol. wt. 54.09, b.p. −4.4°. Supplied in cylinders by Matheson Co., Allied Chem. Corp., KK, and MCB.

D-(−)-Butane-2,3-diol, 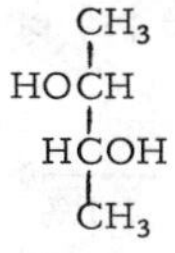
CH3
HOCH
HCOH
CH3

Mol. wt. 90.42, αD −12.9° (neat). Suppliers: British Drug Houses, KK.

DL-Camphor has been resolved by vapor-phase chromatography of the diastereoisomeric ketals prepared with this readily available (by fermentation) diol.[1]

[1]J. Casanova, Jr., and E. J. Corey, *Chem. Ind.*, 1664 (1961)

L(+)-Butane-2,3-dithiol. Mol. wt. 122.25.

The reagent (5) is prepared from D-(−)-butane-2,3-diol (1) by the reactions outlined.[1] In case a synthetic racemic compound contains a carbonyl group which

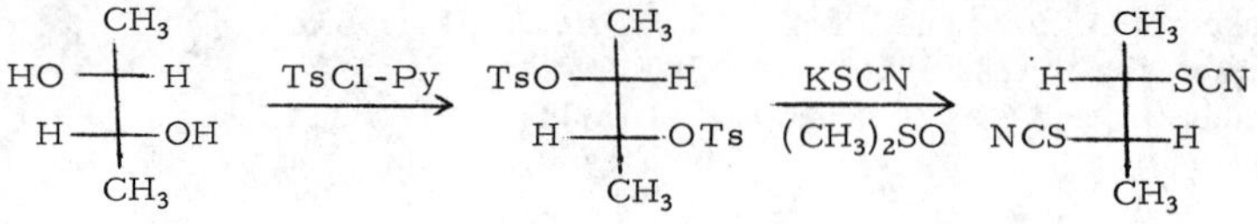

(1) (2) (3)

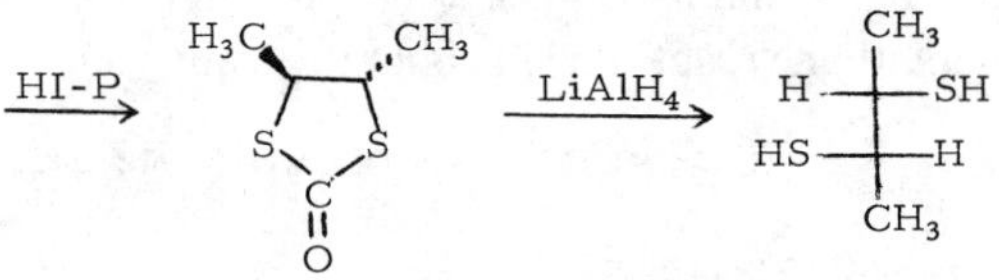

(4) (5)

eventually is to be removed, condensation with the optically active dithiol (5) gives a mixture of diastereoisomeric thioketals, and separation and desulfurization then affords the optically active desoxo compounds.

[1]E. J. Corey and R. B. Mitra, *Am. Soc.*, **84**, 2938 (1962)

Butanone ethyleneketal (2-Methyl-2-ethyl-1,3-dioxolane), 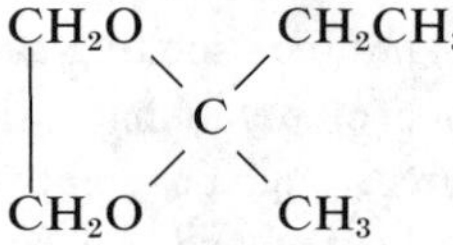 . Mol. wt. 116.16, b.p. 117°. Supplier: Baker.

Preparation and use in exchange ketal formation (dioxolonation).[1] The yields are sometimes higher and the selectivity greater than in direct reaction with ethylene glycol. In the steroid series the order of reactivity is: saturated 3- and 20-ketones > αβ-unsaturated 3- and 20-ketones > 2- or 4-bromo-3-ketones >> (no reaction) 17-ketones, 2,4-dibromo-3-ketones.[2]

[1]H. J. Dauben, B. Löken, and H. J. Ringold, *Am. Soc.*, **76**, 1359 (1954); G. Rosenkranz, M. Velasco, and F. Sondheimer, *ibid.*, **76**, 5024 (1954)

[2]C. Djerassi, "Steroid Reactions," 14–16, Holden-Day (1963)

***t*-Butyl acetate.** $CH_3CO_2C(CH_3)_3$. Mol. wt. 116.16, b.p. 96°, sp. gr. 0.866. Suppliers: B, E, F, KK, MCB. Preparation by three methods.[1]

A procedure for the conversion of N-protected amino acids into their *t*-butyl esters consists in reaction with *t*-butyl acetate in the presence of perchloric acid.[2]

$$\underset{\text{1 mml.}}{PhNHCH(CH_3)CO_2H} + \underset{\text{4 ml.}}{CH_3CO_2C(CH_3)_3} + \underset{\text{0.2 mml.}}{HClO_4 \cdot 2\,H_2O} \xrightarrow[91\%]{} PhNHCH(CH_3)CO_2C(CH_3)_3$$

The mixture is shaken for a few days at room temperature. The reaction is believed

$$CH_3CO_2C(CH_3)_3 \xrightarrow[-CH_3CO_2H]{H^+} (CH_3)_3C^+ \xrightarrow[-H^+]{RCO_2H} RCO_2C(CH_3)_3$$

to involve a carbonium ion intermediate, but the only evidence is that 3-nitrophthalic acid and 2,4,6-trinitrobenzoic acid afford *t*-butyl esters in yields of 85% and 22%. *t*-Butyl esters are regarded as useful carboxyl-masking derivatives because they are resistant to hydrolysis, ammonolysis, and hydrogenolysis. A free amino acid can be converted into the *t*-butyl ester by use of a large excess of *t*-butyl acetate (1 : 100) and a slight excess of perchloric acid.[3]

[1]R. H. Baker and F. G. Bordwell; C. R. Hauser *et al.*; A. Spassow, *Org. Syn., Coll. Vol.*, **3**, 141, 142, 144 (1955)
[2]E. Taschner, C. Wasielewski, and J. F. Biernat, *Ann.*, **646**, 119 (1961)
[3]E. Taschner, A. Chimiak, B. Bator, and T. Sokolowska, *Ann.*, **646**, 134 (1961)

t-Butyl acetoacetate, $CH_3COCH_2CO_2C(CH_3)_3$. Mol. wt. 158.19, b.p. 85°/20 mm. Suppliers: A, E, F, KK.

Prepared by reaction of diketene with *t*-butanol in the presence of sodium ethoxide,[1] the reagent is employed in a novel synthesis of acyloins.[2] The reagent (1) is alkylated as enolate, the product (2) is converted into its enolate (3), and the

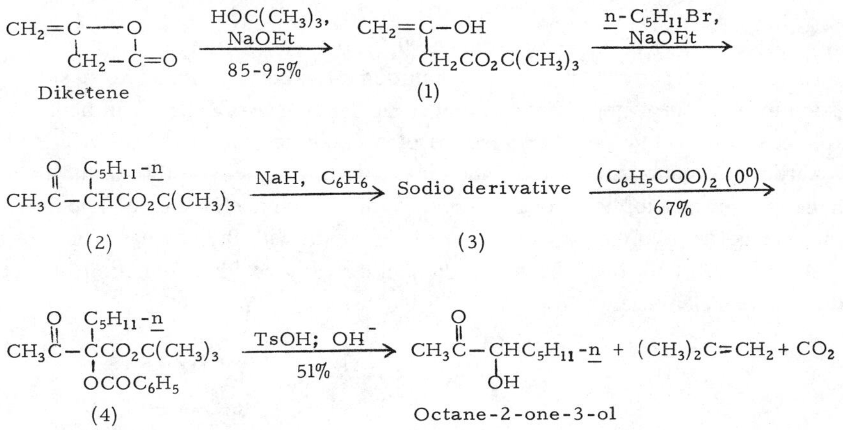

benzoyloxy group is introduced into the α-position (4) by reaction with dibenzoyl peroxide. Elimination of the *t*-butyl group and decarboxylation is accomplished by heating the ester (4) with a catalytic amount of *p*-toluenesulfonic acid; saponification under mild conditions then gives the (base-sensitive) acyloin.

A general synthesis of α,β-unsaturated ketones such as (2)[3] starts with a Knoevenagel condensation of the reagent with an aldehyde. The product (1) when heated

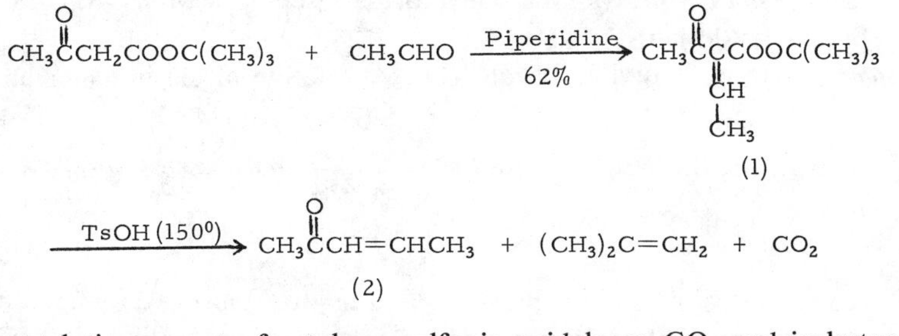

with a catalytic amount of *p*-toluenesulfonic acid loses CO_2 and isobutene with formation of the unsaturated ketone (2).

[1]S.-O. Lawesson, S. Grönwall, and R. Sandberg, *Org. Syn.*, **42**, 28 (1962)

[2]S.-O. Lawesson and S. Grönwall, *Chem. Scand.*, **14**, 1445 (1960)

[3]S.-O. Lawesson, E. H. Larsen, G. Sundström, and H. J. Jakobsen, *Chem. Scand.*, **17**, 2216 (1963)

t-**Butylamine**, $(CH_3)_3CNH_2$. Mol. wt. 73.14, b.p. 45°, sp. gr. 0.698. Suppliers: A, E, F, MCB.

This amine is used to advantage instead of ammonia in the Glaser oxidative coupling of terminal acetylenes.[1]

This hindered amine was used to effect dehydrobromination of dibutyl 1-bromo-3,3,3-trichloropropane-1-boronate where usual reagents failed.[2] Potassium ethoxide

$$Cl_3CCH_2\underset{\underset{Br}{|}}{CH}B(OC_4H_9)_2 \ + \ (CH_3)_3CNH_2 \ \xrightarrow[68\%]{\text{Refl. 1 hr.}} Cl_3CCH{=}CHB(OC_4H_9)_2$$

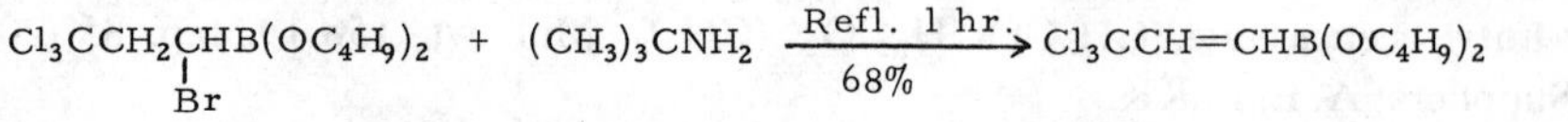

and even sodium *t*-butoxide cause displacement of the bromine atom by alkoxide with no detectable dehydrobromination.

[1]M. D. Cameron and G. E. Bennett, *J. Org.*, **22**, 557 (1957)

[2]D. S. Matteson and R. W. H. Mah, *J. Org.*, **28**, 2174 (1963)

n-**Butyl azide**, $n\text{-}C_4H_9\overset{+}{N}{=}N{=}\overset{-}{N}$. Mol. wt. 99.13, b.p. 106.5°/mm.

This azide, prepared[1] by adding a solution of *n*-butyl bromide to a slurry of of sodium azide in water, cleaves ketones of the type $ArCOCH_2R$ in benzene or nitrobenzene at 70–90° in the presence of a catalytic amount of sulfuric acid to two carbonyl components.[2] Acetophenone gives benzaldehyde (80%) and formaldehyde (83%); propiophenone gives benzaldehyde and acetaldehyde. The reaction is interpreted as involving combination of the azide with the conjugate acid of the ketone, release of nitrogen from the diazonium cation with elimination of water, hydration, and cleavage.

$$C_6H_5\overset{\overset{O}{\|}}{C}CH_3 \ \xrightarrow{H^+} \ C_6H_5\overset{\overset{OH}{|}}{\underset{+}{C}}CH_3 \ \xrightarrow{Bu\overset{+}{N}{=}N{=}\overset{-}{N}} \ C_6H_5\overset{\overset{OH}{|}}{\underset{\underset{Bu\overset{+}{N}{=}N{=}\overset{-}{N}}{|}}{C}}CH_3 \ \xrightarrow[-H_2O]{-N_2}$$

$$C_6H_5\underset{\underset{Bu\overset{+}{N}}{|}}{C}{=}CH_2 \ \longleftrightarrow \ C_6H_5\overset{\overset{+}{C}}{\underset{\underset{BuN}{\|}}{}}{-}\overset{-}{C}H_2 \ \xrightarrow[-H^+]{2\ H_2O} \ C_6H_5\overset{\overset{HO\ \ OH}{|\ \ \ |}}{\underset{\underset{BuNH}{|}}{C}}{-}CH_2 \ \xrightarrow{-BuNH_2} C_6H_5CHO \ + \ CH_2O$$

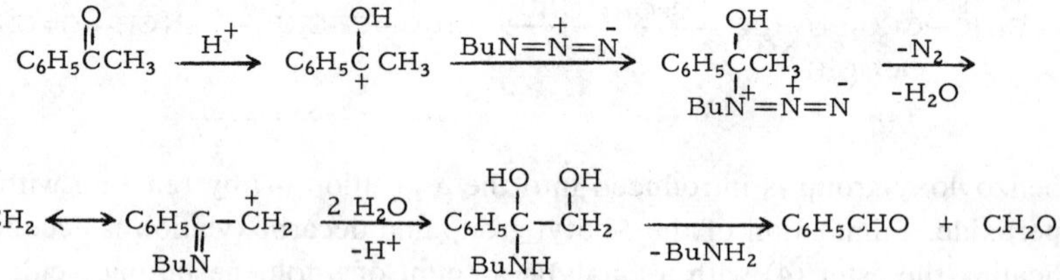

[1]J. H. Boyer and L. R. Morgan, Jr., *Am. Soc.*, **80**, 2020 (1958)

[2]J. H. Boyer and J. Hamer, *ibid.*, **77**, 951 (1955); J. H. Boyer and L. R. Morgan, Jr., *ibid.*, **81**, 3369 (1959)

t-**Butyl azidoformate** (*t*-**Butyloxycarbonyl azide**), see (2). Mol. wt. 143.74, b.p. 73–76°. Supplier: Aldrich.

Preparation from *t*-butyl carbazate (1).[1] A solution of (1) in water and acetic

$$(CH_3)_3CO\overset{\overset{O}{\|}}{C}NHNH_2 \ + \ NaNO_2 \ \xrightarrow[64-82\%]{10-15^0} \ (CH_3)_3CO\overset{\overset{O}{\|}}{C}{-}N{=}\overset{+}{N}{=}\overset{-}{N}$$

(1) 0.62 m. in 72 g. AcOH   0.68 m.   (2)
     and 700 ml. water

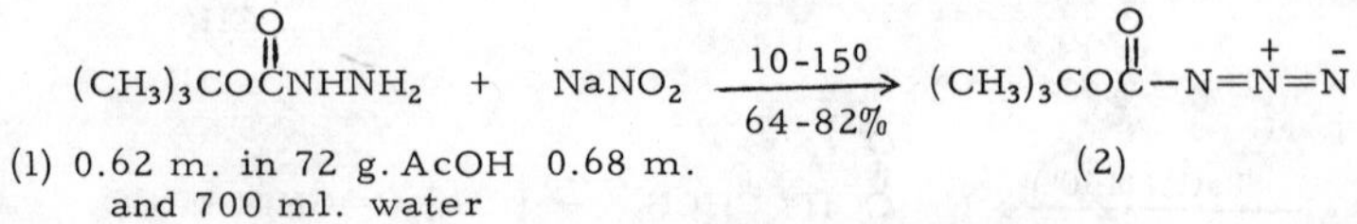

acid is treated with solid sodium nitrite, added in 40–50 min. at a temperature controlled to 10–15° by cooling. Water is added and the product is extracted with ether and distilled.

***N*-Protection of amino acids.** The reagent reacts readily with amino acids and esters (aqueous dioxane, MgO) to give *t*-butyloxycarbonylamino acids, which are not otherwise available, since *t*-butyl chloroformate, $(CH_3)_3COCOCl$, is too un-

$$(CH_3)_3CO\overset{O}{\overset{\|}{C}}N{=}\overset{+}{N}{=}N \; + \; H_2N\overset{R}{\overset{|}{C}}HCO_2H \longrightarrow (CH_3)_3CO\overset{O}{\overset{\|}{C}}-\overset{H}{\overset{|}{N}}-\overset{R}{\overset{|}{C}}HCO_2H$$

stable for convenient manipulation.[2] For these reasons *t*-butyl azidoformate is used widely in peptide synthesis.[3] The group is stable to hydrogenation and to sodium in liquid ammonia and is more resistant to alkali than the carbobenzoxy group. It is readily removed by hydrogen bromide in acetic acid. Selective removal of this group in the presence of a carbobenzoxy group can be accomplished with hydrogen chloride in acetic acid, ether, ethyl acetate, or nitromethane.

[1]L. A. Carpino, B. A. Carpino, P. J. Crowley, C. A. Giza, and P. H. Terry, *Org. Syn.*, **44**, 15 (1964)

[2]A. R. Choppin and J. W. Rogers, *Am. Soc.*, **70**, 2967 (1948)

[3]R. Schwyzer, P. Sieber, and H. Kappeler, *Helv.*, **42**, 2622 (1959); R. A. Boissonnas, *Advances in Org. Chem.*, **3**, 171 (1963); K. Hofmann *et al.*, *Am. Soc.*, **87**, 620 (1965)

***t*-Butyl carbazate**, $(CH_3)_3CO\overset{O}{\overset{\|}{C}}NHNH_2$. Mol. wt. 132.17, m.p. 41–42°.

Preparation. Carpino and co-workers have perfected two methods of preparation. In method I[1] a mixture of *t*-butanol, pyridine, and chloroform is stirred during drop-

$$\text{I.}\quad \underset{5.33\ \text{m.}}{(CH_3)_3COH} + \underset{5.33\ \text{m.}}{\text{Pyridine}} + \underset{1.6\ \text{l.}}{CHCl_3} + \underset{4.85\ \text{m.}}{CH_3SCOCl} \xrightarrow[52\text{-}62\%]{\text{Refl. 24 hrs.}} (CH_3)_3CO\overset{O}{\overset{\|}{C}}SCH_3$$

$$\xrightarrow[70\text{-}80\%]{H_2NNH_2} (CH_3)_3CO\overset{O}{\overset{\|}{C}}NHNH_2$$

wise addition of methyl chlorothioformate (Stauffer Chem. Co.). The reaction product, *t*-butyl-S-methylthiol carbonate, on reaction with hydrazine affords *t*-butyl carbazate.

In method II[2] a mixture of *t*-butanol, quinoline, and methylene chloride is stirred

$$\text{II.}\quad \underset{3.35\ \text{m.}}{(CH_3)_3COH} + \underset{3.33\ \text{m.}}{\text{Quinoline}} + \underset{500\ \text{ml.}}{CH_2Cl_2} + \underset{3.32\ \text{m.}}{C_6H_5OCOCl} \xrightarrow[71\text{-}76\%]{} (CH_3)_3CO\overset{O}{\overset{\|}{C}}OC_6H_5$$

$$\xrightarrow[89\text{-}97\%]{H_2NNH_2} (CH_3)_3CO\overset{O}{\overset{\|}{C}}NHNH_2$$

during dropwise addition of phenyl chloroformate during 4 hrs. The reaction product, *t*-butyl phenyl carbonate, is converted into *t*-butyl carbazate by reaction with hydrazine.

Uses. *t*-Butyl carbazate is a key intermediate for the preparation of *t*-butyl azidoformate and *t*-butyl azidodiformate (see also Ref. 2).

[1]L. A. Carpino, D. Collins, S. Göwecke, J. Mayo, S. D. Thatte, and F. Tibbetts, *Org. Syn.*, **44**, 20 (1964)

[2]L. A. Carpino, B. A. Carpino, C. A. Giza, R. W. Murray, A. A. Santilli, and P. H. Terry, *ibid.*, **44**, 22 (1964)

***t*-Butyl chloride**, $(CH_3)_3CCl$. Mol. wt. 92. 57, b.p. 51°, sp. gr. 0.85. Suppliers: A, B, E, F, KK, MCB.

Preparation. In a separatory funnel shake vigorously a mixture of 50 ml. of *t*-butanol and 250 ml. of concd. hydrochloric acid. In about 10 min. the conversion should be complete. Separate the *t*-butyl chloride layer, dry it over calcium chloride, and distil; yield 45–50 g.

***t*-Butyl chloroacetate**, $ClCH_2CO_2C(CH_3)_3$. Mol. wt. 150.61, b.p. 56°/16 mm. Suppliers: E, F.

Preparation.[1] Useful in glycidic ester condensation because of avoidance of self-condensation.[2]

[1]A. L. McCloskey, G. S. Fonken, R. W. Kluiber, and W. S. Johnson, *Org. Syn., Coll. Vol.*, **4**, 263 (1963)
[2]W. S. Johnson *et al.*, *Am. Soc.*, **75**, 4995 (1953)

***i*-Butyl chloroformate** and ***sec*-Butyl chloroformate.** Mol. wt. 136.58, b.p. 121°, sp. gr. 1.05; b.p. 116°. Suppliers: B, E, F. Preparation.[1]

The reagents are used for making mixed anhydrides that serve as intermediates in peptide synthesis.[2-4] A protected amino acid, such as the N-carbobenzoxy derivative (Cb), is brought into solution in toluene or tetrahydrofurane by addition of enough triethylamine to form the salt and a butyl chloroformate is added at 0°

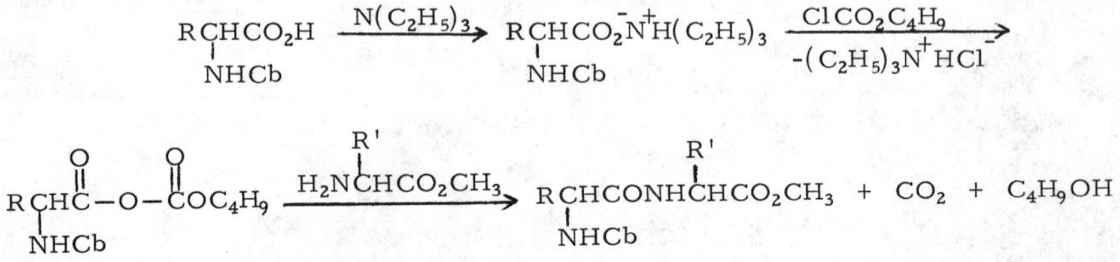
RCHCO2H
NHCb
N(C2H5)3
RCHCO2⁻N⁺H(C2H5)3
NHCb
ClCO2C4H9
-(C2H5)3N⁺HCl⁻
RCHC—O—COC4H9
NHCb
H2NCHCO2CH3
R'
RCHCONHCHCO2CH3 + CO2 + C4H9OH
NHCb

to produce the mixed anhydride. A solution of an amino acid ester (or peptide ester) to be N-acylated is added in an inert solvent and the mixture allowed to come to room temperature; evolution of carbon dioxide begins immediately. The butyl chloroformates are preferred to lower esters for preparation of peptides of moderate or high molecular weight; ethyl chloroformate is preferred for synthesis of dipeptides.[4]

t-Butyl chloroformate has been prepared;[5] it begins to decompose at 10° and is hydrolyzed with extreme ease.

[1]J. R. Vaughan, Jr., and R. L. Osato, *Am. Soc.*, **74**, 676 (1952)
[2]R. A. Boissonnas *et al.*, *Helv.*, **34**, 874 (1951); **35**, 2229, 2237 (1952)
[3]J. R. Vaughan, Jr., *et al.*, *Am. Soc.*, **73**, 3547 (1951); **74**, 676, 6137 (1952); **75**, 5556 (1953); **76**, 2474 (1954)
[4]T. Wieland and H. Bernhard, *Ann.*, **572**, 190 (1951)
[5]A. R. Choppin and J. W. Rogers, *Am. Soc.*, **70**, 2967 (1948)

***t*-Butyl chromate**, $(CH_3)_3COCrO_2(OH)$. Mol. wt. 174.13.

The first description of this reagent reported an excellent yield in the oxidation of cholesteryl acetate to 7-ketocholesteryl acetate;[1] but subsequent reports have not substantiated this claim. In a modified procedure found to be more satisfactory the reagent is prepared in carbon tetrachloride containing acetic acid and acetic anhydride.[2] Some workers[3,4] have found even the modified procedure to give poor yields; others[5-7] have found the reagent as good as or better than conventional reagents (CrO_3, Oppenauer, NBS). Use of the reagent in pyridine offers no advantages over use of chromic anhydride in the same solvent.[8]

[1]R. V. Oppenauer and H. Oberrauch, *Anales asoc. quim. argentina*, **37**, 246 (1949) [*C. A.*, **44**, 3871 (1950)]

[2]K. Heusler and A. Wettstein, *Helv.*, **35**, 284 (1952)

[3]P. N. Rao and P. Kurath, *Am. Soc.*, **78**, 5660 (1956)

[4]R. E. Beyler, A. E. Oberster, F. Hoffman, and L. H. Sarett, *ibid.*, **82**, 170 (1960)

[5]C. Sannié, J.-J. Panouse, and S. Vertalier, *Bull. Soc.*, **22**, 1039 (1955)

[6]N. B. Haynes, D. Redmore, and C. J. Timmons, *J. Chem. Soc.*, 2420 (1963)

[7]A. Katz, *Helv.*, **40**, 487 (1957)

[8]E. Menini and J. K. Norymberski, *Biochem. J.*, **84**, 195 (1962)

t-Butyl cyanoacetate, $N{\equiv}CCH_2CO_2C(CH_3)_3$. Mol. wt. 141.17, b.p. 67–68°/1.5 mm.
Preparation.[1]

$$NCCH_2CO_2H \quad + \quad PCl_5 \xrightarrow{\text{With cooling}} NCCH_2COCl \xrightarrow[\substack{4\text{ m. }C_6H_5N(CH_3)_2 \\ \text{in }600\text{ ml. ether}}]{4\text{ m. }(CH_3)_3COH}$$

4 m. in 2 l. ether   4 m.   (Isolated)

$$NCCH_2CO_2C(CH_3)_3$$

Yield 63-67%

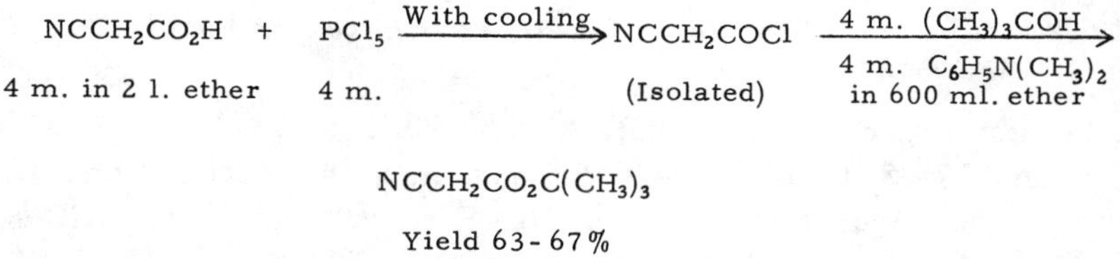

Use in condensations. The reagent can be used in any condensation where ethyl cyanoacetate is used, but it has the added advantage that the carbo-*t*-butoxy group, which may serve in conjunction with the cyano group to activate the α-hydrogens (for cyanomethylation, etc.), can later be removed simply by pyrolysis at 150° in the presence of a trace of p-toluenesulfonic acid. Examples:[2]

$$RCHO + \underset{\overset{|}{CH_2CO_2C(CH_3)_3}}{\overset{CN}{}} \xrightarrow[75\text{-}99\%]{\text{Piperidine}} \underset{\overset{|}{RCH=CCO_2C(CH_3)_3}}{\overset{CN}{}}$$

$$RX + \underset{\overset{|}{CH_2CO_2C(CH_3)_3}}{\overset{CN}{}} \xrightarrow{NaH, THF} \underset{\overset{|}{RCHCO_2C(CH_3)_3}}{\overset{CN}{}}$$

$$2\ CH_2{=}CHCO_2C_2H_5 + \underset{\overset{|}{CO_2C(CH_3)_3}}{CH_2CN} \longrightarrow (CH_3)_3CO_2C\underset{\overset{|}{CH_2CH_2CO_2C_2H_5}}{\overset{CH_2CH_2CO_2C_2H_5}{C}}CN$$

$$\xrightarrow[62\% \text{ over all}]{TsOH\ 150^0} \underset{\overset{|}{CH_2CH_2CO_2C_2H_5}}{\overset{CH_2CH_2CO_2C_2H_5}{CHCN}}$$

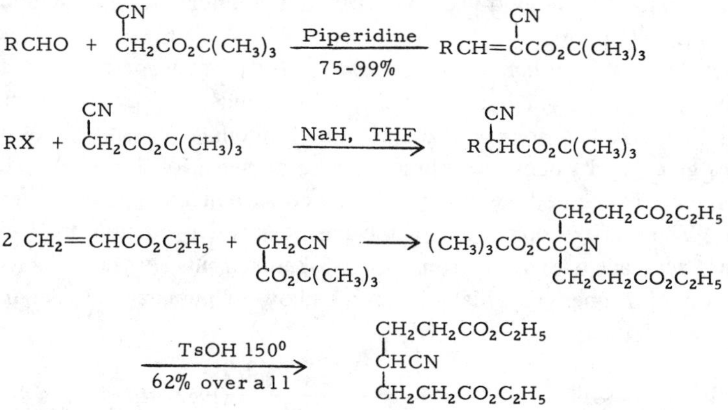

[1]R. E. Ireland and M. Chaykovsky, *Org. Syn.*, **41**, 5 (1961)

[2]S.-O. Lawesson, E. H. Larsen, and H. J. Jakobsen, *Arkiv Kemi*, **23**, 453 (1965)

t-Butyl cyanoformate, $N{\equiv}CCO_2C(CH_3)_3$. Mol. wt. 127.14, b.p. 64–65°/55 mm.
Preparation.[1]

$$CH_3OCH_2CO_2C(CH_3)_3 \xrightarrow[(C_6H_5COO)_2]{NBS, CCl_4} \underset{\overset{|}{Br}}{CH_3OCHCO_2C(CH_3)_3} \xrightarrow[NaHCO_3]{H_2O} O{=}CHCO_2C(CH_3)_3$$

$$\xrightarrow[50\% \text{ overall}]{H_2NOH} HON{=}CHCO_2C(CH_3)_3 \xrightarrow[75\%]{Ac_2O-(C_2H_5)_3N} N{\equiv}CCO_2C(CH_3)_3$$

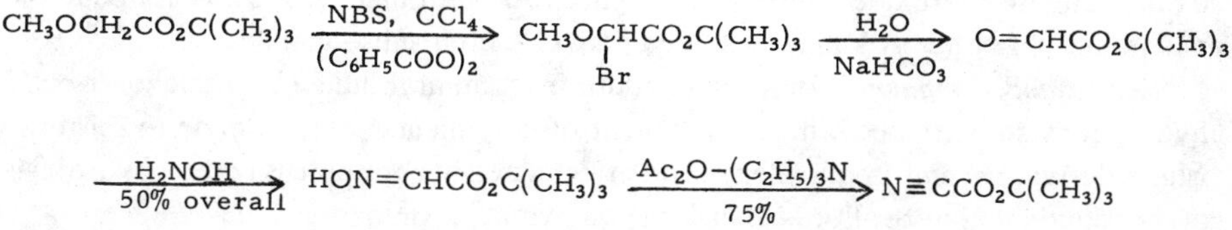

N-Protective group in peptide synthesis.[2] The reagent provides a new and improved method for the preparation of N-*t*-butyloxycarbonyl derivatives of amino acids or peptides, since it reacts with the NH_2 group with elimination of HCN to form the derivative, as illustrated for L-leucine ethyl ester. The reactants are mixed at 0° and agitated at room temperature for 1 hr. The solvent is evaporated and the product isolated from the dried neutral fraction.

$$(CH_3)_3CO\overset{O}{\overset{\|}{C}}-C\equiv N \ + \ \underset{\underset{H}{|}}{H-N}-\underset{\underset{CH_2CH(CH_3)_2}{|}}{CHCO_2C_2H_5} \ \xrightarrow[92\%]{-HCN} \ (CH_3)_3CO\overset{O}{\overset{\|}{C}}-\underset{\underset{H}{|}}{N}\underset{\underset{CH_2CH(CH_3)_2}{|}}{CHCO_2C_2H_5}$$

20 mml.   22 mml.
In hexane   In benzene

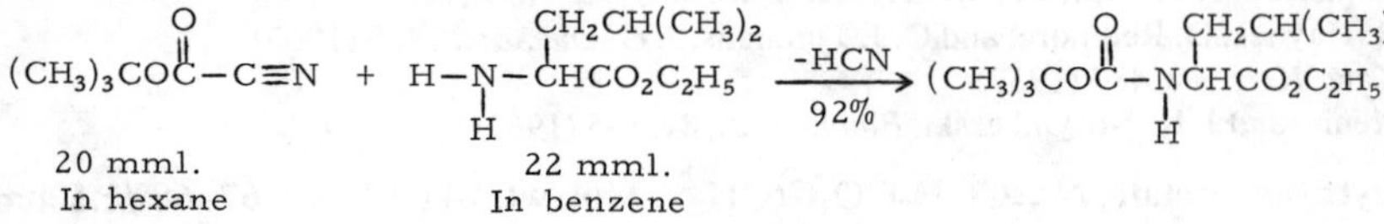

[1] L. A. Carpino, *J. Org.*, **29**, 2820 (1964)
[2] M. Leplawy and W. Stec, *Bull. acad. polon. sci., ser. sci. chim.*, (in English), **12**, 21 (1964) [*C. A.*, **61**, 1933 (1964)]

t-Butyl hydroperoxide, $(CH_3)_3C-OOH$. Mol. wt. 90.12, m.p. 3–4°, b.p. 40°/23 mm.,[1] 42°/18 mm.,[2] 37–38°/16 mm.,[3] 35–36°/8.5 mm.[4] Suppliers: Lucidol Corp., Div. of Wallace Tiernan Inc., Borden Chem. Co., Eastern Chem. Corp., KK, MCB. Preparation.[5] Assay procedure.[6]

Purification. Lucidol material contains 76–90% of active reagent. Purification has been accomplished by fractionation[2,3] and by refluxing in an azeotrope separator until homogeneous and then distillation.[1] Kochi and Mocadlo[3] vacuum distilled Lucidol material several times and obtained a product of greater than 99% purity. Bartlett and McBride[4] extracted Borden material with cold 15% aqueous potassium hydroxide, followed by neutralization with solid ammonium chloride and vacuum distillation (98.4% pure).

Reactions with unsaturated compounds. *t*-Butyl hydroperoxide decomposes smoothly at 95–105° and hydroxylates double bonds, and even α,β-unsaturated ketones; the reactions are catalyzed by small amounts of osmium tetroxide and afford *cis*-glycols.[7] In benzene solution in the presence of Triton B as catalyst it converts α,β-unsaturated ketones into the corresponding epoxides in yields of 50–90%.[8] The steric requirement of the base-catalyzed reaction are indicated by the failure to react with isophorone or Δ^4-3-ketosteroids. Payne[9] effected smooth preparation of β-phenylglycidaldehyde by allowing cinnamaldehyde and *t*-butyl

$$C_6H_5CH{=}CHCHO \ + \ (CH_3)_3COOH \ \xrightarrow[\underset{73\%}{}]{\overset{added\ NaOH}{pH\ 8.5\ 35-40°}} \ C_6H_5CH\overset{\diagdown\ \diagup}{\underset{O}{\rule{0pt}{0pt}}}CHCHO$$

0.5 m.   0.6 m.
$\underbrace{\qquad\qquad\qquad\qquad\qquad\qquad}$
In CH_3OH

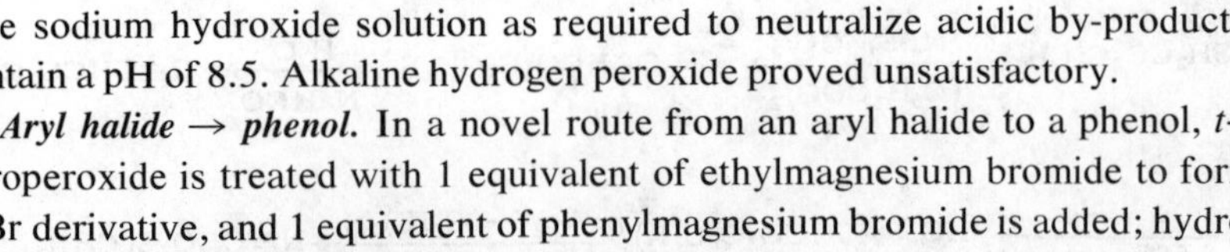

hydroperoxide to react in methanol for 5–6 hrs. during continuous addition of dilute sodium hydroxide solution as required to neutralize acidic by-products and maintain a pH of 8.5. Alkaline hydrogen peroxide proved unsatisfactory.

Aryl halide → phenol. In a novel route from an aryl halide to a phenol, *t*-butyl hydroperoxide is treated with 1 equivalent of ethylmagnesium bromide to form the MgBr derivative, and 1 equivalent of phenylmagnesium bromide is added; hydrolysis of the reaction mixture affords phenol in good yield.[10]

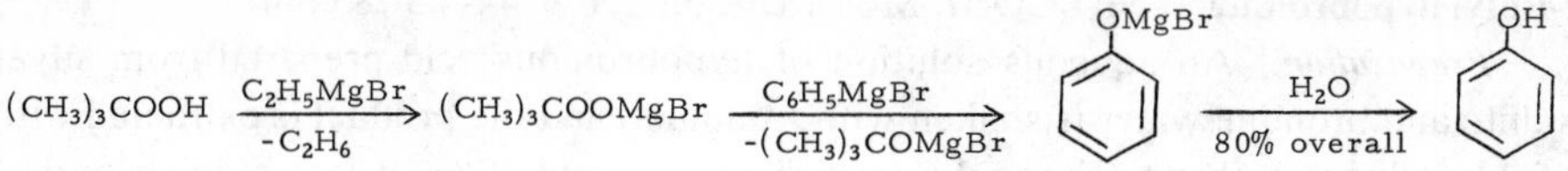

Decarboxylation. Wiberg and co-workers[11] developed an interesting method for decarboxylating bicyclo[2.1.1]hexane-5-carboxylic acid (1) to the parent hydrocarbon (4). The acid chloride (2) was stirred into a mixture of *t*-butyl hydroperoxide

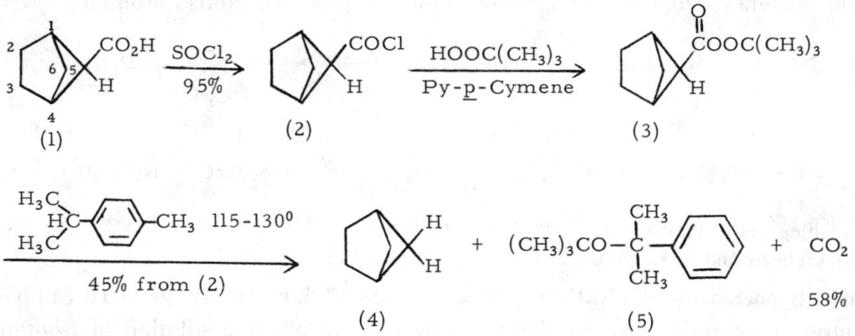

and *p*-cymene with ice cooling to form the peroxycarboxylate (3), which was not isolated but obtained as an acid-free dry solution in *p*-cymene. When this solution was heated, carbon dioxide was evolved and *p*-cymene served as hydrogen donor to form the hydrocarbon (4) and acceptor for the *t*-butoxy radical to form (5). The reaction was applied also to the 1-chloro derivative of (1) but the overall yield was only 26%.

See also N,N′-Carbonyldiimidazole.

α-Oxidation of a ketone. Methyl 7-ketodehydroabietate (1) when heated in acetic acid with *t*-butyl hydroperoxide and a trace of sulfuric acid for 55 hrs. at 50–55° afforded the unsaturated lactone (2) in very high yield.[12]

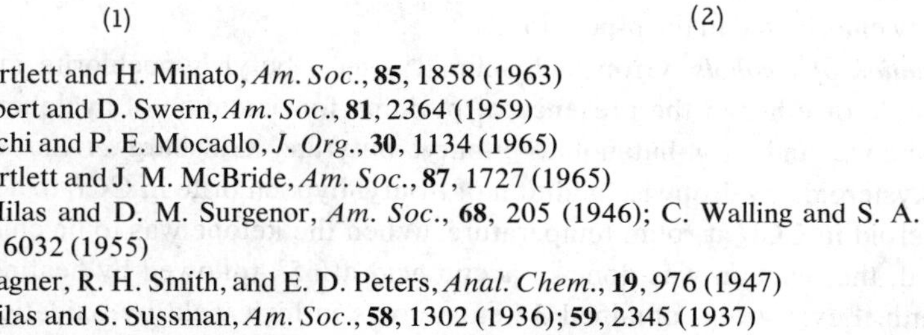

[1]P. D. Bartlett and H. Minato, *Am. Soc.*, **85**, 1858 (1963)
[2]L. S. Silbert and D. Swern, *Am. Soc.*, **81**, 2364 (1959)
[3]J. K. Kochi and P. E. Mocadlo, *J. Org.*, **30**, 1134 (1965)
[4]P. D. Bartlett and J. M. McBride, *Am. Soc.*, **87**, 1727 (1965)
[5]N. A. Milas and D. M. Surgenor, *Am. Soc.*, **68**, 205 (1946); C. Walling and S. A. Buckler, *ibid.*, **77**, 6032 (1955)
[6]C. D. Wagner, R. H. Smith, and E. D. Peters, *Anal. Chem.*, **19**, 976 (1947)
[7]N. A. Milas and S. Sussman, *Am. Soc.*, **58**, 1302 (1936); **59**, 2345 (1937)
[8]N. C. Yang and R. A. Finnegan, *Am. Soc.*, **80**, 5845 (1958)
[9]G. B. Payne, *J. Org.*, **25**, 275 (1960)
[10]S.-O. Lawesson and N. C. Yang, *Am. Soc.*, **81**, 4230 (1959)
[11]K. B. Wiberg, B. R. Lowry, and T. H. Colby, *Am. Soc.*, **83**, 3998 (1961)
[12]E. Wenkert, R. W. J. Carney, and C. Kaneko, *Am. Soc.*, **83**, 4440 (1961)

***t*-Butyl hypobromite**, $(CH_3)_3COBr$. Mol. wt. 153.03, b.p. 44–45°/85 mm.

Preparation.[1] An aqueous solution of hypobromous acid prepared from silver sulfite and bromine water is shaken with *t*-butanol, and the product is extracted with trichlorofluoromethane (Freon 11) and distilled; yield 42%. It is a reddish orange liquid with a bromine-like odor and is stable at 0° in the dark for long periods. It is rapidly destroyed at 85° or on irradiation, or by sodium bicarbonate. The reagent reacts with an alkane and bromotrichloromethane to form an alkyl bromide.

The reagent reacts with an olefin and ethanol to give an α-ethoxy bromide:[2]

$$>C{=}C< \; + \; C_2H_5OH \; + \; (CH_3)_3COBr \longrightarrow \; >\underset{Br}{C}-\underset{OC_2H_5}{C}< \; + \; (CH_3)_3COH$$

$$CH_2{=}CHOC_2H_5 \; + \; C_2H_5OH \; + \; (CH_3)_3COBr \longrightarrow BrCH_2CH(OC_2H_5)_2$$

[1]C. Walling and A. Padwa, *J. Org.*, **27**, 2976 (1962)
[2]J.-M. Geneste and A. Kergomard, *Bull. soc.*, 470 (1963)

***t*-Butyl hypochlorite**, $(CH_3)_3COCl$. Mol. wt. 108.57, b.p. 78°, sp. gr. 0.91. Supplier: Frinton. Preparation[1] by passing chlorine into an alkaline solution of *t*-butanol. *Caution:* an explosion during preparation has been reported;[2] it is recommended that the rate of flow of chlorine be regulated so that the temperature never exceeds 20°, as recorded by a thermometer dipping in the liquid.

Halogenation of hydrocarbons. Walling[3] found *t*-butyl hypochlorite useful for light- or radical-induced chain chlorination of toluene. Thus, in the presence of azobisisobutyronitrile as initiator, the reaction with toluene at 40° affords benzyl chloride (84%), *t*-butanol (97%), and 1–3% each of chlorotoluenes, methyl chloride, and acetone.

$$(CH_3)_2\overset{CN}{C}N{=}N\overset{CN}{C}(CH_3)_2 \xrightarrow[-N_2]{} (CH_3)_2\overset{CN}{C}\cdot \xrightarrow[(CH_3)_2C(CN)Cl]{(CH_3)_3COCl}$$

$$(CH_3)_3CO\cdot \xrightarrow[-(CH_3)_3COH]{C_6H_5CH_3} C_6H_5\dot{C}H_2 \xrightarrow[-(CH_3)_3CO\cdot]{(CH_3)_3COCl} C_6H_5CH_2Cl$$

Olefins react to form allylic chlorides in good yield.[4] Reactivities of allylic C—H bonds are in the order tertiary > secondary > primary. Substitution products arising without shift of the double bond retain their *cis* or *trans* stereochemistry. Phenols are usually chlorinated in the *o*-position.[5]

Oxidation of alcohols. Grob and Schmid[6] used *t*-butyl hypochlorite in carbon tetrachloride or ether in the presence of pyridine for oxidation of cyclohexanol to cyclohexanone and of *n*-butanol to *n*-butyl butyrate. Ginsburg *et al.*[7] oxidized 3-hydroxysteroids by dropwise addition of *t*-butyl hypochlorite in CCl_4 to a solution of the steroid in CCl_4 at room temperature. When the ketone was to be chlorinated as formed, the reaction was done in acetic acid at 65°, followed by heating on the steam bath. Levin *et al.*[8] obtained 3-ketosteroids in high yield by oxidation of the 3-alcohols with *t*-butyl hypochloride in dry *t*-butanol.

t-Butyl hypochlorite has been found particularly useful for the oxidation of a mixture of *cis*- and *trans*-2-phenylcyclohexanol.[9] The oxidation is carried out by titrating the alcohol in carbon tetrachloride containing 1 equivalent of pyridine at

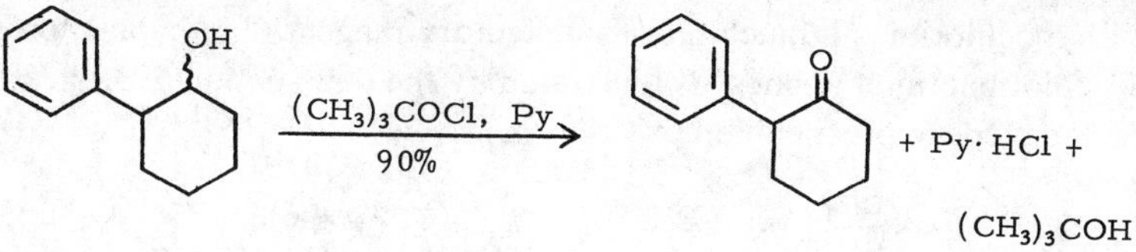

−5 to 0°. The precipitate of pyridine hydrochloride is filtered off and the solution is washed with water, dried, and distilled.

Chlorination of ketones. In some of the examples that follow, the reagent effects both oxidation and chlorination, as in the conversion of cholesterol into 6β-chloro-Δ^4-cholestene-3-one (1) and in the oxidation-chlorination of the multifunctional

1.[10]

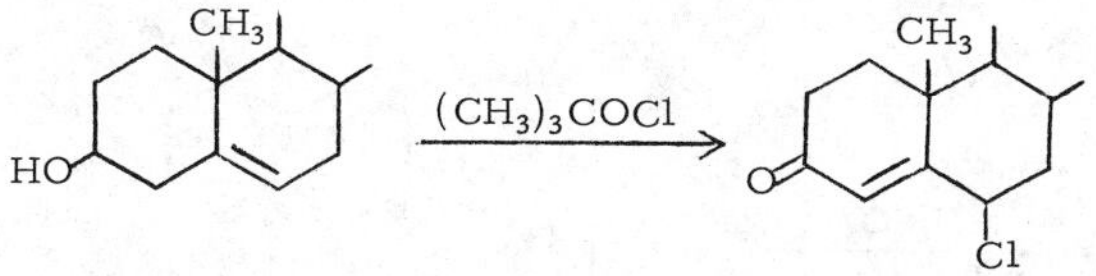

2.[11]

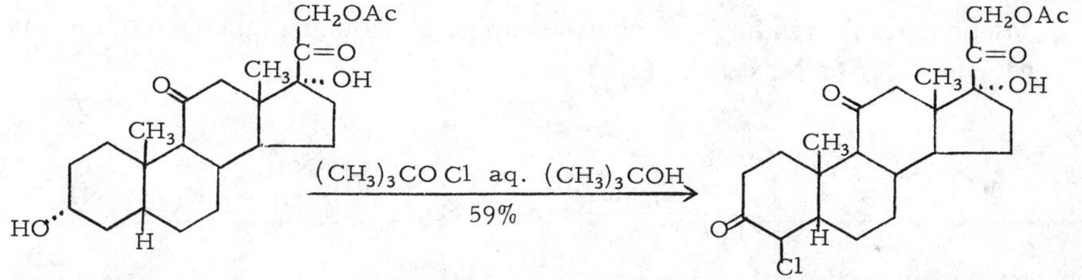

steroid (2). In this case water is needed for effective completion of the chlorination step. Chloro-3-ketosteroids (3 and 4) were obtained both from the ketones and from the corresponding alcohols by reaction with *t*-butyl hypochlorite in acetic acid at steam bath temperature.

3.[7]

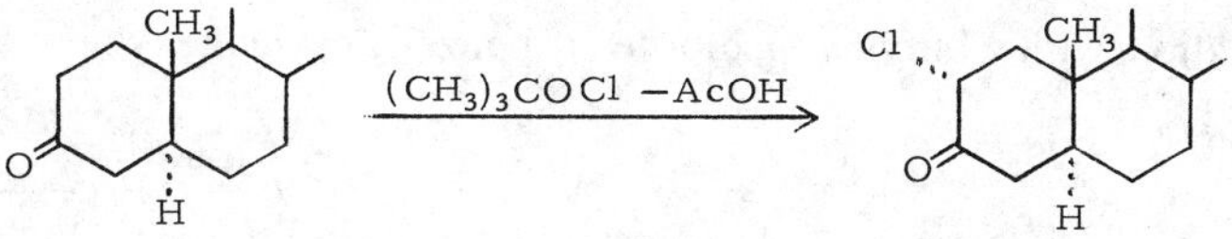

4.[7]

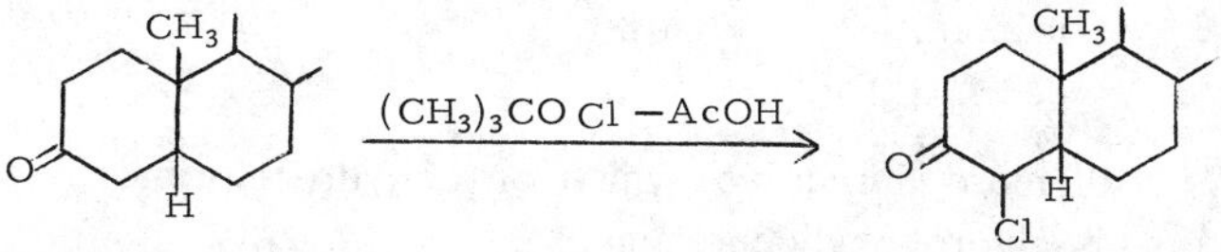

N-Chlorination. N-Acetylamides are converted into N-chloro-N-acetylamides in yields of 90% or more by treatment with excess *t*-butyl hypochlorite in methanol.[12] The preparation of diazaquinones from cyclic hydrazides evidently involves an intermediate N-chloroamide.[13] *See also* 4-Phenyl-1,2,4-triazoline-3,5-dione.

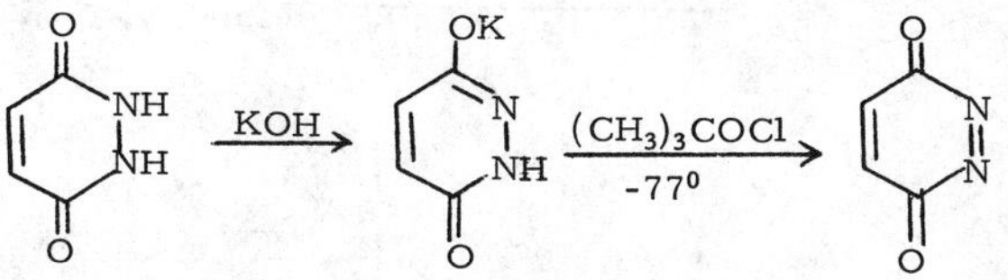

t-Butyl hypochlorite is a much more satisfactory reagent than hypochlorous acid for the N-chlorination of amines, as illustrated by the degradation of 3β-acetoxy-20-amino-Δ^5-pregnene to pregnenolone.[14]

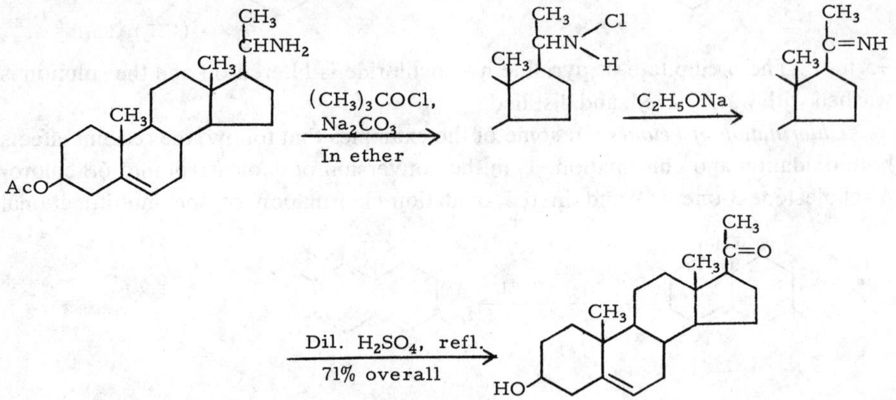
$(CH_3)_3COCl$,
Na_2CO_3
In ether
C_2H_5ONa
Dil. H_2SO_4, refl.
71% overall

Baumgarten's synthesis of phenacylamine and other α-aminoketones[15] involves in the first step an N,N-dichlorination.

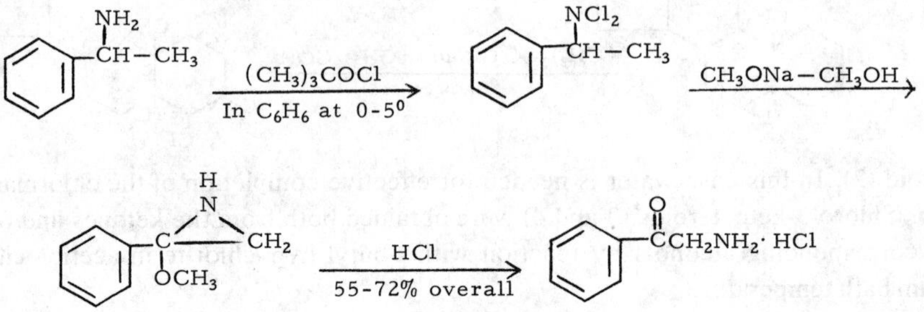
$(CH_3)_3COCl$
In C_6H_6 at 0-5°
$CH_3ONa-CH_3OH$
HCl
55-72% overall

By reaction with *t*-butylhypochlorite in benzene, cyclohexylamine is converted

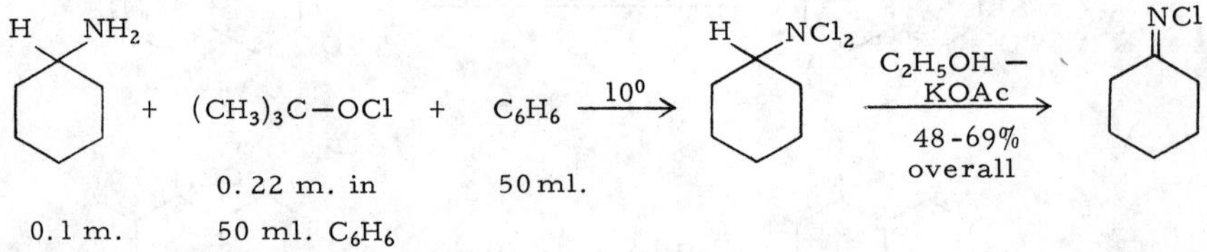
+ $(CH_3)_3C-OCl$ + C_6H_6
10°
0.1 m.
0.22 m. in
50 ml. C_6H_6
50 ml.
C_2H_5OH –
KOAc
48-69%
overall

into the N,N-dichloro derivative, which on dehydrochlorination with potassium acetate affords N-chlorohexylideneimine.[16]

Sulfides → sulfoxides. In methanol solution at −70° the reagent oxidizes 4-substituted thianes almost exclusively to *cis*-sulfoxides.[17]

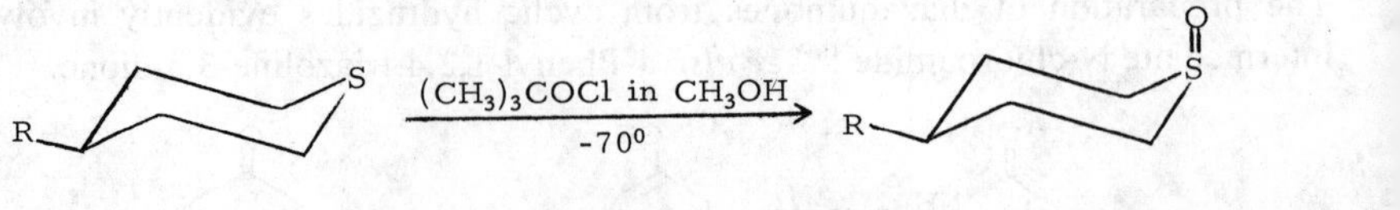
$(CH_3)_3COCl$ in CH_3OH
-70°

R =		% cis
p-ClC_6H_4	98	
$C(CH_3)_3$	100	
CH_3	89	

Alkaloid synthesis. Büchi and Manning[18] transformed ibogaine (**5**) into the naturally occurring ester alkaloid voacangine (**9**) by the four-step process formulated. Treatment of ibogaine with *t*-butyl hypochlorite afforded the chloroindolenine (**6**),

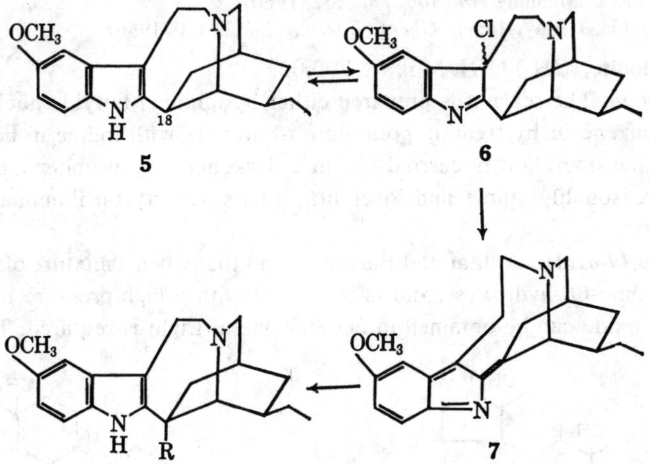
OCH3
N
H
18
Cl
N

5 **6** **7**

8, R = C≡N
9, R = COOCH$_3$

which combined with potassium cyanide to give 18-cyanoibogaine (**8**). Vigorous basic hydrolysis of 8 followed by esterification with diazomethane furnished voacangine (**9**).

Diazoalkanes. The diazoalkane (1) is converted by *t*-butyl hypochlorite in the presence of ethanol into the acetal (3); (2) may be an intermediate.[19]

$$C_6H_5\overset{-}{C}H-\overset{+}{N}{=}N + (CH_3)_3COCl \xrightarrow[-(CH_3)_3COH,\ -N_2]{} \left[C_6H_5CH\begin{matrix}Cl\\ OC_2H_5\end{matrix} \right]$$

(1) (2)

$$\xrightarrow[49\%]{C_2H_5OH} C_6H_5CH(OC_2H_5)_2 + HCl$$

(3)

[1]H. M. Teeter and E. W. Bell, *Org. Syn., Coll. Vol.*, **4**, 125 (1963)
[2]C. P. C. Bradshaw and A. Nechvatal, *Proc. Chem. Soc.*, 213 (1963)
[3]C. Walling and B. B. Jacknow, *Am. Soc.*, **82**, 6108, 6113 (1960)
[4]C. Walling and W. Thaler, *Am. Soc.*, **83**, 3877 (1961); C. A. Grob, H. Kny, and A. Gagneux, *Helv.*, **40**, 130 (1957)
[5]D. Ginsburg, *Am. Soc.*, **73**, 2723 (1951)
[6]C. A. Grob and HJ. Schmid, *Helv.*, **36**, 1763 (1953)
[7]J. J. Beereboom, C. Djerassi, D. Ginsburg, and L. F. Fieser, *Am. Soc.*, **75**, 3500 (1953)
[8]G. S. Fonken, J. L. Thompson, and R. H. Levin, *Am. Soc.*, **77**, 172 (1955)
[9]D. E. Butler (Parke Davis and Co.), private communication
[10]D. Ginsburg, *Am. Soc.*, **75**, 5489 (1953)
[11]A. R. Hanze, G. S. Fonken, A. V. McIntosh, Jr., A. M. Searcy, and R. H. Levin, *Am. Soc.*, **76**, 3179 (1954)
[12]R. C. Petterson and A. Wambsgans, *Am. Soc.*, **86**, 1648 (1964)
[13]T. J. Kealy, *ibid.*, **84**, 966 (1962)
[14]W. E. Bachmann, M. P. Cava, and A. S. Dreiding, *Am. Soc.*, **76**, 5554 (1954); see also M. P. Cava and B. R. Vogt, *Tetrahedron Letters*, 2813 (1964)

[15]H. E. Baumgarten and F. A. Bower, *Am. Soc.*, **76**, 4561 (1954); H. E. Baumgarten and J. M. Petersen, *ibid.*, **82**, 459 (1960); *Org. Syn.*, **41**, 82 (1961)
[16]G. H. Alt and W. S. Knowles, *J. Org.*, **25**, 2047 (1960); *Org. Syn.*, **45**, 16 (1965)
[17]C. R. Johnson and D. McCants, Jr., *Am. Soc.*, **87**, 1109 (1965)
[18]G. Büchi and R. E. Manning, *Am. Soc.*, **88**, 2532 (1966)
[19]H. Baganz and H.-J. May, *Angew. Chem., Internat. Ed.*, **5**, 420 (1966)

***t*-Butyl hypoiodite**, $(CH_3)_3COI$. Mol. wt. 200.02.

Preparation. The reagent is prepared either by mixing *t*-butyl hypochlorite with iodine in benzene or by treating potassium *t*-butoxide with iodine in benzene.[1] In either case the operation is carried out in a darkened vessel under nitrogen. The reagent is reasonably stable and loses little of its activity on illumination at 25° for 30 min.

Steroid 6,19-oxides. Akhtar and Barton[1] found that when a mixture of the reagent and a 5α-bromo-6β-hydroxysteroid is irradiated with a high-pressure mercury arc lamp a 6,19-oxide can be obtained in 60–80% yield. Light is required. Thus, in the

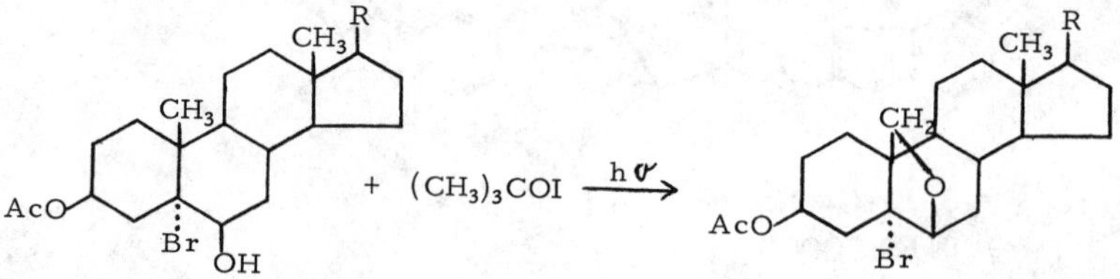
CH3
R
CH3
AcO
Br
OH
+ (CH3)3COI
hν
CH2
O
AcO
Br

dark, 3β-acetoxycholestane-6β-ol is oxidized by the reagent in high yield to the 6-ketone.

Decarboxylation.[2] A benzene solution of a carboxylic acid and *t*-butyl hypoiodite presumably contains an equilibrium amount of *t*-butanol and the acyl hypoiodite, which on illumination decomposes to the alkyl iodide and CO_2:

$$(CH_3)_3COI + R\overset{\overset{O}{\|}}{C}OH \rightleftharpoons (CH_3)_3COH + R\overset{\overset{O}{\|}}{C}OI \xrightarrow{h\nu} RI + CO_2$$

Thus illumination of a solution of the acid and the reagent provides a convenient method of decarboxylation comparable to the Hunsdiecker reaction, as in the example formulated:

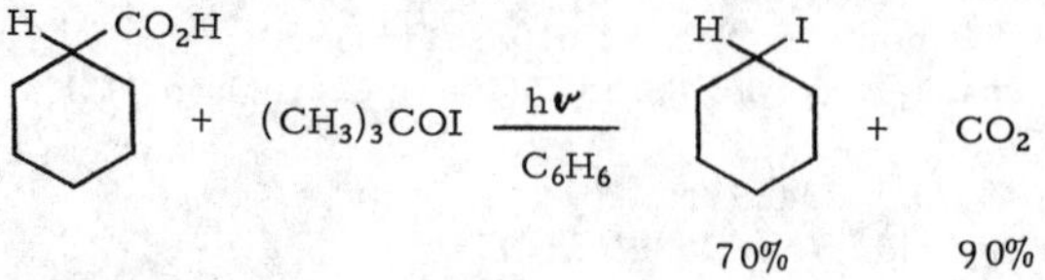
H
CO2H
+ (CH3)3COI
hν
C6H6
H
I
+ CO2
70%
90%

Primary, secondary, and tertiary acids can all be decarboxylated by this method. Adipic acid proved to be too sparingly soluble in benzene for satisfactory reaction but was decarboxylated in a mixture of benzene and sulfolane to give the iodide in 62% yield.

N-Iodoamides and γ-lactones. Barton and co-workers[3] found *t*-butyl hypoiodite to be an excellent reagent for the preparation of N-iodo derivatives of amides (benzamide, succinimide, *n*-butyramide, octadecaneamide). An N-iodoamide of suitable structure (2) is converted on photolysis into the γ-lactone (3).

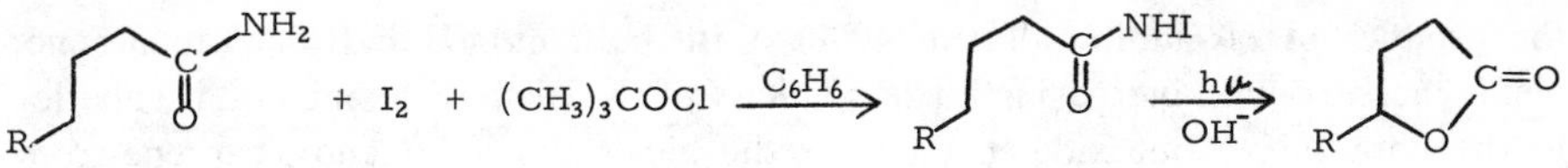

(1) (2) (3)

[1]M. Akhtar and D. H. R. Barton, *Am. Soc.*, **86**, 1528 (1964)

[2]D. H. R. Barton, H. P. Faro, E. P. Serebryakov, and N. F. Woolsey, *J. Chem. Soc.*, 2438 (1965)

[3]D. H. R. Barton, A. L. J. Beckwith, and A. Goosen, *J. Chem. Soc.*, 181 (1965)

***n*-Butyllithium**, $CH_3CH_2CH_2CH_2Li$. Mol. wt. 64.05. Suppliers of solutions in hexane: Alfa, Columbia, Foote Mineral Co., Peninsular ChemResearch.

Preparation. The reagent can be prepared from *n*-butyl bromide and lithium wire in ether at an initial temperature of −10° and then at 0–10° in a total time of about 3 hrs.[1] If *n*-butyl chloride is used, a reflux time of 10 hrs. is required.[2]

Use in the Wittig reaction. The Wittig reagent (2) required for the synthesis of

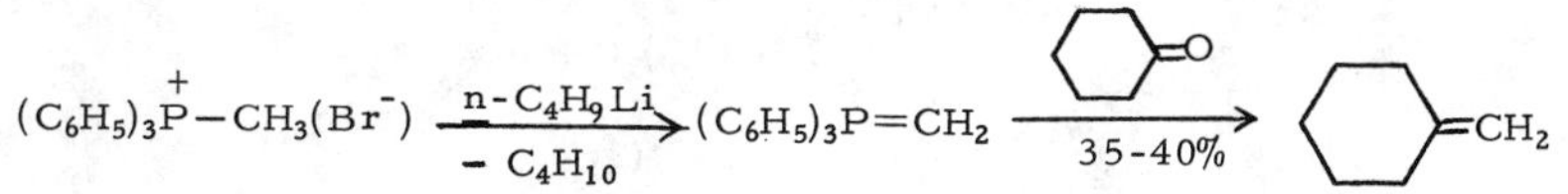

(1) (2) (3)

methylenecyclohexane (3) is prepared by dehydrohalogenation of triphenylmethylphosphonium bromide (1) with *n*-butyllithium.[3]

Metalating agent. *n*-Butyllithium is one of the most useful agents for the replacement of aromatic hydrogen in ethers, usually *ortho* to the ether function; tetrahydrofurane is generally preferred as solvent,[4] but ether is often used. For example, 1,7-dimethoxynaphthalene yields the 6-Li derivative, convertible into the corresponding methyl and carboxy compounds.[2]

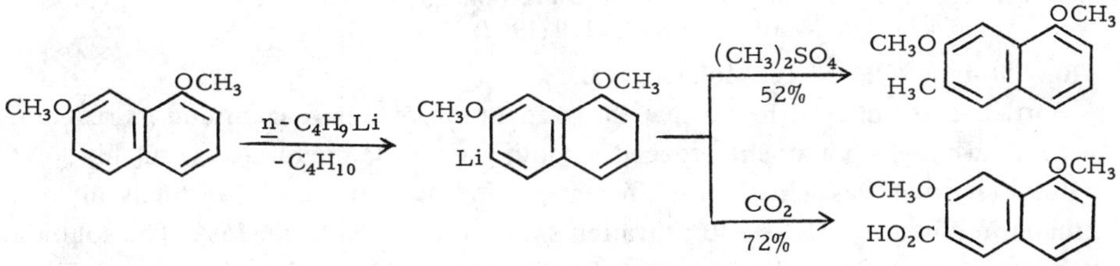

Generation of carbenes. *n*-Butyllithium and methyllithium are both used for production of carbenes.[5] For example, treatment of bromotrichloromethane with *n*-butyllithium in ether at −60° in the presence of cyclohexene gives 7,7-dichloronor-

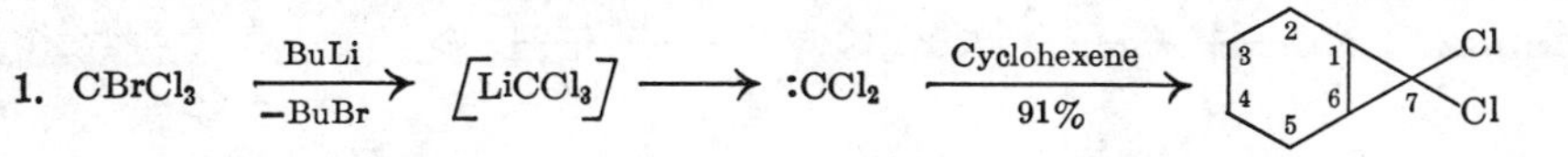

7,7-Dichloronorcarane

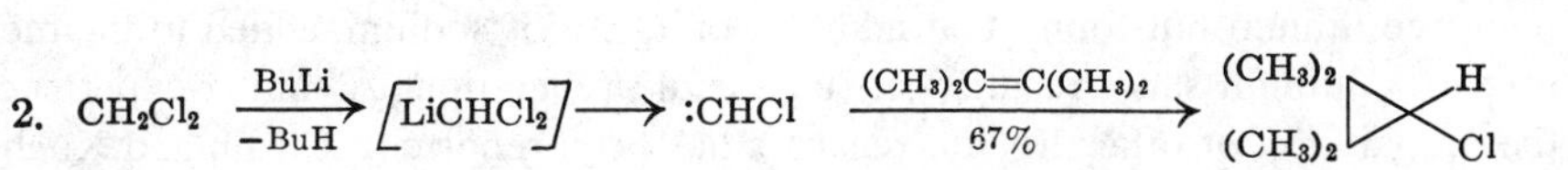

carane (1). Use of methylene chloride (2) leads to generation of chlorocarbene and provides a synthesis of monochlorocyclopropanes. *trans*-Butene-2 gives exclusively the product of *cis*-addition (yield 40%). With both mono- and dichlorocarbene, yields increase with increasing nucleophilicity of the olefin. The method is superior to the potassium *t*-butoxide method for the generation of phenoxycarbene from benzal chloride.[6]

Synthesis of allenes. A reaction first observed by Doering and LaFlamme,[7] conversion of a *gem*-dihalocyclopropane into an allene by reaction with magnesium[7] or with an alkyllithium,[8] is probably the best general method for the synthesis of allenes. For example, decene-1 (1) is converted into the *gem*-dibromide (2), and this is dehalogenated with *n*-butyllithium to give 1,2-undecadiene. In this case the

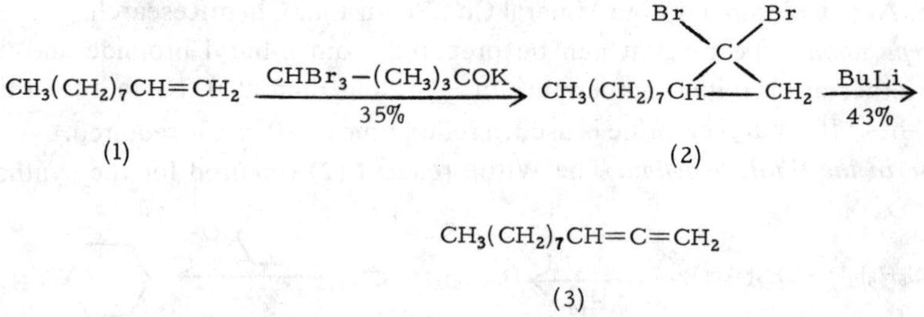
CH3(CH2)7CH=CH2
CHBr3–(CH3)3COK
35%
(1)
Br
Br
C
CH3(CH2)7CH
CH2
(2)
BuLi
43%
CH3(CH2)7CH=C=CH2
(3)

yield in the second step was raised to 68% by use of methyllithium. Dichlorides are inert to methyllithium but react slowly with *n*-butyllithium.

[1]R. G. Jones and H. Gilman, *Org. Reactions*, **6**, 352 (1951)
[2]R. A. Barnes and W. M. Bush, *Am. Soc.*, **81**, 4705 (1959)
[3]G. Wittig and U. Schoellkopf, *Org. Syn.*, **40**, 66 (1960)
[4]H. Gilman and S. Gray, *J. Org.*, **23**, 1476 (1958)
[5]W. T. Miller, Jr., and C. S. Y. Kim, *Am. Soc.*, **81**, 5008 (1959); G. L. Closs and L. E. Closs, *ibid.*, **82**, 5723 (1960)
[6]U. Schöllkopf, A. Lerch, and W. Pitteroff, *Tetrahedron Letters*, 241 (1962)
[7]W. von E. Doering and P. M. LaFlamme, *Tetrahedron*, **2**, 75 (1958)
[8]W. R. Moore and H. R. Ward, *J. Org.*, **27**, 4179 (1962)

***t*-Butyllithium**, $(CH_3)_3CLi$. Mol. wt. 64.05.

Bartlett and Lefferts[1] found that the preparation of this reagent and its reaction with a carbonyl component proceed smoothly if the temperature is suitably low. *t*-Butyl chloride was added with efficient stirring to a suspension of finely divided lithium in ether at −35 to −40°; titration showed the yield to be 75%. The solution was cooled to −60 to −70° and treated with an ethereal solution of hexamethylacetone; workup afforded crystalline tri-*t*-butylcarbinol in high yield.

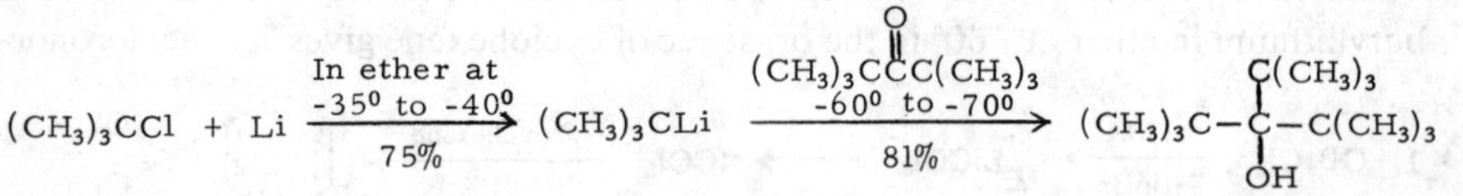
(CH3)3CCl + Li
In ether at
-35° to -40°
75%
(CH3)3CLi
O
(CH3)3CCC(CH3)3
-60° to -70°
81%
C(CH3)3
(CH3)3C–C–C(CH3)3
OH

Stiles and Mayer[2] were unable to prepare *t*-butyllithium by this procedure from sodium-free lithium but found that addition of 1–2% of sodium, added to the melt in preparing lithium sand, resulted in successful preparation. A similar experience in the preparation of other lithium reagents has been reported.[3] Curtin and Koehl[4] found that fine copper bronze can also be used to activate lithium.

[1]P. D. Bartlett and E. B. Lefferts, *Am. Soc.*, **77**, 2804 (1955)
[2]M. Stiles and R. P. Mayer, *Am. Soc.*, **81**, 1501, note 38b (1959)

[3]J. B. Wright and E. S. Gutsell, *Am. Soc.*, **81**, 5193, note 10 (1959); C. W. Kamienski and D. L. Esmay, *J. Org.*, **25**, 1807 (1960)
[4]D. Y. Curtin and W. J. Koehl, Jr., *Am. Soc.*, **84**, 1967 (1962)

***n*-Butyl nitrite**, *n*-C_4H_9ONO. Mol. wt. 103.12, b.p. 77° (24–27°/43 mm.). Suppliers: E, F. Preparation from *n*-BuOH, H_2SO_4, H_2O, $NaNO_2$ at 0°.[1]

Azide peptide synthesis. The azide method of peptide synthesis is very useful because of the absence of racemization;[2] however, a drawback is that on reaction

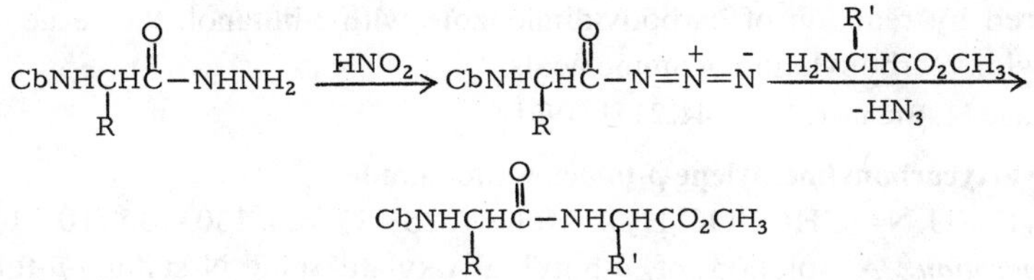

of the hydrazide with nitrous acid to form the azide some of the corresponding amide is also produced. In the Rudinger modification,[3] diazotization is done with *n*-butyl nitrite or nitrosyl chloride and amide formation is at most negligible. With *n*-butyl nitrite slightly more than 2 equivalents are required for optimal yields and tetrahydrofurane serves as solvent. With nitrosyl chloride acid it is not necessary and dioxane is used as solvent.

Nitrosation. An example is the conversion of phenacyl chloride into ω-chloroisonitrosoacetophenone.[4]

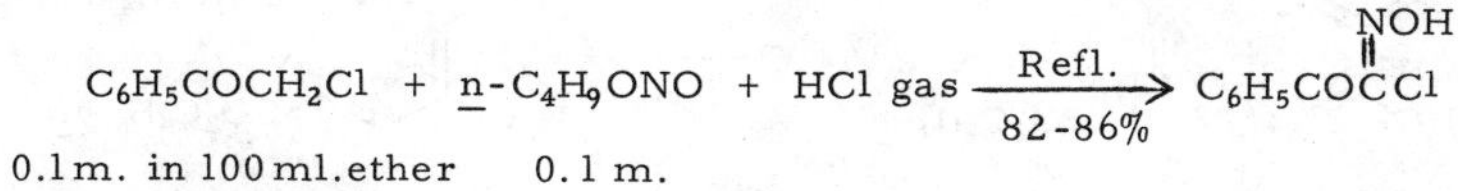

[1]W. A. Noyes, *Org. Syn., Coll. Vol.*, **2**, 108 (1943)
[2]N. A. Smart, G. T. Young, and M. W. Williams, *J. Chem. Soc.*, 3902 (1960); T. Wieland and H. Determann, *Angew. Chem., Internat. Ed.*, **2**, 358 (1963)
[3]J. Honzl and J. Rudinger, *Coll. Czech.*, **26**, 2333 (1961)
[4]N. Levin and W. H. Hartung, *Org. Syn., Coll. Vol.*, **3**, 191 (1955)

***t*-Butyl nitrite**, $(CH_3)_3CONO$. Mol. wt. 103.12, b.p. 63°. Supplier: KK.

Azide peptide synthesis. Hofmann *et al.*[1] used the Rudinger modification of the azide synthesis (*see n*-Butyl nitrite) except that they used *t*-butyl nitrite and hydrogen chloride in dimethylformamide. Comparison with the standard azide procedure showed the modified procedure to be superior in yield and in ease of operation.

[1]K. Hofmann *et al.*, *Am. Soc.*, **87**, 620 (1965)

***t*-Butyl *p*-nitrophenylcarbonate**, *p*-$O_2NC_6H_4OC(=O)OC(CH_3)_3$. Mol. wt. 239.22, m.p. 78.5–79.5°.

Preparation from *p*-nitrophenyl chlorocarbonate and *t*-butanol and use for preparation of *t*-butyloxycarbonylamino acids:[1]

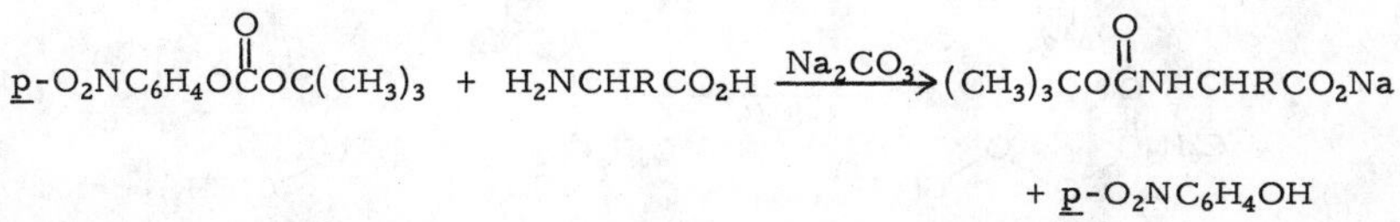

The protected derivatives have satisfactory physical properties and are resistant to hydrogenation and to sodium in liquid ammonia.

[1]G. W. Anderson and A. C. McGregor, *Am. Soc.*, **79**, 6180 (1957)

t-Butyloxycarbonylimidazole, 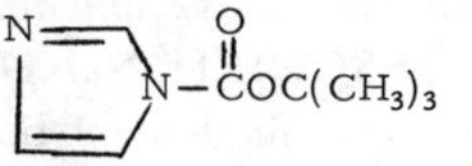

Mol. wt. 168.20, m.p. 46–47°.

Prepared by reaction of carbonyldimidazole with *t*-butanol, the reagent is used for *t*-butyloxycarbonylation of amino acids.[1]

[1]W. Klee and M. Brenner, *Helv.*, **44**, 2151 (1961)

N-*n*-Butyloxycarbonylmethylene-*p*-toluenesulfonamide, *p*-$CH_3C_6H_4SO_2N{=}CHCOOC_4H_9$-*n*. Mol. wt. 283.35, b.p. 130–135°/10^{-3} torr.

Preparation. A solution of *n*-butyl glyoxylate[1] and N-sulfinyl-*p*-toluenesulfonamide in benzene is refluxed and a solution of aluminum chloride in nitrobenzene is added slowly.[2]

$$\text{p-}CH_3C_6H_4SO_2N{=}SO + O{=}CHCO_2C_4H_9\text{-n} \xrightarrow[67\%]{-SO_2} \text{p-}CH_3C_6H_4SO_2N{=}CHCO_2C_4H_9\text{-n}$$

Diels-Alder dienophile.[2] The adduct of the reagent with 2,3-dimethylbutadiene was obtained in unrecorded yield by refluxing the components in benzene for 10

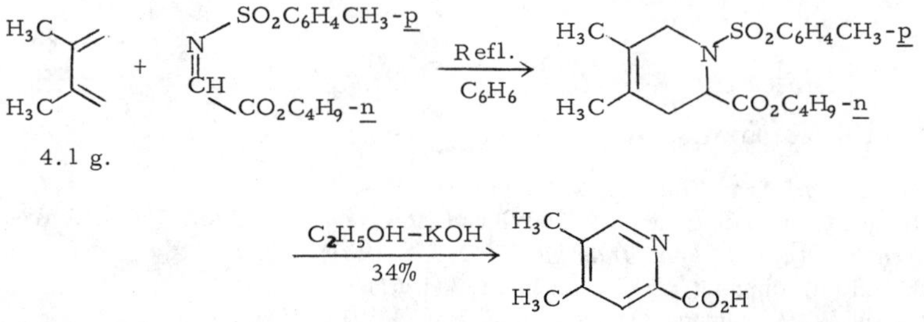

H3C
H3C
4.1 g.
SO2C6H4CH3-p
N
CH
CO2C4H9-n
Refl.
C6H6
H3C
H3C
N
SO2C6H4CH3-p
CO2C4H9-n
C2H5OH-KOH
34%
H3C
H3C
N
CO2H

hrs. Saponification of the adduct gave 4,5-dimethylpyridine-2-carboxylic acid in low yield.

[1]F. J. Wolf and J. Weijlard, *Org. Syn., Coll. Vol.*, **4**, 124 (1963)

[2]R. Albrecht and G. Kresze, *Ber.*, **98**, 1431 (1965)

t-Butyl perbenzoate (t-Butyl peroxybenzoate), $C_6H_5\overset{O}{\overset{\|}{C}}OOC(CH_3)_3$. Mol. wt. 194.22, b.p. 75–76°/0.2 mm., stable at room temperature; half-life at 100°, 18 hrs. Suppliers (pract.): Lucidol Div., Frinton, KK, MCB. Reviews.[1]

Synthesis of t-butyl ethers. Reaction with Grignard reagents provides a convenient synthesis of *t*-butyl ethers, aromatic and aliphatic.[2] Since the *t*-butyl ether

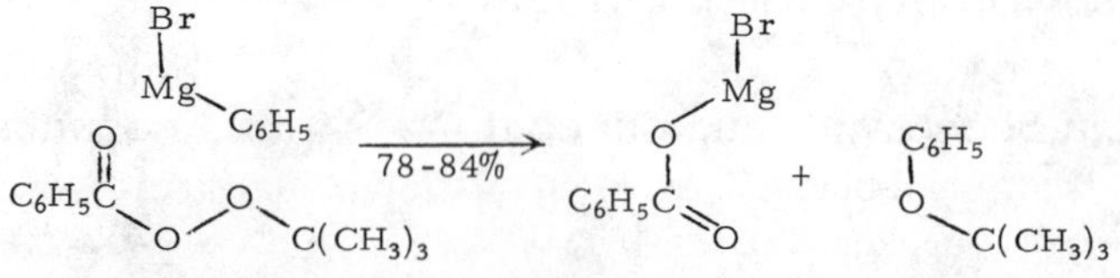

Br
Mg
C6H5
O
C6H5C
O
O
C(CH3)3
78-84%
Br
Mg
O
C6H5C
O
C6H5
O
C(CH3)3

of a phenol is readily hydrolyzed by acid, the reaction provides the means for conversion of an aryl halide into a phenol[3,4] (2-bromothiophene → 2-hydroxythiophene[3]).

Reaction with ethers. Decomposition of *t*-butyl perbenzoate by cuprous chloride in the presence of an ether, $(RCH_2)_2O$, and an alcohol, $R'OH$, leads to the acetal, $RCH(OR')_2$.[5]

$$\underline{n}\text{-}C_3H_7CH_2O\ C_4H_9\text{-}\underline{n} + C_6H_5\overset{O}{\overset{\|}{C}}\text{—OOC}(CH_3)_3 \xrightarrow{Cu^+} \underline{n}\text{-}C_3H_7\overset{H}{\underset{OCOC_6H_5}{\overset{OC_4H_9\text{-}\underline{n}}{C}}}$$

$$\xrightarrow{\underline{n}\text{-}C_6H_{13}OH} \underline{n}\text{-}C_3H_7\overset{H}{\underset{}{C}}(OC_6H_{13}\text{-}\underline{n})_2$$

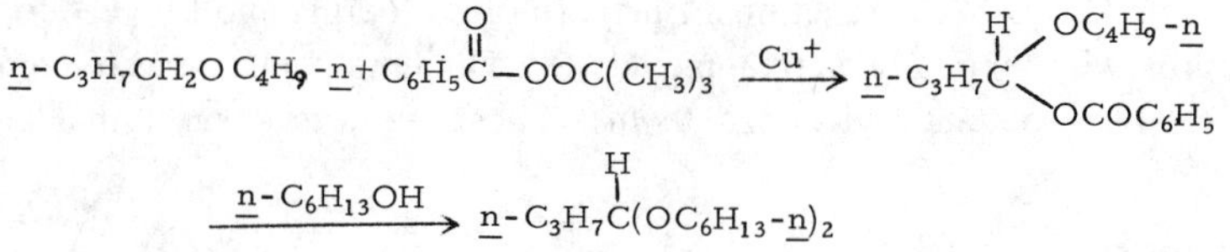

In the presence of a cuprous halide, *t*-butyl perbenzoate reacts at 67–90° with an aliphatic or cyclic ether, for example tetrahydrofurane, to give initially the unstable α-benzoyloxy derivative (1).[5] This ester can be isolated after a reaction period of 4–6 hrs. and converted by mild heat treatment into 2,3-dihydrofurane, which is

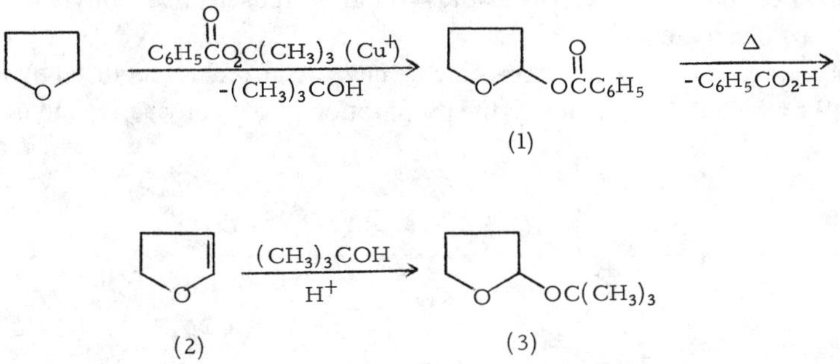

(1)

(2)   (3)

readily obtainable in yield of 66%. If the original reaction mixture is heated for 14–48 hrs., the unsaturated ether (2) which slowly forms adds *t*-butanol under catalysis from the benzoic acid to give as the final product the acetal (3), 2-*t*-butoxytetrahydrofurane.

Reaction with unsaturated compounds (reviews[6]). Kharasch, Sosnovsky, and Yang[7] reported that, in the presence of a catalytic amount of a copper or cobalt salt, *t*-butyl perbenzoate reacts with an olefin to give an allylic benzoate with no allylic rearrangement; with a terminal olefin the only product isolated was the allylic ester with a terminal double bond. A further example is the reaction with 1-methylene-4-*t*-butylcyclohexane.[8] However, in contrast to these early reports, Kochi[9] found by

$$RCH_2CH{=}CH_2 \xrightarrow[\text{CuBr}]{C_6H_5\overset{O}{\overset{\|}{C}}OOC(CH_3)_3} RCHCH{=}CH_2 + (CH_3)_3COH$$
$$\underset{OCOC_6H_5}{|}$$

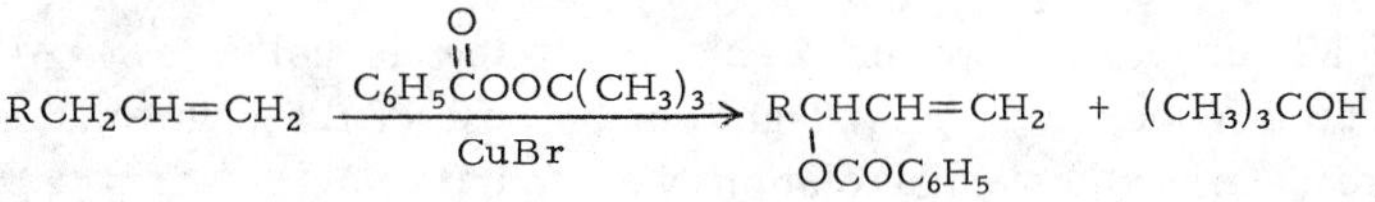

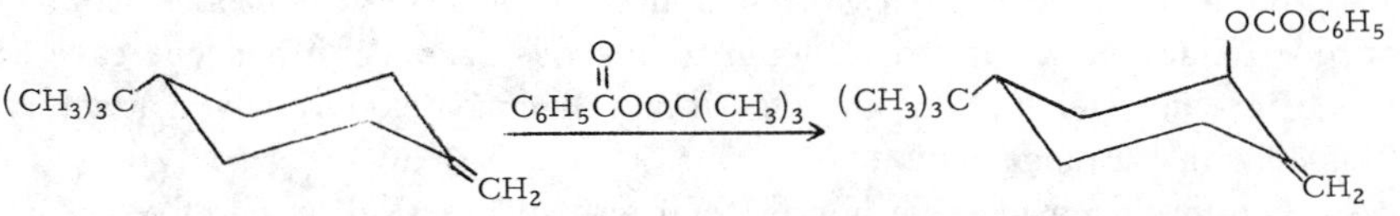

gas-liquid chromatography that the reaction is indeed attended with allylic rearrangement. Thus butene-1 and butene-2 gave essentially the same mixtures of 3-benzoxybutene-1 and crotyl benzoate. The olefins and the esters were shown to be stable

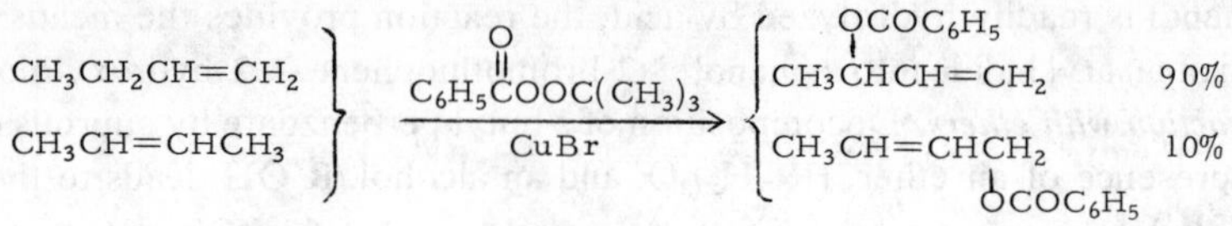

under the conditions of the reactions. Furthermore, Goering and Mayer[10] found that the reaction with optically active bicyclo[3.2.1]octene-2 (1) gives racemic *exo*-bicyclo[3.2.1]-Δ^3-octene-2-yl benzoate (4). These authors suggest that an allylic

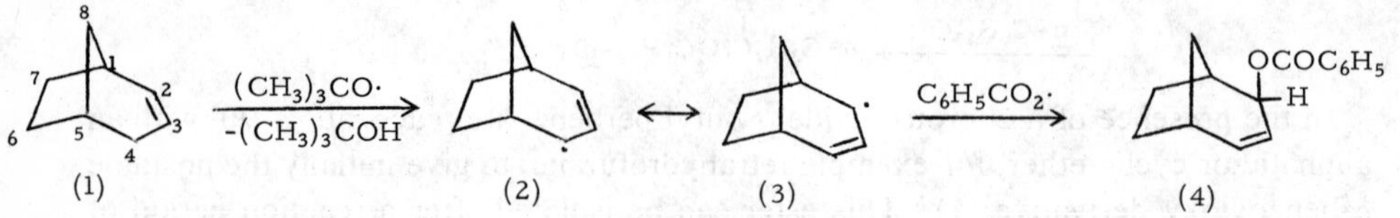

hydrogen at C_4 is abstracted to give a symmetrical radical (2 ↔ 3), with complete randomization of carbon atoms 2 and 4, and stereoselective conversion into the racemic *exo*-benzoate.

Other examples in which rearrangements have occurred are reported by Denney.[11] Denney[12] extended the reaction to the preparation of optically active allylic alcohols.

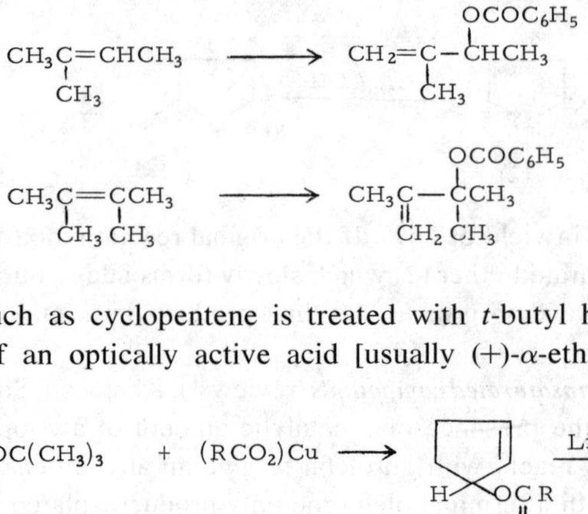

A cyclic olefin such as cyclopentene is treated with *t*-butyl hydroperoxide and the copper salt of an optically active acid [usually (+)-α-ethyl camphorate] to

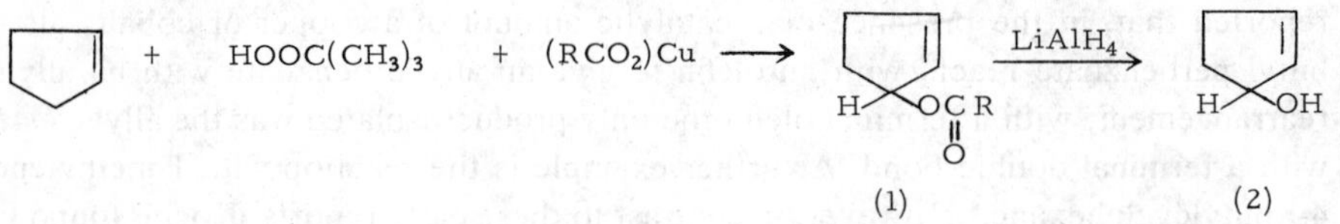

give an optically active allylic ester (1), cleaved to the alcohol (2) by $LiAlH_4$. The total asymmetric induction was not high, but because of the high rotation of the optically pure materials substantial rotations were found.

A preliminary report[13] states that reaction of cholesteryl benzoate with *t*-butyl hydroperoxide and a catalytic amount of cuprous bromide in benzene gave equal parts of Δ^5-cholestene-3β,7α- and -3β,7β-diol dibenzoate. Here allylic attack is not attended with rearrangement.

A reaction of considerable interest and also of practical preparative value discovered by Story[14] is the reaction of *t*-butyl perbenzoate with norbornadiene to give 7-*t*-butoxynorbornadiene. Noteworthy is the production of an ether rather than a benzoate; as in the reactions cited above allylic attack is not possible in this case,

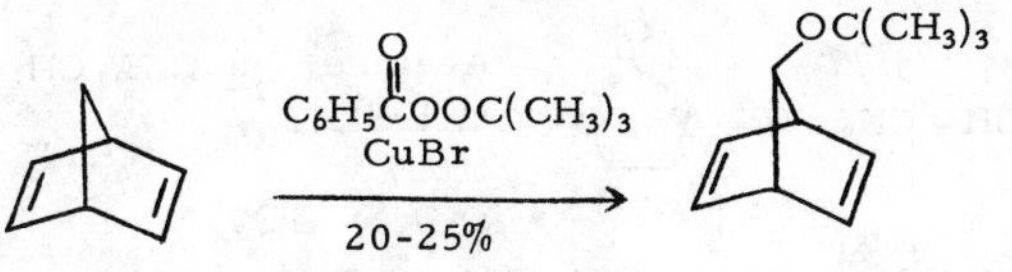
$C_6H_5COOC(CH_3)_3$
CuBr
20-25%
$OC(CH_3)_3$

but perhaps hydrogen abstraction at C_7 is anchimerically assisted by the two double bonds. 7-*t*-Butoxynorbornadiene is cleaved by perchloric acid in acetic acid in 73% yield to 7-norbornadienyl acetate, and this is cleaved by lithium aluminum hydride to give 7-norbornadienol in 95% yield.

[1]S.-O. Lawesson and G. Sosnovsky, *Svensk Kem. Tids.*, **75**, 568 (1963); *Angew. Chem., Internat. Ed.*, **3**, 269 (1964
[2]S.-O. Lawesson and N. C. Yang, *Am. Soc.*, **81**, 4230 (1959); C. Frisell and S.-O. Lawesson, *Org. Syn.*, **41**, 91 (1961)
[3]C. Frisell and S.-O. Lawesson, *Org. Syn.*, **43**, 55 (1963)
[4]S.-O. Lawesson and C. Frisell, *Arkiv Kemi*, **17**, 393 (1961)
[5]S.-O. Lawesson and C. Berglund, *Angew. Chem.*, **73**, 65 (1961)
[6]G. Sosnovsky, *Tetrahedron*, **13**, 241 (1961); S.-O. Lawesson, C. Berglund, and S. Grönwall, *Chem. Scand.*, **14**, 944 (1960)
[7]M. S. Kharasch, G. Sosnovsky, and N. C. Yang, *Am. Soc.*, **81**, 5819 (1959); G. Sosnovsky and N. C. Yang, *J. Org.*, **25**, 899 (1960)
[8]B. Cross and G. H. Whitham, *J. Chem. Soc.*, 1650 (1961)
[9]J. K. Kochi, *Am. Soc.*, **84**, 774 (1962)
[10]H. L. Goering and U. Mayer, *Am. Soc.*, **86**, 3753 (1964)
[11]D. Z. Denney, A. Appelbaum, and D. B. Denney, *Am. Soc.*, **84**, 4969 (1962)
[12]D. B. Denney, R. Napier, and A. Cammarata, *J. Org.*, **30**, 3151 (1965)
[13]A. L. J. Beckwith and G. W. Evans, *Proc. Chem. Soc.*, 63 (1962)
[14]P. R. Story, *J. Org.*, **26**, 287 (1961); P. R. Story and S. R. Fahrenholtz, *Org. Syn.*, **44**, 12 (1964)

γ-Butyrolactone, $\underbrace{CH_2CH_2CH_2CO}_{O}$. Mol. wt. 86.90, b.p. 206° (90–92°/17 mm.), sp. gr. 1.128. Suppliers: General Aniline and Film Corp., Eastern, B, MCB. For the reaction with potassium cyanide, *see* Glutaric acid; for the synthesis of dicyclohexyl ketone, *see* Sodium methoxide, condensation catalyst.

Friedel-Crafts reaction. The Friedel-Crafts reaction of γ-butyrolactone with benzene can be conducted so as to give either of two products by varying the ratio

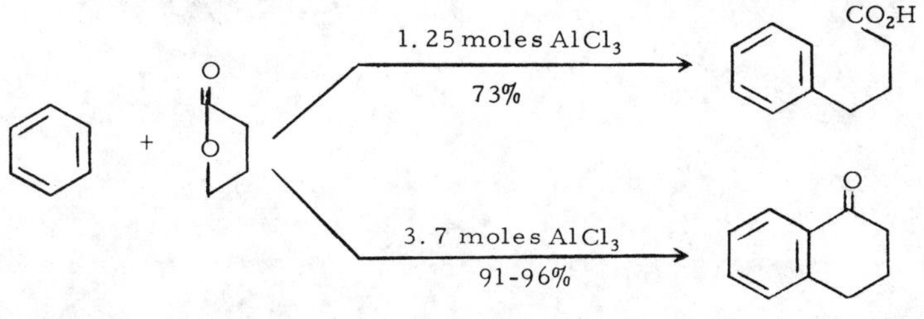
1. 25 moles $AlCl_3$
73%
CO_2H
3. 7 moles $AlCl_3$
91-96%

of aluminum chloride to lactone. With a molar ratio of 1:1.25, γ-phenylbutyric acid is produced in 73% yield and only 11% of α-tetralone is formed. With a ratio of 2.5 the yield of ketone rises to 66%;[1] with a 3.7 molar ratio α-tetralone is formed in nearly quantitative yield.[2]

Addition to terminal olefins. Under the influence of sunlight or a mercury vapor lamp, γ-butyrolactone adds to a terminal olefin (heptene-1, octene-1, decene-1) to give the 2-alkyllactone.[3] The 4-alkyllactone is a minor product.

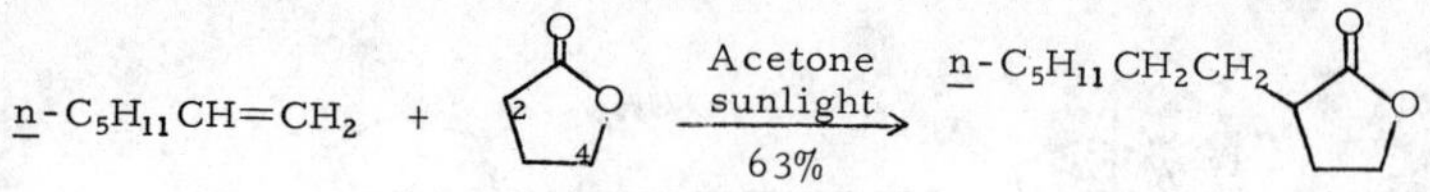
n-C5H11CH=CH2
+
2
O
4
Acetone
sunlight
63%
n-C5H11CH2CH2
O

[1]W. E. Truce and C. E. Olson, *Am. Soc.*, **74**, 4721 (1952)
[2]C. E. Olson and A. R. Bader, *Org. Syn., Coll. Vol.*, **4**, 898 (1963)
[3]D. Elad and R. D. Youssefyeh, *Chem. Comm.*, 7 (1965)

C

Cadmium acetate–Zinc acetate. The catalyst, prepared by heating a mixture of 2.5 g. each of cadmium acetate dihydrate and zinc acetate dihydrate to remove the water of hydration,[1] promotes addition of primary aliphatic amines to acetylene to give ethylidenimines. Thus ethylamine, heated with acetylene and catalyst in an autoclave at 120–140° for 29 hrs., afforded N-ethylethylidenimine (1).[1] Catalyzed reaction

1. $C_2H_5NH_2$ + $CH\equiv CH$ $\xrightarrow[55\%]{\text{Cat.}}$ $CH_3CH=N-C_2H_5$

2. $(CH_3)_2NH$ + $CH_3C\equiv CH$ $\xrightarrow{\text{Cat.}}$ $CH_3\underset{\underset{N(CH_3)_2}{|}}{C}=CH_2$

$\xrightarrow[71\%]{CH_3C\equiv CH}$ $CH_3\underset{\underset{N(CH_3)_2}{|}}{\overset{\overset{CH_3}{|}}{C}}-C\equiv CCH_3$

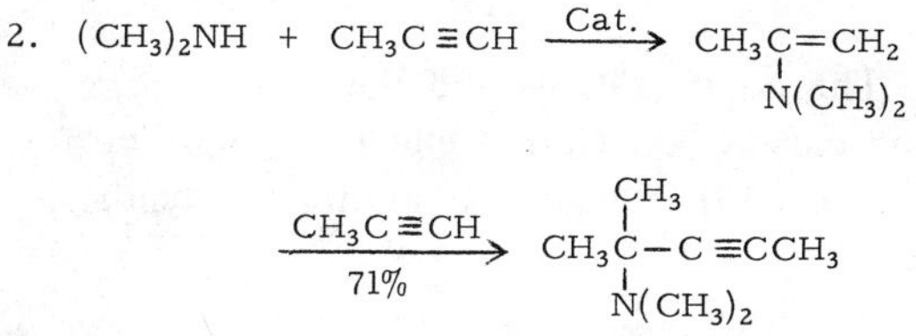

of propyne with dimethylamine in the ratio 2:1 gave N,N,1,1-tetramethyl-2-butynylamine (2).[2]

[1]C. W. Kruse and R. F. Kleinschmidt, *Am. Soc.*, **83**, 213 (1961)
[2]*Idem, ibid.*, **83**, 216 (1961)

Calcium carbonate, $CaCO_3$.

Dehydrohalogenation. Finely divided calcium carbonate has been used for dehydrohalogenation of steroid α-bromoketones.[1,2] Dimethylformamide has also been used as solvent.[3]

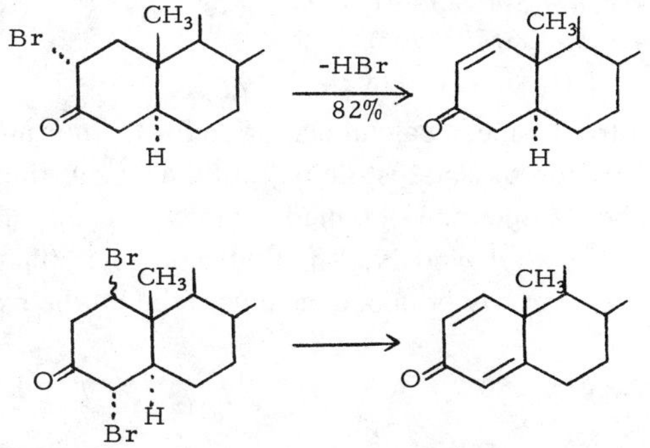

Tosylate rearrangement. Mazur[4] devised an ingenious scheme for the synthesis of perhydroazulenes and first explored the stereochemistry by applying the reaction in the steroid series. Cholestane-5α,6α-diol 6-tosylate (1) is converted quantitatively into the A-homo-B-nor-5-ketosteroid (2) when refluxed with finely divided calcium carbonate–dimethylformamide for 6–8 hrs. or heated with potassium *t*-butoxide in *t*-butanol at 100°. Cholestane-4α,5α-diol 4-tosylate (3) gave A-nor-B-homo-5-

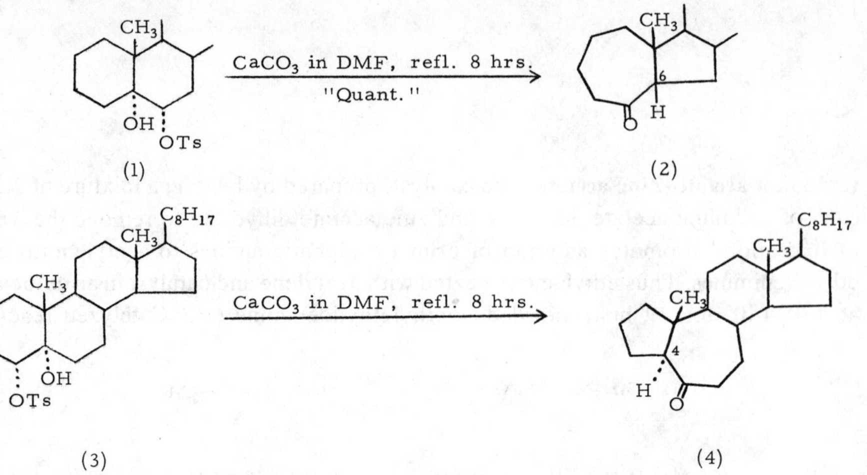

ketocholestane (4). The configurations and the ring juncture were inferred from the rotary dispersion curves: the Cotton effect was positive for (2) and negative for (4). Takeda *et al.*[5] used the Mazur method for the synthesis of 5-hydroxyperhydroazulene-8-one (6).

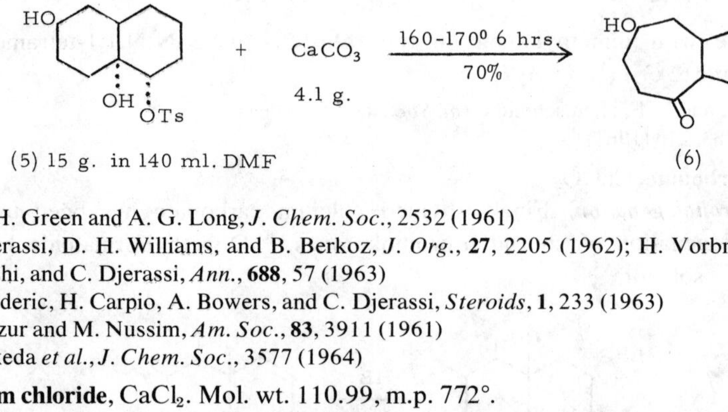

[1]G. F. H. Green and A. G. Long, *J. Chem. Soc.*, 2532 (1961)
[2]C. Djerassi, D. H. Williams, and B. Berkoz, *J. Org.*, **27**, 2205 (1962); H. Vorbrüggen, S. C. Pakrashi, and C. Djerassi, *Ann.*, **688**, 57 (1963)
[3]J. A. Zderic, H. Carpio, A. Bowers, and C. Djerassi, *Steroids*, **1**, 233 (1963)
[4]Y. Mazur and M. Nussim, *Am. Soc.*, **83**, 3911 (1961)
[5]K. Takeda *et al.*, *J. Chem. Soc.*, 3577 (1964)

Calcium chloride, $CaCl_2$. Mol. wt. 110.99, m.p. 772°.

Blanchard[1] substituted 12-mesh calcium chloride for the normally used potassium carbonate or calcium oxide as catalyst-dehydrating agent in the condensation of a ketone with a secondary amine and obtained enamines in generally improved yield. By this method N,N-dimethyl- and N,N-diethylenamines, difficultly accessible by previous procedures, are readily obtained, as illustrated by the example formulated.

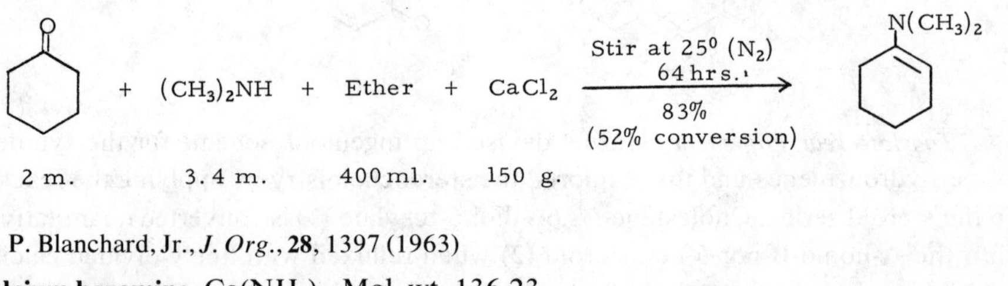

[1]E. P. Blanchard, Jr., *J. Org.*, **28**, 1397 (1963)

Calcium hexamine, $Ca(NH_3)_6$. Mol. wt. 136.23.

This solid complex is formed when ammonia is passed into an ethereal suspension of the ground metal.[1] The reagent reduces double bonds of aromatic hydrocarbons as follows:

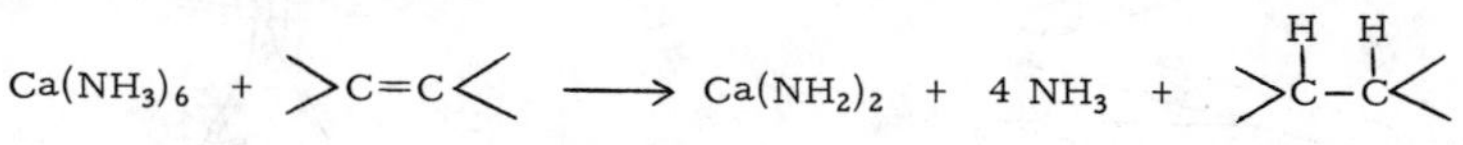

Polycyclic aromatic hydrocarbons are reduced to derivatives with one isolated benzene ring, as shown in the examples. The conversion of phenanthrene to the

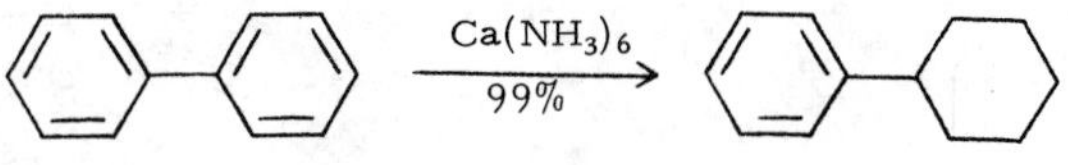

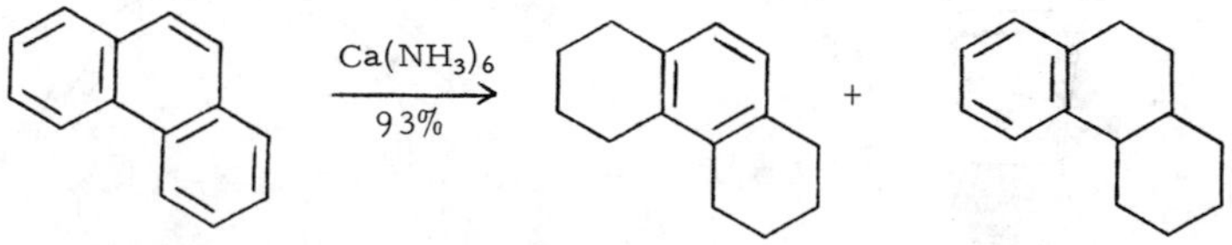

sym- and *as*-octahydrides appears to proceed through initial formation of the 1,2,3,4-tetrahydro- and 9,10-dihydro derivatives. Benzene is reduced to 1,4-cyclohexadiene (30% yield).

The reagent also reduces an alkyl benzyl sulfide to a mercaptan and toluene.[2]

[1]H. Boer and P. M. Duinker, *Rec. trav.*, **77**, 346 (1958)

[2]J. Van Schooten, J. Knotnerus, H. Boer, and P. M. Duinker, *ibid.*, **77**, 935 (1958)

Calcium hydride, CaH_2. Mol. wt. 42.10. Supplier: Ventron Corp.

Drying agent. Calcium hydride reacts irreversibly with water according to the following equation:

$$CaH_2 + 2\ H_2O \longrightarrow Ca(OH)_2 + 2\ H_2$$

Thus 1 g. of calcium hydride reacts quantitatively with 0.85 g. of water to give 1 l. of hydrogen. The reaction provides a convenient source of pure but wet hydrogen.

The hydride is an excellent reagent for drying hydrocarbons, ethers (including dioxane and tetrahydrofurane), amines (including pyridine), esters, the butanols (but not methanol or ethanol, which form alkoxides). Since the hydride is an effective condensation catalyst, it should not be used to dry aldehydes, reactive ketones, acids, or acyl halides. Chlorides and bromides should be dried over calcium hydride only at moderate temperatures.

One method of drying is simply to let the solvent stand over the hydride for several hours with protection from moisture and provision for escape of hydrogen, achieved conveniently by use of a diffusion tube (Fig. C-1) having a spiral of 1 or 2 mm. capillary tubing.[1] The principle was introduced by Grosse[2] for use in high-pressure reactions with small amounts of material. For example, the reaction mixture is charged into the Hershberg-designed assembly[3] of Fig. C-2 and this is placed in a beaker in an autoclave. Nitrogen is admitted very slowly, to avoid breaking the capillary, to give a pressure considerably higher than that anticipated for the reaction mixture. Gases under pressure diffuse through capillaries so very slowly that there is little chance for the escape of a volatile reagent or solvent into the autoclave chamber.

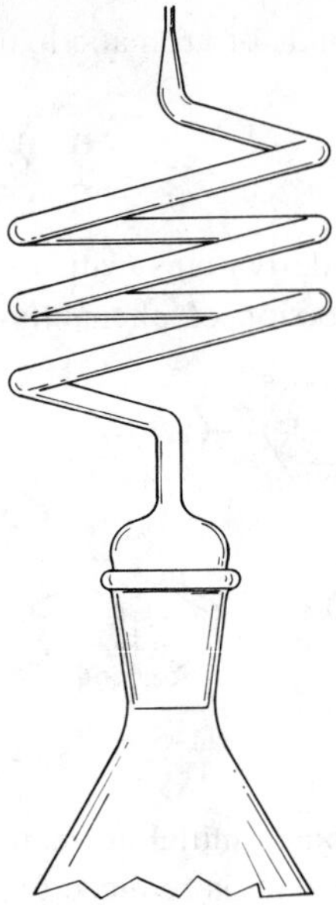

Fig. C-1

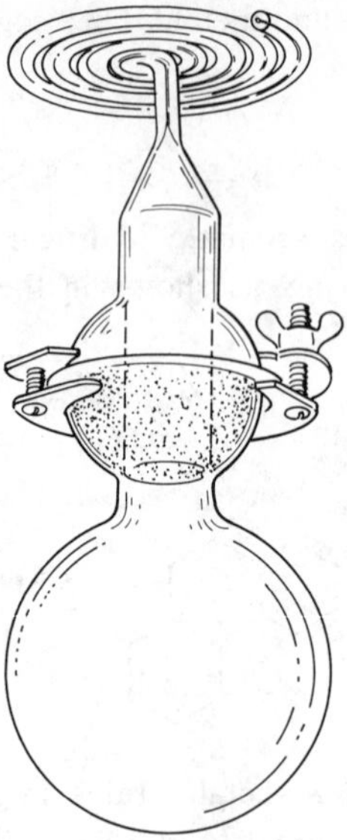

Fig. C-2

Diffusion tubes

Two dynamic methods of drying are described in the *Organic Syntheses* procedure for the preparation of *t*-butyl cyanoacetate.[4] Submitters Ireland and Chaykovsky dried *t*-butanol by refluxing it over calcium hydride overnight and distilling. However, refluxing *t*-butanol may be troublesome because of solidification of the alcohol (m.p. 24–25°) in the condenser. Checkers Tishler and Zambito effected drying by stirring the *t*-butanol with calcium hydride for several hours and then distilled the alcohol.

Condensation catalyst. Cyclization of β-propionylphenylhydrazine to 3-methyloxindole is effected by heating it with either freshly ground calcium hydride or 40-mesh hydride supplied by Ventron.[5]

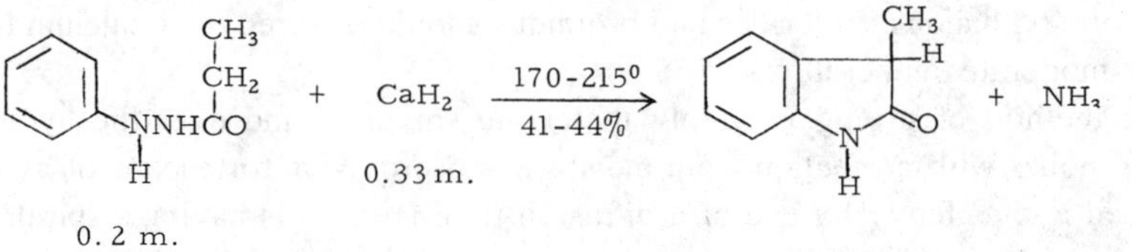
CH3
CH2
NNHCO
H
0. 2 m.
+
CaH2
0.33 m.
170-215°
41-44%
CH3
H
N
H
O
+ NH3

[1]Contributed by M. S. Newman, Ohio State University
[2]A. V. Grosse, *Am. Soc.*, **60**, 212 (1938)
[3]L. F. Fieser, "Experiments in Organic Chemistry," 3rd Ed., 249 D. C. Heath (1957)
[4]R. E. Ireland and M. Chaykovsky, *Org. Syn.*, **41**, 5 (1961)
[5]A. S. Endler and E. I. Becker, *Org. Syn., Coll. Vol.*, **4**, 657 (1963)

Calcium–Liquid NH_3. This combination is effective for the reduction of 12α- or 12β-acetoxy-11-ketosteroids to the 11-ketones.[1] Neither sodium nor potassium

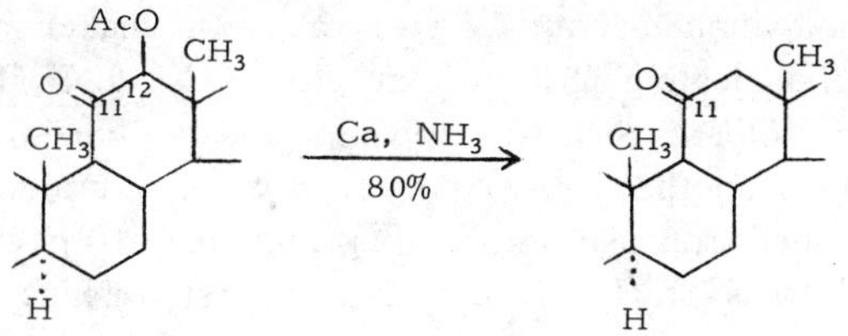
AcO
CH3
O
12
11
CH3
Ca, NH3
80%
H
CH3
O
11
CH3
H

(1) (2)

is satisfactory; barium gives results comparable to calcium. The free ketol corresponding to (1) gave chiefly the 11α,12β-diol. The 17β-acetoxy-17α-aceto steroid (3) is converted by calcium in liquid ammonia into the normal (17β) 20-ketosteroid (5).[2]

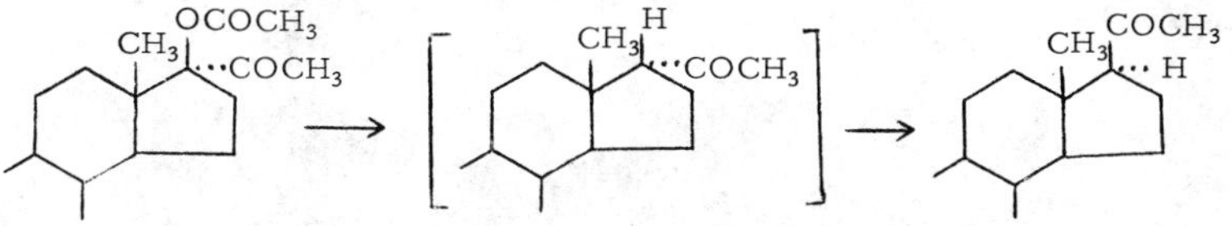
OCOCH3
CH3
COCH3
H
CH3
COCH3
COCH3
CH3
H

(3) (4) (5)

[1]J. H. Chapman, J. Elks, G. H. Phillipps, and L. J. Wyman, *J. Chem. Soc.*, 4344 (1956)
[2]J. S. Mills, H. J. Ringold, and C. Djerassi, *Am. Soc.*, **80**, 6118 (1958)

Calcium sulfate, anhydrous, $CaSO_4$ (Drierite). Suppliers: W. A. Hammond Drierite Co., F, MCB.

This drying agent, called soluble anhydrite to distinguish it from the insoluble mineral of the same composition but different behavior, is prepared according to Hammond and Withrow[1] (Ohio State Univ.) by heating either the hemihydrate or the dihydrate at 230–250° for 2–3 hrs. Exhausted reagent can be regenerated in the same way. In drying an alcohol or other organic solvent, the soluble salt reverts to the hemihydrate, $CaSO_4 \cdot \frac{1}{2}H_2O$, and absorbs 10–12% of its weight of water without danger of shrinking and consequent channeling. Soluble anhydrite is stable, neutral, inert, and insoluble in organic liquids. Added to 95% ethanol, it produces a marked rise in temperature. In completeness of drying air, Drierite stands between phosphorus pentoxide and concd. sulfuric acid and is surpassed only by the former. A comparison with other agents in the drying of liquids (*see* Drying agents) showed that magnesium sulfate and sodium sulfate are slower in their action and less efficient than Drierite.

The Hammond Co. supplies an indicating form which turns from blue to red when exhausted. It is prepared by impregnating the surface of the calcium sulfate granules with 3.5–5% of anhydrous cobalt chloride. The blue cobalt salt is indifferent to aprotic solvents insoluble in water (benzene) or partially soluble in water (diethyl ether), but is leached off by water-miscible solvents, both protonic (ethanol, acetic acid) and aprotic (dimethylformamide, dimethyl sulfoxide, triglyme).

For general use in the drying of ethereal extracts, one of us[2] previously recommended shaking with saturated sodium chloride solution and filtering through a cone of anhydrous sodium sulfate (*which see*), but he is now convinced by the literature cited and from personal trials that calcium sulfate is definitely superior. In one informative experiment, ether was saturated with water at 25° and a 100-ml.

portion, which should contain about 1.2 g. of water, was shaken in a glass-stoppered flask with 10 g. of indicating Drierite (capacity: 1–1.2 g. H_2O). After about 70 seconds no further color change was apparent and the blue was almost entirely replaced by red. When another 100-ml. portion of water-saturated ether was run out of a separatory funnel dropwise onto a filter containing 10 g. of the Drierite, the color change again was practically complete (a first washing with brine is still recommended).

The 1-lb. bottle of Drierite is converted into a handy dispenser for pouring the free-flowing solid onto a filter by drilling a hole in the Bakelite cap, inserting the cylinder of a plastic drying tube, forming a collar around the cylinder on the inside with Plastic Wood, letting this dry, and securing it with glue (Fig. C-3).

[1] W. A. Hammond and J. R. Withrow, *Ind. Eng. Chem.*, **25**, 650, 1112 (1933)
[2] *Org. Expts.*, 53

DRIERITE

Fig. C-3 Drierite dispenser

***d*-10-Camphorsulfonic acid**, (1). Mol. wt. 232.30, m.p. 194° dec., $\alpha_D + 24°$. Suppliers: A, B, E, F, MCB.

Preparation by sulfonation of natural camphor with acetic–sulfuric anhydride.[1]

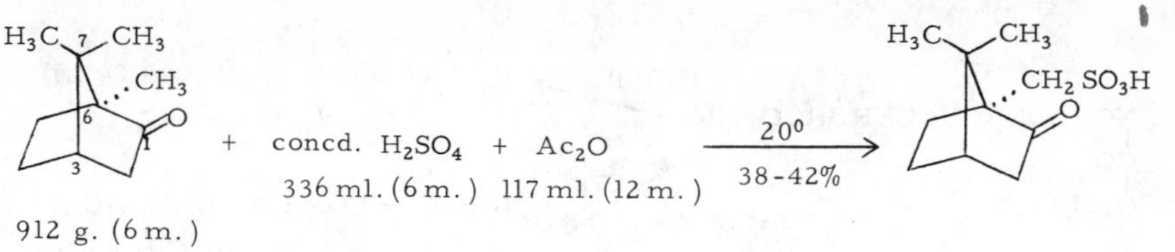

Use. Resolution of (±) bases.

[1]P. D. Bartlett and L. H. Knox, *Org. Syn.*, **45**, 12 (1965)

10-Camphorsulfonyl chloride. Mol. wt. 250.74, m.p. 67–68° (*d*-), 83–84° (*dl*-).

Preparation.[1] A mixture of 2 moles each of the sulfonic acid and PCl_5 is cooled and allowed to liquefy and then stirred mechanically in a cooling bath and then at room temperature. Quenching with ice and washing with water affords satisfactory product in essentially quantitative yield.

Uses. The optically active form is useful for the resolution of (±)-alcohols and amines. The crude *dl*-form (m.p. 81–83°) is suitable for oxidation to (±)-ketopinic acid:[2]

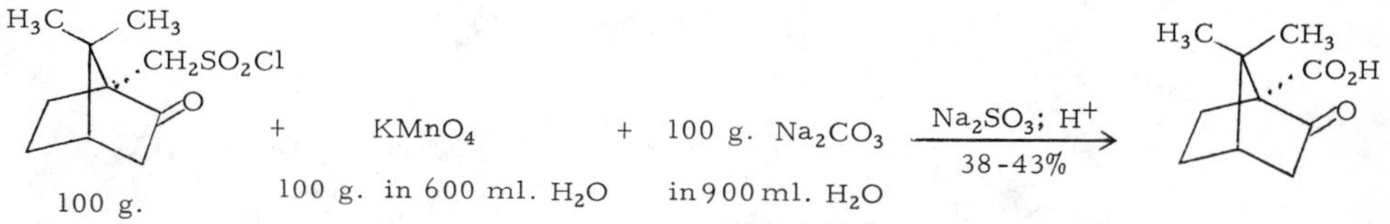

This substance is of interest as a β-keto acid which is not readily decarboxylated.

[1]P. D. Bartlett and L. H. Knox, *Org. Syn.*, **45**, 14 (1965)
[2]*Idem, ibid.*, **45**, 55 (1965)

Capryl alcohol, $CH_3CH_2CH_2CH_2CH_2CH_2CH(OH)CH_3$. Mol. wt. 130.22, b.p. 179°, sp. gr. 0.819. Suppliers: E, F.

Useful for the control of foaming.[1]

[1]W. W. Hartman, J. B. Dickey, and J. G. Stampfli, *Org. Syn.*, *Coll. Vol.*, **2**, 175 (1943)

Carbobenzoxy chloride, $C_6H_5CH_2OCOCl$. Mol. wt. 170.59, m.p. 0°. Supplier: Frinton, KK.

Peptide synthesis. Introduced by Bergmann and Zervas[1] in 1932 for the protection of amino groups in peptide synthesis, the reagent (CbCl) still affords probably the most widely used means of *N*-protection. Preparation from benzyl alcohol and phosgene.[1,2] A procedure stated to give better material than the *Organic Syntheses* procedure is described by Katchalski.[3]

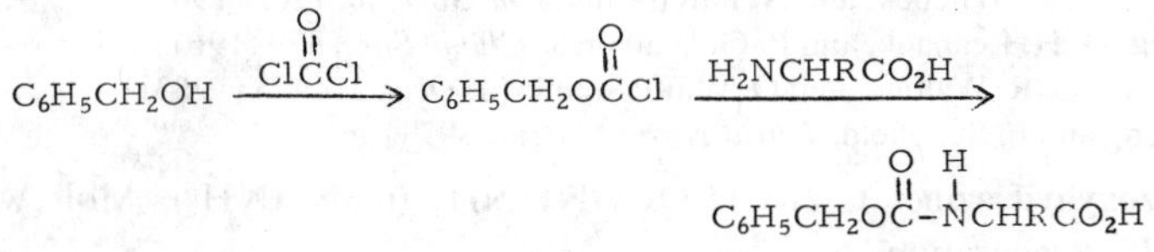

Once a synthetic sequence has been completed, the N-protective Cb group can be removed either by hydrogenolysis or by treatment with hydrogen bromide in acetic acid.[4] Discussion of relative merits of N-protective groups and methods of cleavage.[5]

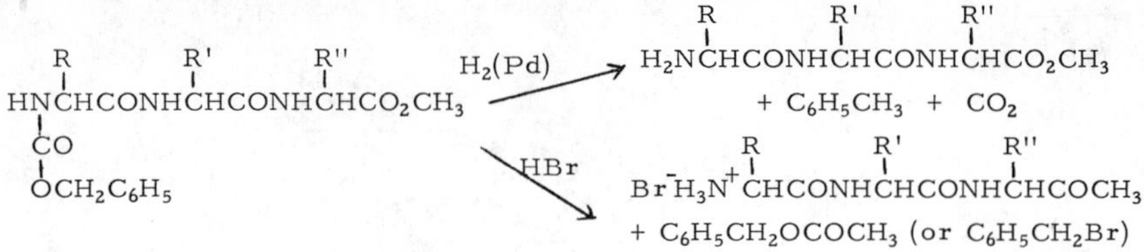

The reagent has been used also for protection of the sulfhydryl group of cysteine.[6] The protective group is removed almost quantitatively by treatment with excess sodium methoxide under nitrogen for 5–10 min. at room temperature.

Isolation of amines. Amino sugars have been isolated and purified as the N-carbobenzoxy derivatives. The group is removed by hydrogenolysis.[7] Thus Wintersteiner and co-workers[8] isolated crude mycosamine hydrochloride from hydrolysis of the antibiotic nystatin, but were unable to induce the material to crystallize. Reaction with carbobenzoxy chloride in aqueous sodium carbonate and extraction with ethyl acetate gave an oil and, when a solution of this in warm ethanol was chilled and scratched, colorless needles in clusters soon deposited.

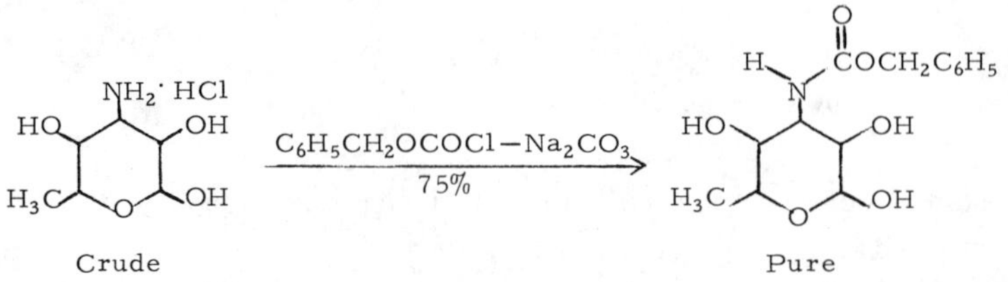

Photolytic cleavage. The Cb group can be cleaved by photolysis; in the case of glycine the yield was 75%.[9] The reaction provides a means for the detection of Cb-amino acids on paper. The chromatogram after drying is irradiated and the paper is developed with ninhydrin in the usual way.

[1]M. Bergmann and L. Zervas, *Ber.*, **65**, 1192 (1932)
[2]A. C. Farthing, *J. Chem. Soc.*, 3213 (1950); H. E. Carter, R. L. Frank, and H. W. Johnston, *Org. Syn., Coll. Vol.*, **3**, 167 (1955)
[3]E. Katchalski, *Methods in Enzymology*, **3**, 541 (1957)
[4]D. Ben-Ishai and A. Berger, *J. Org.*, **17**, 1564 (1952); G. W. Anderson *et al.*, *Am. Soc.*, **74**, 5311 (1952); I. Schumann and R. A. Boissonnas, *Helv.*, **35**, 2240 (1952)
[5]R. A. Boissonnas and G. Preitner, *Helv.*, **36**, 875 (1953); R. A. Boissonnas, *Advances in Organic Chemistry*, **3**, 159 (1963)
[6]M. Sokolovsky, M. Wilchek, and A. Patchornik, *Am. Soc.*, **86**, 1202 (1964)
[7]C. G. Greig, D. H. Leaback, and P. G. Walker, *J. Chem. Soc.*, 879 (1961)
[8]J. D. Dutcher, D. R. Walters, and O. Wintersteiner, *J. Org.*, **28**, 995 (1963)
[9]J. A. Barltrop and P. Schofield, *Tetrahedron Letters*, 697 (1962)

Carbobenzoxyhydrazine, $C_6H_5CH_2OCONHNH_2$ ($CbNHNH_2$). Mol. wt. 166.18, m.p. 69–70°. Preparation.[1]

Peptide synthesis. Hofmann[2] employed the reagent in a novel scheme for peptide synthesis illustrated as follows. An N-protected amino acid chloride (1) is condensed with the reagent to form a peptide bond (2) and the protective phthaloyl group is removed with hydrazine to liberate the free amino group (3) for coupling with a

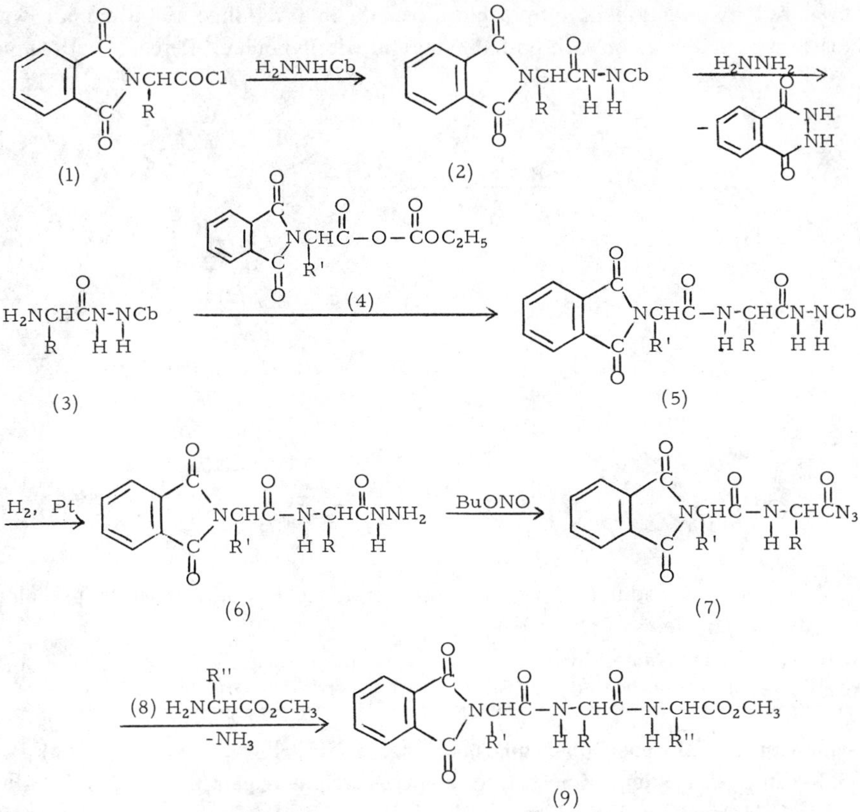
NCHCOCl
R
(1)
H2NNHCb
NCHCN-NCb
R H H
(2)
H2NNH2
NH
NH
NCHC-O-COC2H5
R'
(4)
H2NCHCN-NCb
R H H
(3)
NCHC-N-CHCN-NCb
R' H R H H
(5)
H2, Pt
NCHC-N-CHCNNH2
R' H R H
(6)
BuONO
NCHC-N-CHCN3
R' H R
(7)
R''
(8) H2NCHCO2CH3
-NH3
NCHC-N-CHC-N-CHCO2CH3
R' H R H R''
(9)

mixed anhydride (4) to establish a second peptide link (5). The carbobenzoxy group is then cleaved by hydrogenolysis to the free hydrazide (6), convertible with an alkyl nitrite into the azide (7). Coupling of this with an amino acid ester (8) then forms a third peptide link (9). The method affords the opportunity for extension of the peptide chain at the carboxy end. It played an important role in the synthesis of peptides related to ribonuclease.[3]

[1]N. Rabjohn, *Am. Soc.*, **70**, 1181 (1948)
[2]K. Hofmann *et al.*, *Am. Soc.*, **74**, 470 (1952)
[3]K. Hofmann *et al.*, *Am. Soc.*, **87**, 620 (1965)

Carboethoxymethylenetriphenylphosphorane, $(C_6H_5)_3P{=}CHCO_2C_2H_5$. Mol. wt. 348.37, m.p. 117°. Supplier: Aldrich. *See* Carbomethoxymethylenetriphenylphosphorane.

Carboethoxymethyltriphenylphosphonium bromide, $C_2H_5O_2CCH_2P(C_6H_5)_3Br$. Mol. wt. 429.29. Supplier: Aldrich.

This is the precursor of the preceding Wittig reagent.

N-Carboethoxyphthalimide, 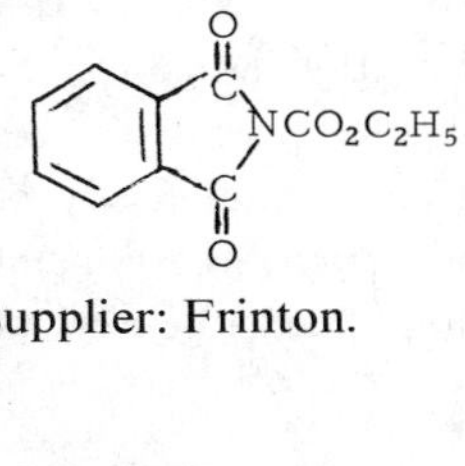
NCO2C2H5

Mol. wt. 219.19, m.p. 80°. Supplier: Frinton.

Prepared by reaction of ethyl chloroformate in N,N-dimethylformamide with potassium phthalimide or with phthalimide and triethylamine.[1] It reacts with amino

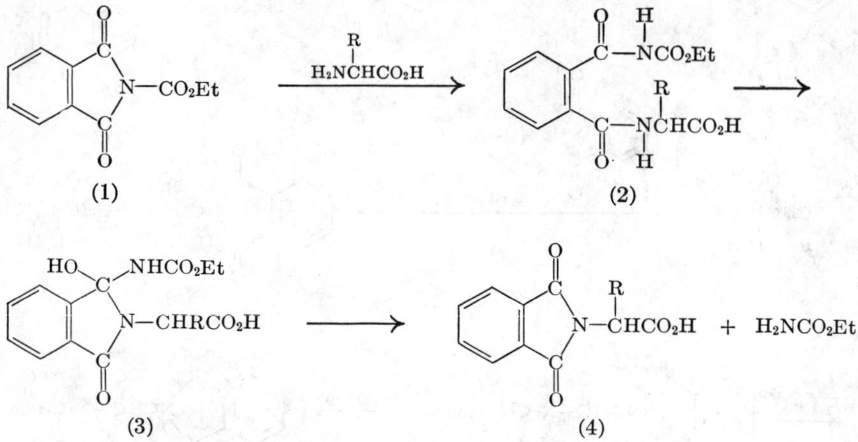

acids in aqueous sodium carbonate solution to give optically pure phthaloyl derivatives (4) in yields of 85–96%.

[1]G. H. L. Nefkens, *Nature*, **185**, 309 (1960); G. H. L. Nefkens, G. I. Tesser, and R. J. F. Nivard, *Rec. trav.*, **79**, 688 (1960); R. A. Boissonnas, *Advances in Organic Chemistry*, **3**, 181 (1963)

Carbomethoxymethylenetriphenylphosphorane, $(C_6H_5)_3P{=}CHCO_2CH_3$ (mol. wt. 333.34, m.p. 163°), and **Carboethoxymethylenetriphenylphosphorane**, $(C_6H_5)_3P{=}CHCO_2C_2H_5$ (mol. wt. 348.37, m.p. 117°).

These Wittig reagents are prepared by reaction of ethyl or methyl bromoacetate with triphenylphosphine and treatment of the adduct with base:

$$RO_2CCH_2Br \xrightarrow{(C_6H_5)_3P} RO_2CCH_2\overset{+}{P}(C_6H_5)_3(\bar{B}r) \xrightarrow{OH^-} RO_2CCH{=}P(C_6H_5)_3$$

Isler[1] used the carboethoxy reagent for the synthesis of diethylnorbixin.

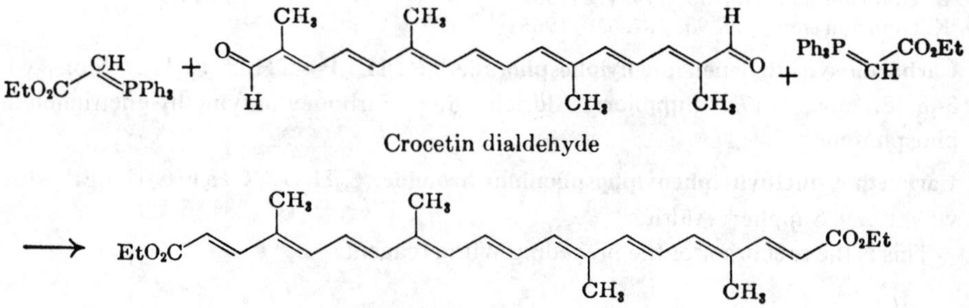

Diethylnorbixin

Bestmann[2] found that the ylide (1) is sufficiently basic to be alkylated with a reactive halide (*e.g.*, allyl bromide, methyl bromoacetate, benzyl bromide) to give a product (2) which on saponification affords the corresponding carboxylic acid (3).

$$CH_2{=}CHCH_2Br + \underset{\underset{P(C_6H_5)_3}{\|}}{CHCO_2CH_3} \longrightarrow CH_2{=}CHCH_2{-}\underset{\underset{+P(C_6H_5)_3(Br^-)}{|}}{CHCO_2CH_3}$$

(1) (2)

$$\xrightarrow{OH^-} CH_2{=}CHCH_2CH_2CO_2H + \overset{-}{O}{-}\overset{+}{P}(C_6H_5)_3$$

(3)

Reaction of the reagent with an α-bromoketone affords an α,β-unsaturated γ-keto ester.[3]

$$RCOCH_2Br + \underset{\underset{P(C_6H_5)_3}{\|}}{CHCO_2CH_3} \xrightarrow[60\text{-}85\%]{} RCOCH{=}CHCO_2CH_3$$

Under suitably drastic conditions, carboethoxymethylenetriphenylphosphorane reacts with an epoxide to give a carboethoxycyclopropane.[4] Trials with other phosphoranes were unsuccessful.

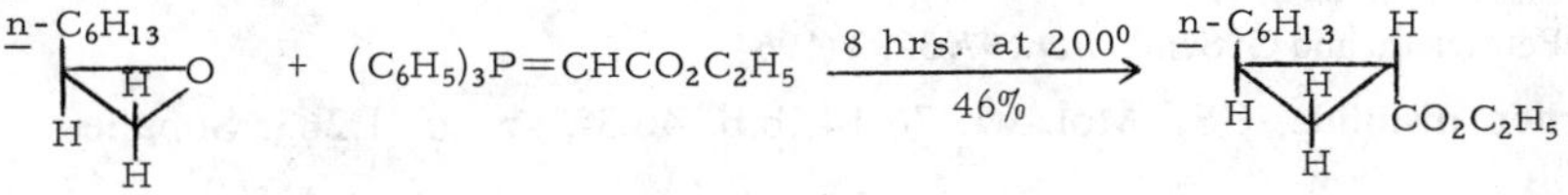

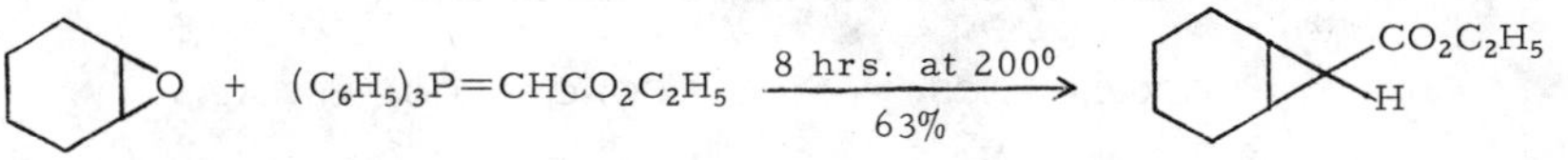

Eiter[5] used the reagent (methyl ester) in one of the more practicable syntheses of "queen substance" isolated from royal jelly.

$$O{=}CH(CH_2)_3CH{=}O \xrightarrow[\text{nearly quant.}]{(C_6H_5)_3P{=}CHCO_2Me} O{=}CH(CH_2)_3CH{=}CHCO_2Me \xrightarrow[60\%]{HC{\equiv}CCH_2Br,\ Zn}$$

(1) (2)

$$HC{\equiv}CCH_2\underset{\underset{OH}{|}}{CH}(CH_2)_3CH{=}CHCO_2Me \xrightarrow[60\%]{20\%\ H_2SO_4\ (Hg^{++})}$$

(3)

$$CH_3CO\overset{8}{C}H{=}\overset{7}{C}H(CH_2)_3CH{=}CHCO_2Me \xrightarrow{H_2,\ Pd} 7,8\text{-Dihydride} \xrightarrow[40\%\ \text{from (4)}]{Na_2CO_3}$$

(4) (5)

$$CH_3CO(CH_2)_5\overset{\overset{H}{|}}{C}{=}\underset{\underset{H}{|}}{C}CO_2H$$

Queen substance

Petragnani and Schill[6] used the ethyl ester in the synthesis of unsaturated fatty acids, for example:

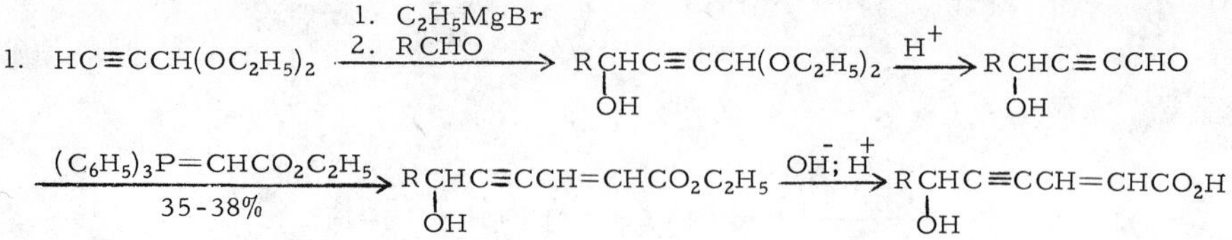

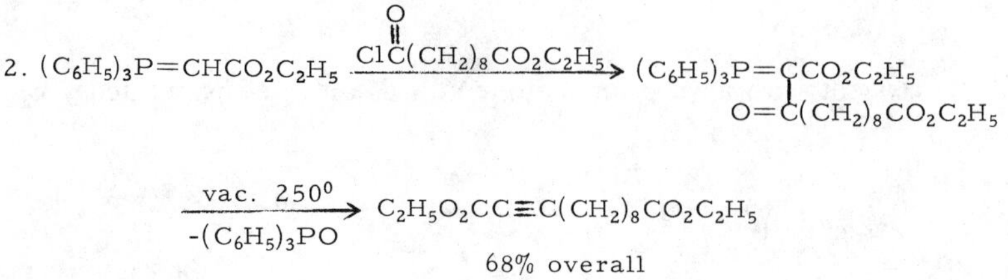

1.　$HC \equiv CCH(OC_2H_5)_2$ $\xrightarrow[\text{2. R CHO}]{\text{1. } C_2H_5MgBr}$ $\underset{\underset{OH}{|}}{R}CHC \equiv CCH(OC_2H_5)_2$ $\xrightarrow{H^+}$ $\underset{\underset{OH}{|}}{R}CHC \equiv CCHO$

$\xrightarrow[35-38\%]{(C_6H_5)_3P=CHCO_2C_2H_5}$ $\underset{\underset{OH}{|}}{R}CHC \equiv CCH=CHCO_2C_2H_5$ $\xrightarrow{OH^-; H^+}$ $\underset{\underset{OH}{|}}{R}CHC \equiv CCH=CHCO_2H$

2.　$(C_6H_5)_3P=CHCO_2C_2H_5$ $\xrightarrow{Cl\overset{\overset{O}{||}}{C}(CH_2)_8CO_2C_2H_5}$ $\underset{O=\overset{|}{C}(CH_2)_8CO_2C_2H_5}{(C_6H_5)_3P=\overset{|}{C}CO_2C_2H_5}$

$\xrightarrow[-(C_6H_5)_3PO]{\text{vac. } 250^0}$ $C_2H_5O_2CC \equiv C(CH_2)_8CO_2C_2H_5$

68% overall

[1] O. Isler, H. Gutmann, M. Montavon, R. Rüegg, G. Ryser, and P. Zeller, *Helv.*, **40**, 1242 (1957)
[2] H. J. Bestmann and H. Schulz, *Ber.*, **95**, 2921 (1962)
[3] H. J. Bestmann, F. Seng, and H. Schulz, *Ber.*, **96**, 465 (1963)
[4] D. B. Denny, J. J. Vill, and M. J. Boskin, *Am. Soc.*, **84**, 3944 (1962)
[5] K. Eiter, *Ann.*, **658.**, 91 (1962)
[6] N. Petragnani and G. Schill, *Ber.*, **97**, 3293 (1964)

Carbon disulfide, CS_2. Mol. wt. 76.14, b.p. 46.3°, sp. gr. 1.263. Suppliers: B, F, MCB.

Silver salts of carboxylic acids are converted into the corresponding acid anhydrides when heated with CS_2 in a sealed tube at 100° for 8 hrs.[1]

$$4 \ RCOOAg + CS_2 \longrightarrow 2 \ (RCO)_2O + CO_2 + 2 \ Ag_2S$$

[1] D. Bryce-Smith, *Proc. Chem. Soc.*, 20 (1957)

Carbonyl chloride, *see* Phosgene.

N,N'-Carbonyldiimidazole, 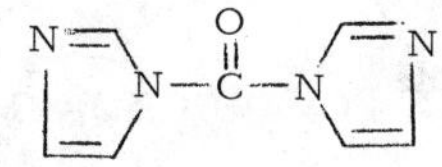

Mol. wt., 162.15, m.p. 116–118°. Supplier: KK. Prepared from phosgene and imidazole in the ratio 1:4 in tetrahydrofurane.[1] Review of uses.[2]

The reagent reacts with a carboxylic acid to form an imidazolide (1),[3] which on reduction affords an aldehyde (2).[4] Under basic catalysis, the reagent can be used

(1)

(2)

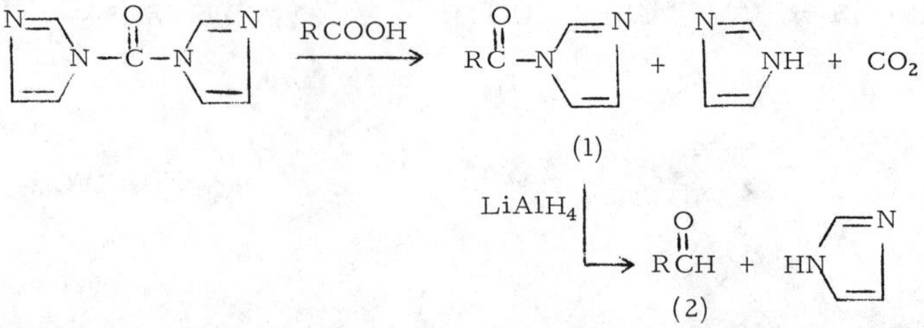

for the synthesis of an ester (3),[5] and it reacts with a primary amine to give a derivative (4), which decomposes at room temperature to imidazole and an isocyanate (5).[6]

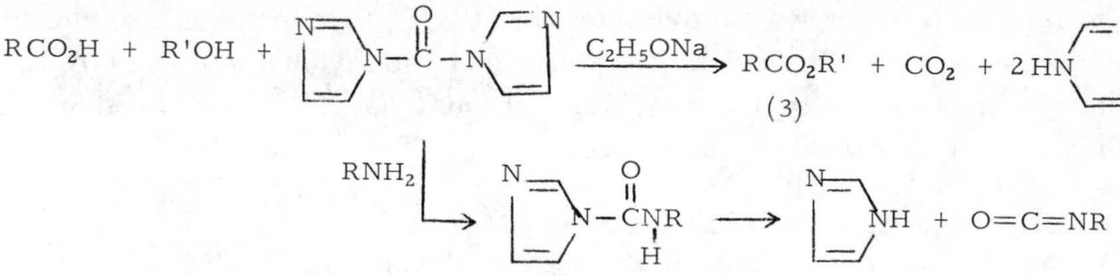

(3) (4) (5)

Paul and Anderson[7] employed the reagent in peptide synthesis as shown in the formulation and stated that racemization was slight, but Weygand[8] noted extensive racemization with the reagent, and Paul[9] reports that in the synthesis of a valyltyrosine peptide the mixed anhydride method is superior in yield and in purity of product.

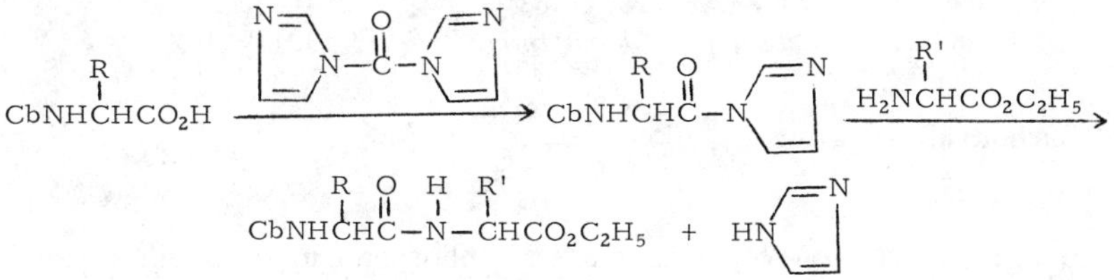

N,N′-Carbonyldiimidazole is used also for the conversion of a carboxylic acid into the *t*-butyl ester of the corresponding peracid.[10] Yields range from 64 to 80%. N,N′-Thionyldiimidazole functions in the same way but yields are lower.

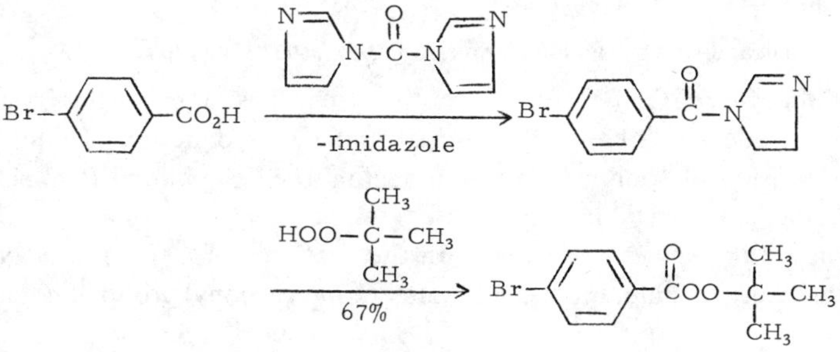

Staab and Datta[11] found that even highly sensitive acid chlorides can be prepared in good yield by passing hydrogen chloride through a cold solution of an imidazolide (6) in chloroform or methylene chloride. The reaction *per se* is reversible, since acid

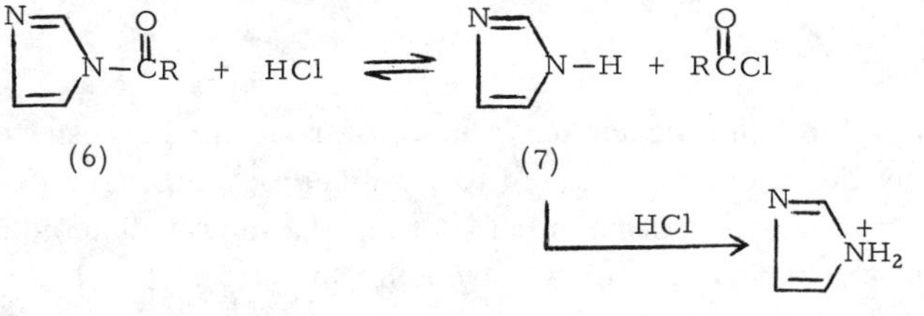

(6) (7) (8)

chlorides react with imidazole to form imidazolides, but the imidazole formed reacts with hydrogen chloride to form imidazolium chloride, which precipitates, and thus the reaction goes to completion.

The previously unknown formyl chloride, HCOCl, was prepared for the first time by passing HCl into a chloroform solution of 1-formylimidazole at −60°. After addition of methanol, methyl formate was isolated in 65–70% yield. On addition of methanol at −40°, the yield of ester was only 9%.

[1]H. A. Staab and K. Wendel, *Ber.*, **93**, 2910 (1960)
[2]H. A. Staab, *Angew. Chem., Internat. Ed.*, **1**, 351 (1962)
[3]H. A. Staab, M. Lüking, and F. H. Dürr, *Ber.*, **95**, 1275 (1962)
[4]H. A. Staab and H. Bräunling, *Ann.*, **654**, 119 (1962)
[5]H. A. Staab and A. Mannschreck, *Ber.*, **95**, 1284 (1962)
[6]H. A. Staab and W. Benz, *Ann.*, **648**, 72 (1961)
[7]R. Paul and G. W. Anderson, *Am. Soc.*, **82**, 4596 (1960); *J. Org.*, **27**, 2094 (1962)
[8]F. W. Weygand, A. Prox, L. Schmidhammer, and W. Konig, *Angew. Chem., Internat. Ed.*, **2**, 183 (1963)
[9]R. Paul, *J. Org.*, **28**, 236 (1963)
[10]R. Hecht and C. Rüchardt, *Ber.*, **96**, 1281 (1963)
[11]H. A. Staab and A. P. Datta, *Angew. Chem., Internat. Ed.*, **3**, 132 (1964)

N,N′-Carbonyldi-s-triazine, (structure uncertain).

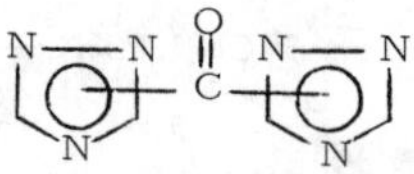

Prepared by reaction of *s*-triazole with phosgene in tetrahydrofurane in the presence of pyridine as HCl-acceptor, the reagent is obtained in 80% yield as a white powder of 95% purity.[1]

Peptide synthesis, *compare* N,N′-Carbonyldiimidazole. Racemization, if any, is slight. Dimethylformamide, a good solvent for small and large peptides, can be used as solvent.

[1]H. C. Beyerman and W. M. Van Den Brink, *Rec. trav.*, **80**, 1372 (1961)

Carbonyl fluoride, COF_2. Mol. wt. 60.01, toxic gas, b.p. −81°. Prepared from carbon monoxide and silver (II) fluoride[1] and by introduction of liquid phosgene into a suspension of sodium fluoride in acetonitrile.[2] Suppliers: Peninsular Chem-Research, Pierce Chem. Co.

Reaction with ketones and amides. In the presence of a trace of N,N-dimethylformamide or of pyridine, the reagent adds to the carbonyl group of cyclohexanone

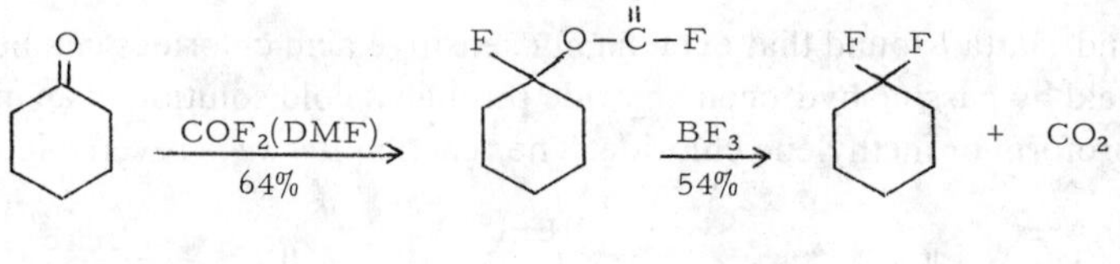
COF2(DMF)
64%
BF3
54%
+ CO2

(1) (2) (3)

to give the distillable 1-fluorocyclohexyl fluoroformate (2).[2] This ester on treatment with boron fluoride etherate is cleaved to 1,1-difluorocyclohexane (3) and carbon dioxide. With N,N-dimethylformamide (4) at 25° the intermediate adduct was not isolated, but the product was α,α-difluorotrimethylamine (5).

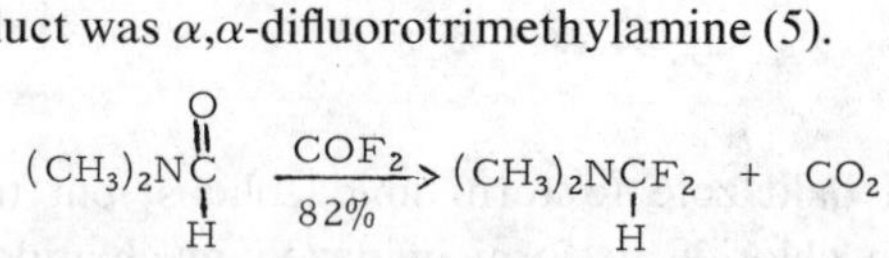
(CH3)2NCH=O
COF2
82%
(CH3)2NCF2H + CO2

(4) (5)

Addition to perfluoroolefins and azaolefins.[2,3] Carbonyl fluoride reacts with perfluoroolefins to give simple adducts, for example, (6)→(7). The addition is

$$\underset{(6)}{CF_3CF{=}CF_2} \xrightarrow[80\%]{COF_2} \underset{(7)}{CF_3CF(COF){-}CF_3}$$

accomplished with excess carbonyl fluoride in acetonitrile at 100–150° in the presence of a catalytic amount of cesium fluoride.

$$\underset{(8)}{CF_3N{=}CF_2} \xrightarrow[56\%]{COF_2} \underset{(9)}{CF_3N(COF)CF_3}$$

Reaction with phenols and alcohols. Carbonyl fluoride reacts with a phenol or alcohol (autoclave, 100–200°) to form the fluoroformate (10). In a second step the fluoroformate on reaction with sulfur tetrafluoride (150–200°) affords the aryl(alkyl) trifluoromethyl ether.[4]

$$Ar(R)OH \xrightarrow[-HF]{COF_2} \underset{(10)}{Ar(R)OC(=O)F} \xrightarrow{SF_4} Ar(R)OCF_3$$

[1]M. W. Farlow, E. H. Man, and C. W. Tullock, *Inorg. Syn.*, **6**, 155 (1960)
[2]F. S. Fawcett, C. W. Tullock, and D. D. Coffman, *Am. Soc.*, **84**, 4275 (1962)
[3]R. D. Smith, F. S. Fawcett, and D. D. Coffman, *Am. Soc.*, **84**, 4285 (1962)
[4]W. A. Sheppard, *J. Org.*, **29**, 1 (1964); P. E. Aldrich and W. A. Sheppard, *ibid.*, **29**, 11 (1964)

N-Carbonylsulfamic acid chloride, $O{=}C{=}NSO_2Cl$. Mol. wt. 141.55, b.p. 106–107°. Preparation from cyanogen chloride and sulfur trioxide.[1]

The reagent reacts with alcohols and phenols to give urethane-N-sulfonyl chlorides, which are hydrolyzed by water to carbamates.[2] It reacts with an olefin to give either a

$$ROH + O{=}C{=}NSO_2Cl \longrightarrow ROCONHSO_2Cl \xrightarrow{H_2O} ROCONH_2$$

β-lactam-N-sulfonyl chloride or an unsaturated acid amido-N-sulfonyl chloride or

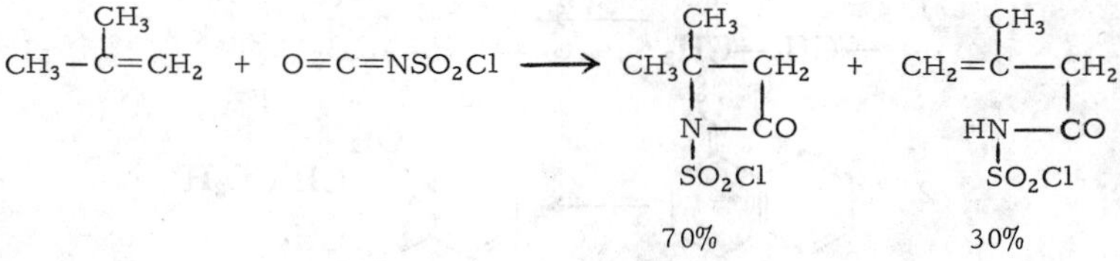
CH3
CH3—C=CH2 + O=C=NSO2Cl
CH3C—CH2
N—CO
SO2Cl
70%
CH2=C—CH2
HN—CO
SO2Cl
30%

both. It reacts with benzaldehyde with elimination of carbon dioxide and formation of N-benzalsulfamidic acid chloride.[3]

$$C_6H_5CHO + O{=}C{=}NSO_2Cl \xrightarrow[-CO_2]{} C_6H_5CH{=}NSO_2Cl$$

It reacts readily at room temperature with conjugated dienes to give amide sulfochlorides.[4]

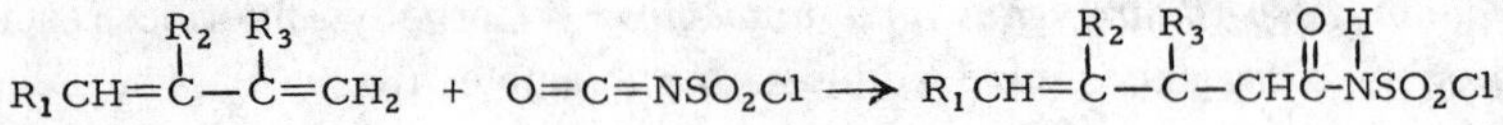

[1]R. Graf, *Ber.*, **89**, 1071 (1956)
[2]R. Graf, *Ber*, **96**, 56 (1963)
[3]R. Graf, *Ann.*, **661**, 111 (1963)
[4]H. Hoffmann and H. J. Diehr, *Tetrahedron Letters*, 1875 (1963)

4-(4-Carboxyphenyl)semicarbazide. Mol. wt. 195.17, m.p. 218–220°.

The reagent is made by condensing *p*-aminobenzoic acid with nitrourea and refluxing the resulting *p*-ureidobenzoic acid with hydrazine in refluxing methanol.[1]

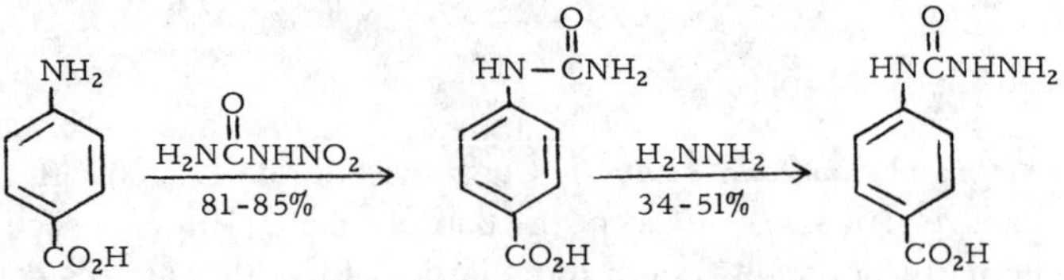

It is used for the resolution of asymmetric carbonyl compounds: the reagent condenses to form a semicarbazone which contains a carboxyl group and which is thus capable of forming diastereoisomeric salts with optically active bases. Used for the resolution of *dl*-3-methylcyclohexanone.

[1]J. K. Shillington, G. S. Denning, Jr., W. B. Greenough, III, T. Hill, Jr., and O. B. Ramsay, *Am. Soc.*, **80**, 6551 (1958)

Caro's acid (Sulfomonoperacid),[1] 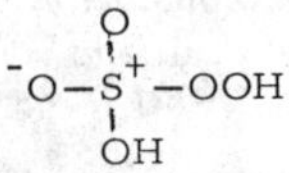

Prepared as required by stirring 10 g. of finely powdered potassium persulfate (mol. wt. 270.33) into 7 ml. of ice-cold concd. sulfuric acid and when the mixture is homogeneous adding 40–50 g. of ice.[2]

Uses. (a) Oxidation of arylamines to nitroso compounds.[1,3] (b) Baeyer-Villiger oxidation of ketones to lactones (1)[2,4] or esters (3).[5] When an aromatic ring is present (2), cleavage occurs adjacent to this ring.[6]

1. 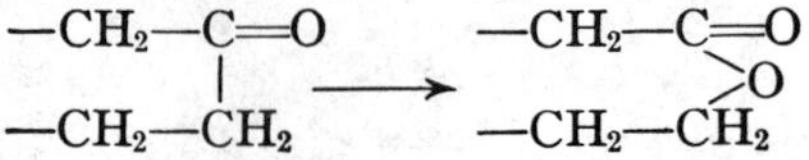

2. 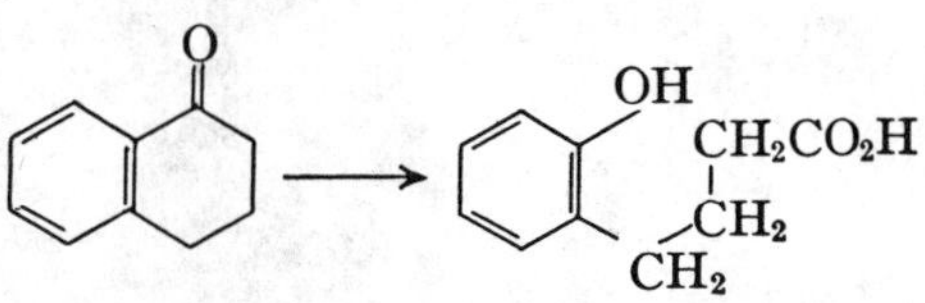

3. 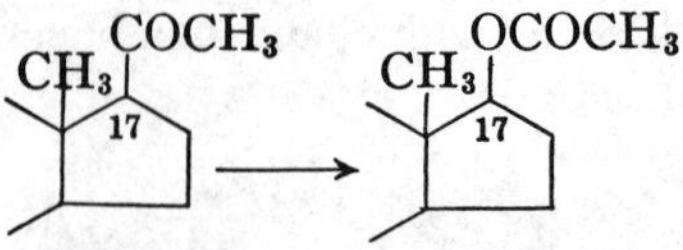

Oxidative cleavage of ketones has been carried out in ligroin at 50–65°[4] with Caro's acid prepared as described, in a solution containing more sulfuric acid, less water, and no organic solvent,[2] by refluxing the ketone with a mixture of potassium persulfate (4 g.) and concd. sulfuric acid (1 ml.) in 90% acetic acid (150 ml.),[7] and with Baeyer and Villiger's dry reagent,[2] prepared by grinding 10 g. of potassium persulfate thoroughly in a mortar with 6 ml. of concd. sulfuric acid, adding 30 g. of potassium sulfate, and grinding the mixture to a dry powder. The powder is added to a cold solution of the ketone in acetic acid and the reaction is allowed to proceed at room temperature for 7–10 days.

A similar reagent for the oxidative cleavage of ketones is prepared by cautiously mixing acetic anhydride (65 g.), concd. sulfuric acid (30 g.), and perhydrol (25 g.).[5] Benzophenone (20 g.) added to this solution is converted, after standing for several days at 0°, into phenylbenzoate in nearly quantitative yield.

The behavior of Δ^4-3-ketosteroids on oxidative cleavage[8,9] is illustrated by the reaction of testosterone propionate (1). Pettit and Kasturi[9] mixed 9 g. of potassium persulfate and 10 g. of concd. sulfuric acid in a mortar, diluted with 150 ml. of acetic acid, added the mixture to a solution of 9 g. of testosterone propionate in 150 ml. of acetic acid, and let the mixture stand with intermittent shaking at room temperature for 7 days. Workup included saponification and reacidification, and the product (2 g.) was characterized as the lactone 3-oxo-17β-hydroxy-4-oxa-5α-androstane (4). A possible reaction sequence is formulated.

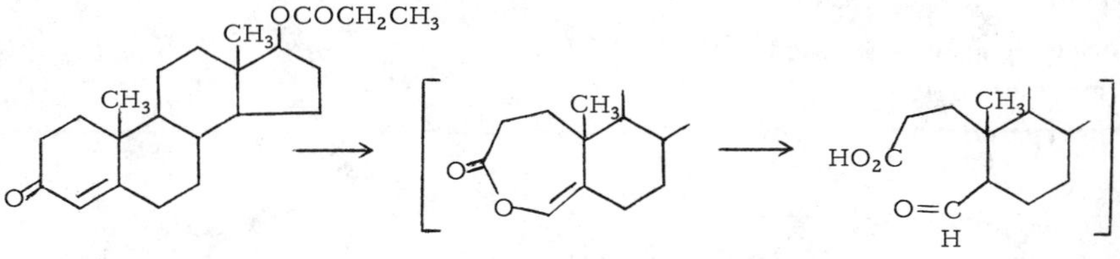
OCOCH₂CH₃
CH₃
CH₃
O
CH₃
O
O
CH₃
HO₂C
O=C
H

(1) (2) (3)

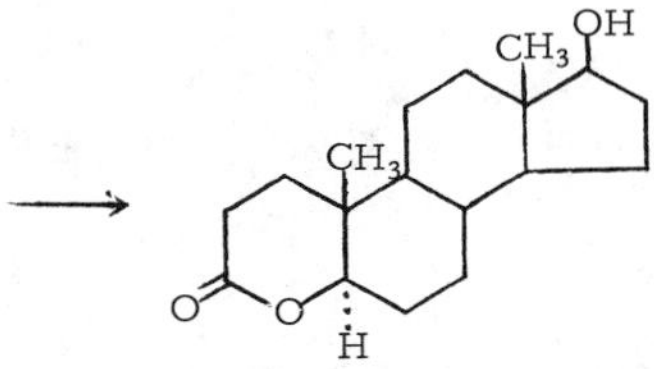
OH
CH₃
CH₃
O
O
H

(4)

[1]H. Caro, *Angew. Chem.*, **11**, 845 (1898)
[2]A. Baeyer and V. Villiger, *Ber.*, **32**, 3625 (1899); **33**, 858 (1900); W. D. Langley, *Org. Syn., Coll. Vol.*, **3**, 334 (1955)
[3]E. Bamberger and R. Hübner, *Ber.*, **36**, 3803 (1903); E. Borel and H. Deuel. *Helv.*, **36**, 805 (1953): 3-nitro-4-aminotoluene→3-nitro-4-nitrosotoluene in 92% yield
[4]L. Ruzicka and M. Stoll, *Helv.*, **11**, 1159 (1928)
[5]R. E. Marker and co-workers, *Am. Soc.*, **62**, 525, 650, 2543, 2621 (1940)
[6]G. Schroeter, German patent 562,827 (1928) [*Chem. Zentr.*, **104**, 127 (1933)]
[7]A. Rollett and K. Bratke, *Monatsh.*, **43**, 685 (1922)
[8]A. Salamon, *Z. physiol. chem.*, **272**, 61 (1942)
[9]G. R. Pettit and T. R. Kasturi, *J. Org.*, **26**, 4557 (1961)

Catechol dichloromethylene ether (3). Mol. wt. 191.01, b.p. 82–89°/12 mm. Supplier: E.

Catechol carbonate (2, m.p. 119°), prepared by shaking a cooled solution of catechol in dilute sodium hydroxide with a 20% solution of phosgene in toluene,[1] reacts with phosphorus pentachloride to give the reagent (3).[2,3] Under Friedel-Crafts conditions, (3) acylates an aromatic hydrocarbon to give a product (4)

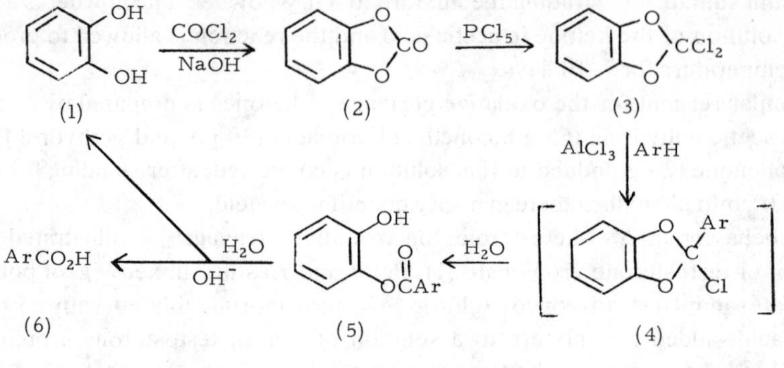
OH
OH
$COCl_2$
NaOH
CO
PCl_5
CCl_2
(1)
(2)
(3)
$AlCl_3$
ArH
Ar
C
Cl
(4)
OH
H_2O
OCAr
(5)
H_2O
OH^-
$ArCO_2H$
(6)

hydrolyzed by water under mild basic catalysis to an aromatic carboxylic acid (6).[3] Yields are in the range 70–100%.

[1]A. Einhorn and E. Lindenberg, *Ann.*, **300**, 141 (1898)

[2]H. Gross, A. Rieche, and E. Höft, *Ber.*, **94**, 544 (1961)

[3]H. Gross, J. Rusche, and M. Mirsch, *Ber.*, **96**, 1382 (1963)

Catechyl phosphorus trichloride (3). Mol. wt. 139.07, m.p. 61–62°, b.p. 132–135°/13 mm. Supplier: A.

Preparation.[1]

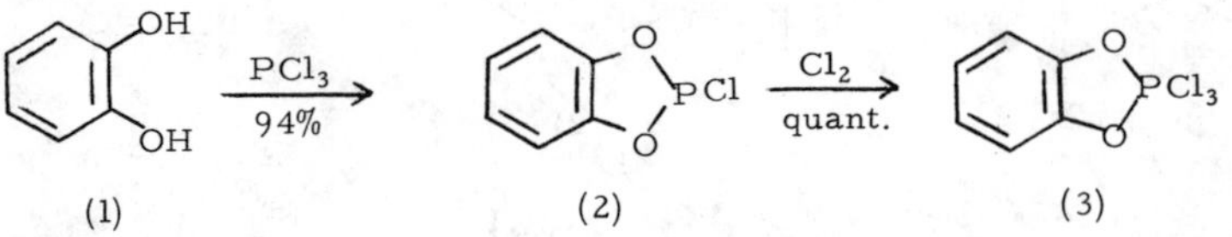
OH
OH
PCl_3
94%
PCl
Cl_2
quant.
PCl_3
(1)
(2)
(3)

The reagent converts carboxylic acids into acid chlorides and carbonyl compounds into *gem*-dihalides:

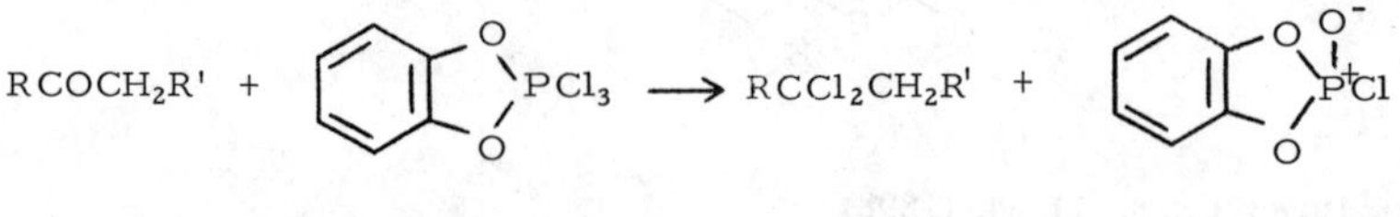
$RCOCH_2R'$ +
PCl_3
$RCCl_2CH_2R'$ +
O^-
PCl

Esters are cleaved to the acid chloride and the alkyl chloride, acid amides are dehydrated to nitriles. Yields for the most part are satisfactory.

[1]H. Gross and J. Gloede, *Ber.*, **96**, 1387 (1963)

Celite, a Johns-Manville registered trademark for a diatomaceous silica product. It is used as a filter aid. Supplier: Fisher. Bernstein[1] recommends that a water-washed benzene extract of a reaction product be filtered through a small pad of Celite and evaporated without further drying.

[1]J. J. Brown, R. H. Lenhard, and S. Bernstein, *Am. Soc.*, **86**, 2185 (1964)

Ceric ammonium nitrate, $(NH_4)_2Ce(NO_2)_6$. Mol. wt. 548.26. Suppliers: MCB, G. F. Smith and Co.

The reagent oxidizes benzyl alcohols to benzaldehydes in over 90% yield.[1] A solution of the oxidant in 50% aqueous acetic acid is added to the alcohol and the solution is warmed on the steam bath until the initial orange color fades to pale yellow.

Toluenes are oxidized to benzaldehydes by the reagent in 50% aqueous acetic acid and in anhydrous acetic acid benzyl acetates are obtained.[2]

[1]W. S. Trahanovsky and L. B. Young, *J. Chem. Soc.*, 5777 (1965)
[2]*Idem*, *J. Org.*, **31**, 2033 (1966)

Cerous hydroxide, $Ce(OH)_3$. Mol. wt. 191.15.

Prepared by adding aqueous sodium hydroxide to a solution of cerous chloride (Aldrich) and collecting and washing the precipitate by centrifugation, the reagent is recommended for the destruction of peroxides in ethers.[1] When the undried white solid is added to diethyl ether, dioxane, or tetrahydrofurane containing peroxide, it turns reddish-brown within a minute or two and the reaction is complete in 15 min., as indicated by a negative test with a slightly acidified solution of potassium iodide containing starch.

[1]J. B. Ramsey and F. T. Aldridge, *Am. Soc.*, **77**, 2561 (1955)

Cesium fluoride, CsF. Mol. wt. 151.91; m.p. 683°; water soluble. Suppliers: Amer. Potash and Chem. Corp., Alfa, KK, MCB.

A study of the Knoevenagel condensation of cyclohexanone and benzaldehyde with three active-methylene components showed that rubidium fluoride and cesium fluoride are more effective catalysts than potassium fluoride and that lithium fluoride

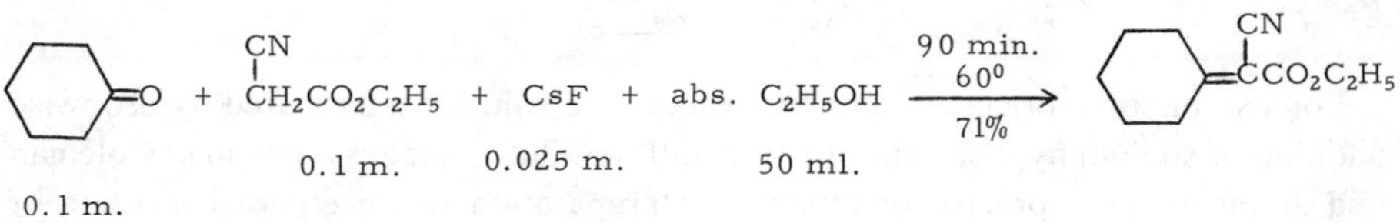

CN
+ $CH_2CO_2C_2H_5$ + CsF + abs. C_2H_5OH
0.1 m.
0.025 m.
50 ml.
0.1 m.
90 min.
60°
71%
CN
$CCO_2C_2H_5$

and sodium fluoride are inactive.[1] Yields obtained by the procedure formulated are shown in the table, along with solubilities.

Fluoride	Yield,%	Soly. (moles/100 ml. C_2H_5OH) at 27°
LiF	0	<0.01
NaF	<5	< .01
KF	52	.06
RbF	77	.16
CsF	71	.25

[1]L. Rand, J. V. Swisher, and C. J. Cronin, *J. Org.*, **27**, 3505 (1962)

Cetyltrimethylammonium bromide, $CH_3(CH_2)_{15}(CH_3)_3\overset{+}{N}\overset{-}{Br}$. Mol. wt. 364.46. Suppliers (techn.): B, E, F, MCB.

The reagent forms water-insoluble cetyltrimethylammonium salts with acidic polysaccharides which are useful for isolation and purification of such substances.[1,2] It is an effective precipitant for sulfated carbohydrates, even of sugar sulfates of low molecular weight. For acidic, nonsulfated polysaccharides molecular weight is important, since the salts of simple uronic acids are water soluble. Wolfrom[3] purified commercial heparin through the cetyltrimethylammonium salt to a product readily convertible into the crystalline barium acid salt.

[1]B. C. Bera, A. B. Foster, and M. Stacey, *J. Chem. Soc.*, 3788 (1955)
[2]Review: J. E. Scott, *Methods of Biochemical Analysis*, **8**, 145 (1960)
[3]M. L. Wolfrom, J. R. Vercellotti, and G. H. S. Thomas, *J. Org.*, **29**, 536 (1964)

Chloral, CCl_3CHO. Mol. wt. 147.39, b.p. 97–98°. Suppliers: A, E, F, MCB. An acid hydrazide when refluxed with an alcohol and chloral (or bromal) is converted into the corresponding ester.[1]

$$ArCH_2CONHNH_2 \xrightarrow[70-80\%]{ROH-CCl_3CHO} ArCH_2COOR$$

[1]T. Kametani and O. Umezawa, *Chem. Pharm. Bull.*, **12**, 379 (1964)

Chloramine, $ClNH_2$. Mol. wt. 51.49.

Preparation. Sisler and co-workers[1] developed a technique for obtaining chloramine by gas-phase reaction of chlorine with excess ammonia and showed that the reagent combines with tertiary amines to form the corresponding 1,1,1-trisubstituted hydrazinium chlorides (1) in good yield, and with tertiary phosphines to form aminophosphonium chlorides (2). The product of reaction (2) can be dehydrohalogenated with magnesium hydride to triphenylphosphine amine.

$$1.\quad (CH_3)_3N + NH_2Cl \longrightarrow (CH_3)_3\overset{+}{N}NH_2Cl^-$$

$$2.\quad (C_6H_5)_3P + NH_2Cl \longrightarrow (C_6H_5)_3P^+NH_2Cl^-$$

$$3.\quad (C_6H_5)_3P^+NH_2Cl^- \xrightarrow{MgH_2} (C_6H_5)_3P{=}NH$$

For use in some organic reactions, chloramine solution is prepared by dropwise addition of sodium hypochlorite solution at 0° to dilute aqueous ammonia.[2] Coleman and Johnson[3] give precise directions for preparation of an ethereal solution of chloramine.

α-Diazoketones. The reaction of α-oximino ketones with chloramine to give α-diazoketones was discovered by Forster[4] in 1915, but was only later resurrected and exploited by Cava[5] and by Meinwald.[6] The reagent is generated by dropwise addition of sodium hypochlorite solution (Chlorox) at 0° to an aqueous solution of the α-oximino ketone in aqueous sodium hydroxide and ammonium hydroxide. Illustrative examples are the preparation of 2-diazo-1-indanone[5] and of 3-diazonopinone.[6] Meinwald[7] suggested that the reaction involves a sequence of steps

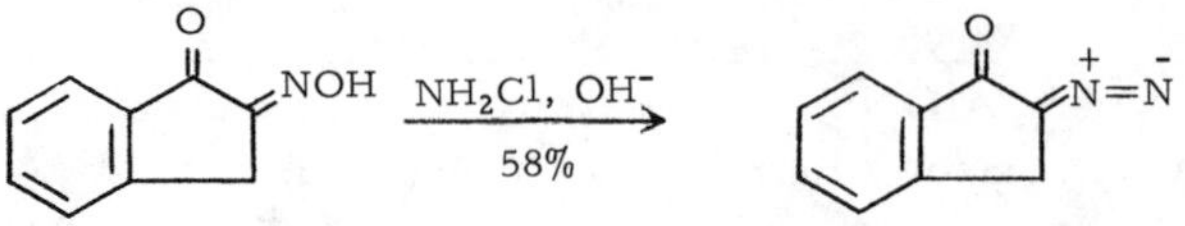
O
NOH
NH₂Cl, OH⁻
58%
O
+ -
N=N

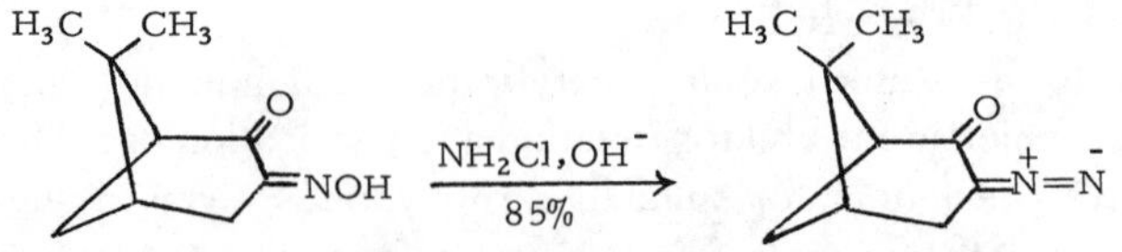
H₃C
CH₃
O
NOH
NH₂Cl,OH⁻
85%
H₃C
CH₃
O
+ -
N=N

initiated by a nucleophilic displacement at the chloramine nitrogen. The carbonyl group plays no part in the sequence outlined, and indeed fluorenone oxime was converted into diazofluorene by the Forster reaction.

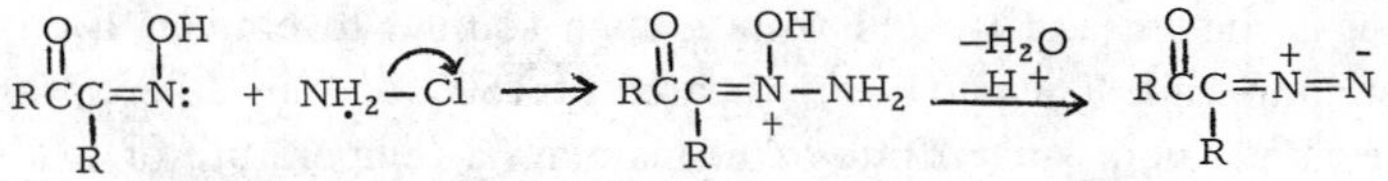

Ring contraction. This efficient route to α-diazoketones acquires particular importance with the finding that a cyclic α-diazoketone on irradiation in an aqueous solvent in the presence of sodium bicarbonate loses nitrogen with formation of a ring-contracted carboxylic acid. In 1962–65 no less than 5 groups reported use of this reaction for the synthesis of ring-contracted nor- and bisnor-steroids.[8-13] Several groups synthesized D-norsteroids, including D-norprogesterone.[10] Cava and Vogt[13] synthesized 1β-carboxy-A-bisnorcholestane by the process formulated. The steps are: conversion into the oximinoketone, formation of the diazoketone, irradiation to

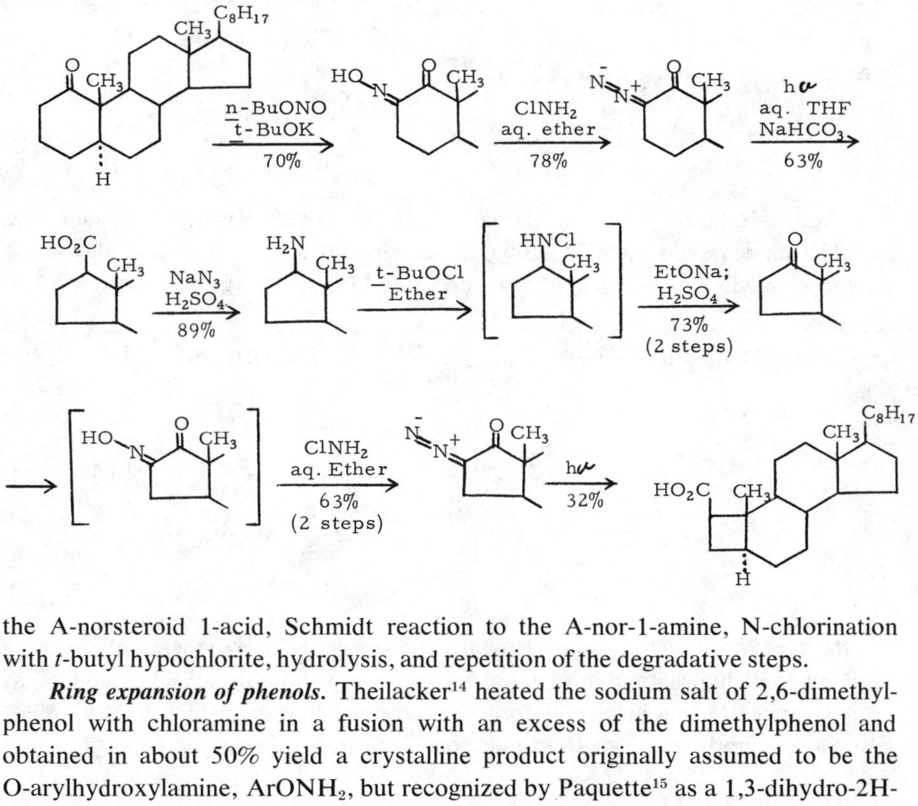

the A-norsteroid 1-acid, Schmidt reaction to the A-nor-1-amine, N-chlorination with *t*-butyl hypochlorite, hydrolysis, and repetition of the degradative steps.

Ring expansion of phenols. Theilacker[14] heated the sodium salt of 2,6-dimethylphenol with chloramine in a fusion with an excess of the dimethylphenol and obtained in about 50% yield a crystalline product originally assumed to be the O-arylhydroxylamine, $ArONH_2$, but recognized by Paquette[15] as a 1,3-dihydro-2H-azepin-2-one (see 3, below).[16] The reaction is generally applicable to 2,6-disubstituted phenols. A procedure by Paquette[17] for the conversion of 2,4,6-trimethylphenol (1) into 1,3-dihydro-3,5,7-trimethyl-2-H-azepine-2-one (3) is as follows.

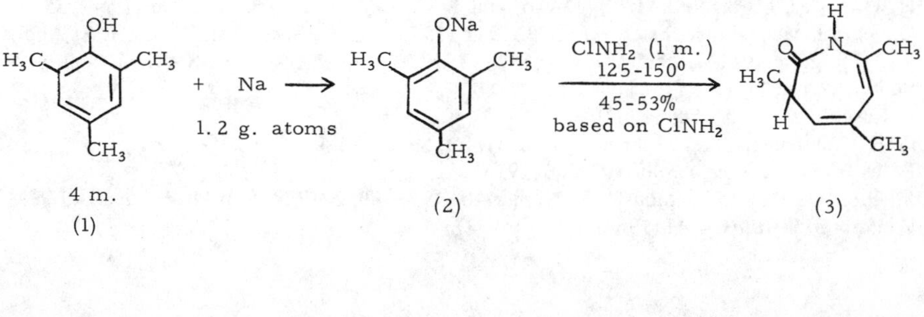

4 m. (1) (2) (3)

The phenol is melted and treated with enough sodium to convert a portion of it into the sodium salt. The mixture is heated at 150°, and an ethereal solution of chloramine is added at such a rate as to maintain a temperature of 125°. Workup included distillation and crystallization of the product, m.p. 131–132°.

Azirines. Schmitz[18] found that chloramine, generated in 80% yield by dropping *t*-butyl hypochlorite into a tenfold excess of 10*M* methanolic ammonia at −40°, reacts with an aldehyde and ammonia to give the diaziridine (a), which was not isolated but which reacted further with 2 moles of the aldehyde to give the triazolidine (b). Schmitz[19] then found that a diaziridine (a) can be dehydrogenated readily

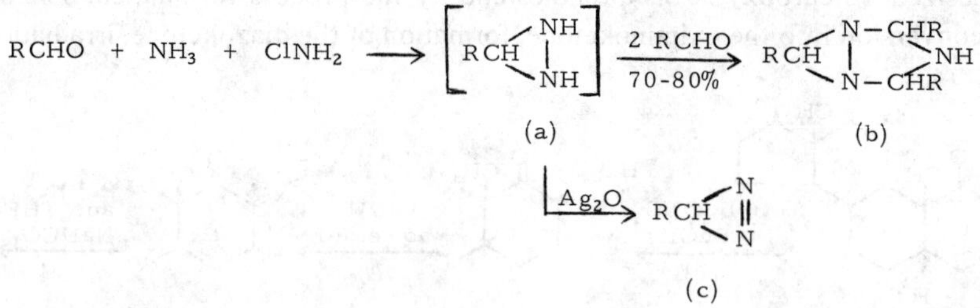
RCHO + NH3 + ClNH2
RCH
NH
NH
2 RCHO
70-80%
N−CHR
NH
N−CHR
(a)
(b)
Ag2O
N
N
(c)

to the diazirine (c) by oxidation. Oxidation of a mixture of formaldehyde, ammonia, and chloramine gave a solution shown to contain diazirine itself (3) by reduction to ammonia and methylamine and by reaction with a Grignard reagent to give (4).

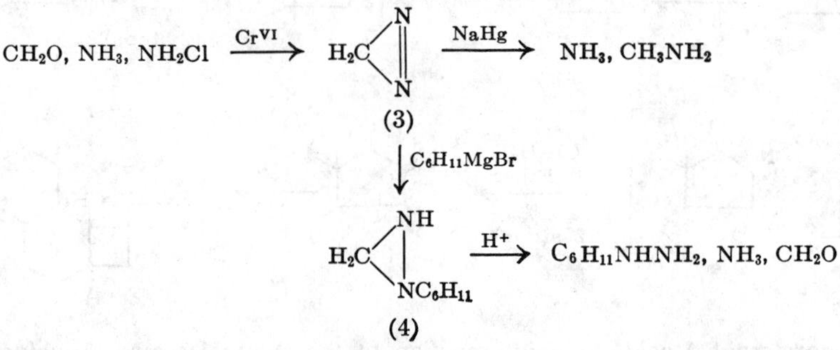
CH2O, NH3, NH2Cl
CrVI
H2C
N
N
NaHg
NH3, CH3NH2
(3)
C6H11MgBr
NH
NC6H11
H+
C6H11NHNH2, NH3, CH2O
(4)

RCH═CH₂ → RCH₂CH₂NH₂. Brown and co-workers[20] found that an alkylborane resulting from hydroboration of an olefin reacts readily with either chloramine or hydroxylamine-O-sulfonic acid to form the corresponding amine in yield of 50–60%. For example, norbornene yields *exo*-norbornylamine.

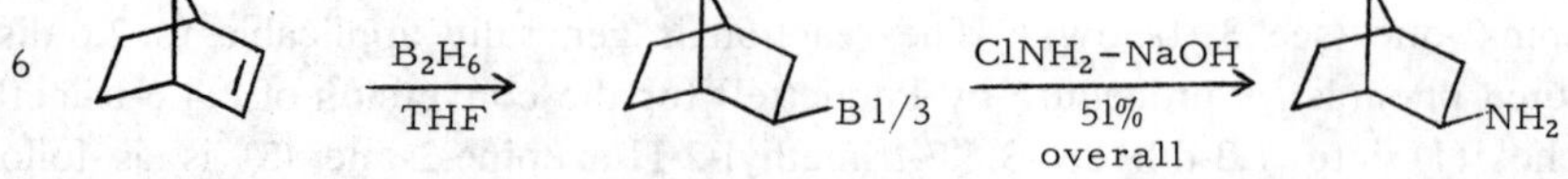
6
B2H6
THF
B 1/3
ClNH2−NaOH
51%
overall
NH2

[1]H. H. Sisler, F. T. Neth, R. S. Drago, and D. Yaney, *Am. Soc.*, **76**, 3906 (1954); G. M. Omietanski and H. H. Sisler, *ibid.*, **78**, 1211 (1956); H. H. Sisler, A. Sarkis, H. S. Ahuja, R. J. Drago, and N. L. Smith, *ibid.*, **81**, 2982 (1959); H. H. Sisler, H. S. Ahuja, and N. L. Smith, *J. Org.*, **26**, 1819 (1961)

[2]F. Raschig, *Ber.*, **40**, 4586 (1907)

[3]G. H. Coleman and H. L. Johnson, *Inorg. Syn.*, **1**, 59 (1939)

[4]M. O. Forster, *J. Chem. Soc.*, **107**, 260 (1915)

[5]M. P. Cava, R. L. Litle, and D. R. Napier, *Am. Soc.*, **80**, 2257 (1958); M. P. Cava and P. M. Weintraub, *Steroids*, **4**, 41 (1964)

[6]J. Meinwald and P. G. Gassman, *Am. Soc.*, **82**, 2857 (1960)
[7]J. Meinwald, P. G. Gassman, and E. G. Miller, *Am. Soc.*, **81**, 4751 (1959)
[8]M. P. Cava and E. Moroz, *Am. Soc.*, **84**, 115 (1962)
[9]J. Meinwald, G. G. Curtis, and P. G. Gassman, *Am. Soc.*, **84**, 116 (1962)
[10]A. Hassner, A. W. Coulter, and W. S. Seese, *Tetrahedron Letters*, 759 (1962)
[11]G. Muller, C. Huynh, and J. Mathieu, *Bull. Soc.*, 296 (1962)
[12]J. L. Mateos, O. Chao, and H. Flores R., *Tetrahedron*, **19**, 1051 (1963)
[13]M. P. Cava and B. R. Vogt, *Tetrahedron Letters*, 2813 (1964)
[14]W. Theilacker and E. Wegner, *Angew. Chem.*, **72**, 127 (1960); W. Theilacker, *ibid.*, **72**, 498 (1960)
[15]L. A. Paquette, *Am. Soc.*, **84**, 4987 (1962); **85**, 3288 (1963)
[16]The correct structure was deduced also (but later) by W. Theilacker, K. Ebke, L. Seidl, and S. Schwerin, *Angew. Chem., Internat. Ed.*, **2**, 154 (1963).
[17]L. A. Paquette, *Org. Syn.*, **44**, 41 (1964)
[18]E. Schmitz, *Ber.*, **95**, 688 (1962)
[19]E. Schmitz and R. Ohme, *Ber.*, **95**, 795 (1962); E. Schmitz and D. Habisch, *Ber.*, **95**, 680 (1962)
[20]H. C. Brown, W. R. Heydkamp, E. Breuer, and W. S. Murphy, *Am. Soc.*, **86**, 3565 (1964)

Chloranil (Tetrachloro-1,4-benzoquinone). Mol. wt. 246.89, m.p. 294°. Suppliers: Baker (purif.), Matheson (techn.). Purified by crystallization from benzene.

Dehydrogenation. Since it is readily available and reasonably stable, chloranil is widely used for dehydrogenation, although quinones of higher redox potential are often more effective.[1] Chloranil dehydrogenates 1,1-dimethyltetralin (1) with migration of a methyl group but without loss of carbon. Perylene (3) condenses

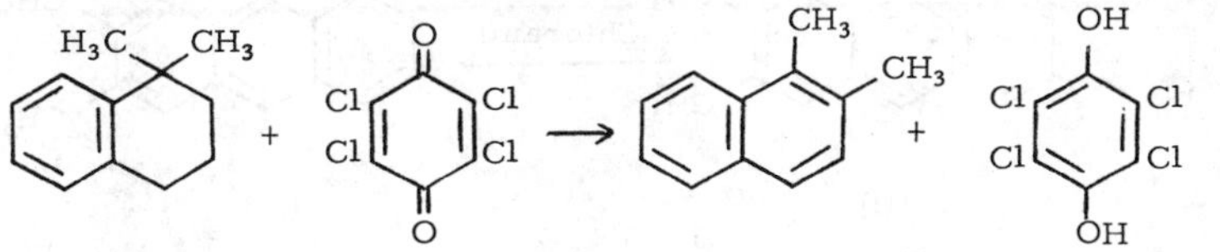

H₃C CH₃
O
Cl Cl
Cl Cl
O
CH₃
CH₃
OH
Cl Cl
Cl Cl
OH
+
+

(1) (2)

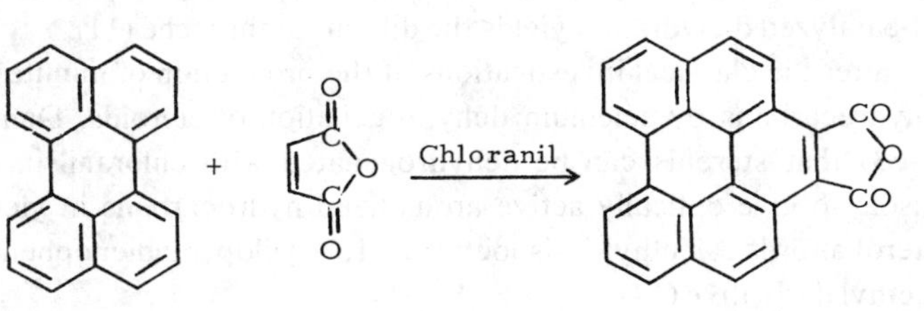

O
C
O
C
O
Chloranil
CO
O
CO
+

(3) (4)

with maleic anhydride in the presence of chloranil for dehydrogenation of the initial Diels-Alder adduct to give the aromatized product (4) in nearly quantitative yield.[2]

Δ^4-3-Ketosteroids (5) are dehydrogenated by chloranil in refluxing *t*-butanol or xylene to the Δ^6-dehydro derivatives (6) in 45–80% yield.[3] Under these conditions,

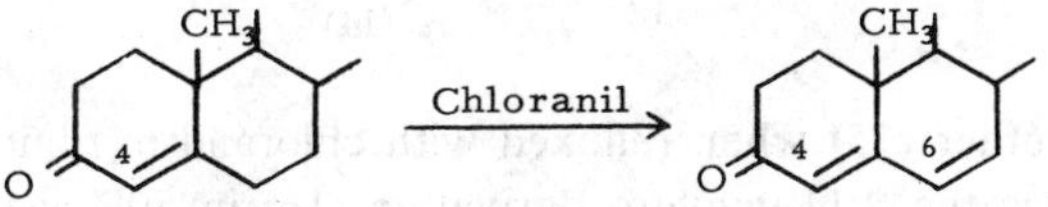

CH₃
O
4
Chloranil
CH₃
O
4
6

(5) (6)

$\Delta^{1,4}$- and $\Delta^{4,6}$-3-ketones are not affected. However, under more vigorous conditions, such as refluxing in *n*- or *sec*-amyl alcohol, Δ^{4}-, $\Delta^{1,4}$-, $\Delta^{4,6}$-, and saturated 3-keto-steroids (7) are all converted into $\Delta^{1,4,6}$-3-ketosteroids (8). For these dehydrogenations chloranil is superior to other readily available quinones. Other examples.[4]

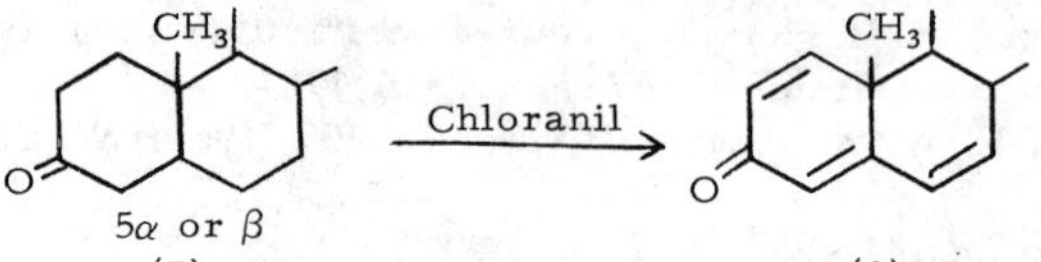
CH3
Chloranil
CH3
5α or β

(7) (8)

Wolthuis[5] employed chloranil in refluxing xylene to effect nearly quantitative dehydrogenation of 1,4-dihydroanthracene derivatives such as (11) obtained by a synthesis starting with (9), the adduct of benzyne with furane. Functioning as an

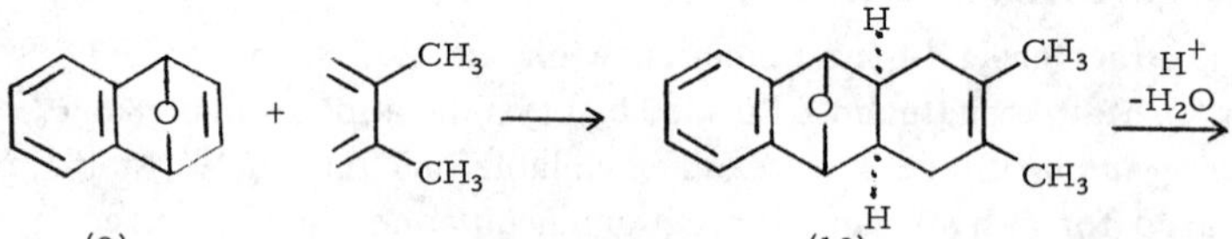
CH3
CH3
H
CH3
CH3
H
H+
-H2O

(9) (10)

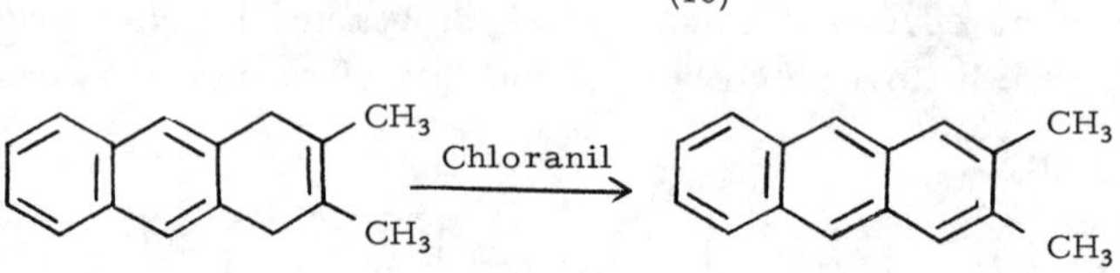
CH3
CH3
Chloranil
CH3
CH3

(11) (12)

active dienophile, (9) reacts with 2,3-dimethylbutadiene to give the adduct (10), which on acid-catalyzed dehydration yields the dihydroanthracene (11).

Many years after the classical investigations of the production of minute amounts of aromatic hydrocarbons by selenium dehydrogenation of steroids, Dannenberg[6] found (1956–63) that steroids can be dehydrogenated with chloranil in refluxing xylene or anisole to give optically active aromatized hydrocarbons in yields up to 50%. Cholesterol affords 3-methyl-3′-isooctyl-$\Delta^{1'}$-1,2-cyclopentadienophenanthrene (13) and its methyl derivative (14).

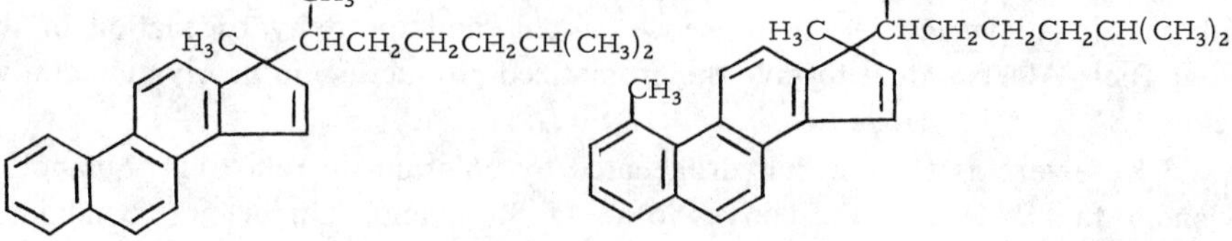
CH3
H3C
CHCH2CH2CH2CH(CH3)2
CH3
H3C
CHCH2CH2CH2CH(CH3)2
CH3

(13) (14)

Estrone methyl ether (15) when refluxed with chloranil in *t*-butanol for 40 hrs. under nitrogen gives the 9,11-dehydro derivative (16) in 50% yield; the dihydrophenanthrene derivative (17) is a by-product.[7]

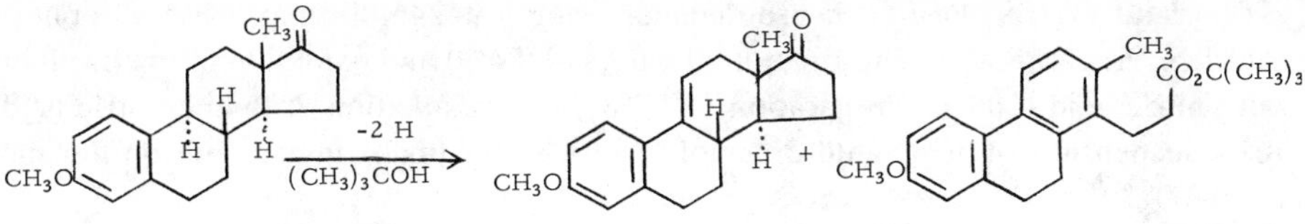

(15) (16) (17)

Veratrole is oxidized by chloranil in 70% (v/v) sulfuric acid at room temperature

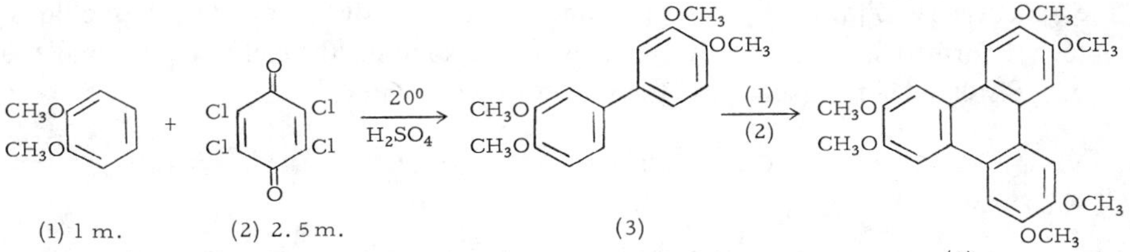

(1) 1 m. (2) 2.5 m. (3) (4)

(10 days) to the hexamethoxytriphenylene (4) in 73% yield.[8] 3,4,3′,4′-Tetramethoxydiphenyl (3) is an intermediate. *o*-Chloranil is almost as effective as chloranil, but bromanil and 2,3-dichloro-5,6-dicyano-1,4-benzoquinone gave somewhat lower yields.

Allylic oxidation. Warren and Weedon[9] used chloranil for allylic oxidation of the carotenoid diol (1). The yield was only 20%, but in this case MnO_2 failed

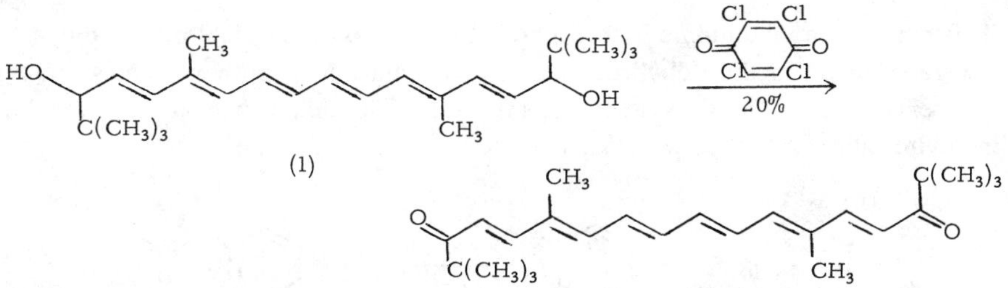

(1) (2)

completely and NBS in chloroform gave the diketone in only 9% yield. *o*-Chloranil destroyed the polyene system. Jensen[10] found that in allylic oxidation of carotenoid alcohols with chloranil yields can be increased by catalysis with iodine and irradiation with artificial Na-light (580–590 mμ), but an obvious disadvantage of this method is that iodine also catalyzes *cis-trans* isomerization of carotenoids.

[1]Review: L. M. Jackman, *Advances in Organic Chemistry, Methods and Results*, **2**, 329 (1960)
[2]E. Clar and M. Zander, *J. Chem. Soc.*, 4616 (1957)
[3]E. J. Agnello and G. D. Laubach, *Am. Soc.*, **82**, 4293 (1960)
[4]J. A. Campbell and J. C. Babcock, *ibid.*, **81**, 4069 (1959); R. M. Dodson and R. C. Tweit, *ibid.*, **81**, 1224 (1959); H. J. Ringold *et al.*, *ibid.*, **81**, 3712 (1959); A. Bowers, L. C. Ibáñez, and H. J. Ringold, *ibid.*, **81**, 5991 (1959)
[5]E. Wolthuis, *J. Org.*, **26**, 2215 (1961)
[6]H. Dannenberg and K.-F. Hebenbrock, *Ann.*, **662**, 21 (1963), and earlier papers cited
[7]A. D. Cross, H. Carpio, and P. Crabbé, *J. Chem. Soc.*, 5539 (1963)
[8]J. M. Matheson, O. C. Musgrave, and C. J. Webster, *Chem. Comm.*, 278 (1965)
[9]C. K. Warren and B. C. L. Weedon, *J. Chem. Soc.*, 3972 (1958)
[10]S. L. Jensen, *Chem. Scand.*, **19**, 1166 (1965)

_o_-Chloranil (Tetrachloro-1,2-benzoquinone). Mol. wt. 245.89, m.p. 133°. Supplier: Aldrich. Preparation by chlorination of catechol (84%) and oxidation of the product with nitric acid (66%).[1] Preparation in 57% yield by addition of fuming nitric acid to a suspension of pentachlorophenol (Aldrich; Eastman) in pentane or hexane at 30–60°.[2]

Dehydrogenation. Allylic, benzylic, and propargylic alcohols are dehydrogenated by *o*-chloranil in carbon tetrachloride at room temperature to the corresponding aldehydes or ketones.[3] The reagent is taken in slight excess and the course of the reaction is followed by disappearance of the yellow color. The tetrachlorocatechol formed is removed by passage of the solution through alumina and the product isolated in the usual way. Examples are formulated.

1. $CH_3CH{=}CHCH{=}CHCH_2OH \xrightarrow[39\%]{15\ min.} CH_3CH{=}CHCH{=}CHCHO$

2. $C_6H_5C{\equiv}C\overset{OH}{\overset{|}{C}}HCH{=}CHC_6H_5 \xrightarrow[91\%]{10\ min.} C_6H_5C{\equiv}C\overset{O}{\overset{\|}{C}}CH{=}CHC_6H_5$

3. $(C_6H_5)_2C{=}\overset{CH_3}{\overset{|}{C}}CH_2OH \xrightarrow[47\%]{4\ days} (C_6H_5)_2C{=}\overset{CH_3}{\overset{|}{C}}CHO$

4.

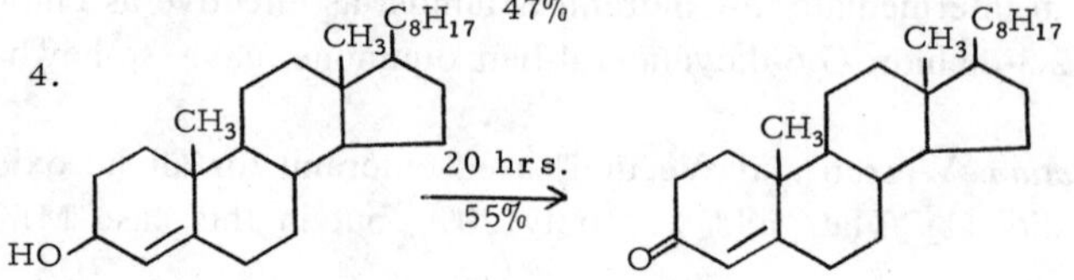

This high-potential quinone is second in effectiveness to 2,3-dichloro-5,6-dicyano-1,4-benzoquinone in the dehydrogenation of tetralin, acenaphthene, and dibenzyl in benzene at 80°.[4] Kinetic studies are reported.[5] The dehydrogenation of the 1,1-dimethyltetralin is attended with Wagner-Meerwein rearrangement.[6]

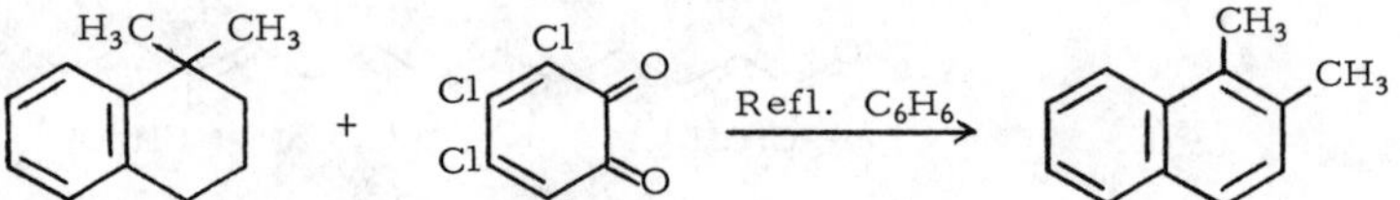

Reid[7] used *o*-chloranil, in combination with perchloric acid in acetic acid, to produce aromatic cationoid salts:

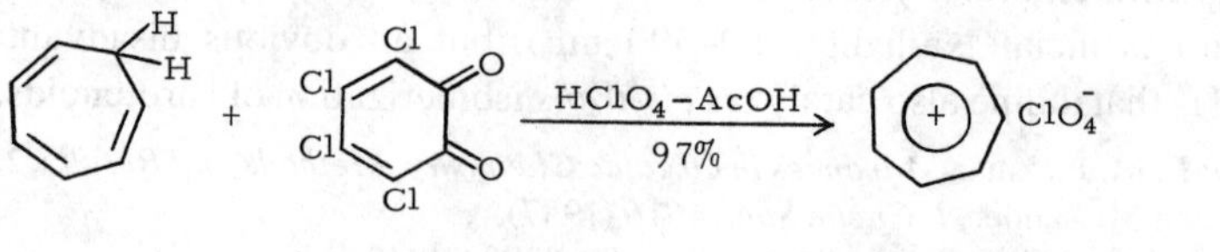

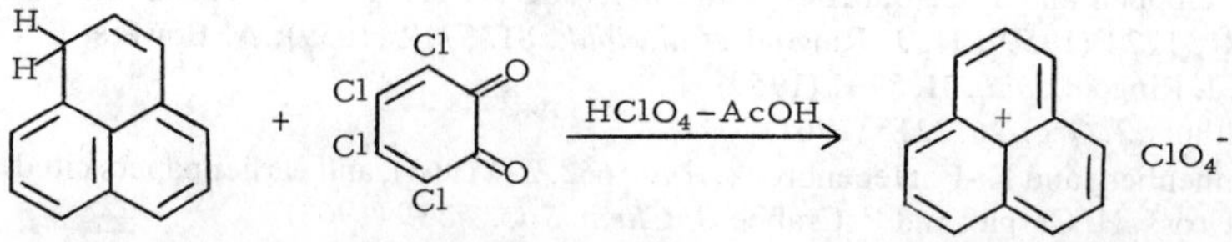

Even though the oxidation-reduction potential of *o*-chloranil (0.87 v.) is less than that of 2,3-dichloro-5,6-dicyanoquinone (1.0 v.), Reid prefers *o*-chloranil because it is more soluble in acetic acid or acetonitrile and because the corresponding hydroquinone is very soluble in ether and hence readily removed.

Oxidation. 1,2,5,6-Tetrahydroxynaphthalenes are oxidized to the amphiquinones in high yield.[8]

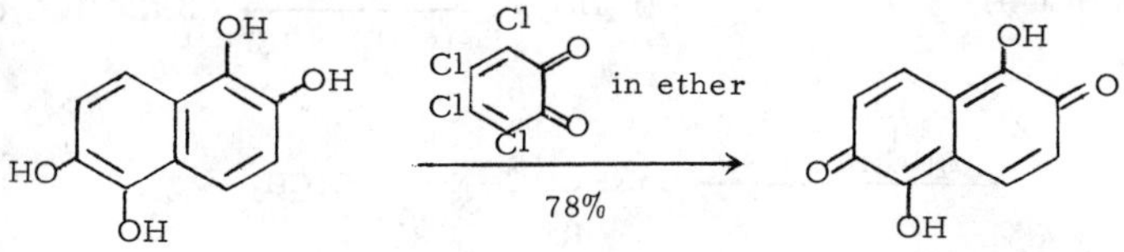
OH
OH
HO
OH
Cl
Cl
Cl
Cl
in ether
78%
OH
O
O
OH

The reagent oxidizes many catechol derivatives at room temperature to the corresponding *o*-quinones in good yield.[8] An example is the oxidation of 1′,2′-dihydroxybenzotropolone (1), which is available by a remarkably simple method:[9]

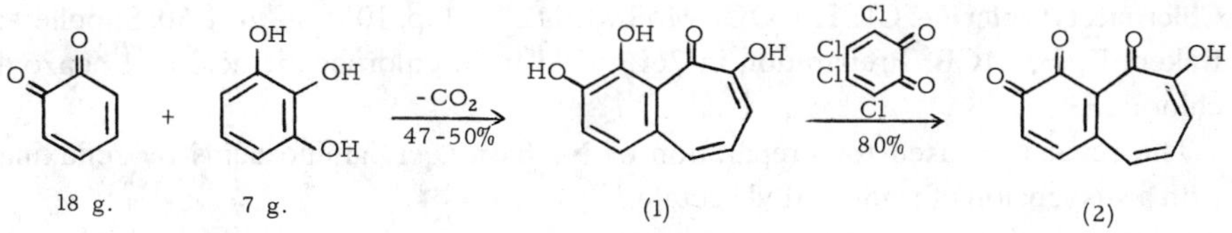
O
O
OH
OH
OH
+
-CO₂
47-50%
OH
O
OH
HO
Cl
Cl
Cl
Cl
80%
O
O
O
OH
18 g.
7 g.
(1)
(2)

a suspension of *o*-benzoquinone in 200 ml. of water precooled to 0° is stirred at 0–5° during addition of a solution of pyrogallol in 70 ml. of water (15 min.); carbon dioxide is evolved and after 2 hrs. excess *o*-benzoquinone is reduced with hydrosulfite and the product (1) is obtained as shiny brown needles, m.p. 186–187°. For oxidation to benzotropolone-1,2-quinone (2) a solution of 408 mg. of (1) in 120 ml. of ether is poured into a solution of 500 mg. of tetrachloro-*o*-benzoquinone in 50 ml. of ether; the solution darkens and the quinone (2) is obtained as dark brown plates with a greenish shimmer.

[1]C. L. Jackson and R. D. MacLaurin, *Am. Chem. J.*, **37**, 7 (1907); C. L. Jackson and P. W. Carleton, *ibid.*, **39**, 493 (1908); A. G. Brook, *J. Chem. Soc.*, 5040 (1952)
[2]W.-H. Chang, *J. Org.*, **27**, 2921 (1962)
[3]E. A. Braude, R. P. Linstead, and K. R. Wooldridge, *J. Chem. Soc.*, 3070 (1956)
[4]E. A. Braude, A. G. Brook, and R. P. Linstead, *J. Chem. Soc.*, 3569 (1954)
[5]E. A. Braude, L. M. Jackman, R. P. Linstead, and J. S. Shannon, *J. Chem. Soc.*, 3116 (1960)
[6]E. A. Braude, L. M. Jackman, R. P. Linstead, and G. Lowe, *J. Chem. Soc.*, 3133 (1960)
[7]D. H. Reid, M. Fraser, B. B. Molloy, H. A. S. Payne, and R. G. Sutherland, *Tetrahedron Letters*, 530 (1961); D. H. Reid and R. G. Sutherland, *J. Chem. Soc.*, 3295 (1963)
[8]L. Horner and W. Dürckheimer, *Z. Naturforschung*, **14b**, 741 (1959); L. Horner and K.-H. Weber, *Ber.*, **98**, 1246 (1965)
[9]L. Horner and K.-H. Weber, *Ber.*, **95**, 1227 (1962)

Chloroacetic anhydride, $(ClCH_2CO)_2O$. Mol. wt. 170.99, m.p. 46°. Supplier: Eastman.

Preferred to chloroacetyl chloride for N-acylation of amino acids in alkaline solution.[1]

[1]S. M. Birnbaum, L. Levintow, R. B. Kingsley, and J. P. Greenstein, *J. Biol. Chem.*, **194**, 455 (1952)

Chloroacetonitrile, $ClCH_2CN$. Mol. wt. 75.50, b.p. 124–125°. Suppliers: A, B, E, F, MCB.

Peptide synthesis. Reaction of an N-protected amino acid with chloroacetonitrile and triethylamine in acetic acid gives an activated cyanomethyl ester capable of reacting with an amino acid ester.[1,2] Yields are improved (75–95%) by use of excess chloroacetonitrile as solvent.

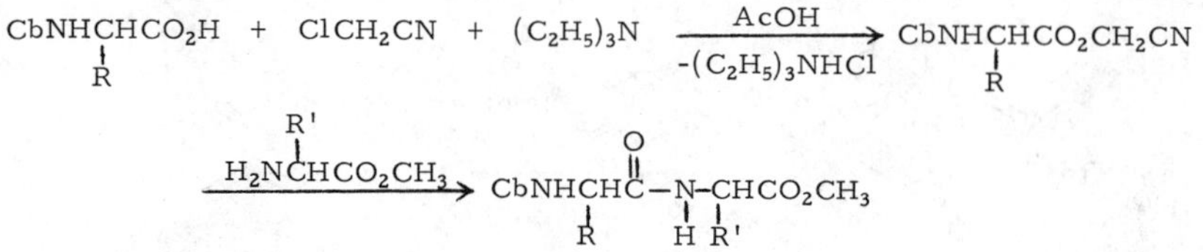
CbNHCHCO₂H + ClCH₂CN + (C₂H₅)₃N
AcOH
-(C₂H₅)₃NHCl
CbNHCHCO₂CH₂CN
R
R'
H₂NCHCO₂CH₃
CbNHCHC—N—CHCO₂CH₃
H R'

[1]R. Schwyzer, B. Iselin, and M. Feurer, *Helv.*, **38**, 69 (1955)
[2]R. Schwyzer, M. Feurer, B. Iselin, and H. Kägi, *Helv.*, **38**, 80 (1955); R. Schwyzer, M. Feurer, and B. Iselin, *Helv.*, **38**, 83 (1955); B. Iselin, M. Feurer, and R. Schwyzer, *Helv.*, **38**, 1508 (1955); R. Schwyzer, B. Iselin, W. Rittel, and P. Sieber, *Helv.*, **39**, 872 (1956)

Chloroacetyl chloride, $ClCH_2COCl$. Mol. wt. 112.95, b.p. 105°, sp. gr. 1.50. Suppliers: Baker, E, F, MCB. Preparation in 76% yield from chloroacetic acid and benzoyl chloride.[1]

The reagent is used for preparation of N-chloroacetylamino acids by refluxing with a suspension of acid in ethyl acetate.[2]

[1]H. C. Brown, *Am. Soc.*, **60**, 1325 (1938)
[2]E. Ronwin, *J. Org.*, **18**, 127 (1953)

Chloroacetylhydrazide hydrochloride, $ClCH_2CONHNH_2 \cdot HCl$. Mol. wt. 145.00, m.p. 150° dec. The reagent is prepared by reaction of the activated *p*-nitrophenyl ester (1) with hydrazine.[1] It is used for the generation of diimide, *which see.*

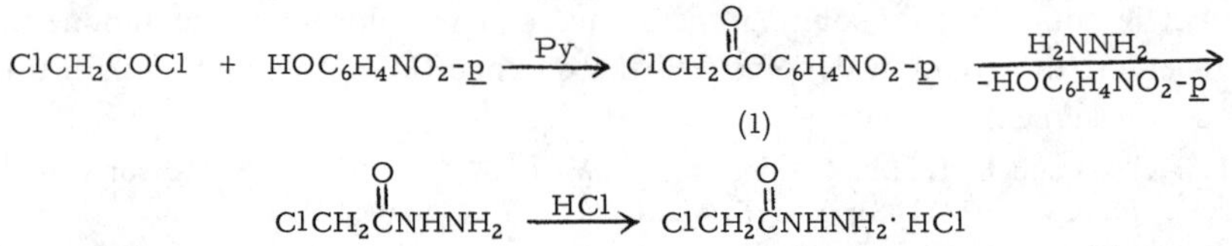
ClCH₂COCl + HOC₆H₄NO₂-p
Py
ClCH₂COC₆H₄NO₂-p
(1)
H₂NNH₂
-HOC₆H₄NO₂-p
ClCH₂CNHNH₂
HCl
ClCH₂CNHNH₂· HCl

[1]R. Buyle, *Helv.*, **47**, 2449 (1964)

Chlorodiazomethane, $ClCHN_2$. Mol. wt. 76.50.

Prepared by reaction of diazomethane and *t*-butyl hypochlorite, the reagent decomposes to give chlorocarbene:[1]

$$ClCH{=}\overset{+}{N}{=}\overset{-}{N} \longrightarrow :CHCl + N_2$$

[1]G. L. Closs and J. J. Coyle, *Am. Soc.*, **84**, 4350 (1962)

Chloroform, $CHCl_3$. Mol. wt. 119.39, b.p. 61.2°. Commercial material is stabilized against conversion to phosgene by 0.7% ethanol.

Chloroform is useful as solvent for Friedel-Crafts reactions not only because it dissolves the aluminum chloride–acetyl chloride complex but also because it has an orienting effect. Thus the reaction of naphthalene with 2 equivalents each of

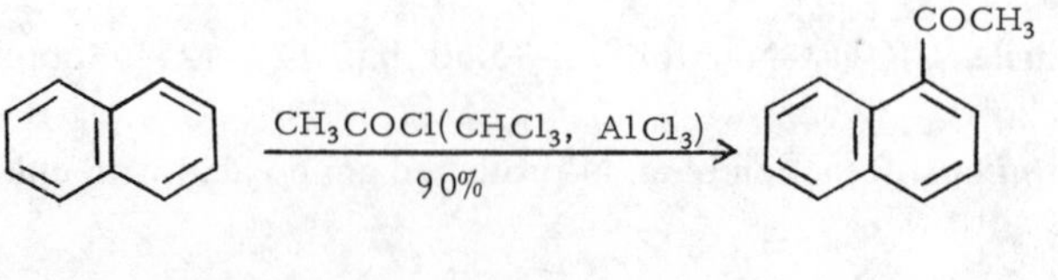
COCH₃
CH₃COCl(CHCl₃, AlCl₃)
90%

acetyl chloride and aluminum chloride in chloroform gives 1-acetylnaphthalene of 99% purity in 90% yield.[1] 9-Acetylanthracene can be obtained in 73% yield with chloroform as solvent; with 3 moles each of acetyl chloride and aluminum chloride, the chief product is 1,5-diacetylanthracene (81.5% yield).[2]

[1]H. F. Bassilios, S. M. Makar, and A. Y. Salem, *Bull. soc.*, **21**, 72 (1954)
[2]H. F. Bassilios, M. Shawky, and A. Y. Salem, *Rec. trav.*, **81**, 679 (1962)

Chloroform and an organic peroxide. Dowbenko[1] has described an interesting radical-induced cycloaddition of chloroform to *cis,cis*-1,5-cyclooctadiene (1) to produce a 2-(trichloromethyl)bicyclo[3.3.0]octane (2) and hydrolysis of the latter to *exo-cis*-bicyclo[3.3.0]octane-2-carboxylic acid (3). In the procedure cited the oxidant is

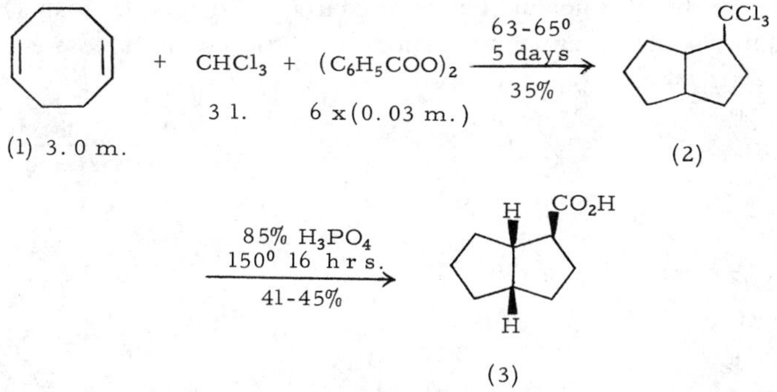
+ CHCl3 + (C6H5COO)2
63-65°
5 days
35%
CCl3
3 l.
6 x(0.03 m.)
(1) 3.0 m.
(2)
85% H3PO4
150° 16 hrs.
41-45%
H
CO2H
H
(3)

dibenzoyl peroxide, added in six portions at the reflux temperature; di-*t*-butyl peroxide also can be used.[2] Similar cycloadditions to *cis,cis*-1,5-cyclooctadiene can be effected in the presence of dibenzoyl peroxide, or di-*t*-butyl peroxide and *t*-butylformamide.[3]

[1]R. Dowbenko, Pittsburgh Plate Glass Co., procedure submitted to *Org. Syn.*
[2]R. Dowbenko, *Am. Soc.*, **86**, 946 (1964); *Tetrahedron*, **20**, 1843 (1964); see also L. Friedman and H. Shechter, *Am. Soc.*, **83**, 3159 (1961)
[3]L. Friedman and H. Shechter, *Tetrahedron Letters*, 238 (1961)

Chloroiridic acid, probably $H_2IrCl_6 \cdot 6\,H_2O$. Supplier: Johnson, Matthey, Ltd.

Henbest *et al.*[1] used this reagent as catalyst in combination with aqueous isopropanol and trimethyl phosphite for the reduction of ketones; the reaction mixture is refluxed for 1–5 days. The main reducing agent is the trivalent phosphorus species; a little of the isopropanol is converted into acetone. The trimethyl phosphite

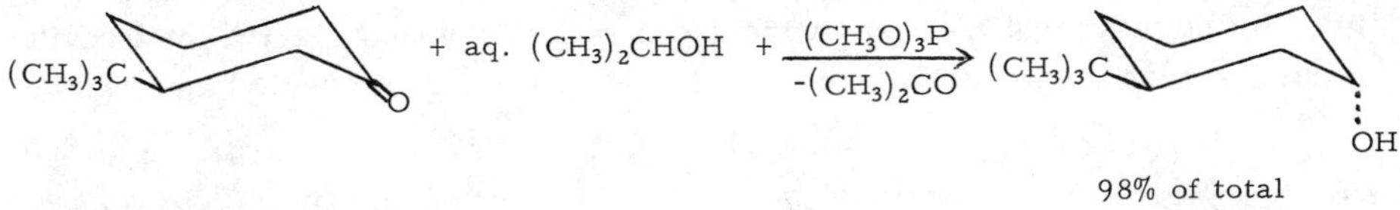
(CH3)3C
+ aq. (CH3)2CHOH +
(CH3O)3P
-(CH3)2CO
(CH3)3C
OH
98% of total

is extensively or wholly hydrolyzed. Tervalent iridium species with one or more phosphorus-containing groups as ligands may be present. The special feature of the method is that it affords an exceptionally high proportion of axial alcohols; most methods for the reduction of unhindered cyclohexanones give equatorial alcohols as the major products. On reduction by the new method 3-*t*-butylcyclohexanone, 3,3,5-trimethylcyclohexanone, and cholestanone afford axial alcohols in yields of 98, 99, and 92%, respectively. Two hundred moles of ketone have been reduced per iridium atom.

The following additional information was kindly supplied by Dr. Henbest: "From our more recent work it appears that any soluble chloro-derivative of tri- or tetravalent iridium will work provided that sufficient phosphorous acid (or alkyl phosphite and water to hydrolyze it) is present to neutralize the salt and effect the reduction. The phosphorous acid acts as the reducing agent probably via an iridium species containing one or more phosphite groups as ligand(s). Soluble iridium compounds include K_2IrCl_6, $(NH_4)_2IrCl_6$, K_3IrCl_6 and iridium trichloride. This last compound is available in the U. S. A. in a soluble form (J. Bishop and Co.); the insoluble form is ineffective."

For the reduction of 4-*t*-butylcyclohexanone (Dow Chem. Co.) to the axial *cis*-alcohol (99% pure), Eliel and Doyle[2] used iridium trichloride (Fisher) and noted that this catalyst could be reused many times with little loss in stereoselectivity.

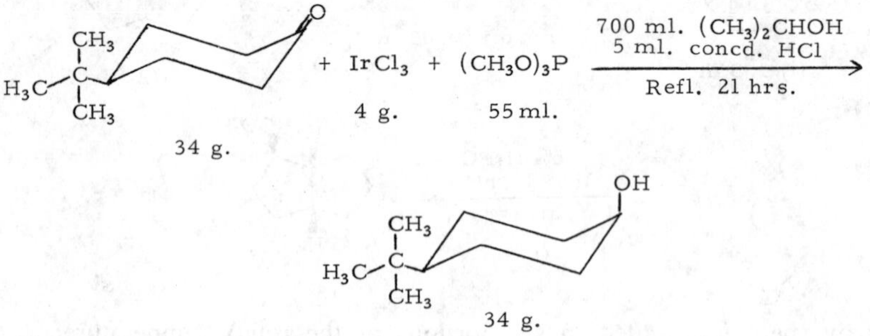
CH3
H3C
CH3
+ IrCl3 + (CH3O)3P
4 g.
55 ml.
700 ml. (CH3)2CHOH
5 ml. concd. HCl
Refl. 21 hrs.
34 g.
OH
CH3
H3C
CH3
34 g.

M. p. 75-77° (pure: 82-83.5°)

[1]Y. M. Y. Haddad, H. B. Henbest, Mrs. J. Husbands, and T. R. B. Mitchell, *Proc. Chem. Soc.*, 361 (1964)

[2]E. L. Eliel and T. W. Doyle, procedure submitted to *Org. Syn.*

Chloromethyl methyl ether, $ClCH_2OCH_3$, Mol. wt. 80.51, b.p. 58–60°. Suppliers: B, E, MCB. The reagent is prepared by passing hydrogen chloride into a mixture of formalin and methanol until saturated, saturating the water layer with calcium

$$CH_2{=}O + CH_3OH + HCl \xrightarrow[86\text{-}89\%]{} ClCH_2OCH_3 + H_2O$$

chloride to salt out the product, and drying and distilling the organic layer.[1]

For conversion to CH_3OCH_2MgCl and CH_3OCH_2Li, *see* Methylal.

Chloromethylation. Having attempted to chloromethylate 5-nitrosalicylaldehyde by conventional methods without success, Taylor and Davis[2] obtained the desired 3-chloromethyl derivative in high yield by using chloromethyl methyl ether

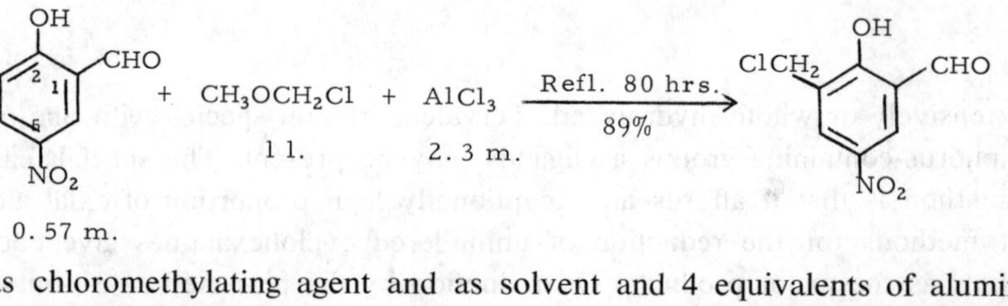
OH
CHO
NO2
0.57 m.
+ CH3OCH2Cl + AlCl3
1 l.
2.3 m.
Refl. 80 hrs.
89%
OH
ClCH2
CHO
NO2

both as chloromethylating agent and as solvent and 4 equivalents of aluminum chloride (one for each oxygen atom) plus 10% catalytic excess. The aluminum chloride was added with stirring at 5° over a 1-hr. period. The slurry was let come to room temperature and then refluxed until evolution of HCl ceased (80 hrs.).

Methoxymethyl ethers of phenols. These ethers are readily prepared by reaction of chloromethyl methyl ether with a suspension of the dry sodium salt of a phenol in benzene or toluene. They are useful as protective groups because they are stable to alkali, to potassium cyanide, to Grignard reagents, and to *n*-butyllithium, but can be hydrolyzed when desired by very gentle treatment with acid.[3] They are cleaved more easily than the corresponding benzyl ethers. Examples of synthetic uses are as follows.

For the synthesis of 2,2′-dihydroxybenzoin (4), LaForge[4] stirred a suspension of the yellow sodium salt of salicylaldehyde in toluene containing chloromethyl

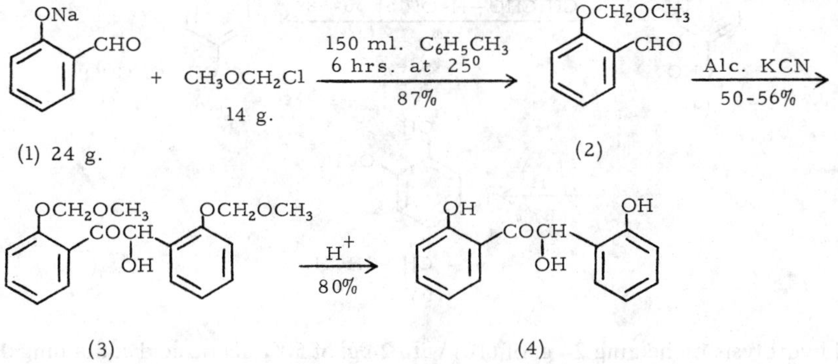

methyl ether at room temperature until the yellow material had completely disappeared. Treatment of (2) with potassium cyanide in ethanol gave the benzoin (3), which was cleaved to (4) by boiling a solution of 15 g. in 100 ml. of 50% acetic acid containing 0.4 g. of concd. sulfuric acid for 12–15 min.

Interested in the production of phenolic polystyrenes, Stern, English, and Cassidy[5] prepared the monomer 2-vinylhydroquinone (8) as follows. Hydroquinone bis-methoxymethyl ether (5) was converted by reaction of the lithium derivative with

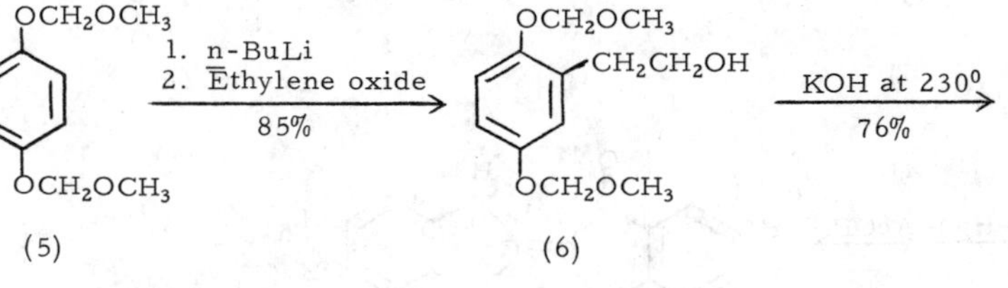

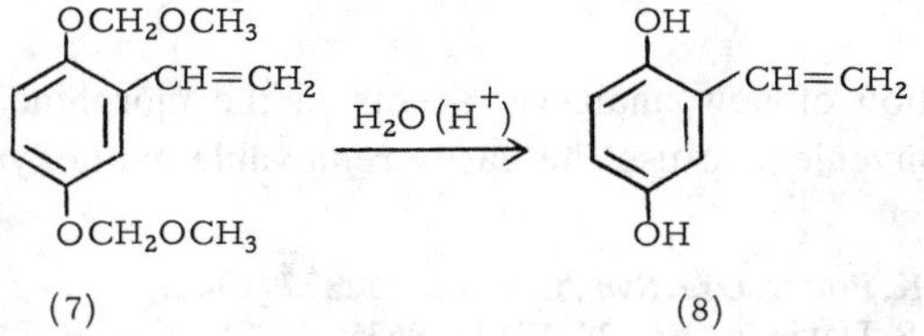

ethylene oxide into the β-ethanol derivative (6). Dehydration of (6) to the vinyl derivative (7) was done by an interesting procedure. A flask containing 70 g. of KOH pellets and a trace of picric acid as polymerization inhibitor was heated at 230° at 0.5 mm. pressure until vapors subsided and the potassium hydroxide was nearly all solid. Then 0.15 mole of the alcohol (6) was added dropwise in 45 min. The vinyl derivative (7) distilled as formed and redistillation gave pure (7) in high yield. Elimination of the protective methoxymethyl groups was accomplished by refluxing for 1 hr. under nitrogen a solution of 0.15 mole of (7) in 20 ml. of methanol containing 1 small drop of concd. hydrochloric acid; the yield is not clear.

Pauly[6] synthesized coniferyl aldehyde (11) by condensing (9), the methoxymethyl ether of vanillin, with acetaldehyde under carefully defined conditions of alkalinity,

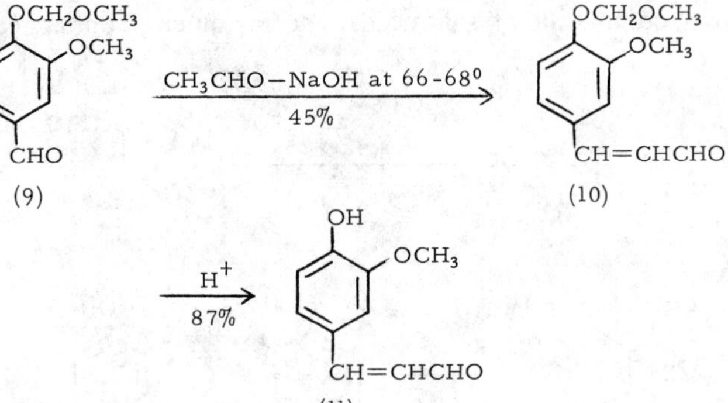
OCH2OCH3
OCH3
CHO
CH3CHO–NaOH at 66-68°
45%
OCH2OCH3
OCH3
CH=CHCHO
OH
OCH3
H+
87%
CH=CHCHO

(9) (10) (11)

and hydrolysis by heating 24 g. of (10) with 24 g. of 50% acetic acid containing 0.3 g. of concd. sulfuric acid.

A synthesis of 5,5′,6,6′-tetrahydroxyindigo (14) by Harley-Mason[7] involves condensation of the O-protected *o*-nitroaldehyde (12) with acetone and alkali and acid hydrolysis of the product (13).

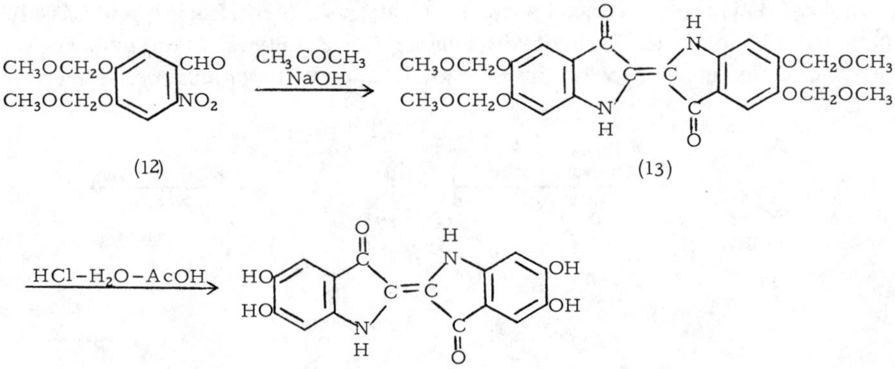
CH3OCH2O
CH3OCH2O
CHO
NO2
CH3COCH3
NaOH
OCH2OCH3
OCH2OCH3
HCl–H2O–AcOH
HO
HO
OH
OH

(12) (13)

In an investigation of new analgesic agents in the morphine series, Rapaport *et al.*[8] found it convenient to use the easily removable methoxymethyl ether as a blocking group.

[1]C. S. Marvel and P. K. Porter, *Org. Syn., Coll. Vol.*, **1**, 377 (1941)
[2]L. D. Taylor and R. B. Davis, *J. Org.*, **28**, 1713 (1963)
[3]See review: J. F. W. McOmie, *Advances in Organic Chemistry*, **3**, 232–233 (1963).
[4]F. B. LaForge, *Am. Soc.*, **55**, 3040 (1933)
[5]R. Stern, J. English, Jr., and H. G. Cassidy, *Am. Soc.*, **79**, 5792 (1957)
[6]H. Pauly and K. Wäscher, *Ber.*, **56**, 603 (1923); H. Pauly and K. Feuerstein, *Ber.*, **62**, 297 (1929)

[7]J. Harley-Mason, *J. Chem. Soc.*, 1244 (1948)
[8]M. A. Abdel-Rahman, H. W. Elliott, R. Binks, W. Küng, and H. Rapaport, *J. Med. Chem.*, **9**, 1 (1966)

Chloromethylphosphoric dichloride, $ClCH_2POCl_2$. Mol. wt. 167.38, b.p. 79–81°/10 mm. Preparation from phosphorus trichloride and paraformaldehyde at 250°.[1] Supplier: Stauffer Chemical Co.

For the preparation of chloromethyl phosphonothioic dichloride, a mixture of the reagent with tetraphosphorus decasulfide (Stauffer) is stirred under nitrogen at 180–190° for 6 hrs.[2]

$$10\,ClCH_2\overset{O}{\overset{\|}{P}}Cl_2 + P_4S_{10} \xrightarrow[66\text{–}72\%]{180\text{–}190°} 10\,ClCH_2\overset{S}{\overset{\|}{P}}Cl_2 + P_4O_{10}$$

[1]R. A. B. Bannard, J. R. Gilpin, G. R. Vavasour, and A. F. McKay, *Can. J. Chem.*, **31**, 976 (1953)
[2]R. Schmutzler, *Org. Syn.*, **46**, 21 (1966)

N-Chloromethylphthalimide, 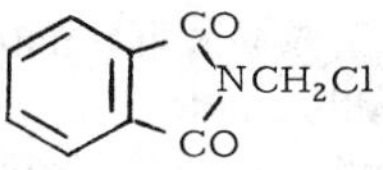

Mol. wt. 195.61, m.p. 132°. Suppliers: British Drug Houses, Frinton. Preparation.[1]

In a solution in ethyl acetate containing 1 equivalent of triethylamine, the reagent reacts with a carboxylic acid or with the carboxyl group of an N-protected amino

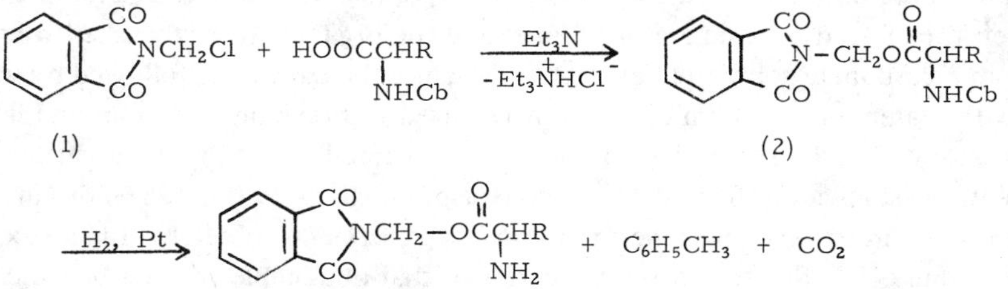

(3)

acid or peptide to form a phthalimidomethyl ester; the solution is filtered from triethylamine hydrochloride and evaporated (yields 70–80%). Since the ester linkage established is stable to hydrogenation, an N-protective carbobenzoxy group can be eliminated if desired to give (3). The carboxy-protective group can be eliminated by treatment with hydrogen chloride in dry dioxane or ethyl acetate (16 hrs.), with hydrogen bromide in acetic acid (10–15 min.), or with hydrazine hydrate in ethanol (3 hrs.).

[1]G. H. L. Nefkens, *Nature*, **193**, 974 (1962)

***m*-Chloroperbenzoic acid**, m-$ClC_6H_4CO_3H$. Mol. wt. 172.57, m.p. 92°. Supplier of 85% pure acid: FMC Corp. (technical bulletin available). The contaminant is *m*-chlorobenzoic acid, which is more strongly acidic than the peracid. Thus peracid of purity of 99+% by iodometric assay can be prepared by washing the 85% material with a phosphate buffer of pH 7.5 and drying the residue at reduced pressure.[1] This white solid peracid is exceptionally stable: less than 1% decomposition per year at room temperature.

Solubility (pure acid, g./100 g. at 25°)

Water	0.154	Methylene chloride	11.2
Hexane	1.4	Ethyl acetate	51.0
Carbon tetrachloride	2.1	*t*-Butanol	69.0
Benzene	8.0	Diethyl ether	89.4
Chloroform	9.8	Ethanol	113.0

Epoxidation. A procedure[2] for the preparation of cholesterol α-epoxide is as follows. A 500-ml. flask equipped with a stirrer, thermometer, condenser (for vent-

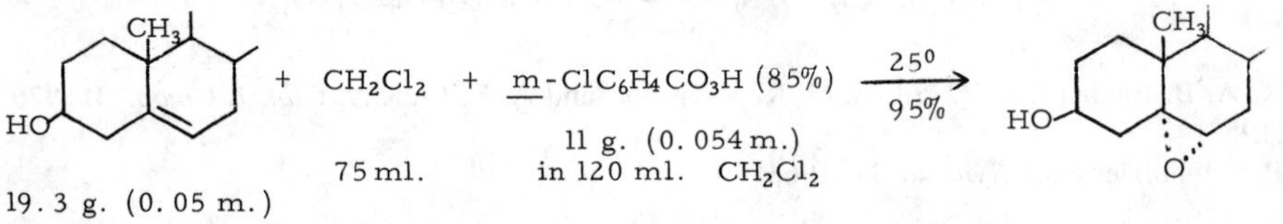

CH3
HO
+ CH2Cl2 + m-ClC6H4CO3H (85%)
25°
95%
11 g. (0.054 m.)
75 ml.
in 120 ml. CH2Cl2
19.3 g. (0.05 m.)

ing oxygen or for emergency cooling), and graduated dropping tube is charged with cholesterol (purified through the dibromide) and methylene chloride. The stirrer is started and the temperature is kept at 25° by cooling as required during addition in 10 min. of a solution of *m*-chloroperbenzoic acid in methylene chloride. Agitation is continued for 20–30 min., when titration of an aliquot should indicate consumption of 1 equivalent of the reagent. Excess peracid is then destroyed by addition of 10% sodium sulfite until a test with starch-iodide paper is negative. The reaction mixture is then transferred to a separatory funnel and the organic layer is washed with 5% sodium bicarbonate solution to extract the *m*-chlorobenzoic acid, followed by washing with water (plus sodium chloride, if required to break an emulsion) and finally with saturated sodium chloride solution. The organic layer is dried and stripped of solvent; crystallization from 88% aqueous acetone gives 19.0 g. (95%) of pure 3β-hydroxy-5α,6α-epoxycholestane, m.p. 141–143°, αD −46° ($CHCl_3$). In the experience of one of us, this result is far superior to that obtainable with perbenzoic acid or perphthalic acid, both of which afford mixtures containing substantial amounts of the β-oxide. *m*-Chloroperbenzoic acid thus appears to be of superior stereoselectivity.

Various attempts to oxidize 7,8- and 8,9-unsaturated tetracyclic triterpenes such as (1) to the corresponding epoxides by reaction with perbenzoic acid afforded mixtures of epoxide with the 7,9(11)-diene. Fried *et al.*[3] found that the difficulty is eliminated by use of *m*-chloroperbenzoic acid in chloroform, which gave exclusively

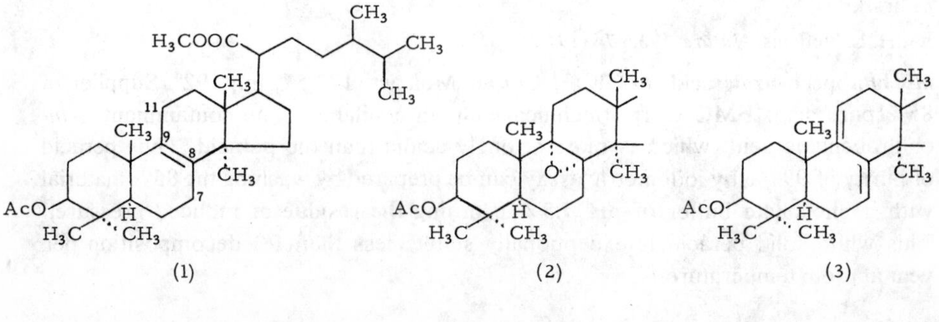

H3COOC
CH3
AcO
H3C
H
11
9
8
7

(1) (2) (3)

the desired epoxide. Thus when the product from (1) was crystallized from methanol containing a few drops of pyridine the $8\alpha,9\alpha$-epoxide was obtained in 95% yield.

Other satisfactory results obtained with the reagent are indicated as follows:[2]

$$n\text{-}C_6H_{13}CH=CH_2 \ + \ m\text{-}ClC_6H_4CO_3H \ \xrightarrow[\text{81\%}]{\text{In } C_6H_6, \ 5 \text{ hrs. at } 25^0} \ n\text{-}C_6H_{13}CH\text{---}CH_2$$

0.1 m.   0.113 m.

$$(CH_3)_2C=CHCOCH_3 \ + \ m\text{-}ClC_6H_4CO_3H \ \xrightarrow[\text{77\%}]{\text{In } CH_2Cl_2, \ 3 \text{ hrs. at } 25^0} \ (CH_3)_2C\text{---}CHCOCH_3$$

0.1 m.   0.11 m.

$$CH_3CH=CHCO_2C_2H_5 \ + \ m\text{-}ClC_6H_4CO_3H \ \xrightarrow[\text{70\%}]{\substack{\text{In } CH_2ClCH_2Cl \\ \text{refl. } 3 \text{ hrs.}}} \ CH_3CH\text{---}CHCO_2C_2H_5$$

0.2 m.   0.25 m.

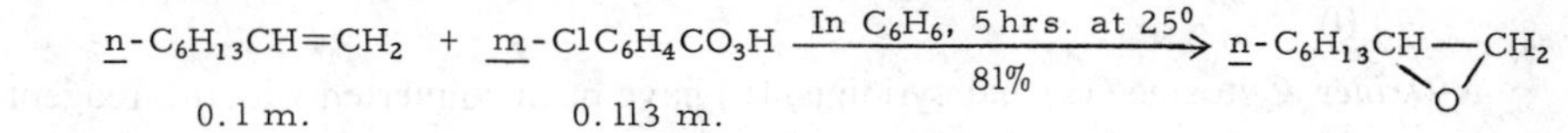

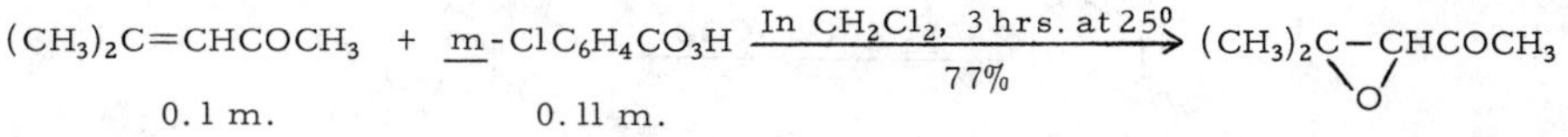

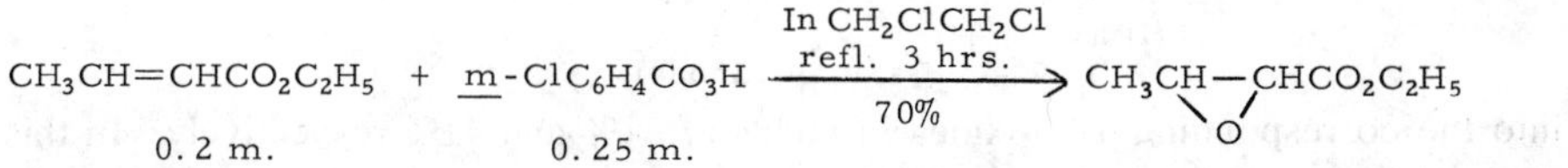

Pasto and Cumbo[4] found the reagent useful for the preparation of volatile epoxides, for example of *cis-* and *trans*-2-butene. The hydrocarbon is distilled into a solution of the peracid in dioxane at 0° and the mixture is stirred under a dry ice–acetone condenser until the hydrocarbon ceases to reflux (about 10 hrs.). Fractionation then affords the epoxide in yield of 52–60%.

Kinetic studies of the epoxidation of ethyl crotonate and of *trans*-stilbene have been reported.[1]

In one step of a synthesis of 2,7-dimethyloxepin (6), the reagent proved superior to other peracids for the selective epoxidation of the more highly substituted double bond of the diene (2).[5] When perbenzoic acid was used the yield of (3) was only 61%.[6]

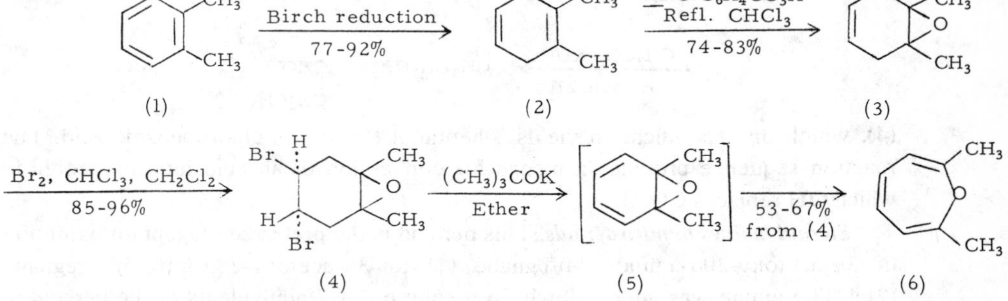

Baeyer-Villiger Oxidation. An example is the oxidation of a 20-ketosteroid and hydrolysis of the product to produce the corresponding 17β-hydroxysteroid.[2]

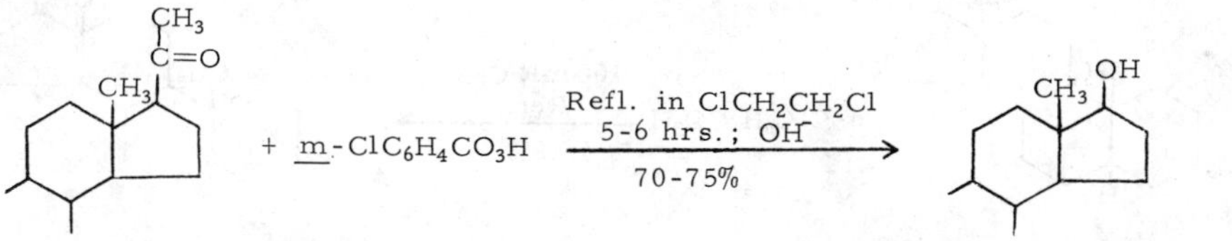

Cleavage of the tetrahydrochromane (2).[7] Reduction of chromane (1) with a six-fold excess of lithium in ethylamine affords tetrahydrochromane (2) in moderate yield. This substance is cleaved by *m*-chloroperbenzoic acid, if in low yield, to the middle-size cyclic ketolactone (3), of interest because of its relationship to macrolide antibiotic lactones.

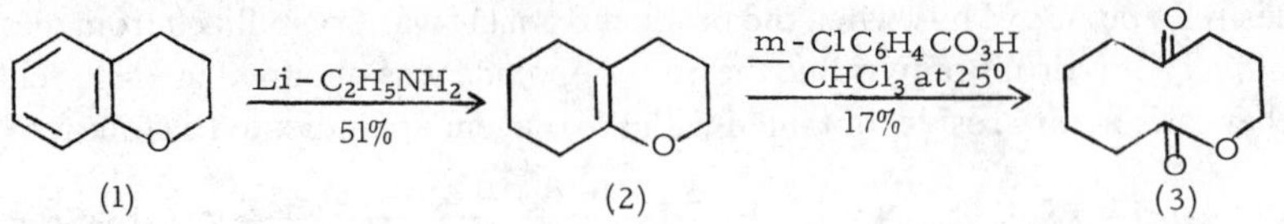

$$\text{(1)} \xrightarrow[51\%]{\text{Li}-\text{C}_2\text{H}_5\text{NH}_2} \text{(2)} \xrightarrow[17\%]{\underline{m}-\text{ClC}_6\text{H}_4\text{CO}_3\text{H} \atop \text{CHCl}_3 \text{ at } 25^0} \text{(3)}$$

N-Oxides. Cytosine (1a) and cytidine (1b) have been converted with this reagent

$$\xrightarrow{\underline{m}-\text{ClC}_6\text{H}_4\text{CO}_3\text{H}}$$

(1) a, R = H
    b, R = β-$\underline{\text{D}}$-ribofuranosyl

(2)

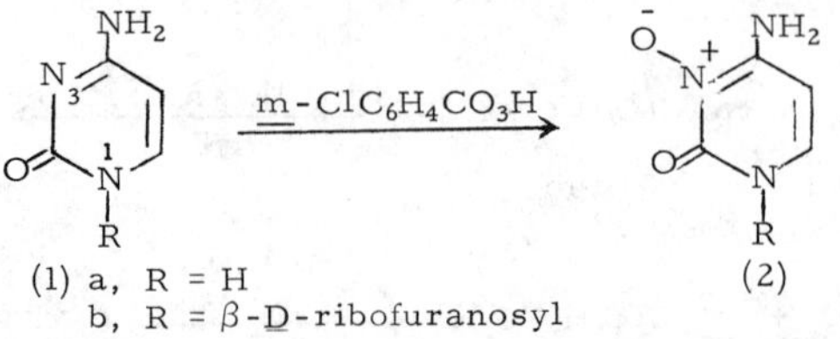

into the corresponding 3-N-oxides in yields of 21% and 41% respectively.[8] In this case neither perphthalic acid nor pertrifluoroacetic acid proved satisfactory.

Carboxy-inversion reaction. Denney and Sherman[9] found that *m*-chloroperbenzoic acid (2) reacts readily with a carboxylic acid chloride (1) in hexane solution at 0° in the presence of pyridine to form a mixed peroxide (3). When the mixture is stirred for several hours at room temperature the peroxide rearranges to the mixed carbonate

$$\underset{\underset{\text{CH}_2\text{CH}_3}{|}}{\text{CH}_3\text{CH}_2\text{CH}_2\text{CH}_2\text{CH}}\overset{\overset{\text{O}}{\|}}{\text{C}}\text{Cl} \;+\; \text{HO}_2\text{CC}_6\text{H}_4\text{Cl}-\underline{m} \;+\; \text{Py} \xrightarrow{\text{Hexane } 0^0}$$

(1)   (2)

$$\underset{\underset{\text{CH}_2\text{CH}_3}{|}}{\text{CH}_3\text{CH}_2\text{CH}_2\text{CH}_2\text{CH}}\overset{\overset{\text{O}}{\|}}{\text{C}}-\text{O}_2-\overset{\overset{\text{O}}{\|}}{\text{C}}\text{C}_6\text{H}_4\text{Cl}-\text{m} \xrightarrow{25^0} \underset{\underset{\text{CH}_2\text{CH}_3}{|}}{\text{CH}_3\text{CH}_2\text{CH}_2\text{CH}_2\text{CH}}-\text{O}\overset{\overset{\text{O}}{\|}}{\text{C}}-\text{O}-\overset{\overset{\text{O}}{\|}}{\text{C}}\text{C}_6\text{H}_4\text{Cl}-\underline{m}$$

(3)   (4)

$$\xrightarrow[73\% \text{ overall}]{\text{CH}_3\text{OH}-\text{KOH}} \underset{\underset{\text{CH}_2\text{CH}_3}{|}}{\text{CH}_3\text{CH}_2\text{CH}_2\text{CH}_2\text{CHOH}} \quad (5)$$

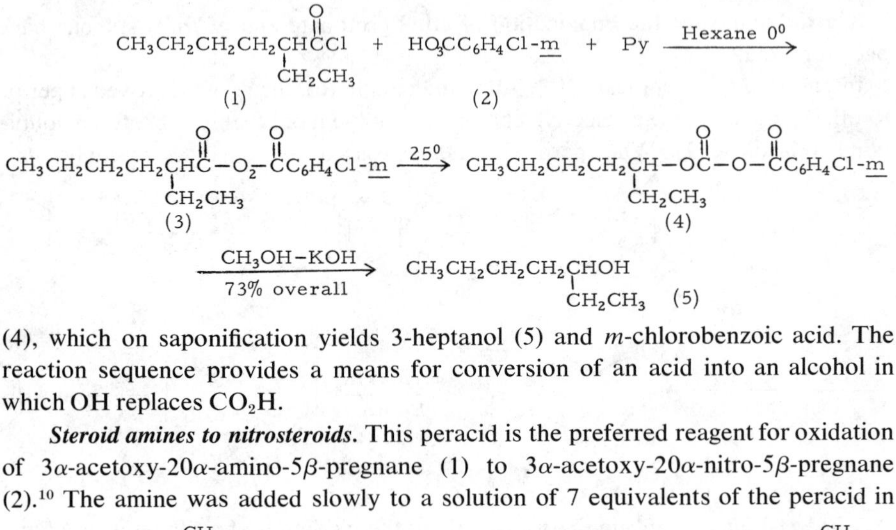

(4), which on saponification yields 3-heptanol (5) and *m*-chlorobenzoic acid. The reaction sequence provides a means for conversion of an acid into an alcohol in which OH replaces CO_2H.

Steroid amines to nitrosteroids. This peracid is the preferred reagent for oxidation of 3α-acetoxy-20α-amino-5β-pregnane (1) to 3α-acetoxy-20α-nitro-5β-pregnane (2).[10] The amine was added slowly to a solution of 7 equivalents of the peracid in

$$+ \; \underline{n}-\text{ClC}_6\text{H}_4\text{CO}_3\text{H} \xrightarrow[58\%]{\substack{100 \text{ ml. CHCl}_3 \\ \text{Refl.}}}$$

12 g.

(1) 3 g.   (2)

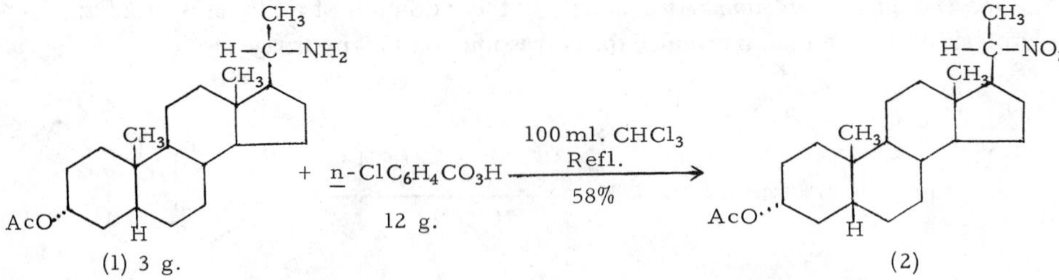

refluxing chloroform. The sequence of events probably is $\text{RNH}_2 \longrightarrow \text{RNHOH} \longrightarrow \text{RNO} \longrightarrow \text{RNO}_2$, and in initial trials the infrared spectrum indicated that some material was lost by conversion of the 20-nitroso compound (RNO) to the dimer. Hence, in the procedure cited, the peracid was taken in large excess to prevent accumulation and dimerization of the nitroso compound.

[1]N. N. Schwartz and J. H. Blumbergs, *J. Org.*, **29**, 1976 (1964)
[2]Procedure worked out in the FMC laboratories
[3]J. Fried, J. W. Brown, and M. Applebaum, *Tetrahedron Letters*, 849 (1965)
[4]D. J. Pasto and C. C. Cumbo, *J. Org.*, **30**, 1271 (1965)
[5]L. A. Paquette and J. H. Barrett, procedure submitted to *Org. Syn.*
[6]W. Hückel and V. Wörffel, *Ber.*, **88**, 338 (1955)
[7]I. J. Borowitz and G. Gonis, *Tetrahedron Letters*, 1151 (1964)
[8]T. J. Delia, M. J. Olsen, and G. B. Brown, *J. Org.*, **30**, 2766 (1965)
[9]D. B. Denney and N. Sherman, *J. Org.*, **30**, 3760 (1965)
[10]C. H. Robinson, L. Milewich, and P. Hofer, *J. Org.*, **31**, 524 (1966)

Chloropicrin, CCl_3NO_2. Mol. wt. 164.39, b.p. 112°, sp. gr. 1.692. Lachrymator and lung irritant. Suppliers: Internat. Minerals and Chem. Co., B, E, F.

The reagent is obtained in high yield by passing chlorine into a well-cooled mixture of picric acid and calcium hydroxide.[1] The product is separated by steam distillation, dried, and distilled.

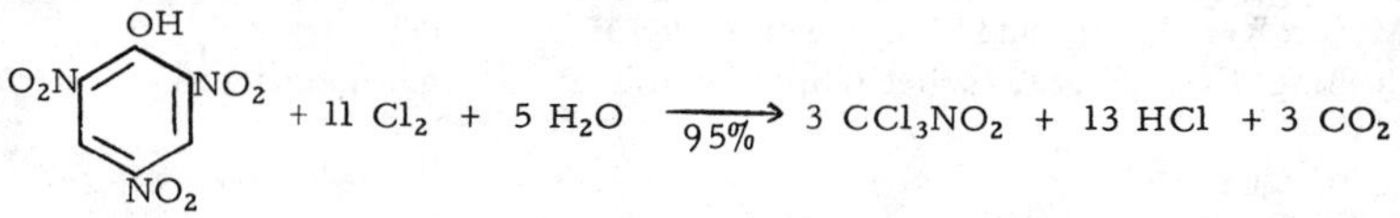
OH
O₂N
NO₂
NO₂
+ 11 Cl₂ + 5 H₂O
95%
3 CCl₃NO₂ + 13 HCl + 3 CO₂

For a use, *see* Tetraethyl orthocarbonate.

[1]E. D. G. Frahm, *Rec. trav.*, **50**, 1125 (1931)

5-Chlorosalicylaldehyde. Mol. wt. 156.57, m.p. 99°. Suppliers: B, E, F.

Whereas benzaldehyde reacts with free amino acids to give Schiff bases which have not been isolated and appear to be unstable, 5-chlorosalicylaldehyde forms yellow Schiff bases which are stabilized by chelation and which are isolable in good yield (F. C. McIntire, 1947). The increased stability of these Schiff bases suggested

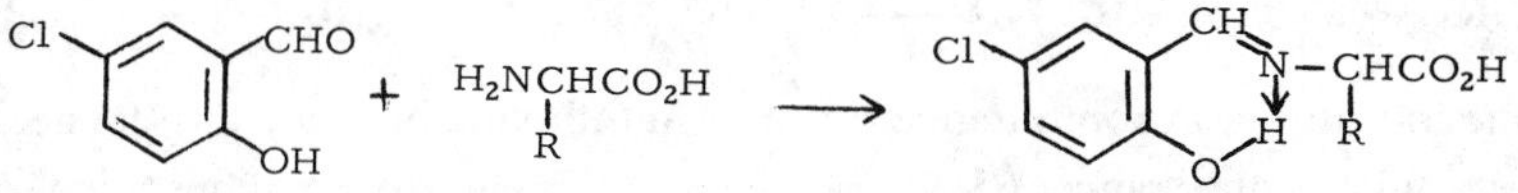
Cl
CHO
OH
+
H₂NCHCO₂H
R
Cl
CH
N−CHCO₂H
H
R
O

their use as N-protective derivatives in peptide synthesis.[1] Removal of the N-arylidene residue can be accomplished with very dilute hydrochloric acid in methanol at room temperature and is evident from decolorization. No racemization of peptide derivatives was observed.

[1]J. C. Sheehan and V. J. Grenda, *Am. Soc.*, **84**, 2417 (1962)

N-Chlorosuccinimide, 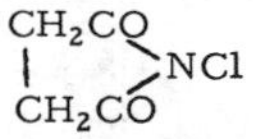
CH₂CO
NCl
CH₂CO

Mol. wt. 133.54, m.p. 148°. Suppliers: B, E, F, MCB.

Use as an oxidizing agent[1] and for the preparation of N-chloroamines.[2] It reacts with aniline and with N-alkylanilines to give *o*- and *p*-chloro derivatives in a ratio of 3.4 : 1.8.[3] N-Chloro intermediates are postulated.

[1]M. F. Hebbelynck and R. H. Martin, *Experientia*, **5**, 69 (1949); *Bull. soc. chim. Belg.*, **60**, 54 (1951); C. A. Grob and HJ. Schmid, *Helv.*, **36**, 1763 (1953)
[2]H. Ruschig, W. Fritsch, J. Schmidt-Thomé, and W. Haede, *Ber.*, **88**, 883 (1955); L. Lábler and F. Sorm, *Coll. Czech.*, **24**, 4010 (1959)
[3]R. S. Neale, R. G. Schepers, and M. R. Walsh, *J. Org.*, **29**, 3390 (1964)

Chlorosulfonic acid, $ClSO_2OH$. Mol. wt. 116.54, b.p. 151°, sp. gr. 1.79. Suppliers pract.): E, F, MCB. Purified by distillation.

For the preparation of sulfate esters, commonly in pyridine solution;[1] other solvents are sulfur dioxide[2] and ethylene dichloride (b.p. 84°).[3] Procedure for the chlorosulfonation of acetanilide.[4]

The reagent reacts readily with a carboxylic anhydride to give the acid chloride, which can be prepared in high yield by distillation at reduced pressure and temperature.[5]

$$
\underset{\substack{\text{n-}C_3H_7C \diagdown O \\ \text{n-}C_3H_7C \diagdown O \\ 0.25 \text{ m.}}}{} + \underset{0.20 \text{ m.}}{ClSO_3H} \xrightarrow{-5°} \underset{}{\text{n-}C_3H_7C \diagdown OSO_3H} + \underset{95\%}{\text{n-}C_3H_7C \diagdown Cl}
$$

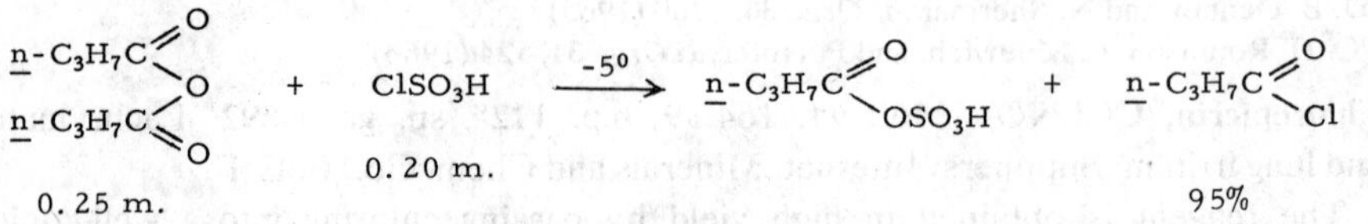

[1]C. Neuberg and L. Liebermann, *Biochem. Z.*, **121**, 326 (1921); R. S. Tipson, "Sulfonic Esters of Carbohydrates," *Advances in Carbohydrate Chemistry*, **8**, 107 (1953)
[2]K. H. Meyer, R. P. Piroué, and M. E. Odier, *Helv.*, **35**, 574 (1952)
[3]I. B. Cushing, R. V. Davis, E. J. Kratovil, and D. W. MacCorquodale, *Am. Soc.*, **76**, 4590 (1954)
[4]*Org. Expts.*, Chapt. 34.4
[5]M. Schmidt and K. E. Pichl, *Ber.*, **98**, 1003 (1965)

Chlorotris(triphenylphosphine)rhodium, $[(C_6H_5)_3P]_3RhCl$. Mol. wt. 913.09

This organometallic complex is obtained as purple-red crystals by interaction of ethanolic solutions of $RhCl_3 \cdot 3H_3O$ and a sixfold molar excess of triphenylphosphine.[1,2] It catalyzes exceedingly rapid hydrogenation of double and triple bonds.[1] Aliphatic aldehydes are decarbonylated to the corresponding paraffins according to the equation:[3]

$$RCHO + RhCl[P(C_6H_5)_3]_3 \longrightarrow RH + RhCl(CO)[(C_6H_5)_3P]_2 + (C_6H_5)_3P$$

The reaction occurs at room temperature or in refluxing benzene or toluene. Yields are generally in the range 65–90%. *See also* Tris-(triphenylphosphine chlororhodium.

The reagent also decarbonylates aroyl chlorides to aryl chlorides (ArCOCl ⟶ ArCl) at relatively low temperatures.[4]

[1]J. F. Young, J. A. Osborn, F. H. Jardine, and G. Wilkinson, *Chem. Comm.*, 131 (1965)
[2]A. Bennett and P. A. Longstaff, *Chem. Ind.*, 846 (1965)
[3]J. Tsuji and K. Ohno, *Tetrahedron Letters*, 3969 (1965)
[4]J. Blum, *Tetrahedron Letters*, 1605 (1966)

β-Chlorovinyl ethyl ether, $ClCH{=}CHOC_2H_5$. Mol. wt. 106.56, *cis*-isomer, b.p. 45–47°/79 mm.; *trans*-isomer, b.p. 59–62°/79 mm.

Preparation.[1]

$$C_2H_5OH + Cl_2 \longrightarrow Cl_2CHCH \underset{OC_2H_5}{\overset{OC_2H_5}{\diagdown}} \xrightarrow[72\%]{\text{Act. Zn}} ClCH{=}CHOC_2H_5$$

Zinc dust is activated by treatment with a solution of ammonium chloride and thereafter with copper sulfate solution. The unsaturated ether obtained is a mixture of the *cis*- and *trans*-isomers in a ratio of about 4:1. Only the *cis*-ether gives ethoxyacetylene in good yield (76%) on being heated with potassium hydroxide.

Uses. See Ethoxyacetylene; Lithium ethoxyacetylene.

[1]D. A. Van Dorp, J. F. Arens, and O. Stephenson, *Rec. trav.*, **70**, 289 (1951)

Cholesterol,

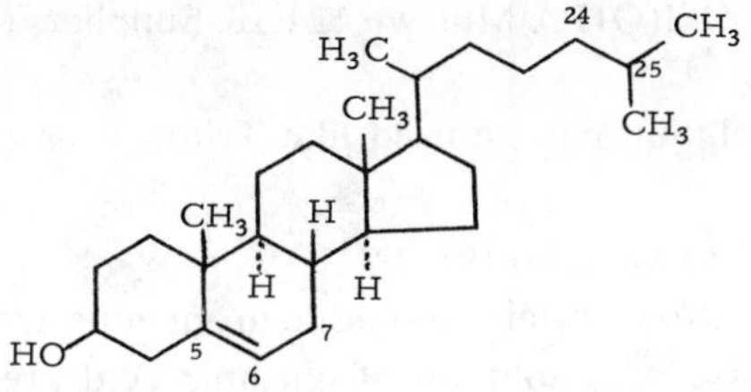
24
CH3
H3C
25
CH3
CH3
CH3
H
H
H
HO
5
6
7

Mol. wt. 386.64, m.p. (evacuated capillary) 149.5–150°. Suppliers: Wilson, A, B, E, F, KK, MCB.

As initially prepared from spinal cord and brain of cattle, commercial cholesterol contains a total of about 3% of the following companions: cholestanol,[1] Δ^7-cholestenol,[1] 7-dehydrocholesterol,[2] and 24-hydroxycholesterol.[3,4] Crystalline material stored for even a few months in contact with air also contains, as products of air oxidation, 25-hydroxycholesterol[4] and 7-ketocholesterol.[4] Purification is accomplished efficiently by conversion to the 5α,6β-dibromide.[5] Addition of a solution of bromine in acetic acid to an ethereal solution of cholesterol precipitates the crystalline dibromide, and the mother liquor retains cholestanol, 7-ketocholesterol, and products of dehydrogenation of Δ^7-cholestenol and 7-dehydrocholesterol. Debromination is accomplished by reaction with zinc dust and a trace of acetic acid in ether, and crystallization from methanol–ether eliminates minute amounts of diols and gives pure cholesterol. The absence of Δ^7-stenols can be established by sensitive color tests carried out by a microtechnique in melting-point capillaries.[6] Purified material can be kept indefinitely if sealed in a vial under nitrogen. If a melting-point determination is done in an open capillary tube, autoxidation may occur rapidly enough to lower the melting-point. The overall yield of pure cholesterol from Wilson cholesterol is 68–69%.

[1]L. F. Fieser, *Am. Soc.*, **75**, 4395 (1953)
[2]A. Windaus and O. Stange, *Z. physiol. Chem.*, **244**, 218 (1936)
[3]A. Ercoli and P. de Ruggieri, *Am. Soc.*, **75**, 3284 (1953)
[4]L. F. Fieser, W.-Y. Huang, and B. K. Bhattacharyya, *J. Org.*, **22**, 1380 (1957)
[5]L. F. Fieser, *Am. Soc.*, **75**, 5421 (1953); *Org. Syn., Coll. Vol.*, **4**, 195 (1963)
[6]*Org. Expts.*, Chapt. 12

Cholesteryl chloroformate, ROCOCl. Mol. wt. 449.10, m.p. 119.5–121°, αD −26.7°. Supplier: Aldrich. Preparation.[1]

The reagent is useful for isolation and identification of primary or secondary amines present in small amounts in aqueous solution.[2] A solution of cholesteryl

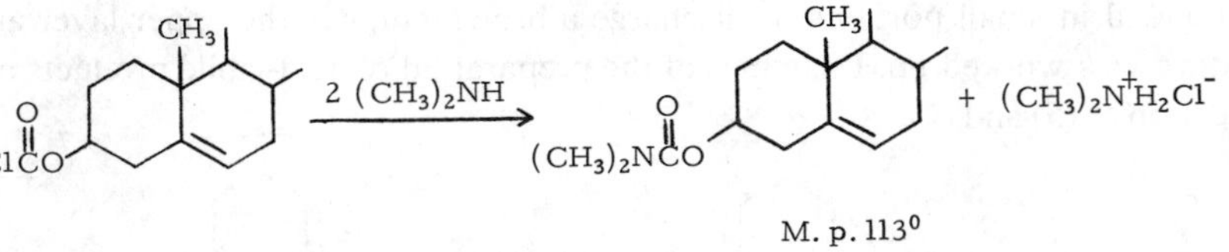
CH3
O
ClCO
2 (CH3)2NH
CH3
O
(CH3)2NCO
+ (CH3)2N+H2Cl−

M. p. 113°

chloroformate (in ether) is shaken with a 1% aqueous solution of the amine, and the carbamate is isolated from the water-washed ethereal solution.

[1]H. Wieland, E. Honold, and J. Vila, *Z. physiol. Chem.*, **130**, 326 (1923)
[2]A. F. McKay and G. R. Vavasour, *Can. J. Chem.*, **31**, 688 (1953)

Choline, $(CH_3)_3N^+CH_2CH_2OH(OH^-)$. Mol. wt. 121.18. Suppliers (50% in methanol): E, F.

The commercial 50% solution can be used like Triton B as catalyst for cyano-ethylation.[1]

[1]S. Pietra, *Boll. Fac. Chim. Ind. Bologna*, **11**, 78 (1953)

Chromic acid, H_2CrO_4. *See also* Chromic anhydride in aqueous acetic acid.

In aqueous solution. Use of a solution of chromic acid prepared from dilute sulfuric acid and sodium dichromate dihydrate is illustrated by the oxidation of hydroquinone to quinone, isolated by a tedious process in 86–92% yield.[1]

Jones reagent in acetone.[2] The reagent is a solution of chromic acid and sulfuric acid in water.[3] An oxidation is carried out by titrating a stirred solution of the alcohol in acetone at 15–20° with the Jones reagent; the mixture separates into a green lower layer of chromium salts and an upper layer which is an acetone solution of the oxidation product. The method has the merit that a secondary alcohol containing a double or triple bond can be oxidized to the corresponding ketone without attack of the center of unsaturation. Thus Djerassi *et al.*[4] attained a high yield in the oxidation of pregnenolone to the nonconjugated Δ^5-pregnene-3,20-dione (2). They prepared a standard solution by dissolving 26.72 g. of chromic trioxide "in 23 ml. of conc'd. sulfuric acid diluted with water to a volume of 100 ml." and for oxidation of 3.0 g. of pregnenolone used 2.75 ml. of this solution.

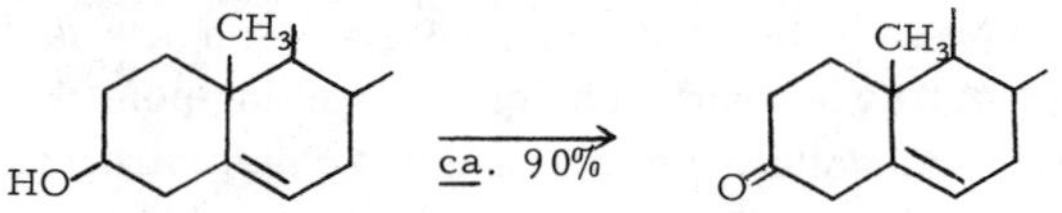
CH3
HO
ca. 90%
CH3
O

(1) (2)

Meinwald *et al.*[5] recommend this method for oxidation of nortricyclanol (3) to

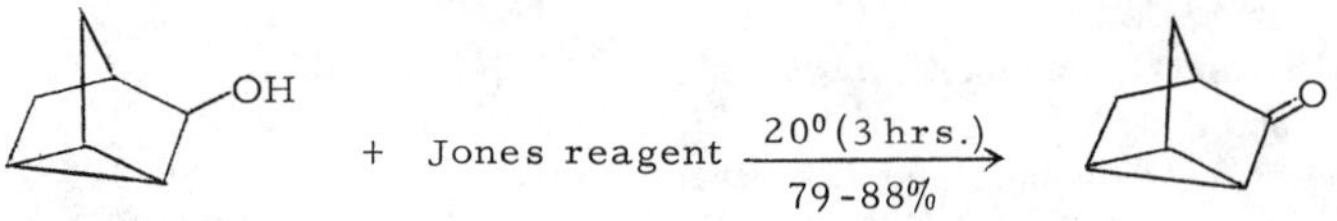
OH
+ Jones reagent
20° (3 hrs.)
79-88%
O

1 m. in 60 ml. acetone 0.7 m.

(3) (4)

the highly strained and reactive ketone (4). They prepared 0.7 mole of Jones reagent by dissolving 70 g. of chromium trioxide in 500 ml. of water and cooling this solution with manual stirring during cautious addition of 61 ml. of concd. sulfuric acid. After addition of the oxidant to a stirred solution of the alcohol in acetone, bisulfite was added in small portions to discharge a brown color in the upper layer and the mixture was worked up. Examples of the preparation of acid-labile products are the oxidation of (5) and (7).[6]

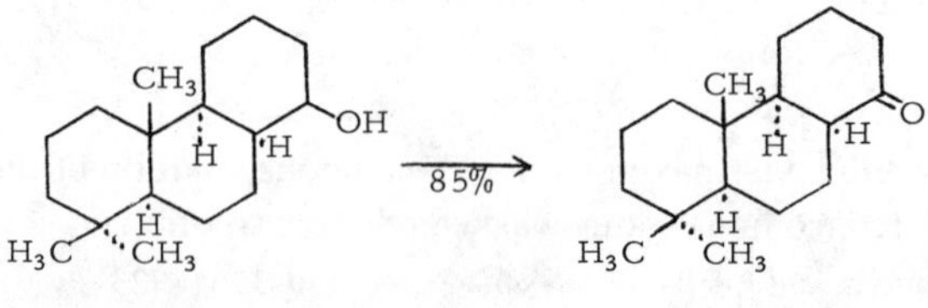
CH3
OH
H
H
H
H3C
CH3
85%
CH3
O
H
H
H
H3C
CH3

(5) (6)

(7) 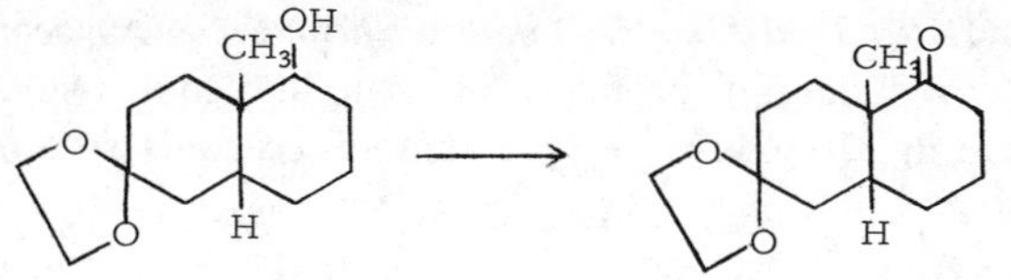 (8)
OH
CH3
H
CH3
O
H

The Jones method is rapid and the yields high. The main limitation is the low solvent power of acetone. However, the method was used successfully by Eisenbraun[7] for oxidation of cyclooctanol to cyclooctanone on a fairly large scale.

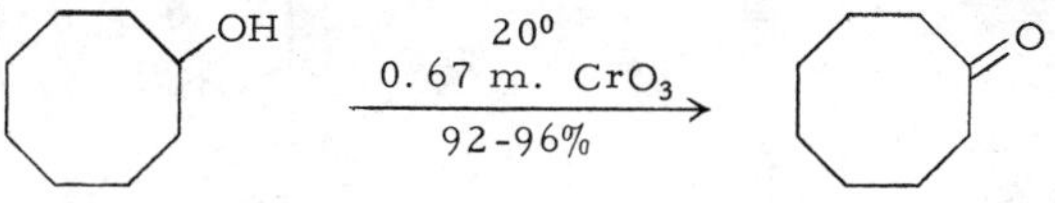
OH
20°
0.67 m. CrO3
92-96%
O

0.5 m.

Two-phase oxidation.[8] A simple procedure for the oxidation of a secondary alcohol to a ketone consists in adding a cold solution of sodium dichromate and sulfuric acid containing a little acetic acid to a stirred solution or suspension of the alcohol in benzene while cooling to 1–6°.[9] Since the ketone as formed is distributed into the benzene layer and protected against secondary reactions, sensitive ketones can be isolated in high yield without appreciable inversion at adjacent centers of asymmetry. Thus a *cis-trans* mixture of 1,5-dihydroxydecalins affords a *cis-trans* mixture of decalin-1,5-diones in 71–76% yield; in the two-phase system the *cis* isomer present is not converted into the more stable *trans* diketone.

The mixture of stereoisomeric α-decalols resulting from hydrogenation of α-naphthol over W-7 Raney nickel or ruthenium oxide is oxidized as follows. A solution of 12 g. of sodium dichromate dihydrate in 190 ml. of water is treated with

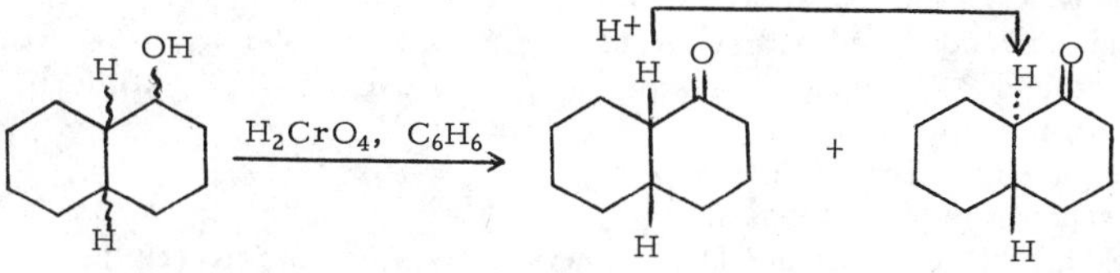
OH
H
H
H2CrO4, C6H6
H+
H
O
H
+
H
O
H

57 ml. of concd. sulfuric acid and 36 ml. of acetic acid, cooled to room temperature, and poured into a flask containing a solution of 54 g. of α-decalol mixture in 300 ml. of benzene. The mixture is stirred at room temperature, and in this case stirring is continued for 24 hrs. to effect conversion of the *cis* to the *trans* isomer. After conventional workup and distillation, the yield of *trans*-α-decalone is 49.5 g. (93%). The same procedure has been used for the preparation of 2-methylcyclohexanone from the alcohol in 85–88% yield.[10]

A similar two-phase procedure[11] involves addition over 15 min. of the theoretical quantity of chromic acid (from $Na_2Cr_2O_7 \cdot 2H_2O$, H_2SO_4, and water) to a solution of the alcohol in ether. After 2 hrs. at 25–30° the ether layer is separated and the product isolated. Oxidation of *l*-menthol gave *l*-menthone in 97% yield with only a trace of *d*-isomenthone, whereas 3–4% of this isomer was present on oxidation with aqueous chromic acid (50–55°), H_2CrO_4 in 90% acetic acid (25°), or H_2CrO_4–acetone (5–10°), and the yields in these oxidations were 90, 71, and 86%. This procedure has been used for the selective oxidation of steroidal 16β,20α-diols to 16-keto-20α-ols.[12]

Kiliani reagent (H_2CrO_4–H_2SO_4–H_2O) in acetic acid. Kiliani[13] prepared his reagent by dissolving 60 g. of sodium dichromate dihydrate in a cooled solution of 80 g. of concd. sulfuric acid in 270 g. of water; the solution contains 10% of chromic acid.

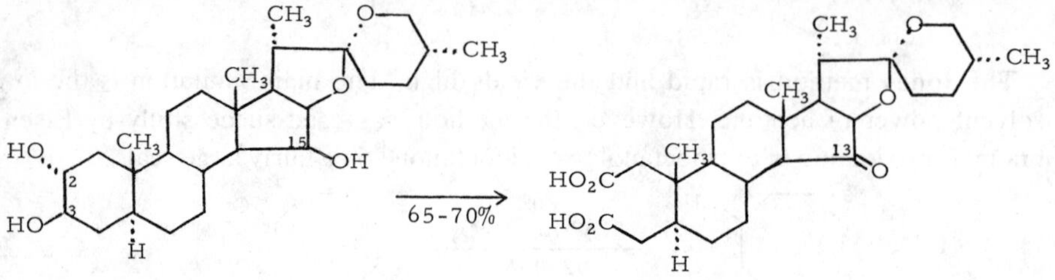

CH3
CH3
CH3
HO
HO
2
3
15
OH
H
O
CH3
65-70%
HO2C
HO2C
13
O

Digitogenin Digitogenic acid

For oxidation of the triol digitogenin to digitogenic acid, a 13-keto-2,3-seco-2,3-dioic acid, 5 g. of the sapogenin was stirred with 50 g. of acetic acid, when it largely dissolved, and 40 g. of the Kiliani reagent was added in portions in 3–4 min. with cooling. A precipitate separated and then dissolved, and soon digitogenic acid began to crystallize. After standing for at least 5 hrs. the product was collected and washed with 30% acetic acid and then, very thoroughly, with water. In a modern use of this method, Pelletier and Locke[14] prepared the Kiliani reagent by dissolving 53 g. of CrO_3 and 80 g. of concd. H_2SO_4 in 400 ml. of water.

[1]E. B. Vliet, *Org. Syn., Coll. Vol.*, **1**, 482 (1941)
[2]Contributed in part by R. E. Ireland
[3]K. Bowden, I. M. Heilbron, E. R. H. Jones, and B. C. L. Weedon, *J. Chem. Soc.*, 39 (1946); P. Bladon, J. M. Fabian, H. B. Henbest, H. P. Koch, and G. W. Wood, *ibid.*, 2402 (1951); A. Bowers, T. G. Halsall, E. R. H. Jones, and A. J. Lemin, *ibid.*, 2555 (1953)
[4]C. Djerassi, R. R. Engle, and A. Bowers, *J. Org.*, **21**, 1547 (1956)
[5]J. Meinwald, J. Crandall, and W. E. Hymans, *Org. Syn.*, **45**, 77 (1965)
[6]R. E. Ireland and P. W. Schiess, *Tetrahedron Letters*, No. 25, 37 (1960); R. F. Church and R. E. Ireland, *ibid.*, 493 (1961)
[7]E. J. Eisenbraun, *Org. Syn.*, **45**, 28 (1965)
[8]Contributed in part by W. S. Johnson
[9]W. S. Johnson, C. D. Gutsche, and D. K. Banerjee, *Am. Soc.*, **73**, 5464 (1951)
[10]E. W. Warnhoff, D. G. Martin, and W. S. Johnson, *Org. Syn., Coll. Vol.*, **4**, 164 (Note 1) (1963)
[11]H. C. Brown and C. P. Garg, *Am. Soc.*, **83**, 2952 (1961)
[12]S. Noguchi, M. Imanishi, and K. Morita, *Chem. Pharm. Bull.*, **12**, 1184 (1964)
[13]H. Kiliani and B. Merk, *Ber.*, **34**, 3562 (1901)
[14]S. W. Pelletier and D. M. Locke, *Am. Soc.*, **87**, 761 (1965); their XX ⟶ XXI

Chromic anhydride, CrO_3. Mol. wt. 100.01; 1.5 oxygen equivalents per mole of anhydride. The oxidizing power of the reagent increases with decreasing water content of the solvent medium.

Aqueous acetic acid (preparative procedures). (a) A solution of CrO_3 in a small volume of water is diluted with acetic acid; the solution is run into a solution of β-methylnaphthalene in acetic acid and the temperature is controlled to 60° by cooling; yield of 2-methyl-1,4-naphthoquinone, 38–42%.[1]

(b) A solution of naphthalene in acetic acid is added to a solution of CrO_3 in 80% acetic acid with stirring and cooling in an ice-salt bath to control the temperature to 10–15° (3 hrs.). The mixture is stirred overnight and let stand for 3 days. The yield of 1,4-naphthoquinone is 18–22%.[2]

(c) A suspension of 90% phenanthrene in an aqueous solution of CrO_3 is stirred during addition of concentrated sulfuric acid at a rate sufficient to maintain boiling. More CrO_3 and water are added and the mixture is refluxed. After separation from anthraquinone and hydrocarbons through the bisulfite addition compound, phenanthrenequinone is obtained in yield of 44–48%.[3]

(d) A solution of 39.6 g. of CrO_3 in a mixture of 50 ml. of water, 120 ml. of acetic acid, and 36 ml. of concd. sulfuric acid is added during 1 hr. at 0° to a solution of 22.9 g. of cholesteryl acetate dibromide (1) in 410 ml. of ethylene dichloride–acetic acid (18:23) with stirring of the mixture at 0° for 3 hrs. After debromination and

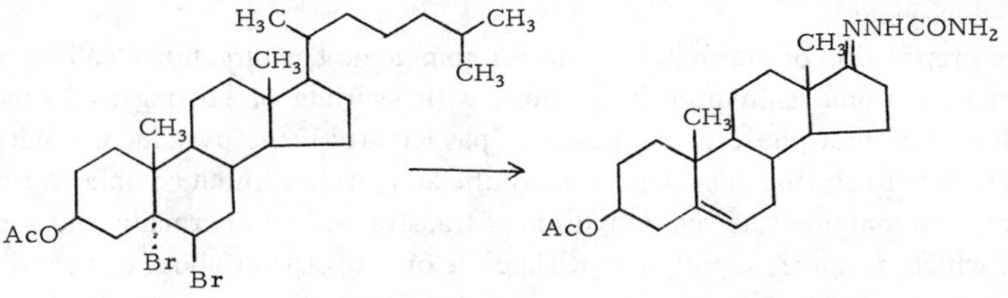

(1) (2)

reaction with semicarbazide, the semicarbazone acetate (2) of androstenolone is obtained in 7.7% yield. The yield from the more stable 5β,6α-dibromide is 15.3%.[4]

Aqueous acetic acid (rate of oxidation). For study of steric acceleration of oxidation of secondary alcohols, the standard basis for comparison is the rate of oxidation at 25° with a solution of chromic acid in 90.9% acetic acid.[5]

Anhydrous CrO_3–acetic acid (Fieser reagent[6]). The oxidation is done by stirring a solution of the alcohol in anhydrous acetic acid with suspended chromic anhydride at a controlled temperature. The oxidant is insoluble in the anhydrous solvent but dissolves as oxidation progresses; addition of water precipitates the product. The reagent oxidizes the side chain of 2-acetoxy-3-alkyl-1,4-naphthoquinones,[6] for example:

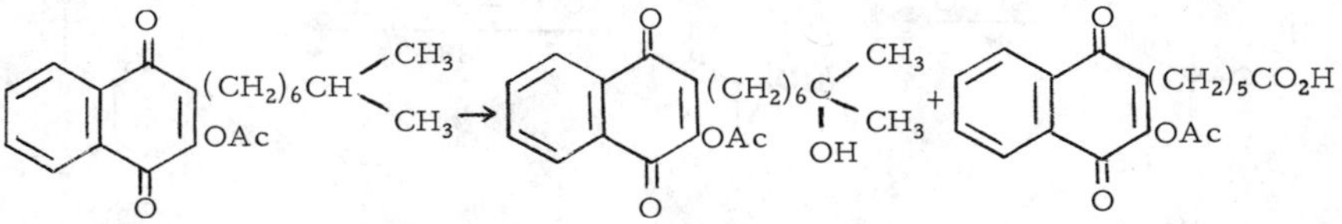

Cholestane-3β-ol, oxidized at 40–60°, affords cholestane-3-one in yield of 83%.[7] The power of the anhydrous reagent is illustrated by its use for smooth oxidation of *t*-alcohols to keto acids, as exemplified by the reactions formulated.[8]

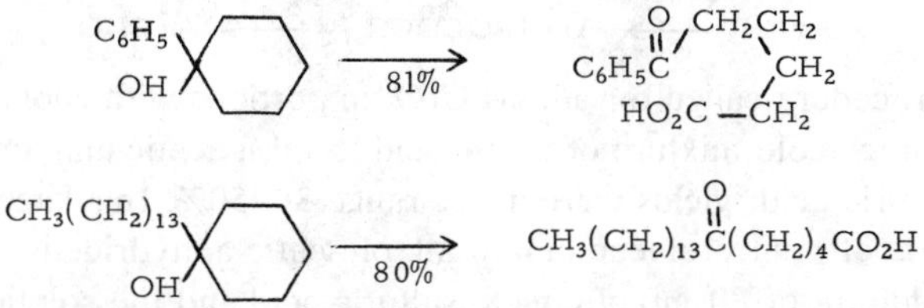

CrO_3–pyridine complex (Sarett reagent[9]). A solution of the complex in pyridine is useful for oxidation of substances containing acid-sensitive groups. With chromic anhydride in pyridine, it is possible to oxidize the alcoholic function of cholesterol α-oxide (1) without disturbance of the epoxide group and obtain the ketoxide (2); the oxide ring of cholesterol β-oxide does not similarly withstand attack but is oxidized by CrO_3-Py to 6β-hydroxy-Δ^4-cholestene-3-one.[10]

(1) 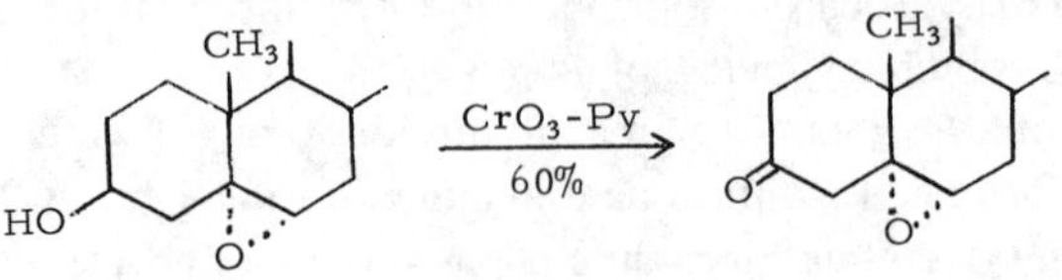 (2)

The CrO_3-Py reagent oxidizes primary allylic and benzylic alcohols to aldehydes in 50–80% yield.[11]

The preparation of the complex calls for comment. The directions[9] call for adding 1 part of chromic anhydride in portions with swirling or stirring to 10 parts of pyridine. The first phase of the reaction, particularly if the pyridine is colder than 15°, appears to consist in slow solution of the anhydride without complex formation. After a few minutes the red anhydride is transformed exothermally into a yellow solid, which dissolves rapidly on swirling (below 30°). After about one third of the anhydride has been added and mostly dissolved, the yellow complex begins to precipitate. When the pyridine was added to the chromic anhydride, the mixture usually inflamed.

CrO_3–pyridine–water (Cornforth reagent[12]). Since preparation of the Sarett reagent is tedious and potentially hazardous, the Cornforth reagent appears to be preferable where applicable. It is prepared by gradual addition of a solution of 50 g. of chromic anhydride in 50 ml. of water to 500 ml. of pyridine with stirring and ice cooling. Cornforth, Cornforth, and Popják[12] used this oxidant in one step of a synthesis of mevalonolactone. The triol (1) was condensed with acetaldehyde to give the 1,3-dioxane (2) for protection of the primary alcoholic group, and (2) was oxidized with the CrO_3–Py–H_2O reagent to give (3) in high yield.

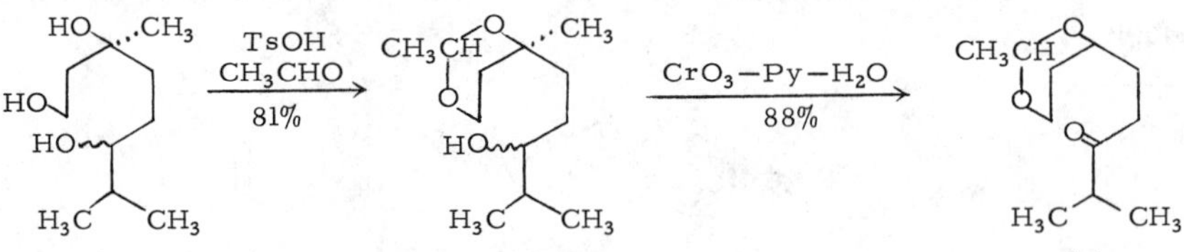

(1) (2) (3)

CrO_3–acetic anhydride–H_2SO_4 (Thiele reagent).[13] This powerful reagent converts an arylmethyl compound into the diacetate of an aldehyde, from which the aldehyde can be recovered by acid hydrolysis:

$$ArCH_3 \longrightarrow ArCH(OCOCH_3)_2 \longrightarrow ArCHO$$

The original procedure called for adding CrO_3 in portions with cooling to a solution of $ArCH_3$ in an ice-cold mixture of acetic acid (5 ml.), acetic anhydride (5–10 ml.), and concd. sulfuric acid; yields were in the range 30–50%. In a later procedure,[14] a solution of 50 g. of p-nitrotoluene in 400 ml. of acetic anhydride is cooled with ice during slow addition of 80 ml. of concd. sulfuric acid and the solution is cooled to 0°. A second solution is prepared by slow addition of 100 g. of CrO_3 in portions with careful cooling to 450 ml. of acetic anhydride (addition of acetic anhydride to solid CrO_3 has resulted in explosions). This second solution is then added slowly with stirring and cooling in a salt-ice bath to the first solution at such a rate that the temperature never exceeds 10° (1.5–2 hrs.). *p*-Nitrobenzaldiacetate is obtained in 65–66% yield.

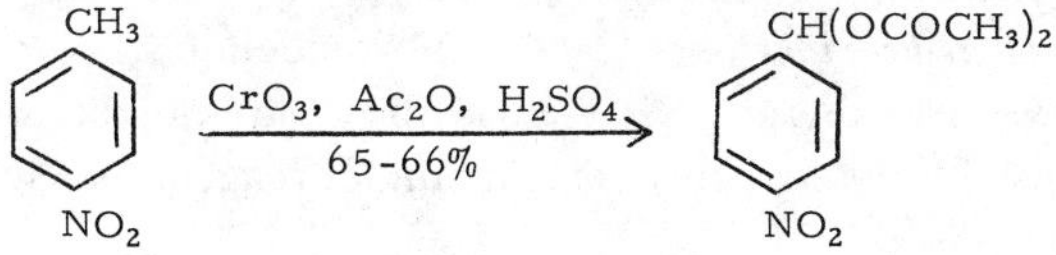
CH3
CH(OCOCH3)2
CrO3, Ac2O, H2SO4
65-66%
NO2
NO2

CrO_3–H_2SO_4. A solution of 36 g. of 3,4-dinitrotoluene in 180 ml. of concd. sulfuric acid at 30° is stirred during slow addition of 60 g. of chromic anhydride in small portions, with cooling to control the temperature to 45–50°; after 4 hrs. the solution is poured onto ice. The yield of 3,4-dinitrobenzoic acid is 89%.[15]

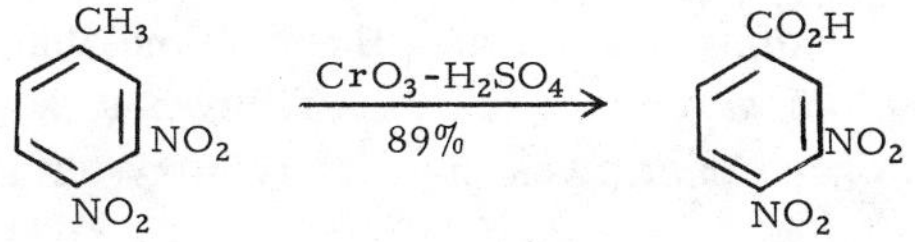
CH3
CO2H
CrO3-H2SO4
89%
NO2
NO2
NO2
NO2

CrO_3–dimethylformamide.[16] Chromic anhydride is stirred with a solution of the alcohol (*e.g.* cholestanol) in dimethylformamide until dissolved and a catalytic amount of sulfuric acid is added. Reaction proceeds at room temperature. Acetonide and ketal groups are not disturbed.

CrO_3–Ac_2O–AcOH. This reagent, diacetyl chromate, oxidizes tetraphenylethylene largely to benzopinacol carbonate and tetraphenylethylene oxide, with a minor amount of benzophenone.[17]

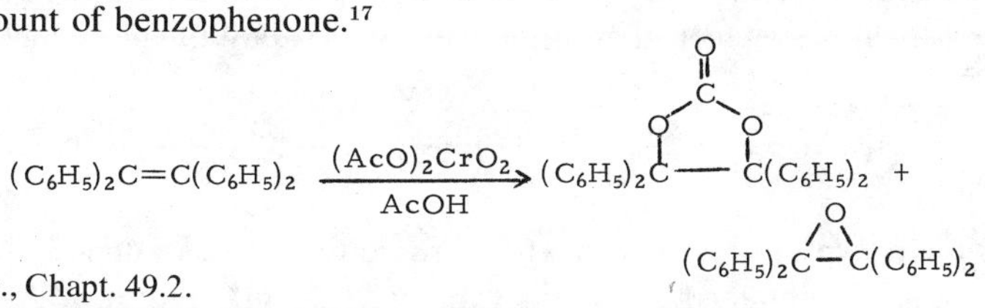
(C6H5)2C=C(C6H5)2
(AcO)2CrO2
AcOH
(C6H5)2C
C(C6H5)2
+
(C6H5)2C
C(C6H5)2

[1]*Org. Expts.*, Chapt. 49.2.
[2]E. A. Braude and J. S. Fawcett, *Org. Syn., Coll. Vol.*, **4**, 698 (1963)
[3]R. Wendland and J. LaLonde, *ibid.*, **4**, 757 (1963)
[4]S. P. J. Maas and J. G. de Heus, *Rec. trav.*, **77**, 531 (1958)
[5]J. Schreiber and A. Eschenmoser, *Helv.*, **38**, 1529 (1955)
[6]L. F. Fieser, *Am. Soc.*, **70**, 3237 (1948); K. Nakanishi and L. F. Fieser, *ibid.*, **74**, 3910 (1952)
[7]L. F. Fieser, *ibid.*, **75**, 4391 (1953)
[8]L. F. Fieser and J. Szmuszkovicz, *ibid.*, **70**, 3352 (1948)
[9]G. I. Poos, G. E. Arth, R. E. Beyler, and L. H. Sarett, *ibid.*, **75**, 422 (1953)
[10]B. Ellis and V. Petrow, *J. Chem. Soc.*, 4417 (1956)
[11]J. R. Holum, *J. Org.*, **26**, 4814 (1961)
[12]R. H. Cornforth, J. W. Cornforth, and G. Popják, *Tetrahedron*, **18**, 1351 (1962)
[13]J. Thiele and E. Winter, *Ann.*, **311**, 353 (1900)
[14]T. Nishimura, *Org. Syn., Coll. Vol.*, **4**, 713 (1963)
[15]E. Borel and H. Deuel, *Helv.*, **36**, 806 (1953)
[16]G. Snatzke, *Ber.*, **94**, 729 (1961)
[17]W. A. Mosher, F. W. Steffgen, and P. T. Lansbury, *J. Org.*, **26**, 670 (1961)

Chromous acetate, $Cr(OCOCH_3)_2$. Mol. wt. 170.10. Deep red powder. It is oxidized much more rapidly by air when moist than when dry, but even the dry material undergoes slow air oxidation.

Preparation.[1,2] In one method[2] a mixture of 9 g. of chromium metal powder (Fisher) and 100 ml. of 6*M* hydrochloric acid was stirred with water cooling to room temperature until the reaction was complete. A deoxygenated solution of

50 g. of sodium acetate in 100 ml. of deoxygenated water was added with stirring and ice cooling. The precipitated chromous acetate was collected and washed with deoxygenated water, ethanol, and ether; yield 12 g., 80–85% purity.

Reduction of an α-bromoketone.[3] 2α,4α-Dibromocholestane-3-one was reduced to 4α-bromocholestane-3-one in 78% yield with freshly prepared chromous acetate taken in large excess. With the theoretical amount of reagent the yield was only 20%.[4]

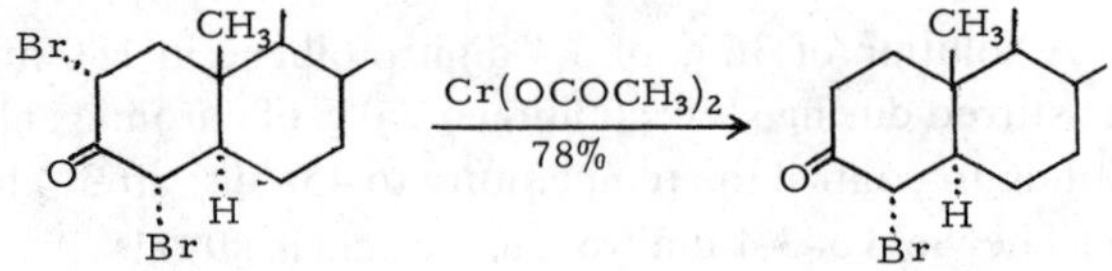

Br
CH_3
O
H
Br
$Cr(OCOCH_3)_2$
78%
CH_3
O
H
Br

Reduction of bromohydrins. Barton and Basu[5] found that reduction of a 9α-bromo-11β-hydroxysteroid with a large excess of chromous acetate in the presence of butanethiol as hydrogen donor gives the 11β-hydroxysteroid in high yield. The

(1) 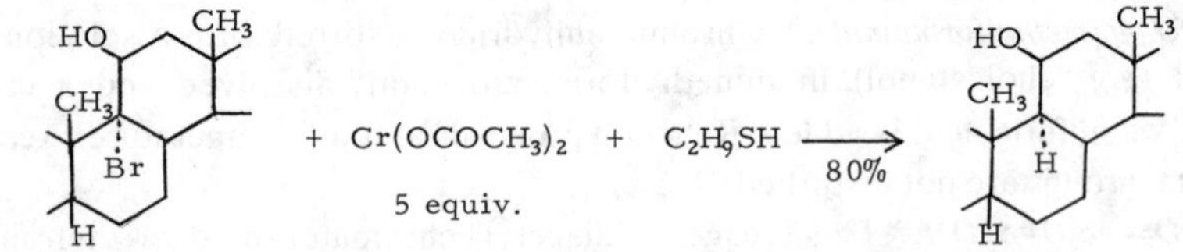 (2)

HO
CH_3
CH_3
Br
H
+ $Cr(OCOCH_3)_2$ + C_4H_9SH
80%
5 equiv.
HO
CH_3
CH_3
H
H

reaction is considered to involve formation of a radical which is attacked by the thiol to give the radical RS· which then dimerizes. In a study of the reduction of 9α-

OH
Br
Cr^{++}
OH
BuSH
OH
H
BuS·
BuSSBu

bromo-11β-hydroxyprogesterone (3) with chromous acetate in dimethyl sulfoxide, Barton, Basu, *et al.*[2] obtained products of three types: 11β-hydroxy-5,9-cyclopregnane-3,20-dione (4), the 11β-hydroxysteroid (5), and the $\Delta^{9(11)}$-olefinic derivative (6), the normal product.

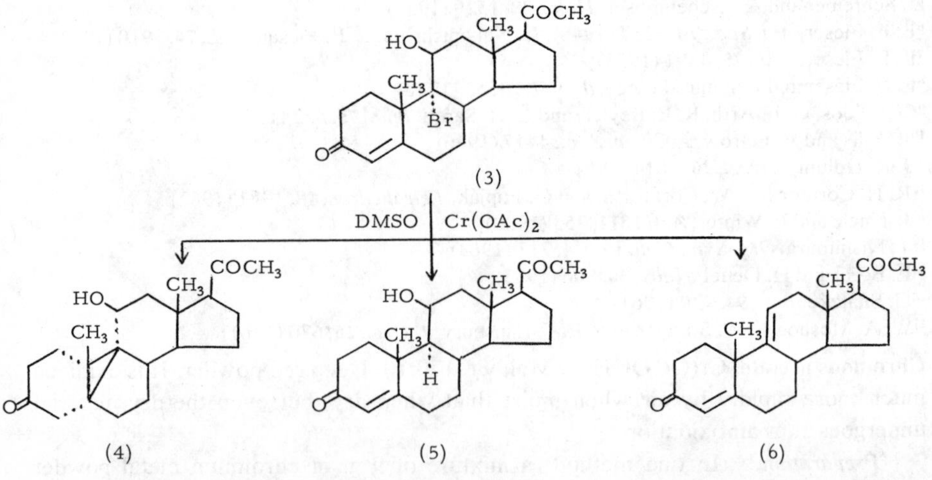

$COCH_3$
CH_3
HO
CH_3
Br
O
(3)
DMSO
$Cr(OAc)_2$
$COCH_3$
CH_3
HO
CH_3
O
$COCH_3$
CH_3
HO
CH_3
H
O
$COCH_3$
CH_3
CH_3
O

(4) (5) (6)

Reductive cleavage of an epoxide. The 16α,17α-epoxysteroid (1) was converted by chromous acetate into the β-hydroxyketone (2) and the unsaturated ketone (3).[6]

(1)

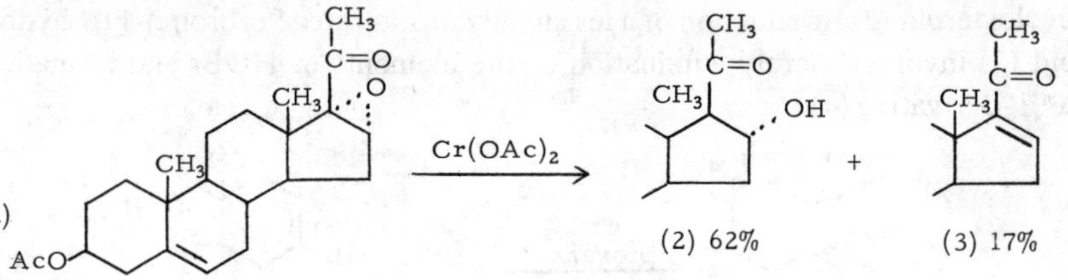

(2) 62% (3) 17%

[1]M. R. Hatfield, *Inorg. Syn.*, **3**, 148 (1950)
[2]D. H. R. Barton, N. K. Basu, R. H. Hesse, F. S. Morehouse, and M. M. Pechet, *Am. Soc.*, **88**, 3016 (1966)
[3]K. L. Williamson and W. S. Johnson, *J. Org.*, **26**, 4563 (1961)
[4]R. M. Evans *et al.*, *J. Chem. Soc.*, 4356 (1956)
[5]D. H. R. Barton and N. K. Basu, *Tetrahedron Letters*, 3151 (1964)
[6]V. Schwarz, *Coll. Czech.*, **26**, 1207 (1961)

Chromous chloride,[1] $CrCl_2$. Mol. wt. 122.92. Supplier of aqueous solution (oxygen absorbent): Fisher.

Preparation.[2] Zinc dust (400 g.) is amalgamated by shaking with a solution of 32 g. of mercuric chloride and 20 ml. of concd. hydrochloric acid in 400 ml. of water. The aqueous phase is decanted, and 800 ml. of water, 80 ml. of concd. hydrochloric acid, and 200 g. of chromic chloride hexahydrate are added. Carbon dioxide from a dry ice generator is bubbled through the mixture to provide agitation and prevent reoxidation by air. When the solution turns light blue it is ready for use. A procedure for preparation of an alcoholic solution using a Jones reductor is also described.[3]

Dehalogenation. The reagent is particularly useful for the dehalogenation of *vic*-dihalides.[4,5] When a solution of the dihalide in acetone is treated with a mixture of chromous chloride solution and acetone, reaction occurs at room temperature and can be followed by the change from the blue color of chromous ion to the green color of chromic ion. A typical workup procedure involves removing the acetone *in vacuo* and filtering the product if solid or extracting if liquid. By this process 5,6-dichlorosteroids are dechlorinated in yields of 90% or better.[5]

Reduction. Chromous chloride is useful also for reduction of α-haloketones to the parent ketones,[3,4] for the reduction of epoxides to olefins,[6] and reduction of

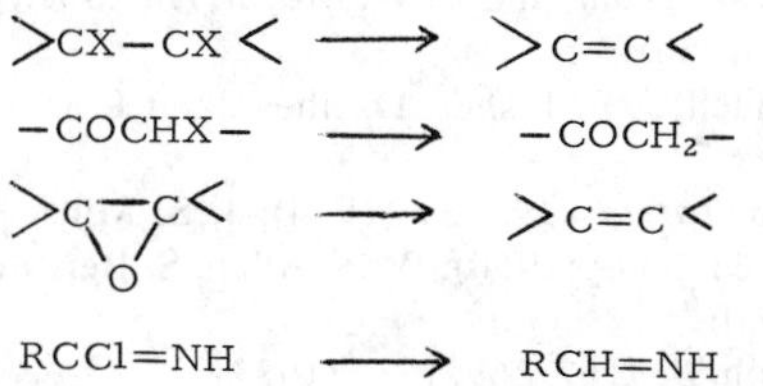

imide chlorides to imines.[7] It reduces aromatic aldehydes bimolecularly but does not reduce aliphatic aldehydes or aromatic ketones.[8] Reduction of the steroid 9α-halo-1,4-diene-3-one (1) with chromous chloride is attended with rearrangement to

(1)

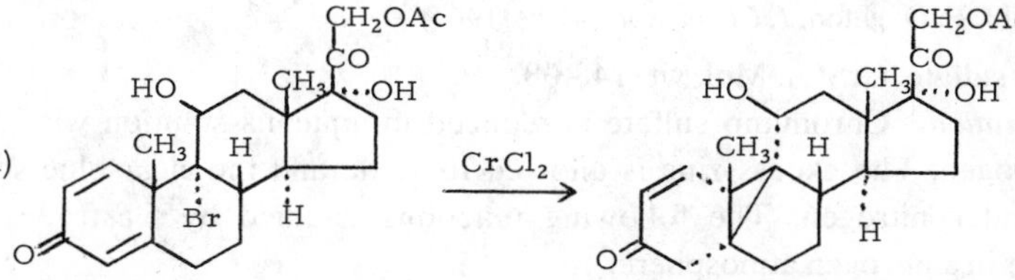

(2)

5,9-cyclosteroid (2).[9] Reduction of a less highly unsaturated 9α-bromo-11β-hydroxysteroid (3) involves merely elimination of the elements of HOBr and formation of the $\Delta^{9(11)}$-derivative (4).[10]

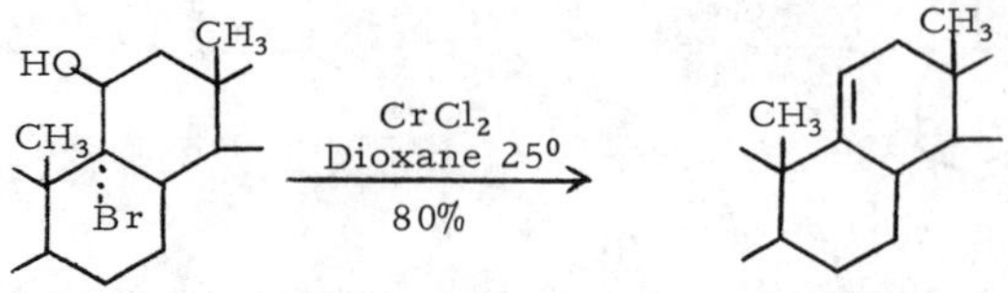

HO
CH3
CH3
Br
CrCl2
Dioxane 25°
80%
CH3
CH3

(3) (4)

Chromous chloride was used to effect reductive cleavage of the γ-lactone (5).[11] In this case reduction with zinc and acetic acid or with calcium and liquid ammonia proceeded very slowly.

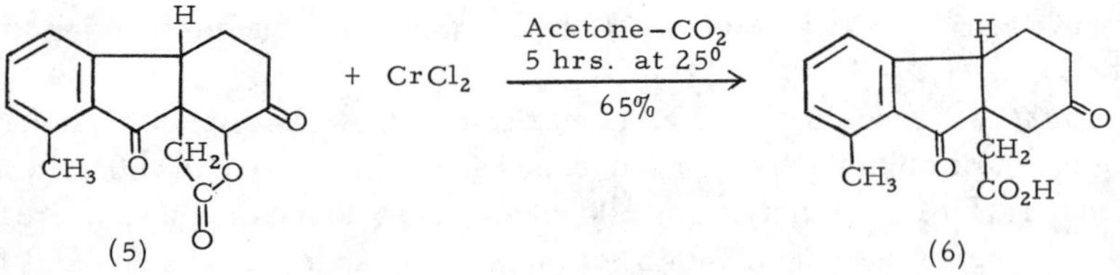

H
CH3
O
CH2
O
C
O
+ CrCl2
Acetone–CO2
5 hrs. at 25°
65%
H
O
CH3
O
CH2
CO2H

(5) (6)

Reductive cleavage of 2,4-dinitrophenylhydrazones.[12] Certain 2,4-dinitrophenylhydrazones of 3-ketosteroids can be cleaved under mild conditions if the nitro groups are reduced to amino groups by chromous chloride. Thus when the 2,4-DNP of 4,5α-dihydrocortisone 21-acetate in methylene chloride was shaken in a separatory funnel with chromous chloride and dilute hydrochloric acid for 15 min., workup of the organic layer afforded the ketone in 94% yield. Use of a water-immiscible solvent protects the released ketone; an inert atmosphere if required can be provided with dry ice. Cortisone 2,4-DNP was cleaved in only 3–5 hrs. and the yield was 60%. DNP derivatives of $\Delta^{1,4}$-diene-3-ones were completely resistant to hydrolysis.

[1]Contributed by Leon Mandell
[2]G. Rosenkranz, O. Mancera, J. Gatica, and C. Djerassi, *Am. Soc.*, **72**, 4077 (1950)
[3]W. F. McGuckin and H. L. Mason, *Am. Soc.*, **77**, 1822 (1955)
[4]P. L. Julian, W. Cole, A. Magnani, and E. W. Meyer, *Am. Soc.*, **67**, 1728 (1945); W. S. Allen *et al.*, *Am. Soc.*, **82**, 3696 (1960)
[5]F. A. Cutler, Jr., L. Mandell, J. F. Fisher, D. Shew, and J. M. Chemerda, *J. Org.*, **24**, 1621, 1629 (1959)
[6]W. Cole and P. L. Julian, *J. Org.*, **19**, 131 (1954); J. S. Mills, A. Bowers, C. Djerassi, and H. J. Ringold, *Am. Soc.*, **82**, 3399 (1960); W. S. Allen, S. Bernstein, L. I. Feldman, and M. J. Weiss, *ibid.*, **82**, 3696 (1960)
[7]J. v. Braun and W. Rudolph, *Ber.*, **67**, 269, 1735 (1934)
[8]J. B. Conant and H. B. Cutter, *Am. Soc.*, **48**, 1016 (1926)
[9]O. Gnoj, E. P. Oliveto, C. H. Robinson, and D. H. R. Barton, *Proc. Chem. Soc.*, 207 (1961)
[10]J. Fried and E. F. Sabo, *Am. Soc.*, **79**, 1130 (1957)
[11]H. O. House and R. G. Carlson, *J. Org.*, **29**, 74 (1964)
[12]J. Elks and J. F. Oughton, *J. Chem. Soc.*, 4729 (1962)

Chromous sulfate, $CrSO_4$. Mol. wt. 148.08.

Preparation.[1] Chromium sulfate is reduced in aqueous solution with zinc dust under nitrogen. The excess zinc is allowed to settle and the clear blue solution is filtered under nitrogen. The following reactions, studied by Castro,[1,2] were all carried out in a nitrogen atmosphere.

Debromination. *vic*-Dibromides are converted into olefins by the reagent in 1:1 water–dimethylformamide solution in good yield. The reaction is complete in 5–30 min.

$$\begin{array}{c} C_6H_5 \\ | \\ HCBr \\ | \\ HCBr \\ | \\ C_6H_5 \end{array} \quad \xrightarrow[\substack{100\%}]{\substack{N_2 \\ CrSO_4 \text{ in aq. DMF}}} \quad \begin{array}{c} C_6H_5CH \\ \| \\ HCC_6H_5 \end{array}$$

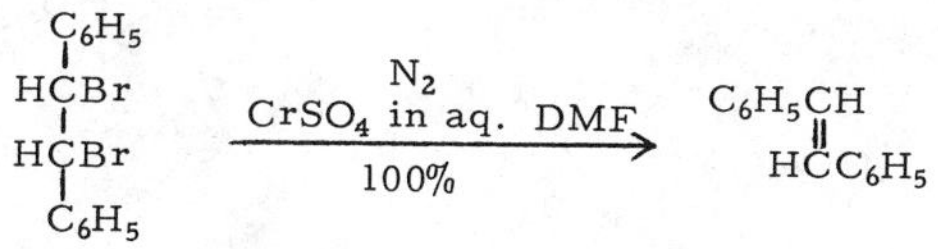

Reduction.[2] Under the same conditions, alkyl bromides are reduced to hydrocarbons only slowly (1–3 days). Allyl bromide and allyl chloride are reduced smoothly

$$CH_3CH_2CH_2Br \quad \xrightarrow[99\%]{CrSO_4 \text{ in aq. DMF}} \quad CH_3CH_2CH_3$$

to propylene. Benzhydryl bromide affords the dimeric product *sym*-tetraphenylethane.

$$(C_6H_5)_2CHBr \quad \xrightarrow[90\%]{CrSO_4 \text{ in aq. acetone}} \quad (C_6H_5)_2CHCH(C_6H_5)_2$$

Acetylenes are reduced by chromous sulfate in water or aqueous DMF to *trans*-olefins in high yield. In the example cited, the reaction is complete at room tempera-

$$HOCH_2C\equiv CCH_2OH \quad \xrightarrow[84\%]{CrSO_4 \text{ in } H_2O} \quad \begin{array}{c} HOCH_2 \\ \diagdown \\ H \diagup \end{array} C=C \begin{array}{c} H \\ \diagdown \\ \diagup CH_2OH \end{array}$$

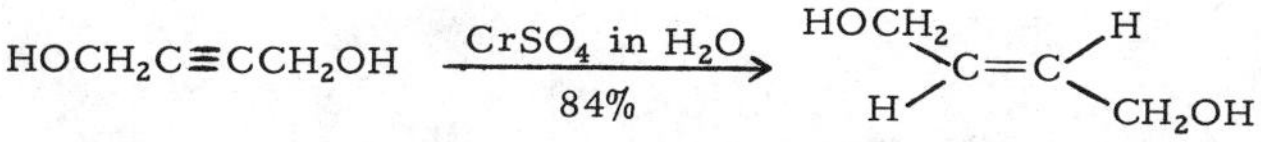

ture in about 10 min. Phenylacetylene is reduced in 2–3 hrs.; diphenylacetylene showed no appreciable reaction in 1 day.

[1] C. E. Castro, *Am. Soc.*, **83**, 3262 (1961)
[2] C. E. Castro and W. C. Kray, Jr., *Am. Soc.*, **85**, 2768 (1963); **86**, 4603 (1964); C. E. Castro and R. D. Stephens, *ibid.*, **86**, 4358 (1964)

Chromyl acetate, $(CH_3CO_2)_2CrO_2$. Mol. wt. 202.10, m.p. 30.5°. The reagent has been isolated[1] by stirring a suspension of 5 g. of CrO_3 in 20 ml. of carbon tetrachloride and 2 g. of acetic anhydride for 5 hrs. (25°) with exclusion of moisture and light, filtering the red solution from CrO_3, and high-vacuum distillation below 20°. It forms red crystals very sensitive to moist air. In a microanalytical study of the oxidation of branched-chain fatty acids and esters, the reagent was generated in carbon tetrachloride in the presence of the substrate.[2] It attacks somewhat selectively a tertiary hydrogen at the γ-position (*compare* Anhydrous CrO_3–acetic acid, p. 145).

[1] H.-L. Krauss, *Angew. Chem.*, **70**, 502 (1958)
[2] J. Cason, P. Tavs, and A. Weiss, *Tetrahedron*, **18**, 437 (1962)

Chromyl chloride, CrO_2Cl_2. Mol. wt. 154.92, b.p. 116°, sp. gr. 1.92. Supplier: Allied Chemical Corp., Baker and Adamson Dept., Alfa, KK. Preparation.[1]

Oxidation. The reagent, a cherry red, volatile liquid, is used in the Étard reaction[2] for oxidation of aromatic methyl derivatives to aldehydes.[3] The reaction is catalyzed by a trace of olefin.[4]

$$C_6H_5CH_3 \quad \xrightarrow{CrO_2Cl_2} \quad C_6H_5CHO$$

Barton *et al.*[5] observed some unusual oxidations by chromyl chloride in the lanosterol series, namely oxidation of $\Delta^{2, 8(9)}$-lanostadiene (1) to the $\Delta^{2, 9(11)}$-7-ketone (2) and of the $\Delta^{7, 9(11)}$-diene (3) to the Δ^8-7-ketone (4). The reactions are particularly

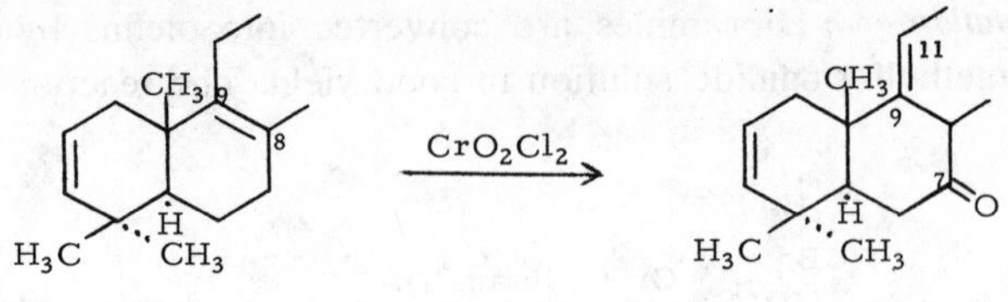

(1) (2)

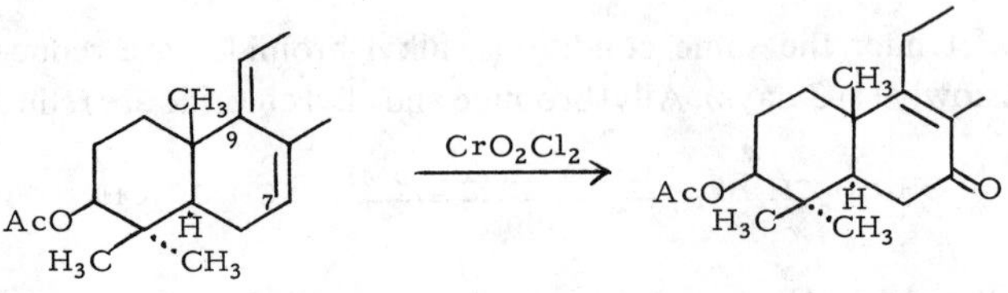

(3) (4)

unexpected in the light of reactions of the reagent with unsaturated steroids described in the next section.

Reaction with olefins. Chromyl chloride reacts with an olefin, for example cyclohexene, in carbon tetrachloride to give a brown solid of the approximate

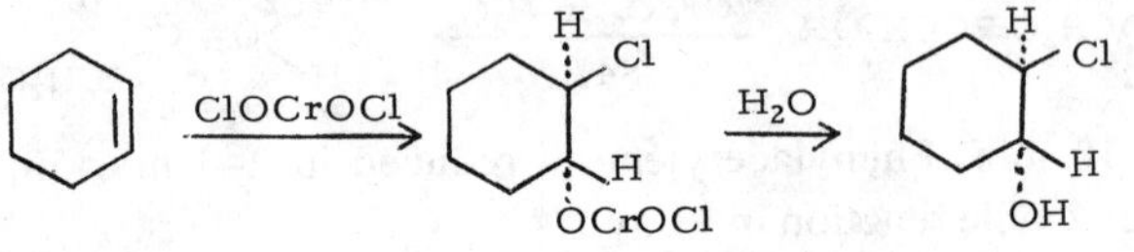

(1) (2) (3)

composition (2), which on hydrolysis gives the *trans*-chlorohydrin (3).[6] The chlorohydrin from an unsymmetrical olefin is positionally isomeric with that obtained by addition of hypochlorous acid; thus a terminal olefin, $RCH{=}CH_2$, yields the β-chloro primary alcohol, $RCHClCH_2OH$ (yields 35–50%). The same reversal has been observed with a Δ^2-steroid derived from hecogenin.[7]

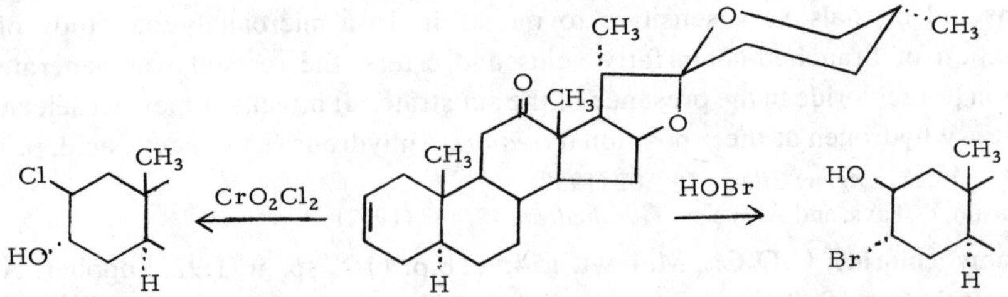

[1]H. D. Law and F. M. Perkin, *J. Chem. Soc.*, **91**, 191 (1907); H. H. Sisler, *Inorg. Syn.*, **2**, 205 (1946)

[2]M. A. Étard, *Compt. rend.*, **90**, 534 (1880); *Ann. chim. phys.*, [5], **22**, 223 (1881)

[3]L. N. Ferguson, *Chem. Rev.*, **38**, 237 (1946)

[4]A. Tillotson and B. Houston, *Am. Soc.*, **73**, 221 (1951)

[5]D. H. R. Barton, P. J. L. Daniels, J. F. McGhie, and P. J. Palmer., *J. Chem. Soc.*, 3675 (1963)

[6]S. J. Cristol and K. R. Eilar, *Am. Soc.*, **72**, 4353 (1950)

[7]H. L. Slates and N. L. Wendler, *Am. Soc.*, **78**, 3749 (1956)

Chromyl trichloroacetate, $CrO_2(OCOCCl_3)_2$. The formation of undesired products of chlorination in the Étard reaction is eliminated by use of this organic ester, prepared as a deep red solution from chromic anhydride and trichloroacetic anhydride in carbon tetrachloride; in the absence of moisture and light, the solution is stable indefinitely at room temperature.[1] It oxidizes cycloalkenes to dialdehydes in yields of 30–70%. The mechanism suggested accounts for the formation of small amounts of epoxides and *trans*-diols as by-products.

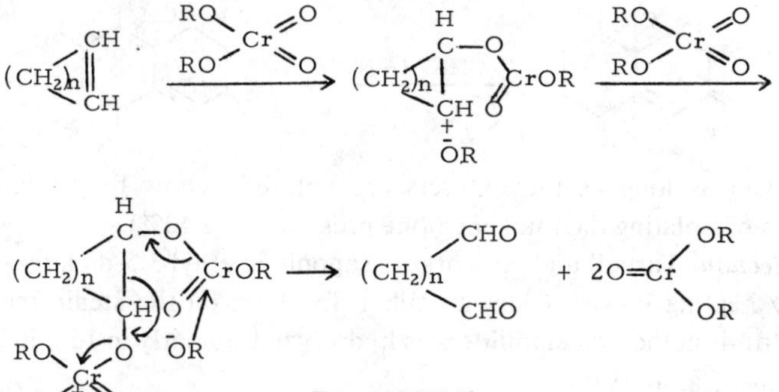

$R = -COCCl_3$

[1]H. Schildknecht and W. Föttinger, *Ann.*, **659**, 20 (1962)

N-*trans*-Cinnamoylimidazole (1). Mol. wt. 198.22, m.p. 133–133.5°. Prepared in high yield by reaction of cinnamoyl chloride with imidazole in benzene at 10–25°.[1] The reagent reacts rapidly and quantitatively with the active site of α-chymotrypsin and hence can be used for the spectrophotometric determination of the normality of an enzyme solution by titration.

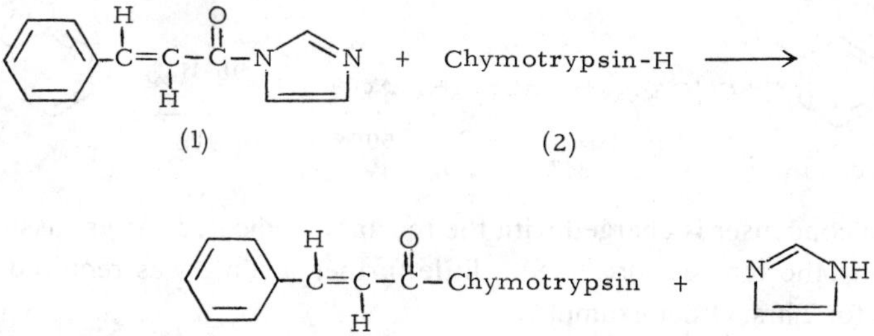
(1)
(2)

[1]G. R. Schonbaum, B. Zerner, and M. L. Bender, *J. Biol. Chem.*, **236**, 2930 (1961)

Claisen's alkali.[1] Preparation: dissolve 35 g. of potassium hydroxide in 25 ml. of water, cool, add 100 ml. of methanol, and cool.

Extraction of cryptophenols and enols. Phenols insoluble in aqueous alkali, for example 2,4,6-triallylphenol,[1] can be extracted from petroleum ether with Claisen's alkali. Vitamin K_1 is easily isolated from a 3–5% alfalfa concentrate by shaking an alcoholic suspension of the oil with aqueous sodium hydrosulfite, extracting with petroleum ether, and extracting the K_1 hydroquinone from petroleum ether with Claisen's alkali. The yellow extract is diluted with water and K_1 hydroquinone

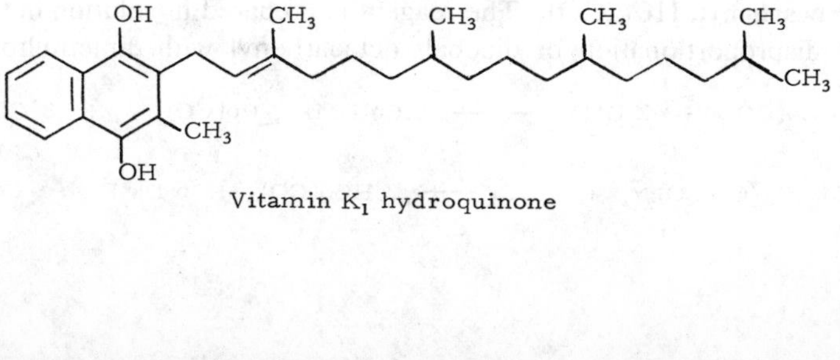

Vitamin K_1 hydroquinone

(colorless) is isolated by ether extraction and crystallization from petroleum ether; oxidation with silver oxide in ether gives pure vitamin K_1.[2]

Exhaustive dichromate oxidation of cholesterol and removal of an extensive acidic fraction leaves a mixture of Δ^4-cholestene-3,6-dione and several monoketones and other neutral products.[3] Repeated extraction of a solution in petroleum ether with

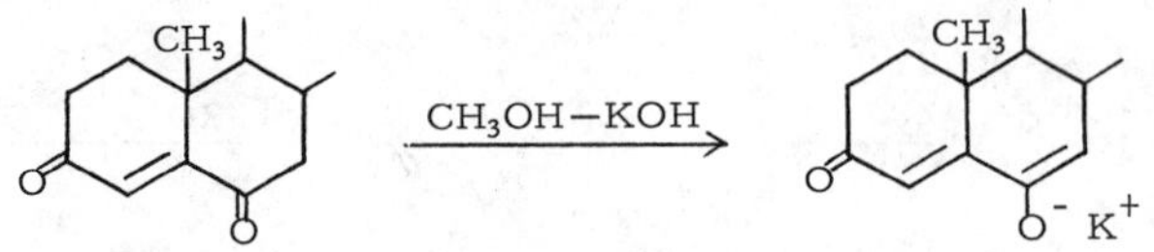
CH3
CH3OH–KOH
CH3
K+

Claisen's alkali as long as the extracts are colored yellow by enolate affords a simple means of isolating the total enedione present (yield 40%).

Saponification. Tarbell and co-workers saponified ethyl 3,5-dichloro-4-hydroxybenzoate by heating it with Claisen's alkali for 1 hr. on the steam bath (93–97% yield);[4] 2-nitro-4-methoxyacetanilide was hydrolyzed similarly in 15 min. of heating.[5]

[1]L. Claisen, *Ann.*, **418**, 96 (1919)
[2]L. F. Fieser, *Am. Soc.*, **61**, 3467 (1939)
[3]L. F. Fieser, *Org. Syn., Coll. Vol.*, **4**, 189 (1963)
[4]D. S. Tarbell, J. W. Wilson, and P. E. Fanta, *ibid.*, **3**, 267 (1955)
[5]P. E. Fanta and D. S. Tarbell, *ibid.*, **3**, 661 (1955)

Cobalt acetate bromide, CoOAcBr. The reagent, prepared *in situ* from cobalt acetate and HBr, is a very active catalyst for the autoxidation of activated methylene or methyl groups. For example,[1] a flask equipped with a Vibromix stirrer, a gas inlet

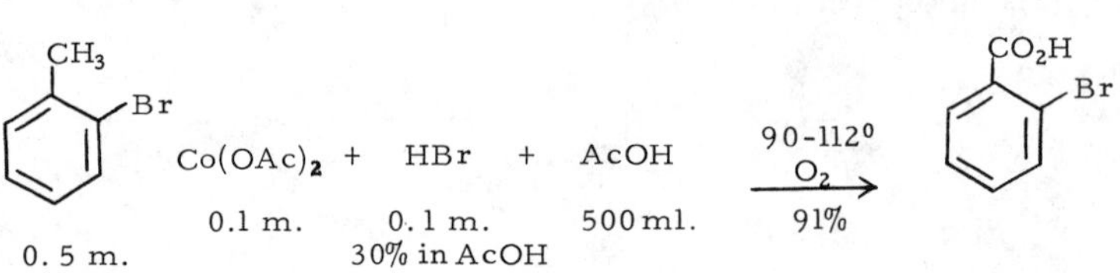
CH3
Br
Co(OAc)2 + HBr + AcOH
0.1 m. 0.1 m. 500 ml.
30% in AcOH
0. 5 m.
90-112°
O2
91%
CO2H
Br

tube, and condenser is charged with the reactants indicated. After passing in oxygen and raising the temperature to 90°, little further heating was required to maintain refluxing for 2 hrs. Other examples:

$$C_6H_5\overset{O}{\overset{\|}{C}}OCH_2CH_2C_6H_5 \xrightarrow[71\%]{CoOAcBr,\ O_2} C_6H_5\overset{O}{\overset{\|}{C}}CH_2\overset{O}{\overset{\|}{C}}C_6H_5$$

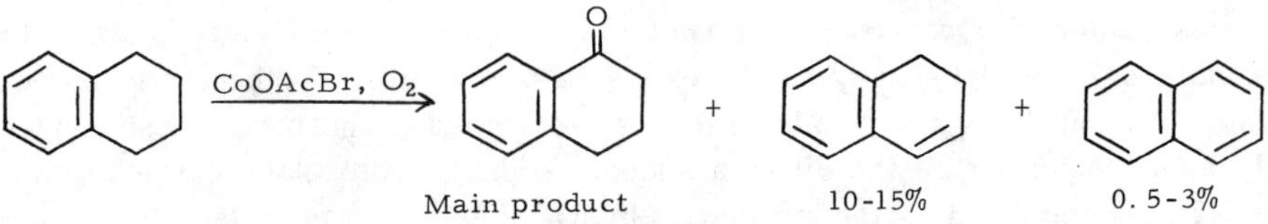
CoOAcBr, O2
+
+
Main product
10-15%
0. 5-3%

[1]A. S. Hay and H. S. Blanchard, *Can. J. Chem.*, **43**, 1306 (1965)

Cobalt hydrocarbonyl, $HCo(CO)_4$. The reagent is prepared in solution in toluene or hexane by disproportionation of dicobalt octacarbonyl with dimethylformamide,

$$3\ Co_2(CO)_8 + 12\ DMF \longrightarrow 2\ Co(DMF)_6[Co(CO)_4]_2 + 8\ CO$$

$$Co(DMF)_6[Co(CO)_4]_2 + 2\ HCl \longrightarrow 2\ HCo(CO)_4 + 6\ DMF + CoCl_2$$

followed by acidification.[1] It is regarded as the true catalyst in the hydroformylation of olefins (oxo reaction);[2,3] *see* Dicobalt octacarbonyl.

[1]L. Kirch and M. Orchin, *Am. Soc.*, **80**, 4428 (1958)
[2]I. Wender, H. W. Sternberg, and M. Orchin, *Am. Soc.*, **75**, 3041 (1953)
[3]R. F. Heck and D. S. Breslow, *Am. Soc.*, **83**, 4023 (1961)

Cobaltous chloride, $CoCl_2$. Mol. wt. 129.85. Prepared from the hexahydrate by dehydration in a stream of hydrogen chloride and dried in vacuum at 100–150°.[1] Suppliers: Alfa, F. KK, MCB.

A catalytic amount of this halide influences the course of Grignard reactions through intermediate formation of RCoCl, which promotes coupling with an organic halide (Kharasch[1, 2]):

1.[1] $C_6H_5MgBr + C_6H_5Br \xrightarrow[86\%]{CoCl_2} C_6H_5-C_6H_5$

2.[3] $C_6H_5MgBr + C_6H_5C{\equiv}CBr \xrightarrow[75\%]{CoCl_2} C_6H_5C{\equiv}CC_6H_5$

[1]M. S. Kharasch and E. K. Fields, *Am. Soc.*, **63**, 2316 (1941), and later papers
[2]Reviewed by D. H. Hey, *Ann. Reports, Chem. Soc.*, **41**, 195 (1944); **45**, 160 (1948)
[3]H. K. Black, D. H. S. Horn, and B. C. L. Weedon, *J. Chem. Soc.*, 1704 (1954)

***s*-Collidine (2,4,6-Trimethylpyridine).** Mol. wt. 121.18, b.p. 171°. Suppliers: B, E, MCB. Redistil before use.

This base is generally superior to pyridine for dehydrohalogenation. Addition of 4% of 3,5-lutidine is advantageous (synthesis of carotinoid polyenes by reaction at 140° under nitrogen).[1]

[1]H. H. Inhoffen, H.-J. Krause, and S. Bork, *Ann.*, **585**, 132 (1954)

Copper (or Cu^{++})–Ascorbic acid. Anbar *et al.*[1] found that *o*-iodobenzoic acid is reduced rapidly to benzoic acid by copper (II) ions and ascorbic acid in neutral aqueous solution but that reduction of *p*-iodobenzoic acid under the same conditions is at least 1,000 times slower. Safe and Moir[2] effected selective reduction of the iodine atom of (1) without disturbance of the nitro group with copper metal and alkaline ascorbic acid. The method failed with the corresponding bromo compound.

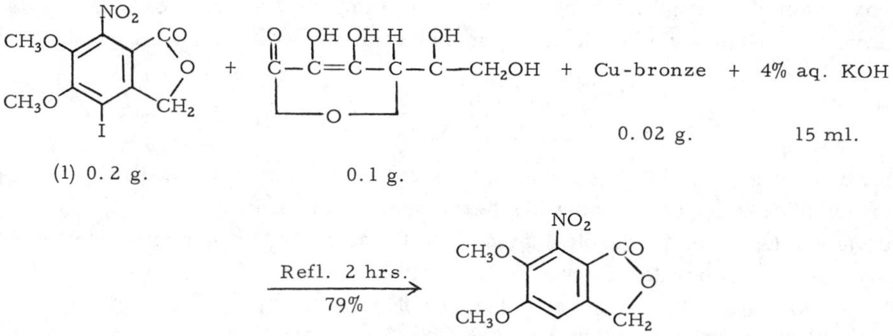
NO2
CH3O
CO
O
CH3O
CH2
I
(1) 0.2 g.
O OH OH H OH
C—C=C—C—C—CH2OH
O
0.1 g.
Cu-bronze
0.02 g.
4% aq. KOH
15 ml.
Refl. 2 hrs.
79%
NO2
CH3O
CO
O
CH3O
CH2

[1]M. Anbar, S. Guttmann, and C. Friedman, *Proc. Chem. Soc.*, 10 (1963)
[2]S. Safe and R. Y. Moir, *Can. J. Chem.*, **43**, 337 (1965)

Copper bronze. Ordinary copper bronze does not give as satisfactory results in the Ullmann reaction as bronze that has been activated with a 2% solution of iodine in acetone.[1]

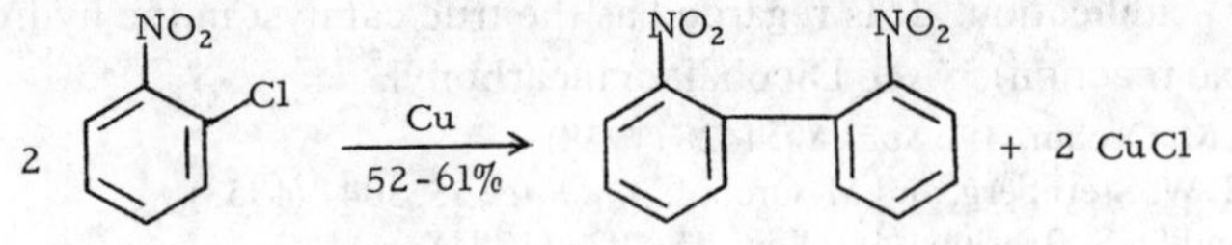

[1]R. C. Fuson and E. A. Cleveland, *Org. Syn., Coll. Vol.*, **3**, 339 (1955)

Copper chromite (Lazier catalyst). Supplier: Harshaw (CU-0202P; 556–002). For preparation of the catalyst[1] an aqueous solution of barium nitrate and cupric nitrate trihydrate is stirred during addition of a solution of ammonium chromate, prepared from ammonium dichromate and aqueous ammonia. The reddish brown precipitate of copper barium ammonium chromate is washed and dried and decomposed by heating in a muffle furnace at 350–450°. The ignition residue is pulverized, washed with 10% acetic acid, dried, and ground to a fine black powder.

Use of the catalyst in hydrogenation is illustrated by the hydrogenation of sebacoin to an easily separated mixture of *cis*- and *trans*-cyclodecane-1,2-diol.[2] The *cis*-diol (m.p. 138°) crystallizes from 1:1 benzene–ethanol, the *trans*-diol (m.p. 54°) crystallizes from pentane.

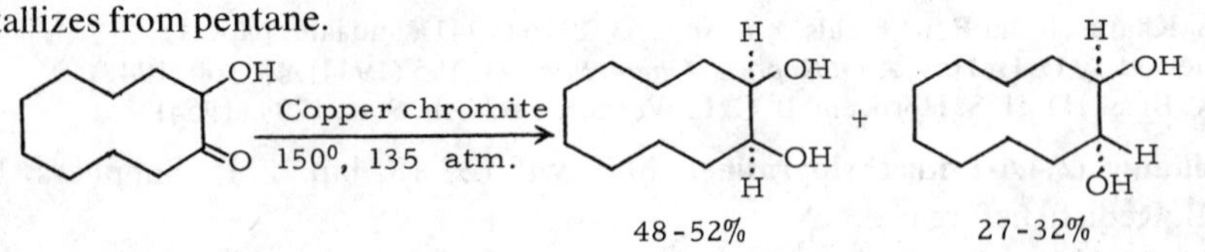

A procedure for preparation of β-methyl-δ-valerolactone from 3-methylpentane-1,5-diol[3] shows that copper chromite is an effective catalyst for dehydrogenation. Presumably the reaction proceeds through the aldehyde and hemiacetal. The evolution of hydrogen is nearly quantitative and the yield of lactone high.

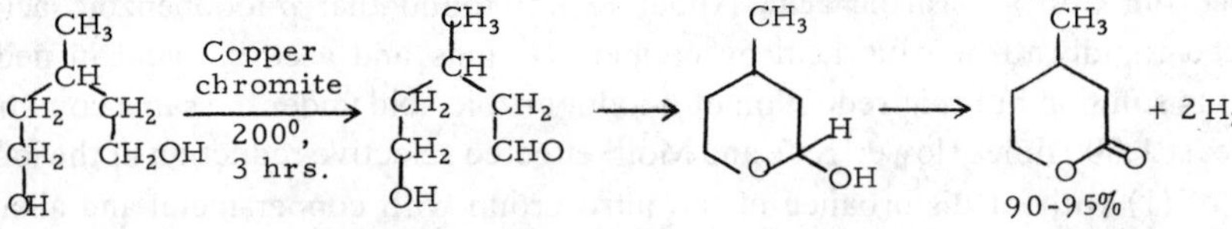

Copper chromite and copper–chromium oxide are effective catalysts for the decarboxylation of acids in quinoline solution. An example is the decarboxylation of *cis*-α-phenylcinnamic acid in refluxing quinoline to *cis*-stilbene; the two catalysts cited

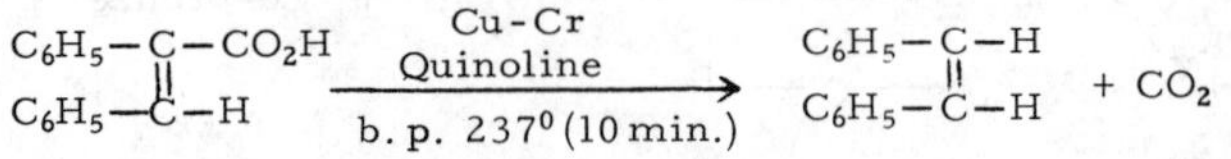

are equally effective.[4] Related catalysts[5,6] have been used in similar decarboxylations in quinoline. Copper chromite has been used without solvent as catalyst for the decarboxylation of imidazole-4,5-dicarboxylic acid[7] and isodehydroacetic acid.[8] *See also* Copper powder; Copper salts.

[1]W. A. Lazier and H. R. Arnold, *Org. Syn., Coll. Vol.*, **2**, 142 (1943)
[2]A. T. Blomquist and A. Goldstein, *Org. Syn., Coll. Vol.*, **4**, 216 (1963)
[3]R. I. Longley, Jr., and W. S. Emerson, *ibid.*, **4**, 677 (1963)
[4]*Org. Expts.*, Chapt. 42.9
[5]C. R. Kinney and D. P. Langlois, *Am. Soc.*, **53**, 2189 (1931)
[6]T. Reichstein, A. Grüssner, and H. Zschokke, *Helv.*, **15**, 1067 (1932)
[7]H. R. Snyder, R. G. Handrick, and L. A. Brooks, *Org. Syn., Coll. Vol.*, **3**, 471 (1955)
[8]N. R. Smith and R. H. Wiley, *ibid.*, **4**, 337 (1963)

Copper–Chromium oxide (Adkins catalyst HJS 2). The catalyst is prepared[1] by stirring a solution of copper nitrate and barium nitrate into an aqueous solution of sodium dichromate and ammonium hydroxide. The orange precipitate is washed, dried, and stirred at 350° to effect decomposition. The cooled material is leached with 10% acetic acid, dried, and pulverized. The catalyst so obtained is brownish black.

With this catalyst, methyl palmitate afforded palmityl alcohol in 78% yield on hydrogenation at 175° for 3 hrs.[1] Aldehydes and ketones were hydrogenated at room temperature with catalyst that had been activated by heating it at 100° under hydrogen pressure.[1] The same activation can be effected by refluxing the catalyst in cyclohexanol for 4 hrs.; during this time 11% of cyclohexanone is formed.[2] Cholestane-3β-ol, refluxed in xylene with three times its weight of Adkins catalyst (HJS 2), affords cholestane-3-one in moderate yield.

This Adkins catalyst is effective for saturation of the reactive 9,10-double bond of phenanthrene,[3] and for hydrogenolysis of tetrahydrofurfuryl alcohol to 1,5-pentanediol.[4]

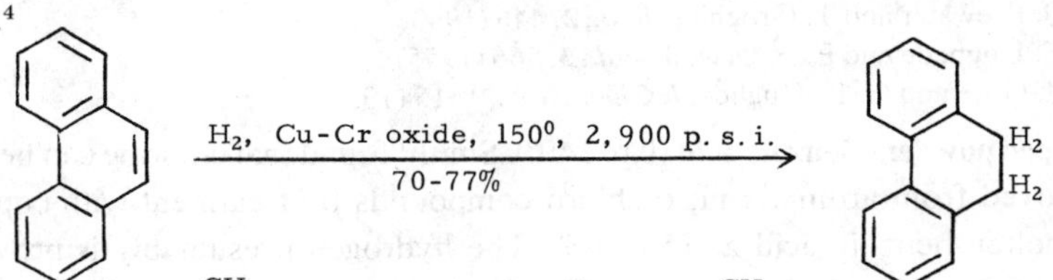
H₂, Cu-Cr oxide, 150°, 2,900 p. s. i.
70-77%
H₂
H₂

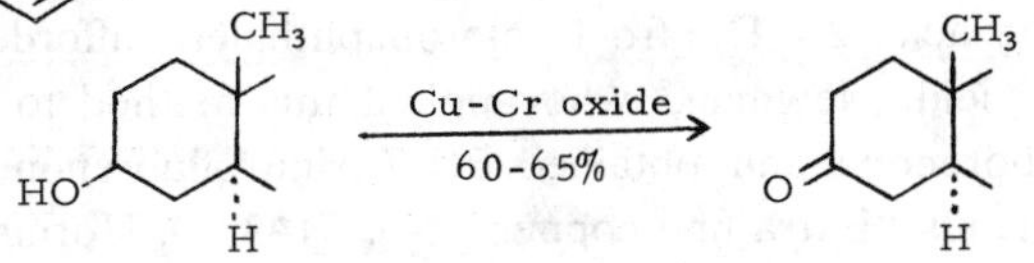
CH₃
HO
H
Cu-Cr oxide
60-65%
CH₃
O
H

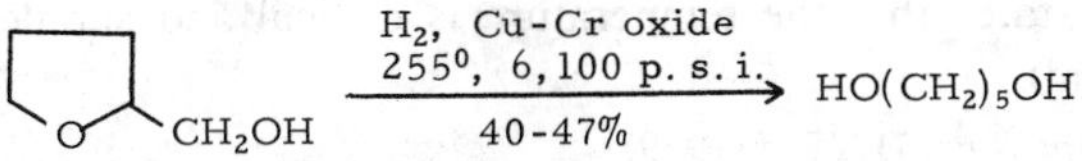
CH₂OH
H₂, Cu-Cr oxide
255°, 6,100 p. s. i.
40-47%
HO(CH₂)₅OH

[1]H. Adkins, E. E. Burgoyne, and H. J. Schneider, *Am. Soc.*, **72**, 2626 (1950); see also Note 11 in a description of the Lazier catalyst, *Org. Syn., Coll. Vol.*, **2**, 142 (1943)
[2]W. R. Nes, *J. Org.*, **23**, 899 (1958)
[3]D. D. Phillips, *Org. Syn., Coll. Vol.*, **4**, 313 (1963)
[4]D. Kaufman and W. Reeve, *ibid.*, **3**, 693 (1955)

Copper powder. Use of commercial copper powder as a decarboxylation catalyst in quinoline solution is illustrated by the decarboxylation of 3-methyl-2-furoic acid[1] and of *m*-nitrocinnamic acid.[2]

When a β-hydroxy acid or its *t*-butyl ester is heated in quinoline solution with a trace of copper powder, decarboxylation and dehydration occur and the product is an exocyclic olefin.[3] For decarboxylation of substituted cinnamic acids in quinoline,

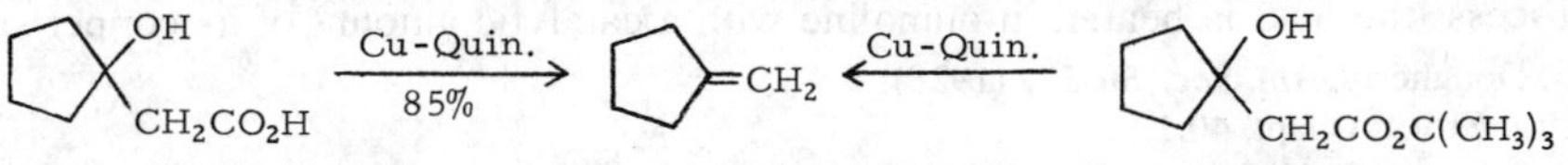
OH
CH₂CO₂H
Cu-Quin.
85%
CH₂
Cu-Quin.
OH
CH₂CO₂C(CH₃)₃

copper powder is a better catalyst than copper sulfate or copper acetate.[4]

Copper powder is used as catalyst in the reaction of sodium with aniline to form the sodium derivative, $C_6H_5\bar{N}HNa^+$.[5] An active copper powder is preferred as catalyst for the synthesis of *p*-nitrodiphenyl ether[6] and *o*-methoxydiphenyl ether.[7] The active catalyst is prepared[6] by reducing copper sulfate in aqueous solution with zinc dust and leaching the precipitate with 5% hydrochloric acid to remove excess

$$p\text{-}NO_2C_6H_4Cl + KOC_6H_5 \xrightarrow[80\text{–}82\%]{Cu(150\text{–}160°)} p\text{-}NO_2C_6H_4OC_6H_5$$

$$C_6H_5Br + KOC_6H_4OCH_3\text{-}o \xrightarrow[54\text{–}61\%]{Cu(200°)} C_6H_5OC_6H_4OCH_3\text{-}o$$

zinc. Active catalyst prepared by essentially the same method[8] is preferred for the Ullmann synthesis of 2,2′-dinitrodiphenyl.[8]

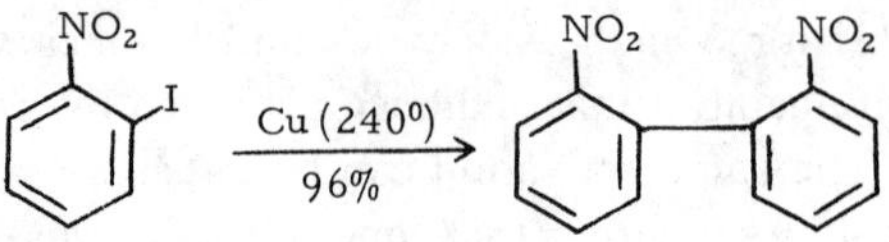
NO2
I
Cu (240°)
96%
NO2
NO2

[1]D. M. Burness, *Org. Syn., Coll. Vol.*, **4**, 628 (1963)
[2]R. H. Wiley and N. R. Smith, *ibid.*, **4**, 731 (1963)
[3]M. Vilkas and N. A. Abraham, *Bull. soc.*, 1196 (1960)
[4]C. Walling and K. B. Wolfstirn, *Am. Soc.*, **69**, 852 (1947)
[5]W. R. Vaughan, *Org. Syn., Coll. Vol.*, **3**, 329 (1955)
[6]R. Q. Brewster and T. Groening, *ibid.*, **2**, 446 (1943)
[7]H. E. Ungnade and E. F. Orwoll, *ibid.*, **3**, 566 (1955)
[8]P. H. Gore and G. K. Hughes, *J. Chem. Soc.*, 1615 (1959)

Copper powder–Benzoic acid (b.p. 249°). Smith[1] found that chlorine can be reductively removed from aromatic nitro-chloro compounds by treatment with copper powder in molten benzoic acid at 150–200°. The hydrogen presumably is provided by the carboxyl group of the acid. 2,4-Dinitro-1-chloronaphthalene afforded 1,3-dinitronaphthalene in 74% yield. Newman,[2] who applied this method to 4-bromo- and 4-iodo-2,5,7-trinitrofluorenone and obtained 2,5,7-trinitrofluorenone in yields of 78 and 22%, recommends "Extra fine copper" No. 5143, O. Hormmel Co., Pittsburgh, Pa., and states that the temperature is critical and should be in the range 170–180° (internal).

[1]W. T. Smith, Jr., *Am. Soc.*, **71**, 2855 (1949)
[2]M. S. Newman and J. Blum, *Am. Soc.*, **86**, 5600 (1964)

Copper salts. *o*-Benzoylbenzoic acid can be decarboxylated in high yield by heating a mixture of the acid with a small amount of its copper salt to 260–270° until evolution of carbon dioxide is at an end.[1] An even simpler procedure is to add a little

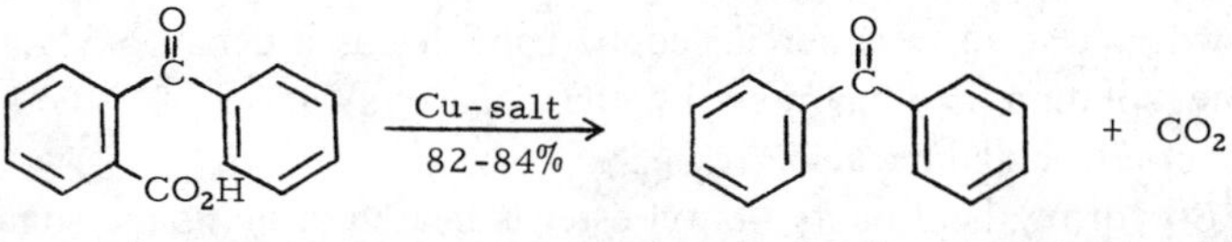
O
C
CO2H
Cu-salt
82-84%
O
C
+ CO2

basic copper carbonate (0.5 g.) to the acid (15 g.) in a distilling flask, melt the mixture and evacuate to eliminate water, and then heat at 260–270°.[2] In another process, the acid is heated in quinoline with a catalytic amount of its copper salt.[3]

[1]G. Dougherty, *Am. Soc.*, **50**, 571 (1928)
[2]*Org. Expts.*, Chapt. 40.1
[3]E. Piers and R. K. Brown, *Can. J. Chem.*, **40**, 559 (1962)

Cotton, absorbent. An excellent drying agent[1] for use in a drying tube placed at the top of a reflux condenser or dropping funnel. The cotton can be dried by heating it in an oven at 100°. On acetolysis it affords α-cellobiose octaacetate in 35–37% yield.[2]

[1]F. P. Pingert, *Org. Syn., Coll. Vol.*, **3**, 19 (1955)
[2]G. Braun, *Org. Syn., Coll. Vol.*, **2**, 124 (1943)

Cupric acetate, anhydrous, $Cu(OCOCH_3)_2$. Mol. wt. 181.63. Supplier: F.

Anhydrous material, prepared by refluxing the hydrate with acetic anhydride and washing the insoluble product with dry ether,[1] is used in methanol–pyridine for the oxidative coupling of terminal acetylenes to diynes.[2-4] Mechanism.[2] Review.[5]

$$2\ RC{\equiv}CH \xrightarrow{Cu(OAc)_2} RC{\equiv}CC{\equiv}CR$$

Intramolecular oxidative coupling of the ω,ω-diacetylenic ester (1) to the diynolide (2) was a key step in a new type of synthesis of the macrocyclic lactone exaltolide.[6]

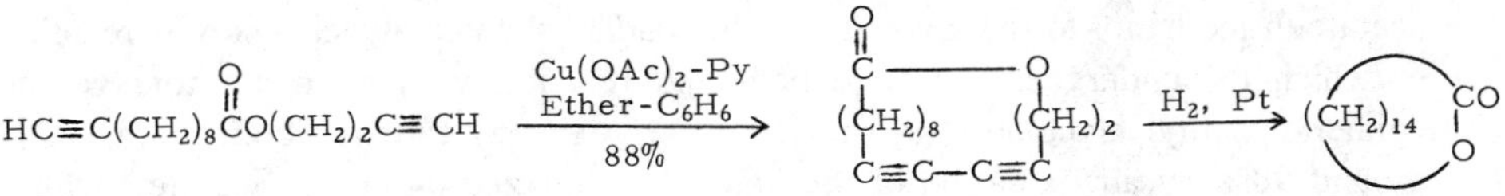
HC≡C(CH2)8CO(CH2)2C≡CH
Cu(OAc)2-Py
Ether-C6H6
88%
(CH2)8
(CH2)2
C≡C—C≡C
H2, Pt
(CH2)14
CO
O

(1) (2) (3)

[1]E. Späth, *Sitzungsber. Akad. Wiss. Wien*, **120**, 117 (1911)
[2]A. A. Clifford and W. A. Waters, *J. Chem. Soc.*, 3056 (1963)
[3]G. Eglinton and A. R. Galbraith, *J. Chem. Soc.*, 889 (1959)
[4]F. Sondheimer and R. Wolovsky, *Am. Soc.*, **81**, 1771 (1959); F. Sondheimer, Y. Amiel, and Y. Gaoni, *ibid.*, **81**, 1771 (1959)
[5]G. Eglinton and W. McCrae, *Advances in Organic Chemistry*, **4**, 225 (1963)
[6]W. McCrae, *Tetrahedron*, **20**, 1773 (1964)

Cupric acetate monohydrate, $Cu(OCOCH_3)_2 \cdot H_2O$. Mol. wt. 199.64. Suppliers: B, MCB.

(a) One use of the reagent is for isolation of 1,3-diketones as the chelated copper derivatives: acetylacetone,[1] diisovalerylmethane,[2] ethyl diacetylacetate.[3] The

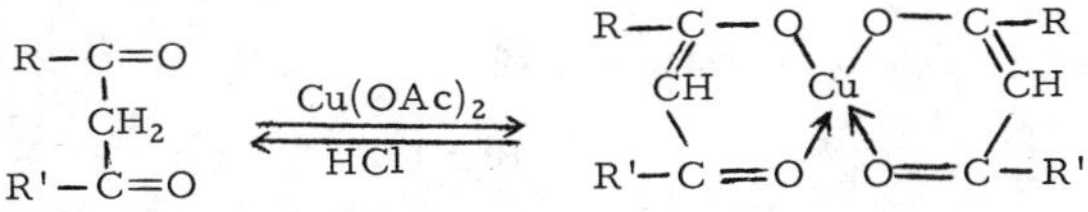
R—C=O
CH2
R'—C=O
Cu(OAc)2
HCl
R—C
CH
R'—C=O
Cu
C—R
CH
C—R'

diketone is treated as such with aqueous cupric acetate[1,3] or a solution in methanol is treated with aqueous cupric acetate.

(b) The reagent is an effective catalyst for the monocyanoethylation of aromatic amines.[4]

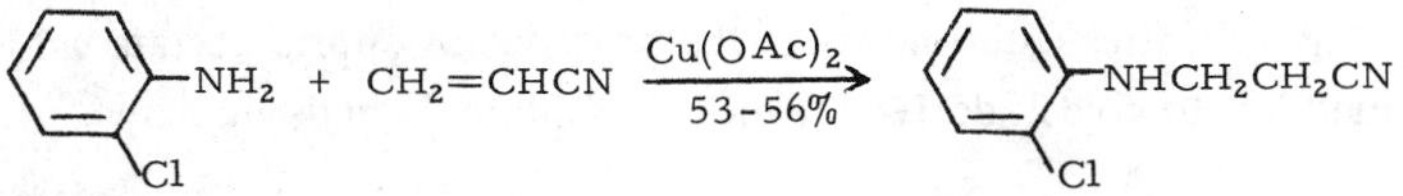
NH2 + CH2=CHCN
Cl
Cu(OAc)2
53-56%
NHCH2CH2CN
Cl

(c) Cupric acetate oxidizes α-ketols to α-diketones, for example, sebacoin (1) to sebacil (2); the reaction is conducted in aqueous acetic acid–methanol; at 75°

(1)

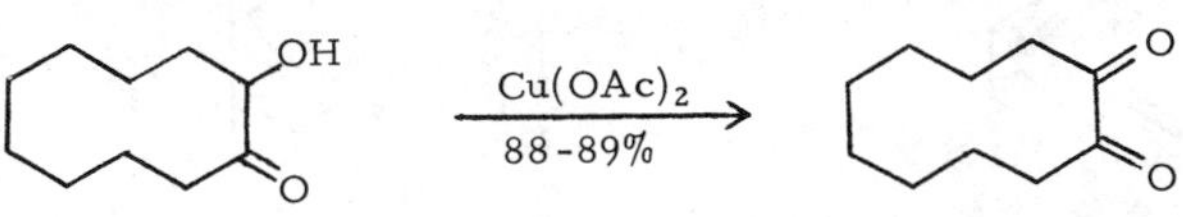
OH
O
Cu(OAc)2
88-89%
O
O

(2)

the color changes from blue to red.[5] Steroid D-homoketols (3) afford diosphenols

(3)

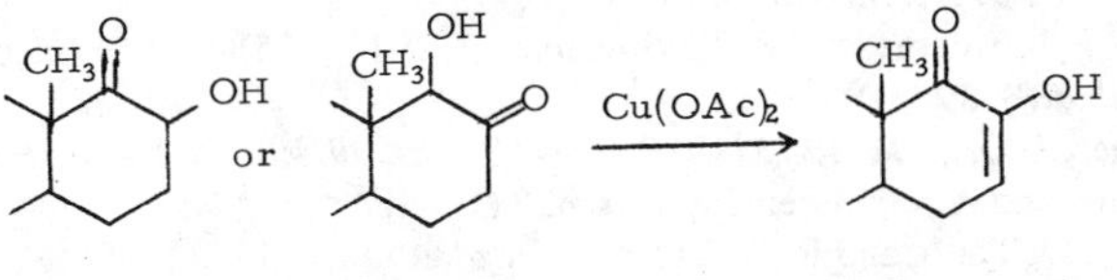
CH3
O
OH
or
CH3
OH
O
Cu(OAc)2
CH3
O
OH

(4)

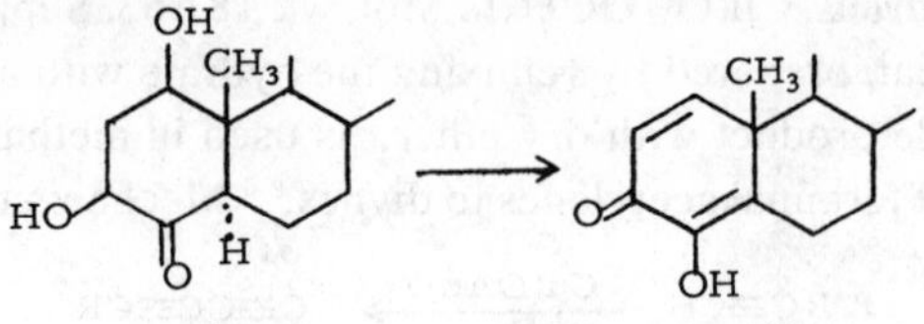
(5) (6)

(4),[6] and the steroid ketodiol (5) yields the 1,4-diene-3-one-4-ol (6).[7] The α-ketol side chain (7) of a steroid such as cortisone is oxidized rapidly (30 min.) by cupric acetate in methanol to the glyoxal (8); the oxidant then catalyzes a slower reaction in which (8) undergoes rearrangement and reaction with methanol to give the epimeric methyl glycolates (9) and (10).[8] Glyoxal (8) can be obtained in yields around 90% by air oxidation of the ketol (7) catalyzed by cupric acetate;[9] with a ratio of cupric acetate to steroid of 1:8, oxidation is complete in an hour or less.

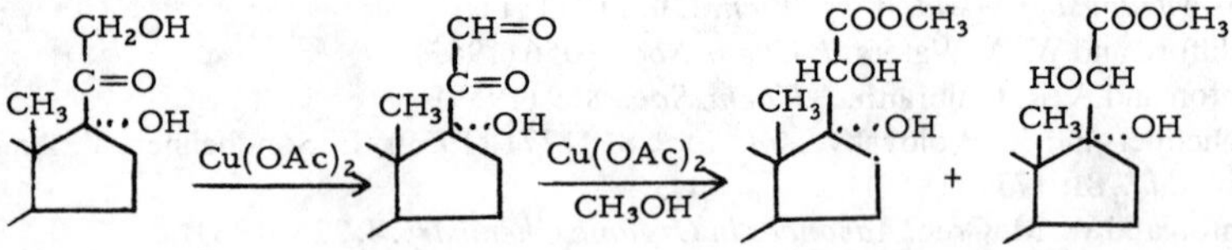
Cu(OAc)₂
Cu(OAc)₂
CH₃OH
(7) (8) (9) (10)

For Glaser oxidative coupling of phenylacetylene to diphenyldiacetylene, Campbell and Eglinton[10] used cupric acetate in pyridine–methanol. The deep blue suspension became green when refluxed (1 hr.). The cooled mixture was acidified with stirring and cooling and the product was collected by ether extraction.

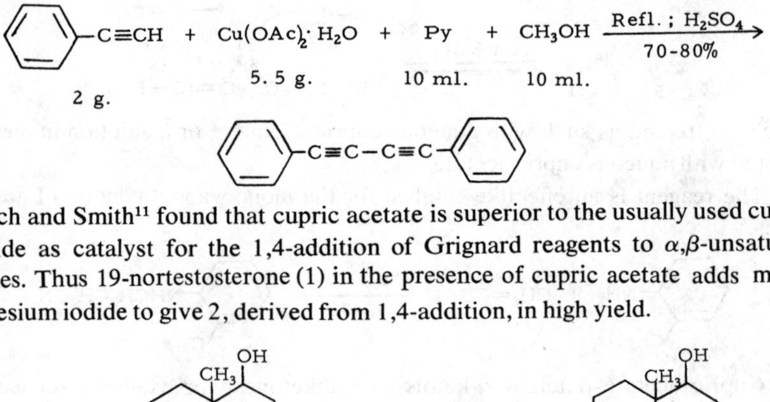
+ Cu(OAc)₂·H₂O + Py + CH₃OH
Refl. ; H₂SO₄
70-80%
2 g.
5. 5 g.
10 ml.
10 ml.

Birch and Smith[11] found that cupric acetate is superior to the usually used cuprous bromide as catalyst for the 1,4-addition of Grignard reagents to α,β-unsaturated ketones. Thus 19-nortestosterone (1) in the presence of cupric acetate adds methylmagnesium iodide to give 2, derived from 1,4-addition, in high yield.

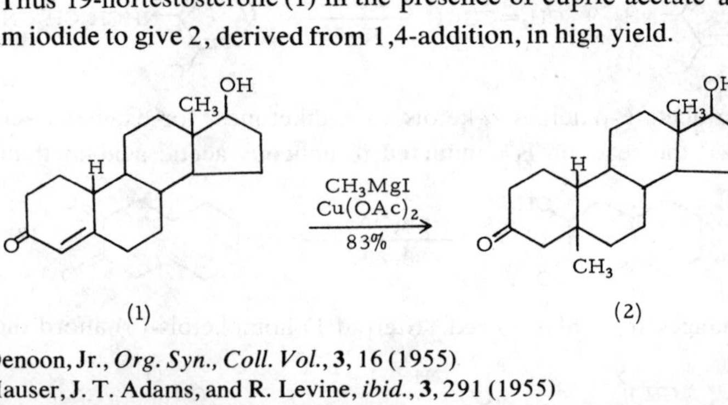
CH₃MgI
Cu(OAc)₂
83%
(1) (2)

[1]C. E. Denoon, Jr., *Org. Syn., Coll. Vol.*, **3**, 16 (1955)
[2]C. R. Hauser, J. T. Adams, and R. Levine, *ibid.*, **3**, 291 (1955)
[3]A. Spassow, *ibid.*, **3**, 390 (1955)
[4]S. A. Heininger, *J. Org.*, **22**, 1213 (1957); *Org. Syn., Coll. Vol.*, **4**, 146 (1963)
[5]A. T. Blomquist and A. Goldstein, *ibid.*, **4**, 838 (1963)
[6]N. L. Wendler, D. Taub, and R. P. Graber, *Tetrahedron*, **7**, 173 (1959)

[7]K. Sasaki, *Chem. Pharm. Bull.*, **9**, 684 (1961)
[8]M. L. Lewbart and V. R. Mattox, *J. Org.*, **28**, 1773, 1779, 2001 (1963)
[9]*Idem, ibid.*, **28**, 2001 (1963)
[10]I. D. Campbell and G. Eglinton, *Org. Syn.*, **45**, 39 (1965)
[11]A. J. Birch and M. Smith, *Proc. Chem. Soc.*, 356 (1962); *see also* J. A. Marshall, W. I. Fanta, and H. Roebke, *J. Org.*, **31**, 1016 (1966)

Cupric bromide, $CuBr_2$. Mol. wt. 223.37, m.p. 498°; black solid, very soluble in water (green solution). Suppliers: Baker and Adamson, B, F, MCB.

Bromination of carbonyl compounds. Kochi[1] observed that cupric chloride reacts with acetone as follows:

$$CH_3COCH_3 + 2\ CuCl_2 \longrightarrow ClCH_2COCH_3 + 2\ CuCl + HCl$$

The more reactive cupric bromide has been used in subsequent work. Isophorone reacts with cupric bromide in methanol at room temperature to give the α-bromo derivative (1), which is unstable in the presence of HBr and methanol and rearranges

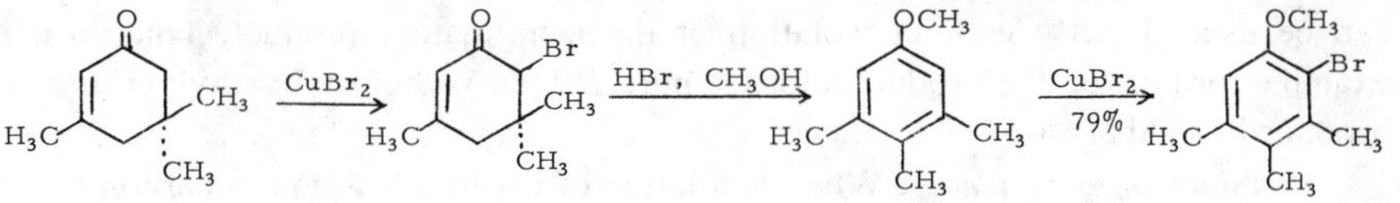

(1) (2) (3)

to give 3,4,5-trimethylanisole (2).[2] Further bromination of (2) affords (3) in good yield.

Cupric bromide reacts with α,β-unsaturated aldehydes in ethanol in the manner illustrated for acrolein.[3] In the reaction with a Δ^4-3-ketosteroid in methanol, a

$$CH_2{=}CHCHO + 3\ EtOH + 2\ CuBr_2 \xrightarrow[67\%]{} EtOCH_2\underset{Br}{C}HCH(OEt)_2 + 2\ CuBr + HBr + H_2O$$

6-bromo derivative is formed, but reacts with solvent methanol to give a 6β-methoxy derivative.[4] The reagent is useful for the bromination of 17-ketosteroids where

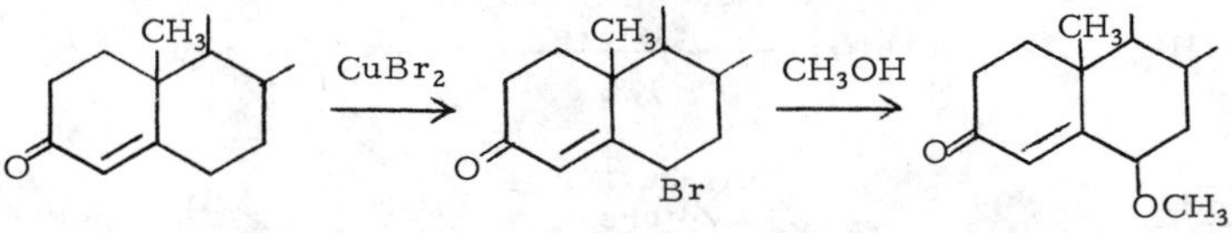

direct bromination gives poor results.[5] One advantage is that the bromination can be conducted in the presence of a 5,6-double bond without attack at this point.[6]

Both Doifode and Marathey[7] and King and Ostrum[8] observed that on reaction with a phenol containing an acetyl group cupric bromide, unlike bromine or NBS, attacks the acetyl group in preference to the aromatic ring, but the experimental conditions and the yields stand in sharp contrast. The former chemists carried out

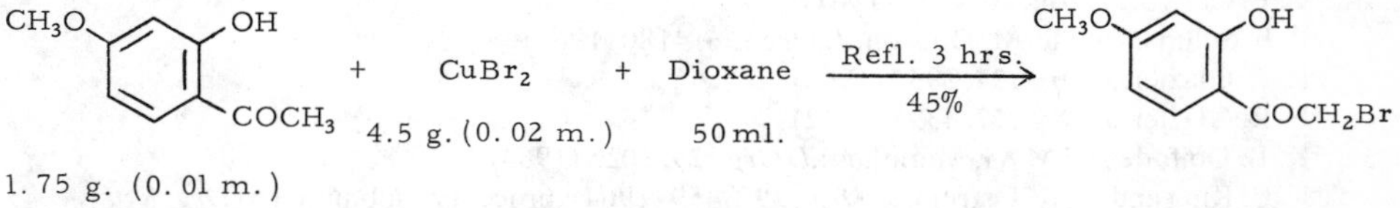

the reaction in homogeneous solution in dioxane and the yields in two examples were 45 and 54%. All other workers had used a homogeneous solution in water, an alcohol, or dimethylformamide. However, King and Ostrum found it highly

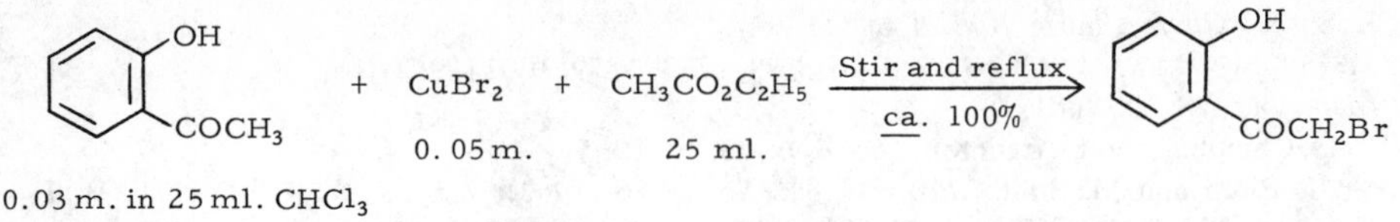
OH
COCH₃
+ CuBr₂ + CH₃CO₂C₂H₅
0.05 m.
25 ml.
Stir and reflux
ca. 100%
OH
COCH₂Br
0.03 m. in 25 ml. CHCl₃

advantageous to use a heterogeneous suspension of finely ground (80 mesh) cupric bromide in chloroform–ethyl acetate. A suspension of the bromide in ethyl acetate is stirred magnetically under reflux and a chloroform solution of the ketone, taken in slight excess to avoid dibromination, is added. The reaction proceeds rapidly with evolution of hydrogen bromide, which is only slightly soluble in the solvent system, and with conversion of black cupric bromide into white cuprous bromide. Completion of the reaction is indicated by cessation of HBr evolution, disappearance of all the black solid, and a change in the color of the solution from green to amber. The cuprous bromide is practically insoluble and is easily removed by filtration. The resulting solution, containing the ω-bromoketone in practically quantitative yield, can be used directly without isolation of the lachrimatory product. Thus in the example formulated the product afforded pure 2′-hydroxybenzoylmethylpyridinium bromide in yield of 85–93%.

Bromination of aromatics. When a solution of the hydrocarbon in carbon tetrachloride is refluxed with the reagent, anthracene affords 9-bromoanthracene in 99% yield and pyrene affords 1-bromopyrene in 94% yield.[9] Bromination of anthracene in nitrobenzene at 210° afforded 9,10-dibromoanthracene in low yield.[10]

Under strictly anhydrous conditions cupric bromide reacts with toluene to give *o*-bromotoluene (63%).[11] If a small amount of water is added the major product is a mixture of phenyltolylmethanes, $C_6H_5CH_2C_6H_4CH_3$.

Reaction with ethylenic and acetylenic compounds. Isolated double bonds are halogenated by the reagent.[12] Internal acetylenes undergo *trans* halogenation, whereas terminal acetylenes are converted into tribromoolefins.

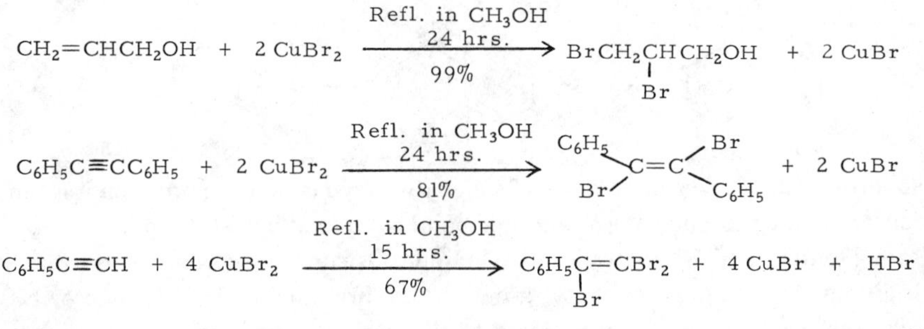
CH₂=CHCH₂OH + 2 CuBr₂
Refl. in CH₃OH
24 hrs.
99%
BrCH₂CHCH₂OH + 2 CuBr
Br
C₆H₅C≡CC₆H₅ + 2 CuBr₂
Refl. in CH₃OH
24 hrs.
81%
C₆H₅
Br
C=C
Br
C₆H₅
+ 2 CuBr
C₆H₅C≡CH + 4 CuBr₂
Refl. in CH₃OH
15 hrs.
67%
C₆H₅C=CBr₂ + 4 CuBr + HBr
Br

[1]J. K. Kochi, *Am. Soc.*, **77**, 5274 (1955)
[2]A. W. Fort, *J. Org.*, **26**, 765 (1961)
[3]C. E. Castro, *J. Org.*, **26**, 4183 (1961)
[4]P. B. Sollman and R. M. Dodson, *J. Org.*, **26**, 4180 (1961)
[5]E. R. Glazier, *J. Org.*, **27**, 2937 (1962)
[6]E. R. Glazier, *J. Org.*, **27**, 4397 (1962)
[7]K. B. Doifode and M. G. Marathey, *J. Org.*, **29**, 2025 (1964)
[8]L. C. King and G. K. Ostrum, *J. Org.*, **29**, 3459 (1964); procedure submitted to *Org. Syn.*
[9]D. C. Nonhebel, *Proc. Chem. Soc.*, 307 (1961)
[10]J. C. Ware and E. E. Borchert, *J. Org.*, **26**, 2263 (1961)
[11]P. Kovacic and K. E. Davis, *Am. Soc.*, **86**, 427 (1964)
[12]C. E. Castro, E. J. Gaughan, and D. C. Owsley, *J. Org.*, **30**, 587 (1965)

Cupric carbonate, basic, $CuCO_3 \cdot Cu(OH)_2$. Mol. wt. 221.11. Suppliers: B, F, MCB. The reagent is ***prepared*** by adding an equivalent amount of a solution of sodium carbonate to a solution of cupric sulfate and washing the precipitate until nearly free from sulfate ion. The wet basic cupric carbonate is used without drying or weighing. In the presence of ammonium hydroxide it effects the ***oxidation:***[1]

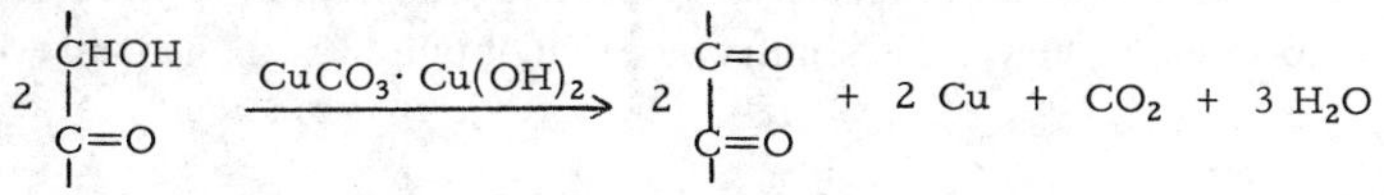

CHOH
2
C=O
CuCO₃· Cu(OH)₂
2
C=O
C=O
+ 2 Cu + CO₂ + 3 H₂O

Decarboxylation. A mixture of 15 g. of benzoylbenzoic acid and 0.5 g. of basic copper carbonate is heated gently to melt the acid, evacuated on the steam bath to remove water, and heated at 265° until evolution of carbon dioxide has ceased (30 min.).[2]

[1]J. R. Totter and W. J. Darby, *Org. Syn., Coll. Vol.*, **3**, 460 (1955)
[2]*Org. Expts.*, Chapt. 40

Cupric chloride, anhydrous, $CuCl_2$. Mol. wt. 134.45. Supplier: Fisher. Preparation by heating the dihydrate (mol. wt. 170.49) in an oven at 110° overnight.[1]

Chlorination of aromatics. By the procedure formulated, Nonhebel[2] chlorinated anthracene to 9-chloroanthracene in very high yield. The progress of the reaction is evident from the conversion of brown cupric chloride to white cuprous chloride.

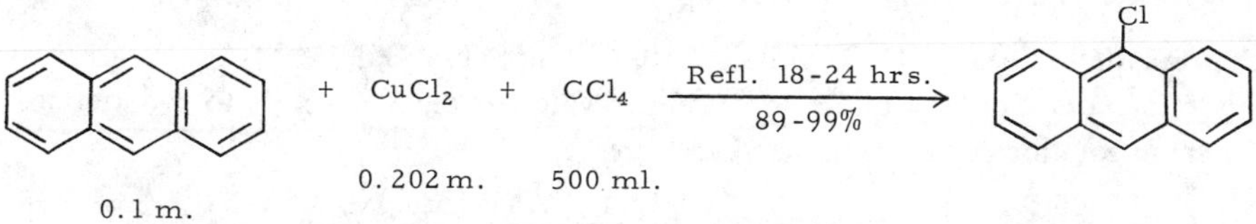

+ CuCl₂ + CCl₄
Refl. 18-24 hrs.
89-99%
Cl
0.202 m.
500 ml.
0.1 m.

At the end of the reaction the cuprous chloride is removed by filtration, and the carbon tetrachloride solution is chromatographed on alumina. The product is obtained as a lemon-yellow solid. By the same procedure pyrene affords 1-chloropyrene in 90% yield; phenanthrene is not chlorinated.

Chlorination of diethyl phosphite to form diethyl phosphorochloridate.[3]

$$(C_2H_5O)_2POH + 2\,CuCl_2 \xrightarrow[74\%]{} (C_2H_5O)_2\overset{O}{\overset{\|}{P}}-Cl + 2\,CuCl + HCl$$

[1]J. C. Ware and E. E. Borchert, *J. Org.*, **26**, 2263 (1961)
[2]D. C. Nonhebel, *Org. Syn.*, **43**, 15 (1963)
[3]T. D. Smith, *J. Chem. Soc.*, 1122 (1962)

Cupric nitrate–Acetic anhydride. For nitration, probably via acetyl nitrate.[1] In the examples formulated[2,3] cupric nitrate was added, either as a solid or suspended in acetic anhydride, to a solution of the substrate in acetic anhydride.

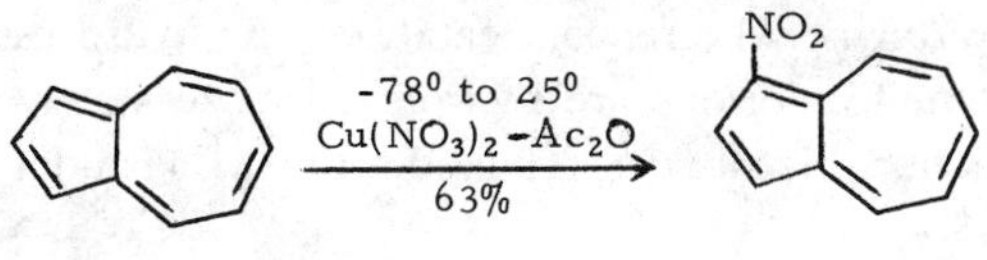

-78° to 25°
Cu(NO₃)₂-Ac₂O
63%
NO₂

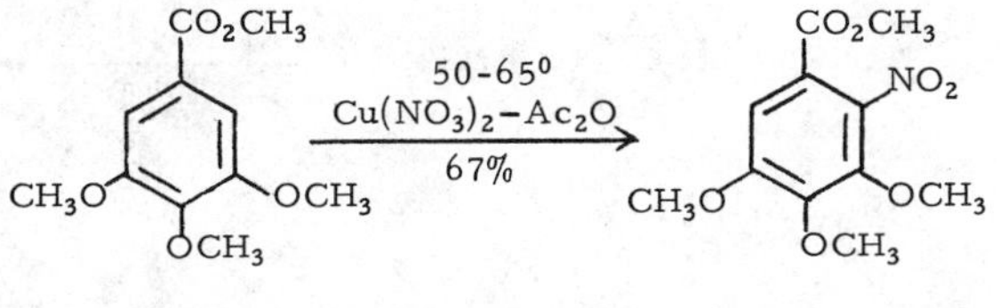

CO₂CH₃
CH₃O
OCH₃
OCH₃
50-65°
Cu(NO₃)₂-Ac₂O
67%
CO₂CH₃
NO₂
CH₃O
OCH₃
OCH₃

[1]G. Bacharach, *Am. Soc.*, **49**, 1522 (1927)
[2]A. G. Anderson, Jr., J. A. Nelson, and J. J. Tazuma, *ibid.*, **75**, 4980 (1953)
[3]K. I. H. Williams, S. E. Cremer, F. W. Kent, E. J. Sehm, and D. S. Tarbell, *ibid.*, **82**, 3982 (1960)

Cupric nitrate–Pyridine. This complex in the presence of triethylamine as base catalyzes the reaction of Δ^5-cholestene-3-one with molecular oxygen to form Δ^4-cholestene-3,6-dione.[1] The yield is about twice that obtained by dichromate oxidation.[2]

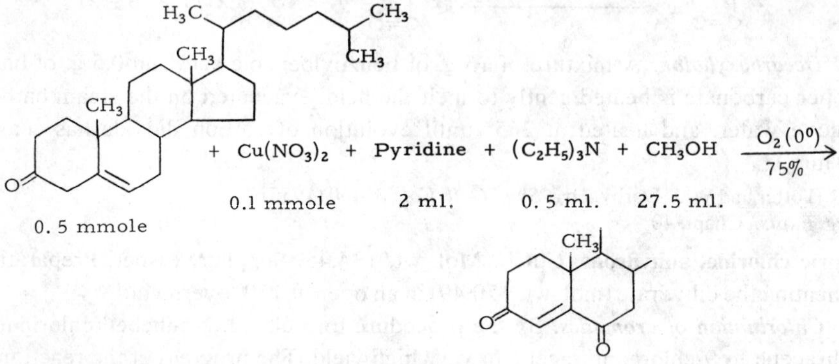
H3C
CH3
CH3
CH3
CH3
+ Cu(NO3)2 + Pyridine + (C2H5)3N + CH3OH
O2 (0°)
75%
0.1 mmole
2 ml.
0. 5 ml.
27.5 ml.
0. 5 mmole
CH3

[1]H. C. Volger and W. Brackman, *Rec. trav.*, **84**, 579 (1965)
[2]L. F. Fieser, *Org. Syn., Coll. Vol.*, **4**, 189 (1963)

Cupric sulfate, $CuSO_4{\cdot}5H_2O$. Mol. wt. 249.69. Suppliers: B, F, MCB.

Oxidation. Copper sulfate in pyridine–water oxidizes α-ketols to α-diketones; the spent solution can be reoxidized with air.[1]

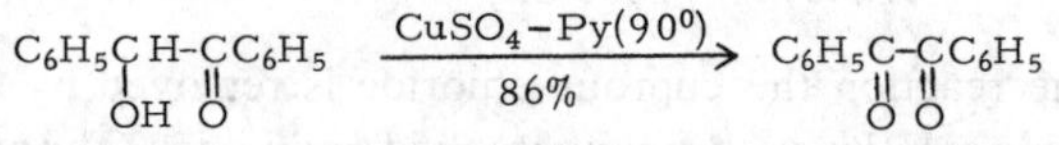
C6H5CH-CC6H5
OH O
CuSO4-Py(90°)
86%
C6H5C-CC6H5
O O

The indanonediol acid (2), isolated as an intermediate in the Hooker oxidation of a 2-hydroxy-3-alkyl-1,4-naphthoquinone (1) to the next lower homolog (3), can be

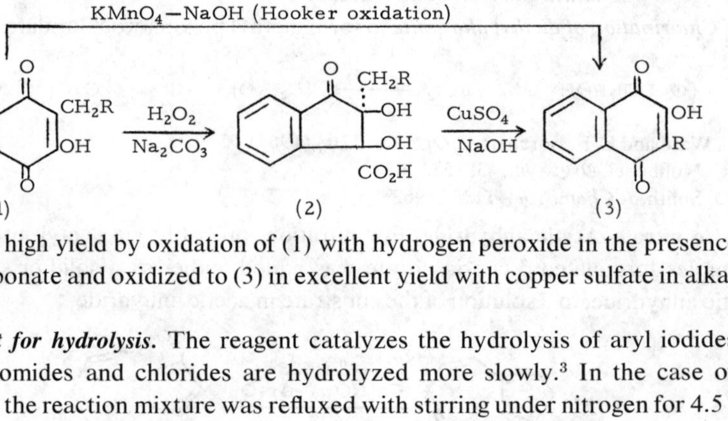
KMnO4–NaOH (Hooker oxidation)
CH2R
OH
H2O2
Na2CO3
CH2R
OH
OH
CO2H
CuSO4
NaOH
OH
R

(1) (2) (3)

prepared in high yield by oxidation of (1) with hydrogen peroxide in the presence of sodium carbonate and oxidized to (3) in excellent yield with copper sulfate in alkaline solution.[2]

Catalyst for hydrolysis. The reagent catalyzes the hydrolysis of aryl iodides to phenols; bromides and chlorides are hydrolyzed more slowly.[3] In the case of 5-iodovanillin the reaction mixture was refluxed with stirring under nitrogen for 4.5 hrs.

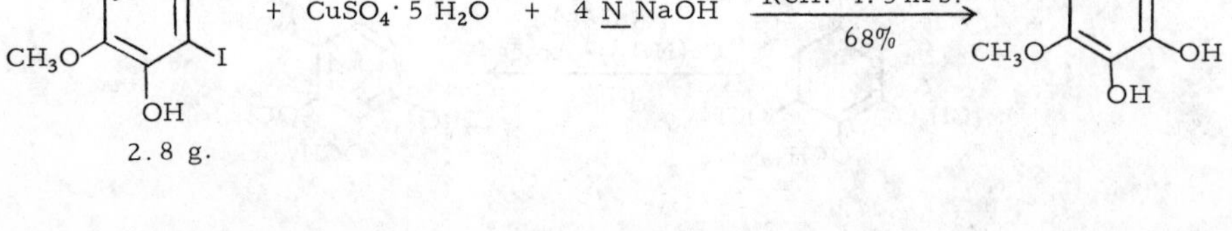
CHO
CH3O
I
OH
2. 8 g.
+ CuSO4 · 5 H2O + 4 N NaOH
Refl. 4. 5 hrs.
68%
CHO
CH3O
OH
OH

[1]H. T. Clarke and E. E. Dreger, *Org. Syn., Coll. Vol.*, **1**, 87 (1941)
[2]L. F. Fieser and M. Fieser, *Am. Soc.*, **70**, 3215 (1948)
[3]S. K. Banerjee, M. Manolopoulo, and J. M. Pepper, *Can. J. Chem.*, **40**, 2175 (1962)

Cuprous ammonium chloride. The combination of cuprous chloride and ammonium chloride in a slightly acidic aqueous solution catalyzes oxidative (air) coupling of terminal acetylenes to diacetylenes.[1,2] The groups NH_2, OH, CO_2H, and CO_2R do not interfere. In the synthesis formulated, cross coupling was accomplished in a mixture of ethanol and $0.08 N$ hydrochloric acid containing the catalyst.

$$CH_2=CH(CH_2)_4C\equiv CH + HC\equiv C(CH_2)_7CO_2H \xrightarrow[30\%]{O_2} CH_2=CH(CH_2)_4C\equiv CC\equiv C(CH_2)_7CO_2H$$

Erythrogenic acid

[1]K. Bowden *et al., J. Chem. Soc.*, 1579 (1947); J. D. Rose and B. C. L. Weedon, *ibid.*, 782 (1949); J. P. Riley, *ibid.*, 2193 (1953)
[2]H. K. Black and B. C. L. Weedon, *ibid.*, 1785 (1953)

Cuprous bromide, Cu_2Br_2. Mol. wt. 286.91, m.p. 504°. Suppliers: E. H. Sargent Co., Alfa, F. MCB.

Preparation and use in a Sandmeyer reaction.[1]

Homologization. The reagent catalyzes the reaction of ketene acetals with diazomethane to give cyclopropanone acetals:[2]

$$CH_2=C\overset{OC_2H_5}{\underset{OC_2H_5}{<}} + CH_2N_2 \xrightarrow[35\%]{Cu_2Br_2} CH_2-C\overset{OC_2H_5}{\underset{CH_2}{<}}OC_2H_5 + N_2$$

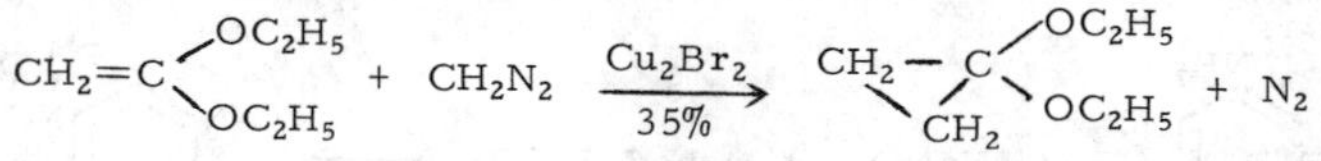

In an improved process for the synthesis of tropilidene (5) by E. Müller,[3] a solution of diazomethane in benzene is added gradually to refluxing benzene containing cuprous bromide as catalyst. Benzene is used in large excess, and the product is isolated most easily by filtering the solution from the catalyst and adding it to a solution of phosphorus pentachloride in carbon tetrachloride. The tropylium chloride which separates is dissolved in water and treated with perchloric acid to afford tropylium perchlorate in 85% yield. The success of the method is attributed to formation of the intermediate (3), a deactivated electrophilic carbon metal complex. Tropilidene

$$\overset{-}{C}H_2-\overset{+}{N}\equiv N \xrightarrow{Br-Cu} Br-\overset{-}{C}u-CH_2-\overset{+}{N}\equiv N \xrightarrow{-N_2}$$

(1)   (2)

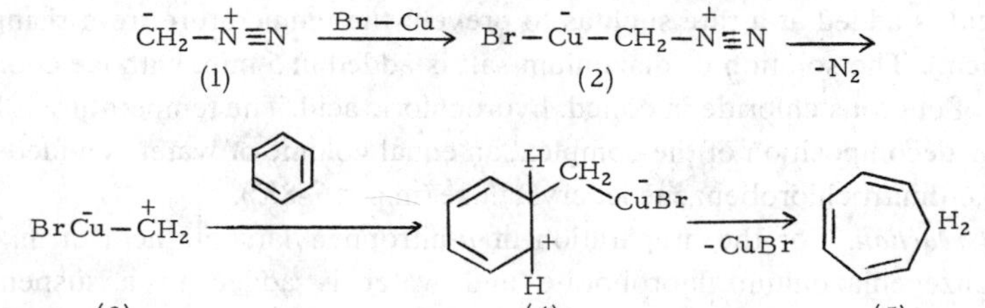

(3)   (4)   (5)

can be recovered from the perchlorate by reduction with sodium borohydride and extraction with ether. As expected, a monoalkylbenzene gives a mixture of three ring-expanded products in similar yield.[4] At temperatures ranging from 120 to 190°,

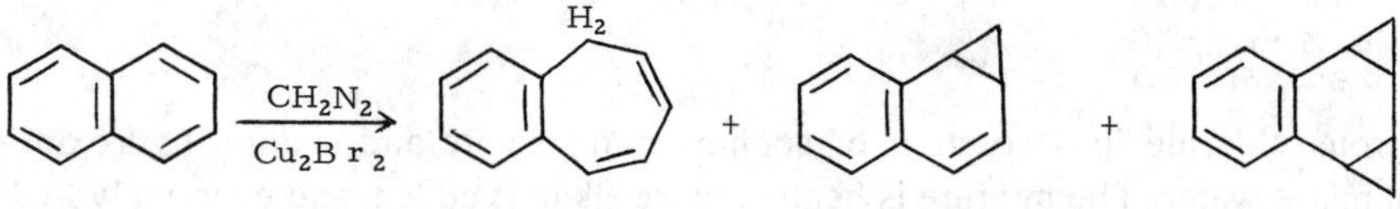

(1)   (2)   (3)

naphthalene gives the hydrocarbons (1)–(3); hydrocarbon (2) can be rearranged to (1) by heating at 260°.[5] Phenanthrene affords two products.

Catalyst for 1,4-Grignard addition. With cuprous bromide as catalyst, methylmagnesium bromide reacts with $\Delta^{1(9)}$-octalone-2 exclusively by 1,4-addition to give 9-methyl-*cis*-decalone-2.[6]

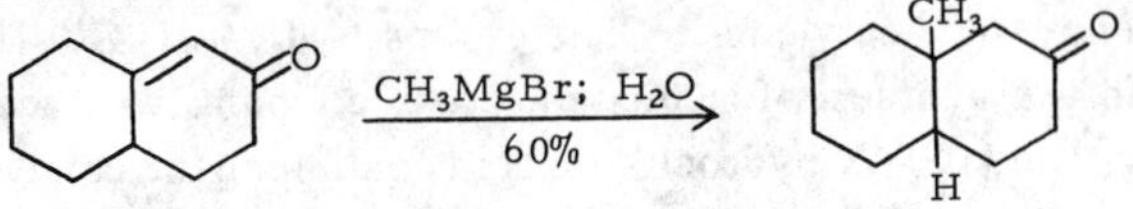
CH₃MgBr; H₂O
60%
CH₃
O
O
H

[1]J. S. Buck and W. S. Ide, *Org. Syn., Coll. Vol.*, **2**, 132 (1943)
[2]M. F. Dull and P. G. Abend, *Am. Soc.*, **81**, 2588 (1959)
[3]E. Müller and H. Fricke, *Ann.*, **661**, 38 (1963)
[4]E. Müller, H. Fricke, and H. Kessler, *Tetrahedron Letters*, 1501 (1963)
[5]*Idem, ibid.*, 1525 (1964)
[6]A. J. Birch and R. Robinson, *J. Chem. Soc.*, 501 (1943); *see also* R. F. Church, R. E. Ireland, and D. R. Shridhar, *J. Org.*, **27**, 707 (1962)

Cuprous chloride, Cu_2Cl_2. Mol. wt. 197.99, m.p. 422°, v. slightly soluble in water. Suppliers: E. H. Sargent, B, F, KK, MCB.

Preparation and use in a Sandmeyer reaction.[1]

Sandmeyer reaction. Application of the reaction to 2,6-dinitroaniline is of particular interest because of the special method employed for diazotization of this

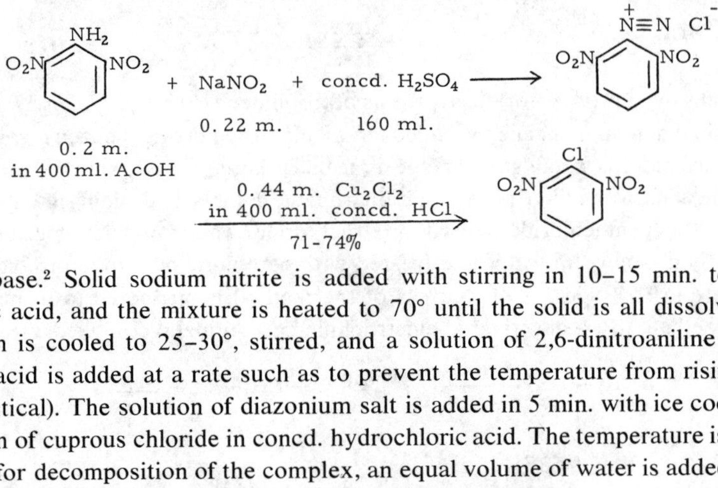
NH₂
O₂N
NO₂
+ NaNO₂ + concd. H₂SO₄
0.22 m.
160 ml.
0.2 m.
in 400 ml. AcOH
N≡N Cl⁻
O₂N
NO₂
0.44 m. Cu₂Cl₂
in 400 ml. concd. HCl
71-74%
Cl
O₂N
NO₂

weak base.[2] Solid sodium nitrite is added with stirring in 10–15 min. to concd. sulfuric acid, and the mixture is heated to 70° until the solid is all dissolved. The solution is cooled to 25–30°, stirred, and a solution of 2,6-dinitroaniline in warm acetic acid is added at a rate such as to prevent the temperature from rising above 40° (critical). The solution of diazonium salt is added in 5 min. with ice cooling to a solution of cuprous chloride in concd. hydrochloric acid. The temperature is brought to 20° for decomposition of the complex, an equal volume of water is added and the yellow 2,6-dinitrochlorobenzene let crystallize (m.p. 86–88°).

Bart reaction. For the preparation of *p*-nitrophenylarsonic acid[3] a mixture of *p*-nitrobenzenediazonium fluoroborate and water is added to a suspension of

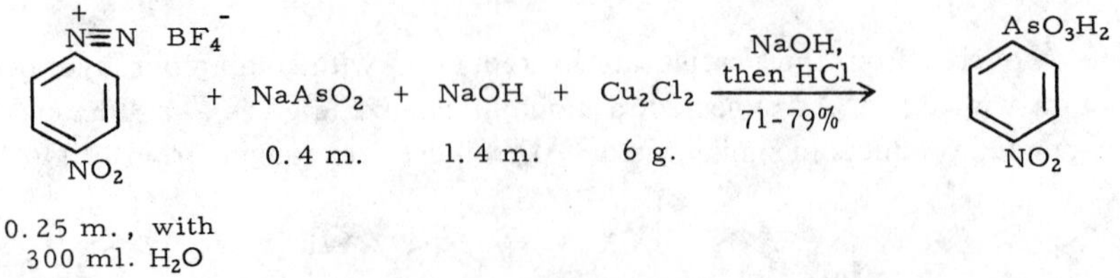
N≡N BF₄⁻
NO₂
0.25 m., with
300 ml. H₂O
+ NaAsO₂ + NaOH + Cu₂Cl₂
0.4 m.
1.4 m.
6 g.
NaOH,
then HCl
71-79%
AsO₃H₂
NO₂

cuprous chloride in a solution of sodium metaarsenite and sodium hydroxide in 600 ml. of water. The mixture is heated, more alkali is added, and eventually hydrochloric acid is added to liberate the arsonic acid from its salt.

Gattermann-Koch synthesis. A procedure for the synthesis of *p*-tolualdehyde[4] calls for stirring a suspension of cuprous chloride and aluminum chloride in toluene

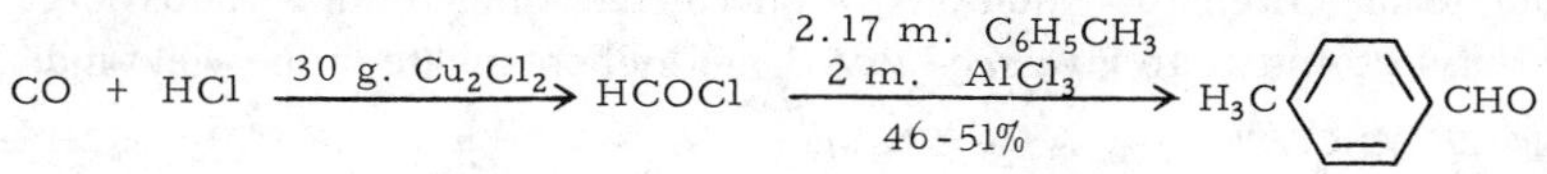
CO + HCl
30 g. Cu2Cl2
HCOCl
2.17 m. C6H5CH3
2 m. AlCl3
46-51%
H3C
CHO

and passing in carbon monoxide and hydrochloric acid and hydrolyzing the resulting complex with ice.

Displacement of aromatic halogen. With both cuprous chloride and copper as catalyst 4-bromo-*o*-xylene can be converted into the corresponding amine.[5] The

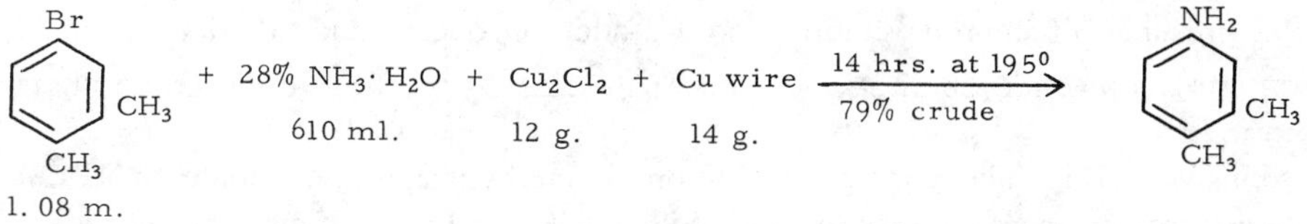
Br
CH3
CH3
1.08 m.
+ 28% NH3·H2O + Cu2Cl2 + Cu wire
610 ml.
12 g.
14 g.
14 hrs. at 195°
79% crude
NH2
CH3
CH3

reactants are placed in a hydrogenation bomb which is closed, heated to 195°, and rocked for 14 hrs.

2-Acetylamino-3-chloroanthraquinone (2) is an intermediate to an important solubilized vat dye, but is available only by a lengthy synthesis. The corresponding 3-bromo compound (1), however, is available in high yield by bromination of 2-aminoanthraquinone and acetylation. Hardy and Fortenbaugh[6] found that (1) reacts with cuprous chloride in α-picoline at 130° to give (2) in almost quantitative yield.

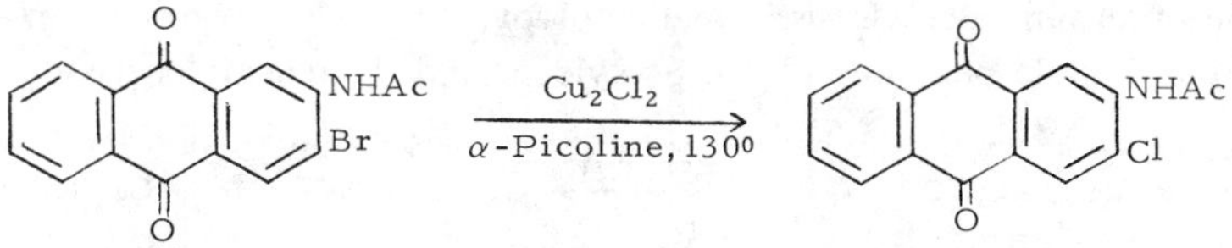
NHAc
Br
Cu2Cl2
α-Picoline, 130°
NHAc
Cl

(1) (2)

Catalyst of Grignard additions. Kharasch[7] discovered that in the reaction of a Grignard reagent with an α,β-unsaturated ketone a catalytic amount of cuprous chloride leads to greatly enhanced 1,4-addition at the expense of 1,2-addition. In the absence of a catalyst, isophorone reacts with methylmagnesium bromide, by either the direct or inverse procedure, to give almost exclusively products of

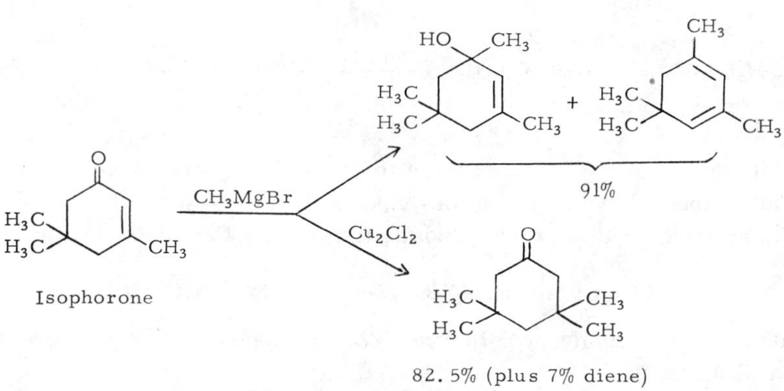
HO
CH3
H3C
H3C
CH3
+
CH3
H3C
H3C
CH3
91%
H3C
H3C
CH3
CH3MgBr
Cu2Cl2
Isophorone
H3C
H3C
CH3
CH3
82.5% (plus 7% diene)

1,2-addition. When the ketone is added to a solution of methylmagnesium bromide containing 1 mole percent of cuprous chloride, addition is almost exclusively 1,4.

Under usual Grignard conditions, *n*-butylmagnesium bromide adds largely 1,4 to *sec*-butyl crotonate to give *sec*-butyl 3-methylheptanoate in the yield indicated.[8]

$$\underset{0.4\ m.}{CH_3CH{=}CHCO_2\underset{\displaystyle CH_3}{\underset{|}{C}}HCH_2CH_3} + \underset{1.3\ m.}{\text{n-}C_4H_9MgBr} \xrightarrow[68-78\%]{} \text{n-}C_4H_9\underset{\displaystyle CH_3}{\underset{|}{C}}HCH_2CO_2\underset{\displaystyle CH_3}{\underset{|}{C}}HCH_2CH_3$$

Addition of 1.4 g. of cuprous chloride in seven portions during addition of the ester to the Grignard reagent improves the yield to 80–85%. The uncatalyzed addition of *n*-butylmagnesium bromide to *sec*-butyl sorbate results in 1,2- and 1,4-addition; in the presence of cuprous chloride no 1,2-addition occurs and addition is almost

$$CH_3CH{=}CHCH{=}CHCO_2\underset{\displaystyle CH_3}{\underset{|}{C}}HCH_2CH_3 + \text{n-}C_4H_9MgBr \xrightarrow[74\%]{Cu_2Cl_2} CH_3\underset{\displaystyle C_4H_9\text{-n}}{\underset{|}{C}}HCH{=}CHCH_2CO_2\underset{\displaystyle CH_3}{\underset{|}{C}}HCH_2CH_3$$

exclusively 1,6.[9] The conjugate addition of methylmagnesium iodide to Δ^{16}-20-keto steroids can be effected in yield of 90% or better by use of cuprous chloride as catalyst and tetrahydrofurane as solvent.[10]

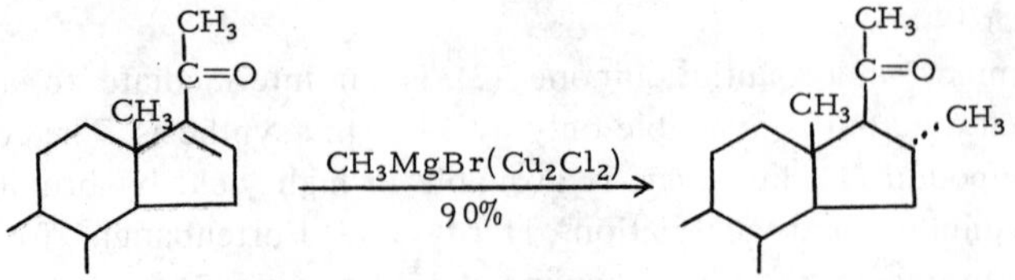

Catalyst for the coupling of terminal acetylenes. Cuprous chloride catalyzes the oxidation of terminal acetylenes in methanol–pyridine by air or oxygen in amounts of only 0.012 mole per mole of the acetylene, and the diacetylene is obtained in

$$2\ CH_3\overset{\displaystyle CH_3}{\underset{\displaystyle OH}{C}}{-}C{\equiv}CH \xrightarrow[90\%]{Cu_2Cl_2 \atop O_2,\ Py\text{-}CH_3OH} CH_3\overset{\displaystyle CH_3}{\underset{\displaystyle OH}{C}}{-}C{\equiv}C{-}C{\equiv}C{-}\overset{\displaystyle CH_3}{\underset{\displaystyle OH}{C}}CH_3$$

high yield after a brief reaction period.[11] A simplified procedure in which oxygen is retained in a balloon has the advantage that a pressure of 10–15 p.s.i. shortens the reaction period.[12] Another procedure[13] employs as catalyst a 1:1 complex of cuprous chloride and N,N,N′,N′-tetramethylethylenediamine. The complex is soluble in many organic solvents, yields are high, and the reaction times short. The reagent also catalyzes the coupling of a terminal acetylene with a 1-bromoacetylene.[14] Yields

$$C_6H_5C{\equiv}CBr + HC{\equiv}C\underset{\displaystyle OH}{\underset{|}{C}}(C_6H_5)_2 \xrightarrow[87\%]{Cu^+ \atop aq.\ C_2H_5NH_2} C_6H_5C{\equiv}CC{\equiv}C\underset{\displaystyle OH}{\underset{|}{C}}(C_6H_5)_2$$

m.p. 86°

are in the range 70–95%. Both of these routes to diacetylenes have been of great value in synthesis (review[15]). Cuprous chloride is a catalyst for the coupling of a terminal acetylene with an allylic chloride to form a 1,4-eneyne.[16] In combination

$$RC{\equiv}CH + ClCH_2CH{=}CH_2 \xrightarrow{Cu^+} RC{\equiv}CCH_2CH{=}CH_2$$

with a basic amine, cuprous chloride catalyzes the coupling of a terminal acetylene with an allenic halide.[17]

$$(CH_3)_2C{=}C{=}CHBr + HC{\equiv}CCH_2OH \xrightarrow[33\%]{Cu_2Cl_2-C_2H_5NH_2,\ DMF} (CH_3)_2C{=}C{=}CHC{\equiv}CCH_2OH$$

Photolytic rearrangement. Irradiation of the π-complex of *cis,cis*-1,5-cyclooctadiene with cuprous chloride produces tricyclo[3.3.0.0^{2,6}]octane, a remarkably stable hydrocarbon.[18]

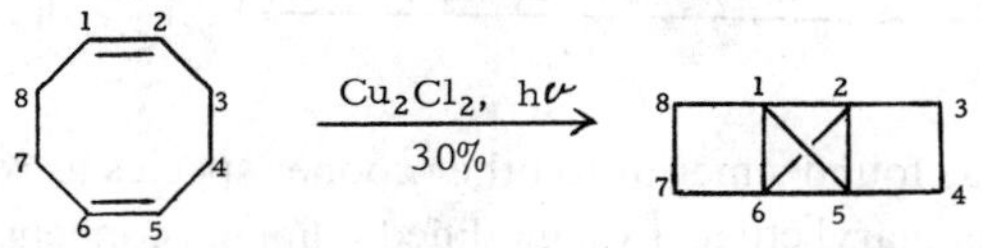
1
2
3
4
5
6
7
8
Cu2Cl2, hν
30%
1
2
3
4
5
6
7
8

[1]C. S. Marvel and S. M. McElvain, *Org. Syn., Coll. Vol.*, **1**, 170 (1941)
[2]F. D. Gunstone and S. H. Tucker, *ibid.*, **4**, 160 (1963)
[3]A. W. Ruddy and E. B. Starkey, *ibid.*, **3**, 665 (1955)
[4]G. H. Coleman and D. Craig, *ibid.*, **2**, 583 (1943)
[5]W. A. Wisansky and S. Ansbacher, *ibid.*, **3**, 307 (1955)
[6]W. B. Hardy and R. B. Fortenbaugh, *Am. Soc.*, **80**, 1716 (1958)
[7]M. S. Kharasch and P. O. Tawney, *Am. Soc.*, **63**, 2308 (1941)
[8]J. Munch-Petersen, *Org. Syn.*, **41**, 63 (1961)
[9]J. Munch-Petersen, C. Bretting, P. M. Jorgensen, S. Refn, V. K. Andersen, and A. Jart, *Chem. Scand.*, **15**, 277 (1961)
[10]K. Heusler *et al.*, *Helv.*, **42**, 2043 (1959)
[11]H. A. Stansbury, Jr., and W. R. Proops, *J. Org.*, **27**, 320 (1962)
[12]*Org. Expts.*, Chapt. 14
[13]A. S. Hay, *J. Org.*, **27**, 3320 (1962)
[14]W. Chodkiewicz, *Ann. chim.*, [13], **2**, 819 (1957)
[15]G. Eglinton and W. McCrae, *Advances in Organic Chemistry*, **4**, 252–274 (1963)
[16]P. Kurtz, *Ann.*, **658**, 6 (1962)
[17]C. S. L. Baker, P. D. Landor, and S. R. Landor, *Proc. Chem. Soc.*, 340 (1963)
[18]R. Srinivasan, *Am. Soc.*, **85**, 3048 (1963)

Cuprous iodide, CuI. Mol. wt. 190.45. Supplier: Fisher.

Wolff rearrangement (Arndt-Eistert reaction). The reagent has been used in acetonitrile–methanol to effect homogeneous rearrangement, for example of α-diazoacetophenone.[1] After standing for several hours at room temperature, methyl phenylacetate is obtained in good yield.

$$C_6H_5\overset{O}{\overset{\|}{C}}CH{=}\overset{+}{N}{=}\overset{-}{N} + CH_3OH-CH_3CN \xrightarrow[75\text{-}85\%]{CuI} C_6H_5CH_2CO_2CH_3 + N_2$$

[1]P. Yates and J. Fugger, *Chem. Ind.*, 1511 (1957)

Cuprous oxide, Cu_2O. Mol. wt. 143.08, Suppliers: B, F, MCB.

Weedon and co-workers[1] found that 1,4-disubstituted vinylacetylenes can be obtained in yields of some 30–60% by dissolving a catalytic amount of cuprous oxide in hot acetic acid and adding a terminal acetylene. The reaction probably involves initial formation of the copper acetylide.

$$2\ RC{\equiv}CH \xrightarrow{CuOAc} RCH{=}CHC{\equiv}CR$$

Bacon and Hill[2] used cuprous oxide in an aprotic solvent (usually DMF, also DMSO) as catalyst for the reaction of aryl halides with various nucleophiles. The displacement is of particular value for the preparation of ethers and thioethers.

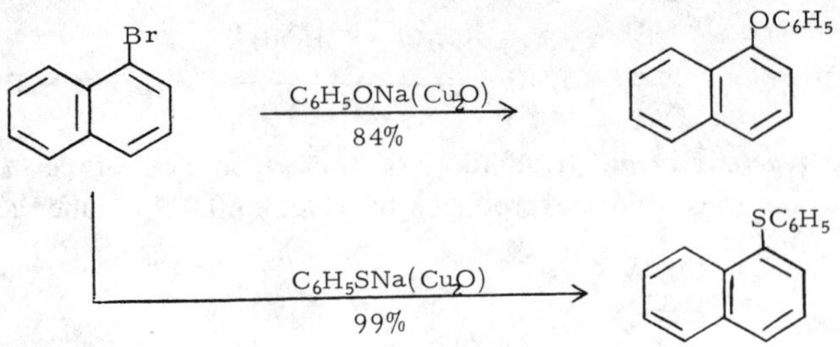
Br
C6H5ONa(Cu2O)
84%
OC6H5
C6H5SNa(Cu2O)
99%
SC6H5

Cuprous oxide was found superior to other copper species as reagent and catalyst for the preparation of diaryl ethers by a modified Ullmann condensation.[3]

[1]R. F. Garwood, E. Oskay, and B. C. L. Weedon, *Chem. Ind.*, 1684 (1962)
[2]R. G. R. Bacon and H. A. O. Hill, *J. Chem. Soc.*, 1108 (1964)
[3]R. G. R. Bacon and O. J. Stewart, *J. Chem. Soc.*, 4953 (1965)

Cyanamide, NH_2CN. Mol. wt. 42.04, m.p. 46°. Suppliers: B, E, F. Preparation.[1,2]

Under catalysis by hydrogen chloride, methanol adds to the triple bond of cyanamide with formation of methylisourea hydrochloride.[2] In strongly alkaline solution

$$H_2NC{\equiv}N \xrightarrow[69-80\%]{CH_3OH-HCl} H_2N\overset{\displaystyle OCH_3}{\overset{|}{C}}{=}NH\cdot HCl$$

cyanamide adds to phenyl isocyanate to give, after acidification, 1-cyano-3-phenylurea.[3]

$$C_6H_5N{=}C{=}O \xrightarrow[62-67\%]{H_2NCN;\ NaOH;\ HCl} C_6H_5\overset{H}{N}-\overset{O}{\overset{\|}{C}}-\overset{H}{N}CN$$

[1]L. A. Pinck and J. M. Salisbury, *Inorg. Syn.*, **3**, 39 (1950)
[2]F. Kurzer and A. Lawson, *Org. Syn., Coll. Vol.*, **4**, 645 (1963)
[3]F. Kurzer and J. R. Powell, *ibid.*, **4**, 213 (1963)

Cyanic acid (N≡COH) ⇌ **Isocyanic acid** (HN=C=O). Mol. wt. 43.03, b.p. 23.5°, pK_a 3.66. Acrid odor, strong lachrymator and vessicant. A dilute solution in ice water contains the species N≡COH and may be kept for several hours. Solutions in ether or benzene contain the iso species HN=C=O and can be kept for several weeks. All the evidence indicates that the only species present in the vapor phase is isocyanic acid.[1]

Preparation of isocyanic acid. Isocyanic acid is prepared by the thermal depolymerization of cyanuric acid, a solid trimer available commercially (B, E, F, MCB) or prepared[2] by hydrolysis of the less expensive cyanuric chloride (B, E, F, MCB)

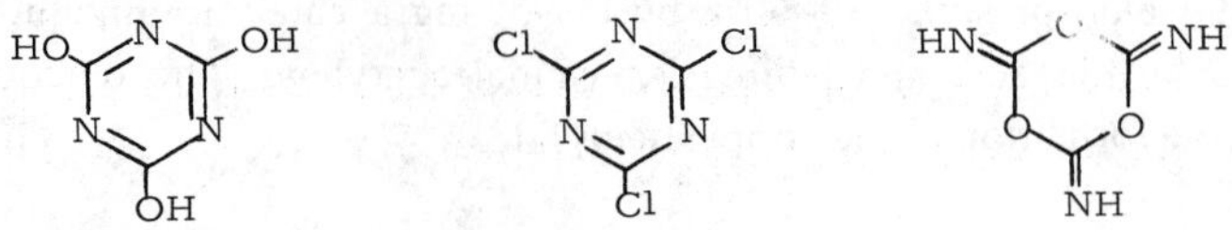
HO
N
OH
N
N
OH
Cl
N
Cl
N
N
Cl
HN
NH
O
O
NH

Cyanuric acid   Cyanuric chloride   Cyamelide

with 1:1 concd. sulfuric acid in water. The liquid reagent polymerizes to the sparingly soluble trimers cyanuric acid and cyamelide.

Allophanates. The chief use for isocyanic acid is for the conversion of primary, secondary, or tertiary alcohols into crystalline allophanates. Depolymerization of cyanuric acid can be done at 360–400° in a slow stream of carbon dioxide and the

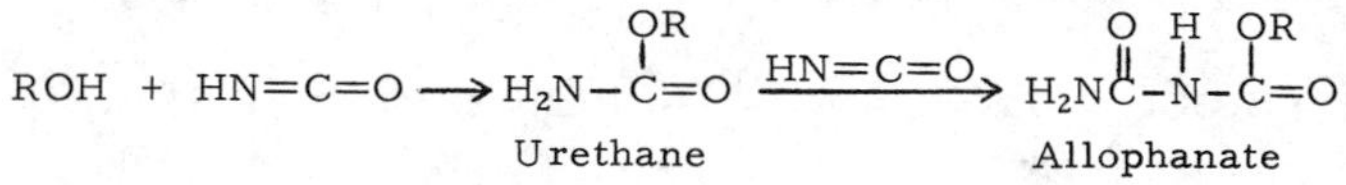
ROH + HN=C=O
H₂N–C=O
OR
HN=C=O
H₂NC–N–C=O
O H OR
Urethane
Allophanate

gas absorbed directly in a liquid alcohol. Zobrist and Schinz,[2] who give a detailed description of an isocyanic acid generator, absorbed the reagent in ether to a concentration of 30–35% and used 0.8 ml. of this solution for reaction with 0.1 g. of substance of molecular weight about 150. Simple alcohols react rapidly and usually an overnight reaction period is adequate; in the case of α-phenylethanol several days are required.[3] If insufficient isocyanic acid is used, the urethane, H_2NCO_2R, becomes the major product. Excess reagent polymerizes to cyanuric acid and cyamelide, which can be eliminated by extraction of the allophanate with benzene.

Except for derivatives of lower alcohols, allophanates are usually high-melting, well crystalline derivatives suitable for characterization and isolation; the alcohol can be regenerated by warming the allophanate with methanolic alkali. That the reaction is applicable even to a highly reactive, acid-sensitive alcohol is illustrated by the isolation of vitamin D_3 by Windaus and co-workers[4] from a complex mixture resulting on irradiation of 7-dehydrocholesterol. Treatment of the crude, oily mixture with isocyanic acid afforded directly a solid product easily purified by

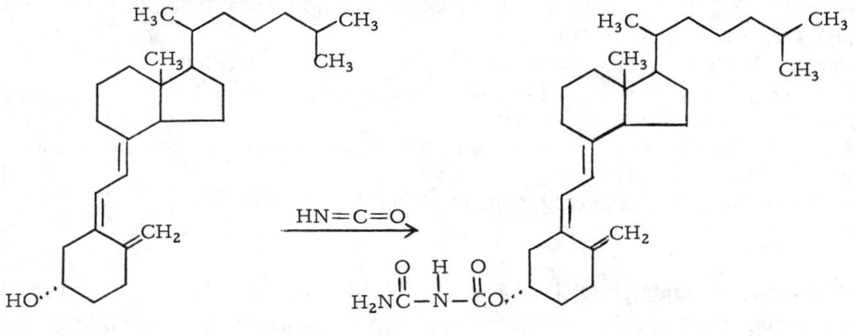
H₃C
CH₃
CH₃
CH₃
CH₂
HO
HN=C=O
H₃C
CH₃
CH₃
CH₃
CH₂
O H O
H₂NC–N–CO

Vitamin D_3 m.p. 82-84° Allophanate m.p. 173-174°

crystallization from acetone and converted into the pure vitamin by hydrolysis. Isolation was accomplished also through the 3,5-dinitrobenzoate, but this lower-melting derivative (m.p. 129°) was obtained in crystalline form only after chromatography of the oily reaction mixture.

Addition to carbonyl compounds. Isocyanic acid adds to the carbonyl group of many aldehydes (formaldehyde, chloral, trifluoroacetaldehyde) and ketones (*s*-dichlorotetrafluoroacetone, perfluorocyclobutanone) to give α-hydroxy isocyanates, a new class of compounds.[5] Isocyanic acid and the carbonyl compound are mixed in

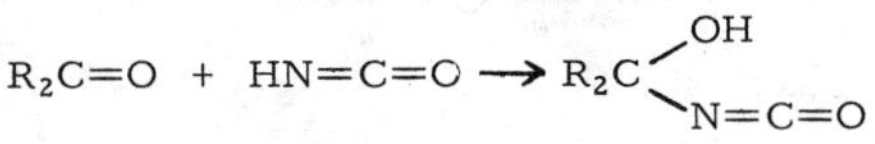
R₂C=O + HN=C=O
R₂C
OH
N=C=O

equimolecular amounts at −78° and the mixture is let warm to a temperature of about 0°, depending upon the stability of the product.

Reaction with acid chlorides. Isocyanic acid reacts with acid chlorides in ether or tetrahydrofurane in the presence of pyridine at −10 to 0° to give acyl isocyanates, usually in good yield.[6]

$$\mathrm{R\overset{O}{\overset{\|}{C}}Cl} + \mathrm{HN{=}C{=}O} \xrightarrow{\mathrm{Py}} \mathrm{R\overset{O}{\overset{\|}{C}}N{=}C{=}O} + \mathrm{Py\cdot HCl}$$

Addition to α,β-unsaturated ethers. In the presence of a catalytic amount of *p*-toluenesulfonic acid, isocyanic acid adds to vinyl *n*-butyl ether to give 1-butoxyethyl isocyanate in high yield.[7]

$$\underset{\text{50 g. in 50 ml. ether}}{\mathrm{\underline{n}\text{-}C_4H_9OCH{=}CH_2}} + \underset{\text{14.9 ml.}}{\mathrm{HN{=}C{=}O}} + \underset{\text{0.1 g.}}{\mathrm{TsOH}} \xrightarrow[\text{90\%}]{38\text{-}40^{\circ}} \mathrm{\underline{n}\text{-}C_4H_9O\overset{CH_3}{\overset{|}{C}}HN{=}C{=}O}$$

Addition to olefins. Isocyanic acid adds to terminal olefins having an electron-releasing group at C_2, but yields are low.[8]

$$\mathrm{C_6H_5\overset{CH_3}{\overset{|}{C}}{=}CH_2} + \mathrm{HN{=}C{=}O} \xrightarrow[41\%]{\mathrm{H^+}} \mathrm{C_6H_5\underset{CH_3}{\underset{|}{\overset{CH_3}{\overset{|}{C}}}}{-}N{=}C{=}O}$$

[1]N. V. Sidgwick, "The Chemical Elements and Their Compounds," **1**, 673, Oxford (1950)
[2]F. Zobrist and H. Schinz, *Helv.*, **35**, 2380 (1952)
[3]M. A. Spielman, J. D. Barnes, and W. J. Close, *Am. Soc.*, **72**, 2520 (1950); H. W. Blohm and E. I. Becker, *ibid.*, **72**, 5342 (1950)
[4]A. Windaus, Fr. Schenck, and F. von Werder, *Z. physiol. Chem.*, **241**, 100 (1936)
[5]F. W. Hoover, H. B. Stevenson, and H. S. Rothrock, *J. Org.*, **28**, 1825 (1963)
[6]P. R. Steyermark, *J. Org.*, **28**, 586 (1963)
[7]F. W. Hoover and H. S. Rothrock, *J. Org.*, **28**, 2082 (1963); J. L. McClanahan and J. L. Harper, *Chem. Ind.*, 1280 (1963)
[8]F. W. Hoover and H. S. Rothrock, *J. Org.*, **29**, 143 (1964)

2-Cyanoethyl phosphate, $\mathrm{(HO)_2\overset{O}{\overset{\|}{P}}OCH_2CH_2CN}$. Mol. wt. 151.06, m.p. 165°. Supplier: British Drug Houses. Prepared by reaction of phosphorus oxychloride with hydracrylonitrile, $HOCH_2CH_2CN$ (Eastman).[1,2]

The reagent is used for the synthesis of phosphomonoesters of alcohols, particularly of nucleotides.[1] It is coupled to the alcoholic component in pyridine using dicyclohexylcarbodiimide (DCC) as condensing agent, as illustrated for the case of 5′-O-tritylthymidine. Acid hydrolysis eliminates the protective trityl group, and very mild

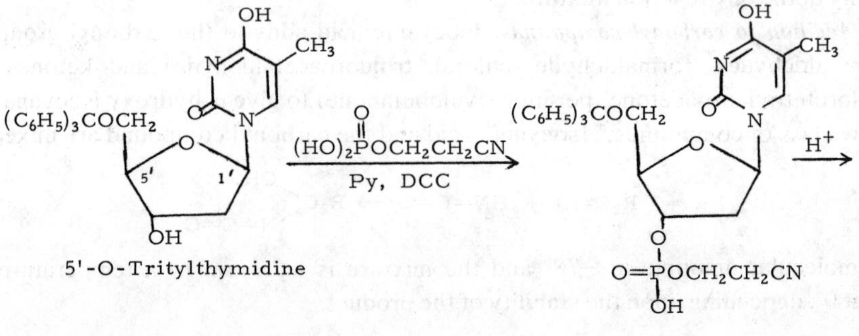
OH
CH3
N
N
(C6H5)3COCH2
O
5'
1'
OH
O
(HO)2POCH2CH2CN
Py, DCC
(C6H5)3COCH2
H+
O=POCH2CH2CN
OH

5'-O-Tritylthymidine

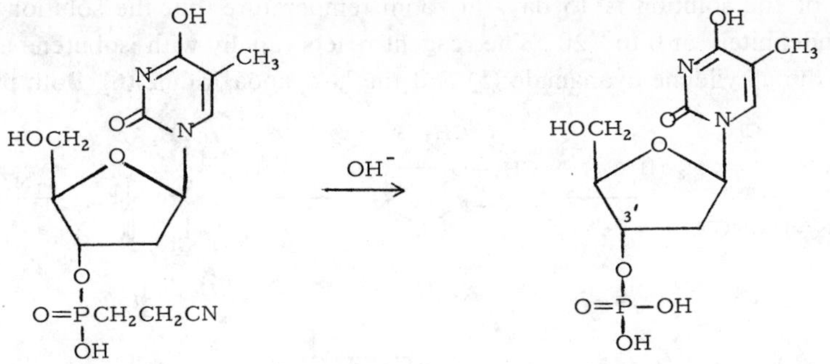

Thymidine-3'-phosphate

alkaline hydrolysis gives the desired phosphate ester. 2-Cyanoethyl phosphate has the advantage of being completely stable to water. The reagent has been used for the preparation of steroid 21-phosphate esters,[3,4] for example, of triamcinolone acetonide (1). A solution of (1), the reagent (2), and dicyclohexylcarbodiimide in pyridine was

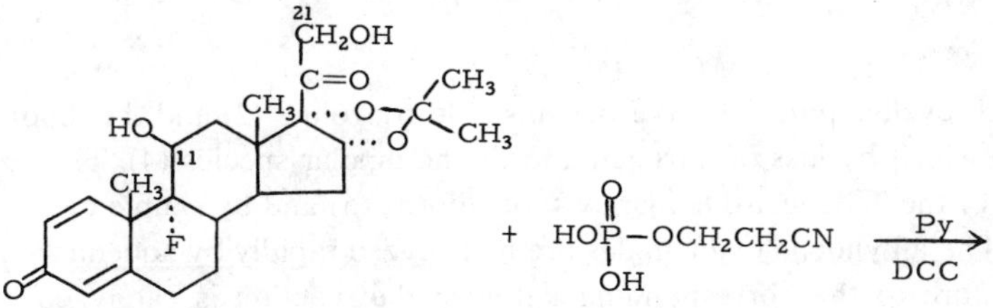

(1) (2)

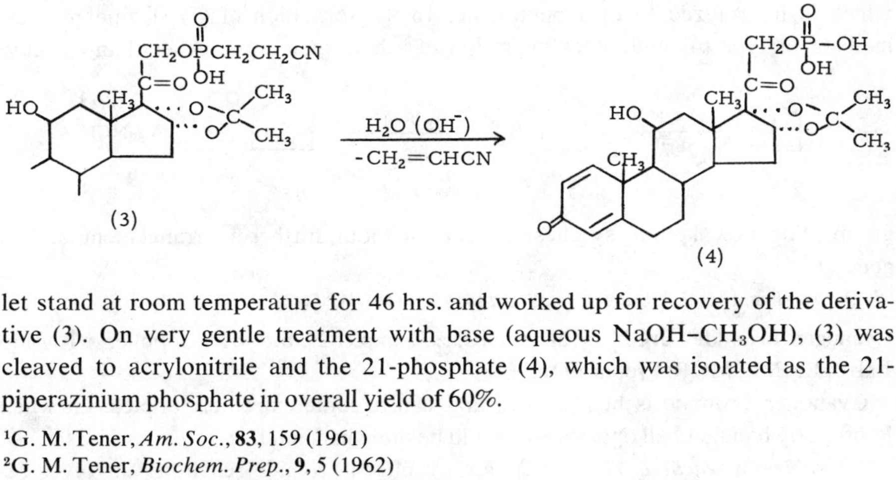

(3) (4)

let stand at room temperature for 46 hrs. and worked up for recovery of the derivative (3). On very gentle treatment with base (aqueous $NaOH-CH_3OH$), (3) was cleaved to acrylonitrile and the 21-phosphate (4), which was isolated as the 21-piperazinium phosphate in overall yield of 60%.

[1]G. M. Tener, *Am. Soc.*, **83**, 159 (1961)
[2]G. M. Tener, *Biochem. Prep.*, **9**, 5 (1962)
[3]J. G. Moffatt, *Am. Soc.*, **85**, 1118 (1963)
[4]R. B. Brownfield, *Steroids*, **2**, 597 (1963)

Cyanogen azide, $\overset{-}{N}=\overset{+}{N}=N-C\equiv N$.[1] Mol. wt. 68.04, highly explosive, colorless oil. For a similar but safer reagent, *see* Picryl azide.

The reagent is prepared by distilling cyanogen chloride into a suspension of sodium azide in acetonitrile at a temperature maintained below 12°, allowing the mixture to warm to room temperature, and filtering to remove sodium chloride. The half-life of the solution is 15 days at room temperature, but the solution can be stored indefinitely at 0 to −20°. The reagent reacts rapidly with isobutene at 0–35° to form the alkylidene cyanamide (5) and the N-cyanoaziridine (6). Both products

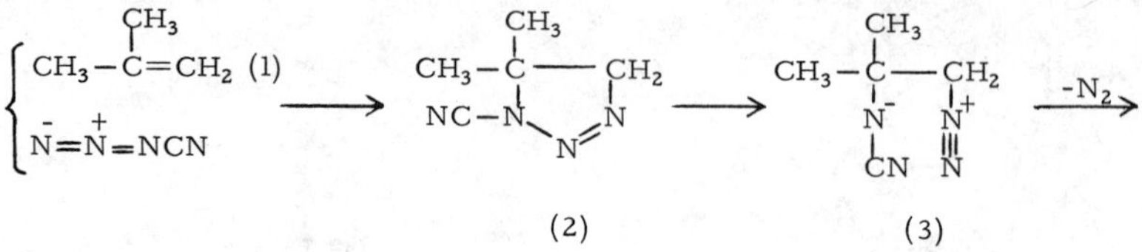

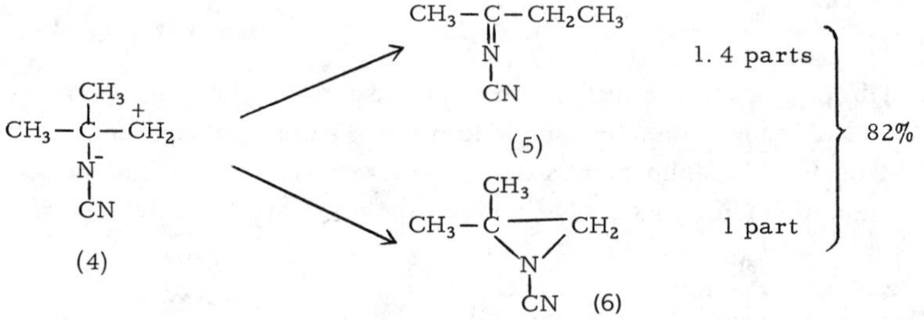

may arise through cycloaddition to give the unstable triazoline (2) and the dipolar intermediate (3), which by loss of nitrogen affords the dipolar species (4). This, by a rearrangement of the Tiffeneau-Demjanov type affords (5) and by simple cyclization affords (6). The alkylidene cyanamides are hydrolyzed rapidly by aqueous acid at room temperature to the corresponding ketones; the reaction is catalyzed by silver ions. Thus cyclopentene gives exclusively the alkylidene cyanamide (7), which is hydrolyzed to cyclopentanone. In the formation of (7), a dipolar intermediate of type (4) undergoes migration of a hydrogen atom rather than an alkyl

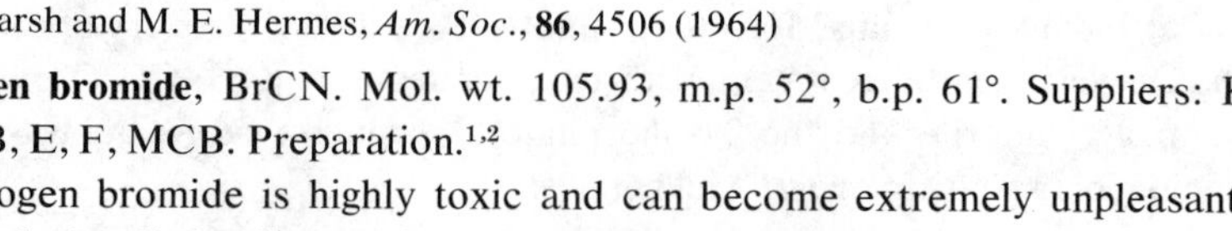

group. The new ketone synthesis occurs without further rearrangement and is general.

[1]F. D. Marsh and M. E. Hermes, *Am. Soc.*, **86**, 4506 (1964)

Cyanogen bromide, BrCN. Mol. wt. 105.93, m.p. 52°, b.p. 61°. Suppliers: Kaplop Labs., B, E, F, MCB. Preparation.[1,2]

Cyanogen bromide is highly toxic and can become extremely unpleasant when improperly handled. All operations should be done in a hood.

von Braun reaction. The von Braun cyanogen bromide reaction[3,4] involves reaction of a tertiary amine with cyanogen bromide to yield a disubstituted cyanamide and an alkyl bromide:

$$R_2NR + BrCN \longrightarrow R_2NCN + RBr$$

A procedure for application of the reaction to N,N-dimethyl-α-naphthylamine[5] calls for refluxing the amine with cyanogen bromide for 16 hrs. The cooled mixture is treated with ether and the solution filtered from a little insoluble α-naphthyltrimethylammonium bromide. The solution is extracted with dilute hydrochloric

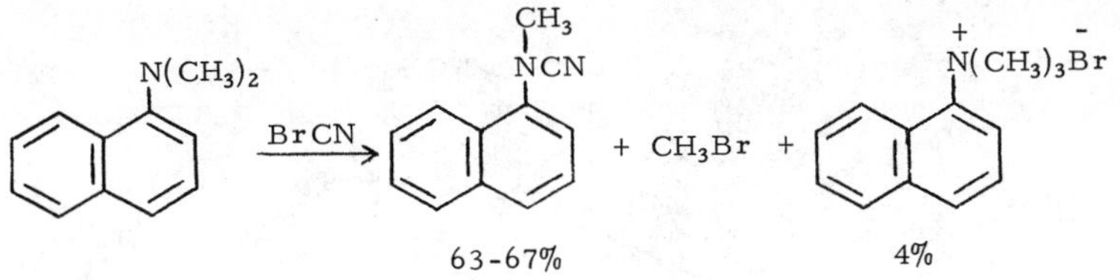

acid to remove starting material, and the N-methyl-1-naphthylcyanamide is recovered and distilled.

The reaction is particularly useful for investigation of the structure of alkaloids. For example, in the study of lupinene, lupinane (1) was cleaved by reaction with cyanogen bromide in boiling benzene to a product shown to be (2) by degradation to quinolinic acid (3).[6]

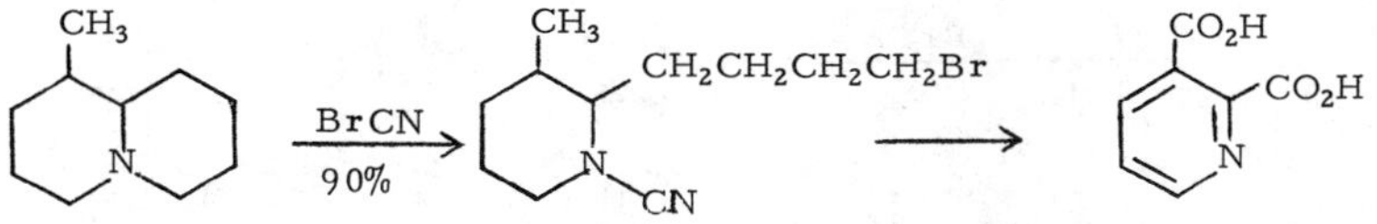

(1) (2) (3)

Cleavage of thioethers. von Braun found[6] that a dialkyl thioether is cleaved by cyanogen bromide at a somewhat elevated temperature (60–70°, sealed tube) to an alkyl thiocyanate and an alkyl bromide. Results for the first three mixed ethers disclose a clear relationship of Haftfestigkeit to size of group:

$$CH_3CH_2SCH_3 \xrightarrow{BrCN} CH_3CH_2SCN + CH_3Br$$

$$CH_3CH_2CH_2SCH_2CH_3 \xrightarrow{BrCN} CH_3CH_2CH_2SCN + CH_3CH_2Br$$

$$CH_3CH_2CH_2CH_2SCH_2CH_2CH_3 \xrightarrow{BrCN} CH_3CH_2CH_2CH_2SCN + CH_3CH_2CH_2Br$$

E. Gross and B. Witkop[7] used this reaction for the selective cleavage of methionine peptides, as illustrated in the formulation:

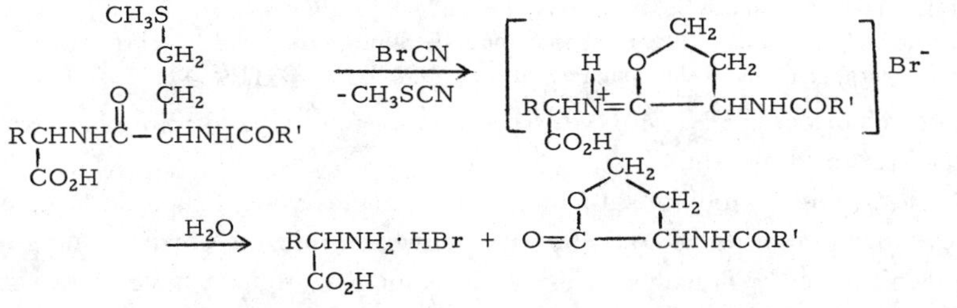

Fragmentation of human pituitary growth hormone (HGH) with cyanogen bromide in 70% formic acid located the two disulfide bridges and confirmed the primary sequence of 188 amino acids deduced from the usual enzymic digestion.[7a]

Reaction with 1,2-amino alcohols. Newhall *et al.*[8] reduced the oximino ketone (1), and treated the resulting mixture of *cis*- and *trans*-amino alcohols, (2) and (3), with cyanogen bromide. A basic fraction was found to contain the crystalline *cis*-amino-

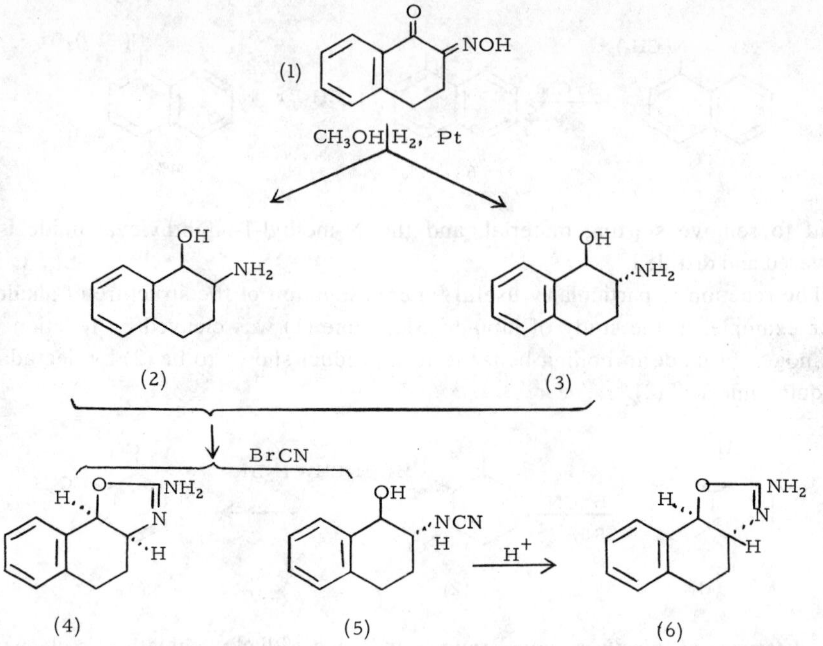

oxazoline (4) and a neutral fraction contained the *trans*-hydroxycyanamide (5), which was cyclized by acid to the *trans*-aminooxazoline (6). Yields are low.

Acylation. When sodium salicylate is refluxed with cyanogen bromide in acetic acid, carbon dioxide is evolved and acetylsalicylic acid is formed; the reaction awaits clarification.[9]

[1]K. H. Slotta, *Ber.*, **67**, 1028 (1934)
[2]W. W. Hartman and E. E. Dreger, *Org. Syn., Coll. Vol.*, **2**, 150 (1943)
[3]J. von Braun, *Ber.*, **40**, 3914 (1907); **42**, 2219 (1909); **44**, 1252 (1911)
[4]H. A. Hageman, "The von Braun Reaction," *Org. Reactions*, **7**, 198 (1953); K. W. Bentley, *Techniques of Organic Chemistry*, **11**, 773 (1963)
[5]H. W. J. Cressman, *Org. Syn., Coll. Vol.*, **3**, 608 (1955)
[6]J. von Braun and P. Engelbertz, *Ber.*, **56**, 1573 (1923)
[7]E. Gross and B. Witkop, *Am. Soc.*, **83**, 1510 (1961); *J. Biol. Chem.*, **237**, 1856 (1962)
[7a]C. H. Li, W.-K. Liu, and J. S. Dixon, *Am. Soc.*, **88**, 2050 (1966)
[8]W. F. Newhall, G. I. Poos, J. D. Rosenau, and J. T. Suh, *J. Org.*, **29**, 1809 (1964)
[9]B. S. Thyagarajan and K. Rajagopalan, *Tetrahedron Letters*, 729 (1965)

Cyanogen chloride, ClCN. Mol. wt. 61.48, m.p. −6.9°, b.p. 13°. Supplier: American Cyanamid Co. Preparation.[1]

Nitrile synthesis. Sodium salts of carboxylic acids when heated in a sealed tube with cyanogen chloride at 200–300° are converted into the corresponding nitriles, often in high yield.[2] Tracer studies[3] with cyanogen bromide have shown that the carbon dioxide liberated is derived from the cyanogen halide and that the bond

$$R\overset{*}{C}O_2Na + BrCN \longrightarrow R\overset{*}{C}N + CO_2 + NaBr$$

between the alkyl group and the carboxyl group is not broken. Sodium dehydroabietate (1) is converted into the nitrile (2) with retention of configuration, and the same is true of sodium O-methylpodocarpate, which has the opposite configuration at C_4.[4]

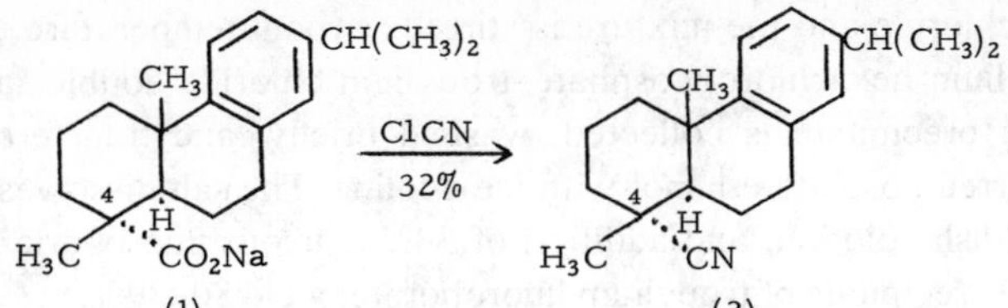

(1) (2)

Alkyl cyanates. Kauer and Henderson[5] found that an alcohol such as neopentyl alcohol can be converted into the corresponding cyanate, if in low yield, by successive treatment with sodium hydride and cyanogen chloride. 1,4-Dihydroxybicyclo-[2.2.2]octane afforded the dicyanate, m.p. 155–156°, in reasonable yield. The

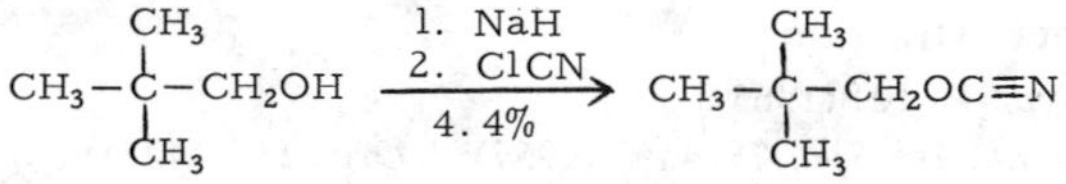
1. NaH
2. ClCN
4.4%

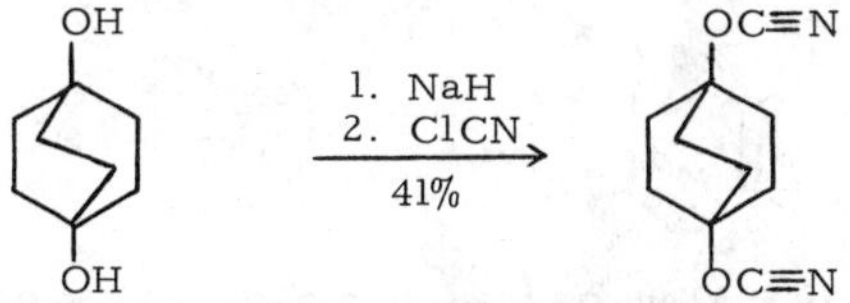
1. NaH
2. ClCN
41%

requirements specified for successful reaction are that the R group be bulky enough to prevent trimerization and that the carbonium ion R^+ have energy high enough to prevent solvolysis under the conditions of synthesis. However, Jensen and Holm[6] prepared ethyl cyanate as a colorless liquid in high yield by spontaneous decomposition of 2-ethoxy-1,2,3,4-thiatriazole in ether.

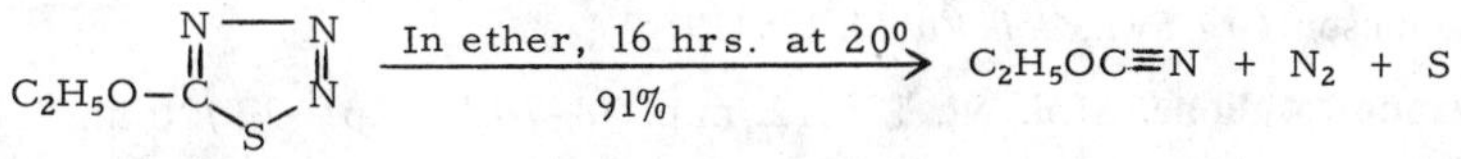
In ether, 16 hrs. at 20°
91%

[1]G. H. Coleman, R. W. Leeper, and C. C. Schulze, *Inorg. Syn.*, **2**, 90 (1946)
[2]E. V. Zappi and O. Bonso, *Anales asoc. quim. argentina*, **35**, 137 (1947) [*C. A.*, **42**, 7704 (1948)]
[3]D. E. Douglas *et al.*, *Can. J. Chem.*, **31**, 1127 (1953); **36**, 1256 (1958)
[4]J. A. Barltrop, A. C. Day, and D. B. Bigley, *J. Chem. Soc.*, 3185 (1961)
[5]J. C. Kauer and W. W. Henderson, *Am. Soc.*, **86**, 4732 (1964)
[6]K. A. Jensen and A. Holm, *Chem. Scand.*, **18**, 826 (1964)

Cycloheptatriene (Tropilidene). Mol. wt. 92.13, b.p. 115.5°, n^{20}D 1.5243. Suppliers: Shell (contains 8.7% toluene); Aldrich. For preparation from benzene and diazomethane catalyzed by cuprous bromide, *see* Cuprous bromide.

The hydrocarbon is conveniently purified and stored as tropylium fluoroborate, which is indefinitely stable, nonhydroscopic, and nonexplosive. The procedure[1] is as follows. Tropilidene is added to a suspension of phosphorus pentachloride in

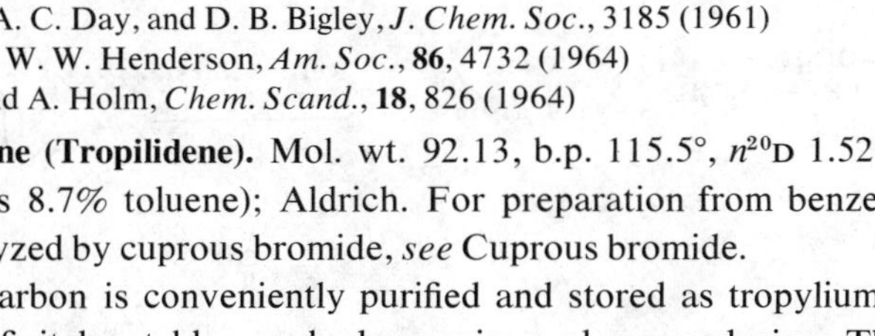
3 PCl5
-2 PCl3, -2 HCl
2 HBF4
BF4
+ PCl5 + 2 HCl

carbon tetrachloride and the mixture is stirred at room temperature until precipitation of tropylium hexachlorophosphate–tropylium chloride double salt is complete (3 hrs.). The precipitate is collected, washed briefly, and transferred quickly to cold, well-stirred absolute ethanol with ice cooling. The salt dissolves exothermally to give a reddish solution, and addition of 50% aqueous fluoroboric acid produces a dense white precipitate of tropylium fluoroborate; yield 80–89%.

A simpler but more costly procedure[2,3] involves dissolving triphenylcarbinol in acetic anhydride, cooling, adding 48% HBF_4, followed by a benzene solution of tropilidene (with cooling) and then ether as precipitant.

For regeneration of the hydrocarbon an aqueous solution of tropylium fluoroborate is treated with an aqueous solution of sodium borohydride and tropilidene is collected by ether extraction.[3]

[1]K. Conrow, *Org. Syn.*, **43**, 101 (1963)
[2]H. J. Dauben, Jr., *et al.*, *Am. Soc.*, **79**, 4557 (1957); *J. Org.*, **25**, 1442 (1960)
[3]E. Müller and H. Fricke, *Ann.*, **661**, 38 (1963)

Cyclohexane-1,3-dione, 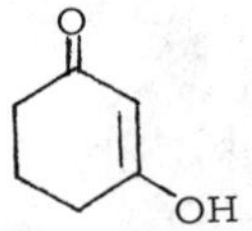
OH

Mol. wt. 112.12, m.p. 105°, λ^{EtOH} 253 mμ (22,300), pKa 5.89. Supplier: Aldrich (this material is stabilized with 20% aqueous sodium chloride solution). This enolic diketone is prepared[1] by hydrogenation of resorcinol in dilute aqueous alkali over a nickel–nickel oxide–kieselguhr catalyst at 50° and a hydrogen pressure of 1,000–1,500 lbs. (yield 85–95%).

For one use, *see* 2-Cyclohexenone.

[1]R. B. Thompson, *Org. Syn., Coll. Vol.*, **3**, 278 (1955)

Cyclohexane-1,4-dione. Mol. wt. 112.12, m.p. 78–79.5°, b.p. 122°/20 mm. Supplier: Columbia.

Preparation.[1] Diethyl succinate is added to a hot solution prepared from sodium and ethanol and the mixture is refluxed for 3–4 hrs. A thick precipitate of the

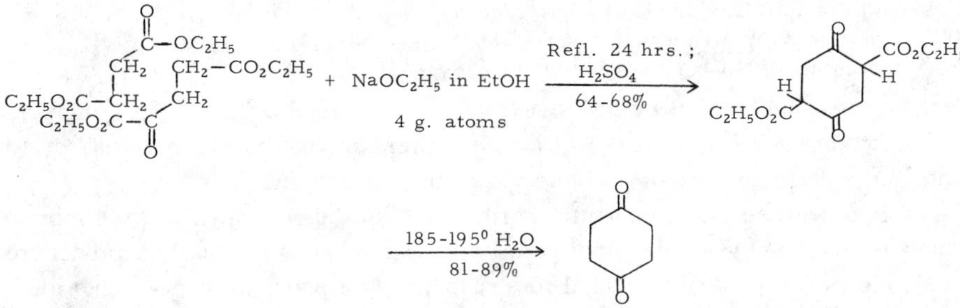
+ NaOC2H5 in EtOH
4 g. atoms
Refl. 24 hrs.; H2SO4
64-68%
185-195° H2O
81-89%

disodium derivative of 2,5-dicarboethoxy-1,4-cyclohexanedione is collected, acidified, and the product crystallized from ethyl acetate. For hydrolysis and decarboxylation, the ester is heated briefly with water in a steel pressure vessel. Cyclohexane-1,4-dione is obtained by distillation as a pale yellow solid, m.p. 77–79°.

[1]A. T. Nielsen and W. R. Carpenter, *Org. Syn.*, **45**, 25 (1965)

Cyclohexanone. Mol. wt. 98.15, b.p. 154–155°, sp. gr. 0.947. Suppliers: B, E, F, MCB.

***H-Acceptor in Oppenauer oxidation**, see* Aluminum isopropoxide.

Cyclohexylidene derivatives of 1,2-diols. Whereas 1,2-diols had traditionally been characterized by reaction with acetone to produce the acetonide or isopropylidene derivatives, Micovic and Stojiljkovic[1] recommended conversion of sugars and other polyols into their cyclohexylidene derivatives because of the ease with which these substances crystallize. Angyal *et al.*[2] explored the behavior of inositol (1) and found that this cyclitol gradually dissolved in a mixture of cyclohexanone and benzene containing *p*-toluenesulfonic acid on boiling under a Dean-Stark water separator.

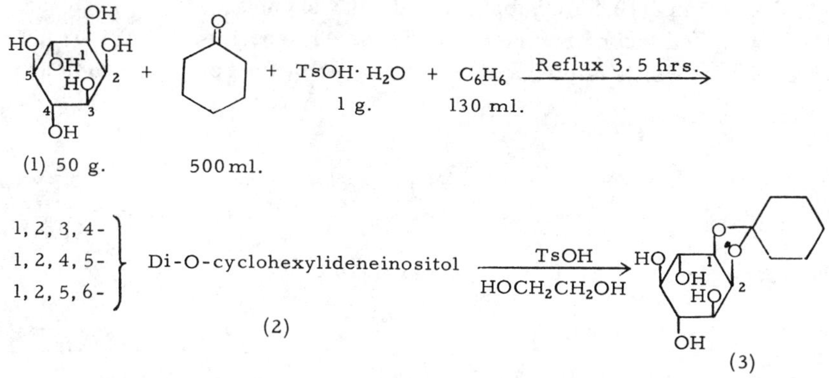
+ TsOH· H₂O + C₆H₆
1 g.
130 ml.
Reflux 3. 5 hrs.
(1) 50 g.
500 ml.
1, 2, 3, 4-
1, 2, 4, 5-
1, 2, 5, 6-
Di-O-cyclohexylideneinositol
(2)
TsOH
HOCH₂CH₂OH
(3)

The water that separated (23 ml.) was considerably more than 1 mole, and the monocyclohexylidene derivative (3) obtained by extraction with water was only a minor product. Most of the inositol had been converted into the three diketols (2), which were separated by chromatography and crystallization. The structures were established by tests with sodium periodate and methylation experiments and by the observation that all three di-derivatives are converted on very mild hydrolysis into the 1,2-mono-derivative (3), which is *cis*. Note that the three di-derivatives all contain one group bridging *trans*-hydroxyl groups (3,4-, 4,5-, and 5,6-). A later paper[3] reports improvements in the preparative procedure and several new derivatives, including 1,2,3,4,5,6-tri-O-cyclohexylideneinositol.

In reporting a new synthesis of (±)-shikimic acid methyl ester (4), Grewe and Hinrichs[4] noted that this substance is convertible in high yield into the cyclohexylidene derivative (5), an oil; (−)-shikimic acid methyl ester afforded a crystalline derivative, m.p. 61°. A mixture of 6 g. of the ester, 6.6 ml. of cyclohexanone,

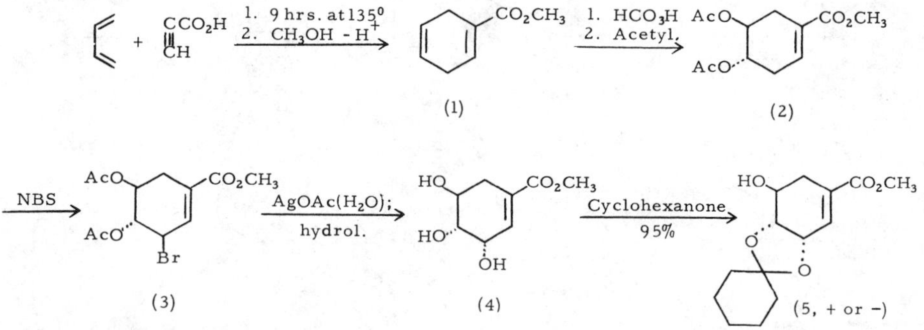
1. 9 hrs. at 135°
2. CH₃OH - H⁺
(1)
1. HCO₃H
2. Acetyl.
(2)
NBS
(3)
AgOAc(H₂O);
hydrol.
(4)
Cyclohexanone
95%
(5, + or −)

4 ml. of dimethylformamide, 50 ml. of benzene, 0.6 g. of Dowex 50 X-4 was refluxed for 100 min. under a water separator, the solution was filtered, washed with methylene chloride, and evaporated in vacuum. Crystallization of the residue from petroleum ether afforded the pure derivative in 95% yield.

[1]V. M. Mićović and A. Stojiljković, *Tetrahedron*, **4**, 186 (1958)
[2]S. J. Angyal, M. E. Tate, and S. D. Gero, *J. Chem. Soc.*, 4116 (1961)
[3]S. J. Angyal, G. C. Irving, D. Rutherford, and M. E. Tate, *J. Chem. Soc.*, 6662 (1965)
[4]R. Grewe and I. Hinrichs, *Ber.*, **97**, 443 (1964)

2-Cyclohexenone. Mol. wt. 96.12, b.p. 56–57.5°/10 mm. or 96–97°/72 mm., $n^{27}D$ 1.458, λ^{EtOH} 226 mμ (10,400). Suppliers: Aldrich, Columbia.

This conjugated ketone is prepared from cyclohexane-1,3-dione (1) via the enol ethyl ether (2), which is prepared by azeotropic distillation of a solution of (1)

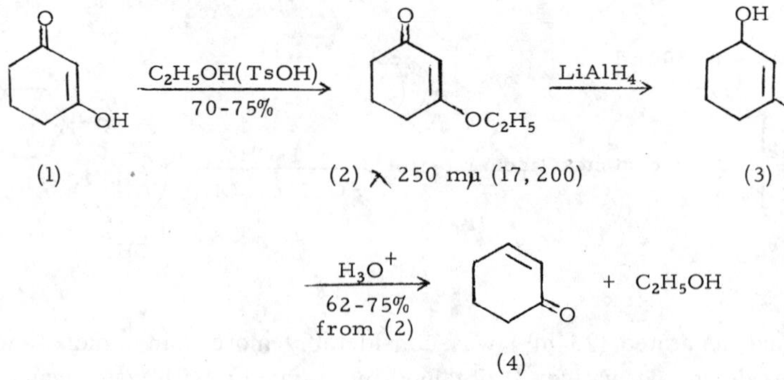
OH
C_2H_5OH(TsOH)
70-75%
OC_2H_5
$LiAlH_4$
OH
OC_2H_5
(1)
(2) λ 250 mμ (17,200)
(3)
H_3O^+
62-75%
from (2)
+ C_2H_5OH
(4)

and *p*-toluenesulfonic acid in absolute ethanol and benzene.[1] Reduction with lithium aluminum hydride in ether to the hydroxy enol ether (3), and acid hydrolysis gives a β-hydroxy ketone which loses water to form 2-cyclohexenone (4).[2] The procedure illustrates a general method for the preparation of α,β-unsaturated aldehydes and ketones from enol ethers of dicarbonyl compounds. Other examples.[3]

[1]W. F. Gannon and H. O. House, *Org. Syn.*, **40**, 41 (1960)
[2]*Idem, ibid.*, **40**, 14 (1960)
[3]H. Born, R. Pappo, and J. Szmuszkovicz, *J. Chem. Soc.*, 1779 (1953); P. Seifert and H. Schinz, *Helv.*, **34**, 728 (1951); H. Favre, B. Marinier, and J. C. Richer, *Can. J. Chem.*, **34**, 1329 (1956)

2-Cyclohexylcyclohexanone, Mol. wt. 180.28, b.p. 137°/12 mm., sp. gr. 0.975. Supplier: National Allied Division, Allied Chem. Corp. The reagent is made by

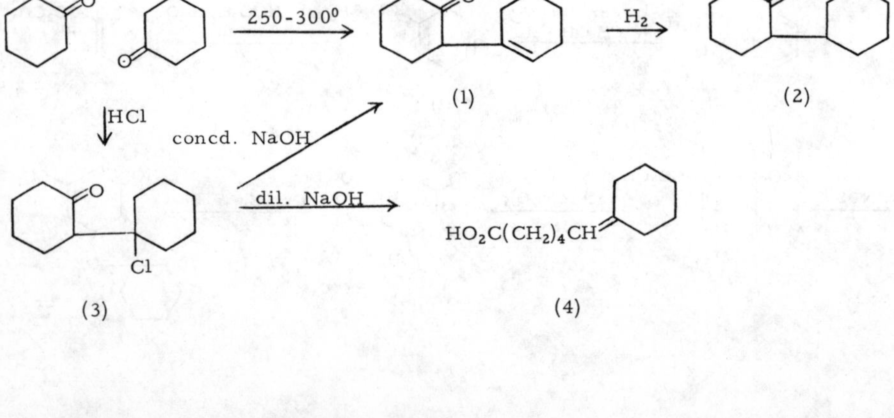
250-300°
H_2
(1)
(2)
HCl
concd. NaOH
dil. NaOH
Cl
$HO_2C(CH_2)_4CH$
(3)
(4)

self-condensation of cyclohexanone at a high temperature (1)[1] or by acid catalysis (3) and elimination of HCl,[2] followed by hydrogenation. Note that dilute alkali opens (3) to the acid (4).

[1]W. Treibs, *Ber.*, **61**, 683 (1928)
[2]J. Plesek, *Coll. Czech.*, **21**, 902 (1956)

1-Cyclohexyl-3-(2-morpholinoethyl)-carbodiimide (3). Mol. wt. 221.33, b.p. 145°/0.2 mm.

Preparation from cyclohexyl isothiocyanate and N-(2-aminoethyl)morpholine (Carbide and Carbon Chemicals Co.).[1]

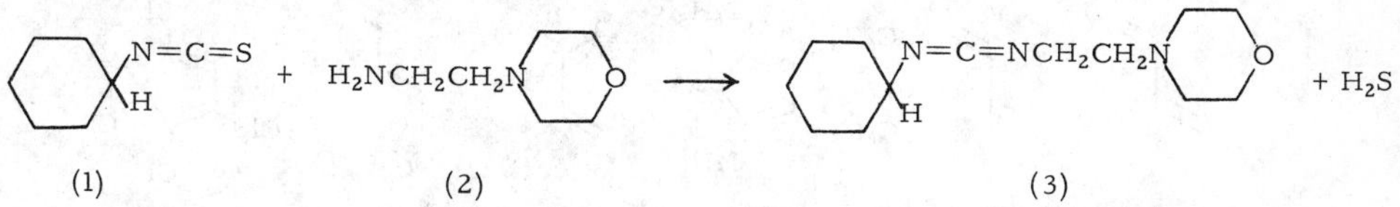

Peptide synthesis. The compound is useful as a coupling agent because the corresponding urea derivative is soluble in dilute acid and hence readily separated from the peptide.[1] Squibb workers[2] found it far superior to DCC for the final cyclization step in the synthesis of a peptide lactone related to the antibiotic vernamycin B_α; methylene chloride was found to be an excellent solvent. The polar character of the urea formed facilitated its separation from the product. Of all other coupling agents tested, only N,N′-carbonyldiimidazole and N-ethyl-5-phenylisoxazolium 3′-sulfonate permitted isolation of the cyclized peptide, if in much lower yield.

[1]J. C. Sheehan and J. J. Hlavka, *J. Org.*, **21**, 439 (1956)
[2]M. A. Ondetti and P. L. Thomas, *Am. Soc.*, **87**, 4373 (1965)

Cyclopentadiene. Mol. wt. 66.10, b.p. 41°, sp. gr. 0.80.

The volatile diene dimerizes readily and is prepared as required by depolymerization of technical dicyclopentadiene (Enjay, high purity, Matheson, purity 85%; b.p. 170°, sp. gr. 0.98). The depolymerization is done by heating the dimer carefully under a fractionating column;[1,2] by heating the dimer under a partial condenser containing methylene chloride;[3] or by adding the dimer at a suitable rate to mineral oil maintained at 240–270°.[4]

Procedures are given for conversion of cyclopentadiene into its sodio derivative[1,5] as an intermediate to ferrocene[1] and ruthenocene.[5]

In view of the widespread interest in the Diels-Alder reaction, the results of a kinetic study of the order of reactivity of dienes toward maleic anhydride[6] is of particular interest. The figures under the formulas give values of rate constants, $10^8 \times k_2$ (liters/mole/sec.) in dioxane at 30°. Where cisoid and transoid forms are both possible, the transoid form is shown, although there is evidence that a bulky group at the 2-position favors the cisoid form required for reaction.[7]

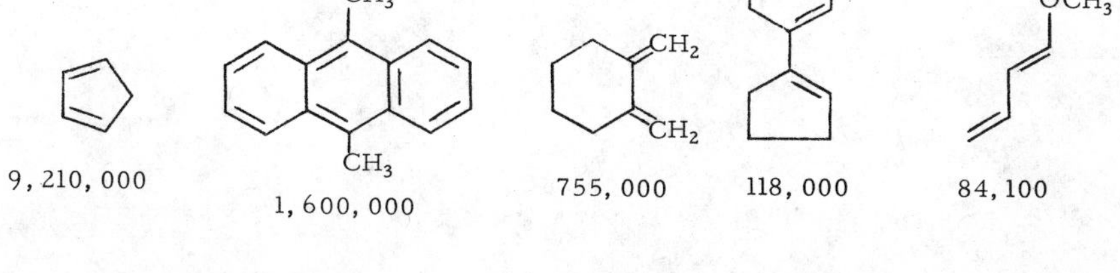
CH3
CH3
CH2
CH2
OCH3
9,210,000
1,600,000
755,000
118,000
84,100

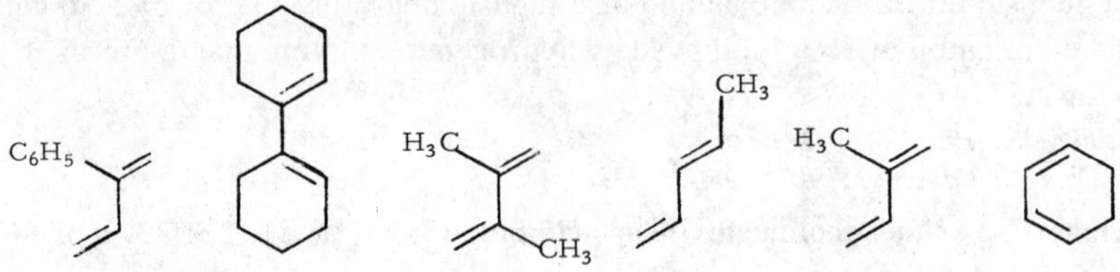

60,000 39,100 33,600 22,700 15,400 13,200

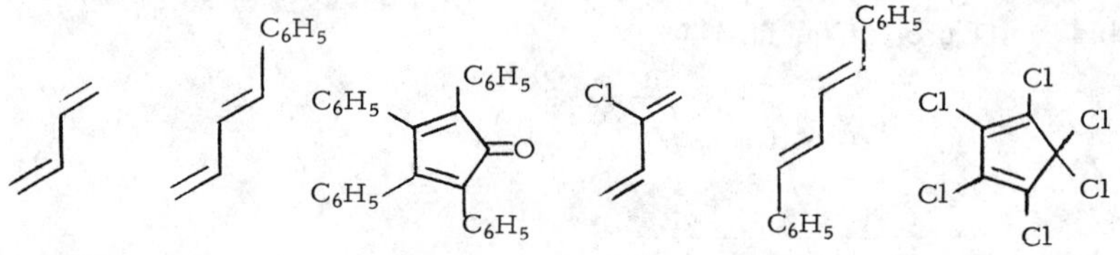

6,830 4,280 1,670 690 296 1.14

[1]G. Wilkinson, *Org. Syn., Coll. Vol.*, **4**, 476 (1963)
[2]*Org. Expts.*, Chapt. 15
[3]R. B. Moffett, *Org. Syn., Coll. Vol.*, **4**, 238 (1963)
[4]M. Korach, D. R. Nielsen, and W. H. Rideout, *Org. Syn.*, **42**, 50 (1962)
[5]D. E. Bublitz, W. E. McEwen, and J. Kleinberg, *ibid.*, **41**, 96 (1961)
[6]J. Sauer, D. Lang, and A. Mielert, *Angew. Chem. Internat. Ed.*, **1**, 268 (1962)
[7]D. Craig, J. J. Shipman, and R. B. Fowler, *Am. Soc.*, **83**, 2885 (1961)

Cyclopentanol. Mol. wt. 86.13, b.p. 134–136°/630 mm. Suppliers: Arapahoe, B, E, F, MCB, Columbia.

Cyclopentanone. Mol. wt. 84.11, b.p. 131°, sp. gr. 0.948. Suppliers: Arapahoe (from adipic acid and barium hydroxide[1]), B, E, F, MCB, Columbia.

[1]J. F. Thorpe and G. A. R. Kon, *Org. Syn., Coll. Vol.*, **1**, 192 (1941)

Cyclopentyl chloroformate, C_5H_9OCOCl. Mol. wt. 148.59, b.p. 69–70.5°/25 mm. The reagent is prepared by the reaction of cyclopentanol with phosgene[1] and has found some use in peptide synthesis for introducing the carbocyclopentyloxy group as

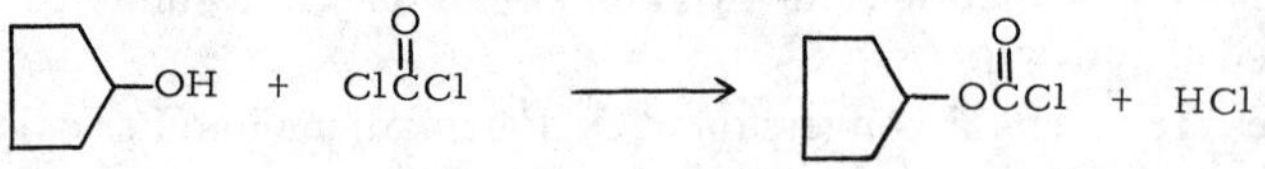

an N-protective group.[2] Derivatives of a few amino acids are sirups. The masking group is resistant to hydrogenolysis but can be cleaved by hydrogen bromide in nitromethane (4–8 min.).

[1]S. Nakanishi, T. C. Myers, and E. V. Jensen, *Am. Soc.*, **77**, 3099 (1955)
[2]F. C. McKay and N. F. Albertson, *Am. Soc.*, **79**, 4686 (1957)

D

Dehydroabietylamine,
Mol. wt. 285.46.

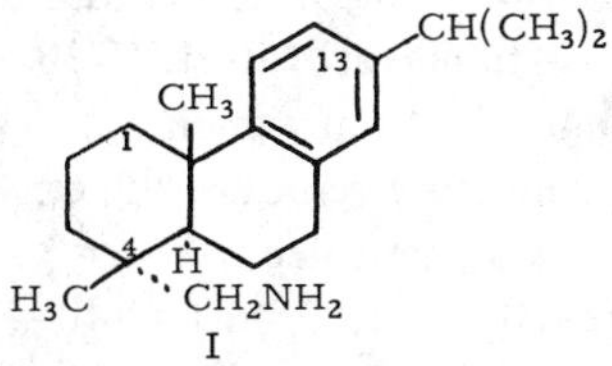
I

In working with semisynthetic penicillins, Sjöberg and Sjöberg[1] observed that dehydroabietylamine (I) gives nicely crystalline salts with various penicillins. They then found this optically active and readily available base to be an excellent reagent for the resolution of acids. Thus several attempts to resolve (±)-γ-phenylpropylsuccinic acid with all common bases met with no success, but resolution was effected easily with I. Further resolutions.[2]

Hercules Powder Company's Amine D contains 50% of I, 20% each of the dihydro and tetrahydro derivatives, and neutral rosin material. For isolation of I[1,3] a solution of 2 kg. of Amine D in 3.3 l. of toluene is treated at 65–70° with 460 g. of acetic acid. Dehydroabietylamine acetate (m.p. 141–143°) is obtained in three crops totaling 960 g. On neutralization of a solution of the acetate in warm water the base separates and is extracted with benzene; evaporation of the dried extract in vacuum gives a viscous oil which slowly crystallizes.

[1]B. Sjöberg and S. Sjöberg, *Arkiv Kemi*, **22**, 447 (1964)
[2]W. J. Gottstein and L. C. Cheney, *J. Org.*, **30**, 2072 (1965)
[3]L. C. Cheney, U. S. patent 2,787,637 (1957) [*C.A.*, **51**, 13,926 (1957)]

***trans,trans*-1,4-Diacetoxybutadiene** (7). Mol. wt. 170.46, m.p. 103–104°.

Preparation. Reppe *et al.*[1] prepared this diene from cyclooctatetraene by the reactions formulated, but regarded it as 1,2-diacetoxy-Δ³-cyclobutene. Criegee *et al.*[2] established the true structure, and Hill and Carlson[3] report a short preparative

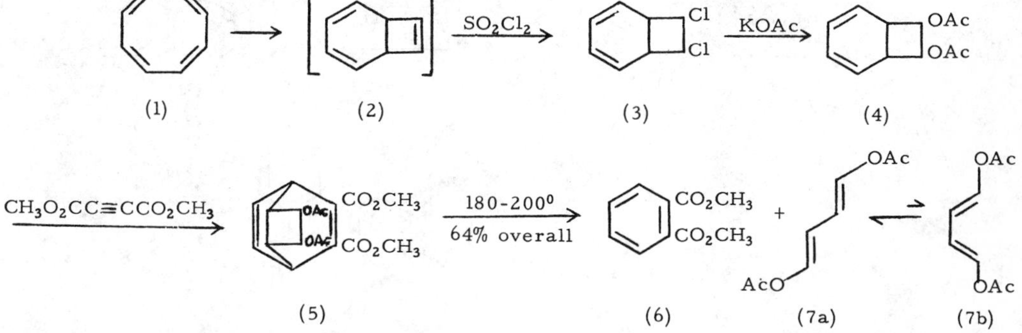

SO₂Cl₂
Cl
Cl
KOAc
OAc
OAc
(1)
(2)
(3)
(4)
CH₃O₂CC≡CCO₂CH₃
CO₂CH₃
CO₂CH₃
180-200°
64% overall
CO₂CH₃
CO₂CH₃
+
OAc
AcO
OAc
OAc
(5)
(6)
(7a)
(7b)

procedure based on Cope's observation that the diacetate (4) can be prepared conveniently by reaction of cyclooctatetraene with mercuric acetate in acetic acid and pyrolysis of the resulting addition compound. Thus 52 g. of cyclooctatetraene was added to a stirred suspension of 160 g. of mercuric acetate in 400 ml. of acetic acid. After 10–15 min. the addition compound which separated was collected and decomposed by heating at 70–75° for 2 hrs. The warm solution was separated from the mercury formed and poured into water. The diacetate (4) solidified on cooling (83 g.) and a solution of this and 54 g. of dimethyl acetylenedicarboxylate in 250 ml. of benzene was refluxed for 6 hrs. The mixture was filtered and distilled at reduced pressure. Material distilling at 140–155°/20 mm. was a mixture of dimethyl phthalate and the diene (7) from which (7) crystallized in the cooled receiver. Trituration with petroleum ether removed dimethyl phthalate, and crystallization of the residue from acetone–petroleum ether gave colorless needles of the diacetoxybutadiene, m.p. 102–104°, yield 26–31 g. (40–48% overall).

Diels-Alder reactions. Hill and Carlson[3] showed that reaction of the reagent with dienophiles in refluxing benzene or without solvent at 100–110° provides a useful one-step synthesis of aromatic compounds. An example other than that formulated

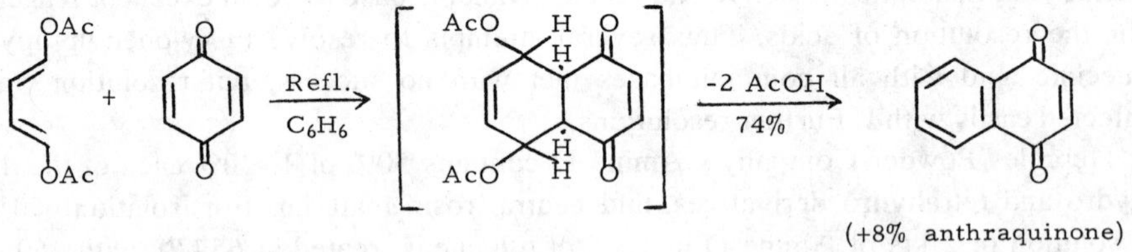
OAc
OAc
+
O
O
Refl.
C6H6
AcO
H
H
AcO
H
H
O
O
-2 AcOH
74%
O
O
(+8% anthraquinone)

is the reaction with β-nitrostyrene to form 2-nitrodiphenyl (20%). This novel route to unsymmetrical biphenyls was utilized by Hill and Carlson[4] for the synthesis of the alkaloid ismine by the reactions formulated.

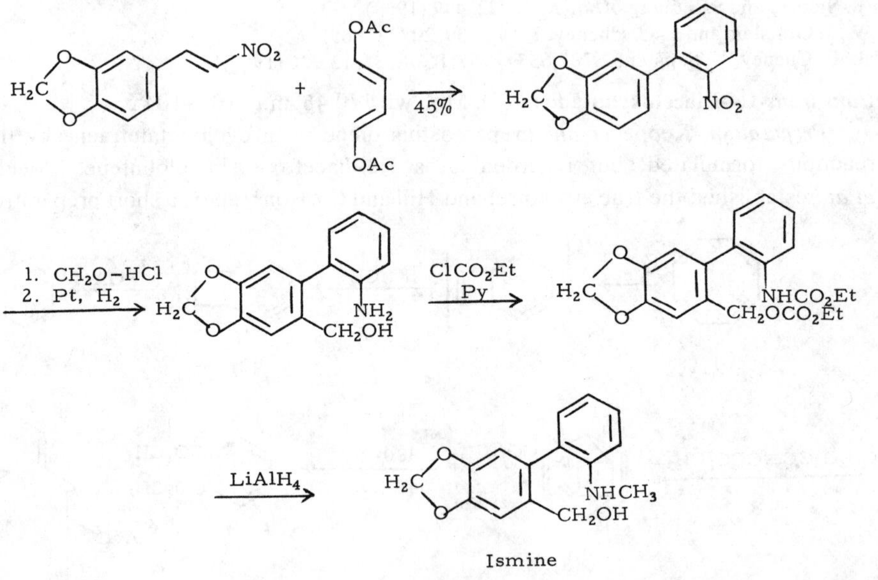
H2C
O
O
NO2
+
OAc
OAc
45%
NO2
1. CH2O-HCl
2. Pt, H2
NH2
CH2OH
ClCO2Et
Py
NHCO2Et
CH2OCO2Et
LiAlH4
NHCH3
CH2OH
Ismine

The synthesis of *dl*-shikimic acid (5) was achieved by both a Glasgow group[5] and a midwestern U.S. group[6]; the formulation refers to the first report since a suggested[6] correction to the stereochemistry of intermediates (1)–(3) has been shown by n.m.r. double resonance experiments to be invalid (letter from R. A.

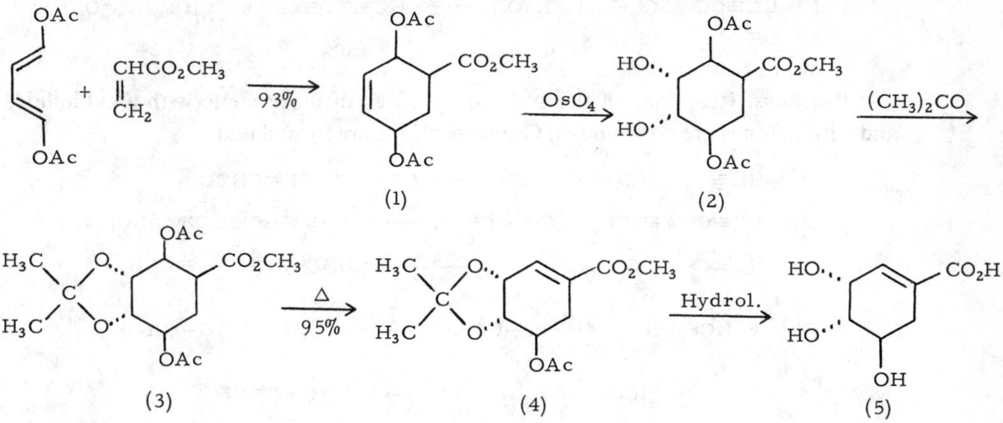
OAc
CHCO2CH3
CH2
93%
OAc
CO2CH3
OAc
(1)
OsO4
HO
HO
OAc
CO2CH3
OAc
(2)
(CH3)2CO
H3C
H3C
C
OAc
CO2CH3
OAc
(3)
Δ
95%
CO2CH3
OAc
(4)
Hydrol.
HO
HO
CO2H
OH
(5)

Raphael). The reaction of *trans,trans*-1,4-diacetoxybutadiene with methyl acrylate gave an adduct (1) which was *cis*-hydroxylated to (2) by rear attack and was then converted into the acetonide (3). Treatement of (3) with base under a variety of conditions failed to effect elimination of acetic acid, but conversion to (4) was effected in high yield by pryolysis.[5] Elimination of acetone (60% acetic acid) and saponification then afforded *dl*-shikimic acid (5), which was resolved.

[1]W. Reppe, O. Schlichting, K. Klager, and T. Toepel, *Ann.*, **560**, 1 (1948)
[2]R. Criegee, W. Hörauf, and W. D. Schellenberg, *Ber.*, **86**, 126 (1953)
[3]R. K. Hill and R. M. Carlson, *Tetrahedron Letters*, 1157 (1964); *J. Org.*, **30**, 2414 (1965)
[4]*Idem, ibid.*, **30**, 1571 (1965)
[5]R. McCrindle, K. H. Overton, and R. A. Raphael, *J. Chem. Soc.*, 1560 (1960)
[6]E. E. Smissman, J. T. Suh, M. Oxman, and R. Daniels, *Am. Soc.*, **84**, 1040 (1962)

Diacetylene, HC≡CC≡CH. Mol. wt. 50.06, m.p. −36 to −35°, b.p. 10.3°, polymerizes rapidly above 0°.

Preparation. The hydrocarbon can be prepared by dehydrochlorination of the readily available 1,4-dichlorobutyne-2 with 10% sodium hydroxide.[1] If either the mono- or di-sodium derivative is required, for example for reaction with an alkyl

$$ClCH_2C{\equiv}CCH_2Cl \xrightarrow{10\%\ NaOH} HC{\equiv}CC{\equiv}CH$$

halide or a ketone, it can be prepared directly by reaction of 1,4-dichlorobutyne-2 in liquid ammonia with 3 or 4 equivalents of sodamide.[2,3]

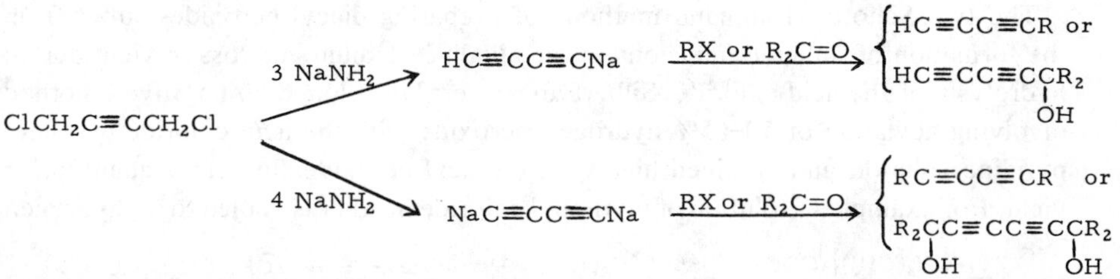
ClCH2C≡CCH2Cl
3 NaNH2
HC≡CC≡CNa
RX or R2C=O
HC≡CC≡CR or
HC≡CC≡CCR2
OH
4 NaNH2
NaC≡CC≡CNa
RX or R2C=O
RC≡CC≡CR or
R2CC≡CC≡CCR2
OH
OH

A new method useful for the preparation of small batches of diacetylene as required involves base-catalyzed cleavage of 2,7-dimethyl-3,5-octadiyne-2,7-diol (from oxidative coupling of the carbinol from acetylene and acetone).

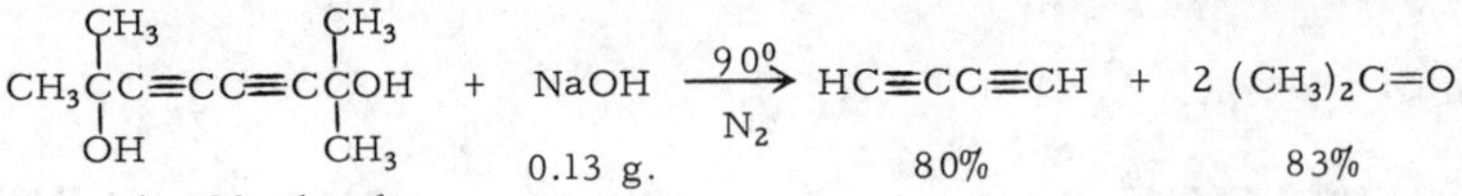

0.1 m. in 100 ml. xylene

Reactions. Reactions of the sodio derivatives of diacetylene with alkyl halides and with ketones are cited above. Other reactions are formulated:

1.[5] $HC{\equiv}C{-}C{\equiv}CH + CH_3OH \longrightarrow HC{\equiv}C{-}CH{=}CHOCH_3$

2.[6] $HC{\equiv}C{-}C{\equiv}CH + CH_2{=}CHCN \longrightarrow HC{\equiv}C{-}C{\equiv}CCH_2CH_2CN$

3.[7] $HC{\equiv}C{-}C{\equiv}CH + 2\ CH_2O \xrightarrow[95^0]{\text{Ag salt}} HOCH_2C{\equiv}C{-}C{\equiv}CCH_2OH$

4.[7] $R_2NCH_2OH + HC{\equiv}C{-}C{\equiv}CH \xrightarrow[-H_2O]{60\text{-}70^0} R_2NCH_2C{\equiv}C{-}C{\equiv}CH$

5.[7] $R_2NH + HC{\equiv}C{-}C{\equiv}CH \xrightarrow[40\text{-}50^0]{\text{Ag salt}} H_2C{=}C(NR_2){-}C{\equiv}CH$

[1]A. W. Johnson, *J. Chem. Soc.*, 1009 (1946)
[2]J. B. Armitage, E. R. H. Jones, and M. C. Whiting, *J. Chem. Soc.*, 44 (1951); 1993 (1952); 3317 (1953)
[3]F. Bohlmann, *Ber.*, **84**, 545 (1951)
[4]R. J. Tedeschi and A. E. Brown, *J. Org.*, **29**, 2051 (1964)
[5]T. Herbertz, *Ber.*, **85**, 475 (1952); A. W. Johnson, *J. Chem. Soc.*, 1009 (1946); W. Franke *et al.*, *Ber.*, **86**, 793 (1953)
[6]W. Franke and H. Meister, German patent 828,572
[7]J. W. Copenhaver and M. H. Bigelow, "Acetylene and Carbon Monoxide Chemistry," 306–308, Reinhold (1949)

Diacyl peroxides, $R\overset{O}{\overset{\|}{C}}OO\overset{O}{\overset{\|}{C}}R$. Preparation:[1] (a) RCOCl + Na_2O_2; (b) RCOCl + H_2O_2 + NaOH. Use for alkylation of 2-hydroxy-1, 4-naphthoquinone.[2] The reaction

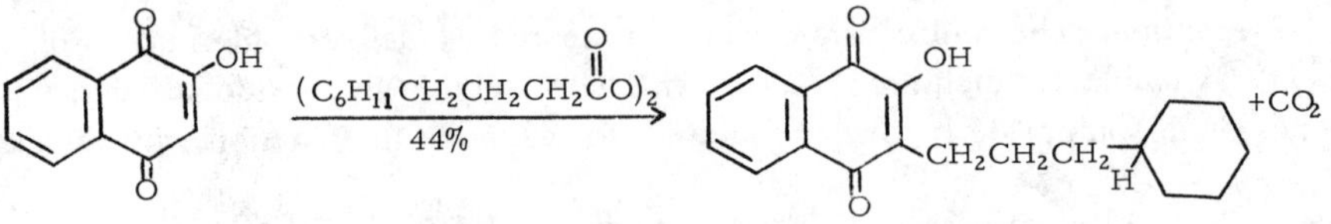

M-1916

is conducted in acetic acid solution at 90–95°. By-products identified are RCO_2H, RH, RR, $ROCOCH_3$, ROH, RCO_2R.

The two Schotten-Baumann methods of preparing diacyl peroxides suffer from the formation of difficult emulsions in the alkaline medium and loss in yield due to hydrolysis of the acid chloride. Silbert and Swern[3] developed an improved method involving acylation of 50–65% hydrogen peroxide with the acid chloride in ether–pyridine solution and obtained highly pure diacyl peroxides in nearly quantitative yield. For example, a solution of myristoyl chloride in ether is cooled to 0°, hydrogen

$$\underset{0.1\text{m. in 175ml. ether}}{2\ \underline{n}\text{-}CH_3(CH_2)_{12}COCl} + \underset{0.75\text{ m.}}{60\%\ H_2O_2} + \underset{0.12\text{ m.}}{\text{Pyridine}} \xrightarrow[98\%]{0\text{-}5^0} [CH_3(CH_2)_{12}COO]_2$$

peroxide is added, followed by dropwise addition of pyridine with stirring and cooling. The diacyl peroxide precipitates as formed. After the addition the ice bath is removed and stirring continued for a total of 1 hr. More ether is added to produce a homogeneous solution, which is washed with dilute hydrochloric acid, 5% potassium bicarbonate, and water, and dried over sodium sulfate. The peroxide content is determined iodometrically.

The improved procedure gave excellent results when applied to the following acid chlorides:[4]

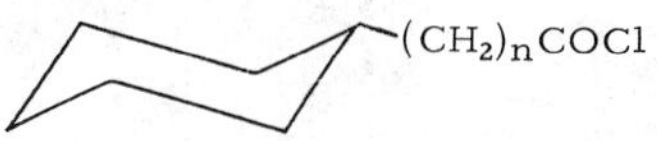
(CH2)nCOCl

(n = 5, 6, 7, 8)

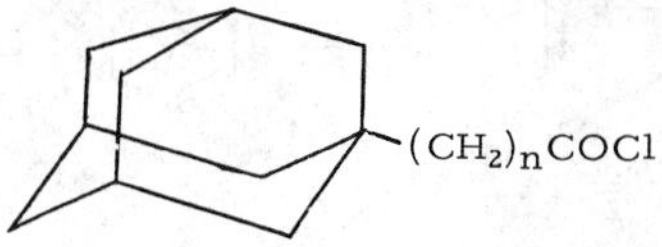
(CH2)nCOCl

(n = 1, 2, 3)

[1]L. F. Fieser, M. Leffler, *et al.*, *Am. Soc.*, **70**, 3178 (1948)
[2]*Idem*, *ibid.*, **70**, 3174–3215 (1948)
[3]L. S. Silbert and D. Swern, *Am. Soc.*, **81**, 2364 (1959)
[4]L. F. Fieser, M. Z. Nazer, and J. Schirmer, unpublished results

Diamylamine, $(C_5H_{11})_2NH$. Mol. wt. 157.29, b.p. of di-*n*-amylamine 189°, of diisoamylamine 183°. Supplier of mixture of isomers: Sharples Chem. Co.

Abietic acid is conveniently isolated from acid-isomerized wood rosin by slowly adding commercial diamylamine to a stirred solution of the material in acetone.[1] The salt which separates (αD −18°) is crystallized four times from acetone and the purified salt (αD −60°) is treated in ethanol with acetic acid and water. After recrystallization from dilute ethanol the yield of abietic acid is 40%.

[1]G. C. Harris and T. F. Sanderson, *Org. Syn., Coll. Vol.*, **4**, 1 (1963)

***dl*-Dianilino-1,2-diphenylethane,** Mol. wt. 364.47, m.p. 146–149°. Supplier: KK. Preparation.[1]

The reagent reacts specifically with most aliphatic aldehydes to give high-melting imidazolidines. Benzaldehyde itself does not react but nitro and halo derivatives do.[2]

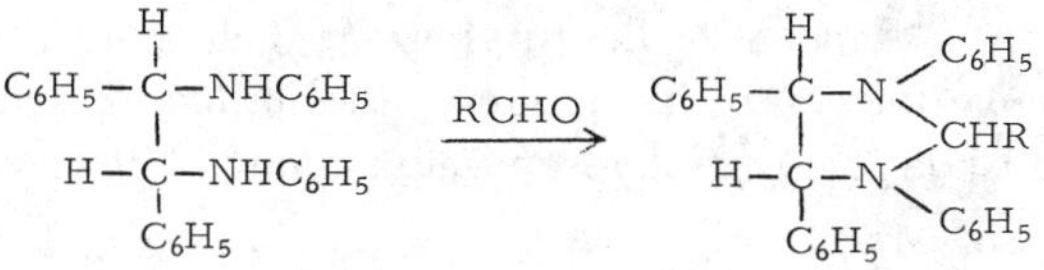
H
C6H5—C—NHC6H5
H—C—NHC6H5
C6H5
RCHO
H
C6H5—C—N
C6H5
CHR
H—C—N
C6H5
C6H5

[1]R. Jaunin, *Helv.*, **39**, 111 (1956)
[2]R. Jaunin and J.-P. Godat, *Helv.*, **44**, 95 (1961)

1,2-Dianilinoethane. Mol. wt. 212.29, m.p. 67.5°. Preparation and use as a specific reagent for aldehydes; ketones do not react.[1] On addition of a catalytic amount of

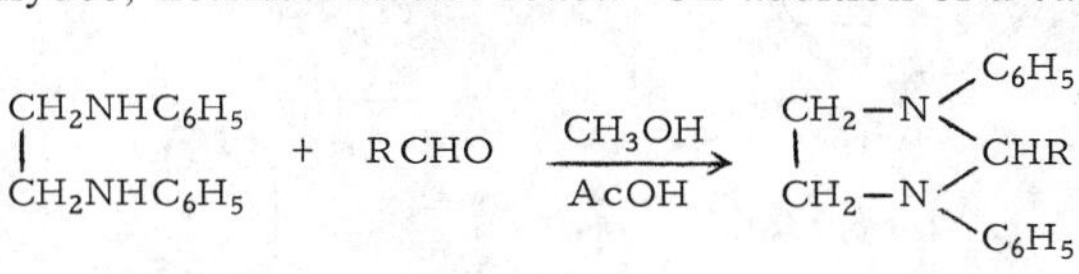
CH2NHC6H5
CH2NHC6H5
+ RCHO
CH3OH
AcOH
CH2—N
C6H5
CHR
CH2—N
C6H5

acetic acid to a solution of the reagent and aldehyde in methanol, the imidazolidine derivative separates in 1–2 min. The formaldehyde derivative melts at 126°. For recovery of the aldehyde, the derivative is shaken with 10% hydrochloric acid and the aldehyde extracted with ether.

[1]H.-W. Wanzlick and W. Löchel, *Ber.*, **86**, 1463 (1953)

3,3′-Dianisole bis-4,4′-(3,5-diphenyl)-tetrazolium chloride. *See* Blue tetrazolium.

2,5-Di-*p*-anisyl-3,4-diphenylcyclopentadienone. Mol. wt. 390.46, m.p. 196°.

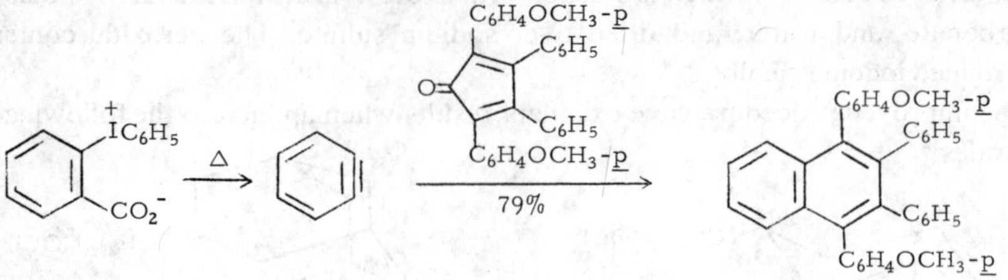

Prepared[1] by the method of Dilthey, the reagent is more efficient than tetraphenyl-cyclopentadienone as a reagent for trapping benzyne generated from diphenyl-iodonium carboxylate.[2] In experiments with equimolar quantities of reactants, the yields of tetraarylnaphthalenes isolated by chromatography were 79 and 57%, respectively.

[1] S. B. Coan, D. E. Trucker, and E. I. Becker, *Am. Soc.*, **77**, 60 (1955)
[2] F. M. Beringer and S. J. Huang, *J. Org.*, **29**, 445 (1964)

Di-*p*-anisylphenylmethyl chloride (2). Mol. wt. 326.81, m.p. 114°.

Preparation.[1]

$$2\ \underline{p}\text{-}CH_3OC_6H_4MgBr\ +\ C_6H_5CO_2CH_3\ \longrightarrow\ \underline{p}\text{-}CH_3OC_6H_4\text{-}\overset{\overset{\displaystyle OH}{|}}{\underset{\underset{\displaystyle C_6H_5}{|}}{C}}\text{-}C_6H_4OCH_3\text{-}\underline{p}$$

(1)

$$\xrightarrow{CH_3COCl}\ \underline{p}\text{-}CH_3OC_6H_4\text{-}\overset{\overset{\displaystyle Cl}{|}}{\underset{\underset{\displaystyle C_6H_5}{|}}{C}}\text{-}C_6H_4OCH_3\text{-}\underline{p}$$

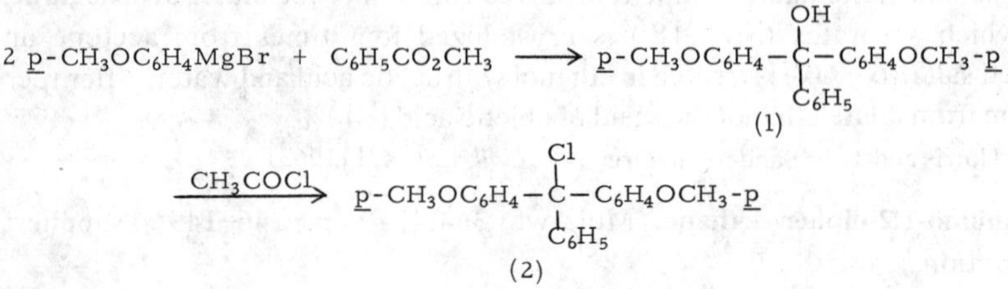

(2)

Protective group. The trityl group is not generally satisfactory as a protective group in the nucleoside field because the glycosyl bonds are labile under the acidic conditions required for removal of the trityl group. Thus Khorana *et al.*[1,2] found it impossible to remove the 5′-O-trityl group of the uridine derivative (3a) without affecting the acid-labile 2′-O-tetrahydropyranyl group. However, substitution in

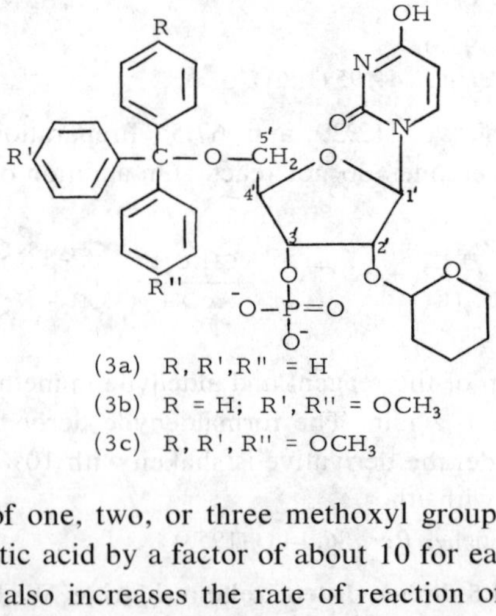

(3a) R, R′, R″ = H
(3b) R = H; R′, R″ = OCH₃
(3c) R, R′, R″ = OCH₃

the protective group of one, two, or three methoxyl groups increases the rate of hydrolysis in 80% acetic acid by a factor of about 10 for each group. Introduction of *p*-methoxyl groups also increases the rate of reaction of the trityl halide with secondary hydroxyl groups and limits the selectivity of reaction with the primary 5′-hydroxyl function. The best balance is achieved with di-*p*-anisylphenylmethyl chloride and the derivative (3b).

[1]M. Smith, D. H. Rammler, I. H. Goldberg, and H. G. Khorana, *Am. Soc.*, **84**, 430 (1962)
[2]H. Schaller, G. Weimann, B. Lerch, and H. G. Khorana, *Am. Soc.*, **85**, 3821 (1963)

1,5-Diazabicyclo[3.4.0]nonene-5, see (2). Mol. wt. 124.19, b.p. 100–102°/12 mm., $n^{20}D$ 1.5190.

Preparation.[1]

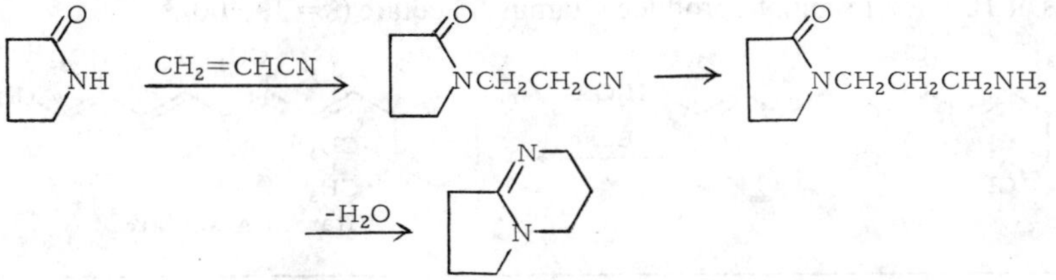
CH₂=CHCN
NH
NCH₂CH₂CN
NCH₂CH₂CH₂NH₂
-H₂O

(2)

Dehydrohalogenation. This base is a superior reagent for effecting dehydrohalogenation. Truscheit and Eiter[2] were unable to dehydrochlorinate the chloro ester (1) with usual tertiary bases (pyridine, quinoline, N,N-dimethylaniline) but by stirring a mixture with the bicyclic amidine base at 90°, adding ice and sulfuric acid, and working up the neutral fraction, they obtained in 85% yield a 70:30 mixture of *cis*- and *trans*-methyl tridecene-10-yne-12-oate (3).

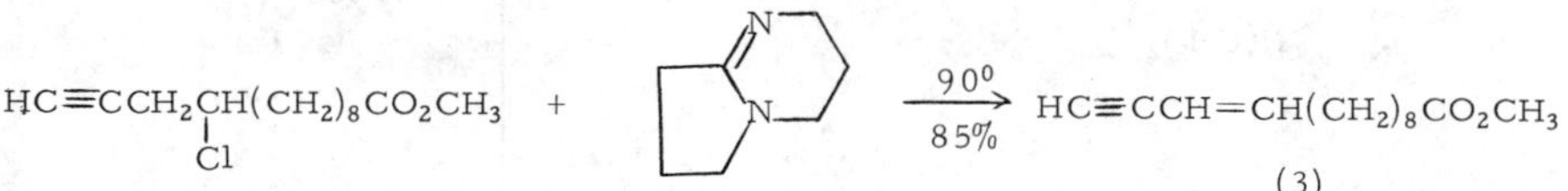
HC≡CCH₂CH(CH₂)₈CO₂CH₃
Cl
90°
85%
HC≡CCH=CH(CH₂)₈CO₂CH₃

(1) 109 g. (2) 117.4 g. (3)

Vogel *et al.*[3] obtained 2,7-dimethyloxepin (7a) as a stable yellow liquid in 60% yield by reaction of the epoxide dibromide (5a) with potassium *t*-butoxide in ether;

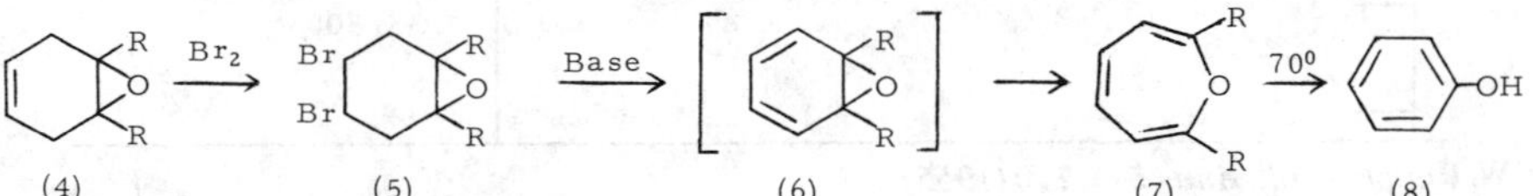
Br₂
Br
Br
Base
70°
OH

(4) (5) (6) (7) (8)
R = H
(a) R = CH₃, (b) R = H

the unsaturated epoxide (6a) is a likely intermediate. Attempts to synthesize oxepin itself by the same method led only to the formation of phenol, but when diazabicyclo-[3.4.0]nonene-5 was used as base, a yellow, unstable product was obtained in good yield in 95% purity. This substance, b.p. 27°/14 mm., begins to rearrange to phenol above 70°. It is assumed to be oxepin because it adds maleic anhydride even at 20° and because the NMR spectrum shows absorption only in the olefinic region.

Eiter and Oediger[4] investigated the synthesis of conjugated eneynes by the Wittig reaction but found the most satisfactory route to be dehydrobromination of 4-bromo-1-alkynes with the cyclic amidine base (see formulation at top of p. 190).

The relative efficiency of the reagent is apparent from the following results for

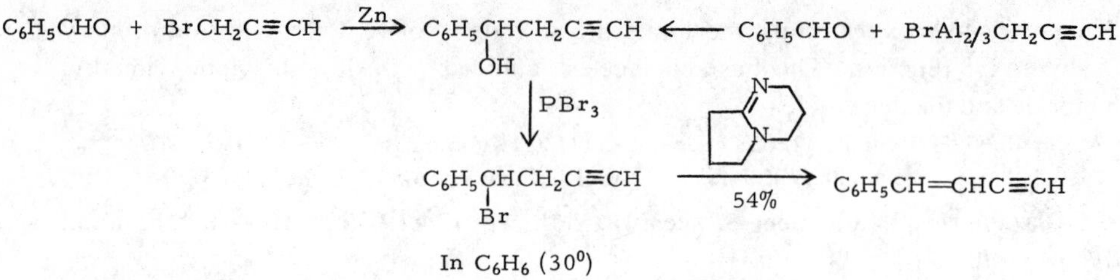
$C_6H_5CHO + BrCH_2C{\equiv}CH \xrightarrow{Zn} C_6H_5CH(OH)CH_2C{\equiv}CH \longleftarrow C_6H_5CHO + BrAl_{2/3}CH_2C{\equiv}CH$
PBr3
$C_6H_5CH(Br)CH_2C{\equiv}CH \xrightarrow{54\%} C_6H_5CH{=}CHC{\equiv}CH$
In C6H6 (30°)

the formation of vitamin A acetate on treatment of the chloride (8) with various bases at 100° for 15 min. to produce vitamin A acetate (ε = 39,400).[5]

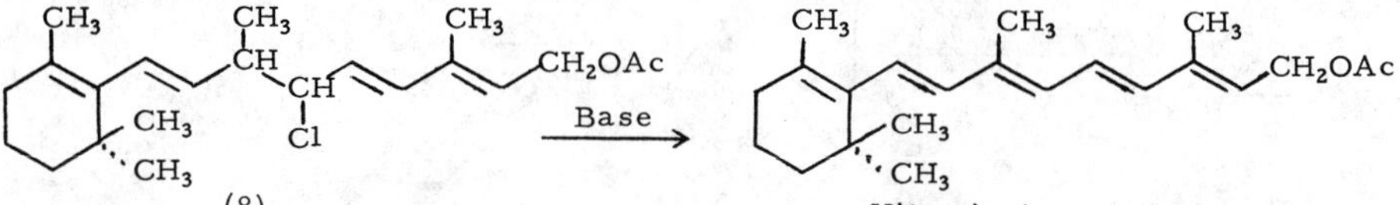
CH3
CH3
CH
CH
Cl
CH3
CH3
CH2OAc
Base
CH2OAc

(8) Vitamin A acetate

Base	Extinction
Pyridine, dimethylaniline, N-methylmorpholine	0
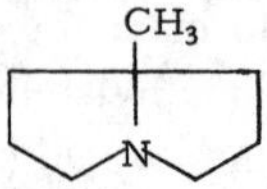	1, 400
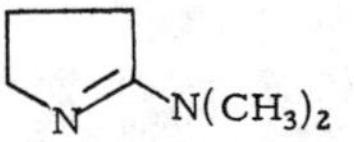	7, 500
 in pyridine	17, 500
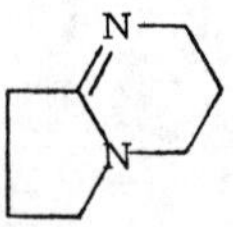	33, 500

[1]W. Reppe, *et al.*, *Ann.*, **596**, 210 (1955)
[2]E. Truscheit and K. Eiter, *Ann.*, **658**, 65 (1962)
[3]E. Vogel, R. Schubart, and W. A. Böll, *Angew. Chem.*, *Internat. Ed.*, **3**, 510 (1964)
[4]K. Eiter and H. Oediger, *Ann.*, **682**, 62 (1965)
[5]H. Oediger, H.-J. Kabbe, F. Möller, and K. Eiter, *Ber.*, **99**, 2012 (1966)

9-Diazofluorene (1). Prepared by oxidation of fluorenone hydrazone,[1] the reagent has been used for the deoxygenation of pyridine N-oxides.[2] With 4 equivalents of reagent, 2-picoline and 2,6-lutidine were obtained in yields of 45 and 62%.

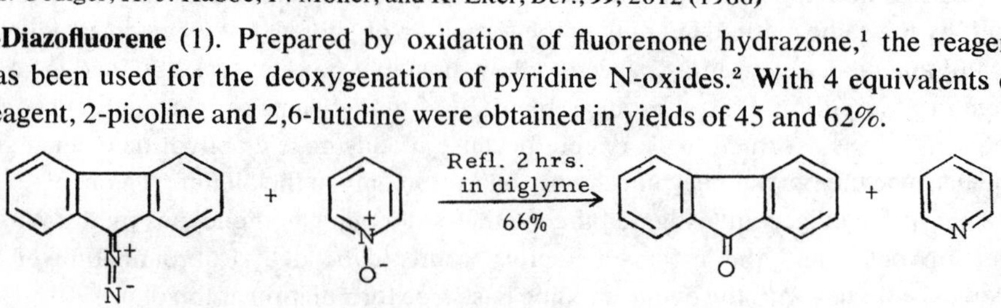
Refl. 2 hrs.
in diglyme
66%

(1) 4 equiv.

[1]C. D. Nenitzescu and E. Solomonica, *Org. Syn., Coll. Vol.*, **2**, 497 (1943)
[2]E. E. Schweizer, G. J. O'Neill, and J. N. Wemple, *J. Org.*, **29**, 1744 (1964)

Diazomethane, $\overset{+}{C}H_2{=}\overset{+}{N}{=}\overset{-}{N} \rightarrow \overset{-}{C}H_2{-}\overset{+}{N}{\equiv}N$. Mol. wt. 42.04, m.p. −145°, b.p. −23°. Since diazomethane is toxic and explosive, operations with sizable quantities should be conducted in a well-ventilated hood behind a safety screen with careful observation of safety precautions of the specific procedure being followed.

Determination. The reagent is prepared and used in ethereal solution of content determined as follows. Dilute an aliquot portion with absolute ether and add a measured quantity of a 0.2 *N* solution of benzoic acid in absolute ether until the solution is colorless and excess acid is present. Add water and titrate the excess benzoic acid with 0.1 *N* sodium hydroxide solution.

Preparation. von Pechmann,[1] discoverer of diazomethane, generated the reagent by the action of alcoholic potassium hydroxide on nitrosomethylurethane, but since this liquid precursor is an active skin irritant a succession of other methods have been proposed. Of these methods we shall list only the four that now appear to have particular merit and these are listed in order of preference.

(a) ***From Bis-(N-methyl-N-nitroso)terephthalamide.***[2]

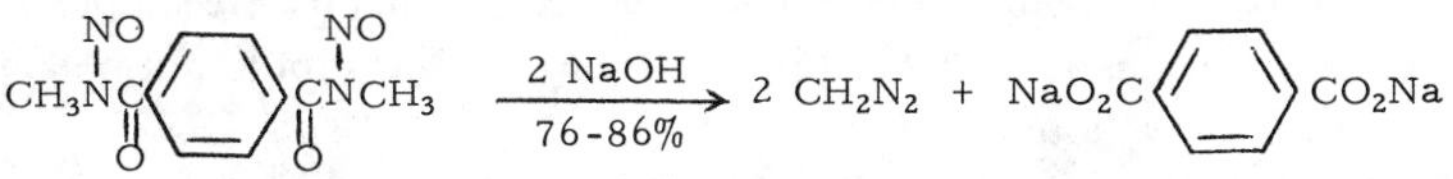

The reagent is available at modest cost from the Explosives Department, E. I. du Pont de Nemours and Co., Gibbstown, J. J., under the trade name EXR-101, which is a mixture of the nitrosamide with 30% white mineral oil as stabilizer, and which may be stored indefinitely at room temperature (bulletin available). Supplied by Aldrich in lots of 100 g. and 500 g.

The charge of nitrosamide is added in one batch to an ice-cold (0°) mixture of ether, diethylene glycol monomethyl ether, and 30% aqueous sodium hydroxide and the ether-diazomethane mixture is distilled through an efficient condenser into an ice-cooled receiver. The distillate of about 2 l. contains about 0.8 mole of diazomethane.

(b) ***From p-Toluenesulfonylmethylnitrosamide.*** Mol. wt. 214.25, m.p. between 55 and 60°. Preparation of reagent.[3] Supplier: Aldrich (Diazald). *Caution:* one laboratory has reported that the substance is a potent skin irritant.[4]

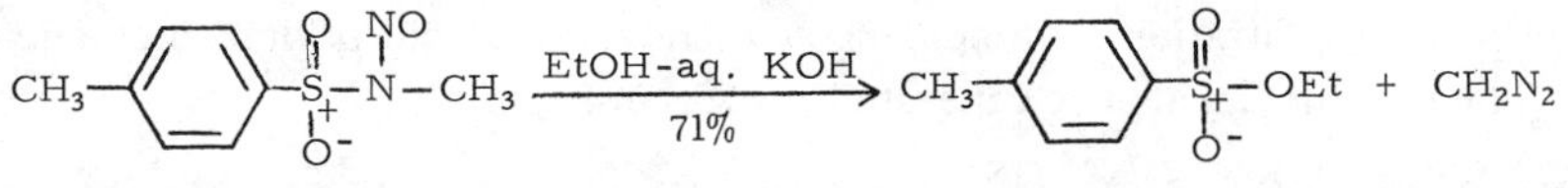

For the preparation of an ethereal–alcoholic solution of diazomethane,[5] a 100-ml. distilling flask is charged with a solution of 5 g. of potassium hydroxide in 8 ml. of water and connected to an efficient condenser delivering into two receiving flasks in series, both cooled in ice. The second receiver contains 25 ml. of ether and the inlet tube dips below the surface of the solvent. The generating flask is heated in a water bath at 65° and a solution of 21.4 (0.1 mole) of Diazald in 130 ml. of ether is added from a dropping funnel in about 25 min. When the dropping funnel is empty, another 20 ml. of ether is added slowly, and the distillation continued until the distilling ether is colorless. The combined distillate contains about 3 g. of diazomethane.

For the preparation of an alcohol-free ethereal solution,[5,6] the generating flask is charged with 35 ml. of diethylene glycol monoethyl ether, 10 ml. of ether, and a solution of 6 g. of potassium hydroxide in 10 ml. of water. The flask is heated in a water bath at 70° and a solution of 21.4 g. of Diazald in 130 ml. of ether is added over 20 min., with occasional shaking of the flask. The distillate is collected as above; yield about 3 g. of diazomethane.

(c) *From N-Methyl-N'-nitro-N-nitrosoguanidine.* Mol. wt. 147.11, m.p. 118°, dec. Preparation.[7] Supplier: Aldrich.

For generation of diazomethane,[8] a mixture of 100 ml. of ether and 30 ml. of

$$CH_3N-\overset{\overset{NH}{\|}}{\underset{\underset{NO}{|}}{C}}-\underset{\underset{H}{|}}{N}NO_2 \xrightarrow[73\%]{KOH} CH_2N_2$$

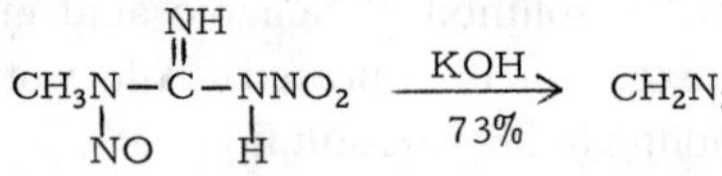

40% aqueous potassium hydroxide solution in a distilling flask is cooled to 0°, 10 g. of crystalline reagent is added in portions with shaking, and the yellow diazomethane solution is distilled from the mixture.

(d) *From N-Nitrosomethylurea*, $CH_3\underset{\underset{NO}{|}}{N}CONH_2$. Mol. wt. 103.09. The reagent is

easily prepared,[9] and diazomethane can be generated from it by the simple procedure of (c).[10] However, it has to be stored in a refrigerator and should not be kept above 20° for more than a few hours.

(e) *From N,N'-Dinitroso-N,N'-dimethyloxamide* (2). The starting material N,N'-dimethyloxamide (1) is obtained in 94% yield by reaction of diamyl oxalate with methylamine[11] and it is supplied by Aldrich. Nitrosation with either N_2O_3 or N_2O_4 in carbon tetrachloride gives the diazomethane precursor (2) as an orange solid,

$$CH_3\underset{\underset{H}{|}}{N}-\underset{\underset{O}{\|}}{C}-\underset{\underset{O}{\|}}{C}-\underset{\underset{H}{|}}{N}CH_3 \xrightarrow[75\%]{\overset{N_2O_3\ or\ N_2O_4}{In\ CCl_4}} CH_3\underset{\underset{NO}{|}}{N}-\underset{\underset{O}{\|}}{C}-\underset{\underset{O}{\|}}{C}-\underset{\underset{NO}{|}}{N}CH_3$$

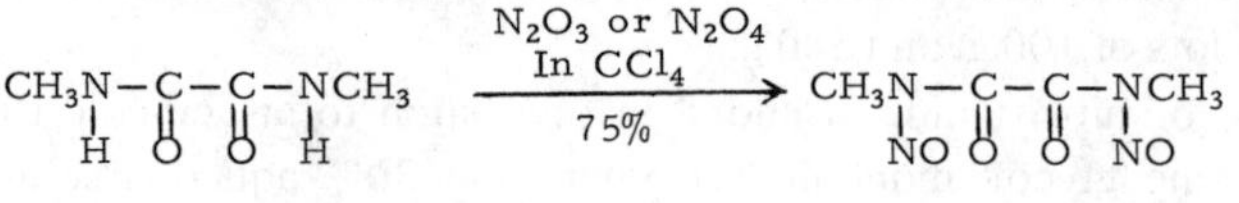

(1)   (2)

m.p. 68°, mol. wt. 174.13[12] This substance is a stable, nonexplosive, nonirritating reagent, which can be stored indefinitely at room temperature in the dark. For the preparation of an ethereal solution of diazomethane, a solution of 8.7 g. of (2) in 250 ml. of ether is added dropwise with stirring in 20 min. to a solution at 60° of 3 g. of sodium in 100 ml. of butanol, and the yellow distillate is collected in a receiver cooled to −30°. Titration with benzoic acid shows the yield of diazomethane to be 3.2 g. The reagent can also be generated as a gas or *in situ.*

$$CH_3\underset{\underset{NO}{|}}{N}-\underset{\underset{O}{\|}}{C}-\underset{\underset{O}{\|}}{C}-\underset{\underset{NO}{|}}{N}CH_3 \xrightarrow[76\%]{2\ C_4H_9ONa} 2\ CH_2{=}\overset{+}{N}{=}\overset{-}{N} + C_4H_9O\underset{\underset{O}{\|}}{C}-\underset{\underset{O}{\|}}{C}OC_4H_9 + 2\ NaOH$$

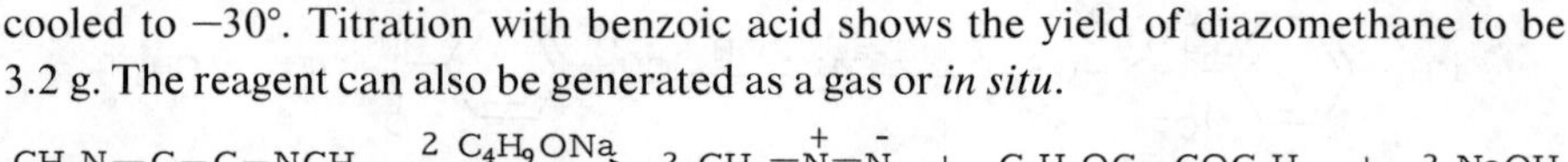

(2)

Esterification of acids. The widely used reaction of diazomethane with carboxylic acids was discovered by von Pechmann,[1] who noted also the effectiveness of the reagent for the methylation of phenols, mineral acids, hydrogen cyanide, and phthalimide. The reagent also methylates tropolones, alkylsulfonic acids,[13] and arylsulfonic acids.[14] The fact that neutral alcohols do not react with diazomethane suggests that an acidic substance supplies a proton required for catalysis of the esterification.

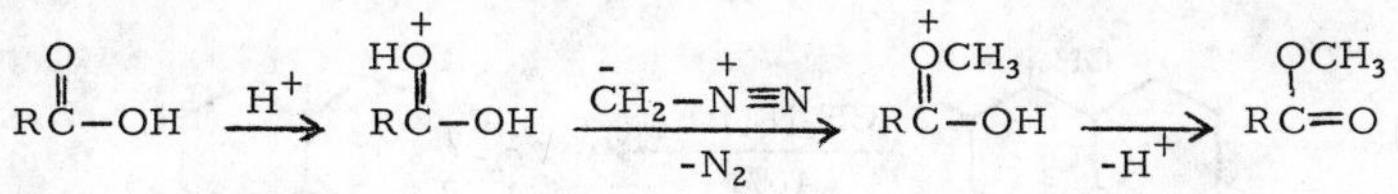

$$\underset{\underset{\text{O}}{\overset{\parallel}{R\overset{}{C}-OH}}}{} \xrightarrow{\overset{+}{H}} \underset{\underset{\text{O}}{\overset{\overset{+}{HO}}{\parallel}}}{R\overset{}{C}-OH} \xrightarrow[-N_2]{\overset{-}{CH_2}-\overset{+}{N}\equiv N} \underset{\underset{\text{O}}{\overset{\overset{+}{OCH_3}}{\parallel}}}{R\overset{}{C}-OH} \xrightarrow[-\overset{+}{H}]{} \underset{\overset{OCH_3}{\overset{|}{R\overset{}{C}=O}}}{}$$

Methylation of alcohols. Although alcohols alone are inert to diazomethane, they can be methylated efficiently by diazomethane under catalysis by boron trifluoride etherate[15] or fluoroboric acid.[16] A mineral acid cannot be used for catalysis because it itself reacts with the reagent, but a Lewis acid performs the function ascribed above to an organic acid. In investigating the structure of estriol D-glucosiduronic acid,

$$ROH \xrightarrow{BF_3} \underset{\overset{|}{H}}{R\overset{+}{O}\overset{-}{B}F_3} \xrightarrow{\overset{-}{CH_2}-\overset{+}{N}\equiv N} ROCH_3 + BF_3$$

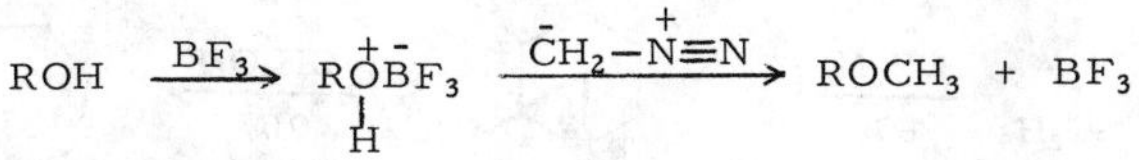

Neeman[17] made use of both uncatalyzed methylation of the phenolic group and BF_3-catalyzed methylation of alcoholic groups.

E. Müller[18] found that aluminum chloride can also be used as catalyst and that in general yields are higher than with boron trifluoride.

Ring Enlargement. In a procedure developed by Kohler and co-workers[19] for the preparation of cycloheptanone, a mixture of cyclohexanone, methanol, and a catalytic amount of finely ground sodium carbonate is stirred during slow addition of

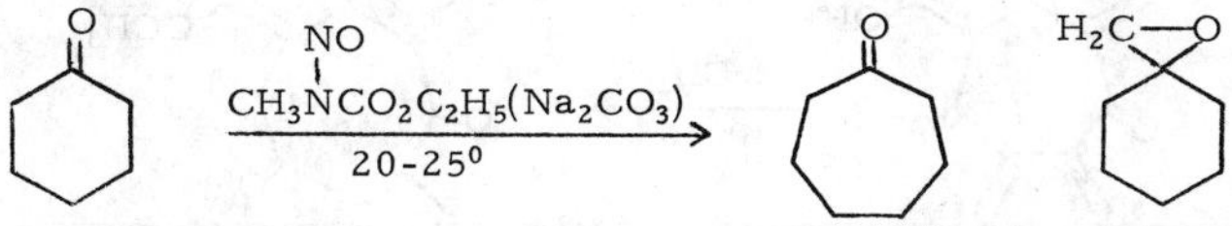

$$\text{(1)} \qquad\qquad\qquad\qquad \text{(2) 63\%} \qquad \text{(3) 15\%}$$

nitrosomethylurethane (4.55 moles). Cycloheptanone is obtained in good yield and the chief companion is methylenecyclohexane oxide. The same procedure was used for the preparation of the succeeding cyclic ketones through C_{10}. de Boer and Backer[20] generated diazomethane from Diazald and potassium hydroxide in aqueous ethanol in the presence of cyclohexanone; slow addition of the solid nitro-samide to the stirred mixture controls the rate of the reaction. The cycloheptanone formed was isolated as the bisulfite addition product and so separated from a little cyclooctanone present, but the yield of cycloheptanone was only 33–36%.

The formation from cyclohexanone of the ring-enlarged ketone (2) and the epoxide (3) indicate that the initial reaction is an addition to the carbonyl group:

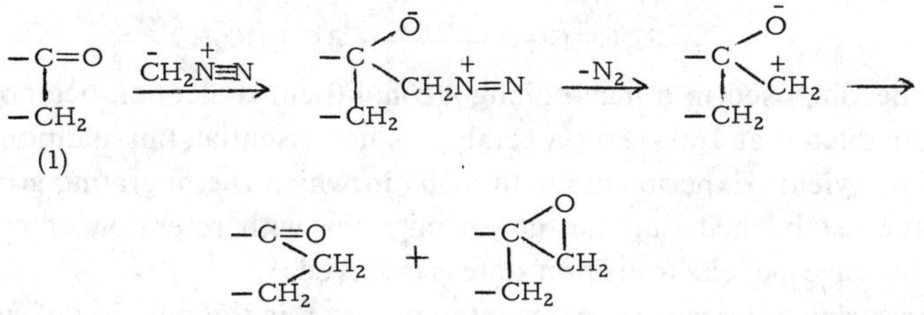

Boron trifluoride etherate markedly accelerates formation of a ketone and so suppresses epoxide formation.[21]

Both fluoroboric acid and boron trifluoride etherate catalyze the homologation of steroid Δ^4-3.ketones.[22] The same transformation can be effected by use of aluminum chloride as catalyst.[23]

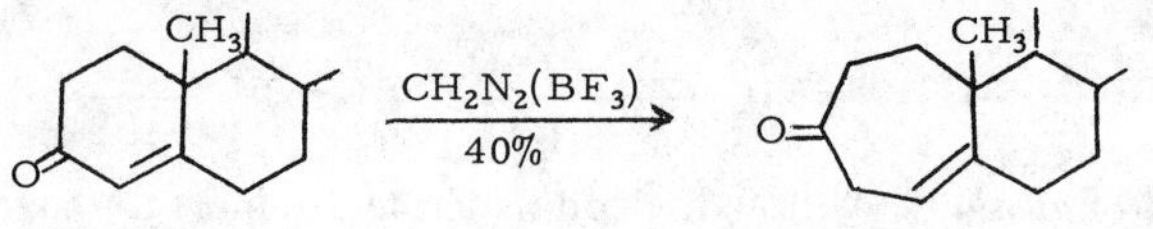
CH$_3$
CH$_2$N$_2$(BF$_3$)
40%
CH$_3$

Cholestenone A-Homocholestenone

Leonard and co-workers[24] found that diazomethane adds to perchlorate salts of enamines to give highly reactive aziridinium salts which are cleaved by water or methanol to give ring-enlarged derivatives.

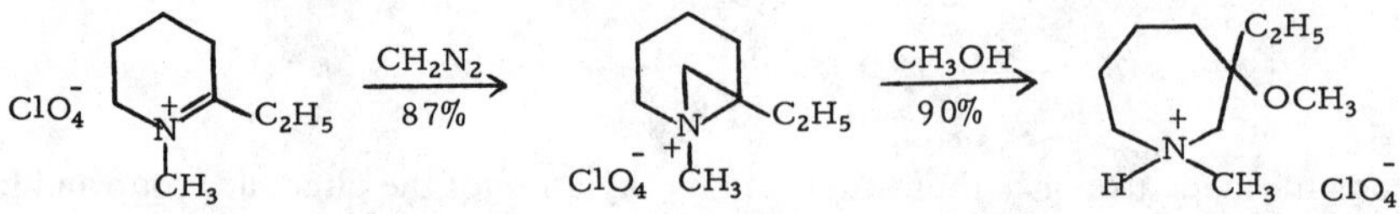
ClO$_4^-$
C$_2$H$_5$
CH$_3$
CH$_2$N$_2$
87%
ClO$_4^-$
CH$_3$
C$_2$H$_5$
CH$_3$OH
90%
C$_2$H$_5$
OCH$_3$
H
CH$_3$
ClO$_4^-$

Fieser[25] explored the action of excess diazomethane on phthaloylnaphthol (I, 1-hydroxy-7,12-dihydropleiadene-7,12 dione) and characterized the product as having an additional methyl and an additional methylene group but did not establish the structure. Some 32 years later Crombie and co-workers[26] raised the yield to

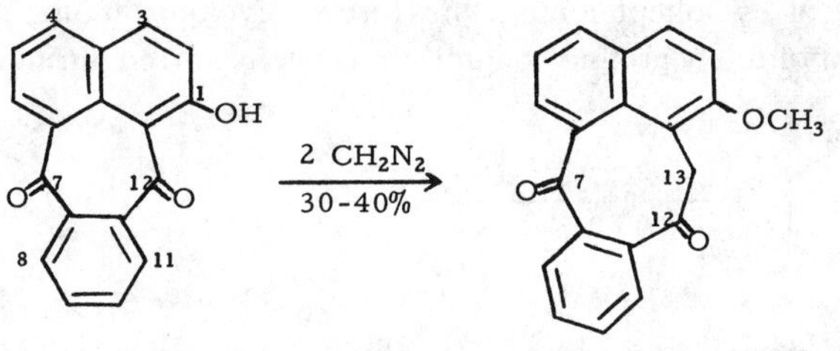
OH
2 CH$_2$N$_2$
30-40%
OCH$_3$

I II

30–40% by using a 7-fold excess of reagent and established the structure II, a product of ring enlargement and methylation (the methyl ether of I is indifferent to diazomethane).

Arndt-Eistert reaction. A few developments since a review of the literature on this reaction of 1942[27] may be mentioned. In an improved procedure,[28] the diazo-

$$RCOCl + 2\ CH_2N_2 \xrightarrow[-CH_3Cl]{} RCOCH{=}\overset{+}{N}{=}\overset{-}{N} \xrightarrow{Ag}$$

$$RCH{=}C{=}O \xrightarrow{H_2O} RCH_2CO_2H$$

ketone is decomposed in a high-boiling alcohol (benzyl alcohol, octanol-2) and the reaction conducted at 160–180°; a catalyst is not essential, but addition of collidine improves the yield. Experiments with acids in which the migrating group is asymmetric have established that the group migrates with retention of configuration, probably because a cyclic transition state is involved.[29]

Buchner ring enlargement. A recent application is the reaction of benzene with diazomethane catalyzed by cuprous bromide to give tropilidene (*see* Cuprous bromide).

Generation of iodomethylzinc iodide. Generation of this active reagent in the Simmons-Smith reaction can be accomplished by addition of a solution of diazomethane to an ethereal solution of zinc iodide and the olefin.[30]

$$ZnI_2 + CH_2N_2 \longrightarrow ICH_2ZnI + N_2$$

Other reactions. Reactions of diazomethane with a variety of carbonyl compounds have been reviewed by Gutsche.[31]

[1]H. von Pechmann, *Ber.*, **27**, 1888 (1894); **28**, 855 (1895)
[2]J. A. Moore and D. E. Reed, *Org. Syn.*, **41**, 16 (1961)
[3]Th. J. de Boer and H. J. Backer, *Org. Syn., Coll. Vol.*, **4**, 943 (1963)
[4]D. Kubik and V. I. Stenberg, *Chem. Ind.*, 248 (1966)
[5]Th. J. de Boer and H. J. Backer, *Rec. trav.*, **73**, 229 (1954)
[6]*Idem, Org. Syn., Coll. Vol.*, **4**, 250 (1963)
[7]A. F. McKay and G. F. Wright, *Am. Soc.*, **69**, 3028 (1947)
[8]A. F. McKay, *ibid.*, **70**, 1974 (1948)
[9]F. Arndt, *Org. Syn., Coll. Vol.*, **2**, 461 (1943)
[10]*Idem, ibid.*, **2**, 165 (1943)
[11]O. C. W. Allenby and G. F. Wright, *Can. J. Research*, **25B**, 295 (1947)
[12]H. Reimlinger, *Ber.*, **94**, 2547 (1961)
[13]F. Arndt and C. Martius, *Ann.*, **499**, 264 (1932)
[14]J. S. Showell, J. R. Russell, and D. Swern, *J. Org.*, **27**, 2853 (1962)
[15]E. Müller and W. Rundel, *Angew. Chem.*, **70**, 105 (1958); E. Müller, M. Bauer, and W. Rundel, *Z. Naturforschung*, **14B**, 209 (1959)
[16]M. Caserio, J. D. Roberts, M. Neeman, and W. S. Johnson, *Am. Soc.*, **80**, 2584 (1958); *Tetrahedron*, **6**, 36 (1959); M. Neeman and W. S. Johnson, *Org. Syn.*, **41**, 9 (1961)
[17]M. Neeman and Y. Hashimoto, *Am. Soc.*, **84**, 2972 (1962)
[18]E. Müller, R. Heischkeil, and M. Bauer, *Ann.*, **677**, 55 (1964)
[19]E. P. Kohler, M. Tishler, H. Potter, and H. T. Thompson, *Am. Soc.*, **61**, 1059 (1939); see also D. J. Cram and R. C. Helgeson, *ibid.*, **88**, 3515 (1966)
[20]Th. J. de Boer and H. J. Backer, *Org. Syn., Coll. Vol.*, **4**, 225 (1963)
[21]H. O. House, E. J. Grubbs, and W. F. Gannon, *Am. Soc.*, **82**, 4099 (1960)
[22]W. S. Johnson, M. Neeman, and S. P. Birkeland, *Tetrahedron Letters*, No. 5, 1 (1960); W. S. Johnson, M. Neeman, S. P. Birkeland, and N. A. Fedoruk, *Am. Soc.*, **84**, 989 (1962)
[23]E. Müller, B. Zeeh, R. Heischkeil, F. Fricke, and H. Suhr, *Ann.*, **662**, 38 (1963)
[24]N. J. Leonard and K. Jann, *Am. Soc.*, **84**, 4806 (1962); N. J. Leonard, K. Jann, J. V. Paukstelis, and C. K. Steinhardt, *J. Org.*, **28**, 1499 (1963)
[25]L. F. Fieser, *Am. Soc.*, **55**, 4963 (1933)
[26]M. E. C. Biffin, L. Crombie, and J. A. Elvidge, *J. Chem. Soc.*, 7500 (1965)
[27]W. E. Bachmann and W. S. Struve, *Org. Reactions*, **1**, 38 (1942)
[28]A. L. Wilds and A. L. Meader, *J. Org.*, **13**, 763 (1948)
[29]K. B. Wiberg and T. W. Hutton, *Am. Soc.*, **78**, 1640 (1956)
[30]G. Wittig and K. Schwarzenbach, *Ann.*, **650**, 1 (1961); H. Hoberg, *Ann.*, **656**, 1, 15 (1962)
[31]C. D. Gutsche, "The Reaction of Diazomethane and Its Derivatives with Aldehydes and Ketones," *Org. Reactions*, **8**, 364 (1954)

***trans*-1,2-Dibenzoylethylene.** Mol. wt. 236.27, m.p. 109–110°. Suppliers: E, F.

Preparation from fumaryl chloride and benzene.[1]

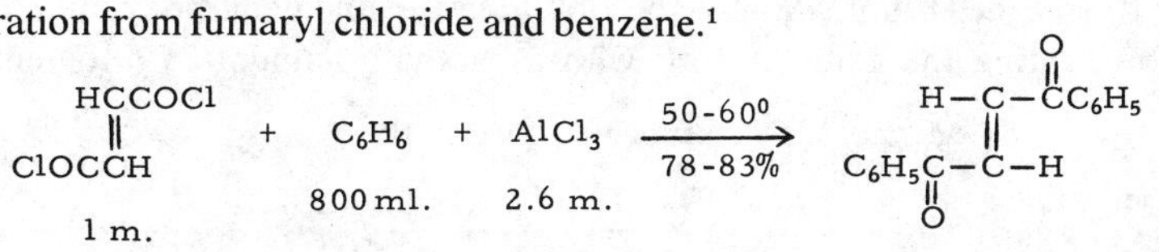
HCCOCl
ClOCCH
1 m.
+
C_6H_6
800 ml.
+
$AlCl_3$
2.6 m.
50-60°
78-83%
O
H—C—CC₆H₅
C₆H₅C—C—H
O

Dienophile. Refluxed with 2,3-diphenylbutadiene in ethanol for 4 hrs., the reagent afforded the adduct in 90% yield.[2]

Conversion of acids to anhydrides. In combination with tri-*n*-butylphosphine (or triphenylphosphine), the reagent converts carboxylic acids into their anhydrides in 50–80% yield.[3] Elimination of the elements of water follows a normal course: the trisubstituted phosphine combines with the oxygen, and dibenzoylethylene accepts hydrogen and affords dibenzoylethane. When tri-*n*-butylphosphine was added to a benzene solution of dibenzoylethylene at room temperature, the solution assumed a red color with evolution of heat. After 10 min. the solution was added to a benzene solution of 2 equivalents of propionic acid and the mixture was refluxed for 2 hrs.; propionic anhydride was obtained in 78% yield along with dibenzoylethane (77%) and tri-*n*-butylphosphine oxide (84%). The suggested pathway is formulated.

$$C_6H_5\overset{O}{\overset{\|}{C}}CH=CH\overset{O}{\overset{\|}{C}}C_6H_5 \xrightarrow{Bu_3P} C_6H_5\overset{\overset{+}{O}P Bu_3}{\underset{|}{C}}=CH-CH=\overset{\overset{-}{O}}{\overset{|}{C}}C_6H_5 \xrightarrow{RCO_2H} C_6H_5\overset{\overset{+}{O}P Bu_3}{\underset{|}{C}}=CHCH_2\overset{O}{\overset{\|}{C}}C_6H_5$$

$$\xrightarrow[-C_6H_5COCH_2CH_2COC_6H_5]{RCO_2H} R\overset{O}{\overset{\|}{C}}-\underset{\underset{\overset{\|}{O}}{OCR}}{\overset{+}{O}P Bu_3} \longrightarrow \begin{matrix} R\overset{O}{\overset{\|}{C}} \\ R\overset{}{\underset{\overset{\|}{O}}{C}} \end{matrix}\rangle O \ + \ O=PBu_3$$

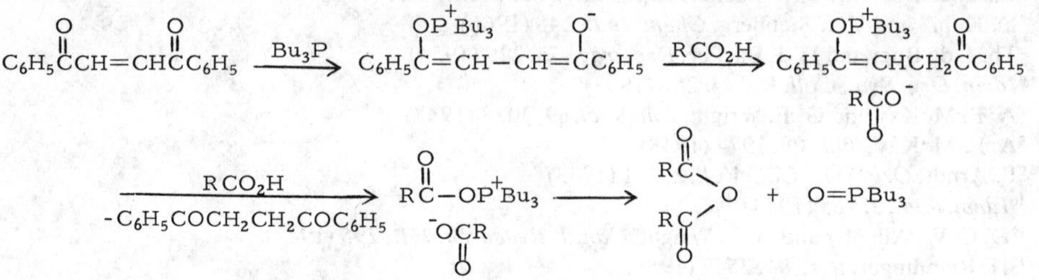

[1]R. E. Lutz, *Org. Syn., Coll. Vol.*, **3**, 248 (1955)
[2]C. F. H. Allen and J. W. Gates, Jr., *Am. Soc.*, **65**, 1283 (1943)
[3]I. Kuwajima and T. Mukaiyama, *J. Org.*, **29**, 1385 (1964)

Dibenzoyl peroxide, $C_6H_5\overset{O}{\overset{\|}{C}}OO\overset{O}{\overset{\|}{C}}C_6H_5$. Mol. wt. 242.22, m.p. 107°. Suppliers: E, F, MCB. Commercial material can be purified by dissolving it in cold chloroform and adding methanol to the point of saturation. Analysis.[1]

Polymerization catalyst. Since the reagent decomposes on moderate heating to radicals,[2] it is useful as a catalyst for radical-induced polymerizations.

Counter-Markownikoff addition of HBr. Kharasch and Mayo[3] discovered that, in the presence of dibenzoyl peroxide, hydrogen bromide adds to unsymmetrical olefins in the counter-Markownikoff direction.

$$(CH_3)_2C=CH_2 \ + \ HBr \xrightarrow{(C_6H_5COO)_2} (CH_3)_2CHCH_2Br$$

Chlorination of alkanes. Kharasch[4] developed a method for the low-temperature radical chlorination of alkanes utilizing sulfuryl chloride catalyzed by dibenzoyl peroxide.

$$RH \ + \ SO_2Cl_2 \xrightarrow[40-80°]{(C_6H_5COO)_2} RCl \ + \ HCl \ + \ SO_2$$

Acyloin synthesis. In a novel synthesis by Lawesson and Grönwall[4a] *t*-butyl acetoacetate is prepared from diketene and *t*-butanol and alkylated, and the benzoyl-oxy group is introduced in the α-position by reaction of the sodio derivative with dibenzoyl peroxide. Elimination of the *t*-butyl group and decarboxylation is accomplished by heating the ester at 160° with a catalytic amount of *p*-toluenesulfonic

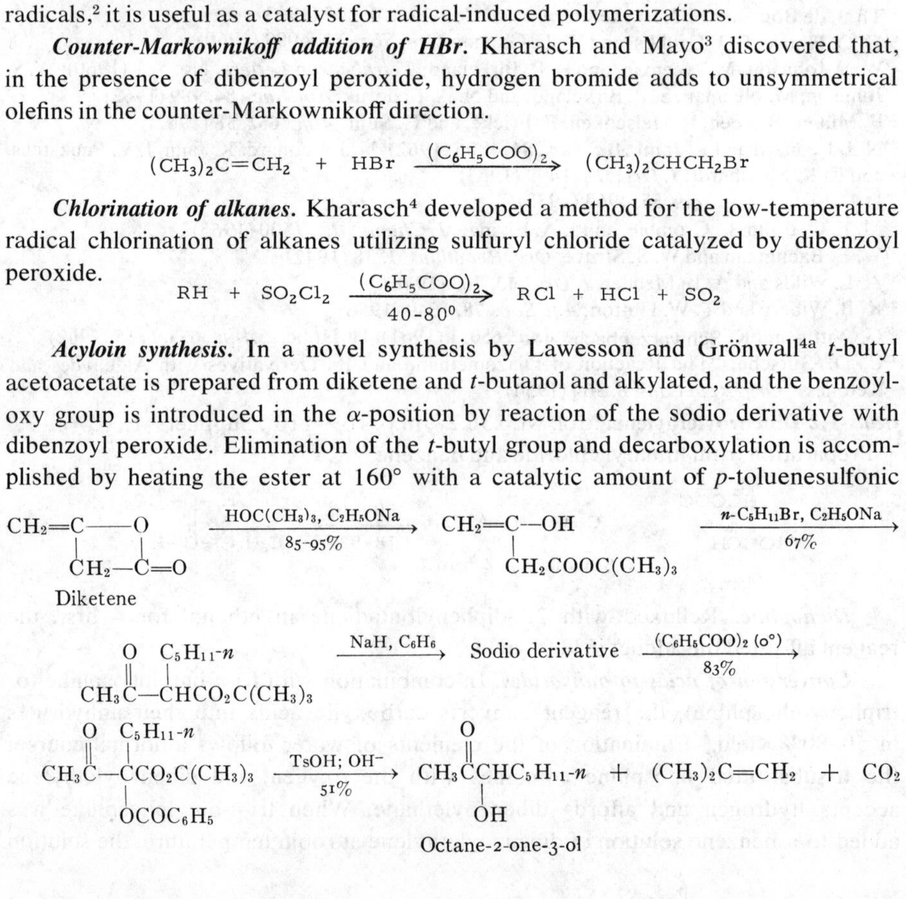

$$\underset{\text{Diketene}}{\overset{\overset{CH_2=C-O}{|}}{\underset{CH_2-C=O}{|}}} \xrightarrow[85-95\%]{HOC(CH_3)_3, C_2H_5ONa} \overset{CH_2=C-OH}{\underset{CH_2COOC(CH_3)_3}{|}} \xrightarrow[67\%]{n\text{-}C_5H_{11}Br, C_2H_5ONa}$$

$$\overset{O \quad C_5H_{11}\text{-}n}{CH_3\overset{\|}{C}-\overset{|}{C}HCO_2C(CH_3)_3} \xrightarrow{NaH, C_6H_6} \text{Sodio derivative} \xrightarrow[83\%]{(C_6H_5COO)_2 \ (0°)}$$

$$\underset{OCOC_6H_5}{\overset{O \quad C_5H_{11}\text{-}n}{CH_3\overset{\|}{C}-\overset{|}{C}CO_2C(CH_3)_3}} \xrightarrow[51\%]{TsOH; \ OH^-} \underset{OH}{\overset{O}{CH_3\overset{\|}{C}CHC_5H_{11}\text{-}n}} \ + \ (CH_3)_2C=CH_2 \ + \ CO_2$$

Octane-2-one-3-ol

acid; saponification of the benzoate group under mild conditions gives the (base-sensitive) acyloin.

Bromination with N-bromosuccinimide. Allylic bromination of 2-heptene is carried out under nitrogen with NBS and a trace of dibenzoyl peroxide; the hydrocarbon is

$$CH_3CH_2CH_2CH_2CH=CHCH_3 \ + \ NBS \ + \ (C_6H_5COO)_2 \xrightarrow[\substack{58-64\% \\ \text{based on NBS}}]{\substack{N_2 \\ \text{Refl. 2 hrs.}}} CH_3CH_2CH_2CHCH=CHCH_3$$

0.41 m. in 250 ml. CCl_4 0.27 m.   0.2 g. $\overset{|}{Br}$

$$+ \ NBS \ + \ (C_6H_5COO)_2 \xrightarrow[\substack{71-79\% \\ \text{based on NBS}}]{\text{Refl. 20 min.}}$$

2.24 m. in 700 ml. C_6H_6 2 m.   4 g.

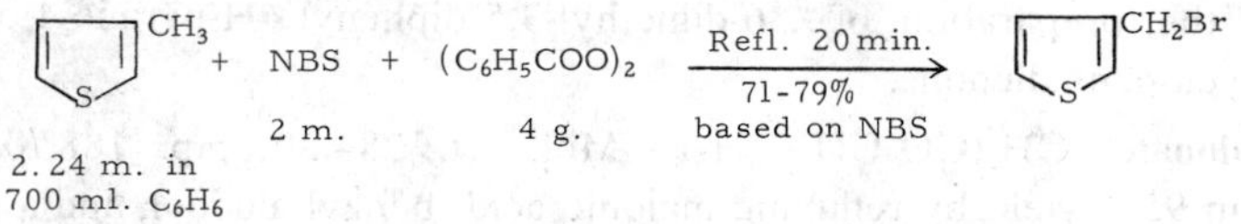

taken in excess and the yield is based on the NBS used.[5] In the benzylic bromination of 3-methylthiophene, this substrate is likewise taken in excess, no nitrogen is used; the proportion of catalyst is higher and the time shorter.[6]

Radical addition of enanthaldehyde to diethyl maleate.[7] The reaction is carried out in a jacketed flask with refluxing trichloroethylene in the jacket to control the temperature, which is critical. The aldehyde is taken in large excess; the peroxide is added in two portions.

$$CH_3(CH_2)_5CHO \ + \ \begin{matrix} HCCO_2C_2H_5 \\ \| \\ HCCO_2C_2H_5 \end{matrix} \ + \ (C_6H_5COO)_2 \xrightarrow[\substack{71-76\% \\ \text{based on ester}}]{\substack{18-24 \text{ hrs.} \\ 84-85^0}}$$

2 m.   1 m.   1 g.

$$\begin{matrix} O \\ \| \\ CH_3(CH_2)_5CCHCOOC_2H_5 \\ | \\ CH_2COOC_2H_5 \end{matrix}$$

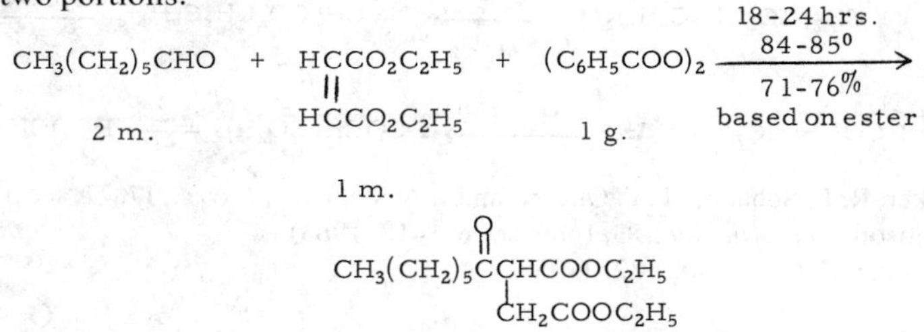

Reaction with enamines. The reaction of dibenzoyl peroxide with an enamine provides an indirect way of introducing a benzoyloxy group into an active methylene compound, such as (1).[8]

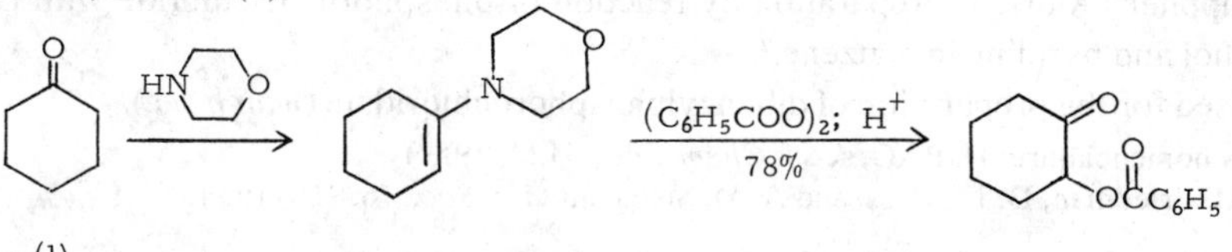

(1)

Catalysis of benzylic bromination, see 1,3-Dibromo-5,5-dimethylhydantoin.

Benzoyloxylation of malonic esters. Dibenzoyl peroxide reacts with the sodium derivative of diethyl ethylmalonate to give diethyl (O-benzoyl)ethyltartronate.[9] This is a general reaction for β-dicarbonyl compounds.

$$C_2H_5CH(CO_2C_2H_5)_2 \xrightarrow[\substack{-H_2 \\ Na^+}]{NaH, \ C_6H_6} C_2H_5\bar{C}(CO_2C_2H_5)_2 \xrightarrow{\substack{O \ O \\ \| \ \| \\ C_6H_5COCOC_6H_5}} \begin{matrix} C_2H_5 \\ \diagdown \\ C_6H_5CO \end{matrix} \begin{matrix} \diagup C(CO_2C_2H_5)_2 \\ \\ \| \\ O \end{matrix}$$

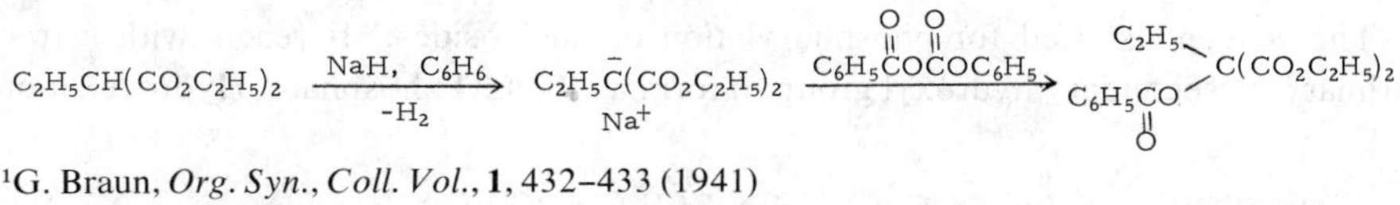

[1]G. Braun, *Org. Syn., Coll. Vol.*, **1**, 432–433 (1941)
[2]C. Walling, "Free Radicals in Solution," 63 ff., 474 ff., Wiley (1957)
[3]M. S. Kharasch and F. R. Mayo, *Am. Soc.*, **55**, 2468 (1933)
[4]M. S. Kharasch and H. C. Brown, *Am. Soc.*, **61**, 2142 (1939)
[4a]S.-O. Lawesson and S. Grönwall, *Chem. Scand.*, **14**, 1445 (1960)

[5]F. L. Greenwood, M. D. Kellert, and J. Sedlak, *Org. Syn., Coll. Vol.*, **4**, 108 (1963)
[6]E. Campaigne and B. F. Tullar, *ibid.*, **4**, 921 (1963)
[7]T. M. Patrick, Jr., and F. B. Erickson, *ibid.*, **4**, 430 (1963)
[8]S.-O. Lawesson, H. J. Jakobsen, and E. H. Larsen, *Chem. Scand.*, **17**, 1188 (1963)
[9]S.-O. Lawesson, T. Busch, and C. Berglund, *Chem. Scand.*, **15**, 260 (1961); E. H. Larsen and S.-O. Lawesson, *Org. Syn.*, **45**, 37 (1965)

Dibenzyl ketone, $C_6H_5CH_2COCH_2C_6H_5$. Mol. wt. 210.26, m.p. 32.5–33°. Suppliers (pract.): Matheson, E, MCB. The commercial material (m.p. 31–34°) can be purified by crystallization from anhydrous ether at −70°.

Used for the preparation of 2,6-dimethyl-3,5-diphenyl-4H-pyrane-4-one and tetraphenylcyclopentadienone.

Dibenzyl malonate, $CH_2(CO_2CH_2C_6H_5)_2$. Mol. wt. 284.30, b.p. 188°/0.2 mm. Preparation in 95% yield by refluxing malonic acid, benzyl alcohol, and a trace of sulfuric acid in toluene under a constant water separator (method A[1]). The ester is sensitive to heat and should be distilled rapidly through a short-path system.[2]

The reagent is used in a general method for the conversion of an acid chloride into an acylmalonic ester which on hydrogenolysis and decarboxylation yields a methyl ketone.[3] the procedure formulated is that used at one stage in the total synthesis of aldosterone.[2]

$$CH_2(CO_2CH_2C_6H_5)_2 \xrightarrow[C_6H_6,\ N_2]{NaH} Na^+\bar{C}H(CO_2CH_2C_6H_5)_2 \xrightarrow{RCOCl}$$

$$RCOCH(CO_2CH_2C_6H_5)_2 \xrightarrow{H_2,\ Pd} RCOCH(CO_2H)_2 \xrightarrow{\Delta} RCOCH_3$$

[1]B. R. Baker, R. E. Schaub, M. V. Querry, and J. A. Williams, *J. Org.*, **17**, 77, see p. 88 (1952)
[2]W. S. Johnson *et al.*, *Am. Soc.*, **85**, 1409, see p. 1418 (1963)
[3]R. E. Bowman, *J. Chem. Soc.*, 325 (1950)

Dibenzyl phosphonate[1] (formerly Dibenzyl phosphite), $(C_6H_5CH_2O)_2\overset{O}{\overset{\|}{P}}H$. Mol. wt. 262.26, m.p. 216.5°.

Supplier: Aldrich. Preparation by reaction of phosphorus trichloride with benzyl alcohol and pyridine in benzene.[2]

Used for the preparation of dibenzylphosphorochloridate (*which see*).

[1]New nomenclature: IUPAC, see *J. Chem. Soc.*, 5122 (1952)
[2]O. M. Friedman, D. L. Klass, and A. M. Seligman, *Am. Soc.*, **76**, 916 (1954)

Dibenzylphosphorochloridate, $(C_6H_5CH_2O)_2\overset{O}{\overset{\|}{P}}Cl$. Mol. wt. 296.68.

Previously known as dibenzyl chlorophosphate or dibenzylchlorophosphonate. New nomenclature.[1] Review.[2] Prepared by chlorination of dibenzyl phosphonate with anhydrous N-chlorosuccinimide.[3]

The reagent is used for phosphorylation of nucleosides.[4] It reacts with a free primary or secondary hydroxyl group to form a dibenzyl phosphate. Hydrogenation

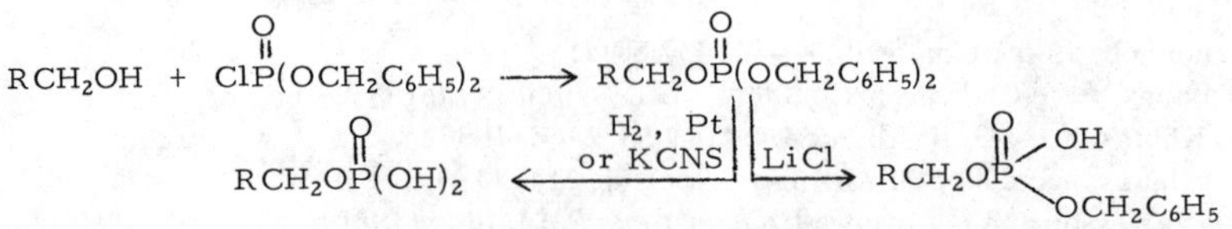
RCH₂OH + ClP(OCH₂C₆H₅)₂
RCH₂OP(OCH₂C₆H₅)₂
H₂ , Pt
or KCNS
LiCl
RCH₂OP(OH)₂
RCH₂OP
OH
OCH₂C₆H₅

of this ester or treatment with potassium thiocyanate in boiling acetonitrile removes both benzyl groups to form a phosphate ester; lithium chloride reacts to remove a single benzyl group and affords the monobenzyl ester.[5]

[1]IUPAC, see *J. Chem. Soc.*, 5122 (1952)
[2]D. M. Brown, *Advances in Organic Chemistry*, **3**, 74 (1963)
[3]G. W. Kenner, A. R. Todd, and F. J. Weymouth, *J. Chem. Soc.*, 3675 (1952); R. H. Hall and H. G. Khorana, *Am. Soc.*, **76**, 5056 (1954)
[4]A. R. Todd, "The Pedler Lecture," *J. Chem. Soc.*, 647 (1946); G. W. Kenner, *Fortsch. Chemie org. Naturstoffe*, **8**, 96 (1951)
[5]V. M. Clark and A. R. Todd, *J. Chem. Soc.*, 2030 (1950); S. M. H. Christie, D. T. Elmore, G. W. Kenner, A. R. Todd, and F. J. Weymouth, *ibid.*, 2947 (1953)

Diborane, B_2H_6. Mol. wt. 27.69, m.p. −165°, b.p. −92.5°, Spontaneously flammable gas. Callery supplies diborane as a compressed gas. Alfa and Ventron supply a solution approximately 1 *M* in BH_3 in tetrahydrofurane. The solution is stored at 0–5° and measured samples are removed by a hypodermic syringe with a long needle.

The gas is highly toxic by inhalation, and the nose cannot be depended upon to warn of unsafe concentrations (diborane temporarily deadens the olefactory senses). Although pure diborane is not spontaneously flammable in air at room temperature, it should always be assumed to contain decomposition products which can make it spontaneously flammable.

Diborane is valued as a fuel for use in rocket propellants. Callery Chemical Co. developed equipment and procedures for the safe handling of diborane and by 1962 had produced about 5 million pounds of the material. It is offered for shipment in chrome–molybdenum steel cylinders at −80° (minimum order 100 g.), and technical bulletins present information on handling procedures and safety precautions.

Diborane is generated conveniently according to equation (1) by dropwise addition of a solution of sodium borohydride in diglyme (diethyleneglycol dimethyl ether,

$$1.\quad 3\ NaBH_4 + 4\ BF_3 \longrightarrow 2\ B_2H_6 + 3\ NaBF_4$$

b.p. 161°, water-miscible) to a stirred mixture of boron trifluoride etherate and diglyme, and the gas generated is swept by a slow stream of nitrogen into a reaction flask containing a solution of an olefin or other reactant, usually dissolved in tetrahydrofurane;[1] a diagram of the apparatus is given on page 32 of a review article.[2] Sodium borohydride is taken in 20% excess, boron trifluoride etherate in 50% excess. An exit tube from the reaction flask dips into a trap of acetone to react with excess

$$2.\quad 4\ (CH_3)_2CO + B_2H_6 \longrightarrow 2[(CH_3)_2CHO]_2BH$$

$$3.\quad [(CH_3)_2CHO]_2BH + 3\ H_2O \longrightarrow 2\ (CH_3)_2CHOH + H_2 + H_3BO_3$$

$$4.\quad B_2H_6 + 6\ H_2O \longrightarrow 6\ H_2 + 2\ H_3BO_3$$

B_2H_6 by formation of diisopropoxyborane (2), which can be destroyed by reaction with water as in (3).

In another method[3] a generating flask charged with mercurous chloride and diglyme is flushed with nitrogen, stirred mechanically, and the pressure is reduced

$$\underset{4\ g.}{2\ NaBH_4} + \underset{23.6\ g.}{Hg_2Cl_2} \xrightarrow[88\%]{Diglyme} 2\ Hg + 2\ NaCl + B_2H_6 + H_2$$

to near that of the solvent vapor. With the outlet tube connected to a series of traps cooled in liquid nitrogen, a solution of sodium borohydride in diglyme is added in small portions from a dropping tube. Reaction is immediate and exothermal and does not require heating. Pumping is continued after the addition to ensure complete transfer of the diborane to the traps. Iodine can be used in place of a metal salt, but in this case it is essential to add the iodine to the borohydride solution.

Reduction. The reaction of diborane with acetone (2) and other simple carbonyl compounds to form dialkoxyboranes[4] may involve the equilibrium $B_2H_6 \rightleftharpoons 2 BH_3$, and a possible formulation of the first step is shown in (5). The monoalkoxyborane

$$5. \quad R_2C=O \xrightarrow{BH_3} \underset{H-\overset{+}{B}H_2}{R_2C=O^-} \longrightarrow \underset{H\cdots BH_2}{R_2C\cdots O} \longrightarrow \underset{H\quad BH_2}{R_2C-O}$$ 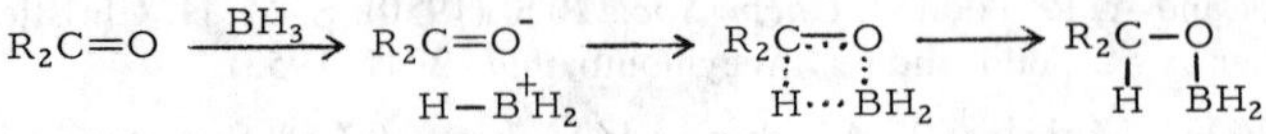

has not been isolated and evidently reacts with a second mole of carbonyl compound to produce the dialkoxyborane. That the reaction stops at this stage, even in the case of acetaldehyde, is presumably due to hindrance. Demonstration of the ready formation and hydrolysis of dialkoxyboranes (reactions 2 and 3) suggested a general method for the reduction of unsaturated compounds which was not exploited at the time because of the rarity of diborane. Later, H. C. Brown and co-workers found that diborane at room temperature rapidly reduces aldehydes, ketones, epoxides, carboxylic acids, nitriles, azo compounds, and *t*-amides.[5] The reductions can be carried out either by passing externally generated diborane into a solution of the organic compound or by adding boron trifluoride etherate to a solution of sodium borohydride and the compound in tetrahydrofurane or diglyme. Because of the much greater solubility of diborane in tetrahydrofurane than in diglyme, the former solvent offers some advantage.

Comparison of the relative reactivity of a number of groups to diborane indicated the following order of decreasing reactivity:[6]

$$-CO_2H \;>\; -\overset{|}{C}=\overset{|}{C}- \;>\; R_2C=O \;>\; RC{\equiv}N \;>\; \overset{O}{\underset{}{\overbrace{>C-C<}}} \;>\; -CO_2R'$$

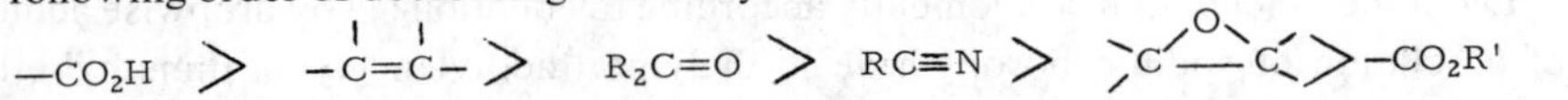

The following groups are not reduced:

$$-COCl, \;-NO_2, \;-SO_2-, \;RX, \;ArX, \;\text{ethers, phenols}$$

The ready reduction of the carboxyl group is particularly noteworthy, since this group is not reduced by $NaBH_4$. The order of reactivity of the nitrile and ester groups to B_2H_6 is the reverse of that in reductions by $LiBH_4$.[7] Acid chlorides are reduced by $NaBH_4$,[7] but not by B_2H_6.

The marked difference in the relative sensitivity of groups to sodium borohydride and to diborane is attributed[5] to the fact that $Na^+BH_4^-$ is essentially a base, reaction occurring through nucleophilic attack of BH_4^- on an electron-deficient center, whereas B_2H_6 is a Lewis acid which preferentially attacks the group at a position of high electron density. By judicious use of diborane and sodium borohydride, it becomes possible to reduce specific groups in the presence of other groups, and to reverse the process. For example, a carboxyl group can be reduced selectively with diborane in the presence of a carbonyl group or a nitro group. The carboxyl group of the anthraquinone I is reduced preferentially to give II in about 60% yield; the

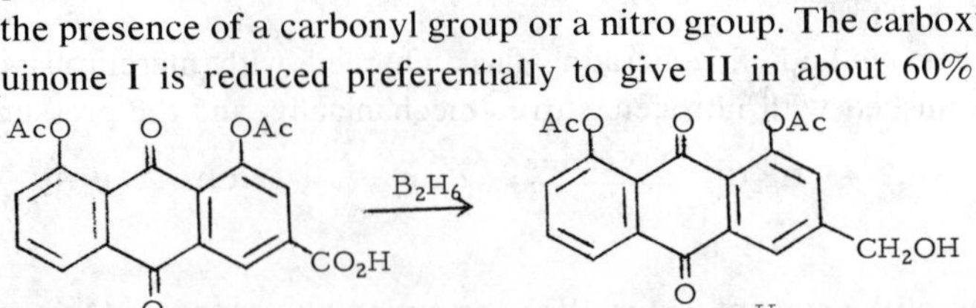

quinone ring is attacked only with use of more reagent.[8] Anthraquinones can be reduced to anthracenes by B_2H_6 generated *in situ*.[8] A solution of anthraquinone (2 g.) in diglyme turns deep red on addition of sodium borohydride, and addition of boron trifluoride etherate and stirring (2 hrs.) affords anthracene (1.2 g.)

Reduction of readily available aldoximes and ketoximes with diborane in tetrahydrofurane provides a facile and convenient method for the preparation of monosubstituted hydroxylamines.[9] Nitro compounds are not reduced by diborane but

$$RR'C{=}NOH \xrightarrow{B_2H_6} RR'CH{-}N(B{<}){-}O{-}B{<} \xrightarrow[\Delta]{aq.\ NaOH} RR'CHNHOH$$

$$R = \text{alkyl or aryl},\ R' = H \text{ or alkyl}$$

salts of nitro compounds are reduced to hydroxylamines.

$$RC(R'){=}NO_2^-Me^+ \xrightarrow{B_2H_6;\ H^+ \text{ or } OH^-} RCH(R')NHOH$$

Under usual conditions, lactones are reduced by lithium aluminum hydride to diols. A procedure for the reduction of δ-lactones with diborane in tetrahydrofurane to cyclic acetals has been applied to the synthesis of several unusual oxasteroids.[10]

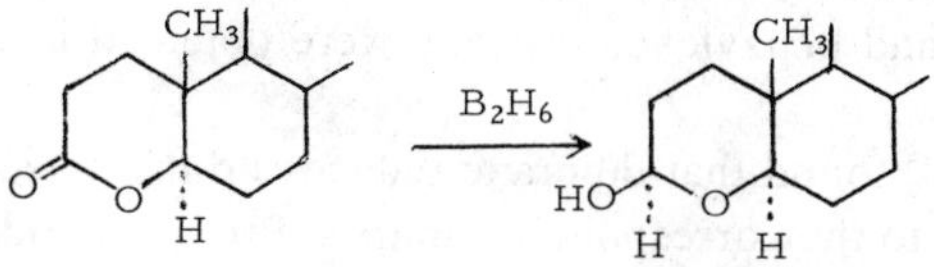
CH3
O
H
B2H6
CH3
HO
O
H
H

An unhindered olefin reacts with diborane with utilization of all three available H—B bonds to form a trialkylborane. These substances are relatively stable to

$$3\ RCH{=}CH_2 + \xrightarrow{1/2\ B_2H_6} (RCH_2CH_2)_3B \xrightarrow{3\ CH_3CO_2H} 3\ RCH_2CH_3 + (CH_3COO)_3B$$

aqueous acids and bases but undergo surprisingly facile protonolysis by acetic acid. Consequently hydroboration followed by protonolysis offers a method for the hydrogenation of double bonds.[11] As applied to olefins, the method can hardly compete with hydrogenation. However, the observations suggested a method for homogeneous hydrogenation catalyzed by an alkylborane.[12] Cyclohexene or caprylene containing 3.8% of tri-*n*-butylborane is hydrogenated quantitatively in 3 hrs. at 220° under 1000 p.s.i. of hydrogen. The reaction is valuable for reduction

$$BR_3 \xrightarrow[-RH]{H_2} R_2BH \xrightarrow{R'CH=CH_2} R'CH_2CH_2BR_2 \xrightarrow{3\ H_2} R'CH_2CH_3$$

of high polymers in solution (*e.g.*, *cis*-polybutadiene).

Alkynes can be reduced by hydroboration and protonolysis. The internal acetylene 3-hexyne reacts with diborane to form an unsaturated boron derivative which is

$$3\ C_2H_5{-}C{\equiv}C{-}C_2H_5 + 1/2\ B_2H_6 \xrightarrow{0^0} C_2H_5{-}CH{=}C(C_2H_5){-}B1/3 \xrightarrow[AcOH]{0^0} C_2H_5{-}CH{=}CH{-}C_2H_5$$

converted by anhydrous acetic acid into *cis*-hexene-3 of high purity.[13] Attempted direct hydroboration of 1-hexyne with diborane in diglyme showed that this terminal acetylene reacted in part by addition of 2 moles of reagent, but smooth reduction to 1-hexene was accomplished with the hindered reagent bis-3-methyl-2-butylborane. (Brown uses the trivial name disiamylborane, since the alkyl group has the only *sec*-isoamyl structure possible). The boron atom of the adduct was shown to be on the

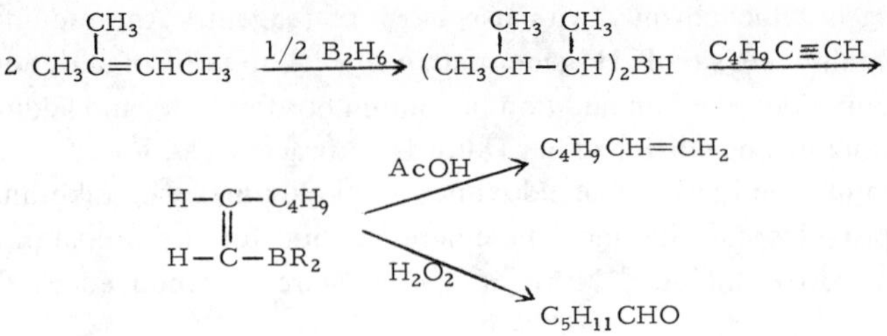

terminal carbon by oxidation with hydrogen peroxide and isolation of *n*-hexaldehyde in 88% yield.

Of further interest is a synthesis of cyclopropanes from allylic chlorides.[14] Addition of diborane to allyl chloride in ether gives a mixture of mono-, di-, and tri (γ-chlo-

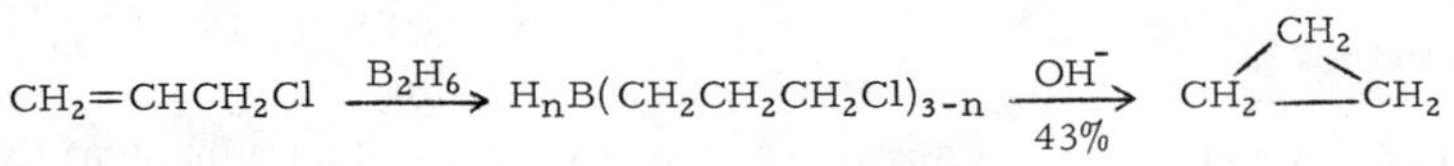

(n = 0, 1, and 2)

ropropyl)boranes. Basic hydrolysis of the mixture affords cyclopropane. Methyl-, *n*-propyl-, phenyl-, and benzylcyclopropane were obtained in this way in yields of 45–71%.

Brown and Heim[15] found that diborane reduces primary, secondary, and tertiary amides in high yield to the corresponding amines. Primary amides are reduced more

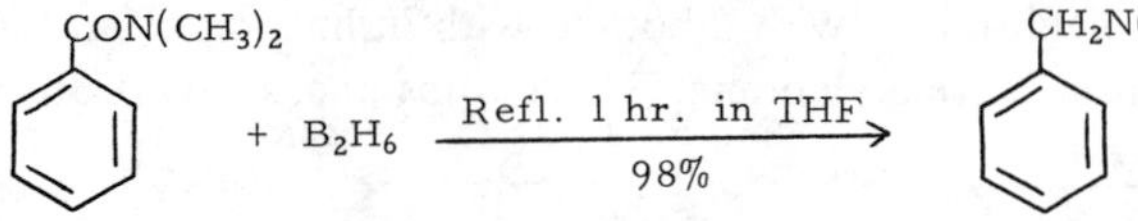

slowly and the reaction may require 8 hrs. of refluxing. Reductions of the same amides with lithium aluminum hydride are slower and the yields lower. Papanastassiou and Bruni[16] found that diborane reduces aliphatic amides to the amines in good yield. The reagent is particularly useful for the reduction of fluoroacetamide derivatives,

$$FCH_2CONHR \xrightarrow{B_2H_6} FCH_2CH_2NHR$$

since $LiAlH_4$ or $LiAlH_4$–$AlCl_3$ cause hydrogenolysis of the C—F bond.

Hydroboration. Johnson and Van Campen[17] noted that tri-*n*-butylborane is oxidized smoothly by alkaline hydrogen peroxide to the borate ester, which is hydrolyzed in the alkaline medium to *n*-butanol and boric acid. Subsequent studies by Brown and co-workers[2,18] established that the oxidation is essentially quantita-

$$(CH_3CH_2CH_2CH_2)_3B \xrightarrow[OH^-]{H_2O_2} (CH_3CH_2CH_2CH_2O)_3B \xrightarrow{HCl} 3\ CH_3CH_2CH_2CH_2OH + H_3BO_3$$

tive at 25°, that the solvents used for formation of the alkylborane do not interfere with the oxidation and hence that the reaction can be performed without isolating the organoborane, and that the reaction is of very wide generality. The sequence of addition of diborane to a double bond and oxidation to an alcohol has become a

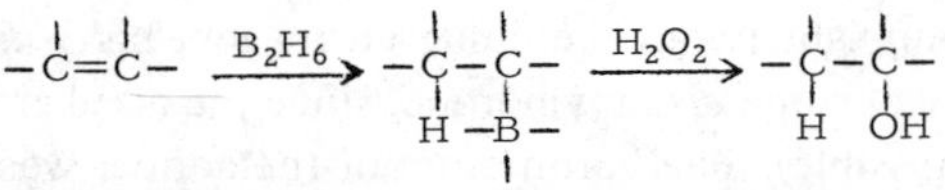

widely used method for accomplishing indirect hydration. We use the term hydroboration for the two-step oxidative process to distinguish it from the process of reduction involving H—B addition and protonolysis. As applied to unsymmetrical olefins, hydroboration differs from hydration by addition of sulfuric acid and hydrolysis in that the addition follows the counter-Markownikoff pattern. Thus terminal alkenes yield almost exclusively the corresponding primary alcohols, for example, 1-octene gives a mixture of 1-octanol and 2-octanol in the ratio 9:1. Addition is

$$RCH{=}CH_2 \xrightarrow{B_2H_6} (RCH_2CH_2)_3B \xrightarrow[OH^-]{H_2O_2} RCH_2CH_2OH$$

thus as expected: the electron-deficient Lewis acid BH_3 becomes linked to the carbon atom richer in hydrogen atoms. 2-Hexene gives about equal parts of the two possible borane derivatives, but if the reaction mixture is refluxed for several hours in diglyme and then oxidized, 1-hexanol is obtained in high yield.[18] The addition of H—B is thus reversible, and boron shifts from an internal position in the carbon chain to a terminal position. Isomerization of an internal olefin, for example 3-ethyl-2-pentene, to a thermodynamically less stable olefin can be accomplished by H—B addition, equilibration, addition of 1-decene as boron acceptor, and slow distillation of the more volatile 3-ethyl-1-pentene resulting by elimination.[19]

Terminal olefins and simple olefins such as 2-butene commonly utilize all three available hydrogens and form the trialkylborane. Trisubstituted olefins such as

$$CH_3CH{=}CHCH_3 \xrightarrow{B_2H_6} CH_3CH\underset{\substack{|\\ B\ 1/3}}{C}HCH_3$$

2-methylbutene-2 and 1-methylcyclohexene utilize only two of the three hydrogen atoms of the borane group to form a dialkylborane. The tetrasubstituted olefin

$$(CH_3)_2C{=}\overset{\substack{CH_3\\ |}}{C}H \xrightarrow{B_2H_6} [(CH_3)_2CH\overset{\substack{CH_3\\ |}}{C}H{-}]_2BH$$

tetramethylethylene reacts rapidly, but utilizes only one of the three available hydrogens and gives the monoalkylborane.

$$(CH_3)_2C{=}\overset{\substack{CH_3\\ |}}{\underset{\substack{|\\ CH_3}}{C}} \xrightarrow{B_2H_6} (CH_3)_2CH\overset{\substack{CH_3\\ |}}{\underset{\substack{|\\ CH_3}}{C}}BH_2$$

The first olefin found not to react with diborane was a 5β-$\Delta^{9(11)}$-steroid.[20] Nussim and Sondheimer[21] investigated other 5β-$\Delta^{9(11)}$-steroids and found them inert to

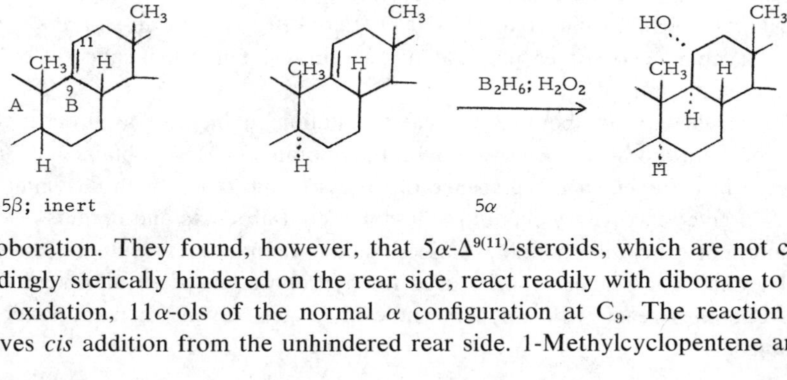
CH3
11
CH3
H
9
A
B
H
CH3
CH3
H
H
B2H6; H2O2
HO
CH3
CH3
H
H
H

5β; inert 5α

hydroboration. They found, however, that 5α-$\Delta^{9(11)}$-steroids, which are not correspondingly sterically hindered on the rear side, react readily with diborane to give, after oxidation, 11α-ols of the normal α configuration at C_9. The reaction thus involves *cis* addition from the unhindered rear side. 1-Methylcyclopentene and 1-

methylcyclohexene have similarly been shown to undergo hydroboration-oxidation to give products of *cis* hydration.[22] Cholesterol reacts with diborane by counter Markownikoff *cis* addition to give a mixture in which the product of rear attack

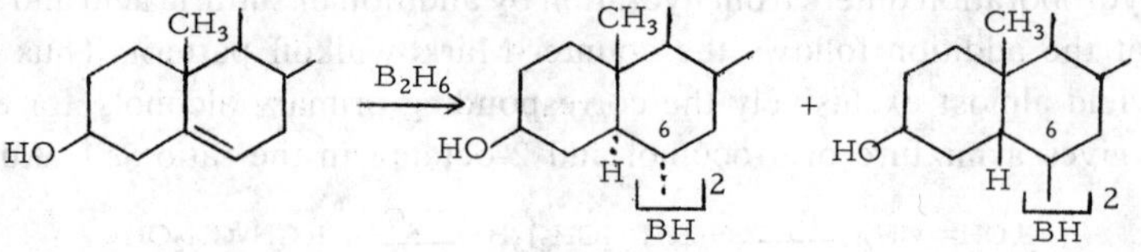

(1) (2)

(1) predominated.[20] The steric course of the reaction and the evident influence of hindrance effects suggest that addition proceeds through a cyclic transition state:

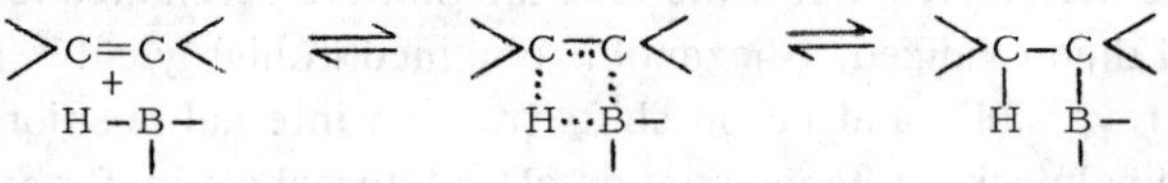

The migration of boron from an internal to a terminal position in a chain at elevated temperatures can be accounted for by a series of eliminations and additions through cyclic transition states.

When hydroboration of an olefin is carried out with the intention of oxidizing an initially formed secondary alcohol to a ketone, hydrogen peroxide oxidation can be dispensed with and the alkylborane oxidized directly with chromic acid. Thus Pappo[23] converted a hydroxyl-free Δ^5-steroid into a mixture of stereoisomeric alkylboranes comparable to (1) and (2), oxidized the mixture with chromic acid, and obtained the 6-keto-5α-steroid in good yield; the initially formed 6-keto-5β-steroid underwent isomerism in the process. H. C. Brown[24] developed an efficient two-phase system for oxidation of either alkylboranes or free alcohols to the ketones using aqueous chromic acid and ether.

The method generally preferred for generation of diborane, either externally or in the presence of an olefin, is that described above (equation 1): addition of a solution of sodium borohydride in diglyme to a solution of boron trifluoride etherate in diglyme. However, the unavailability of diglyme abroad led to alternative procedures.[25,26] Diethyl ether is a satisfactory solvent for use with the combination lithium aluminum hydride–boron trifluoride etherate. For example, a solution of lithium aluminum hydride in ether was added under nitrogen to a stirred solution of boron trifluoride etherate and cholesterol in ether.[26] After H_2O_2 oxidation and acetylation, the products isolated were cholestane-3β,6α-diol diacetate (70%) and coprostane-3β,6β-diol diacetate (15–20%). Later Brown and co-workers[27] investigated a number of other preparative procedures, but none appears to be of outstanding merit.

Use of selected alkylboranes for H—B addition to olefins has broadened the scope of hydroboration. Bis-3-methyl-2-butylborane (BMB), obtained as noted above as in (1) even in the presence of excess olefin, reacts with unsymmetrical olefins to give increased yields of the less hindered products and permits selective reactions not otherwise possible.[28] Since, unlike diborane, it does not reduce carbonyl groups, the reagent can be used for hydroboration of an unprotected acid (2). γ-Lactones react with only one mole of reagent, even when the latter is in excess (3).

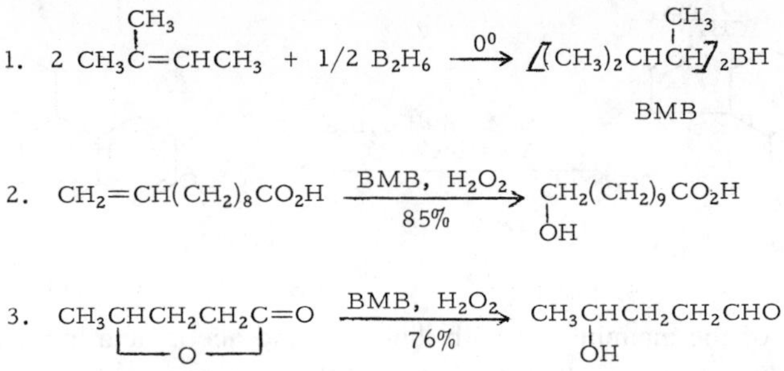

1. $2\ CH_3\overset{\overset{\displaystyle CH_3}{|}}{C}=CHCH_3\ +\ 1/2\ B_2H_6\ \xrightarrow{\ 0^0\ }\ \underset{\underset{\displaystyle BMB}{}}{[(CH_3)_2CH\overset{\overset{\displaystyle CH_3}{|}}{CH}]_2BH}$

2. $CH_2=CH(CH_2)_8CO_2H\ \xrightarrow[85\%]{BMB,\ H_2O_2}\ \underset{\underset{\displaystyle OH}{|}}{CH_2}(CH_2)_9CO_2H$

3. $CH_3CHCH_2CH_2C=O$ $\xrightarrow[76\%]{BMB,\ H_2O_2}$ $CH_3\underset{\underset{\displaystyle OH}{|}}{CH}CH_2CH_2CHO$
     $\underset{\displaystyle O}{\underline{\qquad}}$

Hydroboration of a hindered, optically active olefin (*e.g.*, a terpene or a steroid) proceeds rapidly to give a dialkylborane which can be used for conversion of a suitable olefin, for example *cis*-2-butene, to an optically active alcohol: (−)-2-butanol.[29]

Competitive hydroborations with diborane in diglyme established that the reaction is relatively insensitive to the structure of the olefin.[30] The most reactive olefin studied, 2-methyl-l-butene, is separated by a factor of only 20 or 30 from the least reactive ones, 2,4,4-trimethyl-2-pentene and 2,3-dimethyl-2-butene. In hydroboration with BMB a factor of 10,000 separates reactive 1-octene from cyclohexene, one of the least reactive olefins studied; the study could not be extended to still more inert structures such as 2,4,4-trimethyl-2-pentene. l-Hexyne and 3-hexyne are more reactive than the most reactive olefins studied.

As noted above, the hindered 2,3-dimethyl-2-butene (1) reacts with diborane (1–2 hrs. in diglyme) to give the monoalkyl borane (2), 2,3-dimethyl-2-butylborane

$$CH_3\overset{\overset{\displaystyle CH_3}{|}}{C}=\overset{\overset{\displaystyle CH_3}{|}}{\underset{\underset{\displaystyle CH_3}{|}}{C}}\ \xrightarrow[0^0]{B_2H_6}\ CH_3\overset{\overset{\displaystyle CH_3}{|}}{\underset{\underset{\displaystyle H}{|}}{C}}-\overset{\overset{\displaystyle CH_3}{|}}{\underset{\underset{\displaystyle H}{|}}{C}}-BH_2$$

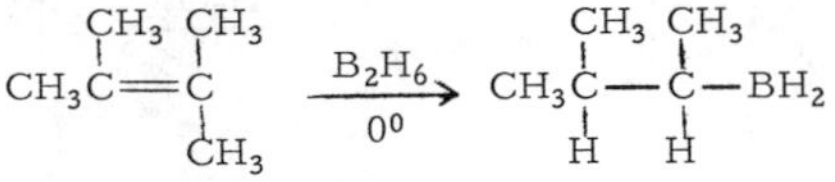

(1)   (2)

("thexylborane"). This reagent has special uses because of its large steric requirement.[31] Special properties and uses are reported for dialkylboranes prepared by controlled hydroboration of cyclohexene, 1-methylcyclohexene, and α-pinene.[32]

In the reduction of a ketone that is either unhindered or of a flexible structure, reduction with sodium borohydride or lithium borohydride gives as the chief product the alcohol of greater stability; thus cholestane-3-one affords the equatorial 3β-ol, which is more stable than the 3α-ol. 2-Methylcyclopentanone and 2-methylcyclohexanone are likewise reduced by metal hydrides to give the more stable *trans*-alcohols in yields of 69–75%. Diborane, although it is a reducing agent of capabilities different from those of the basic metal hydrides, likewise reduces the two cyclanones to the *trans*-alcohols, the yields being 69 and 65%.[33] However, the *cis*-alcohols become the predominant products when an alkylborane of higher steric requirement is used. Thus *cis*-2-methylcyclopentanol and *cis*-2-methylcyclohexanol are formed in yields of 78 and 77% on reduction with bis-3-methyl-2-butylborane, and the yields rise to 94 and 92% when the reagent used is the bulkier diisopinocamphenylborane, from the hydroboration of α-pinene.

Alvarez and Arreguin[34] demonstrated application of the hydroboration reaction to an enol acetate by the conversion of (1) into estriol 3-methyl ether (2).

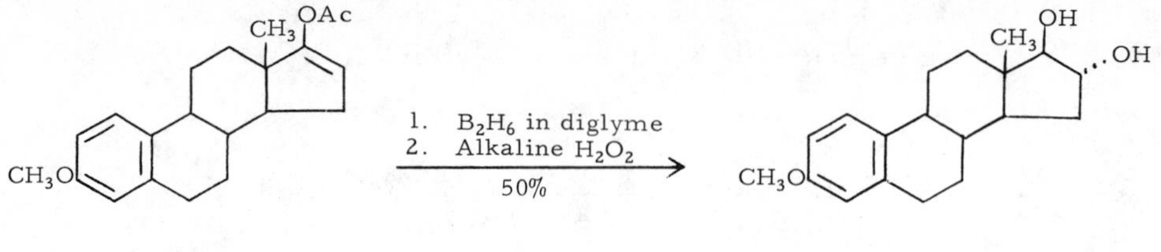
CH3
OAc
1. B2H6 in diglyme
2. Alkaline H2O2
50%
CH3O
CH3
OH
OH
CH3O

(1) (2)

Treatment of the enamine (1) with diborane and acetic acid in tetrahydrofurane gives the borinic acid (2). When (2) is refluxed with acetic acid in diglyme it is converted into cyclohexene in yield of 95% (overall).[35]

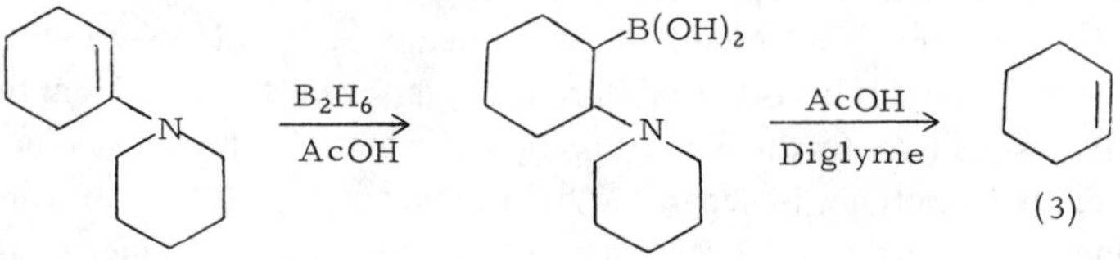
N
B2H6
AcOH
B(OH)2
N
AcOH
Diglyme
(3)

(1) (2)

Cleavage of cyclopropanes. In the absence of a solvent, cyclopropanes are reductively cleaved by diborane.[36] Thus norcarane heated with diborane at 100° for 1–2 hrs. gives the alkylborane (2), identified by peroxide cleavage to cyclohexylmethanol (3). No reaction occurred in tetrahydrofurane.

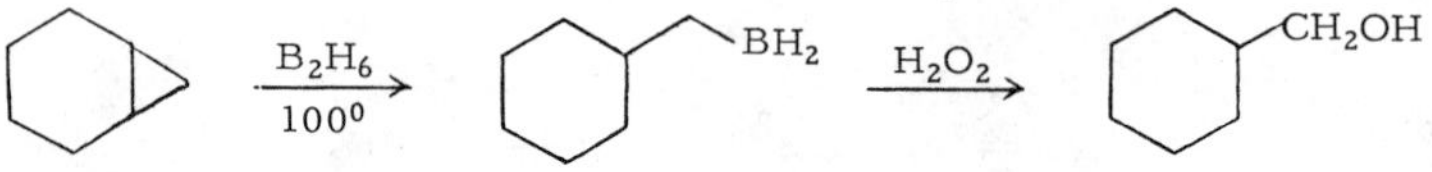
B2H6
100°
BH2
H2O2
CH2OH

(1) (2) (3)

Cleavage of ethers. Long and Freeguard[37] found that the combination of diborane with a halogen, particularly iodine, cleaves an ether readily to give a borate, which is hydrolyzed by water at room temperature within minutes:

$$6\ C_6H_5OCH_3\ +\ 3\ I_2\ +\ B_2H_6\ \longrightarrow\ 2\ B(OC_6H_5)_3\ +\ 6\ CH_3I\ +\ 3\ H_2$$

$$B(OC_6H_5)_3\ +\ 3\ H_2O\ \longrightarrow\ 3\ C_6H_5OH\ +\ H_3BO_3$$

The ether can be symmetrical or unsymmetrical, aliphatic, aromatic, or cyclic. The reaction can be of preparative value; for example the cleavage of tetrahydrofurane with B_2H_6 and I_2 gives 4-iodobutane-1-ol in 90% yield.

[1]H. C. Brown and P. A. Tierney, *Am. Soc.*, **80**, 1552 (1958)
[2]G. Zweifel and H. C. Brown, "Hydration of Olefins, Dienes, and Acetylenes via Hydroboration," *Org. Reactions*, **13**, 1 (1963); *see also* H. C. Brown, "Hydroboration," W. A. Benjamin, Inc., 1962; *Tetrahedron*, **12**, 117 (1961)
[3]G. F. Freeguard and L. H. Long, *Chem. Ind.*, 471 (1965)
[4]H. C. Brown, H. I. Schlesinger, and A. B. Burg, *Am. Soc.*, **61**, 673 (1939)
[5]H. C. Brown and B. C. Subba Rao, *Am. Soc.*, **82**, 681 (1960)
[6]H. C. Brown and W. Korytnyk, *Am. Soc.*, **82**, 3866 (1960)
[7]H. C. Brown and B. C. Subba Rao, *J. Org.*, **22**, 1135 (1957)
[8]D. S. Bapat, B. C. Subba Rao, M. K. Unni, and K. Venkataraman, *Tetrahedron Letters*, No. 5, 15 (1960)

[9]H. Feuer and B. F. Vincent, Jr., *Am. Soc.*, **84**, 3771 (1962); H. Feuer, B. F. Vincent, Jr., and R. S. Bartlett, *J. Org.*, **30**, 2877 (1965); H. Feuer, R. S. Bartlett, B. F. Vincent, Jr., and R. S. Anderson, *J. Org.*, **30**, 2880 (1965)
[10]G. R. Pettit, T. R. Kasturi, B. Green, and J. C. Knight, *J. Org.*, **26**, 4773 (1961)
[11]H. C. Brown and K. Murray, *Am. Soc.*, **81**, 4108 (1959)
[12]E. J. De Witt, F. L. Ramp, and L. E. Trapasso, *Am. Soc.*, **83**, 4672 (1961)
[13]H. C. Brown and G. Zweifel, *Am. Soc.*, **83**, 3834 (1961)
[14]M. F. Hawthorne, *Am. Soc.*, **82**, 1886 (1960)
[15]H. C. Brown and P. Heim, *Am. Soc.*, **86**, 3566 (1964)
[16]Z. B. Papanastassiou and R. J. Bruni, *J. Org.*, **29**, 2870 (1964)
[17]J. R. Johnson and M. G. Van Campen, Jr., *Am. Soc.*, **60**, 121 (1938)
[18]H. C. Brown and B. C. Subba Rao, *Am. Soc.*, **81**, 6423 (1959)
[19]H. C. Brown and M. V. Bhatt, *Am. Soc.*, **82**, 2074 (1960)
[20]W. J. Wechter, *Chem. Ind.*, 294 (1959)
[21]M. Nussim and F. Sondheimer, *Chem. Ind.*, 400 (1960)
[22]H. C. Brown and G. Zweifel, *Am. Soc.*, **83**, 2544 (1961)
[23]R. Pappo, *Am. Soc.*, **81**, 1010 (1959)
[24]H. C. Brown and C. P. Garg, *Am. Soc.*, **83**, 2951 (1961)
[25]R. Dulow and Y. Chrétien-Bessière, *Compt. rend.*, **248**, 416 (1959)
[26]S. Wolfe, M. Nussim, Y. Mazur, and F. Sondheimer, *J. Org.*, **24**, 1034 (1959)
[27]H. C. Brown, K. J. Murray, J. A. Snover, and G. Zweifel, *Am. Soc.*, **82**, 4233 (1960)
[28]H. C. Brown and D. B. Bigley, *Am. Soc.*, **83**, 486 (1961)
[29]H. C. Brown and G. Zweifel, *Am. Soc.*, **83**, 486 (1961)
[30]H. C. Brown and A. W. Moerikofer, *Am. Soc.*, **85**, 2063 (1963)
[31]G. Zweifel and H. C. Brown, *Am. Soc.*, **85**, 2066 (1963)
[32]G. Zweifel, N. R. Ayyangar, and H. C. Brown, *Am. Soc.*, **85**, 2072 (1963)
[33]H. C. Brown and D. B. Bigley, *Am. Soc.*, **83**, 3166 (1961)
[34]F. S. Alvarez and M. Arreguin, *Chem. Ind.*, 720 (1960)
[35]J. W. Lewis and A. A. Pearce, *Tetrahedron Letters*, 2039 (1964)
[36]B. Rickborn and S. E. Wood, *Chem. Ind.*, 162 (1966)
[37]L. H. Long and G. F. Freeguard, *Nature*, **207**, 403 (1965)

Dibromodifluoromethane, CF_2Br_2. Mol. wt. 209.86, b.p. 24.5°. Supplier (99.7% purity) du Pont Co., Organic Chemicals Dept.

The reagent reacts with an alkyllithium or an alkylmagnesium halide to give an olefin according to the following scheme:[1]

$$3\,RCH_2Li(MgBr) + CF_2Br_2 \xrightarrow[40\text{-}70\%]{-70^0} RCH{=}CHCH_2R + RCH_2Br + 2\,LiF$$

$$+ LiBr\ (MgBrF + MgBr_2)$$

The reaction provides a useful method for the preparation of olefins having an uneven number of carbon atoms with a centrally located double bond. The following carbene mechanism is postulated.

$$C_3H_7CH_2Li + CF_2Br_2 \xrightarrow{-70^0} C_3H_7CH_2Br + :CF_2 + LiBr$$

$$C_3H_7CH_2Li + :CF_2 \longrightarrow C_3H_7CH_2\ddot{C}F + LiF$$

$$C_3H_7CH_2\ddot{C}F + C_3H_7CH_2Li \longrightarrow C_3H_7CH_2-\underset{F}{\overset{Li}{\underset{|}{\overset{|}{C}}}}-CH_2C_3H_7$$

$$\xrightarrow[-LiF]{67\%} C_3H_7CH{=}CHCH_2C_3H_7$$

[1]V. Franzen and L. Fikentscher, *Ber.*, **95**, 1958 (1962)

1,3-Dibromo-5,5-dimethylhydantoin (Dibromantin), 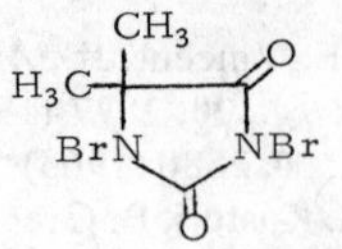

Mol. wt. 285.95, m.p. 195°, dec. Suppliers: Arapahoe, A, B, MCB. A useful reagent comparable to N-bromosuccinimide, it effects allylic bromination, dehydrogenation, substitution of active hydrogen, and oxidation of secondary alcohols. Review.[1]

Attempted side-chain bromination of 2,4-dibenzoylamino-6-methylquinazoline with bromine in chloroform led instead to elimination of the 4-benzoyl group, and reaction with N-bromosuccinimide in chloroform or carbon tetrachloride was

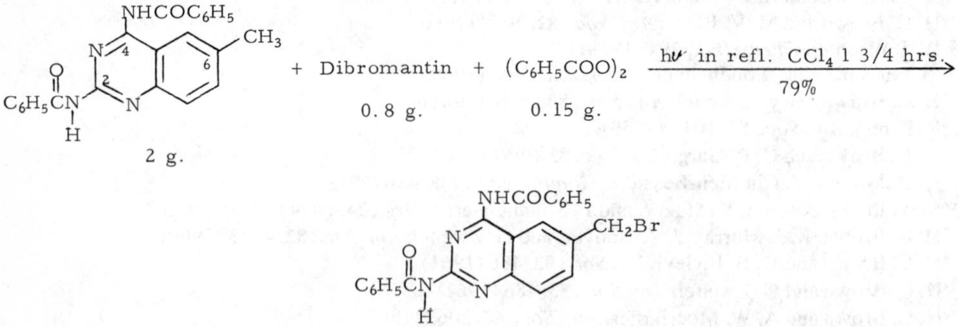

unsatisfactory.[2] The desired 6-bromomethyl derivative was obtained in high yield by bromination in refluxing carbon tetrachloride with dibromantin with catalysis from dibenzoyl peroxide and irradiation by a 500-watt tungsten-filament lamp.

[1]R. A. Reed, *Chem. Prods.*, **23**, 299 (1960) [*C.A.*, **54**, 21059 (1960)]
[2]V. Oakes, H. N. Rydon, and K. Undheim, *J. Chem. Soc.*, 4678 (1962)

Dibromomalonamide, $CBr_2(CONH_2)_2$. Mol. wt. 259.91, m.p. 203°.

Preparation. It is prepared in 94% yield by reaction of malonamide with bromine in the presence of sodium acetate to bind the hydrogen bromide formed.[1]

In phosphorylation. A hydroxy compound is phosphorylated by reaction with 1 mole of dibromomalonamide and 2 moles of triethylphosphite.[2] Thus the reagent was added in portions to a mixture of triethylphosphite and excess *n*-propyl alcohol as solvent. Malonamide promptly separated and, after standing for a time, was

$$2\,ROH + CBr_2(CONH_2)_2 + 2(C_2H_5O)_3P \xrightarrow[72\%]{} 2\,RO\overset{O}{\overset{\|}{P}}(OC_2H_5)_2 + CH_2(CONH_2)_2 + 2\,C_2H_5Br$$

removed by filtration. Tetraethyl pyrophosphate was obtained by the reaction of diethyl hydrogen phosphate with dibromomalonamide and triethylphosphite in ether at 0°.

$$2\,(C_2H_5O)_2\overset{O}{\overset{\|}{P}}OH + CBr_2(CONH_2)_2 + 2\,(C_2H_5O)_3P \xrightarrow[78\%]{0^0}$$

$$2\,(C_2H_5O)_2\overset{O}{\overset{\|}{P}}-O-\overset{O}{\overset{\|}{P}}(OC_2H_5)_2 + CH_2(CONH_2)_2 + 2\,C_2H_5Br$$

[1]J. V. Backes, R. W. West, and M. A. Whiteley, *J. Chem. Soc.*, **119**, 365 (1921)
[2]T. Mukaiyama, T. Hata, and K. Tasaka, *J. Org.*, **28**, 481 (1963)

1,2-Dibromotetrachloroethane, $BrCCl_2CCl_2Br$. Mol. wt. 325.68. No m.p.; the substance decomposes at temperatures above 105°. It is readily purified by vacuum sublimation. Supplier: Dow.

The reagent has been shown to be effective for allylic bromination, the reaction being markedly catalyzed by light.[1] Thus bromination of cyclohexene was effected under illumination with a 275-watt Sylvania sunlamp. The reaction can be carried

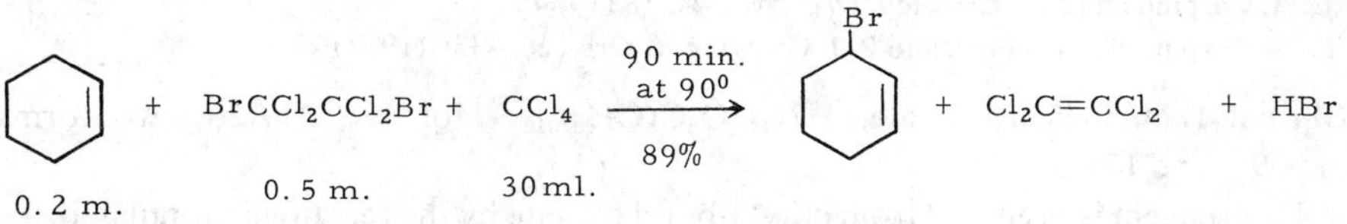

+ BrCCl2CCl2Br + CCl4
90 min.
at 90°
89%
Br
+ Cl2C=CCl2 + HBr
0.2 m.
0.5 m.
30 ml.

out in a homogeneous solution in the reagent or with carbon tetrachloride as diluent. The reagent is less expensive than NBS and, when a mixture of products is formed, the product distribution is different from that obtained with NBS.[1]

[1]E. S. Huyser and D. N. DeMott, *Chem. Ind.*, 1954 (1963)

2,2-Di-*n*-butoxypropane (Acetone di-*n*-butyl ketal), $(CH_3)_2C(OC_4H_9\text{-}n)_2$. Mol. wt. 188.30, b.p. 88–90°/20 mm., sp. gr. 1.4105. Supplier: Eastman. Preparation from *n*-butanol by ketal exchange with 2,2-dimethoxypropane.[1] After removal of a

$$(CH_3)_2C(OCH_3)_2 + 2\ \underline{n}\text{-}C_4H_9OH \xrightarrow[\text{Benzene}]{\text{TsOH}} (CH_3)_2C(OC_4H_9\text{-}\underline{n})_2 + 2\ CH_3OH$$

benzene–methanol azeotrope by distillation, the mixture is cooled, the acid catalyst neutralized with sodium methoxide, and distillation is continued.

[1]N. B. Lorette and W. L. Howard, *Org. Syn.*, **42**, 1 (1962)

Di-*i*-butylaluminum hydride, $Al(i\text{-}Bu)_2H$. For the reduction of α,β-unsaturated γ-lactones to furanes, the reagent is used interchangeably with triisobutylaluminum, *which see.*

Di-*n*-butylamine, $(CH_3CH_2CH_2CH_2)_2NH$. Mol. wt. 129.24, b.p. 159.6°, sp. gr. 0.768, very sparingly soluble in water, pKb 2.7. Suppliers: Pennsalt, A, B, E, F, MCB.

The amine forms salts with abietic acid and related acids which are useful for purification.[1] The acid is regenerated by treatment with acetic acid.

[1]A. W. Burgstahler and L. R. Worden, *Am. Soc.*, **86**, 96 (1964)

Di-*t*-butyl azodiformate, $(CH_3)_3CO\overset{O}{\overset{\|}{C}}N{=}N\overset{O}{\overset{\|}{C}}OC(CH_3)_3$. Mol. wt. 230.26, m.p. 90–92°.

Preparation.[1] A solution of *t*-butyl azidoformate (1) and *t*-butyl carbazate (2) in pyridine is allowed to stand at room temperature for 1 week and diluted with

$$\underset{(1)}{(CH_3)_3CO\overset{O}{\overset{\|}{C}}N{=}\overset{+}{N}{=}\overset{-}{N}} + \underset{(2)}{H_2NNHCO_2C(CH_3)_3} \xrightarrow[77\text{-}80\%]{Py(-HN_3)} \underset{(3)}{(CH_3)_3CO\overset{O}{\overset{\|}{C}}\overset{H}{\overset{|}{N}}{-}\overset{H}{\overset{|}{N}}\overset{O}{\overset{\|}{C}}OC(CH_3)_3}$$

$$\xrightarrow[88\text{-}90\%]{NBS-CH_2Cl_2-Py} \underset{(4)}{(CH_3)_3CO\overset{O}{\overset{\|}{C}}N{=}N\overset{O}{\overset{\|}{C}}OC(CH_3)_3}$$

water. *t*-Butyl hydrazidoformate (3) separates as a snow-white microcrystalline powder and is oxidized by N-bromosuccinimide in pyridine–methylene chloride to di-*t*-butyl azodiformate, obtained by crystallization from ligroin as large lemon-yellow crystals.

Diels-Alder dienophile.[2] It is somewhat less reactive in this respect than diethyl azodiformate.

[1]L. A. Carpino and P. J. Crowley, *Org. Syn.*, **44**, 18 (1964)
[2]L. A. Carpino, P. H. Terry, and P. J. Crowley, *J. Org.*, **26**, 4336 (1961)

Di-*t*-butyl iminodicarboxylate, $HN[CO_2C(CH_3)_3]_2$. Mol. wt. 217.26, two forms, m.p. 90° and 121°.

This reagent is prepared from ethyl *t*-butyl oxalate by the reactions formulated:

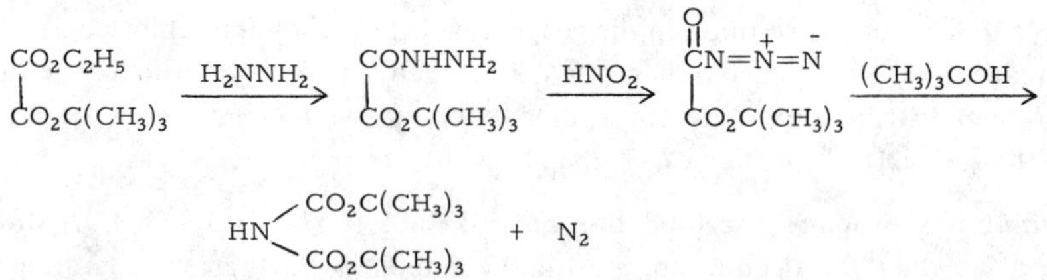

In the last step the azide was warmed with *t*-butanol to effect Curtius rearrangement.[1]

The reagent is used in a Gabriel-type synthesis of primary amines. Thus a suspension of sodium hydride in a solution of the diester in dimethylformamide was stirred at 60° for 6 hrs. to form the sodio derivative, α,α'-dibromo-*o*-xylene was added,

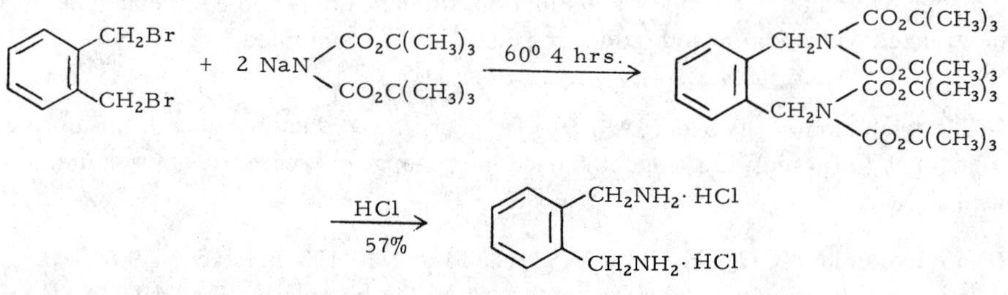

and heating was continued to complete the alkylation. The mixture was diluted with water and the tetraester extracted with methylene chloride and warmed with concd. hydrochloric acid to eliminate the ester groups (as isobutene and CO_2), with formation of *o*-xylylenediamine dihydrochloride.

[1]L. A. Carpino, *J. Org.*, **29**, 2820 (1964)

Di-*t*-butyl malonate, $CH_2[CO_2C(CH_3)_3]_2$. Mol. wt. 216.27, m.p. −6°, b.p. 112–115°/31 mm. Supplier: KK.

Preparation from malonic acid, isobutene, and a catalytic amount of concd. sulfuric acid in ether at a pressure of about 40 p.s.i.[1,2] Preparation via malonyl dichloride;[3] the first step requires heating at 45–50° for 3 days.

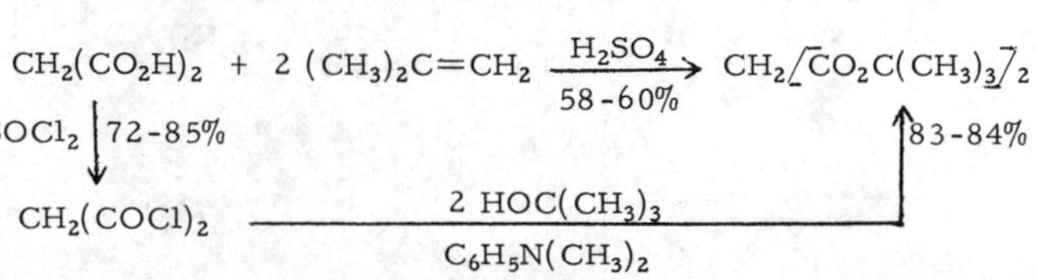

The reagent is useful in the synthesis of ketones because the butyl groups of

$$RCOCl \xrightarrow{CH_2[CO_2C(CH_3)_3]_2} RCOCH[CO_2C(CH_3)_3]_2 \xrightarrow{H^+}$$

$$RCOCH_3 + 2\,(CH_3)_2C{=}CH_2 + 2\,CO_2$$

an acyl derivative can be eliminated pyrolytically; the reaction is done in refluxing toluene or acetic acid with *p*-toluenesulfonic acid as catalyst.[1]

[1]G. S. Fonken and W. S. Johnson, *Am. Soc.*, **74**, 831 (1952)
[2]A. L. McCloskey, G. S. Fonken, R. W. Kluiber, and W. S. Johnson, *Org. Syn., Coll. Vol.*, **4**, 261 (1963)
[3]C. Raha, *ibid.*, **4**, 263 (1963)

Di-*t*-butyl nitroxide, $[(CH_3)_3C]_2N\dot{-}O$ (radical). Mol. wt. 144.23, b.p. 74–75°/35 mm.
Metallic sodium reacts with *t*-nitrobutane in 1,2-dimethoxyethane (glyme) to form a salt, $[(CH_3)_3C]_2NO_2^-Na^+$, which is hydrolyzed to the red liquid radical, stable to oxygen, water, and aqueous alkali.[1] This is a useful radical scavenger, particularly for photolytic decompositions, since it has low absorption in the near ultraviolet.[2]

[1]A. K. Hoffmann, A. M. Feldman, E. Gelblum, and W. G. Hodgson, *Am. Soc.*, **86**, 639 (1964); procedure submitted to *Org. Syn.*
[2]S. F. Nelsen and P. D. Bartlett, *Am. Soc.*, **88**, 143 (1966)

Di-*t*-butyl peroxide, $(CH_3)_3COOC(CH_3)_3$. Mol. wt. 146.22, b.p. 50–52°/90 mm. Suppliers (pract.): Eastern, MCB. Commercial material is purified by distillation at reduced pressure.[1]

Decomposition. This peroxide undergoes smooth thermal decomposition in a first-order reaction to give *t*-butoxy radicals which, in a secondary reaction, decompose to methyl radicals and acetone.[2] The extent of the secondary reaction is dependent upon the reactivity of the solvent toward the *t*-butoxy radical. The

$$(CH_3)_3COOC(CH_3)_3 \longrightarrow 2\,(CH_3)_3CO\cdot \longrightarrow 2\,(CH_3)_2C{=}O + 2\cdot CH_3$$

primary reaction, which is rate-determining, has a half-life of about 20 hrs. at 120° and of about 1 hr. at 150°. The peroxide is thus useful for generation of radicals; hence for initiation of chain-polymerizations and other radical-reactions in the temperature range 120–150°.

Methylation of aromatics. Aromatic substitutions effected by radicals are of more theoretical than preparative interest because the yields are low. Waters and co-workers[3] studied the methylation of monosubstituted benzenes with methyl radicals generated from di-*t*-butyl peroxide, separated the mixture of monomethyl

$$C_6H_5X + (CH_3)_3CO\cdot + \cdot CH_3 \longrightarrow CH_3C_6H_4X + (CH_3)_3COH$$

derivatives from starting material by vapor-phase chromatography, and determined the proportion of the three isomers by infrared spectroscopy. Thus a mixture of chlorobenzene and the peroxide was refluxed and the temperature was kept near 130° by removing the acetone and *t*-butyl alcohol as formed.

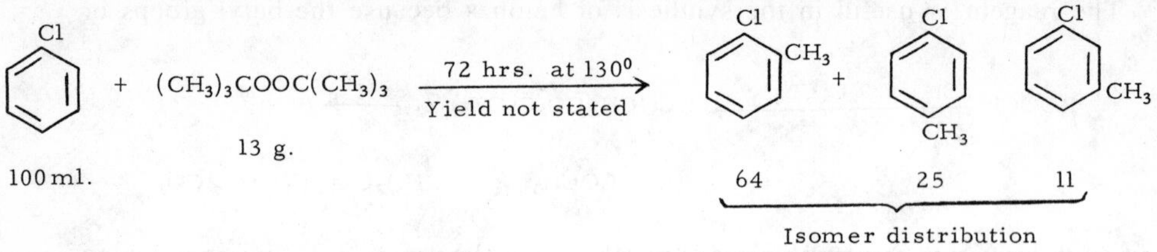
Cl
+ (CH3)3COOC(CH3)3
72 hrs. at 130°
Yield not stated
13 g.
100 ml.
Cl
CH3
+
Cl
CH3
Cl
CH3
64
25
11
Isomer distribution

Addition of amides to olefins. Rieche[4] found that in the presence of decomposing

$$CH_3(CH_2)_3CH{=}CH_2 + HCONH_2 \xrightarrow[150^0]{\cdot R} CH_3(CH_2)_3CH_2CH_2CONH_2$$

$$CH_2{=}CH(CH_2)_8CO_2CH_3 + HCONH_2 \xrightarrow[150^0]{\cdot R} H_2NCO(CH_2)_{10}CO_2CH_3$$

di-*t*-butyl peroxide, radical addition of formamide to hexene-1 can be effected to produce the straight-chain amide in 5–10% yield. Use of the cyclic mesityl oxide

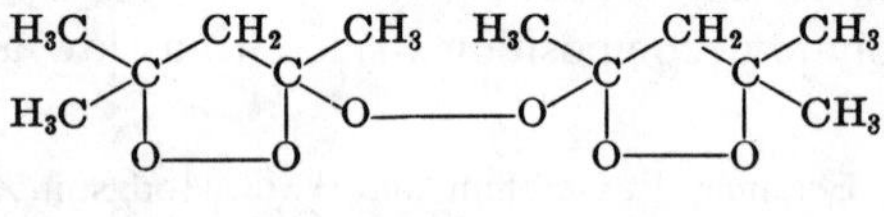

peroxide reduced polymerization and raised the yield to 35–40%. Methyl undecylenate gave dodecanedioic acid methyl ester amide in yield of 40%.

Friedman and Shechter[5] found that the reaction of 1-octene with dimethylformamide and a catalytic amount of di-*t*-butyl peroxide gave a 60:40 mixture of N,N-dimethylnonanamide (1, amidation) and N-methyl-N-nonylformamide (2,

$$CH_3(CH_2)_5CH{=}CH_2 + HCON(CH_3)_2 \xrightarrow[56\%]{\cdot R\ (18\text{ hrs. at }132^0)}$$

$$\underset{(1)}{CH_3(CH_2)_5CH_2CH_2CON(CH_3)_2} + \underset{(2)}{CH_3(CH_2)_5CH_2CH_2CH_2\overset{CH_3}{\overset{|}{N}}CHO}$$

amino alkylation). Under the same conditions N-*t*-butylformamide, N-alkyl-, and N,N-dialkylacetamide react selectively with 1-octene to afford in good yield products of type (1) resulting from attack at the N-alkyl α-carbon atom.

[1]A. L. J. Beckwith and W. A. Waters, *J. Chem. Soc.*, 1108 (1956)
[2]C. Walling and E. S. Huyser, *Org. Reactions*, **13**, 114 (1963)
[3]B. R. Crowley, R. O. C. Norman, and W. A. Waters, *J. Chem. Soc.*, 1799 (1959)
[4]A. Rieche, E. Schmitz, and E. Grundemann, *Angew. Chem.*, **73**, 621 (1961)
[5]L. Friedman and H. Shechter, *Tetrahedron Letters*, 238 (1961)

2,6-Di-*t*-butylpyridine, $C_{13}H_{21}N$. Mol. wt. 191.31, pKa 3.58.[1] Combines with hydrogen chloride, but not with methyl iodide, boron trifluoride, or sulfur trioxide.

[1]H. C. Brown and B. Kanner, *Am. Soc.*, **75**, 3865 (1953)

Di-*t*-butyl succinate, $(CH_3)_3CO\overset{O}{\overset{\|}{C}}CH_2CH_2\overset{O}{\overset{\|}{C}}OC(CH_3)_3$. Mol. wt. 230.30, m.p. 31.5–35°, b.p. 106°/7 mm. Supplier: Baker. Preparation from the acid and isobutene in dioxane (H_2SO_4).[1] Preferable to less hindered esters in the Stobbe condensation because of avoidance of self-condensation.[2]

[1]A. L. McCloskey, G. S. Fonken, R. W. Kluiber, and W. S. Johnson, *Org. Syn., Coll. Vol.*, **4**, 263 (1963)
[2]W. S. Johnson and G. H. Daub, *Org. Reactions*, **6**, 1 (1951)

Di-*n*-butyltin dichloride, *n*-$(C_4H_9)_2SnCl_2$. Mol. wt. 303.83. White crystalline solid, m.p. 43°, b.p. 170°/60 mm. (100°/2 mm.). Suppliers: Metal and Thermit Corp., Alfa, B, E, MCB.

Reaction with vinylmagnesium bromide.[1] In a flask equipped with a dry ice–acetone condenser, mechanical stirrer, and dropping funnel, magnesium turnings are covered with tetrahydrofurane, the stirrer is started, and 5 ml. of vinyl bromide is added.

$$\underset{1.3\text{ m.}}{CH_2{=}CHBr} + \underset{1.2\text{ g. atoms}}{Mg} \xrightarrow{THF} CH_2{=}CHMgBr \xrightarrow[74\text{-}91\%]{0.44\text{ m. }(n\text{-}C_4H_9)_2SnCl_2} (n\text{-}C_4H_9)_2Sn(CH{=}CH_2)_2$$

After the reaction has started, more THF is added, and the rest of the vinyl bromide dissolved in THF is added at such a rate as to maintain gentle refluxing. When the reaction is complete the Grignard solution is cooled, the dry ice–acetone condenser is replaced by a water condenser, and a solution of di-*n*-butyltin dichloride is run in with stirring at gentle reflux. After 20 hrs. of refluxing the mixture is hydrolyzed with saturated ammonium chloride solution (100–120 ml. per mole of Grignard reagent). The organic layer is decanted, the salt residue is washed with ether, the solvents are stripped off, and the di-*n*-butyldivinyltin is distilled at 60°/0.4 mm.

[1]D. Seyferth, *Org. Syn., Coll. Vol.*, **4**, 258 (1963)

Dichloramine, $HNCl_2$. This highly reactive reagent can be generated in aqueous solution by reaction of chlorine and ammonia at pH 3–5; above pH 8 the product is chloramine H_2NCl.[1] Graham[2] generated the reagent from 5% sodium hypochlorite (Chlorox) and ammonium chloride in an aqueous sodium formate–formic acid buffer, and preferred this easily prepared reagent to the less readily available and more hazardous difluoramine for the preparation of diazirine (6). A second reagent required, *t*-octylazomethine (2), was prepared by adding formaldehyde solution with stirring to Rohm and Haas' *t*-octylamine (1, neopentyldimethylcarbinylamine).[3]

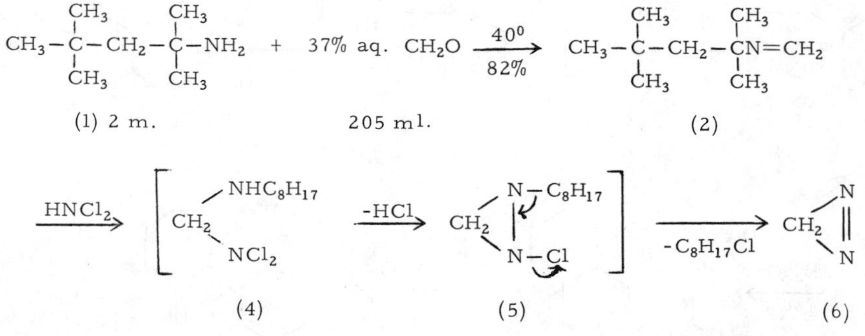
CH3 CH3
CH3–C–CH2–C–NH2 + 37% aq. CH2O
40°
82%
CH3–C–CH2–CN=CH2
(1) 2 m.
205 ml.
(2)
HNCl2
NHC8H17
CH2
NCl2
-HCl
N–C8H17
N–Cl
-C8H17Cl
N
N
(4)
(5)
(6)

For the preparation of diazirine (6) a 5% hypochlorite solution was dropped slowly into a mixture of the formate–formic acid buffer, ammonium chloride, and *t*-octylazomethine maintained at 5–12°, and the volatile contents were removed under vacuum through a series of traps at −35, −80, −142, and −196°. The diazirine, retained largely in the −142° trap, generally contained CO_2 to the extent of 15% or less, and yields of 25–33% were typical.

[1]R. M. Chapin, *Am. Soc.*, **81**, 2112 (1929); R. E. Corbett, W. S. Metcalf, and F. G. Soper, *J. Chem. Soc.*, 1927 (1953)
[2]W. H. Graham, *J. Org.*, **30**, 2108 (1965)
[3]W. H. Graham, private communication

1,3-Dichloro-2-butene, $CH_3C(Cl){=}CHCH_2Cl$. Mol. wt. 125.00, b.p. 127–128°(39°/30 mm.), $n^{20}D$ 1.4740. Physical constants of the pure *cis*- and *trans*-isomers are reported by Hatch and Perry.[1] Supplier: Eastman. According to Toromanoff[2] this Eastman material, a dark brown lachrymatory liquid, can be adequately purified by distillation at normal pressure with an efficient column, careful drying over K_2CO_3 or $CaCl_2$, and redistillation at 30 mm. Vapor-phase chromatographic analysis indicated a 30:70 *cis:trans* mixture. The IR spectrum shows a minute amount of an aromatic impurity.

Preparation. One method is by the cuprous chloride catalyzed hydrochlorination of chloroprene:[3]

$$CH_2{=}C(Cl)CH{=}CH_2 \xrightarrow[58\%]{HCl\,(Cu_2Cl_2)} CH_3C(Cl){=}CHCH_2Cl$$

Another route is from ethyl acetoacetate:[1]

$$CH_3COCH_2CO_2C_2H_5 \xrightarrow{PCl_5} CH_3C(Cl){=}CHCO_2C_2H_5 \xrightarrow{LiAlH_4} CH_3C(Cl){=}CHCH_2OH$$

$$\xrightarrow[Py]{PCl_3} CH_3C(Cl){=}CHCH_2Cl$$

The reagent has been used extensively by Velluz *et al.*[4] for construction of ring A in the total synthesis of 19-norsteroids and estrogens, for example in the sequence

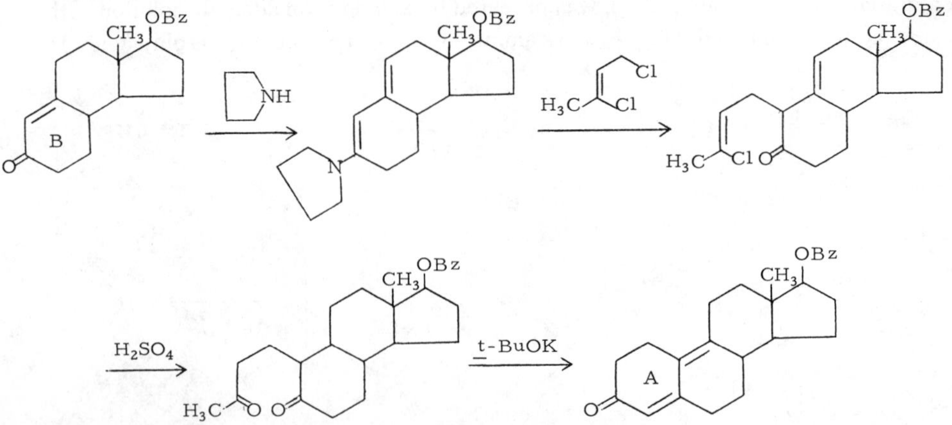
OBz
CH3
B
NH
N
H3C
Cl
Cl
H2SO4
O
t-BuOK
A

formulated. In a model study with a saturated ketone, Julia[5] found that hydrolysis of the vinyl chloride system led mainly to an undesired cyclization.

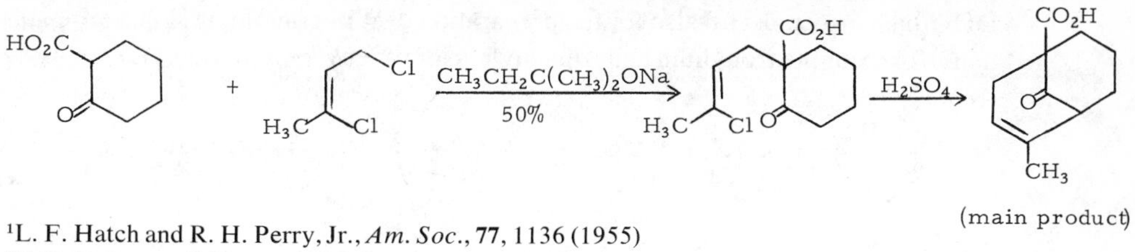
HO2C
+
Cl
H3C
Cl
CH3CH2C(CH3)2ONa
50%
CO2H
H3C
Cl
H2SO4
CO2H
CH3
(main product)

[1]L. F. Hatch and R. H. Perry, Jr., *Am. Soc.*, **77**, 1136 (1955)
[2]E. R. Toromanoff (Roussel Uclaf), private communication
[3]L. F. Hatch and S. G. Ballin, *Am. Soc.*, **71**, 1039 (1949)
[4]L. Velluz, G. Nominé, and J. Mathieu, *Angew. Chem.*, **72**, 725 (1960); L. Velluz, G. Nominé, J. Mathieu, E. Toromanoff, D. Bertin, J. Tessier, and A. Pierdet, *Compt. rend.*, **250**, 1084 (1960); L. Velluz, G. Nominé, R. Bucourt, A. Pierdet, and J. Tessier, *ibid.*, **252**, 3903 (1961),: L. Velluz, G. Nominé, R. Bucourt, A. Pierdet, and Ph. Dufay, *Tetrahedron Letters*, 127 (1961); R. Bucourt, J. Tessier, and G. Nominé, *Bull. soc.*, 1923 (1963)
[5]S. Julia, *Bull. soc.*, **21**, 780 (1954)

2,3-Dichloro-5,6-dicyano-1,4-benzoquinone (DDQ). Mol. wt. 227.01, m.p. 213°. Suppliers: Arapahoe (bulletin available), B, E.

Preparation. The synthesis formulated is due to Thiele.[1] With later improve-

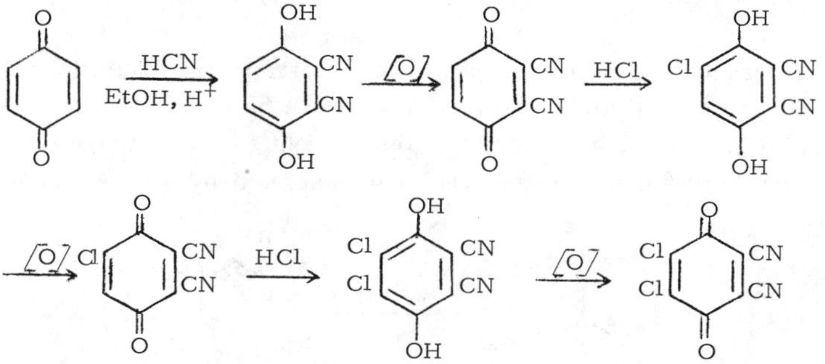
O
O
HCN
EtOH, H+
OH
CN
CN
OH
[O]
O
CN
CN
O
HCl
OH
Cl
CN
CN
OH
[O]
O
Cl
CN
CN
O
HCl
OH
Cl
Cl
CN
CN
OH
[O]
O
Cl
Cl
CN
CN
O

ments[2] the reactions can be completed in 2 days in overall yield from benzoquinone of 70%. More recently Mitchell[3] chlorinated 2,3-dicyanohydroquinone in refluxing acetic acid, obtained the 5,6-dichloro derivative in 40–50% yield, and oxidized this with lead dioxide in aqueous ethanol containing 5% of hydrochloric acid. In the absence of acid no oxidation occurred. However, in a new procedure[4] 2,3-dicyano-hydroquinone is stirred with 1:1 hydrochloric acid–water, and 70% nitric acid is added at 35°, with conversion to DDQ in 90% yield. This shortcut is based on the finding that DDQ is stable in aqueous mineral acid (if not in water) and that 2,3-dicyano-1,4-benzoquinone and 2-chloro-5,6-dicyanobenzoquinone add hydrogen chloride rapidly from aqueous solution.

Aromatization. The high-potential quinone was introduced[2] for use in the dehydrogenation of hydroaromatic compounds; for example, in boiling benzene it converts tetralin into naphthalene and acenaphthene into acenaphthylene. The hydroaromatic compound with a blocking group (1) undergoes aromatization with a 1:2-shift of a methyl group.

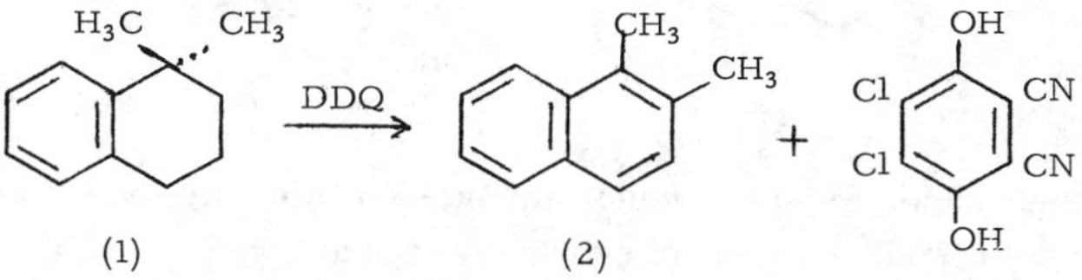
H3C
CH3
DDQ
CH3
CH3
+
OH
Cl
Cl
CN
CN
OH

(1) (2)

DDQ has proved useful also for the preparation of salts containing stable aromatic cations, for example tropylium and triphenylcyclopropenyl perchlorate.[5]

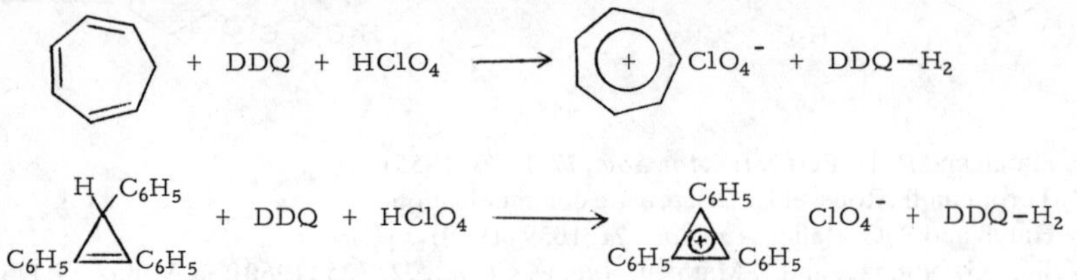

DDQ effects aromatization of ring A of (3) with conversion to equilin (4).[6]

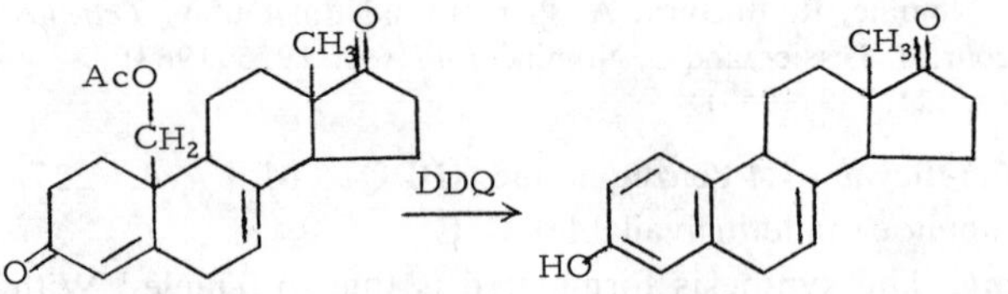

(3) (4)

Dehydrogenation of carbonyl compounds. The reagent is useful for introduction into steroid ketones of double bonds that enhance biological activity. A saturated 3-ketosteroid (5) is dehydrogenated by DDQ in refluxing benzene or dioxane to the $\Delta^{1,4}$-3-ketone (6).[7,8] A Δ^4-3-ketosteroid (7) can undergo dehydrogenation in two ways.

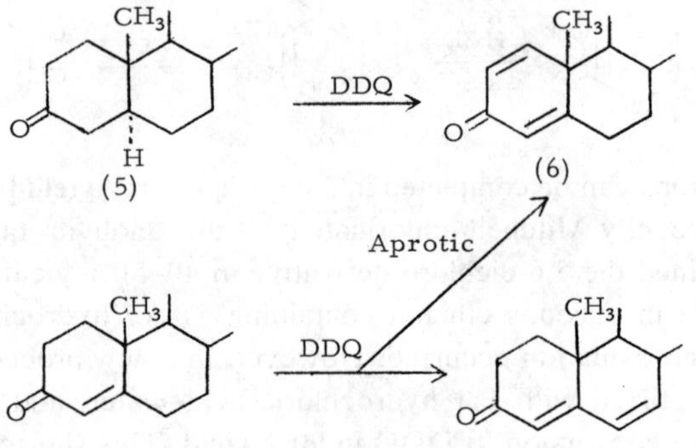

(5) (6) (7) (8)

Under aprotic conditions, or in the presence of a weak proton donor such as *p*-nitrophenol, 1,2-dehydrogenation to give the $\Delta^{1,4}$-3-ketone occurs;[9-12] in the presence of hydrogen chloride, 6,7-dehydrogenation occurs to give the $\Delta^{4,6}$-ketone (8).[12] $\Delta^{4,6}$-3-Ketones (8) undergo 1,2-dehydrogenation only.[9,10] The reaction products are

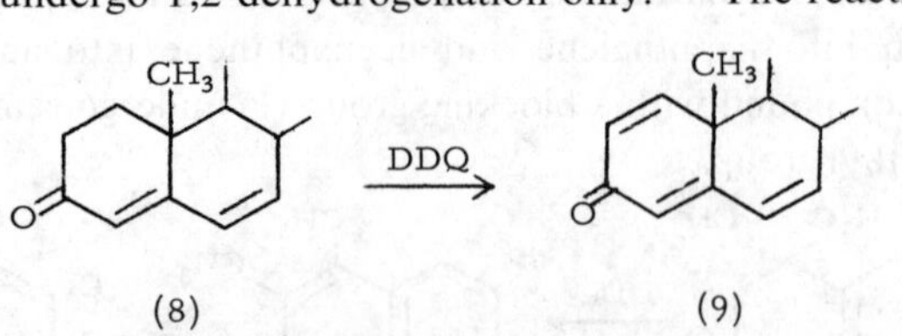

(8) (9)

obtained in high yield, and a particular advantage of the method is that substituents commonly encountered in steroid chemistry are not affected by DDQ.

δ-Lactones are dehydrogenated to the corresponding α,β-unsaturated lactones, but yields are low to fair.[13]

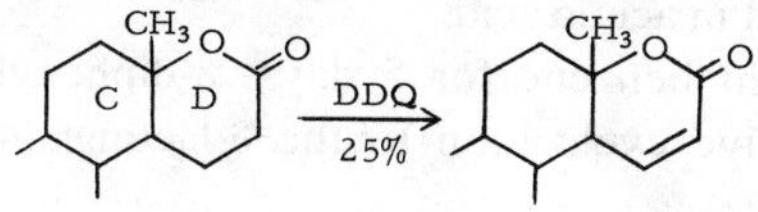
CH3
C
D
DDQ
25%
CH3

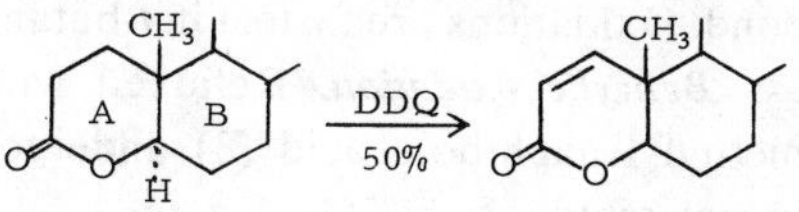
CH3
A
B
H
DDQ
50%
CH3

In dry benzene, DDQ converts the $\Delta^{3,5}$-dienol ether (11) into the 1,4,6-triene-3-one (10); in the presence of water the product is the 4,6-diene-3-one (12).[14] A reaction analogous to (11) → (12) is the dehydrogenation of 17α-acetoxy-6-hydroxymethyl-3-methoxy-$\Delta^{3,5}$-pregnadiene-20-one by DDQ in aqueous acetone to 17α-acetoxy-6-hydroxymethyl-$\Delta^{4,6}$-pregnadiene-3,20-dione.[14a]

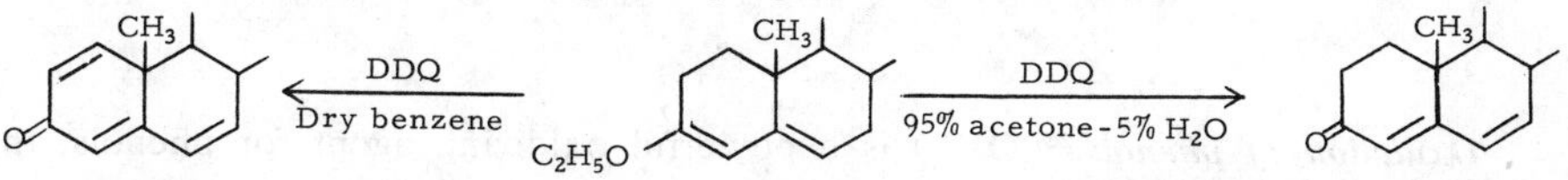
CH3
DDQ
Dry benzene
CH3
C2H5O
DDQ
95% acetone - 5% H2O
CH3

(10) (11) (12)

Oxidation of oxygen functions. DDQ is useful for the selective oxidation of allylic[15-17] and benzylic[18] alcohols. The reagent has the advantage of not showing the variability encountered with active manganese dioxide. It dehydrogenates 2-hydroxymethylene-3-ketosteroids in about 50% yield.[19] The steroid is treated in dioxane solution at room temperature with 1.1–1.5 equivalents of DDQ for 1–5 min.,

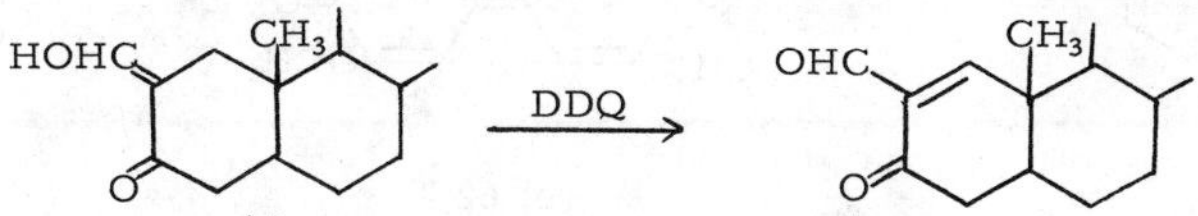
HOHC
CH3
DDQ
OHC
CH3

(1) (2)

and the mixture is diluted with methylene chloride and the product isolated by chromatography.

In a kinetic study of the oxidation of allylic steroid alcohols, Burstein and Ringold[20] observed that 3β-hydroxy-Δ^4-steroids (equatorial) are oxidized about 6 times faster than the 3α-epimers (axial) and that the 3α-hydrogen compound (4) is oxidized 5

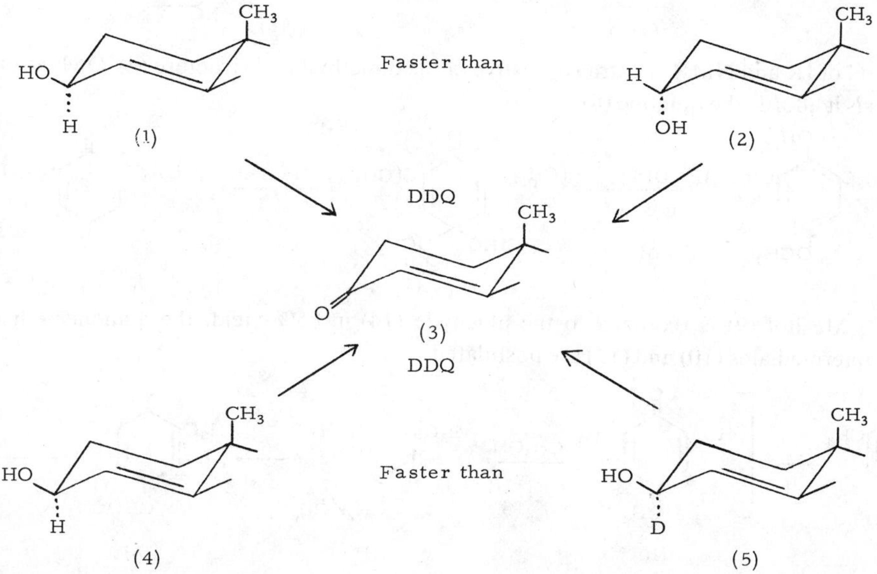
CH3
HO
H
(1)
Faster than
CH3
H
OH
(2)
DDQ
CH3
O
(3)
DDQ
CH3
HO
H
(4)
Faster than
CH3
HO
D
(5)

times as fast as the corresponding 3α-deuterium compound (5). The latter observation indicated that the rate-determining step is the cleavage of the C—H or C—D bond. Oxidations are fastest in *t*-butanol, slowest in acetic acid.

Benzylic oxidation. Refluxed with DDQ in benzene for 5 days, 8-diphenylmethyl-1-naphthoic acid (3) undergoes oxidative cyclization to the δ-lactone (4) in high yield.[21]

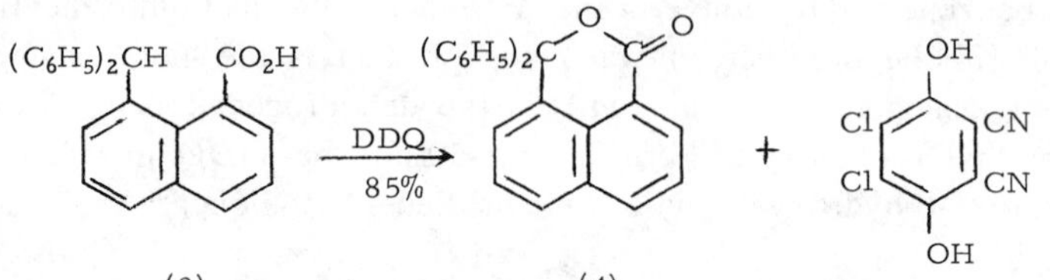

(3) (4)

Oxidation of phenols.[22] DDQ is a powerful oxidizing agent for phenols, the reaction usually proceeding smoothly in methanol solution at room temperature. Usually the chief product is a dimeric substance formed by either C-C or C-O coupling. 2,6-Dimethoxyphenol, with a free *p*-position, gives chiefly the diphenoquinone, with a trace of 2,6-dimethoxyquinone. The hindered phenol (4) affords

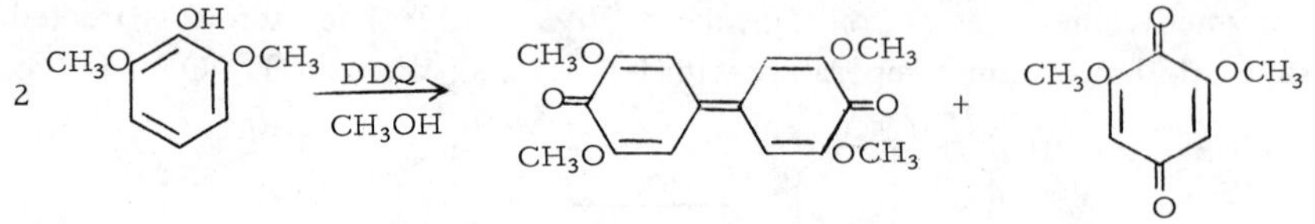

(1) (2) 62% (3) Trace

the C—O dimeric product (5). 2,6-Di-*t*-butyl-4-methoxyphenol (6) gives a yellow

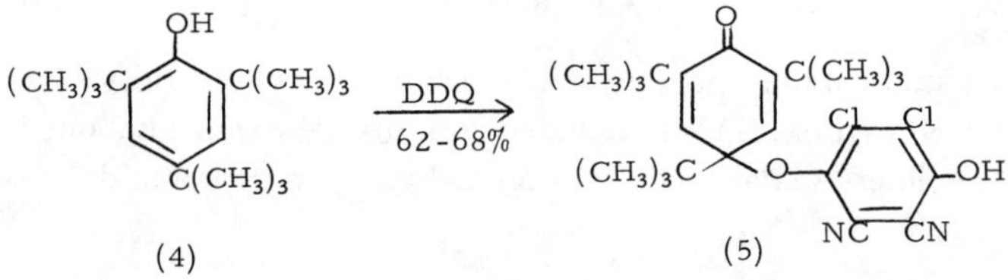

(4) (5)

oil of IR and NMR spectra suggestive of the dimethylketal structure (7). On hydrolysis it affords the quinone (8).

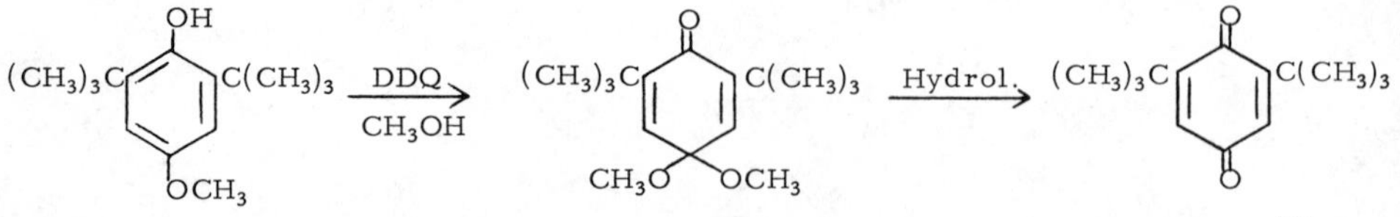

(6) (7) (8)

Mesitol (9) is oxidized to the aldehyde (14) in 85% yield; the quinonemethane intermediates (10) and (12) are postulated.

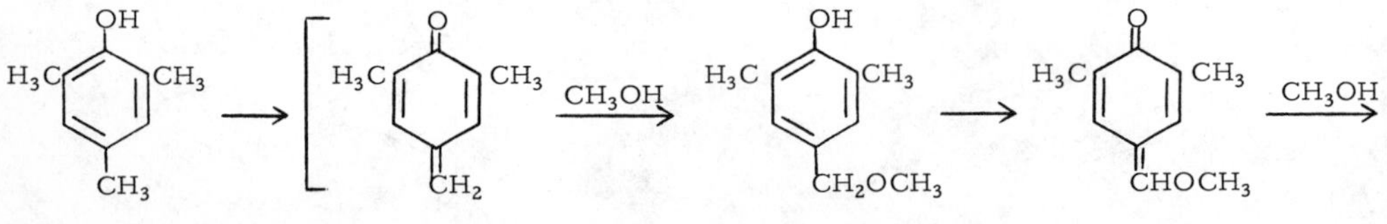

(9) (10) (11) (12)

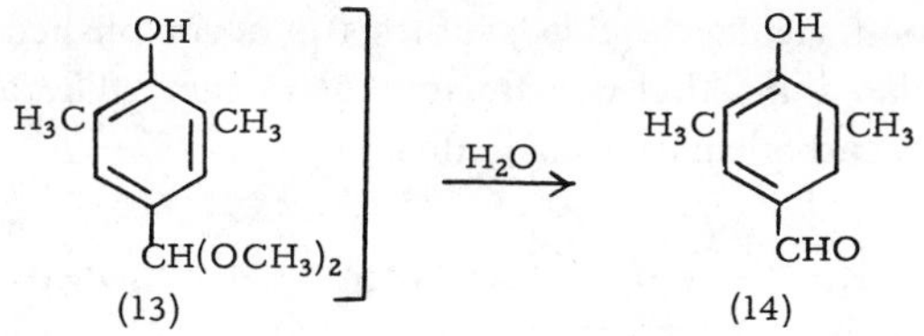

Oxidative dimerization. Enols and enolizable ketones in which α,β-unsaturation is not possible are oxidatively dimerized by DDQ.[23] Thus 2-phenylindane-1,3-dione is converted into the tetrone.

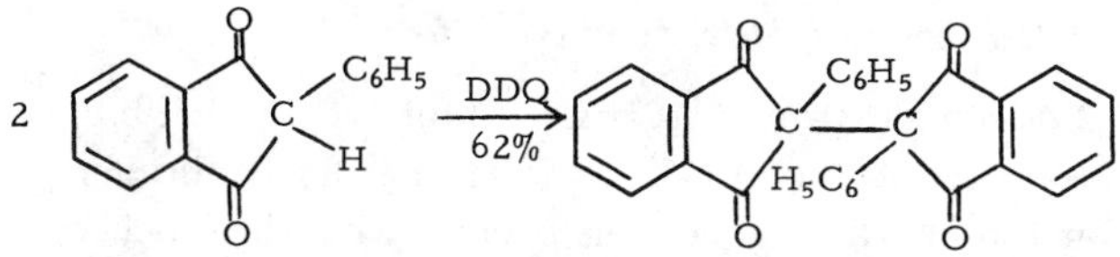

[1]J. Thiele and F. Günther, *Ann.*, **349**, 45 (1906)
[2]E. A. Braude, A. G. Brook, and R. P. Linstead, *J. Chem. Soc.*, 3569 (1954); R. P. Linstead, E. A. Braude, L. M. Jackman, and A. N. Beames, *Chem. Ind.*, 1174 (1954); A. M. Creighton and L. M. Jackman, *J. Chem. Soc.*, 3138 (1960)
[3]P. W. D. Mitchell, *Can. J. Chem.*, **41**, 550 (1963)
[4]D. Walker and T. D. Waugh, *J. Org.*, **30**, 3240 (1965)
[5]D. H. Reid, M. Fraser, B. B. Molloy, H. A. S. Payne, and R. G. Sutherland, *Tetrahedron Letters*, 530 (1961)
[6]J. F. Bagli, P. F. Morand, K. Wiesner, and R. Gaudry, *Tetrahedron Letters*, 387 (1964)
[7]G. Muller, J. Martel, and C. Huynh, *Bull. soc.*, 2000 (1961)
[8]J. A. Zderic, H. Carpio, and D. C. Limon, *J. Org.*, **27**, 1125 (1962)
[9]D. Burn, D. N. Kirk, and V. Petrow, *Proc. Chem. Soc.*, 14 (1960)
[10]D. Burn, V. Petrow, and G. Weston, *J. Chem. Soc.*, 29 (1962)
[11]E. Caspi, P. K. Grover, N. Graven, E. J. Lynde, and Th. Nussbaumer, *ibid.*, 1710 (1962)
[12]H. J. Ringold and A. Turner, *Chem. Ind.*, 211 (1962)
[13]B. Berkoz, L. Cuéllar, R. Grezemkovsky, N. V. Avila, and A. D. Cross, *Proc. Chem. Soc.*, 215 (1964)
[14]S. K. Pradhan and H. J. Ringold, *J. Org.*, **29**, 601 (1964)
[14a]D. Burn, R. V. Coombs, B. Ellis, J. A. H. MacBride, V. Petrow, and Y. P. Yardley, *Chem. Ind.*, 497 (1966)
[15]E. A. Braude, R. P. Linstead, and K. R. Wooldridge, *J. Chem. Soc.*, 3070 (1956)
[16]D. Burn, V. Petrow, and G. O. Weston, *Tetrahedron Letters*, No. 9, 14 (1960)
[17]A. Bowers, P. G. Holton, E. Necoechea, and F. A. Kincl, *J. Chem. Soc.*, 4057 (1961)
[18]H.-D. Becker and E. Adler, *Chem. Scand.*, **15**, 218 (1961)
[19]A. Bowers and co-workers, *J. Org.*, **29**, 3481 (1964); P. Morand, S. Stavrić, and D. Godin, *Tetrahedron Letters*, 49 (1966)
[20]S. H. Burstein and H. J. Ringold, *Am. Soc.*, **86**, 4952 (1964)
[21]A. M. Creighton and L. M. Jackman, *J. Chem. Soc.*, 3138 (1960)
[22]H.-D. Becker, *J. Org.*, **30**, 982 (1965)
[23]*Idem, ibid.*, **30**, 989 (1965)

1,1-Dichlorodiethyl ether, $CH_3CCl_2OC_2H_5$. Mol. wt. 143.01, b.p. 104.5–105.5°. Supplier: KK.

Prepared by addition of two equivalents of dry hydrogen chloride to ethoxyacetylene, the reagent reacts (40°) with a carboxylic acid to form the corresponding

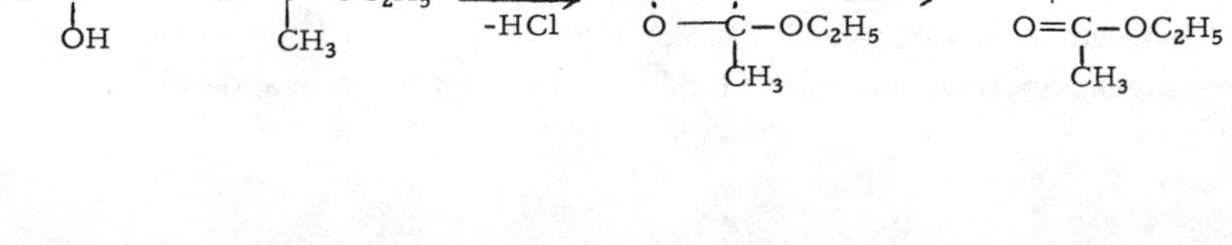

acid chloride in good yield.[1] Peptide synthesis is accomplished by refluxing the components formulated in ethyl acetate for 15–45 min.[2] The amino function is protected by either a carbobenzoxy or a phthaloyl group.

$$\mathrm{CbNH\overset{R}{\overset{|}{C}}HCO_2H + HCl\cdot H_2N\overset{R'}{\overset{|}{C}}HCO_2C_2H_5 + CH_3CCl_2OC_2H_5 \longrightarrow}$$

$$\mathrm{CbNH\overset{R}{\overset{|}{C}}H\overset{O}{\overset{\|}{C}}-\overset{H}{\overset{|}{N}}\,\overset{R'}{\overset{|}{C}}HCO_2C_2H_5 + 3\,HCl + CH_3CO_2C_2H_5}$$

[1]L. Heslinga, G. J. Katerberg, and J. F. Arens, *Rec. trav.*, **76**, 969 (1957)
[2]L. Heslinga and J. F. Arens, *ibid.*, **76**, 982 (1957)

1,1-Dichloro-2,2-difluoroethylene, $Cl_2C{=}CF_2$. Mol. wt. 132.93, b.p. 18.9°/758 mm.[1] Technical grade material, b.p. 19–21°, available from Peninsular ChemResearch on redistillation through a 4-ft. column packed with glass helices gave material homogeneous to V. P. C. and boiling at 18.9°.

Bartlett *et al.*[1] found that in sealed tubes at 80° with excess 1,1-dichloro-2,2-difluoroethylene and a polymerization inhibitor, butadiene, *cis*-piperylene, and *trans*-piperylene each yielded a single 1,2-addition product. Wherever the structures have been established they conform to type I and not II.

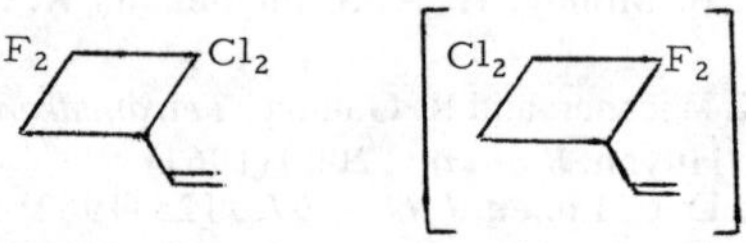
F2
Cl2
Cl2
F2

I II

[1]P. D. Bartlett, L. K. Montgomery, and B. Seidel, *Am. Soc.*, **86**, 616 (1964)

1,1-Dichlorodimethyl ether, CH_3OCHCl_2. Mol. wt. 114.96, b.p. 85–87°.

Prepared by treatment of methyl formate with phosphorus pentachloride,[1] the reagent reacts with a carboxylic acid or anhydride to give the corresponding acid chloride in high yield.[2] Sulfonic acids or their sodium salts afford sulfonyl chlorides;

$$\mathrm{RCO_2H + Cl_2CHOCH_3 \xrightarrow[-HCl]{} RCOCl + HCO_2CH_3}$$

$$\mathrm{(RCO)_2O + Cl_2CHOCH_3 \xrightarrow[-HCl]{} 2\,RCOCl + HCO_2CH_3}$$

carbonyl compounds are converted into *gem*-dichlorides.[2] Under Friedel-Crafts conditions, the reagent reacts with an aromatic hydrocarbon to give an aldehyde.[3]

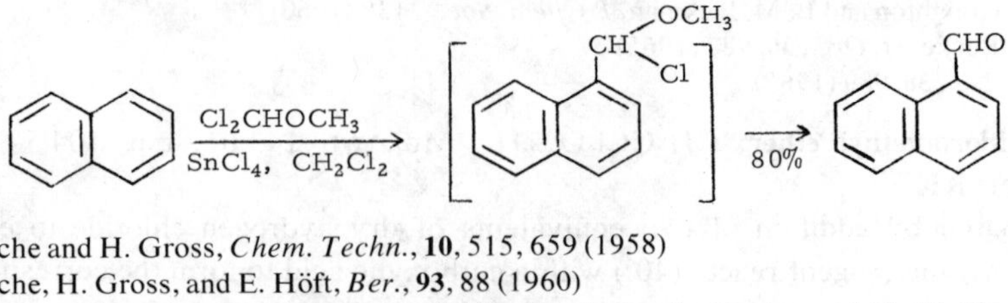
Cl2CHOCH3
SnCl4, CH2Cl2
OCH3
CH
Cl
80%
CHO

[1]A. Rieche and H. Gross, *Chem. Techn.*, **10**, 515, 659 (1958)
[2]A. Rieche, H. Gross, and E. Höft, *Ber.*, **93**, 88 (1960)
[3]A. Rieche and H. Gross, *Ber.*, **92**, 83 (1959); H. Gross, A. Rieche, and G. Matthey, *ibid.*, **96**, 308 (1963)

1,1-Dichloroethylene, $CH_2{=}CCl_2$. Mol. wt. 96.95, b.p. 37°. The technical production of the reagent[1] involves addition of chlorine to vinyl chloride and dehydrochlorination with aqueous alkali (Chemische Werke Hüls AG) or calcium hydroxide (BASF,

$$CH_2{=}CHCl \xrightarrow{Cl_2} ClCH_2CHCl_2 \xrightarrow{Base} CH_2{=}CCl_2$$

Ludwigshafen). The reagent is reasonably stable in storage only if protected with 0.3% of triethylamine or *t*-butylcatechol to inhibit polymerization. The precursor substance, 1,1,2-trichloroethane (mol. wt. 133.41), is available from Baker in purified grade, b.p. 113–115°.

The reagent reacts with an alcohol capable of forming a carbonium ion to give a carboxylic acid:

$$R_3COH \xrightarrow{-OH^-} R_3C^+ \xrightarrow{CH_2{=}CCl_2} R_3CCH_2\overset{+}{C}Cl_2 \xrightarrow{H_2O} R_3CCH_2CO_2H$$

The reaction is conducted in 80–100% sulfuric acid and is greatly accelerated by BF_3.[2] Norbornene (1) is converted in good yield into *exo*-norbornylacetic acid (2), and 1-adamantanol (3) affords 1-adamantylacetic acid.

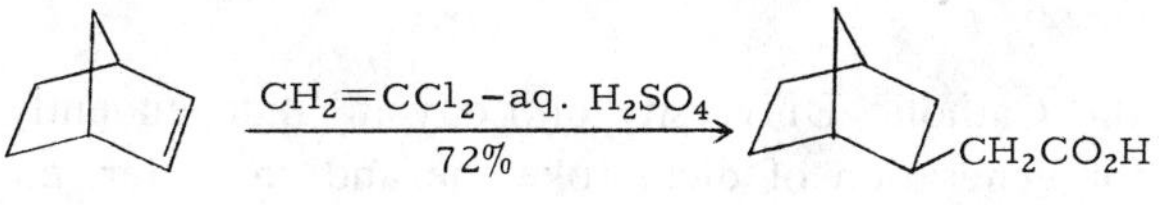

(1) (2)

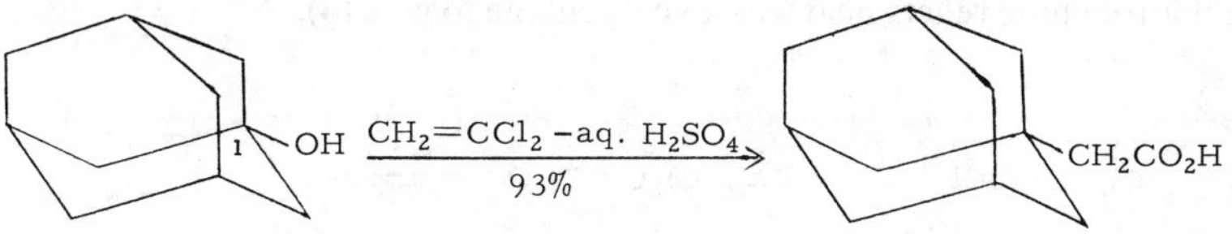

(3) (4)

[1]K. Bott, private communication

[2]K. Bott, *Angew. Chem., Internat. Ed.*, **4**, 956 (1965)

Dichloroformoxime, $Cl_2C{=}NOH$. Mol. wt. 113.94. The reagent is a skin irritant and the vapor is toxic. It is prepared by reduction of chloropicrin with tin and

$$CCl_3NO_2 \xrightarrow{Sn-HCl} Cl_2C{=}NOH$$

concd. hydrochloric acid.[1] It decomposes slowly at room temperature but is stable for months in a refrigerator.

Synthesis of heterocycles. Quilico and co-workers[2] used the reagent for the synthesis of 3-chloroisoxazoles from the Grignard derivatives of terminal acetylenes.

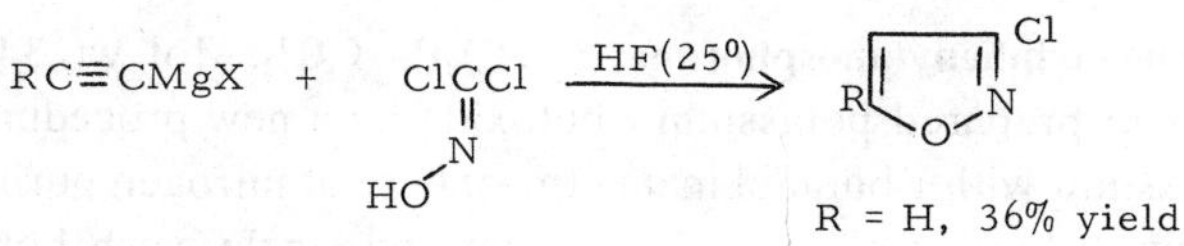

R = H, 36% yield

[1]C. E. Miller, procedure submitted to *Org. Syn.*

[2]P. Bravo, G. Gaudiano, A. Quilico, and A. Ricca, *Gazz.*, **91**, 47 (1961) [*C.A.*, **56**, 12869 (1962)]

Dichloroketene, $Cl_2C{=}C{=}O$ (not isolated).

A Pittsburgh Plate Glass Co. group[1] generated this reagent *in situ* and found that it reacts readily with cyclopentadiene to form a Diels-Alder adduct convertible into tropolone (5). A solution of dichloroacetyl chloride and a ten-fold excess of cyclopentadiene in hexane was treated at 0–5° with a hexane solution of triethylamine. Removal of the solid amine hydrochloride and distillation afforded the adduct (3) in 70–75% yield. This adduct is hydrolyzed by aqueous potassium acetate in acetic acid to tropolone (5), possibly through the mechanism formulated.

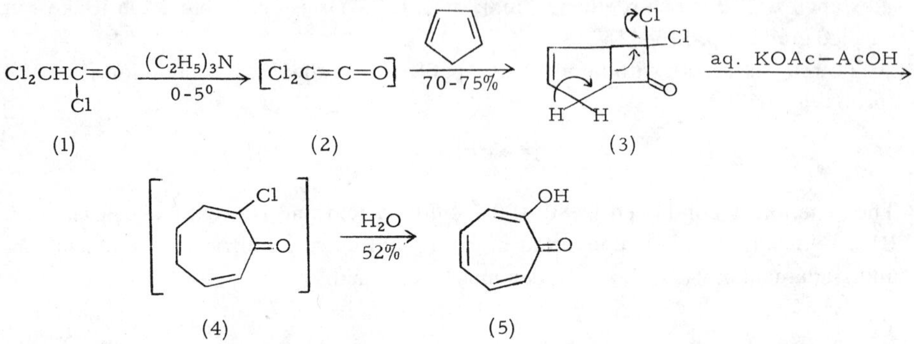

$Cl_2CHC=O$
Cl
$(C_2H_5)_3N$
0-5°
$[Cl_2C=C=O]$
70-75%
Cl
Cl
H
H
aq. KOAc—AcOH
(1)
(2)
(3)
Cl
H_2O
52%
OH
(4)
(5)

A group at the Catholic University of Louvain[2] independently developed the same process for generation of dichloroketene and characterized the product of its addition to cyclopentadiene as the bicyclic dichloroketone (3).[2] They found that dichloroketene reacts also with cyclopentene to give (4).

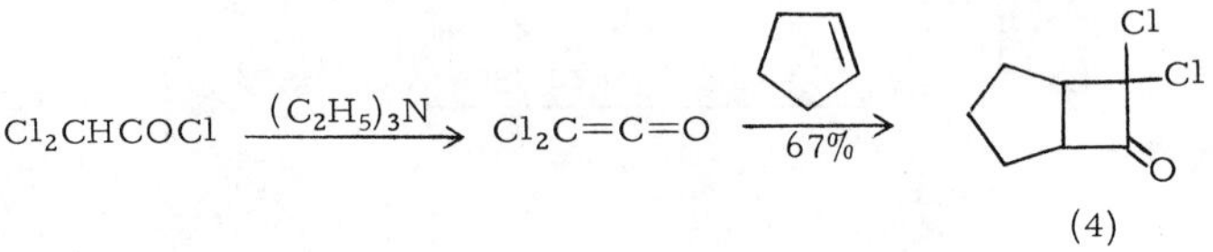

$Cl_2CHCOCl$
$(C_2H_5)_3N$
$Cl_2C=C=O$
67%
Cl
Cl
(4)

Turner and Seden[3] report a low yield synthesis of 4,5-benzotropolone from indene and dichloroketene.

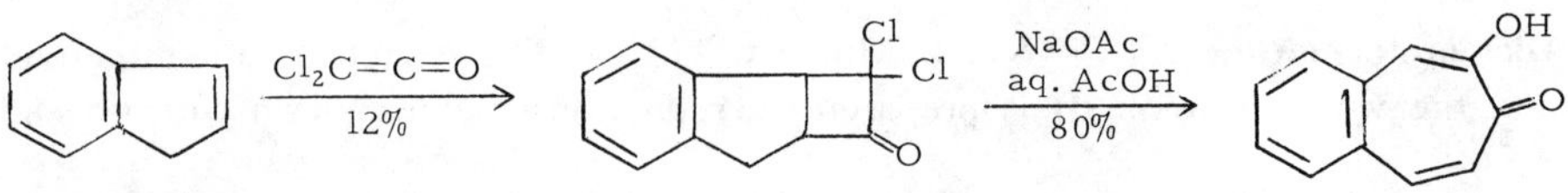

$Cl_2C=C=O$
12%
Cl
Cl
NaOAc
aq. AcOH
80%
OH

[1]H. C. Stevens, D. A. Reich, D. R. Brandt, K. R. Fountain, and E. J. Gaughan, *Am. Soc.*, **87**, 5257 (1965)
[2]L. Ghosez, R. Montaigne, and P. Mollet, *Tetrahedron Letters*, 135 (1966)
[3]R. W. Turner and T. Seden, *Chem. Comm.*, 399 (1966)

Dichloromethylenedioxylenebenzene, *see* Catechol dichloromethylene ether.

Dichloromethylenetriphenylphosphorane, $(C_6H_5)_3P{=}CCl_2$. Mol. wt. 345.20.

Speziale *et al.*[1] prepared potassium *t*-butoxide by a new procedure involving reaction of potassium with *t*-butanol in an atmosphere of nitrogen purified by passage through two wash bottles of Fieser's solution and single wash bottles containing concd. sulfuric acid and anhydrous calcium chloride. Excess *t*-butanol was removed by distillation, pentane was added and the distillation continued, and the potassium *t*-butoxide used in the form of a slurry in pentane. This was cooled to 0–5°, triphenylphosphine was added, and the mixture was stirred well during the dropwise addition of a solution of chloroform in heptane. The reaction of potassium *t*-butoxide with chloroform generates dichlorocarbene, which combines with triphenylphosphine

$$\underset{0.5\ m.}{(C_6H_5)_3P} + \underset{0.5\ m.}{(CH_3)_3COK} + \underset{0.5\ m.}{CHCl_3} + \underset{2\ l.}{\underline{n}\text{-}C_5H_{12}} \xrightarrow[-(CH_3)_2COH,\ -KCl]{0-5^0}$$

$$\underset{(1)}{(C_6H_5)_3P{=}CCl_2} \xrightarrow[68-79\%\ \text{crude}]{\underline{p}\text{-}(CH_3)_2NC_6H_4CHO} \underset{(2)}{\underline{p}\text{-}(CH_3)_2NC_6H_4CH{=}CCl_2}$$

to form a heptane suspension of the phosphorane (1). After addition of *p*-dimethylaminobenzaldehyde in six portions at 0–10° and further stirring, removal of the precipitated triphenylphosphine oxide, evaporation of the filtrate (rotary evaporation), and crystallization of the solid brown residue from methanol gives crude *β*,*β*-dichloro-*p*-dimethylaminostyrene (2). The main impurity is unchanged triphenylphosphine, which can be removed by adding mercuric chloride to a solution of crude (2) in absolute ethanol. An insoluble double salt of mercuric chloride and triphenylphosphine is removed by filtration, and the filtrate concentrated for crystallization.

Dichloromethylenetriphenylphosphorane has been prepared also by the direct reaction of triphenylphosphine with carbon tetrachloride.[2,3] This affords a simple method for preparation of ylides of type (4) and a one-step synthesis of 1,1-dihaloolefins. For example, after a solution of triphenylphosphine and benzaldehyde in carbon tetrachloride has been heated at 60° for 2 hrs., addition of petroleum ether precipitates triphenylphosphine oxide and distillation of the filtrate affords *β*,*β*-

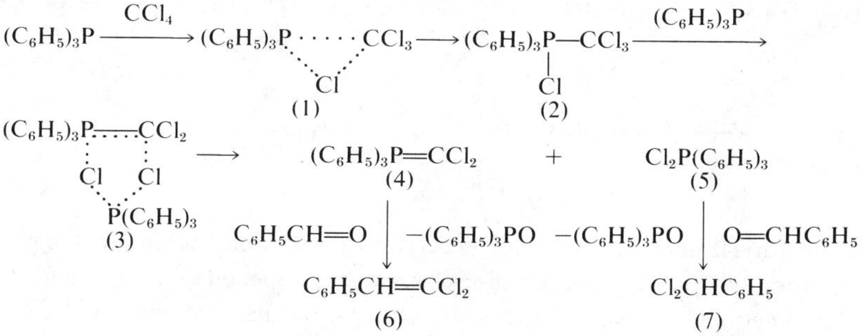

CCl_4
$(C_6H_5)_3P$
$(C_6H_5)_3P \cdots CCl_3$
$(C_6H_5)_3P{-}CCl_3$
Cl
(1)
(2)
$(C_6H_5)_3P{\cdots}CCl_2$
Cl Cl
$P(C_6H_5)_3$
(3)
$(C_6H_5)_3P{=}CCl_2$
(4)
$Cl_2P(C_6H_5)_3$
(5)
$C_6H_5CH{=}O$
$-(C_6H_5)_3PO$
$-(C_6H_5)_3PO$
$O{=}CHC_6H_5$
$C_6H_5CH{=}CCl_2$
(6)
$Cl_2CHC_6H_5$
(7)

dichlorostyrene (6) and benzal chloride (7), both in 72% yield. The reaction does not involve free radicals or a carbene intermediate and is formulated as proceeding through transition states (1) and (3) and the ylide (4). The reaction of carbon tetrabromide with triphenylphosphine and benzaldehyde is complete in a matter of minutes and *β*,*β*-dibromostyrene is obtained in 84% yield.

[1]A. J. Speziale and K. W. Ratts, *Am. Soc.*, **84**, 855 (1962); A. J. Speziale, K. W. Ratts, and D. E. Bissing, *Org. Syn.*, **45**, 33 (1965)

[2]R. Rabinowitz and R. Marcus, *Am. Soc.*, **84**, 1312 (1962)

[3]F. Ramirez, N. B. Desai, and N. McKelvie, *Am. Soc.*, **84**, 1745 (1962)

Dichloromethyllithium, Trichloromethyllithium, $LiCHCl_2$, $LiCCl_3$.

The reagents are prepared by reaction of *n*-butyllithium with methylene chloride or chloroform in tetrahydrofurane–ether–petroleum ether (4:1:1) at −110°.[1] The first reagent is more stable than the second. They react with ketones to give carbinols in good yield.

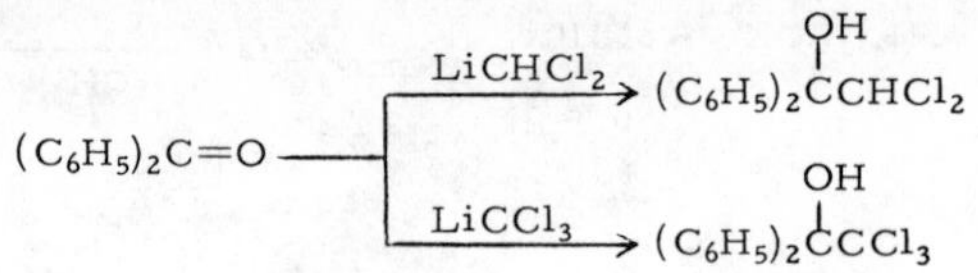

$$(C_6H_5)_2C=O \begin{cases} \xrightarrow{LiCHCl_2} (C_6H_5)_2\overset{\underset{\displaystyle |}{OH}}{C}CHCl_2 \\[2em] \xrightarrow{LiCCl_3} (C_6H_5)_2\overset{\underset{\displaystyle |}{OH}}{C}CCl_3 \end{cases}$$

Deyrup and Greenwald[2] added N-benzylideneaniline in ether to excess dichloro-methyllithium at temperatures below $-70°$ and obtained an orange solution which was allowed to warm to room temperature. Workup afforded crude material in high yield; this first-known monochloroaziridine proved to be too reactive to allow purification for analysis. The reaction proceeds stereoselectively to give the *cis*-isomer.

$$C_6H_5CH=NC_6H_5 + LiCHCl_2 \xrightarrow{-70°} \underset{\overset{\displaystyle |}{Cl}}{\underset{H_5C_6}{\overset{H}{\diagdown}}}NC_6H_5 + LiCl$$

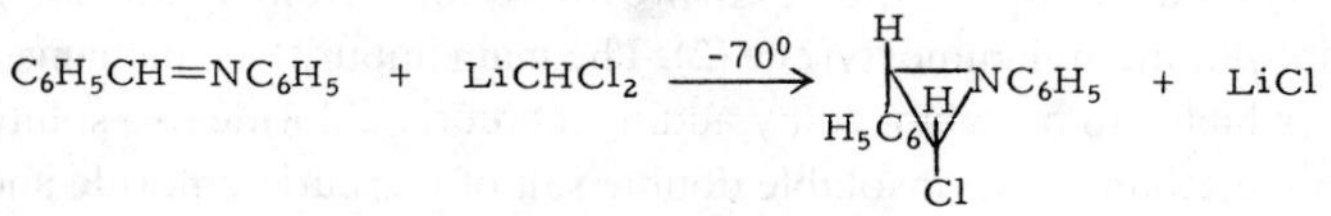

Miller and Whalen[3] generated trichloromethyllithium by adding $CBrCl_3$ to a slurry of CH_3Li in ether at $-115°$ in the presence of 2 equivalents of cyclohexene; after warming to $-100°$ for 1 hr., 7,7-dichloronorcarane was obtained in 77% yield. Breslow and Altman[4] noted that the functioning of the reagent as a dichlorocarbene donor at $-100°$ should favor reaction with gaseous acetylenes, for at $-100°$ these can be obtained as concentrated solutions of the liquid or solid plase. Indeed

$$CH_3CH_2CH_2C\equiv CCH_2CH_2CH_3 + CCl_3CLi \xrightarrow{-95°} \underset{\underset{\displaystyle Cl \quad Cl}{\diagup \diagdown}}{CH_3CH_2CH_2C=\underset{\displaystyle C}{CCH_2CH_2CH_3}}$$

$$\xrightarrow{H_3O^+} \underset{\underset{\displaystyle \overset{\|}{O}}{\diagdown \diagup}}{CH_3CH_2CH_2C=\underset{\displaystyle C}{CCH_2CH_2CH_3}}$$

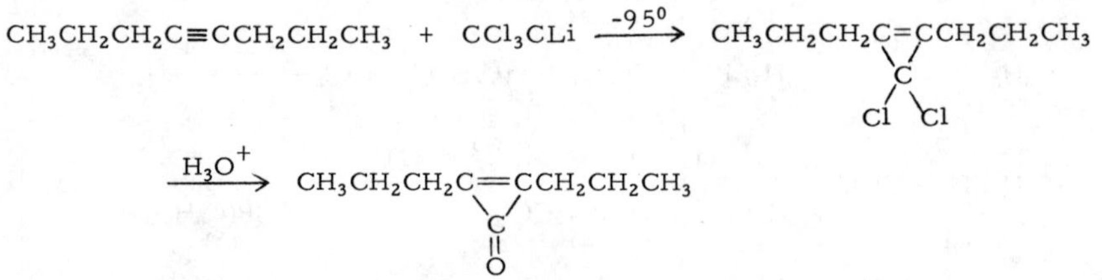

trichloromethyllithium reacted with 4-octyne at $-95°$, and after low-temperature acidification and aqueous workup di-*n*-propylcyclopropenone was obtained in yield of 19%. The reaction succeeded also with 1-pentyne, propyne, and 2-butyne, but not with acetylene itself.

[1] G. Köbrich, K. Flory, and W. Drischel, *Angew. Chem.*, **76**, 536, (1964); *ibid.*, *Internat. Ed.*, **3**, 513 (1964)

[2] J. A. Deyrup and R. B. Greenwald, *Tetrahedron Letters*, 321 (1965); *Am. Soc.*, **87**, 4538 (1965)

[3] W. T. Miller, Jr., and D. M. Whalen, *Am. Soc.*, **86**, 2089 (1964)

[4] R. Breslow and L. J. Altman, *Am. Soc.*, **88**, 504 (1966)

Dicobalt octacarbonyl, $Co_2(CO)_8$. Mol. wt. 342.12, orange red crystals, m.p. $51–52°$. Supplier: Alfa.

Preparation. The reagent is most readily prepared from a Co (II) salt, preferably the carbonate,[1] but also the acetate. It has been prepared in ether[2] or in benzene[3] solution by the reaction of Raney cobalt with carbon monoxide at $150°$ under a pressure of 3,200 p.s.i. It has been prepared also from cobalt (II) carbonate and synthesis gas (1 H_2–1 CO),[4] and by the decomposition of $HCo(CO)_4$:[5]

$$2\ HCo(CO)_4 \longrightarrow Co_2(CO)_8 + H_2$$

Structure. A highly polarized structure such as that formulated does not seem

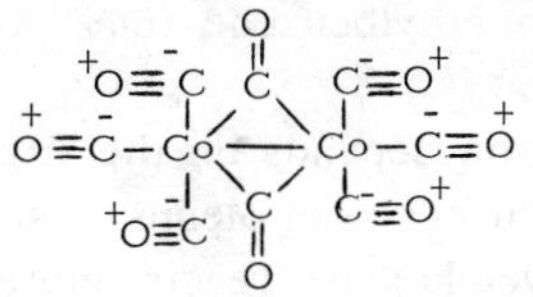

consistent with the low melting point of the complex and its solubility in hydrocarbon solvents. The true nature of the compound is still in doubt.

Oxo reaction. A U.S. patent filed in 1938 by the German Otto Roelen and issued in 1943 disclosed the addition to an olefin of carbon monoxide and hydrogen under catalysis by cobalt, thorium oxide, or kieselguhr. The reaction was at first described as an "oxo synthesis," but after Adkins and Krsek[2,3] had examined the

$$CH_3CH_2CH{=}CH_2 + CO + H_2 \xrightarrow[125\text{-}150^\circ]{125\text{-}200\ \text{atm.}} \underset{70\text{-}80\%}{CH_3CH_2CH_2CH_2CHO} + \underset{20\text{-}30\%}{CH_3CH_2CH(CH_3)CH_2O}$$

reaction more critically and found that use of a solid cobalt catalyst such as dicobalt pentacarbonyl greatly improves the results, the reaction became known as a hydroformylation. A mixture of the two possible aldehydes usually results but the straight-chain aldehyde generally predominates.

Orchin *et al.*[6] have shown experimentally that, under the conditions of the oxo reaction, $Co_2(CO)_8$ is rapidly converted into $HCo(CO)_4$, cobalttetracarbonyl hydride, which is the true catalyst, and that the reaction proceeds by addition of this metal hydride to the olefin. Other examples reported by Adkins and Krsek:[3]

$$CH_3COOCH_2CH{=}CH_2 + CO + H_2 \xrightarrow[69\%]{Co_2(CO)_8} CH_3COOCH_2CH_2CH_2CHO$$

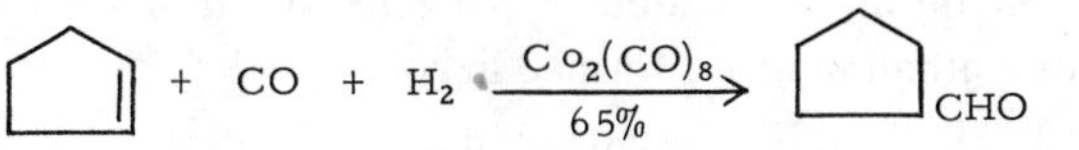

$$\begin{matrix} HCCO_2C_2H_5 \\ \Vert \\ C_2H_5O_2CCH \end{matrix} + CO + H_2 \xrightarrow[51\%]{Co_2(CO)_8} \begin{matrix} CHO \\ | \\ CHCO_2C_2H_5 \\ | \\ CH_2CO_2C_2H_5 \end{matrix}$$

A convenient route to normal paraffinic hydrocarbons involves reversal of the oxo process:

$$\underline{n}\text{-}C_{15}H_{31}CH_2CH_2CH_2OH \xrightarrow{\substack{Ni \\ 240\ \text{atm.}}} [C_{15}H_{31}CH_2CH_2CHO \xrightarrow[-H_2]{-CO}$$
$$\underline{n}\text{-}C_{15}H_{31}CH{=}CH_2] \xrightarrow[95\%]{H_2} \underline{n}\text{-}C_{17}H_{36}$$

Adkins and Krsek[2] noted that certain α,β-unsaturated carbonyl compounds undergo reduction rather than hydroformylation. Thus crotonaldehyde and acrolein are reduced to *n*-butyraldehyde and propionaldehyde, respectively, and methyl vinyl ketone is reduced to methyl ethyl ketone. Orchin and co-workers[4] found that if the reaction is carried out at a higher temperature (180–185°) the carbonyl group is reduced as well. Indeed application of the oxo reaction to an alcohol leads to homologation:

$$(CH_3)_2C(OH)CH_3 + 2\,H_2 + CO \xrightarrow[63\%]{Co_2(CO)_8} (CH_3)_2CHCH_2CH_2OH + H_2O$$

In the aromatic series hydrocarbons are obtained, usually in high yield. Thus acetophenone is reduced to ethylbenzene (64% yield); fluorenone to fluorene (95% yield).

The rate of hydroformylation depends on the structure of the olefin, the order being as follows: straight-chain terminal olefins > straight-chain internal olefins > branched-chain olefins.[7] Cyclohexene reacts more slowly than cyclopentene, cycloheptene, or cyclooctene.

Related reactions. If the oxo reaction is carried out using an alcohol instead of hydrogen, an ester is usually the principal product, although other products are

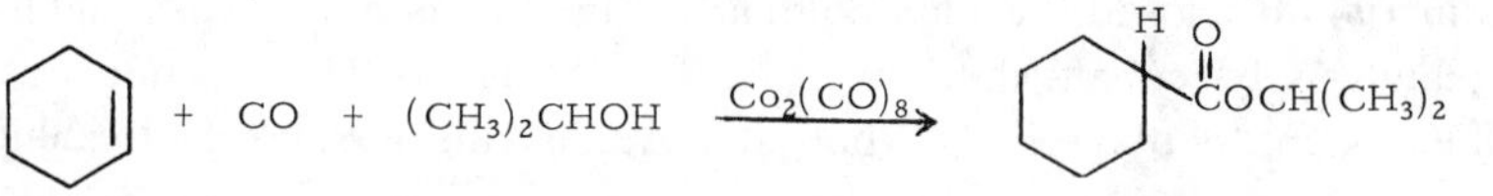

often formed.[8] Application of the reaction to an amine leads to one or two substituted amides, usually in good yield.

$$RCH{=}CH + R'NH_2 + CO \xrightarrow{Co_2(CO)_8} RCH_2CH_2CONHR' + R\underset{\displaystyle CH_3}{\underset{|}{C}H}CONHR'$$

Addition of hydrogen cyanide to an olefin can be accomplished under catalysis with dicobalt octacarbonyl at 130° at a pressure of 25–200 atmospheres.[9] Best results are obtained with terminal olefins, conjugated dienes, which give mainly 1,4-addition products, and with Diels-Alder adducts of cyclopentadiene, which give mixtures and/or position isomers.

$$CH_3CH{=}CH_2 + HCN \xrightarrow[65\%]{Co_2(CO)_8} CH_3\underset{\displaystyle CN}{\underset{|}{C}H}CH_3$$

When benzaldehyde anil is heated in benzene solution with the catalyst under pressure, 2-phenylphthalimidine is formed in good yield.[10]

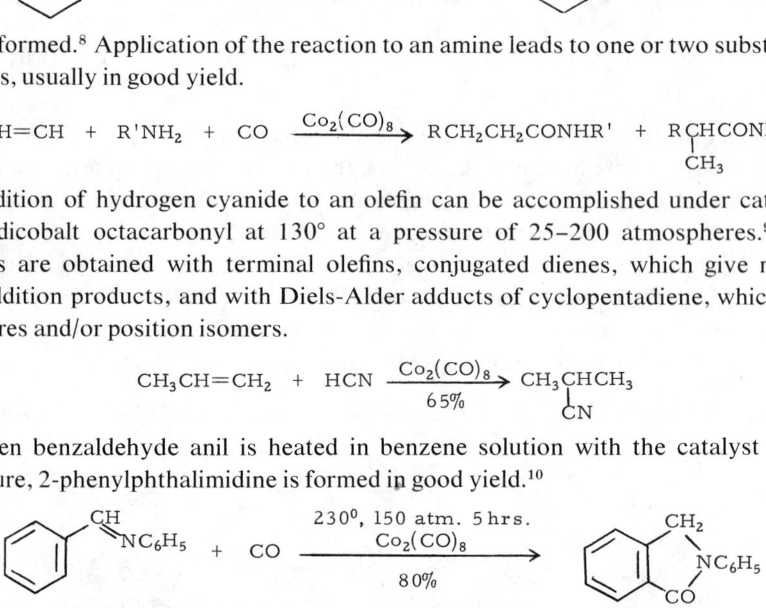

Eisenmann[11] reported that in the presence of the catalyst and under conditions of the oxo reaction propylene oxide reacts with carbon monoxide and methanol to give methyl β-hydroxybutyrate in 40% yield. Later[12] he found that the major

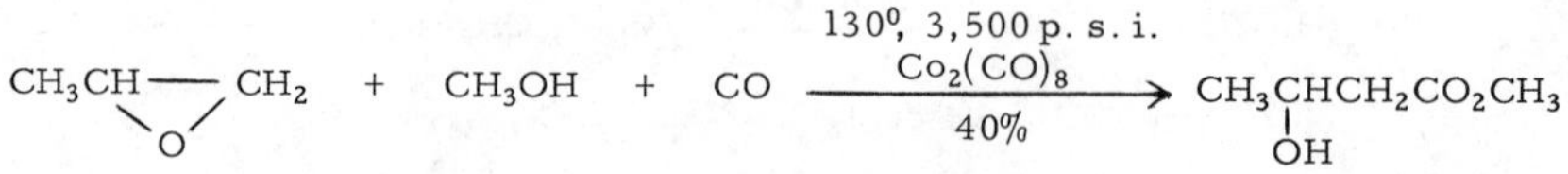

product is acetone and that, in the absence of carbon monoxide, epoxides are isomerized to ketones in good yield in alcohol or benzene as solvent.

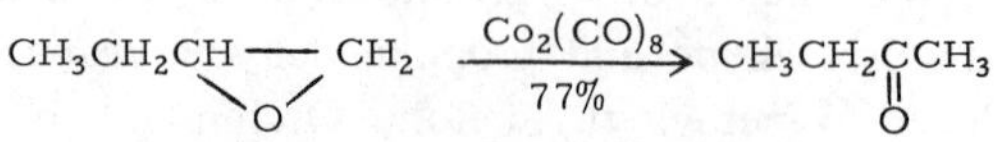

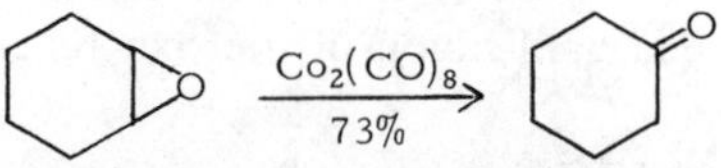

Reaction of acetylene with carbon monoxide. du Pont workers[13] found that acetylene reacts with carbon monoxide in the presence of dicobalt octacarbonyl

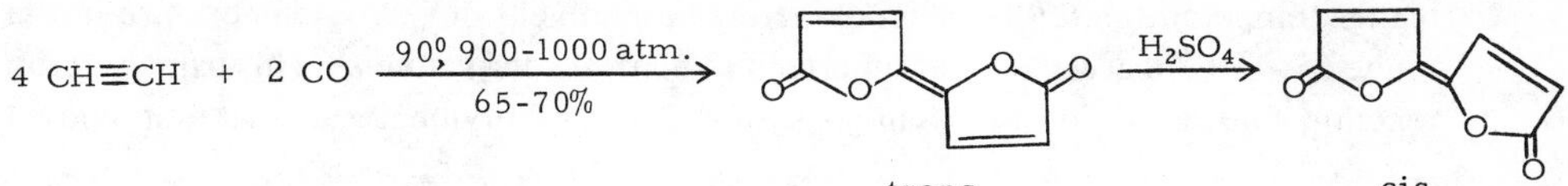
4 CH≡CH + 2 CO
90°, 900-1000 atm.
65-70%
H_2SO_4

trans cis

in an aprotic solvent (nitromethane, acetonitrile, acetone) at 90° and high pressure to give *trans*-bifuranedione. This substance is isomerized quantitatively to the *cis*-isomer in hot concentrated sulfuric acid. The *trans*-isomer is formed in most solvents, but *cis*-bifuranedione is the sole product in tetramethylurea.

Reaction with acetylenes. The formula for dicobalt octacarbonyl written above shows that six of the carbonyl groups are polarized groups singly linked to a cobalt atom but that the other two are bridge groups, each bonded to both atoms of cobalt. Greenfield *et al.*[14] found that the reagent reacts with diphenylacetylene in pentane with displacement of the two bridge carbonyl groups and formation of a highly colored product according to the equation:

$$C_6H_5C{\equiv}CC_6H_5 \;+\; Co_2(CO)_8 \xrightarrow[61\%]{25^0} C_6H_5C_2C_6H_5Co_2(CO)_6 \;+\; 2\ CO$$

The yield cited is for material twice crystallized from methanol. The substance forms dark purple crystals, m.p. 110°, and is characterized by high volatility and high solubility in organic solvents. It has been shown by X-ray analysis[15] to have the structure shown in the formula and in the photograph of a model (Fig. D-1). At

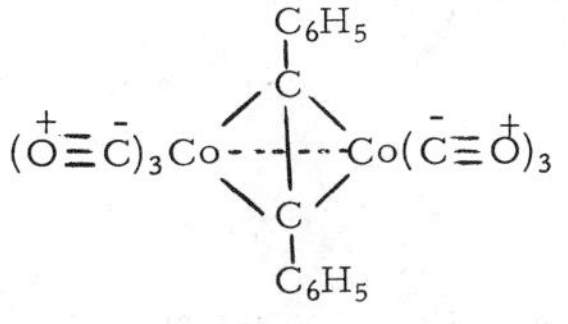
C_6H_5
C
$(\overset{+}{O}{\equiv}\overset{-}{C})_3Co$
$Co(\overset{-}{C}{\equiv}\overset{+}{O})_3$
C
C_6H_5

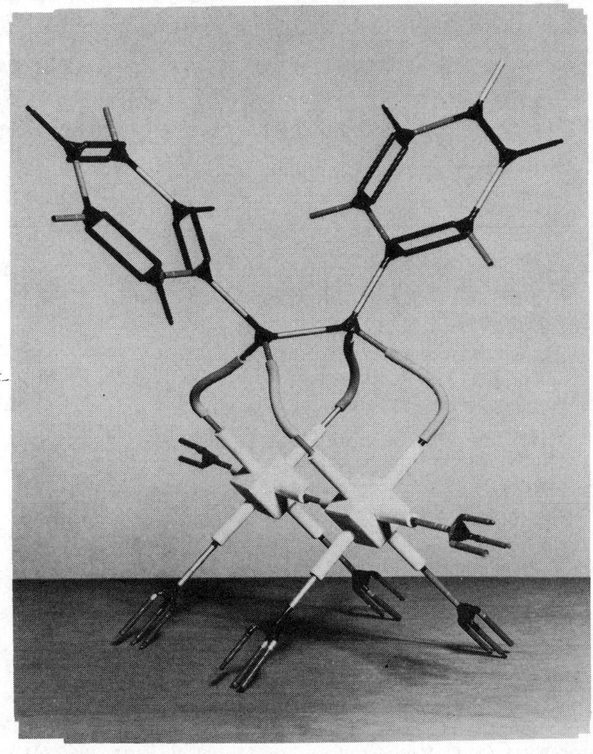

Fig. D-1

higher temperatures (200–280°), or better in a solvent of reflux temperature in the range 65–110°, acetylenes are trimerized to benzene derivatives.[16] This trimerization reaction was used for the synthesis of 1,2,4-tri-*t*-butylbenzene, the first known

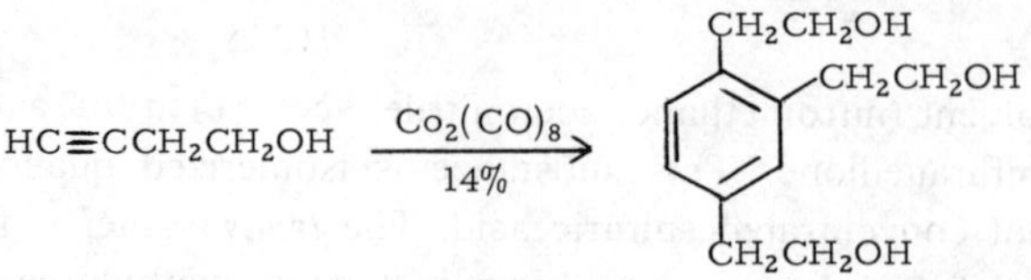
HC≡CCH2CH2OH
Co2(CO)8
14%
CH2CH2OH
CH2CH2OH
CH2CH2OH

benzene derivative having two *ortho* *t*-butyl groups.[17] When *t*-butylacetylene was treated at room temperature with dicobalt octacarbonyl, an organocobaltcarbonyl complex separated in 30% yield. When this was treated with bromine in carbon tetrachloride at 0°, $CoBr_2$ separated and 1,2,4-tri-*t*-butylbenzene was formed in

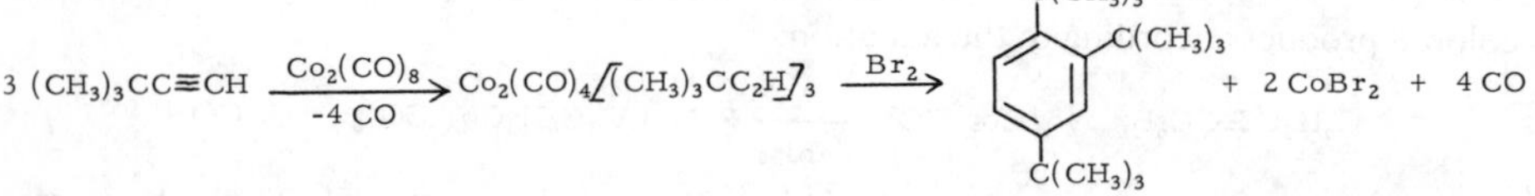
3 (CH3)3CC≡CH
Co2(CO)8
-4 CO
Co2(CO)4[(CH3)3CC2H]3
Br2
C(CH3)3
C(CH3)3
C(CH3)3
+ 2 CoBr2 + 4 CO

70% yield. The complex can also be decomposed to the aromatic hydrocarbon by heating at 150–170° (4 hrs., 56%).

Reduction of aromatics. Certain polycyclic hydrocarbons are partially reduced under conditions of the oxo reaction (H_2, CO, 150–200°, 3,000–3,600 p.s.i.). Whereas benzene, naphthalene, and phenanthrene are resistant, anthracene is reduced to 9,10-dihydroanthracene.[18] Other examples:

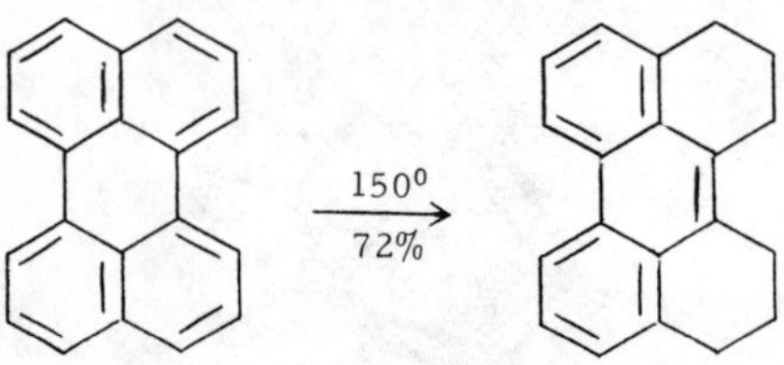
150°
72%

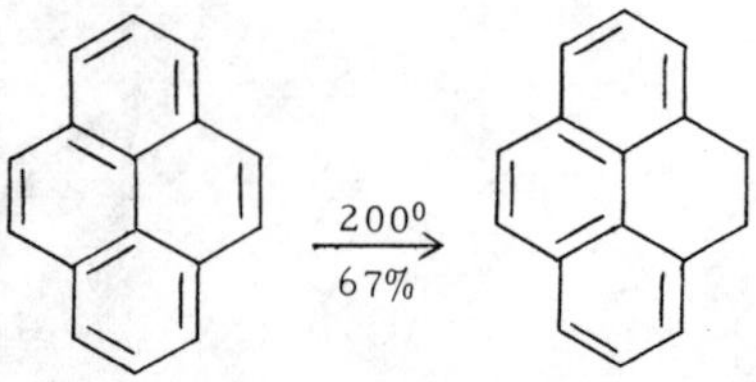
200°
67%

[1]I. Wender, H. W. Sternberg, S. Metlin, and M. Orchin, *Inorg. Syn.*, **5**, 190 (1957)
[2]H. Adkins and G. Krsek, *Am. Soc.*, **70**, 383 (1948)
[3]*Idem, ibid.*, **71**, 3051 (1949)
[4]I. Wender, R. Levine, and M. Orchin, *Am. Soc.*, **72**, 4375 (1950)
[5]P. Gilmont and A. A. Blanchard, *Inorg. Syn.*, **2**, 238 (1946)
[6]I. Wender, R. Levine, and M. Orchin, *Am. Soc.*, **71**, 4160 (1949)
[7]I. Wender, S. Metlin, S. Ergun, H. W. Sternberg, and H. Greenfield, *Am. Soc.*, **78**, 5401 (1956)
[8]G. Natta, P. Pino, and R. Ercoli, *Am. Soc.*, **74**, 4496 (1952)
[9]P. Arthur, Jr., D. C. England, B. C. Pratt, and G. M. Whitman, *Am. Soc.*, **76**, 5364 (1954)
[10]S. Murahashi, *Am. Soc.*, **77**, 6403 (1955)
[11]J. L. Eisenmann, R. L. Yamartino, and J. F. Howard, Jr., *J. Org.*, **26**, 2102 (1961)
[12]J. L. Eisenmann, *J. Org.*, **27**, 2706 (1962)
[13]J. C. Sauer, R. D. Cramer, V. A. Engelhardt, T. A. Ford, H. E. Holmquist, and B. W. Howk, *Am. Soc.*, **81**, 3677 (1959)
[14]H. Greenfield, H. W. Sternberg, R. A. Friedel, J. H. Wotiz, R. Markby, and I. Wender, *Am. Soc.*, **78**, 120 (1956)
[15]W. G. Sly, *Am. Soc.*, **81**, 18 (1959)
[16]W. Hübel and C. Hoogzand, *Ber.*, **93**, 103 (1960)
[17]U. Krüerke, C. Hoogzand, and W. Hübel, *Ber.*, **94**, 2817 (1961)
[18]S. Friedman, S. Metlin, A. Svedi, and I. Wender, *J. Org.*, **24**, 1287 (1959)

"Dicyandiamide" = N-Cyanoguanidine (Ia ⇌ Ib). The reagent is prepared commercially by the dimerization of cyanamide (1) in the presence of a base. Mol. wt. 84.08, m.p. 209–211°. Suppliers: Amer. Cyanamid Co., A, E, F, MCB. The trivial name

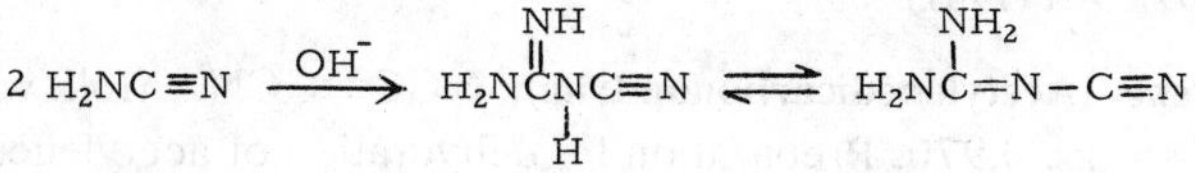

(1) (Ia) (Ib)

"dicyandiamide" is incorrect; the reagent is described most accurately as a mixture of the tautomeric forms of N-cyanoguanidine, Ia ⇌ Ib.

Synthesis of heterocycles. Heated under pressure, dicyandiamide (3 moles) is converted into melamine (2,4,6-triamino-*s*-triazine, 2 moles) discovered by Liebig in 1834:

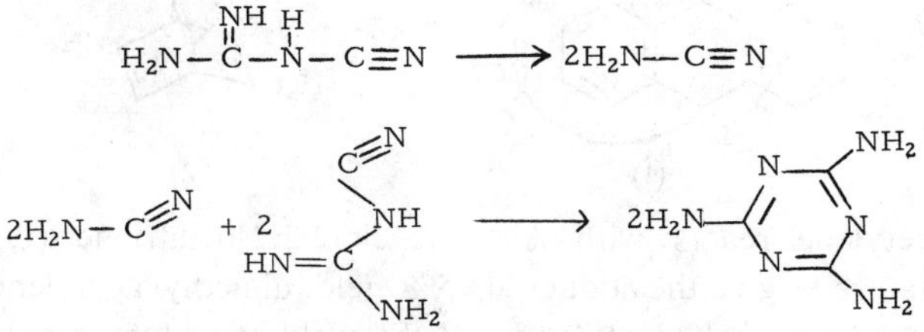

2,4-Diamino-6-phenyl-*s*-triazine (benzoguanamine) is prepared[1] by adding benzonitrile and dicyandiamide to a stirred mixture of potassium hydroxide and methyl

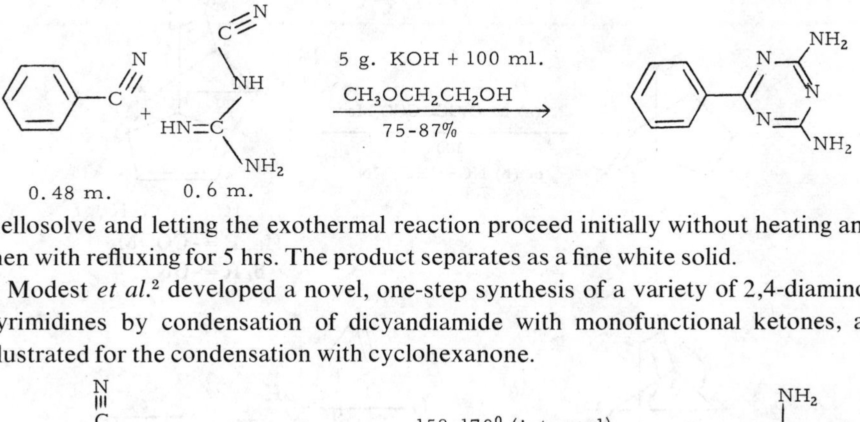

Cellosolve and letting the exothermal reaction proceed initially without heating and then with refluxing for 5 hrs. The product separates as a fine white solid.

Modest *et al.*[2] developed a novel, one-step synthesis of a variety of 2,4-diaminopyrimidines by condensation of dicyandiamide with monofunctional ketones, as illustrated for the condensation with cyclohexanone.

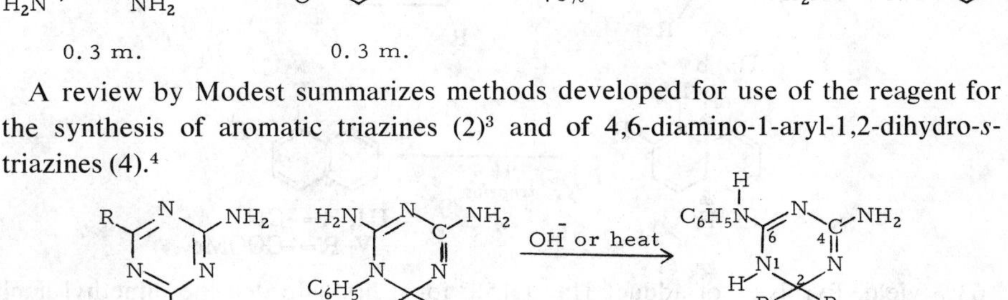

A review by Modest summarizes methods developed for use of the reagent for the synthesis of aromatic triazines (2)[3] and of 4,6-diamino-1-aryl-1,2-dihydro-*s*-triazines (4).[4]

(2) (3) (4)

[1]J. K. Simons and M. R. Saxton, *Org. Syn., Coll. Vol.*, **4**, 78 (1963)
[2]E. J. Modest, S. Chatterjee, and H. K. Protopapa, *J. Org.*, **30**, 1837 (1965)
[3]E. J. Modest in R. C. Elderfield, *Heterocyclic Compounds*, **7**, 650–656 (1961)
[4]*Idem, ibid.*, **7**, 697–701 (1961)

Dicyanoacetylene (Acetylenedicarbononitrile), NCC≡CCN. Mol. wt. 76.06, m.p. 21°, b.p. 76°, sp. gr. 0.970. Preparation by dehydration of acetylenedicarboxamide with P_2O_5.[1]

The acetylene is a highly reactive dienophile which reacts with simple dienes at room temperature to afford Diels-Alder adducts in high yield.[2,3] It reacts with anthracene to give (1) and with durene to give (2).[3]

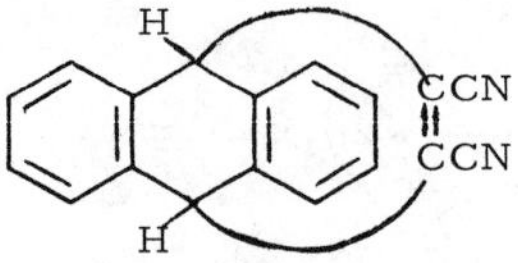
H
CCN
CCN
H

(1)

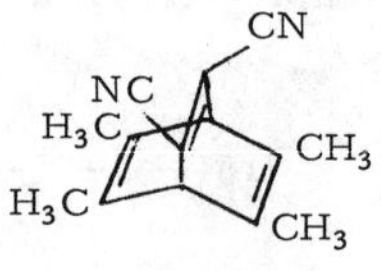
CN
NC
H_3C
CH_3
H_3C
CH_3

(2)

Dicyanoacetylene reacts with $\Delta^{2,5,7}$-bicyclo[2.2.2]octatriene (I, barrelene) at room temperature to give the adduct in 95% yield; dimethyl acetylenedicarboxylate reacts with the hydrocarbon at 100° and the yield of adduct is only 29%.[4] When adduct IIb was heated for 100° for 3.5 hrs. it afforded 1,2-dicyanonaphthalene in

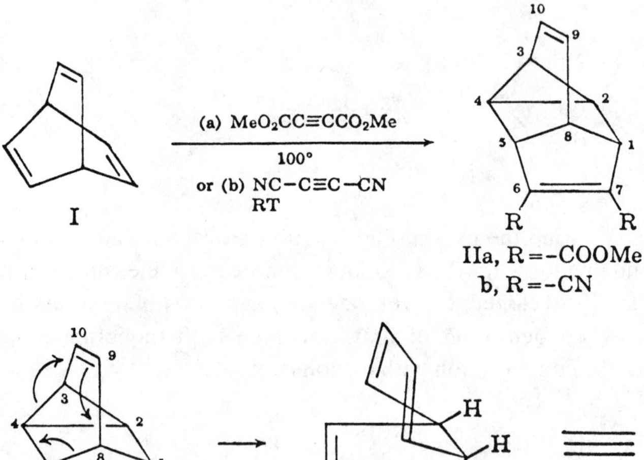
(a) $MeO_2CC{\equiv}CCO_2Me$
100°
or (b) NC−C≡C−CN
RT
10
9
3
4
2
8
5
1
6
7
R
R

I

IIa, R=-COOMe
b, R=-CN

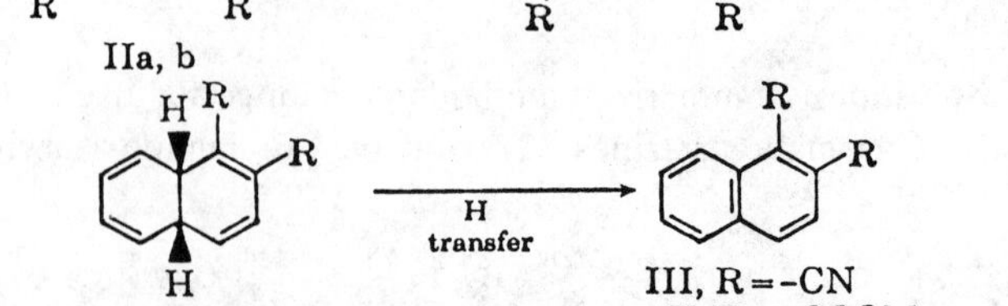
10
9
3
4
2
8
5
1
6
7
R
R
H
H
R
R
IIa, b
H
R
R
H
H
transfer
R
R

III, R=-CN
V, R=-COOMe

63% yield. Pyrolysis of adduct IIa at 150° for 4 hrs. afforded the dimethyl naphthalenedicarboxylate (V) in 33% yield. The reaction is considered to involve the concerted process indicated.

[1]A. T. Blomquist and E. C. Winslow, *J. Org.*, **10**, 149 (1945)
[2]R. C. Cookson and J. Dance, *Tetrahedron Letters*, 879 (1962)
[3]C. D. Weis, *J. Org.*, **28**, 74 (1963)
[4]H. E. Zimmerman and G. L. Grunewald, *Am. Soc.*, **86**, 1434 (1964)

Dicyanomethylene-2,4,7-trinitrofluorene,

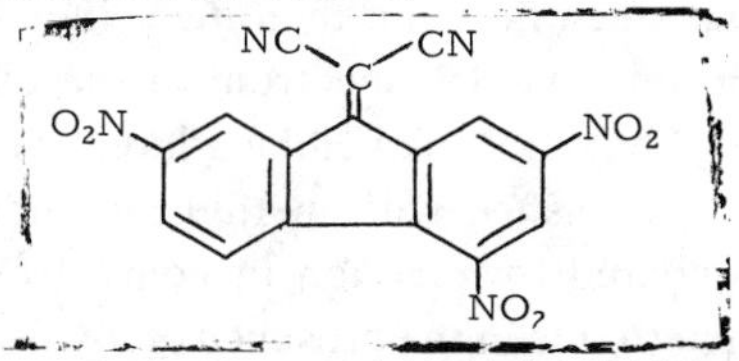
NC
CN
C
O₂N
NO₂
NO₂

Mol. wt. 351.23, m.p. 266–268°. The reagent is prepared in high yield by the condensation of 2,4,7-trinitrofluorenone with malononitrile with piperidine as catalyst.[1] It is intermediate in π-acid strength between tetracyanoethylene and 2,4,7-trinitrofluorenone.

[1]T. K. Mukherjee and L. A. Levasseur, *J. Org.*, **30**, 644 (1965)

Dicyclohexylamine,

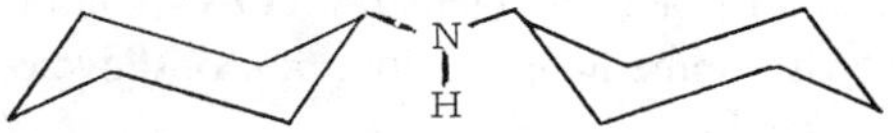
N
H

Mol. wt. 181.31, m.p. 20°, b.p. 256°. Suppliers: Abbott, Monsanto, A, B, E, MCB.

This amine forms salts of outstanding crystallizing properties with a large number of N-protected amino acids.[1] On addition of 1 equivalent of any type in alcohol or ethyl acetate, the salt usually crystallizes at once. After addition of ether, the salt is collected and washed with ether.

[1]E. Klieger, E. Schröder, and H. Gibian, *Ann.*, **640**, 157 (1961)

Dicyclohexylcarbodiimide (DCC), $C_6H_{11}N{=}C{=}NC_6H_{11}$. Mol. wt. 206.33, m.p. 34–35°, b.p. 155°/11 mm. (95–97°/0.2 mm.) Supplier: Upjohn Co. Carwin Org. Chemicals; smallest package, 8 oz. (225 g.).

Handling. The reagent is a potent skin irritant for some individuals. For removal from a storage bottle[1] warm the bottle in hot water or air until the solid has melted, pour liquid into a warm tared flask or a balance pan, and adjust the weight with a warm capillary dropping tube.[2]

Preparation. The reagent is prepared by mercuric oxide oxidation of N,N′-dicyclohexylthiourea, $C_6H_{11}NHC({=}S)NHC_6H_{11}$.[3] A du Pont group[4] developed a general method for the conversion of aliphatic and aromatic isocyanates but did not apply it to the dicyclohexyl derivative.

Dehydrative coupling of hydroxy compounds. Uses of the reagent for coupling two hydroxylic components by elimination of water under mild, neutral conditions have been reviewed.[5,6] A recent example is the preparation of a symmetrical diacyl peroxide by coupling two moles of an acid with one of hydrogen peroxide.[7] An amount of DCC equivalent to the acid (plus 10% excess) is weighed as a liquid into

$$2\ C_6H_5COOH + H_2O_2 + C_6H_{11}N{=}C{=}NC_6H_{11} \xrightarrow[90\%]{0^\circ}$$

$$C_6H_5\overset{O}{\overset{\|}{C}}{-}OO{-}\overset{O}{\overset{\|}{C}}C_6H_5 + C_6H_{11}\overset{H}{\overset{|}{N}}{-}\overset{O}{\overset{\|}{C}}{-}\overset{H}{\overset{|}{N}}C_6H_{11}\ (DCC\text{-}H_2O)$$

a reaction flask equipped with an efficient stirrer and thermometer and cooled in an ice bath, and ether is added and the solution cooled at 0°. Enough dry ethereal hydrogen peroxide to provide a five-fold excess is added. An acid sufficiently soluble in methylene chloride is added as a solution in this solvent at a rate such that the temperature does not rise above 5°. If the acid is not sufficiently soluble, methylene chloride is added first and then the acid is added as a finely divided solid. Stirring is continued until the IR spectrum in the carbonyl region shows the reaction to be complete. The N,N'-dicyclohexylurea which precipitates is removed by suction filtration and washed with methylene chloride. An unsymmetrical diacyl peroxide can be prepared by reaction of equivalent amounts of the two acids with excess hydrogen peroxide in the presence of DCC.

Aryl alkyl ethers are obtained in good yield by heating a phenol (1 equiv.), an alcohol (1.2 equiv.), and DCC (1.1 equiv.) in a sealed tube at 100–110° for 24 hrs.[8] Yields are lower if an inert solvent is used or if the reaction is allowed to proceed at room temperature.

DCC reacts rapidly and nearly quantitatively at 25° in ether solution with mono and di esters of phosphoric acid to yield the corresponding di or tetra esters of pyrophosphoric acid, with immediate precipitation of dicyclohexylurea.[9] It appears

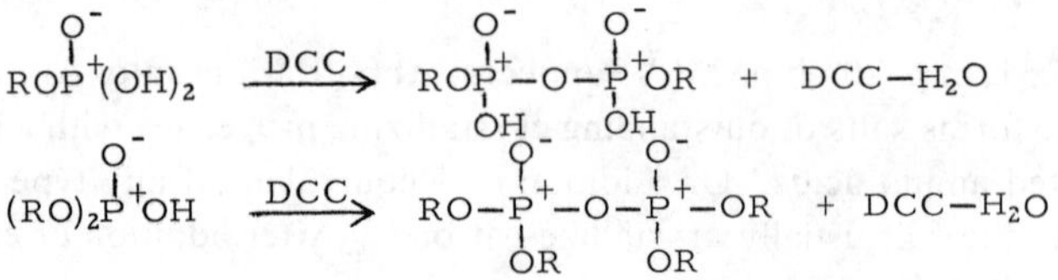

$$ROP^+(OH)_2 \xrightarrow{DCC} ROP^+-O-P^+OR + DCC-H_2O$$

$$(RO)_2P^+OH \xrightarrow{DCC} RO-P^+-O-P^+-OR + DCC-H_2O$$

to be the best reagent for the polymerization of mononucleotides.[10] The reactants are treated with pyridine and the two-phase system is shaken vigorously.

Khorana[11] used DCC for the preparation of symmetrical anhydrides of N-protected amino acids. Thus when N-carbobenzoxy-DL-phenylalanine was stirred at room temperature in ether with DCC the anhydride precipitated along with

$$2 \; C_6H_5CH_2\underset{NHCO_2CH_2C_6H_5}{CHCO_2H} \xrightarrow[90\%]{DCC} (C_6H_5CH_2\underset{NHCO_2CH_2C_6H_5}{CHCO})_2O$$

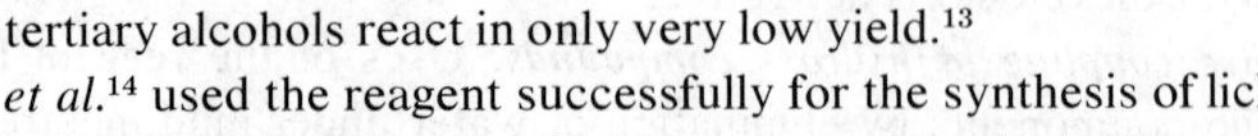

dicyclohexylurea. Extraction of the anhydride with cold ethyl acetate and crystallization afforded pure material, m.p. 125–126°. Schüssler and Zahn[12] used acetonitrile as solvent and obtained anhydrides in somewhat lower yield.

DCC promotes esterification of primary and secondary alcohols under very mild conditions; tertiary alcohols react in only very low yield.[13]

Seshadri et al.[14] used the reagent successfully for the synthesis of lichen depsides, for example (3, methyl evernate). Such esters previously had been prepared by

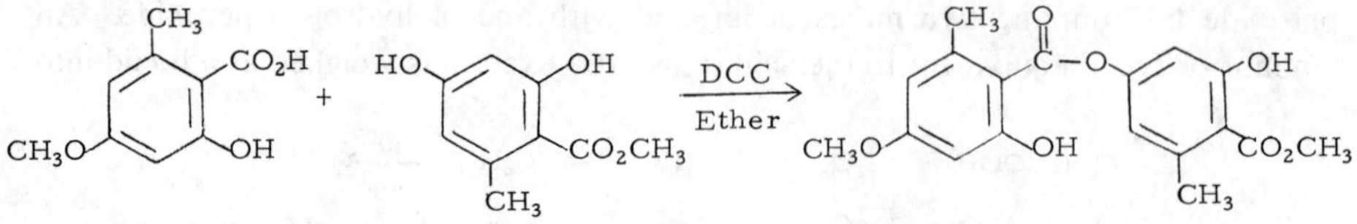

(1)   (2)   (3)

reaction of an acid chloride with a phenol in the presence of aqueous alkali or pyridine. However, a typical depside such as (3) bears a group in the position ortho to the carboxyl group which makes preparation of the acid chloride difficult.

Also, free hydroxyl groups in the acid component must be protected before conversion to the acid chloride. Use of DCC in ether eliminated these difficulties and afforded (3) in good yield. The Indian investigators also used trifluoroacetic anhydride with success.

Peptide synthesis. The dehydrating agent has found wide use in peptide synthesis,[15] although some racemization has been reported.[16] A mixture of acetonitrile

$$RCOOH + H_2NR' \xrightarrow{DCC} RCONHR' + DCC{-}H_2O$$

and tetrahydrofurane is recommended as solvent for minimization of acylurea formation.[17] An active ester for peptide synthesis can be prepared by esterification of an N-acylamino acid with *p*-nitrophenol by dehydration with DCC.[18]

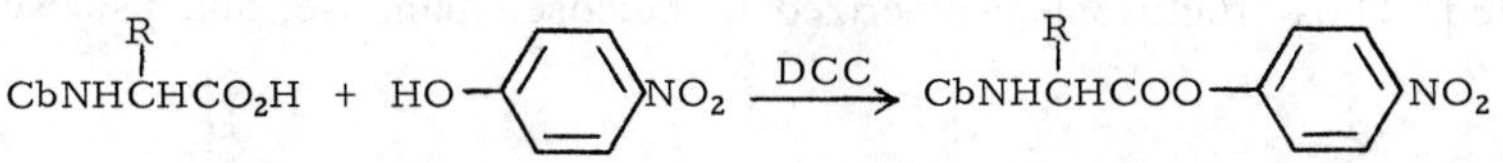

Other N-acylamines. Amines react with carboxylic acids in the presence of DCC at 0° to form amides in 70–90% yield.[19] The reagent has been used for the preparation of N-acyl derivatives of D-glucosamine.

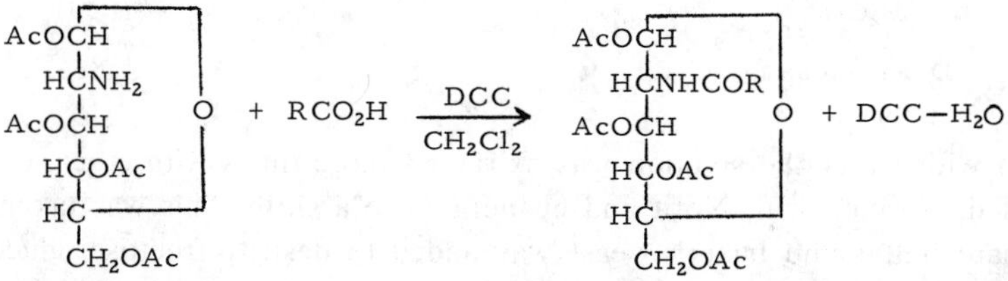

Lactonization. Woodward *et al.*[20] found treatment of the γ-hydroxy acid (1) with DCC in pyridine more effective for lactonization than use of acetic anhydride in pyridine.

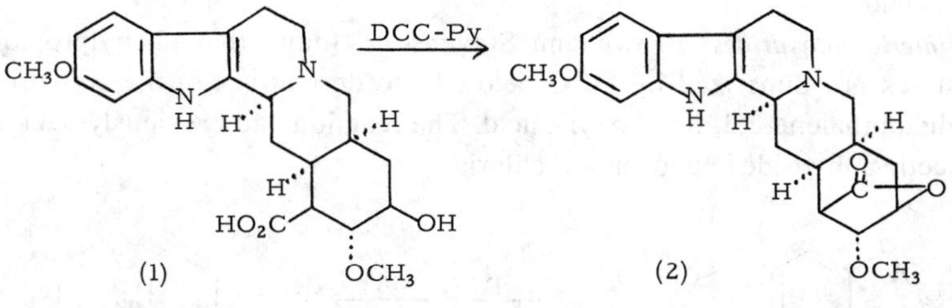

(1) (2)

Johnson *et al.*[21] found the γ-hydroxy acid (3) to resist lactonization by usual methods since the transformation requires a flip of the ring into the boat conformation (4). Lactonization with DCC in pyridine succeeded in some trials but the

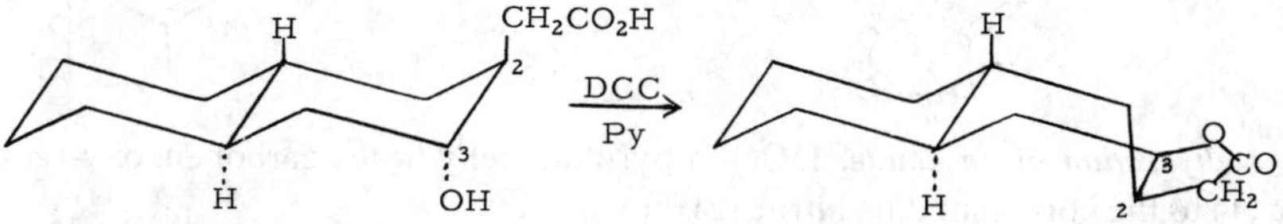

(3) (4)

results were variable. Thus a yield of 55% in one run could not be duplicated. Eventually the lactone was obtained satisfactorily with *p*-toluenesulfonic acid in xylene.

β-Lactams. The synthesis of a penicillin, for example penicillin V, requires cyclization of the corresponding penicilloic acid with formation of the highly strained β-lactam. DCC effects this key step in the synthesis, although in low yield.[22]

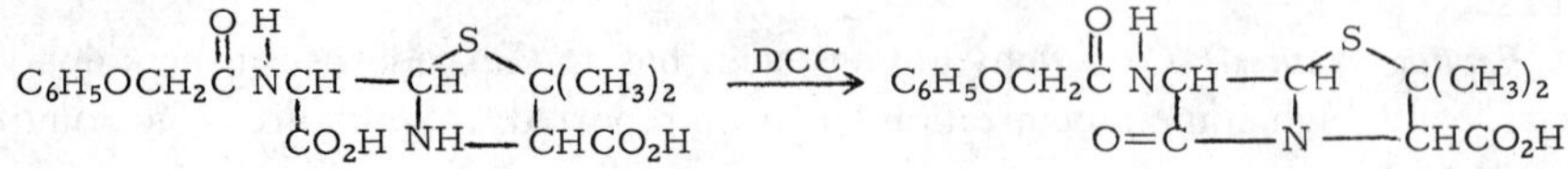

Penicillin V

Lobry de Bruyn-van Ekenstein rearrangement. In the course of experiments aimed at the synthesis of glycosides, Passeron and Recondo[23] discovered that treatment of a reducing sugar with DCC in hot methanol effects the rearrangement indicated. Thus fructose is isomerized to glucose, mannose, and psicose. After

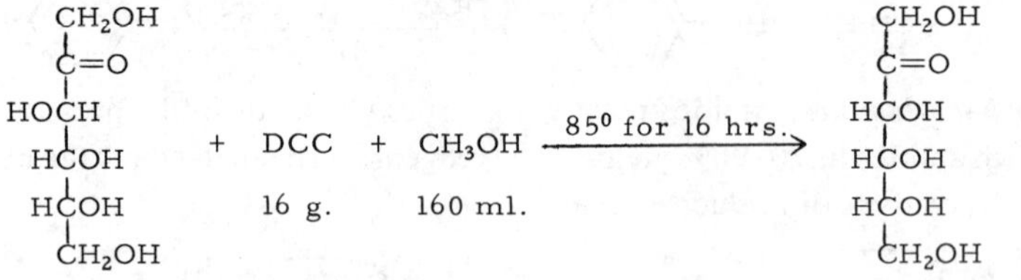

D-Fructose D-Psicose

39 g.

dilution with water the solution was extracted three times with ether, which was discarded, clarified with Norit and evaporated to a sirup. This was taken up in a phosphate buffer and baker's yeast was added to destroy fructose, glucose, and mannose. After filtering off the cells and deionization with Amberlite MB-3, trichloroacetic acid was added to precipitate proteins and the excess removed by either extraction. After concentration to a sirup, psicose was isolated as the phenylosazone in 25% yield.

Dimeric anhydrides. Brown and Stevenson[24] found that phenylpropiolic acid derivatives are dimerized by DCC below 0° to derivatives of the anhydride of 1-phenylnaphthalene-2,3-dicarboxylic acid. The reaction had previously been effected with acetic anhydride in phosphoryl chloride.

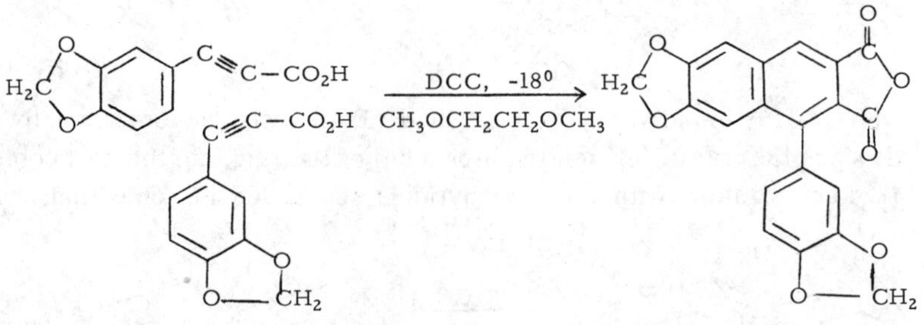

Dehydration of an amide. DCC in pyridine dehydrates carbobenzoxy-L-asparagine (1) to the corresponding nitrile (2).[25]

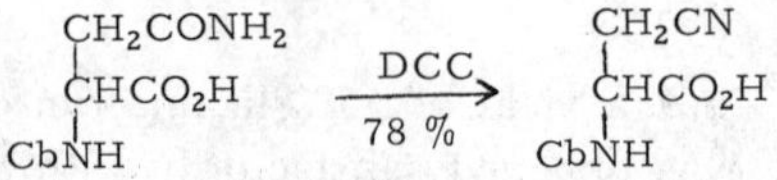

(1) (2)

Synthesis of a heterocycle. An interesting use of the reagent is for the synthesis of barbituric acids.[26]

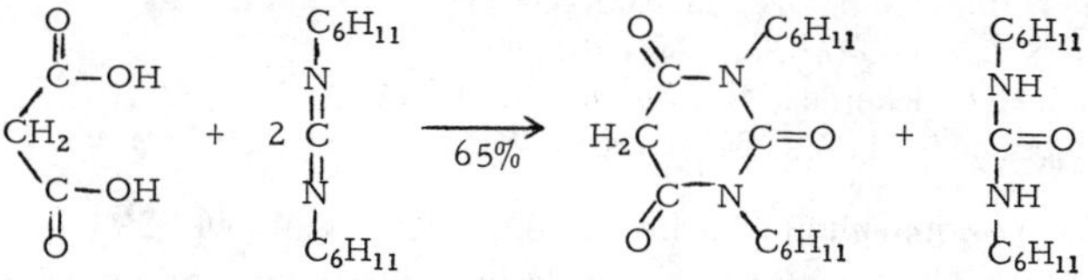

Nucleotides. In an extensive investigation of various reagents for the formation of the 3′,5′-internucleotide linkage, Jacob and Khorana[27] concluded that the combination of DCC with *p*-toluenesulfonyl chloride or mesitylenesulfonyl chloride is the most efficient.

Heterocyclic steroids. 2,3-Seco-5α-cholestane-2,3-dioic acid (1) is converted into the anhydride (2) in high yield by DCC in dioxane at room temperature.[28]

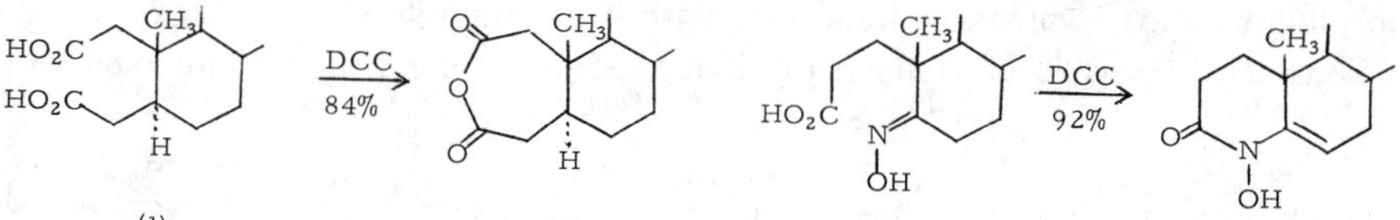

(1) (2) (3) (4)

3,5-Seco-4-norcholestane-5-one-3-oic acid 5-oxime (3) on similar dehydration gives 4-hydroxy-4-aza-Δ^5-cholestene-3-one (4).

[1]M. Bodanszky *et al.*, *Biochem Prep.*, **10**, 122 (1963)

[2]Capillary dropping tubes calibrated for ½ and 1 ml. are available from Macalaster Scientific Co.

[3]E. Schmidt, F. Hitzler, and E. Lahde, *Ber.*, **71**, 1933 (1938)

[4]T. W. Campbell, J. J. Monagle, and V. S. Foldi, *Am. Soc.*, **84**, 3673 (1962)

[5]H. G. Khorana, *Chem. Revs.*, **53**, 145 (1953)

[6]D. M. Brown, *Advances in Organic Chemistry*, **3**, 115 (1963)

[7]F. D. Greene and J. Kazan, *J. Org.*, **28**, 2168 (1963)

[8]E. Vowinkel, *Ber.*, **95**, 2997 (1962); *Angew. Chem., Internat. Ed.*, **2**, 218 (1963)

[9]H. G. Khorana and A. R. Todd, *J. Chem. Soc.*, 2257 (1953); H. G. Khorana, *Am. Soc.*, **76**, 3517 (1954); C. A. Dekker and H. G. Khorana, *ibid.*, **76**, 3522 (1954)

[10]H. G. Khorana and [illegible]lyi, *Am. Soc.*, **83**, 675 (1961); H. G. Khorana, A. F. Turner, and J. P. Vizsolyi, *ibid.*, **83**, 686 (1961); H. G. Khorana, J. P. Vizsolyi, and P. K. Ralph, *ibid.*, **84**, 414 (1962)

[11]D. H. Rammler and H. G. Khorana, *Am. Soc.*, **85**, 1997 (1963)

[12]H. Schüssler and H. Zahn, *Ber.*, **95**, 1076 (1962)

[13]A. Buzas, C. Egnell, and P. Fréon, *Compt. rend.*, **252**, 896 (1961); **255**, 945 (1962); A. Stempel and F. W. Landgraf, *J. Org.*, **27**, 4675 (1962)

[14]S. Neelakantan, R. Padmasani, and T. R. Seshadri, *Tetrahedron*, **21**, 3531 (1965)

[15]J. C. Sheehan and G. P. Hess, *Am. Soc.*, **77**, 1067 (1955); M. Goodman and G. W. Kenner, *Adv. in Protein Chem.*, **12**, 488 (1957); N. F. Albertson, *Org. Reactions*, **12**, 205 (1962)

[16]H. Schwarz and F. M. Bumpus, *Am. Soc.*, **81**, 890 (1949)

[17]H. Zahn and J. F. Diehl, *Z. Naturforsch.*, **12B**, 85 (1957); G. Fölsch, *Chem. Scand.*, **13**, 1407 (1959)

[18]M. Bodanszky and V. du Vigneaud, *Am. Soc.*, **81**, 5688 (1959)

[19]W. A. Bonner and P. I. McNamee, *J. Org.*, **26**, 2554 (1961)

[20]R. B. Woodward, F. E. Bader, H. Bickel, A. J. Frey, and R. W. Kierstead, *Tetrahedron*, **2**, 1 (1958)

[21]W. S. Johnson, V. J. Bauer, J. L. Margrave, M. A. Frisch, L. H. Dreger, and W. N. Hubbard, *Am. Soc.*, **83**, 606 (1961)

[22]J. C. Sheehan and K. R. Henery-Logan, *Am. Soc.*, **81**, 3089 (1959)

[23]S. Passeron and E. Recondo, *J. Chem. Soc.*, 813 (1965)
[24]D. Brown and R. Stevenson, *Tetrahedron Letters*, 3213 (1964)
[25]C. Ressler and H. Ratzkin, *J. Org.*, **26**, 3356 (1961)
[26]A. K. Bose and S. Garratt, *Am. Soc.*, **84**, 1310 (1962)
[27]T. M. Jacob and H. G. Khorana, *Am. Soc.*, **86**, 1630 (1964)
[28]N. L. Doorenbos and M. T. Wu, *Chem. Ind.*, 648 (1965)

N,N′-Dicyclohexyl-4-morpholinocarboxamidine, (3). Mol. wt. 293.44, m.p. 105.5°. Supplier: Aldrich. The reagent is prepared by refluxing morpholine (1) and dicyclohexylcarbodiimide (DCC, 2) in *t*-butanol.[1]

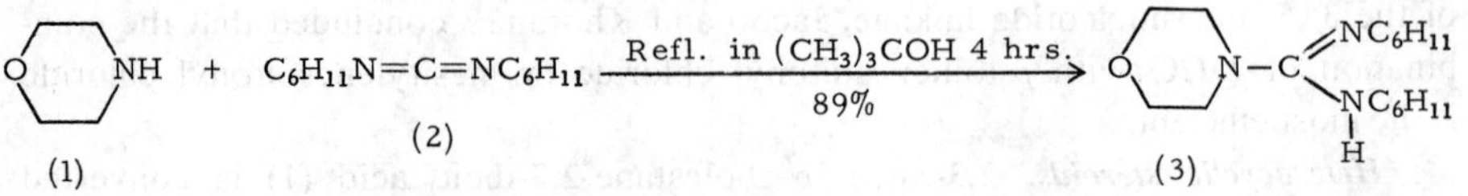

It reacts with a nucleoside-5′-phosphate (4) to give a nucleoside-5′-phosphoromorpholidate (5), which on treatment with DCC in refluxing pyridine gives a nucleoside-3′,5′-cyclic phosphate (6).[1] The reaction of the 5′-phosphate itself with

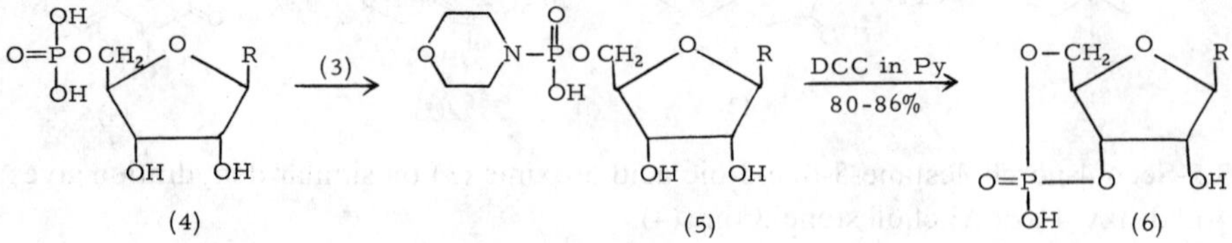

DCC in pyridine is sometimes successful, but in general the phosphates are not sufficiently soluble in pyridine, the solvent of choice.[2]

[1]J. G. Moffatt and H. G. Khorana, *Am. Soc.*, **83**, 649 (1961)
[2]M. Smith, G. I. Drummond, and H. G. Khorana, *Am. Soc.*, **83**, 698 (1961)

Diels-Alder solvents

Introduction. Presented in this section are the results of a study by one of us of the suitability of various solvents, common and uncommon, for carrying out a typical Diels-Alder reaction, namely that between tetraphenylcyclopentadienone and dimethyl acetylenedicarboxylate.[1] Disappearance of the intense purple color

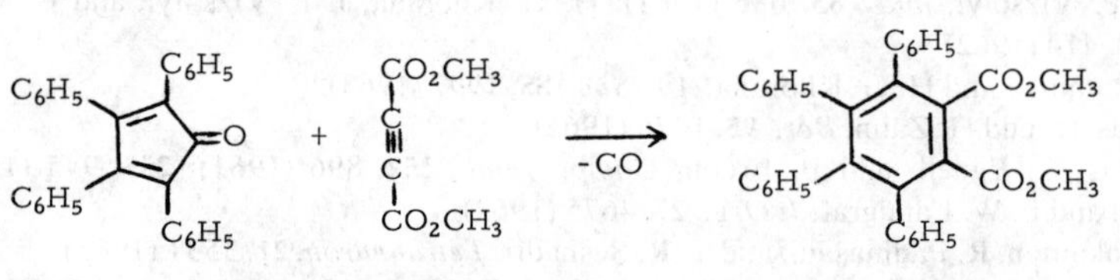

of the dienone signals completion of the reaction. In a student experiment[2] in which the reaction is carried out by refluxing a solution of the components in *o*-dichlorobenzene, b.p. 179°, the purple color disappears dramatically in 5 minutes. In this case the reaction is irreversible, since the initial adduct spontaneously loses carbon monoxide with aromatization of the newly formed ring.

Preliminary trials at room temperature. The interesting finding by Pearson[3] that hindered ketones resistant to oximation at high temperatures can be oximated by allowing a strongly basic solution of the reactants to stand at room temperature for one to six months suggested trial of the above Diels-Alder reaction at room temperature. In a first exploration a suspension of 2 g. of finely ground tetraphenylcyclopentadienone in 10 ml. of a solvent and 2 ml. (3 equiv.) of dimethyl acetylenedicarboxylate was let stand at room temperature with occasional brief stirring with a rod. With *o*-dichlorobenzene as solvent, in 17 days white adduct had separated, and the supernatant liquid was colorless. Crystallization, effected by bringing the solid into solution at about, 80° and adding 15 ml. of 95% ethanol, afforded the pure adduct in 80% yield. With methylene chloride as solvent, reaction was complete in 11 days and afforded a pale tan solution. Evaporation of the solvent and crystallization from 10 ml. of *o*-dichlorobenzene and 15 ml. of 95% ethanol gave the pure adduct in 83% yield. Notice that in both cases the product was identified as dimethyl tetraphenylphthalate; hence carbon monoxide was eliminated from the initial adduct even at 25°.

With acetic acid as solvent, the reaction came to a standstill after about 2 months, but examination of the greyish solid which had separated showed that crystals of adduct had occluded the purple diene component. Similar results invalidated trials in benzene and tetrahydrofurane. It was evident that for comparisons to be significant the suspensions would have to be agitated.

Reaction at reflux temperature. Agitation of a suspension by refluxing (Table 1) often proceeds well but may be attended with troublesome bumping. More serious is the fact that the boiling point of the pure solvent is not necessarily a measure of the reaction temperature. Thus cyclohexane and benzene boil at nearly the same

Table 1 Reaction of 2 g. of tetraphenylcyclopentadienone with 2 ml. of dimethyl acetylenedicarboxylate (3 equiv.) in 10 ml. of refluxing solvent (yield of adduct: 80–97%)

Solvent	B.p.	Time for completion	Terminal color, liq. phase
o-Dichlorobenzene	179°	5 min.	Yellow
Tetrachloroethylene	120.8°	15 min.	Yellow
Acetic acid	118.2°	15 min.	Yellow
Methylcyclohexane	100.9°	40 min.	Yellow
Thiophene	84°	2 hrs.	Orange
1,2-Dimethoxyethane	83°	2 hrs.	Yellow
Cyclohexane	80.8°	2.4 hrs.	Colorless
Benzene	80.1°[a]	2 hrs.	Colorless
Methyl ethyl ketone	79.6°	3.2 hrs.	Colorless
Carbon disulfide	46.3°	23 hrs.	Brown
Methylene chloride	40.7°	24 hrs.	Colorless

[a]Actual liquid temperature: 86–90.5°

temperature but differ greatly in solvent power for the dienone; 10 ml. of solutions in cyclohexane and in benzene saturated at 25° contain 22 mg. and 493 mg. of dienone, respectively. Thus with benzene the elevation in boiling point is such that the liquid temperature of the refluxing mixture at the end of the reaction is about 10° above the boiling point of benzene (see note to Table 1); with cyclohexane the liquid temperature is close to the boiling point. Thus comparison of reaction rates in liquids of the same boiling point is not valid.

Even though the adduct separates in colorless form, a terminal color in the liquid phase is regarded as objectionable. This consideration would eliminate for practical use for the reaction at reflux temperature all solvents listed in the table of b.p.

above 81°, as well as carbon disulfide. The two sulfur-containing solvents give particularly intense colors.

Reaction at 78.8°. For better control the reactions were run in a test tube with stirrer mounted in Dow Corning 702 Fluid heated in an Ace Glass boiler with pure benzene. The inner liquid temperature was 78.8°.

A decision on the molar proportions of reactants was based in part on results obtained with maleic anhydride as dienophile shown in Table 2. The time shown is for complete disappearance of purple or pink color. The bimolecular reaction

Table 2 Reaction of 2 g. of tetraphenylcyclopentadienone in 10 ml. of tetrachloroethylene at 78.8°

Dienophile	Equivalents	Time, hrs.
Maleic anhydride[a]	1.5	8.0
Maleic anhydride	3	2.4
$CH_3O_2CC{\equiv}CCO_2CH_3$[b]	3	3.5

[a]The white solid (2.90 g.) contained maleic anhydride; crystallization from 100 ml. of benzene gave 2.18 g. (87%) of prisms, m.p. 223° dec. (pink).

[b]The white solid, collected and rinsed with methanol, weighed 2.47 g. (100%) m.p. 264–266°. Crystallization from 10 ml. of *o*-dichlorobenzene with addition of 15 ml. of 95% ethanol gave 2.40 g. (97%) of prisms of the same m.p.

becomes extremely slow in the terminal stages but with use of 3 equivalents of dienophile the reaction time falls into a convenient range. Hence this ratio was adopted for the comparisons shown in Table 3 (also 1). Note that maleic anhydride is a more reactive dienophile than the acetylenic ester.

Table 3 gives reaction times at 78.8° observed in solvents of four types. Boiling points and solubilities are included for general interest and in some cases have a bearing on the conduct of the experiment or isolation of the product. Thus tetrahydrofurane has such pronounced solvent power for the dienone (960 mg./10 ml. at 25°) that the reaction can be run at a temperature 13.4° above the boiling point of THF. Other powerful solvents are methylene bromide (888 mg.), bromoform (876 mg.), and *s*-tetrachloroethane (760 mg.). For isolation of the adduct it is often necessary merely to cool the suspension, but where necessary a diluent can be added or solvent distilled: isolation of adduct in close to quantitative yield seldom presents a problem.

Table 3 Reaction of 2 g. of tetraphenylcyclopentadienone with 2 ml. of dimethyl acetylenedicarboxylate (3 equiv.) in 10 ml. of solvent with stirring at 78.8°

No.	Solvent		B.p.[a]	Hrs.[b]	Sol.[c]
	Aprotic solvents insoluble in water				
1	Carbon tetrachloride	CCl_4	76.8°	4.0	164
2	Benzene	C_6H_6	80.1°	5.0	493
3	Cyclohexane	$\overline{CH_2CH_2CH_2CH_2CH_2CH_2}$	80.8°	3.6	22
4	Ethylene dichloride	$ClCH_2CH_2Cl$	83.8°	4.0	400
5	Fluorobenzene	C_6H_5F	84.8°	3.9	350

Table 3 Reaction of 2 g. of tetraphenylcyclopentadienone with 2 ml. of dimethyl acetylenedicarboxylate (3 equiv.) in 10 ml. of solvent with stirring at 78.8° (continued)

No.	Solvent		B.p.[a]	Hrs.[b]	Sol.[c]
	Aprotic solvents insoluble in water				
6	Dimethyl carbonate	$(CH_3O)_2C{=}O$	90°	3.0	65
7	Ethyl chloroformate	$ClCO_2C_2H_5$	92°	4.4	386
8	Methylene bromide	CH_2Br_2	98.2°	3.1	888
9	Trimethyl orthoformate	$HC(OCH_3)_3$	104°	5.5	—
10	Tetrachloroethylene	$Cl_2C{=}CCl_2$	120.8°	3.5	130
11	Paraldehyde	$\overline{OCH(CH_3)OCH(CH_3)OCH(CH_3)}$	124°	3.3	23
12	Dimethyl sulfite	$(CH_3O)_2SO$	126°	5.5	102
13	Ethylene dibromide	$BrCH_2CH_2Br$	131.7°	3.5	352
14	Acetic anhydride	$(CH_3CO)_2O$	139.6°	4.8	40
15	Triethyl orthoformate	$HC(OC_2H_5)_3$	146°	5.3	—
16	*s*-Tetrachloroethane	$CHCl_2CHCl_2$	146.3°	3.9	760
17	Bromoform	$CHBr_3$	149.6°	2.8	876
18	Cyclohexanone	$\overline{CH_2CH_2CH_2CH_2CH_2C}{=}O$	156°	3.7	260
19	Tetraethyl orthocarbonate	$(C_2H_5O)_4C$	158°	5.0	—
20	Dimethyl oxalate	$(CO_2CH_3)_2$	163°	5.5	—
21	*o*-Dichlorobenzene	*o*-ClC_6H_4Cl	179°	3.5	296
22	Diethyl oxalate	$(CO_2C_2H_5)_2$	185.4°	5.0	32
23	Iodobenzene	C_6H_5I	188.5°	3.7	396
24	Dimethyl sulfate	$(CH_3O)_2SO_2$	188.5°	3.5	38
25	Nitrobenzene	$C_6H_5NO_2$	210°	5.0	132
26	Diethyl sulfate	$(CH_3CH_2O)_2SO_2$	210°	10.5	170
27	Hexachlorobutadiene	$CCl_2{=}CClCCl{=}CCl_2$	211°	4.5	—
28	Tetraethyl orthosilicate	$(C_2H_5O)_4Si$	71°/20 mm.	6.0	56
29	Diphenyl ether (m.p. 26.9°)	$C_6H_5OC_6H_5$	257°	3.5	352
	Aprotic solvents partially soluble in water				
30	Ethyl acetate	$CH_3CO_2C_2H_5$ (8.5% at 15°)	77.1°	5.3	94
31	Nitromethane	CH_3NO_2 (9.5% at 20°)	101.3°	4.0	43
32	Chloropicrin	CCl_3NO_2 (about 1% at 20°)	112°	3.2	—
	Aprotic solvents, water-miscible				
33	Tetrahydrofurane	$\overline{CH_2CH_2CH_2CH_2O}$	65.4°	5.1	960
34	Acetonitrile	CH_3CN	81.6°	5.5	45
35	1,2-Dimethoxyethane	$CH_3OCH_2CH_2OCH_3$	83°	5.2	214
36	Dioxane	$\overline{OCH_2CH_2OCH_2CH_2}$	101.3°	4.3[a]	386
37	Dimethylformamide	$HCON(CH_3)_2$	153°	3.2	125
38	Tetramethylurea	$(CH_3)_2NC({=}O)N(CH_3)_2$	175.6°	2.5[a]	100
39	Dimethyl sulfoxide	$(CH_3)_2S^+{-}O^-$	189°	3.0[a]	47
40	Triglyme	$CH_3O(CH_2CH_2O)_3CH_3$	222°	3.3	115
41	Sulfolane	$\overline{CH_2CH_2CH_2CH_2SO_2}$	130°/6.5 mm.	3.0[a]	—
42	Ethylene carbonate (m.p. 39°)	$\overline{OCH_2CH_2OC}{=}O$	238°	6.1	—
	Protonic solvents, water-miscible				
43	Trifluoroacetic acid	CF_3CO_2H	72.5°	3.0[a]	30
44	*t*-Butanol	$(CH_3)_3COH$	82.5°	5.7	130
45	Acetic acid (m.p. 16.6°)	CH_3CO_2H	118°	5.8	30
46	Ethylene bromohydrin	$BrCH_2CH_2OH$	150°	4.0	—
47	Phenol (m.p. 42.5°)	C_6H_5OH	181°	5.2	—

Table 3 Reaction of 2 g. of tetraphenylcyclopentadienone with 2 ml. of dimethyl acetylenedicarboxylate (3 equiv.) in 10 ml. of solvent with stirring at 78.8° (continued)

No.	Solvent		B.p.[a]	Hrs.[b]	Sol.[c]
	Protonic solvents, water-miscible				
48	Chloroacetic acid (m.p. 63°)	CH_2ClCO_2H	189.5°	2.7	—
49	Dichloroacetic acid (m.p. 10°)	$CHCl_2CO_2H$	193.5°	2.1[a]	257
50	Trichloroacetic acid (m.p. 58°)	CCl_3CO_2H	196°	1.2[a]	—
51	Triethyleneglycol	$H(OCH_2CH_2)_3OH$	287°	8.0	23

[a]Taken where available from C. Marsden and S. Mann, *Solvents Guide*, Interscience (1963)
[b]Terminal extraneous color
[c]Solubility of dienone at 25°: mg./10 ml.

None of the 32 aprotic solvents insoluble or partially soluble in water gave rise to an objectionable color. With the exception of No. 26, with a reaction time of 10.5 hrs., the reaction times fall in the range 2.8–6.1 hrs. The ten water-soluble, aprotic solvents are similar in reaction time (average 3.9 hrs.); four of them afford rapid reactions but produce extraneous colors.

Citations in the literature for solvents 6 and 42 present an anomaly: dimethyl carbonate, b.p. 90°, sparingly soluble in water; ethylene carbonate, m.p. 39°, b.p. 238°, readily soluble in water: Vorländer[4] determined the molecular weight of ethylene carbonate in phenol and by vapor density determination and found it normal.

The most significant solvent effects are in the small list of water-soluble, protonic solvents. Trichloroacetic acid, a liquid at the temperature of the benzene boiler, takes the lead with a reaction time of only 1.2 hrs. By contrast, the reaction time for triethylene glycol (No. 51) was 8 hrs. Dichloroacetic, trichloroacetic, and trifluoroacetic acid all promote rapid reaction but they also give rise to extraneous pigments in the reaction mixture.

Effect of additives. A particularly striking solvent effect was one of negative catalysis evident from the results given in Table 4 for reactions carried out with

Table 4 Effect of additives on reactions under the conditions of Table 3

10 ml. Solvent alone (Table 3) or 10 ml. Solvent plus additive	Time, hrs.	Product, g. Th. = 2.47 g.
Abs. Ethanol	6.5	
95% Ethanol[a]	9.3	2.38
Abs. Ethanol and 1 ml. concd. H_2SO_4	9.6	
Acetic acid + 1 ml. Ac_2O	5.0	
Acetic acid + 1 ml. water	9.3	
Boron trifluoride etherate	5.6	1.78

[a]Solubility of dienone at 25°: 10 mg./10 ml.
[b]Solubility of dienone at 25°: 81 mg./10 ml.

stirring at 78.8°. A reaction time of 6.5 hrs. in absolute ethanol is extended to 9.3 hrs. in 95% ethanol or to 9.6 hrs. by addition of 10% by volume of sulfuric acid. A reaction time of 5.8 hrs. in ordinary acetic acid (Table 3) is extended to 9.3 hrs. by addition of 10% of water; added acetic anhydride destroys incidental moisture and speeds up the reaction. Boron trifluoride does not appear to catalyze the reaction.

Oils as solvents. The most promising oils tried are Kadol, a refined mineral oil, and Dow Corning 702 Silicone Fluid, a narrow boiling-range cut of methyl end-blocked polymethylsiloxane:

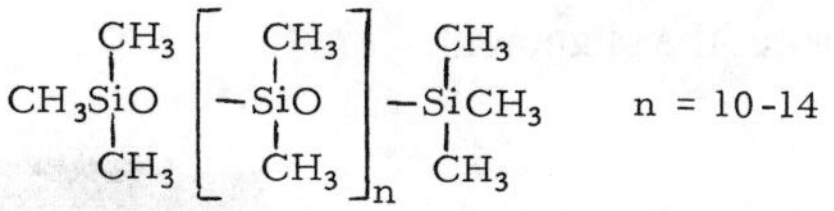

The results recorded in Table 5 show that the reactions are rapid and the yields high.

Table 5 Runs at 78.8° under the conditions of Table 3

Solvent	Time, hrs.	Dienone, mg./10 ml.	Diluent	Yield (Th. = 2.47 g.)	M.p.
Silicone 702	3.2	24.4	CH_3OH	2.47 g.	263–265°
Kadol	5.0	–	C_6H_6	2.22 g.	257–260°

These solvents are so stable that extraneous pigments are not likely to arise, even at much higher temperatures. Solvent containing excess dienophile should be recoverable in usable form, for example by dilution with petroleum ether, removal of adduct, and evaporation.

High-temperature reaction. The reaction of tetraphenylcyclopentadienone with diphenylacetylene with elimination of carbon monoxide and formation of hexaphenylbenzene can be carried out by strong heating of a mixture of 0.5 g. of each component in a test tube,[5] but the method is unsatisfactory on a larger scale because of the high melting point of the product (454–456°). A satisfactory procedure[6] utilizes as solvent benzophenone, m.p. 48°, b.p. 305.4°. A mixture of 40 g. of this ketone, 0.021 mole of tetraphenylcyclopentadienone, and 0.043 mole of diphenylacetylene in a round-bottomed flask fitted with an air condenser is heated over a microburner so that it refluxes briskly (liquid temperature 301–303°). The purple color begins to fade in

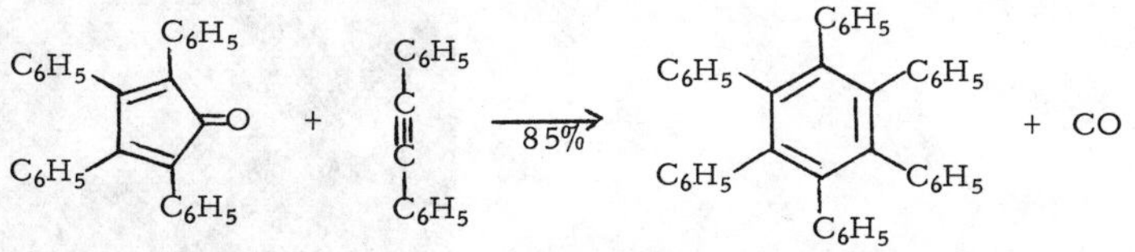

10–15 min. and changes to reddish brown in 25–30 min. When no further lightening in color is observed (45 min.) the burner is removed and 8 ml. of diphenyl ether is added to prevent subsequent solidification of the benzophenone. Crystals of product that separate are brought back into solution by reheating, and the solution is let stand for crystallization. Collected and washed free of brownish solvent with benzene, the product is obtained as colorless plates, m.p. 454–456°; yield 9.4 g. (85%). A satisfactory solvent for recrystallization is diphenyl ether (7 ml./g.).

To prevent air oxidation, the melting point is taken in a capillary evacuated with use of an adapter (Fig. D-2) consisting of a glass tube capped with a rubber vaccine stopper (A. H. Thomas Co. 2310-B) with a hole pierced through it with an awl. The hole is lubricated with glycerol and the capillary is grasped close to the sealed end and thrust through nearly to the full length (identifying wax pencil dots may have to be restored). Connect to the suction pump, make a guiding mark where the tube is to be sealed, hold the point of sealing near the small flame of a microburner, steady your hands on the bench so that neither will move much when the tube is melted, and then hold the tube in the flame until the walls collapse to form a flat seal and remove it at once. Take the tube out of the adapter, and if the two straight parts are not in a line soften the seal and correct the alignment.

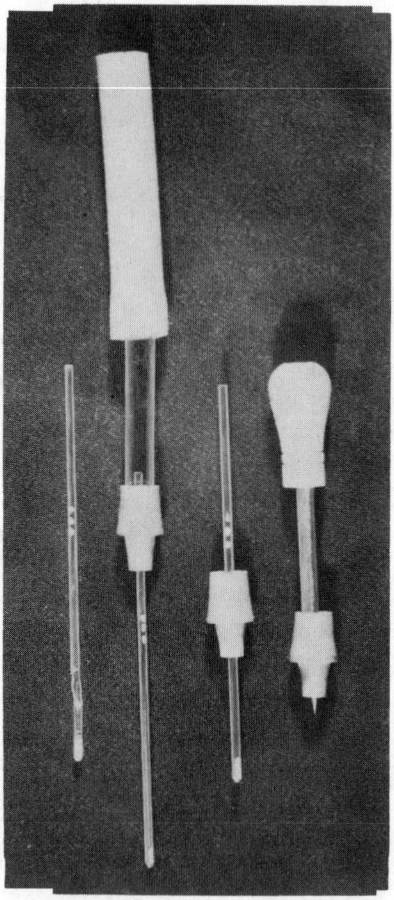

Fig. D-2 Evacuation and sealing of capillary

The melting point is determined conveniently with the Mel-Temp apparatus and 90–510° thermometer supplied by Laboratory Devices, P. O. box 68, Cambridge, Mass. 02139 (Fig. D-3). The evacuated capillary is sealed close to the sample to prevent sublimation, and repeated determinations are made with the same sample. The figure 456° is the average of determinations of the temperature of melting, 454° is the average of observations of the point of solidification. When the amount of diphenylacetylene was reduced to 1.2 times the theory the yield was the same, but the melting point was 450–452°. Other solvents tried and the liquid temperatures are:

stearic acid (340–365°), di-*n*-butyl phthalate (320–325°), phenyl salicylate (290°). The first two solvents are unsatisfactory because of a side reaction consuming some of the dienone, the third because the reaction is too slow.

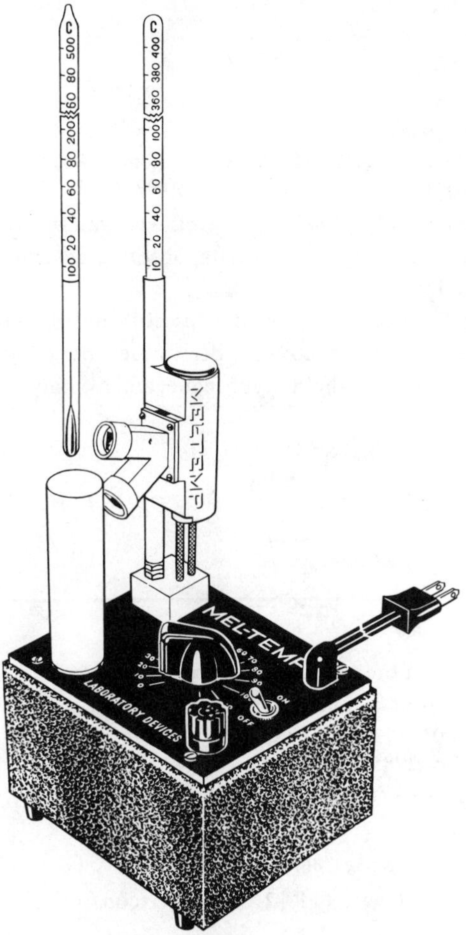
MEL-TEMP
LABORATORY DEVICES
ON
OFF

Fig. D-3 Mel-Temp apparatus

[1]W. Dilthey, I. Thewalt, and O. Trösken, *Ber.*, **67**, 1959 (1934)
[2]*Org. Expts.*, 305
[3]D. E. Pearson and O. D. Keaton, *J. Org.*, **28**, 1557 (1963)
[4]D. Vorländer, *Ann.*, **280**, 187 (1894)
[5]*Org. Expts.*, 307
[6]L. F. Fieser, procedure submitted to *Org. Syn.*

Diethoxymethyl acetate, $CH_3CO_2CH(OC_2H_5)_2$. Mol. wt. 162.18, b.p. 79–80°/24 mm., 85–87°/30 mm.

The reagent is prepared by reaction of acetic anhydride with triethyl orthoformate (41% conversion at equilibrium).[1] The preformed reagent is superior to the

$$(CH_3CO)_2O + HC(OC_2H_5)_3 \rightleftharpoons CH_3CO_2C_2H_5 + CH_3CO_2CH(OC_2H_5)_2$$

combination acetic anhydride–triethyl orthoformate for the conversion of 4,5-diamino-2,6-dichloropyrimidine (1) to 2,6-dichloropurine (2).[2]

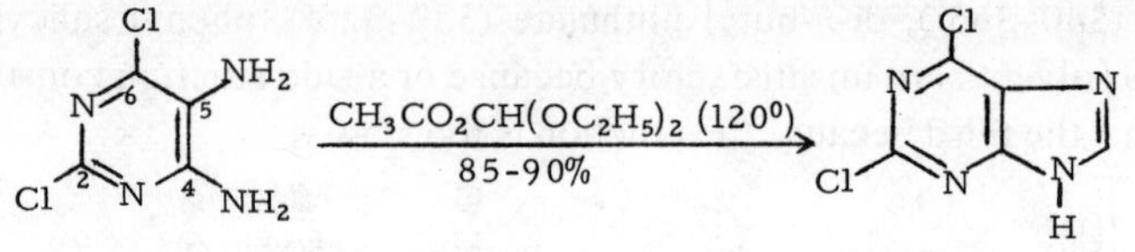

(1) (2)

[1]H. W. Post and E. R. Erickson, *J. Org.*, **2**, 260 (1937)
[2]J. A. Montgomery and C. Temple, Jr., *Am. Soc.*, **79**, 5239 (1957); J. A. Montgomery and L. B. Holum, *ibid.*, **80**, 404 (1958); J. A. Montgomery and C. Temple, Jr., *ibid.*, **80**, 409 (1958)

2,2-Diethoxypropane (Acetone diethyl ketal), $(CH_3)_2C(OC_2H_5)_2$. Mol. wt. 132.20, b.p. 115°. Suppliers: B, E, F. Preparation from triethyl orthoformate, acetone, absolute ethanol, and a trace of *p*-toluenesulfonic acid; yield 75%.[1]

The reagent is higher boiling than 2,2-dimethoxypropane (b.p. 80°) and better adapted to the conversion of cyclitols into polyacetonide derivatives by ketal exchange.[2] A mixture of the powdered cyclitol, 2,2-diethoxypropane, and a catalytic

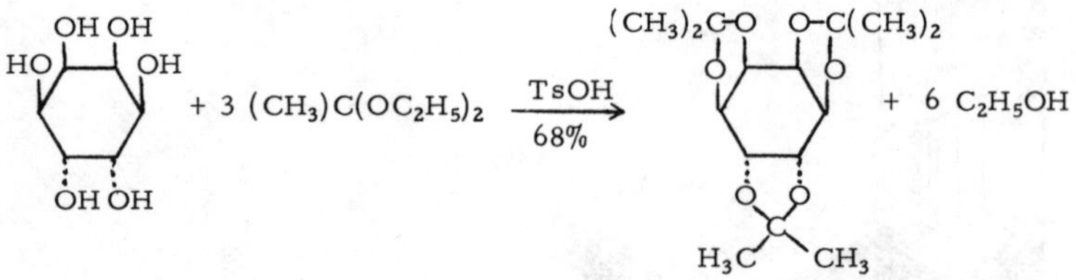

amount of *p*-toluenesulfonic acid is heated on the steam bath for 1–2 hrs. The usual method (acetone + $ZnCl_2$) produces mixtures of di- and triacetonides in 5 hrs.

[1]C. D. Hurd and M. A. Pollack, *Am. Soc.*, **60**, 1905 (1938)
[2]S. J. Angyal and R. M. Hoskinson, *J. Chem. Soc.*, 2985 (1962)

Diethyl acetylenedicarboxylate, $C_2H_5O_2CC{\equiv}CCO_2C_2H_5$. Mol. wt. 170.16, b.p. 96–98°/8 mm. Supplier: Baker. Preparation.[1]

[1]E. H. Huntress, T. E. Lesslie, and J. Bornstein, *Org. Syn., Coll. Vol.*, **4**, 330 (1963)

Diethylaluminum cyanide, $(C_2H_5)_2AlCN$. Mol. wt. 111.12, highly viscous oil, b.p. 150°/0.07 mm.

Preparation.[1] A solution of triethylaluminum in benzene is stirred magnetically with ice cooling during dropwise addition of a solution of hydrogen cyanide in benzene. When the evolution of ethane ceases, the solvent is evaporated and the product distilled.

$$(C_2H_5)_3Al + HCN \xrightarrow[93\%]{C_6H_6\ 0^0} (C_2H_5)_2AlCN + C_2H_6$$

Hydrocyanation. The reagent effects hydrocyanation more efficiently than the combination of hydrogen cyanide with triethylaluminum (*which see*). Thus the reaction with cholestenone in toluene is about 10,000 times as fast as the reaction of the combination reagent. Unlike the combination reagent, it can be used for the cleavage of highly hindered steroid epoxides, such as a 9α,11α-epoxide.

[1]W. Nagata and M. Yoshioka, *Tetrahedron Letters*, 1913 (1966)

Diethylaluminum iodide, $(C_2H_5)_2AlI$. Mol. wt. 212.01. Suppliers: Alfa, KK.

Hoberg[1] found that diethylaluminum iodide (1) reacts with diazomethane in pentane at −80° with evolution of nitrogen and formation of crystalline iodomethylaluminumdiethyl (2). This substance is stable in the temperature range −80 to −50° but at −10 to 0° it decomposes with formation of ethylene. Formation of

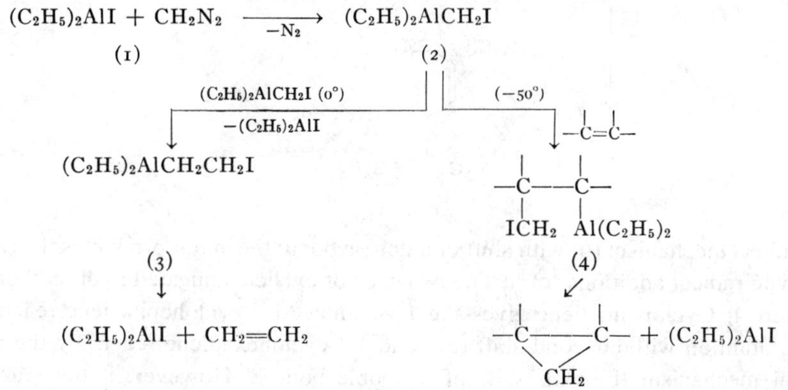

ethylene is interpreted as resulting from reaction of two molecules of (2) to form (3) and elimination of diethylaluminum iodide. Carbene is discounted as an intermediate because at a temperature at which the organometallic compound (2) is stable it reacts rapidly with cyclohexene to give norcarane. A simple addition (4) and elimination accounts for the reaction.

[1]H. Hoberg, *Ann.*, **656**, 1, 15 (1962)

Diethyl azodicarboxylate (DAD), $C_2H_5O_2C—N{=}N—CO_2C_2H_5$. Mol. wt. 174.16, b.p. 93–95°/5 mm., f.p. 6°. Supplier: Aldrich.

Preparation. Two *Organic Syntheses* procedures[1,2] are practically identical in the first step (about 83%); in one[1] oxidation is done with hypochlorous acid in 82% yield, and in the other[2] the oxidant is fuming nitric acid (75%). The second procedure

$$ClCO_2C_2H_5 \xrightarrow[Na_2CO_3]{H_2NNH_2} C_2H_5O_2C\overset{H}{N}-\overset{H}{N}CO_2C_2H_5 \xrightarrow{(a)\ HOCl\ or\ (b)\ HNO_3} C_2H_5O_2CN{=}NCO_2C_2H_5$$

was published because it is safer and gives a product which is purer and keeps better. Material made by hypochlorite oxidation invariably contains about 1% of chlorine and has exploded on attempted distillation (communication from B. C. McKusick, editor of the *Org. Syn.* volume in question.)

Diels-Alder reactions. This dienophile reacts with some conjugated dienes by 1,4-addition, as shown in example 1[3] (*see also* Anthracene-9,10-biimide). Example 2 illustrates two other modes of reaction: the major product arises by a cyclic,

1.

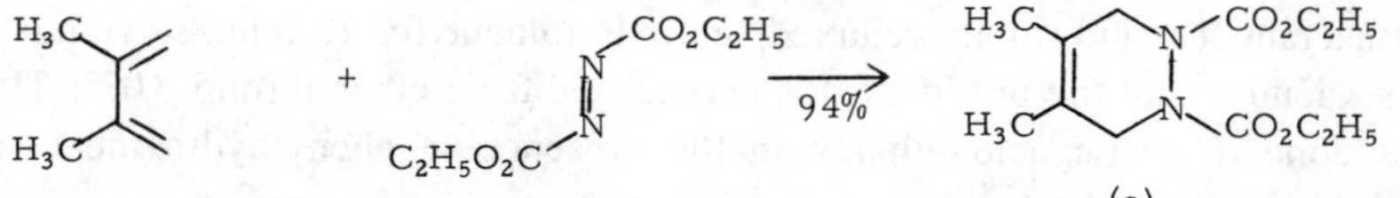

(a)

2. 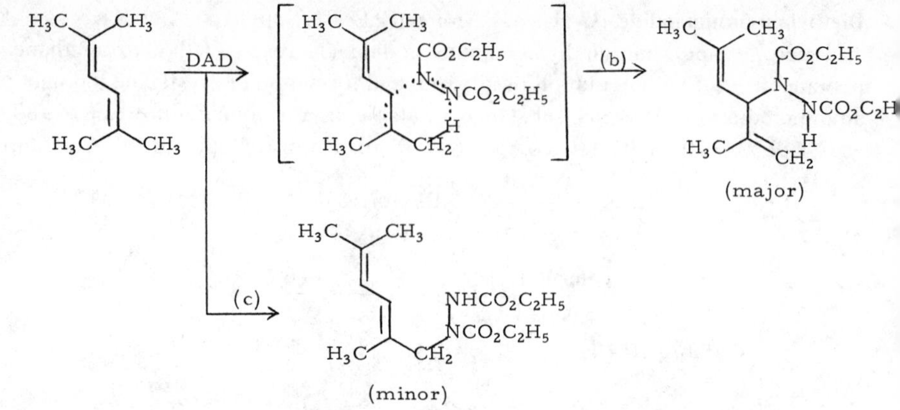

nonradical mechanism (b) with shift of a double bond; the minor product is the result of allylic radical additions (c).[3,4] The behavior of cyclic conjugated hydrocarbons is of interest. Cyclopentadiene gives the 1,4-adduct (d),[5] cycloheptatriene reacts by allylic addition without bond shift (c),[6] and 1,3-cyclohexadiene reacts by the nonradical mechanism (b) with shift of a double bond.[3,4] However, if the reaction mixture of reagent with 1,3-hexadiene is irradiated, the normal Diels-Alder adduct can be obtained in 87% yield.[7] The reagent reacts with steroid $\Delta^{5,7}$-dienes by

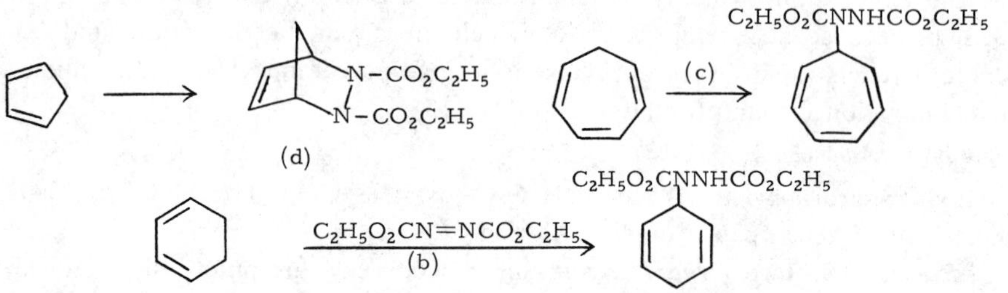

nucleophilic addition at C_7 with abstraction of hydrogen from C_9.[8] Dimethyl acetylenedicarboxylate reacts less readily but also abnormally.[9]

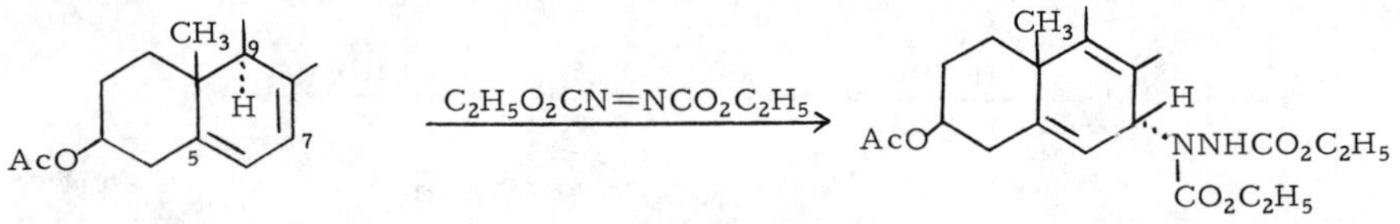

Synthesis of α-ketoaldehydes. An unusual reaction of the reagent with diazoketones provides a route to α-ketoaldehydes, isolated as the phenylosazones.[10]

$$\underset{(1)}{p\text{-}O_2NC_6H_4\overset{O}{\overset{\|}{C}}CH{=}\overset{+}{N}{=}\overset{-}{N}} + \underset{(2)}{C_2H_5O_2CN{=}NCO_2C_2H_5} \xrightarrow[79\%]{-N_2}$$

$$\underset{(3)}{p\text{-}O_2NC_6H_4\overset{O}{\overset{\|}{C}}CH{=}NN(CO_2C_2H_5)_2} \xrightarrow[75\%]{C_6H_5NHNH_2,\ AcOH} \underset{(4)}{p\text{-}O_2NC_6H_4\overset{NNHC_6H_5}{\overset{\|}{C}}CH{=}NNHC_6H_5}$$

Reactants (1) and (2) are refluxed gently in toluene for 15 min., solvent is removed in vacuum, and the product (3) is crystallized from ethanol (m.p. 107°). Hydrolysis is done in acetic acid–ethanol in the presence of phenylhydrazine to form the phenylosazone (4).

Dehydrogenation. Diethyl azodicarboxylate has been shown to effect photochemical dehydrogenation of isopropanol to pinacol[11] and of cyclohexanol to cyclohexanone.[12] Actually the reagent effects nonphotochemical oxidation of alcohols, mercaptans, anilines, and hydrazobenzenes with formation of diethyl hydrazinodicarboxylate.[13]

[1]N. Rabjohn, *Org. Syn., Coll. Vol.*, **3**, 375 (1955). J. C. J. MacKenzie, A. Rodgman, and G. F. Wright, *J. Org.*, **17**, 1666 (1952), recommend that material prepared by Rabjohn's method be washed repeatedly with cold 3% aqueous sodium carbonate and then distilled in small lots just prior to use.

[2]J. C. Kauer, *Org. Syn., Coll. Vol.*, **4**, 411 (1963)

[3]B. T. Gillis and P. E. Beck, *J. Org.*, **27**, 1947 (1962); **28**, 3177 (1963)

[4]B. Franzus and J. H. Surridge, *J. Org.*, **27**, 1951 (1962); B. Franzus, *ibid.*, **28**, 2954 (1963); W. A. Thaler and B. Franzus, *ibid.*, **29**, 2226 (1964)

[5]O. Diels, J. H. Blom, and W. Koll, *Ann.*, **443**, 242 (1925); P. G. Gassman and K. T. Mansfield, procedure submitted to *Org. Syn.*

[6]J. M. Cinnamon and K. Weiss, *J. Org.*, **26**, 2644 (1961)

[7]R. Askani, *Ber.*, **98**, 2551 (1965)

[8]A. van der Gen, J. Lakeman, M. A. M. P. Gras, and H. O. Huisman, *Tetrahedron*, **20**, 2521 (1964)

[9]A. van der Gen, J. Lakeman, U. K. Pandit, and H. O. Huisman, *Tetrahedron*, **21**, 3641 (1965)

[10]E. Fahr and F. Scheckenbach, *Ann.*, **655**, 86 (1962)

[11]G. O. Schwenck and H. Formaneck, *Angew. Chem.*, **70**, 505 (1958)

[12]R. C. Cookson, I. D. R. Stevens, and C. T. Watt, *Chem. Comm.*, 259 (1965)

[13]F. Yoneda, K. Suzuki, and Y. Nitta, *Am. Soc.*, **88**, 2328 (1966)

Diethyl carbonate, $C_2H_5OC(=O)OC_2H_5$. Mol. wt. 118.13, b.p. 127°, sp. gr. 0.975. Suppliers: FMC Corp., A, B, E, F, MCB.

The reagent is useful for the cathylation (carboethoxylation) of compounds having an active methylene group. An ethyl ester of type (1) reacts with diethyl carbonate

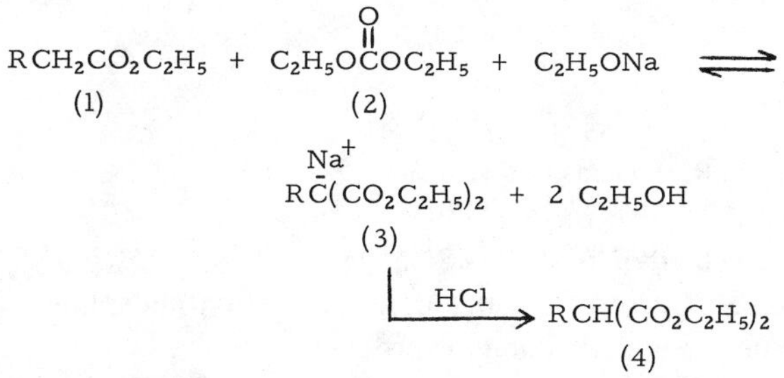

in the presence of sodium ethoxide to give the sodio derivative (3) in a reversible reaction. However, even if the equilibrium is unfavorable, an efficient conversion can be realized by using 4–8 moles of diethyl carbonate and distilling the ethanol as formed.[1] Acidification of (3) gives the substituted malonic ester (4).

Similar condensation of ketones with diethyl carbonate under basic catalysis affords β-keto esters.[2] Ethyl phenylcyanoacetate is prepared[3,4] from dry, freshly prepared sodium ethoxide (0.52 mole), phenylacetonitrile (0.5 mole), diethyl carbonate (2.5 moles), and toluene. As ethanol is removed by distillation, more toluene is added. After acidification of the enolate, ethyl phenylcyanoacetate is obtained in 70–78% yield.

$$C_6H_5CH_2CN + C_2H_5OCOC_2H_5 \xrightarrow[-2\ HOC_2H_5]{NaOC_2H_5} C_6H_5\overset{-}{C}(CN)COC_2H_5(Na^+) \xrightarrow{H^+} C_6H_5CH(CN)CO_2C_2H_5$$

A procedure for the preparation of tridiphenylcarbinol[5] calls for gradual addition of sodium to a vigorously stirred solution of *p*-chlorodiphenyl and diethyl carbonate

$$3\ C_6H_5C_6H_4Cl\text{-}\underline{p} + (C_2H_5O)_2CO + 6\ Na \xrightarrow[35\text{-}40\%]{} (\underline{p}\text{-}C_6H_5C_6H_4)_3CONa + 3\ NaCl + 3\ NaOC_2H_5$$

in refluxing benzene. At the end of the reaction the mixture is acidified and the tridiphenylcarbinol recovered in the usual way.

[1]V. H. Wallingford, A. H. Homeyer, and D. M. Jones, *Am. Soc.*, **63**, 2056 (1941)
[2]*Idem, ibid.*, **63**, 2252 (1941)
[3]*Idem, ibid.*, **64**, 576 (1942)
[4]E. C. Horning and A. F. Finelli, *Org. Syn., Coll. Vol.*, **4**, 461 (1963)
[5]A. A. Morton, J. R. Myles, and W. S. Emerson, *ibid.*, **3**, 831 (1955)

Diethyl chlorophosphite, $(C_2H_5O)_2POCl$. Mol. wt. 172.56, b.p. 64°/6–7 mm.

Preparation in 81% yield by reaction of diethyl phosphite with carbon tetrachloride and triethylamine.[1]

ArOH ⟶ ArH. In a procedure for the removal of the phenolic hydroxyl group of estrone by the method of Kenner and Williams,[2] a G. D. Searle group[3] found it preferable to use diethyl chlorophosphite itself rather than to generate it *in situ* from diethyl phosphonate.

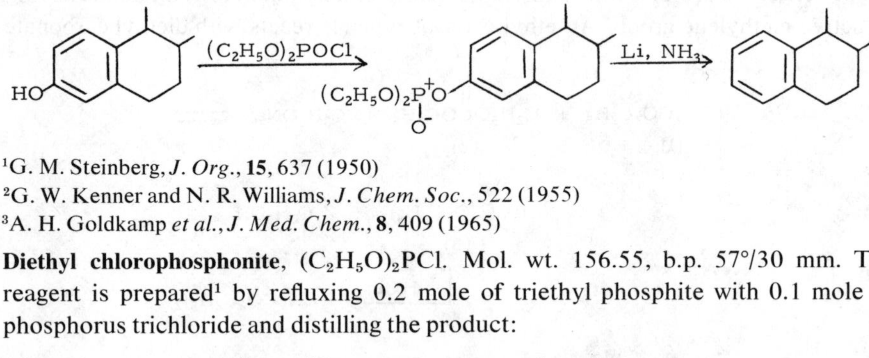

HO
(C₂H₅O)₂POCl
(C₂H₅O)₂P⁺O
O⁻
Li, NH₃

[1]G. M. Steinberg, *J. Org.*, **15**, 637 (1950)
[2]G. W. Kenner and N. R. Williams, *J. Chem. Soc.*, 522 (1955)
[3]A. H. Goldkamp *et al.*, *J. Med. Chem.*, **8**, 409 (1965)

Diethyl chlorophosphonite, $(C_2H_5O)_2PCl$. Mol. wt. 156.55, b.p. 57°/30 mm. The reagent is prepared[1] by refluxing 0.2 mole of triethyl phosphite with 0.1 mole of phosphorus trichloride and distilling the product:

$$2\ (C_2H_5O)_3P + PCl_3 \longrightarrow 3\ (C_2H_5O)_2PCl$$

Its use in peptide synthesis[2] is illustrated as follows:

$$H_2N\overset{R}{C}HCO_2C_2H_5 \xrightarrow[-HCl]{(C_2H_5O)_2PCl} (C_2H_5O)_2PNH\overset{R}{C}HCO_2C_2H_5$$

$$\xrightarrow{CbNH\overset{R'}{C}HCO_2H} CbNH\overset{R'}{C}HCONH\overset{R}{C}HCO_2C_2H_5 + (C_2H_5O)_2POH$$

[1]H. G. Cook, J. D. Ilett, B. C. Saunders, G. J. Stacey, H. G. Watson, I. G. E. Wilding, and S. J. Woodcock, *J. Chem. Soc.*, 2921 (1949)
[2]G. W. Anderson, J. Blodinger, R. W. Young, and A. D. Welcher, *Am. Soc.*, **74**, 5304 (1952)

Diethyl(2-chloro-1,1,2-trifluoroethyl)amine, $(C_2H_5)_2NCF_2CHClF$. Mol. wt. 189.62, b.p. 32–33°/5.5–6.0 mm. Prepared from chlorotrifluoroethylene and diethylamine: $(C_2H_5)_2NH + CF_2{=}CClF$.[1]

The reagent reacts with primary and secondary hydroxysteroids in methylene chloride at 25° to give the corresponding fluoro compounds. The reaction proceeds with essentially complete inversion of configuration except in the case of Δ^5-3β-hydroxysteroids, with which retention of configuration is observed. Thus Δ^5-3β-hydroxyandrostene-17-one gives the corresponding 3β-fluoro derivative in 78% yield.[2] Other workers[3] report that the reaction is sensitive to solvent and temperature effects and that fluorination is accompanied by elimination and rearrangement reactions.

The reagent reacts with 19-hydroxy-Δ^4-androstene-3,17-dione (1) in refluxing acetonitrile to give the 5,19-cyclosteroids (2) and (3).[4]

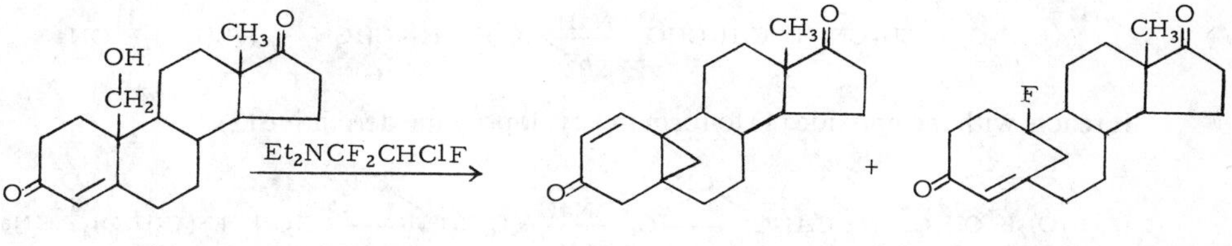
CH3
OH
CH2
Et2NCF2CHClF
CH3
+
CH3
F

(1) (2) 47% crude (3) 38% crude

Smith *et al.*[5] attempted to prepare a 1-fluorosteroid by the action of the reagent on 1α-hydroxy-5α-androstane-3,17-dione but obtained instead three products of retropinacol rearrangement of the C_{19}-methyl group: Δ^4-, $\Delta^{5(10)}$-, and Δ^9-1β-methylestrene-3-ketones.

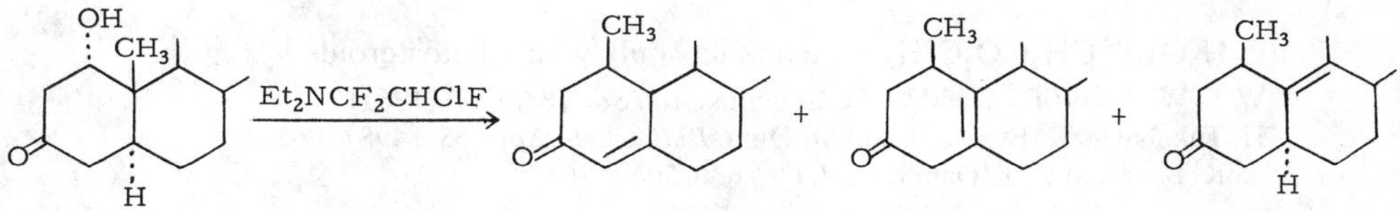
OH
CH3
H
Et2NCF2CHClF
CH3
+
CH3
+
CH3
H

Knox *et al.*[6] found that the reagent reacts with 19-hydroxy-Δ^5-3-acetoxysteroids to give chiefly three products of skeletal rearrangement.

[1] R. L. Pruett *et al., Am. Soc.*, **72**, 3646 (1950)
[2] D. E. Ayer, *Tetrahedron Letters*, **23**, 1065 (1962)
[3] L. H. Knox, E. Velarde, S. Berger, D. Cuadriello, and A. D. Cross *ibid.*, **26**, 1249 (1962); *J. Org.*, **29**, 2187 (1964)
[4] L. H. Knox, E. Velarde, and A. D. Cross, *Am. Soc.*, **87**, 3727 (1965)
[5] L. L. Smith, T. J. Foell, and D. M. Teller, *J. Org.*, **30**, 3781 (1965)
[6] L. H. Knox, E. Velarde, S. Berger, I. Delfín, R. Grezemkovsky, and A. D. Cross, *J. Org.*, **30**, 4160 (1965)

Diethylcyanamide, $(C_2H_5)_2NCN$. Mol. wt. 98.15, b.p. 68°/10 mm. Prepared by the reaction of cyanogen bromide (1 mole) with diethylamide (2 moles).[1] Supplier: KK.

Cyanamide, $N{\equiv}C{-}NH_2$, in the tautomeric form $HN{=}C{=}NH$, is a carbodiimide and can be used in peptide synthesis. Diethylcyanamide, although incapable of tautomerism, functions as a dehydrating agent for the coupling of an N-protected amino acid with an amino acid ester.[2] Best yields are obtained by heating the three components without solvent at 100° for 2 hrs.

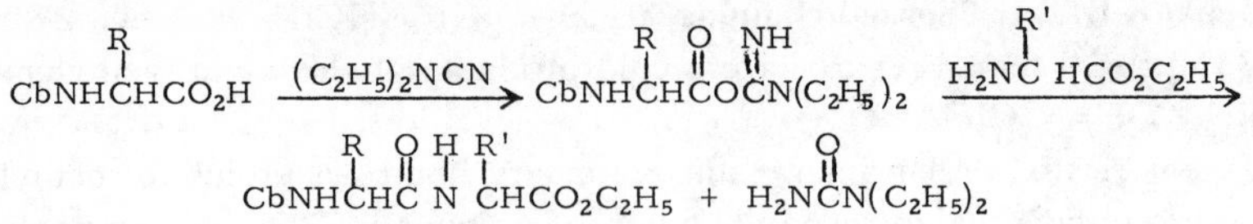

[1]O. Wallach, *Ber.*, **32**, 1872 (1899)
[2]G. Losse and H. Weddige, *Ann.*, **636**, 144 (1960)

Diethyl cyanomethylphosphonate, $(C_2H_5O)_2\overset{O^-}{\overset{|}{P^+}}CH_2CN$. Mol. wt. 177.17. Supplier: Aldrich.

The reagent, useful in a modified Wittig reaction, reacts with aldehydes and ketones under catalysis by a strong base to give α, β-unsaturated nitriles:

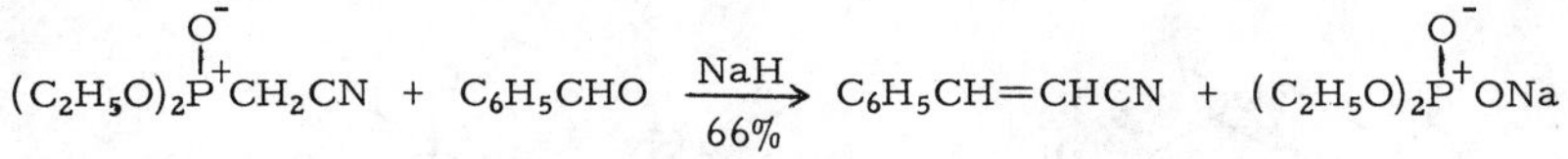

It reacts with the epoxide (1) to form the cyclopropane derivative (2).

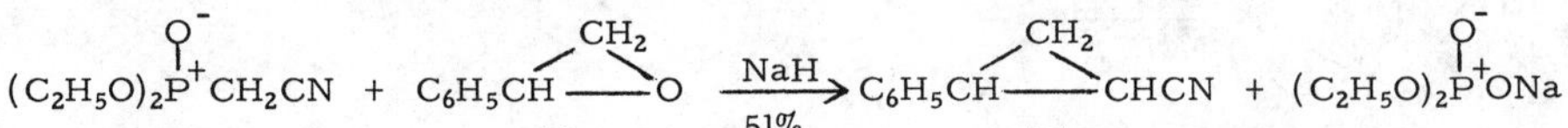

(1) (2)

In the original procedure Wadsworth and Emmons[1] used sodium hydride as base and 1,2-dimethoxyethane as solvent. In a Japanese procedure[2] sodium amide is used as base and tetrahydrofurane as solvent. The reagent reacts with steroids having a keto group at C_3, C_{17}, or C_{20} and is thus less selective than triethylphosphonoacetate, $(C_2H_5O)_2\overset{O^-}{\overset{|}{P^+}}CH_2CO_2C_2H_5$, which reacts only with 3-ketosteroids.[3]

[1]W. S. Wadsworth, Jr., and W. D. Emmons, *Am. Soc.*, **83**, 1733 (1961)
[2]H. Takahashi, K. Fujiwara, and M. Ohta, *Bull. Chem. Soc.*, **35**, 1498 (1962)
[3]A. K. Bose and R. T. Dahill, Jr., *J. Org.*, **30**, 505 (1965)

Diethylene glycol dimethyl ether, *see* Diglyme.

Diethyl ethoxycarbonylmethylphosphonate, *see* Triethyl phosphonoacetate.

Diethyl ethylenepyrophosphite (3). Mol. wt. 228.12, b.p. 68–69°/0.4 mm. The reagent is prepared from ethylene chlorophosphite (1) and diethyl phosphonate (2).[1] It is

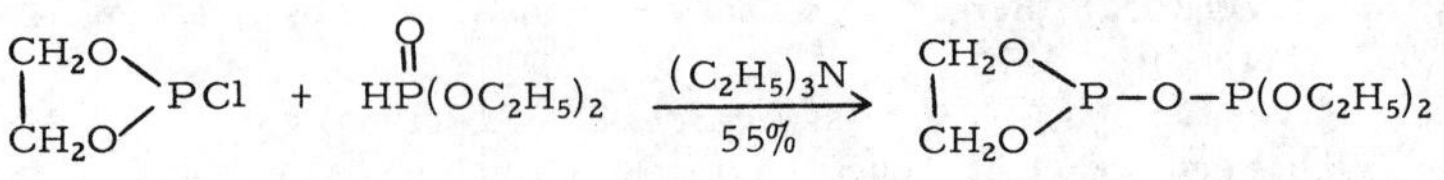

(1) (2) (3)

used for the formation of peptide bonds in the same way that tetraethylpyrophosphite is used; an advantage is that it is more readily prepared.

[1]G. W. Anderson and A. C. McGregor, *Am. Soc.*, **79**, 6180 (1957)

Diethyl oxalate, $(CO_2C_2H_5)_2$. Mol. wt. 146.14, b.p. 83–86°/25 mm. Suppliers: B, E, F, MCB.

Reactions, *see also* Sodium ethoxide, condensation of esters with diethyl oxalate; Sodium methoxide, heterocycles.

Ethoxalyl ketones. The acylation of ketones with diethyl oxalate has been reviewed by Hauser.[1] The reaction product, being an α-keto ester, loses CO on heating to give a β-keto ester. Examples are the acylation of acetone to form ethyl acetopyruvate[2] and of cyclohexanone to give ethyl cyclohexanone-2-glyoxalate.[3]

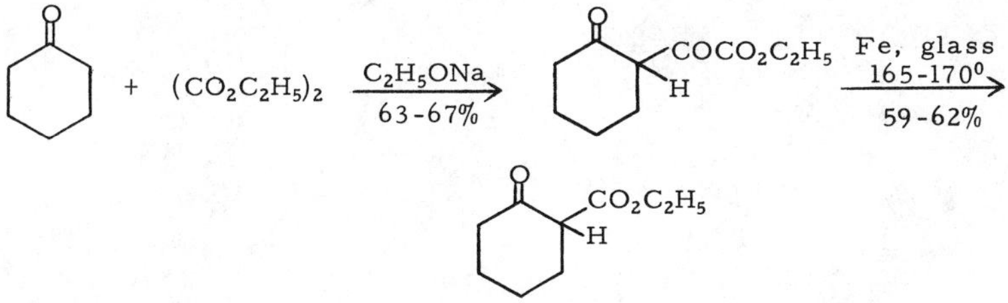
+ (CO2C2H5)2
C2H5ONa
63-67%
COCO2C2H5
H
Fe, glass
165-170°
59-62%
CO2C2H5
H

Bachmann[4] used the sequence in a synthesis of equilenin (in this case dimethyl oxalate was used). Pyrolysis proved to be erratic until it was found that addition of powdered soft glass promotes smooth elimination of CO.

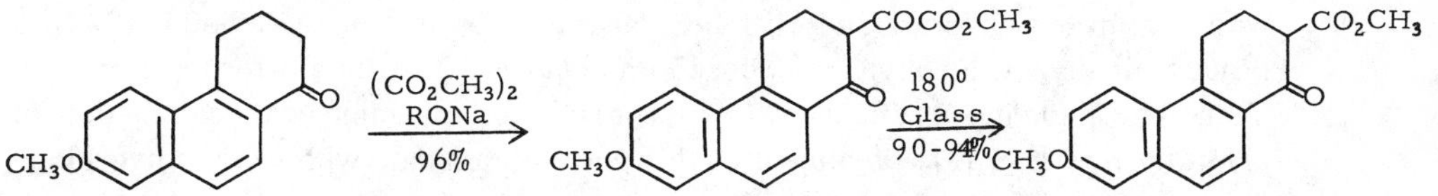
CH3O
(CO2CH3)2
RONa
96%
CH3O
COCO2CH3
180°
Glass
90-94%
CH3O
CO2CH3

Indole synthesis.[5] In the first step of the synthesis *o*-nitrotoluene is condensed with diethyl oxalate in the presence of potassium ethoxide, and the product is isolated as the deep purple potassium enolate. Hydrogenation in acetic acid reduces the nitro group and effects cyclization to ethyl indole-2-carboxylate. This on hydrolysis and decarboxylation affords indole.

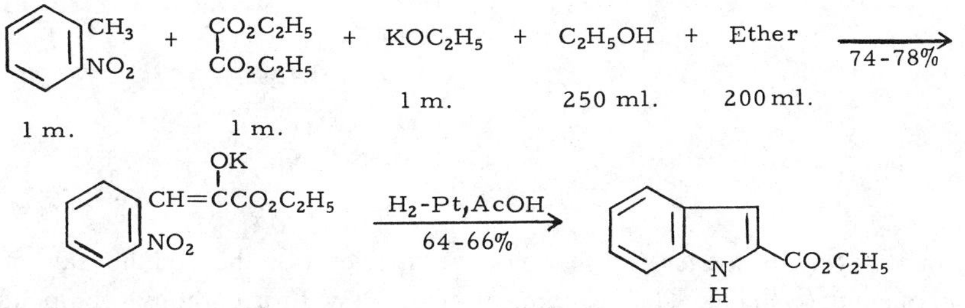
CH3
NO2
1 m.
+
CO2C2H5
CO2C2H5
1 m.
+ KOC2H5
1 m.
+ C2H5OH
250 ml.
+ Ether
200 ml.
74-78%
OK
CH=CCO2C2H5
NO2
H2-Pt,AcOH
64-66%
N
H
CO2C2H5

[1]C. R. Hauser, F. W. Swamer, and J. T. Adams, *Org. Reactions*, **8**, 84, 117 (1954)
[2]C. S. Marvel and E. E. Dreger, *Org. Syn., Coll. Vol.*, **1**, 238 (1941)
[3]H. R. Snyder, L. A. Brooks, and S. H. Shapiro, *ibid.*, **2**, 531 (1943)
[4]W. E. Bachmann, W. Cole, and A. L. Wilds, *Am. Soc.*, **62**, 824 (1940)
[5]W. E. Noland and F. J. Baude, *Org. Syn.*, **43**, 40 (1963)

Diethyl phosphonate (formerly Diethyl phosphite or Diethyl hydrogen phosphite),

$(C_2H_5O)_2\overset{O}{\overset{\|}{P}}H$. Mol. wt. 138.11, b.p. 187–188° (73–74°/14 mm.). Preparation in 93% yield by reaction of ethanol with phosphorus trichloride in either carbon tetrachloride or ether.[1] Suppliers: Virginia-Carolina Chem. Co., A, B, E, MCB.

Reduction of phenols. The reagent provides a means for the reduction of a phenol to the parent hydrocarbon.[2] A solution of the phenol and diethyl hydrogen phosphite in carbon tetrachloride is treated with triethylamine and allowed to stand for 24 hrs. for complete separation of triethylamine hydrochloride (1). The phenol diethyl phosphate ester is collected, dissolved in tetrahydrofurane, and reduced

$$1. \quad ArOH + HP(OC_2H_5)_2 + CCl_4 + (C_2H_5)_3N \longrightarrow$$
$$ArO\overset{O^-}{\underset{+}{P}}(OC_2H_5)_2 + CHCl_3 + (C_2H_5)_3N \cdot HCl$$

$$2. \quad ArO\overset{O^-}{\underset{+}{P}}(OC_2H_5)_2 + 2 Na + NH_3 \longrightarrow ArH + NaNH_2$$
$$+ \ NaO\overset{O^-}{\underset{+}{P}}(OC_2H_5)_2$$

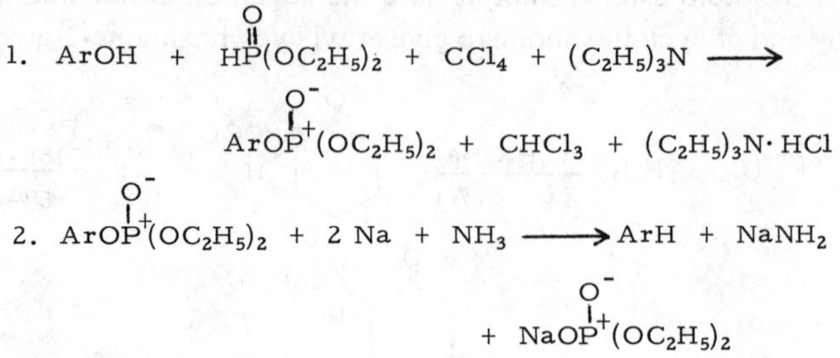

with sodium and liquid ammonia (2). Yields of crude hydrocarbons are 70–80%. Pelletier[3] adapted the reaction to a semimicro procedure (50–200 mg.) suitable for characterization of phenolic degradation products of natural products.

Use in synthesis. Wadsworth and Emmons[4] condensed diethyl phosphonate with carbon tetrachloride and cyclohexylamine according to A. R. Todd (1945) to produce diethyl cyclohexylaminephosphoramidate (3). The initially formed trichloromethylphosphonate (2) reacts with a second mole of amine with elimination of chloroform to give (3). Treatment of this phosphoramidate with sodium hydride in

$$(C_2H_5O)_2\overset{O}{\underset{}{P}}H + CCl_4 \xrightarrow[-HCl]{C_6H_{11}NH_2} (C_2H_5O)_2\overset{O}{\underset{}{P}}CCl_3 \xrightarrow[-CHCl_3]{C_6H_{11}NH_2}$$
$$(1) \hspace{5cm} (2)$$

$$(C_2H_5O)_2\overset{O}{\underset{}{P}}\overset{H}{\underset{}{N}}C_6H_{11} \xrightarrow[-H_2]{NaH} (C_2H_5O)_2\overset{O}{\underset{}{P}}-\overset{-}{N}C_6H_{11}(Na^+)$$
$$(3) \hspace{5cm} (4)$$

$$60\% \downarrow C_6H_5N=C=O$$

$$C_6H_5N=C=NC_6H_{11} + (C_2H_5O)_2\overset{O}{\underset{}{P}}ONa$$
$$(5) \hspace{4cm} (6)$$

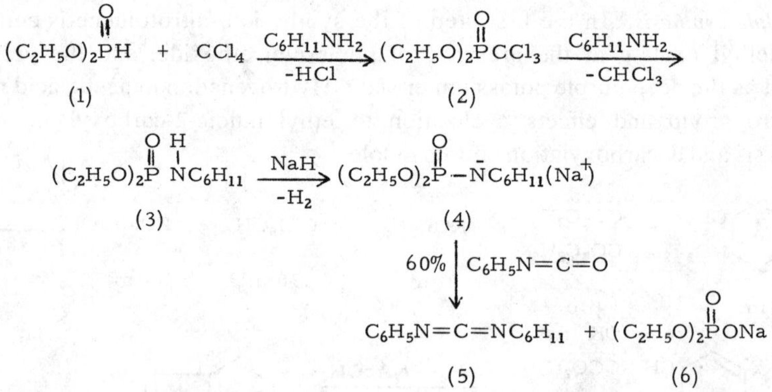

1,2-dimethoxyethane generates the reactive species (4), the sodium salt of the phosphonamidate anion. The reaction of this salt with a carbonyl compound is illustrated for the reaction with phenyl isocyanate to give N-phenyl-N'-cyclohexylcarbodiimide (5). Other primary amines can be used. The carbonyl component can be an aldehyde, a ketene, carbon dioxide, or carbon disulfide.

Special solvent. Diethyl phosphonate is miscible with water and with ordinary organic solvents and has remarkable solvent power. It is recommended as solvent and acid-acceptor for peptide syntheses utilizing tetraethyl pyrophosphite[5] or a reagent such as ethyl dichlorophosphite.[6]

It is a useful solvent for the preparation of arylhydrazones and 2,4-dinitrophenylhydrazones.[7] No catalyst is required because a trace of water hydrolyzes the reagent to the acidic ethyl hydrogen phosphonate, $HP(O)(OC_2H_5)OH$. The small volume of solvent required improves yields. Although the reaction usually is rapid at room temperature, yields may be improved by heating the mixture for 1 hr. on the steam bath.

[1]H. McCombie, B. C. Saunders, and G. J. Stacey, *J. Chem. Soc.*, 380 (1945)
[2]G. W. Kenner and N. R. Williams, *ibid.*, 522 (1955)
[3]S. W. Pelletier and D. M. Locke, *J. Org.*, **23**, 131 (1958)
[4]W. S. Wadsworth, Jr., and W. D. Emmons, *Am. Soc.*, **84**, 1316 (1962)
[5]G. W. Anderson, J. Blodinger, and A. D. Welcher, *Am. Soc.*, **74**, 5309 (1952)
[6]R. W. Young, K. H. Wood, R. J. Joyce, and G. W. Anderson, *Am. Soc.*, **78**, 2126 (1956)
[7]J. A. Maynard, *Australian J. Chem.*, **15**, 867 (1962)

Diethyl sulfate, $(C_2H_5O)_2SO_2$. Mol. wt. 154.19, b.p. 210°, sp. gr. 1.17. Suppliers: A, B, E, F, KK, MCB. *See also* Dimethyl sulfate.

C-Alkylation. A procedure for the preparation of *n*-propylbenzene[1] calls for dropwise addition of freshly distilled diethyl sulfate to benzylmagnesium chloride.

$$\underset{2\text{ m.}}{C_6H_5CH_2Cl} + \underset{2\text{ g. at.}}{Mg} \xrightarrow{\text{Ether}} C_6H_5CH_2MgCl \xrightarrow[70\text{-}75\%]{4\text{ m. }(C_2H_5O)_2SO_2} C_6H_5CH_2CH_2CH_3$$

[1]H. Gilman and W. E. Catlin, *Org. Syn., Coll. Vol.*, **1**, 471 (1941)

N,N-Diethyl-1,2,2-trichlorovinylamine, $Cl_2C{=}C(Cl)N(C_2H_5)_2$. Mol. wt. 202.52, b.p. 67°/6.2 mm.

The formation of this reagent by reaction of N,N-diethyl-2,2,2-trichloroacetamide and triethyl phosphite entails oxidation of the phosphorus atom by the amide with migration of a chlorine atom.[1] The resulting trichloroenamine reacts rapidly with a

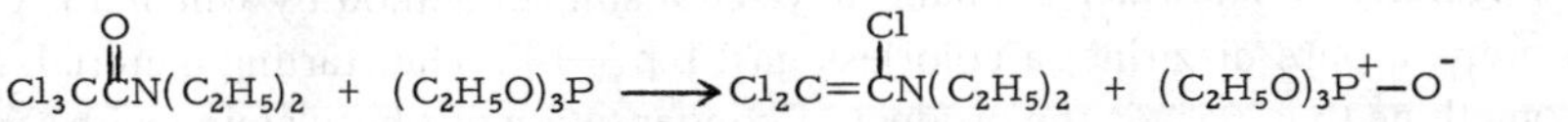

carboxylic acid or an alcohol to give the corresponding acid chloride or alkyl chloride in good yields and it reacts with primary amines to form amidines.[2]

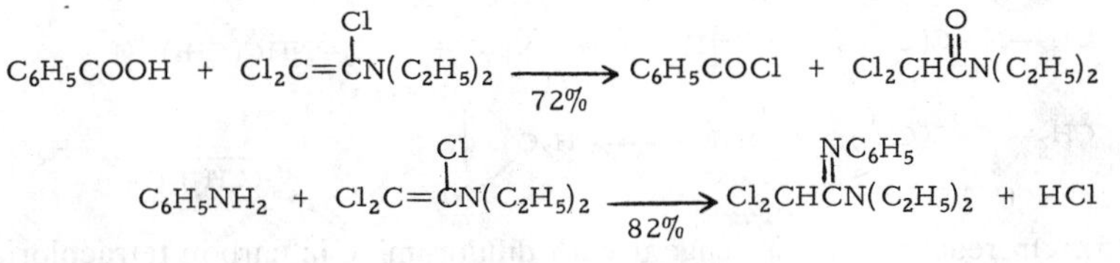

[1]A. J. Speziale and R. C. Freeman, *Am. Soc.*, **82**, 903 (1960)
[2]*Idem, ibid.*, **82**, 909 (1960)

Diethylzinc–Methylene iodide. This combination reacts with cyclohexene to give norcarane in fair yield.[1] The reaction may be related to the Simmons-Smith reaction

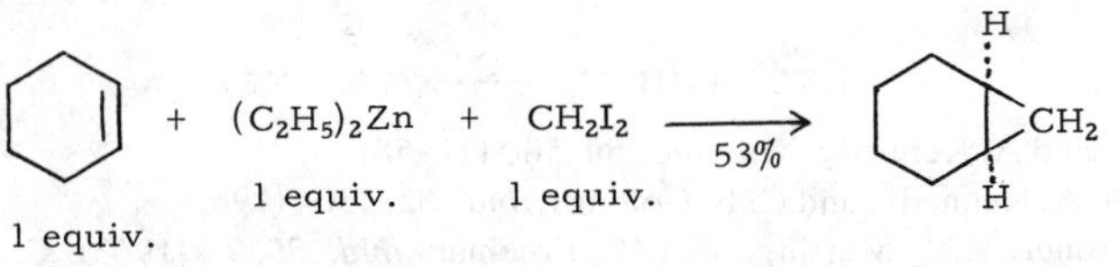

but is much more rapid. Thus the methylene iodide must be added slowly to moderate the reaction. Note that diethylzinc takes fire in air.

[1]J. Furukawa, N. Kawabata, and J. Nishimura, *Tetrahedron Letters*, 3353 (1966)

Difluoramine, HNF_2. Mol. wt. 53.02, b.p. −23°.

Preparation. The starting material for the preparation of this reagent is tetrafluorohydrazine, made by reaction of nitrogen trifluoride with various metals at 375° (1).[1] Conversion into difluoroamine is accomplished by heating tetrafluorohydrazine with thiophenol in an evacuated bulb; yield 74%.[2]

1. $2 NF_3 + M \longrightarrow F_2N-NF_2 + MF_2$

2. $F_2N-NF_2 + C_6H_5SH \longrightarrow 2 HNF_2 + C_6H_5SSC_6H_5$

Deamination. The reagent effects direct deamination of aliphatic and aromatic primary amines, for example:

$$3 \underline{n}-C_4H_9NH_2 + HNF_2 \longrightarrow 2 \underline{n}-C_4H_9\overset{+}{N}H_3\overset{-}{F} + N_2 + \underline{n}-C_4H_{10}$$

Coupling reaction. When treated with difluoramine, dibenzylamine, aziridine, and azetidine release nitrogen and the remaining fragments couple to form dibenzyl, ethylene, and cyclopropane.[3]

$$\begin{array}{c} C_6H_5CH_2 \\ \\ C_6H_5CH_2 \end{array}\!\!\!N-H \xrightarrow[53\%]{HNF_2} C_6H_5CH_2CH_2C_6H_5 + N_2$$

$$\triangleright\!\!N-H \xrightarrow[80\%]{HNF_2} \begin{array}{c} CH_2 \\ \| \\ CH_2 \end{array} + N_2$$

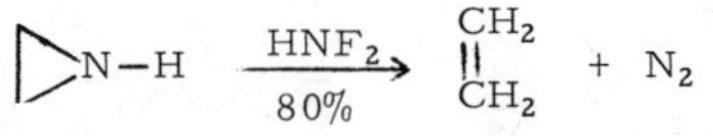

Aziridine

$$\square\!\!-\!N-H \xrightarrow[40\%]{HNF_2} \triangle + N_2$$

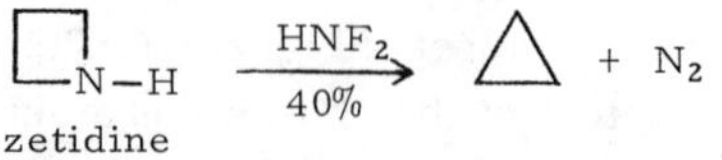

Azetidine

Synthesis of diazirine. Graham[4] devised a simple method by which it is possible to prepare pure diazirine, a colorless gas, b.p. $-14°$. The starting material, *t*-butyl-azomethine (b.p. 65°), is the product of condensation of *t*-butylamine with formalde-

(a) $(CH_3)_3CNH_2 + CH_2O \longrightarrow (CH_3)_3CN=CH_2 + H_2O$

(b) $CH_2=N-C(CH_3)_3 + HNF_2 \longrightarrow :\overset{..}{N}F + CH_2=\overset{+}{N}HC(CH_3)_3\ (\overset{-}{F})$

(c) $CH_2=N-C(CH_3)_3 + :\overset{..}{N}F \longrightarrow H_2C\!\!\begin{array}{c} \diagup N-C(CH_3)_3 \\ | \\ \diagdown NF \end{array} \xrightarrow[-(CH_3)_3CF]{} H_2C\!\!\begin{array}{c} \diagup N \\ \| \\ \diagdown N \end{array}$ 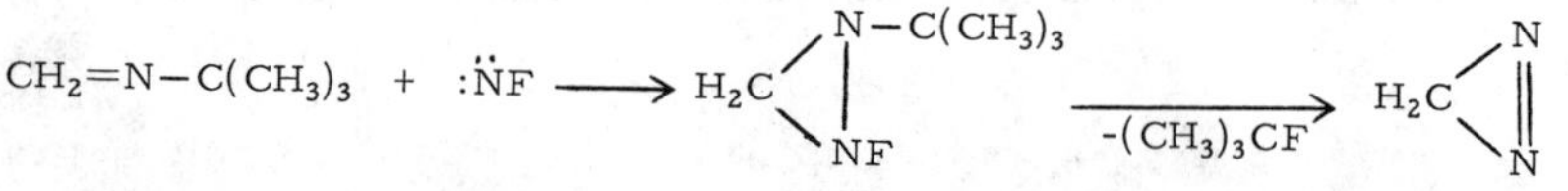

hyde (a). On reaction of this reagent with difluoramine in carbon tetrachloride in a vacuum system, diazirine is formed and is caught in a $-128°$ trap. The probable steps are formulated in (b) and (c).

Reaction with C-nitroso compounds. The reaction affords the corresponding N'-fluorodiimide-N-oxide.[5]

$$RNO + HNF_2 \xrightarrow{Py} R\overset{\overset{\displaystyle O^-}{|+}}{N}=NF$$

[1]C. B. Colburn and A. Kennedy, *Am. Soc.*, **80**, 5004 (1958)
[2]J. P. Freeman, A. Kennedy, and C. B. Colburn, *ibid.*, **82**, 5304 (1960)
[3]C. L. Bumgardner, K. J. Martin, and J. P. Freeman, *ibid.*, **85**, 97 (1963); C. L. Bumgardner and J. P. Freeman, *ibid.*, **86**, 2233 (1964)
[4]W. H. Graham, *Am. Soc.*, **84**, 1063 (1962)
[5]T. E. Stevens and J. P. Freeman, *J. Org.*, **29**, 2279 (1964)

sym-Difluorotetrachloroacetone, $CFCl_2COCFCl_2$. Mol. wt. 231.86, b.p. 124°. Suppliers: Allied Chem. Corp., Gen. Chem. Div., Peninsular ChemResearch.

This reagent affords a ready source of chlorofluorocarbene:[1]

$$Cl_2FC\overset{\overset{\displaystyle O}{\|}}{C}CFCl_2 \xrightarrow{BuO^-} Cl_2FC\overset{\overset{\displaystyle O}{\|}}{C}OBu + :CFCl + Cl^-$$

A suspension of 1 mole of potassium *t*-butoxide in 500 ml. of cyclohexene was stirred at 0° during addition of 0.5 mole of the precursor and then for 3 hrs. at 0–5°. Fractionation of the filtered liquid afforded 7-fluoro-7-chloronorcarane (1) in 36% yield. α-Methylstyrene afforded (2) in 46% yield.

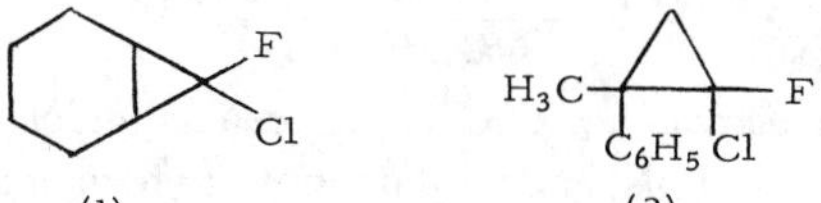
F
Cl
(1)
H_3C
F
C_6H_5 Cl
(2)

[1]B. Farah and S. Hornesky, *J. Org.*, **28**, 2494 (1963)

Diglyme, $CH_3OCH_2CH_2OCH_2CH_2OCH_3$. Mol. wt. 134.17, b.p. 161°, miscible with water. Suppliers: Ansul Chem. Co., (pure) BJ.

Purification.[1] A 1-l portion of diglyme (Ansul Chemical Co.) is stored over 10 g. of small pieces of calcium hydride for 12 hrs. and decanted into a distilling flask. Sufficient lithium aluminum hydride is added to ensure an excess of active hydride (*see* purification of THF), and the solvent is distilled at 62–63°/15 mm. and stored under nitrogen.

[1]Contributed by George Zweifel, University of California, Davis, California

1,4-Dihydronaphthalene-1,4-*endo*-oxide (6). Mol. wt. 144.16, m.p. 56°.

Wittig and Pohmer[1] prepared this reagent by shaking a solution of *o*-fluorobromobenzene in furane with lithium amalgam for four days; they found that it is cleaved by mineral acid to α-naphthol. In a simpler route to the reagent described by Fieser

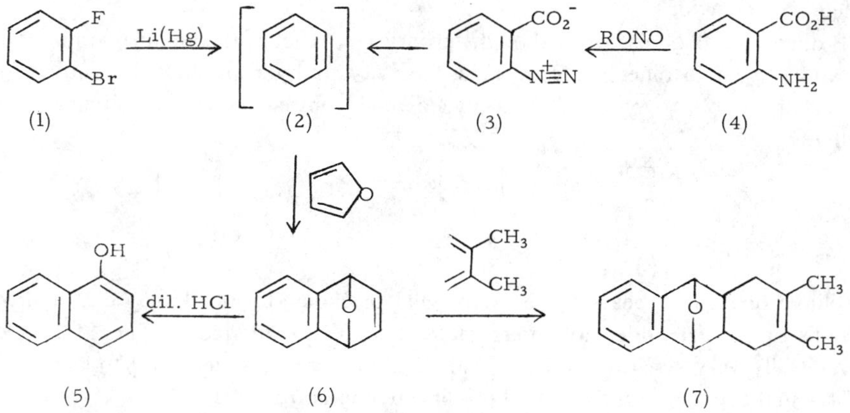
F
Br
(1)
Li(Hg)
(2)
CO_2^-
$N\equiv N$
(3)
RONO
CO_2H
NH_2
(4)
O
CH_3
CH_3
OH
dil. HCl
(5)
O
(6)
O
CH_3
CH_3
(7)

and Haddadin,[2] solutions of 20 ml. of isoamyl nitrite in 20 ml. of 1,2-dimethoxyethane and of 13.7 g. of anthranilic acid in 45 ml. of 1,2-dimethoxyethane were added simultaneously by drops to a refluxing mixture of 50 ml. of 1,2-dimethoxyethane and 50 ml. of furane in a period of 1 hr. The crude, yellowish oxide (7.4 g., 51%) gave colorless, pure material on crystallization from petroleum ether.

Wolthuis found the reagent to be a reactive dienophile;[3] for example it combines with 2,3-dimethylbutadiene to form the adduct (7). For another example[2] *see* Isobenzofurane.

[1]G. Wittig and L. Pohmer, *Ber.*, **89**, 1334 (1956)
[2]L. F. Fieser and M. J. Haddadin, *Can. J. Chem.*, **43**, 1599 (1965)
[3]E. Wolthuis, *J. Org.*, **26**, 2215 (1961)

Dihydropyrane. Mol. wt. 84.11, b.p. 85°. Preparation by dehydration-rearrangement of tetrahydrofurfuryl alcohol over alumina.[1-3] Suppliers: du Pont Electrochemicals Dept., Matheson, A, B, E, F, MCB.

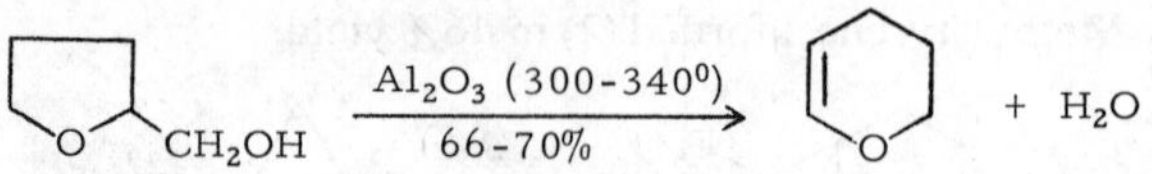

Al_2O_3 (300-340°)
66-70%
CH_2OH
+ H_2O

O-, S-, N-Protective derivatives. The reagent reacts under mild acid catalysis with primary and secondary alcohols, and with phenols, to form tetrahydropyranyl ethers, which are stable to bases, to RMgX, to $LiAlH_4$, to acetic anhydride in pyridine, and to oxidation.[4] When synthetic operations at another site in the molecule have

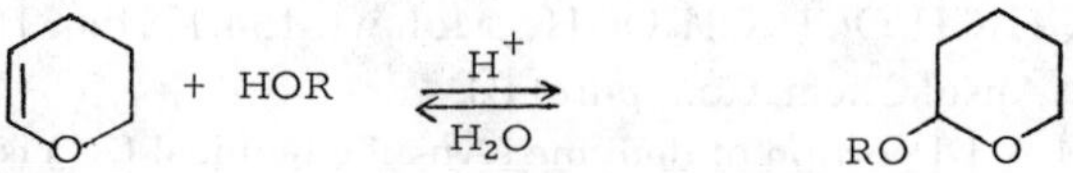

+ HOR
H^+
H_2O
RO

been completed, the ethers are cleaved easily by acids (refluxing with acetic acid gives the corresponding acetate).[5] Dihydropyrane is useful also for the protection of carboxyl groups,[6] of sulfhydryl groups,[7] and of the imidazole hydrogen in purines[8];

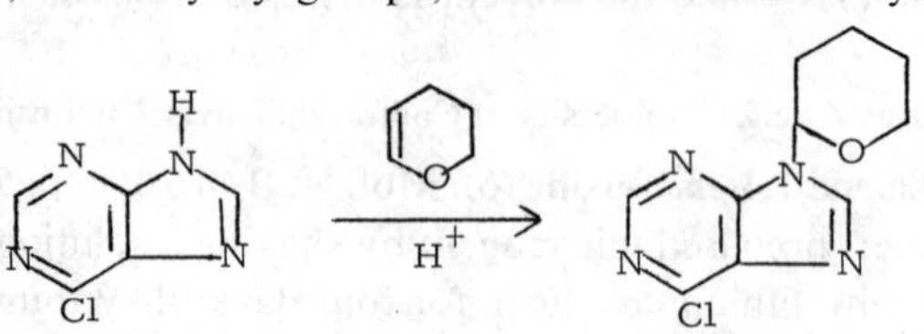

H^+
Cl
Cl

in each case the active-hydrogen compound is regenerated by hydrolysis with dilute mineral acid.

Propargylic alcohols, even though tertiary (1), react with dihydropyrane under catalysis from *p*-toluenesulfonic acid to give tetrahydropyranyl derivatives (2), which are stable to base but hydrolyzed readily by aqueous acid or magnesium sulfate.[9]

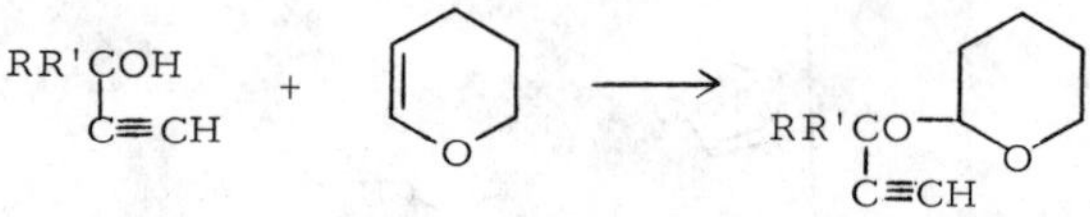

RR'COH
C≡CH
+
RR'CO
C≡CH

(1) (2)

Dihydropyrane appears to be the reagent of choice for blocking the 2′-hydroxy groups of 5′-nucleotide monomers that are to be polymerized.[10] The THP-ethers are usually very sensitive to acids and should be prepared just prior to use or else stored in the dry state at the lowest possible temperature.

Dihydropyrane has been found to react unselectively with axial and equatorial hydroxyl groups of an inositol.[11]

Reactions. Dihydropyrane on hydrolysis with dilute hydrochloric acid affords

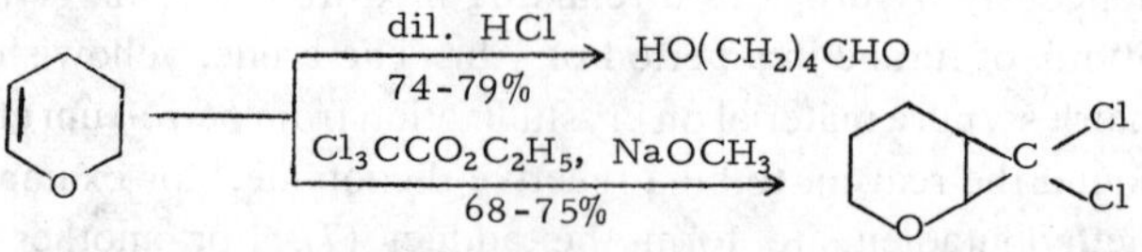

dil. HCl
74-79%
$HO(CH_2)_4CHO$
$Cl_3CCO_2C_2H_5$, $NaOCH_3$
68-75%
Cl
Cl

5-hydroxypentanal.[12] It reacts with dichlorocarbene generated from ethyl trichloroacetate and sodium methoxide in pentane to give 2-oxa-7,7-dichloronorcarane.[13]

Reaction with amides. Dihydropyrane reacts with aliphatic and aromatic amides in benzene or dimethylformamide under catalysis by hydrogen chloride to form

N-(2-tetrahydropyranyl)amides.[14] The products are stable to alkali but are hydrolyzed by concd. phosphoric acid to the amide and 5-hydroxyvaleraldehyde.

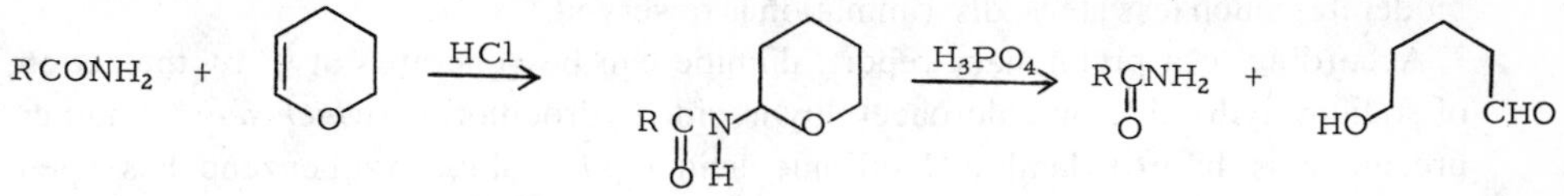

[1]R. Paul, *Bull. soc.*, **53**, 1489 (1933)
[2]L. E. Schniepp and H. H. Geller, *Am. Soc.*, **68**, 1646 (1946)
[3]R. L. Sawyer and D. W. Andrus, *Org. Syn., Coll. Vol.*, **3**, 276 (1955)
[4]R. Paul, *Bull. soc.*, **1**, 971 (1934); G. F. Woods and D. N. Kramer, *Am. Soc.*, **69**, 2246 (1947); W. E. Parham and E. L. Anderson, *ibid.*, **70**, 4187 (1948); A. C. Ott, M. F. Murray, and R. L. Pederson, *ibid.*, **74**, 1239 (1952); R. G. Jones and M. J. Mann, *ibid.*, **75**, 4048 (1953)
[5]W. G. Dauben and H. L. Bradlow, *ibid.*, **74**, 559 (1952)
[6]R. E. Bowman and W. D. Fordham, *J. Chem. Soc.*, 3945 (1952)
[7]W. E. Parham and D. M. DeLaitsch, *Am. Soc.*, **76**, 4962 (1954)
[8]B. K. Robins, E. F. Godefroi, E. C. Taylor, R. L. Lewis, and A. Jackson, *ibid.*, **83**, 2574 (1961)
[9]D. N. Robertson, *J. Org.*, **25**, 931 (1960)
[10]D. B. Straus and J. R. Fresco, *Am. Soc.*, **87**, 1364 (1965)
[11]S. J. Angyal and S. D. Gero, *J. Chem. Soc.*, 5255 (1965)
[12]R. Paul, *Bull. soc.*, **1**, 976 (1934); G. F. Woods, Jr., *Org. Syn., Coll. Vol.*, **3**, 470 (1955)
[13]W. E. Parham, E. E. Schweizer, and S. A. Mierzwa, Jr., *Org. Syn.*, **41**, 76 (1961)
[14]A. J. Speziale, K. W. Ratts, and G. J. Marco, *J. Org.*, **26**, 4311 (1961)

Dihydroresorcinol, *see* Cyclohexane-1,3-dione.

Diimide, HN=NH. The reagent, generated *in situ* by cupric ion-catalyzed oxidation of hydrazine with oxygen (air),[1] hydrogen peroxide,[1] potassium ferricyanide,[2] or mercuric oxide,[2] reduces olefins, alkynes, and azo compounds. Reduction of the

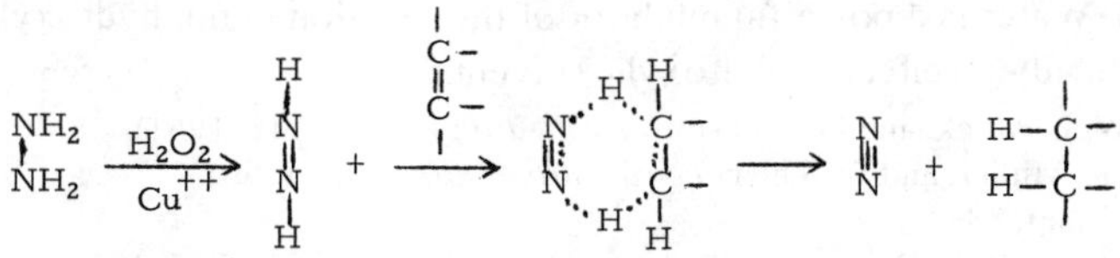

relatively unreactive double bond of cholesterol proceeded in yield of only 20%. In a simple procedure[3] for the reduction of azobenzene to hydrazobenzene, a suspension of the azo compound in ethanol is cooled to 0°, 95% hydrazine and a trace of copper sulfate solution are added, and 30% hydrogen peroxide is added by drops at 0° until the color is discharged.

Dipotassium azodicarboxylate (2), prepared by hydrolysis of azodicarboxamide (1), is a less convenient source since a large excess of reagent is required.[1,4] A

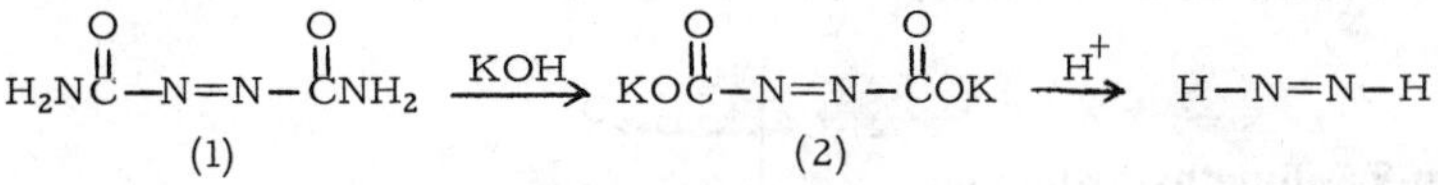

(1) (2)

further source is *p*-toluenesulfonylhydrazide, but again a 100% excess of reagent is required.[2,5] Reduction of an olefin is effected by refluxing the reagent and substrate in diethylene glycol dimethyl ether.

Diimide reduces symmetrical multiple bonds (C=C, C≡C, N=N) more readily than polar bonds (C≡N, C=O, C=N, O=N$^+$—O$^-$, O=S$^+$—O$^-$).[6] The reduction is a stereospecific *cis*-addition.[2,7] A study of the reduction of olefins by generation of diimide in an ethanolic solution of the olefin and hydrazine in an oxygen atmosphere at 45–55° showed that where the bulk effect in an olefin is pronounced, reduction is markedly subject to steric approach control; where the bulk effect is moderate, much less steric discrimination is observed.[8]

According to a preliminary report, diimide can be generated at 0° by the action of sodium hydroxide on chloroacetylhydrazide hydrochloride (*which see*).[9] Another precursor is hydroxylamine-O-sulfonic acid (*which see*). Azobenzene has been reduced quantitatively to hydrazobenzene and allyl alcohol to propanol (70%).

$$ClCH_2CONHNH_2 \cdot HCl + 2\,NaOH \longrightarrow \underbrace{CH_2{=}C{=}O + H_2O}_{\longrightarrow\ CH_3CO_2H} + HN{=}NH + 2\,NaCl$$

In the absence of a substrate, diimide disproportionates to nitrogen and hydrazine, which combines with the ketene formed to give acetylhydrazide.

cis,trans,trans-$\Delta^{1,5,9}$-Cyclododecatriene can be selectively reduced to *cis*-cyclododecene by diimide (hydrazine hydrate, cupric sulfate, air).[10] The result shows that there is considerable difference in reactivity between the *cis* and *trans* double bonds.

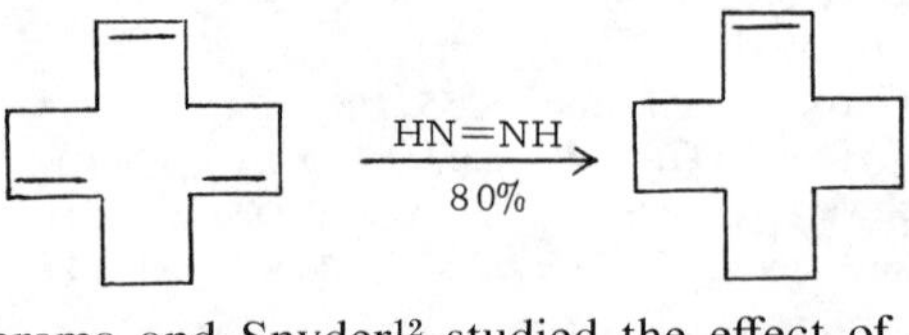
HN=NH
80%

Reviews.[11] Hamersma and Snyder[12] studied the effect of variation in reaction conditions in reductions with diimide generated from dipotassium azodicarboxylate. Air does not have a deleterious effect. Rate of reduction decreases with solvents in the following order: pyridine > dioxane > dimethyl sulfoxide > methanol > ethanol > *n*-butanol. Water is a powerful inhibitor of the reaction in nonhydroxylic solvents but has practically no effect in hydroxylic solvents.

[1]E. J. Corey, W. L. Mock, and D. J. Pasto, *Tetrahedron Letters*, 347 (1961)
[2]S. Hünig, H.-R. Müller, and W. Thier, *Tetrahedron Letters*, 353 (1961)
[3]*Org. Expts.*, Chapt. 38
[4]E. E. van Tamelen, R. S. Dewey, and R. J. Timmons, *Am. Soc.*, **83**, 3725 (1961)
[5]R. S. Dewey and E. E. van Tamelen, *Am. Soc.*, **83**, 3729 (1961)
[6]E. E. van Tamelen, R. S. Dewey, M. F. Lease, and W. H. Pirkle, *Am. Soc.*, **83**, 4302 (1961)
[7]E. J. Corey, D. J. Pasto, and W. L. Mock, *Am. Soc.*, **83**, 2957 (1961)
[8]E. E. van Tamelen and R. J. Timmons, *Am. Soc.*, **84**, 1067 (1962)
[9]R. Buyle, A. Van Overstraeten, and F. Eloy, *Chem. Ind.*, 839 (1964)
[10]M. Ohno and M. Okamoto, *Tetrahedron Letters*, 2423 (1964)
[11]C. E. Miller, *J. Chem. Ed.*, **42**, 254 (1965); S. Hünig, H. R. Müller, and W. Thier, *Angew. Chem., Internat. Ed.*, **4**, 271 (1965)
[12]J. W. Hamersma and E. I. Snyder, *J. Org.*, **30**, 3985 (1965)

1,3-Diiodo-5,5-dimethylhydantoin,

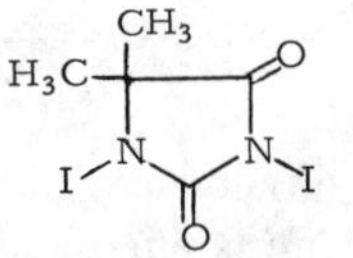
CH3
H3C
N
N
I
I
O
O

Mol. wt. 379.94.

The reagent has been prepared in 74% yield by dropwise addition of a solution of iodine monochloride in carbon tetrachloride to an ice-cold aqueous solution of 5,5-dimethylhydantoin and an equivalent amount of sodium hydroxide.[1] The precipitated product was collected, washed with ice water and with ethyl acetate, dried in vacuum at 60°, and used without further purification. It effects nuclear iodination of only highly reactive aromatic compounds (aniline, phenol; not naphthalene). It does not participate in homolytic reactions such as side-chain iodination of toluene in the presence of dibenzoyl peroxide. Like NIS, it reacts with the enol acetates of saturated and unsaturated ketones to give α-iodoketones.

[1]O. O. Orazi, R. A. Corral, and H. E. Bertorello, *J. Org.*, **30**, 1101 (1965)

Diiron nonacarbonyl (Iron enneacarbonyl), $Fe_2(CO)_9$(1). Mol. wt. 363.79. Preparation.[1]

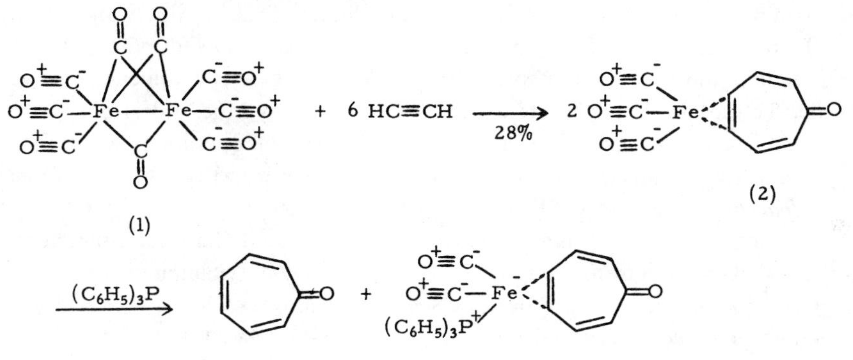
+ 6 HC≡CH
28%
2
$(C_6H_5)_3P$
+

(1) (2) (3) (4)

This complex contains six terminal carbonyl groups and three carbonyl groups bridging the two atoms of iron. When the complex is treated with acetylene at 20–25° and 20–24 atm., acetylene displaces the bridging carbonyls with formation of tropone iron tricarbonyl (2), which is an orange solid of double melting point (63.5–64° and 83–84°). This complex is obtained in only trace amounts with $Fe(CO)_5$ and $Fe(CO)_2$. Treatment of the complex (2) with triphenylphosphine affords free tropone (3, 69% yield) and a new complex (4), tropone iron dicarbonyltriphenylphosphine. Tropone can also be obtained, if in low yield, by oxidation of (2) with ferric chloride:

$$C_7H_6OFe(CO)_3 + 2\,FeCl_3 \xrightarrow[3.0\%]{} \text{Tropone} + 3\,FeCl_2 + 3\,CO$$

Cookson and co-workers[2] treated norbornadiene with diiron nonacarbonyl in the dark and isolated five ketones and four dimeric hydrocarbons. The highest-melting dimer, m.p. 163–164°, was isolated easily in 4% yield because it is the first component eluted from a column of silica gel. Lemal and Shim[3] isolated the same dimer from a complex mixture resulting on irradiation of a mixture of norbornadiene and iron pentacarbonyl. Initial characterization excluding alternative structures indicates

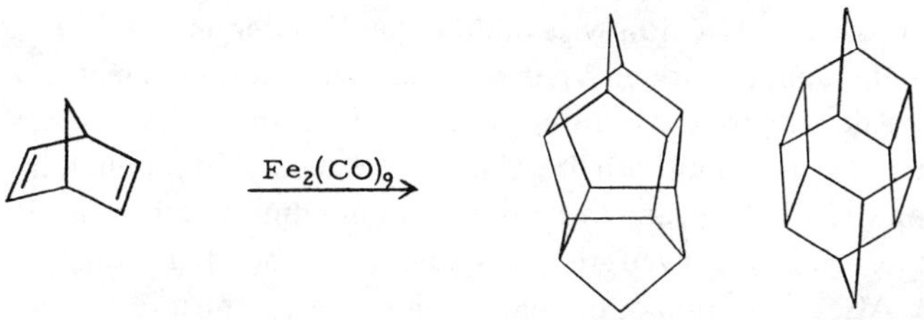
$Fe_2(CO)_9$

I II

that the two molecules of dimer have either united face to face to give II, or at right angles to each other to give I. Structure I is favored by the British group.

[1]E. Weiss, and W. Hübel, *Ber.*, **95**, 1179 (1962)
[2]C. W. Bird, D. L. Colinese, R. C. Cookson, J. Hudec, and R. O. Williams, *Tetrahedron Letters*, 373 (1961)
[3]D. M. Lemal and K. S. Shim, *ibid.*, 368 (1961)

Diisobutylaluminum hydride, $(CH_3)_2CHCH_2\overset{H}{\overset{|}{Al}}CH_2CH(CH_3)_2$. Mol. wt. 142.21, b.p. 140°/4 mm. Supplier ($\frac{1}{2}$ lb. unit in pressure cylinder): Texas Alkyls Inc. (Techn. Bull. Pl 640,928, 1964). Being a liquid, the reagent is dissolved more easily than lithium aluminum hydride. Suitable solvents are ether, benzene, toluene, and cyclohexane. Tetrahydrofurane forms a complex with the hydride and is not suitable. Since the reagent is pyrophoric in concentrated form, all operations with this and other organoaluminum compounds must be conducted in an oxygen-free atmosphere.[1]

Preparation.[2] Gensler and Bruno[3] prepared the reagent by refluxing 160 ml. of a 25% solution of triisobutylaluminum (Hercules Inc., Ethyl Corporation) in heptane for 2–3 hrs. Distillation afforded 27 ml. of colorless diisobutylaluminum hydride, b.p. 80–90°/0.05 mm. For another procedure see Eisch and Kaska.[3a]

Reduction of acetylenes. Since the reagent reacts only very slowly with internal olefins at 70–80° but reduces disubstituted acetylenes at lower temperatures, it is useful for the reduction of disubstituted acetylenes to ethylenes.[4] The *cis*-olefin is obtained almost exclusively.

$$C_2H_5C{\equiv}CC_2H_5 \xrightarrow[N_2\ 45^0]{[(CH_3)_2CHCH_2]_2AlH} \underset{95\%\ \underline{cis}}{C_2H_5CH{=}CHC_2H_5}$$

In a synthesis of isomers of linoleic acid, Gensler and Bruno[3] used the reagent for the reduction of several diynes to *cis*-dienes. For example, 1-chloro-$\Delta^{7,10}$-heptade-

$$\underset{(1)\ 0.052\ m.}{CH_3(CH_2)_5C{\equiv}\overset{10}{C}CH_2C{\equiv}\overset{7}{C}(CH_2)_5CH_2Cl} + \underset{(2)\ 0.15\ m.}{(CH_3)_2CHCH_2\overset{H}{\overset{|}{Al}}CH_2CH(CH_3)_2} \xrightarrow[0^0]{N_2}$$

$$\text{Complex} \xrightarrow[-5^0]{CH_3OH\text{-pet. ether}} \underset{(3)}{CH_3(CH_2)_5\overset{H}{\overset{|}{C}}{=}\overset{H}{\overset{|}{C}}CH_2\overset{H}{\overset{|}{C}}{=}\overset{H}{\overset{|}{C}}(CH_2)_5CH_2Cl} + (CH_3)_3CH + Al(OCH_3)_3$$

cadiyne (1) was treated by dropwise addition at 0° under nitrogen with a large excess of diisobutylaluminum hydride. When the reaction was complete, the organometallic complex was decomposed by dropwise addition with cooling of a mixture of petroleum ether and methanol with liberation of the reduction product, the *cis,cis*-diene (3), evolution of isobutane, and formation of aluminum trimethoxide. Treatment with dilute sulfuric acid dissolved the alkoxide and liberated also the *cis,cis*-diene chloride (3). After repetition of the reduction process to ensure complete conversion of the diyne (1), the pure *cis,cis*-diene (3) was obtained in 82% yield.

Dropwise addition of 0.41 mole of reagent to 0.2 mole of cyclododecyne and gentle heating effected stereoselective reduction to *cis*-cyclododecene, isolated after decomposition with methanol (isobutane is evolved).[5] Gas chromatography indicated that less than 0.4% of the *trans* isomer was formed.

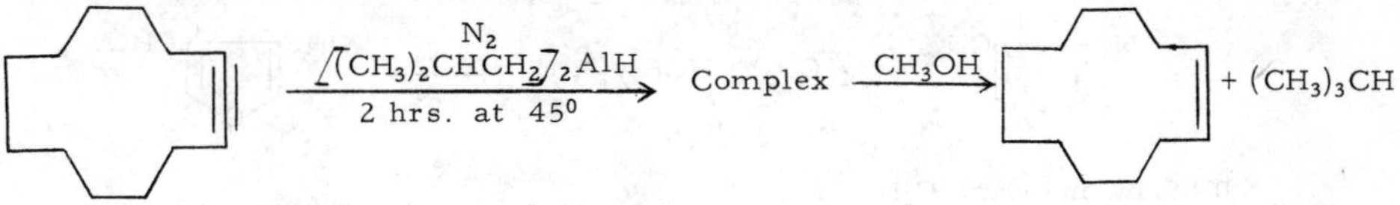
N2
[(CH3)2CHCH2]2AlH
2 hrs. at 45°
Complex
CH3OH
+ (CH3)3CH

Surprisingly, Eisch and Kaska[3a] found that, depending on the experimental conditions, 1-phenylpropyne can be reduced to either *cis*- or *trans*-1-phenylpropene. If a 1:1 ratio is used the *cis* isomer is the exclusive product.

Reduction of quinoline. Neumann[6] found that dropwise addition of quinoline to a stirred mixture of one equivalent of the reagent in ether under nitrogen at 0–10° produces initially a deep red solution. At the end of the addition the mixture turns yellow and the amine (2) separates as long, colorless prisms. This need not be isolated but is hydrolyzed by cautiously dropping in aqueous dioxane. Pure 1,2-dihydroquino-

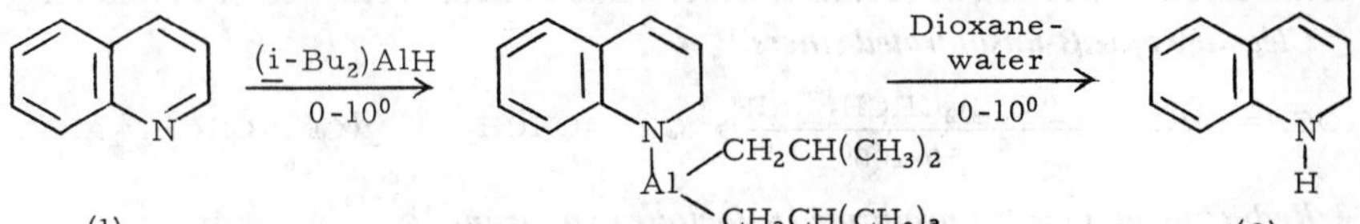
(i-Bu2)AlH
0-10°
Dioxane-water
0-10°
CH2CH(CH3)2
CH2CH(CH3)2

(1) (2) (3)

line is obtained in 95% yield. Further reduction of this dihydride requires more vigorous conditions. 1,2,3,4-Tetrahydroquinoline was obtained in 70% yield by refluxing quinoline with 2 equivalents of reagent in benzene for 24 hrs.

Reduction of a γ-lactone. Schmidlin and Wettstein[7] found the reagent useful for reduction of the γ-lactone (4) to the hemiacetal (5).

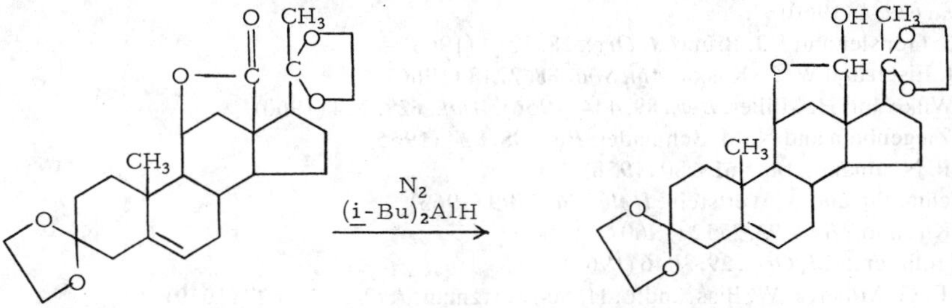
N2
(i-Bu)2AlH

(4) (5)

Isomerization of an epoxide. In xylene solution at 120° the reagent isomerizes the epoxide (6) to the allylic alcohol (7).[8]

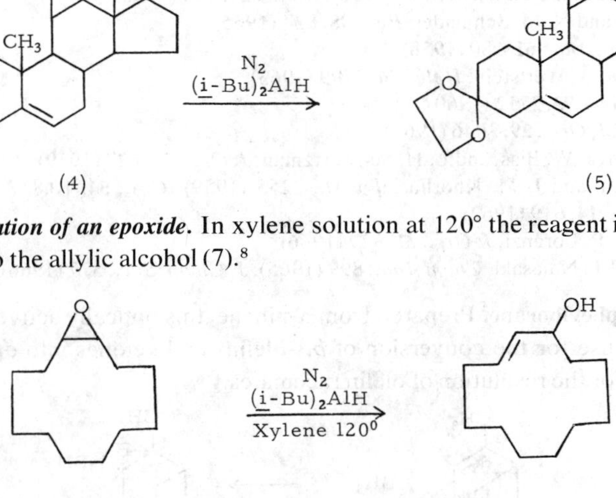
N2
(i-Bu)2AlH
Xylene 120°

(6) (7)

ArCN ⟶ ArCHO. After attempts to reduce 3,4-dicyanofurane to the dialdehyde by conventional methods had failed, Trofimenko[9] (du Pont) stirred a suspension of the dinitrile in benzene under nitrogen and added the metal hydride. The

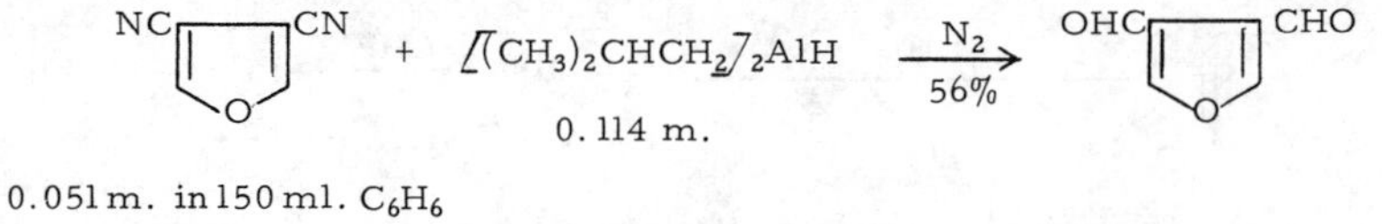
NC
CN
+ [(CH3)2CHCH2]2AlH
0.114 m.
N2
56%
OHC
CHO
0.051 m. in 150 ml. C6H6

mixture warmed up to about 50° and the solid dissolved. Methanol was added to the viscous reaction mixture to decompose the complex and workup afforded the pure aldehyde in moderate yield. The thiophene analog afforded the corresponding dialdehyde in 23% yield. 2,3,4,5-Tetracyanothiophene was not reduced. Benzonitrile has been reduced to benzaldehyde in 90% yield.[10]

Other reductions. Miller *et al.*[10] report reduction of benzoic acid and of methyl benzoate to benzyl alcohol and of *n*-butyl caproate to *n*-butanol and *n*-hexanol. Russian investigators[11] report reduction of aliphatic and aromatic esters to aldehydes, of ortho esters to acetals, of acetals to ethers, and of benzyl ethyl ether to toluene.

Cleavage of α,β-unsaturated ethers:[12]

$$C_2H_5OCH{=}CHCH_3 \xrightarrow[50\text{-}100^0]{[(CH_3)_2CHCH_2]_2AlH} CH_2{=}CHCH_3 + [(CH_3)_2CHCH_2]_2AlOC_2H_5$$

Reduction of an α,β-unsaturated γ-lactone to a furane.[13]

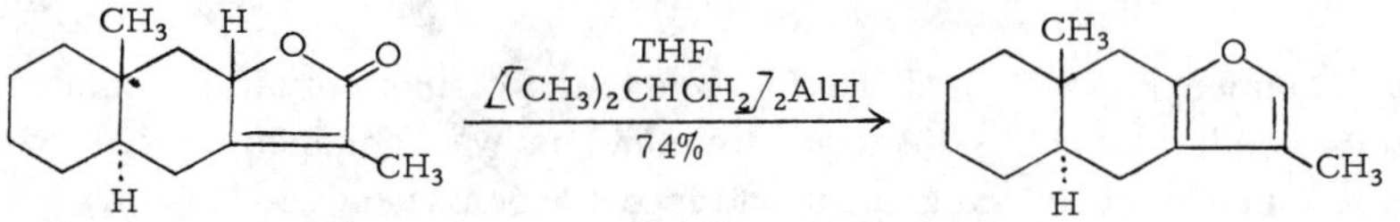
CH3
H
CH3
H
THF
[(CH3)2CHCH2]2AlH
74%
CH3
CH3
H

[1]"Handling and Properties of Triisobutylaluminum." Hercules Powder Co. (now Hercules Inc.), Wilmington, Del.
[2]K. Ziegler *et al.*, *Ann.*, **589**, 91 (1954); British patent 778,098 (1957) [*C.A.*, **51**, 15,081 (1957)]; *Ann.*, **629**, 1 (1960)
[3]W. J. Gensler and J. J. Bruno, *J. Org.*, **28**, 1254 (1963)
[3a]J. J. Eisch and W. C. Kaska, *Am. Soc.* **88**, 2213 (1966)
[4]G. Wilke and H. Müller, *Ber.*, **89**, 444 (1956); *Ann.*, **629**, 224 (1960)
[5]W. Ziegenbein and N. M. Schneider, *Ber.*, **98**, 824 (1965)
[6]W. P. Neumann, *Ann.*, **618**, 80 (1958)
[7]J. Schmidlin and A. Wettstein, *Helv.*, **46**, 2799 (1963)
[8]W. Kirchhof, *Ber.*, **93**, 2712 (1960)
[9]S. Trofimenko, *J. Org.*, **29**, 3046 (1964)
[10]A. E. G. Miller, J. W. Biss, and L. H. Schwartzman, *J. Org.*, **24**, 627 (1959)
[11]L. I. Zakharkin and I. M. Khorlina, *Izvest.*, 2255 (1959) [*C.A.*, **54**, 10,837 (1960)]; *Tetrahedron Letters*, **14**, 619 (1962)
[12]P. Pino and G. P. Lorenzi, *J. Org.*, **31**, 329 (1966)
[13]H. Minato and T. Najasaki, *Chem. Ind.*, 899 (1965); *J. Chem. Soc.*, 377 (1966)

Diisopinocampheylborane. Prepared from α-pinene, this optically active and hindered reagent is of use for the conversion of *bis*-olefins and ketones into optically active alcohols and for the resolution of olefin racemates.[1]

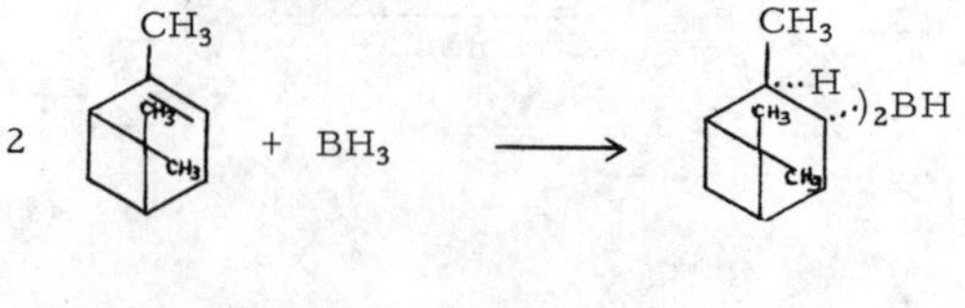
CH3
2
+ BH3
CH3
H
)2BH

In solution in diglyme the reagent reduces flexible monocyclic and rigid bicyclic ketones predominently to the less stable of the two possible alcohols.[2] Comparative results for reduction of four ketones with four reagents are shown in the table.

Reductions, alcohol shown is the less stable epimer:

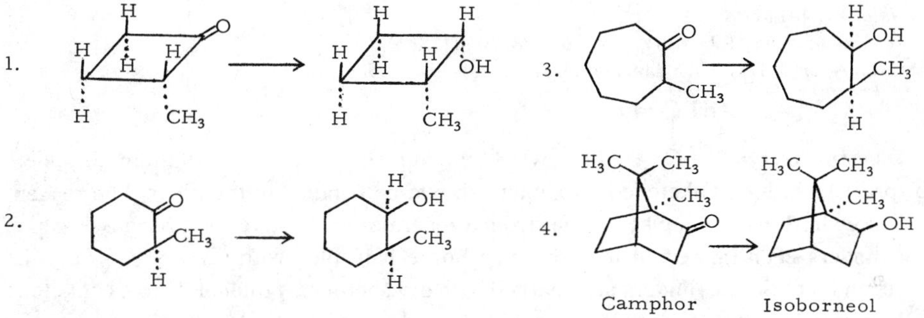
1.
2.
3.
4.
Camphor
Isoborneol

Percent of less stable alcohol

Reduction No.	$LiAlH_4$ THF	B_2H_6 THF	Disiamylborane THF	Diisopinocampheylborane Diglyme
1	25	41	74	83
2	25	26	79	94
3	73	74	64	98
4	91	52	65 (slow)	100 (slow)

[1]H. C. Brown and G. Zweifel, *Am. Soc.*, **83**, 486 (1960); H. C. Brown and D. B. Bigby, *ibid.*, **83**, 3166 (1961); H. C. Brown, N. R. Ayyanger, and G. Zweifel, *ibid.*, **84**, 4341, 4342 (1962); H. C. Brown, N. R. Ayyangar, and G. Zweifel, *ibid.*, **86**, 397, 1071 (1964)

[2]H. C. Brown and V. Varma, *Am. Soc.*, **88**, 2871 (1966)

Diisopropyl peroxydicarbonate, $(CH_3)_2CHO\overset{O}{\overset{\|}{C}}OO\overset{O}{\overset{\|}{C}}OCH(CH_3)_2$. Mol. wt. 206.19, m.p. 8–10°. Suppliers: Pittsburgh Plate Glass Co., Messrs. Novadel Ltd.

The reagent has been used in combination with aluminum chloride as catalyst for the oxygenation of toluene to give in about 50% yield a mixture of cresols of the following isomer distribution: *ortho*, 34%; *meta*, 11%; *para*, 55%.[1]

It is an efficient polymerization catalyst. Since under conditions of heat, friction, or shock it is explosive, the reagent is usually stored and handled in the frozen state.

In the presence of cupric chloride, toluene is converted by the reagent into tolyl isopropyl carbonates, the *ortho* isomer predominating.[2]

$$(CH_3)_2CHO\overset{O}{\overset{\|}{C}}OO\overset{O}{\overset{\|}{C}}OCH(CH_3)_2 + C_6H_5CH_3 \xrightarrow[78-97\%]{CuCl_2\ (60^\circ)} (CH_3)_2CHOH + CO_2$$
$$+ CH_3C_6H_4O\overset{O}{\overset{\|}{C}}OCH(CH_3)_2$$

The reagent is useful as initiator of radical reactions at moderate temperatures, for example the addition of acetyl chloride to heptene-1.[3]

$$CH_3(CH_2)_4CH=CH_2 + Reagent + CH_3COCl \xrightarrow[29\%]{30-70^0} CH_3(CH_2)_4CH_2CH_2CH_2COCl$$

5 g.   2 g. in C_6H_{12} 80 g. $H_2O \downarrow 100\%$

$$CH_3(CH_2)_7CO_2H$$

[1] P. Kovacic and S. T. Morneweck, *Am. Soc.*, **87**, 1566 (1965); P. Kovacic and M. E. Kurz, *ibid.*, **87**, 4811 (1965)

[2] P. Kovacic and M. E. Kurz, *Am. Soc.*, **88**, 2068 (1966)

[3] J. C. Allen, J. I. G. Cadogan, and D. H. Hey, *J. Chem. Soc.*, 1918 (1965)

Diketene, . Mol. wt. 84.07, m.p. $-8°$, b.p. $67–69°/92$ mm. Suppliers (pract.): A, KK. Diketene is toxic and should be handled in the hood. The reagent is prepared[1] by passing ketene gas from a generator into a series of three gas-washing cylinders each immersed in a thermos bottle half filled with acetone–dry ice; the second and third cylinders are charged with dry acetone. A total of 2 moles of ketene is passed into the absorption system in about 4 hrs., during which time the temperature is gradually brought to that of the room. Fractionation of the combined material then affords diketene in yield of 50–55%. Structure.[2]

Reaction with active-H compounds. The reagent is hydrolyzed by water to acetoacetic acid and it reacts with alcohols to form the corresponding esters,[3] as illustrated by a procedure for the preparation of *t*-butyl acetoacetate.[4] The analogous reaction

$$CH_2=C-O + HOC(CH_3)_3 \xrightarrow[75-80\%]{NaOAc\ (80^0)} CH_3\overset{O}{\overset{\|}{C}}CH_2CO_2C(CH_3)_3$$

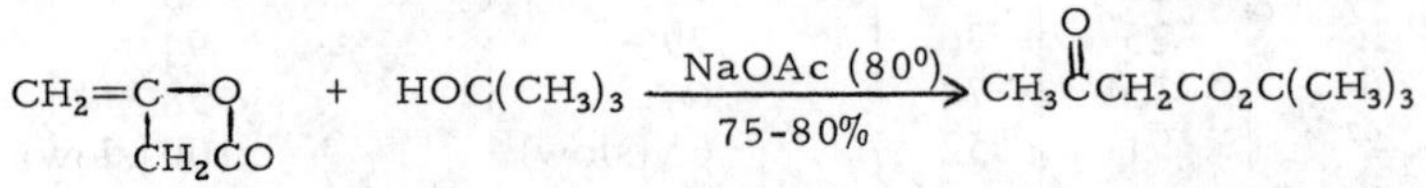

with aromatic amines gives acetoacetanilides: $CH_3COCH_2CONHAr$.[5] Urethane adds to diketene in an analogous manner to give $CH_3COCH_2CONHCO_2C_2H_5$.[6]

Dehydration of alcohols. Acetoacetic esters are useful for effecting dehydration of alcohols.[7] Thus 1-methylcyclohexanol is allowed to react exothermally with

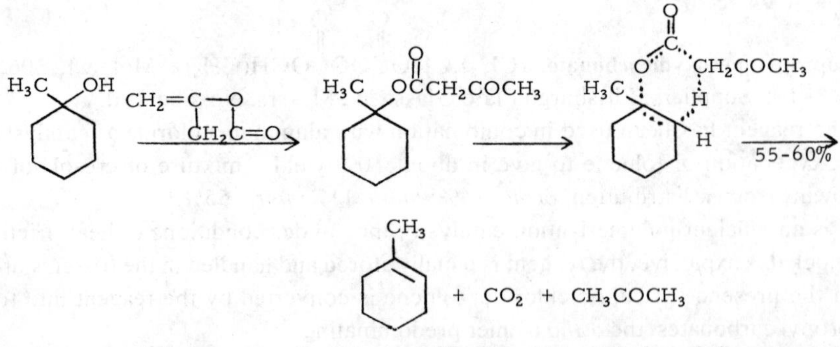

diketene, a catalytic amount of *p*-toluenesulfonic acid is added, and the temperature is raised to effect elimination of acetone and carbon dioxide. 1-Methylcyclohexene of 98% purity is obtained by this procedure, whereas other dehydrative methods give a product containing considerable methylenecyclohexane.

Reaction with hydrogen fluoride. Diketene reacts with anhydrous hydrogen fluoride to give acetoacetyl fluoride.[8]

$$CH_2=C-O \xrightarrow[65\%]{HF} CH_3COCH_2COF$$

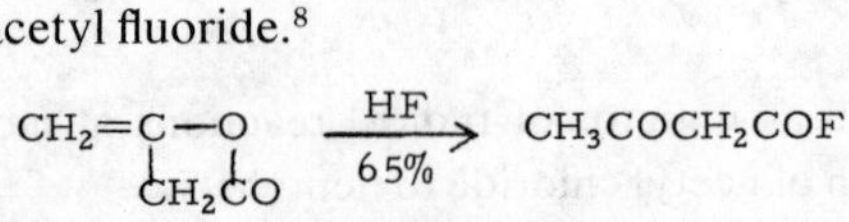

α-Hydroxyketones. A novel modification of the acetoacetic ester synthesis provides a route to α-hydroxyketones.[9] *t*-Butyl acetoacetate (1), prepared in high yield from *t*-butanol and diketene, is alkylated (2) and the benzoyloxy group is

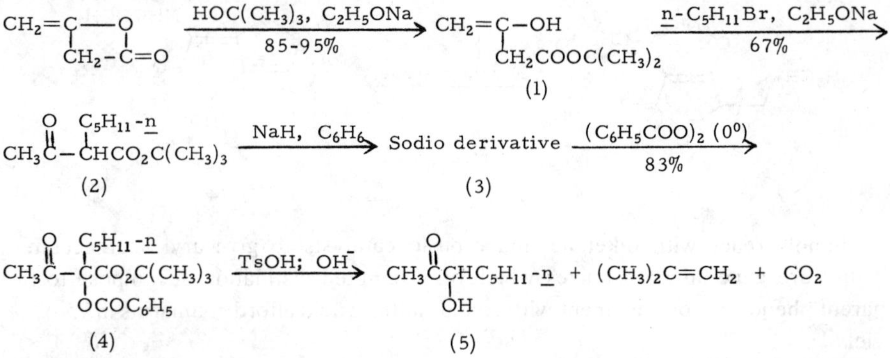

introduced into the α-position by reaction of the sodio derivative (3) with dibenzoyl peroxide in benzene to give (4). Elimination of the *t*-butyl group and decarboxylation is accomplished by heating the ester (4) at 160° with a catalytic amount of *p*-toluenesulfonic acid; saponification under mild condition then gives the (base-sensitive) acyloin (5).

N-Protection in peptide synthesis.[10] Diketene reacts with an amino ester hydrochloride in ethanol at 0–5° in the presence of 1 equivalent of sodium ethoxide or a tertiary amine to give the often crystalline and easily purified N-acetoacetyl derivative (1). After peptide synthesis the protective group is removed by treatment with phenylhydrazine, which affords 1-phenyl-3-methyl-5-pyrazolone.

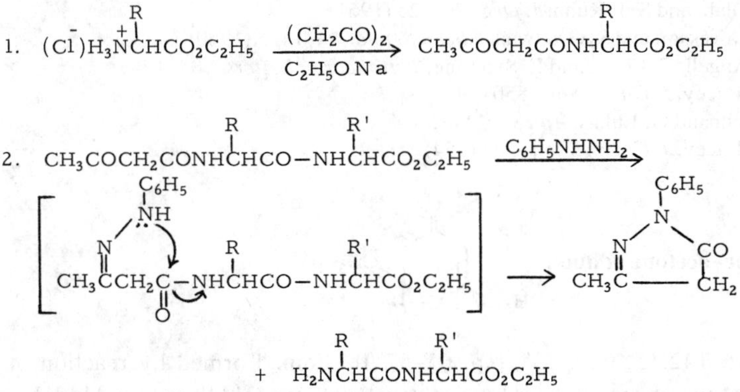

Heterocycles. The N-acetoacetyl derivative of an α-amino ester can be cyclized by base to an α-acetyltetramic acid.[11] Büchi and Lukas[12] used this sequence in the

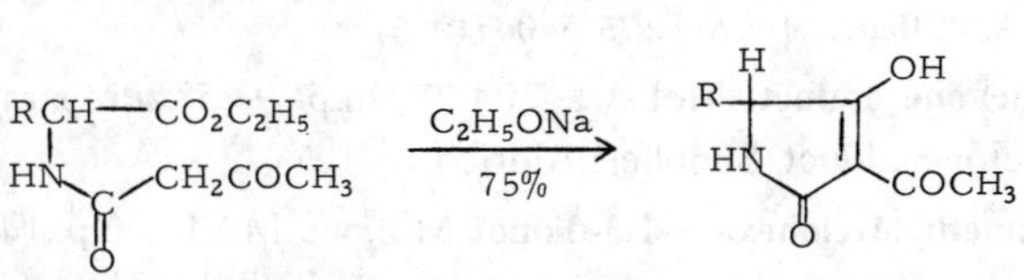

first stage of a synthesis of the antibiotic holomycin (4).

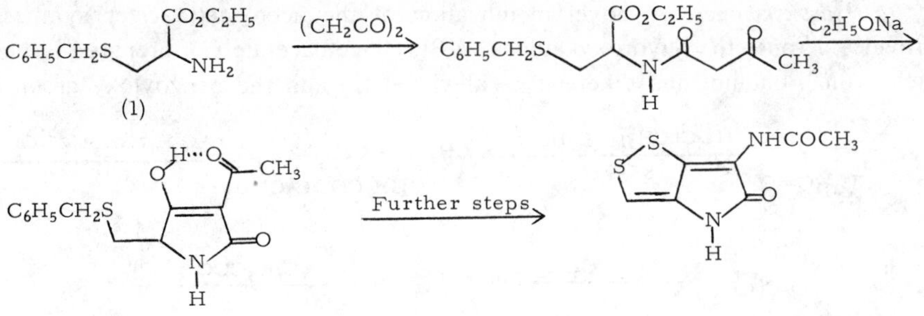

(1) (3) (4)

Phenols react with diketene under basic catalysis to give aryl acetoacetates. These are generally substances which on attempted distillation decompose to the parent phenol, but on treatment with concd. sulfuric acid afford coumarins in 75–90% yield.[13]

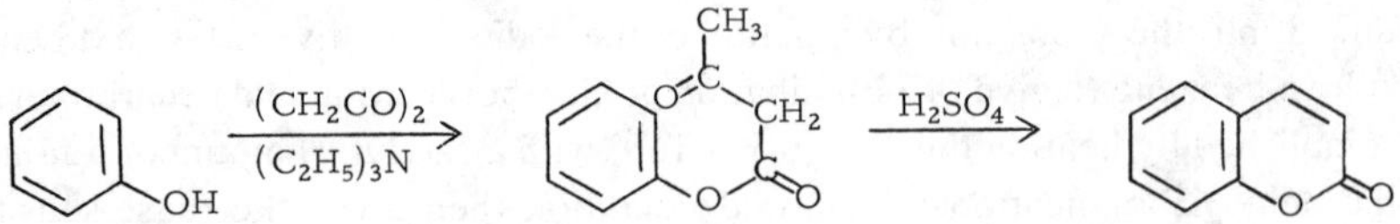

[1]J. W. Williams and J. A. Krynitsky, *Org. Syn., Coll. Vol.*, **3**, 508 (1955)
[2]J. R. Johnson and V. J. Shiner, *Am. Soc.*, **75**, 1350 (1953)
[3]W. Kimel and A. C. Cope, *Am. Soc.*, **65**, 1992 (1943)
[4]S.-O. Lawesson, S. Gronwall, and R. Sandberg, *Org. Syn.*, **42**, 28 (1962)
[5]C. E. Kaslow, and N. B. Sommer, *Am. Soc.*, **68**, 644 (1946)
[6]J. H. Dewar and G. Shaw, *J. Chem. Soc.*, 3254 (1961)
[7]C. Frisell and S.-O. Lawesson, *Arkiv Kemi*, **17**, 401 (1961)
[8]G. A. Olah, and S. J. Kuhn, *J. Org.*, **26**, 225 (1961)
[9]S.-O. Lawesson and S. Gronwall, *Chem. Scand.*, **14**, 1445 (1960)
[10]F. D'Angeli, F. Filira, and E. Scoffone, *Tetrahedron Letters*, 605 (1965)
[11]R. N. Lacey, *J. Chem. Soc.*, 850 (1954)
[12]G. Büchi and G. Lukas, *Am. Soc.*, **86**, 5654 (1964)
[13]R. N. Lacey, *J. Chem. Soc.*, 854 (1954)

Diketene–acetone adduct,

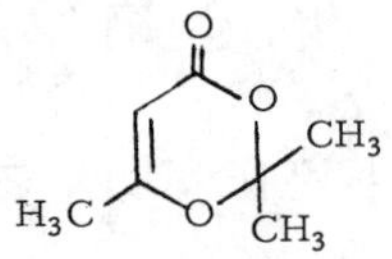

Mol. wt. 142.15, m.p. 13°, b.p. 65–67°/0.2 mm. Formed by reaction of diketene with acetone under catalysis by *p*-toluenesulfonic acid.[1] Supplier: Aldrich.

This pleasant-smelling liquid can be used in many reactions in place of the more volatile diketene. It is stable in the absence of alkali.

[1]M. F. Carroll and A. R. Bader, *Am. Soc.*, **75**, 5400 (1953)

Diketene–acetophenone adduct. Mol. wt. 204.22, m.p. 93.5°. Preparation and use, *see* Diketene–acetone adduct. Supplier: Aldrich.

Dimedone (5,5-Dimethylcyclohexane-1,3-dione). Mol. wt. 140.18, m.p. 149°. Suppliers: E, B, Dajac, Light. Prepared by condensation of diethyl malonate with mesityl oxide, hydrolysis, and decarboxylation.[1] The product is almost entirely enolic.

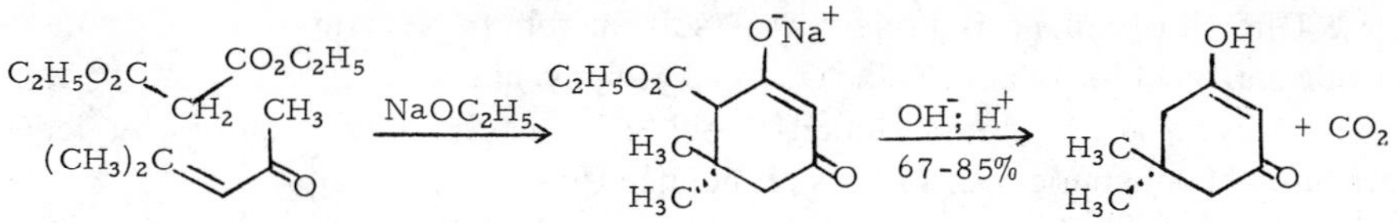

Reaction with aldehydes. Dimedone is particularly useful for isolation and identification of lower aldehydes.[2] It reacts with gaseous formaldehyde without catalyst to give high-melting (188°) methylenebisdimedone with a 10-fold increase in mass.

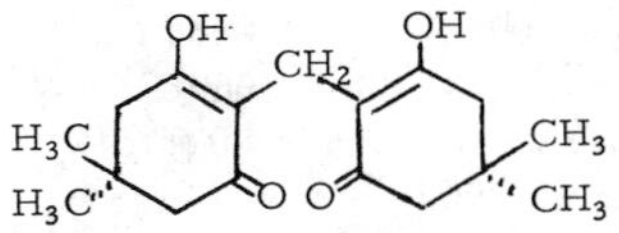

M. p. 188°

N-Protection of amino acids.[3] Dimedone reacts readily with an amino acid ester (but not with the free amino acid) to give an enamine derivative (1), characterized by remarkable stability of the enamine group to acid or base. Thus refluxing (1)

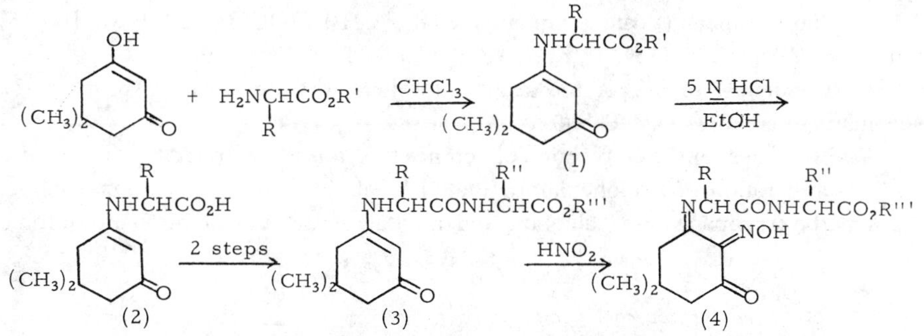

with a mixture of dilute acid and ethanol for 2 hrs. removes the ester group and gives the N-dimedone amino acid (2), which can be used for coupling with an amino acid ester by the DCC method to give the dipeptide derivative (3). The dimedone group can be removed when desired as 2,2-dibromodimedone or, more conveniently, by reaction with nitrous acid to give the deep blue isonitroso compound (4), readily hydrolyzed by dilute HCl.

[1]R. L. Shriner and H. R. Todd, *Org. Syn., Coll. Vol.*, **2**, 200 (1943)

[2]E. C. Horning and M. G. Horning, *J. Org.*, **11**, 95 (1946)

[3]B. Halpern and L. B. James, *Australian J. Chem.*, **17**, 1282 (1964); B. Halpern, *ibid.*, **18**, 417 (1965); B. Halpern and A. D. Cross, *Chem. Ind.*, 1183 (1965)

1,2-Dimethoxyethane (Glyme), $CH_3OCH_2CH_2OCH_3$. Mol. wt. 90.12, b.p. 83 , sp. gr. 0.863, water-miscible. Suppliers: A, E, F, KK, MCB, and (pure) BJ.

Studies of the Favorsky reaction have shown that it is stereospecific in the non-polar solvents ether[1] and 1,2-dimethoxyethane[2] but nonstereospecific in polar methanol.[2]

Wittig[3] found that *o*-dibromobenzene reacts with sodium amalgam in ether to give hexameric phenylenemercury in 52% yield, but that with 1,2-dimethoxyethane as solvent the yield of this product drops to 6% and the main product (41%) is *o*-terphenylenemercury.

1,2-Dimethoxyethane is preferred as solvent for the generation of benzyne by aprotic diazotization of anthranilic acid with isoamyl nitrite.[4]

This solvent is preferred to ether for the Simmons-Smith reaction of acetylenes; formation of polymeric resins is less pronounced.[5]

It is the preferred solvent for the reaction of alkali metals with naphthalene or diphenyl,[6] or with phenol ethers,[7] or for reaction of sodium with *t*-nitrobutane.[8]

1,2-Dimethoxyethane is preferred to dimethyl sulfoxide as solvent for the formylation of ketones with carbon monoxide in the presence of an alkali metal alkoxide.[9] This is due to the fact that DMSO is slightly protic in the presence of a strong base (potassium *t*-butoxide) and hence not inert.

[1]G. Stork and I. J. Borowitz, *Am. Soc.*, **82**, 4307 (1960)
[2]H. O. House and W. F. Gilmore, *Am. Soc.*, **83**, 3980 (1961)
[3]G. Wittig *et al.*, *Ber.*, **95**, 431, 1692 (1962)
[4]*Org. Expts.*, 316
[5]G. Emptoz, L. Vo-Quang, and Y. Vo-Quang, *Bull. soc.*, 2653 (1965)
[6]N. D. Scott, J. F. Walker, and V. L. Hansley, *Am. Soc.*, **58**, 2442 (1936)
[7]A. L. Wilds and N. A. Nelson, *Am. Soc.*, **75**, 5360 (1953)
[8]A. K. Hoffmann and A. T. Henderson, *Am. Soc.*, **83**, 4671 (1961)
[9]A. J. van der Zeeuw and H. R. Gersmann, *Rec. trav.*, **84**, 1535 (1965)

2,2-Dimethoxypropane (Acetone dimethyl ketal), $(CH_3)_2C(OCH_3)_2$. Mol. wt. 104.15, b.p. 79–81°. Suppliers: Dow, B, E, F, MCB, Calbiochem. Preparation.[1]

For an example of use of the reagent for ketal exchange with a primary (or secondary) alcohol, *see* 2,2-Dibutoxypropane.

Ketals. Its use for the conversion of ketones to ketals is illustrated by a procedure for the preparation of cyclohexanone diallyl ketal.[2] A solution of cyclohexanone, 2,2-dimethoxypropane, allyl alcohol, and a catalytic amount of *p*-toluenesulfonic

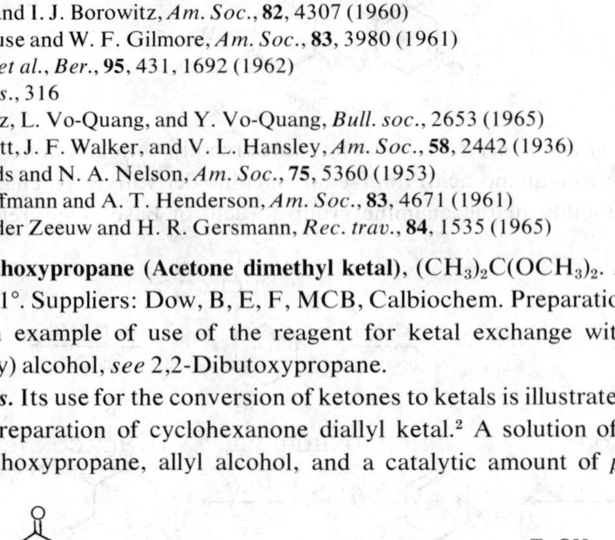
+ CH₂=CHCH₂OH + (CH₃)₂C(OCH₃)₂
TsOH
65-74%
CH₂=CHCH₂O
OCH₂CH=CH₂
+ CH₃COCH₃ + 2 CH₃OH

acid in benzene is distilled through a fractionating column for removal of acetone and a benzene–methanol azeotrope; the solution is cooled, the acid catalyst is neutralized with sodium methoxide, and distillation is resumed.

Acetonides (Isopropylidine derivatives). 2,2-Dimethoxypropane reacts with the dihydroxyacetone side chain of a corticosteroid to give a 17,21-acetonide useful for side-chain protection.[3,4] The blocking group is stable to base but can be removed by

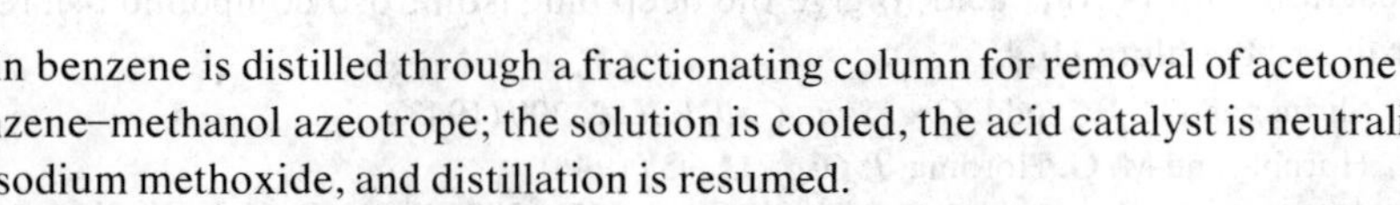
CH₂OH
CO
CH₃
···OH
(CH₃)₂C(OCH₃)₂ (TsOH)
In dimethylformamide
CH₂O
CH₃
CO
C
CH₃
O
CH₃
+ 2 CH₃OH

heating on the steam bath with aqueous acetic or formic acid.

Both *cis*- and *trans*-flavone-3,4-diols give the acetonide derivative of the *cis*-diol on reaction with acetone under conventional conditions, but only the *cis* isomer reacts with 2,2-dimethoxypropane (TsOH catalysis).[5]

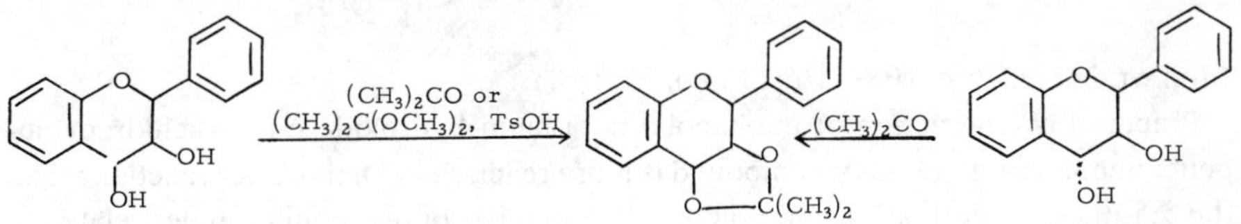
$(CH_3)_2CO$ or
$(CH_3)_2C(OCH_3)_2$, TsOH
OH
OH
$(CH_3)_2CO$
O
O
$C(CH_3)_2$
OH
OH

The reagent has proved useful as water scavenger in the preparation of acetonide derivatives of nucleosides[6] (*see* Di-*p*-nitrophenyl hydrogen phosphate).

Methyl esters. 2,2,-Dimethoxypropane has been used for the preparation of methyl esters of amino acid hydrochlorides.[7] The reagent serves as solvent and as a source of the methoxyl group, and it removes water. The amino acid is suspended in 2,2,-dimethoxypropane, a small volume of concd. hydrochloric acid is added, and

$$H_2NCH(R)CO_2H \xrightarrow[\text{dil. HCl}]{(CH_3)_2C(OCH_3)_2} \overset{-}{Cl}\,H_3\overset{+}{N}CH(R)CO_2CH_3$$

the mixture let stand at room temperature for 18 hrs. A fatty acid can be esterified quantitatively by treatment with methanol, a large excess of 2,2-dimethoxypropane, and a trace of aqueous hydrochloric acid; the reaction is complete in 1 hr. at room temperature.[8]

Enol ethers. The acid-catalyzed reaction of a Δ^4-3-ketosteroid with 2,2-dimethoxypropane in dimethylformamide leads to an enol ether.[9]

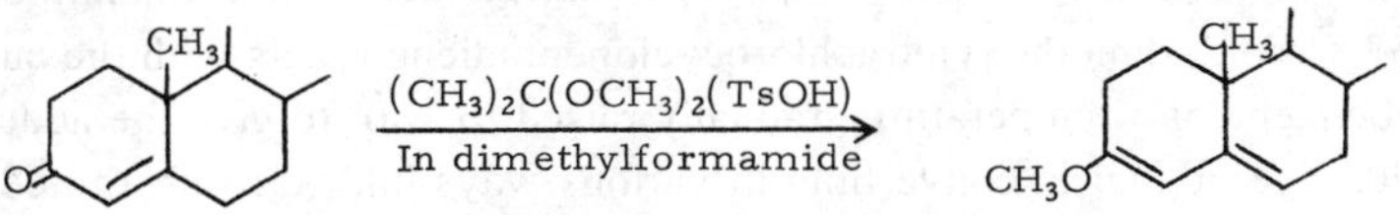
CH_3
$(CH_3)_2C(OCH_3)_2$(TsOH)
In dimethylformamide
CH_3
CH_3O

Reaction with nucleosides. The reagent reacts selectively with the primary 5′-hydroxyl group of a nucleoside (1) to give a 5′-O-1″-methoxyisopropyl derivative

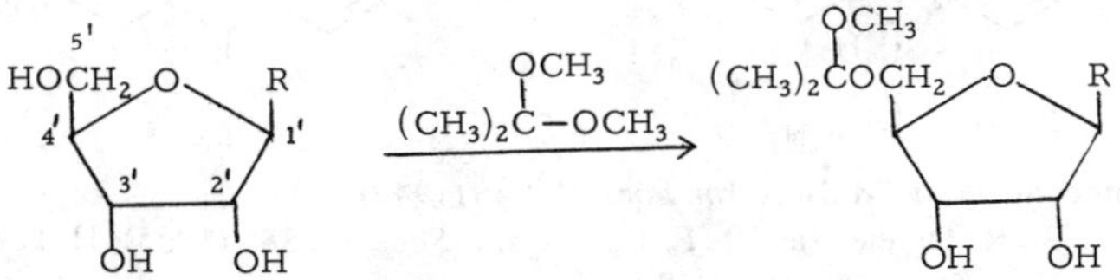
5'
$HOCH_2$
O
R
4'
1'
3'
2'
OH
OH
OCH_3
$(CH_3)_2C-OCH_3$
OCH_3
$(CH_3)_2COCH_2$
O
R
OH
OH

(1) (2)

and hence is a substitute for the commonly used trityl group.[10]

[1]N. B. Lorette, W. L. Howard, and J. H. Brown, Jr., *J. Org.*, **24**, 1731 (1959)
[2]N. B. Lorette and W. L. Howard, *J. Org.*, **25**, 521 (1960); *Org. Syn.*, **42**, 34 (1962)
[3]M. Tanabe and B. J. Bigley, *Am. Soc.*, **83**, 756 (1961)
[4]C. H. Robinson, L. E. Finckenor, R. Tiberi, and E. P. Oliveto, *J. Org.*, **26**, 2863 (1961)
[5]B. R. Brown and J. A. H. MacBride, *J. Chem. Soc.*, 3822 (1964)
[6]A. Hampton, *Am. Soc.*, **83**, 3640 (1961); J. G. Moffatt, *ibid.*, **85**, 1118 (1963)
[7]J. R. Rachele, *J. Org.*, **28**, 2898 (1963)
[8]N. S. Radin, A. K. Hajra, and Y. Akahori, *J. Lipid Res.*, **1**, 250 (1960)
[9]A. L. Nussbaum, E. Yuan, D. Dincer, and E. P. Oliveto, *J. Org.*, **26**, 3925 (1961)
[10]A. Hampton, *Am. Soc.*, **87**, 4654 (1965)

5,5-Dimethoxy-1,2,3,4-tetrachlorocyclopentadiene,

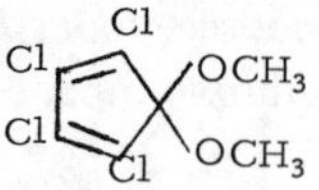

Mol. wt. 263.95, b.p. 108–110°/11 mm.

Prepared by the action of methanolic potassium hydroxide on hexachlorocyclopentadiene,[1] the dimethoxy compound is more reactive in Diels-Alder reactions than the 5,5-dichloro derivative and reacts with a number of dienophiles under relatively mild conditions.[1,2] The adducts are of interest because they can be aromatized by a simple reaction sequence, as illustrated for the acrylic acid adduct, prepared by

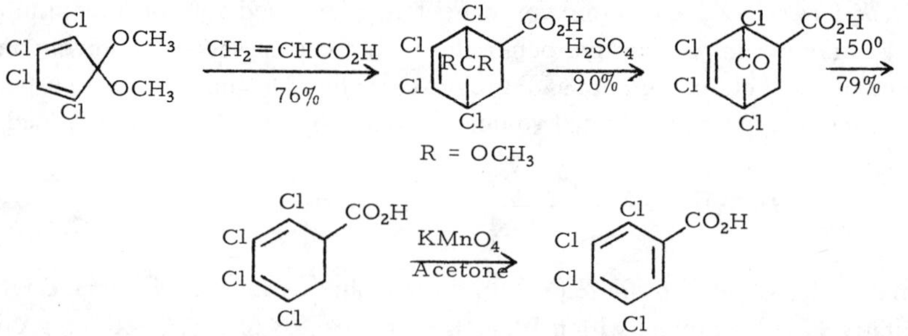

refluxing the components for 48 hrs. In the terminal step, oxidation with permanganate in acetone gave 2,3,4,5-tetrachlorobenzoic acid as the sole product; oxidation with chromic anhydride in acid gave 25% of the aromatic acid and 39% of 1,2,3,4-tetrachlorobenzene.

Whereas hexachlorocyclopentadiene forms an adduct with 1,4-naphthoquinone in only 53% yield, dimethoxytetrachlorocyclopentadiene reacts with the quinone in trichlorobenzene at a temperature gradually raised to 180° to give the adduct (1) in 88% yield. The adduct is convertible in various ways into tetrasubstituted anthraquinones.[3]

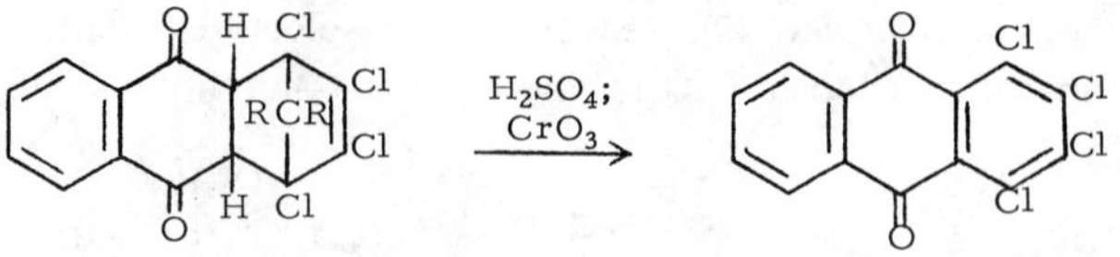

(1) R = OCH_3

[1]J. S. Newcomer and E. T. McBee, *Am. Soc.*, **71**, 946 (1949)
[2]E. T. McBee, W. R. Diveley, and J. E. Burch, *Am. Soc.*, **77**, 385 (1955); H. E. Ungnade and E. T. McBee, *Chem. Rev.*, **58**, 249 (1958)
[3]P. Kniel, *Helv.*, **46**, 492 (1963)

Dimethylacetamide, $CH_3CON(CH_3)_2$. Mol. wt. 87.12, b.p. 165°. Suppliers: B, E, F, MCB. Preparation.[1]

Acetylation. In combination with phosphoryl chloride, the reagent acetylates benzofuranes.[2] (Compare Dimethylformamide–$POCl_3$.)

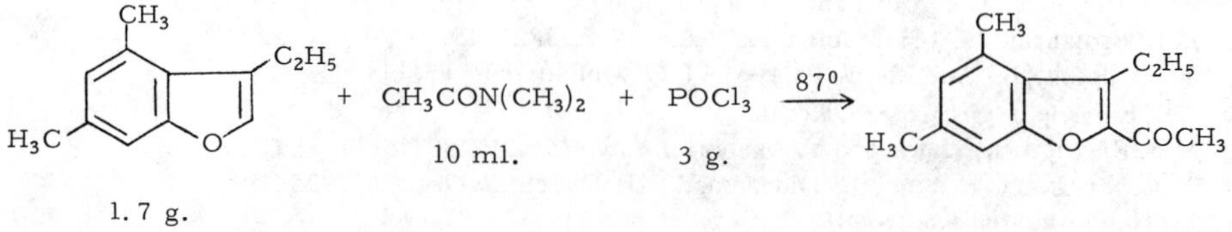

Decarboxylation (solvent effect). Casini and Goodman[3] found that decarboxylation of 7-nitroindole-2-carboxylic acid by heating with a catalytic amount of the copper salt of the acid proceeded in higher yield and with simpler isolation of product when N,N-dimethylacetamide was used as solvent rather than the conventional quinoline.

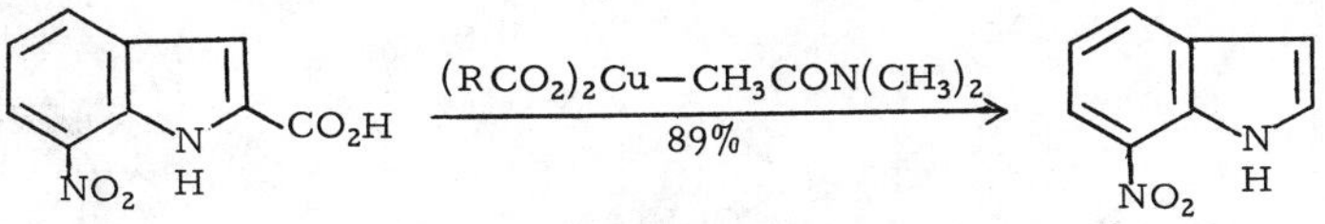

CO2H
NO2
H
(RCO2)2Cu−CH3CON(CH3)2
89%
N
NO2
H

[1]J. R. Ruhoff and E. E. Reid, *Am. Soc.*, **59**, 401 (1937)
[2]D. S. Deorha and P. Gupta, *Ber.*, **97**, 616 (1964)
[3]G. Casini and L. Goodman, *Can. J. Chem.*, **42**, 1235 (1964)

Dimethylacetamide diethyl acetal, $(CH_3)_2NC(CH_3)(OC_2H_5)_2$. Mol. wt. 161.24. A liquid not distillable in vacuum.

Preparation.[1] The method is the same as that described for dimethylformide diethyl acetal; yield 64%.

Reactions with active-methylene and -methyl compounds.[1] The reagent loses ethanol reversibly to form 1-dimethylamino-1-ethoxyethylene:

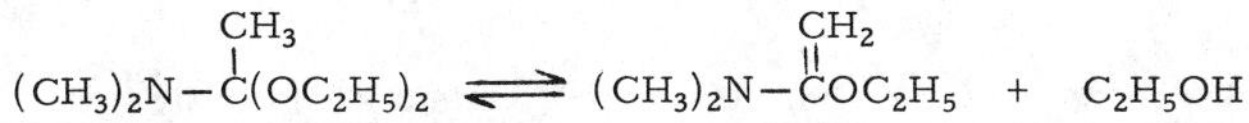

CH3
(CH3)2N−C(OC2H5)2
CH2
(CH3)2N−COC2H5
+
C2H5OH

(1) (2)

Thus either (1), (2), or a mixture serves for condensation with acetophenone, methyl cyanoacetate, malononitrile, nitromethane, cyclopentadiene, and 2,4-dinitrotoluene. The reactions are carried out without solvent and are exothermal. Only in the

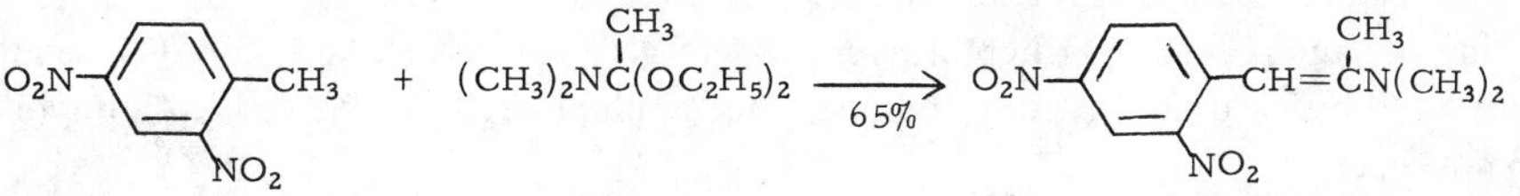

O2N
CH3
NO2
+
CH3
(CH3)2NC(OC2H5)2
65%
O2N
CH3
CH=CN(CH3)2
NO2

condensation with cyclopentadiene was external heating required (refluxing for ½ hr.).

Reaction with allylic alcohols. The reagent (form 2) reacts with allyl alcohol with allylic rearrangement to give allylacetic acid dimethyl amide (4).[1] Eschenmoser and co-workers[2] found this reaction to be stereospecific (the Swiss investigators used

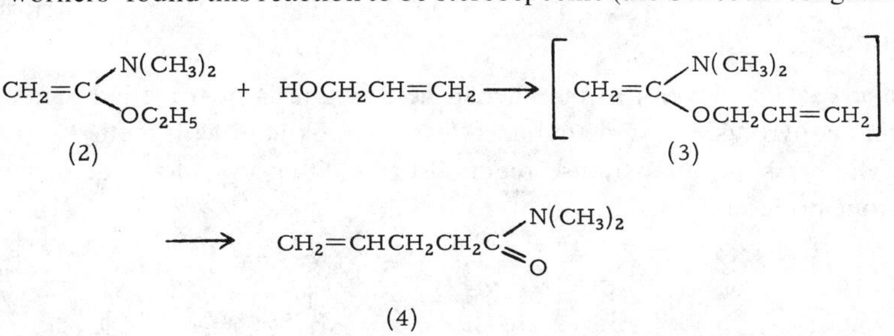

N(CH3)2
CH2=C
OC2H5
(2)
+
HOCH2CH=CH2
N(CH3)2
CH2=C
OCH2CH=CH2
(3)
N(CH3)2
CH2=CHCH2CH2C
O
(4)

a mixture of dimethylacetamide dimethyl acetal and its desmethanol derivative). Thus the allylic alcohol (5), with a 4β-hydroxyl group, gives (6) and the 4α-isomer

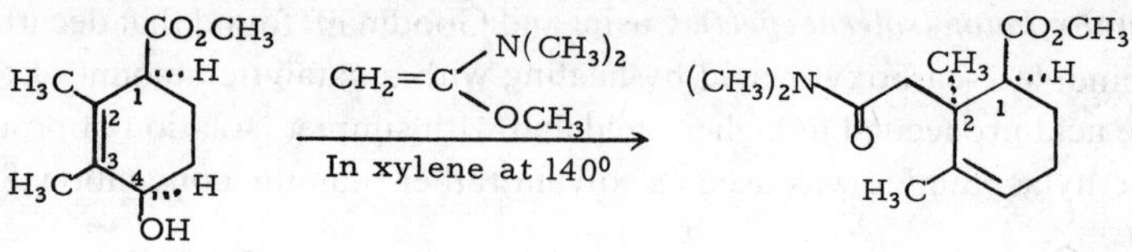

(5) (6)

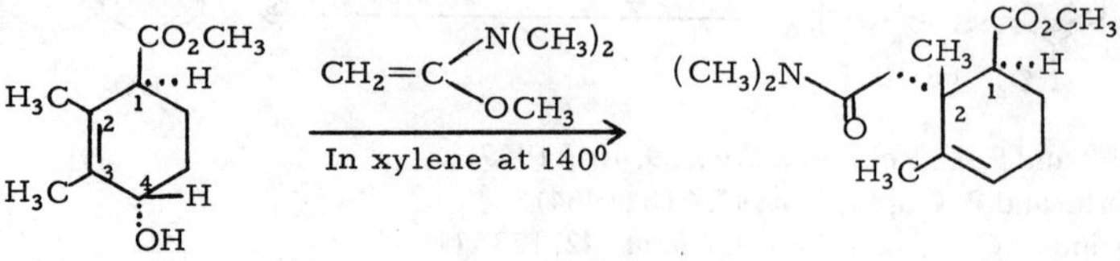

(7) (8)

(7) gives the 2-epimer (8). Benzyl alcohol reacts to give the two products formulated (separated by chromatography).[2]

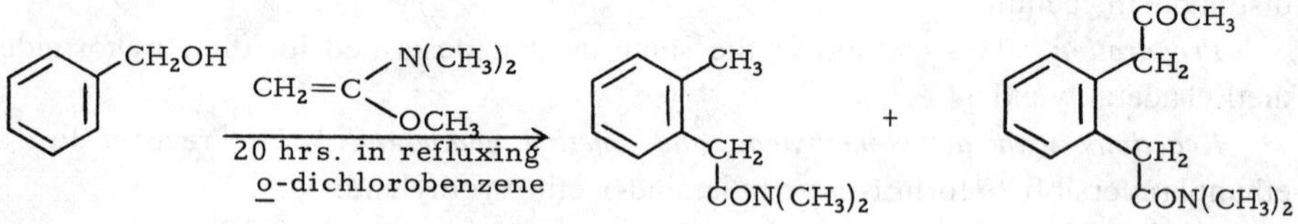

50% 20%

[1]H. Meerwein, W. Florian, N. Schön, and G. Stopp, *Ann.*, **641**, 1 (1961)
[2]A. E. Wick, D. Felix, K. Steen, and A. Eschenmoser, *Helv.*, **47**, 2425 (1964)

Dimethyl acetylenedicarboxylate, $CH_3O_2CC{\equiv}CCO_2CH_3$. Mol. wt. 110.11, b.p. 96–98°/8 mm., sp. gr. 1.10, a powerful lachrymator and vesicant. Suppliers: A, E, F, MCB. Preparation[1] from commercially available potassium acid acetylenedicarboxylate.

The reagent is a useful dienophile: *see* Diels-Alder solvents.

The diester reacts with heptafulvene (1) to give the colorless adduct (2), readily converted by air oxidation into the deep blue azulene (3).[2] Less reactive dienophiles do not react.

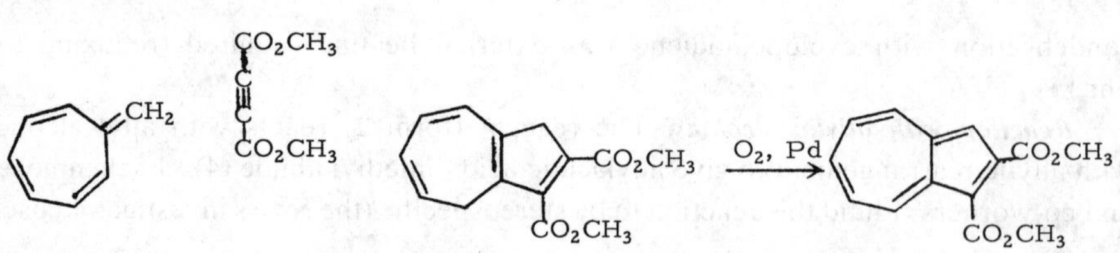

(1) (2) (3)

Wenkert *et al.*[3] developed an efficient new synthesis of trimethyl hemimellitate involving copyrolysis of dimethyl acetylenedicarboxylate with methyl α-pyrone-6-carboxylate. At the temperature required for addition the adduct as formed loses the lactone bridge as CO_2.

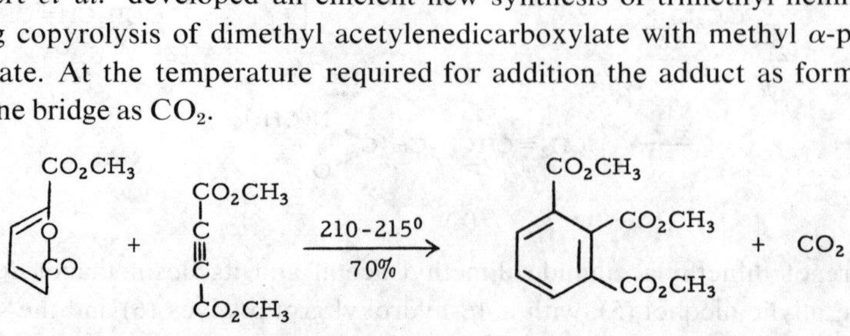

[1]E. H. Huntress, T. E. Lesslie, and J. Bornstein, *Org. Syn.*, *Coll. Vol.*, **4**, 329 (1963)
[2]W. von E. Doering and D. W. Wiley, *Tetrahedron*, **11**, 183 (1960)
[3]E. Wenkert, D. B. R. Johnston, and K. G. Dave, *J. Org.*, **29**, 2534 (1964)

Dimethylamine borane, $(CH_3)_2\overset{+}{N}H\overset{-}{B}H_3$. Mol. wt. 58.93, m.p. 37°, dec. 150°. Preparation.[1]

Schiff bases, for example benzylideneaniline from benzaldehyde and aniline, are reduced rapidly and efficiently by dimethylamine borane in acetic acid solution or suspension.[2] Unlike other borohydrides, dimethylamine borane can be used in

$$3\ C_6H_5CH{=}NC_6H_5 + (CH_3)_2\overset{+}{N}H\overset{-}{B}H_3 + 3\ CH_3CO_2H \xrightarrow[84\%]{} 3\ C_6H_5CH_2NHC_6H_5 + (CH_3)_2NH + (CH_3CO_2)_3B$$

boiling acetic acid with little loss of activity (25% excess reagent is recommended). The following functional groups are not affected: NO_2, Cl, OH, OCH_3, CO_2H, $CO_2C_2H_5$, SO_2NH_2.

[1]H. Nöth and H. Beyer, *Ber.*, **93**, 928 (1960)
[2]J. H. Billman and J. W. McDowell, *J. Org.*, **26**, 1437 (1961)

***p*-Dimethylaminobenzaldehyde**, *p*-$(CH_3)_2NC_6H_4CHO$. Mol. wt. 149.19, m.p. 74°, b.p. 176–177°/17 mm. Unless very pure, the crystals are lemon colored; turns pink on exposure to light. Suppliers: A, B, E, F, MCB.

Preparation:[1]

$$(CH_3)_2NC_6H_5 \xrightarrow{HNO_2} \underline{p}\text{-}(CH_3)_2NC_6H_4NO \xrightarrow[-HCO_2H]{CH_2O\ +\ C_6H_5N(CH_3)_2}$$

$$\underline{p}\text{-}(CH_3)_2NC_6H_4N{=}CHC_6H_4N(CH_3)_2\text{-}\underline{p} \xrightarrow[56-59\%]{CH_2O} \underline{p}\text{-}(CH_3)_2NC_6H_4N{=}CH_2 + \underline{p}\text{-}OHCC_6H_4N(CH_3)_2$$

Color test for pyrroles.[2] One drop of an aqueous, ethereal, or alkaline test solution is mixed on a spot plate with one or two drops of a 5% solution of *p*-dimethylamino-benzaldehyde. If pyrrole or a substance capable of forming pyrrole is present, a violet color will appear. The aldehyde condenses with a tautomeric form of pyrrole (3) to give (4), which is converted by hydrogen ion into the colored quinonoid ion

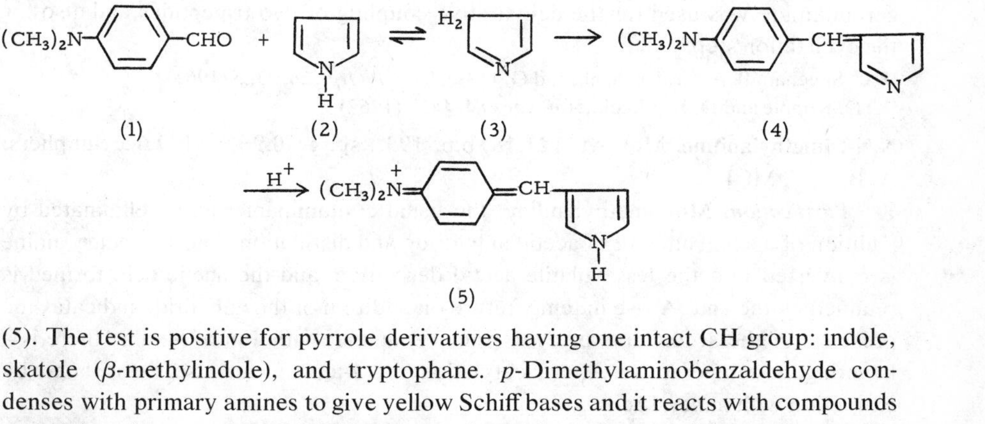

(1) (2) (3) (4) (5)

(5). The test is positive for pyrrole derivatives having one intact CH group: indole, skatole (β-methylindole), and tryptophane. *p*-Dimethylaminobenzaldehyde condenses with primary amines to give yellow Schiff bases and it reacts with compounds containing the grouping—$NHCONH_2$ to give yellow pigments,[3] but only pyrroles give violet pigments.

Test for hydrazine. The reagent condenses with hydrazine to give the product (6), which on prononation forms the yellow-to-orange quinonoid ion (7).[4]

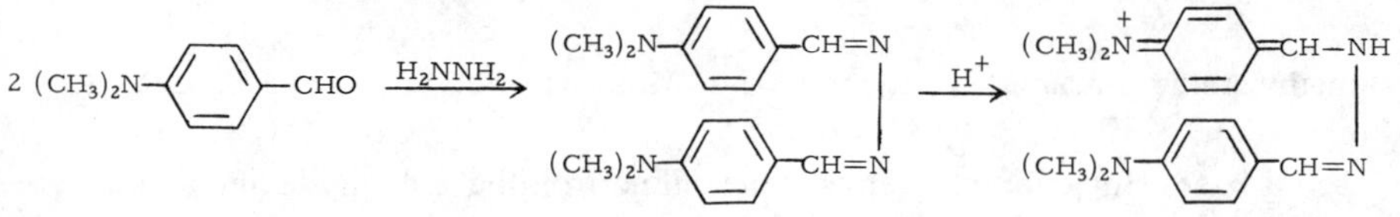

(6) (7)

[1]R. Adams and G. H. Coleman, *Org. Syn., Coll. Vol.*, **1**, 214 (1941)
[2]F. Feigl, *Spot Tests*, **2**, 198, Elsevier (1954)
[3]J. Schreiber and B. Witkop, *Am. Soc.*, **86**, 2441 (1964)
[4]F. Feigl, *loc. cit.*, p. 219

3-Dimethylaminopropylamine, $(CH_3)_2NCH_2CH_2CH_2NH_2$. Mol. wt. 102.18, b.p. 131–134°. Suppliers: A, B, MCB, Light.

The diamine is used to remove excess acid chloride in a novel method for the conversion of steroid secondary and tertiary alcohols into their esters.[1] A solution of the alcohol in ether is treated with ethylmagnesium bromide to form the bromomagnesium alkoxide and this is allowed to react with an acid chloride. In working up a

$$\text{ROH} \xrightarrow{C_2H_5MgBr} \text{ROMgBr} \xrightarrow{R'COCl} \text{ROCOR}' + \text{ClMgBr}$$

reaction mixture from an acid chloride which is hydrolyzed only slowly by dilute alkali, excess 3-dimethylaminopropylamine is added and the resulting basic amide is washed out with dilute acid.

$$R'COCl \xrightarrow{(CH_3)_2NCH_2CH_2CH_2NH_2} (CH_3)_2NCH_2CH_2CH_2NHCOR'$$

[1]D. D. Evans, D. E. Evans, G. S. Lewis, P. J. Palmer, and D. J. Weyell, *J. Chem. Soc.*, 3578 (1963)

N-(3-Dimethylaminopropyl)-N′-ethylcarbodiimide hydrochloride,
$(CH_3)_2NCH_2CH_2CH_2N{=}C{=}NC_2H_5{\cdot}HCl$.
Mol. wt. 175.69. Supplier: Pierce Chem. Co. Preparation.[1]

Peptide synthesis.[2] In the synthesis of a cyclic hexapeptide, this water-soluble carbodiimide was used for the dehydrative coupling of two tripeptides and also for the cyclization step.

[1]J. C. Sheehan, P. A. Cruickshank, and G. L. Boshart, *J. Org.*, **26**, 2525 (1961)
[2]K. D. Kopple and D. E. Nitecki, *Am. Soc.*, **84**, 4457 (1962)

N,N-Dimethylaniline. Mol. wt. 121.18, b.p. 193°, sp. gr. 0.96, pKb 9.62. Suppliers: A, B, E, F, MCB.

Purification. Monomethylaniline, the usual contaminant, can be eliminated by addition of a small amount of acetic anhydride and distillation. The secondary amine is converted into the less volatile acetyl derivative, and the acetic acid formed is retained as the salt. A rise in temperature on addition of the anhydride indicates the presence of the secondary (or even primary) amine; with a pure *t*-amine there is a lowering in temperature. After acetylation the *t*-amine can also be isolated by extraction with mineral acid.

Acylation of t-alcohols. Of three procedures in *Organic Syntheses* for the preparation of *t*-butyl esters by essentially the same method, we have selected the one giving

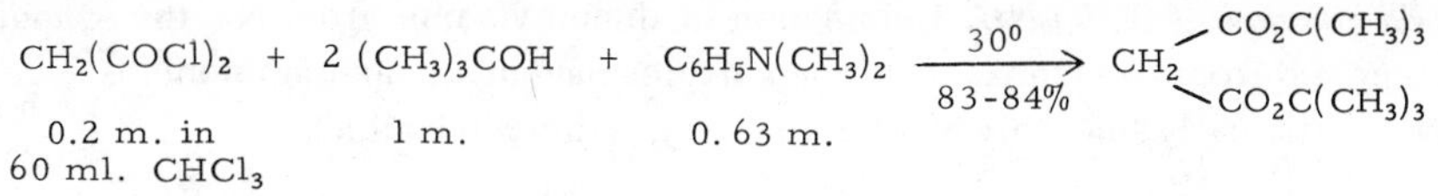
CH2(COCl)2 + 2 (CH3)3COH + C6H5N(CH3)2
30°
83-84%
CH2
CO2C(CH3)3
CO2C(CH3)3
0.2 m. in
60 ml. CHCl3
1 m.
0.63 m.

the best yield: the preparation of di-*t*-butyl malonate.[1] A mixture of the alcohol and dimethylaniline is stirred in a flask cooled in an ice bath, and a solution of malonyl dichloride in chloroform is run in at a rate such that the temperature does not rise above 30°.

The combination of acetyl chloride and dimethylaniline is used for the acetylation of 5α-hydroxysteroids; a 5β-hydroxyl group is unaffected.[2] For example, after a mixture of 3β-acetoxy-5α-hydroxy-$\Delta^{7,9}$-ergostadiene, acetyl chloride, dimethylaniline, and chloroform had been refluxed for 16 hrs., workup, chromatography, and one crystallization afforded the nearly pure 3β,5α-diacetate in excellent yield.[3]

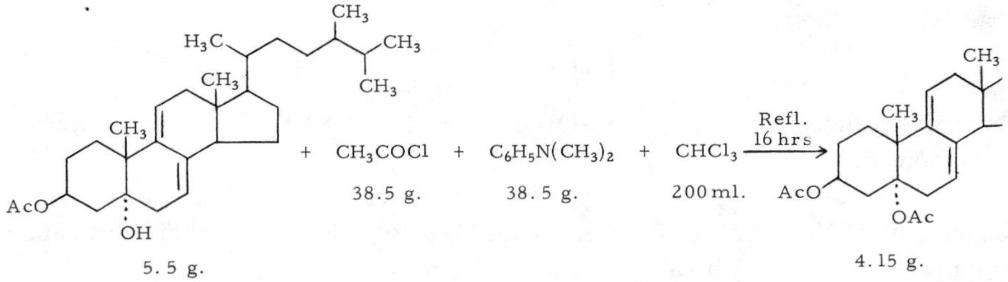
CH3
H3C
CH3
CH3
CH3
CH3
AcO
OH
5.5 g.
+ CH3COCl + C6H5N(CH3)2 + CHCl3
38.5 g.
38.5 g.
200 ml.
Refl.
16 hrs
CH3
CH3
AcO
OAc
4.15 g.

Dehydrohalogenation. Use of the reagent for this purpose is illustrated by a procedure for the preparation of 1,3-cyclohexadiene.[4] The chlorination is done in excess refluxing cyclohexene with *t*-butyl hypochlorite and dibenzoyl peroxide as

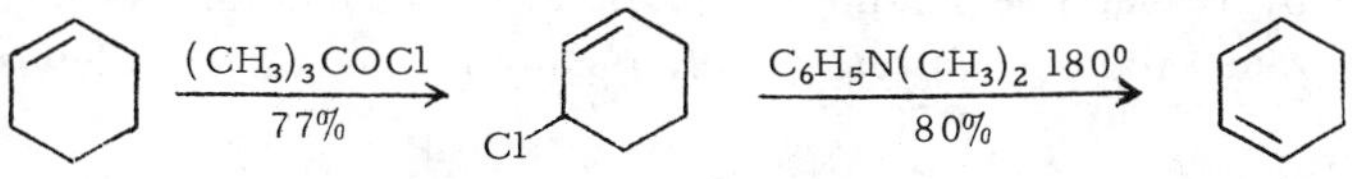
(CH3)3COCl
77%
Cl
C6H5N(CH3)2 180°
80%

catalyst. The next step is carried out at a temperature such that the diene distills as formed.

Decarboxylation. Tarbell *et al.*[5] decarboxylated 3,5-dichloro-4-hydroxybenzoic acid by brief heating of a solution in dimethylaniline.

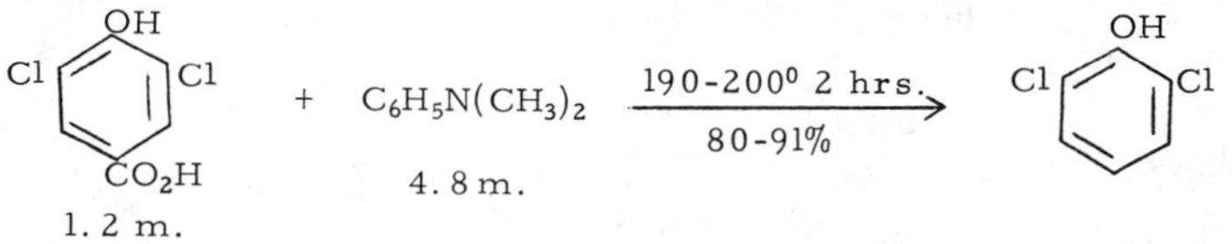
OH
Cl
Cl
CO2H
1.2 m.
+ C6H5N(CH3)2
4.8 m.
190-200° 2 hrs.
80-91%
OH
Cl
Cl

Reaction with tetracyanoethylene. McKusick and Melby[6] added tetracyanoethylene in portions to a stirred solution of dimethylaniline in dimethylformamide while controlling the temperature to 45–50°. *p*-Tricyanovinyl-N,N-dimethylaniline

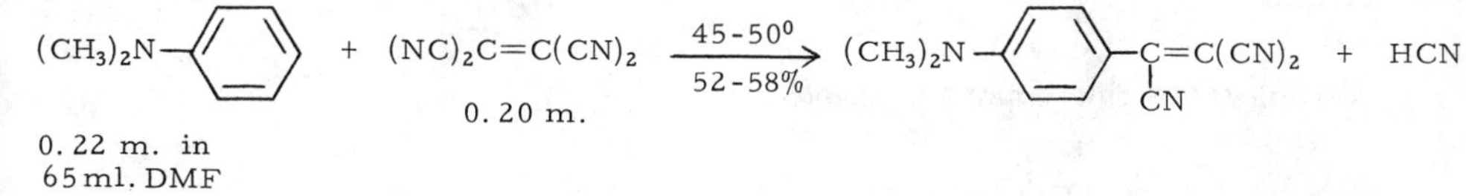
(CH3)2N
0.22 m. in
65 ml. DMF
+ (NC)2C=C(CN)2
0.20 m.
45-50°
52-58%
(CH3)2N
C=C(CN)2
CN
+ HCN

separated as dark blue crystals and was crystallized from acetic acid (deep red solution).

Elimination of $(CH_3)_2NH$. Elimination of dimethylamine from (1), the Mannich base derived from α-phenoxyacetophenone, formaldehyde, and ammonia is accomplished particularly smoothly by brief refluxing in dimethylaniline.[7]

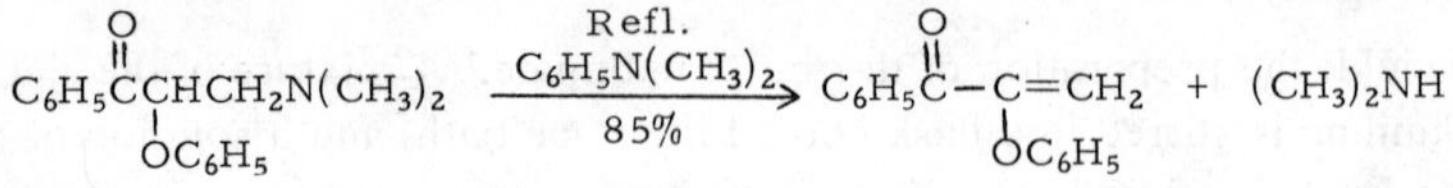

(1)

Other reactions, *see also* *p*-Dimethylaminobenzaldehyde; Oxalyl chloride.

[1]C. Raha, *Org. Syn., Coll. Vol.*, **4**, 263 (1963)
[2]Pl. A. Plattner, Th. Petrzilka, and W. Lang, *Helv.*, **27**, 513 (1944)
[3]P. Bladon *et al.*, *J. Chem. Soc.*, 4883 (1952)
[4]C. A. Grob, H. Kny, and A. Gagneux, *Helv.*, **40**, 130 (1957)
[5]D. S. Tarbell, J. W. Wilson, and P. E. Fanta, *Org. Syn., Coll. Vol.*, **3**, 267 (1955)
[6]B. C. McKusick and L. R. Melby, *ibid.*, **4**, 953 (1963)
[7]J. B. Wright, *J. Org.*, **25**, 1867 (1960)

2,3-Dimethylbutadiene, $CH{=}C(CH_3){-}C(CH_3){=}CH_2$. Mol. wt. 82.14, b.p. 70°. Suppliers: Columbia, B, MCB.

Preparation from pinacol: (a) by slow distillation from a mixture with a small amount of 48% hydrobromic acid (yield 55–60%)[1] and (b) by rapid distillation through a tube packed with alumina and heated in a furnace at 450–470°.[2]

[1]C. F. H. Allen and A. Bell, *Org. Syn., Coll. Vol.*, **3**, 312 (1955)
[2]L. W. Newton and E. R. Coburn, *ibid.*, **3**, 313 (1955)

2,3-Dimethyl-2-butylborane ("Thexylborane"). *See also* Diborane.

Prepared by reaction of 2,3-dimethyl-2-butene with diborane in diglyme or in tetrahydrofurane,[1] the reagent reacts more rapidly with *cis* acyclic olefins than with

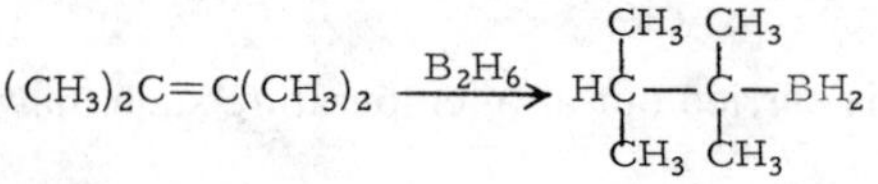

the *trans* isomers, and this difference can be used to effect separation of the *trans olefins* from mixtures. It is a convenient reagent for hydroboration of dienes (diborane itself gives polymeric products):

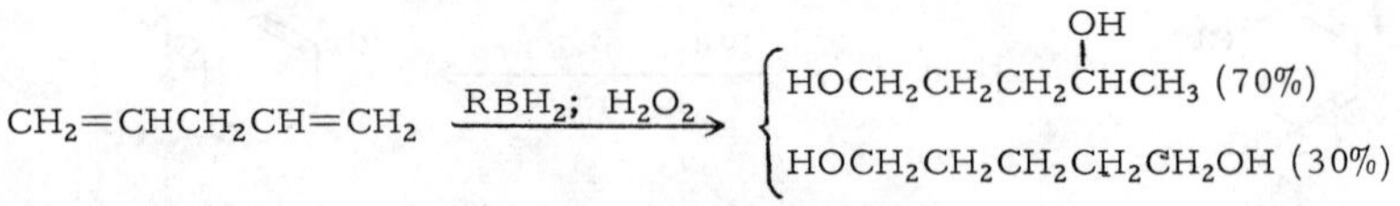

It reacts with an acetylene to give the corresponding vinylborane.

[1]G. Zweifel and H. C. Brown, *Am. Soc.*, **85**, 2066 (1963)

Dimethylchloromethyleneammonium chloride, *see* Dimethylformamide–Thionyl chloride.

3,5-Dimethyl-4-chloromethylisoxazole,

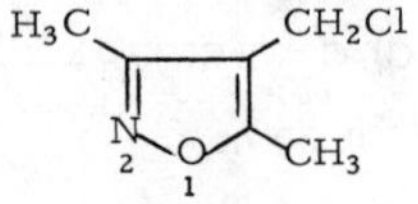

Mol. wt. 145.59, b.p. 67.5°/1 mm.

Preparation. 3,5-Dimethylisoxazole is chloromethylated with paraformaldehyde and hydrogen chloride in the presence of zinc chloride.[1]

Annelation. Prior to publication of details, Stork outlined the synthetic sequence formulated.[2] Alkylation of the enolate of 9-methyloctalin-1,6-dione (1) with the

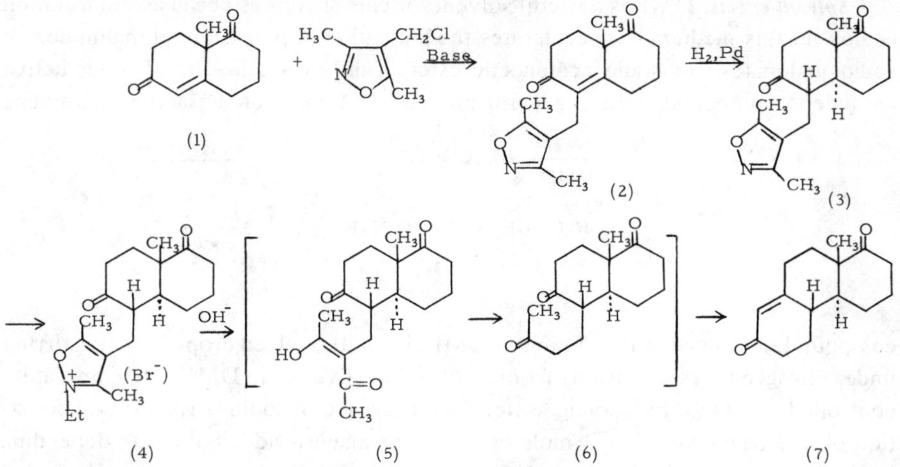
Base
H_2, Pd
OH^-
(1)
(2)
(3)
(4)
(5)
(6)
(7)

reagent gave the product (2) in good yield. The enone double bond was then reduced (3), and there remained the transformation of the isoxazolylmethyl portion into the desired annelated system. This was readily done by addition of triethyloxonium fluoroborate followed by treatment with base. The probable steps are formulated. Intermediate (5) is a symmetrical β-diketone and can undergo cleavage in only one direction. Cyclization of (6) then afforded the desired tricyclic diketone (7).

[1]N. K. Kochetkov, E. D. Khomutova, and M. V. Bazilevskiĭ, *Zhur. Obshcheĭ Khim.*, **28**, 2736 (1958) [*C.A.*, **53**, 9187 (1959)]

[2]G. Stork "The Chemistry of Natural Products," **3**, 131, Butterworths (1964); *Pure and Applied Chem.*, **9**, 131 (1964)

1,2-Dimethyl-4,5-di(mercaptomethyl)benzene (3). Mol. wt. 198.34, m.p. 66–67°, smells similar to *o*-dichlorobenzene.

Preparation. A modification[1] of an earlier method involves refluxing and stirring

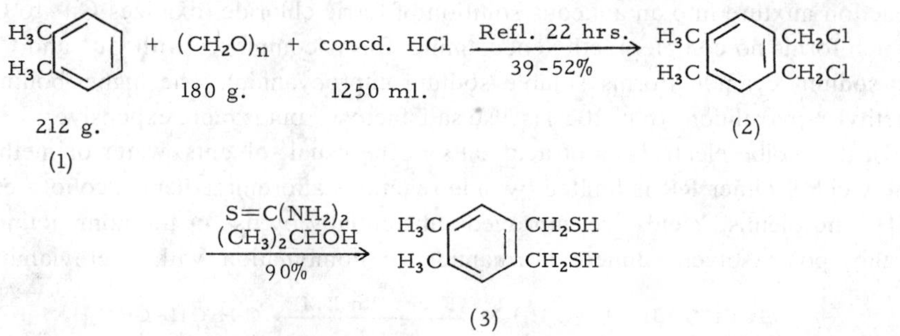
$+ (CH_2O)_n +$ concd. HCl
180 g.
1250 ml.
212 g.
(1)
Refl. 22 hrs.
39-52%
(2)
$S{=}C(NH_2)_2$
$(CH_3)_2CHOH$
90%
(3)

a mixture of *o*-xylene, paraformaldehyde, and hydrochloric acid. The di(chloromethyl) derivative was then refluxed with thiourea in isopropanol.

Characterization of carbonyl compounds. Since it does not possess the evil odor of 1,2-ethanedithiol but condenses readily with aldehydes and ketones under acid catalysis, the reagent is useful for the preparation of characterizing derivatives.[1] Aldehydes and ketones can be differentiated by their rate of reaction with the reagent.

[1]I. Shashak and E. D. Bergmann, *J. Chem. Soc.* C, 1005 (1966)

Dimethylformamide, $HCON(CH_3)_2$. Mol. wt. 73.10, b.p. 153°. Suppliers: du Pont, Matheson, A, B, E, F, MCB.

Solvent effects. DMF is a useful solvent for carbohydrates because mutarotation is slow in this medium.[1] It accelerates the alkylation of potassium phthalimide,[2] of sodio malonates,[3] of sodio acetoacetic esters,[4] and of alkali salts of other active methylene compounds.[4] In a general procedure[4] 1 mole of the active methylene

$$CH_3COCH_2CO_2C_2H_5 \xrightarrow[\text{DMF}]{\text{NaH}} CH_3CO\bar{C}HCO_2C_2H_5(Na^+) \xrightarrow[\text{2 hrs. at } 97^0]{(CH_3)_2CHBr} CH_3COCH(CH(CH_3)_2)CO_2Et + NaBr$$

63% yield

compound (dissolved in the minimum DMF, if solid) is added dropwise with stirring under nitrogen to a suspension of 1 mole of sodium hydride in DMF at a temperature controlled to 40–50° by cooling. After disappearance of sodium hydride and cessation of hydrogen evolution, 1 mole of alkylating agent is added at a rate depending on the heat evolved; in the example formulated the mixture was heated on the steam bath for 2 hrs. Comparison with other metal-solvent systems showed definite advantages for the NaH–DMF system.

In the conversion of an aryl halide to a nitrile the reaction time is shortened and the yield sometimes improved by refluxing the halide and cuprous cyanide in DMF.[5]

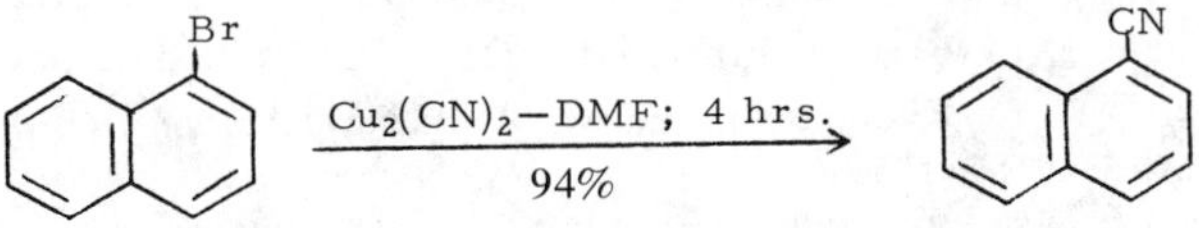
Br
Cu2(CN)2–DMF; 4 hrs.
94%
CN

By this method, 1-bromonaphthalene affords 1-naphthonitrile in 94% yield after refluxing for 4 hrs.; with pyridine as solvent the reaction mixture is heated in an oil bath at 215–225° for 15 hrs., and the yield is 82–90%.[6] Efficient procedures for liberating the nitrile from the cuprous halide complex involve pouring the brown reaction mixture into an aqueous solution of ferric chloride (oxidizes Cu^+ to Cu^{++}, which forms no complex), ethylenediamine (forms complexes with Cu^+ and Cu^{++}), or sodium cyanide (forms soluble sodium cuprocyanide). The higher-boiling N-methyl-2-pyrrolidone (b.p. 202°) is also satisfactory,[7] but is more expensive.

In the Kolbe electrolysis of acid salts in the usual solvents, water or methanol, the yield of dimer RR is limited by side reactions affording ethers, alcohols, esters, RH, and olefins. Yields are improved substantially by use of the nonreacting and highly polar solvent dimethylformamide in combination with triethylamine as

$$C_6H_5CH(CH_3)CO_2H + (C_2H_5)_3N \xrightarrow[41\%]{\text{Elect. in DMF}} C_6H_5CH(CH_3)-CH(CH_3)C_6H_5$$

base.[8] Thus hydratropic acid afforded 2,3-diphenylbutane (*meso* and *dl*) in 41% yield; in a parallel run with triethylamine in methanol the yield of dimer was only 21%, and several products derived from α-methylbenzyl radicals were isolated.

In a procedure for oxidation of alcohols, chromic anhydride is stirred with a solution of the alcohol (*e.g.*, cholestanol) in dimethylformamide until dissolved, and a catalytic amount of sulfuric acid is added; reaction proceeds at room temperature.[9]

The steric course of the reaction of the Wittig reagent (2) with propionaldehyde to give β-ethylstyrene (3) is highly sensitive to the environmental conditions.[10] When the reaction was carried out in DMF as solvent and sodium ethoxide as base and with

$$\underset{(1)}{C_6H_5CH_2\overset{+}{P}(C_6H_5)_3Cl^-} \xrightarrow[\text{DMF}]{NaOC_2H_5} \underset{(2)}{C_6H_5CH{=}P(C_6H_5)_2} \xrightarrow[\text{DMF, LiI}]{CH_3CH_2CHO}$$

$$\underset{(3)}{C_6H_5CH{=}CHCH_2CH_3}$$

added lithium iodide, the *cis*/*trans* ratio was 96:4, as compared to a ratio of 31:69 with the same base but in ether as solvent, with no additive.

In exploring reagents and conditions for efficient synthesis of peptides with activated esters, Pless and Boissonnas[11] found that the reaction of Cb-L-phenylalanine-2,4,5-trichlorophenyl ester (4) with benzylamine is faster in dimethylforma-

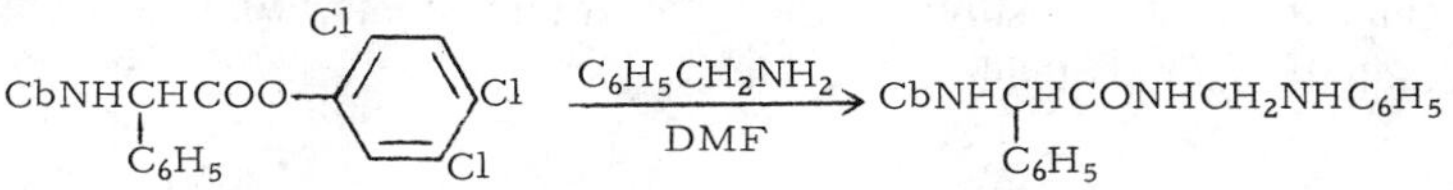

(4) (5)

mide than in other commonly used solvents, a fortunate circumstance since DMF has superior solvent power for large peptides.

DMF is recommended as diluent in the Ullmann synthesis of diaryls from aryl halides[12] and as solvent for the Beckmann rearrangement of steroidal enone oximes.[12a]

The Willgerodt-Kindler reaction (a carbonyl compound with a secondary amine and sulfur to give a thioamide) can be carried out in good yield at 50–60° if DMF is

$$RCOCH_3 + S + R'R'NH \longrightarrow RCH_2CSNR'R' + H_2O$$

used as solvent; pyridine is somewhat less satisfactory.[13] Mayer and Wehl[14] used DMF but added *p*-toluenesulfonic acid as catalyst to accelerate formation of the intermediate enamine and were able to effect reaction at room temperature.

$$RCOCH_3 + (CH_3)_2NH \xrightarrow[\text{DMF}]{TsOH} RC(=CH_2)N(CH_3)_2 \xrightarrow[75\%]{S} RCH_2C(=S)N(CH_3)_2$$

Dehydrohalogenation. Kuhn *et al.*[15] dissolved 2,4-dibromo-1,1,5,5-tetraphenylpentadiene-1,4 in DMF, flushed the system with nitrogen, and slowly added methanolic

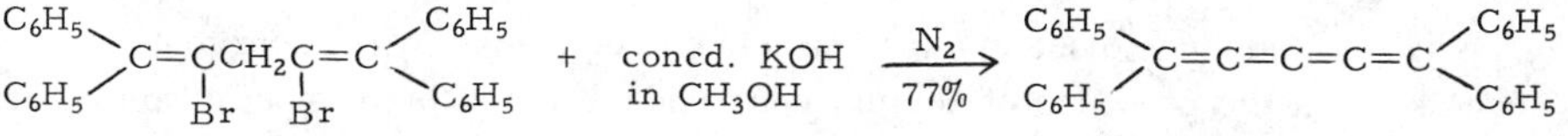

5 ml. 1.6 g.

3 g. in 50 ml. DMF

KOH. After 10 minutes, when the color had changed from blue-green to brown-yellow, methanol (20 ml.) followed by 50% methanol (20 ml.) was added. Tetraphenylpentatetraene separated "in schönen, gelben Prismen."

Bernstein *et al.*[16] were able to effect dehydrohalogenation with DMF alone. When steroid $9\alpha,11\beta$-dichloro-$\Delta^{1,4}$-diene-3-ones (1) or 9α-bromo(chloro)-11β-hydroxy-$\Delta^{1,4}$-diene-3-ones (2) were refluxed in DMF for 0.5 hr. two products, (3) and (4), were obtained in yields of 20–25%. In the aromatization reaction producing (4),

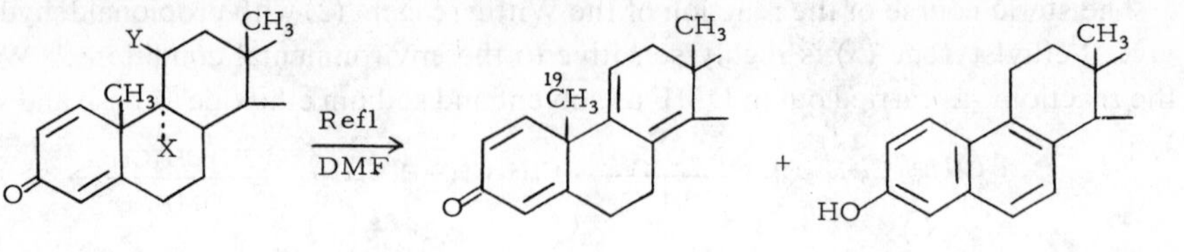

(1) X = Y = Cl
(2) X = Br or Cl, Y = OH

(3)

(4)

the 19-methyl group is evolved as methyl chloride (bromide). When pyridine was used in place of DMF, the reaction required a reflux period of at least 6 hrs.

Reactions (*see also* DMF–$POCl_3$, DMF–$SOCl_2$). In the conversion of an acid chloride into the corresponding N,N-dimethylamide, use of the somewhat objectionable dimethylamine (b.p. 7.4°) can be avoided by substitution of DMF.[17] This permits elimination of any solvent other than the reactants. When an anhydride is used, a catalytic amount of sulfuric acid is added.

$$C_6H_5COCl + HC(=O)N(CH_3)_2 \xrightarrow[97\%]{\text{4 hrs. at } 150^0} C_6H_5CON(CH_3)_2 + HCl + CO$$

$$\text{(succinic anhydride, } CH_2CO{-}O{-}COCH_2\text{)} + HC(=O)N(CH_3)_2 \xrightarrow[90\%]{\text{Trace } H_2SO_4\text{, Refl.}} HO_2CCH_2CH_2CON(CH_3)_2 + CO$$

In the presence of a catalytic amount of di-*t*-butylperoxide at 132°, dimethylformamide adds to octene-1 to give a 60:40 mixture of (1) and (2).[18]

$$CH_3(CH_2)_5CH{=}CH_2 + HCON(CH_3)_2 \xrightarrow{R\cdot} \underset{(1)}{CH_3(CH_2)_5CH_2CH_2CON(CH_3)_2} + \underset{(2)}{CH_3(CH_2)_5CH(CH_3)CON(CH_3)_2}$$

Dimethylformamide reacts with lithium alkyls to give aliphatic aldehydes in yields of 50–80%.[19] Lithium alkenyls react in the same way but in lower yields.[20]

$$\underline{n}\text{-}C_7H_{15}Li + HC(=O)N(CH_3)_2 \longrightarrow \underline{n}\text{-}C_7H_{15}C(OLi)(H)N(CH_3)_2 \xrightarrow[62\%]{H_2O} \underline{n}\text{-}C_7H_{15}CHO$$

Dimethylformamide, used in excess as solvent, condenses with an aromatic amine such as (3) in the presence of sodium methoxide at the reflux temperature to give the corresponding formanilide (4).[21] The same reaction conditions are satisfactory

$$\underset{(3)}{\underline{p}\text{-}FC_6H_4NH_2} + HC(=O)N(CH_3)_2 \xrightarrow[78\%]{NaOCH_3} \underset{(4)}{\underline{p}\text{-}FC_6H_4NHCHO} + (CH_3)_2NH$$

for some but not all substituted anilines. Thus with 2,5-dimethoxyaniline the yield was only 48% with sodium methoxide as base but was raised to 62% by use of sodium hydride.

Use of DMF in the synthesis of α,β-acetylenic aldehydes is illustrated in the formulation.[22]

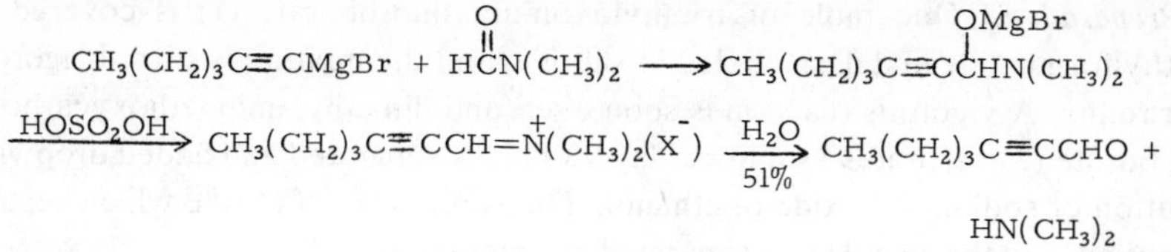

$$CH_3(CH_2)_3C\equiv CMgBr + H\overset{\overset{O}{\|}}{C}N(CH_3)_2 \longrightarrow CH_3(CH_2)_3C\equiv C\overset{\overset{OMgBr}{|}}{C}HN(CH_3)_2$$

$$\xrightarrow{HOSO_2OH} CH_3(CH_2)_3C\equiv CCH=\overset{+}{N}(CH_3)_2(X^-) \xrightarrow[51\%]{H_2O} CH_3(CH_2)_3C\equiv CCHO +$$

$$HN(CH_3)_2$$

Dimethylformamide reacts with a phosphoric acid diester chloride (5) to give the crystalline yellow adduct (6), which is stable in the absence of moisture.[23] This

$$(CH_3)_2N\atop H\overset{}{C}=O + Cl\overset{\overset{O}{\|}}{P}(OR)_2 \longrightarrow \overset{Cl^-}{\underset{H\overset{}{C}-O\overset{\overset{O}{\|}}{P}(OR)_2}{(CH_3)_2\overset{+}{N}}} \qquad \overset{\overset{O}{\|}}{R'O\overset{}{P}OH}\atop \overset{OH\ (8)}{base}$$

$$(5) \qquad\qquad\qquad (6)$$

$$base\Big\downarrow R'OH \qquad\qquad\qquad \Big\downarrow \overset{-(CH_3)_2N\overset{\overset{H}{}}{C}=O}{}$$

$$R'O\overset{\overset{O}{\|}}{P}(OR)_2 \qquad\qquad R'O\overset{\overset{O}{\|}}{P}-O-\overset{\overset{O}{\|}}{P}(OR)_2\atop OH$$

$$(7) \qquad\qquad\qquad (9)$$

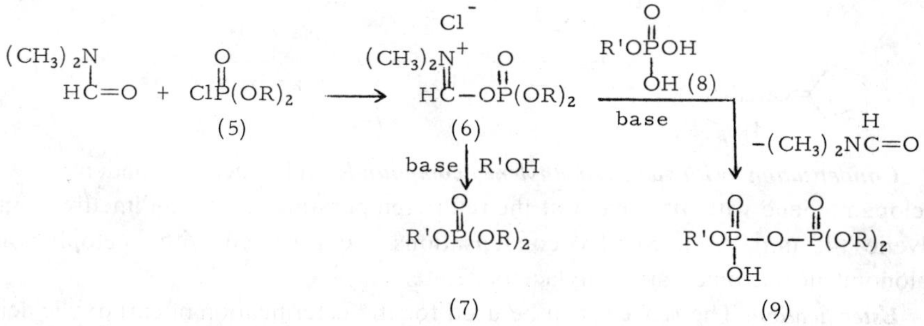

adduct reacts with an alcohol in the presence of base to give a phosphoric acid ester (7), and it reacts with a phosphoric ester (8) to give a pyrophosphate (9). Analogous reactions are reported starting with DMF and a phosphoric acid mono ester dichloride.

[1] R. Kuhn and F. Haber, *Ber.*, **86**, 722 (1953)

[2] J. C. Sheehan and W. A. Bolhofer, *Am. Soc.*, **72**, 2786 (1950)

[3] H. E. Zaugg, B. W. Horrom, and S. Borgwardt, *Am. Soc.*, **82**, 2895 (1960)

[4] H. E. Zaugg, D. A. Dunnigan, R. J. Michaels, L. R. Swett, T. S. Wang, A. H. Sommers, and R. W. DeNet, *J. Org.*, **26**, 644 (1961)

[5] L. Friedman and H. Shechter, *J. Org.*, **26**, 2522 (1961)

[6] M. S. Newman, *Org. Syn., Coll. Vol.*, **3**, 631 (1955)

[7] M. S. Newman and H. Boden, *J. Org.*, **26**, 2525 (1961)

[8] M. Finkelstein and R. C. Petersen, *J. Org.*, **25**, 136 (1960)

[9] G. Snatzke, *Ber.*, **94**, 729 (1961)

[10] L. D. Bergelson and M. M. Shemyakin, *Tetrahedron*, **19**, 149 (1963)

[11] J. Pless and R. A. Boissonnas, *Helv.*, **46**, 1609 (1963)

[12] N. Kornblum and D. L. Kendall, *Am. Soc.*, **74**, 5782 (1952)

[12a] F. Kohen, *Chem. Ind.*, 1378 (1966)

[13] A. Carayon-Gentil, M. Minot, and P. Charbier, *Bull. soc.*, 1420 (1964)

[14] R. Mayer and J. Wehl, *Angew. Chem., Internat. Ed.*, **3**, 705 (1964)

[15] R. Kuhn, Herbert Fischer, and Hans Fischer, *Ber.*, **97**, 1760 (1964)

[16] M. Heller, R. H. Lenhard, and S. Bernstein, *Am. Soc.*, **86**, 2309 (1964)

[17] G. M. Coppinger, *Am. Soc.*, **76**, 1372 (1954)

[18] L. Friedman and H. Shechter, *Tetrahedron Letters*, 238 (1961)

[19] E. A. Evans, *J. Chem. Soc.*, 4691 (1956)

[20] E. A. Braude and E. A. Evans, *J. Chem. Soc.*, 3331 (1955)

[21] G. R. Pettit and E. G. Thomas, *J. Org.*, **24**, 895 (1959); G. R. Pettit, M. V. Kalnins, T. M. Liu, E. G. Thomas, and K. Parent, *ibid.*, **26**, 2563 (1961)

[22] E. R. H. Jones, L. Skattebøl, and M. C. Whiting, *J. Chem. Soc.*, 1054 (1958)

[23] F. Cramer and M. Winter, *Ber.*, **94**, 989 (1961)

Dimethylformamide diethyl acetal, $(CH_3)_2NCH(OC_2H_5)_2$. Mol. wt. 131.22, b.p. 134–136°.

Preparation.[1] One mole of triethyloxonium fluoroborate (1) is covered with dimethylformamide in a flask cooled in salt-ice, and the mixture is stirred vigorously under reflux. A vigorous reaction is soon over, and dimethylaminoethoxycarbonium fluoroborate (2) separates as a lower layer. This is separated and added dropwise to a solution of sodium ethoxide in ethanol. The sodium borofluoride which separates is removed, and the acetal (3) is separated by fractionation.

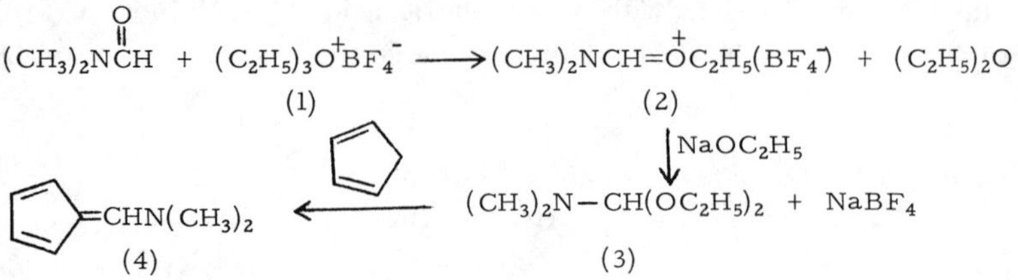

Condensation with active-methylene compounds. The acetal condenses with cyclopentadiene without solvent at the reflux temperature to give 6-dimethylaminofulvene (4, m.p. 67°). Similar condensations are reported with acetophenone, malononitrile, nitromethane, ethylacetoacetate.

Esterification. The reagent can be used for the esterification of carboxylic acids under mild conditions.[2,3] Thus benzoic acid (0.4 mole) reacts with 2 equivalents of the acetal to give ethyl benzoate under the following conditions: in methylene chloride for 5 hrs. at 40°; in benzene for 1 hr. at 80°; in acetonitrile for

$$C_6H_5CO_2H + (CH_3)_2NCH(OC_2H_5)_2 \xrightarrow[90\%]{} C_6H_5CO_2C_2H_5 + C_2H_5OH + (CH_3)_2NCHO$$

36 hrs. at room temperature.[3] It reacts with phenol in ethylene dichloride in 46 hrs. at 84°.[2]

Heterocycles. The reagent condenses with guanidine to give 2,4-diamino-*s*-triazine.[4]

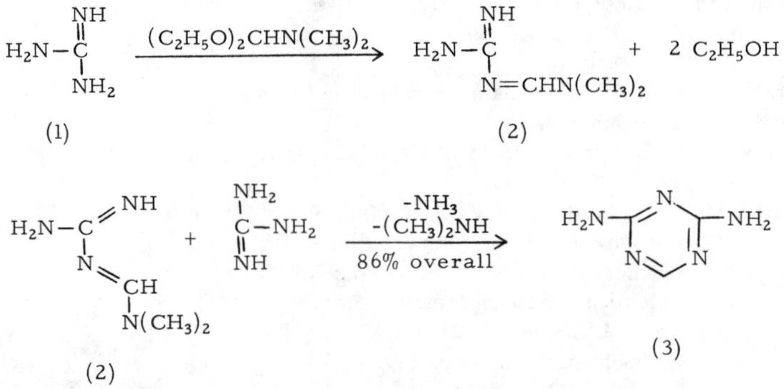

[1]H. Meerwein, W. Florian, N. Schön, and G. Stopp, *Ann.*, **641**, 1 (1961)
[2]H. Vorbrüggen, *Angew. Chem., Internat. Ed.*, **2**, 211 (1963)
[3]H. Brechbühler, H. Büchi, E. Hatz, J. Schreiber, and A. Eschenmoser, *ibid.*, **2**, 212 (1963); *Helv.*, **48**, 1746 (1965)
[4]H. Bredereck, F. Effenberger, and A. Hofmann, *Ber.*, **97**, 61 (1964)

Dimethylformamide–Dimethyl sulfate. Dimethylformamide and dimethyl sulfate react without solvent to form a 1:1 complex (2 days at 25°, 2 hrs. at 60–80°).[1] On

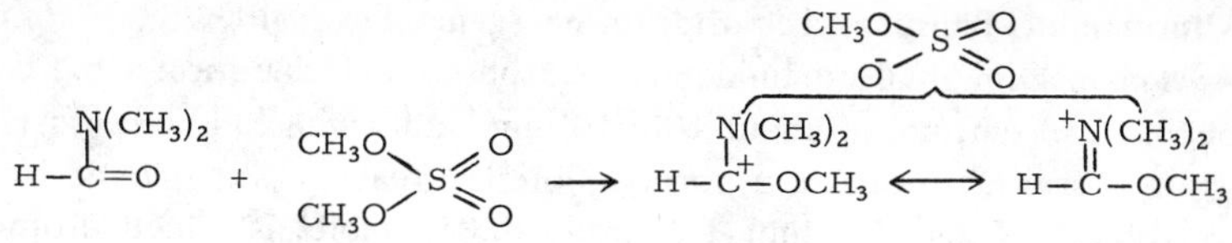

Complex

attempted distillation the complex decomposes to the components. It reacts with cyclopentadienylsodium to give 6-dimethylaminofulvene.[2]

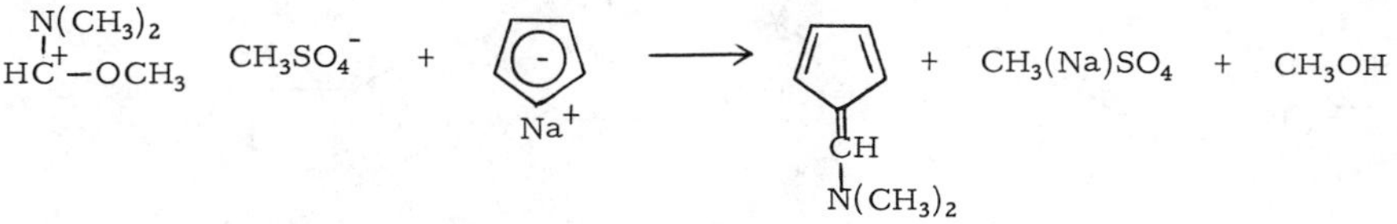

[1]H. Bredereck, F. Effenberger, and G. Simchen, *Angew. Chem.*, **73**, 493 (1961)
[2]K. Hafner *et al.*, *Angew. Chem., Internat. Ed.*, **2**, 123 (1963)

Dimethylformamide dineopentyl acetal (3). Mol. wt. 231.37. *Compare* Dimethylformamide diethyl acetal. Supplier: Fluka.

The acetal (3) mediates the esterification of a carboxylic acid with a benzyl-type alcohol (2).[1] The procedure has the advantage that only one equivalent of the acid

$$\underset{(1)}{RCO_2H} + \underset{(2)}{ArCH_2OH} + \underset{(3)}{(CH_3)_2NCH(OCH_2C(CH_3)_3)_2} \xrightarrow{70\text{-}90\%}$$

$$\underset{(4)}{RCO_2OCH_2Ar} + (CH_3)_2NCHO + 2\ HOCH_2C(CH_3)_3$$

is required and that only volatile by-products are formed. It was tested with a number of N-protected amino acids or peptides using *p*-methoxybenzyl alcohol and *p*-dodecylbenzyl alcohol.

[1]H. Büchi, K. Steen, and A. Eschenmoser, *Angew. Chem., Internat. Ed.*, **3**, 62 (1964); H. Brechbühler, H. Büchi, E. Hatz, J. Schreiber, and A. Eschenmoser, *Helv.*, **48**, 1746 (1965)

Dimethylformamide ethylene ketal, 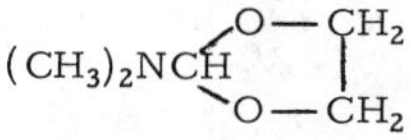

Mol. wt. 117.15, b.p. 142–144°.

Prepared from dimethylformamide diethyl ketal by transacetalization with ethylene glycol.[1] In the presence of a catalytic amount of acetic acid, the reagent reacts selectively with saturated 3-ketosteroids to give the 3-ethyleneketal in the presence of other saturated or unsaturated keto functions.[2] Under the mild conditions cited, Δ^4-3-ketosteroids and 12- and 20-ketosteroids do not react.

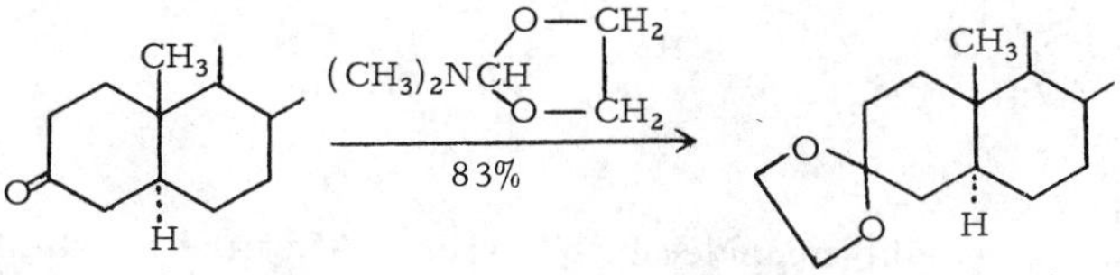

[1]H. Meerwein, W. Florian, G. Schön, and G. Stopp, *Ann.*, **641**, 1 (1961)
[2]H. Vorbrueggen, *Steroids*, **1**, 45 (1963)

Dimethylformamide–Phosphoryl chloride (or phosgene), DMF–POCl$_3$.

Formylation. Dimethylformamide and phosphoryl chloride form a 1:1 complex which begins to dissociate at about 100°/10 mm. and which can be represented as the ion-pair hybrid (1). For use in formylation, the reagent is employed in excess DMF as solvent. Thus phosphoryl chloride (0.055 mole) is added dropwise to

$$(CH_3)_2N-CH \rightleftharpoons (CH_3)_2N=CHOH \xrightarrow[-HCl]{ClP^+Cl_2} (CH_3)_2\overset{+}{N}=CHOP^+Cl_2 \longleftrightarrow (CH_3)_2N-CHOP^+Cl_2$$

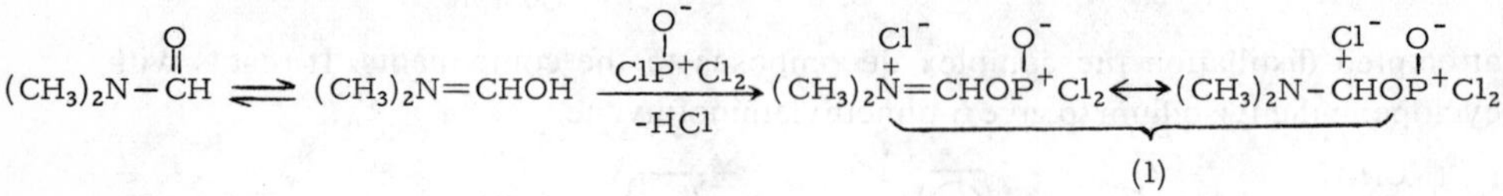

(1)

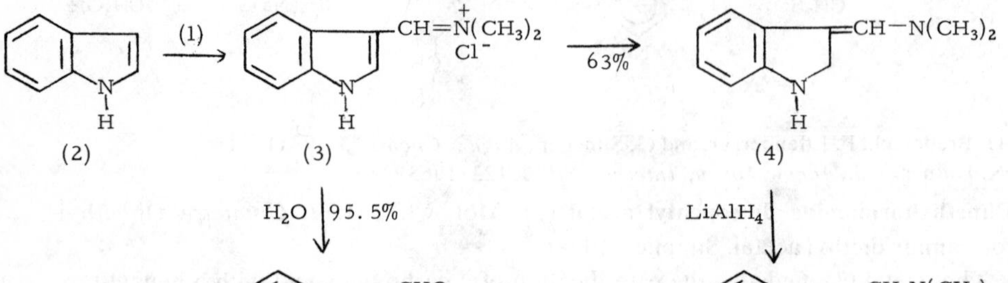

dimethylformamide (0.22 mole) at 10–20° to produce a solution of the complex (1), and indole (0.50 mole) is added in DMF solution; after 45 min. at 35° the solution is poured onto ice-water to give a clean, pale red solution of the salt (3).[1] If the solution is made alkaline and boiled, 3-formylindole (5) separates as an off-white solid in high yield. If the pale red solution of (3) is neutralized to weak acidity at −5°, extraction with chilled chloroform affords 3-dimethylaminomethylenein-dolenine (4). This substance is reduced by LiAlH$_4$ to gramine (6). Application of the same reactions to an aromatic hydrocarbon is reported by Snyder.[2] Application of the reaction sequence to the steroid dienol ether (7) afforded a route to the 6α-methyl Δ^4-3-ketosteroid (11).[3] In this case dimethylformiminium chloride was

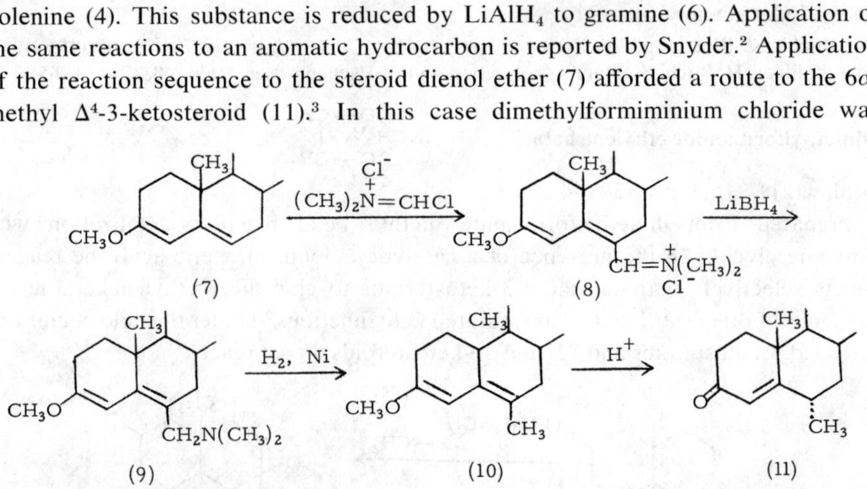

prepared from dimethylformamide and phosgene. After the substitution reaction, the salt (8) was reduced to (9), hydrogenolysis gave (10), and acid hydrolysis gave (11).

Organic Syntheses procedures utilizing DMF–$POCl_3$ reagent afford *p*-dimethylaminobenzaldehyde in 80–84% yield[4] and 2-pyrrolealdehyde in 78–79% yield.[5] Formylation of thiophene derivatives with the reagent is reported.[6]

The DMF–$POCl_3$ combination provides a means for converting even very weakly basic 4,5-diaminopyrimidines into purines under mild conditions.[7] Thus a mixture

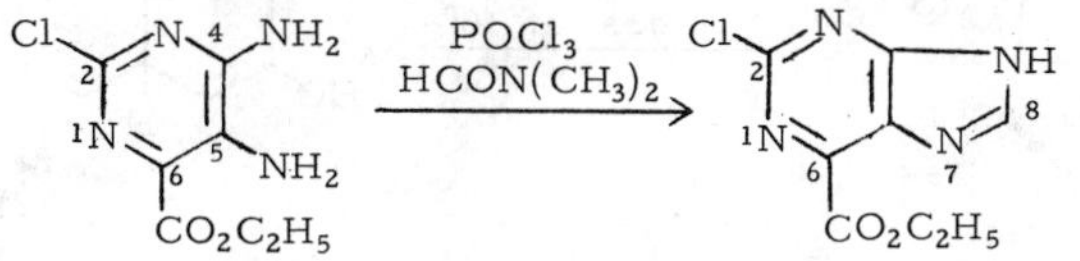

(12) (13)

of 0.2 g. of ethyl 4,5-diamino-2-chloropyrimidine-6-carboxylate (12, pKa 2.6), 10 ml. of dimethylformamide, and 1 ml. of phosphoryl chloride was let stand for 1 hr. and heated on the steam bath for 15 min.; the yield of ethyl 2-chloropurine-6-carboxylate (13) was 0.16 g. Dimethylformamide supplies carbon atom 8.

The reagent prepared from DMF and phosgene reacts with the epoxide (1) in trichloroethylene to give 1-chloro-2-formyloxy-$\Delta^{5,9}$-*cis,cis*-cyclododecadiene (2).[8]

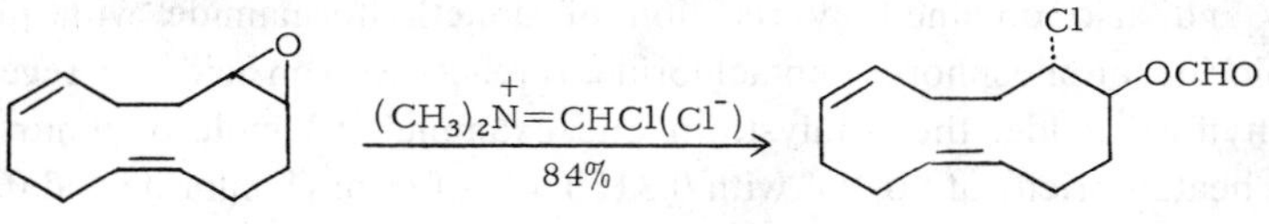

(1) (2)

The DMF–$COCl_2$ reagent reacts with a 3,3-dialkoxy-5α-steroid to give an iminium salt (not isolated), which on hydrolysis with aqueous sodium acetate gives the 2-formyl-3-alkoxy-Δ^2-ene.[9]

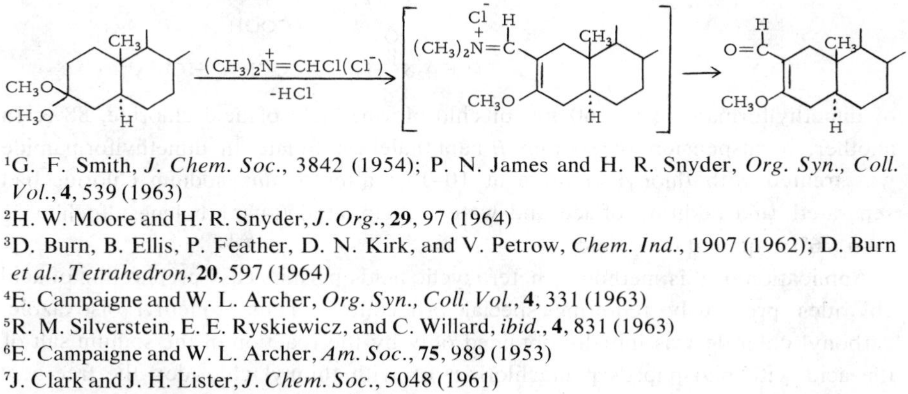

[1] G. F. Smith, *J. Chem. Soc.*, 3842 (1954); P. N. James and H. R. Snyder, *Org. Syn., Coll. Vol.*, **4**, 539 (1963)
[2] H. W. Moore and H. R. Snyder, *J. Org.*, **29**, 97 (1964)
[3] D. Burn, B. Ellis, P. Feather, D. N. Kirk, and V. Petrow, *Chem. Ind.*, 1907 (1962); D. Burn *et al., Tetrahedron*, **20**, 597 (1964)
[4] E. Campaigne and W. L. Archer, *Org. Syn., Coll. Vol.*, **4**, 331 (1963)
[5] R. M. Silverstein, E. E. Ryskiewicz, and C. Willard, *ibid.*, **4**, 831 (1963)
[6] E. Campaigne and W. L. Archer, *Am. Soc.*, **75**, 989 (1953)
[7] J. Clark and J. H. Lister, *J. Chem. Soc.*, 5048 (1961)
[8] W. Ziegenbein and W. Franke, *Ber.*, **93**, 1681 (1960)
[9] D. Burn *et al., Tetrahedron Letters*, **73**, 733 (1964)

Dimethylformamide–Sodium acetate.

Dehydration. Reichstein[1] found this combination the preferred method for effecting the selective dehydration of gitoxigenin (1) to 16-anhydrogitoxigenin (2). A mixture of 3 mg. of (1), 0.3 ml. of DMF, and 1 mg. of anhydrous sodium acetate was heated in a sealed tube at 95° for 4 hrs. Workup by chromatography and crystallization afforded 1.8 mg. of pure (2).

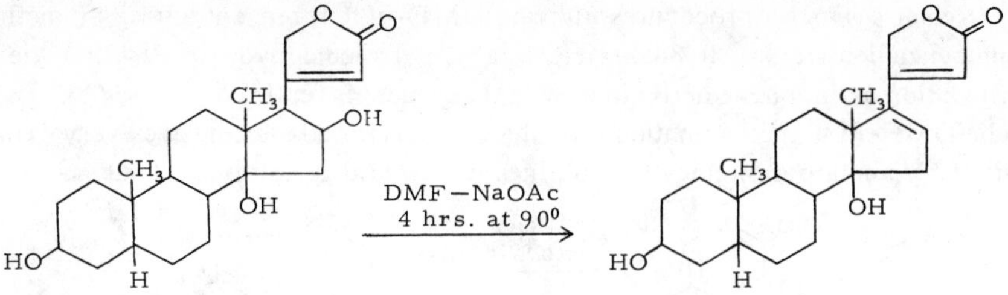
DMF—NaOAc
4 hrs. at 90°

(1) (2)

[1]J. H. Russel, O. Schindler, and T. Reichstein, *Helv.*, **43**, 167 (1960)

Dimethylformamide–Thionyl chloride (Vilsmeier reagent[1]).

Preparation of acid chlorides. The thionyl chloride method of preparing acid chlorides fails with some carboxylic acids (*e.g.*, *p*-$NO_2C_6H_4CO_2H$) and with all sulfonic acids. Bosshard and co-workers[2] found that dimethylformamide catalyzes both reactions, either when used as solvent or when employed in catalytic amount in an inert solvent. The reactive, hygroscopic intermediate dimethylformiminium chloride was isolated from one equivalent each of dimethylformamide and thionyl chloride, and also obtained by reaction of dimethylformamide with phosgene, oxalyl chloride, or phosphorus pentachloride. It reacts with an acid with regeneration of dimethylformamide, the catalyst. In one example, 0.3 mole of *p*-nitrobenzoic acid was heated briefly at 90–95° with 0.315 mole of thionyl chloride and 0.03 mole

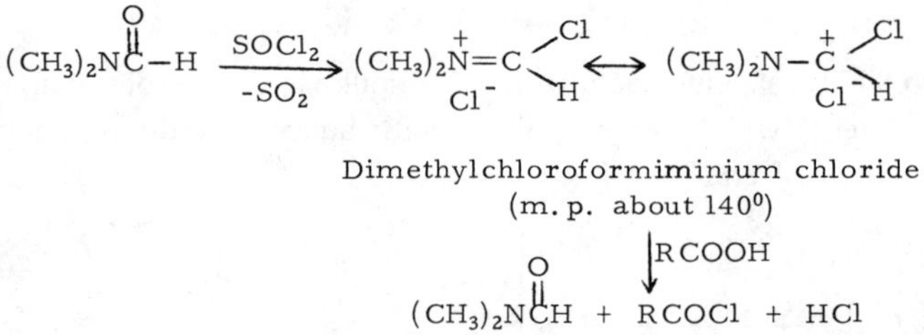
SOCl2
-SO2
Dimethylchloroformiminium chloride
(m. p. about 140°)
RCOOH
RCOCl + HCl

of dimethylformamide in 160 ml. of chloroform; yield of acid chloride, 88%. In another, a suspension of sodium β-naphthalenesulfonate in dimethylformamide was treated with thionyl chloride at 10–15°; after 5 min. sodium chloride had separated, and addition of ice and water precipitated β-naphthalenesulfochloride (m.p. 76°) in quantitative yield.

Application of this method to heterocyclic acids permits easy preparation of acid chlorides previously requiring special procedures. Thus 5-methyl-3-isoxazolecarbonyl chloride was initially obtained only by the reaction of the sodium salt of the acid with phosphorus pentachloride[3] or with thionyl chloride;[4] the free acid does not react with thionyl chloride alone. An improved procedure is as follows.[5]

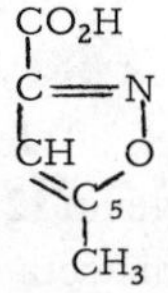

Fifty grams of dimethylformamide was dropped into a stirred suspension of 508 g. of 5-methyl-3-isoxazolecarboxylic acid in 700 g. of thionyl chloride over a period of 10 min. and the suspension was refluxed for 1 hr. (the acid dissolved in 15 min.). After stirring at room temperature overnight, the excess thionyl chloride was removed under vacuum, using a water pump (20 mm.) and a 15-cm. Vigreaux column to facilitate separation. Concentration was continued until the temperature at the head of the column reached 80°, and the residue was then chilled at 4° for several hours. The mass largely crystallized and the flask was inverted into a beaker to drain off a little dimethylformamide; crude yield, 95–100%. Distillation, which often results in violent decomposition,[4] is unnecessary. The crude product containing a little dimethylformamide was used successfully to prepare substituted amides, esters, and ketones ($CdCl_2$ method). The procedure was applied with success to 3-methyl-5-isoxazolecarboxylic acid, pyridine-2,6-dicarboxylic acid, nicotinic acid, isonicotinic acid, and 3-furoic acid. The molar ratio of thionyl chloride to acid specified is 1.47:1.0. With a 1:1 ratio, difficulty was encountered in stirring the thick slurry formed.

Japanese workers[6] report that conversion of the acid (1) to the acid chloride was "much improved" by use of DMF as catalyst. A mixture of the acid and thionyl

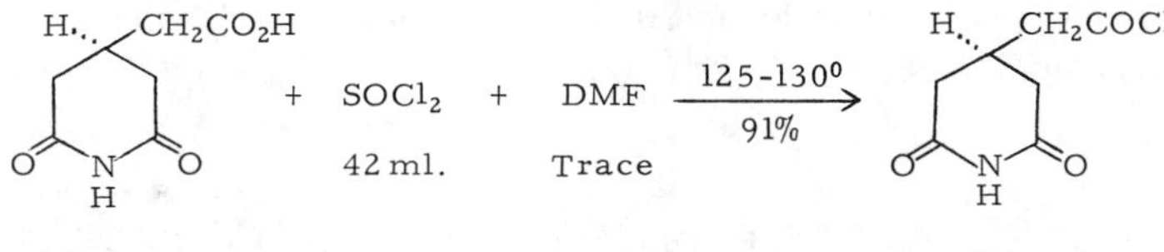
H
CH2CO2H
O
N
H
O
+ SOCl2 + DMF
42 ml.
Trace
125-130°
91%
H
CH2COCl
(1) 10 g.
(2)

chloride was heated at 100° for 15 min. and then cooled to room temperature and treated with a few drops of DMF. The mixture was then refluxed for 15 min.; yield of acid chloride, 91%.

Formylation. The Bosshard group[2] found[7] the reagent useful for the synthesis of *p*-dimethylaminobenzaldehyde:

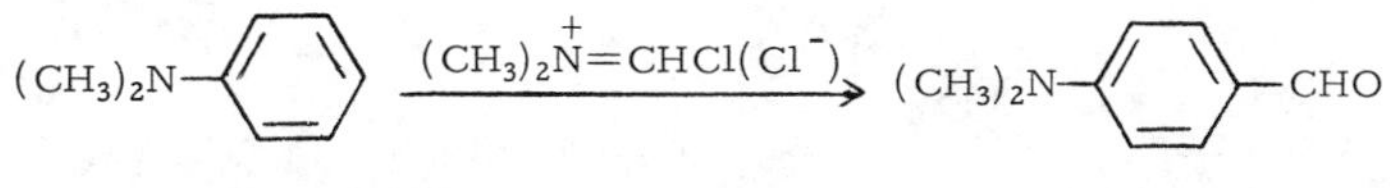
(CH3)2N
(CH3)2N=CHCl(Cl⁻)
(CH3)2N
CHO

The dimethylformamide–thionyl chloride complex reacts rapidly with steroid alcohols to give formyl esters in almost quantitative yield and with retention of configuration;[8] it converts even tertiary alcohols into their formyl esters.[9] It reacts with an epoxide to give the 1-formyloxy-2-chloro compound or the 1,2-dichloro compound.[10]

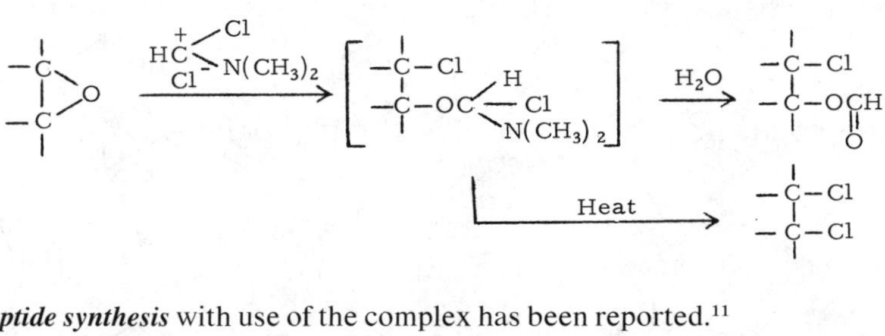
HC
Cl
N(CH3)2
Cl
H
N(CH3)2
H2O
OCH
Heat

Peptide synthesis with use of the complex has been reported.[11]

Dehydration. Lawton and McRitchie[12] tried several conventional methods for dehydration of pyromellitamide (1) to the nitrile (2) without success but finally were able to obtain (2) in unstated yield using thionyl chloride in DMF (60°, 7 hrs.). Bailey *et al.*[13] repeated the preparation and reported a yield of 35%. Thurman[14] found the temperature is critical and was able to increase the yield to 60% by running the reaction initially at 0° and then at room temperature. With this improve-

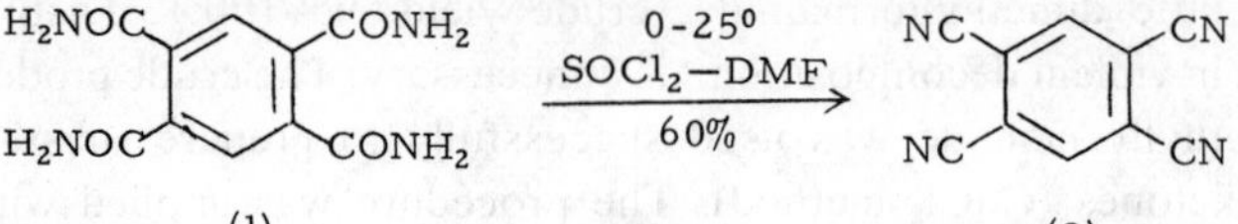
0-25°
$SOCl_2$–DMF
60%

(1) (2)

ment the method became generally applicable, for example for the cyclodehydration of dibenzoylhydrazide.[14]

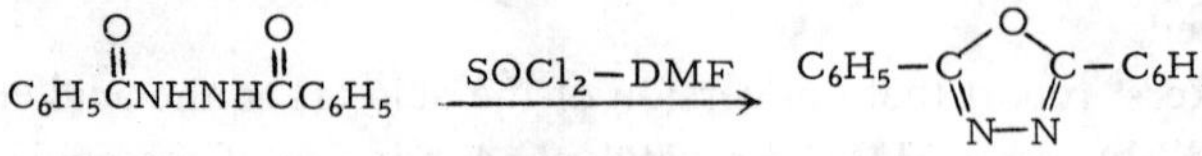
$SOCl_2$–DMF

Chlorination. The reagent can function as a chlorinating agent. Thus picric acid is converted into picryl chloride, and the 1,3,5-triazine (1) yields the corresponding dichloro compound (2).[15]

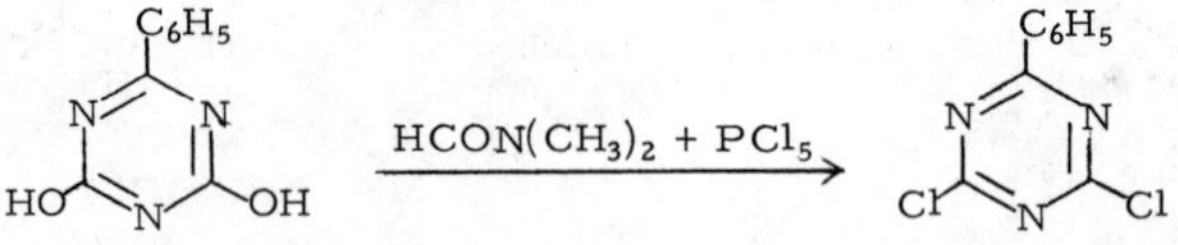
$HCON(CH_3)_2 + PCl_5$

(1) (2)

When the nucleoside (1) was treated with thionyl chloride in the presence of only a catalytic amount of DMF the chloride (2) was formed in almost quantitative yield.[16] However, if (1) is treated with thionyl chloride and excess DMF, the dimethylamino derivative (3) is produced in high yield.

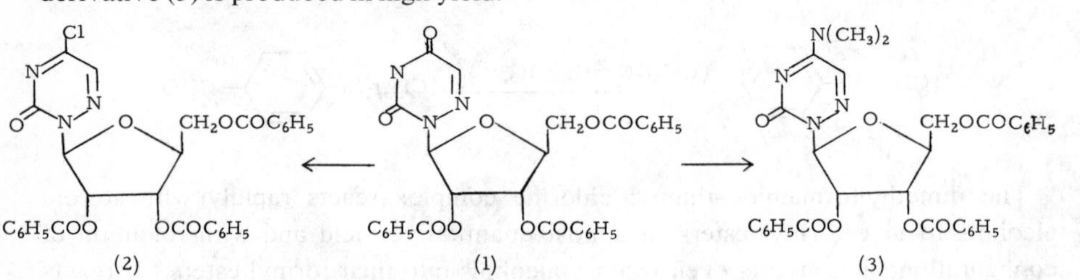

(2) (1) (3)

Dehydrative coupling. Ikehara and Uno[17] showed that the reagent can be used for formation of a 3′,5′-internucleotide linkage.

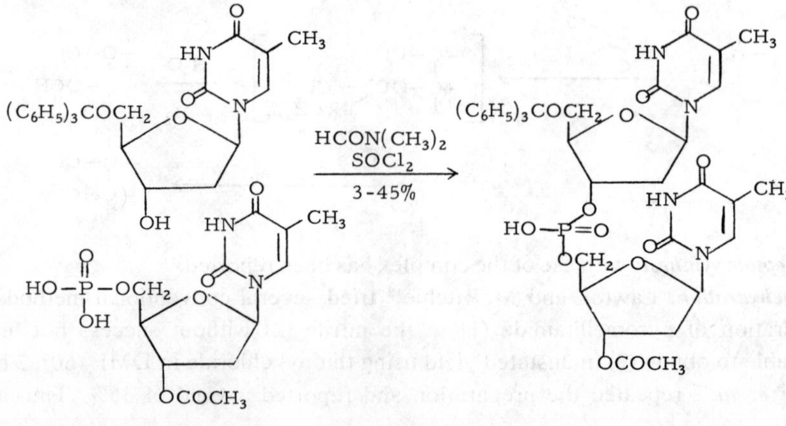
$HCON(CH_3)_2$
$SOCl_2$
3-45%

[1]A. Vilsmeier and A. Haack, *Ber.*, **60B**, 119 (1927), used the roughly equivalent combination of phosphoryl chloride and N-methylformanilide to prepare *p*-alkylaminobenzaldehydes. Phosgene and phosphorus pentachloride have also been used in place of thionyl chloride. Another combination which can be described as a Vilsmeier-type reagent is listed in this book as Dimethylformamide–Phosphoryl chloride.
[2]H. H. Bosshard, R. Mory, M. Schmid, and Hch. Zollinger, *Helv.*, **42**, 1653 (1959)
[3]A. Quilico and L. Panizzi, *Gazz.*, **68**, 625 (1938)
[4]T. S. Gardner, E. Wenis, and J. Lee, *J. Org.*, **26**, 1514 (1961)
[5]Private communication from Thomas S. Gardner and Edward Wenis, Hoffmann-LaRoche, Inc.
[6]Y. Egawa, M. Suzuki, and T. Okuda, *Chem. Pharm. Bull.*, **11**, 589 (1963)
[7]H. H. Bosshard and Hch. Zollinger, *Helv.*, **42**, 1659 (1959)
[8]K. Morita, S. Noguchi, and M. Nishikawa, *Chem. Pharm. Bull.*, **7**, 896 (1959)
[9]Z. Arnold, *Coll. Czech.*, **26**, 1723 (1961)
[10]W. Ziegenbein and W. Franke, *Ber.*, **93**, 1681 (1960); W. Ziegenbein and K.-H. Hornung, *Ber.*, **95**, 2976 (1962)
[11]M. Zaoral and Z. Arnold, *Tetrahedron Letters*, No. 14, 9 (1960)
[12]E. A. Lawton and D. D. McRitchie, *J. Org.*, **24**, 26 (1959)
[13]A. S. Bailey, B. R. Henn, and J. M. Langdon, *Tetrahedron*, **19**, 161(1963)
[14]J. C. Thurman, *Chem. Ind.*, 752 (1964)
[15]H. Eilingsfeld, M. Seefelder, and H. Weidingen, *Angew. Chem.*, **72**, 836 (1960)
[16]J. Žemlička, J. Smrt, and F. Šorm, *Tetrahedron Letters*, 397 (1962); J. Žemlička and F. Šorm, *Coll. Czech.*, **30**, 2052 (1965)
[17]M. Ikehara and H. Uno, *Chem. Pharm. Bull.*, **12**, 742 (1964)

N,N-Dimethylglycinehydrazide hydrochloride, $(CH_3)_2NCH_2CONHNH_2 \cdot HCl$ (Reagent D). Mol. wt. 153. 62, m.p. 181°.

The reagent is analogous to Girard reagents T and P and is prepared in the same way and used for isolation of a ketone from a mixture by conversion to a derivative extractible from ether with dilute hydrochloric acid.[1] It is stated to be more easily purified than reagents T and P.

[1]M. Viscontini and J. Meier, *Helv.*, **33**, 1773 (1950)

N,N-Dimethylhydrazine, $(CH_3)_2NNH_2$. Mol. wt. 60.08, b.p. 63°, sp. gr. 0.784, pKb 6.8, water-miscible. Supplier: FMC Corporation (technical bulletin "Dimazine" available).

Nitrile synthesis. A method for the conversion of aldehydes into nitriles involves formation of the N,N-dimethylhydrazone (1), reaction with methyl iodide (or methyl

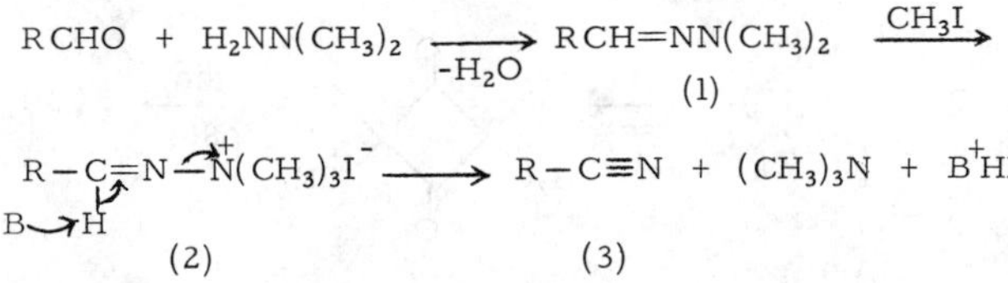
RCHO + H₂NN(CH₃)₂
-H₂O
RCH=NN(CH₃)₂
CH₃I
(1)
R–C=N–N(CH₃)₃I
B
H
R–C≡N + (CH₃)₃N + BHI
(2)
(3)

p-toluenesulfonate) to produce an N,N,N-trimethylhydrazonium salt (2), and *β*-elimination with methanolic sodium hydroxide.[1]

Paquette[2] found this method satisfactory for the conversion of 2-chloro-1-formyl-1-cyclohexene (4) into 2-chloro-1-cyclohexene-1-carbonitrile (7). Formation of the hydrazone (5) was carried out by refluxing the reactants for 16 hrs. under a Dean Stark water separator of the design described by Natelson and Gottfried.[3]

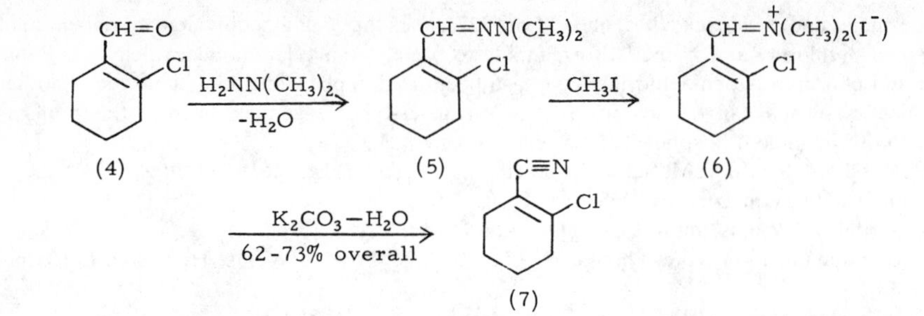

Preparation of hydrazones. It usually is not possible to obtain a hydrazone in good yield by reaction of an aldehyde or ketone with hydrazine because of extensive formation of the azine. This difficulty is circumvented by conversion to the N,N-dimethylhydrazone, followed by exchange with hydrazine.[4] The latter reaction is conducted by refluxing the N,N-dimethylhydrazone with a 2–3 fold excess of anhydrous hydrazine.

[1]R. F. Smith and L. E. Walker, *J. Org.*, **27**, 4372 (1962)
[2]L. A. Paquette, procedure submitted to *Org. Syn.*
[3]S. Natelson and S. Gottfried, *Org. Syn., Coll. Vol.*, **3**, 381 (1955)
[4]G. R. Newkome and D. L. Fishel, *J. Org.*, **31**, 677 (1966)

Dimethylketene, $(CH_3)_2C{=}C{=}O$. Mol. wt. 70.09, b.p. 34°. This reagent combines rapidly with oxygen to form an explosive peroxide; drops of solution allowed to evaporate in air may detonate. Dimerizes on standing.

Preparation. (a) ***From α-bromoisobutyryl bromide.***[1] The α-bromo acid bromide is added dropwise to a mixture of zinc turnings and ethyl acetate in a flask flushed with

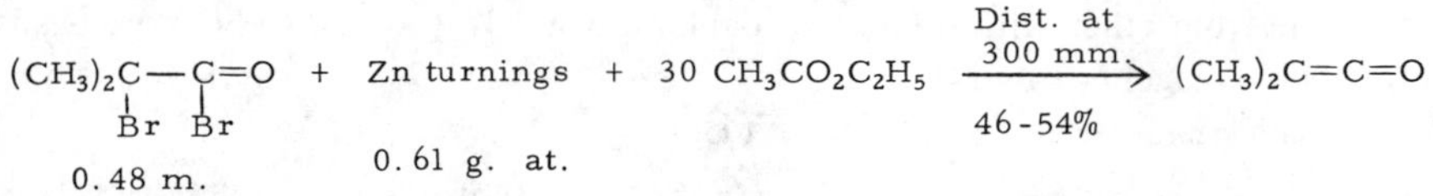

nitrogen and evacuated to a pressure of 300 mm. The dimethylketene distills along with ethyl acetate and is caught in traps cooled in dry ice-acetone. Titration of an aliquot at 0° with 0.1 *N* sodium hydroxide (phenolphthalein) indicated that the distillate is a 9–10% solution of dimethylketene in ethyl acetate.

(b) ***From the dimer.*** Dimethylketene dimer, or tetramethylcyclobutane-1,3-

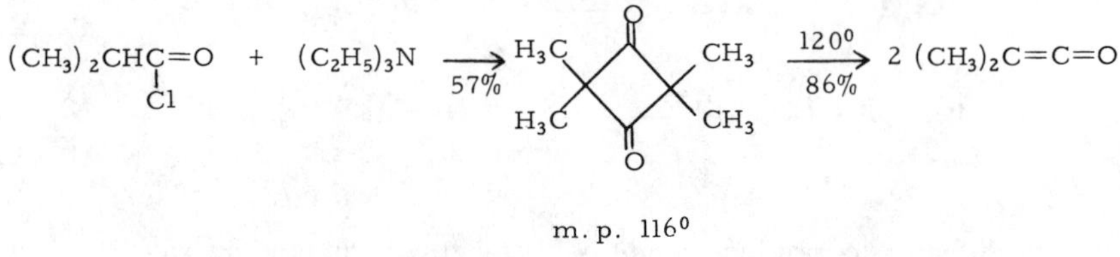

m. p. 116°

dione, is prepared by the action of triethylamine on isobutyryl chloride.[2] Hanford and Sauer[2] describe a pyrolysis lamp for the efficient depolymerization of this comparatively high-melting dimer for generation of dimethylketene.

(c) ***From dimethylmalonic acid.***[2] A mixture of the acid, acetic anhydride, and a trace of sulfuric acid is shaken until the solid dissolves and let stand for 2 days for formation of dimethylmalonic anhydride. A small amount of barium carbonate

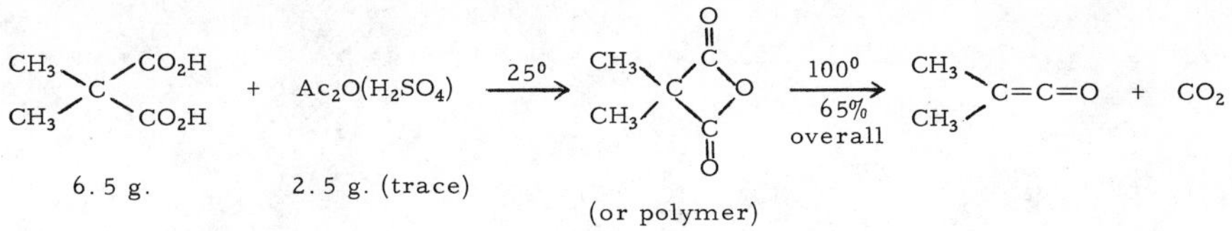
6.5 g.
2.5 g. (trace)
25°
100°
65% overall
(or polymer)

is added and acetic acid and anhydride are removed in vacuum below 60°. The dry anhydride is then decomposed by raising the temperature until vigorous evolution of CO_2 and dimethylketene occurs (about 100°).

(d) ***From dimethylmalonic acid dimethylketeneacylal*** (4). Bestian and Günther[3] heated dimethylmalonic acid with excess acetic anhydride slowly to the boiling

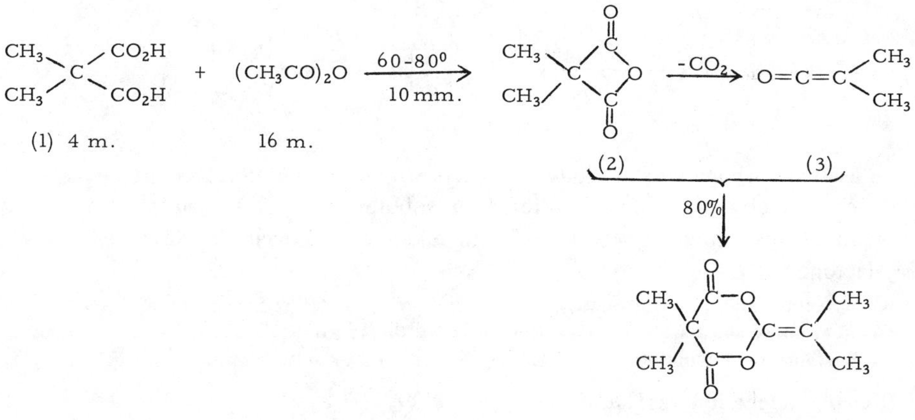
(1) 4 m.
16 m.
60-80°
10 mm.
-CO₂
(2)
(3)
80%

(4) m.p. 80°

point at a pressure of 10 mm. and removed the acetic acid formed continuously through an efficient fractionating column. The residue, a white solid, on crystallization from petroleum ether afforded the pure acylal in 80% yield. This substance is stable to thermal cleavage at 180° but in the presence of potassium carbonate it is cleaved rapidly at 150° to dimethylketene and carbon dioxide. Thus dropwise addition of molten (4) to a small amount of potassium carbonate in a flask heated at 150° produces a gaseous mixture of dimethylketene and carbon dioxide. This gas can be

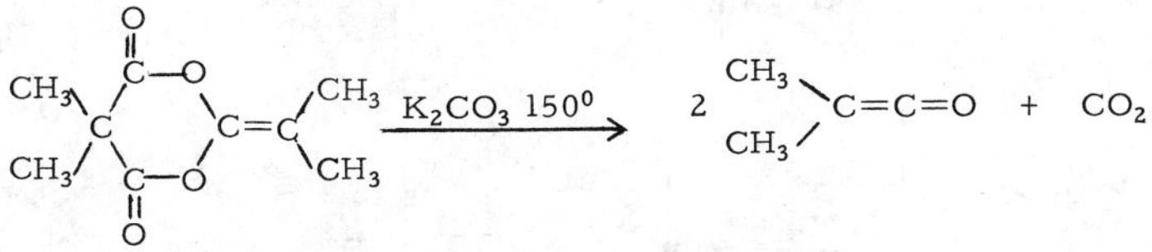
K₂CO₃ 150°

used directly for a reaction requiring dimethylketene, or this reagent can be isolated in liquid form by partial condensation.

The acylal (4) can also be used to carry out reactions with nascent dimethylketene formed *in situ*. It is stable to *t*-butanol at the reflux temperature but with potassium carbonate present to catalyze cleavage it reacts smoothly to form *t*-butyl isobutyrate It reacts with an amine without other catalyst at room temperature to give an isobutyramide. In the presence of an alkali carbonate or the salt of a carboxylic acid, it converts an acid into its anhydride, probably via a mixed anhydride. With

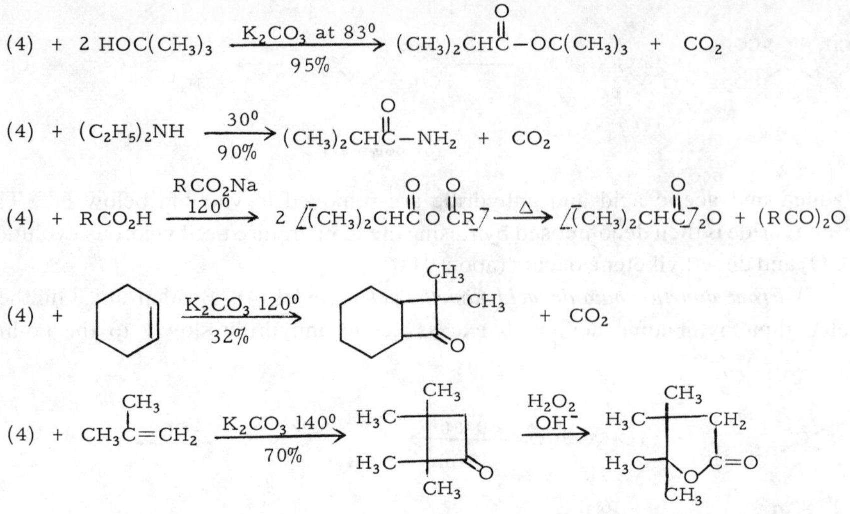

potassium carbonate as a catalyst it reacts with an olefin, taken in large excess, to form a cyclobutanone. The reaction with isobutene gives 2,2,3,3-tetramethylcyclobutanone, as shown by oxidation with alkaline hydrogen peroxide to a known γ-lactone.

[1]C. W. Smith and D. G. Norton, *Org. Syn., Coll. Vol.*, **4**, 348 (1963)
[2]W. E. Hanford and J. C. Sauer, *Org. Reactions*, **3**, 108 (1946)
[3]H. Bestian and D. Günther, *Angew. Chem., Internat. Ed.*, **2**, 608 (1963)

Dimethylmagnesium, $(CH_3)_2Mg$. Mol. wt. 54.39.

A solution of the reagent is prepared[1] in a dry box under nitrogen by dropwise addition of 0.5 mole of dioxane to an ethereal solution of 1 mole of methylmagnesium chloride and filtering the solution from precipitated $MgCl_2$·dioxane. Great care is taken during the operation to prevent the filter cake from becoming partially dry, for diethylmagnesium when dry is extremely pyrophoric. The filtered solution is almost halogen-free and the yield of dimethylmagnesium is 71%.

A Merck group[1] sought to convert 11β,12β-epoxypregnane-3,20-bisethylene ketal (1) into the 11β-hydroxy-12α-methyl steroid (2), but found that use of methyl-

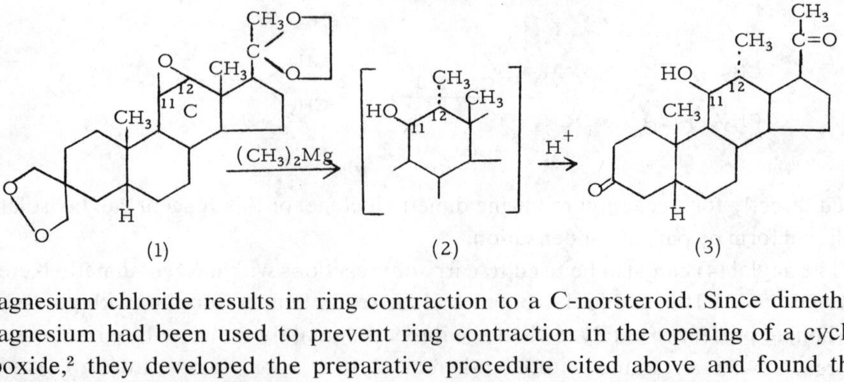

(1) (2) (3)

magnesium chloride results in ring contraction to a C-norsteroid. Since dimethylmagnesium had been used to prevent ring contraction in the opening of a cyclic epoxide,[2] they developed the preparative procedure cited above and found that dimethylmagnesium indeed reacts with (1) to give the *trans*-diaxial product (2), isolated after deketalization as the diketone (3) in 41% yield. The reaction was carried out by displacing the solvent of an ethereal solution of dimethylmagnesium with dioxane, adding the steroid (1), refluxing for 24 hrs., and decomposing excess reagent with saturated ammonium chloride solution (use of acetone, ethyl acetate, or dry ice leads to violent reactions).

[1]B. G. Christensen, R. G. Strachan, N. R. Trenner, B. H. Arison, R. Hirschmann, and J. M. Chemerda, *Am. Soc.*, **82**, 3995 (1960)

[2]P. D. Bartlett and C. M. Berry, *Am. Soc.*, **56**, 2683 (1934)

N,N-Dimethyl-*p*-phenylenediamine, p-$(CH_3)_2NC_6H_4NH_2$. Mol. wt. 136.20, m.p. 38°. Supplier: Eastman. Prepared by reduction of N,N-dimethyl-β-nitrosoaniline with zinc dust in ethanol in the presence of ammonium chloride.[1] The reagent combines with aromatic aldehydes to give crystalline, colored anils suitable for identification. It does not react with ketones; for example acetone can be used as

$$ArCHO \xrightarrow{\underline{p}\text{-}(CH_3)_2NC_6H_4NH_2} ArCH{=}NC_6H_4N(CH_3)_2\text{-}\underline{p}$$

solvent for the preparation of an aldehyde anil. Aliphatic aldehydes form anils but, with few exceptions, these are not crystalline; however, they can be recognized as anils by their solubility in 30% acetic acid.

[1]G. Ed. Utzinger and F. A. Regenass, *Helv.*, **37**, 1901 (1954)

Dimethyl phosphite (Dimethyl phosphonate), $(CH_3O)_2\overset{O}{\overset{\|}{P}}H$. Mol. wt. 110.05, b.p. 64–66°/15 mm. Suppliers: E, F, MCB.

When a suspension of the 4-bromocyclohexadienone (1) in benzene was stirred and treated at room temperature with dimethyl phosphite, no reaction occurred at first but after an induction period of 10 min. the yellow color changed to orange and the mixture became warm.[1] After 3 min. more the solid had dissolved and the color disappeared; workup afforded 2,6-dibromo-4-methylphenol (3) in high yield.

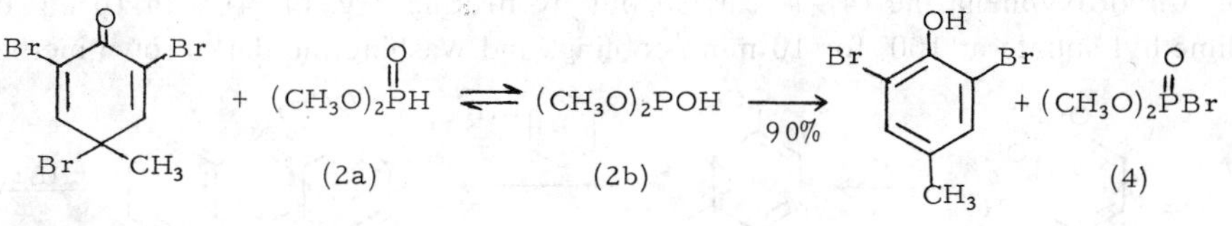

Br
Br
Br
CH3
+ (CH3O)2PH
(CH3O)2POH
90%
OH
Br
Br
CH3
+ (CH3O)2PBr
(2a)
(2b)
(4)

(1) (3)

The autocatalytic nature of the reaction is attributed to slow tautomerism of the pentavalent phosphonate (2a) to the more reactive phosphite (2b), which reacts rapidly with (1). The tautomeric shift is catalyzed by a trace of hydrogen bromide formed by hydrolysis of the phosphoryl bromide (4). With methanol as solvent, methanolysis of (4) with production of acid reduces the induction time to 30 seconds.

[1]B. Miller, *J. Org.*, **28**, 345 (1963)

Dimethyl sulfate, $(CH_3O)_2SO_2$ (*see also* Methyl iodide). Mol. wt. 126.14, b.p. 188.5°, sp. gr. 1.35. Suppliers: A, B, E, F, KK, MCB.

O-Methylation. Quinacetophenone is converted into the dimethyl ether by alternate addition of dimethyl sulfate and aqueous sodium hydroxide to a solution of the substance in 95% ethanol.[1] Anhydrous D-glucose on reaction in aqueous

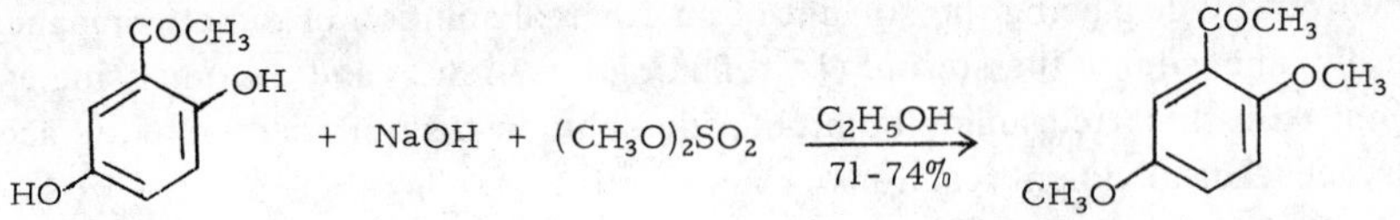

solution with sodium hydroxide and dimethyl sulfate at 55°, followed by heating

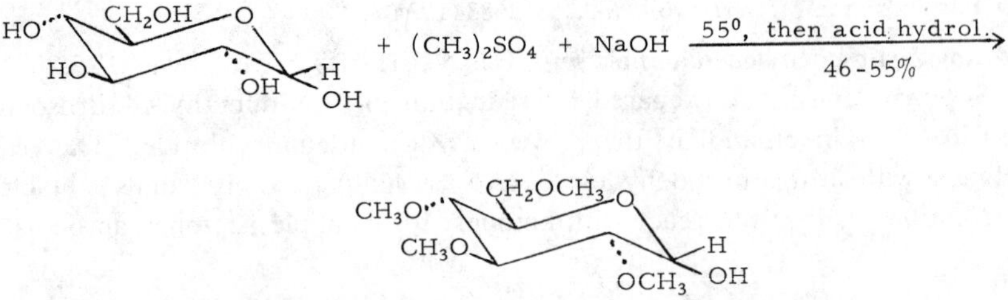

with $2N$ hydrochloric acid at 100° for 1 hr. to hydrolyze the α- and β-methyl glucosides, affords 2,3,4,6-tetra-O-methyl-D-glucose.[2]

ε-Caprolactam is converted into the imino ether, O-methylcaprolactim, by refluxing with dimethyl sulfate in benzene and adding potassium carbonate to decompose the initially formed salt.[3]

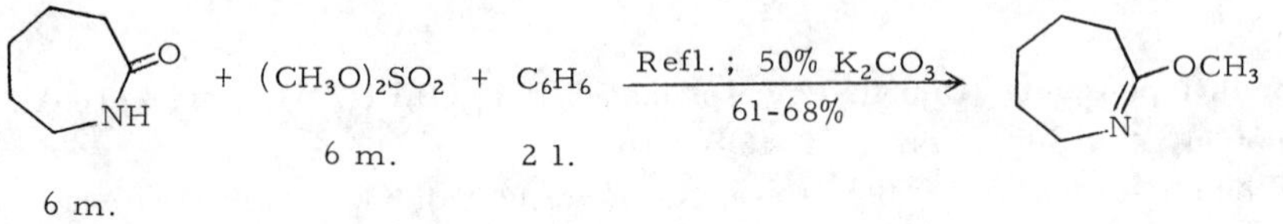

In the method of Bredereck[4] for permethylation of carbohydrates, partially methylated material is dissolved in ether, sodium wire is added followed by dimethyl sulfate diluted with ether. The reaction is complete in about one hour.

***N*-Methylation.** The last step in a synthesis[5] of pyocyanine (6), N-methylation of α-hydroxyphenazine (4), is carried out by heating 2 g. of (4) with 10 ml. of dimethyl sulfate at 100° for 10 min., cooling, and washing the dark brown metho-

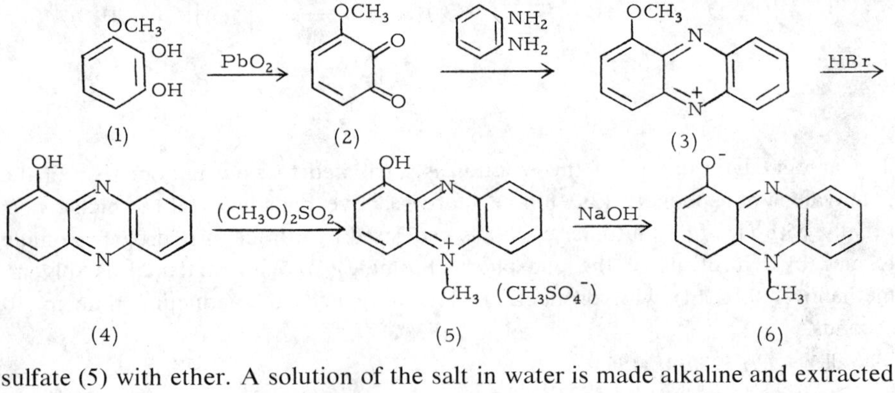

(1) (2) (3) (4) (5) (6)

sulfate (5) with ether. A solution of the salt in water is made alkaline and extracted with chloroform until no more blue color is removed.

A general method for the preparation of unsymmetrical secondary amines[6] involves alkylation of an N-benzylideneamine such as (1) by mixing it with dimethyl sulfate in benzene, heating gently at first and then at reflux, and decomposing the salt (2) with base.

$$C_6H_5CHO \xrightarrow{n\text{-}C_4H_9NH_2} C_6H_5CH{=}NC_4H_9\text{-}n \ (1) \xrightarrow{(CH_3O)_2SO_2}$$

$$C_6H_5CH{=}\overset{+}{N}(CH_3)C_4H_9\text{-}n(CH_3OSO_3^-) \ (2) \xrightarrow[45\text{-}53\%\ \text{overall}]{H_2O;\ NaOH} n\text{-}C_4H_9NHCH_3 \ (3) \quad (+\ C_6H_5CHO\ \text{etc.})$$

Preparation of sulfones.[7] *p*-Toluenesulfonyl chloride is reduced with sodium sulfite and sodium bicarbonate to sodium *p*-toluenesulfinate, which separates as a solid. A mixture of this with dimethyl sulfate, sodium bicarbonate, and water is refluxed for 20 hrs. and the product, methyl *p*-tolyl sulfone, is extracted with benzene.

$$p\text{-}CH_3C_6H_4SO_2Cl \xrightarrow{Na_2SO_3,\ NaHCO_3} p\text{-}CH_3C_6H_4SO_2Na \xrightarrow[69\text{-}73\%]{(CH_3O)_2SO_2\text{-}NaHCO_3} p\text{-}CH_3C_6H_4SO_2CH_3$$

C-Alkylation. *See* Diethyl sulfate.

[1]G. N. Vyas and N. M. Shah, *Org. Syn., Coll. Vol.*, **4**, 837 (1963)
[2]E. S. West and R. F. Holden, *ibid.*, **3**, 800 (1955)
[3]R. E. Benson and T. L. Cairns, *ibid.*, **4**, 588 (1963)
[4]H. Bredereck, G. Hagelloch, and E. Hambsch, *Ber.*, **87**, 35 (1954); H. Bredereck and E. Hambsch, *ibid.*, **87**, 38 (1954)
[5]A. R. Surrey, *Org. Syn., Coll. Vol.*, **3**, 753 (1955)
[6]J. J. Lucier, A. D. Harris, and P. S. Korosec, *Org. Syn.*, **44**, 72 (1964)
[7]L. Field and R. D. Clark, *Org. Syn., Coll. Vol.*, **4**, 674 (1963)

Dimethyl sulfate–Barium oxide, *see also* Methyl iodide. Dimethyl sulfate in combination with barium oxide or barium hydroxide, $Ba(OH)_2 \cdot 8\,H_2O$, provides an effective method for the permethylation of carbohydrates.[1, 2] Dimethyl sulfate is cheaper and sometimes superior to methyl iodide. Dimethylformamide is the usual solvent, but for carbohydrates only slightly soluble in this solvent dimethyl sulfoxide or a mixture of dimethyl sulfoxide and dimethylformamide is used.

[1]R. Kuhn and H. Trischmann, *Ber.*, **96**, 284 (1963)
[2]K. Wallenfels, G. Bechtler, R. Kuhn, H. Trischmann, and H. Egge, *Angew. Chem., Internat. Ed.*, **2**, 515 (1963)

Dimethyl sulfate–Potassium carbonate.[1] This combination is useful for the methylation of plant phenols which tend to decompose or rearrange on methylation with dimethyl sulfate and alkali, for example, 2(3′,4′-dihydroxyphenyl)-6-hydroxybenzofurane-3-carboxylic acid. A solution of the compound in anhydrous acetone is treated with enough dimethyl sulfate and potassium carbonate (10% excess) to methylate all phenolic and carboxylic groups, and the mixture is refluxed for 5 hrs.

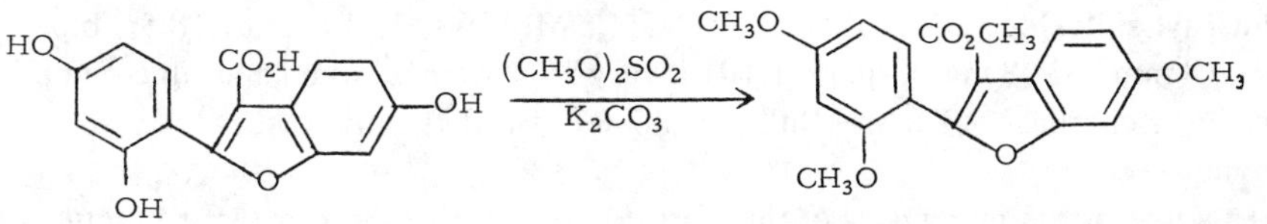
HO
CO2H
OH
OH
(CH3O)2SO2
K2CO3
CH3O
CO2CH3
OCH3
CH3O

with exclusion of moist air. The cooled mixture is centrifuged and the residual inorganic salts washed twice with acetone. Evaporation of the combined acetone solutions gives a sirup which on two crystallizations from ethanol–water or methanol–water affords the methylated phenol-ester in yield of 60–70%. If the compound is not sensitive to alkali, the reaction can be hastened by addition of a few milliliters of 10% methanolic potassium hydroxide.[2]

[1]Contributed by S. G. Sunderwirth, Colorado State University

[2]E. M. Bickoff, R. L. Lyman, A. L. Livingston, and A. N. Booth, *Am. Soc.*, **80**, 3969 (1958)

Dimethyl sulfite, $(CH_3O)_2SO$. Mol. wt. 110.13, b.p. 126–127°. Supplier: Eastman. Preparation by reaction of 1 mole of thionyl chloride with 2.2 moles of methanol.[1]

Esterification of amino acids. Theobald *et al.*[1] heated a mixture of an amino acid, 1.1 equivalents of *p*-toluenesulfonic acid, and a large excess of dimethyl sulfite (5–15 equiv.) to produce the methyl ester *p*-toluenesulfonate. In some cases yields

$$\underset{0.01\ \text{m.}}{C_6H_5CH_2\underset{NH_2}{\underset{|}{C}}HCO_2H} + \underset{0.05\ \text{m.}}{(CH_3O)_2SO} + \underset{0.011\ \text{m.}}{TsOH} \xrightarrow[99\%]{100^0\ 4.5\ \text{hrs.}} C_6H_5CH_2\underset{^+NH_3\ (TsO^-)}{\underset{|}{C}}HCOOCH_3 + SO_2 + CH_3OH$$

were very high, but the methylation of alanine, valine, and lysine could not be carried to completion. The method appears to be less satisfactory than that using thionyl chloride and methanol (*see* Thionyl chloride) except where the *p*-toluenesulfonate is more easily purified than the hydrochloride (this is true of L-proline).

[1]J. M. Theobald, M. W. Williams, and G. T. Young, *J. Chem. Soc.*, 1927 (1963)

Dimethyl sulfone, $(CH_3)_2SO_2$. Mol. wt. 94.14, m.p. 102–109°. Suppliers: Crown Zellerbach, Aldrich. Purified by crystallization from chloroform.

Dropwise addition of ethyl benzoate to potassium *t*-butoxide suspended in a solution of dimethyl sulfone in dimethyl sulfoxide under nitrogen, followed by

$$C_6H_5CO_2C_2H_5 + CH_3SO_2CH_3 \xrightarrow[(CH_3)_2S^+{-}O^-]{(CH_3)_3COK} C_6H_5COCH_2SO_2CH_3$$

stirring at 50–60° for 9 min., and then followed by acidification and workup, afforded ω-(methylsulfonyl)-acetophenone in 91% yield (crude).[1]

Dimethyl sulfone is described as a superior solvent for halogen-exchange reactions, for example, for the conversion of 2-chloropyridine to 2-fluoropyridine.[2]

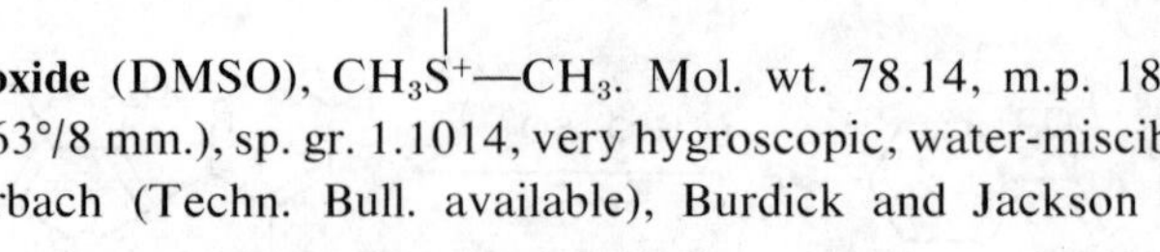
KF in (CH3)2SO2 (110°) 21 days
50%
N
Cl
N
F

[1]H.-D. Becker and G. A. Russell, *J. Org.*, **28**, 1896 (1963)

[2]L. D. Starr and G. C. Finger, *Chem. Ind.*, 1328 (1962); G. C. Finger *et al.*, *J. Org.*, **28**, 1666 (1963)

Dimethyl sulfoxide (DMSO), $CH_3\overset{O^-}{\overset{|}{S^+}}{-}CH_3$. Mol. wt. 78.14, m.p. 18.5°, b.p. 189° (86°/18 mm., 63°/8 mm.), sp. gr. 1.1014, very hygroscopic, water-miscible. Suppliers: Crown Zellerbach (Techn. Bull. available), Burdick and Jackson Labs. (pure). Review.[1]

In some of the many uses of this versatile compound the substance functions like other reagents and undergoes chemical change; such uses are classified below as reactions. In other instances no chemical change of DMSO is involved, or at least none is apparent, and yet the compound has a special solvent power by which it is able to promote a reaction not otherwise realizable, or else it exerts a solvent effect sufficient to accelerate a reaction brought about by another reagent. To be sure, the line of differentiation between solvent power and solvent effect is far from sharp. Striking solvent effects are attributed to the greatly enhanced reactivity of anions in DMSO in contrast to the reduced activity of hydrogen-bonded anions in hydroxylic solvents of comparable dielectric constant. The same property may contribute to solvent power.

Solvent Power

Organic and inorganic compounds. DMSO is not only a potent solvent for many organic compounds but it dissolves certain inorganic salts, as is evident from the table.

Solubility of salts (g./100 g. DMSO at 25°)

Ferric chloride (6 H_2O)	30
Mercuric acetate	100
Sodium dichromate (2 H_2O)	10
Sodium iodide	30
Sodium nitrate	20
Sodium nitrite	1
Stannous chloride (2 H_2O)	40
Zinc chloride	30

Cleavage of digitonides. A new method for splitting a steroid digitonide, for example the 1:1 complex of digitonin with cholesterol, is based upon the observation that DMSO, in which sterols are sparingly soluble, has remarkable solvent power for digitonin and brings about complete dissociation of digitonides at steam-bath temperature.[2] On cooling, the sterol precipitates and is extracted with hexane; digitonin is recovered by evaporation of the DMSO layer to dryness.

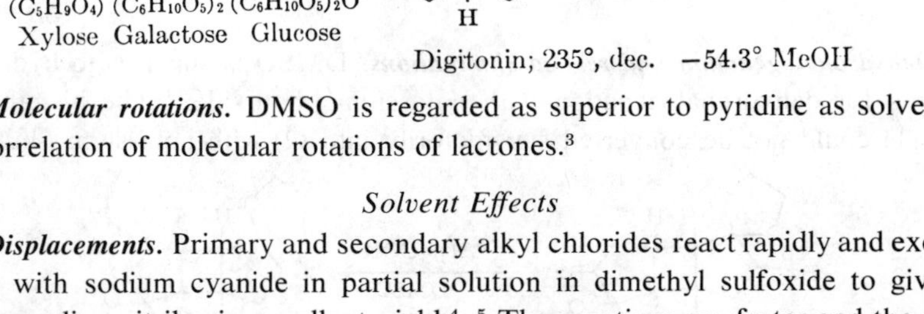

H3C
O
F
CH3
E
CH3
O
C
D
CH3
H
15
HO
OH
2
A
H
H
3
B
(C5H9O4) (C6H10O5)2 (C6H10O5)2O
H
Xylose Galactose Glucose

Digitonin; 235°, dec. −54.3° MeOH

Molecular rotations. DMSO is regarded as superior to pyridine as solvent for the correlation of molecular rotations of lactones.[3]

Solvent Effects

Displacements. Primary and secondary alkyl chlorides react rapidly and exothermally with sodium cyanide in partial solution in dimethyl sulfoxide to give the corresponding nitriles in excellent yield.[4, 5] The reactions are faster and the yields better than in the reactions of the corresponding bromides or iodides in aqueous alcohol.[4] Neopentyl and neophyl halides can be converted by this method to

$$(CH_3)_3CCH_2Cl \xrightarrow[\text{DMSO}]{\text{NaCN}} (CH_3)_3CCH_2CN$$

the nitriles without rearrangement.[5] Solvent DMSO enhances the rate of other bimolecular nucleophilic displacements on carbon: displacement of halogen by azide, thiocyanate, and halide ions,[6] and of tosylate by bromide ion.[7] It also accelerates electrophilic substitutions on carbon. Cram[8] studied the potassium methoxide-catalyzed hydrogen-deuterium exchange of (+)-2-methyl-3-phenylpropionitrile and found that the rate constant for racemization is no less than 10^9 times that observed in methanol.

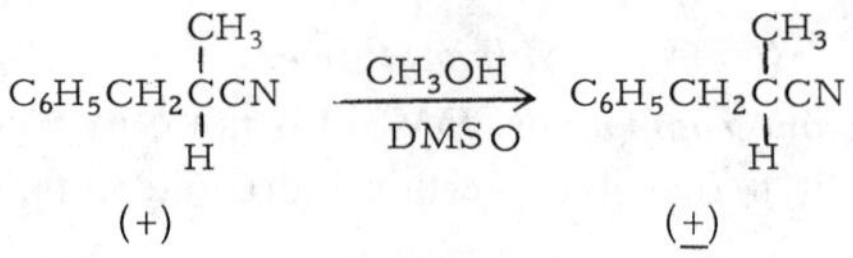

Bromobenzene is converted into *t*-butyl phenyl ether by reaction with potassium *t*-butoxide in DMSO at 25° for 15 hrs. (86% yield); the reaction is much slower in *t*-butanol.[9]

In the reaction of ethyl α-bromobutyrate with sodium nitrite to form ethyl α-nitrobutyrate, Kornblum[10] regards DMSO as superior to DMF as solvent since it has greater solvent power for sodium nitrite (see table).

Cope and Mehta[11] found DMSO to be the best solvent for the reaction of the ditosylate (1) with sodium cyanide.

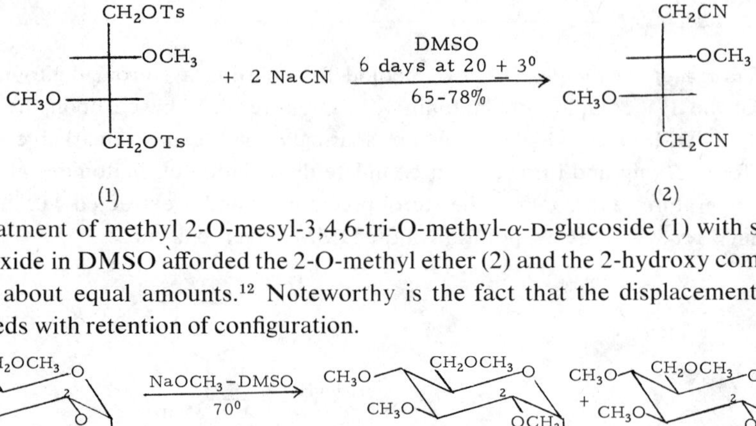

Treatment of methyl 2-O-mesyl-3,4,6-tri-O-methyl-α-D-glucoside (1) with sodium methoxide in DMSO afforded the 2-O-methyl ether (2) and the 2-hydroxy compound (3) in about equal amounts.[12] Noteworthy is the fact that the displacement to (2) proceeds with retention of configuration.

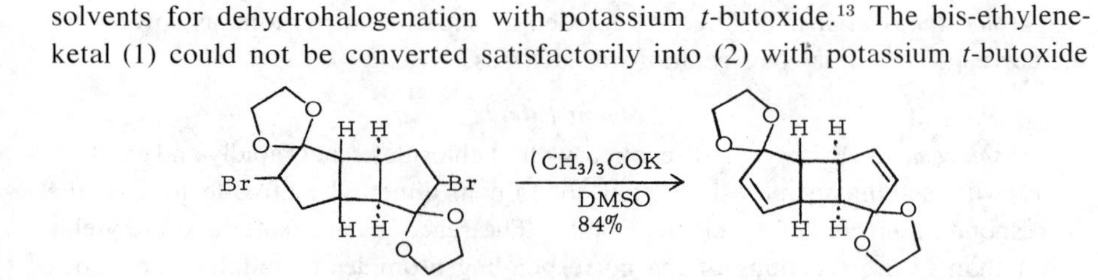

Dehydrohalogenation and related eliminations. DMSO is superior to hydroxylic solvents for dehydrohalogenation with potassium *t*-butoxide.[13] The bis-ethylene-ketal (1) could not be converted satisfactorily into (2) with potassium *t*-butoxide

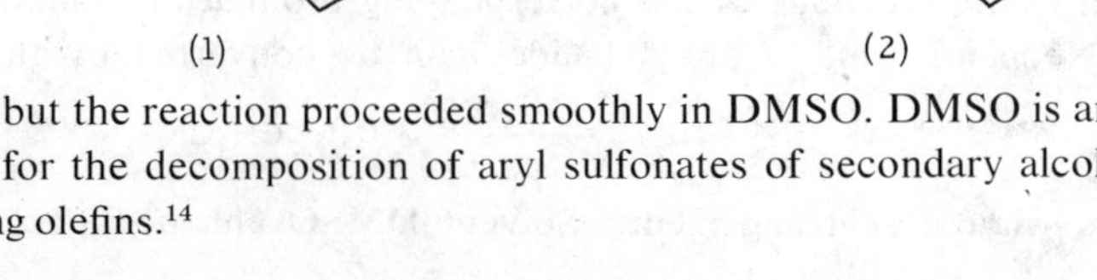

in *t*-butanol, but the reaction proceeded smoothly in DMSO. DMSO is an excellent solvent also for the decomposition of aryl sulfonates of secondary alcohols to the corresponding olefins.[14]

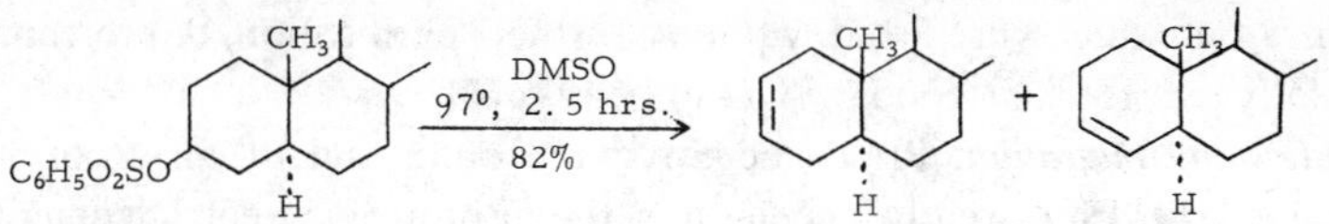

Nace[15] examined the reaction of α-bromo ketosteroids with sodium bicarbonate in DMSO and found that, when the halogen atom and an adjacent hydrogen atom have the *trans*-diaxial relationship, elimination is usually the predominant reaction.

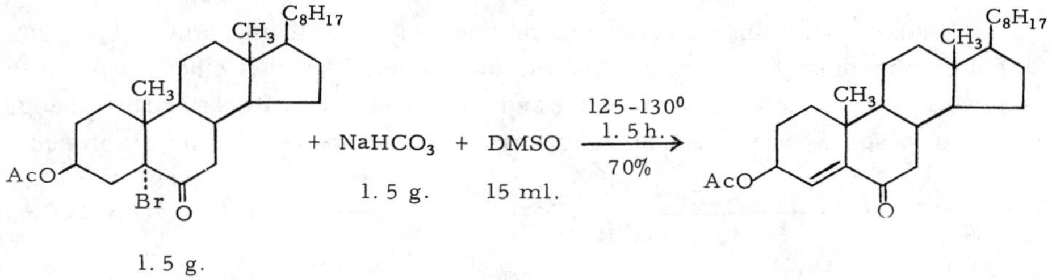

When the bromine atom and an adjacent hydrogen are *cis*-equatorial-axial, hydrolytic oxidation to the diosphenol competes with elimination.

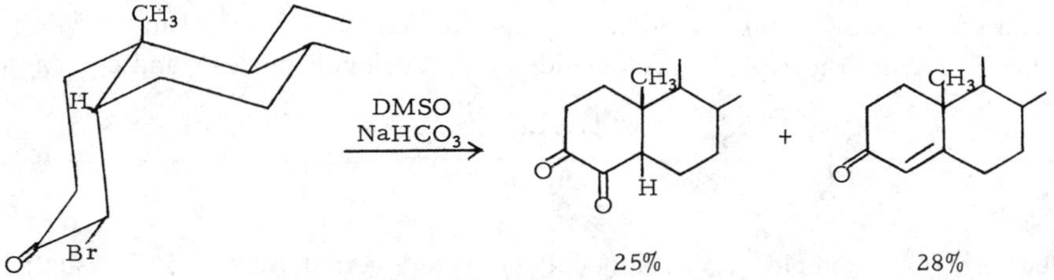

An Esso group[16] studied the β-elimination of isopropyl derivatives to give propylene and arranged the X-groups in the following order of decreasing ease of elimination: Br > SO_2R > SOR > SCN > SR > SH > CN.

$$CH_3CH(X)CH_3 \xrightarrow[\text{at } 55^0]{(CH_3)_3COK-DMSO} CH_3CH{=}CH_2$$

Cram and co-workers[17,18] describe a procedure for the conversion of bromobenzene into phenyl *t*-butyl ether by reaction with potassium *t*-butoxide in dimethyl sulfoxide at 125–130° for 1 minute. Benzyne is assumed to be an intermediate.

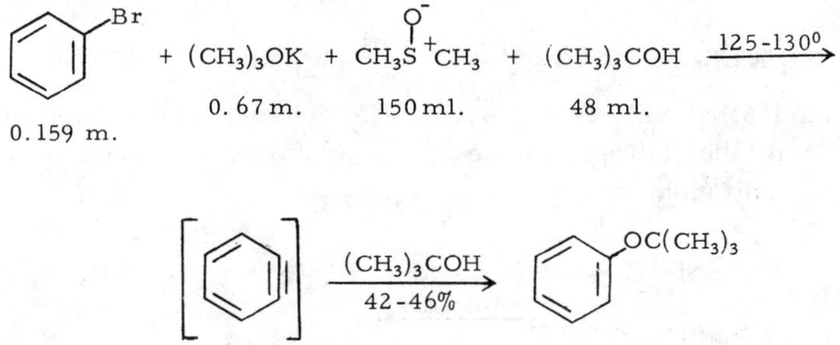

Worthy of note is the statement that commercial DMSO as freshly opened is dry to Karl Fischer reagent and that both this solvent and MSA Research Corporation's potassium *t*-butoxide were used without further purification. Unfortunately the yield, initially reported as "good,"[17] was only 42–46%.

Double bond migration. Price and Snyder[19] found that by virtue of its special solvent effect, DMSO greatly accelerates the prototropic rearrangement of allyl ethers to *cis*-propenyl ethers under catalysis by potassium *t*-butoxide. Cunningham

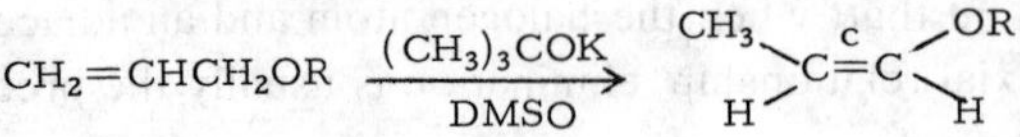

et al.[20] made use of this isomerization in a method for protecting a hydroxyl group of a carbohydrate by conversion to the allyl ether. The allyl ether (1) is easily prepared by reaction with allyl bromide and alkali but is not easily cleaved by aqueous acid or base. However, isomerization by Price's method gives the *cis*-propenyl

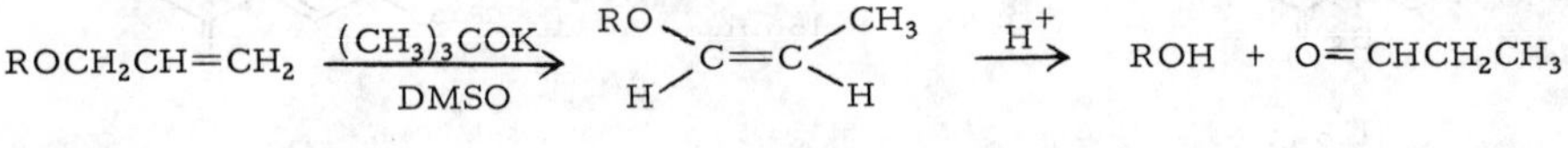

(1) (2)

ether (2) which, being an enol ether, is hydrolyzed readily by dilute acid. The propenyl ether (2) can be cleaved also by permanganate oxidation or by ozonization.

The same base-solvent combination is used extensively for the isomerization of olefins,[21] cyclohexadienes,[22] cyclooctadienes,[23] cyclononadienes[23] and cyclonona-

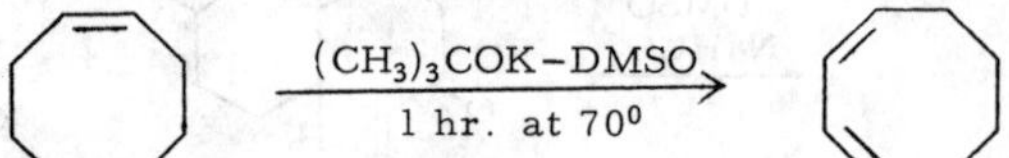

trienes.[23] For example, 1,5-cyclooctadiene is converted rapidly and essentially quantitatively to 1,3-cyclooctadiene under mild conditions.

Alkylation and acylation of ketones and nitriles. DMSO greatly enhances the rate of alkylation of enolate anions.[24,25] Dialkylation of malononitrile and of pentane-2,4-

$$CH_2(CN)_2 + 2\ CH_3I \xrightarrow[NaH]{DMSO} (CH_3)_2C(CN)_2$$

dione by usual methods proceeds poorly but can be accomplished in yields of 60 and 64% by reaction with methyl iodide in DMSO as solvent and sodium hydride as base.[25] The same combination provides an efficient method for acylating a methyl ketone with a methyl ester to produce a symmetrical β-diketone.[26] Best results are obtained with a molecular ratio of sodium hydride to ester to ketone of 2:2:1. The

$$RCOOCH_3 + CH_3COR \xrightarrow{DMSO-NaH} RCOCH_2COR + CH_3OH$$

combination DMSO-NaH is effective for the Thorpe condensation of the 1,4-dinitrile (1) to the β-cyanoenamine (2), and affords reasonable yields in the Dieckmann cyclization.[27]

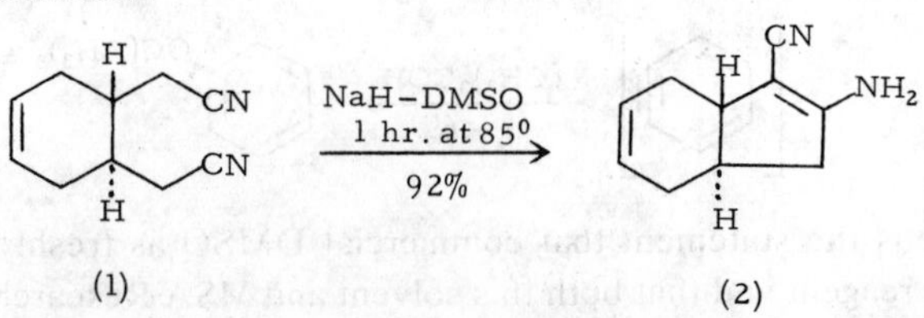

(1) (2)

Thermal decomposition. The usual procedure for carrying out a Wolff-Kishner reduction is to heat a hydrazone with base to a temperature of 200°. Cram *et al.*[28] reported the striking finding that with dimethyl sulfoxide as solvent and potassium *t*-butoxide as base the reaction can be conducted at room temperature. For example,

$$(C_6H_5)_2C{=}NNH_2 \xrightarrow{(CH_3)_3COK,\ DMSO} (C_6H_5)_2CH_2 + N_2$$

1.96 g. of benzophenone hydrazone was added over an 8-hr. period to a stirred mixture of 2 g. of sublimed potassium *t*-butoxide in 5 ml. of DMSO. The procedure lacks manpower efficiency but is of considerable theoretical interest as is the observation[26] that the Cope elimination of an amine oxide, ordinarily done by pyrolysis at temperatures of 120–150°, proceeds smoothly at 25° in an anhydrous mixture of dimethyl sulfoxide and tetrahydrofurane.

$$C_6H_5\underset{CH_3}{CH}\overset{O^-}{\overset{|}{N^+}}(CH_3)_2 \xrightarrow[THF]{DMSO\ 25^0} C_6H_5CH{=}CH_2 + (CH_3)_2NOH$$

Permethylation of polysaccharides. Kuhn and Trischmann[29] found that polysaccharides can be methylated very efficiently in DMSO with dimethyl sulfate and barium oxide and/or barium hydroxide. Srivastava *et al.*[30] used the same method except for the substitution of sodium hydroxide as base. Sodium hydroxide pellets and dimethyl sulfate were added with stirring under nitrogen over 8 hrs. to a solution of undegraded starch in dimethyl sulfoxide. After stirring for another 16 hrs. the mixture was heated to decompose the dimethyl sulfate, cooled, diluted, and neutralized, and the product was extracted with chloroform and precipitated from acetone with ether; yield 91%, $OCH_3 = 42.3\%$. Here the high solvent power of DMSO clearly contributes to the solvent effect.

Hydrolysis. The rate of alkaline hydrolysis of ethyl acetate in DMSO—H_2O mixtures increases with increasing concentration of DMSO.[31]

Reactions

Dehydration. Dehydration of *sec*- and *tert*-benzylic alcohols and *tert*-aliphatic alcohols can be effected by heating the alcohol with 4–8 moles of dimethyl sulfoxide at 160–185° for 9–16 hrs. (Traynelis[32]). 1-Alkylcycloalkanols are dehydrated mainly to the endocyclic olefins as illustrated by the following examples:

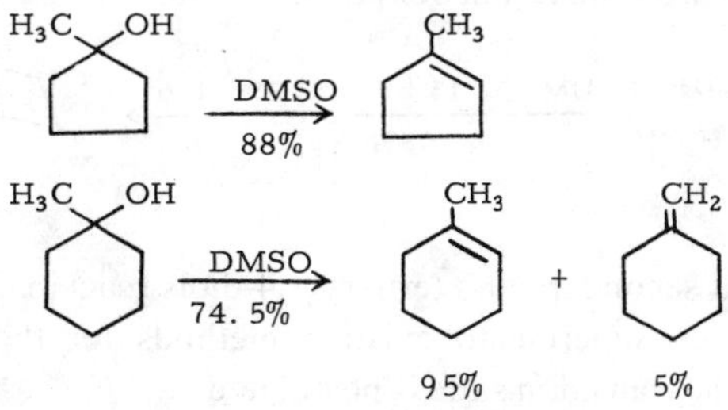
H3C OH
CH3
DMSO
88%
H3C OH
CH3
CH2
DMSO
74. 5%
+
95%
5%

2-Methyl-2-hexanol on dehydration by this method affords about equal amounts of the 1- and 2-olefins, a result consistent with an acid-catalyzed mechanism, but such a mechanism is ruled out by the observation that addition of aniline or sodium *n*-octoxide has no effect on the yield of olefins or the direction of elimination. That a purely thermal elimination is excluded is evident from the comparison of DMSO

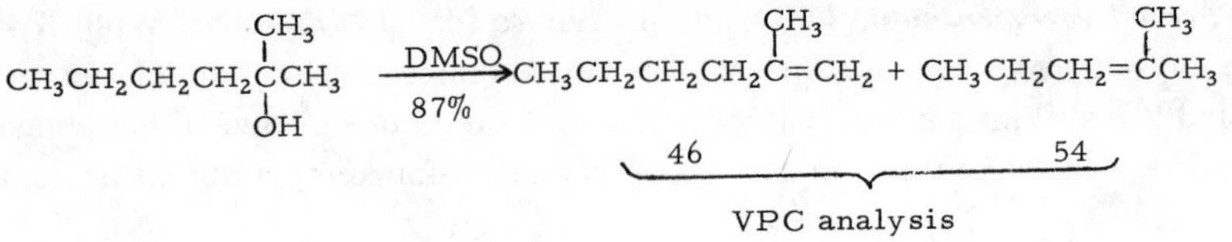
DMSO
87%
46
54
VPC analysis

with other solvents shown in Table. 1. Solvents of higher and of lower dielectric constant do not match the dehydrating action of dimethyl sulfoxide at 160°. There

Table 1 Effect of solvents on 1-phenyl-1-propanol at 160°

Solvent	Dielectric constant, E	Dipole moment, μ	Recovered alcohol, %
Acetamide	59	3.7	100
Dimethyl sulfoxide	48.9	3.95	3
Nitrobenzene	36	3.98	82
Benzonitrile	25	3.94	85
Quinoline	9	2.25	93
Diphenylmethane	2.7	2.95	91

is no correlation between dehydrating action and dipole moment of the solvent. DMSO appears to function as a solvent for promoting ionization of the alcohol to a carbonium ion intermediate. A synthetic potentiality for DMSO dehydration is evident from the conversion of acetonylacetone and 1,4-diphenylbutane-1,4-dione into 2,5-dimethylfurane and 2,5-diphenylfurane in yields of 66 and 60%.

DMSO
190°

R = CH_3 or C_6H_5

In connection with the investigation of the dehydration of alcohols in dimethyl sulfoxide involving reaction at elevated temperatures for substantial periods of time, Traynelis reports identification of a number of dimethyl sulfoxide decomposition products.

Gillis and Beck[33] heated the 1,4-diol (1) with dimethyl sulfoxide in the hope of obtaining the conjugated diene but found the product to be the cyclic oxide (2).

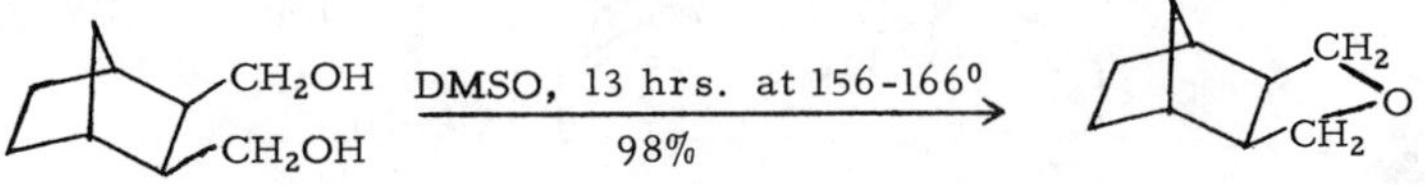
DMSO, 13 hrs. at 156-166°
98%

(1) (2)

They then found that secondary and tertiary 1,4-diols react in the same way and that DMSO-dehydration is superior to existing methods for the synthesis of tetrahydrofuranes. A cyclic transition state is postulated:

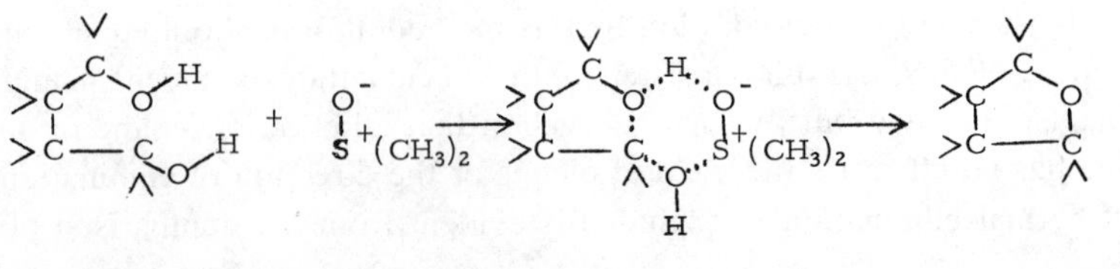

Le Goff[34] treated a mixture of 1,2,3-triphenylcyclopentadiene (3) and 1,2,3-triphenylpropenone (4) with potassium fluoride in dimethyl sulfoxide and so effected Michael-Knoevenagel dehydrative condensation to the bright yellow dihydrohexaphenylpentalene (5).

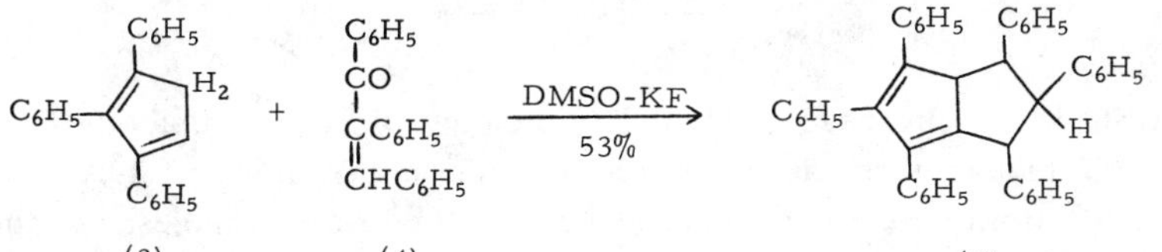

(3) (4) (5)

Oxidation. Kornblum[35] found that phenacyl bromide, as well as its *p*-halo, *p*-nitro, and *p*-phenyl derivatives, can be oxidized to glyoxals by dissolving it in

$$C_6H_5COCH_2Br \xrightarrow[71\%]{DMSO(25^0)} C_6H_5COCHO$$

dimethyl sulfoxide and letting the solution stand at room temperature for about 9 hrs. α-Bromodesoxybenzoin is oxidized smoothly to benzil in 44 hrs. at 45°. The method

$$C_6H_5COCHBrC_6H_5 \xrightarrow[95\%]{DMSO(45^0)} C_6H_5COCOC_6H_5$$

is not satisfactory for benzyl bromides. However, a later investigation[36] disclosed that not only benzyl halides but saturated primary chlorides, bromides, and iodides can be converted into aldehydes by oxidation of the corresponding tosylates with DMSO. For example, 1-iodoheptane (7 g.) is added to a solution of silver tosylate (11 g.) in acetonitrile at 0–5° (protected from light) and the mixture let come to room temperature overnight. The crude tosylate obtained by addition of water and extrac-

$$CH_3(CH_2)_5CH_2I \xrightarrow[CH_3CN]{TsOAg} CH_3(CH_2)_5CH_2OTs \xrightarrow{DMSO} \underset{70\%\ overall}{CH_3(CH_2)_5CHO}$$

tion with ether is added to a solution prepared by heating 150 ml. of DMSO to 150° (some foaming occurs), cooling, adding 20 g. of sodium bicarbonate, and bubbling nitrogen through the solution. The mixture is heated to 150°, kept at this temperature for 3 min., and cooled rapidly; the aldehyde is isolated as the 2,4-dinitrophenylhydrazone. Benzylic tosylates are oxidized smoothly at 100°.

Although Kornblum stated that ordinary halides are not oxidized by his method but must first be converted into the tosylates, Johnson and Pelter[37] found that DMSO

$$CH_3(CH_2)_6CH_2I \xrightarrow[74\%]{DMSO\ 4\ min.\ at\ 150^0} CH_3(CH_2)_6CHO$$

oxidized *n*-octyl iodide rapidly to the aldehyde at 150°. Other primary iodides were oxidized satisfactorily. With primary chlorides and with secondary iodides yields were poor.

An efficient procedure for the preparation of ethyl glyoxalate consists in oxidation of ethyl bromoacetate in 3 equivalents of dimethyl sulfoxide at 70° in the presence of 1,2-epoxy-3-phenoxypropane, a nonalkaline scavenger for hydrogen bromide, and

$$BrCH_2CO_2C_2H_5 + (CH_3)_2\overset{+}{S}-\overset{-}{O} \xrightarrow[70\%]{70^0} O{=}CHCO_2C_2H_5 + HBr + (CH_3)_2S$$

with methyl bromide added to convert the dimethyl sulfide formed into trimethylsulfonium bromide and so suppress side reactions.[38] Epoxides are oxidized smoothly

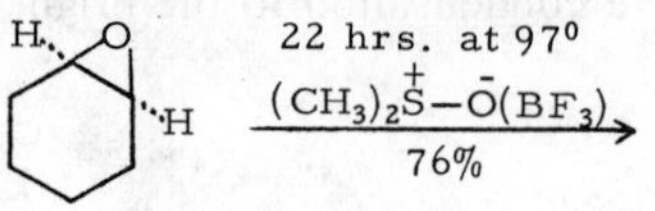
22 hrs. at 97°
(CH3)2S—O(BF3)
76%
OH

by DMSO in the presence of a catalytic amount of boron trifluoride etherate at steam-bath temperature, as illustrated for the preparation of 2-hydroxycyclohexanone.[39] Both 2β,3β-epoxycholestane and 2α,3α-epoxycholestane, the latter

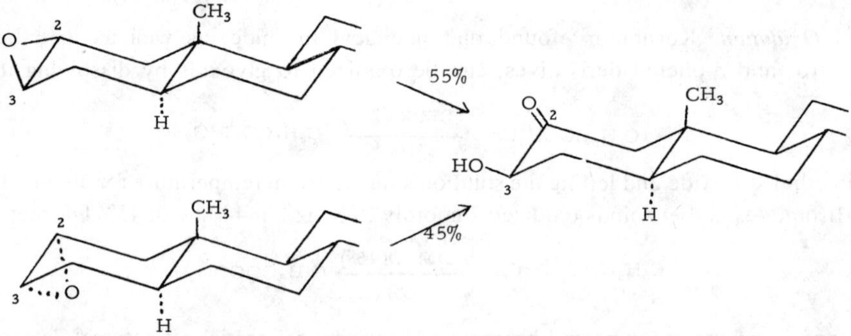
55%
45%
HO

with dioxane as co-solvent, on oxidation by this method afford cholestane-3β-ol-2-one (equatorial OH), along with small amounts of the 2,3-dione and the 2β,3α-diol.

Pfitzner and Moffatt[40] found that 3′-O-acetylthymidine (I) in dimethyl sulfoxide in the presence of anhydrous H_3PO_4 (orthophosphoric acid) and dicyclohexylcar-

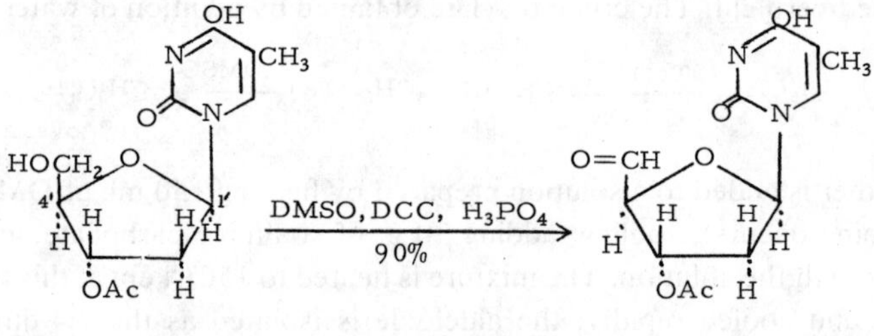
DMSO, DCC, H3PO4
90%

I II

bodiimide (DCC) is oxidized to the aldehyde II at room temperature in high yield. No carboxylic nucleoside derivative could be detected electrophoretically in the reaction mixture. Apparently DMSO is the oxidizing agent and DCC in the presence of the acid reacts with the water formed.

The same investigators[41] found the method generally applicable to the oxidation of primary alcohols to aldehydes and of secondary alcohols to ketones. Thus addition

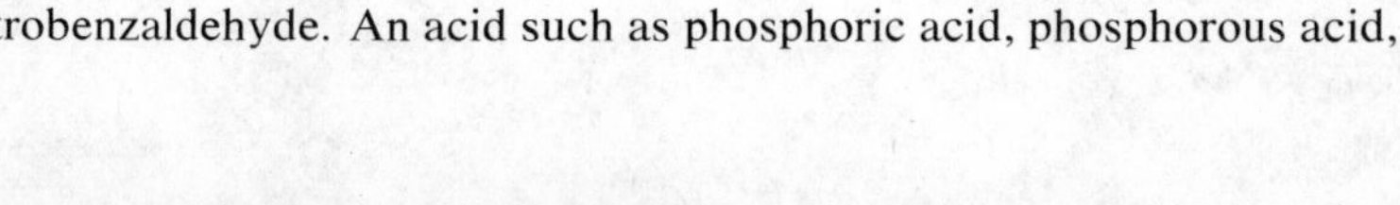
DMSO(excess),DCC (3 mmoles)
H3PO4 (0.5 mmole) 92%
1 mmole

of DCC to a solution of p-nitrobenzyl alcohol and anhydrous phosphoric acid in dry dimethyl sulfoxide led to the quantitative (thin layer chromatography) formation of p-nitrobenzaldehyde. An acid such as phosphoric acid, phosphorous acid, cyanoacetic acid, or pyridinium phosphate is required to promote oxidation within a few hours at room temperature. The stronger acid, trifluoroacetic acid, serves only poorly and hydrogen chloride and sulfuric acid completely inhibit oxidation; however, all three acids function well as their pyridine salts. Other oxidations carried out in DMSO in the presence of DCC and with phosphoric acid or pyridinium trifluoroacetate are as follows:

1-Octanol ⟶ Octylaldehyde (70%)
Testosterone ⟶ Δ^4-Androstene-3,17-dione (92%)
Cholestane-3β-ol ⟶ Cholestane-3-one (68%)
Cholane-24-ol ⟶ Cholane-24-al (85%)

For oxidation of testosterone to the dione it was found best to follow the procedure outlined in the formulation. Oxidation of cholesterol by a similar procedure afforded Δ^5-cholestene-3-one in 66% yield.

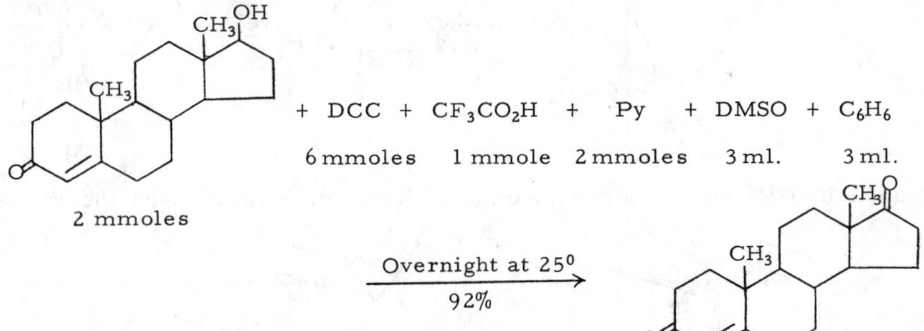
CH3 OH
CH3
+ DCC + CF3CO2H + Py + DMSO + C6H6
6 mmoles 1 mmole 2 mmoles 3 ml. 3 ml.
2 mmoles
Overnight at 25°
92%
CH3 O
CH3

Pfitzner and Moffatt report the interesting finding that an equatorial 11α-hydroxysteroid is oxidized readily by the DMSO–DCC combination whereas the 11β-epimer is inert under the same conditions. The situation is thus the reverse of that found in chromic acid oxidation (Eschenmoser). The authors offer an explanation based on a suggested mechanism for the oxidation.

Dyer *et al.*[42] oxidized a furanoside derivative by the DMSO–DCC–pyridinium phosphate method, and Albright and Goldman[43] applied the Pfitzner-Moffatt procedure successfully for oxidation of yohimbine and other indole alkaloids. Suscep-

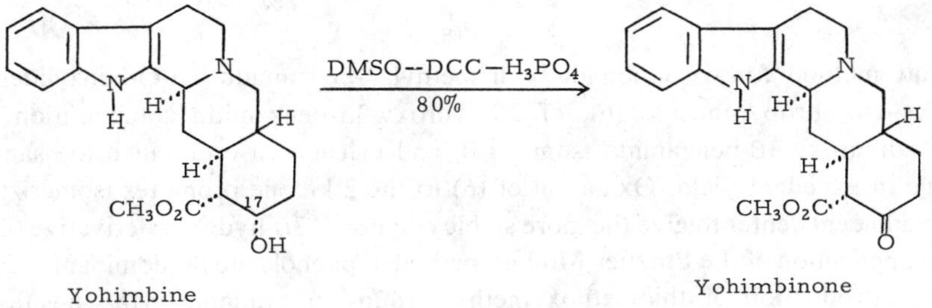
N
H
H
N
H
H
CH3O2C
17
OH
DMSO--DCC-H3PO4
80%
N
H
H
N
H
H
CH3O2C
O

Yohimbine Yohimbinone

tibility of the indole nucleus to oxidation precludes use of most reagents, and the only previously used method, Oppenauer oxidation, gave only moderate yields.

Baker and Buss[44] found the Pfitzner-Moffatt procedure superior for the oxidation of carbohydrate mesylates having a free hydroxyl group, for example the hexitol mesylate (1). In contrast, the Sarrett procedure afforded crystalline material in

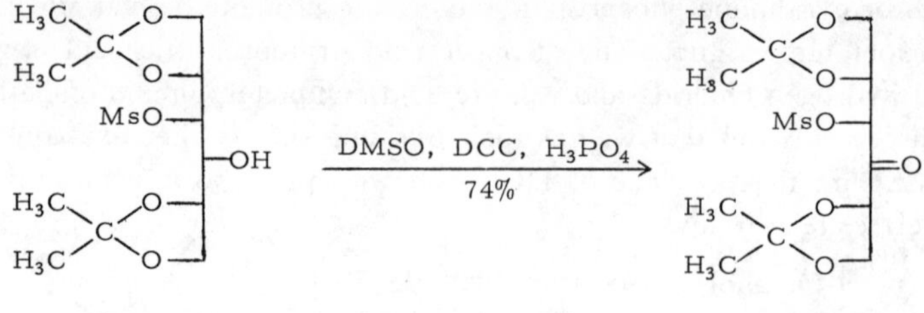

(1) (2)

yields of only 0–5%. However, trials with furanosides and pyranosides showed that the oxidation is subject to steric hindrance. In (3), (4), and (5) the free hydroxyl group is flanked on each side by an ether or acetal group and these compounds all

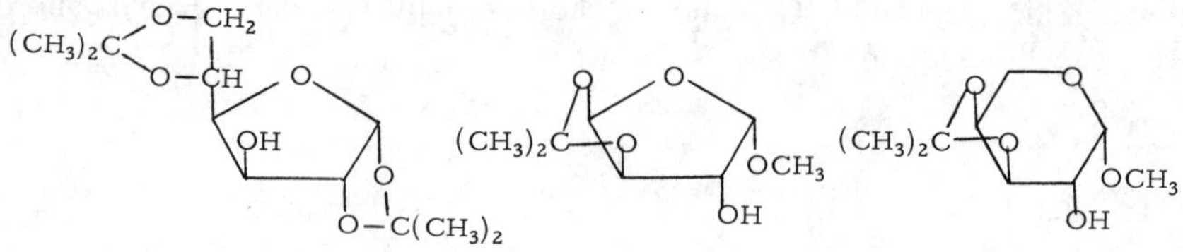

(3) (4) (5)

remained unoxidized at room temperature. Baker and Buss[45] used the Pfitzner-

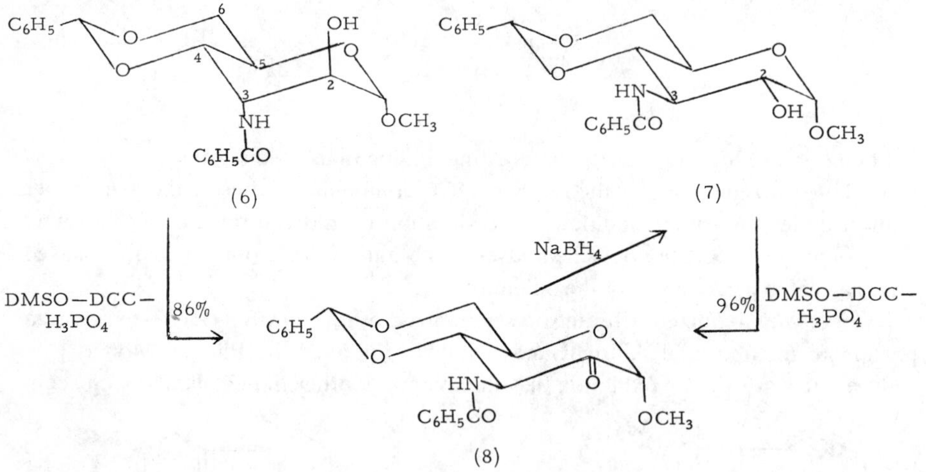

Moffatt method for oxidation of both methyl 3-benzamido-4,6-O-benzylidene-3-desoxy-α-D-altropyranoside (6), of 2β-hydroxy-3α-benzamido configuration, and the 2α-hydroxy-3β-benzamido isomer (7), and in each case obtained the same 2-ketone in excellent yield. Oxidation of (6) to the 2-ketone promotes isomerization at the adjacent center to give the more stable equatorial 3β-hydroxy derivative (4).

On application of the Pfitzner-Moffatt method to phenols the predominant reaction is the introduction of thiomethoxymethyl groups in available *ortho* positions.[46]

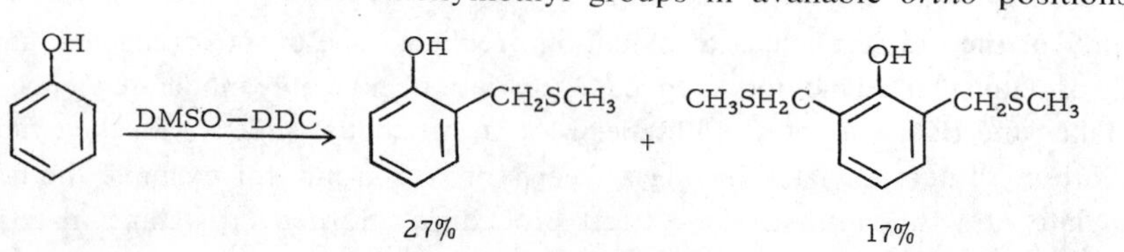

Phenols with no free *ortho* positions give *para* alkylated products. O-Alkylation is observed with more acidic phenols.

The Pfitzner-Moffatt reagent was found to provide the best method for oxidation

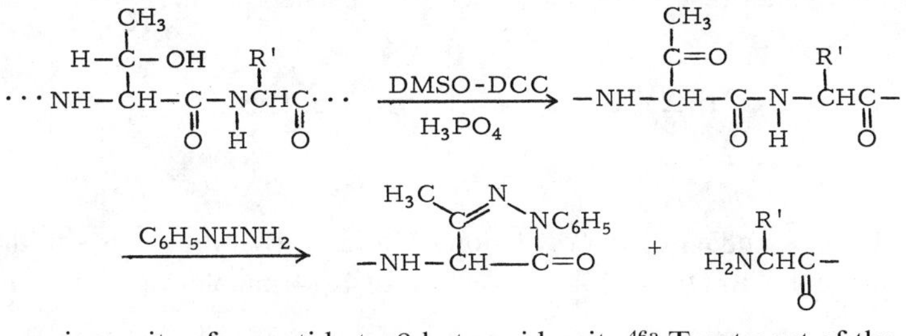

of the threonine units of a peptide to β-keto acid units.[46a] Treatment of the oxidized peptide with phenylhydrazine then effects selective fragmentation, as formulated.

Traynelis and Hergenrother[47] found that benzyl alcohol can be oxidized to benzaldehyde in 80% yield by refluxing a solution of 0.1 mole of the alcohol in 0.7 mole of dimethyl sulfoxide at 190° while bubbling air through the reaction mixture.

$$C_6H_5CH_2OH \xrightarrow[80\%]{\text{DMSO (air)}} C_6H_5CHO$$

In view of the susceptibility of benzaldehyde to autoxidation, it is remarkable that in the presence of DMSO the reaction stops at the aldehyde stage. In fact when a solution of benzaldehyde in DMSO was refluxed for 24 hrs. with air passing through the solution, only 1.6% of benzoic acid was isolated and 87% of benzaldehyde was recovered. Although air facilitates the oxidation, there is no oxygen uptake and isolation of dimethyl sulfide as such or as the mercuric chloride derivative (m.p. 150°) shows that DMSO is the oxidant. Several substituted benzyl alcohols were oxidized by the same method to the aromatic aldehydes. The one allylic alcohol

$$C_6H_5CH{=}CHCH_2OH \xrightarrow[60\%(\text{VPC})]{\text{DMSO(air)}} C_6H_5CH{=}CHCHO$$

$$C_6H_5CH_2CH_2CH_2OH \xrightarrow[26\%]{\text{DMSO(air)}} C_6H_5CH_2CH_2CHO$$

investigated, cinnamyl alcohol, afforded the aldehyde in fair yield. With hydrocinnamyl alcohol the yield was low.

Dimethyl sulfoxide oxidizes certain thio ethers to the sulfoxides in reasonable yield.[48] The ether is heated with a 50% molar excess of DMSO at 160–175° for several hours and the dimethyl sulfide is removed by distillation as formed. The sulfoxides were isolated in high purity by distillation and there was no evidence of sulfone formation. Di-*n*-butyl sulfoxide and tetraethylene sulfoxide were obtained in similar yield, but several sulfides did not react to an appreciable extent. DMSO is a superior reagent for the oxidation of thiols to disulfides.[49] A solution of the thiol in

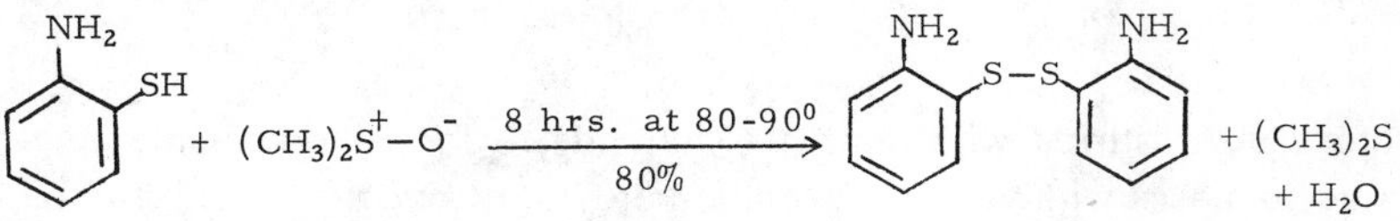

DMSO is stirred at 80–90° for a maximum of 8 hrs., and the disulfide is obtained in excellent yield and purity. Yields of both aliphatic and aromatic disulfides were in

the range 80–100%. Experiments conducted in a nitrogen atmosphere gave the same results as those carried out in air.

In the presence of 5 mole % of bromine, DMSO oxidizes isonitriles to isocyanates, probably by a chain reaction.[50] The reaction is carried out in refluxing chloroform for 24 hrs.

$$(CH_3)_2CHN=C \xrightarrow{Br_2} (CH_3)_2CHN=CBr_2 \xrightarrow{(CH_3)_2\overset{+}{S}-\overset{-}{O}}$$

$$(CH_3)_2CHN=C=O + (CH_3)_2S + Br_2$$

Russell et al.[51] studied the autoxidation of di- and tri-arylmethanes in the system DMSO–t-BuOH–t-BuOK at 25°. A solution of 3.14 mmoles of triphenylmethane,

$$(C_6H_5)_3CH + DMSO-\underline{t}-BuOH(80:20) + \underline{t}-BuOK \xrightarrow{O_2} (C_6H_5)_3COH$$

3.14 mmoles   25 ml.   6 mmoles   3 mmoles

shaken in the system under oxygen, absorbed 2.6 mmoles of gas in 2 min. and 3.2 mmoles in 20 min., and afforded triphenylcarbinol in high yield. Diphenylmethane is oxidized to benzhydrol, benzophenone, and the dimethyl sulfoxide adduct of the ketone:

$$(C_6H_5)_2\overset{\overset{OH}{|}}{C}CH_2\overset{\overset{O^-}{|}}{\underset{}{\overset{+}{S}}}CH_3$$

Ketenes and ketenimines on treatment in DMSO with a few drops of concd. hydrochloric acid undergo oxidative hydrolysis.[52]

$$(C_6H_5)_2C=C=O \xrightarrow[88\%]{aq.\ HCl-DMSO} (C_6H_5)_2\overset{\overset{OH}{|}}{C}-CO_2H$$

$$(C_6H_5)_2C=C=N-\!\!\!\left\langle\!\!\!\bigcirc\!\!\!\right\rangle\!\!-CH_3 \xrightarrow[91.5\%]{aq.\ HCl-DMSO} (C_6H_5)_2\overset{\overset{OH}{|}}{C}-\overset{\overset{O}{\|}}{C}-\overset{\overset{H}{|}}{N}-\!\!\!\left\langle\!\!\!\bigcirc\!\!\!\right\rangle\!\!-CH_3$$

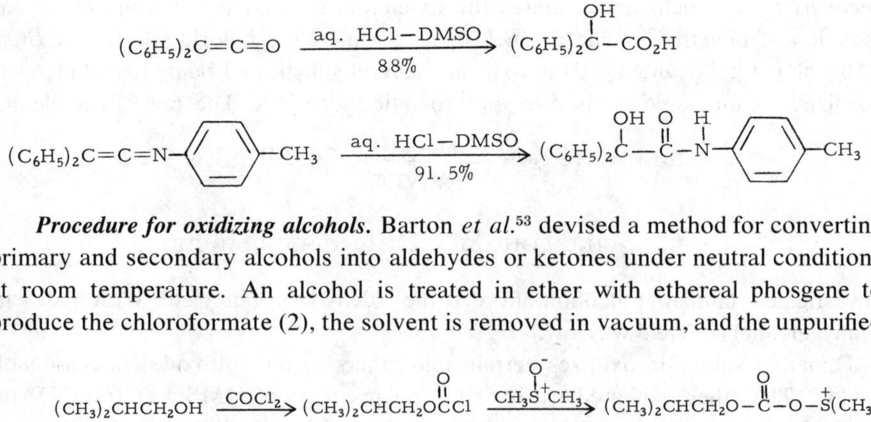

Procedure for oxidizing alcohols. Barton et al.[53] devised a method for converting primary and secondary alcohols into aldehydes or ketones under neutral conditions at room temperature. An alcohol is treated in ether with ethereal phosgene to produce the chloroformate (2), the solvent is removed in vacuum, and the unpurified

$$(CH_3)_2CHCH_2OH \xrightarrow{COCl_2} (CH_3)_2CHCH_2O\overset{\overset{O}{\|}}{C}Cl \xrightarrow{CH_3\overset{+}{S}{}^{\overset{O^-}{|}}CH_3} (CH_3)_2CHCH_2O-\overset{\overset{O}{\|}}{C}-O-\overset{+}{\underset{Cl^-}{S}(CH_3)_2}$$

(1)   (2)   (3)

$$\xrightarrow{-CO_2} (CH_3)_2CHCH-O-\overset{+}{\underset{Cl^-}{S}(CH_3)_2} \xrightarrow[70\%\ overall]{(C_2H_5)_3N} (CH_3)_2CHCH=O + (CH_3)_2S + (C_2H_5)_3\overset{+}{N}HCl^-$$

(4)   (5)

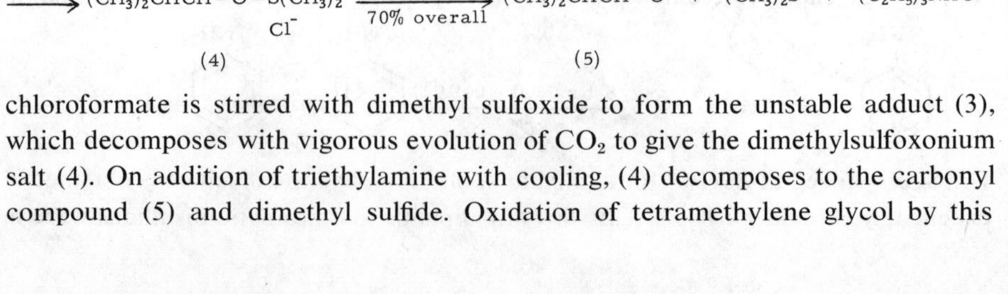

chloroformate is stirred with dimethyl sulfoxide to form the unstable adduct (3), which decomposes with vigorous evolution of CO_2 to give the dimethylsulfoxonium salt (4). On addition of triethylamine with cooling, (4) decomposes to the carbonyl compound (5) and dimethyl sulfide. Oxidation of tetramethylene glycol by this method afforded succinic aldehyde in 80% yield; the compound is not easily accessible by other methods. Yields of (—)-menthone (35%) and of cholestanone (20%) were disappointing.

Methylenation of ketones. Lunn[54] heated Δ^1-5α-androstene-3-one with DMSO containing boron trifluoride etherate in the hope of obtaining the 1,3-dione. Instead he obtained in 30% yield the 4-methylene ketone (2). He then added paraformalde-

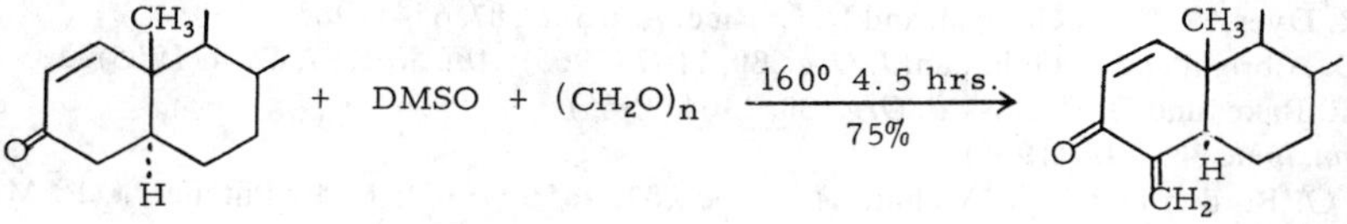
CH3
H
+ DMSO + (CH2O)n
160° 4.5 hrs.
75%
CH3
H
CH2

(1) (2)

hyde to the reaction mixture and raised the yield to 75%. The reagent reacts readily with a saturated 3-ketone but gives a mixture of four products, all in low yield.

[1]C. Agami, *Bull. Soc.*, 1021 (1965)
[2]C. H. Issidorides, I. Kitagawa, and E. Mosettig, *J. Org.*, **27**, 4693 (1962)
[3]B. Witkop, *Experientia*, **12**, 372 (1956)
[4]R. A. Smiley and C. Arnold, *J. Org.*, **25**, 257 (1960)
[5]L. Friedman and H. Shechter, *J. Org.*, **25**, 877 (1960); see also P. A. Argabright and D. H. Hall, *Chem. Ind.*, 1365 (1964)
[6]J. Miller and A. J. Parker, *Am. Soc.*, **83**, 117 (1961)
[7]J. Cason and J. S. Correia, *J. Org.*, **26**, 3645 (1961)
[8]D. J. Cram, B. Rickborn, C. A. Kingsbury, and P. Haberfield, *Am. Soc.*, **83**, 3678 (1961)
[9]D. J. Cram, B. Rickborn, and G. R. Knox, *Am. Soc.*, **82**, 6412 (1960)
[10]N. Kornblum and R. K. Blackwood, *Org. Syn., Coll. Vol.*, **4**, 454 (1963)
[11]A. C. Cope and A. S. Mehta, *Am. Soc.*, **86**, 5626 (1964)
[12]E. D. M. Eades, D. H. Ball, and L. Long, Jr., *J. Org.*, **31**, 1159 (1966)
[13]P. E. Eaton, *Am. Soc.*, **84**, 2344 (1962)
[14]H. R. Nace, *Am. Soc.*, **81**, 5428 (1959)
[15]R. N. Iacona, A. T. Rowland, and H. R. Nace, *J. Org.*, **29**, 3495 (1964); H. R. Nace and R. N. Iacona, *ibid.*, **29**, 3498 (1964)
[16]T. J. Wallace, J. E. Hofmann, and A. Schriesheim, *Am. Soc.*, **85**, 2739 (1963); J. E. Hofmann, T. J. Wallace, and A. Schriesheim, *ibid.*, **86**, 1561 (1964)
[17]D. J. Cram, B. Rickborn, and G. R. Knox, *Am. Soc.*, **82**, 6412 (1960)
[18]M. R. V. Sahyun and D. J. Cram, *Org. Syn.*, **45**, 89 (1965)
[19]C. C. Price and W. H. Snyder, *Am. Soc.*, **83**, 1773 (1961)
[20]J. Cunningham, R. Gigg, and C. D. Warren, *Tetrahedron Letters*, 1191 (1964)
[21]A. Schriesheim, J. E. Hofmann, and C. A. Rowe, Jr., *Am. Soc.*, **83**, 3731 (1961)
[22]A. J. Birch, J. M. H. Graves, and J. B. Siddall, *J. Chem. Soc.*, 4234 (1963)
[23]D. Devaprabhakara, C. G. Cardenas, and P. D. Gardner, *Am. Soc.*, **85**, 1553 (1963)
[24]H. E. Zaugg, *Am. Soc.*, **83**, 837 (1961)
[25]J. J. Bloomfield, *J. Org.*, **27**, 2742 (1962)
[26]J. J. Bloomfield, *J. Org.*, **26**, 4112 (1961)
[27]J. J. Bloomfield and P. V. Fennessey, *Tetrahedron Letters*, 2273 (1964)
[28]D. J. Cram, M. R. V. Sahyun, and G. R. Knox, *Am. Soc.*, **84**, 1734 (1962)
[29]R. Kuhn and H. Trischmann, *Ber.*, **96**, 284 (1963)
[30]H. C. Srivastava, P. P. Singh, S. N. Harshe, and K. Virk, *Tetrahedron Letters*, 493 (1964)
[31]E. Tommila and M.-L. Murto, *Chem. Scand.*, **17**, 1947 (1963)
[32]V. J. Traynelis, W. L. Hergenrother, J. R. Livingston, and J. A. Valicenti, *J. Org.*, **27**, 2377 (1962); V. J. Traynelis, W. L. Hergenrother, H. T. Hanson, and J. A. Valicenti, *ibid.*, **29**, 123 (1964); V. J. Traynelis and W. L. Hergenrother, *ibid.*, **29**, 221 (1964)
[33]B. T. Gillis and P. E. Beck, *J. Org.*, **28**, 1388 (1963)

[34]E. Le Goff, *Am. Soc.*, **84**, 3975 (1962)
[35]N. Kornblum *et al.*, *Am. Soc.*, **79**, 6562 (1957)
[36]N. Kornblum, W. J. Jones, and G. J. Anderson, *Am. Soc.*, **81**, 4113 (1959)
[37]A. P. Johnson and A. Pelter, *J. Chem. Soc.*, 520 (1964)
[38]I. M. Hunsberger and J. M. Tien, *Chem. Ind.*, 88 (1959)
[39]T. Cohen and T. Tsuji, *J. Org.*, **26**, 1681 (1961)
[40]K. E. Pfitzner and J. G. Moffatt, *Am. Soc.*, **85**, 3027 (1963)
[41]*Idem, ibid.*, **85**, 3027–3028 (1963); definitive papers: *ibid.*, **87**, 5661, 5670 (1965)
[42]J. R. Dyer, W. E. McGonigal, and K. C. Rice, *Am. Soc.*, **87**, 654 (1965)
[43]J. D. Albright and L. Goldman, *J. Org.*, **30**, 1107 (1965); *Am. Soc.*, **87**, 4214 (1965)
[44]B. R. Baker and D. H. Buss, *J. Org.*, **30**, 2304 (1965)
[45]*Idem, ibid.*, **30**, 2308 (1965)
[46]M. G. Burdon and J. G. Moffatt, *Am. Soc.*, **87**, 4656 (1965); K. E. Pfitzner, J. P. Marino, and R. A. Olofson, *ibid.*, **87**, 4658 (1965)
[46a]F. D'Angeli, E. Scoffone, F. Filira, and V. Giormani, *Tetrahedron Letters*, 2745 (1966)
[47]V. J. Traynelis and W. L. Hergenrother, *Am. Soc.*, **86**, 298 (1964)
[48]S. Searles, Jr., and H. R. Hays, *J. Org.*, **23**, 2028 (1958)
[49]C. N. Yiannios and J. V. Karabinos, *J. Org.*, **28**, 3246 (1963); T. J. Wallace, *Am. Soc.*, **86**, 2018 (1964)
[50]H. W. Johnson, Jr., and P. H. Daughhetee, Jr., *J. Org.*, **29**, 246 (1964)
[51]G. A. Russell, E. G. Janzen, H.-D. Becker, and F. J. Smentowski, *Am. Soc.*, **84**, 2652 (1962)
[52]I. Lillien, *J. Org.*, **29**, 1631 (1964)
[53]D. H. R. Barton, B. J. Garner, and R. H. Wightman, *J. Chem. Soc.*, 1855 (1964)
[54]W. H. W. Lunn, *J. Org.*, **30**, 2925 (1965)

Dimethyl sulfoxide-derived reagent (a). Sodium methylsulfinylmethide (1). Corey and Chaykovsky[1] prepared this highly reactive reagent in solution by adding powdered sodium hydride to excess dimethyl sulfoxide with stirring under nitrogen at 65–70° until hydrogen was no longer evolved. This salt (1) reacts with ethyltriphenyl-

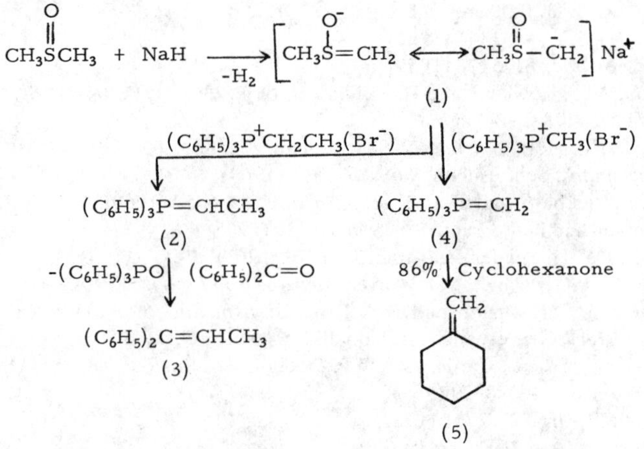
CH₃SCH₃ + NaH
-H₂
CH₃S=CH₂
CH₃S–CH₂
Na⁺
(1)
(C₆H₅)₃P⁺CH₂CH₃(Br⁻)
(C₆H₅)₃P⁺CH₃(Br⁻)
(C₆H₅)₃P=CHCH₃
(2)
-(C₆H₅)₃PO
(C₆H₅)₂C=O
(C₆H₅)₂C=CHCH₃
(3)
(C₆H₅)₃P=CH₂
(4)
86%
Cyclohexanone
CH₂
(5)

phosphonium bromide at room temperature to form the Wittig ylide (2) as shown by reaction with benzophenone to form 1,1-diphenylpropene-1 (3).[2] Addition of methyltriphenylphosphonium bromide to the solution of sodio derivative (1) forms the methylene ylide (4), which converts cyclohexanone, camphor, and cholestane-3-one to the methylene compounds in yields of 86, 73, and 69%. This Corey reagent (1) thus provides a simple method for the preparation of Wittig ylides.

Corey and Chaykovsky[3] and Becker *et al.*[4] independently discovered that condensation of an ester with an alkali methylsulfinylmethide affords a β-ketosulfoxide (6) in high yield. In the first case the ester (neat if a liquid, in THF if a solid) was added at 0° under nitrogen to a solution of 2 equivalents of (1) prepared in DMSO

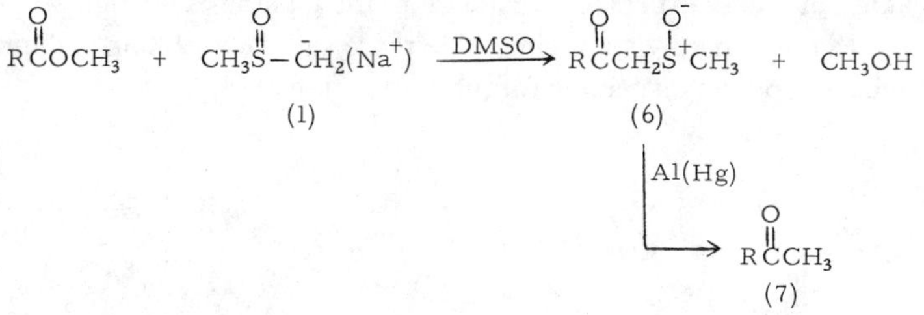

with NaH; yields 70–90%. The Becker group prepared a solution of the potassium salt corresponding to (1) from potassium *t*-butoxide and DMSO and also obtained β-ketosulfoxides in high yields. Corey and Chaykovsky found further that the now readily prepared β-ketosulfoxides (6) can be reduced smoothly with aluminum amalgam to methyl ketones (7) and stressed this feature of the new development. Becker and Russell[5] applied the condensation reaction to diethyl phthalate and developed an efficient new synthesis of ninhydrin using sodium methoxide as base for generation of the active reagent. The ester is added dropwise to a suspension of the base in DMSO which is stirred by a stream of nitrogen. The β-ketosulfoxide (II) initially formed undergoes intramolecular ester condensation with formation

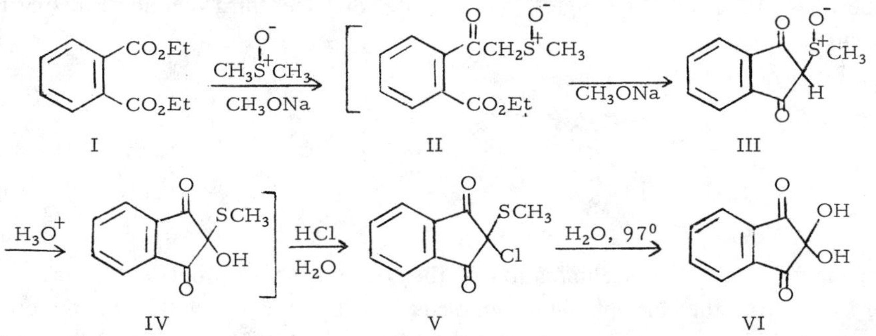

of the 1,3-indanedione system IV, which in the presence of hydrochloric acid undergoes the Pummerer rearrangement to 2-chloro-2-methylmercapto-1,3-indanedione (V). This crystalline product is obtained in yield of 80% and is hydrolyzed in boiling water nearly quantitatively to ninhydrin (VI).

Potassium methylsulfinylmethide evidently is the effective species in a tricarbon condensation of *p*-methoxybenzaldehyde with diphenylmethane and DMSO in the

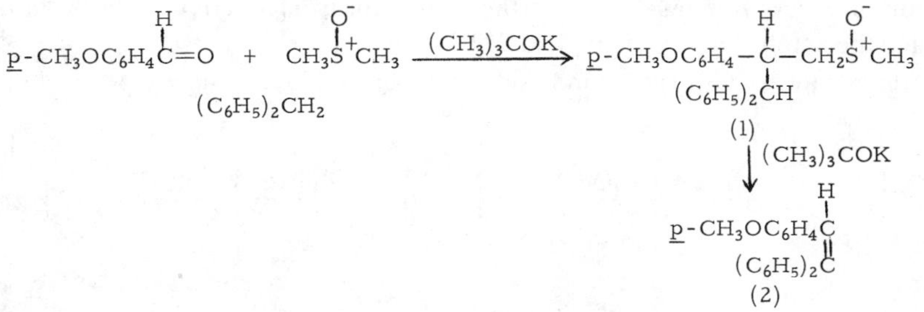

presence of potassium *t*-butoxide to give (1) as the initial product; prolongation of the reaction period or use of a higher concentration of base effects elimination to give (2).[6]

Shemyakin and co-workers,[7] investigating the synthesis of tetracycline antibiotics, found sodium methylsulfinylmethide more satisfactory than sodium hydride, sodium amide, or sodium dispersion for the cyclization of (3) to (4).

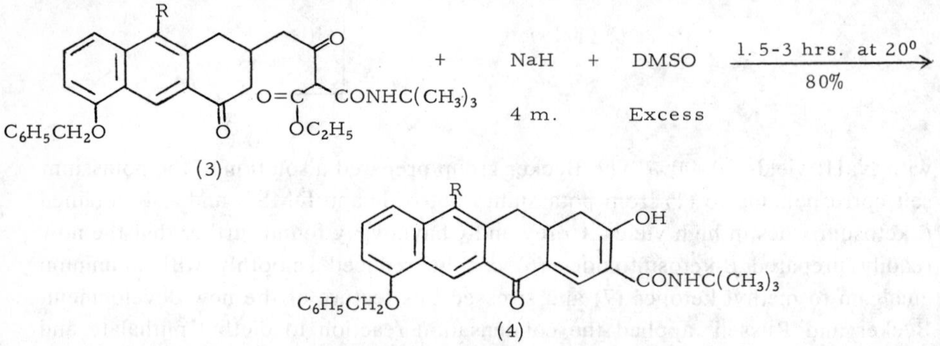
R
$C_6H_5CH_2O$
O=C
$CONHC(CH_3)_3$
OC_2H_5
+ NaH + DMSO
4 m.
Excess
1. 5-3 hrs. at 20°
80%
OH
$CONHC(CH_3)_3$

(3) (4)

In achieving the total synthesis of *dl*-caryophyllene, Corey[8] employed sodium methylsulfinylmethide in three of the steps. One was for Wittig replacement of a carbonyl oxygen by a methylene group. A second was to effect Dieckmann-type conversion of VII into VIII, probably via an intermediate bridged lactone which undergoes attack and cleavage by methoxide ion. The third reaction utilizing the

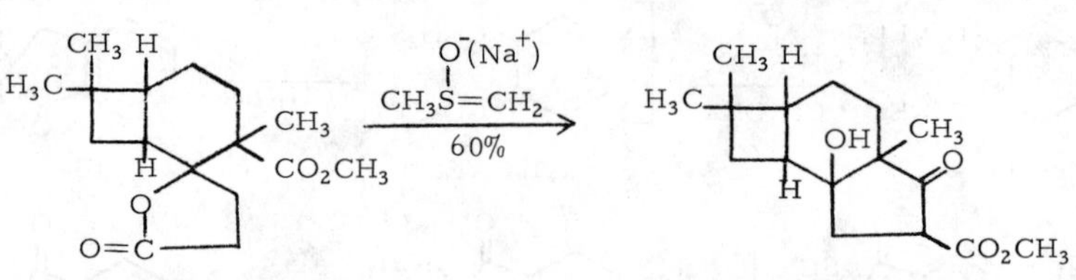
CH_3 H
H_3C
CH_3
CO_2CH_3
O=C
$\bar{O}(Na^+)$
$CH_3S=CH_2$
60%
OH
CO_2CH_3

VII VIII

reagent was the internal elimination of the tosylate IX to produce the ketone X, the key step in the caryophyllene synthesis. Treatment of IX with sodium methylsulfinylmethide (3 equiv.) in dimethyl sulfoxide at 25° for 30 min. followed by

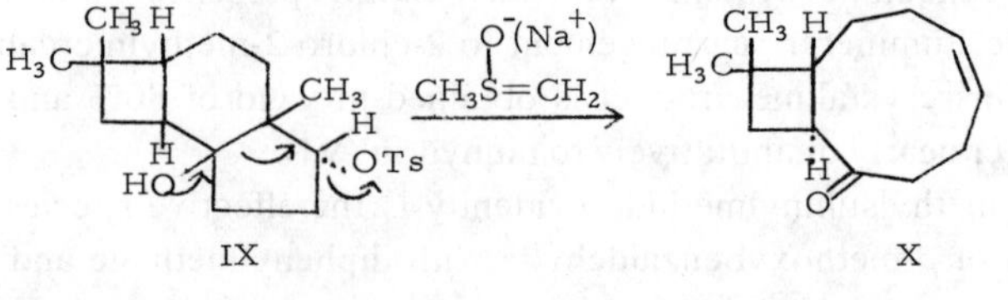
CH_3 H
H_3C
CH_3
HO
OTs
$\bar{O}(Na^+)$
$CH_3S=CH_2$

IX X

addition of excess *t*-butanol and a further 2 hrs. for isomerization of the ring fusion afforded the ketone "in excellent yield."

Gardner and co-workers[9] found sodium methylsulfinylmethide a very effective

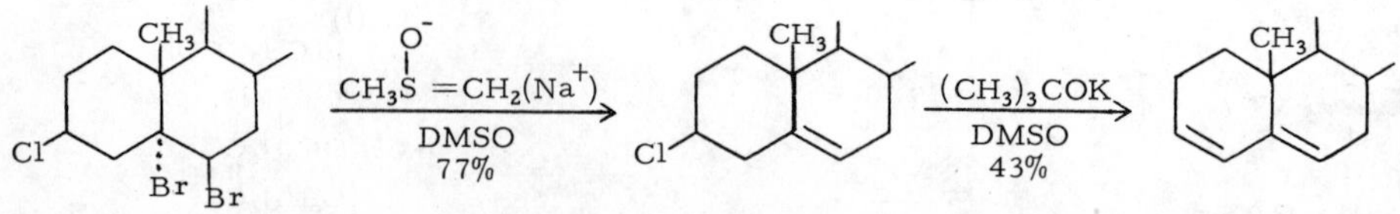
CH_3
Cl
Br
Br
$CH_3S=CH_2(Na^+)$
DMSO
77%
$(CH_3)_3COK$
DMSO
43%

(1) (2) (3)

reagent for dehalogenation of dibromides. Thus 3β-chloro-5α,6β-dibromocholestane, a *vic*-dibromide, is converted by the reagent into 3β-chloro-Δ^5-cholestene (2) as the sole product (yield after crystallization 77%). In contrast, potassium *t*-butoxide in DMSO converts (2) into $\Delta^{3,5}$-cholestadiene (3) and a mixture of unidentified dienes. Thus one reagent is the more effective for dehalogenation and the other for dehydrohalogenation. 7,7-Dibromonorcarane (4), a *gem*-dibromide, reacts with the reagent in DMSO at room temperature to give 7-bromobicyclo[4.1.0]-heptane (5), the *trans*-isomer predominating. Further dehydrohalogenation of (5) does not occur even after rather long reaction times. The *gem*-dibromide (6) with a

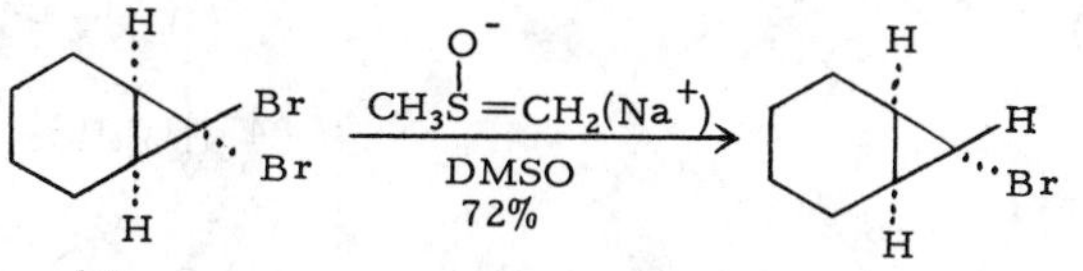
H
Br
Br
H
$CH_3\overset{O^-}{S}=CH_2(Na^+)$
DMSO
72%
H
H
Br
H

(4) (5)

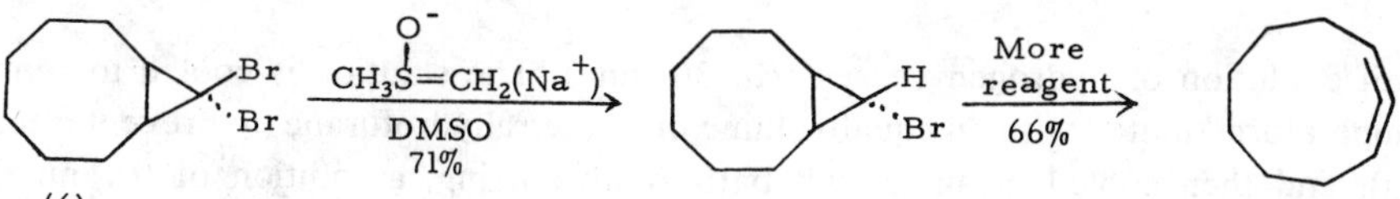
Br
Br
$CH_3\overset{O^-}{S}=CH_2(Na^+)$
DMSO
71%
H
Br
More
reagent
66%

(6) (7) (8)

[6.2.0] system similarly gives the monobromide (7) in the first step but (7) readily reacts further to give the allene (8). The formation of the bromocyclopropanes (5) and (7) from the *gem*-dibromo precursors almost certainly involves nucleophilic displacement on bromine to give an intermediate bromocyclopropylcarbanion which then becomes protonated by solvent. The most obvious route from the monobromide (7) to the allene (8) involves α-elimination of HBr induced by DMSO⁻ playing the role of base, and giving rise to an intermediate cyclopropyl carbene.

Roberts and Whiting[10] developed a method for effecting the anhydrous hydrolysis of esters and nitriles in DMSO. A solution of sodium methylsulfinylmethide is prepared from sodium hydride and DMSO and titrated with a solution of an appropriate amount of water in DMSO, using triphenylmethane as indicator. This produces a fine suspension of sodium hydroxide which hydrolyzes ethyl benzoate very rapidly at room temperature. As compared with reactions in hydroxylic solvents, rates are enhanced by a factor of 10^4–10^5. Benzonitrile is hydrolyzed to benzamide. Methyl and ethyl mesitoates are hydrolyzed readily at 25°.

DMSO in the presence of potassium *t*-butoxide (potassium methylsulfinylmethide is formed) methylates diolefins and reactive aromatic hydrocarbons, for example:[11]

$$CH_2{=}CHCH{=}CH_2 \xrightarrow[50\%]{DMSO-(CH_3)_3COK} CH_2{=}CHCH{=}CHCH_3$$

cis/trans = 80/20

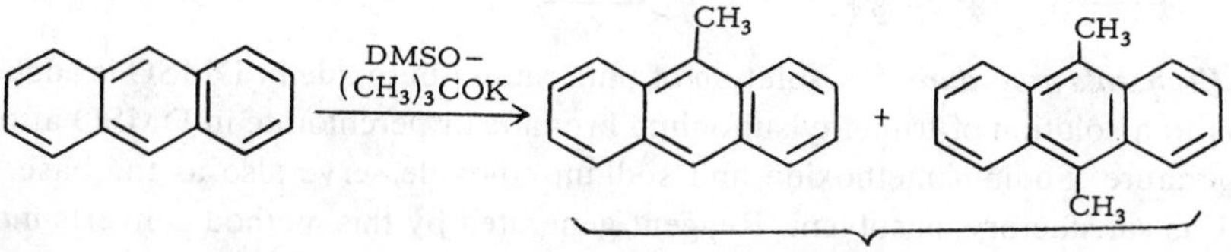
DMSO–
$(CH_3)_3COK$
CH_3
+
CH_3
CH_3

82:18

Dimethyl sulfoxide-derived reagent (b). Dimethylsulfonium methylide, $(CH_3)_2S\!\!=\!\!CH_2$.

This heat-labile reagent is prepared in solution by dehydrohalogenation of trimethyl-sulfonium iodide, bromide, or perchlorate. It is an exceedingly selective methylene-transfer reagent and converts carbonyl compounds into epoxides (oxiranes).

Corey's procedure.[12] The salt is trimethylsulfonium iodide (1), the base is a solution of sodium methylsulfinylmethide (2), prepared by heating a mixture of powdered sodium hydride (0.1 m.) with excess DMSO with stirring under nitrogen at 75°

$$(CH_3)_2\overset{+}{S}-CH_3(I^-) \;+\; CH_3\overset{O^-}{\underset{|}{S}}\!\!=\!\!CH_2(Na^+) \longrightarrow (CH_3)_2S\!\!=\!\!CH_2 \;+\; (CH_3)_2S\!\!=\!\!O$$
$$(1) \qquad\qquad\qquad (2) \qquad\qquad\qquad\qquad (3)$$

$$84\% \downarrow (C_6H_5)_2C\!\!=\!\!O$$

$$(C_6H_5)_2C\overset{\displaystyle O}{\underset{\textstyle (4)}{\diagup\!\!\!\diagdown}}CH_2$$

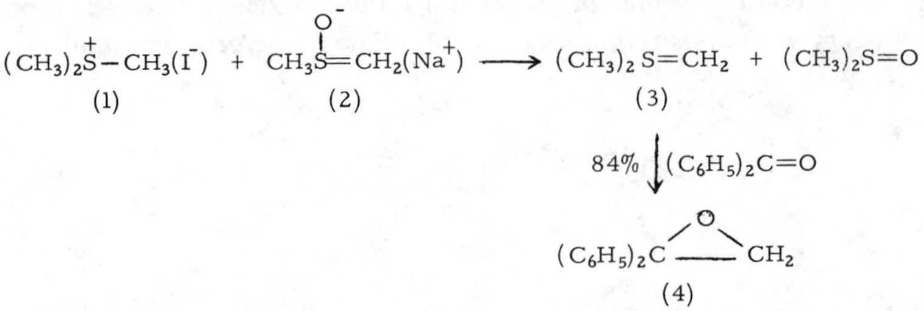

until evolution of hydrogen ceases (*ca.* 30 min.). The solution is cooled to room temperature, diluted with an equal volume of dry tetrahydrofurane (to prevent freezing), and then cooled in an ice-salt bath. With stirring, a solution of 0.1 m. of trimethylsulfonium iodide (1) in 80 ml. of DMSO is added in the course of about 3 min. The mixture is stirred for 1 min. longer and the carbonyl component is added at a moderate rate of injection, neat if a liquid, or as a solution in DMSO or THF. Stirring is continued at salt-ice temperature for 15 min., and then the bath is removed for 30–60 min. The mixture is diluted with 3 volumes of water, and extracted with an appropriate solvent.

Both simple carbonyl compounds and α,β-unsaturated ketones react with this reagent with exclusive formation of epoxides (oxiranes), not cyclopropanes. Yields are as follows: benzaldehyde, 75%; cycloheptanone, 97%; benzalacetophenone, 87%; carvone, 89%; eucarvone, 93%; pulegone, 90%; Δ^4-cholestene-3-one, 90%.

The reagent reacts with butadiene to give vinylcyclopropane, probably by 1,2-addition and elimination of dimethyl sulfide.[12a]

$$CH_2\!\!=\!\!CHCH\!\!=\!\!CH_2 \xrightarrow{(CH_3)_3S\!\!=\!\!CH_2} \underset{\underset{+S(CH_3)_2}{\overset{|}{CH_2}}}{\overset{CH_2}{\underset{|}{}}}\!\!\!\!\!\!\overset{-}{C}HCH\!\!=\!\!CH_2 \longrightarrow \underset{CH_2}{\overset{CH_2}{\diagup\!\!\!\diagdown}}\!\!CHCH\!\!=\!\!CH_2 \;+\; (CH_3)_2S$$

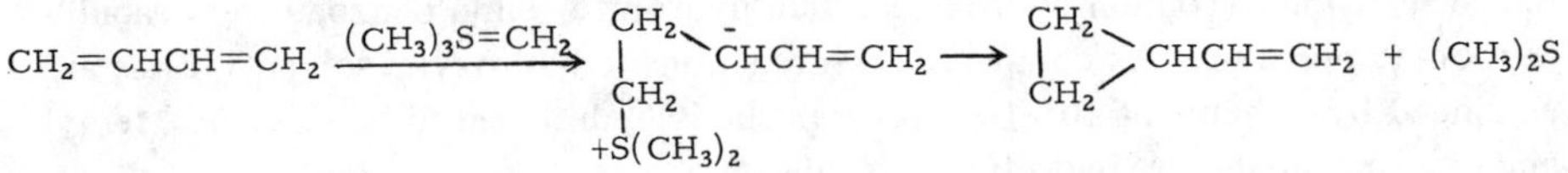

The reagent converts 17-ketosteroids into $17\beta,20$-epoxy-21-norpregnane derivatives reducible to 17β-hydroxy-17α-methylsteroids.[13]

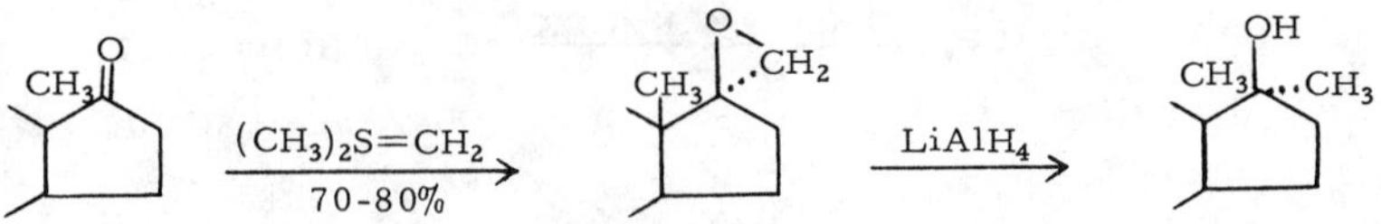

Franzen's procedure.[14] A solution of potassium *t*-butoxide in DMSO is added by drops to a solution of trimethylsulfonium bromide or perchlorate in DMSO at room temperature. Sodium methoxide and sodium ethoxide serve also as the base, and DMF is satisfactory as solvent. Reagent generated by this method converts mono-functional aldehydes and ketones into epoxides in good yield. Franzen initially regarded the product of reaction with benzalacetophenone as a cyclopropane but this inference was later shown to be in error.[15]

Dimethyl sulfoxide-derived reagent (c). Dimethyloxosulfonium methylide, $(CH_3)_2\overset{O}{\overset{\|}{S}}{=}CH_2$. This more stable reagent introduced by Corey[16] reacts with monofunctional carbonyl compounds to give epoxides (oxiranes), but with α,β-unsaturated ketones it gives cyclopropyl ketones. In one procedure a mixture of 0.1 mole each of powdered sodium hydride and trimethyloxosulfonium iodide (1)[17,18] is placed in a three-

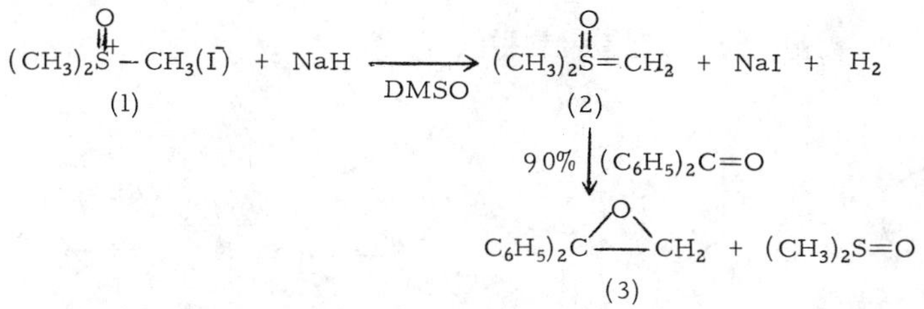

(CH3)2S+–CH3(I-) + NaH
DMSO
(CH3)2S=CH2 + NaI + H2
(1)
(2)
90%
(C6H5)2C=O
C6H5)2C—CH2 + (CH3)2S=O
(3)

necked, round-bottomed flask fitted with a mechanical stirrer, a serum stopper (rubber port), and a reflux condenser carrying at the top a three-way stopcock for making connection either to a dry nitrogen source or to the suction pump. The system is placed under nitrogen by repeated alternate evacuation and admission of nitrogen, the tube leading to the pump is disconnected and joined to a mercury bubbler, and nitrogen is passed through the system at a barely perceptible rate. The flask is surrounded by a water bath at 10–15°, the stirrer is started, and 100 ml. of dimethyl sulfoxide (distilled at reduced pressure from calcium hydride) is introduced at a moderate rate from a hypodermic syringe by injection through the rubber port. A vigorous evolution of hydrogen ensues. After 5 min. the water bath is removed and in another 15–30 min. the evolution of hydrogen stops, and formation of a milky white solution of dimethyloxosulfonium methylide (2) is complete. For reaction with benzophenone, a solid, a solution of 0.08–0.09 mole of the ketone in dimethyl sulfoxide and tetrahydrofurane is introduced fairly rapidly by injection (slight cooling may be necessary). After stirring for 10–15 min. at room temperature, the mixture is warmed to 50° for 1 hr., cooled, diluted with about 3 volumes of water, and the product is extracted with an appropriate solvent and purified by crystallization or distillation. The product from benzophenone is the epoxide (3). Epoxides are produced also from 4-phenylcyclohexanone (72%), cycloheptanone (71%), and benzaldehyde (56%).

In case a sizable batch of reagent is to be prepared and stored, the preferred starting material is trimethyloxosulfonium chloride, prepared by treating a concentrated solution of the corresponding iodide with gaseous chlorine until iodine no longer precipitates. The solution is decanted from the iodine and freed of dissolved iodine by extraction with ether; evaporation of the aqueous solution affords the crystalline chloride. A mixture of equimolar amounts of trimethyloxosulfonium chloride and sodium hydride in tetrahydrofurane is stirred rapidly under reflux until evolution of hydrogen ceases (2–3 hrs.), the mixture is cooled, and the salts removed by rapid suction filtration under nitrogen pressure through a Büchner funnel containing a matting of dried Celite filter-aid into a three-necked, round-bottomed flask. The receiver is flushed with nitrogen, and the openings closed with two glass stoppers and one rubber serum stopper. Samples are withdrawn by syringe, added to water, and titrated with standard acid to the phenolphthalein end point. The colorless solution of ylide can be stored at 0° for months without appreciable decomposition.

When an α,β-unsaturated ketone reacts with dimethyloxosulfonium methylide, the attack is at the carbon–carbon double bond and the product is a cyclopropane derivative, as illustrated for benzalacetophenone. Carvone and eucarvone behave

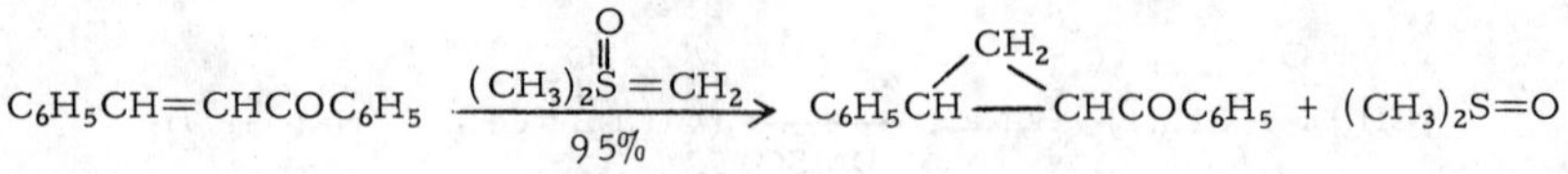

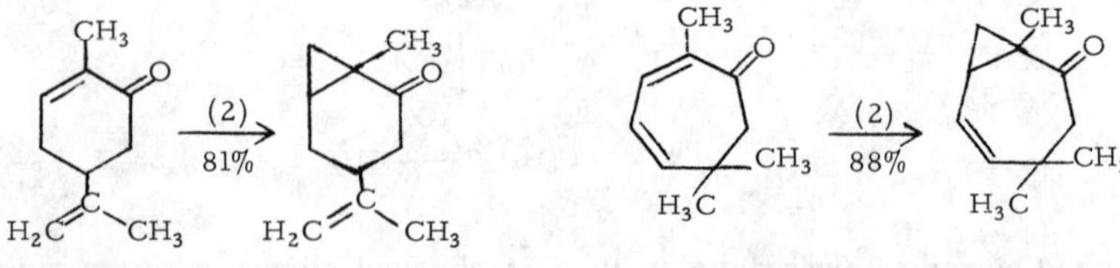

$$C_6H_5CH=CHCOC_6H_5 \xrightarrow[95\%]{(CH_3)_2\overset{O}{\overset{\|}{S}}=CH_2} C_6H_5\overset{CH_2}{\overset{\diagup}{CH}} -CHCOC_6H_5 + (CH_3)_2S=O$$

Carvone

Eucarvone

similarly to give cyclopropanes by reaction of a double bond conjugated with the carbonyl group. The reagent reacts with N-methyl-2-quinolone (1) to give (2).[19]

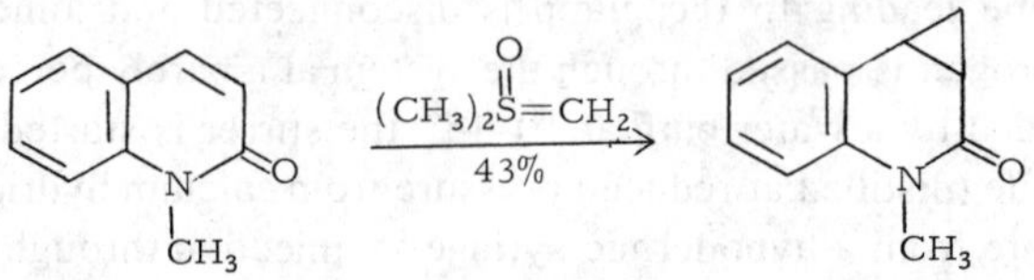

(1)   (2)

It should be noted that dimethyloxosulfonium methylide is more stable than dimethylsulfonium methylide to spontaneous thermal decomposition in inert media and that it is a less reactive nucleophile as shown for the test case of 1,1-diphenyl-ethylene, which reacts with $(CH_3)_2S=CH_2$ but not with $(CH_3)_2S(=O)=CH_2$.

Dimethyloxosulfonium methylide can be acylated by treatment of 2 equivalents of reagent with 1 equivalent of an acid chloride to give a stable β-ketooxosulfonium ylide such as (1, m.p. 120°). These are of interest because upon excitation with 253

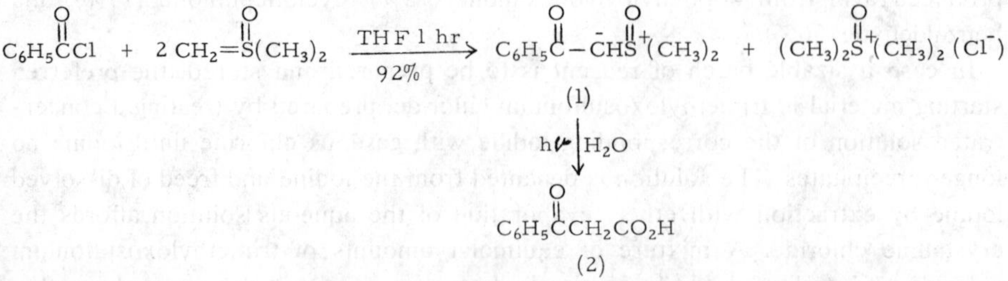

$$\underset{C_6H_5\overset{O}{\overset{\|}{C}}Cl}{} + 2\ \underset{CH_2=\overset{O}{\overset{\|}{S}}(CH_3)_2}{} \xrightarrow[92\%]{THF\ 1\ hr} \underset{C_6H_5\overset{O}{\overset{\|}{C}}-\overset{O}{\overset{\|}{C}}H\overset{+}{S}(CH_3)_2}{} + (CH_3)_2\overset{O}{\overset{\|}{\overset{+}{S}}}(CH_3)_2\ (Cl^-)$$

(1)

$$\downarrow h\nu\ |\ H_2O$$

$$C_6H_5\overset{O}{\overset{\|}{C}}CH_2CO_2H$$

(2)

$m\mu$ irradiation they suffer rupture of the dipolar S—C bond with smooth conversion into the carboxylic acid (2), presumably via ketocarbene and ketene intermediates. The two-step process is analogous to the Arndt-Eistert chain extension procedure. α,β-Unsaturated ethyl esters undergo conjugate addition followed by cyclization, as illustrated for ethyl Δ^1-cyclohexenylcarboxylate. The condensation product (3) on irradiation gives the sulfoxide ester (4), which on peroxide oxidation

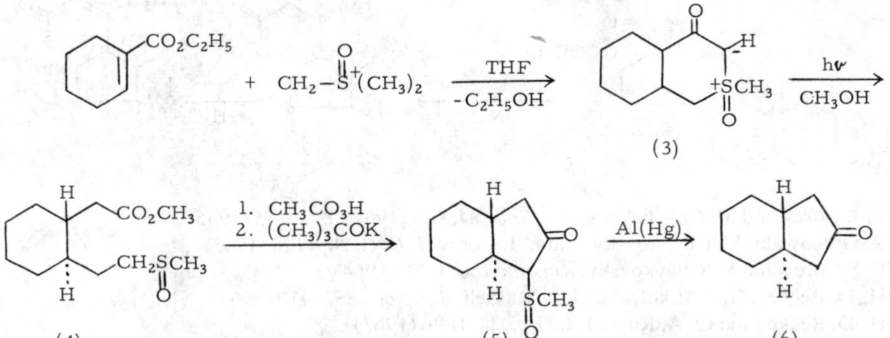

CO2C2H5
CH2−S(CH3)2
THF
-C2H5OH
+SCH3
hν
CH3OH
CO2CH3
CH2SCH3
1. CH3CO3H
2. (CH3)3COK
SCH3
Al(Hg)

(3) (4) (5) (6)

and cyclization with base gives (5); reduction by aluminum amalgam then affords *trans*-hydrindane-2-one (6). The sequence represents a carbocycle synthesis of a new type.[20]

The reagent reacts with an α,β-unsaturated acid amide to give a pyrrolidone or a cyclopropanecarboxylic amide, or both, depending upon the substituents present.[21]

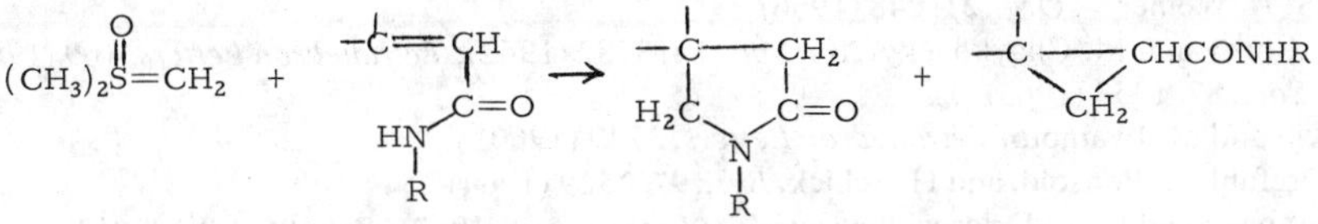

(CH3)2S=CH2
HN
R
CH2
H2C
C=O
N
CHCONHR
CH2

It methylates hydrazones, oximes, acids, and phenols in good yield; it methylates some aromatic hydrocarbons.[22] Examples:

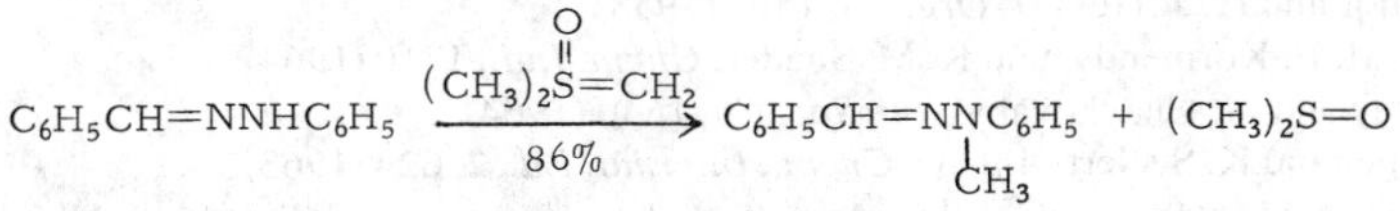

C6H5CH=NNHC6H5
(CH3)2S=CH2
86%
C6H5CH=NNC6H5 + (CH3)2S=O
CH3

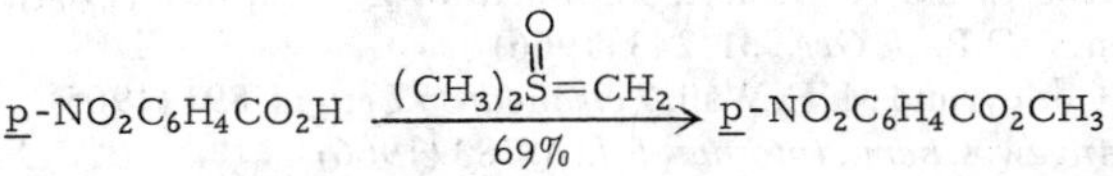

p-NO2C6H4CO2H
(CH3)2S=CH2
69%
p-NO2C6H4CO2CH3

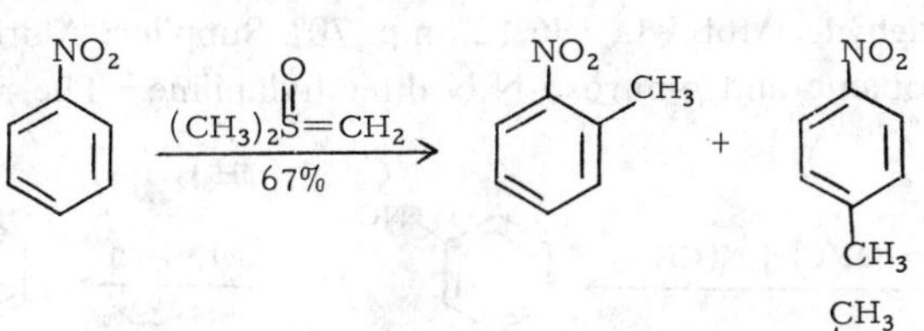

NO2
(CH3)2S=CH2
67%
CH3

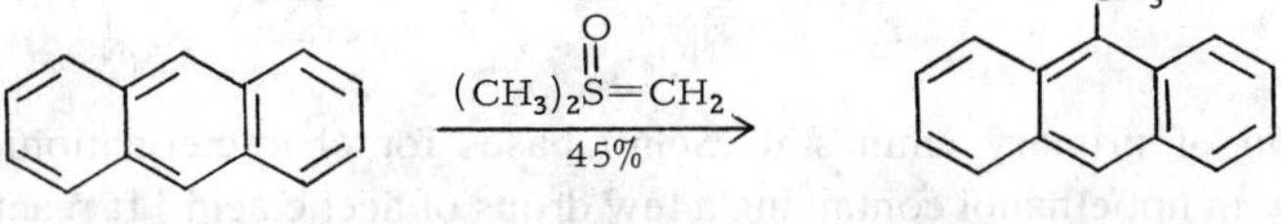

(CH3)2S=CH2
45%
CH3

Dimethylsulfonium methylide and dimethylsulfoxonium methylide react with dihydrotestosterone stereospecifically and in opposite directions.[23]

Dimethylsulfoxonium methylide reacts with tertiary α-acetoxyketones by cycloaddition and dehydration to afford butenolides.[24]

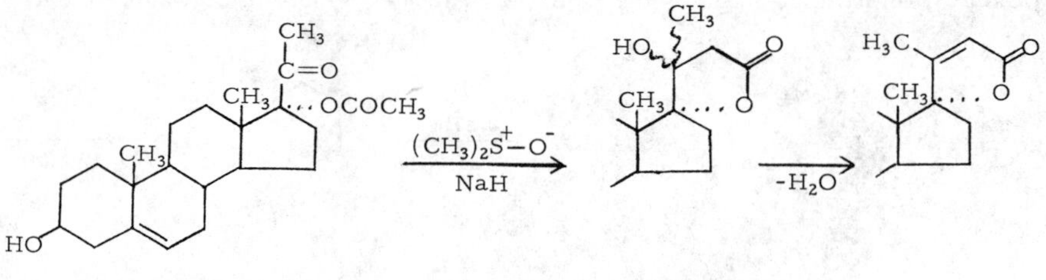

[1]E. J. Corey and M. Chaykovsky, *Am. Soc.*, **84**, 866 (1962); **87**, 1345 (1965)
[2]R. Greenwald, M. Chaykovsky, and E. J. Corey, *J. Org.*, **28**, 1128 (1963)
[3]E. J. Corey and M. Chaykovsky, *Am. Soc.*, **86**, 1639 (1964)
[4]H.-D. Becker, G. J. Mikol, and G. A. Russell, *Am. Soc.*, **85**, 3410 (1963)
[5]H.-D. Becker and G. A. Russell, *J. Org.*, **28**, 1896 (1963)
[6]G. A. Russell and H.-D. Becker, *Am. Soc.*, **85**, 3406 (1963)
[7]A. I. Gurevich, M. M. Shemyakin, *et al.*, *Tetrahedron Letters*, 877 (1964)
[8]E. J. Corey, R. B. Mitra, and H. Uda, *Am. Soc.*, **86**, 485 (1964)
[9]C. G. Cardenas, A. N. Khafaji, C. L. Osborn, and P. D. Gardner, *Chem. Ind.*, 345 (1965); C. L. Osborn, T. C. Shields, B. A. Shoulders, C. G. Cardenas, and P. D. Gardner, *ibid.*, 766 (1965)
[10]W. Roberts and M. C. Whiting, *J. Chem. Soc.*, 1290 (1965)
[11]P. A. Argabright, J. E. Hofmann, and A. Schriesheim, *J. Org.*, **30**, 3233 (1965); G. A. Russell and S. A. Weiner, *J. Org.*, **31**, 248 (1966)
[12]E. J. Corey and M. Chaykovsky, *Am. Soc.*, **84**, 3782 (1962); *Tetrahedron Letters*, 169 (1963); *Am. Soc.*, **87**, 1353 (1965)
[12a]J. Kiji and M. Iwamoto, *Tetrahedron Letters*, 2749 (1966)
[13]G. Drefahl, K. Ponsold, and H. Schick, *Ber.*, **97**, 3529 (1964)
[14]V. Franzen and H.-E. Driesen, *Tetrahedron Letters*, 661 (1962); *Ber.*, **96**, 1881 (1963)
[15]E. J. Corey and M. Chaykovsky, *Tetrahedron Letters*, 169 (1963)
[16]E. J. Corey and M. Chaykovsky, *Am. Soc.*, **84**, 867 (1962); **87**, 1353 (1965)
[17]R. Kuhn and H. Trischmann, *Ann.*, **611**, 117 (1958)
[18]R. T. Major and H.-J. Hess, *J. Org.*, **23**, 1563 (1958)
[19]B. Loev, M. F. Kormendy, and K. M. Snader, *Chem. Ind.*, 1710 (1964)
[20]E. J. Corey and M. Chaykovsky, *Am. Soc.*, **86**, 1640 (1964)
[21]H. Metzger and K. Seelert, *Angew. Chem., Internat. Ed.*, **2**, 624 (1963)
[22]H. Metzger, H. König, and K. Seelert, *Tetrahedron Letters*, 867 (1964); V. J. Traynelis and J. V. McSweeney, O.P., *J. Org.*, **31**, 243 (1966)
[23]C. E. Cook, R. C. Corley, and M. E. Wall, *Tetrahedron Letters*, 891 (1965)
[24]H.-G. Lehmann, *Angew. Chem., Internat. Ed.*, **4**, 783 (1966)

2,4-Dinitrobenzaldehyde. Mol. wt., 196.12, m.p. 70°. Supplier: Aldrich. Preparation from 2,4-dinitrotoluene and *p*-nitroso-N,N-dimethylaniline.[1] The reagent is useful

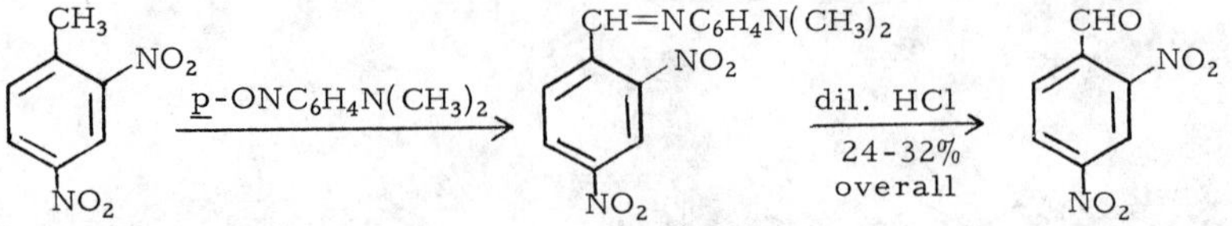

for conversion of primary amines to Schiff bases for characterization; reaction occurs readily in hot ethanol containing a few drops of acetic acid.[2] It reacts in boiling acetic anhydride with α-picoline, 2,6-dimethylquinoline, and other compounds with an activated methyl group to give 2,4-dinitrostyryl derivatives.[3]

[1]G. M. Bennett and E. V. Bell, *Org. Syn., Coll. Vol.*, **2**, 223 (1943)
[2]F. Wild, "Characterization of Organic Compounds," p. 229 Cambridge (1947)
[3]G. M. Bennett and W. L. C. Pratt, *J. Chem. Soc.*, 1465 (1929)

2,4-Dinitrobenzenesulfenyl chloride. Mol. wt. 234.62, m.p. 96°. Preparation from 2,4-dinitrochlorobenzene.[1] Suppliers: A, B, E, F, MCB. Review of uses.[2]

The reagent enters into a number of ionic reactions involving attack on sulfur by a nucleophile with displacement of chlorine, for example, in the reaction with aniline.[3]

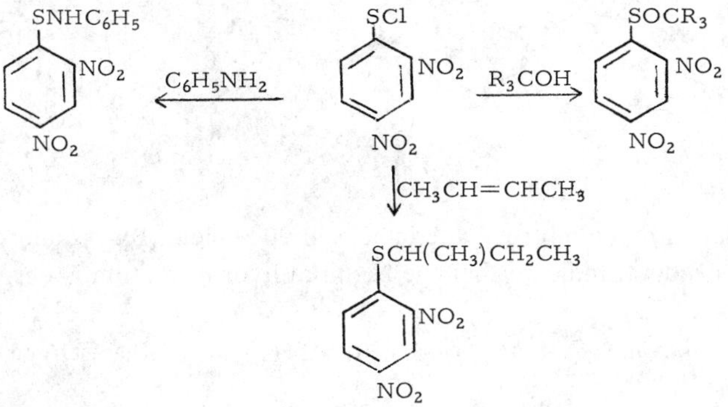
SNHC6H5
NO2
NO2
C6H5NH2
SCl
NO2
NO2
R3COH
SOCR3
NO2
NO2
CH3CH=CHCH3
SCH(CH3)CH2CH3
NO2
NO2

Alcohols, including tertiary alcohols, react readily with the reagent.[2] The reagent adds stereospecifically (*trans*) to unhindered olefins, for example, to the 2-butenes; tetrasubstituted olefins do not react.[4] The reagent can be used for Friedel-Crafts substitution of aromatic compounds,[5] including azulene.[6] The 2,4-dinitrobenzenesulfenyl derivatives formed are colored solids readily purified by crystallization or chromatography and hence are useful for purposes of identification.

Protective group. The 2,4-dinitrobenzenesulfenyl group is potentially useful for blocking hydroxyl groups in nucleosides.[7] It is stable to phosphorylating agents, to dicyclohexylcarbodiimide, and to acids and bases, but can be selectively removed

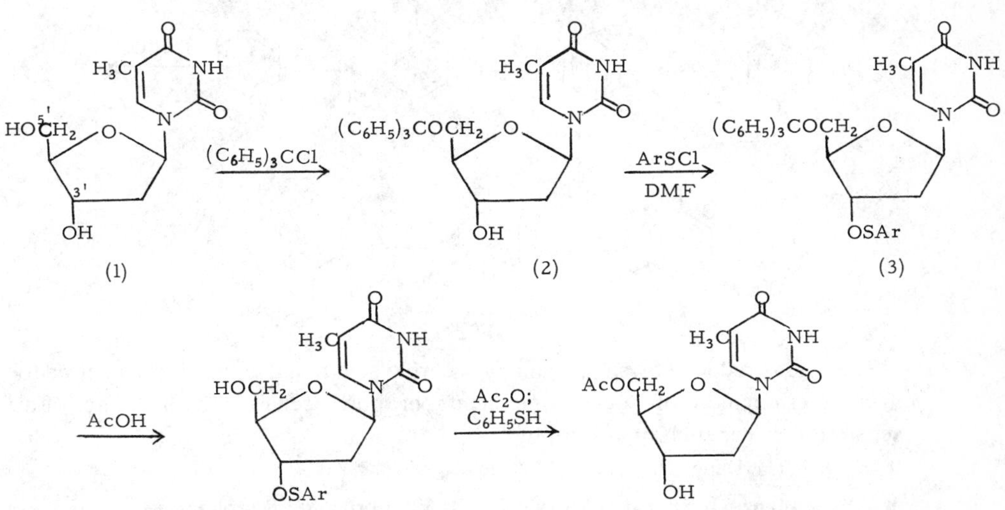
H3C
NH
HOCH2
5'
3'
OH
(C6H5)3CCl
(C6H5)3COCH2
OH
ArSCl
DMF
OSAr
AcOH
HOCH2
OSAr
Ac2O;
C6H5SH
AcOCH2
OH

(1) (2) (3) (4) (5)

by reaction with thiophenol or thiosulfate. Thus thymidine (1) can be transformed into the 5′-acetyl derivative (5) by tritylation of the primary 5′-hydroxyl group (2), reaction of (2) in dimethylformamide with 2,4-dinitrobenzenesulfenyl chloride to give (3), cleavage of the trityl group (4), acetylation, and removal of the blocking group.

Barton *et al.*[8] found that a 2,4-dinitrobenzenesulfenyl ester such as (1) can be cleaved by photolysis under mild, neutral conditions. Thus irradiation of (1) in

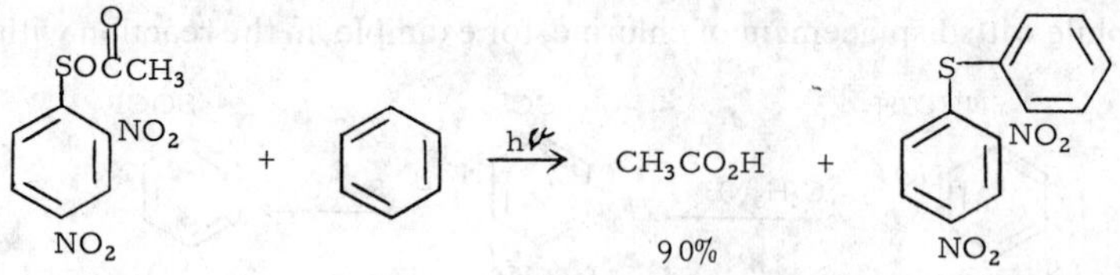
SOCCH3
NO2
NO2
+
hν
CH3CO2H
+
S
NO2
NO2
90%

(1) (2) 73%

benzene under nitrogen afforded acetic acid in 90% yield. By-products such as (2), the chief by-product, indicate that the required hydrogen atom is derived from the solvent.

[1]N. Kharasch *et al.*, *Am. Soc.*, **69**, 1612 (1947); **71**, 2724 (1949); N. Kharasch and R. B. Lanford, *Org. Syn.*, **44**, 47 (1964)
[2]N. Kharasch, S. J. Potempa, and H. L. Wehrmeister, *Chem. Rev.*, **39**, 269 (1946)
[3]J. H. Billman, J. Garrison, R. Anderson, and B. Wolnak, *Am. Soc.*, **63**, 1920 (1941)
[4]N. Kharasch *et al.*, *Am. Soc.*, **71**, 2724 (1949); D. J. Cram, *ibid.*, **71**, 3887 (1949)
[5]N. Kharasch *et al.*, *Am. Soc.*, **72**, 3529 (1950); **75**, 1081, 6035 (1953)
[6]A. G. Anderson, Jr., and R. N. McDonald, *Am. Soc.*, **81**, 5669 (1959)
[7]R. L. Letsinger, F. Fontaine, V. Mahadevan, D. A. Schexnayder, and R. E. Leone, *J. Org.*, **29**, 2615 (1964)
[8]D. H. R. Barton, Y. L. Chow, A. Cox, and G. W. Kirby, *J. Chem. Soc.*, 3571 (1965)

3,4-Dinitrobenzoic acid. Mol. wt. 212.12, m.p. 166°. Suppliers: A, E, F, MCB. Preparation from *p*-toluidine.[1]

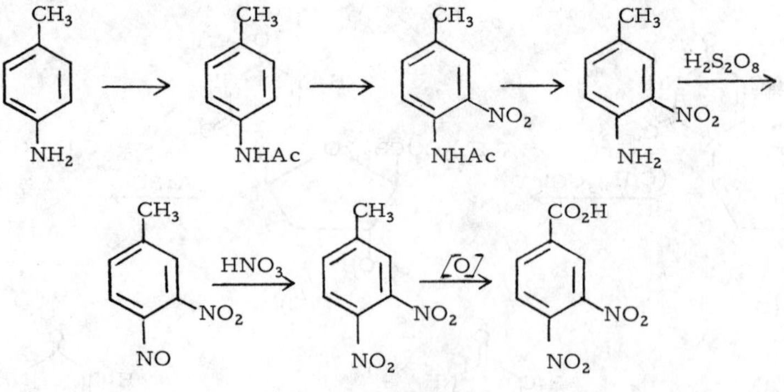
CH3
NH2
CH3
NHAc
CH3
NO2
NHAc
CH3
NO2
NH2
H2S2O8
CH3
NO2
NO
HNO3
CH3
NO2
NO2
[O]
CO2H
NO2
NO2

The acid is reduced by sugars and by ascorbic acid to a dye, and it can be used for microdetermination of sugar components of polysaccharide hydrolyzates after separation by paper chromatography.[1]

[1]E. Borel and H. Deuel, *Helv.*, **36**, 801 (1953)

3,5-Dinitrobenzoyl chloride. Mol. wt. 230.57, m.p. 68°. Suppliers: B, E, F. The corresponding acid is prepared by nitration of benzoic acid;[1] suppliers: Eastman, Light. A process of purification of the acid for use in the determination of creatinine is reported.[2]

The acid chloride is useful for the isolation of low-melting alcohols; reaction in pyridine gives an ester of lower solubility and higher melting point. For example, vitamin D (m.p. 82–84°) from fish liver oil concentrates was isolated by esterification with 3,5-dinitrobenzoyl chloride, and purification by chromatography and crystallization as the 3,5-dinitrobenzoate, m.p. 132°.[3]

[1]R. Q. Brewster, B. Williams, and R. Phillips, *Org. Syn., Coll. Vol.*, **3**, 337 (1955)
[2]A. P. Jansen, W. Sombroek, and E. C. Noyons, *Chem. Weekblad*, **43**, 731 (1947) [*C.A.*, **42**, 2307 (1948)]
[3]H. Brockmann, *Z. physiol.*, **245**, 96 (1937); H. Brockmann and A. Busse, *ibid.*, **249**, 176 (1937)

Di-*p*-nitrobenzylphosphoryl chloridate, $(p\text{-}NO_2C_6H_4CH_2O)_2POCl$. Mol. wt. 386.70, m.p. 107°.

Prepared from *p*-nitrobenzyl phosphate and phosphorus pentachloride, this stable, crystalline reagent is useful for phosphorylation of hydroxy and amino compounds; the substituted benzyl groups are removed by catalytic hydrogenation.[1]

[1]L. Zervas and I. Dilaris, *Am. Soc.*, **77**, 5354 (1955)

2,4-Dinitrofluorobenzene (DNF), 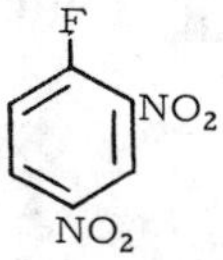
F
NO2
NO2

Mol. wt. 186.10, b.p. 137°/2 mm. Suppliers: Eastern Chem. Corp.; B, E, MCB. Preparation.[1,2]

Peptide analysis. The reagent was introduced by Sanger[3] for identification of the amino-terminal group of a protein or peptide. Condensation occurs under mild conditions to form a 2,4-dinitrophenyl protein; on acid hydrolysis the terminal

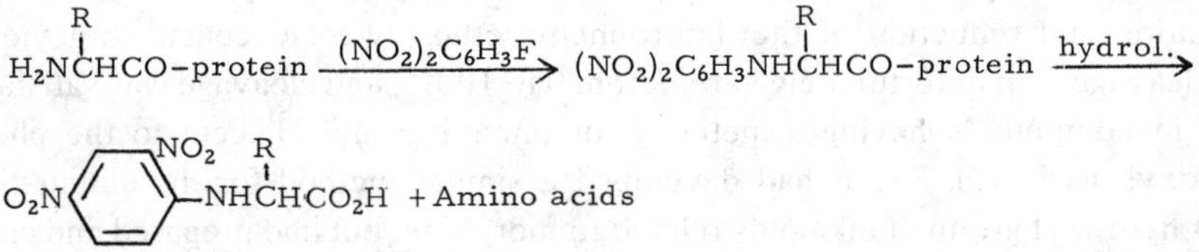
R
H2NCHCO−protein
(NO2)2C6H3F
R
(NO2)2C6H3NHCHCO−protein
hydrol.
NO2 R
O2N
NHCHCO2H
+ Amino acids

amino acid is liberated as the bright yellow 2,4-dinitrophenyl derivative, which is easily separated from the accompanying amino acid mixture and which can be identified by paper chromatography.

Characterization of hydroxy compounds. The reagent reacts with phenols in acetone solution under catalysis by triethylamine to form 2,4-dinitrophenyl ethers useful for identification.[4] Under similar conditions it affords solid derivatives of alcohols.[5]

Degradation of aldoses. It has been used in a Wohl-type degradation in which an aldose oxime is treated with 2,4-dinitrofluorobenzene to form the oxime aryl ether, which decomposes to the next lower aldose, 2,4-dinitrophenol, and hydrogen cyanide.[6]

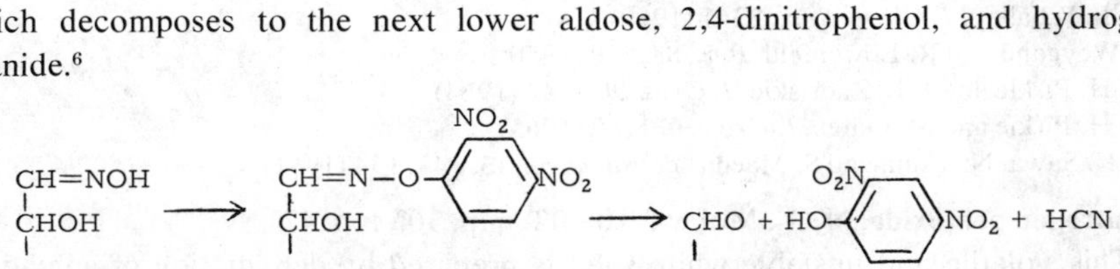
CH=NOH
CHOH
NO2
CH=N−O
NO2
CHOH
O2N
CHO + HO
NO2 + HCN

ArOH ⟶ ArH. Pirkle and Zabriskie[7] reported a method for the replacement of a phenolic hydroxyl group by hydrogen which was employed by Pirkle and Gates[8] in an investigation of potential analgesic agents, for example, for conversion of *cis*-tetrahydrodesoxycodeine (1) into (2). Arylation of (1) with 2,4-dinitrofluorobenzene

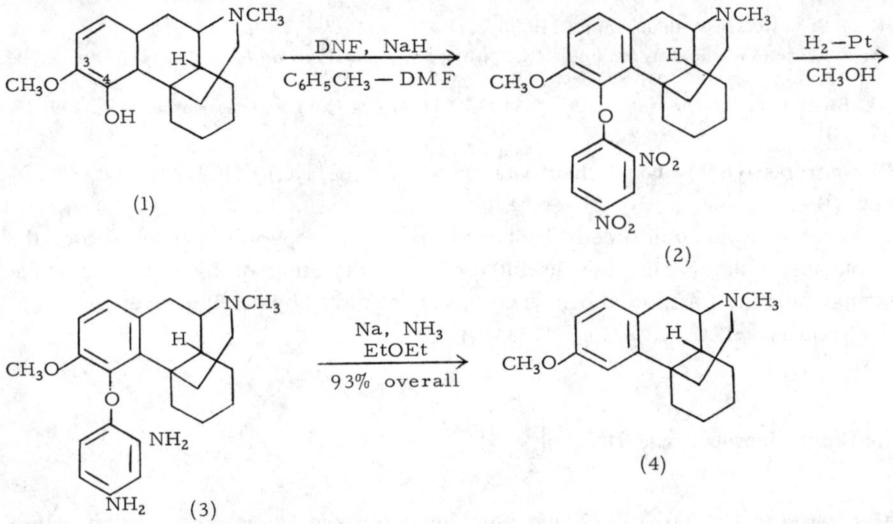
NCH3
DNF, NaH
C6H5CH3—DMF
H2—Pt
CH3OH
CH3O
OH
NO2
Na, NH3
EtOEt
93% overall
NH2

(1) (2) (3) (4)

in toluene–dimethylformamide with sodium hydride as catalyst gave the 2′,4′-dinitrophenyl ether (2) in nearly quantitative yield. This was hydrogenated to the diamino derivative (3), which was not isolated because of facile autoxidation but cleaved with sodium and liquid ammonia in ether to give the desoxy derivative (4) in high overall yield. In the case of model compounds explored[7] yields in the formation and reduction of the dinitrophenyl ethers were excellent, but yields in the cleavage step unfortunately varied from 0 to 100%, and cleavage was satisfactory only in compounds having a methoxyl or phenyl group adjacent to the phenolic hydroxyl, as in (1). Sawa[9] had developed a similar method for the elimination of the 4-hydroxyl group of alkaloids related to morphine, but had prepared and cleaved the 4-phenyl ether. However, preparation of phenyl ethers by the Ullmann reaction is considerably more cumbersome than the smoothly proceeding arylation with DNF, the dinitrophenyl ethers crystallize better, and the diaminophenyl ethers dissolve more easily in the cleavage solvent. Furthermore, the amino groups favor cleavage of the ether in the desired direction.

[1]H. G. Cook and B. C. Saunders, *Biochem. J.*, **41**, 558 (1947)
[2]G. C. Finger and J. L. Finnerty, *Biochem. Preps.*, **3**, 120 (1958)
[3]F. Sanger, *Biochem. J.*, **39**, 507 (1945); **40**, 261 (1946); **45**, 563 (1949); R. R. Porter and F. Sanger, *ibid.*, **42**, 287 (1948)
[4]J. D. Reinheimer, J. P. Douglass, H. Leister, and M. B. Voelkel, *J. Org.*, **22**, 1743 (1957)
[5]W. B. Whalley, *J. Chem. Soc.*, 2241 (1950)
[6]F. Weygand and R. Löwenfeld, *Ber.*, **83**, 559 (1950)
[7]W. H. Pirkle and J. L. Zabriskie, *J. Org.*, **29**, 3124 (1964)
[8]W. H. Pirkle and M. Gates, *J. Org.*, **30**, 1769 (1965)
[9]Y. K. Sawa, N. Tsuji, and S. Maeda, *Tetrahedron*, **15**, 144, 154 (1961)

Dinitrogen pentoxide, N_2O_5. Mol. wt. 108.02, m.p. 30°, b.p. 47°.

This volatile and unstable white solid is prepared by dehydration of anhydrous nitric acid with phosphoric oxide followed by sublimation in a stream of ozonized oxygen; it is stored at $-78°$.[1] It has been prepared also from nitric acid and trifluoroacetic anhydride.[2]

Dinitrogen pentoxide reacts with propylene in methylene chloride at $-25°$ to give 1-nitro-2-propyl nitrate and a mixture of nitroolefins in which 3-nitropropene

$$CH_3CH{=}CH_2 \xrightarrow{N_2O_5} \underset{27\%}{CH_3\overset{ONO_2}{\overset{|}{C}}HCH_2NO_2} + \underset{16\%}{O_2NCH_2CH{=}CH_2}$$

predominates. *trans*-Stilbene reacted to give a mixture of nitratonitro compounds in high yield, but with *cis*-stilbene the yield was very low.[3] In each case the predominant isomer formed was the product of *cis*-addition, probably through a cyclic transition state.[1]

The reagent converts aliphatic secondary amines into nitramines in excellent yield;[4] a solution in carbon tetrachloride is added to a carbon tetrachloride solution of excess amine at $-25°$.

$$[(CH_3)_2CH]_2NH + N_2O_5 \xrightarrow[91\%]{} [(CH_3)_2CH]_2NNO_2 + [(CH_3)_2CH]_2\overset{+}{N}H_2NO_3^-$$

Nitration of diphenyl-2-carboxylic acid with concentrated nitric acid at room temperature gives a 79:21 mixture of 2′-nitrodiphenyl-2-carboxylic acid (2) and 4′-nitrodiphenyl-2-carboxylic acid (3). However, nitration with dinitrogen pentoxide in carbon tetrachloride at 0° (30 min.) gave (2) in good yield as the sole product.[5]

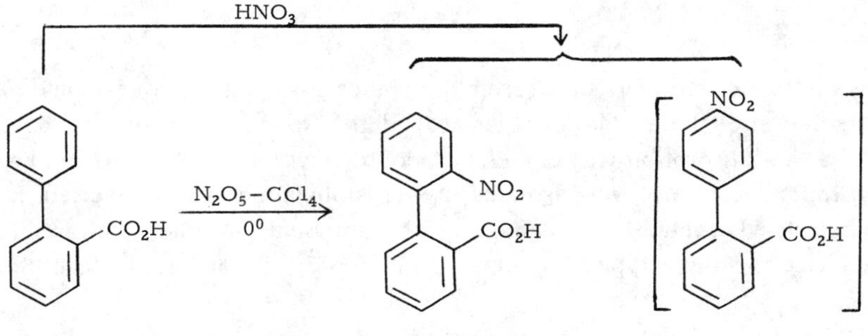
HNO3
N2O5−CCl4
0°
CO2H
NO2
CO2H
NO2
CO2H

(1) (2) (3)

Caesar and Goldfrank[6] describe an apparatus for the generation of 40–60 g. of the reagent per day for use in the synthesis of nitrate esters:

$$ROH + N_2O_5 \longrightarrow RONO_2 + HONO_2$$

Complete nitration of a starch is carried out with a solution of dinitrogen pentoxide in chloroform, and the reaction is promoted by addition of sodium fluoride to precipitate the nitric acid formed as a complex $NaF{\cdot}HONO_2$.

Wolfrom and Rosenthal[7] used this method to prepare D-gluconamide pentanitrate and D-galactonamide pentanitrate.

[1] T. E. Stevens and W. D. Emmons, *Am. Soc.*, **79**, 6008 (1957)
[2] J. H. Robson, *Am. Soc.*, **77**, 107 (1955)
[3] T. E. Stevens, *J. Org.*, **24**, 1136 (1959)
[4] W. D. Emmons, A. S. Pagano, and T. E. Stevens, *J. Org.*, **23**, 311 (1958)
[5] D. H. Hey, J. A. Leonard, and C. W. Rees, *J. Chem. Soc.*, 4579 (1962)

[6]G. V. Caesar and M. Goldfrank, *Am. Soc.*, **68**, 372 (1946); see also N. S. Gruenhut, M. Goldfrank, M. L. Cushing, and G. V. Caesar, *Inorg. Syn.*, **3**, 78 (1950)
[7]M. L. Wolfrom and A. Rosenthal, *Am. Soc.*, **75**, 3662 (1953)

Dinitrogen tetroxide, N_2O_4. Mol. wt. 92.02, m.p. −9.3°, b.p. 21.3°. Supplied in steel cylinders by Matheson ("nitrogen tetroxide"). Purified by oxidation at 0° in a stream of oxygen until the blue color changes to red-brown; after distillation from P_2O_5, the oxide solidifies in the deep freeze to nearly colorless crystals.[1]

Nitration of olefins.[2] The reaction of dinitrogen tetroxide with olefins in ether or ester solvents proceeds particularly well in the presence of added oxygen (to prevent interference by N_2O_3) to form *vic*-dinitroalkanes, nitro-nitrites, or nitro-nitrates.[3] Nitration of cholesteryl acetate was conducted by passing a mixture of

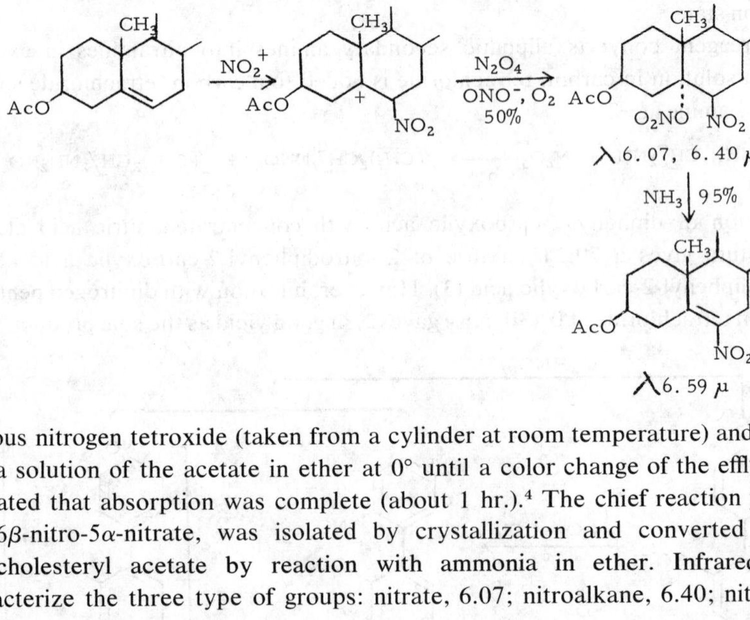
CH3
AcO
NO2+
CH3
AcO
+
NO2
N2O4
ONO⁻, O2
50%
CH3
AcO
O2NO
NO2
λ6.07, 6.40 μ
NH3
95%
CH3
AcO
NO2
λ6.59 μ

gaseous nitrogen tetroxide (taken from a cylinder at room temperature) and oxygen into a solution of the acetate in ether at 0° until a color change of the effluent gas indicated that absorption was complete (about 1 hr.).[4] The chief reaction product, the 6β-nitro-5α-nitrate, was isolated by crystallization and converted into 6-nitrocholesteryl acetate by reaction with ammonia in ether. Infrared bands characterize the three type of groups: nitrate, 6.07; nitroalkane, 6.40; nitroolefin, 6.59 μ.

In a procedure developed by Seifert[5] an olefin, for example *cis*-cyclooctene, is nitrated as usual in ether in the presence of oxygen, and the mixture of adducts,

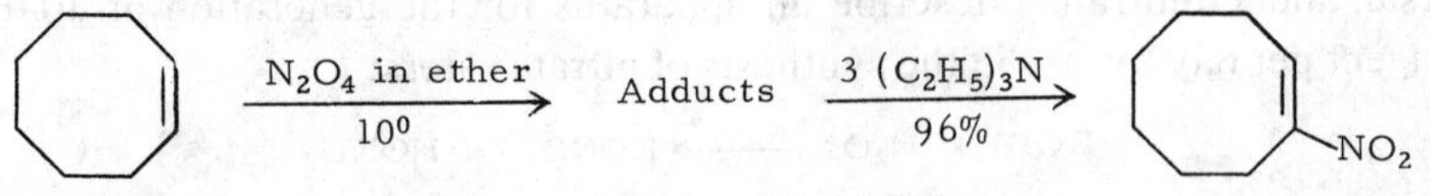
N2O4 in ether
10°
Adducts
3 (C2H5)3N
96%
NO2

without isolation, is treated with 3 moles of triethylamine for ½ hr. Crude 1-nitro-1-cyclooctene is obtained in high yield. 1-Octadecene by the same procedure affords 1-nitro-1-octadecene in 80% yield. Triethylamine is preferred to ammonia, the usual base, because it is a stronger base.

Nitration of acetylenes. The reaction generally yields an unstable, complex mixture of products. Thus Freeman and Emmons[6] identified five products resulting from the nitration of 3-hexyne. Terminal acetylenes give particularly unstable reaction mixtures. Campbell *et al.*[7] isolated three products from the reaction of diphenyl-acetylene with dinitrogen tetroxide; the relative amounts vary with the experimental

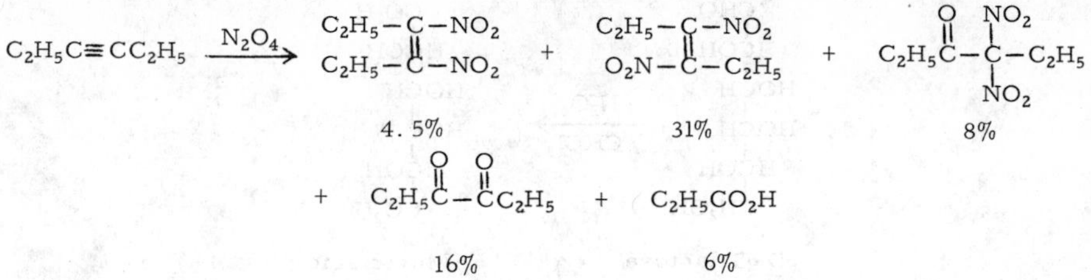

conditions. Usually the main products are *cis*- and *trans*-1,2-dinitrostilbene; a minor product is the highly colored 5-nitro-2-phenylisatogen.

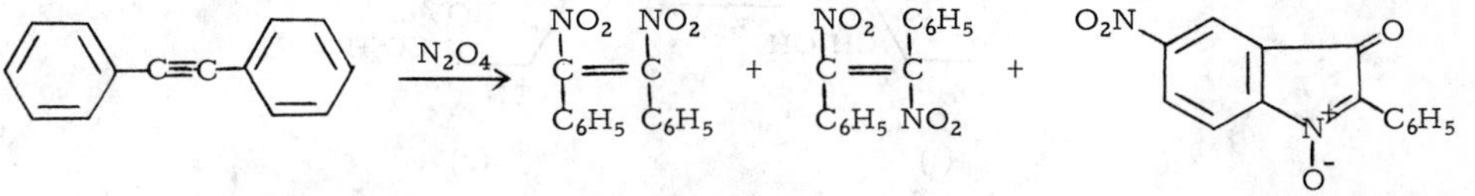

Ponzio reaction. In the Ponzio reaction, dinitrogen tetroxide in ether converts benzaldoxime into phenyldinitromethane.[8] In an improved procedure[9] a fume-off is

$$C_6H_5CH{=}NOH \xrightarrow[38\%]{N_2O_4} C_6H_5CH(NO_2)_2$$

prevented by controlled addition of the oxime in ether to an ethereal solution of dinitrogen tetroxide under reflux.

Oxidation of hydroxyl functions. A mixture rich in dinitrogen tetroxide made by slow distillation of a mixture of nitric acid, sulfuric acid, and arsenious oxide can be kept in a glass-stoppered bottle at 0° and handled by pipet as a liquid.[10] The reagent

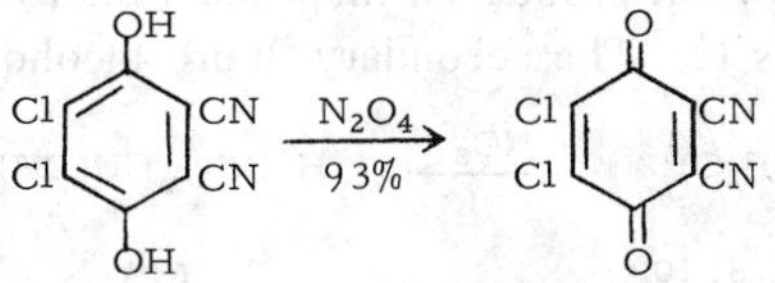

is recommended for oxidation of substituted hydroquinones to quinones, particularly those of high potential. In the example formulated, a suspension of the finely ground hydroquinone (20 g.) in CCl_4 (300 ml.) was treated with 6 ml. of reagent, introduced in 5 min. After stirring for 5 min. longer the quinone was filtered off and crystallized from chloroform–benzene. The method is not suitable for the oxidation of hydroquinone itself.

According to brief literature reports, dinitrogen tetroxide can be used to oxidize primary alcohols either to aldehydes or to acids. Field and Grundy[11] describe oxidation of twelve benzyl alcohols to the aldehydes in yields of 91–98%. A solution of the alcohol (0.1 mole) in chloroform or carbon tetrachloride was treated at 0° with an ice-cold solution of dinitrogen tetroxide in $CHCl_3$ or CCl_4, and after 15 min. at 0° the solution was let stand overnight and then worked up. A stock solution of nitrogen tetroxide in $CHCl_3$ or CCl_4 was prepared either from bulk reagent, or gas generated by thermal decomposition of lead nitrate was passed into the solvent at −15°.

On the other hand, Maurer and Drefahl[12] report use of nitrogen tetroxide for oxidation of D-galactose to mucic acid, and the method was used in two instances

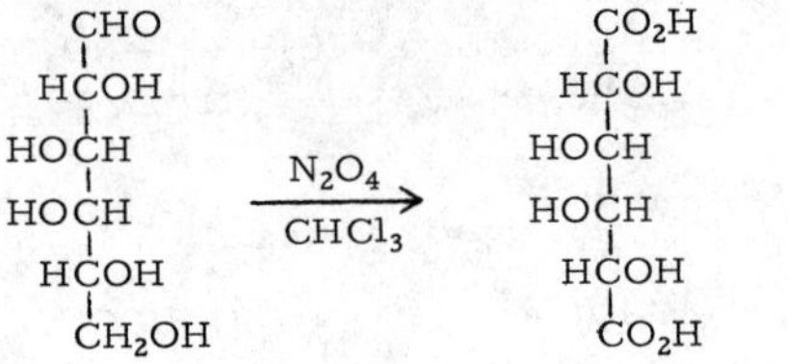

D-Galactose   Mucic acid

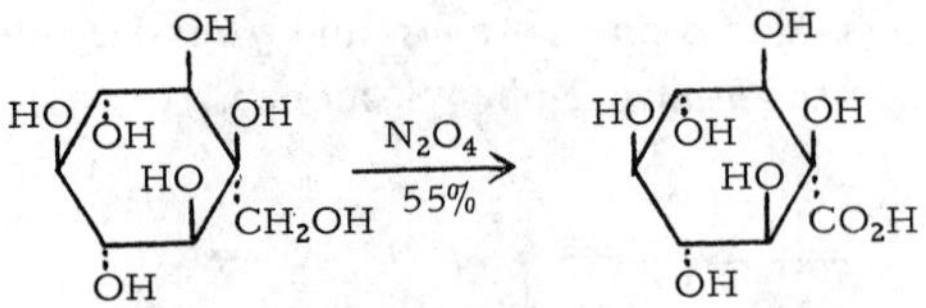

(1)   (2)

by Posternak and Reymond[13] for the preparation of inositol carboxylic acids. In the example formulated, 10 ml. of dinitrogen tetroxide was added with ice cooling to 468 mg. of powdered (1) and the mixture was kept at 0° for 6 days.

Langenbeck and Richter[14] examined the oxidation of a number of primary alcohols and diols and found that, except in the case of benzylic alcohols, the products are mono- or dicarboxylic acids; kinetic studies showed that aldehydes are not intermediates.

Scribner[15] found primary fluoro alcohols of types (1) to be unaffected by dinitrogen tetroxide but that they can be oxidized in good yield by nitrogen dioxide (NO_2), generated by passing nitric oxide and air through a Pyrex tube packed with quartz chips and heated at 400°. The products from primary fluoro alcohols (1) are aldehyde hydrates or aldehydrols (2). The secondary fluoro alcohol (3) afforded the fluoro

$$H(CF_2)_nCH_2OH \xrightarrow{NO_2} H(CF_2)_nCH(OH)_2$$

(1) n = 6, 8, 10   **(2)**

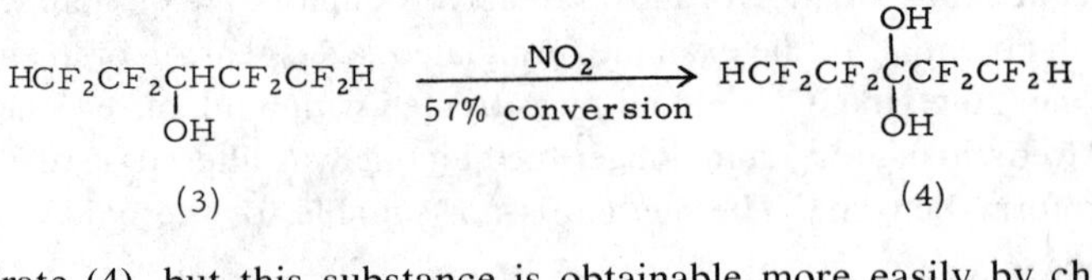

(3)   (4)

ketone hydrate (4), but this substance is obtainable more easily by chromic acid oxidation.

Oxidation of amides. In investigating the reaction of amides with dinitrogen tetroxide, White[16] prepared stock solutions by absorbing gaseous oxide in carbon tetrachloride or acetic acid and determining the titer iodimetrically (under N_2). Reaction with dinitrogen tetroxide proved to be a general method for the conversion of the N-acyl derivatives of primary alkylamines into N-alkyl-N-nitrosoamides, for example (1) → (2). The reaction is carried out in the presence of excess sodium

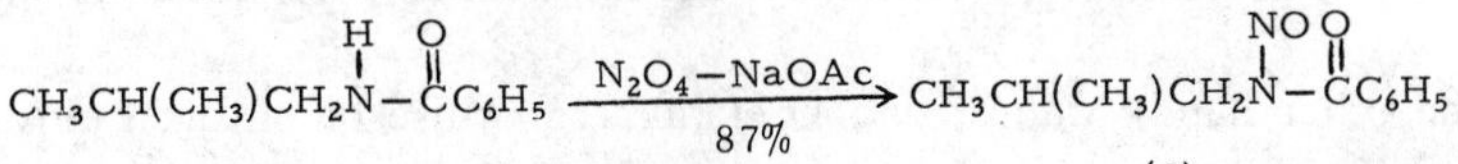

(1)   (2)

acetate to react with the nitric acid liberated, since otherwise the reverse reaction of denitrosation occurs and the equilibrium yield of (2) is only 60%. White found

$$RN(H)-C(=O)R' + N_2O_4 \rightleftharpoons RN(NO)-C(=O)R' + HNO_3$$

$$HNO_3 + NaOAc \longrightarrow NaNO_3 + HOAc$$

further that N-alkyl-N-nitrosoamides decompose smoothly at temperatures in the range 20–80° with elimination of nitrogen and formation of esters. Combination of

$$RN(NO)-C(=O)R' \xrightarrow{\Delta} ROC(=O)R' + N_2$$

the two reactions provides a neat method for the conversion of an aliphatic amine into the corresponding alcohol. The steps involved are acylation of the amine, nitrosation of the amide, thermal elimination of nitrogen to form the corresponding ester, and hydrolysis of the ester.[18] The White procedure also provides a means of converting a lactam, such as that formulated,[19] into the corresponding lactone. In

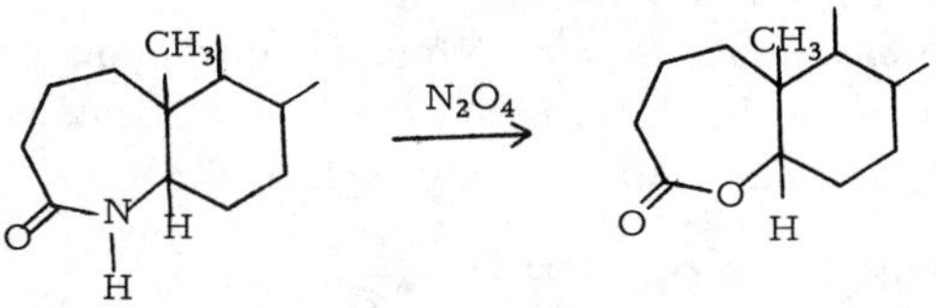
CH3
N2O4
N
H
H
O
CH3
O
H

the example cited and in a related case,[20] heat treatment to effect rearrangement is unnecessary.

The White deamination procedure has been used in the transformation of the steroidal alkaloids tomatidine (1) and solasodine into dihydroneotigogenin (4) and dihydrotigogenin, respectively.[21] Thus (1) was converted into the N,22-dihydride

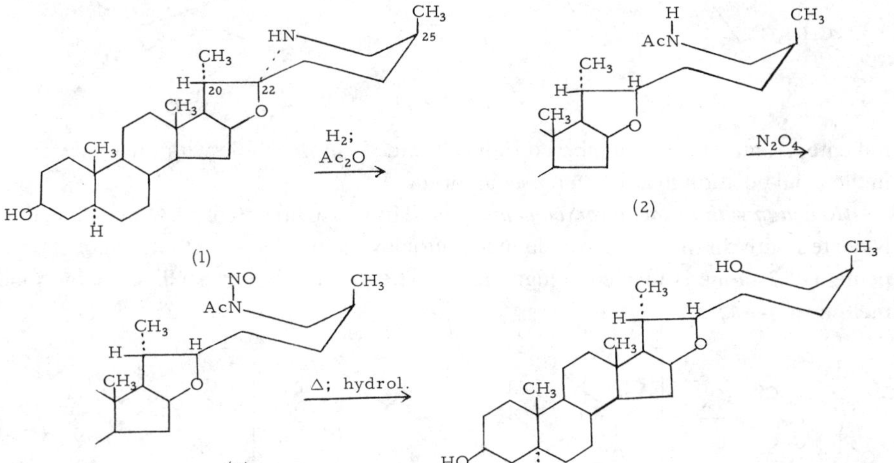
HN
CH3
25
CH3
H
20
22
CH3
O
CH3
HO
H
(1)
H2;
Ac2O
H
AcN
CH3
CH3
H
H
CH3
O
(2)
N2O4
NO
AcN
CH3
CH3
H
H
CH3
O
(3)
Δ; hydrol.
HO
CH3
CH3
H
H
CH3
O
CH3
HO
H
(4)

and its O,N-diacetate (2), and this with dinitrogen tetroxide gave the nitrosoamide (3). Thermal decomposition and hydrolysis then gave dihydroneotigogenin (4). The degradation has been used also for the synthesis of epitestosterone from pregnenolone acetate.[22]

A related reaction is the conversion of a carboxylic acid amide into the corresponding acid. Treatment of an acetylaldonamide with nitrogen tetroxide in acetic acid at 0° afforded the acetylaldonic acid in good yield.[23] Another example is as follows:[24]

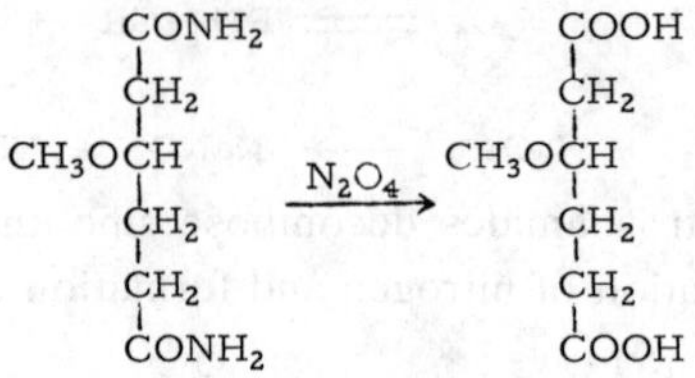
CONH2
CH2
CH3OCH
CH2
CH2
CONH2
N2O4
COOH
CH2
CH3OCH
CH2
CH2
COOH

Oxidation of phosphorus and of sulfur compounds. Dinitrogen tetroxide oxidizes trisubstituted phosphates in methylene chloride at dry ice temperature.[25] It oxidizes

$$(RO)_3P \xrightarrow[75\text{-}85\%]{N_2O_4} (RO)_3\overset{+}{P}-O^-$$

alkyl sulfides to sulfoxides (without further oxidation to sulfones) and trialkyl and triaryl phosphines to phosphine oxides.

Synthesis of diazonium salts. Scribner[26] found that addition of dinitrogen tetroxide to an ethereal solution of the Schiff base benzylidene aniline gives a precipitate of anhydrous benzenediazonium nitrate (highly explosive).

$$C_6H_5N{=}CHC_6H_5 + N_2O_4 \xrightarrow[90\%]{\text{Ether}} C_6H_5\overset{+}{N}{\equiv}N(NO_3^-) + O{=}CHC_6H_5$$

Isomerization of cyclic sulfoxides. Johnson and McCants[27] found that dinitrogen tetroxide equilibrates 4-substituted thiane-l-oxides. Treatment of either the *cis*-isomer (1) or the *trans*-isomer (2) gives an equilibrium mixture of the composition

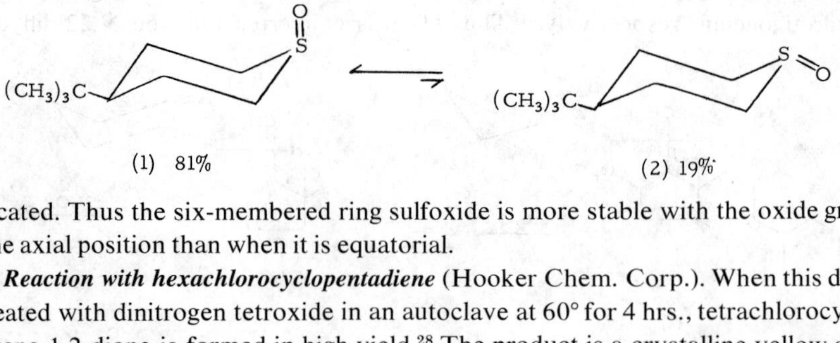
O
S
(CH3)3C
(CH3)3C
S
O

(1) 81% (2) 19%

indicated. Thus the six-membered ring sulfoxide is more stable with the oxide group in the axial position than when it is equatorial.

Reaction with hexachlorocyclopentadiene (Hooker Chem. Corp.). When this diene is heated with dinitrogen tetroxide in an autoclave at 60° for 4 hrs., tetrachlorocyclopentene-1,2-dione is formed in high yield.[28] The product is a crystalline yellow solid melting at 44–45.5°.

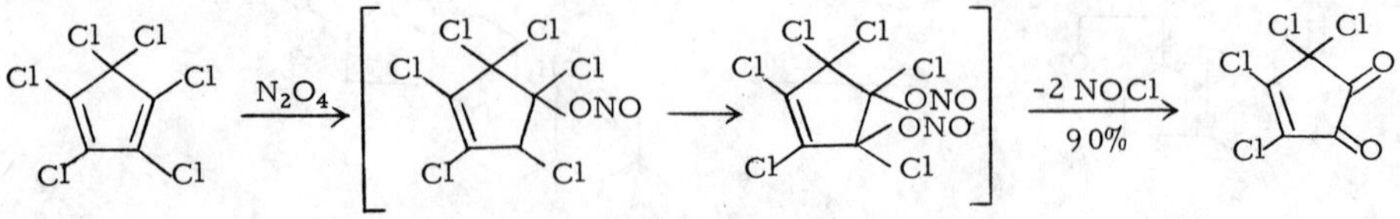
Cl Cl
Cl
Cl
Cl
Cl
N2O4
ONO
ONO
ONO
-2 NOCl
90%
O
O

[1]Review: H. Shechter, *Record of Chemical Progress*, **25**, 55 (1964)
[2]J. R. Cox, Jr., and F. H. Westheimer, *Am. Soc.*, **80**, 5441 (1958)
[3]N. Levy, C. W. Scaife, *et al.*, *J. Chem. Soc.*, 1093, 1096, 1100 (1946); 52 (1948); 2627 (1949)
[4]C. E. Anagnostopoulos and L. F. Fieser, *Am. Soc.*, **76**, 532 (1954)

[5]W. K. Seifert, *J. Org.*, **28**, 125 (1963)
[6]J. P. Freeman and W. D. Emmons, *Am. Soc.*, **79**, 1712 (1957)
[7]K. N. Campbell, J. Shavel, Jr., and B. K. Campbell, *Am. Soc.*, **75**, 2400 (1953)
[8]G. Ponzio, *J. prakt. Chem.*, **73**, 494 (1906)
[9]L. F. Fieser and W. von E. Doering, *Am. Soc.*, **68**, 2252 (1946)
[10]A. G. Brook, *J. Chem. Soc.*, 5040 (1952)
[11]B. O. Field and J. Grundy, *J. Chem. Soc.*, 1110 (1955)
[12]K. Maurer and G. Drefahl, *Ber.*, **75**, 1489 (1942)
[13]Th. Posternak and D. Reymond, *Helv.*, **36**, 1370 (1953)
[14]W. Langenbeck and M. Richter, *Ber.*, **89**, 202 (1956)
[15]R. M. Scribner, *J. Org.*, **29**, 279 (1964)
[16]E. H. White, *Am. Soc.*, **77**, 6008 (1955)
[17]E. H. White, *Am. Soc.*, **77**, 6011, 6014 (1955)
[18]For a further example, see T. Fujii, M. Tashiro, K. Ohara, and M. Kumai, *Chem. Pharm. Bull.*, **8**, 266 (1960)
[19]R. Stevenson, *J. Org.*, **28**, 188 (1963)
[20]P. Bladon and W. McMeekin, *J. Chem. Soc.*, 3504 (1961)
[21]Y. Sato and H. G. Latham, Jr., *J. Org.*, **22**, 981 (1957)
[22]F. Alvarez, *Steroids*, **2**, 393 (1963)
[23]C. D. Hurd and J. C. Sowden, *Am. Soc.*, **60**, 235 (1938)
[24]K. Brenneisen, Ch. Tamm, and T. Reichstein, *Helv.*, **39**, 1233 (1956)
[25]C. C. Addison and J. C. Sheldon, *J. Chem. Soc.*, 2705 (1956)
[26]R. M. Scribner, J. Org., **29**, 3429 (1964)
[27]C. R. Johnson and D. McCants, Jr., *Am. Soc.*, **86**, 2935 (1964); **87**, 1109 (1965)
[28]R. M. Scribner, *J. Org.*, **30**, 3657 (1965)

Dinitrogen tetroxide–Boron trifluoride, $N_2O_4 \cdot BF_3$. Prepared by passing BF_3 gas into a solution of dinitrogen tetroxide in nitromethane at 0°, it is an amorphous white solid insoluble in solvents with which it does not react.[1] Stirred with benzene for a week, it affords nitrobenzene in 39% yield. It differs from nitric acid in the isomer distribution (*o*-substitution favored), but the record of yields is far from impressive.[2]

[1]G. B. Bachman, H. Feuer, B. R. Bluestein, and C. M. Vogt, *Am. Soc.*, **77**, 6188 (1955)
[2]G. B. Bachman and C. M. Vogt, *Am. Soc.*, **80**, 2987 (1958)

Dinitrogen trioxide–Boron trifluoride, $N_2O_3 \cdot BF_3$. The components react at −70° to form a stable white 1:1 complex.[1] It is a weak nitrating agent but a powerful diazotizing agent in nonaqueous solution (benzene, chloroform). The diazonium salt which separates as a solid can be thermally decomposed to form fluoro derivatives (Schiemann reaction) or used in Gomberg-Bachmann arylation.

[1]G. B. Bachman and T. Hokama, *Am. Soc.*, **79**, 4370 (1957)

2,4-Dinitroiodobenzene, 2,4-$(NO_2)_2C_6H_3I$. Mol. wt. 294.01, m.p. 90°. Supplier: KK. The reaction of 2,4-dinitrochlorobenzene with sodium iodide has been carried out in

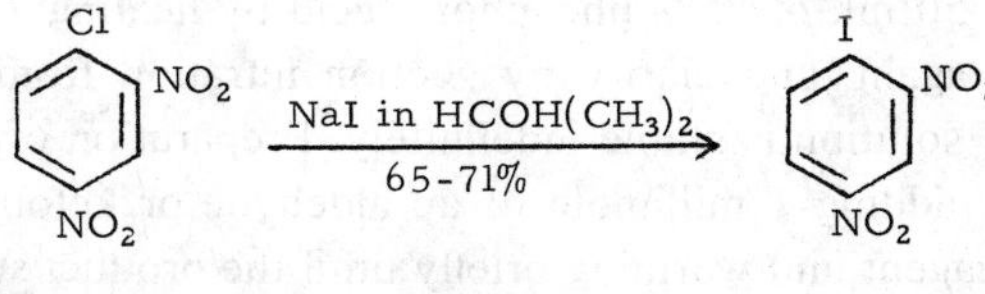
Cl
NO2
NO2
NaI in HCOH(CH3)2
65-71%
I
NO2
NO2

refluxing ethylene glycol,[1] but dimethylformamide is the solvent of choice.[2]

Bunnett and Conner[2] determined velocities and equilibria for the interchange of chlorine, bromine, and iodine atoms in the halogeno-2,4-dinitrobenzenes.

[1]G. M. Bennett and I. H. Vernon, *J. Chem. Soc.*, 1783 (1938)
[2]J. F. Bunnett and R. M. Conner, *Org. Syn.*, **40**, 34 (1960)

Di(p-nitrophenyl)carbonate (1). Mol. wt. 304.21, m.p. 144–145°.

Preparation.[1]

$$O{=}CCl_2 \;+\; 2\;NaOC_6H_4NO_2\text{-}\underline{p} \;\longrightarrow\; O{=}C\Big\langle{}^{OC_6H_4NO_2\text{-}\underline{p}}_{OC_6H_4NO_2\text{-}\underline{p}} \;+\; 2\;NaCl$$

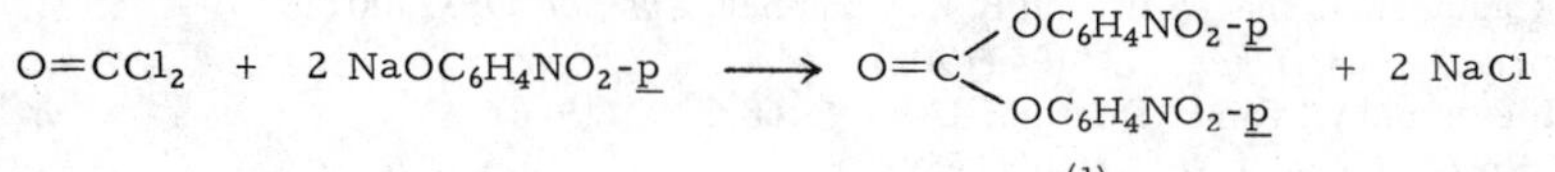

$$(1)$$

Peptide synthesis.[1] This is an excellent reagent for the preparation of activated p-nitrophenyl esters of N-protected amino acids. A solution of the N-protected amino acid (2) in acetic acid containing a little pyridine is treated with 1.5 equivalents

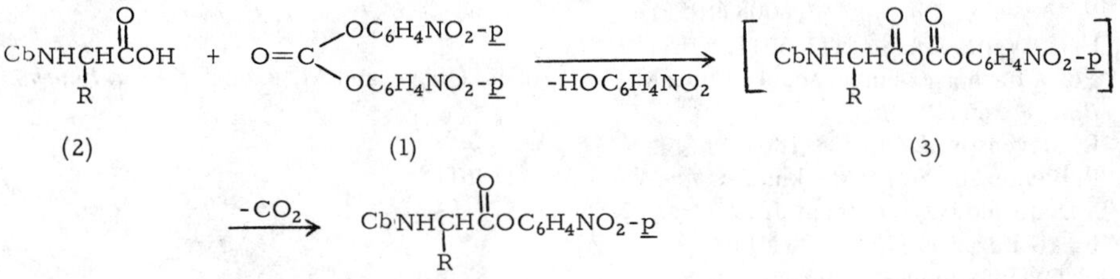

$$\underset{(2)}{Cb NHCHCOH}\;\underset{R}{\overset{O}{\|}} \;+\; \underset{(1)}{O{=}C\Big\langle{}^{OC_6H_4NO_2\text{-}\underline{p}}_{OC_6H_4NO_2\text{-}\underline{p}}}\;\xrightarrow[{-HOC_6H_4NO_2}]{}\; \left[\underset{(3)}{Cb NHCHCOCOC_6H_4NO_2\text{-}\underline{p}}\;\underset{R}{\overset{O\;\;O}{\|\;\;\|}}\right]$$

$$\xrightarrow{-CO_2}\; \underset{(4)}{Cb NHCHCOC_6H_4NO_2\text{-}\underline{p}}\;\overset{O}{\|}\;\underset{R}{}$$

of the reagent and refluxed for 3 hrs. Elimination of p-nitrophenol gives the mixed anhydride (3), which decomposes as formed with loss of carbon dioxide and formation of the activated ester (4).

[1]T. Wieland, B. Heinke, and K. Vogeler, *Ann.*, **655**, 189 (1962); R. Glatthard and M. Matter, *Helv.*, **46**, 795 (1963)

2,4-Dinitrophenylhydrazine. Mol. wt. 198.14, m.p. 197°. Suppliers: Eastern, B, E, F, MCB.

Preparation.[1] Dissolve 100 g. of 2,4-dinitrochlorobenzene in 200 ml. of triethylene glycol, stir mechanically in a salt-ice bath to 15° (some material may crystallize), and add 28 ml. of 64% hydrazine solution by drops while controlling the temperature to $20 \pm 3°$ (25 min.). When the strongly exothermal reaction is over, digest the paste on the steam bath, add 100 ml. of methanol and digest further, then cool, collect, and wash with methanol. Yield 98 g. (100%), m.p. 190–192°.

The reagent, under acid catalysis, reacts with aldehydes and ketones to produce sparingly soluble and high-melting 2,4-dinitrophenylhydrazones well suited for characterization and identification. It is a high-melting red solid sparingly soluble in ethanol, but two procedures can be recommended for preparing solutions of the reagent.

Carbonyl compounds, characterization. (a)[1,2] Dissolve 2.0 g. of 2,4-dinitrophenylhydrazine in 50 ml. of 85% phosphoric acid by heating, cool, add 50 ml. of 95% ethanol, cool again, and clarify by suction filtration from a trace of solid. The resulting $0.1\,M$ solution is stable indefinitely. Preparation of a DNP derivative is accomplished by adding 1 millimole of an aldehyde or ketone to 10 ml. of the $0.1\,M$ solution of reagent and warming briefly until the product starts to crystallize. Although less destructive to sensitive compounds than solutions containing a mineral acid, the phosphoric acid reagent sometimes results in elimination of a tertiary hydroxyl group.[3]

(b)[4] Dissolve 4 g. of 2,4-dinitrophenylhydrazine in 120 ml. of diethylene glycol dimethyl ether (diglyme) by warming, and let cool to room temperature (the solution is said to keep at room temperature for several days). Add 5 ml. of the solution to a solution of 0.1 g. of a carbonyl compound in 1 ml. of ethanol (or diglyme), heat on the steam bath, and add 3 drops of concd. hydrochloric acid. The color changes from red to yellow and the derivative soon separates.

(c) Parrick and Rasburn[5] recommend use of dimethylformamide or dimethyl sulfoxide as solvent. The reagent is very soluble in these solvents and can be used at a concentration such that even fairly soluble derivatives separate spontaneously. The Canadian investigators were able to prepare the DNP (2) only by this method.

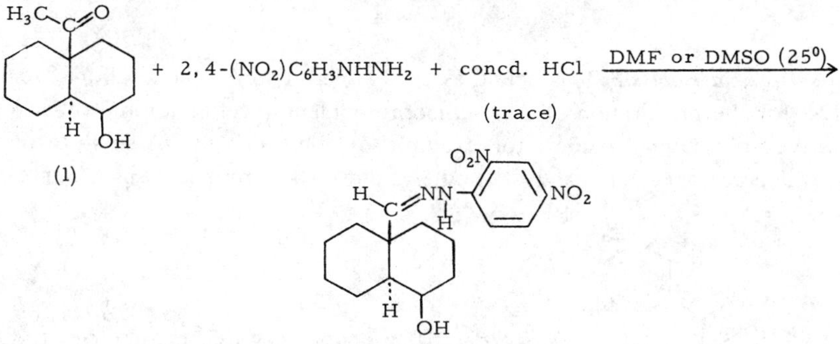
H3C C=O
+ 2,4-(NO2)C6H3NHNH2 + concd. HCl
(trace)
DMF or DMSO (25°)
H
OH
(1)
O2N
H C=NN
H
NO2
H
OH
(2)

Dehydrohalogenation. 2,4-Dinitrophenylhydrazine is used as a dehydrohalogenating agent in the terminal stages of the lengthy bile acid synthesis of cortisone. Dehydrohalogenation of (1) by ordinary methods proceeds poorly because of the *cis* orientation of the 5β-H and the 6β-bromine atom. However, 2,4-dinitrophenylhydrazine condenses with (1) by a nonstereospecific mechanism and effects dehydrohalogenation to give (2) in high yield.[6]

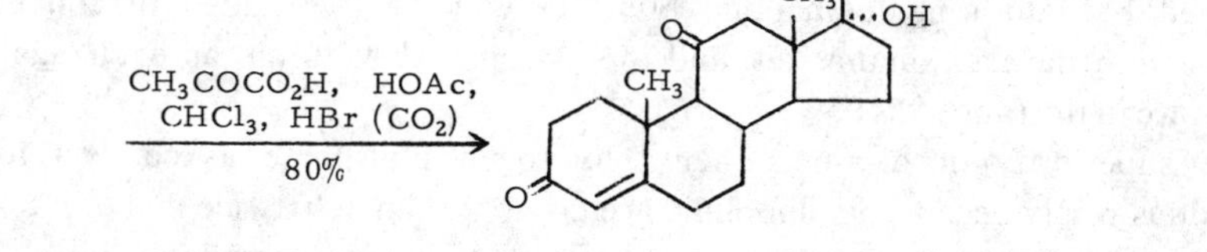
CH3
O
H
Br
(1)
2,4-(NO2)C6H5NHNH2
NaOAc–HOAc (CO2)
78%
CH3
(NO2)2C6H4NHN
(2)
CH3COCO2H, HOAc,
CHCl3, HBr (CO2)
80%
CH2OAc
CO
CH3
OH
O
CH3
O

Cortisone acetate

[1]*Org. Expts.*, Chapt. 16
[2]C. D. Johnson, *Am. Soc.*, **73**, 5888 (1951)
[3]H. Reich, K. F. Crane, and S. J. Sanfilippo, *J. Org.*, **18**, 822 (1953)
[4]H. J. Shine, *J. Org.*, **24**, 252 (1959)
[5]J. Parrick and J. W. Rasburn, *Can. J. Chem.*, **43**, 3453 (1965)
[6]V. R. Mattox and E. C. Kendall, *Am. Soc.*, **72**, 2290 (1950); W. F. McGuckin and E. C. Kendall, *ibid.*, **74**, 3951 (1952)

Di-*p*-nitrophenyl hydrogen phosphate (2). Mol. wt. 340.19, m.p. 175–175.5, pKa 1.7 in 50% ethanol. Supplier: Aldrich.

Preparation.[1]

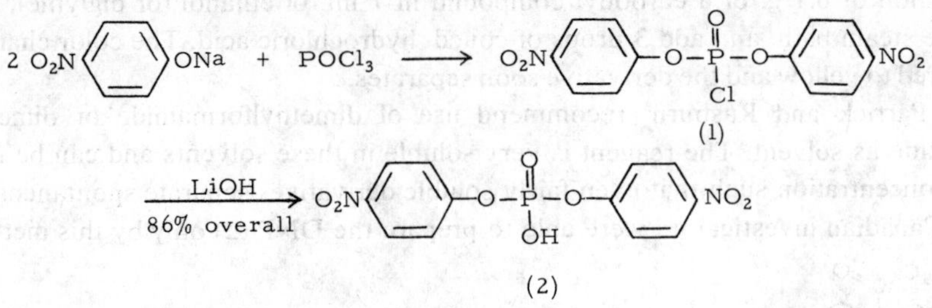

Use as acid catalyst.[2] This acidic ester is superior to *p*-toluenesulfonic acid as catalyst for the preparation of 2′,3′-O-isopropylidene (acetonide) and other ketal derivatives of ribonucleosides, for example of adenosine (3). A mixture of the nucleoside, acetone, the catalyst, and 2,2-dimethoxypropane (which serves as

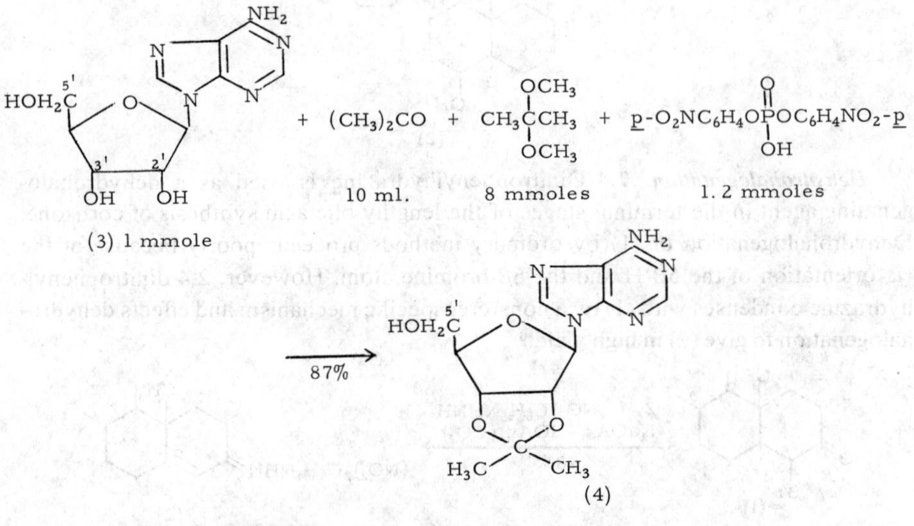

water scavenger) is stirred until a clear solution results (1–3 hrs.). A solution prepared by saturating methanolic NH_4OH with CO_2 is added to convert the catalyst into the ammonium salt, and this is removed with an ion-exchange resin. Yields are in the range 85–95%.

The superiority of di-*p*-nitrophenyl hydrogen phosphate as catalyst for the reaction is partly because its lipophilic area is approximately twice that of *p*-toluenesulfonic acid and hence its salts with nucleosides are correspondingly more soluble in acetone. A further advantage of the procedure is that the formation of a nucleoside acetonide can be coupled with phosphorylation to produce the corresponding nucleotide. Thus, after the preparation of adenosine-2′,3′-acetonide (4) in solution, the acetone and 2,3-dimethoxypropane can be removed by distillation and the residue treated with tetra-*p*-nitrophenyl pyrophosphate for conversion into the nucleotide acetonide.

[1]J. G. Moffatt and H. G. Khorana, *Am. Soc.*, **79**, 3741 (1957)

[2]A. Hampton, *Am. Soc.*, **83**, 3640 (1961); *Biochem. Prepn.*, **10**, 91 (1963); A. Hampton, J. C. Fratantoni, P. M. Carroll, and S. Wang, *Am. Soc.*, **87**, 5481 (1965)

Dioxane, $O(CH_2CH_2)_2O$. Mol. wt. 88.10, m.p. 11.8°, b.p. 101.3°. Suppliers: A, B, BJ, E, F, MCB.

Purification. Benson, McBee, and Rand,[1] in submitting an *Organic Syntheses* procedure for the preparation of N-iodosuccinimide, purify dioxane "only by use of sodium chips and distillation" (reference to Fieser[2]). Checkers B. C. McKusick and T. J. Kealy used a newly opened bottle of "spectroquality Reagent" (MCB). In purifying the solvent for a kinetic study Burstein and Ringold[3] state that they applied the method of Fieser[2] and then stored the solvent over a Linde molecular sieve for at least a week. Actually Fieser cited two methods from the literature, the first of which is designed to eliminate water and glycol acetal by hydrolysis to acetaldehyde. The methods are as follows:

(a)[4] A mixture of 2 l. of commercial dioxane, 27 ml. of concd. hydrochloric acid, and 200 ml. of water is refluxed for 12 hrs., during which time a slow stream of nitrogen is bubbled through the solution to entrain acetaldehyde. The solution is cooled, and potassium hydroxide pellets are added slowly with shaking until they no longer dissolve and a second layer has separated. The dioxane is decanted, treated with fresh potassium hydroxide pellets to remove adhering aqueous liquor, decanted into a clean flask, and refluxed with sodium for 10–12 hrs., when the metal should remain bright. The solvent is then distilled from the sodium and stored out of contact with air.

(b)[5] Peroxides are removed and the aldehyde content decreased by passing the solvent through a column of alumina (80 g. for 100–200 ml. of solvent). The method is applicable to ether and other anhydrous solvents.

Another method recently recommended[6] is distillation of reagent grade dioxane from lithium aluminum hydride.

Solvent effect.[7] "We have had several instances where we wished to convert amides to thioamides:

$$RC(=O)NR'_2 \xrightarrow{P_2S_5} RC(=S)NR'_2$$

(R = alkyl, aryl; R' = H, alkyl, aryl)

Many solvents have been employed in the literature, including pyridine, DMF, toluene, dioxane, chloroform, etc. Although in most cases pyridine or DMF work fine, in other instances the reaction products are intractable tars. We have found that dioxane is unexpectedly the solvent of choice; it has uniformly given high yields (80–90%) of easily purified product even in reactions where *no* product could be obtained using other solvents."

[1]W. R. Benson, E. T. McBee, and L. Rand, *Org. Syn.*, **42**, 73 (1962)
[2]L. F. Fieser, "Experiments in Organic Chemistry," 3rd ed., p. 284, D. C. Heath, Boston, 1955
[3]S. H. Burstein and H. J. Ringold, *Am. Soc.*, **86**, 4952 (1964)
[4]K. Hess and H. Frahm, *Ber.*, **71**, 2627 (1938); see also E. Eigenberger, *J. prakt. Chem.*, **130**, 75 (1931)
[5]W. Dasler and C. D. Bauer, *Ind. Eng. Chem., Anal. Ed.*, **18**, 52 (1946)
[6]R. L. Augustine and J. A. Caputo, *Org. Syn.*, **45**, 80 (1965)
[7]Contributed by Bernard Loev, Smith Kline and French Laboratories

Dioxane dibromide, $OC_4H_8O \cdot Br_2$. Mol. wt. 247.93, m.p. 64°, orange-yellow. Supplier: Frinton.

Bromine is added to dioxane with stirring and cooling, the hot solution is poured into ice water, and the orange precipitate is collected and dried.[1] The complex is very volatile and must be stored in a stoppered container. Anhydrous reagent can be obtained by pouring the solution in dioxane into petroleum ether cooled to −20°.[2]

The reagent is useful for the controlled bromination of phenols. For example, 25 g. of reagent was added with cooling to 9.4 g. of phenol and the mixture poured onto water; extraction with ether gave *p*-bromophenol.[1] Acetylcarbinols can be

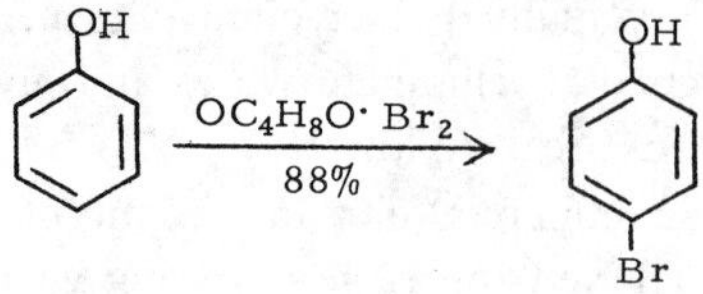
OH
OC4H8O· Br2
88%
OH
Br

brominated by reaction in ether solution with dioxane dibromide under irradiation of the red solution with strong visible light.[2]

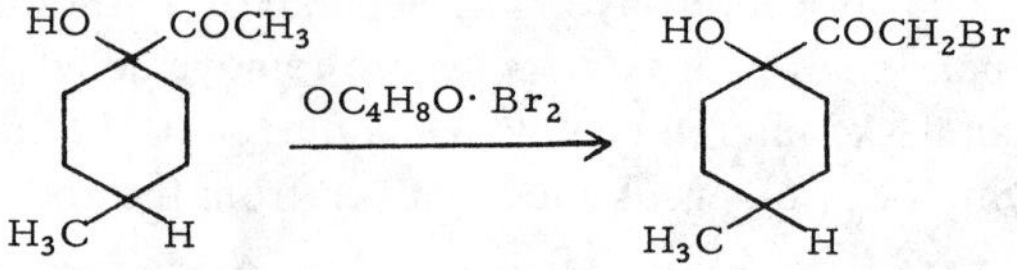
HO
COCH3
H3C
H
OC4H8O· Br2
HO
COCH2Br
H3C
H

Kosolapoff[3] used a solution or suspension of the reagent prepared by adding bromine to excess dioxane for the controlled bromination of reactive compounds: aniline → *p*-bromoaniline, 68% yield.

[1]L. A. Yanovskaya, A. P. Terent'ev, and L. I. Belen'kii, *J. Gen. Chem.*, **22**, 1594 (1952) [*C.A.*, **47**, 8032 (1953)]
[2]J. D. Billimoria and N. F. Maclagan, *J. Chem. Soc.*, 3257 (1954)
[3]G. M. Kosolapoff, *Am. Soc.*, **75**, 3596 (1953)

Dioxane diphosphate,

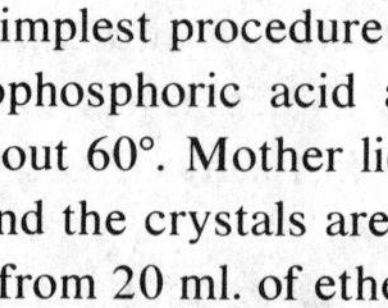
H
OPO(OH)2
O
O
H
OPO(OH)2

Mol. wt. 222.15, m.p. 83–87°, very hygroscopic, decomposed by water; soluble in dioxane, ether, ethanol, ethyl acetate; insoluble in hydrocarbon solvents.

Preparation.[1] The simplest procedure is to mix 1 mole of dioxane with 2 moles of 88% aqueous orthophosphoric acid and stir to induce crystallization as the temperature rises to about 60°. Mother liquor is removed by suction filtration with use of a rubber dam, and the crystals are dried in vacuum over sodium hydroxide. Crystallization of 10 g. from 20 ml. of ether at −60° did not change the melting point. Baer[1] suggested that the reagent might be used for the purification of dioxane or in place of anhydrous phosphoric acid.

Phosphorylation. Chambers *et al.*[2] found the reagent useful for the preparation of nucleoside diphosphoric acids (3, R = adenine, uracil, etc.). A nucleotide (1) is converted into the 5′-phosphoramidate (2) by reaction with ammonia and dicyclohexylcarbodiimide (DCC), and the amidate (2) is treated with dioxane diphosphate in *o*-chlorophenol (3 hrs., 0°). The diphosphoric acid (3) is precipitated with petroleum ether and purified as the ammonium salt. Yields are around 90%, whereas when 85% phosphoric acid was used yields were in the order of 50–60%. However, anhydrous

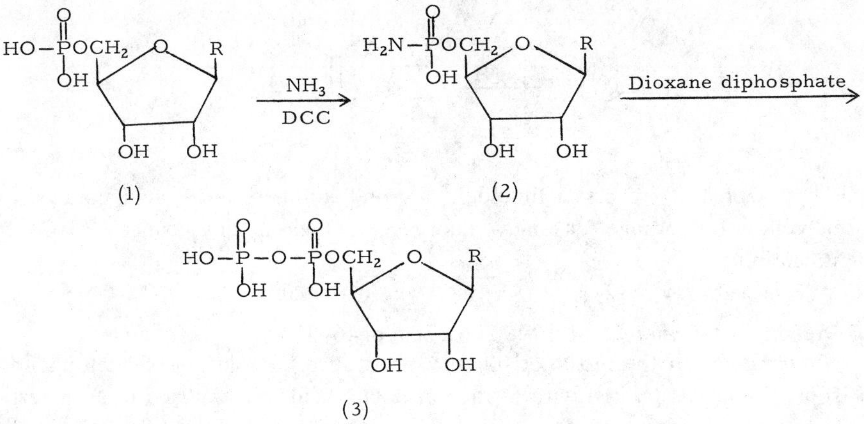

conditions are required and during humid weather the reactions should be run in a dry box.

[1]E. Baer, *Am. Soc.*, **66**, 303 (1944)
[2]R. W. Chambers, P. Shapiro, and V. Kurkov, *Am. Soc.*, **82**, 970 (1960)

Diphenylacetylene, $C_6H_5C{\equiv}CC_6H_5$. Mol. wt. 178.22, m.p. 61°. Suppliers: Farchan, Orgmet, A, B. Preparation from benzil by conversion into the bishydrazone and oxidation with yellow mercuric oxide in benzene (67–73% from benzil).[1]

Preparation from *trans*-stilbene:[2] *trans*-Stilbene (20 g.) is heated with 400 ml. of acetic acid on the steam bath until dissolved, 40 g. of pyridinium hydrobromide

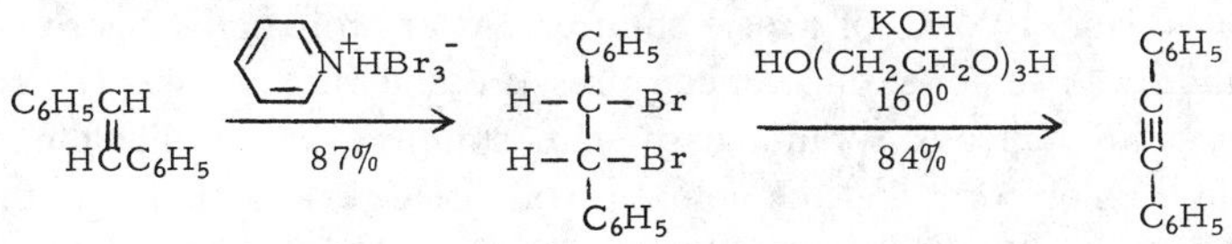

perbromide is added, and the mixture is heated on the steam bath and swirled for 5 min. *meso*-Stilbene dibromide separates at once in pearly plates, and after cooling the product is collected and washed with methanol. A mixture of this material (32.4 g., m.p. 236–237°), 65 g. of potassium hydroxide, and 130 ml. of triethylene glycol is swirled over a free flame to mix the contents and bring the temperature to 160°, when potassium bromide begins to separate. After 5 min. more at 160–170° to complete the reaction, the hot mixture is poured into a beaker, and the flask is rinsed alternately with water and ethanol. After cooling, the product is collected, washed with water, and dried (16.5 g.). Crystallization from 50 ml. of 95% ethanol gives diphenylacetylene in two crops; 11.8 g., m.p. 61.5–62.5°; 2.4 g., m.p. 58–59°.

Use of the reagent as a dienophile is illustrated by a procedure for the synthesis of hexaphenylbenzene described in a note on "Diels-Alder dienophiles."

[1]A. C. Cope, D. S. Smith, and R. J. Cotter, *Org. Syn., Coll. Vol.*, **4**, 377 (1963)
[2]L. F. Fieser, *Org. Syn.*, **46**, 46 (1966)

2-Diphenylacetyl-1,3-indanedione-1-hydrazone. Mol. wt. 354.39, m.p. 305°, dec.

The parent acyl-1,3-indanedione (1) is prepared by condensation of dimethyl phthalate with methyl benzhydryl ketone in the presence of sodium methoxide.[1]

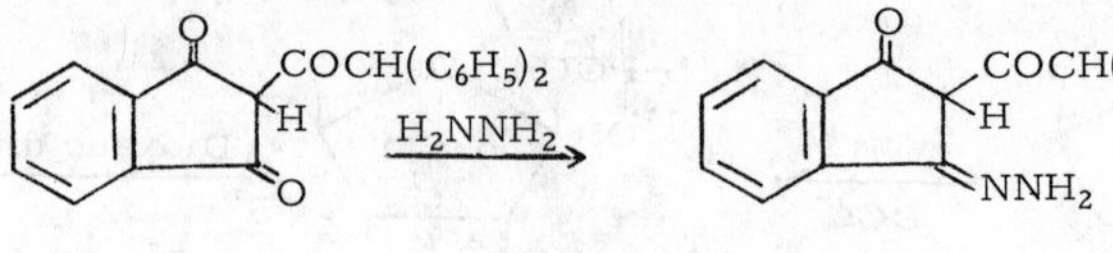

(1) (2)

The hydrazone (2) reacts with aldehydes and ketones under acid catalysis to form yellow to orange, strongly fluorescent, high-melting azines of use for identification.

[1]R. A. Braun and W. A. Mosher, *Am. Soc.*, **80**, 2749, 3048 (1958)

2,3-Diphenylbutadiene, (2). Mol. wt. 206.27, m.p. 46–47°.

Preparation. In the most satisfactory procedure,[1] a solution of acetophenone in isopropanol was treated with a trace of acetic acid and exposed in an inverted flask to sunlight for several months. The resulting 2,3-diphenylbutane-2,3-diol (1, 35 g.) was treated with about 20 mg. of N-phenyl-β-naphthylamine (polymeriza-

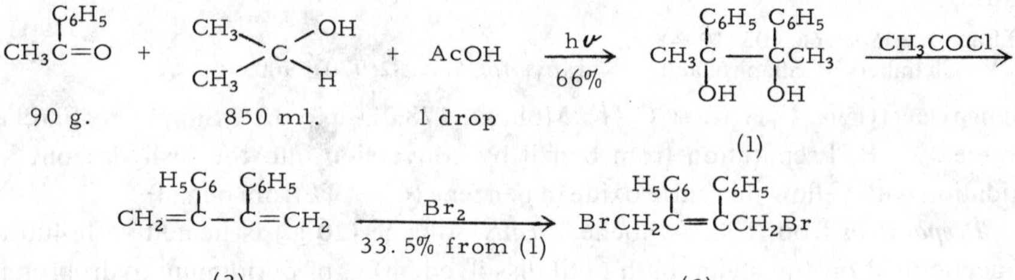

(2) (3)

tion inhibitor) and 105 ml. of acetyl chloride. After an initially vigorous reaction, excess halide was removed at reduced pressure and the mixture was cooled and neutralized with aqueous sodium carbonate solution. The crude diene (2) was taken up into chloroform and the solution was dried, evaporated, and the product converted into the 1,4-dibromo adduct (3), m.p. 144–147°. This derivative is stable in storage. The diene, when required, can be recovered in 92% yield by refluxing the dibromide (3) with zinc dust in acetone for 2 hrs. The diene can be kept in the dark under nitrogen for several weeks, otherwise it quickly becomes greasy and colored.

A shorter method[2] involves addition to dimethyl sulfoxide with stirring under nitrogen of sodium hydride in oil to produce a dark gray solution of sodium methyl-

$$\underset{30\ \text{ml.}}{2\ CH_3SOCH_3} + \underset{2.4\ \text{g.}}{2\ NaH\ (50\%\ \text{in oil})} \xrightarrow[75^0]{} 2\ CH_3SO\overset{-}{C}H_2\overset{+}{Na} + 2\ H_2$$

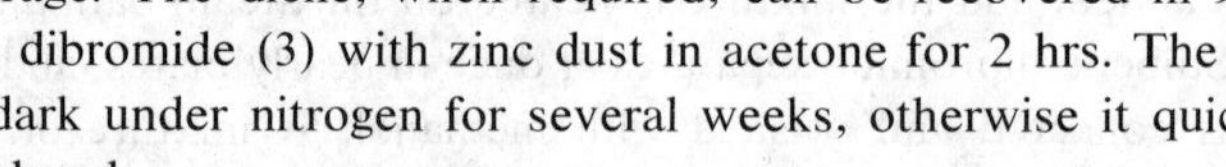

sulfinylmethylide. A solution of diphenylacetylene in dimethyl sulfoxide is then added by drops and the mixture heated at 65° for 2 hrs. Workup and crystallization from methanol gives pure 2,3-diphenylbutadiene, m.p. 47–48° (stable on storage in a refrigerator in the dark for over a year).

[1]C. F. H. Allen, C. G. Eliot, and A. Bell, *Can. J. Research*, **17B**, 75 (1939)

[2]Issei Iwai and Junya Ide, procedure submitted to *Org. Syn.*

trans,trans-**1,4-Diphenylbutadiene.** Mol. wt. 206.27, m.p. 153°. Suppliers: A, B. The hydrocarbon is easily prepared in a short working period as follows:[1]

$$C_6H_5CH_2Cl \xrightarrow{(C_2H_5O)_3P} C_6H_5CH_2\overset{\underset{OC_2H_5}{|}}{P^+}(OC_2H_5)_2(Cl^-) \xrightarrow[-\ C_2H_5Cl]{Refl.\ 1\ hr.}$$

$$C_6H_5CH_2\overset{\underset{|}{O^-}}{P^+}(OC_2H_5)_2 \xrightarrow[NaOCH_3,\ DMF]{C_6H_5CH=CHCHO} \quad C_6H_5\overset{H\ \ \ H}{\underset{H\ \ \ H}{C=C-C=C}}C_6H_5$$

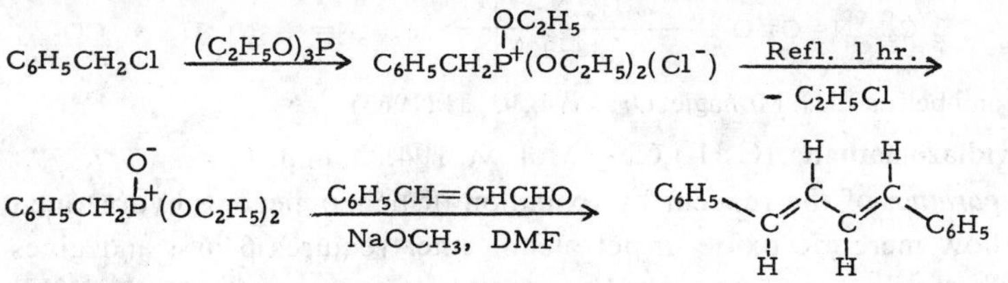

Diels-Alder reaction: synthesis of *p*-terphenyl.[1] Use of the reagent as a Diels-Alder diene is illustrated by the synthesis of *p*-terphenyl.

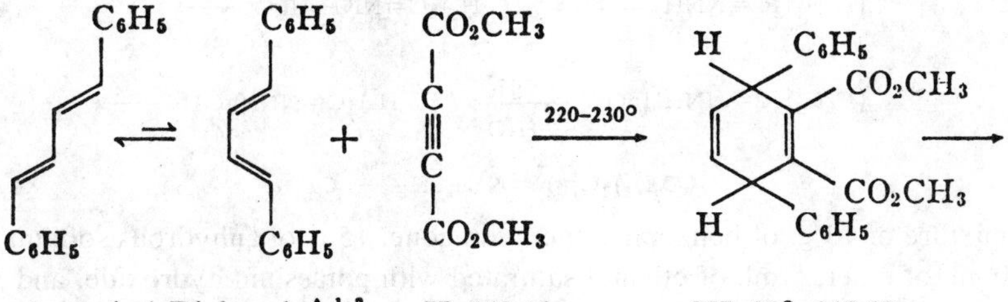

trans, trans-1, 4-Diphenyl-$\triangle^{1,3}$-
butadiene (I, mol. wt. 206.27)   II, 142.11   III, 98°, 348.38

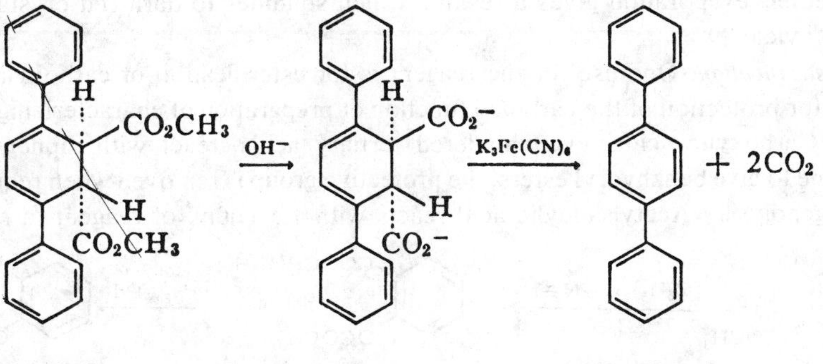

IV, 170°   V   *p*-Terphenyl
(VI, 230.29, 211°)

[1]*Org. Expts.*, Chapt. 24. *See also* R. N. McDonald and T. W. Campbell, *Org. Syn.*, **40**, 36 (1960)

Diphenylcarbamoyl chloride $(C_6H_5)_2NCOCl$. Mol. wt. 231.68, m.p. 86°. Suppliers: B, E.

The reagent can be used for introduction of a carboxyl group into an aromatic compound, for example a polystyrene resin.[1] Friedel-Crafts acylation gives an amide, which on hydrolysis gives the aromatic acid.

$$(C_6H_5)_2N\overset{\underset{|}{O}}{C}Cl + HAr \xrightarrow{AlCl_3} (C_6H_5)_2N\overset{\underset{|}{O}}{C}-Ar \xrightarrow{Hydrol.} (C_6H_5)_2NH + HO\overset{\underset{|}{O}}{C}-Ar$$

[1]R. L. Letsinger and M. J. Kornet, *Am. Soc.*, **85**, 3045 (1963)

Diphenylcarbodiimide, $C_6H_5N=C=NC_6H_5$. Mol. wt. 194.23, b.p. 110–112°/0.2 mm.

An efficient method for preparation of aromatic and aliphatic carbodiimides involves the reaction of an isocyanate with 3-methyl-1-phenyl-3-phospholene-1-oxide, the most readily prepared catalyst of its type.[1]

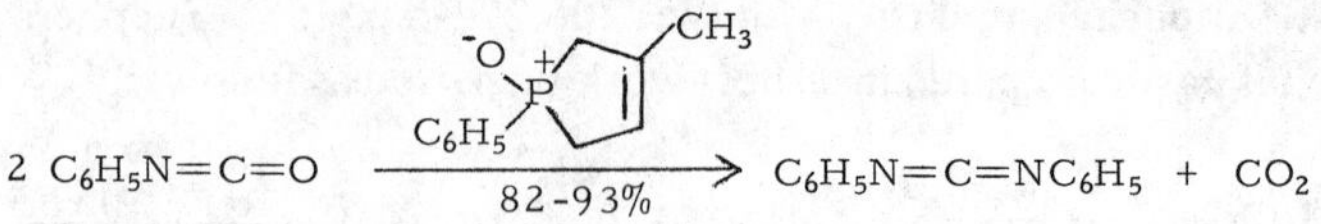
CH3
O
P
C6H5
2 C6H5N=C=O
82-93%
C6H5N=C=NC6H5 + CO2

[1]T. W. Campbell and J. J. Monagle, *Org. Syn.*, **43**, 31 (1963)

Diphenyldiazomethane, $(C_6H_5)_2CN_2$. Mol. wt. 194.23, m.p. 32°.

Preparation of the reagent by oxidation of benzophenone hydrazone (Aldrich) with yellow mercuric oxide in petroleum ether requires 6 hrs. and gives a liquid product.[1] An improved procedure[2] is based upon the finding that the reaction is catalyzed by base, probably as follows:

$$(C_6H_5)_2C{=}NNH_2 \xrightarrow{HgO} (C_6H_5)_2C{=}N\overset{+}{N}H_2Hg\overset{-}{O} \longrightarrow$$

$$(C_6H_5)_2C{=}NNHHgOH \xrightarrow[-H_2O]{OH^-} (C_6H_5)_2C{=}N\overset{-}{N}HgOH \longrightarrow$$

$$(C_6H_5)_2C{=}\overset{+}{N}{=}\overset{-}{N} + Hg + OH^-$$

A mixture of 13 g. of benzophenone hydrazone, 15 g. of anhydrous sodium sulfate, 200 ml. of ether, 5 ml. of ethanol saturated with potassium hydroxide, and 35 g. of yellow mercuric oxide is shaken for 75 min. in a pressure bottle wrapped with a wet towel. The solution is filtered and evaporated and the residue is taken up in petroleum ether; evaporation gives a residue which solidifies to dark red crystals, m.p. 29–32°; yield 89%.

Esterification. One use for the reagent is for esterification of carboxylic acids, either for protection of the carboxyl function or preparation of characterizing derivatives. Carboxylic acids, even hindered tertiary acids, react with diphenyldiazomethane to give benzhydryl esters; the protective group is removed when required by hydrogenolysis.[3] Acetylsalicylic acid reacts with 1.5 equiv. of reagent in refluxing

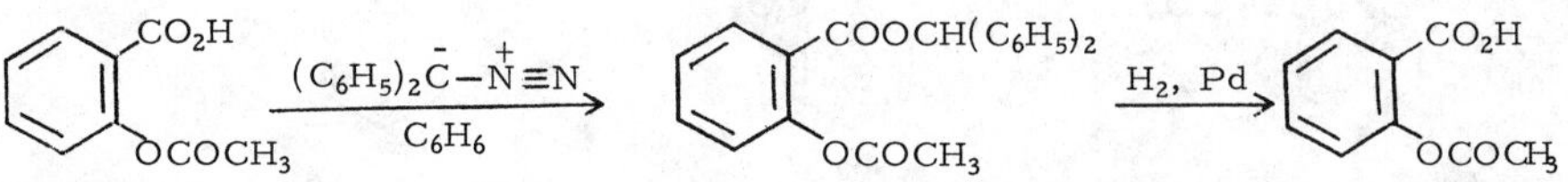
CO2H
OCOCH3
(C6H5)2C−N≡N
C6H6
COOCH(C6H5)2
OCOCH3
H2, Pd
CO2H
OCOCH3

benzene in 20 min. (color change from red-violet to yellow), and the crystalline benzhydryl ester (m.p. 107°) is obtained in quantitative yield. Carboxylic acids can be characterized, and acid mixtures analyzed, by spectrophotometric determination of the rate of reaction with diphenyldiazomethane in alcohol or benzene solution.[4]

Additions. The course of the reaction of diphenyldiazomethane with α,β-unsaturated carbonyl compounds has been clarified by a careful study of the reaction with methyl acrylate (1).[5] When the reaction flask was washed with acid, rinsed three times with distilled water, and dried, addition of a solution of diphenyldiazomethane in pentane or hexane to a stirred solution of methyl acrylate in the same solvent at 0–5° resulted in nearly quantitative evolution of nitrogen and formation of 2,2-diphenylcyclopropanecarboxylic acid methyl ester (4) in yields up to 89%; the cyclopropane derivative evidently arises from decomposition of the initially formed 1-pyrazoline (2). On the other hand, rinsing the flask with trisodium phosphate solution followed by three rinses with distilled water and drying led to high yields of the 2-pyrazoline (3). When, in addition, triethylamine was added to the reaction mixture prior to introduction of diphenyldiazomethane practically no nitrogen was

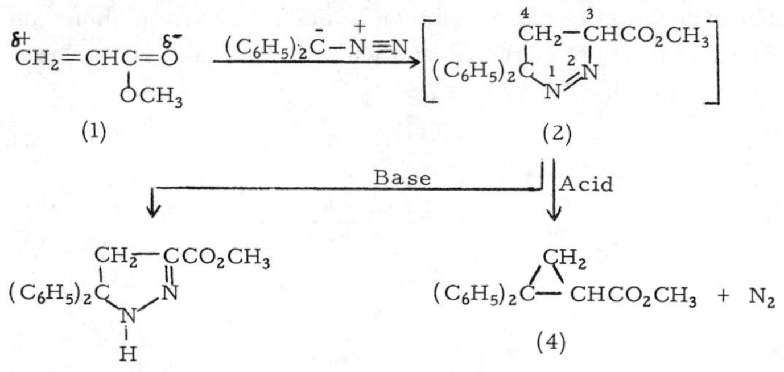

(1) (2) (3) (4)

evolved, and pure 5,5-diphenyl-3-carbomethoxy-2-pyrazoline (3, m.p. 140° dec.) was obtained in 70% yield. In the addition of diphenyldiazomethane to 1,4-naphthoquinone the initially formed adduct evidently enolizes to the hydroquinone before other reactions can occur.[6]

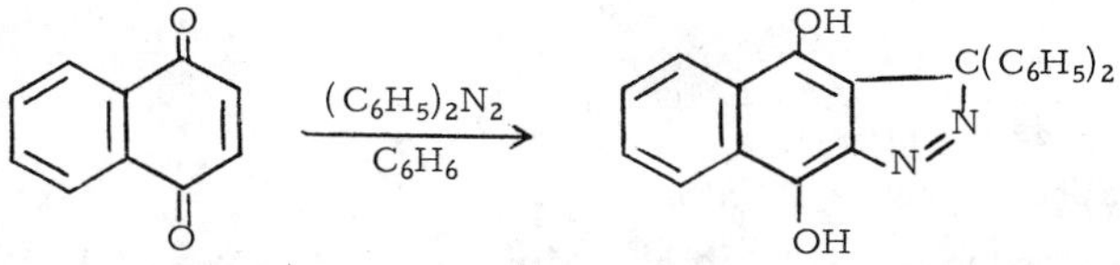

Peptide synthesis. Benzhydryl esters obtained by reaction of the reagent with N-protected amino acids are useful in peptide synthesis, particularly in the case of cysteine peptides.[7] Benzhydryl esters are comparable to *t*-butyl esters in acid lability. These esters cannot be prepared directly but can be made from the *p*-toluenesulfonate salt of the amino acid to be esterified.[8]

[1]L. I. Smith and K. L. Howard, *Org. Syn., Coll. Vol.*, **3**, 351 (1955)
[2]J. B. Miller, *J. Org.*, **24**, 560 (1959)
[3]E. Hardegger, Z. El Heweihi, and F. G. Robinet, *Helv.*, **31**, 439 (1948)
[4]J. D. Roberts and C. M. Regan, *Anal. Chem.*, **24**, 360 (1952)
[5]W. M. Jones, T. H. Glenn, and D. G. Baarda, *J. Org.*, **28**, 2887 (1963)
[6]L. F. Fieser and Mary A. Peters, *Am. Soc.*, **53**, 4080 (1931)
[7]R. G. Hiskey and J. B. Adams, Jr., *Am. Soc.*, **87**, 3969 (1965)
[8]A. A. Aboderin, G. R. Delpierre, and J. S. Fruton, *Am. Soc.*, **87**, 5469 (1965)

N,N′-Diphenylformamidine, $C_6H_5NHCH{=}NC_6H_5$. Mol. wt. 196.25, m.p. 141°. Suppliers: B, E. Preparation from aniline and formic acid:[1]

$$C_6H_5NH_2 + HOCH{=}O + H_2NC_6H_5 \xrightarrow{-2\,H_2O} C_6H_5NHCH{=}NC_6H_5$$

The reagent condenses with a compound having an activated methylene group, for example acetylacetone, with elimination of aniline:[2]

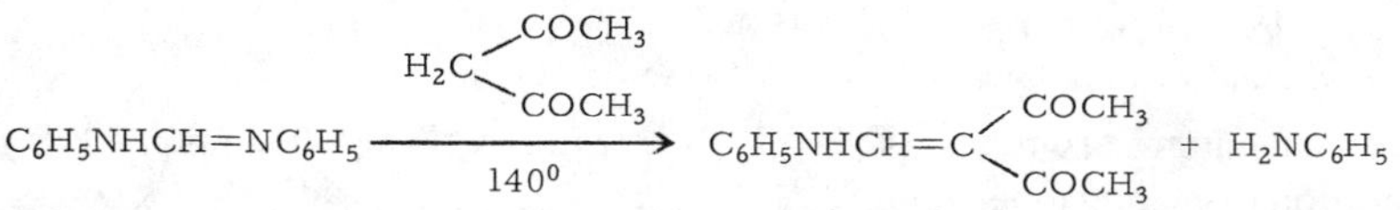

A synthesis of 2,4-dihydroxyisophthalaldehyde (4), investigated by Shoesmith and Haldane[3] and later improved by Kuhn and Staab,[1] involves condensation of the reagent with resorcinol, probably reacting in the diketo form (2). The reagents are heated under evacuation to aid in removal of aniline and to minimize air oxidation, and the red Schiff base (3) is hydrolyzed with dilute alkali to (4). The overall yield of 2,4-dihydroxyisophthalaldehyde is 21–24%.

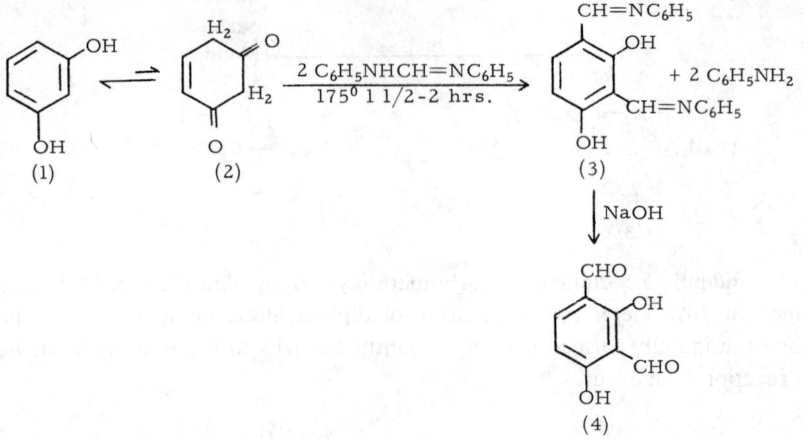
OH
H2
O
2 C6H5NHCH=NC6H5
175° 1 1/2-2 hrs.
H2
CH=NC6H5
OH
+ 2 C6H5NH2
CH=NC6H5
OH
O
OH
(1)
(2)
(3)
NaOH
CHO
OH
CHO
OH
(4)

[1]R. Kuhn and H. A. Staab, *Ber.*, **87**, 272 (1954)
[2]F. B. Dains, *Ber.*, **35**, 2504 (1902)
[3]J. B. Shoesmith and J. Haldane, *J. Chem. Soc.*, 2704 (1923)

1,1-Diphenylhydrazine, $(C_6H_5)_2NNH_2$. Mol. wt. 184.24, m.p. 34.5°. Supplied as the hydrochloride by Eastman. Preparation from diphenylamine by conversion to the N-nitrosoamine and reduction with zinc dust and acetic acid in ethanol.[1,2]

The reagent reacts with aldohexoses to form sparingly soluble and nicely crystalline diphenylhydrazones.[2] In the preparation of D-arabinose by Wohl degradation of D-glucose, 1,1-diphenylhydrazine is useful for isolation of additional D-arabinose from the mother liquor.[3]

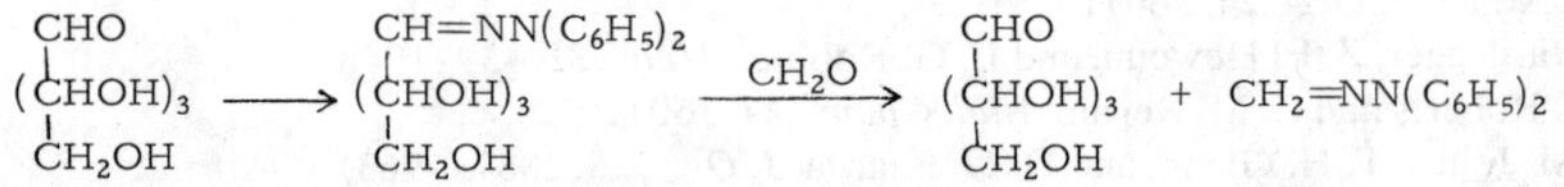
CHO
(CHOH)3
CH2OH
CH=NN(C6H5)2
(CHOH)3
CH2OH
CH2O
CHO
(CHOH)3
CH2OH
+ CH2=NN(C6H5)2

Whereas phenylhydrazine reacts with an osone to form a phenylosazone, 1,1-diphenylhydrazine reacts selectively with the aldehydic group to give a crystalline osone diphenylhydrazone:[4]

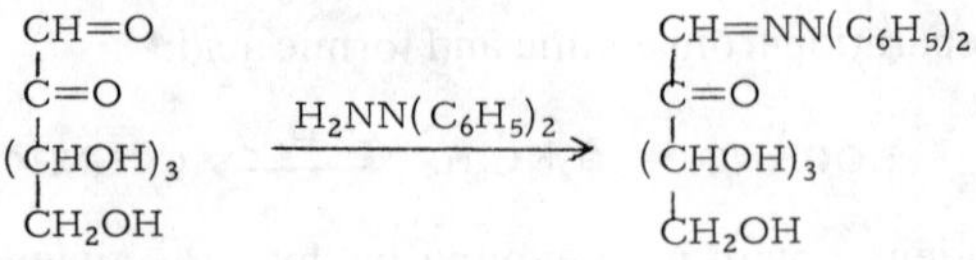
CH=O
C=O
(CHOH)3
CH2OH
H2NN(C6H5)2
CH=NN(C6H5)2
C=O
(CHOH)3
CH2OH

[1]E. Fischer, *Ann.*, **190**, 174 (1877)
[2]R. Stahel, *Ann.*, **258**, 242 (1890)
[3]G. Braun, *Org. Syn., Coll. Vol*, **3**, 102 (1955)
[4]G. Hanseke and W. Liebenow, *Ber.*, **87**, 1068 (1954)

Diphenyliodonium bromide, $(C_6H_5)_2I^+Br^-$. Mol. wt. 361.03, m.p. 208°, dec. Preparation from iodosobenzene.[1]

The reagent is a general phenylating agent; for example, it reacts with potassium phenoxide to give diphenyl ether (76%) and iodobenzene, and effects C-phenylation of carbanions.[2]

[1]F. M. Beringer *et al., Am. Soc.*, **75**, 2705 (1953)
[2]F. M. Beringer *et al., Am. Soc.*, **75**, 7708 (1953); **84**, 2819 (1962); W. Mayer *et al., Ber.*, **93**, 2761 (1960)

Diphenyliodonium-2-carboxylate monohydrate. Mol. wt. 341.13, m.p. 220°, dec.
This stable and safe reagent is useful for the generation of benzyne in an aprotic solvent in the presence of a reactive diene as trapping agent.[1] The original preparative procedure[2] has been improved[3] to a point such that 2 g. of *o*-iodobenzoic acid (Aldrich, Eastman) can be converted into 2.1 g. of the benzyne precursor in 1½-2 hrs. working time. The synthesis of 1,2,3,4-tetraphenylnaphthalene can be

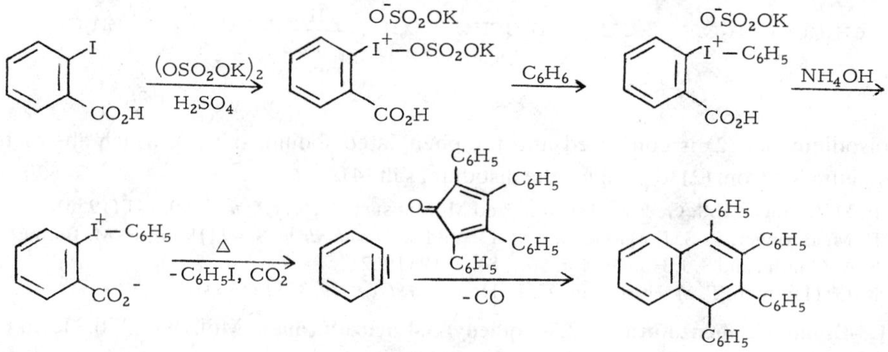

accomplished by heating a mixture of diphenyliodonium-2-carboxylate monohydrate (slight excess) and tetraphenylcyclopentadienone in triethylene glycol dimethyl ether at 200–205° until the purple color is discharged (3–4 min.), or by refluxing the reactants in *mp*-diethylbenzene for 45 min. The yield of hydrocarbon is 92%, with 69% utilization of the benzyne precursor.

[1]E. Le Goff, *Am. Soc.*, **84**, 3786 (1962); see also, F. M. Beringer and S. J. Huang, *J. Org.*, **29**, 445 (1964)
[2]F. M. Beringer and I. Lillien, *Am. Soc.*, **82**, 725 (1960)
[3]*Org. Expts.*, Chapt. 61; L. F. Fieser and M. J. Haddadin, *Org. Syn.*, **46**, 107 (1966)

Diphenyliodonium chloride, $(C_6H_5)_2I^+Cl^-$. Mol. wt. 316.57, m.p. 228–229°.

Preparation.[1] A mixture of benzene, potassium iodate, sulfuric acid, and acetic anhydride is stirred until reaction is complete and then diluted with water. Extraction with ether and treatment with aqueous ammonium chloride gives a precipitate of diphenyliodonium chloride in yield of 63%.

Phenylation of anions. The reagent phenylates acidic di- and tri-ketones in good yield.[2] *t*-Butanol is the preferred solvent and sodium is used to generate the anion. A radical mechanism is proposed.

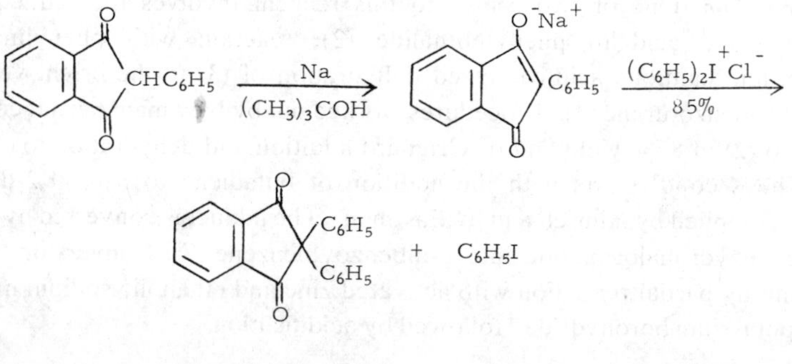

Hauser[3] found the reagent useful for effecting terminal phenylation of acetylacetone after treatment with sodium amide in liquid ammonia to produce (2). When a 2:1 ratio of disodium acetylacetone to diphenyliodonium chloride was used the yield was 92%, but with a 1:1 ratio the yield dropped to 47%. Apparently the

$$CH_3\overset{O}{\overset{\|}{C}}CH_2\overset{O}{\overset{\|}{C}}CH_3 \xrightarrow[\text{NH}_3]{2\ \text{NaNH}_2} CH_3\overset{O}{\overset{\|}{C}}\overset{-}{C}H\overset{O}{\overset{\|}{C}}CH_2 \xrightarrow[-C_6H_5I]{(C_6H_5)_2I^+Cl^-}$$

(1)   (2)

$$CH_3\overset{O}{\overset{\|}{C}}\overset{-}{C}H\overset{O}{\overset{\|}{C}}CH_2C_6H_5 \xrightarrow{(2)} CH_3\overset{O}{\overset{\|}{C}}\overset{-}{C}H\overset{O}{\overset{\|}{C}}\overset{-}{C}HC_6H_5 \xrightarrow[\substack{92\% \\ \text{overall}}]{H_2O} CH_3\overset{O}{\overset{\|}{C}}CH_2\overset{O}{\overset{\|}{C}}CH_2C_6H_5$$

(3)   (4)   (5)

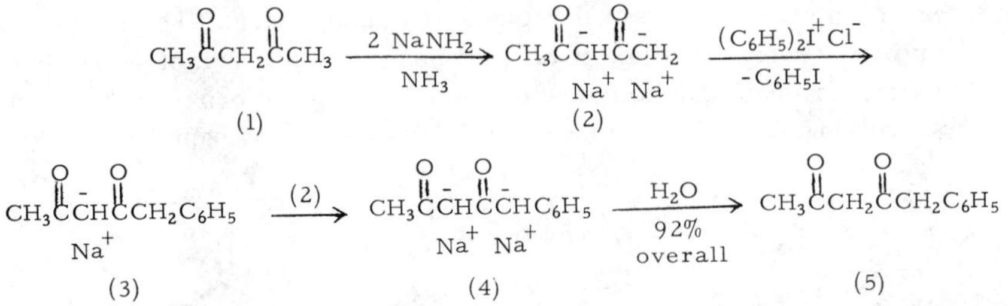

disodium salt (2) is converted into the phenylated sodium salt (3), which abstracts sodium ion from (2) to produce the disodium salt (4).

[1]F. M. Beringer, E. J. Geering, I. Kuntz, and M. Mausner, *J. Phys. Chem.*, **60**, 141 (1956)
[2]F. M. Beringer, P. S. Forgione, and M. D. Yudis, *Tetrahedron*, **8**, 49 (1960); F. M. Beringer, S. A. Galton, and S. J. Huang, *Am. Soc.*, **84**, 2819 (1962)
[3]K. G. Hampton, T. M. Harris, and C. R. Hauser, *J. Org.*, **29**, 3511 (1964)

1,3-Diphenylisobenzofurane (2,5-Diphenyl-3,4-benzofurane). Mol. wt. 270.31, m.p. 131°.

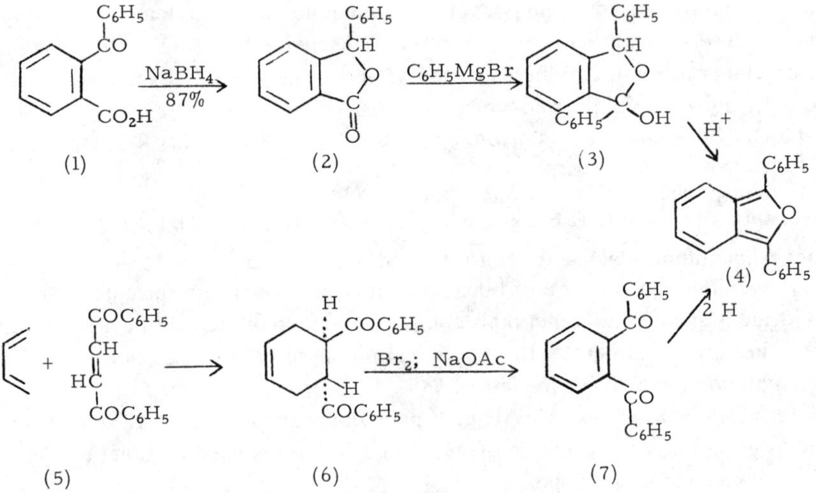

Preparation. One of two routes to this reagent involves the reduction of *o*-benzoylbenzoic acid to phenylphthalide (2), reaction with phenylmagnesium bromide, and cautious acid-catalyzed dehydration of (3) to the bright yellow 1,3-diphenylisobenzofurane (4).[1] Procedures worked out by Newman[2] permit conversion of (1) into (2) in 87% yield and of Grignard addition and dehydration to (4) in 87% yield. The second[3] starts with the addition of butadiene to *trans*-1,2-dibenzoylethylene (supplied by Aldrich and by Eastman). The adduct is converted by bromination and dehydrohalogenation into *o*-dibenzoylbenzene (7). Conversion to (4) has been done by partial reduction with activated zinc and ethanolic sodium hydroxide[3] or with potassium borohydride,[4] followed by acidification.

Trapping agent. The isobenzofurane is highly reactive to dienophiles, since the *o*-quinonoid ring becomes aromatic in the process. It reacts with a number of ethylenic dienophiles.[5] The reaction with acetylenedicarboxylic acid proceeds to the stage of the bis-adduct (8) even in the presence of a large excess of the acid.[6]

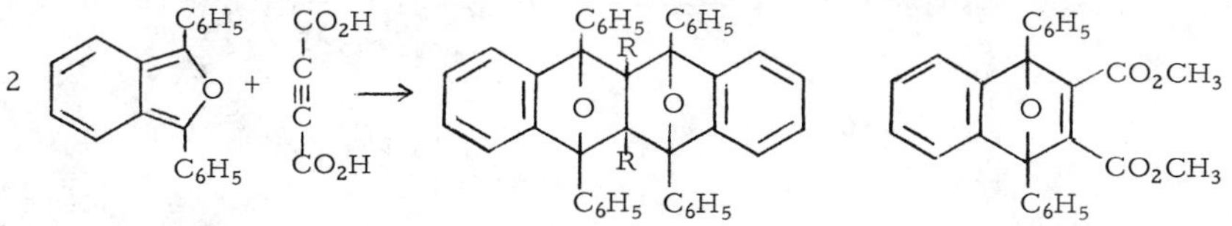

(4) (8, R = COOH) (9)

On the other hand, the reaction with dimethyl acetylenedicarboxylate afforded only the 1:1 adduct (9); both (8) and (9) were obtained in high yield.

The efficiency of diphenylisobenzofurane as a trapping agent is demonstrated by its reaction with benzyne to produce the adduct (10) in 85% yield.[7] The reagent

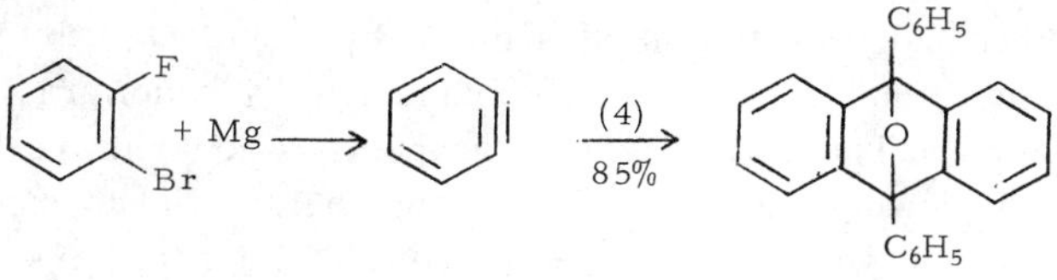

(10)

thus has been used to demonstrate the transient existence of highly strained cycloalkynes[8] and benzocyclobutadienes.[9]

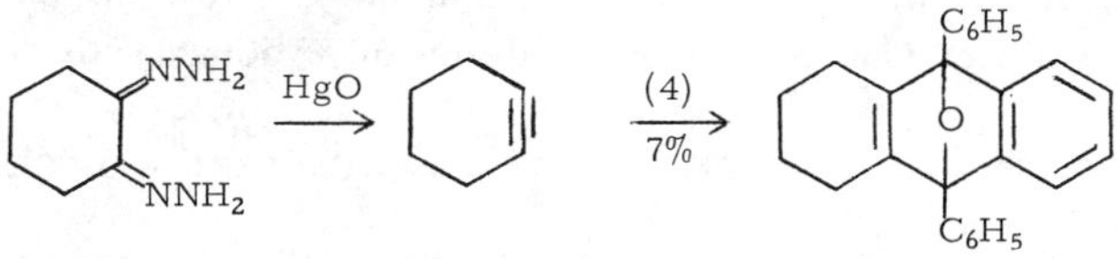

[1]A. Guyot and J. Catel, *Bull soc.*, [3], **35**, 1124 (1906)
[2]M. S. Newman, *J. Org.*, **26**, 2630 (1961)
[3]R. Adams and M. H. Gold, *Am. Soc.*, **62**, 56 (1940)
[4]M. P. Cava, M. J. Mitchell, and A. H. Deana, *J. Org.*, **25**, 1481 (1960)
[5]J. A. Norton, *Chem. Rev.*, **31**, 319 (1942)
[6]J. A. Berson, *Am. Soc.*, **75**, 1240 (1953)
[7]G. Wittig, E. Knauss, and K. Niethammer, *Ann.*, **630**, 10 (1960)
[8]G. Wittig and A. Krebs, *Ber.*, **94**, 3260 (1961); G. Wittig and R. Pohlke, *Ber.*, **94**, 3276 (1961)
[9]M. P. Cava and R. Pohlke, *J. Org.*, **27**, 1564 (1962); C. D. Nenitzescu *et al.*, *Ann.*, **653**, 79 (1962)

Diphenylketene, $(C_6H_5)_2C{=}C{=}O$. Mol. wt. 194.22, b.p. 120°/3.5 mm.

Preparation. Staudinger[1] prepared diphenylketene for the first time by the action of zinc turnings in ether on α-chlorodiphenylacetyl chloride, an intermediate now readily available by the action of phosphorus pentachloride on benzilic acid (1),[2] but evidently the yield was poor. He then tried dehydrohalogenation of

$$1.\quad (C_6H_5)_2\underset{\displaystyle OH}{\underset{|}{C}}COOH \xrightarrow[65-80\%]{PCl_5} (C_6H_5)_2\underset{\displaystyle Cl}{\underset{|}{C}}-\underset{\displaystyle Cl}{\underset{|}{C}}{=}O \xrightarrow{Zn} (C_6H_5)_2C{=}C{=}O$$

$$2.\quad (C_6H_5)_2\underset{\displaystyle H}{\underset{|}{C}}-\underset{\displaystyle Cl}{\underset{|}{C}}{=}O \xrightarrow{Base} (C_6H_5)_2C{=}C{=}O$$

diphenylacetyl chloride (2) with tripropylamine or with quinoline,[3] but later[4] expressed preference for a method introduced by Schroeter.[5] The Schroeter method is described in *Organic Syntheses:*[6] benzil monohydrazone is oxidized in benzene

$$3. \quad C_6H_5C-COC_6H_5 \xrightarrow[\text{CaSO}_4,\ 35^0]{\text{HgO, C}_6\text{H}_6} \underset{+\overset{|}{N}=N^-}{\overset{\overset{\displaystyle C_6H_5}{|}}{C_6H_5C-C=O}} \xrightarrow{-N_2}$$

$$\left[\overset{\overset{\displaystyle C_6H_5}{|}}{C_6H_5\overset{..}{C}-C=O} \right] \xrightarrow{58\%} (C_6H_5)_2C=C=O$$

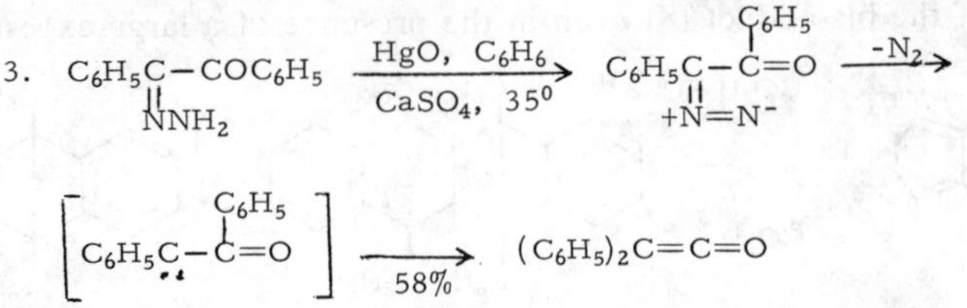

solution with yellow mercuric oxide in the presence of a dehydrating agent, and the filtered solution is dropped into a flask maintained at 100–110° to decompose the diazoketone and remove the benzene. The reagent should be stabilized with a trace of hydroquinone to inhibit polymerization and stored under nitrogen.

Taylor *et al.*[7] prefer a modification of Staudinger's second method. Diphenylacetyl chloride is obtained in 82–94% yield by reaction of the acid with thionyl chloride in ether and a solution in ether of triethylamine, added by drop with ice cooling under nitrogen. Diphenylketene is obtained by distillation at 118–119°/1 mm. in yield of 78–84%. The method utilizes inexpensive, nontoxic starting materials, involves two simple steps, and has the advantage that the diphenylketene, until the final distillation, is never exposed to temperatures greater than 30–35° and hence that polymerization is minimized.

Peptide synthesis. In introducing diphenylketene as a reagent for peptide synthesis, Losse and Demuth[8] used Staudinger's original synthesis.[1] A carbobenzoxy amino acid is treated with the ketene and the amino acid ester in tetrahydrofurane in the presence of triethylamine. Yields are around 50%; optical activity is retained.

$$\underset{\text{CbNHCHCO}_2\text{H}}{\overset{\overset{\displaystyle R}{|}}{}} + O=C=C(C_6H_5)_2 \xrightarrow{(C_2H_5)_3N} \underset{\text{CbNHCHC}-O-\text{CCH}(C_6H_5)_2}{\overset{\overset{\displaystyle R}{|}\ \ \overset{\displaystyle O}{\|}\ \ \overset{\displaystyle O}{\|}}{}}$$

$$\xrightarrow{\underset{\text{H}_2\text{NCHCO}_2\text{C}_2\text{H}_5}{\overset{\overset{\displaystyle R'}{|}}{}}} \underset{\text{CbNHCHCONHCHCO}_2\text{C}_2\text{H}_5}{\overset{\overset{\displaystyle R}{|}\ \ \ \ \ \overset{\displaystyle R'}{|}}{}} + HO_2CCH(C_6H_5)_2$$

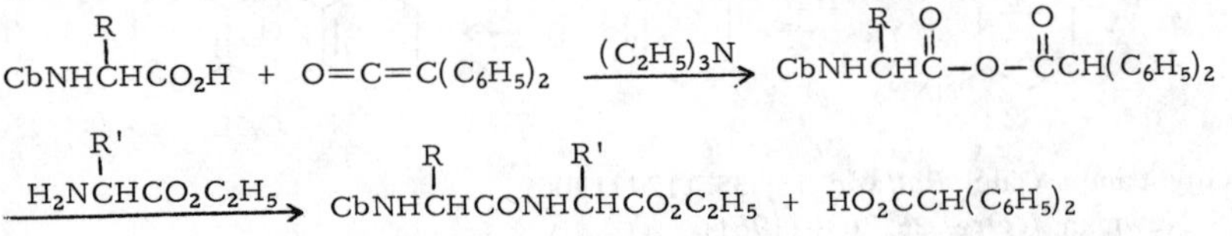

The mixed anhydride obtained from an N-acylamino acid (1) and diphenylketene, for example (2), is attacked by a phenol to give an optically pure aryl ester (3).[9]

$$\underset{\underset{\underset{\text{O}}{\overset{\displaystyle \|}{}}}{\overset{\displaystyle \text{HNCOCH}_2\text{C}_6\text{H}_5}{|}}}{\overset{\overset{\displaystyle \text{O}}{\|}}{H_2NCCH_2CH_2CHCO_2H}} \quad + \quad O=C=C(C_6H_5)_2 \xrightarrow[(C_2H_5)_3N]{\text{THF}}$$

(1)

$$\underset{\underset{\underset{\text{O}}{\overset{\displaystyle \|}{}}}{\overset{\displaystyle \text{HNCOCH}_2\text{C}_6\text{H}_5}{|}}}{\overset{\overset{\displaystyle \text{O}}{\|}\ \ \ \ \ \ \overset{\displaystyle \text{O}}{\|}\ \ \overset{\displaystyle \text{O}}{\|}}{H_2NCCH_2CH_2CHC-O-CCH(C_6H_5)_2}} \xrightarrow[\text{60--65\% overall}]{\text{HO}\langle\ \rangle\text{NO}_2}$$

(2)

$$\underset{\underset{\underset{\text{O}}{\overset{\displaystyle \|}{}}}{\overset{\displaystyle \text{HNCOCH}_2\text{C}_6\text{H}_5}{|}}}{\overset{\overset{\displaystyle \text{O}}{\|}\ \ \ \ \ \ \overset{\displaystyle \text{O}}{\|}}{H_2NCCH_2CH_2CHC-O-}}\langle\ \rangle\text{NO}_2 \quad + \quad \underset{}{\overset{\overset{\displaystyle \text{O}}{\|}}{HOCCH(C_6H_5)_2}}$$

(3)

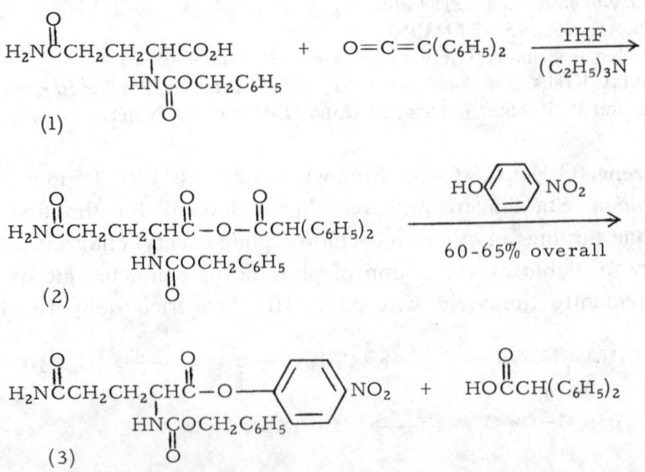

[1]H. Staudinger, *Ber.*, **38**, 1735 (1905)
[2]J. H. Billman and P. H. Hidy, *Am. Soc.*, **65**, 760 (1943)
[3]H. Staudinger, *Ber.*, **40**, 1148 (1907); **44**, 1619 (1911)
[4]H. Staudinger, *Ber.*, **44**, 1623 (1911)
[5]G. Schroeter, *Ber.*, **42**, 2346 (1909)
[6]L. I. Smith and H. H. Hoehn, *Org. Syn., Coll. Vol.*, **3**, 356 (1955)
[7]E. C. Taylor, A. McKillop, and G. H. Hawks, procedure submitted to *Org. Syn.*
[8]G. Losse and E. Demuth, *Ber.*, **94**, 1762 (1961)
[9]D. T. Elmore and J. Smyth, *Proc. Chem. Soc.*, 18 (1963)

Diphenylketene *p*-tolylimine, $(C_6H_5)_2C{=}C{=}N{-}C_6H_4CH_3$-*p*. Mol. wt. 283.36, m.p. 83.5°, yellow. Supplier: Pierce.

Preparation:[1]

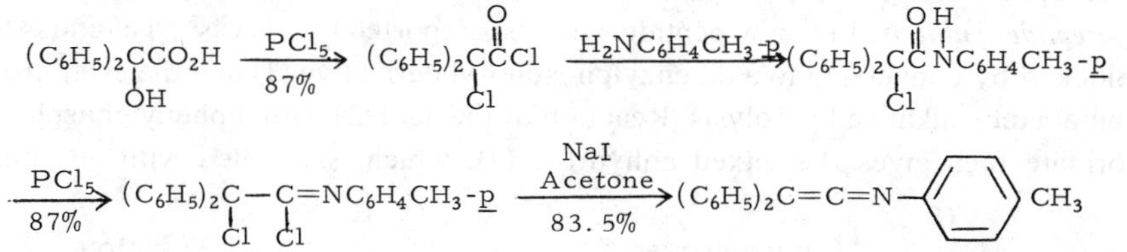

Reactions. The reagent reacts with a carboxylic acid to form an imide.[2] Succinic acid and phthalic acid are converted by the reagent into the anhydrides, but the

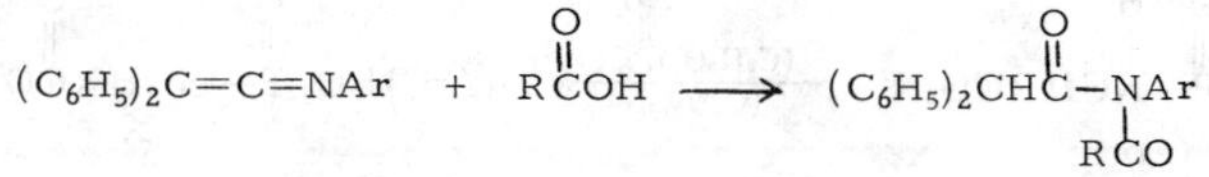

yields are not impressive. Like dicyclohexylcarbodiimide, the keteneimine can be used in peptide synthesis.[3]

[1]C. L. Stevens and J. C. French, *Am. Soc.*, **75**, 657 (1953)
[2]C. L. Stevens and M. E. Munk, *Am. Soc.*, **80**, 4065 (1958)
[3]*Idem, ibid.*, **80**, 4069 (1958)

Diphenylphosphide, lithium salt (2). The reagent is prepared in solution in tetrahydrofurane from diphenylphosphine (1) and *n*-butyllithium.[1] In refluxing THF

$$\underset{(1)}{(C_6H_5)_2PH} + \underline{n}\text{-BuLi} \xrightarrow{THF} \underset{(2)}{(C_6H_5)_2\overset{-}{P}\overset{+}{Li}} + \underline{n}\text{-BuH}$$

the reagent dealkylated anisole in good yield in 4 hrs.[2] In the same reaction period

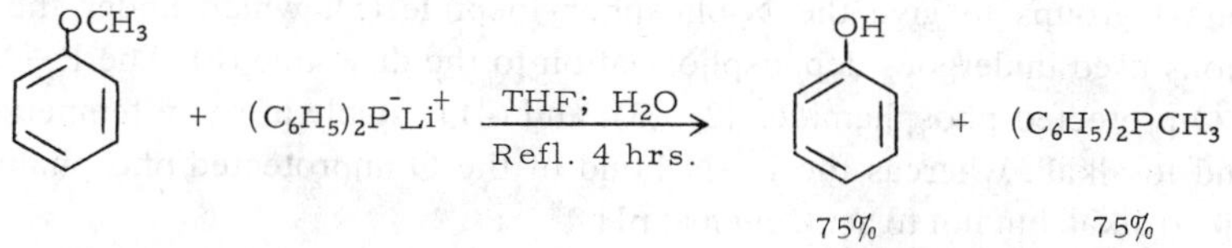

the yield of phenol from phenetole was only 9%. Hence with a mixed ether it should be possible to effect demethylation without deethylation. Benzyl and allyl ethers are cleaved by the reagent.

[1]D. Wittenberg and H. Gilman, *J. Org.*, **23**, 1063 (1958)
[2]F. G. Mann and M. J. Pragnell, *Chem. Ind.*, 1386 (1964)

Diphenylphosphorochloridate, $(C_6H_5O)_2\overset{O}{\overset{\|}{P}}Cl$. Mol. wt. 268.63, b.p. 147–148°/1.3 mm. Supplier: Aldrich.

Prepared by heating a mixture of 1.1 moles of phosphoryl chloride and 2 moles of phenol to 180° and distilling the product.[1] It is used for the preparation of DL-glyceraldehyde 3-phosphoric acid,[1] and dihydroxyacetone phosphate,[2] and diphenyl-phosphoroisothiocyanatidate (see below). This is the reagent of choice for the synthesis of monoalkyl phosphates: it (1) reacts with an alcohol in pyridine solution

$$(C_6H_5O)_2\overset{\overset{O}{\|}}{P}Cl \xrightarrow{HOR} (C_6H_5O)_2\overset{\overset{O}{\|}}{P}-OR \xrightarrow{H_2,\ Pt} (HO)_2\overset{\overset{O}{\|}}{P}-OR$$
$$(1) \qquad\qquad\qquad (2) \qquad\qquad\qquad (3)$$

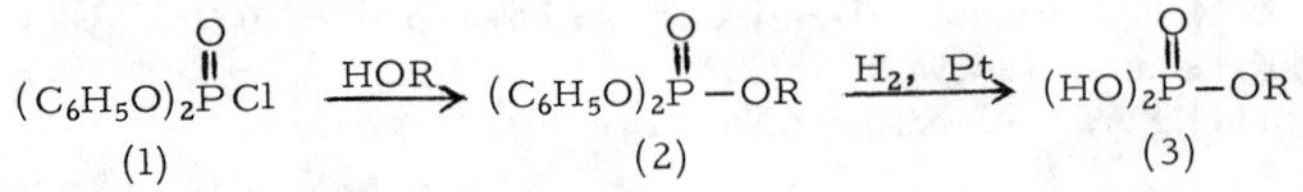

to give a triester (2), from which the phenyl groups can be removed by hydrogenolysis. Brigl and Müller[3] first demonstrated this use of the reagent for the synthesis of glycerol and fructose phosphates; subsequent applications are reviewed by Brown.[4]

Peptide synthesis. In a new peptide synthesis reported by Zervas[5] an amino ester is blocked by conversion to a dibenzylphosphoryl derivative (1) and the ester group is removed by alkaline hydrolysis. Reaction of the acid (2) with diphenylphosphoro-chloridate then gives the mixed anhydride (3), which is coupled with an amino

$$(ArCH_2O)_2\overset{\overset{O}{\|}}{P}-Cl \xrightarrow{H_2NCHRCO_2CH_3} (ArCH_2O)_2\overset{\overset{O}{\|}}{P}-\overset{\overset{H}{|}}{N}CHRCO_2CH_3 \xrightarrow{OH^-}$$
$$[Ar = p\text{-}NO_2C_6H_4] \qquad\qquad\qquad (1)$$

$$(ArCH_2O)_2\overset{\overset{O}{\|}}{P}-\overset{\overset{H}{|}}{N}CHRCO_2H \xrightarrow[\text{(C}_2\text{H}_5)_3\text{N}]{\text{(C}_6\text{H}_5\text{O)}_2\overset{O}{\|}{P}\text{Cl}} (ArCH_2O)_2\overset{\overset{O}{\|}}{P}-\overset{\overset{H}{|}}{N}CHR\overset{\overset{O}{\|}}{C}OP(OC_6H_5)_2$$
$$(2) \qquad\qquad\qquad\qquad\qquad\qquad (3)$$

$$\xrightarrow{H_2NCHR'CO_2CH_2C_6H_5} (ArCH_2O)_2\overset{\overset{O}{\|}}{P}-\overset{\overset{H}{|}}{N}CHR\overset{\overset{O}{\|}}{C}-\overset{\overset{H}{|}}{N}CHR'CO_2CH_2C_6H_5$$
$$(4)$$

$$\xrightarrow[\text{(H}^+)]{H_2,\ Pd} (HO)_2\overset{\overset{O}{\|}}{P}-\overset{\overset{H}{|}}{N}CHR\overset{\overset{O}{\|}}{C}-\overset{\overset{H}{|}}{N}CHR'CO_2H + 2\ CH_3C_6H_4NH_2 + C_6H_5CH_3$$
$$(5)$$
$$\xrightarrow[\text{(H}^+)]{-H_3PO_4 \mid H_2O} H_2NCHRCONHCHR'CO_2H$$
$$(6)$$

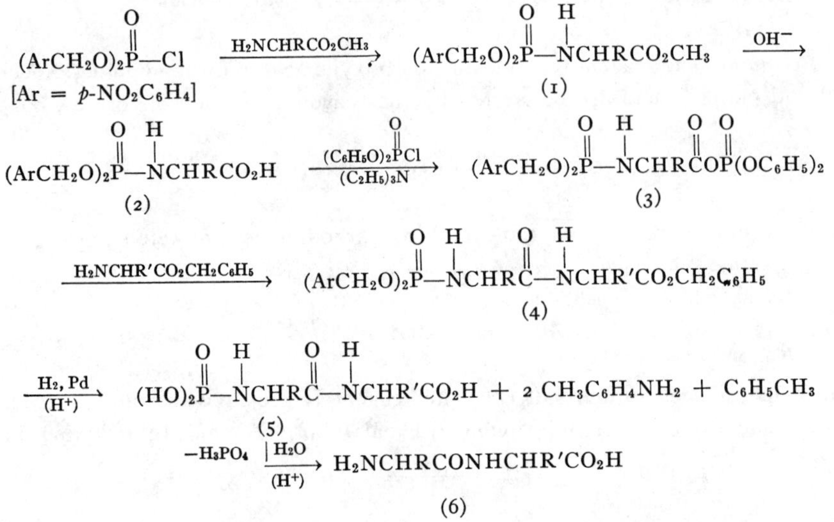

acid benzyl ester to form the dipeptide derivative (4). Hydrogenation removes all the benzyl groups to give the N-phosphorylpeptide (5), which under the acidic conditions used undergoes dephosphorylation to the dipeptide (6). The P—N bond of the O-protected phosphamides (2), (3), and (4) is stable at room temperature to acid and to alkali, whereas the P—N bond of the O-unprotected phosphamide (5) is stable to alkali but not to acid, even at pH 4.

[1] E. Baer, *Biochem. Preps.*, **1**, 50 (1951)
[2] C. E. Ballou, *Biochem. Preps.*, **7**, 45 (1960)
[3] P. Brigl and H. Müller, *Ber.*, **72**, 2121 (1939)
[4] D. M. Brown, *Advances in Organic Chemistry*, **3**, 89–90, Interscience Publishers (1963)
[5] A. Cosmatos, I. Photaki, and L. Zervas, *Ber.*, **94**, 2644 (1961)

Diphenylphosphoroisothiocyanatidate (2). Mol. wt. 291.26, b.p. 210°/0.1 mm., sp. gr. 1.29. Supplier: Aldrich.

Prepared[1] by reaction of diphenylphosphorochloridate (1) with potassium thio-cyanate in acetonitrile solution, from which potassium chloride promptly precipitates:

$$\underset{(1)}{(C_6H_5O)_2P(=O)Cl} + KNCS \longrightarrow \underset{(2)}{(C_6H_5O)_2P(=O){-}N{=}C{=}S} + KCl$$

The reagent reacts with the triethylamine salt of an N-acylated peptide to form a mixed anhydride, which decomposes to an acyl thiohydantoin bearing a substituent group identifying the amino acid at the carboxyl end of the peptide chain.[1]

[1] G. W. Kenner, H. G. Khorana, and R. J. Stedman, *J. Chem. Soc.*, 673 (1953)

Diphenyl phthalate. Mol. wt. 318.31, m.p. 74–76°. Supplier: Eastman. Preparation from phthaloyl chloride and phenol in 97% yield.[1]

The reagent is used for conversion of an amino acid into its phthaloyl derivative without racemization.[2] The acid, the reagent, and triethylamine are heated in phenol

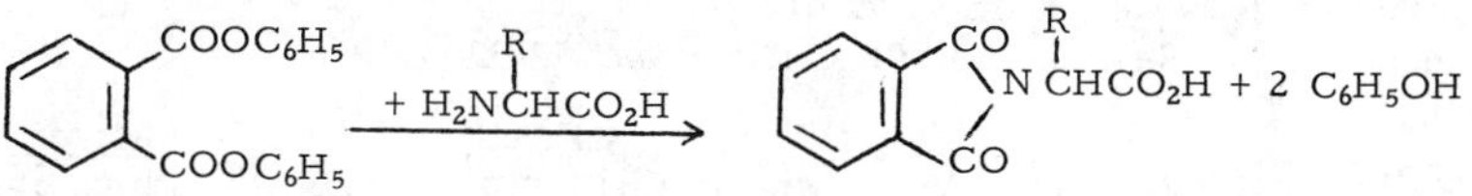

at 120–130° for 5–7 min., and the product is isolated by extraction from ether with sodium bicarbonate, followed by acidification.

[1] F. F. Blicke and O. J. Weinkauff, *Am. Soc.*, **54**, 330 (1932)

[2] F. Weygand and J. Kaelicke, *Ber.*, **95**, 1031 (1962)

Diphenylpicrylhydrazyl (a radical, see 4). Mol. wt. 394.32, m.p. 137–138°, violet. Suppliers: A, E.

Preparation. This stable, deep violet radical is prepared by condensation of 1,1-diphenylhydrazine with picryl chloride[1] and oxidation of the resulting hydrazine with the active lead dioxide of Kuhn and Hammer.[2] Purification.[3]

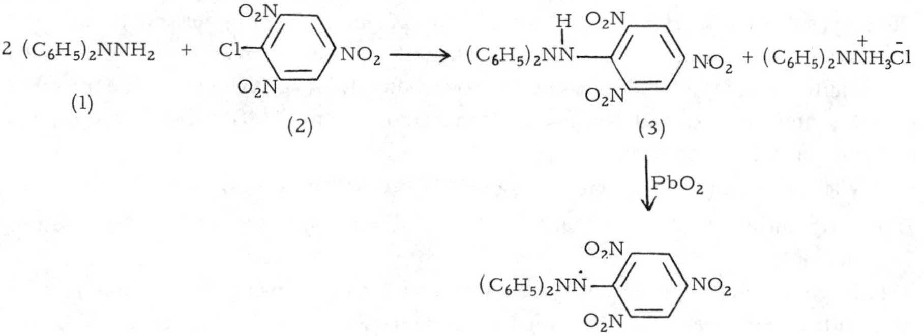

(4)

Dehydrogenation. Linstead[4] found that the radical can be used for hydrogen transfer. The reaction proceeds fairly rapidly in refluxing $CHCl_3$ or CCl_4 and can be

$$XH_2 + 2\,(C_6H_5)_2N\dot{N}Ar \longrightarrow X + 2\,(C_6H_5)_2N\overset{H}{N}Ar$$

followed by the change in color from deep violet to red to pale orange. The hydrazine formed precipitated almost quantitatively. Yields are good and sometimes reach 90–95%.

Bartlett and Kwart[5] regarded this radical as an ideal inhibitor (terminator) of radical chain reactions, but Hammond[6] found that the reaction of this radical with azobisisobutyronitrile is nonstoichiometric and oxygen-sensitive.

Dowd[7] effected dehydrogenation of (5) to (6) by refluxing in benzene for 72 hrs. with excess 2,2-diphenylpicrylhydrazyl.

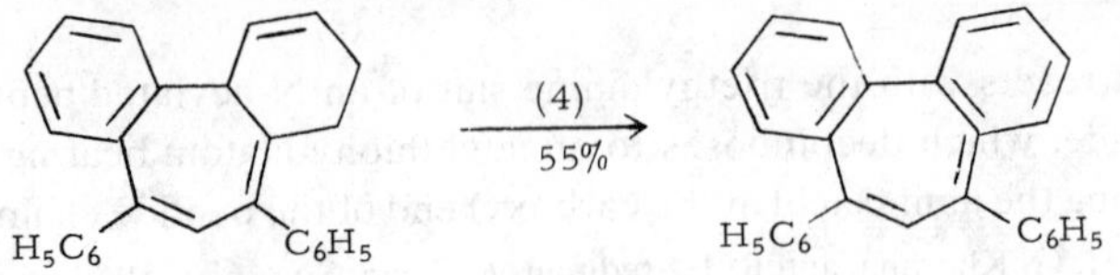

(5) (6)

[1]S. Goldschmidt and K. Renn, *Ber.*, **55**, 628 (1922)
[2]R. Kuhn and I. Hammer, *Ber.*, **83**, 413 (1950)
[3]J. A. Lyons and W. F. Watson, *J. Polymer Sci.*, **18**, 141 (1955)
[4]E. A. Braude, A. G. Brook, and R. P. Linstead, *J. Chem. Soc.*, 3574 (1954)
[5]P. D. Bartlett and H. Kwart, *Am. Soc.*, **72**, 1051 (1950)
[6]G. S. Hammond, J. N. Sen, and C. E. Boozer, *Am. Soc.*, **77**, 3244 (1955)
[7]P. Dowd, *Chem. Comm.*, 568 (1965)

Diphenylsilane, $(C_6H_5)_2SiH_2$. Mol. wt. 184.31, b.p. 75–76°/0.5 mm. Supplier: Peninsular ChemResearch.

The reagent, prepared by reduction of diphenyldichlorosilane (Dow Corning) with lithium aluminum hydride,[1] preferably in tetrahydrofurane,[2] reduces certain diaryl ketones to the corresponding hydrocarbons at the reflux temperature of about 260°. Benzophenone ⟶ diphenylmethane (37%); anthraquinone ⟶ anthracene (45%).[2]

[1]R. A. Benkeser, H. Landesman, and D. J. Foster, *Am. Soc.*, **74**, 648 (1952)
[2]H. Gilman and J. Diehl, *J. Org.*, **26**, 4817 (1961)

Diphenyl sulfoxide, $C_6H_5\overset{\overset{\displaystyle O^-}{|}}{S^+}C_6H_5$. Mol. wt. 202.27, m.p. 71°. Supplier: Eastman.

Although it is a solid at room temperature, the reagent proved useful for a study of the autoxidation of weakly acidic hydrocarbons (toluene, xylenes) in the presence of potassium *t*-butoxide at 100°.[1] Under these conditions DMSO is oxidized rapidly and is not a suitable solvent.

[1]T. J. Wallace, A. Schriesheim, and N. Jacobson, *J. Org.*, **29**, 2907 (1964)

Diphenyltelluride, $(C_6H_5)_2Te$. Mol. wt. 281.81, b.p. 312–320° dec., 174°/10 mm. Supplier: KK.

The reagent (1) is prepared by reaction of tellurium dibromide with phenylmagnesium bromide in ether and purified by conversion to crystalline, yellow diphenyl-

$$2\ C_6H_5MgBr + TeBr_2 \longrightarrow \underset{(1)}{(C_6H_5)_2Te} + 2\ MgBr_2$$

$$(C_6H_5)_2Te + Br_2 \longrightarrow \underset{(2)}{(C_6H_5)_2TeBr_2}$$

tellurium dibromide (2, m.p. 199–202°).[1] Diphenyltelluride is recovered by reaction of dibromide (2) with sodium bisulfite.

Because of its affinity for bromine, diphenyltelluride has been suggested for debromination of *vic*-dibromides,[2] as demonstrated for a few examples:

$$C_6H_5CHBrCHBrCO_2H + (C_6H_5)_2Se \longrightarrow C_6H_5CH{=}CHCO_2H + (C_6H_5)_2SeBr_2$$

However, debromination of cholesterol dibromide proceeded poorly.

[1]K. Lederer, *Ber.*, **48**, 1345 (1915)
[2]M. de Moura Campos, N. Petragnani, and C. Thomé, *Tetrahedron Letters*, **15**, 5 (1960)

Diphenyltin dihydride, $(C_6H_5)_2SnH_2$. Mol. wt. 298.94, m.p. −20°.

The reagent is prepared in ether solution by reduction of diphenyltin dichloride (Metal and Thermit Corp.) with lithium aluminum hydride.[1] It can be isolated by crystallization from petroleum ether–methylene chloride, but solutions in benzene are only moderately stable at room temperature.

Diphenyltin dihydride reduces aldehydes and simple ketones to alcohols in yields of about 80%, and the reaction differs from reduction with a metal hydride in that a

$$>C{=}O + (C_6H_5)_2SnH_2 \longrightarrow >CHOH + (C_6H_5)_2Sn$$

hydrolysis step is not required.[2] The diphenyltin formed, being insoluble in ether, is easily separated.

[1]H. G. Kuivila, A. K. Sawyer, and A. G. Armour, *J. Org.*, **26**, 1426 (1961)
[2]H. G. Kuivila and O. F. Beumel, Jr., *Am. Soc.*, **83**, 1246 (1961)

Diphenylzinc, $(C_6H_5)_2Zn$. Mol. wt. 219.58, m.p. 107°.

Preparation. (a) From diphenylmercury. Diphenylmercury, a highly toxic substance melting at 121–123°, is prepared by reaction of bromobenzene in xylene with 3% sodium amalgam.[1] A mixture of this substance and zinc wool is refluxed

$$2\ C_6H_5Br + Na_2Hg \xrightarrow[32\text{-}37\%]{\text{Xylene}} (C_6H_5)_2Hg + 2\ NaBr$$

$$(C_6H_5)_2Hg + Zn \xrightarrow{\text{Xylene}} (C_6H_5)_2Zn + ZnHg$$

in xylene in an atmosphere of oxygen-free carbon dioxide and the protective gas is used in collection and washing of the crystalline product.[2]

(b) A more convenient method is by reaction of phenyllithium and sublimed zinc bromide in ether.[3]

$$2\ C_6H_5Li + ZnBr_2 \longrightarrow (C_6H_5)_2Zn + 2\ LiBr$$

Reaction with diazonium salts. Curtin and Tveten[3] found that reagent prepared by either method reacts with a benzenediazonium fluoroborate in dimethylformamide to give a *trans*-azo compound in high yield. However, when diphenylzinc prepared

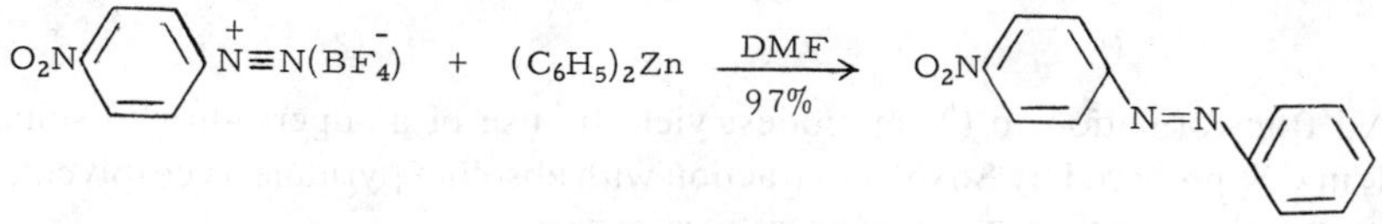
O2N
N≡N(BF4)
+ (C6H5)2Zn
DMF
97%
O2N
N=N

by method (a) was taken in large excess the product was predominantly the *cis*-isomer.

[1]H. O. Calvery, *Org. Syn., Coll. Vol.*, **1**, 228 (1941)
[2]K. A. Kozeschkow, A. N. Nesmejanow, and W. I. Potrosow, *Ber.*, **67**, 1138 (1934)
[3]D. Y. Curtin and J. L. Tveten, *J. Org.*, **26**, 1764 (1961)

Diphosphorus tetraiodide, P_2I_4 (formerly phosphorus diiodide). Mol. wt. 569.59, m.p. 124.5°. Preparation.[1]

Kuhn and Winterstein[2] introduced the reagent for the conversion of 1,2-glycols into olefins. Thus hydrocinnamoin is reduced in high yield to 1,6-diphenylhexatriene. A suspension of the diol in ether is shaken with the reagent until conversion to the

$$\underset{\text{50 g. + 700 ml. ether}}{C_6H_5CH{=}CH\underset{OH}{C}H\underset{OH}{C}HCH{=}CHC_6H_5} + \underset{\text{60 g.}}{P_2I_4} \xrightarrow{90\%} C_6H_5CH{=}CHCH{=}CHCH{=}CHC_6H_5$$

yellow hydrocarbon is complete. The reaction probably proceeds through the unstable *vic*-diiodide.

Kuhn also used the reagent for the transformation of polyacetylene diols to tetraarylcumulenes,[3] for example:

$$(C_6H_5)_2\underset{OH}{C}-C{\equiv}C-C{\equiv}C-\underset{OH}{C}(C_6H_5)_2 \xrightarrow{P_2I_4} (C_6H_5)_2C{=}C{=}C{=}C{=}C{=}C(C_6H_5)_2$$

Bohlmann[4] employed the same reaction sequence to the synthesis of a series of aliphatic cumulenes carrying substituents that block rearrangement to acetylenes.

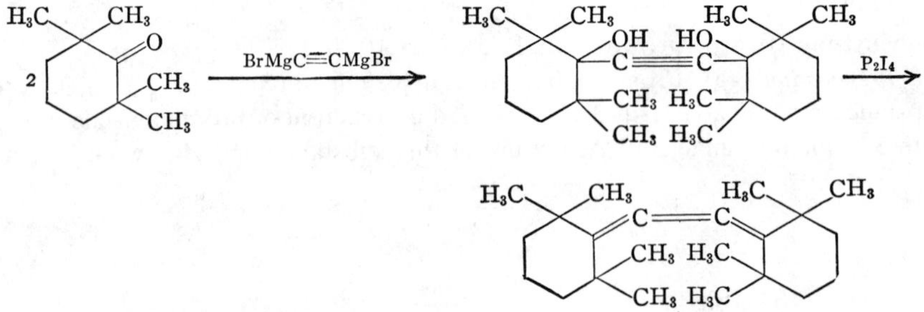
H₃C
CH₃
O
2
CH₃
CH₃
BrMgC≡CMgBr
OH
HO
C
C
P₂I₄

Inhoffen et al.[5] sought to apply the Kuhn-Winterstein reaction to the glycol (1) but found the reagent to be ineffective in ether, THF, benzene, or pyridine, but

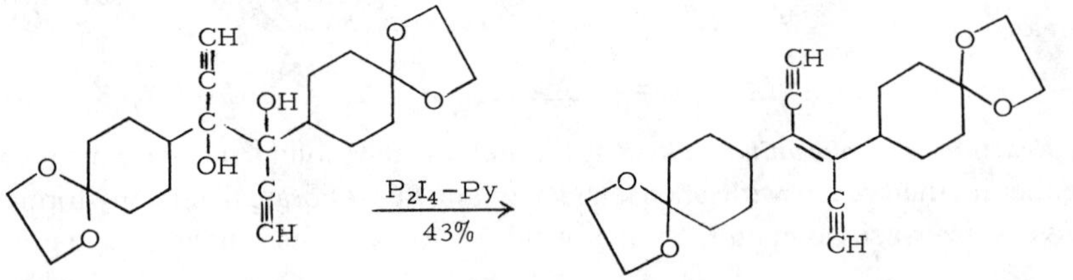
CH
C
OH
P₂I₄–Py
43%

(1) (2)

effected transformation to (2) in modest yield by use of a supersaturated solution of P_2I_4 in CS_2 prepared by Soxhlet extraction with absolute pyridine as cosolvent.

[1]F. E. E. Germann and R. N. Traxler, *Am. Soc.*, **49**, 307 (1927)

[2]R. Kuhn and A. Winterstein, *Helv.*, **11**, 106 (1928)

[3]Paper I: R. Kuhn and K. Wallenfels, *Ber.*, **71**, 1899 (1938); Paper VI: R. Kuhn and K. L. Scholler, *Ber.*, **87**, 598 (1954)

[4]F. Bohlmann, *Ber.*, **85**, 386 (1952)

[5]H. H. Inhoffen, K. Radscheit, U. Stache, and V. Koppe, *Ann.*, **684**, 24 (1965)

Dipotassium tetramethyl osmate, *see* Potassium triacetyl osmate.

Disodium phenanthrene. In a preliminary communication, Vogel *et al.*[1] state that this reagent debrominates (1, "cyclooctatetraene dibromide") to give bicyclo-[4.2.0]octa-2,4,7-triene (2) in good yield. The purest sample contained 95% (2) and the only impurity was cyclooctateraene (3), produced during distillation at −20°,

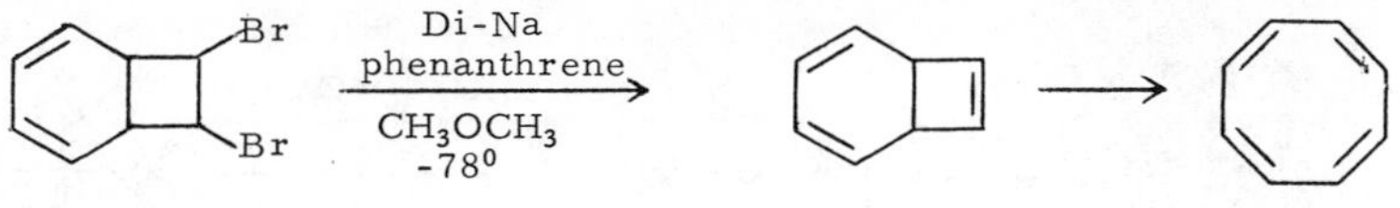

(1) (2) (3)

[1]E. Vogel, H. Kiefer, and W. R. Roth, *Angew. Chem., Internat. Ed.*, **3**, 442 (1964)

5,5′-Dithiobis(2-nitrobenzoic acid). Mol. wt. 368.33, m.p. 238°. Supplier: Aldrich. Preparation.[1]

In aqueous solution at pH 8, the reagent reacts with a sulfhydryl group to form a highly colored anion, λ_{max}412 mμ, ε = 13,600, and is thus of use for the quantitative

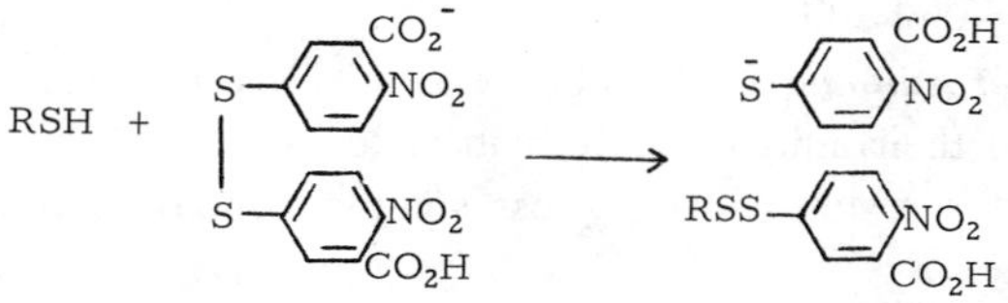

determination of sulfhydryl groups.[1] It has been used in a study of microbiol transsulfurization[2] and in an investigation of the effect of radioprotective agents on tissue sulfhydryl levels.[3]

[1]G. L. Ellman, *Arch. Biochem. Biophys.*, **82**, 70 (1959)
[2]M. Flavin, *J. Biol. Chem.*, **237**, 768 (1962)
[3]B. Sörbo, *Arch. Biochem. Biophys.*, **98**, 342 (1962)

Di-p-toluoyl-D-tartrate,

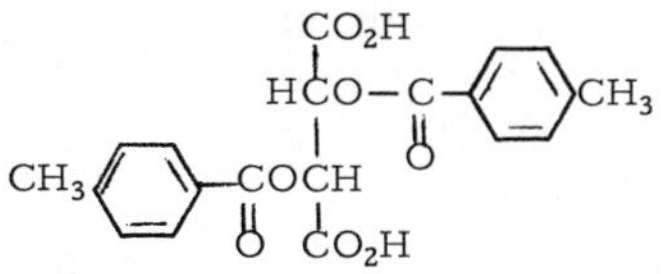

Mol. wt. 386.34, m.p. 169–171°, αD −141° (EtOH). Suppliers: Aldrich, Frinton. The reagent is prepared[1,2] by stirring a mixture of D-tartaric acid and *p*-toluoyl chloride in xylene for 3 hrs., pouring the resulting solution into benzene, collecting the crystalline anhydride formed, and hydrolyzing it to the diacid.

This optically active diacid has been used for the resolution of *dl*-isolysergic acid hydrazide[1] and of a synthetic intermediate.[3] It is useful also for the characterization of amines, since the salts (chiefly acid salts) are stable, crystalline solids of high melting points.[2,4] A salt can be prepared by mixing ethereal solutions of the acid and of the amine.[2] The amine can be regenerated by shaking a solution of the salt in chloroform – methanol with aqueous alkali.[2]

[1]A. Stoll and A. Hofmann, *Helv.*, **26**, 922 (1943)
[2]D. A. A. Kidd, *J. Chem. Soc.*, 4675 (1961)
[3]R. B. Woodward *et al.*, *Tetrahedron*, **2**, 50 (1958)
[4]R. F. Collins, *J. Chem. Soc.*, 2053 (1960)

Di-p-tolylcarbodiimide, *p*-$CH_3C_6H_4N{=}C{=}NC_6H_4CH_3$-*p*. Mol. wt. 222.28, m.p. 56°.

Preparation by oxidation of di-*p*-tolylthiourea with yellow mercuric oxide.[1,2]

Addition of alcohols. The reagent adds alcohols in dioxane in the presence of a base to give N,N′-di-*p*-tolyl-O-alkyl pseudourea esters (2) in quantitative yield.[2] The pseudourea ester (2) reacts with a dialkyl phosphate (3) to give a neutral tertiary

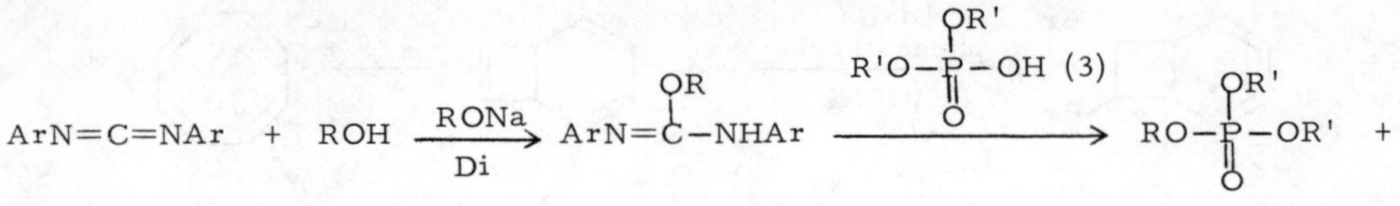

$$\text{ArN=C=NAr} + \text{ROH} \xrightarrow[\text{Di}]{\text{RONa}} \text{ArN=}\overset{\overset{\displaystyle OR}{|}}{C}\text{-NHAr} \xrightarrow{\overset{\displaystyle R'O-\overset{\overset{\displaystyle OR'}{|}}{\underset{\displaystyle O}{\overset{\|}{P}}}-OH \ (3)}{}} \text{RO}-\overset{\overset{\displaystyle OR'}{|}}{\underset{\displaystyle O}{\overset{\|}{P}}}-OR' \ +$$

(1)   (2)   (4)

$$\text{ArN}-\overset{\overset{\displaystyle H}{|}}{\underset{}{\overset{\displaystyle O}{\overset{\|}{C}}}}-\overset{\overset{\displaystyle H}{|}}{N}\text{Ar}$$

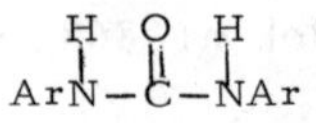

(5)

phosphate (4) and the diarylurea (5). Dicyclohexyldicarbodiimide (DCC) does not undergo the alcohol-addition.

Dehydration of sulfonic acids. The reagent converts aromatic and aliphatic sulfonic acids into their anhydrides in high yield.[3] It is preferred to DCC for this purpose because the resulting urea is less soluble in benzene and easily separated from the anhydride.

Preparation of pyrophosphoric acid esters. The reagent reacts with mono and di esters of phosphoric acid at room temperature to give the corresponding symmetrical di and tetra esters of pyrophosphoric acid.[4] Yields are "well-nigh" quantitative. DCC reacts similarly but isolation of tetra-*p*-nitrophenyl pyrophosphate presented some difficulty owing to similar solubilities of the ester and the substituted urea. With di-*p*-tolylcarbodiimide in dioxane this difficulty was overcome.

Nucleoside phosphites. Reaction of a nucleoside with phosphorous acid and di-*p*-tolylcarbodiimide in pyridine affords the nucleoside phosphite in good yield.[5] Thus

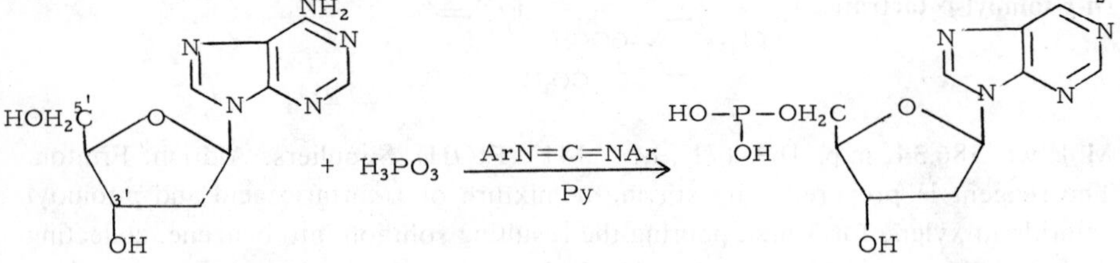

(6)   (7)

a mixture of desoxyadenosine with phosphorous acid is thoroughly dried by evaporation with pyridine thrice at 10^{-2} mm. and then dissolved in dry pyridine along with di-*p*-tolylcarbodiimide. After standing for 3 days at room temperature, the major product, isolated as the ammonium salt, proved to be desoxyadenosine-5'-phosphite (42%). The minor product (29%) was the 3'-phosphite.

[1]F. Zetzsche, H. E. Meyer, H. Overbeck, and W. Nerger, *Ber.*, **71**, 1512 (1938); F. Zetzsche and W. Nerger, *ibid.*, **73**, 467 (1940)
[2]H. G. Khorana, *Can. J. Chem.*, **32**, 227 (1954)
[3]*Idem, ibid.*, **31**, 585 (1953)
[4]H. G. Khorana and A. R. Todd, *J. Chem. Soc.*, 2257 (1953)
[5]J. A. Schofield and A. R. Todd, *ibid.*, 2316 (1961)

Divinylmercury, $Hg(CH=CH_2)_2$. Mol. wt. 254.70, b.p. 48–50°/14 mm.

The reagent is prepared by addition of a solution of mercuric chloride in tetrahydrofurane to a solution of vinylmagnesium bromide in the same solvent.[1] Because of toxicity, it should be handled with extreme care.

If the reagent is warmed on the steam bath with an aliphatic or aromatic carboxylic acid without solvent, a rapid reaction affords a vinyl ester, ethylene, and metallic mercury (1).[2] If the reaction is carried out in an inert solvent to lengthen the reaction

$$1.\quad RCO_2H + Hg(CH{=}CH_2)_2 \longrightarrow RCOOCH{=}CH_2 + CH_2{=}CH_2 + Hg$$

$$2.\quad RCO_2H + Hg(CH{=}CH_2)_2 \longrightarrow RCOOHgCH{=}CH_2 + CH_2{=}CH_2$$

$$3.\quad ArOH + Hg(CH{=}CH_2)_2 \longrightarrow ArOCH{=}CH_2 + CH_2{=}CH_2 + Hg$$

$$4.\quad ArSH + Hg(CH{=}CH_2)_2 \longrightarrow ArSCH{=}CH_2 + CH_2{=}CH_2 + Hg$$

time it is possible to obtain the intermediate vinylmercuric ester (2). These esters are crystalline solids which decompose on heating to give the vinyl ester and mercury. Divinylmercury reacts similarly with phenols (3) and with thiophenols (4) to form vinyl ethers and vinyl thioethers.[2]

[1]G. F. Reynolds, R. E. Dessy, and H. H. Jaffé, *J. Org.*, **23**, 1217 (1958)
[2]D. J. Foster and E. Tobler, *Am. Soc.*, **83**, 851 (1961)

Dowtherm A, a eutectic mixture containing 26.5% of diphenyl and 73.5% of diphenyl ether. M.p. 12°, b.p. 258°.

Spatz[1] found that thermal decomposition of diaryl fumerates to *trans*-stilbene derivatives, originally carried out by dry distillation, often proceeds more satisfactorily by refluxing in Dowtherm for 36–50 hrs. However, the yield of *trans*-

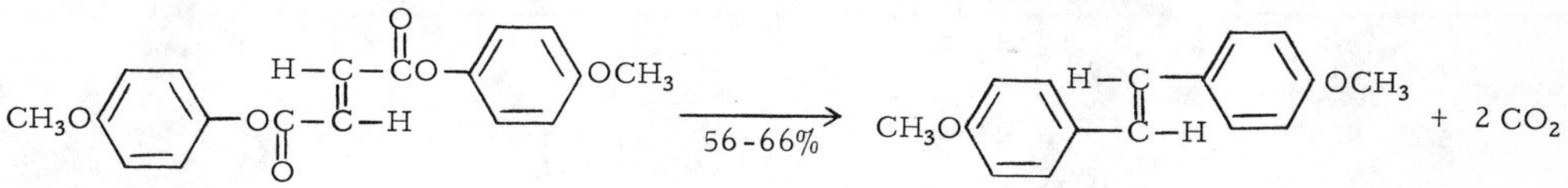
H−C−CO
CH3O
OCH3
OC−C−H
56-66%
H−C
C−H
+ 2 CO2

stilbene itself was only 19%, the reaction with the di-*p*-nitro derivative, and other yields were in the range 10–40%.

Pearlman[2] prepared 3-isoquinuclidone (3) by hydrogenating *p*-aminobenzoic acid in water at 50 p.s.i., adding DMF and cooling, collecting the mixture of *cis*- and

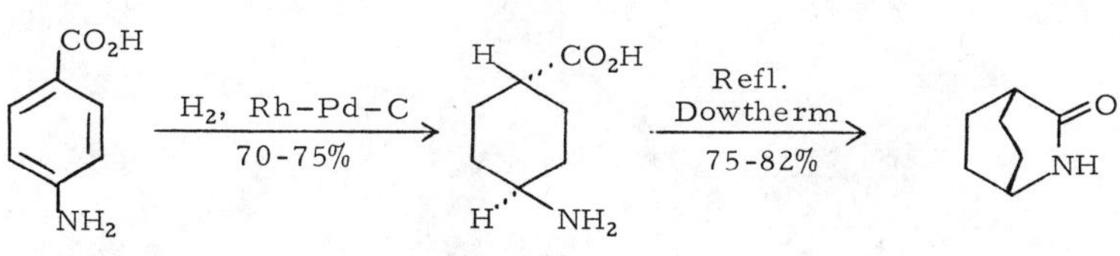
CO2H
NH2
H2, Rh−Pd−C
70-75%
H
CO2H
H
NH2
Refl.
Dowtherm
75-82%
O
NH
+ trans

(1) (2) 208 g. (3) M.p. 195-197°
650 ml.

trans-hexahydrides (2), and refluxing it in Dowtherm for 20 min. 3-Isoquinuclidone is an excellent substitute for camphor for Rast molecular weight determination.[3]

[1]S. M. Spatz, *J. Org.*, **26**, 4158 (1961)
[2]W. M. Pearlman, procedure submitted to *Org. Syn.*
[3]E. Ferber and H. Brückner, *Ber.*, **76**, 1019 (1943)

Drying agents. Pearson and Ollerenshaw[1] state that whereas the determination of the water content of solvents by the Karl Fischer method is tedious and requires large samples, determination by spectrophotometric measurement of the overtone band of water near 5,300 cm.$^{-1}$ is a simple matter requiring no expenditure of sample.

Comparison of the speed and efficiency in the drying of benzene, diethyl ether, and ethyl acetate at room temperature showed that calcium chloride, anhydrous calcium sulfate, and 4A-Molecular Sieves are all fast drying agents which remove considerable amounts of water. In comparison, magnesium sulfate and sodium sulfate are slow drying agents which remove only small amounts of water. With a given agent, diethyl ether was dehydrated considerably faster than either benzene or ethyl acetate.

[1]B. D. Pearson and J. E. Ollerenshaw, *Chem. Ind.*, 370 (1966)

E

Epichlorohydrin, $\overset{O}{CH_2\text{—}CHCH_2Cl}$. Mol. wt. 92.53, b.p. 117°, sp. gr. 1.183. Suppliers: A, B, E, F, MCB.

In the bromination of the enol acetate (1), addition of epichlorohydrin as proton acceptor increases the yield of the axial 6β-bromoketone (2), resulting from kinetic control, to 85%.[1] Bromination in acetic acid in the presence of collidine gave a

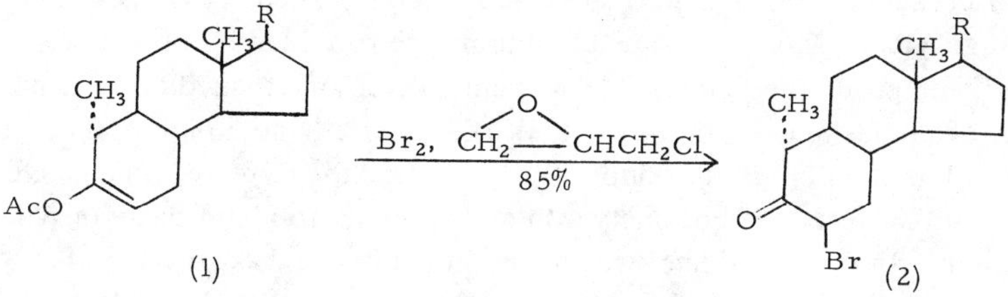

R
CH_3
CH_3
AcO
Br_2, CH_2—$CHCH_2Cl$
O
85%
R
CH_3
CH_3
O
Br

(1) (2)

1:1 mixture of the 6α- and 6β-compounds. Petrow *et al.*[2] used ethylene oxide and propylene oxide for the same purpose; they found that in the bromination of 3-keto-Δ^4-steroids addition of an epoxide increases the yield of 4-bromo-3-keto-Δ^4-steroids to 50% or more; with collidine as additive yields are in the order of 5–30%.

[1]M. P. Hartshorn and E. R. H. Jones, *J. Chem. Soc.*, 1312 (1962)

[2]D. N. Kirk, D. K. Patel, and V. Petrow, *J. Chem. Soc.*, 627 (1956)

1,2-Epoxy-3-phenoxypropane, $\overset{O}{CH_2\text{—}CHCH_2OC_6H_5}$. Mol. wt. 150.18, m.p. 1–3°. Suppliers: B, E, F, MCB.

The reagent is useful as a nonalkaline scavenger for hydrogen bromide.[1]

[1]I. M. Hunsberger and J. M. Tien, *Chem. Ind.*, 88 (1959)

Ethane-1,2-diamineborane, $H_3B\!:\!\underset{H}{\overset{H}{N}}CH_2CH_2\underset{H}{\overset{H}{N}}\!:\!BH_3$. Mol. wt. 87.79, dec. 89°.

This 1:1 complex, a white crystalline solid of considerable stability, is prepared either by heterogeneous absorption of diborane by ethylenediamine in a high vacuum apparatus or by reaction of ethylenediamine with the tetrahydrofurane–borane complex:[1]

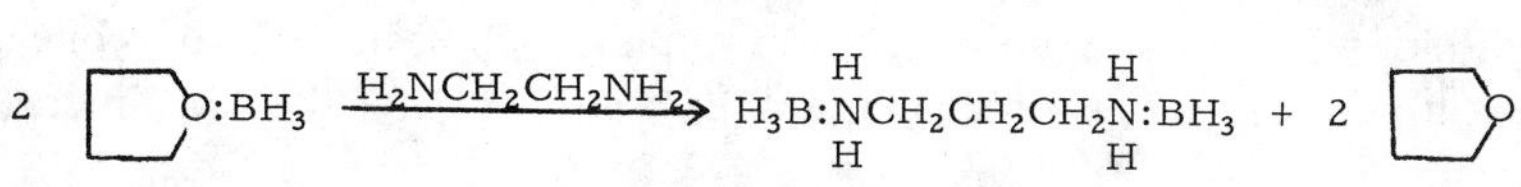

2
O:BH_3
$H_2NCH_2CH_2NH_2$
H H
H_3B:$NCH_2CH_2CH_2N$:BH_3 + 2
H H
O

It is a reducing agent similar in reactivity and selectivity to sodium borohydride. It is readily soluble in tetrahydrofurane, diethylene glycol dimethyl ether, and ethanol, slightly soluble in water or carbon tetrachloride, and insoluble in ether or benzene. Cinnamaldehyde is reduced by the reagent in tetrahydrofurane to cinnamyl alcohol in 94% yield.

[1]H. C. Kelley and J. O. Edwards, *Am. Soc.*, **82**, 4842 (1960)

Ethanedithiol, $HSCH_2CH_2SH$. Mol. wt. 94.20, b.p. 146° (63°/46 mm.), sp. gr. 1.14. Suppliers: Eastern, A, B, E, F, MCB. Prepared from ethylene dibromide and thiourea (55–62% yield)[1] or potassium thiolacetate (56% yield).[2]

The reagent is used for conversion of a ketone into its ethylenethioketal which on desulfurization affords the corresponding desoxo compound. In one efficient

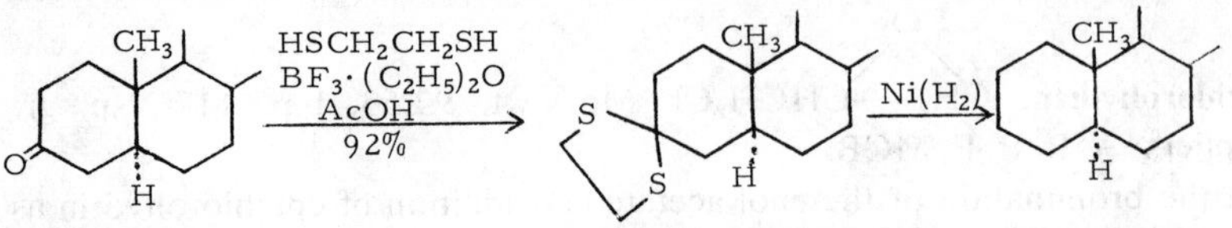

CH3
HSCH2CH2SH
BF3·(C2H5)2O
AcOH
92%
S
S
H
Ni(H2)

procedure (A) a solution of 600 mg. of cholestanone and 1 ml. of ethanedithiol in 15 ml. of hot acetic acid was treated with 1 ml. of boron trifluoride etherate and let stand for crystallization. The derivative separated in excellent needles of high purity and in high yield.[3] More vigorous conditions are provided by procedure B,[3] which takes advantage of the remarkable solvent power of ethanedithiol: a mixture of 180 mg. of cholestane-3,6-dione (1) and 0.2 ml. of ethanedithiol in a test tube was treated with 0.2 ml. of boron trifluoride etherate and the mixture homogenized with a stirring rod. The mixture became warm and soon set to a stiff paste of white solid. After 5 min., 8 ml. of methanol was added, the mixture was stirred and cooled, and the product collected; yield of cholestane-3,6-dione bisethylenethioketal 233 mg. (94%). By applying to the 3,6-diketone (1) the more gentle procedure A and using 1 equivalent of reagent it was possible to prepare the 3-monoethylenethioketal (2).

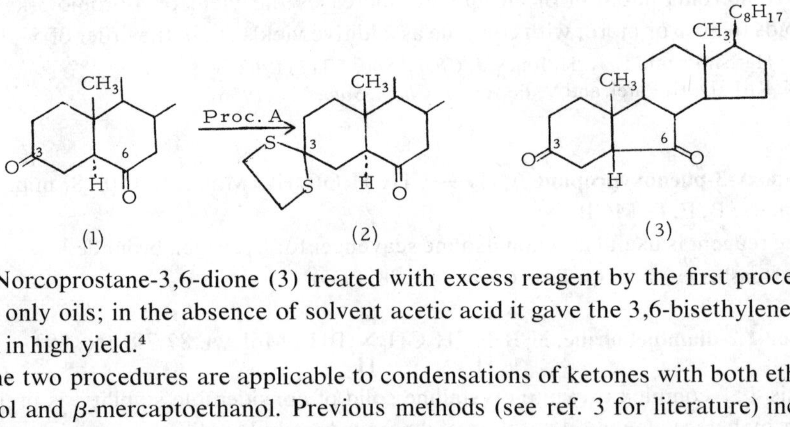

CH3
Proc. A
S
S
C8H17

(1) (2) (3)

(B) Norcoprostane-3,6-dione (3) treated with excess reagent by the first procedure gave only oils; in the absence of solvent acetic acid it gave the 3,6-bisethylenethioketal in high yield.[4]

The two procedures are applicable to condensations of ketones with both ethanedithiol and β-mercaptoethanol. Previous methods (see ref. 3 for literature) include use of zinc chloride and sodium sulfate, hydrogen chloride in ether, *p*-toluenesulfonic acid in benzene with azeotropic distillation, and an exchange method.

Abnormal products have been encountered in the BF_3-catalyzed condensation of ethanedithiol with steroid ketones.[5] For example, Δ^4-cholestene-3β-ol-6-one acetate (4) on treatment with excess ethanedithiol (BF_3–AcOH) gave the C_4-epimers (5) and (6).

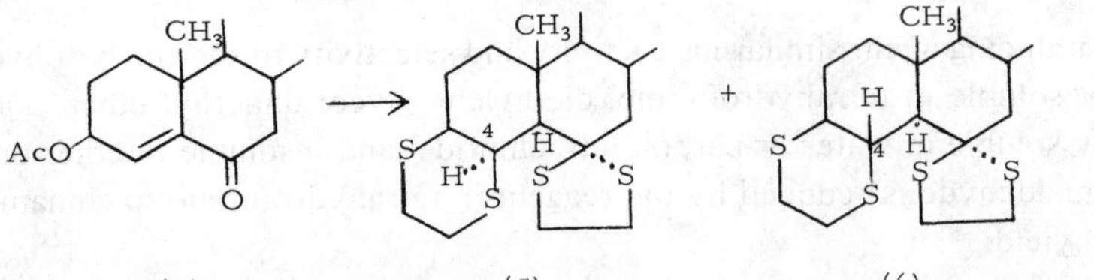

CH3
AcO
O
S
S
S
S
H
+

(4) (5) (6)

[1]A. J. Speziale, *Org. Syn., Coll. Vol.*, **4**, 401 (1963)
[2]L. N. Owen and P. N. Smith, *J. Chem. Soc.*, 2973 (1951)
[3]L. F. Fieser, *Am. Soc.*, **76**, 1945 (1954)
[4]*Idem, ibid.*, **75**, 4386 (1953)
[5]L. F. Fieser, C. Yuan, and T. Goto, *Am. Soc.*, **82**, 1996 (1960)

Ethanolamine, $HOCH_2CH_2NH_2$. Mol. wt. 57.05, b.p. 171°, sp. gr. 1.022. Suppliers: A, B, E, F, MCB.

The reagent has been used for the cleavage of picric acid salts of amines.[1] The finely divided yellow picrate is shaken with a mixture of ether and a concentrated aqueous solution of ethanolamine, and the ether solution is drawn off and extracted repeatedly with ethanolamine solution until it is no longer colored. Hünig and Baron studied the velocity of splitting of quaternary ammonium salts on refluxing with ethanolamine.[2]

$$R_4\overset{+}{N} + H_2NCH_2CH_2OH \longrightarrow R_3N + RNHCH_2CH_2OH + \overset{+}{H}$$

[1]N. Weiner and I. A. Kaye, *J. Org.*, **14**, 868 (1949)
[2]S. Hünig and W. Baron, *Ber.*, **90**, 395, 403 (1957)

Ethoxyacetylene, $HC{\equiv}COC_2H_5$. Mol. wt. 86.09, b.p. 49°. Suppliers: Pfister, KK, Chem. Samples Co.

Preparation. A preparative procedure that has been carefully standardized[1] and subsequently improved[2] involves addition of diethylchloroacetal (1) to a solution of sodamide in liquid ammonia to form the sodio derivative (2), which is cooled to

$$\underset{(1)}{ClCH_2CH(OC_2H_5)_2} \xrightarrow[NH_3]{NaNH_2} \underset{(2)}{NaC{\equiv}COC_2H_5} \xrightarrow{H_2O} \underset{(3)}{HC{\equiv}COC_2H_5}$$

−70° and treated with a solution of sodium chloride precooled to −20° to liberate ethoxyacetylene (3). Since this terminal step is attended with a fire hazard, Stork and Tomasz[3] developed an alternative procedure starting with the addition of bromine to vinyl ethyl ether (4) and eliminating HBr by reaction of (5) with N,N-

$$\underset{(4)}{CH_2{=}CHOC_2H_5} \xrightarrow{Br_2} \underset{(5)}{\underset{Br\ \ Br}{CH_2CHOC_2H_5}} \xrightarrow{(C_2H_5)_2NC_6H_5} \underset{(6)}{BrCH{=}CHOC_2H_5}$$

$$\xrightarrow[NH_3]{LiNH_2} \underset{(7)}{LiC{\equiv}COC_2H_5} \xrightarrow[-70^0]{H_2O} \underset{(3)}{HC{\equiv}COC_2H_5}$$

diethylaniline to give β-bromovinyl ethyl ether (6). Reaction of (6) with lithium amide in liquid ammonia eliminated a second mole of HBr and afforded the lithium derivative (7). Treatment of (7) at −70° with a cold solution of calcium chloride liberated ethoxyacetylene (3).

Synthetic uses. Arens,[4] in reviewing the many reactions of ethynyl ethers and thioethers, states that the first use of ethoxyacetylene for the synthesis of α,β-unsaturated aldehydes was described in a paper published in Russia in 1945,[5] but that the paper was abstracted incorrectly and remained unknown in the western world. van Dorp and Arens[6, 7] rediscovered the method and in 1947 described its use in the synthesis of vitamin A aldehyde. Ethoxyacetylene is treated with ethylmagnesium bromide to form the Grignard reagent, and this is added to the carbonyl group of the "C_{18}-ketone" to produce the acetylenic carbinol (1). Semihydrogenation

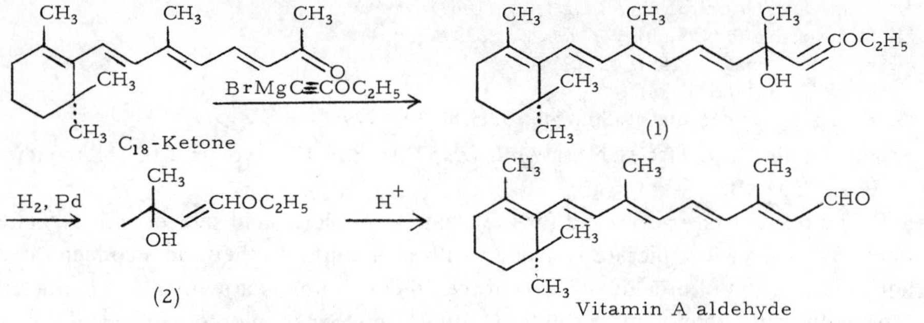

to (2) and acid hydrolysis of this enol ether gives vitamin A aldehyde. Heilbron *et al.*[8] report further syntheses of α,β-unsaturated aldehydes by this method and also a general method for the synthesis of α,β-unsaturated esters and acids involving hydration of an ethoxyacetylenic carbinol such as (3). Brief shaking of the carbinol with 10% sulfuric acid effects conversion into the unsaturated ester, which on hydrolysis affords β-methylcrotonic acid (5) in overall yield of 55%.

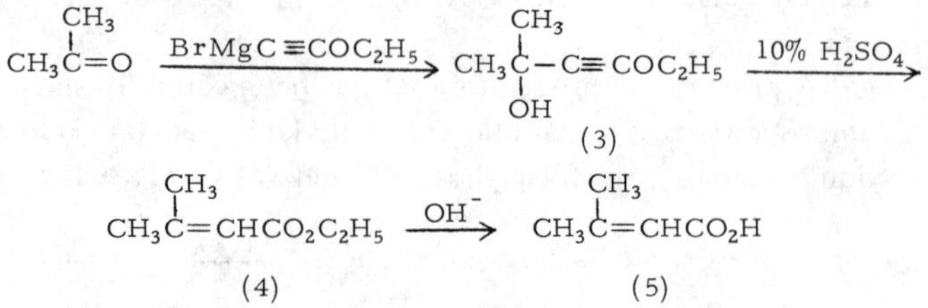

One step in the Sarett total synthesis of cortisone[9] involved addition of ethoxyvinylmagnesium bromide to the reactive carbonyl group of the diketone (6) to produce

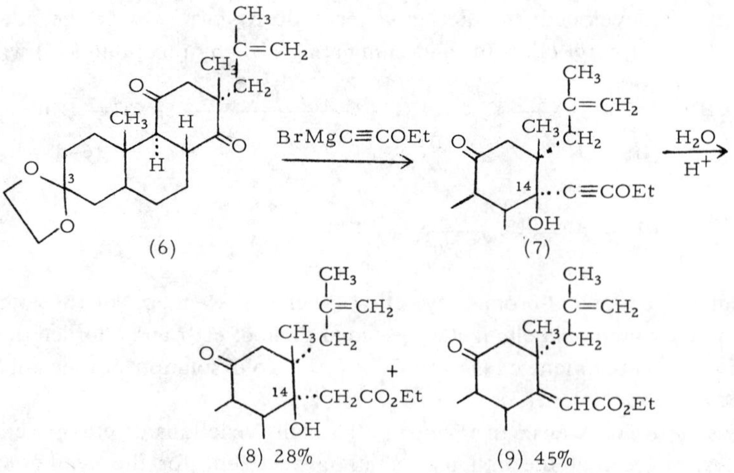

the acetylenic carbinol (7, and its 14-epimer). Acid-catalyzed hydration of the ethynyl group was carried out under very mild conditions in order to avoid elimination of the protective ethyleneketal group at C_3: a solution of (7) in 65 ml. of tetrahydrofurane was treated with 3.9 ml. of 10% sulfuric acid and stirred for 3 hrs. at room temperature. Isolation of the two products (8) and (9) showed that hydration had been accomplished in part with retention of the tertiary hydroxyl at C_{14}.

For effecting addition of the ethoxyethynyl group to the carbonyl group of crotonaldehyde, Stork and Tomasz[3] added ethoxyacetylene in ether to an ethereal
solution of phenyllithium and added the aldehyde to the resulting suspension of
lithium ethoxyacetylide (1). The resulting carbinol is rearranged readily under acid
catalysis to ethyl sorbate (4).

$$C_2H_5OC\equiv CH \xrightarrow{C_6H_5Li} C_2H_5OC\equiv CLi + CH_3CH=CHCHO \longrightarrow$$
$$(1) \qquad\qquad\qquad (2)$$

$$\overset{OH}{\underset{|}{C_2H_5OC\equiv CCHCH=CHCH_3}} \xrightarrow{H^+} C_2H_5O\overset{O}{\overset{||}{C}}CH=CHCH=CHCH_3$$
$$(3) \qquad\qquad\qquad\qquad (4)$$

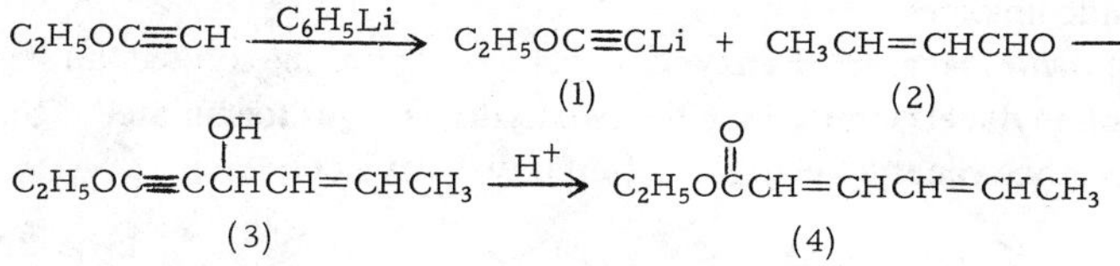

Another route to α,β-unsaturated esters involves reaction of an aldehyde or ketone
with ethoxyacetylene in the presence of boron fluoride etherate.[10]

$$R_2C=O + HC\equiv COC_2H_5 \xrightarrow{BF_3} R_2C=CHCO_2C_2H_5$$

Ethoxyacetylene reacts with diphenylketene in nitromethane at $-20°$ to give the

$$\begin{array}{c} (C_6H_5)_2C=C=O \\ + \\ C_2H_5OC\equiv CH \end{array} \xrightarrow[35\%]{-20°} \begin{array}{c} (C_6H_5)_2C-C=O \\ |\quad\quad| \\ C_2H_5OC=CH \end{array}$$

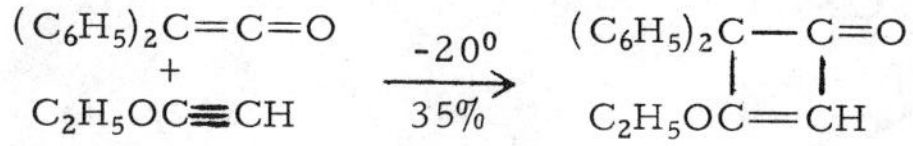

enol ethyl ether of 2,2-diphenylcyclobutane-1,3-dione in modest yield.[11] In benzene
solution the reaction takes a complicated course and affords two unusual products
of rearrangement.[12] Ketene reacts with ethoxyacetylene to give cyclobutane-1,3-
dione enol ethyl ether in 30% yield.[13]

Peptide synthesis. Arens[14] has used ethoxyacetylene as a dehydrating agent in
peptide synthesis. As solvent, ethyl acetate containing 0.5% of added water proved
superior to the anhydrous ester. An N-protected amino acid and an amino acid ester
are refluxed in moist ethyl acetate with 4–5 equivalents of ethoxyacetylene until

$$\underset{\underset{R}{|}}{CbNHCHCO_2H} + \underset{\underset{R'}{|}}{H_2NCHCO_2C_2H_5} + HC\equiv COC_2H_5 \longrightarrow$$

$$\underset{\underset{R}{|}\quad\quad\underset{R}{|}}{CbNHCHCONHCHCO_2C_2H_5} + CH_3CO_2C_2H_5$$

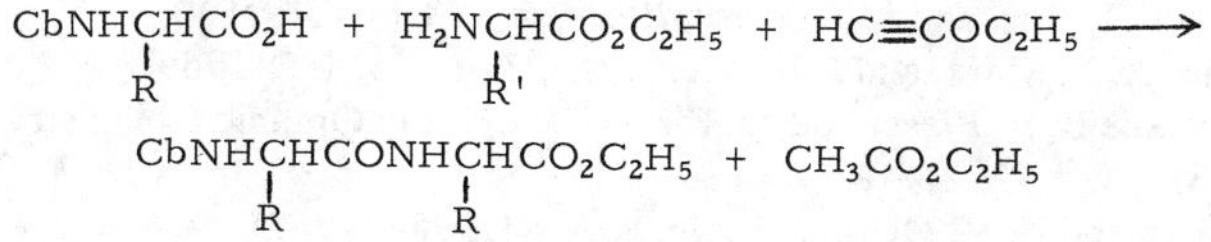

coupling is complete, and the ethyl acetate is removed by evaporation. Yields are
satisfactory and there is no racemization.

Bodanszky and Birkhimer[15] found that in certain cases an N-protected amino
acid reacts with *p*-nitrophenol in the presence of 2 equivalents of ethoxyacetylene
at 50° for 1 hr. and at room temperature for 24 hrs. to give the *p*-nitrophenyl ester.
However, the reaction often stops at the ethoxyvinyl ester stage or at the anhydride
stage.

Phosphorylation. Ethoxyacetylene (2) reacts with a diester of phosphoric acid
(1) to form a 1-ethoxyvinyl ester (3).[16] The reaction is carried out in a chlorinated

$$(C_6H_5CH_2O)_2\overset{O}{\overset{||}{P}}OH + HC\equiv COC_2H_5 \longrightarrow (C_6H_5CH_2O)_2\overset{O}{\overset{||}{P}}O\overset{CH_2}{\overset{||}{C}}OC_2H_5$$
$$(1) \qquad\qquad (2) \qquad\qquad\qquad (3)$$

$$\xrightarrow{C_6H_5CO_2H} (C_6H_5CH_2O)_2\overset{O}{\overset{||}{P}}-O\overset{O}{\overset{||}{C}}C_6H_5$$
$$(4)$$

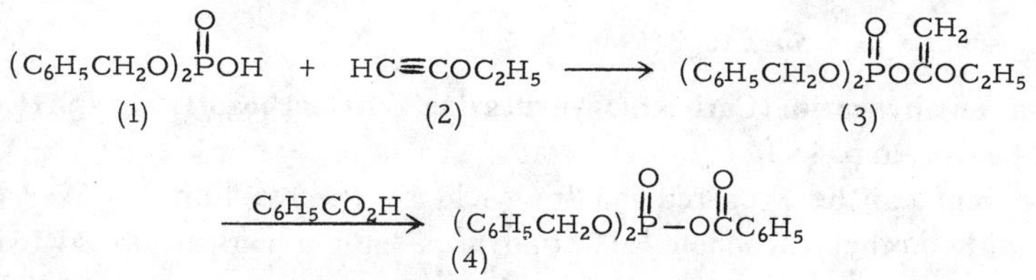

solvent or in CH_2Cl_2-DMF at 0° and affords an enol phosphate (3), which is an active phosphorylating agent. Thus it reacts with benzoic acid to form the mixed anhydride (4). Adenosine-5′-phosphate, as the pyridine salt, reacts with ethoxyacetylene to form an active ester which in methanol solution gives the monomethyl ester of adenosine-5′-monophosphate. The method has been used with success to form internucleoside linkages.

Reaction with benzyne. Benzyne generated from diazotized anthranilic acid reacts with phenylacetylene to give the two hydrocarbons formulated.[17] Surprisingly, it reacts with ethoxyacetylene to form 2-ethoxyphenylacetylene.

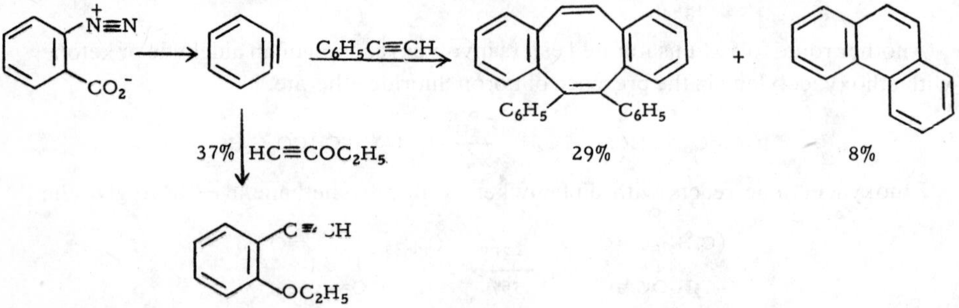
$\overset{+}{N}\equiv N$
CO_2^-
$C_6H_5C\equiv CH$
C_6H_5
C_6H_5
+
37%
$HC\equiv COC_2H_5$
29%
8%
$C\equiv CH$
OC_2H_5

[1]E. R. H. Jones, G. Eglinton, M. C. Whiting, and B. L. Shaw, *Org. Syn., Coll. Vol.*, **4**, 404 (1963)
[2]H. H. Wasserman and P. S. Wharton, *Am. Soc.*, **82**, 661 (1960)
[3]G. Stork and M. Tomasz, *Am. Soc.*, **86**, 471 (1964)
[4]J. F. Arens, *Advances in Organic Chemistry*, **2**, 117 (1960)
[5]N. A. Preobrajensky and V. V. Shokina, *Zhur. Obschcheĭ Khim*, **15**, 65 (1945)
[6]D. A. van Dorp and J. F. Arens, *Nature*, **160**, 189 (1947)
[7]J. F. Arens and D. A. van Dorp, *Rec. trav.*, **67**, 973 (1948)
[8]I. Heilbron, E. R. H. Jones, M. Julia, and B. C. L. Weedon, *J. Chem. Soc.*, 1823 (1949)
[9]G. E. Arth, G. I. Poos, R. M. Lukes, F. M. Robinson, W. F. Johns, M. Feurer, and L. H. Sarett, *Am. Soc.*, **76**, 1715 (1954)
[10]H. Vieregge, H. J. T. Bos, and J. F. Arens, *Rec. trav.*, **78**, 664 (1959)
[11]H. J. Panneman, A. F. Marx, and J. F. Arens, *Rec. trav.*, **78**, 487 (1959)
[12]For a summary, see L. F. Fieser and M. Fieser, "Topics in Organic Chemistry," pp. 504–505, Reinhold (1963)
[13]H. H. Wasserman and E. V. Dehmlow, *Am. Soc.*, **84**, 3786 (1962)
[14]J. Nieuwenhuis and J. F. Arens, *Rec. trav.*, **77**, 1153 (1958)
[15]M. Bodanszky and C. A. Birkhimer, *Chem. Ind.*, 1620 (1962)
[16]H. H. Wasserman and D. Cohen, *J. Org.*, **29**, 1817 (1964)
[17]M. Stiles, U. Burckhardt, and A. Haag, *J. Org.*, **27**, 4715 (1962)

1-Ethoxybutene-1-yne-3, $HC\equiv CCH=CHOC_2H_5$. Mol. wt. 96.13, b.p. 42°/15 mm., 71°/53 mm. The reagent is prepared by the action of sodium ethoxide on 1,4-dichloro-2-butyne, presumably through the intermediate formation of diacetylene.[1] For uses, *see* 1-Methoxybutene-1-yne-3.

$$ClCH_2C\equiv CCH_2Cl \xrightarrow{NaOC_2H_5} HC\equiv CC\equiv CH \xrightarrow{C_2H_5OH} HC\equiv CCH=CHOC_2H_5$$

[1]A. W. Johnson, *J. Chem. Soc.*, 1009 (1946)

Ethoxycarbonylhydrazine (Carboethoxyhydrazine, Ethyl carbazate). $H_2NNHCO_2C_2H_5$. Mol. wt. 104.11, m.p. 45°.

The reagent can be prepared in 90% yield by the reaction of 85% hydrazine hydrate with diethyl carbonate.[1] It condenses with a variety of aldehydes and

$$H_2NNH_2 + C_2H_5OC(=O)OC_2H_5 \longrightarrow H_2NNHC(=O)OC_2H_5 + C_2H_5OH$$

ketones to give crystalline carboethoxyhydrazones of characteristic melting point.[2] Particularly striking is a report by Joly and Nominé[3] that the reagent can be used to convert a 4β-bromo-3-keto-5β-steroid (1) to the Δ^4-3-ketosteroid (3).[3] Bromoketones of type (1) are resistant to dehydrohalogenation by usual methods because

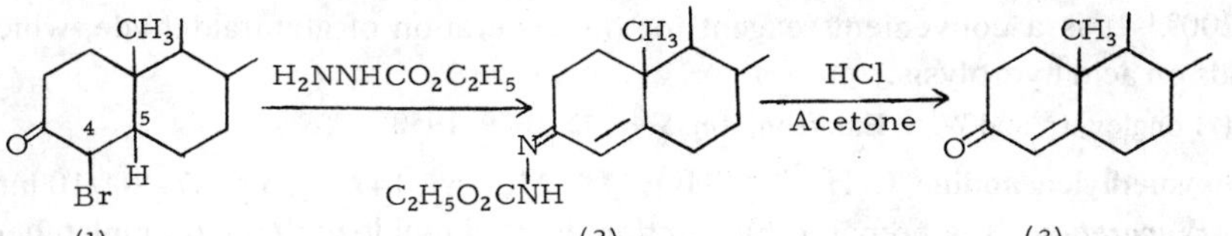
CH3
4
5
Br
H
H2NNHCO2C2H5
CH3
N
C2H5O2CNH
HCl
Acetone
CH3

(1) (2) (3)

of the *cis*-orientation of the groups to be eliminated. 2,4-Dinitrophenylhydrazine and semicarbazide attack (1) by a nonstereospecific mechanism and yield halogen-free derivatives of type (2), which afford (3) on hydrolysis. The carboethoxyhydrazone (2) would appear to have the advantage that it is hydrolyzed very easily by hydrochloric acid in acetone. Successful use of the method is reported by Kirk and Petrow.[4]

[1]O. Diels, *Ber.*, **47**, 2183 (1914)
[2]N. Rabjohn and H. D. Barnstorff, *Am. Soc.*, **75**, 2259 (1953)
[3]R. Joly and G. Nominé, *Bull. Soc.*, 1381 (1956)
[4]D. N. Kirk and V. Petrow, *J. Chem. Soc.*, 1691 (1959)

Ethoxycarbonylmethylenetriphenylphosphorane, $(C_6H_5)_3P{=}CHCO_2C_2H_5$. Mol. wt. 348.37, m.p. 123–125°. Supplier: Aldrich. *See* Carboethoxymethylenetriphenylphosphorane.

2-Ethoxy-1,2-dihydro-1-methyl-6,8-dinitroquinoline, (1). Mol. wt. 279.24, m.p. 124°.

Preparation by Skraup synthesis from 2,4-dinitroaniline and glycerol, followed by treatment with ethanol.[1] The reagent is a specific precipitant for hydroperoxides.[1] Thus Rieche *et al.*,[2] on achieving the synthesis of allylhydroperoxide in preparative quantity by the method formulated, characterized the compound by conversion into the derivative (2), which forms yellow needles, m.p. 101. 5–103°.

$$CH_2{=}CHCH_2MgCl \xrightarrow{ZnCl_2} CH_2{=}CHCH_2ZnCl \xrightarrow{O_2}$$

$$CH_2{=}CHCH_2OOZnCl \xrightarrow{H_2O_2} CH_2{=}CHCH_2OOH$$

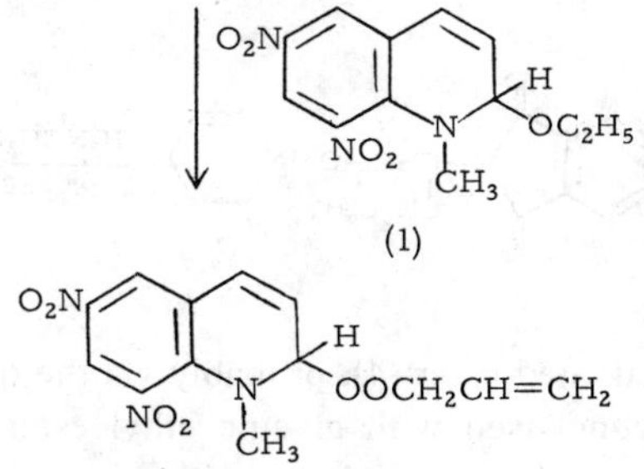
O2N
H
N
OC2H5
NO2
CH3
(1)
O2N
H
N
OOCH2CH=CH2
NO2
CH3

(2)

[1]A. Rieche, E. Schmitz, and P. Dietrich, *Ber.*, **92**, 2239 (1959)
[2]H. E. Seyfarth, J. Henkel, and A. Rieche, *Angew. Chem., Internat. Ed.*, **4**, 1074 (1965)

2-Ethoxy-3,4-dihydro-1,2-pyrane. Mol. wt. 128.17, b.p. 143–145°, 42°/16 mm. Suppliers: Shell Chem. Co., Union Carbide, Baker.

The reagent is prepared by heating acrolein with ethyl vinyl ether in an autoclave

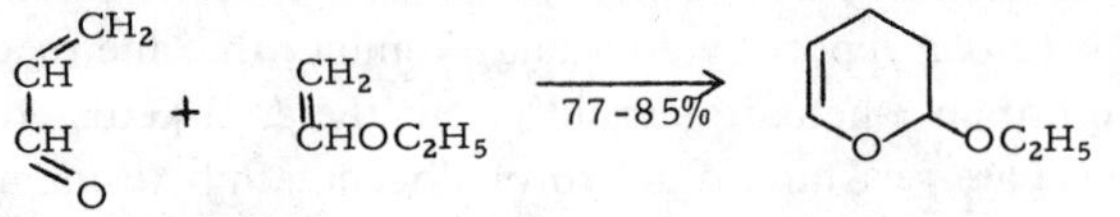

at 200°.[1] It is a convenient reagent for the generation of glutaraldehyde, which it yields on acid hydrolysis.

[1]R. I. Longley, Jr. and W. S. Emerson, *Am. Soc.*, **72**, 3079 (1950)

Ethoxymethyleneaniline, $C_6H_5N{=}CHOC_2H_5$. Mol. wt. 149.19, b.p. 87–88°/10 mm.

Preparation. It is prepared by reaction of ethyl iodide with the (completely dry) silver salt of formanilide and stored in the dark:[1]

$$C_6H_5N(H)CHO \longrightarrow C_6H_5N{=}CHOAg \xrightarrow{C_2H_5I} C_6H_5N{=}CHOC_2H_5$$

Aldehyde synthesis. Monier-Williams[1] showed that the reagent reacts with a Grignard reagent to give an imine, hydrolyzable to an aldehyde:

$$RMgX + C_2H_5OCH{=}NC_6H_5 \longrightarrow RCH{=}NC_6H_5 \xrightarrow{H_2O} RCHO$$

In a comprehensive review of methods for the conversion of Grignard reagents into aldehydes, Smith and Nichols[2] concluded that this is the best route, at least for aromatic aldehydes. The main difficulty lies in the preparation and cost of the reagent.

[1]G. W. Monier-Williams, *J. Chem. Soc.*, **89**, 273 (1906)
[2]L. I. Smith and J. Nichols, *J. Org.*, **6**, 489 (1941)

2-Ethoxypyridine-1-oxide (1). Mol. wt. 123.15, m.p. 72–74°.

Preparation.[1] 2-Ethoxypyridine is heated overnight at 60° with 30% hydrogen peroxide in acetic acid solution. The N-oxide is obtained in 60–65% yield as greyish-white crystals.

Peptide synthesis.[2] The reagent (1) reacted readily with N-phthaloyl-L-phenylalanyl chloride (2) at room temperature with evolution of ethyl chloride and formation

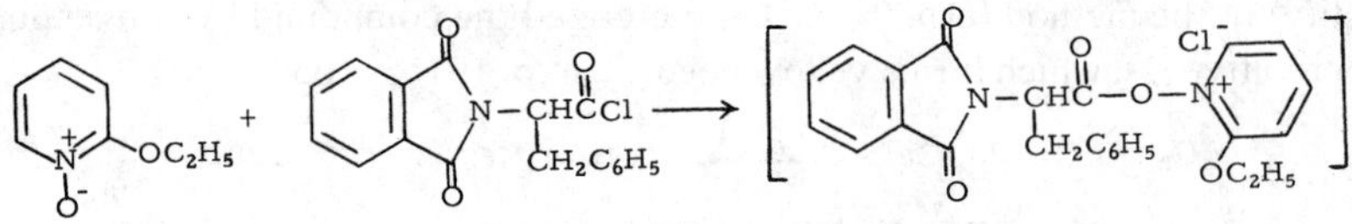

(1) (2) (3)

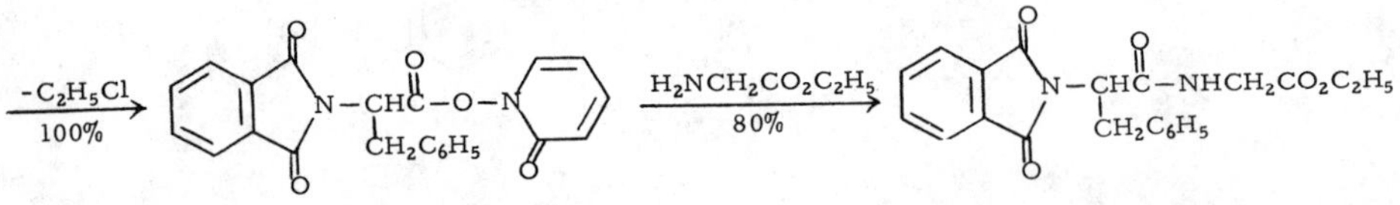

(4) (5)

of the activated ester (4), probably via the quaternary salt (3). Ester (4) in turn was readily condensed with glycine ethyl ester to give optically pure N-phthaloyl-L-phenylalanylglycine ethyl ester (5).

[1]L. A. Paquette, private communication
[2]*Idem, Am. Soc.*, **87**, 5186 (1965)

Ethyl (Methyl) azidoformate, $C_2H_5OC(=O)—N=\overset{+}{N}=\overset{-}{N}$. Mol. wt., 115.09, b.p. 114° (liable to explode), 25°/2 mm., sp. gr. 1.118.

Preparation.[1] Ethyl chloroformate (35 g.) is agitated with an aqueous solution of sodium azide (50 g.) until the pungent odor is gone; the heavy oil is dried and distilled at reduced pressure.

Reaction with benzene. Irradiation of a solution of ethyl azidoformate (1) in benzene affords N-carboethoxyazepine (4) in about 70% yield.[2] In a study of the flash-photolytic decomposition of ethyl azidoformate, Lwowski[3] found spectro-

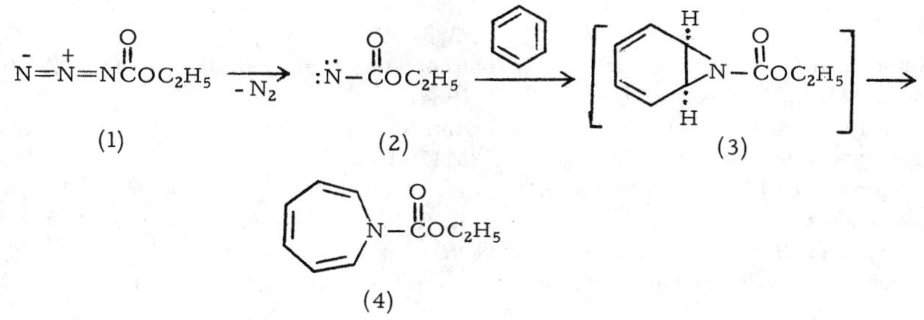

graphic evidence of the formation of the highly reactive carboethoxynitrene (2). Evidence that in the formation of N-carboethoxyazepine (4) the next step is the formation of the unstable aziridine (3) is afforded by isolation of a stable aziridine of this type from the reaction of ethyl azidoformate with cyclohexene.[3]

Cycloaddition to acetylenes and nitriles. Irradiation of a mixture of the reagent and phenylacetylene at 130° gives small amounts of the 1-carboethoxytriazole (1) and

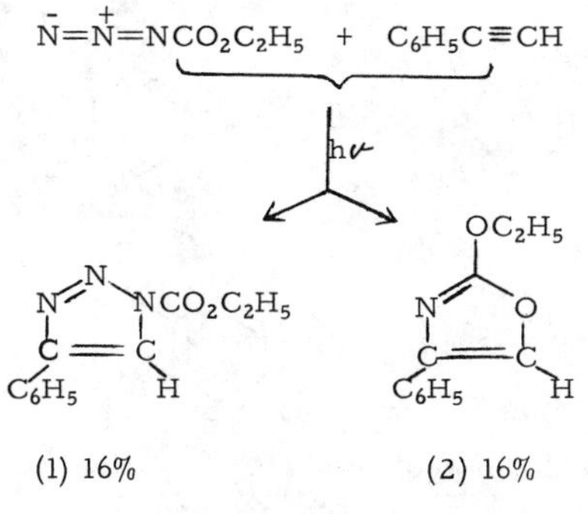

the 2-ethoxyoxazole (2).[4] Irradiation of the reagent in the presence of an aliphatic nitrile gives a 2-alkyl-5-ethoxy-1,3,4-oxadiazole in fair yield.[5]

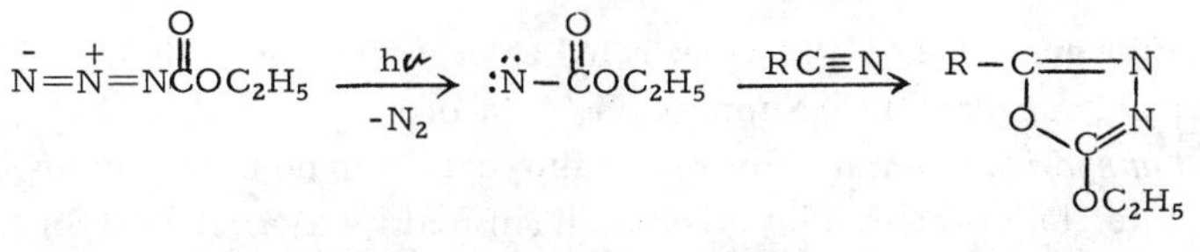

Note, however, that the reaction of methyl azidoformate with the strained olefin norbornene (1) proceeds by a 1,3-cycloaddition to give the unstable triazoline adduct (2), which decomposes in refluxing toluene to give the aziridine (3, 40%), the imide (4, 55%), and *syn*-2-norbornene-7-methylcarbamate (5).[6]

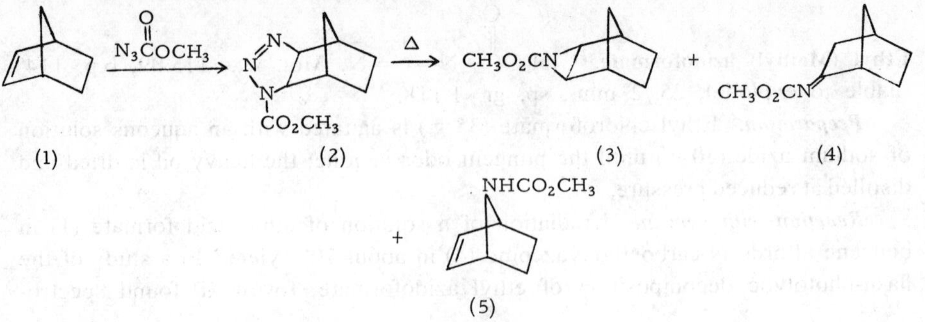

N_3COCH_3
Δ
CO_2CH_3
CH_3O_2CN
CH_3O_2CN
$NHCO_2CH_3$
(1)
(2)
(3)
(4)
(5)

[1]M. O. Forster and H. E. Fierz, *J. Chem. Soc.*, **93**, 81 (1908)

[2]K. Hafner and C. König, *Angew Chem., Internat. Ed.*, **2**, 96 (1963); K. Hafner, D. Zinser, and K.-L. Moritz, *Tetrahedron Letters*, 1733 (1964)

[3]R. S. Berry, D. Cornell, and W. Lwowski, *Am. Soc.*, **85**, 1199 (1963); W. Lwowski, T. J. Maricich, and T. W. Mattingly, Jr., *Am. Soc.*, **85**, 1200 (1963)

[4]R. Huisgen and H. Blaschke, *Tetrahedron Letters*, 1409 (1964)

[5]W. Lwowski, A. Hartenstein, C. DeVita, and R. L. Smick, *Tetrahedron Letters*, 2497 (1964); R. Huisgen and H. Blaschke, *Ann.*, **686**, 145 (1965)

[6]A. C. Oehlschlager, P. Tillman, and L. H. Zalkow, *Chem. Comm.*, 596 (1965)

N-Ethylbenzisoxazolium fluoroborate, 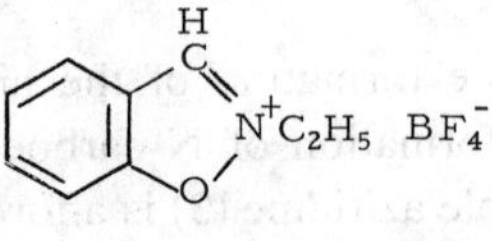

H
C
$\overset{+}{N}C_2H_5$ BF_4^-
O

Mol. wt. 217.99, m.p. 109–110.2°. Prepared by the reaction of benzisoxazole with triethyloxonium fluoroborate.[1]

This is a reagent for peptide coupling through an intermediate active ester: .

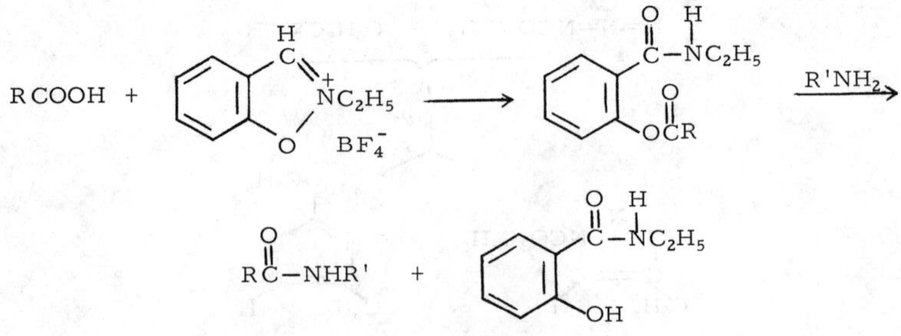

RCOOH +
$\overset{+}{N}C_2H_5$
BF_4^-
$C-NC_2H_5$
OCR
$R'NH_2$
RC—NHR' +
$C-NC_2H_5$
OH

It is also useful for the cyclization of peptides.[2]

[1]D. S. Kemp and R. B. Woodward, *Tetrahedron*, **21**, 3019 (1965)

[2]S. Rajappa and A. S. Akerkar, *Tetrahedron Letters*, 2893 (1966)

Ethyl carbamate, *see* Urethane.

Ethyl chloroformate (Cathyl chloride; Ethyl chlorocarbonate), $ClCO_2C_2H_5$. Mol. wt. 108.53, b.p. 92°, sp. gr. 1.135. Suppliers: FMC Corp., A, E, F, MCB.

Mixed anhydride synthesis. For use of the reagent in peptide synthesis, *see* Butyl chloroformate. The principle involved is illustrated by a procedure for the preparation of diethyl benzoylmalonate (3).[1] Benzoic acid is condensed with cathyl chloride in toluene in the presence of triethylamine to produce the mixed anhydride (1), and an ethereal solution of ethoxymagnesium malonic ester (2), prepared from malonic ester, magnesium, ethanol, and a trace of carbon tetrachloride as catalyst, is added

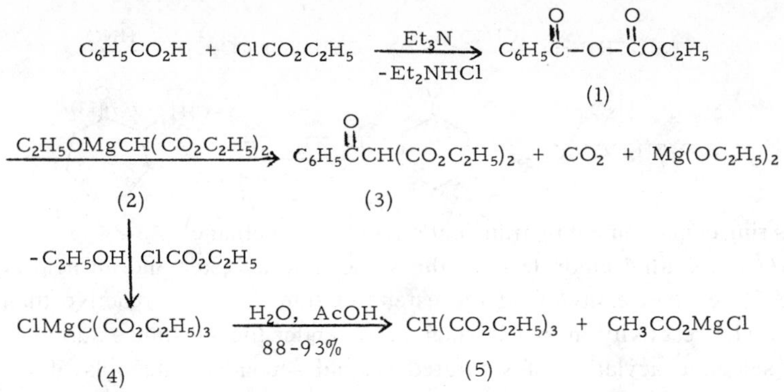

$$C_6H_5CO_2H + ClCO_2C_2H_5 \xrightarrow[-Et_2NHCl]{Et_3N} C_6H_5\overset{O}{\overset{\|}{C}}-O-\overset{O}{\overset{\|}{C}}OC_2H_5$$

(1)

$$\xrightarrow{C_2H_5OMgCH(CO_2C_2H_5)_2} C_6H_5\overset{O}{\overset{\|}{C}}CH(CO_2C_2H_5)_2 + CO_2 + Mg(OC_2H_5)_2$$

(2)   (3)

$$-C_2H_5OH \downarrow ClCO_2C_2H_5$$

$$ClMgC(CO_2C_2H_5)_3 \xrightarrow[88-93\%]{H_2O, \ AcOH} CH(CO_2C_2H_5)_3 + CH_3CO_2MgCl$$

(4)   (5)

at 0°. Ethoxymagnesium malonic ester (2) serves also as intermediate for reaction with cathyl chloride to produce tricarboethoxymethane (5, tricathylmethane).[2]

The mixed anhydride procedure has been applied to the preparation of symmetrical anhydrides of fatty acids,[3] particularly those containing hydroxyl groups and hence sensitive to reagents commonly used for anhydride formation.[4] The acid

$$CH_3(CH_2)_5\underset{OH}{CH}CH_2\overset{H}{\underset{|}{C}}=\overset{H}{\underset{|}{C}}(CH_2)_7\overset{O}{\overset{\|}{C}}OH + ClCO_2C_2H_5 \xrightarrow[THF \ (0^0)]{(C_2H_5)_3N}$$

(1)

$$CH_3(CH_2)_5\underset{OH}{CH}CH_2\overset{H}{\underset{|}{C}}=\overset{H}{\underset{|}{C}}(CH_2)_7\overset{O}{\overset{\|}{C}}O\overset{O}{\overset{\|}{C}}OC_2H_5 \xrightarrow[95\%]{(1) + (C_2H_5)_3N, \ THF \ (0^0)}$$

(2)

$$CH_3(CH_2)_5\underset{OH}{CH}CH_2\overset{H}{\underset{|}{C}}=\overset{H}{\underset{|}{C}}(CH_2)_7\overset{O}{\overset{\|}{C}}O\overset{O}{\overset{\|}{C}}(CH_2)_7\overset{H}{\underset{|}{C}}=\overset{H}{\underset{|}{C}}CH_2\underset{OH}{CH}(CH_2)_5CH_3$$

(3) m. p. 35-35. 5°

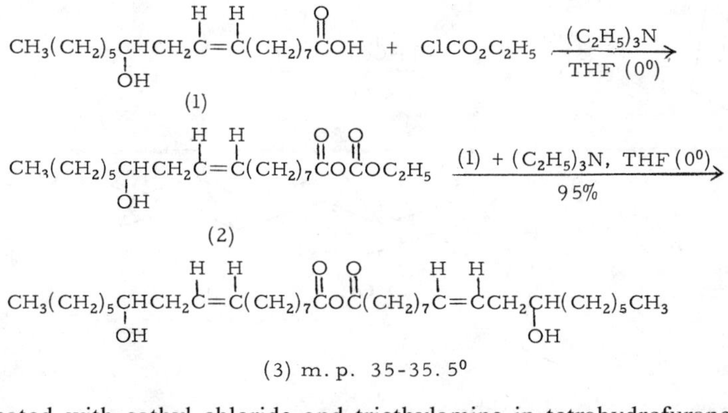

(1) is treated with cathyl chloride and triethylamine in tetrahydrofurane at 0° to form the mixed anhydride (2), and a second mole of the acid is added along with an equivalent amount of triethylamine. After standing overnight at room temperature the triethylamine hydrochloride was removed by filtration, and solvent was removed from the filtrate and washings in a rotary evaporator in vacuum. An ethereal solution of the residue was washed with acid and with carbonate, dried, and evaporated to give pure anhydride in high yield.

Fischer and Grob[5] converted 1-carboxycyclo [2.2.2] octane (4) to the 1-carbamoyl derivative (6) via the mixed anhydride (5).

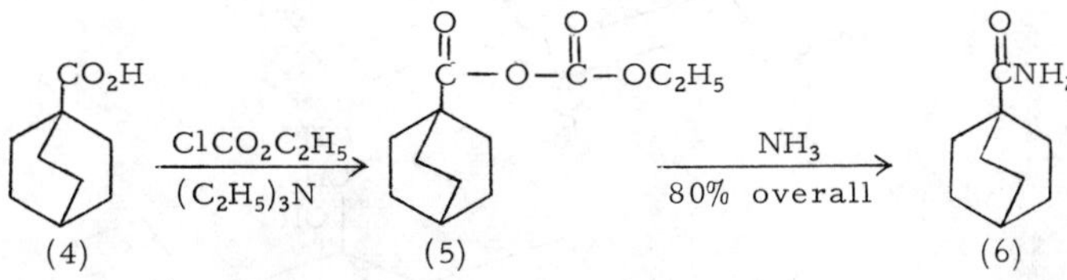

For use of cathyl chloride in the *preparation of a urethane*, *see* Diethyl azidodicarboxylate. Another example is the synthesis of 2-phenylcycloheptanone; the reaction

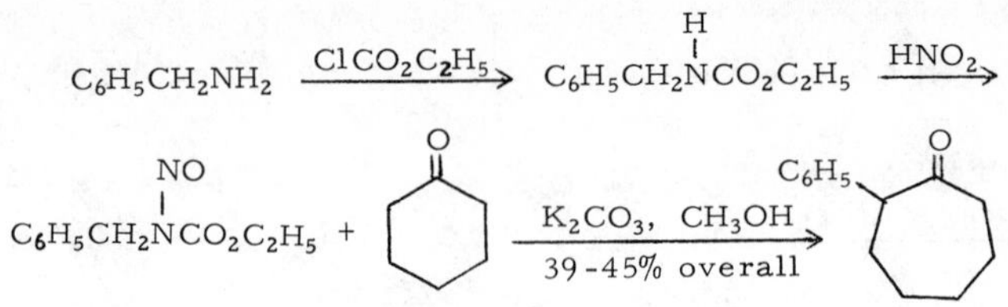

involves ring englargement by transient phenyldiazomethane.[6]

Acylation. Cathyl chloride is at the same time as ester and an acid chloride; because of resonance involving the ester function, it is less reactive than acetyl chloride. The reactivity in fact is such as to render the substance an ideal reagent for the selective acylation of saturated steroid secondary alcohols of equatorial orientation; axial groups if present are untouched. Thus methyl cholate, treated with a large excess of cathyl chloride in pyridine at room temperature, affords the 3-

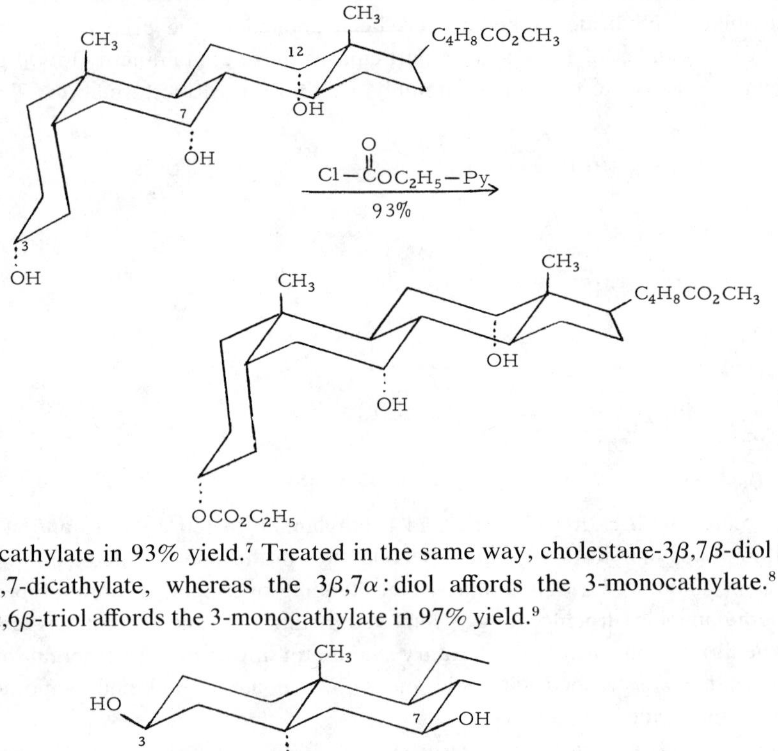

monocathylate in 93% yield.[7] Treated in the same way, cholestane-3β,7β-diol gives the 3,7-dicathylate, whereas the 3β,7α:diol affords the 3-monocathylate.[8] The 3β,5α,6β-triol affords the 3-monocathylate in 97% yield.[9]

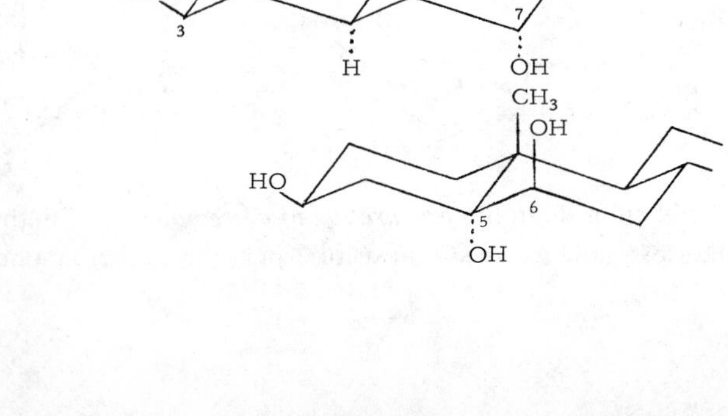

In a procedure for the preparation of 2,4-dimethyl-5-carboethoxypyrrole the reagent serves for introduction of a carboethoxy group into the pyrrole ring.[10]

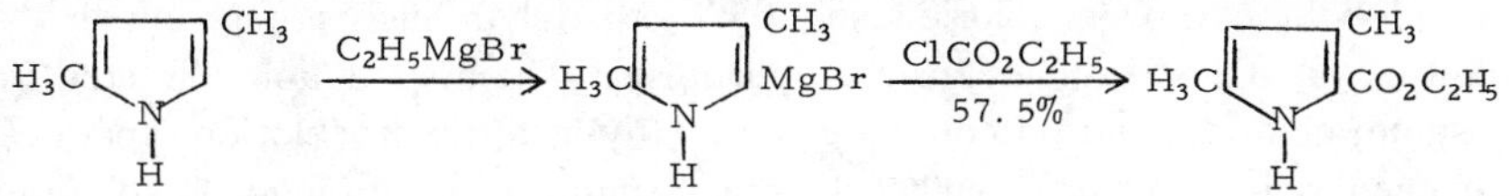

Synthesis of β-cyanoesters. Sauers and Cotter[11] discovered a novel synthesis illustrated for the case of maleamic acid (2), readily prepared from maleic anhydride. When the acid-amide is stirred with 2 equivalents each of cathyl chloride and

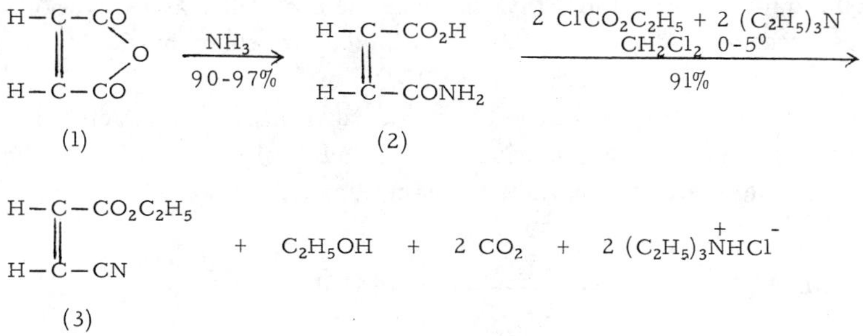

triethylamine in methylene chloride at 0–5°, carbon dioxide is evolved, triethylamine hydrochloride separates, and workup of the filtered solution affords ethyl β-cyanoacrylate in high yield. By the same scheme phthalamic acid afforded ethyl *o*-cyanobenzoate in yield of 84%. Diels-Alder adducts of maleic anhydride are well suited for use in the new reaction sequence. Experiments with model compounds showed that simple amides are not dehydrated to nitriles by cathyl chloride and that simple acids are not esterified and that proximity of the carboxyl and amide functions is essential, but did not distinguish between two possible mechanisms.

[1]J. A. Price and D. S. Tarbell, *Org. Syn., Coll. Vol.*, **4**, 285 (1963)
[2]H. Lund and A. Voigt, *ibid.*, **2**, 594 (1943)
[3]E. Schipper and J. Nichols, *Am. Soc.*, **80**, 5714 (1958)
[4]J. S. Nelson, L. A. Goldblatt, and T. H. Applewhite, *J. Org.*, **28**, 1905 (1963)
[5]H. P. Fischer and C. A. Grob, *Helv.*, **47**, 564 (1964)
[6]C. D. Gutsche and H. E. Johnson, *Org. Syn., Coll. Vol.*, **4**, 780 (1963)
[7]L. F. Fieser and S. Rajagopalan, *Am. Soc.*, **71**, 3938 (1949)
[8]L. F. Fieser, J. E. Herz, M. W. Klohs, M. A. Romero, and T. Utne, *Am. Soc.*, **74**, 3309 (1952)
[9]L. F. Fieser and S. Rajagopalan, *Am. Soc.*, **72**, 5530 (1950); **73**, 118 (1951)
[10]Hans Fischer, *Org. Syn., Coll. Vol.*, **2**, 198 (1943)
[11]C. K. Sauers and R. J. Cotter, *J. Org.*, **26**, 6 (1961)

Ethyl diazoacetate, $\overset{-}{N}{=}\overset{+}{N}{=}CHCO_2C_2H_5$. Mol. wt. 114.10, b.p. 42°/10 mm., sp. gr. 1.09. Suppliers: Aldrich, Columbia.

Preparation. The preparation of the reagent by a combination of *Organic Syntheses* procedures is shown in the formulation. In the first step formalin is

1. $2\ CH_2{=}O + NaCN + NH_4Cl \longrightarrow CH_2{=}NCH_2CN + NaCl + 2\ H_2O$

2. $CH_2{=}NCH_2CN + C_2H_5OH + 2\ HCl + 2\ H_2O \longrightarrow$
$$\overset{-}{Cl}\overset{+}{N}H_3CH_2CO_2C_2H_5 + CH_2O + NH_4Cl$$

3. $\overset{-}{Cl}\overset{+}{N}H_3CH_2CO_2C_2H_5 + NaNO_2 \longrightarrow \overset{-}{N}{=}\overset{+}{N}{=}CHCO_2C_2H_5 + NaCl + 2\ H_2O$

converted into methyleneamino-acetonitrile (61–71%).[1] This is then converted into glycine ethyl ester hydrochloride by refluxing with a mixture of absolute ethanol saturated with hydrogen chloride and 95% ethanol in such proportion that the amount of water just sufficies for the hydrolysis.[2] The hot solution is filtered from ammonium chloride, and on cooling glycine ethyl ester hydrochloride separates in white needles in yield of 87–90%. In the terminal step[3] an aqueous solution of glycine ethyl ester hydrochloride is stirred with methylene chloride at −5° under nitrogen, ice-cold sodium nitrite solution is added, followed by dilute sulfuric acid. Methylene chloride extracts ethyl diazoacetate as formed and protects it from the acid required for its formation. Ethyl diazoacetate is obtained as a yellow oil suitable for use as a reagent in yield of 79–88%. In a similar procedure, in which ether is used for extraction, the yield (unchecked) was 90%.[4]

In a newer procedure[5] the reagent is prepared from N-acetylglycine ethyl ester by nitrosation in carbon tetrachloride and treatment of the resulting N-nitroso-N-acetylglycine ethyl ester with barium oxide–barium hydroxide.

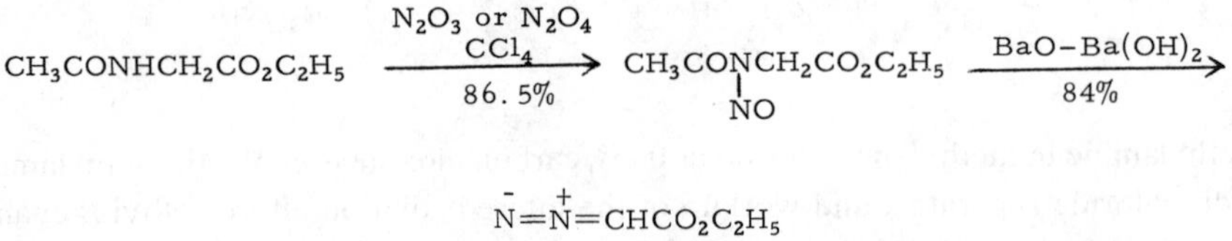
CH₃CONHCH₂CO₂C₂H₅
N₂O₃ or N₂O₄
CCl₄
86.5%
CH₃CONCH₂CO₂C₂H₅
NO
BaO−Ba(OH)₂
84%
N=N=CHCO₂C₂H₅

Buchner reaction. One example of the Buchner method of ring enlargement,[6] the reaction of ethyl diazoacetate with benzene, was reinvestigated by Grundmann and

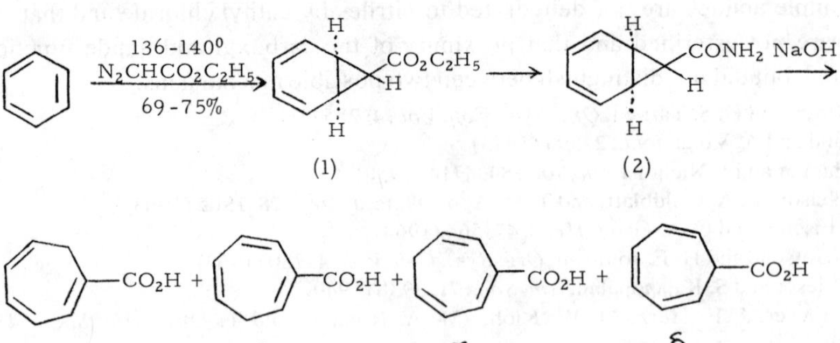
136-140⁰
N₂CHCO₂C₂H₅
69-75%
H
CO₂C₂H₅
H
H
(1)
H
CONH₂
H
H
NaOH
(2)
CO₂H +
CO₂H +
CO₂H +
CO₂H
α
β
γ
δ
M. p. 71⁰
56⁰
Liq.
31⁰

Ottmann[4] with the following results. They found that a trace of heavy metal catalyzes decomposition of the diazo ester and so heated a mixture of the ester with 20 parts of benzene in a glass insert of an autoclave at 136–140° and so obtained 7-carbo-ethoxynorcaradiene (1) in much improved yield. The ester (1) is converted into the amide (2), which on alkaline hydrolysis undergoes rearrangement to a mixture of four cycloheptatrienecarboxylic acids. Characterization of these acids led to assignment of the structures formulated.

Pfau and Plattner[7] employed a Buchner reaction for the synthesis of vetiva-zulene from the indane derivative I. At the high temperature at which the condensation of I with ethyl diazoacetate was carried out the adduct II was in part isomerized to III; saponification of the mixture completed the isomerization, and dehydrogena-

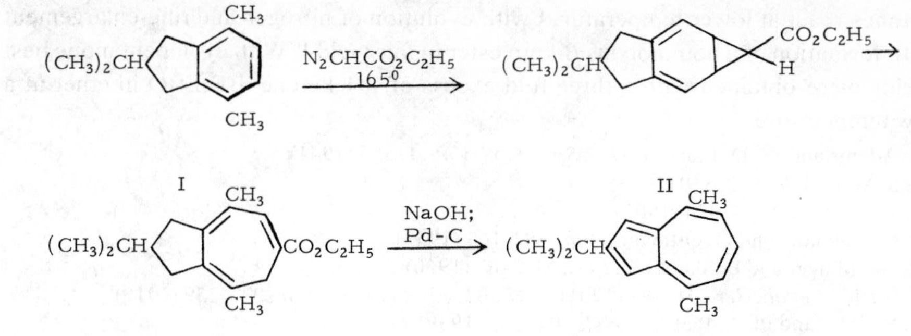

I II III Vetivazulene

tion-decarboxylation completed the synthesis. A general route to tropolones is illustrated by the synthesis of stipitatic acid.[8]

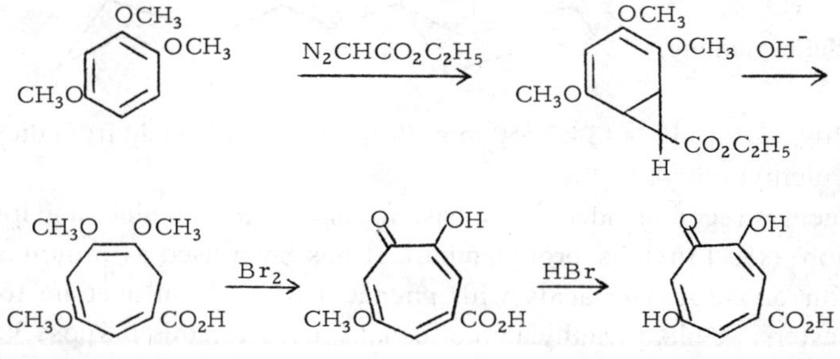

Stipitatic acid

Addition to quinones. Ethyl diazoacetate adds to the more reactive double bond of 1,4-naphthoquinone in warm benzene without loss of nitrogen, possibly by 1,3-dipolar attack of the structure formulated.[9] The products isolated, 3-carboethoxy-

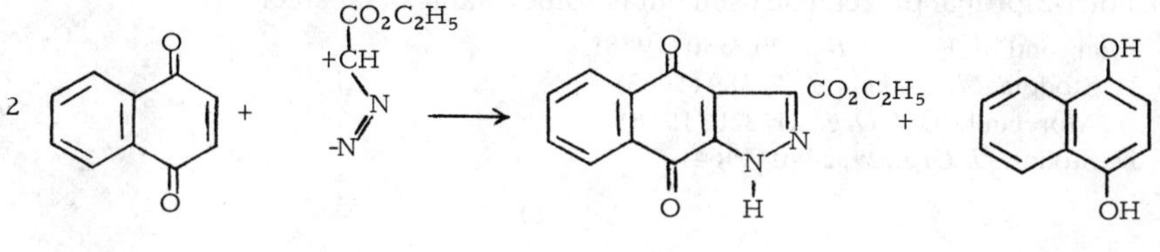

47% 48%

lin-naphthindazole-4,9-quinone and naphthohydroquinone, show that the initial adduct undergoes isomerization and oxidation.

Reaction with ketones. The reagent reacts very sluggishly with ketones, and usually the product is an enol ether. However, in the presence of boron trifluoride,

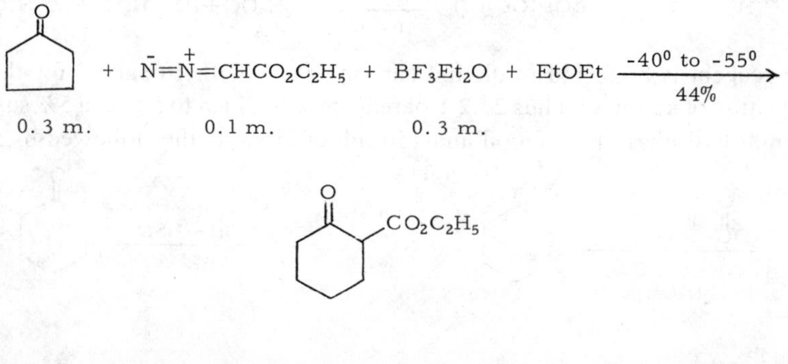

-40° to -55°
44%
0.3 m.
0.1 m.
0.3 m.

ketones react at lower temperatures with evolution of nitrogen and ring-enlargement with formation of a homologous β-keto ester in fair yield.[10] With cyclopentanone best yields were obtained with a three-fold excess of 1:1 ketone–BF_3Et_2O in ether at a low temperature.

[1]R. Adams and W. D. Langley, *Org. Syn., Coll. Vol.*, **1**, 355 (1941)
[2]C. S. Marvel, *ibid.*, **2**, 310 (1943)
[3]N. E. Searle, *ibid.*, **4**, 424 (1963)
[4]C. Grundmann and G. Ottmann, *Ann.*, **582**, 163 (1953)
[5]H. Reimlinger and L. Skatteböl, *Ber.*, **93**, 2162 (1960)
[6]E. Buchner *et al.*, *Ber.*, **34**, 982 (1901); **36**, 3502, 3509 (1903); *Ann.*, **377**, 259 (1910)
[7]A. St. Pfau and Pl. A. Plattner, *Helv.*, **22**, 202 (1939)
[8]R. B. Johns, A. W. Johnson, and M. Tišler, *J. Chem. Soc.*, 4605 (1954)
[9]L. F. Fieser and M. A. Peters, *Am. Soc.*, **53**, 4080 (1931)
[10]W. T. Tai and E. W. Warnhoff, *Can. J. Chem.*, **42**, 1333 (1964)

Ethyldicyclohexylamine, 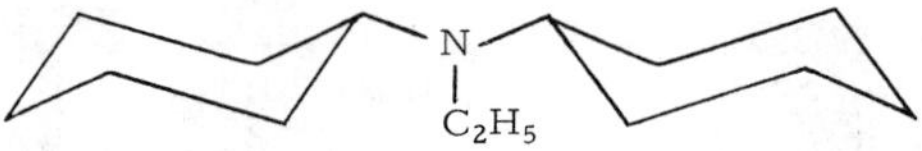

Mol. wt. 209.37, b.p. 132–134.2°/10 mm. Prepared in 91% yield from dicyclohexylamine and diethyl sulfate.[1,2]

The reagent is recommended as the base for alkylation of amines and for dehydrohalogenation[1] (*see* Ethyldiisopropylamine). It has been used as proton acceptor in the reaction of carboxylic acids with phenacyl bromide in acetone to form the phenacyl esters.[2] Unlike triethylamine, usually used for this purpose,[3] the hydrochloride of the reagent is soluble in acetone and other water-miscible solvents. In the presence of ethyldicyclohexylamine, dimethyl and diethyl sulfate convert carboxylic acids into their esters in high yield.[4] The commercially available hindered amine tris(2-hydroxypropyl)amine, $[CH_3CH(OH)CH_2]_3N$, Eastman's "1,1′,1-nitrilotri-2-propanol," can be used but is somewhat less satisfactory.

[1]S. Hünig and M. Kiessel, *Ber.*, **91**, 380 (1958)
[2]F. H. Stodola, *Microchem. J.*, **7**, 389 (1963)
[3]W. T. Moreland, Jr., *J. Org.*, **21**, 820 (1956)
[4]F. H. Stodola, *J. Org.*, **29**, 2490 (1964)

Ethyl diethoxyphosphinyl formate, $C_2H_5O\overset{O}{\overset{\|}{C}}—\overset{O}{\overset{\|}{P}}(OC_2H_5)_2$. Mol. wt. 130.17, b.p. 122.5–123°.

Prepared[1] by heating ethyl chloroformate with triethyl phosphite (Arbusov reaction):

$$C_2H_5O\overset{O}{\overset{\|}{C}}Cl + C_2H_5OP(OC_2H_5)_2 \xrightarrow[50-60\%]{} C_2H_5O\overset{O}{\overset{\|}{C}}-\overset{O}{\overset{\|}{P}}(OC_2H_5)_2 + C_2H_5Cl$$

The reagent is preferred to diethyl carbonate[2] or diethyl oxalate[2] for the carboethoxylation of ketones.[3] Thus 21.2 g. of reagent was added to 5 g. of a 5% suspension of sodium hydride in paraffin oil and 150 ml. of dibutyl ether followed by 9.8 g. of

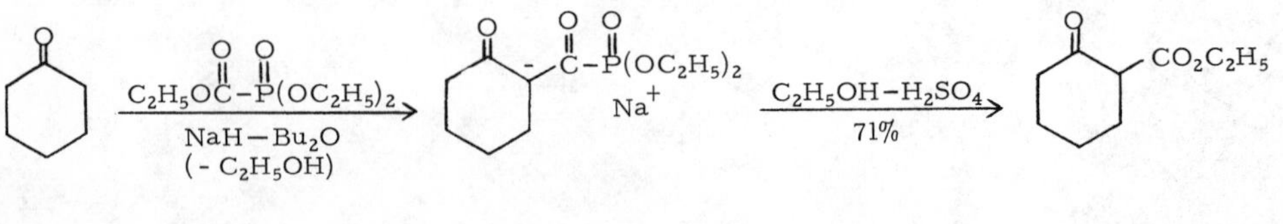
C₂H₅OC–P(OC₂H₅)₂
NaH–Bu₂O
(- C₂H₅OH)
C–P(OC₂H₅)₂
Na+
C₂H₅OH–H₂SO₄
71%
CO₂C₂H₅

cyclohexanone in small portions. The temperature was held below 30° for 1 hr., solvent and ethanol were removed in vacuum, and the mixture was poured into anhydrous ethanol containing sulfuric acid. Workup afforded 2-carboethoxycyclohexanone, b.p. 93–94°, identified by its copper chelate, m.p. 178°.

[1]P. Nylén, *Ber.*, **57**, 1023 (1924); see also T. Reetz, D. H. Chadwick, E. E. Hardy, and S. Kaufman, *Am. Soc.*, **77**, 3813 (1955)
[2]F. W. Swamer and C. R. Hauser, *Am. Soc.*, **72**, 1352 (1950)
[3]I. Shahak, *Tetrahedron Letters*, 2201 (1966)

Ethyldiisopropylamine, $(CH_3)_2CHN(C_2H_5)CH(CH_3)_2$. Mol. wt. 129.24, b.p. 128°. Preparation from diisopropylamine and diethyl sulfate.[1] Supplier: Aldrich ("diisopropylethylamine").

This tertiary amine (R_2NR') is a strong base but it is so hindered that it cannot be alkylated. In the alkylation of an amine it serves as a proton acceptor which cannot combine with the alkyl halide. It is useful also for the capture of acid in a dehydrohalogenation.

$$C_6H_5NH_2 + 2\ C_4H_9Br \xrightarrow[97\%]{R_2NR'} C_6H_5N(C_4H_9)_2$$

$$CH_3CH_2OCH(Cl)CH_3 \xrightarrow{R_2NR'} CH_3CH_2OCH{=}CH_2$$

[1]S. Hünig and M. Kiessel, *Ber.*, **91**, 380 (1958)

1-Ethyl-3-(3'-dimethylaminopropyl)carbodiimide hydrochloride (3). Mol. wt. 177.70. Supplier: Aldrich.

Preparation.[1] Condensation of ethyl isocyanate with N,N-dimethyl-1,3-propanediamine (suppliers: B, E, F, MCB) gives the urea (1), which is dehydrated to (2) with tosyl chloride and triethylamine in methylene chloride. The hydrochloride

$$C_2H_5N{=}C{=}O \xrightarrow{H_2N(CH_2)_3N(CH_3)_2} \underset{(1)}{C_2H_5NHCONH(CH_2)_3N(CH_3)_2} \xrightarrow{TsCl}$$

$$\underset{(2)}{C_2H_5N{=}C{=}N(CH_2)_3N(CH_3)_2} \xrightarrow{Py\cdot HCl} \underset{(3)}{C_2H_5N{=}C{=}N(CH_2)_3\overset{+}{N}H(CH_3)_2\ Cl^-}$$

is prepared by metathesis with pyridine hydrochloride in methylene chloride solution and precipitated with ether.

Peptide synthesis. Sheehan and co-workers[2] used this water-soluble reagent for a simplified and rapid synthesis of tetra- and pentapeptides without isolation of intermediates. The reagent (1.1 equiv.) is added to a solution of the N-carbobenzoxyamino acid (1 equiv.), the amino acid ester hydrochloride or peptide ester hydrochloride (1 equiv.), and triethylamine (1 equiv.) in methylene chloride. After 1 hr. at room temperature the solution was washed successively with water (to remove excess reagent and the urea), dilute hydrochloric acid, and sodium bicarbonate solution. The carbobenzoxy group is removed by hydrogenolysis and the product used directly in the next step.

[1]J. C. Sheehan, P. A. Cruickshank, and G. L. Boshart, *J. Org.*, **26**, 2525 (1961)
[2]J. C. Sheehan, J. Preston, and P. A. Cruickshank, *Am. Soc.*, **87**, 2492 (1965)

Ethylene carbonate, $\begin{matrix} CH_2O \\ | \\ CH_2O \end{matrix} \diagdown C=O$ 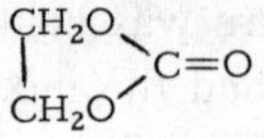

Mol. wt. 88.06, m.p. 39°, b.p. 125°/10 mm. Suppliers; Jefferson Chem. Co., A, B, E, F, MCB.

Ethylene chlorophosphite, $\begin{matrix} CH_2O \\ | \\ CH_2O \end{matrix} \diagdown PCl$ 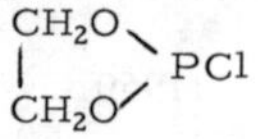

Mol. wt. 126.49, b.p. 56°. The reagent is prepared in 66% yield by the action of phosphorus trichloride on ethylene glycol.[1]

(a) Peptide coupling agent.

(b) Identification of asparaginyl and glutaminyl residues in peptides. Ressler and Kashelikar[2] found this to be the most satisfactory reagent for the dehydration of asparaginyl and glutaminyl residues (1) of peptides to the corresponding cyano-peptides (2). This step is followed by micro Birch reduction (Na, NH_3, CH_3OH) to

$$-NHCHCONH- \atop (CH_2)_{1-2}CONH_2 \qquad \xrightarrow[(C_2H_5O)_3P]{\substack{CH_2O \\ | \diagdown PCl \\ CH_2O}} \qquad -NHCHCONH- \atop (CH_2)_{1-2}C\equiv N \qquad \xrightarrow{Na,\ NH_3,\ CH_3OH}$$

(1)   (2)

$$-NHCHCONH- \atop (CH_2)_{1-2}CH_2NH_2 \qquad \xrightarrow{6\ NHCl} \qquad NH_2CHCO_2H \atop CH_2CH_2NH_2 \qquad or \qquad H_2NCHCO_2H \atop CH_2CH_2CH_2NH_2$$

(3)   (4)   (5)

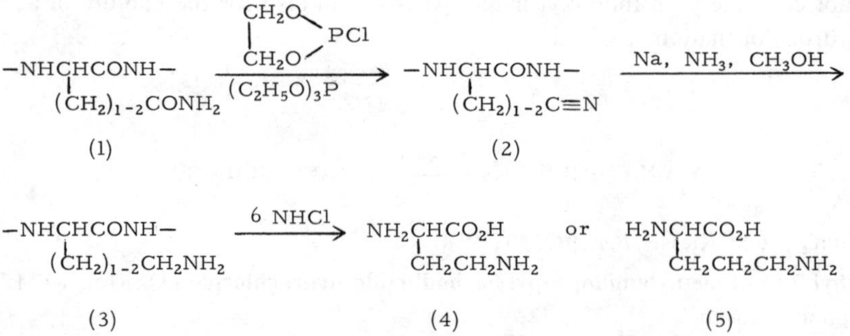

give an amine (3); hydrolysis then liberates either 2,4-diaminobutyric acid (4) or ornithine (5). The method was applied to the polypeptide bacitracin A and established one of the four previously suggested structures.

[1]H. J. Lucas, F. W. Mitchell, Jr., and C. N. Scully, *Am. Soc.*, **72**, 5491 (1950)
[2]C. Ressler and D. V. Kashelikar, *Am. Soc.*, **88**, 2025 (1966)

Ethylenediamine, $H_2NCH_2CH_2NH_2$. Mol. wt. 60.10, m.p. 8.5°, b.p. 117°, sp. gr. 0.90, miscible with water. Suppliers: Dow, Union Carbide, A, B, E, F, MCB. The commercial material contains about 2% of water. For preparation of anhydrous reagent, *see* N-Lithioethylenediamine.

The iodination of phenols and arylamines with KI_3 solution is advantageously carried out in the presence of an HI-acceptor such as ethylamine[1] or ethylene-diamine.[2]

The reagent is used for the preparation of ethylene thiourea:[3]

$$\begin{matrix} CH_2NH_2 \\ | \\ CH_2NH_2 \end{matrix} \xrightarrow{CS_2} \begin{matrix} H \\ CH_2N-C=S \\ | \quad\quad + \quad S^- \\ CH_2NH_3 \end{matrix} \xrightarrow[83-89\%]{HCl} \begin{matrix} CH_2-NH \\ | \quad\quad\quad C=S \ + \ H_2S \\ CH_2-NH \end{matrix}$$

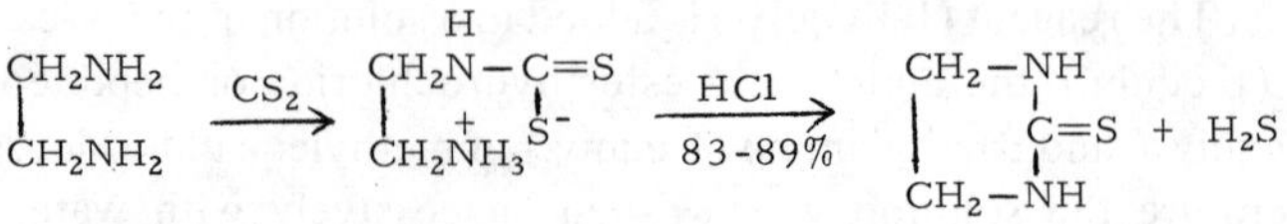

After the reaction of an aryl halide with cuprous cyanide to form a nitrile, added ethylenediamine complexes with cuprous and cupric ions and facilitates isolation of the nitrile[4] (*see* Ferric chloride).

[1]J. C. Clayton and B. A. Hems, *J. Chem. Soc.*, 840 (1950)
[2]K. T. Potts, *J. Chem. Soc.*, 3711 (1953)

[3]C. F. H. Allen, C. O. Edens, and J. VanAllan, *Org. Syn., Coll. Vol.*, **3**, 394 (1955)
[4]L. Friedman and H. Shechter, *J. Org.*, **26**, 2522 (1961)

Ethylenediaminediborane, *see* Ethane-1,2-diaminoborane.

Ethylenediaminetetraacetic acid (EDTA), $(HO_2CCH_2)_2NCH_2CH_2N(CH_2CO_2H)_2$ Mol. wt. 292.24. *Disodium salt* (2 H_2O), mol. wt. 372.24. *Tetrasodium salt*, mol. wt. 380.17. Suppliers: B, E, F, MCB.

Sequestration of zinc ions. Spirocyclopentane (3) was discovered by G. Gustavson (1896) as a product of the action of zinc dust on pentaerythrityl tetrabromide (1) in aqueous ethanol. Later workers, chiefly N. D. Zelinsky (1912) and I. N. Shokhor (1954) investigated the reaction extensively without finding a way to inhibit extensive

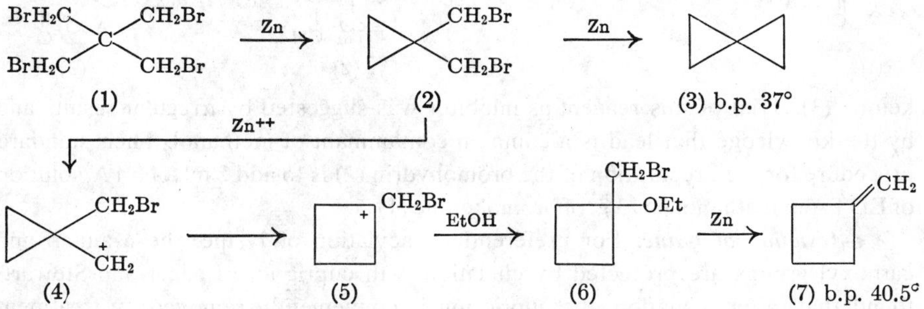
BrH₂C
CH₂Br
BrH₂C
CH₂Br
Zn
CH₂Br
CH₂Br
Zn
(1)
(2)
(3) b.p. 37°
Zn⁺⁺
CH₂Br
CH₂
(4)
CH₂Br
(5)
EtOH
CH₂Br
OEt
(6)
Zn
CH₂
(7) b.p. 40.5°

production of methylenecyclobutane (7), a by-product probably arising by electrophilically induced rearrangement of the intermediate dibromide (2) via the intermediates (4–6). Applequist[1] devised a simple expedient which confirms the mechanism of the rearrangement and raises the yield of spirocyclopentane (94% pure) to 81%: addition of tetrasodium ethylenediaminetetraacetate to sequester zinc ion. The charge was as follows: 0.828 m. of $C(CH_2Br)_4$, 1470 ml. of 95% ethanol, 7.43 m. of NaOH in 510 ml. water, 0.138 m. of NaI, 3.28 g. atoms of zinc dust, 2.57 m. of disodium ethylenediaminetetraacetate. The pentaerythrityl tetrabromide was added slowly to a stirred mixture of the other reagents and a slow stream of nitrogen carried volatile products to cold traps.

Russian workers (1953) had attempted to synthesize spirocyclohexane (9) by reaction of (8) with zinc dust but had isolated only the product of rearrangement, (10). McGreer[2] applied Applequist's sequestration procedure and obtained the

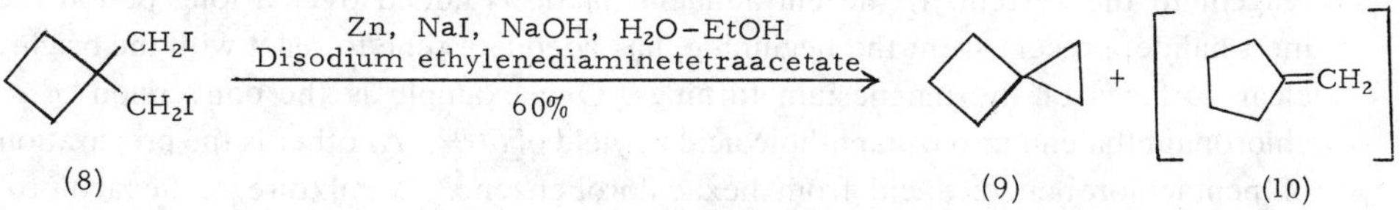
CH₂I
CH₂I
Zn, NaI, NaOH, H₂O–EtOH
Disodium ethylenediaminetetraacetate
60%
+
=CH₂

(8) (9) (10)

spirostane in good yield. Conia *et al.*[3] found that the dibromoketone (11) with zinc dust in the presence of the sequestering agent affords the spiroalkanone (12) as the sole product but that without sequestration the product is a 40:60 mixture of

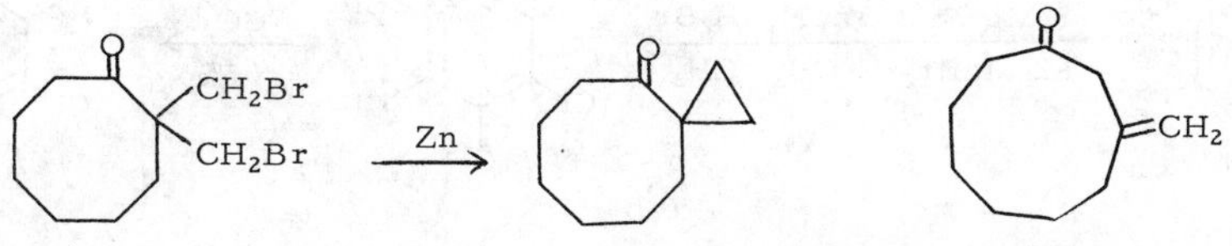
CH₂Br
CH₂Br
Zn
=CH₂

(11) (12) (13)

(12) and (13). With lower homologs of (11) having rings of 5, 6, and 7 carbon atoms, spiroalkanones were obtained as major products with sequestration.

Sequestration of lead ions. Goto and Kishi[4] found that sodium borohydride reduction of the 7α-bromo-3β,5α-diol-6-one diacetate (1) in fully purified methanol–dioxane affords the bromohydrin (2) in excellent yield but that on addition of no more than 4γ of lead acetate/200 mg. of (1) the reaction product is the debrominated

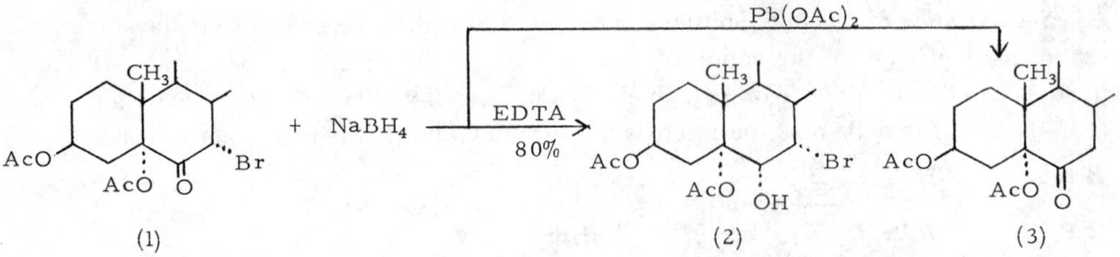
Pb(OAc)$_2$
CH$_3$
+ NaBH$_4$
EDTA
80%
AcO
Br
AcO
O
OH
(1)
(2)
(3)

ketone (3). Trial of this reagent as inhibitor was suggested by irregular results and by the knowledge that lead is a common contaminant of methanol. Their standard procedure for the preparation of the bromohydrin (2) is to add 2 ml. of a 1 *N* solution of EDTA in methanol per 3 g. of bromoketone (1).

ε-Acylation of lysine. For preferential ε-acylation of lysine, the α-amino and carboxyl groups are protected by chelation with cupric ion. Ledger and Stewart[5] found that, after acylation, the cupric ion is conveniently removed by treatment with EDTA.

[1]D. E. Applequist, G. F. Fanta, and B. W. Henrikson, *J. Org.*, **23**, 1715 (1958)
[2]D. E. McGreer, *Can. J. Chem.*, **38**, 1638 (1960)
[3]J.-M. Conia, P. Leriverend, and J.-L. Bouket, *Tetrahedron Letters*, 3189 (1964)
[4]T. Goto and Y. Kishi, *Tetrahedron Letters*, 513 (1961); *Bull. Chem. Soc. Japan*, **35**, 2044 (1962)
[5]R. Ledger and F. H. C. Stewart, *Australian J. Chem.*, **18**, 933 (1965)

Ethylene dibromide, $BrCH_2CH_2Br$. Mol. wt. 187.87, m.p. 9–10°, b.p. 131.7°, sp. gr. 2.180. Suppliers: B, E, F, MCB.

Entrainment reagent. Ethylene dibromide appears to be superior to ethyl bromide as an entrainment reagent for the conversion of an unreactive halide into a Grignard reagent.[1] One advantage is that the dibromide reacts with magnesium to give $MgBr_2$ and ethylene and hence does not introduce a second Grignard reagent to the system. If the entrainment halide is added over a long period the inert halide, present from the beginning, has adequate time to react with the bright, clear surfaces of the magnesium turnings. One example is the conversion of α-chloronaphthalene into α-naphthoic acid in yield of 56%.[1] Another is the preparation of pentachlorobenzoic acid from hexachlorobenzene.[2] A mixture of hexachlorobenzene, magnesium, and ether is refluxed gently with stirring, and a solution of ethylene dibromide in benzene is added through a Hershberg funnel (p. 783) over

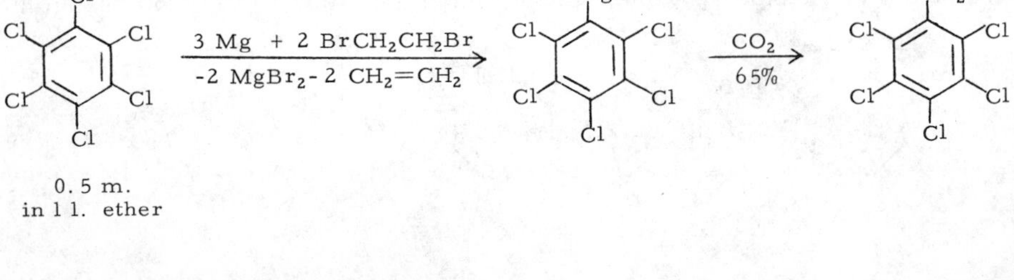
Cl
3 Mg + 2 BrCH$_2$CH$_2$Br
-2 MgBr$_2$ - 2 CH$_2$=CH$_2$
MgCl
CO$_2$
65%
CO$_2$H
0.5 m.
in 1 l. ether

a period of 48 hrs. (little attention is required if the platinum wire fits very snugly). Dried carbon dioxide is passed into the dark brown reaction mixture, 10% hydrochloric acid is added, and the reaction product is purified as the ammonium salt and crystallized from 50% aqueous methanol.

[1]D. E. Pearson, D. Cowan, and J. D. Beckler, *J. Org.*, **24**, 504 (1959)
[2]D. E. Pearson and D. Cowan, *Org. Syn.*, **44**, 78 (1964)

Ethylene glycol, $HOCH_2CH_2OH$. Mol. wt. 62.07, b.p. 199.5–201°, sp. gr. 1.11.

Solvent effects. Ethylene glycol has high solvent power for sodium hydroxide and potassium hydroxide, and the high-boiling point permits raising the reaction temperature when desired. Thus hydrolysis of *t*-butylurea to *t*-butylamine is accomplished by refluxing for 4 hrs. a mixture of 70 g. of *t*-butylurea, 60 g. of sodium hydroxide in 75 ml. of water, and 225 ml. of ethylene glycol.[1]

$$(CH_3)_3C\overset{H}{\overset{|}{N}}-\overset{O}{\overset{\|}{C}}-NH_2 + NaOH \xrightarrow[71-78\%]{} (CH_3)_3CNH_2 + NH_3 + Na_2CO_3$$

Ethylene glycol also serves as an efficient solvent for the production of monovinylacetylene by the dehydrochlorination of 1,3-dichloro-2-butene with potassium hydroxide in a flask fitted with a Friedrichs condenser and a Truebore stirrer carry-

$$\underset{1\ m.}{ClCH_2CH=\overset{Cl}{\overset{|}{C}}CH_3} + \underset{400\ g.}{KOH} + \underset{500\ ml.}{HOCH_2CH_2OH} \xrightarrow[43-48\%]{165-170^0} H_2C=CH-C\equiv CH$$

ing a Teflon paddle and operating in a Truebore bearing (Ace Glass). A mixture of powdered potassium hydroxide flakes and ethylene glycol is stirred under nitrogen at 165–170°, 100 ml. of ethylene glycol mono-*n*-butyl ether is added to control foaming, and 1 mole of the dichloro compound is added in 1 hr.[2]

Since ethylene glycol has the solvent power to dissolve potassium fluoride, it is the preferred solvent for the conversion of *n*-hexyl bromide into *n*-hexyl fluoride.[3]

$$CH_3(CH_2)_4CH_2Br + KF \xrightarrow[40-45\%]{} CH_3(CH_2)_4CH_2F$$

The Fischer indole synthesis can be carried out very simply and without use of an acid catalyst by heating a phenylhydrazone in ethylene glycol for a sufficient period.[4] Acetophenone phenylhydrazone afforded 2-phenylindole in 54% yield after

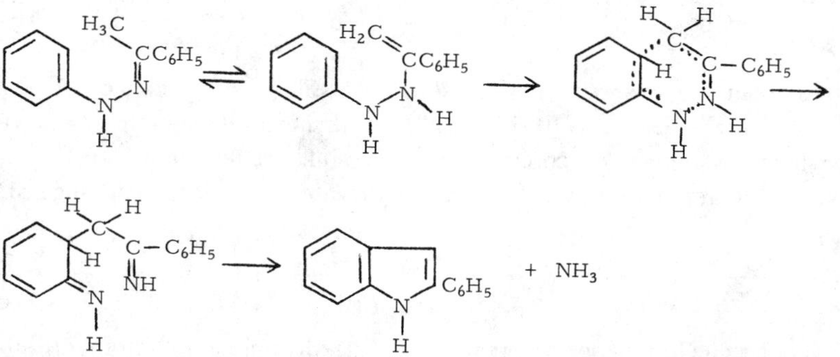

refluxing for 48 hrs.; methyl ethyl ketone phenylhydrazone gave 2,3-dimethylindole in 70% yield in a reflux period of 3 hrs. Use of diethylene glycol shortens the reaction time.

Ethylene ketals (Dioxolone derivatives). Ketalization with ethylene glycol is employed extensively for protection of a carbonyl function during synthetic operations, since the ethyleneketal group is resistant to bases, Grignard reagents, lithium alkyls, and metal hydrides, and since it can be eliminated when desired by acid hydrolysis. A typical procedure[5] is that applied to the tricyclic ketone I, an intermediate in one of the steroid total syntheses by Johnson's group (the formula numbers

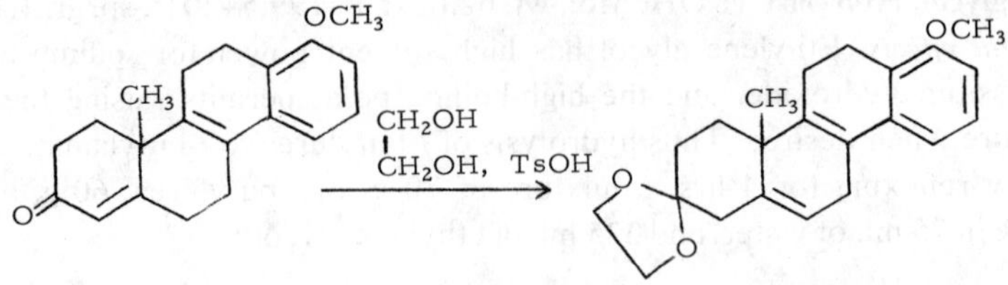
OCH3
CH3
CH2OH
CH2OH, TsOH
OCH3
CH3
O
O

I XXI

are those of the paper cited). A mixture of 1.83 g. of I, 92 ml. of toluene, 0.073 g. of *p*-toluenesulfonic acid monohydrate, and 16 ml. of ethylene glycol was distilled very slowly, with addition of fresh toluene at intervals to maintain the starting volume. Distillation for 6 hrs. afforded about 100 ml. of distillate; workup afforded 1.52 g. (72%) of nearly pure ketal, m.p. 117.5–118.5°. Note that the double bond migrates out of conjugation on ketalization.

Another method of ketalization which has been applied successfully to several steroid ketones consists in heating the ketone with ethylene glycol, triethyl orthoformate, and *p*-toluenesulfonic acid, and distilling off ethanol.[6] Sulfuric acid is less

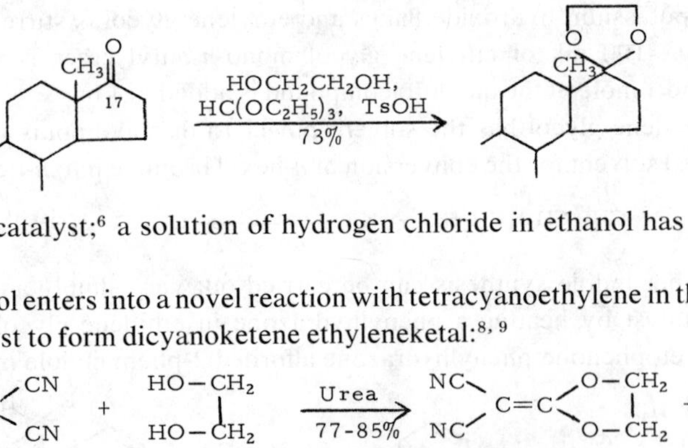
O
CH3
17
HOCH2CH2OH,
HC(OC2H5)3, TsOH
73%
O O
CH3

satisfactory as catalyst;[6] a solution of hydrogen chloride in ethanol has given good results.[7]

Ethylene glycol enters into a novel reaction with tetracyanoethylene in the presence of urea as catalyst to form dicyanoketene ethyleneketal:[8, 9]

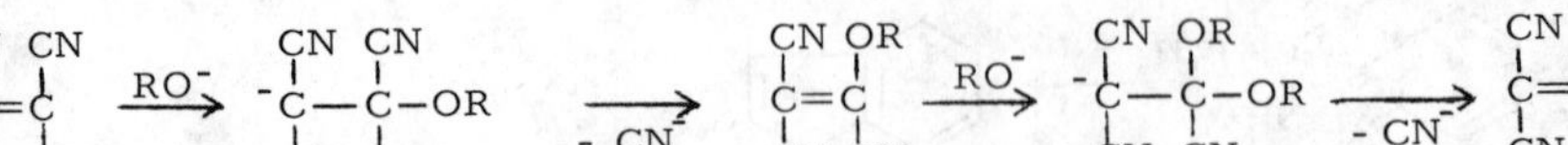
NC
CN
C=C
NC
CN
+
HO−CH2
HO−CH2
Urea
77-85%
NC
C=C
NC
O−CH2
O−CH2
+ 2HCN

Finely divided tetracyanoethylene is added to a solution of urea in ethylene glycol, and the mixture is heated to 70–75° with hand stirring until solution is complete. The ketal which crystallizes on cooling in slightly pink needles, m.p. 116°, is collected and washed with ethylene glycol and then with water. The mechanism suggested is:[10]

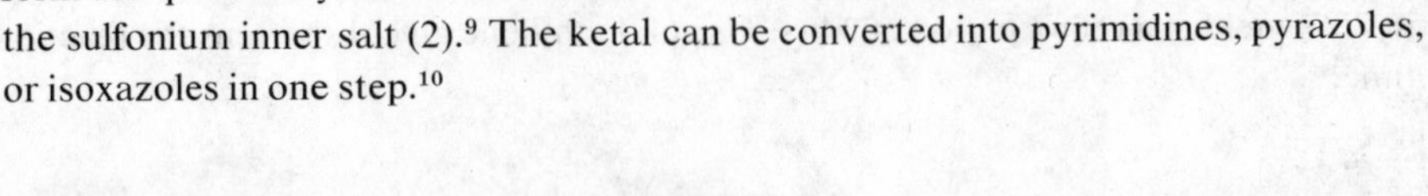
CN CN
C=C
CN CN
RO−
CN CN
−C−C−OR
CN CN
- CN−
CN OR
C=C
CN CN
RO−
CN OR
−C−C−OR
CN CN
- CN−
CN OR
C=C
CN OR

Dicyanoketene ethyleneketal reacts with triethylamine in tetrahydrofurane to form the quaternary ammonium inner salt (1); it reacts with dimethyl sulfide to form the sulfonium inner salt (2).[9] The ketal can be converted into pyrimidines, pyrazoles, or isoxazoles in one step.[10]

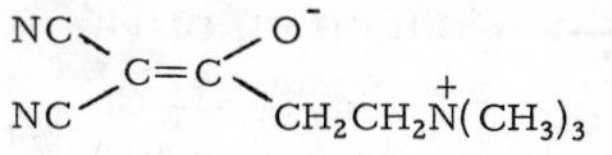

(1) M. p. 212°

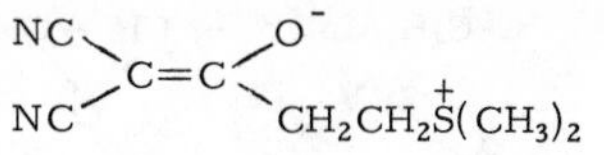

(2) M. p. 171°

Engel and Rakhit[11] found that selective ketalization of the 12-keto group of the 12,20-diketone (1) by reaction with ethylene glycol and boron trifluoride etherate proceeds best with methylene chloride present as co-solvent to provide a homogeneous medium. Without the cosolvent, the yield was 57%.

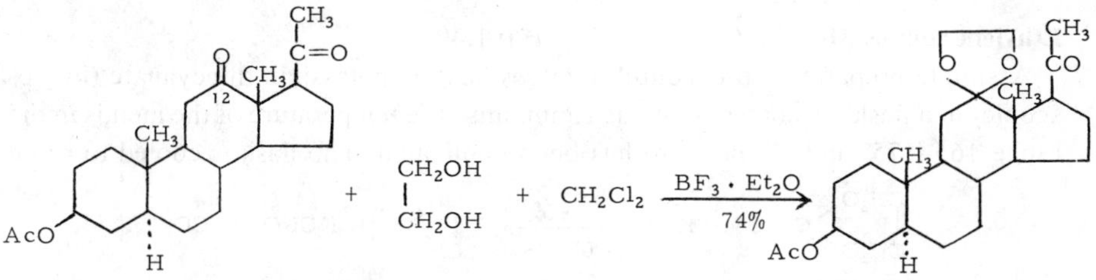

(1) (2)

Removal of the isopropylidene (acetonide) group. Hampton *et al.*[12] found that for a given concentration of acid (0.01 *N*), the conversion of isopropylideneuridine (1) into uridine (2) proceeds ten times more rapidly in ethyleneglycol solution than in water. Methanol and ethanol were significantly less effective.

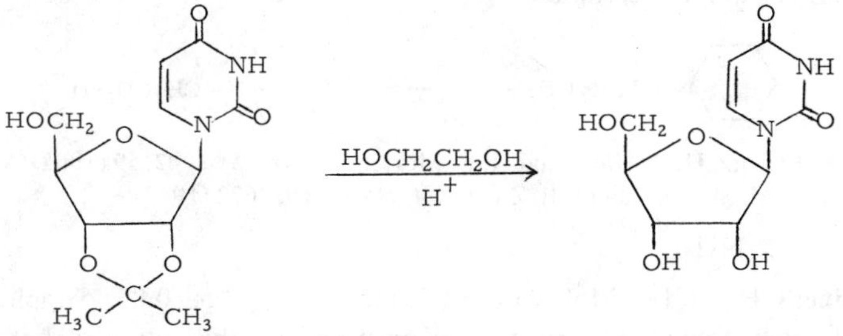

(1) (2)

[1]D. E. Pearson, J. F. Baxter, and K. N. Carter, *Org. Syn., Coll. Vol.*, **3**, 154 (1955)
[2]G. F. Hennion, C. C. Price, and T. F. McKeon, Jr., *ibid.*, **4**, 683 (1963)
[3]A. I. Vogel, J. Leicester, and W. A. T. Macey, *ibid.*, **4**, 525 (1963)
[4]J. T. Fitzpatrick and R. D. Hiser, *J. Org.*, **22**, 1703 (1957)
[5]W. S. Johnson *et al.*, *Am. Soc.*, **78**, 6300 (1956)
[6]A. Marquet, M. Dvolaitsky, H. B. Kagan, L. Mamlok, C. Ouannes, and J. Jacques, *Bull. soc.*, 1822 (1961)
[7]E. Vogel and H. Schinz, *Helv.*, **33**, 116 (1950)
[8]C. L. Dickinson and L. R. Melby, *Org. Syn., Coll. Vol.*, **4**, 276 (1963)
[9]W. J. Middleton and V. A. Englehardt, *Am. Soc.*, **80**, 2788 (1958)
[10]*Idem, ibid.*, **80**, 2829 (1958)
[11]C. R. Engel and S. Rakhit, *Can. J. Chem.*, **40**, 2153 (1962)
[12]A. Hampton, J. C. Fratantoni, P. M. Carroll, and S. Wang, *Am. Soc.*, **87**, 5481 (1965)

Ethylene oxide. Mol. wt. 44.05, b.p. 10.7°, sp. gr. 0.89.

In the synthesis of *n*-hexyl alcohol from *n*-butylmagnesium bromide and ethylene oxide, an ethereal solution of the Grignard reagent is kept at a temperature of 10° or below by stirring in a salt-ice bath during the addition of ethylene oxide (4–6 hrs.).[1]

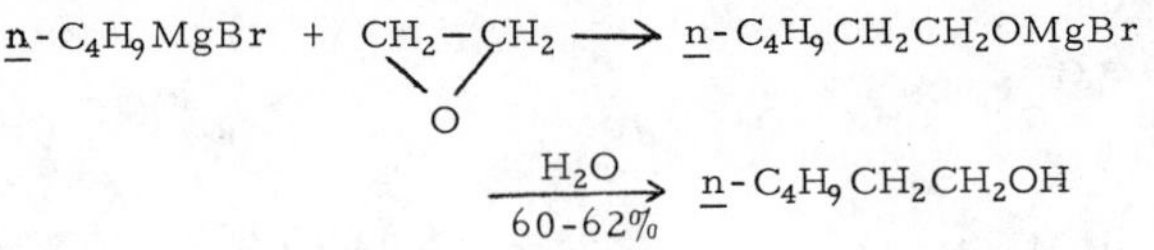

A convenient arrangement is to place the cylinder of condensed gas on a scale so that the decrease in weight can be noted and to have the gas inlet tube extend about 2 cm. above the surface of the liquid so that gaseous ethylene oxide condenses in the cooled reaction mixture.

[1]E. E. Dreger, *Org. Syn., Coll. Vol.*, **1**, 306 (1941)

Ethylene sulfide. Mol. wt. 60.12, b.p. 54°, n^{20}D 1.4960.

A simple preparative procedure[1] involves heating potassium thiocyanate (hygroscopic) in a flask evacuated to about 1 mm. until the temperature of the melt is in the range 165–175° and all moisture has been eliminated. The flask is cooled to room

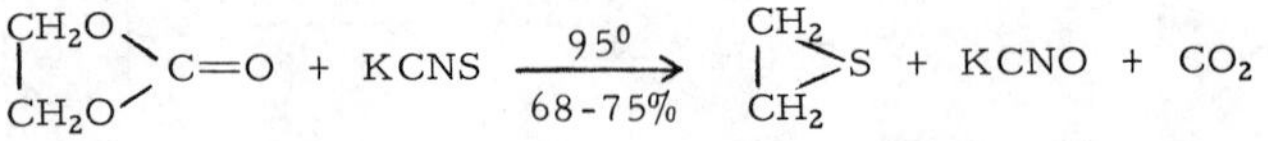

temperature, ethylene carbonate is added, and the reaction flask is heated slowly, with a downward condenser in place to deliver into a dry ice–acetone cooled receiver. Reaction begins when the temperature in the fused potassium thiocyanate layer reaches 90°. Ethylene sulfide distills and is collected in the receiver. The reagent as prepared is suitable for use, for example for reaction with a primary or secondary amine to form an aminomercaptan.[2]

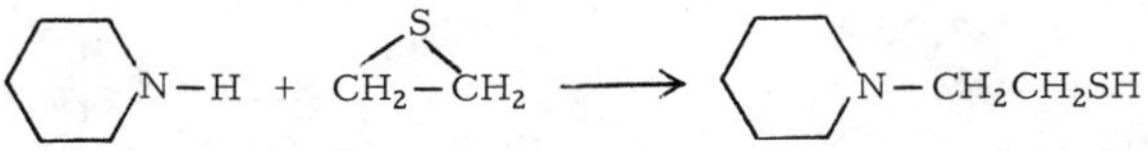

[1]S. Searles, E. F. Lutz, H. R. Hays, and H. E. Mortensen, *Org. Syn.*, **42**, 59 (1962)
[2]H. R. Snyder, J. M. Stewart, and J. B. Ziegler, *Am. Soc.*, **69**, 2672 (1947)

Ethylenimine, CH_2—CH_2 (bridged by NH). Mol. wt. 43.07, b.p. 56°, sp. gr. 0.83. Suppliers: Dow, Light. Relatively toxic.[1] The reagent is prepared by the action of hot alkali on 2-aminoethyl hydrogen sulfate (Aldrich, Eastman). The *Organic Syntheses* procedure[2] should be consulted for references and notes, but the yield (34–37%) has been greatly improved by addition of an alkaline solution of the sulfate to hot alkali at such a rate that the ethylenimine is removed as formed by flash distillation.[3]

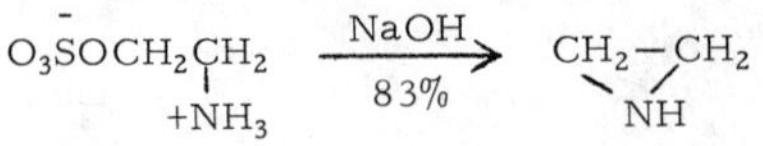

The reagent alkylates primary and secondary amines in the presence of 1 mole of aluminum chloride.[4] For example, di-*n*-butylamine is added with cooling and stirring

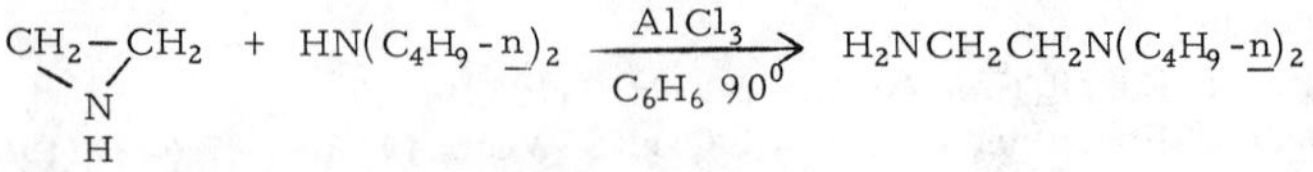

to a suspension of aluminum chloride in benzene, the temperature is raised to 90°, and ethylenimine is introduced as a gas by connecting an ampoule of the reagent to a gas inlet tube and immersing the ampoule in a water bath maintained at 75–85°. Primary amines require a temperature of about 180°. Yields range from 77 to 89%.

References to the following reactions of ethylenimine are to be found in a review:[5] cleavage by acids (reversal of formation); hydrolysis in weakly acidic solution; reaction with thiols, for example with cysteine to give (1); reaction with carbon disulfide to form 2-mercaptothiazoline (2). The reaction of ethylenimine with

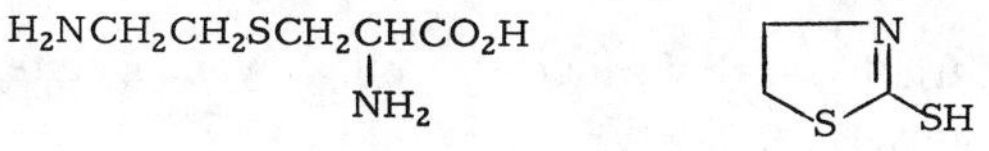

(1) (2)

aldehydes and ketones to give 2-oxazolidines is reported,[6] and several other reactions are described by Bestian.[7]

Ethylenimine proved to be an ideal reagent for the synthesis of S-acetylpantetheine (3) from pantothenic acid.[8] Pantothenic acid, in the form of the trimethylamine salt,

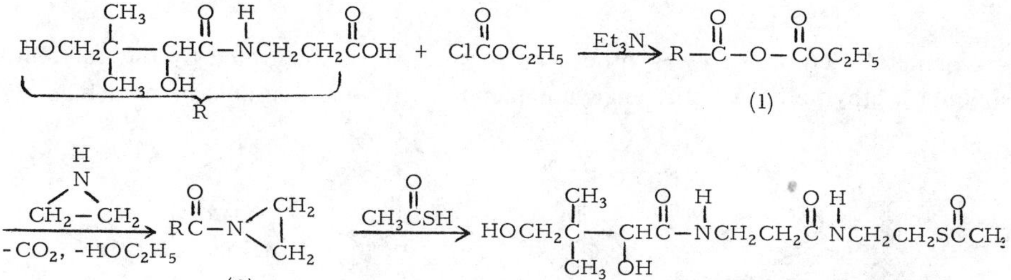

(2) (3)

is condensed in dimethylformamide solution to form the mixed anhydride (1), this is allowed to react with ethylenimine to form the amide (2), and the ring is then cleaved with thiolacetic acid to form S-acetylpantetheine (3).

An aldehyde synthesis utilizing ethylenimine is illustrated by the following example.[9] Reaction of cyclopropanecarboxylic acid chloride with ethylenimine and triethylamine in ether at 0° gives a precipitate of triethylamine hydrochloride and a solution of the 1-acylaziridine (3), which is reduced by lithium aluminum hydride

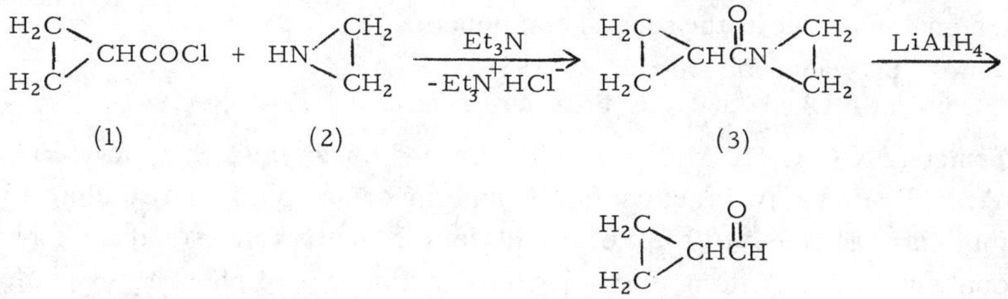

(4)

to give, after hydrolysis, cyclopropanecarboxaldehyde (4) in 60% yield. The acylaziridine (3) is heat-labile but need not be isolated. This procedure is particularly useful for the conversion of aliphatic acid chlorides into aldehydes where direct reduction with lithium tri-*t*-butoxyaluminohydride proceeds in low yield.

[1]J. P. Danehy and D. J. Pflaum, *Ind. Eng. Chem.*, **30**, 778 (1938)
[2]C. F. H. Allen, F. W. Spangler, and E. R. Webster, *Org. Syn., Coll. Vol.*, **4**, 433 (1963)
[3]W. A. Reeves, G. L. Drake, Jr., and C. L. Hoffpauir, *Am. Soc.*, **73**, 3522 (1951)
[4]G. H. Coleman and J. E. Callen, *Am. Soc.*, **68**, 2006 (1946)
[5]J. S. Fruton in Elderfield's "Heterocyclic Compounds," **I**, 61 (1950)
[6]J. B. Doughty, C. L. Lazzell, and A. R. Collett, *Am. Soc.*, **72**, 2866 (1950)

[7]H. Bestian, *Ann.*, **566**, 210 (1950)
[8]R. Schwyzer, *Helv.*, **35**, 1903 (1952)
[9]H. C. Brown and A. Tsukamoto, *Am. Soc.*, **83**, 2016 (1961)

Ethyl ethoxymethylenecyanoacetate (2). Mol. wt. 169.18, m.p. 53°. Supplier: Kay-Fries.

Preparation from triethyl orthoformate (1), ethyl cyanoacetate, acetic anhydride, and zinc chloride (catalyst).[1]

$$\underset{(1)}{C_2H_5OCH(OC_2H_5)_2} + \underset{\displaystyle CN}{CHCO_2C_2H_5} + 2\,(CH_3CO)_2O \xrightarrow{ZnCl_2}$$

$$\underset{(2)}{C_2H_5OCH{=}\underset{\displaystyle CN}{C}CO_2C_2H_5} + 2\,CH_3CO_2C_2H_5 + 2\,CH_3CO_2H$$

A general method for the synthesis of pyrimidines is illustrated by the condensation of ethyl ethoxymethylenecyanoacetate with thiourea in the presence of

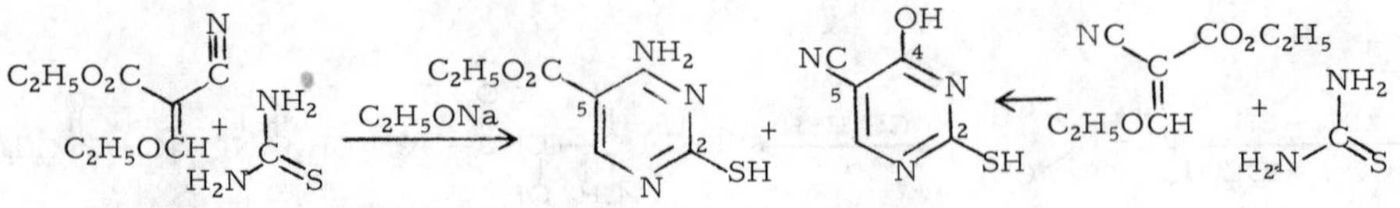

(a) (3) 76-80% (4) 7-12% (b)

sodium ethoxide.[2] The main product (3), 2-mercapto-4-amino-5-carboethoxypyrimidine, is formed by condensation as in (a), the minor product (4), 2-mercapto-4-hydroxy-5-cyanopyrimidine, from condensation as in (b). The major product (3) separates from an aqueous medium at 25°; the minor product separates from the filtrate on cooling overnight at 0°. The synthetic method can be varied by use of guanidine or an amidine and with an alkoxymethylenemalonic ester or an alkoxymethylenemalononitrile as the second component.

[1]E. Grégoire de Bellemont, *Bull. soc.*, **25**, 18 (1901)
[2]T. L. V. Ulbricht, T. Okuda, and C. C. Price, *Org. Syn., Coll. Vol.*, **4**, 566 (1963)

Ethyl formate, $HCO_2C_2H_5$. Mol. wt. 74.08, b.p. 54°, sp. gr. 0.92. Suppliers: Fritzsche Bros., A, B, E, F, MCB. Practical grade material can be purified by adding 15 g. of potassium carbonate per 100 ml. of liquid, letting the mixture stand for 1 hr. with occasional swirling, decanting into a dry flask, adding 5 g. of phosphorus pentoxide, and fractionation with exclusion of moisture.[1]

Grignard synthesis. Use of the reagent for the Grignard synthesis of a secondary alcohol is illustrated by the procedure formulated:[1]

$$\underline{n}\text{-}C_4H_9MgBr \xrightarrow{HCO_2C_2H_5} (\underline{n}\text{-}C_4H_9)_2CHOMgBr \xrightarrow[83\text{-}85\%]{H_2O} (\underline{n}\text{-}C_4H_9)_2CHOH$$

Formylation of ketones. The condensation of cyclohexanone with ethyl formate in the presence of either sodium ethoxide or sodium hydride gives a product which can be described as 2-hydroxymethylenecyclohexanone (1) or the tautomer 2-formylcyclohexanone (2);[2] for simplicity the process is described as a formylation. In the procedure cited, formylation was followed by reaction with hydrazine to

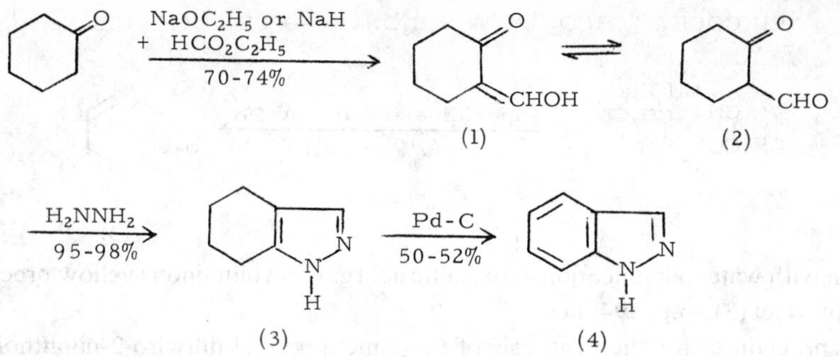

give the tetrahydroindazole (3), which was dehydrogenated to indazole (4). Formylation of a cyclic ketone is often used to block one α-position so that the α'-position can be alkylated; the blocking group is then eliminated by alkaline hydrolysis. W. S. Johnson[3] improved the process by O-alkylation of the hydroxymethylene derivative (6) with isopropyl iodide, methylation of this to give (8), and elimination of the O-alkyl group by acid hydrolysis and the formyl group by basic hydrolysis to (10).

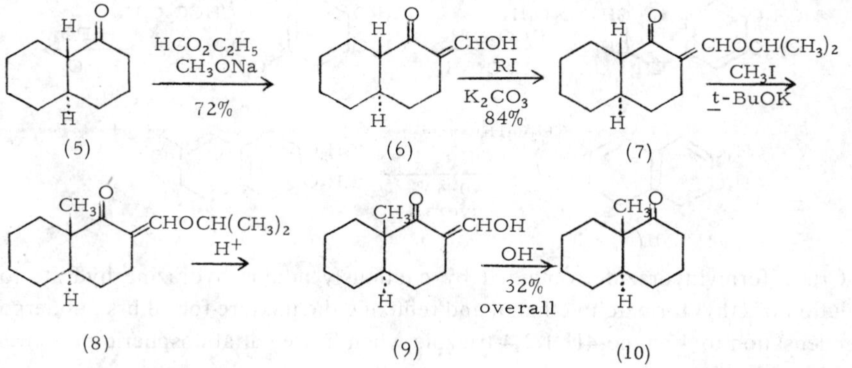

1,3,5-Triacetylbenzene can be synthesized by an interesting process involving formylation of acetone.[4] A solution of sodium ethoxide in absolute ethanol is

$$CH_3COCH_3 + HCO_2C_2H_5 + NaOC_2H_5 \longrightarrow CH_3COCH_2CHO$$

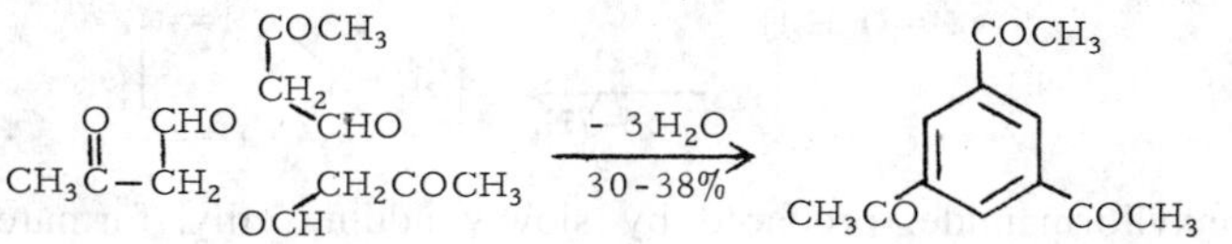

M. p. 163°

diluted with ether, and a mixture of acetone and ethyl formate is dropped into the stirred solution under reflux during 2 hrs. Under the conditions of its formation the formyl derivative condenses to the aromatic triketone.

A procedure for the synthesis of 3-cyano-6-methyl-2(1)-pyridone[5] (3) requires isolation of the sodio derivative of formylacetone, (1), and this was accomplished by stirring a suspension of sodium methoxide in ether with ice cooling, dropping in a mixture of acetone and ethyl formate, and removing the ether at a temperature below 70°. The solid residue was treated with an aqueous solution of cyanoacetamide (Eastman) and piperidine acetate and the solution was refluxed for 2 hrs. After

$$CH_3COCH_3 + HCO_2C_2H_5 + CH_3ONa \longrightarrow CH_3COCH{=}CH\bar{O}\overset{+}{Na}$$

(1)

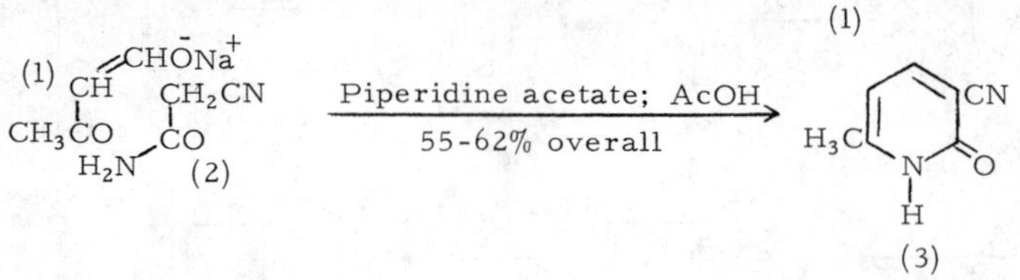

dilution with water, acidification with acetic acid gave a voluminous yellow precipitate of the product (3), m.p. 294° dec.

In a procedure[6] for the synthesis of 6,7-dimethoxy-3,4-dihydro-2-naphthoic acid (IV) by the method of G. P. Crowley and R. Robinson,[7] ethyl γ-veratrylbutyrate (I) is formylated with ethyl formate by dropwise addition of a solution of I and ethyl formate in ether to a suspension of sodium ethoxide in ether at −10°. After acidification, the formyl derivative II is collected by ether extraction and cyclized to III with a mixture of 90% phosphoric acid and concentrated sulfuric acid at 0–10°. Saponification gives the acid IV, m.p. 193°.

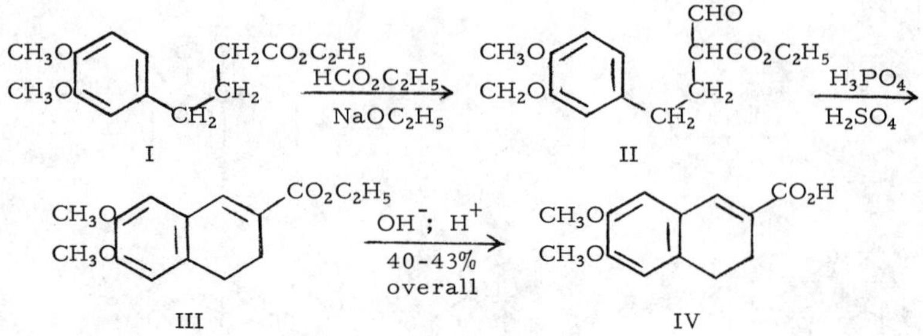

Crude formylhydrazide, obtained by cautiously adding hydrazine hydrate to a solution of ethyl formate in ethanol and refluxing the mixture for 18 hrs., undergoes condensation to 4-amino-4H-1,2,4-triazole when heated at atmospheric pressure to 200° for 3 hrs.[8]

$$HCO_2C_2H_5 \xrightarrow{H_2NNH_2} HCNHNH_2 \ (\overset{O}{\parallel})$$

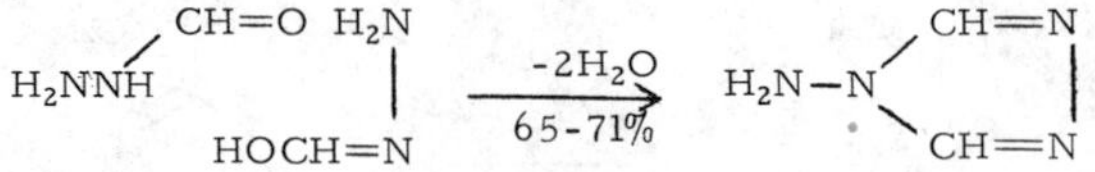

N-Cyclohexylformamide, prepared by slowly adding ethyl formate to cyclohexylamine with stirring and cooling, is an intermediate in an efficient method for conversion of the amine to cyclohexyl isocyanide.[9]

$$C_6H_{11}NH_2 \xrightarrow{HCO_2C_2H_5} C_6H_{11}NHCHO \xrightarrow[67\text{-}72\%]{POCl_3-Py} C_6H_{11}N{=}C$$

Primary and some secondary amines can be formylated in good yield by heating the amine with ethyl formate at 100–110° for 1 hr. in an autoclave.[10] Under the same conditions ethyl acetate does not react.

[1]G. H. Coleman and D. Craig, *Org. Syn., Coll. Vol.*, **2**, 179 (1943)

[2]C. Ainsworth, *ibid.*, **4**, 536 (1963). For application in the steroid series, see R. O. Clinton *et al.*, *Am. Soc.*, **83**, 1478 (1961)

[3]W. S. Johnson and H. Posvic, *Am. Soc.*, **69**, 1361 (1947)
[4]L. Claisen and N. Stylos, *Ber.*, **21**, 1145 (1888); R. L. Frank and R. H. Varland, *Org. Syn., Coll. Vol.*, **3**, 829 (1955)
[5]R. P. Mariella, *ibid.*, **4**, 210 (1963)
[6]H. L. Holmes and L. W. Trevoy, *Org. Syn., Coll. Vol.*, **3**, 300 (1955)
[7]G. P. Crowley and R. Robinson, *J. Chem. Soc.*, 2001 (1938)
[8]C. F. H. Allen and A. Bell, *Org. Syn., Coll. Vol.*, **3**, 96 (1955)
[9]I. Ugi, R. Meyr, M. Lipinski, F. Bodesheim, and F. Rosendhal, *Org. Syn.*, **41**, 13 (1961)
[10]J. Moffat, M. V. Newton, and G. J. Papenmaier, *J. Org.*, **27**, 4058 (1962)

Ethyl isocyanate, $C_2H_5N{=}C{=}O$. Mol. wt. 71.08, b.p. 60°. Preparation.[1] Suppliers: A, B, E, F.

S-Protection. The reagent reacts with cysteine at room temperature in dimethylformamide to form S-ethylcarbamoylcysteine. New syntheses of glutathione and

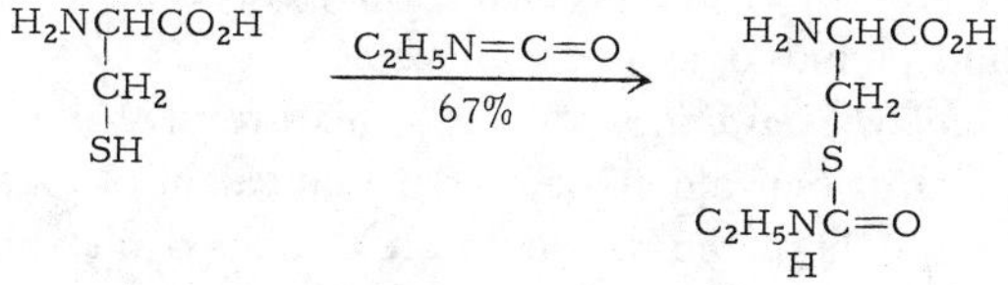
H₂NCHCO₂H
CH₂
SH
C₂H₅N=C=O
67%
H₂NCHCO₂H
CH₂
S
C₂H₅NC=O
H

oxytocin use the ethylcarbamoyl group for protection of the thiol function of cysteine.[2] The group is stable under acid and neutral conditions but is readily cleaved by basic reagents.

[1]L. Gattermann, *Ann.*, **244**, 36 (1888)
[2]St. Guttmann, *Helv.*, **49**, 83 (1966)

Ethyl 2-mercaptoethylcarbonate, (2). Mol. wt. 118.13, b.p. 84°/9 mm.

Preparation[1] in high yield from ethyl 2-hydroxyethylthiol carbonate (1) by catalytic rearrangement with uranyl acetate dihydrate.

$$\underset{(1)}{C_2H_5OCOSCH_2CH_2OH} \xrightarrow{(CH_3COO)_2UO_2\cdot 2\,H_2O} \underset{(2)}{C_2H_5OCO_2CH_2CH_2SH}$$

Mercaptoethylation of primary and secondary amines.[1]

$$R(Ar)NH_2 + C_2H_5OCO_2CH_2CH_2SH \longrightarrow R(Ar)NHCH_2CH_2SH + C_2H_5OH + CO_2$$

[1]D. D. Reynolds, D. L. Fields, and D. L. Johnson, *J. Org.*, **26**, 5125 (1961)

Ethyl metaphosphate, *see* Polyphosphate esters.

N-Ethylmorpholine, 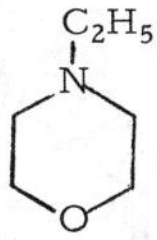
C₂H₅
N
O

Mol. wt. 115.7, b.p. 138–9°. Suppliers: A, B, E, F, MCB.

This solvent is preferred for the determination of active hydrogen by reaction with lithium aluminum hydride.[1] For this use it is dried over calcium hydride, heated with $LiAlH_4$ for 2 hrs. at 90–100°, and then distilled at 20 mm. pressure.

It is a good solvent for both naphthostyrile (1) and lithium aluminum hydride, and proved useful for the reduction of (1) to benz(*cd*)indolenin (2).[2] A solution of 1 g. of (1) in 20 ml. of N-ethylmorpholine was stirred with exclusion of air and treated at 50–60° with 0.5 g. of $LiAlH_4$. The solution turned green and then fluorescent blue, and after 15 min. the mixture was hydrolyzed carefully with water and worked up.

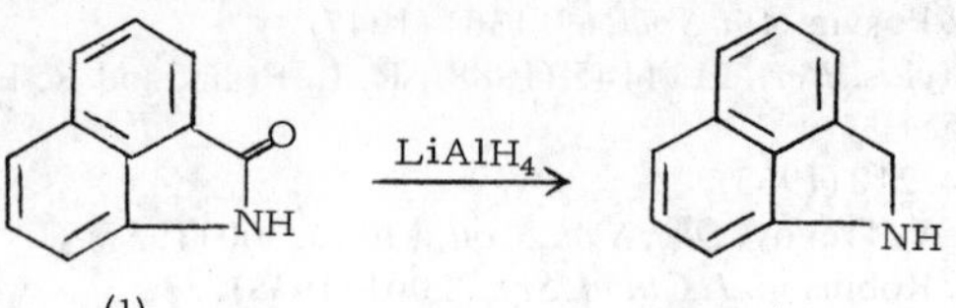

(1)   (2)

[1]F. A. Hochstein, *Am. Soc.*, **71**, 305 (1949)
[2]A. Stoll, Th. Petrzilka, and J. Rutschmann, *Helv.*, **33**, 2254 (1950)

Ethyl orthocarbonate, *see* Tetraethyl orthocarbonate.

Ethyl orthoformate, *see* Triethyl orthoformate.

Ethyl N-phenyliminophosphite, $C_2H_5OP=NC_6H_5$. Mol. wt. 167.15, b.p. 145–147°/0.08 mm. The reagent is prepared in 50% yield from 1 mole of ethyl phosphorodichloridate and 3 moles of aniline in benzene.[1]

Preparation of dialkyl phosphites.[1] When a mixture of 2 equivalents of the reagent (1), 1 equivalent of benzaldehyde, and 1 equivalent of an alcohol in toluene is heated at 100° for 4 hrs., the products are a dialkyl phosphite (5, 75%) and benzylideneaniline (3, 66%).

$$C_2H_5OP=NC_6H_5 \;+\; C_6H_5CH=O \longrightarrow \left[\begin{array}{c} C_2H_5O-P-NC_6H_5 \\ | \quad\quad | \\ O-CHC_6H_5 \end{array} \right] \longrightarrow$$

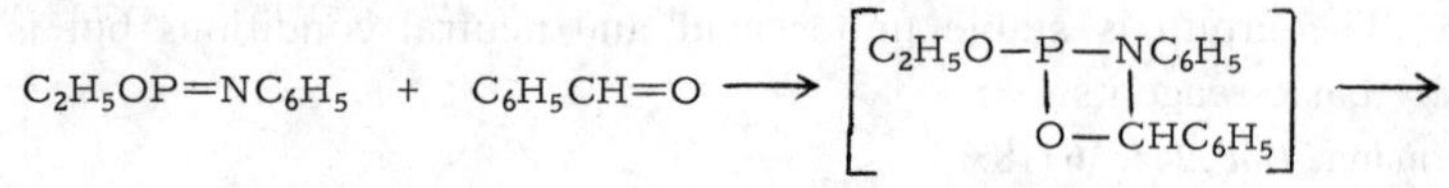

(1)   (2)

$$C_6H_5N=CHC_6H_5 \;+\; C_2H_5O-P=O \xrightarrow{ROH} \begin{array}{c} C_2H_5O \\ RO \end{array}\!\!\!>\!\!P\!\!<\!\!\begin{array}{c} O \\ H \end{array}$$

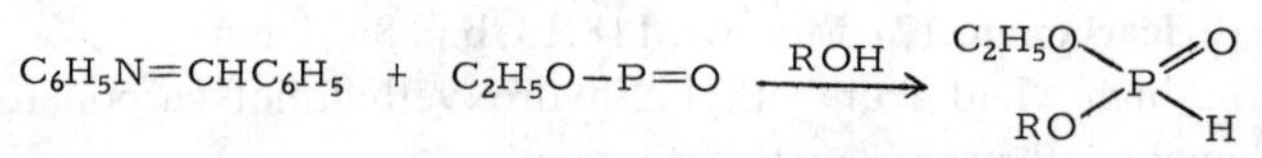

(3)   (4)   (5)

[1]O. Mitsunobu and T. Mukaiyama, *J. Org.*, **29**, 3005 (1964)

N-Ethyl-5-phenylisoxazolium-3′-sulfonate (1). Mol. wt. 253.28. Suppliers: Aldrich, Pilot Chemicals. Preparation and use in peptide synthesis.[1,2] It is recommended that the commercial salt be dissolved in excess 1 N hydrochloric acid and precipitated with acetone to give a fluffy product.[2]

The reagent (1) reacts with an N-protected amino acid or peptide in acrylonitrile or nitromethane in the presence of triethylamine to form an activated ester (2), which

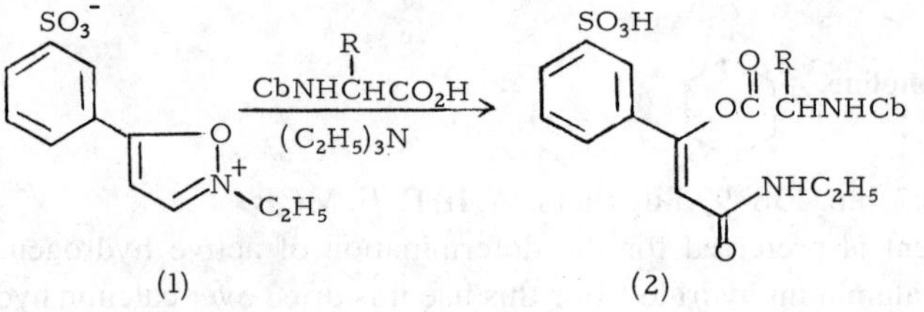

(1)   (2)

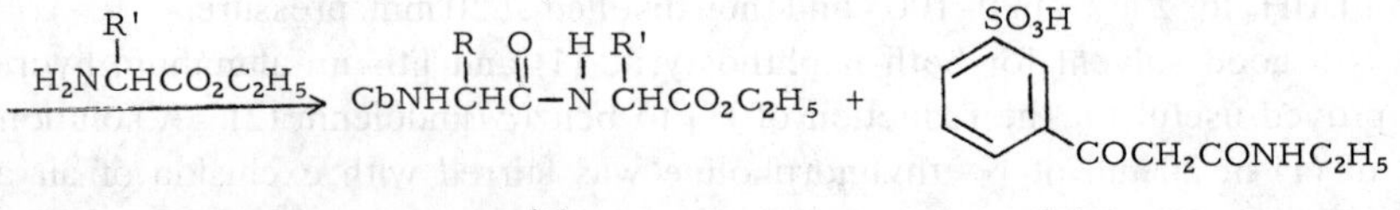

(3)   (4)

on reaction with an amino acid ester affords the peptide derivative (3) and the water-soluble by-product (4). Examples.[3]

In a synthesis of oxytocin requiring coupling of a pentapeptide and a tetrapeptide, Fosker and Law[4] found the reagent superior to DCC or N,N′-carbonyldiimidazole.

[1]R. B. Woodward and R. A. Olofson, *Am. Soc.*, **83**, 1007 (1961); R. B. Woodward, R. A. Olofson, and H. Mayer, *Am. Soc.*, **83**, 1010 (1961)
[2]R. B. Woodward and R. A. Olofson, procedure submitted to *Org. Syn.*
[3]C. H. Li, B. Gorup, D. Chung, and J. Ramachandran, *J. Org.*, **28**, 178 (1963); R. B. Kelly, *ibid.*, **28**, 453 (1963); R. Schwyzer and H. Kappeler, *Helv.*, **46**, 1550 (1963)
[4]A. P. Fosker and H. D. Law, *J. Chem. Soc.*, 4922 (1965)

Ethyl propenyl ether, $CH_3CH{=}CHOC_2H_5$. Mol. wt. 87.13, b.p. 63°. Preparation.[1]

Ethyl vinyl ether is used for effecting the transformation (a), whereas ethyl propenyl ether effects the change (b). Both reactions are employed in the Isler

$$\text{(a)}\quad RCHO \xrightarrow{CH_2=CHOC_2H_5} RCH{=}CHCHO$$

$$\text{(b)}\quad RCHO \xrightarrow{CH_3CH=CHOC_2H_5} RCH{=}\overset{\displaystyle CH_3}{\overset{|}{C}}CHO$$

synthesis of β-carotene under essentially identical reaction conditions.[2] Thus details given under **Ethyl vinyl ketone** apply to the conversion of the C_{16}-aldehyde (1) to the C_{19}-aldehyde (4). The overall yield from (1) was 66–71%.

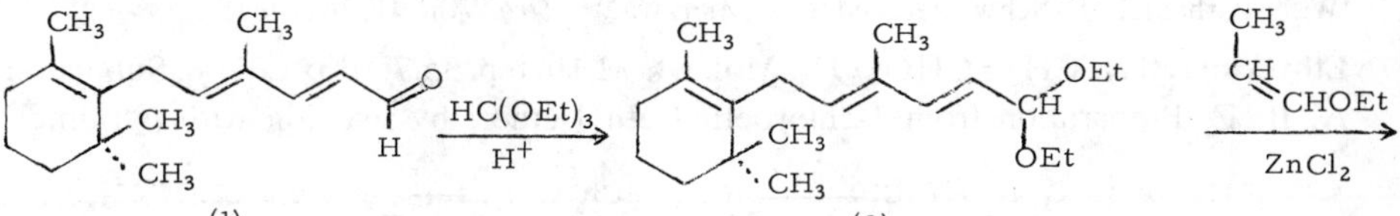
CH3
CH3
CH3
CH3
H
HC(OEt)3
H+
CH3
CH3
OEt
CH
OEt
CH3
CH3
CH3
CH
CHOEt
ZnCl2

(1) (2)

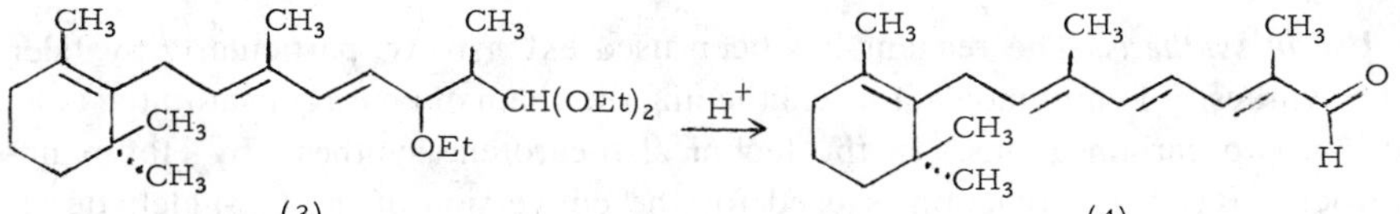
CH3
CH3
CH3
CH(OEt)2
CH3
CH3
OEt
H+
CH3
CH3
CH3
O
CH3
CH3
H

(3) (4)

[1]A. Kirrmann, *Bull. Soc.*, **6**, 841 (1939)
[2]O. Isler, H. Lindlar, M. Montavon, R. Rüegg, and P. Zeller, *Helv.*, **39**, 249 (1956)

Ethyl propionate, $CH_3CH_2COC_2H_5$. Mol. wt. 102.14, b.p. 99°, sp. gr. 0.896.

In the alkoxide-catalyzed condensation of an aromatic aldehyde with phthalide to give a 2-arylindane-1,3-dione the yield is only about 34%. However when ethyl propionate was added as scavenger for the water formed, yields were as high as 100%.[1] The scavenger ester is taken in excess to serve as solvent. Although the

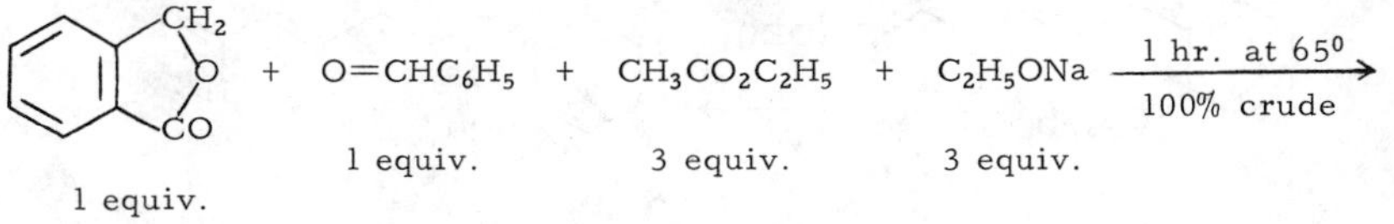
CH2
O
CO
+ O=CHC6H5 + CH3CO2C2H5 + C2H5ONa
1 hr. at 65⁰
100% crude
1 equiv.
3 equiv.
3 equiv.
1 equiv.

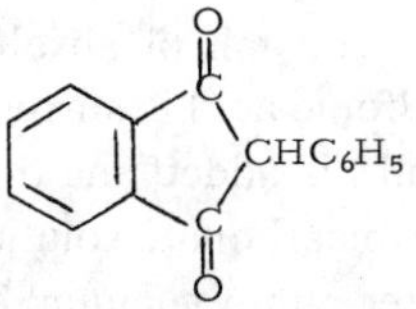
O
C
CHC6H5
C
O

stoichiometry called for 1 equivalent of alkoxide to form the sodium enolate of the product and 1 equivalent to form sodium propionate, it was found preferable in practice to use 3 equivalents. Other unhindered esters serve also as water scavengers.

[1]S. L. Shapiro, K. Geiger, and L. Freedman, *J. Org.*, **25**, 1860 (1960); S. L. Shapiro, K. Geiger, J. Youlus, and L. Freedman, *ibid.*, **26**, 3580 (1961)

Ethyl trichloroacetate, $CCl_3CO_2C_2H_5$. Mol. wt. 191.44, b.p. 55–58°/12 mm. Suppliers: B, E, F.

The reagent is used to generate dichlorocarbene, for example for reaction with dihydropyrane to give 2-oxa-7,7-dichloronorcarane.[1] A flask is flushed with nitrogen

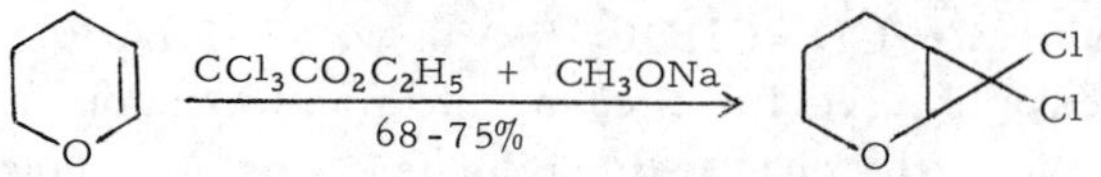

CCl3CO2C2H5 + CH3ONa
68-75%
Cl
Cl
O
O

and charged with 0.92 mole of sodium methoxide, 0.8 mole of dihydropyrane, and 600 ml. of olefin-free pentane. The light yellow solution is stirred for 15 min. in an ice-water bath, and 0.86 mole of ethyl trichloroacetate is added in 3–4 minutes. The mixture is stirred at 6° under nitrogen for 6 hrs. and let stand at room temperature overnight; the color changes from orange-yellow to brown. Water is added, the layers are separated, and the aqueous layer is extracted with pentane. Fractionation affords product of b.p. 74–76°/8 mm.

[1]W. E. Parham, E. E. Schweizer, and S. A. Mierzwa, Jr., *Org. Syn.*, **41**, 76 (1961)

Ethyl vinyl ether, $CH_2{=}CHOC_2H_5$. Mol. wt. 72.10, b.p. 35.7° $n^{20}D$ 1.3768. Suppliers: A, B, E. Preparation from 1-chloroethyl ethyl ether[1] by reaction with pyridine:[2]

$$CH_3CHO + HOC_2H_5 + HCl \xrightarrow[94\%]{} CH_3CHClOC_2H_5 \xrightarrow[43\%]{C_5H_5N} CH_2{=}CHOC_2H_5 + C_5H_5\overset{+}{N}HCl^-$$

Use in synthesis. The reagent has been used extensively, particularly by Isler in the synthesis of carotenoids, for lengthening the chain of an α,β-unsaturated aldehyde by two carbon atoms.[3] In the technical β-carotene synthesis by Hoffmann-La Roche, Basel, the reaction is used for the conversion of the C_{14}-aldehyde (1) into the C_{16}-aldehyde (4). The procedure is as follows.[4] A mixture of 325 g. of (1),

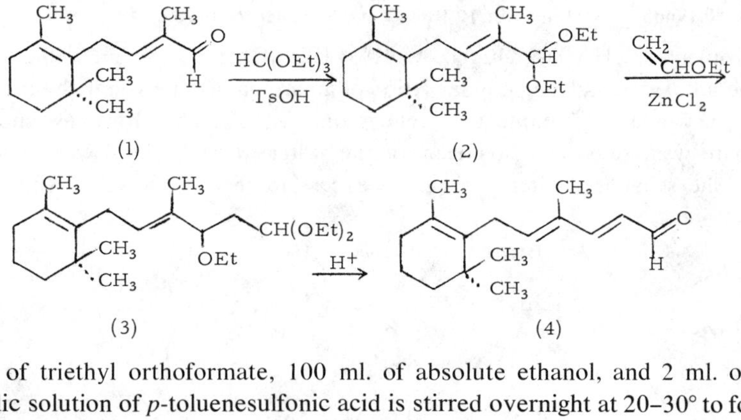

CH3
CH3
O
H
CH3
CH3
HC(OEt)3
TsOH
OEt
CH
OEt
CH2
CHOEt
ZnCl2
(1)
(2)
OEt
CH(OEt)2
H+
(3)
(4)

240 g. of triethyl orthoformate, 100 ml. of absolute ethanol, and 2 ml. of a 1% ethanolic solution of *p*-toluenesulfonic acid is stirred overnight at 20–30° to form the diethyl acetal (2). Pyridine (18 ml.) is added and the solution is poured into 450 ml. of ice-cold 2.5% aqueous sodium bicarbonate solution. The upper layer is separated, and solvent is distilled off at water-pump vacuum. The residual acetal (2) is treated with 2 ml. of a 10% solution of zinc chloride in ethyl acetate and stirred at 30–35° during simultaneous slow addition of 160 ml. of ethyl vinyl ether and 28 ml. of the

10% zinc chloride solution to effect addition to form the β-ethoxydiethylacetal (3). After stirring at room temperature overnight, a solution of 110 g. of sodium acetate in 1 l. of acetic acid and 77 ml. of water is stirred in and the mixture is stirred at 95° for 3 hrs. The moist acetic acid liberates the aldehyde group and eliminates the β-ethoxy group to give the C_{16}-aldehyde (4). The solution is cooled and poured into water, and the crystalline aldehyde that separates is crystallized from 400 ml. of methanol at −10°. The yield of (4) of satisfactory quality for the next step (m.p. 77–78°) is 240 g. [66% from (1)].

An interesting synthesis of citral reported by a Russian group starts with 6-methyl-Δ^5-heptene-2-one (4) prepared by Cope rearrangement of the acetoacetate ester (3).[5] The ketone (4) was converted in very high yield into the diethyl ketal (5) by

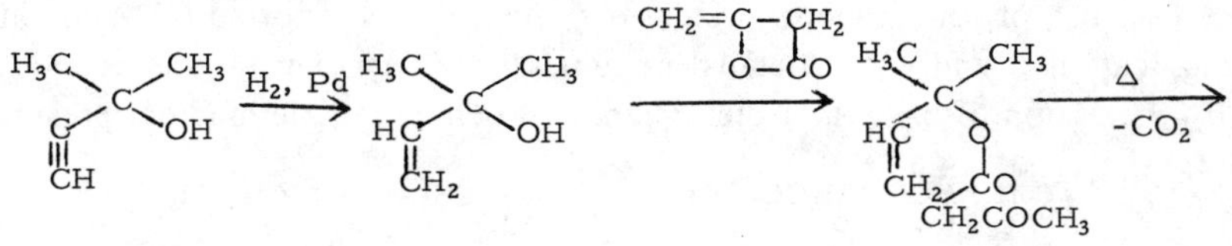
H₂, Pd
Δ
-CO₂

(1) (2) (3)

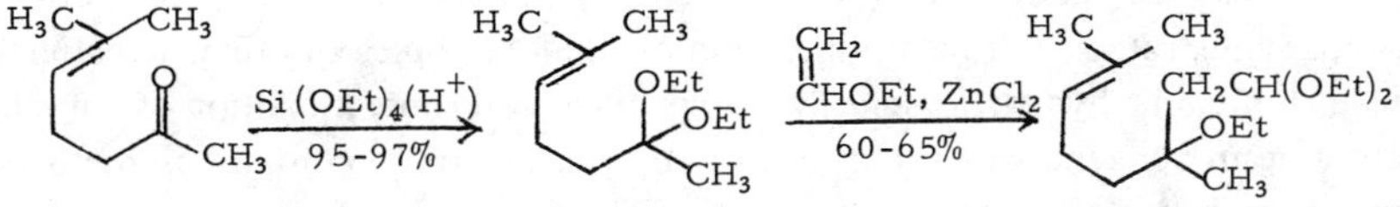
Si(OEt)₄(H⁺)
95-97%
ZnCl₂
60-65%

(4) (5) (6)

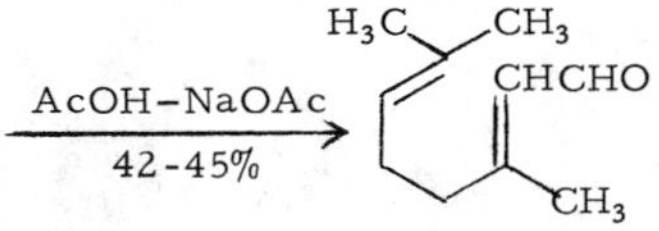
AcOH–NaOAc
42-45%

Citral

reaction with the readily available tetraethoxysilane in the presence of orthophosphoric acid and a small amount of *p*-toluenesulfonic acid and about 0.3 mole of alcohol. Addition of the ketal to ethyl vinyl ketone in acetic acid proceeds better with zinc chloride than with boron trifluoride as catalyst. Hydrolysis of (6) gave citral.[6]

O-Protective group. In the presence of an acid the reagent reacts with the 2′-hydroxyl group of a ribonucleoside 3′-phosphate-5′-acetate to give the α-ethoxyethyl derivative. The α-ethoxyethyl group is preferred to the tetrahydropyranyl

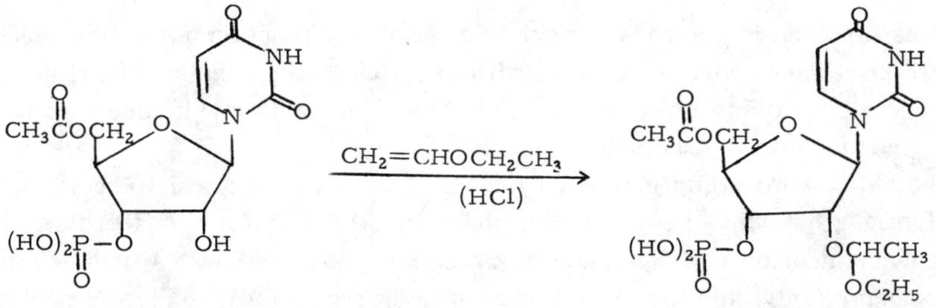
CH₂=CHOCH₂CH₃
(HCl)

group in oligonucleotide synthesis because it can be removed with extreme ease.[7] Thus it is split by 5% acetic acid in 2 hrs. at 20° with no detectable isomerization of the 5'-3'-internucleotide linkage. Under the same conditions a tetrahydropyranyl ether is cleaved to the extent of only 37%, and some isomerization occurs under the more acidic conditions required for full cleavage.

[1]H. R. Henze and J. T. Murchison, *Am. Soc.*, **53**, 4077 (1931)
[2]C. D. Hurd and D. G. Botteron, *Am. Soc.*, **68**, 1200 (1946)
[3]O. Isler and P. Schudel, *Advances in Org. Chem.*, **4**, 128–130 (1963)
[4]O. Isler, H. Lindlar, M. Montavon, R. Rüegg, and P. Zeller, *Helv.*, **39**, 249 (1956)
[5]G. I. Samokhavalov *et al.*, *J. Gen. Chem. U.S.S.R.*, **27**, 2560 (1957)
[6]*Idem, ibid.*, **29**, 2538 (1959)
[7]S. Chládek and J. Smrt, *Chem. Ind.*, 1719 (1964)

Ethyl vinyl ketone, $CH_2{=}CHCOCH_2CH_3$. Mol. wt. 84.11, b.p. 102°; 45–47°/100 mm.; water azeotrope, b.p. 83°. Suppliers: A, KK.

The first use of this reagent in steroid synthesis was reported by Wilds in two communications,[1] and some details are not clear. The reagent was generated *in situ* by the action of base on 1-diethylamino-3-pentanone methiodide, presumably

$$(C_2H_5)_2NCH_2CH_2COCH_2CH_3 \xrightarrow{OH^-} CH_2{=}CHCOCH_2CH_3$$

prepared by the condensation of methyl ethyl ketone with formaldehyde and diethylamine. Wilds' scheme for use of the reagent for the construction of the steroid ring B will be illustrated with more fully documented examples of other investigations. Thus in Johnson's hydrochrysene synthesis of testosterone, a solution of 5-methoxy-2-tetralone in benzene was run into a solution of sodium methoxide in methanol to form the enolate, and a methanol solution of 1-diethylamino-3-pentanone methiodide was added dropwise with stirring at 2–5°.[2] The reaction product, the tricyclic ketone

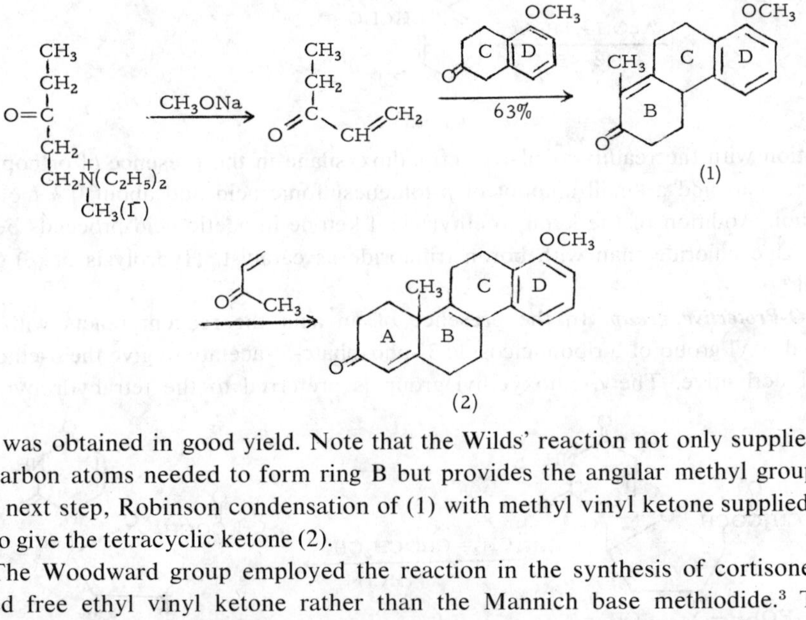
CH3
CH2
O=C
CH2
CH2N(C2H5)2
CH3(I⁻)
CH3ONa
CH3
CH2
C
O
CH
CH2
OCH3
C
D
63%
OCH3
CH3
C
D
B
(1)
CH3
OCH3
CH3
C
D
A
B
(2)

(1) was obtained in good yield. Note that the Wilds' reaction not only supplies the 4 carbon atoms needed to form ring B but provides the angular methyl group. In the next step, Robinson condensation of (1) with methyl vinyl ketone supplied ring A to give the tetracyclic ketone (2).

The Woodward group employed the reaction in the synthesis of cortisone but used free ethyl vinyl ketone rather than the Mannich base methiodide.[3] Their improvement of an early preparative procedure[4] is as follows.[5] Friedel-Crafts reaction of propionyl chloride with ethylene in chloroform affords 1-chloro-3-pentanone, and when a mixture of this with an equivalent amount of diethylamine is heated gradually to 200° ethyl vinyl ketone distills.

$$CH_3CH_2COCl + CH_2{=}CH_2 \xrightarrow[71\text{-}76\%]{AlCl_3\ (0^\circ)} CH_3CH_2COCH_2CH_2Cl$$

$$\xrightarrow[55\%]{(C_2H_5)_2NH} CH_3CH_2COCH{=}CH_2$$

Conversion of the bicyclic ketone (1) to the tricyclic ketone (4) was done by a technique introduced by Shunk and Wilds.[6] The ketone was formylated and the

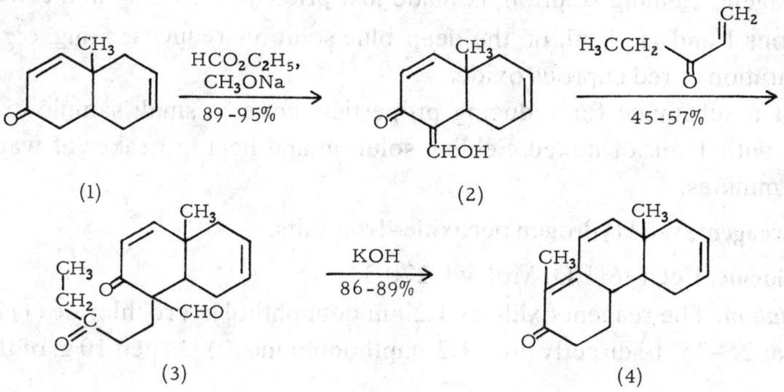

CH3
HCO2C2H5,
CH3ONa
89-95%
CH3
CHOH
CH2
H3CCH2
O
45-57%
(1)
(2)
CH3
CH3
CH2
C
O
CHO
KOH
86-89%
CH3
CH3
(3)
(4)

formyl derivative (2) treated in *t*-butanol under nitrogen with ethyl vinyl ketone and a catalytic amount of potassium *t*-butoxide to give the adduct (3), which on hydrolysis with aqueous potassium hydroxide in dioxane afforded the tricyclic ketone (4).

[1]A. L. Wilds, J. W. Rolls, W. C. Wildman, and K. E. McCaleb, *Am. Soc.*, **72**, 5794 (1950); A. L. Wilds, J. W. Rolls, D. A. Tyner, R. Daniels, S. Kraychy, and M. Narnik, *ibid.*, **75**, 4878 (1956)

[2]W. S. Johnson, J. Szmuszkovicz, E. R. Rogier, H. I. Hadler, and H. Wynberg, *Am. Soc.*, **78**, 6285 (1956)

[3]R. B. Woodward, F. Sondheimer, D. Taub, K. Heusler, and W. M. McLamore, *Am. Soc.*, **74**, 4223 (1952)

[4]E. M. McMahon, J. N. Roper, Jr., H. P. Utermohlen, Jr., R. H. Hasek, R. C. Harris, and J. H. Brant, *Am. Soc.*, **70**, 2977 (1948)

[5]Ref. 3, p. 4239

[6]C. H. Shunk and A. L. Wilds, *Am. Soc.*, **71**, 3946 (1949)

Ethynylmagnesium bromide, HC≡CMgBr.

The reagent is prepared by dropwise addition of a solution of ethylmagnesium bromide in tetrahydrofurane to the same solvent which is kept saturated with acetylene:[1]

$$HC{\equiv}CH + C_2H_5MgBr \longrightarrow HC{\equiv}CMgBr + C_2H_6$$

Inverse addition is essential, since the mono-Grignard reagent can react with ethylmagnesium bromide to form the di-Grignard derivative.

The reagent adds to most carbonyl compounds to form the corresponding ethynylcarbinols in good yield. An example is the reaction with cinnamaldehyde.[1]

$$C_6H_5CH{=}CHCH{=}O \xrightarrow[55\text{-}65\%]{HC\equiv CMgBr} C_6H_5CH{=}CH\underset{\displaystyle OH}{\underset{|}{C}}HC{\equiv}CH$$

[1]L. Skattebøl, E. R. H. Jones, and M. C. Whiting, *Org. Syn., Coll. Vol.*, **4**, 792 (1963)

F

Fehling solution. Solution I: 34.64 g. of $CuSO_4 \cdot 5\,H_2O$ dissolved in water and diluted to 500 ml. Solution II: 173 g. of sodium potassium tartrate (Rochelle salt) and 65 g. of sodium hydroxide dissolved in water and diluted to 500 ml.

The reagent, Fehling solution, is made just prior to use by mixing equal volumes of solutions I and II; 1 ml. of the deep blue solution reduces 5 mg. of D-glucose with separation of red cuprous oxide.

To test a substance for reducing properties, cover a small sample (5 mg.) in a test tube with 1 ml. of mixed Fehling solution and heat in beaker of water at 100° for a few minutes.

Fenton's reagent, *see* Hydrogen peroxide–Iron salts.

Ferric chloride, $FeCl_3 \cdot 6H_2O$. Mol. wt. 270.32.

Oxidation. The reagent oxidizes 1,2-aminonaphthol hydrochloride (1) in aqueous solution at 25–35° to directly pure 1,2-napthoquinone (2).[1] When 10 g. of this yellow-

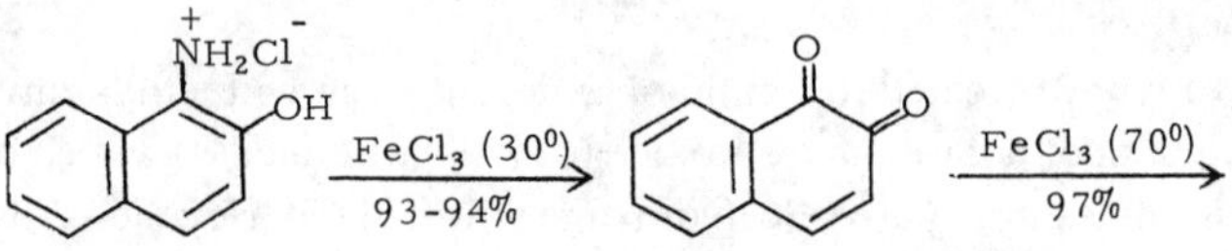
NH2Cl
OH
FeCl3 (30°)
93-94%
O
O
FeCl3 (70°)
97%

(1) (2)

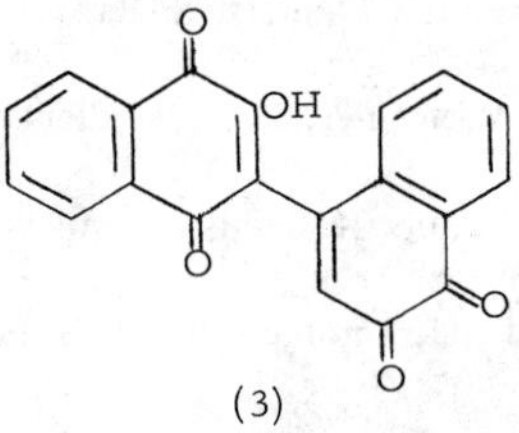
O
OH
O
O
O

(3)

orange quinone (2) is stirred at 65–70° with an aqueous solution of 40 g. of $FeCl_3 \cdot 6H_2O$ in 400 ml. of water for about 1 hr., the quinone partially dissolves, and suddenly the solution clears with separation of the pure yellow oxidation product (3), 2-hydroxy-3,3′-dinaphthyl-1,4,1′,2′-diquinone.[2,3] Preparations of 6-methoxy- and 5,6-dimethoxy-1,2-naphthoquinone by the first method are described by Gates.[4] High yields characterize the oxidation of (4) to 2-amino-1,4-naphthoquinone-4-imine hydrochloride in aqueous solution at room temperature and of (6) in warm acetic acid (1 part)–water (2 parts) to 2-acetylamino-1,4-naphthoquinone (7).[5]

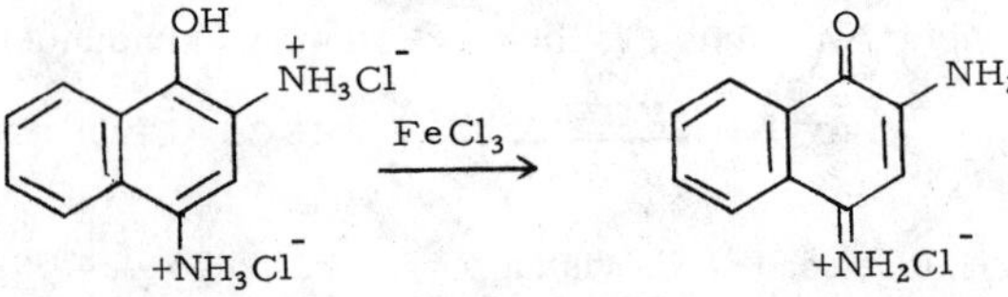
OH
NH3Cl
FeCl3
+NH3Cl
O
NH2
+NH2Cl

(4) (5)

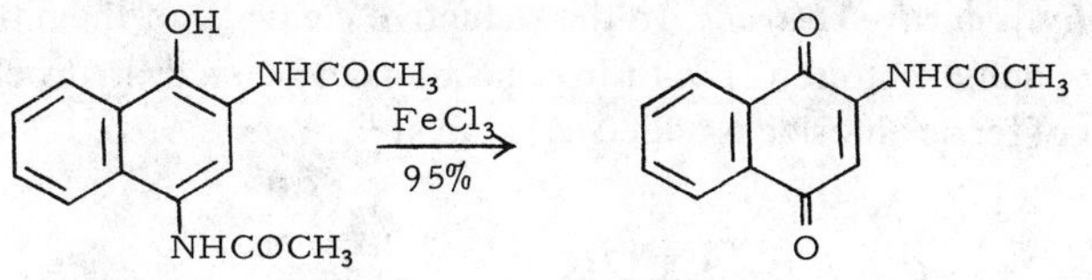
OH
NHCOCH3
FeCl3
95%
NHCOCH3
O
NHCOCH3
O

(6) (7)

Other uses of ferric chloride as an oxidant are for conversion of the dithiol (8) to (±)-thioctic acid (9)[6] and of chavicol (10) to magnolol (11).[7]

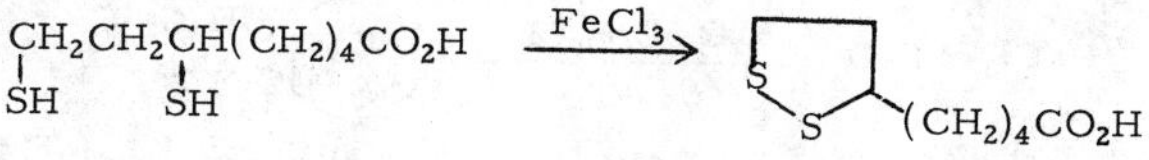
CH2CH2CH(CH2)4CO2H
SH SH
FeCl3
S
S
(CH2)4CO2H

(8) (9)

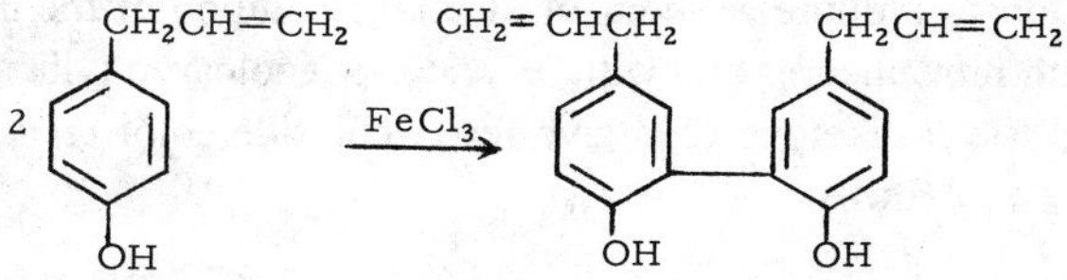
CH2CH=CH2
2
OH
FeCl3
CH2= CHCH2
CH2CH=CH2
OH
OH

(10) (11)

Ferric chloride in chloroform oxidizes the metacyclophane (12) to the bis-dienone (13) in high yield.[8]

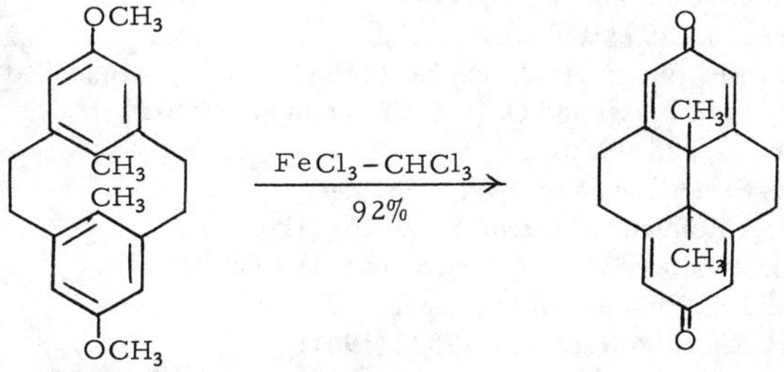
OCH3
CH3
CH3
OCH3
FeCl3-CHCl3
92%
O
CH3
CH3
O

(12) (13)

The reaction of cuprous cyanide with an aryl halide to produce a nitrile gives an intermediate complex of the nitrile with cuprous halide. Friedman and Shechter[9] found that the nitrile can be liberated from the complex very easily by addition of

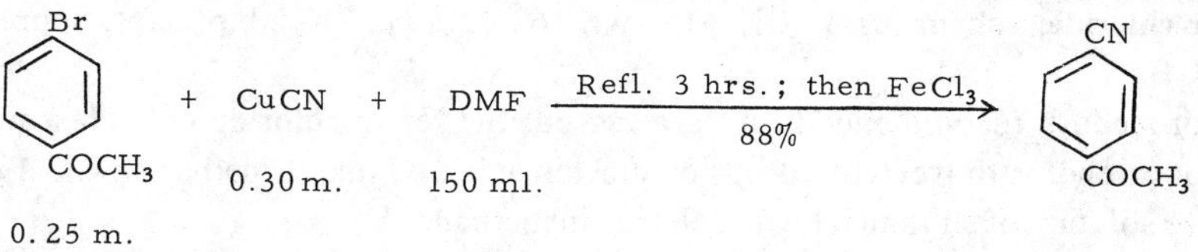
Br
COCH3
0.25 m.
+ CuCN + DMF
0.30 m.
150 ml.
Refl. 3 hrs.; then FeCl3
88%
CN
COCH3

aqueous ferric chloride to oxidize the cuprous to cupric ions, which form no complex with the nitrile.

Of a number of syntheses of cyclohexane-1,2-dione investigated the method preferred is oxidation of 2-hydroxycyclohexanone with ferric chloride in acid solution.[9a]

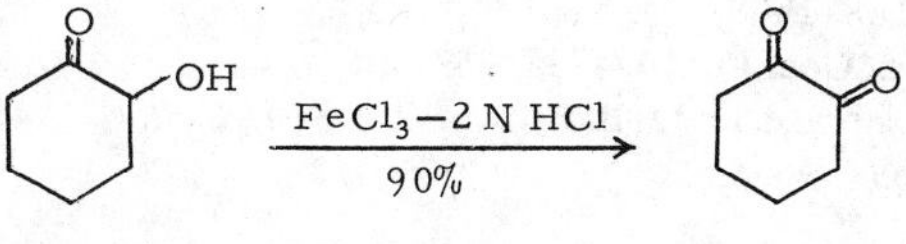
O
OH
FeCl3-2 N HCl
90%
O
O

Catalyst for reductive cleavage. In the reductive cleavage of the nitropropene (1) to *o*-methoxyphenylacetone (2) with iron powder and concd. hydrochloric acid, a small amount of ferric chloride is added as catalyst.[10]

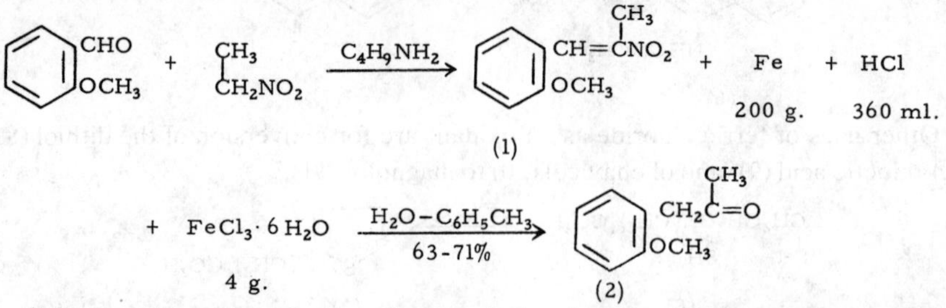

Color test. Some phenols give characteristic colors on addition of a drop of ferric chloride solution to a dilute aqueous or alcoholic solution of the material. The test is negative with nitrophenols and with *m*- and *p*-phenolcarboxylic acids. The β-keto ester (1)[11] and the β-diketone (2)[12] give ferric chloride color tests indicative of the presence of the enol forms.

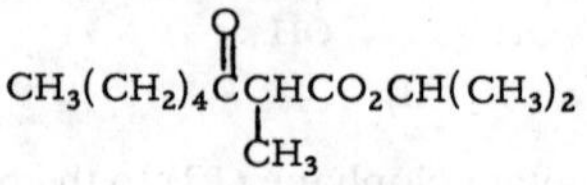
(1)

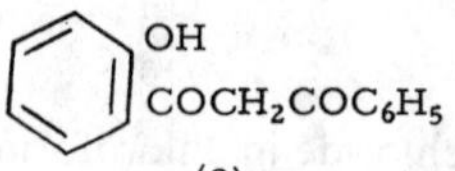
(2)

[1]L. F. Fieser, *Org. Syn., Coll. Vol.*, **2**, 430 (1943)
[2]H. Wichelhaus, *Ber.*, **30**, 2199 (1897)
[3]S. C. Hooker and L. F. Fieser, *Am. Soc.*, **58**, 1216 (1936). The original structure assignment has been revised by L. F. Fieser and D. H. Sachs, unpublished work.
[4]M. Gates, *Am. Soc.*, **72**, 228 (1950)
[5]L. F. Fieser and M. Fieser, *Am. Soc.*, **56**, 1565 (1934)
[6]B. A. Lewis and R. A. Raphael, *J. Chem. Soc.*, 4263 (1962)
[7]H. Erdtman and J. Runeberg, *Acta. Chem. Scand.*, **11**, 1060 (1957)
[8]V. Boekelheide and J. B. Phillips, *Am. Soc.*, **85**, 1545 (1963)
[9]L. Friedman and H. Shechter, *J. Org.*, **26**, 2522 (1961)
[9a]L. De Borger, M. Anteunis, H. Lammens, and M. Verzele, *Bull. Soc. chim. Belg.*, **73**, 73 (1964)
[10]R. V. Heinzelman, *Org. Syn., Coll. Vol.*, **4**, 573 (1963)
[11]K. L. Rinehart, *ibid.*, **4**, 120 (1963)
[12]T. S. Wheeler, *ibid.*, **4**, 478 (1963)

Ferric chloride, anhydrous, $FeCl_3$. Mol. wt. 162.22, m.p. 282°, b.p. 315°. Suppliers: F, MCB.

A procedure recommended for carrying out the ferric chloride color test with a phenol or enol is to treat 1–2 drops of substance in 2–3 ml. of methanol with 1 drop of a 1% solution of sublimed ferric chloride in methanol.[1]

Anhydrous ferric chloride is reduced with iron powder in refluxing tetrahydrofurane under nitrogen for production of the ferrous chloride required for the synthesis of ferrocene.[2] In the preparation of a solution of sodamide in liquid ammonia, a catalytic amount of anhydrous ferric chloride is added to the liquid ammonia followed by enough sodium to convert the iron salt into catalytic iron.[3, 4]

[1]H. Henecka, *Ber.*, **81**, 188 (1948)
[2]G. Wilkinson, *Org. Syn., Coll. Vol.*, **4**, 473 (1963)
[3]N. A. Khan, F. E. Deatherage, and J. B. Brown, *Org. Syn., Coll. Vol.*, **4**, 851 (1963)
[4]N. A. Khan, *ibid.*, **4**, 969 (1963)

Ferrous sulfate. $FeSO_4 \cdot 7H_2O$. Mol. wt. 278.03. Suppliers: B, F, MCB.

Use of ferrous sulfate in an alkaline medium for the reduction of an aromatic nitro group *ortho* to a carbonyl function was introduced by Claisen[1] and extended by Bamberger.[2] A modern version is a procedure for the reduction of *o*-nitrobenzaldehyde to *o*-aminobenzaldehyde.[3]

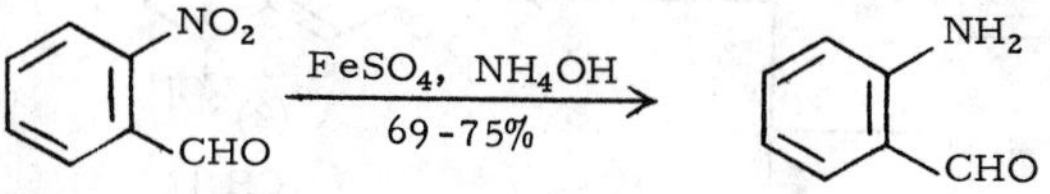

As yet unexplained is the use of ferrous sulfate as moderator to control the otherwise stormy reaction of aniline, glycerol, nitrobenzene, and concentrated sulfuric acid in the Skraup synthesis of quinoline.[4]

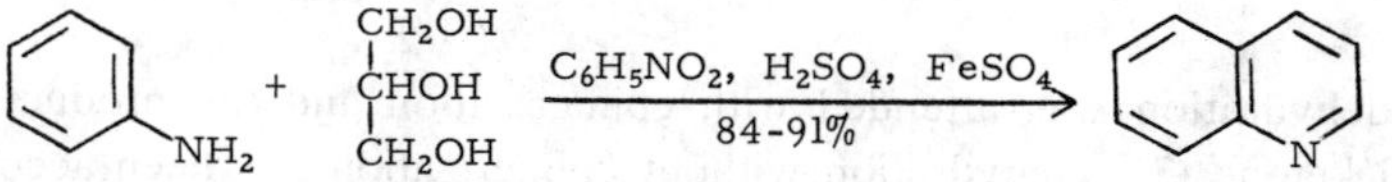

[1]L. Claisen and J. Shadwell, *Ber.*, **12**, 353 (1879); L. Claisen and C. M. Thompson, *Ber.*, **12**, 1947 (1879)
[2]E. Bamberger and Ed. Demuth, *Ber.*, **34**, 1330 (1901)
[3]L. I. Smith and J. W. Opie, *Org. Syn., Coll. Vol.*, **3**, 56 (1955)
[4]H. T. Clarke and A. W. Davis, *ibid.*, **1**, 478 (1941)

Fieser's solution (for absorption of oxygen).[1] The solution is prepared by dissolving 20 g. of potassium hydroxide in 100 ml. of water and adding 2 g. of sodium anthraquinone-β-sulfonate and 15 g. of sodium hydrosulfite ($Na_2S_2O_4$) in the warm solution. The mixture is stirred until a clear, blood-red solution is obtained and cooled to room temperature. Traces of oxygen in tank nitrogen are removed by passage through two or three wash bottles containing the solution and then into a wash bottle of saturated lead acetate solution to absorb a trace of hydrogen sulfide sometimes present. The sulfonated anthrahydroquinone dianion absorbs oxygen with great speed and is continually regenerated by the hydrosulfite. Since the reagent operates on a catalytic principle, the original efficiency is retained to the point of exhaustion of the hydrosulfite and the wash bottles do not require refilling as long as the solution remains clear and bright. When the color changes to dull red or brown or when a precipitate appears, the solution should be replaced. With fresh commercial hydrosulfite the efficient capacity of the above quantity of solution is 788 ml. of oxygen.

This finding, made when I was a graduate student, resulted from taking Professor Baxter's course on gas analysis while doing research with Professor Conant involving measurement of the oxidation-reduction potentials of anthraquinonesulfonates over a range of pH. Reductive titration with sodium hydrosulfite in the alkaline range presented considerable difficulty because the nitrogen available was of poor quality and removal of oxygen by passage over hot copper uncertain. I thus found it expedient to use an already titrated red solution for sweeping the nitrogen before it passed into a fresh solution to be titrated. When the analysis course reached the experiment on the determination of oxygen in air, Professor Baxter allowed me to try my solution in place of pyrogallol. When the first trials seemed promising, he encouraged me to extend the work in the following summer and submit a paper for publication. – L. F. F.

[1]L. F. Fieser, *Am. Soc.*, **46**, 2639 (1924).

Fischer reagent, *see* Karl Fischer reagent.

Florisil. A chromatographic magnesium silicate of analysis: MgO, 15.5%, SiO_2, 84%; Na_2SO_4, 0.5%. Supplier: Floridin Co.

A Searle group[1] found that the 5α-hydroxy-6β-methyl-3-ketosteroid (1) on dehydration by refluxing with florisil in benzene is converted in good yield into the 3-keto-Δ^4-6β-methyl steroid (2). When basic alumina (Woelm-1) was used as

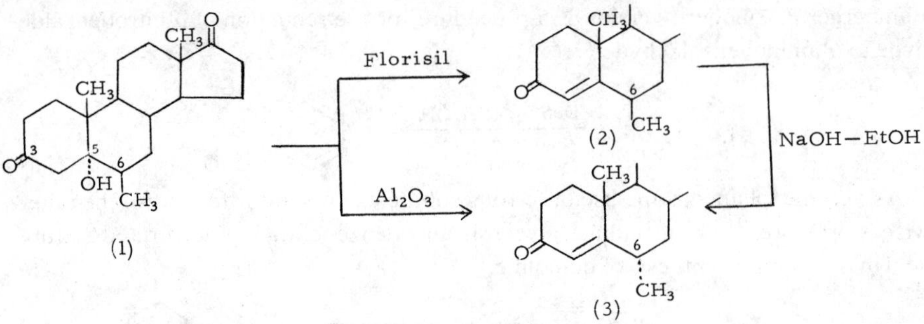
Florisil
Al_2O_3
NaOH—EtOH
(1)
(2)
(3)

catalyst, dehydration was attended with epimerization and the product was the 6α-methyl ketone (3). Dehydration without epimerization has been accomplished also with thionyl chloride–pyridine[2] and with very dilute sodium hydroxide in ethanol.[3] The driving force for epimerization comes from the fact that a 6β-substituent is axial and a 6α-substituent is equatorial. 6β-Acetoxy-3-keto-Δ^4-steroids are epimerized to the 6α-isomers by the action at 0° of hydrogen chloride in chloroform containing 0.7% ethanol (but not in pure chloroform!).[4]

[1] R. H. Bible, Jr. and N. W. Atwater, *J. Org.*, **26**, 1336 (1961)
[2] R. B. Turner, *Am. Soc.*, **74**, 5362 (1952)
[3] J. A. Campbell, J. C. Babcock, and J. A. Hogg, *Am. Soc.*, **80**, 4717 (1958)
[4] P. Th. Herzig and M. Ehrenstein, *J. Org.*, **16**, 1050 (1951); L. F. Fieser, *Am. Soc.*, **75**, 4377 (1953)

Fluorene-9-carboxylic acid. Mol. wt. 210.23. Supplier: Aldrich. In a study of esters of phenols of use as protective derivatives which could be cleaved by photolysis, Barton *et al.*[1] concluded that the esters of fluorene-9-carboxylic acid are the most useful. However, yields of phenol formed on photoylsis were at most 60%.

[1] D. H. R. Barton, Y. L. Chow, A. Cox, and G. W. Kirby, *J. Chem. Soc.*, 3571 (1965)

Fluoroboric acid, HBF_4. Mol. wt. 87.83. Prepared by the slow addition of boric acid to hydrofluoric acid with constant stirring in a copper, lead, silver-plated, or waxed container.[1-3] Supplier (48% aqueous solution): MCB.

Schiemann and related displacements. Addition of an ice-cold solution of fluoroboric acid to a vigorously stirred solution of benzenediazonium chloride at 0° gives a yellowish crystalline precipitate of benzenediazonium fluoroborate, which is collected, washed with ice water, methanol, and ether, and dried.[2] The solid is

$$C_6H_5\overset{+}{N}H_3\overset{-}{Cl} \xrightarrow{HNO_2} C_6H_5\overset{+}{N}{\equiv}N(\overset{-}{Cl}) \xrightarrow{HBF_4} C_6H_5\overset{+}{N}{\equiv}N(\overset{-}{BF_4})$$

16 moles

$$\xrightarrow[51-57\%]{\text{Pyrolysis}} C_6H_5F + N_2 + BF_3$$

placed in a 12-l. flask connected through a long, wide condenser to three 2-l. Erlenmeyers arranged in series and immersed in salt-ice mixtures. An exit tube from the last receiver carries voluminous fumes of BF_3 to the hood. The solid is then heated cautiously and intermittently with a free flame to keep the decomposition in progress.

Workup includes removal of a little phenol. Other examples are the preparation of 4,4'-difluorodiphenyl[4] and *p*-fluorobenzoic acid.

A related reaction is employed for the preparation of *p*-dinitrobenzene.[1] In this case a suspension of the diazonium fluoroborate in water is added with stirring to

$$\underline{p}\text{-}NO_2C_6H_4NH_2 \xrightarrow[95\text{-}99\%]{HNO_2,\ HBF_4} \underline{p}\text{-}NO_2C_6H_4\overset{+}{N}{\equiv}N(\overset{-}{BF_4}) \xrightarrow[67\text{-}82\%]{NaNO_2(Cu)} \underline{p}\text{-}NO_2C_6H_4NO_2$$

copper powder suspended in an aqueous solution of sodium nitrite. Ether is added to break the froth. The product separates as a solid containing copper. It is washed and dried and extracted with benzene.

Catalyst for methylation with diazomethane. Saturated alcohols do not react with diazomethane alone but afford methyl ethers under catalysis by fluoroboric acid or boron trifluoride. Neeman and Johnson[5] give a detailed procedure for the conversion of cholestanol into the methyl ether in 95% yield by reaction in ether–methylene chloride in the presence of fluoroboric acid.

Salts of nonbenzenoid aromatic hydrocarbons. Tropylium fluoroborate (3) is the salt of choice for the preparation of substituted tropilidenes because it is indefinitely stable, nonhygroscopic, and nonexplosive. A convenient procedure for its preparation[6] involves reaction of tropilidene (1) with phosphorus pentachloride in carbon tetrachloride to produce a precipitate of the double salt (2). The salt is stirred into

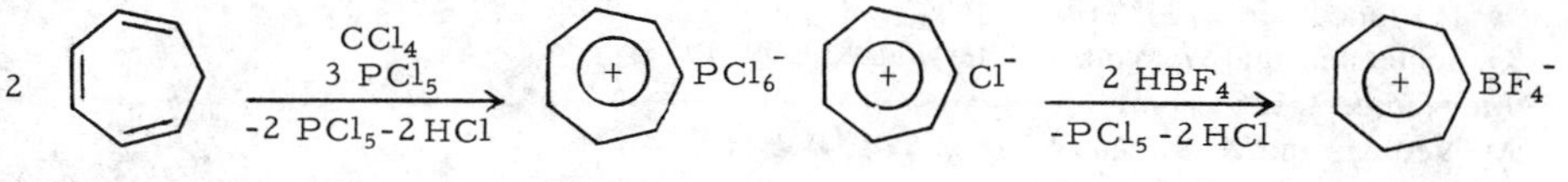
2
CCl4
3 PCl5
-2 PCl5-2 HCl
PCl6-
Cl-
2 HBF4
-PCl5 -2 HCl
BF4-

(1) (2) (3)

ice-cold absolute ethanol to produce a reddish solution, and addition of aqueous fluoroboric acid precipitates white tropylium fluoroborate (3).

Dauben and Bertelli[7] synthesized heptalene (9) starting with reduction of naphthalene-1,5-dicarboxylic acid (4) with sodium and ethanol in liquid ammonia to the

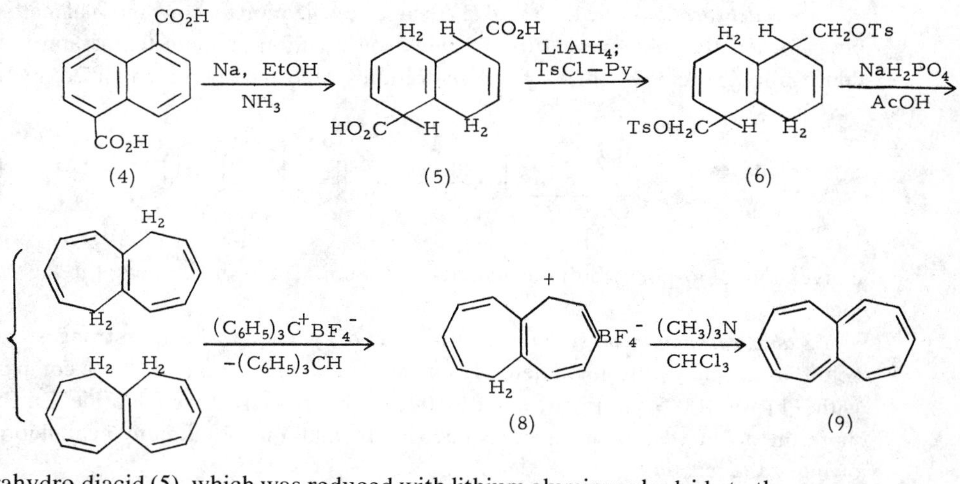
CO2H
CO2H
Na, EtOH
NH3
H2 H CO2H
HO2C H H2
LiAlH4;
TsCl—Py
H2 H CH2OTs
TsOH2C H H2
NaH2PO4
AcOH
H2
H2
H2 H2
(C6H5)3C+BF4-
-(C6H5)3CH
BF4-
H2
(CH3)3N
CHCl3

(4) (5) (6) (7) (8) (9)

tetrahydro diacid (5), which was reduced with lithium aluminum hydride to the corresponding diol. Solvolytic rearrangement of the ditosylate (6) in acetic acid containing sodium dihydrogen phosphate then gave a mixture (7) of 1,6- and 1,10-dihydroheptalene, and hydride abstraction by reaction with triphenylmethylcarbonium fluoroborate in methylene chloride afforded 1-heptalenium fluoroborate (8) as bright yellow cubes, moderately stable in air. Addition of excess trimethylamine to a solution of the salt (8) in chloroform at 0° and removal of trimethylammonium fluoroborate gave a red filtrate from which heptalene was isolated by evaporative distillation and chromatography as a reddish brown viscous liquid.

Hafner *et al.*[8] have described the preparation and use in synthesis of azulenium fluoroborates. Thus when 4,6,8-trimethylazulene (1) in ether solution is treated with an ethereal solution of fluoroboric acid, the violet color fades and colorless 4,6,8-trimethylazulenium fluoroborate (2) separates. On addition of benzaldehyde, the

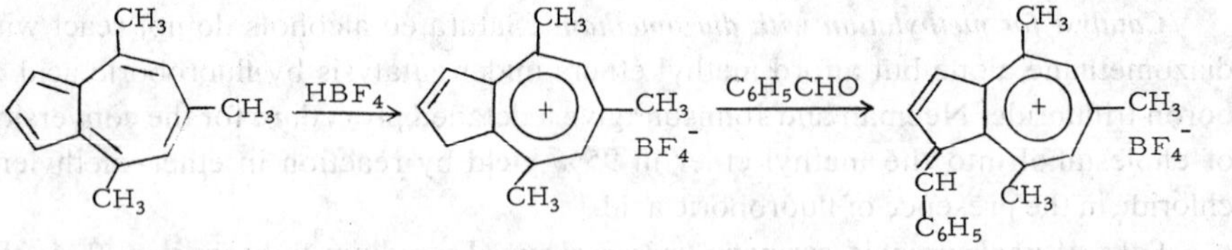
CH3
CH3
CH3
HBF4
CH3
CH3
BF4-
CH3
C6H5CHO
CH3
CH3
BF4-
CH
CH3
C6H5

(1) (2) (3)

colorless salt (2) gradually dissolves and after a few hours the colored 1-benzylidene-4,6,8-trimethylazulenium fluoroborate (3) separates in 97% yield.

[1]E. B. Starkey, *Org. Syn., Coll. Vol.*, **2**, 225 (1943)
[2]D. T. Flood, *ibid.*, **2**, 295 (1943)
[3]G. Schiemann and W. Winkelmüller, *ibid.*, **2**, 299 (1943)
[4]*Idem, ibid.*, **2**, 188 (1943)
[5]M. Neeman and W. S. Johnson, *Org. Syn.*, **41**, 9 (1961)
[6]K. Conrow, *ibid.*, **43**, 101 (1963)
[7]H. J. Dauben, Jr., and D. J. Bertelli, *Am. Soc.*, **83**, 4657, 4659 (1961)
[8]K. Hafner, H. Pelster, and J. Schneider, *Ann.*, **650**, 62 (1961); K. Hafner, H. Pelster, and H. Patzelt, *ibid.*, **650**, 80 (1961)

Fluorosulfonic acid, FSO_2OH. Mol. wt. 100.08, b.p. 165.5°. Supplier: Allied Chem. Corp.

Cyclodehydration. The keto acid (1) resisted cyclization by conventional methods but afforded the diketone (2) in high yield by reaction at room temperature with fluorosulfonic acid.[1] Other cyclodehydrations reported are exemplified by the

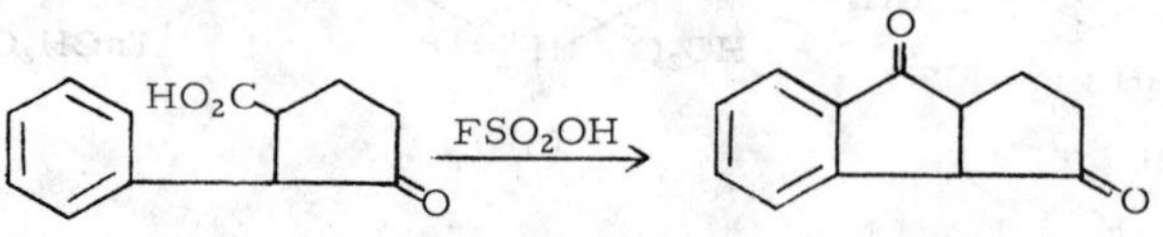
HO2C
FSO2OH
O
O
O

(1) (2)

conversion of γ-phenylbutyric acid to α-tetralone at room temperature in 61% yield.

Stable carbonium ions. Fluorosulfonic acid appears to be the strongest pure acid that has as yet been studied. H_0 for the neat acid is about −12.6, as compared with −11 for 100% sulfuric acid and −10 for anhydrous hydrogen fluoride. Furthermore the acidity can be enhanced somewhat by addition of antimony pentafluoride, owing to the reaction:[2]

$$SbF_5 + 2\ FSO_2OH \longrightarrow [H_2SO_3F]^+[SbF_5(OSO_2F)]^-$$

Olah *et al.*[3] used this system, diluted with liquid sulfur dioxide for better mixing, at −60° for generation of stable carbonium ions from the corresponding alcohols.

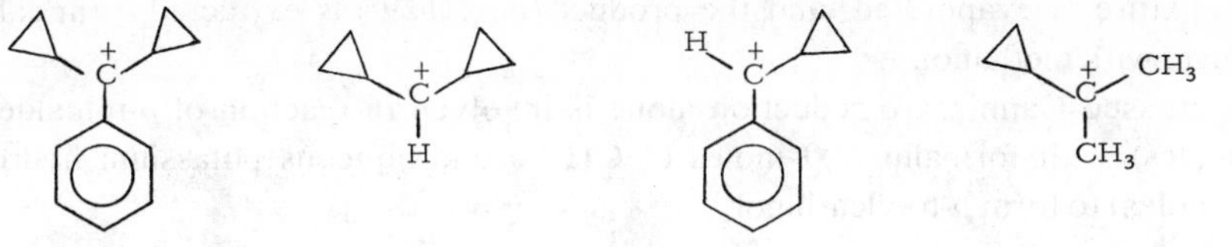

The first known mono- and dicyclopropylcarbonium ions were prepared in this way. Deno *et al.*[4] treated alcohols with fluorosulfonic acid alone at −50° and also obtained stable ions, as evidenced by NMR characterization.

[1]W. Baker, G. E. Coates, and F. Glockling, *J. Chem. Soc.*, 1376 (1951)

[2]R. J. Gillespie in G. A. Olah's "Friedel-Crafts and Related Reactions," **1**, 191, Interscience, N.Y. (1963)

[3]G. A. Olah, M. B. Comisarow, C. A. Cupas, and C. U. Pittman, Jr., *Am. Soc.*, **87**, 2997 (1965); C. U. Pittman, Jr., and G. A. Olah, *ibid.*, **87**, 2998 (1965)

[4]N. C. Deno, J. S. Liu, J. O. Turner, D. N. Lincoln, and R. E. Fruit, Jr., *Am. Soc.*, **87**, 3000 (1965)

Formaldehyde, $CH_2{=}O$. Mol. wt. 30.03, b.p. −21°, sp. gr. 0.81. Commercial formalin is an aqueous solution containing 37% of formaldehyde and 8–10% of methanol; 100 ml. of the solution contains 40 g. of formaldehyde. Where this much water is undesirable but a little can be tolerated, use can be made of methylal, $CH_2(OCH_3)_2$, which is readily hydrolyzed in an acidic medium. Dry gaseous formaldehyde required for a Grignard reaction is generated by heating the solid polymer paraformaldehyde and leading the gas into the reaction mixture.

Some of the many reactions of this versatile reagent are illustrated below.

With diethyl malonate. Two different products are obtainable in moderate yield by reaction of malonic ester with formalin. With a 1:2 ratio of aldehyde to ester and diethylamine as catalyst, the product is the tetraester (1), an intermediate in one synthesis of glutaric acid.[1] With a 2:1 ratio and potassium bicarbonate as catalyst, the product is the crystalline solid diethyl bis(hydroxymethyl)malonate (2).[2]

$$CH_2{=}O + CH_2(CO_2C_2H_5)_2 \xrightarrow[61\%]{(C_2H_5)_2NH} CH_2\begin{cases} CH(CO_2C_2H_5)_2 \\ CH(CO_2C_2H_5)_2 \end{cases} \xrightarrow{78\%}$$

(1)

$$CH_2\begin{cases} CH_2CO_2H \\ CH_2CO_2H \end{cases}$$

$$2\ CH_2{=}O + CH_2(CO_2C_2H_5)_2 \xrightarrow[72\text{-}75\%]{KHCO_3} (HOCH_2)_2C(CO_2C_2H_5)_2$$

(2)

Reduction. The familiar preparation of pentaerythritol from acetaldehyde and formaldehyde involves aldolization and crossed Cannizzaro reduction. In the same way, cyclohexanone reacts with 5 moles of formaldehyde to give the pentaol (3).[3] The reaction is conducted by adding calcium oxide (1.25 moles) to a stirred mixture

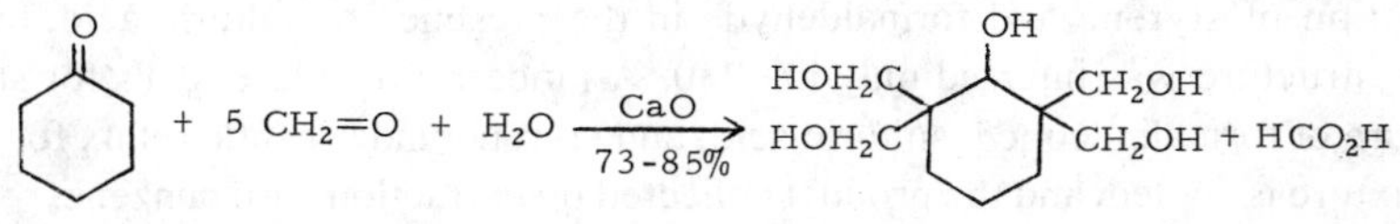

(3)

of cyclohexanone (2 moles), paraformaldehyde (11 moles), and water at 10–15°, with eventual rise to 35°. A little formic acid is added to neutralize particles of lime, the mixture is evaporated, and the product (m.p. 129°) is extracted from calcium formate with methanol.

A crossed Cannizzaro reduction alone is involved in reaction of *p*-tolualdehyde (3 moles) with formalin (3.9 moles of CH_2O) and aqueous potassium hydroxide (7.6 moles) to form *p*-tolylcarbinol.[4]

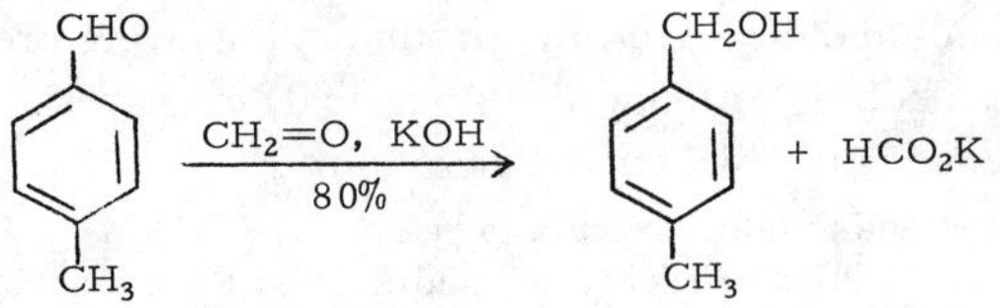

Palladium-on-$BaSO_4$ catalyst is prepared by treating a hot suspension of barium sulfate in an aqueous solution containing palladium chloride with formalin and a little alkali.[5]

A reduction by formaldehyde is involved in the first step of the preparation of *p*-dimethylaminobenzaldehyde (5) from *p*-nitrosodimethylaniline (1), dimethylaniline

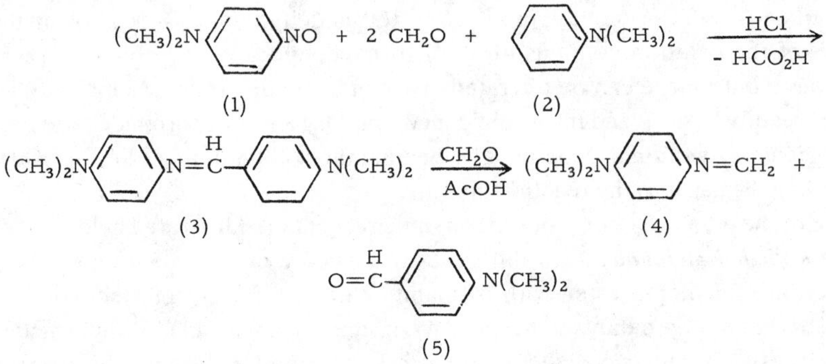

(2), and 2 moles of formaldehyde in hydrochloric acid solution.[6] In the formation of the yellow benzylidene derivative (3), 1 mole of the aldehyde becomes oxidized to formic acid. In the next step the benzylidene derivative (3) is hydrolyzed by oxygen-exchange with formaldehyde. *p*-Dimethylaminobenzaldehyde of satisfactory melting point (73°) but somewhat pigmented is obtained in yield of 56–59%.

Mannich reaction. This reaction is exemplified by the preparation of 5-methylfurfuryldimethylamine from a mixture of 40% aqueous dimethylamine, formalin,

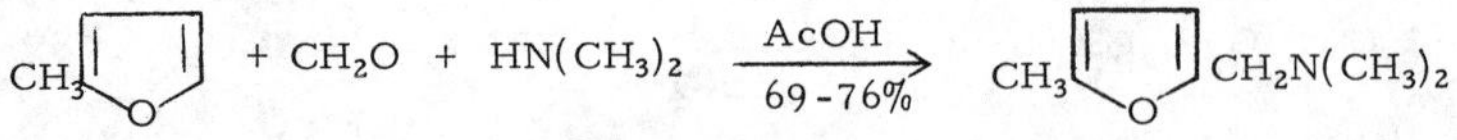

acetic acid, and 2-methylfurane.[7] After an initial exothermal reaction, the mixture is heated briefly on the steam bath, cooled, neutralized with sodium hydroxide, and the product is steam distilled.

Prins reaction. 4-Phenyl-*m*-dioxane was prepared first by H. J. Prins in 1919 by reaction of styrene and formaldehyde in the presence of sulfuric acid, but the correct structure was inferred only in 1930. A modern procedure[8] calls for stirring a mixture of formalin, concd. sulfuric acid, and styrene under gentle reflux for 7 hrs. The mixture is cooled, and the product collected by extraction with benzene.

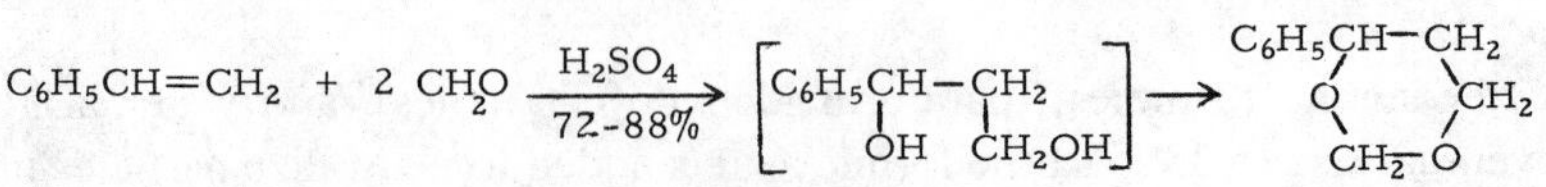

Chloromethylation. 2-Chloromethylthiophene is prepared by passing hydrogen chloride with dry ice cooling into a vigorously stirred mixture of thiophene and concentrated hydrochloric acid and then adding formalin at a rate that will permit

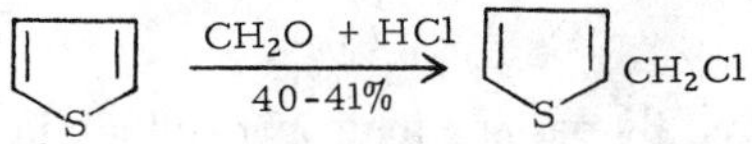

control of the temperature to 5° (4 hrs.).[9] The product is recovered by ether extraction and distillation (73–75°/17 mm.). Mesitylene is chloromethylated by essentially the same procedure in yield of 55–61%.[10] 1-Chloromethylnaphthalene is obtained in 74–77% yield by stirring a mixture of naphthalene, paraformaldehyde, acetic acid, phosphoric acid, and concd. hydrochloric acid at 80–85° for 6 hrs.[11]

Oxidative condensation with a phenol. In a procedure for the preparation of aurin tricarboxylic acid,[12] the first step is addition of solid sodium nitrite in small portions with stirring and ice cooling to concd. sulfuric acid to produce a solution of oxidant

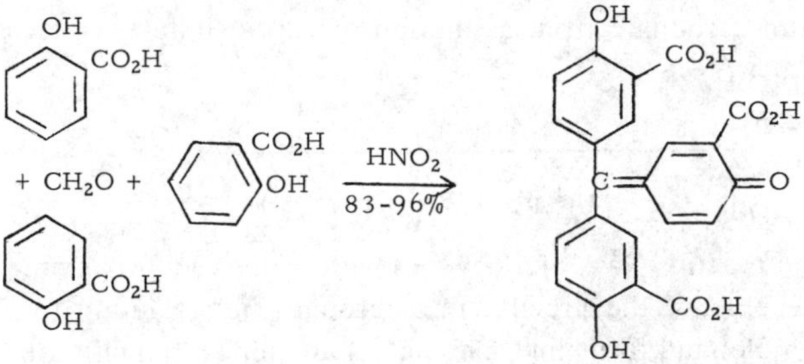

Aurin tricarboxylic acid

oxides of nitrogen. Solid salicylic acid is stirred in small portions, and formalin is added dropwise with vigorous cooling. The reaction product is isolated as the yellowish brown ammonium salt.

Methylation. β-Phenylethylamine is converted into the N,N-dimethyl derivative by adding formalin to a solution of the amine in 90% formic acid; after evolution of carbon dioxide subsides, the mixture is heated on the steam bath for 8 hrs. and worked

$$C_6H_5CH_2CH_2NH_2 + 2\ CH_2O + 2\ HCO_2H \xrightarrow[74\text{-}83\%]{} C_6H_5CH_2CH_2N(CH_3)_2 + 2\ CO_2 + 2\ H_2O$$

up.[13] Another example is the alkylation of ammonium chloride to produce methylamine hydrochloride.[14]

$$NH_4Cl + 2\ CH_2O \xrightarrow[45\text{-}51\%]{} CH_3\overset{+}{N}H_3\overset{-}{Cl} + HCO_2H$$

Synthesis of heterocycles. The first step of a procedure for the synthetic preparation of 2,6-dimethylpyridine (2,6-lutidine) involves condensation of 2 moles of ethyl acetoacetate with 1 mole each of formaldehyde (as formalin) and ammonia to produce the 1,4-dihydropyridine (1).[15] Oxidation with HNO_3–H_2SO_4 to the pyridine (2), saponification (3), and decarboxylation gave 2,6-dimethylpyridine (4).

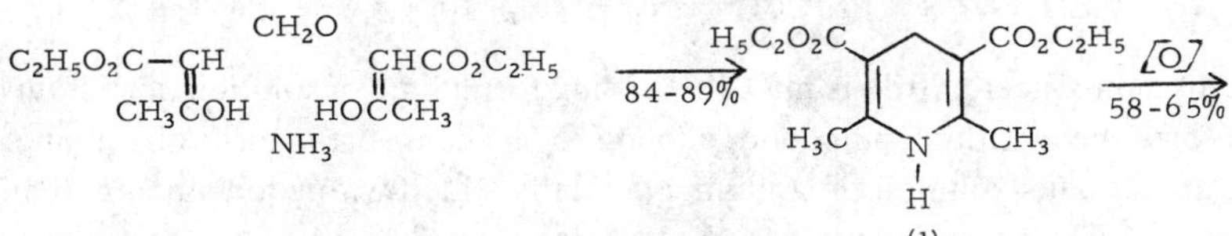

(1)

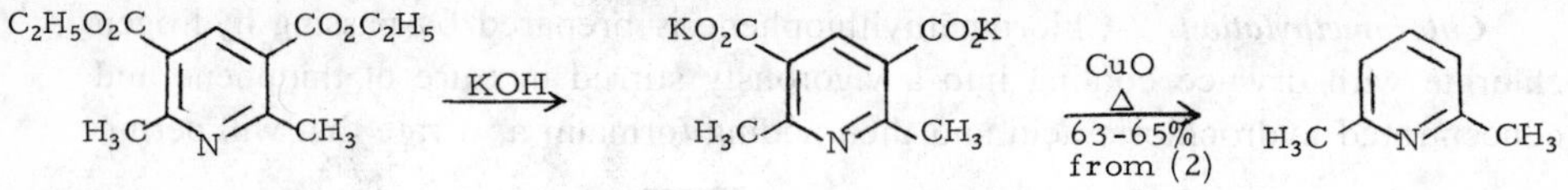

(2)   (3)   (4)

sym-Trithiane is prepared by passing hydrogen sulfide into a mixture of formalin and concd. hydrochloric acid until no more crystalline product separates.[16] The yield

$$3\ CH_2O\ +\ 3\ H_2S\ \xrightarrow[92-94\%]{HCl}\ S\underset{CH_2-S}{\overset{CH_2-S}{\diagdown}}CH_2\ +\ 3\ H_2O$$

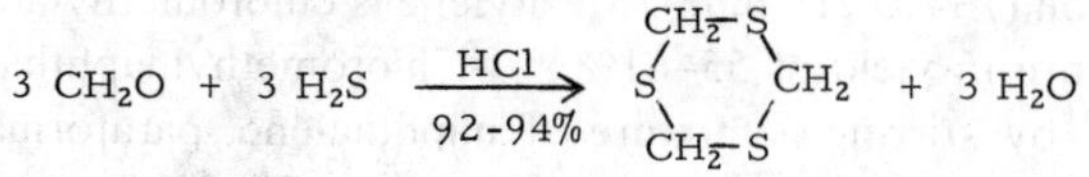

cited is for material crystallized from benzene by an "inverted filtration method."

4-Hydroxymethylimidazole is the end product of a remarkable sequence of reactions which occur when D-fructose is heated with ammonium hydroxide, basic cupric carbonate, and formaldehyde.[17] Basic cupric carbonate (1 mole) is precipitated with sodium carbonate from a solution of cupric sulfate, collected, washed,

$$\begin{array}{c} CH_2OH \\ | \\ C=O \\ | \\ (CHOH)_3 \\ | \\ CH_2OH \end{array} \xrightarrow[\substack{54-60\% \\ \text{(as picrate, m.p. 205}^0)}]{CH_2O,\ NH_3,\ CuCO_3\cdot Cu(OH)_2,\ O_2} \begin{array}{c} CH_2OH \\ | \\ {}^4C=CH \\ | \quad \diagdown N^2 \\ HN-CH \\ {}_5 \quad {}_1 \end{array}$$

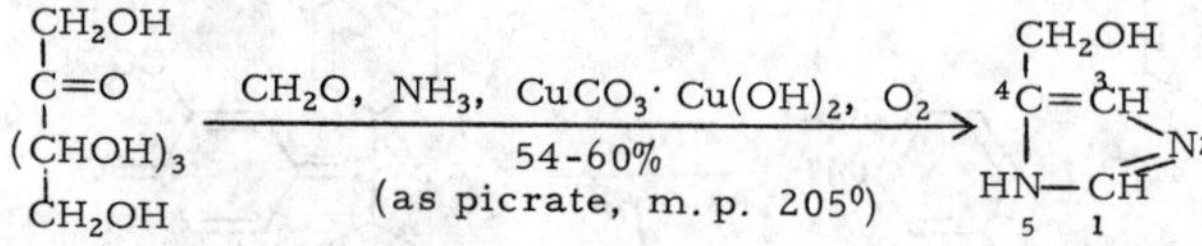

transferred to the reaction flask and covered with water and 28% ammonia solution (12 moles). The bulk of the precipitate is brought into solution by swirling, D-fructose (0.475 mole) and formalin (1.35 moles) are added, the mixture is heated on the steam bath for 30 min., and then air is bubbled through the solution while heating is continued for 2 hrs. longer. 5-Hydroxymethylimidazole is isolated as the picrate and this is converted into the hydrochloride. The chief steps involved seem to be as follows. The ketohexose (1) suffers reverse aldolization to dihydroxyacetone and

$$\begin{array}{cc} CH_2OH & CH_2OH \\ | & | \\ C=O & C=O \\ | & | \\ HOCH & CH_2OH \\ | & \\ HCOH & CH=O \\ | & | \\ HCOH & CHOH \\ | & | \\ CH_2OH & CH_2OH \end{array} \xrightarrow{[O]} \begin{array}{c} CH_2OH \\ | \\ C=O \\ | \\ CH=O \end{array} = \begin{array}{c} O=C-CHO \\ \\ H_3N \quad CH_2=O \end{array} \xrightarrow[NH_3]{} \begin{array}{c} CH_2OH \\ | \\ C=CH \\ | \quad \diagdown N \\ HN-CH \end{array}$$

(1)   (2)   (3)   (4)

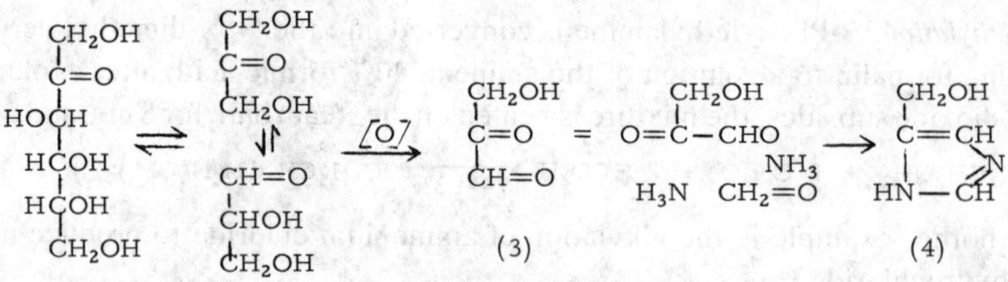

glyceraldehyde, and these on oxidation both afford the osone (3). Condensation of (3) with 1 mole of formaldehyde and 2 moles of ammonia then affords the imidazole (4).

Other reactions with nitrogen compounds. For the preparation of methyleneaminoacetonitrile (m.p. 129°) a mixture of formalin (18.9 moles) and ammonium chloride is stirred at 0° during the addition in 6 hrs. of an aqueous solution of sodium cyanide.[18]

$$2\ CH_2O\ +\ NaCN\ +\ NH_4Cl\ \xrightarrow[61-71\%]{}\ CH_2=NCH_2CN\ +\ NaCl\ +\ 2\ H_2O$$

Diethylaminoacetonitrile is made by adding formalin to a solution of sodium bisulfite to form the addition compound, cooling to 35°, and stirring in diethylamine; after 2 hrs. an aqueous solution of sodium cyanide is added with good stirring to mix the two layers.[19] The upper layer is separated, dried, and distilled.

$$(C_2H_5)_2NH + NaCN + CH_2O + NaHSO_3 \xrightarrow[88-90\%]{} (C_2H_5)_2NCH_2CN + Na_2SO_3 + H_2O$$

In the preparation of glycolonitrile, formalin is added dropwise to a stirred solution of potassium cyanide in water at 0–10°. Dilute sulfuric acid is added for neutralization, the precipitated potassium sulfate is removed by filtration, and the filtrate is extracted with ether in a continuous extractor.[20] The ethereal extract is

$$CH_2O + KCN + H_2O \xrightarrow[76-80\%]{} HOCH_2CN + KOH$$

dried, treated with a little absolute ethanol as polymerization inhibitor, and distilled at 86–88°/8 mm.

Steroid bismethylenedioxy (BMD) derivatives. A Merck group[21] found that on stirring 50 g. of cortisone (1) with chloroform (2 l.), formalin (0.5 l.), and concd. hydrochloric acid (0.5 l.) for 48 hrs., the hormone is converted into the crystalline BMD derivative (2, m.p. 258–261°).[21] Other steroids having the highly sensitive

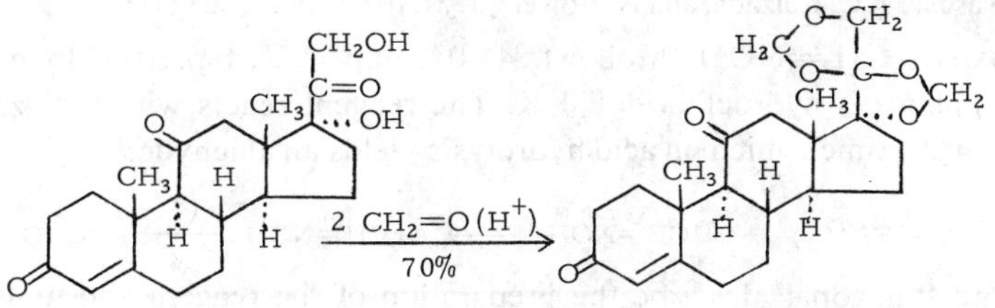
CH₂OH
C=O
CH₃
OH
O
CH₃
H
H
H
2 CH₂=O (H⁺)
70%
O–CH₂
H₂C
O–C–O
CH₃
CH₂
O

(1) (2)

dihydroxyacetone side chain react similarly and are thereby converted into highly crystalline and high-melting derivatives. The BMD protective group is stable to alkylation, acylation, ketalization, bromination, oxidation, reduction, or acid-catalyzed rearrangement. The spiroketal group survives refluxing with 1 N sulfuric acid in 90% methanol for 18 hrs. but can be removed by heating with aqueous formic or acetic acid.

Review: C. Djerassi, "Steroid Reactions," pp. 56–61 (1963).

In the original Merck procedure commercial 30% formalin was used as the source of formaldehyde, and in the case of cortisol (1) it was observed that the 11β-hydroxyl group reacted in part to give the 11β-methoxymethylene ether (2), ascribable to the presence of methanol in the formalin. Syntex workers[22] circumvented this side

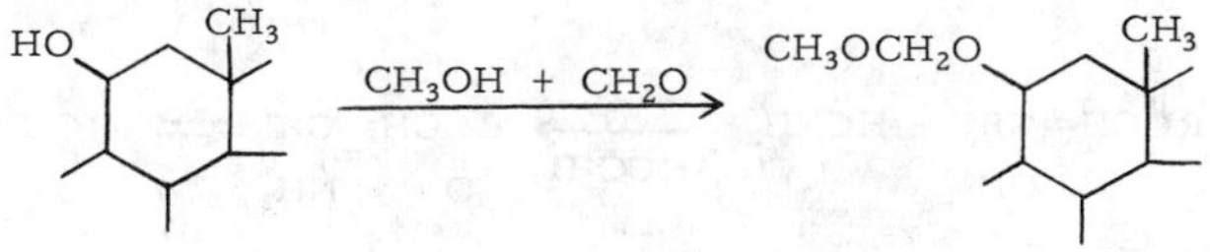
HO
CH₃
CH₃OH + CH₂O
CH₃OCH₂O
CH₃

(1) (2)

reaction by use of paraformaldehyde as a source of methanol-free formaldehyde. Thus cortisol in chloroform solution on treatment with a mixture of paraformaldehyde, water, and concd. hydrochloric acid afforded cortisol-BMD in 80% yield.

[1] T. J. Otterbacher, *Org. Syn., Coll. Vol.*, **1**, 290 (1941)
[2] P. Block, Jr., *Org. Syn.*, **40**, 27 (1960)
[3] H. Wittcoff, *Org. Syn., Coll. Vol.*, **4**, 907 (1963)
[4] D. Davidson and M. Weiss, *ibid.*, **2**, 590 (1943)

[5]R. Mozingo, *ibid.*, **3**, 685 (1955)
[6]R. Adams and G. H. Coleman, *ibid.*, **1**, 214 (1941)
[7]E. L. Eliel and M. T. Fisk, *ibid.*, **4**, 626 (1963)
[8]R. L. Shriner P. R. Ruby, *ibid.*, **4**, 786 (1963)
[9]K. B. Wiberg and H. F. McShane, *ibid.*, **3**, 197 (1955)
[10]R. C. Fuson and N. Rabjohn, *ibid.*, **3**, 557 (1955)
[11]O. Grummitt and A. Buck, *ibid.*, **3**, 195 (1955)
[12]G. B. Heisig and W. M. Lauer, *ibid.*, **1**, 54 (1941)
[13]R. N. Icke and B. B. Wisegarver, *ibid.*, **3**, 723 (1955)
[14]C. S. Marvel and R. L. Jenkins, *ibid.*, **1**, 347 (1941)
[15]A. Singer and S. M. McElvain, *ibid.*, **2**, 214 (1943)
[16]R. W. Bost and E. W. Constable, *ibid.*, **2**, 610 (1943)
[17]J. R. Totter and W. J. Darby, *ibid.*, **3**, 460 (1955)
[18]R. Adams and W. D. Langley, *ibid.*, **1**, 355 (1941)
[19]C. F. H. Allen and J. A. VanAllan, *ibid.*, **3**, 275 (1955)
[20]R. Gaudry, *ibid.*, **3**, 436 (1955)
[21]R. E. Beyler, R. M. Moriarity, F. Hoffman, and L. H. Sarett, *Am. Soc.*, **80**, 1517 (1958); R. E. Beyler, F. Hoffman, R. M. Moriarity, and L. H. Sarett, *J. Org.*, **26**, 242 (1961)
[22]J. A. Edwards, M. C. Calzada, and A. Bowers, *J. Med. Chem.*, **7**, 528 (1964)

Formaldoxime, $CH_2{=}NOH$. Mol. wt. 45.04, m.p. 2.5°, b.p. 109°/15 mm. sp. gr. 1.133. Supplier (as hydrochloride): KK. The reagent reacts with a diazonium salt to give an aryl oxime, which on acid hydrolysis yields an aldehyde:[1]

$$Ar\overset{+}{N}{\equiv}N(\overset{-}{Cl}) + HCH{=}NOH \longrightarrow ArCH{=}NOH \longrightarrow ArCHO$$

Jolad and Rajagopal[2] describe the preparation of the reagent and its use for the synthesis of 2-bromo-4-methylbenzaldehyde.

[1]W. F. Beech, *J. Chem. Soc.*, 1297 (1954)
[2]S. D. Jolad and S. Rajagopal, *Org. Syn.*, **46**, 13 (1966)

Formamide, $HCONH_2$. Mol. wt. 45.04, b.p. 193°, sp. gr. 1.14. Suppliers: du Pont, A, B, E, F, MCB.

This versatile reagent when heated with a halogen compound at 150° for several hours reacts either to give the formylamine (1) or the formate (2).[1] It reacts with the

$$1.\quad RX + 2\,H\overset{O}{\overset{\|}{C}}NH_2 \longrightarrow RNH\overset{O}{\overset{\|}{C}}H + CO + NH_4X$$

$$2.\quad RX + 2\,H\overset{O}{\overset{\|}{C}}NH_2 \longrightarrow RO\overset{O}{\overset{\|}{C}}H + HCN + NH_4X$$

$$3.\quad R\overset{O}{\overset{\|}{C}}CH{=}\underset{OH}{\underset{|}{C}}R' + H\overset{O}{\overset{\|}{C}}NH_2 \xrightarrow[-HCO_2H]{} R\underset{O}{\underset{\|}{C}}CH{=}\underset{NH_2}{\underset{|}{C}}R' \rightleftharpoons R\underset{O}{\underset{\|}{C}}CH_2\underset{NH}{\underset{\|}{C}}R'$$

enolic form of a β-diketone with replacement of the hydroxyl function by NH_2 (3).[2] It reacts with an α-hydroxyketone to form an imidazole, possibly through the steps

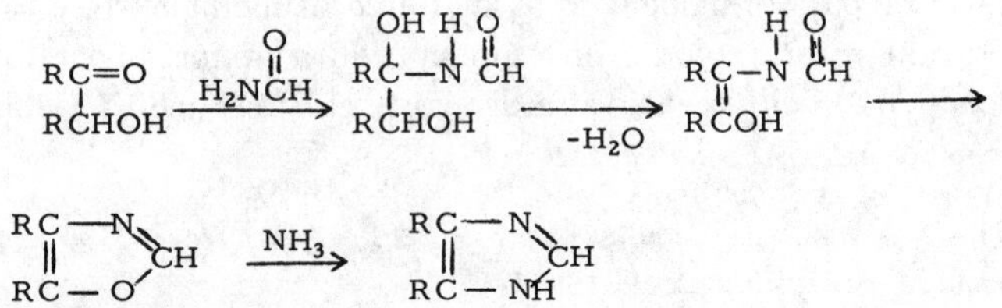

outlined. One intermediate in this scheme is an oxazole, and indeed the conversion of oxazoles into imidazoles has been demonstrated. α-Halo and α-amino ketones react with formamide in the same way as α-hydroxy ketones to form imidazoles.

A one-step synthesis of 4-methylpyrimidine[3,4] involves heating a mixture of formamide, solid ammonium chloride, and a small volume of water to 180–190°, and adding 4,4-dimethoxy-2-butanone in the course of 6 hrs. The ammonium chloride serves as an acidic salt to hydrolyze the ketal bonds.

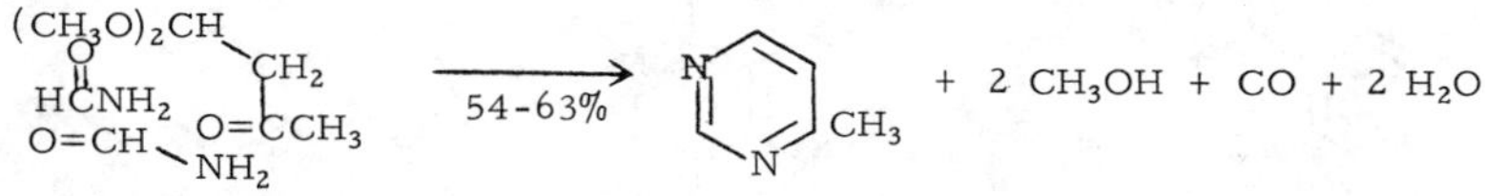
(CH3O)2CH
CH2
HCNH2
O=CCH3
O=CH
NH2
54-63%
N
N
CH3
+ 2 CH3OH + CO + 2 H2O

Addition to olefins. Under the influence of light, formamide adds to terminal olefins in the anti-Markownikoff sense to give crystalline alkylamides.[5] Addition of acetone improves the yield and shortens the irradiation time. Photoaddition under

$$RCH{=}CH_2 + HCONH_2 \xrightarrow[50-90\%]{h\nu} RCH_2CH_2CONH_2$$

these conditions to norbornene is sterospecific and gives exclusively norbornane-*exo*-carboxyamide.[6]

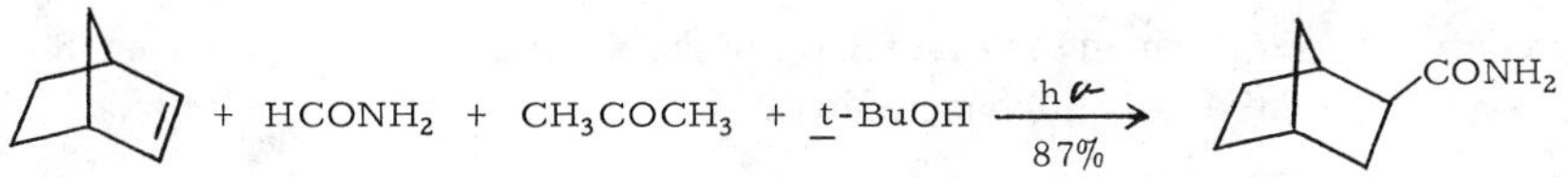
+ HCONH2 + CH3COCH3 + t-BuOH
hν
87%
CONH2

Review.[7]

[1]H. Bredereck, R. Gompper, and G. Theilig, *Ber.*, **87**, 537 (1954)
[2]H. Bredereck, R. Gompper, and G. Morlock, *Ber.*, **90**, 942 (1957)
[3]H. Bredereck and G. Theilig, *Ber.*, **86**, 88 (1953); G. Theilig, *ibid.*, **86**, 96 (1953)
[4]H. Bredereck, *Org. Syn.*, **43**, 77 (1963)
[5]D. Elad and J. Rokach, *J. Org.*, **29**, 1855 (1964)
[6]*Idem*, *J. Chem. Soc.*, 800 (1965)
[7]H. Bredereck, R. Gompper, H. G. v. Schuh, and G. Theilig, *New Methods of Preparative Organic Chemistry*, **3**, 241 (1964)

Formamidine acetate, $H_2NCH{=}\overset{+}{N}H_2(CH_3CO_2^-)$. Mol. wt. 104.11, m.p. 164°. Suppliers: A, B, E, F, MCB.

Preparation.[1,2] Ammonia is passed into a mixture of triethyl orthoformate and acetic acid heated under reflux in an oil bath maintained at 125–130°. Formamidine

$$\underset{90\text{ g.}}{HC(OC_2H_5)_3} + \underset{49.2\text{ g.}}{CH_3CO_2H} + NH_3 \xrightarrow[87-90\%]{125-130^0} H_2NCH{=}\overset{+}{N}H_2(CH_3CO_2^-) + 3\ C_2H_5OH$$

acetate starts to crystallize from the boiling mixture in 20–30 min. When the reaction is at an end, the mixture is cooled, and the product collected and washed with absolute ethanol. The acetate is stable in storage and for many reactions it can be used in place of free formamidine. By contrast, the hydrochloride is extremely deliquescent and cannot be used in most condensations but must be converted into the free base.

Use of an alkylamine (RNH_2) in place of ammonia gives an N,N′-dialkylformamidine acetate: $RNHCH{=}\overset{+}{N}HR(CH_3CO_2^-)$.[3]

Heterocycles. Typical condensations affording 4-aminopyrimido[4,5-d]pyrimidine (1) described by Taylor[1] are formulated in the chart; in each case the reactants

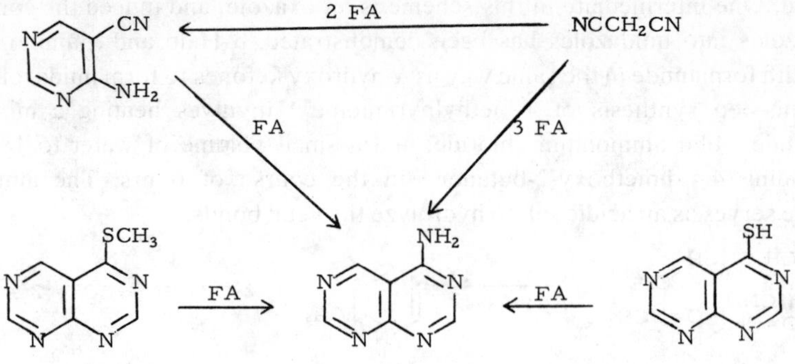
2 FA
NCCH₂CN
CN
NH₂
FA
3 FA
SCH₃
NH₂
SH
FA
FA

(1)

FA = formamidine acetate

are merely refluxed in 2-ethoxyethanol. Condensation of the reagent with *o*-phenylenediamine affords benzimidazole.[1]

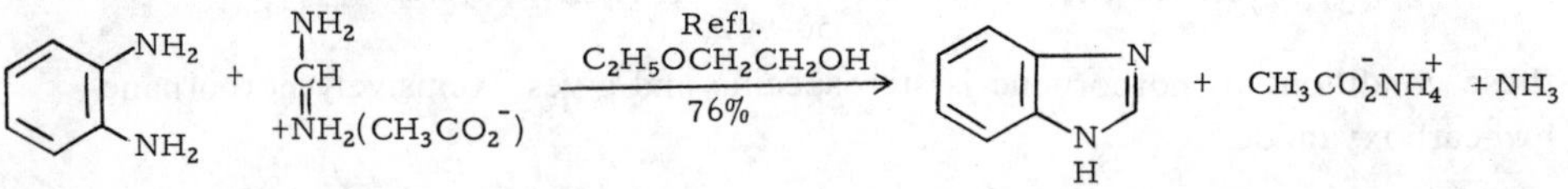
Refl.
C₂H₅OCH₂CH₂OH
76%

Bredereck[4] used formamidine acetate for the synthesis of purines from N-alkyl and N-acyl derivatives of aminoacetonitrile.

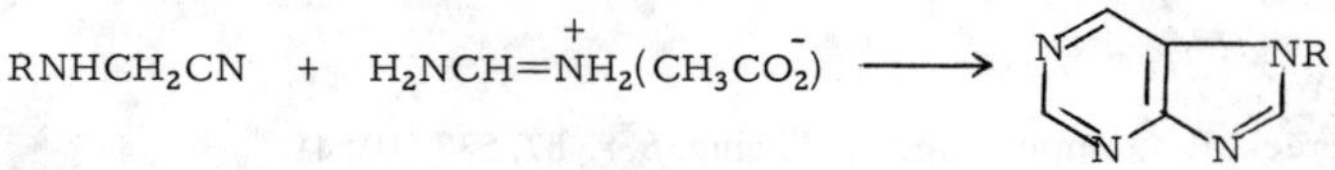

[1]E. C. Taylor and W. A. Ehrhart, *Am. Soc.*, **82**, 3138 (1960)
[2]E. C. Taylor, W. A. Ehrhart, and M. Kawanisi, procedure submitted to *Org. Syn.*
[3]E. C. Taylor and W. A. Ehrhart, *J. Org.*, **28**, 1108 (1963)
[4]H. Bredereck, F. Effenberger, G. Rainer, and H. P. Schosser, *Ann.*, **659**, 133 (1962); H. Bredereck, F. Effenberger, and G. Rainer, *ibid.*, **673**, 82, 88 (1964)

Formic acid, HCO_2H. Mol. wt. 46.03, m.p. 8.4°, b.p. 100.5°, sp. gr. 1,220, pKa 3.77.

Formylation. *See also* Acetic–formic anhydride. Formic acid is a reactive acylating agent for alcohols and amines and sometimes requires no catalyst or other solvent. Cholic acid yields the triformyl derivative on being heated with 87% formic acid

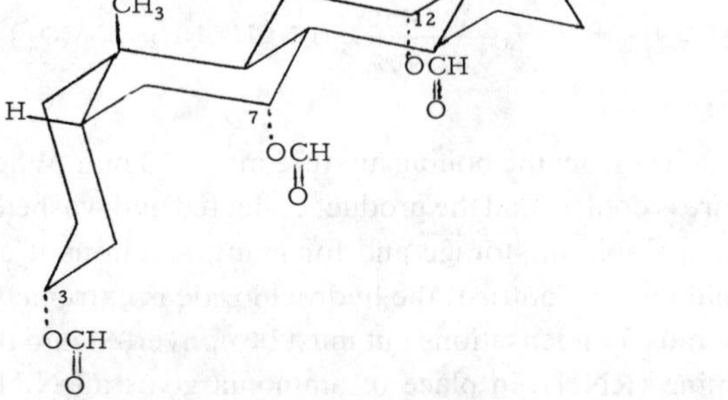

(sp. gr. 1.2, 2 ml./g.) at 50–55° for 5 hrs.[1] Formylation of the 11α-hydroxyl group of 11-epicortisol was accomplished in pyridine at 0° by dropwise addition of a mixture of 1 ml. of 99% formic acid and 0.4 ml. of acetic anhydride.[2] DL-Cystine has been formylated by stirring a mixture of 40 g. of the amino acid and 600 ml. of 87% formic acid and adding 50 ml. of acetic anhydride at such a rate as to maintain a temperature of 60°.[3]

In the first step of the preparation of *o*-tolyl isocyanide,[4] a solution of *o*-toluidine and 98% formic acid in toluene is refluxed under a condenser with water separator for about 3 hrs., toluene and excess formic acid are distilled, and the residual N-

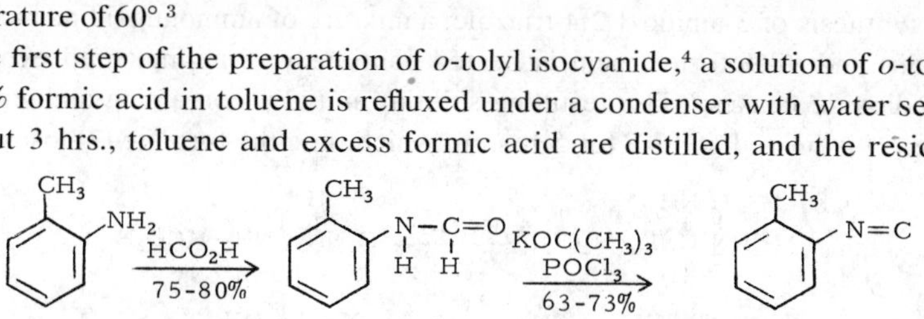

o-tolylformamide is crystallized. On addition of this amide to a hot stirred suspension of potassium *t*-butoxide in *t*-butanol, the alkoxide soon dissolves. The solution is then cooled to 10–20° during addition of phosphoryl chloride in 30–40 min. On completion of the reaction the mixture is added to a cold solution of sodium bicarbonate and the *o*-tolyl isocyanide (an oil) is extracted with petroleum ether and distilled.

A procedure for the preparation of allyl alcohol introduced by B. Tollens (1870) and later improved[5] calls for heating glycerol with three successive portions of formic acid to a temperature of 260° (1½ days) and recovery of the allyl alcohol from the distillate (46% yield). Glycerol monoformate is formed and suffers pyrolysis to the unsaturated alcohol, carbon dioxide, and water.

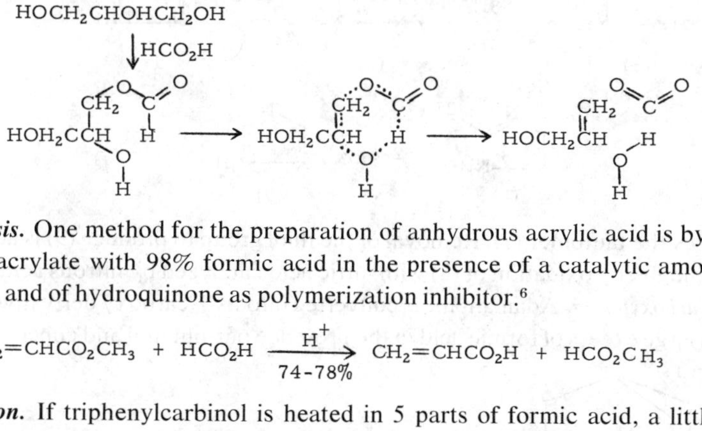

Acidolysis. One method for the preparation of anhydrous acrylic acid is by heating methyl acrylate with 98% formic acid in the presence of a catalytic amount of sulfuric acid and of hydroquinone as polymerization inhibitor.[6]

$$CH_2{=}CHCO_2CH_3 + HCO_2H \xrightarrow[74-78\%]{H^+} CH_2{=}CHCO_2H + HCO_2CH_3$$

Reduction. If triphenylcarbinol is heated in 5 parts of formic acid, a little dissolves to give a yellow solution of the halochromic salt, carbon dioxide is slowly evolved, and the color eventually disappears as reduction to triphenylmethane is complete.[7]

$$(C_6H_5)_3COH + HCO_2H \longrightarrow (C_6H_5)_3CH + CO_2 + H_2O$$

Formic acid appears to be the reducing agent in the alkylation of an amine with formic acid and formaldehyde (*see* Formaldehyde, methylation).

Synthesis of Heterocycles. Benzimidazole is prepared conveniently by heating a mixture of *o*-phenylenediamine (0.5 mole) and 90% formic acid (0.75 mole) at

100° for 2 hrs., cooling, and neutralizing with alkali.[8] The product is collected and crystallized from water (m.p. 170–172°).

In a synthesis of 3-amino-1,2,4-triazole, a mixture of aminoguanidine bicarbonate (Eastman) and 98% formic acid is heated cautiously (foaming) until evolution of carbon dioxide ceases.[9] The resulting solution initially containing aminoguanidine formate (2) is heated at 120° for 5 hrs., in which time the formate (2) loses water to

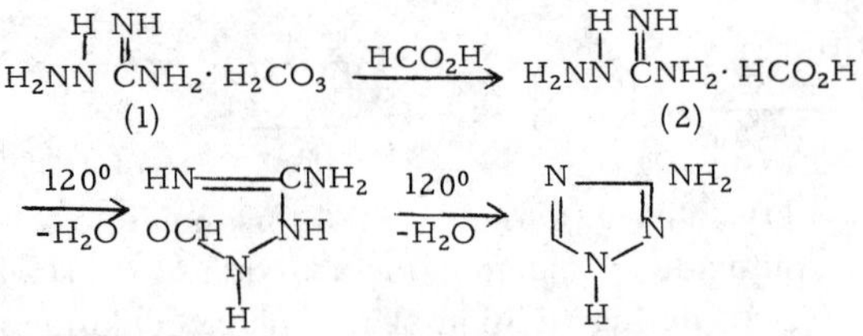
H NH
$H_2NN\ CNH_2 \cdot H_2CO_3$
HCO_2H
H NH
$H_2NN\ CNH_2 \cdot HCO_2H$
120°
$-H_2O$
HN
CNH_2
OCH
NH
N
H
120°
$-H_2O$
N
NH_2
N
N
H

(1) (2) (3) (4)

form the N-formyl derivative (3), which undergoes cyclodehydration to the amino-triazole (4); yield of crystallized product, m.p. 152–153°, 68–70%.

A synthesis of the parent 1,2,4-triazole (9)[9] starts with the conversion of thio-semicarbazide (5) into the formyl derivative (6), accomplished by brief heating with 90% formic acid. Cyclization of (6) by alkali to 3-thiol-1,2,4-triazole (8) probably

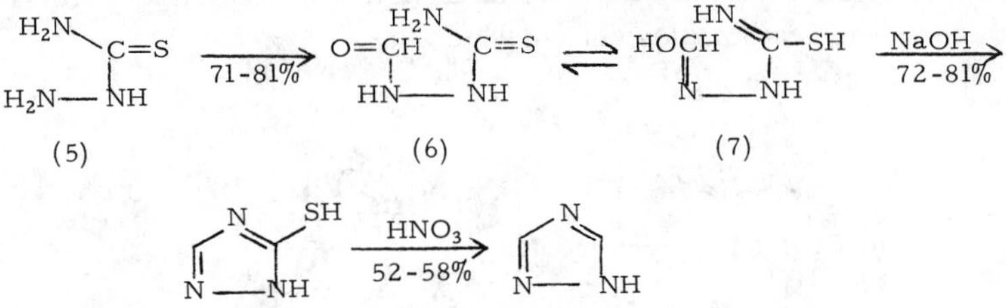
H_2N
C=S
H_2N
NH
71-81%
H_2N
O=CH
C=S
HN
NH
HN
HOCH
C–SH
N
NH
NaOH
72-81%
N
SH
N
NH
HNO_3
52-58%
N
N
NH

(5) (6) (7) (8) (9)

involves the tautomer (7). Removal of the thiol group to produce (9) is accomplished, surprisingly, by oxidation of (8) with nitric acid catalyzed by nitrous acid.

Carboxylation. Adamantane is converted into its 1-carboxy derivative by reaction with a large excess of formic acid in the presence of *t*-butanol and concd. sulfuric acid,

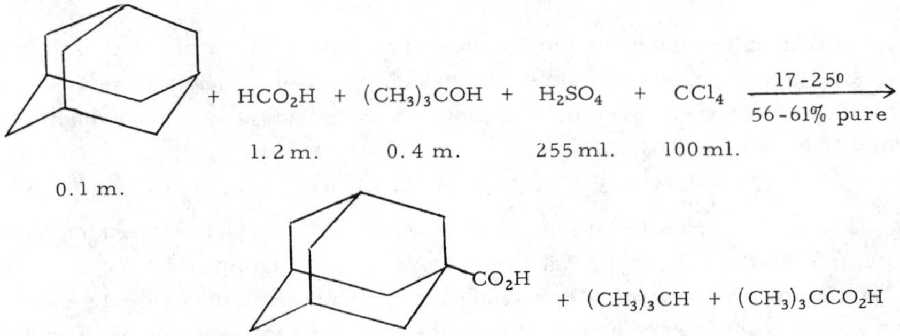
+ HCO_2H + $(CH_3)_3COH$ + H_2SO_4 + CCl_4
17-25°
56-61% pure
1.2 m.
0.4 m.
255 ml.
100 ml.
0.1 m.
CO_2H
+ $(CH_3)_3CH$ + $(CH_3)_3CCO_2H$

with carbon tetrachloride as solvent.[10] A side reaction is reduction of *t*-butanol to isobutane, which is in part carboxylated to trimethylacetic acid. The workup includes crystallization of ammonium adamantanecarboxylate, and the by-product acid salt is retained in the mother liquor.

The procedure illustrates a general method of carboxylating saturated compounds having a tertiary hydrogen, for example isopentane, 2,2-dimethylbutane, and methylcyclohexane.

Another example is the synthesis of 1-methylcyclohexanecarboxylic acid from crude 2-methylcyclohexanol.[10a]

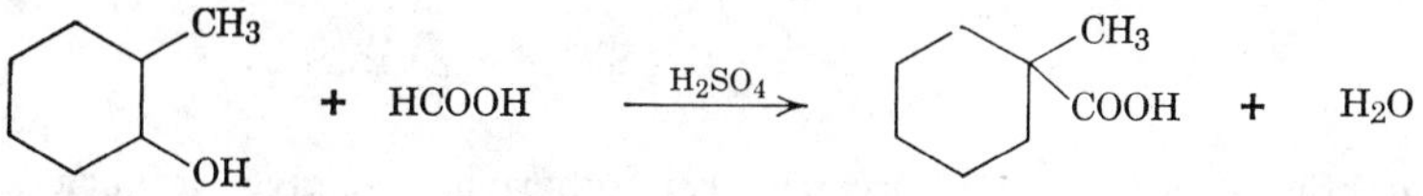
CH3
OH
+ HCOOH
H2SO4
CH3
COOH
+ H2O

Beckmann rearrangement.[11] Ketoximes undergo smooth rearrangement when refluxed in formic acid for 1–6 hrs.[11]

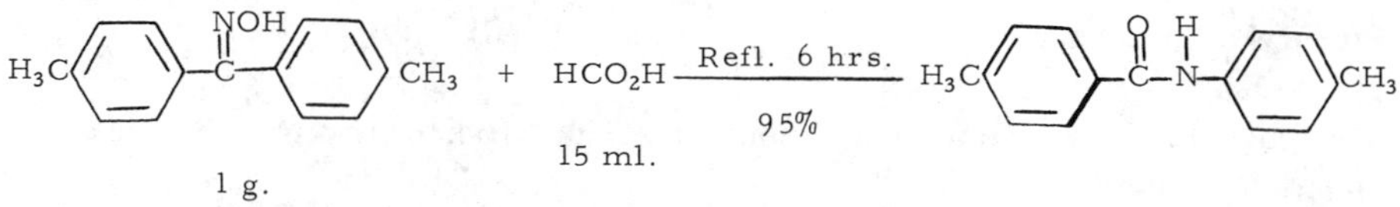
NOH
H3C
CH3
+ HCO2H
Refl. 6 hrs.
95%
15 ml.
1 g.
H3C
O H
C–N
CH3

[1]F. Cortese and L. Bauman, *Am. Soc.*, **57**, 1393 (1935)
[2]F. Reber, A. Lardon, and T. Reichstein, *Helv.*, **37**, 45 (1954)
[3]V. du Vigneaud, R. Dorfman, and H. S. Loring, *J. Biol. Chem.*, **98**, 577 (1932)
[4]I. Ugi and R. Meyr, *Org. Syn.*, **41**, 101 (1961)
[5]O. Kamm and C. S. Marvel, *Org. Syn., Coll. Vol.*, **1**, 42 (1941)
[6]C. E. Rehberg, *ibid.*, **3**, 33 (1955)
[7]H. Kauffmann and P. Pannwitz, *Ber.*, **45**, 766 (1912); A Kovache, *Ann. chim.*, [9], **10**, 184 (1918)
[8]E. C. Wagner and W. H. Millett, *Org. Syn., Coll. Vol.*, **2**, 65 (1943)
[9]C. Ainsworth, *Org. Syn.*, **40**, 99 (1960)
[10]H. Koch and W. Haaf, *ibid.*, **44**, 1 (1964)
[10a]W. Haaf, *Org. Syn.*, **46**, 72 (1966)
[11]T. van Es, *J. Chem. Soc.*, 3881 (1965)

Formic acid–Formamide. 3,4-Dihydroisoquinolines undergo reductive formylation to give N-formyl 1,2,3,4-tetrahydroisoquinolines in good yield when refluxed for 1

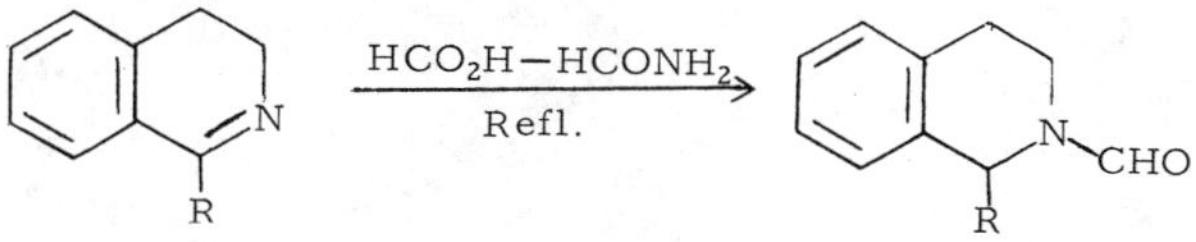
N
R
HCO2H–HCONH2
Refl.
N
CHO
R

hr. with formic acid and formamide.[1] Hydroxyl groups if present prevent the reaction and must be protected as ether groups.

[1]J. Gardent, *Bull. Soc.*, 118 (1960); I. Baxter, L. T. Allan, and G. A. Swan, *J. Chem. Soc.*, 3645 (1965); I. Baxter and G. A. Swan, *ibid.*, 4015 (1965)

Formyl fluoride, HCOF. Mol. wt. 48.02, b.p. –29°.

For the preparation, *see* Acyl fluorides. The reagent formylates aromatic hydrocarbons, alcohols, and phenols.[1]

$$ArH + FCHO \longrightarrow ArCHO + HF \text{ (yields 56-78\%)}$$

$$ROH + FCHO \longrightarrow ROCHO + HF \text{ (yields 69-92\%)}$$

[1]G. A. Olah and S. J. Kuhn, *Am. Soc.*, **82**, 2380 (1960)

Formylimidazole (1). Mol. wt. 96.09, m.p. 53–55°. Hygroscopic, decomposes with loss of CO at 60°.

The reagent, prepared in 85% yield by reaction of N,N′-carbonyldiimidazole with formic acid in tetrahydrofurane, formylates amines at room temperature. It reacts

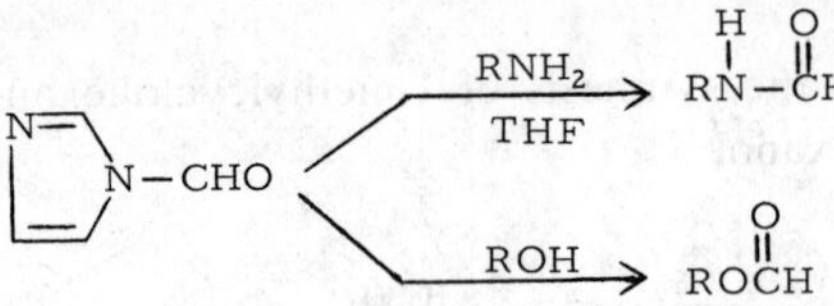
RNH₂
THF
ROH
N–CHO
RN–CH
ROCH

(1)

with alcohols at room temperature, usually exothermally, to give formic acid esters.

[1]H. A. Staab and B. Polenski, *Ann.*, **655**, 95 (1962)

Fremy's salt, *see* Potassium nitrosodisulfonate.

Furfuryl chloroformate, 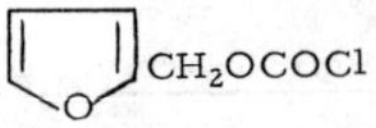 (not isolated).

The reagent is generated in toluene solution at –60° from furfuryl alcohol, phosgene, and triethylamine.[1]

Peptide synthesis. The protective N-furfurylcarbonyl group of an amino acid is very acid-labile and is removed at 0° by HBr–AcOH.

[1]G. Losse, H. Jeschkeit, and E. Willenberg, *Angew. Chem., Internat. Ed.*, **3**, 307 (1964)

G

Galvinoxyl. Mol. wt. 421.62, m.p. 158° dec.

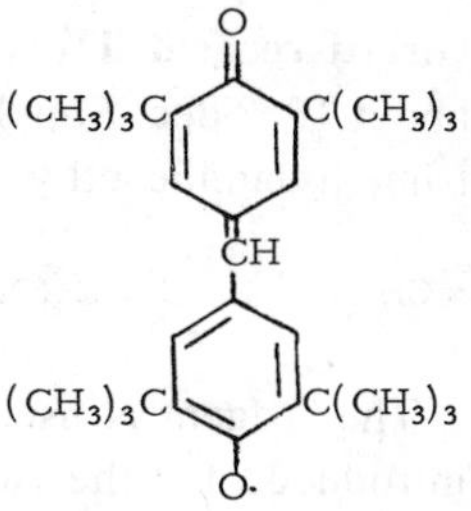

This hindered, deep blue phenoxyl radical is obtained by oxidation of 3,3′,5,5′-tetra-*t*-butyl-4,4′-dihydroxydiphenylmethane with lead dioxide in ether and isooctane[1] or with alkaline ferricyanide.[2] Galvinoxyl is an efficient radical scavenger, reacting quantitatively with radicals with an odd electron on either carbon or oxygen.[3] Contrary to early reports, the radical is not stable to oxygen.[4] The galvinoxyl precursor is available from Frinton Laboratories.

[1]Galvin M. Coppinger, *Am. Soc.*, **79**, 501 (1957)
[2]M. S. Kharasch and B. S. Joshi, *J. Org.*, **22**, 1435 (1957)
[3]P. D. Bartlett and C. Rüchardt, *Am. Soc.*, **82**, 1756 (1960); P. D. Bartlett and T. Funahashi, *ibid.*, **84**, 2596 (1962); P. D. Bartlett, B. H. Gontarev, and H. Sakurai, *ibid.*, **84**, 3101 (1962); F. D. Greene, W. Adam, and J. E. Cantrill, *ibid.*, **83**, 3461 (1961)
[4]F. D. Greene and W. Adam, *J. Org.*, **28**, 3550 (1963)

Gases, inert. An inquiry addressed to Blaine C. McKusick, Central Research Department, du Pont Experimental Station, brought the following response: "You asked about use of inert gases such as nitrogen, helium, and argon. We buy all three

	Molecular weight	Sp. gr. (air = 1)
Helium	4.003	0.137
Nitrogen	28.016	0.967
Argon	39.944	1.38

from the Air Reduction Company. We use large quantities of what is called Seaford Nitrogen. It is oxygen-free (< 20 ppm. O_2) and is used to provide a dry, inert atmosphere in dry boxes, coordination polymerizations, and reactions of Grignard Reagents. It is high quality as it comes from the cylinder and requires no further purification. It is called Seaford Nitrogen because it was originally designed for use in the first nylon plant, located at Seaford, Delaware, and the specifications drawn up for it were those required in this operation. A dry, inert atmosphere is essential in making nylon from the salt of adipic acid and hexamethylenediamine. Seaford nitrogen costs $2.88/220 cu. ft.

"Helium costs $18.02/220 cu. ft. We use a lot of it, principally in gas chromatography. It is used occasionally to sweep a system that includes traps immersed in liquid nitrogen, where an inert gas that will not condense at the temperature of liquid nitrogen is desired. However, argon is used more frequently for this purpose.

"Argon costs $11.88/220 cu. ft. It is used in situations where nitrogen is unacceptable because it reacts. An example is a reaction involving lithium, which forms a nitride with nitrogen. We use it when heating metals that react with nitrogen when hot and in some of our metallurgy operations. Since it is heavier than air, argon is very efficient in purging a system of air. In such a case, it is put in at the bottom of the system and the air is displaced at the top."

Girard reagent P (Pyridine), $Py^+CH_2CONHNH_2(Cl^-)$. Mol. wt. 187.63, m.p. 200°. Suppliers: Arapahoe, Eastern, B, E, F, MCB. *See* Girard reagent T.

Girard reagent T (Trimethylaminoacetohydrazide chloride). Mol. wt. 167.64, m.p. about 185° dec. Suppliers: Arapahoe, Eastern, A, B, E, F, MCB. Preparation from trimethylamine, ethyl chloroacetate, and hydrazine in absolute ethanol.[1]

$$(CH_3)_3N + ClCH_2CO_2C_2H_5 + H_2NNH_2 \xrightarrow[83-89\%]{C_2H_5OH,\ 0-60^\circ} (CH_3)_3\overset{+}{N}CH_2CONHNH_2(\overset{-}{Cl})$$

The reagent is used for separation of ketonic from nonketonic material and was introduced for the isolation of estrone from dark oily concentrates of hydrolyzed urine by conversion to the water-soluble Girard derivative, separation from non-

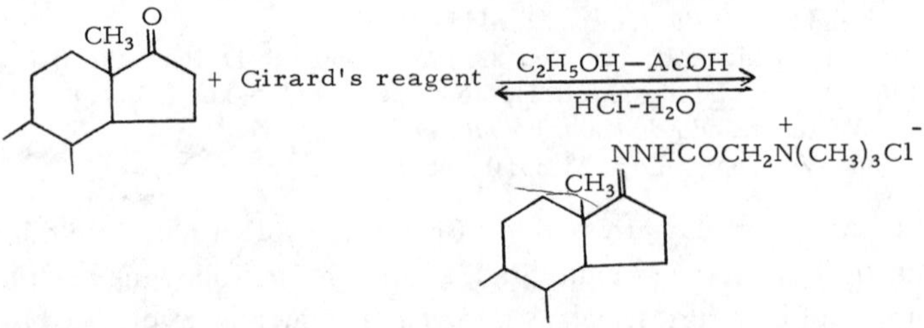

CH3
O
+ Girard's reagent
C2H5OH—AcOH
HCl-H2O
NNHCOCH2N(CH3)3Cl
CH3

ketones, and hydrolysis.[2] In a typical procedure, a solution of 0.5 g. of impure fluorenone (mol. wt. 180), 0.5 g. of reagent T, and 0.5 ml. of acetic acid in 5 ml. of 95% ethanol is refluxed for 30 min. to form the derivative, cooled, and poured into a separatory funnel. After addition of ether, water, and sodium chloride solution (to avoid an emulsion), the layers are separated and the aqueous solution is treated with 1 ml. of concd. hydrochloric acid and heated on the steam bath to hydrolyze the Girard derivative and drive off the ether. Fluorenone separates as a yellow oil which solidifies on cooling.

Saturated ketones such as cholestane-3-one form Girard derivatives which can be hydrolyzed with very dilute acid; Δ^4-3-ketones react more readily with the reagent than saturated ketones, but a considerably higher concentration of acid is required for hydrolysis.[3] Benzophenone reacts only very slowly with reagent T, but fluorenone reacts rapidly. Girard derivatives of aldehydes can be split in high yield.[4]

Isolation of a Girard derivative can be accomplished as follows: a suspension of Girard reagent and the ketone in acetic acid is warmed briefly on the steam bath until the solids are dissolved, the solution is evaporated to dryness at reduced pressure, and the residue is crystallized from methanol–acetone.[5] 17-Ketosteroids can be determined by polarographic analysis of the Girard derivatives in buffered aqueous solution; these derivatives are reduced at the dropping mercury electrode at a half-wave potential of −1.4 v.[5,6] The Girard derivatives of 3-ketosteroids are not reducible under the conditions of the determination, whereas Δ^4-3-ketosteroid Girard derivatives are reduced at a potential of −1.1 v. and are determinable simultaneously in the same specimen. Androstenolone can be determined by microanalytical Oppenauer oxidation to the conjugated ketone and polarography of the

Girard derivative. A number of modifications and improvements of the original methods of analysis have been reported.[7]

Dehydration. Ehrenstein and co-workers[8] find that 3-keto-5β-hydroxysteroids can be dehydrated to the Δ^4-3-ketosteroids by refluxing with the reagent in acetic acid solution.

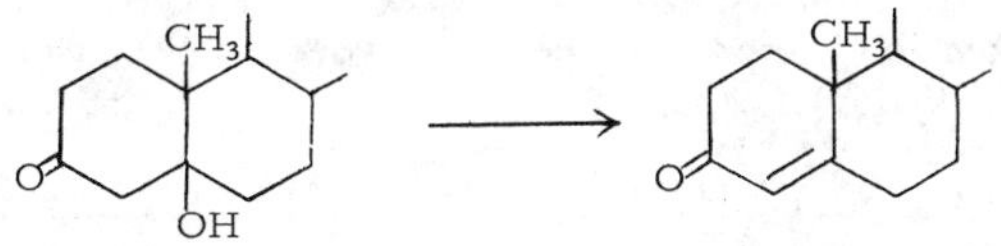
CH3
OH
CH3

See review.[9]

[1]A. Girard, *Org. Syn., Coll. Vol.*, **2**, 85 (1943)
[2]A. Girard and G. Sandulesco, *Helv.*, **19**, 1095 (1936)
[3]T. Reichstein, *Helv.*, **19**, 1107 (1936). For examples of separation of saturated and α,β-unsaturated lactones see A. Zaffaroni, R. B. Burton, and E. H. Keutmann, *J. Biol. Chem.*, **177**, 109 (1949); M. E. Kuehne, *Am. Soc.*, **83**, 1492 (1961)
[4]E. Lederer and G. Nachmias, *Bull. soc.*, **16**, 400 (1949)
[5]J. K. Wolfe, E. B. Hershberg, and L. F. Fieser, *J. Biol. Chem.*, **136**, 653 (1940)
[6]E. B. Hershberg, J. K. Wolfe, and L. F. Fieser, *J. Biol. Chem.*, **140**, 215 (1941)
[7]P. Kabasakalian and J. McGlotten, *J. Electrochem. Soc.*, **105**, 261 (1958)
[8]M. Ehrenstein and M. Dunnenbergen, *J. Org.*, **21**, 774 (1956); M. Ehrenstein and K. Otto, *ibid.*, **24**, 2506 (1959)
[9]O. H. Wheeler, *Chem. Rev.*, **62**, 205 (1962)

Glutaraldehyde, $OCHCH_2CH_2CH_2CHO$. Mol. wt. 100.11, b.p. 75–81°/15 mm. Available as a 30% aqueous solution from Carbon and Carbide Chemicals Co., A, B, E, F.

Glutaraldehyde is conveniently prepared by stirring a mixture of 120 g. of 2-ethoxy-3,4-dihydro-2H-pyrane, 300 ml. of water, and 25 ml. of concd. hydrochloric

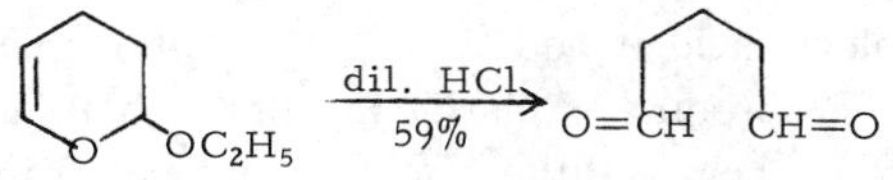
O
OC2H5
dil. HCl
59%
O=CH
CH=O

acid for 22 min., when the temperature rises to 38° and the mixture becomes clear.[1] After 1½ hrs. the mixture is neutralized with sodium bicarbonate, saturated with sodium chloride, and extracted with ether. Distillation gives 55 g. of glutaraldehyde.

The classical synthesis of pseudopelletierine by R. Robinson (1924), as improved by later work of C. Schöpf (1935), K. Ziegler (1950), and A. C. Cope (1951),[2] involves condensation of glutaraldehyde with methylamine and acetonedicarboxylic

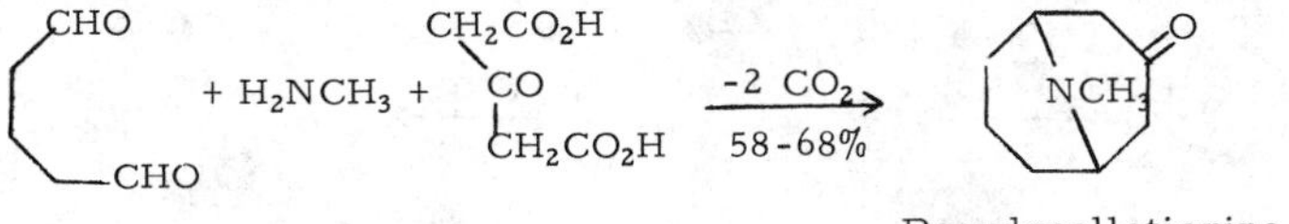
CHO
CHO
+ H2NCH3 +
CH2CO2H
CO
CH2CO2H
-2 CO2
58-68%
NCH3
O
Pseudopelletierine

acid. Best results are obtained when glutaraldehyde is generated in aqueous solution by adding 2-ethoxy-3,4-dihydro-2H-pyrane and concd. hydrochloric acid to distilled water that has been deoxygenated by passing in a stream of nitrogen. The resulting colorless solution of glutaraldehyde is treated under nitrogen with aqueous solutions of methylamine hydrochloride, acetonedicarboxylic acid, $Na_2HPO_4 \cdot 12H_2O$, and sodium hydroxide. Carbon dioxide is evolved, and as the solution is stirred under nitrogen the pH rises from 2.5 to 4.5 in 24 hrs. Concentrated hydrochloric acid is added, and the solution is heated on the steam bath for 1 hr. to effect complete decarboxylation. After neutralization with alkali, repeated extraction with methylene chloride and purification affords pure colorless pseudopelletierine, m.p. 63–64°.

[1]R. I. Longley, Jr., and W. S. Emerson, *Am. Soc.*, **72**, 3079 (1950)

[2]For the procedure cited and references to the earlier work see A. C. Cope, H. L. Dryden, Jr., and C. F. Howell, *Org. Syn., Coll. Vol.*, **4**, 816 (1963)

Glutaric acid, $HO_2CCH_2CH_2CH_2CO_2H$. Mol. wt. 132.11, m.p. 98°, 63.9 g. dis. in 100 g. H_2O at 20°, pKa_1 4.33, pKa_2 5.57. Suppliers: Eastern, A, B, E, F, MCB.

The first[1] of four *Organic Syntheses* procedures starts with the costly trimethylene dibromide and involves conversion to the dinitrile and acid hydrolysis (4 hrs.).

$$BrCH_2CH_2CH_2Br \xrightarrow[77\text{-}86\%]{2\ NaCN} NCCH_2CH_2CH_2CN \xrightarrow[83\text{-}85\%]{HCl} HO_2CCH_2CH_2CH_2CO_2H$$

Extraction with ether and crystallization from ether–benzene affords pure material (m.p. 97–98°) in high yield. A preparation from formaldehyde and diethyl malonate has been cited; *see* Formaldehyde.

In a later procedure,[2] γ-butyrolactone is heated with potassium cyanide at 190–195° for 2 hrs., water is added to dissolve the potassium salt of the cyano acid, and the warm solution is treated with enough hydrochloric acid to liberate the free carboxyl group and to effect partial hydrolysis to give glutaric acid monoamide.

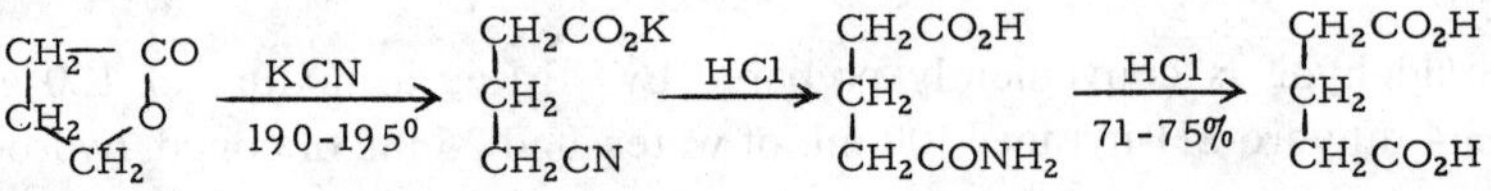

Concentrated hydrochloric acid is added and the solution refluxed 1 hr. to effect complete hydrolysis. The product is clarified with Norit in aqueous solution, the filtrate is evaporated to dryness and the diacid crystallized from chloroform to give material melting at 98–99°.

In a fourth procedure[3] dihydropyrane is hydrolyzed by brief heating with 0.2*N* nitric acid on the steam bath to δ-hydroxyvaleraldehyde, and the solution is added in about 3 hrs. to concd. nitric acid, with stirring and cooling at 10° or below. The aqueous liquor is removed by distillation and the residue crystallized from ether–benzene. Crude glutaric acid, m.p. 90°, is obtained in yield of 70–75%.

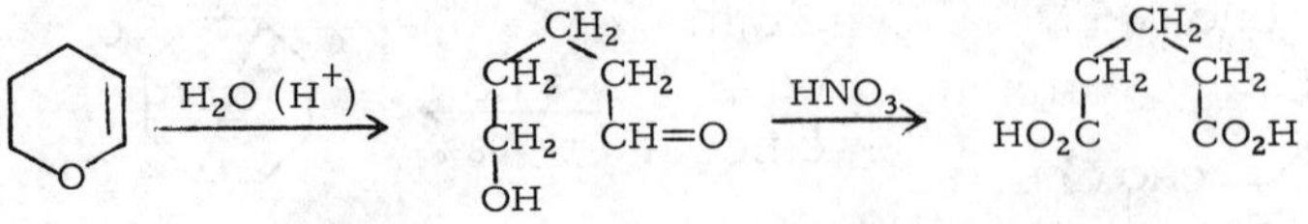

[1]C. S. Marvel and E. M. McColm, *Org. Syn., Coll. Vol.*, **1**, 536 (1941); C. S. Marvel and W. F. Tuley, *ibid.*, **1**, 289 (1941)

[2]G. Paris, L. Berlinguet, and R. Gaudry, *ibid.*, **4**, 496 (1963)

[3]J. English, Jr., and J. E. Dayan, *ibid.*, **4**, 499 (1963)

Glycine, $H_2NCH_2CO_2H$. Mol. wt. 75.07, pK_1 (COOH) 2.4, pK_2 9.8. Preparation from methyleneaminoacetonitrile[1] and from chloroacetic acid.[2] Suppliers: A, B, E, F, MCB.

Glycine, other amino acids, and aminophenols catalyze the Knoevenagel reaction[3, 4] (*compare* β-Alanine). Thus benzalmalononitrile is obtained in quantitative yield by adding a catalytic amount of glycine to a 70% aqueous-alcoholic solution saturated with benzaldehyde and containing 1 equivalent of malononitrile.[4] Glycine catalyzes

$$C_6H_5CHO + CH_2(CN)_2 \xrightarrow[100\%]{H_2NCH_2CO_2H} C_6H_5CH{=}C(CN)_2$$

the aldol condensation of dihydroxyacetone and glycollic aldehyde to give mixtures containing small amounts of ketopentose.[5]

[1]W. K. Anslow and H. King, *Org. Syn., Coll. Vol.*, **1**, 298 (1941)
[2]J. M. Orten and R. M. Hill, *ibid.*, **1**, 300 (1941)
[3]H. D. Dakin, *J. Biol. Chem.*, **7**, 49 (1909); F. S. Prout, *J. Org.*, **18**, 928 (1953)
[4]J. B. Bastús, *Tetrahedron Letters*, 955 (1963)
[5]J. A. Gascoigne, W. G. Overend, and M. Stacey, *Chem. Ind.*, 402 (1959)

Glyoxal, OCHCHO. Mol. wt. 58.04, m.p. 15°, b.p. 50.4°, sp. gr. 1.14. Bisulfite compound, $OCHCHO \cdot 2NaHSO_4 \cdot H_2O$, mol. wt. 316.20. Supplier of bisulfite compound: Union Carbide Chem. Co. Suppliers of 30% aqueous solution of glyoxal: Dow, A, B, E, F, MCB.

Preparation of glyoxal bisulfite by refluxing a mixture of paraldehyde, selenium dioxide, dioxane, and a small volume of 50% acetic acid for 6 hrs.[1] The solution is decanted from selenium, steam distilled to remove paraldehyde and dioxane, and

$$(CH_3CHO)_3 + 3\ H_2SeO_3 \longrightarrow 3\ OCHCHO + 3\ Se + 6\ H_2O$$

treated with a slight excess of 25% lead acetate solution to precipitate selenious acid as lead selenide. The precipitate is removed by filtration, and the solution is saturated with hydrogen sulfide, treated with Norit, and filtered hot. The colorless filtrate is concentrated at reduced pressure and added to a solution of sodium bisulfite in 40% ethanol. The yield of bisulfite compound is 62–64%. It can be crystallized by dissolving it in water and adding enough alcohol to make a 40% solution.

Quinoxaline synthesis. Condensation of glyoxal bisulfite with *o*-phenylenediamine in aqueous solution at about 80° affords quinoxaline in high yield, and this base on permanganate oxidation affords 2,3-pyrazinedicarboxylic acid.[2] In the absence of sodium bisulfite, aqueous glyoxal reacts with *o*-phenylenediamine to give quinoxaline in only about 30% yield along with resinous by-products.

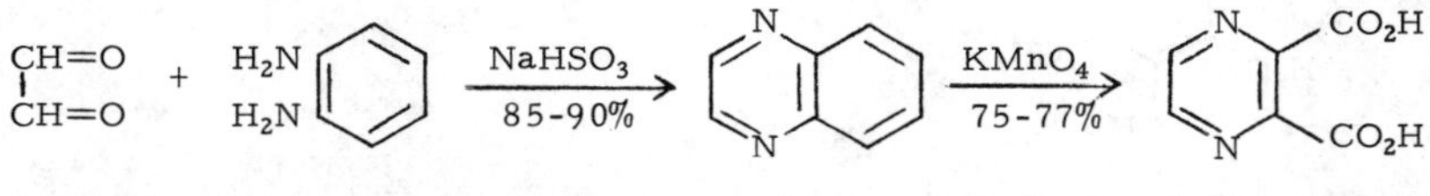
CH=O
CH=O
+
H₂N
H₂N
NaHSO₃
85-90%
N
N
KMnO₄
75-77%
N
N
CO₂H
CO₂H

Synthesis of quinones. In the presence of strong alkali, the dialdehyde disproportionates to glycolic acid:

$$O{=}CHCH{=}O + H_2O \longrightarrow HOCH_2CO_2H$$

However, Homolka[3] found that when air is passed into a suitably buffered aqueous solution, the aldehyde condenses to hexahydroxybenzene, which is oxidized to give the sodium salt of tetrahydroxyquinone. In a procedure by Fatiadi and Sager[4] 600 g. of commercial 30% glyoxal is added to an aqueous solution of sodium sulfite and sodium bicarbonate at 40–45° and a brisk stream of air is drawn through the solution for 1 hr. without application of heat and then for 1 hr. longer at 80–90°. Greenish black crystals of the disodium salt of tetrahydroxyquinone begin to separate in the first few minutes. The salt is collected, washed, and heated with 2 *N* hydrochloric

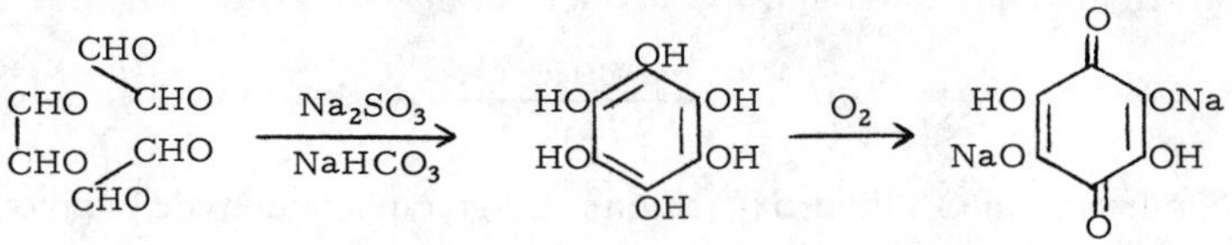

acid; the solution on cooling deposits 11–15 g. (6.2–8.4%) of glistening black crystals of tetrahydroxyquinone. Although the yield is low, the starting material is cheap and the process simple. Tetrahydroxyquinone on reduction with stannous chloride affords hexahydroxybenzene and on further oxidation yields rhodizonic acid and triquinonyl.

Isonaphthazarin can be synthesized by adding a solution of *o*-phthalaldehyde in dioxane to an aqueous solution of glyoxal bisulfite, potassium cyanide, and sodium bicarbonate and stirring the solution in the presence of air at 20°.[5] The solution rapidly acquires an intense violet color and is acidified after 15 min. (since the product is very sensitive to alkali) to precipitate bright red isonaphthazarin.

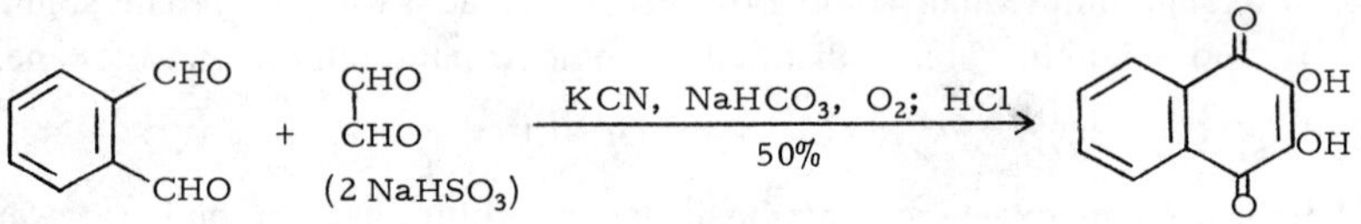

[1]A. R. Ronzio and T. D. Waugh, *Org. Syn., Coll. Vol.*, **3**, 438 (1955)
[2]R. G. Jones and K. C. McLaughlin, *ibid.*, **4**, 824 (1963)
[3]B. Homolka, *Ber.*, **54**, 1393 (1921)
[4]A. J. Fatiadi and W. F. Sager, *Org. Syn.*, **42**, 90 (1962)
[5]F. Weygand, *Ber.*, **75**, 625 (1942)

Gold chloride, $HAuCl_4 \cdot 3\,H_2O$. Mol. wt. 394.08.

This is a specific reagent for the oxidation of α-tocopherol to α-tocopherylquinone.[1] Addition of an aqueous solution of gold chloride to a solution of α-tocopherol in

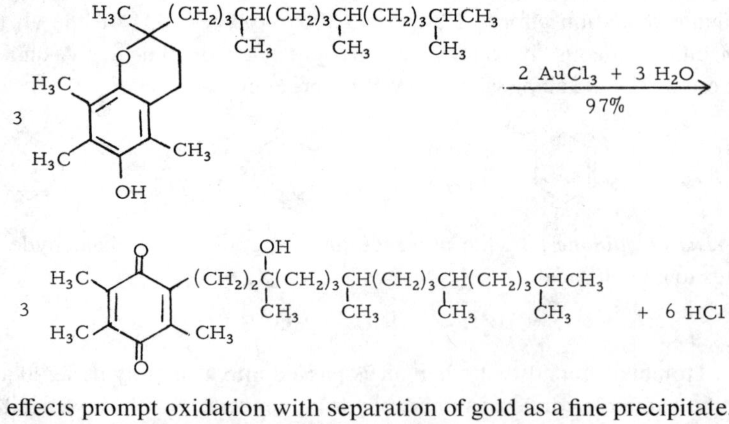

ethanol effects prompt oxidation with separation of gold as a fine precipitate. Evaporation of the filtered solution gives analytically pure α-tocopherylquinone as a golden yellow oil. Oxidation with ferric chloride or with silver nitrate gives highly impure products.

[1]P. Karrer *et al.*, *Helv.*, **21**, 951 (1938); **23**, 455 (1940)

Grignard reagents.

The following reagents are supplied by Arapahoe:

	Mol. wt.	Concn.	Solvent	Sp. gr.
Methylmagnesium bromide	119.26	3 *M*	Ether	1.02
Methylmagnesium bromide	119.26	2 *M*	THF–Benzene	1.06
Methylmagnesium iodide	166.27	3 *M*	Ether	1.09
Zerewitinoff reagent (CH_3MgI)	166.27	1 *M*	Diamyl ether	
Ethylmagnesium chloride	88.84	3 *M*	Ether	0.85
Ethylmagnesium bromide	133.29	3 *M*	Ether	1.01
n-Butylmagnesium chloride	116.89	3 *M*	Ether	0.88
Phenylmagnesium bromide	181.33	3 *M*	Ether	1.14
p-Chlorophenylmagnesium bromide	215.79	1.4 *M*	Ether	0.94

Peninsular ChemResearch offers: methylmagnesium bromide in ether and in di-*n*-butyl ether; methylmagnesium iodide in di-*n*-butyl ether; *n*-hexylmagnesium bromide in ether; tetramethylenedimagnesium bromide in ether; phenylmagnesium bromide in ether.

Metal and Thermit Corp. offers the following reagents:

CH_3MgCl	n-C_4H_9MgCl	n-$C_8H_{17}MgCl$
C_2H_5MgCl	n-$C_5H_{11}MgCl$	C_6H_5MgCl
n-C_3H_7MgCl	n-$C_6H_{13}MgCl$	$CH_2{=}CHMgCl$

Apparatus. An assembly designed by E. B. Hershberg and shown in Fig. G-1 meets usual requirements for the preparation of a Grignard reagent. A three-necked flask with ground-glass openings is mounted on a removable steam cone and provided with a coil-type condenser, a dropping funnel with a pressure-equalizing side tube, and a mercury sealed Hershberg stirrer[1] (Figs. G-2 and G-3) made of tantalum wire. Dry nitrogen is introduced at the top of the condenser and initially can be allowed to sweep through the apparatus and escape at the mouth of the dropping funnel; when the funnel is closed a slight positive pressure is maintained at the nitrogen tank, as indicated by the level of mercury in the escape valve. This gas-trap arrangement[2] is economical of nitrogen and has the advantage over use of a continuous stream of inert gas that it obviates evaporation of solvent. Prepurified grade nitrogen can be used without purification.

One reason for the nitrogen atmosphere is to prevent destruction of the Grignard reagent by air oxidation: $2\,RMgX + O_2 \longrightarrow 2\,ROMgX$.[3] Another is that it facilitates elimination of moisture. Both Newman and Kaugars[4] and Scala and Becker[5] recommend Dow Chemical Company's sublimed magnesium. Turnings made from this special magnesium are available from Dow. The required amount of magnesium turnings is placed in the flask, nitrogen is passed in, and both parts of the empty dropping funnel are flushed with the gas. When the air has been displaced the flask is heated gently with a free flame under continued flow of nitrogen to ensure elimination of any moisture adhering to the surface of the glass or to the metal; use of nitrogen prevents surface oxidation of the warm metal. When the flask has cooled completely, the nitrogen flow is reduced to a barely perceptible rate and small

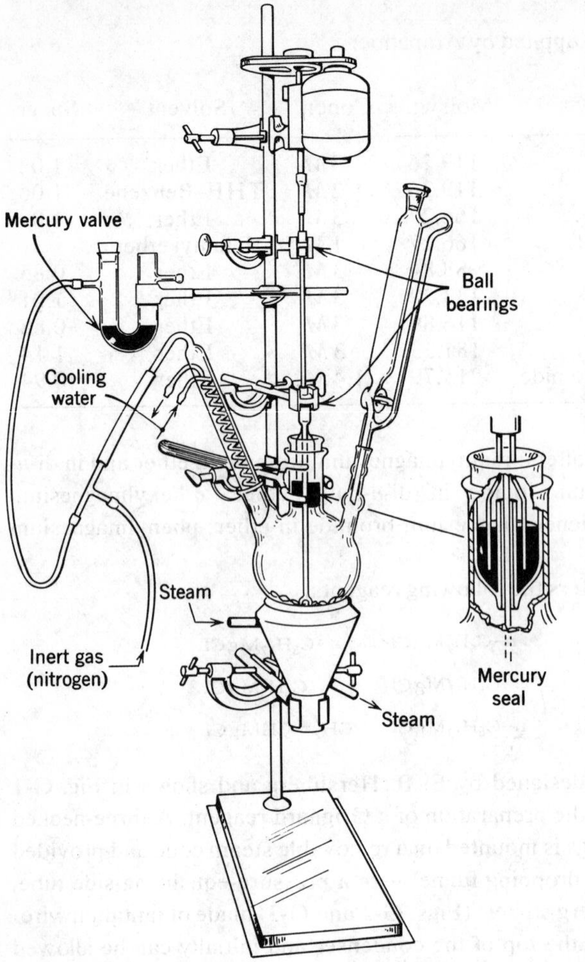
Mercury valve
Ball bearings
Cooling water
Steam
Inert gas (nitrogen)
Mercury seal
Steam

Fig. G-1 Apparatus for the Grignard reaction

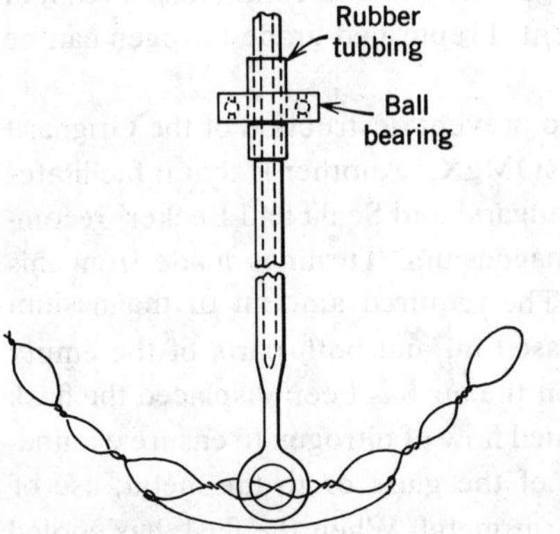
Rubber tubing
Ball bearing

Fig. G-2 Hershberg stirrer

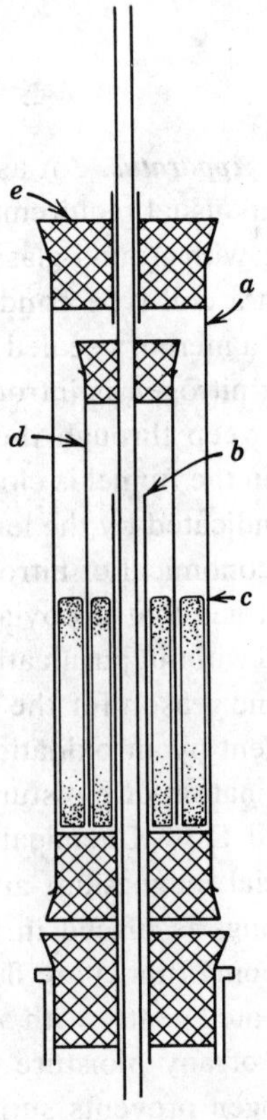
e
a
d
b
c

Fig. G-3 Mercury seal for stirrer

quantities of halogen compound and ether are introduced through the funnel. The stirrer is started even though very little liquid is present, for the crushing of pieces of light magnesium is often effective in initiating reaction. If stirring alone is ineffective, one expedient is to insert a flattened stirring rod and crush a piece of metal with a twisting motion against the bottom of the flask. Another is to add a little methylmagnesium iodide or a small crystal of iodine; according to one report,[6] bromine is more effective as a starter than iodine. Gilman's catalyst[7] is prepared by interaction of magnesium and iodine in ether–benzene.

Analysis of RMgX. Gilman[8] determined yields of Grignard reagents by diluting to a standard volume, adding an aliquot portion to excess 0.2 *N* acid, and back titrating with 0.2 *N* alkali, using phenolphthalein as indicator. Averages of closely agreeing runs and analyses are as follows:

	% RMgX
Ethyl bromide	93.1
n-Amyl bromide	88.0
sec-Amyl bromide	66.8
t-Amyl bromide	23.7
n-Hexyl bromide	92.0
n-Butyl chloride	91.2
n-Butyl iodide	85.6
Benzyl chloride	93.1
Bromobenzene	94.7

Gilman's test[9] for the presence of Grignard reagent is useful: about 0.5 ml. of the ethereal solution is removed with a capillary dropping tube and treated with an equal volume of a 1% solution of Michler's ketone in dry benzene and then, slowly, with 1 ml. of water. Addition of several drops of a 0.2% solution of iodine in acetic acid then produces a characteristic greenish blue color if Grignard reagent was present.

Gaseous halides. For preparation of small amounts of methylmagnesium bromide, a sealed vial of methyl bromide (b.p. 4.6°) is cooled in ice, opened, and fitted with a length of rubber tubing connected through a calcium chloride drying tube to a glass delivery tube leading into the reaction flask. The reagent is then introduced by warming the vial in the hand. Methyl chloride (b.p. −24°) is run directly from a cylinder into the mixture of magnesium and ether. Difficulty sometimes encountered in starting the reaction usually can be overcome by stirring and heating. Reagent can be conserved by providing the flask with a mercury valve or balloon, or with a cold finger condenser inserted in a side tubulature consisting of a test tube filled with dry ice. The latter scheme avoids difficulty from back pressure and makes possible an initial high concentration of methyl chloride favorable for starting the reaction. A highly reactive halide such as allyl bromide is converted into the Grignard reagent in very dilute solution to minimize coupling.

Entrainment procedure. *See also* Ethylene dibromide. A procedure for converting an unreactive halide into its magnesio halide derivative introduced by Grignard[10] consists in adding one equivalent of ethyl bromide to the ethereal solution of the refractory halide and dropping this mixture slowly onto sufficient magnesium to react with both halides. The auxiliary halide keeps the magnesium clean and active, and possibly functions by an exchange reaction. The list of Grignard reactions presented below includes under the heading *Acids and Esters* procedures for the conversion of bromomesitylene into the corresponding carboxylic acid with and without entrainment with ethyl bromide; the results show that the entrainment technique boosts the yield by about 25%. In this case the presence of two Grignard reagents in the solution is not objectionable because carbonation of C_2H_5MgBr gives a liquid acid which does not interfere with isolation of $ArCO_2H$, a crystalline product. For cases where the presence of two Grignard reagents might cause trouble, Pearson[11] introduced use as entraining reagent of ethylene dibromide, which reacts with magnesium to form ethylene and magnesium bromide:

$$BrCH_2CH_2Br + Mg \longrightarrow CH_2{=}CH_2 + MgBr_2$$

It cleanses and activates the magnesium for reaction with the inert halide without introducing a second Grignard reagent. This procedure consistently gave yields higher than those obtained by entrainment with ethyl bromide.

Vinyl halides. Vinyl halides, even the iodide, are not convertible to Grignard reagents by the usual procedure, but Normant[12] found that vinyl chloride and vinyl bromide readily form vinylmagnesium halides in solution in tetrahydrofurane (THF) or in a diether of di- or triethylene glycol. See review.[12]

Co-solvents. Some addition reactions do not proceed satisfactorily at the boiling point of ether and require addition to the ethereal solution of Grignard reagent of a solvent of higher boiling point (benzene, xylene) and removal of ether by distillation. Leigh[13] obtained Grignard reagents in high yield in ligroin, toluene, or xylene by first adding 1 mole of tetrahydrofurane for each mole of magnesium. Ashby and Reed[14] obtained a solution in benzene or toluene by using triethylamine as the complexing agent. A solution of an alkyl or aryl halide in benzene was added to

$$\underset{0.5\text{ m.}}{\underline{n}\text{-}C_4H_9Cl} + \underset{430\text{ ml.}}{C_6H_6} + \underset{0.6\text{ g. at.}}{Mg} + \underset{0.5\text{ m.}}{(C_2H_5)_3N} \xrightarrow[93\%]{\text{2 hrs.}\ 40\text{-}50^\circ} \underline{n}\text{-}C_4H_9MgCl$$

magnesium turnings diluted with benzene and containing triethylamine. About 30 ml. of the solution was added and the reaction was started by gently warming. The remainder of the solution was then added over a 2-hr. period, keeping the temperature at 40–50°. The resulting solution, clear and often colorless, was filtered through a medium sintered-glass funnel. Yields were determined by isolation of the Grignard reagent as a solid by removal of the solvent under vacuum, followed by elemental analysis.

Inverse Grignard. A procedure for conducting an inverse Grignard reaction consists in preparing the reagent in the apparatus of Fig. G-1, replacing the dropping funnel by a siphon tube fitted into a cork and bent to extend to the bottom of the flask, and transferring the solution under nitrogen pressure to a dry dropping funnel flushed with nitrogen. The funnel is then attached to a reaction flask containing a solution of the second component. Filtration, to remove traces of metal that might interfere, can be accomplished by introducing the plug of glass wool into the bottom of the siphon tube.

Hydrolysis. At the end of the reaction the MgX-derivative can be decomposed by adding 25% sulfuric acid or constant boiling hydrochloric acid dropwise with cooling under reflux, or by pouring the reaction mixture onto ice and dilute acid. If the reaction product is sensitive to acid, hydrolysis can be done with an aqueous solution saturated with ammonium chloride at 25°. (To prepare approximately 127 ml. of saturated solution, stir magnetically 38 g. of NH_4Cl with 100 ml. of distilled water until solution is complete and cool to 25°.) One procedure is to add a sufficiently large volume of the solution to bring initially precipitated basic salts into solution (about 3 l. per mole of Mg.)[15] A better method is to add just enough of the solution to precipitate the magnesium salts and leave a nearly anhydrous supernatant solution of the reaction product in ether. The reaction mixture is cooled and stirred under reflux in the original flask, and saturated ammonium chloride solution (25°) is added slowly from a dropping funnel at a rate controlled by the rapidity of refluxing. Usually 150–175 ml. of saturated solution per mole of magnesium is required to reach a point where a clear separation occurs, and this point should not be passed.[16] The solution at first becomes cloudy and opaque and then, when sufficient ammonium chloride has been added, the solution suddenly becomes clear, and a white salt separates as a tough cake that is liable to stop the stirrer. The mixture is allowed to settle for several minutes, the supernatant solution is decanted, and the dense precipitate is washed with one or two portions of fresh ether. The ethereal solution requires no drying and can be evaporated directly for recovery of the product. The procedure is applicable to reaction mixtures in which as much as one third of the solvent is benzene. For a reaction run with tetrahydrofurane as solvent, the amount of ammonium chloride solution recommended is 100–120 ml. per mole of RMgX.[17]

Examples. A summary is given below of the Grignard reactions described in *Organic Syntheses* (O.S.) with abbreviated references in which the boldface figure gives the number of a collective volume or of an individual volume from **40** on. Unless otherwise indicated, the steps of forming the Grignard reagent and of hydrolyzing the addition compound are not shown and the overall yield given is for isolated reaction product of acceptable purity. *Note:* Only the products are listed in the index.

Hydrocarbons and C—C Coupling

O.S. **2**, 478

$$CH_3CH_2CH_2\underset{\displaystyle Br}{CH}CH_3 \xrightarrow[(\underline{n}\text{-}C_4H_9)_2O]{Mg} CH_3CH_2CH_2\underset{\displaystyle MgBr}{CH}CH_3 \xrightarrow[50\text{-}52\%]{H_2O}$$

b. p. 143°

$$CH_3CH_2CH_2CH_2CH_3 \text{ (b. p. 36°)}$$

O.S. **1**, 471

$$C_6H_5CH_2MgCl \xrightarrow[70\text{-}75\%]{(C_2H_5O)_2SO_2} C_6H_5CH_2CH_2CH_3$$

O.S. **2**, 360

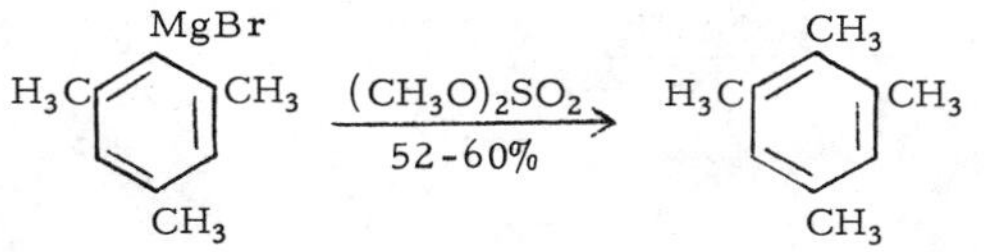
MgBr
H3C
CH3
CH3
(CH3O)2SO2
52-60%
CH3
H3C
CH3
CH3

O.S. **2**, 47

$$C_6H_5CH_2MgCl + 2\,\underline{p}\text{-}CH_3C_6H_4SO_3C_4H_9\text{-}\underline{n} \xrightarrow[50\text{-}59\%]{} C_6H_5CH_2C_4H_9\text{-}\underline{n} + \underline{n}\text{-}C_4H_9Cl + (TsO)_2Mg$$

O.S. 1, 186

$$C_6H_{11}MgBr \xrightarrow[60\text{-}64\%]{BrCH_2C(Br)=CH_2} C_6H_{11}\,CH_2C(Br)=CH_2$$

O.S. 3, 121

$$2\ CH_2=CHCH_2Cl \xrightarrow[55\text{-}65\%]{Mg} CH_2=CHCH_2-CH_2CH=CH_2$$

Alcohols or Dehydration Products

O.S. 1, 188

$$C_6H_{11}MgCl \xrightarrow[64\text{-}69\%]{CH_2=O} C_6H_{11}CH_2OH$$

O.S. 1, 306

$$CH_3CH_2CH_2CH_2MgBr \xrightarrow[60\text{-}62\%]{CH_2-CH_2\ (\text{-O-})} CH_3CH_2CH_2CH_2CH_2CH_2OH$$

O.S. 2, 179

$$2\ n\text{-}C_4H_9MgBr + HCO_2C_2H_5 \xrightarrow[83\text{-}85\%]{} (\underline{n}\text{-}C_4H_9)_2CHOH$$

O.S. 2, 406

$$(CH_3)_2CHMgBr + CH_3CHO \xrightarrow[53\text{-}54\%]{} (CH_3)_2CHCH(OH)CH_3$$

O.S. 41, 49

$$CH_2=CHCH_2MgBr \xrightarrow[57\text{-}59\%]{CH_2=CHCHO} CH_2=CHCH_2CH(OH)CH=CH_2$$

O.S. 3, 696

$$CH_3MgCl + CH_3CH=CHCHO \xrightarrow[81\text{-}86\%]{} CH_3CH=CHCH(OH)CH_3$$

O.S. 3, 200

$$\underline{m}\text{-}ClC_6H_4MgBr \xrightarrow[82\text{-}88\%]{CH_3CHO} \underline{m}\text{-}ClC_6H_4CH(OH)CH_3$$

O.S. 3, 839

$$2\ C_6H_5MgBr + C_6H_5CO_2C_2H_5 \xrightarrow[89\text{-}93\%]{} (C_6H_5)_3COH$$

O.S. 2, 602

$$3\ C_2H_5MgBr \xrightarrow[82\text{-}88\%]{O=C(OC_2H_5)_2} (C_2H_5)_3COH$$

O.S. 1, 226

$$2\ C_6H_5MgBr \xrightarrow{CH_3CO_2C_2H_5} (C_6H_5)_2C(OH)CH_3 \xrightarrow[67\text{-}70\%]{Distil} (C_6H_5)_2C=CH_2$$

O.S. 3, 729

C6H5MgBr +
Ac2O
42-48%
C6H5

O.S. **4**, 792

$$HC{\equiv}CMgBr \xrightarrow[58\text{-}69\%]{C_6H_5CH{=}CHCHO} HC{\equiv}CCH(OH)CH{=}CHC_6H_5$$

O.S. **3**, 237

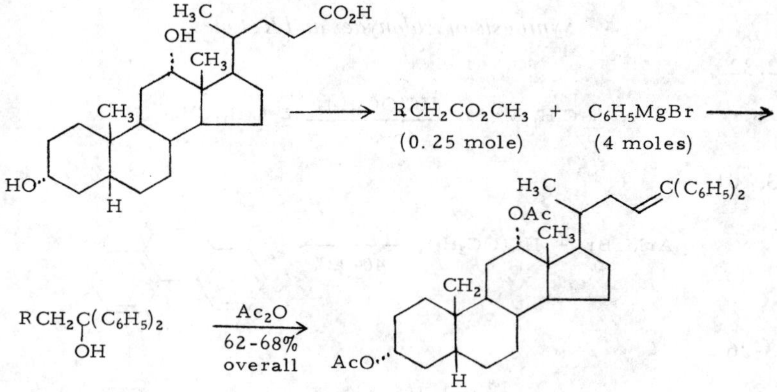

Acids, Esters

O.S. **1**, 524

$$(CH_3)_3CMgCl \xrightarrow[69\text{-}70\%]{CO_2} (CH_3)_3CCO_2H$$

O.S. **1**, 361

$$CH_3CH_2CH(CH_3)MgCl \xrightarrow[76\text{-}86\%]{CO_2} CH_3CH_2CH(CH_3)CO_2H$$

O.S. **2**, 425

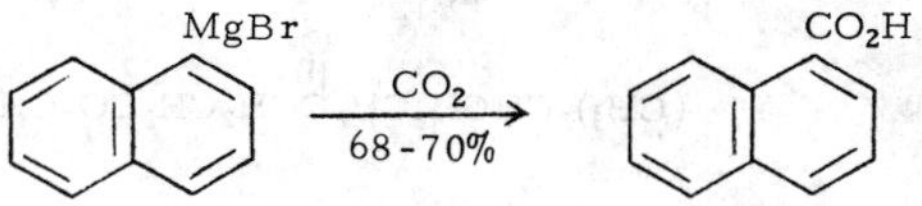

O.S. **3**, 553 (Entrainment method)

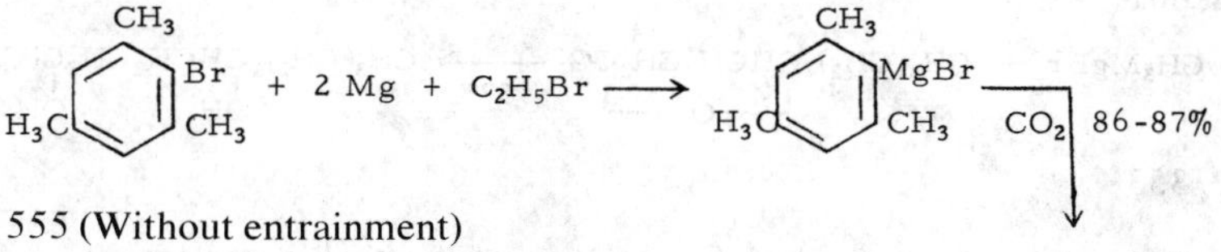

O.S. **3**, 555 (Without entrainment)

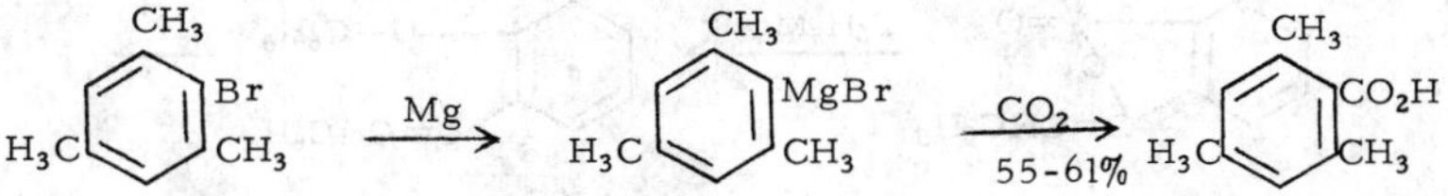

O.S. **2**, 282

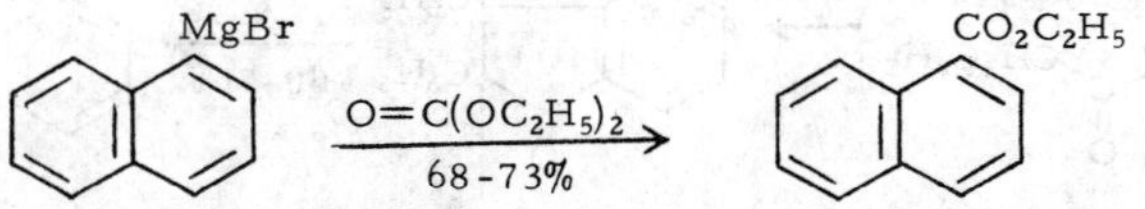

O.S. **2**, 198

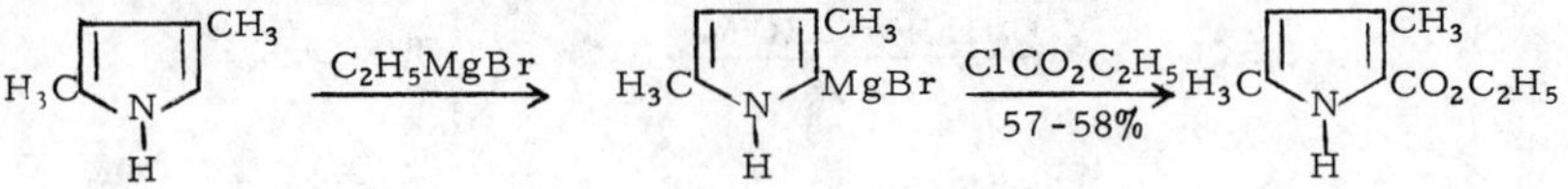

Synthesis of Aldehydes and Ketones

O.S. **2**, 323

$$\underline{n}\text{-}C_5H_{11}MgBr \xrightarrow[45\text{-}50\%]{HC(OC_2H_5)_3} \underline{n}\text{-}C_5H_{11}CHO$$

O.S. **3**, 701

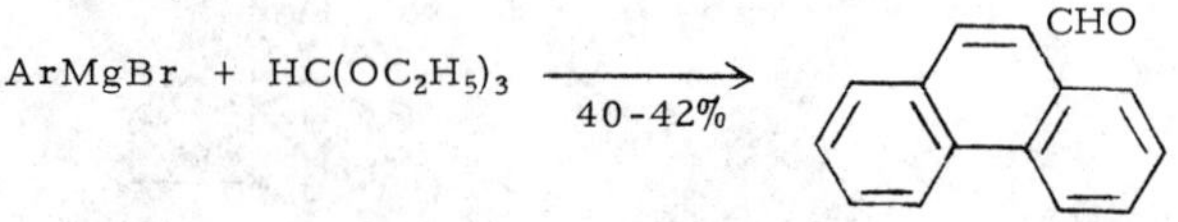

O.S. **3**, 26

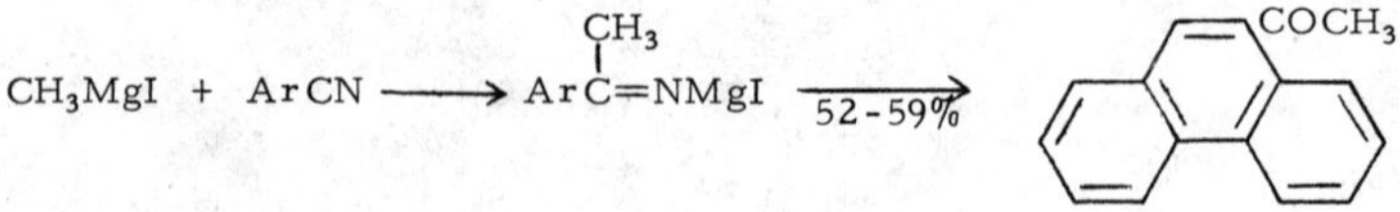

O.S. **3**, 562

$$C_6H_5MgBr \xrightarrow[71\text{-}78\%]{CH_3OCH_2CN} CH_3OCH_2\overset{O}{\overset{\|}{C}}C_6H_5$$

O.S. **3**, 601

$$(CH_3)_2CHCH_2CH_2MgBr \xrightarrow{CdCl_2} R_2Cd \xrightarrow[73\text{-}75\%]{Cl\overset{O}{\overset{\|}{C}}CH_2CH_2CO_2CH_3}$$

$$(CH_3)_2CHCH_2CH_2\overset{O}{\overset{\|}{C}}CH_2CH_2CO_2CH_3$$

Lactones + RMgX

O.S. **4**, 601

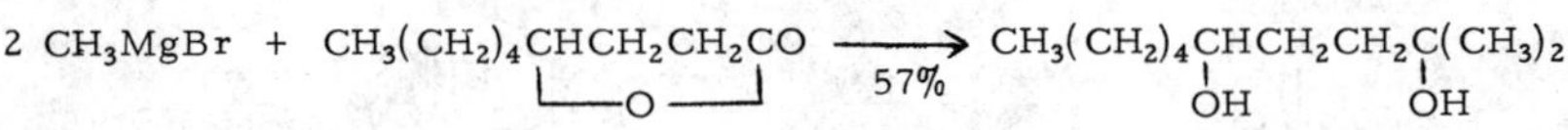

O.S. **3**, 353

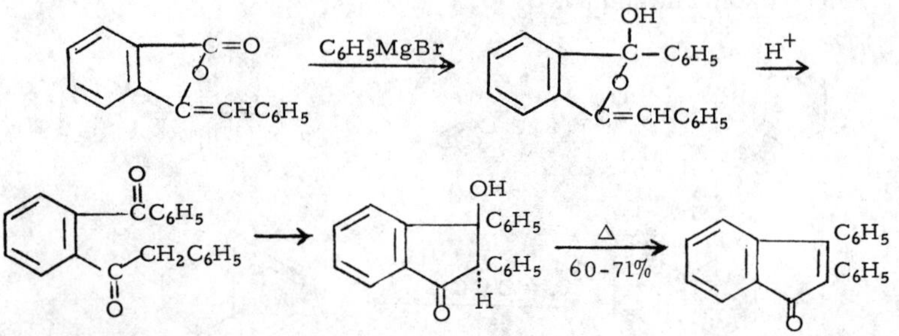

1,4-Additions

O.S. **4**, 93

$$CH_3CH_2C(CH_3){=}C(CN)CO_2C_2H_5 \xrightarrow[92\text{-}95\%]{C_6H_5CH_2MgCl} CH_3CH_2C(CH_3)(CH_2C_6H_5){-}CH(CN)CO_2C_2H_5$$

O.S. **41**, 60

$$\underline{n}\text{-}C_4H_9MgBr + CH_3CH{=}CHCO_2CH(CH_3)CH_2CH_3 \xrightarrow[68\text{-}78\%]{}$$

$$CH_3CH(C_4H_9\text{-}\underline{n})CH_2CO_2CH(CH_3)CH_2CH_3 \xrightarrow[90\text{-}94\%]{KOH} CH_3CH(C_4H_9\text{-}\underline{n})CH_2CO_2H$$

Synthesis of Ethers

O.S. **41**, 91

$$C_6H_5MgBr + C_6H_5C({=}O)OOC(CH_3)_3 \xrightarrow[78\text{-}84\%]{} C_6H_5OC(CH_3)_3$$

O.S. **43**, 55

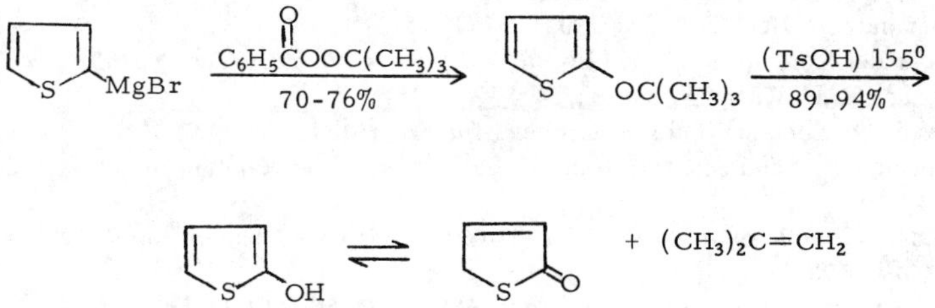

O.S. **4**, 667

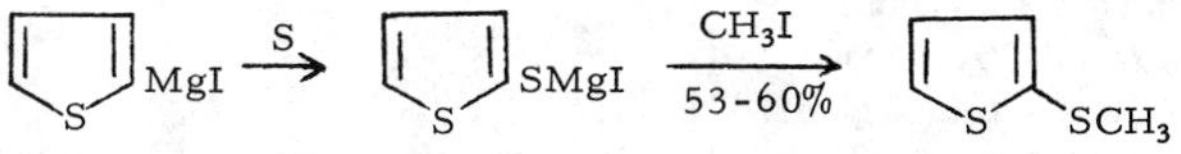

Acceptors other than Carbon

O.S. **1**, 550

$$3\ C_6H_5MgBr \xrightarrow[82\text{-}90\%]{SbCl_3} (C_6H_5)_3Sb$$

O.S. **3**, 771

$$C_6H_5MgBr + Se \xrightarrow[57\text{-}71\%]{} C_6H_5SeH$$

O.S. **4**, 258

$$\underset{(THF)}{CH_2{=}CHMgBr} \xrightarrow[74\text{-}91\%]{(\underline{n}\text{-}C_4H_9)_2SnCl_2} (\underline{n}\text{-}C_4H_9)_2Sn(CH{=}CH_2)_2$$

O.S. **4**, 605

$$C_6H_5CH_2MgCl + C_6H_5CH{=}NCH_3 \xrightarrow[87\text{-}95\%]{} C_6H_5CH(CH_2C_6H_5){-}NCH_3$$

O.S. **4**, 881

$$4\ C_2H_5MgBr + SnCl_4 \xrightarrow[89\text{-}96\%]{} (C_2H_5)_4Sn$$

O.S. **4**, 910

$$C_6H_5MgBr + (C_6H_5)_3AsO \xrightarrow[84\text{-}87\%]{\text{Then HCl}} (C_6H_5)_4AsCl \cdot HCl$$

Recent Examples[18, 19]

$$C_6H_5MgBr + ClCH_2C_6H_4F\text{-}\underline{o} \xrightarrow[85\%]{\text{In benzene at } 73^0} C_6H_5CH_2C_6H_4F\text{-}\underline{o}$$

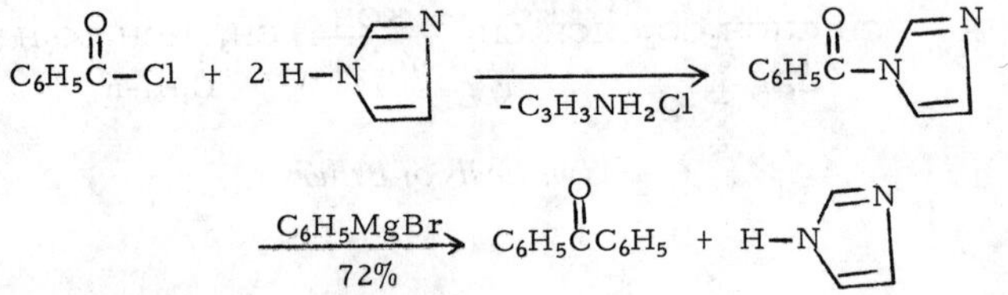

[1]E. B. Hershberg, *Ind. Eng. Chem.*, **8**, 313 (1936)
[2]H. Gilman and A. P. Hewlett, *Rec. trav.*, **48**, 1124 (1929)
[3]H. Gilman and A. Wood, *Am. Soc.*, **48**, 806 (1926)
[4]M. S. Newman and G. Kaugars, *J. Org.*, **30**, 3295 (1965)
[5]A. A. Scala and E. I. Becker, *J. Org.*, **30**, 3491 (1965)
[6]G. J. M. Van der Kerk and J. G. A. Luijten, *Org. Syn.*, *Coll. Vol.*, **4**, 882 (1963)
[7]H. Gilman and R. H. Kirby, *Rec. trav.*, **54**, 577 (1935)
[8]H. Gilman, E. A. Zoellner, and J. B. Dickey, *Am. Soc.*, **51**, 1576 (1929)
[9]H. Gilman and F. Schulze, *Am. Soc.*, **47**, 2002 (1925); H. Gilman and L. L. Heck, *ibid.*, **52**, 4949 (1930)
[10]V. Grignard, *Compt. rend.*, **198**, 625 (1934). See also H. Clement, *ibid.*, **198**, 665 (1934); E. Urion, *ibid.*, **198**, 1244 (1934)
[11]D. E. Pearson, D. Cowan, and J. D. Beckler, *J. Org.*, **24**, 504 (1959)
[12]H. Normant, *Compt. rend.*, **239**, 1510 (1954); *Bull. soc.*, 728 (1957); H. Normant, Advances in Organic Chemistry, **2**, 1 (1960)
[13]T. Leigh, *Chem. Ind.*, 426 (1965)
[14]E. C. Ashby and R. Reed, *J. Org.*, **31**, 971 (1966)
[15]L. Skattebøl, E. R. H. Jones, and M. C. Whiting, *Org. Syn.*, *Coll. Vol.*, **4**, 792 (1963)
[16]E. R. Coburn, *ibid.*, **3**, 696 (1955)
[17]D. Seyferth, *ibid.*, **4**, 258 (1963)
[18]F. A. Vingiello, S.-G. Quo, and J. Sheridan, *J. Org.*, **26**, 3202 (1961)
[19]H. A. Staab and E. Jost, *Ann.*, **655**, 90 (1962); H. A. Staab, *Angew. Chem. Internat. Ed.*, **1**, 351 (1962)

H

Haloform test.[1] A modification of a test devised by Fujiwara[2] permits detection of fluoroform, chloroform, bromoform, or iodoform or of any precursor of a haloform or of a dihalocarbene. A few drops of a mixture suspected of containing a haloform or a precursor is added to a mixture of 3 ml. of 10% sodium hydroxide and 2 ml. of pyridine. If the mixture turns pink to bright blue-red within one or two minutes of vigorous shaking at room temperature, a haloform is present. If the color change occurs not at room temperature but when the solution is heated on the steam bath for one minute, a precursor of a haloform or a dichlorocarbene is indicated. The intensity of the color is proportional to the concentration of haloform or of precursor.

This sensitive test permits use of the readily available sodium hypochlorite in place of sodium hypoiodite for recognition of methyl carbonyl or methylcarbinol structures. It is useful also when detection of a haloform by infrared spectroscopy or gas-liquid chromatography is unsatisfactory, as may happen in certain reactions in which haloform-type cleavage occurs to a slight extent.[3]

[1]Contributed by Madeleine M. Joullié, University of Pennsylvania
[2]K. Fujiwara, *Sitz. Nat. Ges. Rostock*, **6**, 33 (1916) [*C.A.*, **11**, 3201 (1917)]
[3]A. C. Pierce and M. M. Joullié, *J. Org.*, **28**, 658 (1963)

Hexachlorocyclopentadiene (I). Mol. wt. 272.80, b.p. 235–238°. Suppliers: Hooker Electrochemical Co., A, MCB.

This diene reacts with [2.2.1]bicycloheptadiene (norbornadiene, II) to give the Diels-Alder adduct III, useful as an insecticide and named aldrin in honor of K. Alder. The reagent is the precursor of 5,5-dimethoxy-1,2,3,4-tetrachlorocyclopentadiene, *which see*.

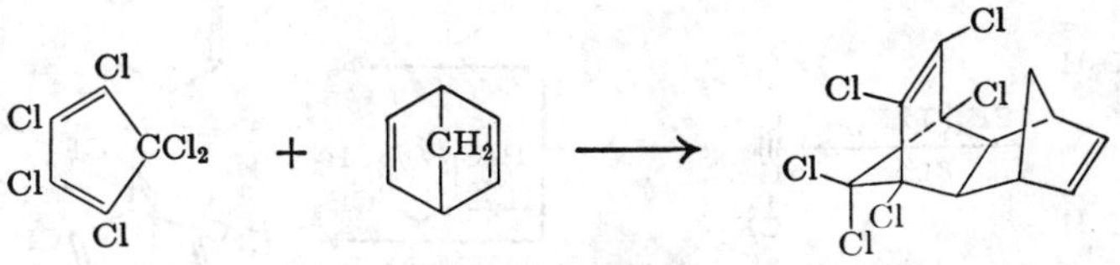

I II (III) Aldrin

Hexaethylphosphorous triamide, $[(C_2H_5)_2N]_3P$.[1,2] Mol. wt. 247.36, b.p. 245–246° (120–122°/10 mm.), $n^{20}D$ 1.4750. For preparative procedure and reactions, *see* Hexamethylphosphorous triamide.

[1]A. Michaelis, *Ann.*, **326**, 129 (1903)
[2]C. Stuebe and H. P. Lankelma, *Am. Soc.*, **78**, 976 (1956)

Hexafluoro-2-butyne, $CF_3C{\equiv}CCF_3$. Mol. wt. 162.04, b.p. −25° to −24°. Supplier: Peninsular ChemResearch (see Fig. H-1).

This reagent has been prepared by dechlorination of $CF_3CCl{=}CClCF_3$ with zinc[1] but is prepared most efficiently by the reaction of acetylenedicarboxylic acid with sulfur tetrafluoride catalyzed by titanium tetrafluoride.[2] The acetylene is an effective Diels-Alder dieneophile.[3] It reacts with butadiene at room temperature to

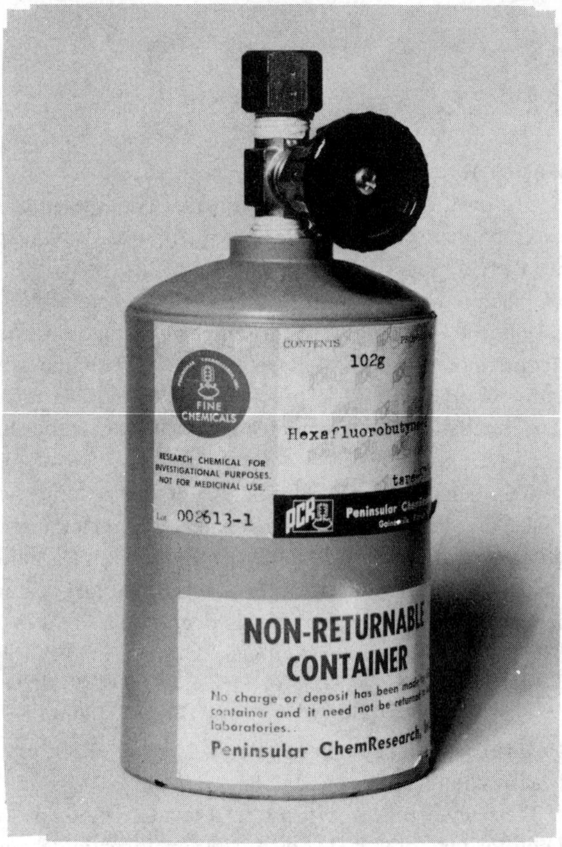
CONTENTS
102g
FINE CHEMICALS
RESEARCH CHEMICAL FOR INVESTIGATIONAL PURPOSES. NOT FOR MEDICINAL USE.
CONTAINER

Fig. H-1 Convenient container for a gaseous reagent

give the substituted 1,4-cyclohexadiene (1). It adds to durene at 200° to form the substituted bicyclooctatriene (2).

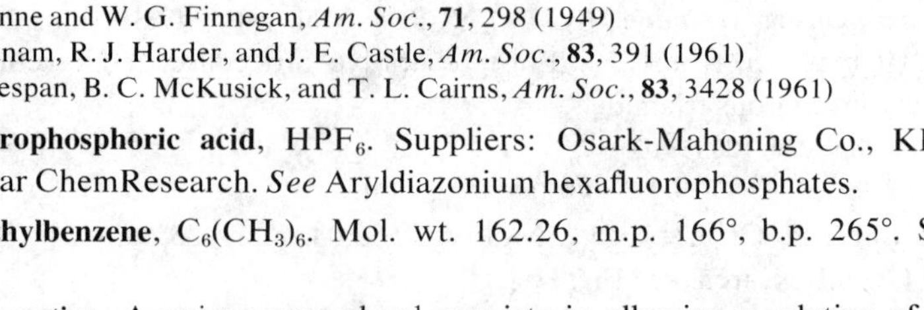

[1]A. L. Henne and W. G. Finnegan, *Am. Soc.*, **71**, 298 (1949)
[2]R. E. Putnam, R. J. Harder, and J. E. Castle, *Am. Soc.*, **83**, 391 (1961)
[3]C. G. Krespan, B. C. McKusick, and T. L. Cairns, *Am. Soc.*, **83**, 3428 (1961)

Hexafluorophosphoric acid, HPF_6. Suppliers: Osark-Mahoning Co., KK, MCB, Peninsular ChemResearch. *See* Aryldiazonium hexafluorophosphates.

Hexamethylbenzene, $C_6(CH_3)_6$. Mol. wt. 162.26, m.p. 166°, b.p. 265°. Suppliers: A, E.

Preparation. A unique procedure[1] consists in allowing a solution of 100 g. of phenol in 1 l. of methanol to drop at the rate of 110 ml./hr. onto an active alumina catalyst heated to 530° in an apparatus provided with a gas-exit tube leading to the hood. A pale yellow solid which collects in the receiver is transferred to a suction funnel with methanol. Crystallization from ethanol affords colorless crystals, m.p. 165–166°, in yield of 55-57%. The mechanism of the reaction is not known.

Reaction. For an interesting oxidative rearrangement, *see* Pertrifluoroacetic acid.

[1]N. M. Cullinane, S. J. Chard, and C. W. C. Dawkins, *Org. Syn., Coll. Vol.*, **4**, 520 (1963)

Hexamethyldisilazane, $(CH_3)_3SiNHSi(CH_3)_3$. Mol. wt. 161.40, b.p. 125.5°. Suppliers: Aldrich, Peninsular ChemResearch. Prepared from trimethylchlorosilane and ammonia in 65% yield.[1]

Peptide synthesis. Review.[2] The reagent reacts with amino acids (reflux, a few drops of H_2SO_4) to form trimethylsilyl esters of N-trimethylsilylamino acids.

$$\underset{\displaystyle NH_2}{RCHCO_2H} + (CH_3)_3SiNHSi(CH_3)_3 \longrightarrow \underset{\displaystyle NHSi(CH_3)_3}{RCHCO_2Si(CH_3)_3} + NH_3$$

Hydroxyl (serine) and sulfhydryl groups (cysteine) are also silylated. These derivatives can be used in peptide synthesis by the phosphoryl chloride, imidazole, mixed anhydride, or *p*-nitrophenyl ester methods. An attractive feature is that the silyl groups are removed merely by treatment with water in the normal workup. Trimethylchlorosilane is not suitable because of the amphoteric nature of the amino acids.

Derivatives of alcohols. Treatment with hexamethyldisilazane in pyridine solution at room temperature converts bile acids[3] and sugars[4] into O-trimethylsilyl ethers, which are much more volatile than usual ethers and well suited for gas-liquid chromatography.

[1]R. O. Sauer, *Am. Soc.*, **66**, 1707 (1944)
[2]L. Birkofer and A. Ritter, *Angew. Chem., Internat. Ed.*, **4**, 417 (1965)
[3]M. Makita and W. W. Wells, *Anal. Biochem.*, **5**, 523 (1963)
[4]C. C. Sweeley, R. Bentley, M. Makita, and W. W. Wells, *Am. Soc.*, **85**, 2497 (1963)

Hexamethylenetetramine (Hexamine), $(CH_2)_6N_4$. Mol. wt. 140.19, white solid. Suppliers: A, B, E, F, MCB. The reagent can be prepared merely by allowing a mixture of formalin and concentrated ammonia solution to evaporate.

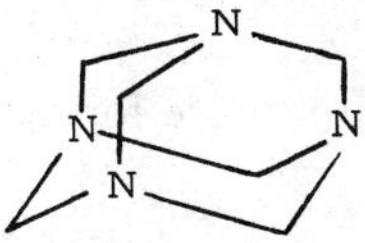

Hexamine is employed in medicine under the name urotropin as a urinary antiseptic, effective because it slowly releases formaldehyde and ammonia and renders urine basic.

Cyclonite. Fuming nitric acid destroys the inner bridge system of hexamine by oxidation and nitrates the peripheral nitrogens to produce cyclonite, an explosive of ballistic strength 150.2 relative to that of TNT taken as 100. Cyclonite was introduced in World War I and played an important part in World War II. Bachmann and Sheehan,[1] working at the University of Michigan under contract with the

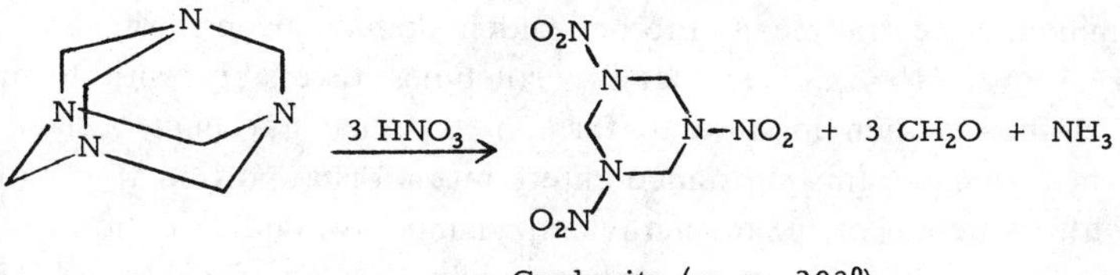
3 HNO₃
N–NO₂ + 3 CH₂O + NH₃

Cyclonite (m.p. 203°)

National Defense Research Committee, were able to double the yield by adding ammonium nitrate and acetic anhydride to the reaction mixture; one mole of hexamine then yields two moles of cyclonite rather than one:

$$C_6H_{12}N_4 + 4\ HNO_3 + 2\ NH_4NO_3 + 6\ (CH_3CO)_2O \xrightarrow[70\%]{} \underset{\text{Cyclonite}}{2C_3H_6N_6O_6} + 12\ CH_3CO_2H$$

Three reactions employing hexamine have found sufficient synthetic use to become known as name reactions:

Delépine reaction.[2] An activated halide reacts with hexamine in a nonaqueous medium to form an iminium salt which is hydrolyzed by ethanolic hydrochloric acid to produce the corresponding amine, as hydrochloride; formaldehyde is removed as the volatile diethyl formal. For example, 2,3-dibromopropene-1 (1) forms the salt (2), which on hydrolysis affords the amine (3).[3] In this example the

$$\underset{(1)}{CH_2{=}C(Br)CH_2Br} + C_6H_{12}N_4 \xrightarrow[86\text{-}91\%]{} \underset{(2)}{CH_2{=}C(Br)CH_2\overset{+}{N}C_6H_{12}N_3(Br^-)}$$

$$\xrightarrow[59\text{-}72\%]{EtOH\text{-}HCl;\ NaOH} \underset{(3)}{CH_2{=}C(Br)CH_2NH_2}$$

halide (1) is allylic. Amine formation has been observed also with comparably activated benzyl halides and phenacyl halides, which form iminium salts that are hydrolyzed with cold alcoholic hydrochloric acid to primary amines, $ArCH_2NH_2$ and $ArCONH_2$.[4]

$$ArCOCH_2Br \xrightarrow{(CH_2)_6CN_4} ArCOCH_2Br\cdot(CH_2)_6N_4 \xrightarrow[HCl]{C_2H_5OH} ArCOCH_2NH_3^+Cl^-$$

Sommelet reaction.[5, 6] This reaction also involves reaction of a halide, usually a benzyl-type halide, with hexamine to form a heximinium salt, which is then hydrolyzed in a nonacidic aqueous medium such as water or aqueous alcohol. In the procedure formulated, the reaction of benzyl chloride with hexamine is carried out

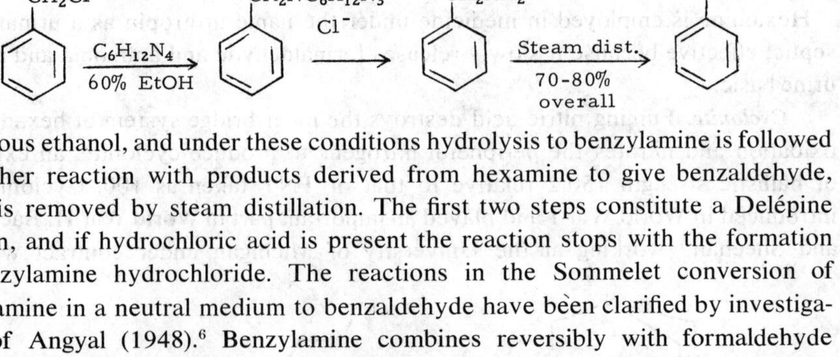
CH₂Cl
C₆H₁₂N₄
60% EtOH
CH₂N⁺C₆H₁₂N₃
Cl⁻
CH₂NH₂
Steam dist.
70-80% overall
CHO

in aqueous ethanol, and under these conditions hydrolysis to benzylamine is followed by further reaction with products derived from hexamine to give benzaldehyde, which is removed by steam distillation. The first two steps constitute a Delépine reaction, and if hydrochloric acid is present the reaction stops with the formation of benzylamine hydrochloride. The reactions in the Sommelet conversion of benzylamine in a neutral medium to benzaldehyde have been clarified by investigations of Angyal (1948).[6] Benzylamine combines reversibly with formaldehyde present in the reaction mixture to form methylenebenzylamine (2) and, in the absence of hexamine, this substance enters into a disproportionation with benzylamine with reduction of (2) to methylbenzylamine (4) and oxidation of (1) to the imine (3), which is hydrolyzed to give benzaldehyde in low yield. At the pH required

$$\underset{(1)}{C_6H_5CH_2NH_2} + \underset{(2)}{C_6H_5CH_2N{=}CH_2} \rightleftharpoons \underset{(3)}{C_6H_5CH{=}NH} + \underset{(4)}{C_6H_5CH_2NHCH_3}$$

$$(3) \xrightarrow{H_2O} \underset{(5)}{C_6H_5CHO + NH_3}$$

for the Sommelet reaction (pH 3.0–6.5), hexamine reacts as the methylene derivative of ammonia, $CH_2{=}NH$, which accepts hydrogen from benzylamine and is reduced to methylamine. This reaction inhibits the disproportionation leading to methylbenzylamine (4) and increases the yield of benzaldehyde. The fundamental process is thus:

$$C_6H_5CH_2NH_2 + CH_2{=}NH + H_2O \longrightarrow C_6H_5CHO + NH_3 + CH_3NH_2$$

The oxidation-reduction process resembles a Cannizzaro reaction and probably proceeds by a similar hydride ion shift involving as acceptor the Schiff base conjugate acid:

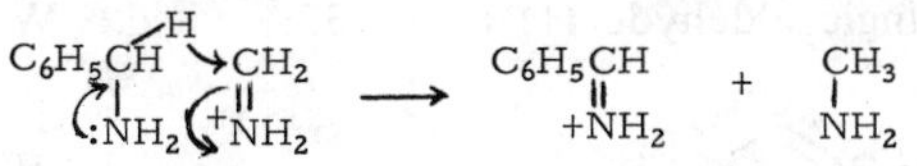

It is also possible that the reaction involves a 1,3-proton shift in (2), followed by hydrolysis:

$$\underset{(2)}{C_6H_5CH_2N{=}CH_2} \longrightarrow C_6H_5CH{=}NCH_3 \xrightarrow{H_2O} C_6H_5CHO + H_2NCH_3$$

In a procedure for the preparation of 2-thiophenealdehyde,[7] the hexaminium salt is prepared in refluxing chloroform and steam distilled to effect the remaining steps

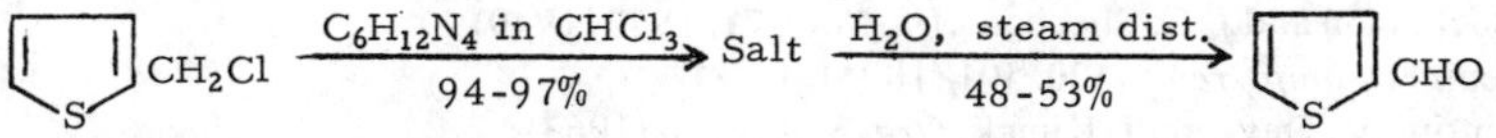

and remove the aldehyde as formed. In the preparation of 3-thiophenealdehyde,[8] the salt is prepared in chloroform and then extracted from chloroform into water. Steam distillation gives a mixture of the aldehyde and methylamine, and the ethereal extract is washed with acid for removal of the amine.

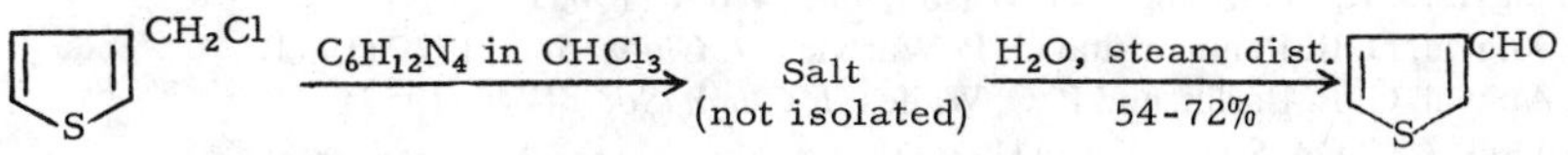

In a Sommelet preparation of α-naphthaldehyde[9] which is noteworthy because of the high yield, the entire reaction is carried out in refluxing 50% acetic acid and the quaternary salt is not isolated. At the end of the reflux period (2 hrs.), hydro-

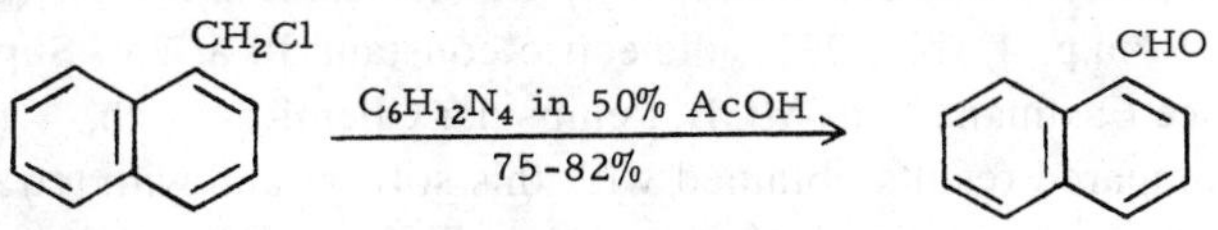

chloric acid is added and refluxing continued for 15 min. to hydrolyze and destroy Schiff bases which otherwise would contaminate the product.

Saturated alkyl halides are convertible into the hexaminium salts in high yield but the yield in the second step is low, for example, 46% in the case of the salt from *n*-heptyl iodide.[10] In the N-heterocyclic series chloromethyl derivatives are mostly inaccessible or unstable, but aminomethyl derivatives available by hydrogenation of the nitriles serve as satisfactory starting materials, for example 3-aminomethylpyridine.[11]

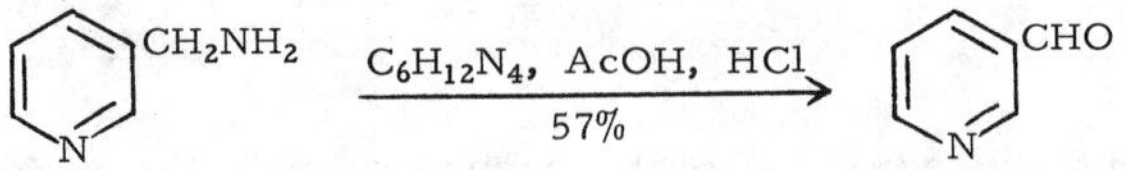
CH2NH2
C6H12N4, AcOH, HCl
57%
CHO

Duff reaction.[12] Hexamine condenses with a phenol in the *ortho* position to give a Schiff base which on hydrolysis yields the *o*-hydroxy aldehyde:

$$\underline{o}\text{-}HOC_6H_4CH{=}NCH_2C_6H_4OH\text{-}\underline{o} \longrightarrow \underline{o}\text{-}HOC_6H_4CHO + H_2NCH_2C_6H_4OH\text{-}\underline{o}$$

Duff heated glycerol with boric acid at 170° until no more water was eliminated, added 25 g. each of a phenol and hexamine, heated at 170° for 15 min., and isolated the chelated *o*-hydroxy aldehyde by steam distillation. For several phenols, yields were in the range 2–8.5 g. Pyrogallol-1,3-dimethyl ether is substituted in the *para* position to give syringic aldehyde (1) in 31–32% yield.[13] W. Baker[14] applied to

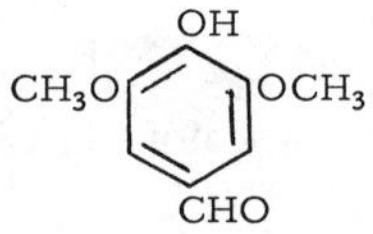
OH
CH3O
OCH3
CHO

(1)

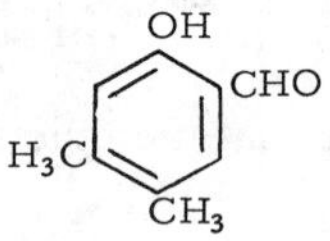
OH
CHO
H3C
CH3

(2)

3,4-dimethylphenol a previous procedure[15] involving heating the phenol with hexamine in acetic acid for 5 hrs. on the steam bath, adding boiling water and hydrochloric acid, and steam distilling the aldehyde (2); yield 32%.

[1]W. E. Bachmann and J. C. Sheehan, *Am. Soc.*, **71**, 1842 (1949)
[2]M. Delépine, *Compt. rend.*, **120**, 501 (1895)
[3]A. T. Bottini, V. Dev, and J. Klinck, *Org. Syn.*, **43**, 6 (1963)
[4]L. M. Long and H. D. Trouliman, *Am. Soc.*, **71**, 2473 (1949)
[5]M. Sommelet, *Compt. rend.*, **157**, 852 (1913)
[6]Review: S. J. Angyal, "Organic Reactions," **8**, 197 (1954)
[7]K. B. Wiberg, *Org. Syn., Coll. Vol.*, **3**, 811 (1955)
[8]E. Campaigne, R. C. Bourgeois, and W. C. McCarthy, *ibid.*, **4**, 918 (1963)
[9]S. J. Angyal, J. R. Tetaz, and J. G. Wilson, *ibid.*, **4**, 690 (1963)
[10]S. J. Angyal, D. R. Penman, and G. P. Warwick, *J. Chem. Soc.*, 1737 (1953)
[11]S. J. Angyal, G. B. Barlin, and P. C. Wailes. *J. Chem. Soc.*, 1740 (1953)
[12]J. C. Duff, *J. Chem. Soc.*, 547 (1941)
[13]C. F. H. Allen and G. W. Leubner, *Org. Syn., Coll. Vol.*, **4**, 866 (1963)
[14]W. Baker, J. F. W. McOmie, and D. Miles, *J. Chem. Soc.*, 820 (1953)
[15]J. C. Duff and E. J. Bills, *J. Chem. Soc.*, 1987 (1932); 1305 (1934)

Hexamethylphosphoric triamide, Hexamethylphosphoramide (HMPA), $[(CH_3)_2N]_3PO$. Mol. wt. 179.19, m.p. 4°, b.p. 232°, dielectric constant 30 at 25°. Suppliers: Monsanto, Tennessee Eastman, A, E, MCB, Peninsular ChemResearch.

Normant[1] compared results obtained with this solvent and with tetrahydrofurane in a large number of reactions of organometallic compounds (both magnesium and sodium).

Fraenkel and co-workers[2] found that sodium, potassium, and lithium dissolve in hexamethylphosphoric triamide to give blue solutions up to 1 *M* which are stable for several hours and then change to red. This aprotic solvent has high solvent power for organometallic compounds, and with it both Grignard reagents and organolithium compounds can be prepared in good yield. The solvent is similar to ammonia but easier to handle.

Schriesheim *et al.*[3] found that the systems $(CH_3)_3COK$–HMPA and even KOH–HMPA are particularly effective in promoting autoxidation of alkyl aromatics with varying side chains to aromatic carboxylic acids at room temperature. With very weakly acidic hydrocarbons (toluene, xylenes, ethylbenzene) yields are in the range 10–50%. HMPA is vastly more active in this reaction than tetramethylene sulfoxide, dimethyl sulfone, or dimethyl sulfoxide. Reasons suggested for the superiority of HMPA are: the hydrogens are completely unreactive to base; the solvent promotes ionization of the hydrocarbon; the low melting point and high boiling point permit use over a wide range of conditions.

Wallace and Schriesheim[4] dried the solvent by distillation under nitrogen over Linde 13X molecular sieve (calcined under N_2 at 350° for 4 hrs.) and stored it in a dry box equipped with a moisture conductivity cell. They found it superior to DMF or tetramethylurea as solvent for base-catalyzed (KOH or NaOH) autoxidation of thiols and disulfides to sulfonic acids. It is superior also for base-catalyzed autoxidation of alkylthiophenes and toluene,[5] and of C_5–C_{12} cyclic ketones to the corresponding dibasic acids.[6]

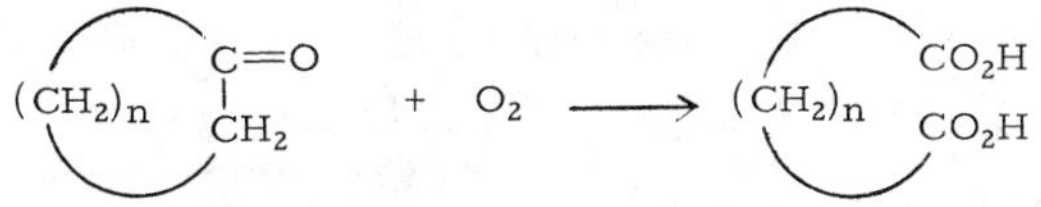
C=O
(CH₂)n
CH₂
+ O₂
(CH₂)n
CO₂H
CO₂H

[1]H. Normant, *Bull. Soc.*, 1888 (1963); 859 (1965); J. Cuvigny and H. Normant, *ibid.*, 2000 (1964); T. Cuvigny, J. Normant, and H. Normant, *Compt. rend.*, **258**, 3502 (1964)
[2]G. Fraenkel, S. H. Ellis, and D. T. Dix, *Am. Soc.*, **87**, 1406 (1965)
[3]J. E. Hofmann, A. Schriesheim, and D. D. Rosenfeld, *Am. Soc.*, **87**, 2523 (1965)
[4]T. J. Wallace and A. Schriesheim, *Tetrahedron*, **21**, 2271 (1965)
[5]T. J. Wallace and F. A. Baron, *J. Org.*, **30**, 3520 (1965)
[6]T. J. Wallace, H. Pobiner, and A. Schriesheim, *J. Org.*, **30**, 3768 (1965)

Hexamethylphosphorous triamide, $[(CH_3)_2N]_3P$. Mol. wt. 163.19, b.p. 162–164° (49–51°/12 mm.), $n^{25}D$ 1.4620.

Preparation[1,2] ***according to Mark.***[3] A solution of 1 mole of phosphorus trichloride in 1.5 l. of ether is stirred in an ice bath under nitrogen to 2° and gaseous dimethyl-

$$PCl_3 + 6(CH_3)_2NH \xrightarrow[97\text{-}100\%]{} [(CH_3)_2N]_3P + 3(CH_3)_2\overset{+}{N}H_2\overset{-}{Cl}$$

amine (b.p. 7.8°, Matheson Co.) is passed in at such a rate that the temperature does not exceed 15° (5–7 hrs.) The white, stirrable slurry of amine hydrochloride which results is let stand overnight at room temperature under nitrogen, filtered, and the filter cake washed with ether. Stripping of solvent from the clear filtrate in a rotating evaporator affords nearly pure triamide. The material can be purified with little loss by charcoal treatment in pentane or by distillation. It is best stored under dry nitrogen.

Epoxidation (Mark[4]). Reaction of the reagent with aromatic aldehydes provides a simple, one-step route to symmetrical and unsymmetrical epoxides, as illustrated by the example formulated.[5] The reagent abstracts one of the aldehydic oxygen atoms and forms hexamethylphosphoric triamide. A solution of the aldehyde in benzene is stirred under reflux, and a solution of the triamide in ether is added at such a rate that the temperature remains at 24–36°. With tap water cooling the

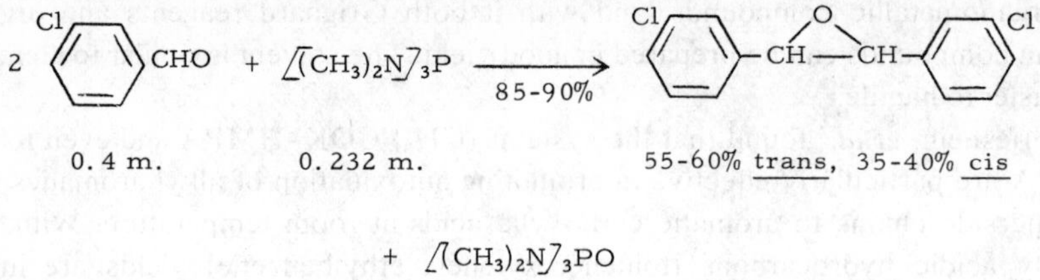

addition requires 30–50 min. The clear solution is heated at 50° for 15 min., the solvent is removed in a rotating evaporator, and the residue is triturated with water and pentane. Recovery by pentane extraction gives an off-white crystalline product, which is a *cis-trans* mixture of the composition indicated (determined by NMR spectrum).

Aliphatic and heterocyclic aldehydes give mainly 1 : 1 adducts, formulated as:[4]

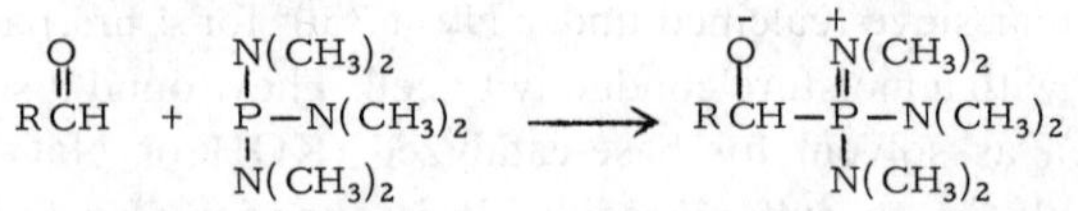

Newman and Blum[6] showed that the reagent can be used to cyclize a dialdehyde such as (1) to the aromatic epoxide (2).

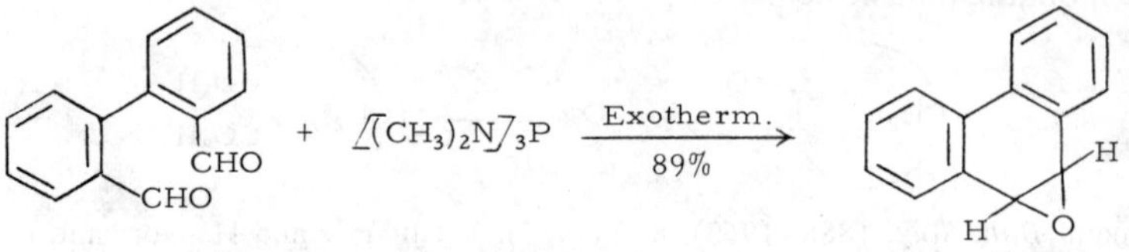

(1) 1 g. (2)

[1]D. R. Carmody and A. Zletz, U.S. patent 2,898,732
[2]A. B. Burg and P. J. Slota, Jr., *Am. Soc.*, **80**, 1107 (1958)
[3]V. Mark, procedure submitted to *Org. Syn.*
[4]V. Mark, *Am. Soc.*, **85**, 1884 (1963)
[5]V. Mark, *Org. Syn.*, **46**, 42 (1966)
[6]M. S. Newman and S. Blum, *Am. Soc.*, **86**, 5598 (1964)

Heyn's catalyst. This 10% platinum catalyst supported on purified charcoal[1] catalyzes the oxidation of primary alcohols by oxygen or air to carboxylic acids.[2] See reviews.[3]

[1]K. Heyns, *Ann.*, **558**, 177 (1947)
[2]K. Heyns *et al.*, *Ann.*, **558**, 187, 192 (1947); *Ber.*, **86**, 110, 833 (1953); **87**, 13 (1954)
[3]K. Heyns and H. Paulsen, "Newer Methods of Preparative Organic Chemistry," **2**, 303 (1963); *Idem*, *Adv. Carbohydrate Chem.*, **17**, 169 (1962)

Hippuric acid, $C_6H_5CONHCH_2CO_2H$. Mol. wt. 179.17, m.p. 188°. Supplier: Eastern Chem. Corp.

Preparation:[1] 1 mole of chloroacetic acid is added to 3 l. of concd. ammonium hydroxide (*ca.* 45 moles) and after 4 days the bulk of the ammonia is removed by distillation. A solution of 1.25 moles of sodium hydroxide is added, and the solution is boiled until the odor of ammonia is completely absent. The solution is filtered, stirred, and kept below 30° by cooling, during simultaneous addition (1 hr.) of

$$ClCH_2CO_2H \xrightarrow{NH_3} H_2NCH_2CO_2H \xrightarrow{NaOH} H_2NCH_2CO_2Na$$

$$\xrightarrow[64-68\%]{C_6H_5COCl} C_6H_5CONHCH_2CO_2H$$

benzoyl chloride (1.1 moles) and a solution of sodium hydroxide (2 moles) at rates such that the solution is always only slightly alkaline. The solution is poured onto hydrochloric acid and the precipitate is crystallized from water.

Azlactone synthesis. Use of the reagent in the azlactone synthesis of amino acids is illustrated by a procedure for the preparation of DL-phenylalanine.[2] A mixture of 0.5 mole each of benzaldehyde, hippuric acid, and anhydrous sodium acetate is treated with 153 g. of acetic anhydride, and the mixture heated and stirred until the

$$C_6H_5CHO + \underset{NHCOC_6H_5}{CH_2CO_2H} + 2(CH_3CO)_2O \xrightarrow[62-64\%]{NaOAc} C_6H_5CH=C\underset{N_{\diagdown C \diagup}^{}O}{\overset{}{\vert}}C=O$$

(1) C_6H_5

$$\xrightarrow[64-67\%]{P, \ (CH_3CO)_2O, \ HI} C_6H_5CH_2\underset{NH_2}{CHCO_2H}$$

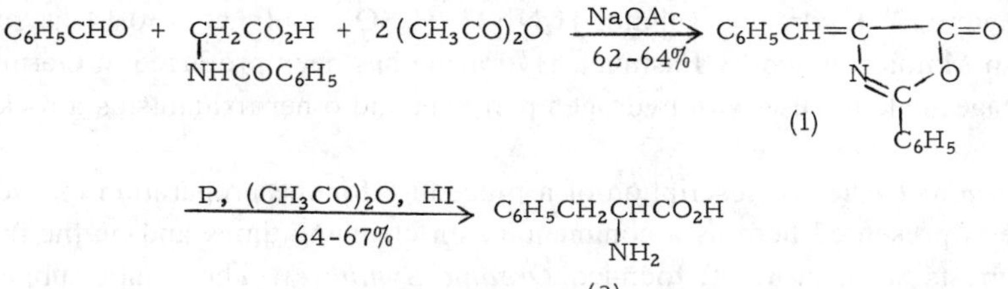

(2)

initially almost solid mass liquefies and turns bright yellow. It is heated on the steam bath for 2 hrs., cooled slightly, diluted with ethanol, and let stand for complete separation of deep yellow crystals of the azlactone (1). A mixture of (1), red phosphorus, and acetic anhydride is stirred during addition of 50% hydriodic acid. After refluxing for 3–4 hrs., the solution is cooled, filtered from phosphorus, and evaporated to dryness at reduced pressure. The residue is taken up in water, and the solution again evaporated to dryness (to eliminate HI). A solution of the residue in water is extracted repeatedly with ether and the extracts discarded. The water solution of phenylalanine is clarified with Norit, heated to expel the ether and then to the boiling point, and neutralized with ammonia to Congo red (the isoelectric point is 5.48). Phenylalanine separates in colorless plates.

Previously limited to condensation with aromatic aldehydes, the reaction of hippuric acid has been applied to aliphatic aldehydes by use of tetrahydrofurane as solvent and substitution of lead acetate for sodium acetate.[3]

[1] A. W. Ingersoll and S. H. Babcock, *Org. Syn., Coll. Vol.*, **2**, 328 (1943)
[2] H. B. Gillespie and H. R. Snyder, *ibid.*, **2**, 489 (1943)
[3] E. Baltazzi and R. Robinson, *Chem. Ind.*, 191 (1954)

Hyamine 1622 and Hyamine 3500. These Rohm and Haas quaternary ammonium germicides are useful in laboratory deodorization in the form of a spray of a dilute aqueous solution:

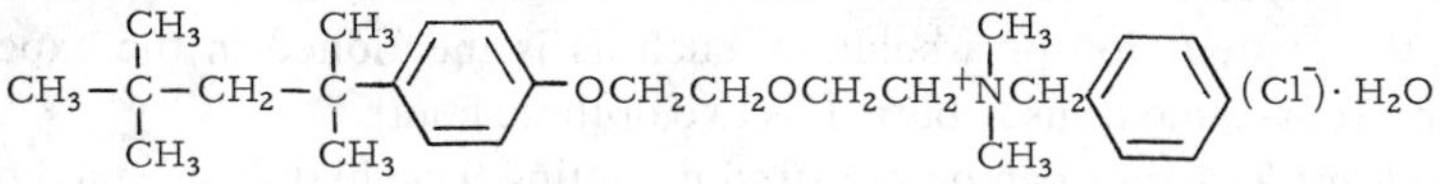

Hyamine 1622 monohydrate, Mol. wt. 466.09

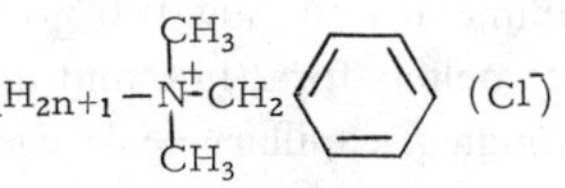

Hyamine 3500, mol. wt. of average ingredient: 358

The reagents are nontoxic but may be irritating if breathed, because of surface activity. A 0.1% aqueous solution of either one should be effective for most deodorization purposes. A stronger solution may be required for a mercaptan odor, a 5% solution for washing up spilled mercaptan.

Hydrazine. Mol. wt. 32.05, b.p. 113.5, sp. gr. 1.01. Suppliers: Olin Mathieson: anhydrous and aqueous solutions; Matheson, Coleman, and Bell: 85% "hydrazine hydrate," the concentration is expressed in terms of $H_2NNH_2{\cdot}H_2O$ (mol. wt. 50.07) even though this entity does not exist as a pure liquid; 100% "hydrazine hydrate" contains 64% N_2H_4. Eastman's hydrazine, 64% in water, is the same as 85% "hydrazine hydrate." Hydrazine sulfate, $H_2NNH_2{\cdot}H_2SO_4$, mol. wt. 130.14, is supplied by Olin Mathieson and by Eastman. Hydrazine has been produced in Germany on a tonnage scale for use with hydrogen peroxide and other oxidants as a rocket and jet fuel.

Historical note. A description of a procedure for the preparation of hydrazine sulfate is presented here as a commentary on changing times and on the fortitude of chemists who, in 1921, founded *Organic Syntheses*. The senior submitter of the procedure[1] was Roger Adams, who had edited Volume 1; his co-worker was B. K. Brown. The senior checker, James B. Conant, was editor of Volume 2 of 1922 in which the procedure was first published; his co-worker was W. L. Hanaway. The basic reactions are simple enough: oxidation of ammonia with hypochlorite and isolation of the product as the sulfate:

$$2\,NH_3 \xrightarrow{NaOCl} H_2NNH_2 \xrightarrow{H_2SO_4} H_2NNH_2{\cdot}H_2SO_4$$

A solution of 300 g. (7.5 moles) of sodium hydroxide in 1.5 l. of water is prepared in a 5-l flask, 1.5 kg. of ice is added, and chlorine is passed in with ice cooling to a gain in weight representing 3 moles. The hypochlorite solution is placed in a 36-cm. evaporating dish together with 1.5 l. of aqueous ammonia, 900 ml. of water, and 375 ml. of 10% gelatin solution as "viscolizer" (increases viscosity) and the solution is boiled down to one third the original volume. After thorough cooling, "the solution is filtered by suction, first through two layers of toweling and then through one thickness of ordinary filter paper over cloth in order to remove finely divided solid impurities. The solution is then placed in a precipitating jar...." (for conversion to the sulfate). The crude hydrazine sulfate, eventually obtained in yield of 34–37%, could be recrystallized to afford 19 g. (12%) of pure material. That the procedure did not at the time seem to present any difficulty is indicated by the following optimistic note: "It is possible for one person, simultaneously evaporating six dishes of the hydrazine mixture, to turn out from 20 to 25 runs in nine hours. The time for the evaporation of a solution, such as is mentioned in the experimental part, with a four-flame Bunsen burner, is two or three hours."

Anhydrous hydrazine can be prepared by letting 95% hydrazine stand overnight with 20% by weight of potassium hydroxide, filtering the resulting gel, and distilling the filtrate under anhydrous conditions.[2]

Day and Whiting[2a] recommend refluxing 100% hydrazine hydrate with an equal weight of sodium hydroxide pellets for 2 hrs. and then distillation in a slow stream of nitrogen introduced through a capillary leak. *Distillation in air can lead to an explosion.*

Reactions. *See also* Diethyl azidodicarboxylate, Diimide, Girard's reagent, Triethyl N-tricarboxylate.

Hydrazine derivatives. N,N′-Dimethylhydrazine can be made by methylation

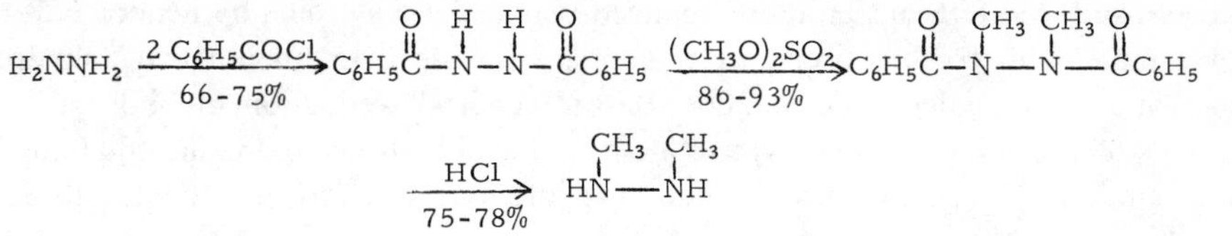

$$H_2NNH_2 \xrightarrow[66-75\%]{2\ C_6H_5COCl} C_6H_5\overset{O}{\overset{\|}{C}}-\overset{H}{\overset{|}{N}}-\overset{H}{\overset{|}{N}}-\overset{O}{\overset{\|}{C}}C_6H_5 \xrightarrow[86-93\%]{(CH_3O)_2SO_2} C_6H_5\overset{O}{\overset{\|}{C}}-\overset{CH_3}{\overset{|}{N}}-\overset{CH_3}{\overset{|}{N}}-\overset{O}{\overset{\|}{C}}C_6H_5$$

$$\xrightarrow[75-78\%]{HCl} HN\overset{CH_3}{\overset{|}{}}-\overset{CH_3}{\overset{|}{N}}H$$

of the N,N'-dibenzoyl derivative and hydrolysis.[3] Thiele[4] discovered a route to monomethylhydrazine which is recorded in *Organic Syntheses.*[5] Benzalazine, obtained as yellow needles by condensation of benzaldehyde with hydrazine,

$$2\ C_6H_5CHO \xrightarrow[91-94\%]{H_2NNH_2} C_6H_5CH=N-N=CHC_6H_5 \xrightarrow{(CH_3O)_2SO_2}$$

Benzalazine

$$\underset{\overset{|}{CH_3}}{\overset{\overset{\bar{O}SO_2OH}{\overset{+}{}}}{C_6H_5CH=N}}-N=CHC_6H_5 \xrightarrow[51-54\%]{3\ H_2O} 2\ C_6H_5CHO + \overset{H}{\overset{|}{C}}H_3N\overset{}{N}H_2 \cdot H_2SO_4 + CH_3OH$$

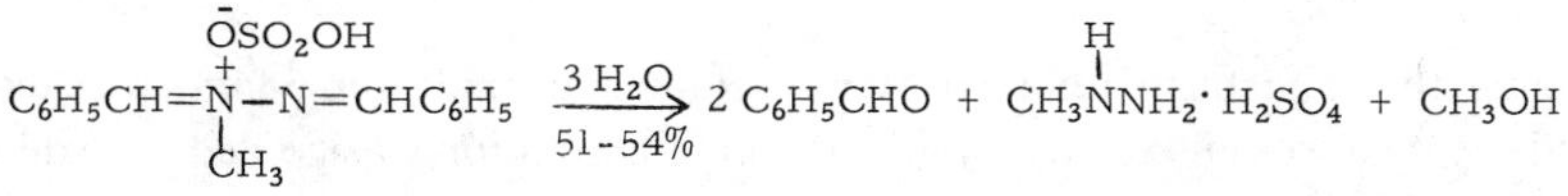

reacts with one equivalent of dimethyl sulfate in refluxing benzene to form a quaternary salt, which separates as a solid. Water decomposes the salt to benzaldehyde, methylhydrazine sulfate, and methanol. Benzaldehyde and benzene are removed by distillation, and methylhydrazine sulfate is obtained by evaporation of the aqueous solution.

Phenylurea, heated with 42% aqueous hydrazine solution for 12 hrs. on the steam bath, gives 4-phenylsemicarbazide in modest yield.[6]

$$C_6H_5NHCONH_2 + H_2NNH_2 \xrightarrow[37-40\%]{} C_6H_5NHCONHNH_2$$

Wolff-Kishner reduction. The literature on the classical Wolff-Kishner procedure has been reviewed by Todd.[7] The modified procedure of Huang-Minlon,[8] which permits reduction on a large scale at atmospheric pressure with efficiency and economy, is illustrated for the case of β-(p-phenoxybenzoyl)-propionic acid. (The story of this discovery is on record.[9]) A mixture of 500 g. (1.85 moles) of the keto acid, 350 g. of potassium hydroxide (3.2 equiv.), 250 ml. of 85% hydrazine (3.6

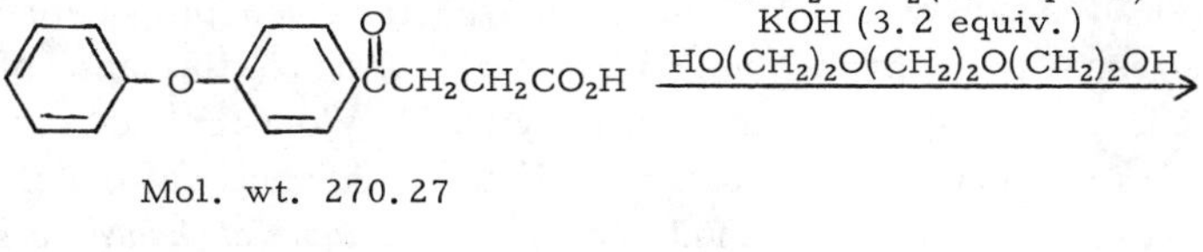

Mol. wt. 270.27

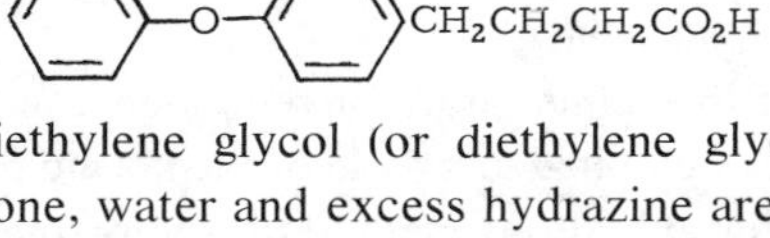

equiv.), and 2.5 l. of triethylene glycol (or diethylene glycol) is refluxed for $1\frac{1}{2}$ hrs. to form the hydrazone, water and excess hydrazine are removed by a take-off condenser until the temperature rises to a point favorable for the decomposition of the hydrazone (195–200°), and refluxing is continued for 4 hrs. more. The cooled solution is diluted with 2.5 l. of water and poured slowly into 1.5 l. of 6 N hydrochloric acid, and the light cream-colored solid is dried. The average yield of material m.p. 64–66° is 451 g. (95%). Pure γ-(p-phenoxyphenyl)-butyric acid melts at 71–72°.

Huang-Minlon showed[10] that steroid keto groups at C_3, C_7, C_{12}, C_{17}, and C_{20} can be reduced satisfactorily by his procedure but that a C_{11} keto group remains unattacked. Moffett and Hunter[11] reduced two 11-keto steroids by heating 0.5–1 g. of sample with freshly prepared sodium methoxide in methanol and anhydrous methanol in a sealed tube at 200°. Barton reports[12] reduction of both carbonyl groups of 7,11-diketolanostanyl acetate by using anhydrous hydrazine and boosting the temperature. A solution of sodium in diethylene glycol was heated to 180°,

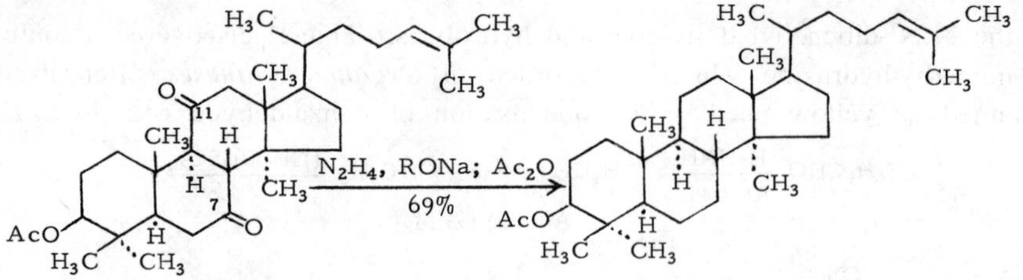

hydrazine that had been refluxed over sodium hydroxide for 3 hrs. was distilled in until the mixture refluxed freely at 180°, and the solution was cooled. The diketone was then added, the solution was refluxed overnight, hydrazine was distilled to raise the liquid temperature to 210°, and the solution was refluxed at this temperature for 24 hrs.

Nagata and Itazaki[13] developed a procedure for the reduction of the highly hindered 11-ketosteroids not requiring anhydrous hydrazine. The ketone (1 m.) is heated with 66 moles of hydrazine hydrate, 8 moles of hydrazine dihydrochloride, and 150 moles of triethylene glycol at 130° for 2.5 hrs. Potassium hydroxide (22 m.) is added as pellets, the temperature is gradually raised to 210° (distillation of hydrazine–water) and held for 2.5 hrs.; yield 90%.

One step in the synthesis of morphine by Gates and Tschudi[14] involved elimination of a keto group from an intermediate having two phenolic methoxyl groups, and under usual Huang-Minlon conditions extensive demethylation occurred. A high yield was achieved by operating at 150–155° for 1 hr.

For the preparation of hydrazones, Schönberg[15] recommends that equivalent amounts of the ketone and hydrazine be refluxed in *n*-butanol (b.p. 117.7°) for 2 hrs. Use of ethanol requires a longer reflux period and use of ethylene or propylene glycol may lead to side reactions. Thus on applying the glycol procedure for conversion of fluorenone to the hydrazone, Baltzly *et al.*[16] observed formation of considerable fluorene (interference by an unusually facile Wolff-Kishner reaction). On refluxing the reactants in *n*-butanol for 4 hrs., the hydrazone was obtained in 67% yield. However, a superior method for the ***preparation of hydrazones*** involves reaction of the ketone with N,N-dimethylhydrazine (*which see*) followed by an exchange reaction with hydrazine.

Henbest[17] found that the temperature of reduction can be lowered substantially and certain side reactions eliminated by use of the strong base potassium *t*-butoxide in boiling toluene. By this anhydrous method the hydrazone (1) affords the reduction

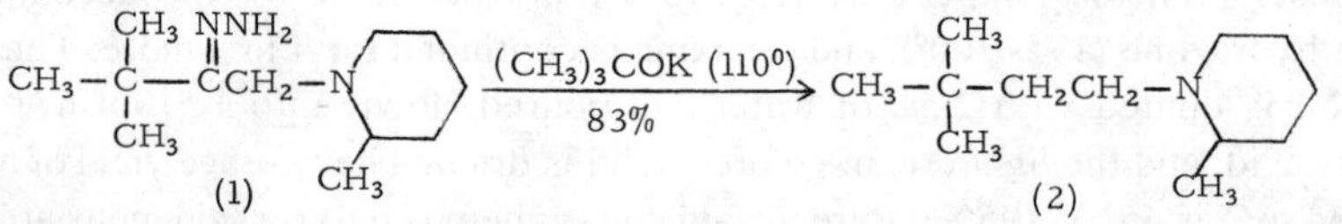

product (2) in high yield, whereas under usual Wolff-Kishner conditions about half of the material suffers β-elimination to give 3,3-dimethylbutene-1. Δ^4-Cholestene-

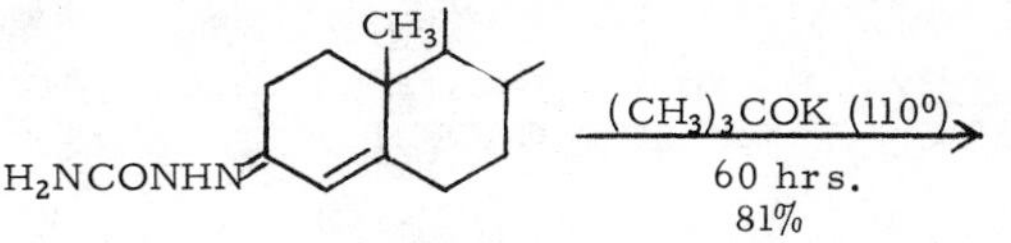

CH_3
$H_2NCONHN$
$(CH_3)_3COK$ (110°)
60 hrs.
81%
CH_3

3-one semicarbazone on being refluxed with potassium *t*-butoxide in toluene until gas evolution ceases (gas buret) reacts only slowly but affords almost pure Δ^4-cholestene in high yield. With the hydrazone the reduction is over in 6 hrs. but the yield is only 65%.

A modified Huang-Minlon reduction was found advantageous in the conversion of methyl cholate to lithocholic acid without purification of intermediates.[18] The 3-hydroxyl group (equatorial) is selectively succinoylated, the hydroxyl functions

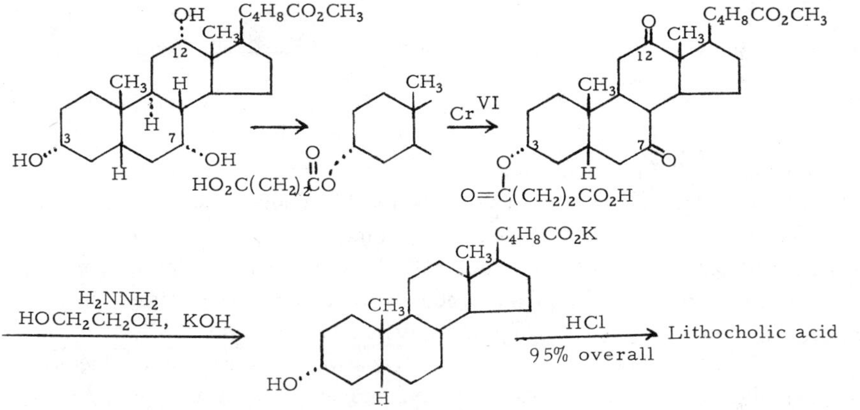

OH
$C_4H_8CO_2CH_3$
CH_3
12
CH_3
H
H
3
7
HO
OH
H
$HO_2C(CH_2)_2CO$
CH_3
Cr VI
$C_4H_8CO_2CH_3$
CH_3
12
CH_3
3
7
H
$O{=}C(CH_2)_2CO_2H$
$C_4H_8CO_2K$
CH_3
H_2NNH_2
$HOCH_2CH_2OH$, KOH
CH_3
HO
H
HCl
95% overall
Lithocholic acid

at C_3 and C_{12} are oxidized, and methyl 3α-succinoxy-7,12-diketocholanate is heated with 85% hydrazine and ethylene glycol at 100° for 1 hr. The resulting clear solution is cooled and potassium hydroxide pellets are added in portions at room temperature to bind the water present. The mixture is heated at 200° for 2 hrs. and then cooled, and the potassium salt that separates is collected by centrifugation. Acidification of a solution of the salt in water gives lithocholic acid in yield of 95% from methyl cholate. Reduction of the diketone under usual Huang-Minlon conditions gives a very impure product in low yield because the diketone is sensitive to the strongly basic medium.

For a procedure for Wolff-Kishner reduction at room temperature, *see* Dimethyl sulfoxide, solvent effects.

A useful modification of the Huang-Minlon procedure was introduced by Gardner[19] in the last step of a neat synthesis of pimelic acid. Furfural is condensed with malonic acid in pyridine (trace of piperidine) to form furylacrylic acid (2), which on treatment with ethanol and hydrogen chloride afforded diethyl γ-ketopimelate (6). This reaction is reported in a patent;[20] it is pictured here as involving 1,4-addition of ethanol to give (3), double bond migration to (4) and then to (5), and ring cleavage to give (6). In reducing the keto diester (6) by the Huang-Minlon procedure in either diethylene glycol or ethylene glycol, Gardner encountered the difficulty that on continuous ether extraction of the acidified reaction mixture the glycol solvents are partially extracted into the ether phase and so turned to the use of triethanolamine as solvent. Thus a mixture of 53.8 g. (0.234 mole) of (6), 230 ml. of 95% triethanolamine, 0.91 mole of 85% KOH, and 37 ml. (0.65 mole) of 85% hydrazine hydrate

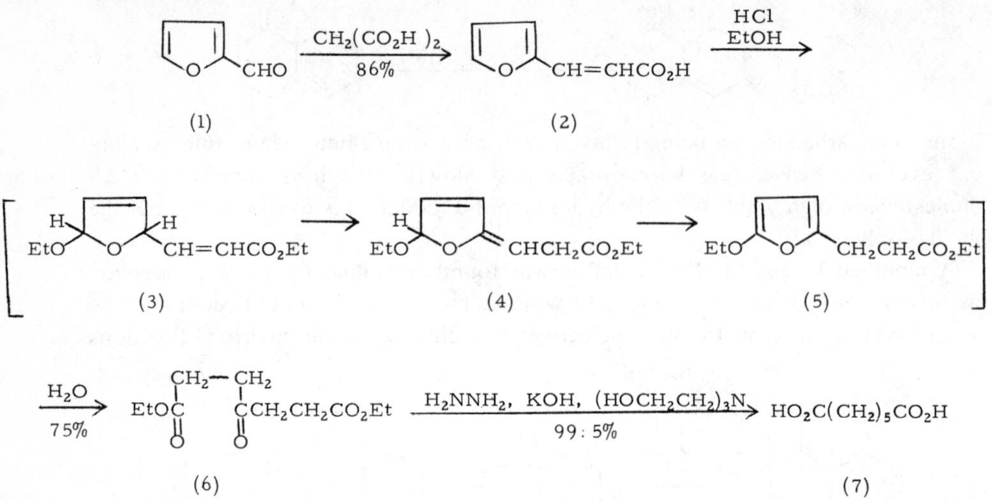

was refluxed for 2.5 hrs., the condenser water was drained until the temperature rose to 195°, and refluxing was continued for 2 hrs. The cooled mixture was treated with 450 ml. of concd. hydrochloric acid and submitted to continuous ether extraction (suspended salts remained suspended entirely in the aqueous phase and offered no problem). One crystallization from concd. hydrochloric acid (several crops) afforded pimelic acid (m.p. 103.5–105°) in practically quantitative yield.

This triethanolamine procedure was used also in a synthesis of docosanedioic acid starting with the condensation of cyclohexanone with morpholine to form 1-morpholino-1-cyclohexene[21] (1), and acylation of this enamine with sebacoyl chloride

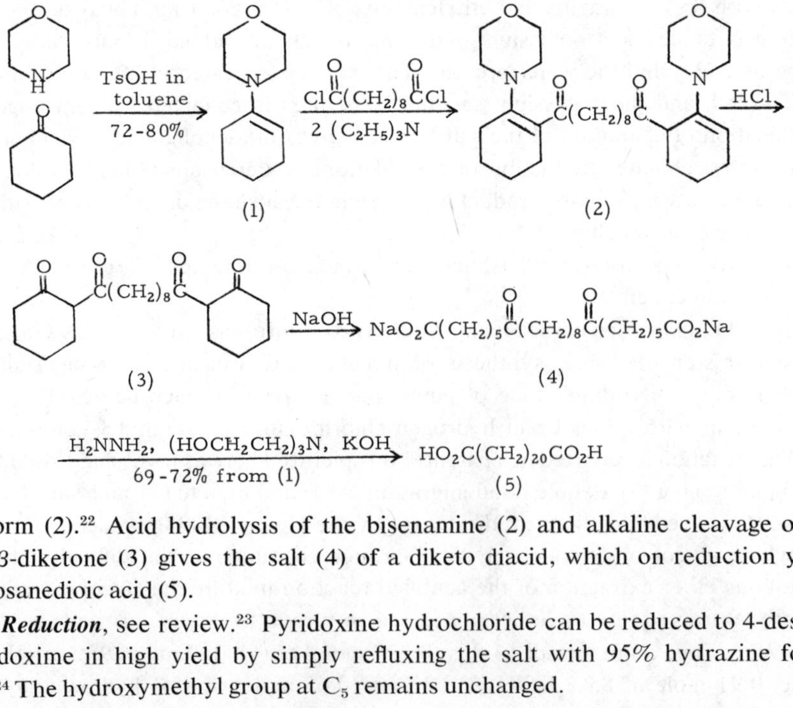

to form (2).[22] Acid hydrolysis of the bisenamine (2) and alkaline cleavage of the bis β-diketone (3) gives the salt (4) of a diketo diacid, which on reduction yields docosanedioic acid (5).

Reduction, see review.[23] Pyridoxine hydrochloride can be reduced to 4-desoxypyridoxime in high yield by simply refluxing the salt with 95% hydrazine for 18 hrs.[24] The hydroxymethyl group at C_5 remains unchanged.

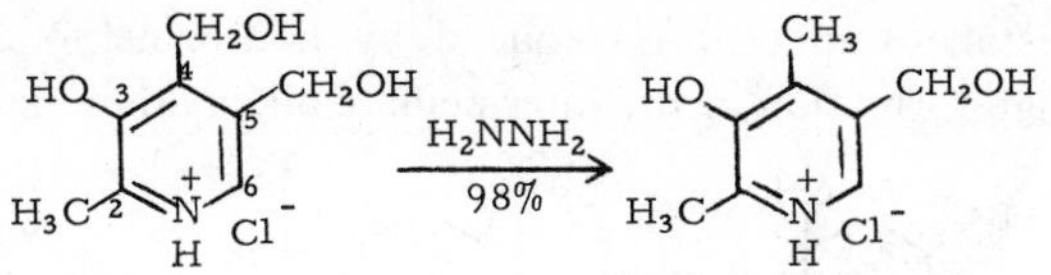

Hydrazine, with or without alkali, has been found to desulfurize cyclic and acyclic ethylenethioketals.[25] The substrate (1 part of weight) in 8–20 parts by volume of

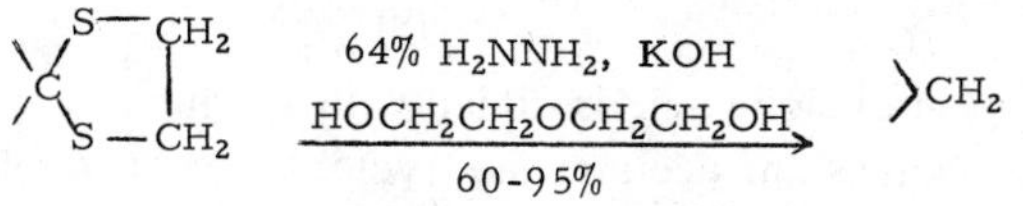

diethylene glycol or triethylene glycol with 3–5 parts by volume of 64% hydrazine ("hydrazine hydrate") and 1.25–2.5 parts by weight of potassium hydroxide, with provision for collecting the evolved gas over water, is heated until gas evolution starts (90–135°), and the temperature is gradually increased to 155–190° to maintain steady gas evolution. Times vary from 30 min. to 3 hrs. Although alkali is not necessary it lowers considerably the temperature of effective reaction.

A novel reaction discovered by Wharton[26] is the reduction of α,β-epoxy ketones by hydrazine to allylic alcohols. When a suspension of the steroid (1) in excess 64%

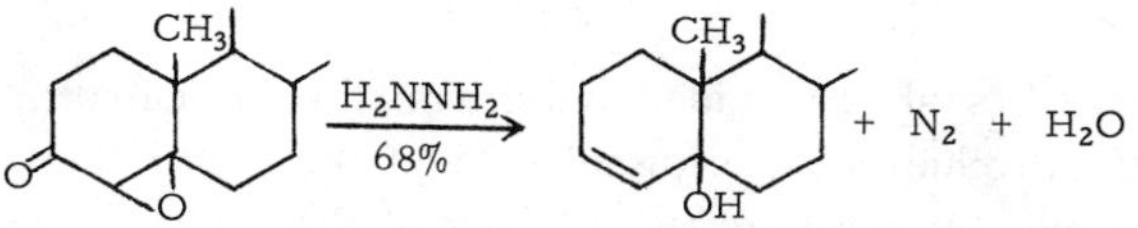

(1) (2)

hydrazine is heated briefly on the steam bath and then refluxed, nitrogen is evolved and the allylic alcohol (2) is produced in good yield. The steroid α,β-oxidoketone (3) reacts similarly but in lower yield.[27]

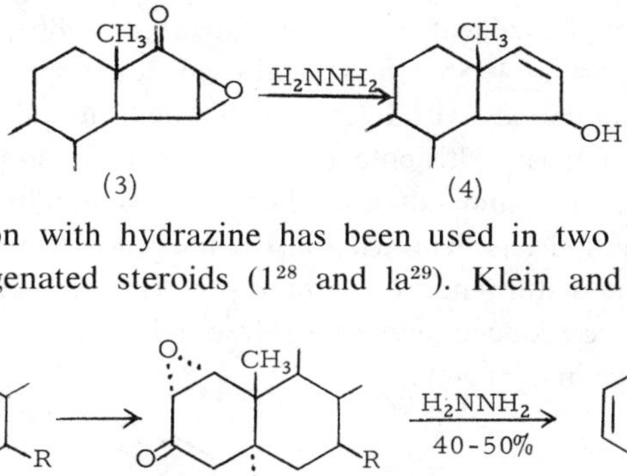

(3) (4)

Wharton reduction with hydrazine has been used in two laboratories for the synthesis of 1-oxygenated steroids (1[28] and la[29]). Klein and Ohloff[30] found the

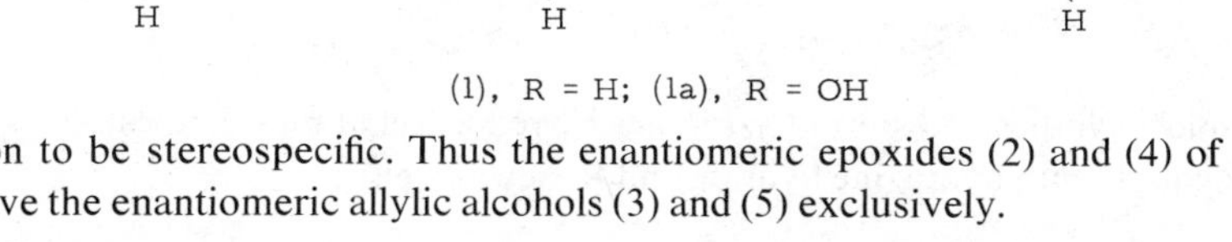

(1), R = H; (la), R = OH

reaction to be stereospecific. Thus the enantiomeric epoxides (2) and (4) of piperitone give the enantiomeric allylic alcohols (3) and (5) exclusively.

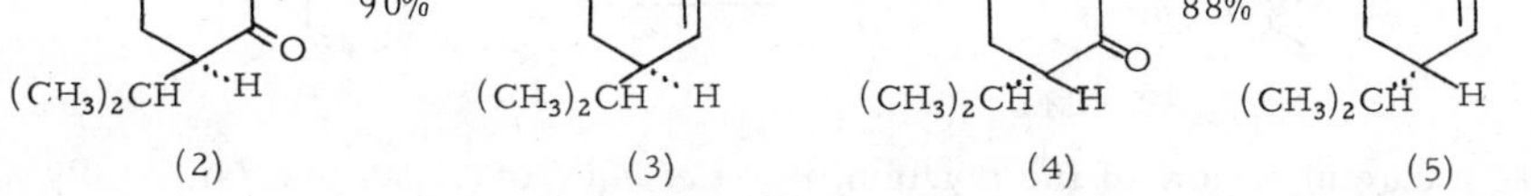

(2) (3) (4) (5)

2α-Bromocholestane-3-one (6) is reduced by hydrazine to Δ^2-cholestene (9), presumably through formation of the intermediate alkenyldiimine (8).[30a] A solution

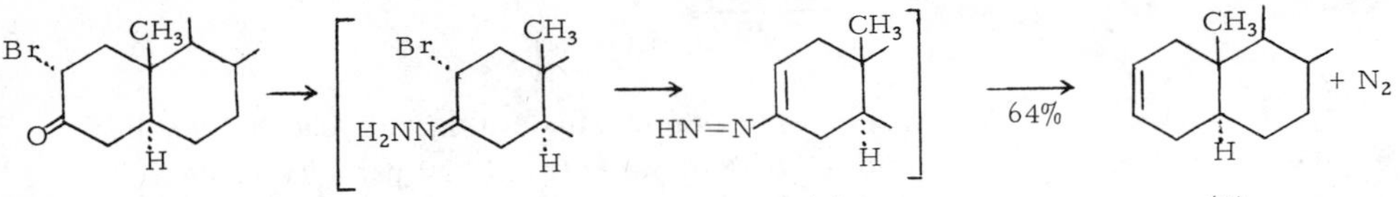

(6) (7) (8) (9)

of the bromoketone in cyclohexene is added with stirring to a refluxing mixture of hydrazine hydrate, potassium acetate and cyclohexene. Cyclohexene is used as solvent to prevent reduction of the product by diimide, should any be formed.

Hydrazine–metal catalyst. Pietra[31] found that in the presence of palladized charcoal hydrazine reduces aromatic nitro compounds to amines in high yield. Thus a suspension of 30 g. of 2-nitrofluorene in 250 ml. of 95% ethanol is warmed with stirring to 50°, 0.1 g. of 10% palladium on charcoal (moistened with alcohol) is added, and 15 ml. of 64% hydrazine is added by drops during 30 min.[32] Another

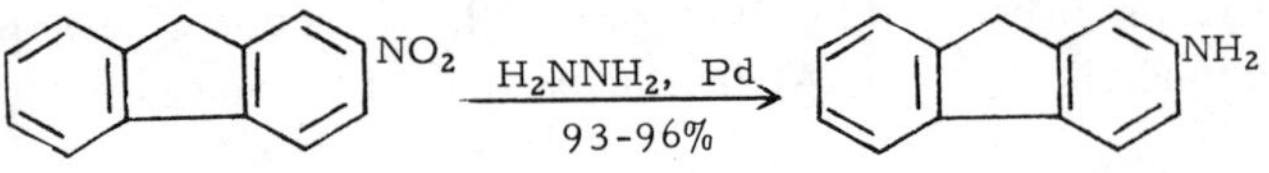

0.1 g. of moistened catalyst is added and the mixture is heated to gentle refluxing. After 1 hr. the nitrofluorene has dissolved and the supernatant liquor is almost colorless. After suitable workup, the product is obtained as colorless crystals of high purity.

The method gives excellent results as applied to the preparation of toluidines, aminodiphenyls, phenylenediamines, aminophenols, *p*-aminobenzoic acid,[31, 33] as well as to the reduction of polycyclic aromatic nitro compounds.[34] When applied to azobenzene or azoxybenzene it gives hydrazobenzene in 80–90% yield.[33]

Prior to Pietra's work of 1955, Furst[35] had used Raney nickel as catalyst for the reduction of nitro compounds with hydrazine. Tarbell[36] found this method successful as applied to a nitrostyryltropolone where other methods were unpromising (ferrous hydroxide, ammonium sulfide, sodium hydrosulfite, hydrogen and platinum in dioxane). The method has been used in the fluorene series for the reduction of nitro groups without disturbance of carbonyl or trifluoroacetylamino functions.[37] Halonitrobenzenes are reduced smoothly to the haloanilines.[38] *o*-Nitrobenzonitrile affords anthranilamide in high yield.[39]

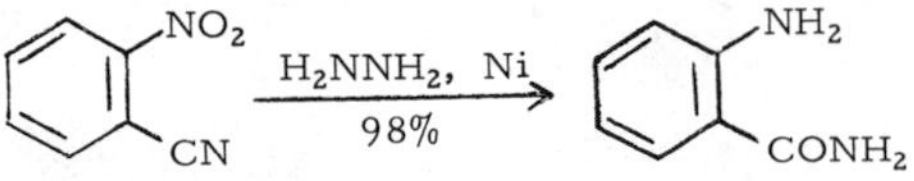

Ponsold[40] synthesized steroid aziridines by reduction of mesylate esters of vicinal azido alcohols with hydrazine hydrate and Raney nickel.

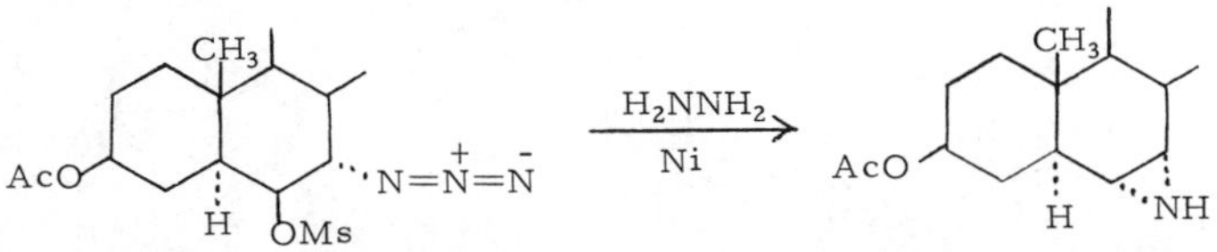

In a careful review of the reaction, Furst *et al.*[41] concluded that Ni, Pd, Pt, and

Ru serve equally well and that this method of reduction is often superior to direct hydrogenation and has the advantage of not requiring special apparatus.

Hjelte[42] showed that α,β-unsaturated acids can be reduced by hydrazine and Raney nickel in an alkaline medium. Thus a solution of cinnamic acid (*trans*) in slightly more than 1 equivalent of aqueous sodium hydroxide was treated with a small amount of W-6 Raney nickel, and 2 moles of hydrazine was added with agitation at 90° during 2 hrs.

$$C_6H_5CH{=}CHCO_2H + 2\ H_2NNH_2 \xrightarrow[85\%]{Ni} C_6H_5CH_2CH_2CO_2H$$

As noted above, Leggetter and Brown[38] found that halonitrobenzenes are reduced smoothly to haloanilines by hydrazine and Raney nickel; with palladium as the catalyst, considerable dehalogenation occurred. A similar observation by Mosby[43] prompted a study[44] which showed that the H_2NNH_2–Pd technique (in C_2H_5OH or CH_3OCH_2OH) provides a general method for the dehalogenation of aromatic halogen compounds (nitro groups if present are reduced).

Synthesis of heterocycles. 3,5-Dimethylpyrazole is prepared from acetylacetone and hydrazine sulfate in aqueous alkali (the reaction with hydrazine hydrate is sometimes violent).[45]

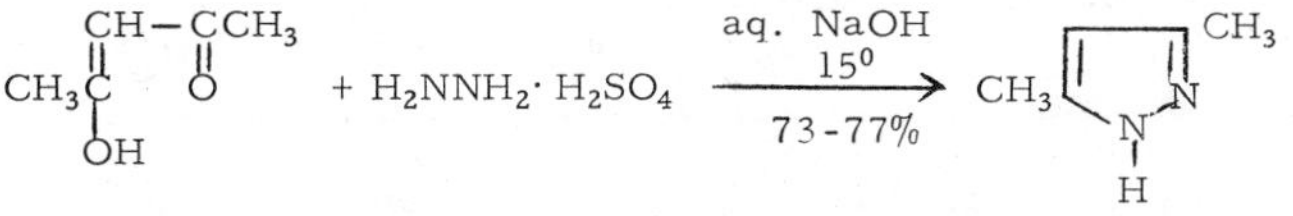
CH—CCH₃
CH₃C
O
OH
+ H₂NNH₂· H₂SO₄
aq. NaOH
15⁰
73-77%
CH₃
CH₃
N
N
H

5-Amino-2,3-dihydro-1,4-phthalazdione (3), known as luminol because oxidation of the substance is attended with a striking chemiluminescence, can be prepared in about 25 min. as follows.[46] A mixture of 1 g. of 3-nitrophthalic acid and 2 ml. of an 8% aqueous solution of hydrazine in a 20 × 150-mm. test tube is heated over a free flame until the solid is dissolved, and 3 ml. of triethylene glycol is added. The solution

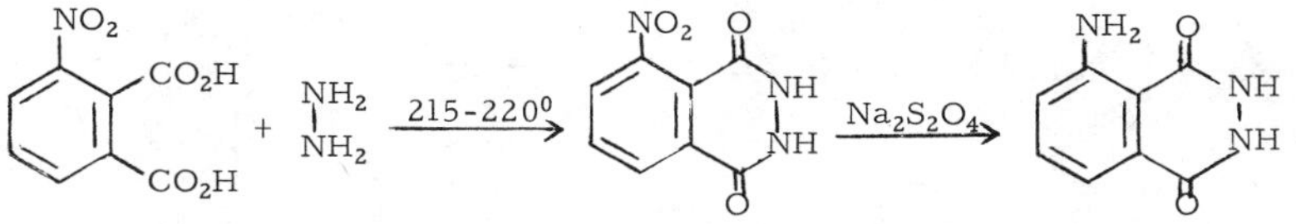
NO₂
CO₂H
CO₂H
+
NH₂
NH₂
215-220⁰
NO₂
O
NH
NH
O
Na₂S₂O₄
NH₂
O
NH
NH
O

(1) (2) (3)

is boiled vigorously under the aspirator tube to distil excess water and then heated at 215–220° for 2 min. to close the heterocyclic ring. The yellow nitro compound (2) separates on cooling and is reduced with sodium hydrosulfite to luminol (3). A dilute solution of (3) in alkali poured into a funnel along with an aqueous solution of potassium ferricyanide and hydrogen peroxide in a dark place produces a beautiful demonstration of chemiluminescence.

For a synthesis of indazole, *see* Ethyl formate.

Other reactions. The preparation of an acid hydrazide from an acid is illustrated by the reaction of *p*-toluenesulfonyl chloride with hydrazine hydrate in tetrahydro-

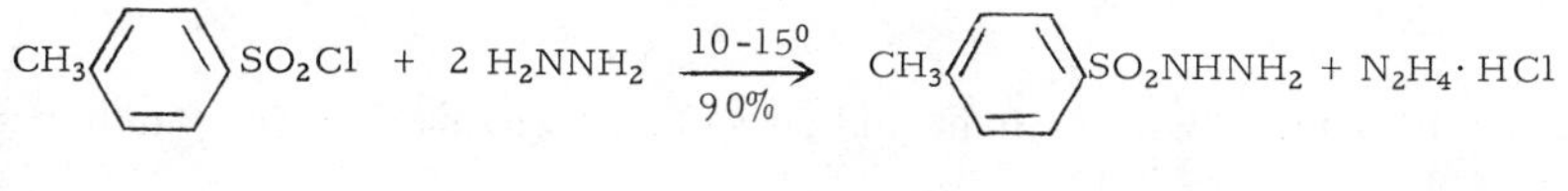
CH₃
SO₂Cl + 2 H₂NNH₂
10-15⁰
90%
CH₃
SO₂NHNH₂ + N₂H₄· HCl

furane.[47] The mixture separates into two layers; the upper THF layer is washed with sodium chloride solution, dried further, filtered, and the product is precipitated with petroleum ether. Use of an ester for hydrazide formation is illustrated by a synthesis of putrescine dihydrochloride.[48]

$$(CH_2)_4(CO_2C_2H_5)_2 \xrightarrow{H_2NNH_2} (CH_2)_4(CONHNH_2)_2 \xrightarrow{HNO_2}$$

$$(CH_2)_4(CON_3)_2 \xrightarrow{87^0} (CH_2)_4(N=C=O)_2 \xrightarrow[\substack{63-67\% \\ \text{overall}}]{H_2O-HCl} (CH_2)_4(\overset{+}{N}H_3\overset{-}{Cl})_2$$

The reaction of cyclohexanone with sodium cyanide and hydrazine sulfate in aqueous solution involves production *in situ* of both hydrogen cyanide and hydrazine.[49]

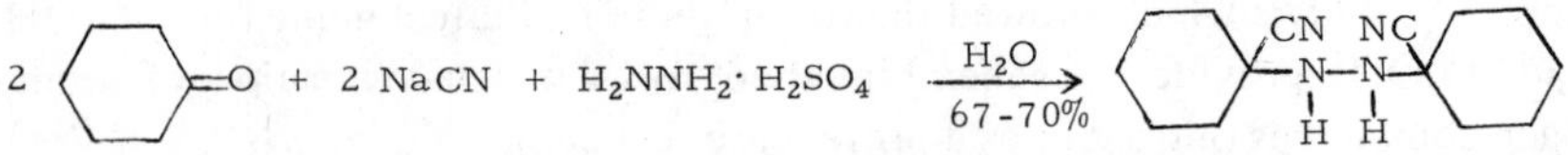

An example of a general synthesis of acetylenes is the conversion of benzil via the dihydrazone (and the unisolated diazide) to diphenylacetylene.[50]

$$C_6H_5COCOC_6H_5 + 2 H_2NNH_2 \xrightarrow[83-87\%]{n\text{-}PrOH} \underset{\substack{\| \\ H_2NN}}{C_6H_5C} - \underset{\substack{\| \\ NNH_2}}{CC_6H_5} \xrightarrow{HgO}$$

$$\left[\begin{array}{c} C_6H_5C - CC_6H_5 \\ \overset{+}{N} \quad \overset{+}{N} \\ \overset{-}{N} \quad \overset{-}{N} \end{array} \right] \xrightarrow[67-73\%]{-2\ N_2} C_6H_5C\equiv CC_6H_5$$

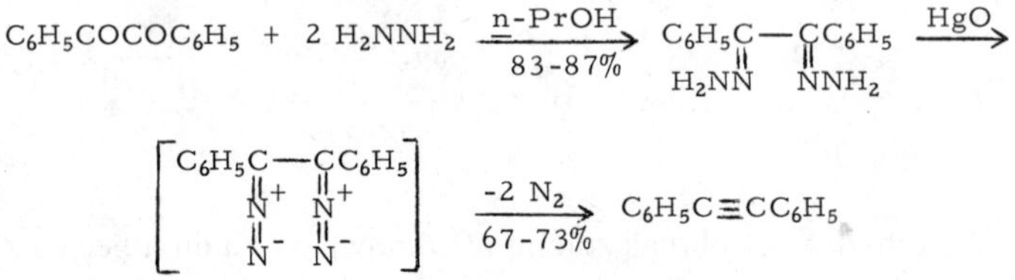

In peptide synthesis involving use of the N-protective phthaloyl group, hydrazine is a useful reagent for removal of the protective group.[51] The protected peptide is

$$+ \ H_2N\overset{R}{\underset{|}{C}}HCONH\overset{R'}{\underset{|}{C}}HCO_2H$$

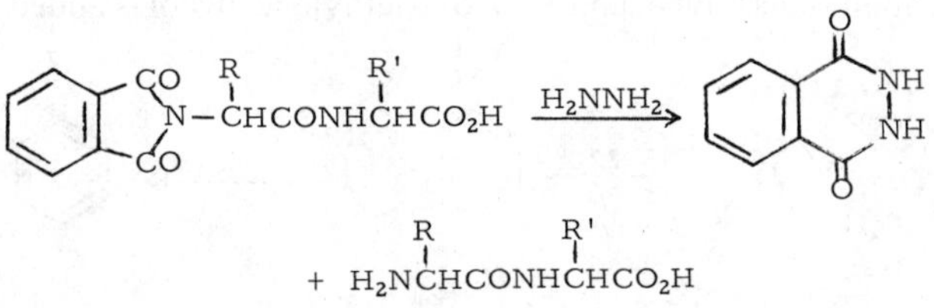

either boiled with hydrazine in ethanol for 2 hrs. or allowed to stand for 1–2 days in an aqueous or alcoholic solution of hydrazine.[52]

A similar hydrazine cleavage is employed in a synthesis of *t*-butylamine, which starts with the condensation of *t*-butanol with urea to form *t*-butylurea.[53] When this

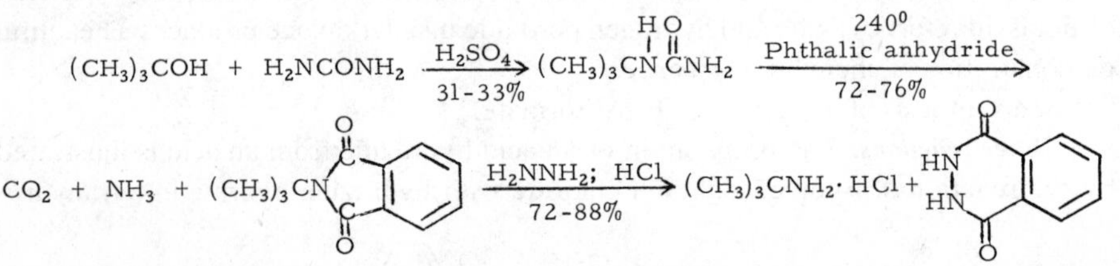

substance is heated with phthalic anhydride, carbon dioxide and ammonia are liberated, and the product is *t*-butylphthalimide. This is cleaved with hydrazine, hydrochloric acid is added, phthalhydrazide is removed by filtration, and the filtrate is worked up for recovery of *t*-butylamine hydrochloride.

Cyclohexane-1,3-dione (largely enolic) reacts with excess hydrazine in ethanol in the presence of air to give the pyridazine (3), which crystallizes in light yellow needles.[54] When the reaction was run in the absence of oxygen, no (3) was formed,

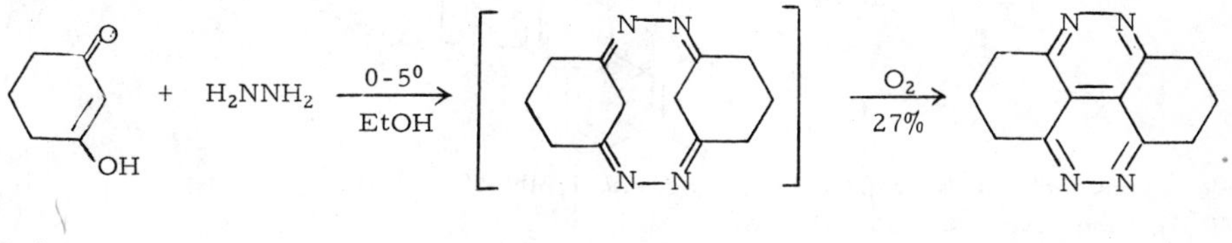

H₂NNH₂
0-5°
EtOH
O₂
27%

(1) (2) (3)

and NMR characterization indicated the intermediate as the diazine (2). This substance is very unstable because the hydrogens on the internal methylene groups are no greater than 1.2 Å apart. There is thus considerable driving force for air oxidation to the aromatic pyridazine (3).

Reaction with phenacyl bromide.[55] Hydrazine reacts with phenacyl bromide (1) in ethanol at 60° according to the following equation to give 2-phenylglyoxal monohydrazone (2):

$$\underset{(1)}{C_6H_5\overset{O}{\overset{\|}{C}}CH_2Br} + 3\ H_2NNH_2 \xrightarrow[60\%]{} \underset{(2)}{C_6H_5\overset{O}{\overset{\|}{C}}CH{=}NNH_2} + N_2H_4Br + 2\ NH_3$$

Isolation of phenacylhydrazine (3) when the reaction was carried out below 5° suggested the following mechanisms:

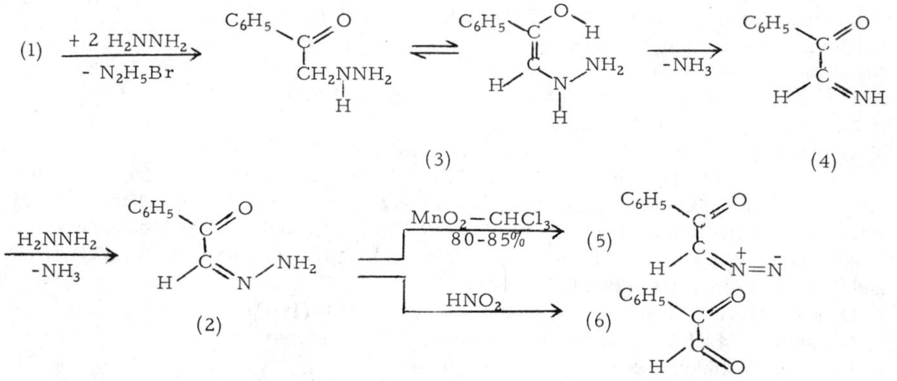

(1)
+ 2 H₂NNH₂
- N₂H₅Br
(3)
-NH₃
(4)
H₂NNH₂
-NH₃
(2)
MnO₂—CHCl₃
80-85%
(5)
HNO₂
(6)

The product (2) on dehydrogenation with manganese dioxide hydrate affords the diazomethyl ketone (5); on treatment with nitrous acid it affords phenylglyoxal (6).

Cleavage of —NCO₂CH₃. Alder and Niklas[56] found that a carbomethoxy group attached to nitrogen in a heterocycle can be eliminated by refluxing with hydrazine hydrate. A University of Amsterdam group[57] made use of this reaction in a total synthesis of 11,12-diazasteroids illustrated in part as follows:

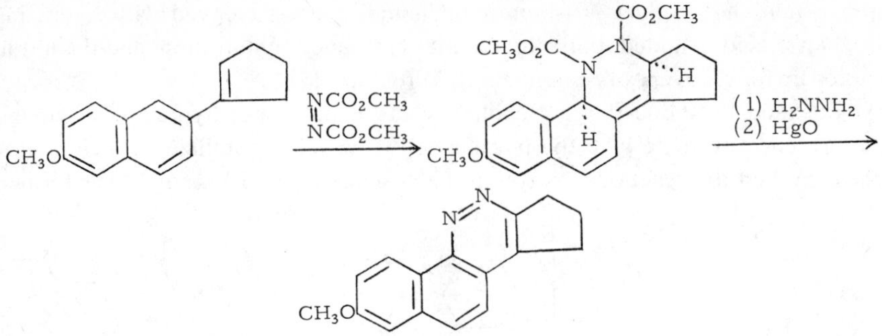

[1]R. Adams and B. K. Brown, *Org. Syn., Coll. Vol.*, **1**, 309 (1941)
[2]R. G. Taborsky, *J. Org.*, **26**, 596 (1961)
[2a]A. C. Day and M. C. Whiting, procedure submitted to *Org. Syn.*
[3]H. H. Hatt, *Org. Syn., Coll. Vol.*, **2**, 208 (1943)
[4]J. Thiele, *Ann.*, **376**, 244 (1910)
[5]H. H. Hatt, *Org. Syn., Coll. Vol.*, **2**, 395 (1943)
[6]A. S. Wheeler, *ibid.*, **1**, 450 (1941)
[7]D. Todd, *Organic Reactions*, **4**, 378 (1948)
[8]Huang-Minlon, *Am. Soc.*, **68**, 2487 (1946)
[9]L. F. Fieser and M. Fieser, "Topics in Organic Chemistry," p. 258, D.C. Heath (1963)
[10]Huang-Minlon, *Am. Soc.*, **71**, 3301 (1949)
[11]R. B. Moffett and J. H. Hunter, *Am. Soc.*, **73**, 1973 (1951)
[12]D. H. R. Barton, D. A. J. Ives, and B. R. Thomas, *J. Chem. Soc.*, 2056 (1955)
[13]W. Nagata and H. Itazaki, *Chem. Ind.*, 1194 (1964)
[14]M. Gates and G. Tschudi, *Am. Soc.*, **78**, 1380 (1956)
[15]A. Schönberg, A. E. K. Fateen, and A. E. M. A. Sammour, *Am. Soc.*, **79**, 6020 (1957)
[16]R. Baltzly, N. B. Mehta, P. B. Russell, R. E. Brooks, E. M. Grivsky, and A. M. Steinberg, *J. Org.*, **26**, 3669 (1961)
[17]M. F. Grundon, H. B. Henbest, and M. D. Scott, *J. Chem. Soc.*, 1855 (1963)
[18]S. Sarel and Y. Yanuka, *J. Org.*, **24**, 2018 (1959)
[19]P. D. Gardner, L. Rand, and G. R. Haynes, *Am. Soc.*, **78**, 3425 (1956)
[20]F. G. Singleton, U.S. patent 2,436,532 (*C.A.*, **42**, 5048, (1946)
[21]S. Hünig, E. Lücke, and W. Brenninger, *Org. Syn.*, **41**, 65 (1961)
[22]*Idem, ibid.*, **43**, 34 (1963)
[23]A. Furst, R. C. Berlo, and S. Hooton, *Chem. Rev.*, **65**, 51 (1965)
[24]R. G. Taborsky, *J. Org.*, **26**, 596 (1961)
[25]V. Georgian, R. Harrisson, and N. Gubisch, *Am. Soc.*, **81**, 5834 (1959)
[26]P. S. Wharton and D. H. Bohlen, *J. Org.*, **26**, 3615 (1961); P. S. Wharton, *ibid.*, **26**, 4781 (1961)
[27]Huang-Minlon and Chung-Tungshun, *Tetrahedron Letters*, 666 (1961); C. Djerassi, D. H. Williams, and B. Berkoz, *J. Org.*, **27**, 2205 (1962); C. Djerassi, G. von Mutzenbecker, J. Fajkos, D. H. Williams, and H. Budzikiewicz, *Am. Soc.*, **87**, 817 (1965); see also W. R. Benn, and R. M. Dodson, *J. Org.*, **29**, 1142 (1964)
[28]C. Djerassi, D. H. Williams, and B. Berkoz, *J. Org.*, **27**, 2205 (1962)
[29]T. Nakano and M. Hasegawa, *Bull. Chem. Pharm.*, **12**, 971 (1964)
[30]E. Klein and G. Ohloff, *Tetrahedron*, **19**, 1091 (1963)
[30a]P. S. Wharton, S. Dunny, and L. S. Krebs, *J. Org.*, **29**, 958 (1964)
[31]S. Pietra, *Ann. Chim. (Rome)*, **45**, 850 (1955)
[32]P. M. G. Bavin, *Org. Syn.*, **40**, 5 (1960)
[33]P. M. G. Bavin, *Can. J. Chem.*, **36**, 238 (1958)
[34]M. J. S. Dewar and T. Mole, *J. Chem. Soc.*, 2556 (1956)
[35]D. Balcom and A. Furst, *Am. Soc.*, **75**, 4334 (1953); A. Furst and R. E. Moore, *Am. Soc.*, **79**, 5492 (1957); R. E. Moore and A. Furst, *J. Org.*, **23**, 1504 (1958)
[36]D. S. Tarbell, R. F. Smith, and V. Boekelheide, *Am. Soc.*, **76**, 2470 (1954)
[37]T. L. Fletcher and M. J. Namkung, *J. Org.*, **23**, 680 (1958)

[38]B. E. Leggetter and R. K. Brown, *Can. J. Chem.*, **38**, 2363 (1960)
[39]K. Butler and M. W. Partridge, *J. Chem. Soc.*, 2396 (1959)
[40]K. Ponsold, *Ber.*, **97**, 3524 (1964)
[41]A. Furst, R. C. Berlo, and S. Hooton, *Chem. Rev.*, **65**, 51 (1965)
[42]N. S. Hjelte, *Chem. Scand.*, **15**, 1200 (1961)
[43]W. L. Mosby, *J. Org.*, **24**, 421 (1959)
[44]*Idem, Chem. Ind.*, 1348 (1959)
[45]R. H. Wiley and P. E. Hexner, *Org. Syn., Coll. Vol.*, **4**, 351 (1963)
[46]*Org. Expts.*, Chapt. 46. For references, see *Org. Syn., Coll. Vol.*, **3**, 69, 656 (1955)
[47]L. Friedman, R. L. Litle, and W. R. Reichle, *Org. Syn.*, **40**, 93 (1960)
[48]P. A. S. Smith, *Org. Syn., Coll. Vol.*, **4**, 819 (1963)
[49]C. G. Overberger, P. Huang, and M. B. Berenbaum, *ibid.*, **4**, 274 (1963)
[50]A. C. Cope, D. S. Smith, and R. J. Cotter, *ibid.*, **4**, 377 (1963)
[51]H. R. Ing and R. F. Manske, *J. Chem. Soc.*, 2348 (1926)
[52]R. A. Boissonnas, *Advances in Org. Chem.*, **3**, 182 (1963)
[53]L. I. Smith and O. H. Emerson, *Org. Syn., Coll. Vol.*, **3**, 151 (1955)
[54]J. K. Stille and R. Ertz, *Am. Soc.*, **86**, 661 (1964)
[55]S. Hauptmann, M. Kluge, K.-D. Seidig, and H. Wilde, *Angew. Chem., Internat. Ed.*, **4**, 688 (1965)
[56]K. Alder and H. Niklas, *Ann.*, **585**, 97 (1954)
[57]A. G. M. Willems, R. R. v. Eck, U. K. Pandit, and H. O. Huisman, *Tetrahedron Letters*, 81 (1966)

Hydrazine acetate.

Peptide synthesis. Hydrazine acetate is an effective reagent for cleaving protective N-phthaloyl groups in a neutral medium and hence can be used in the synthesis of alkali-sensitive peptides such as α-melanotropin.[1] Thus this substance

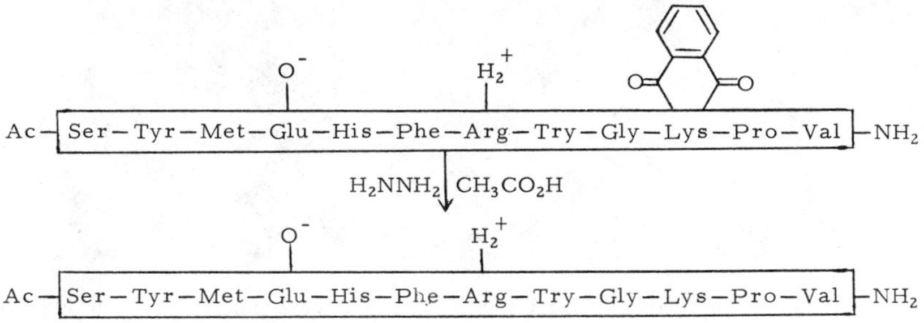
Ac
Ser—Tyr—Met—Glu—His—Phe—Arg—Try—Gly—Lys—Pro—Val
NH₂
H₂NNH₂ CH₃CO₂H
Ac
Ser—Tyr—Met—Glu—His—Phe—Arg—Try—Gly—Lys—Pro—Val
NH₂

was obtained in high yield by heating 440 mg. of the N-phthaloyl precursor with 20 ml. of 2 *M* hydrazine acetate in methanol at 50° for 15 hrs.

[1]R. Schwyzer, A. Costopanagiotis, and P. Sieber, *Helv.*, **46**, 870 (1963)

Hydrazine sulfate, $H_2NNH_2 \cdot H_2SO_4$. Mol. wt. 130.12. Suppliers: E, F, KK, MCB.

Peptide analysis. In the Akabori method for determination of the C-terminal amino acid, the peptide is heated with anhydrous hydrazine at 100–120° for about 8 hrs. to convert internal units into amino acid hydrazides and liberate the free acid from the terminal unit. Unfortunately several amino acids are decomposed under

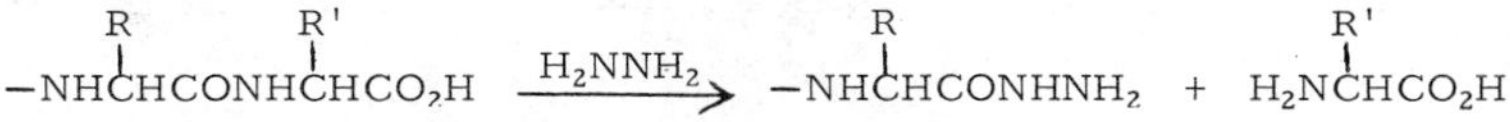
—NHCHCONHCHCO₂H
H₂NNH₂
—NHCHCONHNH₂ + H₂NCHCO₂H

these drastic conditions. Bradbury[1] found that the reaction with hydrazine sulfate is much faster and can be conducted at 60°.

[1]J. H. Bradbury, *Biochem. J.*, **68**, 475 (1958)

Hydrazoic acid, H—N$\overset{+}{=}$N=N⁻. Mol. wt. 43.03, b.p. 37°. *Caution:* very poisonous, use a hood.

Preparation. A solution of hydrazoic acid in benzene or chloroform is prepared[1, 2] in a three-necked flask fitted with an efficient stirrer, a dropping funnel, a thermometer, and a gas-exit tube. A paste is prepared from 65 g. (1 mole) of sodium azide and 65 ml. of warm water, 400 ml. of benzene or chloroform is added, the mixture is cooled to 0°, and 0.5 mole of concd. sulfuric acid is added dropwise with control of the temperature to 0–5°. The organic layer is separated and dried over sodium sulfate. The concentration of the solution of hydrazoic acid is determined by transferring a sample with pipette and pipetter (Fig. H-2) to a glass-stoppered bottle, shaking it with distilled water, and titrating with standard alkali.

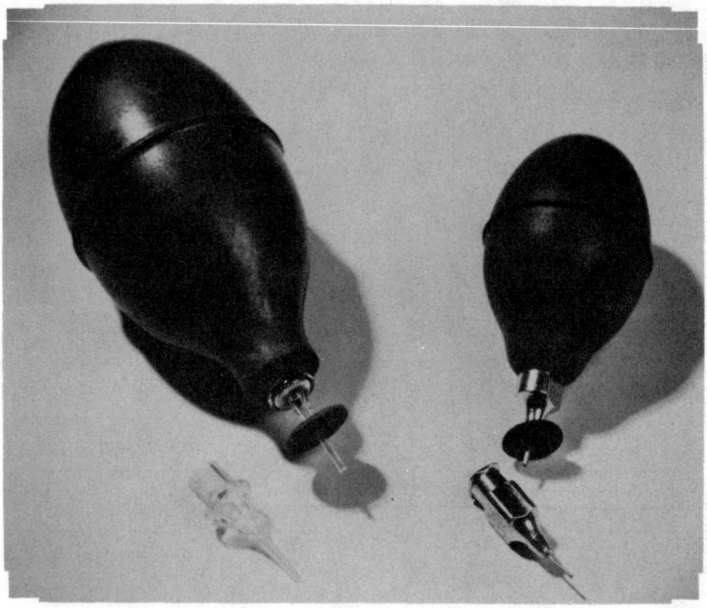

Fig. H-2 Pipetters with needles of steel and of glass (Wilkens-Anderson Co.); a rubber disc is cut out of a piece of Gooch tubing with a corkborer and pierced with an awl

Schmidt reaction[3] (see review[2]). This reaction provides a method for the conversion of a carboxylic acid into an amine in one step. On reaction with hydrazoic acid in benzene in the presence of sulfuric acid, an acid is converted beyond the stage of the acyl azide (a) into the conjugate acid (b), which loses nitrogen more easily than the azide itself, with rearrangement to the isocyanate (d). The one-step

$$\text{R COOH} + \text{HN}_3 \longrightarrow \underset{\text{(a)}}{\text{R}\overset{\text{O}}{\overset{\|}{\text{C}}}-\overset{\cdot\cdot}{\text{N}}\overset{+}{=}\text{N}\overset{-}{=}\text{N}} \overset{\text{H}^+}{\longrightarrow} \underset{\text{(b)}}{\text{R}\overset{\text{O}}{\overset{\|}{\text{C}}}-\overset{\text{H}}{\overset{|}{\text{N}}}-\overset{+}{\text{N}}\equiv\text{N}} \overset{-\text{N}_2}{\longrightarrow}$$

$$\underset{\text{(c)}}{\text{R}\overset{\text{O}}{\overset{\|}{\text{C}}}-\overset{\text{H}}{\overset{|}{\underset{\cdot\cdot}{\text{N}}}}+} \overset{-\text{H}^+}{\longrightarrow} \underset{\text{(d)}}{\text{O}=\text{C}=\text{N}-\text{R}} \overset{\text{H}_2\text{O}}{\longrightarrow} \underset{\text{(e)}}{\text{RNH}_2 + \text{CO}_2}$$

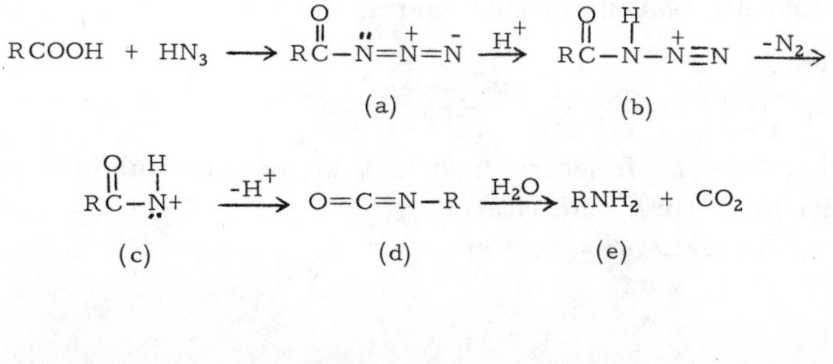

procedure is convenient and the yields are high. For example, a solution of stearic acid (15 g.) in benzene (500 ml.) is treated with concd. sulfuric acid (30 ml.), and the mixture is stirred at 40° during addition of a 5% solution of hydrazoic acid in benzene

$$CH_3(CH_2)_{16}CO_2H \xrightarrow[96\%]{HN_3,\ H_2SO_4;\ H_2O} CH_3(CH_2)_{16}NH_2 + N_2 + CO_2$$

(1.2 equiv.).[4] After gas evolution has ceased, the acid layer is poured into water to precipitate heptadecylamine sulfate, which is collected and crystallized from ethanol.

Benzaldehyde undergoes the Schmidt reaction with formation of benzonitrile in 70% yield;[2] in this case no C→N rearrangement is involved. Ketones, including

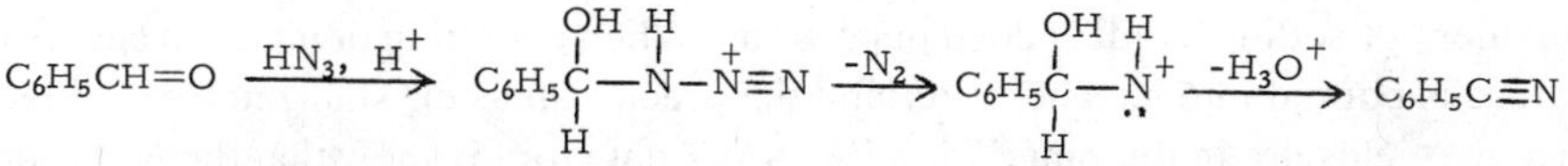

α,β-unsaturated ketones, react with rearrangement to give amides; acetophenone affords acetanilide in 77% yield.[2] As in the Beckmann rearrangement of acetophenone

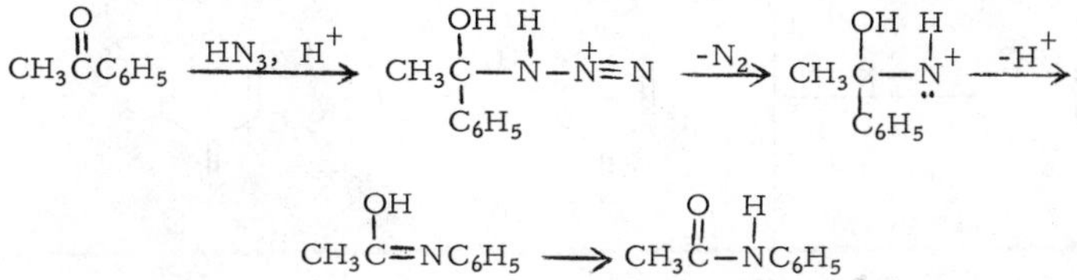

oxime, phenyl is the migrating group. The conversion of acetophenone into acetanilide has been carried out also by drawing a mixture of dry hydrogen azide and air by suction through a mixture of the ketone, benzene, and concd. sulfuric acid. Application of the Schmidt reaction to cyclohexanone effects ring enlargement to ε-caprolactam.[2]

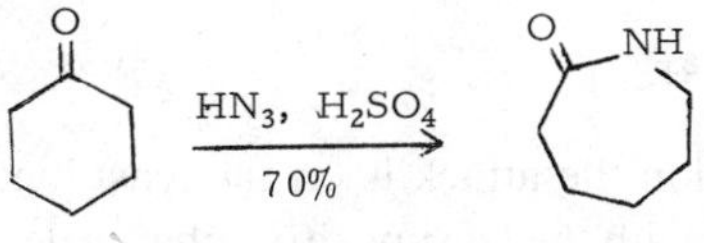
O
HN₃, H₂SO₄
70%
NH

Reaction with quinones. Addition of an aqueous solution of sodium azide to a solution of α-naphthoquinone in acetic acid is attended with rapid evolution of nitrogen and separation of orange-red needles of 2-amino-1,4-naphthoquinone in 87% yield.[5] The reaction is interpreted as involving 1,4-addition of hydrazoic acid

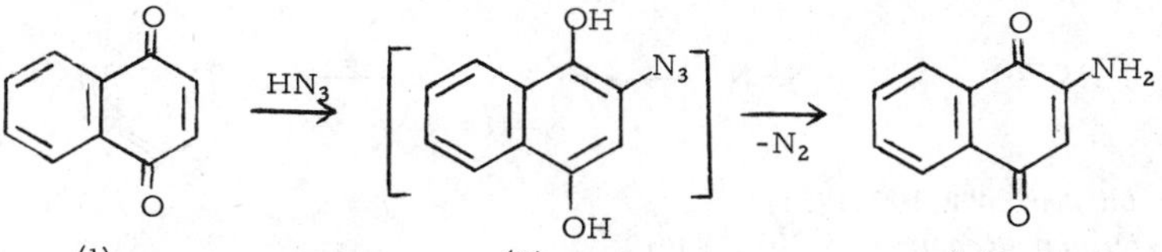
HN₃
OH
N₃
-N₂
NH₂

(1) (2) (3)

to give the azidonaphthohydroquinone (2) and disproportionation, with transference of the hydroquinone hydrogens to the azide group. β-Naphthoquinone and its 3-bromo derivative react similarly to give the corresponding 4-amino-1,2-quinones. 1,4-Benzquinone reacts only to give the adduct, azidohydroquinone.[6] The oxidation reduction potential is lower in the benzene series than with the naphthalene derivatives and evidently not sufficient to bring about a change similar to (2) → (3). Phenanthrenequinone, incapable of a comparable 1,4-addition, reacts with hydrazoic acid in sulfuric acid solutions to give phenanthridone.[7]

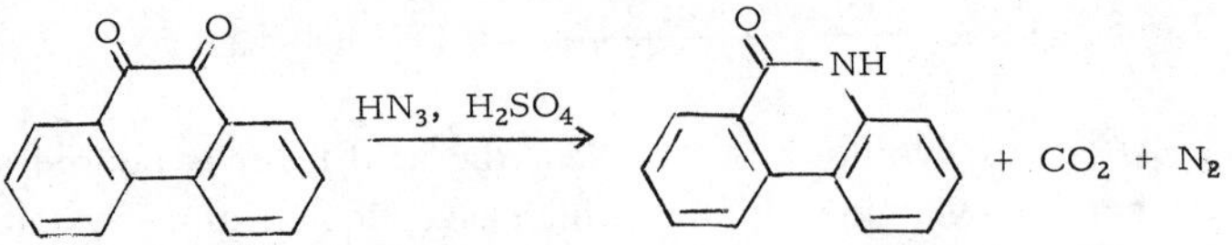

Misiti *et al.*[8] found that in concd. sulfuric acid at 0° hydrazoic acid reacts with some alkylated quinones in an entirely different manner to give 2,5*H*-2,5-azepindiones such as (5). A solution of the quinone (1) in concd. sulfuric acid is treated at 0° with 1 equivalent of sodium azide, added in portions. When evolution of nitrogen ceases the mixture is poured into ice and water and the precipitate is crystallized from aqueous ethanol. Yields are in the range 75–85%. NMR data for (5) show that the NH proton is directly coupled to the vinylic proton and hence that these groups are adjacent.

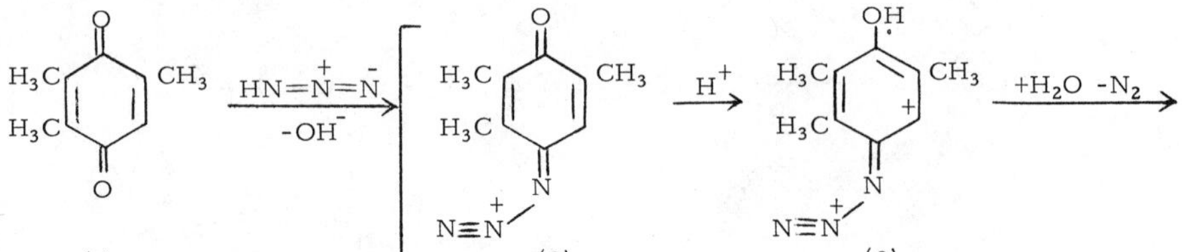

(1) (2) (3)

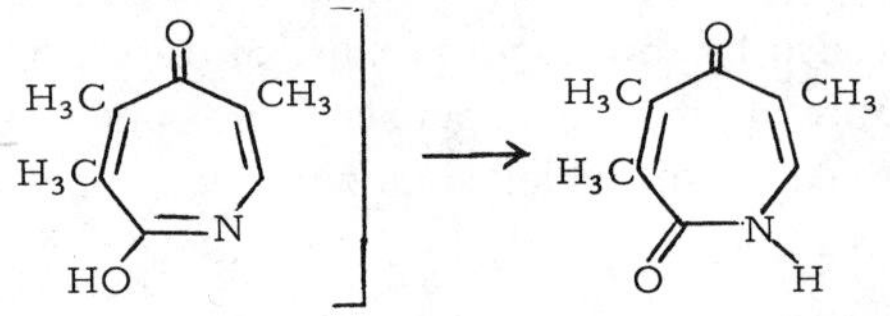

(4) (5)

In this Schmidt-like reaction the attack is on the least hindered carbonyl group and the migrating group is that with the least positive charge density.

Heterocycles. The review cited[2] summarizes reactions of hydrazoic acid leading to heterocycles containing several ring nitrogens.

When an alkyl or aryl nitrile is heated with a benzene solution of hydrazoic acid at 120–150° for 96–120 hrs., a 5-alkyl- or 5-aryltetrazole is produced in excellent yield.[9]

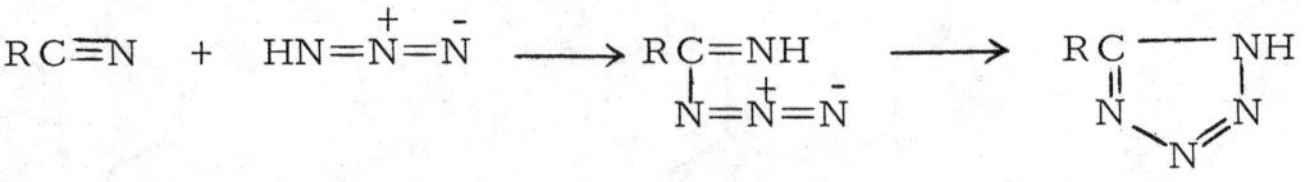

[1]J. von Braun, *Ann.*, **490**, 100 (1931)
[2]H. Wolff, *Organic Reactions*, **3**, 307 (1946)
[3]Karl Friedrich Schmidt, *Angew. Chem.*, **36**, 511 (1923); *Ber.*, **57**, 704 (1924)
[4]L. H. Briggs, G. C. De Ath, and S. R. Ellis, *J. Chem. Soc.*, 61 (1942)
[5]L. F. Fieser and J. L. Hartwell, *Am. Soc.*, **57**, 1482 (1935)
[6]E. Oliveri-Mandalà, *Gazz.*, **52**, II, 139 (1922)
[7]G. Caronna, *Gazz.*, **71**, 481 (1941)
[8]D. Misiti, H. W. Moore, and K. Folkers, *Tetrahedron Letters*, 1071 (1965)
[9]J. S. Mihina and R. M. Herbst, *J. Org.*, **15**, 1082 (1950)

Hydriodic acid. Constant boiling acid is a 57% solution, b.p. 127°, sp. gr. 1.7. Suppliers: F, MCB.

Reduction. Reduction of the azlactone of α-benzoylaminocinnamic acid to phenylalanine is carried out by refluxing with hydriodic acid, red phosphorus, and acetic anhydride.[1] The function of the phosphorus is to destroy iodine as formed. A

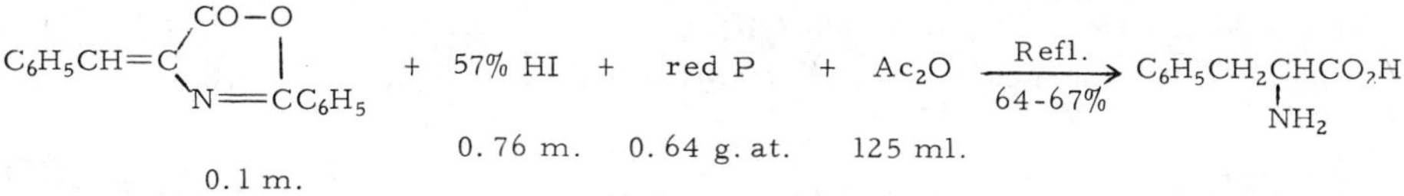

similar mixture is used for the reductive removal of halogen from 3-chloroindole

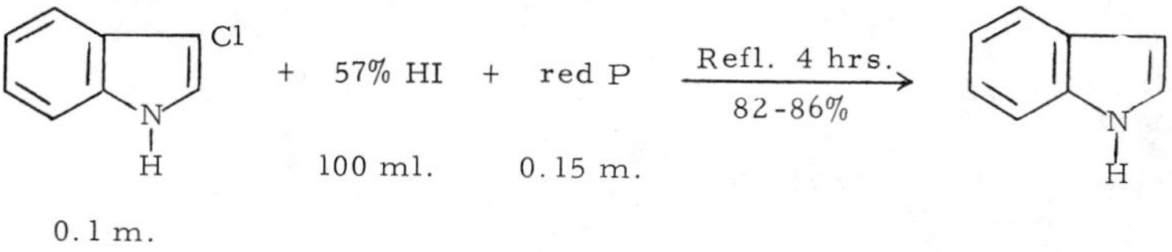

except that no acetic anhydride is added.[2] In the reduction of *m*-nitrobenzenesulfonyl chloride to di-*m*-nitrophenyl disulfide, red phosphorus is not used, but at the end of the reaction solid sodium bisulfite is added to destroy the iodine formed.[3] This

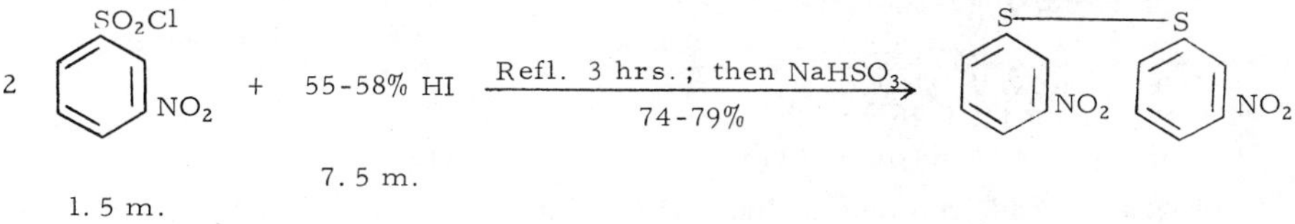

procedure shows that the nitro group is not reduced by hydriodic acid.

In refluxing acetic acid the reagent reduces α-diketones and α-ketols mainly to saturated ketones.[4] Examples:

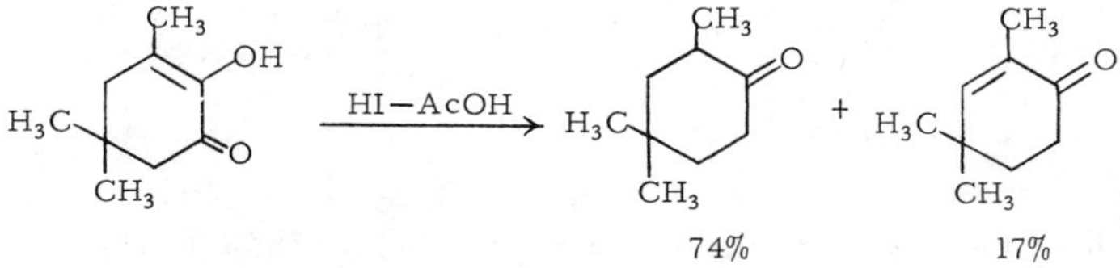

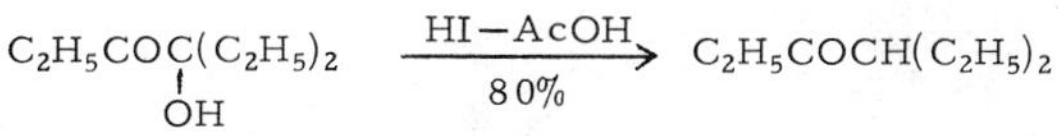

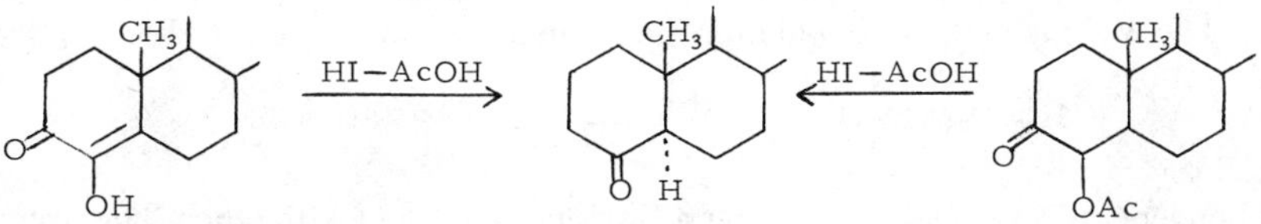

Hydriodic acid is regarded as the most satisfactory reagent for the conversion of a diazoketone into a methyl ketone.[5] Thus 5 ml. of the 57% acid was added to a solution

$$C_6H_5COCH{=}\overset{+}{N}{=}\overset{-}{N} \xrightarrow[96\%]{HI-CHCl_3} C_6H_5COCH_3 + N_2 + I_2$$

of 22 g. of diazoacetophenone in chloroform. After evolution of nitrogen had ceased, water was added, and the chloroform layer separated and shaken with sodium thiosulfate solution to remove iodine.

Ether cleavage. Deulofeu and Guerrero[6] used the combination HI–P–Ac_2O in the last step of the preparation of N-methyl-3,4-dihydroxyphenylalanine.

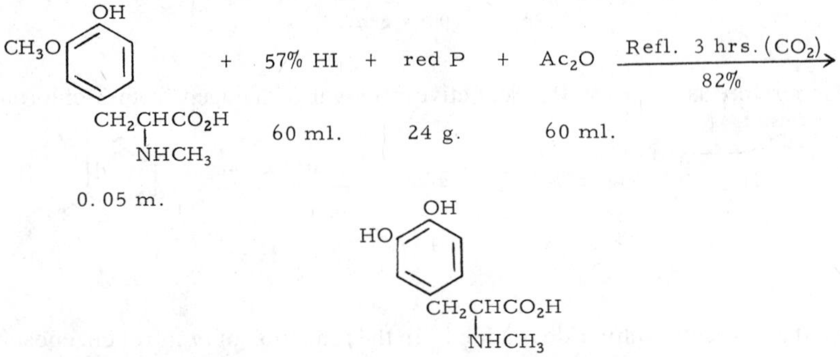

OH
CH₃O
CH₂CHCO₂H
NHCH₃
0.05 m.
+ 57% HI + red P + Ac₂O
60 ml.
24 g.
60 ml.
Refl. 3 hrs. (CO₂)
82%
OH
HO
CH₂CHCO₂H
NHCH₃

[1]H. B. Gillespie and H. R. Snyder, *Org. Syn., Coll. Vol.*, **2**, 489 (1943)
[2]E. F. M. Stephenson, *ibid.*, **3**, 475 (1955)
[3]W. A. Sheppard, *Org. Syn.*, **80**, (1960)
[4]W. Reusch and R. LeMahieu, *Am. Soc.*, **86**, 3068 (1964); W. Reusch, R. LeMahieu, and R. Guynn, *Steroids*, **5**, 109 (1965)
[5]M. L. Wolfrom and R. L. Brown, *Am. Soc.*, **65**, 1516 (1943)
[6]V. Deulofeu and T. J. Guerrero, *Org. Syn., Coll. Vol.*, **3**, 586 (1955)

Hydrobromic acid, constant boiling (48%). B.p. 126°, sp. gr. 1.49, 8.8 moles/l., 70.7 g./100 ml. Suppliers: B, E, F, MCB. Acid that is not colorless can be purified by distillation from stannous chloride.

ROH ⟶ RBr. The preferred procedure for the preparation of primary alkyl bromides calls for use of a small amount of sulfuric acid as catalyst.[1] 2-Bromo-

$$\underset{10\ m.}{(CH_3)_2CHCH_2CH_2OH} + \underset{12.5\ m.}{48\%\ HBr} + \underset{100\ g.}{H_2SO_4} \xrightarrow[88-90\%]{Refl.\ 5-6\ hrs} (CH_3)_2CHCH_2CH_2Br$$

ethylamine hydrobromide is prepared by refluxing 2-ethanolamine with hydrobromic acid, distilling off most of the acid, and precipitating the salt with acetone.[2]

$$\underset{164\ m.}{HOCH_2CH_2NH_2} + \underset{52\ m.}{48\%\ HBr} \xrightarrow[83\%]{} BrCH_2CH_2\overset{+}{N}H_3\overset{-}{Br}$$

Pentaerythritol is converted into the monobromide by refluxing a solution in acetic acid with addition of 48% hydrobromic acid in two batches over a period of about 8 hrs.[3] The workup includes distillation, addition of ethanol and azeotropic distillation

$$HOCH_2C(CH_2OH)_3 \xrightarrow[75-85\%]{HBr-AcOH} BrCH_2C(CH_2OH)_3$$

with benzene, digestion with ether, and Soxhlet extraction with ether. The preparation of β-bromopropionic acid involves both displacement of a hydroxyl group and hydrolysis of a nitrile.[4]

$$HOCH_2CH_2CN + 48\%\ HBr \xrightarrow[82\text{-}83\%]{\text{Refl. 2 hrs.}} BrCH_2CH_2CO_2H$$

4.5 m. 13.6 m.

Hydrolysis of nitriles. The aminonitrile (3) from 3 moles of acetone is refluxed for 2 hrs. with 1 kg. of 48% hydrobromic acid and 600 ml. of water, most of the acid is

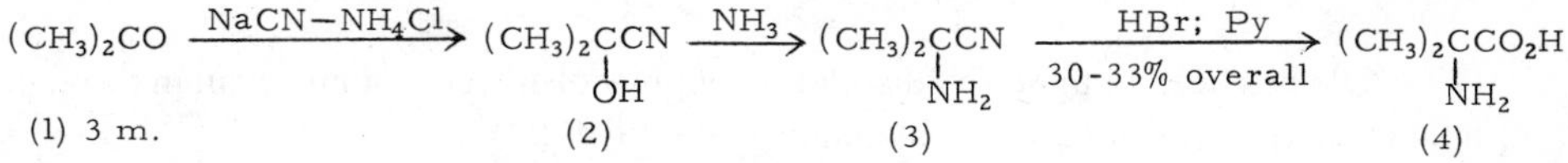

removed by distillation, and a solution of the residue in methanol is treated with pyridine to precipitate the dipolar amino acid (4).[5]

Cleavage of oxides and ethers. The epoxide ring of ethylene oxide is opened

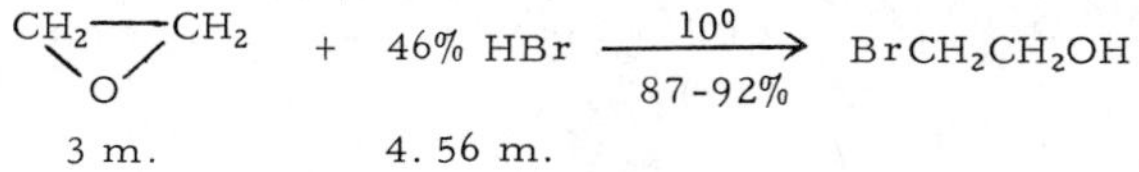

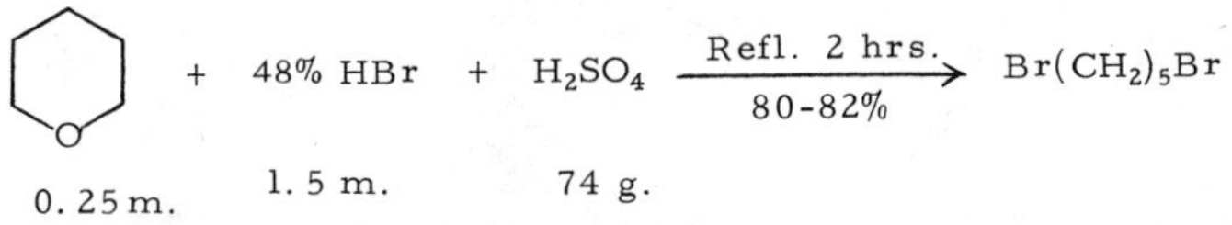

under very mild conditions (10°, 46% acid),[6] whereas cleavage of tetrahydropyrane requires full-strength HBr, addition of H_2SO_4, and refluxing.[7]

Demethylation of α-methoxyphenazine was done with fortified HBr,[8] but the yield was considerably below that obtained by Clarke and Taylor in the demethylation of guaiacol with 48% HBr.[9] In the latter case the mixture was distilled through a

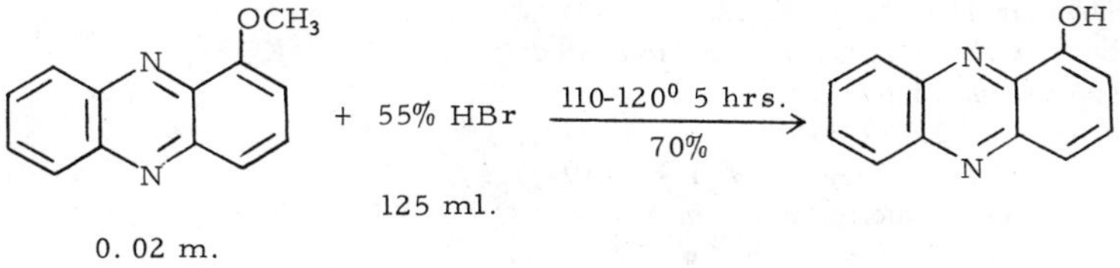

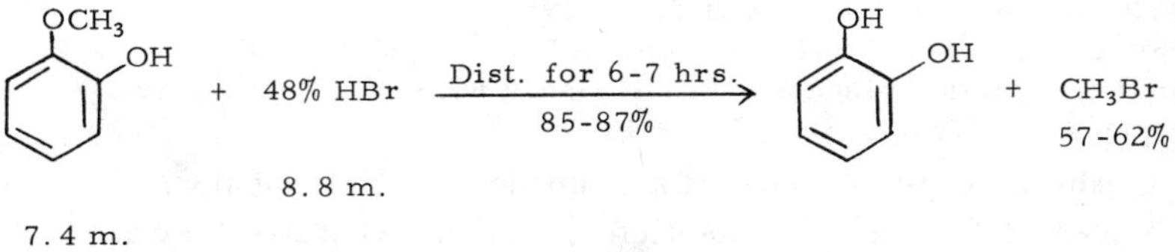

column into an automatic separator which returned the guaiacol to the reaction flask; the methyl bromide formed was absorbed in ice-cold methanol. When the temperature at the head of the column indicated that the guaiacol was all consumed, the cooled mixture was extracted with toluene, and the extract boiled to remove water and let stand for crystallization of colorless catechol.

8-Methoxyquinoline has been demethylated in nearly quantitative yield by refluxing with 48% hydrobromic acid.[10] 6-Methoxy-1-naphthoic acid was demethylated in 90% yield by refluxing the ether in 48% hydrobromic acid and acetic acid for 5 hrs.[11]

Dehydration. 2,3-Dimethylbutadiene has been prepared by addition of a catalytic amount of hydrobromic acid to pinacol and slow distillation.[12] A much better yield

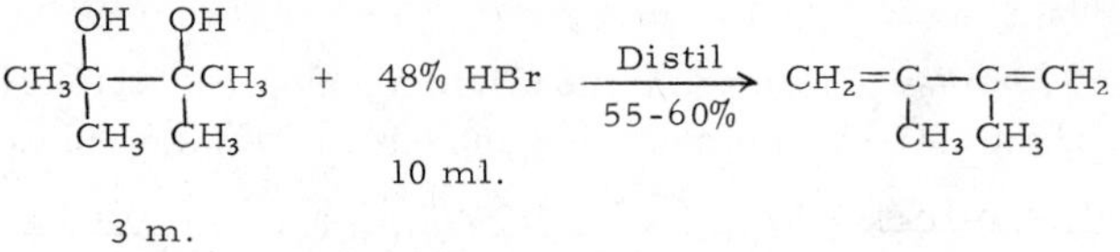

OH OH
CH_3C—CCH_3 + 48% HBr
CH_3 CH_3
10 ml.
3 m.
Distil
55-60%
CH_2=C—C=CH_2
CH_3 CH_3

(79–86%) has been obtained by distillation of pinacol at water-pump vacuum through a tube packed with alumina and maintained at 420–470°.[13]

Reductive cleavage of sulfonamides. The conversion of 2-(*p*-tolylsulfonyl)-dihydroisoindole (1) to 1,3-dihydroisoindole (3) is accomplished by refluxing under

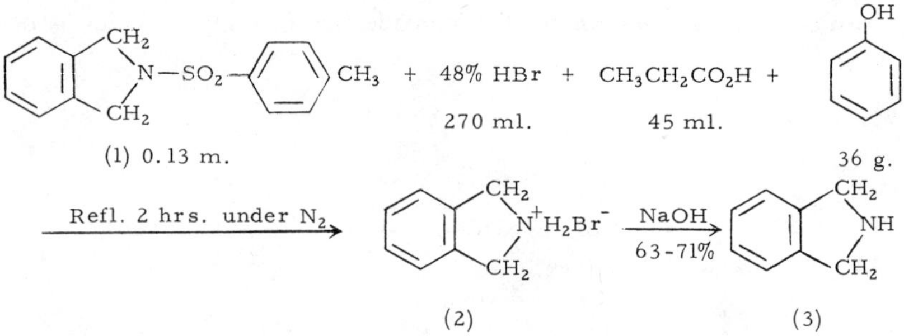

CH_2
N—SO_2
CH_3 + 48% HBr + $CH_3CH_2CO_2H$ +
OH
CH_2
270 ml.
45 ml.
(1) 0.13 m.
36 g.
Refl. 2 hrs. under N_2
$N^+H_2Br^-$
NaOH
63-71%
NH
(2)
(3)

nitrogen a mixture of the amide, hydrobromic acid, propionic acid, and phenol.[14,15] The function of the phenol is to prevent bromination of the amine formed.[16]

[1]O. Kamm and C. S. Marvel, *Org. Syn., Coll. Vol.*, **1**, 25 (1941)
[2]F. Cortese, *ibid.*, **2**, 91 (1943)
[3]S. Wawzonek, A. Matar, and C. H. Issidorides, *ibid.*, **4**, 681 (1963)
[4]E. C. Kendall and B. McKenzie, *ibid.*, **1**, 131 (1941)
[5]H. T. Clarke and H. J. Bean, *ibid.*, **2**, 29 (1943)
[6]F. K. Thayer, C. S. Marvel, and G. S. Hiers, *ibid.*, **1**, 117 (1941)
[7]D. W. Andrus, *ibid.*, **3**, 692 (1955)
[8]A. R. Surrey, *ibid.*, **3**, 753 (1955)
[9]H. T. Clarke and E. R. Taylor, *ibid.*, **1**, 150 (1941)
[10]F. E. King and J. A. Sherred, *J. Chem. Soc.*, 415 (1942)
[11]L. Long, Jr., and A. Burger, *J. Org.*, **6**, 852 (1941)
[12]C. F. H. Allen and A. Bell, *Org. Syn., Coll. Vol.*, **3**, 312 (1955)
[13]L. W. Newton and E. R. Coburn, *ibid.*, **3**, 313 (1955)
[14]J. Bornstein, S. C. Loshua, and A. P. Boisselle, *J. Org.*, **22**, 1255 (1957)
[15]J. Bornstein, J. E. Shields, and A. P. Boisselle, procedure submitted to *Org. Syn.*
[16]S. Searles and S. Nukina, *Chem. Rev.*, **59**, 1077 (1959)

Hydrogen bromide–Acetic acid. Eastman supplies a 30–32% solution.

A procedure for the preparation of 1,2,3,4-tetra-O-acetyl-β-D-glucopyranoside specifies use of acetic acid saturated with hydrogen bromide for cleavage of the 6-trityl group.[1]

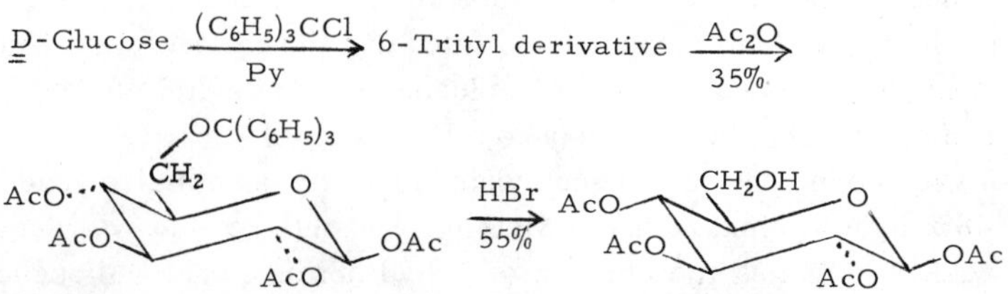

D-Glucose
$(C_6H_5)_3CCl$
Py
6-Trityl derivative
Ac_2O
35%
$OC(C_6H_5)_3$
CH_2
O
AcO
AcO
OAc
AcO
HBr
55%
CH_2OH
O
AcO
AcO
OAc
AcO

[1]D. D. Reynolds and W. L. Evans, *Org. Syn., Coll. Vol.*, **3**, 432 (1955)

Hydrogen bromide, anhydrous. Mol. wt. 80.93, b.p. –66°.

Preparation. The gas is generated by passing hydrogen into a bromine vaporizer and passing the gas mixture through a combustion tube backed with pieces of porous plate and kept at a dull red heat.[1] A feature of the assembly of the bromine vaporizer is shown in Fig. H-3. The inner flask containing bromine is heated by the vapor of refluxing ethyl bromide (b.p. 38.4°). The generator produces 300 g. of hydrogen bromide per hour.

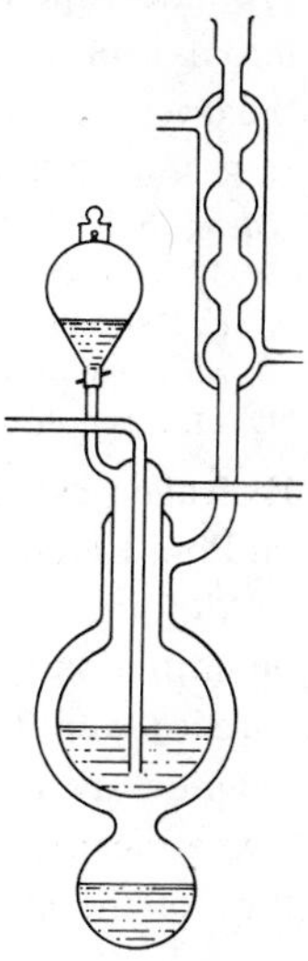

Fig. H-3

ROH(ROAc) ⟶ RBr. Lauryl bromide (dodecyl bromide) is prepared by passing gaseous HBr into the alcohol at 100°,

$$\underline{n}\text{-}C_{12}H_{25}OH \xrightarrow[88\%]{HBr\ (100^0)} \underline{n}\text{-}C_{12}H_{25}Br + H_2O$$

separating the crude bromide from aqueous hydrobromic acid and washing it with concd. sulfuric acid.[2] Decamethylene dibromide is prepared in the same way from the glycol.[3]

In the preparation of 2,3,4,6-tetra-O-acetyl-α-D-glucopyranosyl bromide (α-acetobromoglucose), hydrogen bromide is generated either as described above or by dropping liquid bromine into boiling tetralin.[4]

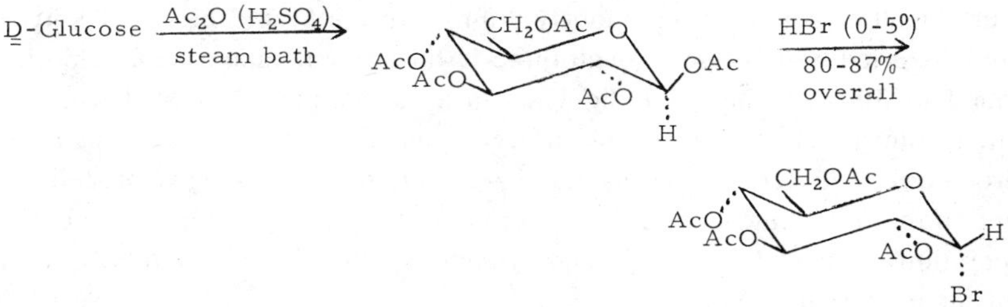

D-Glucose
Ac₂O (H₂SO₄)
steam bath
CH₂OAc
AcO
AcO
AcO
OAc
H
HBr (0-5⁰)
80-87%
overall
CH₂OAc
AcO
AcO
AcO
H
Br

Addition to methyl acrylate.[5] The ester required for this reaction was obtained from a 60% solution in methanol supplied by Rohm and Haas. This solution was treated with hydroquinone and washed repeatedly with 7% aqueous sodium sulfate solution to remove the methanol, and the ester was dried, filtered, and used without distillation. For the preparation of methyl β-bromopropionate, hydrogen bromide

$$\underset{3\text{ m. in 500 ml. ether}}{CH_2{=}CHCO_2CH_3} + \underset{3\text{ m.}}{HBr} \xrightarrow[80-84\%]{0^0} BrCH_2CH_2CO_2CH_3$$

generated by the first method was freed of bromine by bubbling it through a solution of phenol in carbon tetrachloride and was passed into an ethereal solution of methyl acrylate with ice cooling.

Cleavage of a trityl ether.[6] The trityl ether (2) [6-O-trityl-1,2,3,4-tetra-O-acetyl-β-D-glucose] was detritylated by adding a saturated solution of dry hydrogen bromide

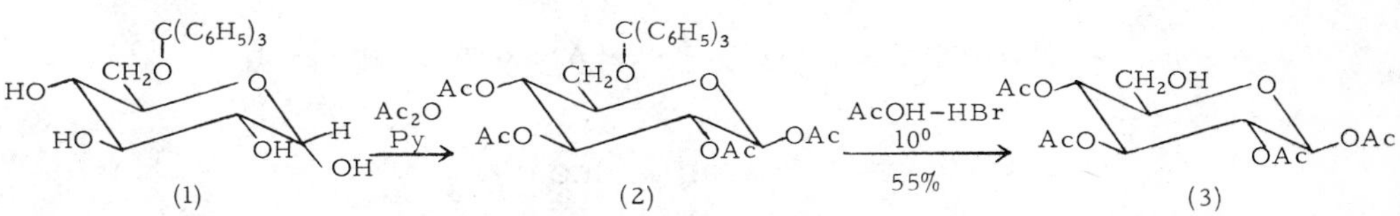

C(C₆H₅)₃
CH₂O
HO
HO
OH
H
OH
Ac₂O
Py
C(C₆H₅)₃
CH₂O
AcO
AcO
OAc
OAc
AcOH–HBr
10⁰
55%
CH₂OH
AcO
AcO
OAc
OAc

(1) (2) (3)

in acetic acid to a solution of (2) in acetic acid and shaking for 45 sec. The trityl bromide which separated was removed quickly by filtration, the filtrate was poured into 1 l. of water, and the mixture was extracted with chloroform. Evaporation of the dried extract gave a sirup which crystallized when rubbed with ether.

[1]J. R. Ruhoff, R. E. Burnett, and E. E. Reid, *Org. Syn., Coll. Vol.*, **2**, 338 (1943)
[2]*Idem, ibid.*, **2**, 246 (1943)
[3]W. L. McEwen, *ibid.*, **3**, 227 (1955)
[4]C. E. Redemann and C. Niemann, *ibid.*, **3**, 11 (1955)
[5]R. Mozingo and L. A. Patterson, *ibid.*, **3**, 576 (1955)
[6]D. D. Reynolds and W. L. Evans, *ibid.*, **3**, 432 (1955)

Hydrogen cyanide. Mol. wt. 27.03, m.p. −13.3°, b.p. 25.7°, sp. gr. 0.688. Suppliers: du Pont Electrochem. Div., McKesson and Robbins.

Although hydrogen cyanide is an active poison, the reagent can be prepared in quantities up to 0.5 kg. without undue danger if a good hood is available and proper care is taken. In the Ziegler procedure,[1] aqueous solutions of sulfuric acid and sodium cyanide are run simultaneously from separatory funnels which deliver into a small funnel secured by copper wire to the upper stopper of a 5-l. flask. The discharge tube of the small funnel is bent to a U shape so that hydrogen cyanide is evolved

$$NaCN + H_2SO_4 \xrightarrow[93\text{-}97\%]{} HCN + NaHSO_4$$

when the two solutions meet; practically all the reaction occurs in the funnel and sodium bisulfate solution continuously drains into the flask. The gaseous hydrogen cyanide is dried by passage through three U-tubes containing calcium chloride and warmed to 30–40°. The gas is liquefied in a coil condenser cooled with ice; the melting point is −15 to −14.5°. Slotta[2] recommends addition of 2 drops of concd. hydrochloric acid per 500 ml. as stabilizer. A note to the Ziegler procedure calls attention to a recommendation by Gattermann[3] that the operator smoke during the preparation since a trace of hydrogen cyanide is sufficient to give tobacco smoke a highly characteristic flavor.

Anhydrous hydrogen cyanide is used in the Gattermann synthesis of aldehydes.[3,4] Thus phenol is treated in dry ether with anhydrous hydrogen cyanide, zinc chloride, and hydrogen chloride gas; the imide separates as the hydrochloride and is hydrolyzed

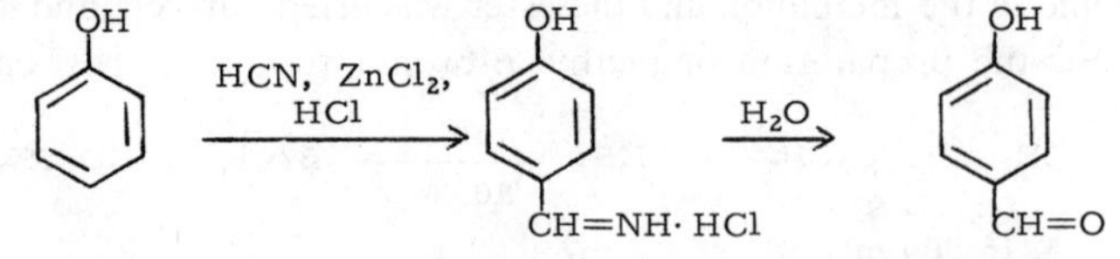
OH
HCN, ZnCl2,
HCl
OH
H2O
OH
CH=NH· HCl
CH=O

by aqueous acid. Phenol ethers and some phenols react with difficulty unless zinc chloride is replaced by the more reactive aluminum chloride. It appears that hydrogen chloride adds to hydrogen cyanide to produce formimino chloride (1), which adds

$$HC{\equiv}N \xrightarrow{HCl} \underset{(1)}{ClCH{=}NH} \xrightarrow{HC{\equiv}N} \underset{(2)}{ClCH{=}NCH{=}NH} \xrightarrow{AlCl_3}$$

$$\underset{(3)}{ClCH{=}NCH{=}NH\cdot AlCl_3} \xrightarrow[-HCl]{ArH} \underset{(4)}{ArCH{=}NCH{=}NH\cdot AlCl_3}$$

$$\xrightarrow{H_2O} \underset{(5)}{ArCH{=}O} + 2\ NH_3 + HCO_2H$$

to a second molecule of hydrogen cyanide to form (2) and that (2) combines with aluminum chloride to form the complex (3). This complex attacks the aromatic component with elimination of hydrogen chloride and formation of the arylmethylene formamidine complex (4), which subsequently is hydrolyzed to the aldehyde.

A modified Gattermann synthesis introduced by R. Adams[5] avoids the use of anhydrous hydrogen cyanide. By passing hydrogen chloride into a mixture of the phenol and zinc cyanide in ether, the effective reagent and catalyst are produced in the reaction mixture in the presence of the substrate.

[1]K. Ziegler, *Org. Syn., Coll. Vol.*, **1**, 314 (1941)
[2]K. H. Slotta, *Ber.*, **67**, 1028 (1934)
[3]L. Gattermann, *Ann.*, **357**, 313 (1907)
[4]*Idem*, *Ber.*, **31**, 1149 (1898)
[5]R. Adams *et al.*, *Am. Soc.*, **45**, 2373 (1923); **46**, 1518 (1924)

Hydrogen fluoride, anhydrous. Mol. wt. 20.01, b.p. 19.4°. Suppliers: Harshaw, Pennsalt, Mathieson. For notes on the handling of the reagent, *see also* Acyl fluorides. Reviews.[1,2]

The liquid reagent has high solvent power, particularly for oxygen-containing substances and for aromatic compounds. It is an effective dehydrating and condensing agent comparable to concd. sulfuric acid but less prone to permit secondary reactions such as enolization and aromatic substitution. Treatment with liquid HF at room temperature provides a simple and efficient method for the cyclization of β-arylpropionic acids and γ-arylbutyric acids to 1-indanones and 1-tetralones, as shown by the yields cited under the formulas of typical ketones.[3] It is far superior

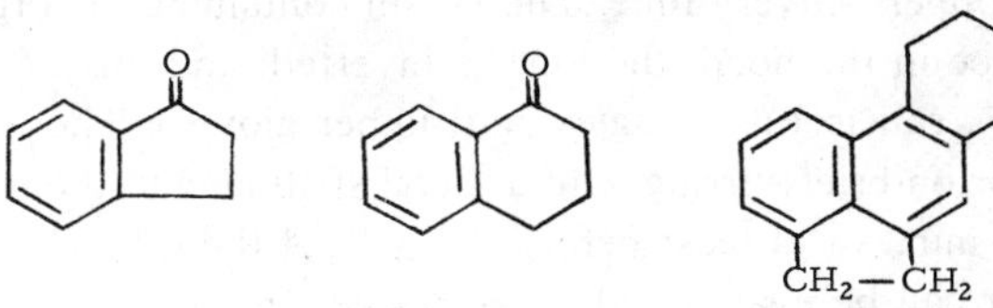

CH_2-CH_2

73% 92% 88%

to other cyclodehydrating agents for application to acids to type (1), for liquid HF gives the product (2) in pure keto-form directly suitable for reaction with Grignard

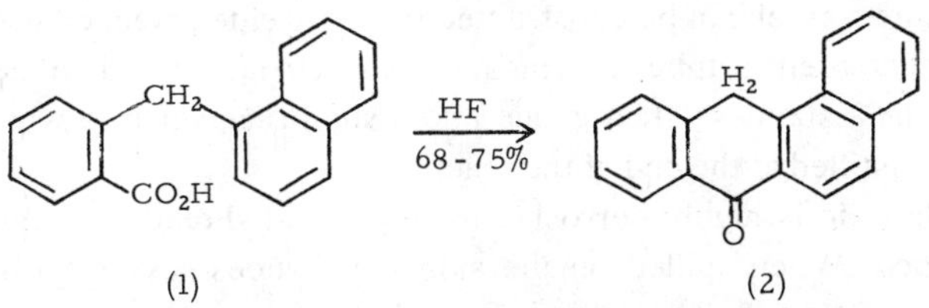

CH_2
CO_2H
HF
68-75%
H_2

(1) (2)

reagents. However, liquid HF does not cyclize *o*-benzoylbenzoic acid. The reagent catalyzes Friedel-Crafts acylation of acenaphthene with free carboxylic acids, but naphthalene and other less reactive hydrocarbons do not react. Acetylation of acenaphthene with acetic acid and HF, and crystallization of the total isomer mixture (91%) affords the less soluble 1-acetoacenaphthene in moderate yield. Acetylation with acetyl chloride and aluminum chloride gives chiefly 3-acetoacenaphthene, but material of satisfactory purity requires purification through the picrate. Reaction of acenaphthene with crotonic acid in liquid HF involves acylation and cyclization to the ketone formulated.

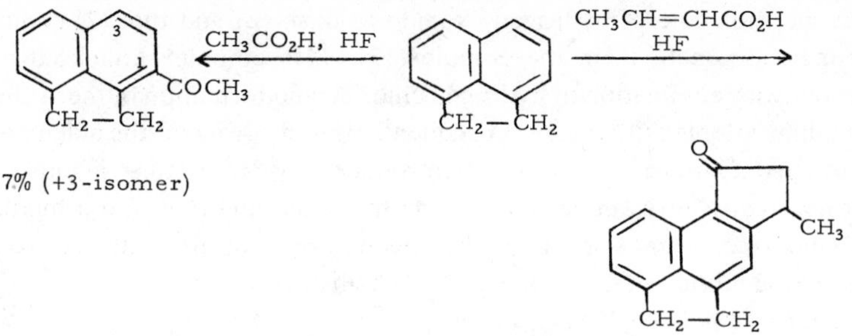

An interesting synthesis reported by Schroeder[4] is the condensation of phthalylacetic acid with naphthalene in liquid HF to form 3,4-benzpyrene-1,5-quinone,

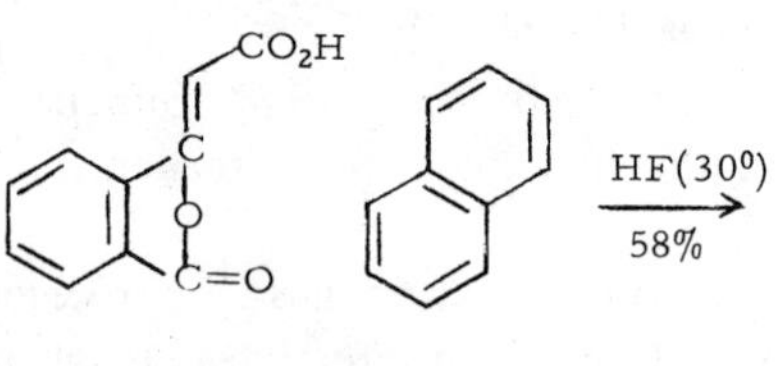

which is convertible into 3,4-benzpyrene by zinc dust distillation. Here the acylating agent enters into a 3-point attack.

Reactions that proceed satisfactorily at room temperature can be carried out in a platinum crucible or, with larger amounts, in an open copper flask or polyethylene vessel. The cylinder is stored in a cold place (5°) until required, when it is removed and fitted with a copper delivery tube. The vessel containing the organic reactants is tared on a balance in the hood, the tank is inverted, and the required weight of hydrogen fluoride is run in (use goggles and rubber gloves). The material usually dissolves at once or on brief stirring with a metal spatula, and the reaction may be complete in 10–20 min., or at least before the bulk of the reagent has evaporated. The excess reagent can be evaporated by gentle heating over a steam bath or in a current of air, or the solution poured into water and a little ice in a beaker and the product quickly collected by suction filtration or by ether or benzene extraction, followed immediately by washing with soda solution. For use at higher temperatures (100°), a pressure vessel can be constructed from a welded steel cylinder fitted with a stainless steel condenser tube carrying a glass water jacket and connected through the condenser to a stainless steel gauge and a steel receiver into which the excess reagent can be distilled at the end of the reaction.

Hydrogen fluoride is highly corrosive to tissue and should be handled with care and not breathed. When spilled on the skin it produces severe burns which only become apparent some 5–8 hrs. later. Parts known or suspected to have been in contact with the reagent should be treated immediately, first by thorough washing with water and then by application of a paste of magnesia, water, and glycerol.

[1]K. Wiechert and J. E. Jones, "Use of Hydrogen Fluoride in Organic Reactions," *Newer Methods of Preparative Organic Chemistry*, p. 315, Interscience (1948)
[2]J. H. Simons, *Ind. Eng. Chem.*, **32**, 178 (1940)
[3]L. F. Fieser and E. B. Hershberg, *Am. Soc.*, **61**, 1274 (1939); **62**, 49 (1940)
[4]H. E. Schroeder, F. B. Stilmar, and F. S. Palmer, *Am. Soc.*, **78**, 446 (1956)

Hydrogen peroxide, H_2O_2. Mol. wt. 64.02. The reagent is available in concentrations from 30 to 100%. The 30% solution is 11.6*M*; 86 ml. = 1 mole. The 90% peroxide supplied by Buffalo Electrochem. Co. is very pure and requires no stabilizer; the rate of decomposition, 1% per year at 30°, is much less than for 30–35% solutions.[1] The papers cited[1] discuss potential hazards in the handling of high-strength hydrogen peroxide. Experiments with sizable amounts of even 30% reagent should be carried out behind an explosion proof safety screen.

For convenience in discussion of a large number of reactions, we have classified them according to reaction conditions: acidic, basic, neutral, or with various additives.

[1]E. S. Shanley and F. P. Greenspan, *Ind. Eng. Chem.*, **39**, 1536 (1947); W. C. Schumb, *ibid.*, **41**, 992 (1949)

Hydrogen peroxide, acidic.

Performic acid. Performic acid is prepared in solution by simply warming hydrogen peroxide of 30% or higher concentration with excess formic acid. The peracid attacks an olefin to form an epoxide, which is then cleaved by formic acid, with opening of the ring in one or both possible directions to give a monoformate or monoformate mixture. In usual practice the material is precipitated with water

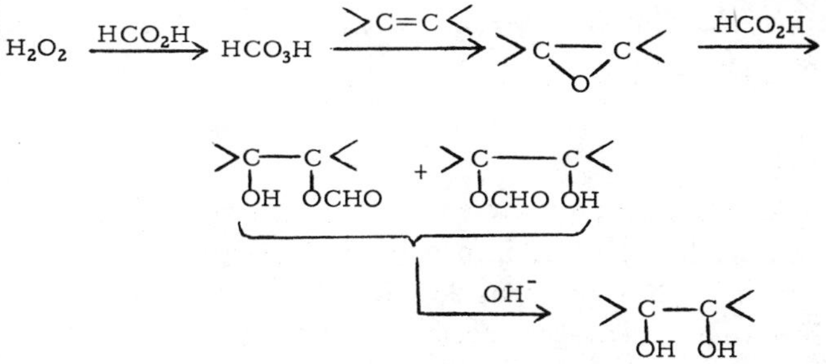

H_2O_2
HCO_2H
HCO_3H
$>C=C<$
HCO_2H
OH
OCHO
OCHO
OH
OH^-
OH
OH

and hydrolyzed with alkali to the diol. An open-chain *cis* olefin affords a *threo* diol, a cyclohexene affords a *trans*-diaxial diol; in each case the reaction can be described as a *trans* hydroxylation.

In one example[1] a mixture of 0.5 mole of oleic acid, 425 ml. of formic acid (90–98%), and 0.5 mole of 30% hydrogen peroxide is heated at 40° for 3 hrs. Excess formic acid is largely removed by distillation at 50°/125 mm. in a stream of nitrogen

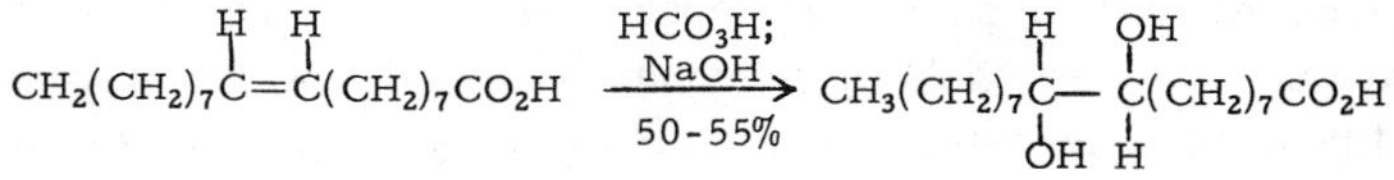

$CH_2(CH_2)_7C=C(CH_2)_7CO_2H$
$HCO_3H;$
NaOH
50-55%
$CH_3(CH_2)_7C—C(CH_2)_7CO_2H$

or carbon dioxide to prevent bumping. The residue is heated with excess 3 *N* sodium hydroxide and the solution is poured into hydrochloric acid. The aqueous solution is decanted and the residue melted under water to remove salts. Three crystallizations from 95% alcohol afford the pure *threo* diol, m.p. 94–95°.

trans-Hydroxylation of cholesterol[2] is accomplished by warming a suspension of 20 g. of the sterol in 200 ml. of 88% formic acid to 70–80° and swirling for about

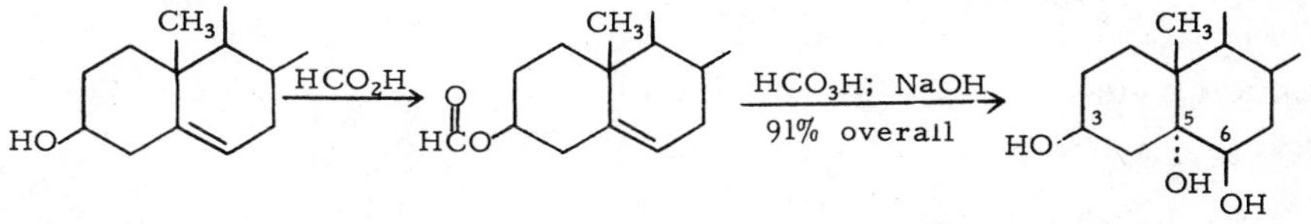

CH_3
HO
HCO_2H
HCO
HCO_3H; NaOH
91% overall
3
5
6
OH
OH

5 min., when cholesteryl formate separates as an oily upper layer. The mixture is cooled to 25° and treated with 20 ml. of 30% hydrogen peroxide. The temperature

slowly rises to 35–40° and then very slowly falls. After a total of 6–15 hrs., 300 ml. of boiling water is added, the mixture is stirred and cooled and the product collected (the 3,6-diformate can be isolated at this point, but in low yield). A solution of the moist material in 600 ml. of methanol is treated with 20 ml. of 25% sodium hydroxide, heated on the steam bath for 10 min., filtered, diluted, and acidified with hydrochloric acid. The precipitated cholestane-3β,5α,6β-triol when fully dry melts at 236–238°. Crystallization from methanol gives needles, m.p. 237–239°. A similar procedure for the reaction of cyclohexene with performic acid and hydrolysis of the product gives *trans*-1,2-cyclohexanediol in 65–73% yield.[3]

The presence of a phenyl group on one or both carbon atoms of an olefin alters the situation by causing cleavage of the epoxide ring with retention of configuration at the phenylated carbon atom with production of the *cis*-glycol. A recent example reported by Rivière[4] is that hydroxylation of 1-phenylcycloheptene (1) with performic acid gives the *cis*-diol (2), identical with the product of hydroxylation with

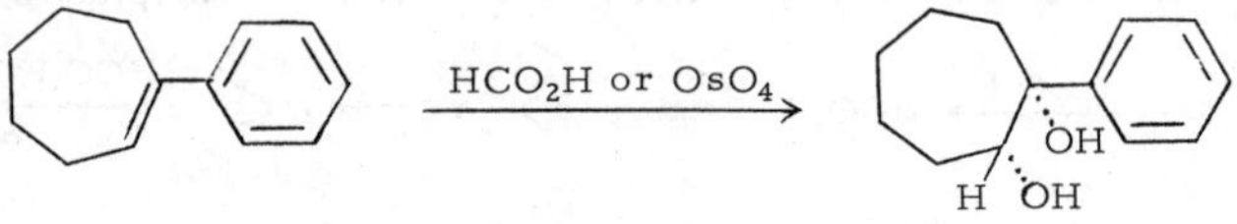
HCO2H or OsO4
OH
H OH

(1) (2)

osmium tetroxide. Rivière cites references to several previous observations of the same phenomenon, for example in papers by Wasserman[5] and by Curtin.[6] The investigators all agree that a *cis*-epoxide is formed, as in an ordinary olefin, but that the phenyl group causes cleavage to occur with retention of configuration.

Preparation of perbenzoic acid (peroxybenzoic acid). In an improved procedure[7] 0.45 mole of 70% hydrogen peroxide is added dropwise behind a safety shield to a

$$C_6H_5CO_2H + H_2O_2 \xrightarrow[85\text{-}90\%]{CH_3SO_3H} C_6H_5CO_3H$$

slurry of 0.30 mole of benzoic acid in 0.9 mole of methanesulfonic acid, with stirring and cooling in ice to maintain a temperature of 25–30° (30 min.). The benzoic acid soon dissolves. After stirring for 2 hrs. longer, the solution is cooled to 15°, ice and cold ammonium sulfate solution are added, and the mixture is extracted with three portions of benzene. The benzene solution is washed twice with ammonium sulfate solution to remove methanesulfonic acid, dried, filtered, and used directly after determination of the concentration by iodometric titration. When required for special uses, pure perbenzoic acid can be obtained from the benzene solution.

Hydroperoxides as intermediates or products. Neopentyl alcohol (3) can be prepared by acid-catalyzed addition of hydrogen peroxide to diisobutylene (1, Texaco)[8] to form the hydroperoxide (2), which in a more strongly acidic medium rearranges to neopentyl alcohol and acetone. Thus 800 g. of 30% hydrogen peroxide

$$\underset{(1)}{(CH_3)_3CCH_2\overset{CH_3}{\overset{|}{C}}{=}CH_2} + H_2O_2 \xrightarrow{H^+} \underset{(2)}{(CH_3)_3CCH_2\underset{OOH}{\underset{|}{\overset{CH_3}{\overset{|}{C}}}}\text{-}CH_3} \xrightarrow[34\text{-}40\%\ \text{overall}]{} \underset{(3)}{(CH_3)_3CCH_2OH} + O{=}C(CH_3)_2$$

is stirred with ice cooling during addition of dilute sulfuric acid, diisobutylene is added, and the mixture is stirred at 25° for 24 hrs. The upper layer of hydroperoxide is separated and added with stirring to 70% sulfuric acid at 15–25° to effect rearrangement. Neopentyl alcohol is separated by total distillation and then fractionated.

Under acid catalysis, hydrogen peroxide adds 1,4 to mesityl oxide (1) to form the hydroperoxide (2); this cyclizes to the peroxide (3), which combines with hydrogen peroxide to give as the final product crystalline mesityl oxide peroxide (4, m.p. 122°).[9]

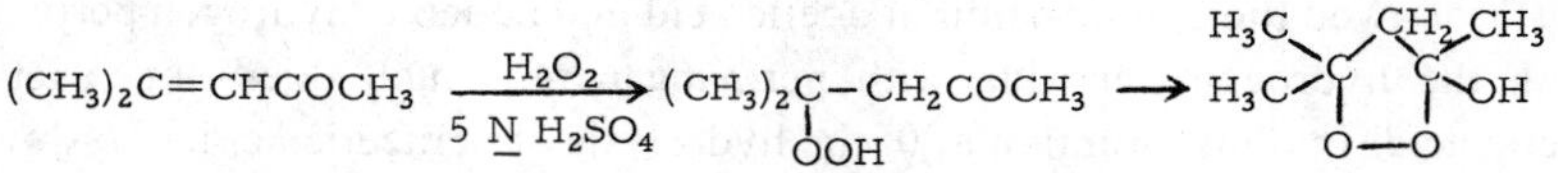

(1) (2) (3)

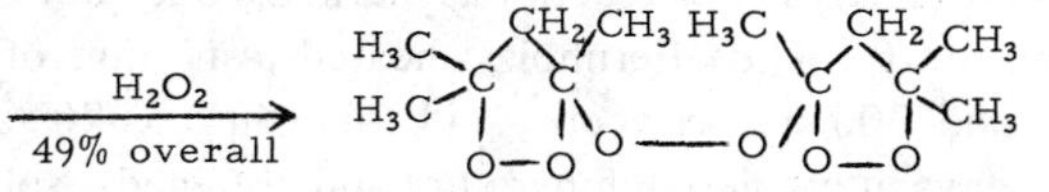

(4)

Saturated 5β- and Δ^4-unsaturated steroid 3-ketones react with hydrogen peroxide in solution in *t*-butanol containing hydrochloric acid to form bishydroperoxides (5, 6) in good yield.[10]

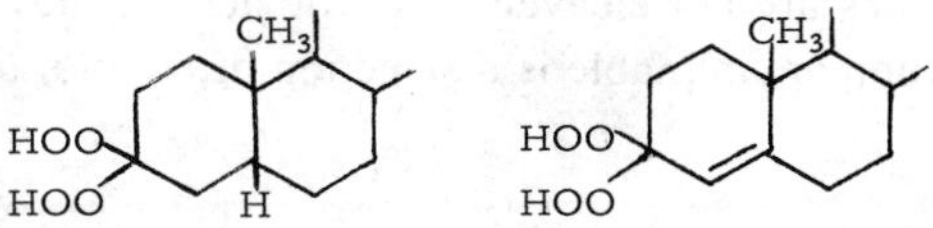

(5) (6)

The reaction of cyclohexanone with hydrogen peroxide in ether has been shown to yield (7) or the mono or bishydroperoxide according to the ratio of the reactants.[11]

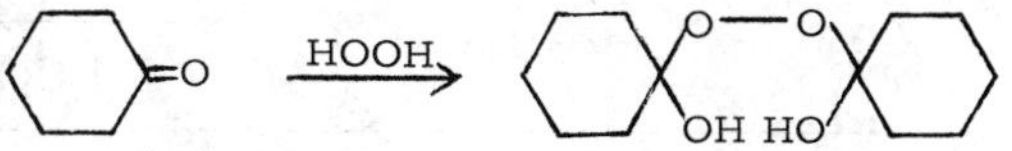

(7)

Presumably (7) results from initial addition of hydrogen peroxide to the carbonyl group to form a hydroperoxide. A similar addition probably is involved in the reaction of 2-acetocyclohexanone (8) with hydrogen peroxide in *t*-butanol containing a trace of sulfuric acid.[12] The initially formed peroxide (9) undergoes acid-catalyzed rearrangement with ring contraction to cyclopentanecarboxylic acid (11), which is obtained in 87% yield.

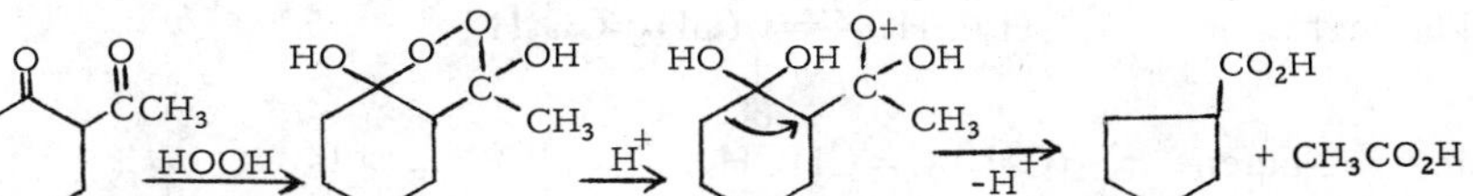

(8) (9) (10) (11)

Hydrogen peroxide–Acetic acid. (*See also* Peracetic acid.) When aqueous hydrogen peroxide, usually 30%, is added to acetic acid at room temperature, a slow reaction occurs and eventually reaches a point of equilibrium:

$$CH_3\overset{O}{\overset{\|}{C}}OH + HOOH \rightleftharpoons CH_3\overset{O}{\overset{\|}{C}}OOH + HOH$$

Reagents prepared in this way (a) contain varying amounts of peracetic acid depending on the strength of the peroxide, the amount of acetic acid taken, the temperature, the time after mixing, and the presence or absence of a catalyst, and should be distinguished from reagents of much higher concentration of CH_3CO_3H which are listed in this book as (b) Peracetic acid, commercial 40%, and (c) Peracetic acid, anhydrous.

Smit[13] followed the equilibration of acetic acid and aqueous hydrogen peroxide by titrating the hydrogen peroxide with permanganate, which does not react with peracetic acid. In dilute solution at 0° the hydrolysis of peracetic acid is so slow that one can titrate the peracid by iodimetry. At room temperature, equilibrium is reached only after several days. The reaction is markedly accelerated by 1% sulfuric acid or by warming to 70°. Thus Fernholz[14] heated a mixture of 30 ml. of 30% hydrogen peroxide and 300 ml. of acetic acid for 5 hrs. at 70° and then let the solution stand for 2 days at ambient temperature and obtained a solution containing 2–2.5% of peracetic acid.

Although aromatic peracids are generally the preferred reagents for the ***epoxidation*** of olefins, the reagent easily prepared by simply mixing 30% hydrogen peroxide with acetic acid can be used for this purpose, since under the mild conditions of formation (25°) epoxides are not cleaved by acetic acid as they are by the stronger formic acid. The accompanying table is a selection of reaction rates determined by Swern.[15]

Rates of Oxidation in Acetic Acid with CH_3CO_3H at 25.8° or with perbenzoic acid at 20–30°

		$k \times 10^3$	
		CH_3CO_3H	$C_6H_5CO_3H$
Ethylene	$CH_2{=}CH_2$	0.19	
Propylene	$CH_3CH{=}CH_2$	4.2	
1-Decene	$CH_3(CH_2)_7CH{=}CH_2$	4.7	
2-Butene	$CH_3CH{=}CHCH_3$	93	
4-Nonene	$CH_3(CH_2)_2CH{=}CH(CH_2)_3CH_3$	105	
Oleic acid	H H $CH_3(CH_2)_7C{=}C(CH_2)_7CO_2H$	67	
Elaidic acid	H $CH_3(CH_2)_7C{=}C(CH_2)_7CO_2H$ H	59	
2-Methyl-2-butene	$(CH_3)_2C{=}CHCH_3$	1240	
Cyclobutene		20.4	
Cyclopentene		195	
Cyclohexene		129	
Cycloheptene		175	
Allylbenzene	$C_6H_5CH_2CH{=}CH_2$	1.9	6-15
Styrene	$C_6H_5CH{=}CH_2$	11.2	35
1,1-Diphenylethylene	$(C_6H_5)_2C{=}CH_2$	48	
Cinnamic acid	$C_6H_5CH{=}CHCO_2H$		0.13

Reactions of another type involve ***cleavage of a carbon–carbon bond***. Phenanthrenequinone is oxidized smoothly to diphenic acid by hydrogen peroxide in warm acetic acid.[16] Use of the reaction in structure elucidation is illustrated as follows.[17] One of five disulfonic acids formed on sulfonation of phenanthrene was converted into the dimethoxyphenanthrene, which on oxidation afforded a dimethoxy-9,10-phenanthrenequinone. Oxidation of this quinone with H_2O_2–AcOH gave an acidic product

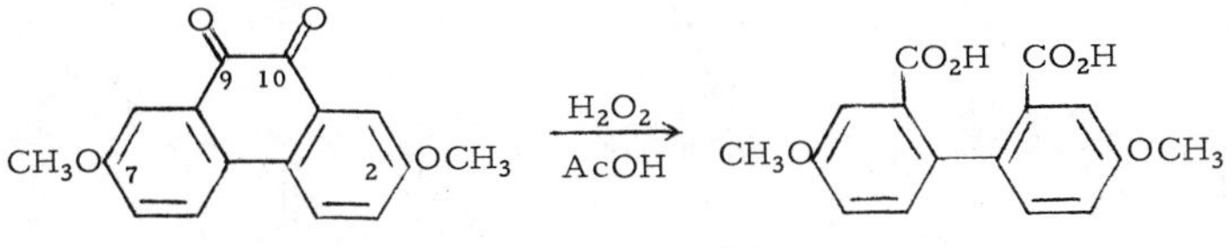
9
10
7
2
CH_3O
OCH_3
H_2O_2
AcOH
CO_2H
CO_2H
CH_3O
OCH_3

(1) (2)

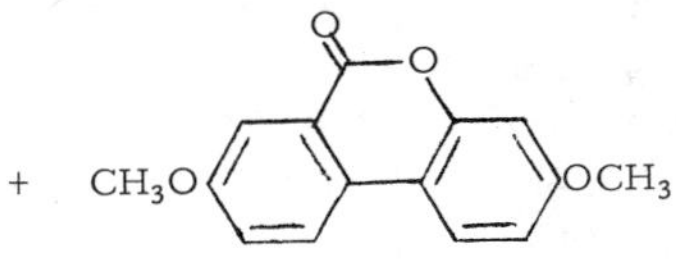
+
CH_3O
OCH_3
O

(3)

identified by synthesis as 3,3′-dimethoxydiphenic acid (2); the product of sulfonation was thus shown to be the 2,7-diacid. The peracid oxidation afforded a substantial amount of a neutral product characterized as the lactone (3); this substance must arise from cleavage of a bond extending from a carbonyl group to a terminal ring.

Diphenic acid itself can be prepared without isolation of phenanthrenequinone by direct oxidation of phenanthrene with hydrogen peroxide in refluxing acetic acid.[18]

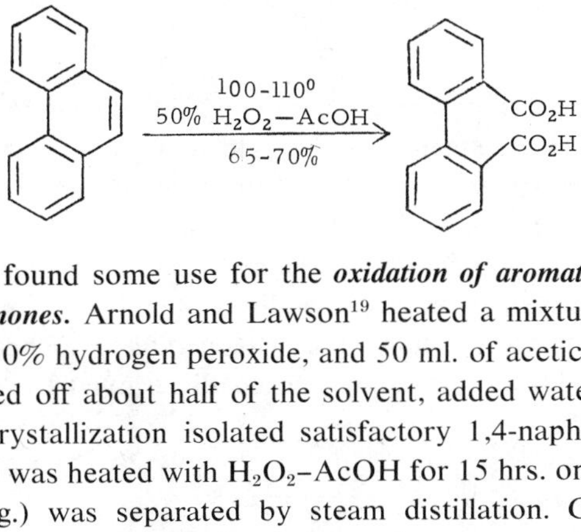
100-110°
50% H_2O_2–AcOH
65-70%
CO_2H
CO_2H

The reagent has found some use for the ***oxidation of aromatic hydrocarbons and phenols to para quinones***. Arnold and Lawson[19] heated a mixture of 10 g. of naphthalene, 25 ml. of 30% hydrogen peroxide, and 50 ml. of acetic acid just above 80° for 45 min., distilled off about half of the solvent, added water to precipitate the product, and by crystallization isolated satisfactory 1,4-naphthoquinone in 20% yield. Durene (5 g.) was heated with H_2O_2–AcOH for 15 hrs. on the steam bath and duroquinone (2.1 g.) was separated by steam distillation. Crude 2-methyl-1,4-naphthoquinone and 2,3-dimethyl-1,4-naphthoquinone were obtained in yields of 30 and 78%. Unchanged hydrocarbon was present at the end of each oxidation.

Cava and Shirley[20] heated a solution of 2 g. of the hydrocarbon (1) in 30 ml. of acetic acid and 6 ml. of 30% hydrogen peroxide at 80–85° for 5 hrs. and isolated the bright yellow quinone (2). Although the yield was low the quinone proved interesting. It reacted with butadiene at 90–100° to give, after 90 min., the colorless adduct (3) in 93% yield. Under the same conditions 2,3-dimethyl-1,4-naphthoquinone was recovered unchanged after 5 days. The enhanced reactivity of (2) as a dienophile is attributed to the decrease in strain which results on conversion of the cyclobutene

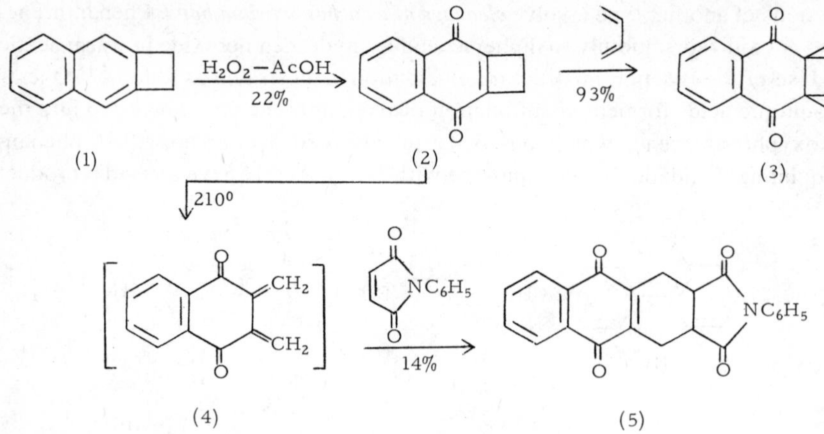
H_2O_2–AcOH
22%
93%
(1)
(2)
(3)
210°
CH_2
CH_2
NC_6H_5
14%
NC_6H_5
(4)
(5)

ring of (2) to the cyclobutane ring of (3). The quinone (2) can also react as a Diels-Alder diene at elevated temperatures, probably by rupture of the cyclobutene ring to produce the transient diene (4). Thus pyrolysis in the presence of N-phenylmaleimide afforded the anthraquinone derivative (5). Dehydrogenation of the initially formed adduct at the expense of the quinone (2) probably accounts for the low yield.

Bryce-Smith[21] found that 2,5- and 2,6-dimethylphenol can be oxidized efficiently to the *p*-benzoquinones with the reagent. Thus a solution of 15 g. of 2,5-dimethylphenol in 100 ml. of acetic acid was treated with 1 ml. of concd. sulfuric acid and

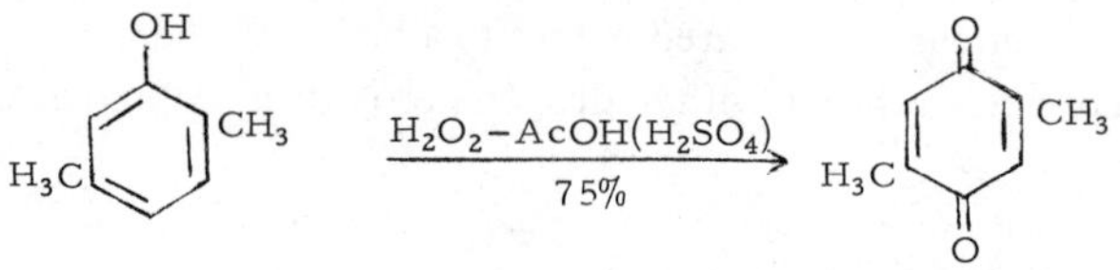
OH
CH_3
H_3C
H_2O_2–AcOH(H_2SO_4)
75%
CH_3
H_3C

cooled to 20° during the dropwise addition of 12 ml. of 80% hydrogen peroxide in the course of 45 min. The temperature was held at 20° for 1 hr. more, let rise to 60°, and checked at 60° for 10 min., when the exothermic reaction was over. On dilution and cooling, the quinone separated in yellow crystals, m.p. 125° (12.5 g.).

Oxidation of arylamines, nitroso compounds, and azobenzenes. Several 2,6-dihaloanilines have been oxidized successfully to the nitroso compounds with 30% hydrogen peroxide in acetic acid at room temperature.[22] When a solution of the reactants is let stand for a time, crystals of the (dimeric) nitroso compound begin to

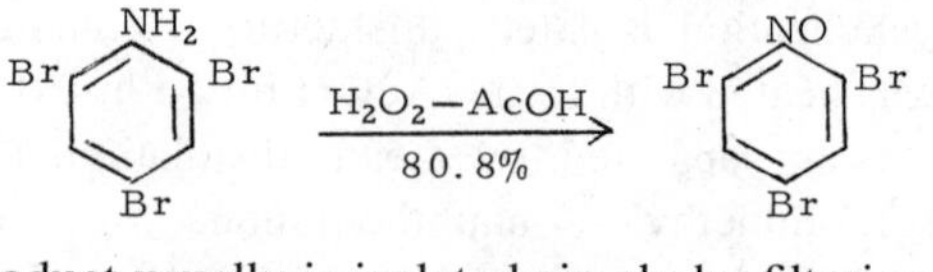
NH_2
Br
Br
Br
H_2O_2–AcOH
80.8%
NO
Br
Br
Br

separate, and the product usually is isolated simply by filtering the crystals. Yields are in the range 20–90%. If the reaction is conducted at steam bath temperature for a more prolonged period oxidation proceeds further to give the nitro compound.

The oxidation to nitroso derivatives apparently requires two bulky *ortho* substituents. Thus *p*-nitroaniline is oxidized under mild conditions to *p,p'*-dinitroazobenzene and *p,p'*-dinitroazoxybenzene.[23]

For oxidation of 1,2-dimethyl-4-nitro-5-nitrosobenzene to the *o*-dinitrobenzene, Kuhn and van Klaveren[24] dissolved 10 g. of nitroso compound in 300 ml. of acetic

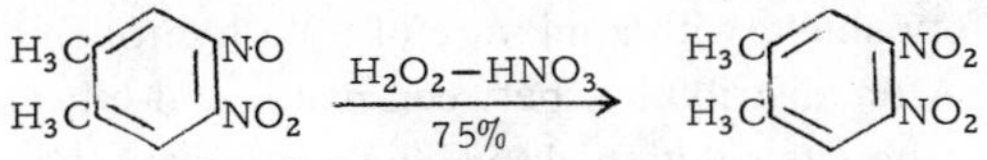

acid and added a mixture of 150 ml. of acetic acid and 150 ml. of 33% hydrogen peroxide, followed by 10 ml. of nitric acid (sp. gr. 1.40). On warming for a few minutes the color changed from dark green to orange, the dilution with water precipitated the dinitroxylene. Three other alkylated *o*-dinitro derivatives were obtained in yields of 62–88%. Other oxidizing agents gave poor results because of attack on the alkyl groups.

The combination of hydrogen peroxide with sulfuric acid serves well for oxidation of aminopyridines to nitropyridines.[25] For example, a solution of 20 g. of 4-amino-2-methylpyridine in 100 ml. of conc. sulfuric acid is dropped into a mixture of 350 ml. of 15% fuming sulfuric acid and 175 ml. of 30% hydrogen peroxide at a temperature kept at 10–20°. The mixture is stirred at 20° for 1 hr., let stand for 2 days, and

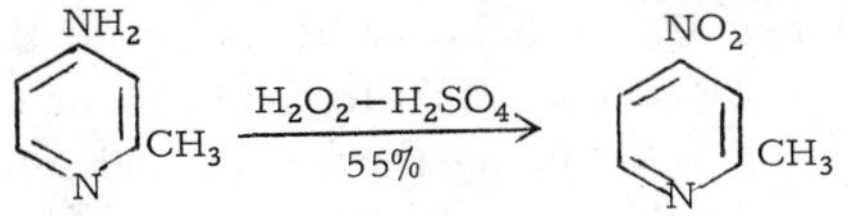

poured onto ice. After neutralization, the nitro compound is extracted with benzene.

Taylor and McKillop[26] found the combination of hydrogen peroxide and trifluoroacetic acid to be the preferred reagent for the oxidation of 5-nitrosopyrimidines (readily available) to the 5-nitropyrimidines.

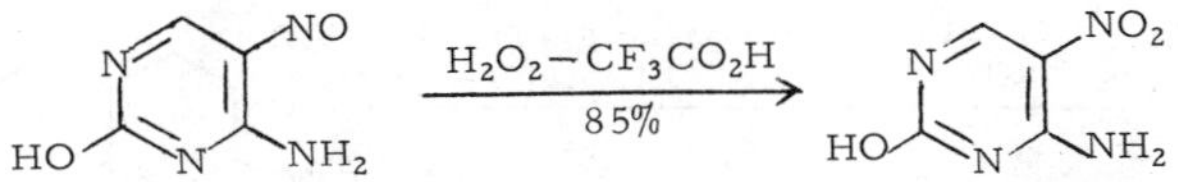

Azobenzenes are oxidized to azoxybenzenes when heated with 30% hydrogen peroxide in acetic acid at 65° or at the reflux temperature.[27] In some instances a reaction period of 24 hrs. is required. Yields are in the range 50–90%.

For generation of halogen. 2,6-Dichloroaniline is prepared by dichlorination of sulfanilamide and hydrolytic removal of the sulfonamide group.[28] The required

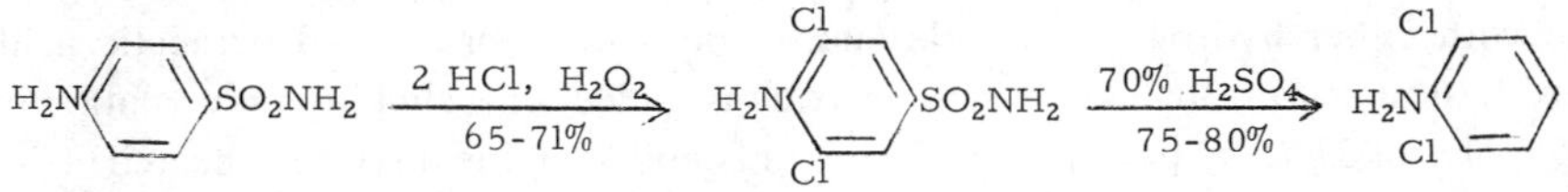

halogen is generated *in situ* by oxidation of hydrochloric acid with hydrogen peroxide. Hydrogen peroxide (0.58 mole) is added dropwise to a stirred solution of 0.29 mole of sulfanilamide in dilute hydrochloric acid and the temperature is checked at 60° as the product begins to separate. Dichlorosulfanilamide is collected and refluxed for 2 hrs. with 70% sulfuric acid. The solution is poured into water and 2,6-dichloroaniline is recovered by steam distillation. 2,6-Dibromoaniline can be prepared by the same procedure.

H_2O_2–BF_3 in the Baeyer-Villiger reaction. Baeyer-Villiger cleavage of simple aliphatic ketones to esters can be effected with a complex of hydrogen peroxide with boron trifluoride etherate.[29] A solution of 2-octanone in ether is treated with ice

$$CH_3CH_2CH_2CH_2CH_2CH_2\overset{O}{\overset{\|}{C}}CH_3 \xrightarrow[60\%]{H_2O_2-BF_3} CH_3CH_2CH_2CH_2CH_2CH_2O\overset{O}{\overset{\|}{C}}CH_3$$

cooling behind a safety shield with a mixture of 90% hydrogen peroxide and boron trifluoride etherate. After distribution between water and ether, the ethereal layer was washed with bicarbonate solution, dried, and evaporated. Fractionation afforded hexanyl acetate as the major product, along with 8% of hexanol and 32% of starting material.

$RSCOCH_3 \longrightarrow RSO_3H$. Swern and co-workers[30] developed an efficient synthesis of alkylsulfonic acids involving photocatalytic addition of thiolacetic acid to a terminal olefin and oxidation of the adduct with 90% hydrogen peroxide in acetic

$$RCH{=}CH_2 + CH_3COSH \xrightarrow[50\text{-}90\%]{h\nu} RCH_2CH_2SCOCH_3 \xrightarrow[\text{quant.}]{H_2O_2,\ AcOH} RCH_2CH_2SO_3H$$

acid. The oxidation is carried out at 65–70° overnight. Oxidation with hydrogen peroxide alone was unsatisfactory. The reaction sequence provides a route to isomer-free *n*-alkylsulfonic acids.

Oxidative rearrangement. Corey and co-workers[31] made the suprising discovery that the 3β,11α-dihydroxy-Δ^{12}-pentacyclic triterpenoid (1) on treatment in methylene chloride with a solution of 30% hydrogen peroxide and *p*-toluenesulfonic acid in *t*-butanol affords, after acetylation, the 11α,12α-epoxide, with a rearranged skeletal

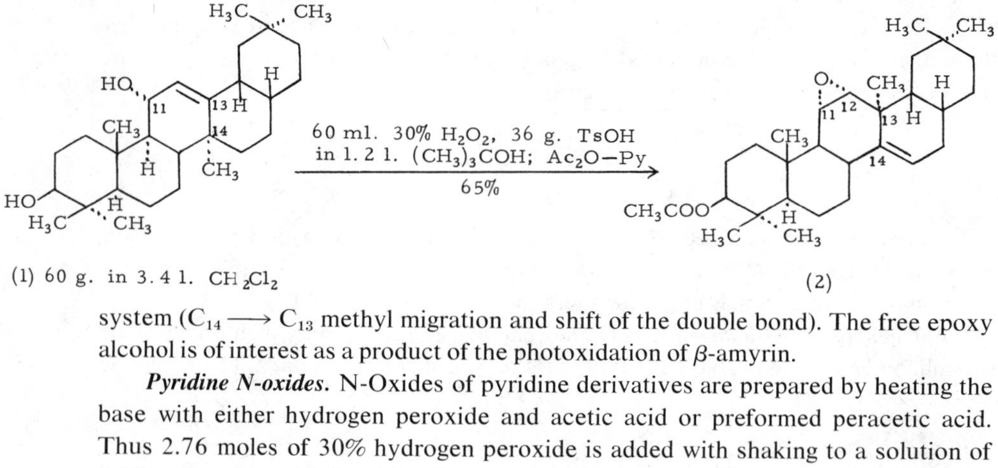
60 ml. 30% H₂O₂, 36 g. TsOH
in 1.2 l. (CH₃)₃COH; Ac₂O–Py
65%

(1) 60 g. in 3.4 l. CH_2Cl_2 (2)

system ($C_{14} \longrightarrow C_{13}$ methyl migration and shift of the double bond). The free epoxy alcohol is of interest as a product of the photoxidation of β-amyrin.

Pyridine N-oxides. N-Oxides of pyridine derivatives are prepared by heating the base with either hydrogen peroxide and acetic acid or preformed peracetic acid. Thus 2.76 moles of 30% hydrogen peroxide is added with shaking to a solution of 2.15 moles of 3-methylpyridine in 600 ml. of acetic acid, the mixture is heated at 70°

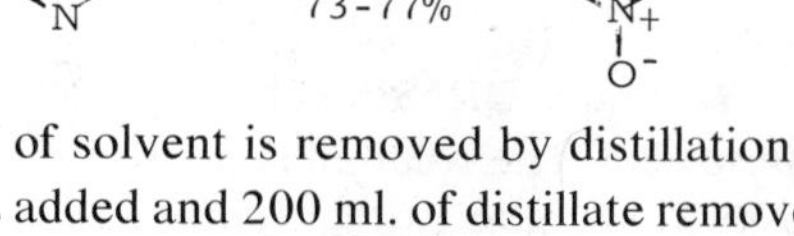
H₂O₂–AcOH
73-77%

for 24 hrs., and 500 ml. of solvent is removed by distillation at reduced pressure.[32] Then 200 ml. of water is added and 200 ml. of distillate removed. The mixture is then made alkaline and the liquid N-oxide is recovered by extraction with chloroform and distilled. Nicotinamide-1-oxide is prepared in the same way and crystallized from water (m.p. 291–293°, yield 73–83%).[33] A procedure for the preparation of pyridine-N-oxide[34] calls for addition with stirring of 1.50 moles of 40% peracetic acid to 1.39 moles of pyridine at such a rate that the temperature does not exceed 40°, solvent is removed, and the product (m.p. 65–66°) is distilled in an apparatus suitable for solids, for example, the two-bulb flask of Fig. H-4.

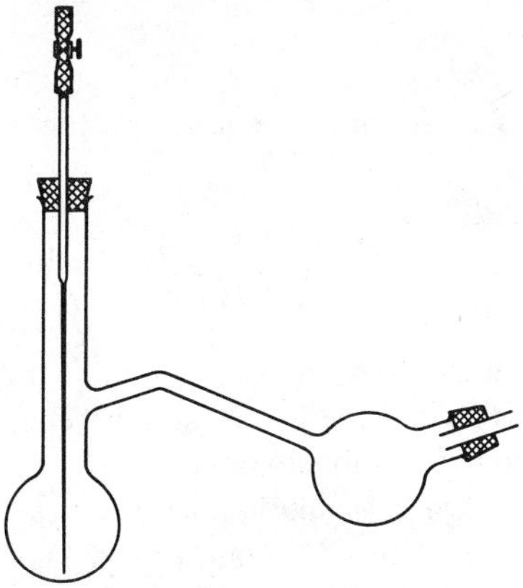

Fig. H-4 Distillation of a solid

[1]D. Swern, J. T. Scanlan, and G. B. Dickel, *Org. Syn., Coll. Vol.*, **4**, 317 (1963)
[2]L. F. Fieser and S. Rajagopalan, *Am. Soc.*, **71**, 3938 (1949)
[3]A. Roebuck and H. Adkins, *Org. Syn., Coll. Vol.*, **3**, 217 (1955)
[4]H. Rivière, *Bull. soc.*, 97 (1964)
[5]H. H. Wasserman and N. E. Aubrey, *Am. Soc.*, **78**, 1726 (1956)
[6]D. Y. Curtin, A. Bradley, and Y. G. Hendrickson, *Am. Soc.*, **78**, 4064 (1956)
[7]L. S. Silbert, E. Siegel, and D. Swern, *Org. Syn.*, **43**, 93 (1963)
[8]J. Hoffman, *Org. Syn.*, **40**, 76 (1960)
[9]A. Rieche, E. Schmitz, and E. Grundemann, *Ber.*, **93**, 2443 (1960)
[10]J. Warnant, R. Joly, J. Mathieu, and L. Velluz, *Bull. soc.*, 331 (1957); L. Velluz, G. Amiard, J. Martel, and J. Warnant, *ibid.*, 879, 1484 (1957)
[11]N. A. Milas, U.S. patent 2,223,807 (1939); R. Criegee, W. Schnorrenberg, and J. Becke, *Ann.*, **565**, 7 (1949)
[12]G. B. Payne, *J. Org.*, **26**, 4793 (1961); L. P. Vinogradova and S. I. Zav'yalov, *Bull. Acad. Sci. USSR, Div. Chem. Sci.*, 2050 (1961)
[13]W. C. Smit, *Rec. trav.*, **49**, 675 (1930)
[14]H. Fernholz, *Ber.*, **84**, 110 (1951)
[15]D. Swern, *Am. Soc.*, **69**, 1692 (1947)
[16]A. F. Holleman, *Rec. trav.*, **23**, 169 (1904)
[17]L. F. Fieser, *Am. Soc.*, **51**, 2471 (1929)
[18]W. F. O'Connor and E. J. Moriconi, *Am. Soc.*, **73**, 4044 (1951); *Ind. Eng. Chem.*, **45**, 277 (1953)
[19]R. T. Arnold and R. Lawson, *J. Org.*, **5**, 250 (1940)
[20]M. P. Cava and R. V. Shirley, *J. Org.*, **26**, 2212 (1961)
[21]D. Bryce-Smith and A. Gilbert, *J. Chem. Soc.*, 873 (1964)
[22]R. R. Holmes and R. P. Bayer, *Am. Soc.*, **82**, 3454 (1964)
[23]H. R. Gutmann, *Experientia*, **20**, 128 (1964)
[24]R. Kuhn and W. van Klaveren, *Ber.*, **71**, 779 (1938)
[25]A. Kirpal and W. Böhm, *Ber.*, **64**, 767 (1931); **65**, 680 (1932); R. H. Wiley and J. L. Hartman, *Am. Soc.*, **73**, 494 (1951); E. V. Brown, *ibid.*, **76**, 3167 (1954)
[26]E. C. Taylor and A. McKillop, *J. Org.*, **30**, 3153 (1965)
[27]B. T. Newbold, *J. Org.*, **27**, 3919 (1962)
[28]M. Seikel, *Org. Syn., Coll. Vol.*, **3**, 262 (1955)
[29]J. D. McClure and P. H. Williams, *J. Org.*, **27**, 24 (1962)
[30]J. S. Showell, J. R. Russell, and D. Swern, *J. Org.*, **27**, 2853 (1962)
[31]I. Agata, E. J. Corey, A. G. Hortmann, J. Klein, S. Proskow, and J. J. Ursprung, *J. Org.*, **30**, 1698 (1965)
[32]E. C. Taylor, Jr., and A. J. Crovetti, *Org. Syn., Coll. Vol.*, **4**, 655 (1963)
[33]*Idem, ibid.*, **4**, 704 (1963)
[34]H. S. Mosher, L. Turner, and A. Carlsmith, *ibid.*, **4**, 828 (1963)

Hydrogen peroxide, basic.

Reaction with enones and quinones. For an α,β-unsaturated ketone to react with hydrogen peroxide a basic medium is required,[1,2] the probable function being to

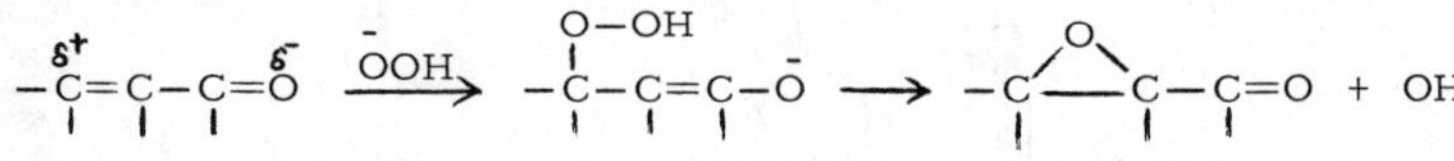

(1) (2)

generate a nucleophilic hydrogen peroxide anion. 1,4-Attack by this species to form (1) and expulsion of hydroxide ion closes the epoxide ring (2). The epoxide function is cleaved by acid but is stable to base.

For conversion of 2-methyl-1,4-naphthoquinone into the 2,3-epoxide,[2] 1 ml. of 30% hydrogen peroxide is added to a solution of 0.2 g. of anhydrous sodium carbonate in 5 ml. of water and the mixture is added to a warm solution of 1 g. of the quinone in 10 ml. of 95% ethanol. The yellow color is discharged at once and on

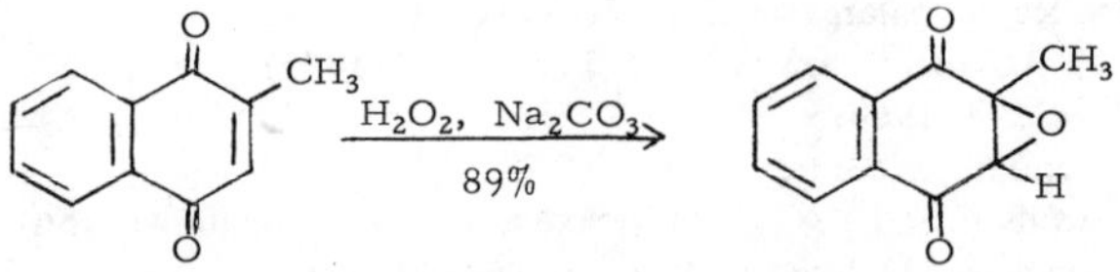
CH3
H2O2, Na2CO3
89%
CH3
O
H

cooling the epoxide separates in colorless crystals (0.97 g.), m.p. 93.5–94.5° (pure 95.5–96.5°). The same procedure is applicable to 1,4-naphthoquinone itself and to its 2,3-dialkyl derivatives, including vitamin K_1.[3] For effecting the first step in a Hooker oxidation of a 2-alkyl-3-hydroxy-1,4-naphthoquinone (3),[4] 0.01 M of a quinone of molecular weight 244–384 is dissolved in 25 ml. of dioxane, by slight warming if necessary, and a solution of 1.2 g. of sodium carbonate in 25 ml. of water is added. The resulting deep red solution or suspension of the sodium salt is heated in a water bath maintained at 70°, the air is largely displaced by passing a slow stream of nitrogen over the surface of the liquid, and 2 ml. of 30% hydrogen peroxide is added. After 20–40 min. the solution becomes colorless. On cooling and acidification the colorless dihydroxyindanonecarboxylic acid (6) separates in high yield. A plausible sequence of events is addition of hydrogen peroxide to the anion (4) to form the hydrate anion (5), and benzilic acid rearrangement to the anion of (6).

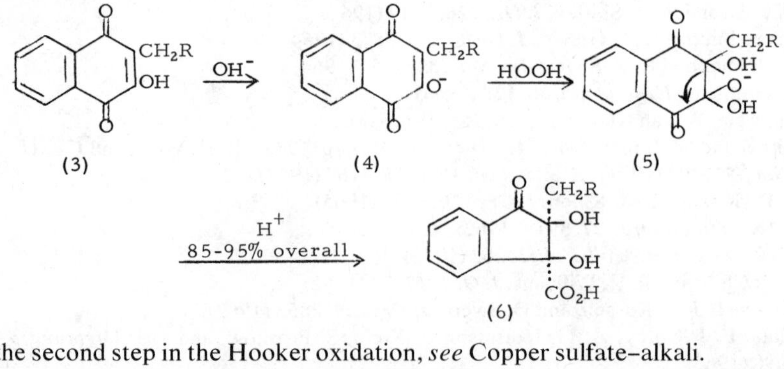
CH2R
OH
OH⁻
CH2R
O⁻
HOOH
CH2R
OH
O⁻
OH
H⁺
85-95% overall
CH2R
OH
OH
CO2H

(3) (4) (5) (6)

For the second step in the Hooker oxidation, *see* Copper sulfate–alkali.

Procedures for the conversion of isophorone[5] (7) and 2,3-diphenylindenone[6] (8) into the corresponding epoxides specify sodium hydroxide as base and methanol or ethanol as organic solvent.

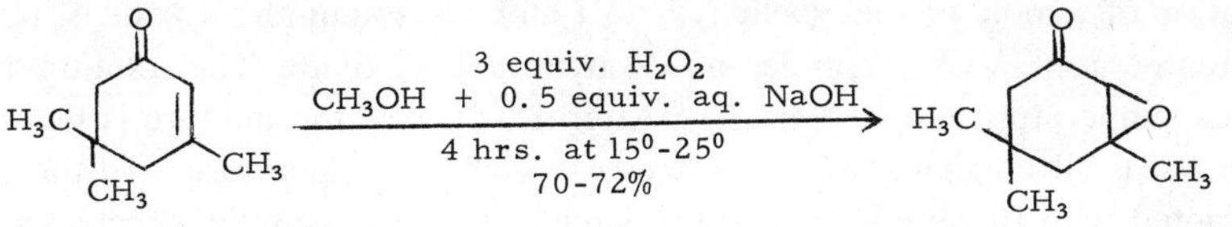

(7)

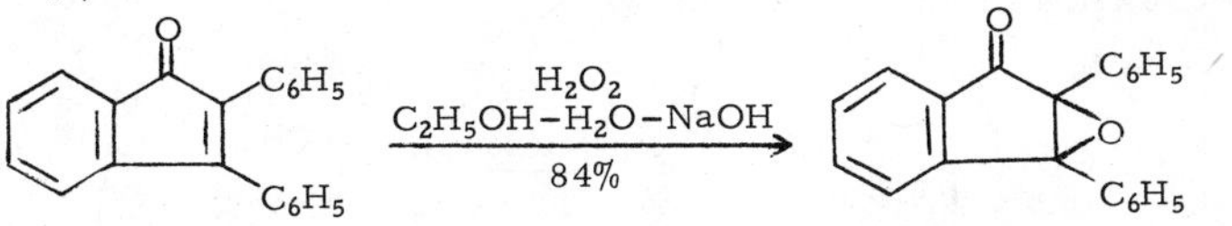

(8)

Early attempts to epoxidize simple α,β-unsaturated aldehydes with hydrogen peroxide in the presence of alkali afforded only acidic products. Payne[7] (Shell Development Co.) found, however, that highly alkali-sensitive aldehydes such as acrolein and methacrolein can be epoxidized successfully by controlling the pH to 8–8.5. Thus acrolein is added to a dilute solution of hydrogen peroxide maintained

$$CH_2=CHCH=O \xrightarrow[75-85\%]{H_2O_2,\ pH\ 8-8.5} CH_2-CHCH=O$$

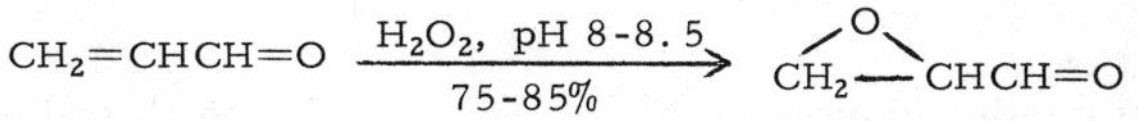

at pH 8–8.5 (pH meter) by the continuous addition of dilute sodium hydroxide. Yields of glycidaldehyde, as determined by titration, are in the range 75–85%. Ethyl ethylidenemalonate was epoxidized satisfactorily by the same procedure in methanol.[8] Later work by Payne[9] suggests that the procedure could be simplified by use of potassium bicarbonate as buffer.

$$CH_3CH=C\begin{smallmatrix}CO_2C_2H_5\\CO_2C_2H_5\end{smallmatrix} \xrightarrow[82\%]{H_2O_2,\ pH\ 7.5-8} CH_3CH-C\begin{smallmatrix}CO_2C_2H_5\\CO_2C_2H_5\end{smallmatrix}$$

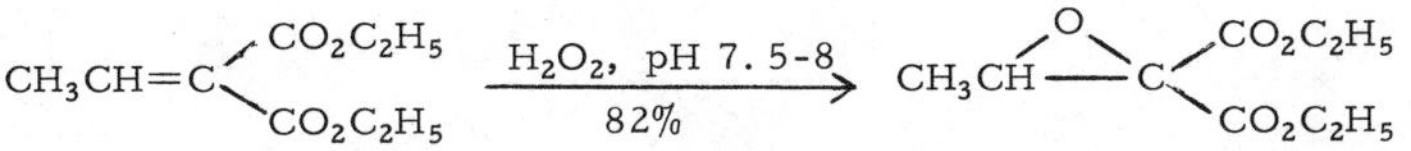

The preparation of 2,5-dihydroxy-1,4-benzoquinone by stirring a solution of hydroquinone in a mixture of 37% hydrogen peroxide and 50% sodium hydroxide solution[10] can be pictured as involving oxidation to quinone, addition of water (or of OH⁻) to form hydroxyhydroquinone (or anion), oxidation to hydroxyquinone, and further addition and oxidation. The product separates as the orange-red disodium salt, which is dissolved in water and acidified.

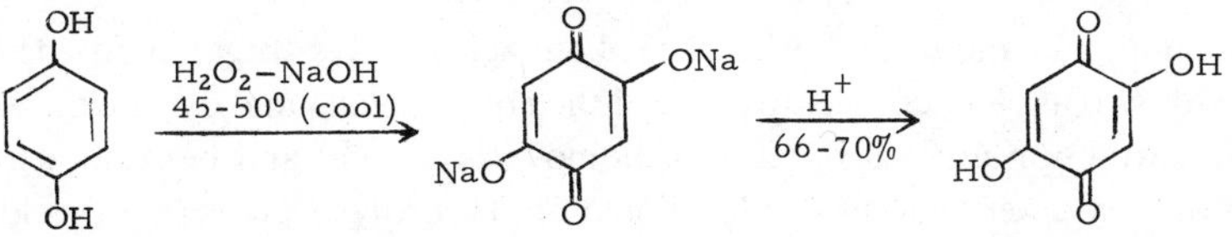

Dakin reaction. This reaction also employs hydrogen peroxide in a basic medium. It is applicable to a phenol carrying in the *ortho* position either an aldehydic or a ketonic group. Thus salicylaldehyde is converted by alkaline hydrogen peroxide into catechol in good yield.[11] The reaction has been interpreted[12] by the mechanism formulated.

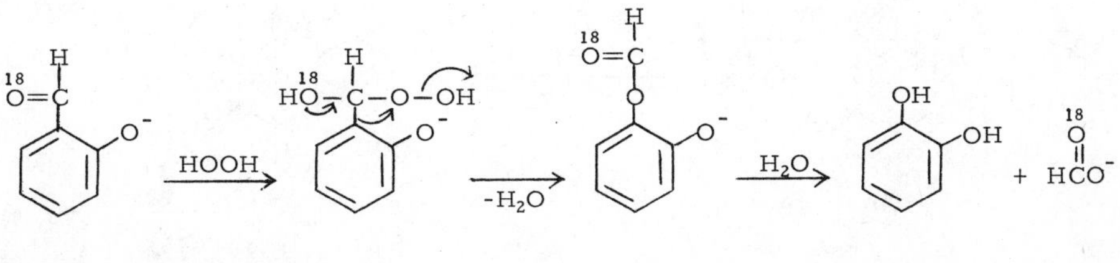

A solution of 1 mole of salicylaldehyde in 1 l. of 1 *N* sodium hydroxide is treated at room temperature with 1.2 moles of 3% hydrogen peroxide. The solution darkens and the temperature rises to 45–50°. After 15–20 hrs. the mixture is neutralized with acetic acid and evaporated to dryness at room temperature, and the residue is extracted with toluene. A procedure later applied to *o*-vanillin specifies that the reaction be run in a stream of nitrogen.[13]

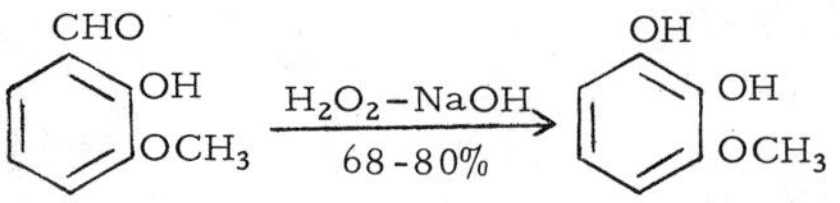

Application of the Dakin reaction to an *o*-acetophenol encounters the difficulty that the starting material forms a chelated sodium salt which is sparingly soluble.

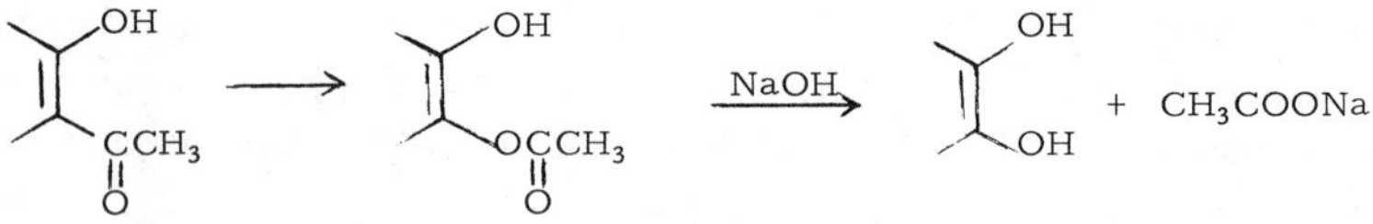

In this case tetramethylammonium hydroxide is preferred to sodium or potassium hydroxide as the base. Triton B is less satisfactory.[14]

Cleavage of α-keto acids and α-diketones. The cleavage of an α-keto acid is the terminal step in a synthesis of homoveratric acid (4) from veratraldehyde (1), via the azlactone (2).[15] Alkali cleaves the azlactone ring with elimination of benzoic

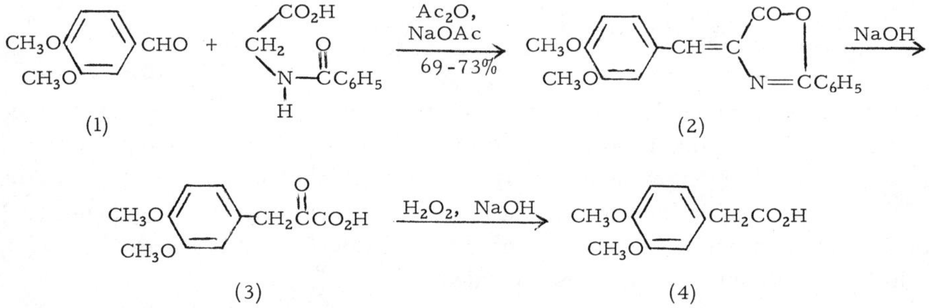

acid and ammonia and formation of the α-keto acid (3). Hydrogen peroxide (30%) is added with stirring and cooling to 15° and the solution let stand overnight. Acidification and extraction gives a mixture of homoveratric acid and benzoic acid, which is separated by esterification and fractionation. Hydrolysis gives (4) in yield of 51% from the azlactone.

Jeger's group[16] used alkaline peroxide to cleave the 11,12-bond of the unsaturated triketone (1) obtained by oxidation of Δ^8-lanostene.

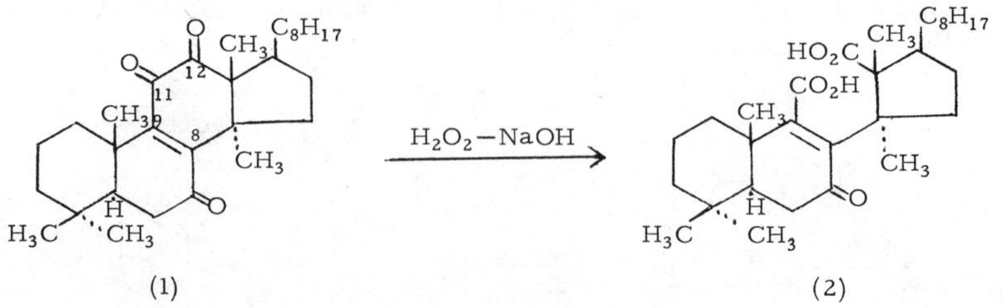

Oxidation of allodunnione (1). Price and Robinson[17] oxidized allodunnione with alkaline peroxide and isolated a crystalline product regarded as a lactonic acid.

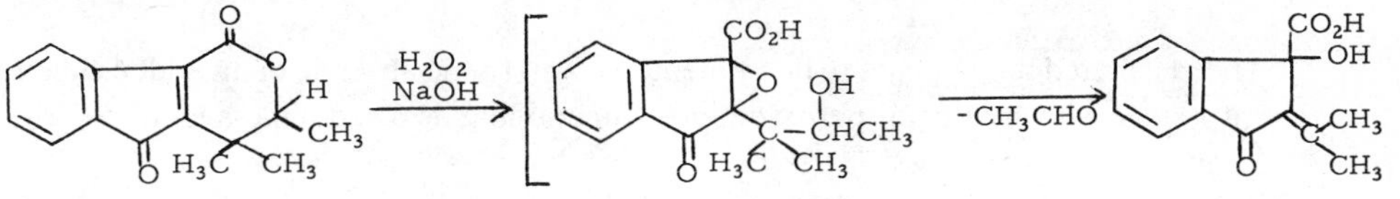

(1) (2) (3)

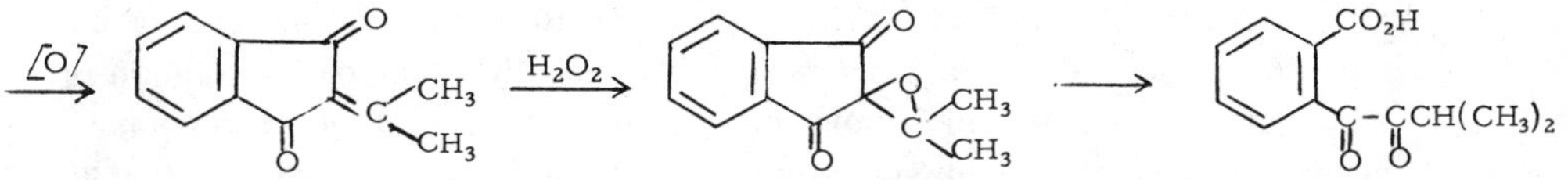

(4) (5) (6)

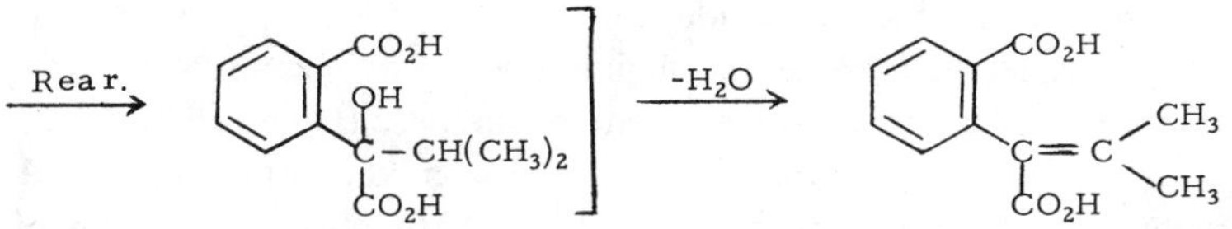

(7) (8)

Later workers[18] showed that the compound is not lactonic and identified it as the known α-isopropylidene-homophthalic acid (8). A possible pathway involving a benzilic acid rearrangement (6 ⟶ 7) is formulated.

Reaction with nitriles. The base-catalyzed reaction of a nitrile with hydrogen peroxide to form the corresponding amide and molecular oxygen was first described by Radziszewski.[19] A mixture of 0.75 mole of *o*-tolunitrile, 2.6 moles (300 ml.)

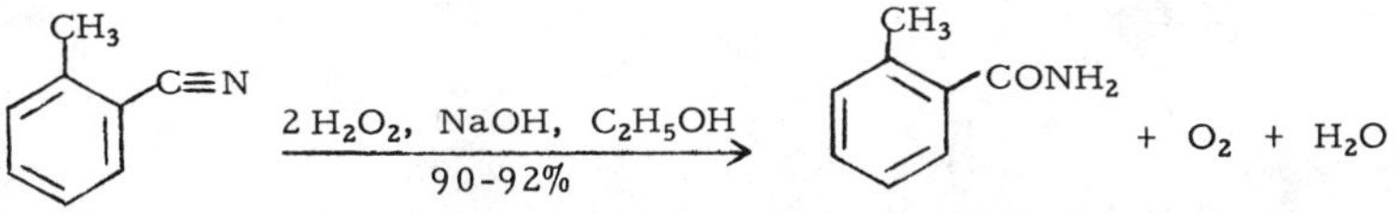

of 30% hydrogen peroxide, 400 ml. of 95% ethanol, and 300 ml. of 6*N* sodium hydroxide forms a homogeneous solution which soon warms up and begins to evolve oxygen.[20] The temperature is controlled at 40–50° by cooling for about 1 hr. and then kept at 50° for 3 hrs. by heating. The mixture is neutralized and steam distilled until 1 l. of distillate is collected and the product let crystallize. A similar procedure affords veratric amide in 87–92% yield.[21] Aromatic nitriles generally afford amides in yields of 80–90%; in the aliphatic series yields are only 50–60%.

Wiberg[22] found that the reaction between benzonitrile and hydrogen peroxide shows first-order dependence on the concentration of the nitrile, hydrogen peroxide, and hydroxide ion; that benzonitrile oxide is not an intermediate; that the oxygen introduced into the nitrile comes from hydrogen peroxide and not from water or hydroxide ion; and that the effect of substituents on the reaction rate parallels that observed in known nucleophilic attacks on the nitrile carbon atom. These results suggested that the initiating reaction is an attack on the nitrile by hydroperoxide anion with formation of the peroxycarboximidic acid (3). This substance has not

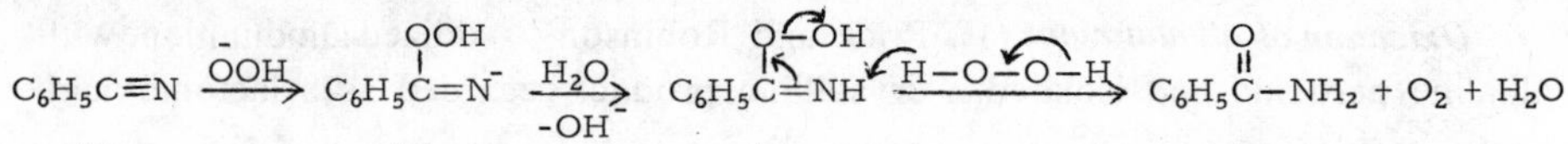

$$C_6H_5C\equiv N \xrightarrow{\overline{O}OH} C_6H_5\overset{OOH}{\underset{}{C}}=N^- \xrightarrow[-OH^-]{H_2O} C_6H_5\overset{O-OH}{\underset{}{C}}=NH \xrightarrow{H-O-O-H} C_6H_5\overset{O}{\underset{}{C}}-NH_2 + O_2 + H_2O$$

(1)   (2)   (3)   (4)

been isolated but appears to be a potent oxidant. In the absence of an added substrate, it reacts with hydrogen peroxide, now functioning as a reducing agent, to form the amide (4) and oxygen.

Payne and co-workers[23] discovered that if the peroxycarboximidic acid (3) is generated in the presence of an olefin, the olefin functions as a reducing agent more powerful than hydrogen peroxide and is converted into the corresponding epoxide; (3) is converted as before into the amide (4). Both benzonitrile and acetonitrile have been used as co-reactants in epoxidations. For example, a solution of 1.5 moles of cyclohexene and 2 moles of acetonitrile in 300 ml. of methanol is stirred at 60° during simultaneous addition of 1 mole of 50% hydrogen peroxide and of 1 N sodium hydroxide, the latter being added at a rate to control the pH to 9.5–10 (pH meter); the amount of cyclohexene epoxide formed corresponded to a yield of 85% based on hydrogen peroxide consumed. Oxidized by this method, pyridine gives pyridine-1-oxide (79%) and aniline gives azobenzene (62%). In a simplified procedure use of potassium bicarbonate as the base (pH 8) eliminates use of a pH meter and controlled addition of alkali.[9] Thus a mixture of 0.6 mole of 2-allylcyclohexanone (1), 0.5 mole of benzonitrile, 300 ml. of methanol, and 10 g. of potassium bicarbonate is stirred during addition of 0.5 mole of 50% hydrogen peroxide and then stirred at room temperature for 40 hrs. The yield of epoxide cited is based on starting

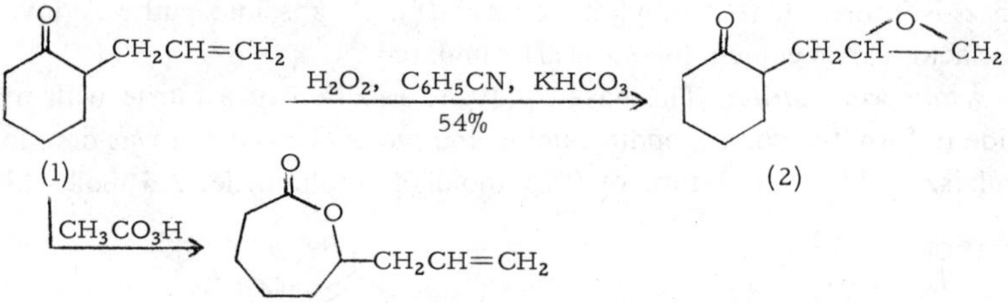

material consumed. Note that the ketone (1) on treatment with peracetic acid undergoes Baeyer-Villiger cleavage to the lactone (3). Oxidized by the bicarbonate-buffered procedure, methylenecyclohexane affords the epoxide in 73% yield.

[1]E. P. Kohler, N. K. Richtmyer, and W. F. Hester, *Am. Soc.*, **53**, 205 (1931); Pl. A. Plattner *et al.*, *Helv.*, **31**, 1822 (1948)

[2]L. F. Fieser, *J. Biol. Chem.*, **133**, 391 (1940)

[3]L. F. Fieser, M. Tishler, and W. L. Sampson, *Am. Soc.*, **62**, 1628 (1940)

[4]L. F. Fieser and M. Fieser, *Am. Soc.*, **70**, 3215 (1948)

[5]R. L. Wasson and H. O. House, *Org. Syn.*, *Coll. Vol.*, **4**, 552 (1963)

[6]E. F. Ullman and J. E. Milks, *Am. Soc.*, **84**, 1315 (1962); E. Weitz and A. Scheffer, *Ber.*, **54**, 2341 (1921)

[7]G. B. Payne, *Am. Soc.*, **81**, 4901 (1959)

[8]G. B. Payne, *J. Org.*, **24**, 2048 (1959)

[9]G. B. Payne, *Tetrahedron*, **18**, 763 (1962)

[10]R. G. Jones and H. A. Shonle, *Am. Soc.*, **67**, 1034 (1945)

[11]H. D. Dakin, *Org. Syn.*, *Coll. Vol.*, **1**, 149 (1941)

[12]C. A. Buton in J. O. Edwards, "Peroxide Reaction Mechanisms," 14–15, Interscience (1962)

[13]A. R. Surrey, *Org. Syn., Coll. Vol.*, **3**, 759 (1955)
[14]W. Baker *et al.*, *J. Chem. Soc.*, 1825 (1952); 1615 (1953)
[15]H. R. Snyder, J. S. Buck, and W. S. Ide, *Org. Syn., Coll. Vol.*, **2**, 333 (1943)
[16]E. Kyburz, B. Riniker, H. R. Schenk, H. Heusser, and O. Jeger, *Helv.*, **36**, 1891 (1953)
[17]J. R. Price and R. Robinson, *J. Chem. Soc.*, 1522 (1939); 1493 (1940)
[18]M. A. Oxman, M. G. Ettlinger, and A. R. Bader, *J. Org.*, **30**, 2051 (1965)
[19]Br. Radziszewski, *Ber.*, **18**, 355 (1885)
[20]C. R. Noller, *Org. Syn., Coll. Vol.*, **2**, 586 (1943)
[21]J. S. Buck and W. S. Ide, *ibid.*, **2**, 44 (1943)
[22]K. B. Wiberg, *Am. Soc.*, **75**, 3961 (1953)
[23]G. B. Payne and P. H. Williams, *J. Org.*, **26**, 651 (1961); G. P. Payne, P. H. Deming, and P. H. Williams, *ibid.*, **26**, 659 (1966)

Hydrogen peroxide, neutral.

Amines oxides. Procedures[1] for the preparation of an aliphatic amine (4), its conversion to the amine oxide (5), and pyrolysis of this to methylenecyclohexane (6, Cope reaction) are shown in the formulation. A homogeneous solution of 0.35

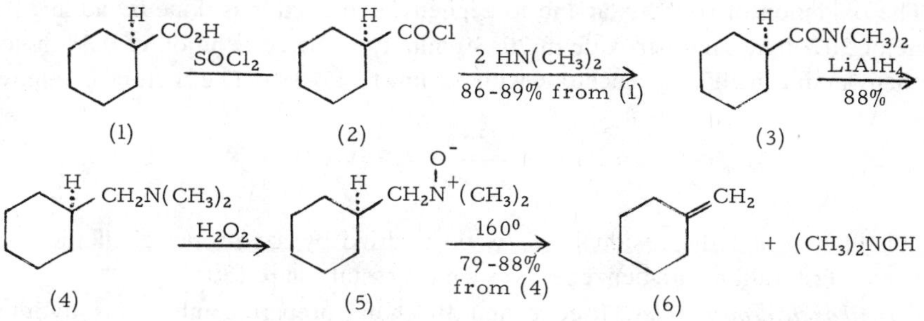
CO2H
SOCl2
COCl
2 HN(CH3)2
86-89% from (1)
CON(CH3)2
LiAlH4
88%
(1)
(2)
(3)
CH2N(CH3)2
H2O2
CH2N+(CH3)2
160°
79-88% from (4)
CH2
+ (CH3)2NOH
(4)
(5)
(6)

mole of N,N-dimethylcyclohexylmethylamine (4) and 0.35 mole of 30% hydrogen peroxide in 45 ml. of methanol is let stand at room temperature for 36 hrs., with addition after 2 hrs. and again after 5 hrs., of a further 0.35 mole of hydrogen peroxide (with other amines, cooling may be necessary). The oxidation is recognized as complete when a drop of solution gives no color when tested with phenolphthalein. Excess hydrogen peroxide is destroyed by adding a small amount of platinum black and stirring until evolution of oxygen has ceased. The solution is filtered and concentrated at 60° under reduced pressure until the amine oxide hydrate solidifies. Heating is continued at 100° at a pressure of 10 mm. until the material melts and then resolidifies. Then, at a temperature of 160°, the Cope reaction is completed in about 2 hrs. The distillate is washed with water and the hydrocarbon layer is separated, dried, and distilled; the methylenecyclohexane is very pure and 1-methylcyclohexane is completely absent. N,N-Dimethylhydroxylamine is recovered from the aqueous washings as the pure hydrochloride in yield of 78–90%.

Oxidation of S and As compounds. Thioethers can be oxidized to sulfoxides with hydrogen peroxide in acetone; with hydrogen peroxide in 50% acetic acid they yield the corresponding sulfones.[2] Thus 16.1 ml. of 30% hydrogen peroxide is added with

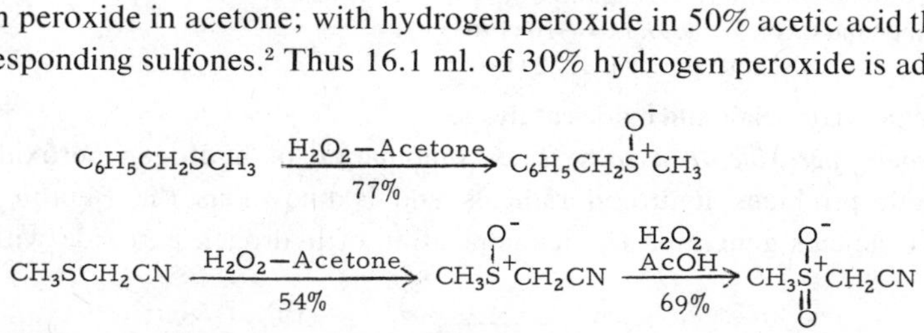
C6H5CH2SCH3
H2O2–Acetone
77%
C6H5CH2S+CH3
CH3SCH2CN
H2O2–Acetone
54%
CH3S+CH2CN
H2O2
AcOH
69%
CH3S+CH2CN

cooling to a solution of 21 g. of benzyl methyl sulfide in 60 ml. of acetone and the solution is let stand overnight. The sulfoxide produced is purified by crystallization (m.p. 54°). Hydrogen peroxide is the preferred oxidant in the final step of the preparation of di-*p*-aminophenyl disulfide.[3] The initial reactions of condensation and

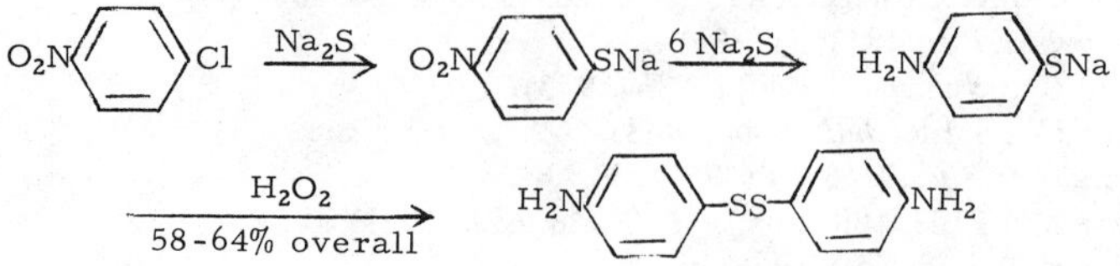
Na₂S
6 Na₂S
H₂O₂
58-64% overall

reduction are done in aqueous solution and the solution filtered to remove insoluble material, chiefly *p*-chloroaniline. The filtrate is concentrated, and 30% hydrogen peroxide is added at 65–70° over a period of 2 hrs. The disulfide separates on cooling as a solid and is purified by crystallization (m.p. 75–76°).

The oxidation of triphenylarsine to triphenylarsine oxide is done by adding 0.41 mole of 30% hydrogen peroxide in 20–30 min. to a stirred solution of 0.33 mole of triphenylarsine in 200 ml. of acetone with cooling to 25–30°.[4] The acetone is removed

$$(C_6H_5)_3As \xrightarrow[84\text{–}87\%]{H_2O_2} (C_6H_5)_3AsO$$

by distillation and the residual yellow oil is dried by azeotropic distillation with benzene. Trituration with benzene gives white crystals, m.p. 189°.

1-Alkoxyhydroperoxides. Rieche and Bischoff[5] prepared anhydrous hydrogen peroxide by treating a solution of 60 g. of 84% peroxide in ether with anhydrous sodium sulfate and magnesium carbonate. After 2 days the mixture was filtered and an aliquot titrated. One equivalent of an acetal was added to 1 equivalent of the ethereal solution, and the mixture was heated cautiously at 70°. The ether was allowed to distil to give a mixture of the acetal and the anhydrous reagent. After a reaction period of about 15 hrs. at 70°, the 1-alkoxyhydroperoxide was isolated in yields in the range 40–70%.

$$\underset{0.1\text{ m.}}{RCH(OC_2H_5)_2} + \underset{0.1\text{ m.}}{HOOH} \xrightarrow{70^0} RCH(OC_2H_5)(OOH) + C_2H_5OH$$

[1] A. C. Cope and E. Ciganek, *Org. Syn., Coll. Vol.*, **4**, 339, 612 (1963).
[2] M. Gazdar and S. Smiles, *J. Chem. Soc.*, **93**, 1833 (1908); S. Hünig and O. Boes, *Ann.*, **579**, 23 (1953).
[3] C. C. Price and G. W. Stacy, *Org. Syn., Coll. Vol.*, **3**, 86 (1955)
[4] R. L. Shriner and C. N. Wolf, *ibid.*, **4**, 910 (1963)
[5] A. Rieche and C. Bischoff, *Ber.*, **94**, 2722 (1961)

Hydrogen peroxide–Salt and oxide catalysts.

Hydrogen peroxide–Iron salts. The combination of hydrogen peroxide with a ferrous salt produces hydroxyl radicals and is known as the Fenton reagent.[1] Hydroxyl radicals generated by decomposition of hydrogen peroxide with ferrous

$$Fe^{2+} + H_2O_2 \xrightarrow{H_2SO_4} Fe^{3+} + OH^- + \cdot OH$$

sulfate attack *t*-butanol to form a carbon radical which dimerizes to produce 2,5-di-methylhexane-2,5-diol.[2] A flask with creased sides and equipped with a high-speed

$$
\underset{\underset{\text{CH}_3}{|}}{\overset{\overset{\text{CH}_3}{|}}{\text{CH}_3-\text{C}}}\text{OH} \xrightarrow{\cdot\text{OH}} \cdot\underset{\underset{\text{CH}_3}{|}}{\overset{\overset{\text{CH}_3}{|}}{\text{CH}_2-\text{C}}}\text{-OH} \xrightarrow[40-46\%]{} \text{HO}-\underset{\underset{\text{CH}_3}{|}}{\overset{\overset{\text{CH}_3}{|}}{\text{C}}}\text{CH}_2\text{CH}_2\underset{\underset{\text{CH}_3}{|}}{\overset{\overset{\text{CH}_3}{|}}{\text{C}}}\text{-OH}
$$

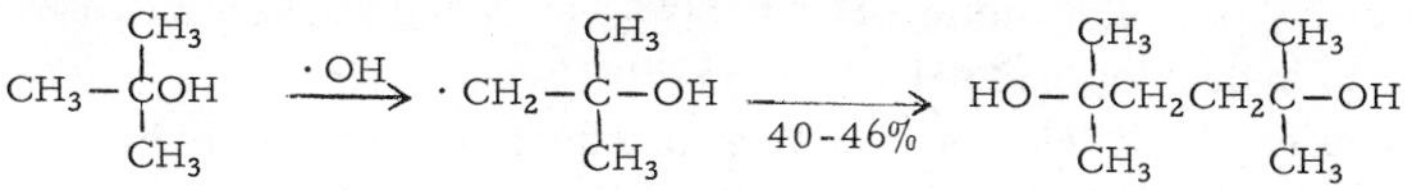

stirrer is charged with 900 ml. of *t*-butanol (9.5 moles) and a solution of 0.5 mole of sulfuric acid in 1.5 l. of water. One dropping buret is charged with 1 mole of 35% hydrogen peroxide and another with a solution of 1 mole of ferrous sulfate in 1 mole of dilute sulfuric acid. The reaction mixture is swept with nitrogen and kept at 10° by ice cooling while the two solutions are run in equivalently in 20 min. The mixture is then neutralized with 1 mole of sodium hydroxide and treated with 450 g. of sodium sulfate (which does not all dissolve) and the organic layer is separated. The aqueous layer is extracted with three 400-ml. portions of *t*-butanol. The combined material is distilled to remove the bulk of the solvent, and the residue extracted with ether. Evaporation gives a solid product which on digestion with ether and cyclohexane affords satisfactory diol, m.p. 87–88°.

Hydroxyl radicals generated from ferrous sulfate and hydrogen peroxide at room temperature attack nitrobenzene to form small amounts of *o*-, *m*-, and *p*-nitrophenol;[3] benzene yields small amounts of phenol and diphenyl.[4]

Dehydrogenation of 2-amino-3,4-dihydroquinoxaline to 2-aminoquinoxaline is effected smoothly by reaction with hydrogen peroxide in combination with a trace amount of ferrous chloride, ferric sulfate, or lead dioxide "as carrier."[5] Details are

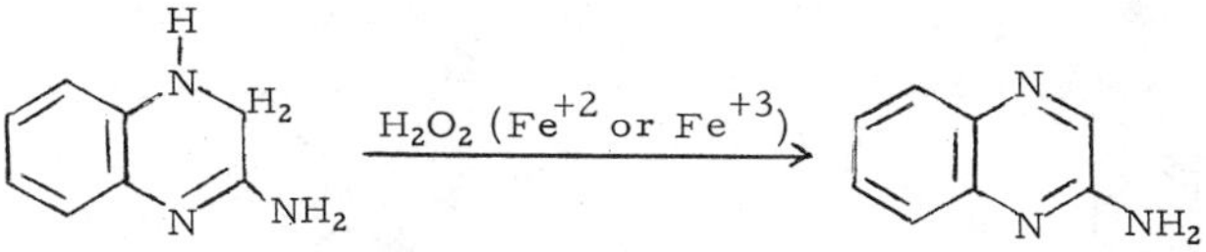

not given, and the nature of the effective oxidant is not clear. Also unclear is the nature of the oxidant in the oxidation of an α-hydroxy acid with hydrogen peroxide in combination with a ferric salt:

$$
\underset{\underset{\text{OH}}{|}}{-\text{CHCO}_2\text{H}} \longrightarrow -\text{CH}{=}\text{O} + \text{CO}_2
$$

Thus in the Ruff degradation,[6] for example of D-glucose to D-arabinose, the aldose is oxidized electrolytically to the aldonic acid and this on treatment with hydrogen peroxide and ferric acetate affords the 2-ketoaldonic acid (aldosulose), which yields D-arabinose by loss of carbon dioxide. The yield of aldopentose from the aldonic acid

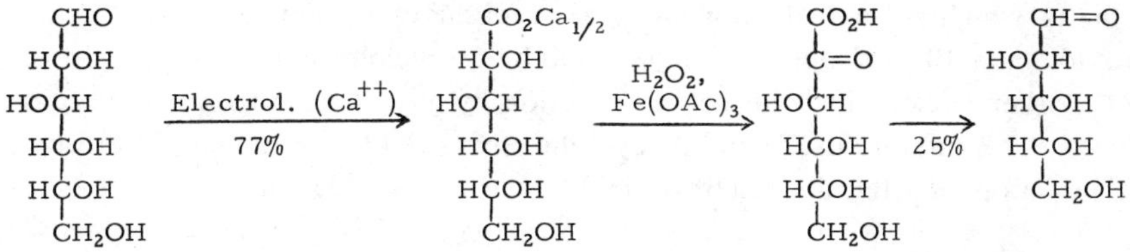

D-Glucose   D-Arabinose

is increased to 44% by addition of ion-exchange resins.[7] However, both the oxidation of the aldose to the aldonic acid and the oxidation of this to the aldopentose can be accomplished with hypochlorite, and a two-stage but single-batch hypochlorite process[8] appears to be superior to the Ruff degradation. The first stage is done with 3 equivalents of hypochlorite at pH 11 (80% yield) and the second stage with 1.4 equivalents of hypochlorite at pH 4.5–5 (65–70% yield).

Udenfriend *et al.*[9] developed a system for effecting aromatic hydroxylation consisting of ferrous sulfate, oxygen, ascorbic acid, and ethylenediaminetetraacetic acid (EDTA) in a phosphate buffer and noted that, in the cases tried, oxygen could be replaced by hydrogen peroxide in a nitrogen atmosphere. EDTA is not essential but greatly increases the reaction rate. Dihydroxymaleic acid and diethyl diketosuccinate were about as active as ascorbic acid; alloxan and ninhydrin were less active. Examples:

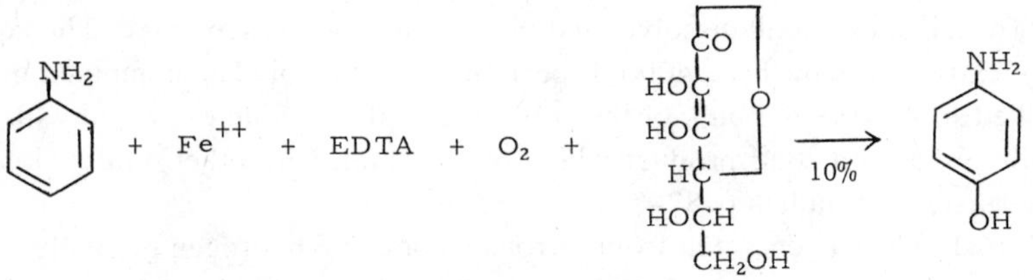
NH2
+ Fe++ + EDTA + O2 +
CO
HOC
HOC
HC
O
HOCH
CH2OH
10%
NH2
OH

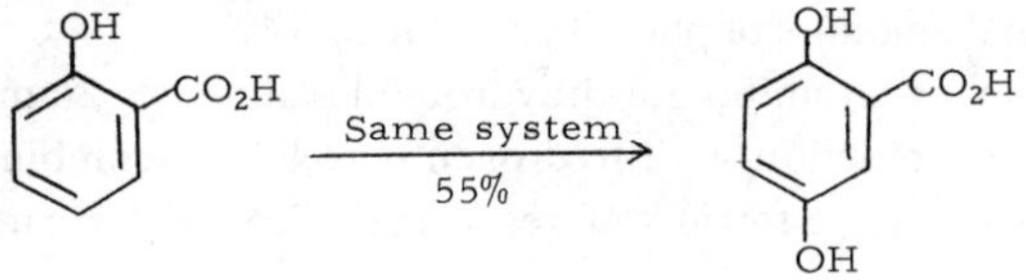
OH
CO2H
Same system
55%
OH
CO2H
OH

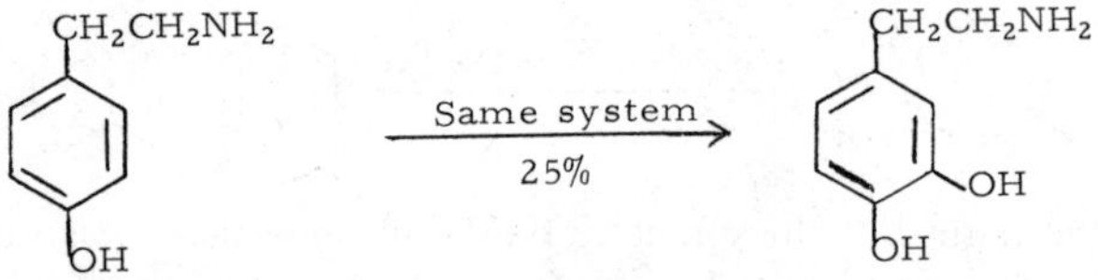
CH2CH2NH2
OH
Same system
25%
CH2CH2NH2
OH
OH

In these and five other instances, the product of oxidation was identical with that produced metabolically from the same substrate *in vivo*.

Vanadium catalysis. Treibs *et al.*[10] prepared a pervanadic acid catalyst by adding 1 g. of vanadium pentoxide with cooling to 10 ml. of 30% hydrogen peroxide, shaking to produce a yellow-red solution, diluting with 100 ml. of acetone, and filtering. The catalyst solution was added to 500 g. of cyclohexene in 5 l. of acetone, 100 g. of 30% hydrogen peroxide was added with stirring at 30°, and the mixture was refluxed for 1 hr. Workup afforded 32 g. of *trans*-cyclohexane-1,2-diol and 41 g. of adipic acid, products which arise from noncatalytic reaction of cyclohexene with hydrogen peroxide, and 40 g. of a product believed to be Δ^2-cyclohexene-1-one (2). However, later workers[11] examined the volatile fraction by gas chromatography and found it to be a 1:2.3 mixture of (2) and Δ^2-cyclohexene-1-ol (3). Tetralin, oxidized by this method, afforded α-tetralone in 65% yield.[10]

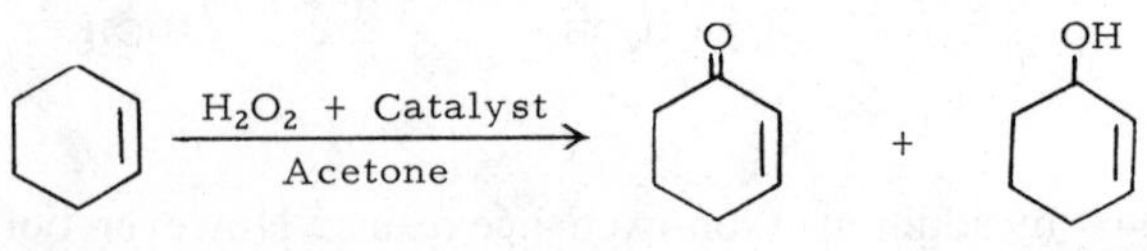
H2O2 + Catalyst
Acetone
O
+
OH

(1) (2) (3)

After a report of an explosion in an oxidation with catalyst prepared by his procedure, Treibs[12] published more detailed instructions and stressed the point that traces of mineral acid must be excluded.

Tungsten catalysis. A primary amine with the group CH_2NH_2 on oxidation with 35% hydrogen peroxide in the presence of a catalytic amount of sodium tungstate ($Na_2WO_4 \cdot 2H_2O$, supplied by Fisher) is oxidized *via* the hydroxylamine to the

$$\underline{n}\text{-}C_3H_7CH_2NH_2 \xrightarrow{H_2O_2(Na_2WO_4)} \underline{n}\text{-}C_3H_7CH_2NHOH \xrightarrow[57\%]{} \underline{n}\text{-}C_3H_7CH{=}NOH$$

aldoxime.[13] The reaction is carried out in water or in aqueous alcohol at room temperature.

Payne and Williams[14] found that maleic, fumaric, and crotonic acid are very resistant to attack by peracetic or perbenzoic acid but are converted into their epoxides in yields of 77, 50, and 50% by hydrogen peroxide–sodium tungstate at pH 4–5.5. Addition of the peroxide initiates an exothermal reaction which is

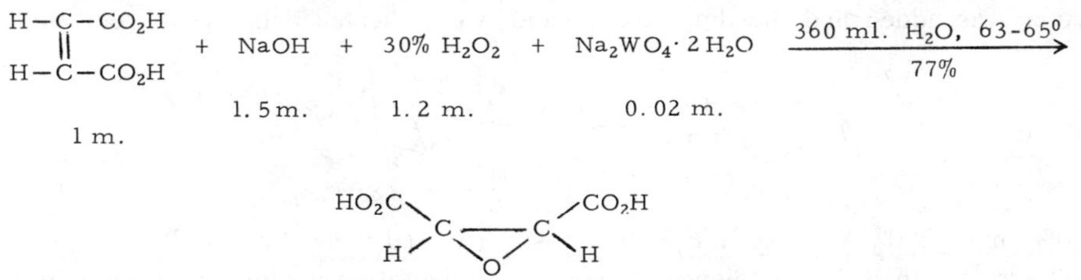

checked at 63–65°. Alkali is run in as required to maintain the proper pH (pH meter). When titration shows a consumption of 1.02 equivalents of hydrogen peroxide, the solution is concentrated in vacuum and treated with acetone to precipitate the epoxide as the sodium salt.

Schultz *et al.*[15] found that oxidation of sulfides to sulfones by hydrogen peroxide in acetic acid requires extended refluxing and gives poor yields but that addition of a tungsten (or vanadium) catalyst improves the result. The catalyst was prepared by stirring a mixture of tungstic anhydride (WO_3, supplied by Fisher) and distilled water and adding sodium hydroxide to effect solution; the pH is then adjusted to 5–6 with acetic acid. In the oxidation of 2-phenylmercaptoethanol this sulfide is added to an aqueous solution of the catalyst and 30% hydrogen peroxide is added

$$C_6H_5SCH_2CH_2OH + H_2O_2\ (H_2WO_4) \xrightarrow[94.5\%]{H_2O\ 63\text{-}75^0} C_6H_5SO_2CH_2CH_2OH$$

at a rate required for maintaining a temperature of 63–75°, and the sulfone is obtained in high yield. In the uncatalyzed oxidation the main product (65%) is 2-phenylsulfinylethanol. If the sulfide is sparingly soluble in water, ethanol or dioxane can be used.

H_2O_2–OsO_4. Olefins can be hydroxylated to *cis* diols in 30–60% yield by treatment with anhydrous hydrogen peroxide in *t*-butanol in the presence of a catalytic

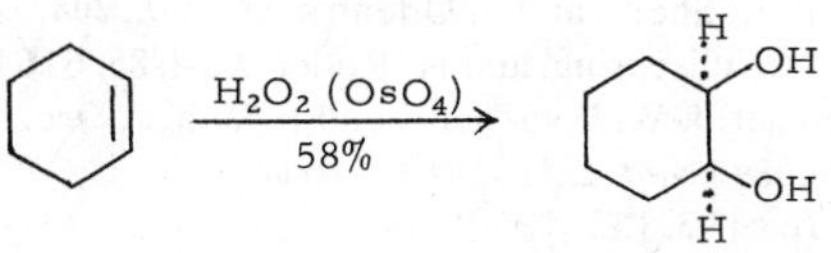
H₂O₂ (OsO₄)
58%
H
OH
OH
H

amount of osmium tetroxide.[16] The olefin reacts with osmium tetroxide to form the osmate ester which is converted by peroxide into the *cis* glycol with regeneration of osmium tetroxide. The combination has been used to oxidize a steroid 17,20-ene to the 17α-ol-20-one.[17] Anhydrous conditions are not necessary; 30% aqueous

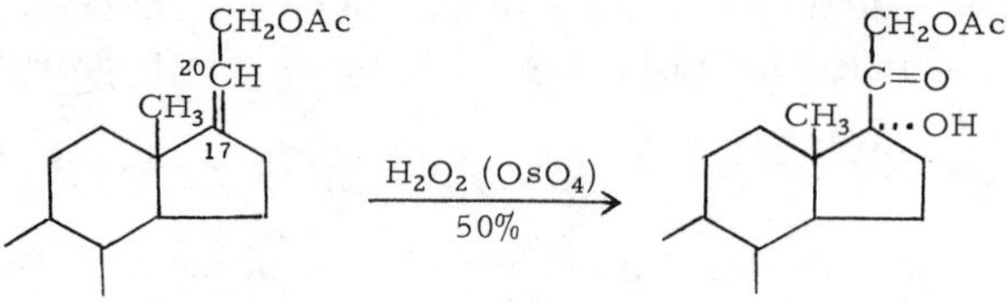

CH_2OAc
20CH
CH_3
17
H_2O_2 (OsO_4)
50%
CH_2OAc
C=O
CH_3
OH

hydrogen peroxide can be used, with acetone or acetone–ether as solvent.[18] Thus a solution of 1 g. of the maleic anhydride–furane adduct (1) in 8 ml. of acetone and 2 ml. of ether was treated with 1 ml. of a solution of 1 g. of OsO_4 in 200 ml. of *t*-butanol followed by 3 drops of 30% hydrogen peroxide.[19] The anhydride suffered hydrolysis in this mixture and the diacid began to separate. After 24 hrs. at 30° ether was added and the dihydroxy diacid was collected and crystallized; yield

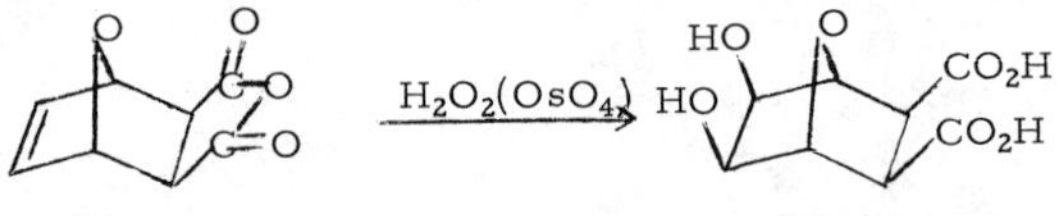

H_2O_2(OsO_4)
HO
HO
CO_2H
CO_2H

(1) (2)

50%, m.p. 200°. Whereas the adduct easily isolated from the reaction of maleic anhydride with cyclopentadiene has the *endo* configuration, addition of the anhydride to furane gives the *exo*-product (1). The configuration shown for the diol is probable and not rigidly established.

Oxidation of Δ[4]-cholestene-3-one in ether solution with 30% hydrogen peroxide and a catalytic amount of OsO_4 gives both possible *cis* diols, which were isolated in the yields indicated.[20]

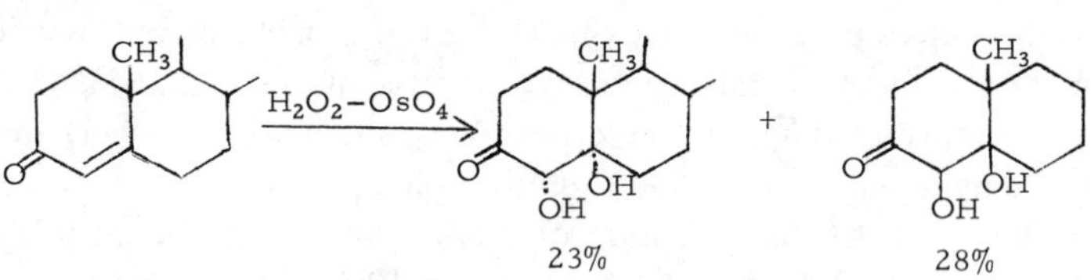

CH_3
H_2O_2–OsO_4
CH_3
OH
OH
+
CH_3
OH
OH
23%
28%

[1]H. S. H. Fenton, *J. Chem. Soc.*, **65**, 899 (1894)
[2]E. L. Jenner, *Org. Syn.*, **40**, 90 (1960)
[3]H. Loebl, G. Stein, and W. Weiss, *J. Chem. Soc.*, 2074 (1949)
[4]J. R. L. Smith and R. O. C. Norman, *J. Chem. Soc.*, 2897 (1963)
[5]K. Pfister, 3rd, A. P. Sullivan, Jr., J. Weijlard, and Max Tishler, *Am. Soc.*, **73**, 4955 (1951)
[6]O. Ruff, *Ber.*, **31**, 1573 (1898); **32**, 550, 3677 (1899); **34**, 1362 (1901); O. Ruff and G. Ollendorff, *ibid.*, **33**, 1798 (1900)
[7]H. G. Fletcher, Jr., H. W. Diehl, and C. S. Hudson, *Am. Soc.*, **72**, 4546 (1950)
[8]R. L. Whistler and R. Schweiger, *Am. Soc.*, **81**, 5190 (1959); R. L. Whistler and K. Yagi, *J. Org.*, **26**, 1050 (1961)
[9]S. Udenfriend, C. T. Clark, J. Axelrod, and B. B. Brodie, *J. Biol. Chem.*, **208**, 731 (1954); B. B. Brodie, J. Axelrod, P. A. Shore, and S. Udenfriend, *ibid.*, **208**, 741 (1954)
[10]W. Treibs, G. Franke, G. Leichsenring, and H. Röder, *Ber.*, **86**, 616 (1953)
[11]E. J. Eisenbraun, A. R. Bader, J. W. Polacheck, and E. Reif, *J. Org.*, **28**, 2057 (1963)
[12]W. Treibs, *Angew. Chem., Internat. Ed.*, **3**, 812 (1964)
[13]K. Kahr and C. Berther, *Ber.*, **93**, 132 (1960)

[14]G. B. Payne and P. H. Williams, *J. Org.*, **24**, 54 (1959); see also Y. Liwschitz, Y. Rabinsohn, and D. Perera, *J. Chem. Soc.*, 1116 (1962)
[15]H. S. Schultz, H. B. Freyermuth, and S. R. Buc, *J. Org.*, **28**, 1140 (1963)
[16]N. A. Milas and S. Sussman, *Am. Soc.*, **58**, 1302 (1936); **59**, 2345 (1937)
[17]K. Miescher and J. Schmidlin, *Helv.*, **33**, 1840 (1950)
[18]M. Mugdan and D. P. Young, *J. Chem. Soc.*, 2988 (1949)
[19]R. Daniels and J. L. Fischer, *J. Org.*, **28**, 320 (1963)
[20]J. F. Eastham, G. B. Miles, and C. A. Krauth, *Am. Soc.*, **81**, 3114 (1959)

Hydrogen peroxide–Selenium dioxide. A. Stoll[1] effected olefin hydroxylations with hydrogen peroxide in combination with a stochiometric amount of selenium dioxide in

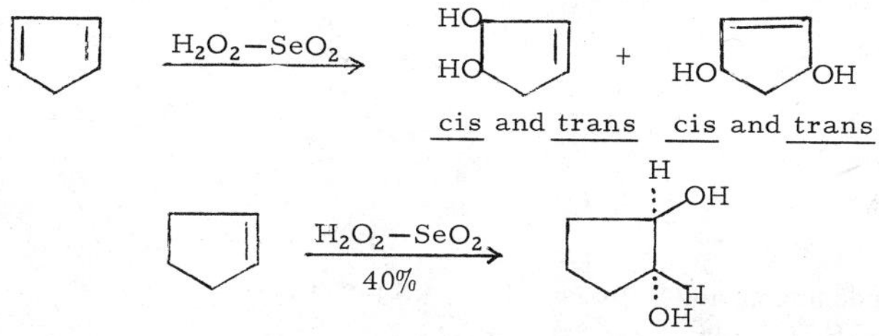
$H_2O_2-SeO_2$
HO
HO
+
HO
OH
cis and trans
cis and trans
$H_2O_2-SeO_2$
40%
H
OH
H
OH

t-butanol at 0° (65 hrs.). Cyclopentadiene reacted by 1,2- and 1,4-addition to give a mixture of the *cis*- and *trans*-cyclopentane-1,2- and -1,3-diols, and cyclopentene gave *trans*-cyclopentane-1,2-diol.

C. W. Smith[2] investigated oxidations by hydrogen peroxide promoted by a purely catalytic amount of selenium dioxide. Acrolein is oxidized smoothly to monomeric acrylic acid with a 15% solution of hydrogen peroxide in *t*-butanol in the presence of 5 g. of selenium dioxide per mole of peroxide. Curiously, the same

$$CH_2{=}CHCHO \xrightarrow[90\%]{H_2O_2\ (SeO_2)} CH_2{=}CHCO_2H$$

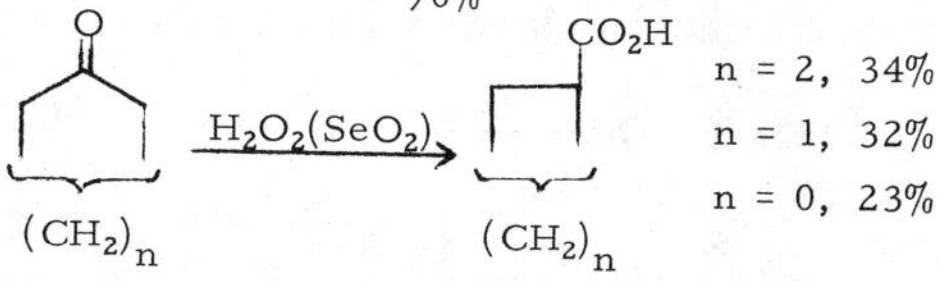
O
$H_2O_2(SeO_2)$
CO_2H
$(CH_2)_n$
$(CH_2)_n$
n = 2, 34%
n = 1, 32%
n = 0, 23%

combination oxidizes cycloheptanone, cyclohexanone, and cyclopentanone with ring contraction to give cyclohexane-, cyclopentane-, and cyclobutanecarboxylic acid in the yields cited.[3] The contraction of a 5-membered ring is unusual. Aside from the finding that cyclohexane-1,2-dione is not oxidized by the reagent to the ring-contracted acid, nothing is known of the mechanism of the reaction. The reagent oxidizes cyclododecanone to cycloundecanecarboxylic acid in 32% yield.[4] Dodecane-1, 12-dioic acid was isolated as a by-product in 17% yield.

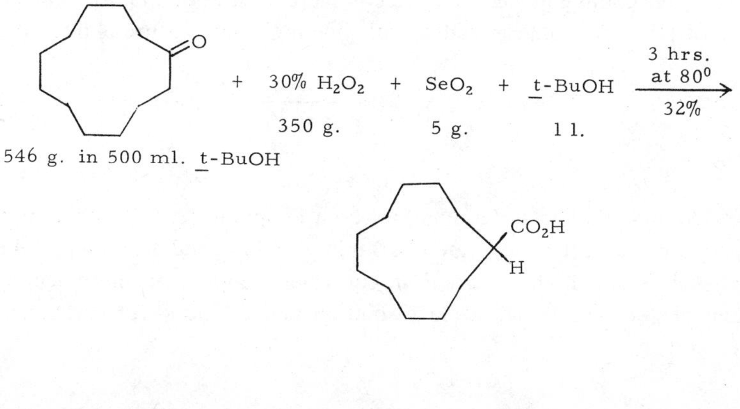
O
546 g. in 500 ml. t-BuOH
+ 30% H_2O_2 + SeO_2 + t-BuOH
350 g.
5 g.
1 l.
3 hrs.
at 80°
32%
CO_2H
H

The oxidation of steroid ketones by H_2O_2–SeO_2 has been investigated by Caspi.[5] Oxidation of cholestane-3-one gives a mixture of ring-contracted acids and a

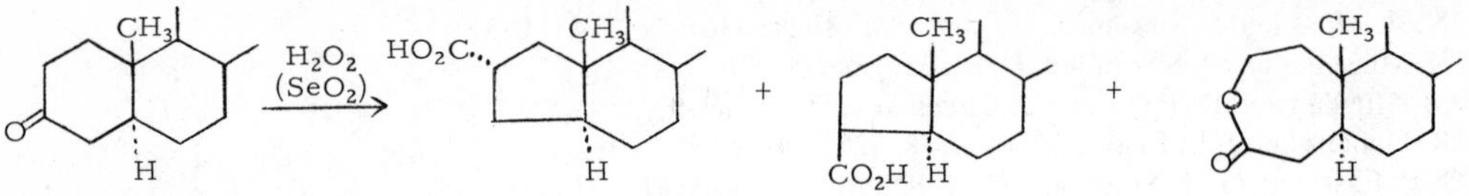

product of Baeyer-Villiger fission. The Δ^4-3-ketosteroid (1) yields the ε-lactone (2), which is easily rearranged to the γ-lactone (3).

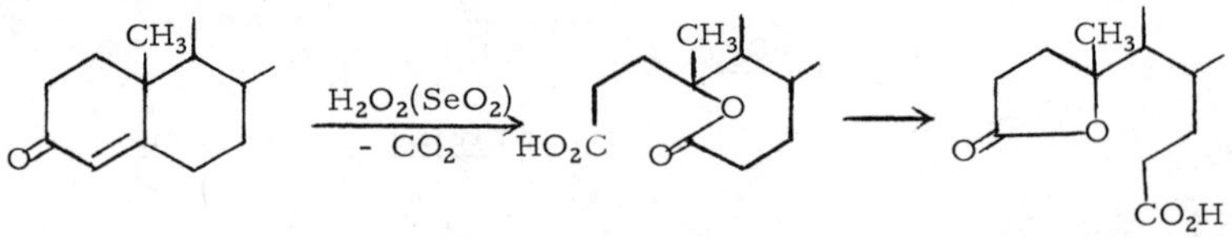

(1) (2) (3)

[1]A. Stoll, A. Lindenmann, and E. Jucker, *Helv.*, **36**, 268 (1953)
[2]C. W. Smith and R. T. Holm, *J. Org.*, **22**, 746 (1957)
[3]G. B. Payne and C. W. Smith, *J. Org.*, **22**, 1680 (1957)
[4]W. D. Dittmann, W. Kirchoff, and W. Stumpf, *Ann.*, **681**, 30 (1965)
[5]E. Caspi and S. N. Balasubrahmanyam, *Tetrahedron Letters*, 745 (1963); Experientia, **19**, 396 (1963); *J. Org.*, **28**, 3383 (1963); E. Caspi, Y. Shimizu, and S. N. Balasubrahmanyam, *Tetrahedron*, **20**, 1271 (1964); H. M. Hellman and R. A. Jerussi, *ibid.*, **20**, 741 (1964)

Hydroxylamine hydrochloride, $HONH_2 \cdot HCl$. Mol. wt. 69.50, m.p. 151°, 100 g. $H_2O^{17°}$ dissolves 83.3 g. Suppliers, Baker, Eastman.

Preparation.[1] Sodium hydroxylamine disulfonate (1), prepared by passing sulfur dioxide into an aqueous solution of sodium nitrite and sodium bisulfite, reacts with acetone to form acetoxime (2); acid hydrolysis affords hydroxylamine hydrochloride.

$$NaNO_2 + NaHSO_3 + SO_2 \longrightarrow \underset{(1)}{HON(SO_3Na)_2} \xrightarrow{O=C(CH_3)_2}$$

$$\underset{(2)}{HON=C(CH_3)_2} \xrightarrow[70\%\ overall]{HCl-H_2O} \underset{(3)}{HONH_2 \cdot HCl} + O=C(CH_3)_2$$

Derivatives. For the preparation of benzohydroxamic acid,[2] isolated as the potassium salt, separate solutions of 46.7 g. (0.67 mole) of hydroxylamine hydrochloride in 240 ml. of methanol and of 1 mole of potassium hydroxide in 140 ml. of methanol are prepared at the boiling points, cooled to 30–40°, and the second solution is added to the first. After cooling in ice to ensure complete separation of potassium chloride, 0.33 mole of ethyl benzoate is added with shaking. The mixture is filtered quickly

$$C_6H_5CO_2C_2H_5 + HONH_2 \cdot HCl + KOH \xrightarrow[57-60\%]{} C_6H_5\overset{O}{\overset{\|}{C}}-NHOK \xrightarrow[91-95\%]{AcOH} C_6H_5\overset{O}{\overset{\|}{C}}-NHOH$$

(more KCl?), and the filtrate let stand at room temperature. Crystals of potassium benzohydroxamate start to separate in 20 min. to 3 hrs., and after 48 hrs. the salt is collected. A mixture of the salt and an equivalent amount of dilute acetic acid is stirred and heated to produce a clear solution and let stand for crystallization at room temperature and then at 0°. Benzohydroxamic acid separates in white crystals; recrystallization from ethyl acetate gives pure material, m.p. 125–128°.

Hydroxyurea (m.p. 137–141°) is obtained[3] by preparing a solution of hydroxylamine hydrochloride, urethane, and sodium hydroxide in water, letting it stand at room temperature for 3 days, carefully neutralizing with hydrochloric acid, extracting

$$H_2NCO_2C_2H_5 + H_2NOH \cdot HCl \xrightarrow[53\text{-}73\%]{NaOH} H_2N\overset{O}{\overset{\|}{C}}-\overset{H}{\overset{|}{N}}-OH$$

with ether, and evaporating the aqueous phase to dryness at reduced pressure. The residual mixture of hydroxyurea and sodium chloride is extracted with absolute ethanol, from which hydroxyurea crystallizes.

Oximes. In one procedure[4] a mixture of 0.55 mole of benzophenone, 0.86 mole of hydroxylamine hydrochloride, 200 ml of 95% ethanol, and 40 ml. of water is

$$(C_6H_5)_2C{=}O + H_2NOH \cdot HCl \xrightarrow[98\text{-}99\%]{NaOH} (C_6H_5)_2C{=}NOH$$

treated with 2.75 moles of powdered sodium hydroxide, added in portions. The mixture is refluxed for 5 min. and poured into a dilute hydrochloric acid. The precipitated oxime is directly pure, m.p. 141–142°.

For the preparation of heptaldoxime,[5] a suspension of 4 moles of heptaldelyde in a solution of 5 moles of hydroxylamine hydrochloride in 600 ml. of cold water is stirred with cooling to 45° during addition of a solution of 2.5 moles of sodium carbonate in 500 ml. of water. After stirring for 1 hr. longer, the oily upper layer is

$$CH_3(CH_2)_5CHO + H_2NOH \cdot HCl \xrightarrow[81\text{-}93\%]{Na_2CO_3} CH_3(CH_2)_5CH{=}NOH$$

separated, washed twice with water, and distilled at 6 mm. The oxime so obtained (m.p. 44–46°) can be used directly for reduction with sodium and absolute ethanol to *n*-heptylamine. The pure oxime melts at 53–55°.

Nitriles. Veratronitrile is prepared[6] by heating a mixture of the oxime with acetic anhydride cautiously until a vigorous reaction is over, refluxing for 20 min., and then

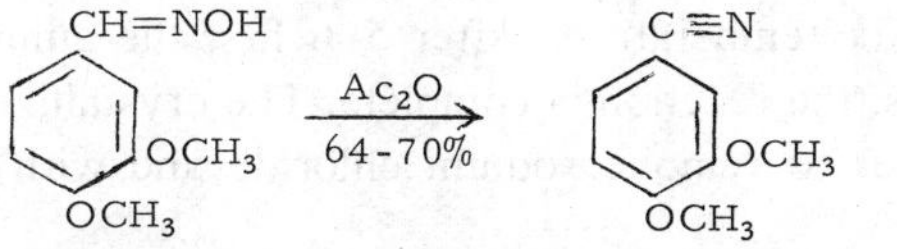

pouring the solution with stirring into cold water. Veratronitrile separates in nearly colorless crystals, m.p. 66–67°.

A simpler procedure dispenses with preformation of the oxime and utilizes acetic acid as the solvent.[7] A mixture of 5 g. of anisaldehyde with 1.2 equivalents each of

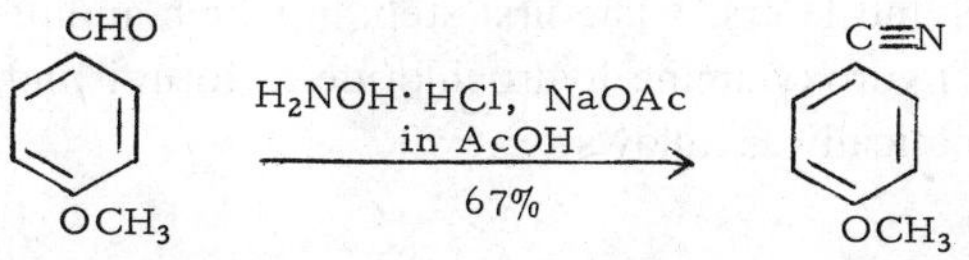

hydroxylamine hydrochloride and anhydrous sodium acetate in 20 ml. of acetic acid is refluxed overnight and filtered from sodium chloride. The solvent is removed by distillation at reduced pressure and the product recovered by ether extraction and crystallized from petroleum ether (m.p. 59–60°). Vanillin and piperonal afforded nitriles by this method in yields of 80 and 50%. Trials were unpromising with pyridine-3-aldehyde (yield 7%) and *n*-butyraldehyde.

A procedure for the conversion of D-glucose monohydrate via the oxime to pentaacetyl D-gluconitrile is lengthy and the overall yield 50%.[8]

α-Keto acids are converted into nitriles, with decarboxylation, when refluxed

$$\mathrm{R\overset{O}{\overset{\|}{C}}CO_2H + H_2NOH \longrightarrow \left[R\overset{NOH}{\overset{\|}{C}}CO_2H\right] \xrightarrow{H^+} RCN + CO_2 + H_2O}$$

under nitrogen with 1 equivalent of aqueous hydroxylamine hydrochloride.[9] Yields are good, although some hindered keto acids do not react.

Nucleophilic substitution. The interesting reaction formulated[10] is carried out by dissolving 20 g. of 1-nitronaphthalene and about 6 equivalents of hydroxylamine hydrochloride in 1.2 l. of 95% ethanol in a flask heated in a bath at 50–60°, and

NO2
+ NH2OH
KOH
55-60%
NO2
NH2
+ H2O

stirring the solution during addition in 1 hr. of 100 g. of potassium hydroxide (about 10 equiv.) in 630 ml. of methanol.

1,4-Addition-reduction. The synthesis of DL-β-amino-β-phenylpropionic acid (3) from cinnamic acid (1) involves 1,4-addition of hydroxylamine to the conjugated system followed by reduction of the hydroxylamino acid (2) by the hydroxylamine.[11]

$$\underset{(1)}{\mathrm{C_6H_5CH{=}CHCO_2H}} \xrightarrow{\mathrm{H_2NOH}} \underset{(2)}{\left[\mathrm{C_6H_5\underset{NHOH}{\underset{|}{C}H}CH_2CO_2H}\right]} \xrightarrow[34\%]{\mathrm{H_2NOH}} \underset{(3)}{\mathrm{C_6H_5\underset{NH_2}{\underset{|}{C}H}CH_2CO_2H}}$$

A hot solution of 2 moles of hydroxylamine hydrochloride in 100 ml. of hot water is added to a hot solution prepared from 2 gram atoms of sodium and 1.6 l. of absolute ethanol and, after cooling, the sodium chloride formed is removed by suction filtration. Addition of 1 mole of cinnamic acid produces a voluminous precipitate of (2), which dissolves at the reflux temperature. After 5–6 hrs. the amino acid (3) begins to separate and in 9 hrs. the reaction is complete. The crystallized product is washed alternately with water to remove sodium chloride and with absolute alcohol for drying.

Reduction. In the Cadiot-Chodkiewicz acetylene coupling reaction, hydroxylamine hydrochloride is generally added as a reducing agent to maintain the copper salt in the cuprous stage.[12]

Synthesis of isatin. This synthesis originated with Sandmeyer[13] and was further improved by Marvel and Hiers.[14] The first step, in which aniline is condensed with chloral hydrate and hydroxylamine hydrochloride to form isonitrosoacetanilide (1) is represented schematically as follows:

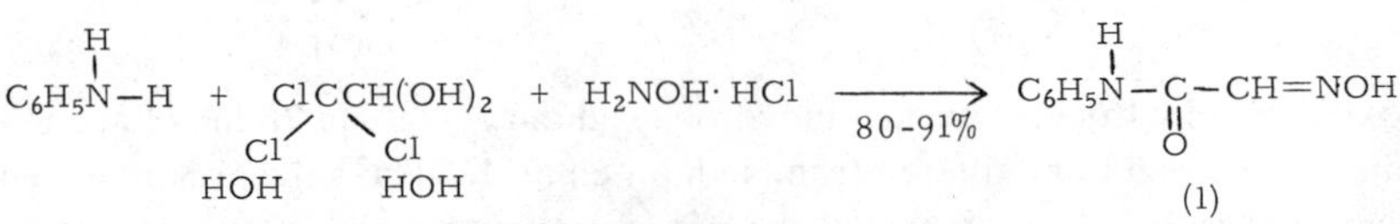

H
C6H5N—H + ClCCH(OH)2 + H2NOH· HCl
Cl Cl
HOH HOH
80-91%
H
C6H5N—C—CH=NOH
O

(1)

When the reactants are heated together in water solution the crystalline product

(1) soon separates in high yield. In the next step, (1) is added slowly with stirring to concd. sulfuric acid at 60–70° and the solution is poured onto ice. By cyclization and hydrolysis, (1) is converted in good yield into isatin (2).

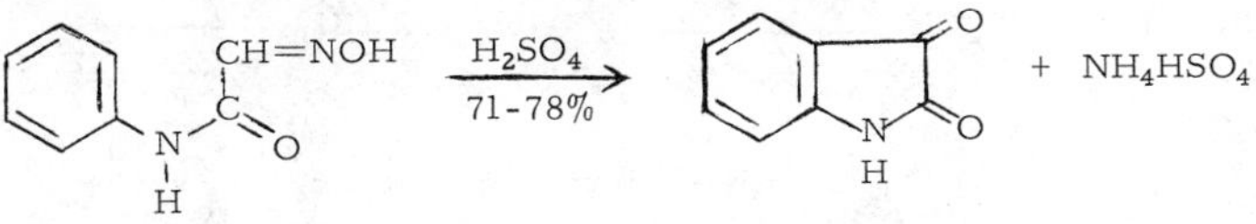

(1) (2)

[1]W. L. Semon, *Org. Syn., Coll. Vol.*, **1**, 318 (1941)
[2]C. R. Hauser and W. B. Renfrow, Jr., *ibid.*, **2**, 67 (1943)
[3]R. Deghenghi, *Org. Syn.*, **40**, 60 (1960)
[4]A. Lachman, *Org. Syn., Coll. Vol.*, **2**, 70 (1943)
[5]E. W. Bousquet, *ibid.*, **2**, 313 (1943)
[6]J. S. Buck and W. S. Ide, *ibid.*, **2**, 622 (1943)
[7]J. H. Hunt, *Chem. Ind.*, 1873 (1961)
[8]H. T. Clarke and S. M. Nagy, *Org. Syn., Coll. Vol.*, **3**, 690 (1955)
[9]A. Ahmad and I. D. Spencer, *Can. J. Chem.*, **39**, 1340 (1961)
[10]C. C. Price and S.-T. Voong, *Org. Syn., Coll. Vol.*, **3**, 664 (1955)
[11]R. E. Steiger, *ibid.*, **3**, 91 (1955)
[12]W. Chodkiewicz, *Ann. chim.*, [13], **2**, 819 (1957)
[13]T. Sandmeyer, *Helv.*, **2**, 237, 239 (1919)
[14]C. S. Marvel and G. S. Hiers, *Org. Syn., Coll. Vol.*, **1**, 327 (1941)

Hydroxylamine sulfate, $HONH_2 \cdot \frac{1}{2}H_2SO_4$. Mol. wt. 82.07. Suppliers: Alfa, E, F, MCB.

The reaction of 1,3-dicarbonyl compounds with hydroxylamine, usually as the hydrochloride, is widely used for the synthesis of isoxazoles.[1] The reaction is

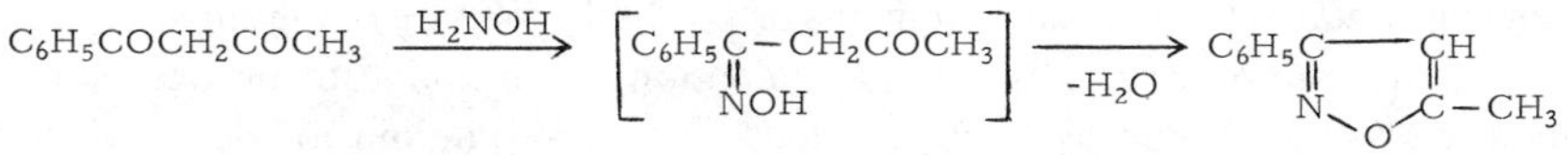

usually carried out in aqueous alcohol, with or without addition of base. In a procedure for the preparation of 3,5-dimethylisoxazole, hydroxylamine sulfate is used rather than the hydrochloride since it is cheaper.[2] A mixture of 2,4-pentanedione, hydroxylamine sulfate, and potassium carbonate is heated under reflux on the steam bath for 1.5 hrs. until a ferric chloride test is negative.

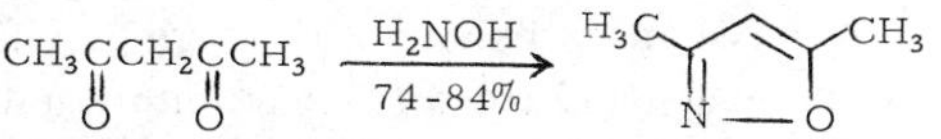

[1]A. Quilico in A. Weissberger's "The Chemistry of Heterocyclic Compounds," **17**, 6 (1962)
[2]C. E. Miller, procedure submitted to *Org. Syn.*

Hydroxylamine-O-sulfonic acid, $\overset{+}{N}H_3OSO_3^-$. Mol. wt. 113.09, m.p. 210° dec., hygroscopic. Soluble in water and methanol in the cold; decomposes slowly in water at 25°. Suppliers: Allied Chem. Co., Alfa, E, F. The reagent is prepared[1] by adding

$$(\overset{+}{N}H_3OH)_2SO_4^{=} + 2\,H_2SO_4 \cdot SO_3 \xrightarrow[98-99\%]{} 2\,\overset{+}{N}H_3OSO_3^- + 3\,H_2SO_4$$

dry bis(hydroxylammonium)sulfate to 30% fuming sulfuric acid with stirring, cooling to 0°, and adding ether dropwise to precipitate the product, which is washed 7 times with ether to remove sulfuric acid and dried in vacuum; purity 92–98%, as determined by reaction with excess iodide ion and titration of the iodine formed with sodium thiosulfate solution.

Deamination. The reagent can be used to effect deamination of a primary amine via the methanesulfonamide or an arylsulfonamide.[2] For example, benzylamine

$$C_6H_5CH_2NHSO_2CH_3 \xrightarrow[-H_2SO_4]{\overset{+}{N}H_3OSO_3^-} C_6H_5CH_2\underset{NH_2}{\underset{|}{N}}SO_2CH_3 \xrightarrow[-CH_3SO_2H]{}$$

$$C_6H_5CH_2N{=}NH \xrightarrow[-N_2]{} C_6H_5CH_3$$

methanesulfonamide is treated in aqueous–alcoholic alkali with 15–25 equivalents of the alkali-unstable reagent. The formulation shows the pathway postulated.

Generation of diimide. Hydroxylamine-O-sulfonic acid reacts with cyclohexanone in an alkaline medium to give 1,1-dihydroxyazocyclohexane (3).[3] This substance decomposes rapidly even at room temperature to cyclohexanone, nitrogen, and

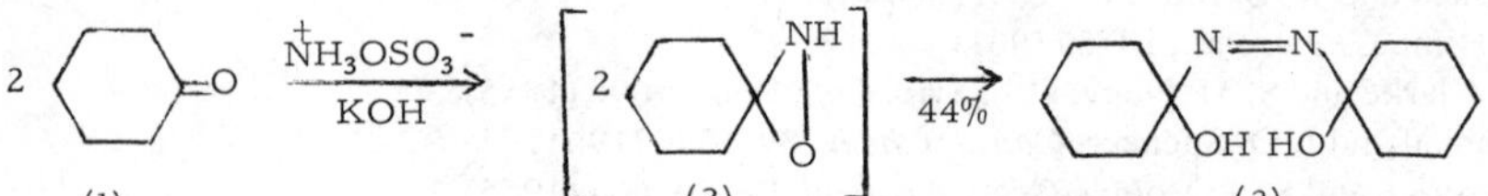
2
$\overset{+}{N}H_3OSO_3^-$
KOH
2
NH
O
44%
N=N
OH HO

(1) (2) (3)

hydrazine; that the decomposition proceeds by way of diimide follows from the observation that decomposition in the presence of quinone or of azobenzene gave hydroquinone or hydrazobenzene. Hydroxylamine-O-sulfonic acid itself can be used in alkaline solution for generation of diimide:[4]

$$2\ \overset{+}{N}H_3OSO_3^- + 4\ OH^- \longrightarrow HN{=}NH + 2SO_4^{=}$$

Thus ethylene, cyclohexene, fumaric acid, and cinnamic acid are all reduced if the reagent is added to an alkaline aqueous or alcoholic solution of the substrate.

Amination. Another use is for the amination of an amine to produce an unsymmetrically substituted hydrazine.[5] Thus 1-aminopyridinium iodide is prepared by adding 0.3 mole of pyridine to a freshly prepared solution of 0.1 mole of hydroxylamine-O-sulfonic acid in cold water, heating for 20 min. on the steam bath, cooling,

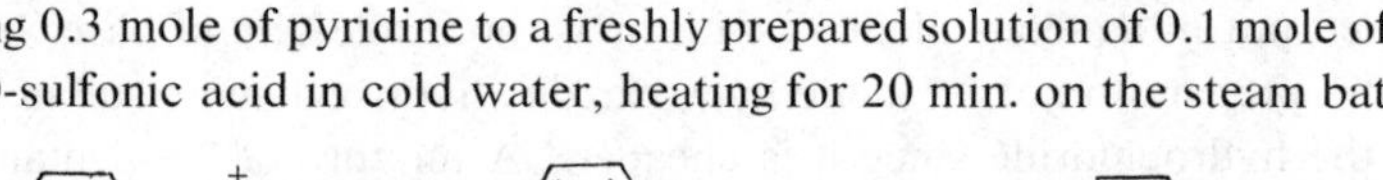
N
$\overset{+}{N}H_3OSO_3^-$
K_2CO_3
$\overset{+}{N}-\overset{-}{N}H$
HI
63-72%
$\overset{+}{N}-NH_2\ I^-$

and adding 0.1 mole of potassium carbonate with stirring. The water and excess pyridine are removed in an evacuated rotatory evaporator, and the organic matter is extracted from potassium sulfate with ethanol. After addition of 0.1 mole of hydriodic acid and cooling to −20°, 1-aminopyridinium iodide separates and is recrystallized from absolute ethanol. The method is generally applicable to primary, secondary, and tertiary amines; yields are in the range 30–80%.

An interesting reaction is the N-amination by the reagent of benzotriazole (1) to give 1-aminobenzotriazole (2),[6] a compound of particular interest as a benzyne precursor (*see* Lead tetraacetate, miscellaneous oxidations).

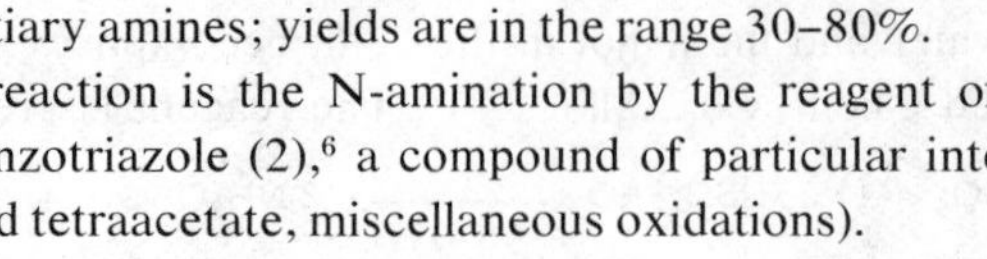
N
N
N
H
aq. KOH (60°)
$\overset{+}{N}H_3OSO_3^-$
N
N
N
NH_2

(1) (2)

Attempts to use the reagent for the conversion of carboxylic acids into amides were not promising.[7] In the presence of aluminum chloride (2 equiv.) the reagent reacts with an aromatic hydrocarbon with introduction of an amino group, but yields are very low: benzene → aniline (28%); toluene → toluidine (50%), *o*; *m*; *p* = 51; 13; 36.[8]

H. C. Brown[9] showed that the reagent reacts readily with an alkylborane to produce the corresponding amine. Examples:

$$CH_3(CH_2)_5CH{=}CH_2 \xrightarrow{B_2H_6} [CH_3(CH_2)_5CH_2CH_2]_3B \xrightarrow[\text{64\% overall}]{\overset{+}{N}H_3OSO_3^-} CH_3(CH_2)_5CH_2CH_2NH_2$$

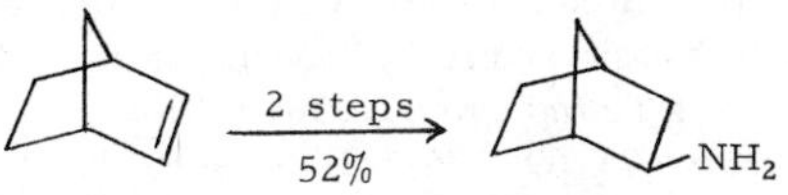

Brown *et al.*[9a] found that the reagent is soluble in diglyme and that in this solvent it converts the organoboranes from unhindered and hindered olefins into the corresponding amines, for example:

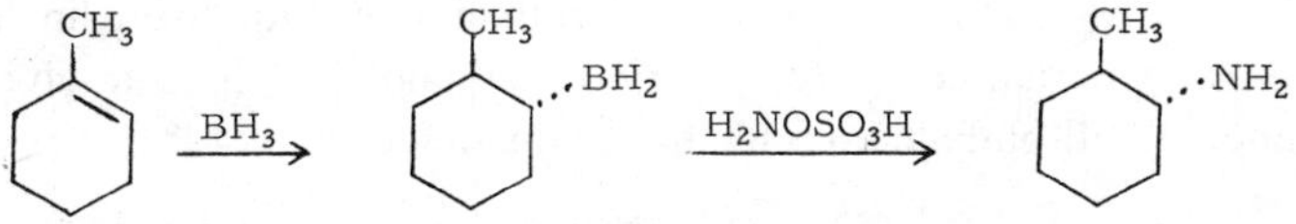

Benzisoxazoles. The reagent is used in a general synthesis of benzisoxazoles from salicylaldehydes.[10] Salicylaldehyde is treated in ether solution with an excess of the reagent and with anhydrous sodium sulfate; the sodium salt of the oxime

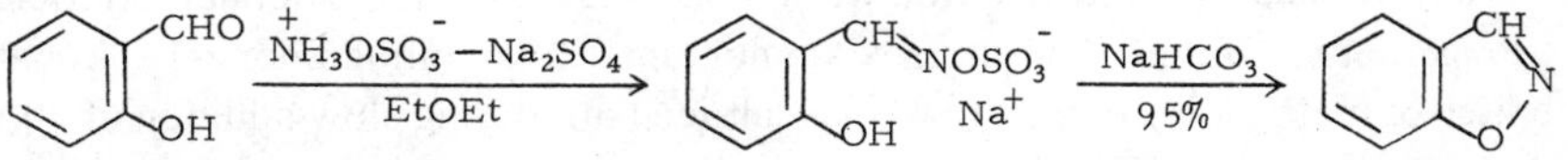

sulfonate separates in a few minutes and is treated with sodium bicarbonate to effect cyclization.

Diaziridines and diazirines. Hydroxylamine-O-sulfonic acid and ammonia react with cyclohexanone to form 3,3-pentamethylenediaziridine, which can be oxidized with silver oxide to 3,3-pentamethylenediaziridine.[11] The product, a liquid, is distilled

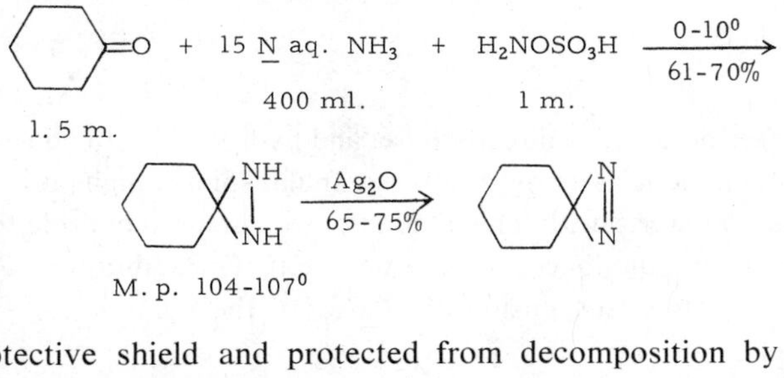

behind a protective shield and protected from decomposition by dilution with ether and storage in a refrigerator.

2-Hydroxycyclohexanone reacts with the reagent and ammonia to give the diaziridine in 45% yield; oxidation with silver oxide gave the diazirine in 70% yield.[12]

[1]H. J. Matsuguma and L. F. Audrieth, *Inorg. Syn.*, **5**, 122 (1957)
[2]A. Nickon and A. Sinz, *Am. Soc.*, **82**, 753 (1960); A. Nickon and A. S. Hill, *ibid.*, **86**, 1153 (1964)
[3]E. Schmitz and R. Ohme, *Angew. Chem., Internat. Ed.*, **2**, 157 (1963); *Ber.*, **97**, 2521 (1964)
[4]E. Schmitz, R. Ohme, and S. Schramm, *Angew. Chem.*, **73**, 807 (1961); R. Appel and W. Büchner, *ibid.*, **73**, 807 (1961); *Ann.*, **654**, 1 (1962)
[5]R. Gösl and A. Meuwsen, *Org. Syn.*, **43**, 1 (1963)
[6]C. D. Campbell and C. W. Rees, *Chem. Comm.*, 192 (1965); see also C. W. Rees and R. C. Storr, *ibid.*, 193 (1965)
[7]G. B. Bachman and J. E. Goldmacher, *J. Org.*, **29**, 2576 (1964)
[8]P. Kovacic and R. P. Bennett, *Am. Soc.*, **83**, 221 (1961); P. Kovacic, R. P. Bennett, and J. L. Foote, *ibid.*, **84**, 759 (1962)
[9]H. C. Brown, W. R. Heydkamp, E. Breuer, and W. S. Murphy, *Am. Soc.*, **86**, 3565 (1964)
[9a]M. W. Rathke, N. Inoue, K. R. Varma, and H. C. Brown, *Am. Soc.*, **88**, 2870 (1966)
[10]D. S. Kemp and R. B. Woodward, *Tetrahedron*, **21**, 3019 (1965)
[11]H. J. Abendroth, *Angew. Chem.*, **73**, 69 (1961); E. Schmitz, R. Ohme, and R.-D. Schmidt, *Ber.*, **95**, 2714 (1962); E. Schmitz and R. Ohme, *Org. Syn.*, **45**, 83 (1965)
[12]E. Schmitz, A. Stark, and C. Hörig, *Ber.*, **98**, 2509 (1965)

N-Hydroxymethylphthalimide. Mol. wt. 177.16, m.p. 138–141°, Supplier: British Drug Houses.

Prepared in 90% yield by reaction of phthalimide with formalin, the reagent reacts with aromatic amines in refluxing 80% ethanol ($\frac{1}{2}$–2 hrs.) to give crystalline N-(arylaminomethyl)-phthalimides melting in the range 112–242°.[1]

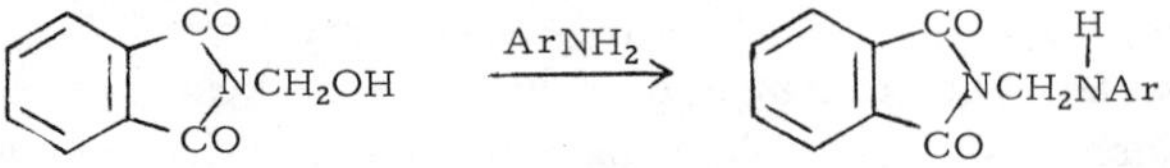
CO
NCH₂OH
CO
ArNH₂
CO
H
NCH₂NAr
CO

[1]M. B. Winstead and H. W. Heine, *Am. Soc.*, **77**, 1913 (1955)

2-Hydroxy-1,4-naphthoquinone. Mol. wt. 174.15, 195–196°, dec. Supplier: Aldrich.

Preparation. One method utilizes ammonium 1,2-naphthoquinone-4-sulfonate.[1] One liter of methanol is cooled to 0° in a salt-ice bath during slow addition of 80 ml. of concd. sulfuric acid. The cooling bath is removed, 1 mole of the ammonium salt is added and stirred to an even paste, and the temperature is gradually raised to boiling, with separation of the ether (2), more methanol is added, and the mixture is

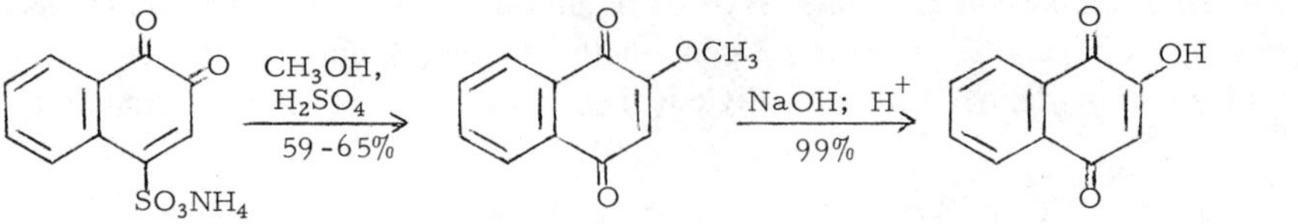
O
O
SO₃NH₄
CH₃OH,
H₂SO₄
59-65%
O
OCH₃
O
NaOH; H⁺
99%
O
OH
O

(1) (2) (3)

cooled. The ether obtained is directly pure, and hydrolysis with aqueous alkali gives hydroxynaphthoquinone as a bright yellow granular solid of high purity.

A better method starts with Thiele addition of acetic anhydride to 1,4-naphthoquinone[2] (1,2-naphthoquinone can also be used). On addition of 2 ml. of boron trifluoride etherate to a suspension of 15.8 g. of the quinone in 40 ml. of acetic

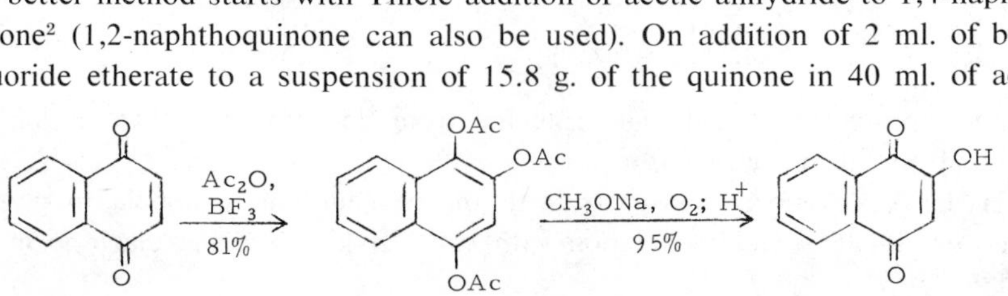
O
O
Ac₂O,
BF₃
81%
OAc
OAc
OAc
CH₃ONa, O₂; H⁺
95%
O
OH
O

anhydride the quinone slowly dissolves with a slight temperature rise. Workup affords pure, colorless 1,2,4-triacetoxynaphthalene. For hydrolysis the triacetate is stirred with sodium methoxide in methanol at 0°; the hydroquinone liberated is oxidized by air with separation of the bright red sodium salt of hydroxynaphthoquinone. Acidification of an aqueous solution affords bright yellow hydroxynaphthoquinone, m.p. 195–196°.

[1]L. F. Fieser and E. L. Martin, *Org. Syn., Coll. Vol.*, **3**, 465 (1955)
[2]L. F. Fieser, *Am. Soc.*, **70**, 3165 (1948)

2-Hydroxy-5-nitrobenzyl bromide,

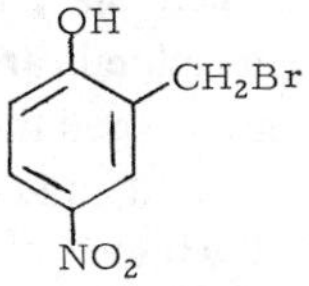
OH
CH_2Br
NO_2

Mol. wt. 232.05, m.p. 145°. Prepared by bromomethylation of *p*-nitrophenol.[1] The reagent has high selectivity for reaction with the tryptophane residues of proteins. The only other amino acid that reacts is cysteine, and this reacts more slowly.[1]

[1]D. E. Koshland, Jr., Y. D. Karkhanis, and H. G. Latham, *Am. Soc.*, **86**, 1448 (1964); H. R. Horton and D. E. Koshland, Jr., *ibid.*, **87**, 1126 (1965)

N-Hydroxyphthalimide (4). Mol. wt. 163.13, m.p. 237–240°. Supplier: Pierce Chem. Co.

The preparation of the reagent in 70% yield by reaction of N-carboethoxyphthalimide (1) with hydroxylamine and triethylamine in boiling absolute ethanol appears to involve cleavage of the heterocyclic ring (2) and its re-formation (3) with elimination of urethane; the solution turns red with formation of the triethylammonium salt of N-hydroxyphthalimide (4), and after acidification and dilution with water this separates in nearly colorless needles.[1] The reagent is used in peptide

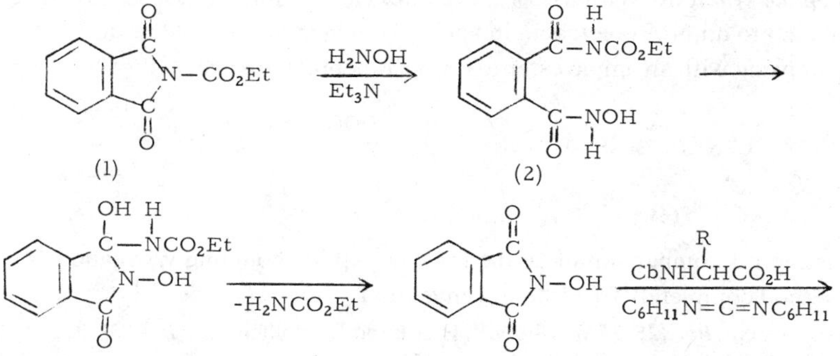
$N-CO_2Et$
H_2NOH
Et_3N
$C-NCO_2Et$
$C-NOH$
$-H_2NCO_2Et$
N–OH
$CbNHCHCO_2H$
$C_6H_{11}N=C=NC_6H_{11}$

(1) (2) (3) (4)

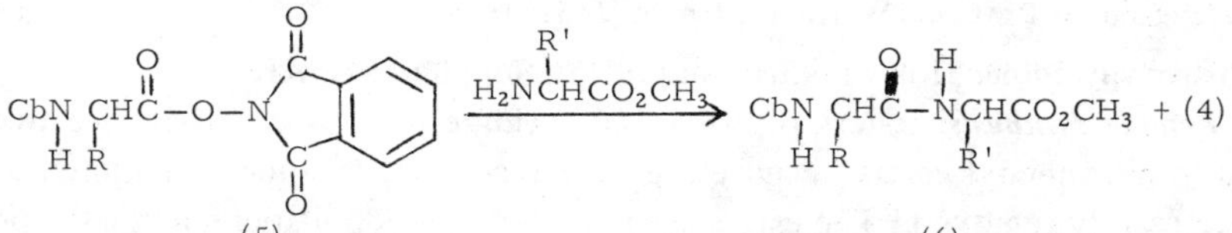
$H_2NCHCO_2CH_3$
$CbN\ CHC-NCHCO_2CH_3$ + (4)

(5) (6)

synthesis as follows. Condensation of (4) with an N-protected amino acid in the presence of dicyclohexylcarbodiimide gives the activated ester (5), which reacts readily with an amino ester to give the protected dipeptide (6); the N-hydroxyphthalimide formed is removed by shaking with aqueous bicarbonate solution.

The synthesis of N-hydroxyphthalimide by Nefkens (1⟶4) is indirect. A direct synthesis worked out later by Mazur and Plume[2] is shorter and the yield is higher (91%). A solution of 1.1 moles of hydroxylamine hydrochloride in 1.5 l. of pyridine in a 1-l. one-necked round-bottomed flask was cooled to 30°, 1.0 mole of phthalic anhydride was added in one portion, and the flask was swirled until a clear solution resulted (42°). The solution was heated at an internal temperature of 90° for 15 min. and the pyridine was distilled on a rotating evaporator at water-pump pressure. The hot viscous residue was added rapidly to 1 l. of 1 *N* acetic acid and the resulting precipitate was collected and washed thoroughly with 0.01 *N* acetic acid.

[1]G. H. L. Nefkens and G. I. Tesser, *Am. Soc.*, **83**, 1263 (1961); G. H. L. Nefkens, G. I. Tesser, and R. J. F. Nivard, *Rec. trav.*, **81**, 683 (1962)
[2]R. H. Mazur and G. Plume, procedure submitted to *Org. Syn.*

N-Hydroxypiperidine (2 ⇌ 3). Mol. wt. 101.15, m.p. 39°, b.p. 110°/55 mm.

Preparation by oxidation of piperidine with 3% hydrogen peroxide.[1] A more satisfactory method is by oxidation of N-ethylpiperidine to the N-oxide followed by Cope elimination.[2]

Structure. The reagent reacts as the hydroxylamine (2) with phenylisocyanate

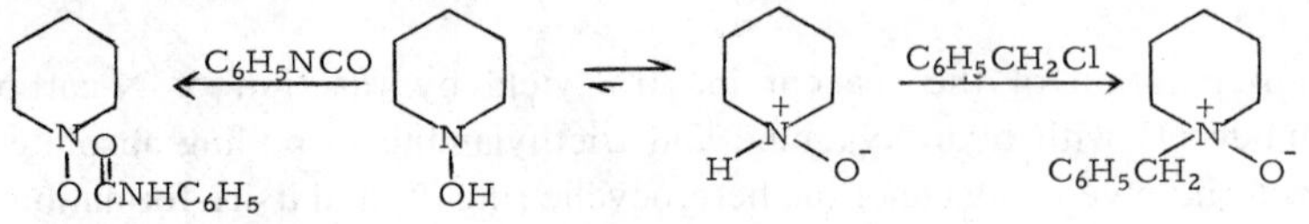

C_6H_5NCO
$C_6H_5CH_2Cl$
N
O
$CNHC_6H_5$
OH
H
$C_6H_5CH_2$

(1) (2) (3) (4)

to give (1) and as the amine oxide (3) to give the N-benzyl derivative (4).[1]

Peptide synthesis. With use of dicyclohexylcarbodiimide (DCC), the reagent can be coupled to an N-protected amino acid (5) to give an activated ester[2,3] (6) capable of condensing with an amino ester to form a dipeptide linkage.

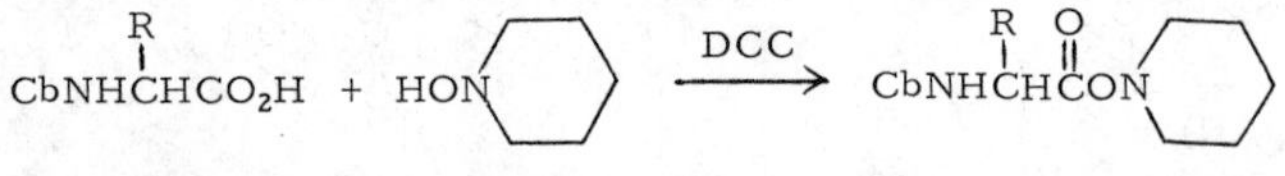

R
$CbNHCHCO_2H$ + HON
DCC
R O
CbNHCHCON

(5) (2) (6)

In a careful comparison of 16 methods of peptide coupling Weygand *et al.*[4] found this method completely free from racemization.

[1]R. Wolffenstein, *Ber.*, **25**, 2777 (1892); F. Haase and R. Wolffenstein, *Ber.*, **37**, 3228 (1904)
[2]B. O. Handford, J. H. Jones, G. T. Young, and T. F. N. Johnson, *J. Chem. Soc.*, 6814 (1965)
[3]S. M. Beaumont, B. O. Handford, J. H. Jones, and G. T. Young, *Chem. Comm.*, 53 (1965)
[4]F. Weygand, A. Prox, and W. König, *Ber.*, **99**, 1451 (1966)

3-Hydroxypyridine. Mol. wt. 95.10, m.p. 126°. Suppliers: A, KK.

Peptide synthesis. Esters of 3-hydroxypyridine offer some advantage over the usual *p*-nitrophenyl esters because the unreacted ester is soluble in dilute acid and hence readily removed.[1] The esters are readily prepared by coupling the Cb-peptide and 3-hydroxypridine with dicyclohexylcarbodiimide in ethyl acetate in the presence of triethylamine.

[1]E. Taschner and B. Rzeszortarska, *Angew. Chem., Internat. Ed.*, **4**, 594 (1965)

N-Hydroxysuccinimide, 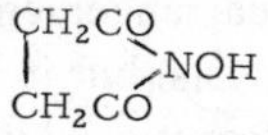

Mol. wt. 115.09, dec. 175°. Supplier: Aldrich. Preparation.[1,2]

The reagent is used in peptide synthesis in the same way that N-hydroxyphthalimide is used, and it has the advantage that the by-product of coupling an activated ester with an amino ester is the water-soluble N-hydroxysuccinimide.

[1]R. Wegler, F. Grewe, and K. Mehlose, U.S. patent 2,816, 111 (1957)
[2]G. W. Anderson, J. E. Zimmerman, and F. M. Callahan, *Am. Soc.*, **86**, 1839 (1964)

Hypobromous acid, aqueous, *see* N-Bromoacetamide.

Hypochlorous acid in ether. The hypochlorous acid generated by passing chlorine into sodium bicarbonate solution can be extracted with ether and the solution dried quickly and the titer determined.[1] The Ruschig process[1] for a modified Curtius degradation of 3β-acetoxy-Δ^5-bisnorcholenic acid (1) to pregnenolone (7) involves

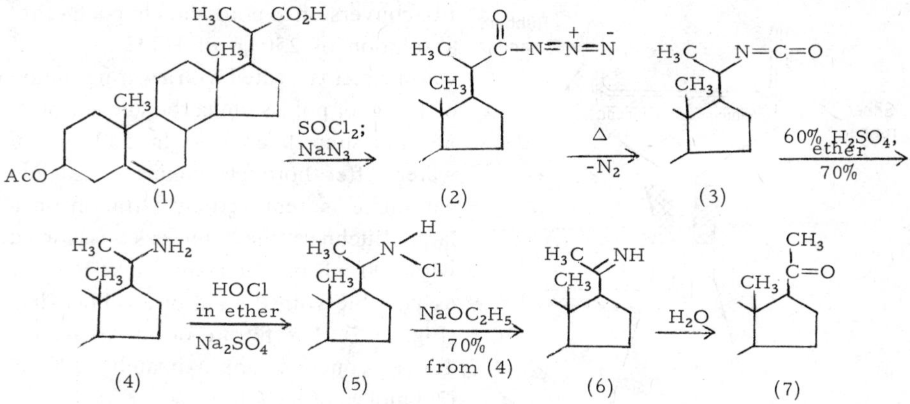
SOCl2; NaN3
Δ
-N2
60% H2SO4, ether
70%
HOCl in ether
Na2SO4
NaOC2H5
70% from (4)
H2O
(1)
(2)
(3)
(4)
(5)
(6)
(7)

conversion through the acid chloride to the azide (2) and the isocyanate (3), which was hydrolyzed to the amine (4) by stirring an ethereal solution with 60% sulfuric acid. An ethereal solution of the amine was added at 0° to an ethereal solution of hypochlorous acid in the presence of anhydrous sodium sulfate to retain the water formed on production of the chloroamine (5). This substance was obtained as a crystalline solid, and on reaction with sodium ethoxide it afforded a solid ketimine (6) which was hydrolyzed easily to pregnenolone (7).

In an efficient degradation of pregnenolone acetate (8) to androstenolone (14), the first step[2] consists in reaction of the oxime (9) with phosphoryl chloride in

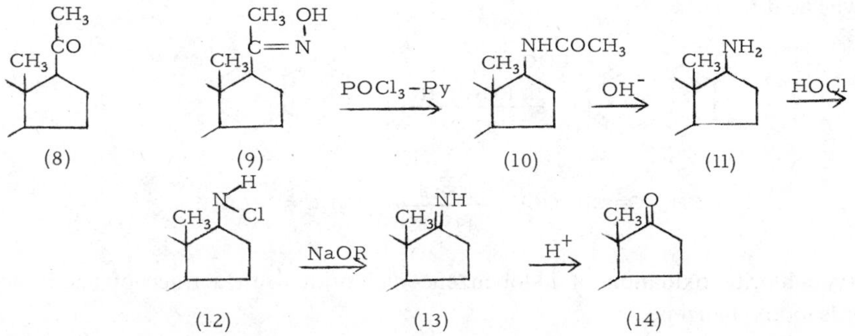
POCl3-Py
OH⁻
HOCl
NaOR
H⁺
(8)
(9)
(10)
(11)
(12)
(13)
(14)

pyridine, which effects quantitative Beckmann rearrangement to the 17-acetylamine (10). The amide is very resistant to acid hydrolysis, but it is hydrolyzed smoothly by alcoholic alkali at 170° to the amine (11). Reaction with hypochlorous acid in ether effects conversion to the chloroamine (12), which then yields the ketimine (13) and the 17-ketone (14).[3] Androstenolone is obtained from pregnenolone in overall yield of 70%.[4]

[1]H. Ruschig, W. Fritsch, J. Schmidt-Thomé, and W. Haede, *Ber.*, **88**, 883 (1955)
[2]J. Schmidt-Thomé, *Ber.*, **88**, 895 (1955)
[3]G. Ehrhart, H. Ruschig, and W. Aumüller, *Angew. Chem.*, **52**, 363 (1939)
[4]J. Schmidt-Thomé, *Ann.*, **603**, 43 (1957)

Hypohalite solution (*see also* Sodium hypobromite, sodium hypochlorite). Sodium hypochlorite solution can be prepared by passing chlorine into dilute alkali with ice cooling.[1] Calcium hypochlorite is available commercially as a stable, water-soluble solid (trade names: HTH, Perchloron). Suppliers: Harshaw, McKesson and Robbins, Olin Mathieson, Pennsalt. For conversion to potassium hypochlorite[2] a solution of 250 g. of HTH in 1 l. of warm water is treated with a warm solution of 175 g. of potassium carbonate and 50 g. of potassium hydroxide in 500 ml. of water. After thorough shaking the calcium carbonate is removed by filtration on a large Büchner; the filter cake is stirred with 200 ml. of water and sucked as dry as possible with the aid of a rubber dam (Fig. H-5). The filtrate of approximately 1.5 l. contains approximately 200 g. (2.3 moles) of KOCl.

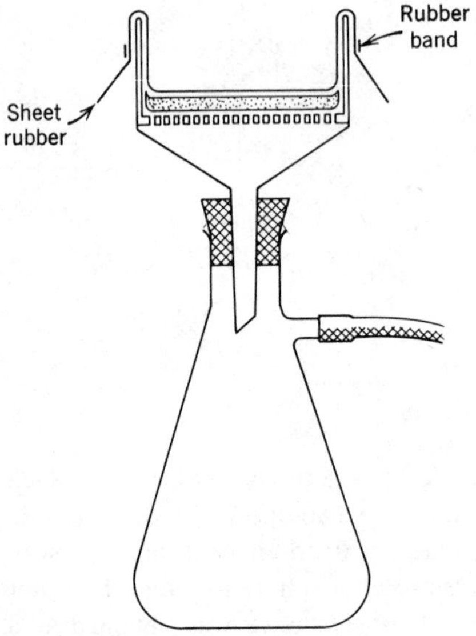
Rubber
band
Sheet
rubber

Fig. H-5 Rubber dam for pressing the filter cake

"Chlorox," one of the commercially available preparations of sodium hypochlorite, is a stablized solution containing 5.25% of NaOCl by weight.

For the oxidation of ammonia to hydrazine, *see* Hydrazine.

Two examples of the preparation of an acid by hypochlorite oxidation of a methyl ketone are: 2-acetonaphthalene → β-naphthoic acid[2] and mesityl oxide → dimethylacrylic acid.[3]

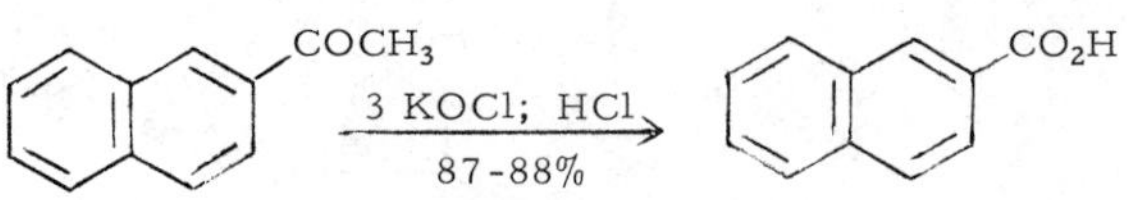
COCH3
3 KOCl; HCl
87-88%
CO2H

$$(CH_3)_2C{=}CHCOCH_3 \xrightarrow[49-53\%]{KOCl} (CH_3)_2C{=}CHCO_2H$$

Hypochlorite oxidation of iodobenzene dichloride with a trace of acetic acid affords iodoxybenzene.[4]

$$C_6H_5\overset{+}{I}Cl(Cl^-) + NaOCl + H_2O \xrightarrow[87-92\%]{65-75^\circ} C_6H_5\overset{O^-}{\overset{|}{I}}{}^{+}{=}O + NaCl + 2\,HCl$$

An interesting reaction of the early steroid literature is the oxidation of cholesterol suspended in sodium hypobromite solution to the Diels acid.[5] In describing a supposedly improved procedure, Shoppee and Summers[6] do not state the yield. The

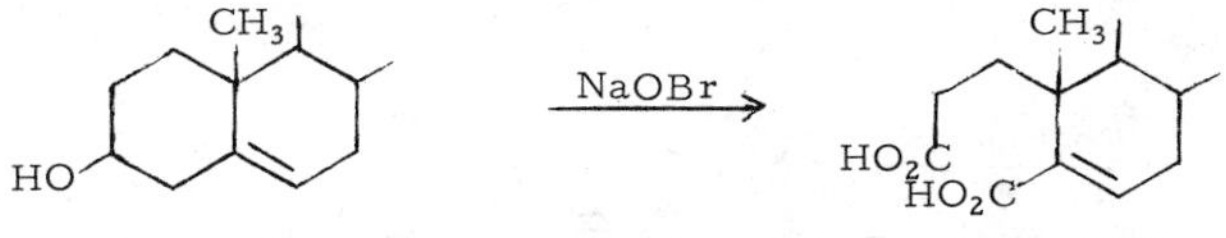
CH3
HO
NaOBr
CH3
HO2C
HO2C

Diels acid

Diels acid is high melting and sparingly soluble and can be isolated easily even though present in very small amounts. One of us[7] oxidized cholesterol with sodium dichromate dihydrate in benzene–acetic acid and isolated the Diels acid in yield of 2.97% along with 6 neutral oxidation products. One of these, Δ^5-cholestene-3-one, might be but is not a precursor of the Diels acid. Cholesteryl acid chromate would seem to offer an attractive possibility for intramolecular attack of the allylic β-hydrogen at C_4. The active species in the hypobromite oxidation may be the hypobromite of cholesterol.

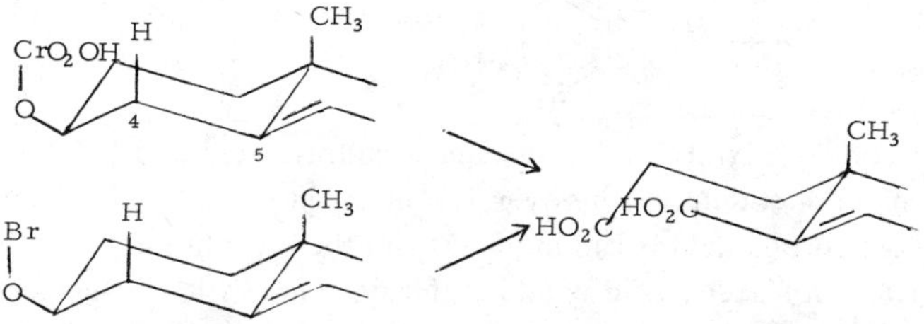
CrO2OH
H
CH3
4
5
Br
H
CH3
CH3
HO2C
HO2C

Moore and Link[8] developed a method for the oxidation of aldoses with hypoiodite in methanol to produce aldonic acids, which are isolated as the potassium or barium aldonates in 90% yield. The procedure was worked out for the purpose of preparing

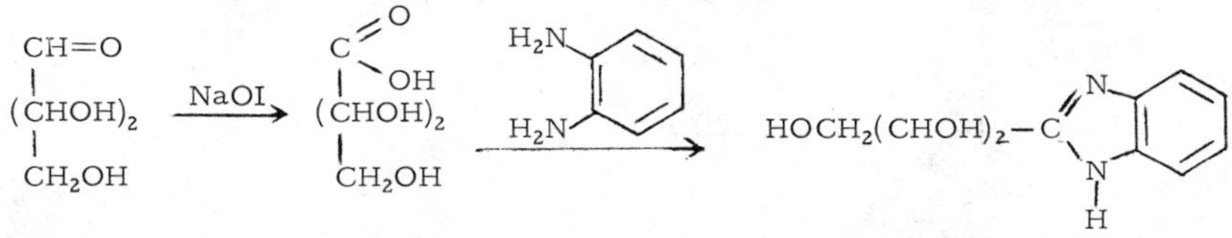
CH=O
(CHOH)2
CH2OH
NaOI
O
C
OH
(CHOH)2
CH2OH
H2N
H2N
HOCH2(CHOH)2–C
N
N
H

intermediates suitable for condensation with *o*-phenylenediamine to form benzimidazole derivatives, since these have properties ideal for characterization and identification (m.p. and αD, m.p. of hydrochlorides and picrates).

[1]R. Adams and B. K. Brown, *Org. Syn., Coll. Vol.*, **1**, 309 (1941)
[2]M. S. Newman and H. L. Holmes, *ibid.*, **2**, 428 (1943)
[3]L. I. Smith, W. W. Prichard, and L. J. Spillane, *ibid.*, **3**, 302 (1955)
[4]M. W. Formo and J. R. Johnson, *ibid.*, **3**, 486 (1955)
[5]O. Diels and E. Abderhalden, *Ber.*, **36**, 3177 (1903); **37**, 3092 (1904)
[6]C. W. Shoppee and G. H. R. Summers, *J. Chem. Soc.*, 2528 (1952)
[7]L. F. Fieser, *Am. Soc.*, **75**, 4377, 4386, 4395 (1953)
[8]S. Moore and K. P. Link, *J. Biol. Chem.*, **133**, 293 (1940)

Hypophosphorous acid, H_3PO_2. Mol. wt. 65.99. Suppliers of 50% aqueous solution (sp. gr. 1.22): Baker, Fisher.

This reducing agent reagent was introduced by Mai[1] for deamination, that is, for replacement of a diazonium salt group by hydrogen. For the reduction of tetrazotized *o*-tolidine, Kornblum[2] used about six times the theory of 30% hypophosphorous acid;

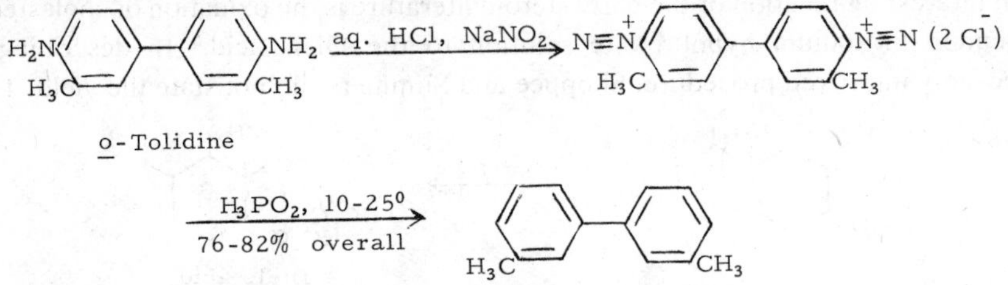

aq. HCl, NaNO2
o-Tolidine
H3PO2, 10-25°
76-82% overall

this is cooled in ice and treated with the filtered, clear red solution of diazotized amine. Immediate evolution of nitrogen occurs. The flask is placed in a refrigerator for 8–10 hrs. and let stand at room temperature for 8–10 hrs. before workup.

In the preparation of 2,4,6-tribromobenzoic acid from *m*-aminobenzoic acid,[3] this substance (0.2 mole) is brominated in solution in dilute hydrochloric acid and the tribromo compound which precipitates is collected and washed and used in the next step without drying. A diazotizing solution is prepared by stirring with ice

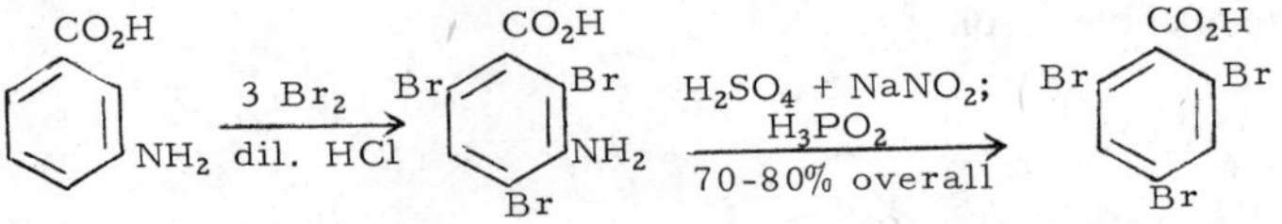

3 Br2
dil. HCl
H2SO4 + NaNO2;
H3PO2
70-80% overall

cooling a precooled mixture of 1 l. of concd. sulfuric acid and 0.5 l. of water and adding 0.54 mole of sodium nitrite over 15 min. A large excess (1.8 moles) of cold 50% hypophosphorous acid is run in at −5° and then a solution of the bromination product in 1.85 l. of acetic acid is added during 1 hr. with the temperature maintained at −10° to −15°. After 2 hrs. the temperature is allowed to rise to 5° and the mixture is let stand in a refrigerator for 36 hrs.

Another use for reagent is for reductive elimination of halogen from at least some nitroaryl halides, for example picryl iodide:[4]

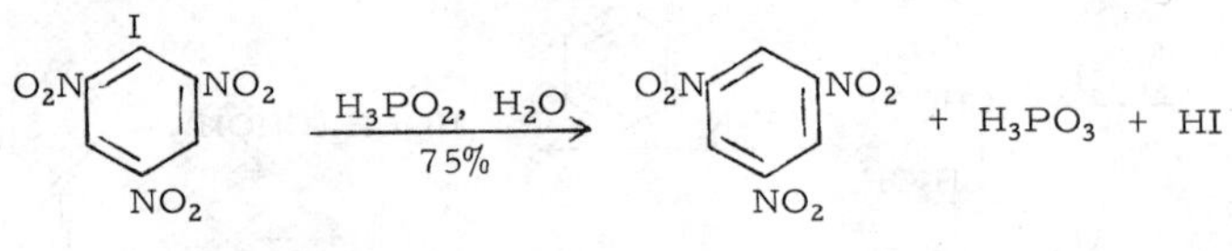

H3PO2, H2O
75%
+ H3PO3 + HI

Neither 2,4-dinitrothiophene nor the 2,5-dinitro isomer has been nitrated satisfactorily, but the 2,3,4-trinitro compound[5] is obtainable by the following reaction

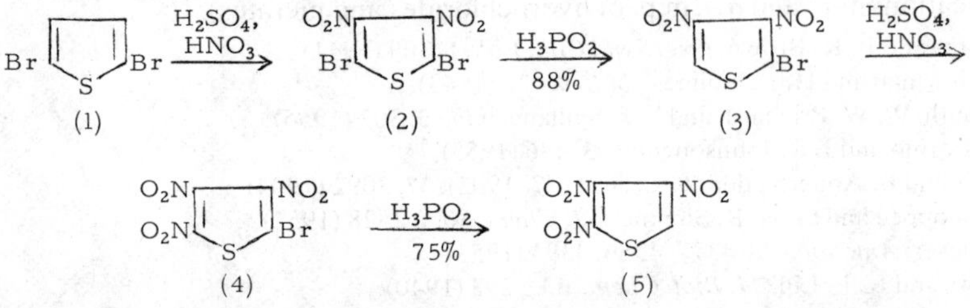

H2SO4,
HNO3
H3PO2
88%
H2SO4,
HNO3
(1)
(2)
(3)
H3PO2
75%
(4)
(5)

sequence. 2,5-Dibromothiophene (1) is converted into the 3,4-dinitro derivative (2), which on treatment in acetone with hypophosphorous acid in 5-molar excess suffers elimination of one of the bromine atoms with conversion into (3) in high yield. A third nitro group can now be introduced to give (4), in which the remaining bromine atom is active enough for reductive elimination.

[1]J. Mai, *Ber.*, **35**, 162 (1902)
[2]N. Kornblum, *Org. Syn., Coll. Vol.*, **3**, 295 (1955)
[3]M. M. Robison and B. L. Robison, *ibid.*, **4**, 947 (1963)
[4]A. H. Blatt and N. Gross, *J. Org.*, **22**, 1046 (1957)
[5]A. H. Blatt and N. Gross, E. W. Tristram, *ibid.*, **22**, 1588 (1957)

I

Imidazole, 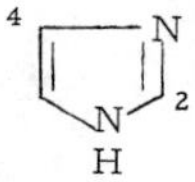

Mol. wt. 68.08, m.p. 90°, b.p. 257°, pKb 7.0; picrate, m.p. 212°. Suppliers: A, B, E, F, MCB, Chem. Intermed. and Res. Labs.

Preparation. Imidazole can be prepared[1] by a remarkable method discovered by Maquenne:[2] D-tartaric acid is converted into the dinitrate, and the ester on reaction with ammonium hydroxide and formaldehyde affords imidazole-4,5-dicarboxylic acid, which is decarboxylated to the parent heterocycle. Diketosuccinic

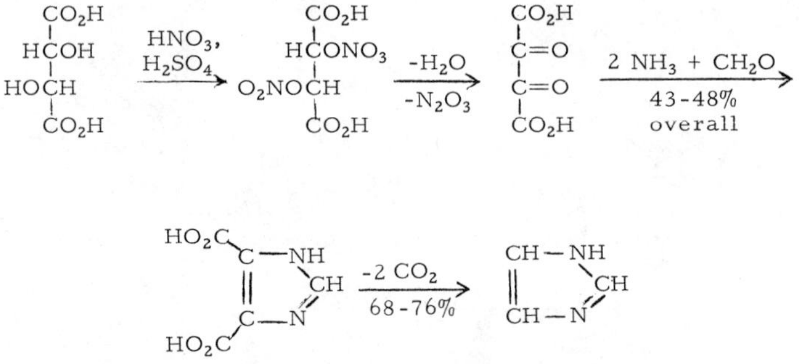

acid is an intermediate but is not isolated. Powdered D-tartaric acid (200 g.) is stirred with 432 ml. of concd. nitric acid and 432 ml. of fuming nitric acid, and 800 ml. of concd. sulfuric acid is slowly added. The temperature is checked at 38°, and after 3 hrs. the tartaric acid dinitrate which separates is collected on a glass filter cloth and pressed nearly dry "by means of a large flat glass stopper or the bottom of a 125-ml. Erlenmeyer flask" (we recommend the stainless steel spatula shown in Fig. I-1). It is not necessary to cite further details to indicate that the procedure is neither easy nor elegant. Other procedures employ glyoxal, ammonia, and formaldehyde.[3]

A simple new synthesis described by Bredereck *et al.*[4] starts with bromoacetaldehyde diethyl acetal (4), readily available according to Bedoukian[5] by bromination of vinyl acetate and treatment of the product with ethanol. The German workers

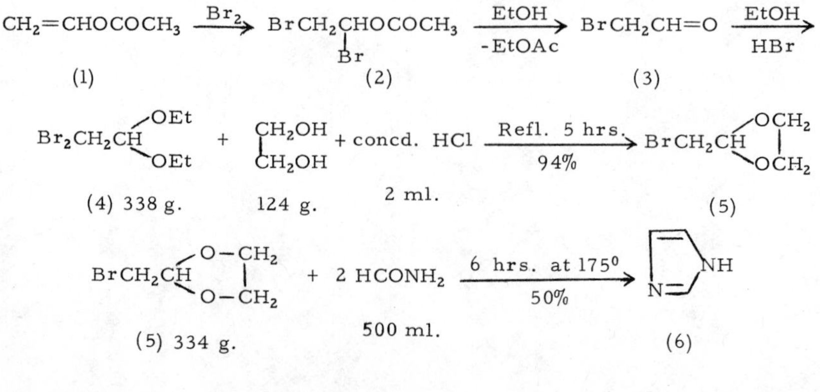

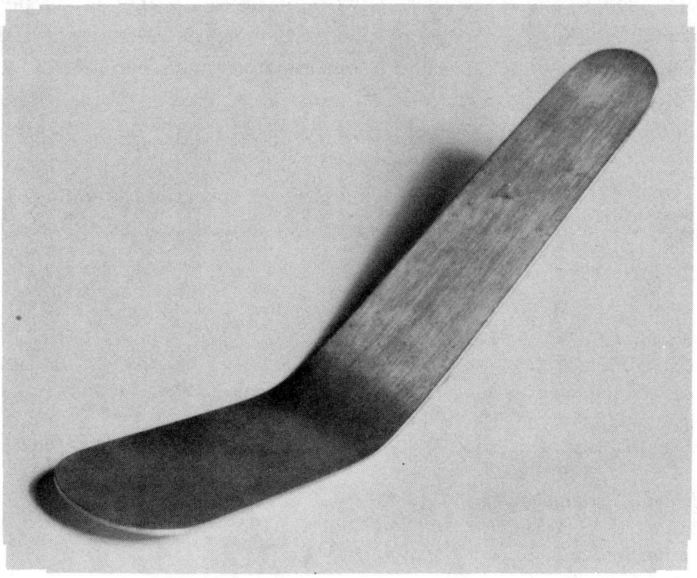

Fig. I-1 Wide-blade 5-inch spatula (Wilkens-Anderson Co.)

found it convenient to first convert the diethyl acetal (4) into the ethylene ketal (5) by acid-catalyzed exchange with ethylene glycol. They then heated the cyclic acetal (5) with excess formamide while bubbling in ammonia to prevent decomposition of acid-sensitive imidazole, which was obtained in yield of 50%.

Synthesis of carbonyl compounds. Staab developed interesting syntheses of ketones[6] and of aldehydes[7] using imidazole. An aliphatic or aromatic acid is convertible via the acid chloride into the imidazolide, the carbonyl group of which is susceptible to nucleophilic attack. Thus ketones are obtainable, often in good yield, by reaction of an imidazolide with a Grignard reagent. The imidazolide is reduced

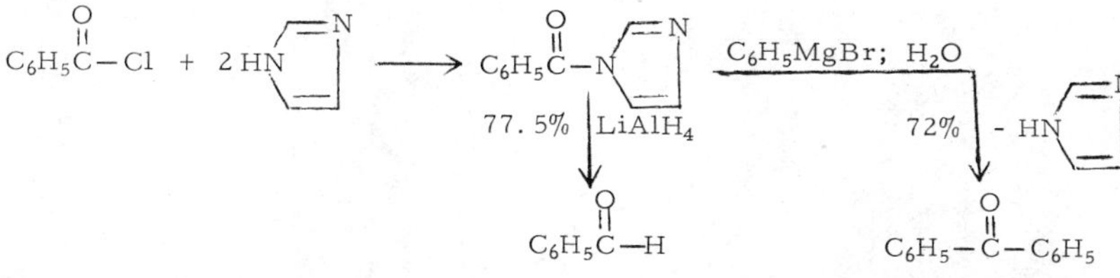

smoothly by lithium aluminum hydride to the corresponding aldehyde. The selectivity of the reaction makes possible the synthesis of ester-aldehydes from mono esters of dicarboxylic acids and of acylamino aldehydes from acylamino carboxylic acids.

See also N,N'-Carbonyldiimidazole, N,N-Thionyldiimidazole, and N,N'-Thiocarbonyldiimidazole.

Catalytic activity. Growing evidence that the imidazole ring of the histidine residue of several hydrolytic enzymes is responsible for the proteolytic acitivty of these enzymes led to the discovery in two laboratories[8,9] that imidazole has the power to catalyze the hydrolysis of phenyl acetate and of substituted phenyl acetates. The effective catalyst is the nonprotonated imidazole species. Imidazole also markedly accelerates the reaction of an amino acid ester with a *p*-phenyl ester or other activated ester; the mechanism probably involves an acyl imidazole derivative.[10,11]

Note, however, that Beyerman, Weygand, *et al.*[12] observed that imidazole caused racemization in a number of peptide couplings and is less effective than 1,2,4-triazole.

[1]H. R. Snyder, R. G. Handrick, and L. A. Brooks, *Org. Syn., Coll. Vol.*, **3**, 471 (1955)
[2]M. Maquenne, *Ann. Chim.*, [6], **24**, 525 (1891)
[3]B. Radziszewski, *Ber.*, **15**, 1493 (1882); R. Behrend and J. Schmitz, *Ann.*, **277**, 338 (1893)
[4]H. Bredereck, R. Gompper, R. Bangert, and H. Herlinger, *Ber.*, **97**, 827 (1964)
[5]P. Z. Bedoukian, *Am. Soc.*, **66**, 651 (1944)
[6]H. A. Staab and E. Jost, *Ann.*, **655**, 90 (1962); H. A. Staab, *Angew. Chem., Internat., Ed.*, **1**, 351 (1962)
[7]H. A. Staab and H. Bräunling, *Ann.*, **654**, 119 (1962)
[8]M. L. Bender and B. W. Turnquest, *Am. Soc.*, **79**, 1652 (1957)
[9]T. C. Bruice and G. L. Schmir, *Am. Soc.*, **79**, 1663 (1957)
[10]R. H. Mazur, *J. Org.*, **28**, 2498 (1963)
[11]T. Wieland and K. Vogeler, *Angew. Chem., Internal. Ed.*, **2**, 42 (1963) *Ann.*, **680**, 125 (1964)
[12]H. C. Beyerman, F. Weygand, *et al.*, *Rec. trav.*, **84**, 213 (1965)

Indicators. Velluz[1] gives a table listing over 100 indicators, with notes and references.

Mallinckrodt Chemical Works supplies a kit of eight dry indicators covering the pH range 3–10, each with the color change and pH marked on the label. Each bottle

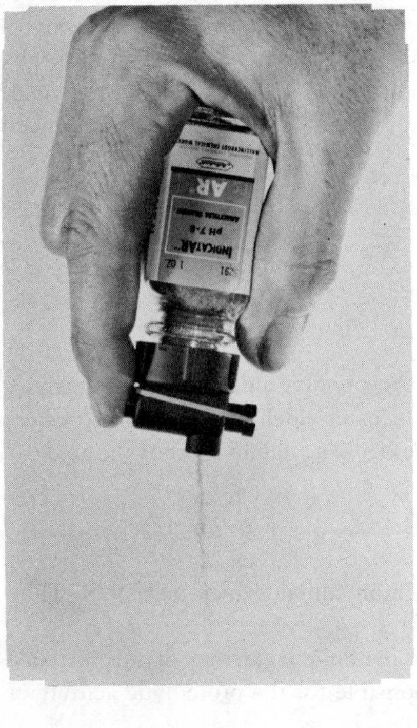
AR

Fig. I-2

has a push-button dispenser which automatically meters approximately 0.1 g. (equal to 1 or 2 drops of standard indicating solution). (See Fig. I-2.)

[1]L. Velluz, *Substances Naturelles de Synthese*, **7**, Part 3 (1953)

Indicators, Hammett acidity. With a set of 17 basic indicators any acidity lying in the range from a dilute aqueous mineral acid to that of pure sulfuric acid may be determined in any solvent.[1] The range is from +2.8 (4-aminoazobenzene) to −9.3 (2,4,6-trinitroaniline). Aldrich supplies a set of 500 mg. of each analytically pure indicator.

[1]L. P. Hammett and A. J. Deyrup, *Am. Soc.*, **54**, 2721 (1932); L. P. Hammett and M. A. Paul, *ibid.*, **56**, 827 (1934)

Iodine. Mol. wt. 253.82, m.p. 113.6°, b.p. 184°.

Aromatic iodination. Aniline can be iodinated by adding iodine to a stirred mixture of the amine and aqueous sodium bicarbonate at 12–15°.[1] The reaction is complete in about 30 min. and the product is collected and crystallized.

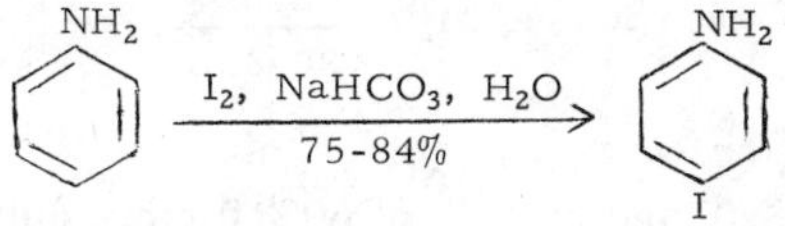
NH2
I2, NaHCO3, H2O
75-84%
NH2
I

The preferred procedure for iodination of veratrole employs silver trifluoroacetate. An aqueous solution of the salt is prepared from the acid and silver oxide, filtered, and evaporated to dryness; the salt is purified by Soxhlet extraction with ether.[2] A solution of iodine in chloroform is added with stirring under reflux to a mixture of

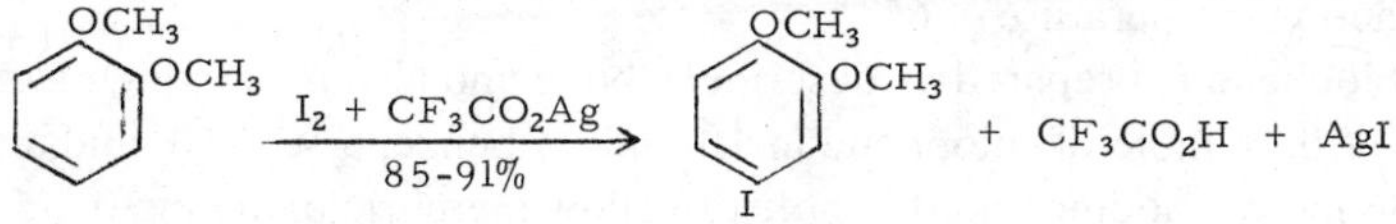
OCH3
OCH3
I2 + CF3CO2Ag
85-91%
OCH3
OCH3
I
+ CF3CO2H + AgI

veratrole and silver trifluoroacetate, silver iodide is removed by filtration, the solution is evaporated, and the product distilled. Here silver trifluoroacetate functions as base as does sodium bicarbonate in the iodination of aniline. Iodination of veratrole in the presence of mercuric oxide as base yields only 40–55% of product contaminated with mercury salts.

Two methods have been described for effecting the smooth reaction of tyrosine with iodine to form 3,5-diiodotyrosine. In one (a)[3] a solution of iodine (320 g.) and

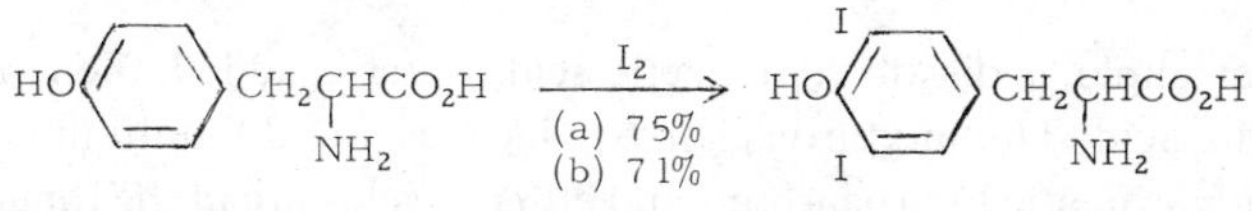
HO
CH2CHCO2H
NH2
I2
(a) 75%
(b) 71%
I
HO
CH2CHCO2H
NH2
I

sodium iodide (400 g.) in water (1.3 l.) is added dropwise to a stirred solution of L-tyrosine (100 g.) in 20% aqueous ethylamine (1 l.), and the mixture is stirred for 30 min. longer. Excess iodine is removed by reduction, the pH is adjusted to 5–6, and the amino acid which separates is dissolved in *N*-hydrochloric acid and reprecipitated with ammonia. In a second procedure (b)[4] powdered iodine is suspended in a solution of tyrosine in acetic acid and hydrochloric acid and 30% hydrogen peroxide is added slowly with shaking.

In the absence of activating substituents, direct introduction of iodine can be accomplished with use of nitric acid or sulfur trioxide to oxidize HI to I_2 and so displace the equilibrium. For the preparation of iodobenzene[5] a mixture of 1.5 moles of iodine and excess benzene (455 ml.) is treated at 50° with 6.15 moles of nitric

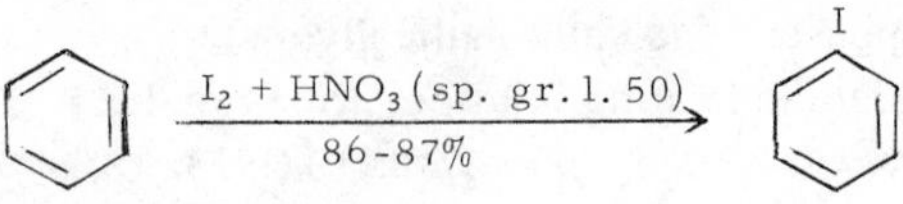
I
I₂ + HNO₃ (sp. gr. 1.50)
86-87%

acid, added in 1¾ hrs. The temperature rises to a gentle boil and the mixture is then refluxed for 15 min. The oil is separated and steam distilled from sodium hydroxide, and yellow nitro compounds are removed by stirring with dilute hydrochloric acid and iron filings. Iodobenzene is then steam distilled, dried, and distilled. By essentially the same procedure, thiophene affords 2-iodothiophene in 68–72% yield.[6]

The more vigorous conditions required when the aromatic ring contains deactivating substituents are achieved by using fuming sulfuric acid as oxidant and raising the temperature. Thus a mixture of 1 mole of phthalic anhydride, a first charge of

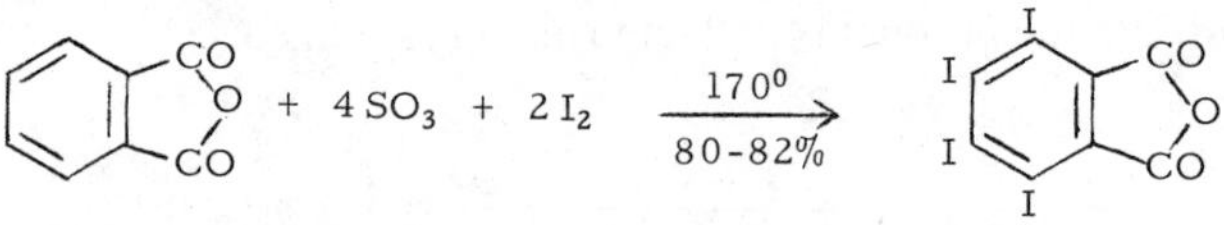
CO
O
CO
+ 4 SO₃ + 2 I₂
170°
80-82%
I
I
I
I
CO
O
CO

iodine (total of 2.12 moles), and 600 ml. of 60% fuming sulfuric acid is heated continuously to 65°, the remainder of the iodine is added in portions, and eventually the temperature is raised to 170° for 2 hrs.[7] After cooling, tetraiodophthalic anhydride is collected on a glass filter cloth and washed with concd. sulfuric acid. *m*-Dinitrobenzene affords 1,3-dinitro-5-iodobenzene in 67–70% yield by a similar procedure but at a reaction temperature of 65°.[8]

2-Iodothiophene is prepared as described above and also by stirring mechanically a solution of 0.84 mole of thiophene in 100 ml. of benzene at 5–10° and alternately adding 0.86 mole of iodine and 0.7 mole of yellow mercuric oxide in 30–35 min. and allowing the temperature to rise to about 45°.[9]

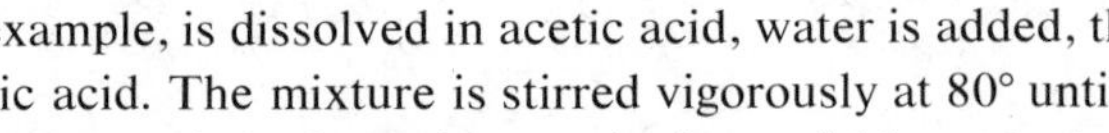
2
S
2 I₂ + HgO
60-62%
2
S
I
+ HgI₂ + H₂O

Wirth[10] has described a high-yield procedure for iodination with iodic acid as oxidant and sulfuric acid as catalyst:

$$5\ ArH + 2\ I_2 + HIO_3 \longrightarrow 5\ ArI + 3\ H_2O$$

p-Xylene, for example, is dissolved in acetic acid, water is added, then sulfuric acid, iodine, and iodic acid. The mixture is stirred vigorously at 80° until the iodine color disappears. With aromatic hydrocarbons and ethers, yields are in the range 75–90%.

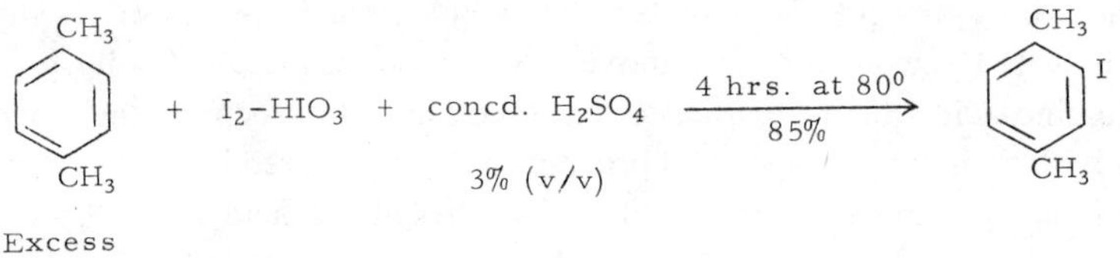
CH₃
CH₃
Excess
+ I₂-HIO₃ + concd. H₂SO₄
3% (v/v)
4 hrs. at 80°
85%
CH₃
I
CH₃

Ogata[11] reports iodination of aromatic compounds in fair yield by reaction with iodine and peracetic acid in acetic acid solution; the effective reagent may be acetyl hypoiodite, CH_3COOI. The reagent reacts with cyclohexene to give 1-iodo-2-acetoxycyclohexane, but in low yield.[12] The Prévost reaction is far better.

Indirect iodination. Two examples illustrate introduction of aromatic iodine by displacement of the HgCl group. In one,[13] mercuric acetate is dissolved in hot phenol to form the *o*-acetoxymercuric derivative, the solution is poured into hot water, and

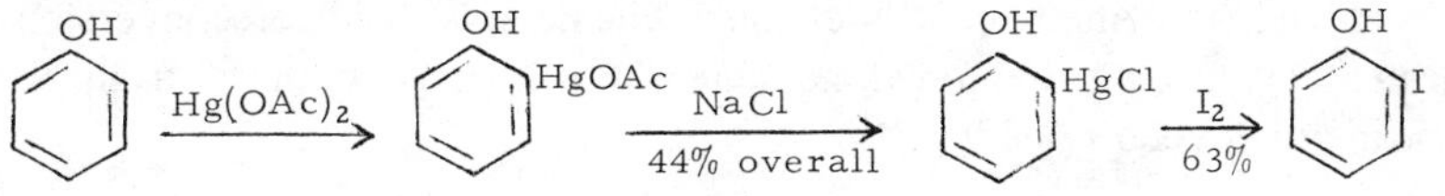
OH
Hg(OAc)2
OH
HgOAc
NaCl
44% overall
OH
HgCl
I2
63%
OH
I

a hot solution of sodium chloride is added to precipitate the *o*-chloromercuri derivative. This product on reaction with iodine in chloroform solution affords *o*-iodophenol. In another,[14] the chloromercuri derivative is obtained from a sulfonate by displacement. The reaction with iodine is conducted in refluxing ethanol.

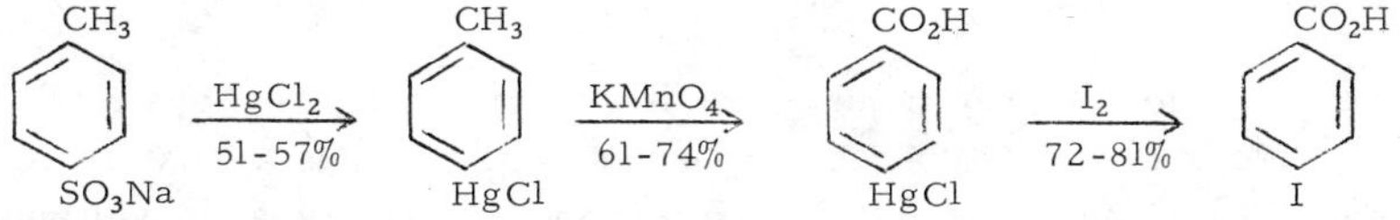
CH3
SO3Na
HgCl2
51-57%
CH3
HgCl
KMnO4
61-74%
CO2H
HgCl
I2
72-81%
CO2H
I

Displacement of OH. The *Organic Syntheses* preparation of methyl iodide[15] is noteworthy for the size of the run, the yield being 4150–4250 g. (this procedure was

$$CH_3OH \xrightarrow[93\text{-}95\%]{P\ +\ I_2} CH_3I$$

checked by L. F. F. single handed). Methanol, and with it methyl iodide as it is formed, distils into a long reflux condenser and the condensate flows into a chamber containing 2 kg. of iodine and leaches iodine into a second chamber charged with 200 g. each of red and yellow phosphorus to form PI_3 and from it CH_3I. When the first batch of iodine is exhausted a second 2-kg. charge is introduced. The methyl iodide produced is of highest purity.

Cetyl iodide is prepared by stirring a mixture of 1 mole of the alcohol, 0.32 gram atom of red phosphorus, and 1.06 gram atoms of iodine at 145–150° for 5 hrs.[16]

$$\underline{n}\text{-}C_{16}H_{33}OH \xrightarrow[78\%]{P\ +\ I_2} \underline{n}\text{-}C_{16}H_3I$$

Oxidation. The use of iodine as an oxidant is exemplified by a procedure for the preparation of dibenzoyl disulfide.[17] A solution of potassium hydroxide in absolute ethanol is saturated with hydrogen sulfide and treated with benzoyl chloride at 15°. The potassium chloride formed is removed by filtration, and iodine is added to the filtrate in the amount required to complete the oxidation.

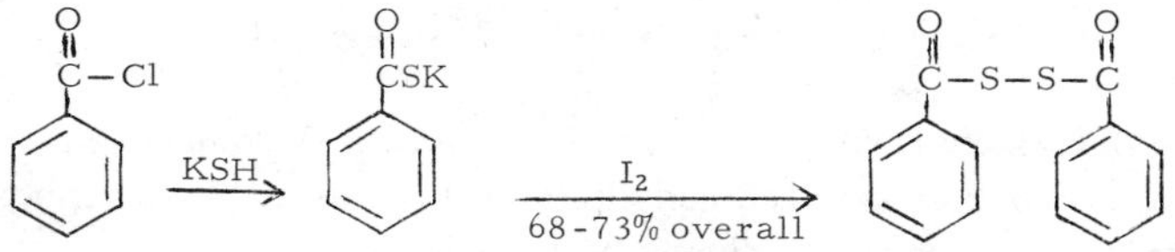
O
C—Cl
KSH
O
CSK
I2
68-73% overall
O
C—S—S—C
O

An oxidation of a different type is involved in the conversion of acetylacetone via the sodium enolate into tetraacetylethane.[18] Treatment of the 1,3-diketone with

$$CH_3COCH_2COCH_3 \xrightarrow{NaOH,\ H_2O,\ CH_3OH} CH_3\overset{ONa}{\overset{|}{C}}{=}CHCOCH_3 \xrightarrow[41\text{-}59\%\ \text{overall}]{I_2}$$

$$\begin{array}{c} CH_3COCHCOCH_3 \\ | \\ CH_3COCHCOCH_3 \end{array}$$

aqueous-methanolic sodium hydroxide affords the crystalline sodium enolate, which is collected and dried. A suspension of the sodium enolate in ether is then treated with a solution of iodine in ether; the solvent is allowed to evaporate and the mixture is crystallized from methanol. The same scheme was used for conversion of the tetraester (1) into the quinonedimethane (3), which crystallized from benzene-hexane in bright yellow needles.[19]

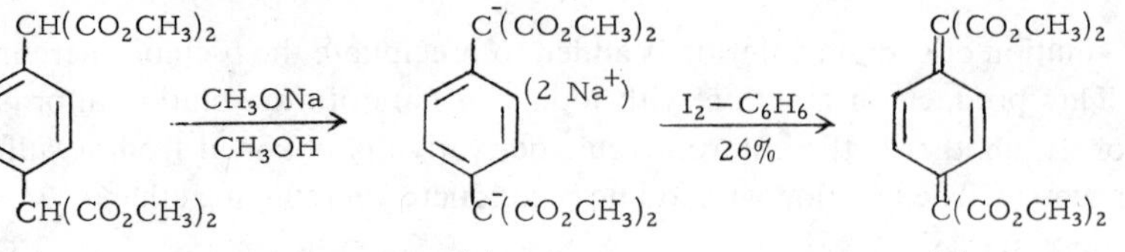
CH(CO2CH3)2
CH3ONa
CH3OH
C(CO2CH3)2
(2 Na+)
I2−C6H6
26%
C(CO2CH3)2

(1) (2) (3)

The combination of oxygen and iodine is the most satisfactory system found for the photocyclization of stilbenes to phenanthrenes.[20] Thus a mixture of *trans*-stilbene (0.01 m.) and iodine (0.005 m.) in cyclohexane on irradiation with a mercury

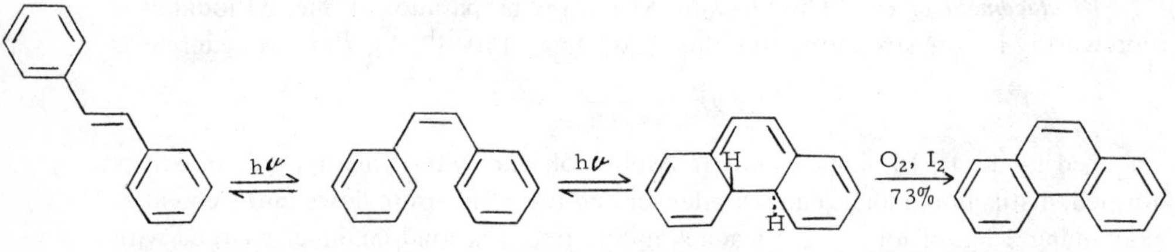
hν
hν
H
H
O2, I2
73%

lamp in the presence of oxygen gave pure phenanthrene in 73% yield. With several substituted stilbenes, addition of one equivalent of cupric bromide or chloride improves the yield significantly.[21]

Dehydration. In one example, illustrating the efficiency of iodine as catalyst for dehydration, a solution of 100 g. of benzopinacol in 500 ml. of acetic acid is treated

$$(C_6H_5)_2\underset{OH}{C}-\underset{OH}{C}(C_6H_5)_2 \xrightarrow[95\text{-}96\%]{I_2} (C_6H_5)_3C-\underset{O}{\overset{\|}{C}}C_6H_5$$

with 1 g. of iodine and refluxed for about 5 min.[22] The solution turns red and on cooling deposits very pure benzopinacolone in nearly quantitative yield. The filtrate can be used for the processing of further batches until 500 g. of benzopinacol has been rearranged. Dehydration without rearrangement is involved in the conversion of diacetone alcohol into mesityl oxide.[23] The procedure calls for treating 1100 g. of

$$(CH_3)_2\underset{OH}{C}-CH_2-COCH_3 \xrightarrow[65\%]{(I_2)} (CH_3)_2C{=}CHCOCH_3$$

crude diacetone alcohol with 0.1 g. of iodine. Since a rearrangement is involved in one of these two iodine-catalyzed dehydrations it is reasonable to suppose that both reactions proceed through carbonium ions, for example:

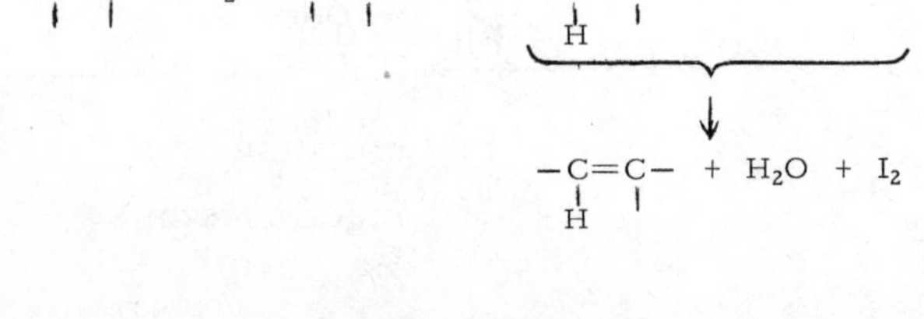
H OH
δ+ δ−
H HO···I···I
+ I2
+ HOI + I−
+ H2O + I2

The usual laboratory procedure for the preparation of β-myrcene (2) involves heating linalool (1) with a trace of iodine in vacuum at 150–160°.[24] A recent study[25] showed that β-myrcene is accompanied by *trans*-β-ocimene (3) and a trace of the *cis* isomer. The main products are readily separated by fractionation and obtained in the yields indicated.

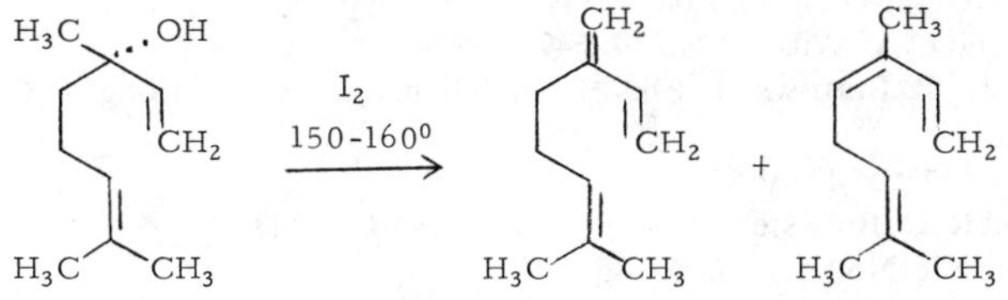
H3C
OH
CH2
H3C
CH3
I2
150-160°
CH2
CH2
H3C
CH3
+
CH3
CH2
H3C
CH3

(1) (2) 45% (3) 16%

Catalysis. A procedure for the preparation of 4-bromo-*o*-xylene[26] calls for addition of bromine over 3 hrs. to a stirred mixture of 500 g. of *o*-xylene with 12 g. of iron filings and a crystal of iodine. A note states that neither catalyst suffices by itself.

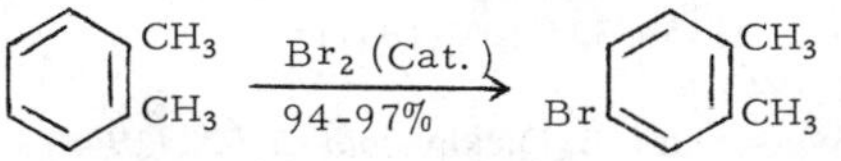
CH3
CH3
Br2 (Cat.)
94-97%
Br
CH3
CH3

The use of iodine as starter for the preparation of Grignard reagents has been mentioned. Work of Zechmeister has abundantly demonstrated the efficiency of iodine as a catalyst for the *cis-trans* isomerization of carotenoid pigments.[27]

Iodine in trace amounts has been used as catalyst in Friedel-Crafts acylations of furane and thiophene[28] and of more active members of the benzene series such as anisole and acetanilide.[29] Oddly enough, it is not effective for benzoylation of anthracene.[30]

Iodine has been used to activate zinc for the Reformatsky reaction;[31] an instance is reported where the Reformatsky reaction was erratic unless the metal was activated by iodine.[32]

Generation of HI. Benzilic acid can be reduced in high yield to diphenylacetic acid as follows.[33] A mixture of 250 ml. of acetic acid, 15 g. of red phosphorus, and

$$(C_6H_5)_2C(OH)CO_2H \xrightarrow[94\text{-}97\%]{HI} (C_6H_5)_2CHCO_2H$$

5 g. of iodine is let stand for 15–20 min. until the iodine has been converted to hydrogen iodide; then 5 ml. of water and 0.44 mole of benzilic acid are added and the mixture is refluxed for 2½ hrs.

Iodination of ketones. In a Syntex procedure for the 21-acetoxylation of 20-ketopregnanes, the ketone is treated with iodine in tetrahydrofurane in the presence of calcium oxide, followed by displacement of the iodine function by reaction with

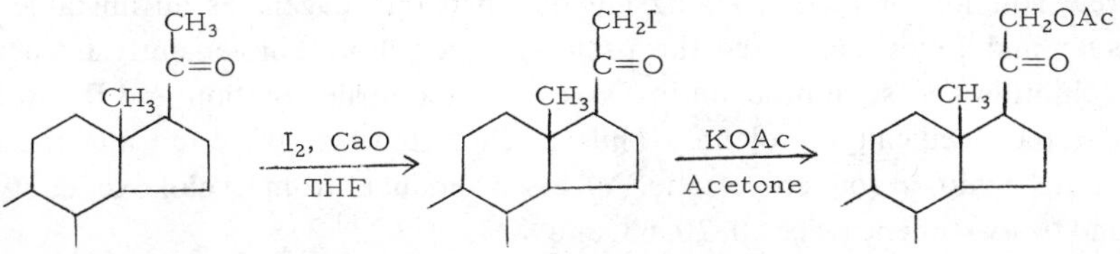
CH3
C=O
CH3
I2, CaO
THF
CH2I
C=O
CH3
KOAc
Acetone
CH2OAc
C=O
CH3

potassium acetate in acetone.[34] Halpern and Djerassi,[35] who used the procedure, state that the THF should contain some peroxide. Wall and co-workers,[36] on obtaining somewhat irregular results, postulated that the reaction involves radicals, and found that by adding azobisisobutyronitrile as initiator they could obtain consistently good results.

[1]R. Q. Brewster, *Org. Syn., Coll. Vol.*, **2**, 347 (1943)
[2]D. E. Janssen and C. V. Wilson, *ibid.*, **4**, 547 (1963)
[3]J. H. Barnes, E. T. Borrows, J. Elks, B. A. Hems, and A. G. Long, *J. Chem. Soc.*, 2824 (1950)
[4]L. Jurd, *Am. Soc.*, **77**, 5747 (1955)
[5]F. B. Dains and R. Q. Brewster, *Org. Syn., Coll. Vol.*, **1**, 323 (1941)
[6]H. Y. Lew and C. R. Noller, *ibid.*, **4**, 545 (1963)
[7]C. F. H. Allen and H. W. J. Cressman, *ibid.*, **3**, 796 (1955)
[8]T. L. Fletcher, M. J. Namkung, W. H. Wetzel, and H.-L. Pan, *J. Org.*, **25**, 1342 (1960)
[9]W. Minnis, *Org. Syn., Coll. Vol.*, **2**, 357 (1943); K. E. Miller and C. G. Lex, procedure submitted to *Org. Syn.*
[10]H. O. Wirth, O. Königstein, and W. Kern, *Ann.*, **634**, 84 (1960)
[11]Y. Ogata and K. Nakajima, *Tetrahedron*, **20**, 43 (1964)
[12]Y. Ogata, K. Aoki, and Y. Furuya, *Chem. Ind.*, 304 (1965)
[13]F. C. Whitmore and E. R. Hanson, *Org. Syn., Coll. Vol.*, **1**, 326 (1941)
[14]F. C. Whitmore *et al.*, *ibid.*, **1**, 159, 325, 519 (1941)
[15]H. S. King, *ibid.*, **2**, 399 (1943)
[16]W. W. Hartman, J. R. Byers, and J. B. Dickey, *ibid.*, **2**, 322 (1943)
[17]R. L. Frank and J. R. Blegen, *ibid.*, **3**, 116 (1955)
[18]R. G. Charles, *ibid.*, **4**, 869 (1963)
[19]D. S. Acker and W. R. Hertler, *Am. Soc.*, **84**, 3370 (1962)
[20]F. B. Mallory, C. S. Wood, and J. T. Gordon, *Am. Soc.*, **86**, 3094 (1964); C. S. Wood and F. B. Mallory, *J. Org.*, **29**, 3373 (1964)
[21]D. J. Collins and J. J. Hobbs, *Chem. Ind.*, 1725 (1965)
[22]W. E. Bachmann, *Org. Syn., Coll. Vol.*, **2**, 73 (1943)
[23]J. B. Conant and N. Tuttle, *ibid.*, **1**, 345 (1941)
[24]B. A. Arbusow and W. S. Abramow, *Ber.*, **67**, 1942 (1934)
[25]Y.-R. Naves and F. Bondavalli, *Helv.*, **48**, 563 (1965)
[26]W. A. Wisansky and S. Ansbacher, *Org. Syn., Coll. Vol.*, **3**, 138 (1955)
[27]L. Zechmeister, *Progress in the Chemistry of Organic Products*, **18**, 223 (1960)
[28]H. D. Hartough and A. I. Kosak, *Am. Soc.*, **68**, 2639 (1946)
[29]S. Chodroff and H. C. Klein, *Am. Soc.*, **70**, 1647 (1948); I. A. Kaye, H. C. Klein, and W. J. Burlant, *ibid.*, **75**, 745 (1953)
[30]P. H. Gore and J. A. Hoskins, *J. Chem. Soc.*, 5744 (1965)
[31]W. E. Bachmann, W. Cole, and A. L. Wilds, *Am. Soc.*, **62**, 824 (1940)
[32]J. D. Hardstone and K. Schofield, *J. Chem. Soc.*, 5194 (1965)
[33]C. S. Marvel, F.D. Hager and E. C. Caudle, *Org. Syn., Coll. Vol.*, **1**, 224 (1941)
[34]H. J. Ringold and G. Stork, *Am. Soc.*, **80**, 250 (1958)
[35]O. Halpern and C. Djerassi, *Am. Soc.*, **81**, 439 (1959)
[36]E. S. Rothman, T. Perlstein, and M. E. Wall, *J. Org.*, **25**, 1966 (1960)

Iodine azide, $I{-}N{=}\overset{+}{N}{=}\overset{-}{\bar{N}}$ (not isolated).

Generation. By the reaction of an aqueous suspension of silver azide and an ethereal solution of iodine, Hantzsch[1] obtained the reagent as an unstable solid. Hassner and Levy[2] generated the pseudohalogen more conveniently from iodine monochloride and sodium azide in DMF or acetonitrile solution and found that it adds stereospecifically to olefins. Thus Δ^2-cholestene is converted into the *trans*-diaxial 2β-azido-3α-iodocholestane. Yields of adducts from cyclohexene, styrene, *cis*- and *trans*-stilbene were 80, 70, 63, and 80%.

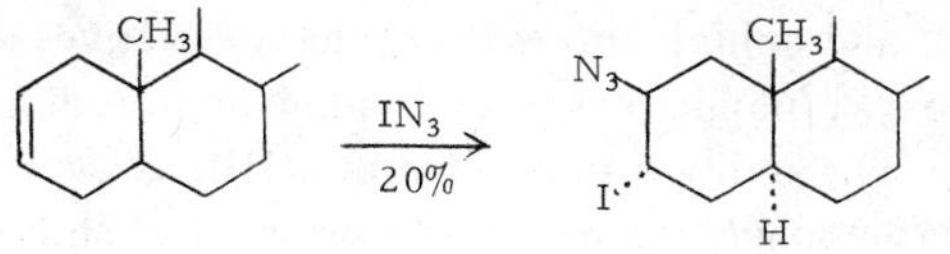

The adduct (1) of iodine azide and styrene can be converted into either the aziridine (3) or the azirine (4):[3]

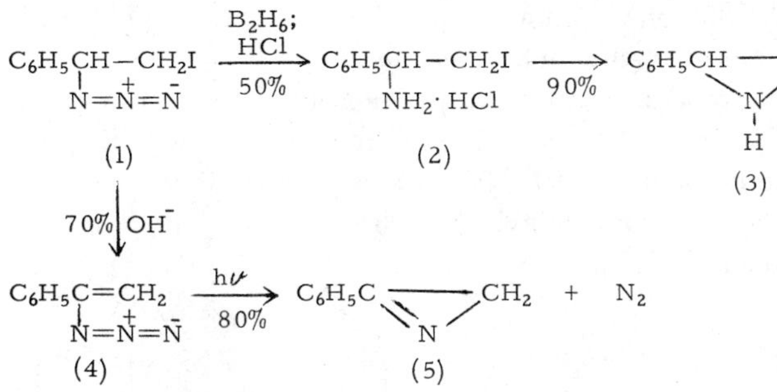

[1]A. Hantzsch, *Ber.*, **33**, 524 (1900)
[2]A. Hassner and L. A. Levy, *Am. Soc.*, **87**, 4203 (1965)
[3]Private communication from A. Hassner, Univ. Colorado

Iodine isocyanate, IN=C=O. Mol. wt. 168.93. Preparation from silver cyanate and iodine.[1] The reagent adds to olefins in the *trans* manner to give iodo-isocyanates,[2] of interest because they can be transformed into ethyleneimines.[3] The reaction is

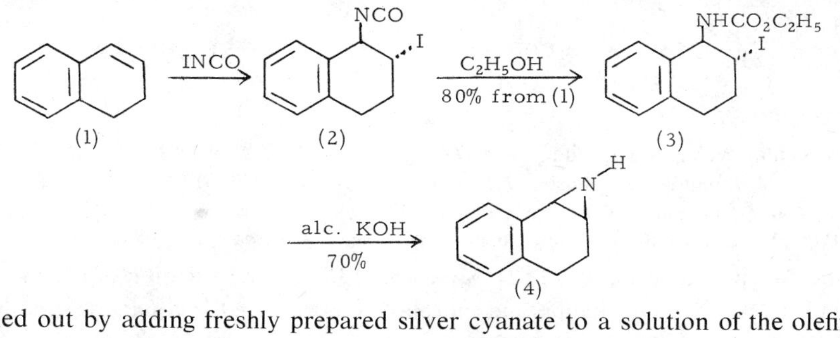

carried out by adding freshly prepared silver cyanate to a solution of the olefin in anhydrous ether; the slurry is stirred and cooled, and solid iodine is added. The reagent adds to Δ^2-cholestene to give the *trans*-diaxial 3α-iodo-2β-cholestanyliso-cyanate.[4]

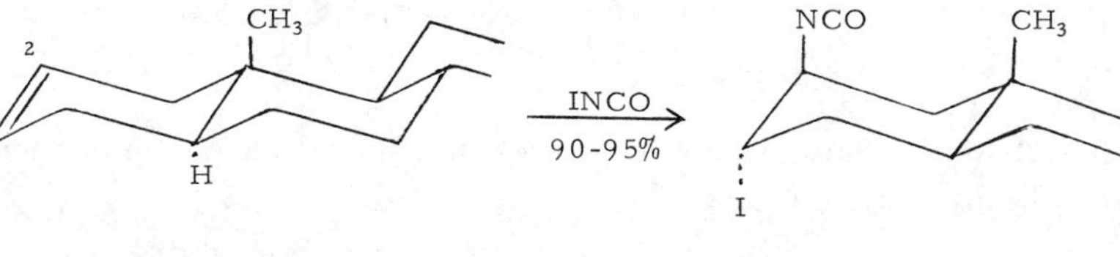

[1]L. Birckenbach and M. Lindhard, *Ber.*, **64**, 961 (1931)
[2]G. Drefahl and R. Ponsold, *Ber.*, **93**, 519 (1960)
[3]A. Hassner and C. C. Heathcock, *Tetrahedron*, **20**, 1037 (1964)
[4]*Idem*, *Tetrahedron Letters*, 1125 (1964); *J. Org.*, **30**, 1748 (1965)

Iodine monobromide. Mol. wt. 206.83, m.p. 41°, b.p. 116.° In contrast to the chloride, which is an iodinating agent, iodine monobromide is a mild brominating agent capable of converting phenol and aniline into the *p*-bromo derivatives.[1] A solution suitable for use is prepared by warming a mixture of iodine and bromine in acetic acid until solution is complete. On addition of a substrate, HBr is evolved and I_2 begins to separate. Yields: α-bromonaphthalene, 55%; 4-bromo-1-naphthol (crude), 63%.

[1]W. Militzer, *Am. Soc.*, **60**, 256 (1938)

Iodine monochloride. Mol. wt. 162.38, m.p. 14°, 27°; b.p. 97°. Suppliers: Alfa, E, F, KK, MCB. A solution of iodine monochloride in acetic acid (Wijs' solution) is used to determine iodine values of fats and oils.

The reagent is prepared by passing chlorine into 254 g. of iodine with shaking until the gain in weight is 71 g.[1, 2] The reagent is either used directly[1] or distilled; the yield of material boiling at 97–105° is 87%. It is used for iodination of *p*-nitroaniline, anthranilic acid, and salicylic acid in acetic acid solution to the following products in the yields indicated:

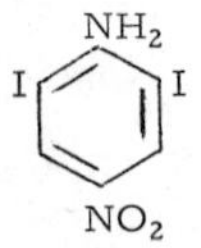

56-64%[1]

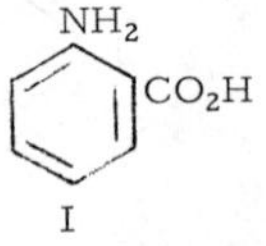

67-74%[3]

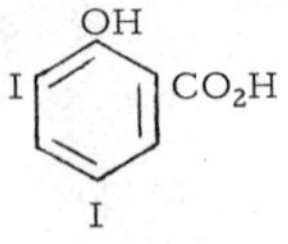

91-92%[2]

Phenolic acids of types (1) and (2) yield triiodo derivatives on reaction with iodine monochloride in acetic acid solution.[4]

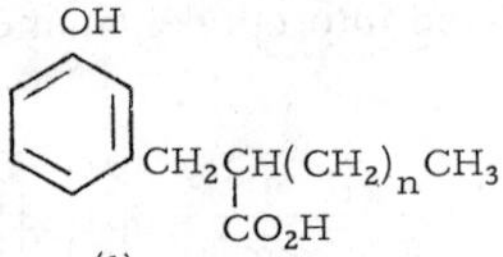

(1)

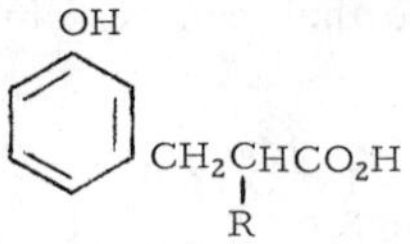

(2)

[1]R. B. Sandin, W. V. Drake, and F. Leger, *Org. Syn.*, *Coll. Vol.*, **2**, 196 (1943)
[2]G. H. Woollett and W. W. Johnson, *ibid.*, **2**, 343 (1943)
[3]V. H. Wallingford and P. A. Krueger, *ibid.*, **2**, 349 (1943)
[4]D. Papa, H. F. Ginsberg, I. Lederman, and V. DeCamp, *Am. Soc.*, **75**, 1107 (1953)

Iodine–Morpholine complex (1, 2). Mol. wt. 213.03. The formation of a crystalline, orange 1:1 complex of iodine and morpholine is reported in a patent.[1] Southwick and Kirchner[2] who regard the substance as the charge-transfer complex (2) found that, with excess morpholine, the complex is an effective reagent for the iodination

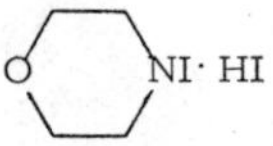

(1)

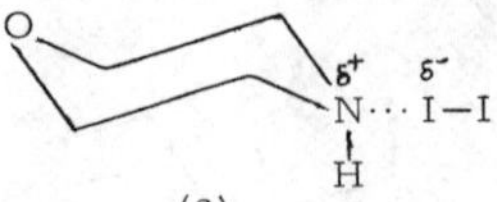

(2)

of terminal acetylenes. Thus it was used successfully for the iodination of phenylacetylene and 1-ethynylcyclohexanol.

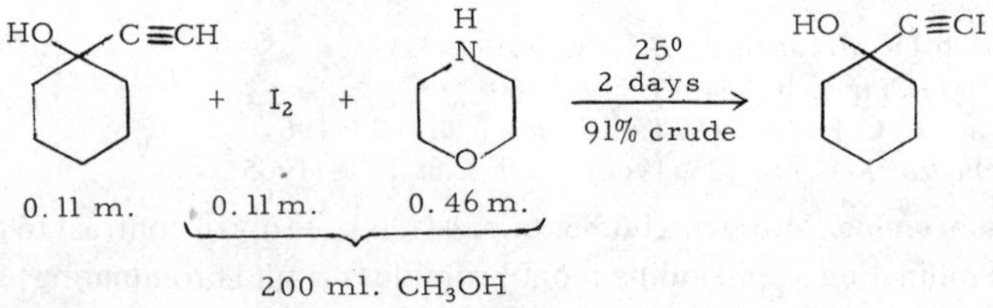

Iodination of the benzofurane (1) was largely unsuccessful except by use of the iodine–morpholine reagent.[3]

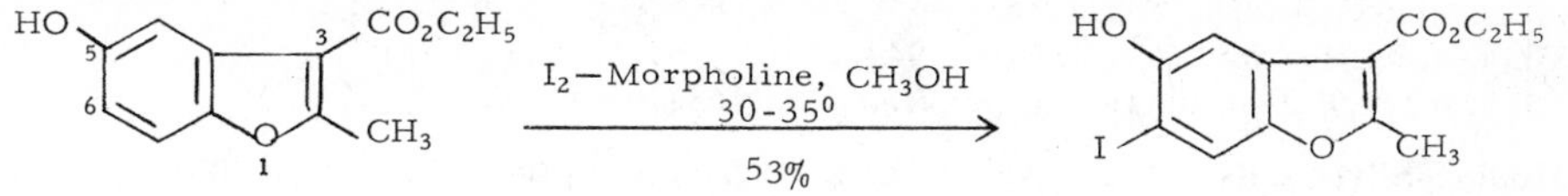

(1) (2)

Chabrier *et al.*[4] found that phenols and arylamines can be iodinated in good yield by reaction of the substrate, iodine, and morpholine in the molecular ratio 1:1:3 in an anhydrous solvent (ethanol, ether, benzene). For example, phenol afforded 2,4,6-triiodophenol in 90% yield.

[1]R. V. Rice and G. D. Beal, U.S. patent 2,290,710 [*C. A.*, **37**, 502 (1943)]
[2]P. L. Southwick and J. R. Kirchner, *J. Org.*, **27**, 3305 (1962)
[3]C. A. Giza and R. L. Hinman, *J. Org.*, **29**, 1453 (1964)
[4]P. Chabrier, J. Seyden-Penne, and A.-M. Fouace, *Compt. rend.*, **245**, 174 (1957)

Iodine pentafluoride, IF_5. Mol. wt. 221.91, m.p. −8°, b.p. 97° dec., sp. gr. 3.29. Supplier: Matheson Co.

A primary amine having an α-methylene group is oxidized by the reagent in methylene chloride in part to the nitrile and in part to the aldehyde, both in very low yield.[1] The reagent oxidizes hydrazobenzene to azobenzene and effects Beckmann rearrangement of oximes. The most significant use is for oxidation of *t*-butylamine

$$\underset{0.040\text{ m.}}{C_6H_5CH_2NH_2} + \underset{0.043\text{ m.}}{IF_5} \xrightarrow{\text{Stir at }15^0\text{ 2 hrs.; ice}} \underset{20\%}{C_6H_5CN} + \underset{9\%}{C_6H_5CHO}$$

to azoisobutane, for the yield is fair and the product less easily accessible by other methods.

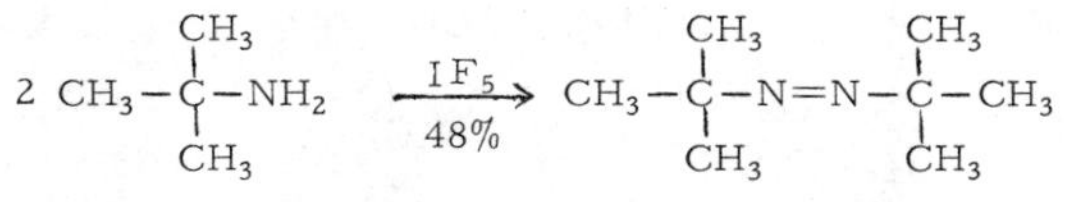

Alkyl and aryl isothiocyanates are converted by IF_5 in pyridine into thiobis-N-(trifluoromethyl)amines.[2]

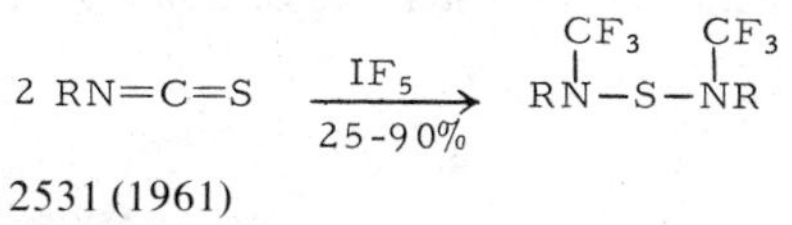

[1]T. E. Stevens, *J. Org.*, **26**, 2531 (1961)
[2]T. E. Stevens, *J. Org.*, **26**, 3451 (1961)

Iodine–Pyridine. Methyl ketones of the types $ArCOCH_3$ and $R_2CHCOCH_3$ react with iodine in pyridine solution to give the pyridinium iodides in yields sometimes

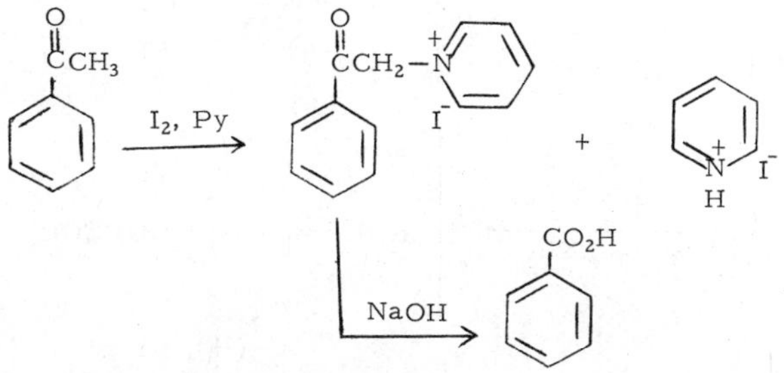

as high as 95%.[1,2] The salts are cleaved by alkali to the corresponding acids. 2-Methylchromone, a vinylogous methyl ketone, is similarly substituted.[3]

[1]F. Kröhnke, *Ber.*, **66**, 1386 (1933)
[2]L. C. King, *Am. Soc.*, **66**, 894, 1612 (1944)
[3]J. Schmutz, R. Hirt, and H. Lauener, *Helv.*, **35**, 1168 (1952)

Iodine–Silver salts (CF_3CO_2Ag or AgF). Haszeldine[1] prepared silver trifluoroacetate by mixing aqueous solutions of sodium trifluoroacetate and silver nitrate, filtering from a little silver chloride, and extracting continuously with ether. This silver salt combines with iodine in nitrobenzene to give a solution of a complex of which the active species is the hypoiodite CF_3COOI.[2] The complex iodinates benzene and halobenzenes in yields of 60–80%. However, 2-naphthol afforded the 1-iodo compound in only 20% yield and 1-naphthol gave a intractable product.[3] The complex oxidizes primary alcohols to aldehydes in variable yield (25–80%) and oxidizes ketones to α-diketones in low yield (33-50%).[3] The combination I_2 + AgF iodinates simple aromatic compounds in low yield (10–50%).[3]

[1]R. N. Haszeldine, *J. Chem. Soc.*, 584 (1951)
[2]R. N. Haszeldine and A. G. Sharpe, *J. Chem. Soc.*, 993 (1952)
[3]E. D. Bergmann and I. Shahak, *J. Chem. Soc.*, 1418 (1959)

Iodoacetamide, ICH_2CONH_2. Mol. wt. 184.97, m.p. 95°. Suppliers: A, B.

Witkop[1] achieved selective cleavage of methionyl peptides by a method based on the observation that sulfonium salts of methionine (9) decompose with elimination of the sulfur function and formation of homoserine lactone (10). Highest yields are obtained with use of iodoacetamide as the alkylating agent.

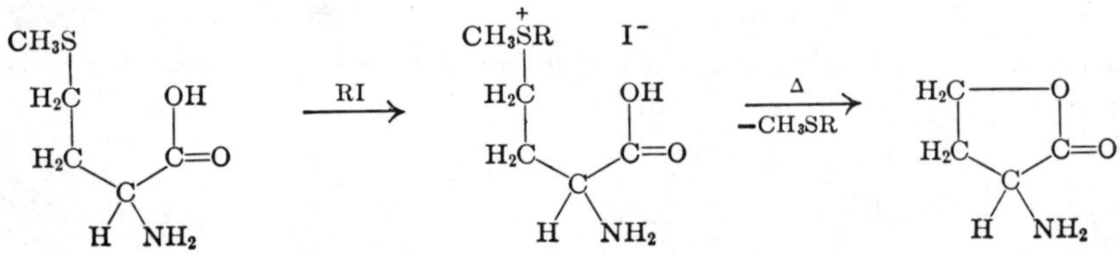

CH3S
H2C
OH
H2C
C=O
C
H
NH2
RI
+
CH3SR
I⁻
Δ
−CH3SR
H2C
O

(8) (9) (10)

Application of the reaction to the cleavage of a methionine-containing peptide is shown in the formulation.

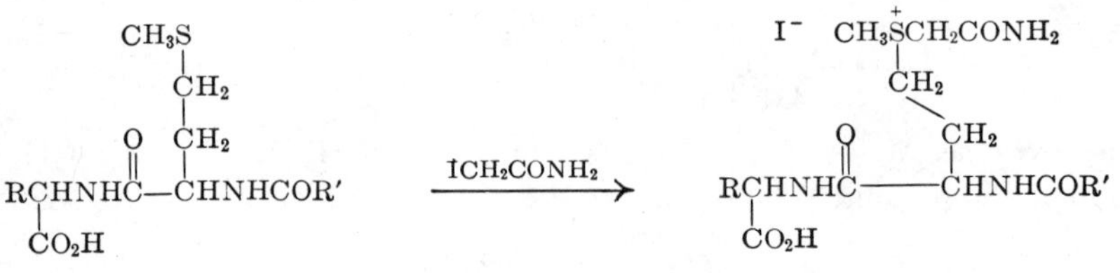

CH3S
CH2
O
RCHNHC—CHNHCOR'
CO2H
ICH2CONH2
I⁻ CH3SCH2CONH2
RCHNHC——CHNHCOR'

(11) (12)

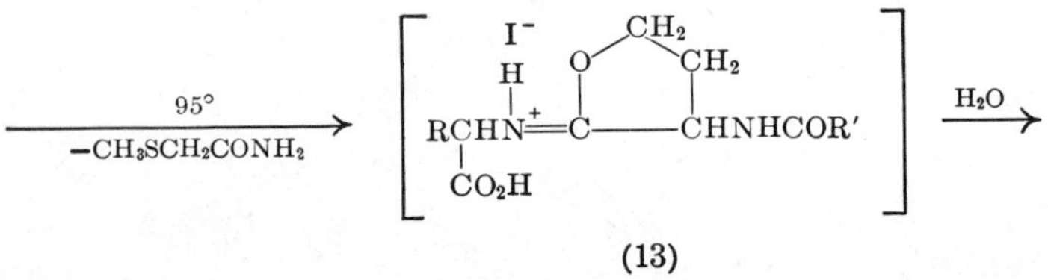

95°
−CH3SCH2CONH2
I⁻
H
O
CH2
RCHN=C——CHNHCOR'
CO2H
H2O

(13)

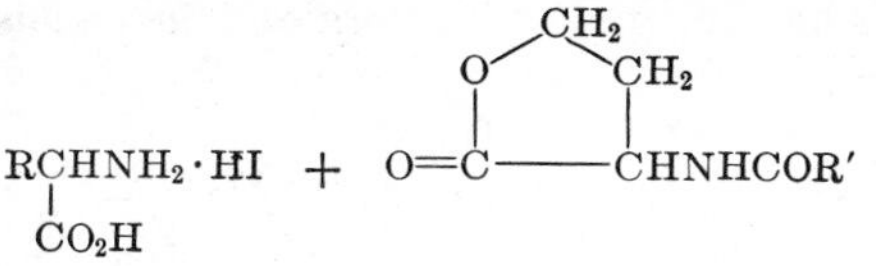

(14) (15)

[1]W. B. Lawson, E. Gross, C. M. Foltz, and B. Witkop, *Am. Soc.*, **84**, 1715 (1962)

Iodobenzene, C_6H_5I. Mol. wt. 204.01, m.p. −29°, b.p. 188.5° (b.p. 77–78°/30 mm.), sp. gr. 1.824. Suppliers: A, B, E, F, MCB. Reaction of benzene with iodine and nitric acid affords in 86–87% yield iodobenzene containing traces of nitro compounds.[1] A purer product is obtained by the Sandmeyer reaction in 74–76% yield.[2] In a more recent procedure[3] a mixture of 0.079 mole of iodine, 10 g. of concd. sulfuric acid, and

$$2\ C_6H_6 + I_2 + CH_3CO_3H \xrightarrow[67-77\%]{70^0} 2\ C_6H_5I + CH_3CO_2H + H_2O$$

100 g. of benzene is stirred at 70° and treated in the course of 1.5 hrs. with 122 ml. of a 1.5*M* solution of peracetic acid prepared by stirring a mixture of 40 g. of 30% hydrogen peroxide and 1 g. of concd. sulfuric acid in a bath at 30° and dropping in 180 g. of acetic anhydride in the course of 3.5 hrs. (the solution is stable for at least a week in a cool place in the dark). When the addition is complete and the brown solution has become almost colorless, the mixture is diluted with water, and the organic layer is separated, suitably washed, dried, and distilled.

Triphenylamine can be prepared by reaction of diphenylamine with iodobenzene in refluxing nitrobenzene in the presence of 1 equivalent of potassium carbonate and a trace of copper powder.[4] Unchanged diphenylamine is precipitated as the hydro-

$$2\ (C_6H_5)_2NH + C_6H_5I + K_2CO_3 \xrightarrow[82-85\%]{C_6H_5NO_2\,(Cu)} 2\ (C_6H_5)_3N + 2\ KI + CO_2$$

chloride by passing hydrogen chloride gas into a benzene solution of the crude product.

[1]F. B. Dains and R. Q. Brewster, *Org. Syn., Coll. Vol.*, **1**, 323 (1941)
[2]H. J. Lucas and E. R. Kennedy, *ibid.*, **2**, 351 (1943)
[3]Y. Ogata and K. Nakajima, *Tetrahedron*, **20**, 43 (1964); procedure submitted to *Org. Syn.*
[4]F. D. Hager, *Org. Syn., Coll. Vol.*, **1**, 544 (1941)

Iodobenzene dichloride (Phenyliodo dichloride), $C_6H_5I^+Cl(Cl^-)$. Mol. wt. 274.92, m.p. 115–120°, dec., yellow, decomposes on storage. Supplier: KK.

The reagent is prepared by passing dry chlorine into a solution of iodobenzene in chloroform cooled in a salt-ice bath.[1] The product separates as yellow crystals (87–94%).

Uses (*see also* Iodosobenzene; Iodoxybenzene). An interesting use is for the conversion of cholesterol into the 5α,6α-dichloride.[2, 3] Molecular chlorine, reacting

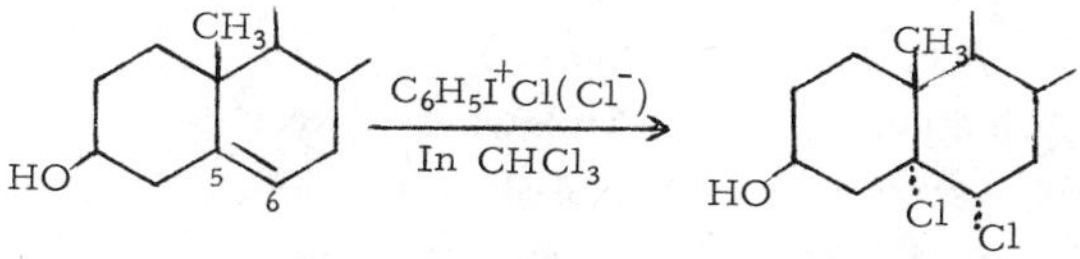
CH3
HO
5
6
C6H5I+Cl(Cl-)
In CHCl3
CH3
HO
Cl
Cl

by an ionic mechanism, gives the 5α,6β-dichloride, the product of *trans* addition. If water is present, iodobenzene dichloride reacts in part by *trans* addition, but in the complete absence of moisture the *cis*-dichloride is the exclusive product. A cyclic transition state is postulated.

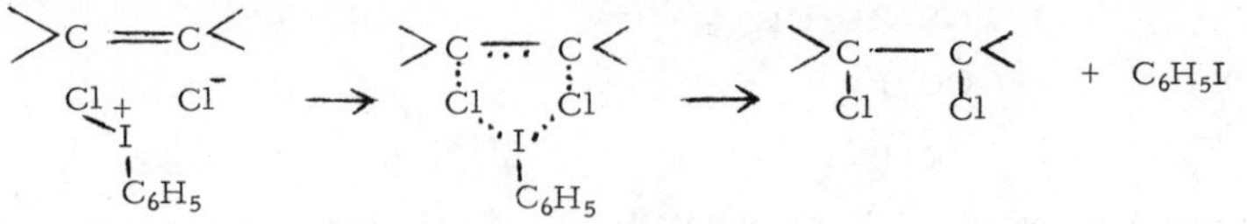

Under photocatalysis iodobenzene dichloride reacts with saturated hydrocarbons to form chlorinated hydrocarbons, iodobenzene, and hydrogen chloride.[4]

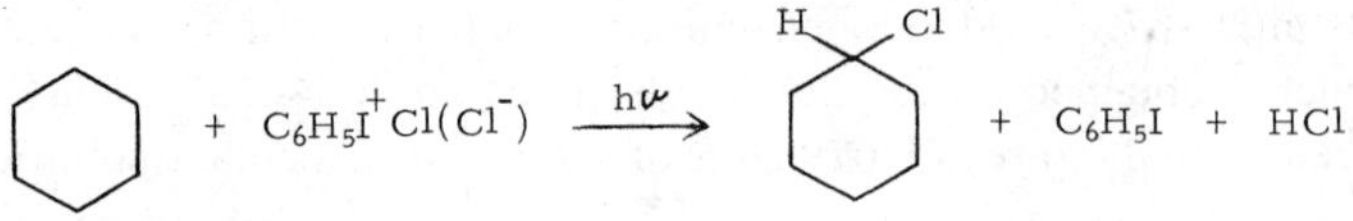

[1]C. Willgerodt, *J. prakt, Chem.* (2), **33**, 155 (1886); H. J. Lucas and E. R. Kennedy, *Org. Syn., Coll. Vol.*, **3**, 482 (1955)
[2]C. J. Berg and E. S. Wallis, *J. Biol. Chem.*, **162**, 683 (1946)
[3]D. H. R. Barton and E. Miller, *Am. Soc.*, **72**, 370 (1950)
[4]D. F. Banks, E. S. Huyser, and J. Kleinberg, *J. Org.*, **29**, 3692 (1964)

***o*-Iodobenzoic acid**, *o*-$IC_6H_4CO_2H$. Mol. wt. 216.02, m.p. 162°. Suppliers: A, B, E, F, MCB. The reagent is obtained in "almost quantitative" yield by diazotization of anthranilic acid in dilute sulfuric acid and addition of a solution of potassium iodide in dilute sulfuric acid.[1] It can be crystallized from hot water.

In a modern procedure[2] anthranilic acid is diazotized in dilute sulfuric acid and the filtered solution is added to potassium iodide, also in the dilute acid. The crude brown acid (92%) is converted into the ethyl ester, b.p. 150–151°/13 mm. (71%), which on hydrolysis with ethanolic potassium hydroxide gives the pure acid, m.p. 163°, in almost theoretical yield.

For uses of the reagent *see* Diphenyliodonium-2-carboxylate; *o*-Iodosobenzoic acid.

[1]W. Wachter, *Ber.*, **26**, 1744 (1893)
[2]G. P. Baker, F. G. Mann, N. Sheppard, and A. J. Tetlow, *J. Chem. Soc.*, 3721 (1965)

5-Iodo-2,4-dinitrophenylhydrazine (2). Mol. wt. 328.47, m.p. 248°.

Preparation.[1]

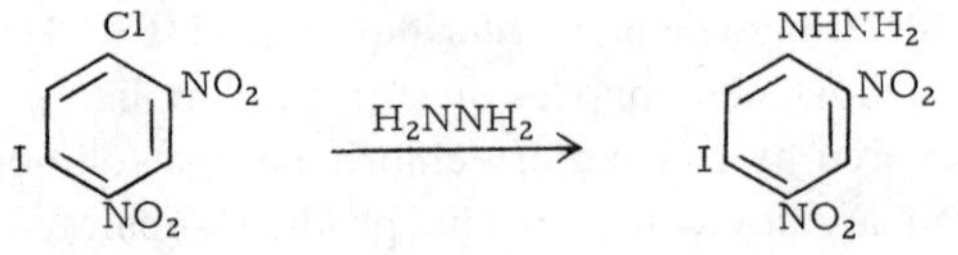

(1) (2)

Use.[1] The reagent reacts with ketosteroids to form crystalline, high-melting derivatives.

[1]P. Karlson and H. Hoffmeister, *Ann.*, **662**, 1 (1963)

Iodomethylmercuric iodide, ICH_2HgI. Mol. wt. 468.46.

This mercurial can be prepared in satisfactory yield by the reaction of excess methylene iodide with finely divided mercury under ultraviolet irradiation.[1,2] It is a methylene-transfer reagent like the Simmons-Smith reagent. Thus when the reagent was refluxed with diphenylmercury and excess cyclohexene in benzene for 8 days, norcarane and phenylmercuric iodide were obtained in the yields indicated.[2] Despite the long reaction time, this method may be preferable in large-

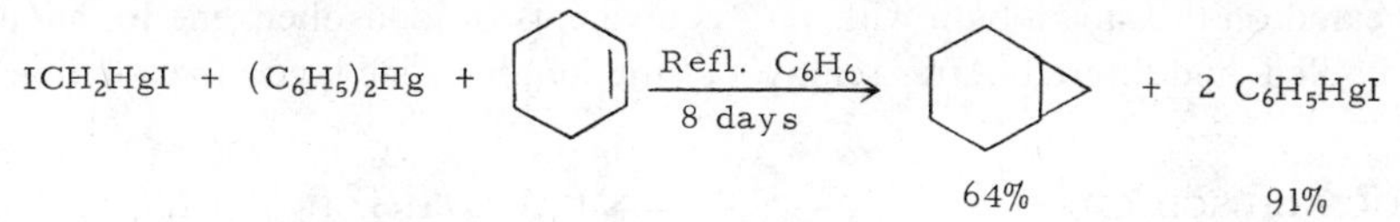

scale runs to procedures involving diazomethane. It may be preferable to the Simmons-Smith method for compounds sensitive to zinc iodide, a Lewis acid.

[1] H. E. Simmons and R. D. Smith, *Am. Soc.*, **81**, 4256 (1959)
[2] D. Seyferth and M. A. Eisert, *Am. Soc.*, **86**, 121 (1964)

Iodosobenzene. $C_6H_5\overset{+}{I}-\overset{-}{O}$. Mol. wt. 220.01, dec. (*explodes*) about 210°, yellow. Supplier: Aldrich.

Preparation. The reagent is prepared by stirring iodosobenzene diacetate with 3 *N* sodium hydroxide, collecting and drying the solid, and macerating it with chloroform to remove a little iodobenzene.[1] This procedure is preferable to an older one[2] involving alkaline hydrolysis of iodosobenzene dichloride because iodoso-

$$C_6H_5I^+-OCOCH_3(CH_3CO_2^-) \xrightarrow[85-93\%]{NaOH} C_6H_5I^+-O^-$$

benzene diacetate is more stable and more easily prepared than the dichloride and the overall yield is greater (75% versus 54%). Crude, wet iodosobenzene is satisfactory for use in some reactions and can be analyzed as follows.[2] A sample is added to a mixture of dilute sulfuric acid, potassium iodide, and chloroform; the mixture is shaken for about 15 min. and then titrated with 0.1 *N* sodium thiosulfate. Iodoxybenzene can be analyzed by the same method; the two reagents can be distinguished from the fact that iodosobenzene reduces iodide ion in a saturated solution of sodium borate whereas iodoxybenzene does not.

Diaryliodonium salts. Iodosoarenes are useful for the preparation of diaryliodonium salts. One method involves Friedel-Crafts type condensation with an aromatic compound in the presence of acids.[3] For example, a cold solution of 5 g.

$$C_6H_5\overset{+}{I}-\overset{-}{O} + C_6H_6 \xrightarrow[74\%]{H_2SO_4;\ NaBr} C_6H_5-\overset{+}{I}-C_6H_5\ (Br^-)$$

of iodosobenzene, 12.5 ml. of benzene, 65 ml. of acetic acid, and 12.5 ml. of acetic anhydride is stirred during dropwise addition of 5 ml. of concd. sulfuric acid. After 24 hrs. water is added and the solution is extracted twice with ether, clarified with decolorizing carbon, and filtered. On addition of sodium bromide, diphenyliodonium bromide precipitates.

Diphenyliodonium iodide can be prepared by stirring a mixture of 0.1 mole each of iodosobenzene and iodoxybenzene with 1 *N* sodium hydroxide for 24 hrs.[4] The

$$C_6H_5I^+-O^- + C_6H_5\overset{\overset{O}{\|}}{I^+}-O^- \xrightarrow{OH^-} (C_6H_5)_2I^+(IO_3^-) \xrightarrow[70-72\%]{KI} (C_6H_5)_2I^+(I^-)$$

resulting brown slurry is stirred with water and the supernatant solution of diphenyliodonium iodate is decanted through a filter and treated with potassium iodide to precipitate the iodonium iodide.

Oxidation. Iodosobenzene has been suggested as a reagent for the oxidation of sulfides to sulfones.[5] Thus a solution of 12 g. of the thiodiglycol (1) in 50 ml. of water was heated on the steam bath with 1.15 equivalents of iodosobenzene for about 15 min., cooled, and filtered. After removal of the iodobenzene layer, the solution was

$$\underset{(1)}{HOCH_2CH_2SCH_2CH_2OH} + C_6H_5\overset{+}{I}-O^- \xrightarrow[81\%]{} \underset{(2)}{HOCH_2CH_2\overset{O^-}{\overset{|}{S^+}}CH_2CH_2OH} + C_6H_5I$$

evaporated to dryness and the 2,2′-dihydroxydiethyl sulfoxide (2) was purified by crystallization.

[1]H. Saltzman and J. G. Sharefkin, *Org. Syn.*, **43**, 60 (1963)
[2]H. J. Lucas, E. R. Kennedy, and M. W. Formo, *Org. Syn., Coll. Vol.*, **3**, 483 (1955)
[3]F. M. Beringer *et al.*, *Am. Soc.*, **75**, 2705 (1953); **81**, 342 (1959)
[4]H. J. Lucas and E. R. Kennedy, *Org. Syn., Coll. Vol.*, **3**, 355 (1955)
[5]A. H. Ford-Moore, *J. Chem. Soc.*, 2126 (1949)

Iodosobenzene diacetate (Phenyliodoso diacetate), $C_6H_5\overset{+}{I}OCOCH_3(CH_3CO_2^-)$. Mol. wt. 322.10, m.p. 158°.

Preparation. The reagent is prepared by stirring 0.1 mole of iodobenzene in a bath at 30° during dropwise addition of 0.24 mole (0.48 equiv.) of commercial 40% peracetic acid.[1] The product crystallizes and is collected after cooling and washed with water. The purity is determined by treatment of a sample with dilute

$$C_6H_5I + CH_3CO_3H + CH_3CO_2H \xrightarrow[97\text{-}98\%]{30^0} C_6H_5\overset{+}{I}OCOCH_3(CH_3CO_2^-) + H_2O$$

sulfuric acid, potassium iodide, and chloroform and titration with sodium thiosulfate.[2] Iodosobenzene diacetate is the preferred precursor for the preparation of iodosobenzene, but its preparation from iodosobenzene and acetic acid[3] demonstrates the reverse reaction.

Comparison with lead tetraacetate. Iodosobenzene diacetate cleaves *vic*-glycols in acetic acid at 50–80° but the rate constant is about one hundredth that of lead tetraacetate.[4,5] Further analogies to lead tetraacetate are that the reagent is capable of converting olefins into glycol diacetates[6] and that on decomposition in refluxing acetic acid containing 2,4,6-trinitrotoluene this substrate is methylated to trinitro-*m*-xylene to the extent of about 20%.[3]

Oxidation. The reagent oxidizes primary aromatic amines in benzene solution at room temperature to azo compounds (1–2 days), but yields are variable; toluidines, 42, 56, and 6%; *p*-nitroaniline (53%); α-naphthylamine, 3%.[6] Oxidations in acetic

$$2\ C_6H_5NH_2 \xrightarrow[95\%]{C_6H_5\overset{+}{I}OCOCH_3(CH_3COO^-)} C_6H_5N{=}NC_6H_5$$

acid solutions have also been studied.[7] Some N-arylacetamides undergo ring acetylation.[8] In benzene solution *o*-nitroaniline is oxidized to benzofurazane oxide in high yield.[6]

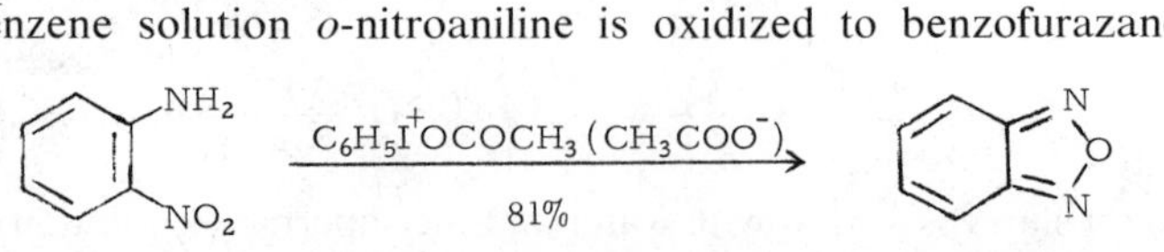

Szmant[9] found that the reagent oxidizes the diamino sulfone (1) to the cyclic azo compound (2). A solution of the reactants is stored at room temperature for 3 days.

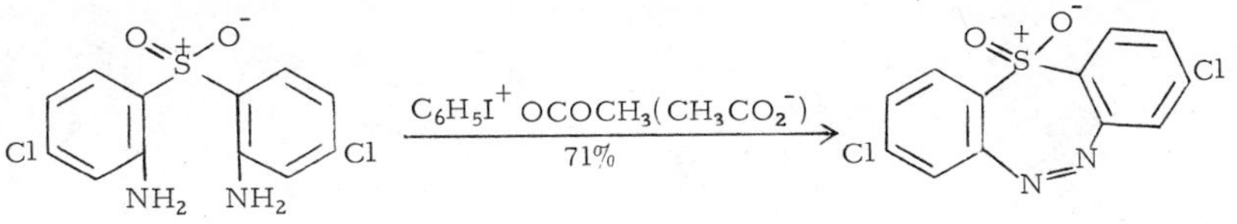
$C_6H_5\overset{+}{I}OCOCH_3(CH_3CO_2^-)$
71%
Cl
Cl
NH2
NH2
Cl
Cl
N=N

(1) (2)

The halogen-free sulfone corresponding to (1) afforded the parent cycloazo compound in yield of 95%.[10]

In combination with a catalytic amount of osmium tetroxide in *t*-butanol-pyridine, the reagent effects oxidative hydroxylation of double bonds.[11]

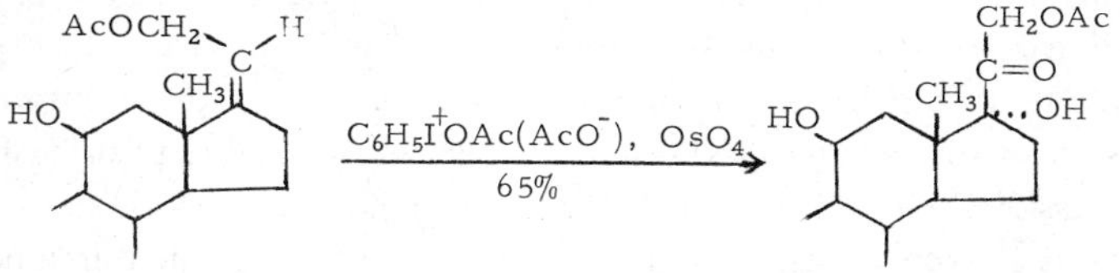
AcOCH2
H
CH3
HO
$C_6H_5\overset{+}{I}OAc(AcO^-)$, OsO_4
65%
CH2OAc
C=O
CH3
OH
HO

Iodosobenzene diacetate has been employed as a coupling agent in the preparation of diaryliodonium salts:[12]

$$Ar\overset{+}{I}OCOCH_3(CH_3CO_2^-) + Ar'H + H_2SO_4 \longrightarrow Ar\overset{+}{I}Ar'(HOSO_3^-) + 2\ CH_3CO_2H$$

[1]J. G. Sharefkin and H. Saltzman, *Org. Syn.*, **43**, 62 (1963)
[2]H. J. Lucas, E. R. Kennedy, and M. W. Formo, *Org. Syn.*, *Coll. Vol.*, **3**, 483 (1955)
[3]R. B. Sandin and W. B. McCormack, *Am. Soc.*, **67**, 2051 (1945)
[4]R. Criegee and H. Beucker, *Ann.*, **541**, 218 (1939)
[5]S. J. Angyal and R. J. Young, *Am. Soc.*, **81**, 5251 (1959)
[6]K. H. Pausacker, *J. Chem. Soc.*, 1989 (1953)
[7]G. B. Barlin, K. H. Pausacker, and N. V. Riggs, *J. Chem. Soc.*, 3122 (1954)
[8]G. B. Barlin and N. V. Riggs, *J. Chem. Soc.*, 3125 (1954)
[9]H. H. Szmant and R. Infante, *J. Org.*, **26**, 4173 (1961)
[10]H. H. Szmant and R. L. Lapinski, *Am. Soc.*, **78**, 458 (1956)
[11]J. A. Hogg *et al.*, *Am. Soc.*, **77**, 4436, 4438, 6401 (1955)
[12]F. M. Beringer *et al.*, *Am. Soc.*, **81**, 342 (1959)

***o*-Iodosobenzoic acid** (1⇌2). Mol. wt. 264.02, dec. above 200°. Supplier: Aldrich. In chemical behavior and infrared spectrum, the compound differs so markedly from the *m*- and *p*-isomers that it is regarded as existing largely as the cyclic tautomer (2).[1]

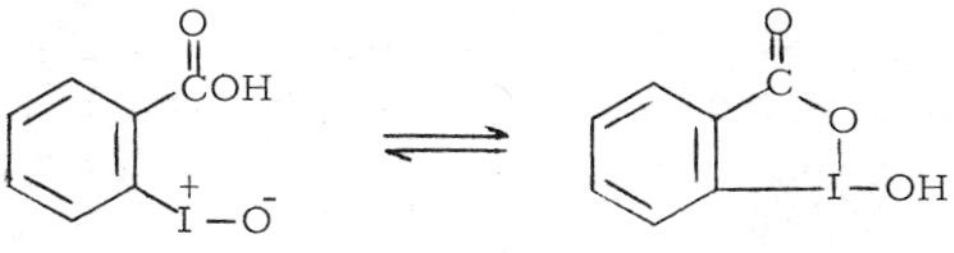
COH
I–O
C
O
I–OH

(1) (2)

Preparation by oxidation of *o*-iodobenzoic acid with fuming nitric acid at 50°.[2]

Determination of SH groups. An approximately 0.02 *N* solution is standardized by iodimetry and employed in weakly alkaline solution for the determination of cysteine, glutathione, and sulfhydryl groups of proteins.[3]

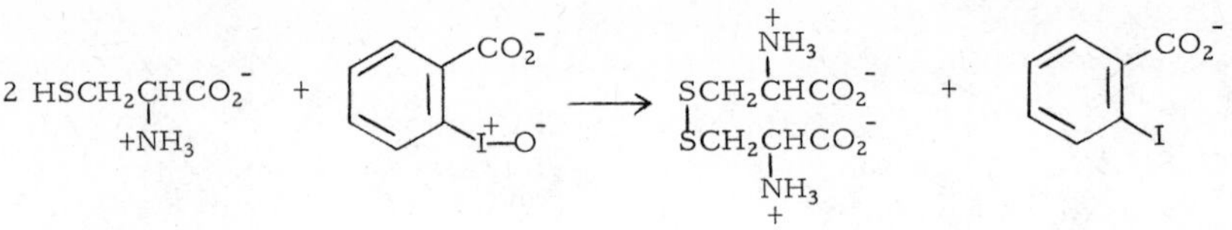

[1]G. P. Baker, F. G. Mann, N. Sheppard, and A. J. Tetlow, *J. Chem. Soc.*, 3721 (1965)
[2]V. Meyer and W. Wachter, *Ber.*, **25**, 2632 (1892); P. Askenasy and V. Meyer, *Ber.*, **26**, 1354 (1893)
[3]L. Hellerman, F. P. Chinard, and P. A. Ramsdell, *Am. Soc.*, **63**, 2551 (1941)

N-Iodosuccinimide $(CH_2CO)_2NI$. Mol. wt. 224.99, m.p. 201°. dec. Suppliers: Arapahoe, B, KK.

Preparation.[1, 2] N-Silver succinimide is prepared by adding freshly precipitated, moist silver oxide to a boiling solution of succinimide in water (with protection from light). The resulting suspension is filtered by suction and the filtrate let stand for separation of the silver salt. This is dried in air, ground, and dried in vacuum at 110°; yield 47%. The salt is added to a mixture of iodine and dioxane in a brown bottle which is shaken occasionally for one hour and then warmed in a bath at 50°. The silver iodide is removed by filtration and the filtrate is diluted with carbon tetrachloride and cooled for crystallization of N-iodosuccinimide. Material melting at 193–199° is obtained in yield of 81–85%.

Iodination of enol acetates. Djerassi and Lenk[1] found that treatment of heptane-2-one with either acetic anhydride or isopropenyl acetate affords the enol acetate (2) and that this reacts with N-iodosuccinimide (NIS) without solvent or in dioxane to give the pure α-iodoketone (3, 82%) and N-acetylsuccinimide (4, 90%). The formation of (4) and the fact that NIS does not share the radical reactivity of NBS suggest that the reaction follows an ionic course, probably through the intermediate

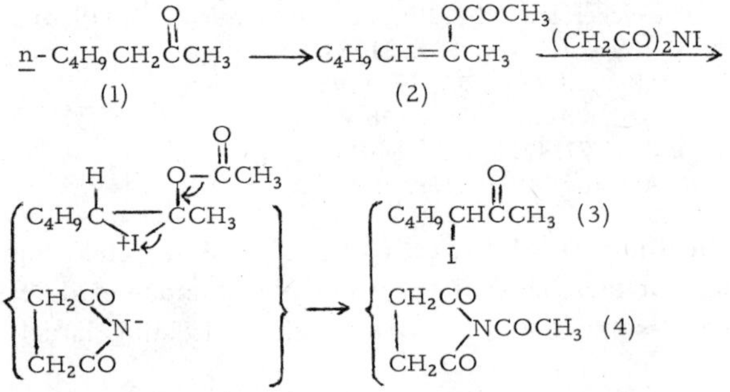

iodonium complex. In the steroid series NIS reacts selectively with an enol acetate group without touching a 5,6-double bond. Thus the enol acetate (6), readily available from pregnenolone and isopropenyl acetate, reacts with NIS in dioxane at 85°

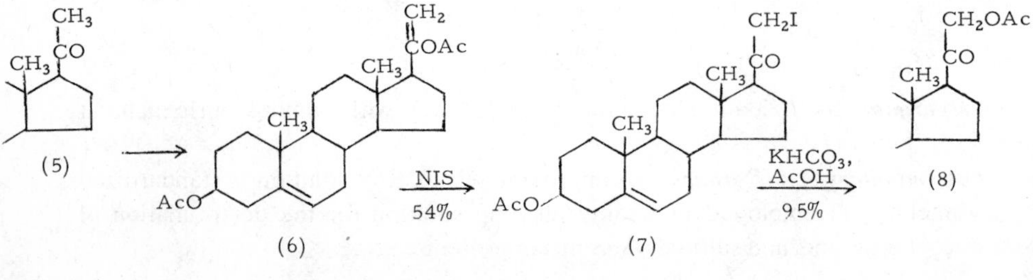

to give the 21-iodo-20-one (7), convertible into the 3,21-diacetate (8). Enol acetates of Δ^4-3-ketosteroids are more reactive, and the reaction with NIS proceeds at room temperature to produce the corresponding 6-iodo-Δ^4-3-ketones.[3]

[1]C. Djerassi and C. T. Lenk, *Am. Soc.*, **75**, 3494 (1953)
[2]W. R. Benson, E. T. McBee, and L. Rand, *Org. Syn.*, **42**, 73 (1962)
[3]C. Djerassi, J. Grossman, and G. H. Thomas, *Am. Soc.*, **77**, 3826 (1955)

Iodoxybenzene, $C_6H_5\overset{O^-}{\overset{|}{I^+}}{=}O$. Mol. wt. 236.02, m.p. 230° (*explodes*), yellow. Solubility in 1 l. H_2O: 2.8 g. at 12°, 12 g. at 100°. Supplier: Aldrich.

Preparation. One standardized procedure[1] involves disproportionation of iodosobenzene. A thin paste of 110 g. of iodosobenzene in water is steam distilled in a 5-1. flask until removal of iodobenzene is complete; the recovery of pure halide is about 90%. The hot mixture is cooled and the white solid is collected, dried, and ground in a mortar. The purity as determined by iodimetric titration is 99%.

$$2\ C_6H_5\overset{+}{I}{-}\overset{-}{O} \xrightarrow[92\text{-}95\%]{\text{Steam distil.}} C_6H_5\overset{O^-}{\overset{|}{I^+}}{=}O + C_6H_5I$$

A second method is[2] by hypochlorite oxidation of freshly prepared, pulverized iodobenzene dichloride. A mixture of 0.4 mole of dichloride and 1 mole of commercial sodium hypochlorite solution (1.15 l. of Chlorox) and 2 ml. of acetic acid is

$$C_6H_5\overset{+}{I}Cl(Cl^-) + NaOCl + H_2O \xrightarrow[87\text{-}92\%]{} C_6H_5\overset{O^-}{\overset{|}{I^+}}{=}O + NaCl + 2\ HCl$$

stirred vigorously with a Hershberg stirrer in a bath at 65–75° until the mixture becomes frothy and the color of the solid changes from yellow to white. The workup is as before except that the dried product is washed with chloroform to remove iodobenzene.

A third method[3] has the advantage over the two cited in that it is a simple one-step process and that the yield is higher. Iodobenzene (0.1 mole) is stirred at 35° during addition of 0.5 mole of 40% peracetic acid (65 ml.) over a 30-min. period, during which time iodosobenzene diacetate separates. After addition of 80 ml. of

$$C_6H_5I \xrightarrow[72\text{-}80\%]{2\ CH_3CO_3H} C_6H_5\overset{O^-}{\overset{|}{I^+}}{=}O$$

water the temperature is slowly raised to 100° and kept at 100° for 45 min. for hydrolysis of the diacetate to iodosobenzene and further oxidation to iodoxybenzene, which is collected after cooling, dried, and extracted with chloroform (purity 99.0–99.9%).

[1]H. J. Lucas and E. R. Kennedy, *Org. Syn., Coll. Vol.*, **3**, 485 (1955)
[2]M. W. Formo and J. R. Johnson, *ibid.*, **3**, 486 (1955)
[3]J. G. Sharefkin and H. Saltzman, *Org. Syn.*, **43**, 65 (1963)

Ion-exchange resins. Most of the commercially available ion-exchange resins have a polystyrene matrix cross-linked with 3–5% of divinylbenzene. Cation exchangers usually contain sulfonic acid groups introduced by sulfonation, whereas anion exchangers contain quaternary amino groups introduced by chloromethylation followed by amination. Exchangers of both types are supplied by the Rohm and

Haas Co. under the trade name Amberlite resins and by the Dow Chemical Co. as Dowex resins. The firms Fisher and Matheson, Coleman and Bell identify the resins offered as Rexyns and as Permutits, respectively.

Use of an exchange resin as catalyst for a reaction has the important advantage that the resins are insoluble in water and in organic solvents and can be removed by filtration without leaving any unwanted ion in the solution. A spent cation exchanger can be rejuvenated by washing with a mineral acid and a spent anion exchanger by washing with alkali.

In the following examples of uses of resins, a Rohm and Haas Amberlite Ion-exchange Resin is designated by the abbreviation IR.

Acidic Resins

Hydrolysis. Examples are the hydrolysis of the enamine (1)[1] and the amide (2).[2] The same resin, IR-120 (H^+), is superior to sulfuric acid for effecting hydrolytic

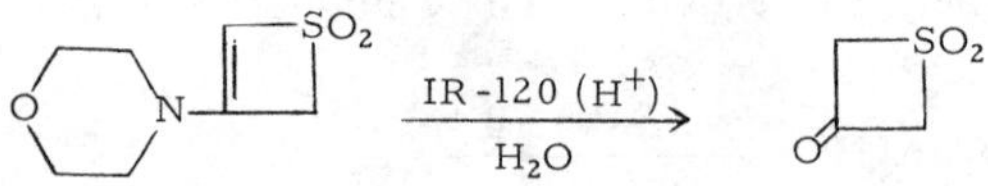

SO2
N
O
IR-120 (H+)
H2O
SO2

(1)[1]

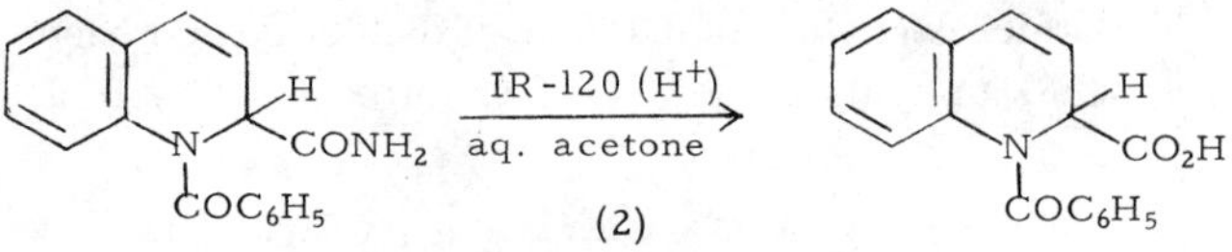

H
N
CONH2
COC6H5
IR-120 (H+)
aq. acetone
H
N
CO2H
COC6H5

(2)

decarboxylation of esters of acetoacetic acid.[3] Amberlite IR-100 (H^+) is superior to hydrochloric acid for hydrolysis of ordinary esters, and the advantage increases with increasing molecular weight of the ester.[4] The same resin hydrolyzes N-(2-O-methyl-D-glucosyl)-piperidine (3) in 80% yield, whereas with 1 *N* sulfuric acid (90 min., steam bath) the yield was only 61%.[5]

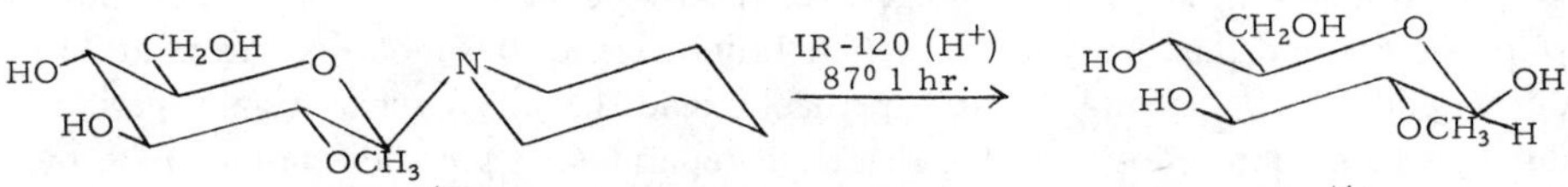

HO
CH2OH
O
HO
OCH3
N
IR-120 (H+)
87° 1 hr.
HO
CH2OH
O
HO
OH
OCH3
H

(3) (4)

Acetonide formation. An acidic resin was used to advantage for the conversion of orotidine methyl ester (1) into the acetonide (2).[6] A small amount of 2,2-dimethoxy-propane was added as water scavenger to a suspension of the ester and the resin in acetone and the mixture was shaken at room temperature for 3 hrs.

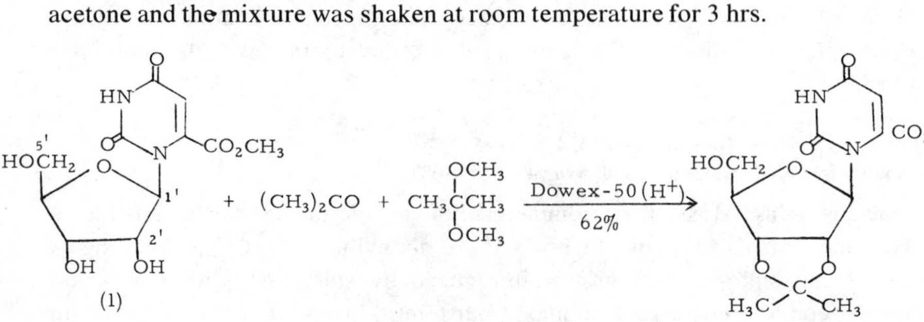

O
HN
O
N
CO2CH3
5'
HOCH2
O
1'
2'
OH
OH
+ (CH3)2CO + CH3CCH3
OCH3
OCH3
Dowex-50 (H+)
62%
HN
CO2CH3
HOCH2
H3C
C
CH3

(1) (2)

Methyl glycosides. The same resin, Dowex-50 (H^+) has been used for the preparation of methyl glycosides.[7]

Amide formation. An acidic resin proved to be an excellent catalyst for the dehydrative reaction of the amine (1) with the acid (2) to form the amide (3).[8] The

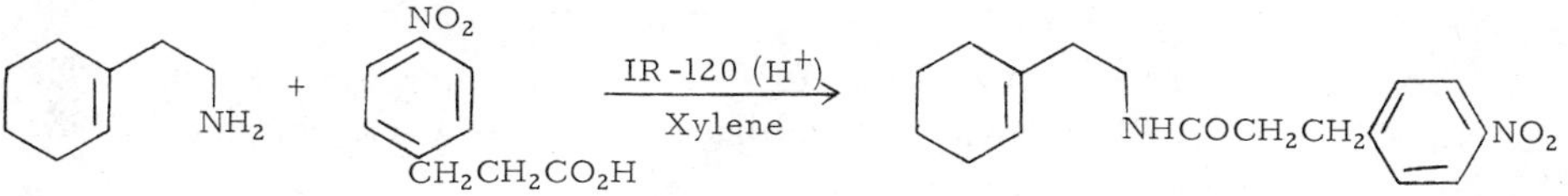

(1) (2) (3)

components were refluxed in xylene in the presence of an amount of resin representing 3% of the weight of (2) until 1 mole of water had been collected in a water separator. The yield was almost quantitative.

Esterification.[9] Methyl esters of amino acids are conveniently prepared by use of a strong cation-exchange resin as catalyst: IR-120 (H^+) or Zeo-Karb 222 (H^+). The resin is treated with 2–4 volumes of 2 *N* hydrochloric acid, washed free of acid, and dried. A mixture of the resin, amino acid, and methanol is then refluxed and stirred for 3 hrs. and the resin is removed by filtration.

Rearrangement. Newman[10] used Dowex-50 to rearrange ethynylcarbinols to α,β-unsaturated ketones. For example, a mixture of 1-ethynylcyclohexanol, acetic acid, water, and the resin is refluxed for 45 mins.

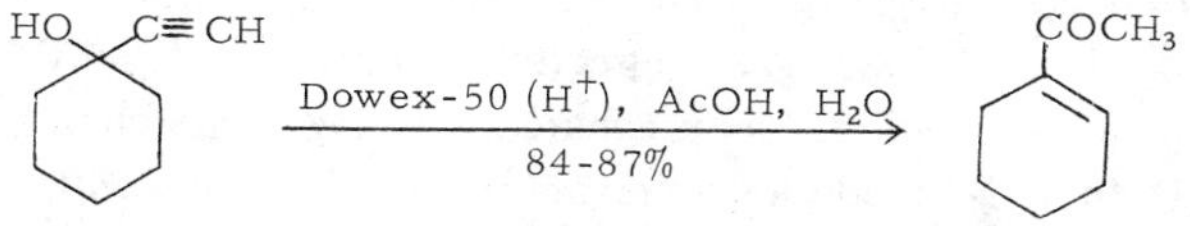

Ketone condensations. Lorette[11] used the same resin to effect self-condensation of ketones to α,β-unsaturated ketones. The reactions are very slow but good yields are obtained.

$$(CH_3)_2CO + CH_3COCH_3 \xrightarrow[87.4\%]{\text{Dowex-50 }(H^+)} (CH_3)_2C{=}CHCOCH_3$$

Phenol condensations. A cation-exchange resin is an excellent catalyst for the alkylation of phenol with isobutene.[12] A mixture of phenol and the resin was stirred at 80° and isobutene was passed in over a period of 3 hrs. The temperature was

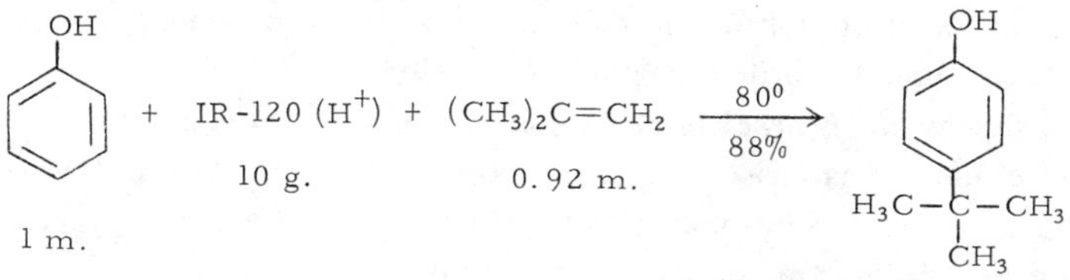

raised to 120° for 4 hrs. and the resin was filtered off. The yield of *p*-*t*-butylphenol was 88% (57.5% conversion). Benzene and xylene are not alkylated.

An acidic resin can be used in place of sulfuric acid for the von Pechmann[13] coumarin synthesis. For condensations such as that of resorcinol with ethyl acetoacetate, the two resins indicated appear to be equally satisfactory (Zeo-Karb is a British polystyrenesulfonic acid).[14] Use of a resin simplified purification of the product.

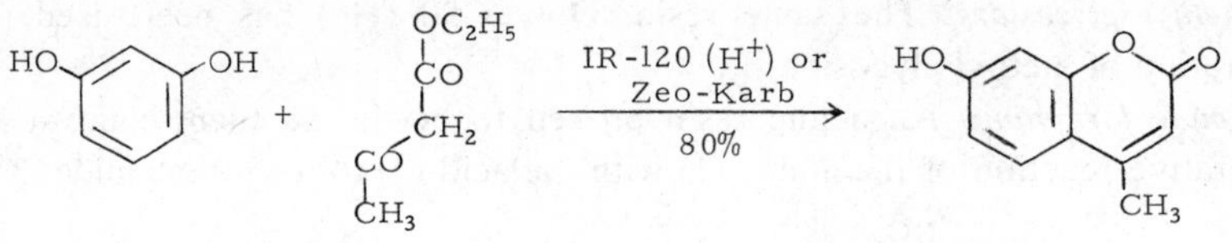

The same resins have been used for C-acylation of phenols with anhydrides, probably via O-acylation and Fries rearrangement.[15] However, yields are low even with 1,3-dihydroxybenzenes.

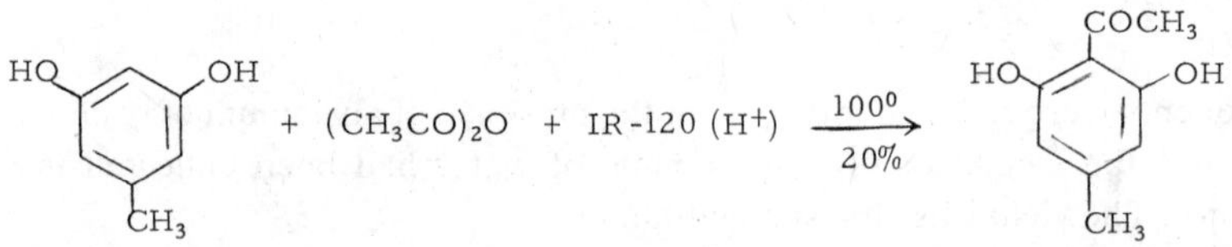

Hydration of acetylenes. Dowex-50 itself is not an efficient catalyst for hydration of an acetylene, but Newman[10] prepared an excellent catalyst for this reaction by impregnating the resin with 1% mercuric sulfate (by treatment with mercuric oxide

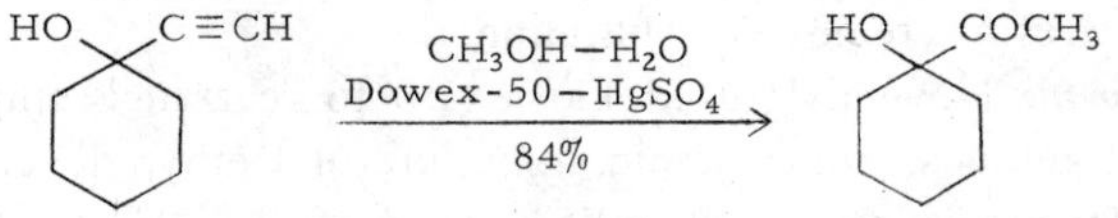

in dil. H_2SO_4). This method was used in one step of a synthesis of 19-norprogesterone.[16] Billimoria and Maclagan[17] effected hydration of 1-ethynylcyclohexanol satisfactorily with Zeo-Karb 225 impregnated with 1% of mercuric sulfate. In some cases (*e.g.*, 1-ethynyl-2-methylcyclohexanol) pretreatment with 20% mercuric sulfate was necessary.

Basic Resins

Hydrolysis. A solution of 5 g. of 1-methyl-3-carbomethoxypyridinium iodide in 50 ml. of water was passed through a 2 × 15-cm. column of a strongly basic resin in

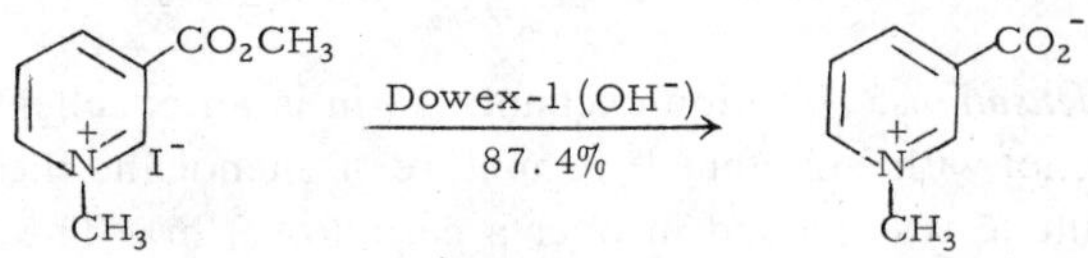

the hydroxide form; elution of the column with 50 ml. of water, evaporation, and treatment with ethanol afforded the pure betaine in high yield.[18]

Free base from the hydrochloride. A solution of an alkyl guanidine hydrochloride in absolute ethanol was passed through a column of basic resin previously washed with absolute ethanol. Elution with ethanol afforded the alkyl guanidine uncontaminated with inorganic material in quantitative yield.[19]

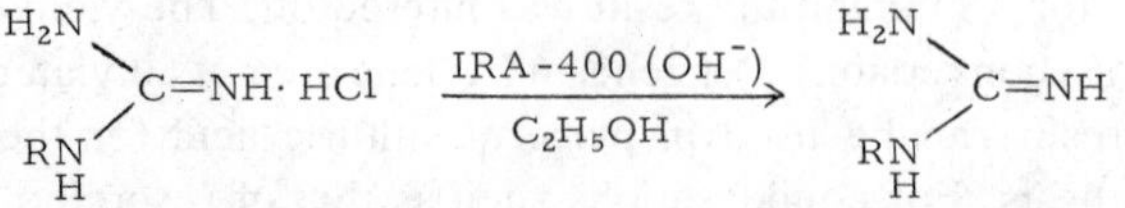

Dehydrohalogenation. Hinman and Lang[20] found usual reagents [NH_3, $(C_2H_5)_3N$] unsatisfactory for dehydrochlorination of 3-chloro-1,1-dimethylindolinium chloride (1) since 1-methylindole was formed in appreciable amounts by demethylation. They

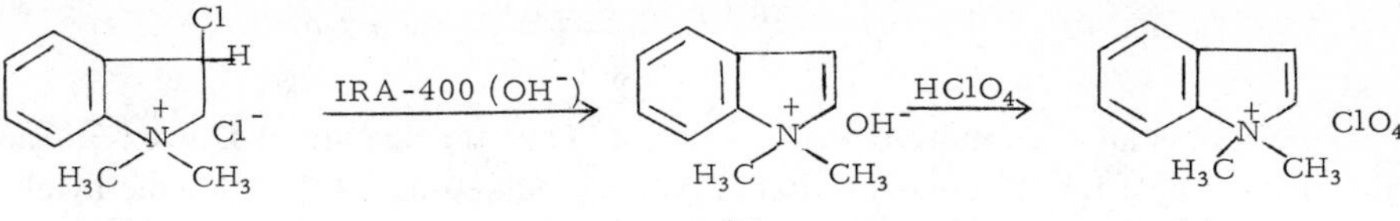

IRA-400 (OH⁻)
HClO₄

(1) (2) (3)

then proved that the desired reaction can be effected smoothly with a basic resin and isolated the product as the perchlorate (3) in 83% overall yield.

The same resin proved to be an excellent reagent for the cyclization of (3) and (5) to the lactams.[21] The precursor is dissolved in absolute ethanol, the resin is added, and the mixture is stirred at room temperature for one hour.

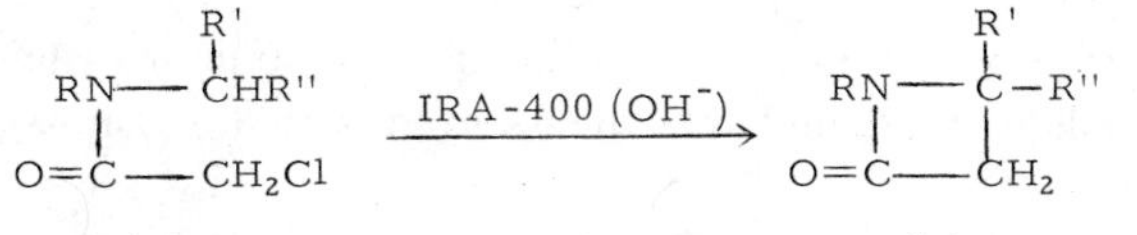

IRA-400 (OH⁻)

(3) (4)

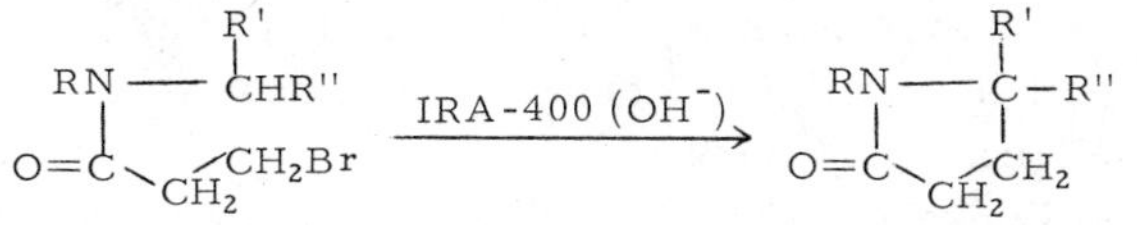

IRA-400 (OH⁻)

(5) (6)

Hydration. Galat[22] found a basic resin useful for the hydration of nicotinonitrile to nicotinamide. Use of alkali itself leads to formation of the acid as a by-product. In another case the yield of amide was 83%.[23]

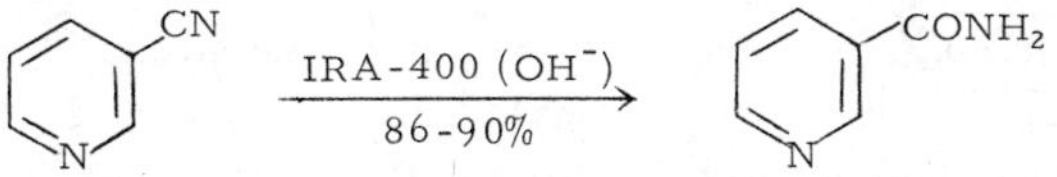

IRA-400 (OH⁻)
86-90%

Condensations. The aldol condensation of an aldehyde with a nitroalkane can be effected with the strongly basic IRA-400.[24]

$$CH_3CH_2CHO + CH_3NO_2 + \text{IRA-400 } (OH^-) \xrightarrow[68\%]{50 \text{ hrs. at } 25^0} CH_3CH_2CH(OH)CH_2NO_2$$

Bergmann[25] found resins IR-400 (OH^-) and IR-410 (OH^-) satisfactory catalysts for the Michael reaction.

$$(CH_3)_2CHNO_2 + CH_2{=}CHCN \xrightarrow[90\%]{\text{IR-400 } (OH^-)} (CH_3)_2C(NO_2)CH_2CH_2CN$$

The strongly basic Dowex 1-X10 (a quaternary ammonium hydroxide resin) has been used to condense carbonyl compounds with cyclopentadiene to produce fulvenes.[26] The system has the advantage that the reaction time can be controlled by regulating the time of contact of the catalyst with the reactants.

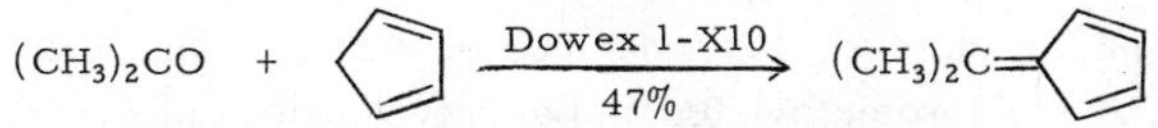

(CH₃)₂CO +
Dowex 1-X10
47%
(CH₃)₂C=

Astle and co-workers[27] have used the weakly basic IR-4B or Dowex-3 resin as catalysts for the Knoevenagel condensation of aldehydes with cyanoacetic acid. Actually the salt obtained by treatment of a basic resin with an excess of acetic or

$$\mathrm{RCHO} + \underset{\displaystyle |\atop \mathrm{CN}}{\mathrm{CH_2CO_2H}} \xrightarrow{\text{IR-4B or Dowex-3}} \mathrm{RCH{=}}\underset{\displaystyle |\atop \mathrm{CN}}{\mathrm{CCO_2H}}$$

benzoic acid is an even more effective catalyst. Thus the acetate of Dowex-3 is an effective catalyst for the condensation of unhindered ketones with active methylene compounds containing a cyano group. Dowex-3, a polystyrene-polyamine resin containing primary, secondary, and tertiary amino groups, affords an acetate which is as effective as piperidine acetate.

Displacements. For the displacement $ArCH_2X \rightarrow ArCH_2CN$, a strongly basic anion exchanger (Amberlite IRA-400 or Dowex 21K) was first converted into

$$\mathrm{C_6H_5CH_2Br} + \mathrm{IRA\text{-}400(CN^-)} \xrightarrow[69\%]{\mathrm{C_2H_5OH}\ 3\ \text{hrs.}\ (65^0)} \mathrm{C_6H_5CH_2CN}$$

the cyanide form by washing with 20% aqueous sodium cyanide.[28] A solution of the benzyl halide in ethanol was then stirred with 1.2 equivalents of the resin for 3 hrs.

A similar process has been used to prepare benzyl ethers of phenols, a strongly basic exchanger being washed with a solution of the phenol in water and in dilute alkali and then with alcohol.[29] The modified resin is then stirred with a solution of a benzyl halide until reaction is complete.

Weinstock and Boekelheide[30] introduced Amberlite IRA-400 (OH^-) for conversion of alkyltrimethylammonium iodides into the corresponding quaternary ammonium hydroxides. The procedure is simpler than the conventional silver oxide method and avoids undesirable side reactions sometimes encountered. In the case of a derivative of β-erythroidine, this technique raised the yield in a Hofmann degradation from 40 to 78%.

Epoxide formation. Australian workers[31] used the strongly basic Deacidite FF for conversion of 1-O-*p*-toluenesulfonylinositol (1) into 1,2-anhydro-(±)-inositol (2). Use of aqueous alkali led to 1,2-anhydromyoinositol (3) by "epoxide migration."

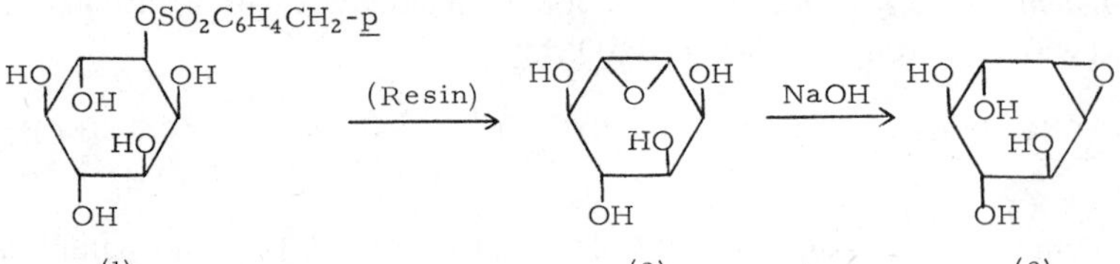
OSO2C6H4CH3-p
HO
OH
OH
HO
OH
(Resin)
HO
O
OH
HO
OH
NaOH
HO
O
OH
HO
OH

(1) (2) (3)

Solid-phase peptide synthesis. A new principle of peptide synthesis described by Merrifield[32] involves attachment of the first amino acid of the chain to a solid polymer by a covalent bond, stepwise addition of succeeding units until the desired sequence is assembled, and removal of the peptide from the solid support. When the growing peptide chain is attached to an insoluble solid particle, it can be filtered and washed free of reagents and by-products. The resin selected, a copolymer of styrene and divinylbenzene, was partially chloromethylated to provide a point for attachment of an N-protected amino acid through reaction with its triethylammonium salt (1). To increase the resistance of the resulting benzyl ester group to acid cleavage, the chloromethylated polymer was nitrated (2). After reaction of (1) and (2) to product (3), the N-protecting group could be removed with hydrogen bromide in acetic acid with little ester cleavage. Neutralization with excess triethylamine then gave (4), with a free N-terminal amino group available for coupling with a second N-protected amino acid by dehydration with N,N'-dicyclohexylcarbodiimide in dimethylformamide to give (5). The steps described complete one cycle. Further cycles can be carried out by alternately deprotecting and coupling with an appropriate carbobenzoxyamino acid. Finally, a completely protected peptide such as (5) is decarbobenzoxylated with HBr and the free peptide (6) is liberated from the polymer by saponification. The feasibility of the idea was demonstrated

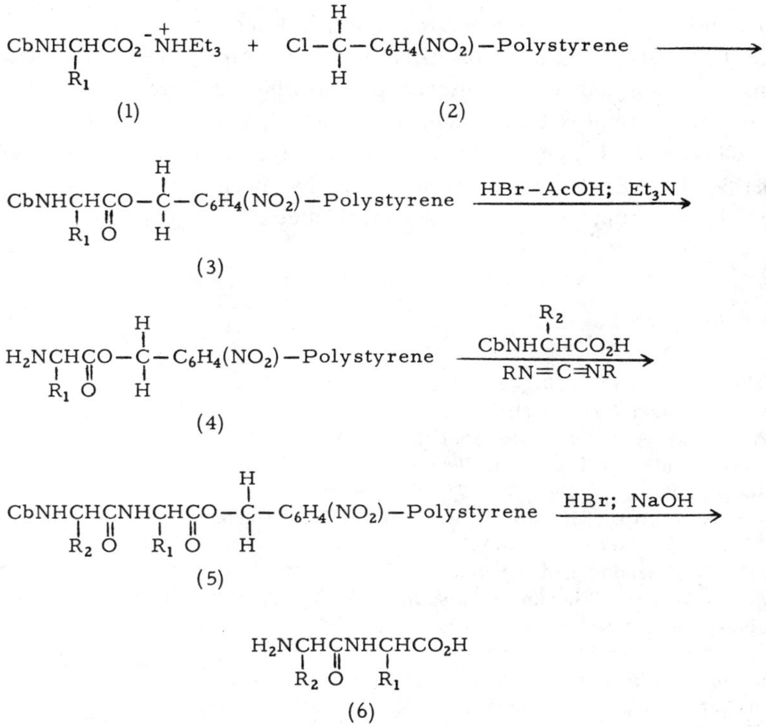

$$
\underset{(1)}{\underset{\underset{R_1}{|}}{CbNHCHCO_2}\overset{+}{}NHEt_3} \quad + \quad \underset{(2)}{Cl-\underset{\underset{H}{|}}{\overset{\overset{H}{|}}{C}}-C_6H_4(NO_2)-Polystyrene} \quad\longrightarrow
$$

$$
\underset{(3)}{\underset{R_1\ \ O}{\underset{|\quad\ ||}{CbNHCHCO}}-\underset{\underset{H}{|}}{\overset{\overset{H}{|}}{C}}-C_6H_4(NO_2)-Polystyrene} \quad\xrightarrow{HBr-AcOH;\ \ Et_3N}
$$

$$
\underset{(4)}{\underset{R_1\ \ O}{\underset{|\quad\ ||}{H_2NCHCO}}-\underset{\underset{H}{|}}{\overset{\overset{H}{|}}{C}}-C_6H_4(NO_2)-Polystyrene} \quad\xrightarrow[RN=C=NR]{\overset{\overset{R_2}{|}}{CbNHCHCO_2H}}
$$

$$
\underset{(5)}{\underset{R_2\ \ O\quad R_1\ \ O}{\underset{|\quad ||\quad\ |\quad ||}{CbNHCHCNHCHCO}}-\underset{\underset{H}{|}}{\overset{\overset{H}{|}}{C}}-C_6H_4(NO_2)-Polystyrene} \quad\xrightarrow{HBr;\ \ NaOH}
$$

$$
\underset{(6)}{\underset{R_2\ O\quad R_1}{\underset{|\quad ||\quad\ |}{H_2NCHCNHCHCO_2H}}}
$$

by the synthesis of L-leucyl-L-alanylglycyl-L-valine, identical with a sample prepared by the *p*-nitrophenyl ester procedure.

Letsinger[33] reported investigation of solid-phase peptide synthesis utilizing a styrene–divinylbenzene polymer supplied with carboxyl groups by reaction with diphenylcarbamyl chloride and aluminum chloride and hydrolysis. Reduction with lithium aluminum hydride to the hydroxymethyl polymer, reaction with phosgene to form the chloroformyl derivative, condensation with L-leucine ethyl ester, and alkaline hydrolysis gave polymer–leucine. This intermediate on successive treatment with isobutyl chlorocarbonate and triethylamine in toluene and with glycine benzyl ester *p*-toluenesulfonate and triethylamine in dimethylformamide gave polymer-leucylglycyl benzyl ester. Treatment with 15% hydrobromic acid in acetic acid eliminated the benzyl group and effected cleavage from the polymer, and addition of ether to the filtered solution gave a precipitate of the dipeptide hydrobromide.

In more recent work, Letsinger[34] prepared a functionalized polymer directly by copolymerization of styrene and divinylbenzene with *p*-vinylbenzyl alcohol or *p*-vinylbenzoic acid. He recommends that the polymer contain only about 0.2–0.5% of divinylbenzene, since a low degree of cross linkage reduces diffusion problems.

Ion-exchange chromatography. Moore and Stein[35] developed the chromatography of peptides and amino acids on ion-exchange resins into a powerful tool for the analysis of protein hydrolyzates. However, the method is not generally applicable to proteins or polynucleotides because large polymers are not tightly held and because the resins are of low capacity. In this area absorbents based on cellulose are promising (see next section).

Anion-Exchange Celluloses

Since the usual ion-exchange resins have very limited use for chromatography of proteins, Peterson and Sober[36] investigated adsorbents derived from cellulose, of which the most useful is DEAE-cellulose. It is prepared by treating a high-purity wood cellulose with 2-chloroethyldiethylamine hydrochloride to give a *die*thyl-*a*mino*e*thyl (DEAE) derivative. This material has been used successfully for the fractionation of serum proteins[37] and oligonucleotides.[38]

$$\underset{\text{(cellulose)}}{R{-}OH} + ClCH_2CH_2N(C_2H_5)_2 \longrightarrow ROCH_2CH_2\overset{+}{\underset{H}{N}}(C_2H_5)_2(Cl^-)$$

[1]R. H. Hasek *et al.*, *J. Org.*, **28**, 2496 (1963)
[2]R. F. Collins, *Chem. Ind.*, 736 (1957)
[3]M. J. Astle and J. A. Oscar, *J. Org.*, **26**, 1713 (1961)
[4]C. W. Davies and G. G. Thomas, *J. Chem. Soc.*, 1607 (1952)
[5]J. E. Hodge and C. E. Rist, *Am. Soc.*, **74**, 1498 (1952)
[6]J. G. Moffatt, *Am. Soc.*, **85**, 1118 (1963); see also K. Erne, *Chem. Scand.*, **9**, 893 (1955)
[7]D. F. Mowery, Jr., *J. Org.*, **26**, 3484 (1961)
[8]M. Walter, H. Besendorf, and O. Schnider, *Helv.*, **44**, 1546 (1961)
[9]P. J. Mill and W. R. C. Crimmin, *Biochim. Biophys. Acta*, **23**, 432 (1957)
[10]M. S. Newman, *Am. Soc.*, **75**, 4740 (1953)
[11]N. B. Lorette, *J. Org.*, **22**, 346 (1957)
[12]B. Loev and J. T. Massengale, *J. Org.*, **22**, 988 (1957)
[13]H. von Pechmann and C. Duisberg, *Ber.*, **16**, 2119 (1883)
[14]E. V. O. John and S. S. Israelstam, *J. Org.*, **26**, 240 (1961)
[15]P. Price and S. S. Israelstam, *J. Org.*, **29**, 2800 (1964)
[16]R. Bucourt, J. Tessier, and G. Nominé, *Bull. soc.*, 1923 (1963)
[17]J. D. Billimoria and N. F. Maclagan, *J. Chem. Soc.*, 3257 (1954)
[18]E. M. Kosower and J. W. Patton, *J. Org.*, **26**, 1318 (1961)
[19]R. Greenhalgh and R. A. B. Bannard, *Can. J. Chem.*, **39**, 1017 (1961); C. Y. Meyers and L. E. Miller, *Org. Syn., Coll. Vol.*, **4**, 39 (1963) describe the conversion of aminocaproic acid hydrochloride to the dipolar amino acid with a column of Amberlite IR-4B resin.
[20]R. L. Hinman and J. Lang, *J. Org.*, **29**, 1449 (1964)
[21]B. G. Chatterjee, V. V. Rao, and B. N. G. Mazumdar, *J. Org.*, **30**, 4101 (1965)
[22]A. Galat, *Am. Soc.*, **70**, 3945 (1948): see also J. M. Bobbitt and D. A. Scola, *J. Org.*, **25**, 560 (1960)
[23]J. M. Bobbitt and R. E. Doolittle, *J. Org.*, **29**, 2298 (1964)
[24]M. J. Astle and F. P. Abbott, *J. Org.*, **21**, 1228 (1956)
[25]E. D. Bergmann and R. Corett, *J. Org.*, **21**, 107 (1956); **23**, 1507 (1958)
[26]G. H. McCain, *J. Org.*, **23**, 632 (1958)
[27]M. J. Astle and W. C. Gergel, *J. Org.*, **21**, 493 (1956); R. W. Hein, M. J. Astle, and J. R. Shelton, *J. Org.*, **26**, 4874 (1961)
[28]M. Gordon, M. L. DePamphilis, and C. E. Griffin, *J. Org.*, **28**, 698 (1963)
[29]E. J. Rowe, K. L. Kaufman, and C. Piantadosi, *J. Org.*, **23**, 1622 (1958)
[30]J. Weinstock and V. Boekelheide, *Am. Soc.*, **75**, 2546 (1953)
[31]S. J. Angyal, V. Bender, and J. H. Curtin, *J. Chem. Soc.*, 798 (1966)

[32]R. B. Merrifield, *Am. Soc.*, **85**, 2149 (1963). For references to later papers, see R. B. Merrifield and J. M. Stewart, *Nature*, **207**, 522 (1965); J. M. Stewart and D. W. Wooley, *ibid.*, **206**, 619 (1965).
[33]R. L. Letsinger and M. J. Kornet, *Am. Soc.*, **85**, 3045 (1963)
[34]R. L. Letsinger and V. Mahadevan, *Am. Soc.*, **87**, 3526 (1965)
[35]S. Moore and W. H. Stein, *Advances in Protein Chemistry*, **11**, 191 (1956)
[36]E. A. Peterson and H. A. Sober, *Am. Soc.*, **78**, 751 (1956)
[37]H. A. Sober *et al.*, *Am. Soc.*, **78**, 756 (1956)
[38]C. Coutsogeorgopoulos and H. G. Khorana, *Am. Soc.*, **86**, 2926 (1964), and earlier papers by H. G. Khorana

Iron, Fe. At. wt. 55.85. Suppliers of iron powder: B, F, MCB.

Reduction. Kornblum[1] found that reduction of (—)-2-nitrooctane to the amine by iron and acetic acid proceeds with at least 82% retention of optical purity. Platinum hydrogenation in acetic acid was somewhat less satisfactory, and platinum hydrogenation in absolute ethanol led to 90% racemization. Reduction with $LiAlH_4$ was attended with complete racemization.

Koopman[2] found that in the reduction of the nitro group with iron and concd.

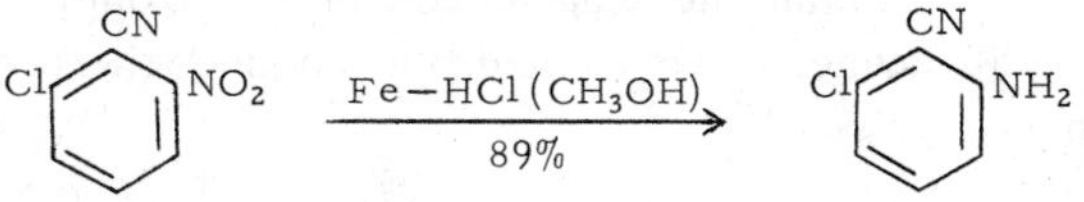
CN
Cl
NO₂
Fe—HCl (CH₃OH)
89%
CN
Cl
NH₂

hydrochloric acid addition of methanol is beneficial. Without methanol, reduction is incomplete.

Effect of traces of iron. Slaunwhite and Neely[3] observed that a trace of iron exerts a striking effect on the course of the bromination of estrone in acetic acid. Iron-free bromine (Mallinckrodt, sealed glass ampoule) gives 2,4-dibromoestrone,

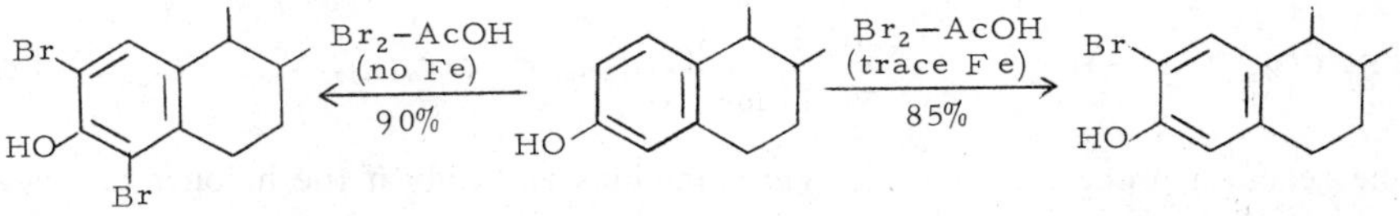
Br
HO
Br
Br₂—AcOH
(no Fe)
90%
HO
Br₂—AcOH
(trace Fe)
85%
Br
HO

but a trace of iron blocks 4-substitution and leads to 2-bromoestrone in high yield. Water (15–20%) has the opposite effect and leads to the production of 4-bromoestrone.

For the effect of a trace of iron on the reduction of aromatic compounds with Li or Na, ammonia, and an alcohol, *see* Birch reduction.

[1]N. Kornblum and L. Fishbein, *Am. Soc.*, **77**, 6266 (1955)
[2]H. Koopman, *Rec. trav.*, **80**, 1075 (1961)
[3]W. R. Slaunwhite, Jr., and L. Neely, *J. Org.*, **27**, 1749 (1962)

Iron pentacarbonyl, $Fe(CO)_5$. Mol. wt. 195.90, yellow, m.p. —20°, b.p. 103°. Suppliers: Alfa, KK, Mond Nickel Co., Ltd., England.

Reaction with acetylenes. Reppe and Vetter[1] found that acetylenic hydrocarbons react with iron pentacarbonyl in moist ethanol at 50–80° and a pressure of 40 atm. to form hydroquinones. At higher temperatures the main product is the ethyl ester of an acrylic acid (*see* Nickel carbonyl). Some insight into this remarkable reaction was provided by the observation of Sternberg *et al.*[2] that exposure to sunlight of a mixture of dimethylacetylene and $Fe(CO)_5$ causes separation of orange crystals of the composition $Fe(CO)_5(CH_3C{\equiv}CCH_3)_2$. This substance is formulated as a

π-complex formed by union of the two alkyne molecules with two carbonyl groups, since on exposure to air the complex affords duroquinone and on treatment with acid it yields durohydroquinone and carbon monoxide quantitatively.

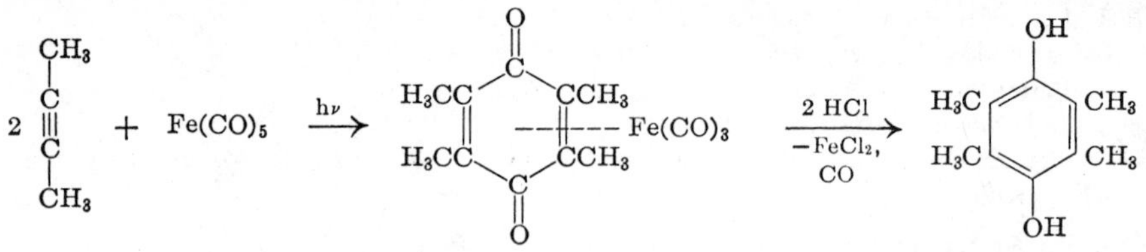

Complex

From the reaction of iron tetracarbonyl with diphenylacetylene, Schrauzer[3] isolated a π-complex incorporating only one molecule of carbon monoxide, a tetraphenylcyclopentadienone derivative.

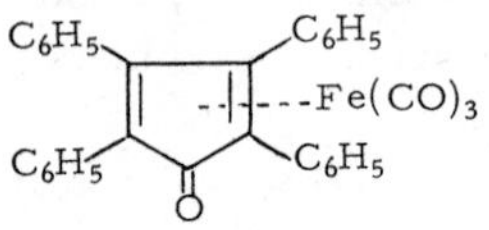

Trimerization of benzonitrile. Kettle and Orgel[4] found that the reagent converts benzonitrile into the trimer, 2,4,6-triphenyl-1,3,5-triazine, in good yield. On cooling, the product crystallizes from the reaction mixture.

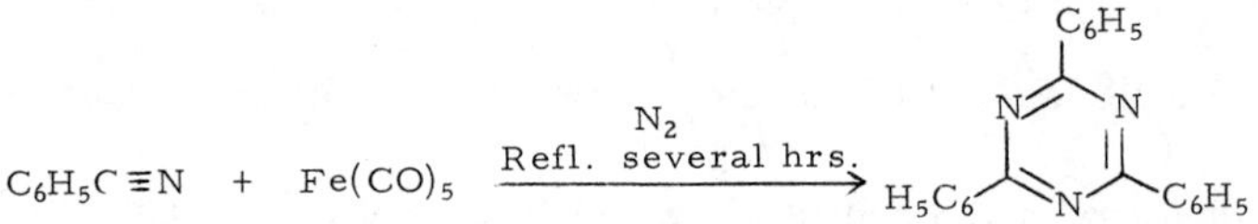

Coupling of gem-dihalides. Iron pentacarbonyl effects dechlorinative coupling of dichloro- or dibromodiphenylmethane to give tetraphenylethylene in good yields.[5]

$$2\,Fe(CO)_5 + 2\,(C_6H_5)_2CCl_2 \xrightarrow[\text{95\% crude}]{\text{Refl. in } C_6H_6} (C_6H_5)_2C{=}C(C_6H_5)_2 + 2\,FeCl_2 + 10\,CO$$

The reaction proceeds only with *gem*-dihalides and only if the halogen atoms are activated by groups such as phenyl, cyano, or carboalkoxy groups.

[1]W. Reppe and H. Vetter, *Ann.*, **582**, 133 (1953)
[2]H. W. Sternberg, R. Markby, and I. Wender, *Am. Soc.*, **80**, 1009 (1958)
[3]G. N. Schrauzer, *Chem. Ind.*, 1403 (1958)
[4]S. F. A. Kettle and L. E. Orgel, *Proc. Chem. Soc.*, 307 (1959)
[5]C. E. Coffey, *Am. Soc.*, **83**, 1623 (1961)

Isoamyl nitrite–Trifluoroacetic acid. For the diazotization of the weakly basic aminophenol (1), which is unstable in the presence of a trace of water, a solution

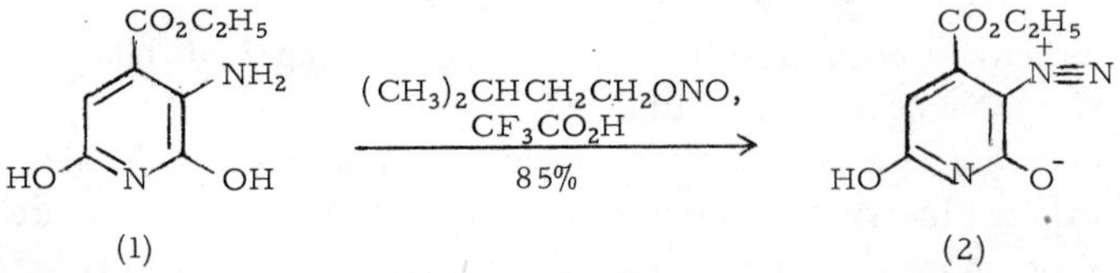

(1) (2)

of isoamyl nitrite in ethyl acetate is added at 0° to a stirred solution of the aminophenol in ethyl acetate containing trifluoroacetic acid.[1] The yellow diazo compound (2) was obtained after concentration of the solution and washed with ether.

[1]Z. B. Papanastassiou, A. McMillan, V. J. Czebotar, and T. J. Bardos, *Am. Soc.*, **81**, 6056 (1959)

Isobenzofurane (VIIIa, not isolated).

The reagent is generated most conveniently by thermal decomposition of the adduct IIIa of 1,4-dihydronaphthalene-1,4-endo-oxide (Ia) and tetraphenylcyclopentadienone (II). Heated alone at 165°, this adduct loses carbon monoxide and affords 1,2,3,4-tetraphenylbenzene (IX) as the only isolated product.[1] The transient

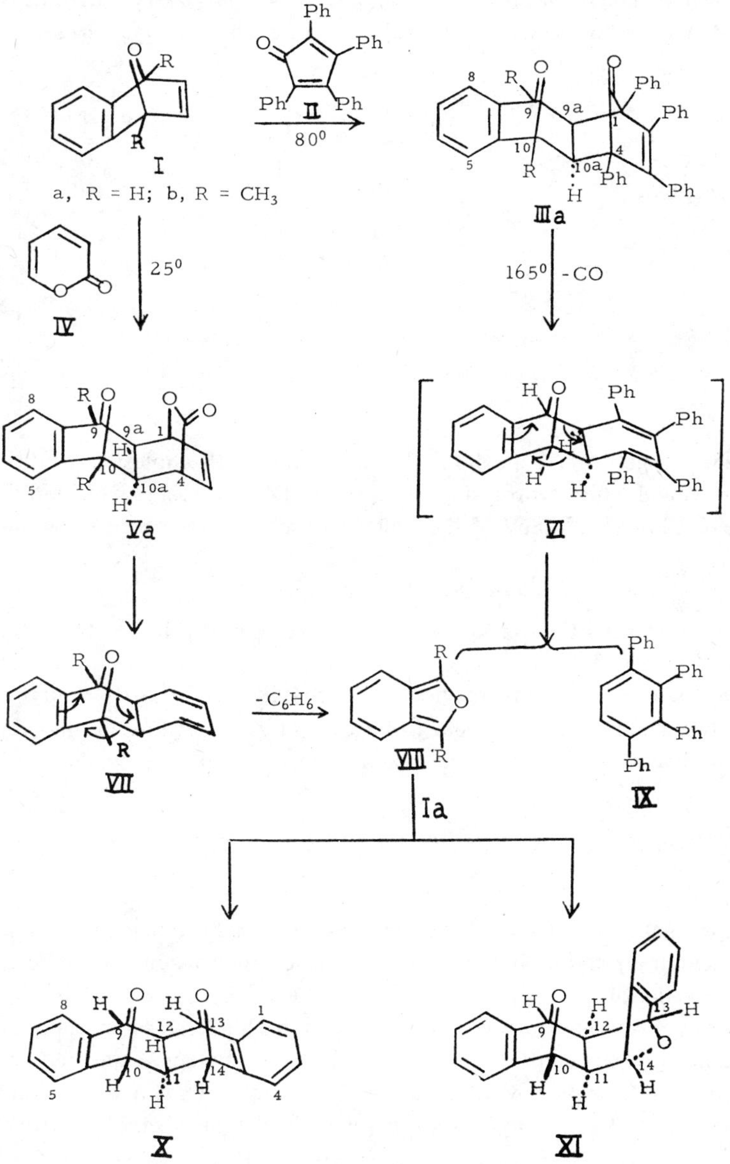
Ph
Ph
Ph
Ph
80°
I
a, R = H; b, R = CH3
IIIa
IV
25°
165°
-CO
Va
VI
- C6H6
VIII
IX
Ia
X
XI

existence of isobenzofurane was established by trapping it with the 1,4-*endo*-oxide (Ia) and isolation of the adducts X and XI. Isobenzofurane was generated also by pyrolysis of the adduct Va from the 1,4-*endo*-oxide (Ia) and α-pyrone (IV), and identified by conversion into X and XI.

[1] L. F. Fieser and M. J. Haddadin, *Can. J. Chem.*, **43**, 1599 (1965)

Isobornyloxyaluminum dichloride (1).

Preparation. The reagent is prepared by treating an ethereal solution of 0.1 mole of isoborneol with 0.09 mole of aluminum chloride and 0.025 mole of lithium aluminum hydride; a small amount of *t*-butanol is added and let react until hydrogen evolution ceases.[1]

Reduction.[1] The dichloride is a reducing agent similar to mixed hydride ($HAlCl_2$) and it is fairly stereoselective in the reduction of cyclohexanone derivatives in that

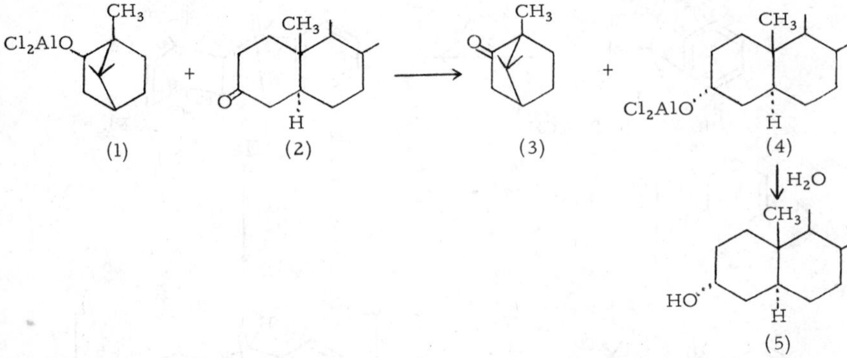

CH3
Cl2AlO
CH3
H
CH3
Cl2AlO
CH3
H
H2O
CH3
HO
H
(1)
(2)
(3)
(4)
(5)

it produces largely axial alcohols. Thus it reacts with cholestanone (2) to give camphor (3) and, after hydrolysis and crystallization, cholestane-3α-ol (5), in about 70% yield. Chromatography of the crude alcohol indicated it to be an 86:14 α:β mixture.

[1]E. L. Eliel and D. Nasipuri, *J. Org.*, **30**, 3809 (1965)

Isobutene, $(CH_3)_2C{=}CH_2$. Mol. wt. 56.10, b.p. −6.9°, sp. gr. 0.59. Suppliers: Matheson Co., MCB.

t-Butyl esters. The conversion of a carboxylic acid into its *t*-butyl ester by acid-catalyzed addition to isobutene is illustrated by a procedure for the preparation of di-*t*-butyl malonate.[1] Isobutene, taken from a cylinder, is condensed in a dry

$$\underset{0.48\ \text{m.}}{CH_2(CO_2H)_2} + \underset{1.5\ \text{m.}}{CH_2{=}C(CH_3)_2} + \underset{100\ \text{ml.}}{\text{Ether}} + \underset{5\ \text{ml.}}{\text{concd. } H_2SO_4} \xrightarrow[58-60\%]{} CH_2\begin{cases} CO_2C(CH_3)_3 \\ CO_2C(CH_3)_3 \end{cases}$$

ice–acetone bath and added to a cooled mixture of malonic acid, ether, and concd. sulfuric acid in a pressure bottle, and this is closed and shaken at room temperature until the acid dissolves.

The value of *t*-butyl esters in synthesis is that this hindered ester group is resistant to saponification but can be removed when desired by acid-catalyzed reversal of the reaction of formation with liberation of isobutene. Thus a general method for the synthesis of β-keto ethyl esters[2] involves acylation of ethyl *t*-butyl malonate

$$CH_2\begin{cases} CO_2C_2H_5 \\ CO_2C(CH_3)_3 \end{cases} \xrightarrow{Mg\ +\ C_2H_5OH} C_2H_5OMgCH\begin{cases} CO_2C_2H_5 \\ CO_2C(CH_3)_3 \end{cases} \xrightarrow{RCOCl}$$

$$RCOCH\begin{cases} CO_2C_2H_5 \\ CO_2C(CH_3)_3 \end{cases} \xrightarrow[\text{heat}]{TsOH} RCOCH_2CO_2C_2H_5 + (CH_3)_2C{=}CH_2 + CO_2$$

via the ethoxymagnesium derivative and simultaneous elimination of isobutene and carbon dioxide. The ethyl *t*-butyl malonate required for this synthesis is obtained by treating diethyl malonate with one equivalent of potassium hydroxide in ethanol to produce monoethyl malonate and treating this with isobutene and sulfuric acid.[3]

A synthesis of ketones of the type $R'COCH_2R$ involves alkylation of di-*t*-butyl malonate (as sodio derivative) to (2), acylation to (3), and elimination of isobutene and carbon dioxide.[4]

$$\underset{(1)}{CH_2[CO_2C(CH_3)_3]_2} \xrightarrow[EtONa]{RX} \underset{(2)}{RCH[CO_2C(CH_3)_3]_2} \xrightarrow[NaH]{R'COCl}$$

$$\underset{(3)}{R'COC(R)[CO_2C(CH_3)_3]_2} \xrightarrow[Heat]{TsOH} \underset{(4)}{R'COCH_2R} + 2\ (CH_3)_2C{=}CH_2 + 2\ CO_2$$

α-Acetylaminoketones are obtained by a similar process starting with the acylation of the acetylamino derivative of di-*t*-butyl malonate.[5]

$$C_6H_5COCl + CH_3CONHCH[CO_2C(CH_3)_3]_2 \xrightarrow{NaH}$$

$$C_6H_5COC(NHCOCH_3)[CO_2C(CH_3)_3]_2 \xrightarrow[Heat]{TsOH} C_6H_5COCH_2NHCOCH_3 + 2\,(CH_3)_2C{=}CH_2 + 2\ CO_2$$

***t*-Butyl ethers** are similarly prepared by reaction of an alcohol or phenol with isobutene in the presence of an acid catalyst, and the blocking group can be split with trifluoroacetic acid. An example of the use of the sequence is a synthesis of testosterone benzoate[6] from androstenolone. Of several acid catalysts investigated for ether formation, a 1:1 complex of boron trifluoride and phosphoric acid proved to be the most satisfactory. It can be purchased (Badische Anilin und Soda-Fabrik) or prepared by introducing BF_3 gas into anhydrous phosphoric acid. The latter reagent was prepared from 80% phosphoric acid by addition of the calculated amount of phosphorus pentoxide.

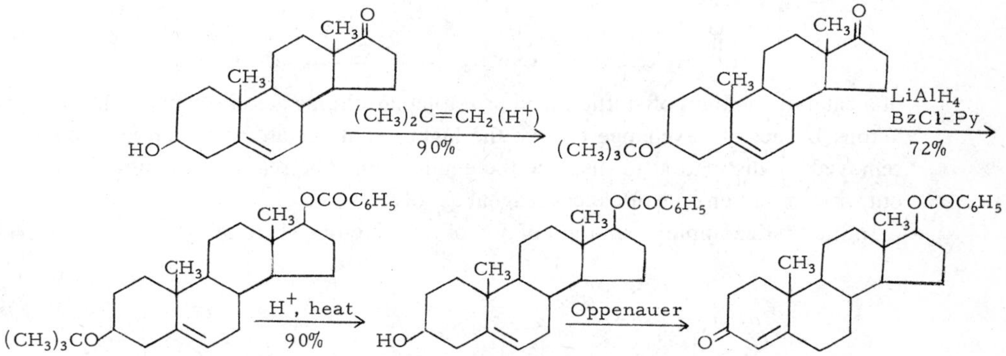
CH3
CH3
HO
(CH3)2C=CH2 (H+)
90%
(CH3)3CO
LiAlH4
BzCl-Py
72%
OCOC6H5
H+, heat
90%
HO
Oppenauer
O

Peptide synthesis. An N-protected amino acid (1) is easily converted into the *t*-butyl ester (2) by reaction with excess isobutene in the presence of sulfuric acid or *p*-toluenesulfonic acid.[7] Removal of the N-protective group by hydrogenolysis

$$\underset{(1)}{CbNHCH(R)CO_2H} \xrightarrow[H_2SO_4]{CH_2{=}C(CH_3)_2} \underset{(2)}{CbNHCH(R)CO_2C(CH_3)_3} \xrightarrow{H_2-Pd} \underset{(3)}{H_2NCH(R)CO_2C(CH_3)_3}$$

gives an amino acid ester more stable than the methyl or ethyl ester because the hindered character of the *t*-butyl group inhibits intermolecular condensation with an amino group. The *t*-butyl group can be removed when desired by elimination of isobutene by refluxing in benzene with an acid catalyst (H_2SO_4, TsOH) or with HBr–AcOH. A free amino acid can also be esterified by reaction with isobutene in 10:1 dioxane–sulfuric acid.[8]

The *t*-butyl group is useful also for the protection of the hydroxyl group of serine or tyrosine.[9] The ethers are cleaved by HBr–AcOH or by HCl–$CHCl_3$. The group is not useful for protection of sulfhydryl groups because drastic conditions are required for cleavage.

[1]A. L. McCloskey, G. S. Fonken, R. W. Kluiber, and W. S. Johnson, *Org. Syn., Coll. Vol.*, **4**, 261 (1963)
[2]D. S. Breslow, E. Baumgarten, and C. R. Hauser, *Am. Soc.*, **66**, 1286 (1944)
[3]R. E. Strube, *Org. Syn., Coll. Vol.*, **4**, 417 (1963)
[4]G. S. Fonken and W. S. Johnson, *Am. Soc.*, **74**, 831 (1952); W. H. Puterbaugh, F. W. Swamer and C. R. Hauser, *ibid.*, **74**, 3438 (1952)
[5]A. W. Schrecker and M. M. Trail, *Am. Soc.*, **80**, 6077 (1958)
[6]H. C. Beyerman, G. J. Heiszwolf, *J. Chem. Soc.*, 755 (1963); *Rec. trav.*, **84**, 203 (1965)
[7]G. W. Anderson and F. M. Callahan, *Am. Soc.*, **82**, 3359 (1960)
[8]R. Roeske, *J. Org.*, **28**, 1251 (1963)
[9]H. C. Beyerman and J. S. Bontekoe, *Rec. trav.*, **81**, 691 (1962); F. M. Callahan, G. W. Anderson, R. Paul, and J. E. Zimmerman, *Am. Soc.*, **85**, 201 (1963); E. Wünsch and J. Jentsch, *Ber.*, **97**, 2490 (1964); K. Podůska and M. I. Titov, *Coll. Czech.*, **30**, 1611 (1965)

Isocyanic acid, *see* Cyanic acid.

Isopropenyl acetate, $CH_3C(OCOCH_3){=}CH_2$. Mol. wt. 100.11, b.p. 96°/750 mm., sp. gr. 0.93. Suppliers: Carbon and Carbide Chem. Co., Tennessee Eastman, Monomer-Polymer, B, MCB.

Enol acetylation. The reagent is made commercially by sulfuric acid-catalyzed reaction of acetone with ketene.[1] It is the enol acetate of acetone and in the presence

$$CH_3\overset{O}{\overset{\|}{C}}CH_3 \rightleftharpoons CH_3\overset{OH}{\overset{|}{C}}{=}CH_2 \xrightarrow{CH_2=C=O} CH_3\overset{OCOCH_3}{\overset{|}{C}}{=}CH_2$$

of a catalytic amount of sulfuric acid or *p*-toluenesulfonic acid it reacts with a higher ketone by acetate exchange to give the higher enol acetate and acetone, which is removed by distillation to displace the equilibrium. The reaction usually is carried out with either benzene or excess reagent as solvent.

In the first example[2] a solution of 3 g. of the triketone (1) in 50 ml. of isopropenyl

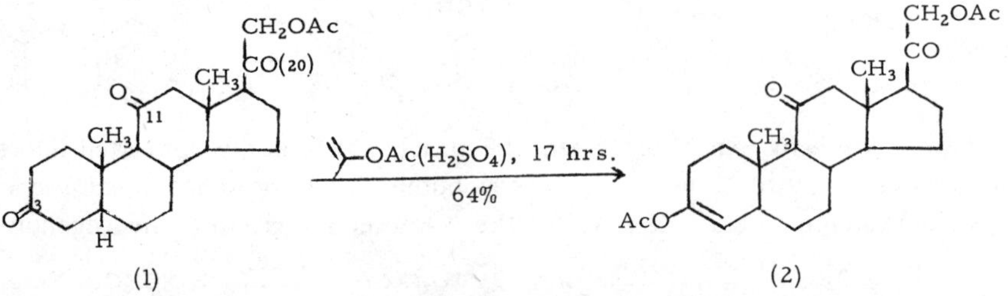

CH2OAc
CO(20)
CH3
O
11
CH3
3
H
OAc(H2SO4), 17 hrs.
64%
CH2OAc
CO
CH3
O
CH3
AcO

(1) (2)

acetate and 2 drops of concd. sulfuric acid was refluxed for 17 hrs., and the enol acetate (2) was isolated by chromatography on Woelm alumina (activity III). Note that even in the presence of a large excess of reagent the 11- and 20-keto groups remain unattacked; also that in the 5β-series a 3-ketone gives the Δ^3-enol acetate. A similar procedure applied to (3), a 3-ketone of the 5α-series, affords the Δ^2-enol diacetate (4).[3]

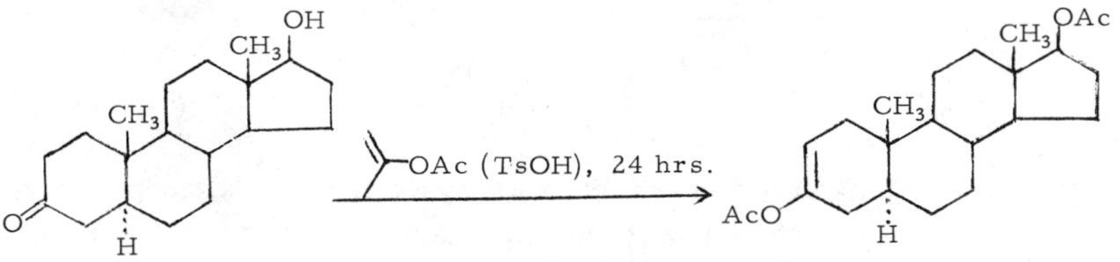
OH
CH3
CH3
OAc (TsOH), 24 hrs.
H
OAc
CH3
CH3
AcO
H

(3) (4)

Progesterone (5) presents an interesting case. Moffett and Weisblat[4] heated progesterone with isopropenyl acetate and *p*-toluenesulfonic acid monohydrate (TsOH) and obtained (6), in which the Δ^4-3-keto group has afforded a $\Delta^{3,5}$-dienol acetate

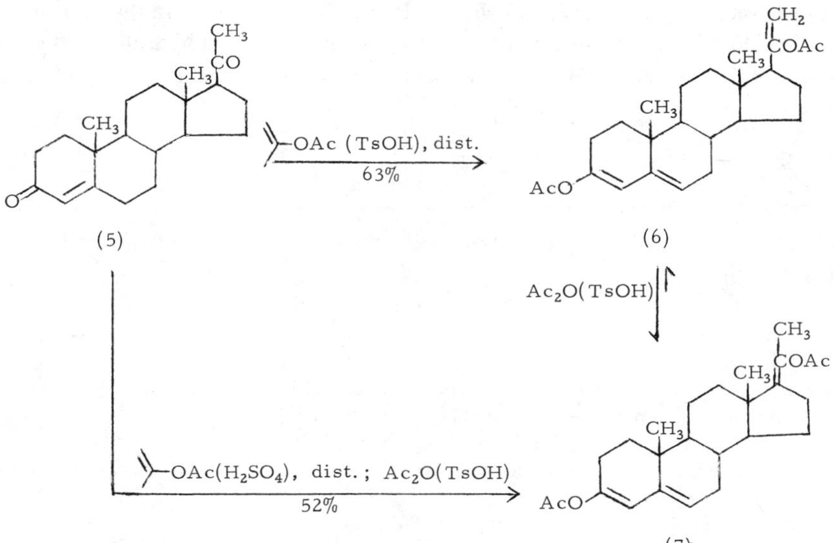
CH3
CO
CH3
CH3
OAc (TsOH), dist.
63%
CH2
COAc
CH3
CH3
AcO
Ac2O(TsOH)
OAc(H2SO4), dist.; Ac2O(TsOH)
52%
CH3
COAc
CH3
CH3
AcO

(5) (6) (7)

group and which also possesses a Δ^{20}-enol acetate group.[4] Djerassi[5] obtained the isomeric Δ^{17}-enol acetate (7) as the main product by an only slightly different procedure: progesterone is heated with isopropenyl acetate with sulfuric acid as catalyst and the crude product is refluxed for 2 hrs. with acetic anhydride containing TsOH, a procedure known to be effective for the isomerization of Δ^{20}- to Δ^{17}-enol acetates.[6] By the isopropenyl acetate–TsOH procedure. Moffett and Weisblat[4] converted several 20-keto steroids, including Δ^5-pregnene-3β-ol-20-one, into their Δ^{20}-enol acetates. Using the older acetic anhydride–TsOH procedure, Fieser and Huang-Minlon[7] obtained from pregnenolone the two geometrically isomeric Δ^{17}-enol acetates.

Fajkos and Šorm[8] converted androsterone acetate (8) into the only possible enol acetate (9) by heating 3 g. of (8) with 25 ml. of isopropenyl acetate and 1 ml. of a catalyst solution of 0.02 ml. of concd. sulfuric acid in isopropenyl acetate and collected 10 ml. of distillate in 2 hrs.; they then added 25 ml. more of reagent and 1 ml. of catalyst solution and collected 25 ml. in 2 hrs.

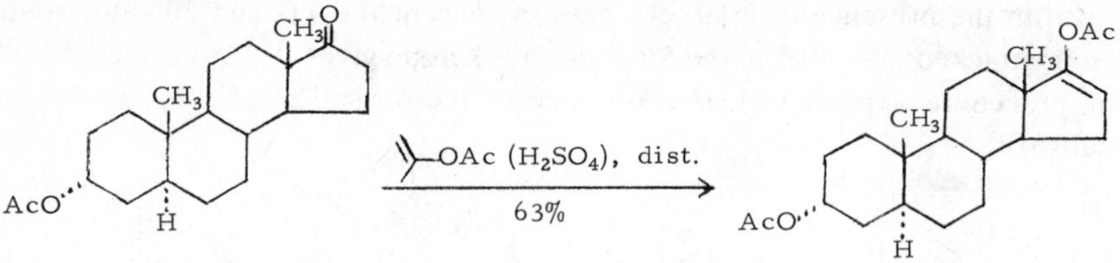
CH3
O
CH3
AcO
H
OAc (H2SO4), dist.
63%
OAc
CH3
CH3
AcO
H

(8) (9)

11β-Hydroxysteroids, resistant to acetylation by usual methods, can be acetylated by treatment overnight at room temperature with isopropenyl acetate and *p*-toluenesulfonic acid.[9]

House and Trost[10] state that enol acetylation with isopropenyl acetate yields chiefly the less highly substituted isomer whereas the reverse is true of enol acetylation with acetic anhydride, but we fail to find in the paper experimental evidence in support of these generalizations.

Condensation with succinic anhydride.[11] In 1,2-dichloroethane in the presence of aluminum chloride, isopropenyl acetate reacts with succinic anhydride (or the acid chloride) to give 2-acetyl-1,3-cyclopentanedione in moderate yield:

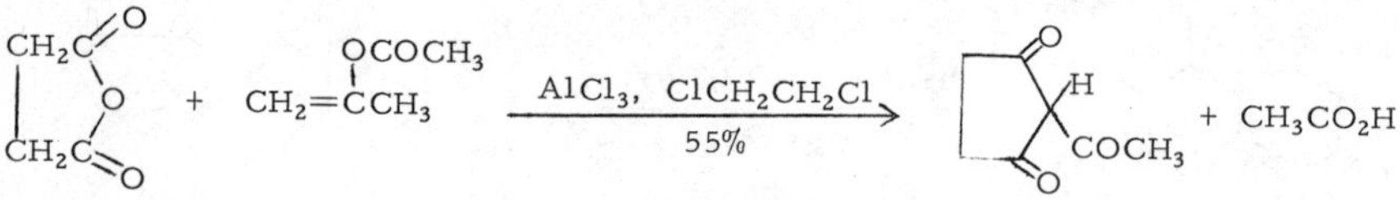
CH2C
O
O
CH2C
O
+
OCOCH3
CH2=CCH3
AlCl3, ClCH2CH2Cl
55%
O
H
COCH3
O
+ CH3CO2H

Yields are lower in the condensation with glutaric anhydride (40%) and maleic anhydride (16%).

[1]H. J. Hagemeyer, Jr., and D. C. Hull, *Ind. Eng. Chem.*, **41**, 2920 (1949)
[2]R. Deghenghi and C. R. Engel, *Am. Soc.*, **82**, 3201 (1960)
[3]R. Villotti, H. J. Ringold, and C. Djerassi, *Am. Soc.*, **82**, 5693 (1960); the experimental data are cited by C. Djerassi, *Steroid Reactions*, p. 41, Holden-Day, Inc. (1963)
[4]R. B. Moffett and D. I. Weisblat, *Am. Soc.*, **74**, 2183 (1952)
[5]C. Djerassi, J. Grossman, and G. H. Thomas, *Am. Soc.*, **77**, 3826 (1955)
[6]H. Vanderhaeghe, E. R. Katzenellenbogen, K. Dobriner, and T. F. Gallagher, *Am. Soc.*, **74**, 2810 (1952)
[7]L. F. Fieser and Huang-Minlon, *Am. Soc.*, **71**, 1840 (1949)
[8]J. Fajkos and F. Šorm, *Coll. Czech.*, **24**, 766 (1959)
[9]E. P. Oliveto *et al.*, *Am. Soc.*, **75**, 5486 (1953); see also T. G. Halsall, D. W. Theobald, and K. B. Walshaw, *J. Chem. Soc.*, 1029 (1964)
[10]H. O. House and B. M. Trost, *J. Org.*, **30**, 2502 (1965)
[11]F. Merényi and M. Nilsson, *Chem. Scand.*, **17**, 1801 (1963); **18**, 1368 (1964); M. Nilsson, *ibid.*, **18**, 441 (1964)

Isopropylidene malonate. Mol. wt. 144.12, m.p. 97°, pKa 5.2.

The reagent is prepared in about 50% yield by reaction of malonic acid and acetone with either acetic anhydride[1, 2] or isopropenyl acetate,[2] in each case in the presence of a catalytic amount of sulfuric acid. This acetonide is a strong acid and can be hydrolyzed under very mild conditions; since the methylene group is

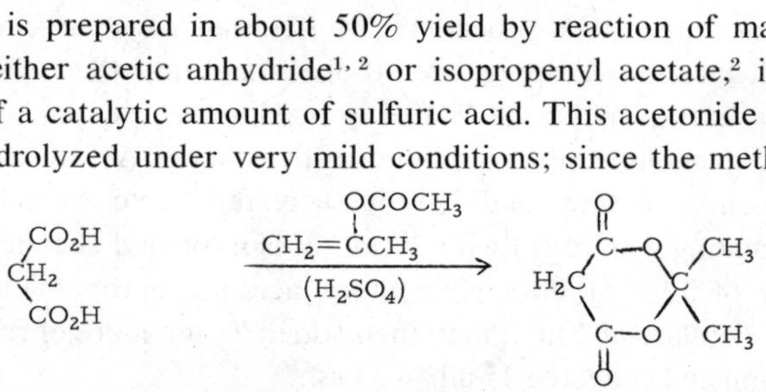
CO2H
CH2
CO2H
OCOCH3
CH2=CCH3
(H2SO4)
O
C
O
CH3
H2C
C
C
O
CH3
O

very reactive, the compound provides an attractive alternative to malonic ester in synthesis. Methylation,[3] best done by treating isopropylidene malonate with silver oxide and methyl iodide in acetonitrile,[2] has a strong tendency to proceed to the dialkylation stage. Alkylation with benzyl chloride in methanol, ethanol, or dimethylformamide gave only isopropylidene dibenzylmalonate.[4]

Corey[5] found the reagent useful for the synthesis of half-esters of α,β-unsaturated malonic esters. Isopropylidene malonate condenses readily with carbonyl compounds, even with the highly hindered mesitaldehyde (1). Thus the condensation of (1) and (2) was conducted in pyridine solution, and the product (3) on being refluxed with absolute ethanol and a trace of hydrogen chloride afforded ethyl hydrogen mesitylidene malonate (4).

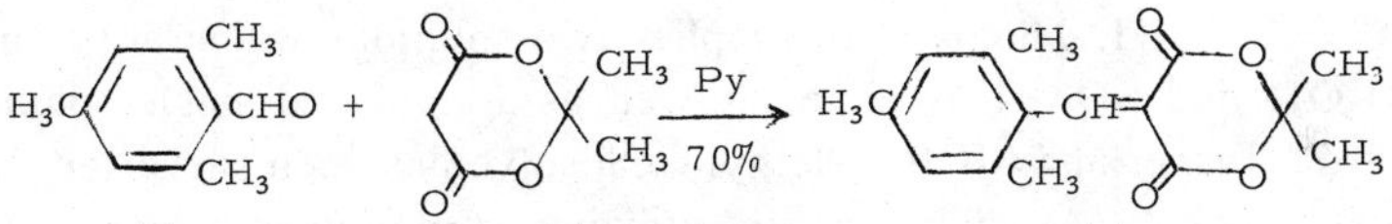
CH3
H3C
CHO
CH3
O
CH3
CH3
Py
70%
CH3
H3C
CH
CH3
O
CH3
CH3

(1) (2) (3)

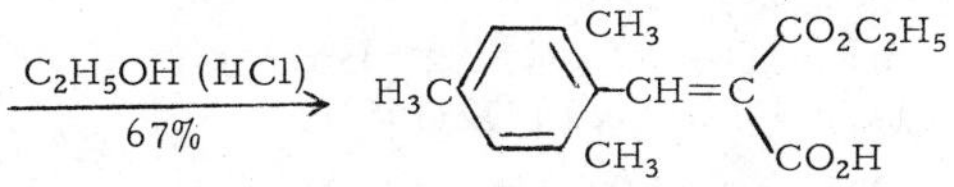
C2H5OH (HCl)
67%
CH3
H3C
CH=C
CH3
CO2C2H5
CO2H

(4)

[1]A. N. Meldrum, *J. Chem. Soc.*, **93**, 598 (1908)
[2]D. Davidson and S. A. Bernhard, *Am. Soc.*, **70**, 3426 (1948)
[3]E. Ott, *Ann.*, **401**, 159 (1913)
[4]J. A. Hedge, C. W. Kruse, and H. R. Snyder, *J. Org.*, **26**, 992 (1961)
[5]E. J. Corey, *Am. Soc.*, **74**, 5897 (1952)

K

Karl Fischer reagent. This reagent for determination of small amounts of water in inert solvents is a solution of iodine, sulfur dioxide, and pyridine in methanol.[1] The reaction with water is:

$$I_2 + SO_2 + H_2O \longrightarrow 2\ HI + SO_3$$

Pyridine combines with both products. A Karl Fischer titration unit is available (A.H. Thomas Co.) for detection of the end point by means of platinum electrodes and a pH meter. A.H. Thomas Co. supplies two solutions which are to be mixed (I_2 in CH_3OH and SO_2 in pyridine) as well as a stabilized single solution (also available from Fisher and MCB). Methyl Cellosolve has been suggested as solvent in place of methanol.[2] A similar reagent utilizing a solution of bromine and sulfur dioxide in chloroform has been described.[3]

[1]Karl Fischer, *Angew. Chem.*, **48**, 394 (1935)
[2]E. D. Peters and J. L. Jungnickel, *Anal. Chem.*, **27**, 450 (1955)
[3]R. Belcher and T. S. West, *J. Chem. Soc.*, 1772 (1953)

Ketene, $CH_2{=}C{=}O$. Mol. wt. 42.04, b.p. −48°. Highly toxic.

Preparation. An early procedure[1] involving distillation of acetone through a combustion tube packed with broken pieces of porous porcelain and heated in a

$$CH_3\overset{CH_3}{\overset{|}{C}}O \xrightarrow{\Delta} CH_2{=}C{=}O + CH_4$$

gas or electric combustion furnace at 695–705° afforded ketene in yield of 25–29%. Then Williams and Hurd[2] described an improved generator ("ketene lamp") which raises the yield to 80–90% and which has an output of 0.45 mole of ketene per hr. E. B. Hershberg's adaptation of the Williams-Hurd generator is shown in Fig. K-1. Refluxing acetone comes in contact with a glowing grid of resistance wire and is cracked to ketene and methane; unchanged acetone is condensed and returned to the boiling flask and ketene evolved in the gas phase along with methane is absorbed directly by a liquid reactant or in a solution, or else condensed in a dry ice trap. Another generator has a pyrolyzing coil submerged in acetone in a 1-l. flask submerged in a water bath.[3] Another method involves pyrolysis of acetic anhydride at 500–510°[4] in an apparatus that achieves rapid separation of ketene from the

$$CH_3\overset{OCOCH_3}{\overset{|}{C}}{=}O \xrightarrow{500\text{-}510^0} CH_2{=}C{=}O + CH_3COOH$$

by-product acetic acid and so prevents recombination. The method has the advantage that there are no by-product gases. Processes used industrially are discussed in reviews.[5]

A laboratory procedure which rivals the acetone–hot wire process consists in the pyrolysis of diketene.[6] Since this material is available as starting material, the reliable diketene method now appears to be the method of choice.

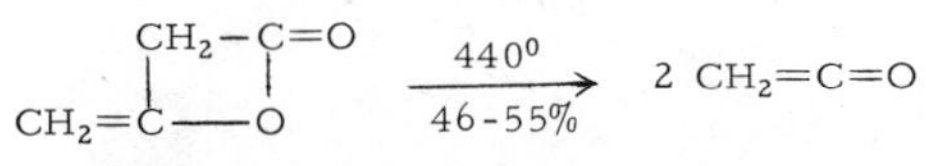

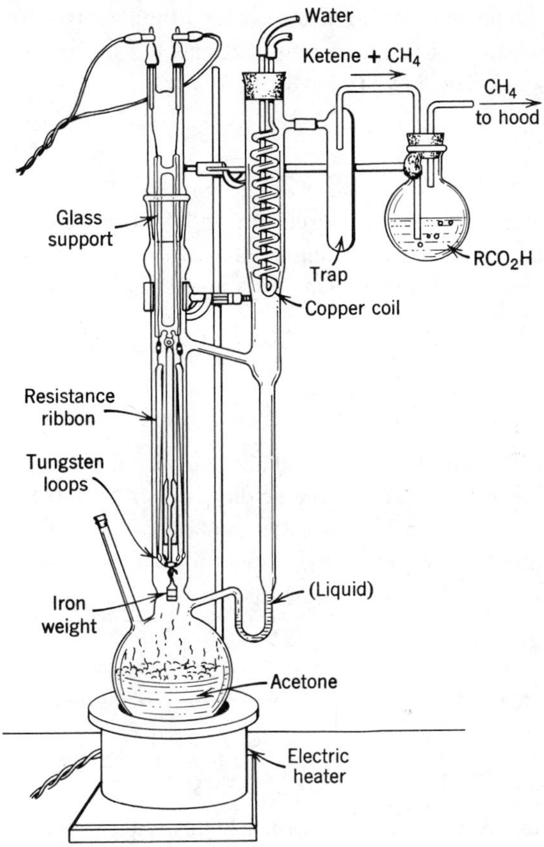
Water
Ketene + CH_4
CH_4
to hood
Glass support
Trap
RCO_2H
Copper coil
Resistance ribbon
Tungsten loops
Iron weight
(Liquid)
Acetone
Electric heater

Fig. K-1 Ketene generator

Reactions. Ketene is stable to some extent at –80°; at 0° it rapidly dimerizes. The usual procedure for conducting a reaction is to pass gaseous ketene into a liquid reactant or a solution. Thus for the preparation of *n*-caproic anhydride[7] 1 mole of the acid is placed in a gas-washing bottle cooled in ice and 0.5–0.55 mole of

$$CH_3(CH_2)_4CO_2H + CH_2{=}C{=}O \xrightarrow[80\text{-}87\%]{} \begin{matrix} CH_3(CH_2)_4C{=}O \\ \quad\quad\quad\quad\;\; O \\ CH_3(CH_2)_4C{=}O \end{matrix} + CH_3CO_2H$$

gaseous ketene generated from acetone is passed in. The resulting mixture is fractionated slowly to allow for the complete conversion of the initially formed mixed anhydride; low-boiling fractions contain acetone, ketene, acetic acid, and acetic anhydride.

The reaction of ketene with primary amines proceeds rapidly and in high yield.

$$C_6H_5NH_2 + CH_2{=}C{=}O \longrightarrow C_6H_5N(H){-}C({=}O){-}CH_3$$

The reaction with water is slow and water is not a good absorbent for ketene. The reaction is catalyzed by acids, and ketene is absorbed readily by dilute acetic acid. Hydrogen peroxide is more readily acetylated to give diacetyl peroxide.[8]

$$2\ CH_2{=}C{=}O + HOOH \longrightarrow CH_3C({=}O){-}OO{-}C({=}O)CH_3$$

The reaction of ketene with *t*-butanol requires catalysis by a strong acid.[5] A mixture of 74 g. of *t*-butanol and 0.5 ml. of concd. sulfuric acid is heated to 60° and ketene is passed in for 1½ hrs.

$$(CH_3)_3COH + CH_2{=}C{=}O \xrightarrow[60\%]{(H_2SO_4)} (CH_3)_3COCOCH_3$$

Taking advantage of the low rate of reaction of ketene with water in neutral or alkaline solution, Bergmann and Stern[9] were able to effect the N-acetylation of amino acids by passing ketene into an alkaline solution of the acid at room temperature. Jackson and Cahill[10] studied the acetylation of L-tryptophane and found that

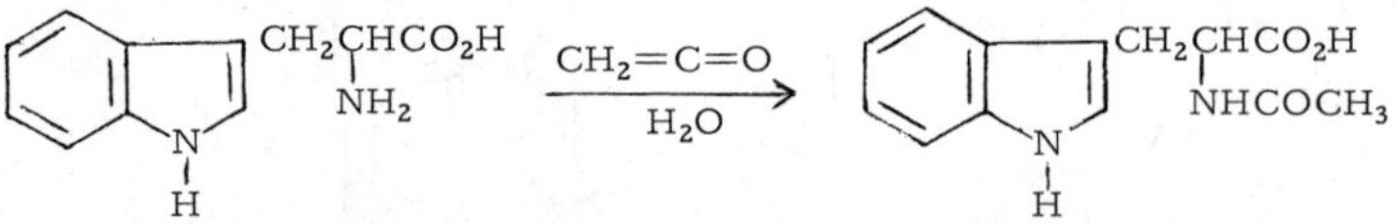

if the solution is kept alkaline to phenolphthalein while passing in ketene no racemization occurs. If the acidity is such that acetic acid is generated in the solution, racemization ensues within an hour. Cysteine reacts readily with ketene in slightly alkaline solution to give the N,S-diacetyl derivative.[11]

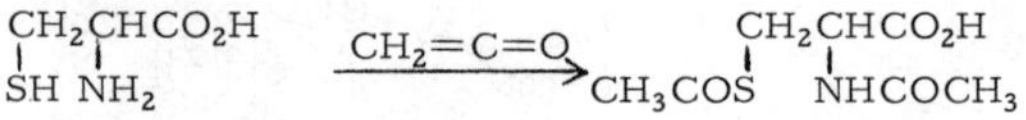

Küng[12] discovered that under properly defined conditions ketene adds formaldehyde smoothly to form β-propiolactone. The process can be conducted on a

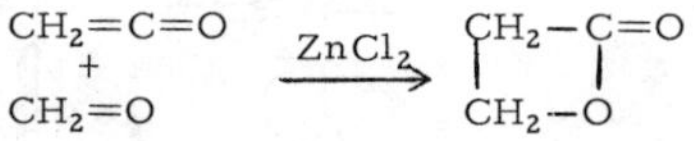

technical scale and affords a highly reactive reagent of promise for many syntheses.[13] Observations on the formation of β-lactones by catalyzed reaction of ketene with ketones are summarized by Zaugg.[14]

6-Hydroxycoumarane-3-one reacts with acetic anhydride to give both the O-acetate (2) and the enol acetate (3); use of ketene permits selective O-acetylation.[15]

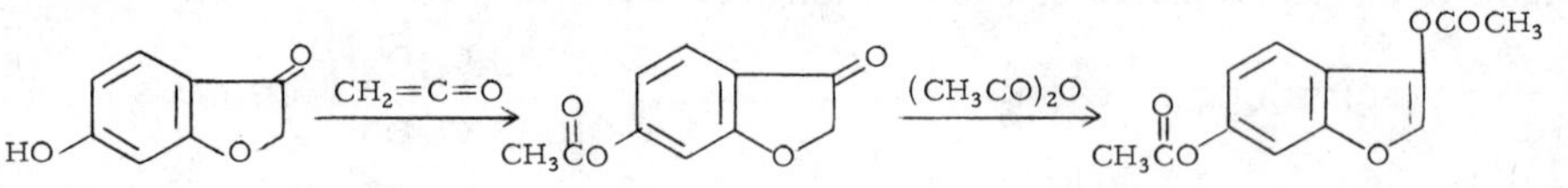

(1) (2) (3)

Ried and Mengler[16] found that ketene reacts with diazoketones at room temperature (toluene solution, 2 days) to form β,γ-butenolides (1) which are hydrolyzed to γ-keto acids (2) reducible in turn to the corresponding acids (3). The sequence

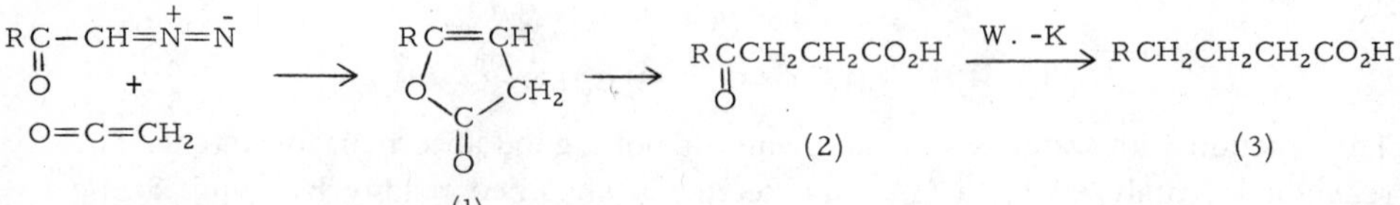

(1) (2) (3)

provides a means for lengthening an acid side chain by three carbon atoms, but a serious limitation is that yields in the first step are in the range 30–50%.

Addition of a cold (−78°) methylene chloride solution of diazomethane to a methylene chloride solution of excess ketene and removal of the excess by vacuum distillation at −78° affords a solution of pure cyclopropanone.[17]

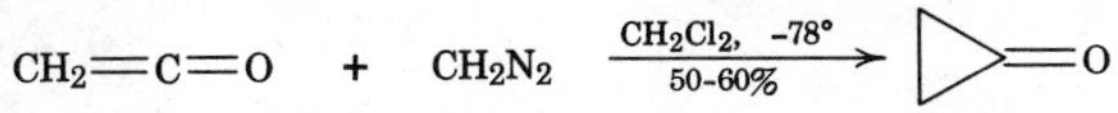

[1]C. D. Hurd, *Org. Syn., Coll. Vol.*, **1**, 330 (1941)
[2]J. W. Williams and C. D. Hurd, *J. Org.*, **5**, 122 (1940)
[3]C. Hamalainen and J. D. Reid, *Ind. Eng. Chem.*, **41**, 1018 (1949)
[4]G. J. Fischer A. F. MacLean, and A. W. Schnizer, *J. Org.*, **18**, 1055 (1953)
[5]R. N. Lacey, *Adv. in Org. Chem.*, **2**, 213 (1960); G. Quabeck, *Newer Methods of Preparative Org. Chem.*, **2**, 133 (1963)
[6]A. B. Boese, Jr., U.S. patent 2,218,066 [*C.A.*, **35**, 1072 (1941)]; S. Andreades and H. D. Carlson, *Org. Syn.*, **45**, 50 (1965)
[7]J. W. Williams and J. A. Krynitsky, *Org. Syn., Coll. Vol.*, **3**, 164 (1955)
[8]J. d'Ans and W. Frey, *Ber.*, **45**, 1845 (1912)
[9]M. Bergmann and F. Stern., **63**, 437 (1930)
[10]R. W. Jackson and W. M Cahill, *J. Biol. Chem.*, **126**, 37 (1938)
[11]A. Neuberger, *Biochem. J.*, **32**, 1452 (1938)
[12]F. E. Küng, U.S. patent 2,356,459 [*C.A.*, **39**, 88 (1945)]
[13]T. L. Gresham, J. E. Jansen, *et al.*, *Am. Soc.*, **70**, 998, 999, 1001, 1003, 1004 (1948)
[14]H. E. Zaugg, *Organic Reactions*, **8**, 305 (1954)
[15]W. Logemann, G. Cavagna, and G. Tosolini, *Ber.*, **96**, 1680 (1963)
[16]W. Ried and H. Mengler, *Ann.*, **678**, 113 (1964)
[17]N. J. Turro and W. B. Hammond, *Am. Soc.*, **88**, 3673 (1966)

$$\begin{array}{l} COCO_2H \\ | \\ CH_2CH_2CO_2H \end{array}$$

α-Ketoglutaric acid, $CH_2CH_2CO_2H$ (with $COCO_2H$), Mol. wt. 146.09. m.p. 110–111°. Preparation by ester condensation of diethyl succinate and diethyl oxalate in the presence of potassium ethoxide and hydrolysis.[1] Suppliers: Aldrich, MCB.

The reagent is recommended for the cleavage of 2,4-dinitrophenylhydrazones of steam-volatile ketones by an exchange reaction.[2]

[1]L. Friedman and E. Kosower, *Org. Syn., Coll. Vol.*, **3**, 510 (1955)
[2]H. R. Harrison and E. J. Eisenbraun, *J. Org.*, **31**, 1294 (1966)

Kiliani reagent, *see* Chromic acid.

L

Lead acetate trihydrate, $Pb(OCOCH_3)_2 \cdot 3\,H_2O$. Mol. wt. 379.35, solubility in g. per 100 ml. H_2O: $45.6^{15°}$, $200^{100°}$. Suppliers: B, MCB.

Neutralization of HCl. Lead hydroxide obtained by hydrolysis of the acetate appears to be superior to lead oxide for the liberation of an α-amino acid from its hydrochloride. In the synthesis of α-aminodiethylacetic acid formulated[1] the crude

$$(C_2H_5)_2CO \xrightarrow{NaCN + NH_4Cl} (C_2H_5)_2\underset{OH}{CCN} \xrightarrow{NH_3} (C_2H_5)_2\underset{NH_2}{CCN} \xrightarrow[\text{(cold)}]{HCl}$$

$$(C_2H_5)_2\underset{\overset{+}{N}H_3Cl^-}{CCONH_2} \xrightarrow[\text{(hot)}]{HCl} (C_2H_5)_2\underset{\overset{+}{N}H_3Cl^-}{CCO_2H} \xrightarrow[\substack{39-43\% \\ \text{overall}}]{Pb(OH)_2} (C_2H_5)_2\underset{NH_2}{CCO_2H}$$

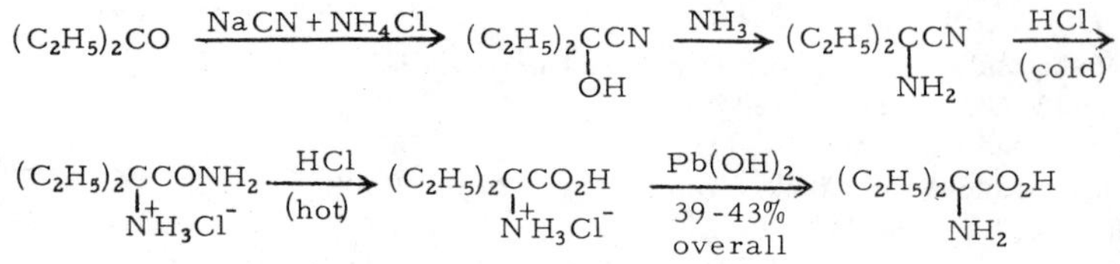

hydrochloride from 1 mole of diethyl ketone is heated in 2 l. of water with a moist paste of lead hydroxide prepared by adding 1.5 l. of $2N$ sodium hydroxide to a stirred solution of 1.5 moles of lead acetate in 1.3 l. of water and filtering the mixture. For a reason not stated, the suspension is diluted with water to a volume of about 3.5 l. and then concentrated at reduced pressure to a volume of about 2 l. before filtration from lead salts. The filtrate is saturated with hydrogen sulfide to precipitate lead sulfide, which is removed by suction filtration. The filtrate is concentrated at reduced pressure nearly to dryness and the amino acid crystallized from ethanol.

Removal of selenium and selenious acid. Use of selenium dioxide or selenious acid as an oxidant is attended with the difficulty of freeing the reaction mixture of colloidal selenium and of excess oxidant. A procedure[2] for the oxidation of paraldehyde to glyoxal and isolation of the product as the bis-bisulfite addition compound specifies use of lead acetate as more satisfactory for the removal of selenious acid than sulfur dioxide, provided that the solution is kept cool and a large excess is avoided. A

$$(CH_3CHO)_3 \xrightarrow{H_2SeO_3} O{=}CHCH{=}O \xrightarrow[\substack{72-74\% \\ \text{overall}}]{NaHSO_3} NaO_3S{-}\underset{OH}{CH}{-}\underset{OH}{CHSO_3Na}$$

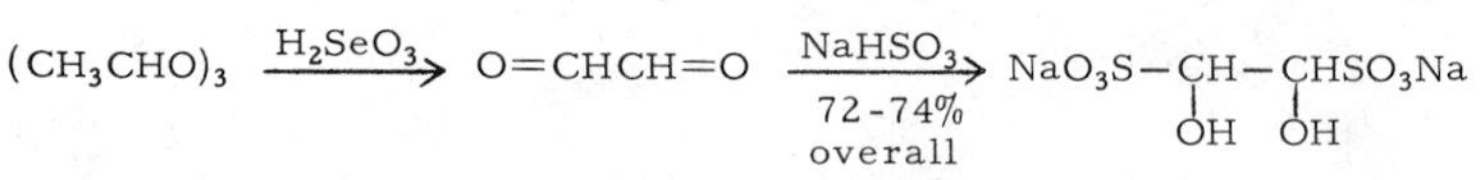

mixture of 1.72 moles of selenious acid, a large excess of paraldehyde, 540 ml. of dioxane, and 40 ml. of 50% acetic acid is refluxed for 6 hrs. and decanted from a deposit of selenium. The solution is then distilled for removal of paraldehyde and dioxane, decanted from a further small amount of selenium and titrated with 25% lead acetate solution until a filtered test portion gives no further precipitate of lead selenite. The filtered solution is saturated with hydrogen sulfide for removal of excess lead ion, treated with Norit and filtered. The colorless filtrate is concentrated at reduced pressure and stirred for 3 hrs. with a filtered solution of sodium bisulfite in 40% ethanol. Glyoxal bis-bisulfite separates and is washed with ethanol and then with ether.

Elimination of H_2S. The reagent has been used for the conversion of the thiourea (1) to the cyanamide (2) by elimination of the elements of H_2S.[3] A suspension of 0.2 mole of *o*-chlorophenylthiourea (1) in boiling water is treated with a boiling

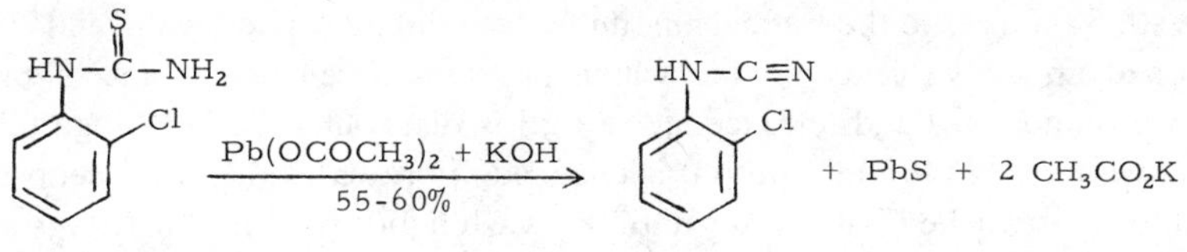

(1) (2)

solution of potassium hydroxide and then with a hot saturated solution of lead acetate trihydrate. After cooling to 0° the lead sulfide is removed by filtration and the colorless filtrate is acidified with acetic acid. *o*-Chlorophenylcyanamide (2) precipitates and is crystallized from benzene–petroleum ether.

[1]R. E. Steiger, *Org. Syn., Coll. Vol.*, **3**, 66 (1955)
[2]A. R. Ronzio and T. D. Waugh, *ibid.*, **3**, 438 (1955)
[3]F. Kurzer, *ibid.*, **4**, 172 (1963)

Lead dioxide, PbO_2. Mol. wt. 289.21, dec. 290°. Suppliers: F, MCB.

***$HOArCH_3 \longrightarrow HOArCO_2H$*.** In extending preliminary observations of others, Graebe and Kraft[1] showed that the methyl group of a methylated phenol can be transformed into carboxyl by a process describable as an oxidative alkali fusion, the oxidant being lead dioxide. They applied the method successfully to the three cresols, the three toluic acids, 2,4-dimethylphenol, and *o*-cresotinic acid (1, available from Eastman). A preparative procedure[2] for the conversion of (1) into 2-hydroxyisophthalic acid (2) is as follows. A mush of 240 g. of potassium hydroxide pellets and

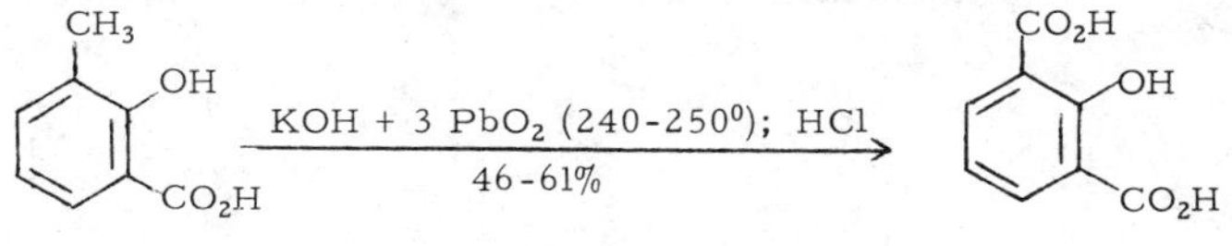

(1) (2)

50 ml. of water in a stainless steel beaker is let cool and stirred during addition of 40 g. (0.263 mole) of *o*-cresotinic acid. The beaker is clamped in a cold oil bath and 240 g. (1 mole) of lead dioxide is stirred in all at once. With steady manual stirring the temperature of the bath is slowly brought to 240°, when the lumpy brown mass quickly turns to a bright orange melt containing heavy crystals of lead monoxide. After 15 min. at 240° the temperature is raised briefly to 250°, the melt is let cool, and the congealing mass is distributed over the walls of the beaker. After addition of water and of enough sodium sulfide to precipitate all lead ion present, the solution is filtered and acidified. The amount of lead dioxide specified is 1.25 times the theory; if just the theoretical amount is used the yield is lowered by about 20%.

***Hydroquinones $\longrightarrow$ quinones*.** Willstätter and Parnas[3] introduced use of a suspension of lead dioxide in benzene for oxidation of 2,6-dihydroxynaphthalene to

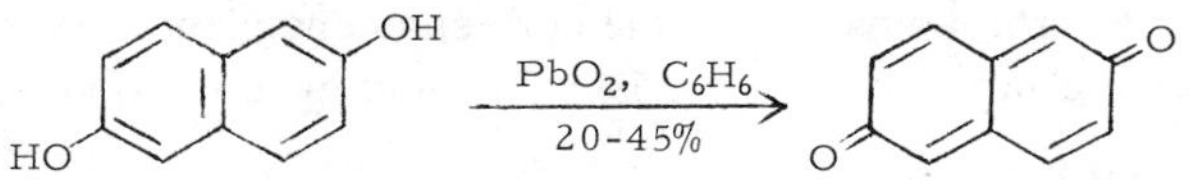

amphinaphthoquinone, a quinone of high oxidation potential and great sensitivity; it is decomposed even by water. They noticed considerable variations in the quality of commercial samples of the oxidant; some were wholly unsatisfactory, others could be used if taken in large excess. Later workers failed to obtain the quinone even with 50 times the theoretical amount of lead dioxide. Then Kuhn and Hammer[4] developed an active lead dioxide which provides a reliable means of preparing sensitive quinones. Lead tetraacetate (50 g.) is distributed in centrifuge tubes and rubbed with water (460 ml.) until it is changed to brown oxide. After centrifugation the material is washed with water until the wash liquor is neutral to litmus and then collected and washed four times with acetone and four times with ether. Yield 23 g. (92%). For oxidation, 2 g. of finely pulverized 2,6-dihydroxynaphthalene is added to a boiling suspension of 4 g. of active lead dioxide, the mixture is shaken for 1.2 min. and filtered. The orange filtrate is concentrated at reduced pressure and amphinaphthoquinone is obtained in 2 crops in yield of 45%, over twice the earlier yield.[3] *o*-Benzoquinone was obtained satisfactorily by oxidation of catechol with this reagent in benzene in the presence of anhydrous sodium sulfate. The active lead dioxide is very satisfactory for oxidation of 2,6-naphthylenedibenzenesulfonamide to 2,6-naphthoquinone dibenzosulfonimide.[5]

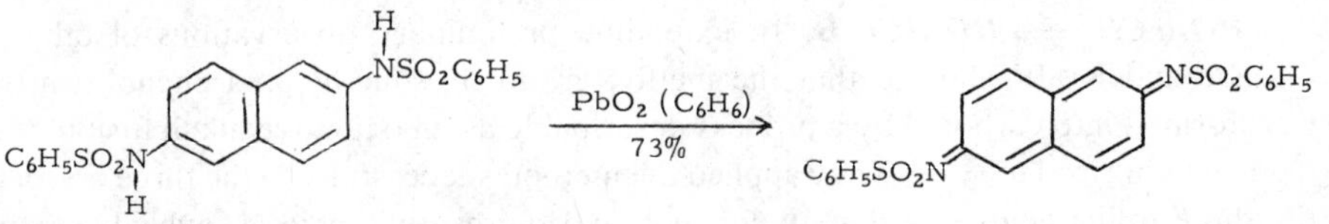
H
$NSO_2C_6H_5$
$C_6H_5SO_2N$
H
PbO_2 (C_6H_6)
73%
$NSO_2C_6H_5$
$C_6H_5SO_2N$

The oxidation of a catechol derivative to an *o*-benzoquinone is the first step in a synthesis of pyocyanine (6) introduced by Wrede and Strack[6] and further developed by Surrey.[7] Pyrogallol monomethyl ether (1) is oxidized by shaking a solution in

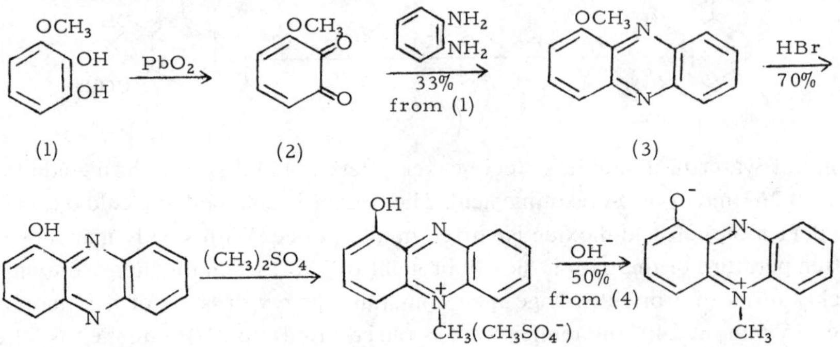
OCH_3
OH
OH
PbO_2
OCH_3
NH_2
NH_2
33%
from (1)
OCH_3
N
N
HBr
70%

(1) (2) (3)

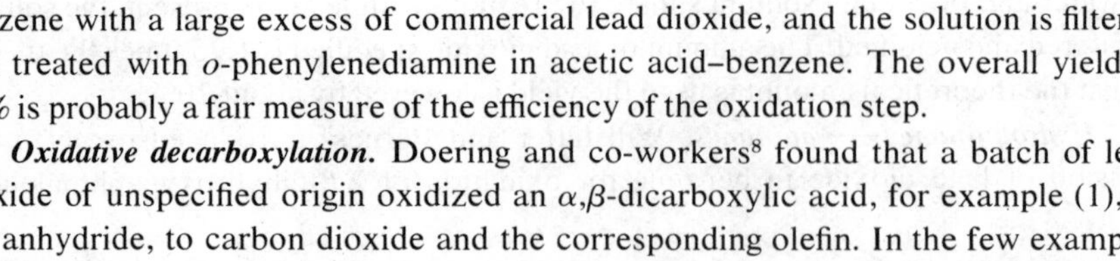
OH
N
N
$(CH_3)_2SO_4$
OH
N
N
$CH_3(CH_3SO_4^-)$
OH^-
50%
from (4)
O^-
N
N
CH_3

(4) (5) (6)

benzene with a large excess of commercial lead dioxide, and the solution is filtered and treated with *o*-phenylenediamine in acetic acid–benzene. The overall yield of 33% is probably a fair measure of the efficiency of the oxidation step.

Oxidative decarboxylation. Doering and co-workers[8] found that a batch of lead dioxide of unspecified origin oxidized an α,β-dicarboxylic acid, for example (1), or the anhydride, to carbon dioxide and the corresponding olefin. In the few examples studied yields were in the range 20–35%. The acid or anhydride was ground with

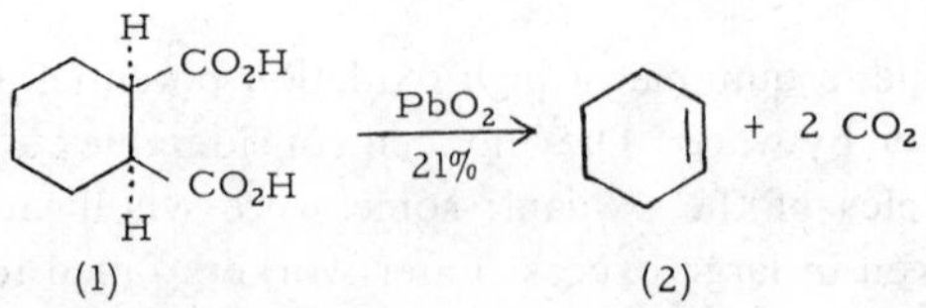
H
CO_2H
CO_2H
H
PbO_2
21%
+ 2 CO_2

(1) (2)

the oxidant in a ball mill and the fine powder was heated carefully either alone or mixed with powdered glass. Later, Doering and Finkelstein[9] reported difficulty in reproducing the yields using various samples of lead dioxide, including material prepared according to Kuhn and Hammer.[4] They then found that lead dioxide prepared from lead acetate trihydrate by reaction with alkali and oxidation with calcium hypochloride gave consistent yields of about 19%. Lead tetraacetate appears to be a superior reagent for oxidative decarboxylations of this type.

McElvain and Eisenbraun[10] applied the Doering oxidation procedure to nepetonic acid (1) and found that this γ-keto acid suffers elimination of the elements of formic acid and affords the α,β-unsaturated ketone (2) in 34% yield.

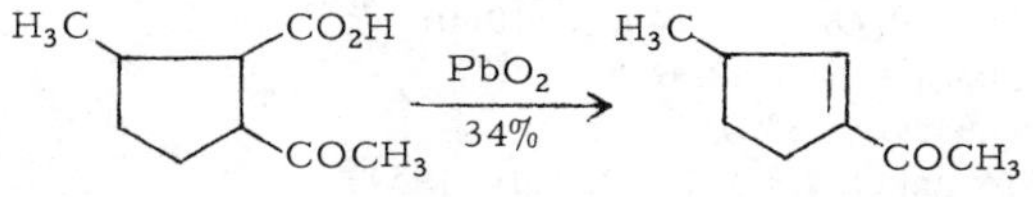
H3C
CO2H
PbO2
34%
COCH3
H3C
COCH3

(1) (2)

Oxidative cyclization. The active lead dioxide of Kuhn and Hammer was used in the three examples formulated below. The first two are reported by Hassall

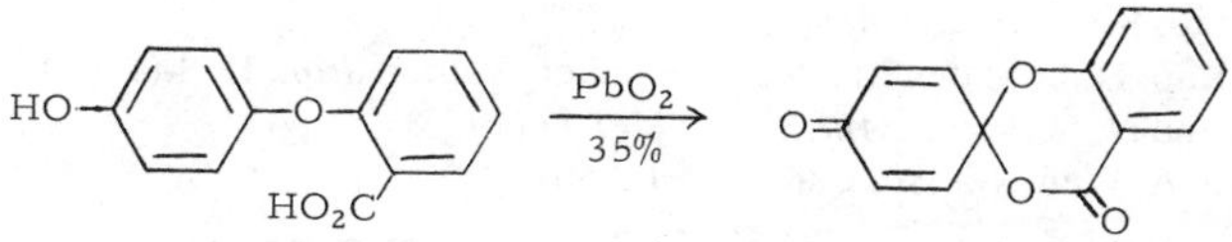
HO
O
HO2C
PbO2
35%
O
O
O
O

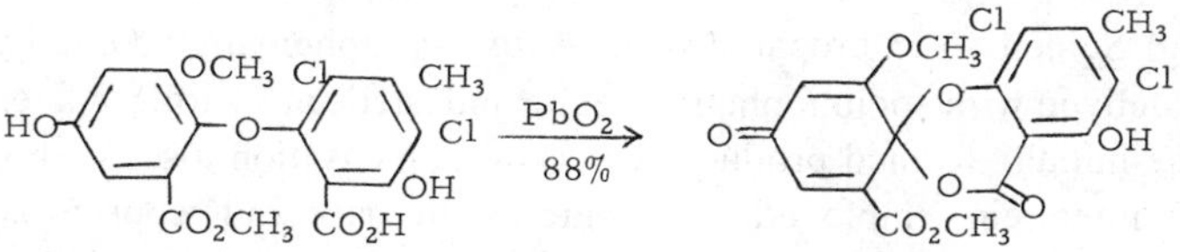
OCH3
Cl
CH3
HO
O
Cl
OH
CO2CH3
CO2H
PbO2
88%
Cl
CH3
OCH3
Cl
O
OH
O
CO2CH3

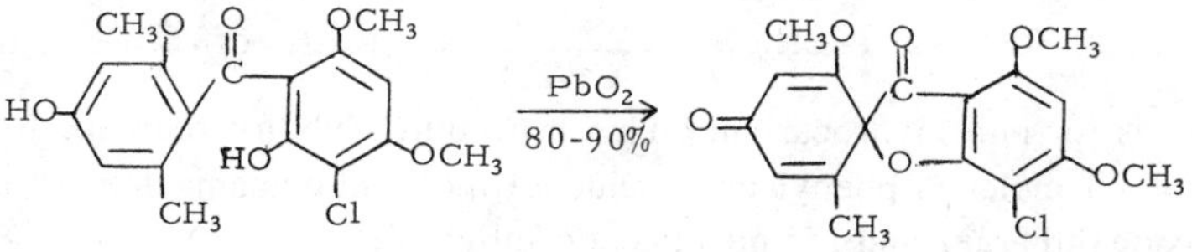
CH3O
O
OCH3
HO
C
HO
OCH3
CH3
Cl
PbO2
80-90%
CH3O
O
OCH3
O
C
O
OCH3
CH3
Cl

and Lewis.[11] With ferricyanide as oxidant the products were obtained in only very low yield. The third reaction, reported by Taub, Wendler, *et al.*[12] of Merck, is a high-yield oxidative ring closure of a *p*-hydroxybenzophenone to dehydrogriseofulvin. In this case oxidation by shaking an ether-acetone solution with Kuhn-Hammer lead dioxide was preferred to active manganese dioxide. With silver oxide the yield was only 5–10%.

Wilmarth and Schwartz[13] describe a form of lead dioxide which they regard as more active than that of Kuhn and Hammer. It is prepared (footnote 9 of the paper cited) by adding water to a solution of lead tetraacetate in acetic acid and chloroform and washing the precipitated lead dioxide with water and then with the solvent with which it is to be used. It has been used for oxidation of 1,1,4,4-tetraphenyl-2,3-dibenzoyltetrazane (1) to the hydrazyl radical (2) and for the preparation of the very unstable dihydrazyl radical (4).[14]

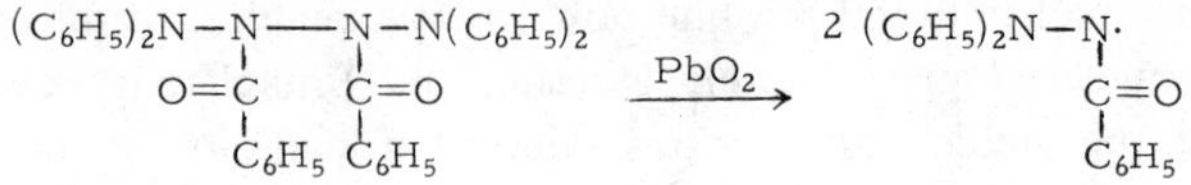

(1) (2)

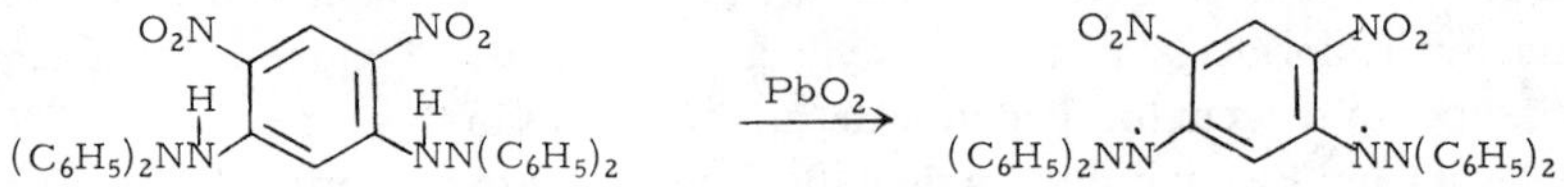

(3) (4)

[1]C. Graebe and H. Kraft, *Ber.*, **39**, 799 (1906)
[2]D. Todd and A. E. Martell, *Org. Syn.*, **40**, 28 (1960)
[3]R. Willstätter and J. Parnas, *Ber.*, **40**, 1406 (1907)
[4]R. Kuhn and I. Hammer, *Ber.*, **83**, 413 (1950)
[5]R. Adams and R. A. Wankel, *Am. Soc.*, **73**, 2219 (1951)
[6]F. Wrede and E. Strack, *Ber.*, **62**, 2051 (1929); *Z. physiol.*, **181**, 74 (1929)
[7]A. R. Surrey, *Org. Syn., Coll. Vol.*, **3**, 753 (1955)
[8]W. von E. Doering, M. Farber, and A. Sayigh, *Am. Soc.*, **74**, 4370 (1952)
[9]W. von E. Doering and M. Finkelstein, *J. Org.*, **23**, 141 (1958)
[10]S. M. McElvain and E. J. Eisenbraun, *Am. Soc.*, **77**, 1599 (1955)
[11]C. H. Hassall and J. R. Lewis, *J. Chem. Soc.*, 2312 (1961)
[12]D. Taub, C. H. Kuo, H. L. Slates, and N. L. Wendler, *Tetrahedron*, **19**, 1 (1963)
[13]W. K. Wilmarth and N. Schwartz, *Am. Soc.*, **77**, 4543 (1955)
[14]J. Heidberg and J. A. Weil, *Am. Soc.*, **86**, 5173 (1964)

Lead oxide (yellow), PbO. Mol. wt. 223.21. Suppliers: B, F, MCB.

Thiele and Schleussner[1] prepared *trans,trans*-1,4-diphenylbutadiene by reaction of cinnamaldehyde with sodium phenylacetate and acetic anhydride at a temperature such that the initially formed product of Perkin condensation loses carbon dioxide. Kuhn and Winterstein[2] employed lead oxide as the base and reported an increase in crude yield from 20 to 34%. The lower yield of 23–25% reported in *Organic*

$$C_6H_5CH{=}CHCHO + \underset{}{\overset{CO_2H}{|}}\!\!\!CH_2C_6H_5 \xrightarrow[23\text{-}25\%]{Ac_2O,\ PbO} C_6H_5CH{=}CHCH{=}CHC_6H_5$$

Syntheses is for pure hydrocarbon.[3] This procedure calls for refluxing for 5 hrs. a mixture of 1.1 moles of phenylacetic acid, 1.1 moles of cinnamaldehyde, 0.55 mole of lead oxide (litharge), and 155 ml. of acetic anhydride.

Lead oxide is used in an early procedure[4] for the conversion of an amino acid hydrochloride into the free amine. A solution of alanine hydrochloride in water is

$$CH_3CHO \xrightarrow{NH_4Cl\ +\ NaCN} \underset{NH_2}{CH_3CHCN} \xrightarrow{HCl} \underset{+NH_3Cl^-}{CH_3CHCO_2H} \xrightarrow[52\text{-}60\%\ \text{overall}]{PbO} \underset{NH_2}{CH_3CHCO_2H}$$

boiled with lead oxide for 1 hr. and the mixture is cooled, filtered from lead chloride, boiled with a fresh batch of lead oxide, cooled, and filtered again. Then the chloride ion content is reduced further with lead hydroxide precipitated from lead acetate solution. Finally the chloride ion content is determined by titration and an equivalent amount of silver nitrate is added.

For other methods of liberating a neutral amino acid from its hydrochloride, *see* Lead acetate; Ion-exchange resins; Pyridine; Silver carbonate.

[1]J. Thiele and K. Schleussner, *Ann.*, **306**, 198 (1899)
[2]R. Kuhn and A. Winterstein, *Helv.*, **11**, 103 (1928)

[3]B. B. Corson, *Org. Syn.*, *Coll. Vol.*, **2**, 229 (1943)
[4]E. C. Kendall and B. F. McKenzie, *ibid.*, **1**, 21 (1941)

Lead tetraacetate. $Pb(OCOCH_3)_4$. Mol. wt. 443.39. Suppliers: Arapahoe; Matheson, Coleman and Bell; G. Frederick Smith Co. Commercial material contains 96–97% of small crystals of lead tetraacetate moistened with acetic acid and a trace of acetic anhydride to prevent hydrolysis. The paste is best stored at 5°. To obtain dry material, scrape a batch of paste onto a suction funnel, wash it with acetic acid to form an even cake, suck the cake thoroughly, and let it dry at room temperature in a dark place in the hood. Usually the presence of acetic acid in the reaction mixture is not objectionable and in this case the paste can be used as received after iodimetric analysis. A weighed sample of about 0.5 g. is dissolved in 5 ml. of acetic acid by gentle warming, and a solution of 12 g. of anhydrous sodium acetate and 1 g. of potassium iodide in 100 ml. of water is added. After swirling for several minutes the flask wall is rinsed with water, and the liberated iodine is titrated with 0.1 *N* sodium thiosulfate to a starch end point.

$$\% \ Pb(OAc)_4 = 2.217 \times (\text{ml. thiosulfate/wt. of sample})$$

After a reaction, excess lead tetraacetate can be destroyed with ethylene glycol prior to addition of water; otherwise brown lead dioxide precipitates.

Preparation.[1,2] A mixture of 600 ml. of acetic acid and 400 ml. of acetic anhydride in a wide-mouthed or three-necked flask is heated to 55° with mechanical

$$Pb_3O_4 + 8\ CH_3CO_2H \longrightarrow Pb(OCOCH_3)_4 + 2\,Pb(OCOCH_3)_2 + 4\,H_2O$$

stirring, and 700 g. (1.03 mole) of red lead oxide (Pb_3O_4) is added in portions of 15–20 g. A fresh addition is made only after the color due to the preceding portion has largely disappeared, and the temperature is kept between 55 and 80°. At the end of the reaction the thick and somewhat dark solution is cooled, and the crystalline lead tetraacetate which separates is collected and washed with acetic acid. The crude product without being dried is dissolved in hot acetic acid, and the solution is clarified with Norit, filtered, and cooled. The colorless crystalline product is dried in a vacuum desiccator over potassium hydroxide in the dark and stored in the desiccator; yield 320–350 g. (70–77%).

Reviews.[3]

Hydroquinones ⟶ quinones. Introduction of lead tetraacetate as a reagent for the oxidation of organic compounds was an outcome of the classical investigations by Otto Dimroth of the *Coccus* insect pigments carminic acid, kermesic acid, and laccaic acid. The three pigments were found to be convertible by mild oxidation into labile products which at first could not be isolated but which reverted readily to the starting materials on gentle reduction. Having characterized carminic acid as

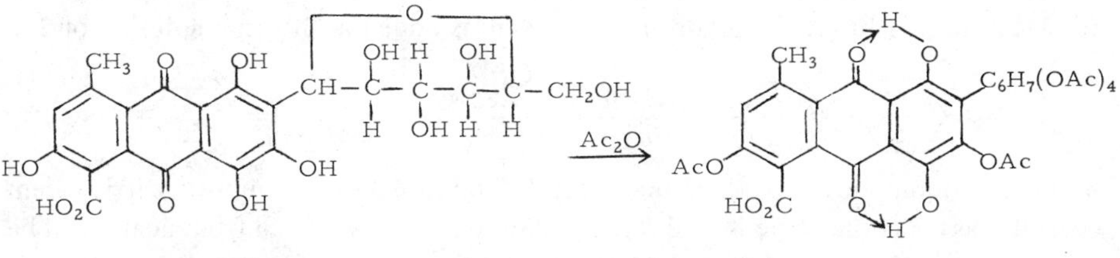
CH3
OH
CH–C–C–C–C–CH2OH
OH H OH
H OH H H
HO
OH
HO2C
OH
Ac2O
H
CH3
C6H7(OAc)4
AcO
OAc
HO2C
H

Carminic acid   Hexaacetate

an anthraquinone containing eight hydroxyl groups, Dimroth suspected that the labile oxidation product is an anthradiquinone and so explored the oxidation of polyhydroxyanthraquinones such as quinizarin and alizarin.[4] Oxidation of quinizarin with lead dioxide in benzene (method of Willstätter and Kalb[5]) afforded quinizarin-quinone in pure form. However, later irregularities in repetition and extension of the reaction led Dimroth to try oxidation with lead tetraacetate in acetic acid solution and this reagent soon became the oxidant of choice.[6,7] For preparation of quinizarin-quinone,[7] a mixture of 20 g. of quinizarin, 40 g. of lead tetraacetate, and 50 ml. of

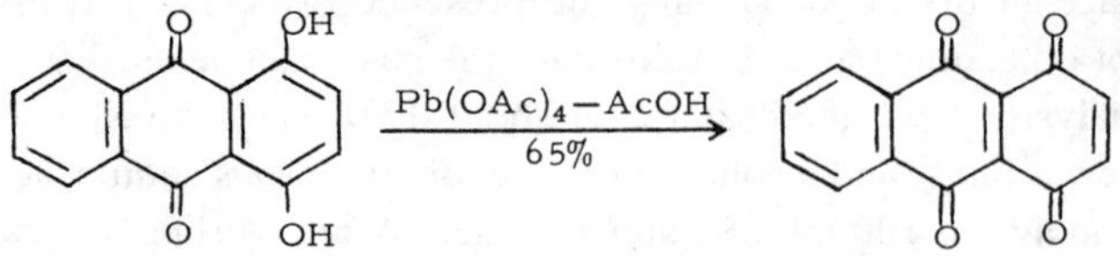
OH
O
OH
Pb(OAc)₄—AcOH
65%
O

acetic acid is ground in a mortar. In about 5 min. the mixture sets to a mass of needles of the diquinone, and after stirring for 10 min. more, or until a test portion no longer gives a red color when diluted with water, the product is collected, washed with water, and dried. For crystallization it is dissolved in 200 ml. of nitrobenzene and the solution is filtered on a prewarmed suction funnel and diluted with 2 volumes of carbon disulfide; the 1,4,9,10-diquinone separates in yellow needles.

Dimroth[6] found that carminic acid on reaction with acetic anhydride containing a trace of sulfuric acid affords the fully acetylated octaacetate, but that, when a mixture of the pigment with acetic anhydride without catalyst is heated on the steam bath until a test portion on dilution with water gives no red color, the product is a hexa-acetate (see formula above). The finding that this hexaacetate on oxidation with lead tetraacetate affords an anthradiquinone was a key step in elucidation of the structure of carminic acid. A study of model polyhydroxyanthraquinones showed that α-hydroxyl groups invariably are acylable with considerable difficulty as compared to β-hydroxyl groups. The modern interpretation of the Dimroth observations is as follows. Carminic acid on reaction with acetic anhydride alone affords only the hexaacetate because the two α-hydroxyl groups are hydrogen-bonded to the quinone carbonyl oxygen atoms. A catalytic amount of sulfuric acid protonates the carbonyl oxygens and hence liberates the α-hydroxyl groups for acetylation to give the octa-acetate. In a further investigation[8] Dimroth demonstrated the formation of 1,4,5,-8,9,10-anthratriquinone on lead tetraacetate oxidation, but the product was not isolated in pure form.

König *et al.*[9] found the preparation of the highly sensitive diphenoquinone by oxidation of 4,4′-dihydroxydiphenyl with lead dioxide[5] unsatisfactory, even when the active lead dioxide prepared according to Kuhn and Hammer by hydrolysis of lead tetraacetate was used. However, they found that the oxidation can be accomplished very satisfactorily with lead tetraacetate itself. A solution of 7.14 g. of lead tetraacetate in 140 g. of anhydrous acetic acid is added at 30° to a solution of 2 g.

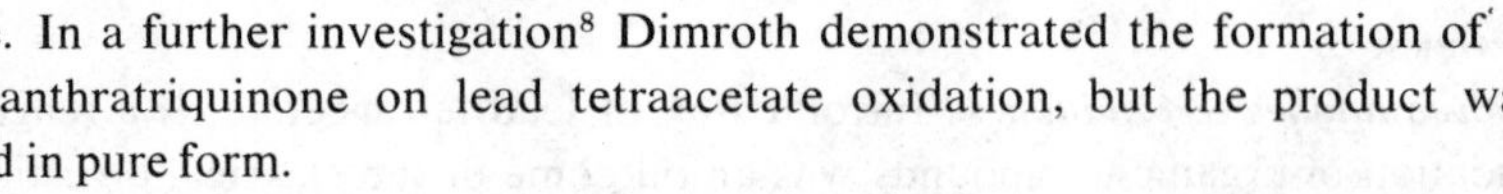
HO
OH
Pb(OAc)₄
Dioxane—AcOH
45%
O
O

of the hydroquinone in 80 ml. of dioxane in 1–2 min. In 3–5 min. more the irridescent crystal mass is cooled briefly and rapidly filtered and crystallized from acetone. The quinone forms brown-violet crystals, m.p. *ca.* 165° dec.

Diacyldiimides are intensely colored compounds, generally red in solution, and are crystalline substances stable in storage; for example dibenzoyldiimide, $C_6H_5CON=NCOC_6H_5$, melts at 119.5–121.5′. Clement[10] found that these compounds can be prepared more conveniently than by previous methods by oxidation of the corresponding diacylhydrazide with lead tetraacetate. He then tried oxidation of

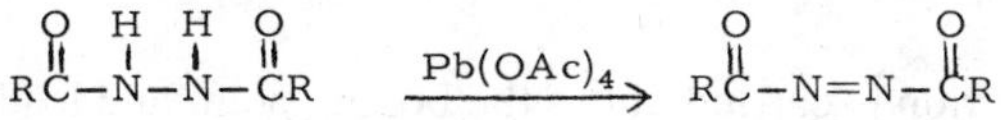

O H H O
RC—N—N—CR
Pb(OAc)4
O O
RC—N=N—CR

phthalhydrazide (1) with lead tetraacetate in acetonitrile at 0° and obtained a green solution which rapidly lost color and deposited an intractable white polymer. That the green color was due to the diazaquinone (2) was proved by carrying out the oxidation in the presence of butadiene as trapping agent and characterizing the reaction product as the adduct (3). For comparison, the cyclic dihydrazides (4) and

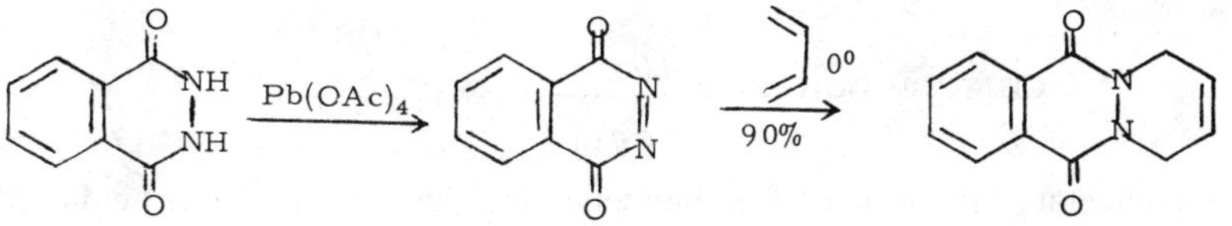

Pb(OAc)4
0°
90%

(1) (2) (3)

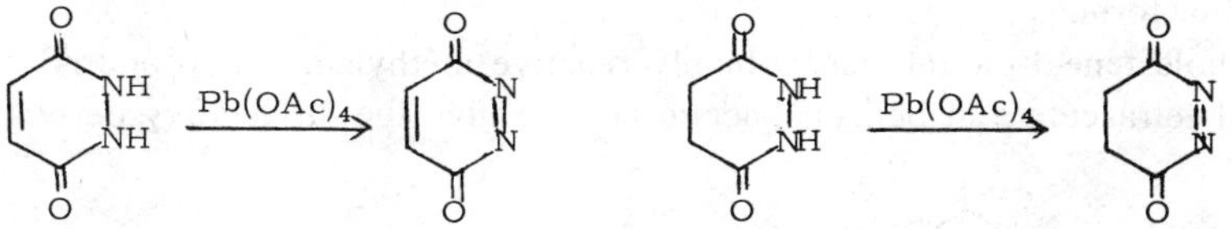

Pb(OAc)4
Pb(OAc)4

(4) (5) (6) (7)

(6) were similarly oxidized with lead tetraacetate in the presence of butadiene, and the transient existence of the cyclic diacyl diimides (5) and (7) was demonstrated by characterization of the adducts. The three cyclic diacyldiimides show extraordinary reactivity as Diels-Alder dienophiles. The butadiene adducts of (2), (5), and (7) were obtained in yields of 90, 76, and 54%, respectively. The preferred procedure is to stir a mixture of the cyclic hydrazide (10 mmoles), the diene (at least 10 mmoles) in about 50 ml. of methylene chloride containing 1 ml. of acetic acid (to suppress hydrolysis of the oxidant) at room temperature and add 1 equivalent of lead tetraacetate in small increments as fast as it is consumed. A convenient test for lead tetraacetate involves spotting solution onto a filter paper, allowing the solvent to evaporate, and moistening with water; a brown spot (lead dioxide) indicates unchanged reagent (alternatively the reaction solution is spotted on moistened starch iodide paper). Anthracene is a relatively unreactive diene, but all three cyclic diacylimides afforded anthracene adducts; the yields being (2), 71%; (5), 23%; (7), 3%. The oxidations are all rapid, and the decrease in yield is due to decreasing rate of addition in the order (2) > (5) > (7). In experiments on a 10-mmole scale the reaction mixture was treated with 7 g. of Woelm neutral alumina and the whole evaporated to dryness at reduced pressure. The resulting solid was packed on top of a chromatographic column of 50 g. of Woelm alumina deactivated by 10% of water, and the chromatogram was then developed (CCl_4; CH_2Cl_2).

α-Acetoxylation of carbonyl compounds. Early in his investigations of the use of lead tetraacetate, Dimroth[1] noticed that on boiling a solution of lead tetraacetate in acetic anhydride the oxidant is rapidly reduced. Investigation showed that the reagent effects acetoxylation of the two methyl groups to give O-acetylglycolic

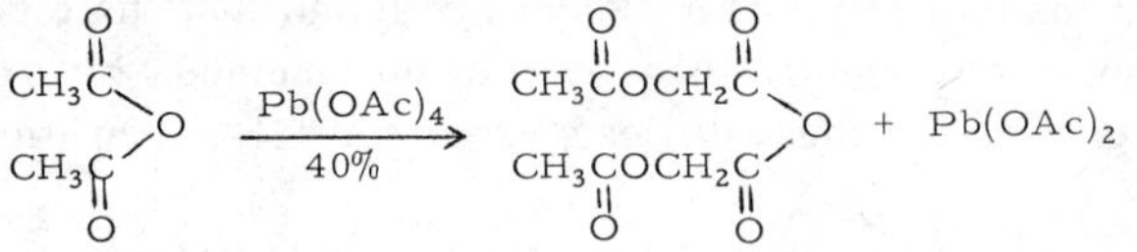

anhydride. The reaction is carried out at the boiling point and lead acetate separates as oxidation proceeds. Acetone reacts in acetic acid at the boiling point to give a mixture of the α-acetoxy and α,α'-diacetoxy compounds. The reaction is particularly favorable for preparative purposes as applied to compounds with a reactive methylene group, for oxidation proceeds at a lower temperature and side reactions are avoided. Ethyl acetoacetate reacts particularly well and in this case the preferred solvent is benzene. Lead tetraacetate (0.84 equiv.) is added in 4 portions to a solution of

$$CH_3COCH_2CO_2C_2H_5 \xrightarrow[C_6H_6]{Pb(OAc)_4} CH_3CO\underset{}{\overset{OCOCH_3}{\overset{|}{C}}}HCO_2C_2H_5$$

ethyl acetoacetate in thiophene-free benzene, and the temperature is held at 35° by cooling. Lead acetate is insoluble in benzene and separates in flocculent, then crystalline form.

Δ^5-Cholestene-3-one (1) has a highly reactive methylene group and is oxidized by lead tetraacetate in benzene–acetic acid to give the 4α-acetoxy derivative (2)

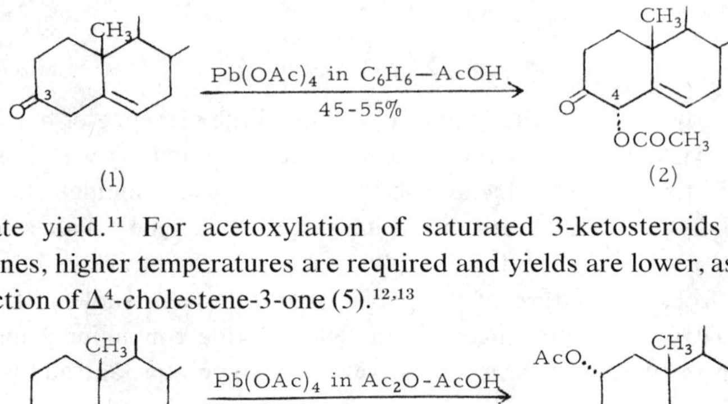

(1) (2)

in moderate yield.[11] For acetoxylation of saturated 3-ketosteroids and steroid Δ^4-ene-3-ones, higher temperatures are required and yields are lower, as exemplified by the reaction of Δ^4-cholestene-3-one (5).[12,13]

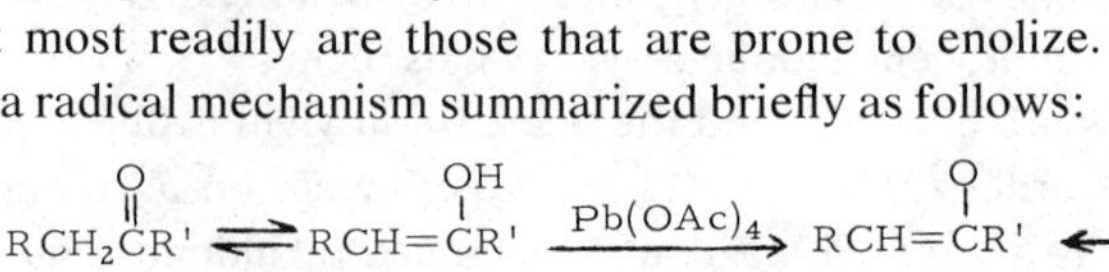

(5) (6)

The results cited are consistent with the postulate that enolization is the rate-determining step in the acetoxylation of a carbonyl compound, since the compounds that react most readily are those that are prone to enolize. Cavill and Solomon[14] proposed a radical mechanism summarized briefly as follows:

$$RCH_2\overset{O}{\overset{\|}{C}}R' \rightleftharpoons RCH{=}\overset{OH}{\overset{|}{C}}R' \xrightarrow{Pb(OAc)_4} RCH{=}\overset{O\cdot}{\overset{|}{C}}R' \longleftrightarrow R\dot{C}H{-}\overset{O}{\overset{\|}{C}}R'$$

$$\xrightarrow{OAc} R\overset{OAc}{\overset{|}{C}}H{-}\overset{O}{\overset{\|}{C}}R'$$

Henbest *et al.*[15] concede that by-products arising from dehydrogenative coupling at high temperatures are indicative of a competing radical reaction, but suggest that the normal pathway of acetoxylation is conversion of the enol (7) into the ester (8) and intramolecular donation of an acetoxy group to the adjacent carbon to give (10). We suggest participation of the transition state (9). With the idea that a Lewis acid

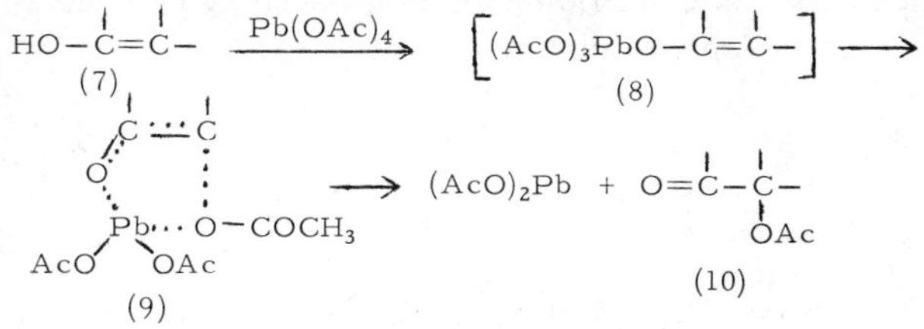

should speed up the enolization of a ketone and hence promote its acetoxylation, Henbest tried reaction of cholestane-3-one (I) with lead tetraacetate in acetic acid

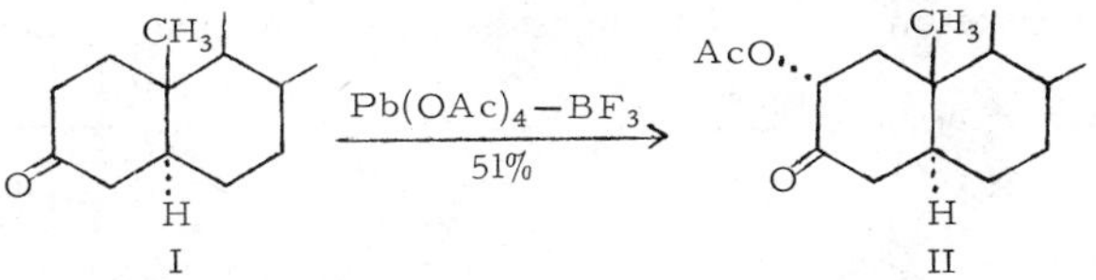

containing boron trifluoride etherate under nitrogen and found that the reaction is complete at 25° in 145 min. Best yields in the 21-acetoxylation of 20-ketosteroids were obtained with 5% methanol in benzene as solvent. Acetoxylation of a 11-keto-pregnane was slower, but at 50° the 9α-acetoxy-11-ketone was obtained in 15% yield.

A half-ester of malonic acid such as III, prepared in 70% yield by partial saponification of the di-ester, is oxidized by lead tetraacetate in benzene at 50° to the α-acetoxy derivative V, which loses carbon dioxide at 200° to give the α-acetoxy ester VI.[16] Yields are in the range 35–80%. That the half-esters are oxidized much

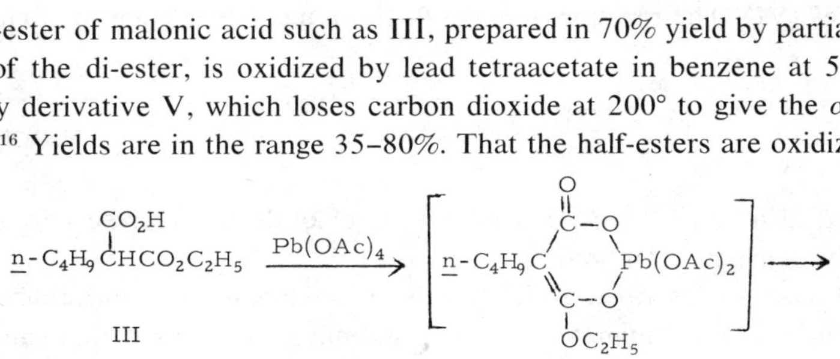

more readily than neutral malonates is attributed to formation of the cyclic intermediate IV, which promotes enolization.

Oxidation of a β,γ-unsaturated acid of type (VII) with lead tetraacetate gives either of the allylic acetoxy derivatives (VIII) or (IX), or a mixture.[17]

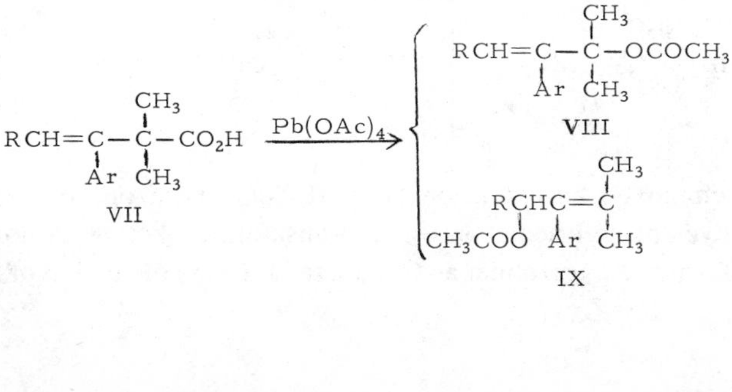

Although acetoxylation of ketones with lead tetraacetate and with mercuric acetate generally afford the same product, Zalkow[18] reports that pulegone reacts with the former reagent to give an epimeric mixture of 4-acetoxy derivatives and with the latter to give a 2-acetoxy epimer mixture.

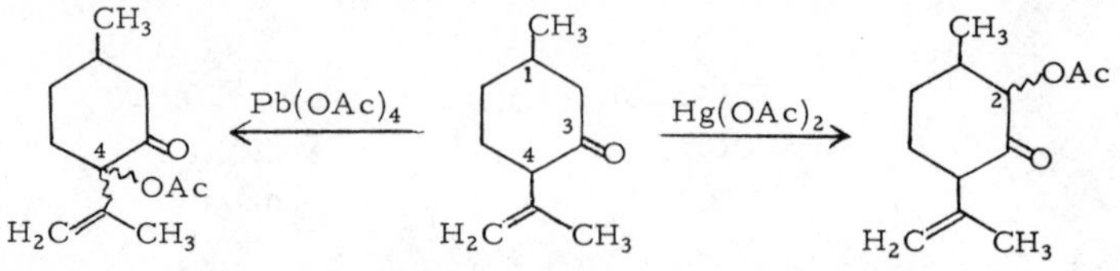

Related acetoxylations. Iffland[19] found that lead tetraacetate oxidizes ketoximes (I) smoothly to nitrosoacetates (III). The Henbest mechanism cited above for acetoxylation of a ketone (7 → 10) suggests the intermediate II and the transition

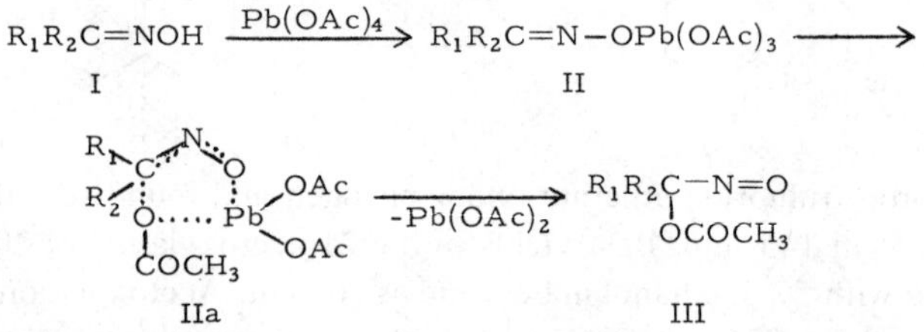

state IIa. A similar mechanism accounts for the oxidation of ketohydrazones (IV) to azoacetates (VI). The reaction is run at 0–10° in methylene chloride. Benzene and

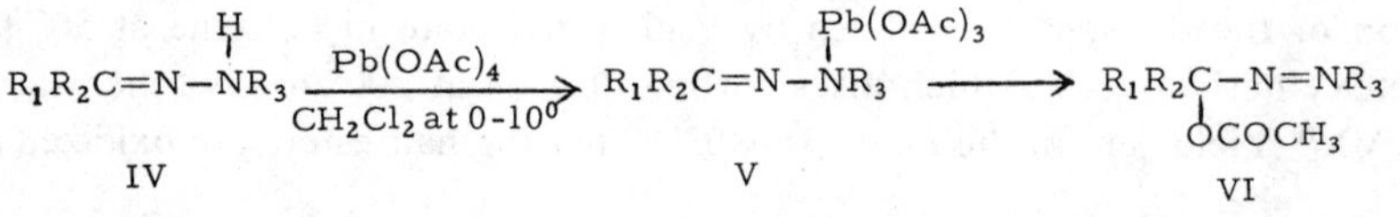

acetic acid also can be used, but methylene chloride is preferred because it has higher solvent capacity for both reactants.

A third reaction described by Iffland[20] is the oxidation of a semicarbazone (1) to a carbamate (4) with elimination of elemental nitrogen. The reaction can be interpreted as involving initial formation of (2) via a tetravalent lead ester and donation of an acetoxyl group, as in the formation of VI above. Elimination of nitrogen gives the ionic species (3), which rearranges by a 1,2-shift to the carbamate (4).

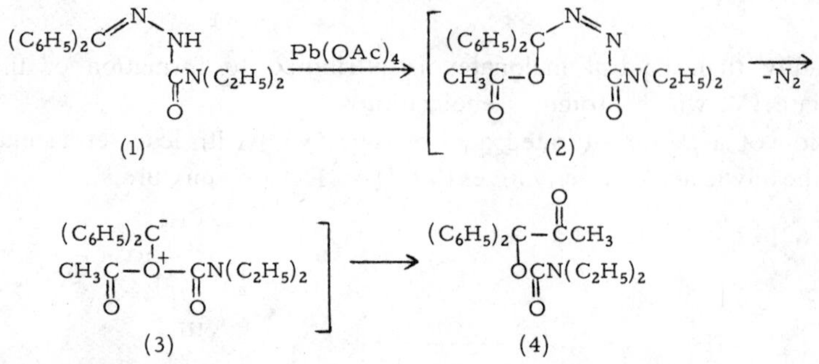

Freeman[21] employed an extension of the Iffland hydrazone oxidation in a new synthesis of cyclopropyl acetates. An α,β-unsaturated ketone is condensed with hydrazine to form a 2-pyrazoline, as formulated for the conversion of mesityl oxide

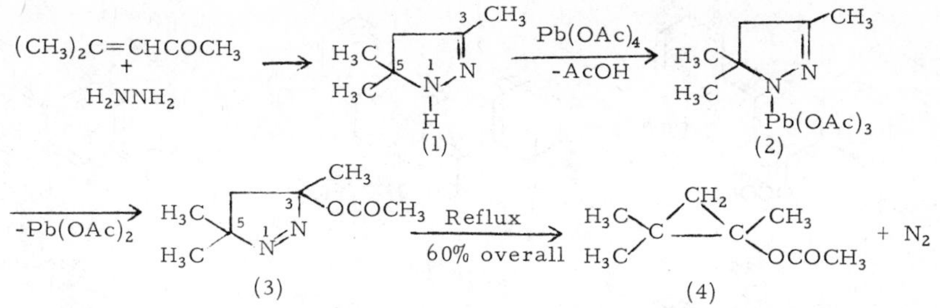

into 3,5,5-trimethyl-2-pyrazoline (1). The crude material is taken up in methylene chloride and a solution of lead tetraacetate in the same solvent is added at 10–15° in the course of 15 min. The formation of 3,5,5-trimethyl-3-acetoxy-1-pyrazoline (3) in this oxidation is pictured as proceeding through the lead ester (2), as in the Iffland oxidations. Decomposition of (3) with loss of nitrogen is effected by merely refluxing crude (3) until gas evolution stops. Twelve examples of the synthesis are reported.

Acetoxylation of hydrocarbons. In a paper of 1923,[1] Dimroth reported preliminary observations on the oxidation of aromatic hydrocarbons and olefins with lead tetraacetate. Toluene, he found, affords benzyl acetate in very low yield; oxidation of diphenylmethane and triphenylmethane proceeded more readily but offered nothing of preparative promise. Dimroth observed also that anethole reacts to give in small yield a product of addition of two acetoxyl groups to the olefin linkage.

Rudolf Criegee worked for the doctorate with Dimroth at Würzburg on the mechanism of diazo coupling, but the results were initially recorded only in his dissertation (1925). He then undertook a comprehensive investigation of the reaction of lead tetraacetate with olefins and aromatic hydrocarbons and in 1930 at the age of 28 presented the results in his first publication, a 40-page *Annalen* paper.[22] The results were hardly commensurate with the effort and experimental skill evident from the paper. Cyclohexene reacted in part by addition of two acetoxyl groups, as noted by Dimroth, to give a mixture of *cis* and *trans* products (2), and in part by

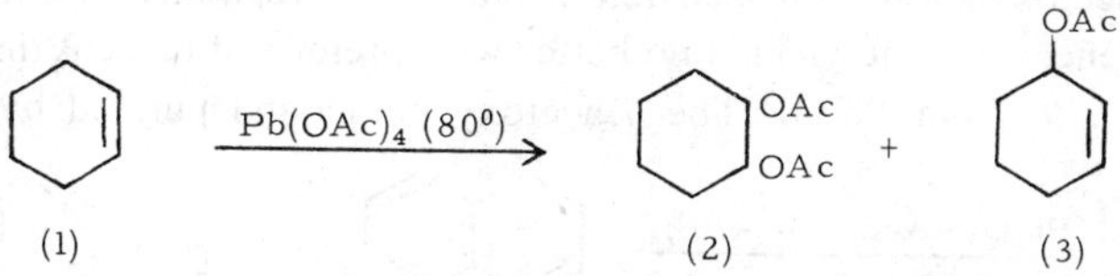

allylic acetoxylation (3), but the yields were very low and considerable high-boiling material remained unidentified. Indene reacted similarly. Cyclopentadiene and 2,3-dimethylbutadiene gave chiefly small amounts of 1,2-diacetoxy derivatives.

In 1911, over a decade before Dimroth's introduction of lead tetraacetate as an organic reagent, K. H. Meyer[23] found that addition of lead dioxide to a solution of anthracene in acetic acid results in formation of 9-acetoxyanthracene in 40–50% yield and a small amount of oxanthrone acetate. The observation attracted no attention until 1938, when Hershberg[24] found that certain polynuclear hydrocarbons on oxidation with lead tetraacetate in acetic acid afford acetoxy derivatives as the sole products, in some cases with surprising ease. Oxidized in acetic acid at steam-bath temperature, 1,2-benzanthracene gives the pure 10-acetoxy derivative (1) in moderate yield, while 10-methyl-1,2-benzanthracene reacts less cleanly to give the 10-acetoxymethyl compound (2) in low yield. 1,2,5,6-Dibenzanthracene (3) is not attacked, and

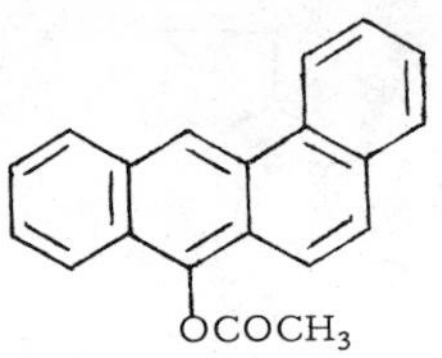

(1) 53%

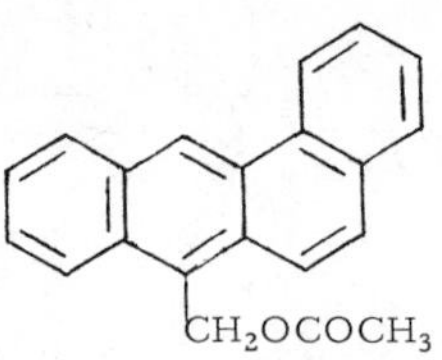

(2) 17%

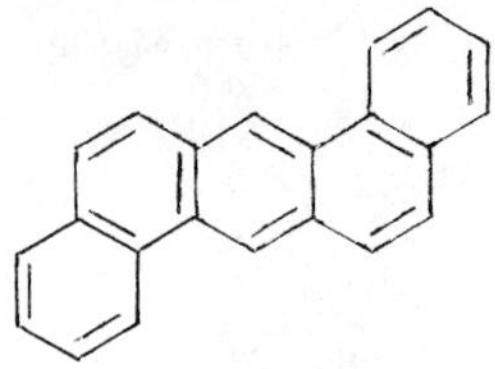

(3) No reaction

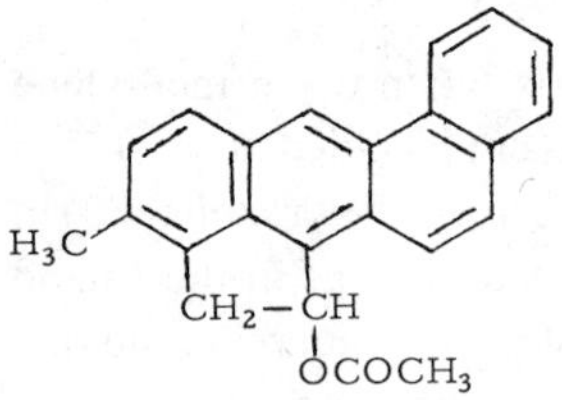

(4) 46%

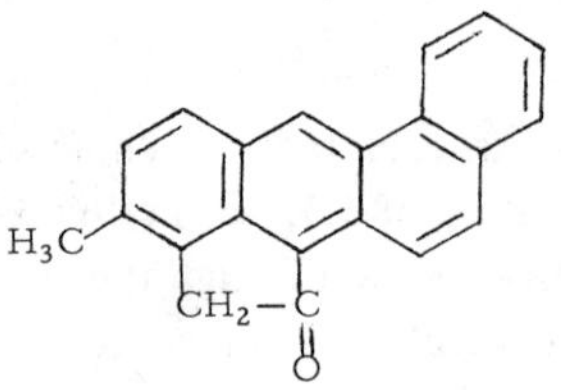

(5) 7%

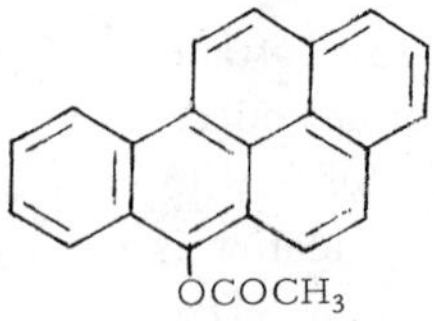

(6) 85%

the yellow impurity accompanying hydrocarbon prepared by the Elbs reaction can be removed effectively by oxidation with lead tetraacetate. Methylcholanthrene was oxidized by slow dropwise addition of a solution of lead tetraacetate in acetic acid to a stirred solution of the hydrocarbon in benzene with ice cooling. The chief product was 15-acetoxy-20-methylcholanthrene (4), a product of benzylic oxidation analogous to (2). A small amount of the 15-ketone (5) is formed, probably through the 15,15-diacetoxy compound. 3,4-Benzpyrene was oxidized by mixing solutions of the oxidant in acetic acid and of the hydrocarbon in benzene at room temperature; the reaction was complete in half an hour, and the yield of nearly pure 5-acetoxy-3,4-benzpyrene (6) was 94%. Methylcholanthrene and 3,4-benzpyrene are among the most potent known carcinogens, but note that 1,2-benzanthracene is noncarcinogenic and that 1,2,5,6-dibenzanthracene is a carcinogen of considerable potency.

Benzylic acetoxylation with lead tetraacetate has preparative value as applied to acenaphthene,[25] and the yield is no better with preformed reagent than with that generated *in situ* from Pb_3O_4. The 7-acetoxy compound purified by distillation

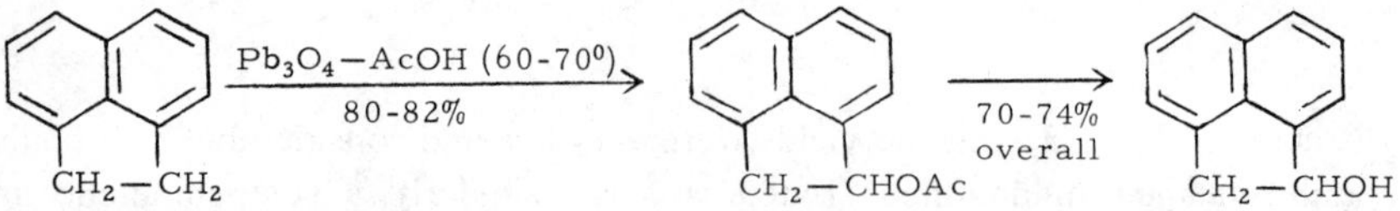
Pb3O4—AcOH (60-70°)
80-82%
70-74%
overall
CH2—CH2
CH2—CHOAc
CH2—CHOH

contains traces of acenaphthene and acenaphthenone, but after saponification 7-acenaphthenol is obtained pure by crystallization.

Putnam[26] reinvestigated the oxidation of anthracene with lead tetraacetate with results of interest. Oxidation in acetic acid at 35–40° or in benzene–acetic acid at 50–55° give a small amount of a new compound characterized as a 9,10-diacetoxy-9,10-dihydroanthracene (2), m.p. 173°. Oxidation in refluxing benzene for 45 hrs. gave substantial amounts of both this compound and its geometrical isomer, m.p. 127°. When a solution of either isomer in acetic acid is warmed, 9-acetoxyanthracene (3) is produced in good yield. The acetoxylation of anthracene in acetic acid to (3) is thus the result of 1,4-addition of acetoxyl groups to the diene system and 1,4-elimination of acetic acid. The isomeric addition products (2) are reasonably stable

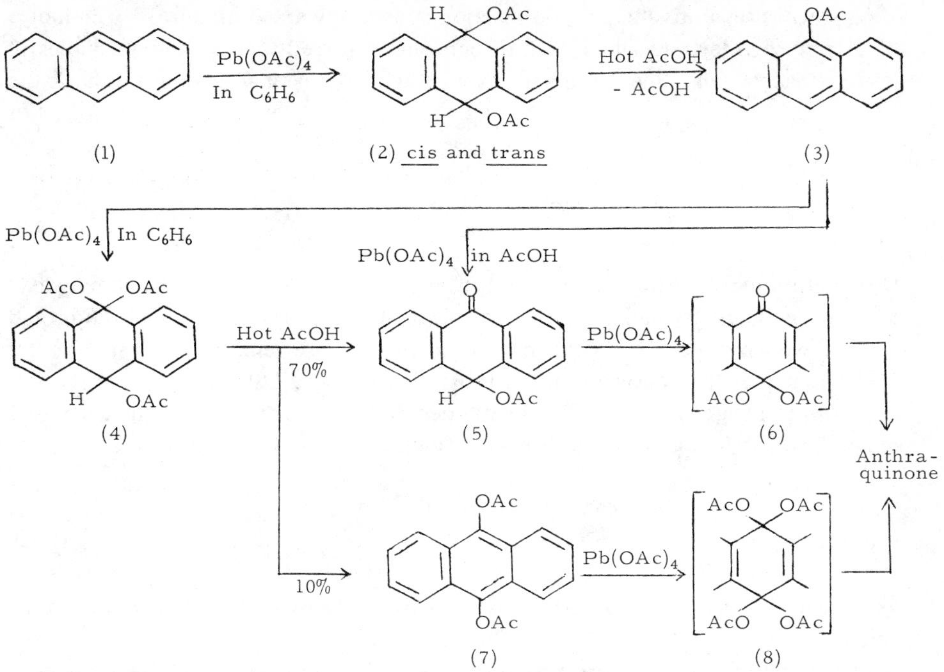

in hot benzene, but are sensitive to acetic acid. The acetoxylation of 3,4-benzpyrene evidently is a direct substitution, since this hydrocarbon has no additive power, for example for maleic anhydride.

The second product identified by Meyer, "oxanthrone acetate" (5, better 10-acetoxy-9-anthrone), was obtained in moderate amount by oxidation of 9-acetoxyanthracene (3) with lead tetraacetate in acetic acid. Oxidation of (3) in refluxing benzene resulted in 1,4-addition to give the triacetoxy compound (4). This substance when heated in acetic acid is converted largely into 10-acetoxy-9-anthrone (5) by loss of acetic anhydride and to a lesser extent into 9,10-diacetoxyanthracene (7) by loss of acetic acid. If (4) is an intermediate in the oxidation of (3) in acetic acid to (5), the acetoxy group in the product (5) must be attached to a different *meso* carbon atom (5) than in (3), and this inference was shown to be correct by oxidation of 2-methyl-9-acetoxyanthracene and identification of the product as 2-methyl-10-acetoxy-9-anthrone by synthesis. Both (5) and (7) on further oxidation with lead tetraacetate in acetic acid yield anthraquinone, probably via the products of acetoxylation of (5) and 1,4-addition to (7).

***vic*-Glycol cleavage.** Criegee's second paper[27] (1931), also based on his own experimental work at Würzburg, described discovery of the now classical cleavage of *vic*-glycols by lead tetraacetate in acetic acid at room temperature to aldehydes, ketones, or both, according to the structure of the glycol. Criegee isolated the carbonyl components as the *p*-nitrophenylhydrazones in yields of 80–90% and showed that a *vic*-glycol can be titrated quantitatively with 0.1 *N* lead tetraacetate in acetic acid.

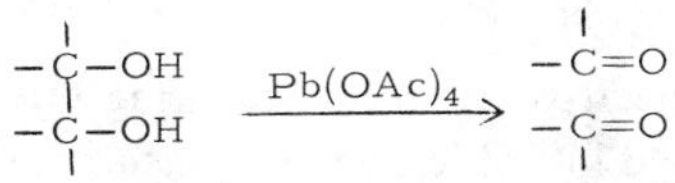

A typical preparative use of the reaction is as follows.[28] A mixture of 1.24 moles of di-*n*-butyl D-tartrate and 1.25 l. of benzene is stirred rapidly, and 1.3 moles of lead tetraacetate is added in about 25 min. with occasional cooling to check the

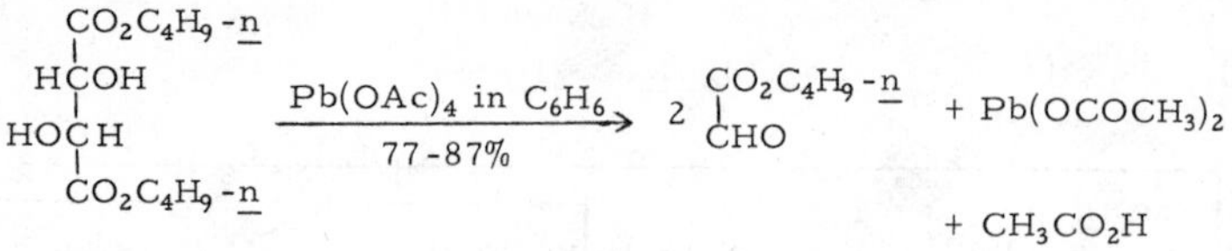

temperature at 30°. After stirring for 1 hr. more, during which time the gummy lead acetate becomes crystalline, the inorganic material is removed by filtration and washed with benzene. The benzene and acetic acid are removed by distillation at 65°/40 mm. and the *n*-butyl glyoxalate is distilled at 65–79°/20 mm.

Glycol cleavage is characterized by its specificity, speed of reaction, and generality of application. Compounds readily cleaved include those with the groups:

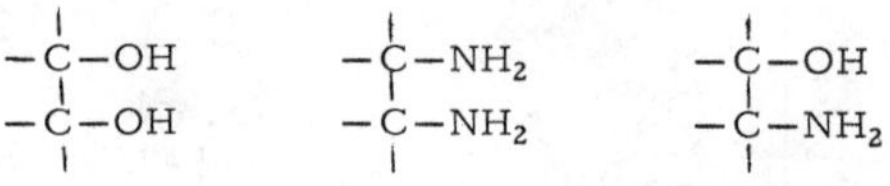

as well as α-hydroxy acids, α-hydroxy aldehydes, α-hydroxy ketones, and oxalic acid.

In the first of three comprehensive studies of the relative rates of glycol cleavage by lead tetraacetate in acetic acid,[29-31] Criegee noted that with the six pairs of *cis*-*trans* isomers or near isomers listed in Table 1 the *cis* isomer reacts much faster than the *trans*. On the assumption that the relationship is general, and that the

Table 1 Rates of glycol cleavage (k^{20})

	cis	*trans*
Indane-1,2-diol	27,800	0.47
Tetralin-1,2-diol	40.2	1.86
Cyclohexane-1,2-diol	5.0	0.22
D-Mannonic acid γ-lactone	39	
D-Arabonic acid γ-lactone		0.01
α-Methyl-D-mannofuranoside	900	
Ethyl D-glucofuranoside-5,6-carbonate		0.01

hydroxyl groups are closer together in a *cis*- than in a *trans*-glycol, and noting that the reaction is bimolecular, Criegee suggested that cleavage proceeds through an intermediate cyclic ester (I). However, L. P. Kuhn[32] applied to several pairs of diols

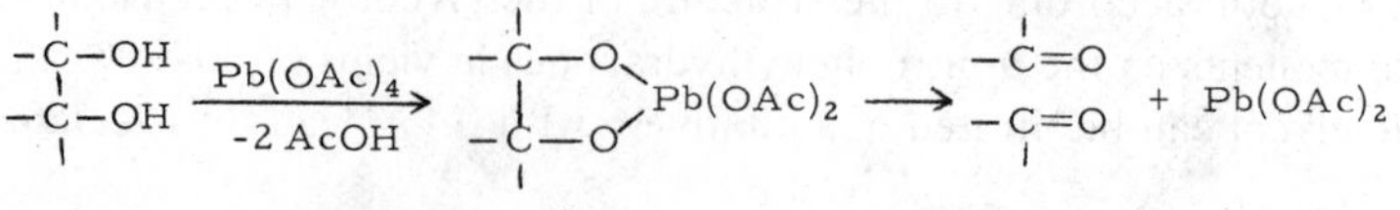

a technique of infrared spectroscopy for measuring OH:OH separation distances with the results summarized in Table 2. The measurements give values for $\Delta\nu$, the band separation distance, which is inversely proportional to the distance between the two hydroxyl groups. The rate constants for glycol cleavage are those summarized by Kuhn from the literature. The $\Delta\nu$ values show that in rings containing less than 10 carbon atoms the *cis* hydroxyls are closer than the *trans*, but that the situation is

Table 2 IR band separation and rates of $Pb(OAc)_4$ cleavage

Cyclic 1,2-diol	$\Delta\nu$, cm^{-1}		k^{20}	
	cis	*trans*	*cis*	*trans*
C_5-*cis*	61		40,000	
C_5-*trans*		0		12.8
C_6-*cis*	38		5	
C_6-*trans*		33		0.2
C_7-*cis*	44			
C_7-*trans*		37		
C_8-*cis*	51			
C_8-*trans*		43		
C_9-*cis*	49		2.9	
C_9-*trans*		45		20.7
C_{10}-*cis*	44		2.6	
C_{10}-*trans*		45		100
C_{12}-*cis*	38		1.3	
C_{12}-*trans*		51		73.6
C_{16}-*cis*	—		7.8	
C_{16}-*trans*		50		91.2

reversed in rings containing 10 or more carbons. There is a reversal also in the order of relative reactivity, but this occurs at C_9 rather than at C_{10}. Comparison of the data shows that there is no simple correlation between reaction rates and $\Delta\nu$, and this indicates that the distance between the OH groups is not the main factor that determines reaction rate. A later paper by Criegee[31] reports rate constants for a series of bicyclic pinacols as well as $\Delta\nu$ values determined by Kuhn. Again, a rough

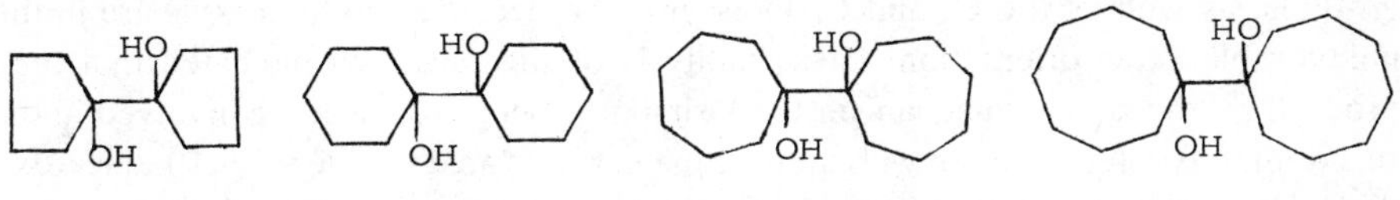
HO
OH
HO
OH
HO
OH
HO
OH

k^{20}	0.04	11.8	2,390	3,000
$\Delta\nu$	36	46	60	65

parallelism is evident. However, a glaring discrepancy emerged from a study of the isomeric 9,10-dimethyldihydrophenanthrene-9,10-diols. In 1940, when only the 164° isomer derived from phenanthrenequinone was known, the substance was thought to be the *cis*-isomer because of the high rate of cleavage, k^{20} 192.[30] However,

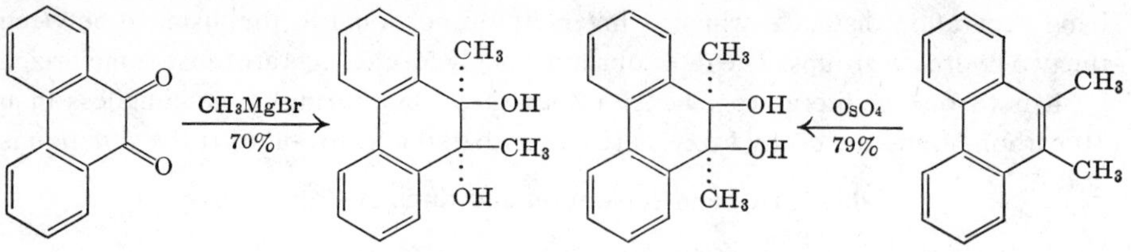
CH₃MgBr
70%
CH₃
OH
CH₃
OH
CH₃
OH
OH
CH₃
OsO₄
79%
CH₃
CH₃

M.p. 164° M.p. 104°
k²⁰ 192, Δν 0 k²⁰ 7.5, Δν 27cm⁻¹

the finding that dipotassium tetramethylosmate, a sensitive reagent developed (1942) for the characterization of *cis*-diols,[33] does not react with the fast-acting diol prompted investigation of the osmylation of 9,10-dimethylphenanthrene.[31] The synthesis gave a second isomer (104°) and this reacted at a lower rate (k^{20} 7.5) even though, from the method of preparation, it must be the *cis*-diol. Criegee thus established beyond question that the fast-acting dimethyldihydrophenanthrenediol is *trans* and the slow-acting one *cis*. The rate relationship is the same with the two 9,10-dihydrophenanthrene-9,10-diols; Booth and Boyland[34] proved the isomer obtained by reduction of phenanthrenequinone to be *trans* by resolving it, and found that the other isomer alone forms an acetonide.

The results cited invalidated the postulate that glycol cleavage proceeds through a cyclic ester, since such an ester could hardly be formed from a *trans*-dihydrophenanthrenediol, in which the hydroxyl groups are held rigidly in the diaxial, antiparallel orientation. Furthermore, the mechanism gives no account of a striking basic catalysis by water, methanol, or potassium acetate.[35] Thus the rate constant for *cis*-dimethyldihydrophenanthrenediol rises from 7.5 in acetic acid to 202 in 75:25 AcOH–CH_3OH and to 2430 in 25:75 AcOH–CH_3OH. Water is more effective than methanol, but the proportion of water cannot be increased beyond 15% without hydrolysis of the reagent. A concerted mechanism[32] involving participation of a Lewis base (HO^-, CH_3O^-) in formation of transition state II accounts for both the catalytic effect and for the rapid cleavage of the *trans*-dihydrophenanthrenediols,

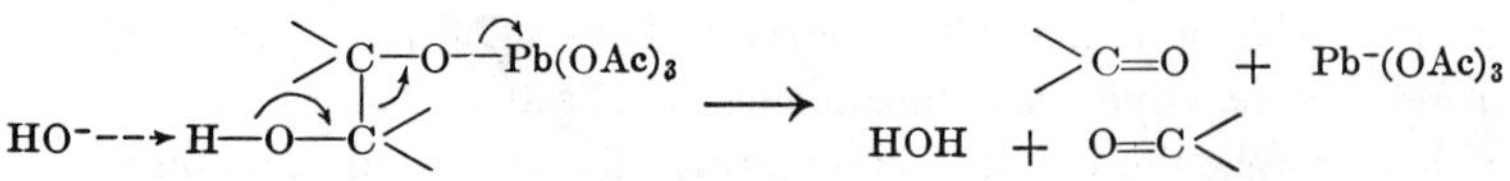
C—O—Pb(OAc)₃
HO⁻---→ H—O—C
C=O + Pb⁻(OAc)₃
HOH + O=C

II

since the diaxial antiparallel hydroxyls are coplanar and hence ideally oriented for formation and collapse of the transition state. The much slower cleavage of isomeric *cis*-diols, as well as the C_5 and C_6 diols, may be because the hydroxyls are in the unfavorable skew orientation; basic catalysis of the reactions indicated that they proceed by the acyclic mechanism II. A similar process seems to be involved in the reaction of lead tetraacetate with anthracene in acetic acid to give the 9,10-diacetoxy derivative, since the reaction is markedly catalyzed by water, methanol, or potassium acetate.[36] The addition of only 0.6% of water shortens the time required for consumption of 0.5 mole of reagent at 25° from 40 to 14 hrs.

Bell[37] found that glycol cleavage by lead tetraacetate is markedly catalyzed by trichloroacetic acid; the acid catalysis is accounted for by the transition state III. A practical application of acid-catalyzed glycol cleavage is reported by Grob.[38] The diaxial IV could not be cleaved satisfactorily under ordinary conditions to 1,6-

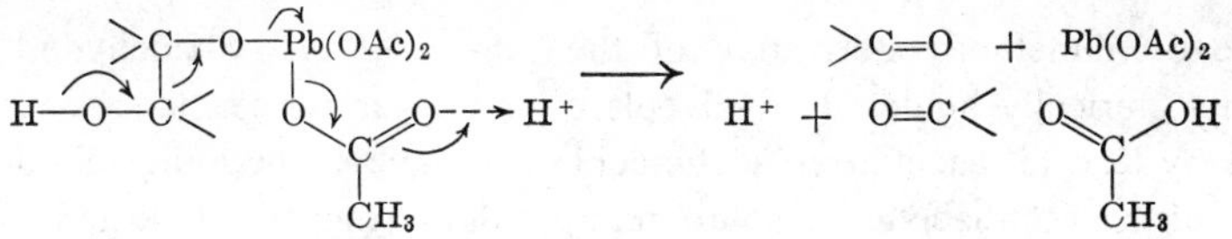

III

dioxo-$\Delta^{3,8}$-decadiene (V) but reacted in methanol in the presence of trichloroacetic acid to give the high-melting tetramethyl ketal VI in 75% yield.

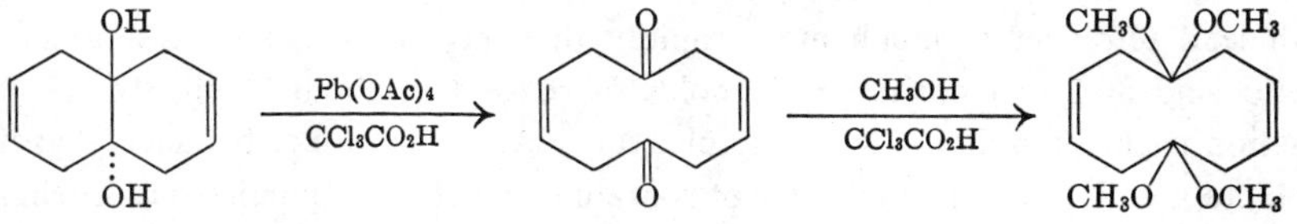

IV m.p. 83° V m.p. 185° VI m.p. 202°

The interesting case of four stereoisomeric ring B *vic*-diols derived from cholesterol has been explored by Angyal and Young.[39] Ring B, being locked between two other rings, is rigid and not free to flip. Rate constants for lead tetraacetate cleavage in

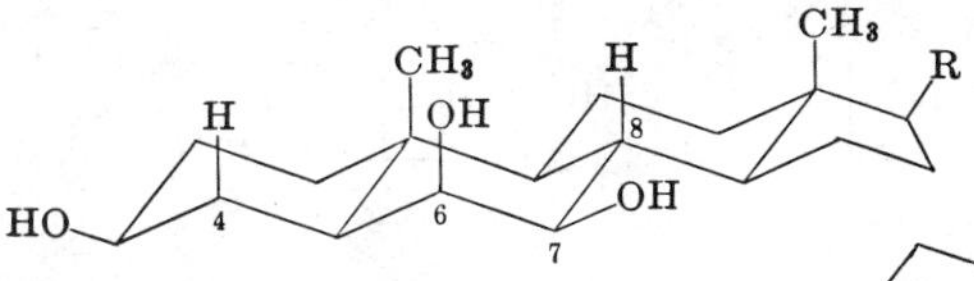

(8) $k^{25°} = 160$

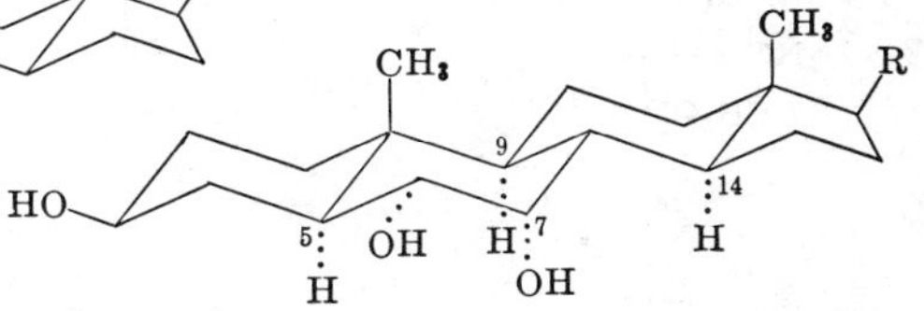

(9) $k^{25°} = 42$

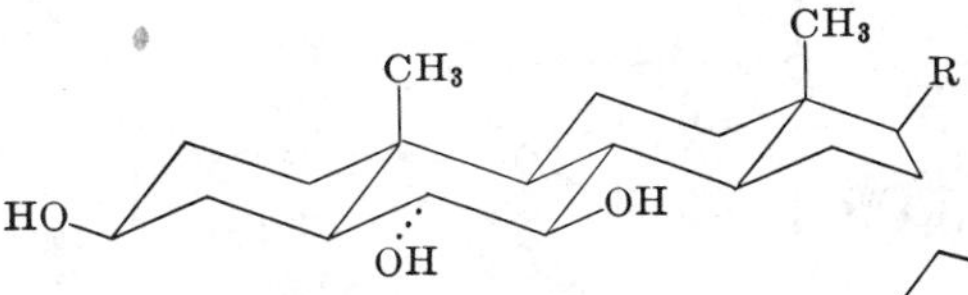

(10) $k^{25°} = 2.1$

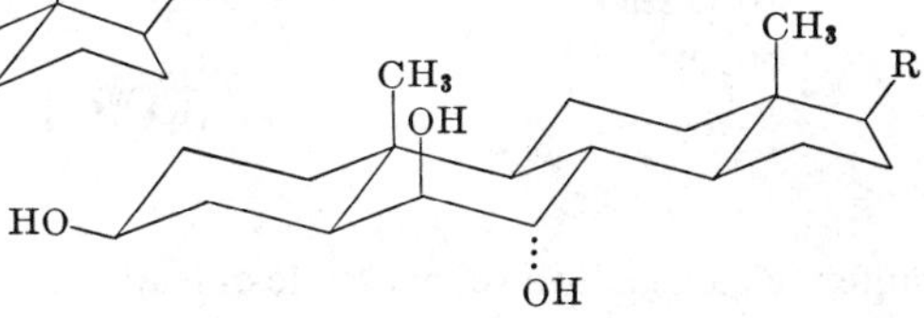

(11) $k^{25°} = 0$

acetic acid are given under the formulas. The greater reactivity of the 6β,7β-diol (8) over the 6α,7α-diol (9) may be in part because repulsion of the 6β-hydroxyl group by the axial methyl group facilitates stronger bonding with the equatorial OH and in part because cleavage of (8) relieves the steric strain imposed by the CH_3:OH interaction. The *trans*-diequatorial diol (10) is less reactive than either of the two *cis*-diols. The diaxial diol (11) resisted attack under the conditions tried, but acid catalysis was not investigated. Cholestane-3β,5α,6β-triol (13), in which the *vic*-glycol group is diaxial, is oxidized rapidly by the Grob procedure with formation of

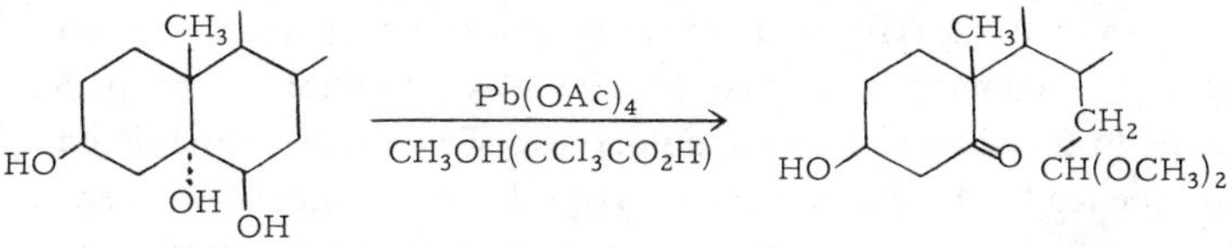

(13) (14)

(14), the 6,6-dimethoxy derivative of the 5,6-seco-5-keto-6-aldehyde.[40] Perlin[41] found that sterically hindered *vic*-glycols of the sugar series that are resistant to cleavage by lead tetraacetate in acetic acid or to aqueous periodic acid often react rapidly with lead tetraacetate in solution in pyridine. Angyal's cholestane-3β,6β,7α-triol consumes 0.8 mole of reagent in pyridine at 0° in 6 hrs.

In an exploratory experiment of 1932, Criegee[42] noted that D-glucose rapidly consumes lead tetraacetate without concurrent production of formaldehyde or formic acid. In a careful study of the reaction, Perlin and Brice[43] found that D-glucose reacts with lead tetraacetate much more rapidly than *cis*-cyclohexane-1,2-diol; in 98% acetic acid at 25° it consumes 2 moles of reagent in about 3 min. In the initial reaction D-glucopyranose (I) suffers cleavage of the α-hydroxy hemiacetal group at C_1-C_2 to give II, the 4-formyl ester of D-arabinose, which is stabilized by cyclization to the pyranose III. The formyl group may then migrate through the cyclic ester IV to give the 3-formyl ester of D-arabinose, V. The α-hydroxy hemiacetal group of V

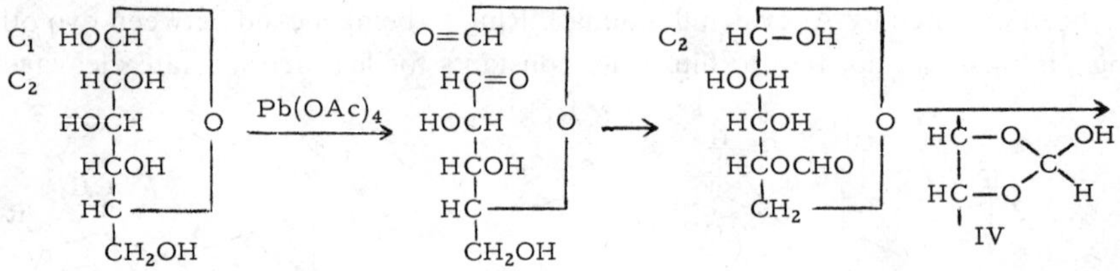
C₁ HOCH
C₂ HCOH
HOCH
HCOH
HC
CH₂OH
O
Pb(OAc)₄
O=CH
HC=O
HOCH
HCOH
HC
CH₂OH
C₂ HC—OH
HOCH
HCOH
HCOCHO
CH₂
HC—O
HC—O
OH
C
H

I II III IV

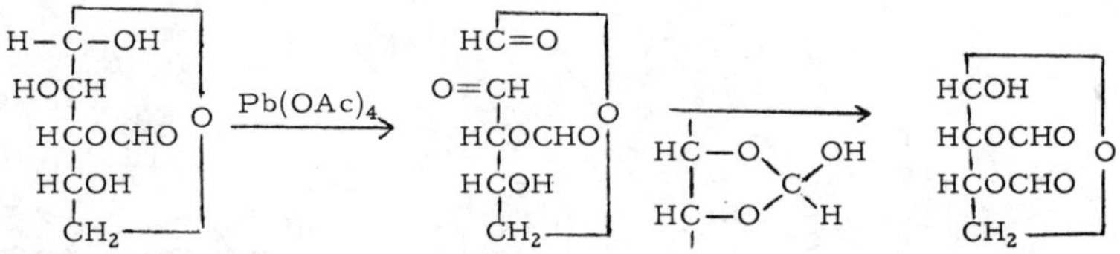
H—C—OH
HOCH
HCOCHO
HCOH
CH₂
O
Pb(OAc)₄
HC=O
O=CH
HCOCHO
HCOH
CH₂
HC—O
HC—O
OH
C
H
HCOH
HCOCHO
HCOCHO
CH₂

V VI VII

suffers cleavage by a second mole of lead tetraacetate with production of D-erythrose 2,4-diformate (VI) which, by acyl migration and cyclization, affords the end product, D-erythrose 2,3-diformate (VII). The locations assigned to the formyl groups are probable but not certain; the esters are all very labile and are hydrolyzed by water alone. Glycol cleavage by lead tetraacetate has been useful in the determination of the structures and configurations of a natural octulose[44] and a natural nonulose.[45]

Oxidation of phenols. Bamberger[46] prepared the dienone *p*-tolulquinol (2) by acid-catalyzed rearrangement of *p*-tolylhydroxylamine (1). In a modern procedure (Goodwin[47]) the hydroxylamine is added to partially frozen dilute sulfuric acid, and the mixture is shaken and allowed to come to room temperature; a complicated mixture results from which (2) is isolated by chromatography in 33% yield. A less satisfactory route to (2) is by oxidation of *p*-cresol with a peracid in a neutral medium (Bamberger[48]). Kenner[49] found that 2,6-dialkyl-4-nitrophenols are oxidized by lead tetraacetate to 2,6-dialkyl-1,4-benzoquinones. Then Wessely initiated a series of studies on the lead tetraacetate oxidation of substituted phenols to *p*- and *o*-arenones of types (3) and (8).[50] Thus *p*-cresol on oxidation affords a mixture of *p*-toluquinol acetate (3) and the diacetoxy-*o*-arenone (8). Goodwin and Witkop[47] used this method

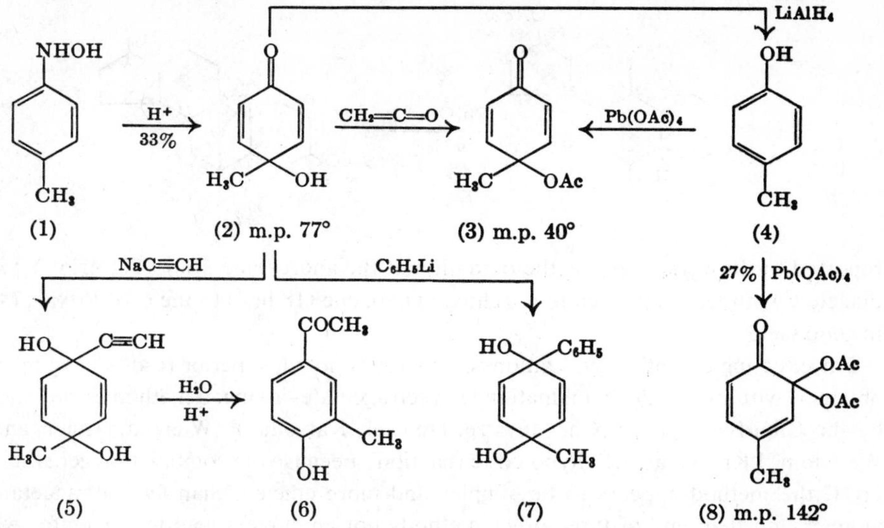
LiAlH₄
NHOH
H⁺
33%
CH₂=C=O
Pb(OAc)₄
OH
H₃C
OH
H₃C
OAc
CH₃
CH₃
(1)
(2) m.p. 77°
(3) m.p. 40°
(4)
NaC≡CH
C₆H₅Li
27%
Pb(OAc)₄
COCH₃
HO
C≡CH
HO
C₆H₅
OAc
OAc
H₂O
H⁺
CH₃
H₃C
OH
HO
CH₃
OH
CH₃
(5)
(6)
(7)
(8) m.p. 142°

to obtain the *ortho* product (8) in 27% yield but did not record the yield of the *para* product (3) isolated from the mixture. Evidently the route to (3) via (1) and (2) is the preferred preparatory route even though the yield in the first step is very low. Note the analogy of these products to those formed on lead tetraacetate oxidation of anthracene. *o*-Toluquinol probably arises as follows:

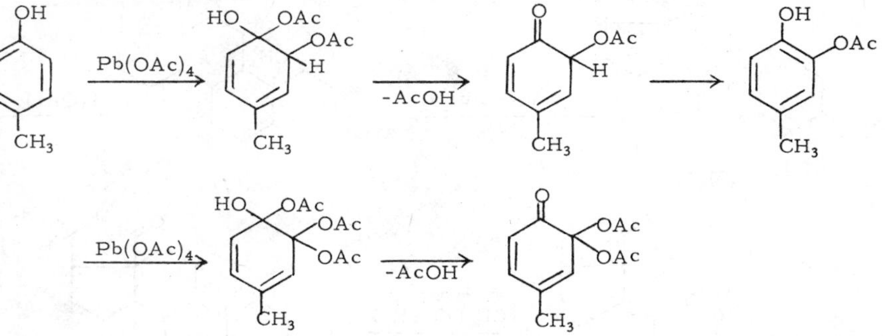
OH
HO
OAc
OAc
Pb(OAc)₄
H
-AcOH
OAc
H
OH
OAc
CH₃
CH₃
CH₃
CH₃
HO
OAc
OAc
OAc
Pb(OAc)₄
-AcOH
OAc
OAc
CH₃
CH₃

Oxidation of estrone with lead tetraacetate affords a dienone in about 20% yield.[51]

Substitution of steroid angular methyl groups. Lead tetraacetate alone has been used to oxidize a steroid alcohol having a hydroxyl group strategically located for attack of an angular methyl group, such as the 3β-acetoxy-20β-hydroxy-5α-pregnane

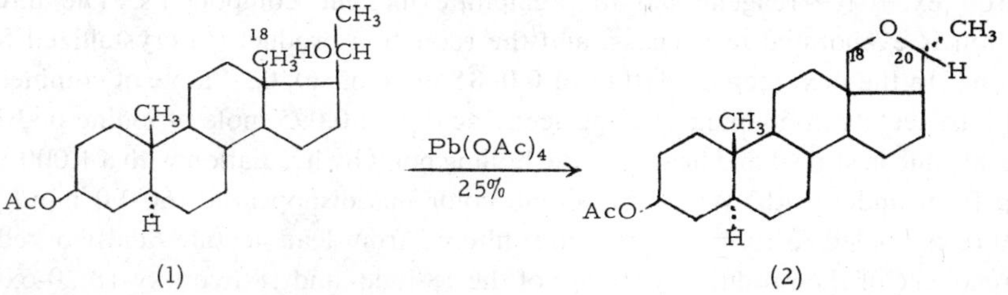
CH₃
18
CH₃HOCH
CH₃
AcO
H
Pb(OAc)₄
25%
O
CH₃
18
20
H
CH₃
AcO
H

(1) (2)

(1). Arigoni, Jeger, and co-workers[52] refluxed (1) in benzene overnight with 2 g. of freshly dried lead tetraacetate, chromatographed the reaction mixture, and isolated about 200 mg. of starting material and 250 mg. of the 18,20-oxide (2). An example

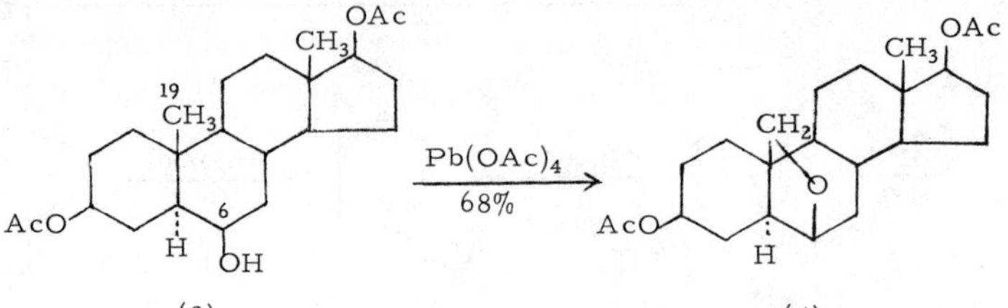
OAc
CH3
19
CH3
6
AcO
H
OH
Pb(OAc)4
68%
CH2
O
AcO
H
OAc
CH3

(3) (4)

reported by Bowers *et al.*[53] is the oxidation of the androstane-3β,6β,17β-triol-3,17-diacetate with lead tetraacetate in refluxing benzene (18 hrs.) to the 6β,19-oxide (4) in good yield.

For effecting oxidative cyclizations of the types noted, superior results have been obtained with use of the combination lead tetraacetate–iodine, a method introduced by the Ciba-Basel group: Ch. Meystre, Heusler, Kalvoda, P. Wieland, Anner, and Wettstein.[54] Known as the "hypoiodite reaction" because the combination generates HOI, the method appears to be simpler and more efficient than lead tetraacetate alone,[55] and than any of three other methods not employing lead tetraacetate. An improved procedure[54] is as follows. Pregnenolone acetate (5) is first reduced with lithium tri-*t*-butoxyaluminum hydride in tetrahydrofurane to 3β-acetoxy-20β-hydroxy-Δ^5-pregnene (6); the 20α-isomer is formed in minor amount. At the end

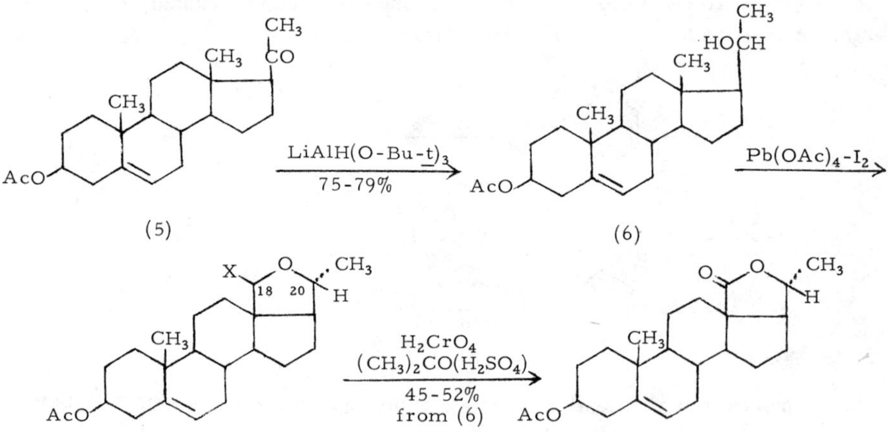
CH3
CH3
CO
CH3
AcO
LiAlH(O-Bu-t)3
75-79%
CH3
HOCH
CH3
CH3
AcO
Pb(OAc)4-I2

(5) (6)

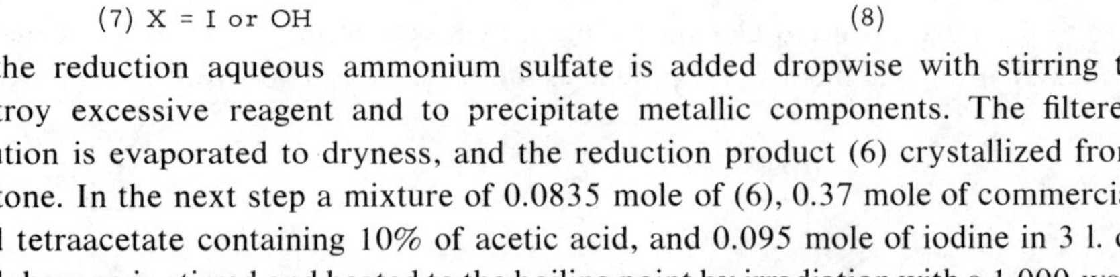
X
O
CH3
18
20
H
CH3
AcO
H2CrO4
(CH3)2CO(H2SO4)
45-52%
from (6)
AcO
O
O
CH3
H
CH3

(7) X = I or OH (8)

of the reduction aqueous ammonium sulfate is added dropwise with stirring to destroy excessive reagent and to precipitate metallic components. The filtered solution is evaporated to dryness, and the reduction product (6) crystallized from acetone. In the next step a mixture of 0.0835 mole of (6), 0.37 mole of commercial lead tetraacetate containing 10% of acetic acid, and 0.095 mole of iodine in 3 l. of cyclohexane is stirred and heated to the boiling point by irradiation with a 1,000-watt lamp from underneath. When the iodine color has disappeared (60–90 min.), the mixture is cooled to room temperature, filtered from lead acetate, and worked up for recovery of the product, a mixture of the 18-iodo- and 18-hydroxy-18,20-oxides (7). In the completing step this mixture is oxidized by the Jones procedure to 3β-acetoxy-20β-hydroxy-Δ^5-pregnene-18-oic acid 18 → 20 lactone (8). For further examples of the hypoiodite reaction see Paper VII[56] of the Ciba series.

In studying the oxidative cyclization of 20-hydroxysteroids to 18,20-oxides, Jeger and co-workers observed the occurrence of some fragmentation to give 17-acetoxy compounds.[57] They then investigated the oxidation of various monohydroxysteroids by lead tetraacetate in refluxing benzene with the following results. The 19-hydroxyketosteroid (9) is converted in high yield into the 10-acetoxy

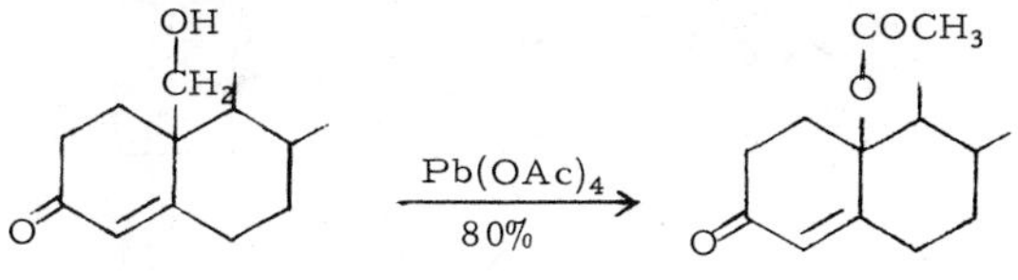

(9) (10)

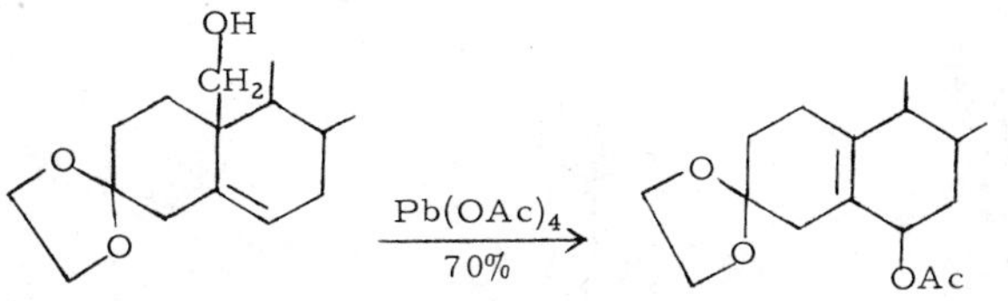

(11) (12)

steroid (10). The 19-hydroxy-Δ^5-steroid (11) affords the 6β-acetoxy-19-norsteroid (12). Other fragmentation reactions observed are formulated. Iodine is not favorable to fragmentation.

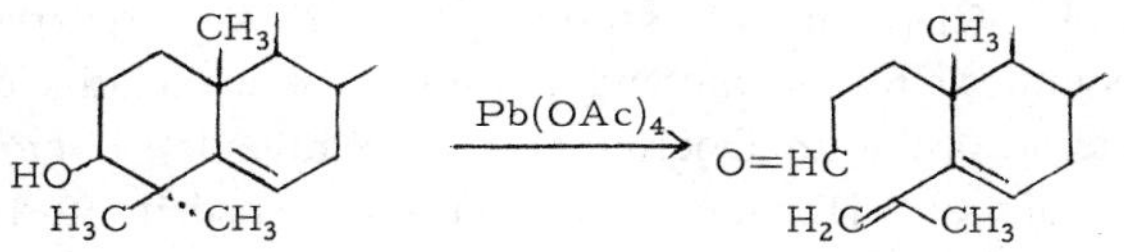

(13) (14)

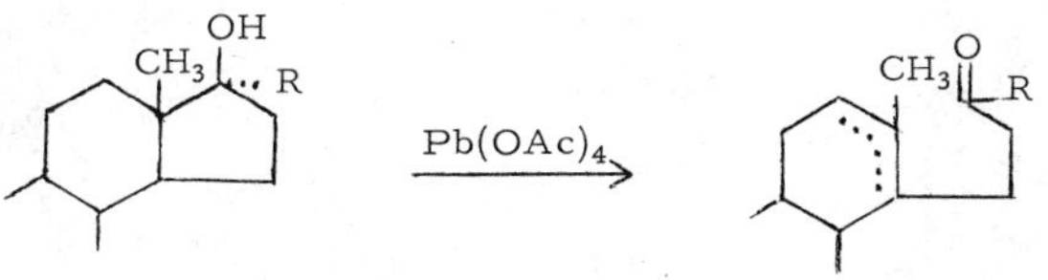

(15) (16)

Barton made two interesting applications of the lead tetraacetate–iodine method. One effects smooth decarboxylation of primary and secondary carboxylic acids to the corresponding iodides, as in the examples formulated.[58] A 5% suspension of lead tetraacetate in carbon tetrachloride is stirred and irradiated with a tungsten lamp. The carboxylic acid is added, followed by iodine until the color persists. If

$$CH_3(CH_2)_5\underset{\text{OAc}}{CH}(CH_2)_{10}CO_2H \xrightarrow[82\%]{Pb(OAc)_4\text{-}I_2} CH_3(CH_2)_5\underset{\text{OAc}}{CH}(CH_2)_9CH_2I$$

$$\text{Cyclohexanecarboxylic acid} \xrightarrow[91\%]{Pb(OAc)_4\text{-}I_2} \text{Cyclohexyl iodide}$$

$$HO_2C(CH_2)_4CO_2H \xrightarrow[33\%]{Pb(OAc)_4\text{-}I_2} ICH_2CH_2CH_2CH_2I + CO_2$$

no carboxylic acid is added the following reaction ensues: $Pb(OCOCH_3)_4 + I_2 \rightarrow 2CH_3I + 2CO_2 + Pb(OAc)_2$. The second application was to the conversion of the steroid 20-carboxamide (17) through the N-iodoamide (18) to the lactone (19).[59]

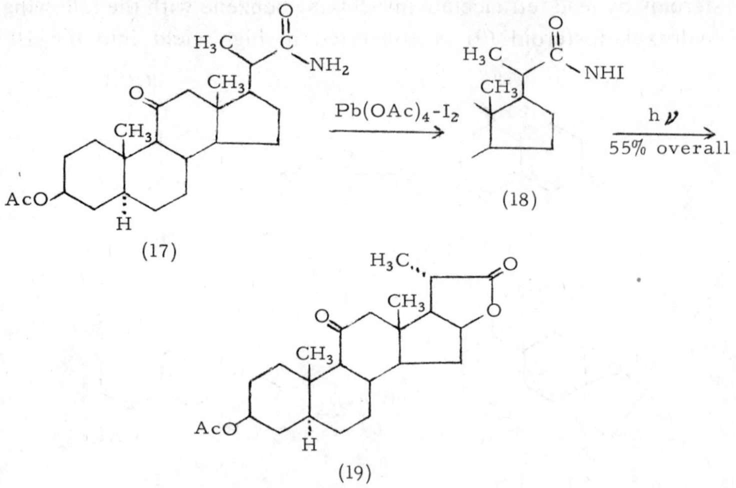
H3C
NH2
CH3
O
CH3
AcO
H
(17)
Pb(OAc)4-I2
H3C
NHI
CH3
(18)
hν
55% overall
H3C
CH3
CH3
AcO
H
(19)

A mixture of the amide (17) with 3 moles of lead tetraacetate and 4 moles of iodine in chloroform is irradiated at 15° for 5 hrs.

Oxidative decarboxylation. Having occasion to prepare a quantity of 1-carboethoxy-Δ^2-bicyclo[2.2.2]-octene (8), Grob and co-workers[60] prepared the anhydride (6) by the neat synthesis formulated and investigated the oxidative decarboxylation with lead dioxide according to Doering (*see* Lead dioxide). Even with the active lead dioxide of Kuhn and Hammer, the reaction proceeded poorly and the yield of (8) was only 30–37%. The Swiss workers then found that the yield can be more

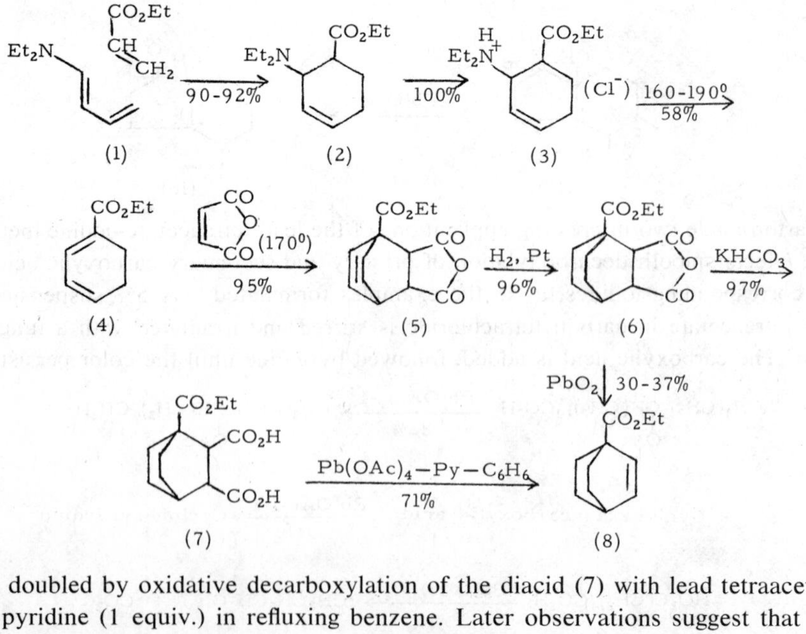
CO2Et
Et2N
CH
CH2
(1)
90-92%
CO2Et
Et2N
(2)
100%
H
Et2N+
CO2Et
(Cl⁻)
(3)
160-190°
58%
CO2Et
(4)
CO
O
CO
(170°)
95%
CO2Et
CO
O
CO
(5)
H2, Pt
96%
CO2Et
CO
O
CO
(6)
KHCO3
97%
PbO2
30-37%
CO2Et
CO2H
CO2H
(7)
Pb(OAc)4-Py-C6H6
71%
CO2Et
(8)

than doubled by oxidative decarboxylation of the diacid (7) with lead tetraacetate and pyridine (1 equiv.) in refluxing benzene. Later observations suggest that the reagent is still more effective in dimethyl sulfoxide containing 1 equivalent of pyridine.[61]

Büchi[62] effected oxidative decarboxylation of the monocarboxylic acid (9) by refluxing it with lead tetraacetate in benzene under nitrogen for 14 hrs. The olefinic isomers (10) and (11) were separated by gas chromatography and isolated in small

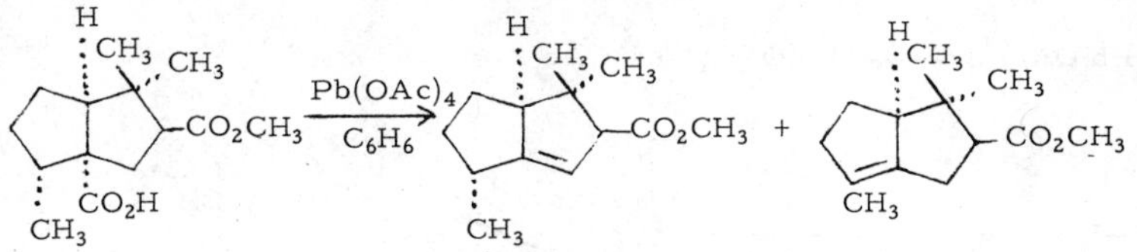
Pb(OAc)4
C6H6
H
CH3
CO2CH3
CO2H
+

(9) (10) (11)

amounts. Büchi's group has encountered other instances in which the product of oxidation of a monocarboxylic acid is not an olefin but an ester in which the carboxyl group is replaced by an acetoxy group.[63] Such a case was encountered by Cope[64] on investigating the oxidation of 4-cycloheptene-1-carboxylic acid (6), prepared by a novel synthesis reported by Stork in a communication.[65] The pyrrolidine enamine (1) of cyclopentanone reacts with 1 equivalent of acrolein in dioxane, initially at

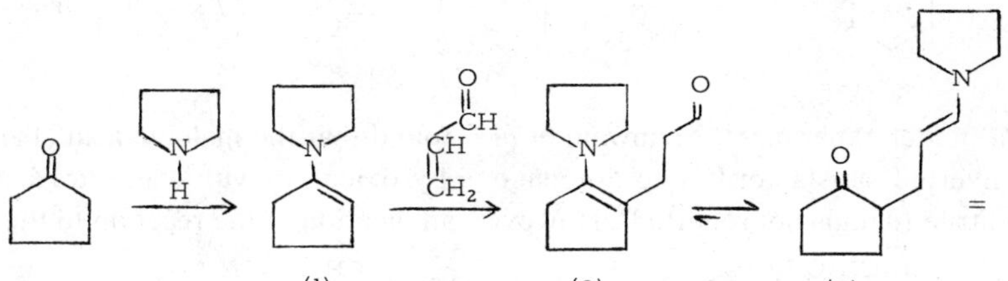

(1) (2) (3)

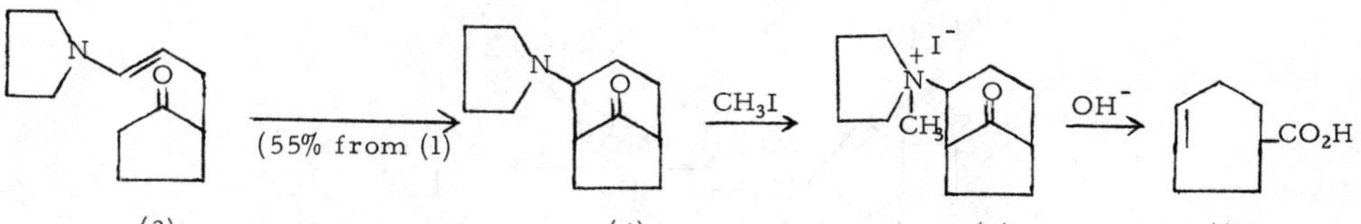
(55% from (1))
CH3I
OH-
CO2H

(3) (4) (5) (6)

0°, to give the bicyclic aminoketone (4) by Michael addition to (2), and probably exchange of the enamine group from the keto carbonyl to the more reactive aldehyde carbonyl group to give (3), and cyclization to (4). The corresponding methiodide (5) is cleaved by base to 4-cycloheptene-1-carboxylic acid (6). Put through the same process, the enamine of cyclohexanone afforded the bicyclic aminoketone analogous to (4) in 75% yield and this afforded 4-cyclooctene-1-carboxylic acid.

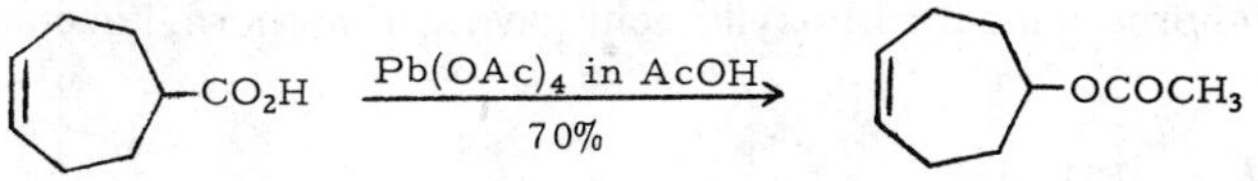
CO2H
Pb(OAc)4 in AcOH
70%
OCOCH3

(6) (7)

Cope's group found that the cycloheptene acid (6) on oxidation with lead tetraacetate in acetic acid affords 1-acetoxy-Δ^4-cycloheptene (7) in good yield.

Thus oxidation of an acid with the carbonyl group on a saturated carbon atom can give rise either to an olefin or to an acetoxy compound. Possible pathways are as follows:

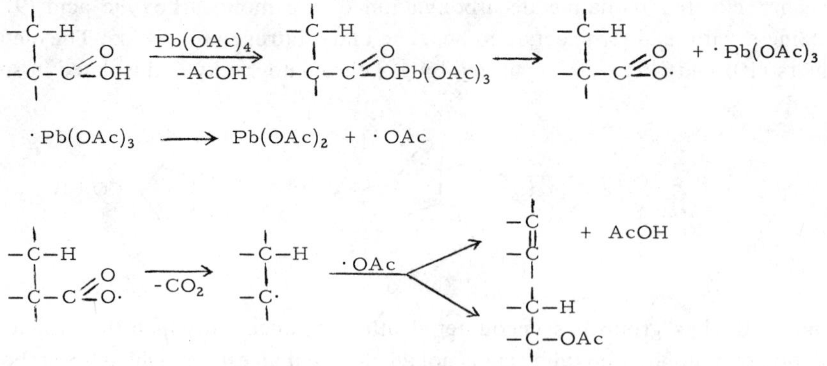

In their synthesis of Dewar benzene (10) van Tamelen and Pappas[66] effected oxidative decarboxylation of the anhydride (9) with lead tetraacetate in pyridine at 43–45°.

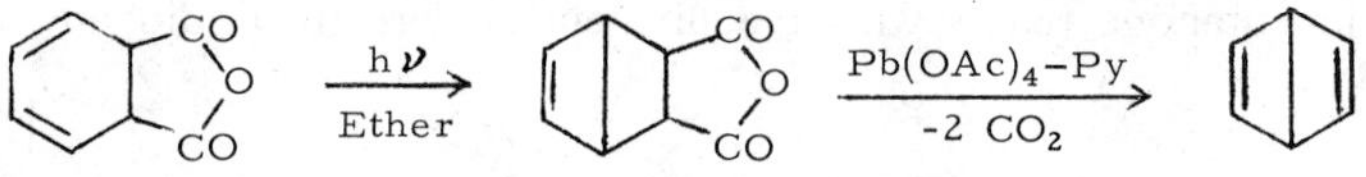

(8) (9) (10)

In a model experiment, Meinwald *et al.*[67] found that the malonic acid (11) could be converted satisfactorily into 5-nonanone by oxidation with lead tetraacetate in acetonitrile (details not reported). However, application of the reaction to the cyclic

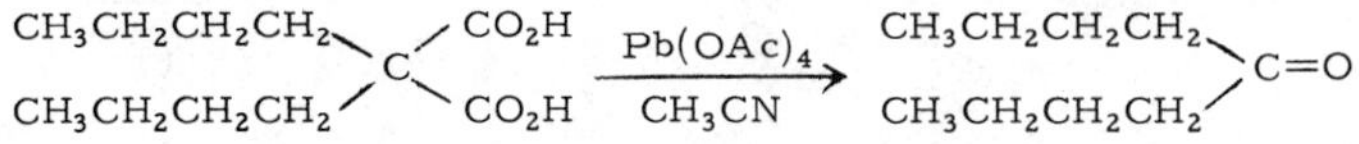

(11) (12)

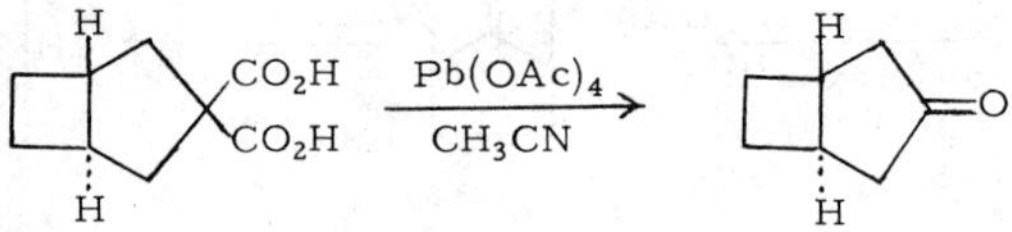

(13) (14)

malonic acid (13) afforded only a trace of the desired ketone (14), as determined by gas chromatography and spectral analysis.

Corey and Casanova[68] found that oxidation of either *meso-* or *dl*-1,2-diphenylsuccinic acid with lead tetraacetate in pyridine afforded *trans*-stilbene (40–45% yield); *cis*-stilbene was shown *not* to be an intermediate. Similar oxidation of either *endo-* or *exo*-norbornane-2-carboxylic acid gave *exo*-norbornyl acetate (24–67%).

They presented evidence that an intermediate carbonium ion is involved. On application of the reaction to the acid (1), the only product which could be isolated was the lactone (2).[69]

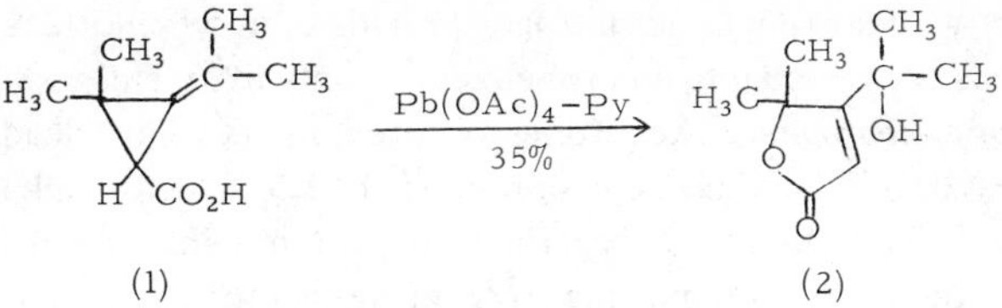
Pb(OAc)$_4$–Py
35%

(1) (2)

Oxidative decarboxylation was a key step in elucidation of the structure of the photochemical adduct of benzene with 2 moles of maleic anhydride (3).[70] The oxidation product (4) is the maleic anhydride adduct of cyclooctatetraene.

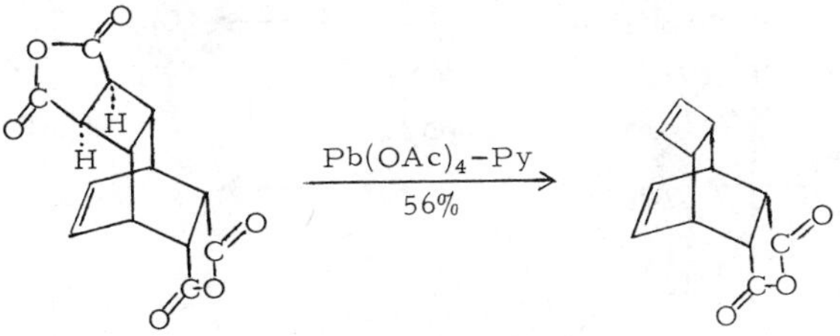
Pb(OAc)$_4$–Py
56%

(3) (4)

Winstein *et al.*[70a] oxidized the acid I to the acetate II in about 60% yield with lead tetraacetate and pyridine in benzene.

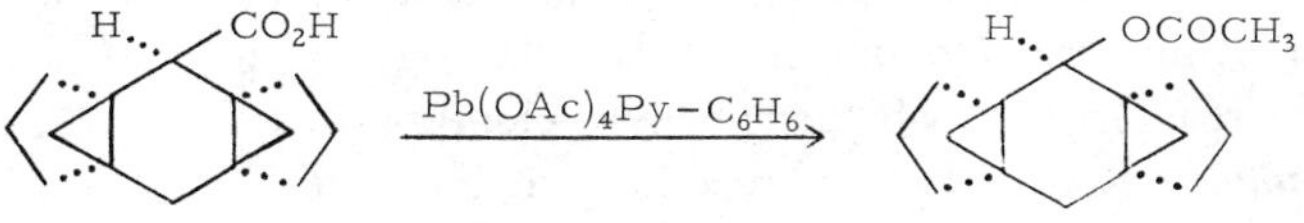
Pb(OAc)$_4$Py–C$_6$H$_6$

I II

Oxidative demethylation. In benzene solution, 1 mole of lead tetraacetate oxidizes 21-desoxyajmaline-17-acetate (1) to the 2-hydroxy derivative (2).[71] Excess oxidant gives the 1-demethyl-Δ^1-compound (3), an indolenine.

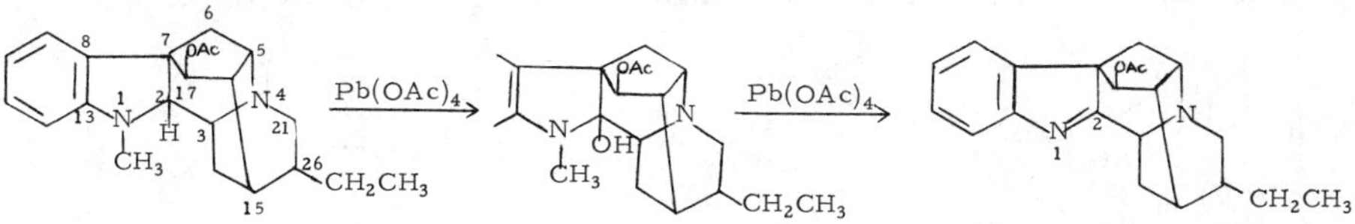
Pb(OAc)$_4$
Pb(OAc)$_4$

(1) (2) (3)

Oxidative rearrangement. A reaction related to the Hofmann rearrangement of N-bromoamides occurs on oxidation of an aromatic amide with lead tetraacetate to give the isocyanate or a derived product in moderate to excellent yield.[72] Simple aliphatic amides react rapidly in dimethylformamide containing triethylamine.

$$RC(=O)-NH_2 \xrightarrow{Pb(OAc)_4} RC(=O)-\ddot{N}: \longrightarrow RN=C=O$$

Halodecarboxylation. Kochi[73] found that treatment of a carboxylic acid with lead tetraacetate and an ionic halide, usually lithium chloride, results in the reaction

$$RCO_2H + Pb(OAc)_4 + LiCl \longrightarrow RCl + CO_2 + Pb(OAc)_2 + LiOAc + AcOH$$

formulated in which the carboxyl group is displaced by the halogen atom. The salt is added to a solution of the acid and lead tetraacetate in benzene and the mixture is heated to 80° with stirring under nitrogen. Yields of alkyl chlorides were high, but benzoic acid was converted into chlorobenzene in only 8% yield.

Cleavage of cyclopropanes. According to a preliminary report, lead tetraacetate in acetic acid cleaves alkyl and aryl cyclopropanes to 1,3-diacetoxyalkanes and mono-acetoxy olefins.[74] The reaction is regarded as involving electrophilic attack of lead tetraacetate on the cyclopropane ring to form a γ-acetoxy organolead ester which decomposes to the products. The results cited for the cleavage of phenylcyclopropane at 75° include product analysis as determined by vapor-phase chromatography.

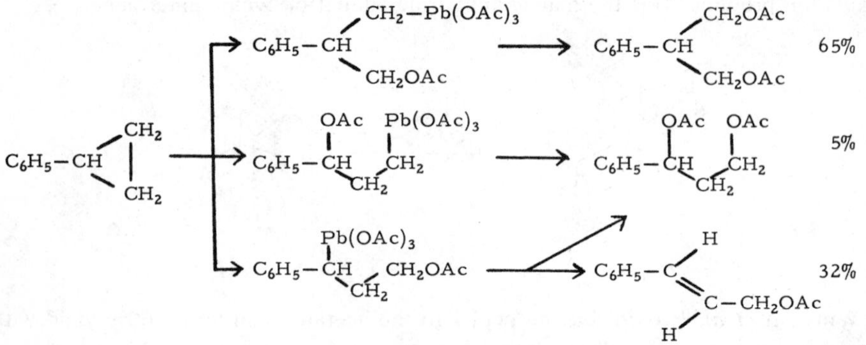
CH₂–Pb(OAc)₃
C₆H₅–CH
CH₂OAc
CH₂OAc
C₆H₅–CH
CH₂OAc
65%
CH₂
C₆H₅–CH
CH₂
OAc Pb(OAc)₃
C₆H₅–CH CH₂
CH₂
OAc OAc
C₆H₅–CH CH₂
CH₂
5%
Pb(OAc)₃
C₆H₅–CH CH₂OAc
CH₂
H
C₆H₅–C
C–CH₂OAc
H
32%

p-Bromophenylcyclopropane reacts about half as rapidly as phenylcyclopropane, and the ratio of symmetrical to unsymmetrical cleavage is higher. Monoalkylcyclo-propanes also react more slowly than phenylcyclopropane.

Methylation. Discovery of this reaction was the result of a chance observation made by one of us in attempting to improve a known procedure for converting butadiene–toluquinone (I) into 2-methyl-1,4-naphthoquinone (IV).[75] Oxidation of the isomerized product II with silver oxide stops at the stage of the highly sensitive quinone III. Chromic acid carried the oxidation to the desired stage but affords IV in yield of only 50%. Lead tetraacetate might either acetoxylate one of the activated

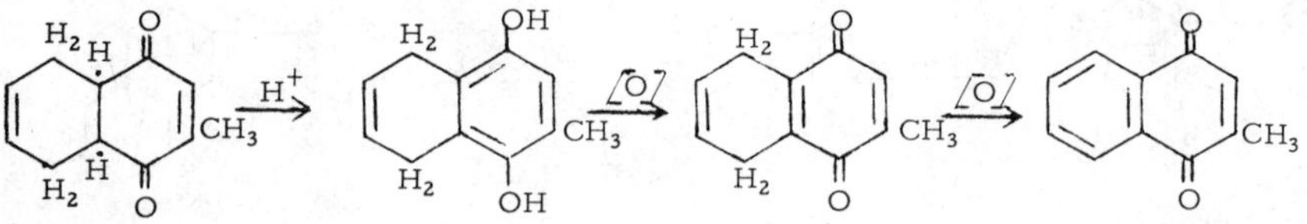
H₂ H
H
H₂
CH₃
H⁺
H₂
OH
OH
H₂
CH₃
[O]
H₂
H₂
CH₃
[O]
CH₃

I II III IV

methylene groups of III or add to the double bond to give a glycol diacetate, and either intermediate should lose acetic acid readily with formation of IV. When the hydroquinone II was warmed on the steam bath with 2 moles of lead tetraacetate in acetic acid, a reaction occurred and afforded a crystalline yellow product resembling IV but melting slightly above the melting point of pure IV (106–107°). In a repetition of the experiment with 3 moles of oxidant the product melted sharply at 126.5–127.5° and was identified as 2,3-dimethyl-1,4-naphthoquinone. Fieser and Chang[75] then found that methylation of quinones with lead tetraacetate is a general reaction and that higher alkyl groups can be introduced by gradual addition of Pb_3O_4 to a hot solution of a quinone in a given carboxylic acid. Substitution of a diacyl peroxide for a tetravalent lead ester as a source of alkyl radicals led to a highly useful method of alkylating quinones.[76]

Although the reaction is of no preparative value, it is of interest that aromatic nitro compounds can be *op*-methylated by radicals generated by the decomposition of lead tetraacetate.[77]

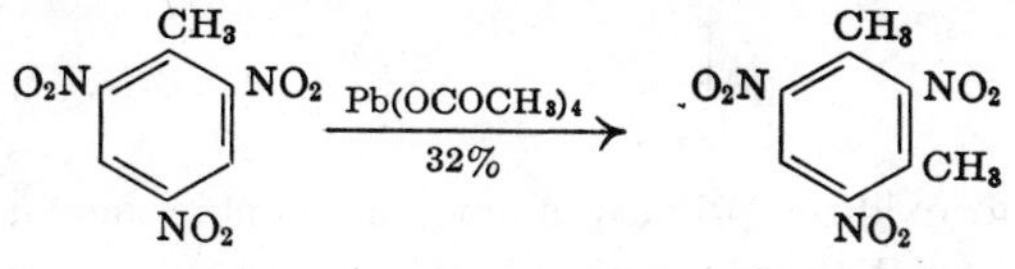

M.p. 80.6° M.p. 182°

Miscellaneous oxidations. Lead tetraacetate in acetic acid at 35° oxidizes 1-benzenesulfonamidonaphthalene (1) to N-benzenesulfonyl-1,4-naphthoquinone-4-imine (2) and it oxidizes 2-benezenesulfonamidonaphthalene (3) to 2-benzene-sulfonamido-1,4-naphthoquinone (4).[78]

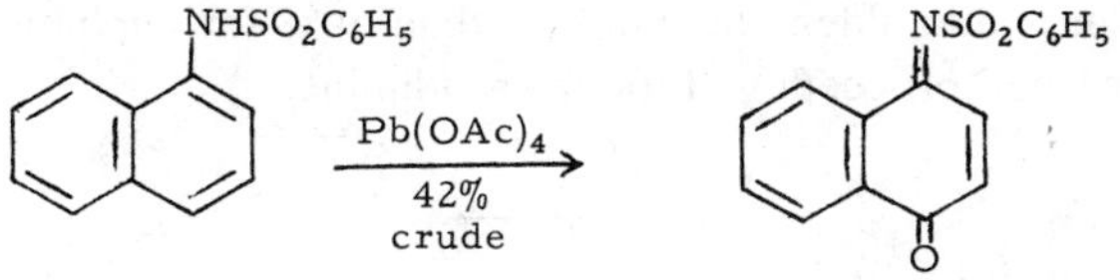

(1) (2)

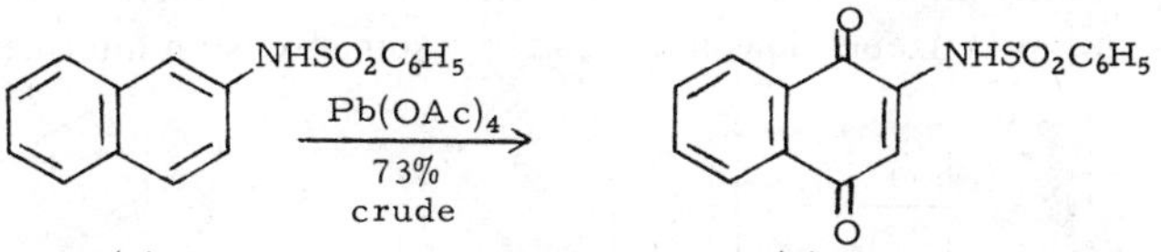

(3) (4)

N,N-Dimethylaniline on treatment with the reagent in acetic anhydride and chloroform at room temperature is dealkylated and affords N-methylacetanilide and formaldehyde.[79] The reaction is general for tertiary amines having one aromatic group.

$$C_6H_5N(CH_3)_2 \xrightarrow{Pb(OAc)_4} C_6H_5N(COCH_3)CH_3 + CH_2O \quad (83\%,\ 61\%)$$

Field[80] found that lead tetraacetate is an efficient reagent for oxidation of thiols to disulfides; the thiols can be alkyl, aryl, benzyl, or heterocyclic. He later found that a

$$2\,RSH + Pb(OAc)_4 \longrightarrow RSSR + Pb(OAc)_2 + 2\,AcOH$$

disulfide can be oxidized further with lead tetraacetate in methanol–chloroform to a methyl sulfinate,[81] as in a procedure for the preparation of methyl benzenesulfinate.[82] A solution of 0.25 mole of diphenyl disulfide in chloroform–methanol is stirred under

$$(C_6H_5S)_2 + 3\,Pb(OAc)_4 + 4\,CH_3OH \longrightarrow 2\,C_6H_5\overset{O^-}{\overset{|}{S^+}}OCH_3 + 3\,Pb(OAc)_2 + 4\,AcOH + 2\,AcOCH_3$$

reflux during addition at hourly intervals of eight 0.125-mole portions of lead tetraacetate, each freshly dissolved in 250 ml. of chloroform (the chloroform solution decomposes on standing).

Fuson *et al.*[83] oxidized diarylvinyl alcohols (1) with the reagent to the acetates of the corresponding diarylglycolaldehydes (3). Presumably the adduct (2) is an intermediate.

$$(\text{Mesityl})_2\text{C}=\text{CHOH} \xrightarrow[40^\circ]{\text{Pb(OAc)}_4 \text{ in AcOH}} \left[(\text{Mesityl})_2\overset{|}{\underset{\text{OAc}}{\text{C}}}-\overset{|}{\underset{\text{OAc}}{\text{CHOH}}}\right] \xrightarrow{-\text{AcOH}} (\text{Mesityl})_2\overset{|}{\underset{\text{OAc}}{\text{C}}}-\text{CHO}$$

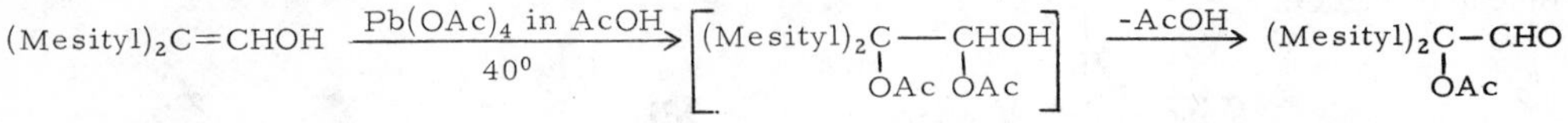

(1)   (2)   (3)

Lead tetraacetate oxidizes primary amines containing an α-methylene group to nitriles, probably through the aldimine.[84] Yields are nearly doubled when the ratio of amine to oxidant is 1:2, but even so the highest yield reported is 61%. A perhaps

$$\text{RCH}_2\text{NH}_2 \xrightarrow{\text{Pb(OAc)}_4} \langle\text{RCH}=\text{NH}\rangle \longrightarrow \text{RCN}$$

related reaction is the conversion of an aldehyde into a nitrile by treatment with lead tetraacetate (2–3 fold excess) in benzene while passing in a stream of ammonia.[84a] Yields are high for aromatic aldehydes but less than 50% for aliphatic aldehydes. The reaction is regarded as proceeding through an aldimine.

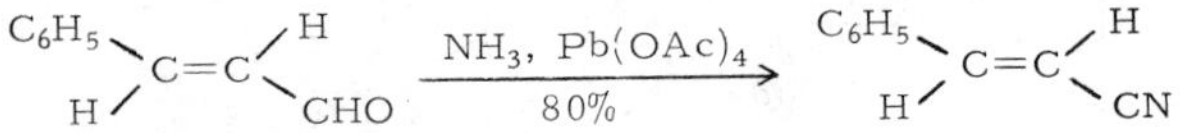

Campbell and Rees[85] report that benzyne can be generated rapidly and almost quantitatively under mild conditions by oxidation of 1-aminobenzotriazole with lead tetraacetate in benzene. The nitrene (2) presumably is an intermediate. When

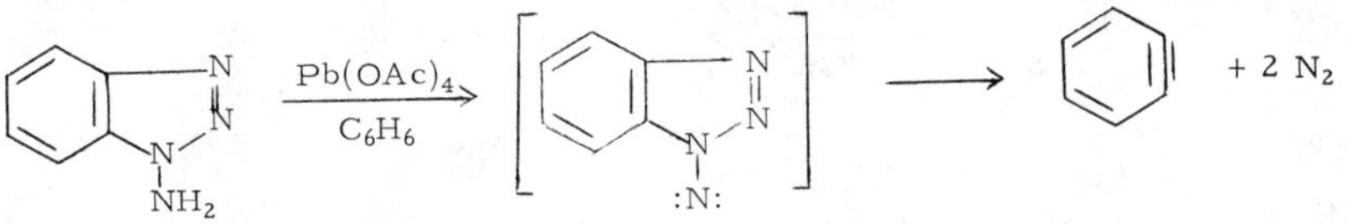

(1)   (2)   (3)

the oxidation was carried out in the presence of tetraphenylcyclopentadienone, 1,2,3,4-tetraphenylnaphthalene was obtained in 95% yield.

Lead tetraacetate oxidizes 2-aminobenzotriazole (4) to *cis,cis*-muconitrile (7) in 64% yield, presumably through the nitrenes (5) and (6).[86] The same product is obtained in 50% yield by lead tetraacetate oxidation of *o*-phenylenediamine (8).[87]

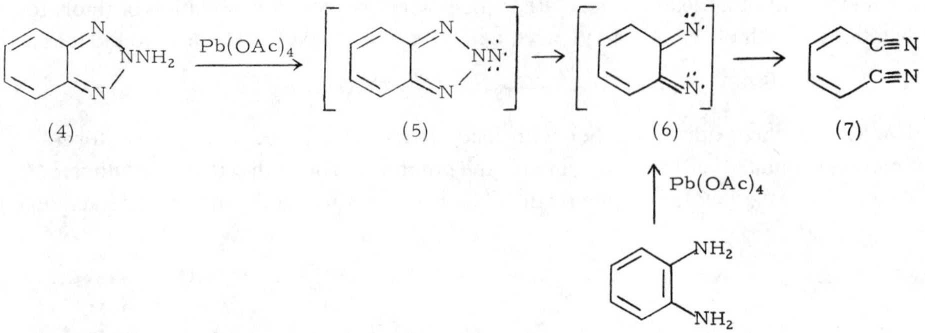

(4)   (5)   (6)   (7)

(8)

Partch[88] studied the oxidation of primary and secondary alcohols with lead tetraacetate in pyridine.

Ciganek[89] at du Pont found that carbonyl cyanide hydrazone (1) can be obtained

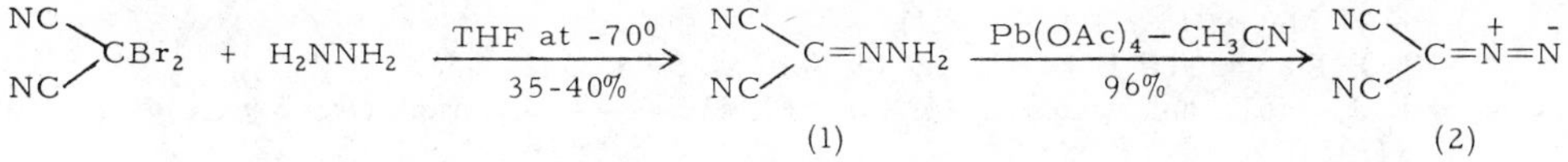

$$NC\diagdown CBr_2 + H_2NNH_2 \xrightarrow[\substack{35-40\%}]{THF\ at\ -70^0} NC\diagdown C=NNH_2 \xrightarrow[\substack{96\%}]{Pb(OAc)_4-CH_3CN} NC\diagdown C=\overset{+}{N}=\overset{-}{N}$$

(1)   (2)

very easily by reaction of dibromomalononitrile with hydrazine in tetrahydrofurane, but attempts to oxidize the hydrazone to the diazomethane (2) with the usual reagents all failed. He then found that dicyanodiazomethane (2) can be obtained in almost quantitative yield by oxidation of (*l*) with lead tetraacetate in acetonitrile. Dicyanodiazomethane, a highly electrophilic diazoalkane, is of interest as a precursor of dicyanocarbene. The oxidation procedure, limited to the preparation of diazo compounds which are stable to acetic acid (a reaction product), has been used also for the preparation of bis-(trifluoromethyl)diazomethane.[90]

Lead tetraacetate reacts with a primary amide (1) in refluxing benzene or benzene–acetic acid to give an acylamine (5) and a trace of the dialkylurea (6).[91] Spectroscopic

$$RCONH_2 \xrightarrow{Pb(OAc)_4} R\overset{\overset{O}{\|}}{C}\overset{..}{N}: + Pb(OAc)_2 + AcOH$$

(1)   (2)

$$RN=C=O \xrightarrow{AcOH} R\overset{H}{N}\overset{\overset{O}{\|}}{C}OAc \longrightarrow R\overset{H}{N}Ac + R\overset{H}{N}\overset{O}{C}\overset{H}{N}R$$

(3)   (4)   (5)   (6)

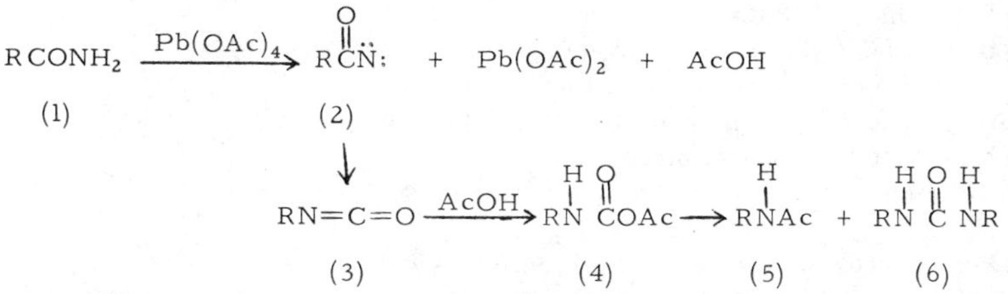

detection of the alkyl isocyanate (3) suggested that the reaction involves formation and rearrangement of the acyl nitrene (2). Yields of the aryl amines are 40–80%. The reaction should be useful for degradation.

[1]O. Dimroth and R. Schweizer, *Ber.*, **56**, 1375 (1923)

[2]This essentially the method of Dimroth and Schweizer[1] as modified by R. Hellmuth, Dissertation, Würzburg (1930).

[3]R. Criegee, *Angew. Chem.*, **70**, 173 (1958); *Newer Methods of Preparative Organic Chemistry*, **2**, 368 (1963)

[4]O. Dimroth and E. Schultze, *Ann.*, **411**, 345 (1916)

[5]R. Willstätter and L. Kalb, *Ber.*, **38**, 1235 (1905)

[6]O. Dimroth and H. Kämmerer, *Ber.*, **53**, 471 (1920)

[7]O. Dimroth, O. Friedemann, and H. Kämmerer, *Ber.*, **53**, 481 (1920)

[8]O. Dimroth and V. Hilcken, *Ber.*, **54**, 3050 (1921)

[9]K.-H. König, W. Schulze, and G. Möller, *Ber.*, **93**, 554 (1960)

[10]R. A. Clement, *J. Org.*, **25**, 1724 (1960); **27**, 1115 (1962)

[11]L. F. Fieser and R. Stevenson, *Am. Soc.*, **76**, 1728 (1954)

[12]S. Seebeck and T. Reichstein, *Helv.*, **27**, 948 (1944)

[13]L. F. Fieser and M. A. Romero, *Am. Soc.*, **75**, 4716 (1953)

[14]G. W. K. Cavill and D. H. Solomon, *J. Chem. Soc.*, 4426 (1955)

[15]H. B. Henbest, D. N. Jones, and G. P. Slater, *J. Chem. Soc.*, 4472 (1961); J. D. Cocker, H. B. Henbest, G. H. Phillipps, G. P. Slater, and D. A. Thomas, *ibid.*, 6 (1965)

[16]M. Vilkas and M. Rouhi-Laridjani, *Compt. rend.*, **251**, 2544 (1960); M. Rouhi-Laridjani and M. Vilkas, *ibid.*, **254**, 1090 (1962)

[17]J. Jacques, C. Weidmann, and A. Horeau, *Bull. soc.*, 424 (1959)

[18]L. H. Zalkow, J. W. Ellis, and M. R. Brennan, *J. Org.*, **28**, 1705 (1963); L. H. Zalkow and J. W. Ellis, *ibid.*, **29**, 2626 (1964)

[19]D. C. Iffland, and G. X. Criner, *Chem. Ind.*, 176 (1956); D. C. Iffland, L. Salisbury, and W. R. Schafer, *Am. Soc.*, **83**, 747 (1961)

[20]D. C. Iffland and T. M. Davies, *Am. Soc.*, **85**, 2182 (1963)

[21]J. P. Freeman, *J. Org.*, **29**, 1379 (1964)

[22]R. Criegee, *Ann.*, **481**, 263 (1930)

[23]K. H. Meyer, *Ann.*, **379**, 37 (1911)
[24]L. F. Fieser and E. B. Hershberg, *Am. Soc.*, **60**, 1893, 2542 (1938)
[25]L. F. Fieser and J. Cason, *Am. Soc.*, **62**, 432 (1940); J. Cason, *Org. Syn.*, *Coll. Vol.*, **3**, 3 (1955)
[26]L. F. Fieser and S. T. Putnam, *Am. Soc.*, **69**, 1038 (1947)
[27]R. Criegee, *Ber.*, **64**, 260 (1931)
[28]F. J. Wolf and J. Weijlard, *Org. Syn.*, *Coll. Vol.*, **4**, 124 (1963)
[29]R. Criegee, L. Kraft, and B. Rank, *Ann.*, **507**, 159 (1933)
[30]R. Criegee, E. Büchner, and W. Walther, *Ber.*, **73**, 571 (1940)
[31]R. Criegee, E. Höger, G. Huber, P. Kruck, F. Marktscheffel, and H. Schellenberger, *Ann.*, **599**, 81 (1956)
[32]L. P. Kuhn, *Am. Soc.*, **76**, 4323 (1954)
[33]R. Criegee, B. Marchand, and H. Wannowius, *Ann.*, **550**, 99 (1942)
[34]J. Booth and E. Boyland, *Biochem. J.*, **44**, 361 (1949); J. Booth, E. Boyland, and E. E. Turner, *J. Chem. Soc.*, 1188, 2808 (1950)
[35]R. Criegee and E. Büchner, *Ber.*, **73**, 563 (1940)
[36]L. F. Fieser and S. T. Putnam, *Am. Soc.*, **69**, 1041 (1947)
[37]R. P. Bell, V. G. Rivlin, and W. A. Waters, *J. Chem. Soc.*, 1696 (1958)
[38]C. A. Grob and P. W. Schiess, *Helv.*, **43**, 1546 (1960)
[39]S. J. Angyal and R. J. Young, *Am. Soc.*, **81**, 5251 (1959)
[40]L. F. Fieser and F. Mukawa, unpublished
[41]H. R. Goldschmid and A. S. Perlin, *Can. J. Chem.*, **38**, 2280 (1960)
[42]R. Criegee, *Ann.*, **495**, 211 (1932)
[43]A. S. Perlin and C. Brice, *Can. J. Chem.*, **34**, 541 (1956)
[44]A. J. Charlson and N. K. Richtmyer, *Am. Soc.*, **81**, 1512 (1959); **82**, 3428 (1960)
[45]H. H. Sephton and N. K. Richtmyer, *J. Org.*, **28**, 2388 (1963)
[46]E. Bamberger, *Ber.*, **34**, 61 (1901)
[47]S. Goodwin and B. Witkop, *Am. Soc.*, **79**, 179 (1957)
[48]E. Bamberger, *Ber.*, **36**, 2028 (1903)
[49]E. Jones and J. Kenner, *J. Chem. Soc.*, 1851 (1931); J. Kenner and F. Morton, *ibid.*, 679 (1934)
[50]Paper I: F. Wessely, G. Lauterbach-Keil, and F. Sinwel, *Monatsh.*, **81**, 811 (1950). Paper IX: F. Wessely, E. Zbiral, and J. Jörg, *ibid.*, **94**, 227 (1963)
[51]A. M. Gold and E. Schwenk, *Am. Soc.*, **80**, 5683 (1958); German patent 1,015,802
[52]G. Cainelli, M. Lj. Mihailović, D. Arigoni, and O. Jeger, *Helv.*, **42**, 1124 (1959)
[53]A. Bowers, E. Denot, L. Cuéllar Ibáñez, Ma. Elena Cabezas, and H. J. Ringold, *J. Org.*, **27**, 1862 (1962)
[54]Ch. Meystre, K. Heusler, J. Kalvoda, P. Wieland, G. Anner, and A. Wettstein, *Experientia*, **17**, 475 (1961); *Helv.*, **45**, 1317 (1962); K. Heusler, P. Wieland, and Ch. Meystre, *Org. Syn.*, **45**, 57 (1965)
[55]J. F. Bagli, P. F. Morand, and R. Gaudry, *J. Org.*, **28**, 1207 (1963)
[56]Ch. Meystre, J. Kalvoda, G. Anner, and A. Wettstein, *Helv.*, **46**, 2844 (1963)
[57]M. Amorosa, D. Arigoni, O. Jeger, *et al.*, *Helv.*, **45**, 2674 (1962)
[58]D. H. R. Barton and E. P. Serebryakov, *Proc. Chem. Soc.*, 309 (1962); D. H. R. Barton, H. P. Faro, E. P. Serebryakov, and N. F. Woolsey, *J. Chem. Soc.*, 2438 (1965)
[59]D. H. R. Barton and A. L. J. Beckwith, *Proc. Chem. Soc.*, 335 (1963)
[60]C. A. Grob, M. Ohta, E. Renk, and A. Weiss, *Helv.*, **41**, 1191 (1958); further examples in this series are reported by N. B. Chapman, S. Sotheeswaran, and K. J. Toyne, *Chem. Comm.*, 214 (1965)
[61]Private communication from C. A. Grob cited by R. Criegee.[3b]
[62]G. Büchi, R. E. Erickson, and N. Wakabayashi, *Am. Soc.*, **83**, 927 (1961)
[63]Private communication from G. Büchi
[64]A. C. Cope, C. H. Park, and P. Scheiner, *Am. Soc.*, **84**, 4862 (1962)
[65]G. Stork and H. K. Landesman, *Am. Soc.*, **78**, 5129 (1956)
[66]E. E. van Tamelen and S. P. Pappas, *Am. Soc.*, **85**, 3297 (1963)
[67]J. Meinwald, J. J. Tufariello, and J. J. Hurst, *J. Org.*, **29**, 2914 (1964)
[68]E. J. Corey and J. Casanova, Jr., *Am. Soc.*, **85**, 165 (1963)

[69]J. Meinwald, J. W. Wheeler, A. A. Nimetz, and J. S. Liu, *J. Org.*, **30**, 1038 (1965)
[70]E. Grovenstein, Jr., D. V. Rao, and J. W. Taylor, *Am. Soc.*, **83**, 1705 (1961)
[70a]L. Birladeanu, T. Hanafusa, and S. Winstein, *Am. Soc.*, **88**, 2315 (1966)
[71]M. F. Bartlett, B. F. Lambert, and W. I. Taylor, *Am. Soc.*, **86**, 729 (1964)
[72]H. E. Baumgarten and A. Staklis, *Am. Soc.*, **87**, 1141 (1965)
[73]J. K. Kochi, *Am. Soc.*, **87**, 2500 (1965)
[74]R. J. Ouellette and D. L. Shaw, *Am. Soc.*, **86**, 1651 (1964)
[75]L. F. Fieser and F. C. Chang, *Am. Soc.*, **64**, 2043 (1942)
[76]L. F. Fieser and A. E. Oxford, *Am. Soc.*, **64**, 2060 (1942)
[77]L. F. Fieser, R. C. Clapp, and W. H. Daudt, *Am. Soc.*, **64**, 2052 (1942)
[78]H. J. Richter and R. L. Dressler, *J. Org.*, **27**, 4066 (1962)
[79]L. Horner, E. Winkelmann, K. H. Knapp, and W. Ludwig, *Ber.*, **92**, 288 (1959)
[80]L. Field and J. E. Lawson, *Am. Soc.*, **80**, 838 (1958)
[81]L. Field, C. B. Hoelzel, J. M. Locke, and J. E. Lawson, *Am. Soc.*, **83**, 1256 (1961); **84**, 847 (1962)
[82]L. Field and J. M. Locke, *Org. Syn.*, **46**, 62 (1966)
[83]R. C. Fuson *et al.*, *Am. Soc.*, **79**, 1938 (1957)
[84]M. Lj. Mihailović, A. Stojiljković, and V. Andrejević, *Tetrahedron Letters*, 461 (1965). For other papers by this Yugoslavian group on "Reactions with Lead Tetraacetate," *see* M. Lj. Mihailović *et al.*, *Tetrahedron*, **21**, 955 (1966). *See also* ref. 52.
[84a]K. N. Parameswaran and O. M. Friedman, *Chem. Ind.*, 988 (1965)
[85]C. D. Campbell and C. W. Rees, *Proc. Chem. Soc.*, 296 (1964)
[86]C. D. Campbell and C. W. Rees, *Chem. Comm.*, 192 (1965)
[87]K. Nakagawa and H. Onoue, *Chem. Comm.*, 396 (1965)
[88]R. E. Partch, *J. Org.*, **30**, 2498 (1965)
[89]E. Ciganek, *J. Org.*, **30**, 4198 (1965)
[90]D. M. Gale, W. J. Middleton, and C. G. Krespan, *Am. Soc.*, **87**, 657 (1965)
[91]B. Acott and A. L. J. Beckwith, *Chem. Comm.*, 161 (1965)

Lead tetrabenzoate, $Pb(OCOC_6H_5)_4$. Mol. wt. 691.65. Preparation.[1,2] The reagent is an oxidant similar to lead tetraacetate.[1]

[1]R. Criegee, *Ann.*, **481**, 263 (1930)
[2]C. D. Hurd and P. R. Austin, *Am. Soc.*, **53**, 1543 (1931)

Lead tetrafluoride, PbF_4. In following up his classical work on lead tetraacetate, Dimroth[1] made an exploratory study of the reaction of lead tetraacetate in chloroform solution with liquid hydrogen fluoride to generate lead tetrafluoride:

$$Pb(OAc)_4 + 4\,HF \rightleftharpoons PbF_4 + 4\,AcOH$$

He found that the reaction is reversible and reaches a point of equilibrium, and so investigated reactions of PbF_4 only with olefins which do not react with lead tetraacetate, 1,1-Diphenylethylene (1) is such an olefin, and it was found to react with *in situ* generated lead tetrafluoride to give in moderate yield a crystalline product

$$\underset{(1)}{(C_6H_5)_2C{=}CH_2} \xrightarrow[42\%]{PbF_4\ (0^{\circ})} \underset{(2)}{[(C_6H_5)_2\underset{F}{C}-\underset{F}{CH_2}]} \longrightarrow \underset{(3)}{C_6H_5CF_2CH_2C_6H_5}$$

assumed to be the 1,2-adduct (2). Dimroth's adduct was later shown to be a product of rearrangement, 1,2-diphenyl-1,1-difluoroethane (3).[2]

Discovery of the surprising effect of fluorine substituents on the biological activities of adrenocortical hormones prompted a Syntex group[3] to investigate the reaction of pregnenolone acetate (4) in methylene chloride with lead tetrafluoride generated from lead tetraacetate and hydrogen fluoride at low temperatures. The best result, obtained with a reaction period of 15 min. at −75°, afforded in 27% yield

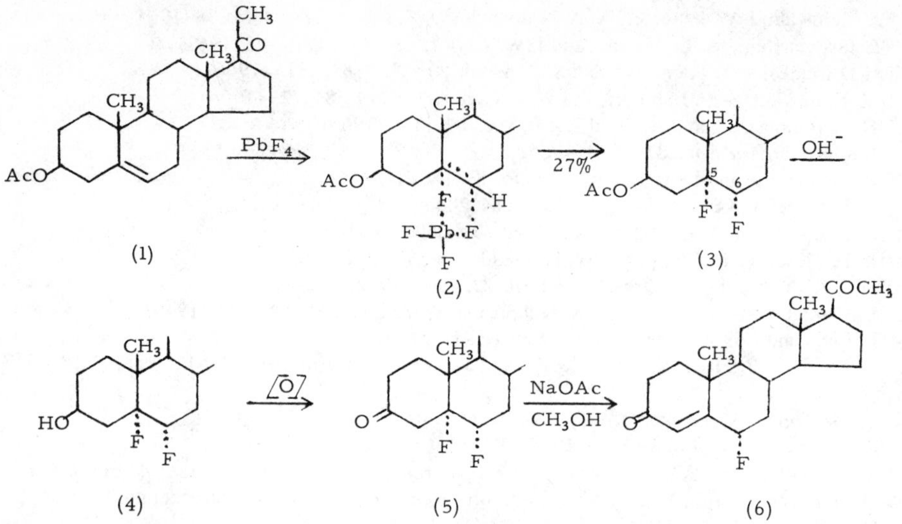

the 5α,6α-difluoro derivative (3), probably arising through the transition state (2). Saponification to the 3β-ol (4), oxidation to the 3-ketone (5), and dehydrohalogenation with sodium acetate in methanol gave the already known 6α-fluoroprogesterone (6). Lead tetrafluoride generated *in situ* from lead dioxide and sulfur tetrafluoride has been used for the addition of fluorine to the double bonds of certain halogenated olefins.[4]

[1]O. Dimroth and W. Bockemüller, *Ber.*, **64**, 516 (1931)
[2]J. Bornstein and M. R. Borden, *Chem. Ind.*, 441 (1958)
[3]A. Bowers, P. J. Holton, E. Denot, M. C. Loza, and R. Urquiza, *Am. Soc.*, **84**, 1050 (1962)
[4]E. R. Bissell and D. B. Fields, *J. Org.*, **29**, 1591 (1964)

Lemieux-Johnson oxidation, *see* Periodate–Osmium tetroxide oxidation.

Levulinic acid, $CH_3COCH_2CH_2CO_2H$. Mol. wt. 116.11, m.p. 33–35°. Suppliers: Quaker Oats, A, B, E, F, MCB.

Preparation. Tollens[1] discovered that levulinic acid can be obtained by the action of dilute mineral acids on carbohydrates; starting materials include sucrose,[1] starch,[2] glucose,[3] furfural,[4] and 4-hydroxymethylfurfural.[4] In the *Organic Syntheses* procedure[5] a solution of 500 g. of sucrose in 1 l. of water is treated with 250 ml. of

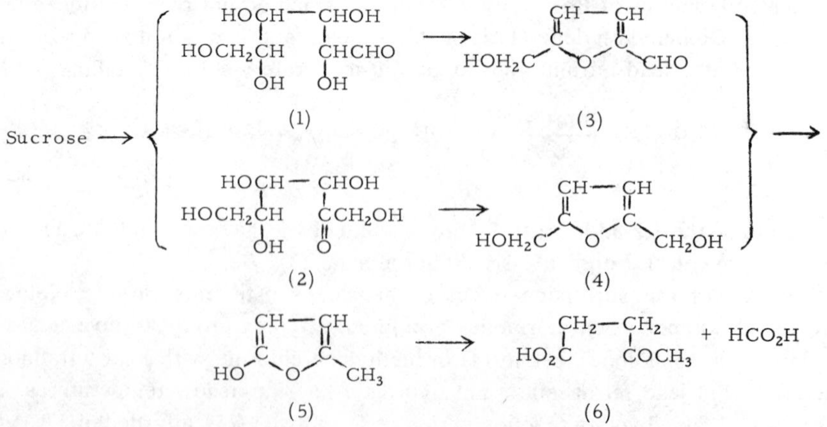

concd. hydrochloric acid and heated on the steam bath for 24 hrs. A black solid resulting from carbonation is removed by filtration, and the filtrate is evaporated to dryness. Extraction with ether (6–7 hrs.) and distillation affords material that solidifies almost completely at 30°. The yield of 72–76 g. is 21–22% of the theory based on utilization of both components of the disaccharide. Intermediate furane derivatives of types (3) and (4) evidently eliminate formaldehyde and suffer reduction by formaldehyde. A paper which appeared after publication of the procedure cited states that glucose is a better starting material than sucrose; there is less humus material and the yield is slightly higher.[3]

***Cleavage of 2,4-dinitrophenylhydrazones*[6] *and oximes.*[7]** Keeney[6] devised a semi-micro colorimetric procedure for estimation of 2,4-dinitrophenylhydrazones which involved use of levulinic acid as acceptor molecule for regeneration of ketones from the derivatives by exchange. He found that hydrazones of nonconjugated ketones are split much more rapidly than are those of conjugated or aromatic ketones.

DePuy[7] experienced difficulty in hydrolyzing the dioxime I (the dimer of cyclopentadienone oxime) with acid, either alone or in combination with formaldehyde since I, II, and III are all highly sensitive to strong acids. He then tried levulinic acid as acceptor and found that, when I is stirred at room temperature with levulinic acid to which 10 volume per cent of 1 *N* hydrochloric acid had been added, the dioxime dissolved in about 3 hrs. and from the solution the keto-oxime II was

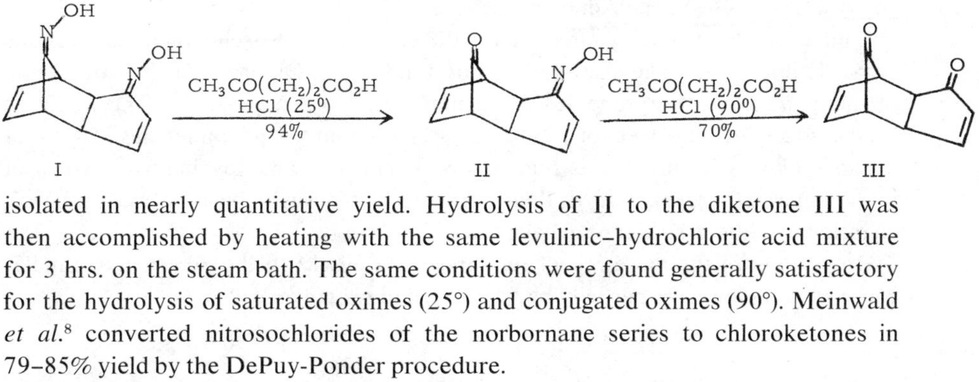

OH
OH
$CH_3CO(CH_2)_2CO_2H$
HCl (25°)
94%
O
N
OH
$CH_3CO(CH_2)_2CO_2H$
HCl (90°)
70%
O
O

I II III

isolated in nearly quantitative yield. Hydrolysis of II to the diketone III was then accomplished by heating with the same levulinic–hydrochloric acid mixture for 3 hrs. on the steam bath. The same conditions were found generally satisfactory for the hydrolysis of saturated oximes (25°) and conjugated oximes (90°). Meinwald *et al.*[8] converted nitrosochlorides of the norbornane series to chloroketones in 79–85% yield by the DePuy-Ponder procedure.

A procedure developed by Mattox and Kendall[9] for the liberation of cortisone acetate from the C_3 2,4-dinitrophenylhydrazone utilizes a mixture of pyruvic acid, acetic acid, chloroform, and hydrogen bromide and affords the parent compound in 80% yield. For comparison, DePuy[7] prepared a mixture of 1 g. of Δ^4-cholestene-3-one 2,4-DNP with 100 ml. of chloroform and 100 ml. of the usual levulinic–hydrochloric acid and heated it under reflux for 3 hrs. Chromatography afforded pure Δ^4-cholestene-3-one in high yield.

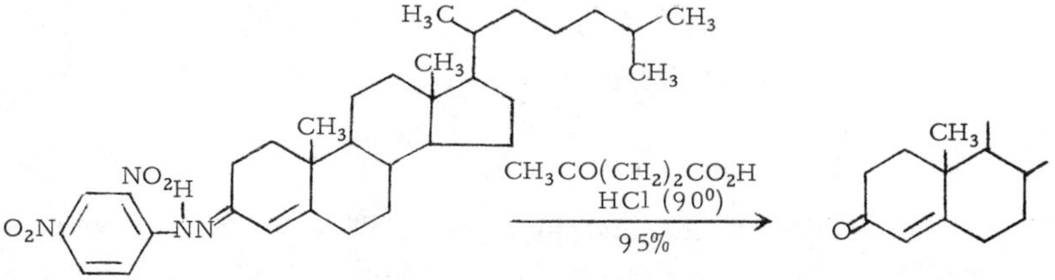

H_3C
CH_3
CH_3
CH_3
CH_3
NO_2 H
O_2N
NN
$CH_3CO(CH_2)_2CO_2H$
HCl (90°)
95%
CH_3

[1]A. von Grote and B. Tollens, *Ann.*, **175**, 181 (1875); **206**, 226 (1880)
[2]P. Rischbieth, *Ber.*, **20**, 1773 (1887)
[3]P. P. T. Sah and S.-Y. Ma, *Am. Soc.*, **52**, 4881 (1930)
[4]H. P. Teunissen, *Rec. trav.*, **50**, 1 (1931)
[5]B. F. McKenzie, *Org. Syn., Coll. Vol.*, **1**, 335 (1941)
[6]M. Keeney, *Anal. Chem.*, **29**, 1489 (1957)
[7]C. H. DePuy and B. W. Ponder, *Am. Soc.*, **81**, 4629 (1959)
[8]J. Meinwald, Y. C. Meinwald, and T. N. Baker, III, *Am. Soc.*, **86**, 4074 (1964)
[9]V. R. Mattox and E. C. Kendall, *Am. Soc.*, **70**, 882 (1948); *J. Biol. Chem.*, **188**, 287 (1951)

Lindlar catalyst, Pd–$CaCO_3$–PbO. Lindlar and Dubuis[1] have described in detail a slight modification of Lindlar's original procedure.[2] A solution of palladium chloride in dilute hydrochloric acid is neutralized to pH 4–4.5 and added to a suspension of precipitated calcium carbonate and the suspension is stirred and heated on the

$$PdCl_2 \xrightarrow{HCl} H_2PdCl_2 \xrightarrow{H_2O} PdO \xrightarrow{HCO_2H} Pd$$

$$Pb(OAc)_2 \xrightarrow{H_2O} PbO$$

steam bath to precipitate the red-brown oxide. Sodium formate solution is added to the stirred, warm suspension, and after about 40 min. reduction is complete and the color is black. The precipitate is collected and washed on a Büchner funnel and partially deactivated by stirring it in a slurry of 7.7% aqueous lead acetate at 80° for 45 min. The poisoned catalyst is washed and dried at 60–70°. The yield of catalyst from 1.48 g. of palladium chloride is 19.5 g.

Lindlar catalyst is highly effective for the selective hydrogenation of triple bonds to *cis*-double bonds. The example cited by Lindlar and Dubuis is the hydrogenation of phenylacetylene to styrene. A hydrogenation flask is charged with 2.04 g. (0.02 m.) of phenylacetylene, 0.10 g. of Lindlar catalyst, 1.0 ml. of quinoline, and 150 ml. of hexane. The air in the flask is removed by evacuating and flushing with hydrogen three times. Hydrogenation is then performed at a slight positive pressure with agitation by either a shaker or a magnetic stirrer.

Only a few of the more recent references are cited.[3] In the presence of Lindlar catalyst, cumulenes are partially hydrogenated to *cis*-polyenes.[4]

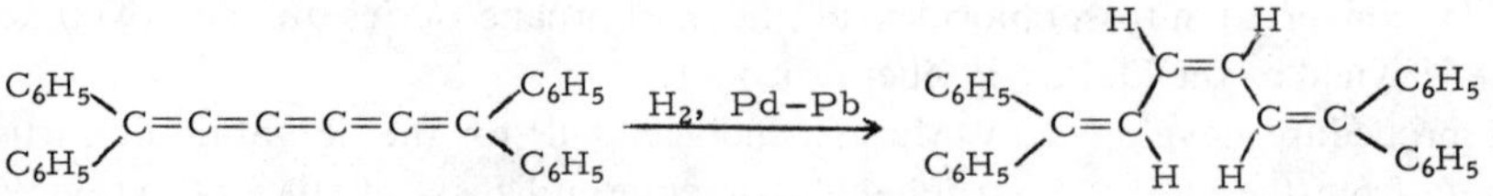
C_6H_5
C_6H_5
C=C=C=C=C=C
C_6H_5
C_6H_5
H_2, Pd–Pb
C_6H_5
C_6H_5
C=C
H
H
C=C
H
H
C=C
C_6H_5
C_6H_5

Cram and Allinger[5] employed in place of the Lindlar catalyst a palladium-on-barium sulfate catalyst poisoned by synthetic quinoline (quinoline from coal tar was unsatisfactory). Thus on hydrogenation of dimethyl 5-decynedioate (1) in methanol the mildly exothermal reaction ceased abruptly after 20 min. with uptake

$$CH_3O_2C(CH_2)_3C{\equiv}C(CH_2)_3CO_2CH_3 \quad + \quad 5\%\ Pd{-}BaSO_4 \quad + \quad H_2 \xrightarrow[97\%]{}$$

(1) 19. 2 g. in 100 ml. CH_3OH — 0. 4 g.

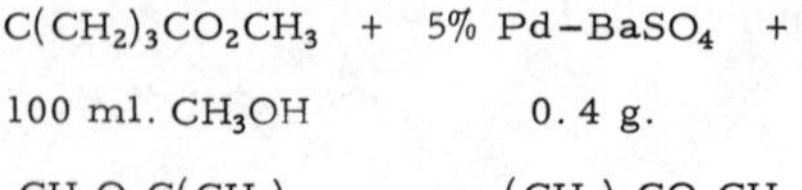
$CH_3O_2C(CH_2)_3$
$(CH_2)_3CO_2CH_3$
C=C
H
H

(2)

of exactly 1 mole of hydrogen, and dimethyl *cis*-5-decenedioate (2) was obtained in 97% yield. Schneider[6] states that he has found this a satisfactory and convenient substitute for the Lindlar catalyst but considers an even better catalyst to be 5% palladium-on-barium sulfate, used with pyridine as solvent. This reduces acetylenes to *cis*-olefins and stops sharply at the olefin stage.

[1]H. Lindlar and R. Dubuis, procedure submitted to *Organic Syntheses*
[2]H. Lindlar, *Helv.*, **35**, 446 (1952)
[3]R. T. Arnold and G. Smolinsky, *Am. Soc.*, **82**, 4918 (1960); R. Rüegg, U. Gloor, A. Langemann, M. Kofler, C. von Planta, G. Ryser, and O. Isler, *Helv.*, **43**, 1745 (1960); J. D. Surmatis and A. Ofner, *J. Org.*, **26**, 1171 (1961); J. M. Osbond, P. G. Philpott, and J. C. Wickens, *J. Chem. Soc.*, 2779 (1961)
[4]R. Kuhn and H. Fischer, *Ber.*, **93**, 2285 (1960)
[5]D. J. Cram and N. L. Allinger, *Am. Soc.*, **78**, 2518 (1956)
[6]W. P. Schneider, private communication

N-Lithioethylenediamine, $H_2NCH_2CH_2NH_2$ (mol. wt. 60.10) + Li (at. wt. 6.94) → $H_2NCH_2CH_2NHLi + H_2$. N-Lithioethylenediamine (98% pure) is available from Foote Mineral Co. (1 mole per bottle).

Reggel, Friedman, and Wender[1] discovered that this reagent has remarkable power for isomerization of olefins and for dehydrogenation of dihydrobenzenoid systems. The anhydrous ethylenediamine required is prepared either by heating commercial material with sodium for a day or two and distilling[1] or by azeotropic distillation with thiophene-free benzene, followed by distillation from potassium hydroxide pellets and then from sodium.[2] For preparation of the reagent, purified ethylenediamine is stirred under nitrogen at 90–110° and lithium is added in portions; an initial deep blue color disappears at the end of the addition, and a pale yellow solution results.[1]

Reggel *et al.*[1] observed that N-lithioethylenediamine in excess ethylenediamine as solvent rapidly isomerizes terminal alkenes to internal alkenes. Thus 1-octene on being refluxed with the reagent for 2 hrs. affords 90% of internal alkenes, mainly 2-octene. More striking is dehydrogenation, effected under very mild conditions. On addition of 4-vinylcyclohexene (1) to a solution of N-lithioethylenediamine

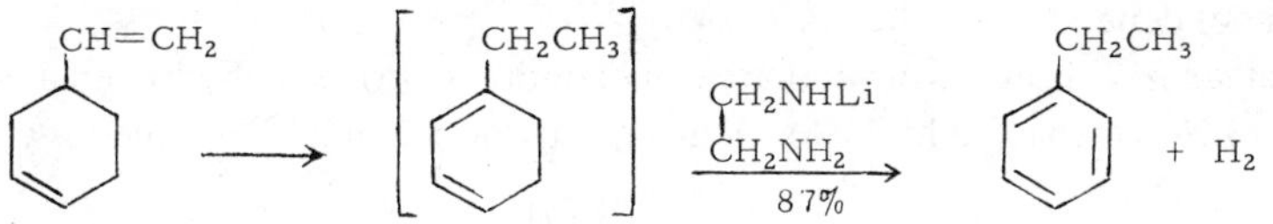
CH=CH2
CH2CH3
CH2NHLi
CH2NH2
87%
CH2CH3
+ H2

(1) (2) (3)

at room temperature, hydrogen was evolved, and ethylbenzene (3) was isolated as the sole organic product. The first step, evidently, is migration of the terminal double bond, probably to give the conjugated diene (2). The next step, dehydrogenation of (2) to (3) at room temperature, is surprising. In another example a solution of N-lithioethylenediamine is prepared from 2 moles of lithium and 375 ml. of ethylenediamine, a 2-mole portion of *d*-limonene is added with stirring at 100–115° during 1.5 hrs. (vigorous gas evolution), and the mixture is refluxed for 0.5 hr.

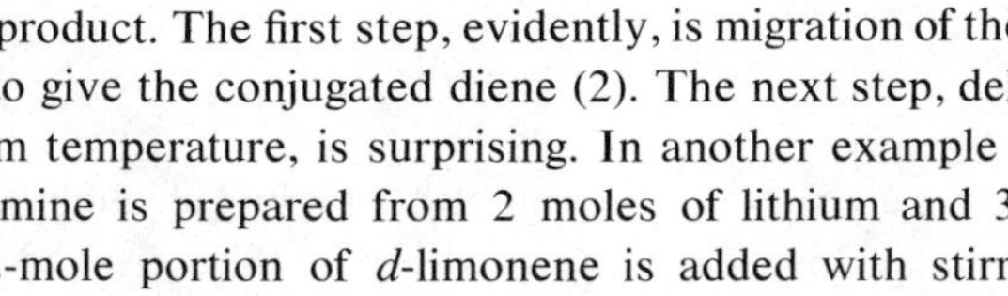
CH3
H3C CH2
H
H2NCH2CH2NLi
92%
CH3
+ H2
CH
H3C CH3

d-Limonene p-Cymene

The product consisted entirely of *p*-cymene, b.p. 170–173°. In a test experiment, the reagent from 0.25 mole of lithium and excess ethylenediamine was treated with 1 mole of *d*-limonene, and the mixture was heated for 1 hr. No hydrogen was evolved, and the product consisted entirely of *d*-limonene, b.p. 169–172°, recovery 92%. This observation seems to us to rule out a concerted mechanism suggested by Reggel involving a nine-membered cyclic transition state and implying that the reagent plays a catalytic role; another argument against this mechanism is that lithium triethylenediamine is also an effective dehydrogenating agent. Our co-worker, Makhluf Haddadin, has suggested the following mechanism. The strongly basic lithium derivative attacks the hydrocarbon (2) with displacement of one of the acidic hydrogens by lithium (4). In the second step, reaction of (4) with ethylenediamine to form (3) and lithium hydride, the latter in turn reacts with ethylenediamine with formation of hydrogen and lithioethylenediamine. The reaction

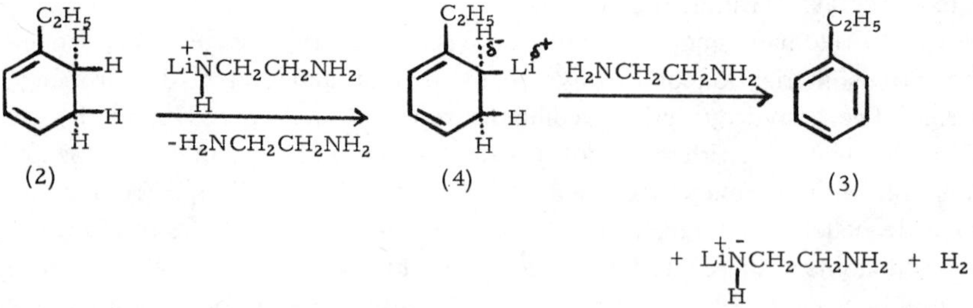
C_2H_5
H
$Li\overset{+}{N}^{-}CH_2CH_2NH_2$
$-H_2NCH_2CH_2NH_2$
$H_2NCH_2CH_2NH_2$
$+ Li\overset{+}{N}^{-}CH_2CH_2NH_2 + H_2$

(2) (4) (3)

requires basic conditions and does not proceed as long as an appreciable amount of the acidic hydrocarbon (2) is still present. Only when the hydrocarbon has been completely neutralized does the solution become sufficiently basic for occurrence of the concluding step. Lithioethylenediamine is thus required in full molar amount (and possibly in slight excess) and, although it is regenerated and not consumed, it is not a catalyst because it is formed in one reaction and regenerated in another reaction not dependent on the first.

Reggel *et al.* observed also that α-phellandrene (5) is dehydrogenated by the reagent to *p*-cymene (yield 74%). A group at Poona, India,[2] obtained *p*-cymene in

(5)

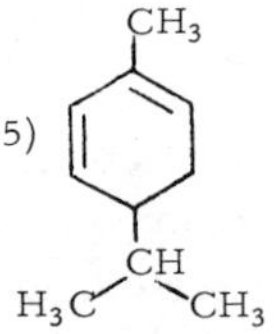
CH_3
CH
H_3C
CH_3

(6)

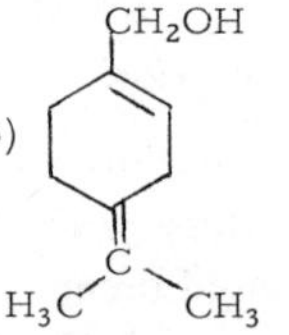
CH_2OH
C
H_3C
CH_3

(7)

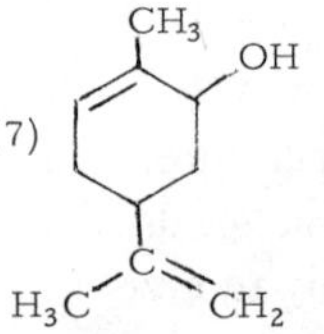
CH_3
OH
C
H_3C
CH_2

"high yield" by similar reactions of terpinolene, perillyl alcohol (6), and carveol (7). They found that the cyclopropane ring of Δ^3-carene (8) is cleaved in both directions to give a mixture of about equal parts of *p*-cymene (9) and *m*-cymene (10). The

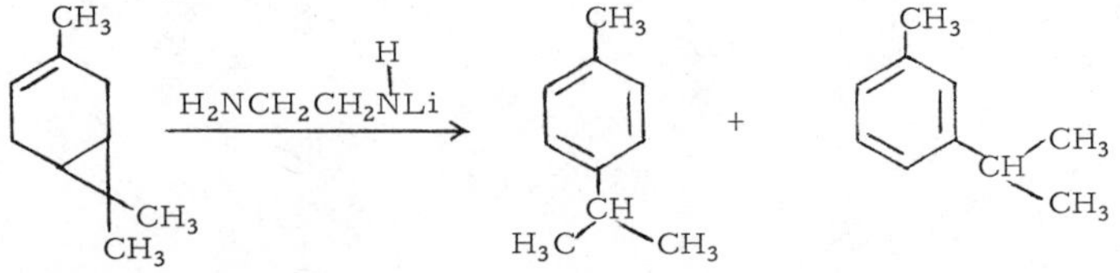
CH_3
CH_3
CH_3
H
$H_2NCH_2CH_2NLi$
CH_3
CH
H_3C
CH_3
+
CH_3
CH
CH_3
CH_3

(8) (9) (10)

sesquiterpene γ_1-cadinene (11) is converted into calamenene (12), presumably by migration of the exocyclic double bond to the disubstituted ring and aromatization.

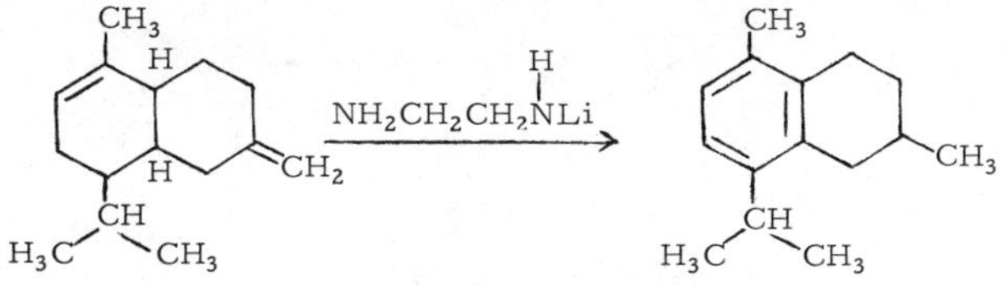

(11) (12)

Aromatization of the second ring could not be effected by repeating the treatment. This demonstration of partial aromatization opens some doors and closes others. For one thing it discouraged one of us, for a time, from attempting to use the reagent for aromatization of guaiene to guaiazulene (an eventual trial was negative). Aromatization of a hydroaromatic ring evidently requires the presence of two double bonds or of a cyclopropane ring and a double bond.[2] The observation that the conjugated ketone *dl*-piperitone (13) remained unaltered on treatment with the reagent

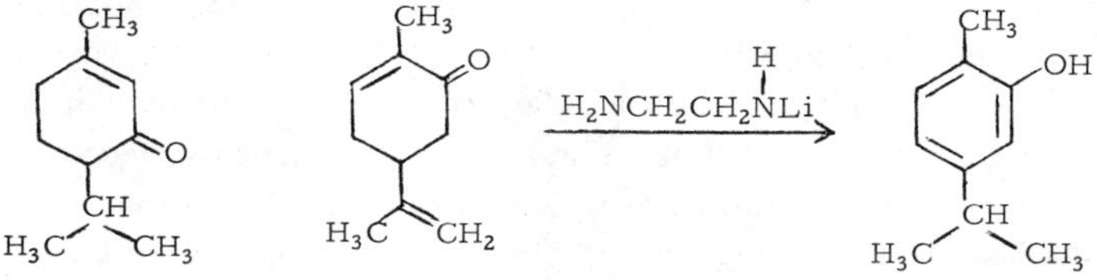

(13) (14) (15)

shows that an enolic double bond does not function as one of the two required double bonds. Carvone (14), containing two double bonds, affords carvacrol (15) in "quantitative yield." Geraniol (16) is isomerized by the reagent to the conjugated diene (17). Although excess reagent was used and refluxing extended for 8 hrs., no dehydrogenation to the triene occurred.

$$(CH_3)_2C{=}CHCH_2CH_2\overset{CH_3}{C}{=}CHCH_2OH \xrightarrow{H_2NCH_2CH_2\overset{H}{N}Li} (CH_3)_2C{=}CHCH{=}CH\overset{CH_3}{C}HCH_2CH_2OH$$

(16) (17)

According to a brief note,[3] lithioethylenediamine dehydrogenates cyclohexanol and 2-methylcyclohexanol to cyclohexanone and 2-methylcyclohexanone in yields of 54 and 38%. Saturated primary alcohols afford aldehydes in low to moderate yield (25–66%) and benzyl alcohol gives benzaldehyde in 80% yield.

Although lithioethylenediamine had been prepared and used extensively by Reggel and others, isolation of the compound was first reported by Beumel and Harris[4] (Foote Mineral Co.).

Further examples of isomerization and partial dehydrogenations are shown in the formulations.[5]

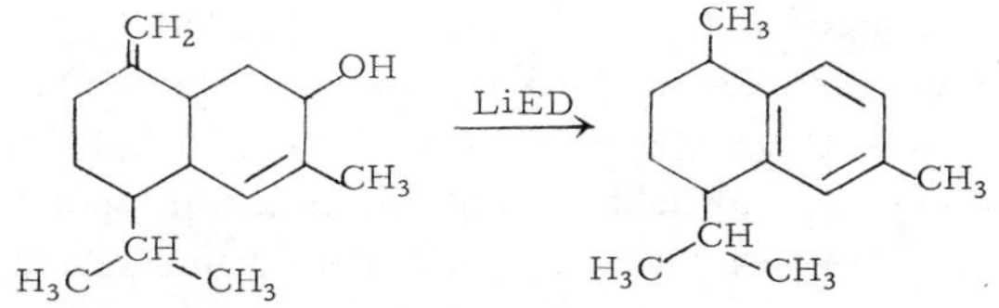

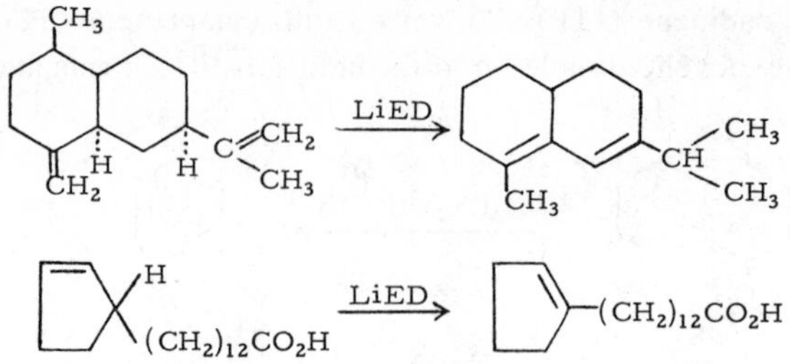

[1]L. Reggel, S. Friedman, and I. Wender, *J. Org.*, **23**, 1136 (1958)
[2]B. S. Tyagi, B. B. Ghatge, and S. C. Bhattacharyya, *J. Org.*, **27**, 1430 (1962)
[3]S. Smith, R. Thompson, and E. O. Woolfolk, *J. Org.*, **27**, 2662 (1962)
[4]O. F. Beumel, Jr., and R. F. Harris, *J. Org.*, **28**, 2775 (1963); **30**, 814 (1965)
[5]B. S. Tyagi, B. B. Ghatge, and S. C. Bhattacharyya, *Tetrahedron*, **19**, 1189 (1963)

Lithium. At. wt. 6.94, m.p. 180.5°, b.p. 1332°, sp. gr. $0.534^{20°}$, solubility: 10.9 g./100 g. NH_3 at −33.2°; 36.5 g./l. CH_3NH_2 at −22.8°. Supplier of lithium ribbon ($\frac{1}{5}$ in. ribbon $\frac{1}{16}$ in. thick on spool in oil): Foote. Suppliers of wire, shot, and dispersions (see below): Lithium Corp. Amer. now Gulf Resources and Chem. Corp., Foote.

Techniques of handling. Most procedures in the literature were worked out when lithium was available only in the form of large rods or slugs coated with paraffin oil. Such material can be hammered into thin foil and the pieces washed free of oil with ether and cut with scissors into chips which are allowed to fall directly into ether in the reaction flask.[1, 2] Hershberg[3] put lithium through a sodium press equipped with a 3-mm. die to produce wire. Bartlett and Lefferts[4] prepared fine shot by melting lithium in mineral oil, adding a few drops of oleic acid to inhibit coalescence, and shaking vigorously for a few seconds. The mixture was cooled and filtered through a pad of glass wool, and the silvery gray particles about 1 mm. in diameter were washed with ether and transferred to the reaction flask. Burgstahler[5] recommends cautious addition of ligroin to the still hot mineral oil suspension before the contents becomes cool enough to set to a gel (caused by the oleic acid). The shot rises cleanly into the neck of the flask and is easily removed with a spoon-type spatula.

The lithium ribbon now available can be wiped to remove protective oil and placed in petroleum ether.[6] Pieces are cut with scissors, argon-dried to remove solvent, and added to the reaction flask.

Lithium Corporation of America now offers lithium of typical sodium content 0.02% in several convenient forms. ***Wire***, as recommended by Hershberg,[3] is $\frac{1}{8}$ in. in diameter and comes in spools of $\frac{1}{4}$ lb., 1 lb., and 2 lbs. Hershberg's technique as applied to the commercial wire is as follows. The length of wire to be taken is calculated from the relationship that 1 cm. of dry wire weighs 0.0423 g. (12 cm. weighs 0.508 g.). The petrolatum coated wire is cut into the proper number of 12-cm. lengths, and a fraction thereof, and the rods are placed in a test tube having a stopcock at the bottom and an argon inlet near the rubber stopper. The rods are washed free of oil with successive portions of benzene and ether under argon, and the wash liquor is drained out through the stopcock under argon pressure. For introduction of lithium, the reaction flask is flushed with argon and the gas allowed to stream out of the widest opening of the flask. The lithium container is flushed with argon while being opened and held at right angles to the opening of the flask, a length of wire is withdrawn with a forceps, and short cylinders are cut with scissors and allowed to drop into the flask. Fresh, silvery surfaces are thus exposed and protected.

The surface area of the $\frac{1}{8}$-in. wire is 1,660 sq. in./lb. The Lithium Corp. lithium ***shot*** has a particle size of approximately 4 to 16 mesh. Lithium ***dispersions*** available in mineral oil, petrolatum, or paraffin wax have a mean diameter of 15 microns and a surface area of 5.3×10^5 sq. in./lb. The compositions are as follows:

Wt.% of Li	Mineral oil	Petrolatum	Oleic acid	Wax
30	69	—	1	—
30	61	8	1	—
30	—	70	—	—
30	—	—	2	68

The dispersion in paraffin wax (m.p. *ca.* 52°) is supplied in rods 1 in. diameter × 8 in. length; each rod contains 22.7 g. of lithium metal. Rods are sealed in individual polyethylene tubes and packed 20 rods to a can. The dispersion in wax has exceptional shelf life, does not segregate, is easily and safely handled in air, and the required amount of metal is obtained by cutting and weighing the calculated amount of dispersion. The dispersion is converted into dry lithium metal powder by washing away the paraffin wax matrix with pentane or hexane in a closed vessel under dry argon, followed by removal of solvent and drying in a stream of argon. Dry lithium powder is extremely reactive to air, water vapor, and nitrogen.

Dispersions in mineral oil will segregate on storage but uniformity of composition can be restored by stirring without agglomerating the particles. The dispersion is then spooned from the container for weighing. With dispersions in wax, segregation is not a problem.

Fire control. Lithium fires can be controlled by smothering with anhydrous lithium chloride or with Ansul Plus 50 fire-extinguishing compound.

Alkyl and aryl derivatives. *See also* *n*-Butyllithium; Lithium–Sodium alloy; Methyllithium; Phenyllithium.

Alkyl and aryl derivatives of lithium are made by treating a halide with 2 equivalents of lithium metal in dry ether by essentially the technique employed in preparing

$$RBr + 2\ Li \longrightarrow RLi + LiBr$$

a Grignard reagent, and yields are excellent.[7, 8] The reaction with lithium starts more readily and proceeds at a greater rate than that with magnesium. Differences are that lithium should not be warmed in a stream of nitrogen because it tends to form the nitride and that lithium floats on ether. Lithium alkyls can be prepared not only in the presence of ether but also with low-boiling petroleum ether as sole solvent;[9] the hydrocarbon solvent makes possible the preparation of organolithium compounds (*e.g.*, isopropyllithium) otherwise difficultly obtainable.

Lithium derivatives are somewhat more reactive than the corresponding Grignard reagents[10] and, in the reaction with an α,β-unsaturated ketone, differ from the magnesiohalides in giving rise to a much higher proportion of the 1,2-addition product.[11] An important application is in utilization of aromatic chloro compounds in syntheses which cannot be accomplished by the Grignard method. Alkyl- and aryllithiums react readily with ethylene oxide without the need for heating, as is the case with a Grignard reagent. Care must be taken to use exactly 1 mole of ethylene oxide per mole of R(Ar)Li to avoid formation of products of the type $ROCH_2CH_2(OCH_2CH_2)_xOH$.

Not all organic halides react satisfactorily with lithium, but often the desired lithium compound can be prepared indirectly by exchange reaction with a readily available lithium compound such as *n*-butyllithium:

$$\underline{n}\text{-}C_4H_9Li \;+\; RX \qquad RLi \;+\; \underline{n}\text{-}C_4H_9X$$

The exchange reaction was discovered in 1938 by Gilman[12] and by Wittig[13] and has since found wide use.[14] The preparation of *n*-butyllithium (phenyllithium, though less reactive, is also used), the interchange step, and a reaction with a carbonyl reactant can be carried out in succession in the same flask, and only 30–60 min. need be allowed for formation of the intermediate RLi compound. The lithium atom can be replaced by MgBr, and occasionally this is the most expedient route to a given Grignard reagent.[15]

The yield of an alkyl- or aryllithium can be determined by the acid-titration method of analysis used for assay of Grignard reagents, or by adding benzophenone to an aliquot portion and determining the weight of tertiary alcohol formed.[16] The double titration method of Gilman and Haubein[17] has long been used. An oxidimetric V_2O_5 method[18] is a simple and accurate procedure for the determination of alkyllithiums in hydrocarbon solution. However, neither of the latter two methods can be used for the determination of aryllithiums.

In exploratory studies preparatory to eventual attempts to synthesize plasmalogens such as (1), Craig *et al.*[19] found that dehydrobromination of the glycerol α-bromo-

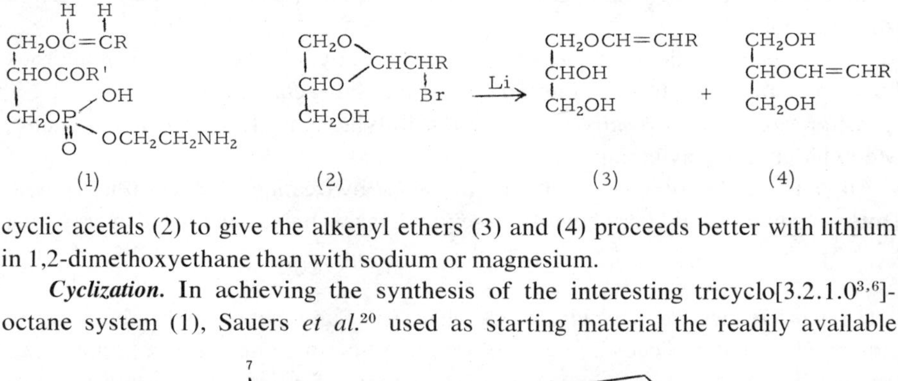
Li

(1) (2) (3) (4)

cyclic acetals (2) to give the alkenyl ethers (3) and (4) proceeds better with lithium in 1,2-dimethoxyethane than with sodium or magnesium.

Cyclization. In achieving the synthesis of the interesting tricyclo[3.2.1.0^{3,6}]-octane system (1), Sauers *et al.*[20] used as starting material the readily available

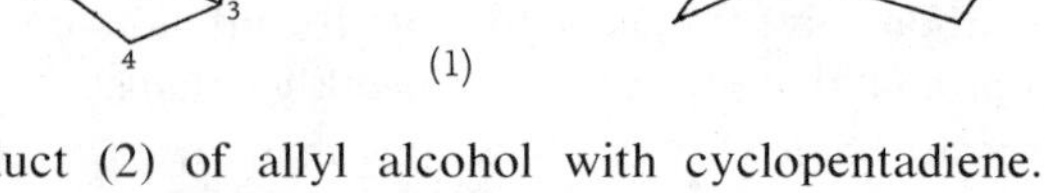
7
1
8
2
5
6
3
4

(1)

Diels-Alder adduct (2) of allyl alcohol with cyclopentadiene. The unsaturated alcohol was converted into 6-chloromethyl-3-oxatricyclo[3.2.1.0^{2,4}]octane (3). For cyclization of (3) to the tricyclic alcohol (4), lithium in tetrahydrofurane proved superior to sodium in toluene. The alcohol (4) has a comparatively strained cyclobutane ring, but was converted easily into the parent hydrocarbon by oxidation (CrO_3–Py) and Wolff-Kishner reduction.

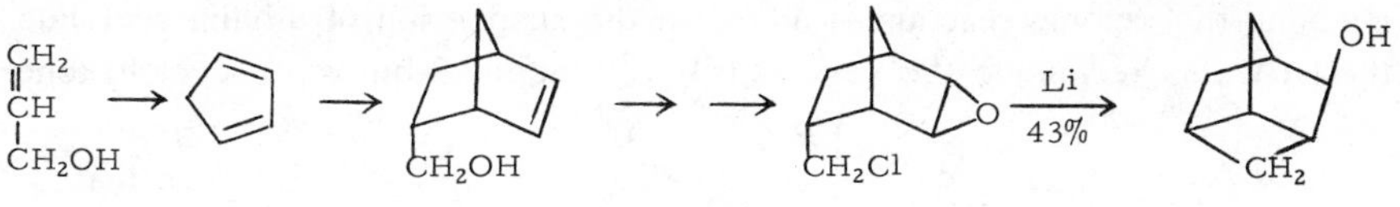

(2) (3) (4)

[1]R. B. Woodward and E. C. Kornfeld, *Org. Syn., Coll. Vol.*, **3**, 413 (1955)
[2]H. Heaney and I. T. Millar, *Org. Syn.*, **40**, 105 (1960)
[3]L. F. Fieser and E. B. Hershberg, *Am. Soc.*, **59**, 396 (1937)
[4]P. D. Bartlett and E. B. Lefferts, *Am. Soc.*, **77**, 2804 (1955)
[5]A. W. Burgstahler, private communication
[6]C. D. Gutsche and H. H. Peter, *Org. Syn., Coll. Vol.*, **4**, 887 (1963)
[7]K. Ziegler and H. Colonius, *Ann.*, **479**, 135 (1930)
[8]H. Gilman *et al.*, *Am. Soc.*, **54**, 1957 (1932)
[9]H. Gilman, W. Langham, and F. W. Moore, *Am. Soc.*, **62**, 2327 (1940)
[10]H. Gilman and R. H. Kirby, *Am. Soc.*, **55**, 1265 (1933)
[11]A. Lüttringhaus, *Ber.*, **67**, 1602 (1934)
[12]H. Gilman and A. L. Jacoby, *J. Org.*, **3**, 108 (1938)
[13]G. Wittig, U. Pockels, and H. Dröge, *Ber.*, **71**, 1903 (1938)
[14]R. G. Jones and H. Gilman, *Org. Reactions*, **6**, 339 (1951)
[15]H. Gilman and C. E. Arntzen, *Am. Soc.*, **72**, 3823 (1950)
[16]H. Gilman, E. A. Zoellner, and W. M. Selby, *Am. Soc.*, **54**, 1957 (1932)
[17]H. Gilman and A. H. Haubein, *Am. Soc.*, **66**, 1515 (1944)
[18]P. F. Collins, C. W. Kamienski, D. L. Esmay, and R. B. Ellestad, *Anal. Chem.*, **33**, 468 (1961)
[19]J. C. Craig, D. P. G. Hamon, H. W. Brewer, and H. Härle, *J. Org.*, **30**, 907 (1965)
[20]R. R. Sauers, R. A. Parent, and S. B. Damle, *Am. Soc.*, **88**, 2257 (1966)

Lithium acetylide, LiC≡CH. Mol. wt. 31.97.

Procedures for the preparation of sodamide in liquid ammonia and for its reaction with acetylene to form sodium acetylide are applicable also to the preparation of

$$Li \xrightarrow[-H_2]{NH_3(Fe^{+++})} LiNH_2 \xrightarrow[-NH_3]{HC\equiv CH} LiC\equiv CH$$

lithium amide and lithium acetylide.[1] Although sodium acetylide is generally used for the conversion of a ketone into an ethynylcarbinol (Nef reaction), lithium acetylide is often superior, particularly in the case of α,β-unsaturated ketones. For example, β-ionone (1) reacts with lithium acetylide to give the carbinol (2) with a conversion

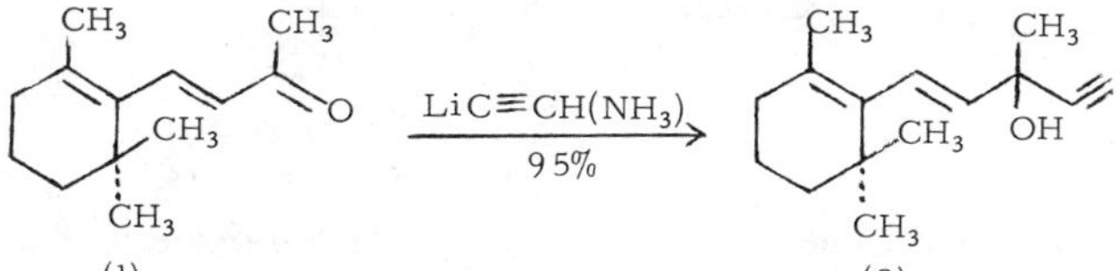

(1) (2)

of 79% and yield of 95%.[2] With sodium acetylide under identical conditions (3 hrs. at −34°) the conversion was 27% and the yield 74%. $\Delta^{2,4,6}$-Octatrienal (3) reacts with lithium acetylide to give the ethynylcarbinol (4) in 72% yield, whereas with sodium acetylide the yield is less than one third as good.[3]

$$\underset{(3)}{CH_3CH{=}CHCH{=}CHCH{=}CHCHO} \xrightarrow[73\%]{LiC\equiv CH(NH_3)} \underset{(4)}{CH_3CH{=}CHCH{=}CHCH{=}CH\underset{OH}{\underset{|}{C}}HC\equiv CH}$$

If the ketone is sensitive to liquid ammonia, lithium acetylide can be prepared in ammonia in the usual way and precipitated with ether. After displacement of all the ammonia, the ketonic reactant is added to the suspension of lithium acetylide in ether. By this technique the very sensitive 2-methyl-3-butoxy-Δ^2-cyclohexene-1-

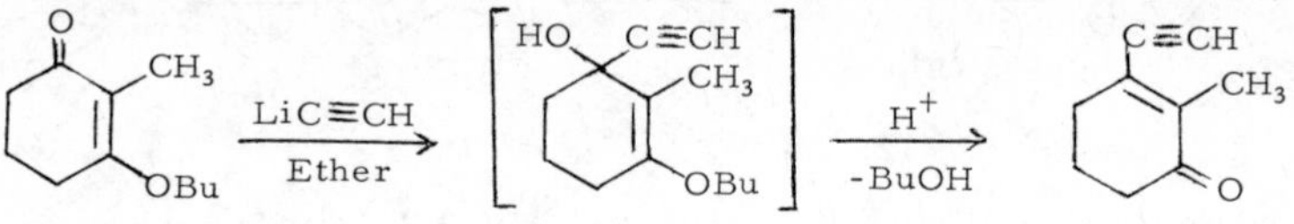

(5) (6) (7)

one (5) is converted into the ethynylcarbinol (6), which rearranges with loss of butanol to give (7).[4]

[1] R. A. Raphael, "Acetylenic Compounds in Organic Syntheses," 193–194, Academic Press (1955)
[2] W. Oroshnik and A. D. Mebane, *Am. Soc.*, **71**, 2062 (1949)
[3] R. A. Raphael, *loc. cit.*,[1] p. 197
[4] A. Eschenmoser, J. Schreiber, and S. A. Julia, *Helv.*, **36**, 482 (1953)

Lithium acetylide–Ethylenediamine, $LiC{\equiv}CH \cdot H_2NCH_2CH_2NH_2(LiC{\equiv}CH \cdot EDA)$. The stabilized form of lithium acetylide,[1] available in 100-g. bottles from Foote Mineral Co., need not be handled and used in liquid ammonia but is handled as a dry powder and can be used in solvents such as diglyme, dioxane, dimethylformamide, and dimethyl sulfoxide.

Solubilities in g./100 g. at 25°: hexane, 0; benzene, 0.008; xylene, 0.04; diethyl ether, 0.05; THF, 0.4; dioxane, 0.6; *n*-propylamine, 17.0; *n*-butylamine, 12.1; di-*n*-butylamine, 0.03.

Reaction with halides. A typical halogen displacement carried out by Pattison and Dear[2] in dimethyl sulfoxide afforded 7-fluoro-1-heptyne in excellent yield.

$$F(CH_2)_5Cl + LiC{\equiv}CH \cdot EDA \xrightarrow[92\%]{DMSO\ 25^0} F(CH_2)_5C{\equiv}CH$$

Addition to ketones. Huffman and Arapakos[3] were able to convert the ketone (1) into the ethynylcarbinol (2) in only very low yield with either sodium acetylide

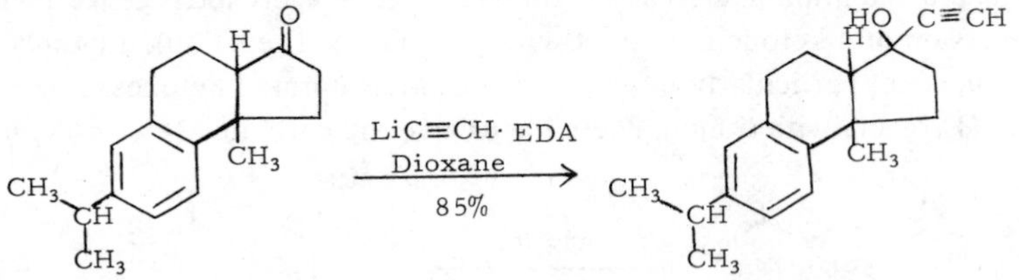

(1) (2)

in liquid ammonia or acetylene and potassium 2-methylbutoxide, but with lithium acetylide–EDA the reaction proceeded smoothly and in high yield.

[1] O. F. Beumel, Jr., and R. F. Harris, *J. Org.*, **28**, 2775 (1963); **29**, 1872 (1964)
[2] F. L. M. Pattison and R. E. A. Dear, *Can. J. Chem.*, **41**, 2600 (1963)
[3] J. W. Huffman and P. G. Arapakos, *J. Org.*, **30**, 1604 (1965)

Lithium–Alkylamine reduction.

		B.P.	Sp. Gr.
Li +	Methylamine	−6.5°	0.699
	Dimethylamine	7.4°	0.680
	Ethylamine	16.6°	0.689
	i-Propylamine	34°	0.694
	n-Propylamine	48.7°	0.719
	Morpholine	128°	

Benkeser[1] discovered that the combination of lithium with an alkylamine of low molecular weight is a powerful reducing agent for aromatic hydrocarbons and certain other systems. The amine most frequently used in ethylamine. An example of the reaction is the reduction of naphthalene by lithium and ethylamine to give in high yield a mixture of $\Delta^{9,10}$- and $\Delta^{1,9}$-octalin. In reporting another reduction to be cited below, Burgstahler[2] states that use of nonredistilled tank ethylamine inhibits the reaction, presumably because traces of iron compounds from the tank catalyze amide formation at the expense of reduction. Lithium wire washed with ether and cut into small pieces usually is satisfactory, although Burgstahler recommends fine lithium shot prepared according to Bartlett (*see* Lithium).

A reduction can be carried out by dissolving the substance to be reduced in a sufficiently large excess of redistilled, anhydrous ethylamine in a flask fitted with a Dewar-type dry ice condenser (Ace Glass 6121), adding the calculated amount of lithium, and stirring magnetically; a deep blue color which develops initially is discharged when reduction is complete. In the absence of a substrate, lithium slowly dissolves in refluxing ethylamine without evolution of hydrogen to give a deep blue solution which apparently contains Li^+ bound to a negatively charged solvent complex. This reaction, although irreversible, is usually carried out in the presence of a substrate which reacts with the blue complex almost as rapidly as it is formed, probably by 1,4-addition of lithium. We have formulated a possible sequence of events in the reduction of naphthalene to illustrate the type of process that appears

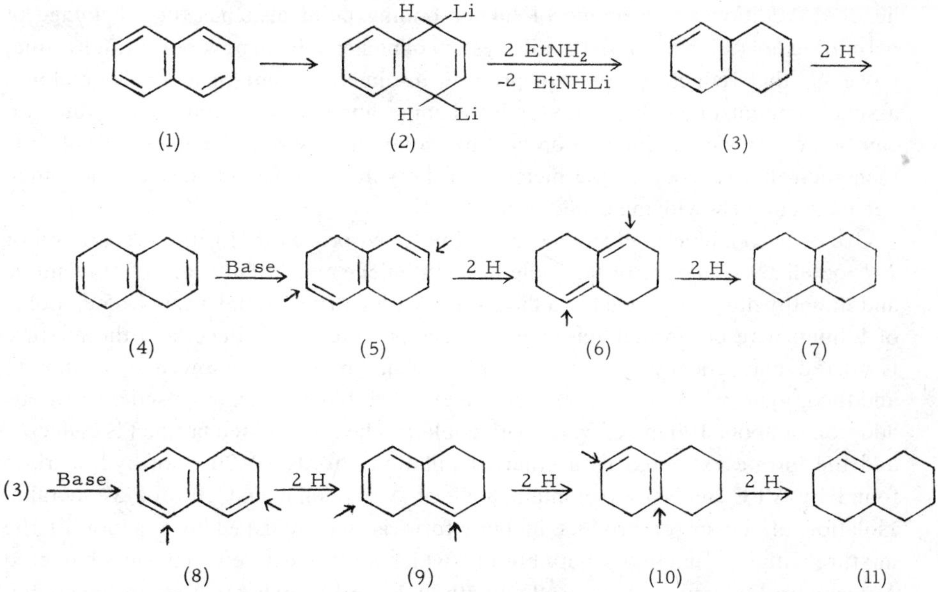
H
Li
2 EtNH2
-2 EtNHLi
H
Li
2 H
(1)
(2)
(3)
Base
2 H
2 H
(4)
(5)
(6)
(7)
(3)
Base
2 H
2 H
2 H
(8)
(9)
(10)
(11)

to be involved, but it is completely hypothetical. The 1,4-dilithio adduct (2) reacts with ethylamine to form Δ^2-dialin (3) which in part suffers a second 1,4-addition across a benzene ring to give, after metal exchange with solvent, isotetralin (4). This hydrocarbon seems a likely intermediate since it was obtained by Grob[3] in 82% yield by reduction of naphthalene with sodium in liquid ammonia–ether–absolute ethanol at −70°. At the higher temperature of the Li–$EtNH_2$ reaction, isotetralin may suffer base-catalyzed isomerization to the conjugated triene (5), convertible by a 1,6-addition (6) followed by a 1,4-addition into $\Delta^{9,10}$-octalin (7), the main end product. A possible route to $\Delta^{1,9}$-octalin (11), the minor product, involves base-catalyzed isomerization of Δ^{2}-dialin (3) to the conjugated Δ^{1}-dialin (8), followed by 1,8-addition (9), 1,6-addition (10), and 1,4-addition (11).

Treatment of toluene in methylamine with 4 equivalents of lithium effects reduction to a mixture of methylcyclohexenes found by VPC analysis to contain 59% of 1-methylcyclohexene (V) and 41% of 3- and 4-methylcyclohexene.[4] The reactions

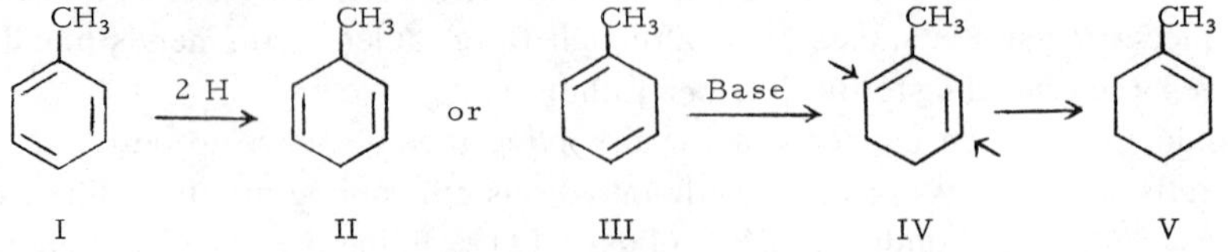

again are accounted for on the assumption of 1,4-addition, isomerization of nonconjugated to conjugated dienes, and 1,4-addition. Benkeser and co-workers[5] found that greater selectivity can be achieved by replacing a considerable portion of the methylamine or ethylamine by dimethylamine, *n*- or *i*-propylamine, or morpholine. The added amine increases the proportion of the more thermodynamically stable reduction product. Thus reduction of toluene with 4 equivalents of lithium in a mixture of 10% (by volume) of methylamine and 90% of dimethylamine afforded a mixture of 82% of 1-methylcyclohexene (V) and 18% of 3- and 4-methylcyclohexene. Addition of an amine of higher boiling point also has the advantage of permitting operation in a less volatile system of higher solvent power. It is interesting, however, that reduction does not proceed in dimethylamine or morpholine alone; a small amount of methylamine or ethylamine appears to be required. Reduction can be accomplished with *n*-propylamine alone, but the reflux temperature (49°) is then such that reaction of the metal with the solvent to form hydrogen and amide begins to compete with metal addition.

Use of a co-amine is embodied in an improved procedure for the preparation of $\Delta^{9,10}$-octalin.[6] A mixture of 0.2 mole of naphthalene and 250 ml. each of ethylamine and dimethylamine is placed in a flask fitted with a dry ice condenser, 1.65 g.-atoms of lithium wire cut in half-centimeter pieces is added all at once, and the mixture is stirred magnetically for 14 hrs. The solvent mixture is allowed to evaporate and the grayish white residue (containing excess lithium) is decomposed by cautious addition of about 100 ml. of water with cooling. The precipitated product is collected and the filtrate extracted with ether; distillation affords 19–20 g. of hydrocarbon found by VPC analysis to contain 80% of $\Delta^{9,10}$-octalin and 20% of $\Delta^{1,9}$-octalin. Isolation of the major product in pure form is accomplished by reaction of the mixture with bis-3-methyl-2-butylborane, which adds selectively to the less hindered $\Delta^{1,9}$-isomer. Oxidation of the product with hydrogen peroxide to convert the adduct into an easily separated alcohol, followed by distillation, affords $\Delta^{9,10}$-octalin of 99% purity in yield from naphthalene of 50–54%.

Before the technique of VPC analysis was available, Benkeser[1] observed that ethylbenzene is reduced by excess lithium at the boiling point of ethylamine (16.6°) to ethylcyclohexene and ethylcyclohexane, while at −78° 1-ethylcyclohexene is the sole product. The later quantitative work already cited[4] established that reduction of toluene in methylamine with 4 equivalents of lithium gives 1-methylcyclohexene

(V) in 59% yield and 37% of a mixture of 3-methylcyclohexene (VI) and 4-methylcyclohexene (VII). Possible routes to these isomers are suggested in the formulation. In the presence of an adequate amount of lithium, these isomers are reduced fairly

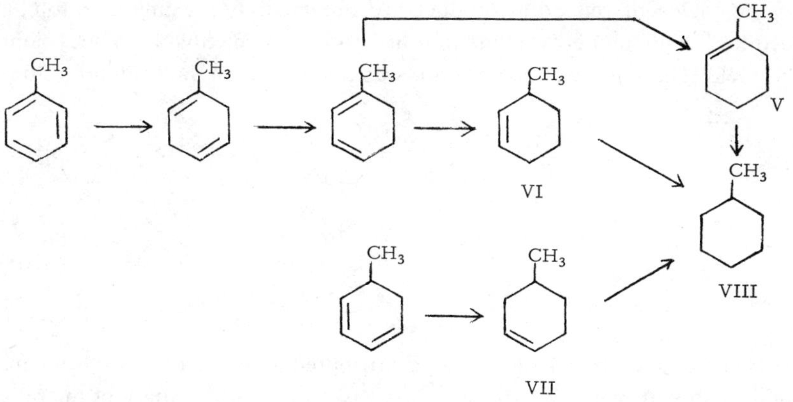
CH_3
V
VI
VII
VIII

readily to methylcyclohexane, evidently by 1,2-addition of lithium and metal exchange with the amine. 1-Methylcyclohexene (V) is reduced much more slowly, for example to the extent of only 4% in 3 hrs. (Li + CH_3NH_2) in comparison with 59% reduction of 4-methylcyclohexene under the same conditions. The inductive effect of the methyl group evidently slows down the uptake of electrons by the double bond.

Acetylenic bonds are reduced readily by a process of strictly *trans*-addition.[7] Thus 5-decyne (0.1 mole) is reduced by lithium (0.25 mole) in ethylamine at −78° in fair yield to *trans*-5-decene, and 3-octyne affords *trans*-3-octene (52%). However,

$$CH_3(CH_2)_3C{\equiv}C(CH_2)_3CH_3 \xrightarrow[52\%]{Li-C_2H_5NH_2} CH_3(CH_2)_3\overset{H}{C}{=}\underset{H}{C}(CH_2)_3CH_3$$

reduction carried out in ethylamine at the boiling point (16.6°) afforded appreciable amounts of the alkanes, particularly when lithium was taken in excess. The result is in contrast to reduction of alkynes by sodium in liquid ammonia, which stops cleanly at the olefin stage, and confirms the finding with the cyclohexene derivatives that a nonterminal, disubstituted double bond is fairly susceptible to reduction by the Li-RNH_2 system.

Benkeser and co-workers[8] found that the reduction of aromatic nitro compounds by the lithium–amine reagent stops rather cleanly at the aromatic amine even though arylamines can be reduced by excess lithium in ethylamine to cyclohexene derivatives. This surprising finding is accounted for as follows. Reduction of the nitro group is attended with generation of alkyl amide ions, and as the concentration of these ions increases an equilibrium is established between the arylamine and the

$$ArNO_2 \xrightarrow{Li-EtNH_2} \underbrace{ArNH_2 + Et\bar{N}H(Li^+)}$$
$$\rightleftharpoons$$
$$Ar\bar{N}H(Li^+) + EtNH_2$$

anilide ions. Since an alkyl amide ion is a much stronger base than the aryl amide ion, the equilibrium lies well in the direction of the aryl amide ion. Because of its negative charge, the aryl amide ion is not prone to take on electrons and reduction stops at this stage. On the other hand, if the starting material is the arylamine the ring is reduced rapidly at first, since little or no aryl amide ion is present, and reasonably good yields of reduction product are obtained. For example, *p*-toluidine is reduced by lithium and ethylamine in 6 hrs. to *trans*-1-methyl-4-aminocyclohexane in 50% yield. That the reaction does not stop at the stage of the 1-aminocyclohexene

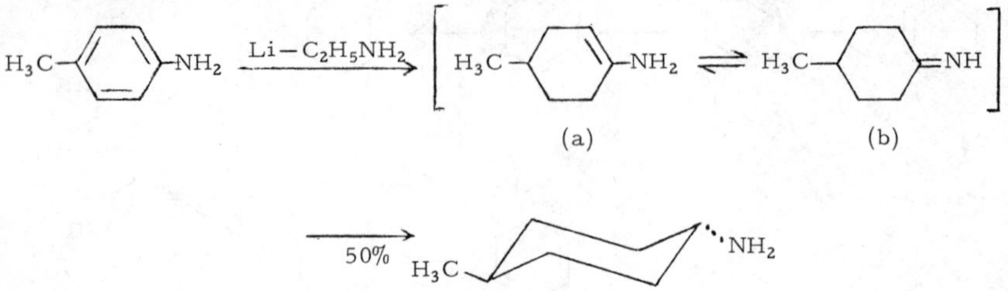
H₃C
NH₂
Li–C₂H₅NH₂
H₃C
NH₂
H₃C
NH
(a)
(b)
50%
H₃C
NH₂

(a), as in the case of the alkylbenzene, is attributed to the fact that (a) is an enamine in equilibrium with the imine (b), which is easily reduced. A study of the reduction of N-methylaniline and N,N-dimethylaniline afforded results supporting this postulate.[4] Notice that reduction of *p*-toluidine affords a product in which both substituents are equatorial. Reduction of *o*- and *m*-toluidine also gave in each case the most stable diequatorial cyclohexane derivative. The reaction thus should be useful for the stereospecific synthesis of cyclohexane derivatives.

In seeking to synthesize abietic acid (3) from dehydroabietic acid (1), Burgstahler and Worden[2] first tried without success to reduce the aromatic ring with sodium or lithium and ethanol in liquid ammonia under the usual Birch conditions, even in the presence of ether as cosolvent. They then turned to the more powerful reducing

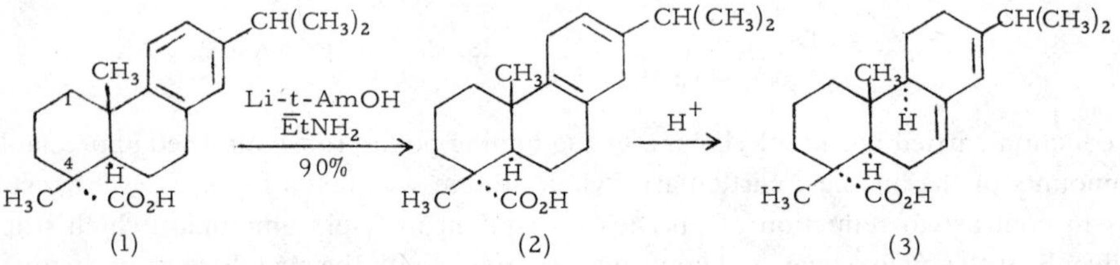
CH(CH₃)₂
CH₃
H₃C
H
CO₂H
Li-t-AmOH
EtNH₂
90%
CH(CH₃)₂
CH₃
H₃C
H
CO₂H
H⁺
CH(CH₃)₂
CH₃
H
H₃C
H
CO₂H

(1) (2) (3)

system developed by Benkeser. Lithium and ethylamine reduced the aromatic ring but also reduced the carboxyl group to an aldehyde group (reduction of other acids to aldehydes has been observed, but yields are generally low[9]). However, in the presence of a very weakly acidic proton source, such as *t*-amyl alcohol or *t*-butanol, selective reduction of dehydroabietic acid to $\Delta^{8,12}$-abietadienoic acid (2) was achieved in yield of 90%. Acid isomerization furnished a mixture from which abietic acid (3) was isolated as the diisoamylamine salt. Alcohols more acidic than the two mentioned, for example ethanol or isopropanol, almost completely inhibited the reduction. The reaction was carried out with 300 mg. (1 mmole) of dehydroabietic acid and 4.3 ml. (40 mmoles) of *t*-amyl alcohol in a flask equipped with a dry ice condenser and a high-speed stirrer (slow-speed stirring resulted in incomplete reduction). Redistilled anhydrous ethylamine (30 ml.) was added, followed by 240 mg. (35 mg. -atoms) of finely divided lithium shot (Bartlett, *see* Lithium). After 15 min., when the lithium had dissolved and a pale blue cast pervaded the solution, a second 240-mg. charge of lithium was added. When the characteristic deep blue color first persisted, *t*-amyl alcohol was added until the mixture remained colorless (*ca.* 5 ml.) and the mixture was worked up.

Henbest[10] found lithium–ethylamine an effective reagent for the reduction of allylic esters or ethers to olefins. For example a solution of 0.4 g. of the benzoate of Δ^4-cholestene-3β-ol in 20 ml. of ethylamine on reaction with 0.1 g. of lithium

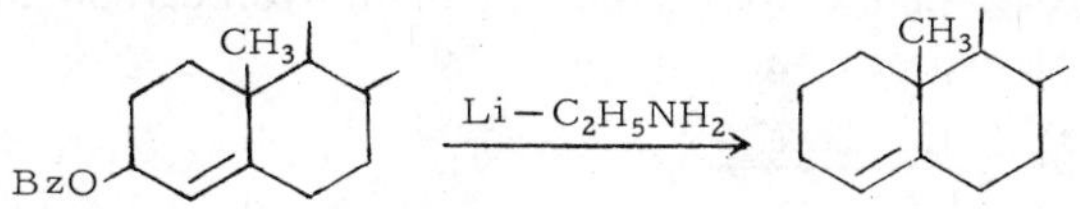
CH3
BzO
Li–C2H5NH2
CH3

afforded 0.224 g. of pure Δ^4-cholestene. For the reduction of the 7α,8α-epoxide derived from 3β-acetoxy-Δ^7-lanostene to the 7α-ol, Fried *et al.*[11] added *t*-butanol as a proton source. One advantage of the reagent over Li–NH_3 is that most organic compounds are more soluble in ethylamine than in liquid ammonia. For the reduction of steroid epoxides to axial alcohols, the reagent is more powerful and more specific than lithium aluminum hydride.[12] Thus 7α,8α- and 9α,11α-epoxides are not reduced

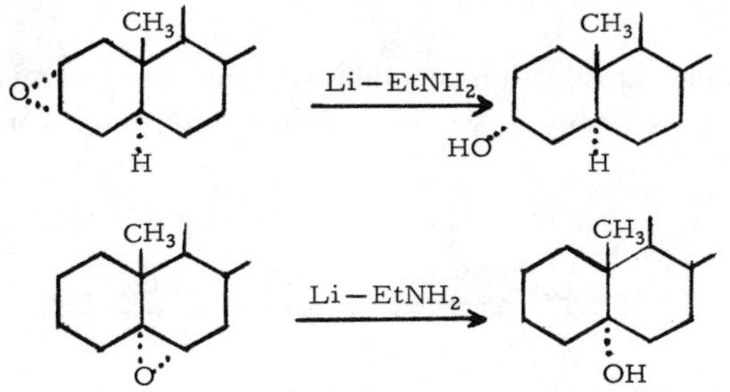
CH3
O
H
Li–EtNH2
CH3
HO
H
CH3
O
Li–EtNH2
CH3
OH

by the metal hydride but react with lithium–ethylamine to give the 8α- and 9α-alcohols. If the molecule contains a hydroxyl group as, for example, in the case of cholesterol α-oxide (1), reduction with lithium–ethylamine affords both the axial alcohol (2) and the olefinic cholesterol (3).[13] With the β-epoxide, only starting material and olefin (3) are formed. The interpretation advanced is that the anion from the

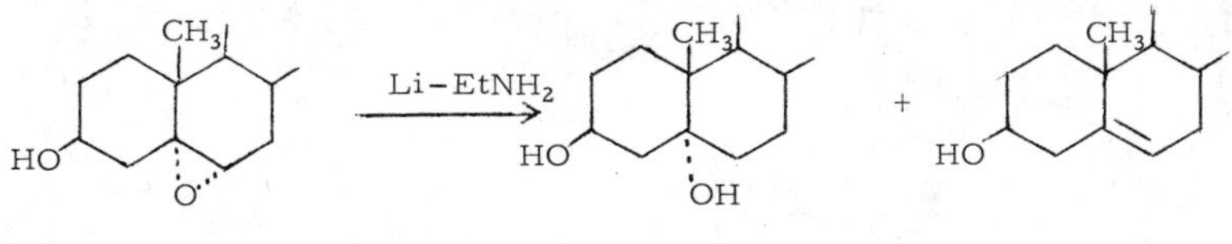
CH3
HO
O
Li–EtNH2
CH3
HO
OH
+
CH3
HO

(1) (2) (3)

hydroxyl group inhibits formation of an anionic intermediate that leads to the axial alcohol.

Henbest found that lithium–ethylamine also is an effective reagent for the desulfurization of ethylenethioketals.[14] The Diels-Alder adduct of benzoquinone with 2 moles of butadiene, for which Henbest deduced the configuration (4), was converted

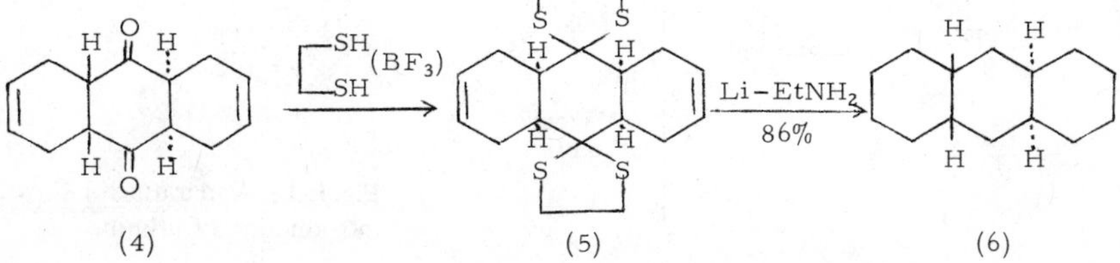
H O H
H O H
SH
SH
(BF3)
S S
H H
H H
S S
Li–EtNH2
86%
H H
H H

(4) (5) (6)

into the bis-ketal (5) and this when treated with lithium and ethylamine at −20° suffered desulfurization and reduction of the two isolated double bonds to give the perhydroanthracene (6) in excellent yield. This efficient and stereospecific method has been employed for the reduction of diones of the cyclohexane series.[15]

Truce[16] found that lithium in ethylamine is superior to sodium in liquid ammonia for the cleavage of a sulfide to a mercaptan and a hydrocarbon and of a sulfone to a sulfinic acid and a hydrocarbon.

$$RSR' \xrightarrow{Li-EtNH_2} RSLi + RLi \xrightarrow{EtNH_2} RSH + RH$$

Examples are as follows:

$$(\underline{n}\text{-}C_{10}H_{21})_2S \xrightarrow{Li-EtNH_2} \underset{86\%}{\underline{n}\text{-}C_{10}H_{21}SH} + \underset{85\%}{\underline{n}\text{-}C_{10}H_{22}}$$

$$\underline{n}\text{-}C_{10}H_{21}SC_6H_5 \xrightarrow{Li-EtNH_2} \underset{87\%}{C_6H_5SH} + \underset{71\%}{\underline{n}\text{-}C_{10}H_{22}} + \underline{n}\text{-}C_{20}H_{42}$$

$$(\underline{n}\text{-}C_{10}H_{21})_2SO_2 \longrightarrow \underset{82\%}{\underline{n}\text{-}C_{10}H_{21}SO_2Li} + \underline{n}\text{-}C_{10}H_{22}$$

Finding that gradual introduction of lithium to the reaction mixture is beneficial, Truce designed an assembly in which refluxing ethylamine condenses and drops

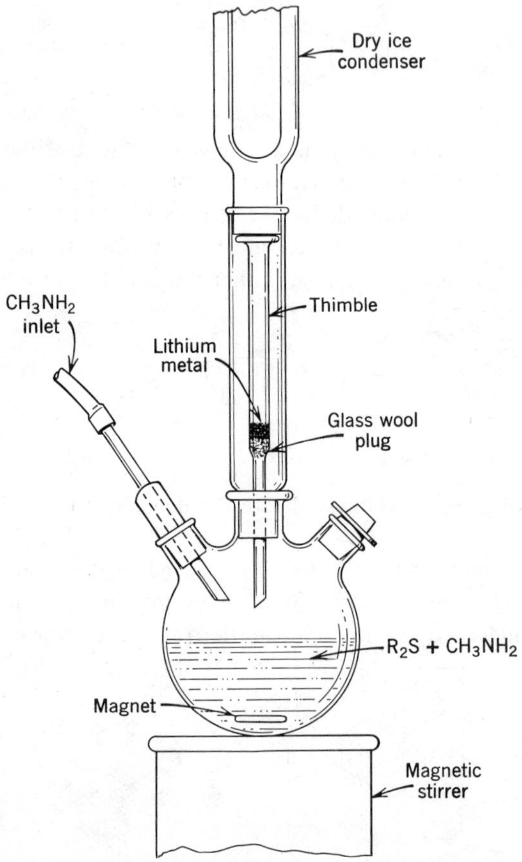
Dry ice
condenser
Thimble
CH3NH2
inlet
Lithium
metal
Glass wool
plug
R2S + CH3NH2
Magnet
Magnetic
stirrer

Fig. L-1 Apparatus for slow introduction of lithium

through a thimble containing the calculated amount of lithium wire cut in small pieces. A drawing kindly supplied by Dr. Truce shows the arrangement (Fig. L-1) The thimble is simply a test tube drawn out at the bottom to a taper and connected to a length of 6-mm. o.d. glass tubing.

It is well known that treatment of a tertiary amide with lithium aluminum hydride effects reductive cleavage to an aldehyde and a secondary amine.[17] Patchornik[18]

$$RC(=O)-NR'_2 \xrightarrow{LiAlH_4} RCH(=O) + HNR'_2$$

found that the N-proline peptide bond is similarly cleaved by reduction with lithium and methylamine. If a peptide contains a C-terminal proline unit, reduction liberates free proline, which can be determined colorimetrically:

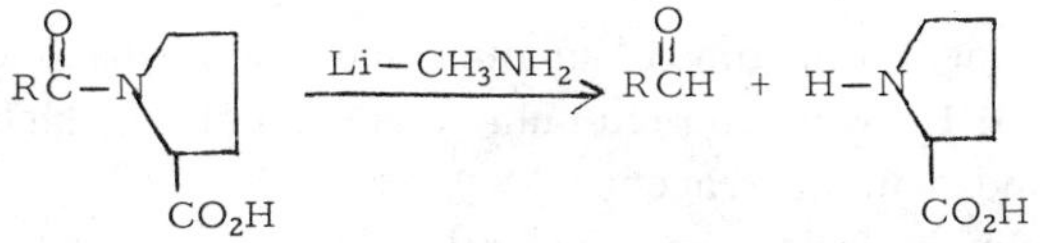

For example, L-alanyl-L-proline is acetylated and the N-acetyl derivative dissolved in methylamine and treated with excess lithium at −70°. After 1 hr. ammonium chloride is added to discharge the blue color, the solvent is allowed to evaporate,

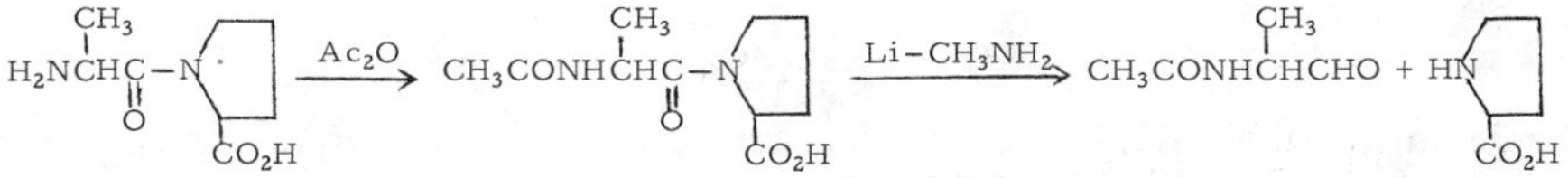

the residue is dissolved in water and the proline determined colorimetrically. A peptide containing an internal proline unit affords on reduction a new peptide having an N-terminal proline unit from which proline can be liberated by digestion with the specific exoenzyme L-proline imino peptidase.

[1]R. A. Benkeser, R. E. Robinson, D. M. Sauve, and O. H. Thomas, *Am. Soc.*, **77**, 3230 (1955)
[2]A. W. Burgstahler and L. R. Worden, *Am. Soc.*, **86**, 96 (1964)
[3]C. A. Grob and P. W. Schiess, *Helv.*, **43**, 1546 (1960)
[4]R. A. Benkeser, J. J. Hazdra, R. F. Lambert, and P. W. Ryan, *J. Org.*, **24**, 854 (1959)
[5]R. A. Benkeser, R. K. Agnihotri, and M. L. Burrous, *Tetrahedron Letters*, 1 (1960); R. A. Benkeser, R. K. Agnihotri, M. L. Burrous, E. M. Kaiser, J. M. Mallan, and P. W. Ryan, *J. Org.*, **29**, 1313 (1964)
[6]R. A. Benkeser and E. M. Kaiser, *J. Org.*, **29**, 955 (1964)
[7]R. A. Benkeser, G. Schroll, and D. M. Sauve, *Am. Soc.*, **77**, 3378 (1955)
[8]R. A. Benkeser, R. F. Lambert, P. W. Ryan, and D. G. Stoffey, *Am. Soc.*, **80**, 6573 (1958)
[9]A. W. Burgstahler, L. R. Worden, and T. B. Lewis, *J. Org.*, **28**, 2918 (1963)
[10]A. S. Hallsworth, H. B. Henbest, and T. I. Wrigley, *J. Chem. Soc.*, 1969 (1957)
[11]J. Fried, J. W. Brown, and L. Borkenhagen, *Tetrahedron Letters*, 2499 (1965)
[12]A. S. Hallsworth and H. B. Henbest, *J. Chem. Soc.*, 4604 (1957)
[13]A. S. Hallsworth and H. B. Henbest, *J. Chem. Soc.*, 3571 (1960)
[14]N. S. Crossley and H. B. Henbest, *J. Chem. Soc.*, 4413 (1960)
[15]R. D. Stolow and M. M. Bonaventura, *Tetrahedron Letters*, 95 (1964)
[16]W. E. Truce, D. P. Tate, and D. N. Burdge, *Am. Soc.*, **82**, 2872 (1960)
[17]For an example, see L. Birkofer and E. Frankus, *Ber.*, **94**, 216 (1961)
[18]A. Patchornik, M. Wilchek, and S. Sarid, *Am. Soc.*, **86**, 1457 (1964)

Lithium aluminum hydride, $LiAlH_4$. Mol. wt. 37.95. Solubility, g./100 g. at 25°: diethyl ether, 35–40 (to attain this degree of solubility it is necessary to concentrate a more dilute solution); tetrahydrofurane, 13; di-*n*-butyl ether, 2; dioxane, 0.1. Supplier: Metal Hydrides, now Ventron Corp.

Reviews. Metal Hydrides Technical Bulletin No. 401, October, 1963; U. Solms, *Chimia*, **5**, 25 (1951); W. G. Brown, *Org. Reactions*, **6**, 649 (1951); N. G. Gaylord, "Reduction with Complex Metal Hydrides," Interscience, N. Y. (1956); N. G. Gaylord, *J. Chem. Ed.*, **34**, 367 (1957); H. Hörmann, ' Newer Methods in Preparative Organic Chemistry," **2**, 213 (1963)

Determination. A procedure and apparatus for assay by hydrogen evolution are described in Metal Hydrides Technical Bulletin No. 401.

History. Discovery of the reagent by Finholt, Bond, and Schlesinger[1] was announced in 1947. It is made by gradual addition of aluminum chloride to a slurry of lithium hydride in ether and filtering from precipitated lithium chloride and

$$4\ LiH + AlCl_3 \longrightarrow LiAlH_4 + 3\ LiCl$$

excess lithium hydride. At the time of announcement of the discovery of the reagent, Nystrom and W. G. Brown[2] reported numerous examples of high-yield reductions of organic compounds with the reagent.

Lithium aluminum hydride reacts instantly with excess water or ethanol with liberation of 4 moles of hydrogen. Hochstein[3] in a quantitative study of lithium

$$LiAlH_4 + 4\ HOH \longrightarrow LiOH + Al(OH)_3 + 4\ H_2$$

$$LiAlH_4 + 4\ C_2H_5OH \longrightarrow LiOC_2H_5 + Al(OC_2H_5)_3 + 4\ H_2$$

aluminum hydride reduction found that compounds which liberate methane from methylmagnesium iodide in the Zerewitinoff test liberate hydrogen from the hydride, although there are some differences in degree and in the response of enolizable

$$LiAlH_4 + \begin{cases} 4\ RCOOH \\ 4\ C_6H_5OH \\ 4\ R_2NH \\ 4\ RSH \\ 4\ RC{\equiv}CH \end{cases} \longrightarrow 4\ H_2 + \begin{cases} RCOOLi + Al(OCOR)_3 \\ C_6H_5OLi + Al(OC_6H_5)_3 \\ LiNR_2 + Al(NR_2)_3 \\ LiSR + Al(SR)_3 \\ LiC{\equiv}CR + Al(C{\equiv}CR)_3 \end{cases}$$

compounds. Thus a primary amine generates only 1 mole of methane from the Grignard reagent but liberates 2 moles of hydrogen from the hydride. Hochstein[3] developed an analytical procedure for determining both active hydrogen and total hydride consumed.

The formula of the reagent is usually written as $LiAlH_4$ but interpreted as $Li^+[AlH_4]^-$ or Li^+AlH_4. Thus in ether solution it exists as aggregates of solvated lithium ions and aluminohydride ions. The first step in the reaction with a ketone is transfer of a hydride ion to give the complex (1); since this has three more hydrogen

$$R_2C{=}O + H{-}\bar{A}lH_3(Li^+) \longrightarrow \underset{(1)}{R_2\underset{H}{C}{-}O\bar{A}lH_3(Li^+)} \xrightarrow{3\ R_2C=O}$$

$$\underset{(2)}{(R_2\underset{H}{C}{-}O)_4\bar{A}l(Li^+)} \xrightarrow{H_2O} 4\ R_2\underset{H}{C}OH + Al(OH)_3 + LiOH$$

atoms available it can react with three more moles of ketone to give the alcoholate (2), which on decomposition with water affords the secondary alcohol, aluminum hydroxide, and lithium hydroxide. Because of its low molecular weight and because

1 mole reduces 4 moles of an aldehyde or ketone, lithium aluminum hydride has a very favorable ratio of reducing capacity to mass.

Technique of handling. The reagent as supplied by Ventron is in the form of large lumps of microcrystalline aggregates. Lots of 100 g., 1 lb., and 8 lbs. are shipped in polyethylene bags within sealed metal cans. This material can be transferred to glass-stoppered bottles, preferably flushed with nitrogen and kept sealed with paraffin when not in frequent use.

The main hazards associated with the reagent arise from its extreme sensitivity to water and to abrasion. Contact with liquid water, such as in inadequately dried apparatus, fingerprints, perspiration-soaked gloves, etc., may cause ignition. Lumps are best crushed by cautious pounding with a hard rubber hammer in a tray with an aluminum foil liner. Grinding with a mortar and pestle should be done in a dry box in an atmosphere of nitrogen or argon and the fine hydride protected thereafter from air or it will deteriorate rapidly. If ground too vigorously in the presence of air the hydride may take fire. Furthermore, the hydride dust is caustic and irritating. However, if the amount required is small, an operator aware of the hazard and suitably protected can crush a lump or two with mortar and pestle and grind the material lightly without a blanketing atmosphere. Note further that in a procedure described below for reduction of an amide, A. C. Cope[25] used lumps of reagent as supplied and found that magnetic stirring of a suspension in ether effects adequate disintegration.

A pail of powdered limestone and a long-handled shovel are recommended by Ventron for smothering a lithium aluminum hydride fire. The only commercial extinguishers recommended are those of the nitrogen-propelled dry powder type, such as Ansul. In no case use water, carbon dioxide, or a chemical extinguisher.

A solution of the hydride in diethyl ether is best prepared by gentle refluxing and stirring under nitrogen for several hours in a flask protected from moisture with a drying tube. Usually an insoluble precipitate is formed, probably owing to atmospheric exposure during handling. This precipitate appears very bulky, but it is easily removed by filtration and found to amount to less than 1% of the hydride. If the solution is exposed to air, the hydride reacts slowly with atmospheric oxygen with liberation of hydrogen. Solutions of the hydride are no safer than the solid and should be handled under dry nitrogen. Spillage may ignite spontaneously after brief exposure.

Techniques of reduction. Early workers investigating the scope of reduction by the new reagent used the hydride in the form of clarified solutions in ether, but the preparation of such solutions requires prolonged refluxing, a troublesome filtration, transfer of solution under protection from air, and sludge disposal. The present practice is to place the solvent, ether or tetrahydrofurane, in a flask fitted with a stirrer, reflux condenser, and dropping tube, add lithium aluminum hydride, usually powdered but sometimes in lump form, and run in a solution of the substance to be reduced at a rate sufficient to maintain gentle boiling.

Destruction of excess reagent. Although the practice is not always justified on the basis of pilot experiments, lithium aluminum hydride is often taken in large excess of the theory (2- to 4-fold excess). If the amount of hydride to be destroyed is considerable, even slow and cautious addition of water with stirring may be attended with difficulty (and danger) due to vigorous frothing from the hydrogen evolved. Use of a 10% solution of sodium hydroxide or ammonium chloride has the advantage that aluminum oxide separates in granular and easily filterable form. Ethyl acetate is often preferred since no hydrogen is evolved and the reduction product, ethanol, is not likely to interfere with isolation of the product.

$$4\ CH_3CO_2C_2H_5 + LiAlH_4 \longrightarrow LiOC_2H_5 + Al(OC_2H_5)_3 + 4\ C_2H_5OH$$

During workup of a reduction a voluminous, gelatinous precipitate of hydroxides often makes separation of the product difficult, and attempted solution of the precipitate with excess acid or base leads to large volumes and to emulsions. Steinhardt[4] states that these difficulties can be avoided by following a procedure[5] in which the stirred reduction mixture from n grams of lithium aluminum hydride is treated by successive dropwise addition of n ml. of water, n ml. of 15% sodium hydroxide solution, and $3n$ ml. of water. This produces a dry granular precipitate which is easy to filter and wash.

Reduction of aldehydes and ketones.[6] The typical reductions formulated were carried out with standardized solutions of lithium aluminum hydride and demonstrated the requirement of 0.25 mole of reagent per mole of carbonyl compound. The modern

$$CH_3(CH_2)_5CHO \xrightarrow[86\%]{0.25\ LiAlH_4} CH_3(CH_2)_5CH_2OH$$

$$CH_3CH_2COCH_3 \xrightarrow[80\%]{0.25\ LiAlH_4} CH_3CH_2\underset{\displaystyle OH}{\underset{|}{C}}HCH_3$$

$$CH_3CH{=}CHCHO \xrightarrow[70\%]{0.25\ LiAlH_4} CH_3CH{=}CHCH_2OH$$

technique is exemplified by a procedure[7] for the reduction of the carbonyl group of 3-ethoxy-2-cyclohexenone (1), an intermediate in the preparation of 2-cyclohexenone. A flask equipped with a reflux condenser, a mechanical stirrer, and a dropping funnel

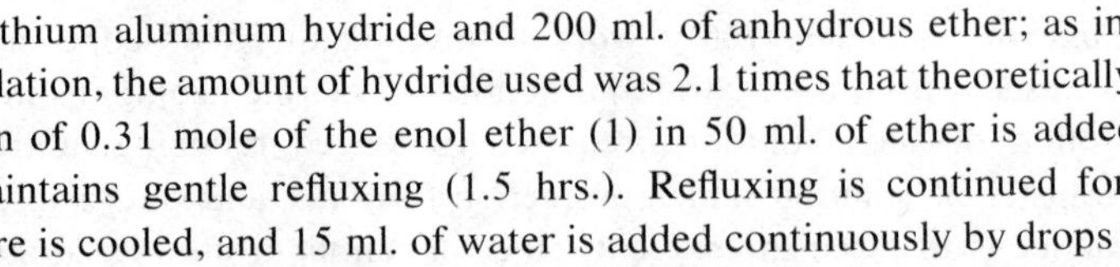
LiAlH4
(2.1 x Theory)
Ether
OC2H5
OH
OC2H5
H+
62-75%
from (1)

(1) (2) (3)

and protected from atmospheric moisture with drying tubes is charged with 0.16 mole of lithium aluminum hydride and 200 ml. of anhydrous ether; as indicated in the formulation, the amount of hydride used was 2.1 times that theoretically required. A solution of 0.31 mole of the enol ether (1) in 50 ml. of ether is added at a rate which maintains gentle refluxing (1.5 hrs.). Refluxing is continued for 30 min., the mixture is cooled, and 15 ml. of water is added continuously by drops (foaming). The mixture is then worked up, and the enol ether (2) is hydrolyzed with acid to 2-cyclohexenone (3).

Carboxylic acids and derivatives. A particularly striking property of lithium aluminum hydride is its ability to reduce carboxylic acids to primary alcohols. Displacement of the acidic hydrogen uses up $\frac{1}{4}$ mole of the reagent and $\frac{1}{2}$ mole more is required for the reduction. Thus good yields are obtained with 0.75 mole of hydride per mole of acid.[8] Triphenylacetic acid is not reduced under ordinary conditions

$$HOOC(CH_2)_8COOH \xrightarrow[97\%]{1.5\ LiAlH_4} HOCH_2(CH_2)_8CH_2OH$$

(ether) but can be converted into the carbinol in good yield either by reduction in the higher-boiling tetrahydrofurane or by conversion into the acid chloride, which is reduced readily in ether.[9] The 4β-carboxyl group of podocarpic acid is axial and

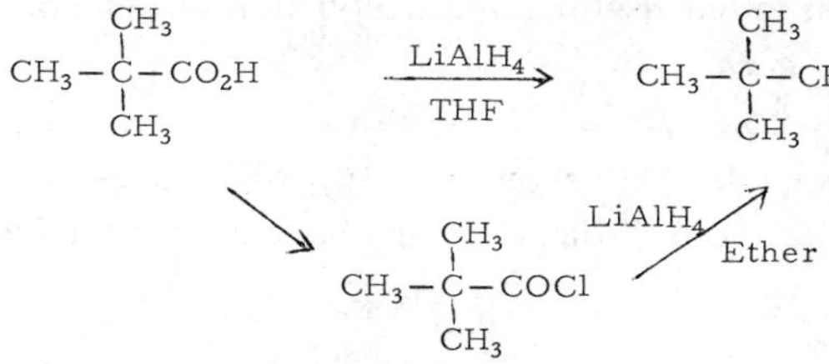

strongly hindered by the axial β-methyl group at C_{10}. Furthermore, the active hydrogens of the carboxyl and phenolic hydroxyl groups consume reagent and give rise to metal oxide groups which decrease the solubility. Thus reduction of

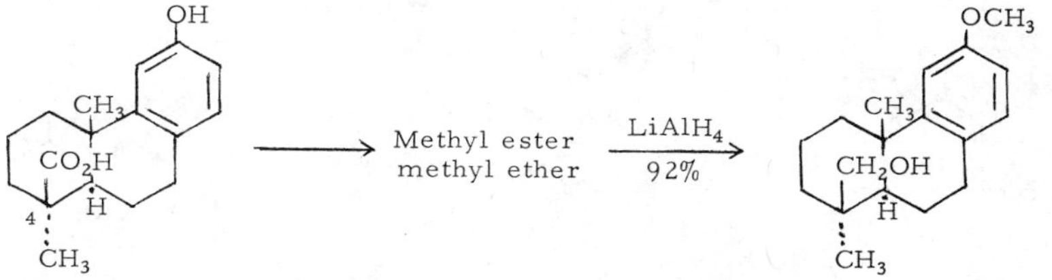

the free acid over a four-day period at room temperature afforded podocarpinol in only 56% yield; reduction of the methyl ester methyl ether proceeded much more satisfactorily.

The reduction of an ester requires 0.5 mole of reagent and probably involves formation and reduction of the corresponding aldehyde. A detailed procedure for

$$C_6H_5\overset{O}{\overset{\|}{C}}OC_2H_5 \xrightarrow{\;0.25\ LiAlH_4\;} \left[C_6H_5\overset{O}{\overset{\|}{C}}H \right] \xrightarrow[90\%]{\;0.25\ LiAlH_4\;} C_6H_5CH_2OH$$

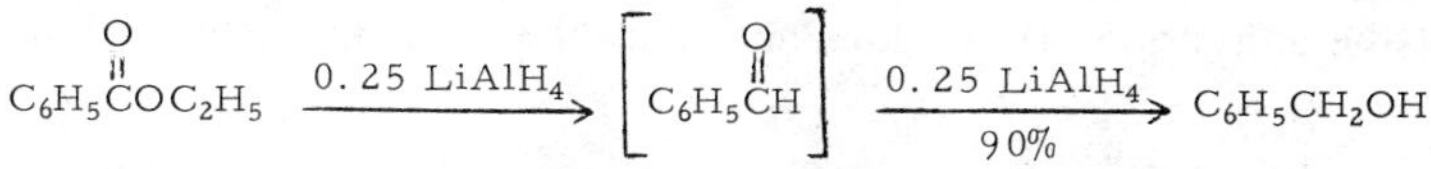

the reduction of ethyl α-(1-pyrrolidyl)propionate (1)[10] calls for heating and stirring 0.56 mole of pulverized lithium aluminum hydride with 300 ml. of ether under

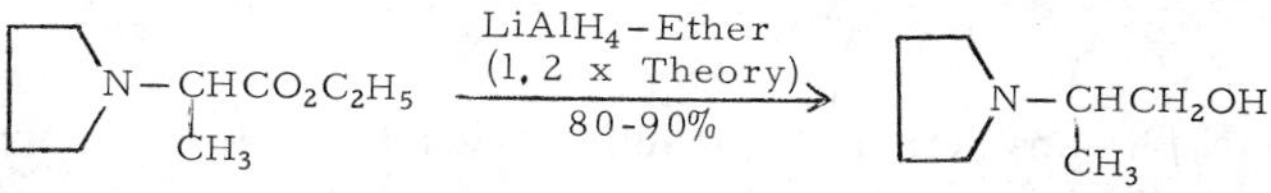

(1)   (2)

reflux until most of the hydride has dissolved, and dropping in 0.92 mole of ester in 200 ml. of ether at such a rate that the solvent refluxes gently. After stirring and refluxing for 30 min. more, 50 ml. of ethyl acetate is added slowly, followed by 600 ml. of 4 N hydrochloric acid.

In describing the preparation of 2,2-dichloroethanol from dichloroacetyl chloride, Sroog and Woodburn[11] state that the reaction appears to be smoother than with the free acid or the ethyl ester. The procedure is essentially the same as for the ester (1);

$$CHCl_2COCl \xrightarrow[64\text{-}65\%]{\begin{array}{c}LiAlH_4\\(1.0\ x\ Theory)\end{array}} CHCl_2CH_2OH$$

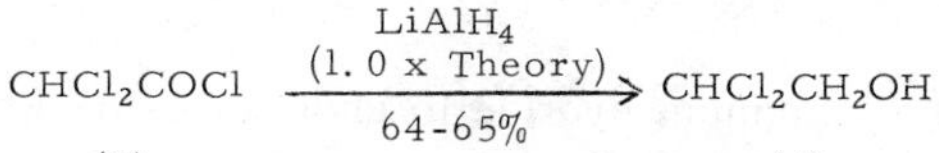

(3)   (4)

0.6 mole of acid chloride was added in ether to a slurry of 0.3 mole of hydride in ether.

Since an acetyl derivative of an alcohol is easily cleaved by lithium aluminum hydride, the reagent is often useful for deacetylation of compounds sensitive to

$$CH_3\overset{O}{\overset{\|}{C}}OR \xrightarrow{0.5\ LiAlH_4} CH_3CH_2OH + HOR$$

acids or bases. For example, $\Delta^{8(14)}$-cholestene-3β,7α-diol diacetate (1) is converted very easily by acid into the $\Delta^{8,14}$-diene (2) and gave a poor product on attempted

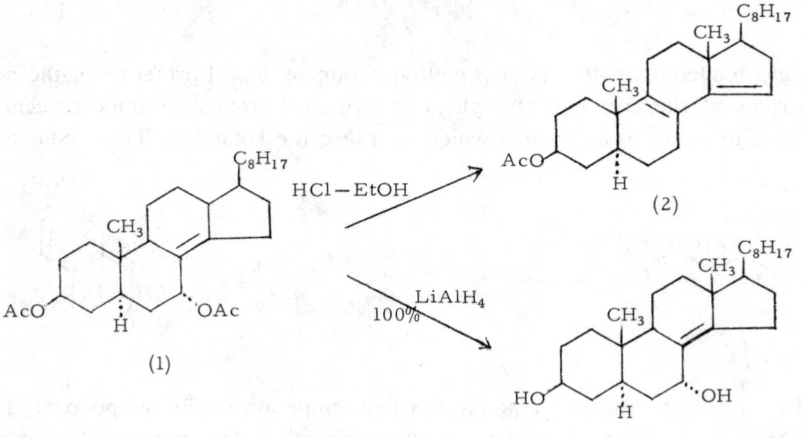

C8H17
CH3
AcO
H
HCl–EtOH
LiAlH4
100%
OAc
HO
OH

(1) (2) (3)

hydrolysis with base, but on brief treatment with lithium aluminum hydride in ether it afforded the almost pure diol (3) in quantitative yield.[12]

Buchta found[13] that lithium aluminum hydride reduces one of the two carbonyl groups of the anhydride (4) but does not touch the second. Evidently hindrance of

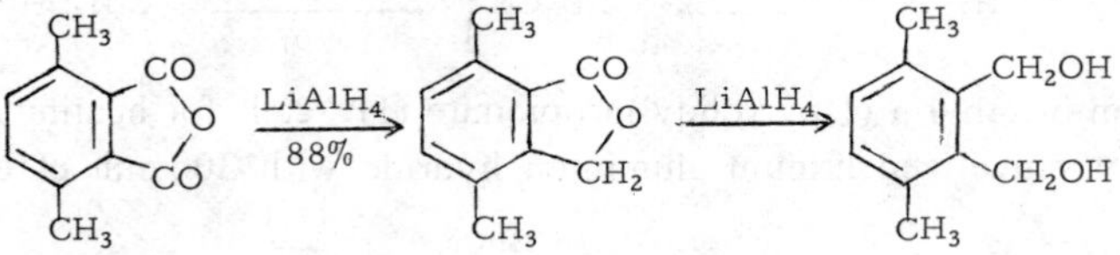

CH3
CO
O
LiAlH4
88%
CH2
CH2OH

(4) (5) (6)

the two *o*-methyl groups permits formation of a metal complex with only one of the carbonyl groups. Reduction of the phthalide to the diol (6) requires production of only one complex and is accomplished under the same experimental conditions.

Epoxides. A synthetic sequence which has been of particular value in the steroid field consists in conversion of an olefin into the epoxide and reductive fission to one of the two possible alcohols. Δ^1-Cholestene (1) gives the 1α,2α-epoxide (2) and this

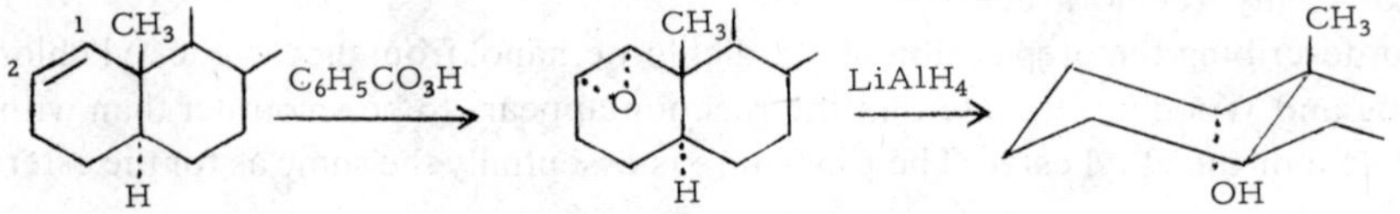

1
2
CH3
C6H5CO3H
LiAlH4
O
H
OH

(1) (2) (3)

is reduced by lithium aluminum hydride to cholestane-1α-ol.[14] The pattern is the same with Δ^2-cholestene (4).[15] In each case the alcohol formed is axial, whereas

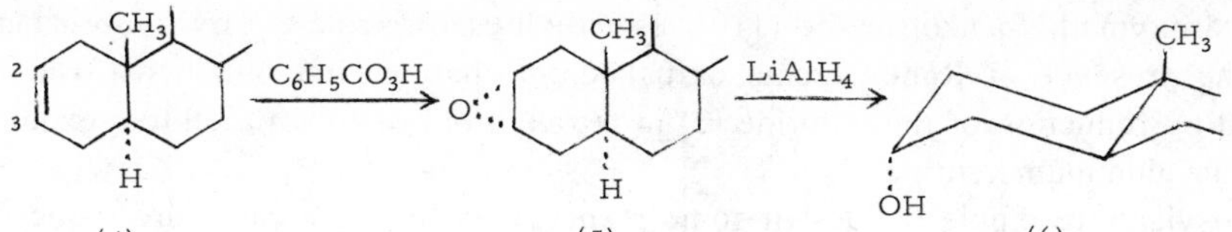

(4) (5) (6)

opening of the oxide ring in the alternative direction would have given an equatorial alcohol. Many other cases have been investigated, particularly by Plattner and coworkers,[16] and the almost invariable rule is that an oxide is cleaved in such a direction as to give an axial alcohol. Before the advent of lithium aluminum hydride, epoxides were reduced by catalytic hydrogenation. The metal hydride is now the reagent of choice and has been found to give good results where hydrogenation has failed or given poor results.

Halides and tosylates. Trevoy and W. G. Brown[17] found that lithium aluminum hydride reduces benzyl iodide and benzyl bromide in high yield at 35° in either diethyl ether or tetrahydrofurane. In tetrahydrofurane at 65°, benzyl chloride and 1-bromodecane are reduced to toluene and to *n*-decane, both in 72% yield. Johnson, Blizzard, and Carhart[18] found that lithium aluminum hydride reacts more sluggishly than in other reductions and presented experimental evidence that not all four hydrogen atoms possess adquate reactivity toward alkyl halides. Thus the reaction probably proceeds in at least two steps, the first of which is much more rapid than the others:

$$LiAlH_4 + RX \longrightarrow RH + LiX + AlH_3$$

$$AlH_3 + 3\ RX \longrightarrow AlX_3 + 3\ RH$$

1-Bromooctane refluxed in tetrahydrofurane for 1 hr. with a 1.4-fold excess of lithium aluminum hydride over the theory afforded *n*-octane in yield of 64%, but with a 3.2-fold excess the yield rose to 96%. The investigators then developed an improved procedure in which hydrogenolysis is effected by means of lithium hydride with a small amount of lithium aluminum hydride as the hydrogen carrier. Thus 1-bromodecane on being refluxed in tetrahydrofurane for 0.5 hr. with 0.13 mole of $LiAlH_4$ and 1.5 moles of AlH_3 per mole of RX afforded *n*-octane in 96% yield.

Buchta[13] made use of hydrogenation with lithium aluminum hydride in a synthesis

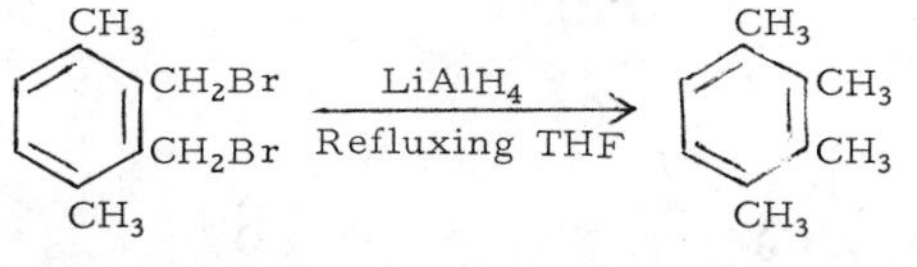

of prehnitene. Breslow[19] used it for the reduction of *sym*-triphenylcyclopropenyl bromide to triphenylcyclopropene. Gaylord[20] tried without success to reduce

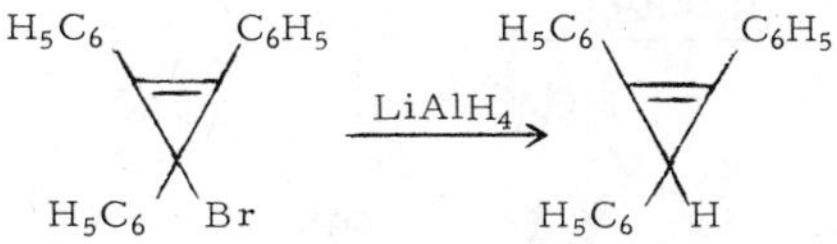

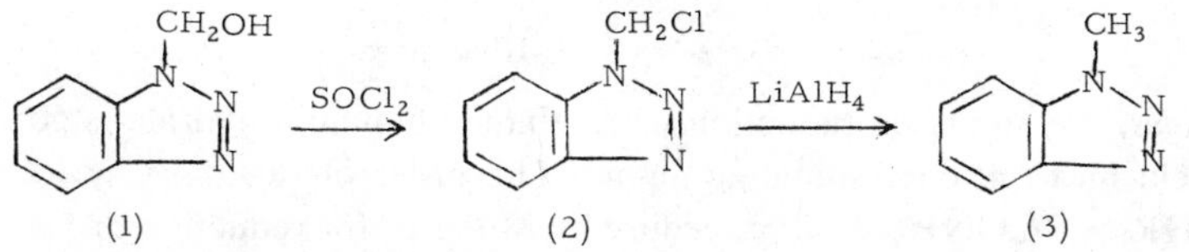

(1) (2) (3)

1-hydroxymethylbenzotriazole (1) to 1-methylbenzotriazole (3) by hydrogenation in the presence of Raney nickel or palladium charcoal but obtained (3) in high yield by reduction of the chloride (2) in tetrahydrofurane with a 10-fold excess of lithium aluminum hydride.

Tosylates resemble halides in some respects and in some cases are reduced as follows:

$$R-OSO_2C_6H_4CH_3\text{-}\underline{p} \xrightarrow{LiAlH_4} RH + HO_3SC_6H_4CH_3\text{-}\underline{p}$$

Schmid and Karrer[21] found that *l*-menthyl tosylate is reduced mainly to *l*-menthane and that the 6-tosylate of di-acetone-D-galactopyranoside (1) affords di-acetone-D-

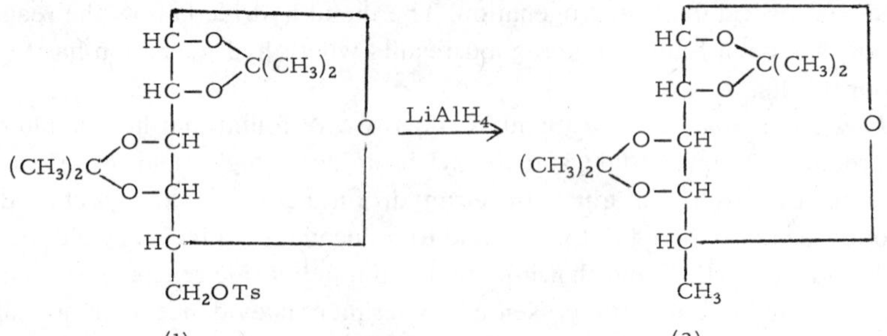
HC–O
C(CH3)2
HC–O
O–CH
(CH3)2C
O–CH
HC
CH2OTs
O
LiAlH4
HC–O
C(CH3)2
HC–O
O–CH
(CH3)2C
O–CH
HC
CH3
O

(1) (2)

fucopyranoside (2). However, two other sugar tosylates, one a primary tosylate, suffered fission as follows:

$$RO{-}SO_2C_6H_4CH_3\text{-}p \xrightarrow{LiAlH_4} ROH + HO_2SC_6H_4CH_3\text{-}p$$

Zorbach[22] used the reaction in a synthesis of D-rhamnose, although the yield in the reduction step was low.

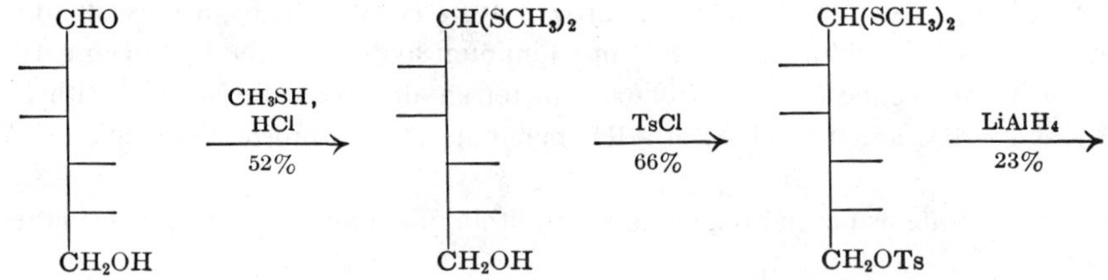
CHO
CH2OH
CH3SH,
HCl
52%
CH(SCH3)2
CH2OH
TsCl
66%
CH(SCH3)2
CH2OTs
LiAlH4
23%

D-Mannose (1) (2)

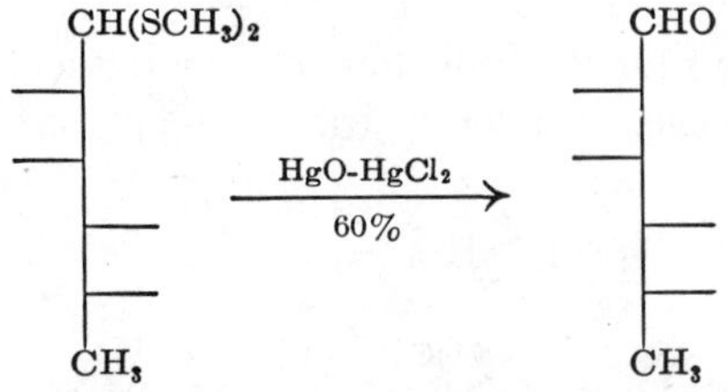
CH(SCH3)2
CH3
HgO-HgCl2
60%
CHO
CH3

(3) D-Rhamnose

Amides. An amide on reduction with lithium aluminum hydride is converted in good yield into the corresponding amine. The order of reactivity is $RCONR_2 >$ $RCONHR > RCONH_2$. In a procedure by Moffett[23] for reduction of 5,5-dimethyl-

$$RCONH_2 \xrightarrow{0.5\ LiAlH_4} RCH_2NH_2$$

2-pyrrolidone (1), 1 mole of pulverized lithium hydride is stirred with tetrahydrofurane under reflux during gradual addition of 0.8 mole of (1) in tetrahydrofurane.

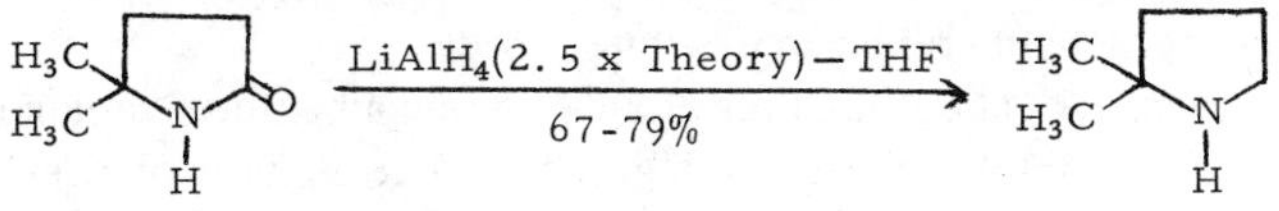

(1) (2)

For reduction of N-methyllauramide (3), which is sparingly soluble in ether, Wilson and Stenberg[24] placed 1 mole of finely divided lithium aluminum hydride in a flask

$$\underset{(3)}{CH_3(CH_2)_{10}CONHCH_3} \xrightarrow[81\text{-}95\%]{LiAlH_4\,(2.7 \times Theory)-Ether} \underset{(4)}{CH_3(CH_2)_{10}CH_2NHCH_3}$$

fitted with a stirrer and a Soxhlet extractor and placed 0.75 mole of the amide in the thimble of the extractor. Extraction was complete in about 3 hrs. and the reaction period was extended for 2 hrs. more. A procedure by Cope and Ciganek[25] for the reduction of N,N-dimethylcyclohexanecarboxamide (5) follows the usual pattern of addition of an ethereal solution of the amide (0.86 mole) to a suspension in ether of lithium aluminum hydride in large excess (0.85 mole) but presents an interesting point of novelty: the metal hydride is not pulverized but employed in

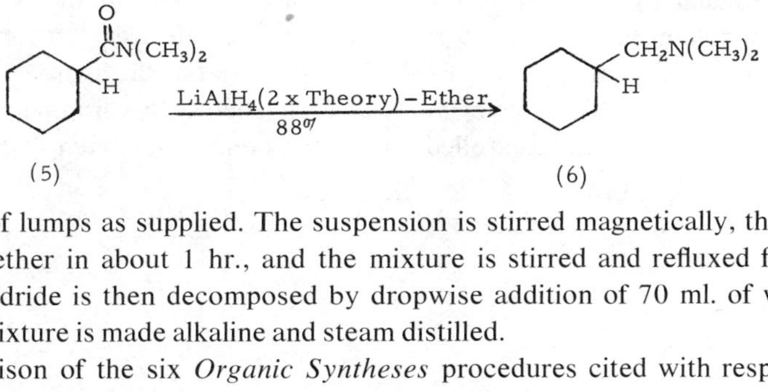

(5) (6)

the form of lumps as supplied. The suspension is stirred magnetically, the amide is added in ether in about 1 hr., and the mixture is stirred and refluxed for 15 hrs. Excess hydride is then decomposed by dropwise addition of 70 ml. of water, and then the mixture is made alkaline and steam distilled.

Comparison of the six *Organic Syntheses* procedures cited with respect to the amount of hydride used reveals a point of significance:

Type of compound reduced	Hydride used/theory	Median yield, %
Ketone	2.1	63.5
Ester	1.2	85
Acid chloride	1.0	64.5
RCONHR′	2.5	73
RCONHR′	2.7	88
$RCONR'_2$	2.0	88

The results lend no support to the practice followed in 4 of the 6 cases of using a large excess of reagent. This practice not only wastes costly reagent but extends the working period by the time required for decomposition of excess hydride, particularly when water is used. Note that the work of Hochstein[3] demonstrated that reductions of unsaturated compounds with lithium aluminum hydride in dilute solution are quantitative. Use of 1.2 times theory allows adequately for slight destruction of reagent by air oxidation in a prolonged reaction period and should be adequate. Cope's simple procedure of using lump reagent and a magnetic stirrer[25] seemed to us particularly attractive and a trial was most impressive. When hydride as supplied was stirred magnetically with ether at room temperature for about 1 hr. the lumps were completely replaced by a fine powder.

Aldehyde synthesis. The reduction of an N-monosubstituted amide (1) probably proceeds by addition to the carbonyl group and formation of a geminal amino

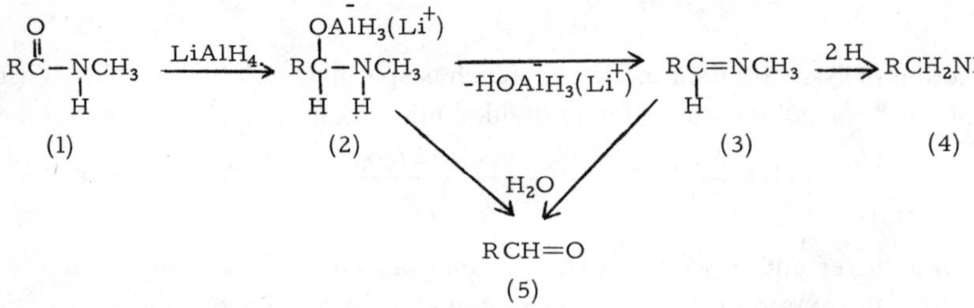

alcohol derivative (2), which by an elimination affords the imine (3), reducible to the final product (4). With an N,N-disubstituted amide the intermediate corresponding to (3) would be an iminium salt. With proper control of the amount of hydride, hydrolysis of (2) or (3) should afford an aldehyde (5). However, the structure of the amide seems to be of prime importance, and satisfactory results have been obtained with amides of only three types. Weygand[26] devised a synthesis of aromatic aldehydes involving reaction of N-methylaniline with phosgene to form methylphenylcarbamyl chloride (1), which reacts with an aromatic hydrocarbon under Friedel-Crafts conditions to give the N-methylanilide of a carboxylic acid; this derivative is reduced by a controlled amount of lithium aluminum hydride to the

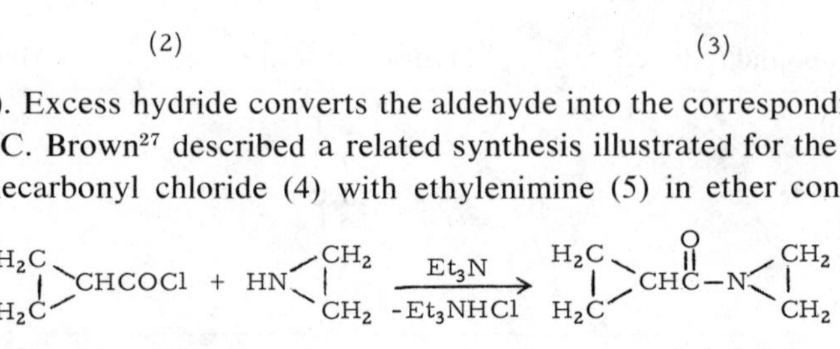

aldehyde (3). Excess hydride converts the aldehyde into the corresponding primary alcohol. H. C. Brown[27] described a related synthesis illustrated for the reaction of cyclopropanecarbonyl chloride (4) with ethylenimine (5) in ether containing one

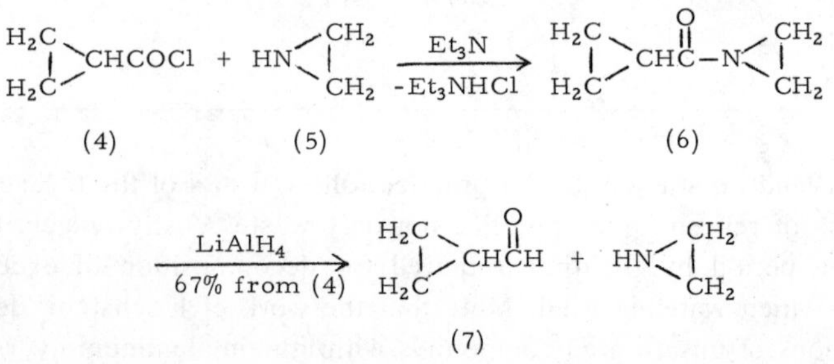

equivalent of triethylamine at 0°. Triethylamine hydrochloride precipitates, and the filtered solution of the 1-acylaziridine (6) on reduction with lithium aluminum hydride affords, after hydrolysis, cyclopropanecarboxaldehyde (7). *n*-Butyryl, *n*-caproyl, and pivaloyl chloride were converted by this synthesis into the corresponding aldehydes in yields of 75, 81, and 79%, and in two trials use of lithium aluminum hydride in 100% excess of the acid chloride did not decrease the yield.

An aldehyde synthesis introduced by Staab[28] involves condensation of an acid chloride with imidazole to form an imidazolide (1) and reductive cleavage with lithium aluminum hydride to the aldehyde (2) and imidazole. In a one-step process

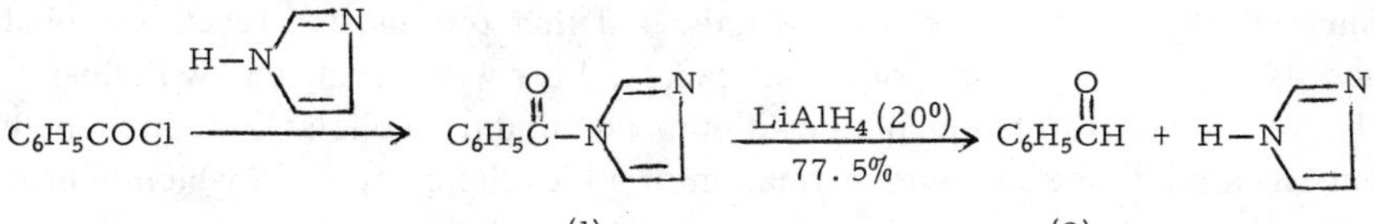
H—N
C_6H_5COCl
C_6H_5C—N
$LiAlH_4$ (20°)
77.5%
C_6H_5CH
+ H—N
N

(1) (2)

the carboxylic acid is treated in tetrahydrofurane or ether with N,N′-carbonyldiimidazole to form the imidazolide, the carbon dioxide is removed by evacuation (since it is reducible), and lithium aluminum hydride is added in amount sufficient to react with the Zerewitinoff active imidazole and to effect reduction. Yields range from 30 to 83%.

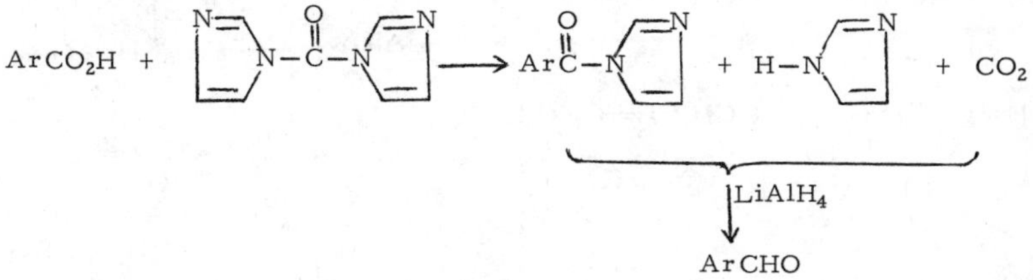
$ArCO_2H$ +
N—C—N
ArC—N
+ H—N
+ CO_2
$LiAlH_4$
ArCHO

In seeking to synthesize methionine-α-C^{14}, Claus and Morgenthau[29] explored the possibility of converting β-methylmercaptopropionitrile into β-methylmercaptopropionaldehyde, but Stephen reduction and direct reduction with lithium aluminum

$$CH_3SCH_2CH_2CN \longrightarrow CH_3SCH_2CH_2CHO$$

hydride both gave unsatisfactory results. They then developed the following general method for the conversion of a nitrile into the aldehyde. Treatment of the nitrile

$$\underset{(1)}{RC\equiv N} \xrightarrow[HCl]{EtOH} \underset{(2)}{R\overset{\overset{+}{N}H_2}{\overset{\|}{C}}{-}OEt(Cl^-)} \xrightarrow{2\ EtOH} \underset{(3)}{RC(OEt)_3}$$

$$\xrightarrow{LiAlH_4} \underset{(4)}{RCH(OEt)_2} \xrightarrow{H_2O} RCHO$$

(1) with ethanol and hydrogen chloride affords the imino ester hydrochloride (2), which on further reaction with ethanol is converted into the ortho ester (3). Reduction of the ortho ester (3) to the diethyl acetal (4) was effected by adding 0.25 mole of a 1 *M* solution of lithium aluminum hydride in ether to a boiling solution of 1 mole of the ortho ester in benzene and refluxing for 4 hrs. Decomposition of the complex with a 30% solution of Rochelle salt afforded the diethyl acetal of β-methylmercapto-

propionaldehyde in 73% yield; the yield of the dimethyl acetal was 97%. Acid hydrolysis afforded the desired β-methylmercaptopropionaldehyde.

Reduction of allylic and propargylic alcohols. Although lithium aluminum hydride does not reduce ethylenic or acetylenic hydrocarbons, it reduces allylic alcohols and their acetylenic counterparts.[30] The reaction was first encountered by Nystrom and W. G. Brown in the reduction of cinnamic acid to dihydrocinnamyl alcohol. Investigation showed that the anomaly is not in the reduction of the carboxyl group,

$$C_6H_5CH{=}CHCO_2H \xrightarrow[85\%]{LiAlH_4} C_6H_5CH_2CH_2CH_2OH$$

but that cinnamyl alcohol is the precursor of the saturated alcohol. Experimentation eliminated alternative pathways and indicated that the unusual reaction probably proceeds as follows. The hydride first reacts rapidly with the alcohol with liberation of hydrogen and formation of the lithium aluminum alcoholate. Reduction then occurs at a moderate rate with formation of a bicyclic complex in which hydrogen has become bonded to one originally unsaturated carbon and aluminum to the other. The second hydrogen atom required for saturation of the original double bond is supplied in the terminal step of hydrolysis of the complex, when water cleaves the

$$4\ C_6H_5CH{=}CHCH_2OH + LiAlH_4 \longrightarrow (C_6H_5CH{=}CHCH_2O)_4AlLi + 4\ H_2$$

Cinnamyl alcohol

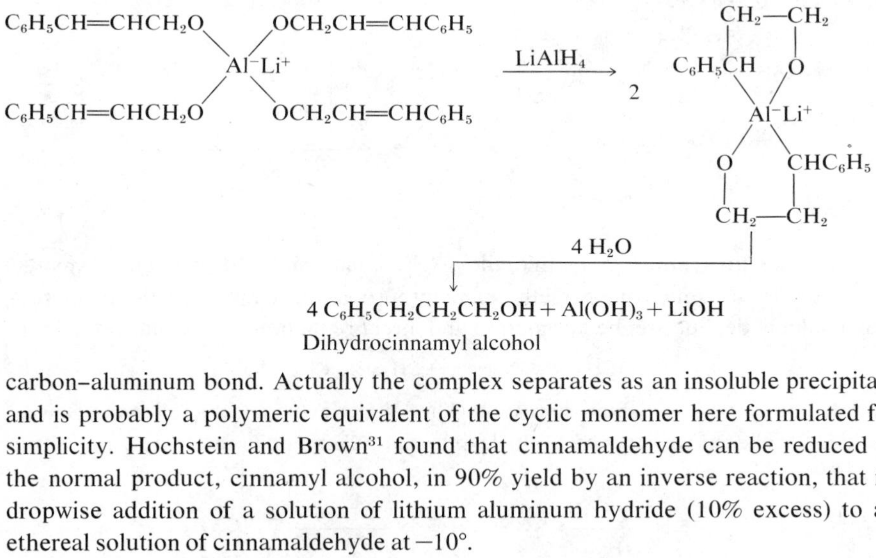

Dihydrocinnamyl alcohol

carbon–aluminum bond. Actually the complex separates as an insoluble precipitate and is probably a polymeric equivalent of the cyclic monomer here formulated for simplicity. Hochstein and Brown[31] found that cinnamaldehyde can be reduced to the normal product, cinnamyl alcohol, in 90% yield by an inverse reaction, that is, dropwise addition of a solution of lithium aluminum hydride (10% excess) to an ethereal solution of cinnamaldehyde at −10°.

Jorgensen *et al.*[32] found that reduction of an allylic alcohol by lithium aluminum hydride can be carried beyond the stage of the saturated alcohol to give a cyclopropane. Thus a cinnamyl acid, ester, aldehyde, or ketone on reduction with 100% excess $LiAlH_4$ in refluxing tetrahydrofurane or dimethoxyethane affords a phenylcyclopropane in yield of 45–80%. The reaction complements the Simmons-Smith synthesis.

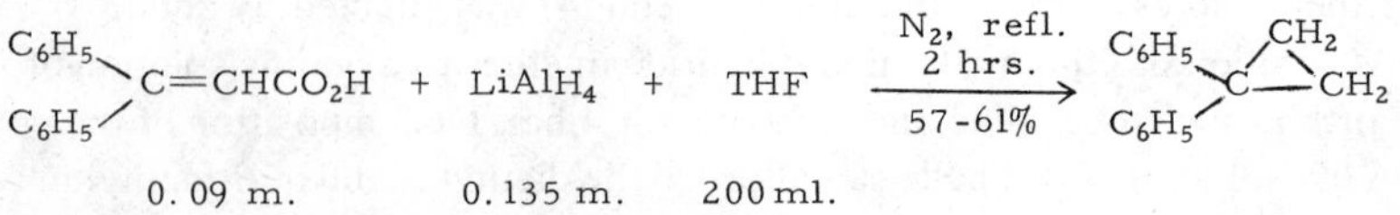

A group at the Glaxo Laboratories[33] found the reaction applicable to propargylic alcohols and used it in a synthesis of vitamin A, and further examples were explored by E. R. H. Jones and co-workers.[34] Phenylpropiolic acid (1) is reduced first to phenylpropargyl alcohol, and this affords a complex which on decomposition with

$$\text{1. } C_6H_5C{\equiv}CCO_2H \xrightarrow[-70^\circ]{LiAlH_4} C_6H_5C{\equiv}CCH_2OH \xrightarrow{LiAlH_4} C_6H_5CH{=}CHCH_2OH$$

Phenylpropiolic acid — Phenylpropargyl alcohol — Cinnamyl alcohol

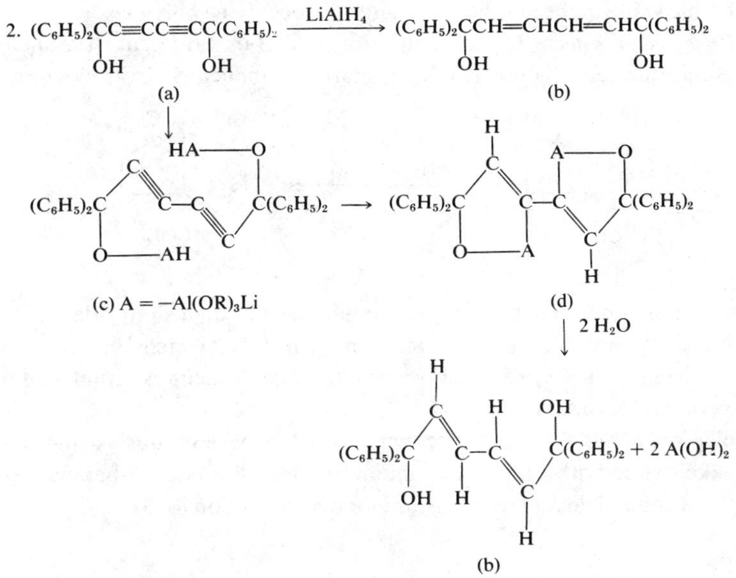
2. (C6H5)2CC≡CC≡CC(C6H5)2
OH OH
(a)
LiAlH4
(C6H5)2CCH=CHCH=CHC(C6H5)2
OH OH
(b)
(c) A = –Al(OR)3Li
(d)
2 H2O
+ 2 A(OH)2
(b)

water affords cinnamyl alcohol. In example 2 a diacetylenic glycol (a) is reduced to a diethylenic glycol (b) in which the double bonds are both *trans*. Possibly the intermediate glycolate (c) cyclizes by *trans* additions to the triple bonds with

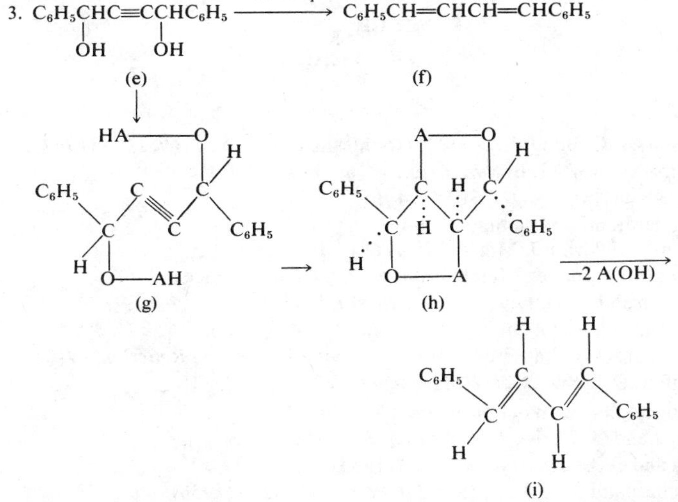
3. C6H5CHC≡CCHC6H5
OH OH
(e)
LiAlH4
C6H5CH=CHCH=CHC6H5
(f)
(g)
(h)
–2 A(OH)
(i)

establishment of carbon–aluminum bonds (d), the cleavage of which introduces a second pair of hydrogens *trans* to those introduced in the first step. The mono-acetylenic glycol (3e) presents a different situation, for reduction results in elimination of both hydroxyl functions and formation of a *trans,trans* diene (i). The interpretation suggested in the formulas is that intramolecular *trans* addition of the glycolate (g) gives an intermediate (h) of stereochemistry favorable for *trans* eliminations involving the first pair of hydrogen atoms as another pair become attached on severance of the carbon–aluminum bonds. This variation of the reduction reaction has been of considerable value in the synthesis of naturally occurring polyenes.

α,β-Unsaturated esters. The reduction of an α,β-unsaturated ester such as (1) to the α,β-unsaturated alcohol (2) is ordinarily complicated by reduction of the

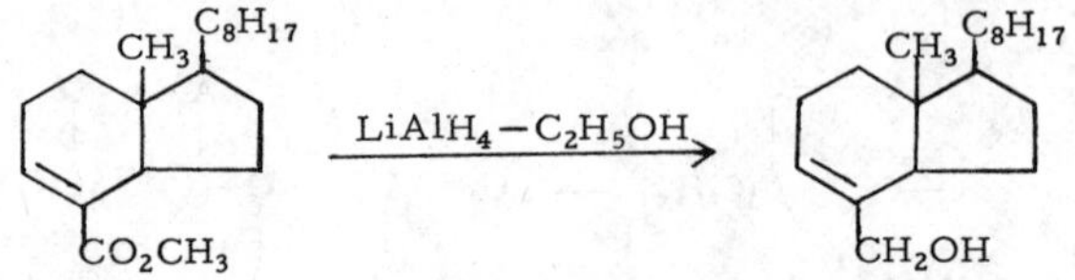

C_8H_{17}
CH_3
$LiAlH_4-C_2H_5OH$
CO_2CH_3
CH_2OH

(1) (2)

conjugated double bond. Lythgoe *et al.*[35] found that this undesired side reaction can be suppressed by use of lithium aluminum hydride to which one equivalent of ethanol has been added. Presumably the effective reagent is lithium aluminum monoethoxyhydride, $LiAlH_3(OC_2H_5)$.

Aziridine syntheses. A recent discovery by a Japanese group[36] is that reduction of dibenzylketoxime with lithium aluminum hydride affords *cis*-2-benzyl-3-phenylaziridine (2) in good yield and with only minor overreduction to (3).

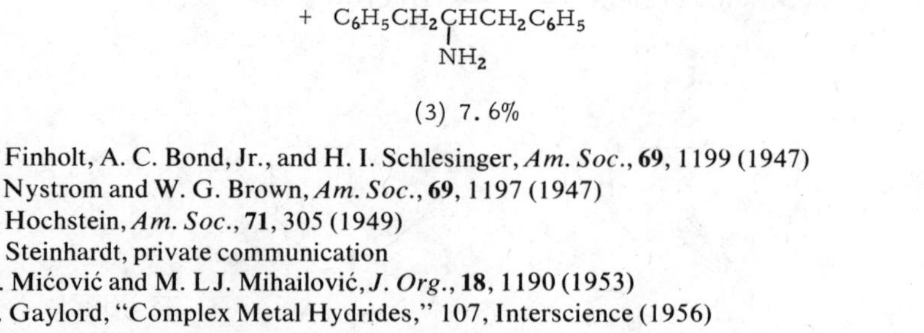

$C_6H_5CH_2CCH_2C_6H_5$ + $LiAlH_4$ + THF
NOH
$C_6H_5CH_2$
$CH_2C_6H_5$
H N H
H
(1) 0.05 m.
80 ml.
(2) 75%
+ $C_6H_5CH_2CHCH_2C_6H_5$
NH_2

(3) 7.6%

[1]A. E. Finholt, A. C. Bond, Jr., and H. I. Schlesinger, *Am. Soc.*, **69**, 1199 (1947)
[2]R. F. Nystrom and W. G. Brown, *Am. Soc.*, **69**, 1197 (1947)
[3]F. A. Hochstein, *Am. Soc.*, **71**, 305 (1949)
[4]C. K. Steinhardt, private communication
[5]V. M. Mićović and M. LJ. Mihailović, *J. Org.*, **18**, 1190 (1953)
[6]N. G. Gaylord, "Complex Metal Hydrides," 107, Interscience (1956)
[7]W. F. Gannon and H. O. House, *Org. Syn.*, **40**, 14 (1960)
[8]R. F. Nystrom and W. G. Brown, *Am. Soc.*, **69**, 2548 (1947)
[9]Unpublished observations cited in Brown's review in *Organic Reactions*, **VI**, 469 (1951)
[10]R. B. Moffett, *Org. Syn., Coll. Vol.*, **4**, 834 (1963)
[11]C. E. Sroog and H. M. Woodburn, *ibid.*, **4**, 271 (1963)
[12]L. F. Fieser and G. Ourisson, *Am. Soc.*, **75**, 4404 (1953)
[13]E. Buchta and G. Loew, *Ann.*, **597**, 123 (1955)
[14]H. B. Henbest and R. A. L. Wilson, *J. Chem. Soc.*, 3289 (1956)

[15]A. Fürst and Pl. A. Plattner, *Helv.*, **32**, 275 (1949)
[16]Pl. A. Plattner, H. Heusser, and A. B. Kulkarni, *Helv.*, **31**, 1822, 1885 (1948); **32**, 265, 1070 (1949); Pl. A. Plattner, H. Heusser, and M. Feurer; *ibid.*, **32**, 587 (1949)
[17]L. W. Trevoy and W. G. Brown, *Am. Soc.*, **71**, 1675 (1949)
[18]J. E. Johnson, R. H. Blizzard, and H. W. Carhart, *Am. Soc.*, **70**, 3664 (1948)
[19]R. Breslow and D. Dowd, *Am. Soc.*, **85**, 2729 (1963)
[20]N. G. Gaylord, *Am. Soc.*, **76**, 285 (1954)
[21]H. Schmid and P. Karrer, *Helv.*, **32**, 1371 (1949)
[22]W. W. Zorbach and C. O. Tio, *J. Org.*, **26**, 3543 (1961)
[23]R. B. Moffett, *Org. Syn.*, *Coll. Vol.*, **4**, 355 (1963)
[24]C. V. Wilson and J. F. Stenberg, *ibid.*, **4**, 564 (1963)
[25]A. C. Cope and E. Ciganek, *ibid.*, **4**, 339 (1963)
[26]F. Weygand and R. Mitgau (Paper VI), *Ber.*, **88**, 301 (1955)
[27]H. C. Brown and A. Tsukamoto, *Am. Soc.*, **83**, 4549 (1961)
[28]H. A. Staab and H. Bräunling, *Ann.*, **654**, 119 (1962)
[29]C. J. Claus and J. L. Morgenthau, Jr., *Am. Soc.*, **73**, 5005 (1951); U.S. patent 2,786,872 (1957) [*C.A.*, **51**, 12130 (1957)]
[30]R. F. Nystrom and W. G. Brown, *Am. Soc.*, **69**, 2548 (1947); **70**, 3738 (1948); R. T. Gilsdorf and F. F. Nord, *J. Org.*, **15**, 807 (1950)
[31]F. A. Hochstein and W. G. Brown, *Am. Soc.*, **70**, 3484 (1948)
[32]M. J. Jorgensen and A. W. Friend, *Am. Soc.*, **87**, 1815 (1965); M. J. Jorgensen and A. F. Thacher, procedure submitted to *Org. Syn.*
[33]J. Attenburrow, A. F. B. Cameron, J. H. Chapman, R. M. Evans, B. A. Hems, A. B. A. Jansen, and T. Walker, *J. Chem. Soc.*, 1094 (1952)
[34]E. B. Bates, E. R. H. Jones, and M. C. Whiting, *J. Chem. Soc.*, 1854 (1954)
[35]R. S. Davidson, W. H. H. Günther, S. M. Waddington-Feather, and B. Lythgoe, *J. Chem. Soc.*, 4907 (1964)
[36]K. Kotera and K. Kitahonoki, procedure submitted to *Org. Syn.*; *see also* K. Kitahonoki *et al.*, *Tetrahedron Letters*, 1059 (1965)

Lithium aluminum hydride–Aluminum chloride, *see also* Aluminum hydride. Mol. wts.: $LiAlH_4$, 37.95; $AlCl_3$, 133.35.

Reviews. M. N. Remick, "The Mixed Hydrides," Metal Hydrides, Inc. (1959); E. L. Eliel, *Record of Chemical Progress*, **22**, 129 (1961).

Nature of the mixed hydrides. Depending upon the molar ratio of the two components, various species are possible:

$$LiAlH_4 + AlCl_3 \longrightarrow LiCl + 2\ AlH_2Cl$$

$$LiAlH_4 + 3\ AlCl_3 \longrightarrow LiCl + 4\ AlHCl_2$$

$$3\ LiAlH_4 + AlCl_3 \longrightarrow 3\ LiCl + 4\ AlH_3$$

The reactions are carried out in ether, since tetrahydrofurane is cleaved by mixed hydrides to *n*-butanol,[1] and the 1:1, 1:3, and 3:1 mixtures all give solutions of a reducing agent or agents. Although lithium chloride is insoluble in ether, the chloride formed in the above reactions does not precipitate and presumably is present as a complex. Aluminum hydride generated by the third reaction appears to be stabilized by complexing. Conductivity measurements indicate the presence of ionic species such as $Al_2Cl_5^+$, AlH_4^- and $Al_2H_2Cl_3^+AlH_4^-$. Although different investigators have used different ratios of hydride to halide, and variation probably exists in the identity and concentration of the active species, the mixed hydrides used all appear to be somewhat less active reducing agents than lithium aluminum hydride itself but to have a greater tendency to bring about reductive fission, equivalent to hydrogenolysis with hydrogen and a catalyst.

Reductive fission. Use of a mixed hydride was first reported by B. R. Brown,[2] who investigated the action of a 1:1.5 mixture on cholestenone enol acetate (1). The result was a mixture, and the only point of interest is that one product, Δ^4-cholestene (2), results from reduction and hydrogenolysis. Reduction of cholestenone

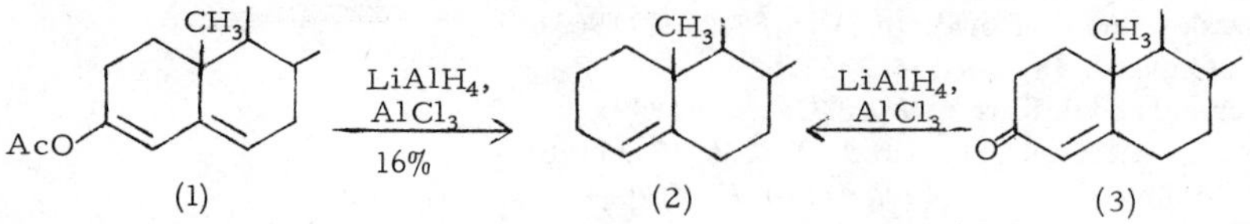

(3) itself with the mixed hydride (1:2) gave Δ^4-cholestene (2) in "good yield."[3] By contrast, reduction of cholestenone enol acetate (1) with lithium aluminum hydride alone gives a mixture of 5 products all of which retain an oxygen function at C_3.[4]

The mixed hydride is the reagent of choice for the reduction of diaryl ketones, alkyl aryl ketones, and certain aryl alcohols to the corresponding hydrocarbons.[5] Usually best yields are obtained if an equimolecular mixture of the ketone and aluminum chloride in ether is added to the 1:1 $LiAlH_4$:$AlCl_3$ reagent.

Marker's method of opening ring F of a steroid sapogenin to form the dihydrosapogenin by hydrogenation in a strongly acid medium is not applicable to diosgenin because the 5,6-double bond is the first point of attack. Doukas and Fontaine[6] found that both saturated and unsaturated sapogenins can be converted into the dihydro

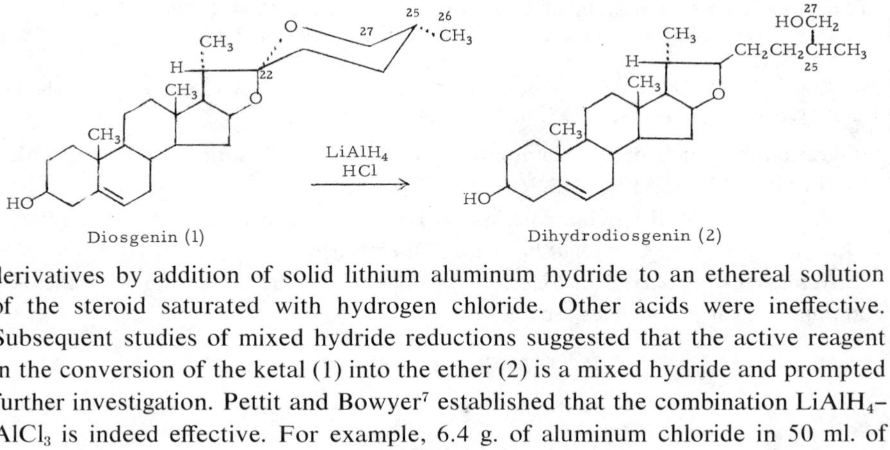

Diosgenin (1) Dihydrodiosgenin (2)

derivatives by addition of solid lithium aluminum hydride to an ethereal solution of the steroid saturated with hydrogen chloride. Other acids were ineffective. Subsequent studies of mixed hydride reductions suggested that the active reagent in the conversion of the ketal (1) into the ether (2) is a mixed hydride and prompted further investigation. Pettit and Bowyer[7] established that the combination $LiAlH_4$–$AlCl_3$ is indeed effective. For example, 6.4 g. of aluminum chloride in 50 ml. of ether is added to an ice-cold mixture of 0.45 g. of lithium aluminum hydride and a solution of 0.5 g. of tigogenin in 75 ml. of ether.

Eliel and co-workers[8] investigated several acetals and ketals and found reduction in ether by mixed hydride to be a general reaction. Best yields are obtained when the ratio of $LiAlH_4$ to $AlCl_3$ is 1:4 and the reagent is taken in 100% excess. Typical examples are:

$$C_6H_5CH(OCH_3)_2 \xrightarrow[88\%]{LiAlH_4-AlCl_3} C_6H_5CH_2OCH_3$$

$$(C_6H_5)(CH_3)C(OC_2H_5)_2 \xrightarrow[81\%]{LiAlH_4-AlCl_3} (C_6H_6)(CH_3)CHOC_2H_5$$

A detailed procedure is given for the reductive cleavage of cyclohexanone ethyleneketal to cyclohexyloxyethanol.[9] A mixture of 1.76 moles of aluminum chloride and

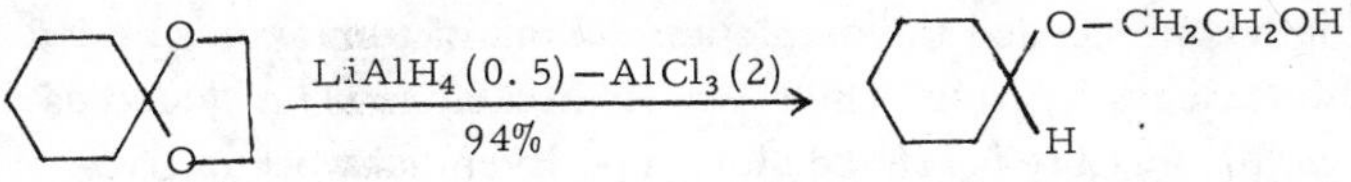
LiAlH4 (0. 5)−AlCl3 (2)
94%
O−CH2CH2OH
H

500 ml. of ether is stirred in an ice-salt bath until a light gray solution results. Lithium aluminum hydride (0.44 mole) is pulverized in a dry box, placed in an Erlenmeyer flask, cooled in an ice bath, and covered with 100 ml. of ether. The flask is swirled to dissolve as much hydride as possible, and the solution is added through a dropping funnel to the reaction flask. Swirling with ether is repeated five times until all the hydride has been added and the gray slurry is stirred for 1 hr. A solution of 0.88 mole of cyclohexanone ethyleneketal in 200 ml. of ether is added at a rate sufficient for gentle refluxing. The ice bath is removed, and the mixture is refluxed for 3 hrs. Excess hydride is destroyed by the dropwise addition of about 12 ml. of water, dilute sulfuric acid is run in, the mixture worked up, and the product distilled.

Enamines are cleaved to alkenes by refluxing in ether for 5–24 hrs. with an equimolecular mixture of lithium aluminum hydride and aluminum chloride, as illustrated for the pyrrolidine enamine of cyclopentanone.[10]

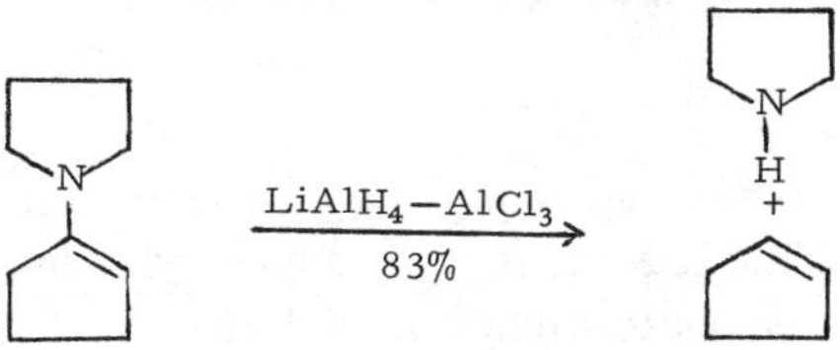
LiAlH4−AlCl3
83%

Reduction of ketones. Wheeler and Mateos[11] found that reduction of 3-cholestanone with mixed hydride gives essentially pure cholestane-3β-ol, with equatorial hydroxyl, whereas reduction with $LiAlH_4$ alone was known to give the 3β- and 3α-epimers in a ratio of approximately 9:1. Eliel and Rerick[12] and Eliel, Martin, and Nasipuri,[13] seeking a route to pure *trans*-4-*t*-butylcyclohexanol (4), found that

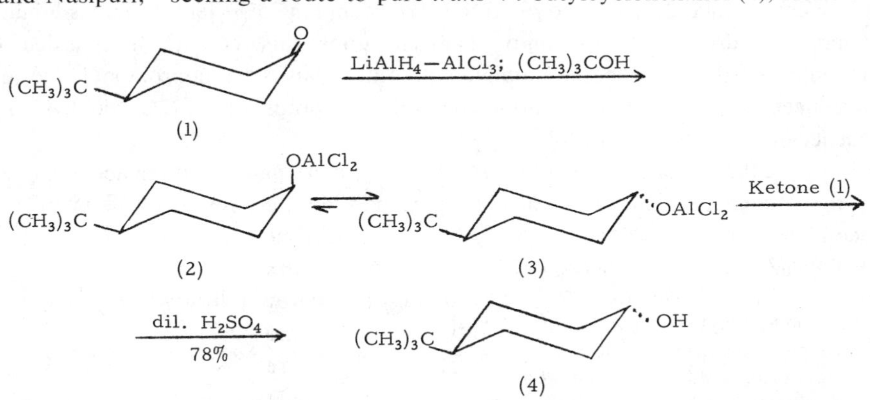
(CH3)3C
LiAlH4−AlCl3; (CH3)3COH
(1)
OAlCl2
(2)
OAlCl2
(3)
Ketone (1)
dil. H2SO4
78%
OH
(4)

mixed hydride reduction of 4-*t*-butylcyclohexanone (1) under normal conditions of kinetic control gives the *trans* and *cis* alcohols in a ratio of 80:20. However, they noted that the intermediate $AlCl_2$-derivatives (2) and (3) possess two bulky groups and hence are ideal for axial→equatorial isomerization. In a standardized procedure,[13] a mixed hydride is prepared as above from 0.5 mole of $AlCl_3$ and 0.145 mole of $LiAlH_4$ in ether, 0.5 mole of the ketone (1) is added in ether. After the reaction has proceeded to completion, 10 ml. of *t*-butanol is added and allowed to react with excess hydride. Then 3 g. of the ketone (1) is added as equilibration catalyst, and the mixture is refluxed for 4 hrs. Eventual workup gives material found by gas chromatographic analysis to contain 96% of *trans* alcohol, 0.8% of *cis* alcohol, and 3.2% of ketone. Crystallization gives the pure *trans* alcohol, m.p. 75–78°, in 78% yield.

Reduction of epoxides. Eliel's group[14] found that the reductive cleavage of triphenylethylene oxide (1) is altered in a striking way by use of mixed hydrides. With lithium aluminum hydride alone, attack is at the least substituted carbon to

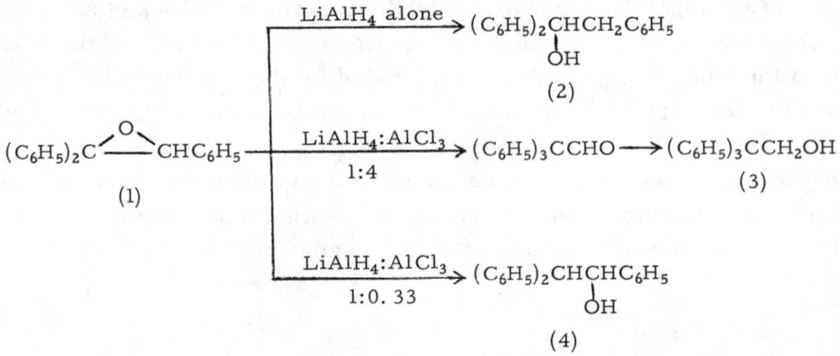

LiAlH4 alone
(C6H5)2CHCH2C6H5
OH
(2)
(C6H5)2C
O
CHC6H5
(1)
LiAlH4:AlCl3
1:4
(C6H5)3CCHO
(C6H5)3CCH2OH
(3)
LiAlH4:AlCl3
1:0.33
(C6H5)2CHCHC6H5
OH
(4)

give the most highly substituted carbinol (2). With a mixed hydride of high aluminum chloride content (1:4), the product is triphenylmethylcarbinol (3), shown to result from an initial rearrangement to triphenylacetaldehyde, which can be effected with $AlCl_3$ alone or with BF_3. With mixed hydride of low $AlCl_3$ content (1:0.33) the product is phenylbenzhydrylcarbinol.

Other reductions. A mixed hydride of ratio 1:1 is superior to lithium aluminum hydride alone for the reduction of nitriles to amines.[15] Unlike lithium aluminum hydride, the mixed reagent does not reduce nitro groups. Thus *p*-nitrobenzaldehyde is reduced by mixed hydride to *p*-nitrobenzyl alcohol in 75% yield. Whereas lithium aluminum hydride tends to remove halogen atoms, this reduction is retarded or prevented by the presence of aluminum chloride. Thus β-bromopropionyl chloride is reduced by a 1:1 mixed hydride to 3-bromo-1-propanol in 77% yield; in the absence of $AlCl_3$ the yield is 44%.[15, 16]

In the reduction of an oxime by $LiAlH_4$, the primary amine is often accompanied by the secondary amine. Use of the mixed hydride increases the amount of the secondary amine formed.[17]

[1]W. J. Bailey and F. Marktscheffel, *J. Org.*, **25**, 1797 (1960)

[2]B. R. Brown, *J. Chem. Soc.*, 2756 (1952). For other examples of hydrogenolysis *see* A. J. Birch and M. Slaytor, *Chem. Ind.*, 1524 (1956).

[3]J. Broome, B. R. Brown, A. Roberts, and A.M.S. White, *J. Chem. Soc.*, 1406 (1960)

[4]W. G. Dauben and J. F. Eastham, *Am. Soc.*, **73**, 3260 (1951)

[5]R. F. Nystrom and C. R. A. Berger, *Am. Soc.*, **80**, 2896 (1958); B. R. Brown and A. M. S. White, *J. Chem. Soc.*, 3755 (1957); J. Blackwell and W. J. Hickinbottom, *ibid.*, 1405 (1961)

[6]H. M. Doukas and T. D. Fontaine, *Am. Soc.*, **73**, 5917 (1951)

[7]G. R. Pettit and W. J. Bowyer, *J. Org.*, **25**, 84 (1960)

[8]E. L. Eliel, V. G. Badding, and M. N. Rerick, *Am. Soc.*, **84**, 2371 (1962)

[9]R. A. Daignault and E. L. Eliel, procedure submitted to *Organic Syntheses*

[10]J. W. Lewis and P. P. Lynch, *Proc. Chem. Soc.*, 19 (1963)

[11]O. H. Wheeler and J. L. Mateos, *Can. J. Chem.*, **36**, 1431 (1958)

[12]E. L. Eliel and M. N. Rerick, *Am. Soc.*, **82**, 1367 (1960)
[13]E. L. Eliel, R. J. L. Martin, and D. Nasipuri, procedure submitted to *Organic Syntheses*
[14]E. L. Eliel and D. W. Delmonte, *Am. Soc.*, **80**, 1744 (1958); E. L. Eliel and M. N. Rerick, *ibid.*, **82**, 1362 (1960); M. N. Rerick and E. L. Eliel, *ibid.*, **84**, 2356 (1962)
[15]R. F. Nystrom, *Am. Soc.*, **77**, 2544 (1955)
[16]R. F. Nystrom, *Am. Soc.*, **81**, 610 (1959)
[17]M. N. Rerick *et al.*, *Tetrahedron Letters*, 629 (1963)

Lithium aluminum hydride–Boron trifluoride etherate. Pettit[1] found that this combination reduces esters to the corresponding ethers, for example 3β-acetoxycholestane affords 3β-ethoxycholestane. 5α-Lanosteryl 3β-formate similarly afforded

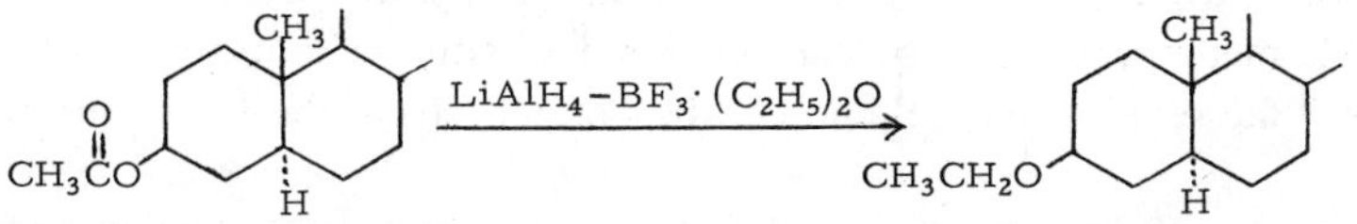
CH3
O
CH3CO
H
LiAlH4–BF3·(C2H5)2O
CH3
CH3CH2O
H

3β-methoxy-5α-lanostane, in this case by reduction with sodium borohydride–boron trifluoride etherate. A solution of the ester in boron trifluoride etherate is added to a cooled suspension of the neutral hydride in ether. Lactones of various types are reduced smoothly by $LiAlH_4$–$BF_3·(C_2H_5)_2O$ to cyclic oxides. Thus the γ-, δ- and ε-lactones formulated were reduced in the yields indicated under the formulas.

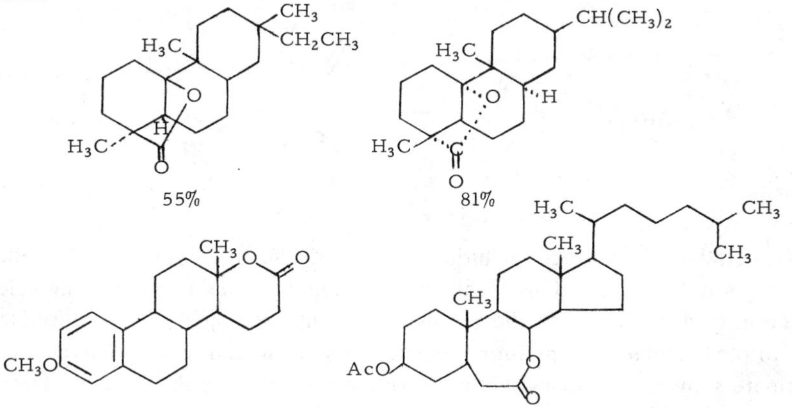
CH3
CH2CH3
H3C
O
H
H3C
CH(CH3)2
H3C
O
H
H3C
C
O
CH3
O
O
CH3O
H3C
CH3
CH3
CH3
CH3
AcO
O
O

55% 81% 71% 60%

[1]G. R. Pettit and T. R. Kasturi, *J. Org.*, **25**, 875 (1960); G. R. Pettit, U. R. Ghatak, B. Green, T. R. Kasturi, and D. M. Piatak, *ibid.*, **26**, 1685 (1961)

Lithium aluminum hydride–Pyridine.[1] Lansbury[2] introduced use of pyridine as a solvent of greater solvent power than tetrahydrofurane for the lithium aluminum hydride reduction of carbonyl compounds. A small-scale reduction is carried out by adding finely divided hydride to a solution of the substrate in pyridine, swirling the solution without cooling for a few minutes until the mildly exothermal reaction is over, adding methanol to decompose excess reagent, pouring the solution into 5% hydrochloric acid, and isolating the product by ether extraction. Use of pyridine as solvent sometimes enhances the reactivity of the metal hydride. Thus benzopinacolone is reduced by lithium aluminum hydride in ether to the corresponding alcohol, but in pyridine solution it suffers cleavage to triphenylmethyllithium and benzyl alcohol. The presence of triphenylmethyl carbanions in the solution can be

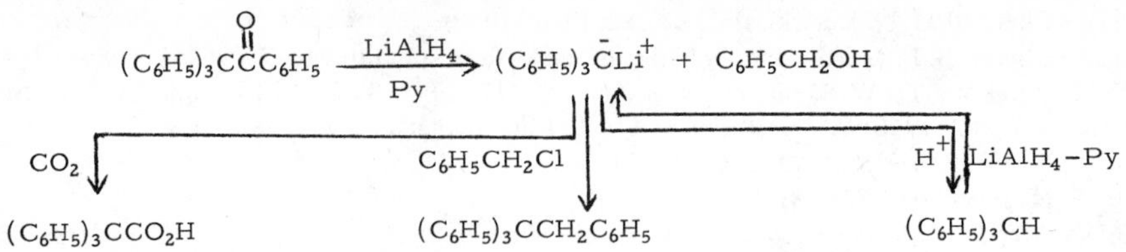
$(C_6H_5)_3CCOC_6H_5$
LiAlH4
Py
$(C_6H_5)_3\bar{C}Li^+$ + $C_6H_5CH_2OH$
CO2
C6H5CH2Cl
H+
LiAlH4-Py
$(C_6H_5)_3CCO_2H$
$(C_6H_5)_3CCH_2C_6H_5$
$(C_6H_5)_3CH$

demonstrated by carbonation and by reaction with benzyl chloride. Quenching with acid produces triphenylmethane, and this reaction can be reversed by treatment of the hydrocarbon with lithium aluminum hydride in pyridine. The still more acidic hydrocarbon fluorene is readily metalated with the hydride in pyridine. By this simple method, followed by carbonation, half-gram quantities of triphenylacetic acid and 9-fluorenecarboxylic acid can be obtained from the hydrocarbons in one hour.

In the course of the study of use of pyridine as solvent, Lansbury[3] discovered that on dissolving powdered lithium aluminum hydride (0.5 g.) in pyridine (50 ml.) and letting the orange solution stand in a stoppered bottle for at least 24 hrs. one obtains a solution of a new, milder reducing agent. NMR and IR data show that the substance lacks Al-H bonds and contains both 1,2- and 1,4-dihydropyridine groups bound to aluminum. It is regarded as tetrakis-(N-dihydropyridyl)-aluminate:

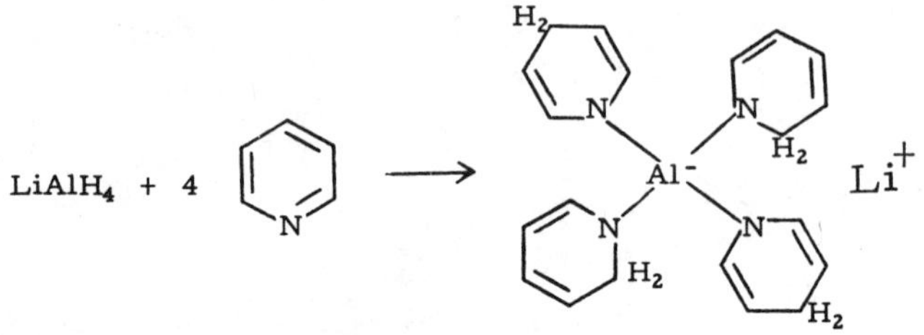
LiAlH4 + 4
N
Al
H2
Li+

A ketone added to the aged solution is reduced effectively, but a carboxylic acid or ester is not reduced. This weak hydride donor is thus useful for the selective reduction of a keto acid to the corresponding hydroxy acid. Both intermolecular and intramolecular competition experiments with tetrakis-(N-dihydropyridyl)-aluminate showed that diaryl ketones are more reactive to this reagent than either dialkyl or aralkyl ketones. This relationship is the opposite of that found by H. C. Brown for reduction with sodium borohydride in isopropyl alcohol, where the order of reactivity is acetone > acetophenone > benzophenone.

[1]Contributed by Peter T. Lansbury, University of Buffalo

[2]P. T. Lansbury, *Am. Soc.*, **83**, 429 (1961); P. T. Lansbury and R. Thedford, *J. Org.*, **27**, 2383 (1962)

[3]P. T. Lansbury and J. O. Peterson, *Am. Soc.*, **83**, 3537 (1961); **85**, 2236 (1963)

Lithium amide, $LiNH_2$. Mol. wt. 22.97. Suppliers: Alfa, Lithium Corp. The contents of a 1-lb. tin should be transferred to 6-oz. narrow-necked amber bottles with screw-caps which are kept tightly stoppered and sealed with paraffin. If left open for minimal periods during weighing, the titer remains fairly constant (initial titer 98%).[1]

Preparation and use in the aldol condensation of esters with aldehydes and ketones. Hauser and Puterbaugh[2] added a small portion of a 1.1-g. batch of lithium shot to 300 ml. of stirred liquid ammonia, followed by a catalytic amount of ferric nitrate.

When the initial blue color faded, the remainder of the lithium was added in small portions. After about 20 min. the resulting gray suspension of lithium amide was treated with an ethereal solution of 0.2 mole of t-butyl acetate, added dropwise.

$$CH_3\overset{O}{\overset{\|}{C}}OC(CH_3)_3 \xrightarrow[\text{2. } C_6H_5COCH_3, \text{ ether}]{\text{1. } LiNH_2, NH_3} C_6H_5\overset{CH_3}{\underset{OLi}{\overset{|}{C}}}CH_2\overset{O}{\overset{\|}{C}}OC(CH_3)_3 \xrightarrow[76\%]{H_2O} C_6H_5\overset{CH_3}{\underset{OH}{\overset{|}{C}}}CH_2\overset{O}{\overset{\|}{C}}OC(CH_3)_3$$

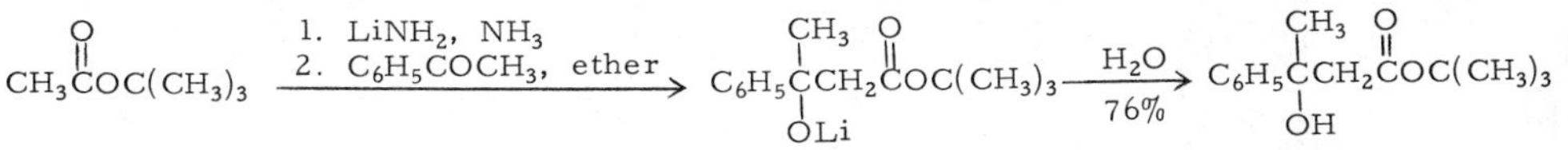

The liquid ammonia was then driven off, an equivalent of the ketone was added, and the mixture refluxed for completion of the aldolization. Lithium amide proved to be superior to sodium amide for these condensations. t-Butyl esters were preferred because they are readily cleaved by pyrolysis or by acid. Dehydration to the α,β-unsaturated esters could be effected with thionyl chloride and pyridine. Hauser and Lindsay[3] succeeded in effecting aldol condensation with ethyl acetate, which undergoes self-condensation very readily; two equivalents of base were used. Sisido[4] has reported further alkylations of t-butyl esters.

Lithium amide has been used in place of sodamide for the preparation of heterocyclic amines with antihistaminic activity.[5] For example, a mixture of 2-aminopyridine, β-dimethylaminoethyl bromide hydrobromide, and 2 equivalents of lithium amide is refluxed in toluene for 22 hrs. Kaye,[1] who notes that lithium amide

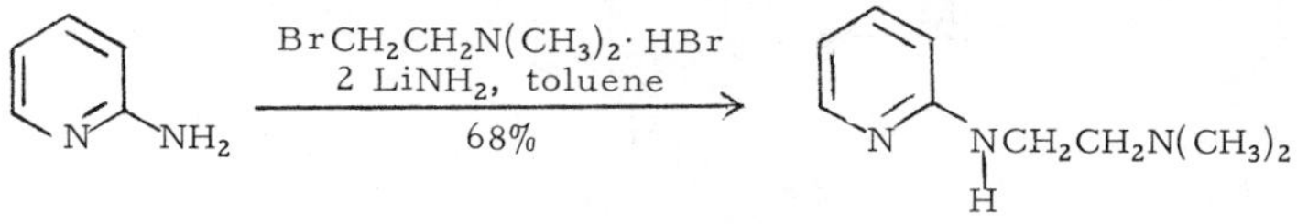

is more easily handled than sodamide, obtained a number of heterocyclic secondary amines of the type formulated in generally fair yield with use of lithium amide as base and found that they can be converted into the corresponding tertiary amines by the same procedure in excellent yield.

[1] I. A. Kaye, *Am. Soc.*, **71**, 2322 (1949); I. A. Kaye and I. C. Kogan, *ibid.*, **73**, 5891 (1951); J. J. Ferraro, I. A. Kaye, and U. Weiss, *J. Chem. Soc.*, 2813 (1964)

[2] C. R. Hauser and W. H. Puterbaugh, *Am. Soc.*, **75**, 1068 (1953)

[3] C. R. Hauser and J. K. Lindsay, *Am. Soc.*, **77**, 1050 (1955). *See also* C. R. Hauser and W. J. Chambers, *J. Org.*, **21**, 1524 (1956).

[4] K. Sisido, Y. Kazama, H. Kodama, and H. Nozaki, *Am. Soc.*, **81**, 5817 (1959); K. Sisido, K. Sei, and H. Nozaki, *J. Org.*, **27**, 2681 (1962)

[5] C. P. Huttrer, C. Djerassi, W. L. Beears, R. L. Mayer, and C. R. Scholz, *Am. Soc.*, **68**, 1999 (1946)

Lithium–Ammonia, *see also* Birch reduction.

Stork[1] deduced that the octalone (1) should be convertible into the less stable of two possible enolates by reaction with lithium in liquid ammonia, and experimentation showed this prediction to be correct. Reduction of the unsaturated ketone (1) afforded only the *trans-β*-decalone (3) as the initial product. The reaction is considered to involve addition of an electron from lithium to ketone (1) to produce a hybrid intermediate (2) with carbanion character at the β-carbon, and this intermediate abstracts a proton from ammonia to form the product (3). These observations suggested a new method of alkylation.[2] Alkylation of the *trans*-2-decalone (4) in the presence of base proceeds through the more stable enolate and gives (5). However, the less stable enolate (2) can be generated by Li–NH$_3$ reduction of the unsaturated ketone (1) and alkylated to give the isomeric methylated product (6).

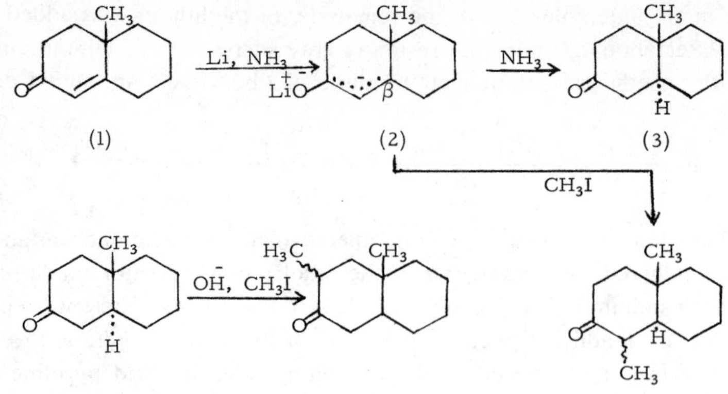
CH3
Li, NH3
LiO
β
NH3
H
CH3I
H3C
OH, CH3I
CH3

(1) (2) (3)

(4) (5) (6)

Schaub and Weiss[3] alkylated testosterone (7) by reducing it with lithium in liquid ammonia and adding methyl iodide in ether. The product was shown to be the

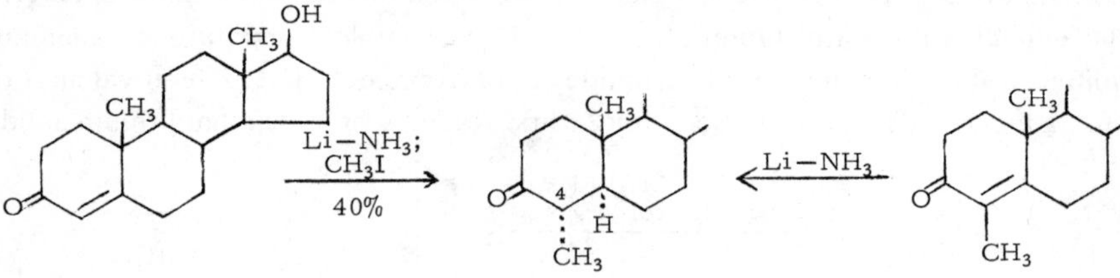
OH
CH3
Li–NH3;
CH3I
40%
4
H
Li–NH3

(7) (8) (9)

4α-methyl ketone (8) from NMR data and from formation of an identical product on reduction of 4-methyltestosterone (9). Schaub and Weiss also showed that $\Delta^{4,6}$-3-dienones (10) are reduced smoothly by Li–NH_3 to Δ^5-3-ketosteroids (12). A solution

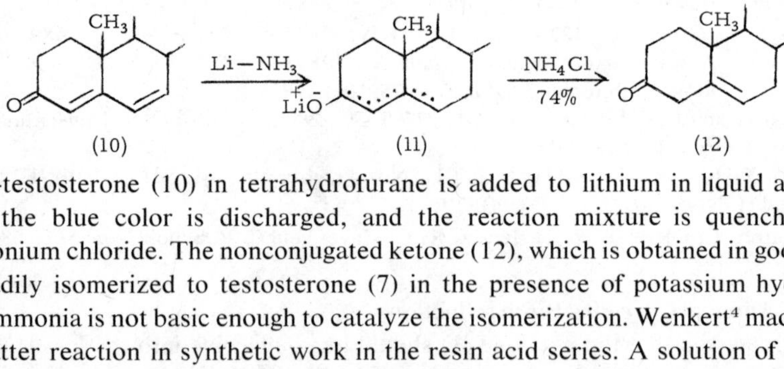
CH3
Li–NH3
LiO
NH4Cl
74%

(10) (11) (12)

of Δ^6-testosterone (10) in tetrahydrofurane is added to lithium in liquid ammonia until the blue color is discharged, and the reaction mixture is quenched with ammonium chloride. The nonconjugated ketone (12), which is obtained in good yield, is readily isomerized to testosterone (7) in the presence of potassium hydroxide, but ammonia is not basic enough to catalyze the isomerization. Wenkert[4] made use of the latter reaction in synthetic work in the resin acid series. A solution of 100 mg. of the dienone (13) and 100 mg. of lithium in 5 ml. of tetrahydrofurane and 80 ml.

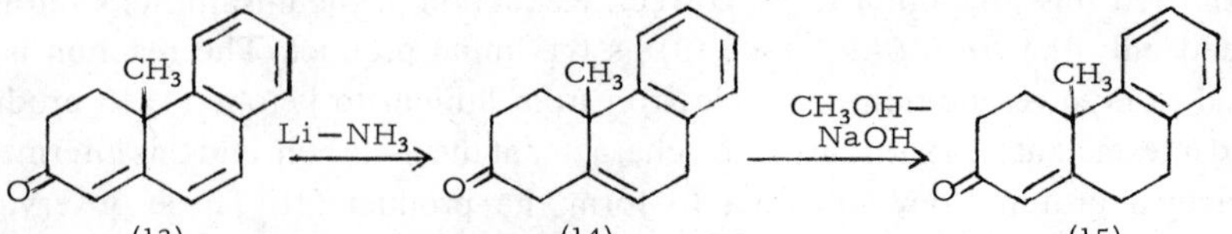
CH3
Li–NH3
CH3OH–
NaOH

(13) (14) (15)

of liquid ammonia was stirred for 3 min., ammonium chloride was added, and the mixture allowed to evaporate under nitrogen. Extraction with chloroform, washing, and evaporation gave 90 mg. of ketone (14) as an oil; this was refluxed with 10%

methanolic sodium hydroxide for 30 min. and crystallization of the product afforded 71 mg. of the conjugated ketone (15).

Dear and Pattison[5] regard lithium–ammonia as superior to sodium–ammonia for the stereospecific *trans*-reduction of alkynes. The reaction is carried out in tetrahydrofurane under pressure in an autoclave at room temperature.

$$CH_3(CH_2)_7C\equiv C(CH_3)_7CO_2H \xrightarrow[97.5\%]{Li-NH_3 \text{ in } THF} CH_3(CH_2)_7\underset{H}{\overset{}{C}}=\underset{(CH_2)_7CO_2H}{\overset{H}{C}}$$

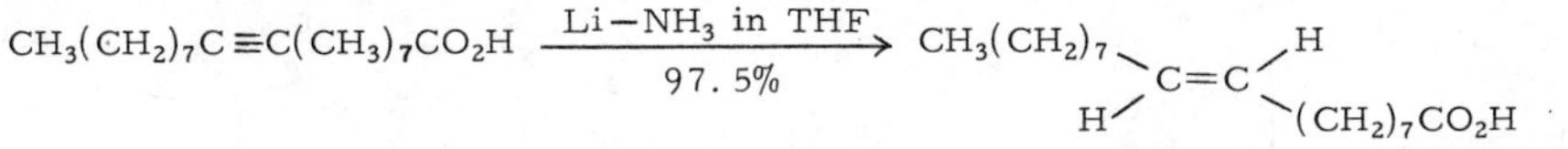

Stearolic acid   Elaidic acid

Methyl desoxypodocarpate (1), which has a highly hindered axial methyl ester group, reacts with Li–NH$_3$ to give the corresponding acid in about 75% yield and the expected alcohol is only a minor product.[6] Methyl dehydroabietate (3), in

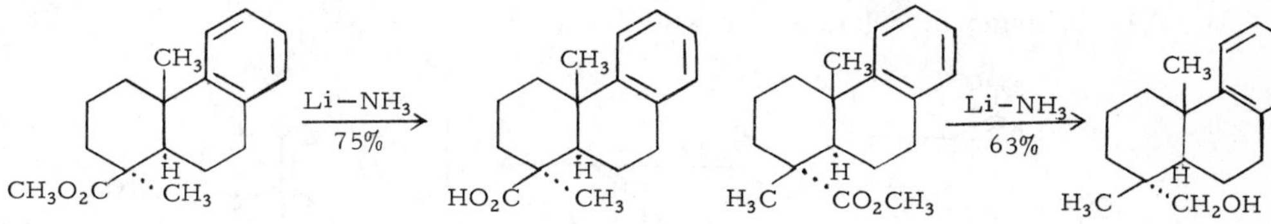

which the ester group is equatorial, was largely reduced to the alcohol (4). A related example is reported by Meyer and Levinson.[7]

[1] G. Stork and S. D. Darling, *Am. Soc.*, **82**, 1512 (1960); G. Stork and J. Tsuji, *Am. Soc.*, **83**, 2783 (1961)

[2] G. Stork, P. Rosen, and N. L. Goldman, *Am. Soc.*, **83**, 2965 (1961)

[3] R. E. Schaub and M. J. Weiss, *Chem. Ind.*, 2003 (1961)

[4] E. Wenkert, A. Afonso, J. B-son Bredenberg, C. Kaneko, and A. Tahara, *Am. Soc.*, **86**, 2038 (1964)

[5] R. E. A. Dear and F. L. M. Pattison, *Am. Soc.*, **85**, 622 (1963)

[6] E. Wenkert and B. G. Jackson, *Am. Soc.*, **80**, 217 (1958)

[7] W. L. Meyer and A. S. Levinson, *J. Org.*, **28**, 2184 (1963)

Lithium–Ammonia–ROH, *see* Birch reduction.

Lithium borohydride, LiBH$_4$. Mol. wt. 21.79, m.p. 284° dec., extremely hygroscopic, soluble in lower primary amines and ethers (diethyl ether, 4 g./100 g.; tetrahydrofurane, 21 g./100 g.); soluble (with reaction) in lower alcohols; soluble in water with slow decomposition: LiBH$_4$ + 2 H$_2$O → LiBO$_2$ + 4 H$_2$ + heat. Supplier: Alfa.

The reagent can be prepared in solution by reaction of sodium borohydride with lithium chloride in absolute ethanol,[1] isopropylamine,[2] and diglyme.[3] It is more

$$NaBH_4 + LiCl \longrightarrow LiBH_4 + NaCl$$

reactive as a reducing agent than sodium borohydride and less reactive than lithium aluminum hydride. Thus it reduces aldehydes, ketones, acid chlorides, oxides, esters, and lactones, but does not reduce carboxylic acids, nitriles, or nitro compounds.[4] The chief advantage over sodium borohydride is the high solubility in ether solvents.

[1] J. Kollonitsch, O. Fuchs, and V. Gábor, *Nature*, **173**, 125 (1954)

[2] H. I. Schlesinger, H. C. Brown, and E. K. Hyde, *Am. Soc.*, **75**, 209 (1953)

[3] H. C. Brown, E. J. Mead, and B. C. Subba Rao, *Am. Soc.*, **77**, 6209 (1955)

[4] H. C. Brown, "Hydroboration," p. 245, Benjamin (1962)

Lithium bromide, LiBr. Mol. wt. 86.86. Supplier: KK. Elimination of nitrogen from the diazoalkene (1) is effected by lithium bromide in ether at 0° and affords equivalent amounts of the methyltricyclohexane (2) and 1-methyl-$\Delta^{1,3}$-cyclohexadiene (3), each in yield of about 25%.[1] An intermediate carbene seems to be involved in the formation of (2).

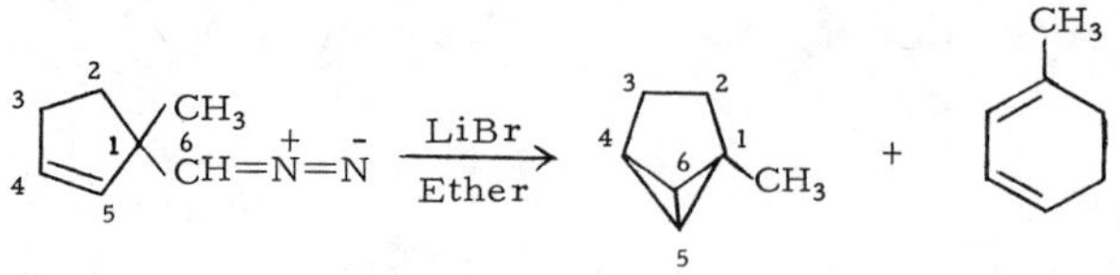

(1) (2) (3)

Lithium bromide in large excess in refluxing acetonitrile slowly (12 hrs.) converts the quinolinium perchlorate (4) into the base (5).[2]

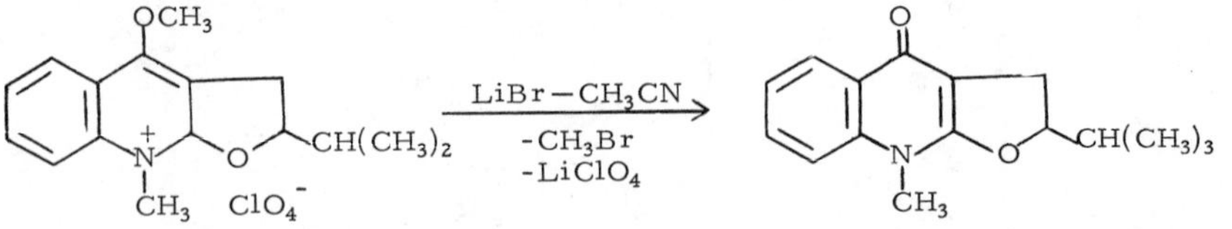

(4) (5)

[1]G. L. Closs and R. B. Larrabee, *Tetrahedron Letters*, 287 (1965)
[2]S. Goodwin and E. C. Horning, *Am. Soc.*, **81**, 1908 (1959); J. W. Huffman and L. E. Browder, *J. Org.*, **29**, 2598 (1964)

Lithium bromide–Boron trifluoride etherate. Aliphatic ethers can be cleaved by reaction with lithium bromide and boron trifluoride etherate in acetic anhydride at room temperature for 30 hrs.[1] Methoxycyclohexane, for example, is converted into a 7:1 mixture of acetoxycyclohexane and cyclohexene. Saturated steroid ethers are cleaved to mixtures of enes and acetates under these conditions; cholesteryl methyl ether gave about equal parts of cholesteryl acetate and cholesteryl bromide. However, Narayanan[2] reports that the lithium halide is not essential and indeed often detrimental. Thus cholesteryl methyl ether treated with boron trifluoride etherate and acetic anhydride in ether at 0° (14 hrs.) gave cholesteryl acetate in 93% yield.

[1]R. D. Youssefyeh and Y. Mazur, *Tetrahedron Letters*, 1287 (1962)
[2]C. R. Narayanan and K. N. Iyer, *Tetrahedron Letters*, 759 (1964)

Lithium–*t*-Butanol–Tetrahydrofurane. The insecticides aldrin and isodrin became attractive starting materials for the synthesis of compounds of interesting ring systems with the discovery by Bruck, Thompson, and Winstein[1] of a simple method for effecting their dechlorination, that is, replacement of all the chlorine atoms by hydrogen. The polyhalogen compound is treated with lithium ribbon (3 g. atoms per

I II (III) Aldrin

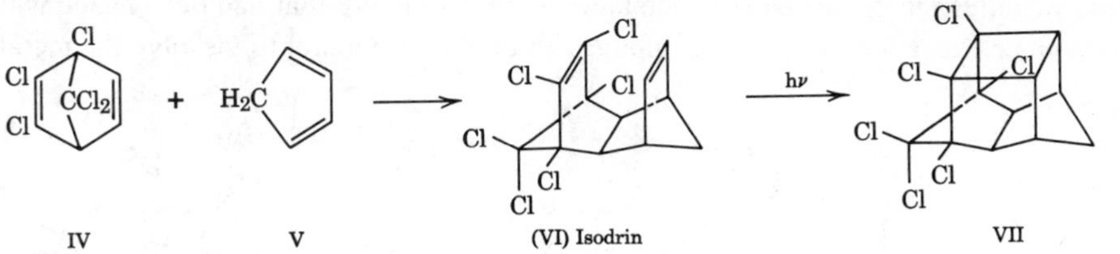

halogen) and *t*-butanol (2 molar equivalents per halogen) in refluxing tetrahydrofurane in an inert atmosphere. In this series yields of hydrocarbons are around 60%. Note that halogens displaced are vinylic, allyic, geminal, and even bridgehead types. Displacement of aromatic halogen is effected efficiently in the conversion of 3′,4′,5′,6′-tetrachlorobenzonorbornene (1) into benzonorbornene (2).[2]

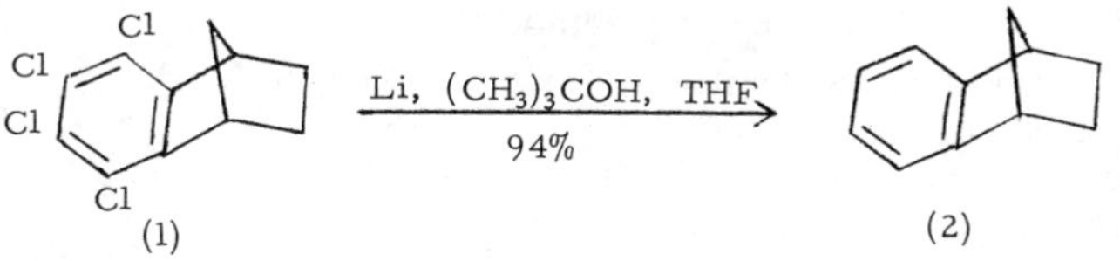

The reaction has been applied[3] also to the adduct (5), readily available by the reactions formulated. This dechlorination with lithium and *t*-butanol in tetrahydrofurane was attended with some reduction, and the product was a mixture of the

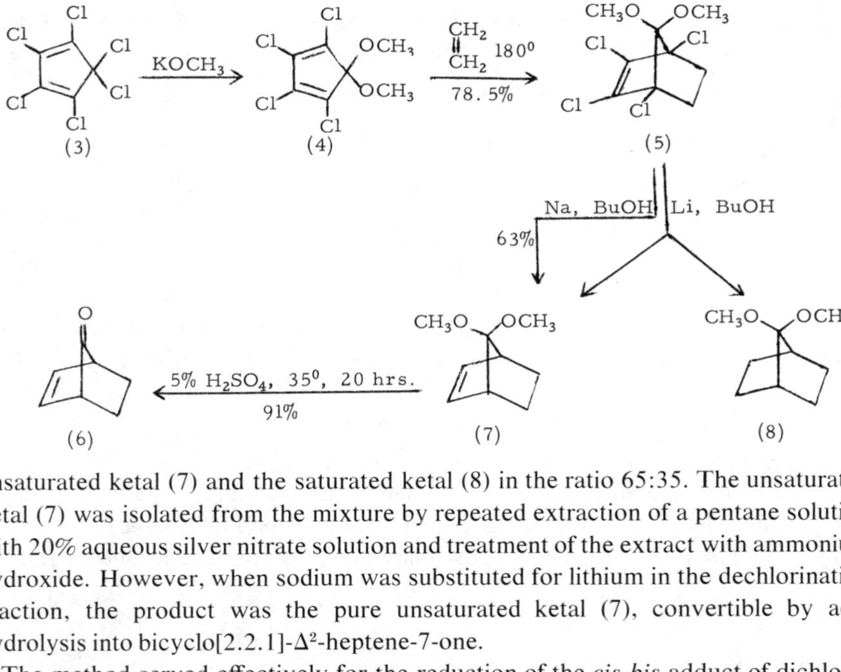

unsaturated ketal (7) and the saturated ketal (8) in the ratio 65:35. The unsaturated ketal (7) was isolated from the mixture by repeated extraction of a pentane solution with 20% aqueous silver nitrate solution and treatment of the extract with ammonium hydroxide. However, when sodium was substituted for lithium in the dechlorination reaction, the product was the pure unsaturated ketal (7), convertible by acid hydrolysis into bicyclo[2.2.1]-Δ^2-heptene-7-one.

The method served effectively for the reduction of the *cis-bis*-adduct of dichlorocarbene to *cis,cis*-1,5-cyclooctadiene.[4] A solution of the halide in tetrahydrofurane was stirred magnetically over an ice bath, and lithium wire that had been flaked with a hammer was added in portions along with enough *t*-butanol to dissolve the metal.

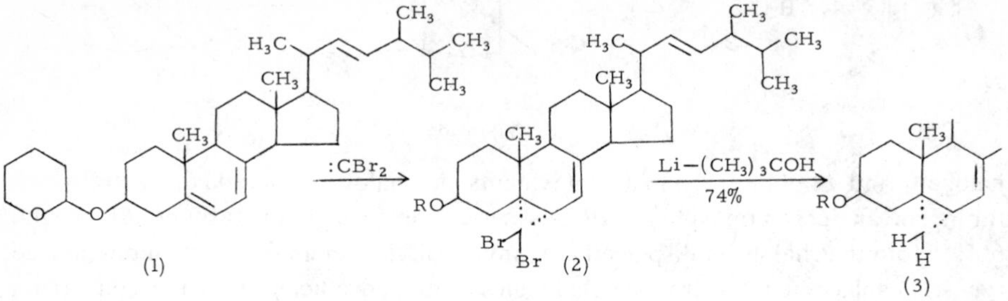

The adduct (2) from ergosteryl tetrahydropyranyl ether (1) and dibromocarbene was reduced successfully by the same method.[5,6]

[1]P. Bruck, D. Thompson, and S. Winstein, *Chem. Ind.*, 405 (1960)
[2]P. Bruck, *Tetrahedron Letters*, 449 (1962)
[3]P. G. Gassman and P. G. Pape, *J. Org.*, **29**, 160 (1964)
[4]L. F. Fieser and D. H. Sachs, *J. Org.*, **29**, 1113 (1964)
[5]M. Z. Nazer, *J. Org.*, **30**, 1737 (1965)
[6]For further examples, see D. I. Schuster and F.-T. Lee, *Tetrahedron Letters*, 4119 (1965)

Lithium carbonate, Li_2CO_3. Mol. wt. 73.89, m.p. 618°. Suppliers: F, MCB.

Dehydrohalogenation. With use of this reagent in refluxing dimethylformamide, House and Bashe[1] were able to effect dehydrobromination of the bromoketone (1) to the unrearranged product (2). Use of the less basic lithium chloride led to

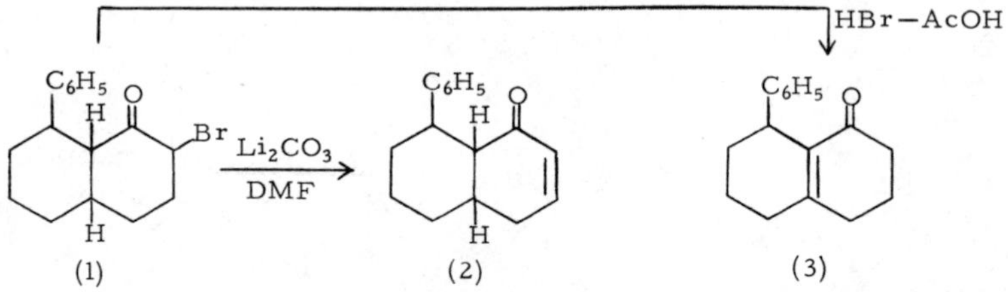

mixtures of (2) and the rearranged ketone (3), and this substance was the sole product formed when dehydrohalogenation was effected by refluxing (1) with HBr in acetic acid.

[1]H. O. House and R. W. Bashe, II, *J. Org.*, **30**, 2942 (1965)

Lithium carbonate–Lithium bromide. Mol. wts. 73.89, 86.86.

Use of this combination in dimethylformamide for effecting dehydrohalogenation was introduced by Joly[1] as an improvement of the method of Holysz using lithium chloride in DMF. As applied to 2β,4β-dibromo-17β-acetoxyetiocholane (1), the

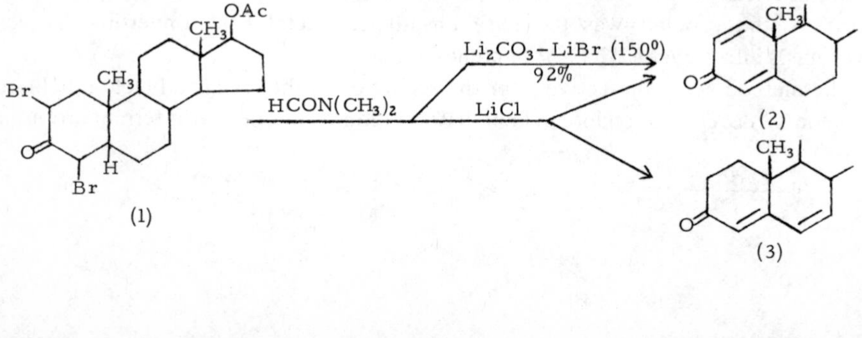

Holysz procedure affords a mixture of the 1,4-dienone (2) and the 4,6-dienone (3). The French investigators heated 4.9 g. of the dibromide 1 with 4.5 g. of lithium carbonate and 5.2 g. of lithium bromide in 150 ml. of dimethylformamide under nitrogen at the reflux temperature (about 150°) for 1 hr. and isolated the pure 1,4-dienone (2) in 92% yield. Under the same conditions, but with lithium carbonate alone, the yield of (2) was only 42%. Three other 2β,4β-dibromo-3-ketones of the 5β series on dehydrogenation with Li_2CO_3–LiBr–DMF at the reflux temperature afforded 1,4-dienones in yields of 71, 90, and 93%. It was then found possible to isolate the intermediate 4β-bromo-Δ^1-ene-3-one (4) by moderating the conditions.[2] Thus (1), treated as before but at 100° for 30 min., afforded (4) in moderate yield.

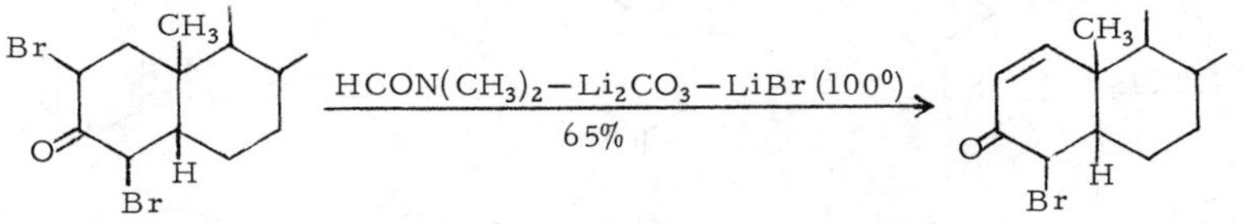

Br
CH3
HCON(CH3)2–Li2CO3–LiBr (100°)
65%
H
Br
CH3
H
Br

(1) (4)

Although results reported are conflicting, the reaction appears to be less satisfactory as applied to 2α,4α-dibromo-3-ketones of the 5α series. Pelc *et al.*[3] treated 17β-acetoxy-2α,4α-dibromo-3-ketoandrostane with Li_2CO_3–LiBr in DMF and obtained a mixture of the 1,4-dienone and the 4,6-dienone in which the latter predominated. However, P. Wieland, Heusler, and Wettstein[4] obtained the 1,4-dienone (6) in satisfactory yield from (5) as follows. A solution of (5) in acetic acid containing

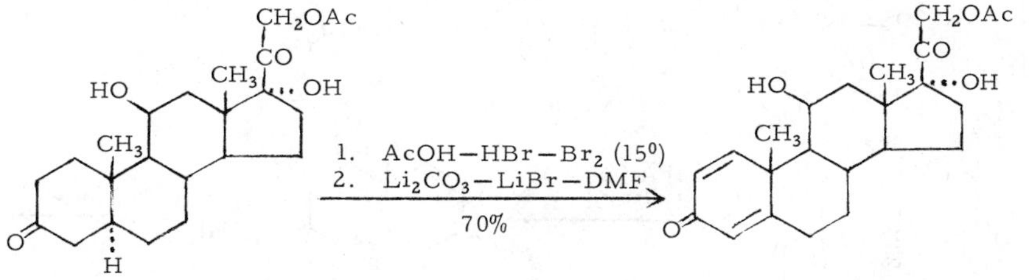

CH2OAc
CO
CH3
OH
HO
CH3
H
1. AcOH–HBr–Br2 (15°)
2. Li2CO3–LiBr–DMF
70%
CH2OAc
CO
CH3
OH
HO
CH3

(5) (6)

a trace of hydrogen bromide was treated with bromine at 15–16°, the solid which precipitated on dilution was collected and the filtrate extracted with ether. The total material when heated in the DMF with Li_2CO_3–LiBr afforded prednisolone 21-acetate (6) in 70% overall yield.

Application of the reaction to 3α-bromocholestane-2-one (9) resulted in part in dehydrohalogenation to the 3-ene-2-one (10) and in part in epimerization at C_3, the axial 3α-bromine giving way to the equatorial 3β-bromoketone (11).[5]

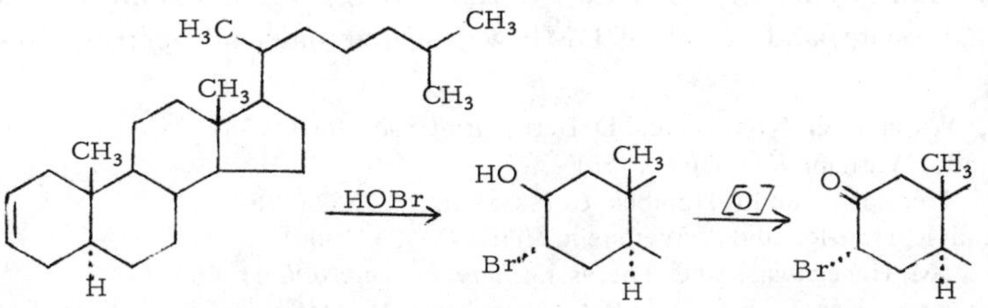

H3C
CH3
CH3
CH3
CH3
H
HOBr
HO
CH3
Br
H
[O]
CH3
Br
H

(7) (8) (9)

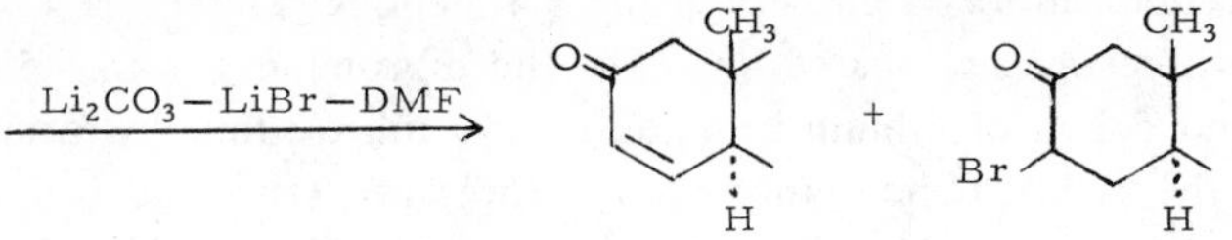

(10) (11)

Plieninger *et al.*[6] found the method advantageous for conversion of the enone (12) into the dienone (15). The yield in the dehydrohalogenation of (14) was over twice that obtained with collidine (40%).

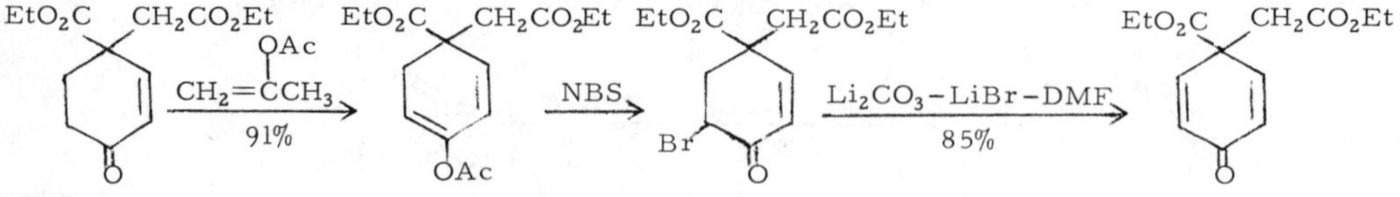

(12) (13) (14) (15)

Corey and Hortmann[7] achieved the synthesis of dihydrocostunolide (6) from santonin (1) by a process indicated in outline in the formulation. For the dehydrobromination of (2) to (3) the method of Joly proved much superior (yield 80–90%) to

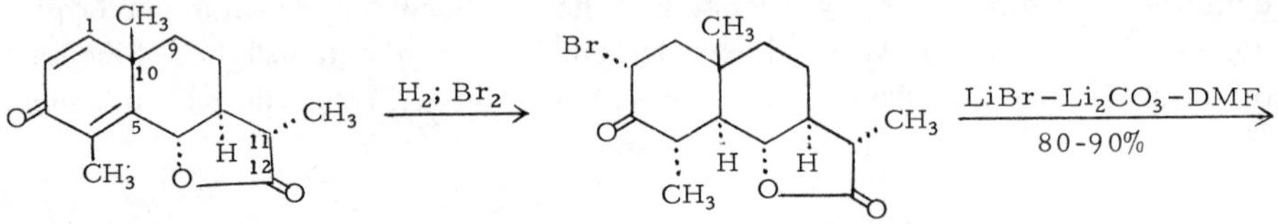

(1) (2)

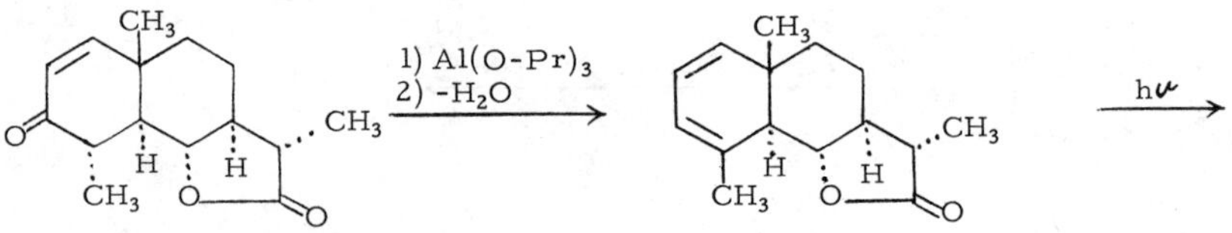

(3) (4)

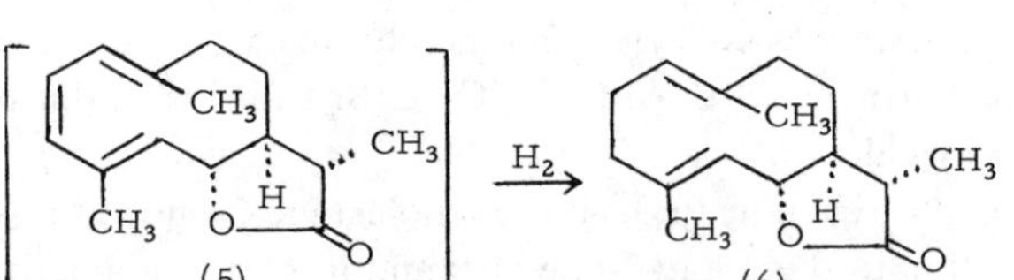

(5) (6)

use of collidine (yield 30%). A mixture of 7.26 g. of (2), 3 g. of lithium bromide, 4 g. of lithium carbonate, and 50 ml. of DMF was heated under nitrogen at 120–125° for 75 min.

[1]R. Joly, J. Warnant, G. Nominé, and D. Bertin, *Bull. soc.*, 366 (1958)
[2]R. Joly and J. Warnant, *ibid.*, 367 (1958)
[3]B. Pelc, S. Heřmánek, and J. Holubek, *Coll. Czech.*, **26**, 1852 (1961)
[4]P. Wieland, K. Heusler, and A. Wettstein, *Helv.*, **43**, 523 (1960)
[5]T. Nakano, M. Hasegawa, and C. Djerassi, *Chem. Pharm. Bull.*, **11**, 465 (1963)
[6]H. Plieninger, G. Ege, H. J. Grasshoff, G. Keilich, and W. Hoffmann, *Ber.*, **94**, 2115 (1961)
[7]E. J. Corey and A. G. Hortmann, *Am. Soc.*, **87**, 5736 (1965)

Lithium chloride, LiCl. Mol. wt. 42.40, m.p. 613°, Suppliers: B, MCB.

The fact that the last step in the original synthesis of cortisone required *cis* elimination of the elements of hydrogen bromide from 4β-bromodihydrocortisone 21-acetate (I), and hence presented difficulty, stimulated the search for better methods of dehydrohalogenation. Holysz[1] found lithium chloride (or bromide) in dimethylformamide to be an effective reagent. The reaction, conducted at 100° in a nitrogen atmosphere, afforded cortisone 21-acetate (II) in good yield.

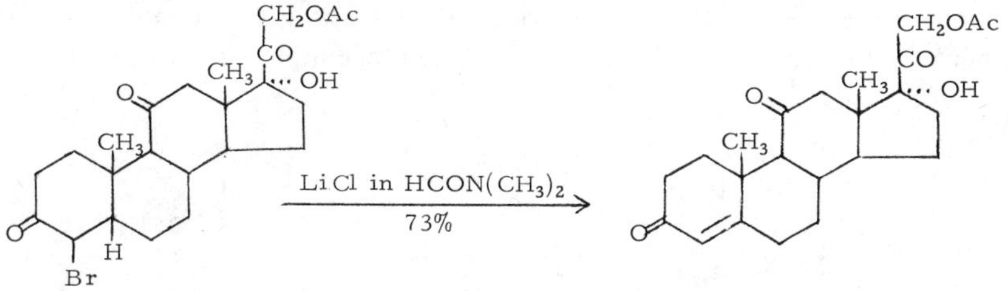
CH2OAc
CO
CH3
OH
CH3
H
Br
LiCl in HCON(CH3)2
73%
CH2OAc
CO
CH3
OH
CH3

I II

The case of 2-chloro-2-methylcyclohexanone (2) presents no comparable steric difficulty, and here dehydrohalogenation by the Holysz method and with collidine both proceed rather poorly and in comparable yield.[2]

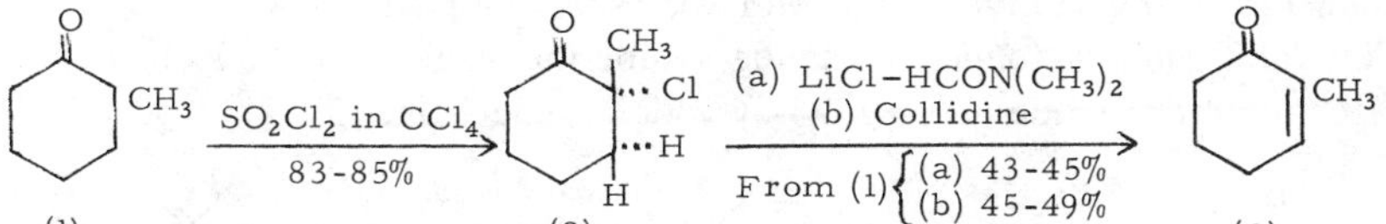
CH3
SO2Cl2 in CCl4
83-85%
CH3
Cl
H
H
(a) LiCl-HCON(CH3)2
(b) Collidine
From (1) (a) 43-45% (b) 45-49%
CH3

(1) (2) (3)

Lithium chloride serves as an electrolyte in the electrolytic reduction of alkylbenzenes in methylamine in a cell which can be operated either with or without an asbestos divider for separation of the anode and cathode compartments.[3] Without the divider, excellent conversions into 1,4-dihydro products result. When the same hydrocarbon is reduced with the divider in place, equally excellent conversions into tetrahydro products are obtained.

Griseofulvic "acid" (4, actually the 2′- or 4′-enol) reacts with phosphoryl chloride in DMF (but not in dimethylacetamide) to give the 4′-chloride (5) in high yield.[4]

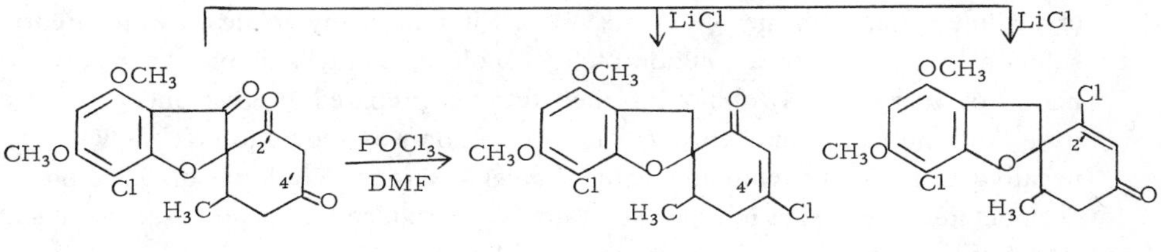
LiCl
LiCl
OCH3
CH3O
Cl
2′
4′
H3C
POCl3
DMF
OCH3
CH3O
Cl
4′
H3C
Cl
OCH3
Cl
2′
CH3O
Cl
H3C

(4) (5) (6)

In the presence of lithium chloride (no solvent) the product is a mixture of the 4′-chloride (5, 40%) and the 2′-chloride (6, 60%).

[1]R. P. Holysz, *Am. Soc.*, **75**, 4432 (1953); see also W. S. Johnson, W. A. Vredenburgh, and J. E. Pike (preparation of XIV), *Am. Soc.*, **82**, 3414 (1960)

[2]E. W. Warnhoff, D. G. Martin, and W. S. Johnson, *Org. Syn.*, *Coll. Vol.*, **4**, 162 (1963)

[3]R. A. Benkeser and E. M. Kaiser, *Am. Soc.*, **85**, 2858 (1963)

[4]L. Stephenson, T. Walker, W. K. Warburton, and G. B. Webb, *J. Chem. Soc.*, 1282 (1962)

Lithium diethoxyaluminum hydride, $LiAlH_2(OC_2H_5)_2$.

The reagent is prepared in solution by addition of 2 moles of ethanol or 1 mole of ethyl acetate to 1 mole of $LiAlH_4$ in ether solution at 0°.[1] It reduces N,N-dimethylamides to aldehydes in yields of 70–90%.[1] Tarbell's group[2] found that 3,4,5-trimethoxybenzaldehyde can be prepared far more readily by this method than by

$$CH_3CH_2CH_2CON(CH_3)_2 \xrightarrow[90\%]{} CH_3CH_2CH_2CHO$$

Rosenmund reduction of the acid chloride. A suspension of 0.26 mole of the amide in ether was treated at 0° with a solution of the reagent prepared from 0.17 mole of

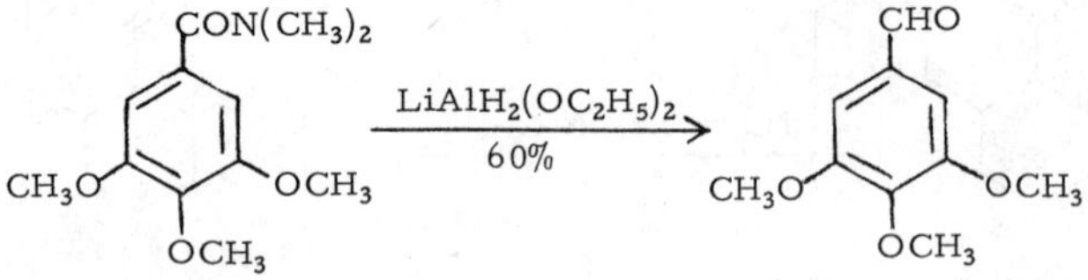
CON(CH3)2
CH3O
OCH3
OCH3
LiAlH2(OC2H5)2
60%
CHO
CH3O
OCH3
OCH3

$LiAlH_4$ (1.3 × theory), added in 2 hrs. at 0°. Workup and crystallization afforded the pure aldehyde in 60% yield.

[1]H. C. Brown and A. Tsukamoto, *Am. Soc.*, **81**, 502 (1959)

[2]T. J. Perun, L. Zeftel, R. G. Nelb, and D. S. Tarbell, *J. Org.*, **28**, 2937 (1963)

Lithium diethylamide, $LiN(C_2H_5)_2$. Mol. wt. 79.07. Supplier: Alfa.

Cope and Tiffany[1] found this amide a superior reagent for the isomerization of cyclooctatetraene oxide (1) to 1,3,5-cyclooctatriene-7-one (3). They postulate that

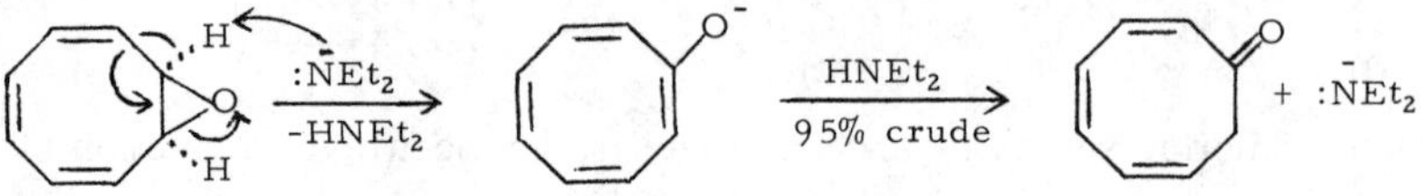
H
H
:NEt2
-HNEt2
HNEt2
95% crude
+ :NEt2

(1) (2) (3)

base-catalyzed removal of a proton from (1) gives an anion which rearranges to the enolate anion (2). The ketone (3) was first obtained from (1) by reaction with phenyllithium, but competing reactions limited the yield to 8%. The sterically hindered base mesityllithium gave an improved yield (39%) of (3), but the weaker base lithium diethylamide gave a still better result; the overall yield from (1) of the crystalline semicarbazone of (3) was 71%. Lithium diethylamide was prepared in solution by addition of a solution of 0.60 mole of diethylamine in ether during a period of 15 min. to a solution of phenyllithium prepared in ether under nitrogen from 0.66 mole of bromobenzene. The conversion was shown to be complete by a negative test for phenyllithium with Michler's ketone.[2] Addition of 0.3 mole of cyclooctatetraene oxide in ether at −8 to −12° produced a deep orange color and after 5 min. the mixture was acidified and worked up.

In a later study[3] lithium diethylamide prepared from *n*-butyllithium was shown to effect the following transformations:

cis-Stilbene oxide ⟶ Desoxybenzoin (70%)
trans-Stilbene oxide ⟶ Diphenylacetaldehyde (66%)
Triphenylethylene oxide ⟶ Benzhydryl phenyl ketone (80%)
1,l-Diphenyl-2-*p*-tolylethylene oxide ⟶ Benzhydryl
p-tolyl ketone (41%)

Tetraphenylethylene oxide – no reaction
1,1-Diphenylethylene oxide – no reaction

Burgstahler[4] prepared the reagent by stirring lithium shot in ethylamine until the blue color was completely discharged and used it for isomerization of the non-conjugated diene (1) to (2). Mineral acids failed to effect isomerization. Lithium di-

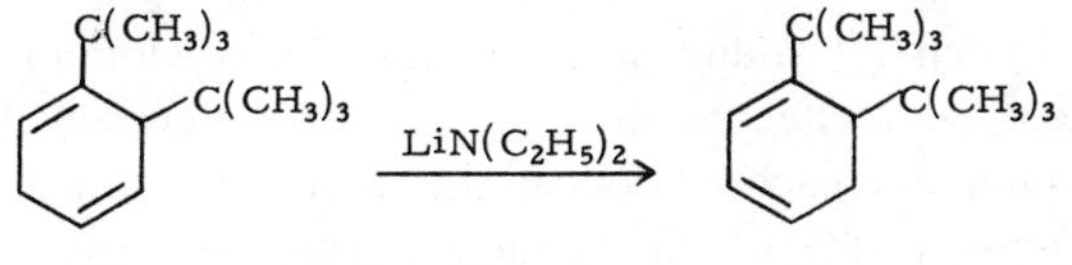

(1) (2)

ethylamide also isomerized the acid (3) to palustric acid (4), but in this case refluxing with potassium hydroxide in ethylene glycol proved to be even more satisfactory.[5] Benkeser *et al.*[6] found that the base rapidly isomerizes 2,5-dihydrocumene, probably to give a mixture of all possible conjugated dienes.

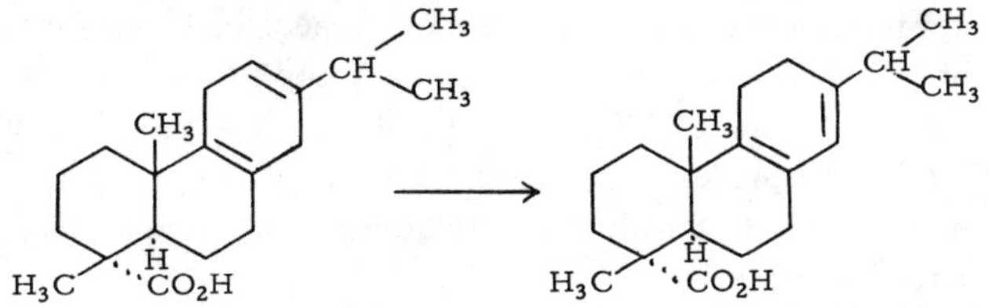

(3) (4)

Ziegler[7] used this base in the alkylation of aliphatic nitriles. Cason[8] found this the best route to 2,2-dimethyloctadecanoic acid.

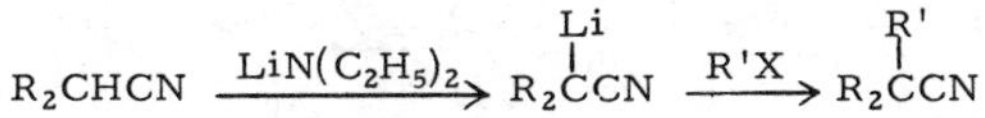

Norbornene epoxide (1) is isomerized by this strongly basic reagent cleanly if in only moderate yield to nortricyclanol (2).[9] Acid catalysts act on (1) to give mixtures of products.

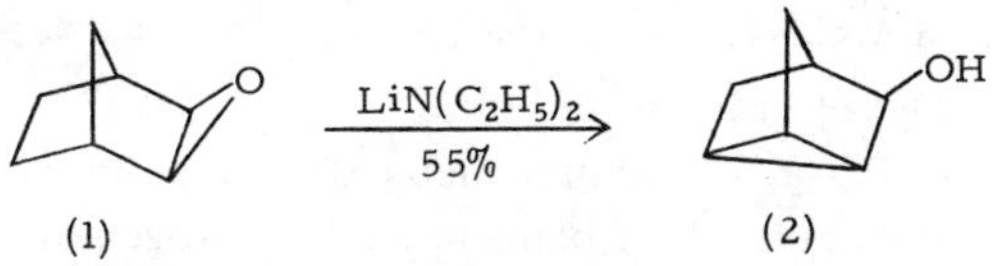

(1) (2)

[1]A. C. Cope and B. D. Tiffany, *Am. Soc.*, **73**, 4158 (1951)
[2]H. Gilman and F. Schulze, *Am. Soc.*, **47**, 2002 (1925)
[3]A. C. Cope, P. A. Trumbull, and E. R. Trumbull, *Am. Soc.*, **80**, 2844 (1958)
[4]A. W. Burgstahler, P.-L. Chien, and M. O. Abdel-Rahman, *Am. Soc.*, **86**, 5281 (1964)
[5]A. W. Burgstahler and L. R. Worden, *Am. Soc.*, **86**, 96 (1964)
[6]R. A. Benkeser, M. L. Burrous, J. J. Hazdra, and E. M. Kaiser, *J. Org.*, **28**, 1094 (1963)
[7]K. Ziegler and H. Ohlinger, *Ann.*, **495**, 84 (1932)
[8]J. Cason *et al.*, *J. Org.*, **15**, 850 (1950)
[9]J. K. Crandall, *J. Org.*, **29**, 2830 (1964)

Lithium dimethylaminosulfonylmethide, $(CH_3)_2NSO_2\bar{C}H_2Li^+$. Mol. wt. 129.08. Prepared[1] in tetrahydrofurane by the action of *n*-butyllithium on N,N-dimethyl-methanesulfonamide. It reacts with ketones to give β-hydroxysulfonamides in high yields.

$$(C_6H_5)_2C{=}O + (CH_3)_2NSO_2\overset{-}{C}H_2\overset{+}{Li} \xrightarrow[85\%]{} (C_6H_5)_2C(OH)CH_2SO_2N(CH_3)_2$$

[1]E. J. Corey and M. Chaykovsky, *Am. Soc.*, **87**, 1345 (1965)

Lithium–Diphenyl, 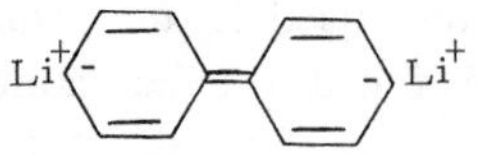
Mol. wt. 166.06.

The deep blue-green 2:1 adduct which lithium forms with diphenyl in solution in tetrahydrofurane is more effective than lithium itself for certain cleavage reactions.[1] One advantage is that the reaction proceeds in a homogeneous system.[1] For example, triphenylamine affords 58% of diphenylamine with the complex but only 8% of

$$(C_6H_5)_3N \xrightarrow[10\ hrs.\ 66^0]{LiC_6H_5C_6H_5Li-THF} \underset{58\%}{(C_6H_5)_2NH} + \underset{1.6\%}{C_6H_5NH_2}$$

the secondary amine with lithium alone. Anisole and diphenyl ether are cleaved in yields of 80 and 96%.

Estrone, a key intermediate in the preparation of medicinally useful 19-norsteroids, can now be prepared in high yield at the remarkably low temperature of 35° from $\Delta^{1,4}$-androstadiene-3,17-dione 17-ethyleneketal (1) by reaction with lithium–diphenyl in THF in the presence of a suitably acidic hydrocarbon such as diphenylmethane to intercept the by-product methyllithium and prevent its addition to the potential 17-carbonyl group.[2]

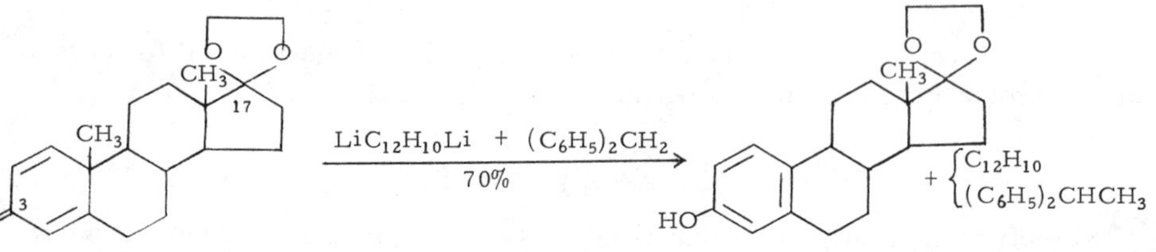
CH3
17
CH3
3
LiC12H10Li + (C6H5)2CH2
70%
CH3
HO
+ C12H10
(C6H5)2CHCH3

(1) (2)

[1]J. J. Eisch, *J. Org.*, **28**, 707 (1963)
[2]H. L. Dryden, Jr., G. M. Webber, and J. J. Wieczorek, *Am. Soc.*, **86**, 742 (1964)

Lithium ethoxide, $LiOC_2H_5$. Mol. wt. 52.00.

Preparation. (1)[1] 3.5 g. of lithium metal (0.5 g. atom) and absolute ethanol (0.5 mole) are added to 500 ml. of benzene under nitrogen, and when the reaction is complete the benzene is removed by distillation, and lithium ethoxide is obtained as a white solid. (2)[2] Lithium wire (3.5 g.) is added to 100 ml. of absolute ethanol under reflux, and when the initial reaction has subsided, the mixture is refluxed for 8 hrs. Evaporation under reduced pressure gives 42 g. of a white solid presumed to be $LiOC_2H_5{\cdot}C_2H_5OH$ (theory 46 g.). (3)[3] A 0.2*M* solution is prepared by dissolving 1.40 g. of lithium wire in 1 l. of absolute ethanol.

Uses. Lithium ethoxide (method 1) or crude lithium ethoxide monoethylate (method 2) is used[2] for reaction with the nitrosourea (1) to generate the carbene 2,2-diphenylcyclopropylidene (3) in the presence of an acceptor olefin such as isobutene. The intermediate diazocylcopropane (2) in part collapses to 1,1-diphenylallene (5), either in a concerted manner or via the carbene (3). Although other bases are effective in this reaction, lithium ethoxide gives "by far the cleanest product."

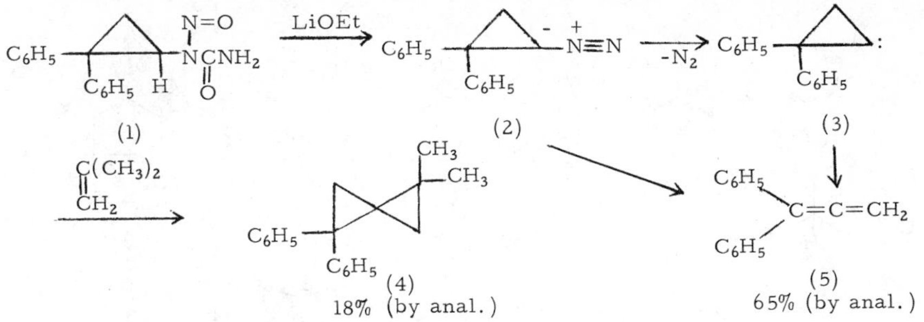
LiOEt
$-N_2$
$C(CH_3)_2$ = CH_2
(1)
(2)
(3)
(4)
18% (by anal.)
(5)
65% (by anal.)

The 0.2 *M* solution of lithium ethoxide in ethanol (method 3) has been used as base in the synthesis of 1,4-diphenyl-1,3-butadiene by the Wittig reaction.[3]

$$C_6H_5CH{=}CHCH_2Cl \xrightarrow[91\text{-}93\%]{(C_6H_5)_3P} C_6H_5CH{=}CHCH_2\overset{+}{P}(C_6H_5)_3(\overset{-}{Cl})$$

$$\xrightarrow[60\text{-}67\%]{C_6H_5CHO\ +\ LiOC_2H_5} C_6H_5CH{=}CHCH{=}CHC_6H_5 + (C_6H_5)_3\overset{+}{P}{-}\overset{-}{O}$$

$$+\ C_2H_5OH\ +\ LiCl$$

[1]T. L. Brown, D. W. Dickerhoof and D. A. Bafus, *Am. Soc.*, **84**, 1371 (1962)
[2]W. M. Jones, M. H. Grasley, and W. S. Brey, Jr., *ibid.*, **85**, 2754 (1963)
[3]R. N. McDonald and T. W. Campbell, *Org. Syn.*, **40**, 36 (1960)

Lithium ethoxyacetylide, $LiC{\equiv}COC_2H_5$.

The reagent can be prepared *in situ* by treating *β*-chlorovinyl ethyl ether with lithium amide in liquid ammonia and used in the synthesis of ethoxyethynylcarbinols of type (1).[1] This substance (1) is convertible by selective hydrogenation (Lindlar

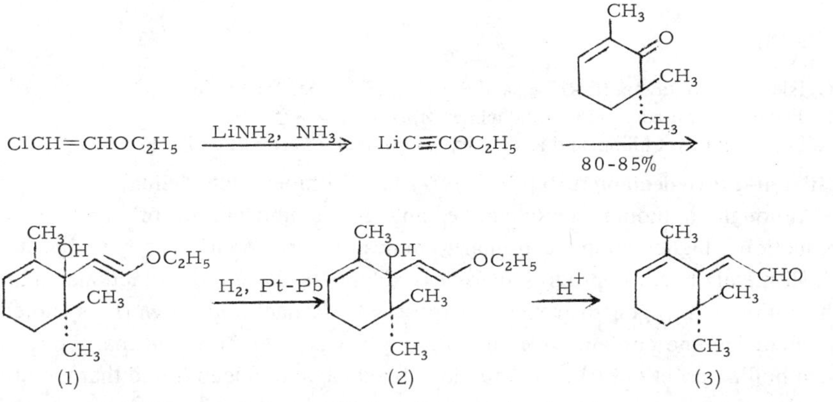
ClCH=CHOC₂H₅
LiNH₂, NH₃
LiC≡COC₂H₅
80-85%
H₂, Pt-Pb
H⁺
(1)
(2)
(3)

catalyst) and hydrolysis of the enol ether (2) into the doubly unsaturated aldehyde (3), of use in the synthesis of carotinoids.

Synthesis of cardenolide aglycones. The reagent has been used for the synthetic construction of the *α,β*-unsaturated *γ*-lactone ring characteristic of the cardiotonic glycosides. Thus an Israeli group[2] synthesized digitoxigenin (4) starting with addition of the reagent to the 20-ketone (1) to give (2). Acid-catalyzed hydration-dehydration gave the olefinic ester (3), which was hydroxylated at C_{21} by allylic oxidation with selenium dioxide. Treatment with aqueous-methanolic hydrochloric acid hydrolyzed the ester group with lactonization to digitoxigenin (4). A Canadian

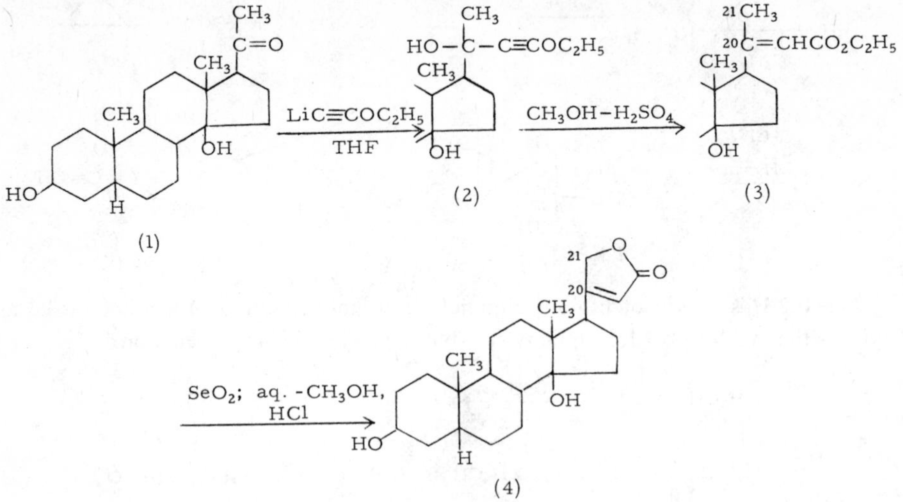

group[3] added the reagent to the 20-keto-3β,5β,14β,21-tetraol 3,21-diacetate (5) to give (6) and on acid hydrolysis obtained periplogenin (7).

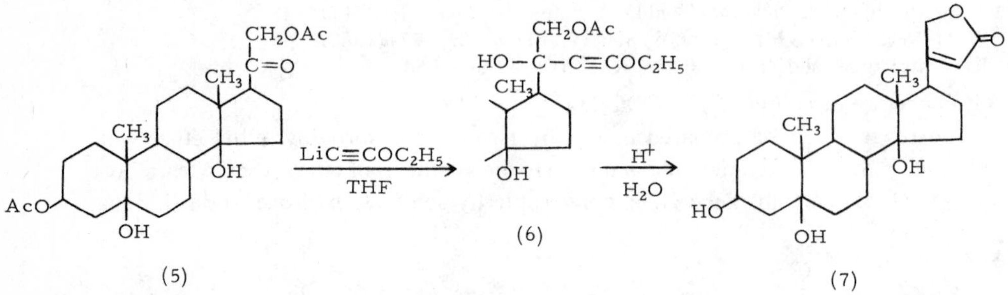

[1]O. Isler, M. Montavon, R. Rüegg, and P. Zeller, *Helv.*, **39**, 259 (1956)
[2]N. Danieli, Y. Mazur, and F. Sondheimer, *Am. Soc.*, **84**, 875 (1962)
[3]R. Deghenghi, A. Philipp, and R. Gaudry, *Tetrahedron Letters*, 2045 (1963)

Lithium–Ethylenediamine (b.p. 117°). *See also* Lithioethylenediamine.

Although ammonia, methylamine, and ethylamine are useful for metal-amine reductions, higher aliphatic primary amines perform poorly, perhaps because the alkali metals become progressively less soluble in these monofunctional amines as the ratio of nitrogen to carbon decreases. L. Reggel and co-workers[1] noted that ethylenediamine contains one amino group per carbon atom and that the relatively high boiling point (117°) is a potential advantage and indeed found that the lithium–ethylenediamine system at 90–100° is an effective combination for reduction. It reduces tetralin (1) to $\Delta^{9(10)}$-octalin (2) and the latter to *trans*-decalin (3). A flask

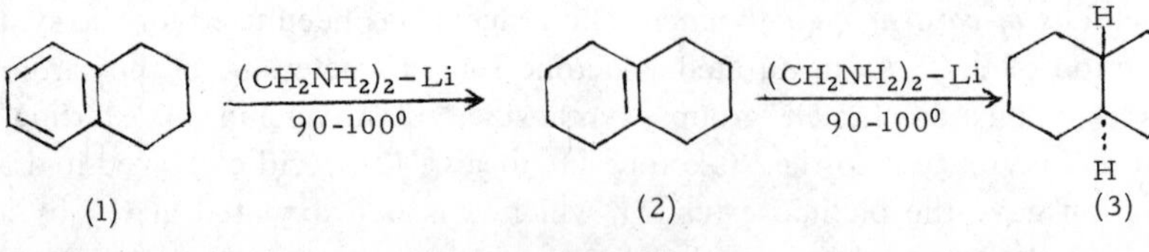

fitted with a stirrer, a condenser, and thermometer is flushed with nitrogen and charged with ethylenediamine (purified by standing over sodium and then distilling) and the substance to be reduced, and lithium is added in 4–8 pieces (0.5–1 g.) through the condenser at 90–100° in 1.5–3 hrs. The rate of addition is controlled by the hydrogen evolved; fresh metal is added before the blue color from a previous addition is discharged. Treatment of tetralin in the diamine with lithium in only 10% excess of the amount of the theory for conversion into (2) afforded (2) in only 32% yield and 46% of tetralin was recovered. Increasing the lithium to 100% excess raised the yield of (2) to 68%, with 31% recovery of starting material. Other examples briefly explored are characterized for the most part by low yields, incomplete conversion, and formation of mixtures.

Corey and Cantrall[2] found the reagent useful for reduction on a microscale of the methylene groups of 3-methylenecholestane (4) and of the triterpenoid derivative

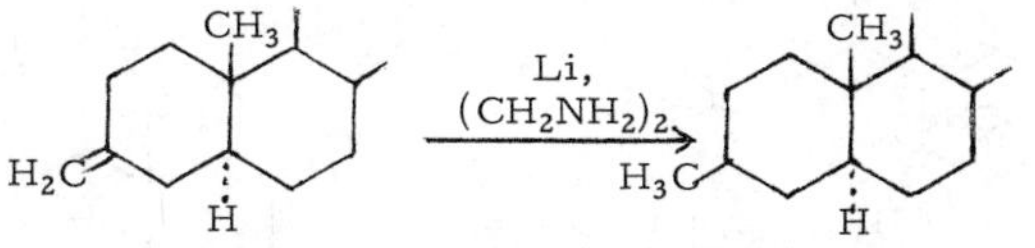

CH3
Li,
(CH2NH2)2
H2C
H3C
H

(4) (5)

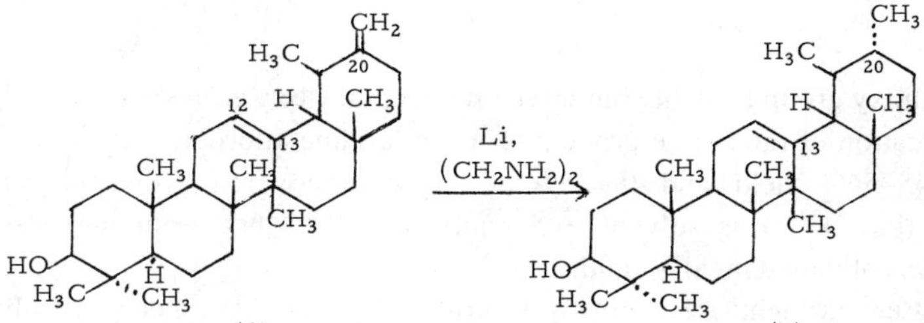

CH2
CH3
H3C
20
12
H
13
Li,
(CH2NH2)2
HO

(6) (7)

(6). In each case the reaction afforded in good yield a product in which the methyl group generated is equatorial. Note that the trisubstituted 12,13-double bond of (7, α-amyrin) remained untouched.

[1]L. Reggel, R. A. Friedel, and I. Wender, *J. Org.*, **22**, 891 (1957)

[2]E. J. Corey and E. W. Cantrall, *Am. Soc.*, **81**, 1745 (1959)

Lithium iodide, anhydrous, LiI. Mol. wt. 133.85, m.p. 446°. Suppliers: Alfa, KK.

Taschner and Liberek[1] found that ester and N-carbomethoxy groups can be cleaved efficiently with anhydrous lithium iodide in refluxing pyridine. N-Acyl and peptide bonds do not react. Having experienced difficulty in hydrolyzing the

$$RCOOCH_3 + LiI \longrightarrow RCOOLi + CH_3I$$

tricyclic ester I, because of sensitivity of the product to both acids and bases, Eschenmoser[2] tried the new method and found that I on being refluxed in pyridine

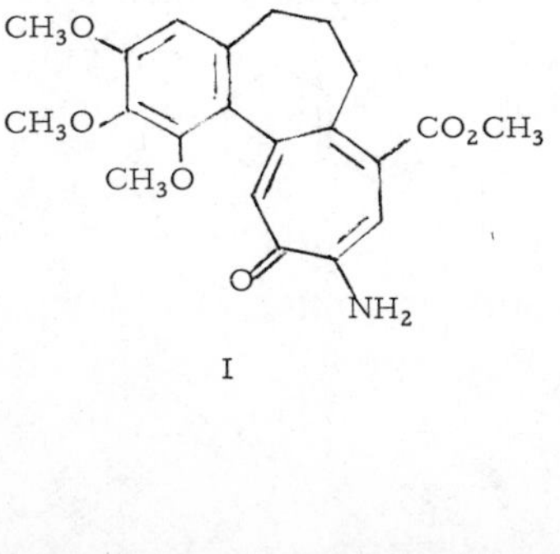

CH3O
CO2CH3
NH2
O

I

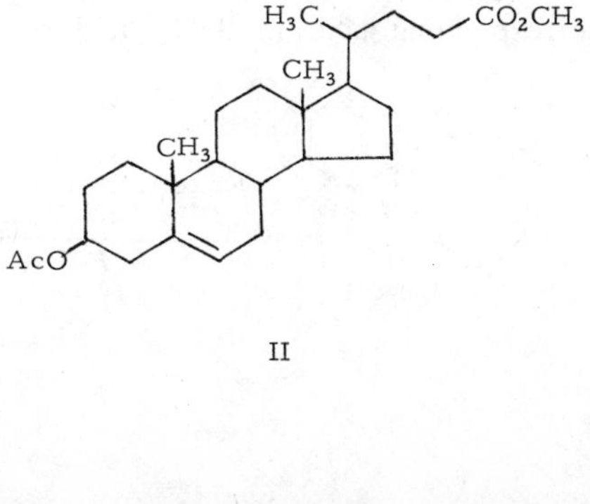

H3C
CO2CH3
CH3
AcO

II

for 2 hrs. with a large excess of lithium iodide is cleaved in high yield to the corresponding acid. In the case of the steroid acetoxy ester II, the boiling point of pyridine (115°) proved to be too low and this solvent was replaced by 2,6-lutidine, b.p. 143°. After refluxing II for 8 hrs. with 6.5 equivalents of lithium iodide in 20 ml. of solvent per mmole of ester, the yield of acetoxy acid was 68–74%, starting material recovered amounted to 3–6%, and only 5–7% of the hydroxy acid was formed. The highly selective cleavage of the methyl group with very little attack of the acetoxy group is in sharp contrast to hydrolysis; the equatorial 3β-hydroxyl group is hydrolyzed with great ease. In the case of O-acetyloleanolic acid methyl ester (III) the 17-

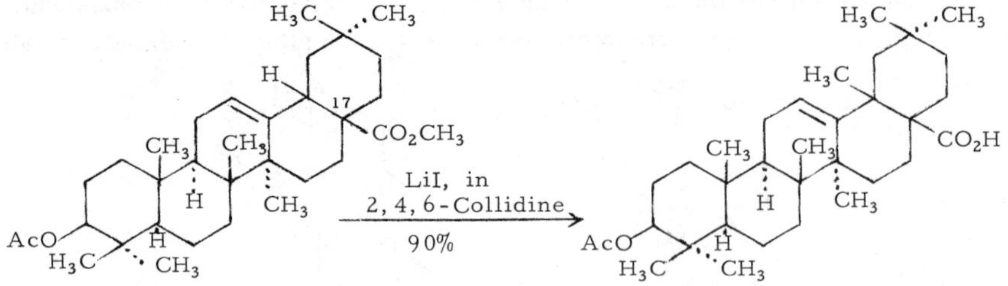
H3C
CH3
H
17
CO2CH3
CH3
CH3
H
CH3
LiI, in
2, 4, 6-Collidine
90%
AcO
H
H3C
CH3
H3C
CH3
H3C
CO2H
CH3
CH3
H
CH3
AcO
H
H3C
CH3

III IV

carbomethoxy group is highly hindered and the 3-acetoxy group somewhat hindered, and application of the above procedure using lutidine afforded the acetoxy acid IV in only 45–56% yield, and 28–35% of III was recovered. However, with 2,4,6-collidine (b.p. 172°) as solvent (6.5 equiv. of LiI, 8 hrs. refluxing) the acetoxy acid IV was obtained in 90% yield.

Several esters highly resistant to hydrolysis by base have been hydrolyzed successfully by refluxing with anhydrous lithium iodide in collidine. The reaction is often slow, for example, hydrolysis of the ester (1) required refluxing under nitrogen

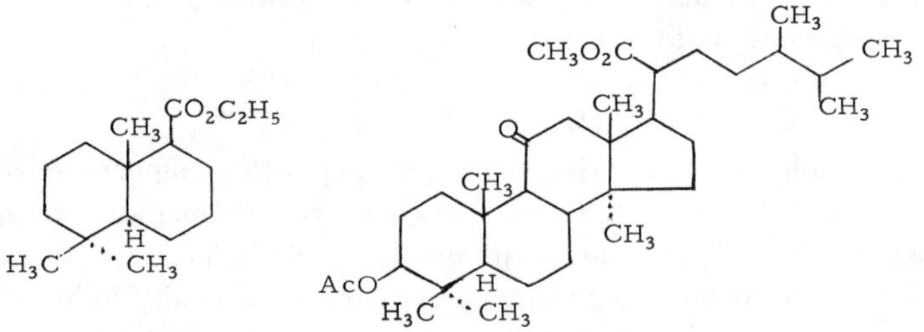
CH3
CO2C2H5
H
H3C
CH3
CH3
CH3O2C
CH3
CH3
CH3
O
CH3
CH3
AcO
H
H3C
CH3

(1) (2)

for 72 hrs., but the corresponding acid was obtained in 94% yield.[3] Alkaline hydrolysis in refluxing diethylene glycol proceeded in only moderate yield. Another example is the strongly hindered, alkali-resistant eburicoic ester derivative (2), which afforded the corresponding acid in 95% yield by the LiI–collidine method.[4] Herz and

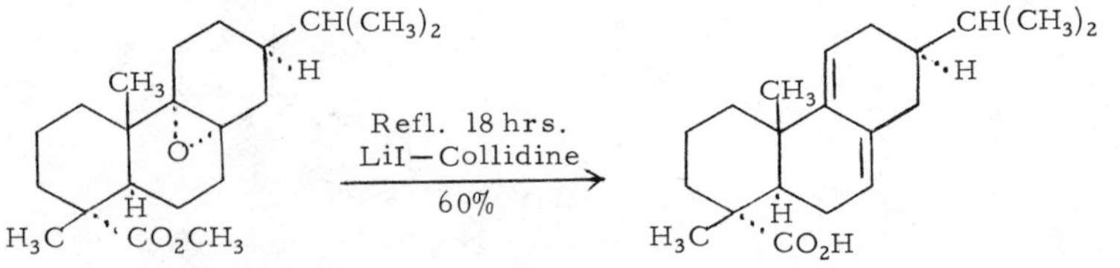
CH(CH3)2
H
CH3
O
H
H3C
CO2CH3
Refl. 18 hrs.
LiI—Collidine
60%
CH(CH3)2
H
CH3
H
H3C
CO2H

(3) (4)

Wahlborg[5] found that treatment of the resin acid derivative (3) with the reagent in refluxing collidine not only hydrolyzed the hindered ester function but eliminated the oxide function to produce a diene system. Treatment with Lewis acids had given unpromising mixtures.

Dean[6] studied the hydrolysis (halogenolysis) of methyl glycyrrhetate, a hindered triterpenoid ester, in a number of solvents and found dimethylformamide superior to pyridine, lutidine, collidine, or N-methyl-2-pyrrolidone in both speed and extent of reaction. In addition, use of DMF facilitated workup; the reaction mixture can be poured onto water and the mixture is acidified and extracted with ether.

[1]E. Taschner and B. Liberek, *Roczniki Chemii*, **30**, 323 (1956) [*C.A.*, **51**, 1039 (1957)]
[2]F. Elsinger, J. Schreiber, and A. Eschenmoser, *Helv.*, **43**, 113 (1960)
[3]W. L. Meyer and A. S. Levinson, *J. Org.*, **28**, 2184 (1963)
[4]G. W. Krakower, J. W. Brown, and J. Fried, *J. Org.*, **27**, 4710 (1962)
[5]W. Herz and H. J. Wahlborg, *J. Org.*, **30**, 1881 (1965)
[6]P. D. G. Dean, *J. Chem. Soc.*, 6655 (1965)

Lithium iodide dihydrate, $LiI \cdot 2H_2O$. Mol. wt. 169.88. Supplier: Fluka.

Whereas anhydrous lithium iodide in a refluxing base of suitable boiling point cleaves esters to the corresponding acids, lithium iodide dihydrate effects cleavage and concomitant decarboxylation, as illustrated for the case of 2-benzyl-2-carbomethoxycyclopentanone. Elsinger[1] refluxed a mixture of this keto ester, lithium

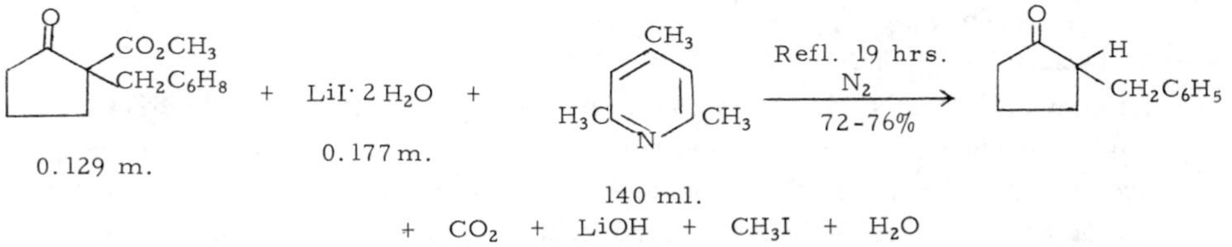
CO2CH3
CH2C6H5
0.129 m.
LiI· 2 H2O
0.177 m.
CH3
H3C
CH3
N
140 ml.
Refl. 19 hrs.
N2
72-76%
H
CH2C6H5
+ CO2 + LiOH + CH3I + H2O

iodide dihydrate, and 2,4,6-collidine under nitrogen and obtained 2-benzylcyclopentanone in good yield. He notes that methyl esters react more rapidly than ethyl esters, which in turn react faster than esters of secondary alcohols. *t*-Butyl esters are cleaved very readily with a catalytic amount of lithium iodide. The checkers prepared lithium iodide dihydrate by azeotropic distillation of the trihydrate with collidine under a Dean-Stark trap[2] (Fig. L-2) until 1 mole of water had been removed.

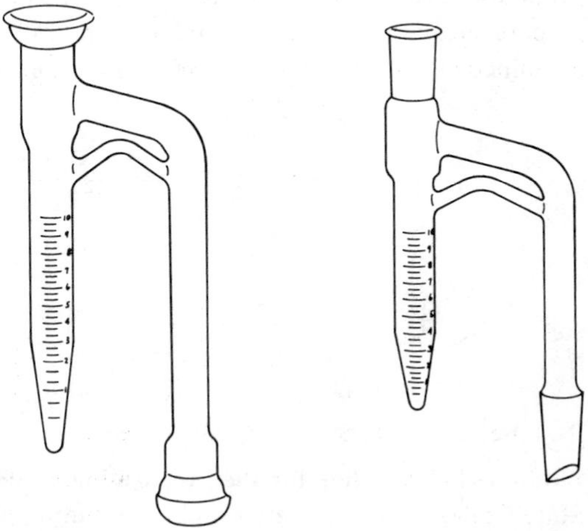

Fig. L-2 Dean-Stark separator[3]

[1]F. Elsinger, *Org. Syn.*, **45**, 7 (1965)
[2]E. W. Dean and D. D. Stark, *Ind. Eng. Chem.*, **12**, 486 (1920)
[3]Ace Glass 7725

Lithium-N-Methylaniline. This combination in ether effects 1,4-reduction of conjugated dienes to olefins:[1]

$$RCH{=}CHCH{=}CHR' + 2\ Li + 2\ C_6H_5NHCH_3 \longrightarrow RCH_2CH{=}CHCH_2R' + 2\ C_6H_5\underset{\displaystyle Li}{\underset{|}{N}}CH_3$$

Examples: butadiene ⟶ butene-2; naphthalene ⟶ 1,4-dihydronaphthalene.

[1]K. Ziegler *et al.*, *Ann.*, **511**, 64 (1934); **528**, 101 (1937); **567**, 1 (1950)

Lithium monocyanoborohydride, $LiB(CN)H_3$·Dioxane. Mol. wt. 134.90.

The reagent is prepared by reaction of lithium borohydride with liquid hydrogen cyanide in ether and isolated as the dioxane complex.[1] It is remarkably stable to

$$\overset{+}{Li}\overset{-}{B}H_4 + HCN \longrightarrow H_3\overset{-}{B}CN(\overset{+}{Li}) + H_2$$

water and to acids and can be refluxed for hours in aqueous dioxane without decomposition. This hydride is a weak reducing agent; it reduces aliphatic and aromatic aldehydes and α-hydroxy ketones but does not reduce ketones, aromatic nitro compounds, carboxylic acids or esters, or azoxy compounds.[2] Pyrene-3-aldehyde (2 mmoles) on being heated in the steam bath for 15 hrs. with 2 mmoles of the dioxane complex afforded the corresponding alcohol in 85% yield.

[1]G. Wittig and P. Raff, *Ann.*, **573**, 202, 209 (1951)
[2]G. Drefahl and E. Keil, *J. pr. Chem.*, **6**, 80 (1958)

Lithium nitride, Li_3N. Mol. wt. 34.83. Preparation.[1] Supplier: Foote Mineral Co.

This substance is almost completely insoluble in most nonreacting solvents and was thought to be unreactive. Koenig *et al.*[2] found, however, that dioxane, diglyme, and other solvents strongly enhance the reactivity of the nitride, possibly by partially dissolving it. Fluorene would not react with lithium nitride in ether or without a solvent at the melting point of the hydrocarbon. But when fluorene was refluxed with lithium nitride overnight in diglyme, carbonation of the solution afforded fluorene-9-carboxylic acid in 31% yield.

[1]E. Masdupuy and F. Gallais, *Inorg. Syn.*, **4**, 1 (1953)
[2]P. E. Koenig, J. M. Morris, E. J. Blanchard, and P. S. Mason, *J. Org.*, **26**, 4777 (1961)

Lithium-*n*-Propylamine. Leonard *et al.*[1] have used this reducing combination to prepare enamines which are stable to hydrolysis because the functional group is contained in a bi- or tricyclic system.[1] Examples:

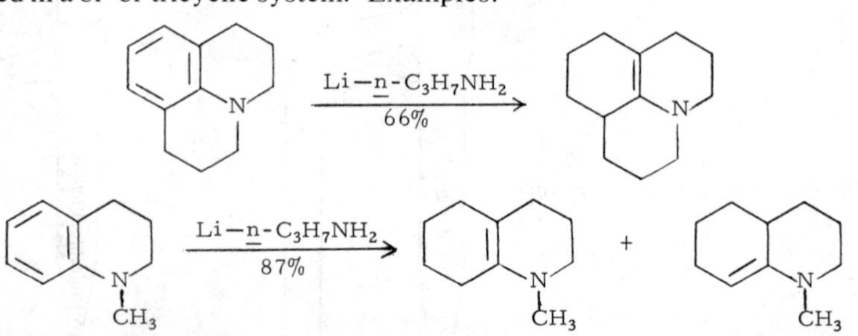
Li−n-C3H7NH2
66%
Li−n-C3H7NH2
87%
+
N
CH3

[1]N. J. Leonard, C. K. Steinhardt, and C. Lee, *J. Org.*, **27**, 4027 (1962)

Lithium-Sodium alloy for the preparation of alkyllithiums.[1] Beel *et al.*[2] at Colorado State College had routinely prepared *n*-butyllithium in 80–95% yield by a procedure of Gilman, Beel, *et al.*[3] from *n*-butyl bromide, diethyl ether, and metallic lithium stated by the supplier (Lithium Corp. of America) to contain 0.05% of sodium. When a change in the manufacturing process afforded purer lithium containing only 0.005% of sodium, the Beel group tested this material and found that it afforded *n*-butyllithium in maximum yield of only 48% (yields determined by the double titration method[4]). They then found that the difficulty can be overcome by melting the high-purity lithium under oil and alloying it with at least 0.5% of sodium. This concentration will ensure rapid reaction and good yield of alkyllithium. It is not sufficient to add an equivalent amount of sodium to the reaction mixture, as the sodium must be alloyed with at least a part of the lithium for realization of the catalytic effect. The sodium content may range up to 5.0% without promoting the side reaction:

$$2\,RCl + 2\,Na \longrightarrow RR + 2\,NaCl$$

These observations were confirmed and expanded by McArthur *et al.*[5] of Olin Mathieson in the course of the preparation of large batches of a variety of alkyllithiums. In the preparation of C_5 and lower alkyllithiums, as little as 0.5% of sodium in the alloy suffices for reasonably rapid reaction. With higher alkyls (C_{10}–C_{24}) the optimum sodium content varies but is somewhat higher. Thus with decyl chloride and octadecyl chloride the optimum content is approximately 1.33% of sodium. The particle size of the lithium is unimportant; if the sodium content is optimal, chunks, shot, shavings, ribbon, foil, and powder react equally well.

In a typical experiment 1.18 moles of lithium containing 1.33% of sodium was slurried in 200 ml. of ether at 7–10° in a three-necked, round-bottomed flask previously purged with dry nitrogen, and 0.5 mole of octadecyl chloride in 60 ml. of ether was added in the course of about 1 hr. Stirring was continued at 7–10° for an additional 1.5 hrs. and the solution was filtered through a loose plug of glass wool under a blanket of nitrogen. The yield of octadecyllithium was 72–88%. Decyllithium was obtained similarly in yield of 67–92%.

Diethyl ether is the best solvent found for the reaction; the reagent grade solvent was noticeably better after being refluxed with lithium for an hour prior to addition of the alkyl halide. After formation of the alkyllithium, the diethyl ether can be replaced by another solvent if the temperature is kept below 10° during the exchange.

In the preparation of large batches of alkyllithium compounds (10–1,000 gallons) it is desirable to increase the quantity of sodium by about 25% and to activate the diethyl ether by prolonged refluxing with lithium in order to avoid long induction periods (15–24 hrs.) and dangerous surges of reaction.

[1]Contributed by Richard E. McArthur, Olin Mathieson Chemical Corporation, New Haven, Connecticut

[2]J. A. Beel, W. G. Koch, G. E. Tomasi, D. E. Hermansen, and P. Fleetwood, *J. Org.*, **24**, 2036 (1959)

[3]H. Gilman, J. A. Beel, C. G. Brannen, M. W. Bullock, G. E. Dunn, and L. S. Miller, *Am. Soc.*, **71**, 1499 (1949)

[4]H. Gilman and A. H. Haubein, *Am. Soc.*, **66**, 1515 (1944)

[5]S. J. Chiras, F. F. Frulla, A. O. Minklei, C. W. Kaufman, H. Hugos, G. T. Motcock, and R. E. McArthur, unpublished work at Olin Mathieson Chemical Corporation under Supply Contracts to U.S. Air Force; AF (33-600)-36839 (1957); AF (33-601) 57–10186 (1957); and AF (33-601) 57-573 (1957)

Lithium tri-*t*-butoxyaluminum hydride, $LiAlH[OC(CH_3)_3]_3$. Mol. wt. 254.30, sublimes at 280° under vacuum, solubility at 25° in g./100 g.: diglyme, 41; tetrahydrofurane; 36, ether, 2; 1,2-dimethoxyethane, 4; *t*-BuOH, 0. Supplier: Ventron Corp. The reagent currently costs 1.5 as much as $LiAlH_4$ and the active hydrogen per gram is only 3.7% that for $LiAlH_4$.

H. C. Brown and co-workers[1] found that lithium aluminum hydride in ether solution reacts with 4 moles of methanol, ethanol, or isopropanol but with only 3 moles of *t*-butanol. Dropwise addition of 1 mole of *t*-butanol at room temperature to a stirred solution of 0.31 mole of $LiAlH_4$ in ether produces a white precipitate of lithium tri-*t*-butoxyaluminum hydride in essentially quantitative yield. The new reagent proved to be a milder reducing agent than $LiAlH_4$, since it reduces aldehydes, ketones, and acid chlorides in diethyl ether or diglyme at 0° but fails to react with esters and nitriles.

Brown and Subba Rao[2] found that addition of lithium tri-*t*-butoxyaluminum hydride in diglyme to a solution of an acid chloride at −78° provides a convenient synthesis of aldehydes. Aromatic acid chlorides with *m*- or *p*-substituents form aldehydes in yields of 60–90%. Substituent nitro-, cyano-, and carboethoxy groups are not affected. *o*-Substituents tend to reduce the yield. Yields are 5–10% lower at −40° than at −78°.

Weissman and Brown[3] found that, in contrast to alkyl esters, which are inert to the reagent, phenyl esters of aliphatic and alicyclic acids react to form aldehydes. However, since the phenyl esters are generally prepared from the acid chlorides, it is preferable to proceed through the acid chloride directly.

Wheeler and Mateos[4] in a preliminary note reported that reduction of cholestane-3-one with either lithium tri-*t*-butoxyaluminum hydride or with lithium aluminum hydride–aluminum chloride affords 99% of the equatorial cholestane-3β-ol, whereas reduction with lithium aluminum hydride alone was known to give only 88–91% of the 3β-ol.[5,6]

Fajkoš[7] studied the reduction of a number of steroid ketones with the tri-*t*-butoxy reagent, but did not check the results of Wheeler and Mateos. Reduction of 3β-formyloxy-Δ^5-androstene-17-one (1) to the 17β-ol (2) in high yield (pure) demonstrates high

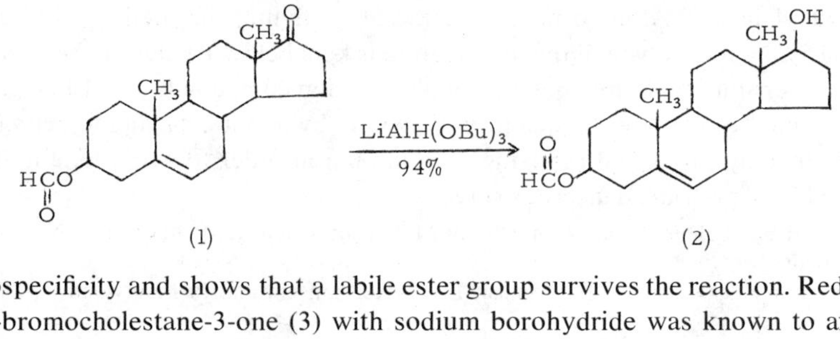
CH3
O
CH3
HCO
O
LiAlH(OBu)3
94%
OH
CH3
CH3
O
HCO

(1) (2)

stereospecificity and shows that a labile ester group survives the reaction. Reduction of 2α-bromocholestane-3-one (3) with sodium borohydride was known to afford a

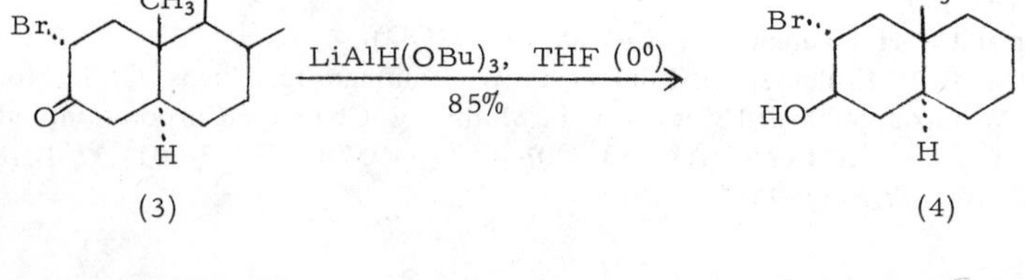
CH3
Br
O
H
LiAlH(OBu)3, THF (0°)
85%
CH3
Br
HO
H

(3) (4)

mixture containing substantial amounts of both epimeric alcohols; reduction with tri-*t*-butoxy reagent gave only the 3β-ol (4); this observation was later confirmed in another laboratory.[8] Fajkoš compared rates of reduction at room temperature of ketones of the 5α- or Δ^5-series and established the order:

$$\text{3-one} > \text{17-one} > \Delta^4\text{-3-one}$$

He then demonstrated the selective reduction of the more reactive group of the dione (5). For reduction of the less reactive of the two keto groups of the dione (7),

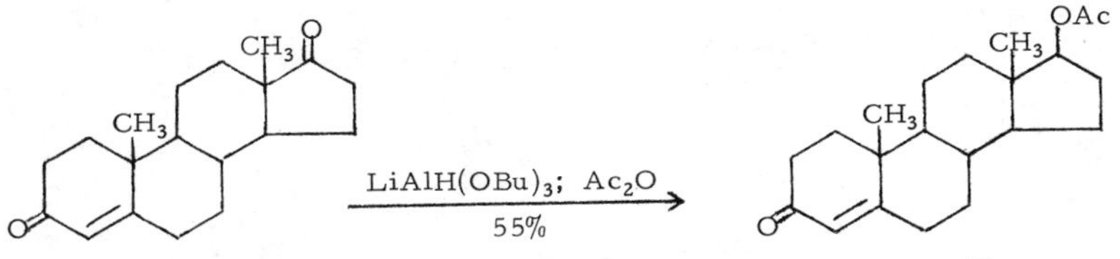
CH3
CH3
LiAlH(OBu)3; Ac2O
55%
OAc
CH3
CH3

(5) (6)

the Czech workers[9] protected the 3-keto group by conversion into the pyrrolidine enamine (8), reduced the 17-keto group, and removed the protecting block. Note that enamines are reduced by $LiAlH_4$–$AlCl_3$ in ether.

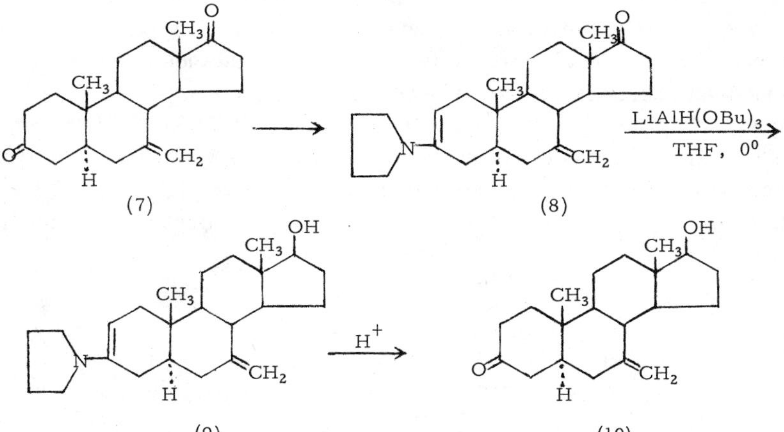
CH3
CH3
CH2
H
CH3
CH3
CH2
H
LiAlH(OBu)3
THF, 0°
OH
CH3
CH3
CH2
H
H+
OH
CH3
CH3
CH2
H

(7) (8)

(9) (10)

Tamm[10] found the reagent useful for reduction of carbonyl groups in the bufadienolide series. Thus reduction of the ketone (11) with sodium borohydride suffered from the fact that the unsaturated lactone ring is attacked to some extent and the yield of (12) was only 40–50%. Reduction with lithium tri-*t*-butoxyaluminum hydride in tetrahydrofurane at 0° proceeded rapidly (15 min.), the lactone ring was

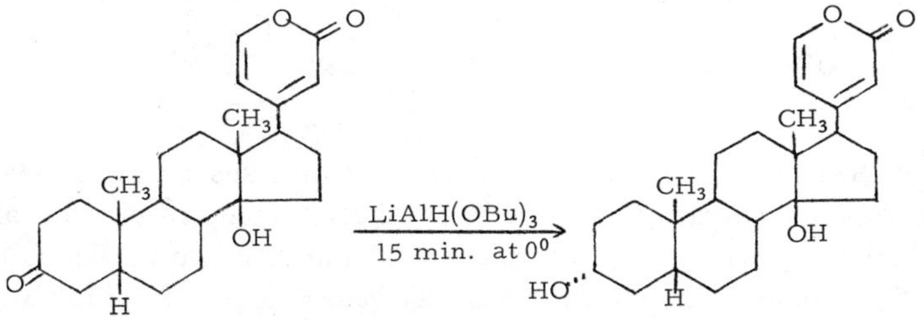
CH3
CH3
OH
H
LiAlH(OBu)3
15 min. at 0°
HO
CH3
CH3
OH
H

(11) (12)

completely resistant to attack, and the yield of (12) was doubled (details not reported). Exploratory reductions on a micro scale with paper-chromatographic control showed that the cardenolide ring of 3-O-acetylstrophanthidin (13) is likewise resistant to attack, that the hindered aldehyde group is reduced at 0° only in 1 hr., and that in this period the acetyl group remains intact. In another case a formyloxy group re-

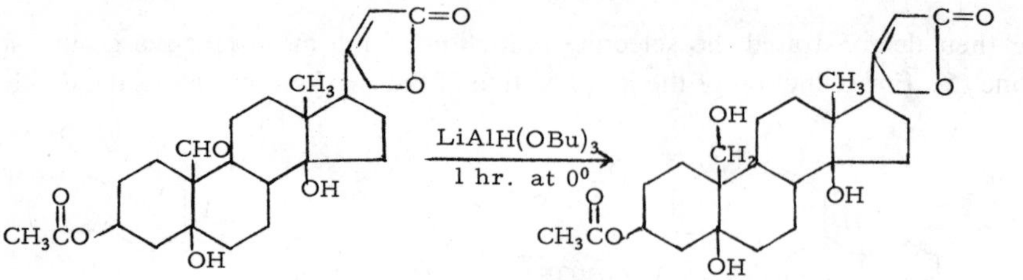
C=O
CH3
O
CHO
OH
CH3CO
OH
LiAlH(OBu)3
1 hr. at 0°
OH
CH2
CH3
O
C=O
OH
CH3CO
OH

(13) (14)

sisted attack at 0° but was reduced at 20°. The 14,15-epoxide group of marinobufagin is reduced by sodium borohydride but not by lithium tri-*t*-butoxyaluminum hydride.

20-Ketosteroids are reduced by the reagent mainly to the 20β-ols, as shown by a Ciba-Basel group.[11] A suspension of 0.4 mole of lithium tri-*t*-butoxyaluminum hydride in 750 ml. of THF is stirred under reflux in a nitrogen atmosphere, cooled to 2°, and 0.2 mole of pregnenolone acetate (15) is added and rinsed in with 50 ml. of solvent. The mixture is stirred at 0–5° for 1.5 hrs. and then a solution of 100 g. of ammonium sulfate in 150 ml. of water is added dropwise in 15–20 min. with vigorous stirring and cooling below 10° (vigorous evolution of hydrogen). One crystallization of the product from acetone affords pure 3β-acetoxy-20β-hydroxy-Δ^5-pregnene in 75–79% yield. Here the reducing agent was taken in 100% excess, but no evidence is presented to show that this large amount is required. Indeed

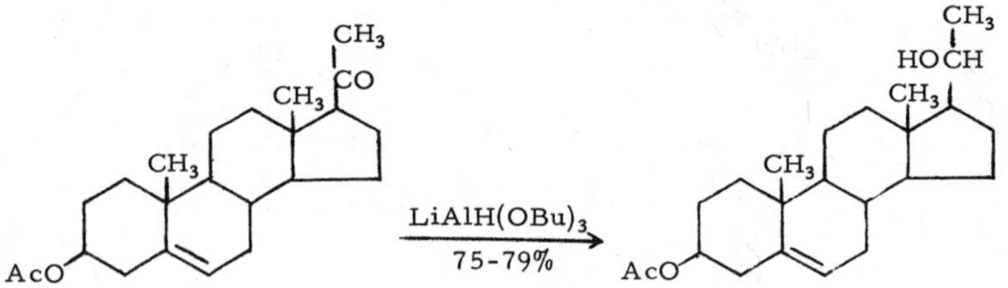
CH3
CO
CH3
CH3
AcO
LiAlH(OBu)3
75-79%
CH3
HOCH
CH3
CH3
AcO

(15) (16)

evidence to the contrary is provided by an experiment by Burgstahler and Nordin.[12] After dropwise addition of a cold (0°) solution of 0.039 mole of Δ^4-cholestene-3-

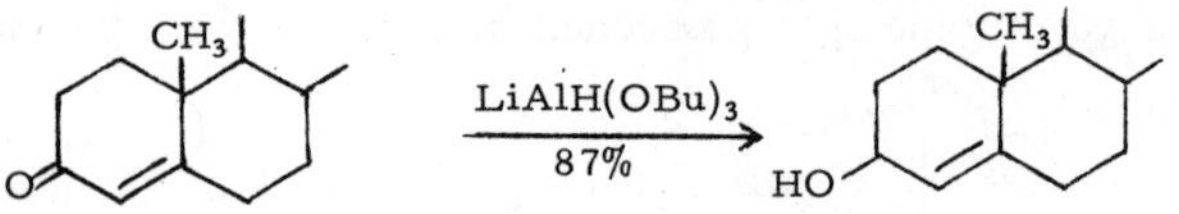
CH3
O
LiAlH(OBu)3
87%
CH3
HO

(17) (18)

one (17) in 200 ml. of ether–benzene (8:1) to a stirred suspension of 0.05 mole of lithium tri-*t*-butoxyaluminum hydride (1.3 × theory) in ether–diglyme at −40 to −50°, the mixture was let stand overnight at 0° and then hydrolyzed by treatment with ice, 5*N* sodium hydroxide, and Rochelle salt. Evaporation of the washed and dried ethereal extract and crystallization from ethyl acetate afforded 13 g. (87%) of

Δ^4-cholestene-3β-ol, m.p. 126–129°. One recrystallization gave pure material, m.p. 131–132°.

In effecting the stereospecific reduction of 6-ketotigogenin acetate (19) to the 6β-ol (20), a Syntex group[13] outdid the Swiss industrialists in liberal use of a costly reagent. To a stirred solution of 10 g. of (19) in 400 ml. of THF at room temperature

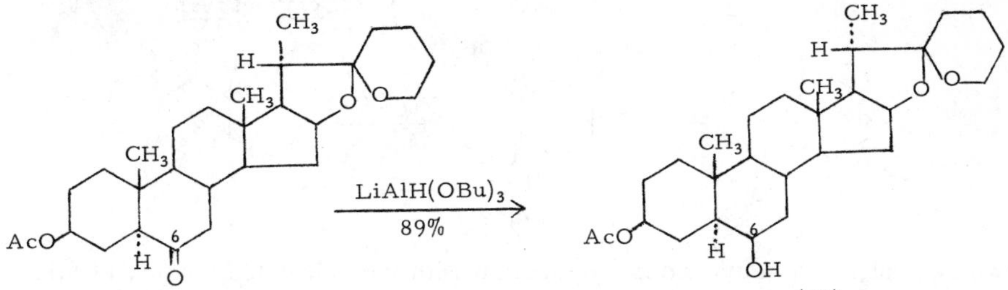
CH3
H
CH3
CH3
AcO
H
6
O
LiAlH(OBu)3
89%
OH

(19) (20)

they added 32 g. of lithium tri-t-butoxyaluminum hydride, or 6 times the amount theoretically required. After 60 hrs. at room temperature the complex was hydrolyzed, and the excess reagent destroyed by pouring the mixture onto 3 l. of 5% acetic acid.

Syntex chemists Zderic and Iriarte[14] explored the reduction of the bismethylenedioxy derivative of prednisone, or prednisone–BMD, in the hope that the bulky reagent would selectively reduce the carbonyl group at C_3. Reaction with excess

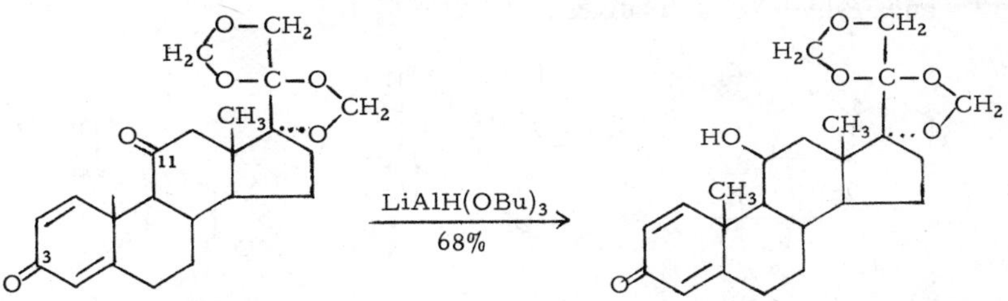
O–CH2
H2C
O–C–O
CH2
CH3
11
3
LiAlH(OBu)3
68%
HO

Prednisone-BMD Prednisolone-BMD

lithium tri-t-butoxyaluminum hydride at room temperature surprisingly effected reduction alone of the hindered 11-keto group and gave prednisolone–BMD in good yield.

Fajkoš and Joska[15] found that reduction of a 16α-bromo-17-ketosteroid (21) with a metal hydride in a polar solvent is attended with considerable inversion at C_{16}, but that reduction with lithium tri-t-butoxyaluminum hydride in a nonpolar solvent affords the 17-epimeric 16α-bromo alcohols (22) and (23).

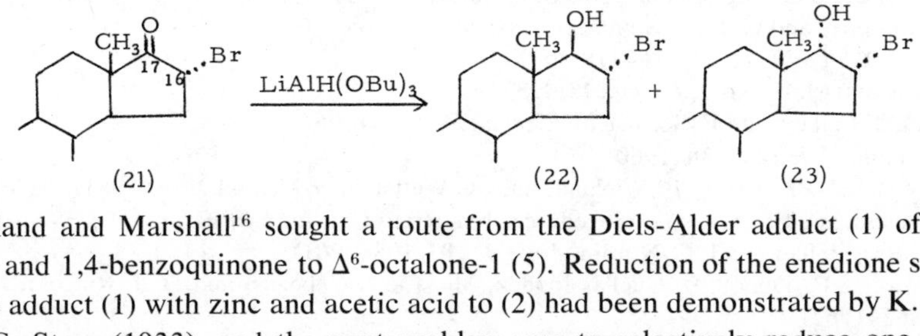
O
CH3
17
16
Br
LiAlH(OBu)3
OH
+

(21) (22) (23)

Ireland and Marshall[16] sought a route from the Diels-Alder adduct (1) of butadiene and 1,4-benzoquinone to Δ^6-octalone-1 (5). Reduction of the enedione system of the adduct (1) with zinc and acetic acid to (2) had been demonstrated by K. Alder and G. Stern (1933), and the next problem was to selectively reduce one of the

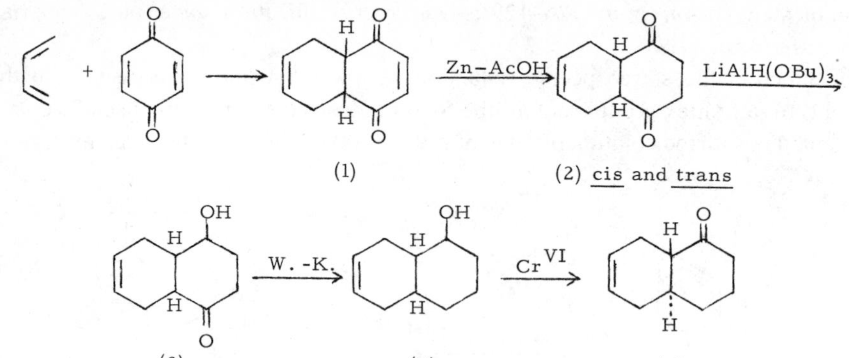

(1) (2) cis and trans

(3) (4) (5)

two equivalent carbonyl groups. Reduction with the calculated amount of lithium aluminum hydride proceeded in the desired manner but in disappointingly low yield, probably because of the spread in reactivity between the four available hydrogens. Reduction with the theoretical amount of lithium tri-*t*-butoxyaluminum hydride in dilute solution in THF proceeded satisfactorily to (3), and Wolff-Kishner reduction followed by chromic acid oxidation gave Δ^6-*trans*-octalone-1 (5) in overall yield from benzoquinone of 52%.

In contrast to the usual reduction of 3-ketosteroids to 3β-hydroxysteroids, $\Delta^{5(10)}$-estrene-3,17-dione (6) is reduced by lithium tri-*t*-butoxyaluminum hydride to the 3α- and 3β-hydroxy derivatives in the ratio 15:1.[17]

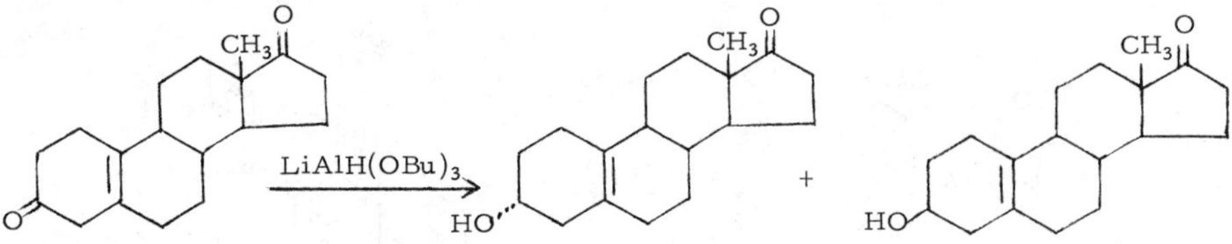

(6) (7) (8)

Engel *et al.*[18] used the reagent to reduce 17α-hydroxy-Δ^4-pregnene-3,20-dione to a mixture of Δ^4-pregnene-3β,17α,20-triols epimeric at C_{20}. The 17,20-glycol was then cleaved by lead tetraacetate to the 17-ketone.

[1] H. C. Brown and R. F. McFarlin, *Am. Soc.*, **80**, 5372 (1958); H. C. Brown and C. J. Shoaf, *ibid.*, **86**, 1079 (1964)
[2] H. C. Brown and B. C. Subba Rao, *Am. Soc.*, **80**, 5377 (1958)
[3] P. M. Weissman and H. C. Brown, *J. Org.*, **31**, 283 (1966)
[4] O. H. Wheeler and J. L. Mateos, *Chem. Ind.*, 395 (1957)
[5] H. R. Nace and G. L. O'Connor, *Am. Soc.*, **73**, 5824 (1951)
[6] C. W. Shoppee and G. H. R. Summers, *J. Chem. Soc.*, 687 (1950)
[7] J. Fajkoš, *Coll. Czech.*, **24**, 2284 (1959)
[8] R. Kwok and M. E. Wolff, *J. Org.*, **28**, 423 (1963)
[9] J. Joska, J. Fajkoš, and F. Šorm, *Coll. Czech.*, **26**, 1646 (1961)
[10] Ch. Tamm, *Helv.*, **43**, 338 (1960)
[11] K. Heusler, J. Kalvoda, P. Wieland, and A. Wettstein, *Helv.*, **44**, 179 (1961); K. Heusler, P. Wieland, and Ch. Meystre, procedure submitted to *Organic Syntheses*
[12] A. W. Burgstahler and I. C. Nordin, *Am. Soc.*, **83**, 198 (1961)
[13] A. Bowers, E. Denot, L. Cuéllar Ibáñez, Ma. Elena Cabezas, and H. J. Ringold, *J. Org.*, **27**, 1862 (1962)
[14] J. A. Zderic and J. Iriarte, *J. Org.*, **27**, 1756 (1962)

[15]J. Fajkoš and J. Joska, *Coll. Czech.*, **27**, 1849 (1962)

[16]R. E. Ireland and J. A. Marshall, *J. Org.*, **27**, 1620 (1962)

[17]W. F. Johns, *J. Org.*, **29**, 1490 (1964)

[18]M. G. Ward, J. C. Orr, and L. L. Engel, *J. Org.*, **30**, 1421 (1965)

Lithium trichloromethide, *see* Trichloromethyllithium.

Lithium triethoxyaluminum hydride, $LiAlH(OC_2H_5)_3$.

The reagent is generated *in situ* by adding either 3 moles of ethanol or 1.5 moles of ethyl acetate to 1 mole of $LiAlH_4$ in 1.3 M ether solution at $0°$.[1] It reduces both

$$RC{\equiv}N \xrightarrow{LiAlH(OEt)_3} \overset{\overset{\displaystyle H}{|}}{RC}{=}N\bar{A}l(OEt)_3(Li^+) \xrightarrow{H_2O} RCHO$$

aliphatic nitriles (70–80%) and aromatic nitriles (80–90%) in ether solution at $0°$. It also reduces acyl dimethylamides to aldehydes in high yield.[2]

$$RCON(CH_3)_2 \xrightarrow{LiAlH(OEt)_3} \overset{\overset{\displaystyle N(CH_3)_2}{|}}{\underset{\underset{\displaystyle H}{|}}{RC}}{-}O\bar{A}l(OEt)_3(Li^+) \xrightarrow{H_2O} RCHO$$

[1]H. C. Brown and C. P. Garg, *Am. Soc.*, **86**, 1085 (1964)

[2]H. C. Brown and A. Tsukamoto, *Am. Soc.*, **86**, 1089 (1964)

Lithium trimethoxyaluminum hydride, $LiAlH(OCH_3)_3$.

Preparation of standard solution. Brown and Weissman[1] found that solutions of this reagent are not stable; for example, a solution found to be 0.67 M in hydride decreased in concentration to 0.57 M within one week. They therefore decided to utilize solutions freshly prepared by the addition of methanol to a standard solution of lithium aluminum hydride in tetrahydrofuran. Details are given for stirring $LiAlH_4$ in THF under nitrogen for 2 hrs., and filtering the solution under nitrogen pressure through a bed of Celite on a sintered-glass funnel and storing the crystal-clear solution in a 1-l. flask with a rubber septum syringe inlet. A hypodermic syringe with a long needle is used for transfer of samples. The solution is stable in storage.

Reduction. In contrast to lithium tri-*t*-butoxyaluminum hydride, the trimethoxy derivative is about as active a reducing agent as $LiAlH_4$.[1] However, it reduces epoxides (particularly internal epoxides) more slowly than $LiAlH_4$. Cinnamaldehyde is reduced to hydrocinnamaldehyde. It shows higher stereoselectivity than $LiAlH_4$ in reduction of bicyclic ketones.[2] Thus camphor is reduced to isoborneol (*exo*) by $LiAlH(OCH_3)_3$ in 98% yield and by $LiAlH_4$ in 91% yield. The reagent also appears more selective in the reduction of unsymmetrical oxides.[3] Thus $LiAlH_4$ reduces styrene oxide (1) to give the secondary alcohol (2) in 96% yield and the primary alcohol (3) in 4% yield, but the amount of (3) is decreased to 1% with

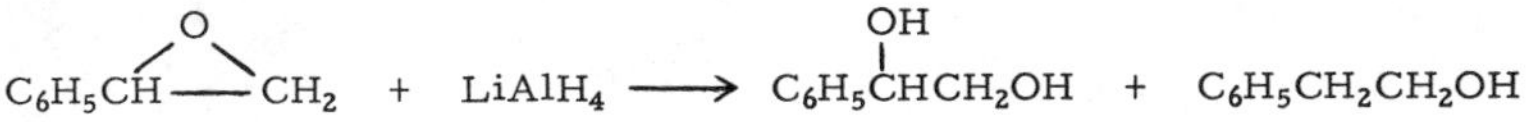

$$\underset{(1)}{C_6H_5\overset{\displaystyle O}{\overset{\displaystyle \diagup \diagdown}{CH}{-}CH_2}} + LiAlH_4 \longrightarrow \underset{(2)}{C_6H_5\overset{\overset{\displaystyle OH}{|}}{CH}CH_2OH} + \underset{(3)}{C_6H_5CH_2CH_2OH}$$

lithium trimethoxyaluminum hydride. In general the reduction of epoxides, oximes, azobenzene, and cyclohexyl tosylate is slower with lithium trimethoxyaluminum hydride than with $LiAlH_4$.

[1]H. C. Brown and P. M. Weissman, *Am. Soc.*, **87**, 5614 (1965)

[2]H. C. Brown and H. R. Deck, *Am. Soc.*, **87**, 5620 (1965)

[3]H. C. Brown, P. M. Weissman, and N. M. Yoon, *Am. Soc.*, **88**, 1458 (1966)

2,6-Lutidine, 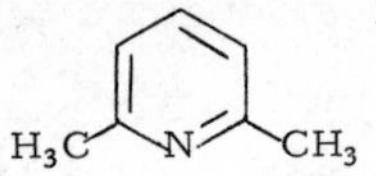

Mol. wt. 107.15, sp. gr. 0.923, b.p. 142–143°. Suppliers: A, B, E, F, KK, MCB. Mazurek and Perlin[1] found this pyridine homolog superior to silver carbonate, the usual acid acceptor, for the preparation of orthoesters (3) from 2,3,4,6-tetra-O-acetyl-α-D-mannopyranosyl bromide (1). Of considerable interest is the fact that

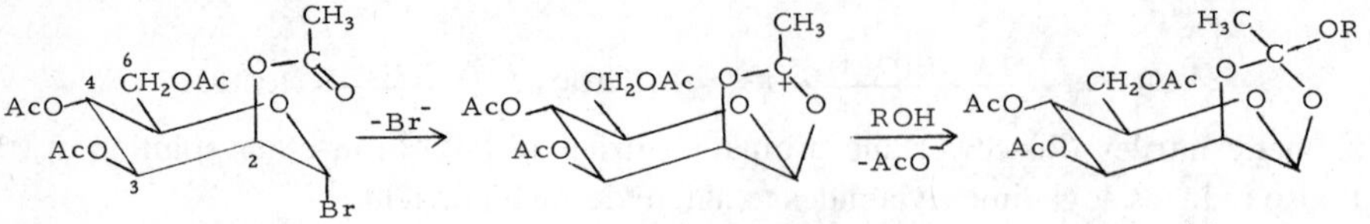

(1) (2) (3)

the OR-*exo* isomer is produced almost exclusively. The reaction is improved by use of chloroform as solvent rather than excess base.

[1]M. Mazurek and A. S. Perlin, *Can. J. Chem.*, **43**, 1918 (1965)

2,6-Lutidine-3,5-dicarboxylic hydrazide, 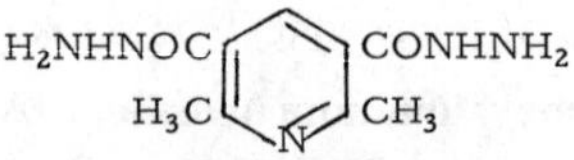

Mol. wt. 223.23. Velluz and Rousseau[1] describe preparation of the reagent and its use for the precipitation of progesterone as the 1:1 derivative in over 95% yield. Progesterone is recovered by displacement with benzaldehyde.

[1]L. Velluz and G. Rousseau, *Bull. soc.*, **13**, 288 (1946)

M

Magnesium, at. wt. 24.32, m.p. 650°. *See also* Grignard reagents.

Absolute methanol and ethanol of grades available commercially in 1931 were conveniently dried by refluxing over magnesium turnings (1 g./100 ml.) for 4 hrs.[1,2]

Reduction. Magnesium amalgam is the reducing agent in the classical reduction of acetone to pinacol.[3] A solution of mercuric chloride in acetone is added gradually under reflux to a flask containing magnesium covered with benzene. The yield of pinacol hydrate based on magnesium used is 43–50%.

An efficient method for the reduction of alkyl and aryl halides is by reaction with magnesium and isopropanol.[4] 1-Bromonaphthalene is reduced at the boiling point of isopropanol (82°). With a less reactive halide, for example chlorobenzene, a

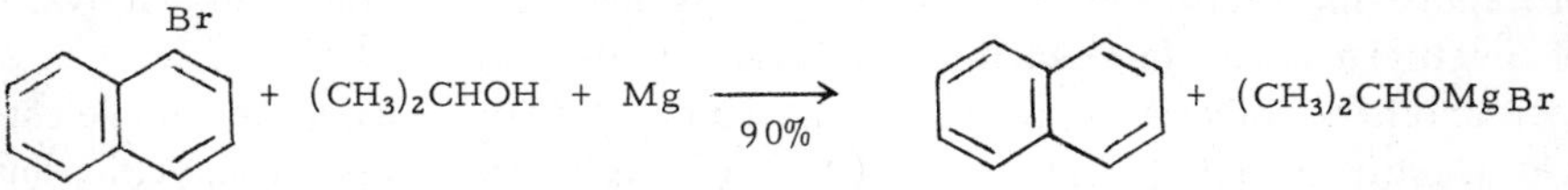
Br
+ (CH3)2CHOH + Mg
90%
+ (CH3)2CHOMgBr

mixture of 0.25 g. atom of magnesium powder and 50 ml. of decalin (b.p. *ca.* 190°) is heated under reflux without stirring, a small crystal of iodine is added, and about one fifth of a mixture of 0.1 mole of chlorobenzene and 0.15 mole of isopropanol is added from a dropping funnel.[4] Reaction is almost immediately apparent in the region of the iodine, and it becomes more vigorous when the stirrer is started and heating is reduced. The remainder of the halide solution is then added in about 30 min., a further 25 ml. of decalin is added to facilitate stirring, and the mixture is refluxed for 1 hr. more. Dilute hydrochloric acid is added dropwise with stirring to the cooled mixture until the solid is dissolved, and the organic layer is washed well with water, dried, and distilled; the yield of pure benzene is 89%.

In the case of a still less reactive halide, or one with a tendency to undergo dehydrohalogenation, it is advantageous to add a reactive halide such as 1-bromonaphthalene or *n*-butyl bromide for entrainment. Thus the reaction of 0.05 mole of cyclohexyl chloride and 0.05 mole of 1-bromonaphthalene with 0.33 g. atom of magnesium and isopropanol (0.3 mole) in 50 + 20 ml. of decalin afforded a mixture of 83% of cyclohexane and 10% of cyclohexene (removable with sulfuric acid). By this procedure cyclohexyl fluoride gives cyclohexane (33%) and benzotrifluoride gives toluene (10%). Fluorobenzene is inert.

In this work magnesium powder Grade 4 from Magnesium Elektron Ltd., Manchester, England, was employed within 6 months of the date of grinding. Use of older or coarser material may lengthen the induction period.

Syntheses via ethoxymagnesiummalonic ester. Four *Organic Syntheses* procedures start with the reaction of absolute ethanol, malonic ester, and magnesium in the presence of a catalytic amount of carbon tetrachloride to produce ethoxymagnesium-

$$C_2H_5OH + CH_2(CO_2C_2H_5)_2 + Mg \xrightarrow{CCl_4} C_2H_5OMgCH(CO_2C_2H_5)_2 + H_2$$

malonic ester. The reaction is started by treating the magnesium (1 mole) and catalyst (1 ml.) with a small portion of a solution of malonic ester (1 mole) in ethanol;

once a vigorous reaction is in progress, the remainder of the solution is added as required to maintain boiling. When crystalline reaction product starts to separate, the mixture is cooled, ether is added to dissolve the ethoxymagnesium derivative, and refluxing is continued until the metal is all consumed. The next step, an acylation, usually is carried out with the ethereal solution as prepared. The formation of the ethoxymagnesium derivative appears to be essentially quantitative. Thus in a synthesis of tricarboethoxymethane described by Lund and Voigt[5] by addition of a solution of ethyl chloroformate to the ethereal solution of the ethoxymagnesium derivative, the overall yield from malonic ester is 88–93%. The superiority of the

$$C_2H_5OMgCH(CO_2C_2H_5)_2 \xrightarrow[-HOC_2H_5]{ClCO_2C_2H_5} ClMgC(CO_2C_2H_5)_3$$

$$\xrightarrow[88\text{-}93\%\text{ from malonic ester}]{AcOH} CH(CO_2C_2H_5)_3$$

ethoxymagnesium derivative as a reactant is evident from comparison with a procedure for preparation of tricarbomethoxymethane via the sodio derivative of dimethyl malonate in which the overall yield is 40–42%.[6]

In another synthesis the product of acylation of ethoxymagnesiummalonic ester with *o*-nitrobenzoyl chloride is converted by acid hydrolysis-decarboxylation into

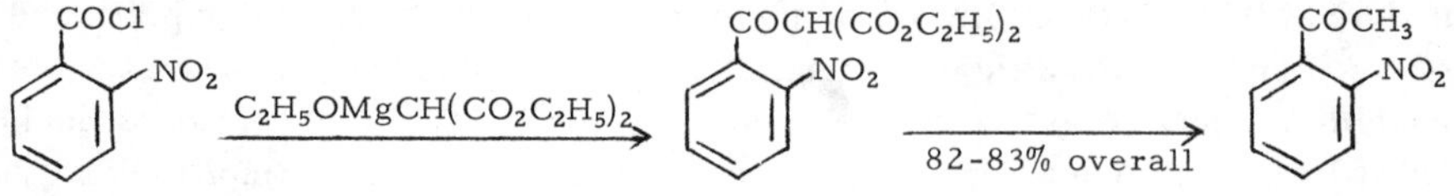
COCl
NO2
C2H5OMgCH(CO2C2H5)2
COCH(CO2C2H5)2
NO2
82-83% overall
COCH3
NO2

o-nitroacetophenone.[7] Acylation of the ethoxymagnesium derivative with phenylacetyl chloride gives a product convertible by the steps shown into naphthoresorcinol.[8]

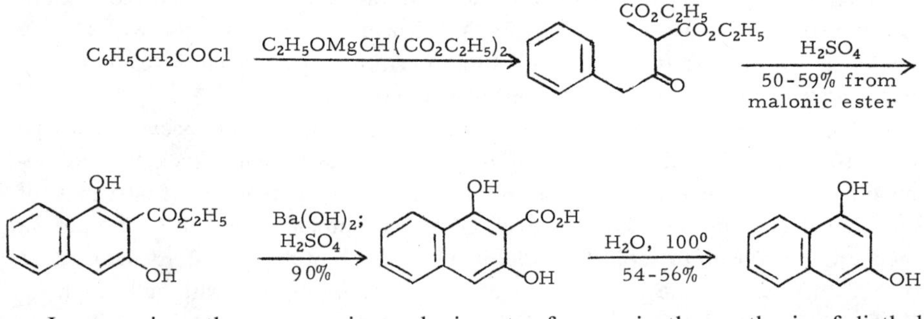
C6H5CH2COCl
C2H5OMgCH(CO2C2H5)2
CO2C2H5
CO2C2H5
H2SO4
50-59% from
malonic ester
OH
CO2C2H5
OH
Ba(OH)2;
H2SO4
90%
OH
CO2H
OH
H2O, 100°
54-56%
OH
OH

In preparing ethoxymagnesiummalonic ester for use in the synthesis of diethyl benzoylmalonate, Price and Tarbell[9] prepared the reagent as usual from malonic ester (0.2 mole), magnesium (0.2 mole), carbon tetrachloride (0.2 ml.), and a slight excess of ethanol (0.275 mole) in ether, and then removed the solvent by evaporation, eventually at reduced pressure. The partially crystalline residue was taken up in benzene and the solvent again removed; this process eliminated a trace of ethanol which might interfere in the next step. This step requires preparation of the mixed benzoic-carbonic anhydride. Benzoic acid and triethylamine are dissolved in toluene to form the amine salt, and the solution is cooled below 0° and treated with ethyl chloroformate. The triethylamine hydrochloride that precipitates does not interfere with the subsequent acylation, carried out by dropwise addition of a solution of ethoxymagnesiummalonic ester in dry ether.

$$C_6H_5CO\overset{-}{O}\overset{+}{N}HEt_3 + ClCOEt \xrightarrow[-Et_3\overset{+}{N}H\overset{-}{Cl}]{\text{Toluene, }0^0} C_6H_5C(=O){-}O{-}C(=O)OEt$$

$$\xrightarrow[68\text{-}75\%\text{ overall}]{C_2H_5OMgCH(CO_2Et)_2} C_6H_5C(=O)CH(CO_2Et)_2 + CO_2 + Mg(OEt)_2$$

Acylation of t-alcohols and of ethyl acetoacetate. *t*-Butyl acetate can be prepared in moderate yield by reaction of *t*-butanol with acetyl chloride and magnesium in

$$2\ CH_3C(=O)Cl + 2\ HOC(CH_3)_3 + Mg \xrightarrow[45\text{-}55\%]{\text{Ether}} 2\ CH_3C(=O)OC(CH_3)_3 + MgCl_2 + H_2$$

ether (Spassow[10]). Acetylation of ethyl acetoacetate with acetyl chloride and magnesium in refluxing benzene (2 hrs.) and purification of the product as the blue copper derivative afforded pure ethyl diacetylacetate in moderate yield (Spassow[11]). However, Viscontini and Merckling[12] carried out the same reaction in absolute

$$CH_3COCH_2CO_2Et \xrightarrow[46\text{-}52\%]{CH_3COCl\ +\ 1/2\ Mg} CH_3COCH(COCH_3)CO_2Et$$

ethanol at room temperature and obtained ethyl diacetyl acetate (ethyl acetyl-acetoacetate) in 73% yield. By this procedure the Swiss workers prepared eleven C-acyl derivatives of ethyl acetoacetate in average yield of 73% and cleaved several of them to the β-keto esters in good yield with either ammonia in ethanol or sodium methoxide.

$$RCOCl + CH_2(COCH_3)CO_2Et \xrightarrow[\text{av. }73\%]{Mg{-}EtOH} RC(=O){-}CH(COCH_3)CO_2Et \xrightarrow[\text{av. }61\%]{EtOH{-}NH_3\text{ or }CH_3ONa} RCOCH_2CO_2Et$$

Reformatsky reaction. This reaction can be carried out conveniently and with improved yield with magnesium in place of the less reactive zinc provided that the *t*-butyl halo ester is used to retard self-condensation of the α-halo ester.[13]

$$C_6H_5C(CH_3){=}O + BrCH_2CO_2C(CH_3)_3 \xrightarrow[89\%]{Mg,\ Ether;\ H_2O} C_6H_5C(CH_3)(OH){-}CH_2CO_2C(CH_3)_3$$

[1]H. Lund and J. Bjerrum, *Ber.*, **64**, 210 (1931)
[2]H. E. Baumgarten and J. M. Petersen, *Org. Syn.*, **41**, 82 (1961)
[3]R. Adams and E. W. Adams, *Org. Syn., Coll. Vol.*, **1**, 459 (1941); *Org. Expts.*, 112
[4]D. Bryce-Smith, B. J. Wakefield, and E. T. Blues, *Proc. Chem. Soc.*, 219 (1963); procedure submitted to *Org. Syn.*
[5]H. Lund and A. Voigt, *Org. Syn., Coll. Vol.*, **2**, 594 (1943)
[6]B. B. Corson and J. L. Sayre, *ibid.*, **2**, 596 (1943)
[7]G. A. Reynolds and C. R. Hauser, *ibid.*, **4**, 708 (1963)
[8]K. Meyer and H. S. Bloch, *ibid.*, **3**, 637 (1955)
[9]J. A. Price and D. S. Tarbell, *ibid.*, **4**, 285 (1963)
[10]A. Spassow, *ibid.*, **3**, 144 (1955)
[11]A. Spassow, *ibid.*, **3**, 390 (1955)
[12]M. Viscontini and N. Merckling, *Helv.*, **35**, 2280 (1952)
[13]T. Moriwake, *J. Org.*, **31**, 983 (1966)

Magnesium bromide etherate.

Preparation. A mixture of 0.3 g. of magnesium dust, 60 ml. of absolute ether, and 30 ml. of benzene is treated with 2.16 g. of mercuric bromide, refluxed for 2 hrs., and the solution is filtered and used immediately.[1] It was known[2] that (1), the 9(11)-monoepoxide of ergosteryl D acetate, is isomerized by boron trifluoride

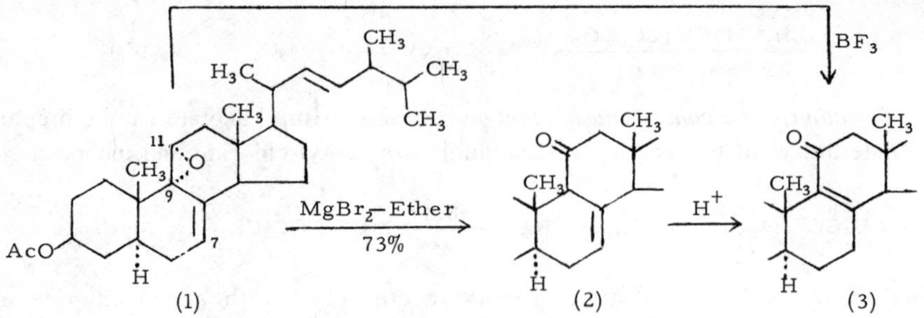

etherate to the conjugated ketone (3). Bachmann[1] found that isomerization to the nonconjugated ketone (2) can be accomplished by adding 75 ml. of the freshly prepared magnesium bromide etherate solution to a solution of 1.23 g. of (1) in 140 ml. of ether and refluxing for 2 hrs. Isomerization to (3) is effected by a trace of hydrochloric acid in ethanol or by passing a benzene–petroleum ether solution through a column of activated alumina.

House[3] showed that both *cis*- and *trans*-2,3-epoxybutane are isomerized by magnesium bromide in ether solution to butane-2-one. The *cis*-epoxide also gave butane-2-one in the presence of boron trifluoride, and the *trans*-epoxide gave both the ketone and isobutyraldehyde. A bromohydrin has been isolated also from the low-temperature isomerization of cyclohexene oxide.[4]

[1]W. E. Bachmann, J. P. Horwitz, and R. J. Warzynski, *Am. Soc.*, **75**, 3268 (1953)
[2]H. Heusser *et al.*, *Helv.*, **34**, 2106 (1951)
[3]H. O. House, *Am. Soc.*, **77**, 5083 (1955)
[4]S. M. Naqvi, J. P. Horwitz, and R. Filler, *Am. Soc.*, **79**, 6283 (1957)

Magnesium–Iodine–Ether.[1] A suspension of metallic magnesium in an ethereal solution of magnesium iodide is a convenient reagent for the removal of vicinal

$$Mg + I_2 \longrightarrow MgI_2$$

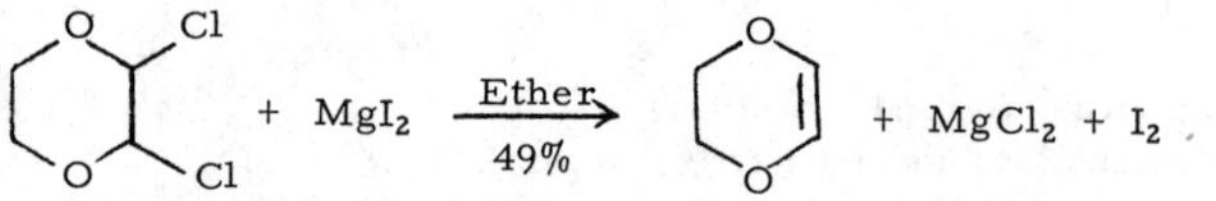

halogens to form a double bond.[2] For dehalogenation of 2 moles of 2,3-dichlorodioxane, a mixture of 3.4 moles of magnesium turnings and 1.2 l. of dry ether is stirred under reflux, and 0.4 mole of iodine is added in small portions. When the mixture becomes colorless a mixture of 2 moles of dichlorodioxane with 200 ml. of ether is added at such a rate that the solution is never darker than light brown (9 hrs.). A decrease in rate as the reaction proceeds is probably due to coprecipitation of iodine ion with the insoluble magnesium chloride formed; this faster reaction accounts also for the requirement of more than a catalytic amount of iodine.[3] The reaction mixture is poured onto ice and water and the organic layer separated, dried, and fractionated; the yield of dioxene, b.p. 93–95°, is 84 g. (49%). The higher-boiling di-*n*-butyl ether is the solvent of choice for dehalogenating less reactive compounds.[4] The reagent is particularly useful for preparing products sensitive to aqueous iodine, such as vinyl ethers.[5]

[1]Contributed by R. K. Summerbell, Northwestern Univ.
[2]R. K. Summerbell and R. R. Umhoefer, *Am. Soc.*, **61**, 3016 (1939)
[3]L. K. Roche, Ph.D. Thesis, Northwestern University (1940)
[4]R. K. Summerbell and R. R. Umhoefer, *Am. Soc.*, **61**, 3020 (1939)
[5]R. K. Summerbell and D. K. A. Hyde, *J. Org.*, **25**, 1809 (1960)

Magnesium methyl carbonate (MMC), $CH_3OMgOCO_2CH_3 + XCO_2$.

Szarvasy[1] passed carbon dioxide into a suspension of magnesium methoxide in methanol until the solid had dissolved, removed the solvent, and obtained a white solid of analysis and properties suggesting the following formulation of the reaction:

$$Mg(OCH_3)_2 + 2\,CO_2 \longrightarrow CH_3O\overset{O}{\overset{\|}{C}}OMgO\overset{O}{\overset{\|}{C}}OCH_3$$

Stiles and Finkbeiner[2] discovered interesting synthetic uses for the reagent in 1959 and later[3] recognized that the original formulation is incorrect. Actually the term "magnesium methyl carbonate" denotes a solution prepared by saturating magnesium methoxide with carbon dioxide, and the active reagent is best represented by the formula $CH_3OMgOCO_2CH_3 + XCO_2$, where X varies widely with solvent and temperature. Detailed, nearly identical procedures for preparing a stock solution on a large scale are presented by Finkbeiner and Stiles[3] and by Finkbeiner and Wagner.[4] In essence this involves adding magnesium turnings over a period of several hours to dry methanol with stirring under reflux until the metal is all converted into magnesium methoxide. The methanol is partly removed at 50° at reduced pressure, dimethylformamide is added, and carbon dioxide is passed in as rapidly as it is absorbed. The remaining methanol is distilled and the result is a pale yellow solution of MMC in dimethylformamide.

Although the chelate magnesium salt (1) of the aci-form of nitroacetic acid had not been characterized, its existence in solution had been suggested from the striking effect of magnesium ions on the rate of decarboxylation of nitroacetic acid.

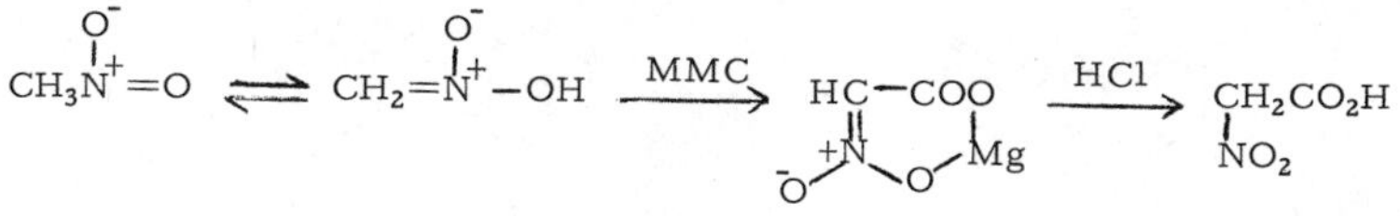
MMC
HCl

(1)

Stiles and Finkbeiner[2] found that treatment of nitromethane with 4 molar equivalents of a 2*M* solution of MMC in dimethylformamide at 50° for 4–5 hrs. resulted in quantitative conversion to magnesium nitroacetate (1), as determined spectrophotometrically. Hydrolysis with ice and hydrochloric acid afforded pure nitroacetic acid in 63% yield. The result shows that the decarboxylation of nitroacetic acid is a reversible reaction and that the position of equilibrium is completely shifted by chelation. Thus chelation provides the driving force for a new route to α-nitro acids and hence, by catalytic hydrogenation, to α-amino acids (Pd-C). Thus nitroethane, 1-nitropropane, and 1-nitrobutane were converted, in the overall yields indicated, into DL-alanine (46%), DL-α-aminobutyric acid (34%), and DL-norvaline (42%). 2-Nitropropane, which is incapable of forming a chelate of type (1), failed to undergo reaction. α-Carboxylation of eight other nitroalkanes and esterification afforded α-nitro esters in average yield of 45%.[4]

Stiles[5] extended use of the reagent to the α-carboxylation of ketones to give β-keto acids (3). The intermediate magnesium salt (2) can be alkylated *in situ*

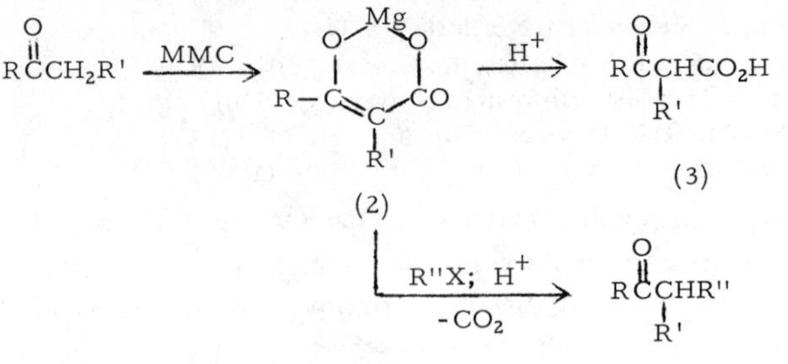

(2) (3) (4)

and then decarboxylated to (4), thus effecting overall α-alkylation. Acetophenone, heated with 3–4 moles of MMC at 110–120° for 1 hr., afforded benzoylacetic acid in 68% yield.

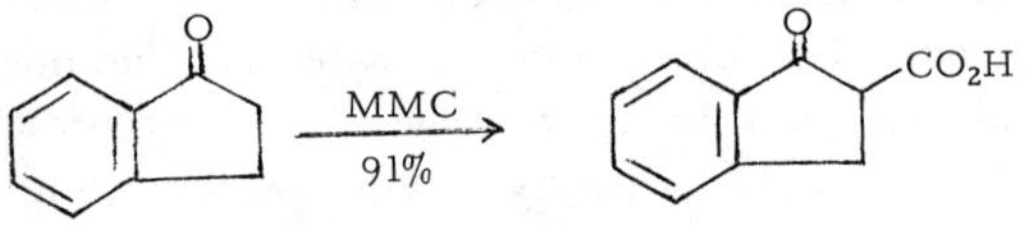

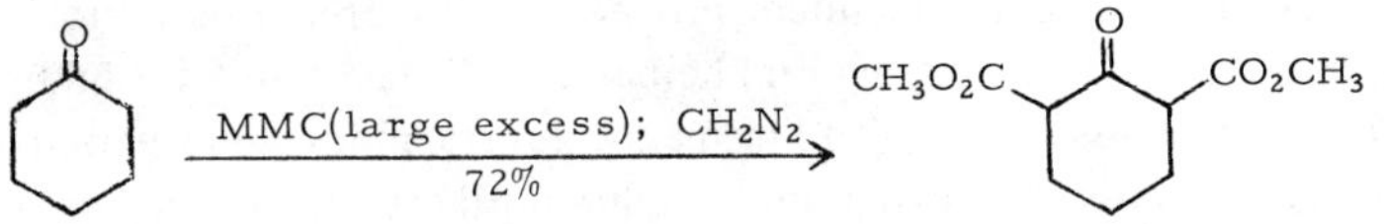

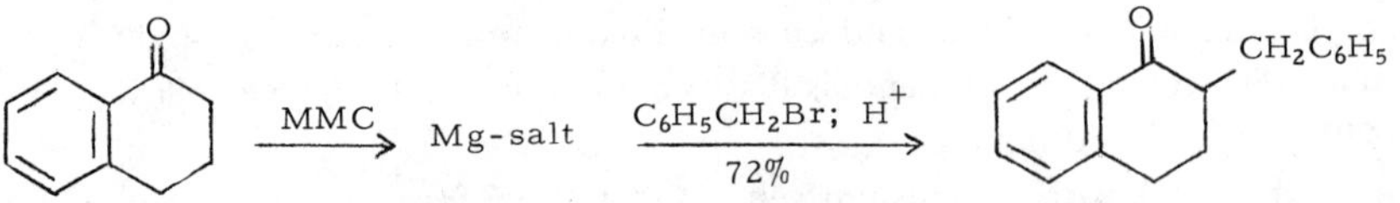

Pelletier[6] applied the new reaction to 5-methoxytetralone-2 (5) and established that the sole product is the 3-carboxylic acid (6). Evidently the steric requirement

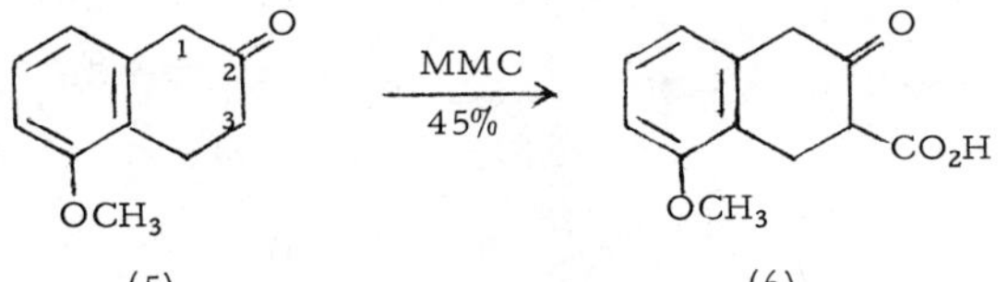

(5) (6)

of the chelate magnesium salt favors equilibration to the 3-position rather than to the more hindered 1-position.

Finkbeiner[7] extended the sequence of α-carboxylation and alkylation-decarboxylation to 3-phenylhydantoin (7). Reaction with MMC, followed by esterification

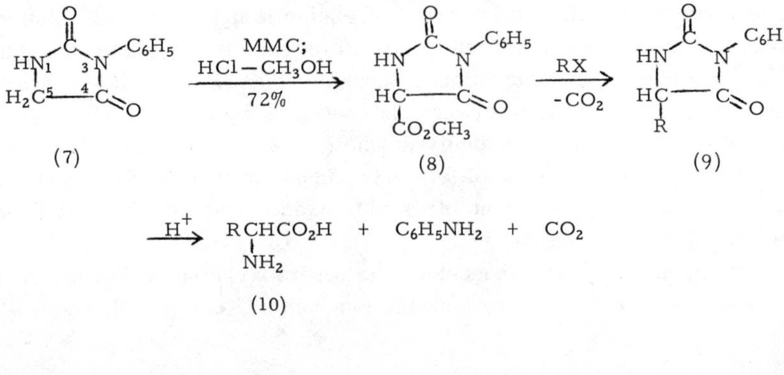

(7) (8) (9) (10)

(CH_3OH–HCl), afforded 3-phenyl-5-carbomethoxyhydantoin (8) in 72% yield. Alkylation of this substance with benzyl chloride and excess methanolic sodium methoxide, followed by reaction with aqueous hydrochloric acid to effect hydrolysis and decarboxylation, afforded 3-phenyl-5-benzylhydantoin. For the synthesis of natural amino acids in DL-form, Finkbeiner treated 3-phenylhydantoin (7) with MMC to produce the chelate magnesium salt, treated this with an appropriate alkylating agent, and poured the mixture on ice and hydrochloric acid to produce the 3-phenyl-5-alkylhydantoin. Hydrolysis to the free amino acids was accomplished according to Gaudry[8] by heating with barium hydroxide under pressure at 160° for 15–30 min., precipitating barium ion with ammonium carbonate, evaporating to dryness, and heating on the steam bath to eliminate excess ammonium carbonate. In the synthesis of DL-Val, Leu, Try, Pro, and Lys, yields of 3-phenyl-5-alkylhydantoin averaged 53% and in the hydrolysis the average yield was 45%. In the synthesis of DL-phenylalanine the yields in the two steps were 98 and 97%.

A recent example of this method of α-carboxylation is reported by Griffin *et al.*[9]

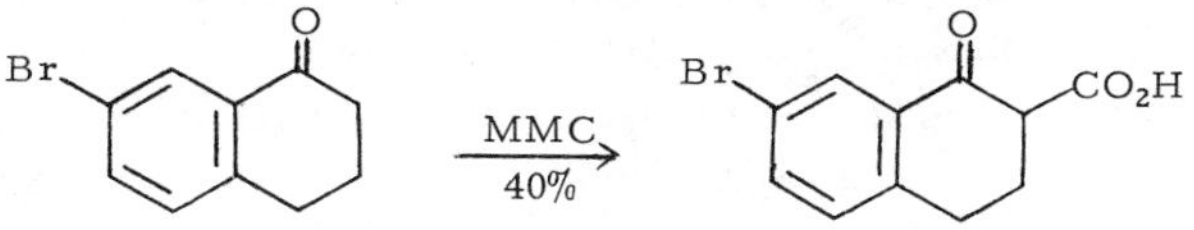
Br
O
MMC
40%
Br
O
CO_2H

[1]E. Szarvasy, *Ber.*, **30**, 1836 (1897)
[2]M. Stiles and H. L. Finkbeiner, *Am. Soc.*, **81**, 505 (1959)
[3]H. L. Finkbeiner and M. Stiles, *Am. Soc.*, **85**, 616 (1963)
[4]H. L. Finkbeiner and G. W. Wagner, *J. Org.*, **28**, 215 (1963)
[5]M. Stiles, *Am. Soc.*, **81**, 2598 (1959)
[6]S. W. Pelletier and P. C. Parthasarathy, *Tetrahedron Letters*, 103 (1964); S. W. Pelletier, R. L. Chappell, P. C. Parthasarathy, and N. Lewin, *J. Org.*, **31**, 1747 (1966)
[7]H. L. Finkbeiner, *Am. Soc.*, **86**, 961 (1964)
[8]R. Gaudry, *Can. J. Res.*, **26B**, 387 (1948)
[9]R. W. Griffin, Jr., J. D. Gass, M. A. Berwick, and R. S. Shulman, *J. Org.*, **29**, 2109 (1964)

Magnesium oxide, MgO. Mol. wt. 40.32, m.p. 2800°.

The preparation of pyrimidine (2) by the hydrogenative dechlorination of 2,4-dichloropyrimidine (1) and of 2,5-dichloropyrimidine (3) was reported at about the

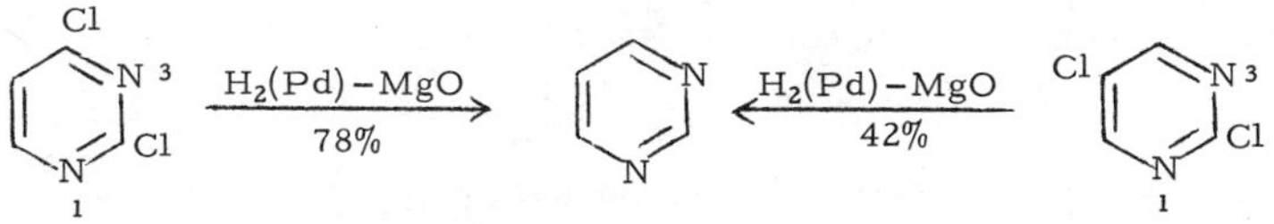
Cl
N 3
Cl
N
1
H_2(Pd)–MgO
78%
N
N
H_2(Pd)–MgO
42%
Cl
N 3
Cl
N
1

(1) (2) (3)

same time by Whittaker[1] (Wellcome Labs.) and by Lythgoe and Rayner[2] (Cambridge Univ.). In each case magnesium oxide was found superior to other bases as hydrogen chloride acceptor, and in each case the crude product was precipitated as pyrimidine mercurichloride, which was split by distillation with sodium sulfide. In an improved procedure[3] a suspension of 84 g. of 2,4-dichloropyrimidine in 420 ml. of ethanol and 840 ml. of water is treated with 50 g. of magnesium oxide and 21 g. of 3% palladium charcoal and agitated under hydrogen at 30–40 lb. pressure until the uptake of hydrogen is complete (80 min.). The yield of pyrimidine purified as described is 78%.

The reagent has been used as hydrogen chloride acceptor in the N-protection of amino acids by reaction with carbobenzoxy chloride.[4]

[1]N. Whittaker, *J. Chem. Soc.*, 1565 (1951)
[2]B. Lythgoe and L. S. Rayner, *J. Chem. Soc.*, 2323 (1951)
[3]N. Whittaker, *J. Chem. Soc.*, 1646 (1953)
[4]V. du Vigneaud and G. L. Miller, *Biochem. Preps.*, **2**, 79 (1952)

Magnesium sulfate, $MgSO_4$; $MgSO_4 \cdot 7\,H_2O$.

Anhydrous magnesium sulfate is sometimes used as a drying agent but may prove unsatisfactory because of its acidity. Robertson[1] developed an efficient procedure for the addition of a tertiary ethynylcarbinol to dihydropyrane to provide a protected derivative consisting in adding a trace of *p*-toluenesulfonic acid to a mixture of the reactants with a slight excess of the reagent and heating for 1 hr. on the steam bath

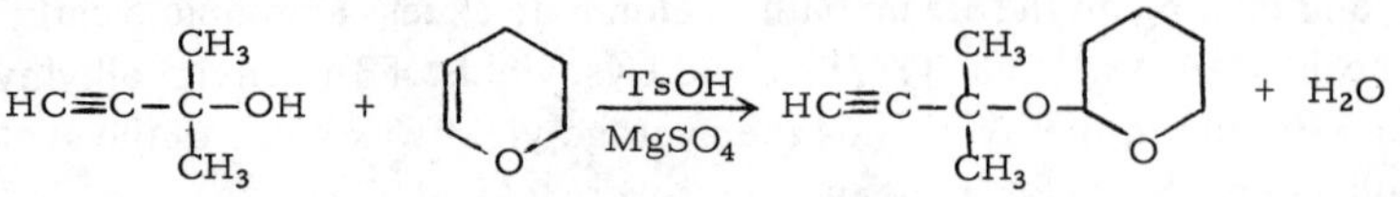
CH3
HC≡C–C–OH
CH3
+
TsOH
MgSO4
CH3
HC≡C–C–O–
CH3
O
+ H2O

(yields 60–84%). He notes that if ether extraction is used in the workup the drying agent should be neutral or basic; magnesium sulfate is sufficiently acidic to reverse the addition in a few hours.

Bernstein's group made deliberate use of this property of the reagent to effect selective hydrolysis of the Δ^5-3,20-bisethyleneketal I.[2] When a solution of I in benzene saturated with water was shaken with anhydrous magnesium sulfate for 1 hr. the 3-ethyleneketal group was removed quantitatively to give the 20-monoethyleneketal II.

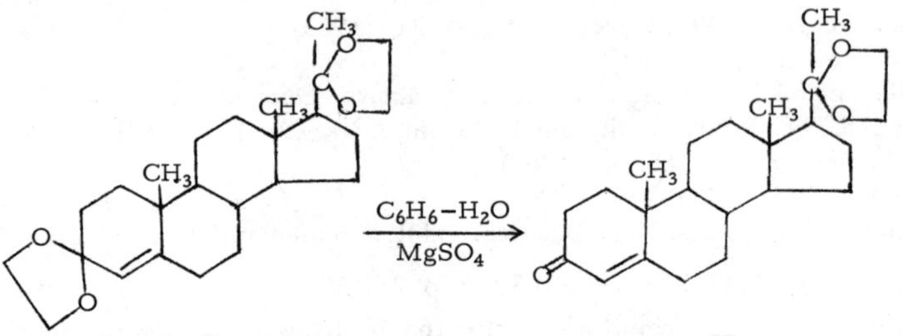
CH3
CH3
CH3
C6H6–H2O
MgSO4
CH3
CH3
CH3

I II

[1]D. N. Robertson, *J. Org.*, **25**, 931 (1960)
[2]J. J. Brown, R. H. Lenhard, and S. Bernstein, *Am. Soc.*, **86**, 2183 (1964)

H—C—CHO
‖
Malealdehyde, H—C—CHO. Mol. wt. 86.09, yel. oil, b.p. 56–59°/9.5 mm., soluble in acetic acid, acetone, or chloroform; sparingly soluble in ether, benzene, carbon tetrachloride; polymerizes rapidly; yellow solutions in water and methanol become colorless on standing; disemicarbazone, needles from water, m.p. 246–247° dec.

In spite of the unfavorable properties noted, Wohl and co-workers[1] succeeded in isolating the highly sensitive dialdehyde in substantially pure form and in characterizing it as crystalline derivatives such as the disemicarbazone. The synthesis involved reaction of acetylenedimagnesium bromide (1) with triethyl orthoformate to produce the bisdiethylacetal of acetylenedialdehyde (2), selective hydrogenation in the presence of Lindlar catalyst to (3), and hydrolysis, carefully conducted at a low temperature. The structure and configuration follow from permanganate oxidation of the bisdiethylacetal (3) and hydrolysis to a product characterized as *meso*-tartaric dialdehyde (5).

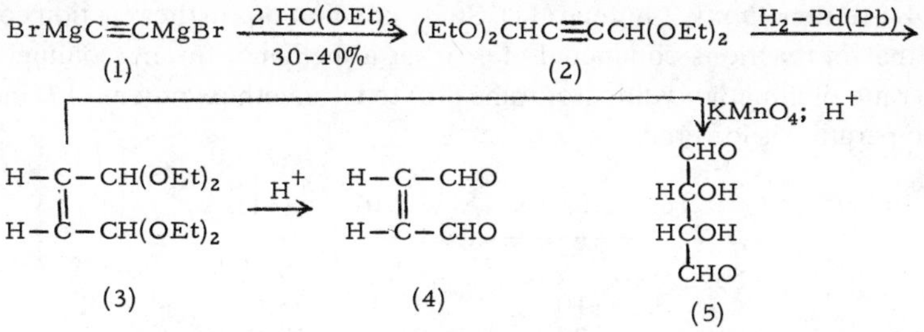
BrMgC≡CMgBr
2 HC(OEt)3
30-40%
(EtO)2CHC≡CCH(OEt)2
H2-Pd(Pb)
KMnO4; H+
(1)
(2)
H−C−CH(OEt)2
H−C−CH(OEt)2
H+
H−C−CHO
H−C−CHO
CHO
HCOH
HCOH
CHO
(3)
(4)
(5)

Wohl's synthesis now appears to be the second best route to a stable precursor which can be hydrolyzed to malealdehyde. Ranking third is a process described by Marquis[2] in which furane is nitrated in acetic anhydride at a low temperature to give a noncrystalline unstable product which yields 2-nitrofurane on treatment with pyridine. J. R. Johnson[3] presented arguments for regarding the labile intermediate as the product of 1,4-addition of acetyl nitrate (7), and hence presumably convertible into malealdehyde.

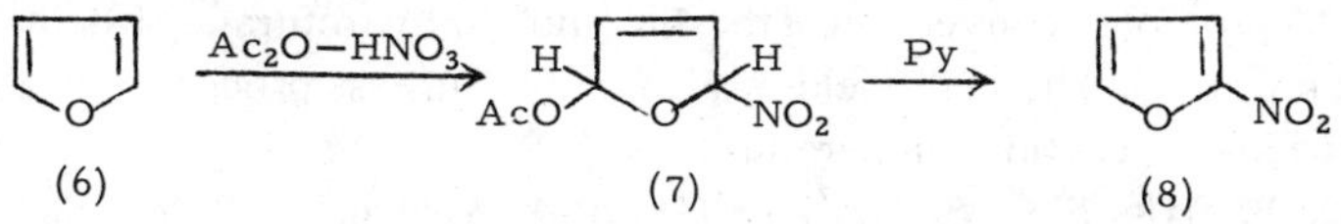
Ac2O−HNO3
H
AcO
H
NO2
Py
NO2
(6)
(7)
(8)

It now appears that the precursor of choice is 2,5-dimethoxy-2,5-dihydrofurane (10), a reagent supplied by Aldrich and by Eastman, and prepared by a procedure

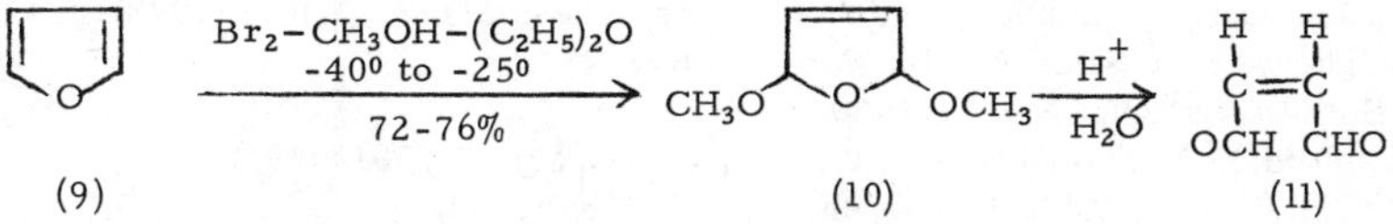
Br2−CH3OH−(C2H5)2O
-40° to -25°
72-76%
CH3O
OCH3
H+
H2O
H H
C=C
OCH CHO
(9)
(10)
(11)

introduced by a Swedish group[4] and further perfected by Burness.[5] A cold (−25°) solution of bromine in methanol is added to a cold (−40°) solution of furane in methanol and ether and the temperature kept below −25°. Gaseous ammonia is then passed in at a temperature below −15° until the mixture is just basic to pH paper. The solution is stirred for 1.5 hrs. and let warm to −5°, when it should become colorless. The ammonium bromide is removed by filtration, the filtrate is concentrated and refiltered, and the product is isolated by distillation, b.p. 50–51°/12 mm.

Acid hydrolysis of 2,5-dimethoxy-2,5-dihydrofurane affords malealdehyde together with a small amount of fumaric dialdehyde.[6,7]

A Russian group[8] investigated the reaction of furane with bromine and methanol with interesting results. Under conditions very similar to those of the Burness

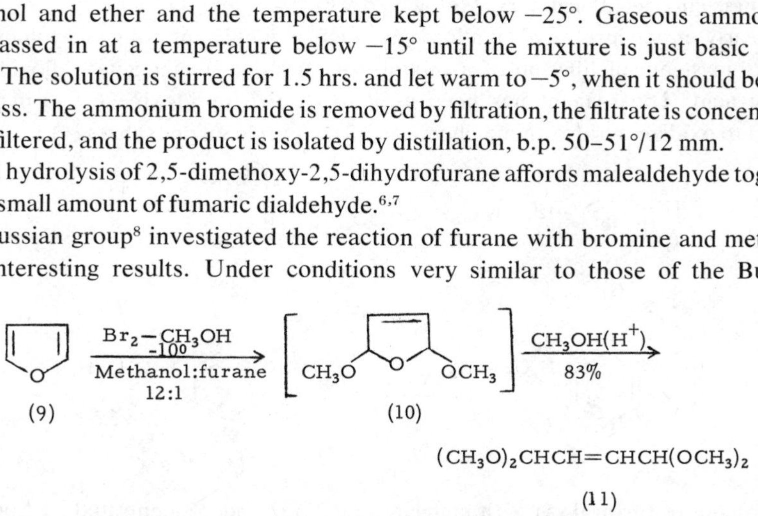
Br2−CH3OH
-10°
Methanol:furane
12:1
CH3O
OCH3
CH3OH(H+)
83%
(9)
(10)
(CH3O)2CHCH=CHCH(OCH3)2
(11)

procedure except that the temperature was higher and that no ether was used as co-solvent, the reaction product, isolated in good yield after neutralization with ammonia, was 1,1,4,4-tetramethoxy-2-butene (11). Study of variations in the reaction condition showed that in reactions conducted at −10° at a methanol:furane volume ratio of 21:1 the ratio of dimethoxydihydrofurane (10) to tetramethoxybutene (11) increases as the temperature is lowered:

Temperature, °C	% (10)	% (11)
0	2	19
−10	18.4	53
−20	36.6	19.6
−30	56.5	3

Another determining factor is the concentration of hydrogen bromide formed in the bromination step. The results just cited show that at a methanol:furane ratio of 21:1 and a temperature of −10° the yield of tetramethoxybutene was 53%; the first experiment cited was conducted at the same temperature (−10°) but with a methanol: furane ratio of 12:1 and the yield of tetramethoxybutene rose to 83%. Hydrogen bromide thus promotes conversion of the dimethoxydihydrofurane (10) into the tetramethoxybutene (11). The high yield of (10) in the Burness procedure is attributable largely to the lower reaction temperature.

[1]A. Wohl and B. Mylo, *Ber.*, **45**, 322, 1746 (1912); A. Wohl and E. Bernreuther, *Ann.*, **481**, 11, 29 (1930)
[2]R. Marquis, *Ann. chim.*, [8], **4**, 216 (1905)
[3]B. T. Freure and J. R. Johnson, *Am. Soc.*, **53**, 1142 (1931)
[4]N. Clauson-Kaas, F. Limborg, and J. Fakstorp, *Chem. Scand.*, **2**, 109 (1948); J. Fakstorp, D. Raleigh, and L. E. Schniepp, *Am. Soc.*, **72**, 869 (1950)
[5]D. M. Burness, *Org. Syn.*, **40**, 29 (1960)
[6]D. L. Hufford, D. S. Tarbell, and T. R. Koszalka, *Am. Soc.*, **74**, 3014 (1952)
[7]K. Alder, H. Betzing, and K. Heimbach, *Ann.*, **638**, 187 (1960)
[8]S. M. Makin and N. I. Telegina, *Journal General Chemistry USSR*, **32**, 1082 (1962) [translated from the Russian Journal, **32**, 1104 (1962)]

Manganese dioxide, (a) ordinary. Mol. wt. 86.93.

An early procedure[1] which affords a quick if inefficient route to toluquinone consists in steam distillation of a mixture of *o*-toluidine, manganese dioxide, and sulfuric acid. The reagent has been used to oxidize pyridoxine (1) to pyridoxal (2)[2] and to oxidize pyridoxamine phosphate (3) to pyridoxal phosphate (4).[3]

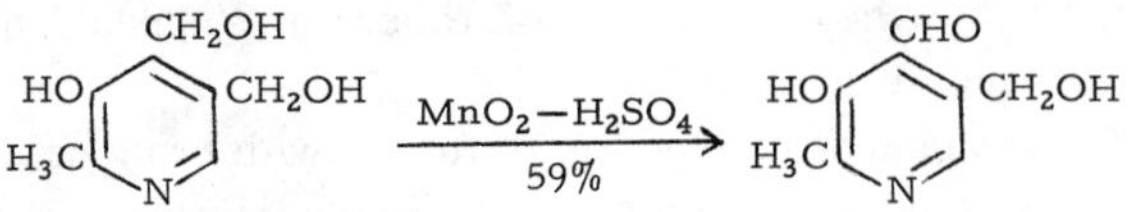
CH₂OH
HO
CH₂OH
H₃C
N
MnO₂—H₂SO₄
59%
CHO
HO
CH₂OH
H₃C
N

(1) (2)

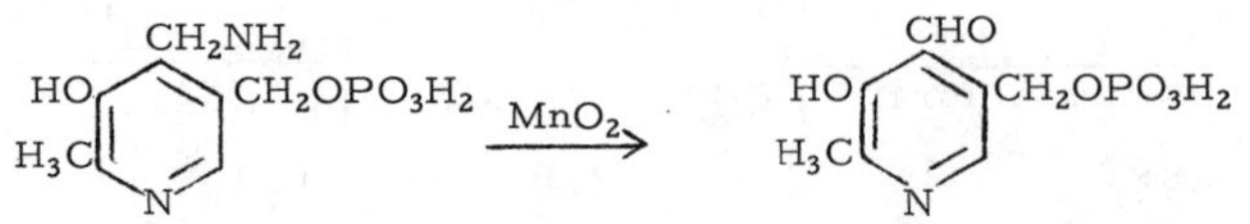
CH₂NH₂
HO
CH₂OPO₃H₂
H₃C
N
MnO₂
CHO
HO
CH₂OPO₃H₂
H₃C
N

(3) (4)

Treatment of furfural (5) with manganese dioxide and concentrated hydrochloric acid affords a compound known as mucochloric acid (7) in 75% yield.[4] An intermediate has been identified as the *β*-chloro-*γ*-hydroxylactone (6).[5]

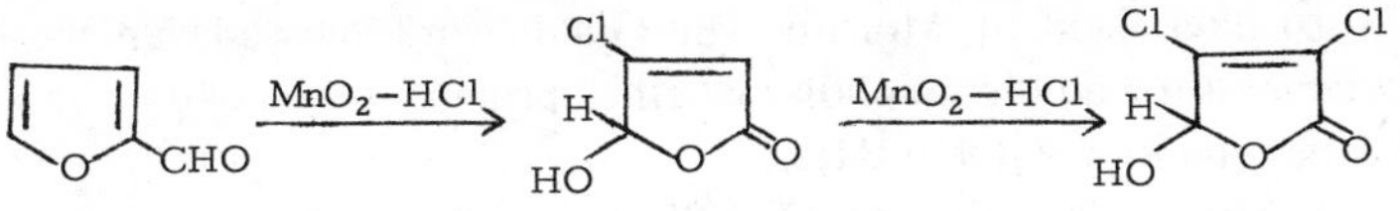

(5) (6) (7)

[1]T. H. Clark, *Am. Chem. J.*, **14**, 565 (1892)
[2]D. Heyl, *Am. Soc.*, **70**, 3434 (1948)
[3]A. N. Wilson and S. A. Harris, *Am. Soc.*, **73**, 4693 (1951)
[4]M. Yanagita, *J. Pharm. Soc. Japan* **72**, 1383 (1952)
[5]Y. Hachihama, T. Shono, and S. Ikeda, *J. Org.*, **29**, 1371 (1964)

Manganese dioxide, (b) active. Mol. wt. 86.93. Supplier: Beacon Chem. Ind. The commercial material is prepared by a method different from those cited below; it is essentially MnO_2 containing about 5% of water of hydration.

Preparation. Morton and co-workers[1] in 1948 prepared an activated form of the oxide by mixing aqueous solutions of equivalent amounts of manganous sulfate and potassium permanganate and washing and drying the precipitate, and found it an effective reagent for oxidation of vitamin A_1 to retinene, whereas ordinary

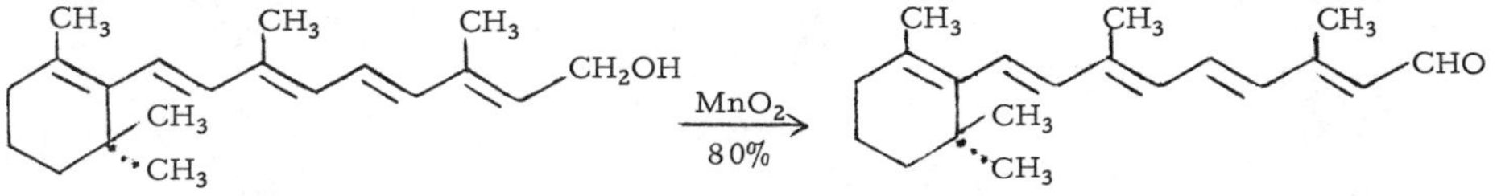

Vitamin A_1 Retinene (A_1 aldehyde)

manganese dioxide is not. The oxidation was carried out by placing a solution of 0.2 g. of vitamin A_1 in 50 ml. of petroleum ether and 10 g. of active MnO_2 in a stoppered flask and letting the mixture stand for 6–10 days with occasional shaking. Other primary allylic alcohols oxidized successfully to aldehydes by the Morton procedure include 2β-ionylidene–ethanol[2] and vitamin A_2.[3]

In working out a synthesis of vitamin A_1, Attenburrow *et al.*[4] of the Glaxo Laboratories (1952) had occasion to investigate the oxidation of five allylic alcohols and developed an active MnO_2 of greater reactivity than that of Morton. Solutions of manganous sulfate and sodium hydroxide are added simultaneously to a stirred solution of potassium permanganate, and the precipitate is collected by centrifugation and washed until the wash liquor is colorless. The solid is dried at 100–120° and ground to a fine powder. This active material is a hydrated oxide containing 3–4% of firmly bound water. In a typical oxidation a solution of 1.0 g. of the secondary

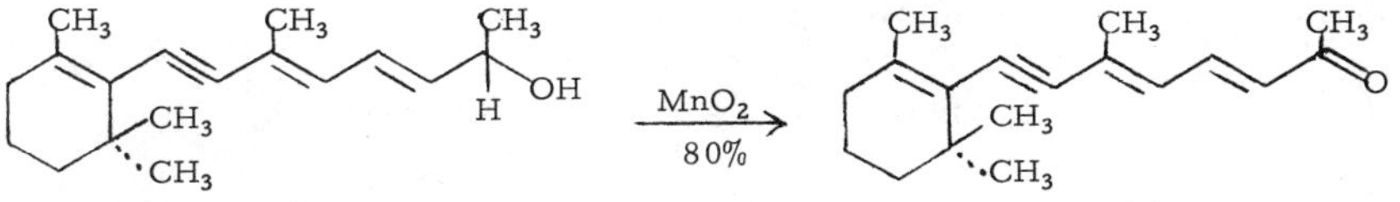

(1) (2)

alcohol (1) in 50 ml. of petroleum ether was shaken with 1 g. of the active MnO_2 for 1 hr.; isolation and distillation gave 0.8 g. of directly pure ketone (2). By comparison, Oppenauer oxidation of (1) and purification through the semicarbazone gave (2) in yield of about 10%. The Attenburrow oxide was satisfactory in all five of the synthetic reactions, as well as in trials with allyl alcohol, cinnamyl alcohol, Δ^3-octyne-2-ol, and 3-dehydro-β-ionol. Yields are as high as with the Morton oxide, and oxidation usually is complete in 0.5–1 hr. at room temperature, as compared with the 6–10 days used by Morton. The Glaxo work showed that the oxidation method is general for allylic alcohols, whether primary or secondary, and also for alcohols of the type RC≡CCH(OH)R′.

Other preparations of active MnO_2 have subsequently been reported from the Syntex[5] and Pfizer[6] laboratories. Gritter and Wallace[7] compared samples of oxide prepared by all the procedures mentioned above, as well as by modified procedures, without finding any striking differences. The majority of investigators have used the Attenburrow oxide and found it fully satisfactory. The material now available commercially is regarded as still more active.

Oxidation of alcohols. Examples already cited illustrate the oxidation of allylic and propargylic primary and secondary alcohols to aldehydes and ketones. An interesting use of the reaction is in a synthesis of 4,5,6,7-tetrahydroindanone-1

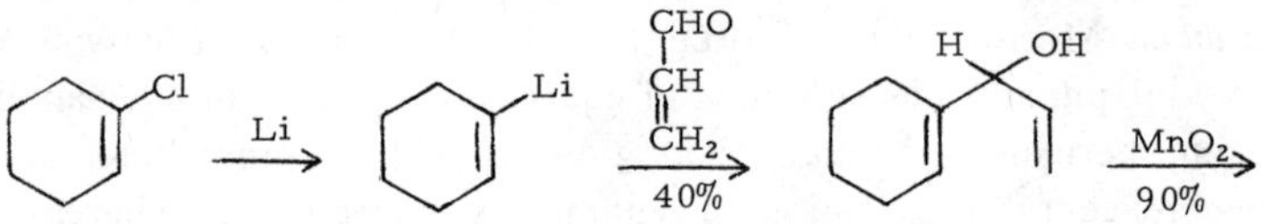

Cl
Li
Li
CHO
CH
CH2
40%
H
OH
MnO2
90%

(1) (2) (3)

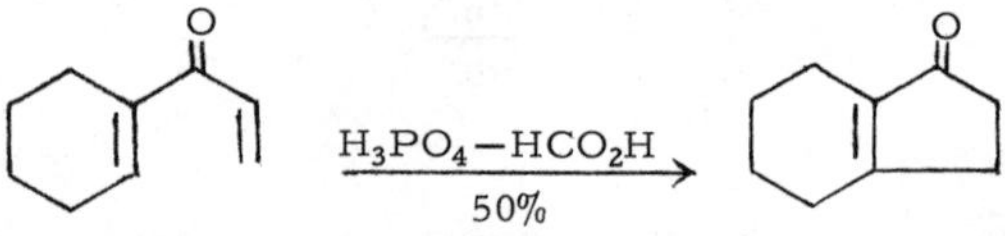

O
H3PO4−HCO2H
50%
O

(4) (5)

(5) described by Braude and Coles.[8] Crombie and Crossley[9] found that α-cyclopropyl alcohols can be oxidized satisfactorily with a suspension of active manganese

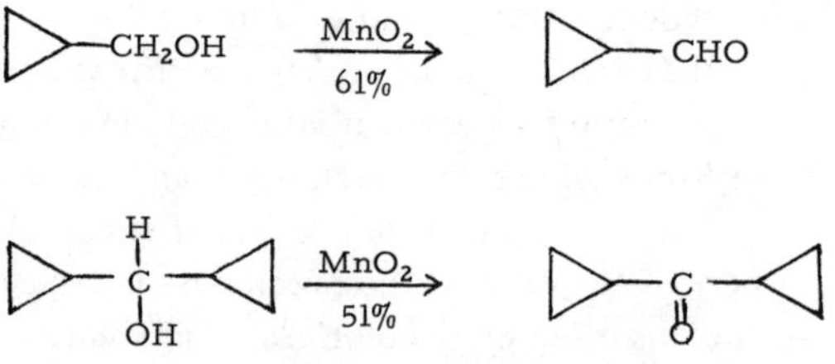

CH2OH
MnO2
61%
CHO
H
C
OH
MnO2
51%
C
O

dioxide in *n*-pentane. Using carefully neutralized oxide in methylene chloride, Stork and Tomasz[10] were able to oxidize the highly unstable carbinol (3) to the ketone (4), the first known vinyl ethynyl ketone.

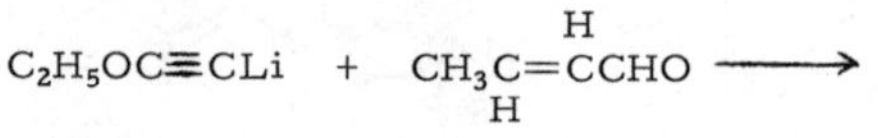

C2H5OC≡CLi
+
H
CH3C=CCHO
H

(1) (2)

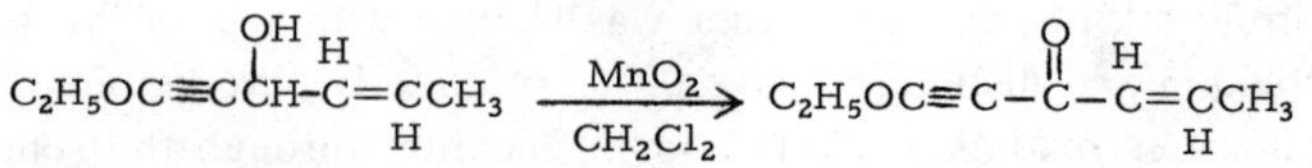

OH
H
C2H5OC≡CCH−C=CCH3
H
MnO2
CH2Cl2
O
H
C2H5OC≡C−C−C=CCH3
H

(3) (4)

Pratt and Van de Castle[11] studied the oxidation of a variety of phenylcarbinols with active MnO_2. After refluxing the reagent with benzene until water no longer collected in a Dean-Stark trap, the carbinol was added and refluxing was continued until evolution of water ceased. Results are summarized as follows: $ArCH_2OH \longrightarrow ArCHO$ (75–85%); $Ar_2CHOH \longrightarrow Ar_2CO$ (88–96%); $ArCHOHR \longrightarrow ArCOR$ (72–95%).

Selective oxidation of an allylic secondary alcohol group without attack of saturated secondary alcohol groups was demonstrated by Mancera, Rosenkranz, and Sondheimer.[5] Adrenosterone (I) on reduction with lithium aluminum hydride

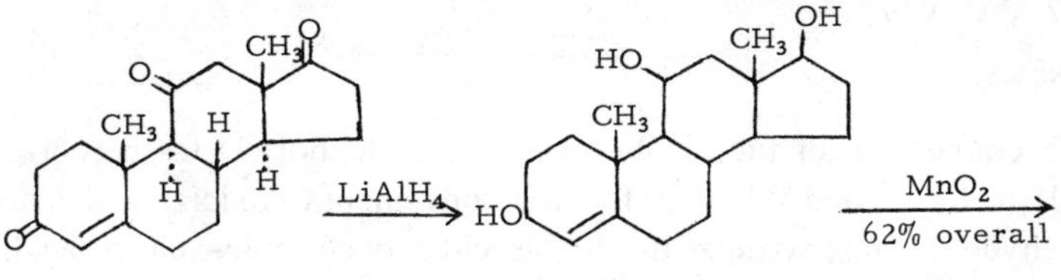
CH3
O
CH3
H
H
H
LiAlH4
HO
CH3
OH
CH3
HO
MnO2
62% overall

I II (3α + 3β)

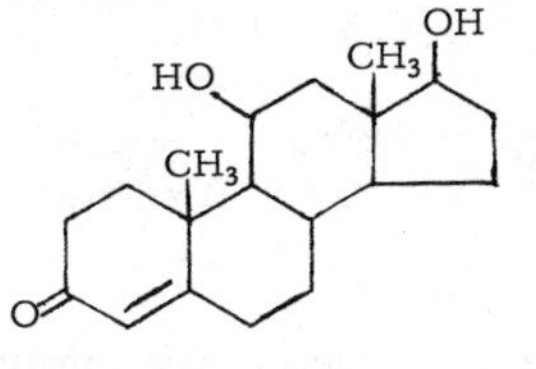
HO
CH3
OH
CH3
O

III

yielded a mixture of the $3\alpha,11\beta,17\beta$- and $3\beta,11\beta,17\beta$-triols (II), and oxidation of the mixture with active MnO_2 in chloroform at room temperature attacked only the allylic alcohol function and gave 11β-hydroxytestosterone (III).

Sondheimer, Amendolla, and Rosenkranz[12] found that cholesterol and other homoallylic Δ^5-3-hydroxysteroids are oxidized by active MnO_2 in refluxing benzene for 8 hrs. to $\Delta^{4,6}$-3-ketosteroids, but the yields are too low for preparative purposes.

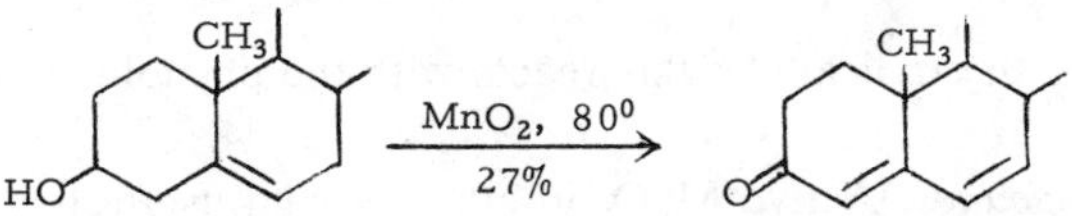
CH3
HO
MnO2, 80°
27%
CH3
O

Δ^4-3-Ketosteroids also afford $\Delta^{4,6}$-dienones in low yield. Harrison[13] of Syntex found that, given sufficient reagent and purified solvents, both primary and secondary saturated alcohols can be oxidized in high yield, but slowly. Thus 100 mg. of 5α-androstane-17β-ol stirred at room temperature in hexane or acetonitrile (20 ml.)

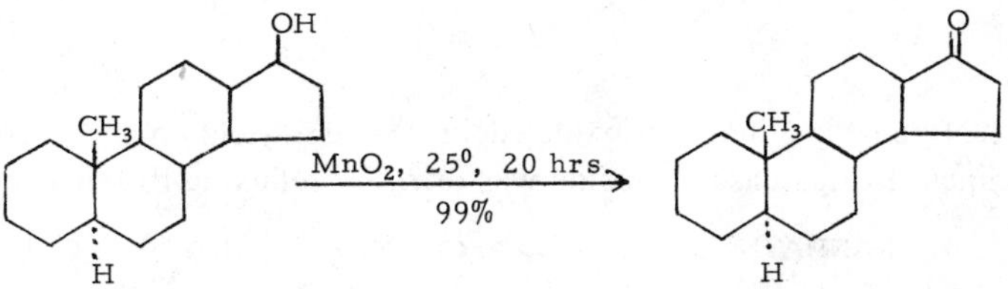
OH
CH3
H
MnO2, 25°, 20 hrs.
99%
O
CH3
H

with 2 g. of Attenburrow oxide for 20 hrs. gave pure 5α-androstane-17-one in practically quantitative yield. The rate of the reaction varies with the solvent used. Thus in dimethyl sulfoxide the oxidation was still incomplete in 7 days. With the same ratio of oxidant to alcohol, 4-methylcyclohexanol in acetonitrile afforded 4-methylcyclohexanone in 71% in 3 days. *n*-Butyraldehyde was obtained in 70% yield by filtering a benzene solution of *n*-butanol through a column of active MnO_2.

Nickon[14] noted that in the activated-MnO_2 oxidation of steroid allylic alcohols (Δ^4-3-ols and Δ^5-7-ols) the quasi-equatorial isomer is oxidized in one tenth to one third the time required for the quasi-axial isomer.

Benzylic alcohols can be oxidized selectively by the Attenburrow reagent as illustrated in the example formulated.[15]

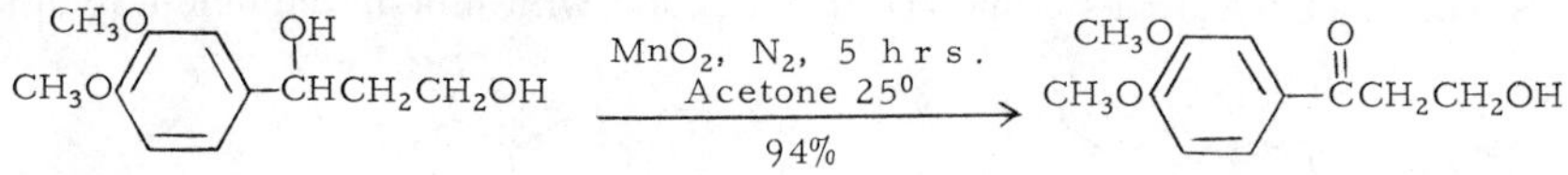

For the conversion of the highly unsaturated alcohol (1) to the corresponding acid (3), Bell, Jones, and Whiting[16] found it convenient to oxidize with active MnO_2 to the aldehyde (2) and to oxidize this to the acid with chromic–sulfuric acid.

$$CH_3CH_2CH_2C{\equiv}CC{\equiv}CCH\overset{c}{=}CHCH_2OH \xrightarrow[90\%]{MnO_2} CH_3CH_2CH_2C{\equiv}CC{\equiv}CCH\overset{c}{=}CHCHO$$

(1) (2)

$$\longrightarrow CH_3CH_2CH_2C{\equiv}CC{\equiv}CCH\overset{c}{=}CHCO_2H$$

(3)

Becker[17] used Attenburrow MnO_2 for the conversion of hindered phenols into quinol ethers. For example, 2,4,6-tri-*t*-butylphenol is oxidized in benzene to give

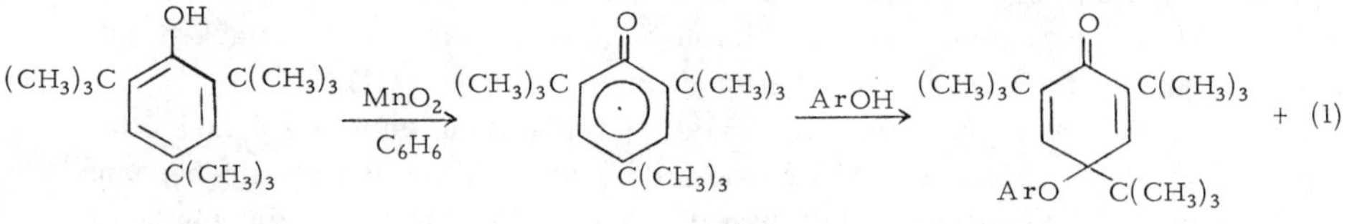

(1) (2) (3)

a blue solution of the radical (2); this reacts with the phenol (1) to give the quinol ether (3).

Nitrogen compounds. Active MnO_2 in ether proved useful for oxidation of the bishydrazone (1) to the purple bisdiazo compound (2), for yellow mercuric oxide, the usual reagent for an oxidation of this type, gave low yields.[18] Pratt and McGovern[19]

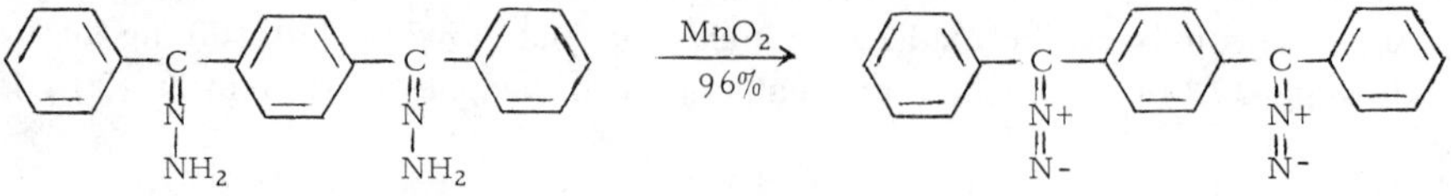

(1) (2)

found that N-benzylanilines are oxidized in 75–90% yield to the corresponding benzal anilines. In this case the oxide was dried by refluxing in benzene for 5 hrs.

$$ArCH_2NHAr' + MnO_2 \longrightarrow ArCH{=}NAr' + MnO + H_2O$$

under a water trap, and the progress of the oxidation was then followed by measuring the water formed. Hydrazobenzenes are oxidized efficiently by this method to azobenzenes. Aniline is oxidized to azobenzene but in this case the reagent should not be dried. Independent papers from the University of Maryland[19] and from Puerto Rico[20] report oxidation of several substituted anilines to azobenzenes in good yield; nitroanilines and aminobenzoic acids are not attacked.

Yates[21] found active MnO_2 superior to mercuric oxide for oxidation of 1-mesitylglyoxal-2-hydrazone to the α-diazoketone; with mercuric oxide the yield was only

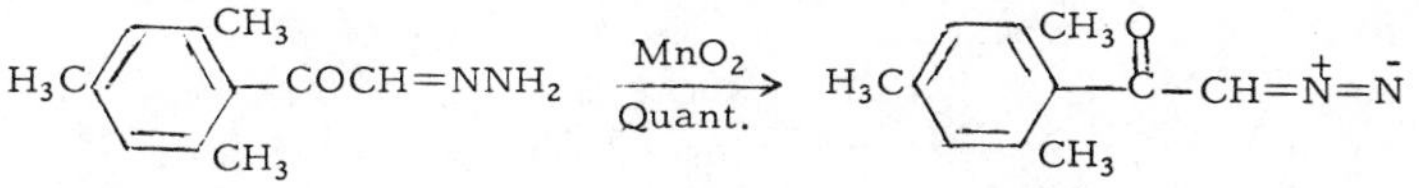

75%. Allinger[22] employed this procedure in the synthesis of 4-carboxy-[8]paracyclophane (3) from 4,5-diketo-[9]paracyclophanemonohydrazone (1).

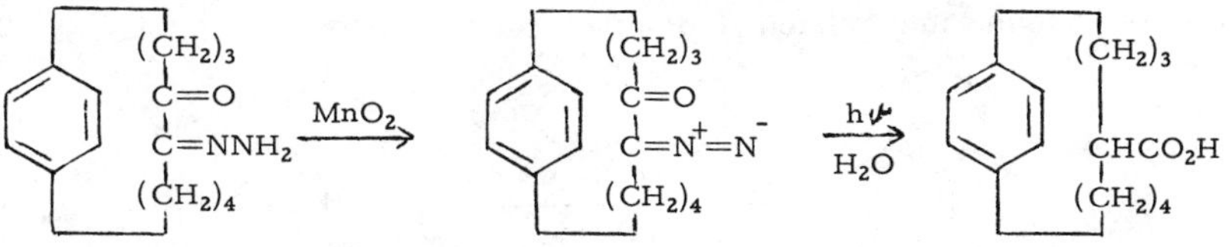

(1) (2) (3)

Kelly[23] found that active MnO_2 oxidizes phenylhydrazides smoothly in aqueous acetic acid at room temperature to give the corresponding acids in good yield; aromatic reaction products are benzene (30%), phenol (27%), and phenyl acetate

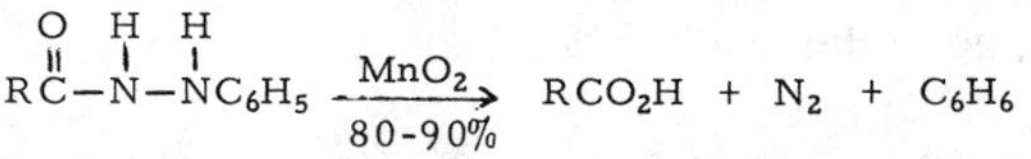

(41%). The reaction was applied to model dipeptides exemplified by the γ-phenylhydrazide of N-carbobenzoxy-α-L-glutamyl-L-methionine methyl ester (4) and found to afford the carboxylic acid (5) in good yield without disturbance of the carbobenzoxy and ester protective groups. The results suggest use of the phenylhydrazide group for protection of carboxyl groups in peptide chemistry.

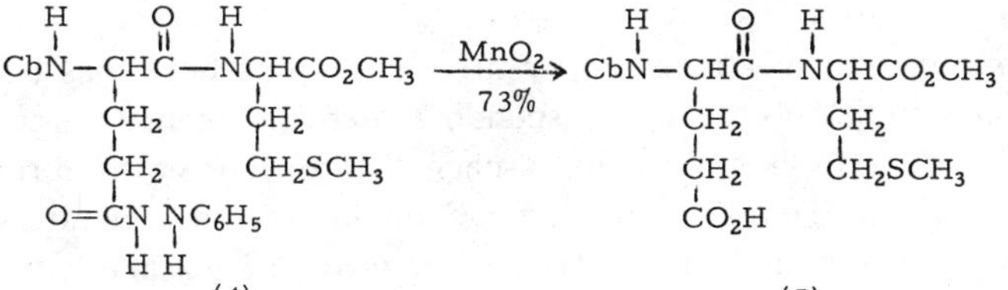

(4) (5)

Henbest[24] found that active MnO_2, used in large amounts, oxidizes N-alkyl- and N,N-dialkylanilines in three ways:

$$>NCH_3 \longrightarrow >NCHO$$

$$>NCH_2R \longrightarrow >NH + O{=}CHR$$

$$>NCH_2CH_2R \longrightarrow [>NCH{=}CHR] \longrightarrow >NCHO + O{=}CHR$$

With many amines the reaction follows a single course; for example, dimethylaniline affords N-methylformanilide in 80% yield. Tri-*n*-alkylamines are oxidized to N-formyldialkylamines:[25]

$$(CH_3CH_2CH_2CH_2)_3N \xrightarrow[40\%]{MnO_2} (CH_3CH_2CH_2CH_2)_2NCHO$$

N,N′-Diphenylpyrazolidine (2) is obtained by oxidation of 1,3-dianilinopropane with active MnO_2. Wittig and co-workers,[26] with Attenburrow oxide, obtained (2) in 31% yield; Daniels and Martin[27] employed MnO_2 prepared according to Henbest, Jones, and Owen[28] and raised the yield to 45%.

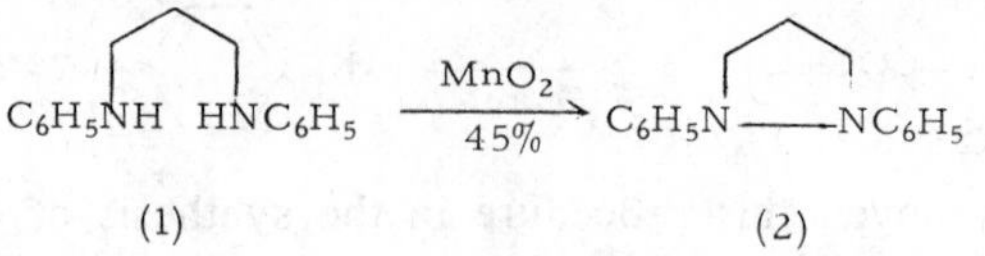

Jansen *et al.*[29] found Attenburrow MnO_2 to be the best reagent tried for dehydrogenation of indoline and related derivatives such as (1) to the indoles (2). With

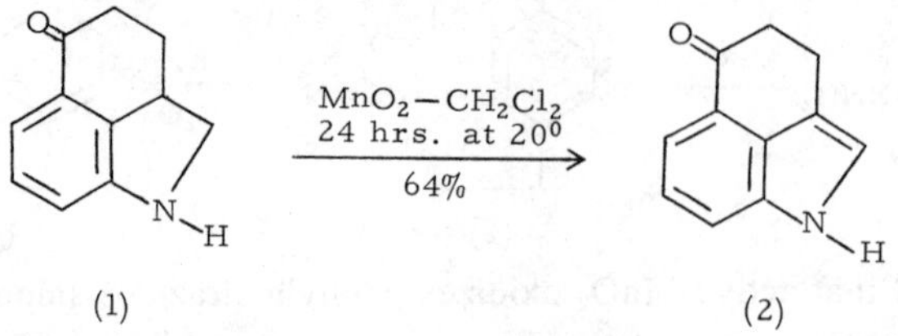

nickel peroxide, the next best reagent, the yield was only 33%. Reimlinger[30] regards active MnO_2 and silver oxide as the best reagents for oxidation of diarylhydrazones to diaryldiazomethanes; other oxidants give azines as by-products.

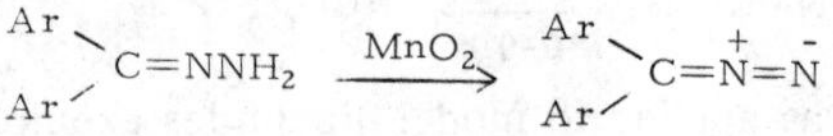

Alkyl sulfides. Di-*n*-butyl sulfide and dibenzyl sulfide have been oxidized to the sulfoxides by prolonged shaking of a petroleum ether solution with active MnO_2.[31] Diallyl sulfoxide was obtained by this method in 13% yield.

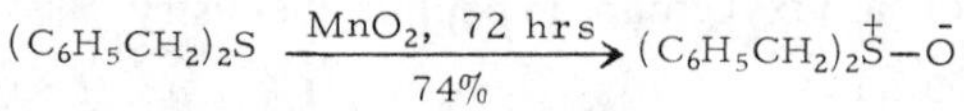

Miscellaneous. Reagent prepared by the Syntex procedure[5] has been used to oxidize pyridine methanols to the corresponding aldehydes, mercaptans to disulfides, aliphatic α-ketols to α-diketones, and N-phenylhydroxylamine to nitrosobenzene. A suspension of the dioxide in an ethereal solution of the substrate was stirred vigorously for 5–6 hrs. and the oxide was removed, and washed with ether. The filtrate and washings were concentrated under reduced pressure and the product isolated by distillation or crystallization.

[1]S. Ball, T. W. Goodwin, and R. A. Morton, *Biochem. J.*, **42**, 516 (1948)
[2]N. L. Wendler, H. L. Slates, and M. Tishler, *Am. Soc.*, **71**, 3267 (1949)
[3]K. R. Farrar, J. C. Hamlet, H. B. Henbest, and E. R. H. Jones, *Chem. Ind.*, 49 (1951)
[4]J. Attenburrow, A. F. B. Cameron, J. H. Chapman, R. M. Evans, B. A. Hems, A. B. A. Jansen, and T. Walker, *J. Chem. Soc.*, 1094 (1952)
[5]O. Mancera, G. Rosenkranz, and F. Sondheimer, *J. Chem. Soc.*, 2189 (1953); F. Sondheimer, O. Mancera, M. Urquiza, and G. Rosenkranz, *Am. Soc.*, **77**, 4145 (1955)
[6]M. Harfenist, A. Bavley, and W. A. Lazier, *J. Org.*, **19**, 1608 (1954)
[7]R. L. Gritter and T. J. Wallace, *J. Org.*, **24**, 1051 (1959)
[8]E. A. Braude and J. A. Coles, *J. Chem. Soc.*, 2014 (1950); 1430 (1952)
[9]L. Crombie and J. Crossley, *J. Chem. Soc.*, 4983 (1963)
[10]G. Stork and M. Tomasz, *Am. Soc.*, **86**, 471 (1964)
[11]E. F. Pratt and J. F. Van de Castle, *J. Org.*, **26**, 2973 (1961)
[12]F. Sondheimer, C. Amendolla, and G. Rosenkranz, *Am. Soc.*, **75**, 5932 (1953)
[13]I. T. Harrison, *Proc. Chem. Soc.*, 110 (1964)
[14]A. Nickon and J. F. Bagli, *Am. Soc.*, **83**, 1498 (1961); A. Nickon, N. Schwartz, J. B. DiGiorgio, and D. A. Widdowson, *J. Org.*, **30**, 1711 (1965)

[15]E. Adler and H.-D. Becker, *Chem. Scand.*, **15**, 849 (1961); *see also* E. P. Papadopoulos, A. Jarrar, and C. H. Issidorides, *J. Org.*, **31**, 615 (1966)
[16]I. Bell, E. R. H. Jones, and M. C. Whiting, *J. Chem. Soc.*, 1313 (1958)
[17]H.-D. Becker, *J. Org.*, **29**, 3068 (1964)
[18]R. W. Murray and A. M. Trozzolo, *J. Org.*, **26**, 3109 (1961)
[19]E. F. Pratt and T. P. McGovern, *J. Org.*, **29**, 1540 (1964)
[20]O. H. Wheeler and D. Gonzalez, *Tetrahedron*, **20**, 189 (1964)
[21]H. Morrison, S. Danishefsky, and P. Yates, *J. Org.*, **26**, 2617 (1961)
[22]N. L. Allinger, L. A. Freiberg, R. B. Hermann, and M. A. Miller, *Am. Soc.*, **85**, 1171 (1963)
[23]R. B. Kelly, *J. Org.*, **28**, 453 (1963); R. B. Kelly, G. R. Umbreit, and W. R. Liggett, *ibid.*, **29**, 1273 (1964)
[24]H. B. Henbest and A. Thomas, *J. Chem. Soc.*, 3032 (1957)
[25]H. B. Henbest and M. J. W. Stratford, *Chem. Ind.*, 1170 (1961)
[26]G. Wittig, W. Joos, and P. Rathfelder, *Ann.*, **610**, 180 (1957)
[27]R. Daniels and B. D. Martin, *J. Org.*, **27**, 178 (1962)
[28]H. B. Henbest, E. R. H. Jones, and T. C. Owen, *J. Chem. Soc.*, 4909 (1957)
[29]A. B. A. Jansen, J. M. Johnson, and J. R. Surtees, *J. Chem. Soc.*, 5573 (1964)
[30]H. Reimlinger, *Ber.*, **97**, 3493 (1964)
[31]D. Edwards and J. B. Stenlake *J. Chem. Soc.*, 3272 (1954)

β-Mercaptoethanol, $HOCH_2CH_2SH$. Mol. wt. 78.13, b.p. 155°. Suppliers: Trylon Chem., A, E, MCB.

Djerassi[1] introduced this reagent for conversion of steroid ketones into the ethylenehemithioketals, derivatives similar to ethylenethioketals but more easily

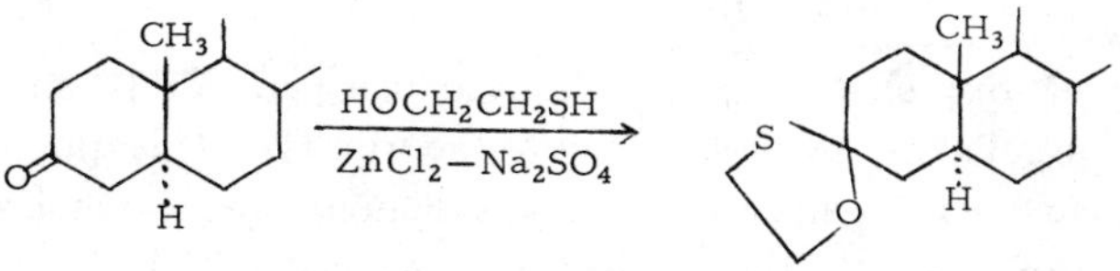
CH3
HOCH2CH2SH
ZnCl2—Na2SO4
H
CH3
S
O
H

reconverted to the ketone (by Raney nickel).[2] In the original procedure for saturated ketones a solution of the ketone in dioxane is treated with freshly fused zinc chloride and anhydrous sodium sulfate and let stand for about 20 hrs. at room temperature. In the case of Δ^4-3-ketones the more powerful catalyst *p*-toluenesulfonic acid is used, but yields are low.

A simpler procedure is to use boron trifluoride etherate as catalyst, either alone or in acetic acid solution.[3] Thus a solution of 300 mg. of androstenolone

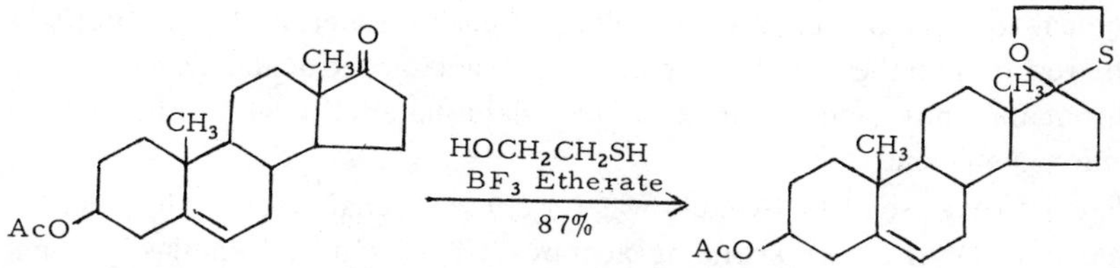
O
CH3
CH3
AcO
HOCH2CH2SH
BF3 Etherate
87%
O
S
CH3
CH3
AcO

acetate, 0.3 ml. of β-mercaptoethanol, and 0.3 ml. of boron trifluoride etherate in acetic acid at 25° soon deposited crystals. A little water was added and the crystals were collected and washed with methanol: 310 mg. of product, m.p. 166–170°. One crystallization from 95% ethanol gave pure material, m.p. 183–184°.

[1]J. Romo, G. Rosenkranz, and C. Djerassi, *Am. Soc.*, **73**, 4961 (1951)
[2]Review: C. Djerassi, "Steroid Reactions," 21–34, Holden-Day (1963)
[3]L. F. Fieser, *Am. Soc.*, **76**, 1945 (1954)

Mercuric acetate, $Hg(OCOCH_3)_2$. Mol. wt. 318.70. Solubility, g. in 100 g. H_2O: $25^{10°}$, $100^{100°}$. Suppliers: B, F, MCB.

Dehydrogenation. Windaus introduced use of mercuric acetate as a reagent for the dehydrogenation of specific unsaturated compounds early in his classical investigations of ergosterol and the D vitamins. He had converted ergosterol (I)

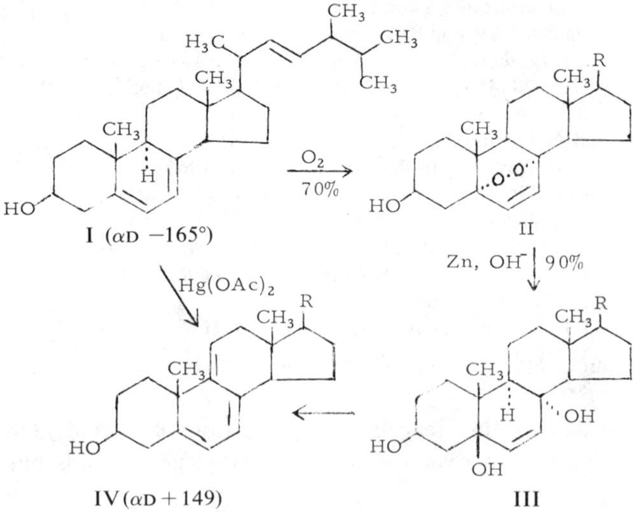

I (αD −165°)
O_2
70%
II
Zn, OH⁻
90%
Hg(OAc)₂
IV (αD + 149)
III

by photochemical oxygenation into the transannular peroxide II[1] and reduced this with zinc dust in refluxing alcoholic alkali to the triol III.[2] This triol has two allylic tertiary alcoholic functions and on vacuum distillation it lost two moles of water and afforded dehydroergosterol (IV). This triene seemed of interest for studies of irradiation, acid isomerization, and other reactions, but the route through the peroxide was not suitable for operation on a large scale. A trial to prepare it directly from ergosterol by reaction with mercuric acetate evidently was undertaken without much hope of success, for the report states that "Mercuri acetat wirkt merkwürdigerweise dehydrierend auf Ergosterin. . . " A solution of mercuric acetate in ethanol containing a little acetic acid was added to a refluxing solution of 25 g. of ergosterol in 1250 ml. of ethanol, and the mixture was refluxed for 40 min. Purification without benefit of chromatography afforded in 40% yield material judged to be pure from the optical rotatory power (see αD values in chart). Later, an ETH group[3] carried out the reaction on the acetate in refluxing dioxane–acetic acid, purified the product by chromatography, and obtained in 37% yield material judged to be pure by the extinction coefficient.

Windaus' next use of the reagent was for dehydrogenation of 5-dihydroergosterol to ergosterol-D; the corresponding acetates are shown in formulas (1) and (4).[4] In the later work on the cortisone problem, ergosterol-D and $\Delta^{7,9(11)}$-dienes of the cholesterol, bile acid, and sapogenin series acquired importance as key intermediates in several routes to 11-oxygenated steroids and considerable work was done to improve the procedure for conversion of Δ^7-enes to $\Delta^{7,9(11)}$-dienes. A Merck group[5] found evidence that the reactions involve a mercurated steroid such as (2), but that a part of the diene (4) comes from a thermolabile mercury-free intermediate isolable under certain reaction conditions and characterized as (5). In their procedure a

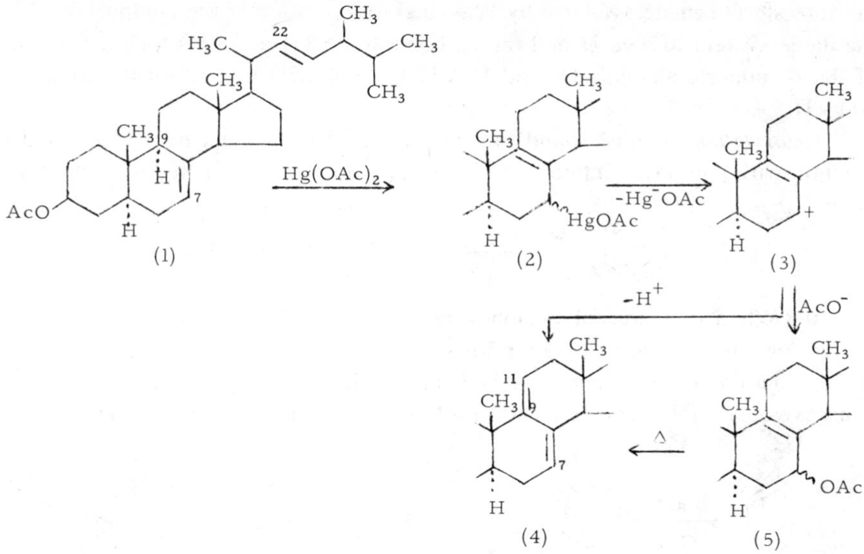

solution of 200 g. (0.44 mole) of $\Delta^{7,22}$-ergostadiene-3β-ol acetate (1) in 1.1 l. of chloroform is treated with a solution of 0.91 mole of mercuric acetate in 1670 ml. of acetic acid, and the suspension is stirred at 25–30° for 23 hrs. The mixture is filtered from mercurous acetate (98% yield) and concentrated to 600 ml. at 70–80° in vacuum; the long distillation promotes complete conversion of (5) to (4). Crystallization at 25° affords diene (4) of 76.5% purity in yield of 78%. This material is suitable for most purposes. Further purification is extremely wasteful.

An ETH group[6] carried out the same reaction by refluxing (1) with mercuric acetate in dioxane–acetic acid under nitrogen. The reaction was complete in 10 min. and afforded in 70% yield crude material of 84% purity and suitable for the purpose at hand.

A third reaction studied by Windaus[7] was the dehydrogenation of the *p*-nitrobenzoate (I) of isodehydrocholesterol ($\Delta^{6,8}$-cholestadiene-3β-ol), a by-product in the commercial production of the $\Delta^{5,7}$-diene required for conversion to vitamin D_3. The reaction was later reinvestigated by Barton and Rosenfelder[8] with a sample of

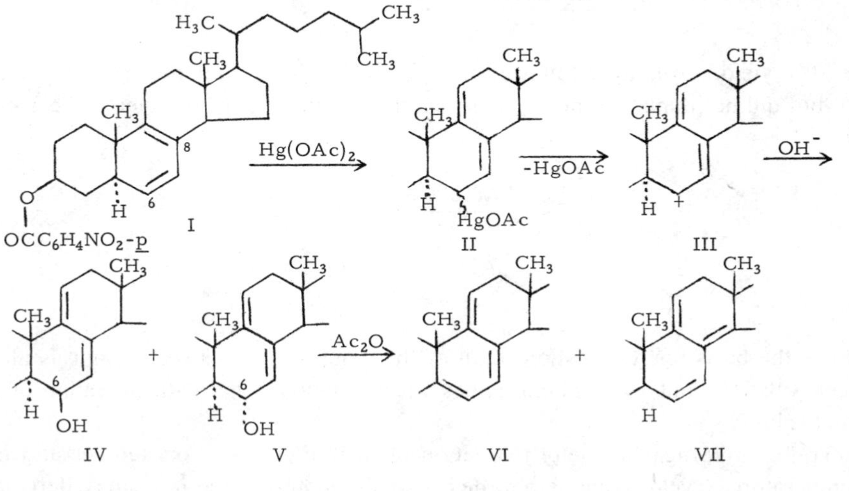

the rare steroid made available by Windaus. Mercuration at the unhindered end of the diene system to give II and the carbonium ion III accounts for the formation of the 6-epimeric alcohols IV and V, which on dehydration afford the trienes VI and VII.

Acetoxylation. Treibs[9] found that olefins and ketones can be acetoxylated by reaction with mercuric acetate. Initially the reaction was conducted without solvent

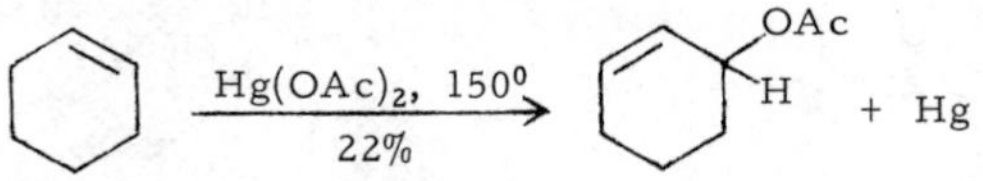

at 150°. When 0.2 mole of cyclohexene and 0.2 mole of mercuric acetate were heated together in a bomb tube for 2 hrs. metallic mercury separated in 100% yield, and distillation of the organic material from considerable tar afforded a little benzene and gave Δ^1-cyclohexenyl-3-acetate in 22% yield. The low yield is characteristic;

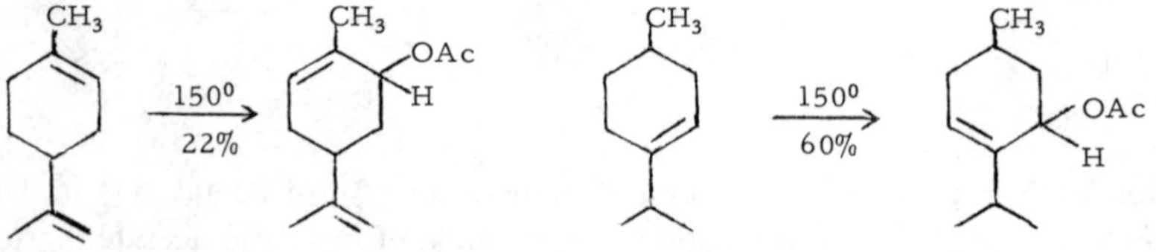

(1) (2) (3) (4)

with unsymmetrical olefins, results are variable. (+)-Limonene (1) yields the (+)-acetoxy derivative,[9] but (+)-menthene-3 (3) is oxidized to the (±)-ester.[10]

The oxidation of (3) was found to proceed in two stages, only the second of

1. $C_{10}H_{18} + 2\,Hg(OCOCH_3)_2 \longrightarrow C_{10}H_{17}OCOCH_3 + 2\,HgOCOCH_3 + CH_3CO_2H$

2. $C_{10}H_{18} + 2\,HgOCOCH_3 \longrightarrow C_{10}H_{17}OCOCH_3 + 2\,Hg + CH_3CO_2H$

which requires a temperature as high as 150°. Thus oxidation of 1-phenylcyclohexene was carried out by stirring 0.1 mole of hydrocarbon with 0.2 mole of mercuric acetate in acetic acid at 95°, and the acetoxy derivative was obtained

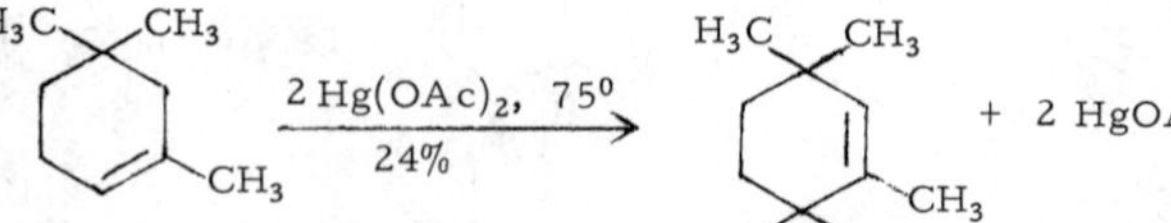

in 70% yield.[11] Alkonyi[12] followed this procedure for oxidation of α-cyclogeraniol (5) but did not duplicate the yield; his characterization of the product as the acetate

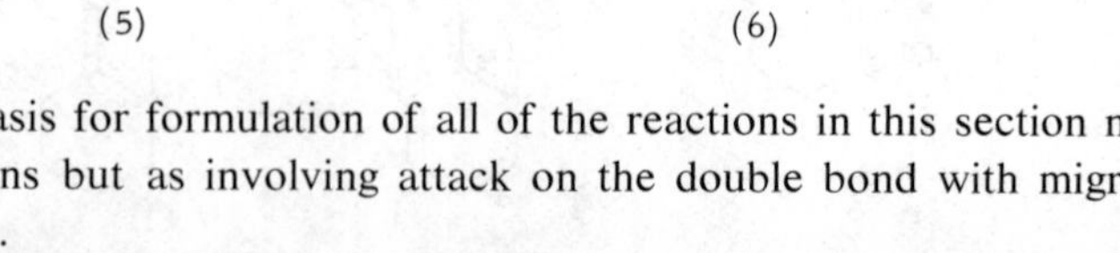

(5) (6)

(6) is the basis for formulation of all of the reactions in this section not as allylic acetoxylations but as involving attack on the double bond with migration of the double bond.

Treibs[9] investigated briefly the action of mercuric acetate on ketones at a high temperature. Cyclohexanone afforded a small amount of the α-acetoxy derivative

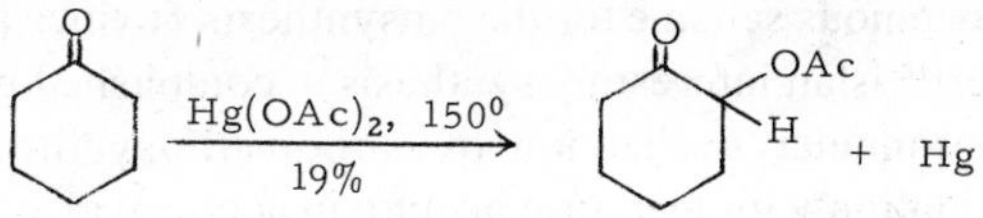
Hg(OAc)2, 150°
19%
OAc
H
+ Hg

along with considerable tar and some phenyl acetate. Under the same conditions tetralin and indane were not attacked.

Dehydrogenation of tertiary amines. Leonard[13-15] found that cyclic tertiary amines can be oxidized to enamines with 4 mole equivalents of mercuric acetate in 5% aqueous acetic acid at steam-bath temperature. When quinolizidine was treated in

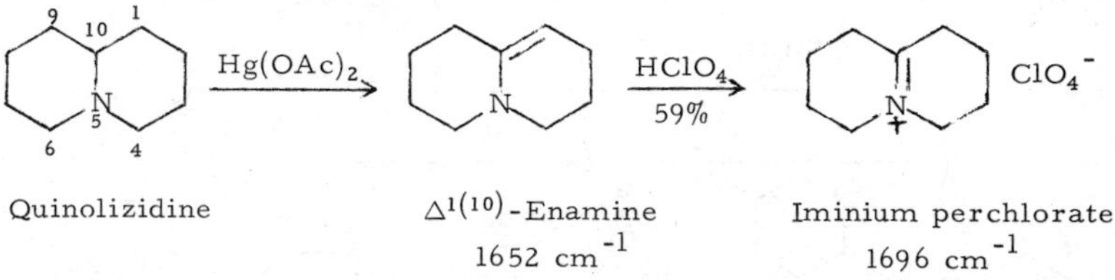
9
10
1
N
5
6
4
Hg(OAc)2
N
HClO4
59%
N
+
ClO4-
Quinolizidine
Δ1(10)-Enamine
1652 cm-1
Iminium perchlorate
1696 cm-1

this way for 1.5 hrs., the mercurous acetate that separated corresponded to 92% of that calculated for removal of two hydrogen atoms, and the yield of $\Delta^{1(10)}$-dehydroquinolizidine, isolated as the iminium perchlorate, was 59%.[13] Note that the enamine and the iminium ion are differentiated sharply by the IR spectrum. Leonard postulates initial formation of the mercurated complex (2) through the π-electrons on

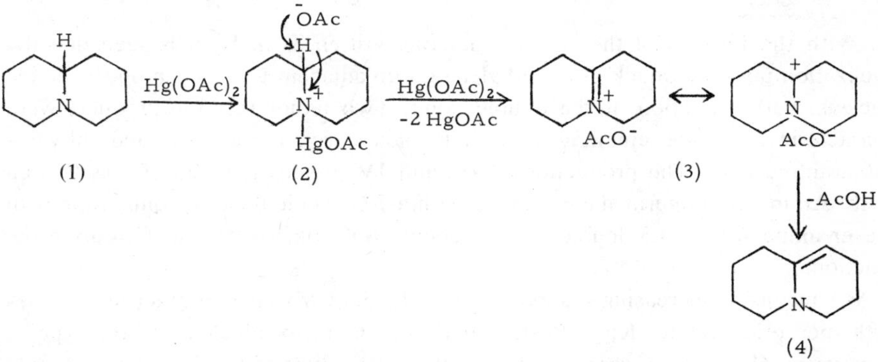
H
N
Hg(OAc)2
OAc
H
N
+
HgOAc
Hg(OAc)2
-2 HgOAc
N
+
AcO-
N
+
AcO-
-AcOH
N

(1) (2) (3) (4)

the nitrogen. Abstraction of the proton from the tertiary carbon is a concerted process with cleavage of the nitrogen-mercury bond and separation of HgOAc, which reacts very rapidly with mercuric acetate to give insoluble mercurous acetate. The amine salt remains in solution as the hybrid (3), and on basification a proton is abstracted from C_1 to give the $\Delta^{1(10)}$-enamine (4).

In the case of *trans*-1-methyldecahydroquinoline (5), reaction with 4 moles of mercuric acetate in 5% acetic acid (30 min. at 90°) dehydrogenation was followed by hydroxylation, and the product obtained after basification was 10-hydroxy-1-methyl-Δ^8-octahydroquinoline (6).[15] The *cis* isomer of (5) also gave (6).

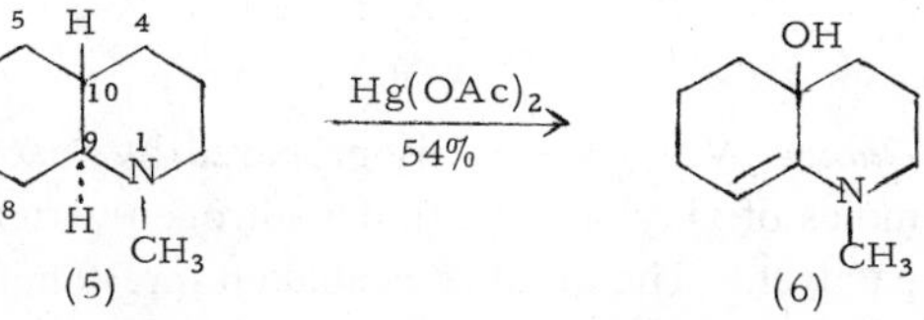
5
H
4
10
9
1
N
8
H
CH3
Hg(OAc)2
54%
OH
N
CH3

(5) (6)

In support of an ingenious scheme for the biosynthesis of vinca and iboga alkaloids proposed by Wenkert[16] is an interesting synthesis accomplished by Kutney, Brown, and Piers[17] by transannular cyclization of carbomethoxydihydrocleavamine (I), accomplished by oxidation with mercuric acetate in acetic acid at room temperature. Chromatography afforded as the main product the vinca alkaloid vincadifformine,

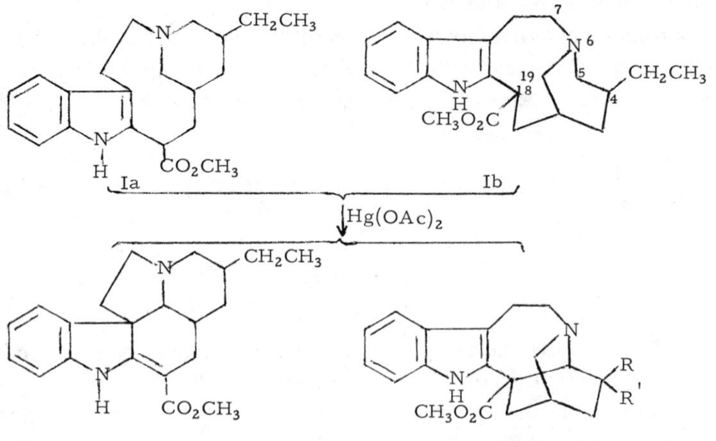
CH2CH3
N
H
CO2CH3
Ia
Ib
7
6
19
18
5
4
CH3O2C
Hg(OAc)2
R
R'

II

III R = C_2H_5, R' = H
IV R = H, R' = C_2H_5

II. With the formula of the starting material written as in Ia, it is seen that the cyclization involves attack of a methylene group adjacent to nitrogen on the indole nucleus, with migration of the double bond. Two minor reaction products were characterized as the epimeric iboga alkaloids coronaridine (III) and dihydrocatharantine (IV). The production of III and IV implies reaction of I as written in Ib. An iminium intermediate from Ib with a 5,6-double bond in equilibrium with the enamine with a 4,5-double bond accounts for isomerization at C_4 during the reaction.

As a means of decreasing side reactions in the dehydrogenation of tertiary amines with mercuric acetate, Knabe[18] studied the oxidation of alkaloids of the type of papaverine (1) in the presence of an equivalent amount of the sequestering agent ethylenediamine tetraacetic acid (EDTA) as the disodium salt. The products were the papaverinium salt (2) and the dihydro salt (3) in the yields noted. In the absence of the complexing agent, papaverine is cleaved at the methylene bridge.

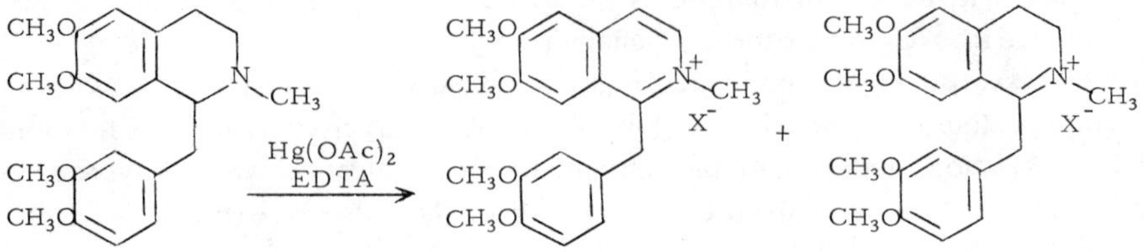
CH3O
N
CH3
Hg(OAc)2
EDTA
X-
+

(1) (2) 80% (3) 20%

Vinyl ester interchange. Vinyl laurate is prepared by reaction of 0.4 mole of lauric acid with 2.4 moles of vinyl acetate under nitrogen with addition of 1.6 g. of mercuric acetate as catalyst.[19] The mixture is shaken for 30 min., 0.15 ml. of 100%

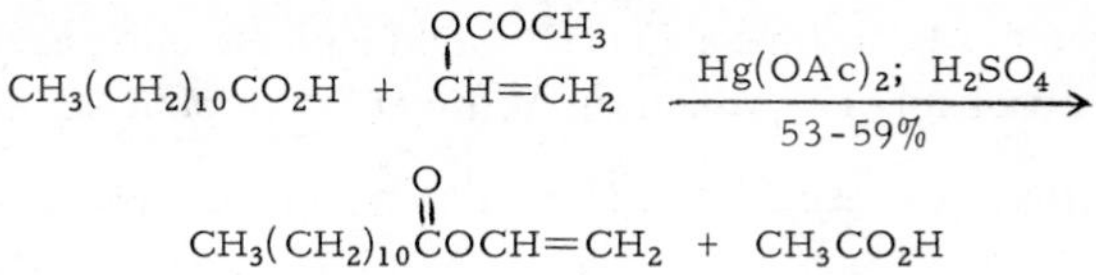

$$CH_3(CH_2)_{10}CO_2H \ + \ \overset{\overset{\displaystyle OCOCH_3}{|}}{CH}=CH_2 \xrightarrow[53-59\%]{Hg(OAc)_2; \ H_2SO_4}$$

$$CH_3(CH_2)_{10}\overset{\overset{\displaystyle O}{\|}}{C}OCH=CH_2 \ + \ CH_3CO_2H$$

sulfuric acid is added, and the mixture is refluxed for 3 hrs. The exchange reaction was studied in some detail by Watanabe and Conlon,[20] who found that only mercuric salts of weak acids, acetic and benzoic acid, are specific catalysts for the reaction. Burgstahler states[21] that the purity of the catalyst is extremely critical and recommends that the material be crystallized from absolute ethanol and then dried thoroughly. Burgstahler uses the exchange reaction as one step in a method for the introduction of an angular substituent into a cyclic system. Thus the vinyl ether (3) on Claisen rearrangement gives the unsaturated aldehyde (4). The reaction, which

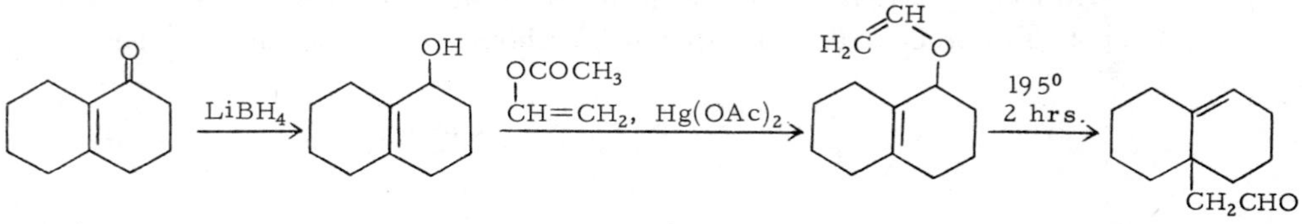

(1)　　　　　　　　　(2)　　　　　　　　　　　(3)　　　　　　　　(4)

is stereospecific, has been applied also to Δ^4-cholestene-3β-ol[21] and to the synthesis of a product of degradation of atisine.[22] Büchi and White[23] used the exchange reaction for the preparation of β-cyclogeranyl vinyl ether (6), an intermediate in the total synthesis of ($\pm$)-thujopsene.

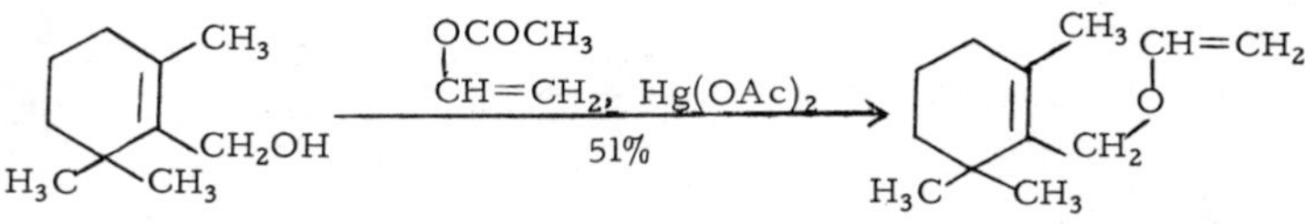

(5)　　　　　　　　　　　　　　　　　　　　　　(6)

Mercurated aromatics. *o*-Chloromercuriphenol is prepared by heating phenol with mercuric acetate in boiling water and adding sodium chloride to cleave the initially formed *o*-acetoxymercuri compound (Whitmore *et al.*[24]).

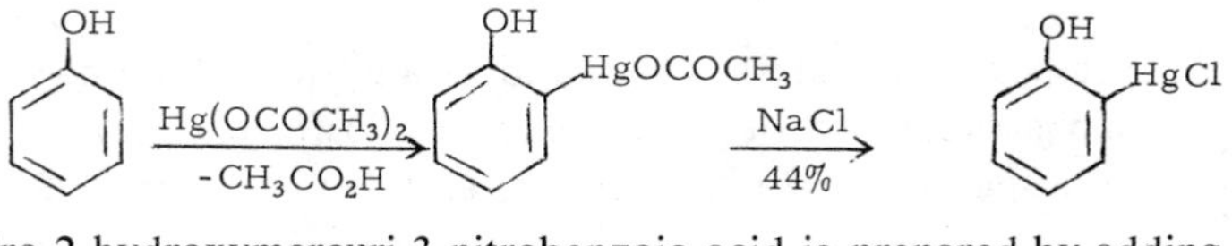

Anhydro-2-hydroxymercuri-3-nitrobenzoic acid is prepared by adding a solution of 1.1 moles of mercuric acetate in dilute acetic acid to an aqueous solution of the disodium salt of 3-nitrophthalic acid, gradually raising the temperature, and heating at 165–175° for about 70 hrs.[25]

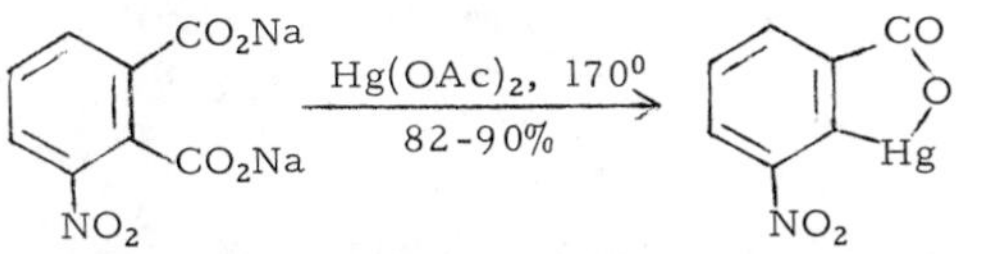

Reaction with α,β-unsaturated acids. A reaction introduced by Abderhalden and Heyns[26] and developed by Carter[27] is of particular value for the synthesis of serine[28] and threonine.[29] Crotonic acid (1), treated with mercuric acetate in methanol, forms an adduct, probably (2), which on treatment with potassium bromide to give

$$CH_3CH=CHCO_2H \xrightarrow{Hg(OAc)_2-CH_3OH} \begin{bmatrix} CH_3CH-CHCO_2H \\ \quad | \qquad | \\ \quad OCH_3 \ \ HgOAc \end{bmatrix} \xrightarrow{KBr}$$

(1)   (2)

$$\begin{bmatrix} CH_3CH-CHCO_2H \\ \quad | \qquad | \\ \quad OCH_3 \ \ HgBr \end{bmatrix} \xrightarrow[75-85\%]{Br_2} \begin{array}{c} CH_3CH-CHCO_2H \\ \quad | \qquad | \\ \quad OCH_3 \ \ Br \end{array}$$

(3)   (4)

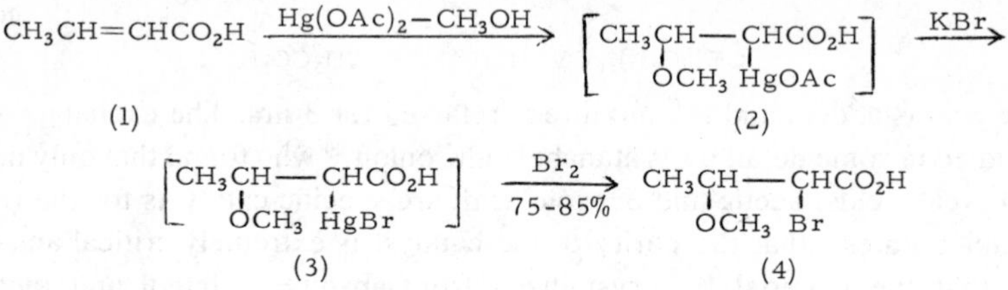

(3), followed by bromination, yields α-bromo-β-methoxy-n-butyric acid (4), convertible into DL-threonine. Methyl acrylate is the starting material for a similar synthesis of a precursor of DL-serine.

Hydration of acetylenes. Although the direct addition of water to an acetylenic compound to produce the corresponding carbonyl compound can be achieved in a

$$\begin{array}{c} | \\ C \\ \| \| \\ C \\ | \end{array} \xrightarrow{HOH(H^+)} \begin{array}{c} | \\ C-OH \\ \| \\ CH \\ | \end{array} \longrightarrow \begin{array}{c} | \\ C=O \\ | \\ CH_2 \\ | \end{array}$$

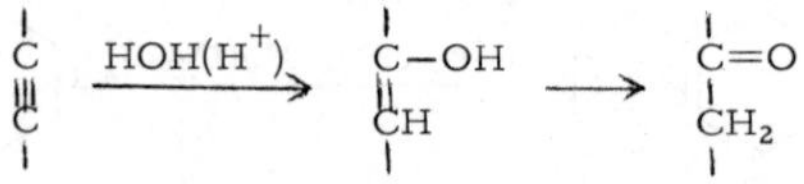

few cases by treatment with hot acid alone, the preferred procedures use either a catalytic amount of mercuric sulfate in dilute sulfuric acid or 2 moles of mercuric acetate. In the latter case an intermediate mercury complex is formed and is cleaved to the carbonyl compound by hydrochloric acid or by hydrogen sulfide. Myddleton[30] isolated some of the complexes and from their properties deduced the following formulation:

1. $RC\equiv CR + 2\,Hg(OCOCH_3)_2 + H_2O \longrightarrow \begin{array}{c} OHgOCOCH_3 \\ | \\ R-C=C-R \\ | \\ HgOCOCH_3 \end{array} + 2\,CH_3CO_2H$

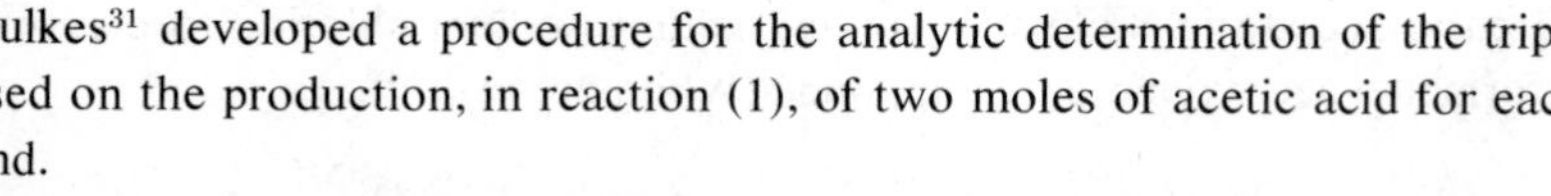

2. $\begin{array}{c} OHgOCOCH_3 \\ | \\ R-C=C-R \\ | \\ HgOCOCH_3 \end{array} \xrightarrow{H_2S} \begin{bmatrix} OH \\ | \\ RC=CHR \longrightarrow RCCH_2R \\ \qquad\qquad\quad \| \\ \qquad\qquad\quad O \end{bmatrix} + 2HgS + 2\,CH_3CO_2H$

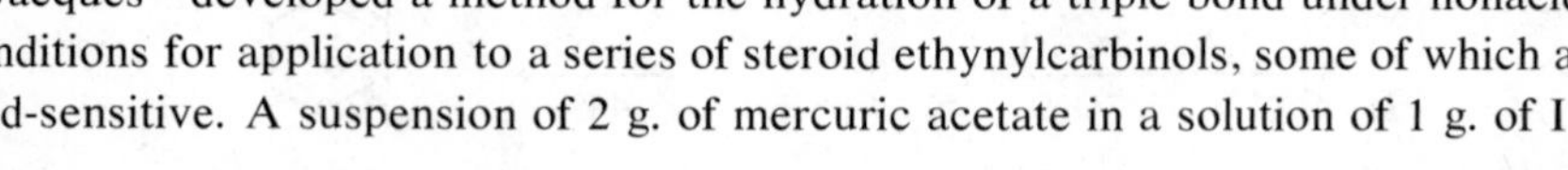

Koulkes[31] developed a procedure for the analytic determination of the triple bond based on the production, in reaction (1), of two moles of acetic acid for each triple bond.

Jacques[32] developed a method for the hydration of a triple bond under nonacidic conditions for application to a series of steroid ethynylcarbinols, some of which are acid-sensitive. A suspension of 2 g. of mercuric acetate in a solution of 1 g. of I in

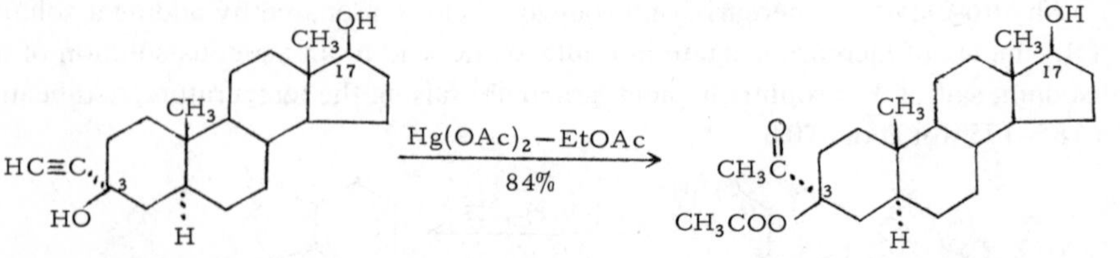

100 ml. of ethyl acetate was stirred at 20° for 24 hrs., hydrogen sulfide was introduced to precipitate mercury compounds, and working up of the filtrate afforded the 3-acetyl-3-acetoxy derivative II in 84% yield. The method proved applicable to

the acid-sensitive 17-ethyleneketal of the 17-ketone corresponding to I. Note that the tertiary alcoholic function at C_3 becomes acetylated, whereas the secondary 17β-hydroxyl of I is unattacked. Intermediate formation of the mercury complex (1) possibly accounts for the result, for migration of an acetyl group can occur via cyclization to (2), with ultimate formation of the *gem*-acetyl-acetoxy compound (4).

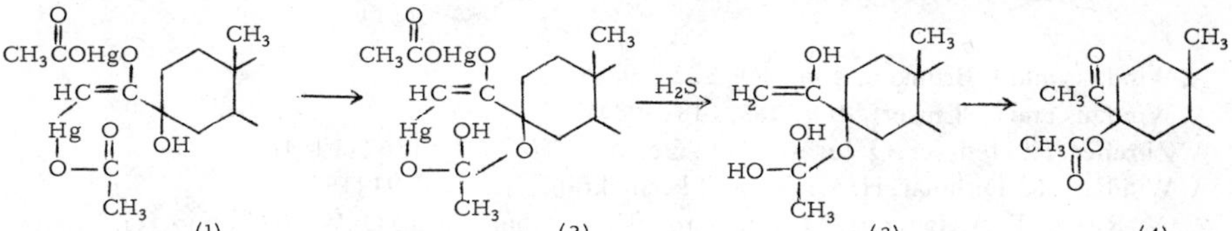
H₂S

(1) (2) (3) (4)

Bromination catalyst. Griseofulvin (1) in acetic acid solution does not react with bromine under ordinary conditions. However, when treated with bromine in the

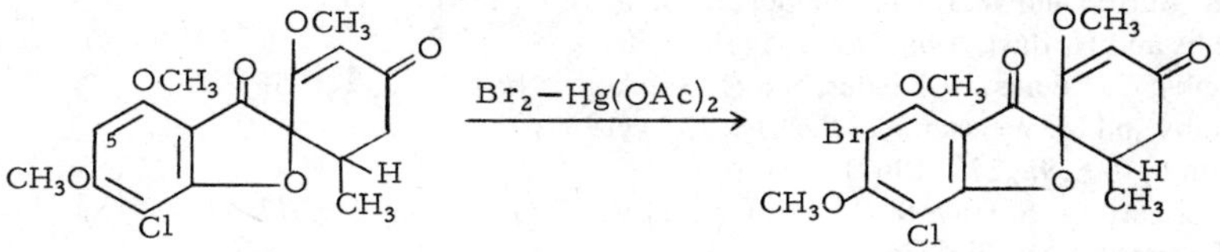
Br₂–Hg(OAc)₂

(1) (2)

presence of 0.5 mole of mercuric acetate it affords the 5-bromo derivative (2).[33] Presumably (1) suffers displacement to the 5-acetoxymercuri derivative, followed by displacement by bromine.

Replacement of halogen in sugar derivatives. Wolfrom *et al.*[34] describes mercuric acetate as an "excellent reagent" for replacement of halogen in acetylated glycopyranosyl halides. Since α-D-glycosyl halides and β-D-glycosyl halides both give β-D-acetates, the steric course appears to be controlled by the ortho ester effect.[35]

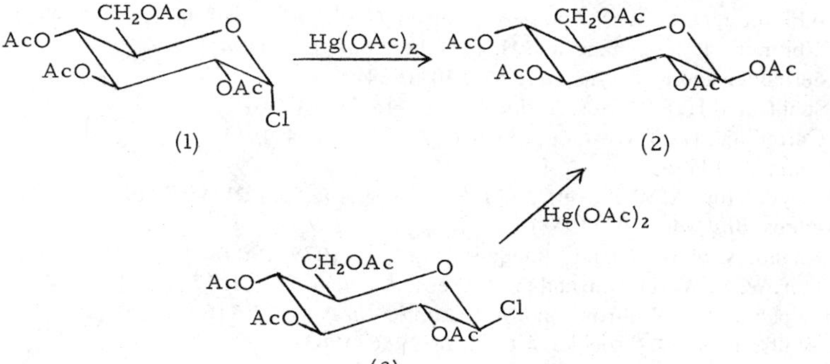
Hg(OAc)₂
Hg(OAc)₂

(1) (2) (3)

Addition to olefins. Mercuric acetate reacts with olefins in alcoholic solution by typical electrophilic addition:[36]

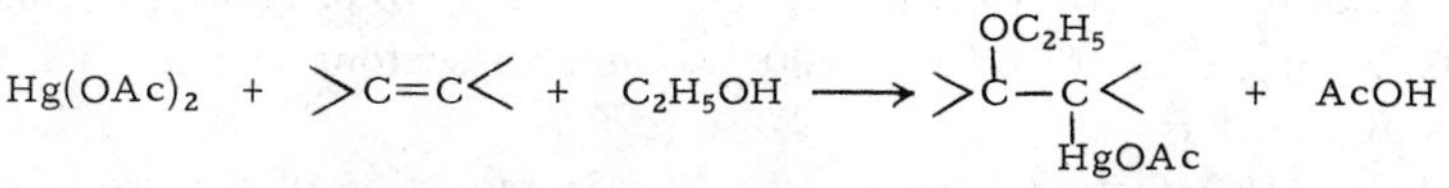

In the case of simple olefins the reaction proceeds by *trans* addition, as illustrated for cyclohexene (1). However, Traylor[37] showed that in the case of norbornene (2) and other strained systems the reactions proceeds by *cis* addition.

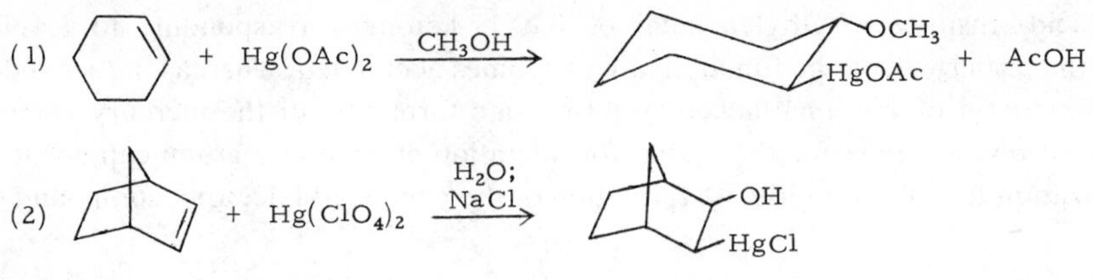

[1]A. Windaus and J. Brunken, *Ann.*, **460**, 225 (1928)
[2]A. Windaus and O. Linsert, *Ann.*, **465**, 148 (1928)
[3]A. Zürcher, H. Heusser, O. Jeger, and P. Geistlich, *Helv.*, **37**, 1562 (1954)
[4]A. Windaus, K. Dithmar, H. Mürke, and F. Suckfüll, *Ann.*, **488**, 91 (1931)
[5]W. V. Ruyle, T. A. Jacob, J. M. Chemerda, E. M. Chamberlin, D. W. Rosenburg, G. E. Sita, R. L. Erickson, L. M. Aliminosa, and M. Tishler, *Am. Soc.*, **75**, 2604 (1953)
[6]G. Saucy, P. Geistlich, R. Helbling, and H. Heusser, *Helv.*, **37**, 250 (1954)
[7]A. Windaus, U. Riemann, and G. Zühlsdorff, *Ann.*, **552**, 135, 142 (1942)
[8]D. H. R. Barton and W. J. Rosenfelder, *J. Chem. Soc.*, 2381 (1951)
[9]W. Treibs and H. Bast, *Ann.*, **561**, 165 (1949)
[10]W. Treibs, G. Lucius, H. Kögler, and H. Breslauer, *Ann.*, **581**, 59 (1953)
[11]W. Treibs and M. Weissenfels, *Ber.*, **93**, 1374 (1960)
[12]I. Alkonyi, *Ber.*, **95**, 279 (1962)
[13]N. J. Leonard, A. S. Hay, R. W. Fulmer, and V. W. Gash, *Am. Soc.*, **77**, 439 (1955)
[14]N. J. Leonard, R. W. Fulmer, and A. S. Hay, *Am. Soc.*, **78**, 3457 (1956)
[15]N. J. Leonard, L. A. Miller, and P. D. Thomas, *Am. Soc.*, **78**, 3463 (1956)
[16]E. Wenkert, *Am. Soc.*, **84**, 98 (1962)
[17]J. P. Kutney, R. T. Brown, and E. Piers, *Am. Soc.*, **86**, 2286, 2287 (1964)
[18]J. Knabe, *Arch. Pharm.*, **292**, 416 (1959); **293**, 121 (1960); J. Knabe and G. Grund, *ibid.*, **296**, 854 (1963); J. Knabe and H. Roloff, *Ber.*, **97**, 3452 (1964)
[19]D. Swern and E. F. Jordan, Jr., *Org. Syn., Coll. Vol.*, **4**, 977 (1963)
[20]W. H. Watanabe and L. E. Conlon, *Am. Soc.*, **79**, 2828 (1957)
[21]A. W. Burgstahler and I. C. Nordin, *ibid.*, **83**, 198 (1961)
[22]A. Ogiso and I. Iwai, *Chem. Pharm. Bull.*, **12**, 820 (1964)
[23]G. Büchi and J. D. White, *Am. Soc.*, **86**, 2884 (1964)
[24]F. C. Whitmore, H. D. West, and E. R. Hanson, *Org. Syn., Coll. Vol.*, **1**, 161 (1941)
[25]F. C. Whitmore, P. J. Culhane, and H. T. Neher, *ibid.*, **1**, 56 (1941)
[26]E. Abderhalden and K. Heyns, *Ber.*, **67**, 530 (1934)
[27]L. R. Schiltz and H. E. Carter, *J. Biol. Chem.*, **116**, 793 (1936)
[28]H. E. Carter, and H. D. West, *Org. Syn., Coll. Vol.*, **3**, 774 (1955)
[29]*Idem, ibid.*, **3**, 813 (1955)
[30]W. W. Myddleton, A. W. Barrett, and J. H. Seager, *Am. Soc.*, **52**, 4405 (1930)
[31]M. Koulkes, *Bull. soc.*, 402 (1953)
[32]H. B. Kagan, A. Marquet, and J. Jacques, *Bull. soc.*, 1079 (1960)
[33]T. Walker, W. K. Warburton, and G. B. Webb, *J. Chem. Soc.*, 1277 (1962)
[34]A. Thompson, M. L. Wolfrom, and M. Inatome, *Am. Soc.*, **77**, 3160 (1955)
[35]M. L. Wolfrom and W. Groebke, *J. Org.*, **28**, 2986 (1963)
[36]J. Chatt, *Chem. Rev.*, **48**, 7 (1951)
[37]T. G. Traylor and A. W. Baker, *Am. Soc.*, **85**, 2746 (1963); T. G. Traylor, *ibid.*, **86**, 245 (1964)

Mercuric chloride, $HgCl_2$. Mol. wt. 271.52, m.p. 277°, b.p. 304°, solubility, g. in 100 parts H_2O: 3.6 at 0°, 61.3 at 100°; soluble in acetone and in 95% ethanol. Suppliers: B, F, MCB.

Amalgams. In effecting Clemmensen-Martin reduction of β-benzoylpropionic acid to γ-phenylbutyric acid. Martin[1] shook a mixture of 120 g. of mossy zinc, 12 g. of mercuric chloride, 200 ml. of water, and 5–6 ml. of concd. hydrochloric acid for 5 min., decanted the liquid from the ***amalgamated zinc***, and added 75 ml. of

$$C_6H_5COCH_2CH_2CO_2H \xrightarrow[82\text{-}89\%]{Zn(Hg),\ HCl,\ Toluene} C_6H_5CH_2CH_2CH_2CO_2H$$

water, 175 ml. of concd. hydrochloric acid, 100 ml. of toluene, and 0.28 mole of keto acid. The mixture was refluxed vigorously for 25–30 hrs. with addition of three 50-ml. portions of concd. hydrochloric acid, and the toluene layer was worked up for isolation of the product. Procedures for Clemmensen reduction of vanillin to creosol[2] and of benzoin to *trans*-stilbene[3] omit use of hydrochloric acid in preparation of amalgam and are slower. In one, granulated zinc is let stand with 5% aqueous mercuric chloride solution for 2 hrs. with occasional shaking; in the other a mixture of zinc dust, water, and mercuric chloride is stirred mechanically for 20–30 min. That the yields of reduction products (60–67% and 53–57%) are lower than obtained by Martin is probably attributable to Martin's use of toluene.

In the reduction of acetone to pinacol,[4] ***magnesium amalgam*** is generated *in situ* by dropwise addition of a solution of 90 g. of mercuric chloride in 505 ml. of acetone to a suspension of 80 g. of magnesium turnings in 800 ml. of benzene.

Aluminum amalgam is prepared *in situ* for reaction with absolute ethanol[5] or *t*-butanol[6] to produce aluminum ethoxide or aluminum *t*-butoxide. In the first case 27 g. of aluminum filings is treated with 276 g. of absolute alcohol, 0.2 g. of mercuric chloride, and "several crystals of iodine," and the mixture is refluxed for several hours. With the higher-boiling alcohol, iodine catalysis is not necessary. A mixture of 64 g. of aluminum shavings, 254 ml. of *t*-butanol and 5–10 g. of aluminum *t*-butoxide is heated to boiling, and 0.4 g. of mercuric chloride is added. As heating is continued the mixture gradually becomes black, and heating is discontinued for 1 hr. Then 309 ml. of *t*-butanol and 200 ml. of benzene are added; eventually the mixture is refluxed for about 10 hrs.

Organomercurials. Nesmajanow[7] prepared β-naphthylmercuric chloride by diazotizing β-naphthylamine hydrochloride and adding to the solution a solution of 1 mole of mercuric chloride in concd. hydrochloric acid to precipitate a yellow complex salt. The complex was collected, washed with water and with acetone, air dried (*explosive*), and a suspension of the complex and copper powder in acetone

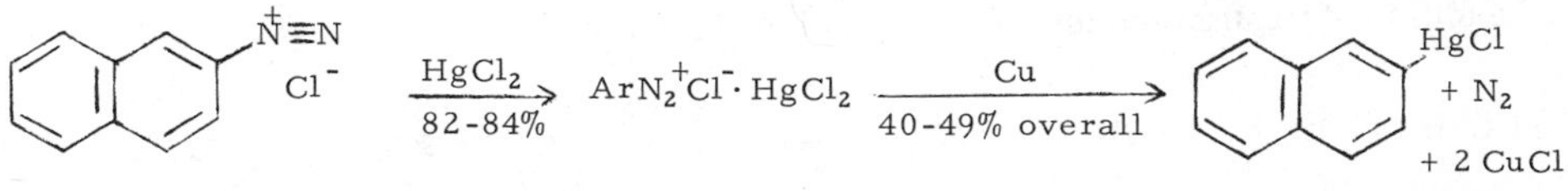
N≡N
Cl⁻
HgCl2
82-84%
ArN2+Cl−·HgCl2
Cu
40-49% overall
HgCl
+ N2
+ 2 CuCl

was stirred at 70° to effect smooth decomposition with liberation of nitrogen. The resulting solid was collected, and the product (m.p. 267°) was extracted with boiling xylene.

Whitmore and co-workers[8] prepared *p*-tolylmercuric chloride by heating an equimolecular mixture of sodium *p*-toluenesulfinate and mercuric chloride in boiling water until evolution of sulfur dioxide ceased (2 hrs.).

$$\text{p-}CH_3C_6H_4SO_2Na\cdot 2\,H_2O + HgCl_2 \xrightarrow[51\text{-}57\%]{} \text{p-}CH_3C_6H_4HgCl + SO_2 + NaCl + 2\,H_2O$$

Isolation of histidine. Histidine forms a sparingly soluble complex with mercuric chloride, whereas other amino acids from proteins do not, and can be precipitated in the form of this derivative.[9] Thus 1.4 kg. of commercial dried blood corpuscle paste is refluxed with 4.5 l. of concd. hydrochloric acid for 18–20 hrs. and most of the excess

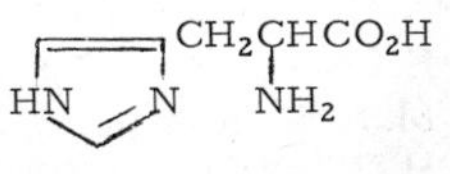
CH2CHCO2H
HN
N
NH2

acid is removed by distillation. The residue is taken up in 8 l. of water and the mixture is neutralized to pH 4.5 and let stand for precipitation of pigment. After filtration and clarification of the filtrate with Norit, the pale yellow solution is diluted to 25 l. and treated with a solution of 600 g. of mercuric chloride in 2 l. of hot 95% ethanol. The pH is adjusted to 7.0–7.5 by addition of sodium carbonate solution and the mixture let stand for separation of the complex. The liquor is siphoned off and the crock filled to the original volume with wash water. After one further washing the complex is collected, and a suspension in water is stirred during saturation with hydrogen sulfide. Mercuric sulfide is removed, the filtrate is concentrated to 1 l., clarified with Norit, and treated with 3 volumes of ethanol. After standing in an ice chest for 3–4 days, a crystallizate of L-histidine monohydrochloride is collected (85–90 g.). One recrystallization gives 75–80 g. of optically pure material.

Correction. An *Organic Syntheses* procedure[10] for converting 1.5 moles of cyclohexene into *trans*-2-chlorocyclohexanol calls for reaction with hypochlorous acid prepared by passing chlorine into iced sodium hydroxide solution after addition of 0.1 mole of mercuric chloride to produce a precipitate of mercuric oxide. The function of the mercuric oxide is not explained either in the procedure or in an early paper by Fortey[11] on which it is based. Fortey states that he obtained a yellowish oil which decomposes on attempted distillation at atmospheric pressure. Osterberg and Kendall[12] prepared a 2% hypochlorite solution by passing CO_2 into a suspension of bleaching powder and filtering off the calcium carbonate. Cyclohexene, when taken in excess and shaken with this solution, afforded *trans*-2-chlorocyclohexanol as a water-white oil which distilled at 760 mm. with only slight decomposition. We conclude that mercuric oxide serves no useful purpose and is actually deleterious.

Cleavage of diethyl mercaptals. E. Fischer[13] introduced use of mercuric chloride for the cleavage of penta-O-acetyl DL-glucose diethylmercaptal, and a slightly modified procedure by Wolfrom[14] has been widely adopted.[15] For example,[16] the diethylmercaptal is dissolved in aqueous acetone, an excess of washed cadmium carbonate is added with vigorous stirring, and then an excess of mercuric chloride in acetone is added gradually. The mixture is stirred at 25° for 24 hrs. with occasional addition of fresh cadmium carbonate and then refluxed briefly.

$$RCH(SC_2H_5)_2 + 2\ HgCl_2 + CdCO_3 \longrightarrow RCHO + 2\ ClHgSC_2H_5 + CdCl_2 + CO_2$$

[1] E. L. Martin, *Org. Syn., Coll. Vol.*, **2**, 499 (1943)
[2] R. Schwarz and H. Hering, *ibid.*, **4**, 203 (1963)
[3] R. L. Shriner and A. Berger, *ibid.*, **3**, 786 (1955)
[4] R. Adams and E. W. Adams, *ibid.*, **1**, 459 (1941)
[5] W. Chalmers, *ibid.*, **2**, 599 (1943)
[6] W. Wayne and H. Adkins, *ibid.*, **3**, 48 (1955)
[7] A. N. Nesmajanow, *ibid.*, **2**, 432 (1943)
[8] F. C. Whitmore, F. H. Hamilton, and N. Thurman, *ibid.*, **1**, 519 (1941)
[9] G. L. Foster and D. Shemin, *ibid.*, **2**, 330, (1943)
[10] G. H. Coleman and H. F. Johnstone, *ibid.*, **1**, 158 (1941)
[11] E. C. Fortey, *J. Chem. Soc.*, **73**, 932 (1898)
[12] A. E. Osterberg and E. C. Kendall, *Am. Soc.*, **42**, 2621 (1920)
[13] E. Fischer, *Ber.*, **27**, 673 (1894)
[14] M. L. Wolfrom, *Am. Soc.*, **51**, 2188 (1929)
[15] H. W. Arnold and W. L. Evans, *Am. Soc.*, **58**, 1950 (1936)
[16] J. English, Jr., and P. H. Griswold, Jr., *Am. Soc.*, **67**, 2040 (1945)

Mercuric cyanide, $Hg(CN)_2$. Mol. wt. 252.65. Supplier: KK.

Helferich[1] reported that mercuric cyanide is sometimes superior to silver carbonate as base in the synthesis of glycosides and disaccharides. Thus the condensation of tetra-O-benzoylbromoglucose (1 equiv.) with benzyl alcohol (1 equiv.) in nitromethane in the presence of mercuric cyanide (1 equiv.) gives tetra-O-benzoylbenzyl-β-D-glucose in 90% yield. Zorbach[2] used this general procedure, except that 1,2-dichloroethane was used as the solvent, in a successful synthesis of the cardiac-active

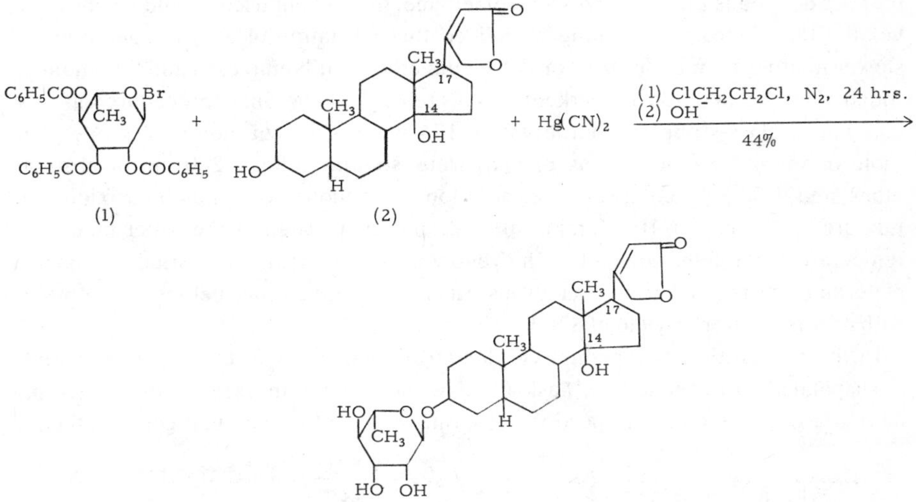

(1) (2) (3)

principle evomonoside (3) from the bromide (1) and digitoxigenin (2). After saponification, evomonoside (3) was isolated in 44% yield. An attempted synthesis with silver carbonate as base was attended with extensive elimination of the tertiary hydroxyl group at C_{14}.

[1]B. Helferich and K. Weis, *Ber.*, **89**, 314 (1956); B. Helferich and R. Steinpreis, *Ber.*, **91**, 1794 (1958)

[2]W. W. Zorbach, G. D. Valiaveedan. and D. V. Kashelikar, *J. Org.*, **27**, 1766 (1962)

Mercuric oxide (yellow, red), HgO. Mol. wt. 216.61, m.p. 630°.

Oxidation of hydrazones. Staudinger's[1] procedure for the preparation of diphenyldiazomethane was adapted with minor changes by *Organic Syntheses*.[2] A pressure

$$(C_6H_5)_2C{=}O \xrightarrow[87\%]{H_2NNH_2} (C_6H_5)_2C{=}NNH_2 \xrightarrow[89\text{-}96\%]{HgO} (C_6H_5)_2C{=}\overset{+}{N}{=}\overset{-}{N}$$

bottle charged with 0.1 mole of benzophenone hydrazone, 0.1 mole of yellow mercuric oxide, and 100 ml. of petroleum ether (b.p. 30–60°) is closed, wrapped in a wet towel, and shaken at room temperature for 6 hrs. The mixture is filtered to remove the mercury formed and a little benzophenone azine, and evaporated to dryness at reduced pressure at room temperature. The material so obtained melts at room temperature but is suitable for use as a reagent; the pure compound melts at 29–30°, but purification by crystallization is difficult.

A procedure for preparing diphenylketene[3] is by thermal decomposition of phenylbenzoyldiazomethane prepared by stirring a mixture of 0.25 mole of benzil monohydrazone, 0.38 mole of mercuric oxide, 35 g. of anhydrous calcium sulfate,

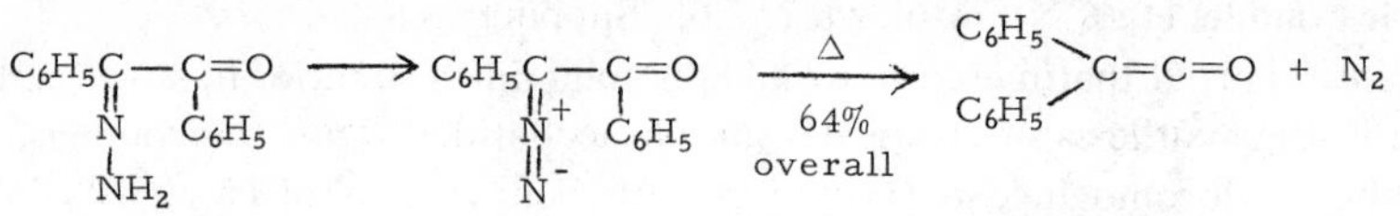

and 200 ml. of benzene at 25–35° for 4 hrs. The benzene solution is added dropwise to a distilling flask heated in a Wood's metal bath at 100–110°; the benzene flash distills, and the diazo compound is transformed into diphenylketene. In case the product desired is phenylbenzoyldiazomethane, the solvent used should be one more volatile than benzene. Staudinger[1] noticed that oxidation of a hydrazone is much slower in ether than in benzene or petroleum ether, but Nenitzescu and Solomonica[4] found that the reaction is markedly catalyzed by alkali. Their procedure calls for charging a glass-stoppered bottle with 0.134 mole of benzil monohydrazone, 0.28 mole of mercuric oxide, 15 g. of anhydrous sodium sulfate, 200 ml. of absolute ether, and 4 ml. of a cold, saturated solution of alcoholic potassium hydroxide. The mixture is shaken for 10–15 min., filtered, and evaporated at the water pump at a temperature not higher than 40°. The yellow crystalline material is dried on a porous plate and crystallized from anhydrous ether to give material melting at about 79° with decomposition in yield of 87–94%.

In the preparation of diphenylacetylene from benzil dihydrazone, this derivative is suspended in benzene in a flask fitted with a stirrer and reflux condenser, and mercuric oxide is added in portions at a rate sufficient to maintain gentle refluxing

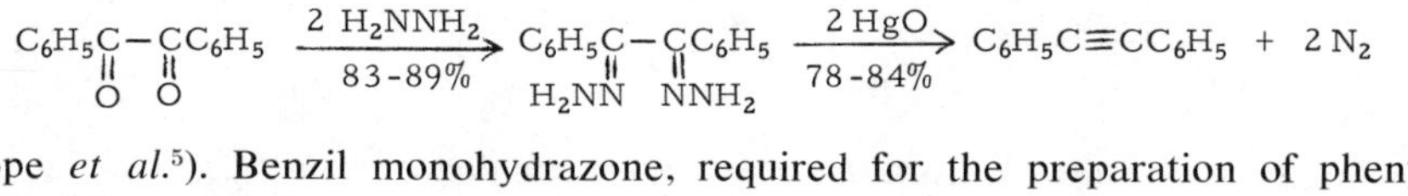

(Cope *et al.*[5]). Benzil monohydrazone, required for the preparation of phenylbenzoyldiazomethane, is prepared by slowly adding hydrazine hydrate to a hot solution of benzil in ethanol; refluxing for 5 min. completes the reaction. To prepare the dihydrazone, the Cope group refluxed benzil with 2 equivalents of hydrazine hydrate in *n*-propanol (b.p. 97°) for 60 hrs.

Catalyst for addition to acetylenes. The conversion of 1-ethynylcyclohexanol to 1-acetylcyclohexanol by hydration of the triple bond is accomplished by dissolving 5 g. of Mallinckrodt red mercuric oxide in a solution of 8 ml. of concd. sulfuric

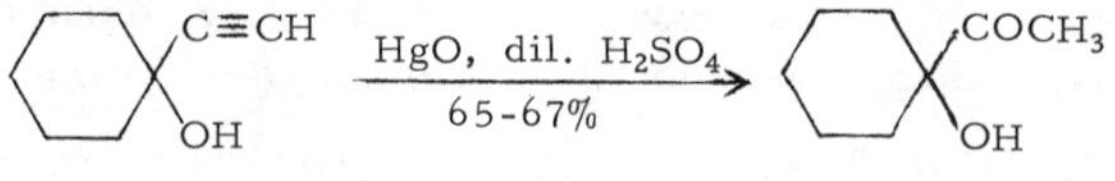

acid in 190 ml. of water, warming to 60°, and dropping in 0.4 mole of 1-ethynylcyclohexanol in 1.5 hrs.[6]

Vinyl chloroacetate is obtained by catalyzed addition of chloroacetic acid to acetylene.[7] A mixture of 2.12 moles of chloroacetic acid (m.p. 63°), 0.2 g. of hydroquinone (polymerization inhibitor), and 20 g. of yellow mercuric oxide is stirred and

$$ClCH_2CO_2H + HC{\equiv}CH \xrightarrow[42-49\%]{HgO} ClCH_2CO_2CH{=}CH_2$$

heated until the chloroacetic acid just melts, and acetylene is passed in through a gas-inlet tube extending to the bottom. After about 30 min. the temperature can be lowered to 40–50° without solidification. The absorption of acetylene, very rapid at first, becomes very slow after about three hours, and the reaction is stopped.

Hunsdiecker reaction. In the original Hunsdiecker reaction, the silver salt of an acid is prepared in one step and treated with bromine in another. Cristol[8] found that isolation of a metal salt can be dispensed with: addition of bromine to a refluxing suspension of red mercuric oxide in a solution of the acid in carbon tetrachloride affords the corresponding alkyl halide in good yield:

$$2\ CH_3(CH_2)_{16}CO_2H + 2\ Br_2 + HgO \xrightarrow[93\%]{} 2\ CH_3(CH_2)_{16}Br + 2\ CO_2 + HgBr_2$$

Iodine can be used in place of bromine.[9] A procedure for the preparation of bromocyclopropane is as follows.[10] A solution of 0.2 mole of cyclopropanecarboxylic

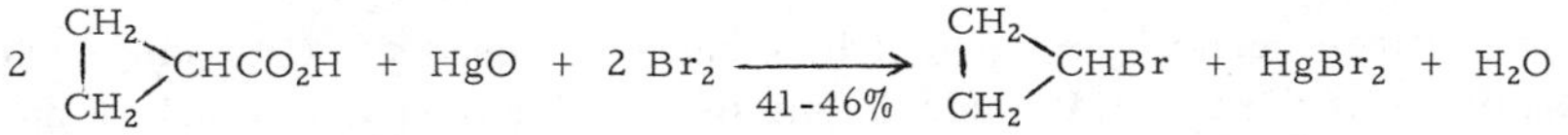

acid and 0.2 mole of bromine in 50 ml. of 1,1,2,2-tetrachloroethane is added dropwise to a stirred suspension of 0.11 mole of red mercuric oxide in 60 ml. of tetrachloroethane in a flask kept in a water bath at 30–35° over a period of 45 min., and stirring is continued until the evolution of carbon dioxide ceases. Checkers F. S. Fawcett and B. C. McKusick (du Pont) followed the reaction with a wet test meter (Albany Meter Co. stainless steel No. Al-18-1) presaturated with carbon dioxide and found the evolution of CO_2 to be 52–60% of the theory. They filtered the mixture on a sintered glass pressure filter (Corning 34020) in order to minimize evaporation losses (see Fig. M-1).

Fig. M-1

When the Cristol procedure was applied to the bridgehead acid (1), bicyclo[2.2.2]octane-1-carboxylic acid, with bromine and mercuric oxide in carbon tetrachloride, the product was a mixture of the expected bromide (2) with an even larger proportion of the unexpected chloride (3).[11] The chloride evidently arises by abstraction of chlorine from the solvent by an intermediate bridgehead radical. The pure bromide was obtained with use as solvent of either bromotrichloromethane or 1,2-dibromoethane.

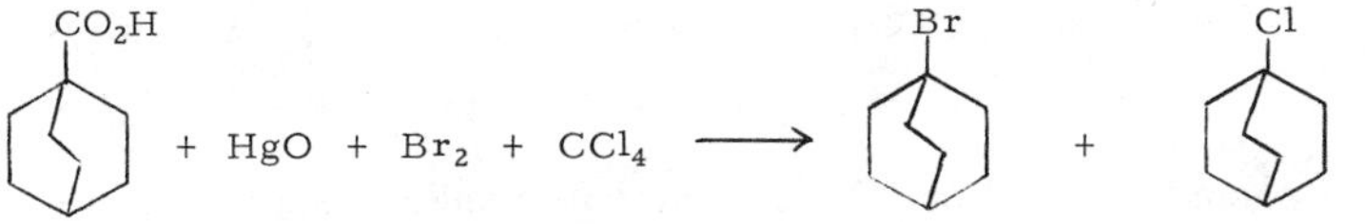

(1) (2) 32% (3) 68%

Davis *et al.*[12] found the Cristol-Firth method most satisfactory for carrying out the Hunsdiecker reaction and suggest procedural details for improvement of results.

Koenigs-Knorr synthesis. The reaction of a poly-O-acylglycosyl halide with a hydroxylic component to form a glucopyranoside[13] is usually effected by use of silver oxide or silver carbonate. The much less expensive yellow mercuric oxide, together with a small amount of mercuric bromide, serves equally well and is not sensitive to light.[14] The acid acceptor, by reaction with the hydrogen bromide formed, gives more catalyst.

[1]H. Staudinger, E. Anthes, and F. Pfenninger, *Ber.*, **49**, 1928 (1916); for possible improvements in the oxidation procedure, see R. Baltzly *et al.*, *J. Org.*, **26**, 3672 (1961)

[2]L. I. Smith and K. L. Howard, *Org. Syn.*, *Coll. Vol.*, **3**, 351 (1955)

[3]L. I. Smith and H. H. Hoehn, *ibid.*, **3**, 356 (1955)

[4]C. D. Nenitzescu and E. Solomonica, *ibid.*, **2**, 496 (1943)

[5]A. C. Cope, D. S. Smith, and R. J. Cotter, *ibid.*, **4**, 377 (1963)
[6]G. W. Stacy and R. A. Mikulec, *ibid.*, **4**, 13 (1963)
[7]R. H. Wiley, *ibid.*, **3**, 853 (1955)
[8]S. J. Cristol and W. C. Firth, Jr., *J. Org.*, **26**, 280 (1961)
[9]S. J. Cristol, L. K. Gaston, and T. Tiedeman, *J. Org.*, **29**, 1279 (1964)
[10]J. S. Meek and D. T. Osuga, *Org. Syn.*, **43**, 9 (1963)
[11]F. W. Baker, H. D. Holtz, and L. M. Stock, *J. Org.*, **28**, 514 (1963); H. D. Holtz and L. M. Stock, *Am. Soc.*, **86**, 5183 (1964)
[12]J. A. Davis, J. Herynk, S. Carroll, J. Bunds, and D. Johnson, *J. Org.*, **30**, 415 (1965)
[13]M. L. Wolfrom and D. R. Lineback, "Methods in Carbohydrate Chemistry," II, 342 (1963)
[14]L. R. Schroeder and J. W. Green, *J. Chem. Soc.*, 530 (1966)

Mercuric oxide–Iodine. This combination has been used as a possibly superior alternative to the lead tetraacetate–iodine reagent for oxidative attack of angular methyl groups. Thus a 6β-hydroxysteroid (1) reacts with mercuric oxide and iodine in a light-induced reaction to give the 6,19-oxide (2) through an intermediate hypoiodite.[1]

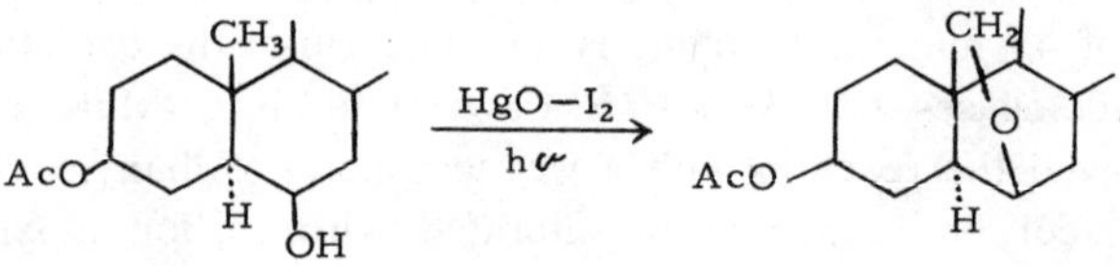
CH3
AcO
H
OH
HgO–I2
hν
CH2
O
AcO
H

(1) (2)

[1]M. Akhtar and D. H. R. Barton, *Am. Soc.*, **86**, 1528 (1964)

Mercuric sulfate, $HgSO_4$. Mol. wt. 296.68.

The reagent, usually in combination with a strong acid, is used as catalyst for effecting additions to acetylenes. Hydration of the acetylenic glycol (1) is effected by adding 25 g. of the finely powdered material within 20 min. to a vigorously

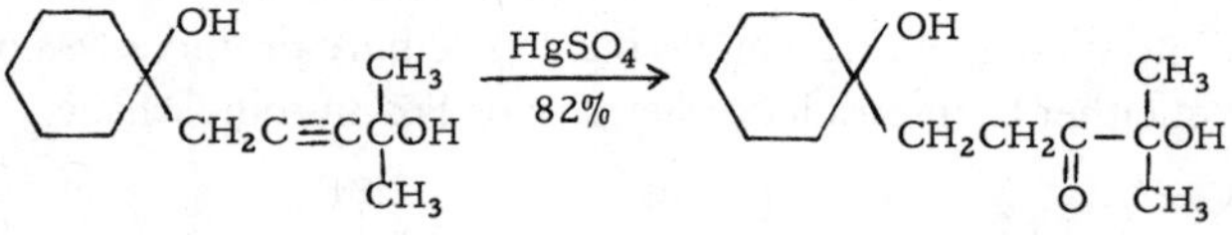
OH
CH3
CH2C≡CCOH
CH3
HgSO4
82%
OH
CH3
CH2CH2C–COH
O CH3

(1) (2)

stirred, ice-cooled solution of 1.25 g. of mercuric sulfate in 125 ml. of 85% formic acid.[1] On each addition the temperature rises 2–3°. After the addition, stirring is continued for 30 min., when finely divided crystals appear. Saturated ammonium sulfate solution is added and the product is extracted with benzene and crystallized from pentane. Hydration of an acetylenic intermediate is a key step in an interesting and highly efficient synthesis of histamine (7) devised by Fraser and Raphael.[2] Commercially available 2-butyne-1,4-diol (1) is converted via the dichloride (2) into the diphthaloylamino compound (3). Hydration to the ketone (4) is accomplished by treating a solution of 3 g. of (3) in 140 ml. of 90% acetic acid with 0.75 g. of mercuric sulfate and 0.5 ml. of concd. sulfuric acid and refluxing for 4 hrs. On addition of water, the high-melting ketone (4) separates and is collected and washed. Acid hydrolysis affords the keto diamine (5), which on treatment with potassium thiocyanate is cyclized to 2-mercaptohistamine (6). The last step, replacement of SH by H, is accomplished in high yield by oxidation with ferric chloride. This surprising reaction is analogous to the desulfurization of mercaptopyrimidines by

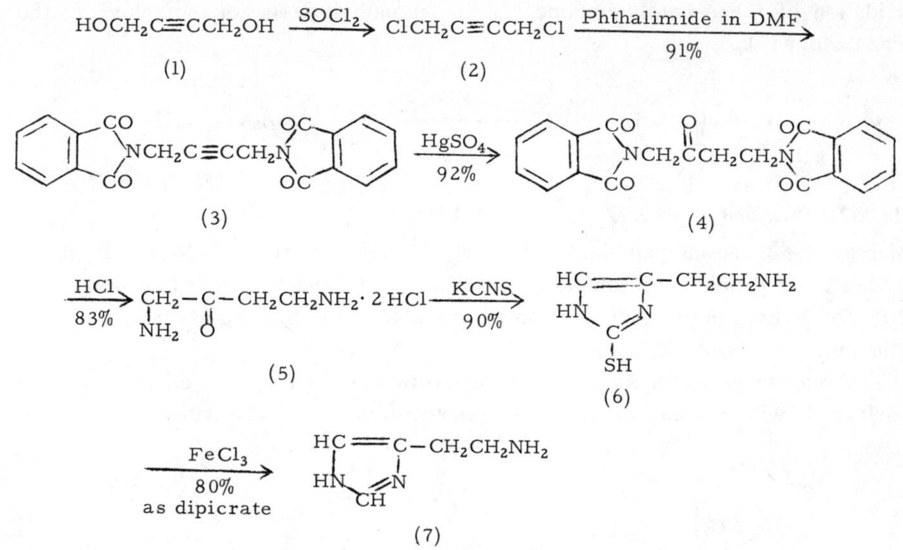

$$\text{HOCH}_2\text{C} \equiv \text{CCH}_2\text{OH} \xrightarrow{\text{SOCl}_2} \text{ClCH}_2\text{C} \equiv \text{CCH}_2\text{Cl} \xrightarrow[\text{91\%}]{\text{Phthalimide in DMF}}$$

$$\qquad\qquad (1) \qquad\qquad\qquad\qquad (2)$$

$$(3) \qquad\qquad\qquad\qquad\qquad\qquad (4)$$

$$(5)$$

$$(6)$$

$$(7)$$

oxidation with hydrogen peroxide or nitric acid, attributed to formation of the sulfonic acid and hydrolysis.[3]

Newman[4] introduced a new hydration catalyst prepared by impregnating Dowex 50, a sulfonated polystyrene resin, with mercuric sulfate. A stirred mixture of 39 g. of 1-ethynylcyclohexanol (1), 100 ml. of acetic acid, 10 ml. of water, and 20 g. of

$$(1) \qquad\qquad (2) \qquad\qquad (3) \qquad\qquad (4)$$

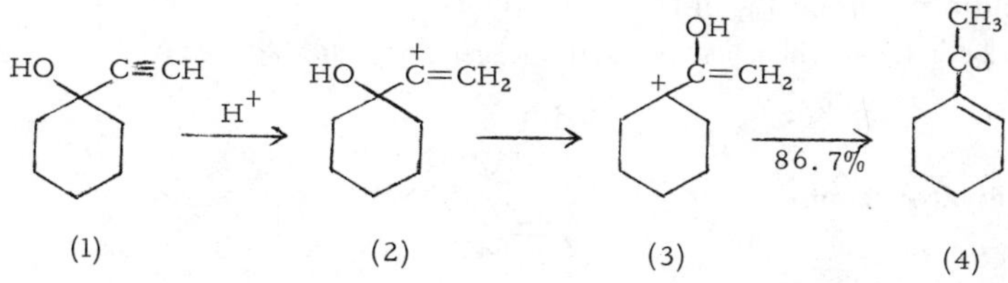

Hg-resin was heated to reflux for 45 min. Filtration from the resin and suitable workup afforded 1-acetylcyclohexene (4) in 86.7% yield.

Two moles of toluene can be added to the triple bond of acetylene in the presence of mercuric sulfate–sulfuric acid.[5] Acetylene is passed into a mixture of 606 g. of

$$2\ \text{C}_6\text{H}_5\text{CH}_3 + \text{CH} \equiv \text{CH} \xrightarrow[\text{60-64\%}]{\text{HgSO}_4} \begin{array}{c} \text{CH}_3\text{C}_6\text{H}_4 \\ \diagdown \\ \diagup \\ \text{CH}_3\text{C}_6\text{H}_4 \end{array} \text{CHCH}_3$$

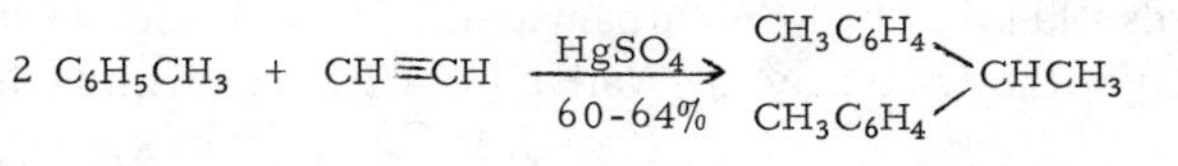

toluene, 70 ml. of concd. sulfuric acid, and 7 g. of mercuric sulfate maintained at 10–15° by ice cooling.

[1]A. Mondon, *Ann.*, **585**, 43 (1954); for a further example of the reaction, *see* Z. G. Hojas, K. J. Doebel, and M. W. Goldberg, *J. Org.*, **29**, 2527 (1964)
[2]M. M. Fraser and R. A. Raphael, *J. Chem. Soc.*, 227 (1952)
[3]R. C. Elderfield, "Heterocyclic Compounds," **6**, 284 (1957)
[4]M. S. Newman, *Am. Soc.*, **75**, 4740 (1953); for a further example of the reaction, see Z. G. Hajos, K. J. Doebel, and M. W. Goldberg, *J. Org.*, **29**, 2527 (1964)
[5]J. S. Reichert and J. A. Nieuwland, *Org. Syn., Coll. Vol.*, **1**, 229 (1941)

Mercuric trifluoroacetate, $\text{Hg(OCOCF}_3)_2$. Newman and Arkell[1] prepared the reagent by dissolving mercuric oxide in anhydrous trifluoroacetic acid and used it for oxidation of the α-ketohydrazone (1), the monohydrazone of dipivaloyl, to the diazoketone (2).

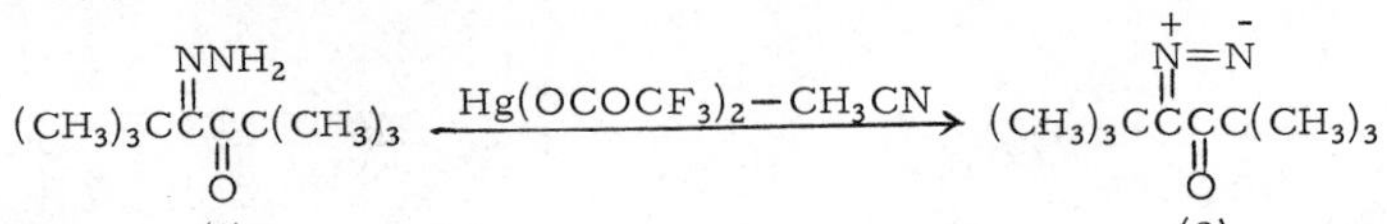

(1) (2)

[1]M. S. Newman and A. Arkell, *J. Org.*, **24**, 385 (1959)

Mercury *p*-toluenesulfonamide, $(p\text{-}CH_3C_6H_4SO_2NH)_2Hg$. Mol. wt. 241.03. Prepared by heating 2.1 moles of *p*-toluenesulfonamide and 1 mole of mercuric oxide at 195–200° for 2½ hrs., pulverizing the cooled product, and extracting excess amide with ethanol.[1]

The reagent is useful for effecting the hydration of steroid 17-ethynylcarbinols such as I, where usual methods give poor results.[1] Thus a mixture of 1 g. of I,

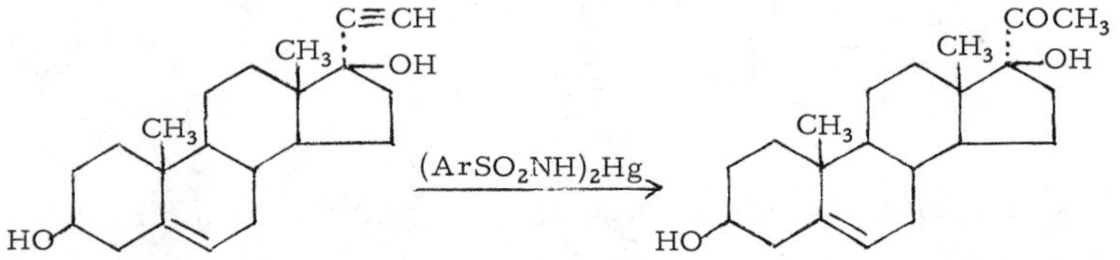

I II

1.86 g. of the reagent, and 50 ml. of 95% ethanol was refluxed for 72 hrs. and hydrogen sulfide was passed in to precipitate mercury. Suitable processing of the filtered solution afforded 1 g. of II.

[1]M. W. Goldberg, R. Aeschbacher, and E. Hardegger, *Helv.*, **26**, 680 (1943)

O-Mesitoylhydroxylamine, 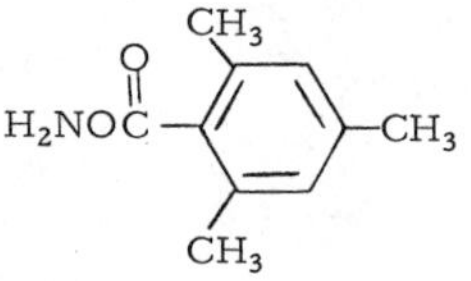

Mol. wt. 179.21, m.p. 31–32°.

Preparation. The reagent is obtained by a general method developed by Carpino[1,2] for the synthesis of O-aroylhydroxylamines as illustrated for O-benzoylhydroxylamine (5). *t*-Butylazidoformate (1) on treatment with hydroxylamine hydrochloride and alkali affords *t*-butyl N-hydroxycarbamate (2). Benzoylation in the presence of triethylamine gives the O-benzoyl derivative (3). Cleavage with hydrogen chloride

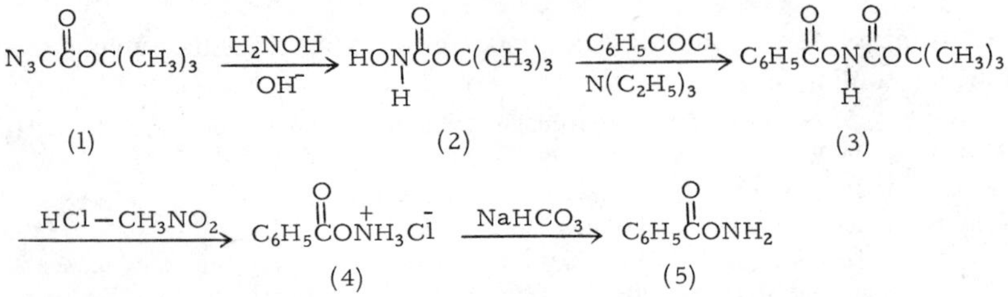

in nitromethane affords the hydrochloride (4), which on neutralization yields O-benzoylhydroxylamine (5),[1] a rather unstable liquid. Use of the hindered mesitoryl chloride afforded the relatively stable solid O-mesitoylhydroxylamine.[2]

Amination. On gentle heating of a mixture of the hindered O-mesitoylhydroxyl-amine with dibenzylamine, the amino group is transferred from the reagent with

$$(CH_3)_3C_6H_2\overset{O}{\overset{\|}{C}}ONH_2 + HN\overset{CH_2C_6H_5}{\underset{CH_2C_6H_6}{\big\langle}} \xrightarrow{58\%} (CH_3)_3C_6H_2\overset{O}{\overset{\|}{C}}OH + H_2NN\overset{CH_2C_6H_5}{\underset{CH_2C_6H_5}{\big\langle}}$$

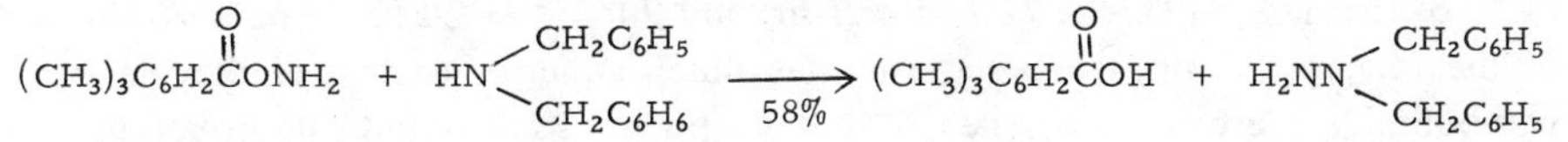

formation of 1,1-dibenzylhydrazine.[2] With unhindered O-benzoylhydroxylamine, amination is only a minor reaction.

The reagent effects amination of the sodium salt of di-*t*-butyliminodicarboxylate

$$(CH_3)_3C_6H_2\overset{O}{\overset{\|}{C}}ONH_2 + Na\underset{}{N}CO_2C(CH_3)_3]_2 \xrightarrow{35-40\%} (CH_3)_3C_6H_2\overset{O}{\overset{\|}{C}}ONa + H_2NN\underset{}{CO_2}C(CH_3)_3]_2$$

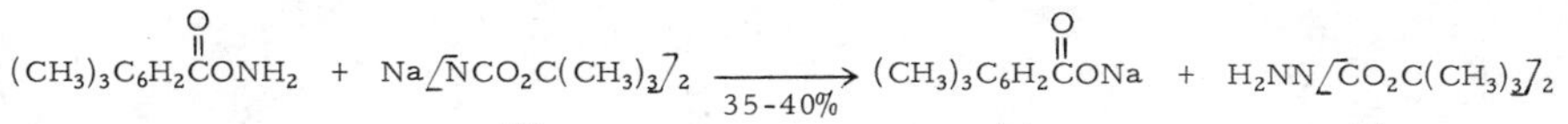

$$(6) \qquad\qquad\qquad (7) \qquad\qquad\qquad (8)$$

(6), at least in modest yield.[3] The reagent is generally useful for the amination of sulfonamides, amides, imides, and pyrroles.[4]

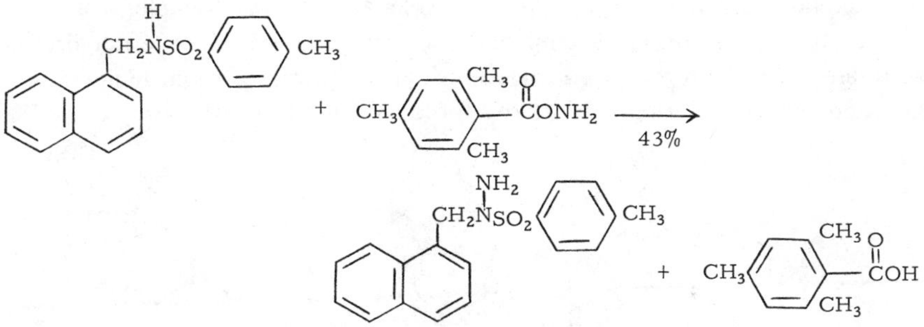

[1] L. A. Carpino, C. A. Giza, and B. A. Carpino, *Am. Soc.*, **81**, 955 (1959)
[2] L. A. Carpino, *Am. Soc.*, **82**, 3133 (1960)
[3] L. A. Carpino, *J. Org.*, **29**, 2820 (1964)
[4] L. A. Carpino, *J. Org.*, **30**, 321 (1965)

Mesitylenesulfonyl chloride, Mol. wt. 218.71, m.p. 57°. Supplier of the free acid: Aldrich.

Khorana[1] found this an efficient reagent for the synthesis of the internucleotide $C_3'-C_5'$ bond. When the protected nucleotide (1) and the protected nucleoside (2)

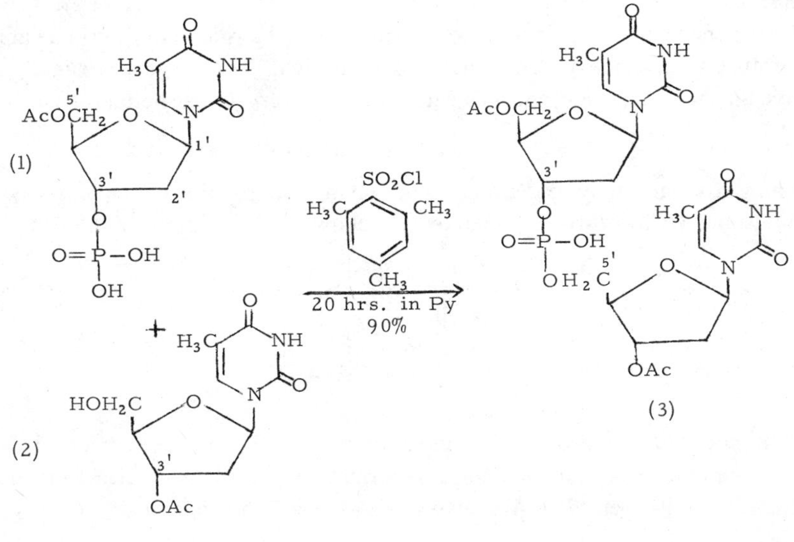

were treated in pyridine solution with 100% excess of mesitylenesulfonyl chloride and the solution let stand at room temperature, the yield of O-diacetylthymidyl-(3′ → 5′)thymidine (3) was 71% after 1 hr. and 90% after 20 hrs. The performance of the reagent in this test synthesis was much superior to that of dicyclohexylcarbodiimide, ethoxyacetylene, N-ethyl-5-phenylisoxazolium fluoroborate; p-toluenesulfonyl chloride was somewhat less effective.

[1]T. M. Jacob and H. G. Khorana, *Am. Soc.*, **86**, 1630 (1964); S. A. Narang and H. G. Khorana, *ibid.*, **87**, 2981 (1965)

Mesityl oxide ethyleneketal,

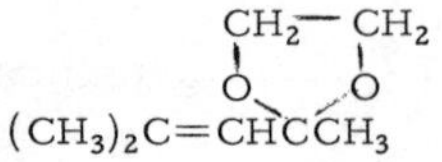

Mol. wt. 142.19, b.p. 58°/25 mm. Preparation.[1]

The reagent has been used for the conversion of Δ^4-3-ketosteroids into the 3-ethyleneketal derivatives by exchange dioxolanation.[2] For example, a mixture of 1 g. of cortisone acetate (1) and 2 ml. of the reagent in 7.5 ml. of dry tetrahydrofurane was treated with 0.06 ml. of concd. sulfuric acid, stirred at room temperature for 2 hrs., and let stand overnight. Workup afforded 844 mg. of nearly pure (2).

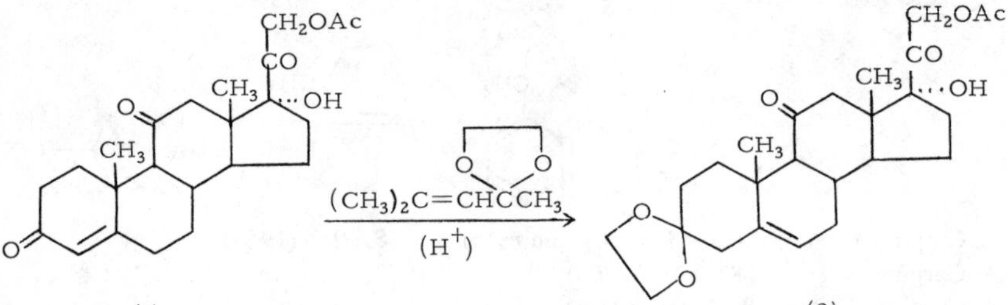

(1) (2)

[1]E. J. Salmi and V. Rannikko, *Ber.*, **72**, 600 (1939)
[2]J. M. Constantin, A. C. Haven, Jr., and L. H. Sarett, *Am. Soc.*, **75**, 1716 (1953)

Mesyl chloride, CH_3SO_2Cl. Mol. wt. 114.56, b.p. 70°/20 mm., sp. gr. 1.47. Suppliers: Aldrich, Eastman. Preparation from the acid (technical) and thionyl chloride (71–83%).[1]

Primary and secondary alcohols react with mesyl chloride in dry pyridine at room temperature to give mesylates, usually in good yield. The reagent is slightly more reactive than tosyl chloride, and mesylates sometimes crystallize better than tosylates.

$$ROH + CH_3SO_2Cl + Py \longrightarrow ROSO_2CH_3 + \overset{+}{Py}H\overset{-}{Cl}$$

Both mesyloxy and tosyloxy are efficient leaving groups. For example, mesylation of cholestane-3β,5α,6β-triol 3-acetate (1) followed by reaction of the 6-mesylate

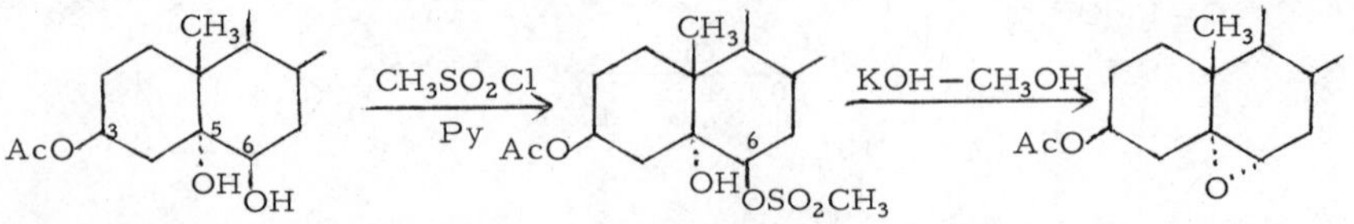

(1) (2) (3)

(2) with potassium hydroxide in methanol affords the 5α,6α-epoxide (3).[2] This reaction sequence has been used for transformation of cholestane-2α,3α-epoxide (4) into the 2β,3β-isomer (7).[3] Acetolysis of (4) gives the diaxial 2β-acetoxy-3α-ol

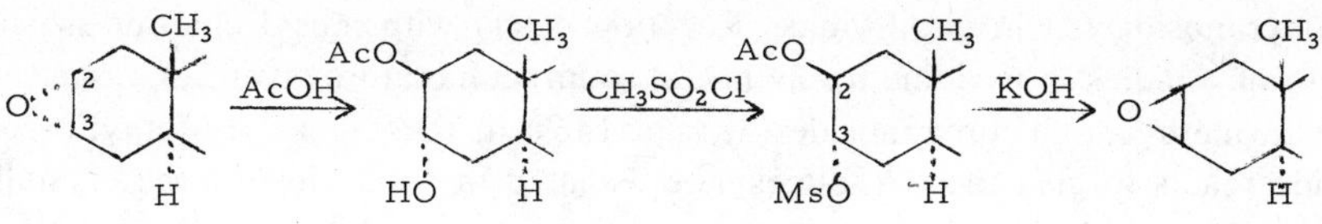

(4) (5) (6) (7)

(5), convertible into the acetate mesylate (6). That this reacts with base with inversion at C_3 to give the β-epoxide (7) demonstrates that mesyloxy is an efficient leaving group whereas the acetoxy group is not. Similarity of the mesyloxy group to a bromine atom is shown by the behavior of two of the four known 5α-spirostane-12-one-2,3-diol dimesylates. Both (8) and (10) on reaction with sodium iodide in acetone at 100° give the Δ^2-olefin (9).[4] Under the same conditions, the dimesylates of the 2β,3β- and 2β,3α-diol are essentially unchanged.

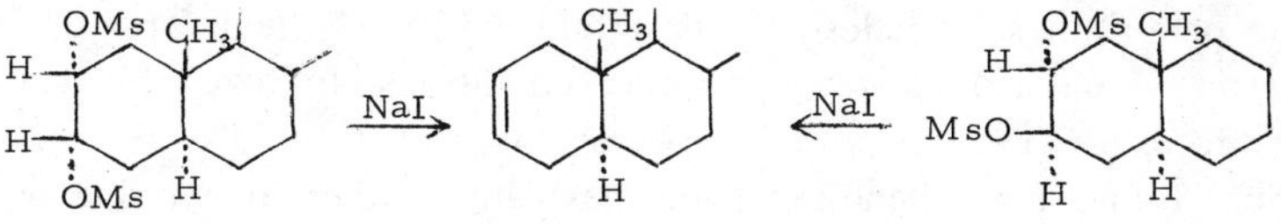

(8) (9) (10)

A mesylate incapable of undergoing *trans* elimination to an olefin may react with base with skeletal rearrangement. One[5] of two cases[5,6] encountered by Wendler in the sapogenin series is the transformation of (11) to the ring-contracted aldehyde (12).

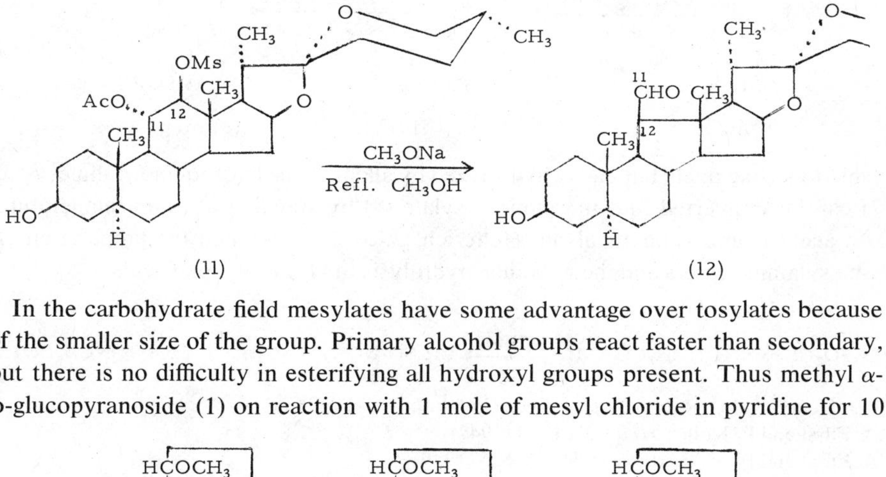

(11) (12)

In the carbohydrate field mesylates have some advantage over tosylates because of the smaller size of the group. Primary alcohol groups react faster than secondary, but there is no difficulty in esterifying all hydroxyl groups present. Thus methyl α-D-glucopyranoside (1) on reaction with 1 mole of mesyl chloride in pyridine for 10

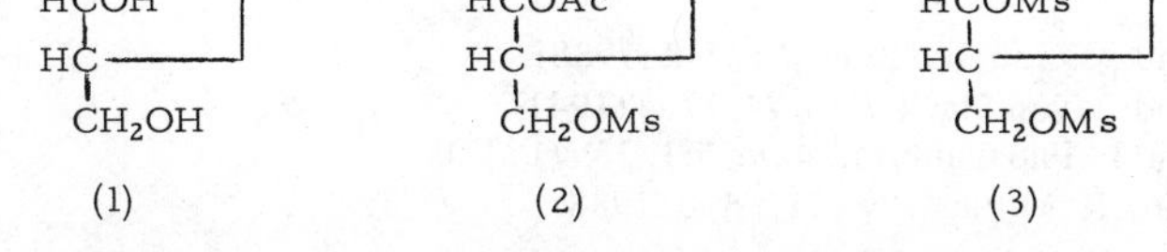

(1) (2) (3)

hrs. at −20° and then for 14 hrs. at 0°, followed by reaction with acetic anhydride for 20 hrs. at 25°, affords crystalline 6-O-mesyl-2,3,4-tri-O-acetyl-α-methyl-D-glucopyranoside (2) in good yield.[7] Reaction of (1) with mesyl chloride in slight excess of 4 moles in pyridine for 24 hrs. at room temperature gives 2,3,4,6-tetra-O-mesyl-α-methyl-D-glucopyranoside (3, m.p. 146°) in 63% yield. β-Methyl-D-cellobioside reacts to give the 6,6′-dimesylate, isolated in good yield as the crystalline

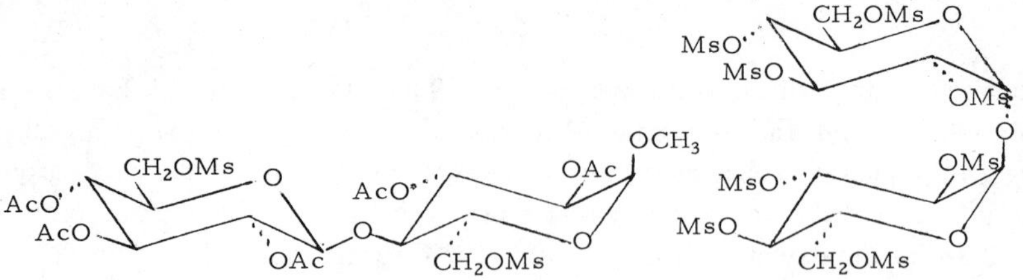
CH₂OMs
AcO
AcO
OAc
AcO
OAc
OCH₃
CH₂OMs
MsO
MsO
CH₂OMs
OMs
OMs
MsO
MsO
CH₂OMs

(4) (5)

pentaacetyl derivative (4).[8] All the hydroxyl groups of trehalose can be mesylated; crystalline octa-O-mesyltrehalose (5) is obtained in 90% yield. Treated with 1.1 moles of mesyl chloride under very mild conditions, followed by acetylation, trehalose affords 6-O-mesylhepta-O-acetyltrehalose.

Helferich[9] found that phenols can be mesylated, either in pyridine or by the Schotten-Baumann procedure, and found that the mesylates are extraordinarily

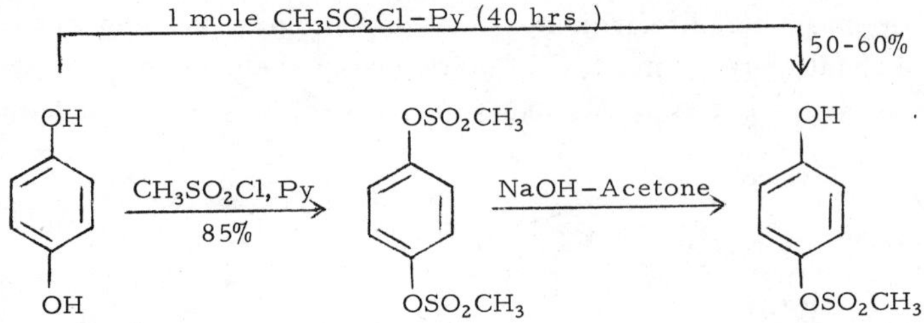
1 mole CH₃SO₂Cl–Py (40 hrs.)
50-60%
OH
OH
OSO₂CH₃
OSO₂CH₃
OH
OSO₂CH₃
CH₃SO₂Cl, Py
85%
NaOH–Acetone

(6) (7) m.p. 167° (8) m.p. 76°

stable to strong acids but easily hydrolyzed by alkali. Thus hydroquinone dimesylate (7) can be converted into the monomesylate (8) by standing at room temperature with acetone and aqueous alkali. Helferich[10] also demonstrated the preparation of N-mesylamino esters and their alkaline hydrolysis to N-mesylamino acids.

$$CH_3SO_2Cl + 2\,H_2NCH_2CO_2C_2H_5 \xrightarrow[80\%]{\text{Ether}} CH_3SO_2N(H)CH_2CO_2C_2H_5 + HCl\cdot H_2NCH_2CO_2C_2H_5$$

[1] P. J. Hearst and C. R. Noller, *Org. Syn., Coll. Vol.*, **4**, 571 (1963)
[2] A. Fürst and F. Koller, *Helv.*, **30**, 1454 (1947)
[3] A. Fürst and Pl. A. Plattner, *Helv.*, **32**, 275 (1949)
[4] H. L. Slates and N. L. Wendler, *Am. Soc.*, **78**, 3749 (1956)
[5] N. L. Wendler, R. F. Hirschmann, H. L. Slates, and R. W. Walker, *Am. Soc.*, **77**, 1632 (1955)
[6] R. Hirschmann, C. S. Snoddy, Jr., C. F. Hiskey, and N. L. Wendler, *Am. Soc.*, **76**, 4013 (1954)
[7] B. Helferich and A. Gnüchtel, *Ber.*, **71**, 712 (1938)
[8] B. Helferich and F. von Stryk, *Ber.*, **74**, 1794 (1941)
[9] B. Helferich and P. Papalambrou, *Ann.*, **551**, 235 (1942)
[10] B. Helferich and R. Mittag, *Ber.*, **71**, 1480 (1938)

Mesyl chloride–Sulfur dioxide. Merck workers Hazen and Rosenburg[1] found mesyl chloride that had been distilled in vacuum to be ineffective in dehydrating an

11β-hydroxy steroid to the $\Delta^{9(11)}$-ene in dimethylformamide containing collidine but that mesyl chloride distilled at atmospheric pressure effected reaction at 25–30° in a few minutes. The substance produced on distillation was identified as sulfur

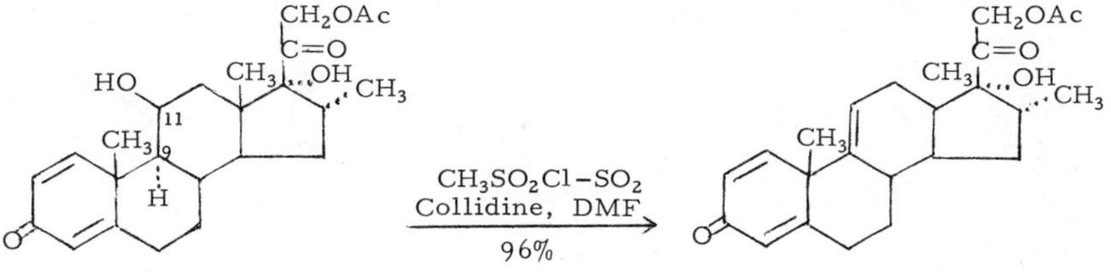

CH₂OAc
C=O
HO
CH₃
OH
CH₃
11
CH₃
9
H
CH₃SO₂Cl–SO₂
Collidine, DMF
96%
CH₂OAc
C=O
CH₃
OH
CH₃
CH₃

I II

dioxide, and indeed mesyl chloride which had failed to react was rendered effective by addition of small amounts of sulfur dioxide.

The efficient dehydration procedure thus evolved is as follows. A solution of 0.04 mole of I in 33 ml. of natural collidine and 100 ml. of dimethylformamide was stirred and cooled to 10°, the cooling bath was removed and, in the course of 1–2 min., 10 ml. (0.128 mole) of mesyl chloride containing 5% of anhydrous sulfur dioxide was added to the clear solution. The temperature was controlled to 25–35° by cooling. After a period of 5 min., during which a light-colored precipitate separated and the solution assumed a reddish hue, the cooling bath was replaced, and the excess mesyl chloride was decomposed by the slow addition of 17 ml. of water. Eventual precipitation with water gave 15.3 g. (96%) of II, m.p. 203–211°. Crystallization from ethanol gave pure material, m.p. 216.5–218°.

The Merck chemists suggest that mesyl chloride and sulfur dioxide react reversibly to give a mixed anhydride acid chloride which has only one branch on the sulfur of the acid chloride function, whereas in mesyl chloride there are two branches, and

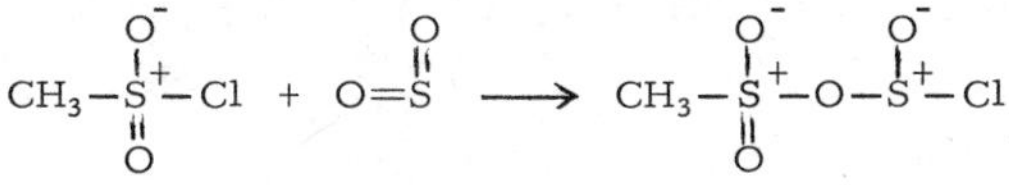

which is better able to attack the hindered β-hydroxyl group. Decomposition ot the resulting ester may proceed through a cyclic transition state.

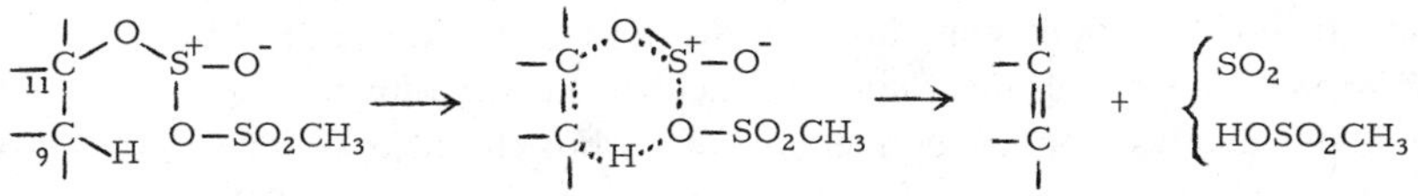

11
9
H
O–SO₂CH₃
SO₂
HOSO₂CH₃

[1]G. G. Hazen and D. W. Rosenburg, *J. Org.*, **29**, 1930 (1964)

Metaphosphoric acid. The reagent is prepared by heating 85% orthophosphoric acid until a clear liquid results.[1] It has been used as shown in the formulation for the

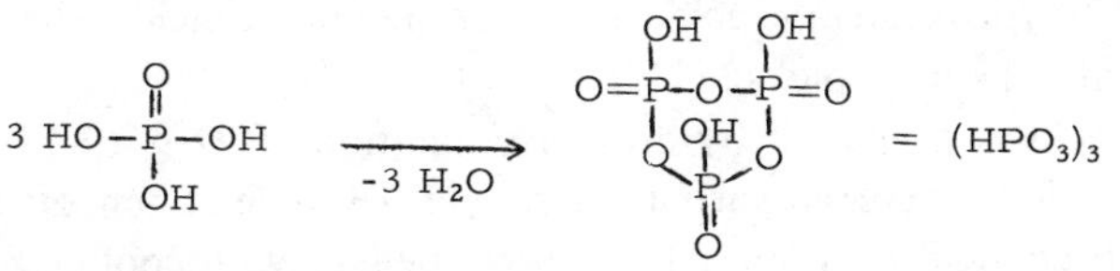

3 HO–P–OH
OH
–3 H₂O
OH
OH
= (HPO₃)₃

synthesis of pyridoxal 5′-phosphate (3), isolated as the sparingly soluble yellow calcium salt.[2] Nitric acid is used in the acid hydrolysis of (2) to destroy N,N-dimethylglycylhydrazine.

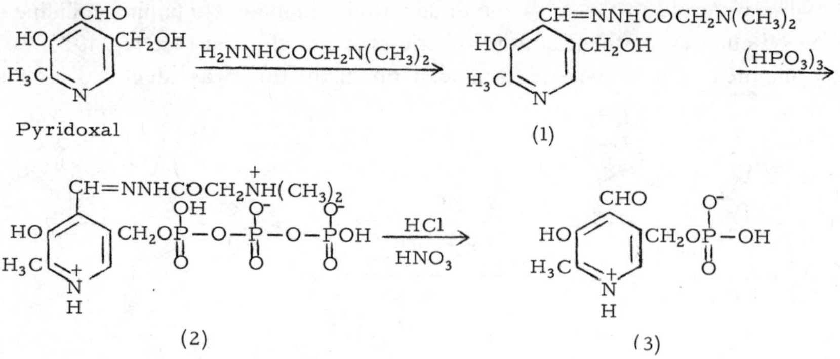
Pyridoxal
(1)
(2)
(3)

[1]M. Viscontini, G. Bonetti, and P. Karrer, *Helv.*, **32**, 1482 (1949)
[2]M. Viscontini, C. Ebnöther, and P. Karrer, *Helv.*, **34**, 1834, 2198 (1951)

Methanephosphonyl dichloride, 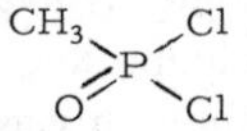

Mol. wt. 132.92, m.p. 32°. A. W. Hofmann[1] prepared the reagent by chlorination of methylphosphonic acid with two equivalents of phosphorus pentachloride. Kinnear and Perren[2] prepared it by reaction of methyl chloride with phosphorus trichloride and aluminum chloride and cautious decomposition of the complex with water.

McKay[3] found that the reagent reacts with at least some 1,2-, 1,3-, and 1,4-glycols to form cyclic esters which are cleaved by acid.

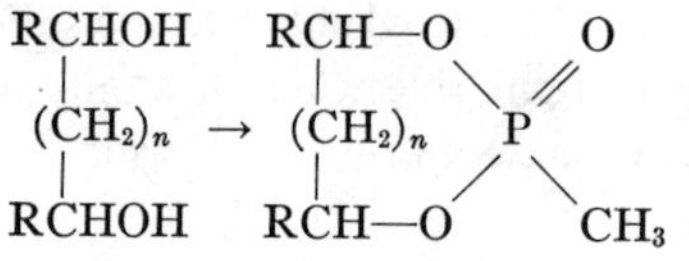

$n = 0, 1, 2$

[1]A. W. Hofmann, *Ber.*, **6**, 303 (1873)
[2]A. M. Kinnear and E. A. Perren, *J. Chem. Soc.*, 3437 (1952)
[3]A. F. McKay *et al.*, *Am. Soc.*, **74**, 5540 (1952); **76**, 3546 (1954)

Methanesulfonic acid, CH_3SO_3H. Mol. wt. 96.11. Suppliers: Standard Oil Co. of Indiana (techn., 95% pure, contains 2% water), Eastern, A, B, E, F, MCB.

Peracids. Swern's group[1] found this acid to be superior to sulfuric acid as the solvent and catalyst for the conversion of carboxylic acids into peroxy acids. It can be used even with some acids containing acid-labile groups. It has good solvent power for carboxylic acids, but the peroxy acid usually precipitates from the medium as it is formed. Peracids are obtained in 75–98% yield from *p-t*-butyl-, *o*- and *p*-nitro-, and *p*-cyanobenzoic acid, terephthalic acid, lauric acid, palmitic acid, stearic acid, 12-hydroxystearic acid, and α-bromostearic acid. The method failed with *m*- and *p*-methoxybenzoic acid.

Perbenzoic acid is prepared[2] by dropwise addition of 22 g. (0.45 mole) of 70% hydrogen peroxide to a stirred partial solution of 0.3 mole of benzoic acid in 86.5 g. of methanesulfonic acid in a 500-ml. tall-form beaker with cooling to 25–30°. The addition takes about 30 min., during which time the benzoic acid all dissolves. After 2 hrs. more the solution is cooled to 15°, diluted with ice and cold saturated ammonium sulfate solution (to decrease the solubility), and extracted with benzene.

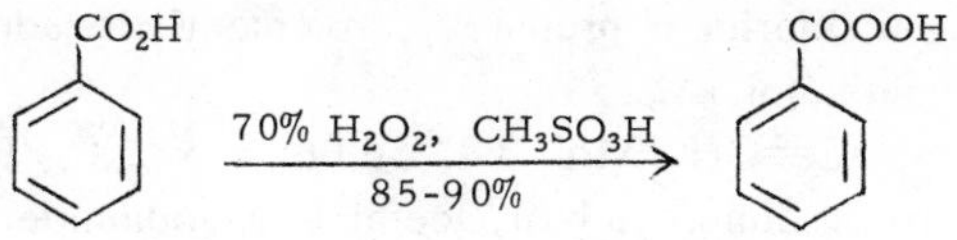
CO2H
70% H2O2, CH3SO3H
85-90%
COOOH

The extract is washed with ammonium sulfate solution, dried, and filtered. The solution is used directly for an epoxidation or other reaction. The yield cited is based upon iodimetric titration.

"Hydrolysis" of esters.[3] A method for cleavage of an ester under nonalkaline conditions consists in refluxing 0.1 mole of ester and 0.1 mole of methanesulfonic acid in 100 ml. of 90% formic acid for 5 hrs. The acid is obtained in 64–97% yield; the alcohol portion of the ester is converted into the formate. Methanesulfonic acid is the acid of choice; sulfuric acid gave much poorer yields, *p*-toluenesulfonic acid still lower, and trifluoroacetic acid and phosphoric acid essentially none.

[1]L. S. Silbert, E. Siegel, and D. Swern, *J. Org.*, **27**, 1336 (1962)
[2]*Idem, Org. Syn.*, **43**, 93 (1963)
[3]B. Loev, *Chem. Ind.*, 193 (1964)

Methanesulfonic anhydride, $(CH_3SO_2)_2O$. Mol. wt. 174.20, m.p. 70°, b.p. 138°/10 min. Prepared by heating the acid with thionyl chloride (80% yield)[1] or with phosphorus pentoxide (55%)[2]. Suppliers: E, F.

The reagent reacts with benzene in the presence of aluminum chloride more rapidly than mesyl chloride and affords methyl phenyl sulfone in 77% yield.[2] It acylates anisole and some alkylbenzenes without catalyst.[3] It forms methane-

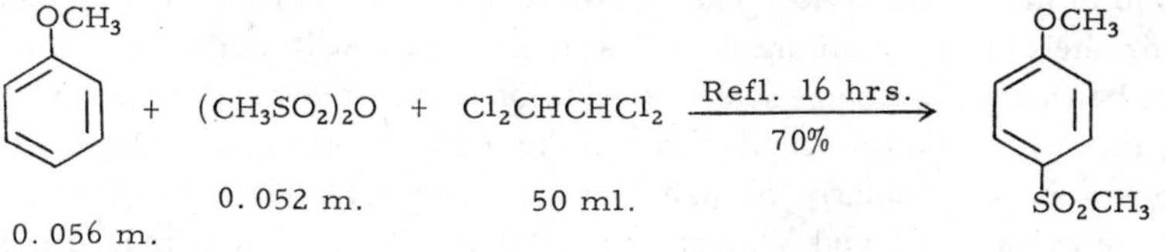
OCH3
0. 056 m.
+ (CH3SO2)2O + Cl2CHCHCl2
0. 052 m.
50 ml.
Refl. 16 hrs.
70%
OCH3
SO2CH3

sulfonyl isocyanate in better yield (38%) than does mesyl chloride (5%), and it resembles trifluoroacetic anhydride in promoting esterification of acetic and benzoic acid.[2] Linstead[4] found that treatment of mannosaccharodilactone (1) with mesyl chloride, tosyl chloride, or thionyl chloride under a variety of conditions gave only tars, but that with methanesulfonic anhydride an almost quantitative yield of the crystalline 2,5-dimesylate (2) was obtained.

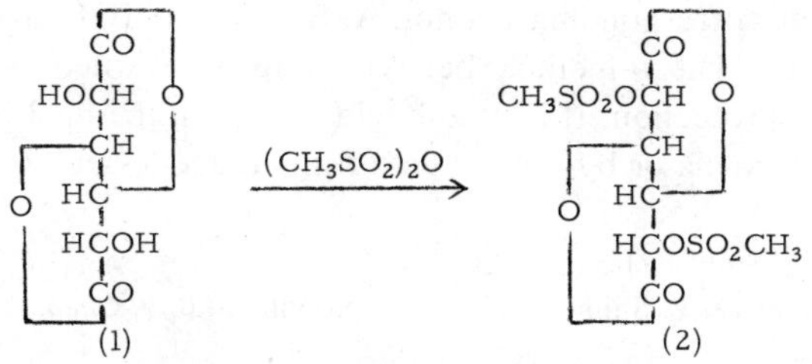
CO
HOCH
CH
HC
HCOH
CO
(1)
(CH3SO2)2O
CO
CH3SO2OCH
CH
HC
HCOSO2CH3
CO
(2)

[1]L. N. Owen and S. P. Whitelaw, *J. Chem. Soc.*, 3723 (1953)
[2]L. Field and P. H. Settlage, *Am. Soc.*, **76**, 1222 (1954)
[3]E. E. Gilbert, *J. Org.*, **28**, 1945 (1963)
[4]R. P. Linstead, L. N. Owen, and R. F. Webb, *J. Chem. Soc.*, 1225 (1953)

Methanesulfonyl chloride, *see* Mesyl chloride.

3% Methanolic hydrogen chloride is prepared conveniently by adding 5 ml. of acetyl chloride to 100 ml. of methanol.

Methoxyacetylene, $CH_3OC{\equiv}CH$. Mol. wt. 56.06, b.p. 23°. The procedure for preparing the ether from dimethylchloroacetal and sodamide is similar to that for ethoxyacetylene; an improved preparation on a 2 mole scale is reported by Wasserman.[1]

Arens[2] found (1950) that an acid, for example benzoic acid, can be converted into its anhydride under very mild conditions by reaction with one-half equivalent of methoxyacetylene. Wasserman[1] found that the reaction of benzoic acid with one

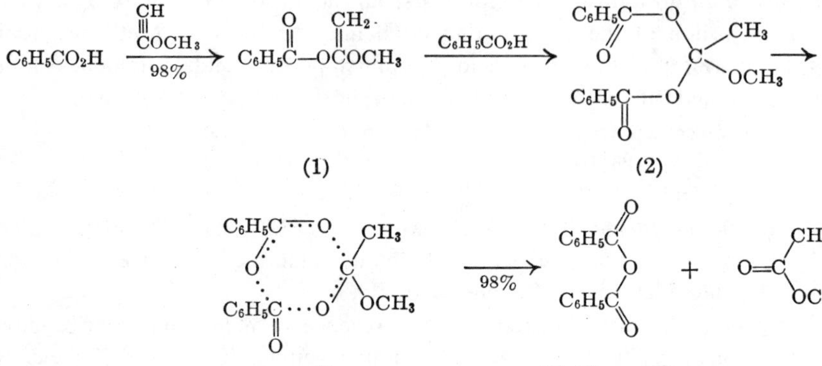

(1) (2) (3) (4) M. p. 45°

equivalent of methoxyacetylene can be controlled to permit isolation of 1-methoxyvinylbenzoate (1) by operating at a low temperature with addition of a trace of mercuric benzoate to catalyze selective addition to the triple bond. Jones[3] postulated that the further reaction proceeds through the unstable ortho ester (2) and that this decomposes intramolecularly through the cyclic transition state (3) to the anhydride (4) and methyl acetate, and Wasserman[4] established that this is indeed the path: reaction of the monoadduct (1) with benzoic acid-O^{18} resulted in equal distribution of isotope between the anhydride and the ethyl acetate.

[1]H. H. Wasserman and P. S. Wharton, *Am. Soc.*, **82**, 661 (1960)
[2]J. F. Arens and T. Doornbos, *Rec. trav.*, **74**, 79 (1955), and earlier papers cited.
[3]G. Eglinton, E. R. H. Jones, B. L. Shaw, and M. C. Whiting, *J. Chem. Soc.*, 1860 (1954)
[4]H. H. Wasserman and P. S. Wharton, *Am. Soc.*, **82**, 1411 (1960)

***p*-Methoxybenzyl chloride**, *p*-$CH_3OC_6H_5CH_2Cl$. Mol. wt. 157.62, b.p. 101–103°/8–10 mm. Suppliers of the corresponding alcohol: A, B, E, F, MCB. Preparation.[1]

Peptide synthesis. The *p*-methoxybenzyl group is regarded as superior to the benzyl group for S-protection. It is readily cleaved from the final peptide either by sodium in liquid ammonia or by boiling with trifluoroacetic acid[2] or with anhydrous hydrogen fluoride.[3]

[1]R. L. Shriner and C. J. Hull, *J. Org.*, **10**, 228 (1945)
[2]S. Akabori, S. Sakakibara, Y. Shimonishi, and Y. Nobuhara, *Bull. Chem. Soc. Japan*, **37**, 433 (1964)
[3]S. Sakakibara, Y. Nobuhara, Y. Shimonishi, and R. Kiyoi, *ibid.*, **38**, 120 (1965)

***p*-Methoxybenzyloxycarbonyl azide**, *p*-$CH_3OC_6H_4CH_2O\overset{O}{\overset{\|}{C}}N{=}N^+{=}N^-$. Mol. wt. 207.18, m.p. 32°.

This stable, crystalline reagent is prepared as shown in the formulation.[1] It reacts with amino acids and peptides to give p-methoxycarbobenzoxy derivatives. This

$$\underline{p}\text{-}CH_3OC_6H_4CH_2OH \xrightarrow[C_6H_5N(CH_3)_2]{ClCOOC_6H_5} \underline{p}\text{-}CH_3OC_6H_4CH_2O\overset{\overset{O}{\|}}{C}OC_6H_5 \xrightarrow{H_2NNH_2}$$

$$\underline{p}\text{-}CH_3OC_6H_4CH_2O\overset{\overset{O}{\|}}{C}NNH_2 \xrightarrow{HNO_2} \underline{p}\text{-}CH_3OC_6H_4CH_2O\overset{\overset{O}{\|}}{C}N=\overset{+}{N}=\overset{-}{N}$$

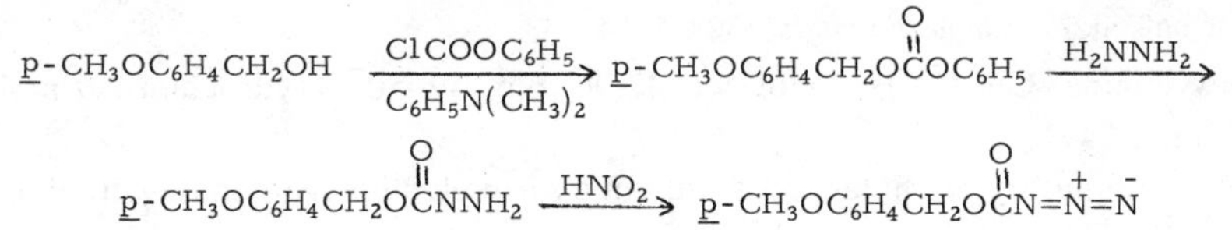

protective group is cleaved by trifluoroacetic acid at 0°, whereas the carbobenzoxy group under the same conditions is stable.

[1]F. Weygand and K. Hunger, Ber., **95**, 1 (1962)

1-Methoxybutene-1-yne-3, $HC{\equiv}CCH{=}CHOCH_3$. Mol. wt. 82.10, b.p. 52°/28 mm. Supplier: Aldrich.

1-Methoxy- and 1-ethoxybutene-1-yne-3 are used to effect the transformation $RCHO \longrightarrow R(CH{=}CH)_2CHO$. Inhoffen[1] converted the ethoxy compound into the lithium derivative, added this to an aldehyde to give (1), selectively hydrogenated

$$RCHO + LiC{\equiv}CCH{=}CHOC_2H_5 \longrightarrow R\underset{\underset{OH}{|}}{C}HC{\equiv}CCH{=}CHOC_2H_5$$

(1)

$$\xrightarrow{H_2, Pd} R\underset{\underset{OH}{|}}{C}HCH{=}CHCH{=}CHOC_2H_5 \xrightarrow{H^+} RCH{=}CHCH{=}CHCHO$$

(2)   (3)

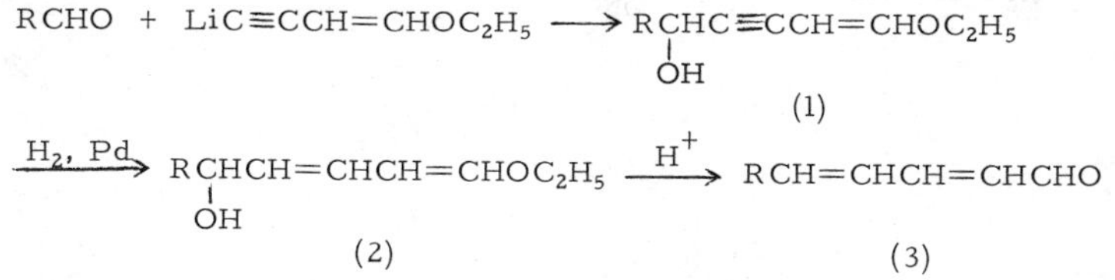

the triple bond, and hydrolyzed the enol ester (2) to produce the polyene aldehyde (3). Marshall and Whiting[2] converted 1-methoxybutene-1-yne-3 into the bromomagnesium derivative and added this to benzaldehyde to produce (4), a propargylic

$$C_6H_5CHO + BrMgC{\equiv}CCH{=}CHOCH_3 \longrightarrow C_6H_5\underset{\underset{OH}{|}}{C}H-C{\equiv}CCH{=}CHOCH_3$$

(4)

$$\xrightarrow{LiAlH_4} C_6H_5\underset{\underset{OH}{|}}{C}HCH{=}CHCH{=}CHOCH_3 \xrightarrow{H^+} C_6H_5CH{=}CHCH{=}CHCHO$$

(5)   (6)

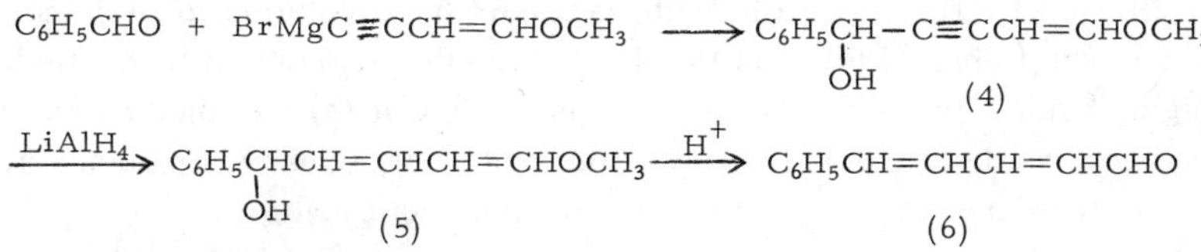

alcohol reducible with lithium aluminum hydride to the diene (5). Acid hydrolysis gave the dienic aldehyde (6, m.p. 43°) in 75% overall yield.

[1]H. H. Inhoffen, F. Bohlmann, and G. Rummert, Ann., **569**, 226 (1950)
[2]D. Marshall and M. C. Whiting, J. Chem. Soc., 4082 (1956)

α-Methoxyethylenetriphenylphosphorane.[1] This Wittig reagent (4) is prepared by quaternization of triphenylphosphine (1) with α-chloroethyl methyl ether (2) in benzene, suspending the solid white salt (3) in glyme under nitrogen at −40° and stirring in potassium t-butoxide in 5 min. This affords a red solution of the unstable but highly reactive ylide (4), useful for the synthesis of methyl ketones from aldehydes

$$(C_6H_5)_3P + Cl\overset{\overset{CH_3}{|}}{C}HOCH_3 \xrightarrow[88\%]{C_6H_6} (C_6H_5)_3\overset{+}{P}-\overset{\overset{CH_3}{|}}{C}HOCH_3(Cl^-) \xrightarrow[-40° \ N_2]{KOC(CH_3)_3 \atop CH_3OCH_2CH_2OCH_3}$$

(1)   (2)   (3)

$$(C_6H_5)_3P{=}\overset{\overset{CH_3}{|}}{C}OCH_3 \xrightarrow{C_6H_5CH=O} C_6H_5CH{=}\overset{\overset{CH_3}{|}}{C}OCH_3 \xrightarrow[88\%]{H^+} C_6H_5CH_2\overset{\overset{CH_3}{|}}{C}{=}O$$

(4)   (5)   (6)

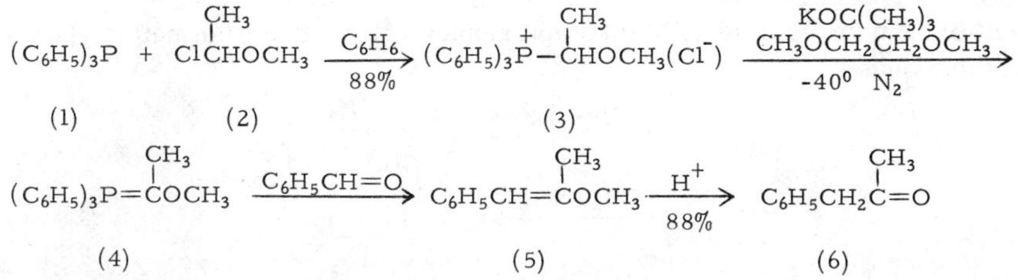

and ketones. Thus benzaldehyde adds to the ylide (4) to give the enol ether (5), converted by acid hydrolysis into the methyl ketone (6).

[1]D. R. Coulson, *Tetrahedron Letters*, 3323 (1964)

Methoxylamine, CH_3ONH_2. Mol. wt. 47.06, b.p. 49–50°. Hydrochloride, mol. wt. 83.53, m.p. 149°.

Preparation[1] by methylation of sodium hydroxylaminedisulfonate with dimethyl sulfate.

Protection of keto groups. Fried and Nutile[2] used the methoxyimino derivative (2) for protection of the $\Delta^{1,4}$-diene-3-one system of the steroid (1). Conversion to

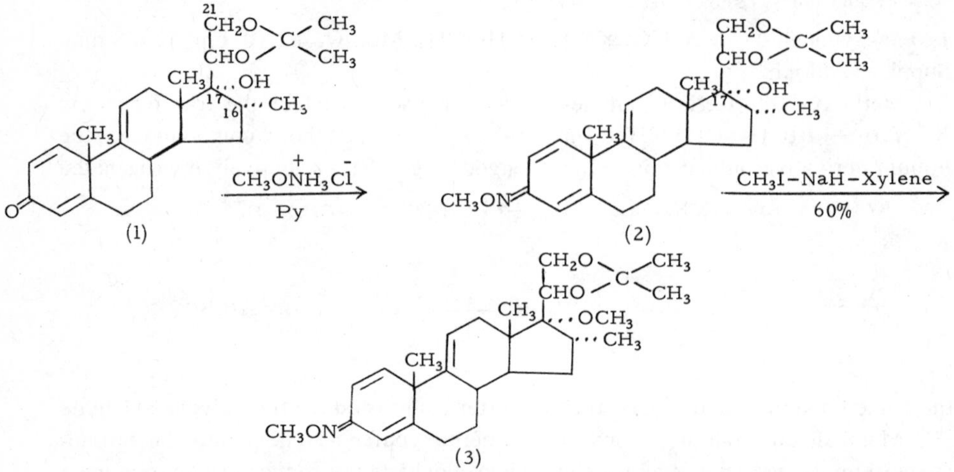

(2) was effected by reaction with methoxylamine hydrochloride in pyridine. The methoxime group of (2) withstood the drastic conditions required for methylation of the highly hindered tertiary 17α-hydroxyl group to give (3). The methoxime group is surprisingly resistant to acid hydrolysis but can be removed by conversion to the semicarbazone followed by hydrolysis with aqueous acetic acid.

Bernstein's group[3] used the methoximino group to protect 20-ketosteroids, and Eardley and Long[4] found that treatment with thionyl chloride in pyridine causes elimination of the 17α-hydroxyl group from 17α,21-dihydroxy-20-methoximino-steroids (1) to give Δ^{16}-derivatives (2).

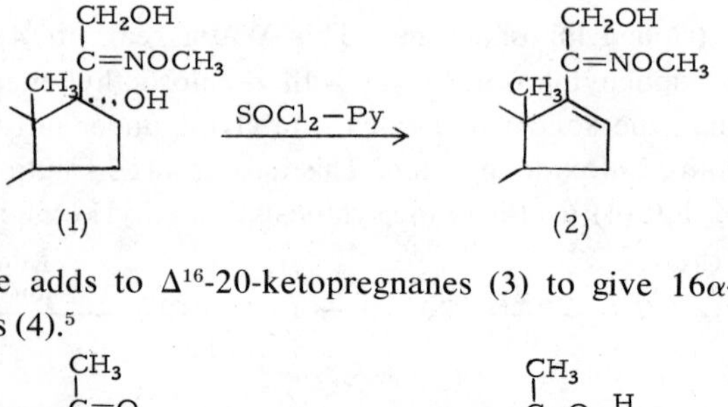

Methoxylamine adds to Δ^{16}-20-ketopregnanes (3) to give 16α-methoxylamino-20-ketopregnanes (4).[5]

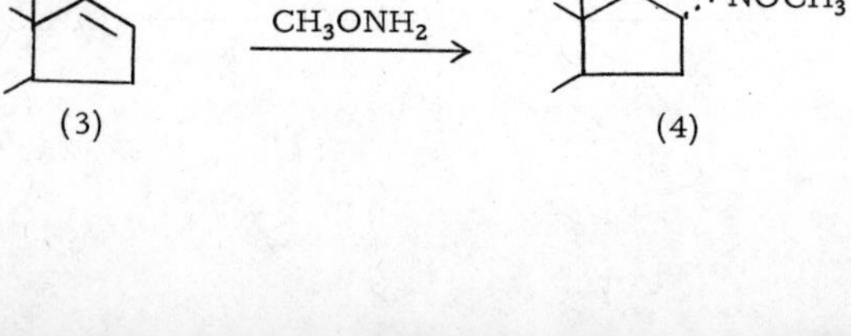

[1]A. R. Goldfarb, *Am. Soc.*, **67**, 1852 (1945); H. Hjeds, *Chem. Scand.*, **19**, 1764 (1965)
[2]J. H. Fried and A. N. Nutile, *J. Org.*, **27**, 914 (1962)
[3]M. Heller, F. J. McEvoy, and S. Bernstein, *J. Org.*, **28**, 1523 (1963)
[4]S. Eardley and A. G. Long, *J. Chem. Soc.*, 130 (1965)
[5]G. Drefahl, K. Ponsold, B. Schönecker, and U. Rott, *Ber.*, **99**, 186 (1966)

Methoxymethylenetriphenylphosphorane, $(C_6H_5)_3P{=}CHOCH_3$ (in ether solution). This Wittig reagent is prepared by reaction of triphenylphosphine with chloromethyl methyl ether to form triphenyl(methoxymethyl)phosphonium chloride (1) and stirring a suspension of the finely powdered salt in ether under nitrogen during gradual addition of one equivalent of ethereal phenyllithium.[1] Levine[1] employed the resulting deep red solution of methoxymethylenetriphenylphosphorane (2) in two-fold excess for reaction with tigogenone (3) to form the methoxymethylene derivative (4), which was hydrolyzed by brief treatment with ether saturated with 72% perchloric acid

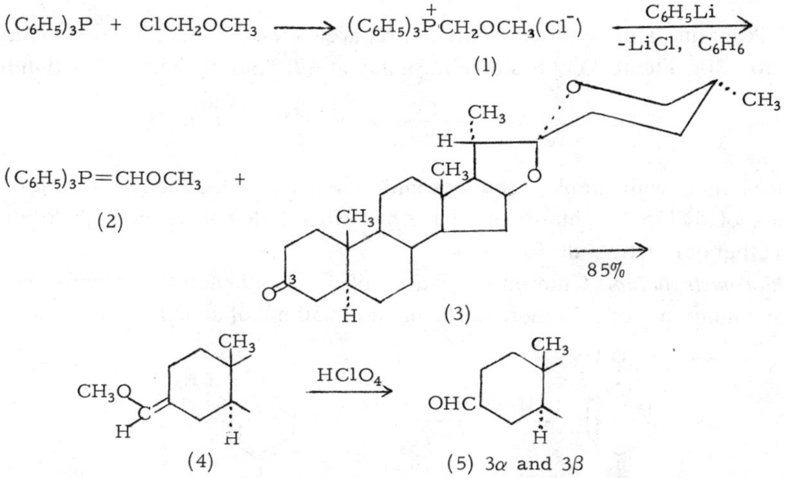
(C6H5)3P + ClCH2OCH3
(C6H5)3P+CH2OCH3(Cl−)
C6H5Li
-LiCl, C6H6
(1)
(C6H5)3P=CHOCH3 +
(2)
CH3
H
O
85%
(3)
CH3O
HClO4
OHC
(4)
(5) 3α and 3β

in nearly quantitative yield to a mixture of the 3-epimeric aldehydes (5). Reaction of the reagent in 100% excess with cyclohexanone and with acetophenone afforded the expected aldehydes in yields of only about 40%.

Wittig[2] later reported reaction of this ylide with several aldehydes, ketones, and cyclic ketones. He found that vinyl ethers are obtainable in higher yield by using the reagent (8) prepared in the same way starting with *p*-tolyl chloromethyl ether (6).

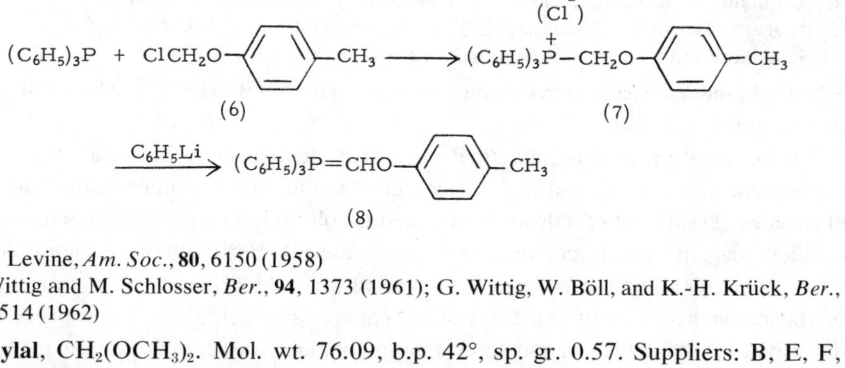
(C6H5)3P + ClCH2O
CH3
(Cl−)
(C6H5)3P+−CH2O
(6)
(7)
C6H5Li
(C6H5)3P=CHO
(8)

[1]S. G. Levine, *Am. Soc.*, **80**, 6150 (1958)
[2]G. Wittig and M. Schlosser, *Ber.*, **94**, 1373 (1961); G. Wittig, W. Böll, and K.-H. Krück, *Ber.*, **95**, 2514 (1962)

Methylal, $CH_2(OCH_3)_2$. Mol. wt. 76.09, b.p. 42°, sp. gr. 0.57. Suppliers: B, E, F, MCB. Preparation from formalin and concd. hydrochloric acid.[1]

Solvent effect. Runge[2] found it possible to convert chloromethyl methyl ether into its Grignard derivative in methylal or tetrahydrofurane as solvent. Methylal

$$CH_3OCH_2Cl \xrightarrow[CH_2(OCH_3)_2]{Mg} CH_3OCH_2MgCl$$

is preferred because of greater stability of the Grignard reagent in this solvent. No reaction between the halide and magnesium occurred in diethyl ether, di-*i*-amyl ether, anisole, dioxane, benzene, or petroleum ether. Runge[2] found that methoxymethylmagnesium chloride reacts with cyclohexene to give a trace of norcarane; however, the yield as determined by gas chromatography, is only about 1%.

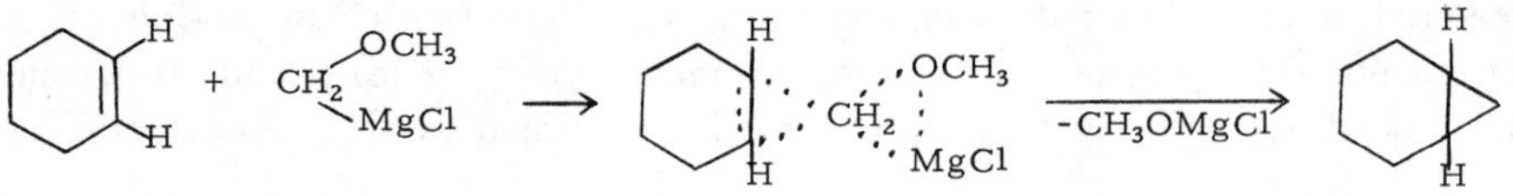
H
H
OCH3
+ CH2
MgCl
H
OCH3
CH2
H
MgCl
-CH3OMgCl
H
H

Schöllkopf and Küppers[3] were able to prepare methoxymethyllithium in methylal at −25 to −30°. The reagent is stable for a day at −70° but decomposes within hours

$$CH_3OCH_2Cl \xrightarrow[CH_2(OCH_3)_2]{2\ Li} CH_3OCH_2Li + LiCl$$

at 0°. Reactions with simple aldehydes and ketones afforded the expected products in yields of 54–75%. Chloromethyl methyl ether failed to react with lithium in diethyl ether or in tetrahydrofurane.

Chloromethylation. Chloromethylation of *p*-nitrophenol is accomplished by preparing a mixture of 0.36 mole of the phenol, 650 ml. of concd. hydrochloric acid,

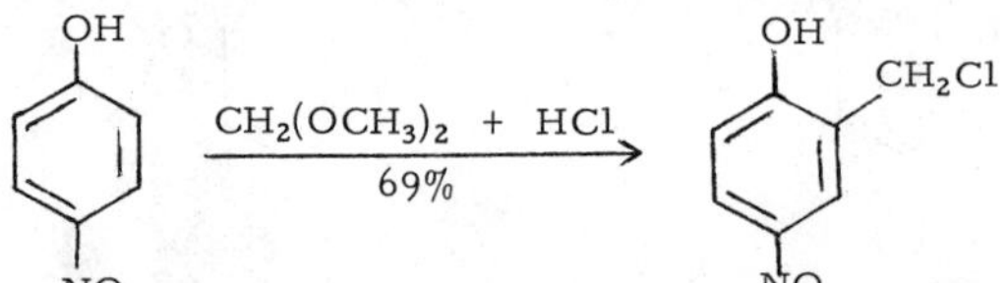
OH
NO2
CH2(OCH3)2 + HCl
69%
OH
CH2Cl
NO2

5 ml. of concd. sulfuric acid, and 1 mole of methylal. Hydrogen chloride is bubbled into the mixture while it is stirred at 70±2° for 4–5 hrs. The chloromethyl derivative begins to crystallize during the first hour. The reaction mixture is cooled in ice and the acid liquor is decanted. Crystallization from benzene affords the pure chloromethyl derivative, m.p. 129–130°, in 69% yield.[1]

[1]C. A. Buehler, F. K. Kirchner, and G. F. Deebel, *Org. Syn., Coll. Vol.*, **3**, 468 (1955)
[2]F. Runge, E. Taeger, C. Fiedler, and E. Kahlert, *J. prakt. Chem.*, **19**, 37 (1963)
[3]U. Schöllkopf and H. Küppers, *Tetrahedron Letters*, 1503 (1964)

3-Methyl-2-benzothiazolonehydrazone hydrochloride (MBTH·HCl). M.p. 240–260°, dec. Supplier: Aldrich.

The preparation of this analytical reagent is described by Sawicki *et al.*,[1] who developed a sensitive test for formaldehyde and other water-soluble aliphatic aldehydes. Treatment of a drop of an aqueous solution of formaldehyde with excess basified reagent[1] effects conversion to the azine (2). Ferric chloride then oxidizes (1) to (3), which condenses with the azine to form the brilliant blue cation (4, one of the resonance structures). Spot plate, paper, silica gel, and column procedures described for the detection and determination of aldehydes are particularly useful for determination of formaldehyde in auto exhaust fumes and polluted air.

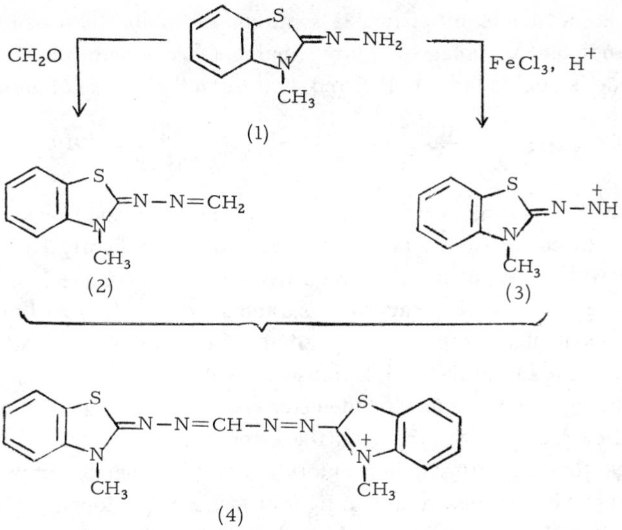
CH₂O
FeCl₃, H⁺
(1)
(2)
(3)
(4)

The reagent reacts with most aniline derivatives to give, after oxidation with ferric chloride, pigments showing ultraviolet absorption at 545–675 mμ.[2] The reagent can be used also for the characterization and determination of carbazoles.[3]

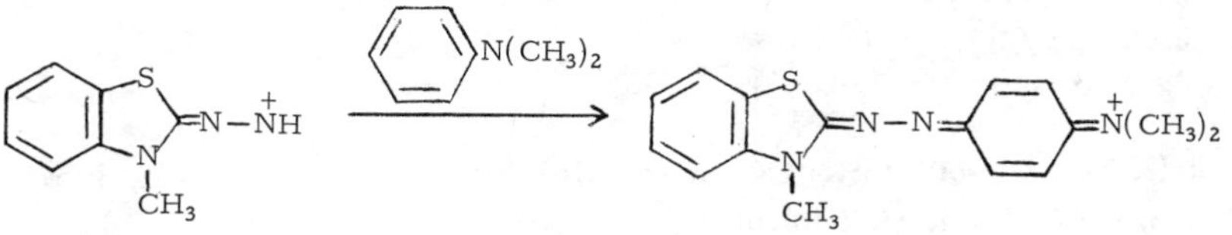

(3)

[1]E. Sawicki, T. R. Hauser, T. W. Stanley, and W. Elbert, *Anal. Chem.*, **33**, 93 (1961)
[2]E. Sawicki, T. W. Stanley, T. R. Hauser, W. Elbert, and J. L. Noe, *ibid.*, **33**, 722 (1961)
[3]E. Sawicki, T. R. Hauser, T. W. Stanley, W. Elbert, and F. T. Fox, *ibid.*, **33**, 1574 (1961)

Methyl borate, $(CH_3O)_3B$. Mol. wt. 103.92, b.p. 68.7°. Suppliers: KK, MSA Res. Corp. Small amounts can be prepared readily from boron trioxide and methanol.[1]

$$2\ B_2O_3 + 3\ CH_3OH \longrightarrow 3\ HBO_2 + B(OCH_3)_3$$

One use is for the transformation of an aryl halide into a phenol where direct hydrolysis is inapplicable.[2] Slow addition of phenylmagnesium bromide solution to

$$C_6H_5MgBr + (CH_3O)_3B \xrightarrow{H_2O(H^+)} \left\{ \begin{matrix} C_6H_5B(OH)_2 \\ (C_6H_5)_2BOH \end{matrix} \right\} \xrightarrow[60-78\%]{H_2O_2} C_6H_5OH$$

methyl borate at −80° and hydrolysis of the adduct affords a mixture of phenylboronic acid and diphenylboronic acid, and this mixture is cleaved by hydrogen peroxide in ether to give phenol in good yield.

[1]W. Gerrard, "The Organic Chemistry of Boron," p. 6, Academic Press (1961)
[2]M. F. Hawthorne, *J. Org.*, **22**, 1001 (1957)

Methyl bromide, CH_3Br. Mol. wt. 94.95, b.p. 4.6°, sp. gr. 1.732.

Methyl bromide is supplied in glass vials (E, F, MCB) or can be generated from sodium bromide, methanol, and sulfuric acid.[1]

For methylation of malonic ester,[2] 2 g. atoms of sodium is dissolved in 1 l. of absolute ethanol and 2 moles of diethyl malonate is added. Then 2.1 moles of methyl bromide is bubbled into the stirred mixture in the course of about 4 hrs. The

$$CH_2(CO_2C_2H_5)_2 \xrightarrow{NaOC_2H_5} \overset{+}{Na}\overset{-}{C}H(CO_2C_2H_5)_2 \xrightarrow[79-83\%]{CH_3Br} CH_3CH(CO_2C_2H_5)_2$$

reaction proceeds smoothly with separation of sodium bromide and with enough heat evolution to cause some boiling. The pale orange, slightly basic suspension is neutralized with acetic acid and filtered from sodium bromide (which is saved). The filtrate is returned to the reaction flask and stirred to prevent bumping during distillation of most of the ethanol. A solution of the sodium bromide in water is added to the residue in the flask to decrease the solubility of the ester and to make the water layer sufficiently dense for the ester layer to float on it. The water layer is separated and extracted with ether and the ester and ether extracts are combined, dried by quick shaking with calcium chloride, and the ether is removed. Any unchanged malonic ester is then removed by a procedure developed by Michael:[3] the ester is shaken for exactly 1 minute with a cold solution of 10 g. of sodium hydroxide in 30 ml. of water (the alkylated malonic ester is attacked only superficially). The ester is then dried again and distilled.

[1]N. Weiner, *Org. Syn., Coll. Vol.*, **2**, 280 (1943)
[2]*Idem, ibid.*, **2**, 279 (1943)
[3]A. Michael, *J. pr.*, (2) **72**, 537 (1905)

3-Methyl-1-butyne-3-ol, $(CH_3)_2\overset{\overset{OH}{|}}{C}C{\equiv}CH$. Mol. wt. 84.11, m.p. 2–3°, b.p. 104°. Suppliers: Air Reduction Chemical and Carbide Co., A, B, E, F, MCB.

Uses. See Cuprous chloride, Diacetylene.

Methyl chloride, CH_3Cl. Mol. wt. 50.49, b.p. −23.7°, sp. gr. 0.920. Supplier: MCB. A generator (NaCl, H_2SO_4, CH_3OH) is described by L. I. Smith in a procedure for the preparation of durene from coal tar xylene.[1] A 5-l. flask is charged with 3.7 l. of

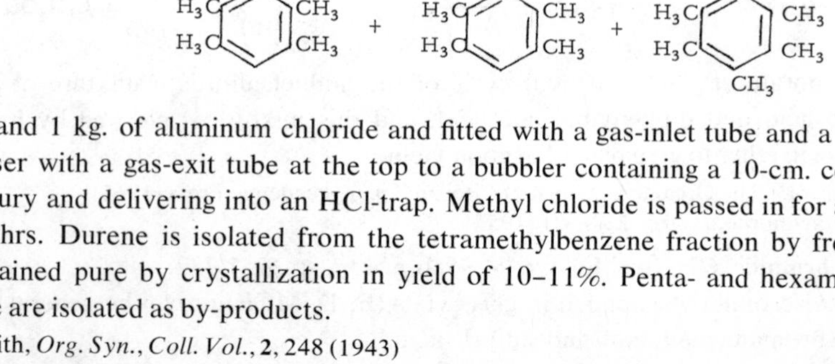
CH3
CH3
H3C
CH3
CH3
H3C
2 CH3Cl AlCl3
H3C
CH3
H3C
CH3
+
CH3
H3C
CH3
H3C
CH3
+
CH3
H3C
CH3
H3C
CH3
CH3

xylene and 1 kg. of aluminum chloride and fitted with a gas-inlet tube and a reflux condenser with a gas-exit tube at the top to a bubbler containing a 10-cm. column of mercury and delivering into an HCl-trap. Methyl chloride is passed in for a total of 100 hrs. Durene is isolated from the tetramethylbenzene fraction by freezing and obtained pure by crystallization in yield of 10–11%. Penta- and hexamethylbenzene are isolated as by-products.

[1]L. I. Smith, *Org. Syn., Coll. Vol.*, **2**, 248 (1943)

3-Methyl-4-chloromethylisoxazole, 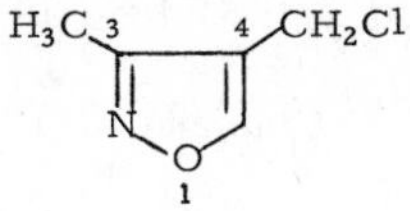

Mol. wt. 119.56, b.p. 62°/1 mm.

The reagent is prepared by chloromethylation of 3-methylisoxazole and used, like 3,5-dimethyl-4-chloromethylisoxazole (*which see*), for annelation.[1] However, the yields in the alkylation step are often poor.

[1]N. K. Kochetkov, E. D. Khomutova, and M. V. Bazilevskiĭ, *Zhur. Obshcheĭ Khim.*, **28**, 2736 (1958) [*C.A.*, **53**, 9187 (1959)]

Methyl chlorosulfite, $CH_3OS^{+}(O^{-})Cl$. Mol. wt. 111.54, b.p. 35–36°/65 mm. Preparation from methanol and thionyl chloride.[1, 2]

$$CH_3OH + SOCl_2 \xrightarrow[88\%]{} CH_3OS^{+}(O^{-})Cl + HCl$$

Berti[2] explored use of the reagent for the conversion of secondary alcohols into olefins by reaction of the alcohol with methyl chlorosulfite in ether containing pyridine to produce the methyl alkyl sulfites and pyrolysis of these esters, but the results were not impressive.

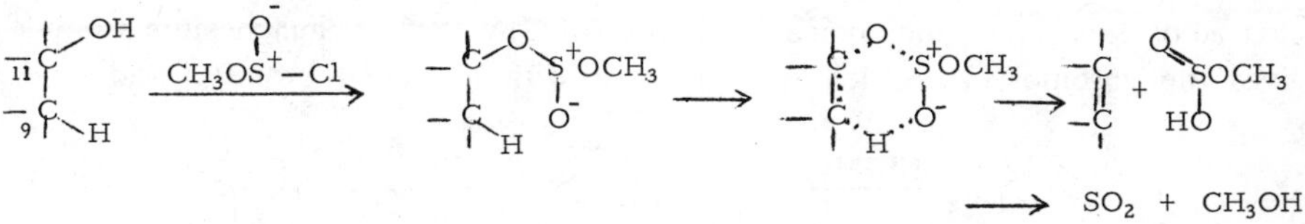

A Merck group[3] later was confronted with the problem of selectively eliminating the 11β-hydroxyl group of cortisol 21-acetate (1). Standard methods of dehydration

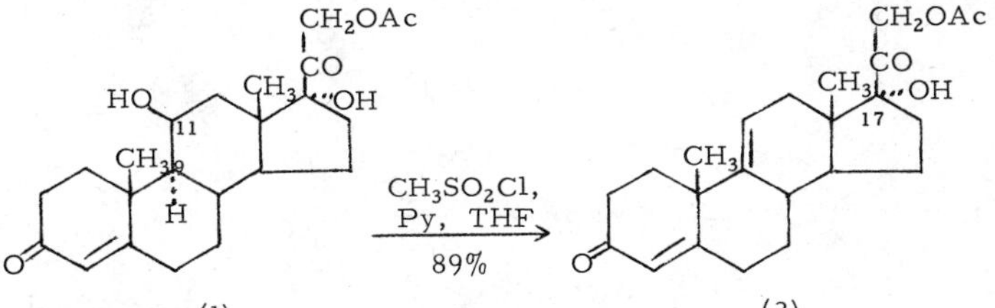

(1) (2)

were precluded by either the unsaturation in ring A or the tertiary 17α-hydroxyl group, but methyl chlorosulfite in the presence of pyridine proved effective. A solution of 0.01 mole of (1) and 8 ml. of pyridine in 150 ml. of tetrahydrofurane was stirred at −10 to −5° during dropwise addition of 0.088 mole of methyl chlorosulfite. The mixture was allowed to come to room temperature and the product precipitated with water. Crude (2), m.p. 226–230°, was obtained in 89% yield; the pure compound melts at 232.5–236.5°. With dimethylacetamide as solvent the yield was 65%. Note that the Merck dehydration done at a low temperature involves *trans* elimination and evidently does not involve the cyclic transition state formulated above for the pyrolytic *cis* elimination.

[1]P. Carré and D. Libermann, *Bull. soc.*, [4], **53**, 1050 (1933)
[2]G. Berti, *Am. Soc.*, **76**, 1213 (1954)
[3]E. M. Chamberlin, E. W. Tristram, T. Utne, and J. M. Chemerda, *J. Org.*, **25**, 295 (1960)

Methyl cyclopropyl ketone, 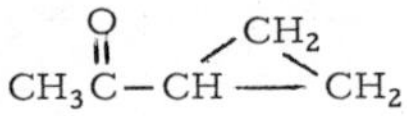

Mol. wt. 84.11, b.p. 110–112°. Suppliers: Aldrich, Columbia, Light.

The starting material for preparation of the reagent is α-acetyl-γ-butyrolactone (1), available from U. S. Industrial Chemicals. One patented route to (1) is that

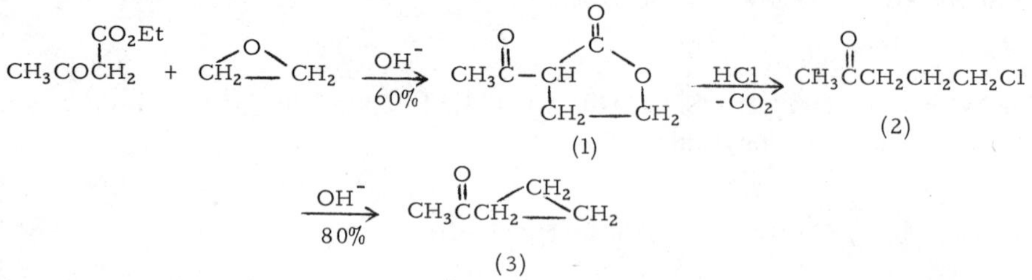

formulated. Cleavage of (1) with hydrochloric acid eliminates carbon dioxide and gives the keto chloride (2), which is cyclized by base to methyl cyclopropyl ketone (3).[1]

This ketone is the key reagent in a general method for the synthesis of terpenoids devised by Julia, Julia, and Guégan.[2] Reaction of (3) with methylmagnesium bromide gives the carbinol (4), which on treatment with hydrogen bromide undergoes

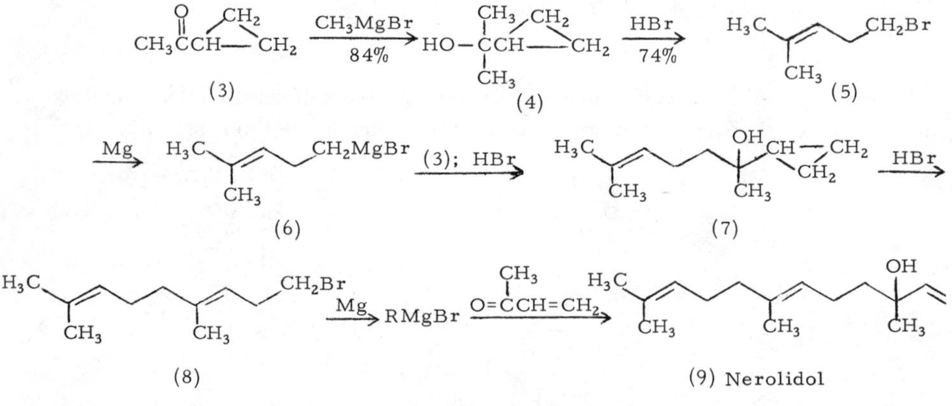

homoallylic rearrangement to give the C_6-bromide (5). Conversion to the Grignard reagent (6) and repetition of the process give the C_{11}-bromide (8). The synthesis can be terminated at this stage by conversion to the Grignard derivative and reaction with methyl vinyl ketone to give nerolidol. The synthesis can also be extended to higher terpenoids.

[1]G. W. Cannon, R. C. Ellis, and J. R. Leal, *Org. Syn., Coll. Vol.*, **4**, 597 (1963)
[2]M. Julia, S. Julia, and R. Guégan, *Bull. soc.*, 1072 (1960)

Methylene chloride, CH_2Cl_2. B.p. 40.8°, sp. gr. 1.34.

Methylene chloride is a useful solvent for Friedel-Crafts acylations, and in one case alters the direction of substitution. Acetylation of chrysene in nitrobenzene or

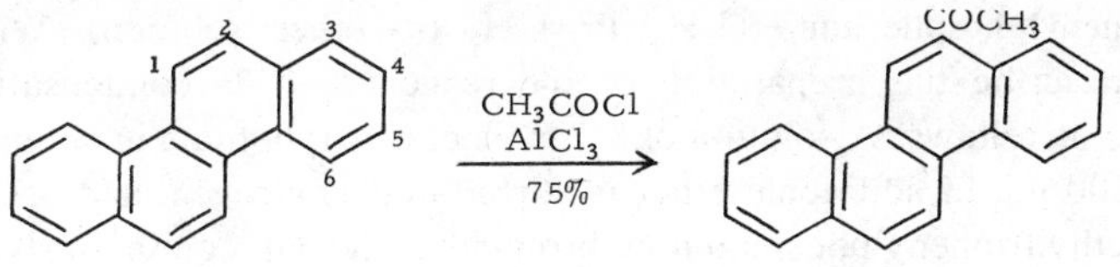
CH3COCl
AlCl3
75%
COCH3

carbon disulfide gives 2-, 4-, and 5-acetylchrysene, wnereas in methylene chloride the sole product is 2-acetylchrysene.[1]

Methylene chloride has high solvent power for magnesium halide etherates, even those derived from terminal acetylenes, which are sparingly soluble in ether, and is recommended for use in Grignard reactions (comparable to the more expensive tetrahydrofurane).[2] Thus a reagent is prepared in ether as usual but the solvent is replaced by methylene chloride prior to subsequent Grignard reaction.

Curtin[3] used a combination of methylene chloride and ether for the lithium aluminum hydride reduction of esters insoluble in ether.

[1]W. Carruthers, *J. Chem. Soc.*, 3486 (1953)
[2]H. G. Viehe and M. Reinstein, *Ber.*, **95**, 2557 (1962)
[3]D. Y. Curtin, J. A. Kampmeier, and M. L. Farmer, *Am. Soc.*, **87**, 874 (1965)

Methylene diurethane, $CH_2(NHCO_2C_2H_5)_2$. Mol. wt. 190.20, m.p. 131°.

Preparation by condensation of urethane with 40% formaldehyde solution in the presence of concd. hydrochloric acid.[1] On treatment with boron trifluoride the reagent (1) loses urethane (2) and affords the unstable methylene urethane (3,

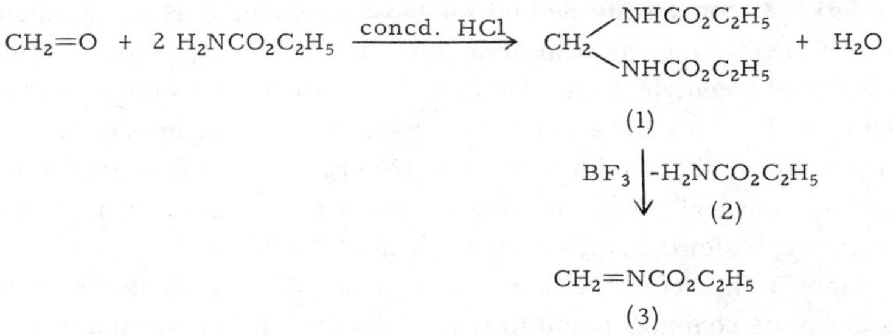
CH2=O + 2 H2NCO2C2H5
concd. HCl
CH2
NHCO2C2H5
NHCO2C2H5
+ H2O
(1)
BF3
-H2NCO2C2H5
(2)
CH2=NCO2C2H5
(3)

probably complexed with BF_3). The latter reagent is a reactive dienophile. Thus it reacts with isoprene (4) to give the tetrahydropyridine derivative (5)[2] and with cyclohexadiene-1,3 (6) to give the isoquinuclidine derivative (7).[3]

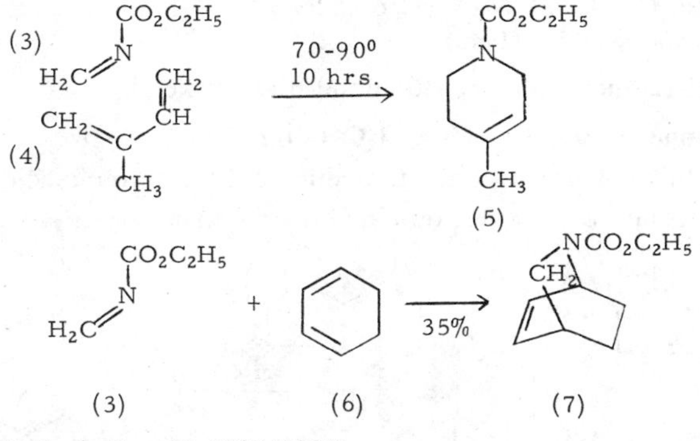
CO2C2H5
(3)
N
H2C
CH2
CH2
CH
C
(4)
CH3
70-90°
10 hrs.
CO2C2H5
N
CH3
(5)
CO2C2H5
N
H2C
+
35%
NCO2C2H5
CH2

(3) (6) (7)

[1]M. Conrad and K. Hock, *Ber.*, **36**, 2206 (1903)
[2]R. Merten and G. Müller, *Angew. Chem.*, **74**, 866 (1962)
[3]M. P. Cava and C. K. Wilkins, Jr., *Chem. Ind.*, 1422 (1964)

Methylenetriphenylphosphorane, $(C_6H_5)_3P{=}CH_2$ (in ether solution). Wittig and Schoellkopf[1] describe the preparation of the reagent and its condensation with cyclohexanone as follows. A solution of 0.1 mole of *n*-butyllithium in about 100 ml. of ether and 200 ml. of additional ether is stirred under nitrogen, and 0.1 mole of crystalline methyltriphenylphosphonium bromide is added continuously over a 5-minute period (evolution of butane causes frothing). Stirring for 4 hrs. at room

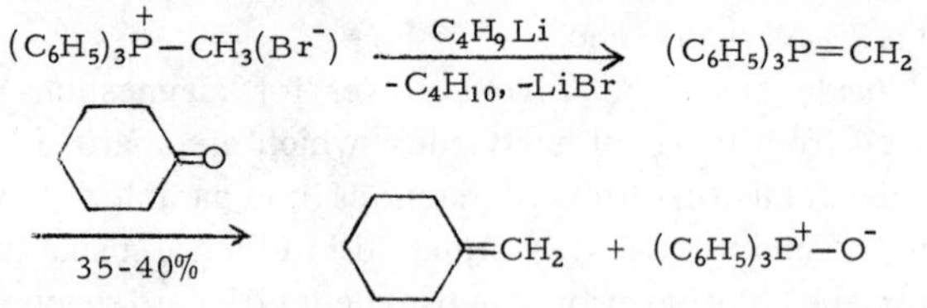

temperature gives an orange solution from which a small amount of the Wittig reagent separates. Cyclohexanone (0.11 mole) is added dropwise. The solution becomes colorless, and a white precipitate of triphenylphosphine oxide separates. The mixture is refluxed overnight, cooled, and the precipitate is removed by suction filtration and washed with ether. The ethereal solution is washed with water, dried, the ether is distilled through an 80-cm. column packed with glass helices, and the methylenecyclohexane is fractioned through a spinning-band column. The product is 99% pure.

[1]G. Wittig and U. Schoellkopf, *Org. Syn.*, **40**, 66 (1960)

Methyl esters.* A convenient method for the conversion of acids into their methyl esters involves refluxing the acid (1 mole) with methanol (3 moles), ethylene dichloride (300 ml., methylene chloride can also be used), and sulfuric acid (3 ml. for an aliphatic acid, 15 ml. for aromatic) for 6–15 hrs. (overnight is convenient). The reaction mixture is diluted with water, and the organic layer is washed with sodium carbonate solution and dried, and the solvent is removed. A number of aliphatic and aromatic acids afford methyl esters in yields of 87–98°%.[1]

The method is a useful alternative to that of Weissberger and Kibler,[2] which requires elaborate equipment, and to that of Baker,[3] which is satisfactory for small-scale operations but inconvenient for larger runs.

*Contributed by M. S. Newman, Ohio State University

[1]R. O. Clinton and S. C. Laskowski, *Am. Soc.*, **70**, 3135 (1948)

[2]A. Weissberger and C. J. Kibler, *Org. Syn., Coll. Vol.*, **3**, 610 (1955)

[3]B. R. Baker, *Am. Soc.*, **65**, 1577 (1943)

2-Methyl-2-ethyl-1,3-dioxolane, *see* Butanone ethylene ketal.

Methyl ethyl ketone (MEK), $CH_3COCH_2CH_3$, b.p. 79.6°.

Knof[1] found this solvent superior to acetic acid for chromic acid oxidation of a mixture of the α- and β-epoxides obtained from 3β-acetoxy-Δ^5-androstene-17-one.

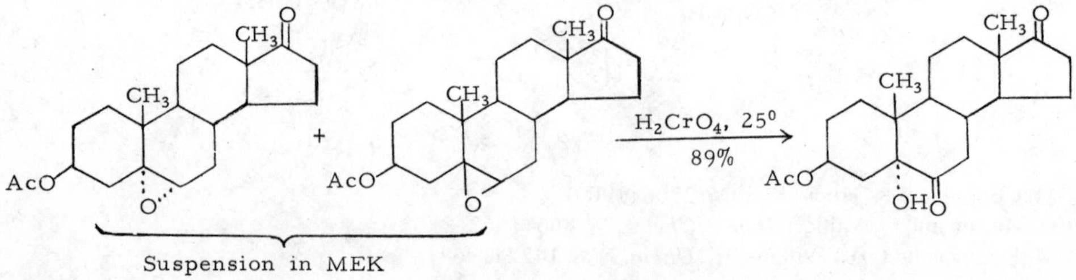
CH3
AcO
+
H2CrO4, 25°
89%
OH
Suspension in MEK

A solution of 18.38 g. of the mixture in 180 ml. of hot MEK was cooled to room temperature to produce a suspension of fine crystals, and 9 ml. of a 75% aqueous solution of CrO_3 was added all at once. After 10 min. the mixture was diluted with 1 l. of water and the produce collected, washed and dried to give 17.13 g. (89%) of material melting at 198–199° (recrystallized: 200–201°).

[1]L. Knof, *Ann.*, **657**, 171 (1962)

3-Methyl-1-ethyl-3-phospholene-1-oxide (I). Mol. wt. 144.15, b.p. 115–119°/1.2–1.3 mm.

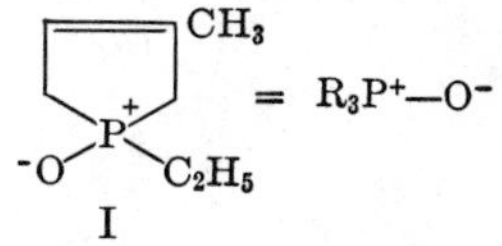

I

du Pont chemists[1] found a novel route to carbodiimides involving dimerization of an isocyanate under catalysis by this phospholine oxide with elimination of carbon dioxide: 2 ArN=C=O ⟶ ArN=C=NAr + CO_2. Kinetic evidence indicates that the reaction is reversible, has a low activation energy, and proceeds in two steps. A suggested sequence of events is shown in the formulation.

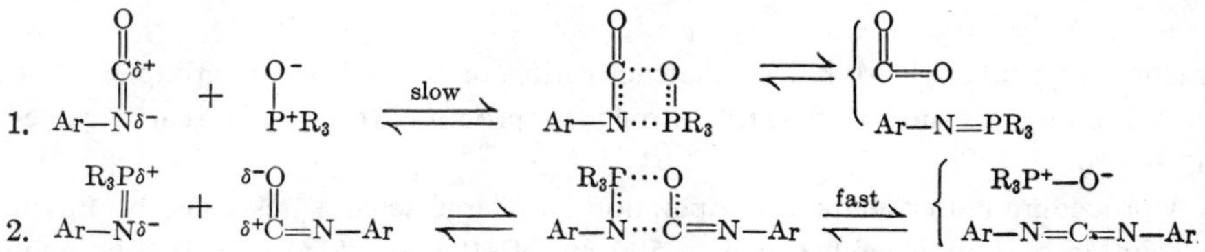

[1]T. W. Campbell, J. J. Monagle, and V. S. Foldi, *Am. Soc.*, **84**, 3673 (1962); J. J. Monagle, T. W. Campbell, and H. F. McShane, Jr., *ibid.*, **84**, 4288 (1962)

Methyl fluorene-9-carboxylate. Mol. wt. 224.25, m.p. 64–65°. Supplier: Eastman.

Preparation. A novel reaction affording crude fluorene-9-carboxylic acid in almost quantitative yield discovered by Vorländer[1] is the basis for a procedure developed by Arnold, Parham, and Dodson[2] and by Richter.[3] A suspension of

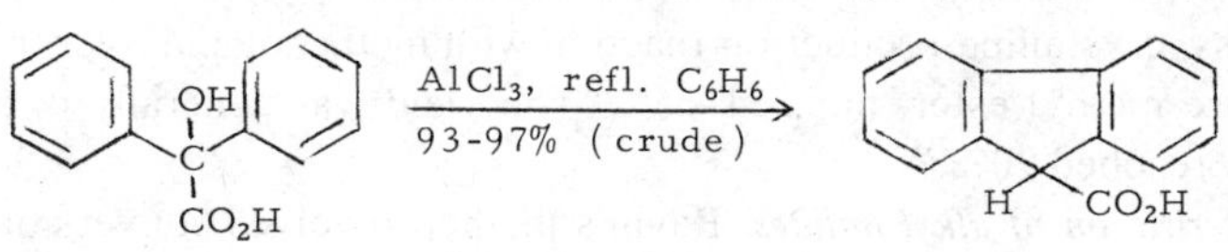
AlCl3, refl. C6H6
93-97% (crude)

I II

0.2 mole of benzilic acid in 700 ml. of benzene is stirred to an even slurry, 0.6 mole of aluminum chloride is added, and the mixture is stirred under reflux for 3 hrs. Hydrogen chloride is evolved, and the initially yellow solution becomes deep red. The complex is decomposed with a little ice, then with water and hydrochloric acid, and the benzene is removed by steam distillation. Repeated extraction with 10% sodium carbonate solution, clarification with Norit, and acidification gives 39–41 g. (93–97%) of fluorene-9-carboxylic acid, m.p. 215–222°. Almost colorless material, m.p. 219–222°, is obtained by stirring the crude acid with 200 ml. of benzene at 45° and collecting the residual solid (30–34 g.).

Vorländer showed that benzene plays no role in the reaction (CS_2 also can be used but is less satisfactory) and noted that diphenylchloroacetic acid is not an intermediate, since it reacts with benzene and aluminum chloride to give triphenylacetic acid. He noted also that diphenylacetic acid does not react with aluminum chloride, that diphenylcarbinol gives products other than fluorene derivatives, and that triphenylcarbinol withstands refluxing with benzene and aluminum chloride. The last observation argues against a mechanism involving intermediates (b) and (c), although it is possible that nonplanarity of the rings interferes with a comparable

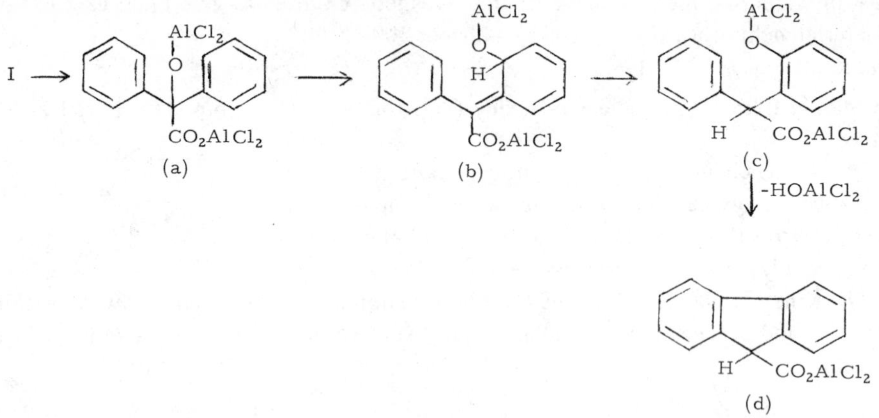
I
$AlCl_2$
O
CO_2AlCl_2
(a)
$AlCl_2$
O
H
CO_2AlCl_2
(b)
$AlCl_2$
O
H
CO_2AlCl_2
(c)
$-HOAlCl_2$
H
CO_2AlCl_2
(d)

reaction of triphenylcarbinol or that activation of the —$OAlCl_2$ group by three phenyl groups is not sufficiently strong to promote the allylic rearrangement (a)⟶(b).

A procedure better adapted to operation on a large scale is described by Bavin.[4] A solution of 1 mole of fluorene in 500 ml. of ether is added with stirring under reflux to an ethereal solution of phenyllithium prepared from 1.5 moles of bromobenzene. After 1 hr. more the orange solution of metallated hydrocarbon is poured as rapidly as possible into powdered dry ice which has been slurried with ether. The mixture is acidified, the solvent removed by steam distillation, and the solid is dissolved in aqueous potassium carbonate solution. Clarification with Norit gives a pale yellow solution which is poured into excess 30% hydrochloric acid. The colorless, crystalline product on reaction with methanol and hydrogen chloride gave the pure methyl ester, m.p. 64–65°. "The yield was generally over 70% and occasionally reached 90%."

Characterization of alkyl halides. Bavin's method involves conversion of methyl fluorene-9-carboxylate by reaction with methanolic sodium methoxide into the yellow anion and reaction of this with an alkyl halide; discharge of the color indicates completion of the reaction. The 9-alkyl esters are crystalline and usually high

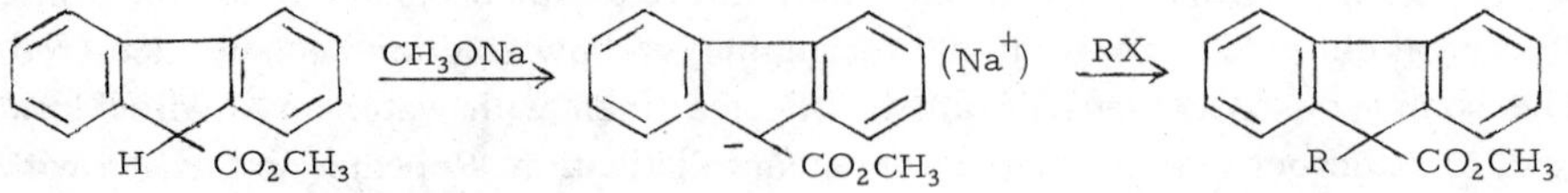
CH_3ONa
(Na^+)
RX
H
CO_2CH_3
CO_2CH_3
R
CO_2CH_3

melting. Hydrolysis to the acid and determination of the equivalent weight by titration establishes the size of the alkyl groups.

[1]D. Vorländer and A. Pritzsche, *Ber.*, **46**, 1793 (1913)
[2]R. T. Arnold, W. E. Parham, and R. M. Dodson, *Am. Soc.*, **71**, 2439 (1949)
[3]H. J. Richter, *Org. Syn., Coll. Vol.*, **4**, 482 (1963)
[4]P. M. G. Bavin, *Anal. Chem.*, **32**, 554 (1960)

N-Methylformanilide, $C_6H_5N(CH_3)CHO$. Mol. wt. 135.16, m.p. 14°. b.p. 131°/22 mm. Suppliers: Aldrich, Eastman. The reagent is prepared by azeotropic distillation of a mixture of N-methylaniline, formic acid, and toluene for removal of the water formed in the condensation and then vacuum distillation of the product; yield 93–97%.[1]

Vilsmeier and Haack[2] found that the reagent forms a 1 : 1 complex with phosphoryl chloride and that this attacks an aromatic compound of adequate reactivity to give a product hydrolyzed by water to an aromatic aldehyde. They prepared *p*-dimethylaminobenzaldehyde, but did not record full details:

$$C_6H_5N(CH_3)CHO + POCl_3 \longrightarrow \text{Complex} \xrightarrow{C_6H_5N(CH_3)_2} \text{Intermediate} \xrightarrow{H_2O}$$

$$C_6H_5NHCH_3 + (CH_3)_2NC_6H_4CHO\text{-}\underline{p}$$

The first finished procedure was a report in a classical paper on pyrene by an I. G. Farbenindustrie group[3] on the preparation of pyrene-3-aldehyde. N-Methylformanilide (135 g.) was mixed with 100 ml. of *o*-dichlorobenzene (to promote

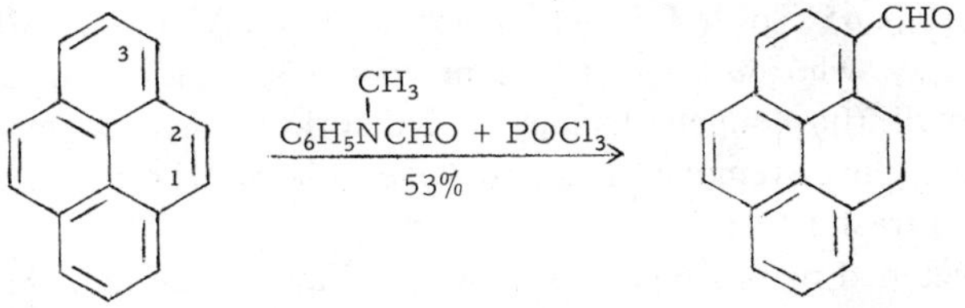
3
2
1
CH3
C6H5NCHO + POCl3
53%
CHO

solution of the hydrocarbon), and the liquid was stirred and maintained at 20–25° during addition in 2 hrs. of 135 g. of $POCl_3$. Pyrene (100 g.) was then added, and the mixture stirred at 90–95° for 2 hrs. The resulting deep red solution on cooling deposited a complex which was collected, washed with benzene, and decomposed with water. Crystallization from ethanol afforded long, bright yellow needles of pyrene-3-aldehyde.

A procedure for the preparation of 9-anthraldehyde is similar, but the reactants and solvent are mixed at once and the mixture stirred mechanically on the steam bath for 20 min.[4] An aqueous solution of sodium acetate is added to the red complex

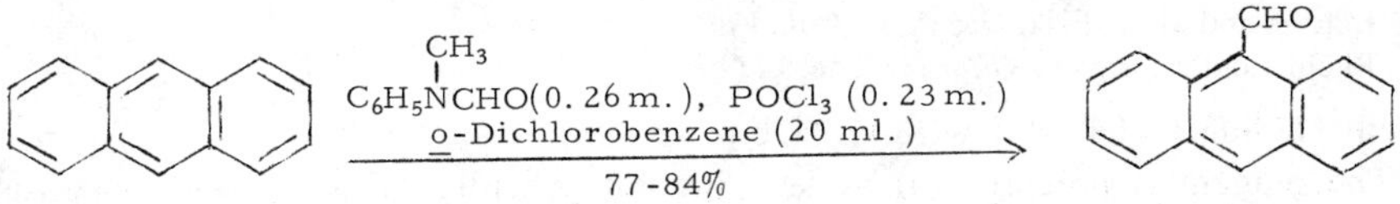
CH3
C6H5NCHO(0.26 m.), POCl3 (0.23 m.)
o-Dichlorobenzene (20 ml.)
77-84%
CHO

to destroy a product of condensation of N-methylaniline with $POCl_3$. Steam distillation and crystallization from acetic acid affords bright yellow crystals of the aldehyde. By the same procedure 3,4-benzpyrene affords the 5-aldehyde in 90% yield.[5] With the less reactive 1,2-benzanthracene, about half of the hydrocarbon

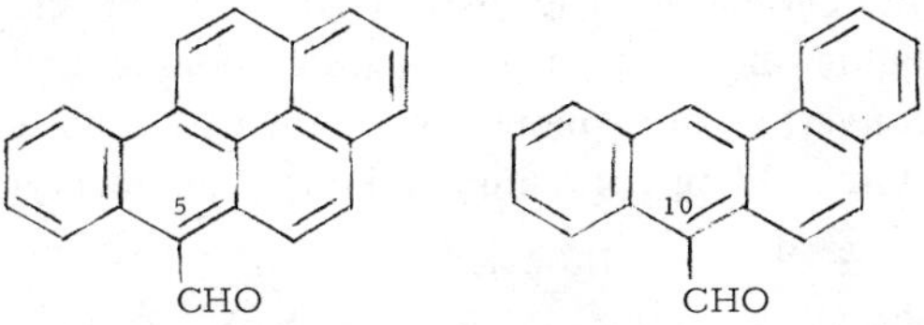
5
CHO
10
CHO

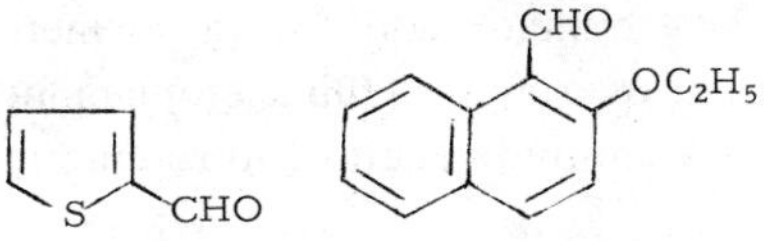
S
CHO
CHO
OC2H5

is recovered and the yield based on material consumed is 63%.[6] The reactivity of thiophene is such as to allow operation at a lower temperature.[7] A mixture of 1 mole each of the reagent and $POCl_3$ is let stand for 30 min. to form the complex, during which time the temperature rises slightly and the color changes from yellow to red. The mixture is then stirred at 25–30° during addition of 1.1 moles of thiophene in 2 hrs. 2-Thenaldehyde is obtained in yield of 71–74%. Conversion of 2-ethoxynaphthalene to the 1-aldehyde is accomplished in 74–84% yield by heating a mixture of the components on the steam bath for 6 hrs.[4]

[1]L. F. Fieser and J. E. Jones, *Org. Syn., Coll. Vol.*, **3**, 590 (1955)
[2]A. Vilsmeier and A. Haack, *Ber.*, **60**, 119 (1927)
[3]H. Vollmann, H. Becker, M. Corell, and H. Streeck, *Ann.*, **531**, 1 (1937)
[4]L. F. Fieser, J. L. Hartwell, and J. E. Jones, *Org. Syn., Coll. Vol.*, **3**, 98 (1955)
[5]L. F. Fieser and E. B. Hershberg, *Am. Soc.*, **60**, 2542 (1938)
[6]L. F. Fieser and J. L. Hartwell, *Am. Soc.*, **60**, 2555 (1938)
[7]A. W. Weston and R. J. Michaels, Jr., *Org. Syn., Coll. Vol.*, **4**, 915 (1963)

2-Methylfurane, 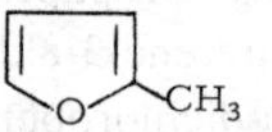

Mol. wt. 82.10, b.p. 65°, sp. gr. 0.916. Suppliers: duPont, A, B, E, F, MCB.

Eliel and Fisk,[1] who describe the Mannich condensation with the reagent to produce 5-methylfurfuryldimethylamine, state that the commercial material contains a stabilizer which can be removed by storage over solid KOH for 24 hrs., decantation, and storage over fresh KOH.

Buchta[2] developed an interesting synthesis of undecylic acid (5) starting with the condensation of 2-methylfurane (1) with vinyl β-carbomethoxyethyl ketone (2).

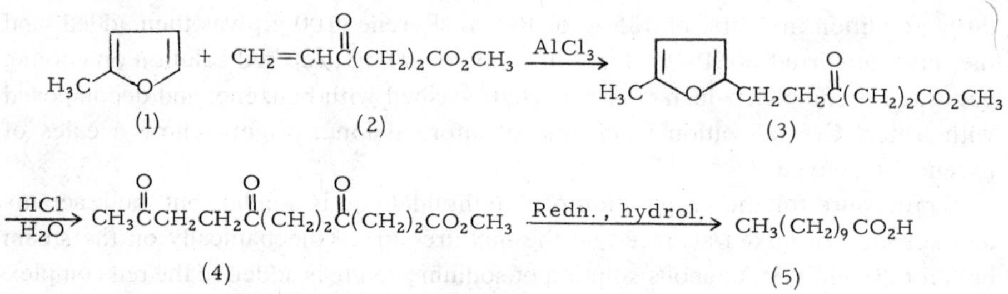

[1]E. L. Eliel and M. T. Fisk, *Org. Syn., Coll. Vol.*, **4**, 626 (1963)
[2]E. Buchta and W. Bayer, *Naturwiss.*, **46**, 14 (1959)

Methyl iodide, CH_3I. Mol. wt. 141.94, b.p. 42.3°, sp. gr. 2.279.

The reagent is prepared (a) by leaching iodine with CH_3OH–CH_3I condensate into a chamber containing a mixture of red and yellow phosphorus, yield 93–95%;[1] and (b) by addition of dimethyl sulfate to a stirred suspension of calcium carbonate in an aqueous solution of potassium iodide; methyl iodide distills as formed and is obtained in yield of 90–94%.[2]

O-Methylation. For esterification of a carboxylic acid: dissolve 0.1 mole of acid in methanol, add 15.5 g. of methyl iodide and 14 g. of potassium carbonate, and stir overnight.[3] Quinacetophenone is converted into a monomethyl ether by cooling a solution in acetone to room temperature, adding potassium carbonate and methyl

$$KI + (CH_3)_2SO_4 \longrightarrow CH_3I \quad CH_3SO_4K$$

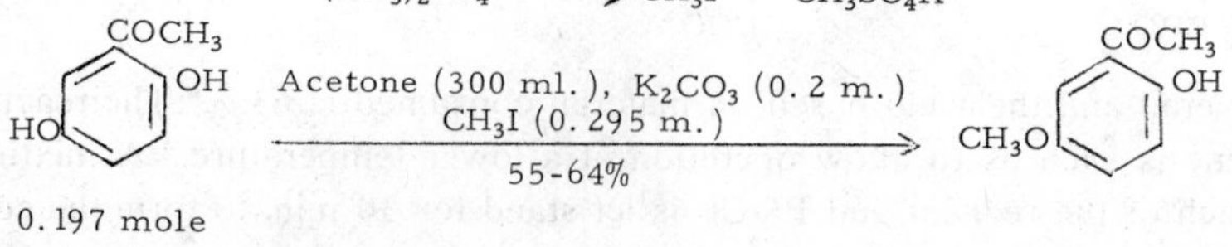

iodide, and refluxing for 6 hrs.[4] Acetone is evaporated and the product isolated by steam distillation.

Purdie[5] introduced the combination methyl iodide–silver oxide for exhaustive methylation of carbohydrates. In an improved procedure[6] a solution of the carbohydrate in methanol is shaken at 25° with methyl iodide and silver oxide, and four further additions are made at 12-hr. intervals, with occasional filtration from silver salts.

Kuhn[7] found that carbohydrates can be completely methylated in a single step by silver oxide and methyl iodide if dimethylformamide is used as solvent. The carbohydrate is dissolved in 10–25 times its weight of DMF and treated at room temperature with methyl iodide (3 equiv. per OH group) and silver oxide (2 equiv. per OH group). Barium oxide can also be used.[8] 1,2-Diketones are converted by this procedure into monoketals. Thus benzil is converted into (1), phenanthrenequinone into (2), and ninhydrin into (3).[9]

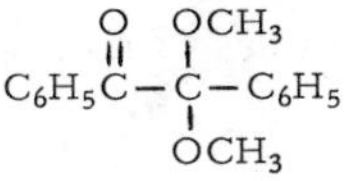
O
OCH3
C6H5C–C–C6H5
OCH3

(1)

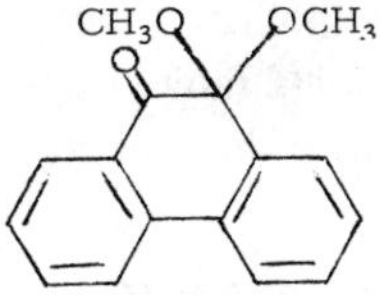
CH3O
OCH3

(2)

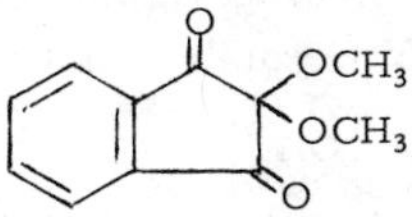
O
OCH3
OCH3
O

(3)

Walker *et al.*[10] found the CH_3I–Ag_2O–DMF method effective for conversion of reducing carbohydrates into completely methylated glycosides without protection of the reducing group. The reaction proceeds smoothly with a variety of aldoses, ketoses, and uronic acids; completely methylated products are obtained in 80–90% yield in a reaction period of 16 hrs.

Kuhn and Trischmann[11] found (1963) that methylation in DMF is advantageously conducted with dimethyl sulfate, rather than with methyl iodide. Substances sparingly soluble in dimethylformamide can be methylated in dimethyl sulfoxide or in a mixture of DMSO and DMF. Inositol was converted by this method into the hexamethyl ether (m.p. 18°).

Srivastava[12] found that a polysaccharide such as maize starch can be permethylated by dissolving it in dimethyl sulfoxide and stirring at 25° for 48 hrs. with methyl iodide and barium oxide. However, a later report[13] states that a higher degree of methylation of starch is obtained by using dimethyl sulfate and sodium hydroxide in dimethyl sulfoxide.

The strongly chelated *peri*-hydroxyl group of juglone can be methylated with methyl iodide and silver oxide in chloroform, but not in dimethyl sulfoxide.[14]

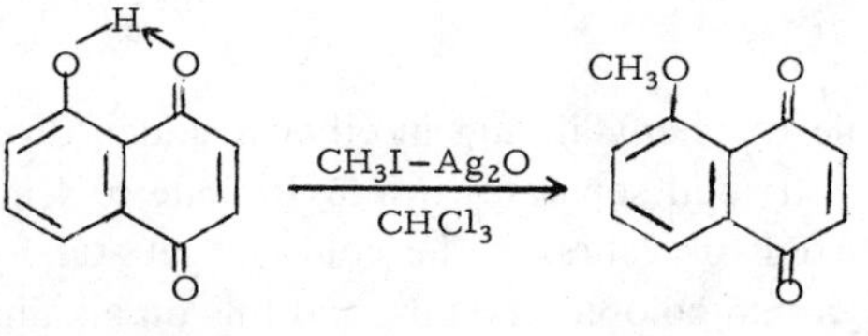
H
O
O
O
CH3I–Ag2O
CHCl3
CH3O
O
O

C-Methylation. For methylation of ethyl (1-ethylpropylidine)cyanoacetate (1, from diethyl ketone and ethyl cyanoacetate) to ethyl (1-ethylpropenyl)methylcyanoacetate (3),[15] 0.4 mole of (1) is added dropwise to a solution of 0.4 mole of sodium ethoxide in ethanol at −5° to form the enolate (2), and 0.44 mole of methyl iodide is added at a rate sufficient to produce a gentle temperature rise. The mixture

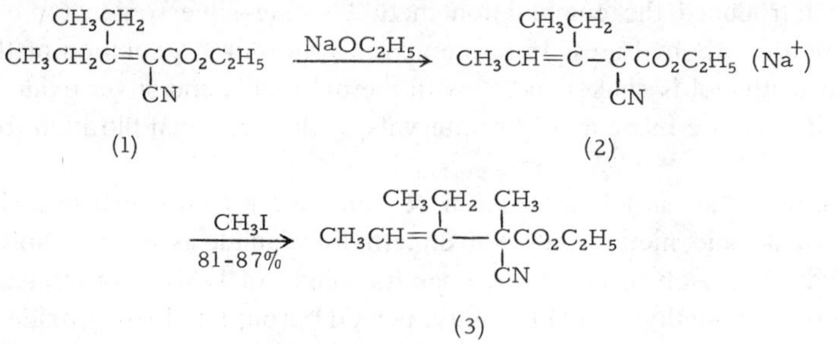

is then refluxed until a piece of red litmus paper dipped into the solution and then moistened shows a neutral reaction. Water is added and the organic layer separated, dried, and distilled.

The C-methylation of dihydroresorcinol (2) is carried out with unisolated material obtained by hydrogenation of resorcinol (2 moles) in alkaline solution.[16] The solution is filtered from catalyst and treated with dioxane and 2.64 moles of methyl iodide,

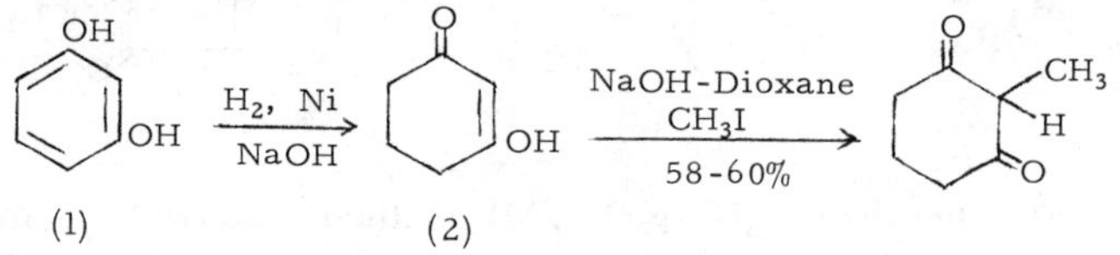

and the mixture is refluxed for 12–14 hrs. On thorough ice cooling, 2-methylcyclohexane-1,3-dione separates in nearly pure form. Crystallization from 95% ethanol gives colorless crystals, m.p. 208–210 dec., with only minor losses.

The synthesis of 1-methylisoquinoline from the parent heterocycle is accomplished by methylation of the Reissert compound (2).[17] This solid (0.32 mole) is dried and stirred under nitrogen with dioxane–ether until dissolved, the solution is cooled

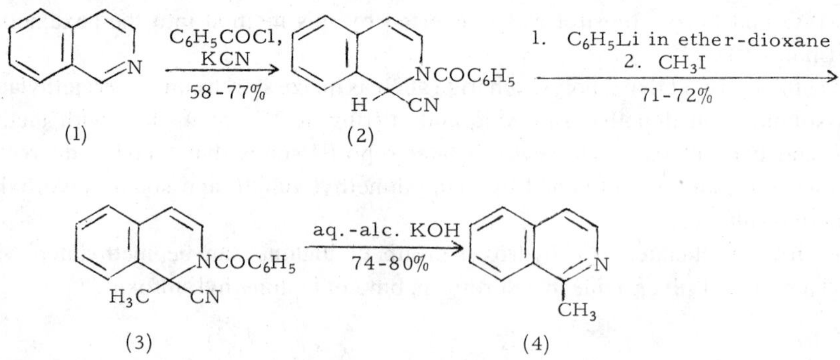

to −10°, and 0.35 mole of phenyllithium in ether is added dropwise; the solution turns deep red and a red solid separates. Methyl iodide (0.4 mole) is then added, and the mixture is stirred for 2 hrs. in the cold and let stand overnight. Suitable workup affords (3) as cream-colored crystals, and this on alkaline hydrolysis yields 1-methylisoquinoline (4).

Addition to trivalent N and P. In the preparation of benzyltrimethylammonium iodide (2), a solution of 1 mole of the tertiary amine (1) in 200 ml. of absolute ethanol is stirred during addition of 1.34 moles of methyl iodide at a rate sufficient to promote gentle refluxing (30 min.).[18] Considerable methiodide (2) separates during the addition; addition of 1 l. of absolute ether precipitates the remainder.

$$\underset{(1)}{C_6H_5CH_2N(CH_3)_2} + CH_3I \xrightarrow[94-99\%]{} \underset{(2)}{C_6H_5CH_2\overset{+}{N}(CH_3)_3(I^-)}$$

Diisopropyl methylphosphonate (4) is prepared by reaction of 2 moles of triisopropyl phosphate and 2 moles of methyl iodide.[19] The methyl iodide is placed in a flask with an efficient condenser and about 50 ml. of triisopropyl phosphate is run in. The flask is heated briefly until the reaction starts and the flame is then withdrawn. The remainder of the reagent is added at a rate sufficient to maintain gentle

$$\underset{(3)}{[(CH_3)_2CHO]_3P} + CH_3I \xrightarrow[85-95\%]{} \underset{(4)}{[(CH_3)_2CHO]_2\overset{O^-}{\overset{|}{P^+}}-CH_3} + (CH_3)_2CHI$$

refluxing. After refluxing the mixture for 1 hr. more, the mixture is worked up by distillation. The total isopropyl iodide recovered amounts to 91%. The phosphonate ester distills at 51°/ 1 mm. as a colorless liquid.

Methylation of carotenoid alcohols. Work in this area has been reviewed and extended by Karrer.[20]

[1]H. S. King, *Org. Syn., Coll. Vol.*, **2**, 399 (1943)
[2]W. W. Hartman, *ibid.*, **2**, 404 (1943)
[3]L. H. Sarett, private communication
[4]G. N. Vyas and N. M. Shah, *Org. Syn., Coll. Vol.*, **4**, 836 (1963)
[5]T. Purdie and J. C. Irvine, *J. Chem. Soc.*, **83**, 1021 (1903)
[6]A. S. Anderson, G. R. Barker, J. M. Gulland, and M. V. Lock, *J. Chem. Soc.*, 369 (1952)
[7]R. Kuhn, H. Trischmann, and I. Löw, *Angew. Chem.*, **67**, 32 (1955); R. Kuhn and H. Baer, *Ber.*, **88**, 1537 (1955); **89**, 504 (1956)
[8]R. Kuhn, H. Baer, and A. Seeliger, *Ann.*, **611**, 236 (1958)
[9]R. Kuhn and H. Trischmann, *Ber.*, **94**, 2258 (1961)
[10]H. G. Walker, Jr., M. Gee, and R. M. McCready, *J. Org.*, **27**, 2100 (1962)
[11]R. Kuhn and H. Trischmann, *Ber.*, **96**, 284 (1963). Review: K. Wallenfels, G. Bechtler, R. Kuhn, H. Trischmann, and H. Egge, *Angew. Chem., Internat. Ed.*, **2**, 515 (1963)
[12]H. C. Srivastava, S. N. Harshe, and P. P. Singh, *Tetrahedron Letters*, 1869 (1963)
[13]H. C. Srivastava, P. P. Singh, S. N. Harshe, and K. Virk, *Tetrahedron Letters*, 493 (1964)
[14]J. F. Garden and R. H. Thomson, *J. Chem. Soc.*, 2483 (1957)
[15]E. M. Hancock and A. C. Cope, *Org. Syn., Coll. Vol.*, **3**, 397 (1955)
[16]A. B. Mekler, S. Ramachandran, S. Swaminathan, and M. S. Newman, *Org. Syn.*, **41**, 56 (1961)
[17]J. Weinstock and V. Boekelheide, *Org. Syn., Coll. Vol.*, **4**, 641 (1963)
[18]W. R. Brasen and C. R. Hauser, *ibid.*, **4**, 585 (1963)
[19]A. H. Ford-Moore and B. J. Perry, *ibid.*, **4**, 325 (1963)
[20]H. Müller and P. Karrer, *Helv.*, **48**, 291 (1965)

Methylketene diethylacetal, $CH_3CH{=}C(OC_2H_5)_2$. Mol. wt. 130.18, b.p. 133–134°.

Preparation[1] by dropwise addition of triethyl α-bromoorthopropionate to a stirred suspension of powdered sodium in refluxing benzene.

$$CH_3CHBrC(OC_2H_5)_3 + 2\ Na \xrightarrow[80\%]{} CH_3CH{=}C(OC_2H_5)_2 + NaBr + NaOC_2H_5$$

Dehydration of aldoximes.[2] The reagent forms an adduct (1) with a *syn*-aldoxime which is decomposed by catalytic amounts of boron trifluoride and mercuric oxide to the isonitrile (2) in good yield; the adduct (3) of an *anti*-oxime decomposes to give the nitrile (4).

$$HON\!=\!CHC_6H_4CH_3 \xrightarrow{CH_3CH=C(OC_2H_5)_2}$$

$$CH_3CH_2\underset{\substack{| \\ O \\ \diagdown N=C \diagup H \\ \diagdown C_6H_4CH_3}}{C}(OC_2H_5)_2 \longrightarrow CH_3CH_2CO_2C_2H_5 \;+\; C_2H_5OH \;+\; CH_3C_6H_4N\!=\!C$$

(1)   (2)

$$CH_3CH_2\underset{\substack{| \\ O \\ \diagdown N=C \diagup C_6H_4CH_3 \\ \diagdown H}}{C}(OC_2H_5)_2 \longrightarrow CH_3CH_2CO_2C_2H_5 \;+\; C_2H_5OH \;+\; CH_3C_6H_4C\!\equiv\!N$$

(3)   (4)

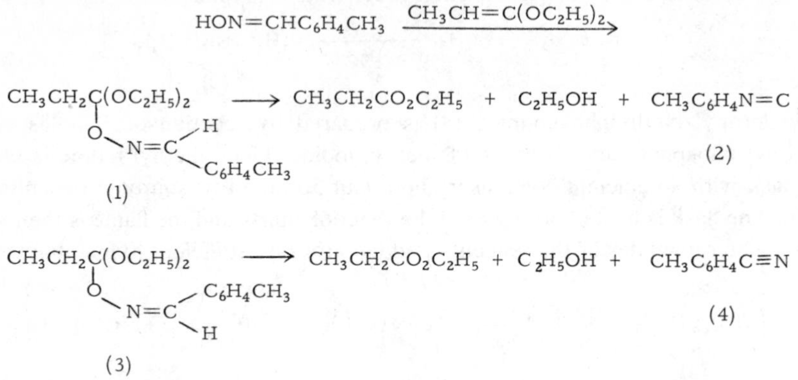

[1]P. M. Walters and S. M. McElvain, *Am. Soc.*, **62**, 1482 (1940)
[2]T. Mukaiyama, K. Tonooka, and K. Inoue, *J. Org.*, **26**, 2202 (1961)

Methyllithium, CH_3Li. Suppliers: Lithium Corp. Amer. (*ca.* $2\,M$ solution in ether), Alfa.

Allene synthesis. Doering discovered a novel synthesis of allenes involving addition of dibromocarbene to an olefin[1] and dehalogenation with magnesium with opening of the three-membered ring.[2] Yields in the second step were low (16–35%), but the products were of high purity. Then both Moore[3] and Skattebøl[4] found that the

$$-CH\!=\!CH- \xrightarrow{:CBr_2} -\underset{\diagdown}{CH}\!\overset{\overset{\displaystyle Br \diagdown C \diagup Br}{|}}{\underset{}{\!-\!-\!}}CH- \xrightarrow{Mg} -CH\!=\!C\!=\!CH- \;+\; MgBr_2$$

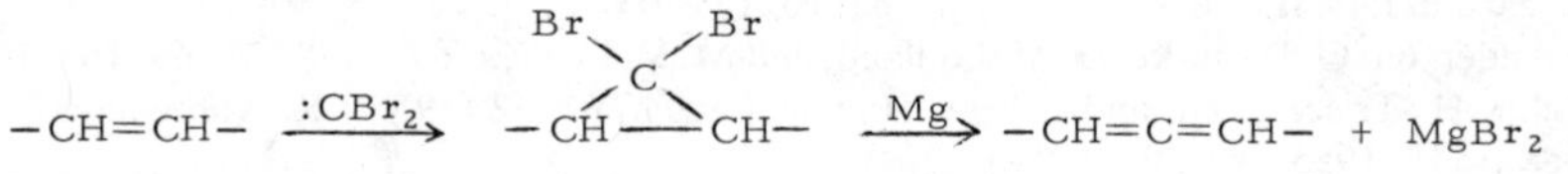

step of dehalogenation-rearrangement can be conducted much more effectively with an alkyllithium than with magnesium:

$$-\underset{\diagdown}{CH}\!\overset{\overset{\displaystyle Br \diagdown C \diagup Br}{|}}{\!-\!-\!}CH- \;+\; CH_3Li \longrightarrow -CH\!=\!C\!=\!CH- \;+\; CH_3Br \;+\; LiBr$$

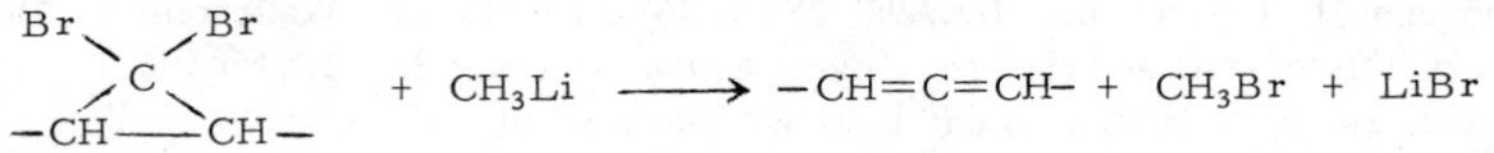

An example is the synthesis of 1,2-cyclononadiene,[3, 4] as described in a detailed procedure.[5] The starting material, *cis*-cyclooctene (2) is available commercially or prepared by partial hydrogenation of *cis,cis*-1,5-cyclooctadiene.[6] A flask fitted with

$$(1) \xrightarrow{H_2,\ Pd} (2) \xrightarrow[65-76\%]{\substack{CHBr_3, \\ t-BuOK}} (3) \xrightarrow{CH_3Li} (4) \xrightarrow[85\%]{Na,\ NH_3} (5)$$

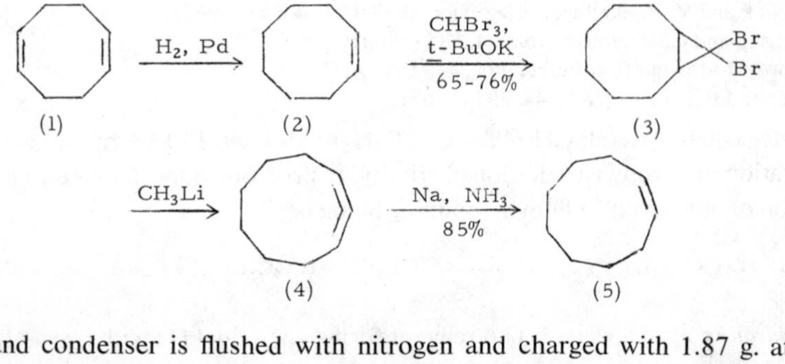

stirrer and condenser is flushed with nitrogen and charged with 1.87 g. atoms of potassium and 2 l. of anhydrous *t*-butanol, and the mixture is stirred under reflux and a slight positive pressure of nitrogen until the metal is dissolved (hydrogen escapes through a mercury trap). About 1.5 l. of *t*-butanol is distilled at atmospheric pressure under nitrogen into a predried flask (and is suitable for reuse). The remaining solvent is then removed at reduced pressure and eventually the white potassium *t*-butoxide is heated at 150° under a pressure of 1–0.1 mm. for 2 hrs. Nitrogen is again introduced, 1.62 moles of *cis*-cyclooctene and 200 ml. of pentane are added (the diluent facilitates stirring), a dropping funnel is charged with 1.66 moles of bromoform, and the flask is cooled in an ice-salt bath. The bromoform is added dropwise to the stirred slurry over a period of 6–7 hrs., the color of the reaction mixture changing gradually from light yellow to brown. The mixture is then allowed to come to room temperature and left stirring overnight. Addition of water, neutralization, extraction with pentane, and distillation affords pure 9,9-dibromobicyclo[6.1.0]-nonane (3, b.p. 62°/0.4 mm.).

The next step is carried out in a flask equipped with a stirrer, pressure-equalized dropping funnel, and a nitrogen inlet connected to a mercury valve (*see* Grignard reagents, Fig. G-3). The flask is charged with 0.66 mole of the dibromide (3) and 0.85 mole of methyllithium in 450 ml. of ether and cooled in a dry ice–acetone bath maintained at −30 to −40°. The methyllithium is added dropwise with stirring during 1 hr. After stirring for 30 min. more, excess methyllithium is destroyed by dropwise addition of 100 ml. of water. More water is then added, and the product is extracted with ether and distilled. The 1,2-cyclooctadiene, obtained in yield of 90–93%, was found by gas chromatography to be 99% pure. Gardner[6] had obtained the hydrocarbon from (3) by the magnesium procedure in 59% yield and reduced it with sodium and ammonia to *cis*-cyclononene (5).

Methyllithium is the generally preferred reagent for the allene synthesis but is less reactive than *n*-butyllithium. Thus Moore[3] found that 1,1-dichloro-2-hexylcyclopropane was recovered unchanged after treatment with methyllithium in reflux-

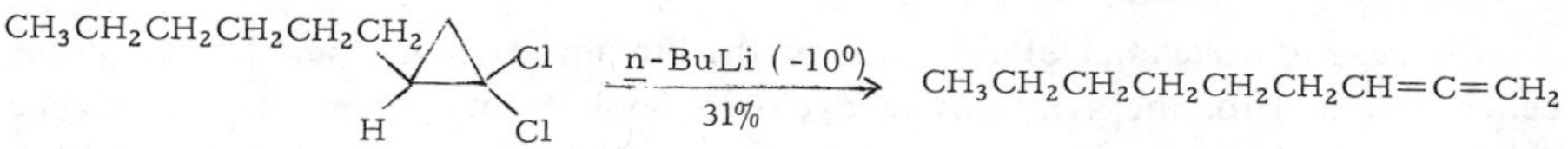

CH3CH2CH2CH2CH2CH2
Cl
H
Cl
n-BuLi (-10°)
31%
CH3CH2CH2CH2CH2CH2CH=C=CH2

ing ether but that it reacted with *n*-butyllithium in ether at −10° to give 1,2-nonadiene in 31% yield (52% conversion).

Although the reaction of methyllithium with *gem*-dibromocyclopropanes generally affords allenes, the reaction with the strained 8,8-dibromocyclo[5.1.0]octane leads to the shower of products formulated.[6a]

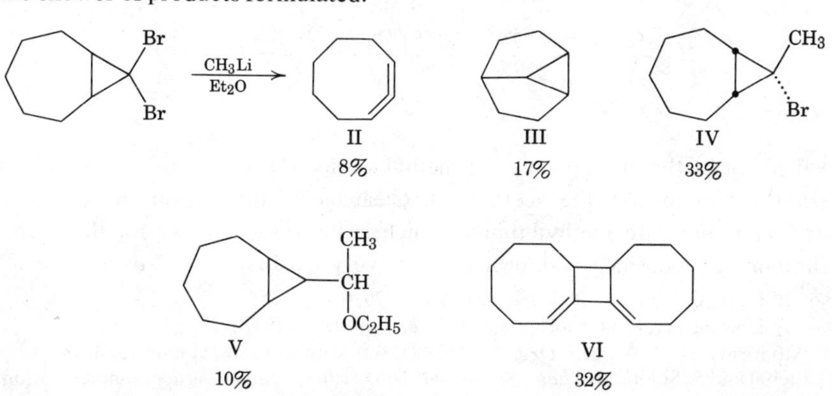

Br
Br
CH3Li
Et2O
II
8%
III
17%
CH3
Br
IV
33%
CH3
CH
OC2H5
V
10%
VI
32%

Untch *et al.*[7] describe an olefin-to-allene conversion involving the following steps:

$$>C=C< + CBr_4 + CH_3Li \longrightarrow >C=C< + :CBr_2 \longrightarrow \overset{Br\quad Br}{\underset{>C-C<}{\diagup C}} \xrightarrow{CH_3Li} >C=C=C<$$

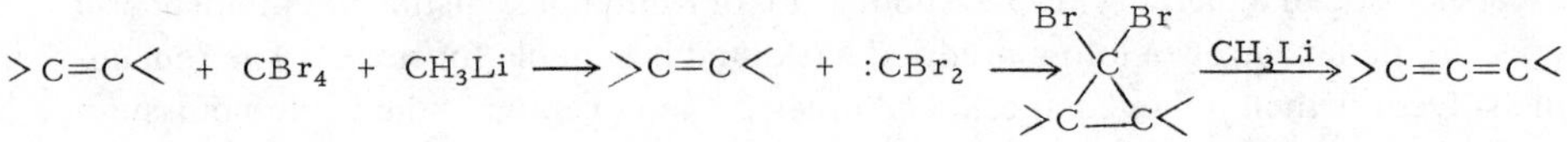

The olefin, taken in 4-fold excess, is treated with 1 equivalent of carbon tetrabromide and 2 equivalents of methyllithium in ether at $-65°$. The yield was twice that obtained with *n*-butyllithium.

Skattebøl[8] employed the reaction for the synthesis of cumulenes from allenes. However, in the first step the yield of the desired bis adduct was only 6%, the major product being the mono adduct.

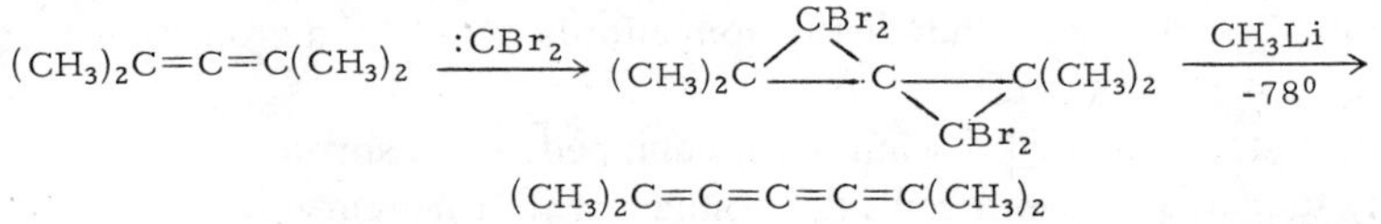

$$(CH_3)_2C=C=C(CH_3)_2 \xrightarrow{:CBr_2} (CH_3)_2C\overset{CBr_2}{\underset{CBr_2}{\diagup C \diagdown}}C(CH_3)_2 \xrightarrow[-78°]{CH_3Li}$$

$$(CH_3)_2C=C=C=C=C(CH_3)_2$$

CH₃Li vs. CH₃MgBr. In the conversion of ketols of type (1) into the *vic*-glycols (2), yields with the Grignard reagent were about 20% whereas with methyllithium yields were about 58%.[9]

$$\underset{\overset{|}{OH}\ \overset{||}{O}}{R-\overset{\overset{R}{|}}{C}-CCH_3} \xrightarrow{CH_3Li} \underset{\overset{|}{OH}\ \overset{|}{OH}}{R-\overset{\overset{R}{|}}{C}-\overset{\overset{CH_3}{|}}{C}CH_3}$$

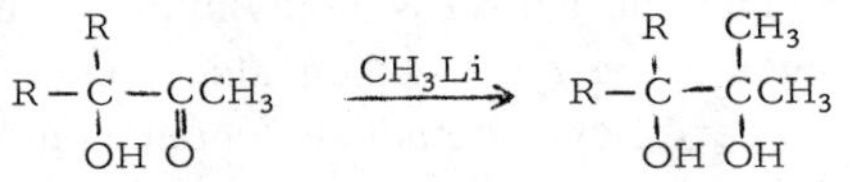

$$(1) \qquad\qquad (2)$$

Methyl ketones from acids. A carboxylic acid or its lithium salt reacts with an alkyllithium to form a ketone.[10] See next paragraph for an example.

$$RCO_2H + 2\ R'Li \longrightarrow R\overset{O}{\overset{||}{C}}R' + Li_2O + R'H$$

Cleavage of acetates. DePuy[11] used methyllithium in the first and last step of a general method for the synthesis of cyclopropanols from cyclopropanecarboxylic acids, for example *trans*-2-phenylcyclopropanecarboxylic acid (1). Reaction with 2

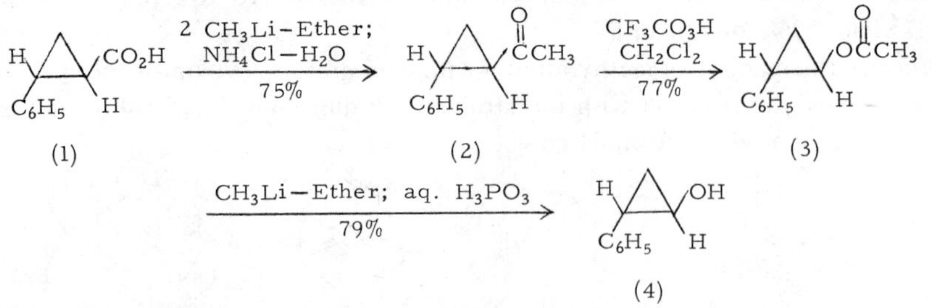

equivalents of methyllithium gave the methyl ketone (2), which underwent Baeyer-Villiger reaction to give the acetate (3). Cleavage to the cyclopropanol (4) was effected smoothly with methyllithium, which is more satisfactory for the purpose than lithium hydride, since cyclopropanols are very sensitive to this strong base.

[1] W. von E. Doering and A. K. Hoffmann, *Am. Soc.*, **76**, 6161 (1954)

[2] W. von E. Doering and P. M. LaFlamme, *Tetrahedron*, **2**, 75 (1958)

[3] W. R. Moore and H. R. Ward, *J. Org.*, **27**, 4179 (1962), with an earlier communication

[4] L. Skattebøl and S. Solomon, *Chem. Scand.*, **17**, 1683 (1963), with an earlier communication

[5]*Idem*, procedure submitted to *Org. Syn.*
[6]P. D. Gardner and M. Narayana, *J. Org.*, **26**, 3518 (1961)
[6a]E. T. Marquis and P. D. Gardner, *Tetrahedron Letters*, 2793 (1966)
[7]K. G. Untch, D. J. Martin, and N. T. Castellucci, *J. Org.*, **30**, 3572 (1965)
[8]L. Skattebøl, *Tetrahedron Letters*, 2175 (1965)
[9]W. J. Hickinbottom, A. A. Hyatt, and M. B. Sparke, *J. Chem. Soc.*, 2533 (1954)
[10]H. Gilman and P. R. Van Ess, *Am. Soc.*, **55**, 1258 (1933); H. Gilman, W. Langham, and F. W. Moore, *ibid.*, **62**, 2327 (1940); C. Tegnér, *Chem. Scand.*, **6**, 782 (1952); J. F. Arens and D. A. Van Dorp, *Rec. trav.*, **65**, 338 (1946); **66**, 759 (1947)
[11]C. H. DePuy, G. M. Dappen, K. L. Eilers, and R. A. Klein, *J. Org.*, **29**, 2813 (1964)

Methylmagnesium bromide. (For suppliers, *see* Grignard reagents.)

In solution in tetrahydrofurane, this reagent reduces a *gem*-dibromocyclopropane at room temperature to afford the monobromocyclopropane in good yield.[1] The

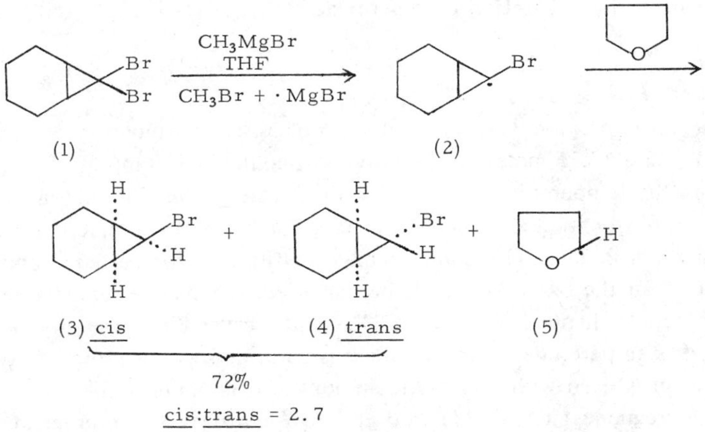
CH₃MgBr
THF
Br
Br
CH₃Br + · MgBr
(1)
(2)
H
Br
H
(3) cis
(4) trans
(5)
72%
cis:trans = 2.7

reaction appears to involve a radical process. That the *cis*-isomer predominates is not surprising: the 7α-bromine atom is removed preferentially because it is less shielded than the 7β-bromine atom of (1).

[1]D. Seyferth and B. Prokai, *J. Org.*, **31**, 1702 (1966)

Methylmagnesium iodide. (For suppliers, *see Grignard reagents.*) Wilds and McCormack[1] encountered difficulties in attempted demethylation of the ketone (2). The usual, strongly acidic reagents lead to self-condensation; even pyridine hydrochloride was unsatisfactory. They then found that the alcoholic precursor (1) can

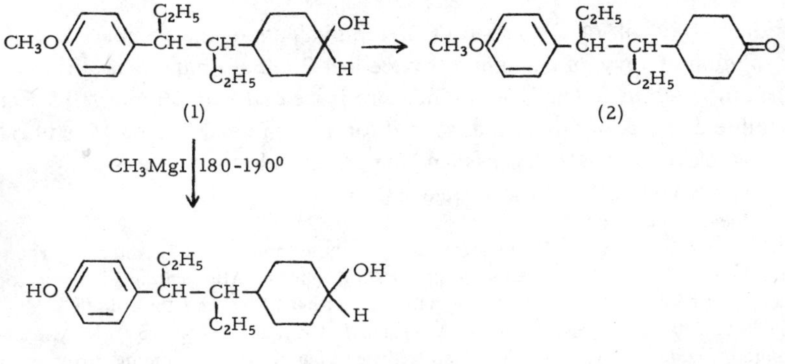
CH₃O
C₂H₅
CH—CH
OH
H
(1)
(2)
CH₃MgI 180-190°
HO
(3)

be demethylated efficiently by heating with methylmagnesium iodide. The ether (1) is treated with an ethereal solution of the Grignard reagent, the solvent is evaporated, and the residual material is heated to 180–190° for 2 hrs.

[1]A. L. Wilds and W. B. McCormack, *Am. Soc.*, **70**, 4127 (1948)

N-Methylmorpholine, 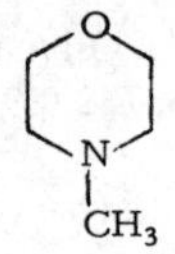

Mol. wt. 101.15, b.p. 113–115°. Suppliers: Union Carbide, A, B, E, MCB.

N-Methylmorpholine oxide–Hydrogen peroxide,* 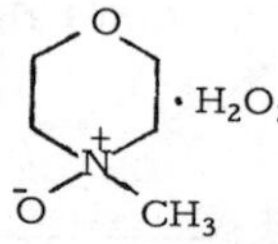

Mol. wt. 151.16, m.p. 73–75°.

The reagent is prepared[1] by slow addition of 34 g. (0.5 mole) of 50% hydrogen peroxide to 26 g. (0.25 mole) of N-methylmorpholine in 100 ml. of *t*-butanol while maintaining the temperature at 30–35° with a water bath. The mixture is diluted with 170 ml. of *t*-butanol and allowed to stand for 48 hrs. to complete the oxidation of N-methylmorpholine. The solution may be titrated[2] for peroxide content and used as such for the oxidation of olefins, or dried with magnesium sulfate and the volatile materials distilled *in vacuo* to leave the crystalline N-methylmorpholine oxide–hydrogen peroxide complex, which is triturated with acetone and collected.

The reagent is used with a catalytic amount of osmium tetroxide for conversion of a $\Delta^{17(20)}$-pregnene such as (1) into the 17α-hydroxy-20-ketopregnane (2).[3] A solution of 3.24 g. of (1) in 110 ml. of *t*-butanol, 30 ml. of methylene chloride, 4.1 ml.

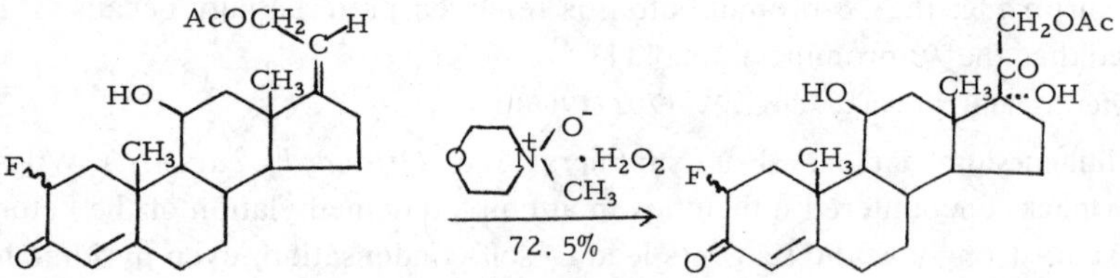
AcOCH₂
H
C
CH₃
HO
CH₃
F
O
• H₂O₂
CH₃
72. 5%
CH₂OAc
CO
CH₃
OH
HO
CH₃
F
O

(1) (2)

of pyridine, 10.9 ml. of a solution of 2.1 equivalents of the complex in *t*-butanol, and a solution of 5 mg. of osmium tetroxide in 1.7 ml. of *t*-butanol is stirred at room temperature overnight. The reaction mixture is treated with 30 ml. of 0.5% sodium hydrosulfite and 2 g. of a filter-aid, stirred for 30 min., and filtered. The filtrate and washings are evaporated to dryness, and the product chromatographed.

*Contributed by William P. Schneider, Upjohn Co.

[1]W. P. Schneider and A. R. Hanze, U. S. patent 2,769,823

[2]A 2-ml. aliquot is added to 20 ml. of acetic acid and about 1 g. of sodium bicarbonate (to keep out air) and 2 ml. of saturated potassium iodide solution. After standing loosely stoppered in the dark for 5 min., the resulting iodine is titrated with 0.1 *N* sodium thiosulfate.

[3]A. H. Nathan, B. J. Magerlein, and J. A. Hogg, *J. Org.*, **24**, 1517 (1959). K. Miescher and J. Schmidlin, *Helv.*, **33**, 1840 (1950), used hydrogen peroxide and osmium tetroxide to effect this transformation.

Methyl nitrate, CH_3ONO_2. Mol. wt. 74.02, b.p. 65°, sp. gr. 1.20. Supplier: K, K.

Preparation.[1] Concd. nitric acid (300 ml.) is chilled in ice and treated with cooling with 300 ml. of concd. sulfuric acid. In a second flask 150 ml. of methanol is cooled in ice to keep the temperature below 10° during cautious addition of 50 ml. of concd. sulfuric acid. One third of the cold nitric–sulfuric acid is placed in each of three 500-ml. Erlenmeyers, and each portion is treated with one third of the methanol–sulfuric acid mixture, added in 2–3 min. with constant swirling. Methyl nitrate separates as an almost colorless oily upper layer. After standing for 15 min. the lower layer of spent acid is separated and quenched with a large volume of water to avoid vigorous decomposition. The combined ester is washed with 25 ml. of ice-cold 22% sodium chloride solution, and the process is repeated with addition of enough alkali to produce a faintly alkaline reaction. The ester is washed free of alkali with ice-cold salt solution, then washed twice with 15 ml. of ice water, dried over calcium chloride, decanted, and used directly. Yield 190–230 g. (66–80%). Distillation is not recommended; the crude ester is satisfactory for synthetic purposes.

Preparation of phenylnitromethane.[2] A solution of sodium ethoxide prepared from 2 g. atoms of sodium and 400 ml. of absolute ethanol is treated with 100 ml. of absolute ethanol and cooled to 0°, when the sodium ethoxide separates as a hard cake. A second 100 ml. portion of ethanol is added, and a chilled mixture of 2 moles of benzyl cyanide and 2.8 moles of methyl nitrate is added with shaking at 5–15°. The reaction proceeds smoothly, with separation of the *aci*-nitro sodium salt (1). The salt is collected (300–320 g.), placed in a 4-l. beaker, and treated with a solution

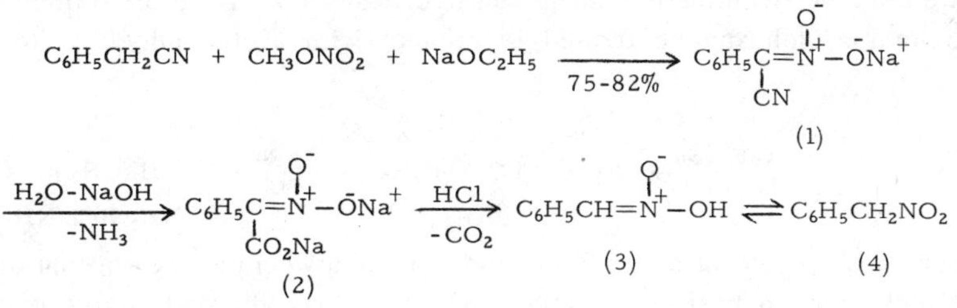

of 300 g. of sodium hydroxide in 1.5 l. of water. The mixture is boiled for hydrolysis of the nitrile group, and when ammonia is no longer evolved (3 hrs.) the beaker is placed in a salt-ice bath and stirred mechanically until the temperature drops to 30°. Then 500 g. of ice is added and concd. hydrochloric acid (850 ml.) is stirred in at 0–10°. Phenylnitromethane, isolated by ether extraction and distilled at 92–94°/4 mm., is obtained in yield of 56–60% (based on the benzyl cyanide used).

[1] A. P. Black and F. H. Babers, *Org. Syn., Coll. Vol.*, **2**, 412 (1943)
[2] *Idem, ibid.*, **2**, 512 (1943)

Methyl nitrite, CH_3ONO. Mol. wt. 61.04, b.p. −12°, sp. gr. 0.99.

A procedure[1] for the conversion of propiophenone into an α-oximino derivative,

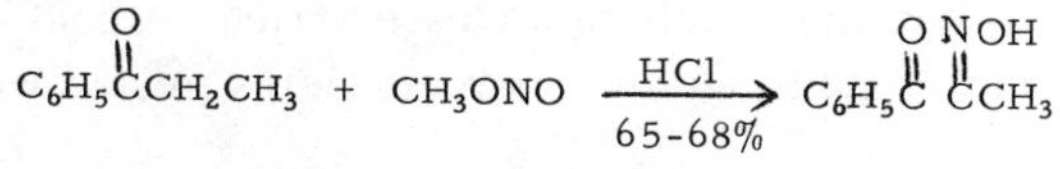

isonitrosopropiophenone, calls for passing in hydrogen chloride and methyl nitrite generated by dropping a 1:2 (by volume) mixture of sulfuric acid and water into a mixture of sodium nitrite, methanol, and water into a solution of propiophenone in ether. After about $4\frac{1}{2}$ hrs. the brown-red solution turns yellow, and the product is isolated by repeated extraction with 10% aqueous alkali; acidification of the combined extracts gives a product melting at 111–113° in 65–68% yield. Crystallization from toluene gives snow-white isonitrosopropiophenone, m.p. 112–113°.

Ferris[2] developed an interesting synthesis of α-amino acids illustrated in the formulation for the synthesis of DL-phenylalanine from 1-phenyl-3-butanone (1).

$$C_6H_5CH_2CH_2COCH_3 \xrightarrow[\text{HCl}]{\text{CH}_3\text{ONO}} C_6H_5CH_2C\underset{\|}{\overset{}{\,}}COCH_3 \xrightarrow[\text{NaOH}]{(CH_3)_2SO_4} C_6H_5CH_2C\underset{\|}{\overset{}{\,}}COCH_3$$
$$\qquad\qquad\qquad\qquad\qquad\qquad\qquad\qquad\text{NOH}\qquad\qquad\qquad\qquad\qquad\qquad\text{NOCH}_3$$

$$\text{(1)}\qquad\qquad\qquad\qquad\qquad\qquad\text{(2)}\qquad\qquad\qquad\qquad\qquad\qquad\text{(3)}$$

$$\xrightarrow[\text{Di}]{\text{NaOCl; H}^+} C_6H_5CH_2C\underset{\|}{\overset{}{\,}}CO_2H \xrightarrow{\text{H}_2,\ \text{Pd/C}} C_6H_5CH_2CHCO_2H$$
$$\qquad\qquad\qquad\qquad\text{NOC}_2\text{H}_5\qquad\qquad\qquad\qquad\text{NH}_2$$

$$\qquad\qquad\qquad\qquad\text{(4)}\qquad\qquad\qquad\qquad\qquad\text{(5)}$$

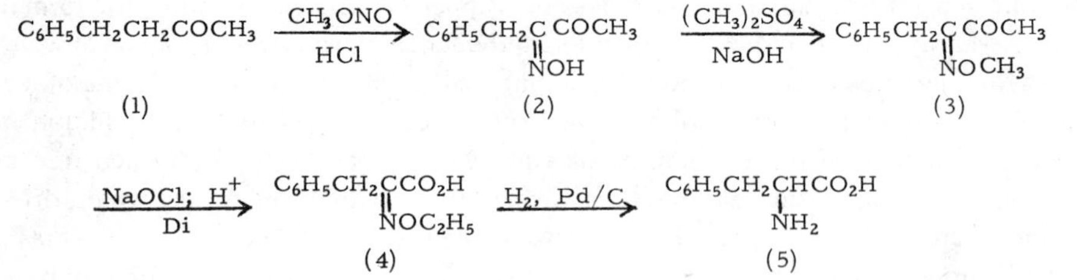

A solution of 1.8 moles of the ketone in 800 ml. of ether is treated with the methyl nitrite generated from 2.58 moles of methanol, 2.28 moles of 95% sodium nitrite in 100 ml. of water, and 2.86 moles of concd. sulfuric acid diluted with 145 ml. of water. The oximino ketone (2) is converted into the O-methyl ether (3) with dimethyl sulfate and alkali, and hypochlorite cleavage in aqueous dioxane affords the acid (4). Hydrogenation completes the synthesis of DL-phenylalanine (5) in overall yield of 50% from (1).

Ferris' FMC group[3] worked out a synthesis of lysine in which cyclohexanone is nitrosated in ether with methyl nitrite and hydrochloric acid and the sodium salt of 2,6-dioximinocyclohexanone treated in ethanol with acetic anhydride to effect

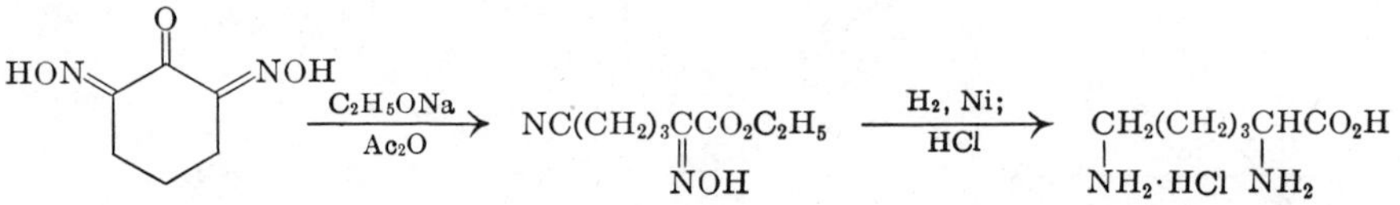

Beckmann rearrangement to ethyl 5-cyano-2-oximinovalerate. Hydrogenation over Raney nickel and a basic co-catalyst and hydrolysis afforded DL-lysine monohydrochloride in 63% overall yield from cyclohexanone. A less fully studied application of the sequence to cyclopentanone affords DL-ornithine in overall yield of 21%.

[1]W. H. Hartung and F. Crossley, *Org. Syn., Coll. Vol.*, 2, 363 (1943)

[2]A. F. Ferris, *J. Org.*, 24, 1726 (1959)

[3]A. F. Ferris, *J. Org.*, 25, 12 (1960); A. F. Ferris, G. S. Johnson, F. E. Gould, and H. K. Latourette, *ibid.*, 25, 492 (1960); A. F. Ferris, G. S. Johnson, and F. E. Gould, *ibid.*, 25, 496 (1960); A. F. Ferris, G. S. Johnson, F. E. Gould, and H. Stange, *ibid.*, 25, 1302 (1960); 26, 2602 (1961)

Methyl oxocarbonium hexafluoroantimonate, $CH_3CO^+SbF_6^-$.[1] Mol. wt. 278.80, m.p. 173–175°.

Preparation.[2] This extremely stable oxocarbonium salt is prepared with provision for exclusion of moisture, preferably in a closed system. A cold solution of acetyl fluoride in Freon 113 (1,1,2-trifluorotrichloroethane) is treated with an equivalent

$$CH_3COF\ +\ Cl_2CFCClF_2\ +\ SbF_5 \xrightarrow{-5\ \text{to}\ 0^0} CH_3CO^+SbF_6^-$$

$$\text{0. 2 m.}\qquad\qquad\text{70 ml.}\qquad\qquad\text{0. 2 m.}$$

Freon solution of freshly distilled antimony pentafluoride, and after stirring for 1 hr. the white, crystalline precipitate is collected, washed with cold Freon 113, and dried in vacuum.

Acetylation. The reagent, neat or in nitromethane solution, acylates aromatic compounds in good yield (1). It is an effective reagent for O-acetylation of alcohols (2), for S-acetylation of mercaptans, and for N-acetylation of primary and secondary amides.

$$1.\quad C_6H_5CH_3(Cl) + CH_3CO^+SbF_6^- \xrightarrow[96\ (87\%)]{CH_3NO_2} \underline{p}\text{-}CH_3\overset{O}{\overset{\|}{C}}C_6H_4CH_3(Cl) + HF + SbF_5$$

$$2.\quad CH_3(CH_2)_6CH_2OH + CH_3CO^+SbF_6^- \xrightarrow[87\%]{CH_3NO_2} CH_3(CH_2)_6CH_2OCOCH_3 + HF + SbF$$

[1]Contributed by George A. Olah, Case Institute of Technology
[2]G. A. Olah, S. J. Kuhn, W. S. Tolgyesi, and E. B. Baker, *Am. Soc.*, **84**, 2733 (1962)

2-Methylpentane-2,4-diol, $CH_3CH(OH)CH_2C(CH_3)(OH)CH_3$

Mol. wt. 118.17, b.p. 135–136°/40 mm., water soluble. Suppliers: Shell Chemical Co. Aldrich, Eastman.

The diol is useful as a reagent for characterization and isolation of aldehydes.[1] A mixture of the aldehyde and diol and an acid catalyst in benzene is refluxed under a

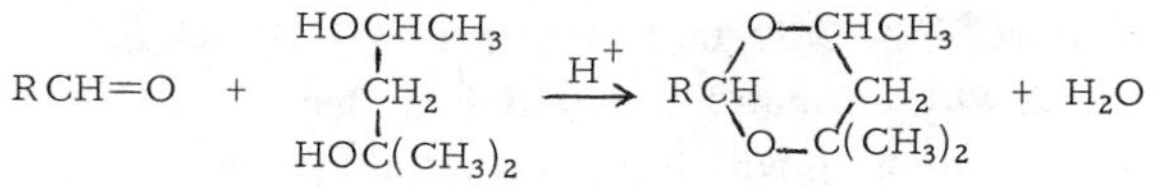

phase-separating head until water evolution ceases. The catalyst is usually *p*-toluenesulfonic acid or a solid catalyst of sulfuric acid absorbed on silica gel, prepared by adding a solution of 0.5 g. of concd. sulfuric acid in 10 ml. of water to 100 g. of 28–200 mesh silica gel. The cyclic six-membered acetals are formed in high yield; for example that from acrolein is obtained in 98% yield, whereas the yield of acrolein ethyleneglycol acetal (five-membered) is only 58%. The cyclic acetals probably are stabilized by the equatorial orientation of three of the substituent groups.

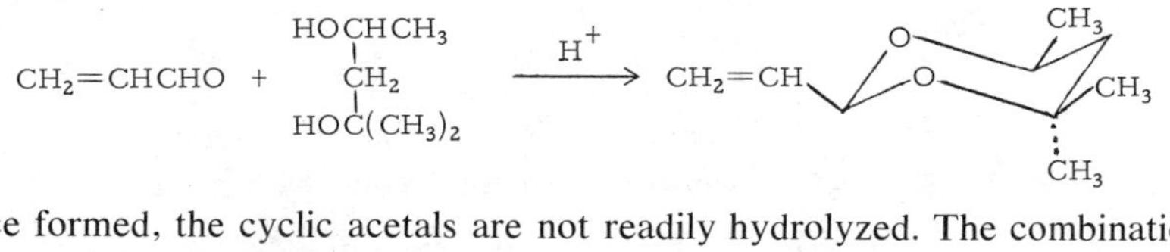

Once formed, the cyclic acetals are not readily hydrolyzed. The combination of ready formation, stability, and sparing solubility of the acetals in water makes it possible to form an acetal in aqueous solution and extract it with benzene. Thus the concentration of an aqueous solution of formaldehyde was reduced from 3 to 0.8% by simply stirring the solution at pH 3 with the diol and benzene; the acetal, 4,4,6-trimethyl-1,3-dioxane, was isolated in 65% yield. Another example is α-hydroxyadipaldehyde (2) the product of acid hydrolysis of (1), the thermal dimer of acrolein. The hydroxydialdehyde is stable only in the aqueous acidic solution of formation and attempts to distil it or otherwise separate it from water have all led to condensations. If, however, 2-methylpentane-2,4-diol and benzene are added to the aqueous solution, the half acetal (3) is obtained in 90–95% yield. Since the half acetal can be

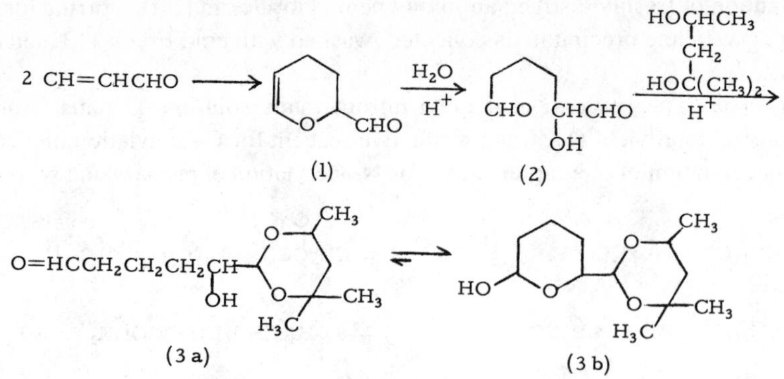

distilled and stored without decomposition, and since it does not react further with the diol, it evidently exists in the cyclic form (3b).

[1]R. F. Fischer and C. W. Smith, *J. Org.*, **25**, 319 (1960); **28**, 594 (1963)

N-Methyl-N-phenylcarbamoyl chloride, $C_6H_5N(CH_3)COCl$

Mol. wt. 169.61, m.p. 85–86°.

Weygand[1] prepared the reagent by saturating 500 ml. of ethyl acetate with phosgene and adding in 1½–2 hrs. a solution of 150 g. of N-methylaniline in 1.5 l. of ethyl acetate while passing in phosgene. Distillation of the solvent and crystallization from ethanol afforded the pure acid chloride in 75–85% yield.

The reagent reacts with aromatic compounds under Friedel-Crafts conditions to give amides, which on reduction with lithium aluminum hydride afford aldehydes.

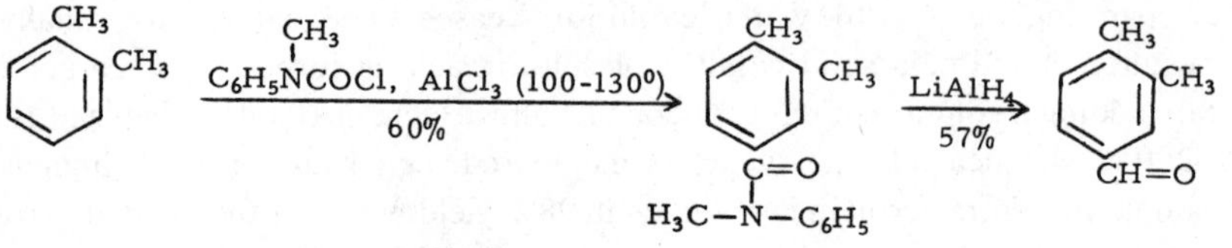

[1]F. Weygand and R. Mitgau, *Ber.*, **88**, 301 (1955)

1-Methylphenylhydrazine, $C_6H_5N(CH_3)NH_2$

Mol. wt. 122.17, b.p. 117–118°/21 mm. Suppliers: Aldrich, Eastman.

Preparation by reduction of N-nitroso-N-methylaniline with zinc dust.[1]

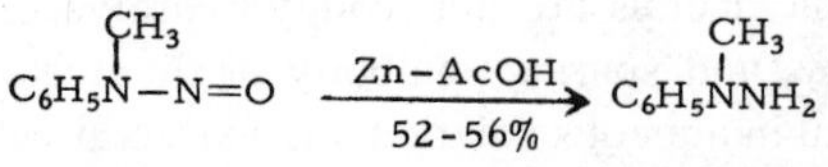

Alkazones. That oxidation of sugars by phenylhydrazine stops with phenylozazone formation was attributed by us to chelate ring formation,[2] and the suggestion has since received experimental support.[3] Chapman[4] reasoned that if the interpretation is correct the reaction with 1-methylphenylhydrazine should proceed beyond the ozazone stage, and so discovered a new class of compounds, described as alkazones. Thus on treatment with excess reagent in water–ethanol–acetic acid, 1,3-dihydroxyacetone, erythrose, and arabinose react with 3, 4, and 5 moles of 1-methylphenylhydrazine to give alkazones melting at 127°, 151° and 159° (poly-

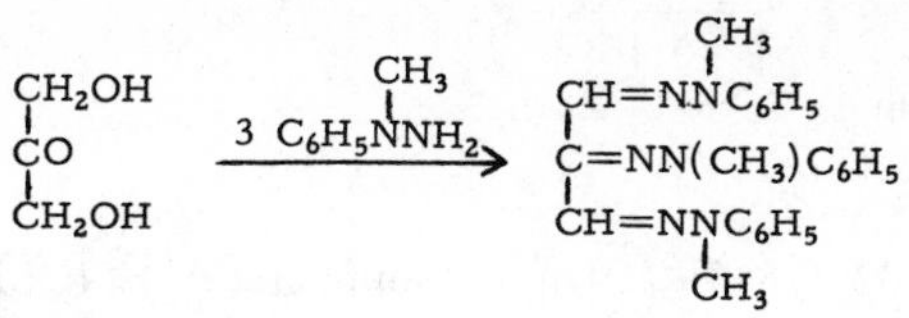

morphs), and 152° in yields of 94, 55, and 30%. The same product is obtained from xylose (39%) as from arabinose. Fructose reacted in poor yield to give an impure alkazone.

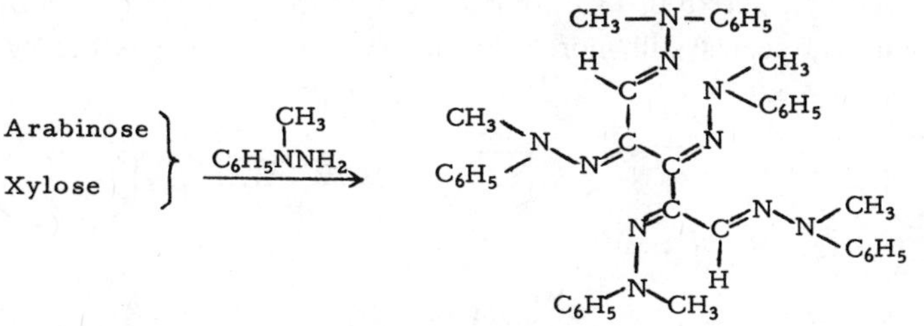

[1]W. W. Hartman and L. J. Roll, *Org. Syn., Coll. Vol.*, **2**, 418 (1943)
[2]L. F. Fieser and M. Fieser, "Organic Chemistry," 1st edition, 351–353, D. C. Heath and Co. (1944)
[3]L. Mester, *Am. Soc.*, **77**, 4301 (1955); *idem, Angew. Chem., Internat. Ed.*, **4**, 574 (1965); L. Mester, E. Moczar, and J. Parello, *Am. Soc.*, **87**, 596 (1965); G. Henseke and H. Kohler, *Ann.*, **614**, 105 (1958)
[4]O. L. Chapman, W. J. Welstead, Jr., T. J. Murphy, and R. W. King, *Am. Soc.*, **86**, 732, 4968 (1964)

3-Methyl-1-phenyl-3-phospholene-1-oxide (3). Mol. wt. 192.19, m.p. 60–65°, b.p. 163–168°/0.65 mm.

Preparation.[1] A 1-l. suction flask is charged with 1 mole of dichlorophenylphosphine (Victor Chemical Works), 300 ml. of commercial isoprene (about 3

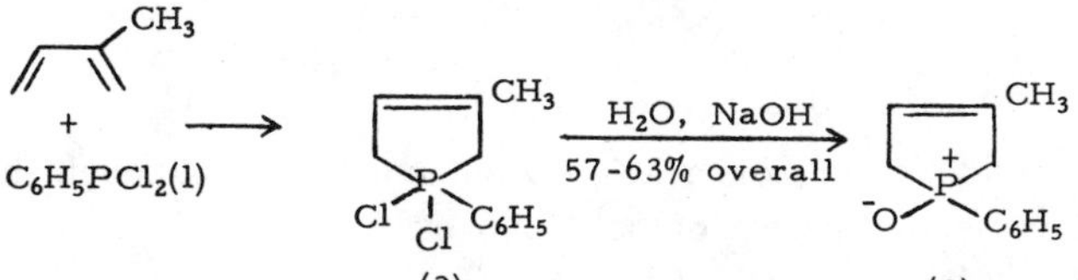

(2) (3)

moles), and 2 g. of the commercial antioxidant Ionol (2,6-di-*t*-butyl-*p*-cresol). The flask is stoppered, the side arm is sealed with tubing and a clamp, and the homogenous solution is let stand at room temperature. White solid is usually apparent within 2–4 hrs., and after 5–7 days the liquid phase is full of the white crystalline adduct (2). The granular adduct is washed, slurried with petroleum ether, and collected in a sintered glass Büchner funnel. The dichloride (2) is hydrolyzed to the oxide (3) with ice water, and the solution is neutralized with alkali and extracted with chloroform.

Use. The reagent is a catalyst for the conversion of isocyanates into carbodiimides.[2] It is comparable to 3-methyl-1-ethyl-3-phospholene-1-oxide but somewhat less reactive; on the other hand, it is more readily accessible.

[1]W. B. McCormack, *Org. Syn.*, **43**, 73 (1963)
[2]T. W. Campbell and J. J. Monagle, *Org. Syn.*, **43**, 31 (1963)

1-Methyl-2-pyrrolidone, 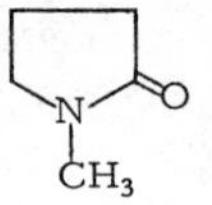
N
O
CH3

Mol. wt. 99.13, b.p. 202° (78–79°/10 mm.). Suppliers: A, E, F, MCB.

Special solvent. Newman[1,2] recommends N-methyl-2-pyrrolidone as solvent for the reaction of an aryl halide with cuprous cyanide to produce the corresponding nitrile. Dimethylformamide (b.p. 153°) has been suggested[3] as a better solvent than originally used pyridine (b.p. 115°), but 1-methyl-2-pyrrolidone has an even higher boiling point and seems still more satisfactory. An example is the synthesis of

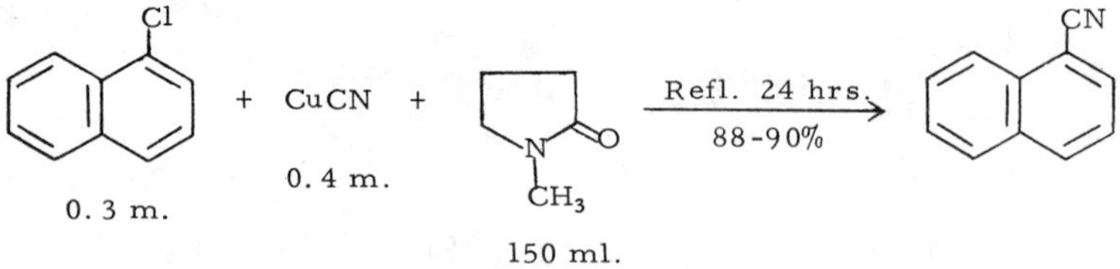
Cl
+ CuCN +
0.3 m.
0.4 m.
CH3
150 ml.
Refl. 24 hrs.
88-90%
CN

1-cyanonaphthalene from 1-chloronaphthalene.[2] Isolation of the nitrile is facilitated by pouring the cooled, dark reaction mixture into a critical amount of ethylenediamine (60 g.) in water (800 ml.). After standing, the mixture is extracted with ether–benzene and the extract cooled, dried, and the product distilled. In the reaction of 1-bromonaphthalene a 3-hr. period of refluxing sufficed.

Henbest[4] found 1-methyl-2-pyrrolidone superior to other aprotic solvents tried for nucleophilic displacement of the tosylate group. In the conversion of the tosylate (1) of cholestane-3β-ol into 3α-cyanocholestane (2) the best yield was obtained in

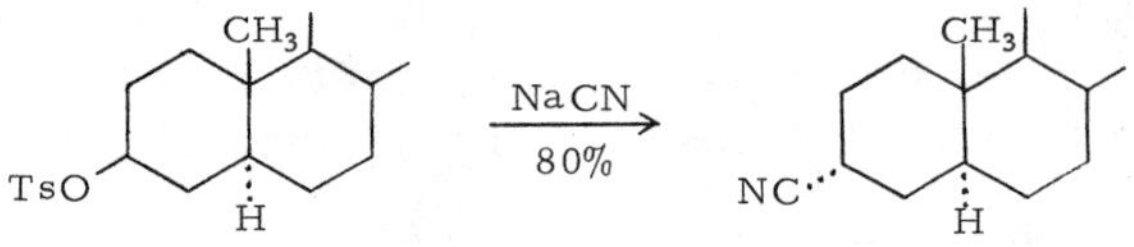
CH3
TsO
H
NaCN
80%
CH3
NC
H

(1) (2)

1-methyl-2-pyrrolidone containing 5% of *t*-butanol at 90°, although the yield was only slightly lower in the absence of the alcohol.

[1]M. S. Newman and H. Boden, *J. Org.*, **26**, 2525 (1961)
[2]M. S. Newman and S. Ramachandran, procedure submitted to *Org. Syn.*
[3]L. Friedman and H. Shechter, *J. Org.*, **26**, 2522 (1961)
[4]H. B. Henbest and W. R. Jackson, *J. Chem. Soc.*, 954 (1962)

1-Methyl-3-*p*-tolyltriazene, *p*-$CH_3C_6H_4N{=}NNHCH_3$. Mol. wt. 149.19, m.p. 78–81°.

Preparation.[1] *p*-Toluidine hydrochloride is diazotized, and the solution of the diazonium chloride is brought to pH 6.8–7.2 at 0° with a cold, concd. solution of sodium carbonate. The resulting deep red solution is added slowly to a vigorously

$$\underline{p}\text{-}CH_3C_6H_4\overset{+}{N}H_3\overset{-}{Cl} \xrightarrow{HNO_2} \underline{p}\text{-}CH_3C_6H_4\overset{+}{N}{\equiv}N(\overset{-}{Cl}) \xrightarrow[53\%]{CH_3NH_2} \underline{p}\text{-}CH_3C_6H_4N{=}N{-}NHCH_3$$

stirred mixture of sodium carbonate, aqueous methylamine, and ice. Extraction with ether gives crude triazene, which on sublimation at 50°/0.1 mm. affords a yellow product m.p. 77–80°. Crystallization from hexane or ether–hexane affords pure, colorless material.

Esterification of acids. Use of the reagent is illustrated by the methylation of 3,5-dinitrobenzoic acid. An ethereal solution of the acid is added slowly to an

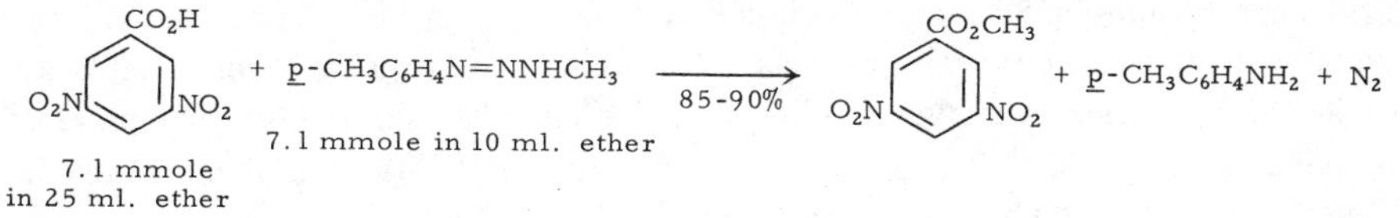
CO2H
O2N
NO2
+ p-CH3C6H4N=NNHCH3
85-90%
CO2CH3
+ p-CH3C6H4NH2 + N2
7.1 mmole in 10 ml. ether
7.1 mmole
in 25 ml. ether

ethereal solution of the reagent; nitrogen is evolved and the solution turns red. When the reaction is complete (about 1 hr.), the ethereal solution is washed with hydrochloric acid to remove *p*-toluidine and then washed with soda solution, dried, and evaporated.

[1]E. H. White, A. A. Baum, and D. E. Eitel, procedure submitted to *Org. Syn.*

Methyltriphenylphosphonium bromide, $(C_6H_5)_3P^+CH_3(Br^-)$. Mol. wt. 257.23, m.p. 232–233°. Supplier: Aldrich.

Preparation.[1] A solution of 0.21 mole of triphenylphosphine in 45 ml. of benzene is placed in a pressure bottle cooled in ice-salt, and 0.29 mole of previously condensed methyl bromide is added. The bottle is sealed, allowed to stand at room temperature for 2 days, opened, and the white solid is collected by suction filtration with the aid of about 500 ml. of hot benzene and dried in a vacuum oven at 100° over phosphorus pentoxide. The yield is 74 g. (99%).

$$(C_6H_5)_3P + CH_3Br \longrightarrow (C_6H_5)_3\overset{+}{P}CH_3(Br^-)$$

[1]G. Wittig and U. Schoellkopf, *Org. Syn.*, **40**, 66 (1960)

Methyl vinyl ether, $CH_3OCH{=}CH_2$. Mol. wt. 58.08, b.p. 13°. Suppliers: General Aniline and Film Corp., Matheson Chem. Co., K and K Laboratories.

In a synthesis of 3-methylpentane-1,5-diol (5), a mixture of 4.08 moles of crotonaldehyde, 5.06 moles of methyl vinyl ether, and 1.1 g. of hydroquinone (to inhibit polymerization) is heated in an autoclave at 200° for 12 hrs. to produce the methylmethoxydihydropyrane (3).[1] This is hydrolyzed to β-methylglutaraldehyde (4) by

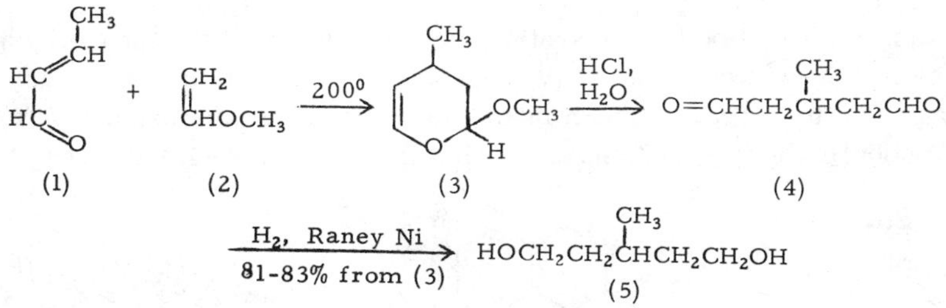
CH3
CH
HC
HC
O
+
CH2
CHOCH3
200°
CH3
OCH3
O
H
HCl,
H2O
CH3
O=CHCH2CHCH2CHO
(1)
(2)
(3)
(4)
H2, Raney Ni
81-83% from (3)
CH3
HOCH2CH2CHCH2CH2OH
(5)

stirring with dilute hydrochloric acid at 50°, solid sodium bicarbonate is added until the solution is neutral to pH indicator paper, and the entire reaction mixture is placed in a rocking hydrogenation autoclave, which is operated at 125° for 4 hrs.[2] A procedure is included for isolation of β-methylglutaraldehyde in 90% yield.

[1]R. I. Longley, Jr., W. S. Emerson, and A. J. Blardinelli, *Org. Syn., Coll. Vol.*, **4**, 311 (1963)
[2]R. I. Longley, Jr., and W. S. Emerson, *ibid.*, **4**, 660 (1963)

Methyl vinyl ketone, $CH_3COCH{=}CH_2$. Mol. wt. 70.09, b.p. 81°, 35–36°/140 mm., n^{20}D 1.4086. Suppliers: du Pont; Chas. Pfizer Co.; Matheson, Coleman and Bell Co.;

Aldrich (Aldrich also supplies 1-diethylamino-3-butanone). Methyl vinyl ketone is made commercially by hydration of vinylacetylene. The du Pont material, an azeotrope containing 85% methyl vinyl ketone and 15% water, can be dried by treatment with excess anhydrous potassium carbonate; the organic layer is separated, chilled in ice, dried first over calcium chloride and then potassium carbonate, and distilled: b.p. 30–32°/120 mm., or 35–36°/140 mm.[1] Material from Matheson, Coleman and Bell can be used as received.

Steroid synthesis. In developing a method of cyclization which has been of immeasurable value in the total synthesis of steroids, Robinson[2] at first experimented with the condensation of ketones with methyl vinyl ketone itself, but encountered difficulties associated with the tendency of the ketone to polymerize. He then turned to possible precursors and met with some success with methyl β-chloroacetyl ketone, $CH_3COCH_2CH_2Cl$. Still better, however, was the methiodide (2) of the Mannich base (1), 1-diethylamino-3-butanone. Treated with a strong base such as sodamide in

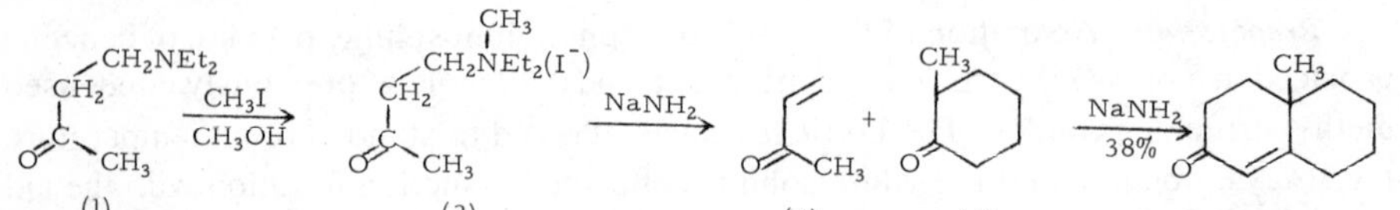
CH2NEt2
CH2
CH3I
CH3OH
CH3
CH3
CH2NEt2(I⁻)
CH2
CH3
NaNH2
CH3
+
CH3
NaNH2
38%
CH3

(1) (2) (3) (4) (5)

the presence of 2-methylcyclohexanone, the methiodide is cleaved to methyl vinyl ketone (3), which reacts with the ketone by a combination of Michael addition and aldolization to give the bicyclic ketone (5). This reaction later became a key step in the Cornforth-Robinson[3] synthesis of epiandrosterone: conversion of the bicyclic ketone (7) into the tricyclic ketone (8).

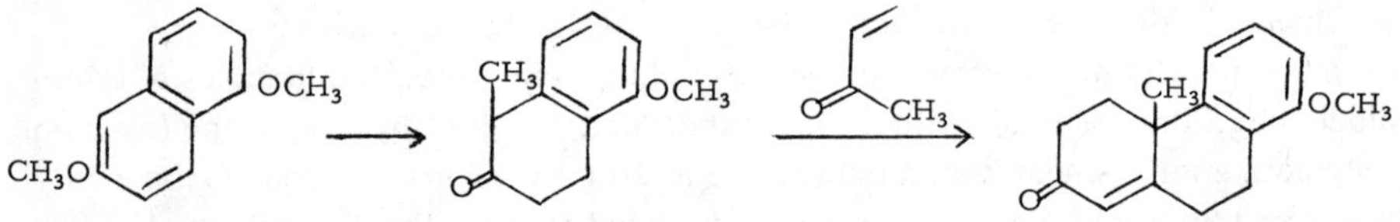
OCH3
CH3O
CH3
OCH3
CH3
CH3
OCH3

(6) (7) (8)

Johnson, in his hydrochrysene synthesis of steroids,[1] used the same scheme for the construction of ring A, as in (9) ⟶ (10). He converted 1-diethylamino-3-butanone into the methiodide with methyl iodide in ether with addition of a little benzyl alcohol to catalyze quaternization. The ether was removed, and the crystalline

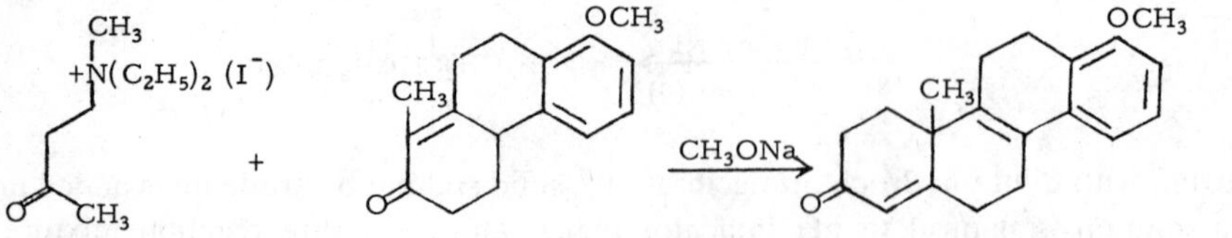
CH3
+N(C2H5)2 (I⁻)
CH3
+
OCH3
CH3
CH3ONa
OCH3
CH3

(9) (10)

salt dissolved in methanol along with the ketone (9) and treated with a solution of sodium methoxide in methanol. At another stage of the synthesis, Johnson used preformed methyl vinyl ketone to effect a Robinson cyclization. Sarett's synthesis of cortisone involved in one step the addition of ring A to (11) to produce (12).[4] The procedure described utilized the 85% aqueous methyl vinyl ketone azeotrope with

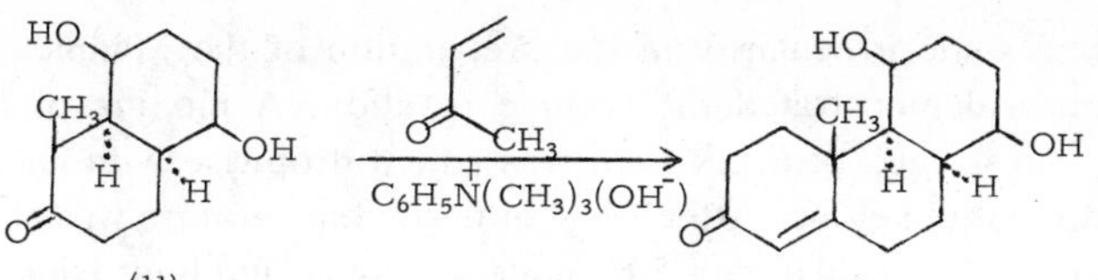

(11) (12)

Triton B as catalyst, but a note indicates that the Mannich base methiodide (with an added equivalent of base) was equally satisfactory.

Ramachandran and Newman,[5] in the synthesis of the methyloctalindione III, effected Robinson cyclization in two steps. Michael addition to II was carried out by

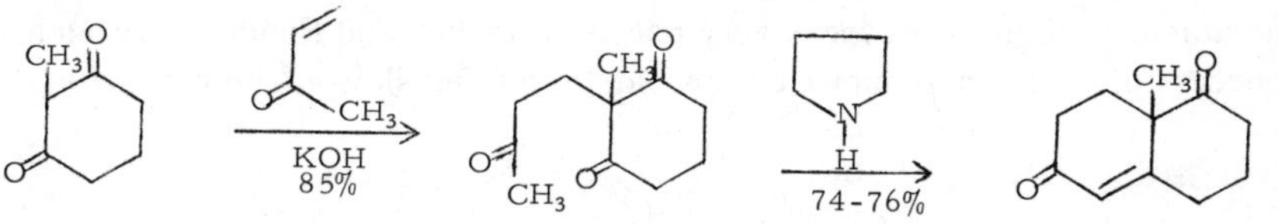

I II III

refluxing for 3 hrs. a mixture of 0.5 mole of I, 0.75 mole of methyl vinyl ketone, 0.25 g. of potassium hydroxide, and 250 ml. of ethanol. The intermediate adduct II was isolated as a crude liquid and heated with benzene and 3 ml. of pyrrolidine under a Dean-Stark phase-separating head for removal of water. Suitable workup and crystallization from ether afforded the bicyclic diketone, III, m.p. 47–50°.

A method for the preparation of 1-diethylamino-3-butanone, first reported by Wilds and Shunk[6] in 1943, is described in more detail in *Organic Syntheses*.[7] A

$$CH_3COCH_3 + CH_2{=}O + (C_2H_5)_2\overset{+}{N}H_2Cl^- \xrightarrow[-H_2O]{} CH_3COCH_2CH_2\overset{+}{N}H(C_2H_5)_2(Cl^-)$$

$$\xrightarrow[\text{62-70\% overall}]{KOH} CH_3COCH_2CH_2N(C_2H_5)_2$$

mixture of 1.6 moles of diethylamine hydrochloride, 2.26 moles of paraformaldehyde, 600 ml. (8.2 moles) of acetone, 80 ml. of methanol, and 0.2 ml. of concd. hydrochloric acid is heated for 12 hrs. under reflux, and the light yellow solution is cooled and neutralized with a solution of 65 g. of sodium hydroxide in 300 ml. of water, and the Mannich base is extracted with ether and distilled. Hagemeyer[8] (1949), apparently unaware of the work of Wilds, described a procedure in which a mixture of 6 moles of diethylamine hydrochloride in 30% aqueous solution, 12 moles of acetone, 8 moles of paraformaldehyde, and 150 ml. of isopropanol is refluxed for 6 hrs. and then concentrated at reduced pressure from a water bath. Neutralization with 50% sodium hydroxide solution, separation of the tertiary amine, drying, and distillation afforded the Mannich base in 80.5% yield.

Hagemeyer found that 1-diethylamino-3-butanone can be cleaved to methyl vinyl ketone by suspending the *t*-amine in Dowtherm, an inert heat-exchange liquid,

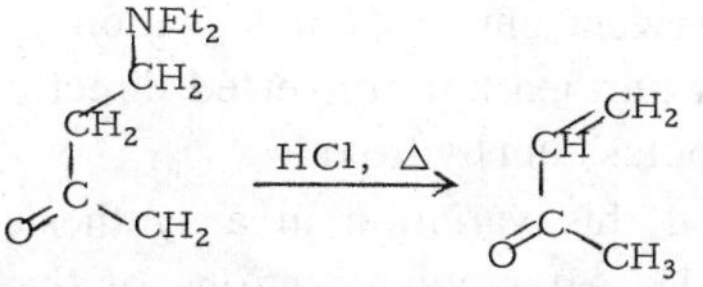

passing in hydrogen chloride to form the amine hydrochloride, and heating at atmospheric pressure; pure methyl vinyl ketone distills in quantitative yield. Newman,[9] on experiencing some irregularity in the preparation of the Mannich base by the Wilds procedure, demonstrated the reverse reaction. A mixture of 2.2 moles of diethylamine and 4.5 ml. of acetic acid was added dropwise with ice cooling to 2 moles of methyl vinyl ketone. After 2 hrs. at 0° the temperature was allowed to rise to 27°, the liquid was washed with 50% aqueous potassium hydroxide solution and with water, dried, and distilled at 69–73°/16 mm.; yield 68%.

The work by Ramachandran and Newman[5] cited above shows that 2-methylcyclohexane-1,3-dione (I) reacts smoothly with methyl vinyl ketone by Michael addition and aldolization to give the methyloctalindione III in good yield. In contrast, the condensation of 2-methylcyclohexanone (2) with methyl vinyl ketone (1) under comparable conditions proceeds very poorly. Marshall and Fanta,[10] after numerous unsuccessful attempts to improve the yield, found that slow addition of methyl vinyl

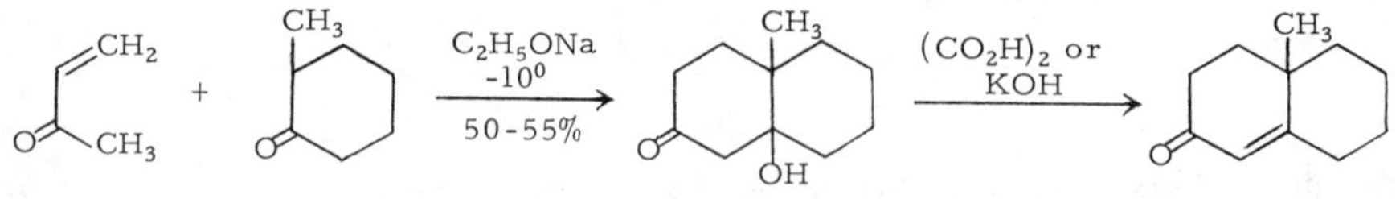
CH2
CH3
CH3
C2H5ONa
-10°
50-55%
CH3
OH
(CO2H)2 or
KOH
CH3

(1) (2) (3) (4)

ketone to an equivalent quantity of 2-methylcyclohexanone at −10° in the presence of a catalytic amount of sodium methoxide effected cyclization to the *cis*-ketol (3) in fairly good yield. This was dehydrated to (4) by steam distillation from oxalic acid or from potassium hydroxide.

In preparation for a Robinson annelation, it is often advantageous to increase the reactivity of the position adjacent to the carbonyl group by introduction of a substituent that can later be removed. Thus Turner *et al.*[11] found direct condensation of 7-methoxy-1-tetralone (1) with methyl vinyl ketone or the corresponding Mannich base to show "little promise." They then converted (1) into the 2-hydroxymethylene

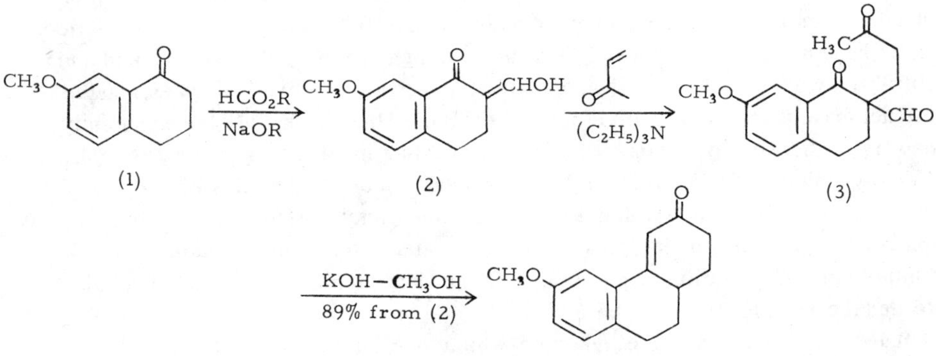
CH3O
HCO2R
NaOR
CH3O
CHOH
(C2H5)3N
H3C
CH3O
CHO
(1)
(2)
(3)
KOH−CH3OH
89% from (2)
CH3O

(4)

derivative (2), which underwent smooth condensation with methyl vinyl ketone to give (3), which was not purified but converted directly into the product (4) by treatment with methanolic potassium hydroxide.

Corey and Nozoe[12] used this variation in a synthesis of helminthosporal (6) from (−)-carvomenthone (1). After condensation of the 3-formyl derivative (2) with methyl vinyl ketone, deformylation of (3) with 2% ethanolic potassium carbonate did not lead to aldol condensation, but cyclization to (5) was accomplished

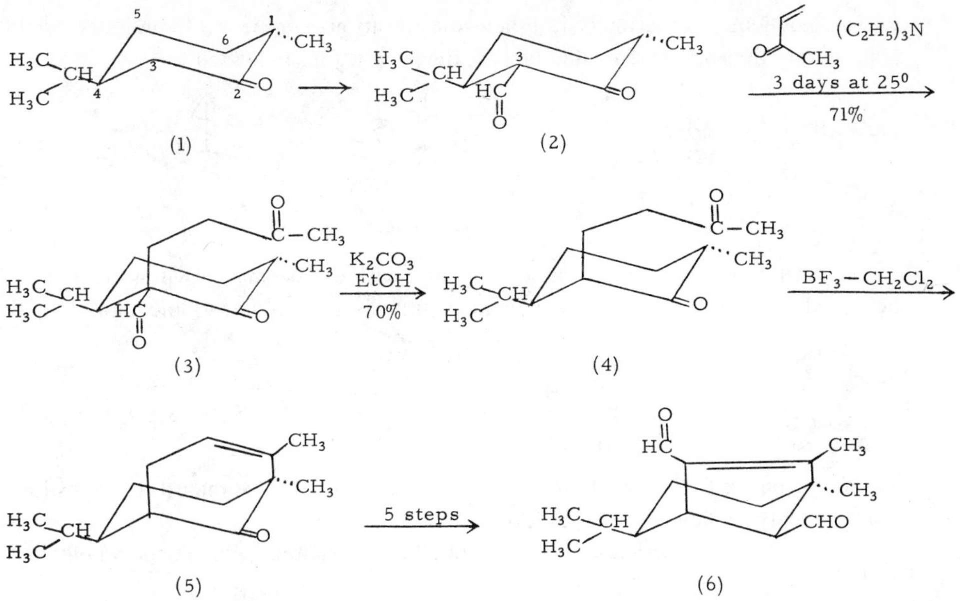

satisfactorily with boron trifluoride. Conversion in five more steps to helminthosporal (6) confirmed the structure and stereochemistry assigned to this crop-destroying toxin and established the absolute configuration.

Vitamin A synthesis. The addition of acetylene to methyl vinyl ketone is a key step in Isler's technical synthesis of vitamin A.[13] The resulting carbinol (2) under

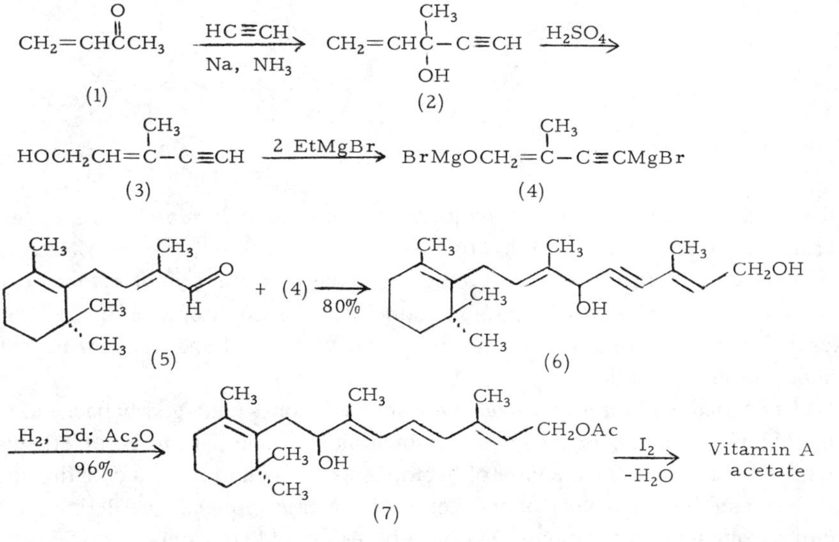

acid catalysis undergoes allylic rearrangement to the primary alcohol (3), which is converted into the di-MgBr derivative (4) for addition to the aldehyde (5). Selective hydrogenation of the triple bond and acetylation to (7), followed by iodine-catalyzed dehydration, completes the synthesis.

Diels-Alder reactions. As a reactive dienophile, methyl vinyl ketone reacts with cyclopentadiene exothermally in ether solution to give in nearly quantitative yield a mixture of the *endo*- and *exo*-adducts in the proportions indicated.[14]

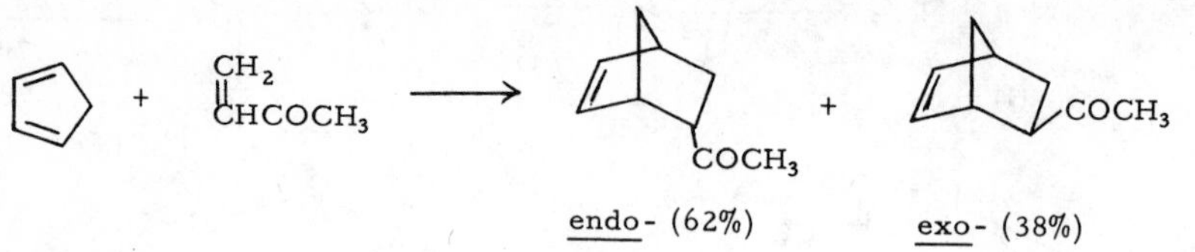
CH2
CHCOCH3
COCH3
COCH3
endo- (62%)
exo- (38%)

Zimmerman and Paufler[15] heated α-pyrone with excess methyl vinyl ketone and achieved decarboxylative double diene synthesis of the diacetylbicyclo[2.2.2]-2-

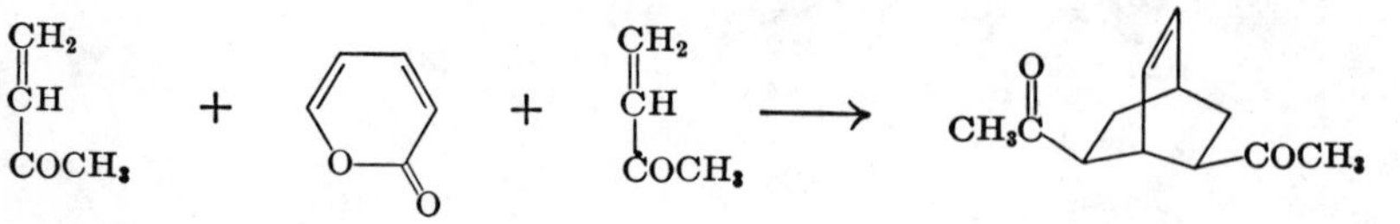
CH2
CH
COCH3
CH2
CH
COCH3
O
O
CH3C
O
COCH3

octene formulated. The diketone, a solid, was converted by standard reactions into the hydrocarbon bicyclo[2.2.2]-2,5,7-octatriene, m.p. 16°.

Synthesis of 4-methylquinolines. Campbell and Schaffner[16] developed an efficient

$$CH_2{=}CH{-}C{\equiv}CH \xrightarrow{CH_3OH(Hg^{++})} CH_3OCH_2CH_2\overset{OCH_3}{\underset{OCH_3}{C}}{-}CH_3 \xrightarrow{H^+}$$

$$CH_3OCH_2CH_2COCH_3 \xrightarrow{H^+} CH_2{=}CHCOCH_3$$

synthesis of lepidine from aniline hydrochloride and either methyl vinyl ketone or one of two more stable precursors, 1,3,3-trimethoxybutane or 4-methoxy-2-butanone:

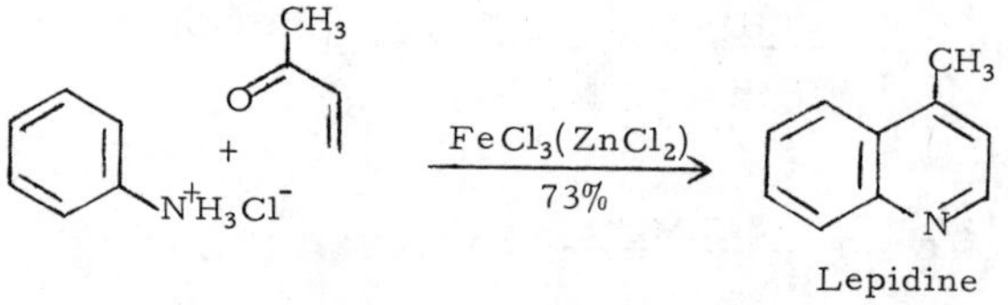
CH3
N+H3Cl−
FeCl3(ZnCl2)
73%
CH3
N
Lepidine

A flask was charged with 0.625 mole of aniline hydrochloride, 1 mole of ferric chloride hexahydrate, 10 g. of anhydrous zinc chloride, and 450 ml. of 95% ethanol. The mixture was stirred at 60–65°, and 0.5 mole of methyl vinyl ketone, 1,3,3-trimethoxybutane, or 4-methoxy-2-butanone was added dropwise in $1\frac{1}{2}$–2 hrs. After refluxing for 2 hrs., the alcohol was largely removed and the mixture made alkaline and steam distilled.

Acetoethylation of carbonyl compounds. Simple ketones undergo Michael addition to methyl vinyl ketone in the presence of lithium amide, Triton B, or ethanolic KOH.[17] As illustrated for the case of acetophenone, the initial product is the open-chain α-acetoethyl derivative of the ketone (1) or a cyclohexanone derivative (2) resulting from aldol cyclization of (1). Several esters, β-keto esters, malonic esters,

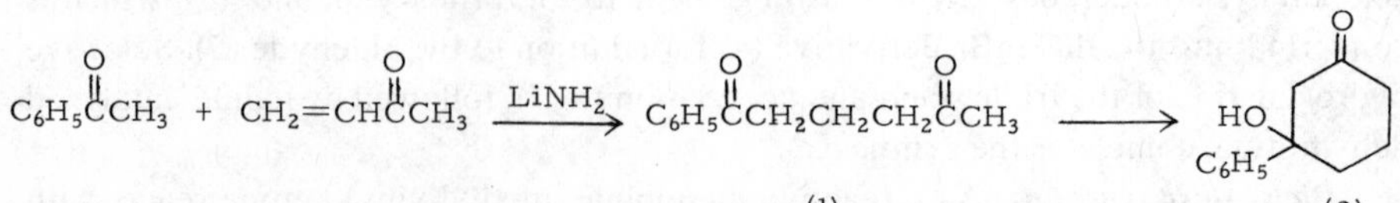
C6H5CCH3 + CH2=CHCCH3
LiNH2
C6H5CCH2CH2CH2CCH3
HO
C6H5

(1) (2)

and amines were acetoethylated, often in good yield, by reaction with methyl vinyl ketone and a basic catalyst.[18]

Synthesis of pyrroles.[19] A key step in a new synthesis of pyrroles is the base-catalyzed condensation of methyl vinyl ketone with an N-tosyl amino acid ester, for example N-tosylglycine ethyl ester (1). Dehydration of the pyrrolidine (2)

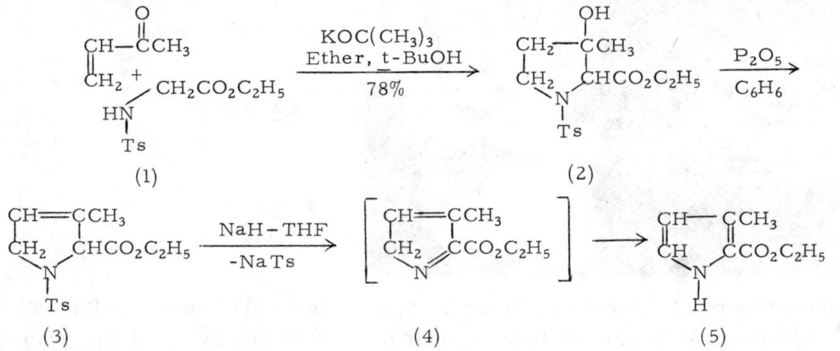

with phosphorus pentoxide in boiling benzene afforded the Δ^3-pyrroline (3), and treatment with sodium hydride in boiling tetrahydrofurane eliminated sodium *p*-toluenesulfonate and gave 2-carboethoxy-3-methylpyrrole (5).

[1]W. S. Johnson, J. Szmuszkovicz, E. R. Rogier, H. I. Hadler, and H. Wynberg, *Am. Soc.*, **78**, 6285 (1956)

[2]E. C. du Feu, F. J. McQuillin, and R. Robinson, *J. Chem. Soc.*, 53 (1937)

[3]J. W. Cornforth and R. Robinson, *J. Chem. Soc.*, 1855 (1949)

[4]G. I. Poos, G. E. Arth, R. E. Beyler, and L. H. Sarett, *Am. Soc.*, **75**, 422 (1953)

[5]S. Ramachandran and M. S. Newman, *Org. Syn.*, **41**, 38 (1961)

[6]A. L. Wilds and C. H. Shunk, *Am. Soc.*, **65**, 469 (1943)

[7]A. L. Wilds, R. M. Nowak, and K. E. McCaleb, *Org. Syn.*, *Coll. Vol.*, **4**, 281 (1963)

[8]H. J. Hagemeyer, Jr., *Am. Soc.*, **71**, 1119 (1949)

[9]S. Swaminathan and M. S. Newman, *Tetrahedron*, **2**, 88 (1958)

[10]J. A. Marshall and W. I. Fanta, *J. Org.*, **29**, 2501 (1964)

[11]R. B. Turner, D. E. Nettleton, Jr., and R. Ferebee, *Am. Soc.*, **78**, 5923 (1956)

[12]E. J. Corey and S. Nozoe, *Am. Soc.*, **87**, 5728 (1965)

[13]O. Isler, W. Huber, A. Ronco, and M. Kofler, *Helv.*, **30**, 1911 (1947)

[14]J. G. Dinwiddie, Jr., and S. P. McManus, *J. Org.*, **30**, 766 (1965)

[15]H. E. Zimmerman and R. M. Paufler, *Am. Soc.*, **82**, 1514 (1960)

[16]K. N. Campbell and I. J. Schaffner, *Am. Soc.*, **67**, 86 (1945)

[17]N. C. Ross and R. Levine, *J. Org.*, **29**, 2341 (1964)

[18]*Idem, ibid.*, **29**, 2346 (1964)

[19]A. H. Jackson, G. W. Kenner, and W. G. Terry, *Tetrahedron Letters*, 921 (1962)

Molecular sieves.[1] Natural zeolites (alumino-silicates) known for some time have a three-dimensional framework with channels and interconnecting cavities forming pores which normally contain water of hydration. A property unique to a zeolite crystal is that the water of hydration can be driven off by heating without causing collapse of the crystal, which remains unchanged and robust with a network of empty cavities or pores. A dehydrated natural zeolite would have potentiality for adsorption of gases or liquids were it not for lack of uniformity in pore size. Research in Union Carbide's Linde Division led to the development in 1954 of synthetic zeolites of completely regular crystal structure and uniform pore size, and these have become of distinct value in adsorption processes.[2] A synthetic zeolite of a given type (see p. 704) adsorbs molecules of only a narrow range in size and hence can be used to effect highly selective separations; by sorbing small molecules and excluding molecules larger than its pores, it functions as a molecular sieve. Figure M-2 shows a sieve

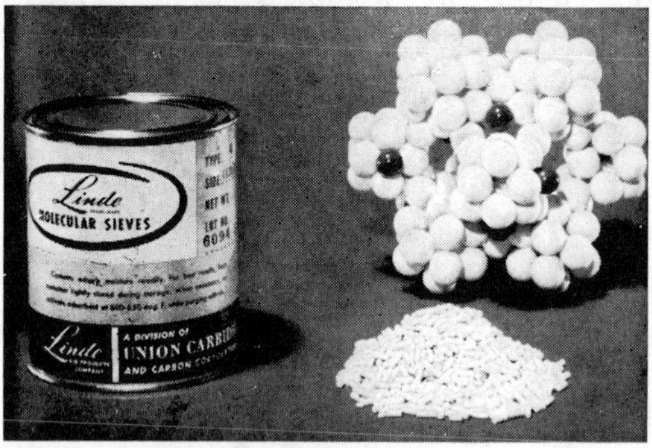
Linde
MOLECULAR SIEVES
A DIVISION OF
UNION CARBI

Fig. M-2

crystal and a pile of cylindrical pellets $\frac{1}{16}$ in. in diameter. The sieve crystal is sturdy and retains its structure on being heated to 320° for recovery of the adsorbed material, regeneration of the sieve, or both.

Suppliers. Union Carbide, F, MCB, Davison Division of W. R. Grace and Co.

Specifications. The sieves most widely used, Types 3A, 4A, 5A, and 13X, are all available as powder, $\frac{1}{16}$-in. pellets, and $\frac{1}{8}$-in. pellets; type 4A sieves are available also in beads of three sizes.

Type	Formula	Nominal pore diameter
3A	$K_9Na_3[(AlO_2)_{12}(SiO_2)_{12}]\cdot 27\,H_2O$	3Å
4A	$Na_{12}[(AlO_2)_{12}(SiO_2)_{12}]\cdot 27\,H_2O$	4Å
5A	$Ca_{4.5}Na_3[(AlO_2)_{12}]\cdot 30\,H_2O$	5Å
13X	$Na_{86}[(AlO_2)_{86}(SiO_2)_{106}]\cdot x\,H_2O$	10Å

Sieve 3A adsorbs water and ammonia and excludes ethane; it dries polar liquids and gases such as methanol, ethanol, propylene. Sieve 4A adsorbs ethanol, H_2S, CO_2, SO_2, C_2H_4, C_2H_6, C_3H_6 and excludes propane and larger hydrocarbons; it dries nonpolar liquids and gases to minimum effluent water content with maximum capacity. Sieve 5A adsorbs n-C_4H_9OH, n-C_4H_{10}, and excludes iso compounds and rings of 4 or more carbons. Type 13X, which corresponds to the natural zeolite faujasite, has a nominal pore diameter of 10 Å; it adsorbs $(n\text{-}C_4H_9)_2NH$ and excludes $(n\text{-}C_4H_9)_3N$. A specific use is for drying hexamethylphosphoric triamide, $[(CH_3)_2N]_3PO$.

Uses. Molecular sieves of types 4A and 5A are highly useful for the drying of gases (air, hydrogen, chemical streams) and are more efficient than any other absorbents. The main use to the organic chemist is for extremely effective drying of solvents. Mere standing of the solvent over powder or pellets may suffice provided that enough time is allowed before distillation or filtration, but it is probably more efficient to provide agitation or to use a Soxhlet as in the esterification procedure given below. Another scheme is to percolate solvent slowly through a 2-ft. column packed with $\frac{1}{16}$-in. pellets of a molecular sieve. Schriesheim *et al.*[3,4] used a 10:1 volume ratio of sieve to solvent.

Specific sieves used for drying various solvents are as follows:

Type 3A: methanol; *see* Esterification, below
Type 4A: acetone,[5] DMSO,[5,6] pyridine,[6] nitromethane,[5] caprylic acid, DMF[7,8]
Type 5A: THF,[9] dioxane[10]
Type 13X: hexamethylphosphoric triamide, $[(CH_3)_2N]_3PO$[4]

Esterification.[11] For the preparation of methyl nonanoate the charge of nonanoic acid, methanol, and concd. sulfuric acid is placed in a flask fitted with a thermometer

$$\underset{1\text{ m.}}{CH_3(CH_2)_7CO_2H} + \underset{4\text{ m.}}{CH_3OH} + \underset{0.05\text{ m.}}{H_2SO_4} \xrightarrow[96\%]{\text{Refl. 15 hrs. 3A Mol. sieve}} CH_3(CH_2)_7CO_2CH_3$$

and Soxhlet extractor charged with 160 g. of conditioned Linde 3A molecular sieve $\frac{1}{16}$-in. pellets and provided with a large-capacity condenser. The extractor is filled with methanol, the cooling water is turned on, and the flask is heated with an electric mantle and stirred magnetically. A reflux temperature in the range 65–67° is maintained for 15 hrs. After cooling, 5.3 g. of anhydrous sodium carbonate is added to neutralize the sulfuric acid, the methanol is removed by distillation at atmospheric pressure, and the product is distilled at 45°/0.2 mm. Fresh molecular sieve is conditioned by heating at 320° for 3 hrs. If recovered material is to be used again it should be dried thoroughly in air to reduce the possibility of fire or explosion and then heated in a well-ventilated furnace at 320° for 12 hrs.

Enamines. Szmuszkovicz[12] states that enamines difficult to prepare by usual methods often can be obtained by use of $\frac{1}{16}$-in pellets of Type 4A sieve.

[1]Bulletin: "Union Carbide Molecular Sieves for Selective Absorption," The British Drug Houses Ltd.; D. W. Breck, "Crystalline Molecular Sieves," *J. Chem. Ed.*, **41**, 678 (1964)
[2]D. W. Breck *et al.*, *Am. Soc.*, **78**, 5963, 5972 (1956)
[3]J. E. Hofmann, T. J. Wallace, and A. Schriesheim, *Am. Soc.*, **86**, 1561 (1964)
[4]T. J. Wallace, H. Pobiner, and A. Schriesheim, *J. Org.*, **30**, 3768 (1965)
[5]S. G. Smith, A. H. Fainberg, and S. Winstein, *Am. Soc.*, **83**, 618 (1961)
[6]H. G. Khorana *et al.*, *Am. Soc.*, **87**, 350 (1965)
[7]A. B. Thomas and E. G. Rochow, *Am. Soc.*, **79**, 1843 (1957)
[8]G. R. Pettit, M. V. Kalnins, T. M. Liu, E. G. Thomas, and K. Parent, *J. Org.*, **26**, 2563 (1961)
[9]M. E. Cain, *J. Chem. Soc.*, 3532 (1964)
[10]S. H. Burstein and H. J. Ringold, *Am. Soc.*, **86**, 4952 (1964)
[11]H. R. Harrison, W. M. Haynes, P. Arthur, and E. J. Eisenbraun (Oklahoma State Univ.), procedure submitted to *Org. Syn.*
[12]J. Szmuszkovicz, *Advances in Organic Chemistry*, **4**, 11 (1963)

Morpholine, $O(CH_2CH_2)_2NH$. Mol. wt. 87.12, b.p. 128°. Suppliers: Union Carbide (Div. Union Carbide Corp.), Aldrich, Eastman. The reagent is made technically by dehydration of diethanolamine.

Willgerodt reaction. Schwenk[1] introduced use of morpholine in a modified Willgerodt reaction which does not require a sealed tube or autoclave. A methyl

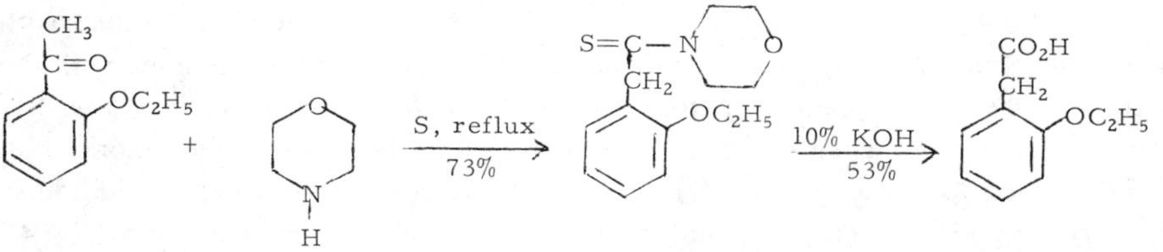
CH3
C=O
OC2H5
+
H
S, reflux
73%
S=C—N
CH2
OC2H5
10% KOH
53%
CO2H
CH2
OC2H5

ketone, refluxed with sulfur and morpholine, is converted into a thioamide, which is hydrolyzed to the arylacetic acid.

Enamines. Like other secondary amines, morpholine reacts with an aldehyde having an α-hydrogen to form an enamine; the derivative can be reduced with formic acid.[2]

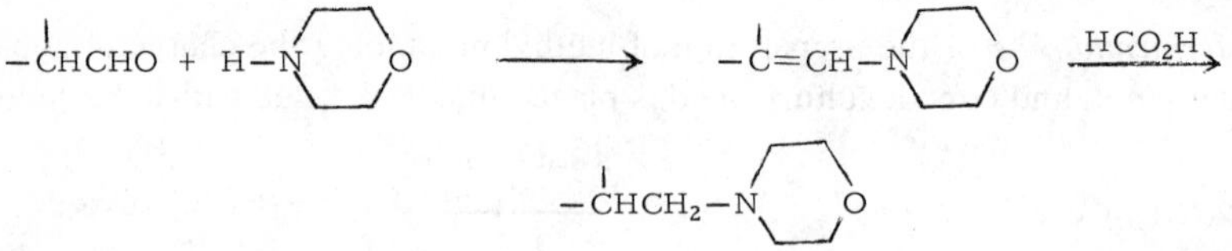

The morpholine enamine of cyclohexanone, 1-morpholino-1-cyclohexene, is prepared by refluxing a mixture of 1.5 moles of cyclohexanone, 1.8 moles of morpholine, 1.5 g. of *p*-toluenesulfonic acid, and 300 ml. of toluene in a flask fitted with a water separator under a reflux condenser.[3] Separation of water commences at once

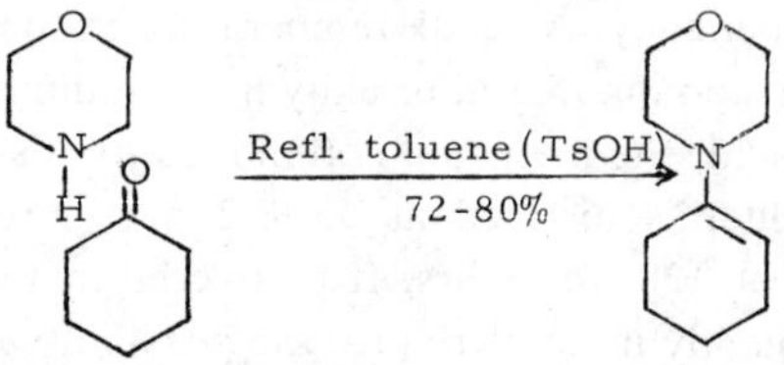

and ceases in 4–5 hrs. Most of the toluene is removed at atmospheric pressure, and the enamine distills at 118–120°/10 mm. Alkylation of enamines of cyclohexanones by alkyl halides[4-6] or electrophilic olefins,[4,5] followed by hydrolysis, is a good route to α-monoalkylcyclohexanones. Acylation of enamines of cyclopentanone and cyclohexanone is the first step in a general procedure for increasing the chain length of a carboxylic acid by 5 or 6 carbon atoms and of a dicarboxylic acid by 10 or 12 carbon atoms.[7]

N-Nitromorpholine.[8] Acetone cyanohydrin nitrate (2) is prepared by stirring acetic anhydride in an ice bath and adding white fuming nitric acid dropwise at 3–5° followed by dropwise addition of acetone cyanohydrin at 5–10° in 45 minutes.

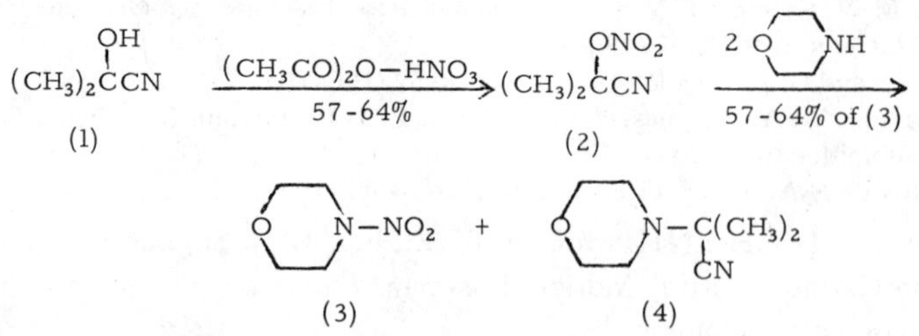

The mixture is quenched with ice and the acetone cyanohydrin nitrate is collected by extraction with methylene chloride (b.p. 62–65°/10 mm., moderately explosive). When a mixture of 0.4 mole of morpholine and 0.2 mole of acetone cyanohydrin nitrate is heated slowly, eventually to 110°, N-nitromorpholine (3) is formed with liberation of acetone and hydrogen cyanide, and these reactants combine with the second mole of morpholine to form α-morpholinoisobutyronitrile (4). This by-product is basic and when the mixture is cooled, poured into 10% hydrochloric acid, and extracted with methylene chloride the by-product is retained in the acid liquor. N-Nitromorpholine separates from absolute ethanol in white crystals, m.p. 57–64°.

The synthesis illustrates a general method for the preparation of both primary and secondary nitramines.

Aldehydes from gem-dihalides.[9] Morpholine is used in the two-step hydrolysis of *gem*-dihalides to aldehydes. The dihalide is warmed with excess morpholine on

$$RCHX_2 + 4\,HN(CH_2CH_2)_2O \longrightarrow RCH[N(CH_2CH_2)_2O]_2 \xrightarrow[HCl]{H_2O} RCHO$$

the steam bath for 2–3 hrs. until precipitation of morpholine hydrochloride is complete. The mixture is added with vigorous stirring to hydrochloric acid, ice, and water. Yields are 60–90%.

gem-Dithiols can be prepared in good yield by passing H_2S into a solution of a ketone and morpholine in ethanol, DMF, or DMSO at room temperature.[10] Undoubtedly the enamine is first formed, since morpholine enamines react with H_2S in DMF under mild conditions to give *gem*-dithiols.[11]

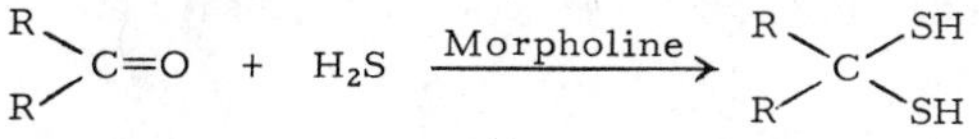
R
C=O + H₂S
R
Morpholine
R
SH
C
R
SH

[1]E. Schwenk and E. Bloch, *Am. Soc.*, **64**, 3051 (1942)
[2]P. L. DeBenneville and J. H. Macartney, *Am. Soc.*, **72**, 3073 (1950)
[3]S. Hünig, E. Lücke, and W. Brenninger, *Org. Syn.*, **41**, 65 (1961)
[4]G. Stork, R. Terrell, and J. Szmuszkovicz, *Am. Soc.*, **76**, 2029 (1954)
[5]G. Stork, A. Brizzolara, H. Landesman, J. Szmuszkovicz, and R. Terrell, *Am. Soc.*, **85**, 207 (1963)
[6]D. M. Locke and S. W. Pelletier, *Am. Soc.*, **80**, 2588 (1958)
[7]S. Hünig *et al.*, *Ber.*, **91**, 129 (1958); **92**, 652 (1959); **93**, 909, 913 (1960)
[8]J. P. Freeman and I. G. Shepard, *Org. Syn.*, **43**, 83 (1963)
[9]M. Kerfanto, *Compt. rend.*, **252**, 3457 (1961); **254**, 493 (1962); *Angew. Chem., Internat. Ed.*, **1**, 459 (1962)
[10]J. Jentzsch, J. Fabian, and R. Mayer, *Ber.*, **95**, 1764 (1962); R. Mayer, G. Hiller, M. Nitzschke, and J. Jentzsch, *Angew. Chem., Internat. Ed.*, **2**, 370 (1963)
[11]C. Djerassi and B. Tursch, *J. Org.*, **27**, 1041 (1962)

1-Morpholino-1-cyclohexene (1). Mol. wt. 167.25, b.p. 118–120°/10 mm. Supplier: Aldrich.

Preparation from a mixture of cyclohexanone, morpholine, toluene, and a trace of *p*-toluenesulfonic acid is described on page 706.

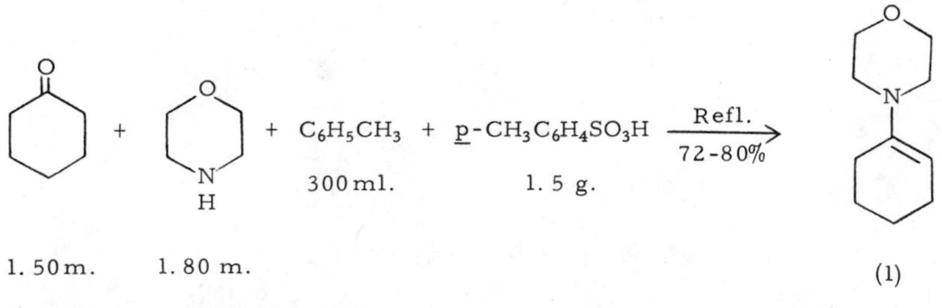
+
O
N
H
+ C₆H₅CH₃ + p-CH₃C₆H₄SO₃H
300 ml.
1. 5 g.
Refl.
72-80%
N
O
1. 50 m.
1. 80 m.

(1)

Synthesis of β-diketones and keto acids. In an efficient synthesis of ω-phenylnonanoic acid[1] a solution of hydrocinnamoyl chloride was added slowly to a solution of the reagent and triethylamine in chloroform at 35°. After 20 hrs. the mixture was refluxed with 10% hydrochloric acid to eliminate the enamine group and give the

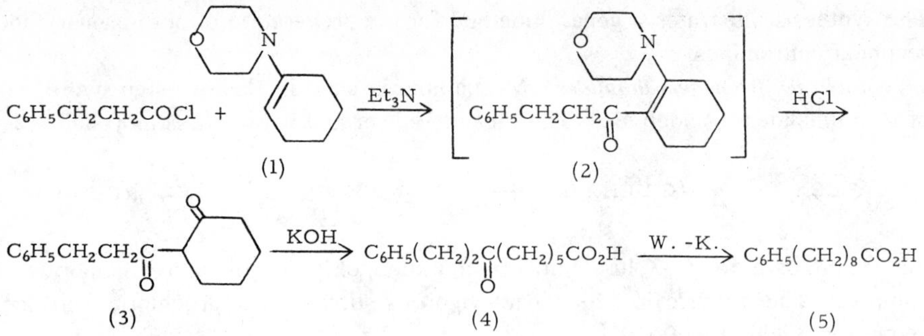

1,3-diketone (3). Alkaline cleavage to the ketoacid (4) and Wolff-Kishner reduction afforded ω-phenylnonanoic acid (5). By a similar process, docosanedioic acid can be obtained from sebacoyl chloride.[2]

[1]L. F. Fieser, J. P. Schirmer, Jr., S. Archer, R. R. Lorenz, and P. I. Pfaffenbach, paper submitted to *J. Med. Chem.*

[2]S. Hünig, E. Lücke, and W. Brenninger, *Org. Syn.*, **43**, 34 (1964)

N

Naphthacene-9,10,11,12-diquinone (4). Mol. wt. 288.24, m.p. 330–333°, dec.

Preparation. The precursor, 11,12-dihydroxy-9,10-quinone (3), has been prepared by two methods, neither of which is very well documented. In one[1] phthalic anhydride is condensed with succinic acid and potassium acetate at 200–220° for 2 hrs. to give α,β-diphthylidene-ethane (1, "ethinediphthalide"), which on treatment with sodium methoxide and methanol, followed by acidification, is isomerized in part to the dihydroxynaphthacenequinone (3, "isoethinediphthalide"), and in part to bis-diketohydrindene (2). The yield of the desired (3) is very low. A second

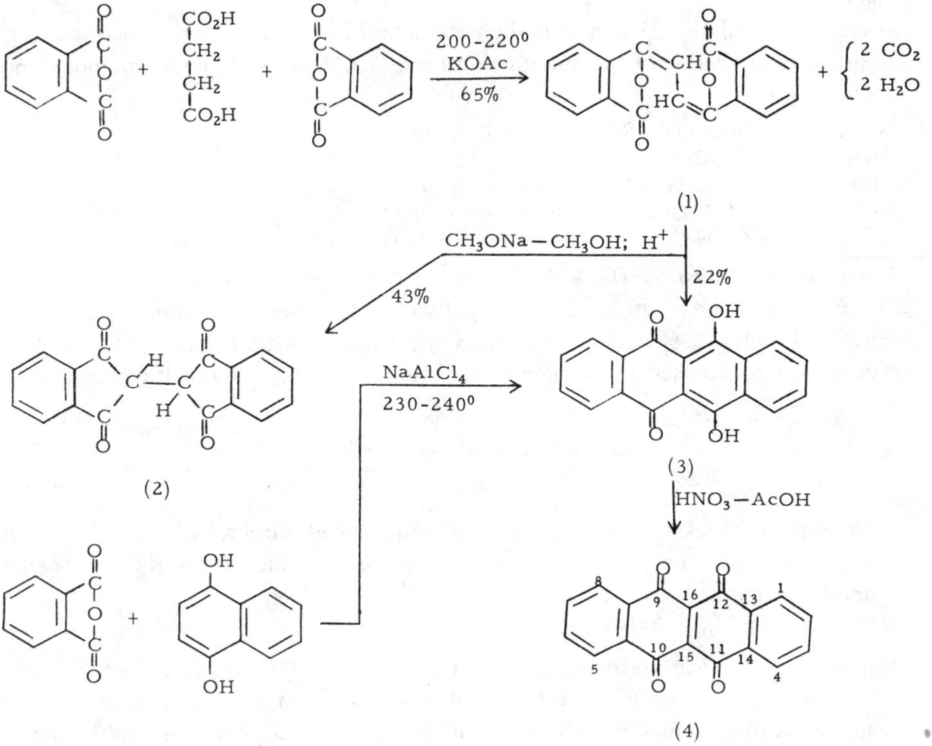
200-220°
KOAc
65%
2 CO_2
2 H_2O
(1)
$CH_3ONa-CH_3OH$; H^+
22%
43%
(2)
$NaAlCl_4$
230-240°
(3)
HNO_3-AcOH
(4)

method which appears more promising although the yield is not recorded is the condensation of phthalic anhydride with 1,4-naphthohydroquinone in a sodium aluminum chloride melt.[2] (Condensation with H_2SO_4—H_3BO_3 might be useful, *see* Boric acid.) The dihydroxyquinone (3), which can be sublimed in vacuum to give dark red needles, m.p. 349°, on oxidation with fuming nitric acid in acetic acid affords naphthacene-9,10,11,12-diquinone in good yield.

Diels-Alder dienophile. The diquinone has a reactive 15,16-double bond, as is evident from the formation of a dichloride and an epoxide.[3] It reacts with 2,3-dimethylbutadiene in acetic acid at 100° to form the colorless adduct (5), m.p. 255–256°.[4] Like other 1,3-diketones which are incapable of enolization,[5] the adduct

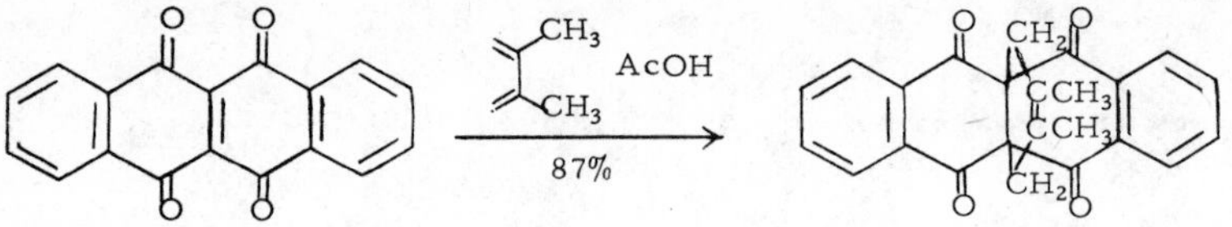

(5)

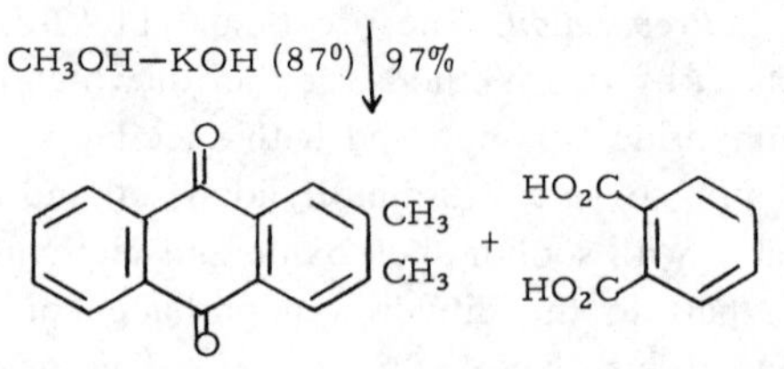

(6)

is cleaved readily by 25% methanolic potassium hydroxide to give, after isomerization and air-oxidation of the initial neutral product, 2,3-dimethylanthraquinone and phthalic acid.

[1]S. Gabriel and E. Leupold, *Ber.*, **31**, 1159, 1272 (1898)
[2]H. Raudnitz, *Ber.*, **62**, 509 (1929)
[3]H. Voswinckel, *Ber.*, **38**, 4015 (1905); **42**, 458 (1909)
[4]L. F. Fieser and J. T. Dunn, *Am. Soc.* **58**, 1054 (1936)
[5]L. J. Beckham and H. Adkins, *Am. Soc.*, **56**, 2676 (1932)

Naphthalene-2,3-dicarboxylic acid. Mol. wt. 216.18, m.p. 239–241°.

Preparation.[1] An autoclave is charged with 1.28 moles of 2,3-dimethylnaphthalene, 3.14 moles (23% excess) of sodium dichromate dihydrate, and 1.8 l. of water. The autoclave is closed and shaken continuously at 250° for 18 hrs. It is then cooled

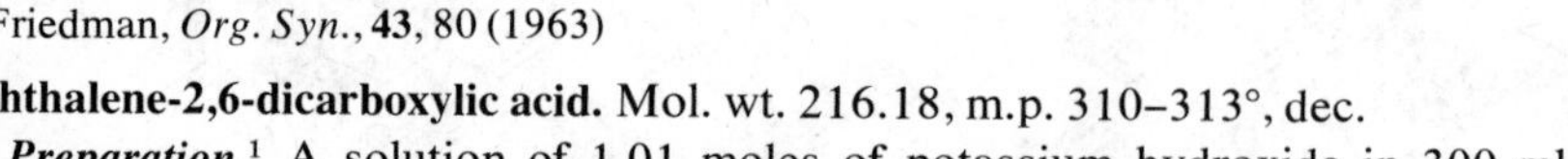

with continual shaking, opened, emptied, and washed. Green hydrated chromium oxide is separated on a large Büchner funnel, and acidification of the filtrate and workup precipitates the diacid.

[1]L. Friedman, *Org. Syn.*, **43**, 80 (1963)

Naphthalene-2,6-dicarboxylic acid. Mol. wt. 216.18, m.p. 310–313°, dec.

Preparation.[1] A solution of 1.01 moles of potassium hydroxide in 300 ml. of water is heated to 60–70°, and 0.505 mole of technical naphthalic anhydride is stirred in. The pH of the deep brown solution is adjusted to 7, and the solution is twice clarified with Norit. The filtrate is concentrated to about 180 ml., cooled, treated with 800 ml. of methanol with stirring, and cooled to 0–5°. The precipitated

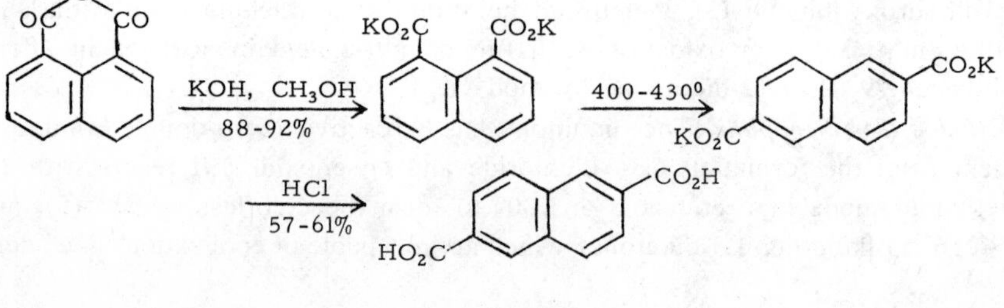

dipotassium salt is collected, washed with methanol, dried at 150°/150 mm., ground with 4% of anhydrous cadmium chloride in a ball mill for 4 hrs., and placed in an autoclave (the function of the cadmium chloride is not disclosed). The autoclave is evacuated to remove oxygen, filled with carbon dioxide at a pressure of about 30 atm., and heated with agitation at an internal temperature of 400–430° for 1½ hrs. In this period the carboxylate groups migrate to less hindered positions, and the product liberated on acidification of the cooled melt is naphthalene-2,6-dicarboxylic acid.

[1]B. Raecke and H. Schirp, *Org. Syn.*, **40**, 71 (1960)

Naphthalene–Magnesium. A mixture of 60 mmoles of magnesium powder and 60 mmoles of naphthalene in 150 ml. of liquid ammonia is stirred, 50 ml. of ether is added, and stirring is continued until nearly all the magnesium has dissolved to give a deep green solution (1½ hrs.)[1] The green complex probably has the composition $C_{10}H_8Mg$. Addition of 120 mmoles of absolute ethanol discharges the color and workup affords 1,4-dihydronaphthalene, m.p. 24–25°. The complex metalates benzyl cyanide in ether to give a yellow precipitate; replacement of the ammonia with ether and refluxing with ethyl iodide affords α-phenylbutyronitrile in 34%

$$C_6H_5CH_2CN \xrightarrow{C_{10}H_8Mg} \text{Precipitate} \xrightarrow{C_2H_5I} C_6H_5\underset{\displaystyle CH_2CH_3}{\underset{|}{C}}HCN$$

yield. Attempted metalation of ethyl phenylacetate led instead by ester condensation to ethyl α,γ-diphenylacetoacetate.

$$C_6H_5CH_2COOC_2H_5 + \overset{\displaystyle C_6H_5}{\overset{|}{C}}H_2CO_2C_2H_5 \xrightarrow{C_{10}H_8Mg} C_6H_5CH_2CO\overset{\displaystyle C_6H_5}{\overset{|}{C}}HCO_2C_2H_5$$

[1]Chr. Ivanoff and P. Markov, *Naturwissenschaften*, **50**, 688 (1963)

Naphthalene–Sodium, $[C_{10}H_8]^-Na^+$.

Naphthalene does not react with sodium in ether, but Scott[1] found that when sodium is introduced to a solution of naphthalene in dimethyl ether or in ethylene glycol dimethyl ether the metal dissolves to give a deep green solution. Carbonation of the green solution gave a mixture of naphthalene, Δ^1-dialin-3,4-dicarboxylic acids, and Δ^2-dialin-1,4-dicarboxylic acids, and the result led Scott to infer that the solution contains the complex $(C_{10}H_8Na_2)C_{10}H_8$. However, a quantitative study of the reaction in tetrahydrofurane by Paul[2] indicated that the metal dissolves by

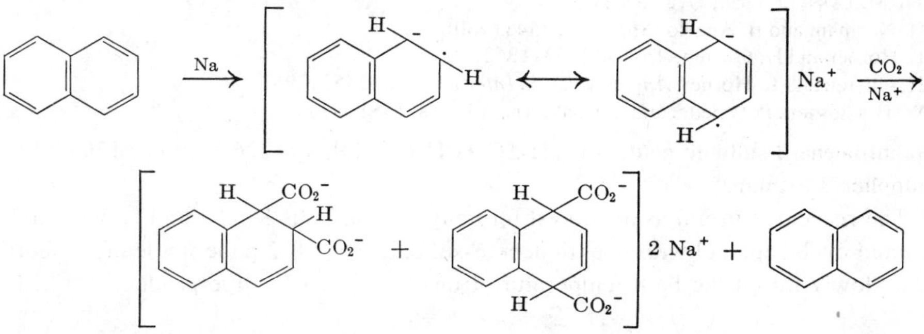
Na
H
H
H
H
Na+
CO2
Na+
H
CO2−
H
CO2−
H
CO2−
H
CO2−
2 Na+

(1/2 mole) (1/2 mole)

transferring an electron to naphthalene to give the green ion radical formulated. Although naphthalene adds only 1 atom of sodium, carbonation of 1 mole of naphthalene–sodium at −70° is attended with disproportionation to give ½ mole each of naphthalene and a mixture of the 1,4- and 3,4-dicarboxylic acids. Lyssy[3] isolated and established the configurations of several components of the mixture.

Normant[4] generated naphthalene–sodium in tetrahydrofurane and showed that it is better than sodium alone in THF for metalation of terminal acetylenes and active-methylene compounds, but yields of acids obtained after carbonation are not impressive:

$$\left.\begin{array}{l} \underline{n}\text{-}C_4H_9C{\equiv}CH \\ HC{\equiv}CH \\ \text{Indene} \\ \text{Fluorene} \\ C_6H_5CH_2CN \end{array}\right\} \text{Naphth.}-\text{Na} \xrightarrow{CO_2} \left\{\begin{array}{l} \underline{n}\text{-}C_4H_9C{\equiv}CCO_2H\ (50\%) \\ HC{\equiv}CCO_2H\ (20\%) \\ \text{Indene-1-COOH}\ (62\%) \\ \text{Fluorene-9-COOH}\ (44\%) \\ C_6H_5CH(CN)CO_2H\ (40\%) \end{array}\right.$$

Horner[5] investigated α-metalation of nitriles, followed by alkylation; typical results are formulated:

$$C_6H_5CH_2CN \xrightarrow[61\%]{\text{Naphth.}-\text{Na};\ \text{cyclohexyl iodide}} C_6H_5\underset{\substack{|\\C_6H_{11}}}{C}HCN$$

$$(C_6H_5)_2CHCN \xrightarrow[77\%]{\text{Naphth.}-\text{Na};\ C_2H_5Br} (C_6H_5)_2\underset{\substack{|\\C_2H_5}}{C}CN$$

$$CH_3CN \xrightarrow[35\%]{\text{Naphth.}-\text{Na};\ C_2H_5Br} CH_3CH_2CH_2CN$$

Wurtz coupling of benzyl halides proceeds particularly well.[6]

$$2\ C_6H_5CH_2Cl \xrightarrow[80\%]{2\ \text{Naphthalene-sodium}} C_6H_5CH_2CH_2C_6H_5$$

The anion-radical in tetrahydrofurane is described as an "almost ideal" reagent for the reductive cleavage of a tosylate to the alcohol.[7] The reagent is prepared by stirring clean pieces of sodium with a slight excess of naphthalene in tetrahydrofurane under nitrogen or argon at room temperature.

[1]N. D. Scott, J. F. Walker, and V. L. Hansley, *Am. Soc.*, **58**, 2442 (1936); J. F. Walker and N. D. Scott, *ibid.*, **60**, 951 (1938)
[2]D. E. Paul, D. Lipkin, and S. I. Weissman, *Am. Soc.*, **78**, 116 (1956)
[3]Th. M. Lyssy, *J. Org.*, **27**, 5 (1962)
[4]H. Normant and B. Angelo, *Bull. soc.*, 354 (1960)
[5]L. Horner and H. Güsten, *Ann.*, **652**, 99 (1962)
[6]H. Güsten and L. Horner, *Angew. Chem., Internat. Ed.*, **1**, 455 (1962)
[7]W. D. Closson, P. Wriede, and S. Bank, *Am. Soc.*, **88**, 1581 (1966)

Naphthalene-β-sulfonic acid, β-$C_{10}H_7SO_3H \cdot H_2O$. Mol. wt. 226.24, m.p. 120–122°. Supplier: Eastman.

The reagent is useful as a catalyst for dehydration. Thus Kohler and co-workers[1] treated cycloheptanol with naphthalene-β-sulfonic acid (1–2 g. per mole of alcohol) and slowly raised the bath temperature from 180 to 250° as the product distilled.

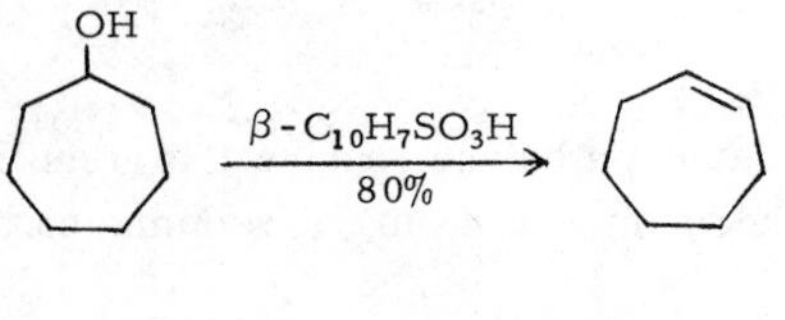
OH
β-C10H7SO3H
80%

Braude and Evans[2] found the ester (1) remarkably resistant to dehydration by usual methods (heating with $KHSO_4$ or with iodine, treatment with $SOCl_2$ in pyridine or with P_2O_5 and triethylamine), but effected dehydration, with rearrangement to (3), by heating 1 g. with 20 mg. of naphthalene-β-sulfonic acid at 130°/100 mm.

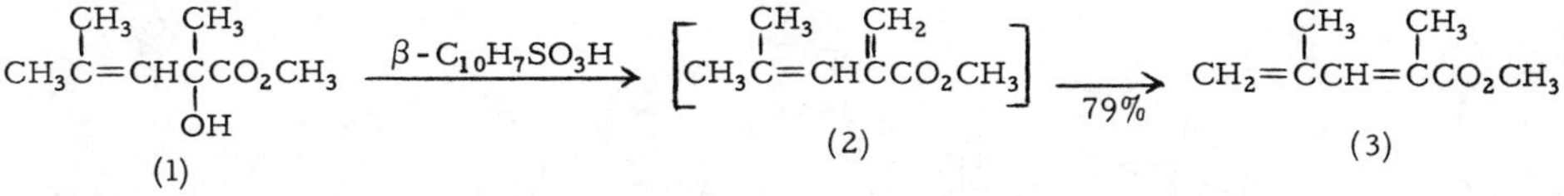
CH3 CH3
CH3C=CHCCO2CH3
OH
(1)
β-C10H7SO3H
CH3 CH2
CH3C=CHCCO2CH3
(2)
79%
CH3 CH3
CH2=CCH=CCO2CH3
(3)

[1]E. P. Kohler, M. Tishler, H. Potter, and H. T. Thompson, *Am. Soc.*, **61**, 1057 (1939)
[2]E. A. Braude and E. A. Evans, *J. Chem. Soc.*, 3324 (1955)

1,2-Naphthalic anhydride. Mol. wt. 198.27, m.p. 166–167°.

Preparation.[1] Ester condensation of ethyl γ-phenylbutyrate with diethyl oxalate gives crude ethyl α-ethoxalyl-γ-phenylbutyrate, which is cyclized by concd. sulfuric

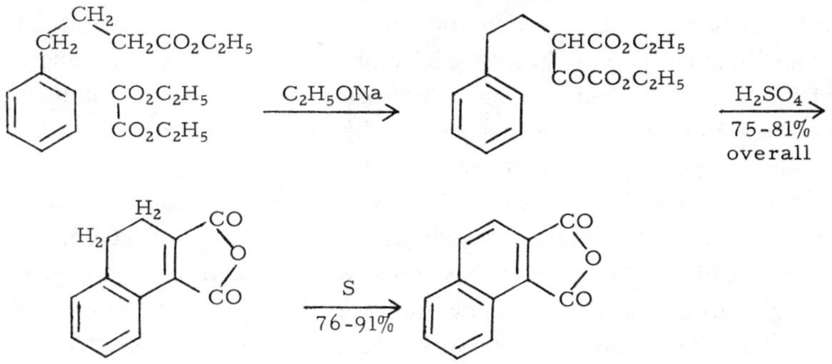
CH2
CH2 CH2CO2C2H5
CO2C2H5
CO2C2H5
C2H5ONa
CHCO2C2H5
COCO2C2H5
H2SO4
75-81%
overall
H2
H2
CO
O
CO
S
76-91%
CO
O
CO

acid at 20–25° to 3,4-dihydro-1,2-naphthalic anhydride. For dehydrogenation a mixture of 0.1 mole of this material and 0.1 mole of sulfur is heated in a bath[2] preheated to 230–235°, the temperature is raised to 250° for 30 min., and the product is distilled at reduced pressure and crystallized from benzene–ligroin.

[1]E. B. Hershberg and L. F. Fieser, *Org. Syn., Coll. Vol.*, **2**, 194, 423 (1943)
[2]Wood's metal or a 10:7.5 mixture of KNO_3 and $NaNO_2$

1,8-Naphthalic anhydride. Mol. wt. 198.17, m.p. 274°. Suppliers: Coaltar Chemicals Corp., Aldrich, Eastman.

Permanganate oxidation of a solution in alkali affords hemimellitic acid.[1]

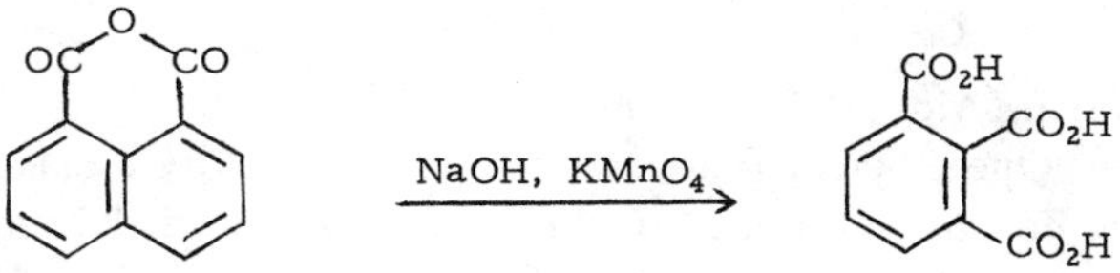
O
OC CO
NaOH, KMnO4
CO2H
CO2H
CO2H

[1]C. Graebe and M. Leonhardt, *Ann.*, **290**, 217 (1896)

1,2-Naphthoquinone (5). Mol. wt. 158.15, m.p. 145–147°, dec., Supplier: Eastman (practical).

Preparation.[1] A mixture of 0.5 mole of sulfanilic acid and 0.25 mole of sodium carbonate is dissolved in 500 ml. of water and the solution is cooled to 15°, treated with a solution of 0.54 mole of sodium nitrite in 100 ml. of water, and poured into a mixture of 1.25 moles of hydrochloric acid and 600 g. of ice. The solution, from which *p*-benzenediazonium sulfonate soon separates, is let stand for 15–25 min.

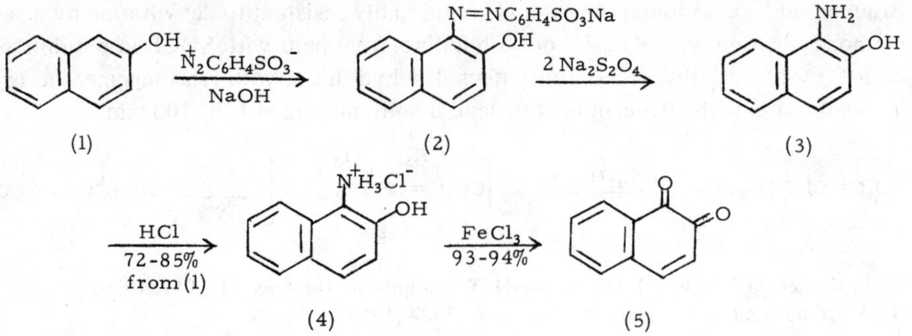

while 0.5 mole of β-naphthol is dissolved in a warm solution of 2.75 moles of sodium hydroxide in 600 ml. of water. The solution is cooled to about 5° by addition of 400 g. of ice, the suspension of diazotized sulfanilic acid is added, and the mixture allowed to stand for 1 hr. The azo compound (2), Orange II, separates to form a stiff paste. The solid is brought into solution by warming to 45–50°, and about one tenth of 230 g. (1.1 moles) of sodium hydrosulfite is added continuously until the froth subsides; the remainder is then added without delay. To give an easily filterable product, the mixture is heated strongly until it begins to froth and then cooled to 25°, and the cream-colored product (3) collected and washed. The crude amino-naphthol is washed into a beaker containing a solution at 30° of 0.63 mole of concd. hydrochloric acid and 2 g. of stannous chloride (antioxidant) in 1 l. of water. When the amine dissolves, the solution is clarified with 10 g. of Norit, filtered by suction, heated to the boiling point, treated with 100 ml. of concd. hydrochloric acid, and then cooled. 1-Amino-2-naphthol hydrochloride (4) separates as completely colorless needles. For oxidation to the quinone (5), a solution of 0.89 mole of ferric chloride in a mixture of 90 ml. of concd. hydrochloric acid and 200 ml. of water is cooled by addition of 250 g. of ice and filtered by suction. A mixture of 0.41 mole of the hydrochloride (4), 5 ml. of concd. hydrochloric acid, and 3 l. of water preheated to 35° is swirled for 1–2 minutes until the solid is dissolved and filtered by suction. The oxidizing agent is added rapidly with swirling, and the quinone, which separates as a voluminous, microcrystalline orange-yellow precipitate, is collected and washed well with water. The highly sensitive product is directly pure; recrystallization lowers its temperature of decomposition.

[1] L. F. Fieser, *Org. Syn., Coll. Vol.*, **2**, 35, 430 (1943)

1,4-Naphthoquinone. Mol. wt. 158.15, m.p. 124–125°.

Preparation. One efficient route to this compound is by a diazo coupling procedure parallel to that described for the preparation of 1,2-naphthoquinone.[1] Another, a diene synthesis, is as follows.[2] A suspension of 1 mole of 1,4-benzoquinone in 500 ml. of acetic acid is chilled until the solvent begins to crystallize and treated with 1.2 moles of condensed butadiene. The flask is closed with a wired-on stopper, cooled in running water, and shaken occasionally for a few hours until the quinone has all dissolved. After 40–48 hrs. the solution is filtered through a pad of decolorizing carbon, and the filtrate and washings heated on the steam bath. A solution of 100 ml. of concd. hydrochloric acid and 15 g. of stannous chloride in 500 ml. of water is added, and the solution heated for 15 min. on the steam bath to effect isomerization to (3), which soon begins to separate in heavy white needles.

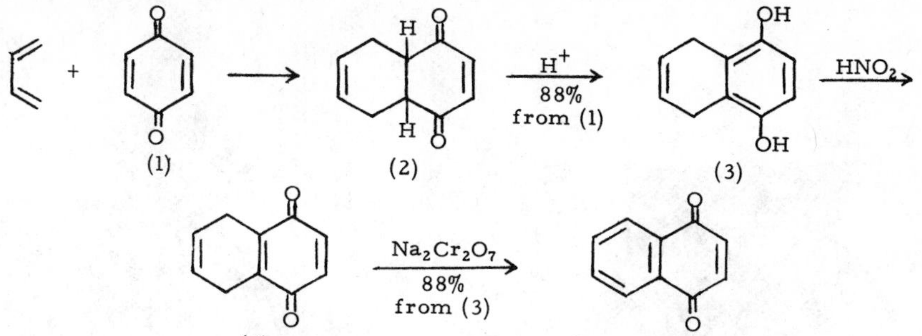
H
H
H^+
88%
from (1)
OH
OH
HNO_2
$Na_2Cr_2O_7$
88%
from (3)

(1) (2) (3) (4) (5)

This dihydronaphthohydroquinone is dissolved in hot acetic acid, the solution is let cool to about 100°, and an aqueous solution of sodium nitrite is run in rapidly with swirling. Nitrous acid effects smooth oxidation to 5,8-dihydro-1,4-naphthoquinone (4, which can be isolated in 91–97% yield). The temperature is adjusted to 65°, and a warm (65°) solution of sodium dichromate containing a little sulfuric acid is added. The temperature is checked at 65–70° for 15 min., and after 45 min. addition of ice and water precipitates bright yellow naphthoquinone, m.p. 124–125°.

Commercial preparations. Current offerings of 1,4-naphthoquinone are: Baker, technical, m.p. 95–120°, 500 g./\$6.05; Eastman, practical, m.p. 121–124°, 500 g./\$7.15; Matheson, Coleman and Bell, pure, m.p. 124–125°, 100 g./\$9.40. The Baker and Eastman materials can hardly be made by either process cited above, since the starting materials, 1-naphthol and 1,4-benzoquinone, cost \$16.40 and \$5.80 per 500 g. The technical material, possibly produced in Japan as a by-product of the air oxidation of naphthalene to phthalic anhydride, seemed to offer a cheap source of 2-hydroxy-1,4-naphthoquinone (*which see*) required for a current research in this laboratory. Pure 1,4-naphthoquinone[1,2] is bright yellow and reacts with acetic anhydride and boron fluoride etherate to give pure, colorless 1,2,4-triacetoxynaphthalene in two crops: 66.5% + 16.6% = 83.1%. Trial of the dark red Eastman material showed it to be completely unsatisfactory for the reaction but also brought a reminder of a property of 1,4-naphthoquinone appropriately emphasized on the label of the MCB material (Fig. N-1). The quinone has an insidious propensity to contact the hands and face, where it produces painful burns and blisters.[3] A hazard in handling the material, even with rubber gloves, is that a trace of dust may cause sneezing and unavoidable spillage. With care it is possible to effect considerable purification by crystallization from acetone (500 ml. per 200 g.) or tetrahydrofurane (300 ml. per 200 g.) without use of decolorizing carbon and filtration, which presents a difficult handling problem. The mixture is best refluxed in a closed system (Fig. N-2), and the solution is stirred while being cooled in an ice bath in order to produce small crystals that tend not to occlude dark impurities. Collected and washed with great care, the material was obtained as a bright yellow solid which, however, darkened somewhat in diffuse light. The recovery was 65–75%. Recrystallization did not eliminate the light sensitivity unless decolorizing carbon was used. In the Thiele reaction, twice crystallized Eastman material afforded a first crop of colorless 1,2,4-triacetoxynaphthalene in yield of 62.5%.

The material offered by Matheson, Coleman and Bell is assuredly of high quality. A 70-g. batch was stirred with acetic anhydride–boron fluoride ethereate at 0° and

1,4-Naphthoquinone
MP 124-125°C

Fig. N-1 Label reads: Warning. May cause burns. Do not breathe dust. Avoid contact with skin, eyes, or clothing.

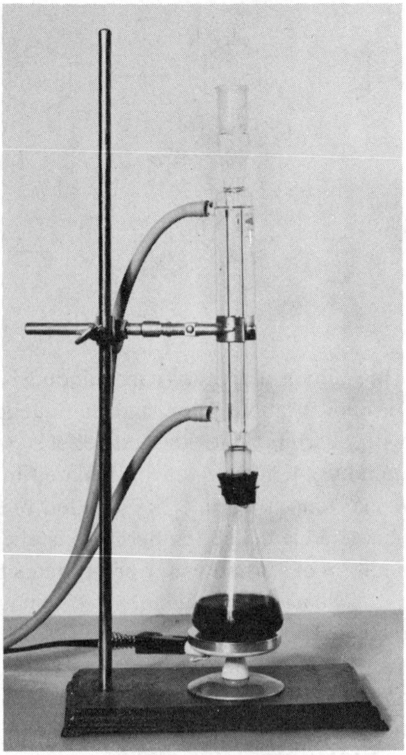

Fig. N-2 Dissolving 30 g. of 1,4-naphthoquinone in 75 ml. of acetone in a 250-ml. Erlenmeyer (short Liebig condenser, Neoprene adapter, 70-watt Wilkens-Anderson hot plate)

then with warming and afforded pure triacetoxynaphthalene in 2 crops: 66% + 9.5% = 75.5%. Although this result nearly matches the best yield as yet recorded (81%),[2] we decided to try use of perchloric acid, since this is the only catalyst found capable of effecting Thiele addition of acetic anhydride to 2-methyl-1,4-naphthoquinone.[3] Indeed the yield was raised to 80% (2 crops) by adding 10 ml. of 71% perchloric acid dropwise to 200 ml. of acetic anhydride which was stirred mechanically in an ice bath, pouring the resulting solution onto 79 g. of Eastman naphthoquinone which had been crystallized twice from tetrahydrofurane, and stirring with ice cooling until reaction was complete. The most satisfactory technique found for weighting out the noxious 1,4-naphthoquinone, for example from a 1-lb. bottle of commercial material, is to use a powder funnel (Fig. N-3) fashioned from an 18 × 28-cm. sheet of cellulose acetate and Scotch tape, rubber gloves, and a dust mask.[4]

[1] L. F. Fieser, *Org. Syn., Coll. Vol.*, **1**, 383 (1941); **2**, 39 (1943)

[2] L. F. Fieser, *Am. Soc.*, **70**, 3165 (1948)

[3] H. Burton and P. F. G. Praill, *J. Chem. Soc.*, 755 (1952)

[4] A comparison of practical and pure quinone was made by dusting small and approximately equal amounts of the materials on bandaids and applying these to the inner right and left forearms of a volunteer. The lesions produced in 20 hrs. were indistinguishable in size and severity.

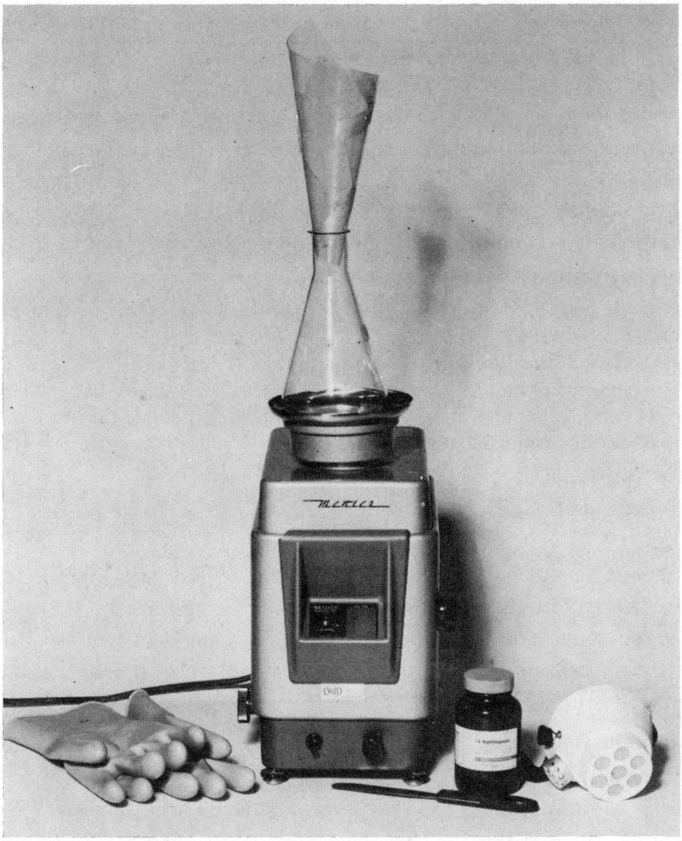
Waring

Fig. N-3 Manipulation of a noxious solid

1,2-Naphthoquinone-4-sulfonate, ammonium and potassium. Mol. wts. 255.25 and 276.31. Supplier of sodium salt: Eastman.

Preparation. A solution of 2.1 moles of β-naphthol in alkali is treated at 0° with sodium nitrite, and then dilute sulfuric acid is run in slowly.[1] 1-Nitroso-2-naphthol separates, the moist cake is transferred to a wide crock, and the funnel rinsed with a cold solution of 5.8 moles of sodium bisulfite and 100 ml. of 6*N* sodium hydroxide in 2 l. of water.[2,3] The mixture is diluted with water to 4–4.5 l. and stirred vigorously

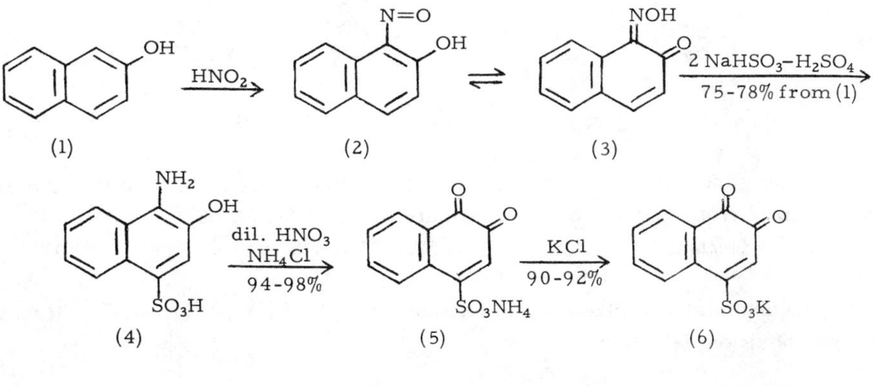
OH
HNO2
N=O
OH
NOH
O
2 NaHSO3–H2SO4
75-78% from (1)
(1)
(2)
(3)
NH2
OH
SO3H
dil. HNO3
NH4Cl
94-98%
O
O
SO3NH4
KCl
90-92%
O
O
SO3K
(4)
(5)
(6)

with a paddle so that all soluble product is dissolved in 3–4 min. The solution is filtered as rapidly as possible, and the clear, golden-yellow filtrate is acidified (H_2SO_4) immediately. 1-Amino-2-naphthol-4-sulfonic acid[4] which separates is light gray, but when washed with warm ethanol until the filtrate is colorless and then with ether (absence of light), it is obtained as a pure white, dust-dry product which weighs 370–380 g. (75–78% based on β-naphthol[5]). Oxidation is accomplished with dil. nitric acid at 25–30°, and the ammonium salt is salted out with ammonium chloride solution. It is obtained as bright orange microcrystals of high purity and is convertible into the potassium salt in high yield.

[1] C. S. Marvel and P. K. Porter, *Org. Syn., Coll. Vol.*, **1**, 411 (1941)
[2] L. F. Fieser, *ibid.*, **2**, 42 (1943)
[3] E. L. Martin and L. F. Fieser, *ibid.*, **3**, 635, note 1 (1955)
[4] Technical 1-amino-2-naphthol-4-sulfonic acid (du Pont Co., Dyes and Chemicals Div.; Allied Chem. Corp., National Aniline Division) is rendered suitable for use in the next step by washing with warm ethanol and then with ether.
[5] E. L. Martin and L. F. Fieser, *Org. Syn., Coll. Vol.*, **3**, 633 (1955)

α-Naphthylisocyanate, $C_{10}H_7N{=}C{=}O$. Mol. wt. 169.18, m.p. 3–5°. Supplier: Eastman.

The reagent is useful for the characterization of hydroxy compounds by conversion into the naphthylurethanes. For example a mixture of 0.2 g. of phenol, 0.2 g. of the isocyanate, and 1 drop of pyridine is heated over a small flame. The mixture generally becomes red within a few minutes and is then cooled, and the solid crystallized from ligroin.

α-Naphthylisothiocyanate, α-$C_{10}H_7N{=}C{=}S$. Mol. wt. 185.25, m.p. 56–57°. Suppliers: A, E, F. Preparation.[1]

End-group analysis of proteins. The reagent is used in the same way that phenylisocyanate is used (*which see*), and has the advantage for spectrophotometric analysis that the naphthyl derivatives have a maximum at 222 mμ of high intensity.[2]

[1] G. M. Dyson and R. F. Hunter, *Rec. trav.*, **45**, 421 (1926)
[2] S. Katsuki, J. E. Scott, and I. Yamashina, *Biochem. J.*, **97**, 25C (1965)

Nickel acetate tetrahydrate, $Ni(OCOCH_3)_2 \cdot 4H_2O$. Mol. wt. 248.87. Supplier: Baker and Adamson, B, F, MCB.

Aldoximes → Amides.[1] The isomerization of piperonaldoxime to piperonylamide[2] is accomplished by refluxing a solution in xylene with catalytic amounts of nickel

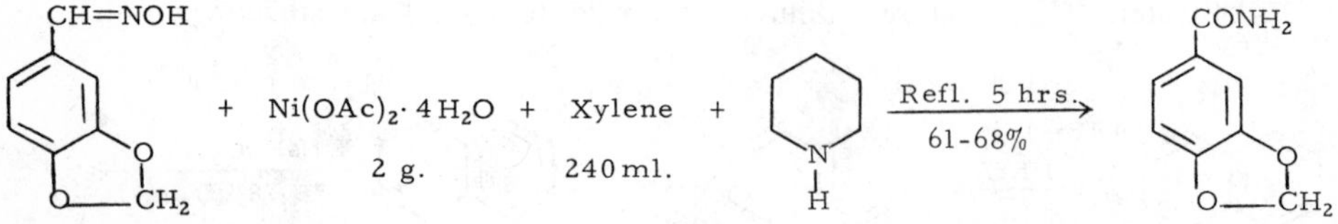
CH=NOH
O
CH2
+ Ni(OAc)2· 4 H2O + Xylene +
2 g.
240 ml.
N
H
Refl. 5 hrs.
61-68%
CONH2
O
O
CH2

0.4 m.

acetate and piperidine. The crude product that separates (m.p. 160–165°) is recrystallized from isopropanol to give pure material (167–168°).

[1] L. Field, P. B. Hughmark, S. H. Shumaker, and W. S. Marshall, *Am. Soc.*, **83**, 1983 (1961)
[2] J. D. Buckman and L. Field, procedure submitted to *Org. Syn.*

Nickel-aluminum alloy (1:1 Raney nickel alloy). Suppliers: W. R. Grace Co., Raney Catalyst Division; Harshaw.

On treatment of this alloy with aqueous sodium hydroxide the aluminum dissolves, with liberation of hydrogen, and the residue is a black sponge of nickel atoms

$$NiAl_2 + 6\,NaOH \longrightarrow Ni + 2\,Na_3AlO_3 + 3\,H_2$$

interspersed with holes left by departed aluminum atoms. The sponge nickel is effective as a hydrogenation catalyst, as described in the next section. Chemical reduction in some cases can be effected by the reaction of Raney alloy with alkali. Papa, Schwenk, *et al.*[1,2] showed that by this method benzyl alcohol can be reduced to toluene (70%), acetophenone to ethylbenzene (70%), and 2-methylcyclohexanone to 2-methylcyclohexanol (80%). Best results are obtained when the substrate is soluble in alkali, as in the case of the diacid (2) resulting from oxidation of β-naphthol with peracetic acid (Page and Tarbell[3]). A solution of the diacid in 500 ml. of 10%

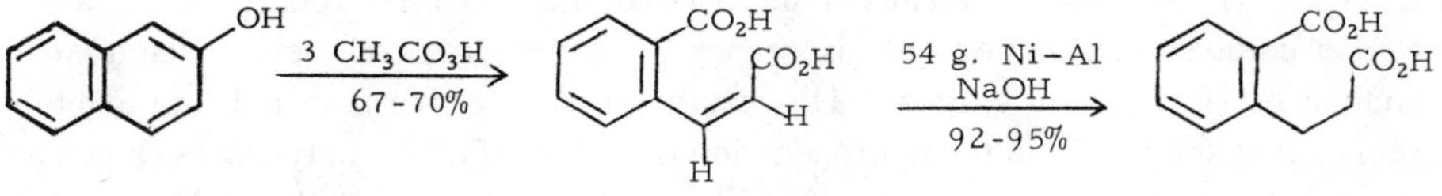
OH
3 CH₃CO₃H
67-70%
CO₂H
CO₂H
H
H
54 g. Ni-Al
NaOH
92-95%
CO₂H
CO₂H

0.14 m. (1) 0.094 m. (2) (3)

sodium hydroxide was warmed to about 87° and 54 g. of Raney alloy was added with stirring over a period of 50 min. After stirring and heating on the steam bath for 1 hr. more, the mixture was filtered by suction with care to keep the residual metal moist, for it becomes pyrophoric when dry. The metal is destroyed by careful treatment with dilute nitric acid (vigorous reaction). If the carbonyl group is attached directly to an aromatic ring the extent of reduction is dependent upon the temperature: at 80–90° Clemmensen-type reduction occurs,[1,2] whereas at 10–20° the product

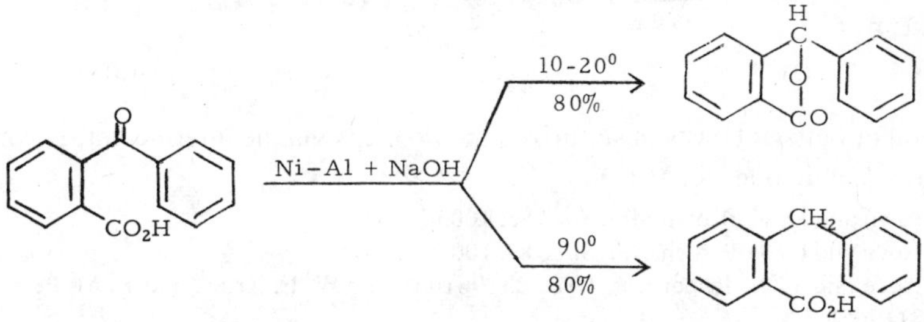
CO₂H
Ni-Al + NaOH
10-20°
80%
H
C
O
CO
90°
80%
CH₂
CO₂H

is the carbinol, as illustrated for the case of *o*-benzoylbenzoic acid.[4] Salicylaldehyde similarly affords saligenin (77%) at 10–20° and *o*-cresol (76%) at 90°.

Raney alloy in an alkaline medium reduces aromatic halides to hydrocarbons in good yield.[5,6] It has been used for reduction of the chlorinated polyene acid (4) to δ-phenyl-*n*-valeric acid (5).[7]

$$\underset{(4)}{C_6H_5CCl{=}CClCCl{=}CClCO_2H} \xrightarrow[86\%]{Ni\text{-}Al\ NaOH} \underset{(5)}{C_6H_5(CH_2)_4CO_2H}$$

Oximes and nitriles are reduced by Raney alloy–sodium hydroxide without external heating to amines in generally satisfactory yields.[8]

[1]D. Papa, E. Schwenk, and B. Whitman, *J. Org.*, **7**, 587 (1942)
[2]E. Schwenk, D. Papa, H. Hankin, and H. Ginsberg, *Org. Syn.*, *Coll. Vol.*, **3**, 742 (1955)
[3]G. A. Page and D. S. Tarbell, *ibid.*, **4**, 136 (1963)
[4]P. L. Cook, *J. Org.*, **27**, 3873 (1962)
[5]E. Schwenk, D. Papa, B. Whitman, and H. Ginsberg, *J. Org.*, **9**, 1 (1944)
[6]G. Märkl, *Ber.*, **96**, 1441 (1963)
[7]A. Roedig, G. Märkl, and V. Schaal, *Ber.*, **95**, 2844 (1962)
[8]B. Staskun and T. van Es, *J. Chem. Soc.*, 531 (1966)

Nickel boride, Ni_2B. H. C. Brown and his son have described two active hydrogenation catalysts. A nickel boride catalyst designated P-1 is prepared by reduction of nickel acetate with sodium borohydride in aqueous solution.[1] Comparative hydrogenations of olefins ranging from very reactive safrole to unreactive cyclohexene showed that the P-1 catalyst is considerably more active than Raney nickel and has less tendency to isomerize olefins in the course of the hydrogenation. A second nickel boride catalyst, designated P-2, is prepared by reduction of nickel acetate with sodium borohydride in ethanol.[2] This catalyst is less active than P-1 and exhibits remarkable selectivity in the hydrogenation of olefins of different types, as shown by the following figures in minutes for half-hydrogenation: safrole, 3.5 (with P-1 catalyst, 6); 1-octene, 7; 3-methyl-1-butene, 15; 3.3-dimethyl-1-butene, 56; cyclohexene, *ca.* 200; *trans*-2-pentene, *ca.* 360; 2-methyl-1-butene, *ca.* 400. The P-2 catalyst reduces 3-hexyne to the *cis*-olefin of 98–99% purity and is regarded as superior to the Lindlar catalyst for such reductions.

Truce[3] has reported that a reagent prepared from nickel (II) chloride hexahydrate in ethanol by reduction with sodium borohydride and evidently similar to Brown and Brown's nickel boride can be used to effect desulfurization. Thus the diphenyl-

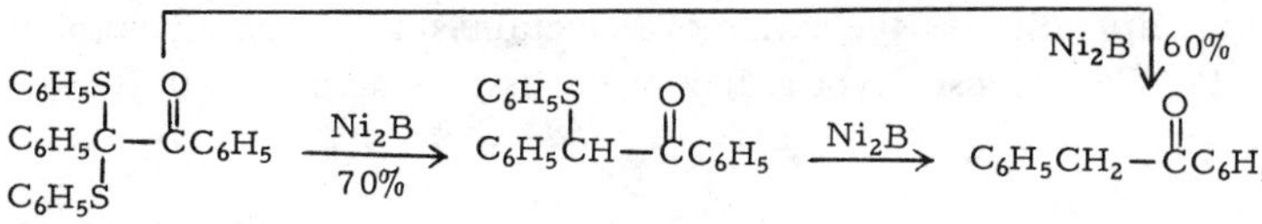
Ni₂B 60%
C₆H₅S
C₆H₅C—CC₆H₅
C₆H₅S
Ni₂B
70%
C₆H₅S
C₆H₅CH—CC₆H₅
Ni₂B
C₆H₅CH₂—CC₆H₅

(1) (2) (3)

mercaptal of benzil (1) was desulfurized in two steps via the intermediate (2) or, with excess reagent, in one step to (3).

[1]C. A. Brown and H. C. Brown, *Am. Soc.*, **85**, 1003 (1963)
[2]H. C. Brown and C. A. Brown, *Am. Soc.*, **85**, 1005 (1963)
[3]W. E. Truce and F. E. Roberts, *J. Org.*, **28**, 961 (1963); W. E. Truce and F. M. Perry, *ibid.*, **30**, 1316 (1965)

Nickel carbonyl, $Ni(CO)_4$. Mol. wt. 170.73, an extremely poisonous, colorless liquid, m.p. −25°, b.p. 43°, sp. gr. 1.31. Suppliers: Alfa, KK.

Nature of the present survey. Although study of the interaction of nickel carbonyl with unsaturated organic compounds has been under way for little more than two decades, reviews by Bird[1] (1962) and by Schrauzer[2] (1964) bear evidence to the extensive literature already on record in documentation of a highly specialized field of organometallic chemistry. Our account will then be limited to a very brief survey of types of reactions now clearly defined, if not as yet fully understood.

Hydrocarboxylation of acetylenes. In 1953 Reppe[3] published details of his "carbonylation" reaction, effected with use of nickel carbonyl and described more appropriately as a hydrocarboxylation (addition of the elements of formic acid).

The reaction, now recognized as more complicated, was originally represented by the equation:

$$4\ CH\equiv CH\ +\ 4\ H_2O\ +\ Ni(CO)_4\ +\ 2\ HCl \longrightarrow 4\ CH_2=CHCO_2H\ +\ NiCl_2\ +\ H_2$$

In the meantime, E. R. H. Jones et al.[4] had examined the reaction for laboratory procedures and found that it often can be carried out at temperatures of 60–70° at atmospheric pressure and that yields are about 40–50%. Other examples[1,2] are as follows:

$$C_6H_5C\equiv CH \xrightarrow{Ni(CO)_4} C_6H_5\overset{\underset{CO_2H}{|}}{C}=CH_2$$

$$C_6H_5C\equiv CCH_3 \xrightarrow{Ni(CO)_4} C_6H_5\overset{\underset{CO_2H}{|}}{C}=CHCH_3\ +\ C_6H_5CH=\overset{\underset{CH_3}{|}}{C}CO_2H$$

$$AcOH_2CC\equiv CCH_2OAc \xrightarrow{Ni(CO)_4} \underset{HO_2C}{\overset{AcOH_2C}{>}}C=C\underset{H}{\overset{CH_2OAc}{<}}$$

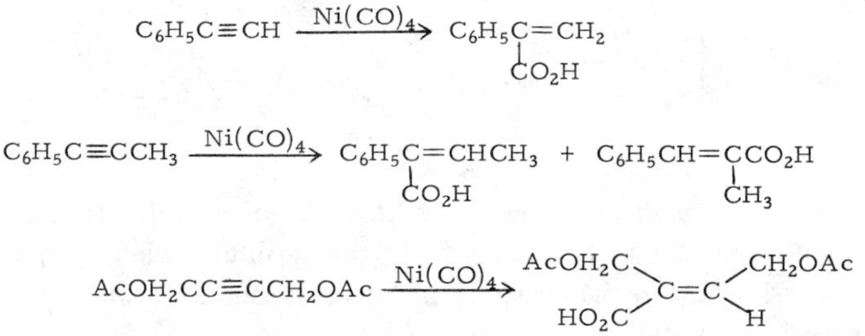

The Reppe hydrocarboxylation of acetylenes was initially carried out in the presence of an acid, but little was known about the function of the acid or the nature of the carbon monoxide transfer agent. Sternberg et al.[5] found that diphenylacetylene can be hydrocarboxylated in alkaline solution and that in this case a nickel carbonyl anion is the source of carbon monoxide. When the hydrocarbon was shaken with a saturated solution of sodium hydroxide in methanol in the presence of excess nickel carbonyl under helium the reaction mixture turned dark red. After 80 hrs., acidification and workup afforded α-phenyl-*trans*-cinnamic acid and 1,2,3,4-tetraphenyl-butadiene in the yields indicated. Note that the cinnamic acid results from *cis* addition of the elements of formic acid.

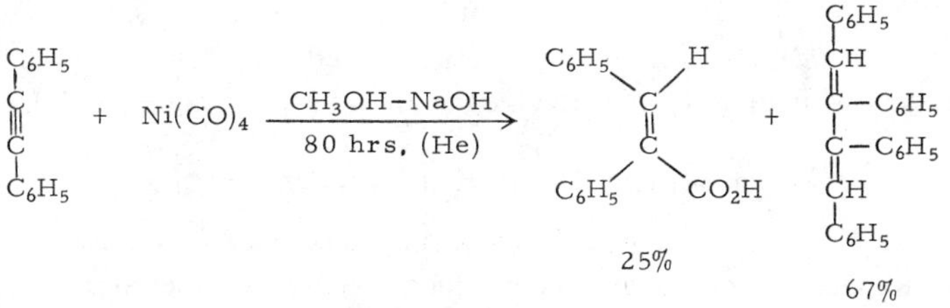

Formation of allenic acids. Chlorosubstituted acetylenes of the type (1) react with nickel carbonyl to form allenic acids (2).[6] Yields are very low (5–15%),

$$R_2\overset{\underset{Cl}{|}}{C}-C\equiv CR'\ +\ Ni(CO)_4\ +\ H_2O \xrightarrow{60°} R_2C=C=\overset{\underset{CO_2H}{|}}{C}R'$$

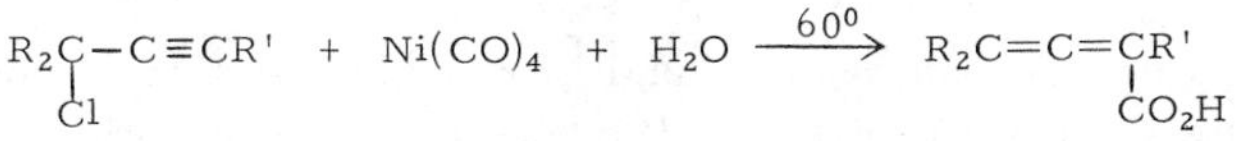

$$(1) \qquad\qquad\qquad\qquad (2)$$

probably because of the instability of the products, α-Hydroxyacetylenes (3) afford allenic acids on treatment with nickel carbonyl, butanol, and dry hydrogen chloride; in this case the reaction involves initial conversion to the chloride.

$$(CH_3)_2\overset{\underset{OH}{|}}{C}-C\equiv CH \xrightarrow[65°]{HCl-BuOH} (CH_3)_2\overset{\underset{Cl}{|}}{C}-C\equiv CH \xrightarrow{Ni(CO)_4} (CH_3)_2C=C=CHCO_2H$$

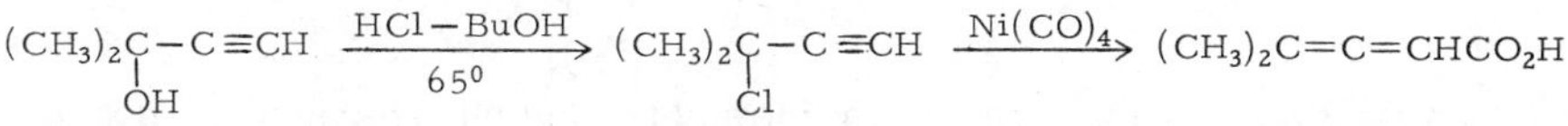

$$(3) \qquad\qquad\qquad (4) \qquad\qquad\qquad (5)$$

Hydrocarboxylation of olefins. Whereas acetylenes usually undergo hydrocar-boxylation at atmospheric pressure and temperature, olefins usually require more drastic conditions. For example, Reppe and Kröper[7] heated cyclohexene with nickel carbonyl at 270°/100 atm. to obtain cyclohexanecarboxylic acid. Bird *et al.*[8] reasoned that a strained olefin should react more readily and found indeed that norbornene (6) reacts with nickel carbonyl in ethanol–water–acetic acid at atmospheric pressure at 50° to give in good yield a mixture of the *exo*-acid (7), the corresponding ester (8), and a trace of the ketone (9). The reaction involves *cis* addition of HCO_2H from the less hindered side.

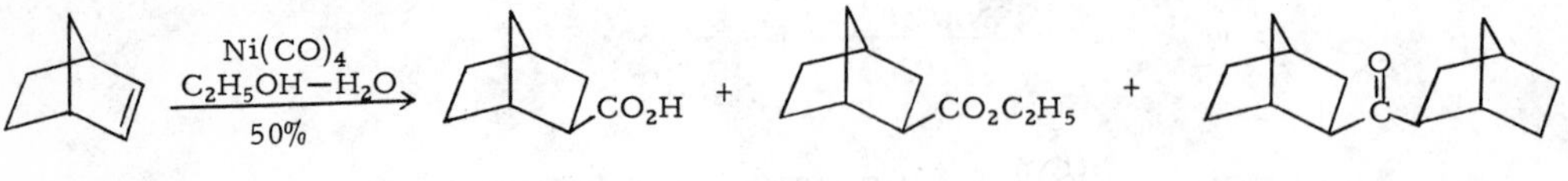

(6) (7) 10% (8) 72% (9)

Reaction with diazoalkanes.[9] Catalytic amounts of nickel carbonyl decompose diazoalkanes to products evidently formed from an intermediate carbene. Use of a large excess of reagent in the presence of ethanol leads to formation of carboxylic acid esters in yields of 20–25%.

$$R_2\overset{-}{C}-\overset{+}{N}\equiv N + Ni(CO)_4 \xrightarrow{-CO} R_2\underset{\underset{N_2^+}{|}}{\overset{}{C}}-\overset{-}{Ni}(CO)_3 \xrightarrow{-N_2}$$

$$R_2\overset{+}{C}-\overset{-}{Ni}(CO)_3 \xrightarrow[-Ni(CO)_2]{} R_2C{=}C{=}O \xrightarrow{C_2H_5OH} R_2CHCO_2C_2H_5$$

With diazonium fluoroborates.[10] In the presence of acetone or acetic acid, nickel carbonyl reacts with a diazonium fluoroborate to form the carboxylic acid in yields in the range 5–70%.

$$Ar\overset{+}{N}\equiv N(BF_4^-) \xrightarrow[AcOH-H_2O]{Ni(CO)_4} ArCO_2H$$

With aryl iodides.[11] In refluxing methanol, ethanol, or isopropanol, nickel carbonyl reacts with an aryl iodide to give the aroate ester in good yield. Under these

$$ArI + Ni(CO)_4 + C_2H_5OH \longrightarrow Ar\overset{O}{\overset{\|}{C}}OC_2H_5$$

conditions, aryl bromides, aryl chlorides, and alkyl halides do not react. When tetrahydrofurane is used as solvent the product is the aril (α-diketone).

$$C_6H_5I + Ni(CO)_4 \xrightarrow[80\%]{Tetrahydrofurane} C_6H_5\overset{O}{\overset{\|}{C}}-\overset{O}{\overset{\|}{C}}C_6H_5$$

Coupling of allylic compounds. Bauld[12] found that allyl and cinnamyl acetates react with nickel carbonyl to give biallyl and bicinnamyl in moderate yields.

$$2\,RCH{=}CHCH_2OAc + Ni(CO)_4 \xrightarrow[THF]{45\text{-}65^0\ 2\ hrs.} RCH{=}CHCH_2CH_2CH{=}CHR + Ni(OAc)_2 + 4\,CO$$

$$\left.\begin{array}{l} R = H \quad 50\% \\ R = C_6H_5 \quad 31\% \end{array}\right\}\text{yield}$$

Corey and Hamanaka[13] employed intramolecular allylic coupling in a novel synthesis of medium-size carbocyclic systems. The doubly allylic α,ω-dihalide (1) on cyclization with nickel carbonyl gave as the chief product the all-*trans*-dimethylcyclododecatriene (2), accompanied by an isomer, probably *cis, trans, trans*.

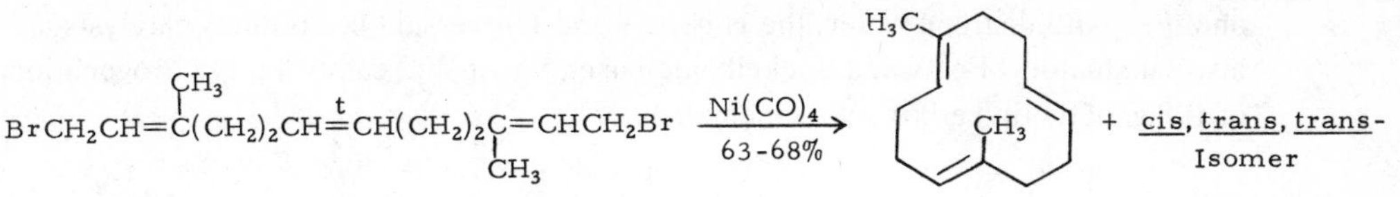
CH3
BrCH2CH=C(CH2)2CH=CH(CH2)2C=CHCH2Br
CH3
Ni(CO)4
63-68%
H3C
CH3
+ cis, trans, trans-
Isomer

(1) (2)

[1]C. W. Bird, *Chem. Rev.*, **62**, 283 (1962)
[2]G. N. Schrauzer, *Advances in Organometallic Chemistry*, **2**, 1 (1964)
[3]W. Reppe, *Ann.*, **582**, 1 (1953)
[4]E. R. H. Jones, T. Y. Shen, and M. C. Whiting, *J. Chem. Soc.*, 230 (1950); 763, 766 (1951)
[5]H. W. Sternberg, R. Markby, and I. Wender, *Am. Soc.*, **82**, 3638 (1960)
[6]E. R. H. Jones, G. H. Whitham, and M. C. Whiting, *J. Chem. Soc.*, 4628 (1957)
[7]W. Reppe and H. Kröper, *Ann.*, **582**, 38 (1953)
[8]C. W. Bird, R. C. Cookson, J. Hudec, and R. O. Williams, *J. Chem. Soc.*, 410 (1963)
[9]C. Rüchardt and G. N. Schrauzer, *Ber.*, **93**, 1840 (1960)
[10]J. C. Clark and R. C. Cookson, *J. Chem. Soc.*, 686 (1962)
[11]N. L. Bauld, *Tetrahedron Letters*, 1841 (1963)
[12]*Idem, ibid.*, 859 (1962)
[13]E. J. Corey and E. Hamanaka, *Am. Soc.*, **86**, 1641 (1964)

Nickel catalysts (a), Raney type. The Raney Catalyst Division and the Davison Chemical Division of W. R. Grace and Co. supply an identical 50% sponge nickel catalyst prepared by the method of Raney by leaching 1:1 nickel–aluminum alloy with alkali. The catalyst is similar to the W-catalysts described in the next section. Harshaw Chem. Co. supplies similar nickel catalysts. Universal Oil Products Co. catalyst kieselguhr pellets are reduced with hydrogen and pulverized before use (*see Org. Syn., Coll. Vol.*, **3**, 278 (1955).

Preparation. Procedures for the preparation of catalyst from Raney nickel–aluminum alloy differ in the method of adding the alloy, in the concentration of sodium hydroxide, in the temperature and duration of digestion, and in the method of washing the catalyst free of sodium aluminate and alkali. Catalysts prepared by the Adkins group with identifying designations (W for Wisconsin) and references are as follows: W-1,[1] W-2,[2] W-3,[3,4] W-4,[3,4] W-5,[5] W-6,[5,6] and W-7[5,6] (the numbers reflect merely the chronology of development). Adkins and Krsek[7] compared the activity of these catalysts in the hydrogenation of β-naphthol. The catalysts most widely used are W-2 and W-6; W-5, although somewhat less active, is rated as an excellent catalyst. The activity is said to be enhanced by addition of a small amount of triethylamine chloroplatinate.[8]

Typical hydrogenations are conducted at moderately high pressures of hydrogen. By heating the 50% sponge nickel with 10% sodium hydroxide, Dominguez[9] obtained a very active catalyst designated T-1, which can be used at low temperatures (40–60°) and low pressures (50–60 p.s.i.). Examples: phenol ⟶ cyclohexanol (92%); coumarin ⟶ 3,4-dihydrocoumarin (83%).

Reduction of C═C and C≡C. Newman *et al.*[10] found that resorcinol can be hydrogenated satisfactorily in alkaline solution in the presence of W-2 catalyst after this has been freed of aluminum by careful washing with 5% alkali solution

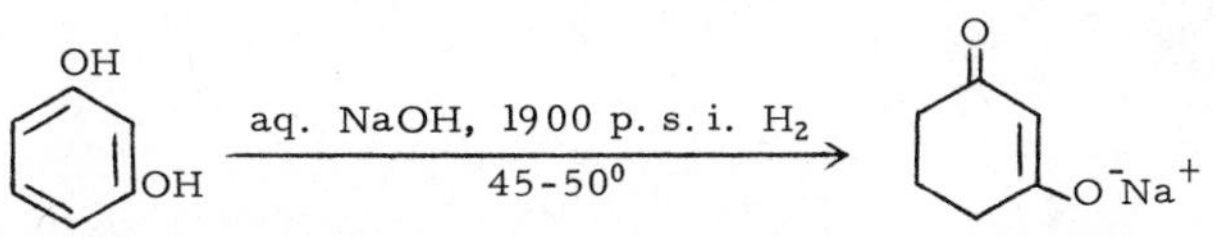
OH
OH
aq. NaOH, 1900 p. s. i. H2
45-50°
O
O⁻Na⁺

and then with distilled water; the Harshaw and Universal Oil Products catalysts are also satisfactory. Feely and Boekelheide[11] used 5 g. of W-2 catalyst for hydrogenation of 108 g. of diethyl ethoxyethylenemalonate.

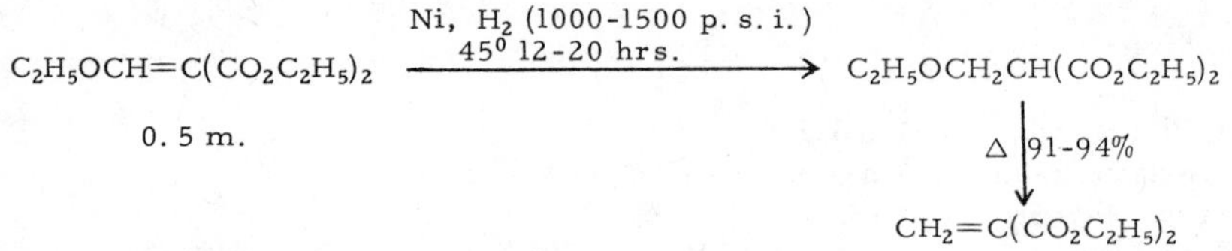

Raney nickel deactivated with piperidine and zinc acetate has been used for semihydrogenation of the acetylenic linkage.[12]

Erman and Flautt[13] tried hydrogenation of *p*-anisylbornylene (1) with Pt and Pd catalysts in various solvents but invariably obtained mixtures of the *endo* and *exo*

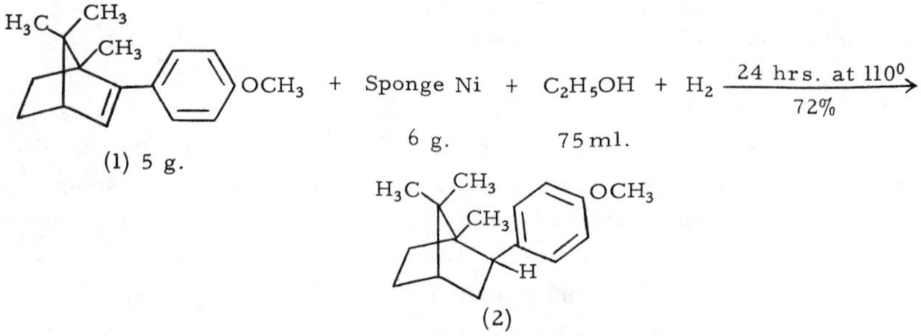

products. Use of W-5 Raney nickel led to reduction of the aromatic ring as well as the olefinic linkage. However, the Davison sponge nickel catalyst afforded homogeneous *exo*-dihydride in good yield.

Aldehydes. The reaction mixture containing the dialdehyde formulated was neutralized with solid sodium bicarbonate and placed in an autoclave and hydrogenated in the presence of 39 g. of W-2 Raney nickel.[14]

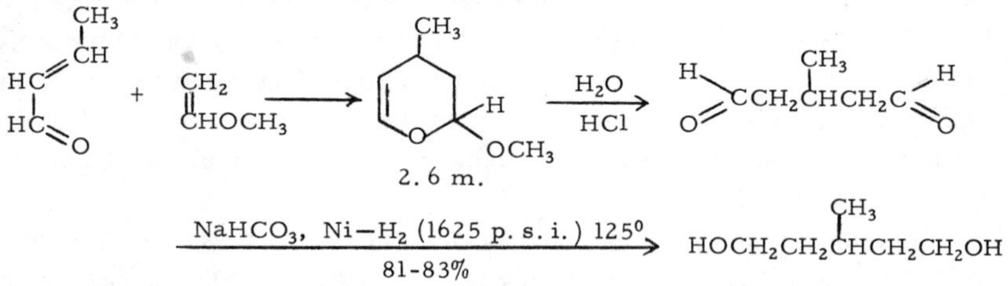

RNO_2 ⟶ RNH_2. Dauben *et al.*[15] effected hydrogenation of 1-(nitromethyl)-cyclohexanol in 450 ml. of acetic acid in the presence of 3 heaping teaspoonfuls of

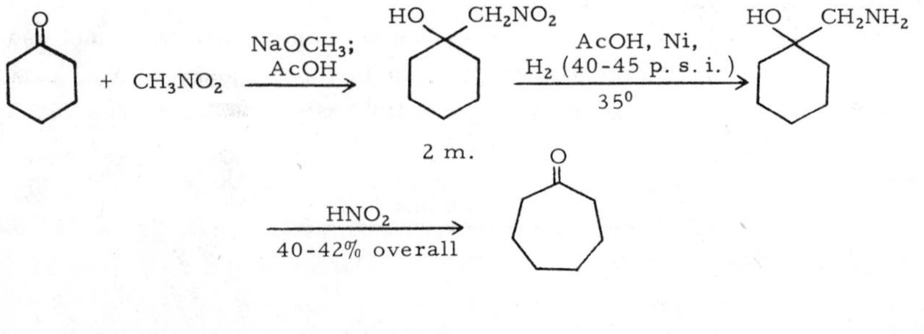

W-4 catalyst at a very slight positive pressure and with the temperature kept at 35° by internal cooling.

For preparation of an intermediate required for the synthesis of 5,5-dimethylpyrrolidone, Moffett[16] used absolute ethanol as solvent and the W-5 catalyst.

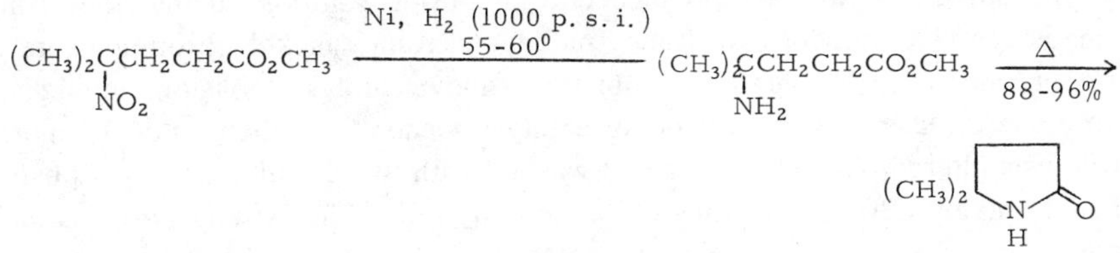

Amines. Primary amines, RNH_2, where R is alkyl, arylalkyl, or cycloalkyl, when boiled with Raney nickel (no H_2) are converted into the secondary amine (80% yield) and some tertiary amine (10%).[17] *n*-Hexylamine ⟶ di-*n*-hexylamine (70%) and tri-*n*-hexylamine (9%). The reaction is considered to involve dehydrogenation to the aldimine followed by addition of the starting amine, elimination of ammonia, and hydrogenation.

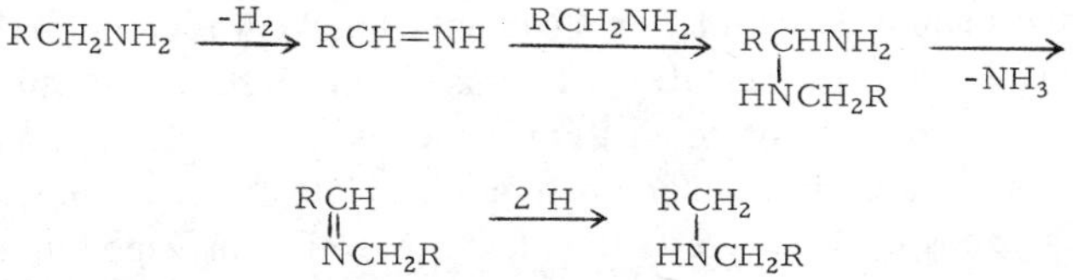

Alkylation of primary amines can be accomplished with Raney nickel (W-2) in the presence of ethanol, as in the preparation of N,N′-diethylbenzidine.[18] The mixture is stirred vigorously during the reflux period.

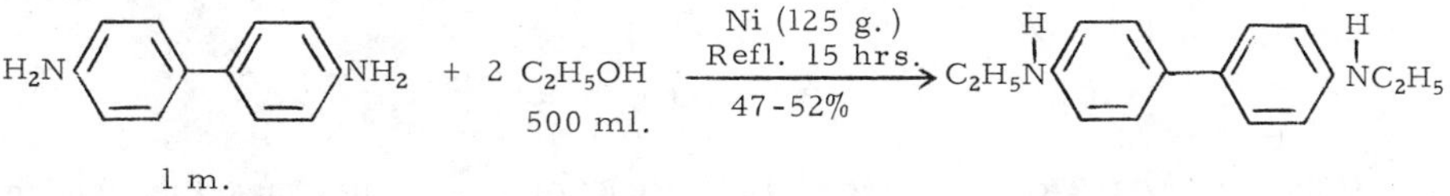

Balcom and Noller[19] condensed methylamine with an aldehyde under hydrogenating conditions and so prepared the secondary amine. Catalysts W-2 and W-6 both proved to be satisfactory. Methylamine was taken in 3:1 molar ratio; with a 1:1 ratio the yield dropped to 75%.

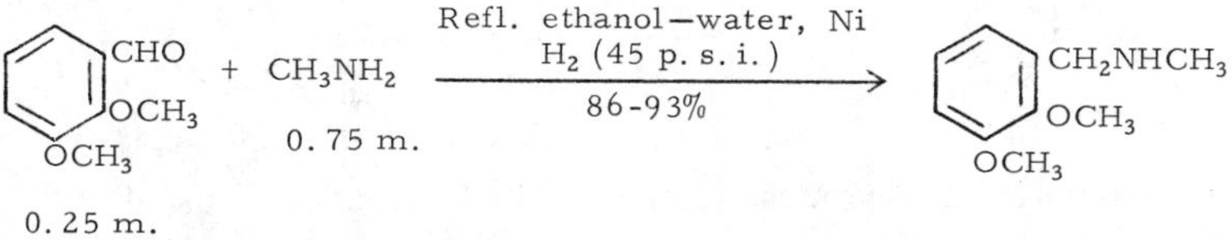

Nitriles. Aromatic[20] and aliphatic[21] nitriles on Raney nickel hydrogenation in ethanol containing phenylhydrazine are reduced to the aldehydes, which are trapped as the phenylhydrazones (8 hrs. at atmospheric pressure or 5 hrs. under pressure). Yields are around 30% if 1 equivalent of phenylhydrazine is used but are increased to about 90% by use of 4 equivalents of the reagent.

Hydrogenation of a nitrile under ordinary conditions suffers from the difficulty that the primary amine formed condenses with the intermediate aldimine to form the secondary amine. Gould, Johnson, and Ferris[22] eliminated this side reaction by

$$RC{\equiv}N \xrightarrow{2\,H} RCH{=}NH \xrightarrow{2\,H} RCH_2NH_2$$
$$\longrightarrow (RCH_2)_2NH$$

carrying out the hydrogenation in acetic anhydride solution so that the primary amine is acetylated as formed. Raney nickel and Raney nickel–chromium catalysts were purchased in active form from the Raney Catalyst Division; immediately before use the 2–3 g. (wet weight) of catalyst needed for reduction of 0.1 mole of nitrile was filtered from the water and washed with two 20-ml. portions of absolute ethanol and two 20-ml. portions of acetic anhydride. The Ni and Ni-Cr catalysts were about equally effective. Sodium acetate usually was used as basic cocatalyst; with sodium hydroxide as cocatalyst the reaction is very rapid, and the metal

$$\underset{0.1\text{ m.}}{CH_3(CH_2)_{11}CN} + \underset{2\text{-}3\text{ g.}}{Ni(H_2)} + \underset{12\text{ g.}}{NaOAc} \xrightarrow[100\%]{Ac_2O\ 50^0} CH_3(CH_2)_{11}CH_2NHAc$$

catalyst can be reused repeatedly. With some nitriles such as benzonitrile no cocatalyst is necessary. The reduction formulated (tridecanenitrile) was conducted at an initial hydrogen pressure of 50 p.s.i. and the uptake was complete in about 1 hr. The amine is isolated as the acetate or else this is hydrolyzed with hydrochloric acid.

Raney nickel hydrogenation of nitriles is aqueous acetic acid or in aqueous acetic acid–pyridine in the presence of sodium hypophosphite (NaH_2PO_2), which becomes oxidized to the phosphate, converts the nitriles into the corresponding aldehydes.[23] The reaction takes place at room temperature and pressure. Yields are in the range 50–90%. Thus a solution of 1 g. of benzonitrile and 2 g. of hydrated sodium hypophosphite in 29 ml. of a 1:1:2 water–acetic acid–pyridine mixture was stirred with 0.3–0.4 g. of Raney nickel at 40–45° for 1 hr.; benzaldehyde was isolated as the 2,4-dinitrophenylhydrazone.

$$C_6H_5CN + NaH_2PO_2 + Ni(H_2) \xrightarrow[90\%]{} C_6H_5CHO$$

***Halides and tosylates* ⟶ *hydrocarbons*.** Reaction with Raney nickel in alkaline solution effects reductive displacement of fluoro-, chloro-, bromo-, and iodo-substitutents in aromatic and aliphatic acids and in phenols (60–70% yields).[24] If the reaction is carried out in D_2O, deuterated derivatives are obtained. Under these

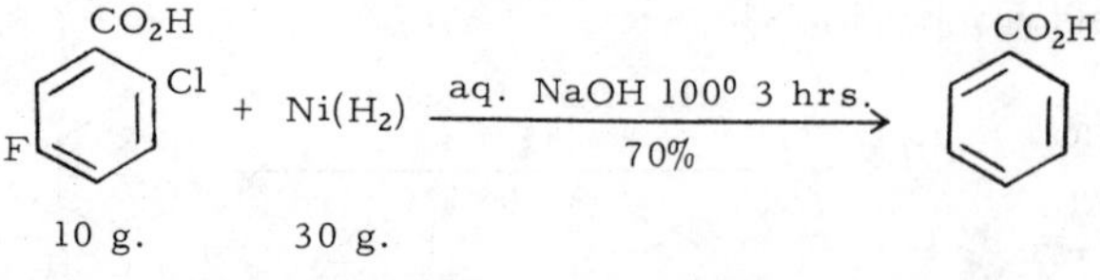
CO2H
Cl
F
+ Ni(H2)
aq. NaOH 100° 3 hrs.
70%
CO2H

10 g. 30 g.

conditions, trifluoromethyl groups, which resist hydrogenation by usual methods, are reduced to methyl groups.[25]

Kenner and Murray found that Raney nickel hydrogenation reduces phenyl tosylates to the corresponding hydrocarbons, at room temperature and pressure.[26] Other examples.[27] On the other hand, tosylates of aliphatic alcohols are merely hydrolyzed to the corresponding alcohols.[26]

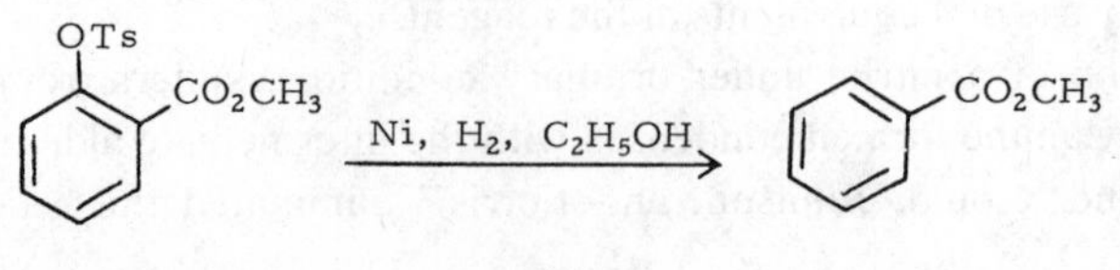
OTs
CO2CH3
Ni, H2, C2H5OH
CO2CH3

Desulfurization. In reviewing the subject of desulfurization with Raney nickel, Pettit and van Tamelen[28] credit Bougault[29] in 1940 with the first demonstration of a reaction which has become of considerable value in both synthesis and degradation. In laying a groundwork for the investigation of sulfur-containing biotin, Merck chemists[30] found that sulfides, disulfides, sulfones, and sulfoxides on reaction with Raney nickel catalyst undergo hydrogenolysis with the replacement of the sulfur atom by two atoms of hydrogen. No added hydrogen is required, for an abundant quantity is adsorbed on, or dissolved in, the catalyst. The W-2 catalyst was found to contain between 170 and 460 ml. of hydrogen per 4 g., depending upon the final temperature used in its preparation. Pyrophoric Raney nickel had been known to be effective for the removal from substrates of impurities, probably containing sulfur, which poison other hydrogenation catalysts.[31] Desulfurization of biotin (1) to desthiobiotin[32] played a key role in the elucidation of structure of the vitamin.

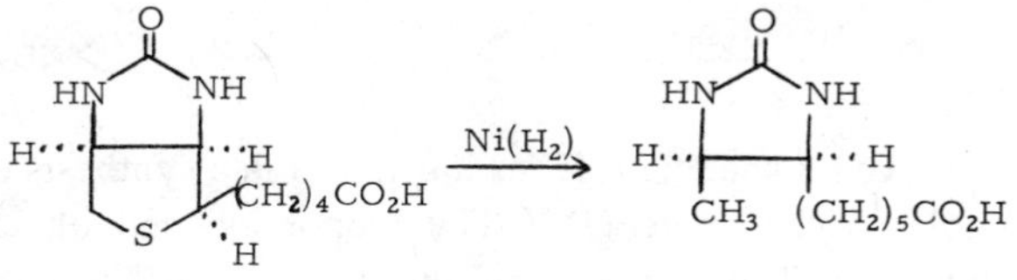
HN
NH
H
H
(CH2)4CO2H
S
H
Ni(H2)
HN
NH
H
H
CH3
(CH2)5CO2H
(1)
(2)

Badger[33] and Buu-Hoï[34] have made extensive use of nickel desulfurization in synthesis. One general scheme employs thiophene derivatives.[33] For example γ-2-thienylbutyric acid (1, succinoylation of thiophene and reduction) is acetylated (as ester) at the 5-position (2) and converted by Huang-Minlon reduction and

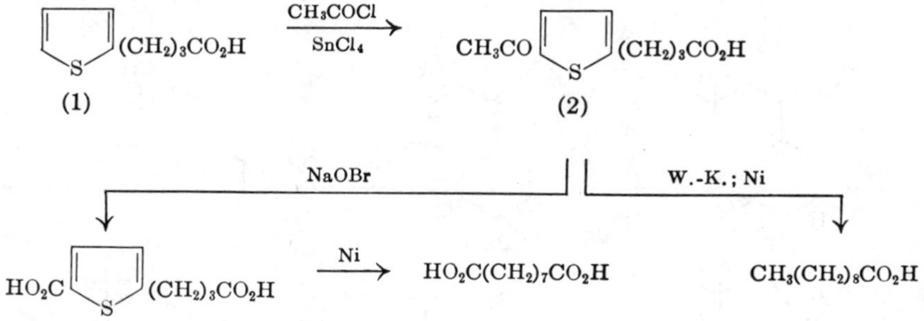
(CH2)3CO2H
S
(1)
CH3COCl
SnCl4
CH3CO
(CH2)3CO2H
S
(2)
NaOBr
W.-K.; Ni
HO2C
(CH2)3CO2H
S
(3)
Ni
HO2C(CH2)7CO2H
(4)
CH3(CH2)8CO2H
(5)

desulfurization with W-6 Raney nickel into decanoic acid (5). Higher homologs can be obtained by acylation of (1) with appropriate acid chlorides. The aceto acid (2) also can be oxidized to (3), which on desulfurization affords azelaic acid (4). Desulfurization of these acids is carried out conveniently in aqueous sodium carbonate solution, and yields usually are very high. Other examples.[35] Badger and co-workers[36] applied the method also to the synthesis of 2,3,10,11-dibenzperylene (3). Tetraphenylthiophene (1, below), readily available by heating tetraphenylcyclopentadienone with sulfur, when heated with molten sodium aluminum chloride undergoes dehydrogenative cyclization to flavophene (2). Desulfurization, accomplished by refluxing flavophene with deactivated Raney nickel in mesitylene, afforded the heptacyclic hydrocarbon (3).

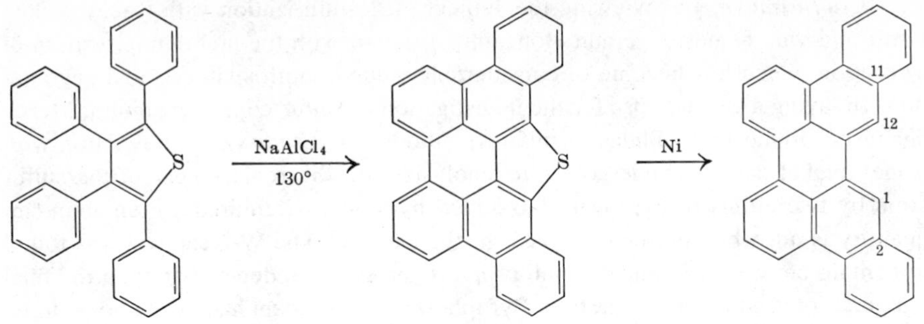

(1) (2) (3)

The sequence of conversion of a ketone to the ethylenethioketal and nickel desulfurization sometimes is useful for effecting selective removal of a carbonyl

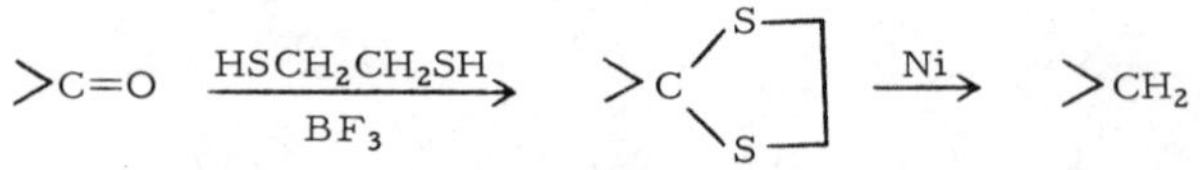

group after it has served a specific role, as illustrated in a synthesis of $\Delta^{9(11)}$-cholestenol from Δ^7-cholestenyl acetate (1).[37] The monoethylenethioketal (8) was desulfurized by refluxing 147 mg. of material in 60 ml. of methanol for 22 hrs. with 3 g. of Raney nickel prepared from 6 g. of alloy.

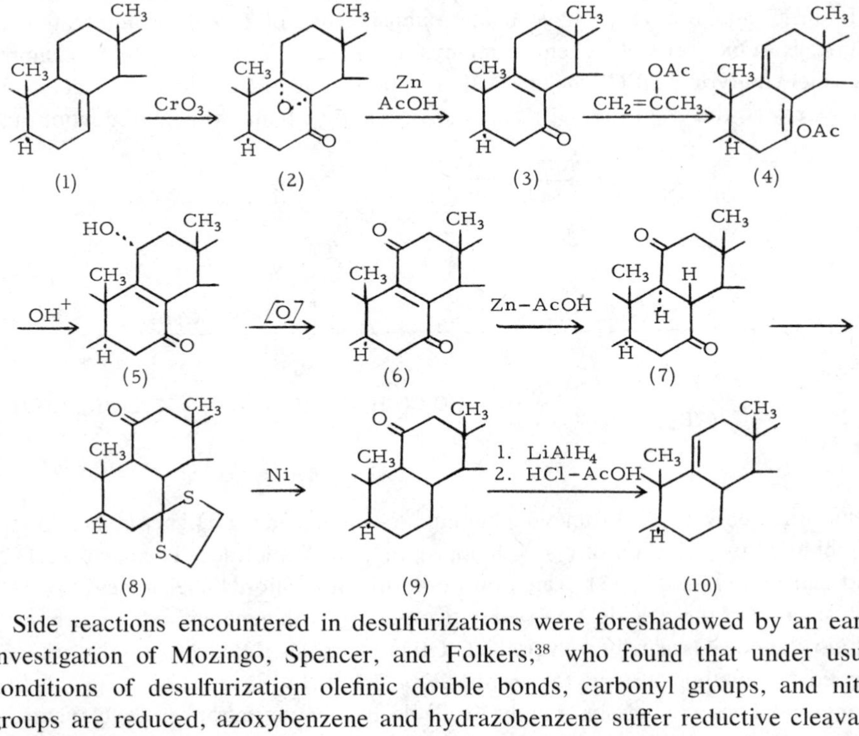

Side reactions encountered in desulfurizations were foreshadowed by an early investigation of Mozingo, Spencer, and Folkers,[38] who found that under usual conditions of desulfurization olefinic double bonds, carbonyl groups, and nitro groups are reduced, azoxybenzene and hydrazobenzene suffer reductive cleavage of the N—N bond, and benzyl alcohol yields toluene. For avoidance of such side reactions, subsequent workers have deactivated the catalyst by refluxing it with acetone.[39] However, this expedient is not always effective and side reactions continue to present a difficulty. Thus Djerassi and Williams[40] prepared the labeled

5α-androstane-11-one (2) by desulfurization of the ethylenethioketal (1) with Raney nickel containing active deuterium but obtained the Δ^2-olefin as the main product when the catalyst was deactivated. Similarly, 17-ethylenethioketals on reaction with

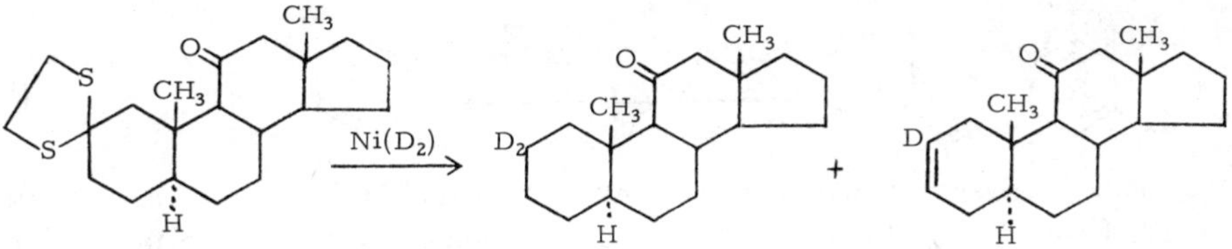

CH₃
CH₃
Ni(D₂)
D
H
+

(1) (2) (3)

the moderately active W-2 catalyst afford mainly the Δ^{16}-steroids, isolable in yields of the range 65–75%.[41]

In a synthesis of *o*-di-*t*-butylbenzene by Burgstahler and Abdel-Rahman[42] the last step, a desulfurization, was carried out originally with freshly prepared W-6 Raney

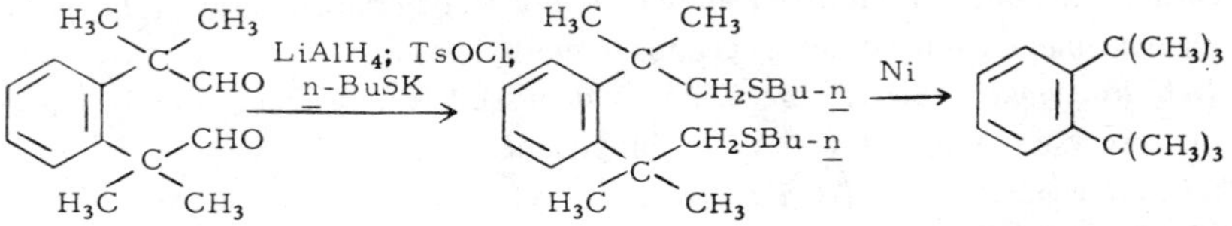

H₃C CH₃
CHO
LiAlH₄; TsOCl;
n-BuSK
CH₂SBu-n
Ni
C(CH₃)₃

nickel in ethanol with stirring at 45–50°, but the yield was low because of extensive reduction of the aromatic ring. In this particular instance, catalyst deactivated by refluxing with acetone was ineffective for desulfurization. Subsequent work led to the development of the following ***simplified preparation*** of a Raney nickel which effects clean desulfurization in 10 min. in refluxing ethanol without appreciable attack of an easily reduced aromatic ring.[43] The alloy powder (150 g.) is added in the course of 30 min. to a stirred solution of 195 g. of sodium hydroxide in 750 ml. of water maintained at 75°. Digestion at a gradually decreasing temperature is continued for an additional 30 min. The supernatant liquid is decanted and the catalyst transferred with distilled water to a 1-l. (or 2-l.) graduated cylinder placed in the sink. A 7- or 8-mm. glass tube reaching to the bottom of the cylinder is connected to the distilled water tap, and the flow of water is adjusted so that the metal rises to within 2–3 inches of the overflowing top of the cylinder. After 10–15 min. the washings are found to be neutral by pH Hydrion paper. The tube is then withdrawn and the water is decanted and replaced with ethanol.

Desulfurization is useful in the synthesis of heterocycles where the desired ring system can be constructed most readily in the form of a sulfhydryl derivative. An example is the synthesis of 4-methyl-6-hydroxypyrimidine via 2-thio-6-methyluracil.[44] The crude thio derivative was desulfurized with 45 g. (wet paste) of Raney nickel.

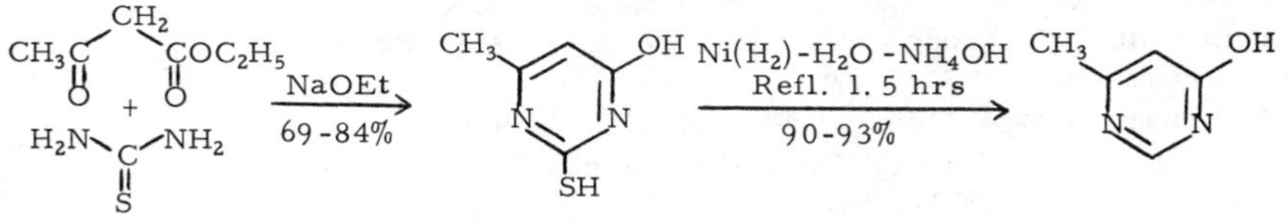

CH₂
CH₃C
COC₂H₅
O
+
H₂N
C
NH₂
S
NaOEt
69-84%
CH₃
OH
N
SH
Ni(H₂)-H₂O-NH₄OH
Refl. 1.5 hrs
90-93%
0.07 m.

Disproportionation. Kleiderer and Kornfeld[45] found that Raney nickel (W-2) in the presence of a hydrogen acceptor (cyclohexanone) catalyzes the dehydrogenation of secondary alcohols. Cholesterol, refluxed with 2 parts of catalyst and cyclohexanone in toluene, afforded cholestenone; fluorenone was converted into fluorenol in 76% yield. In the presence of a hydrogen donor (cyclohexanol, diethylcarbinol),

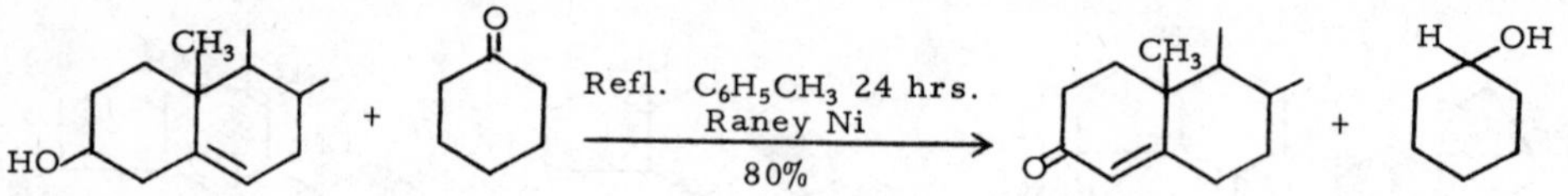

the metal catalyzes reduction of carbonyl groups and activated double or triple bonds. Examples: *trans*-stilbene → dibenzyl; cholestanone → cholestanol (50% yield).

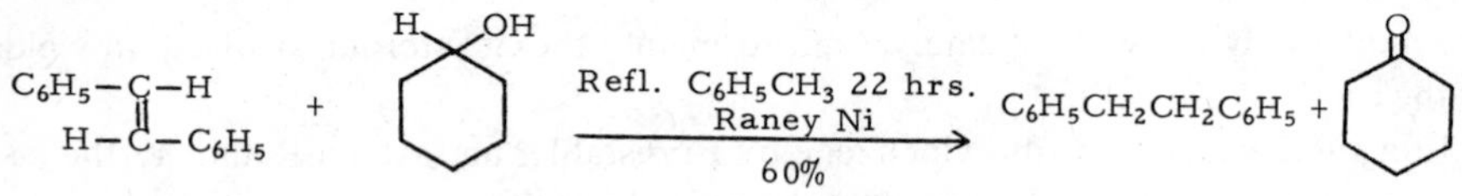

Dehalogenation. With either Raney nickel or Raney cobalt catalyst it is possible to reduce aromatic trifluoromethyl groups to methyl groups.[46]

Dehydrogenation. Sasse[47] describes a degassed Raney nickel catalyst and its use for conversion of pyridine into 2, 2′-bipyridine.

[1] L. W. Covert and H. Adkins, *Am. Soc.*, **54**, 4116 (1932)
[2] R. Mozingo, *Org. Syn., Coll. Vol.*, **3**, 181 (1955)
[3] A. A. Pavlic and H. Adkins, *Am. Soc.*, **68**, 1471 (1946)
[4] H. Adkins and A. A. Pavlic, *Am. Soc.*, **69**, 3039 (1947)
[5] H. Adkins and H. R. Billica, *Am. Soc.*, **70**, 695 (1948)
[6] H. R. Billica and H. Adkins, *Org. Syn., Coll. Vol.*, **3**, 176 (1955)
[7] H. Adkins and G. Krsek, *Am. Soc.*, **70**, 412 (1948)
[8] D. R. Levering and E. Lieber, *Am. Soc.*, **71**, 1515 (1949)
[9] X. A. Dominguez, I. C. Lopez, and R. Franco, *J. Org.*, **26**, 1625 (1961)
[10] A. B. Mekler, S. Ramachandran, S. Swaminathan, and M. S. Newman, *Org. Syn.*, **41**, 56 (1961)
[11] W. Feely and V. Boekelheide, *Org. Syn., Coll. Vol.*, **4**, 298 (1963)
[12] W. Oroshnik, G. Karmas, and A. D. Mebane, *Am. Soc.*, **74**, 295 (1952)
[13] W. F. Erman and T. J. Flautt, *J. Org.*, **27**, 1526 (1962)
[14] R. I. Longley, Jr., and W. S. Emerson, *Org. Syn., Coll. Vol.*, **4**, 660 (1963)
[15] H. J. Dauben, Jr., H. J. Ringold, R. H. Wade, D. L. Pearson, and A. G. Anderson, Jr., *Org. Syn., Coll. Vol.*, **4**, 221 (1963)
[16] R. B. Moffett, *ibid.*, **4**, 357 (1963)
[17] K. Kindler, G. Melamed, and D. Matthies, *Ann.*, **644**, 23 (1961)
[18] R. G. Rice and E. J. Kohn, *Org. Syn., Coll. Voll.*, **4**, 283 (1963)
[19] D. M. Balcom and C. R. Noller, *ibid.*, **4**, 603 (1963)
[20] A. Gaiffe and R. Pollaud, *Compt. rend.*, **252**, 1339 (1961)
[21] *Idem, ibid.*, **254**, 496 (1962)
[22] F. E. Gould, G. S. Johnson, and A. F. Ferris, *J. Org.*, **25**, 1658 (1960); A. F. Ferris, F. E. Gould, G. S. Johnson, and H. Stange, *ibid.*, **26**, 2602 (1961)
[23] O. G. Backeberg and B. Staskun, *J. Chem. Soc.*, 3961 (1962)
[24] N. P. Buu-Hoï, N. D. Xuong, and N. van Bac, *Bull. soc.*, 2442 (1963)
[25] *Idem, Compt. rend.*, **257**, 3182 (1963)
[26] G. W. Kenner and M. A. Murray, *J. Chem. Soc.*, S178 (1949)
[27] P. S. Sarin and T. R. Seshadri, *Tetrahedron*, **8**, 64 (1960): E. Caspi, E. Cullen, and P. K. Grover, *J. Chem. Soc.*, 212 (1963)
[28] G. R. Pettit and E. E. van Tamelen, *Org. Reactions*, **12**, 356 (1962)
[29] J. Bougault, E. Cattelain, and P. Chabrier, *Bull. soc.*, [5], **7**, 781 (1940)
[30] R. Mozingo, D. E. Wolf, S. A. Harris, and K. Folkers, *Am. Soc.*, **65**, 1013 (1943)
[31] H. Adkins, "Reactions of Hydrogen," The University of Wisconsin Press, Madison, Wisconsin, p. 28 (1937)

[32]V. du Vigneaud and D. B. Melville; K. Folkers, D. E. Wolf, R. Mozingo, J. C. Keresztesy, and S. A. Harris, *J. Biol. Chem.*, **146**, 475 (1942)
[33]G. M. Badger, H. J. Rodda, and W. H. F. Sasse, *J. Chem. Soc.*, 4162 (1954)
[34]N. P. Buu-Hoï *et al.*, *J. Chem. Soc.*, 1975 (1954); *Compt. rend.*, **240**, 442, 785 (1955); *Bull. soc.*, 1175, 1583 (1955)
[35]K. E. Miller, C. R. Haymaker, and H. Gilman, *J. Org.*, **26**, 5217 (1961); J. F. McGhie, W. A. Ross, D. Evans, and J. E. Tomlin, *J. Chem. Soc.*, 350 (1962)
[36]G. M. Badger, B. J. Christie, J. M. Pryke, and W. H. F. Sasse, *J. Chem. Soc.*, 4417 (1957)
[37]L. F. Fieser and W.-Y. Huang, *Am. Soc.*, **75**, 5356 (1953)
[38]R. Mozingo, C. Spencer, and K. Folkers, *Am. Soc.*, **66**, 1859 (1944)
[39]G. B. Spero *et al.*, *Am. Soc.*, **70**, 1907 (1948); G. Rosenkranz *et al.*, *ibid.*, **71**, 3689 (1949); L. B. Barkley *et al.*, *ibid.*, **76**, 5017 (1954)
[40]C. Djerassi and D. H. Williams, *J. Chem. Soc.*, 4046 (1963)
[41]J. Fishman, M. Torigoe, and H. Guzik, *J. Org.*, **28**, 1443 (1963)
[42]A. W. Burgstahler and M. O. Abdel-Rahman, *Am. Soc.*, **85**, 173 (1963)
[43]Private communication from Albert W. Burgstahler
[44]H. M. Foster and H. R. Snyder, *Org. Syn., Coll. Vol.*, **4**, 638 (1963)
[45]E. C. Kleiderer and E. C. Kornfeld, *J. Org.*, **13**, 455 (1948)
[46]N. P. Buu-Hoï, N. D. Xuong, and N. V. Bac, *Compt. rend.*, **257**, 3182 (1963)
[47]W. H. F. Sasse, *Org. Syn.*, **46**, 5 (1966)

Nickel catalysts (b), non-Raney types. Tyman[1] prepared a catalyst of the approximate composition Ni/Al_2O_3 by treatment of 1:1 nickel–aluminum alloy with water at 70°. Extensive washing is not required, and the catalyst is not subject to influence by a trace of alkali.

Lapporte and Schuett[2] prepared a catalyst which is extremely effective for the hydrogenation of aromatics by adding triethylaluminum to a solution of nickel (II) 2-ethylhexanoate in benzene or heptane. An exothermic reaction occurred with evolution of a gas which is greater than 95% ethane and production of a black, nonpyrophoric solution of a catalytically active complex.

[1]J. H. P. Tyman, *Chem. Ind.*, 404 (1964)
[2]S. J. Lapporte and W. R. Schuett, *J. Org.*, **28**, 1947 (1963)

Nickel peroxide. Supplier: British Drug Houses.

This is a black, hydrous mixture of higher oxides of nickel made by treating nickel sulfate in alkaline solution with sodium hypochlorite.[1] In alkaline solution, saturated aliphatic primary alcohols of adequate water-solubility are oxidized to the carboxylic acids (30°, 3–5 hrs.). Benzyl alcohol is oxidized by the alkaline reagent to benzoic acid, but with benzene as solvent and a slight excess of nickel peroxide it is oxidized to benzaldehyde. Allylic alcohols also can be oxidized to aldehydes by this method.

A suspension of the reagent oxidizes aniline at 80° to azobenzene in 80% yield and it oxidizes benzylamine at 60° to benzonitrile in 78% yield.[2] Nickel peroxide oxidatively cleaves α-glycols and α-hydroxy acids in aprotic solvents (benzene, ether).[3] In an aqueous alkaline solution α-ketols and α-keto acids are cleaved to carboxylic acids. The reagent thus resembles lead tetraacetate.

The reagent oxidizes a Schiff base of type (1) to the benzoxazole (2) in benzene or ether at room temperature.[4] It oxidizes *o*-phenylenediamines to *cis,cis*-1,4-dicyano-1,3-butadienes, but yields are low (9–26%).[5]

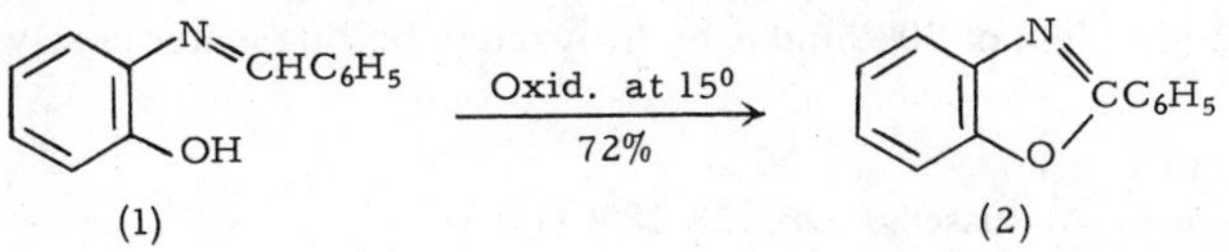
N
CHC6H5
OH
Oxid. at 15°
72%
N
CC6H5
O

(1) (2)

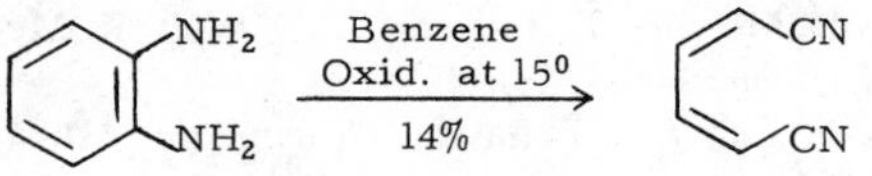

The reagent has been used to oxidize 4-cyanocatechol to the *o*-quinone.[6] Silver oxide was not effective.

Aromatic and allylic aldehydes are converted into amides by oxidation at −20° with nickel peroxide in the presence of ammonia.[7] At higher temperatures nitriles are formed.

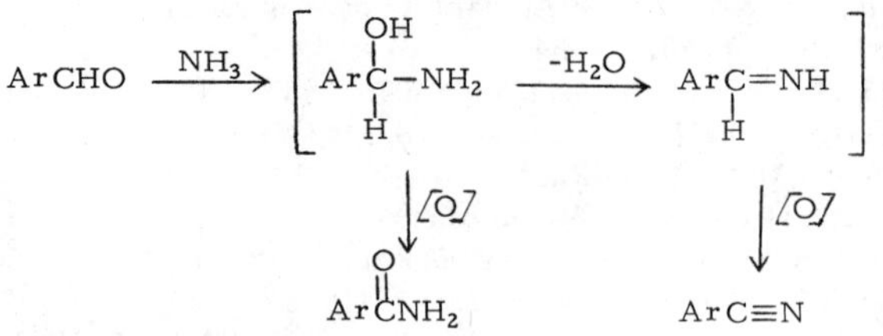

[1]K. Nakagawa, R. Konaka, and T. Nakata, *J. Org.*, **27**, 1597 (1962)
[2]K. Nakagawa and T. Tsuji, *Chem. Pharm. Bull.*, **11**, 296 (1963)
[3]K. Nakagawa, K. Igano, and J. Sugita, *ibid.*, **12**, 403 (1964)
[4]K. Nakagawa, H. Onoue, and J. Sugita, *ibid.*, **12**, 1135 (1964)
[5]K. Nakagawa and H. Onoue, *Tetrahedron Letters*, 1433 (1965)
[6]M. F. Ansell and A. F. Gosden, *Chem. Comm.*, 520 (1965)
[7]K. Nakagawa, H. Onoue, and K. Minami, *Chem. Comm.*, 17 (1966)

Ninhydrin (V). Mol. wt. 178.14, m.p. 253° dec. Suppliers: A, E, F, MCB.

A six-step synthesis of ninhydrin starting with a double ester condensation of dimethyl phthalate with ethyl acetate is presented as a student experiment.[1] A new synthesis[2] involves in the first step dropwise addition of diethyl phthalate to a suspension of sodium methoxide in dimethyl sulfoxide which is stirred with a

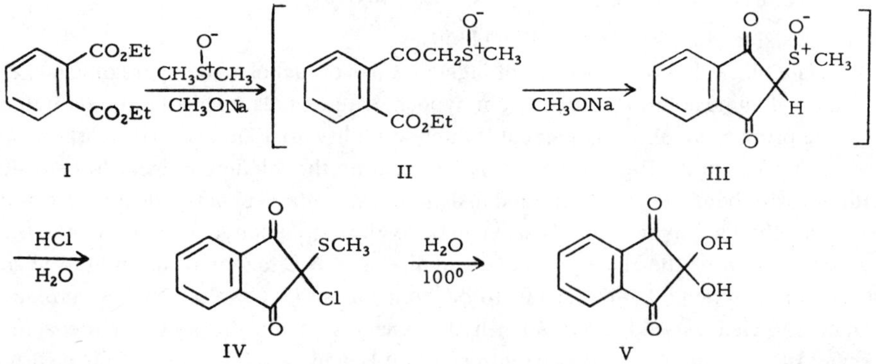

stream of nitrogen. The initial product, the β-keto sulfoxide II, undergoes intramolecular ester condensation with formation of the 1,3-indanedione system, III, which in the presence of hydrochloric acid undergoes Pummerer rearrangement to 2-chloro-2-methylmercapto-1,3-indanedione (IV). This crystalline product (m.p. 63°) is obtained in yield of 80% and is hydrolyzed in boiling water nearly quantitively to ninhydrin (V).

[1]*Org. Expts.*, 138
[2]H.-D. Becker and G. A. Russell, *J. Org.*, **28**, 1896 (1963)

Nitric acid, HNO_3.

For removal of oxides of nitrogen, 100 ml. of fuming nitric acid (90%) is treated with 0.5 g. of urea, and air is passed through the acid for about 20 min., when it should be colorless.[1]

Anhydrous nitric acid (sp. gr. 1.53) is made by distilling fuming nitric acid from an equal volume of concd. sulfuric acid.[2]

Oxidation. The chief use for nitric acid other than in nitration is as an oxidizing agent. As a survey of types of compounds oxidized, conditions, and yields, we present below a summary of procedures appearing in *Organic Syntheses*, with citation of the collective or annual volume number and page. "Concentrated nitric acid" (sp. gr. 1.42) is indicated by its composition, 71% HNO_3.

O.S. **1**, 18

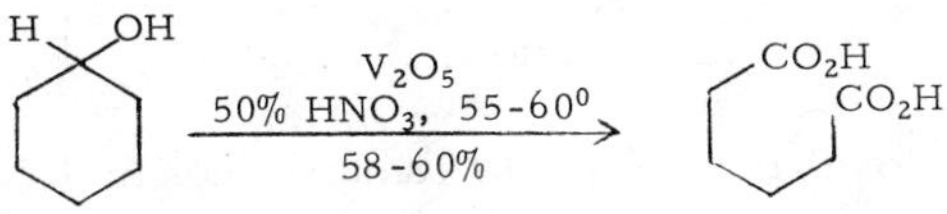
H
OH
V_2O_5
50% HNO_3, 55-60°
58-60%
CO_2H
CO_2H

O.S. **1**, 290

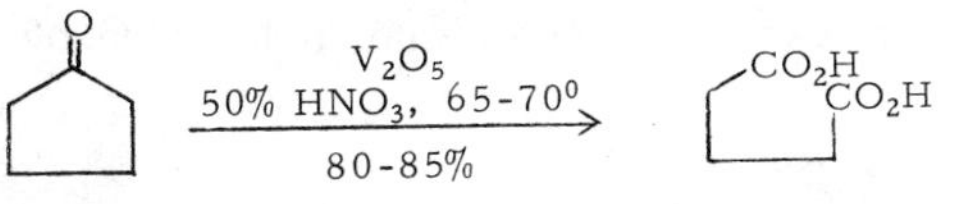
O
V_2O_5
50% HNO_3, 65-70°
80-85%
CO_2H
CO_2H

O.S. **1**, 168

$$ClCH_2CH_2CH_2OH \xrightarrow[78-79\%]{71\%\ HNO_3,\ 25-30^0} ClCH_2CH_2CO_2H$$

O.S. **4**, 499

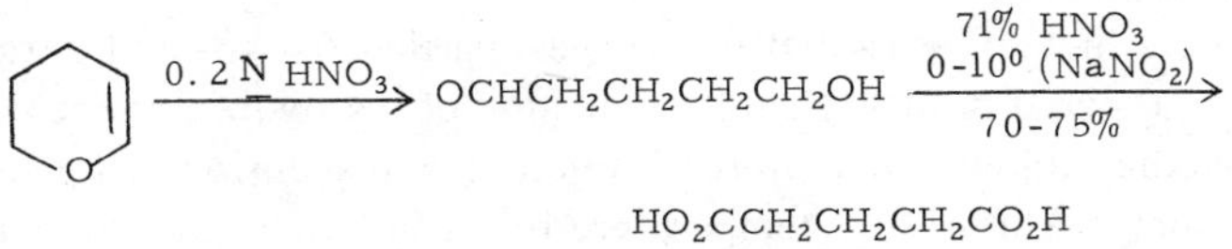
0.2 N HNO_3
$OCHCH_2CH_2CH_2CH_2OH$
71% HNO_3
0-10° ($NaNO_2$)
70-75%
$HO_2CCH_2CH_2CH_2CO_2H$

O.S. **1**, 385

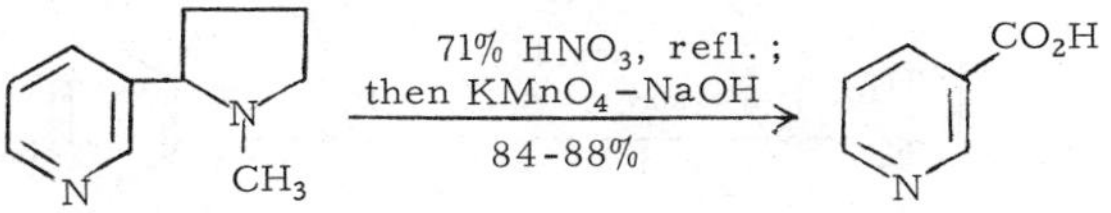
N
N
CH_3
71% HNO_3, refl.;
then $KMnO_4$–NaOH
84-88%
CO_2H
N

O.S. **3**, 791

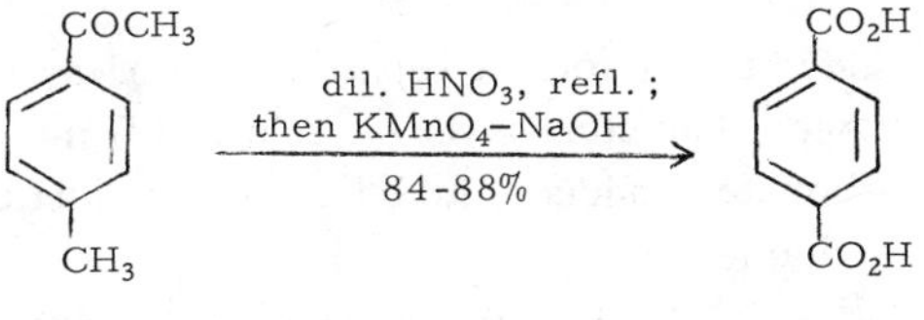
$COCH_3$
CH_3
dil. HNO_3, refl.;
then $KMnO_4$–NaOH
84-88%
CO_2H
CO_2H

O.S. **3**, 820

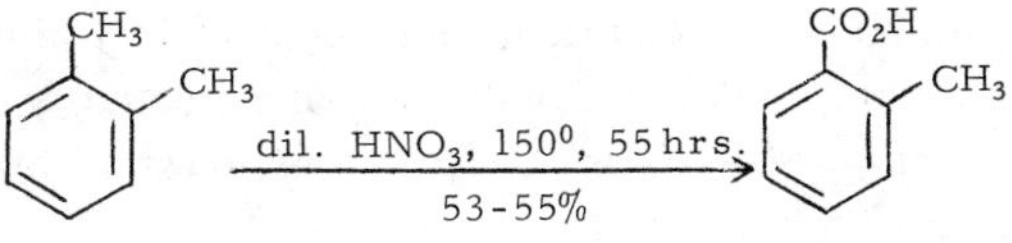
CH_3
CH_3
dil. HNO_3, 150°, 55 hrs.
53-55%
CO_2H
CH_3

O.S. **3**, 822

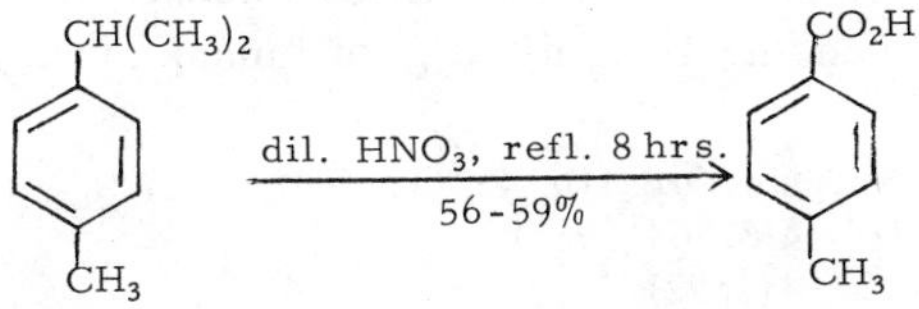
$CH(CH_3)_2$
CH_3
dil. HNO_3, refl. 8 hrs.
56-59%
CO_2H
CH_3

O.S. **2**, 214

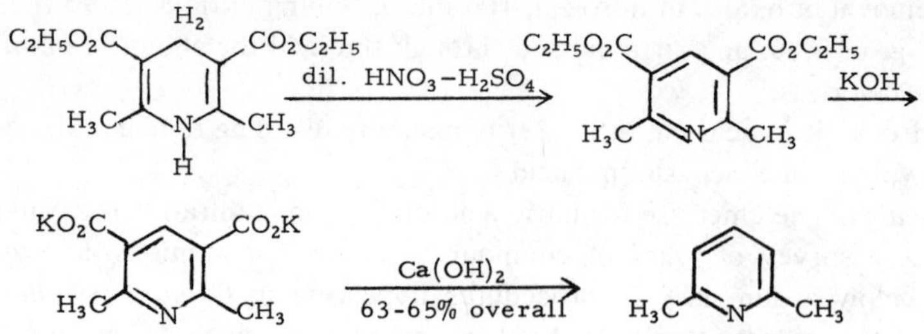

O.S. **40**, 99

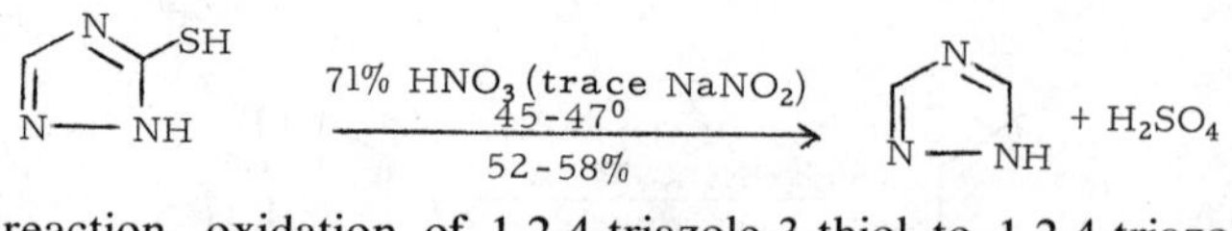

The last reaction, oxidation of 1,2,4-triazole-3-thiol to 1,2,4-triazole, requires comment. The replacement of the sulfhydryl group by hydrogen by oxidation with dilute nitric acid was discovered by Marckwald[3] in the imidazole series:

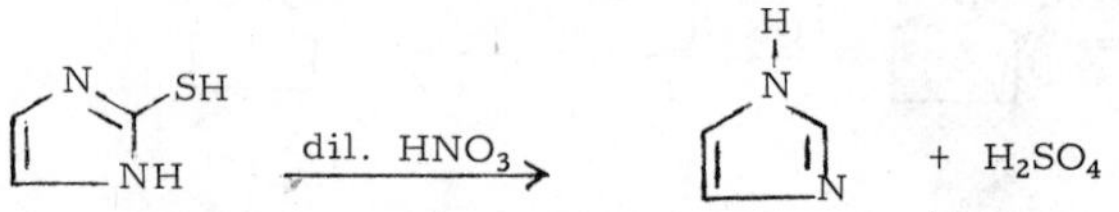

Since the sulfur appears as sulfuric acid, the oxidation probably produces the sulfonic acid, which is hydrolyzed.

Ansell[4] found that some 1,4-benzoquinones, particularly tri- and tetrasubstituted ones, can be prepared in high yield by oxidation of the hydroquinones with concd. nitric acid in ether at low temperature. A solution or suspension of 1 g. of the hydroquinone in ether is stirred at a temperature between 0 and −20°, and concd. nitric

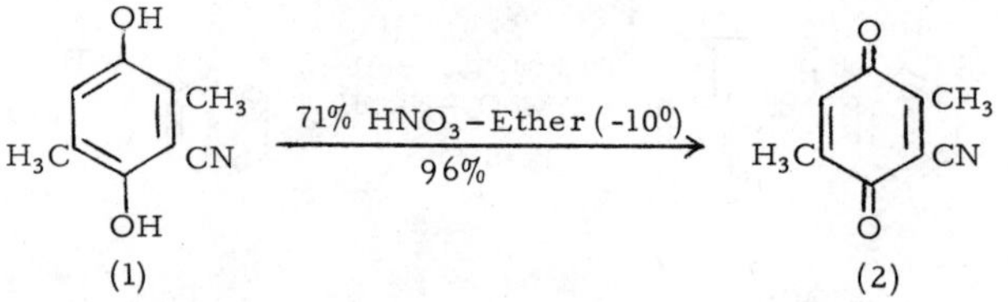

(1) (2)

acid (up to 10 ml.) is added until it forms a separate layer or until the ether layer does not deepen further in color. The ether layer is separated from any nitric acid layer, and solid sodium bicarbonate is added until effervescence just ceases. Evaporation to dryness affords the quinone.

A convenient oxidation procedure not included in the above list is as follows.[5] A mixture of benzoin, 20 ml. of acetic acid, and 10 ml. of 71% nitric acid is heated on the steam bath for 2 hrs., cooled under the tap, and treated with 75 ml. of water. The yield of benzil, m.p. 90–92°, is 93–98%; one crystallization from methanol gives pure product, m.p. 95–96°.

Borel and Deuel[6] oxidized 3-nitro-4-nitrosotoluene to 3,4-dinitrotoluene in 96% yield (crude) by adding 10 g. to 50 g. of fuming nitric acid and controlling the temperature to 25°.

[1] J. P. Freeman and I. G. Shepard, *Org. Syn.*, **43**, 84 (1963)
[2] P. Liang, *Org. Syn., Coll. Vol.*, **3**, 804 (1955)
[3] W. Marckwald, *Ber.*, **25**, 2354 (1892)

[4]M. F. Ansell, B. W. Nash, and D. A. Wilson, *J. Chem. Soc.*, 3028 (1963)
[5]*Org. Expts.*, 215
[6]E. Borel and H. Deuel, *Helv.*, **36**, 806 (1953)

4′-Nitroazobenzene-4-carboxylic acid chloride (4). Mol. wt. 289.67, m.p. 163°. The reagent is prepared by the reactions formulated.[1] In the first step a suspension of 6.2 g. of *p*-nitroaniline in 125 ml. of water is treated with 16.2 g. of 51% peracetic

$$O_2N-\langle\text{ }\rangle-NH_2 \xrightarrow[33\%]{CH_3CO_3H} O_2N-\langle\text{ }\rangle-NO \xrightarrow[85-90\%]{p\text{-}H_2NC_6H_4CO_2H\ (AcOH,\ 60°)}$$

(1)   (2)

$$O_2N-\langle\text{ }\rangle-N=N-\langle\text{ }\rangle-CO_2H \xrightarrow[94\%]{SOCl_2,\ C_6H_5CH_3,\ Na_2CO_3} O_2N-\langle\text{ }\rangle-N=N-\langle\text{ }\rangle-COCl$$

(3)   (4)

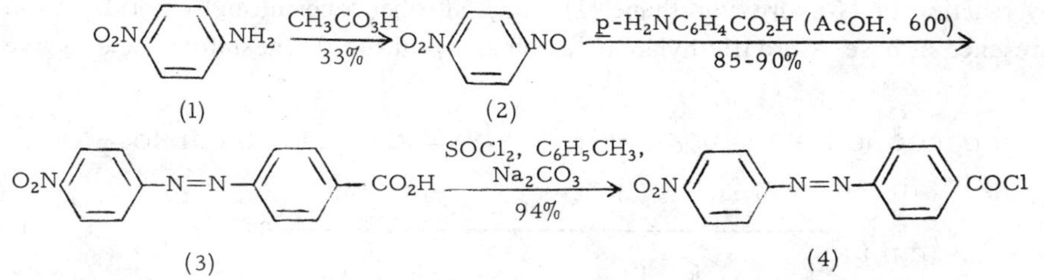

acid, shaken for 72 hrs., and let stand for 48 hrs. Fractional steam distillation affords *p*-nitronitrosobenzene of satisfactory purity in the first two fractions; later fractions contain *p*-dinitrobenzene and 4,4′-dinitroazobenzene.

Esters, prepared by reaction of the reagent with a primary or secondary alcohol in benzene containing pyridine, are highly colored (λ 297–320; ϵ *ca.* 30,000), crystalline, and high melting; methyl ester, 187°, ethyl ester 163°, cetyl ester, 11°; cyclohexyl ester, 165°; crotyl ester, 136°.[1] A study of model esters showed that mixtures are easily separated by countercurrent distribution and chromatography.[2]

The colored acid chloride was developed by Butenandt's group[3] for use in an investigation of the sex attractant principle secreted by the female silkworm moth, bombykol, which indeed was isolated as the 4′-nitroazobenzene-4-carboxylic ester (2). Twelve milligrams of pure ester, m.p. 95–96°, was isolated from 500,000 pairs of scent glands. Oxidation of 1 mg. of this ester with permanganate in acetone, separation of the colored fragment, and esterification with diazomethane gave an

$$HO(CH_2)_9\underset{10}{\overset{H}{C}}=\underset{H}{\overset{H}{C}}-\underset{12}{\overset{H}{C}}=C(CH_2)_2CH_3 \xrightarrow{ArCOCl} \text{Ester} \xrightarrow[acetone]{KMnO_4\ in}$$

(1) Silkworm moth attractant (female)   (2)

$$O_2N-\langle\text{ }\rangle-N=N-\langle\text{ }\rangle-\overset{O}{\overset{\|}{C}}-O(CH_2)_9CO_2H\ +\ HO_2CCO_2H\ +\ HO_2C(CH_2)_2CH_3$$

(3)   (4)   (5)

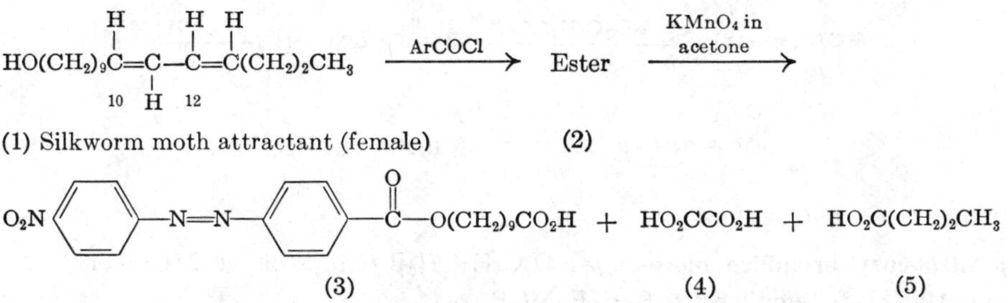

ester identified by paper chromatography as the methyl ester of (3). Steam distillation of the remaining material left a residue of oxalic acid (4) and gave a distillate containing butyric acid (5); both acids were identified by paper chromatography. Bombykol was thus found to be $\Delta^{10,12}$-hexadienol-1.

Amin[4] investigated the colored derivatives found on reaction of the acid chloride with primary and secondary amines and with thiols. Neurath and Doerk[5] prepared and characterized 45 amides derived from reaction of the reagent with primary and secondary amines. The amides have excellent crystallizing properties and absorb at 336–337 mμ with high extinction coefficients.

[1]E. Hecker, *Ber.*, **88**, 1666 (1955)
[2]E. S. Amin and E. Hecker, *Ber.*, **89**, 695 (1956)
[3]A. Butenandt, T. Beckmann, and E. Hecker, *Z. physiol. Chem.*, **324**, 71 (1961); A. Butenandt, D. Stamm, and E. Hecker, *Ber.*, **94**, 1931 (1961)
[4]E. S. Amin, *J. Chem. Soc.*, 3764 (1957); 4769 (1958); 1619 (1959)
[5]G. Neurath and E. Doerk, *Ber.*, **97**, 172 (1964)

***p*-Nitrobenzenesulfonoxyurethane** (3). Mol. wt. 290.26, m.p. 116.5°. Prepared[1] by reaction of N-hydroxyurethane (1) and *p*-nitrobenzenesulfonyl chloride in the presence of base. More triethylamine eliminates *p*-nitrobenzenesulfonic acid to give

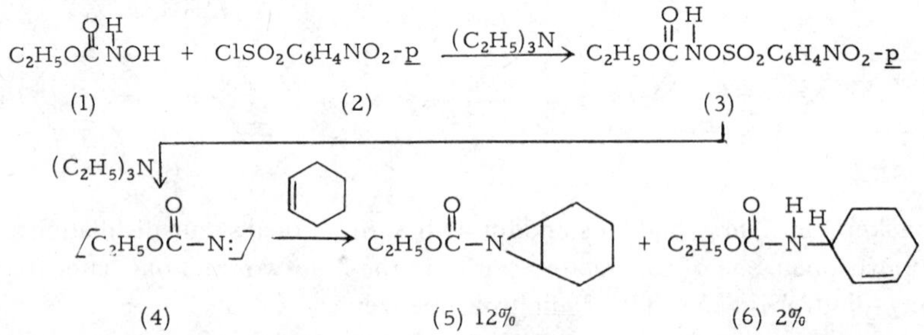

C₂H₅OCNOH + ClSO₂C₆H₄NO₂-p
(C₂H₅)₃N
C₂H₅OCNOSO₂C₆H₄NO₂-p
(1)
(2)
(3)
(C₂H₅)₃N
[C₂H₅OC–N:]
C₂H₅OC–N
+ C₂H₅OC–N
(4)
(5) 12%
(6) 2%

the highly reactive carboethoxynitrene (4), which, on being generated in the presence of cyclohexene, affords the carboethoxyaziridine (5) and 3-cyclohexenylurethane (6).

[1]W. Lwowski, T. J. Maricich, and T. W. Mattingly, Jr., *Am. Soc.*, **85**, 1200 (1963)

***p*-Nitrobenzoyl chloride**, *p*-$NO_2C_6H_4COCl$. Mol. wt. 185.57, m.p. 72–74°. Suppliers: E, F, MCB.

Kiss[1] reported that *p*-benzoylaminobenzoyl esters of sugars usually crystallize extraordinarily well and can be hydrolyzed under conditions (0.1 *N* NaOH in H_2O–CH_3OH) to which most sugars are stable. They are prepared by acylation with *p*-nitrobenzoyl chloride, reduction of the nitro group with Raney nickel, and benzoylation.

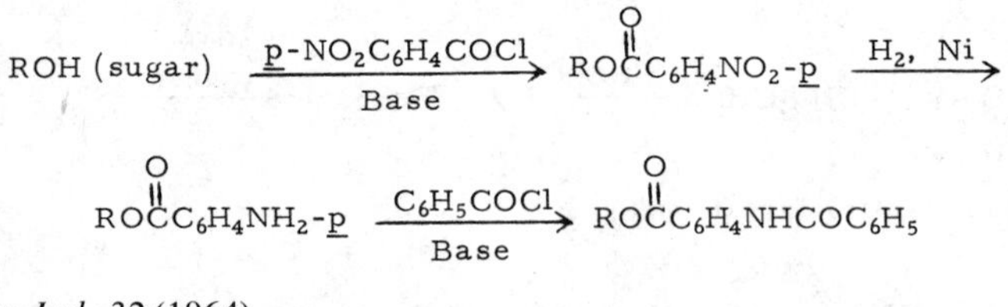

ROH (sugar)
p-NO₂C₆H₄COCl
Base
ROCC₆H₄NO₂-p
H₂, Ni
ROCC₆H₄NH₂-p
C₆H₅COCl
Base
ROCC₆H₄NHCOC₆H₅

[1]J. Kiss, *Chem. Ind.*, 32 (1964)

***p*-Nitrobenzyl bromide (chloride)**, *p*-$NO_2C_6H_4CH_2Br(Cl)$. Mol. wt. 216.04 (171.72), m.p. 100° (71°). Suppliers: A, B, E, F, MCB.

Protective group. The *p*-nitrobenzyl ester group has been used in peptide synthesis for protection of the carboxyl group.[1] An ester is prepared by treatment of an N-protected amino acid with triethylamine and *p*-nitrobenzyl bromide or chloride. The group is entirely resistant to HBr in AcOH, the reagent usually used for cleaving the carbobenzoxy group. It is removed by hydrogenolysis.

[1]R. Schwyzer and P. Sieber, *Helv.*, **42**, 972 (1959); A. P. Fosker and H. D. Law, *J. Chem. Soc.*, 4922 (1965)

***p*-Nitrobenzyl tosylate**, *p*-$CH_3C_6H_4SO_2OCH_2C_6H_4NO_2$-*p*. Mol. wt. 307.32, m.p. 103°.

The reagent is prepared by reaction of silver *p*-toluenesulfonate with *p*-nitrobenzyl bromide (88% crude, 80% pure).[1] It reacts with the sodium or trialkylammonium salt of a carbobenzoxy amino acid or peptide in acetone or dimethylformamide to give the *p*-nitrobenzyl ester in high yield. The protective group is stable to acid and readily removed by catalytic hydrogenation.

[1]D. Theodoropoulos and J. Tsangaris, *J. Org.*, **29**, 2272 (1964)

***p*-Nitrocarbobenzoxy chloride**, *p*-$NO_2C_6H_4CH_2OCOCl$. Mol. wt. 215.59, m.p. 34°. Supplier: Eastman. Preparation.[1]

For use in peptide synthesis the reagent has the advantage over carbobenzoxy chloride of being a solid and hence more easily purified.[1,2] The nitro group permits detection by spectroscopy. The protective group is more readily removed by hydrogenolysis than the carbobenzoxy group;[3] on the other hand the group is more stable to acid hydrolysis (HBr in AcOH) than the Cb group.[4] Shields and Carpenter[5] exploited this difference in stability in the synthesis of a heptapeptide sequence of insulin by coupling a tripeptide to a tetrapeptide. The tripeptide contained an amino-terminal threonine unit, which was carbobenzoxy protected, and a carboxy-terminal lysine unit in which the ϵ-amino group was protected by a *p*-nitrocarbobenzoxy group. The carbobenzoxy group was selectively cleaved, the N^{ϵ}-protected tripeptide coupled with the tetrapeptide, and then all the protective groups were removed by palladium-catalyzed hydrogenolysis.

[1]F. H. Carpenter and D. T. Gish, *Am. Soc.*, **74**, 3818 (1952)
[2]*Idem, ibid.*, **75**, 950 (1953)
[3]C. Berse, R. Baucher, and L. Piche, *J. Org.*, **22**, 805 (1957)
[4]D. T. Gish and V. du Vigneaud, *Am. Soc.*, **79**, 3579 (1957)
[5]J. E. Shields and F. H. Carpenter, *Am. Soc.*, **83**, 3066 (1961)

Nitrogen.* Traces of oxygen in commercial nitrogen can be removed by passage through Fieser's solution, but the gas is then wet and not suitable for use as a protective atmosphere, for example in Grignard and Wittig reactions. A solution which effectively removes traces of oxygen and moisture from a nitrogen stream is prepared by adding 0.5 g. of lithium aluminum hydride to a solution of 10 g. of benzopinacolone in 50 ml. of pyridine. The ketone is rapidly cleaved[1] to give a blood-red solution containing the triphenylmethide anion which, together with the

$$(C_6H_5)_3C\overset{\overset{\displaystyle O}{\|}}{C}C_6H_5 \xrightarrow{LiAlH_4-Py} (C_6H_5)_3\bar{C}Li^+ + HOCH_2C_6H_5$$

metal hydride, reacts with traces of oxygen and moisture. Triphenylmethane can also be used as the carbanion source.[2] The solution is effective as long as it remains red. A trace of pyridine carried over in the gas stream can be removed by passing the gas through a wash bottle containing concentrated sulfuric acid. However, pyridine is an excellent solvent for carbanion reactions,[3] and in this case the presence of a trace of pyridine in the nitrogen is not objectionable.

*Contributed by Peter T. Lansbury, State University of New York at Buffalo
[1]P. T. Lansbury, *Am. Soc.*, **83**, 429 (1961)
[2]P. T. Lansbury and R. Thedford, *J. Org.*, **27**, 2383 (1962)
[3]M. Avramoff and Y. Sprinzak, *Am. Soc.*, **82**, 4953 (1960)

Nitrogen oxides (chiefly N_2O_4, some N_2O_3 and NO_2). Supplied in cylinders by Matheson.

Dox[1] prepared the reagent from arsenious oxide and concd. nitric acid in a rather elaborate generator and absorbed the dried gas in ice-cooled diethyl malonate (400 g.)

$$CH_2(CO_2C_2H_5)_2 \xrightarrow[74-76\%]{N_2O_4} O{=}C(CO_2C_2H_5)_2$$

until the weight had increased by about 200 g. The liquid becomes dark green. After standing in ice for several hours and then at room temperature (red fumes) for 2 days, gas and water are removed at the water pump, and diethyl oxomalonate is distilled at 50 mm. and redistilled at 103–108°/15 mm. as a golden-yellow liquid.

Brook[2] prepared the nitrogen oxides reagent by cautious heating of a mixture of 83 ml. of fuming (98%) nitric acid, 33 ml. of concd. sulfuric acid, and 100 g. of arsenious oxide, and condensed the red vapor boiling in the range 20–30° by salt-ice. The reagent was kept in a glass-stoppered bottle at 0° and handled by pipet as a liquid for oxidation of hydroquinones to quinones. For example, a suspension of

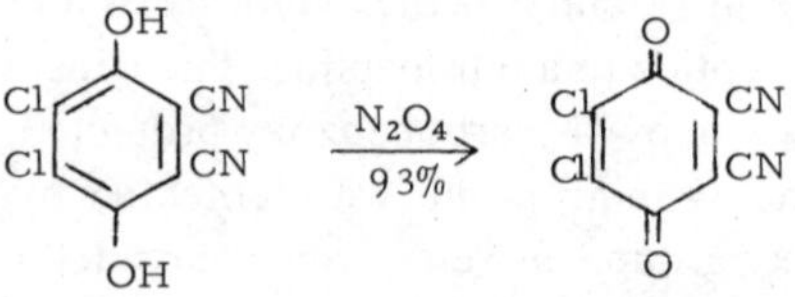
OH
Cl
Cl
CN
CN
OH
N_2O_4
93%
O
Cl
Cl
CN
CN
O

20 g. of finely ground 2,3-dichloro-5,6-dicyanohydroquinone in 300 ml. of carbon tetrachloride was stirred rapidly at room temperature and 6 ml. of the reagent was introduced in 5 min. Stirring was continued for 5 min., and the quinone was filtered off and crystallized from chloroform-benzene. The method was applied with equal success to the oxidation of 2,3-dicyano-, 5-chloro-2,3-cyano-, tetrachloro-, and 2,5-dimethylhydroquinone, of tetrabromo- and tetrachlorocatechol, and of 3,3′,5,5′-tetrabromo-4,4′-dihydroxydiphenyl. It is not suitable for oxidation of hydroquinone itself.

[1]A. W. Dox, *Org. Syn., Coll. Vol.*, **1**, 266 (1941)
[2]A. G. Brook, *J. Chem. Soc.*, 5040 (1952)

Nitrogen tetroxide, *see* Dinitrogen tetroxide.

1-Nitroguanyl-3,5-dimethylpyrazole (4). Mol. wt. 171.16, m.p. 125–126°.

Preparation.[1] Condensation of nitroguanidine (1) with hydrazine gives nitroaminoguanidine (2), and this reacts with acetylacetone (3, enol) with cyclization to form the reagent, 1-nitroguanyl-3,5-dimethylpyrazole (4).

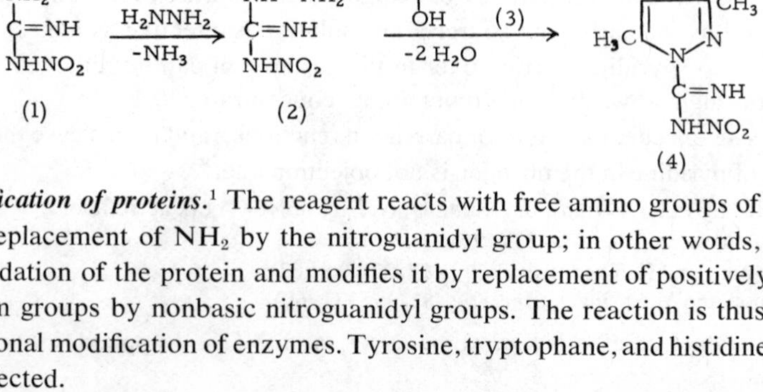
NH_2
C=NH
$NHNO_2$
(1)
H_2NNH_2
$-NH_3$
$NH-NH_2$
C=NH
$NHNO_2$
(2)
CH—C—CH_3
H_3CC
O
OH
(3)
$-2\ H_2O$
CH_3
H_3C
N
N
C=NH
$NHNO_2$
(4)

Modification of proteins.[1] The reagent reacts with free amino groups of a protein (7) with replacement of NH_2 by the nitroguanidyl group; in other words, it effects nitroguanidation of the protein and modifies it by replacement of positively charged ammonium groups by nonbasic nitroguanidyl groups. The reaction is thus of value for intentional modification of enzymes. Tyrosine, tryptophane, and histidine residues are not affected.

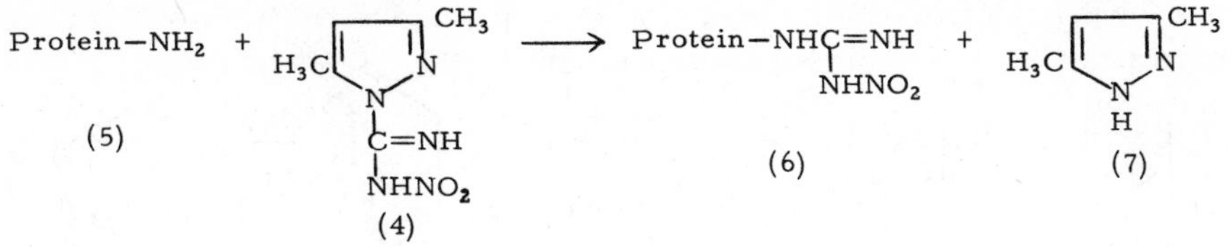

[1]A. F. S. A. Habeeb, *Biochem. biophysica Acta*, **93**, 533 (1964)

Nitromethane, CH_3NO_2. Mol. wt. 61.04, b.p. 101°, sp. gr. 1.31. Suppliers: Commercial Solvents Corp., McKesson and Robbins, Eastman. An early *Organic Syntheses* procedure[1] describes preparation of nitromethane according to Kolbe (1909):

$$ClCH_2CO_2Na \xrightarrow{NaNO_2} NO_2CH_2CO_2Na \xrightarrow[35\text{-}38\%]{H_2O} CH_3NO_2 + NaHCO_3$$

Nitromethane is now made by vapor-phase nitration of petroleum hydrocarbons.

For use as solvent in a kinetic study (Olah[2]), Eastman spectroscopic grade nitromethane was washed three times with a solution containing 25 g. of $NaHCO_3$ and 25 g. of $NaHSO_3$ per liter, then with water, 5% sulfuric acid, water, aqueous $NaHCO_3$, dried overnight with Drierite, then passed through a 2-ft. column of $\frac{1}{16}$ in. type 4A Linde molecular sieves, and distilled at 58°/160 mm. from a small amount of sieve powder to yield solvent neutral to bromphenol blue in ethanol and dry to Karl Fischer reagent (n^{25}D 1.3790).

Solvent effect. Because of its high dielectric constant (39), nitromethane is recommended as solvent for Koenig-Knorr condensation of glycosyl bromides with ethanol in the presence of silver carbonate to yield β-glycosides.[3] In less polar solvents the α-glycoside can be the major product.

Condensation with carbonyl compounds. Nitromethane condenses with benzaldehyde in methanol–aqueous sodium hydroxide at 10–15° to give a sodium enolate,

$$C_6H_5CHO + CH_3NO_2 \xrightarrow{CH_3OH-NaOH} C_6H_5\underset{\displaystyle OH}{CH}CH{=}\overset{\displaystyle O^-}{N^+}ONa \xrightarrow[80\text{-}83\%]{HCl} C_6H_5CH{=}CHNO_2$$

which on acidification affords β-nitrostyrene in good yield.[4] Alkali-catalyzed condensation with formaldehyde to produce 2-nitroethanol[5] is carried out by treating a mixture of 4.2 moles of paraformaldehyde and 2.5 l. (46.6 moles) of nitromethane with aqueous potassium hydroxide. When the reaction is over, the mixture is

$$O_2NCH_3 + CH_2O \xrightarrow[46\text{-}49\%]{1.\ KOH\quad 2.\ H_2SO_4} O_2NCH_2CH_2OH$$

neutralized with acid, and 2.3 l. of nitromethane is recovered by distillation. The residual golden-yellow liquid is treated with an equal volume of diphenyl ether as a heat-dispensing agent to reduce the chance of explosion, and the 2-nitroethanol is distilled (some diphenyl ether codistills.).

A synthesis of cycloheptanone (5) starts with the base-catalyzed addition of nitromethane to cyclohexanone;[6] the procedure calls for a ratio of 3.25 moles of nitromethane to 2.5 moles of ketone. The sodium enolate (2) separates and is collected and acidified to liberate the nitro alcohol (3), which is hydrogenated to the amino alcohol (4). Nitrous acid then effects deaminative rearrangement with ring expansion to cycloheptanone (5).

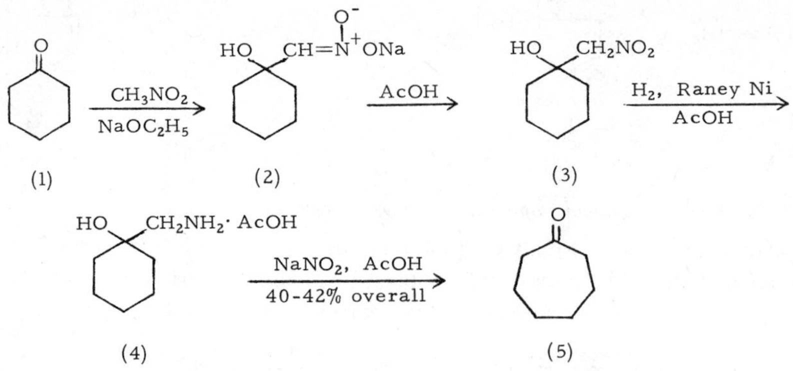

Nitromethane condenses with suitable dialdehydes in an alkaline medium to give cyclic products in which the methyl groups of two nitromethane molecules are incorporated into a ring.[7] Thus the reagent reacts with glyoxal to give a mixture of isomeric inositol derivatives, one of which (1) can be obtained in pure form due to

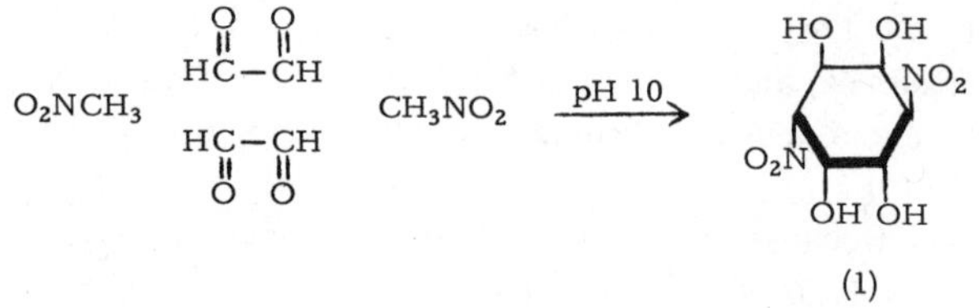

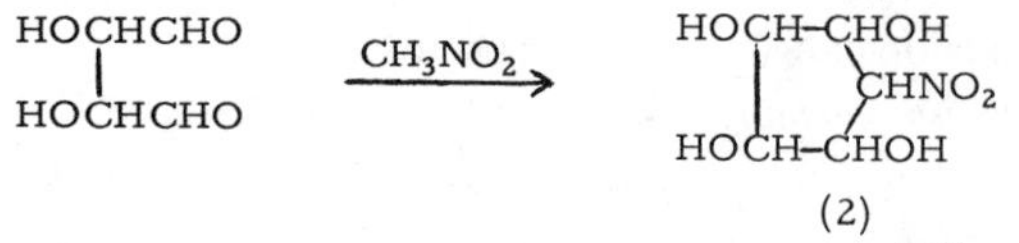

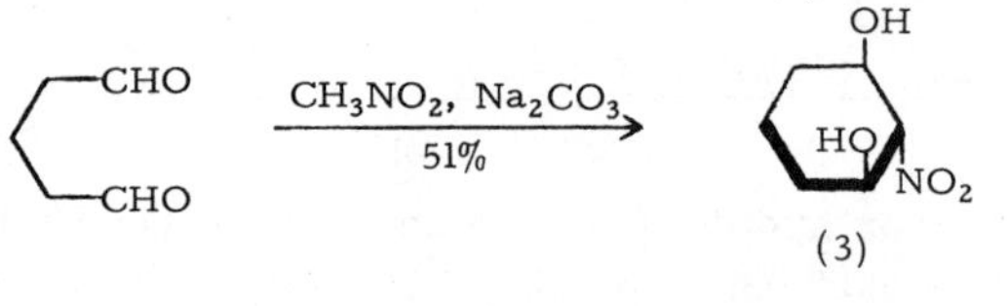

insolubility in water. Tartaraldehyde reacts to give a mixture of stereoisomeric nitrocyclopentane-2,3,4,5-tetraols (2). Glutaraldehyde condenses with nitromethane in sodium carbonate solution to give a mixture of 2-nitrocyclohexane-1,3-diols from which one isomer can be isolated in 51% yield by extraction with ether. The substance was assigned the *trans* configuration (3) on the basis of NMR evidence. *o*-Phthalaldehyde reacts with nitromethane in alcoholic alkali to give, after acidification yellow 2-nitro-3-hydroxyindene (4, m.p. 148°).[7,8]

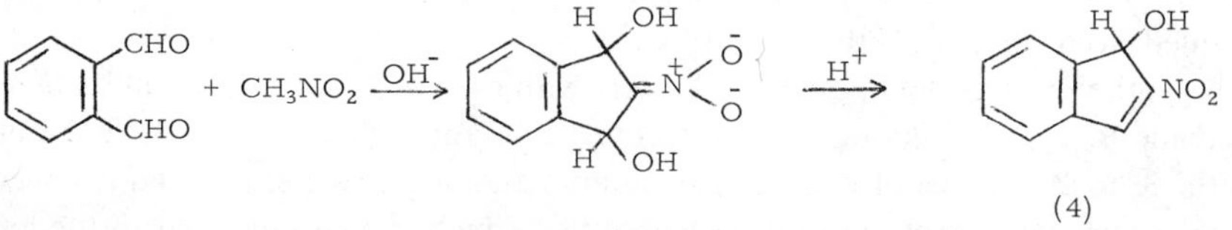

The condensation reaction has been used for the synthesis of 2-amino-3-desoxy-β-D-glucopyranosylhypoxanthin (8) starting with inosine (5).[3]

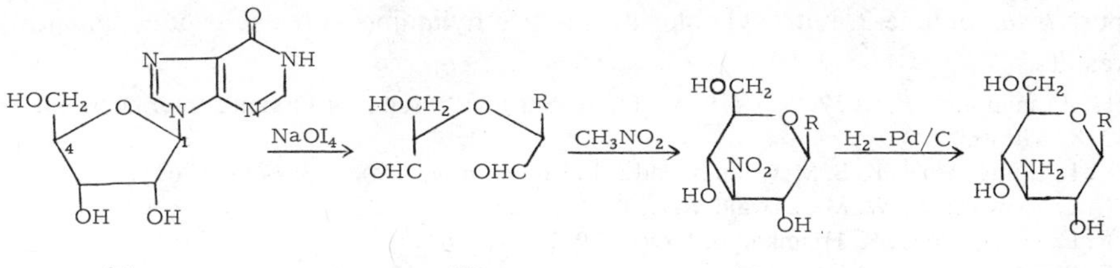

(5) (6) (7) (8)

[1]F. C. Whitmore and M. G. Whitmore, *Org. Syn., Coll. Vol.*, **1**, 401 (1941)
[2]G. A. Olah, S. J. Kuhn, S. H. Flood, and B. A. Hardin, *Am. Soc.*, **86**, 1043 (1964)
[3]P. F. Lloyd and G. P. Roberts, *J. Chem. Soc.*, 2962 (1963); F. W. Lichtenthaler and H. P. Albrecht, *Ber.*, **99**, 575 (1966)
[4]D. E. Worrall, *Org. Syn., Coll. Vol.*, **1**, 413 (1941)
[5]W. E. Noland, *Org. Syn.*, **41**, 67 (1961)
[6]H. J. Dauben, Jr., H. J. Ringold, R. H. Wade, D. L. Pearson, and A. G. Anderson, Jr., *Org. Syn., Coll. Vol.*, **4**, 221 (1963)
[7]Review: F. W. Lichtenthaler, *Angew. Chem., Internat. Ed.*, **3**, 211 (1964)
[8]H. H. Baer and B. Achmatowicz, *ibid.*, **3**, 224 (1964)

3-Nitro-N-nitrosocarbazole (1). Mol. wt. 241.20. Preparation from carbazole.[1]

The reagent can be used to effect transnitrosation of secondary amines.[2] Thus N-methylaniline on being refluxed with the reagent in benzene for 20 min. afforded

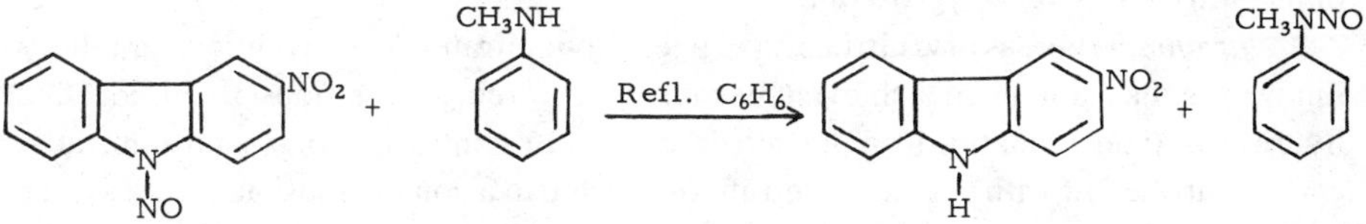

(1) (ArAr'NNO) 78% 55%

N-nitrosoaniline in fair yield. Bumgardner[2] attempted to convert aziridine into the N-nitroso derivative by this method but isolated only ethylene and nitrous oxide.

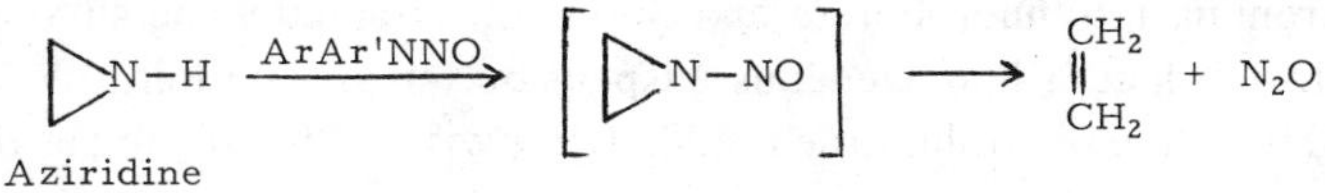

Aziridine

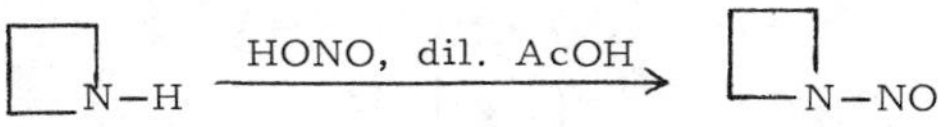

Azetidine

N-Nitrosoazetidine, a stable compound boiling at 195°, has been prepared in direct nitrosation.[3]

Clark and Helmkamp[4] found that deamination of *meso*- and *dl*-2,3-dimethylaziridine by reaction with 3-nitro-N-nitrosocarbazole is highly stereospecific. The *meso*-isomer (1) afforded *cis*-butene-2 of 99.9% purity and the *dl*-isomer gave 99%

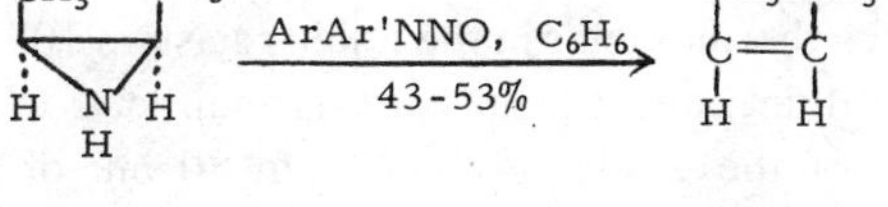

(1)

pure *trans*-butene-2. Nitrosyl chloride and triethylamine were used with comparable results.[5]

[1] H. Lindemann, *Ber.*, **57**, 557 (1924); D. A. Shirley, "Prepn. of Organic Intermediates," p. 218, Wiley (1951)
[2] C. L. Bumgardner, K. S. McCallum, and J. P. Freeman, *Am. Soc.*, **83**, 4417 (1961)
[3] C. C. Howard and W. Marckwald, *Ber.*, **32**, 2032 (1899)
[4] R. D. Clark and G. K. Helmkamp, *J. Org.*, **29**, 1316 (1964)
[5] E. Bertile *et al.*, *Angew. Chem., Internat. Ed.*, **3**, 490 (note 13) (1964)

Nitronium tetrafluoroborate, $NO_2^+BF_4^-$. Mol. wt. 112.83, dec. 170°.

Preparation.[1,2] An ice-cold solution of 1 mole of anhydrous hydrogen fluoride and 1 mole of 95% fuming nitric acid (sp. gr. 1.50) in 200 ml. of nitromethane is stirred in a fused silica or polyethylene flask and kept at −20 to 0° while passing in

$$HNO_3 + HF + 2\,BF_3 \longrightarrow NO_2^+BF_4^- + H_2OBF_3$$

boron trifluoride gas to saturation. The precipitated white crystalline nitronium salt is filtered and washed three times each with 30 ml. of nitromethane and 30 ml. of 1,1,2-trifluoro-1,2,2-trichloroethane or methylene chloride. The moist salt is transferred to a round-bottomed flask and dried in vacuum at 40–50°. The yield of colorless, crystalline nitronium tetrafluoroborate of purity better than 95% is 89–100 g. (88–99%). The salt can be stored indefinitely at room temperature with exclusion of moisture but is very hygroscopic.

Nitration. Whereas aryl nitriles have been mononitrated by conventional methods, dinitration has not been achieved because the forcing conditions required effect hydrolysis (and oxidation) of the nitrile group. In contrast, mono- and dinitration can be carried out with nitronium tetrafluoroborate in a nonaqueous, acid-free system where the only acid originates from the proton elimination during nitration. The solvent used, tetramethylene sulfone or acetonitrile (both water-soluble), is sufficiently basic to keep the acid at a concentration below the level needed to effect detectable hydrolysis (or oxidation).

Nitration of *o*-tolunitrile in two stages is effected as follows.[1,2] A mixture of 0.55 mole of nitronium tetrafluoroborate and 300 g. of tetramethylene sulfone is stirred at 10–20° to produce a homogeneous suspension (about one third of the salt dissolves), and 0.5 mole of *o*-tolunitrile is added dropwise in 20–30 min.; as the reaction

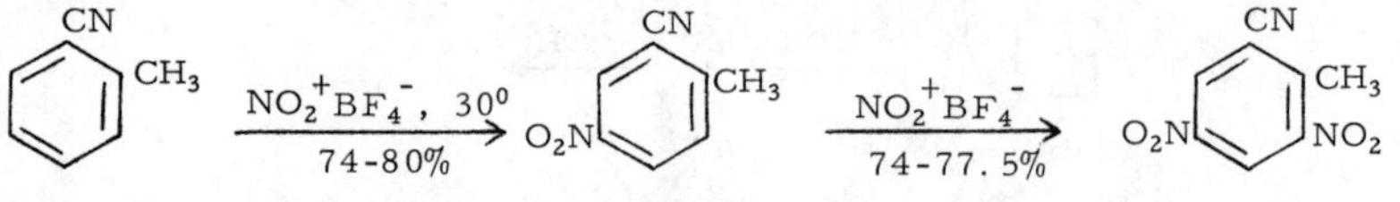
CN
CH3
NO2+BF4-, 30°
74-80%
CN
O2N
CH3
NO2+BF4-
74-77.5%
CN
CH3
O2N
NO2

proceeds the salt dissolves, and the product begins to separate. The cooling bath is removed, and stirring continued at 30–35° for 15 min. The mixture is then poured into ice water and the product collected, washed, dried, and crystallized from 100 ml. of methanol. The yield of 5-nitro-*o*-tolunitrile, m.p. 105–106°, is 74–80%.

In the next step a mixture of 0.43 mole of 5-nitro-*o*-tolunitrile, 0.56 mole of nitronium tetrafluoroborate, and 350 g. of tetramethylene sulfone is stirred under reflux and heated until a thermometer in the liquid registers 100–115°. This temperature is maintained for 1 hr., and the mixture is then let cool and poured into ice water. Crystallization of the crude product from 80 ml. of methanol gives pure 3,5-dinitro-*o*-tolunitrile, m.p. 85.5–86.5°, in 74–77.5% yield.

[1]S. J. Kuhn and G. A. Olah, *Am. Soc.*, **83**, 4564 (1961)
[2]G. A. Olah and S. J. Kuhn, procedure submitted to *Org. Syn.*

p-Nitroperbenzoic acid, p-$O_2NC_6H_4CO_3H$. Mol. wt. 183.12, m.p. 137°.

The reagent is prepared by gradual addition of *p*-nitrobenzoyl chloride to a stirred suspension of sodium peroxide in tetrahydrofurane at −20 to −5° in the presence of a catalytic amount of frozen water.[1] This peracid, a crystalline solid completely stable at room temperature, is 7–20 times as reactive as perbenzoic acid in the epoxidation of olefins.

[1]M. Vilkas, *Bull. soc.*, 1401 (1959)

p-Nitrophenol, p-$O_2NC_6H_4OH$. Mol. wt., 139.11, m.p. 114°, pKa 7.16. Suppliers: du Pont, Dyes and Chem. Div.; Monsanto; Eastman; Aldrich.

Peptide synthesis. *p*-Nitrophenol reacts with a carbobenzoxy amino acid in

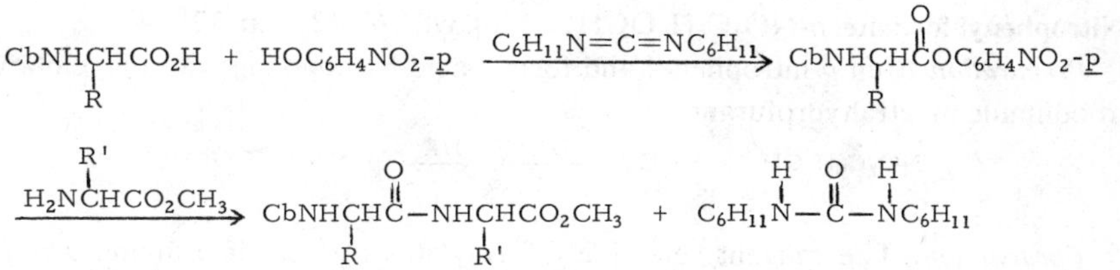

CbNHCHCO2H
R
+ HOC6H4NO2-p
C6H11N=C=NC6H11
CbNHCHCOC6H4NO2-p
R
O
R'
H2NCHCO2CH3
CbNHCHC—NHCHCO2CH3
R
R'
O
+ C6H11N—C—NC6H11
H O H

ethyl acetate in the presence of dicyclohexylcarbodiimide to give a reactive *p*-nitrophenyl ester, which reacts with an amino acid ester at room temperature and without catalyst to form a peptide bond.[1] This procedure is one of the most useful for peptide synthesis.

[1]M. Bodanszky and V. du Vigneaud, *Am. Soc.*, **81**, 5688 (1959); M. Bodanszky, J. Meienhofer, and V. du Vigneaud, *ibid.*, **82**, 3195 (1960); M. Bodanszky, G. S. Denning, Jr., and V. du Vigneaud, *Biochemical Preparations*, **10**, 122 (1963)

1-o-Nitrophenylbutadiene-1,3, (4). Mol. wt. 175.18, pale yellow needles, m.p. 67°.

Preparation. The precursor, 1-chloro-4-(*o*-nitrophenyl)-2-butene (3), is prepared by the Meerwein reaction of diazotized *o*-nitroaniline (2) with butadiene in acetone in the presence of cupric chloride and sodium acetate.[1,2] Dehydrohalogenation with methanolic potassium hydroxide then gives the diene (4).[2]

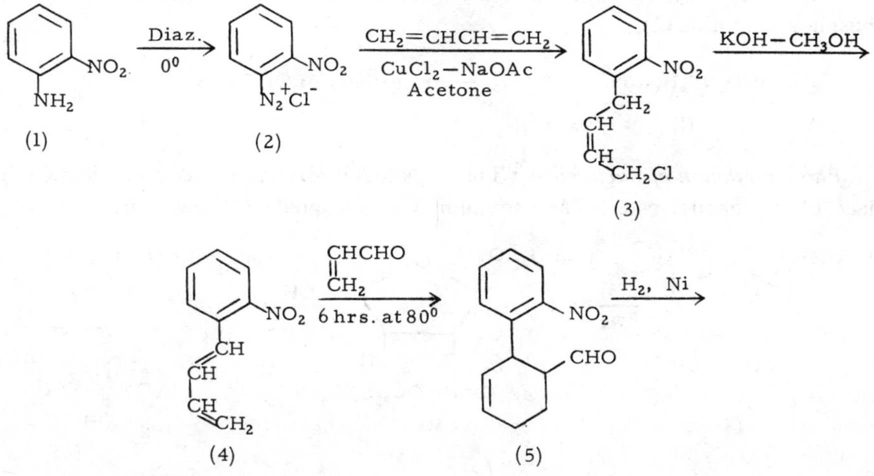

NO2
NH2
(1)
Diaz.
0°
NO2
N2+Cl-
(2)
CH2=CHCH=CH2
CuCl2—NaOAc
Acetone
NO2
CH2
CH
CH
CH2Cl
(3)
KOH—CH3OH
NO2
CH
CH
CH
CH2
(4)
CHCHO
CH2
6 hrs. at 80°
NO2
CHO
(5)
H2, Ni

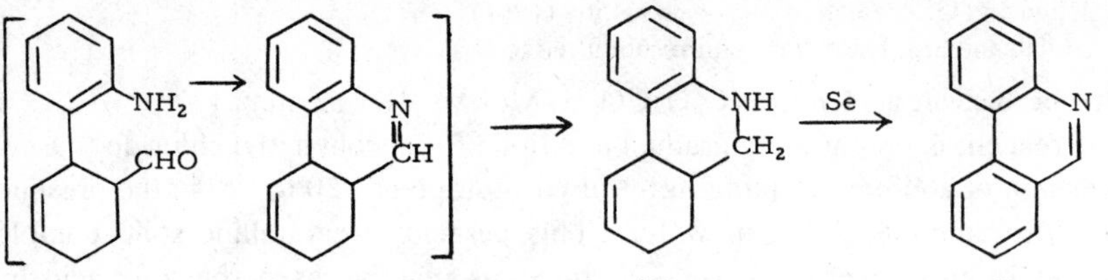
NH₂
CHO
N
CH
NH
CH₂
Se
N

(6) (7) (8) (9)

Diels-Alder reaction. The diene is more reactive than phenylbutadiene. Braude and Fawcett[2] used it for the interesting synthesis of phenanthridine (9) formulated. The overall yield from *o*-nitroaniline was about 15%.

[1]J. H. Sellstedt and W. E. Noland, procedure submitted to *Org. Syn.*
[2]E. A. Braude and J. S. Fawcett, *J. Chem. Soc.*, 3113 (1951)

p-Nitrophenyl formate, $p\text{-}NO_2C_6H_4OCHO$. Mol. wt. 167.12, m.p. 72–74°.

Preparation from *p*-nitrophenol and formic acid by coupling with dicyclohexylcarbodiimide in tetrahydrofurane.[1]

$$p\text{-}NO_2C_6H_4OH + HCO_2H \xrightarrow[63\%]{DCC-THF} p\text{-}NO_2C_6H_4OCHO$$

Formylation. The reagent selectively formylates the terminal amino group of ornithine or lysine.[1] Sheehan's method of formylation with formic acid and acetic anhydride[2] formylates both amino groups.

$$H_2N(CH_2)_{3\text{ or }4}\underset{\displaystyle NH_2}{CH}CO_2H + H\overset{\displaystyle O}{\overset{\|}{C}}OC_6H_4NO_2\text{-}p \xrightarrow[35-50\%]{THF,\ 25^{\circ},\ 6\ hrs.}$$

$$H\overset{\displaystyle O}{\overset{\|}{C}}-\overset{\displaystyle H}{N}(CH_2)_{3\text{ or }4}\underset{\displaystyle NH_2}{CH}CO_2H + HOC_6H_4NO_2$$

[1]K. Okawa and S. Hase, *Bull. Chem. Soc., Japan*, **36**, 754 (1963)
[2]J. C. Sheehan and D.-D. H. Yang, *Am. Soc.*, **80**, 1154 (1958)

p-Nitrophenylphosphorodichloridate, (3). Mol. wt. 255.99, m.p. 43.5–44.5°.

Preparation[1] by reaction of anhydrous sodium *p*-nitrophenoxide (1) with phosphoryl chloride (2).

$$\underset{(1)}{p\text{-}O_2NC_6H_4ONa} + \underset{(2)}{POCl_3} \longrightarrow \underset{(3)}{p\text{-}O_2NC_6H_4O\overset{\displaystyle O}{\overset{\|}{P}}Cl_2} + NaCl$$

Phosphorylation of nucleosides.[1] This crystalline reagent is used to advantage in place of the phenyl ester. Thus thymidine 3′-phosphate (7) was synthesized by

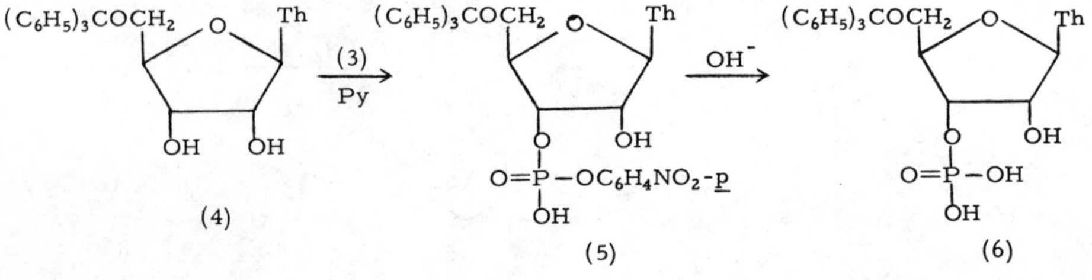
(C₆H₅)₃COCH₂
O
Th
OH
OH
(4)
(3)
Py
(C₆H₅)₃COCH₂
O
Th
O
OH
O=P—OC₆H₄NO₂-p
OH
(5)
OH⁻
(C₆H₅)₃COCH₂
O
Th
O
OH
O=P—OH
OH
(6)

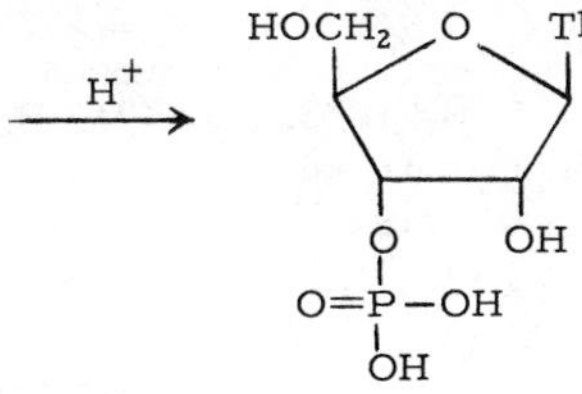

(7)

reaction of the reagent with 5′-O-tritylthymidine (4), removal of the *p*-nitrophenyl group by treatment with alkali, followed by removal of the trityl group by mild acid cleavage. The overall yield was 75%.

[1]A. F. Turner and H. G. Khorana, *Am. Soc.*, **81**, 4651 (1959)

o-Nitrophenylsulfenyl chloride, *o*-$NO_2C_6H_4SCl$. Mol. wt. 189.63, m.p. 74.5–75°.

Preparation:[1]

$$\underline{o}\text{-}NO_2C_6H_4SSC_6H_4NO_2\text{-}\underline{o} + Cl_2 + CCl_4 \xrightarrow[96-97\%]{(I_2)} 2\,\underline{o}\text{-}NO_2C_6H_4SCl$$

N-Protection in peptide synthesis. Zervas *et al.*[2] used the reagent to confer N-protection on amino acids and esters. The esters react with the reagent in the presence of triethylamine; with free amino acids the reaction is carried out in the presence of dioxane and 2*N* sodium hydroxide. The protective group is cleaved easily by acids, even by acetic acid, but preferably with the theoretical amount of hydrogen chloride in ethanol. The wide applicability of the method is demonstrated by the synthesis of peptides available only with great difficulty by conventional syntheses, for example L-phenylalanyl-L-glutaminyl-L-glutamyl-L-glutamine.[3]

The *o*-nitrophenylsulfenyl protective group proved very satisfactory in a large-scale synthesis of L-leucyl-L-methionine amide;[4] an added advantage was that on cleavage of the protective group the reagent could be recovered readily and used again.

[1]M. H. Hubacher, *Org. Syn., Coll. Vol.*, **2**, 455 (1943)
[2]L. Zervas, D. Borovas, and E. Gazis, *Am. Soc.*, **85**, 3660 (1963)
[3]L. Zervas and C. Hamalidis, *Am. Soc.*, **87**, 99 (1965)
[4]P. H. Bentley, H. Gregory, A. H. Laird, and J. S. Morley, *J. Chem. Soc.*, 6130 (1964)

p-Nitrophenyl trifluoroacetate, *p*-$O_2NC_6H_4OCOCF_3$. Mol. wt. 235.12, m.p. 37–39°. Supplier: Aldrich.

Prepared in quantitative yield by refluxing *p*-nitrophenol with trifluoroacetic anhydride, the reagent reacts rapidly with a carbobenzoxyamino acid at room temperature to give the *p*-nitrophenyl ester.[1] The reaction with a carbobenzoxy peptide is slower and requires heat or a long reaction period. *p*-Nitrophenyl trichloroacetate is active only in refluxing pyridine and gives *p*-nitrophenyl esters in poor yield.

[1]S. Sakakibara and N. Inukai, *Bull. Chem. Soc. Japan*, **37**, 1231 (1964)

1-Nitropropane, $CH_3CH_2CH_2NO_2$. Mol. wt. 89.09, b.p. 130°. Suppliers: A, E, F, MCB.

The reaction of indole-3-carboxaldehyde with diammonium hydrogen phosphate and 1-nitropropane to form indole-3-carbonitrile illustrates a general method for the conversion of aromatic aldehydes into the corresponding nitriles.[1, 2]

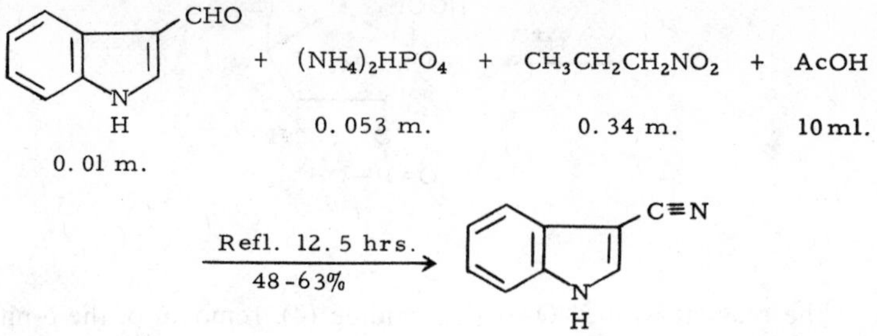
CHO
+ (NH4)2HPO4 + CH3CH2CH2NO2 + AcOH
H
0. 053 m.
0. 34 m.
10 ml.
0. 01 m.
Refl. 12. 5 hrs.
48-63%
C≡N
H

[1]H. M. Blatter, H. Lukaszewski, and G. de Stevens, *Am. Soc.*, **83**, 2203 (1961)
[2]*Idem*, *Org. Syn.*, **43**, 58 (1963)

***p*-Nitroso-N,N-dimethylaniline,** $(CH_3)_2NC_6H_4NO$. Mol. wt. 150.18, m.p. 86°. The reagent is prepared by nitrosation of N,N-dimethylaniline and isolated as the hydrochloride (mol. wt. 185.66).[1, 2]

In one aldehyde synthesis the carbon atom required for the aldehyde group is introduced as formaldehyde. Thus *p*-dimethylaminobenzaldehyde is prepared as follows.[1] A mixture of N,N-dimethylaniline, formalin, and concd. hydrochloric is

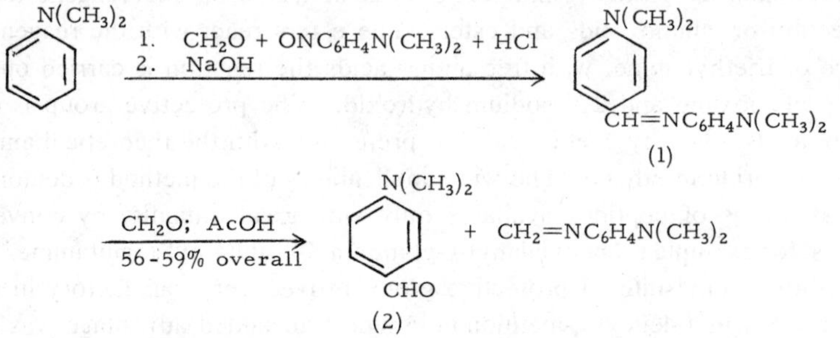
N(CH3)2
1. CH2O + ONC6H4N(CH3)2 + HCl
2. NaOH
N(CH3)2
CH=NC6H4N(CH3)2
(1)
N(CH3)2
CH2O; AcOH
56-59% overall
+ CH2=NC6H4N(CH3)2
CHO
(2)

heated for 10 min. on the steam bath and *p*-nitroso-N,N-dimethylaniline hydrochloride is added rapidly. A vigorous reaction occurs and is complete in about 5 min. The mixture is diluted with water and made alkaline to precipitate the yellow Schiff base (1). Cleavage of the Schiff base is accomplished by reaction with more formaldehyde in dilute acetic acid solution.

In another synthesis an aromatic methyl group activated by an *ortho* or *para* nitro substituent is converted into an aldehyde group, as illustrated by the preparation of 2,4-dinitrobenzaldehyde.[2]

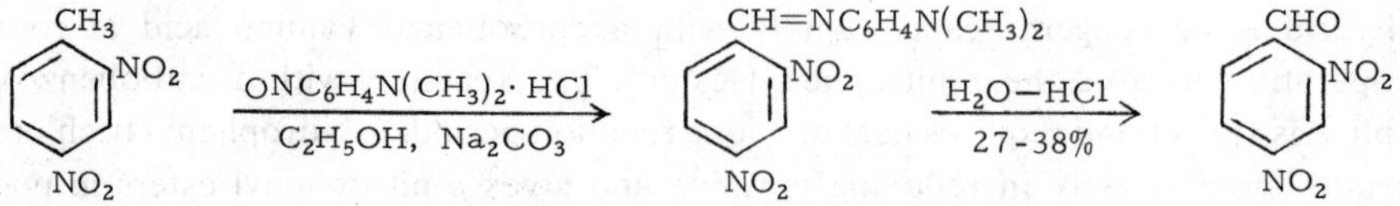
CH3
NO2
ONC6H4N(CH3)2· HCl
C2H5OH, Na2CO3
NO2
CH=NC6H4N(CH3)2
NO2
NO2
H2O—HCl
27-38%
CHO
NO2
NO2

Kröhnke reactions.[3] A halogen compound reacts with pyridine in ethanol to form a pyridinium salt (1), which is treated with *p*-nitroso-N,N-dimethylaniline and alkali to produce a nitrone (2), which on acid hydrolysis affords an aldehyde.

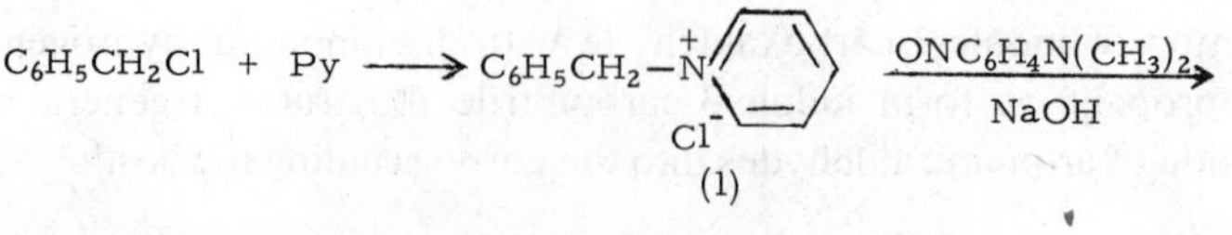
C6H5CH2Cl + Py
C6H5CH2—N
Cl
(1)
ONC6H4N(CH3)2
NaOH

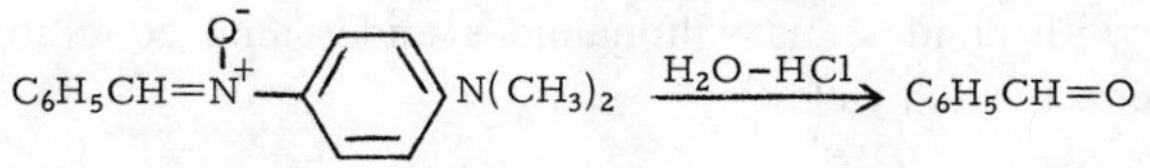

(2)

Ruzicka, Plattner, and Furrer[4] used the reaction for the conversion of cholestanone into the two enol forms of cholestane-2,3-dione: bromination to the 2-bromo derivative, conversion into the 2-pyridinium bromide, and condensation of this with

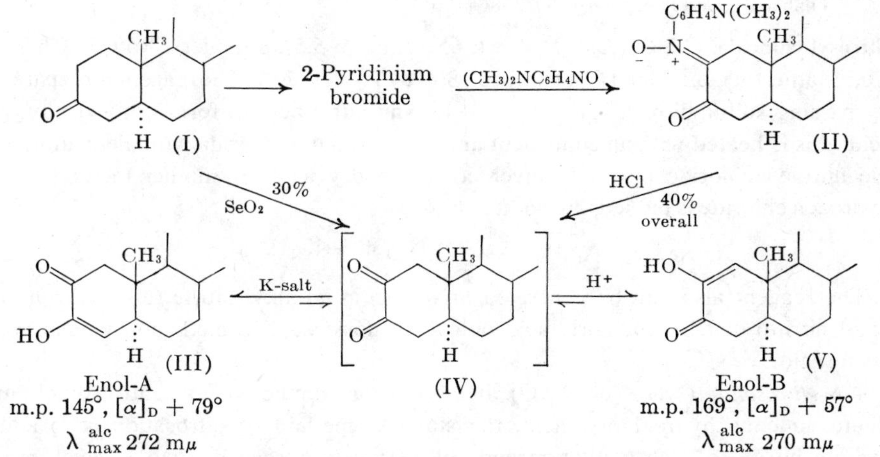

p-nitrosodimethylaniline gave the nitrone II, which on acid hydrolysis gave the dione enol mixture.

[1]R. Adams and G. H. Coleman, *Org. Syn., Coll. Vol.*, **1**, 214 (1941)

[2]G. M. Bennett and E. V. Bell, *ibid.*, **2**, 223 (1943)

[3]F. Kröhnke, *Ber.*, **71**, 2583 (1938)

[4]L. Ruzicka, Pl. A. Plattner, and M. Furrer, *Helv.*, **27**, 524 (1944)

N-Nitrosomethylurea. Generation of diazomethane; *see* page 192.

Nitrosonium hexafluorophosphate (Nitrosyl hexafluorophosphate). $NO^+PF_6^-$. Mol. wt. 174.99. Supplier: Ozark Mahoning Corp.

A novel procedure[1] for studying the competitive alkylation of benzene and toluene involved stirring a solution of 0.125 mole of each hydrocarbon and 0.05 mole of nitrosonium hexafluorophosphate in 50 g. of nitromethane at 25° and adding a solution of 0.025 mole of ethylamine in 20 g. of nitromethane over a period of 20 min. The results indicate the relative reactivities of toluene (T) and benzene (B) to

$$ArH + C_2H_5NH_2 + NO^+PF_6^- \xrightarrow[CH_3NO_2]{25^0} ArC_2H_5 + N_2 + H_2O + HPF_6$$

be: $k_T/k_B = 1.51$. Gas-liquid partition chromatography indicated the following % isomer distribution: *o*, 35.8; *m*, 48.5; *p*, 15.7. The results were similar with acetonitrile as solvent. The hexafluorophosphate salt is preferred to the tetrafluoroborate because of greater solubility.

[1]G. A. Olah, N. A. Overchuk, and J. C. Lapierre, *Am. Soc.*, **87**, 5785 (1965)

Nitrosonium tetrafluoroborate, $NO^+BF_4^-$. Mol. wt. 116.83. Supplier: Ozark-Mahoning Corp. Preparation.[1]

The reagent reacts with secondary amines to give nitrosoamines in good yield.[1] It reacts with amides and sulfonamides under mild conditions to give the corresponding acids in good yield.[2]

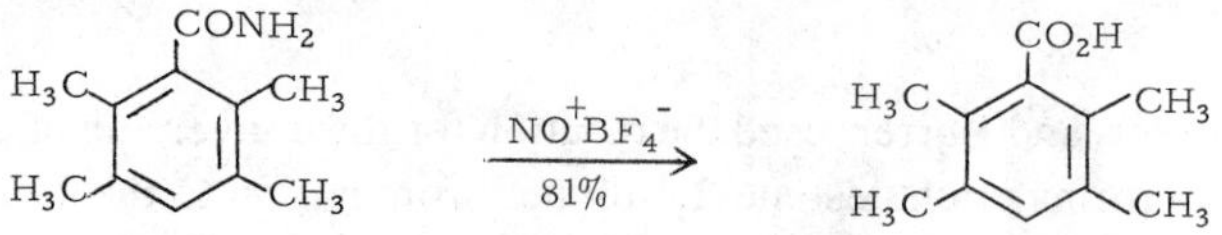

[1]G. Oláh, L. Noszkó, S. Kuhn, and M. Szelke, *Ber.*, **89**, 2374 (1956)
[2]G. A. Olah and J. A. Olah, *J. Org.*, **30**, 2386 (1965)

Nitrosyl chloride, ClNO (gas). Mol. wt. 65.47, b.p. −5.5°. Suppliers: Bowers Chem. Co.; Matheson Co.; Allied Chem. Corp., Solvay Process Div. The reagent is prepared by passing sulfur dioxide gas into cold fuming nitric acid to form nitrosylsulfuric acid; this is heated with an equivalent amount of sodium chloride with distillation of the nitrosyl chloride into a receiver cooled in dry ice.[1] In another method[2] dry hydrogen chloride is passed into sodium nitrite:

$$NaNO_2 + 2\ HCl \longrightarrow ClNO + NaCl + H_2O$$

The reagent also can be generated *in situ* from an alkyl nitrite (ethyl nitrite or amyl nitrite) and hydrochloric acid or hydrochloric acid in methanol, ethanol, or acetic acid.[3]

N-Nitrosation (NH ⟶ NNO). If a secondary amine is sufficiently soluble in dilute aqueous hydrochloric acid, the simplest method of nitrosation is to add sodium nitrite to such a solution and collect the N-nitroso derivative which precipitates. Use of nitrosyl chloride in a nonaqueous medium is indicated for amines, amines, or lactams not soluble in aqueous acid or in case the nitroso compounds are highly reactive.

Heilbron and co-workers[4] dissolved 5 g. of acetanilide in 35 ml. of acetic acid

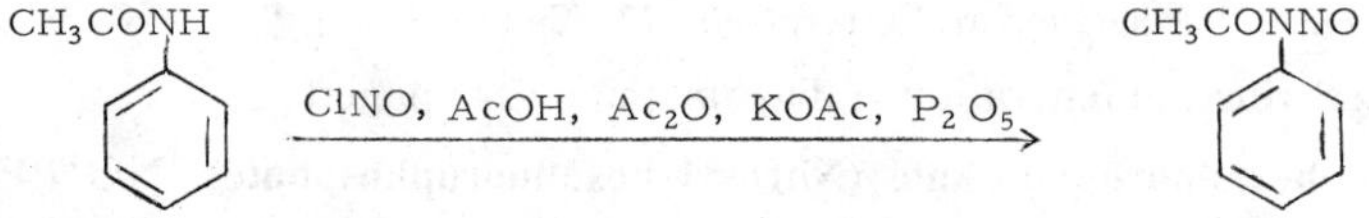

and 15 ml. of acetic anhydride, added 5 g. of fused potassium acetate and 0.5 g. of phosphorus pentoxide, stirred the mixture at 8° and slowly added 3 g. of a 25% solution of nitrosyl chloride in acetic anhydride. In 15 min. the solution was poured into ice and water, and the crystalline, yellow N-nitrosoacetanilide collected (4.7 g.). The method was satisfactory with *p*-nitroacetanilide, which could not be nitrosated with nitrous fumes. Huisgen[5] applied the procedure with success to the preparation of highly reactive N-nitrosolactams of type I. Baker *et al.*[6] modified the Heilbron

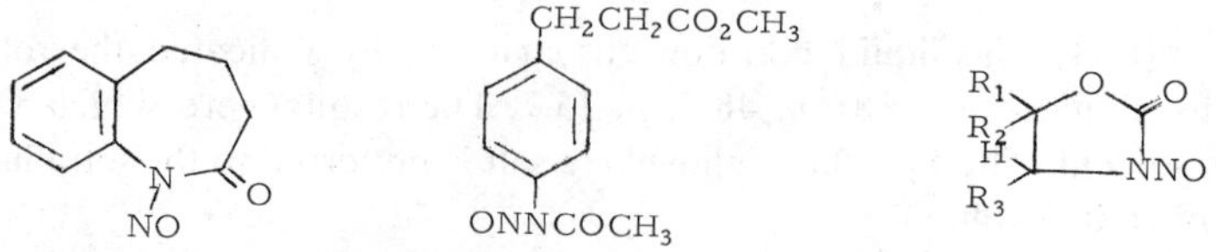

I II III

procedure to the extent of omitting the phosphorus pentoxide and obtained the N-nitroso compound II in 90% yield. Newman and Kutner[7] converted a series of 2-oxazolidones into the 3-nitroso-2-oxazolidones (III) by adding a solution of nitrosyl chloride in acetic anhydride to a solution of the oxazolidone in pyridine at 15°. Chen *et al.*[8] found the pyridine method applicable to a series of diarylamines lacking adequate solubility .n aqueous acid. Diphenylamine afforded N-nitrosodiphenylamine in 95% yield.

Rundel and Müller[9] was able to prepare N-nitrosoaziridine in ethereal solution by reaction of aziridine, nitrosyl chloride, and triethylamine (all in equimolecular amounts) at −70°; at room temperature the product decomposes to ethylene and

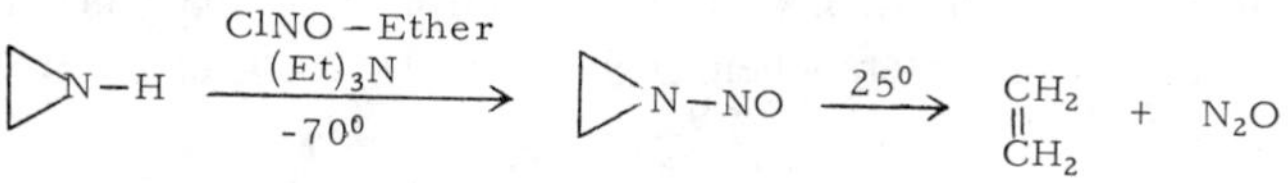
ClNO − Ether
(Et)$_3$N
-70°
N−H
N−NO
25°
CH$_2$
CH$_2$
+ N$_2$O

nitrous oxide. By the same method Clark and Helmkamp[10] prepared N-nitroso-*trans*-2,3-dimethylaziridine as an unstable yellow oil.

Nitroso chlorides. The first known method of transforming terpene hydrocarbons into crystalline derivatives was discovered by Tilden[11] in 1877: addition of nitrosyl chloride to an olefinic double bond to produce a nitroso chloride. His method was to cool a solution of a terpene in chloroform to −10° and saturate the solution with gaseous nitrosyl chloride. The solution becomes bright green, and when a test indicates that the reaction is complete the solution is poured into 2–3 volumes of methanol; the nitroso chloride separates in microcrystalline form. The terpenes investigated by Tilden all contained di- or trisubstituted double bonds and they all give colorless nitroso chlorides. Later investigation of tetrasubstituted olefins showed that they invariably give deep blue nitroso chlorides (1):[3]

R$_2$C=CR$_2$ + ClNO
R$_2$C−CR$_2$
Cl N=O
(blue)

(1)

We suggest that the color is associated with participation of the ionic resonance structure (a):

(1) (a)

A trisubstituted olefin is unsymmetrical, and nitrosyl chloride invariably adds to give the nitroso chloride in which the chlorine atom is attached to the more highly substituted carbon atom, as in (2). Unlike (1), however, the product is colorless. The interpretation now generally accepted is that the less hindered nitroso compound (2) dimerizes to the more stable form (3), which is colorless. Heating some-

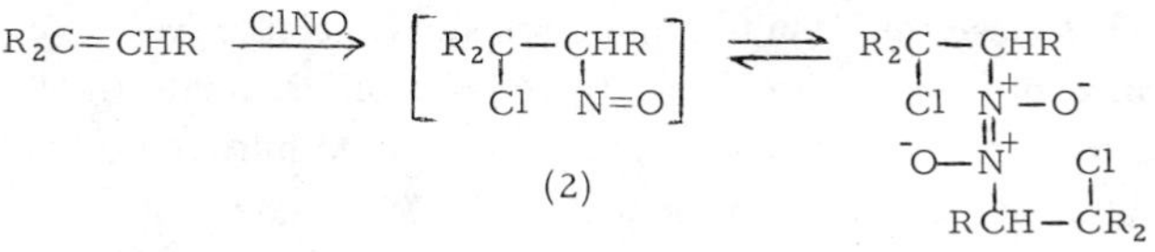
R$_2$C=CHR
ClNO
R$_2$C−CHR
Cl N=O
(2)
R$_2$C−CHR
Cl
RCH−CR$_2$

(3)

times effects partial dissociation to the blue monomer (2). *d*-Limonene, with a trisubstituted endocyclic double bond and a disubstituted exocyclic double bond reacts perferentially at the first of these functions.[3]

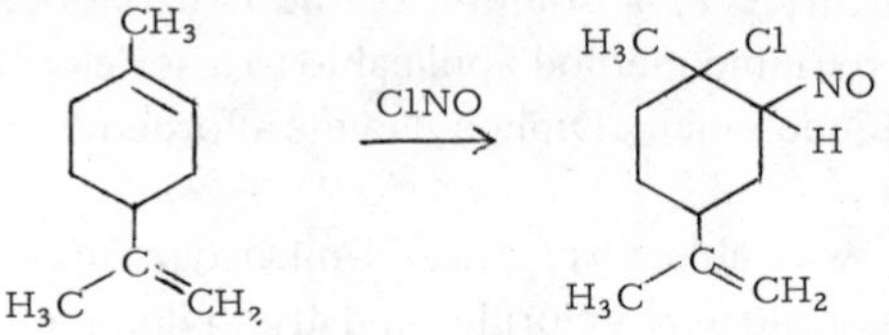

Meinwald and co-workers[12] investigated the stereochemistry of the dimeric nitroso chlorides from norbornene (1) and norbornadiene and concluded that the products have the *exo-cis* stereochemistry as in (3). Lack of rearrangement, *cis*

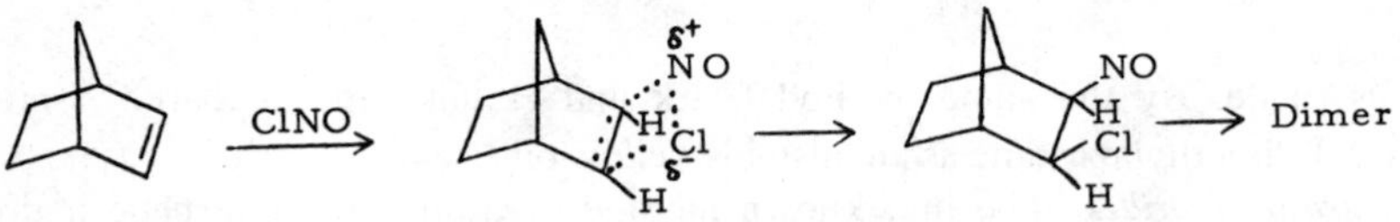

(1) (2) (3, d or l)

addition, and lack of incorporation of a nucleophilic solvent (EtOH, AcOH) into the products all speak against an ionic addition mechanism, and a free radical mechanism initiated by ·NO is unlikely since nitric oxide is unreactive to norbornadiene. Meinwald suggests a four-center mechanism with little development of carbonium ion character in the transition state (2).

In a later paper[13] Meinwald concluded that nitrosyl chloride adds *cis* to strained double bonds, such as that in norbornene, but that it adds *trans* to unstrained olefins (Δ^9-octalin). In accordance with this view of the addition is a report by Hassner[14] on the reaction of cholesteryl acetate (4). Nitrosyl chloride was passed into methylene chloride or carbon tetrachloride until a deep burgundy color developed, the steroid

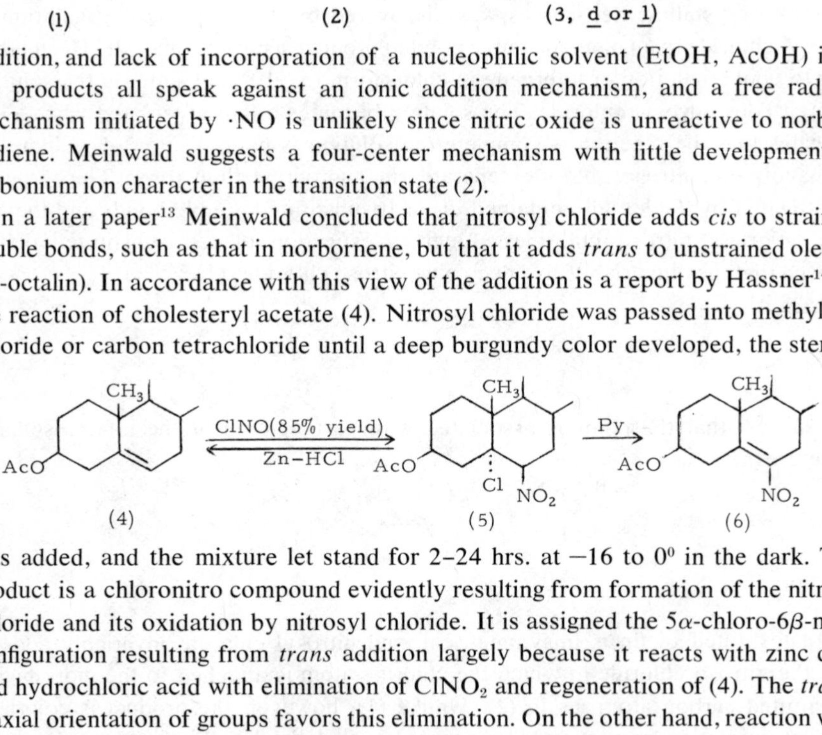

(4) (5) (6)

was added, and the mixture let stand for 2–24 hrs. at −16 to 0° in the dark. The product is a chloronitro compound evidently resulting from formation of the nitroso chloride and its oxidation by nitrosyl chloride. It is assigned the 5α-chloro-6β-nitro configuration resulting from *trans* addition largely because it reacts with zinc dust and hydrochloric acid with elimination of $ClNO_2$ and regeneration of (4). The *trans*-diaxial orientation of groups favors this elimination. On the other hand, reaction with pyridine at room temperature to form 6-nitrocholesteryl acetate (6) would have to be described as an unusually facile *cis* elimination.

The E. R. H. Jones group[15] noted that cholesteryl acetate reacts very slowly with carefully purified nitrosyl chloride and traced the difference to the presence in the ordinary reagent of nitrogen dioxide. They then added nitrogen dioxide to the pure reagent and obtained the nitro chloride in 2 hrs. in 74% yield.

Ohno *et al.*[16] found that the steric course of the addition of nitrosyl chloride is dependent upon the solvent. Thus addition of the reagent to cyclohexene in sulfur dioxide gives the *trans*-adduct, whereas the *cis*-adduct is formed in methylene chloride, chloroform, or trichloroethylene.

Regeneration of an olefin from the nitrosyl chloride adduct is an important step in the preparation of pure Δ^9-octalin. One of two efficient routes to crude hydrocarbon is by hydrogenation of β-naphthol to 2-decalol and dehydration (H_3PO_4 or H_3BO_3), and a second is reduction of tetralin by adding lithium to a solution of tetralin in ethylenediamine.[17] Material obtained by the first method is equilibrated by heating with phosphorus pentoxide with some increase in the proportion of the Δ^9-isomer. Applying and extending improvements reported by others, W. G.

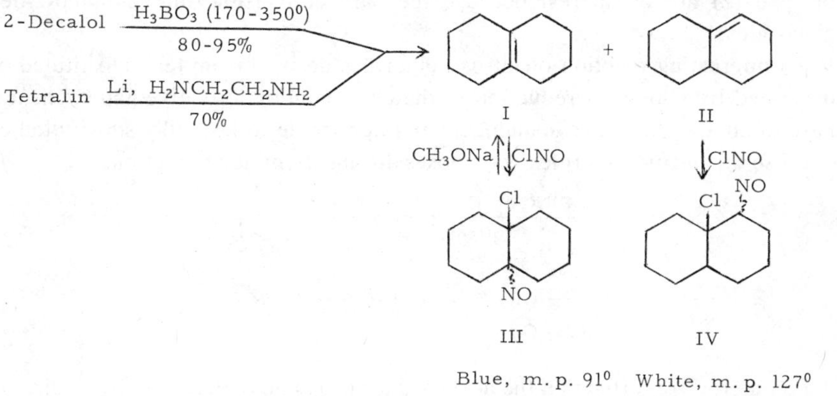
2-Decalol
H_3BO_3 (170-350°)
80-95%
Tetralin
Li, $H_2NCH_2CH_2NH_2$
70%
+
I
II
CH_3ONa
ClNO
ClNO
Cl
NO
Cl
NO
III
IV
Blue, m.p. 91°
White, m.p. 127°

Dauben[17] compared the two methods and found that each affords material shown by NMR to consist largely of Δ^9-octalin (I) but containing 10–20% of hydrocarbon containing di- or trisubstituted double bonds. Benkeser[18] had shown that the principal companion of I is $\Delta^{1(9)}$-octalin (II), and had characterized both isomers by reaction with nitrosyl chloride to form the blue nitroso chloride III and the higher-melting (dimeric) derivative IV. (III and IV are probably both *trans*.) Since Δ^9-octalin predominates in the mixture, the blue nitroso chloride III is easily obtained is pure form by crystallization. Benkeser regenerated pure Δ^9-octalin by treatment of the blue derivative with sodium methoxide. Dauben[17] stirred a solution of III in ether with zinc dust and added hydrochloric acid dropwise; the yield of Δ^9-octalin was 59%, comparable to that obtained with sodium methoxide. Hussey[19] prepared the octalin mixture by dehydration of 2-decalol (H_3PO_4), but effected conversion into the blue nitroso chloride III by mixing 0.5 mole of isoamyl nitrite with 0.5 mole of concd. hydrochloric acid at −10° and adding 0.5 mole of the octalin mixture. After 1.5 hrs. the blue precipitate was collected and washed with iced ethanol (crude yield 75%). Crystallization from acetone–ether gave blue prisms of pure nitroso chloride, m.p. 91–92°, in 60% yield. For regeneration of the hydrocarbon a solution of 0.094 mole of the nitroso chloride in 30 ml. of N,N-dimethylaniline was warmed to 70°, when gas evolution (ClNO) started. Heating at 85° for 2.5 hrs. completed the reaction, and dilution with water and extraction with pentane afforded pure Δ^9-dialin in 85–95% yield.

Although the rate of reaction of olefins with nitrosyl chloride decreases with decreasing number of alkyl substituents, crystalline nitroso chlorides (dimeric) have been obtained even from ethylene and propylene.[3]

The evidence at hand indicates that nitrosyl chloride adds *cis* to the double bond of a dihydropyrane. Serfontein *et al.*[20] found that 3,4,6-tri-O-acetyl-D-glucal (1) gives a crystalline adduct characterized by NMR as the 1-chloro-2-nitroso-1,2-didesoxy sugar (2). Similar results have been reported by Lemieux *et al.*[21] Compounds

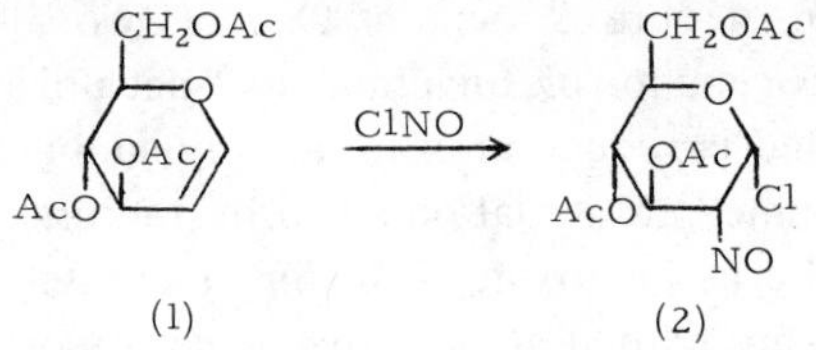

of type (2) are of interest because they are convertible into 2-amino-2-desoxy-glycosides.

An interesting reaction of nitroso chlorides derived from tetrasubstituted olefins described by Closs[22] is reduction to the chloroamino derivative and cyclization by base to an aziridine. The sequence is not applicable to less fully substituted olefins because the nitroso chloride decomposes during attempted reduction.

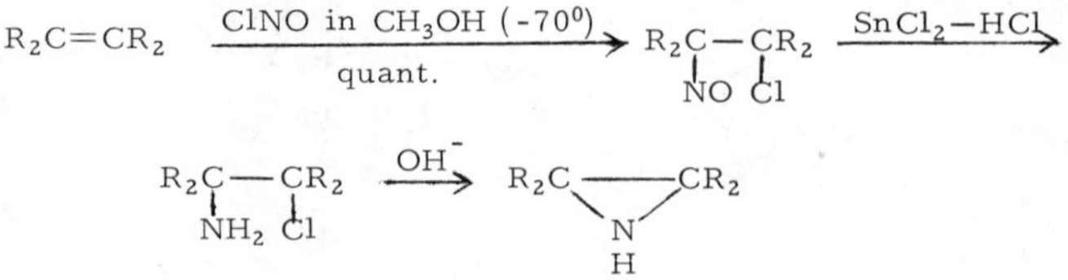

Pummerer[23] reported that the nitroso chloride (2) on refluxing with pyridine affords the α,β-unsaturated oxime (3).

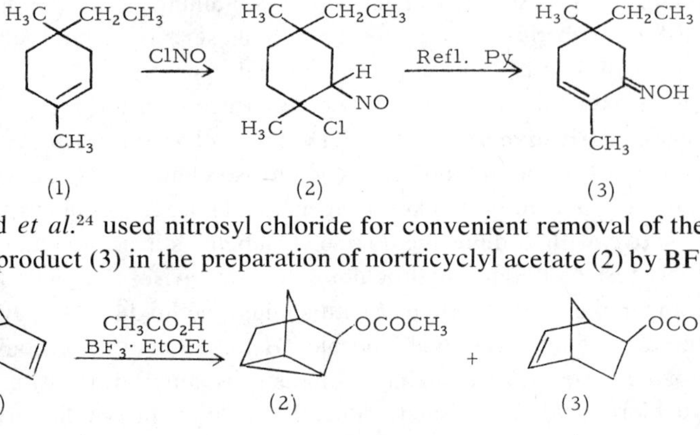

Meinwald *et al.*[24] used nitrosyl chloride for convenient removal of the unwanted olefinic by-product (3) in the preparation of nortricyclyl acetate (2) by BF_3-catalyzed

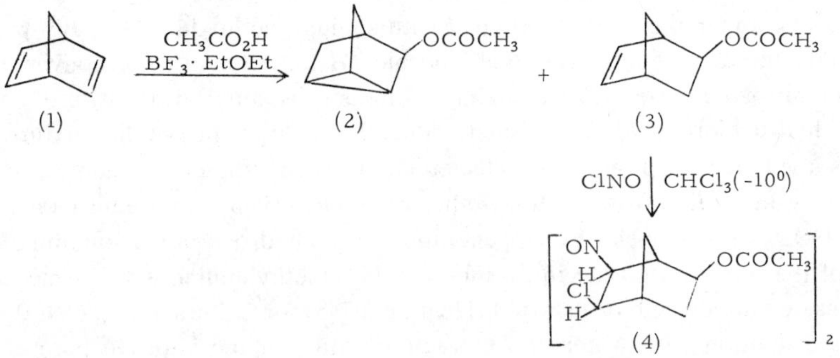

addition of acetic acid to norbornadiene (1). The reaction mixture containing 10–15% of the olefinic acetate (3) was dissolved in chloroform, and nitrosyl chloride was bubbled in at −10° until the color changed through bright green to brownish green. The dimeric adduct (4) separated as a white precipitate, and workup of the filtered solution afforded nortricyclyl acetate (2) as a faintly green liquid in yield of 52–66%.

Deamination of amides. Wolfrom[25] found that penta-O-acetylmannonamide can be deaminated smoothly to the corresponding aldonic acid pentaacetate by reaction with nitrosyl in chloroform.

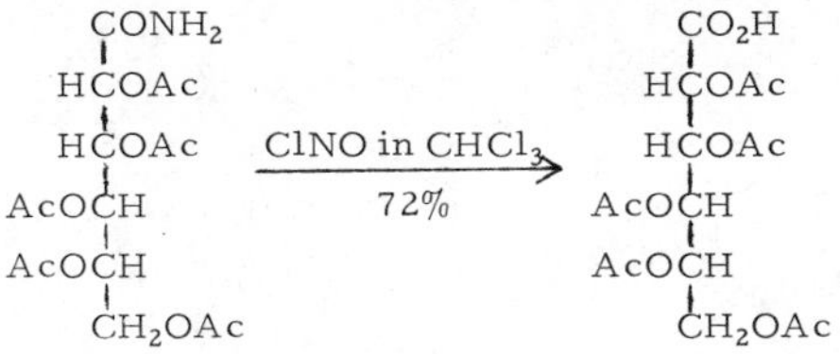

Alkyl nitrites. An efficient method for the preparation of 2-octyl nitrite described by Kornblum and Oliveto[26] is as follows. A solution of 0.6 mole of 2-octanol in 240 ml. of pyridine is stirred at 0–10°, and 0.95 mole of nitrosyl chloride is allowed to evaporate into the solution in the course of about 3 hrs. Petroleum ether and water are run into the mixture, and the organic layer is washed, dried, and distilled. Reaction of the alcohol with sodium nitrite and sulfuric acid afforded the ester in only 59% yield.

$$CH_3CH_2CH_2CH_2CH_2CH_2CH(OH)CH_3 \xrightarrow[80\%]{ClNO} CH_3CH_2CH_2CH_2CH_2CH_2CH(ONO)CH_3$$

Oximation of hydrocarbons. Naylor and Anderson[27] irradiated a mixture of cyclohexane and nitrosyl chloride and obtained cyclohexanone oxime in 71% yield. The reaction is interpreted as a radical reaction represented in part as follows:

$$ClNO \xrightarrow{h\nu} Cl\cdot + \cdot NO$$

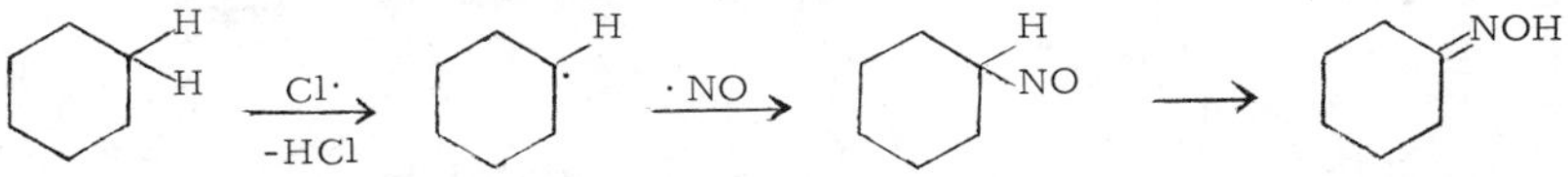

Addition to C═N. Nitrosyl chloride reacts (0°, 40 hrs.) with oximino esters (1) in methylene chloride or ether to give α-chloro-α-nitroso esters (3),[28] which are monomeric blue liquids sensitive to light and heat.

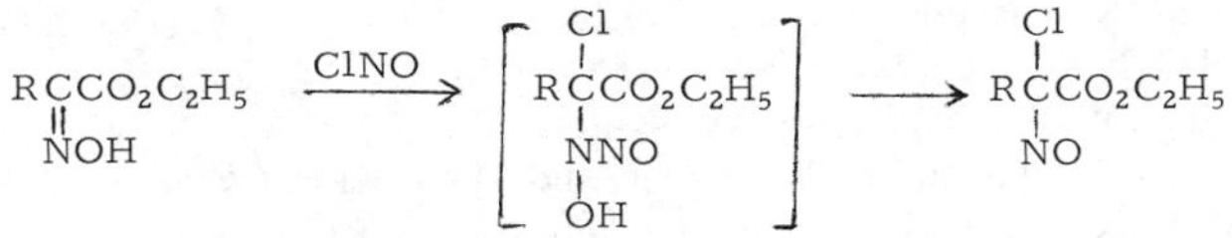

(1) (2) (3)

Oxidation of ketones. Manning and Stansbury[29] treated 1 mole of acetophenone in ethanol at 23–30° with 1.6 moles of nitrosyl chloride and warmed the mixture gradually to 60°. At the end of the process, by-product esters were saponified with alkali and phenylglyoxal diethyl acetal was isolated easily in 53% yield.

$$C_6H_5COCH_3 \xrightarrow[53\%]{ClNO,\ C_2H_5OH} C_6H_5COCH(OC_2H_5)_2$$

Hydrazides ⟶ Azides. The acid azide method of peptide synthesis, for example of the protected dipeptide (3), has the advantage that no racemization of the peptide components occurs.[30] However, conversion of N-carbobenzoxy-S-benzyl-L-cysteine hydrazide (1) into the azide by reaction with nitrous acid under normal conditions affords considerable amide as by-product ($RCONHNHNO \longrightarrow RCONH_2 + \overset{-}{N}{=}\overset{+}{N}{=}O$). Amide formation can be suppressed by operating at high acidity and

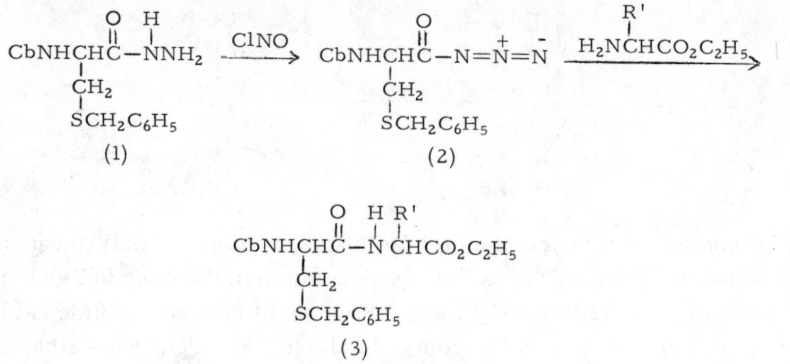

$$
\underset{(1)}{\underset{\underset{SCH_2C_6H_5}{\overset{\displaystyle CH_2}{|}}}{\overset{\overset{\displaystyle O\ \ H}{\|\ \ |}}{CbNHCHC-NNH_2}}} \xrightarrow{ClNO} \underset{(2)}{\underset{\underset{SCH_2C_6H_5}{\overset{\displaystyle CH_2}{|}}}{\overset{\overset{\displaystyle O}{\|}}{CbNHCHC-N=\overset{+}{N}=\overset{-}{N}}}} \xrightarrow{\underset{}{\overset{\overset{\displaystyle R'}{|}}{H_2NCHCO_2C_2H_5}}}
$$

$$
\underset{(3)}{\underset{\underset{SCH_2C_6H_5}{\overset{\displaystyle CH_2}{|}}}{\overset{\overset{\displaystyle O\ \ H\ R'}{\|\ \ |\ \ |}}{CbNHCHC-N\ CHCO_2C_2H_5}}}
$$

low temperature, but these conditions favor another side reaction leading to the sulfoxide. Honzl and Rudinger[31] found that nitrosyl chloride converts the hydrazide into the azide under a variety of conditions without formation of either the amide or the sulfoxide. Thus a suspension of 1 g. of (1) in tetrahydrofurane is treated at $-20°$ with nitrosyl chloride in tetrahydrofurane and dioxane and agitated with a Vibromix for about 6 min., when the hydrazide dissolves. Alternatively, a solution of 1 g. of (1) in tetrahydrofurane $2.2 N$ in hydrogen chloride is treated at $-20°$ with n-butyl nitrite.

ArNH$_2$ ⟶ *ArOH.* Nitrosyl chloride proved preferable to other reagents for the hydrolysis of adenosine-N-oxide to inosine-N-oxide.[32]

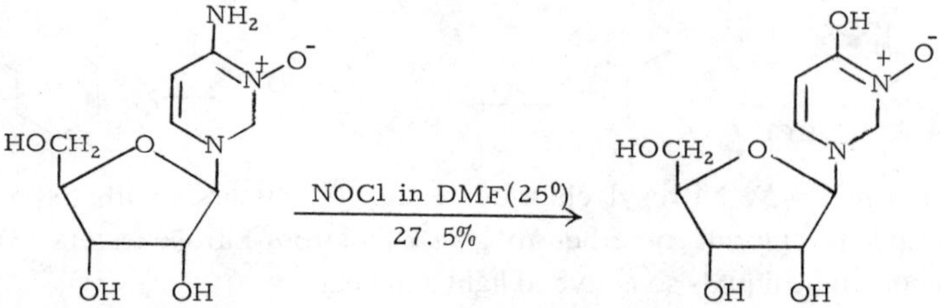

[1]M. L. Wolfrom, M. Konigsberg, and D. I. Weisblat, *Am. Soc.,* **61**, 574 (1939); G. H. Coleman, G. A. Lillis, and G. E. Goheen, *Inorg. Syn.,* **1**, 55 (1939)

[2]J. R. Morton and H. W. Wilcox, *Inorg. Syn.,* **4**, 48 (1953)

[3]See review by L. J. Beckham, W. A. Fessler, and M. A. Kise, *Chem. Rev.,* **48**, 324 (1951).

[4]H. France, I. M. Heilbron, and D. H. Hey, *J. Chem. Soc.,* 369 (1940)

[5]R. Huisgen, *Ann.,* **574**, 180 (1951)

[6]W. A. Skinner, H. F. Gram, and B. R. Baker, *J. Org.,* **25**, 777 (1960)

[7]M. S. Newman and A. Kutner, *Am. Soc.,* **73**, 4199 (1951)

[8]M. M. Chen, A. F. D'Adamo, Jr., and R. I. Walter, *J. Org.,* **26**, 2721 (1961)

[9]W. Rundel and E. Müller, *Ber.,* **96**, 2528 (1963)

[10]R. D. Clark and G. K. Helmkamp, *J. Org.,* **29**, 1316 (1964)

[11]W. A. Tilden and W. A. Shenstone, *J. Chem. Soc.,* **31**, 554 (1877)

[12]J. Meinwald, Y. C. Meinwald, and T. N. Baker, III, *Am. Soc.,* **85**, 2513 (1963)

[13]J. Meinwald, Y. C. Meinwald, and T. N. Baker, III, *Am. Soc.,* **86**, 4074 (1964)

[14]A. Hassner and C. Heathcock, *J. Org.,* **29**, 1350 (1964); K. Tanabe and R. Hayashi, *Chem. Pharm. Bull. (Tokyo),* **10**, 1177 (1962)

[15]W. A. Harrison, E. R. H. Jones, G. D. Meakins, and P. A. Wilkinson, *J. Chem. Soc.,* 3210 (1964)

[16]M. Ohno, M. Okamoto, and K. Nukada, *Tetrahedron Letters,* 4047 (1965)

[17]W. G. Dauben, E. C. Martin, and G. J. Fonken, *J. Org.,* **23**, 1205 (1958)

[18]R. A. Benkeser, R. E. Robinson, D. M. Sauve, and O. H. Thomas, *Am. Soc.,* **77**, 3230 (1955)

[19]A. S. Hussey, J.-F. Sauvage, and R. H. Baker, *J. Org.,* **26**, 256 (1961)

[20]W. J. Serfontein, J. H. Jordaan, and J. White, *Tetrahedron Letters*, 1069 (1964)
[21]R. U. Lemieux, T. L. Nagabhushan, and I. K. O'Neill, *ibid.*, 1909 (1964)
[22]G. L. Closs and S. J. Brois, *Am. Soc.*, **82**, 6068 (1960)
[23]R. Pummerer and F. Graser, *Ann.*, **583**, 207 (1953)
[24]J. Meinwald, J. Crandall, and W. E. Hymans, *Org. Syn.*, **45**, 74 (1965)
[25]M. L. Wolfrom and H. B. Wood, *Am. Soc.*, **73**, 730 (1951)
[26]N. Kornblum and E. P. Oliveto, *Am. Soc.*, **69**, 465 (1947)
[27]M. A. Naylor and A. W. Anderson, *J. Org.*, **18**, 115 (1953)
[28]L. W. Kissinger and H. E. Ungnade, *J. Org.*, **23**, 1517 (1958)
[29]D. T. Manning and H. A. Stansbury, Jr., *J. Org.*, **26**, 3755 (1961)
[30]T. Wieland and H. Determann, *Angew. Chem., Internat. Ed.*, **2**, 358 (1963)
[31]J. Honzl and J. Rudinger, *Coll. Czech.*, **26**, 2333 (1961)
[32]H. Sigel and H. Brintzinger, *Helv.*, **48**, 433 (1965)

Nitrosyl fluoride, FNO. Of several methods for the preparation of this highly reactive gas reviewed by Andreades,[1] that preferred by du Pont workers is pyrolysis of a mixture of nitrosyl fluoroborate and sodium fluoride. Boswell[2] of du Pont found that, when a stream of gas generated in this way was passed into a solution of cholesteryl acetate in methylene chloride at 0°, the solution turned deep blue and workup afforded the 5α-fluoro-6-nitrimine (2) in high yield. This was hydrolyzed by neutral

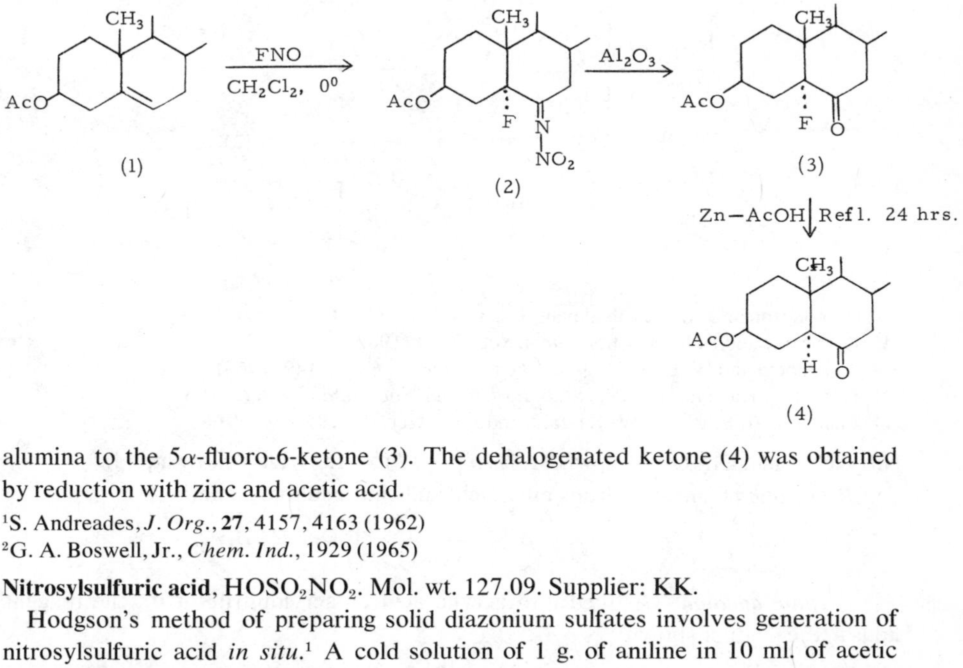
CH3
AcO
FNO
CH2Cl2, 0°
(1)
CH3
AcO
F
N
NO2
(2)
Al2O3
CH3
AcO
F
O
(3)
Zn—AcOH
Refl. 24 hrs.
CH3
AcO
H
O
(4)

alumina to the 5α-fluoro-6-ketone (3). The dehalogenated ketone (4) was obtained by reduction with zinc and acetic acid.

[1]S. Andreades, *J. Org.*, **27**, 4157, 4163 (1962)
[2]G. A. Boswell, Jr., *Chem. Ind.*, 1929 (1965)

Nitrosylsulfuric acid, $HOSO_2NO_2$. Mol. wt. 127.09. Supplier: KK.

Hodgson's method of preparing solid diazonium sulfates involves generation of nitrosylsulfuric acid *in situ*.[1] A cold solution of 1 g. of aniline in 10 ml. of acetic acid is stirred into a solution of 1 g. of sodium nitrite in 5 ml. of concd. sulfuric acid at 0°. After 10–30 min. at 0°, 50 ml. of ether is stirred in to precipitate the solid diazonium sulfate, which is collected and washed with acetic acid–ether and then with ether; yield 74%. The method was applied to a variety of amines but failed with amines containing two or more nitro groups. Furthermore the method gives salts containing a little sodium hydrogen sulfate.

Both shortcomings are eliminated in a modified procedure[2] which utilizes solid nitrosylsulfuric acid prepared by passing sulfur dioxide into nitric acid (sp. gr. 1.5) with cooling until the whole appears to be completely solid. The solid is collected on a sintered glass funnel and washed free of nitric acid with a small amount of acetic acid and then with carbon tetrachloride; yield 70%. A solution or suspension of 2 g. of the amine in 10 ml. of acetic acid was added to a solution of solid nitrosylsulfuric acid (10% excess) in 10 ml. of acetic acid. After 30 min., 100 ml. of ether is added with stirring and cooling to precipitate the diazonium sulfate.

Lactams are obtained from cycloalkanecarboxylic acids by treatment in chloroform solution at 65–70° with nitrosylsulfuric acid in 15–30% fuming sulfuric acid.[3]

$$(CH_2)_n \overset{H}{\underset{CO_2H}{C}} \xrightarrow[\substack{76.5\% \ (n=10) \\ 88.5\% \ (n=11)}]{\substack{HOSO_2NO_2-CHCl_3 \\ 65-70^0}} (CH_2)_n \overset{C=O}{\underset{NH}{|}} + CO_2$$

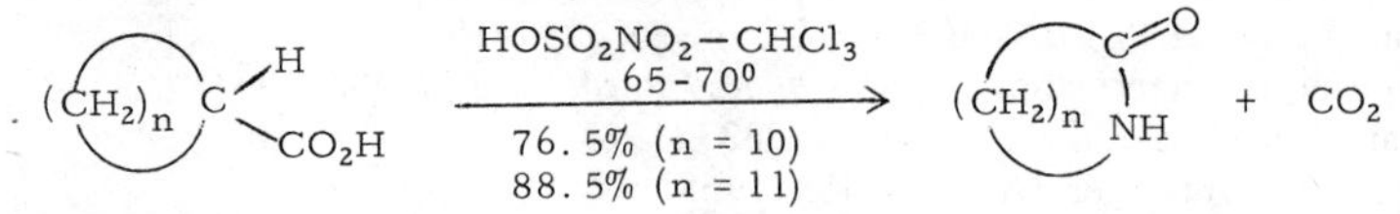

ε-Caprolactam is obtained in 60% yield from cyclohexanecarboxylic acid by treatment with liquid sulfur dioxide, nitric acid, and sulfuric acid.[4] Presumably nitrosylsulfuric acid is generated *in situ*.

Under catalysis by mercuric sulfate, the reagent adds 1,4 to a conjugated diene to give a C-nitroso sulfate (1) which cyclizes spontaneously to a 6H-1,2-oxazine (3) and this on pyrolysis gives a pyrrole (4).[5]

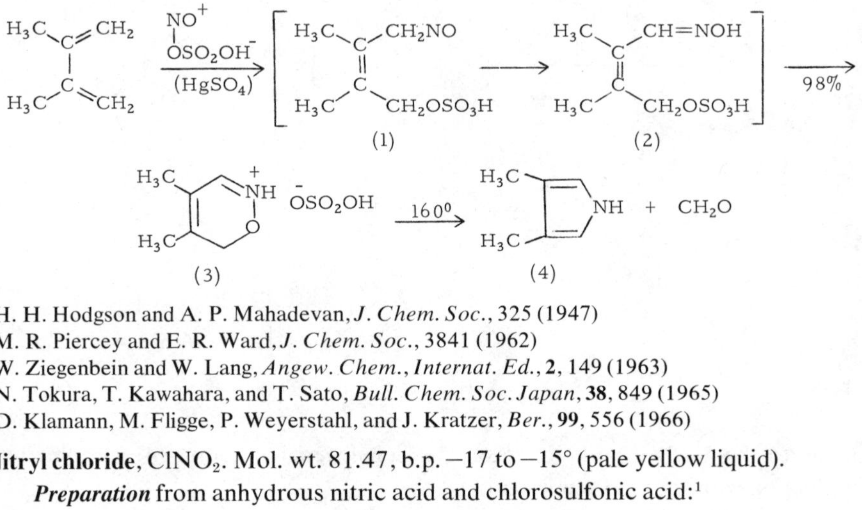

[1] H. H. Hodgson and A. P. Mahadevan, *J. Chem. Soc.*, 325 (1947)
[2] M. R. Piercey and E. R. Ward, *J. Chem. Soc.*, 3841 (1962)
[3] W. Ziegenbein and W. Lang, *Angew. Chem., Internat. Ed.*, **2**, 149 (1963)
[4] N. Tokura, T. Kawahara, and T. Sato, *Bull. Chem. Soc. Japan*, **38**, 849 (1965)
[5] D. Klamann, M. Fligge, P. Weyerstahl, and J. Kratzer, *Ber.*, **99**, 556 (1966)

Nitryl chloride, $ClNO_2$. Mol. wt. 81.47, b.p. −17 to −15° (pale yellow liquid).

Preparation from anhydrous nitric acid and chlorosulfonic acid:[1]

$$HNO_3 + ClSO_3H \xrightarrow[80-90\%]{} ClNO_2 + H_2SO_4$$

Olefinic addition. Distilled into acrylic acid or acrylonitrile at 0°, the reagent adds in the manner shown:

$$CH_2{=}CHCO_2H \xrightarrow[71\%]{ClNO_2} \underset{NO_2 \ Cl}{CH_2 \ CHCO_2H}$$

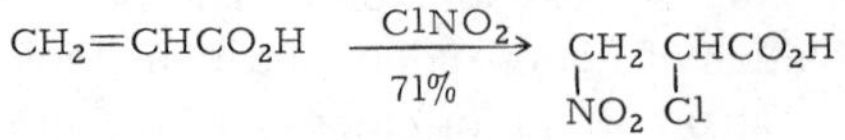

$$CH_2{=}CHCN \xrightarrow{ClNO_2} \underset{NO_2 \ Cl}{CH_2 \ CHCN}$$

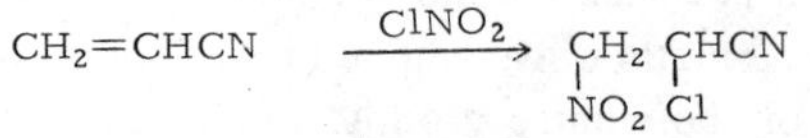

The reagent adds to an unsymmetrical terminal olefin to give the 1-nitro-1-chloroalkane:

$$RCH{=}CH_2 \xrightarrow{ClNO_2} \underset{Cl}{RCHCH_2NO_2}$$

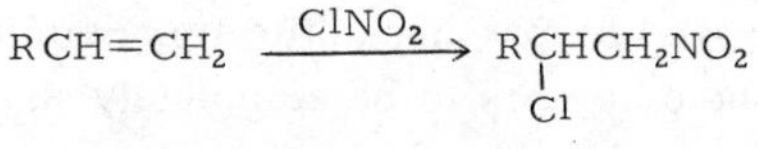

Nitration of ketones. Direct nitration proceeded poorly and α-nitroketones were obtained in only low yield by use of the enol acetate.[2]

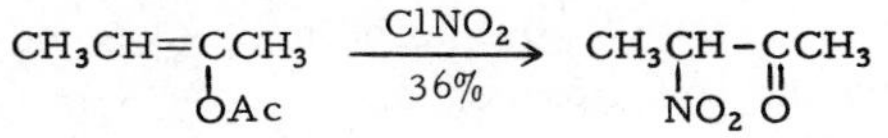

CH3CH=CCH3
OAc
ClNO2
36%
CH3CH−CCH3
NO2 O

[1]H. Shechter, F. Conrad, A. L. Daulton, and R. B. Kaplan, *Am. Soc.*, **74**, 3052 (1952)
[2]G. B. Bachman and T. Hokama, *J. Org.*, **25**, 178 (1960)

Nitryl iodide, INO_2. Mol. wt. 172.92. Hassner and co-workers[1] generated this reagent *in situ* in ether solution at room temperature from one equivalent of silver nitrite and iodine in a nitrogen atmosphere. They found that it adds to some olefins, for example, to Δ^2-cholestene:

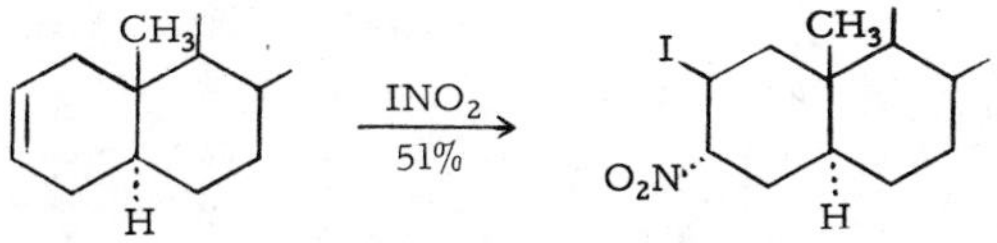

CH3
H
INO2
51%
I
CH3
O2N
H

The adducts are sometimes unstable, but they can be dehydrohalogenated to unsaturated nitro compounds:

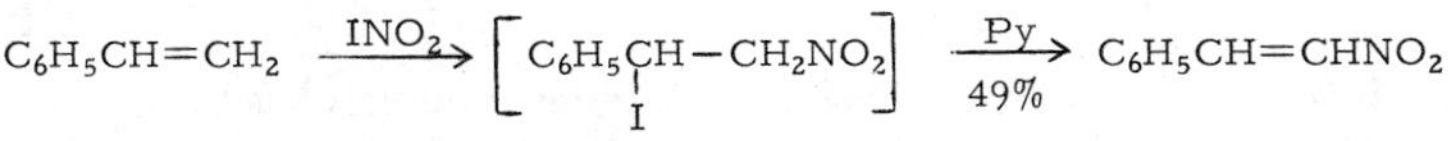

C6H5CH=CH2
INO2
C6H5CH−CH2NO2
I
Py
49%
C6H5CH=CHNO2

[1]A. Hassner, Univ. Colorado, private communication

Norbornene, 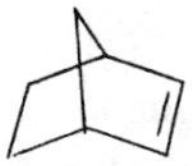

Mol. wt. 94,15, m.p. 46°, b.p. 96°, sp. gr. (30°) 0.8589. Suppliers: Roberts Chem. Co.; Aldrich. Prepared by the reaction of dicyclopentadiene with 2 equivalents of ethylene at 200°;[1] the essential reaction involves equimolecular amounts of monomeric cyclopentadiene and ethylene:

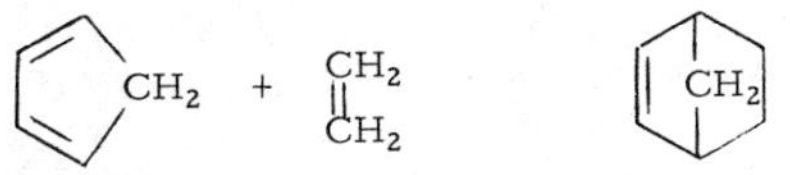

CH2
+
CH2
CH2
CH2

exo-2-Norborneol and 2-norbornanone. Kleinfelter and Schleyer[2] effected smooth addition of formic acid to norbornene to give *exo*-2-norbornyl formate (2) in excellent

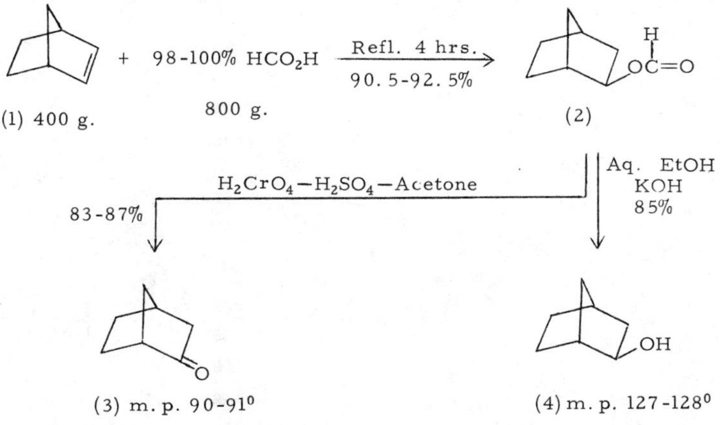

+ 98-100% HCO2H
Refl. 4 hrs.
90.5-92.5%
H
OC=O
(1) 400 g.
800 g.
(2)
H2CrO4−H2SO4−Acetone
83-87%
Aq. EtOH
KOH
85%
O
OH
(3) m.p. 90-91°
(4) m.p. 127-128°

yield. These authors do not mention an earlier paper[3] in which boron trifluoride etherate was employed as catalyst but the yield of (2) was only 71%. The formate on saponification affords pure *exo*-2-norborneol (4), and it was oxidized to 2-norbornanone (3) by the Jones procedure.

Bicyclo[3.2.1]octanone-3.[4] The first step in this synthesis, the addition of dibromocarbene to norbornene, gives the dibromocyclopropyl derivative (1), which

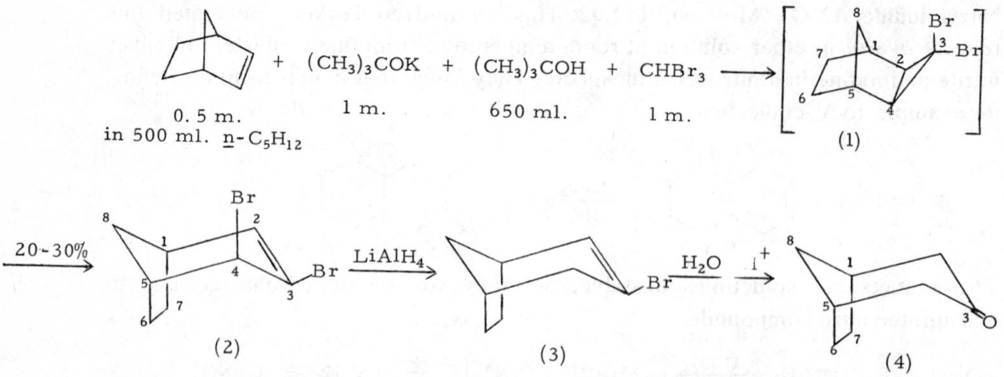
+ $(CH_3)_3COK$ + $(CH_3)_3COH$ + $CHBr_3$
0. 5 m.
in 500 ml. n-C_5H_{12}
1 m.
650 ml.
1 m.
Br
Br
(1)
20-30%
Br
Br
(2)
$LiAlH_4$
Br
(3)
H_2O
H^+
O
(4)

rearranges to the olefinic dibromide (2); the initial *exo*-adduct has been isolated when the halogen is chlorine but not when it is bromine. The allylic 4β-bromine atom of (2) is removed by metal hydride reduction, and acid hydrolysis then gives the 3-ketone (4).

exo-2-Norbornylamine. For the preparation of this amine from norbornene by hydroboration and reaction with $ClNH_2$–NaOH *see* Chloramine (H. C. Brown and co-workers[20]).

Norbornyl-2-exo-carboxylic acid. For the preparation of this acid by the hydrocarboxylation of norbornene at 50°, *see* Nickel carbonyl.

Olefinic reactivity, *see* Phenyl azide.

[1] J. Meinwald and N. J. Hudak, *Org. Syn., Coll. Vol.*, **4**, 738 (1963)
[2] D. C. Kleinfelter and P. von R. Schleyer, *Org. Syn.*, **42**, 79 (1962)
[3] L. Schmerling, J. P. Luvisi, and R. W. Welch, *Am. Soc.*, **78**, 2819 (1956)
[4] C. W. Jefford and J. A. Gunsher (Temple Univ.), procedure submitted to *Org. Syn.*

O

Oleic acid, $\begin{matrix} HC(CH_2)_7CH_3 \\ \| \\ HC(CH_2)_7CO_2H \end{matrix}$ Mol. wt. 282.47, m.p. 13.0–13.2°, 16.0–16.3° (polymorphic forms), b.p. 200–201°/1.2 mm. Suppliers: F (purified); MCB (U.S.P.); E (Techn.).

Purification. Brown *et al.*[1] cite references to four methods of purification of oleic acid. Swern's group[2] describe a procedure involving low-temperature crystallization, distillation, and distillation of the ester. However, in a procedure for the preparation of *threo*-9,10-dihydroxystearic acid by performic acid hydroxylation of oleic acid, Swern *et al.*[3] state that their method of purifying oleic acid is lengthier and less convenient than purification of the dihydroxy acid.

Hydrogen acceptor. Silverwood and Orchin[4] found that in the dehydrogenation of 2.5 mmoles of guaiene with 7.5 mmoles of selenium at 290° for 1 hr. the yield is improved from 14 to 28.6% by addition of 15 mmoles of oleic acid as hydrogen acceptor.

[1]N. A. Khan, F. E. Deatherage, and J. B. Brown, *Org. Syn., Coll. Vol.*, **4**, 851 (1963)
[2]H. B. Knight, E. F. Jordan, Jr., and D. Swern, *Biochem. Preparations*, **2**, 100 (1952)
[3]D. Swern, J. T. Scanlan, and G. B. Dickel, *Org. Syn., Coll. Vol.*, **4**, 317 (1963)
[4]H. A. Silverwood and M. Orchin, *J. Org.*, **27**, 3401 (1962)

Osmium tetroxide, OsO_4. Mol. wt. 254.20, m.p. 40°, b.p. 135°. Soluble in ether, benzene. Solubility in water at 15°: 5.88 g./100 g. Suppliers: J. Bishop and Co., Merck.

This expensive reagent is supplied in sealed glass ampoules; each contains 1.0 g. and is packed in a wooden case with a taped-on scorer. The ampoule is scored in the middle, broken, and the crystals of oxide are scraped out and dissolved in ether. On addition of an olefinic reactant, an osmate ester (usually black) is deposited slowly. After about 24 hrs. the ether is evaporated, and the osmate ester is reductively

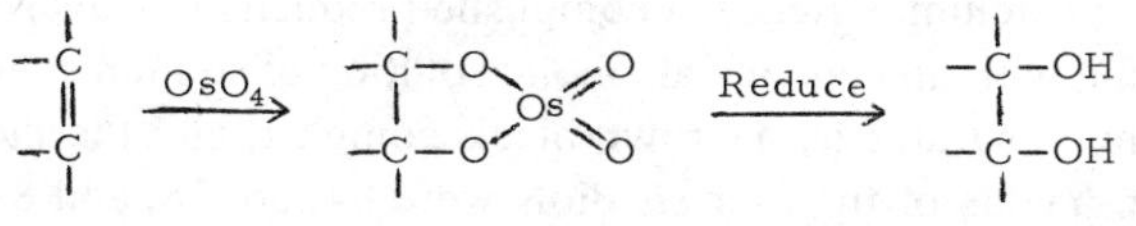

OsO4
Os
Reduce
OH
OH

cleaved by one of the methods described below. The product, usually obtained in high yield, is the *vic*-glycol; where isomerism is possible, the product, since it is derived from the *cis*-osmate ester without disturbance of the C–O bonds, is the pure *cis*- or *erythro*-isomer.

In reporting discovery of this useful method of preparing *cis*-glycols, Criegee[1] recorded his reason for undertaking the research. K. A. Hofmann[2] had found that organic and inorganic compounds completely stable to sodium or potassium chlorate in neutral aqueous solution are oxidized readily on addition of a catalytic amount of osmium tetroxide. For example, a solution of 2 g. each of hydroquinone and sodium chlorate in 100 ml. of water undergoes no change at 20°, but on addition of a solution of 7.5 mg. of OsO_4 in 1.5 ml. of water the solution at once becomes deep brown, and green needles of quinhydrone soon separate. Apart from this example, oxidation of organic compounds by the $NaClO_3$–OsO_4 combination seemed specific for unsaturated compounds, which afforded *vic*-glycols. Furthermore, oxidation of fumaric acid to DL-tartaric acid and of maleic acid to *meso*-tartaric acid established that the reaction is a stereospecific *cis* addition, exactly like permanganate oxidation. Böeseken[3] had postulated that the permanganate reaction gives the *cis*-diol because it proceeds through a necessarily *cis* cyclic intermediate. We formulate the hypothetical intermediate as a manganate ester and postulate that it is water-soluble and hence too

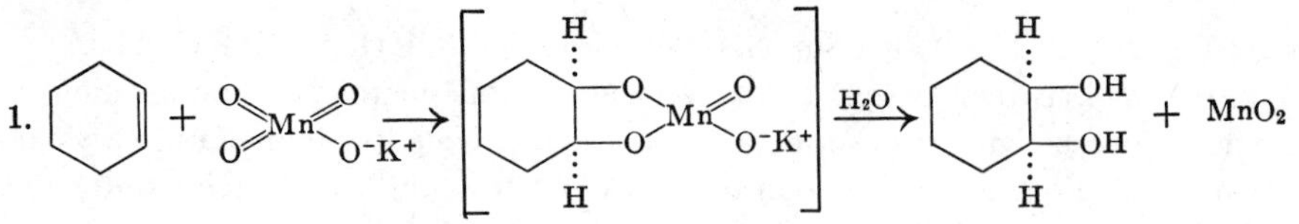

easily hydrolyzed to be isolable. Böeseken postulated that oxidation with $NaClO_3$–OsO_4 in aqueous solution involves formation of a similar intermediate and oxidative cleavage by chlorate with regeneration of the osmium tetroxide.

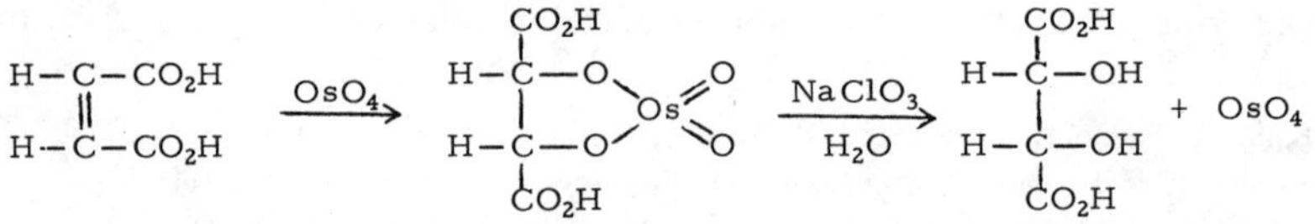

On exploring the reaction of osmium tetroxide with olefins in a nonaqueous solvent (ether, also benzene, cyclohexane), Criegee[1] promptly confirmed the Böeseken hypothesis. From acenaphthylene, indene, Δ^2-ditralin, and Δ^9-octalin he obtained nicely crystalline, bright colored osmate esters in yields of 89–99%; no

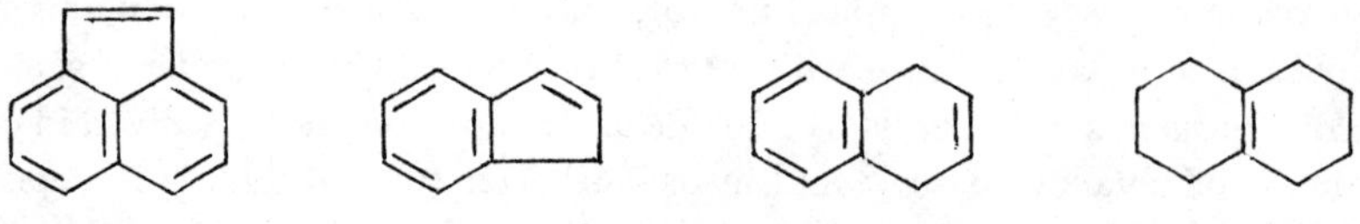

solvent for crystallization was found, but analysis of the samples as prepared gave correct values for osmium. Criegee accomplished reductive cleavage of the osmate esters by refluxing with an aqueous-alcoholic solution of sodium sulfite; osmium, in unidentified form, separates as a brown-black complex, and the diol is recovered from the filtrate. Yields of the four *cis*-diols were 84, 66, 78, and 81%. Except for the four examples cited, osmate esters are usually microcrystalline or amorphous dark brown or black solids.

In a second paper[4] Criegee reported that pyridine markedly catalyzes the reaction of osmium tetroxide with an olefin and that an osmate ester combines with 2 molecules of pyridine to form a complex which can be crystallized from methylene

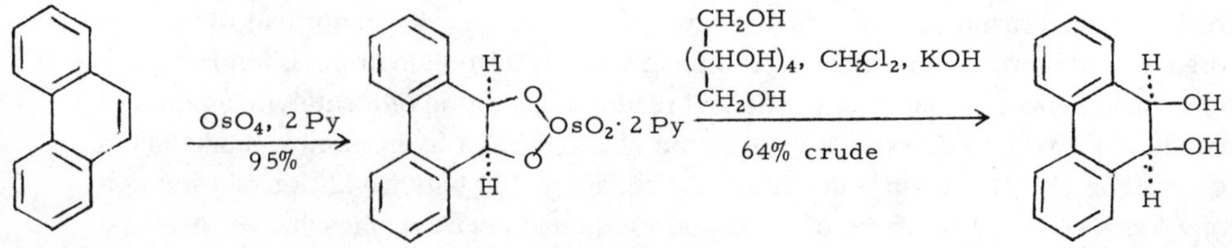

(1) (2) (3)

chloride–petroleum ether. By carrying out the reaction in benzene containing pyridine, even tetraphenylethylene, which does not add bromine, adds OsO_4 to form the dipyridine–osmate ester. Addition of 0.005 mole of phenanthrene to a solution of 0.005 mole of osmium tetroxide in 15 ml. of benzene produced at once an orange color (molecular complex). On addition of 0.01 mole of pyridine the color changed to pure yellow and then gradually brown. After 7 days red-brown crystals of the complex (2) amounted to 2.80 g. (95%). For cleavage, a solution of 2.4 g. of (2) in 25 ml. of methylene chloride was shaken with a solution of 10 g. of mannitol and 1 g. of potassium hydroxide in 100 ml. of water until the organic layer became colorless (about 1 hr.). Evaporation of the methylene chloride left a residue of 0.55 g., which on crystallization from dilute ethanol (Norit) and then from toluene afforded colorless needles of the pure diol (3), m.p. 178–179°. Potassium hydroxide cleaves the osmate ester to a point of equilibrium, and mannitol, by transesterification, displaces the equilibrium.

Criegee established that dienes form 1,2-, not 1,4-adducts.

The alkali-mannitol method has found little subsequent use, and the same is true of other suggested methods: sodium sulfite and zinc dust,[5] alkaline formaldehyde,[6] and ascorbic acid.[6] Baran[7] worked out an effective modification of Criegee's pyridine procedure, as illustrated for androstenolone. A solution of 1.14 g. (3.9 mmoles) of

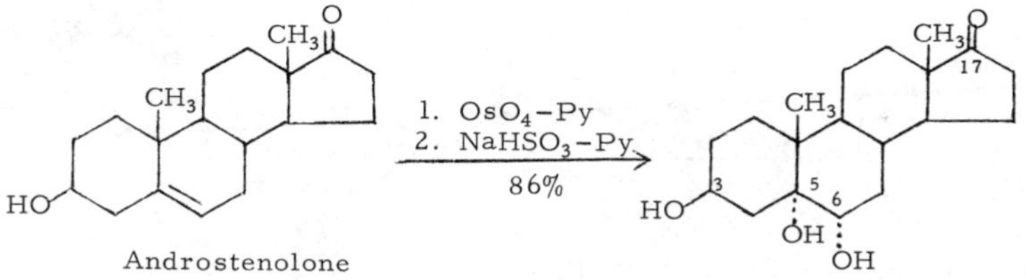
CH3
CH3
HO
Androstenolone
1. OsO4–Py
2. NaHSO3–Py
86%
CH3
17
CH3
HO
3
5
6
OH
OH

androstenolone and 1.0 g. (0.394 mmole) of osmium tetroxide in 15 ml. of pyridine was stirred for 2 hrs. at room temperature, and a solution of 1.8 g. of sodium bisulfite in 30 ml. of water and 20 ml. of pyridine was added with continued stirring. Within 5 min. the complex was cleaved to give an orange solution, which was extracted with 100-, 50-, and 50-ml. portions of methylene chloride. Evaporation of the dried extract and trituration of the residue with ethyl acetate afforded 1.05 g. (86%) of satisfactory product, m.p. 240–243°.

Barton and Elad[8] in an investigation of a natural product introduced a new method of osmate cleavage. A solution of the olefinic product and osmium tetroxide in dioxane was let stand for 48 hrs. and hydrogen sulfide was passed in to precipitate black osmium dioxide; this was removed by filtration and the diol recovered from the filtrate. An application of this procedure impressive for the scale on which costly chemicals were expended is reported by Hirschmann *et al.*[9] of Merck in the synthesis from prednisone (1) of A-norcortisol (9), which proved to be devoid of physiological activity. A solution of 100 g. of prednisone BMD (2) in 720 ml. of pyridine was cooled to 5° and treated with a solution of 69.9 g. of osmium tetroxide (1.1 × theory!) in 408 ml. of pyridine. Black material began to separate in 5 min. After 5 days at room temperature the reaction mixture was stirred into 13.4 l. of petroleum ether and the precipitate was collected, washed, and dissolved in 8 l. of dioxane. The solution was cooled in an ice bath and saturated with hydrogen sulfide. The precipitated osmium dioxide was removed, and the bulk of the solvent removed

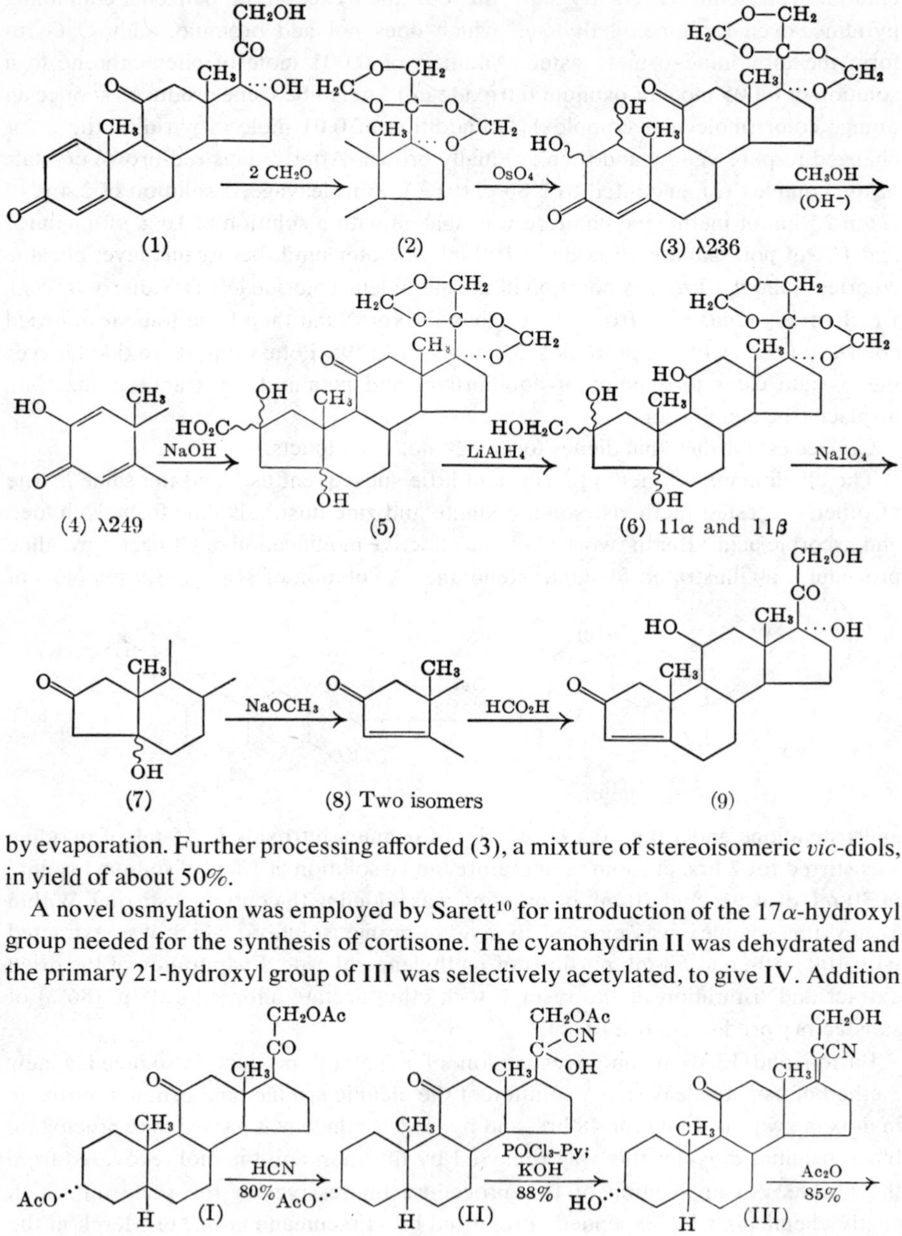

by evaporation. Further processing afforded (3), a mixture of stereoisomeric *vic*-diols, in yield of about 50%.

A novel osmylation was employed by Sarett[10] for introduction of the 17α-hydroxyl group needed for the synthesis of cortisone. The cyanohydrin II was dehydrated and the primary 21-hydroxyl group of III was selectively acetylated, to give IV. Addition

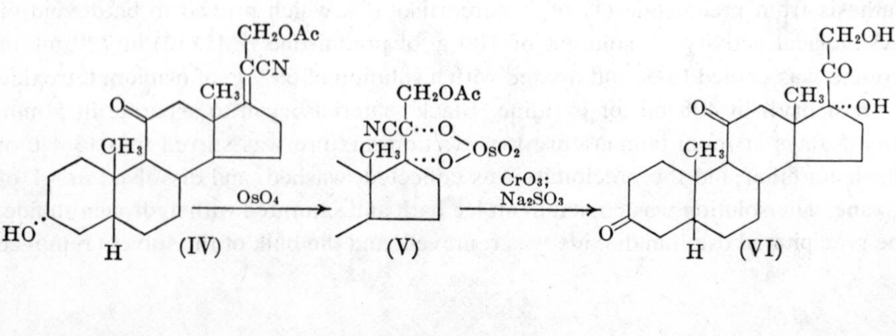

of osmium tetroxide gave the osmate ester V, which Sarett found to be a suitable derivative for oxidation at C_3. Cleavage of the osmate ester with sodium sulfite liberated the 17α-hydroxyl group and was attended with reversal of the cyanohydrin reaction with restoration of the 20-keto group to give VI, which has the desired dihydroxyacetone side chain.

Osmylation of an α,β-unsaturated ketone was a key step in the synthesis of triamcinolone by Bernstein's group[11] at the Lederle Laboratories. The starting material, cortisol 21-acetate 3,20-bisethyleneketal (1), reacted with thionyl chloride in pyridine

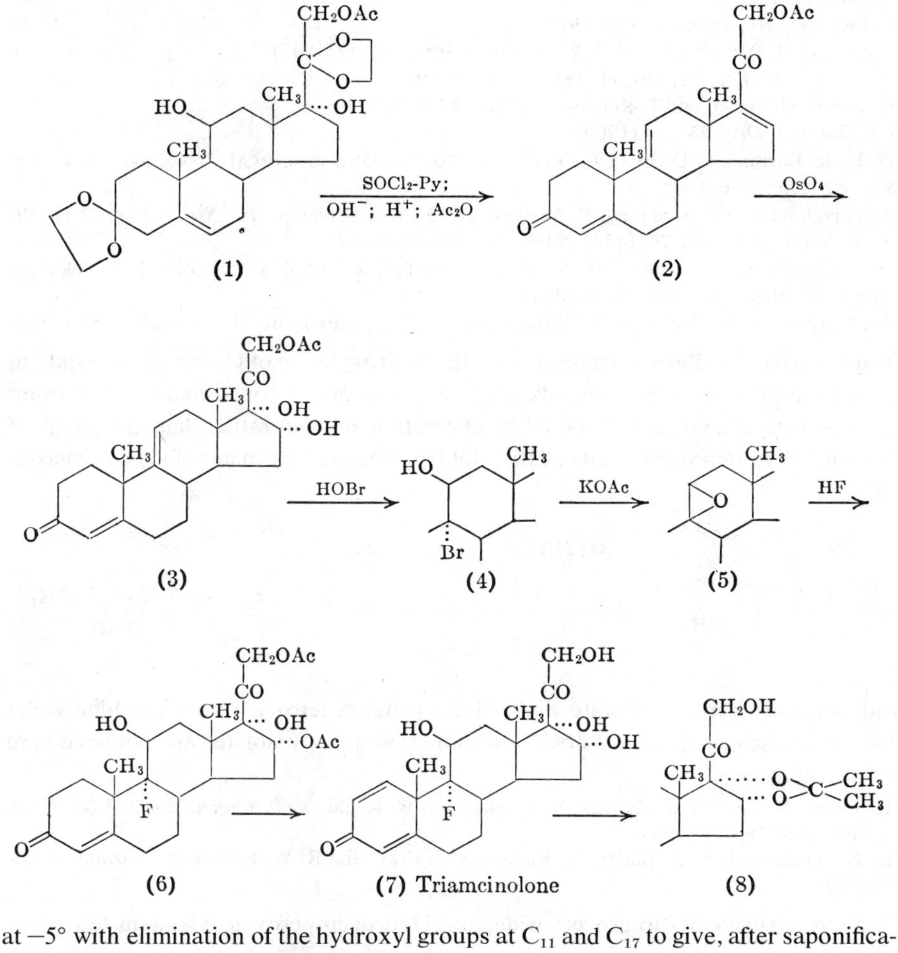
CH2OAc
CH3
HO
OH
O
C
CH3
SOCl2-Py;
OH−; H+; Ac2O
(1)
CO
(2)
OsO4
OH
(3)
HOBr
Br
(4)
KOAc
(5)
HF
OAc
F
(6)
CH2OH
(7) Triamcinolone
(8)

at −5° with elimination of the hydroxyl groups at C_{11} and C_{17} to give, after saponification, deketalization with methanolic sulfuric acid, and reacetylation, the 21-acetoxytriene-3,20-dione (2). The 16,17-double bond was then selectively hydroxylated by reaction with osmium tetroxide in benzene containing pyridine; cleavage of the osmate ester was effected under mild conditions (Na_2SO_3–K_2CO_3 in water–methanol–benzene) to avoid rearrangement with expansion of ring D.

A novel reaction was encountered by Johnson and co-workers[12] in characterizing the nonconjugated ketone (2) obtained from Δ^4-cholestene-3-one (1) on reaction with diazomethane in the presence of boron fluoride. Treatment of (2) with osmium tetroxide gave the hemiketal (3), which readily forms a methyl ether.

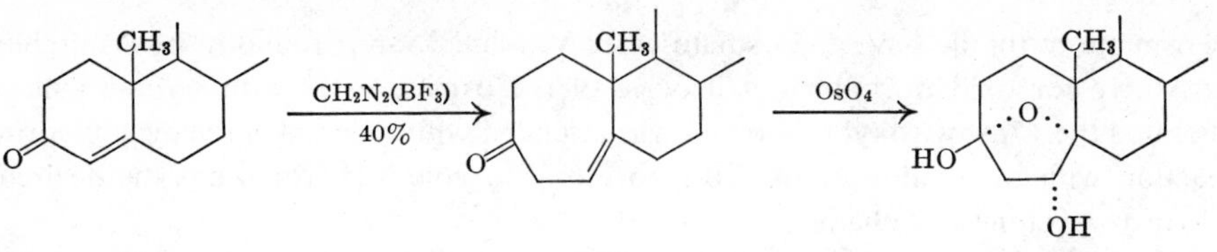

(1) (2) (3)

[1]R. Criegee, *Ann.*, **522**, 75 (1936)
[2]K. A. Hofmann, *Ber.*, **45**, 3329 (1912); K. A. Hofmann, O. Ehrhart, and O. Schneider, *Ber.*, **46**, 1657 (1913)
[3]J. Böeseken, *Rec. trav.*, **41**, 199 (1922)
[4]R. Criegee, B. Marchand, and H. Wannowius, *Ann.*, **550**, 99 (1942)
[5]A. Serini *et al.*, *Ber.*, **71**, 1362 (1938); **72**, 391 (1939)
[6]H. Reich, M. Sutter, and T. Reichstein, *Helv.*, **23**, 170 (1940)
[7]J. S. Baran, *J. Org.*, **25**, 257 (1960)
[8]D. H. R. Barton and D. Elad, *J. Chem. Soc.*, 2085 (1956); see also M. Uskoković *et al.*, *Am. Soc.*, **82**, 4965 (1960)
[9]R. Hirschmann, G. A. Bailey, R. Walker, and J. M. Chemerda, *Am. Soc.*, **81**, 2822 (1959)
[10]L. H. Sarett, *Am. Soc.*, **70**, 1454 (1948); **71**, 2443 (1949)
[11]S. Bernstein, R. H. Lenhard, W. S. Allen, M. Heller, R. Littell, S. M. Stolar, L. I. Feldman, and R. H. Blank, *Am. Soc.*, **78**, 5693 (1956)
[12]W. S. Johnson, M. Neeman, S. P. Birkeland, and N. A. Fedoruk, *Am. Soc.*, **84**, 989 (1962)

Osmium tetroxide–Barium chlorate. For the hydroxylation of (1), an intermediate in the synthesis of reserpine-type alkaloids, a Czech group[1] found that use of barium chlorate with osmium tetroxide, taken in catalytic amount rather than in equivalent amount,[2] gave improved yields of the diol (2). The starting material (1) was shaken

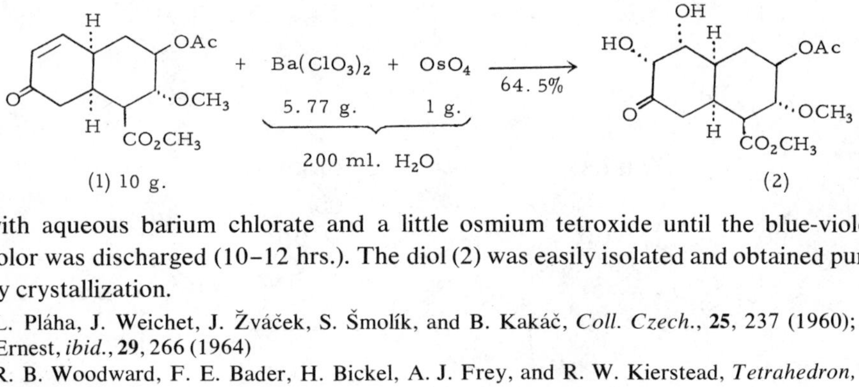

(1) 10 g. (2)

with aqueous barium chlorate and a little osmium tetroxide until the blue-violet color was discharged (10–12 hrs.). The diol (2) was easily isolated and obtained pure by crystallization.

[1]L. Pláha, J. Weichet, J. Žváček, S. Šmolík, and B. Kakáč, *Coll. Czech.*, **25**, 237 (1960); I. Ernest, *ibid.*, **29**, 266 (1964)
[2]R. B. Woodward, F. E. Bader, H. Bickel, A. J. Frey, and R. W. Kierstead, *Tetrahedron*, **2**, 1 (1958)

Osmium tetroxide–Hydrogen peroxide, *see* Hydrogen peroxide–Osmium tetroxide.

Oxalic acid, $(CO_2H)_2$. Mol. wt. 90.04, m.p. 189°, pKa 1.46, pKa_2 4.40, solubility at 20°: 10.2 g./100 g. water. Dihydrate, mol. wt. 126.07, m.p. 101°. Suppliers of dihydrate: B, E, F, MCB; of anhydrous acid: E, F, MCB.

Preparation of anhydrous oxalic acid. In one procedure[1] a mixture of 2 kg. of oxalic acid dihydrate and 3l. of carbon tetrachloride is stirred in a flask fitted with an upright condenser heated with steam and delivering into a cooled downward condenser provided with a gravity separation which collects water and returns carbon tetrachloride to the boiling flask. Dehydration is complete in 18–24 hrs. and the yield is 90%. Alternatively,[2] pulverized hydrate is placed in trays in an oven heated at

98–99° for about 2 hrs.; the material is then crushed and bottled (purity 99.5+%).

Dehydration. The alcohol (1) was not dehydrated by potassium bisulfate at 190–200° or by refluxing acetic anhydride, but afforded the olefin smoothly when heated with anhydrous oxalic acid for 1 hr.[3] Dehydration of Reformatsky carbinols

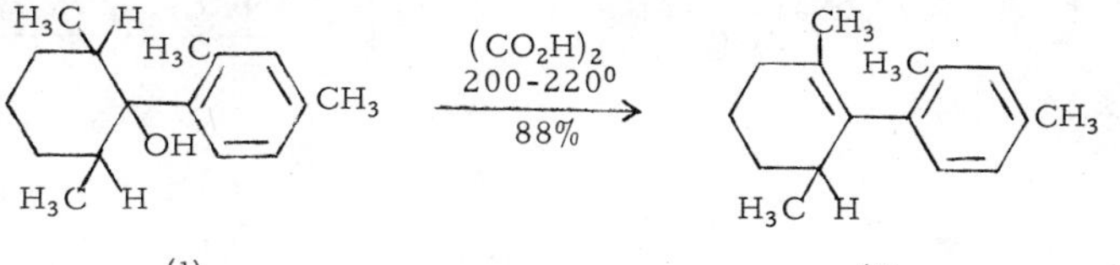
H3C H
H3C
CH3
OH
H3C H
(CO2H)2
200-220°
88%
CH3
H3C
CH3
H3C H

(1) (2)

in the thiophene series with strong acids proceeded poorly, and treatment with thionyl chloride in pyridine afforded tars.[4] However, refluxing for 3–6 hrs. with 6% aqueous oxalic acid effected dehydration in "nearly quantitative" yield.

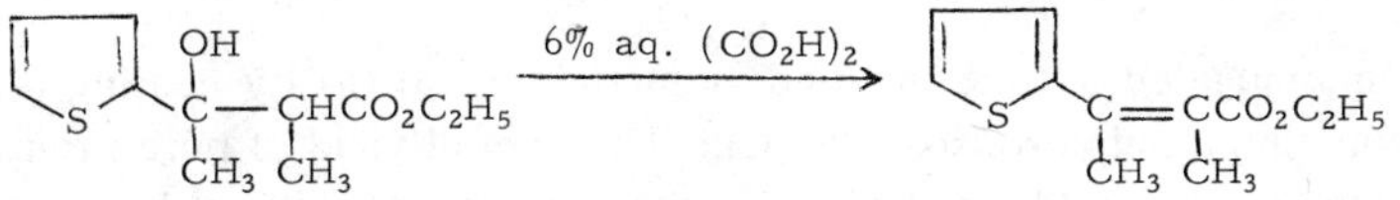
OH
S
C—CHCO2C2H5
CH3 CH3
6% aq. (CO2H)2
S
C=CCO2C2H5
CH3 CH3

Elimination of R_2NH. Cardwell[5] showed that the Mannich reaction of acetone with formaldehyde and dimethylamine gives the unsymmetrical product (1), and that

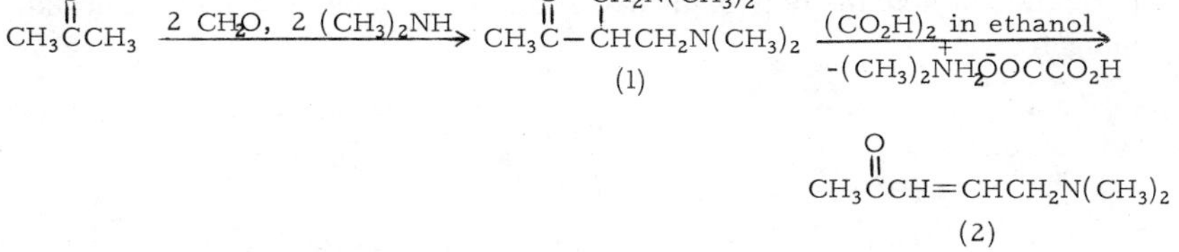
CH3CCH3
2 CH2O, 2 (CH3)2NH
CH3C—CHCH2N(CH3)2
CH2N(CH3)2
(1)
(CO2H)2 in ethanol
-(CH3)2NH2 OOCCO2H
CH3CCH=CHCH2N(CH3)2
(2)

this reacts with oxalic acid in ethanol at 25° with elimination of dimethylamine as the acid oxalate and formation of unsaturated aminoketone (2). The mixture is cooled to 0°, the salt removed by filtration, and the product recovered from the filtrate.

Isolation of amines. Diacetoneamine, prepared by addition of ammonia to mesityl oxide, is isolated by addition of the required weight of oxalic acid to a solution of the amine in ethanol.[6]

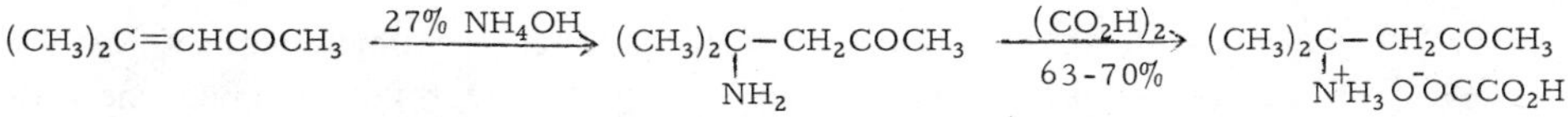
(CH3)2C=CHCOCH3
27% NH4OH
(CH3)2C—CH2COCH3
NH2
(CO2H)2
63-70%
(CH3)2C—CH2COCH3
NH3 OOCCO2H

Acidic condensing agent. In the first synthesis of vitamin K_1, a solution of 5 g. of 2-methyl-1,4-naphthohydroquinone, 1.48 g. of phytol, and 1 g. of anhydrous oxalic acid in 15 ml. of dioxane was heated at 75° for 36 hrs.[7]

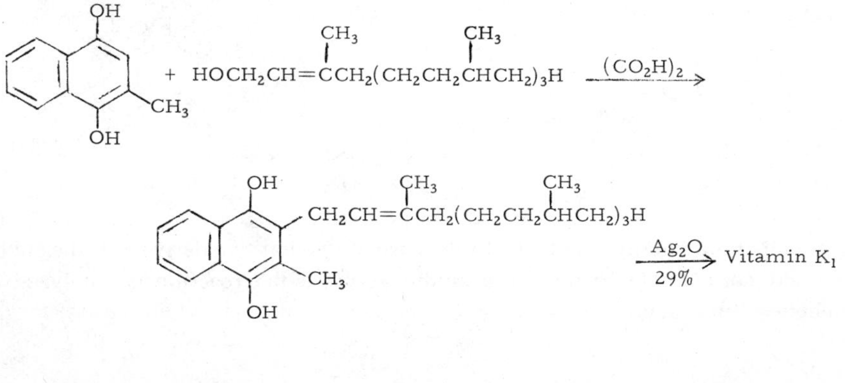
OH
CH3
OH
CH3 CH3
+ HOCH2CH=CCH2(CH2CH2CHCH2)3H
(CO2H)2
OH
CH3 CH3
CH2CH=CCH2(CH2CH2CHCH2)3H
CH3
OH
Ag2O
29%
Vitamin K1

Mild acid catalyst. A procedure[8] for the preparation of Δ^4-cholestene-3-one (5) from cholesterol (1, unpurified) involves oxidation of the dibromide (2), debromination

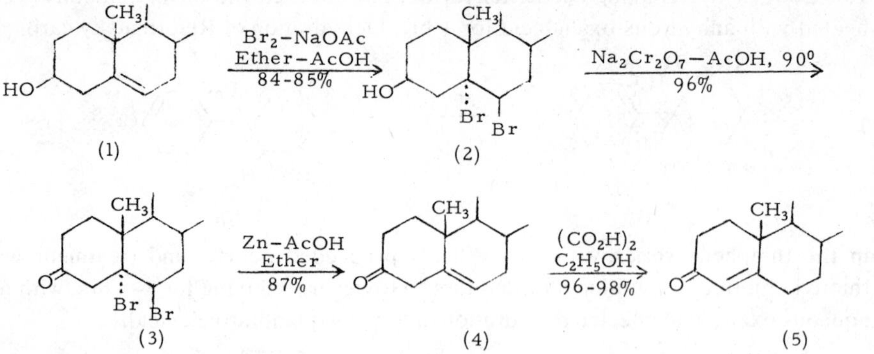
CH3
HO
Br2–NaOAc
Ether–AcOH
84-85%
CH3
HO
Br
Br
Na2Cr2O7–AcOH, 90°
96%
CH3
O
Br
Br
Zn–AcOH
Ether
87%
CH3
O
(CO2H)2
C2H5OH
96-98%
CH3
O

(1) (2) (3) (4) (5)

to the nonconjugated ketone (4), and isomerization to (5) by heating (4) in 95% ethanol containing anhydrous oxalic acid. The overall yield of pure product is 68–69%. Isomerization of (4) in 100-g. batches in ethanol with either hydrochloric acid or sodium hydroxide as catalyst gave yellowish, low-melting material.

Oxalic acid is recommended for the hydrolysis of the enol ethers obtained by Birch reduction to give the nonconjugated ketone; with a mineral acid the product is the conjugated ketone.[9,10]

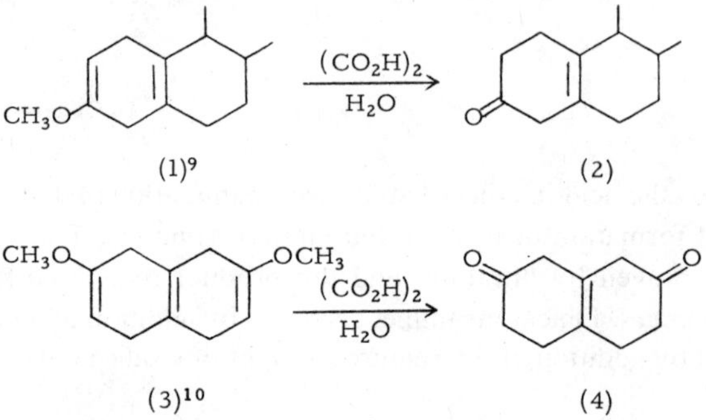
CH3O
(CO2H)2
H2O
O
CH3O
OCH3
(CO2H)2
H2O
O
O

(1)[9] (2)

(3)[10] (4)

Both oxalic acid and the still weaker catalyst adipic acid are sometimes useful for the selective ketalization of steroid diketones.[11] Progesterone, refluxed in benzene with ethylene glycol and oxalic acid, gave a mixture of the 3,20-bisethyleneketal (2) and

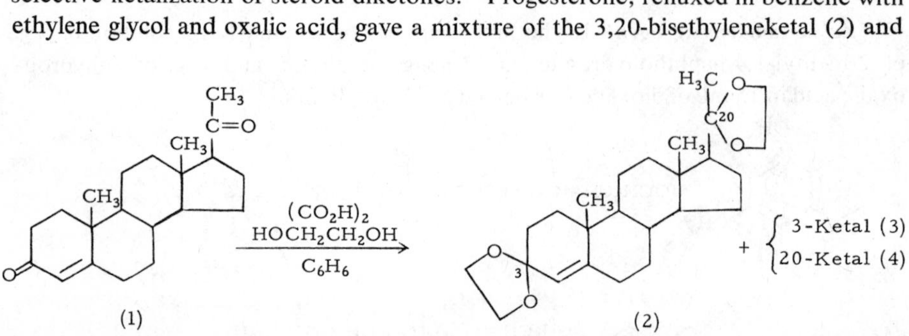
CH3
C=O
CH3
CH3
O
(CO2H)2
HOCH2CH2OH
C6H6
H3C
O
C20
O
CH3
CH3
O
3
O
+
3-Ketal (3)
20-Ketal (4)

(1) (2)

the 3- and 20-monoethyleneketals. In the case of the first two derivatives the double bond did not migrate to the 5,6-position, as when the reaction is catalyzed by *p*-toluenesulfonic acid.

[1]H. T. Clarke and A. W. Davis, *Org. Syn., Coll. Vol.*, **1**, 421 (1941)
[2]E. Bowden, *ibid.*, **1**, 424 (1941)
[3]R. B. Carlin and D. A. Constantine, *Am. Soc.*, **69**, 50 (1947)
[4]R. E. Miller and F. F. Nord, *J. Org.*, **15**, 89 (1950)
[5]H. M. E. Cardwell, *J. Chem. Soc.*, 1056 (1950)
[6]P. R. Haeseler, *Org. Syn., Coll. Vol.*, **1**, 196 (1941)
[7]L. F. Fieser, *Am. Soc.*, **61**, 3467 (1939)
[8]L. F. Fieser, *Am. Soc.*, **75**, 5421 (1953)
[9]D. Burn and V. Petrow, *J. Chem. Soc.*, 364 (1962)
[10]B. Weinstein and A. H. Fenselau, *J. Org.*, **30**, 3209 (1965)
[11]J. J. Brown, R. H. Lenhard, and S. Bernstein, *Am. Soc.*, **86**, 2183 (1964)

Oxalyl bromide, $(COBr)_2$. Mol. wt. 215.85; b.p. 102–103°. Supplier: Kaplop Labs. Preparation.[1]

Treibs found[2] that the reagent reacts with an aliphatic acid bromide in carbon tetrachloride, at a bath temperature raised from 100 to 135°, with introduction of COBr into the α-position. The reaction can be followed by absorbing the hydrogen

$$RCH_2COBr \xrightarrow{(COBr)_2} RCH(COBr)_2 + CO + HBr$$

bromide and measuring the carbon monoxide evolved. Yields are in the range 38–68%. Olefins of types $RCH{=}CH_2$ and $R_2C{=}CH_2$ react poorly, but bromocarbonylation is effected in modest yield with styrene and 1,1-diphenylethylene.[3]

$$C_6H_5CH{=}CH_2 \xrightarrow[40\%]{(COBr)_2} C_6H_5CH{=}CHCOBr$$

$$(C_6H_5)_2C{=}CH_2 \xrightarrow[50\%]{(COBr)_2} (C_6H_5)_2C{=}CHCOBr$$

Highly reactive aromatics react with the reagent: anthracene⟶9-anthroic acid bromide (52%). Ketones undergo α-bromocarbonylation in yields of the order of 10–40%.[4]

[1]H. Staudinger and E. Anthes, *Ber.*, **46**, 1431 (1913)
[2]W. Treibs and H. Orttmann, *Ber.*, **91**, 297 (1958)
[3]*Idem, Ber.*, **93**, 545 (1960)
[4]W. Treibs, J. Riemer, and H. Orttmann, *Ber.*, **93**, 551 (1960)

Oxalyl chloride, $(COCl)_2$. Mol. wt. 126.93, b.p. 62°, sp. gr. 1.50. Suppliers: A, B, E, F, MCB, Kaplop Labs. Preparation from finely ground, anhydrous oxalic acid and phosphorus pentachloride.[1, 2]

Preparation of acid chlorides. Adams and Ulich,[3] in introducing use of the reagent for the conversion of carboxylic acids into the acid chlorides, described two main procedures: (a) reaction of the acid with oxalyl chloride in benzene; (b) conversion of the acid into the sodium salt and treatment of the salt with oxalyl chloride, with or without addition of pyridine as catalyst. Wilds and Shunk[4] investigated the steps in the synthesis of desoxycorticosterone acetate (5) from the acid (1) and found that reaction of (1) with thionyl chloride to form the acid chloride (3) proceeds poorly because of attack of ring A by the reagent. Much better results were obtained by dissolving the acid in one equivalent of aqueous alkali, freezing the solution, and evaporating to dryness at reduced pressure (lyophilization). The dried salt (2) was then suspended in benzene containing a trace of pyridine and treated at 0° with oxalyl chloride. Reaction of the acid chloride with diazomethane afforded (4) in

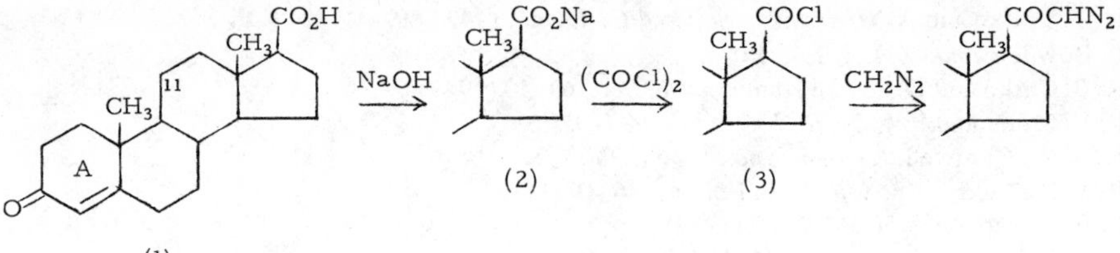

(1)   (2)   (3)

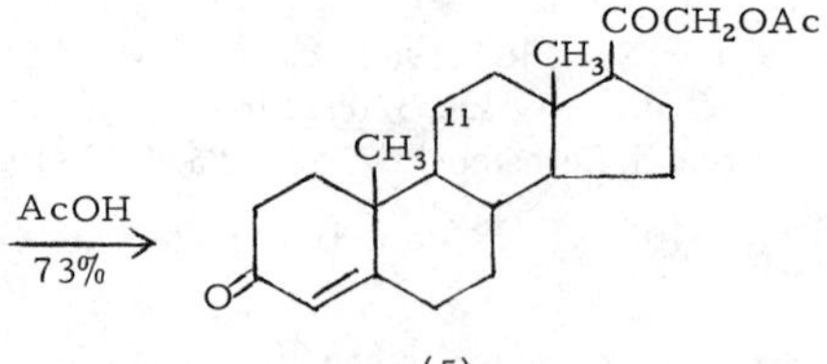

(5)

81% yield from (1), and reaction of (4) with acetic acid completed the synthesis of desoxycorticosterone acetate (5). Reichstein's group,[5] in applying the same reaction sequence for the synthesis of 11-epicorticosterone, repeated the work of Wilds and Shunk on (1) as a model compound but with variable results; the diazoketone (4) was invariably accompanied by the methyl ester of (1). Better results were obtained by treating the free acid (1) with oxalyl chloride in benzene at 20° without adding pyridine. No discoloration occurred in 1 hr., and the pure diazoketone (4), free from ester, was obtained in 75% yield. This method served well for the synthesis of 11-epicorticosterone.

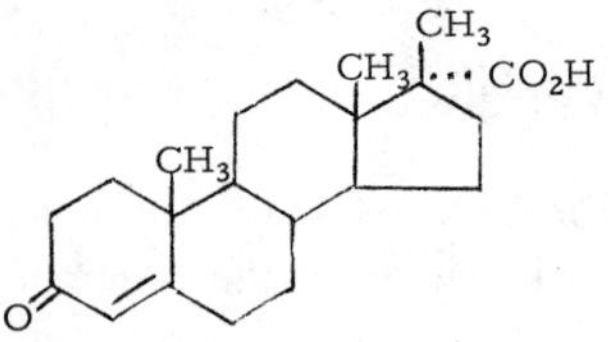

(6)

Engel and Just[6] used the oxalyl chloride–benzene method with success for the conversion of (6) into the acid chloride; in this case yields were unsatisfactory when the sodium salt was treated with oxalyl chloride in the presence of pyridine.

Cason[7] found that the reaction of the half-ester of an unsymmetrical diacid such as (7) with thionyl chloride proceeds with some rearrangement, probably involving the anhydride, and gives a mixture of the two possible ester-acid chlorides. Ställberg-Stenhagen[8] studied

$$HO_2CCH_2CH_2\overset{\overset{\displaystyle C_4H_9}{|}}{\underset{\underset{\displaystyle C_2H_5}{|}}{C}}CO_2CH_3$$

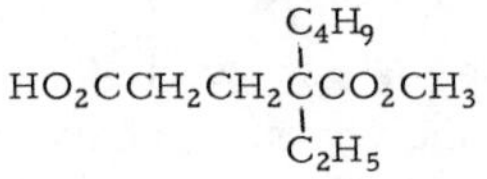

(7)

$$HO_2CCH_2\overset{\overset{\displaystyle CH_3}{|}}{\underset{\underset{\displaystyle H}{|}}{C}}CH_2CO_2CH_3$$

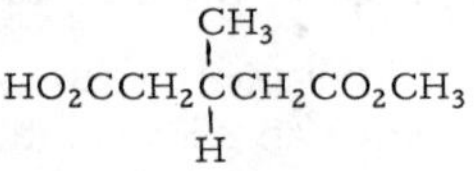

(8)

the reaction of thionyl chloride with the two enantiomers of the ester-acid (8), where rearrangement leads to racemization. With highly purified thionyl chloride no racemization occurred when the reaction was conducted at 30° and the excess reagent removed at a temperature below 50°. At a higher temperature, or with less pure reagent, racemization occurred. Oxalyl chloride gave better results, for with

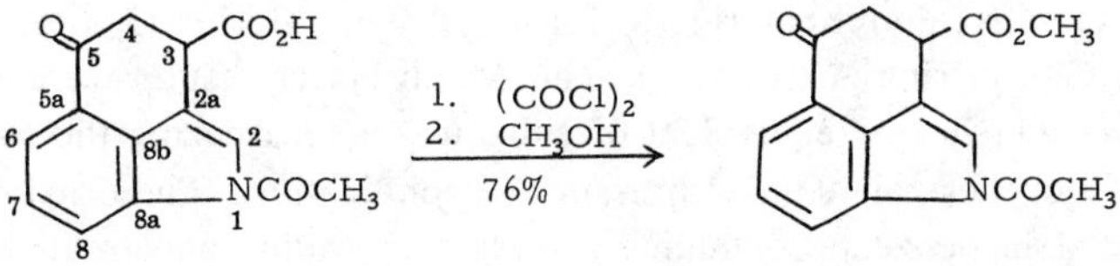

(9)   (10)

this reagent in benzene at 50° no racemization was noted. Linstead[9] found that this procedure effectively avoided rearrangements in the conversion of half-esters of succinic, glutaric, and phthalic acid into the ester-acid chlorides. Szmuszkovicz[10] tried esterification of the acid (9) with diazomethane but obtained the ester (10) in only 55% yield. He then treated the acid with oxalyl chloride in benzene, evaporated to dryness, and refluxed the residue with methanol and so obtained (10) in high yield.

In one step of a total synthesis of rotenone, Miyano[11] needed to convert tubaic acid into the acid chloride, but usual methods were precluded by the extreme

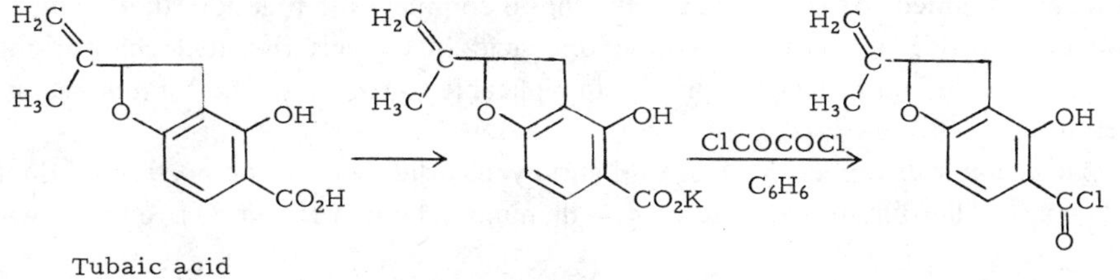
ClCOCOCl
C_6H_6
Tubaic acid

lability of the unsaturated side chain. The problem was solved by conversion of the acid to the dry potassium salt (procedure not recorded) and reaction of this with oxalyl chloride in benzene.

Introduction of COCl (Chlorocarbonylation). Kharasch and Brown[2] found that some saturated hydrocarbons react with oxalyl chloride in the presence of light or a peroxide with replacement of hydrogen by COCl. Thus a mixture of 0.3 mole of cyclohexane, 0.2 mole of oxalyl chloride, and 1.2 g. of dibenzoyl peroxide when

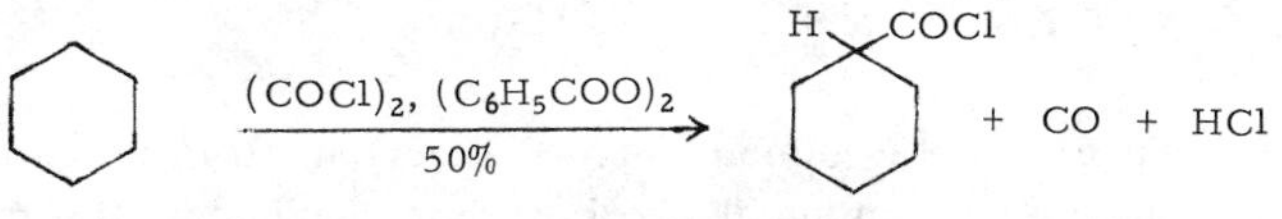
$(COCl)_2$, $(C_6H_5COO)_2$
50%
+ CO + HCl

refluxed for 24 hrs. afforded cyclohexanecarboxylic acid chloride in modest yield. In some instances an olefinic hydrogen can be displaced by COCl.[12] Thus 1,1-diphenylethylene (6.5 g.), refluxed with 4.5 g. of oxalyl chloride for 3–4 hrs., afforded an acid chloride in 50% yield. Bergmann *et al.*[13] raised the yield in this reaction to 95% by using 5 moles of oxalyl chloride per mole of olefin. Neither light nor peroxides have any effect on this reaction, which apparently is ionic. Cyclohexene

$$(C_6H_5)_2C{=}CH_2 \xrightarrow[50\%]{(COCl)_2} (C_6H_5)_2C{=}C\begin{matrix} H \\ COCl \end{matrix}$$

and trimethylethylene failed to react. Anthracene undergoes substitution without catalysis.[14] A solution of 5 g. of anthracene and 30 ml. of oxalyl chloride in 150 ml.

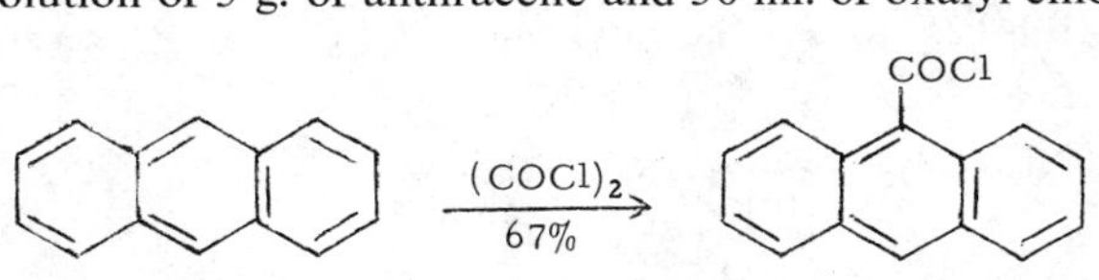
COCl
$(COCl)_2$
67%

of nitrobenzene was heated to 120°, and the temperature was raised to 240° in the course of 5–6 hrs.

Under Friedel-Crafts conditions ($AlCl_3$, CS_2) oxalyl chloride reacts with aromatic hydrocarbons to give carbonyl chlorides hydrolyzed by water to carboxylic acids.

An example is a one-step procedure[15] for the preparation of mesitoic acid in which the yield is as good as in the two-step synthesis via bromomesitylene and carbonation of the Grignard derivative. Oxalyl chloride is added to a suspension of aluminum

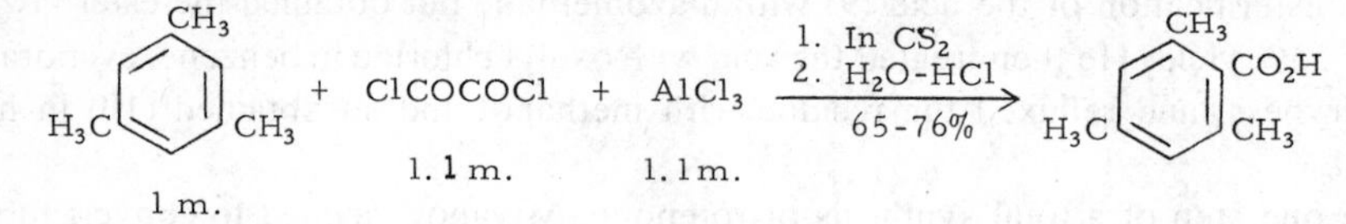

chloride in carbon disulfide at 10–15°, and then a solution of mesitylene in carbon disulfide is added. After eventual refluxing to complete the reaction, the mixture is treated with ice and concd. hydrochloric acid to convert the acid chloride into mesitoic acid. The procedure proved inapplicable to ferrocene, hydroquinone, and veratrole.

Reaction with amines. McDonald[16] surveyed earlier work and reported additional examples. The dihydrochloride of 4,4′-diaminodiphenylmethane (1) refluxed with

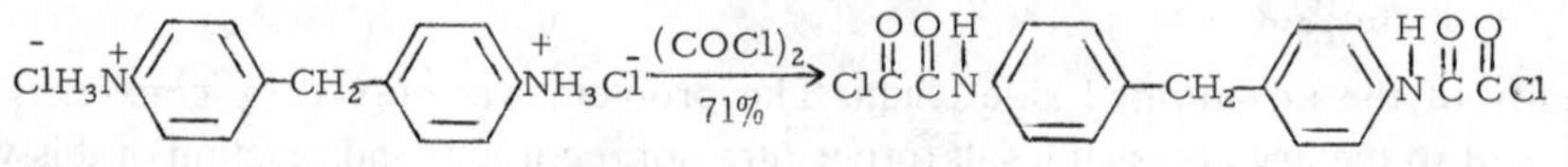

(1) (2)

excess oxalyl chloride afforded the bisoxamic acid chloride (2). Ethylamine hydrochloride (0.06 mole), heated under reflux for 50 hrs. with 1.25 moles of oxalyl chloride, was converted into N-ethyloximidic acid chloride.

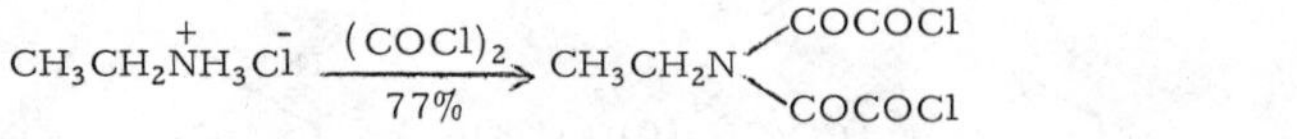

As noted on page 29, N,N-dimethylaniline reacts to form an α-diketone.[17] A suspension of 1 mole of aluminum chloride in carbon disulfide is stirred at 0° and 1.5 moles of the amine is run in with stirring to produce an even slurry, and 0.25 mole of

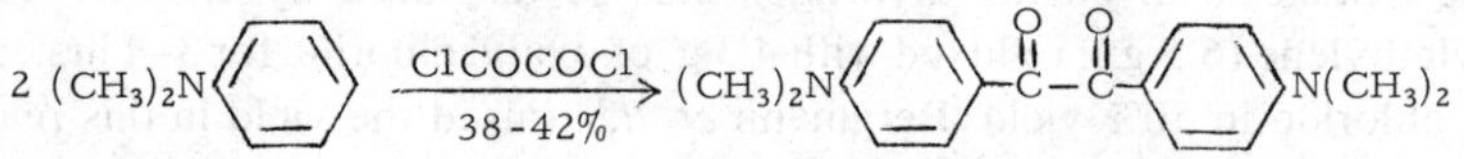

oxalyl chloride is added in carbon disulfide solution. A lengthy workup and crystallization from acetone affords 4,4′-bis(dimethylamine)benzil, m.p. 200–202°. The procedure is particularly important, even though the yield is low, because 4-dimethylaminobenzaldehyde cannot be converted into the benzoin.

ROH → RCl. Oxalyl chloride reacts with an alcohol at room temperature or slightly above to form the alkyl chloroglyoxalate; in the presence of pyridine, this decomposes at 100–125° to give the alkyl chloride.[18]

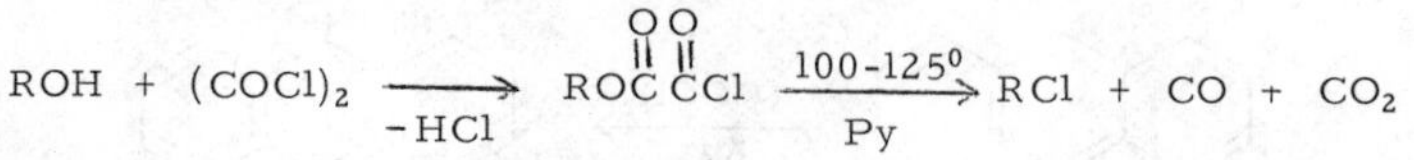

Amide → Isocyanate. A general procedure for the conversion of amides into acylisocyanates[19] provides a route to these compounds more convenient than the reaction of acid chlorides with silver cyanate. In the example formulated a mixture of α-chloroacetamide and 100 ml. of ethylene dichloride is chilled to about 2° and stirred while 0.6 mole of oxalyl chloride is added all at once. The mixture is removed

$$ClCH_2\overset{O}{\overset{\|}{C}}NH_2 + ClCOCOCl \xrightarrow[65\%]{} ClCH_2\overset{O}{\overset{\|}{C}}-N=C=O + CO + 2\ HCl$$

from the ice bath, stirred for 1 hr., and then refluxed at 83° with stirring for 5 hrs. The solution is chilled to 0–10°, the condenser is replaced by a 120-mm. column packed with glass helices, and the solvent is removed *in vacuo* with stirring. The product is then collected at 68–70°/70 mm. Further study[20] showed that aromatic amides give excellent results, but that with aliphatic amides yields are not satisfactory unless there are no α-hydrogens or unless the α-carbon carries an electron-withdrawing group.

Reaction with unsaturated ketones. Deghenghi and Gaudry[21] found that oxalyl chloride reacts with Δ^4-3-ketosteroids to give 3-chloro-$\Delta^{3,5}$-steroids, as illustrated

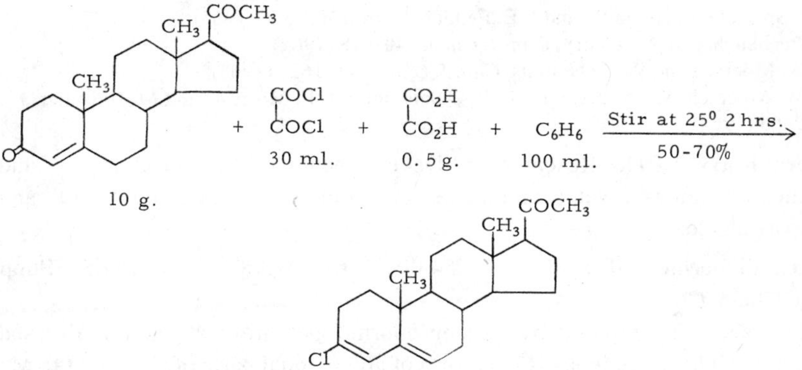
COCH3
CH3
CH3
COCl
COCl
CO2H
CO2H
+
C6H6
Stir at 25° 2 hrs.
50-70%
30 ml.
0. 5 g.
100 ml.
10 g.
COCH3
CH3
CH3
Cl

for progesterone. The reaction is carried out in benzene at room temperature with oxalic acid as catalyst. Phosphorus pentachloride, phosphoryl chloride, and acetyl chloride can also be used but are less effective.[22]

The reagent reacts with the $\Delta^{1,4}$-3-ketosteroid (1) in benzene at room temperature to give the 3-chloro-$\Delta^{1,3,5}$-triene (2), which can be isolated without difficulty but

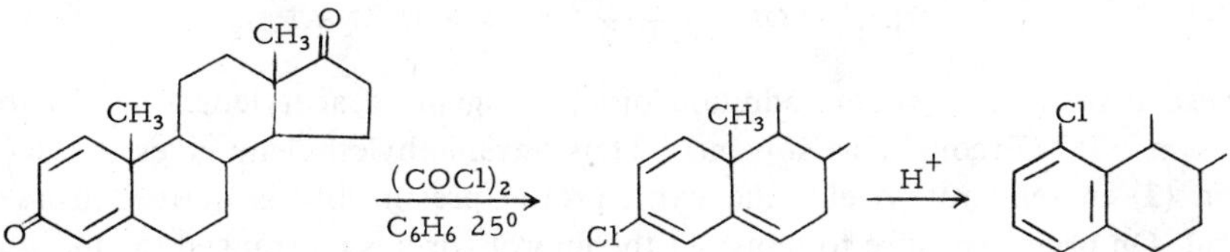
CH3
CH3
(COCl)2
C6H6 25°
CH3
Cl
H+
Cl
CH3

(1) (2) (3)

which is labile and can be rearranged by acid to the aromatic 1-chloro-4-methyl-1,3,5(10)-triene (3).[23] The rearrangement product (3) can be obtained directly from (1) by extending the reaction time, addition of oxalic acid, and use of chloroform as solvent.

[1]H. Staudinger, *Ber.*, **41**, 3563 (1908)
[2]M. S. Kharasch and H. C. Brown, *Am. Soc.*, **64**, 329 (1942)
[3]R. Adams and L. H. Ulich, *Am. Soc.*, **42**, 599 (1920)
[4]A. L. Wilds and C. H. Shunk, *Am. Soc.*, **70**, 2427 (1948)
[5]F. Reber, A. Lardon, and T. Reichstein, *Helv.*, **37**, 45 (1954)
[6]Ch. R. Engel and G. Just, *Can. J. Chem.*, **33**, 1515 (1955)

[7]J. Cason, *Am. Soc.*, **69**, 1548 (1947)
[8]S. Ställberg-Stenhagen, *Am. Soc.*, **69**, 2568 (1947)
[9]J. E. H. Hancock and R. P. Linstead, *J. Chem. Soc.*, 3490 (1953)
[10]J. Szmuszkovicz, *J. Org.*, **29**, 843 (1964)
[11]M. Miyano, *Am. Soc.*, **87**, 3958 (1965)
[12]M. S. Kharasch, S. S. Kane, and H. C. Brown, *Am. Soc.*, **64**, 333 (1942)
[13]F. Bergmann, M. Weizmann, E. Dimant, J. Patai, and J. Szmuszkovicz, *Am. Soc.*, **70**, 1612 (1948)
[14]H. G. Latham, Jr., E. L. May, and E. Mosettig, *Am. Soc.*, **70**, 1079 (1948)
[15]P. E. Sokol, *Org. Syn.*, **44**, 69 (1964)
[16]R. N. McDonald, *J. Org.*, **24**, 1580 (1959)
[17]C. Tüzün, M. Ogliaruso, and E. I. Becker, *Org. Syn.*, **41**, 3 (1961)
[18]S. J. Rhoads and R. E. Michel, *Am. Soc.*, **85**, 585 (1963)
[19]A. J. Speziale and L. R. Smith, *J. Org.*, **27**, 3742 (1962); **28**, 1805 (1963); *Org. Syn.*, **46**, 16 (1966)
[20]A. J. Speziale, L. R. Smith, and J. E. Fedder, *J. Org.*, **30**, 4306 (1965)
[21]R. Deghenghi and R. Gaudry, *Can. J. Chem.*, **40**, 818 (1962)
[22]G. W. Moersch and W. A. Neuklis, *Can. J. Chem.*, **41**, 1627 (1963)
[23]G. W. Moersch, W. A. Neuklis, T. P. Culbertson, D. F. Morrow, and M. E. Butler, *J. Org.*, **29**, 2495 (1964)

Oxygen, heavy. YEDA Research and Development Co., Ltd., produces O^{17} and O^{18} enriched oxygen gas, water, and labeled compounds. Water depleted in O^{17} and O^{18} is also available.

Oxygen difluoride, OF_2. Mol. wt. 54.00, m.p. −223.8°, b.p. −144.8°. Supplier: Allied Chem. Co.

The reagent is prepared by passing fluorine gas into a 2% solution of sodium hydroxide.[1] The effluent gas is a mixture of about equal parts of OF_2 and O_2. Merritt and Ruff[2] trapped OF_2 out of the fluorine stream at liquid oxygen temperature, distilled off oxygen and fluorine, and so obtained material about 99% pure, which was stored at room temperature in large-volume glass bulbs. They found that oxygen difluoride oxidizes primary aliphatic amines smoothly at −78° in Freon 11 as solvent to the nitroso derivatives:

$$3\ RNH_2 + OF_2 \longrightarrow RN{=}O + 2\ R\overset{+}{N}H_3\overset{-}{F}$$

These investigators report addition of the reagent to acetylenes and to some olefins at −78° (Freon 11 as solvent).[3] Thus tetramethylethylene gives the fluorohydrin (2) in moderate yield; the extra proton presumably is derived from the solvent. On brief exposure to moist air the fluorohydrin is rearranged to pinacolone

$$\underset{(1)}{(CH_3)_2C{=}C(CH_3)_2} \xrightarrow[65\text{-}70\%]{OF_2\text{ at }-78^0} \underset{(2)}{(CH_3)_2\underset{OH}{\underset{|}{C}}{-}\underset{F}{\underset{|}{C}}(CH_3)_2} \xrightarrow{H_2O} \underset{(3)}{(CH_3)_3CCOCH_3} + HF$$

(3) and hydrogen fluoride. The only significant product obtained from 1,1-diphenylethylene was ω-fluoroacetophenone (4). Acetylenes react to give α,α-difluoroketones (5).

$$(C_6H_5)_2C{=}CH_2 \xrightarrow{OF_2} C_6H_5\overset{O}{\overset{\|}{C}}CH_2F \quad (4)$$

$$RC{\equiv}CR \xrightarrow{OF_2} RCOCF_2R' \quad (5)$$

[1]D. M. Yost, *Inorg. Syn.*, **1**, 109 (1939)
[2]R. F. Merritt and J. K. Ruff, *Am. Soc.*, **86**, 1392 (1964)
[3]*Idem*, *J. Org.*, **30**, 328 (1965)

Ozone, O_3, a blue, unstable gas; m.p. −192.5° (to a dark blue liquid); b.p. −111.9°.

Commercial equipment. The Welsbach Corp., Ozone Process Division, offers Laboratory Ozonators which produce ozone concentrations up to 4% by weight in air and up to 8% by weight in oxygen. Accessory units are as follows. An ozone concentration meter based on the principle of the thermal conductivity cell continuously measures ozone concentration in the carrier gas. A strip-chart recorder is designed specifically for use with the ozone meter. A portable air dryer is a self-contained unit consisting of a compressor and dryer on a common portable base; no attendance is needed.

A 49-page Welsbach bulletin, "Organic Ozone Reactions and Techniques" (4th Revision, Jan. 1, 1962), with 91 references presents much useful information on mechanism, scope, and experimental conditions, and describes laboratory reactors available from Ace Glass, Inc., and Labglass, Inc. Welsbach's "Basic Manual of Applications and Laboratory Ozonization Techniques" describes properties and applications of ozone and suggestions for laboratory studies and analysis.

Construction of ozone generators. Smith *et al.*[1] give a detailed description with diagrams of a laboratory ozonizer of the type introduced by Harries with three Berthelot tubes in series. Since the maximum concentration of ozone tolerated is 0.15 to 1 part per million in air, Smith passes the effluent gas through a catalytic ozone destroyer consisting of two 30-in. towers in series containing broken glass moistened with 5% sodium hydroxide solution. Henne and Hill[2] used a generator containing three Berthelot tubes and passed a stream of 6% ozonized oxygen into 0.5 mole of 6-methyl-1-heptene in 200 ml. of methylene chloride at −78° at the rate of 20 l. per hr. for 12 hrs. The resulting solution was added gradually to 32.5 g. of zinc dust and 300 ml. of 50% acetic acid; the decomposition of ozonide was exothermal and the methylene chloride distilled. After refluxing and stirring, the mixture was extracted with ether, peroxides were destroyed by washing with potassium iodide solution, and fractionation afforded 5-methylhexanol (62%), 5% of 5-methylhexanoic acid and 9% of polymer. Because of undue heating of the electrodes and because of the fragility of the Berthelot tubes, Henne and Perilstein[3] designed a new ozonizer sturdily built of Pyrex glass and featuring a water-cooled electrode. Decomposition of the ozonide was accomplished by hydrogenation in the presence of 1% palladium supported on calcium carbonate. To minimize handling, ozonization and hydrogenation of the ozonide were carried out in the same solvent and the same container. Methylene chloride could not be used because it prevents hydrogenation. Ethyl acetate and ethanol were preferred as solvents. Yields of simple aldehydes and ketones from olefins were in the range 30–60%.

Dessy and Newman[4] used the generator of Henne and Perilstein for ozonization of technical pyrene. A solution of 25 g. of the hydrocarbon in 100 ml. of dimethylformamide was treated with ozone in 50% excess over a period of about 6 hrs. The solution of ozonide was added with stirring to 500 ml. of 1% aqueous acetic acid to produce a fine suspension which on standing overnight coagulated to a granular brown solid. This was collected and extracted repeatedly with 10% aqueous potassium hydroxide, and the dark brown filtrate was treated with potassium hypochlorite

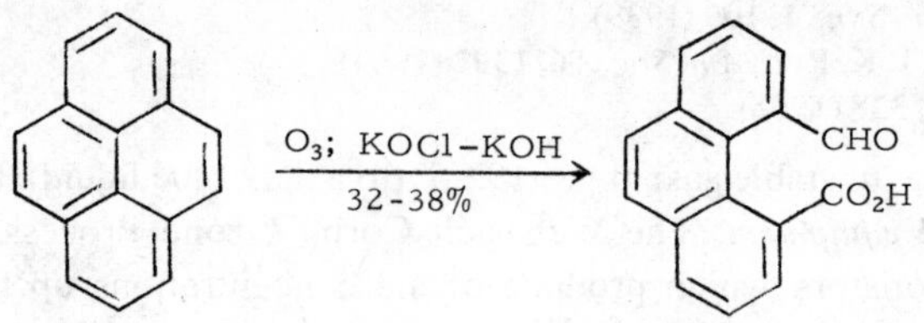
O3; KOCl–KOH
32–38%
CHO
CO2H

solution and let stand overnight. On being heated on the steam bath for 6 hrs. the color changed to orange. 5-Formyl-4-phenanthroic acid was then precipitated as the potassium salt, liberated from the salt, and crystallized.

For ozonization of phenanthrene to produce diphenyl-2,2′-dialdehyde, Bailey and Erickson[5] dried oxygen with a Pittsburgh Laboratory-Lectrodryer and passed it through a Welsbach ozonator. The checkers dried the oxygen by passage through a 30-cm. column of silica gel and used a simple generator capable of producing 3.8% ozone by weight at a flow of 20 l. per hr. A solution of 10 g. of phenanthrene

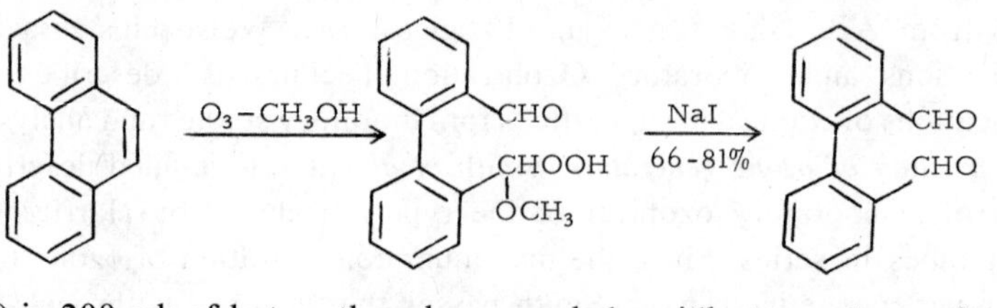
O3–CH3OH
CHO
CHOOH
OCH3
NaI
66–81%
CHO
CHO

(0.056 m.) in 200 ml. of hot methanol was cooled rapidly to produce a fine suspension, and this was transferred to a long, cylindrical gas-absorption vessel with an inlet tube extending over the bottom and cooled to about −30° in a Dewar flask containing dry ice and acetone. Ozonization was continued until all the suspended hydrocarbon had reacted, which required 1.1–1.3 equivalents of ozone. A solution of 0.17 mole of sodium iodide in 30 ml. of acetic acid was added to reduce the peroxidic product of ozonolysis and methanolysis, and after 1 hr. the released iodine was reduced with sodium thiosulfate solution. On evaporation of the methanol under an air blast the product separated. Water was added, and the product collected and crystallized from ether–ligroin to give pale yellow crystals, m.p. 62–63°.

Schmidt and Grafen[6] found that ozonization of the enol ethers of readily available cycloalkanones provides a convenient route to ω-aldehyde acids. The ketone (1) is converted via the diethyl acetal (2) into the enol ether (3), this is ozonized at 0°, and the ozonide decomposed by hydrogenation. The sensitive aldehyde ester

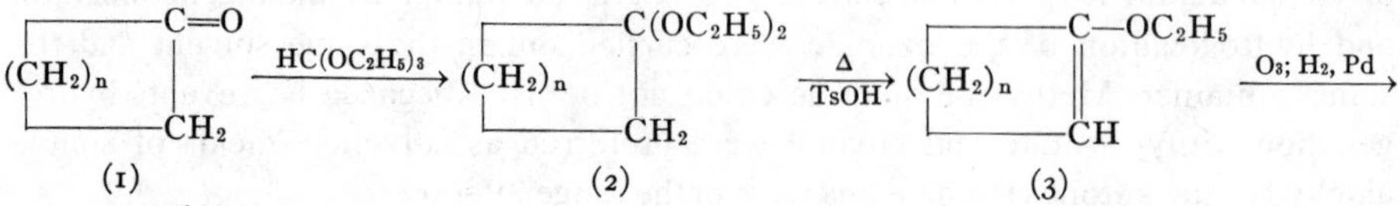
C=O
(CH2)n
CH2
HC(OC2H5)3
C(OC2H5)2
(CH2)n
CH2
Δ
TsOH
C—OC2H5
(CH2)n
CH
O3; H2, Pd

(1) (2) (3)

n = 3, 4, 5, 6, 10

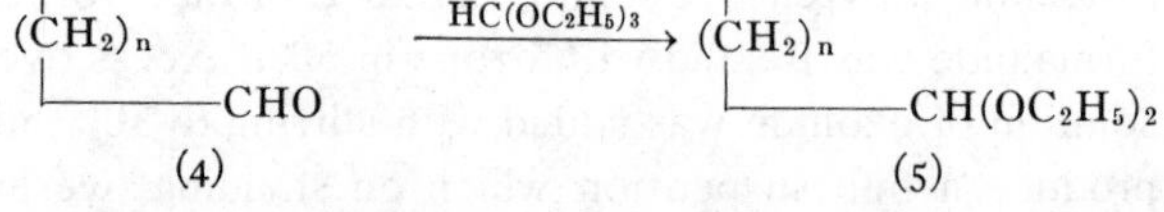
CO2C2H5
(CH2)n
CHO
HC(OC2H5)3
CO2C2H5
(CH2)n
CH(OC2H5)2

(4) (5)

(4) is isolated as the diethyl acetal (5). A competitive experiment in which a mixture of one equivalent each of an enol ether (1-ethoxycyclohexene) and an analogous

olefin (1-*n*-propylcyclohexene) was ozonized at −80° with 0.75 equivalent of oxidant showed that 80% of the material oxidized came from the enol ether component.

Fremery and Fields[7] developed a process for reaction of a cycloolefin with ozone in emulsion with aqueous alkaline hydrogen peroxide to give the diacid in one step, generally in good yield. The emulsifying agent, a polyoxyethylated lauryl alcohol (Atlas Co.), is dissolved in the olefin, and the solution is added to an aqueous solution of sodium hydroxide and hydrogen peroxide in an indented flask fitted with a high-speed stirrer; oxygen containing about 3% of ozone is then run in. Typical examples are as follows:

Cyclooctene ⟶ suberic acid
cis, cis-1,5-Cyclooctadiene (large excess) ⟶ Δ^3-hexene-1,6-dicarboxylic acid
Indene ⟶ homophthalic acid
Butadiene–maleic anhydride adduct ⟶ 1,2,3,4-tetracarboxybutane

In one step in the degradation of biosynthetic cholesterol from labeled acetate to establish the pattern of distribution in ring A, Cornforth, Hunter, and Popják[8] succeeded in converting Δ^5-cholestene into the ozonide in 80% yield by passing ozone into a solution of the hydrocarbon in dry *n*-hexane until a dilute solution of

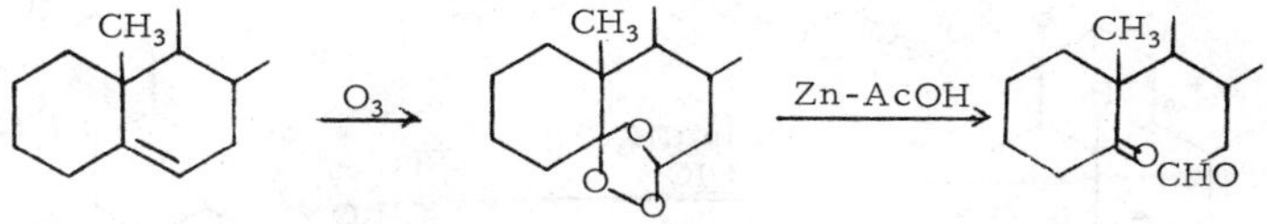

CH3
O3
CH3
Zn-AcOH
CH3
CHO

bromine in acetic acid was no longer rapidly decolorized on mixing with a few drops of the *n*-hexane solution. Centrifugation, washing, and drying gave a glassy ozonide which was reduced in good yield to the corresponding keto aldehyde by shaking a solution in acetic acid with zinc dust for several hours at 25° until iodine was no longer liberated on addition of a crystal of potassium iodide to a sample of the solution.

In developing a highly efficient route from stigmasterol to progesterone (4),

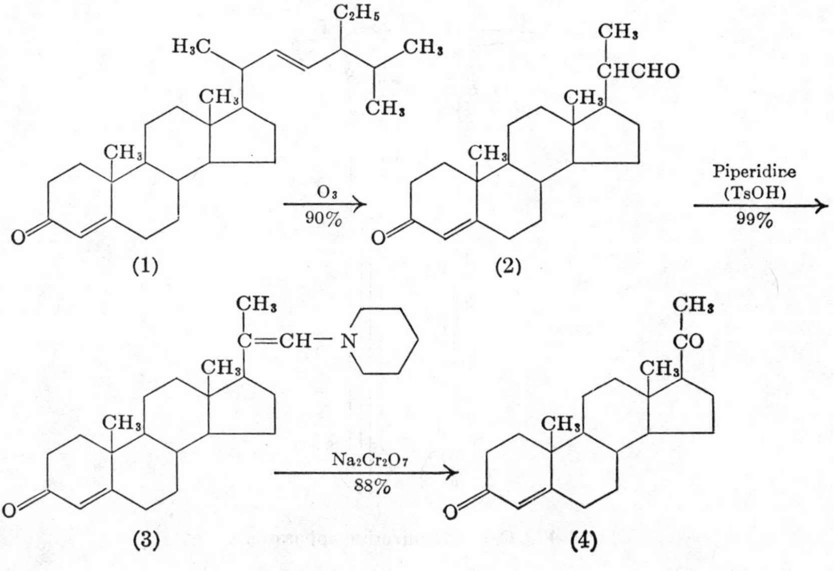

C2H5
H3C
CH3
CH3
CH3
CH3
O3
90%
CH3
CHCHO
CH3
CH3
Piperidine
(TsOH)
99%
(1)
(2)
CH3
C=CH—N
CH3
CH3
Na2Cr2O7
88%
CH3
CO
CH3
CH3
(3)
(4)

Heyl and Herr[9] of Upjohn found that stigmastadienone (1) prepared by Oppenauer oxidation of stigmasterol is attacked by ozone preferentially in the side chain to produce the bisnoraldehyde (2). Slomp and Johnson[10] studied the ozonolysis of stigmastadienone by measuring the rate of decrease in intensity of the band of 10.26 μ due to the side-chain double bond; the rate of reaction of the nuclear double bond was shown by the change in absorption of the Δ^4-ene-3-one system at 6.0 μ. They found that pyridine, added in amount of 1% by volume to a solution of (1) in methylene chloride, markedly increases the selectivity obtainable, and they defined conditions permitting production of the aldehyde (2) in 90% yield. The aldehyde is condensed with piperidine with *p*-toluenesulfonic acid as catalyst, and the enamine (3) is oxidized to progesterone with sodium dichromate in anhydrous acetic acid–benzene.[11]

The use of tetracyanoethylene as buffer in the ozonization of olefins is described in the section on this reagent.

In initial attempts by Johnson and co-workers[12] to convert the furfurylidene ketone (5) into the dibasic acid (6) by ozonization, a reaction which with saturated steroids proceeds in excellent yield, yields were low and variable, probably because excess

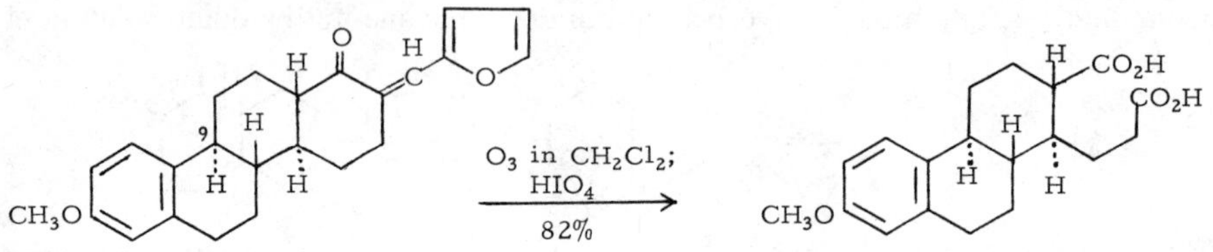
CH₃O
O₃ in CH₂Cl₂;
HIO₄
82%
CO₂H
CO₂H

(5) (6)

ozone attacks the benzylic position C_9. Indeed the situation was remedied by saturating an appropriate volume of methylene chloride with ozone at −78° in chamber B of the special ozonization apparatus[13] of Fig. O-1 and forcing the solution under nitrogen pressure into chamber A containing a magnetic stirring bar

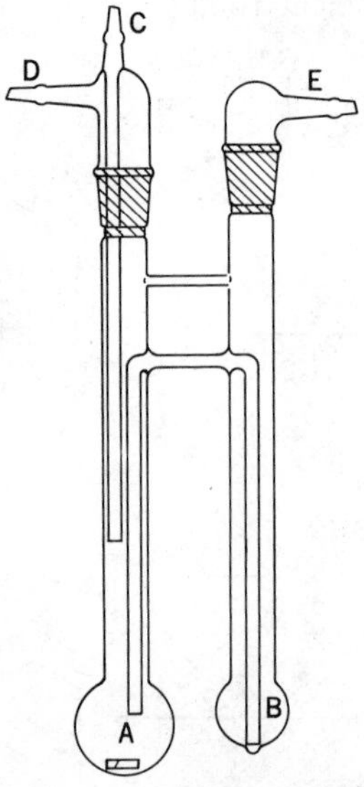
C
D
E
A
B

Fig. O-1 Ozonization apparatus

and a solution of the substrate in methylene chloride, the quantities being adjusted to avoid an excess of ozone. Discharge of the blue color of ozone was instantaneous, and after decomposition of the ozonide with periodic acid, the diacid (6) was isolated in high and reproducible yield. A dialdehyde was obtained satisfactorily from a diolefin by the same technique.[14]

Stille and Foster[15] studied the ozonization of acenaphthylene (1) to naphthalene-1,8-dialdehyde hydrate (2) and found that with ozone in a stream of nitrogen a yield of 73.5% can be realized, whereas with the usual oxygen–ozone mixture the yield was only 16.5%.

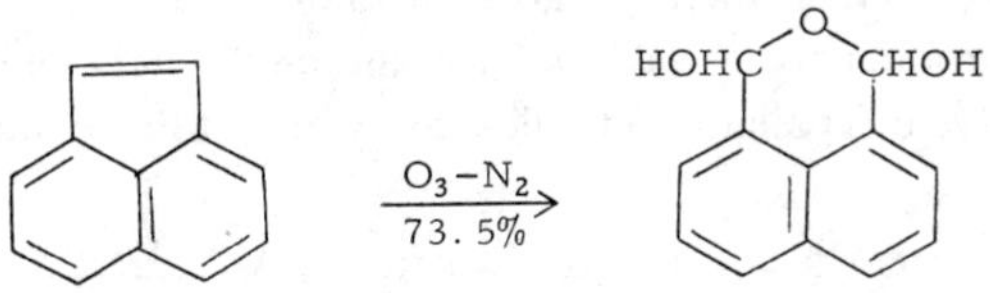
HOHC
CHOH
O
O_3-N_2
73.5%

(1) (2)

[1]L. I. Smith, F. L. Greenwood, and O. Hudrlik, *Org. Syn., Coll. Vol.*, **3**, 673 (1955)
[2]A. L. Henne and P. Hill, *Am. Soc.*, **65**, 752 (1943)
[3]A. L. Henne and W. L. Perilstein, *Am. Soc.*, **65**, 2183 (1943)
[4]R. E. Dessy and M. S. Newman, *Org. Syn., Coll. Vol.*, **4**, 484 (1963)
[5]P. S. Bailey and R. E. Erickson, *ibid.*, **41**, 41 (1961)
[6]U. Schmidt and P. Grafen, *Ann.*, **656**, 97 (1962)
[7]M. I. Fremery and E. K. Fields, *J. Org.*, **28**, 2537 (1963)
[8]J. W. Cornforth, G. D. Hunter, and G. Popják, *Biochem. J.*, **54**, 590 (1953)
[9]F. W. Heyl and M. E. Herr, *Am. Soc.*, **72**, 2617 (1950)
[10]G. Slomp, Jr., and J. L. Johnson, *Am. Soc.*, **80**, 915 (1958)
[11]M. E. Herr and F. W. Heyl, *Am. Soc.*, **74**, 3627 (1952); D. A. Shepherd *et al.*, *ibid.*, **77**, 1212 (1955)
[12]W. L. Meyer, D. D. Cameron, and W. S. Johnson, *J. Org.*, **27**, 1130 (1962)
[13]M. B. Rubin, *J. Chem. Ed.*, **41**, 388 (1964)
[14]W. S. Johnson, M. B. Rubin, *et al.*, *Am. Soc.*, **85**, 1409 (1963)
[15]J. K. Stille and R. T. Foster, *J. Org.*, **28**, 2703 (1963)

P

Palladium acetate, $Pd(OCOCH_3)_2$. Brown crystals from palladium nitrate and acetic acid or by dissolving Pd sponge in hot acetic acid containing a deficiency of HNO_3.[1] Supplier: Engelhard Industries.

The reagent oxidizes a Δ^1-olefin to give the enol acetate as the main product, and the predominant product from a Δ^2-olefin is the allylic acetate.[2] The reactions are considered to involve oxypalladation followed by loss of the elements of HPdOAc.

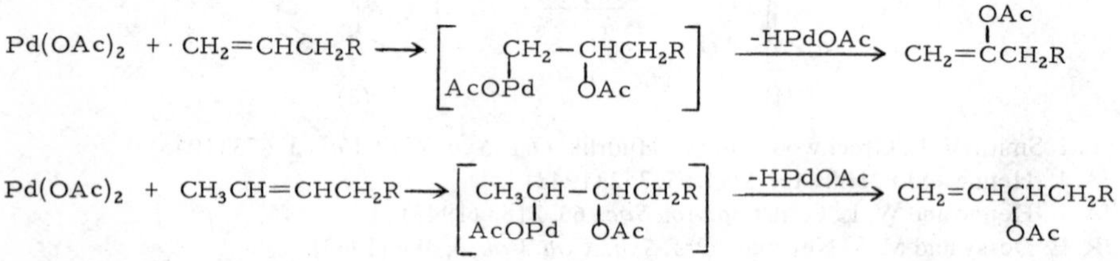

[1]S. M. Morehouse, A. R. Powell, J. P. Heffer, T. A. Stephenson, and G. Wilkinson, *Chem. Ind.*, 544 (1964).
[2]W. Kitching, Z. Rappoport, S. Winstein, and W. G. Young, *Am. Soc.*, **88**, 2054 (1966)

Palladium catalysts, *see also* Lindlar catalyst. Engelhard Industries supplies 5% palladium catalyst on C, Al, Al_2O_3, $BaCO_3$, $BaSO_4$, or $CaCO_3$; also pelleted, granular catalysts, and unsupported palladium oxide and palladium black. MacKay supplies 10% Pd-C catalyst.

Mozingo[1] describes the preparation of four catalysts:

A 5% Pd on $BaSO_4$
B 5% Pd on C
C $PdCl_2$ on C (5% Pd)
D 10% Pd on C

Catalysts A and B are prepared by reduction with alkaline formaldehyde, carry no adsorbed hydrogen, and are less pyrophoric than D, prepared by reduction with

$$Na_2PdCl_4 + CH_2O + 3\,NaOH \longrightarrow Pd + HCO_2Na + 4\,NaCl + 2\,H_2O$$

hydrogen ($PdCl_2 + H_2 \longrightarrow Pd + HCl$). With catalyst C the palladium salt is reduced by hydrogen as needed and there is no loss in activity during storage.

Cram and Allinger[2] reported that a Pd-$BaSO_4$ catalyst partially poisoned with quinoline is superior in reproducibility to the Lindlar catalyst for partial hydrogenation of the triple bond.

Hydrogenolysis. Rosenmund[3] found that Pd-$BaSO_4$ catalyst can be markedly activated for the hydrogenolysis of alcohols of the type ArCH(OH)— by addition of

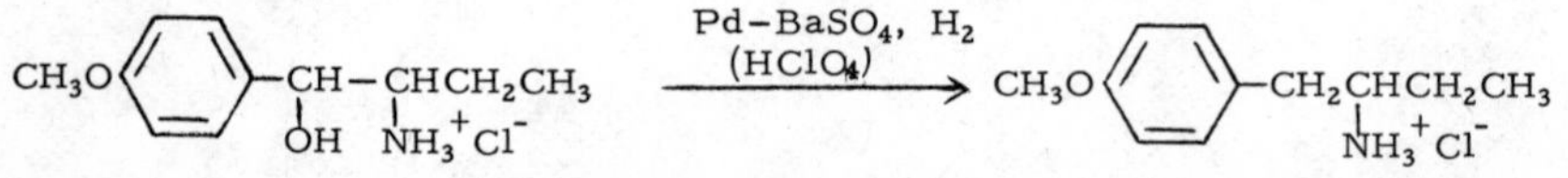

(1) (2)

a little perchloric acid. Thus hydrogenation of (1) to (2) in acetic acid was greatly accelerated by addition of a small volume of a solution of 70% perchloric acid in acetic acid. Kindler[4] found sulfuric acid to be an effective catalyst for the conversion of acetophenone into ethylbenzene. Palladium black was washed with dilute sulfuric

$$C_6H_5COCH_3 \xrightarrow{H_2,\ Pd\ (H_2SO_4)} C_6H_5CHOHCH_3 \longrightarrow C_6H_5CH_2CH_3$$

acid and then washed with water and with methanol until the wash solvent was no longer acidic and so obtained as a highly activated catalyst.

Dehydrogenation. Linstead[5] found that dehydrogenation of hydroaromatic hydrocarbons with palladium charcoal in the liquid phase is promoted by passing carbon dioxide into the liquid to entrain and remove the evolved hydrogen from the equilibrium mixture and also by vigorous boiling, which helps to dislodge the hydrogen from the active surface of the catalyst. The course of the reaction can be followed by absorbing the carbon dioxide and measuring the hydrogen. Tetralin, b.p. 208°, is dehydrogenated in the liquid phase only when it is actually boiling. Dehydrogenation can be effected at 185° when the liquid is made to boil by reducing the pressure or by adding a lower-boiling diluent, but no reaction occurs in the tranquil liquid at 200°. Both *cis*- and *trans*-decalin are dehydrogenated only very slowly at the boiling point (*ca.* 190°), but can be dehydrogenated satisfactorily in the vapor phase at 300°.

Fried and co-workers[6] at Squibb, in the process of degrading eburicoic acid (1) to 14-methylpregnane, converted the acid as acetate into the norketo acid (2) and

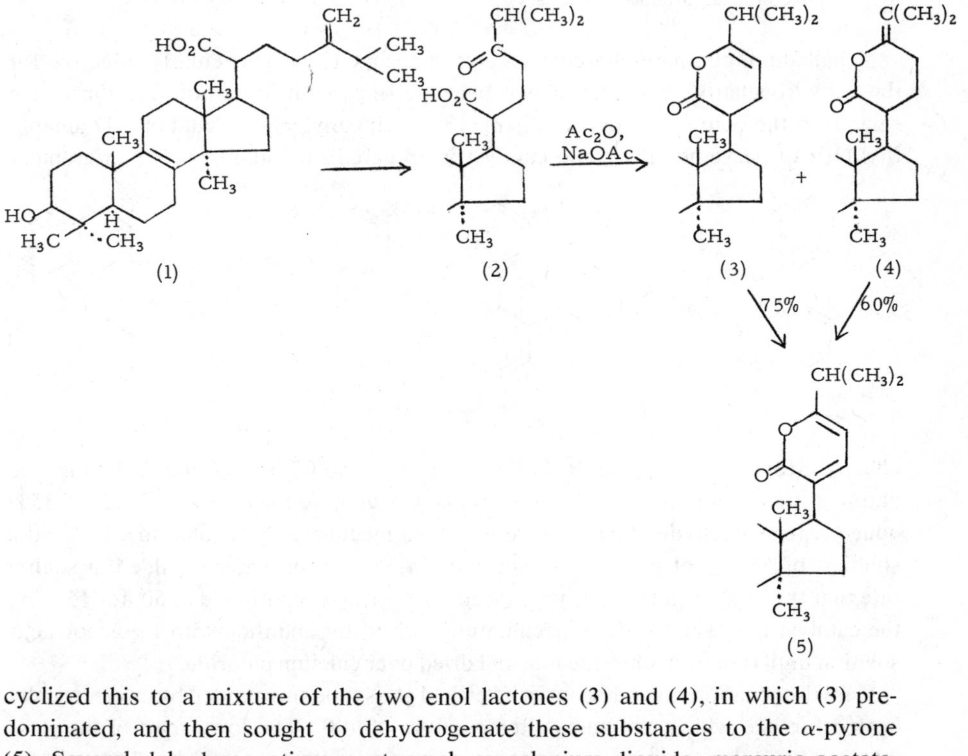
CH2
HO2C
CH3
CH3
HO
H3C
H
CH3
CH(CH3)2
HO2C
Ac2O,
NaOAc
CH(CH3)2
C(CH3)2
(1)
(2)
(3)
(4)
75%
60%
CH(CH3)2
(5)

cyclized this to a mixture of the two enol lactones (3) and (4), in which (3) predominated, and then sought to dehydrogenate these substances to the α-pyrone (5). Several dehydrogenating agents such as selenium dioxide, mercuric acetate, and bromine were tried; it was then found that the reaction can be performed cleanly and in good yield with 10% palladium-on-charcoal in boiling *p*-cymene (b.p. 177°). The endocyclic lactone (3) afforded the pure α-pyrone (5) in 75% yield. The exocyclic lactone (4) could likewise be dehydrogenated to (5), but in this case the yield was only 60% and the time had to be extended from 2 to 6 hrs.; presumably the slow step is migration of the double bond to give (3). Wintersteiner and Moore[7] of Squibb found the method useful in effecting dehydrogenation of (6), a degradation product of isojervine, to the aromatic veratramine derivative (7).

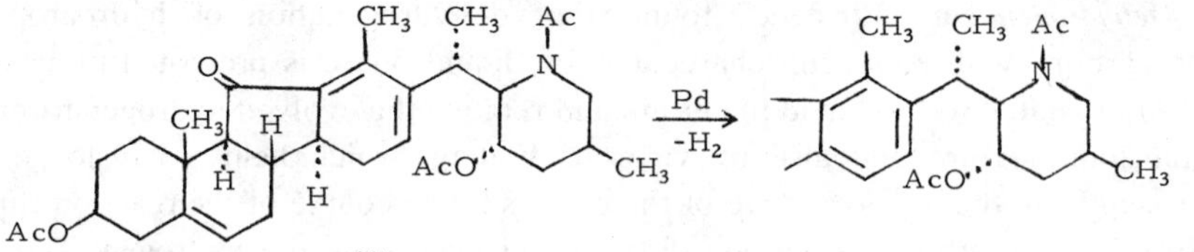
Pd
-H₂

(6) (7)

1-Phenanthrol can be prepared satisfactorily by refluxing the hydroaromatic ketone with palladium catalyst in naphthalene for 24 hrs.[7a]

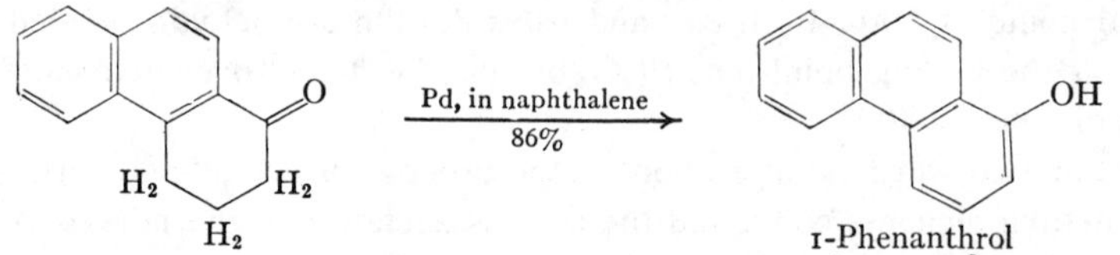
Pd, in naphthalene
86%

1-Ketotetrahydrophenanthrene (m.p. 96°) 1-Phenanthrol (m.p. 157°)

A palladium-platinum-charcoal catalyst appears to be particularly effective for the dehydrogenative coupling of two benzene rings with formation of a third such ring, as in the conversion of *o*-terphenyl (8) to triphenylene (9),[8] and of 1,1′-dinaphthyl (10) to perylene (11).[9] The catalyst is prepared[10] by adding 400 g. of granular

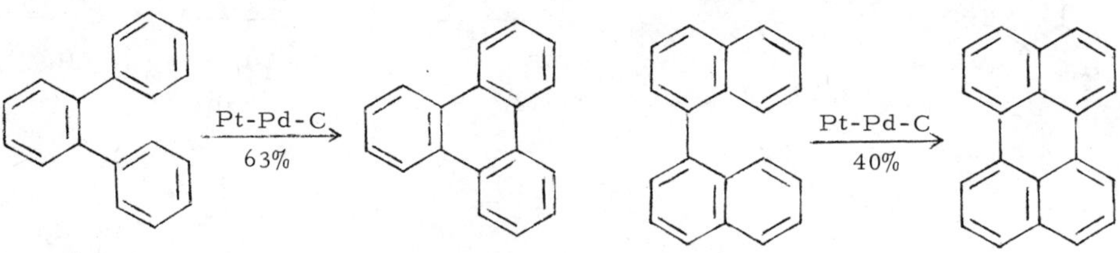
Pt-Pd-C
63%
Pt-Pd-C
40%

(8) (9) (10) (11)

charcoal (for gas absorption, B. D. H.) to a solution of 7.3 g. of chloroplatinic acid and 4 g. of palladium chloride in 400 ml. of water, followed by 635 ml. of 35% aqueous formaldehyde. The mixture is stirred mechanically, cooled to −10°, and a solution of 580 g. of potassium hydroxide in 580 ml. of water is added at such a rate that the temperature does not exceed 5°. Stirring is continued at 60° for 15 min., the catalyst is washed well by decantation with water and then with 1% acetic acid solution until free from chloride ion, and dried over calcium chloride.

Dehydrogenation is carried out in a vertical glass tube surrounded by an electrically heated furnace which maintains a temperature of 490°. The lower part of the glass tube contains the catalyst; the upper part, packed with porcelain beads, serves as preheater and vaporizer. A solution of 4.5 g. of *o*-terphenyl in 15 ml. of purified decalin is passed over the hot catalyst together with hydrogen during 3 hrs. The reaction mixture is chromatographed.

Crawford and co-workers[10a] discovered a surprising combination for effecting dehydrogenation: palladium charcoal and sulfur. The Diels-Alder adduct (3) was obtained by pinacol reduction of the tetralone (1), dehydration to (2) by refluxing

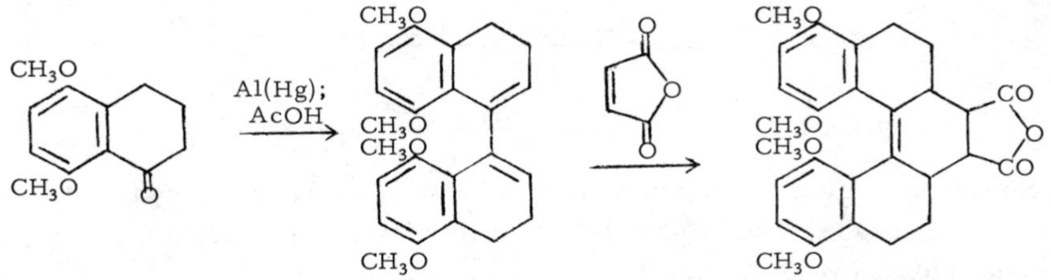

CH_3O
CH_3O
Al(Hg);
AcOH
CH_3O
CH_3O
CH_3O
CH_3O
CH_3O
CH_3O
CH_3O
CH_3O
CO
O
CO

(1) (2) (3)

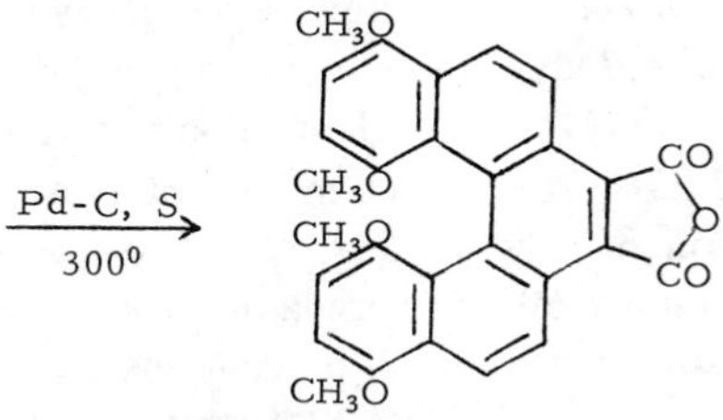

Pd-C, S
300°
CH_3O
CH_3O
CH_3O
CH_3O
CO
O
CO

(4)

with acetic acid, and reaction with maleic anhydride. Attempts to dehydrogenate (3) with palladium charcoal or sulfur alone were unsuccessful. However when an intimate mixture of 0.5 g. of (3), 0.05 g. of 30% Pd-C, and 0.15 g. of sulfur was heated in an oil bath at 300° for 5 min. hydrogen sulfide was evolved, and extraction with chloroform and crystallization from acetic anhydride afforded 0.22 g. of deep red prisms of (4), m.p. 323.5°.

Sulfur or Pd-C alone dehydrogenates (5) to (6) at 300–310°, but at this temperature the combination of the two affords a mixture of equal parts of (6) and (7). Since

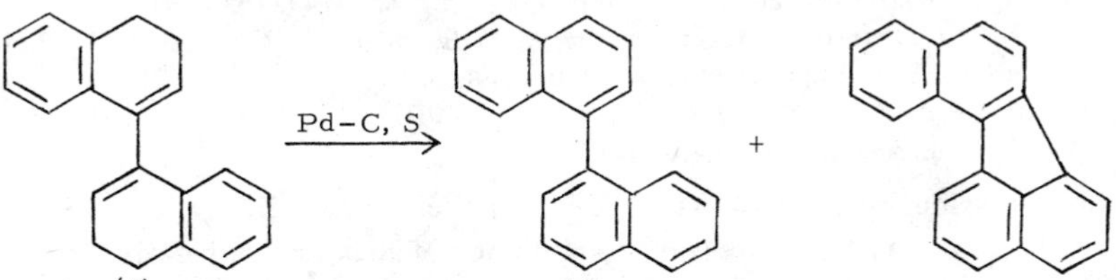

Pd-C, S
+

(5) (6) (7)

1,1′-dinaphthyl (6) is not dehydrogenated to (7) under the conditions of the experiment, cyclodehydrogenation leading to (7) must occur prior to complete aromatization. A tentative explanation of the operation of the catalyst is that palladium abstracts hydrogen atoms, leaving a diradical which cyclizes. Without the sulfur the palladium would rapidly become saturated with hydrogen, but sulfur removes hydrogen from the metal and gives hydrogen sulfide. Hydrogenations can be effected by hydrogen transfer in the presence of a metal catalyst from a hydrogen-rich donor;

the present reaction represents the converse: dehydrogenation by a substance poor in hydrogen, *i.e.*, sulfur.

Decarbonylation of aldehydes. Some aromatic aldehydes are decarbonylated when heated with 5% palladium on charcoal, for example:[11]

$$C_6H_5CHO \xrightarrow[78\%]{\text{Pd-C, 1 hr. at } 179^\circ} C_6H_6 + CO$$

$$\underline{o}\text{-}CH_3OC_6H_4CHO \xrightarrow[94\%]{\text{Pd-C, 1 hr. at } 243^\circ} C_6H_5OCH_3 + CO$$

$$\underline{p}\text{-}O_2NC_6H_4CHO \xrightarrow[77\%]{\text{Pd-C, 0.5 hr. at } 205^\circ} C_6H_5NO_2 + CO$$

The reaction does not proceed satisfactorily with aliphatic aldehydes and with some aryl aldehydes. Thus 1-naphthaldehyde is converted into naphthalene at 210°, but 2-naphthaldehyde, even at 250–255°, gives only a trace of gas.

Reduction with hydrazine and Pd-C, *see* Hydrazine–Metal catalyst.

Palladium hydroxide on carbon.[12] This nonpyrophoric, dry hydrogenation catalyst described by Pearlman (Parke, Davis and Co.) is prepared by stirring a mixture of 100 g. of $PdCl_2$, 240 g. of Darco G–60, and 2 l. of deionized water at 80°, adding all at once a solution of 50 g. of $LiOH \cdot H_2O$ in 200 ml. of water, stopping the heating, but continuing the stirring overnight. The solid is filtered, washed liberally with dilute acetic acid, and dried in vacuum at 60°. The preparation is wetted with solvent before use; even with an organic solvent no special precautions are required. The catalyst formed on reduction of the metal hydroxide is extremely active.

[1]R. Mozingo, *Org. Syn.*, *Coll. Vol.*, **3**, 685 (1955)
[2]D. J. Cram and N. L. Allinger, *Am. Soc.*, **78**, 2518 (1956)
[3]K. W. Rosenmund and E. Karg, *Ber.*, **75**, 1850 (1942)
[4]K. Kindler, E. Schärfe, and P. Henrich, *Ann.*, **565**, 51 (1949)
[5]R. P. Linstead, A. F. Millidge, S. L. S. Thomas, and A. L. Walpole, *J. Chem. Soc.*, 1146 (1937); R. P. Linstead and S. L. S. Thomas, *ibid.*, 1127 (1940)
[6]D. Rosenthal, P. Grabowich, E. F. Subo, and J. Fried, *Am. Soc.*, **85**, 3971 (1963)
[7]O. Wintersteiner and M. Moore, *Tetrahedron Letters*, 795 (1962)
[7a]E. Mosettig and H. M. Duvall, *Am. Soc.*, **59**, 387 (1937)
[8]P. G. Copeland, R. E. Dean, and D. McNeil, *J. Chem. Soc.*, 1687 (1960)
[9]*Idem*, *ibid.*, 1689 (1960)
[10]W. Baker, W. K. Warburton, and L. J. Breddy, *J. Chem. Soc.*, 4149 (1953)
[10a]H. S. Blair, M. Crawford, J. M. Spence, and V. R. Supanekar, *J. Chem. Soc.*, 3313 (1960); M. Crawford and V. R. Supanekar, *ibid.*, 2380 (1964)
[11]J. O. Hawthorne and M. H. Wilt, *J. Org.*, **25**, 2215 (1960)
[12]W. M. Pearlman, procedure submitted to *Org. Syn.*

Palladium chloride, $PdCl_2$. Mol. wt. 177.32. Suppliers: F, MCB.

A new route to an isocyanate is by combination of an amine with carbon monoxide in the presence of palladium chloride:[1]

$$\underline{n}\text{-}C_4H_9NH_2 + CO + PdCl_2 \xrightarrow[49.2\%]{65^\circ,\ 1\text{ atmosph.},\ 48\text{ hrs.}} \underline{n}\text{-}C_4H_9N{=}C{=}O + Pd + 2\,HCl$$

[1]E. W. Stern and M. L. Spector, *J. Org.*, **31**, 596 (1966)

Pentachlorophenol, C_6Cl_5OH. Mol. wt. 266.34, m.p. 189–191°. Suppliers: A, B, E, MCB.

Polypeptide synthesis. Coupling an amino acid with pentachlorophenol by the dicyclohexylcarbodiimide method gives an active ester which readily polymerizes in dimethylformamide solution in the presence of a tertiary amine to give a polyamino acid.[1] The polymers obtained by this method have higher molecular weights than those produced by other methods. This active-ester method is useful for the synthesis of polypeptides having a repeating sequence of amino acids. For example, a dipeptide is converted into the active ester and this is polymerized.

[1]J. Kovacs and A. Kapoor, *Am. Soc.*, **87**, 118 (1965); J. Kovacs, R. Ballina, R. L. Rodin, D. Balasubramanian, and J. Applequist, *ibid.*, 119 (1965); J. Kovacs and M. Q. Ceprini, *Chem. Ind.*, 2100 (1965); J. Kovacs and B. J. Johnson, *J. Chem. Soc.*, 6777 (1965); J. Kovacs, R. Giannotti, and A. Kapoor, *Am. Soc.*, **88**, 2282 (1966)

1,4-Pentadiene, $CH_2{=}CHCH_2CH{=}CH_2$. Mol. wt. 68.11, b.p. 26–27.5°. Suppliers: Aldrich, Baker.

Preparation (a).[1] 1,5-Pentanediol (Suppliers: Union Carbide, Eastman) is acetylated, and the diacetate is dropped into a glass reaction tube mounted vertically in an electric furnace. The heated section of the tube is packed with glass rings

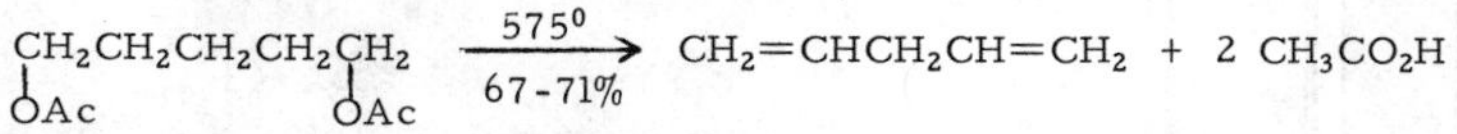
CH₂CH₂CH₂CH₂CH₂
OAc
OAc
575⁰
67-71%
CH₂=CHCH₂CH=CH₂ + 2 CH₃CO₂H

held in place by a plug of glass wool supported by indentations in the tube. Attached at the top of the tube are a nitrogen inlet and a Hershberg dropping tube[2] modified by addition of a pressure-equalizing arm (Fig. P-2 or Fig. P-3). The delivery tube is immersed in a 1-l. round-bottomed flask immersed in an ice bath and having a side arm from which vapors pass successively through a trap immersed in an ice bath and a trap immersed in dry ice–acetone. The temperature at the middle of the tube is raised to 575° while nitrogen is passed through a flowmeter into the tube; then 645 ml.

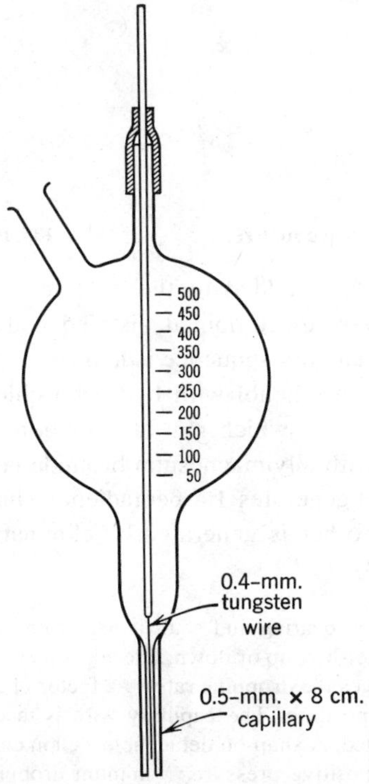
500
450
400
350
300
250
200
150
100
50
0.4-mm. tungsten wire
0.5-mm. × 8 cm. capillary

Fig. P-1 Hershberg dropping tube[2]

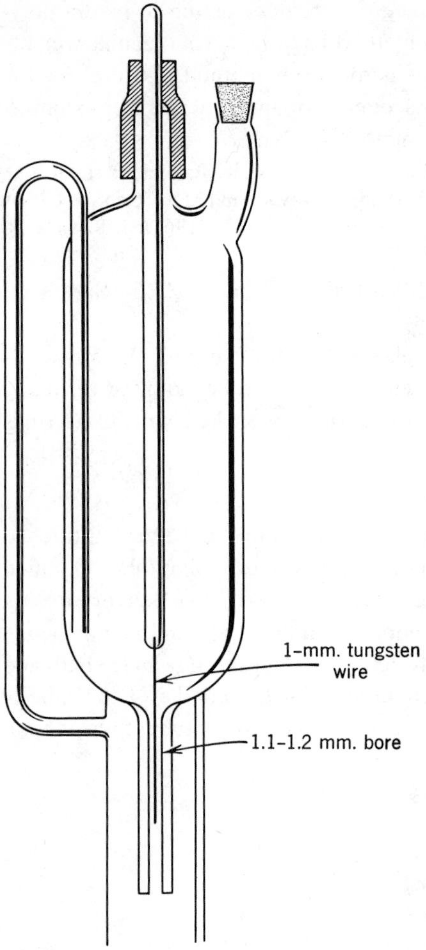
1-mm. tungsten wire
1.1–1.2 mm. bore

Fig. P-2 The same, with pressure equalizer[1]

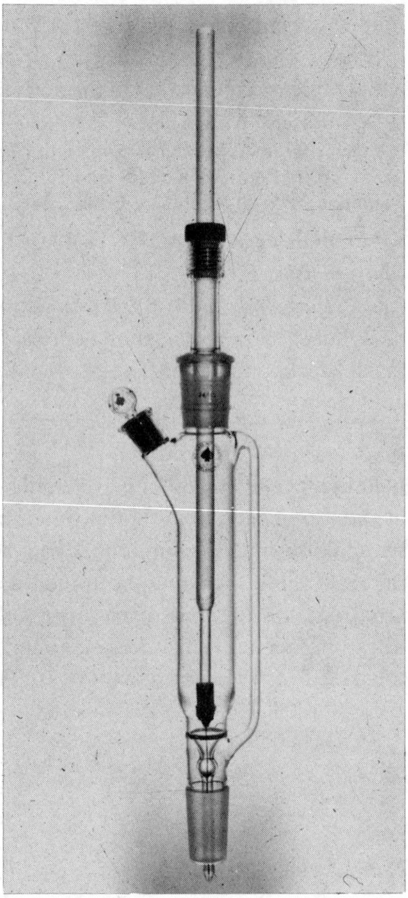

Fig. P-3 Ace Glass 7340*

of 1,5-pentanediol diacetate is added to the tube over a period of 3.5 hrs. The contents of the three receivers are combined, distilled, and redistilled.

Preparation (b).[3] The reaction sequence is shown in the formulation. Treatment of paraldehyde and absolute ethanol with hydrogen chloride gas at −5° produces α-chloroethyl ethyl ether (1), which on bromination affords α,β-dibromoethyl ethyl ether (2). Coupling with allylmagnesium bromide gives (3), which on reaction with zinc dust in *n*-butanol generates 1,4-pentadiene. Thus one double bond is that of allyl bromide, and the other is generated by elimination of $BrOC_2H_5$ from the bromohydrin ethyl ether (3).

*In this new model a precision bearing and shaft is used to maintain adjustment. Coarse adjustment is made by sliding the shaft up or down, fine adjustment by turning the threaded nylon shaft holder. Four turns changes the dropping rate by a factor of 2 near the open position and a factor of $\frac{1}{10}$ near the closed position. The capillary wire is made of platinum so that highly corrosive materials may be used. A snap-on detachable Teflon cone with a more resiliant Viton or silicone O-ring provides positive pressure. Minimum dropping rate about 10 drops/min. Maximum dropping rate about 6 ml./min.

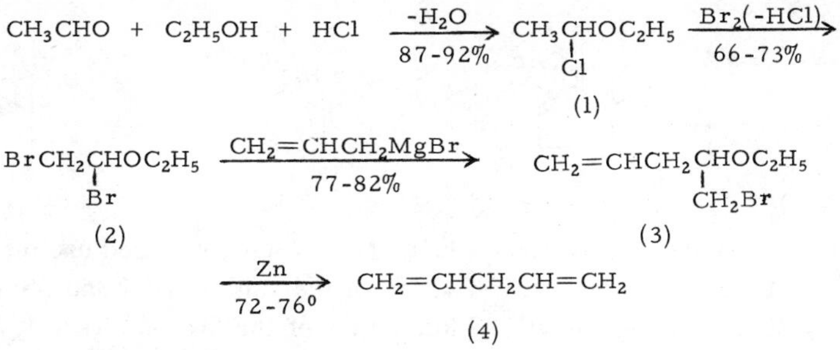

$$CH_3CHO + C_2H_5OH + HCl \xrightarrow[87-92\%]{-H_2O} CH_3\underset{Cl}{CH}OC_2H_5 \xrightarrow[66-73\%]{Br_2(-HCl)}$$

(1)

$$Br CH_2\underset{Br}{CH}OC_2H_5 \xrightarrow[77-82\%]{CH_2=CHCH_2MgBr} CH_2=CHCH_2\underset{CH_2Br}{CH}OC_2H_5$$

(2)   (3)

$$\xrightarrow[72-76°]{Zn} CH_2=CHCH_2CH=CH_2$$

(4)

[1]R. E. Benson and B. C. McKusick, *Org. Syn., Coll. Vol.*, **4**, 746 (1963)
[2]L. F. Fieser and E. B. Hershberg, *ibid.*, **2**, 129 (1943)
[3]O. Grummitt, E. P. Budewitz, and C. C. Chudd, *ibid.*, **4**, 748 (1963)

Peracetic acid (a), *see* Hydrogen peroxide–Acetic acid. Although this solution contains peracetic acid, and is commonly called peracetic acid, it differs substantially in composition and properties from reagents (b) and (c).

Peracetic acid (b), commercial 40% peracetic acid. Material available from the Buffalo Electro Chemical Co. (Becco) has the composition: 40% peracetic acid, 5% hydrogen peroxide, 39% acetic acid, 1% sulfuric acid, and 15% water. The density is 1.15 g. per ml. A fresh sample contains 0.77 mole of peracetic acid per 100 ml. Since the peracid content decreases somewhat on standing, an old sample should be analyzed by treating an aliquot with potassium iodide and titrating the liberated iodine with standard sodium thiosulfate. For certain uses it is advisable to neutralize the sulfuric acid present by addition of sodium acetate.

Caution. The reagent produces severe skin burns.

Epoxidation of olefins. Reif and House[1] worked out the following procedures for the preparation of *trans*-stilbene oxide. Titration of the commercial "40%" peracetic acid showed it to contain 0.497 g. of peracid per ml. For reaction with 0.3 mole of *trans*-stilbene, 65 ml. (0.425 mole) of reagent was used after addition of 5 g. of NaOAc·3 H$_2$O to neutralize the sulfuric acid present. A solution of the hydrocarbon

$$\underset{H}{\overset{C_6H_5}{>}}C=C\underset{C_6H_5}{\overset{H}{<}} \xrightarrow[70-75\%]{CH_3CO_3H} \underset{H}{\overset{C_6H_5}{>}}C\overset{O}{\underset{}{\diagup\diagdown}}C\underset{C_6H_5}{\overset{H}{<}}$$

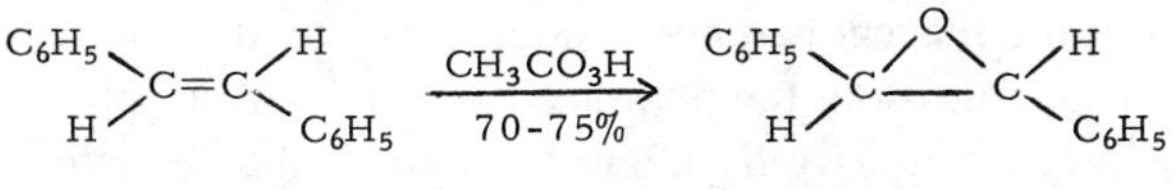

in 400 ml. of methylene chloride was cooled to 20°, the cooling bath was removed, and the solution of peracetic acid was run in with stirring in 15 min. The mixture was then stirred for 15 hrs.; the temperature rose to 32–35° in about 2 hrs. and then gradually fell. After the time specified the optical density of the reaction mixture at 295 mμ should have dropped to less than 3% of the initial value. If more than this amount of *trans*-stilbene remains, it cannot be removed by repeated crystallization, and hence more peracetic acid should be added. When the reaction was complete, suitable workup and crystallization from methanol and then from hexane gave the pure oxide in 70–75% yield.

T. F. Gallagher's method for introducing a 17α-hydroxy group into a 20-keto-steroid consists in formation of the 17-enol acetate (2), epoxidation to (3) with perbenzoic acid, and alkaline hydrolysis. An Upjohn group[2] carried out the epoxi-dation step with commercial 40% peracetic acid after addition of sodium acetate for neutralization and found that the cheaper reagent behaved like perbenzoic acid

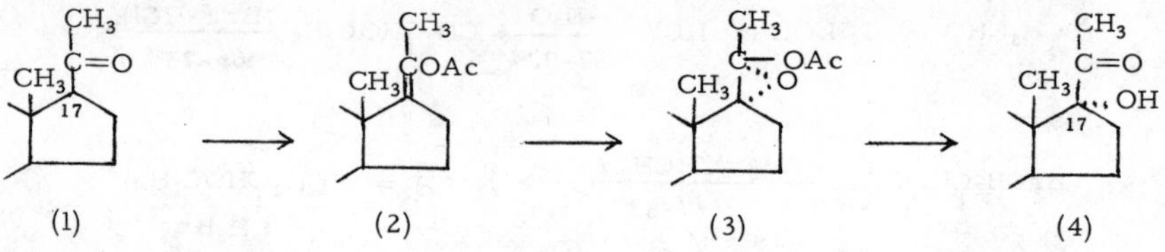

and gave the epoxide (3). Oliveto and Hershberg[3] (Schering) tried use of the commercial peracetic acid without neutralization of the sulfuric acid and obtained the 17α-hydroxy-20-ketone (4) directly. Elimination of the alkaline hydrolysis of (3) makes possible retention of acetyl groups elsewhere in the molecule.

Oxidation. Van Duuren *et al.*[4] oxidized 1,2,5,6-dibenzanthracene to the 3,4-seco-diacid by adding 10 g. of 40% peracetic acid saturated with sodium acetate to

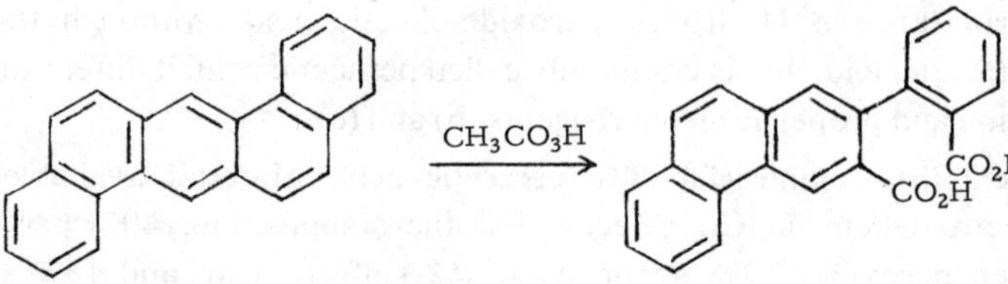

a suspension of 2 g. of the hydrocarbon in 75 ml. of chloroform at 4° and stirring at 4° for 8 hrs., when a clear solution resulted. On continued stirring at 4° the product started to separate and was collected after a reaction period of 8 days.

Page and Tarbell[5] oxidized β-naphthol to *o*-carboxycinnamic acid with 1.1 times the theoretical amount of 40% Becco peracetic acid without addition of sodium acetate. The peracid (0.14 mole) was stirred in a bath at 25–30°, and a solution of 0.14 mole of β-naphthol in 100 ml. of acetic acid was added in 4 hrs. (temperature

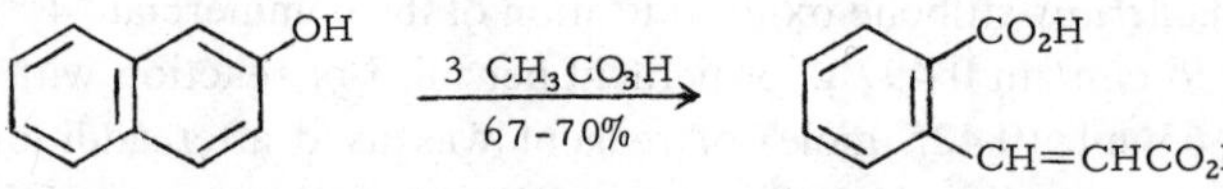

not to exceed 40°). Solid product started separating from the orange solution after about one third of the reagent had been added. The mixture was let stand at 25–30° for 6–8 hrs. and then at room temperature for 4 days. Acid of satisfactory purity was obtained by precipitation from a filtered solution in bicarbonate.

For conversion of pyridine to the N-oxide (H. S. Mosher[6]), 1.5 moles of 40% peracid is added to 1.39 moles of pyridine with stirring and at a rate sufficient to maintain a temperature of 80° (50–60 min.), and the mixture is stirred until the temperature drops to 40°. The solvent is evaporated on the steam bath at the water

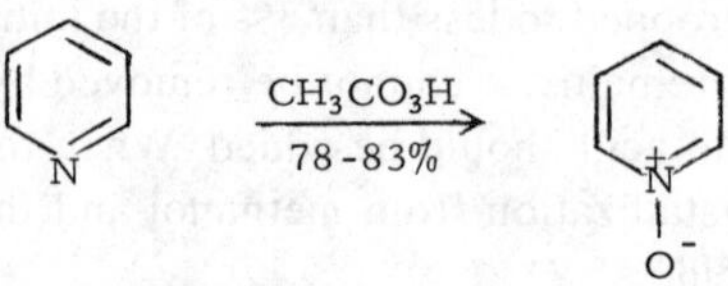

pump and the residual pyridine-N-oxide acetate placed in a flask suitable for collecting a solid distillate (see Fig. H-4). The vacuum pump is protected by a dry ice trap to collect the acetic acid formed by dissociation of the acetate at a low pressure and the oxide is distilled at 95–98°/0.5 mm. The deliquescent base is obtained as a colorless solid, m.p. 65–66° (sealed capillary).

Iodobenzene (0.1 mole) is oxidized to iodosobenzene diacetate by adding 0.24 mole of 40% peracetic acid to iodobenzene with stirring in a bath kept at 30° over

$$C_6H_5I \xrightarrow{CH_3CO_3H} C_6H_5I^{+}OCOCH_3(CH_3COO^{-})$$

a period of 30–40 min. (Sharefkin and Saltzman[7]). On stirring for another 20 min. at 30° a yellow solution is formed and begins to deposit crystals of iodosobenzene diacetate. After thorough cooling the product is collected and washed with cold water; m.p. 158–159° dec., purity 97–98%.

Peracetic acid oxidizes 1,1-dialkylhydrazines to tetrazines (1) in 70–80% yield, and cleaves the benzoylhydrazone of acetophenone as shown (2).[8] Peracetic acid

1. $2\ (C_2H_5)_2NNH_2 \xrightarrow{CH_3CO_3H} (C_2H_5)_2NN{=}NN(C_2H_5)_2$

2. 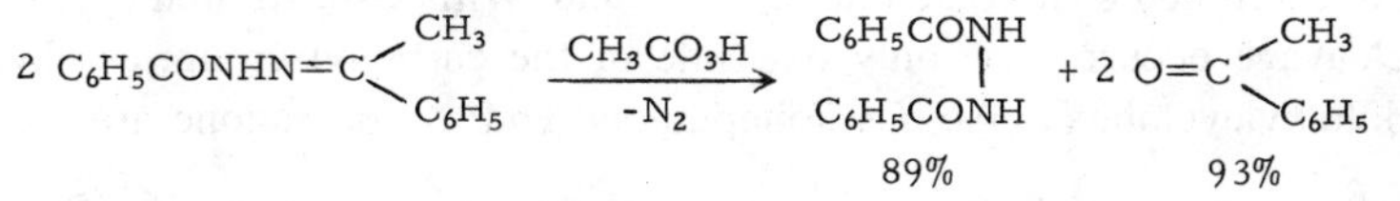

is better for these reactions than yellow mercuric oxide and is soluble in organic solvents (chloroform, aqueous acetic acid).

[1]D. J. Reif and H. O. House, *Org. Syn., Coll. Vol.*, **4**, 860 (1963)
[2]H. V. Anderson, E. R. Garrett, F. H. Lincoln, Jr., A. H. Nathan, and J. A. Hogg, *Am. Soc.*, **76**, 743 (1954)
[3]E. P. Oliveto and E. B. Hershberg, *Am. Soc.*, **76**, 5167 (1954)
[4]B. L. Van Duuren, I. Bekersky, and M. Lefar, *J. Org.*, **29**, 686 (1964)
[5]G. A. Page and D. S. Tarbell, *Org. Syn., Coll. Vol.*, **4**, 136 (1963)
[6]H. S. Mosher, L. Turner, and A. Carlsmith, *ibid.*, **4**, 828 (1963)
[7]J. G. Sharefkin and H. Saltzman, *Org. Syn.*, **43**, 62 (1963)
[8]L. Horner and H. Fernekess, *Ber.*, **94**, 712 (1961)

Peracetic acid (c), anhydrous. Phillips, Frostick, and Starcher[1] (Union Carbide) describe an efficient two-step synthesis of a water-free solution of peracetic acid in ethyl acetate or acetone. In the first step acetaldehyde is autoxidized at 0° under catalysis by ultraviolet light, 0.01% of cobaltous acetate, or ozone to acetaldehyde monoperacetate (AMP). This substance forms crystals (m.p. 20–22°) which explode

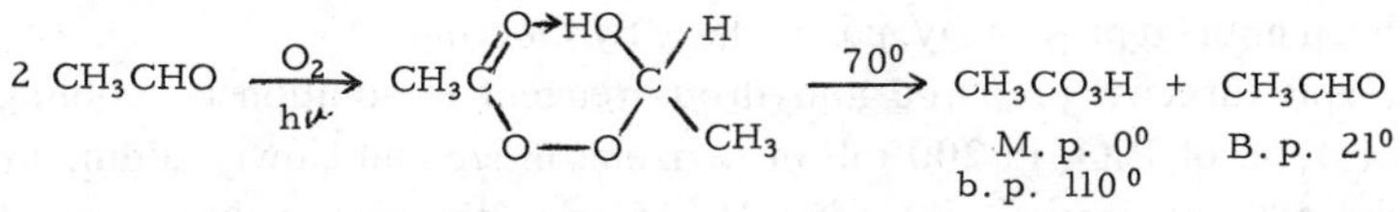

AMP

easily, but solutions can be handled safely. Thus an acetone solution containing 48.5% of AMP, 7.7% of acetic acid, and 0.1% of a stabilizer is fed through a steam-heated coil vaporizer into the middle of a fractionating column, while acetone vapor is fed into the column about 1 inch from the bottom. In this way AMP is pyrolyzed at about 70° to peracetic acid and acetaldehyde and the more volatile component passes into the distillate. The liquid collected from the bottom of the column is an acetone solution containing 27.9% of peracetic acid and 9.1% of acetic acid.

This solution can be used for the epoxidation of α,β-unsaturated esters to glycidic esters.[2] For example, 24 moles of ethyl crotonate was allowed to react with 4130 g. (12 moles) of a 22.2% solution of peracetic acid in ethyl acetate. After addition and

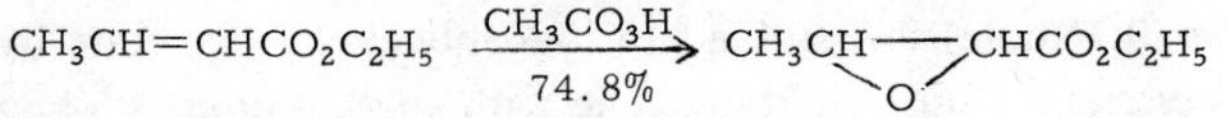
CH3CH=CHCO2C2H5
CH3CO3H
74.8%
CH3CH—CHCO2C2H5
O

reaction periods totaling 5.5 hrs. at 85°, fractionation gave 1167 g. of ethyl 2,3-epoxybutyrate. The excess ethyl crotonate facilitated separation of the by-product acetic acid from the epoxide.

The anhydrous reagent gives improved results in the Baeyer-Villiger oxidation of cyclic ketones to lactones.[3] Thus cyclopentanone afforded δ-valerolactone in

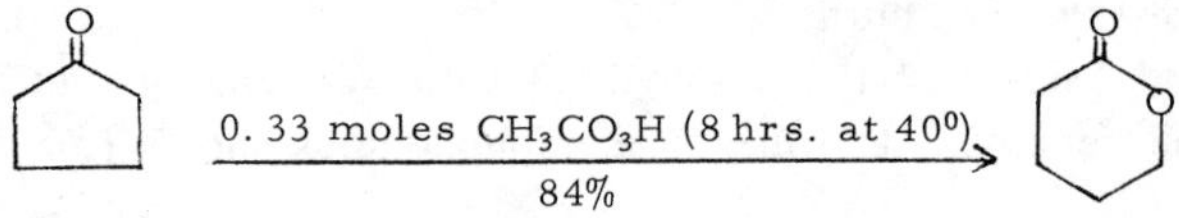
0.33 moles CH3CO3H (8 hrs. at 40°)
84%

high yield under the conditions indicated; in a reaction period of 6.25 hrs., cyclohexanone afforded ε-caprolactone in 85% yield. With 2-substituted cyclohexanones, ring cleavage occurred on only one side of the carbonyl group, as illustrated for 2-cyclohexylcyclohexanone. Cycloheptanone and cyclooctanone are less reactive;

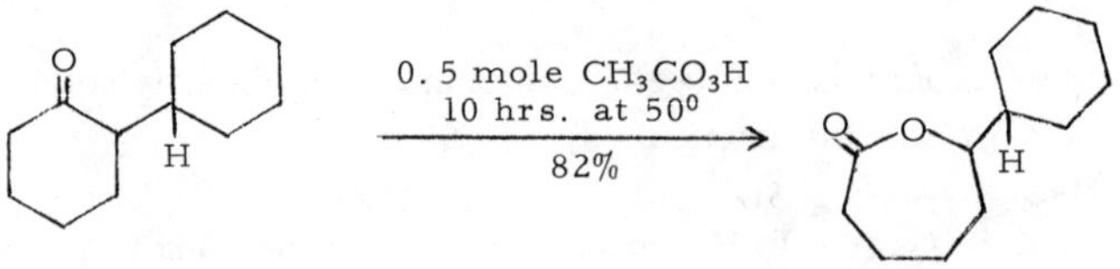
0.5 mole CH3CO3H
10 hrs. at 50°
82%

at 80° cycloheptanone gives about equal parts of the lactone and the diacid and cyclooctanone gives chiefly the diacid.

In the oxidation of cyclododecanone (1), the lactone (2) was obtained in yield as high as 50% by carrying out the reaction in the presence of concd. sulfuric

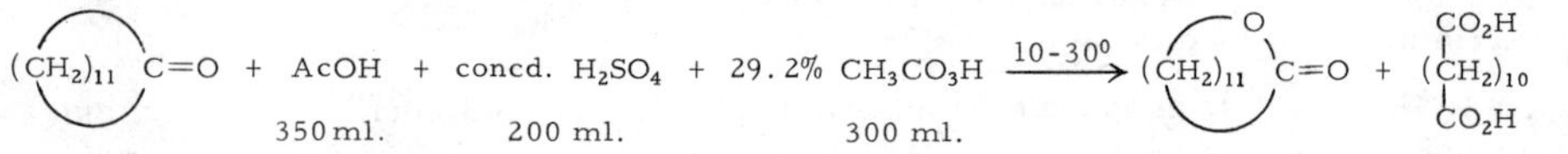
(CH2)11 C=O + AcOH + concd. H2SO4 + 29.2% CH3CO3H
350 ml.
200 ml.
300 ml.
10-30°
(CH2)11 C=O + (CH2)10
CO2H
CO2H

(1) 50 g. (2) 50% (3) 25%

acid and use of a large excess of reagent.[4] Dodecanedioic acid was formed in significant amount. At 20° in the absence of sulfuric acid the reaction was complete only after 36 days. The marked catalysis of the Baeyer-Villiger reaction by sulfuric acid had been noticed previously, particularly by Doering.[5]

Horner and Jürgens[6] prepared anhydrous reagent in solution by cooling a suspension of 45 g. of P_2O_5 in 200 ml. of benzene in ice and slowly adding 60 ml. of commercial 40% peracetic acid. After 10–15 min., the clear solution was filtered, titrated, and stored in the cold out of the light. The anhydrous reagent was used for epoxidation of Schiff bases (azomethines) to isonitrones (oxaziranes). Thus reaction of cyclohexylidenecyclohexylamine (1) with a benzene solution of peracetic

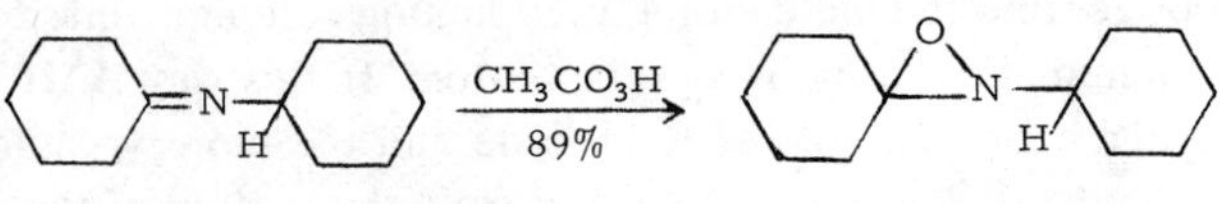
CH3CO3H
89%

(1) (2)

acid (2% excess) at room temperature for 5 hrs. afforded pentamethylene-N-cyclohexylisonitrone (2). A few months before the appearance of the Horner paper Emmons[7] (Rohm and Haas) had reported oxidation of a few azomethines to oxaziranes "in good yield with anhydrous peracetic acid in methylene chloride" without identifying the method used to prepare the anhydrous reagent. Later he reported[8] preparation of anhydrous peracetic acid *in situ* for use in the oxidation of amines, for example *p*-toluidine. Chloroform (60 ml.) was stirred vigorously in an ice bath and treated with 0.5 mole of 90% hydrogen peroxide and 2 drops of sulfuric acid catalyst. Acetic anhydride (0.6 mole) was added dropwise in 30 min. to react with the water present in the peroxide and convert some of the peroxide into diacetyl peroxide. After 15 min. at room temperature the clear solution was diluted with 40 ml. of chloroform and heated to boiling. A solution of 0.1 mole of *p*-toluidine

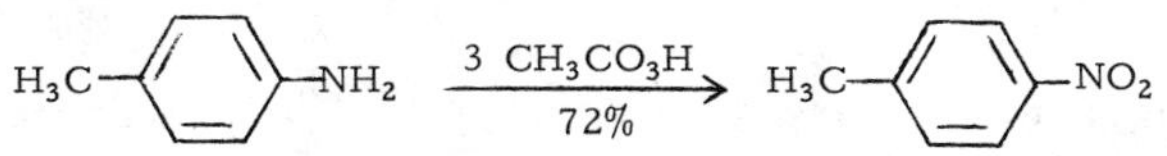
H3C
NH2
3 CH3CO3H
72%
H3C
NO2

in 25 ml. of chloroform was added to the refluxing solution in 20 min. and the yellow solution was refluxed for 1 hr. and poured onto water. Workup and crystallization from ligroin afforded pure *p*-nitrotoluene in 72% yield. Seven other nitrobenzenes were obtained in yields of 62–83%. *sec*-Butylamine, oxidized by this method, gave 2-nitrobutane in 65% yield.

An interesting use for the anhydrous reagent, at least that prepared by the Union Carbide process, is for generation of hydroxyl radicals. Prior to publication of the method by Heywood, Phillips, and Stansbury,[9] Schleyer and Nicholas[10] reported an experiment in which Dr. Heywood of Union Carbide carried out the oxidation and they worked up the reaction mixture. A mixture of 150 g. of adamantane

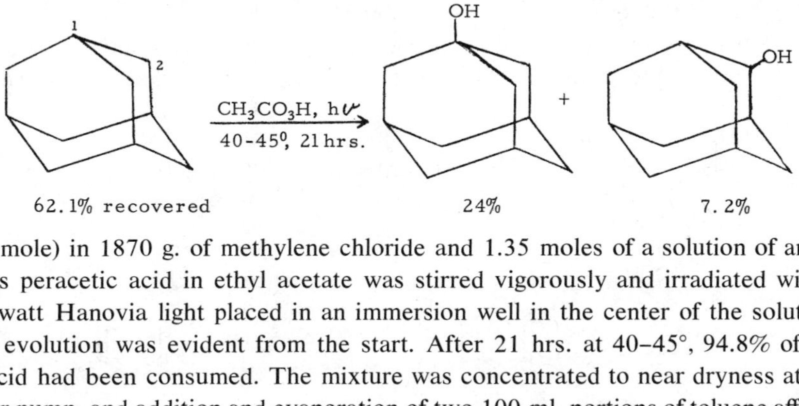
1
2
CH3CO3H, hν
40-45°, 21 hrs.
OH
+
OH
62.1% recovered
24%
7.2%

(1.1 mole) in 1870 g. of methylene chloride and 1.35 moles of a solution of anhydrous peracetic acid in ethyl acetate was stirred vigorously and irradiated with a 100-watt Hanovia light placed in an immersion well in the center of the solution. Gas evolution was evident from the start. After 21 hrs. at 40–45°, 94.8% of the peracid had been consumed. The mixture was concentrated to near dryness at the water pump, and addition and evaporation of two 100-ml. portions of toluene afforded 155 g. of white solid. A part of the material was worked up by chromatography, and a part was oxidized to a more easily separated mixture containing 1-adamantanol and 2-adamantanone. The estimated yields of unchanged hydrocarbon, 1-adamantanol, and 2-adamantanol are given under the formulas. Polyoxygenated material amounted to 1.8% and an unidentified carbonyl compound to 4.5%.

In one of two other oxidations with the Union Carbide reagent in ethyl acetate, Heywood and Phillips[11] found that α,β-unsaturated cyclic acetals such as (1)

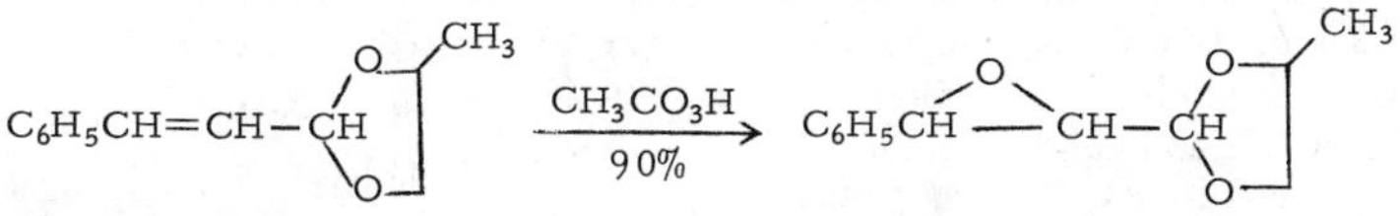
CH3
C6H5CH=CH—CH
CH3CO3H
90%
C6H5CH—CH—CH
CH3

(1) (2)

are oxidized smoothly to the epoxides (2). Surprisingly, the noncyclic unsaturated acetal (3) reacted only at a somewhat higher temperature and afforded no epoxide but the ester (4). The reaction was found to be catalyzed by sulfuric acid, and in the

$$\underset{(3)}{CH_3CH{=}CHCH(OC_4H_9)_2} \xrightarrow[73\%]{CH_3CO_3H} \underset{(4)}{CH_3CH{=}CHCO_2C_4H_9} + C_4H_9OH + CH_3CO_2H$$

$$\underset{(5)}{\underline{n}\text{-}C_3H_7CH(OC_2H_5)_2} \xrightarrow[69\%]{CH_3CO_3H\ (H_2SO_4)} \underset{(6)}{\underline{n}\text{-}C_3H_7CO_2C_2H_5} + C_2H_5OH + CH_3CO_2H$$

presence of this catalyst the unsaturated cyclic acetal (1) was converted into unidentified material and afforded no epoxide. Saturated acetals such as (5) react with peracetic acid in the presence of sulfuric acid to give esters (6). The reaction may proceed through the hemiacetal peracetate (8).

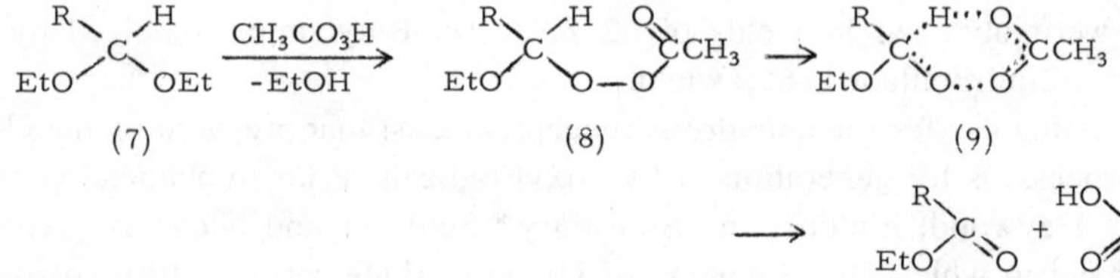

(7) (8) (9)

Crosby[12] devised an interesting synthesis of scopoletin (V) which involves in the first step oxidation of commercial isovanillin (I, 1.28 moles) in ethyl acetate at 40° with 1.40 moles of peracetic acid in the same solvent. The product, 3-hydroxy-4-methoxyphenyl formate (II), distilled at 111–115°/1 mm. and, after two weeks at

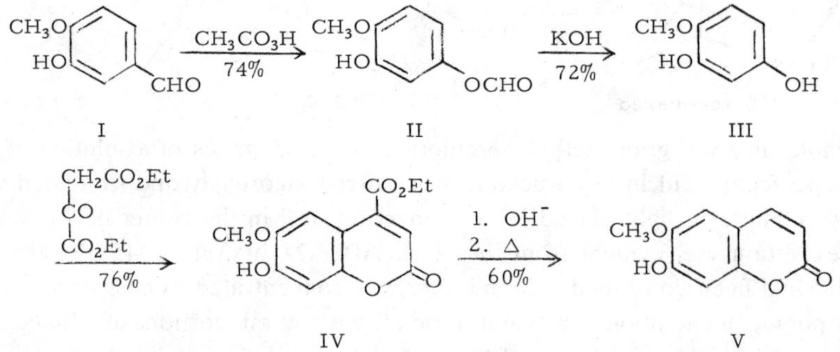

I II III IV V

room temperature, solidified (m.p. 57–58°). Hydrolysis with alcoholic potassium hydroxide and condensation of the phenol III with sodium ethyl oxalacetate in the presence of 85% phosphoric acid afforded the ester IV, which on saponification and decarboxylation (Cu, quinoline) yielded scopoletin, m.p. 198–200°.

[1] B. Phillips, F. C. Frostick, Jr., and P. S. Starcher, *Am. Soc.*, **79**, 5982 (1957); B. Phillips, P. S. Starcher, and B. D. Ash, *J. Org.*, **23**, 1823 (1958)
[2] D. L. MacPeek, P. S. Starcher, and B. Phillips, *Am. Soc.*, **81**, 680 (1959)
[3] P. S. Starcher and B. Phillips, *Am. Soc.*, **80**, 4079 (1958)
[4] K. Kosswig, W. Stumpf, and W. Kirchhof, *Ann.*, **681**, 28 (1965)
[5] W. von E. Doering and L. Speers, *Am. Soc.*, **72**, 5515 (1950)
[6] L. Horner and E. Jürgens, *Ber.*, **90**, 2184 (1957)
[7] W. D. Emmons, *Am. Soc.*, **78**, 6208 (1956)

[8]W. D. Emmons, *Am. Soc.*, **79**, 5528 (1957)
[9]D. L. Heywood, B. Phillips, and H. A. Stansbury, Jr., unpublished
[10]P. von R. Schleyer and R. D. Nicholas, *Am. Soc.*, **83**, 182 (1961)
[11]D. L. Heywood and B. Phillips, *J. Org.*, **25**, 1699 (1960)
[12]D. G. Crosby, *J. Org.*, **26**, 1215 (1961)

Perbenzoic acid, $C_6H_5CO_3H$. Mol. wt. 138.12.

Preparation. An early procedure[1] involved cooling a solution of sodium methoxide in methanol to −5°, adding a cold solution of dibenzoyl peroxide in chloroform,

$$(C_6H_5COO)_2 \xrightarrow[-C_6H_5CO_2CH_3]{CH_3ONa} C_6H_5CO_3Na \xrightarrow[82.5\text{-}86\%]{H_2SO_4} C_6H_5CO_3H$$

extracting the sodium perbenzoate with ice water, washing the extract with cold chloroform to remove ethyl benzoate, liberating the perbenzoic acid by acidification, and extracting it with chloroform. Commercial dibenzoyl peroxide should be crystallized by dissolving it in hot chloroform and adding methanol to the point of crystallization.

For an improved preparation from 70% hydrogen peroxide and benzoic acid in methanesulfonic acid and extraction of the product with benzene (Swern), *see* Hydrogen peroxide, Moyer and Manley[2] (Dow) apparently were unaware of the Swern method on presenting a note on "An Improved Synthesis of Peroxybenzoic Acid." In their procedure, based on one by Vilkas,[3] a solution of 8 g. sodium peroxide in 135 ml. of water at 20° is filtered and stirred magnetically during addition of 175 ml. of denatured ethanol and a solution of 0.5 g. of $MgSo_4 \cdot 7\,H_2O$ in 15 ml. of water; the salt inhibits the catalytic effect of traces of metal ions and makes it possible to run the reaction at room temperature with yield equal to that obtained at −5° without addition of magnesium sulfate. The solution is brought to room temperature and stirred during the addition of 11.6 ml. of benzoyl chloride in 10–12 min. The solution is filtered to remove any dibenzoyl peroxide, acidified with 20% sulfuric acid, and extracted with six 75-ml. portions of carbon tetrachloride, chloroform, or benzene; yield about 75%. About 25 ml. of ethanol is extracted into the organic phase.

Reaction with olefins. The reaction of perbenzoic acid with an olefin usually proceeds smoothly at a low temperature (0–25°) and affords an epoxide in high yield. The olefin is dissolved at room temperature in a solution of perbenzoic acid in chloroform or benzene, with use of more solvent if required, and the solution is let stand at room temperature or below for several hours. Iodimetric titration can be used to follow the course of a reaction or, by using an excess of reagent, to determine the total uptake of reagent and hence the number of double bonds present.

In his classical investigations of ergosterol and vitamin D, Windaus encountered all the now known variations in the reaction of perbenzoic acid with unsaturated compounds. The finding of Windaus and Lüttringhaus that in the presence of excess perbenzoic acid ergosterol consumes exactly three moles of peracid was the first direct proof of the presence of three double bonds. Reaction of the sterol with one mole of perbenzoic acid gave a product later established to be the $3\beta,5\alpha,6\alpha$-triol 6-benzoate. The *trans* 22,23-double bond is notably unreactive. The cisoid 5,8-diene system in ergosterol is highly reactive to dienophiles such as maleic anhydride and, under photosensitization, oxygen adds 1,4 to give the $5\alpha,8\alpha$-transannular peroxide. Thus the preferential addition to the 5,6-double bond showed that perbenzoic acid has no tendency to add 1,4. Evidently the reaction initially affords the α-epoxide

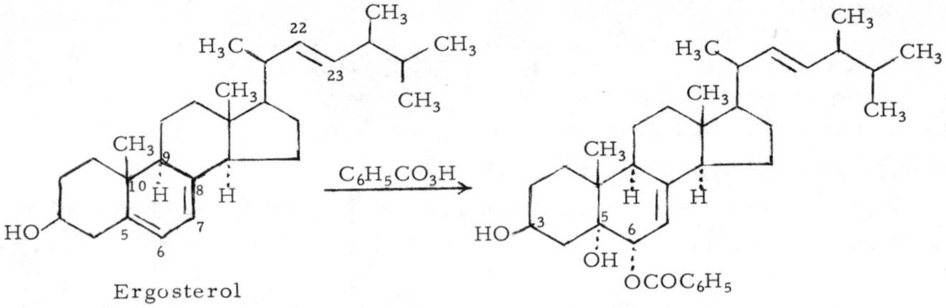

Ergosterol
$C_6H_5CO_3H$

which, being allylic, is easily cleaved to the 5,6-diol 6-benzoate. Whereas saturated epoxides are cleaved to *trans* products, an allylic epoxide can suffer cleavage with either inversion or retention, depending upon the structure and the reaction conditions. That in this case the 6α-configuration is retained might seem attributable to the equatorial orientation of the 6α-benzoate group; a bulky 6β-axial group would set up severe interaction with the axial 10β-methyl group. However, the behavior of lumisterol argues against this interpretation.[5] Lumisterol differs from ergosterol in the configurations at both C_9 and C_{10}, and the rear side is shielded by the angular methyl

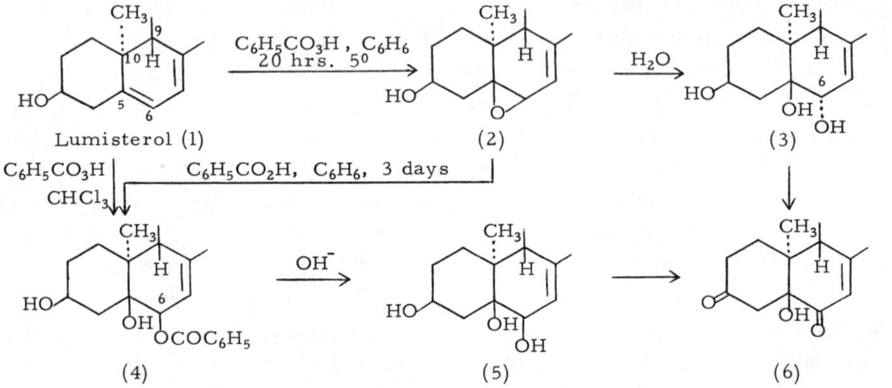

$C_6H_5CO_3H$, C_6H_6
20 hrs. 5°
H_2O
Lumisterol (1)
(2)
(3)
$C_6H_5CO_3H$
$CHCl_3$
$C_6H_5CO_2H$, C_6H_6, 3 days
OH^-
(4)
(5)
(6)

group. Thus whereas perbenzoic acid attacks ergosterol from the rear side, the attack on lumisterol is frontal, and the product of brief reaction with perbenzoic acid in benzene is the β-oxide (2). This oxide can be hydrolyzed by brief heating of a suspension in water; the 3-acetate 5,6-β-epoxide can be hydrolyzed in the same way without disturbance of the acetyl group. The reaction proceeds with normal inversion at C_6, even though the axial 6α-hydroxyl group is strongly hindered by the axial 10α-methyl group. If lumisterol is treated with perbenzoic acid in chloroform in the usual way, or if the solution in benzene is allowed to stand for 3 days, the product is the 3β,5β,6β-triol 6-benzoate (4), convertible on saponification into the triol (5). The 6-epimers, (3) and (5), were correlated by oxidation to the diketone (6). The abnormal reaction in the series, the cleavage of the β-oxide (2) by benzoic acid with retention of configuration at C_6, is attributed to the allylic nature of the oxide and

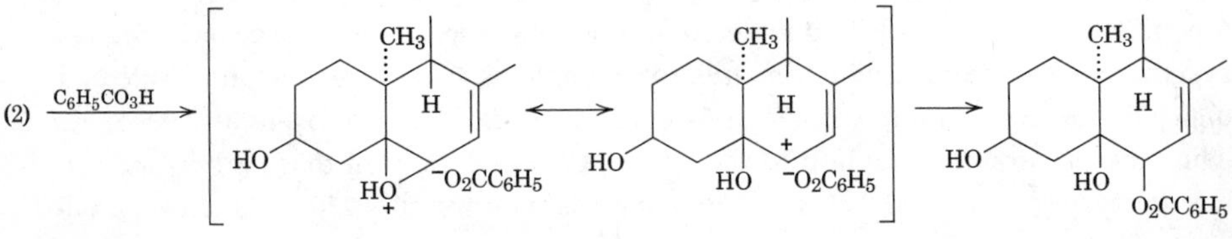

(2)
$C_6H_5CO_3H$

the acidic character of the reagent. The 7,8-double bond promotes development of partial carbonium ion character at C_6 sufficient for attraction of the nucleophile with *cis* opening of the oxide ring.

In the steroid series tetrasubstituted double bonds are hydrogenated with difficulty or not at all, but they nevertheless consume perbenzoic acid. The product isolated, however, is not an epoxide but an allylic alcohol or a conjugated diene. The first case was encountered by Windaus and Lüttringhaus,[4] but a better documented example was reported by Windaus, Linsert, and Eckhardt.[6] Reaction of Δ^8-cholesteryl acetate (1) with perbenzoic acid in chloroform afforded an unsaturated

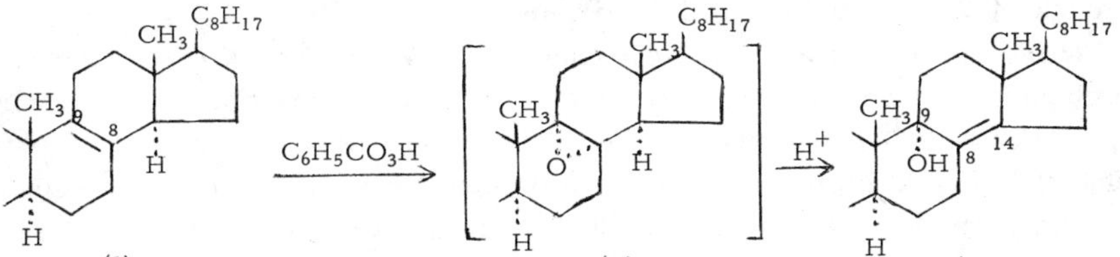

C8H17
CH3
CH3
9
8
H
C6H5CO3H
O
H
H+
14
OH
H

(1) (2) (3)

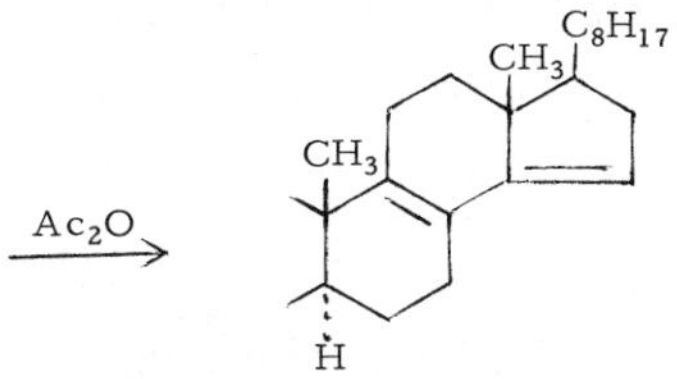

C8H17
CH3
CH3
Ac2O
H

(4)

diol 3-acetate regarded as an allylic tertiary alcohol (3) because on treatment with acetic anhydride it afforded the diene (4). Since Henbest and Wrigley[7] found Δ^8-ergosteryl acetate to react with perbenzoic acid in benzene to give the $8\alpha,9\alpha$-epoxide in good yield, and in view of the supporting evidence cited below, the production of the allylic alcohol (3) is interpreted as involving formation of the epoxide (2) and cleavage to (3) under the influence of a trace of acid present in the chloroform. The supporting evidence, presented by Fieser and Goto[8] in clarifying results reported by Wintersteiner and Moore,[9] is as follows. Δ^7-cholesteryl acetate (I) reacts with

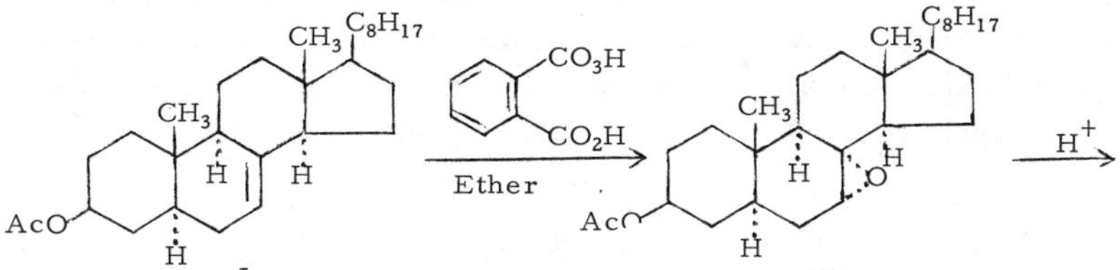

C8H17
CH3
CH3
H
AcO
CO3H
CO2H
Ether
O
H+

I II

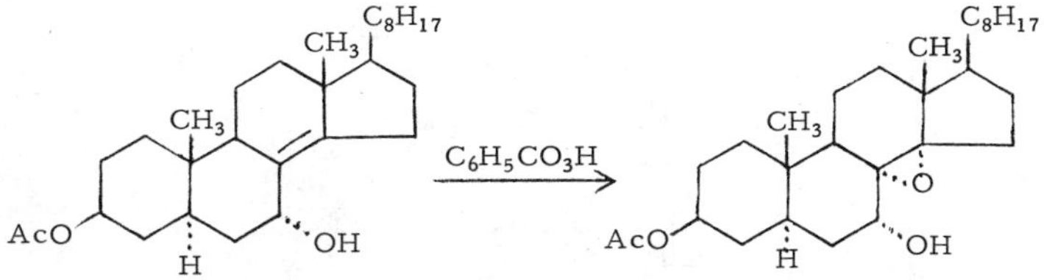

C8H17
CH3
CH3
AcO
OH
H
C6H5CO3H
O

III IV

perphthalic acid in ether to give the 7α,8α-epoxide (II). This oxide does not react further on treatment with perphthalic acid in ether, but in chloroform solution it consumed one mole of perbenzoic acid and gave the oxide alcohol IV. The oxide II alone is stable in dry or wet chloroform containing benzoic acid, and hence this acid is not responsible for the cleavage and further reaction. In the procedure for preparing perbenzoic acid usual at the time, a solution of the sodium salt was acidified with sulfuric acid and extracted with chloroform. Chloroform, shaken with 10% sulfuric acid and dried, was found to be acidic enough to isomerize the oxide II to the allylic alcohol III, isolated as the diacetate. Titration showed that the chloroform contained 0.2 mg. of sulfuric acid per 100 ml.

Oxidation of α-diketones, o-quinones, and ***phenol derivatives*** (peracetic and perphthalic acids were used in some of the reactions described in this section). Karrer,[10] having had experience with the conversion of carotenoids into epoxides, some of which occur in nature, explored the reaction of the unsaturated α-diketone (1) with perphthalic acid in ether. The crystalline reaction product had the composition but not the properties of a monoepoxide and proved to be the anhydride (2). The peracid thus cleaves the bond between the carbonyl groups and inserts an

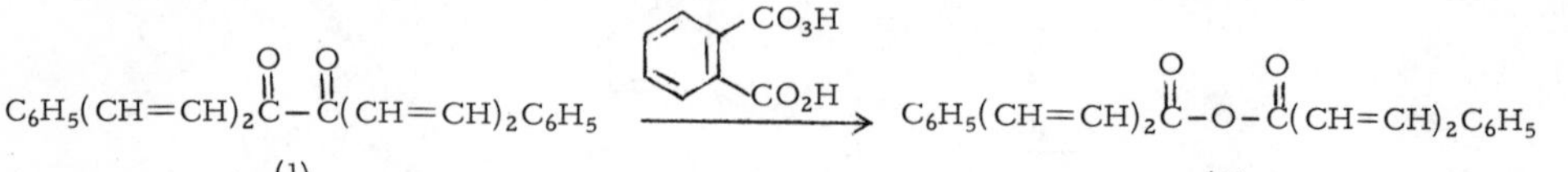

(1) (2)

atom of oxygen. Böeseken[11] had oxidized β-naphthoquinone with peracetic acid and characterized the product as *o*-carboxyallocinnamic acid (5). Karrer and Schneider[12] dissolved 3.5 g. of β-naphthoquinone in 70 ml. of chloroform, and added

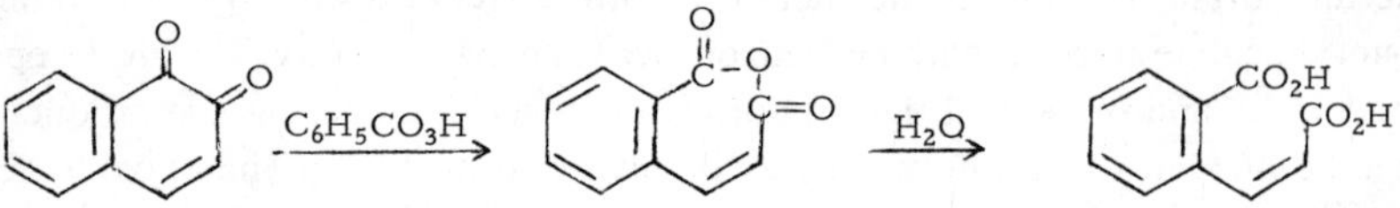

(3) (4) (5)

a solution of 3.4 g. (1.1 equiv.) of perbenzoic acid in 50 ml. of chloroform at room temperature. The solution soon turned deep red, but after 5 days it became bright yellow. Acids were removed by extraction with bicarbonate solution, and evaporation of the dried chloroform and several crystallizations of the residue afforded 0.85 g. of the pure anhydride (4), m.p. 115°. Other examples of the reaction:[13–15]

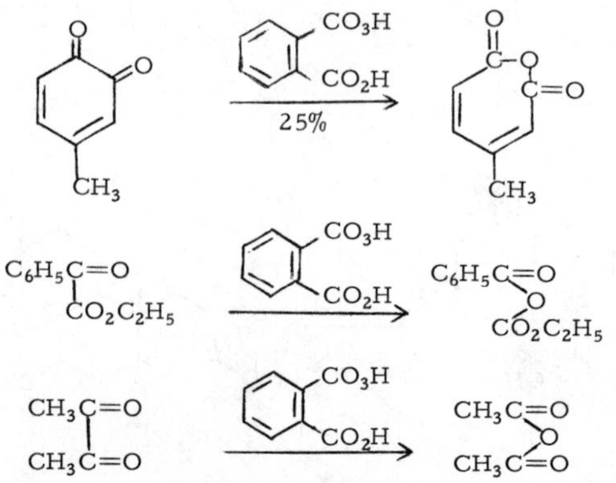

H. Fernholz[16] studied the oxidation of β-naphthol methyl ether with perbenzoic acid and presented convincing evidence of the course of a remarkable reaction: conversion into the monomethyl ester of *o*-carboxyallocinnamic acid (6) in 35% yield. Six grams of the ether (1) was added with cooling to 692 ml. of a 2.5% solution

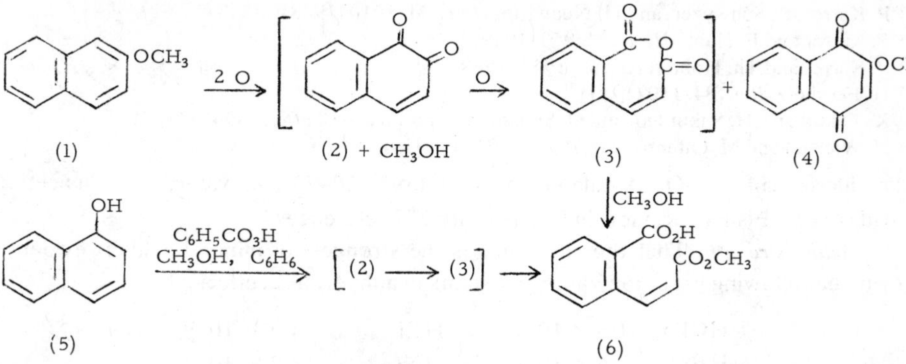
OCH_3
2 O
O
(1)
(2) + CH_3OH
(3)
(4)
OCH_3
CH_3OH
OH
$C_6H_5CO_3H$
CH_3OH, C_6H_6
[(2) → (3)]
CO_2H
CO_2CH_3
(5)
(6)

of perbenzoic acid in benzene, and the solution, soon red, was let stand for 14 days at 10°. Extraction with bicarbonate removed the acidic fraction from which the half-ester (6) was isolated, left a small neutral fraction from which 2-methoxy-1,4-naphthoquinone (4) was isolated in yield of 2.5%. The postulated sequence of events in the main reaction is oxidation to β-naphthoquinone (2) with elimination of methanol, cleavage to the Karrer anhydride (3), and reaction of this with the methanol formed in the first step to give the half-ester (6). When the oxidation was done with a solution of peracetic acid prepared from 300 ml. of acetic acid and 30 ml. of 30% hydrogen peroxide (14 days at 25°), the free diacid corresponding to (6) was obtained in 65% yield; evidently water, present in large excess of the methanol formed, reacted preferentially with the anhydride. Oxidation of 3 g. of the methyl ether (1) with perbenzoic acid in benzene with addition of 30 g. of absolute ethanol affords 0.9 g. of the half ethyl ester of the diacid. Oxidation of α-naphthol in benzene–methanol afforded the half methyl ester (6), which likewise must be formed via the quinone (2) and the anhydride (3).

An interesting observation was that, under the conditions used in these oxidations with perbenzoic and peracetic acid, perphthalic acid reacted to only a negligible extent.

Oxidation of diarylamines. Tokumaru *et al.*[17] found that addition of perbenzoic acid in ether to an ethereal solution of 4,4′-dimethoxydiphenylamine (1) gives a

$$(\underline{p}\text{-}CH_3OC_6H_4)_2NH + C_6H_5CO_3H \xrightarrow[70\%]{} (\underline{p}\text{-}CH_3OC_6H_4)_2N{=}O$$

(1) 0.6 g. in ether   0.66% in ether   (2) m.p. 150–151°

crystalline precipitate of 4,4′-dimethoxydiphenylnitroxide (2), a substance of a type regarded by Wieland as having tetravalent oxygen.[18]

[1] G. Braun, *Org. Syn., Coll. Vol.*, **1**, 431 (1941)
[2] J. R. Moyer and N. C. Manley, *J. Org.*, **29**, 2099 (1964)
[3] M. Vilkas, *Bull. soc.*, 1501 (1959)
[4] A. Windaus and A. Lüttringhaus, *Ann.*, **481**, 119 (1930)
[5] For references, see L. F. Fieser and M. Fieser, "Steroids," pp. 141–143, Reinhold (1959)
[6] A. Windaus, O. Linsert, and H. J. Eckhardt, *Ann.*, **534**, 22 (1938)
[7] H. B. Henbest and T. I. Wrigley, *J. Chem. Soc.*, 4596 (1957)

[8]L. F. Fieser, T. Goto, and B. K. Bhattacharyya, *Am. Soc.*, **82**, 1693 (1960)
[9]O. Wintersteiner and M. Moore, *Am. Soc.*, **65**, 1507 (1943)
[10]P. Karrer, Ch. Cochand, and N. Neuss, *Helv.*, **29**, 1836 (1946)
[11]J. Böeseken and G. Slooff, *Rec. trav.*, **49**, 100 (1930)
[12]P. Karrer and L. Schneider, *Helv.*, **30**, 859 (1947)
[13]P. Karrer, R. Schwyzer, and H. Neuwirth, *Helv.*, **31**, 1210 (1948)
[14]P. Karrer and F. Haab, *Helv.*, **32**, 950 (1949)
[15]P. Karrer and Th. Hohl, *Helv.*, **32**, 1932 (1949)
[16]H. Fernholz, *Ber.*, **84**, 110 (1951)
[17]K. Tokumaru, H. Sakurago, and O. Simamura, *Tetrahedron Letters*, 3945 (1964)
[18]H. Wieland and M. Offenbäcker, *Ber.*, **43**, 2111 (1914)

Perchloric acid, $HClO_4$. Available concentrations: 20–73% in water; 0.1 *N* in acetic acid (Baker, Fisher). Review in English with 277 references.[1]

Acidic strength. That the substance is the strongest of mineral acids is evident from the following list of ionization constants in anhydrous acetic acid.[1]

$HClO_4$	1.6×10^{-4}	HCl	1.4×10^{-7}
HBr	4×10^{-5}	HNO_3	4.2×10^{-8}
H_2SO_4	6×10^{-7}	CCl_3CO_2H	2.3×10^{-10}

Cleavage of rings containing oxygen. Cleavage of an epoxide to a diol with hydrochloric or sulfuric acid in aqueous acetone or dioxane may proceed poorly because of by-product conversion into a chlorohydrin or a sulfate ester. Since perchloric acid is incapable of comparable reactions, it is the reagent of choice. For example, a solution of 0.17 g. of the 5α,6α-epoxide (1) in 13 ml. of tetrahydrofurane was treated

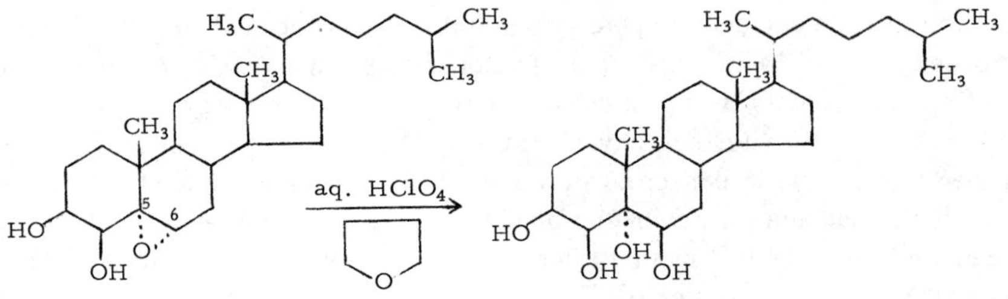
H3C
CH3
CH3
CH3
CH3
5
6
HO
O
OH
aq. HClO4
O
H3C
CH3
CH3
CH3
CH3
HO
OH
OH
OH

(1) (2)

with 0.8 ml. of 30% perchloric acid and let stand at 22° for 7 hrs.[2] Dilution with water and two crystallizations afforded the pure tetraol (2).

In exploring, as a model case, the conversion of Δ^4-cholestene-3-one (3) into cholestane-3,6-dione (7), a Merck group[3] converted 750 mg. of the 3-ethyleneketal (4) into the 5α,6α-epoxide (5) by reaction with perbenzoic acid and treated a solution of (5) in tetrahydrofurane with 3 *N* aqueous perchloric acid solution at room temperature to effect hydrolysis of both the ethyleneketal group at C_3 and the 5α,6α-epoxide group. Treatment of (6) with base effected dehydration and isomerization to the

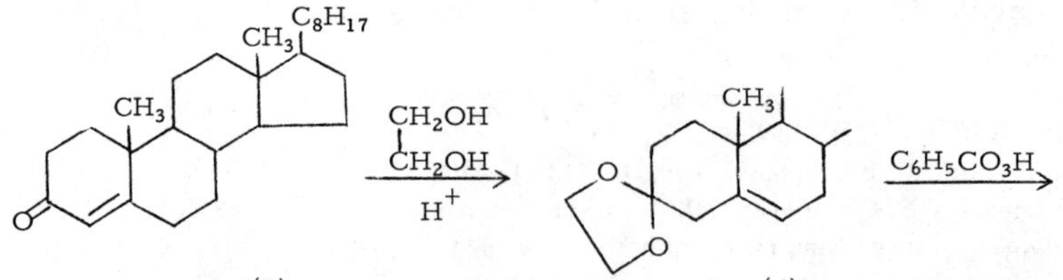
C8H17
CH3
CH3
O
CH2OH
CH2OH
H+
CH3
O
O
C6H5CO3H

(3) (4)

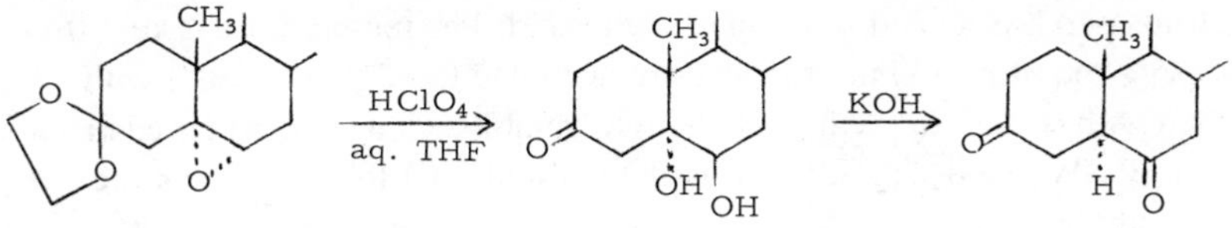
HClO4
aq. THF
KOH

(5) (6) (7)

3,6-dione. Zderic and Limon[4] cleaved the 3,20-bisethyleneketal (8) to the diketone (9) by treatment with 3 *N* aqueous perchloric acid in tetrahydrofurane at room temperature for 3 hrs. In a parallel run with hydrochloric acid in acetic acid the yield was only 65%.

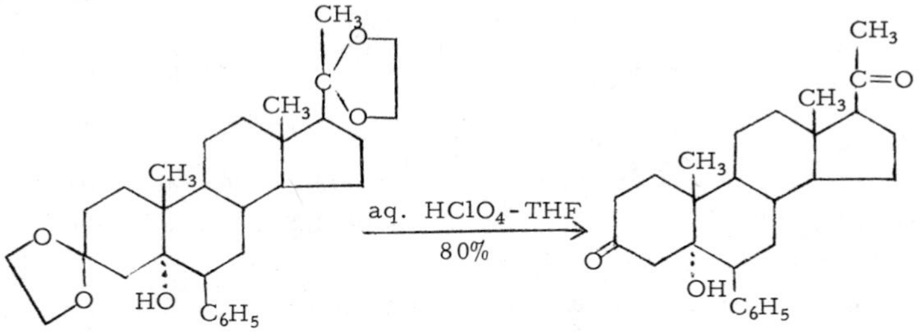
aq. HClO4-THF
80%

(8) (9)

Schmidlin and Wettstein[5] investigated the deketalization of the 3,20-bisethyleneketal (10) under acid catalysis. One complication is that an acid catalyst may

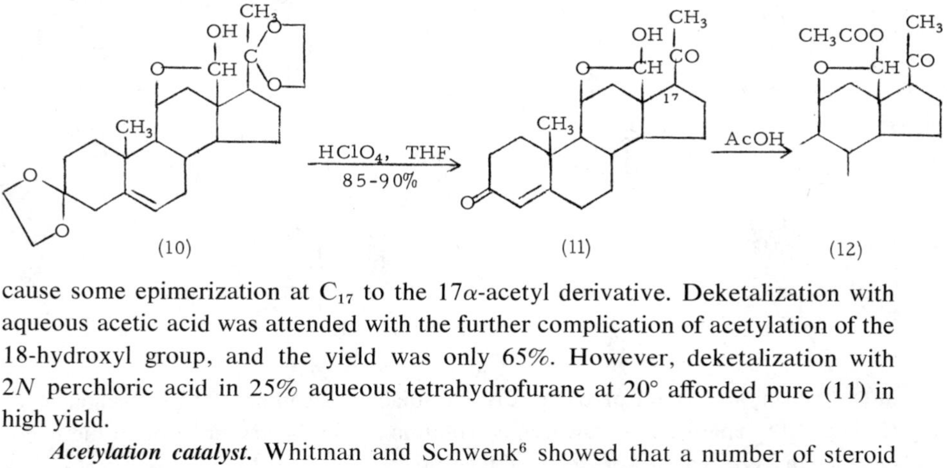
HClO4, THF
85-90%
AcOH

(10) (11) (12)

cause some epimerization at C_{17} to the 17α-acetyl derivative. Deketalization with aqueous acetic acid was attended with the further complication of acetylation of the 18-hydroxyl group, and the yield was only 65%. However, deketalization with 2*N* perchloric acid in 25% aqueous tetrahydrofurane at 20° afforded pure (11) in high yield.

Acetylation catalyst. Whitman and Schwenk[6] showed that a number of steroid alcohols can be acetylated in high yield by suspending 1 g. of steroid in 10 ml. of acetic acid and 3 ml. of acetic anhydride, cooling to 18°, and adding 0.1 ml. of 5 *N*

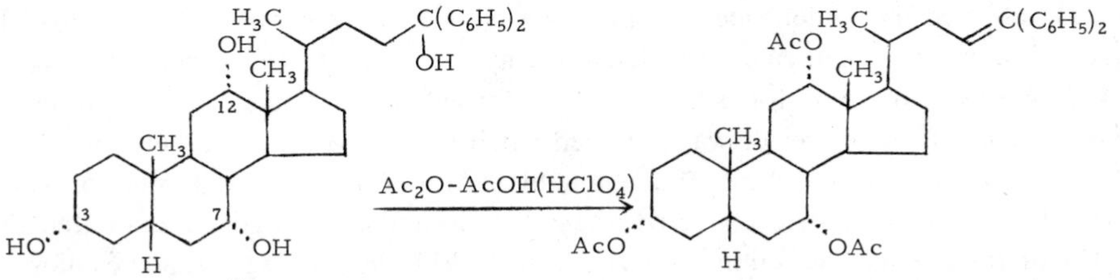
Ac2O-AcOH(HClO4)

(1) (2)

anhydrous perchloric acid solution in acetic acid. The temperature is kept from rising above 35°, and after 30 min. the mixture is cooled to 18° and treated with a little ice and then with water. A bile acid diphenylcarbinol such as (1), prepared in a first step of a Barbier-Wieland degradation, is dehydrated, and the product is the acetylated diphenylethylene (2).

The usual method of acetylation in pyridine is not applicable to compounds such as 5α-bromo-6β-chlorocholestane-3β-ol (3). Ziegler and Shabica[7] cooled a suspension of 0.5 g. of needles of (3) in 10 ml. of acetic acid and 2.5 ml. of acetic

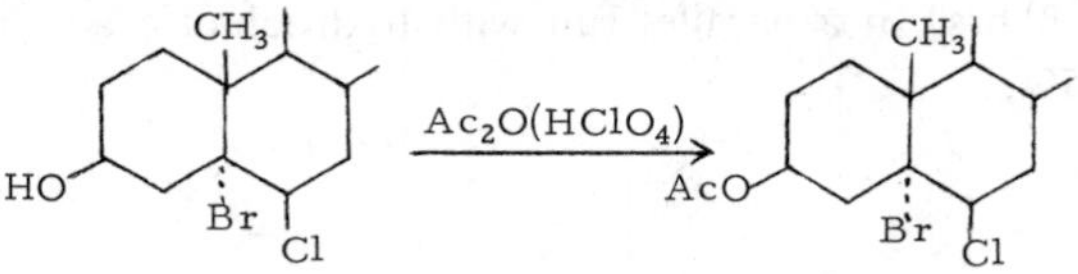
CH3
HO
Br
Cl
Ac2O(HClO4)
CH3
AcO
Br
Cl

(3) (4)

anhydride with 5 drops of 70% perchloric acid. The temperature rose to 20°, the needles of (3) were rapidly replaced by plates of the acetate (4), and after 0.5 hr. workup afforded 0.55 g. of (4).

Acetylation of cevadine (5) with acetic anhydride in pyridine gives the 4,16-diacetate, but a further reaction occurs under catalysis by perchloric acid. Stoll and Seebeck[8] stirred a suspension of 20 g. of cevadine (5) in 120 ml. of acetic anhydride and slowly added 4 ml. of 70% perchloric acid. The temperature rose to 60–70°, and

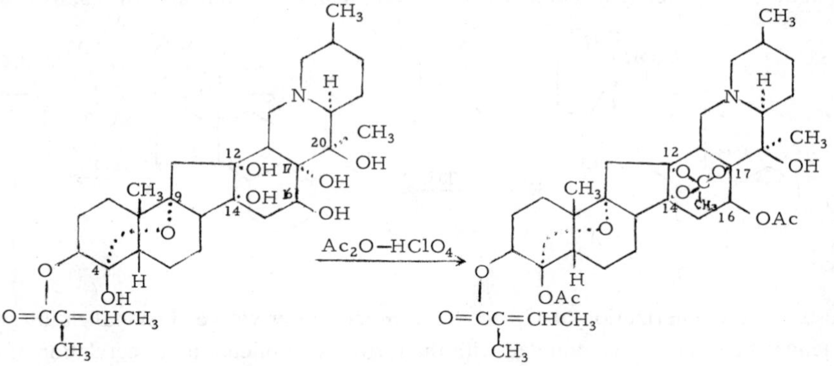
CH3
H
N
20
CH3
OH
12
OH
17
OH
OH
16
OH
CH3
9
14
O
4
H
OH
O=CC=CHCH3
CH3
Ac2O–HClO4
CH3
H
N
CH3
OH
12
17
CH3
14
16
OAc
H
OAc
O=CC=CHCH3
CH3

(5) (6)

after 2 hrs. crystals of the perchlorate of 4,16-diacetylcevadine 12,14,17-orthoacetate (6) began to separate from the yellow solution. After 24 hrs. addition of methanol afforded in 2 crops a total of 24.6 g. of the salt.

Reichstein and co-workers[9] investigated the acetylation of methyl 3α,12β-dihydroxyetianate (7) by four procedures. Reaction with acetic anhydride in pyridine at 20° afforded the 3-monoacetate; the 3α-group is equatorial, but the 12β-hydroxyl group is axial and hindered by the angular methyl group and by the ester group at C_{17}. Acetylation in pyridine for 3 hrs. at 100° afforded the 3,12-diacetate, but in poor yield. A much better result was obtained when a suspension of 510 mg. of (7) in 6 ml. of acetic acid and 2 ml. of acetic anhydride was treated with 0.6 ml. of 60% perchloric acid; diacetate collected after 40 min. and recrystallized amounted to 93% of the theory. Reaction with acetic anhydride containing *p*-toluenesulfonic acid for 26 hrs. at 20° and recrystallization afforded the diacetate in 90% yield.

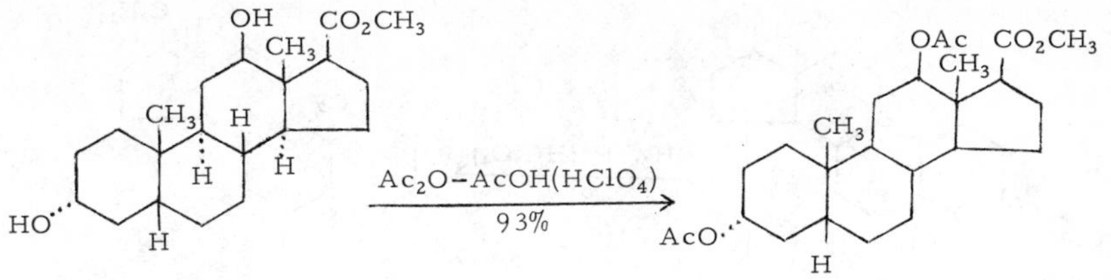

(7) (8)

Fritz and Schenk[10] effected acetylation with acetic anhydride and perchloric acid in ethyl acetate at room temperature, and this combination proved effective for conversion of the δ-keto acid (1) into the lactone (2).[11] In the original procedure of

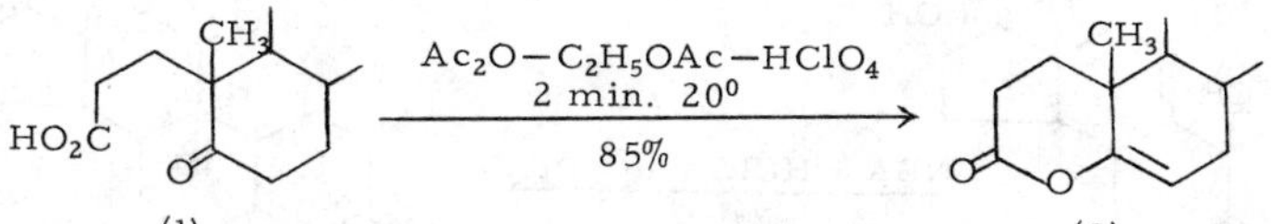

(1) (2)

Turner[12] this lactonization was effected in comparable yield by refluxing the keto acid with acetic anhydride and acetyl chloride for 40 hrs.

Activation of hydrogenation catalysts. An isolated steroid 5,6-double bond possesses only low-order reactivity, for example, it adds difluorocarbene but not dichloro- or dibromocarbene.[13] Hydrogenation of cholesterol with ordinary catalysts stops far short of completion, but Hershberg *et al.*[14] found that perchloric acid has a powerful activating effect. Cholesterol (1250 g.), which had been purified by careful crystallization from ethyl acetate, was dissolved in 17 l. of ethyl acetate, 25 g. of platinum oxide catalyst and 2.0 ml. of 71% perchloric acid were added, and hydrogenation was conducted at 40–50° and 15 p.s.i. pressure. The reaction was complete in 30 min., and pure cholestane-3β-ol was obtained in total yield of 88%. By-products, accounted for in the yields indicated, were coprostanol (β-addition), cholestane (hydrogenolysis), and cholestanyl acetate (ester interchange with the solvent).

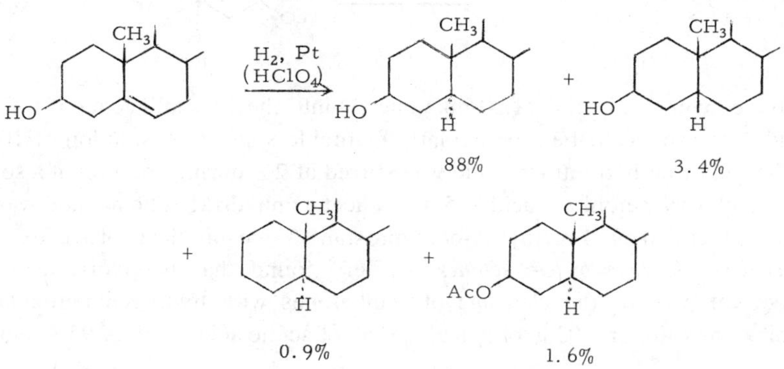

For activation of palladium for hydrogenolysis, *see* Palladium catalyst, hydrogenolysis.

Edward and Ferland[15] found that hydrogenolysis of δ-lactones to cyclic ethers in the presence of Adams catalyst in methanol is greatly accelerated by perchloric acid (0.03 ml. of 70% perchloric acid per 20 mg. of lactone). Surprisingly, several γ- and ε-lactones showed no uptake of hydrogen under the same conditions. No explanation of the unique reactivity of δ-lactones has been presented.

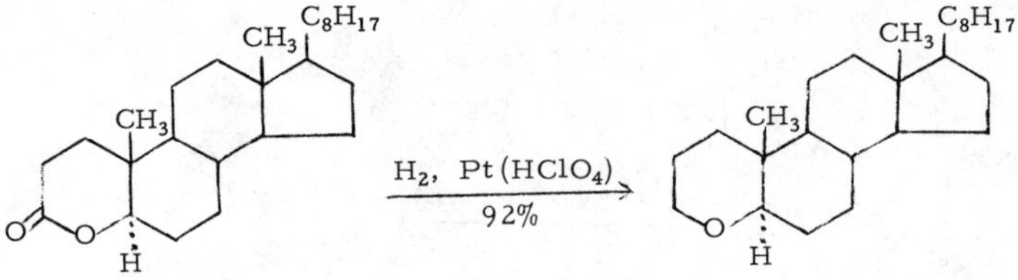

Generation of HOBr. For effecting addition of HOBr to the 9(11)-double bond of (1) to produce the *trans*-axial bromohydrin (2), Fried and Sabo[16] generated the

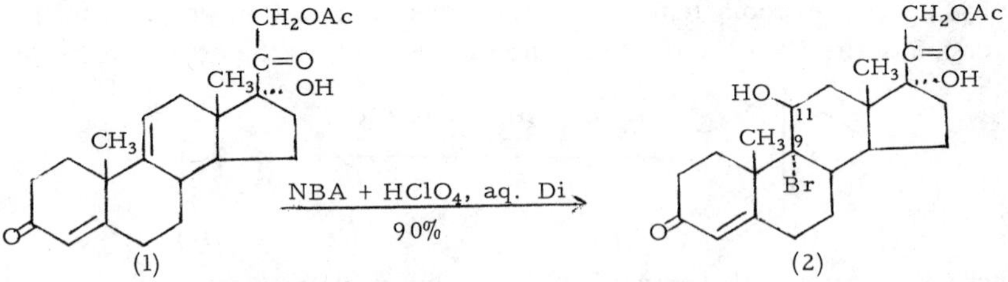

reagent from N-bromoacetamide and perchloric acid in aqueous dioxane and obtained (2) in yield of 90%. With sulfuric acid in place of perchloric acid the yield was only 45%. Barkley *et al.*[17] accomplished two comparable additions to $\Delta^{4,9(11)}$-diene-3-ones with hypobromous acid generated from N-bromosuccinimide in combination with perchloric acid in one case and with sulfuric acid in the other, and the yields were 95 and 84%.

Preparation of enol acetates. Barton *et al.*[18] used perchloric acid as catalyst for

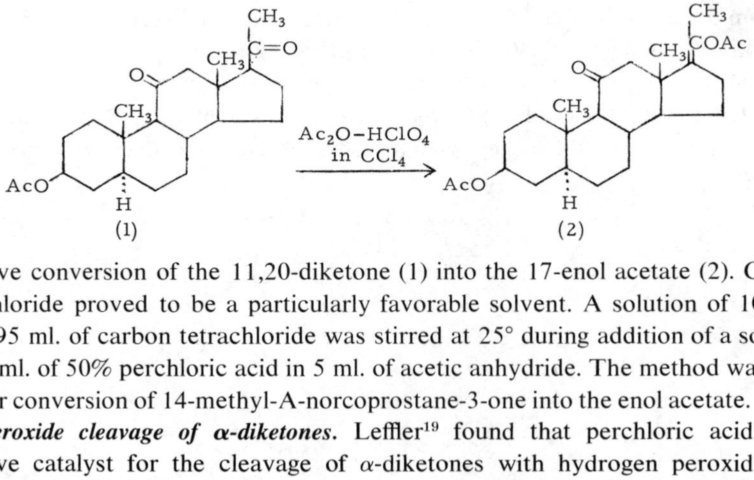

selective conversion of the 11,20-diketone (1) into the 17-enol acetate (2). Carbon tetrachloride proved to be a particularly favorable solvent. A solution of 10 g. of (1) in 95 ml. of carbon tetrachloride was stirred at 25° during addition of a solution of 0.2 ml. of 50% perchloric acid in 5 ml. of acetic anhydride. The method was used also for conversion of 14-methyl-A-norcoprostane-3-one into the enol acetate.

Peroxide cleavage of α-diketones. Leffler[19] found that perchloric acid is an effective catalyst for the cleavage of α-diketones with hydrogen peroxide. For example, a mixture of 1.02 g. of benzil, 25 ml. of acetic acid, 1 ml. of 95% hydrogen

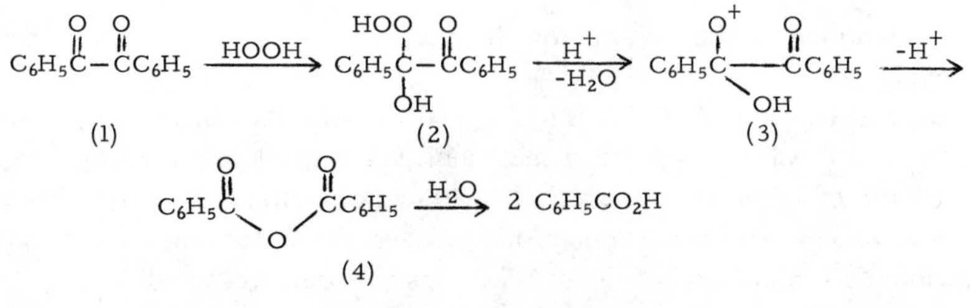

peroxide, and 1 ml. of 70% perchloric acid is refluxed for 15 min. (the color is discharged in 5 min.). The cleavage is known to yield the anhydride (4) as the initial product (*see* Perbenzoic acid, oxidation of α-diketones). A mechanism suggested by Leffler accounts for the catalytic effect of perchloric acid.

Other catalytic effects. Pettit and Bowyer[20] treated 0.14 g. of tigogenone in 3 ml. of ethanedithiol with 1 drop of 72% perchloric acid and noted separation of crystals after several minutes. The ethylenethioketal was isolated by chromatography, and the yield of crude product was 0.15 g.

Rammler and Dekker[21] used perchloric acid to catalyze the benzoylation of a secondary alcoholic group of a carbohydrate. Benzoylation of D-arabitol (1) in

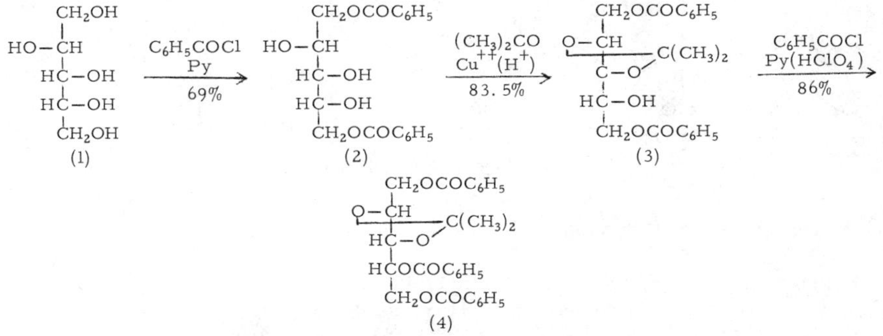

pyridine gives 1,5-di-O-benzoyl-D-arabitol (2), which was converted into the 2,3-acetonide (2,3-O-isopropylidene derivative), and a solution of 1.5 g. of this substance (3) in pyridine was treated with benzoyl chloride and a drop of 60% perchloric acid. After 18 hrs. at room temperature, workup afforded the 1,4,5-tribenzoyl derivative (4).

Burton and Praill[22] found perchloric acid to be an effective catalyst for the Thiele reaction. Whereas 2-methyl-1,4-naphthoquinone is inert to acetic anhydride containing sulfuric acid, it reacts in the presence of perchloric acid to give 2-methyl-1,3,4-triacetoxynaphthalene in moderate yield. The quinone (0.02 mole) is dissolved in 34 ml. of acetic anhydride, 0.01 mole of 72% perchloric acid is added by drops, and the solution is let stand at room temperature for 20 hrs. Workup and crystallization from methanol and then from benzene–ligroin gives colorless prisms.

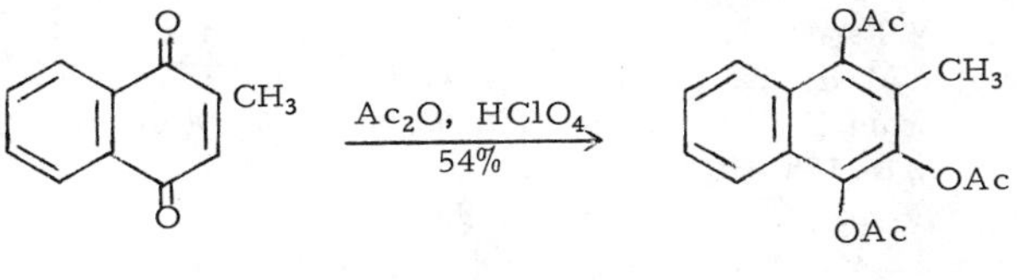
O
CH3
Ac2O, HClO4
54%
OAc
CH3
OAc
OAc

Zderic *et al.*[23] converted 11β,12β-dihydroxysteroids into their acetonides in

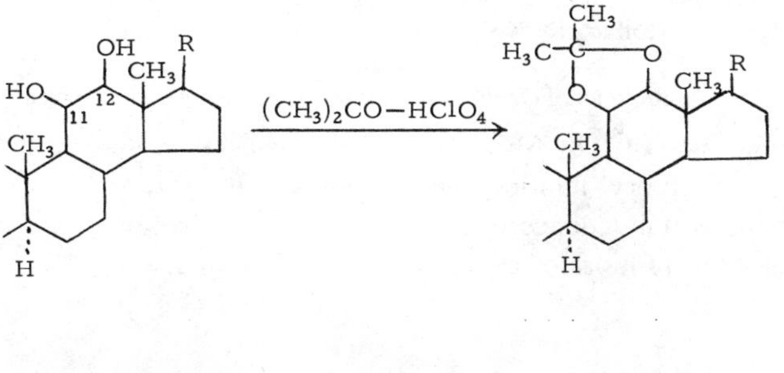
OH
R
CH3
HO
12
11
CH3
H
(CH3)2CO–HClO4
CH3
H3C
O
R
CH3
CH3
H

almost quantitative yield by treatment with acetone and a few drops of 70% perchloric acid. Use of perchloric acid as catalyst was also found to greatly improve the condensation of acetone with the riboside thioinosine to form the 2′,3′-acetonide.[24]

Cyclodehydration. Ghera and Sondheimer[25] found treatment with perchloric acid and acetic anhydride in benzene–acetic acid at room temperature the most effective method for effecting the cyclodehydration of the diol (1) to the hydrocarbons (2) and (3).

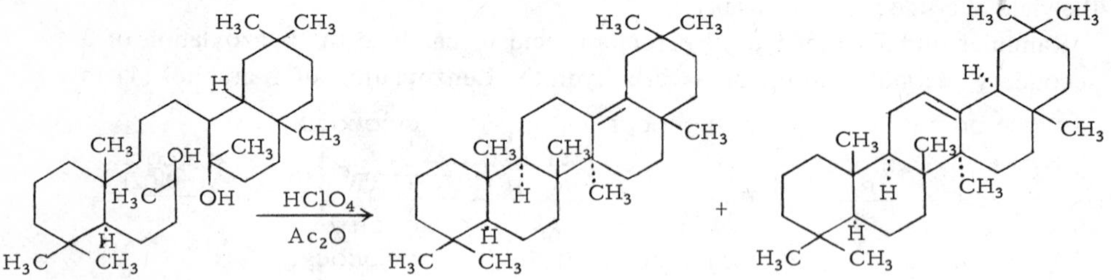

(1) (2) (3)

[1] G. N. Dorofeenko, S. V. Krivun, V. I. Dulenko, and Yu. A. Zhdanov, *Russian Chem. Rev.*, **34**, 88 (1965)
[2] L. F. Fieser and T. Goto, *Am. Soc.*, **82**, 1693 (1960)
[3] G. I. Poos, G. E. Arth, R. E. Beyler, and L. H. Sarett, *Am. Soc.*, **75**, 422 (1953)
[4] J. A. Zderic and D. C. Limon, *Am. Soc.*, **81**, 4570 (1959)
[5] J. Schmidlin and A. Wettstein, *Helv.*, **46**, 2799 (1963)
[6] B. Whitman and E. Schwenk, *Am. Soc.*, **68**, 1865 (1946)
[7] J. B. Ziegler and A. C. Shabica, *Am. Soc.*, **74**, 4891 (1952)
[8] A. Stoll and E. Seebeck, *Helv.*, **35**, 1942 (1952)
[9] S. Pataki, K. Meyer, and T. Reichstein, *Helv.*, **36**, 1295 (1953)
[10] J. S. Fritz and G. H. Schenk, *Anal. Chem.*, **31**, 1808 (1959)
[11] P. N. Rao and L. R. Axelrod, *J. Chem. Soc.*, 1356 (1965); B. E. Edwards and P. N. Rao, *J. Org.*, **31**, 324 (1966)
[12] R. B. Turner, *Am. Soc.*, **72**, 579 (1950)
[13] L. H. Knox, E. Velade, S. Berger, D. Cuadriello, P. W. Landis, and A. D. Cross, Am. Soc., **85**, 1851 (1963)
[14] E. B. Hershberg, E. Oliveto, M. Rubin, H. Staeudle, and L. Kuhlen, *Am. Soc.*, **73**, 1144 (1951)
[15] J. T. Edward and J. M. Ferland, *Chem. Ind.*, 975 (1964)
[16] J. Fried and E. F. Sabo, *Am. Soc.*, **75**, 2273 (1953); *see also* H. B. Henbest, E. R. H. Jones, A. A. Wagland, and T. I. Wrigley, *J. Chem. Soc.*, 2977 (1955)
[17] L. B. Barkley, M. W. Farrar, W. S. Knowles, and H. Raffelson, *Am. Soc.*, **76**, 5017 (1954)
[18] D. H. R. Barton, R. M. Evans, J. C. Hamlet, P. G. Jones, and T. Walker, *J. Chem. Soc.*, 747 (1954); D. H. R. Barton, D. A. J. Ives, and B. R. Thomas, *ibid.*, 903 (1954)
[19] J. E. Leffler, *J. Org.*, **16**, 1785 (1951)
[20] G. R. Pettit and W. J. Bowyer, *J. Org.*, **25**, 84 (1960)
[21] D. H. Rammler and C. A. Dekker, *J. Org.*, **26**, 4615 (1961)
[22] H. Burton and P. F. G. Praill, *J. Chem. Soc.*, 755 (1952)
[23] J. A. Zderic, H. Carpio, and C. Djerassi, *Am. Soc.*, **82**, 446 (1960)
[24] J. A. Zderic, J. G. Moffatt, D. Kau, K. Gerzon, and W. E. Fitzgibbon, *J. Med. Chem.*, **8**, 275 (1965)
[25] E. Ghera and F. Sondheimer, *Tetrahedron Letters*, 3887 (1964)

Perchloryl fluoride, $FClO_3$, Mol. wt. 102.46, gas, b.p. −46.8°. Supplier: Pennsalt Chem. Corp. Perchloryl fluoride itself is stable, but mixtures with oxidizable substances are potentially dangerous. Details for safe handling are described in a bulletin "Perchloryl Fluoride" available from the supplier.

Perchlorylation. Pennsalt chemists[1] found that when perchloryl fluoride is bubbled into a suspension of aluminum chloride in benzene a reaction analogous to a Friedel-Crafts reaction occurs and gives perchlorylbenzene, a colorless liquid,

$$C_6H_6 + FClO_3 + AlCl_3 \xrightarrow[10-20^0]{} C_6H_5ClO_3 + HCl + AlCl_2F$$

b.p. 232°. This substance on alkaline hydrolysis yields phenol; on nitration it affords 3-nitroperchlorobenzene. The perchlorylation reaction is not applicable to nitrobenzene and similarly deactivated aromatics.

Fluorination of active-methylene compounds. The Pennsalt group[2] treated diethyl malonate with two equivalents of sodium ethoxide in ethanol and passed in perchloryl fluoride with cooling and reported isolation of diethyl difluoromalonate in

$$CH_2(CO_2C_2H_5)_2 \xrightarrow{2\ NaOC_2H_5} Na_2C(CO_2C_2H_5)_2 \xrightarrow[10-15^0]{FClO_3} F_2C(CO_2C_2H_5)_2$$

high yield. However, a reinvestigation[3] showed that the reaction is far more complicated and that the following five products can be isolated: diethyl difluoromalonate, diethyl ethylmalonate, diethyl malonate, diethyl ethylfluoromalonate, and diethyl fluoromalonate.

Preparation of α-fluoroketones and vinylogs. Gabbard and Jensen[4] of the Ben May Laboratory of Cancer Research converted cholestanone into the pyrrolidylenamine and bubbled perchloryl fluoride into a benzene solution of this derivative until an initial orange color was discharged (30 sec.). Workup and crystallization afforded 2α-fluorocholestane-3-one in good yield. Joly and Warnant[5] studied the

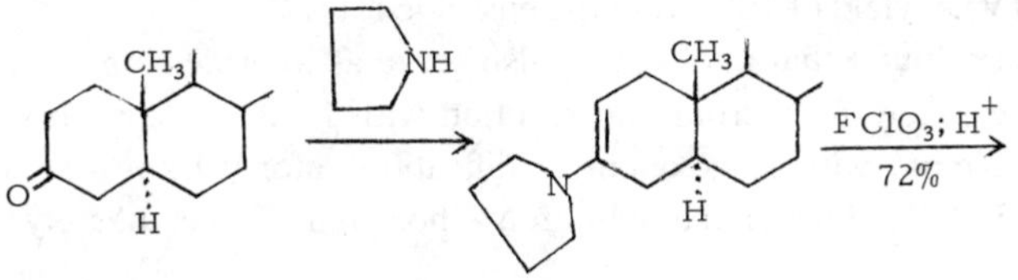
CH3
NH
CH3
FClO3; H+
72%
CH3
F
O
H
N
H
O
H

action of the reagent on the enamines of seven Δ^4-3-ketosteroids (1). A solution of the ketone in 8 volumes of methanol was refluxed under nitrogen and treated with 1 volume of pyrrolidine. The enamine, collected after cooling, was dissolved in 10 volumes of methanol containing 10% of water, and the solution was kept at

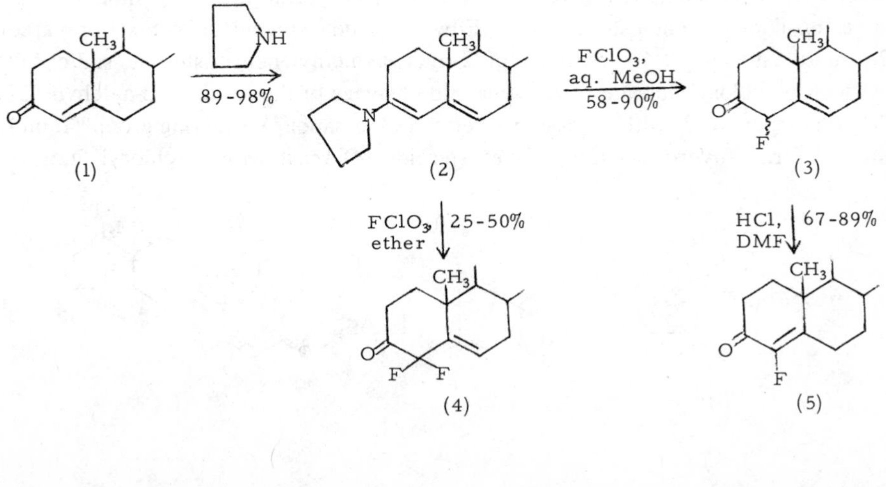
CH3
NH
89-98%
CH3
FClO3,
aq. MeOH
58-90%
CH3
O
N
O
F
(1)
(2)
(3)
FClO3, ether
25-50%
HCl, DMF
67-89%
CH3
CH3
O
F F
O
F
(4)
(5)

−20° by cooling while passing in perchloryl fluoride for about 20 min. Excess reagent was removed with a stream of nitrogen and the product, the 4-fluoro-Δ^5-3-ketone (3), was precipitated with ice water. Isomerization to the conjugated ketone (5) was accomplished by dissolving crude (3) in 10 volumes of dimethylformamide, adding 1 volume of concd. hydrochloric acid, and allowing the solution to stand at room temperature for 20 hrs. The Ben May group[6] investigated the reaction of $\Delta^{3,5}$-dienamines of type (2) with perchloryl fluoride in ether at 0° and obtained a mixture from which the 4,4-difluoro-Δ^5-ketone (4) was isolated in yields of 25–50%. When the solvent used was ether containing 4% of pyridine, the product was a 3:1 mixture of the difluoro ketone (4) and the conjugated 4-fluoroketone (5).

The enamine method is not applicable to 20-ketosteroids, since these ketones only rarely have been found capable of conversion into enamines. However, the Ben May group[7] found that the enamide III, prepared from a Δ^{16}-20-ketone (I)

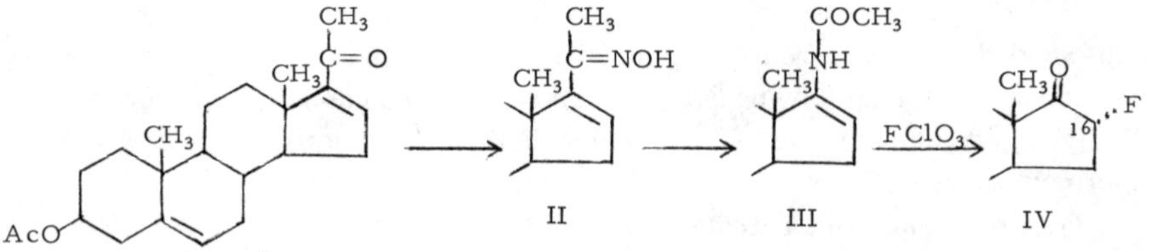
CH3
C=O
CH3
CH3
AcO
CH3
C=NOH
CH3
COCH3
NH
CH3
FClO3
CH3
F
16

I II III IV

by Beckmann rearrangement of the oxime (II), reacts rapidly with perchloryl fluoride in pyridine at room temperature. Removal of one third of the pyridine, acidification to pH 2, extraction with ether, chromatography, and crystallization gave the pure 16α-fluoro-17-ketone (IV) in yield of 70% from the enamide III.

The Ben May workers[8] found that enol ethers also serve as intermediates. Thus the enol ethers of three 3-keto-5α-steroids on reaction with $FClO_3$ in pyridine at room temperature for 2 min., with subsequent acidification, afforded 2α-fluoro-3-ketones in yields of 75–90%. The enol ether of Δ^4-cholestene-3-one, namely 3-

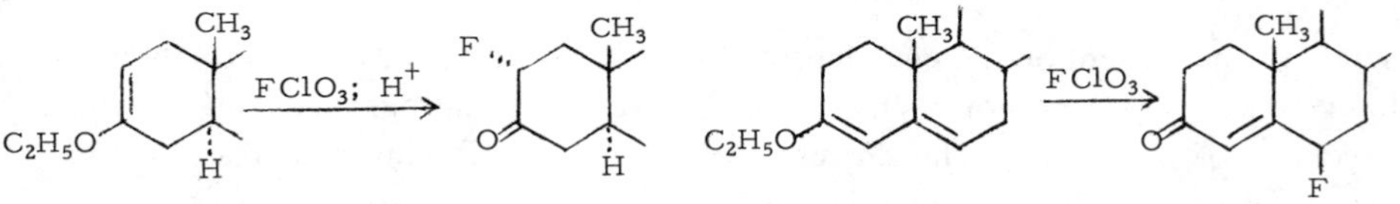
CH3
C2H5O
H
FClO3; H+
F
CH3
O
H
CH3
C2H5O
FClO3
CH3
O
F

ethoxy-$\Delta^{3,5}$-cholestadiene, on reaction with perchloryl fluoride in pyridine at −20° gave 6β-fluoro-Δ^4-cholestene-3-one. Edwards and Ringold[9] (Syntex) prepared 2α-fluorotestosterone by reaction of 2-hydroxymethylenetestosterone sodio salt (1) with perchloryl fluoride in benzene and cleavage of the resulting 2-aldehydo-2-fluoro compound (2) with potassium acetate in methanol. A Schering group[10] found that 16-formyl(hydroxymethyl)-17-ketosteroids (4) react with perchloryl fluoride

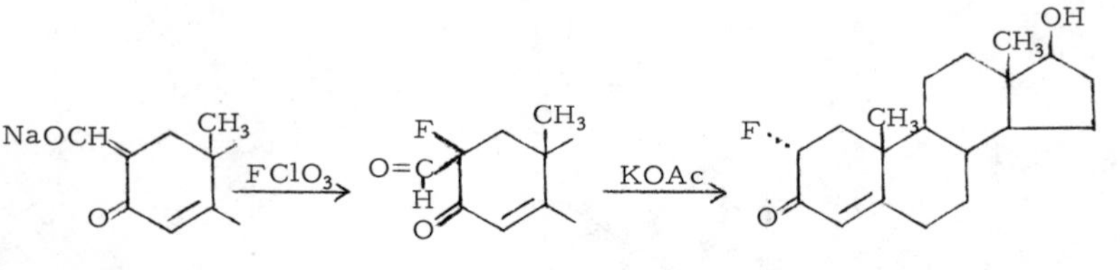
NaOCH
CH3
O
FClO3
F
CH3
O=C
H
O
KOAc
OH
CH3
CH3
F
O

(1) (2) (3)

in *t*-butanol in the presence of potassium *t*-butoxide with formation of the 16,16-difluoro-17-ketosteroids (7) in yields of 25–30%. The mechanism suggested is shown in the formulation:

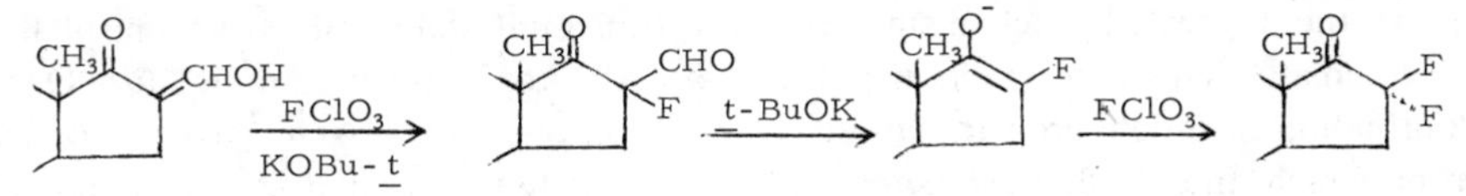
CHOH
FClO3
KOBu-t
CHO
t-BuOK
FClO3

(4) (5) (6) (7)

A Pfizer group[11] developed a synthesis of 6-fluorosteroids from the dienol acetates of Δ^4-3-ketones. Under the conditions of enol acetylation, the 17α-hydroxyl group of the pregnadienedionediol monoacetate (I) was acetylated as well, giving II.

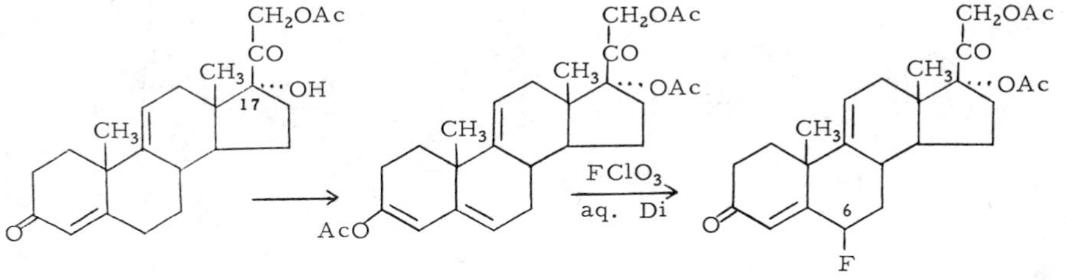
CH2OAc
CO
OH
OAc
AcO
FClO3
aq. Di

I II III

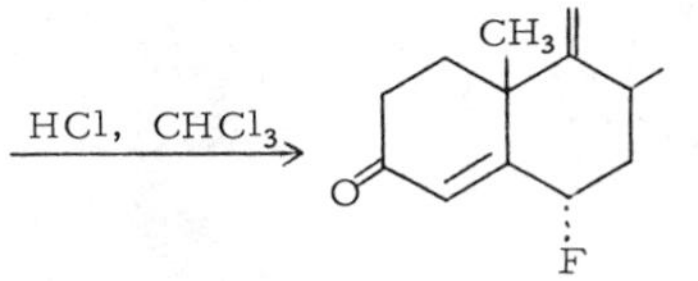
HCl, CHCl3

IV

Reaction of II with perchloryl fluoride in 65% aqueous dioxane gave a 6-epimer mixture rich in the 6β-fluoro compound III but difficult to separate. However, treatment of the mixture in chloroform with hydrogen chloride effected epimerization to the equatorial 6α-fluoro compound IV, which was thus obtainable from II in overall yield of 45%.

Upjohn chemists[12] converted 3,11-diketosteroids into the 3-enol acetates, 3-enol ethers, and 3-enamines and, by reaction of these derivatives with perchloryl fluoride, obtained the 4- and 6-fluoro derivatives.

Oxofluorination. This is the name given by Neeman[13] to a new electrophilic ambident reaction of perchloryl fluoride with olefins of the type of indene (1). A solution of indene in aqueous dioxane is kept saturated with $FClO_3$ at room

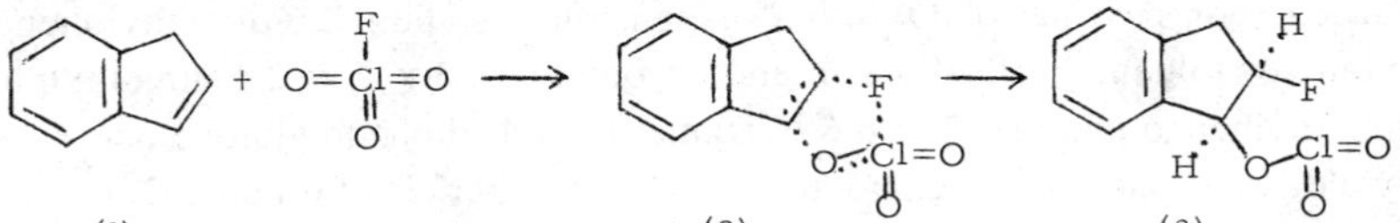

(1) (2) (3)

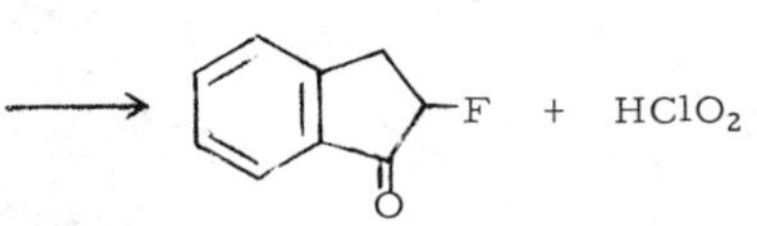
+ HClO2

(4)

temperature for 4 hrs. Column chromatography, fractional crystallization, and gas chromatography afforded five products, the most abundant of which accounted for 46% of the total and was characterized as 2-fluoroindanone (4). The formulation shows the suggested mechanism: (2) is the transition state and (3) an unisolated intermediate. The new reaction afforded a route to 7α-fluoroestradiol (9). Oxofluorination of 6-dehydroestradiol diacetate (5) in aqueous dioxane gave a mixture from which the 7α-fluoro-6-ketone (6) was isolated. Borohydride reduction at −5° gave the *cis*-fluorohydrin (7) as 3,17-diacetate. The free hydroxyl group was eliminated by reduction of the mesylate (8), and hydrolysis afforded 7α-fluoroestradiol (9).

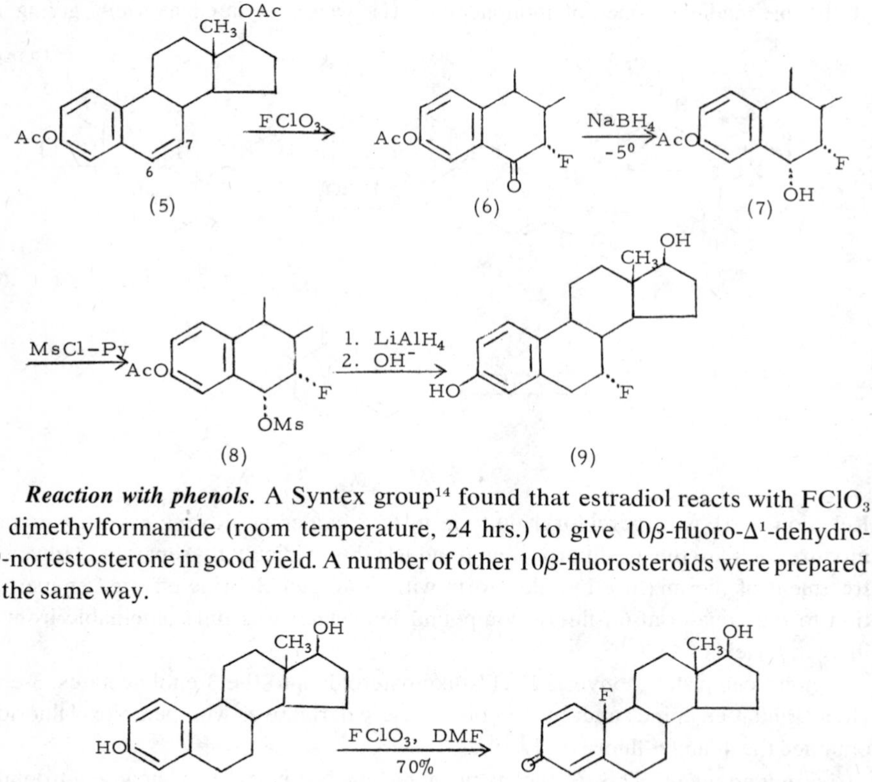
OAc
CH_3
AcO
7
6
$FClO_3$
AcO
F
O
$NaBH_4$
-5°
AcO
F
OH
(5)
(6)
(7)
OH
CH_3
MsCl−Py
AcO
F
OMs
1. $LiAlH_4$
2. OH^-
HO
F
(8)
(9)

Reaction with phenols. A Syntex group[14] found that estradiol reacts with $FClO_3$ in dimethylformamide (room temperature, 24 hrs.) to give 10β-fluoro-Δ¹-dehydro-19-nortestosterone in good yield. A number of other 10β-fluorosteroids were prepared in the same way.

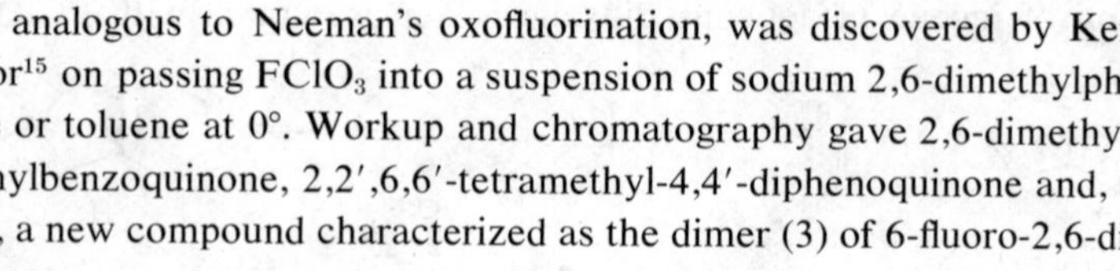
OH
CH_3
HO
$FClO_3$, DMF
70%
OH
CH_3
F
O

A reaction in which perchloryl fluoride functions as an ambident electrophile, somewhat analogous to Neeman's oxofluorination, was discovered by Kende and MacGregor[15] on passing $FClO_3$ into a suspension of sodium 2,6-dimethylphenoxide in pentane or toluene at 0°. Workup and chromatography gave 2,6-dimethylphenol, 2,6-dimethylbenzoquinone, 2,2′,6,6′-tetramethyl-4,4′-diphenoquinone and, in about 20% yield, a new compound characterized as the dimer (3) of 6-fluoro-2,6-dimethyl-

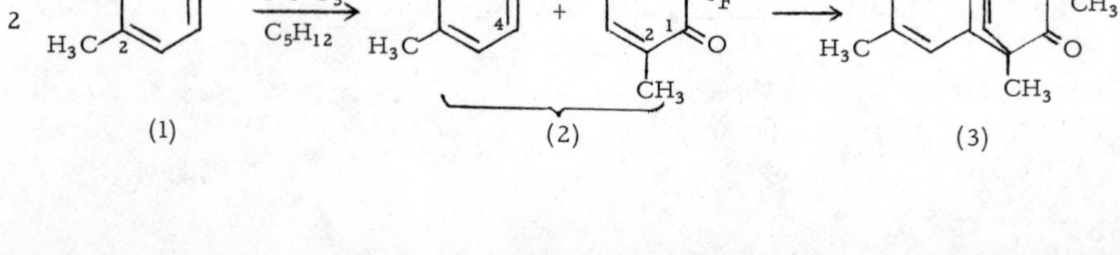
CH_3
Na^+O^-
2
H_3C
$FClO_3$
C_5H_{12}
F
CH_3
O
H_3C
+
CH_3
F
O
CH_3
F
CH_3
O
F
CH_3
H_3C
O
CH_3
(1)
(2)
(3)

2,4-cyclohexadienone (2). Dimerization involves the 4,5-double bond of one molecule and the 2,5-diene system of another.

Another reaction affording dienones was discovered by Taub.[16] 3,5-Dimethoxyphenol (4) reacts with $FClO_3$ in pyridine at 20° to give in good yield a 1:1 mixture of dienones (5) and (6). The 2- and 4-fluoro derivatives of (4) were identified as intermediates.

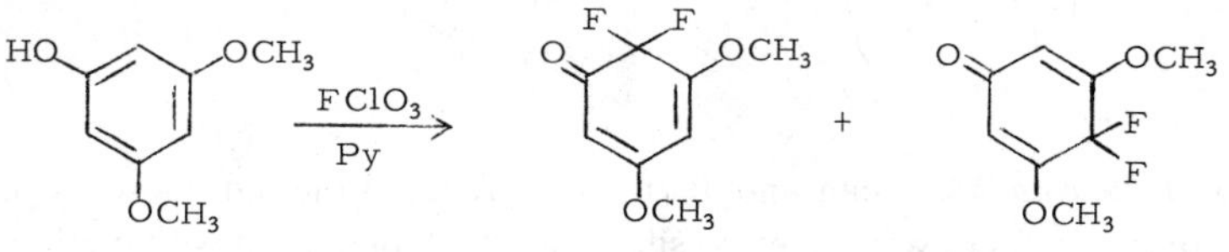

(4) (5) (6)

Reactions with nitroalkanes and oximes. Shechter and Roberson[17] found that the sodium salts of secondary nitroalkanes are convertible by reaction of the anions with

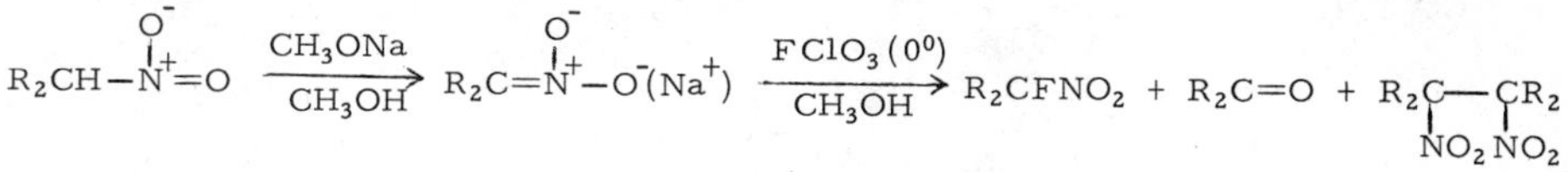

perchloryl fluoride into fluoronitro compounds in moderate yields (36–42%); the corresponding ketones are also formed (32–55%), along with smaller amounts of *vic*-dinitro compounds. Primary nitroalkanes give aldehydes as the principal products. Among other examples of this and related reactions, Freeman[18] reported conversion of 5-nitro-6-phenylnorbornene (1), as anion, into 2-fluoro-2-nitro-3-phenylnorbornene (2) in good yield, along with a small amount of the ketone (3).

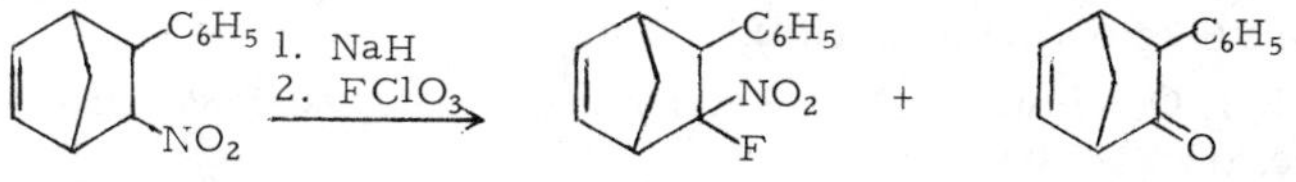

(1) (2) 70% (3) 3.5%

Simple ketoximes are converted into the parent ketones on treatment with base and then with perchloryl fluoride.

$$(C_6H_5)_2C{=}NOH \xrightarrow{NaH} (C_6H_5)_2C{=}NO^-(Na^+) \xrightarrow{FClO_3} (C_6H_5)_2C{=}O$$

Fluorinated heterocycles. Schuetz *et al.*[19] found that 2-fluorothiophene and 2-fluorothianaphthene can be prepared in yields of 49 and 70% by the exothermal reactions of gaseous perchloryl fluoride with the corresponding organolithium compound in ether. Phenyllithium gave a negligible amount of fluorobenzene.

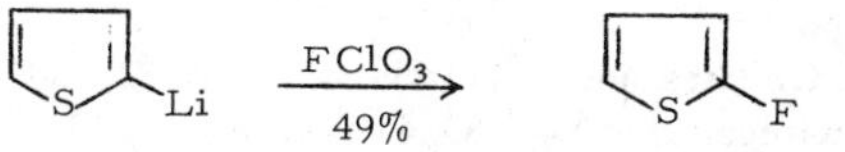

Reaction with β-keto esters.[20] A β-keto ester (1), as the carbanion, reacts with the reagent to give the α-fluoro-β-keto ester (2), which on "ketone splitting" with methanolic sodium hydroxide gives the 3-fluoro-2-ketone (3). On treatment with a catalytic amount of sodium ethoxide, (2) undergoes a retro ester-condensation to give the α-fluorocarboxylic acid ester (4).

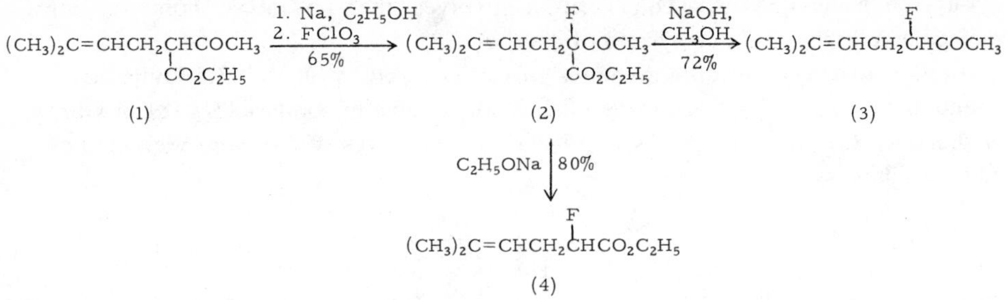

$$(CH_3)_2C=CHCH_2\underset{\underset{CO_2C_2H_5}{|}}{CH}COCH_3 \xrightarrow[\underset{65\%}{2.\ FClO_3}]{1.\ Na,\ C_2H_5OH} (CH_3)_2C=CHCH_2\underset{\underset{CO_2C_2H_5}{|}}{\overset{\overset{F}{|}}{C}}COCH_3 \xrightarrow[72\%]{\underset{CH_3OH}{NaOH,}} (CH_3)_2C=CHCH_2\overset{\overset{F}{|}}{C}HCOCH_3$$

$$(1) \hspace{5cm} (2) \hspace{5cm} (3)$$

$$C_2H_5ONa \downarrow 80\%$$

$$(CH_3)_2C=CHCH_2\overset{\overset{F}{|}}{C}HCO_2C_2H_5$$

$$(4)$$

Indirect fluorination. Dean and Pattison[21] prepared diethyl fluorosuccinate (4) in the following indirect way. Diethyl succinate was condensed with diethyl oxalate

$$\begin{matrix}\underset{\underset{\underset{\underset{CO_2C_2H_5}{|}}{CH_2}}{\underset{|}{CH_2}}}{\overset{\overset{\overset{\overset{CO_2C_2H_5}{|}}{CH_2}}{|}}{CO_2C_2H_5}}\end{matrix} + (CO_2C_2H_5)_2 \xrightarrow{C_2H_5OK} \begin{matrix}\underset{\underset{\underset{\underset{CO_2C_2H_5}{|}}{K\overset{|}{C}COCO_2C_2H_5}}{|}}{\overset{\overset{\overset{CO_2C_2H_5}{|}}{CH_2}}{|}}\end{matrix} \xrightarrow{FClO_3} \begin{matrix}\underset{\underset{\underset{\underset{CO_2C_2H_5}{|}}{F\overset{|}{C}COCO_2C_2H_5}}{|}}{\overset{\overset{\overset{CO_2C_2H_5}{|}}{CH_2}}{|}}\end{matrix}$$

$$(1) \hspace{4cm} (2) \hspace{4cm} (3)$$

$$\xrightarrow{KHCO_3} \begin{matrix}\underset{\underset{\underset{CO_2C_2H_5}{|}}{CHF}}{\overset{\overset{\overset{CO_2C_2H_5}{|}}{CH_2}}{|}}\end{matrix}$$

$$(4)$$

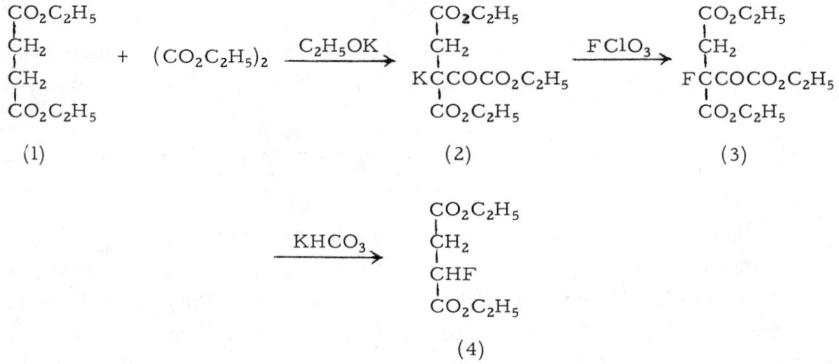

and potassium ethoxide to form (2), the ethoxalyl group of which promotes fluorination by perchloryl fluoride to give (3). The fluoride atom in (3) permits mild alkaline cleavage of the β-keto ester to give (4).

[1]C. E. Inman, R. E. Oesterling, and E. A. Tyczkowski, *Am. Soc.*, **80**, 5286 (1958)

[2]*Idem, ibid.*, **80**, 6533 (1958)

[3]H. Gershon, J. A. A. Renwick, W. K. Wynn, and R. D'Ascoli, *J. Org.*, **31**, 916 (1966)

[4]R. B. Gabbard and E. V. Jensen, *J. Org.*, **23**, 1406 (1958)

[5]R. Joly and J. Warnant, *Bull. Soc.*, 569 (1961)

[6]S. Nakanishi, R. L. Morgan, and E. V. Jensen, *Chem. Ind.*, 1136 (1960)

[7]S. Nakanishi and E. V. Jensen, *J. Org.*, **27**, 702 (1962); S. Nakanishi, *J. Medicinal Chem.*, **7**, 108 (1964)

[8]S. Nakanishi, Ken-ichi Morita, and E. V. Jensen, *Am. Soc.*, **81**, 5259 (1959)

[9]J. Edwards and H. J. Ringold, *Am. Soc.*, **81**, 5262 (1959)

[10]C. H. Robinson, N. F. Bruce, and E. P. Oliveto, *J. Org.*, **28**, 975 (1963)

[11]B. M. Bloom, V. V. Bogert, and R. Pinson, Jr., *Chem. Ind.*, 1317 (1959)

[12]B. J. Magerlein, J. E. Pike, R. W. Jackson, G. E. Vandenberg, and F. Kagan, *J. Org.*, **29**, 2982 (1964)

[13]M. Neeman and Y. Osawa, *Am. Soc.*, **85**, 232 (1963); *idem, Tetrahedron Letters*, 1987 (1963)

[14]J. S. Mills, J. Barrera, E. Olivares, and H. Garcia, *Am. Soc.*, **82**, 5882 (1960)

[15]A. S. Kende and P. MacGregor, *Am. Soc.*, **83**, 4197 (1961)

[16]D. Taub, *Chem. Ind.*, 558 (1962)

[17]H. Shechter and E. B. Roberson, *J. Org.*, **25**, 175 (1960)

[18]J. P. Freeman, *Am. Soc.*, **82**, 3869 (1960)

[19]R. D. Schuetz, D. D. Taft, J. P. O'Brien, J. L. Shea, and H. M. Mork, *J. Org.*, **28**, 1420 (1963)

[20]H. Machleidt, *Ann.*, **667**, 24 (1963); **676**, 67 (1964)

[21]F. H. Dean and F. L. M. Pattison, *Can. J. Chem.*, **41**, 1833 (1963)

Performic acid, *see* Hydrogen peroxide.

Periodates. Sodium periodate (metaperiodate), $NaIO_4$ (213.90); sodium paraperiodate, $Na_3H_2IO_6$ (293.90); potassium periodate, KIO_4 (230.01).

Cyclic 1,3-diketones. Wolfrom and Bobbitt[1] found that several cyclic 1,3-diketones reduce sodium periodate in aqueous solution and are oxidized to dibasic acids. Thus cyclohexane-1,3-dione, in which the diketo form (1) is practically nonexistent, on oxidation with excess periodate (5.7 equiv.) afforded pure glutaric acid in 86.5% yield. The reaction is regarded as involving intermediates (3)–(5),

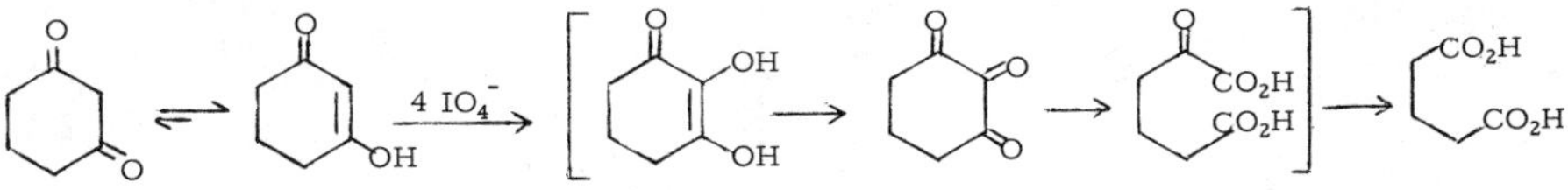

(1) (2) (3) (4) (5) (6)

and indeed (3) and (4) were synthesized and found to consume 3 and 2 equivalents of periodate with production of quantitative amounts of glutaric acid and carbon dioxide. The enediolic, or reductone structure (3) permits further oxidation without immediate cleavage of a carbon–carbon bond. Periodate oxidation of a similar reductone system has been observed in the case of sodium 1,2-dihydroxyanthraquinone-3-sulfonate.[2] 5-Methyl- and 5.5-dimethylcyclohexane-1,3-dione afford

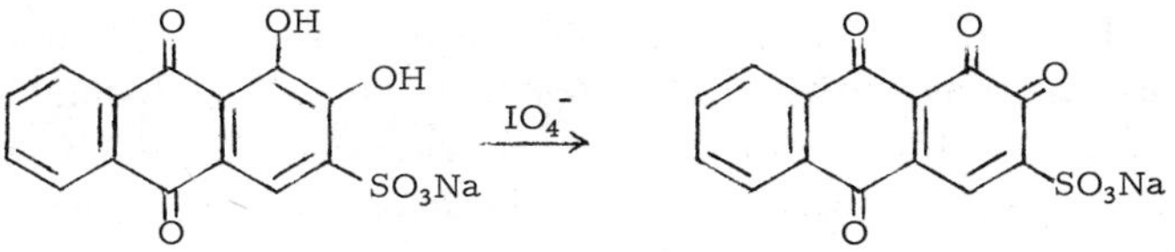

3-methyl- and 3,3-dimethylglutaric acid in yields of 90 and 92%. 2-Substituted cyclohexane-1,3-diones are oxidized to glutaric acid and the monocarboxylic acid.

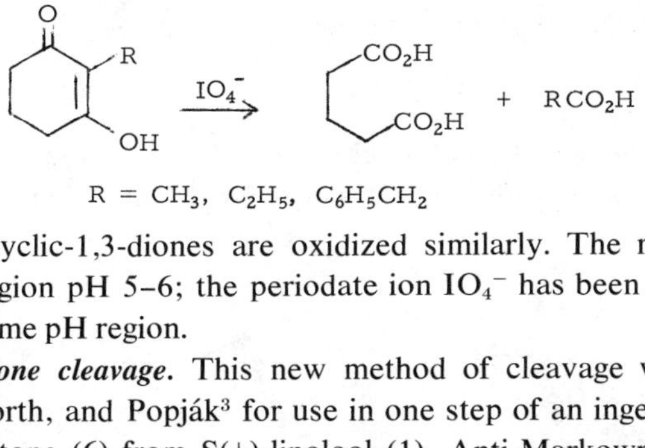

$R = CH_3,\ C_2H_5,\ C_6H_5CH_2$

Five-membered cyclic-1,3-diones are oxidized similarly. The rate of reaction is maximal in the region pH 5–6; the periodate ion IO_4^- has been shown to be most prevalent in the same pH region.

Cornforth ketone cleavage. This new method of cleavage was developed by Cornforth, Cornforth, and Popják[3] for use in one step of an ingenious synthesis of R(–)-mevalonolactone (6) from S(+)-linolool (1). Anti-Markownikoff hydration of both double bonds was accomplished by hydroboration–oxidation to give the triol (2). Acid-catalyzed condensation with acetaldehyde in ether gave the 1,3-dioxane for protection of the primary alcoholic function. For oxidation to the ketone (4), a solution of 37.5 g. of (3) in 100 ml. of pyridine was added to a batch of the CrO_3–H_2O–Py reagent prepared from 60 g. of CrO_3, and the mixture was let stand at room temperature for 3 days, diluted with water, and extracted with ether. Note that the oxidation gave material satisfactory for the next step in 88% yield. It was then necessary to find a method for effecting the oxidative fission: $-COCH_2- \longrightarrow -CO_2H + HO_2C-$. A method was suggested by work of Huebner *et al.*,[4] who have

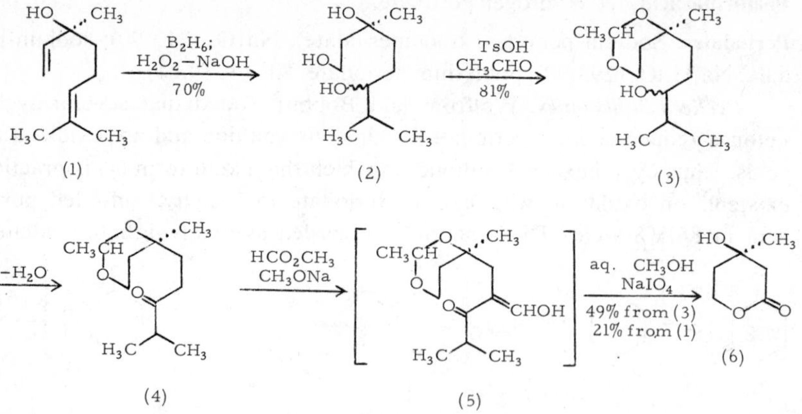

found that several 1,3-dicarbonyl compounds are cleaved by periodate in aqueous solution. An α-hydroxymethylene ketone should be the equivalent of a 1,3-diketone or ketoaldehyde, and hence (4) was condensed with methyl formate to give (5), which was not isolated but immediately oxidized by addition of aqueous methanolic sodium periodate. The acetal group was hydrolyzed during workup, and the product was R(−)-mevalonolactone. Similarly S(+)-mevalonolactone was obtained from (−)-linolool. This work incidentally established the absolute configurations of the two naturally occurring linolools ($\alpha_D + 16.9°$ and $-16.9°$). The sodium mevalonates were compared as substrates for the enzyme mevalonic kinase prepared from rat liver. Utilization of R-mevalonate was substantially complete, whereas that of S-mevalonate was almost undetectable.

Sulfides ⟶ sulfoxides. Leonard and Johnson[5] found that sulfoxides uncontaminated with either sulfides or sulfones are obtainable in high yield by addition of the sulfide to a slight excess of 0.5 *M* aqueous $NaIO_4$. The method was employed by

$$R_2S + NaIO_4 \longrightarrow R_2\overset{+}{S}-\overset{-}{O} + NaIO_3$$

two groups[6,7] for the oxidation of 4-substituted thianes to the sulfoxides. Oxidation of (1)[7] gives a mixture of the *cis* (2) and *trans* (3) sulfoxides, most stable in the

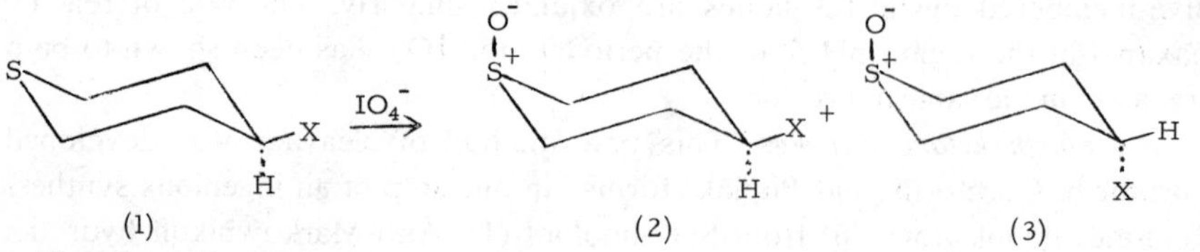

X = Cl, OH, OTs

conformation with the sulfoxide oxygen axial. The mixtures are separated by chromatography, the *trans* isomers being eluted first.

Periodate–permanganate oxidation (Lemieux-von Rudloff).[8–11] In this catalytic method an aqueous solution 0.019 *M* in sodium periodate and 0.0034 *M* in potassium permanganate (pH 7–8) cleaves an olefinic double bond rapidly at 25°. The permanganate is reduced only to the manganate stage, from which it is regenerated by the periodate, which by itself does not attack the olefin. A symmetrically disubstituted olefin (1) or a trisubstituted olefin (2) is converted into an α-ketol, which sometimes

1. $\begin{matrix} RCH \\ \| \\ R'CH \end{matrix} \xrightarrow{KMnO_4-NaIO_4} \begin{matrix} RCHOH \\ | \\ R'CHOH \end{matrix} \longrightarrow \begin{matrix} RCHOH \\ | \\ R'C{=}O \end{matrix} \longrightarrow \begin{matrix} RCO_2H \\ R'CO_2H \end{matrix}$

2. $R_2C{=}CHR' \xrightarrow{KMnO_4-NaIO_4} R_2\underset{OH}{C}-\underset{OH}{CHR'} \longrightarrow R_2\underset{OH}{C}-\underset{O}{\overset{\|}{C}R'} \longrightarrow R_2C{=}O + HO_2CR'$

3. $RCH{=}CH_2 \xrightarrow{KMnO_4-NaIO_4} RCH{=}O + CH_2{=}O$

4. $RCH{=}C(CH_3)_2 \xrightarrow{KMnO_4-NaIO_4} RCH{=}O + O{=}C(CH_3)_2$

can be isolated; further oxidation gives acids or ketones. An olefin with a terminal double bond (3) gives formaldehyde in high yield. Those with an isopropylidine group (4) are cleaved quantitatively to acetone. For oxidation of 2-methylene-5,5-dimethylbicyclo[2.1.1]hexane (I) to the ketone II, Meinwald and Gassman[12] shook a mixture of 2.2 g. of I (a liquid), 14 g. of potassium periodate, 0.5 g. of potassium permanganate, and 20 g. of potassium carbonate in 200 ml. of water for 16 hrs. A solid

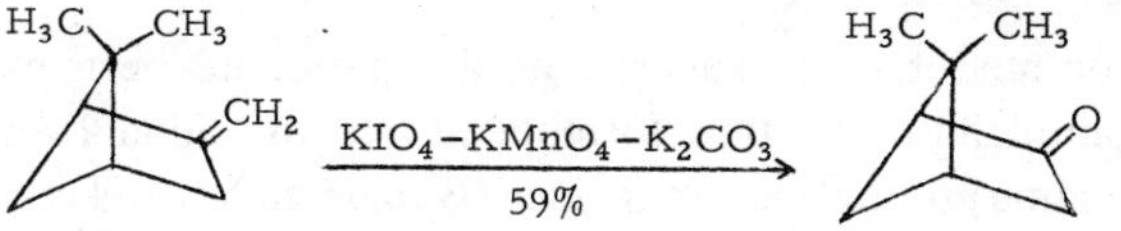
H3C CH3
CH2
KIO4-KMnO4-K2CO3
59%
H3C CH3
O

I II

insoluble in water can be oxidized in aqueous *t*-butanol,[11,13] pyridine,[11] or dioxane.[13]

The reagent has proved useful for elucidation of the structures of unsaturated

$$\underset{(1)}{CH_3CH_2\overset{16}{\underset{OH}{CH}}\overset{15}{\underset{OH}{CH}}CH_2CH{=}\overset{12}{C}HCH_2CH{=}\overset{9}{C}H(CH_2)_7CO_2H} \xrightarrow[\text{aq. t-BuOH}]{KIO_4-KMnO_4-K_2CO_3} \underset{(2)}{HO_2C(CH_2)_7CO_2H}$$

natural products. Thus Gunstone and Morris[14] effected oxidation of 15,16-dihydroxylinoleic acid (1) in water (300 ml.)–*t*-butanol (90 ml.) with enough potassium carbonate added to give a pH of 8–9 and obtained azelaic acid (2). Jacobson and co-workers[15] found evidence of structure (3) for the sex attractant of the gypsy moth by similar oxidation conducted with 4 mg. of attractant. The reagent attacked only

$$\underset{(3)}{CH_3(CH_2)_5\underset{OCOCH_3}{CH}CH_2\overset{H}{C}{=}\overset{H}{C}(CH_2)_5CH_2OH} \xrightarrow[\text{aq. t-BuOH}]{KIO_4-KMnO_4-K_2CO_3}$$

$$\underset{(4)}{CH_3(CH_2)_5\underset{OCOCH_3}{CH}CH_2CO_2H} + \underset{(5)}{HO_2C(CH_2)_5CH_2OH}$$

the double bond, leaving the acetoxyl and hydroxyl groups unchanged. 3-Acetoxy-1-nonanonoic acid (4) was isolated in 92% yield, together with an ω-hydroxy acid (5), which was oxidized with alkaline permanganate to pimelic acid ın 71% yield.

Tschesche and co-workers[16] employed the Lemieux-Rudloff procedure for oxidation of the carbinol phosphate (6) to DL-mevalonic acid 5-phosphate (7). At the end of the reaction excess periodate was destroyed with ethylene glycol, iodate ion was precipitated as barium iodate, and cations were removed with Dowex-50. Mevalonic acid phosphate (7) was then isolated as the crystalline tricyclohexylammonium

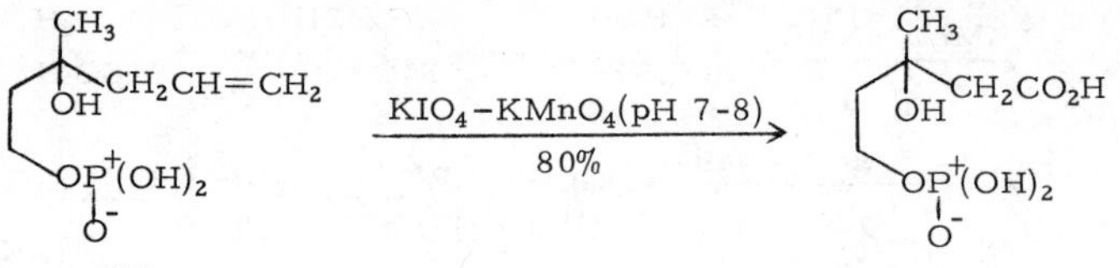
KIO_4–$KMnO_4$(pH 7–8)
80%

(6) (7)

salt. Dried in vacuum, this salt lost one mole of cyclohexylamine with formation of the dicyclohexylammonium salt.

Periodate–osmium tetroxide oxidation (Lemieux-Johnson).[17, 18] The reagent, periodate catalyzed by osmium tetroxide, operates as follows. Osmium tetroxide adds to a double bond to form an osmate ester, and this is oxidized by periodate with cleavage to carbonyl compounds and regeneration of osmium tetroxide. Discovery of the

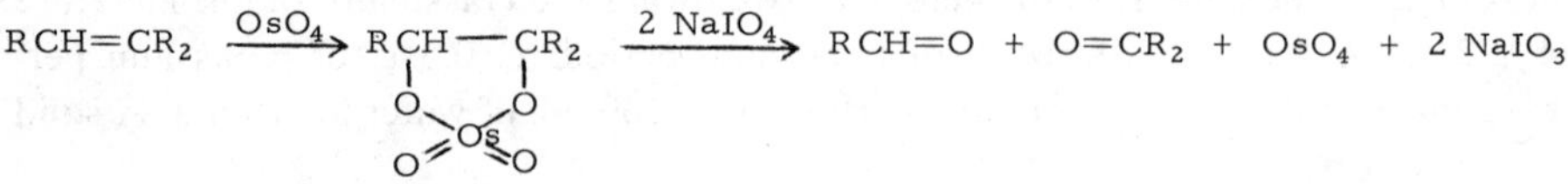
$RCH{=}CR_2$
OsO_4
$RCH—CR_2$
2 $NaIO_4$
$RCH{=}O$ + $O{=}CR_2$ + OsO_4 + 2 $NaIO_3$

method was prompted by the success of the periodate–permanganate method, which converts an olefin of the type formulated into an acid and a ketone. The new method gives the same products as are formed by ozonization and reductive cleavage. Note that the method bears some resemblance to the catalylic $NaClO_3$–OsO_4 method of K. A. Hofmann which lead to Criegee's discovery of the osmate esters, although the earlier method stops with oxidation of an olefin to a *vic*-glycol.

The Lemieux-Johnson group[18] describe two procedures:

(a) A mixture of 5 ml. of water, 15 ml. of dioxane, 0.77 g. of 1-dodecene, and 11.3 mg. of OsO_4 is stirred for 5 min., when it becomes deep brown. While the temperature is kept at 24–26°, 2.06 g. of finely ground sodium periodate ($NaIO_4$) is added in portions over 30 min. Stirring at 24–26° is continued for 1.5 hrs., when the tan-colored slurry has become pale yellow. The mixture is extracted with ether, and the extract dried and treated with a solution of 1 g. of 2,4-dinitrophenylhydrazine in 5 ml. of concd. sulfuric acid, 7.5 ml. of water, and 35 ml. of 95% ethanol. The two-phase system is stirred for 70 min., evaporated to a volume of 50 ml., and the yellow 2,4-dinitrophenylhydrazone is collected in two crops; yield 1.09 g. (68%).

(b) A mixture of 15 ml. of ether, 15 ml. of water, 0.405 g. of cyclohexene, and 65.4 mg. of OsO_4 is stirred at 24–26° during the addition of 2.32 g. of finely ground sodium periodate over a period of 40 min. Stirring at 24–26° is continued for 80 min., by which time the color has changed to yellow and considerable sodium iodate has separated. Conversion into adipaldehyde bis-2,4-dinitrophenylhydrazone is accomplished as in (a); yield 77%.

Dvornik and Edwards[19] used the reagent with success in 80% acetic acid for oxidation of the alkaloid derivative (1) to the α-ketal acetate (2), and Vorbrueggen and

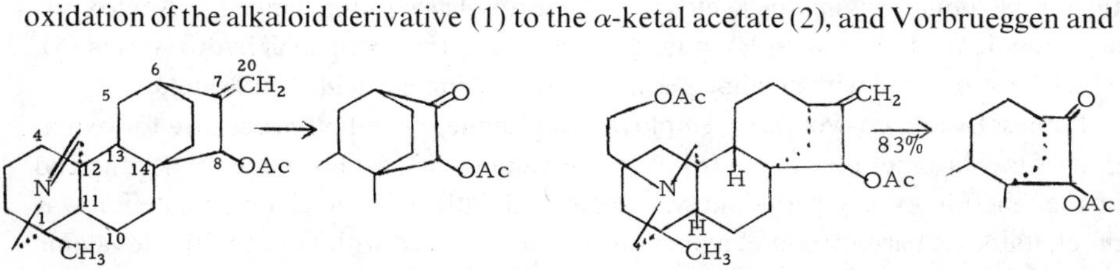
83%

(1) (2) (3) (4)

Djerassi[20] oxidized F-dihydrogarryfoline diacetate (3) to (4), also in 80% acetic acid (at 4°). Note that these two oxidations proceed without attack of the nitrogen atom. When 80% acetic acid is used as solvent the catalytic amount of osmium tetroxide can be removed conveniently by distillation at 40°/20 mm.; osmium can be recovered from the distillate by addition of sodium bisulfate. With dioxane–water as solvent, the crude reaction product obtained after removal of solvent at reduced pressure is dissolved in benzene and ether, and the solution is filtered over deactivated alumina (activity II or III).[20]

An interesting application of the method by Jeger's group[21] is to 8α,20-oxidomanool (5). On oxidation in aqueous dioxane the initially formed α-hydroxyaldehyde (6) suffered reverse aldolization to give the methyl ketone (7) and formaldehyde.

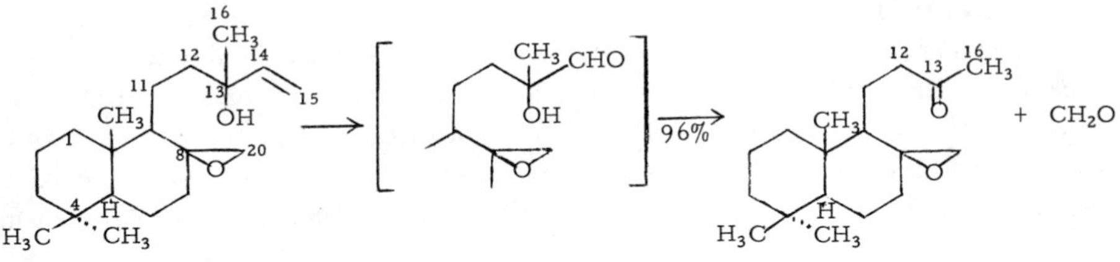
96%
+ CH₂O

(5) (6) (7)

Tarbell and co-workers,[22] seeking a route to 4-carboxytropolone (10), attempted to oxidize the styryl double bond of 4-styryltropolone (8) by ozonization, oxidation

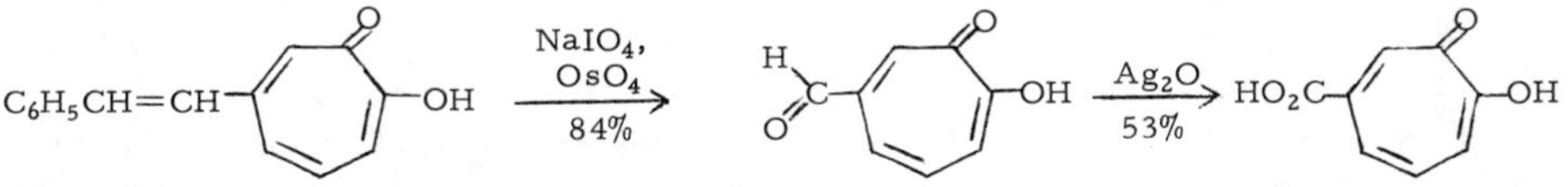
NaIO₄, OsO₄
84%
Ag₂O
53%

(8) (9) (10)

with peracids, nitric acid, chromic acid, permanganate–periodate, or by Prévost oxidation, but all lead to preferential oxidation of the tropolone ring. However, Lemieux-Johnson oxidation in aqueous dioxane and sublimation of the product afforded yellow 4-formyltropolone, m.p. 154–157°, in 84% yield. Oxidation of the aldehyde with silver oxide in basic solution afforded 4-carboxytropolone in 53% yield.

A feature of the catalytic method is that the rate of formation of the dark osmate ester provides an indication of the relative reactivity of the olefin.[20] Compounds having an unhindered methylene group give a coloration in 10–30 min., whereas hindered olefins may require 4–12 hrs. for visual perception of oxidation; in such a case the periodate should be added over a period of 1–4 days at 4°.

Periodate–ruthenium tetroxide oxidation. This reaction, attributed to Pappo and Becker,[23] is illustrated by a procedure of Stork *et al.*[24] for effecting the oxidative

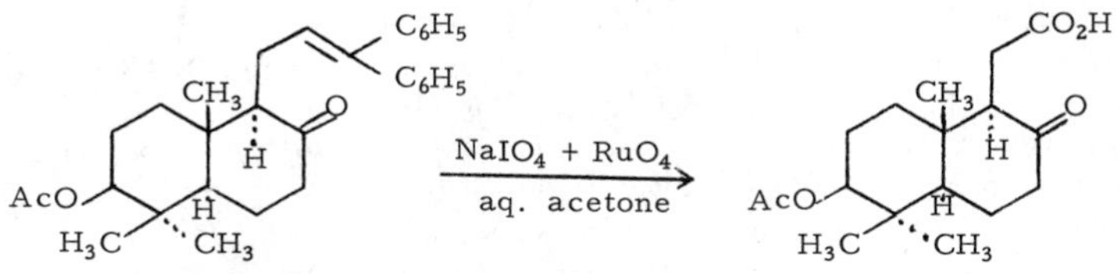
NaIO₄ + RuO₄
aq. acetone

(3) (4)

step of Barbier-Wieland degradation. Oxidation of the keto diphenylethylene (3) to the keto acid (4), "could be done by ozonolysis but was best carried out, in excellent yield, by treatment with ruthenium tetroxide and sodium periodate in aqueous acetone solution." The periodate is the prime oxidant, ruthenium tetroxide functions as catalyst. A stirred solution of 2.22 g. of (3) in 100 ml. of acetone was treated with a carbon tetrachloride solution of 500 mg. of ruthenium tetroxide and then with a solution of 2 g. of sodium periodate in 40 ml. of water. When precipitation of black RuO_2 seemed complete, 3 g. of periodate was added to dissolve it. The addition was repeated twice more in the course of 10 hrs. Then propanol was added to destroy excess reagent, the mixture was filtered to remove black RuO_2, and the solvent was removed by evaporation. This step was combined with others, and the yield was not specifically determined.

Sondheimer *et al.*[25] used the catalytic method with success in another oxidative fission of a Barbier-Wieland diphenylethylene with which ozonolysis and chromic acid oxidation proceeded poorly. Sarel and Yanuka[26] found the method successful as applied to the diphenylethylene (2) with which the periodate–osmium tetroxide method was completely ineffective. The olefinic compound (2) was dissolved in

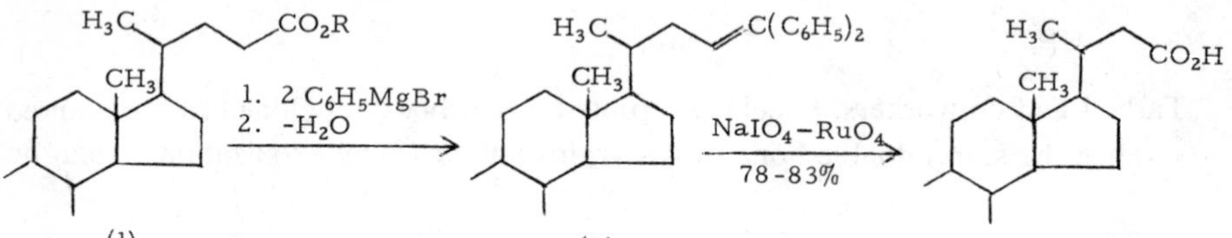
H3C
CO2R
CH3
1. 2 C6H5MgBr
2. -H2O
H3C
C(C6H5)2
CH3
NaIO4-RuO4
78-83%
H3C
CO2H
CH3

(1) (2) (3)

aqueous acetone (80–85%), and 5 mole % of RuO_4 and 140 mole % of $NaIO_4$ were added at 15–25°.

[1]M. L. Wolfrom and J. M. Bobbitt, *Am. Soc.*, **78**, 2489 (1956)
[2]L. K. Ramachandran and P. S. Sarma, *J. Sci. Ind. Research* (*India*), **10B**, 147 (1951) [*C.A.*, **47**, 2160 (1953)]
[3]R. H. Cornforth, J. W. Cornforth, and G. Popják, *Tetrahedron*, **18**, 1351 (1962)
[4]C. F. Huebner, S. R. Ames, and E. C. Bubl, *Am. Soc.*, **68**, 1621 (1946)
[5]N. J. Leonard and C. R. Johnson, *J. Org.*, **27**, 282 (1962); *Am. Soc.*, **84**, 3701 (1962)
[6]C. R. Johnson and D. McCants, Jr., *Am. Soc.*, **86**, 2935 (1964)
[7]J. C. Martin and J. J. Uebel, *Am. Soc.*, **86**, 2936 (1964)
[8]R. U. Lemieux and E. von Rudloff, *Can. J. Chem.*, **33**, 1701 (1955)
[9]*Idem, ibid.*, **33**, 1710 (1955)
[10]E. von Rudloff, *ibid.*, **33**, 1714 (1955)
[11]*Idem, ibid.*, **34**, 1413 (1956)
[12]J. Meinwald and P. G. Gassman, *Am. Soc.*, **82**, 2857 (1960)
[13]M. E. Wall and S. Serota, *J. Org.*, **24**, 741 (1959)
[14]F. D. Gunstone and L. J. Morris, *J. Chem. Soc.*, 2127 (1959)
[15]M. Jacobson, M. Beroza, and W. A. Jones, *Am. Soc.*, **83**, 4819 (1961)
[16]H. Machleidt, E. Cohnen, and R. Tschesche, *Ann.*, **655**, 70 (1962); **672**, 215 (1964)
[17]Contributed in part by H. Vorbrueggen, Stanford University
[18]R. Pappo, D. S. Allen, Jr., R. U. Lemieux, and W. S. Johnson, *J. Org.*, **21**, 478 (1956)
[19]D. Dvornik and O. E. Edwards, *Can J. Chem.*, **35**, 860 (1957)
[20]H. Vorbrueggen and C. Djerassi, *Tetrahedron Letters*, 119 (1961); *Am. Soc.*, **84**, 2990 (1962)
[21]U. Scheidegger, K. Schaffner, and O. Jeger, Helv., **45**, 400 (1962)
[22]D. S. Tarbell, K. I. H. Williams, and E. J. Sehm, *Am. Soc.*, **81**, 3443 (1959)
[23]R. Pappo and A. Becker, *Bull. Res. Council Israel*, **5A**, 300 (1956)
[24]G. Stork, A. Meisels, and J. E. Davies, *Am. Soc.*, **85**, 3419 (1963)

[25]F. Sondheimer, R. Mechoulam, and M. Sprecher, *Tetrahedron*, **20**, 2473 (1964)
[26]S. Sarel and Y. Yanuka, *J. Org.*, **24**, 2018 (1959)

Periodic acid, H_5IO_6. Mol. wt., 227.95, m.p. 122°, dec. 130–140°; soluble in water; pKa_1 1.6, pKa_2 5.7.

vic-Glycol cleavage. Use of periodic acid for the cleavage of *vic*-glycols to carbonyl compounds (1) was introduced by Malaprade[1] in 1928. Compounds with a

$$1. \quad \underset{\underset{OH}{|}}{R}\underset{\underset{OH}{|}}{CH}\underset{|}{C}R_2 \xrightarrow{H_5IO_6} RCH{=}O \ + \ O{=}CR_2$$

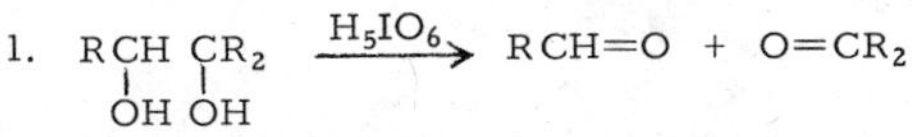

$$2. \quad \underset{\underset{OH}{|}}{R}CH{-}\underset{\underset{NH_2}{|}}{C}HR \xrightarrow{H_5IO_6} RCH{=}O \ + \ O{=}CHR$$

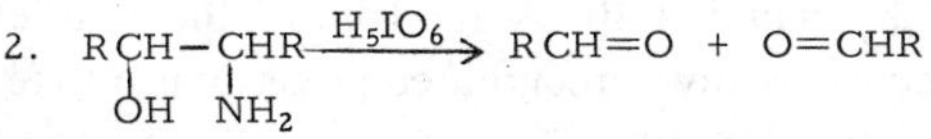

hydroxyl and an amino group on adjacent carbon atoms are also cleaved (2); see review by Jackson.[2] Unlike lead tetraacetate, periodic acid does not oxidize oxalic acid, and it oxidizes α-hydroxy acids only slowly. The reaction is done in aqueous solution or in a solvent mixture containing water, usually at room temperature. Kinetic studies[3, 4] indicate that the reactive species is the ion II and that the reaction proceeds through a cyclic intermediate, either the hydrated ion III or the dehydrated ion IV.

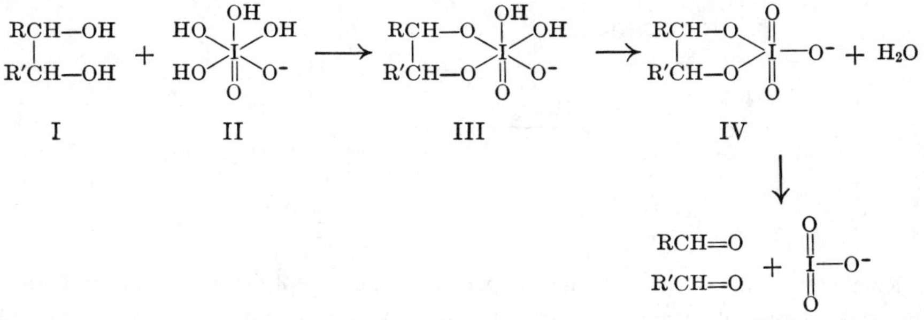

Analytical use of the reaction is illustrated by the fact that glucose, mannose, and galactose on oxidation afford one mole of formaldehyde and five moles of formic acid.[2] It is useful also in the correlation of sugars. Thus oxidation of α-methyl-D-glucopyranoside (1) is carried out by dissolving 12.5 g. of (1) in aqueous solution with 2.1 equivalents of aqueous periodic acid and letting the solution stand at 20–25° for 24 hrs. The oxidation destroys the asymmetric centers at C_2, C_3, and C_4 and gives the dialdehyde (2), and the same dialdehyde is obtained from all the seven

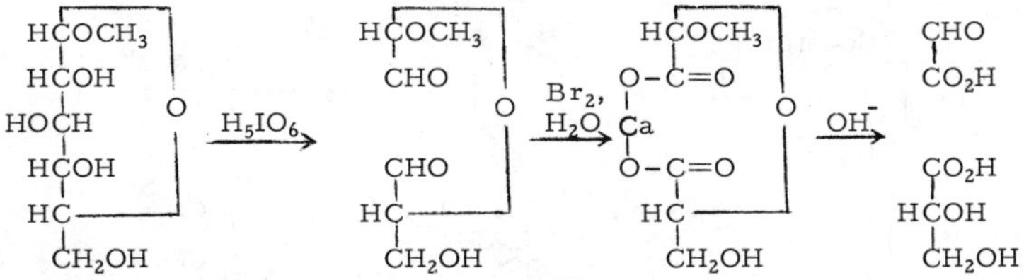

other α-methyl-D-aldopyranosides.[5] The dialdehyde (2) can be oxidized with bromine in a solution kept neutral with a carbonate and the diacid isolated as the calcium salt (3). This salt is easily hydrolyzed and the process provides the best known method for the preparation of optically pure D-glyceric acid.

Cleavage of a water-insoluble diol is exemplified by the oxidation of *erythro*-9,10-dihydroxystearic acid (4, m.p. 132°).[6] A solution of 6 g. of potassium periodate

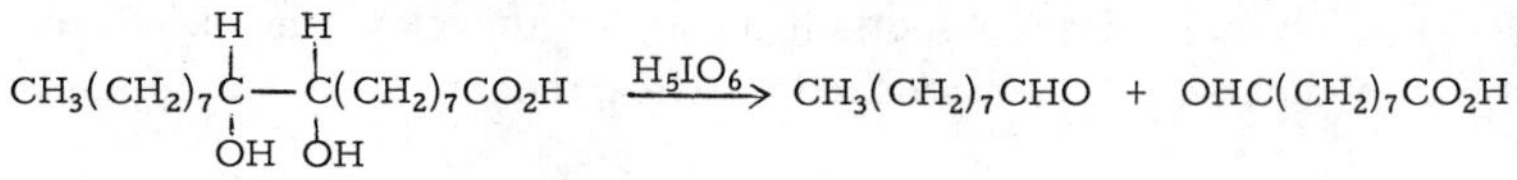

H_5IO_6

(4) (5) (6)

(KIO_4) in 300 ml. of *N* sulfuric acid at 20° is added rapidly to a solution of 8 g. of (4) in 400 ml. of 95% ethanol at 40°. After 10 min. the colorless solution is cooled to 15°, water is added to dissolve precipitated potassium sulfate, and extraction with ether and steam distillation affords 3.3 g. (76%) of pelargonic aldehyde. The aqueous solution is filtered, and cooled in ice and crude azelaic half-aldehyde (6) is collected (3.3 g., 76%). Extraction with petroleum ether leaves a residue of trimer (0.5 g.), and purification through the semicarbazone affords 1.5 g. of the pure aldehyde-acid (m.p. 38°).

Preparative application of the reaction to a partially protected sugar derivative is illustrated by the cleavage of D-galactose-4,5-monoacetonide dimethyl acetal (7) to D-threose-2,3-monoacetonide (8) and glyoxal dimethyl acetal (9).[7]

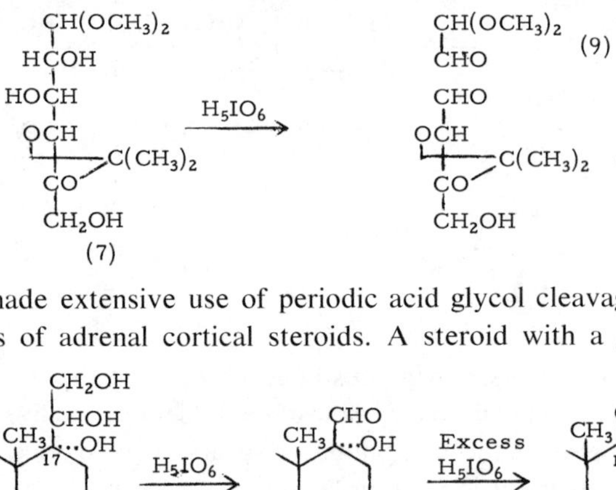

H_5IO_6

(7) (8) (9)

Reichstein made extensive use of periodic acid glycol cleavage in investigating the side chains of adrenal cortical steroids. A steroid with a glycerol-type side

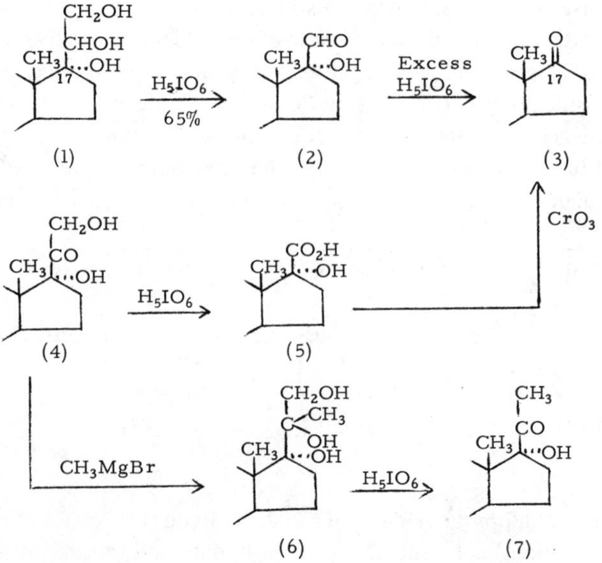

H_5IO_6
65%
(1)
(2)
Excess H_5IO_6
(3)
CrO_3
H_5IO_6
(4)
(5)
CH_3MgBr
H_5IO_6
(6)
(7)

chain (1) yields first an α-hydroxyaldehyde (2), and then the 17-ketone (3).[8] A dihydroxyacetone side chain (4) is cleaved to an α-hydroxyacid (5) oxidizable with chromic acid to the 17-ketone.[9] Periodic acid oxidation was used also in an ingenious process for elimination of a 21-hydroxyl group from (4): addition of methyl Grignard reagent (6) and glycol cleavage.[10]

Cleavage with ethereal periodic acid.[11] In the case of water-insoluble compounds, or where a cleavage product is sensitive to aqueous acid, cleavage of a *vic*-glycol or an epoxide of the type formulated can be carried out with a solution of periodic acid in ether or tetrahydrofurane. A solution containing about 16 mg./ml. can be prepared by stirring an excess of the powdered peracid with dry ether or THF for

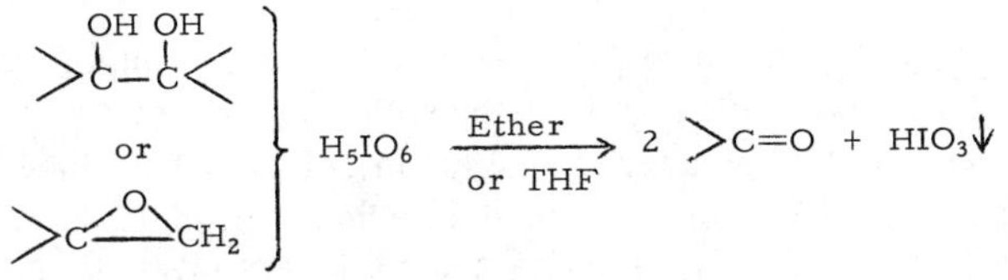

1 hr., allowing the solid to settle, and decanting the supernatant solution. The solution is titrated and the required amount added to a solution of the diol in ether or THF. Iodic acid separates immediately and is so slightly soluble that the reaction mixture can be worked up by simply filtering and evaporating the solution. This procedure proved useful in the conversion of isopimaradiene (1) into the aldehyde semicarbazone (3).[12] Osmylation gave the diol (2), an oil which was not purified

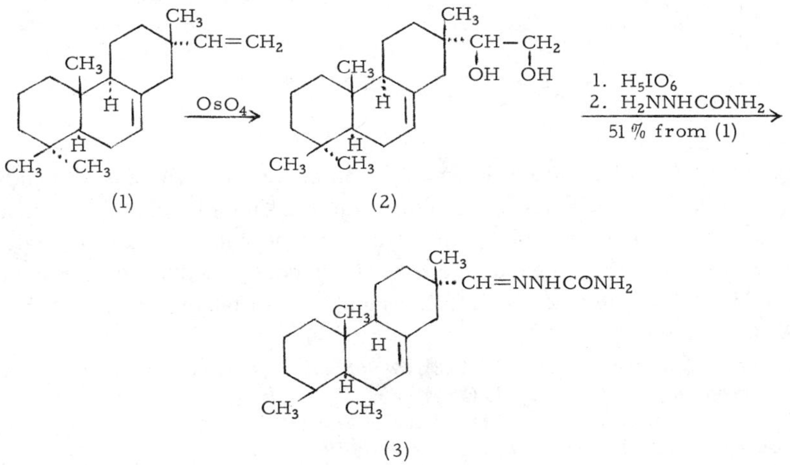

but cleaved with periodic acid in THF. The aldehyde was isolated in the form of the semicarbazone (3), which was obtained in overall yield from (1) of 51%.

Since periodic acid is a hydrate (H_5IO_6), the reagent can be used to effect hydrolytic cleavage of an epoxide. The extremely acid-sensitive γ-bromobutyraldehyde (5) was prepared satisfactorily in this way from the epoxide (4).[13]

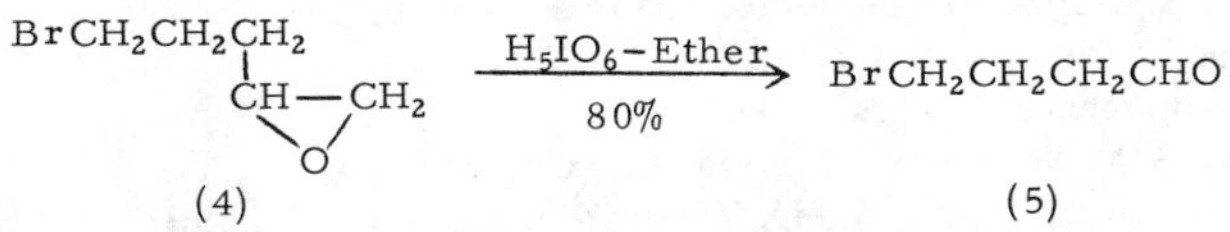

Hydrolysis of 1,3-dioxolanes and epoxides.[14] A carbonyl function is often protected by conversion into a 1,3-dioxolane (ethyleneacetal or ketal) with ethylene glycol. In case quantitative removal of the blocking group is difficult, due to an unfavorable equilibrium, complete hydrolysis can be accomplished in aqueous dioxane solution by addition of an equivalent amount of periodic acid. This destroys the ethylene glycol formed and drives the equilibrium to completion. A procedure employed in a total synthesis of linoleic acid[15] is as follows. A mixture of 1.6 g. (0.0070 mole) of

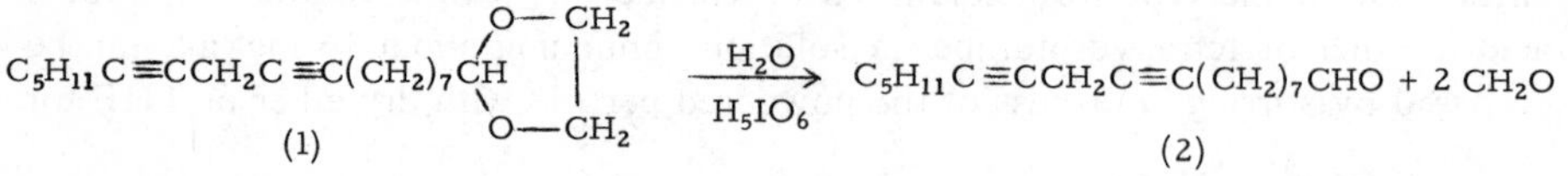

potassium periodate,[16] 2 ml. of 6 *N* sulfuric acid, and 100 ml. of water is warmed briefly to effect solution, cooled, and treated with a solution of 2 g. (0.0066 mole) of the ethyleneacetal of $\Delta^{9,12}$-octadecadiynal (1) in 240 ml. of dioxane. The resulting cloudy mixture is heated rapidly to boiling, when it becomes clear. The flask is stoppered tightly and allowed to stand for 3 hrs. Water (300 ml.) is added and enough sodium bicarbonate to give a weakly basic mixture, and the product is taken up in petroleum ether. The yield of crude aldehyde is quantitative. The aldehyde was not isolated, but was oxidized immediately to $\Delta^{9,12}$-octadecadiynoic acid.

Hydrolysis of an epoxide with sulfuric or hydrochloric acid as catalyst may proceed poorly because of formation of a sulfuric acid ester or a chloride, and use of an acid incapable of forming an ester is indicated. The reagent of choice is aqueous perchloric acid in tetrahydrofurane, but the more costly periodic acid is also effective.

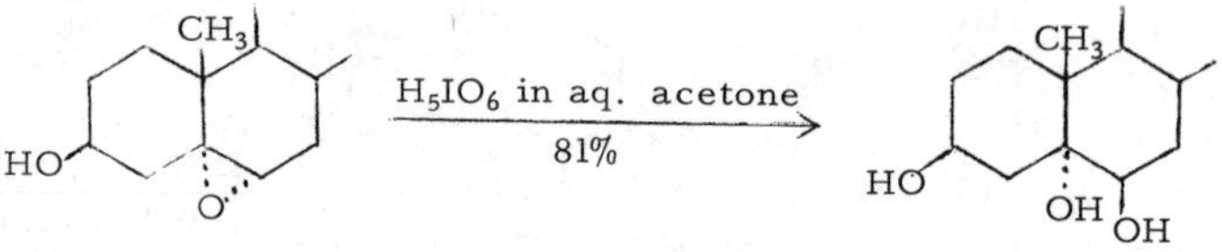
CH3
HO
O
H5IO6 in aq. acetone
81%
CH3
HO
OH
OH

A solution of 1 g. of cholesterol α-oxide in 30 ml. of hot acetone is treated with a solution of 0.625 g. of periodic acid dihydrate in 10 ml. of water.[17] Before all the precipitated oxide has redissolved, thin plates of cholestane-3β, 5α,6β-triol begin to separate. The mixture is refluxed for one half hour, cooled, and the product collected; yield 0.83 g. (81%), m.p. 231–232°. A second crop of material (0.14 g.) melted at 225–226°.

[1]L. Malaprade, *Bull. soc.*, [4], **43**, 683 (1928); *Compt. rend.*. **186**, 382 (1928)
[2]E. L. Jackson, *Org. Reactions*, **2**, 341 (1944)
[3]C. C. Price *et al., Am. Soc.*, **60**, 2726 (1938); **64**, 552 (1942)
[4]G. J. Buist and C. A. Bunton, *J. Chem. Soc.*, 1406 (1954)
[5]E. L. Jackson and C. S. Hudson, *Am. Soc.*, **59**, 994 (1937)
[6]G. King, *J. Chem. Soc.*, 1826 (1938)
[7]E. Pacsu, S. M. Trister, and J. W. Green, *Am. Soc.*, **61**, 2444 (1939)
[8]D. A. Prins and T. Reichstein, *Helv.*, **24**, 396, 945 (1941)
[9]T. Reichstein, Ch. Meystre, and J. von Euw, *Helv.*, **22**, 1107 (1939)
[10]J. von Euw and T. Reichstein, *Helv.*, **24**, 408 (1941); H. G. Fuchs and T. Reichstein, *Helv.*, **24**, 804 (1941)
[11]Contributed by Robert E. Ireland, University of Michigan
[12]R. E. Ireland and J. Newbould, *J. Org.*, **28**, 23 (1963)
[13]R. E. Ireland, unpublished results
[14]Contributed by H. M. Walborsky, Florida State University
[15]H. M. Walborsky, R. H. Davis, and D. R. Howton, *Am. Soc.*, **73**, 2590 (1951)

[16]Preformed periodic acid is recommended in place of periodate and sulfuric acid because of the reduced amounts of water and dioxane required to produce a homogeneous solution.

[17]L. F. Fieser and S. Rajagopalan, *Am. Soc.*, **71**, 3938 (1949)

Permaleic acid, $HO_2CC(H){=}C(H)CO_3H$. Mol. wt. 132.07.

A solution of the reagent is prepared by adding 39.2 g. (0.4 mole) of freshly crushed maleic anhydride to an ice-cold mixture of 11.6 g. of 90% hydrogen peroxide and 150 ml. of methylene chloride and stirring until the solid dissolves.[1] The reagent is somewhat less reactive than trifluoroperacetic acid and much more reactive than the usual peracids. At room temperature, the titer of the solution falls off about 5% in 6 hrs. A reaction is carried out by adding the substance to be oxidized to the methylene chloride solution and eventually filtering the maleic acid which separates.

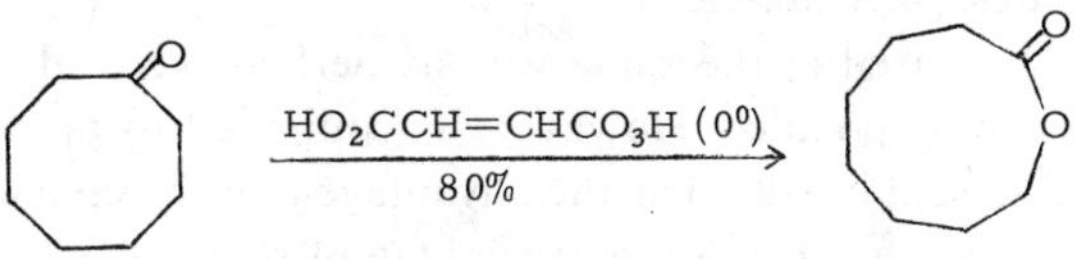
HO₂CCH=CHCO₃H (0°)
80%

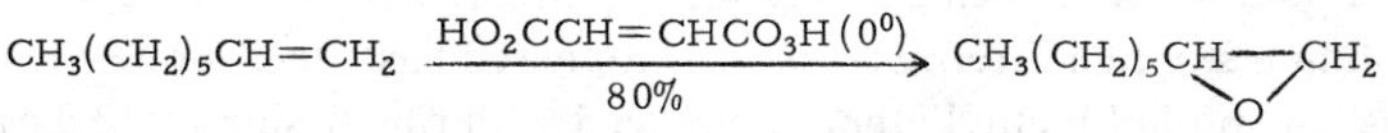
CH₃(CH₂)₅CH=CH₂
HO₂CCH=CHCO₃H (0°)
80%
CH₃(CH₂)₅CH—CH₂
O

Permaleic acid can be used in the Baeyer-Villiger reaction, as illustrated for the case of cyclooctanone. It epoxidizes terminal olefins efficiently at a low temperature; 1-octene affords the epoxide in 80% yield at 0° but at 25° the yield is only 40%. In the case of the more reactive internal olefins the epoxide ring is cleaved even at 0° to give the diol monomaleate. Aromatic amines with electron-withdrawing substituents are oxidized to nitro compounds.

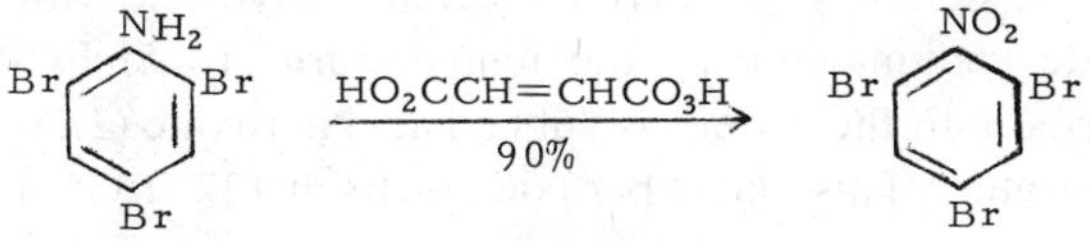
NH₂
Br
Br
Br
HO₂CCH=CHCO₃H
90%
NO₂
Br
Br
Br

[1]R. W. White and W. D. Emmons, *Tetrahedron*, **17**, 31 (1962)

Perphthalic acid (Monoperphthalic acid), $C_6H_4(CO_2H)CO_3H$. Mol. wt. 182.13.

Preparation. An *Organic Syntheses* procedure[1] specifies use of phthalic anhydride pulverized to pass a 14-mesh sieve. A solution of 0.6 mole of sodium carbonate in 250 ml. of water is stirred in an ice-salt bath, cooled to 0°, and treated

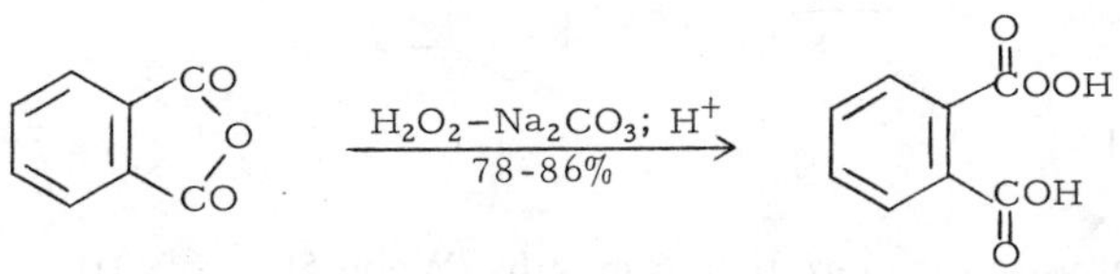
CO
O
CO
H₂O₂–Na₂CO₃; H⁺
78-86%
O
COOH
COH
O

with 0.6 mole of 30% hydrogen peroxide. With the temperature at −5 to 0°, 0.5 mole of pulverized phthalic anhydride is added, and the mixture is stirred vigorously at −5 to 0° for 30 min. The resulting solution or suspension is shaken with 350 ml. of ether and carefully acidified with an ice-cold solution of 30 ml. of concd. sulfuric acid in 150 ml. of water. The peracid is extracted into the ether, and the water layer is extracted twice more with ether. The extract is washed with two 200-ml. portions of 40% ammonium sulfate solution and dried in a refrigerator over anhydrous magnesium sulfate. A 2-ml. portion of the solution is treated with 30 ml. of 20% potassium iodide solution, and after 10 min. the iodine liberated is titrated with 0.1*N* thiosulfate; yield 78–86%.

In our experience pulverization of phthalic anhydride is tedious and we prefer crystallization, which provides material in a form suitable for ready reaction and which eliminates any phthalic acid, usually present in a sample taken from a previously opened bottle. The anhydride is boiled with a liberal quantity of benzene, and the hot solution is filtered from any undissolved phthalic acid and diluted with ligroin. A modification of a published method[2] is as follows. A mixture of 66 g. of the crystallized anhydride, 92 g. of sodium perborate, and 280 ml. of water is stirred at 0° for 2 hrs., shaken with ether, and acidified with 120 ml. of ice-cold 30% sulfuric acid. Extraction is done as described above.

Uses. The reagent is used in the same way as perbenzoic acid, mainly for the conversion of olefins into epoxides. However, as compared to perbenzoic acid in chloroform, perphthalic acid in ether has the advantage that the course of a reaction can be followed by noting the size of the precipitate of phthalic acid formed, or by filtering the solution and seeing if any more phthalic acid separates. One can note also if a solution stored in a refrigerator in a stoppered flask deteriorates on standing.

Whereas β-naphthol methyl ether is oxidized with ring fission by perbenzoic acid in benzene and by peracetic acid in acetic acid at room temperature, it is attacked by perphthalic acid in ether to only a negligible extent.[3]

[1]G. B. Payne, *Org. Syn.*, **42**, 77 (1962)
[2]M. A. Stahmann and M. Bergmann, *J. Org.*, **11**, 589 (1946)
[3]H. Fernholz, *Ber.*, **84**, 110 (1951)

Persuccinic acid (Monoperoxysuccinic acid, 3). Mol. wt. 134.09, m.p. *ca.*, 107°.

On stirring 50 g. of finely powdered succinic anhydride with 85 ml. of 30% hydrogen peroxide and maintaining the temperature at 25° by cooling, the anhydride is converted into the water-insoluble diacyl peroxide (2), which is collected and washed with water.[1] This diacyl peroxide melts at 132–133°, dec., and is stable

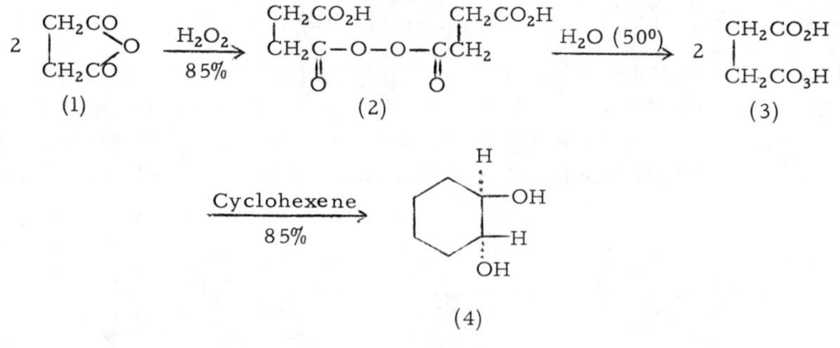

CH_2CO
CH_2CO
H_2O_2
85%
CH_2CO_2H
$CH_2C-O-O-CCH_2$
CH_2CO_2H
H_2O (50°)
CH_2CO_2H
CH_2CO_3H
Cyclohexene
85%
H
OH
H
OH

at room temperature for more than 6 months. When 50 g. of (2) is stirred with 150 ml. of water at 0° for 1 hr. the solid dissolves to give a solution of persuccinic acid (3); about 5% of (3) is hydrolyzed in the process to succinic acid and hydrogen peroxide. The aqueous solution of (3) can be used directly for *trans* hydroxylation of an olefin. Thus the solution obtained from 50 g. of (2) is treated with 0.175 mole of cyclohexene and a surface-active agent (Cémulsol) and agitated with a vibrator for about 70 min. The mixture is neutralized with sodium carbonate, saturated sodium chloride solution is added, and the *trans*-1,2-cyclohexanediol is extracted with ether and ethyl acetate. Crystallization from ethyl acetate gives pure material, m.p. 104°.

[1]R. Lombard and G. Schroeder, *Bull. soc.*, 2800 (1963)

Pertrifluoroacetic acid (Peroxytrifluoroacetic acid), CF_3CO_3H.

This peracid, characterized by effortless preparation and remarkable oxidizing power, was discovered by W. D. Emmons,[1] Rohm and Haas Co., in 1953. Initial observations are described in the section that follows.

—NH_2 ⟶ —NO_2. A first communication[1] notes that trifluoroacetic acid, a very strong acid (pKa 0.3) reacts with aqueous hydrogen peroxide much more rapidly than formic or acetic acid, and that the equilibrium probably is much more favorable to the peracid. For oxidation of 0.05 mole of *p*-aminobenzonitrile to *p*-nitrobenzonitrile, a solution of peracid is prepared by adding 5.1 ml. (0.2 mole) of

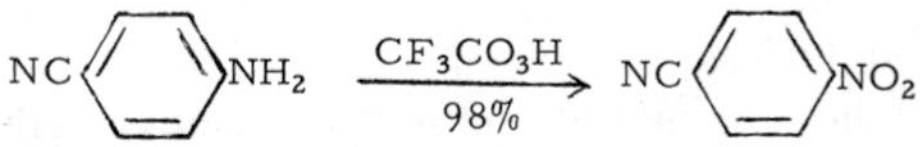
NC
NH2
CF3CO3H
98%
NC
NO2

90% hydrogen peroxide to 40 ml. of trifluoroacetic acid at 20°. The amine is added, the temperature is let rise to 50° and kept there by intermittent cooling for 1 hr. Water precipitates the product in high yield. Aniline afforded nitrobenzene in 79% yield; oxidation with peracetic acid gives 11% of nitrobenzene and 71% of azoxybenzene. The high acidity of trifluoroacetic acid inhibits such secondary condensations. Diethylnitrosoamine by this procedure afforded diethylnitramine in 76% yield.

$$(C_2H_5)_2NNO \xrightarrow[76\%]{CF_3CO_3H} (C_2H_5)_2NNO_2$$

However, Emmons[2] in a later investigation improved the procedure by using anhydrous pertrifluoroacetic acid made by mixing 90% hydrogen peroxide with an appropriate amount of trifluoroacetic anhydride and found methylene chloride (b.p. 41°) to be an exceptionally satisfactory solvent. A solution of the anhydrous reagent in methylene chloride showed no drop in active oxygen in a reflux period of 24 hrs.; the reagent is superior to performic acid in stability. Reactions are easily controlled by operating at the reflux temperature.

In a typical oxidation, 4.1 ml. (0.15 mole) of 90% H_2O_2 is suspended in 100 ml. of methylene chloride, the mixture is stirred and cooled in ice, and 26 ml. (0.18 mole) of trifluoroacetic anhydride is added in one portion. After 5 min. (exothermal reaction), the mixture is allowed to warm to room temperature and 0.1 mole of diethylnitrosoamine in 10 ml. of methylene chloride is added in the course of 30 min. (brisk boiling). After refluxing for 1 hr., the organic layer is separated, washed with water, dried, and evaporated. The residual diethylnitramine corresponded to a yield of 91%; the yield of distilled material was 76%.

The anhydrous reagent in methylene chloride or chloroform as solvent served well for the oxidation of anilines to nitrobenzenes, particularly those with negative substituents.[3] In the example cited the solvent was chloroform, and the yield is

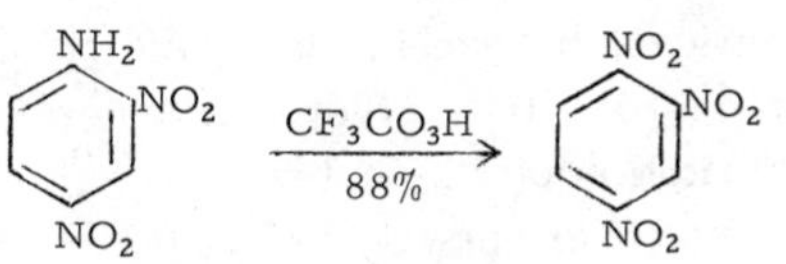
NH2
NO2
NO2
CF3CO3H
88%
NO2
NO2
NO2

for pure material crystallized from ethanol. Both *o*- and *p*-dinitrobenzene and the 2,4,6-trihalonitrobenzenes were all obtained in high yield. Picramide was not oxidized; *p*-anisidine and β-naphthylamine were oxidized too extensively.

Oximes which yield resonance-stabilized acinitroalkanes can be oxidized by the procedure used for oxidation of anilines and nitrosoamines.[4]

$$HON{=}C(CO_2C_2H_5)_2 \xrightarrow[66\%]{CF_3CO_3H,\ CH_2Cl_2} O_2NCH(CO_2C_2H_5)_2$$

Oxidation of the monofunctional *n*-octyl aldoxime in methylene chloride gave a mixture of products, one of which was the nitrile, probably derived from elimination of trifluoroacetic acid from the oxime trifluoroacetate:

$$n\text{-}C_7H_{15}CH{=}NOH \xrightarrow{CF_3CO_3H} n\text{-}C_7H_{15}CH{=}NOCOCF_3 \longrightarrow n\text{-}C_7H_{15}C{\equiv}N + CF_3CO_2H$$

Addition of solid disodium hydrogen phosphate as a buffer to scavenge trifluoroacetic acid from the system did prevent the side .eaction leading to the nitrile but very little nitroalkane was formed. However, furthe: study of the reaction conditions led to the finding that nitroalkanes can be obtained in good yield by peracid oxidation with acetonitrile in the presence of a buffer (Na_2HPO_4, $NaHCO_3$, $NaCO_3$). For example,

$$C_6H_5CH{=}NOH \xrightarrow[\text{In } CH_3CN \\ 77\%]{CF_3CO_3H,\ Na_2HPO_4,\ CO(NH_2)_2} C_6H_5CH_2NO_2$$

benzaldoxime was oxidized with anhydrous reagent in acetonitrile in the presence of disodium hydrogen phosphate and a little urea as scavenger for nitrogen oxides.

Olefins. Emmons and co-workers[5] found that the anhydrous reagent in methylene chloride reacts instantly at 0° with all simple olefins tried; it is much more reactive than performic acid. Higher terminal olefins react rapidly, whereas with performic acid at 40° the reaction takes 8–24 hrs. The initially formed epoxide is cleaved by CF_3CO_2H to the diol monoester, and this is converted into the diol by methanolysis. In initial experiments the monoester reacted to some extent with the epoxide, but this side reaction is inhibited by adding triethylammonium trifluoroacetate to increase the effective concentration of trifluoroacetate ion. Example: a solution

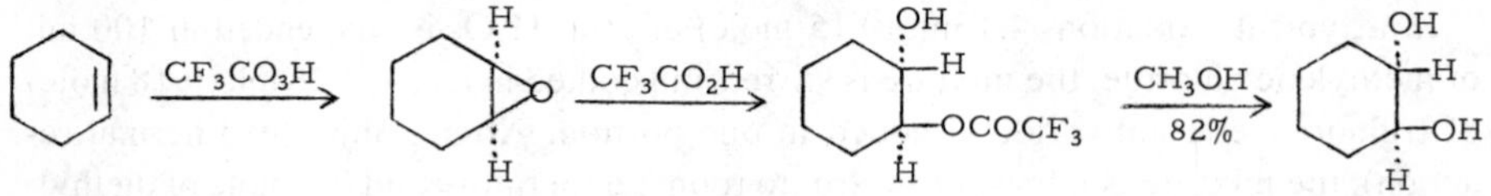
CF₃CO₃H
H
H
CF₃CO₂H
OH
H
OCOCF₃
H
CH₃OH
82%
OH
H
OH
H

prepared from 50 ml. of methylene chloride, 37.2 ml. (0.264 mole) of trifluoroacetic anhydride, and 6 ml. of 90% hydrogen peroxide (0.264 mole) was added in 20 min. to a solution of 0.2 mole of cyclohexene and 0.1 mole of triethylammonium trifluoroacetate in 50 ml. of methylene chloride. The solution was refluxed during the addition, and the mixture was then stirred for 15 min. After removal of volatile solvents at reduced pressure, distillation gave 46.7 g. of crude monoester. This was refluxed with 300 ml. of methanol for 20 hrs., and evaporation gave 19.0 g. (82%) of colorless crystalline *trans*-cyclohexane-1,2-diol. One crystallization from acetone gave pure material, m.p. 103–104°. Hydrogen chloride markedly accelerates methanolysis. In the preparation of 2,3-pentanediol, refluxing with 3% methanolic hydrogen chloride was complete in 2 hrs. (yield 74%).

Emmons and Pagano[6] devised an ingenious scheme for isolation of the initially formed epoxide based on the reasonable assumption that pertrifluoroacetic acid is a much weaker acid than trifluoroacetic acid. The acidity constant could not be determined, but constants for two other acid-peracid pairs suggests a pKa of about 3.7. In the presence of a solid buffer (Na_2CO_3, $NaHCO_3$, Na_2HPO_4) trifluoroacetic

Acid	pKa	Peracid	pKa
CH_3CO_2H	4.8	CH_3CO_3H	8.2
HCO_2H	3.7	HCO_3H	7.1
CF_3CO_2H	0.3	CF_3CO_3H	(3.7 estimated)

acid is neutralized as added, but the peracid reacts much more rapidly with the olefin than with the buffer. When the olefin is consumed, the buffer destroys any excess peracid, and the only substance present in the solution is the epoxide. For epoxidation of 0.2 mole of pentene-1, a mixture of 8.2 ml. of 90% hydrogen peroxide (0.3 mole) and 50 ml. of methylene chloride was stirred in ice during addition of 50.8 ml.

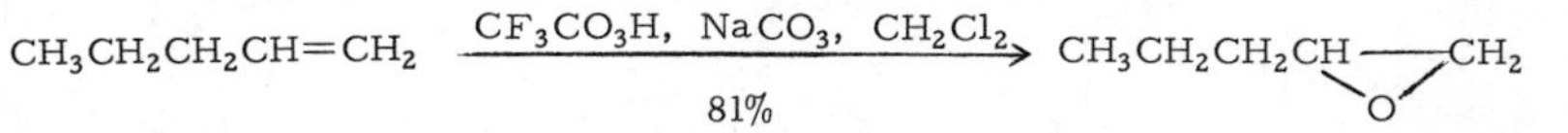

$CH_3CH_2CH_2CH{=}CH_2$
CF_3CO_3H, $NaCO_3$, CH_2Cl_2
81%
$CH_3CH_2CH_2CH$
CH_2
O

(0.76 mole) of trifluoroacetic anhydride and for another 15 min. The solution was than added in the course of 30 min. to a stirred slurry of 0.9 mole of solid sodium carbonate in a solution of 0.2 mole of pentene-1 in 200 ml. of methylene chloride (vigorous boiling). After refluxing for 30 min., salts were removed by centrifigation, and fractionation afforded pentene-1,2-epoxide, b.p. 89–90°, in 81% yield. Epoxides were prepared by this method from methyl methacrylate (84%) and ethyl crotonate (73%).

Micheli[7] applied this "epoxide" procedure to acetyl methyl ursolate, which contains a highly hindered double bond, and isolated the ketone formulated in high yield.

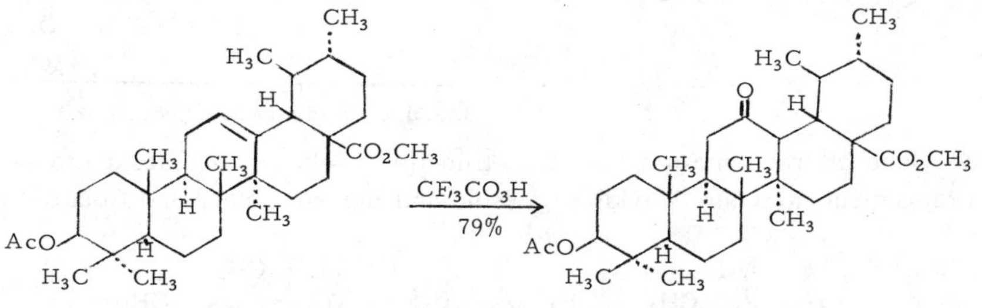

CH_3
H_3C
H
CO_2CH_3
CH_3
CH_3
H
CH_3
AcO
H
H_3C
CH_3
CF_3CO_3H
79%
CH_3
H_3C
O
H
CO_2CH_3
CH_3
CH_3
H
CH_3
AcO
H
H_3C
CH_3

Baeyer-Villiger reaction. Emmons and Lucas[8] found that pertrifluoroacetic acid, in contrast to other peracids, oxidizes most acyclic ketones smoothly and rapidly to esters in excellent yields. In initial trials some 5–10% of material was lost

$$RCOR' \xrightarrow{CF_3CO_3H} RCO{-}OR'$$

by transesterification of the ester with CF_3CO_2H, but this side reaction was eliminated by adding disodium hydrogen phosphate as buffer to remove most of the trifluoroacetic acid as formed. A noteworthy application of the reaction is the oxidation of methyl cyclopropyl ketone to cyclopropyl acetate.[8] Some unreacted ketone

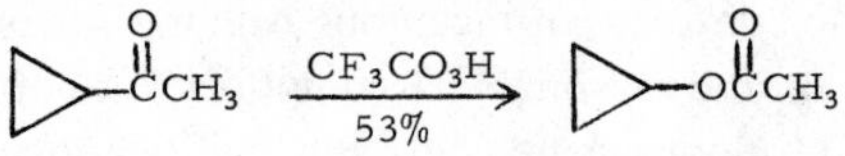
CF3CO3H
53%

present in the reaction mixture had to be removed with Girard's reagent P, and the yield was not high. However, methyl cyclopropyl ketone is completely inert to perbenzoic acid. Trifluoroacetic acid probably functions as catalyst because of its strong acidity:

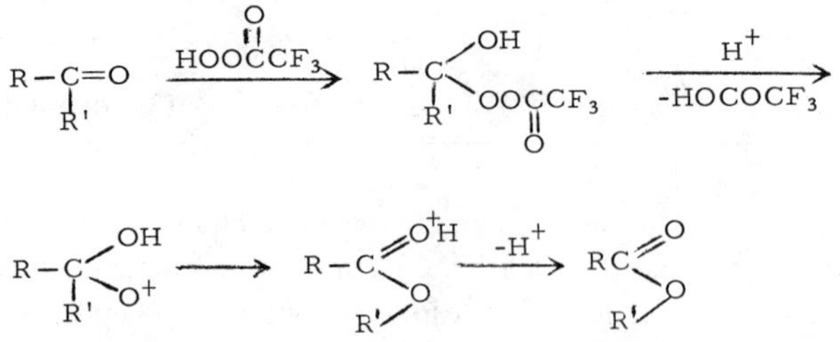
HOOCCF3
H+
-HOCOCF3
-H+

High yields are obtained also in the oxidation of cyclic ketones to the ring-expanded lactones.[9]

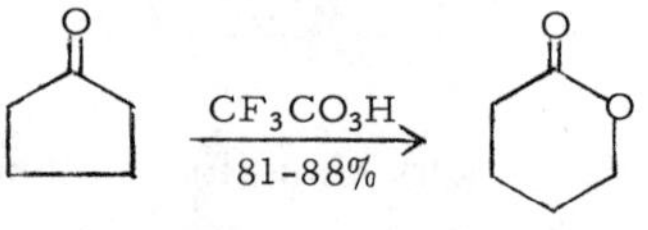
CF3CO3H
81-88%

Oxidation of aromatic compounds. Chambers *et al.*[10] explored oxidation of aromatic hydrocarbons with anhydrous pertrifluoroacetic acid in methylene chloride and, except in the case of benzene and toluene, obtained mixtures shown to contain small amounts of phenols and quinones, as exemplified by the case of *m*-xylene:

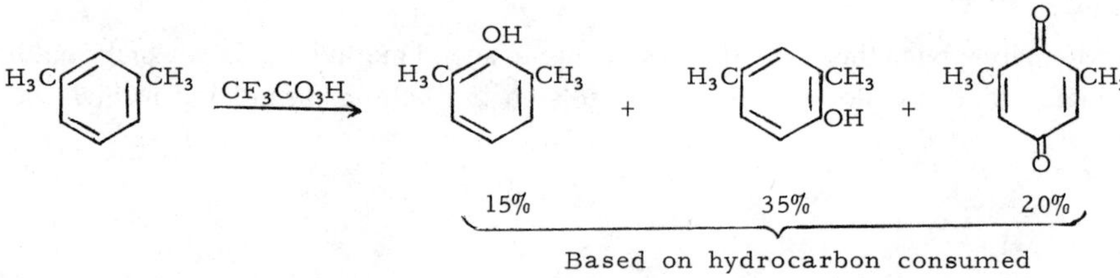
CF3CO3H
15%
35%
20%
Based on hydrocarbon consumed

Mesitylene afforded mesitol and 2,3,5-trimethyl-1,4-benzoquinone, a product of rearrangement. McClure[11] oxidized anisole and diphenyl ether and isolated small

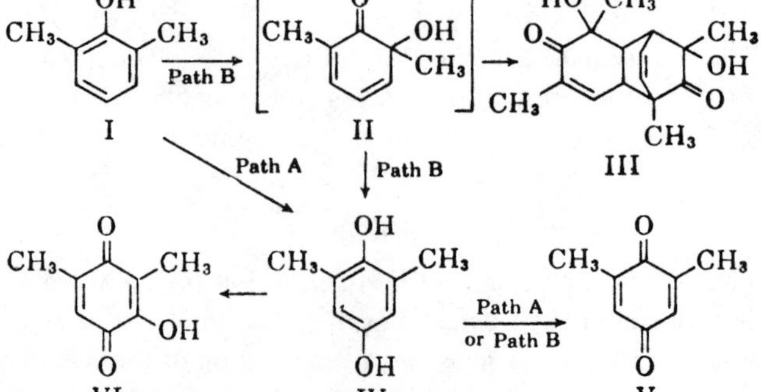
Path B
Path A
Path B
Path A
or Path B
I
II
III
IV
V
VI

amounts (7–35%) of the *o*- and *p*-hydroxy derivatives. He found that slow addition of hydrogen peroxide to a solution of 2,6-dimethylphenol and trifluoroacetic acid in methylene chloride affords primarily a product of *ortho* oxidation, the dimer III of 2,6-dimethyl-*o*-quinol (II, compare oxidation with perchloryl fluoride). However, when the reagents were mixed together all at once the quinone V was obtained in 72% yield. The interpretation advanced is that the rate equation for the formation of V contains at least one term having a higher order in oxidizing agent than the corresponding equation for the formation of III. When the concentration of oxidant is high, V predominates, but when the oxidant is kept at a low concentration by addition over a 20-hr. period, the rate of formation of V is reduced and III becomes the major product. The formation of the *o*-quinol II is attributed to an intermediate VII, in which hydrogen bonding anchors the peracid in a position favorable for attack

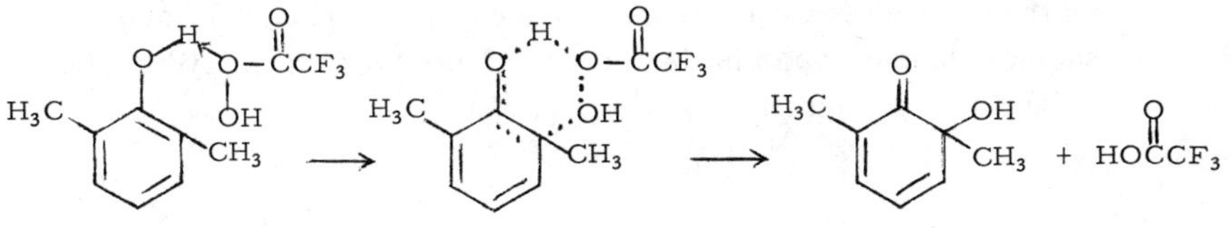

VII VIII II

on the *ortho* carbon atom by the electrophilic peracid oxygen, with formation of the transition state VIII and the product II.

Buehler and Hart[12] reasoned that coordination of a Lewis acid with an organic peracid might facilitate departure of ionic electrophilic hydroxyl and furnish an oxidant potent under mild conditions. Indeed, use of boron trifluororide gave a greatly improved procedure for the preparation of mesitol. The hydrocarbon was

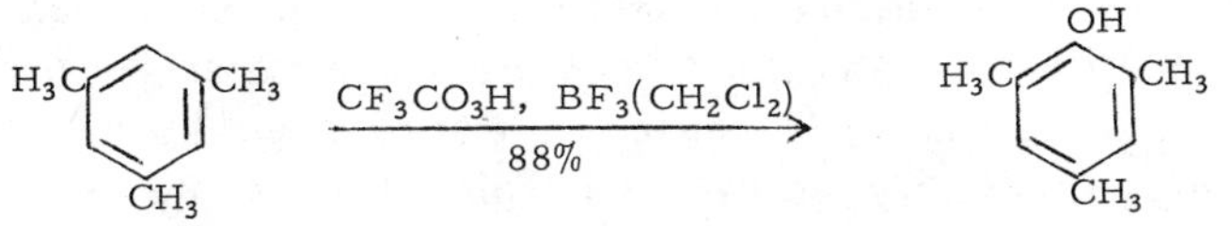

taken in excess in order to inhibit further oxidation of the primary product. A solution was prepared at 0° from 35 g. (0.167 mole) of trifluoroacetic anhydride, 50 ml. of methylene chloride, and 4 ml. (0.148 mole) of 90% hydrogen peroxide and let warm to room temperature. In the course of 2.5 hrs., with ice cooling to keep the temperature below 7°, and while boron trifluoride was bubbled through the reaction mixture, a solution of 56.1 g. (0.468 mole) of mesitylene in 100 ml. of methylene

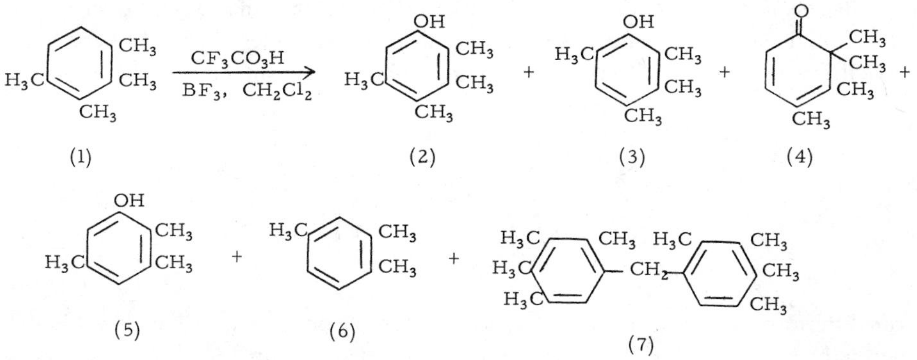

(1) (2) (3) (4)

(5) (6) (7)

chloride was added. Workup afforded 32 g. of mesitylene and 17.7 g. (88%) of mesitol.

Oxidation of prehnitene (1) gave only 9% of the expected product (2), along with lesser amounts of the products (3)–(7), evidently arising from rearrangements. The formation of the dienone (4) is attributed to attack by OH^+ and methyl migration:

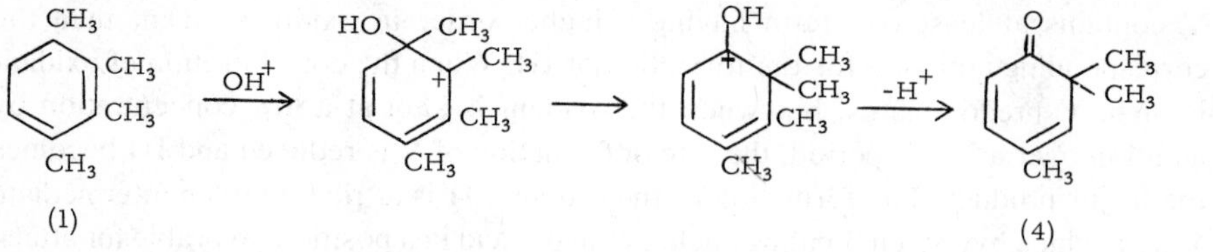

Reasoning that, since a hexaalkylbenzene is incapable of phenol formation, alkyl migration should become important Waring and Hart[13] applied the BF_3-catalyzed reaction to hexamethyl- and hexaethylbenzene at 0° and obtained dienones in

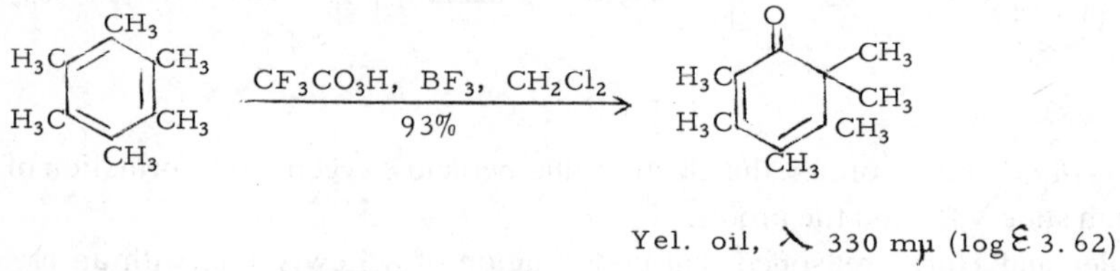

excellent yield. 2,3,4,5,6,6-Hexamethyl-2,4-cyclohexadienone does not dimerize but it forms a maleic anhydride adduct (m.p. 140°) in refluxing ether.

2,6-Dibromopyridine-N-oxide. Because of the inductive effect of the two bromine atoms in 2,6-dibromopyridine, the unshared pair of electrons on nitrogen is relatively inaccessible for coordination, and the N-oxide cannot be prepared by oxidation with perbenzoic acid or peracetic acid, but the more strongly electrophilic pertrifluoroacetic acid gives satisfactory results.[14] A mixture of 3 g. of the amine, 37 ml. of

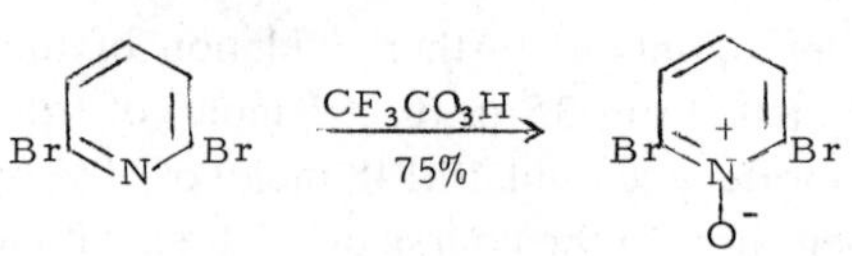

trifluoroacetic acid, and 4 ml. of 30% hydrogen peroxide was refluxed on the steam bath for 3 hrs., and the product crystallized once from methanol.

Benzimidazole synthesis. Nair and Adams[15] found that the reagent oxidizes 2-substituted aniline derivatives of type (1) to benzimidazoles (2) under very mild conditions in yields of 60–90%, depending upon the substituents in the aromatic

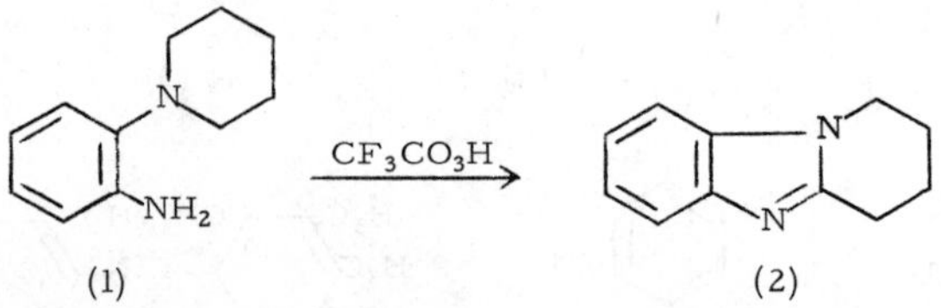

ring. Performic acid is regarded as satisfactory.[16] Both reagents cyclize N-acyl derivatives of types (3) in high yield with loss of the acyl group.[13]

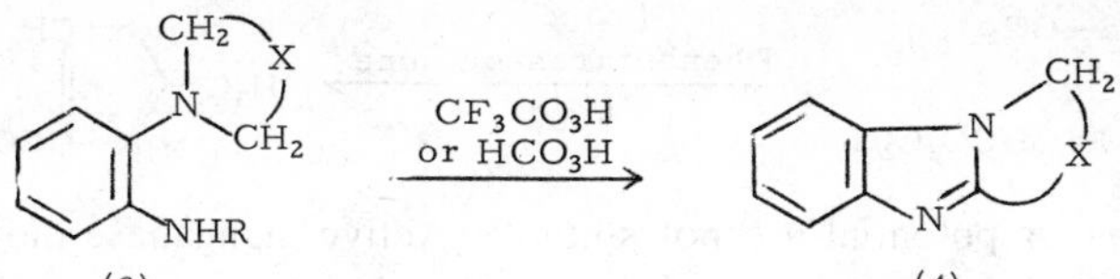

(3) (4)

R = CHO, $COCH_3$, COC_6H_5

X = $(CH_2)_2$, $(CH_2)_3$, CH_2OCH_2, $(CH_2)_4$

Oxidation of Δ⁴-3-ketosteroids. The reagent oxidizes Δ^4-cholestene-3-one to the aldehyde-lactone (3), possibly by formation and rearrangement of (2), an epoxide of an enol lactone.[17]

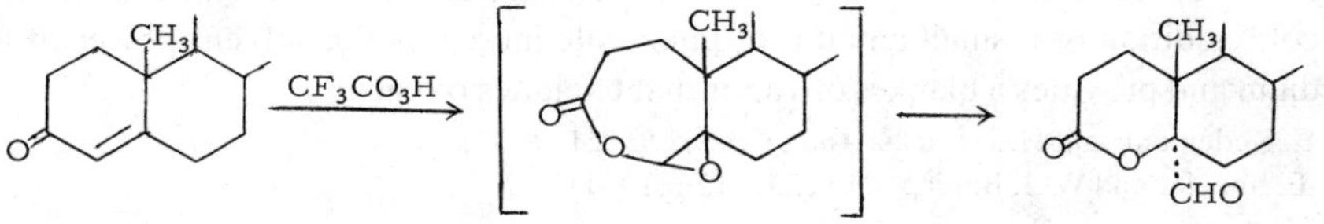

(1) (2) (3)

[1]W. D. Emmons and A. F. Ferris, *Am. Soc.*, **75**, 4623 (1953)
[2]W. D. Emmons, *Am. Soc.*, **76**, 3468 (1954)
[3]*Idem, ibid.*, **76**, 3470 (1954)
[4]W. D. Emmons and A. S. Pagano, *Am. Soc.*, **77**, 4557 (1955)
[5]W. D. Emmons, A. S. Pagano, and J. P. Freeman, *Am. Soc.*, **76**, 3472 (1954)
[6]W. D. Emmons and A. S. Pagano, *Am. Soc.*, **77**, 89 (1955)
[7]R. A. Micheli, *J. Org.*, **27**, 666 (1962)
[8]W. D. Emmons and G. B. Lucas, *Am. Soc.*, **77**, 2287 (1955)
[9]W. F. Sager and A. Duckworth, *Am. Soc.*, **77**, 188 (1955); R. Huisgen and H. Ott, *Tetrahedron*, **6**, 253 (1959); E. E. Smissman, J. F. Muren, and N. A. Dahle, *J. Org.*, **29**, 3517 (1964)
[10]R. D. Chambers, P. Goggin, and W. K. R. Musgrave, *J. Chem. Soc.*, 1804 (1959)
[11]J. D. McClure and P. H. Williams, *J. Org.*, **27**, 627 (1962); J. D. McClure, *ibid.*, **28**, 69 (1963)
[12]C. A. Buehler and H. Hart, *Am. Soc.*, **85**, 2177 (1963); H. Hart *et al.*, *J. Org.*, **29**, 2397 (1964), **30**, 331 (1965)
[13]A. J. Waring and H. Hart, *Am. Soc.*, **86**, 1454 (1964)
[14]R. F. Evans, M. Van Ammers, and H. J. Den Hertog, *Rec. trav.*, **78**, 408 (1959)
[15]M. D. Nair and R. Adams, *Am. Soc.*, **83**, 3518 (1961)
[16]O. Meth-Cohn and H. Suschitzky, *J. Chem. Soc.*, 4666 (1963)
[17]J. T. Pinhey and K. Schaffner, *Tetrahedron Letters*, 601 (1965)

Pfitzner-Moffatt reagent, *see* Dimethyl sulfoxide.

Phenanthrenequinone. Mol. wt. 208.22, m.p. 206–208°. Suppliers (pure): B, E, F, MCB; (technical, 80%): Aldrich.

Preparation. By oxidation of phenanthrene with CrO_3 in acetic acid[1] or with K_2CrO_7 in dil. H_2SO_4.[2-4]

Purification. Yields reported are not strictly comparable because of variations in the purity of the phenanthrene used, particularly with respect to the content of anthracene. Crude phenanthrenequinone can be separated easily from anthraquinone and other impurities through the water-soluble bisulfite addition compound.[1,2] Thus a simple route to pure quinone at moderate cost is by bisulfite purification of the 80% technical material.

Dehydrogenation.[5] Thiazolines are dehydrogenated smoothly by phenanthrenequinone to thiazoles. In acetic acid solution at 100° the reaction is complete in ½ hr.

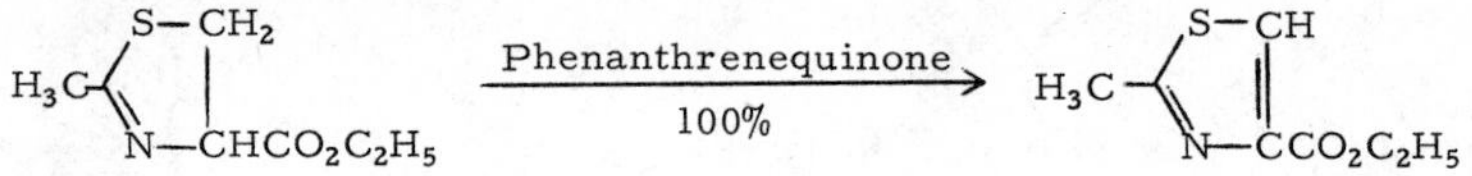
Phenanthrenequinone
100%

Quinones of higher potential are not suitable. Active manganese dioxide gives at most yields of 70%.

[1]C. Graebe, *Ann.*, **167**, 131 (1873)
[2]R. Anschütz and G. Schultz, *Ann.*, **196**, 32 (1879)
[3]L. Oyster and H. Adkins, *Am. Soc.*, **43**, 208 (1921)
[4]H. W. Underwood and E. L. Kochmann, *Am. Soc.*, **46**, 2069 (1924)
[5]Moira A. Barton, G. W. Kenner, and R. C. Sheppard, *J. Chem. Soc.* C, 1061 (1966)

Phenetole, $C_6H_5OC_2H_5$. Mol. wt. 122.16, b.p. 170°, sp. gr. 0.97. Suppliers: B, E, F.

In the saponification of a difficulty split ester with potassium hydroxide in ethylene glycol,[1] addition of a small amount of phenetole increases the solvent power of the medium and provides a blanket of vapor that excludes oxygen.[2]

[1]C. E. Redemann and H. J. Lucas, *Anal. Chem.*, **9**, 521 (1937)
[2]W. E. Shaefer and W. J. Balling, *ibid.*, **23**, 1126 (1951)

Phenol, C_6H_5OH. Mol. wt. 94.11, m.p. 43°, b.p. 181°, pKa 10.0. Suppliers: A, B, E, F.

Bouveault-Blanc reduction. Whereas the reduction of amino acid esters with sodium and ethanol affords the amino alcohols in yield of only 23–30%, Enz[1] found that the yield can be doubled by using sodium and an ethanolic solution of phenol. The optimum molar ratio of ester:phenol:sodium varies between 1:6:8 and 1:7.5:12. The reaction is conducted in an oil bath in the range 160–190°. The method also gives improved yields in the reduction of aromatic esters; in this case addition of quinoline or tetrahydroquinoline as catalyst is advantageous.

Bromine scavenger. When phenol is added as bromine scavenger in the pyrolysis of the tetrabromide (1), the yield of product (2) is 92%.[2] In the absence of phenol, (2) is obtained in yield of 48% along with unidentified products.

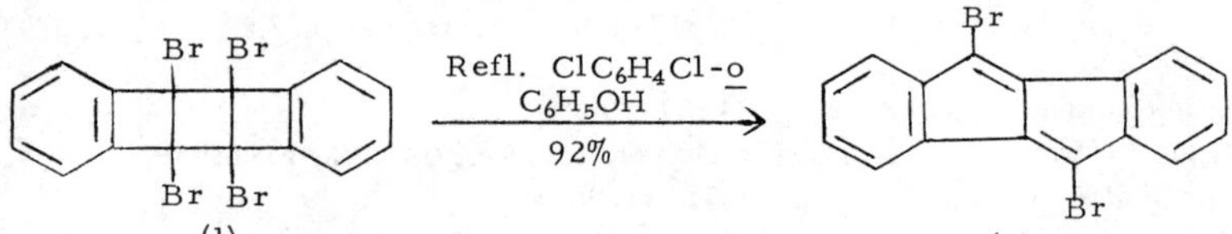
Br Br
Br Br
Refl. ClC6H4Cl-o
C6H5OH
92%
Br
Br

(1) (2)

Synthesis of hexamethylbenzene.[3] In this curious reaction phenol is perhaps described more accurately as a starting material than as a reagent. A solution of phenol in methanol is allowed to drop at a rate of 110 ml. per hr. onto a column of

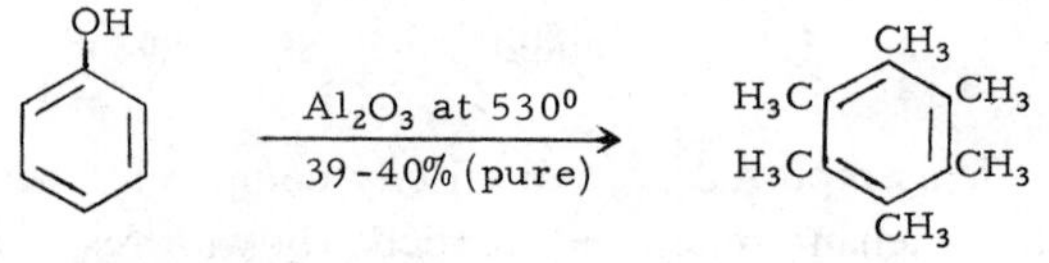
OH
Al2O3 at 530°
39-40% (pure)
CH3
H3C
CH3
H3C
CH3
CH3

100 g. in 1 l. CH_3OH

300 g. of alumina heated to 530°. One crystallization of the crude product gives colorless hexamethylbenzene, m.p. 165–166°, in modest yield.

[1]W. Enz, *Helv.*, **44**, 206 (1961)
[2]M. P. Cava, R. Pohlke, and M. J. Mitchell, *J. Org.*, **28**, 1861 (1963)
[3]N. M. Cullinane, S. J. Chard, and C. W. C. Dawkins, *Org. Syn., Coll. Vol.*, **4**, 520 (1963)

Phenyl acetate, $CH_3CO_2C_6H_5$. Mol. wt. 136.15, b.p. 78–80°/10 mm., suppliers: A, B, E, F.

Pappo, Bloom, and Johnson[1] effected acetylation of the 3β-hydroxyl group of the steroid dienic acid (1) by ester interchange with excess phenyl acetate in the presence of sodium hydride. Conventional acetylation is complicated by the formation of a mixed anhydride and the necessity for a step of selective hydrolysis.

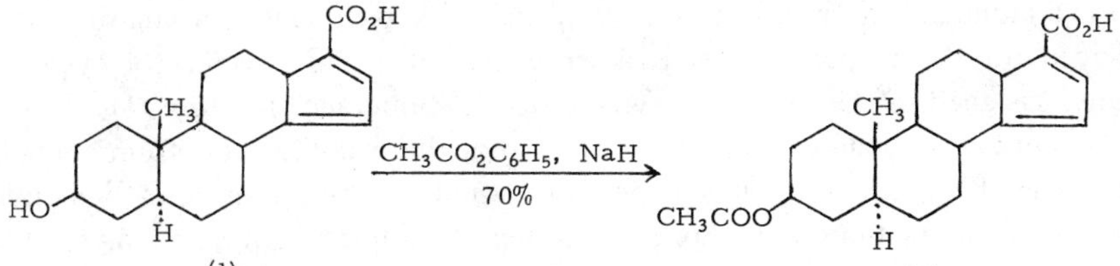
CO2H
CH3
HO
H
CH3CO2C6H5, NaH
70%
CO2H
CH3
CH3COO
H

(1) (2)

[1]R. Pappo, B. M. Bloom, and W. S. Johnson, *Am. Soc.*, **78**, 6347 (1956)

Phenyl azide, $C_6H_5N{=}\overset{+}{N}{=}\overset{-}{N}$. Mol. wt. 119.12, b.p. 41–43°/5 mm., 66–68°/21 mm.

Preparation.[1] A mixture of 55.5 ml. of concd. hydrochloric acid and 300 ml. of water is stirred in an ice-salt bath, and 35.5 g. of phenylhydrazine is added dropwise in 5–10 min.; phenylhydrazine hydrochloride separates as fine white plates.

$$C_6H_5NHN\overset{+}{H_3}\overset{-}{Cl} \xrightarrow[65-68\%]{NaNO_2} C_6H_5N{=}\overset{+}{N}{=}\overset{-}{N}$$

After the temperature has fallen to 0°, 100 ml. of ether is added, and a solution of 25 g. of sodium nitrite in 30 ml. of water is added from a dropping funnel at such a rate that the temperature *never* rises above 5° (25–30 min.). The mixture is steam distilled until about 400 ml. of distillate has been collected. The ethereal layer containing most of the product is separated, the water layer is extracted once with ether, and the total ethereal solution is dried over calcium chloride and placed in a Claisen flask for vacuum distillation. Since phenyl azide explodes when heated at ordinary pressure and occasionally at a lower pressure, the distillation must be carried out behind an explosion-proof screen. The flask is immersed in a water bath at 25–30°, and the ether is removed under reduced pressure. Then the temperature of the water bath is raised to 60–65°, and the phenyl azide is distilled at reduced pressure. The substance is obtained as a pungent, pale yellow oil.

Doering and Odum[2] found that removal of by-product phenol increases the stability of phenyl azide in storage. Thus the ethereal extract of the steam distillate was extracted with 2 *N* sodium hydroxide until no phenol could be detected by the ferric chloride test. Concentration and vacuum distillation afforded phenyl azide in 31% yield.

Addition of cycloolefins. Wolff[3] in 1912 explored the reaction of phenyl azide with a number of olefinic compounds and in the case of styrene, dimethyl fumarate, and 1,4-benzoquinone was able to carry out the reaction at a moderate temperature and establish that the initial product is a 1-phenyl-4,5-dihydro-1,2,3-triazole. The three compounds mentioned have one or two activating groups flanking the double

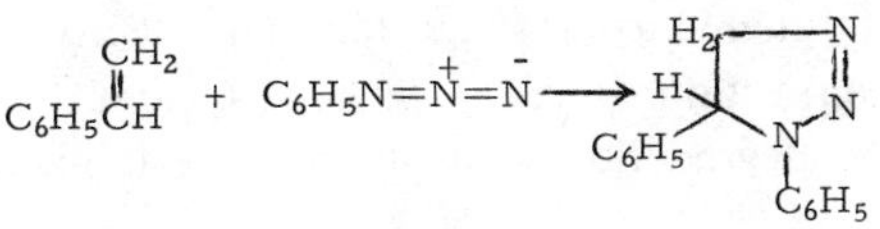
CH2
C6H5CH
+ C6H5N=N=N
H2
N
H
N
C6H5
N
C6H5

bond. The reaction of phenyl azide with styrene, which has only one weakly activating group, is complete at room temperature only after several weeks. Other compounds studied by Wolff reacted only at elevated temperatures such that addition products if formed decomposed with loss of nitrogen.

With this background of information, Alder and Stein[4] were surprised to find that phenyl azide reacts exothermally at room temperature with dicyclopentadiene to form a monoadduct, m.p. 130–131°, in quantitative yield. Two formulas had been advanced for dicyclopentadiene. Krämer and Spilker,[5] discoverers of cyclopentadiene, assigned to the dimer the symmetrical cyclobutane structure (1), resulting from dimerization analogous to the formation of the truxillic acids from cinnamic acid. Wieland[6] suggested the unsymmetrical formula (2), resulting from 1,4-addition of one molecule to the double bond of another. Two lines of evidence favored the

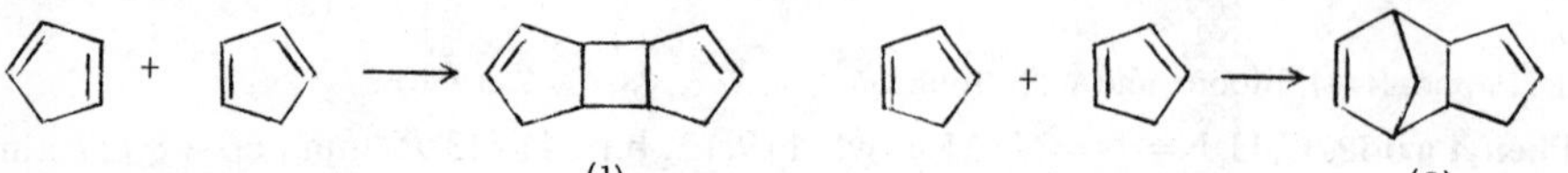

(1) (2)

symmetrical structure (1), but this seemed to Alder and Stein[7] inconsistent with their observation that in the presence of excess phenyl azide the hydrocarbon forms only a monoadduct. Furthermore, model compounds analogous to (1) failed to react with the reagent, whereas the bridged-ring compounds (3), analogous to (2), all reacted readily with phenyl azide at room temperature. On the other hand,

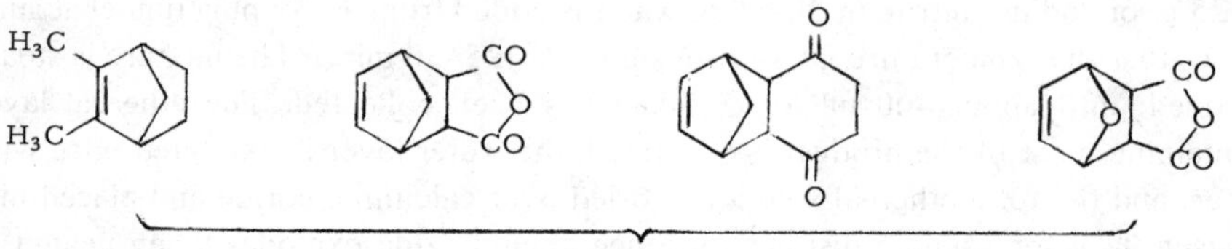
H₃C
H₃C
CO
CO
O
O
CO
CO

(3)

Reactive to $C_6H_5N_3$

the unsaturated anhydride (4), which lacks a one-atom bridge across the six-membered ring, failed to react with phenyl azide even after several weeks. Also unreactive is the anhydride (5), in which the cyclohexane is spanned by a two-carbon bridge. Alder and Stein inferred that a double bond in a strain-free system (4, 5)

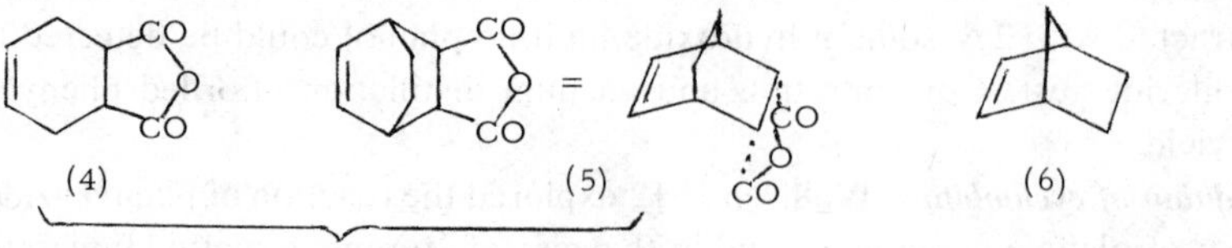
CO
CO
CO
CO
CO
CO

(4) (5) (6)

Strain-free, unreactive Strained, reactive

is inert to phenyl azide whereas one in a strained system possesses special reactivity in consequence of the strain. The selective reactivity of one of the two double bonds of dicyclopentadiene to phenyl azide suggested that selective hydrogenation should be possible. Indeed the reaction slowed down sharply after the uptake of one mole of hydrogen, and oxidative degradation of the dihydride established the unsymmetrical structure (2) as correct.[7]

In another investigation, Alder and Stein[8] found that cyclopentene forms a phenyl azide adduct but is considerably less reactive than the bridged compounds of type (6); cyclohexene formed no adduct. Ziegler[9] later found the reaction with phenyl azide in ether very useful for the characterization of *cis*- and *trans*-cycloalkenes of medium-size rings. Thus *cis*- and *trans*-cyclooctene form crystalline adducts melting at 87 and 111°, respectively. Approximate relative reaction rates of *trans*-cycloalkenes with phenyl azide are shown under the formulas:

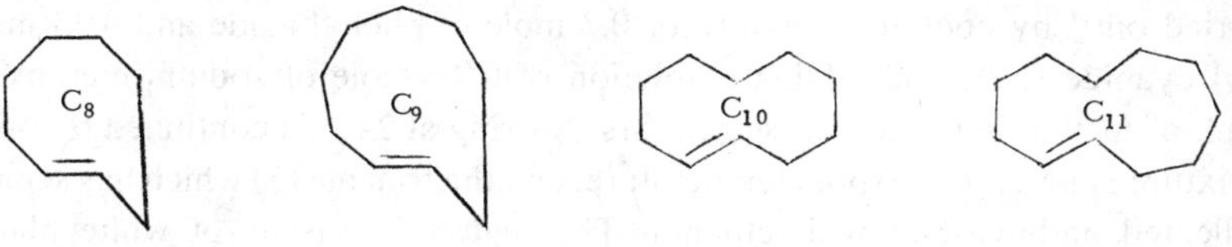
C8
C9
C10
C11

20,000 2,000 10 1

Trapping of cycloalkynes and of benzyne. Wittig[10] demonstrated the transient formation of the C_5–C_7 cycloalkynes by oxidation of the appropriate 1,2-dione bishydrazones in the presence of either of two highly reactive Diels-Alder components, phenyl azide and 1,3-diphenylisobenzofurane. Thus oxidation of the C_7-bishydrazone with mercuric oxide in refluxing benzene in the presence of powdered potassium hydroxide and 1 equivalent of the azide afforded the crystalline adduct in 29% yield. The cyclohexyne adduct was obtained in yield of 8%.

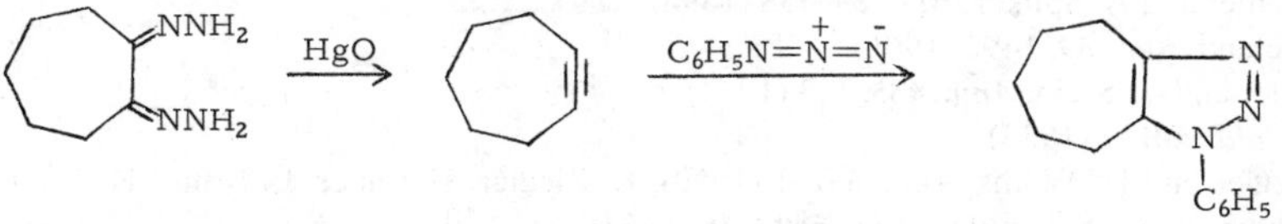
NNH2
NNH2
HgO
C6H5N=N=N
N
N
N
C6H5

Wittig[11] found phenyl azide useful also as a reagent for trapping benzyne generated from a new precursor prepared from sodium *o*-aminobenzenesulfinate. A solution of this salt and sodium nitrite in the least amount of water was stirred at −15° and

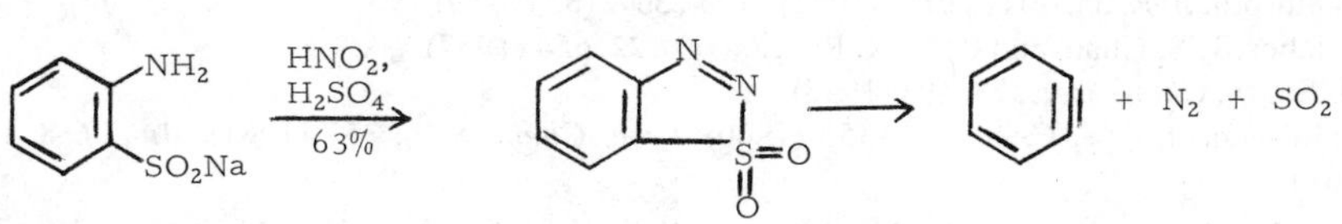
NH2
SO2Na
HNO2,
H2SO4
63%
N
N
S=O
O
+ N2 + SO2

treated with a mixture of dilute sulfuric acid and glycerol. After stirring at −15° for 2 hrs. the mixture was extracted with successive portions of ether at −6°. Drying at −20° and evaporation at 0° gave a yellow, explosive product regarded as the heterocyclic azo-sulfone, rather than a diazonium salt, because it is insoluble in water and soluble in ether. A solution of freshly prepared reagent in tetrahydrofurane was treated with phenyl azide, and the temperature allowed to rise to 10°, at which temperature evolution of nitrogen and sulfur dioxide proceeded briskly for about 2 hrs. Workup, the next day, afforded 1-phenylbenzotriazole, m.p. 89°, in 47% yield.

1,2,3-Triazoles. Procedures[12] for the preparation of 1,4-diphenyl-5-amino-1,2,3-triazole (3) and its rearrangement to 4-phenyl-5-anilino-1,2,3-triazole (4) exemplify general reactions discovered by Otto Dimroth.[13] Under basic catalysis, phenyl azide reacts with benzyl cyanide, ethyl acetoacetate, malonic ester, and similar

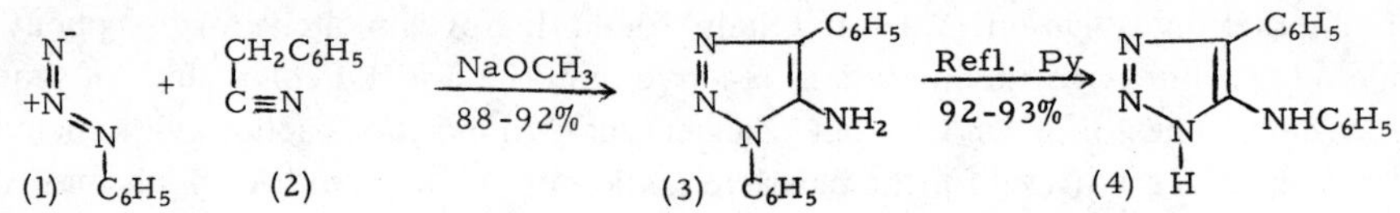
NaOCH3
88-92%
Refl. Py
92-93%
(1)
(2)
(3)
(4)

reagents to form 1,2,3-triazoles in high yield. One possible mechanism for the reaction is suggested by Lieber, Chao, and Rao;[14] a 1,3-dipolar cycloaddition of the type postulated by Smith[15] and by Huisgen[16] seems to us attractive. The reaction is carried out[11] by cooling a mixture of 0.3 mole of phenyl azide and 0.33 mole of benzyl cyanide to 2° and adding a solution of 0.45 mole of sodium methoxide in 150 ml. of methanol in the course of 2 hrs. Stirring at 2–5° is continued for 48 hrs., the mixture is let come to room temperature, and the triazole (3) which has separated is collected and washed with ethanol. The material consists of white platelets, m.p. 169–171°, unchanged on recrystallization from benzene. The basic diphenylaminotriazole (3) is isomerized to the acidic phenylanilinotriazole (4, soluble in alkali) by refluxing in pyridine for 24 hrs. Dilution with ice water gives a milky oil which on stirring and scratching changes to white needles, m.p. 168–169°; the substance is soluble in hot water and in ether, but sparingly soluble in benzene.

[1] R. O. Lindsay and C. F. H. Allen, *Org. Syn., Coll. Vol.*, **3**, 710 (1955)
[2] W. von E. Doering and R. A. Odum, *Tetrahedron*, **22**, 81 (1966)
[3] L. Wolff, *Ann.*, **394**, 68 (1912)
[4] K. Alder and G. Stein, *Ann.*, **485**, 211 (1931)
[5] G. Krämer and A. Spilker, *Ber.*, **29**, 558 (1896)
[6] H. Wieland, *Ber.*, **39**, 1492 (1906)
[7] K. Alder and G. Stein, *Ann.*, **485**, 223 (1931)
[8] *Idem, ibid.*, **501**, 1 (1933)
[9] K. Ziegler and H. Wilms, *Ann.*, **567**, 1 (1950); K. Ziegler, H. Sauer, L. Bruns, H. Froitzheim-Kühlhorn, and J. Schneider, *Ann.*, **589**, 136 (1954)
[10] G. Wittig and A. Krebs, *Ber.*, **94**, 3260 (1961)
[11] G. Wittig and R. W. Hoffmann, *Ber.*, **95**, 2718 (1962)
[12] E. Lieber, T. S. Chao, and C. N. R. Rao, *Org. Syn., Coll. Vol.*, **4**, 380 (1963)
[13] O. Dimroth, *Ber.*, **35**, 4041, 4058 (1902); *Ann.*, **364**, 182 (1909)
[14] E. Lieber, T. S. Chao, and C. N. R. Rao, *J. Org.*, **22**, 654 (1957)
[15] L. I. Smith, *Chem. Rev.*, **23**, 193 (1938)
[16] R. Huisgen, *Angew. Chem.*, **72**, 359 (1960); *Proc. Chem. Soc.*, 357 (1961); *Ann.*, **658**, 169 (1962)

***p*-Phenylazobenzenesulfonyl chloride**, *p*-$ClSO_2C_6H_4N{=}NC_6H_5$. Mol. wt. 280.74, m.p. 125°. Preparation.[1]

The reagent reacts with primary and secondary amines in pyridine solution to form highly crystalline orange to red sulfonamides easily purified by chromatography. The derivatives are useful for identification and separation of amines, and the amines can be recovered by hydrolysis with concd. hydrochloric acid–dioxane.[2]

[1] R. D. Desai and C. V. Mehta, *Indian J. Pharm.*, **13**, 211 (1951)
[2] E. O. Woolfolk, W. E. Reynolds, and J. L. Mason, *J. Org.*, **24**, 1445 (1959)

***p*-Phenylazobenzoyl chloride**, $C_6H_5N{=}NC_6H_4COCl$-*p*. Mol. wt. 244.68, m.p. 92–94°. Suppliers: E, F.

The reagent forms colored esters with alcohols which are useful for chromatographic separation and identification.[1] *Compare* 4′-Nitroazobenzene-4-carboxylic acid chloride.

[1] E. O. Woolfolk, F.-E. Beach, and S. P. McPherson, *J. Org.*, **20**, 391 (1955)

p-Phenylazomaleinanil, 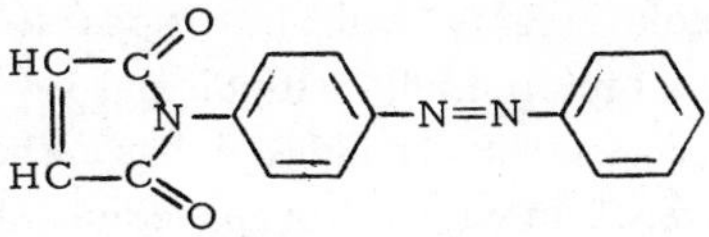

Mol. wt. 173.17, orange needles, m.p. 162°. Preparation.[1] Supplier: Aldrich.

The reagent is a dienophile useful for characterizing dienes and for the investigation of diene mixtures. The adducts from butadiene, cyclopentadiene, $\Delta^{1,3}$-hexadiene, and cosmene, $CH_2{=}CH(CH{=}CH)_2CH{=}CH_2$, are obtainable under mild conditions (20–50°), and they are high-melting, colored compounds which crystallize well. They can be separated from mixtures readily by chromatography and have identifying IR and UV spectra.

[1] P. Naylor and M. C. Whiting, *J. Chem. Soc.*, 2970 (1955)

N-Phenyl-N′-benzoyldiimide, $C_6H_5N{=}NCOC_6H_5$. Mol. wt. 210.23, m.p. 135°. Preparation.[1]

This substance decomposes in the presence of dilute acid or base at room temperature to phenyl radicals, presumably via phenyldiimide, $C_6H_5N{=}NH$.[2] The ease of production of phenyl radicals in a homogeneous medium makes this an appealing precursor.

[1] G. Ponzio and G. Charrier, *Gazz.*, **39**, 1, 596 (1909)
[2] S. G. Cohen and J. Nicholson, *Am. Soc.*, **86**, 3892 (1964)

Phenylboronic acid, $C_6H_5B(OH)_2$. Mol. wt. 119.94, m.p. 215–217°. Supplier: L. Light.

Preparation (50–60% yield) from phenylmagnesium bromide and *n*-butyl borate.[1]

Cyclic esters of diols. The reagent reacts with 1,2-, 1,3-, and 1,4-diols to form cyclic esters.[2] 1,5-Diols give no crystalline products. Both *cis*- and *trans*-cyclo-

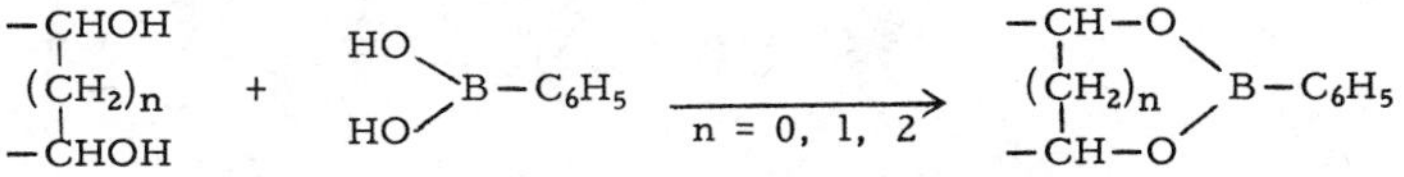

−CHOH
(CH2)n
−CHOH
+
HO
HO
B−C6H5
n = 0, 1, 2
−CH−O
(CH2)n
−CH−O
B−C6H5

pentane-1,2-diol and *cis*- and *trans*-cyclohexane-1,2-diol give esters, but the esters of the *trans*-isomers have much higher melting points. The reagent (as well as the

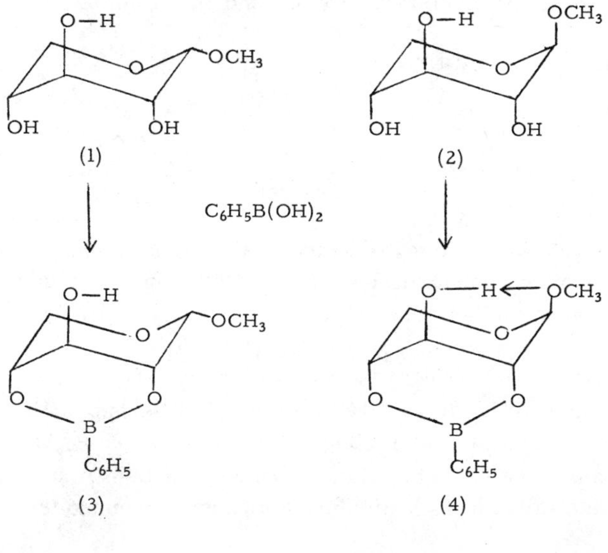

O−H
OCH3
OH
OH
(1)
O−H
OCH3
OH
OH
(2)
C6H5B(OH)2
O−H
OCH3
O
O
B
C6H5
(3)
O—H←OCH3
O
O
B
C6H5
(4)

anhydride, triphenylboraxole) reacts with hexosides to give crystalline cyclic esters in high yield. It reacts preferentially with diaxial 1,3-diol groups, as illustrated for methyl α- and β-xylopyranosides, (1) and (2).[3] The carbohydrate can be liberated from the boronic ester by reaction with 1,3-propanediol. The reagent has been used to provide a protective group in disaccharide synthesis.[4]

[1]F. R. Bean and J. R. Johnson, *Am. Soc.*, **54**, 4415 (1932)
[2]J. M. Sugihara and C. M. Bowman, *Am. Soc.*, **80**, 2443 (1958)
[3]R. J. Ferrier, *J. Chem. Soc.*, 2325 (1961); R. J. Ferrier, D. Prasad, A. Rudowski, and I. Sangster, *ibid.*, 3330 (1964)
[4]R. J. Ferrier and D. Prasad, *J. Chem. Soc.*, 7429 (1965)

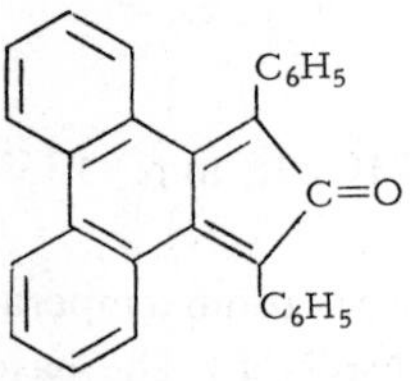

C_6H_5
C=O
C_6H_5

Phenylcyclone. Mol. wt. 382. 43, m.p. 273°, dark green. The reagent is prepared in 90% yield by the condensation of phenanthrenequinone with dibenzyl ketone.[1] Mackenzie[2] found it more reactive than tetraphenylcyclopentadienone in Diels-Alder reactions.

[1]W. Dilthey, I. ter Horst, and W. Schommer, *J. prakt. Chem.*, **143**, 189 (1935)
[2]K. Mackenzie, *J. Chem. Soc.*, 473 (1960)

Phenyldiazomethane, $C_6H_5CH{=}\overset{+}{N}{=}\overset{-}{N}$. Mol. wt. 118.14.

The reagent can be generated in ether solution as required by reaction of the stable precursor N-nitroso-N-benzyl-*p*-toluenesulfonamide (2) with sodium methoxide.[1]

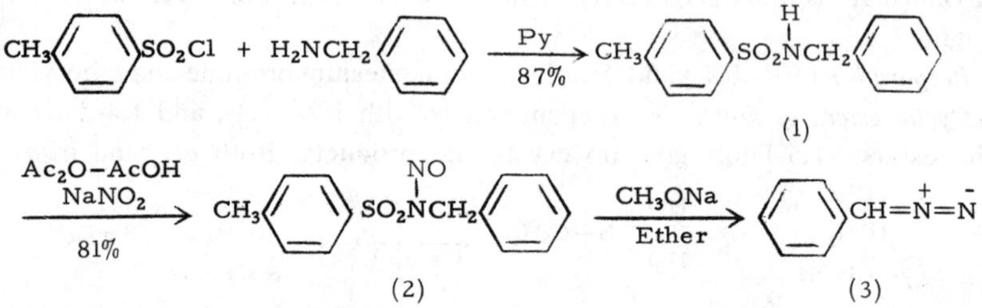

CH_3 SO_2Cl + H_2NCH_2
Py
87%
CH_3 SO_2NCH_2
H
(1)
$Ac_2O-AcOH$
$NaNO_2$
81%
CH_3 SO_2NCH_2
NO
(2)
CH_3ONa
Ether
$CH{=}\overset{+}{N}{=}\overset{-}{N}$
(3)

In an alternative procedure which apparently is preferred,[2] benzaldehyde is condensed with excess hydrazine in ether and the resulting benzalhydrazone is

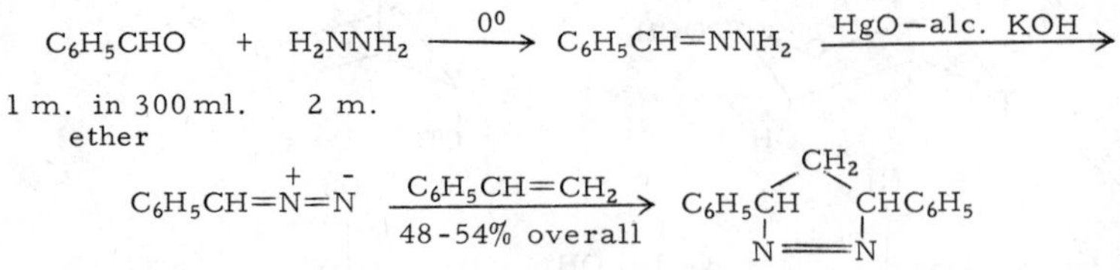

C_6H_5CHO + H_2NNH_2
0^0
$C_6H_5CH{=}NNH_2$
HgO—alc. KOH
1 m. in 300 ml. ether
2 m.
$C_6H_5CH{=}\overset{+}{N}{=}\overset{-}{N}$
$C_6H_5CH{=}CH_2$
48-54% overall
CH_2
C_6H_5CH
CHC_6H_5
N
N

oxidized with yellow mercuric oxide in an alkaline medium. The deep wine-red solution of phenyldiazomethane obtained on filtration on reaction with styrene at room temperature for 24 hrs. affords colorless 3,5-diphenyl-1-pyrazoline.

[1]C. G. Overberger and J.-P. Anselme, *J. Org.*, **28**, 592 (1963)
[2]*Idem, Am. Soc.*, **86**, 658 (1964); procedure submitted to *Org. Syn.*

***o*-Phenylenediamine**, *o*-$H_2NC_6H_4NH_2$. Mol. wt. 108.14, m.p. 101° b.p. 257°, pKb 9.48. Suppliers (practical grade): Ciba, Koppers, A, B, E, F, MCB. Commercial material is liable to be badly discolored from air oxidation. A small sample can be purified by sublimation or distillation in a horizontally mounted test tube evacuated at the water pump. A larger lot is best purified as described in the next section.

Preparation.[1] A mixture of 0.5 mole of *o*-nitroaniline, 40 ml. of 20% sodium hydroxide solution, and 200 ml. of 95% ethanol is stirred vigorously and heated on the steam bath until the solution boils gently. The steam is turned off, and 2 g. atoms

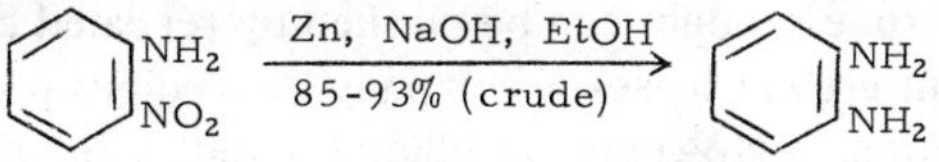
NH2
NO2
Zn, NaOH, EtOH
85-93% (crude)
NH2
NH2

of zinc dust (162 g. of 80% pure metal) is added in 10-g. portions (cautiously at first) frequently enough to maintain boiling. When the addition is complete the mixture is refluxed with continued stirring for 1 hr.; the initially deep red solution becomes nearly colorless. The hot solution is filtered by suction, the zinc is extracted thoroughly with hot ethanol, and the solution is concentrated at water-pump vacuum to a volume of 125–150 ml. After thorough cooling in ice-salt, the faintly yellow crystals are collected, washed with a little ice water, and dried in a vacuum dessicator. The yield of material, m.p. 97–100°, is 46–50 g. (85–93%).

For purification, 50 g. of the above material is dissolved in 175 ml. of hot water containing 1–2 g. of sodium hydrosulfite, and the solution is clarified with Norit. After cooling, eventually in salt-ice, the colorless crystals are collected and washed with a little ice water. The recovery is 46 g. (85%), m.p. 99–101°.

Preparation of heterocycles. Benzimidazole is prepared[2] by heating a mixture of 0.5 mole of *o*-phenylenediamine and 0.75 mole of 90% formic acid in a water

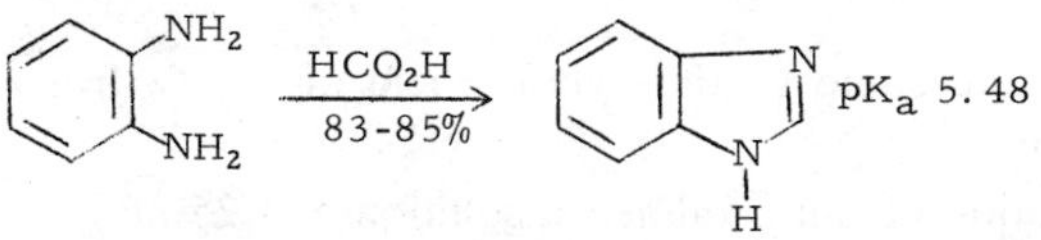
NH2
NH2
HCO2H
83-85%
N
N
H
pKa 5.48

bath at 100° for 2 hrs. After cooling, 10% sodium hydroxide solution is added slowly with thorough mixing until the mixture is just alkaline to litmus. The crude, yellowish benzimidazole is collected, washed with cold water, and purified without being dried (dry weight 97–99%, m.p. 167–168°). A solution of the moist material in 750 ml. of boiling water is clarified with Norit, filtered through a well-heated funnel, and let stand for crystallization, eventually at 10–15°. The yield of colorless product, m.p. 170–172°, is 49–50.5 g. (83–85%).

With use of acetic acid (45 g.) in place of formic acid, 2-methylbenzimidazole, m.p. 172–174°, is obtained in 68% yield. Conversion of aliphatic acids into 2-alkylbenzimidazoles has been proposed as a general method for preparing solid derivatives for identification.[3]

1,2,3-Benzotriazole (supplied by Aldrich and by Eastman) can be prepared by the following one-step process.[4] One mole of *o*-phenylenediamine is dissolved by warming in 2 moles of acetic acid and 300 ml. of water, the solution is cooled to 5°, and a

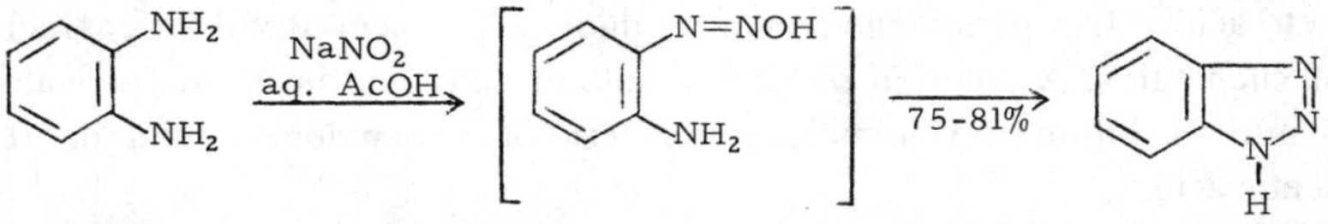
NH2
NH2
NaNO2
aq. AcOH
N=NOH
NH2
75-81%
N
N
N
H

cold solution of 1.09 moles of sodium nitrite in 120 ml. of water is added all at once with gentle stirring. The solution turns dark green and the temperature is allowed to rise rapidly to 80°, when the color changes to clear orange-red (it is essential that the temperature rise to 80°). The cooling bath is removed and, as the solution cools, benzotriazole separates as an oil, which solidifies on ice cooling. The tan-colored product is dried and distilled at 156–159°/2 mm., and the distillate melted and poured into 250 ml. of benzene. The yield of colorless benzotriazole, m.p. 96–97°, is 90–97 g. (75–81%). "The crude product can be purified by repeated crystallizations from benzene or water, but greater losses accompany this tedious process of purification than a single distillation."[4] However, distillation is not without hazard. Attempted distillation from a glass-lined kettle at 160°/2 mm. of about 2,000 lbs. of crude material prepared by the procedure outlined at Maumee Chemical Co., Toledo, Ohio, resulted in an explosion with a property damage of $200,000.[5]

A process which is longer but which gives pure material easily and without distillation is outlined in the formulation.[6] All the intermediates were obtained in the yields reported in pure, colorless form. The benzotriazole was obtained as colorless crystals of higher melting point than that of the direct procedure. The overall yield

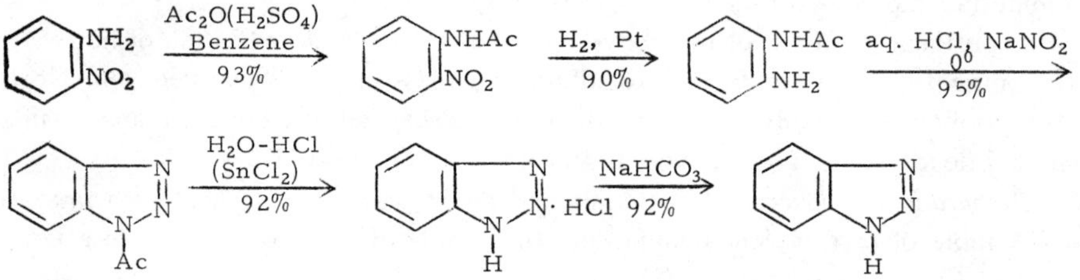
NH2
NO2
Ac2O(H2SO4)
Benzene
93%
NHAc
NO2
H2, Pt
90%
NHAc
NH2
aq. HCl, NaNO2
0°
95%
N
N
N
Ac
H2O-HCl
(SnCl2)
92%
N
N· HCl
N
H
NaHCO3
92%
N
N
N
H

of 67% is nearly the same as the overall yield of 69.5% from *o*-nitroaniline via *o*-phenylenediamine.

For the preparation of quinoxaline,[7] a solution of 1.25 moles of *o*-phenylenediamine in 2l. of water is stirred at 70°, and a solution of 1.29 moles of glyoxal–sodium bisulfite in 1.5l. of hot water (80°) is run in. The mixture is allowed to stand for 15

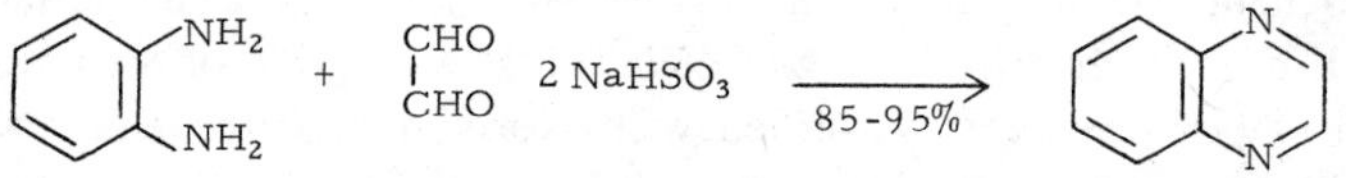
NH2
NH2
+
CHO
CHO
2 NaHSO3
85-95%
N
N

min., cooled to 25°, and 500 g. of $Na_2CO_3 \cdot H_2O$ is added. Quinoxaline separates as an oil or, at a lower temperature, as a crystalline solid (m.p. 29–30°). The product is extracted with ether and distilled at 108–111°/12 mm. 6,7-Diphenylquinoxaline can be prepared by heating a mixture of 210 mg. of benzil and 108 mg. of *o*-phenylenediamine in a test tube on the steam bath until the melt changes to a light tan solid. Crystallization from methanol gives colorless needles, m.p. 126°.[8] Phenanthrenequinone reacts with *o*-phenylenediamine in acetic acid solution to form the quinoxaline (phenanthrazine), which crystallizes in bright yellow needles, m.p. 217°.[9] Morrison[10] found the 3-alkyl-2-hydroxy-quinoxalines readily prepared by heating an α-keto acid with *o*-phenylenediamine in dilute acetic acid useful for characterization of such acids. A solution of 9.85 mmole of keto acid in 25 ml. of water was stirred into a solution of 37 mmoles (an excess) of *o*-phenylenediamine in 50 ml of 10% acetic acid.

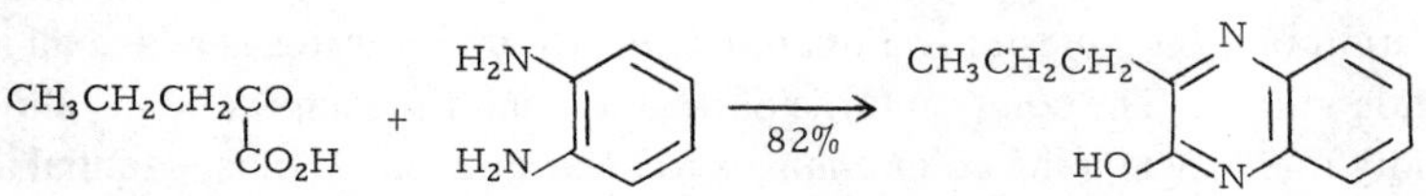
CH3CH2CH2CO
CO2H
+
H2N
H2N
82%
CH3CH2CH2
HO
N
N

The synthesis of 2-mercaptobenzimidazole is accomplished by refluxing for 3 hrs. a mixture of 0.3 mole of *o*-phenylenediamine, 0.33 mole of potassium ethyl xanthate, and 45 ml. of water.[11] The solution is clarified with Norit, heated to 60–70°,

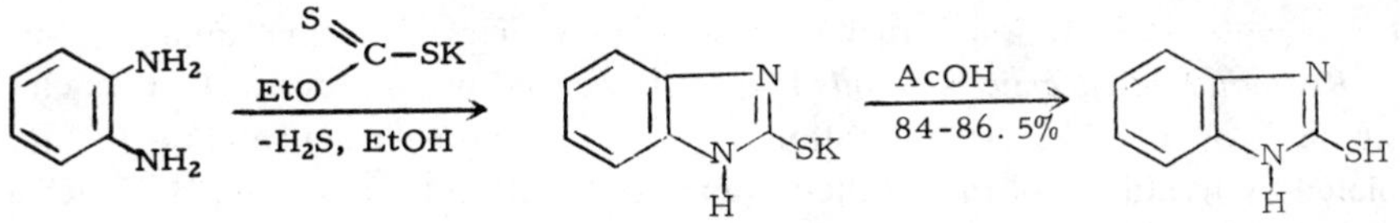
NH2
NH2
S
C-SK
EtO
-H2S, EtOH
N
N
H
SK
AcOH
84-86.5%
N
N
H
SH

diluted with 300 ml. of water at 60–70°, and stirred during the addition of 25 ml. of acetic acid in 50 ml. of water. The product separates in glistening crystals, m.p. 303–304°.

[1]E. L. Martin, *Org. Syn., Coll. Vol.*, **2**, 501 (1943)
[2]E. C. Wagner and W. H. Millett, *ibid.*, **2**, 65 (1943)
[3]R. Seka and R. H. Müller, *Monatsh.*, **57**, 97 (1931): W. O. Pool, H. J. Harwood, and A. W. Ralston, *Am. Soc.*, **59**, 178 (1937)
[4]R. E. Damschroder and W. D. Peterson, *Org. Syn., Coll. Vol.*, **3**, 107 (1955)
[5]*Chem. Eng. News*, **34**, 2450 (1956)
[6]L. F. Fieser and E. L. Martin, *Am. Soc.*, **57**, 1835 (1935)
[7]R. G. Jones and K. C. McLaughlin, *Org. Syn., Coll. Vol.*, **4**, 824 (1963)
[8]*Org. Expts.*, 214 (1964)
[9]O. Hinsberg, *Ann.*, **237**, 340 (1887)
[10]D. C. Morrison, *Am. Soc.*, **76**, 4483 (1954)
[11]J. A. Van Allan and B. D. Deacon, *Org. Syn., Coll. Vol.*, **4**, 569 (1963)

***o*-Phenylene phosphorochloridate**, (2). Mol. wt. 190.53, b.p. 91°/0.9 mm.

Preparation.[1,2] Catechol reacts readily with phosphorus pentachloride to give (1), and when this is warmed with acetic anhydride, acetyl chloride is evolved nearly quantitatively with formation of the reagent (2) in 88% yield.

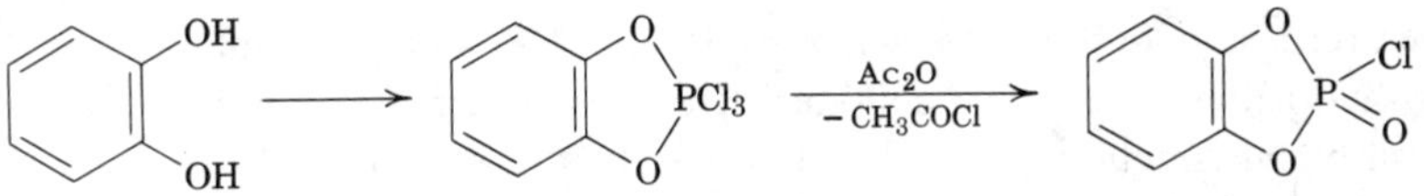
OH
OH
O
PCl3
O
Ac2O
- CH3COCl
O
Cl
P
O
O

(1) (2)

Phosphorylation. The reagent (2) reacts readily with a primary aliphatic alcohol in the presence of base (pyridine) to give (3), which is easily hydrolyzed to the *o*-hydroxyphenyl phosphate ester (4). When this substance is treated in neutral

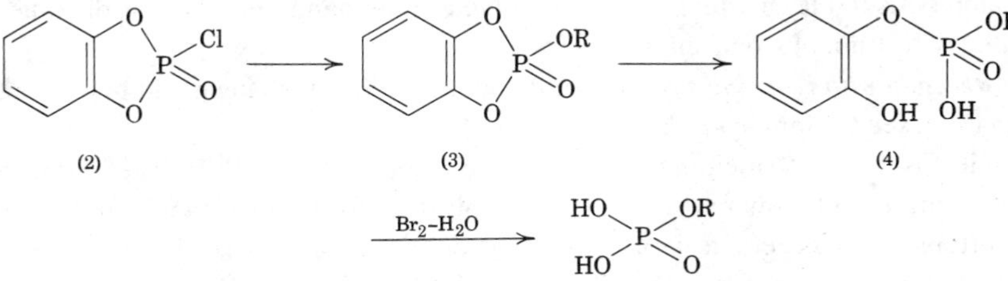
O
Cl
P
O
O
O
OR
P
O
O
O
OR
P
O
OH OH
Br2-H2O
HO
OR
P
HO
O

(2) (3) (4) (5)

aqueous buffer with excess bromine water it is converted into the monoester (5) in good yield.[2] "Except for unsaturated alcohols, as complications may then occur in the oxidation step, *o*-phenylene phosphorochloridate appears to be one of the most convenient and helpful phosphorylating agents available."[2]

[1]W. S. Reich, *Nature*, **157**, 133 (1946)
[2]T. A. Khawaja and C. B. Reese, *Am. Soc.*, **88**, 3446 (1966)

***d*- and *l*-α-Phenylethylamine**, $C_6H_5CH(NH_2)CH_3$. Mol. wt. 121.18, m.p. 184.5°, αD + or −40°. Preparation of racemic amine;[1,2] suppliers of racemic amine: E, MCB.

Resolution of racemic material. Ingersoll[3] worked out a procedure in which the *d*-amine is obtained by crystallization as the salt with *l*-malic acid and the *l*-amine is isolated by treatment of the mother liquor amine with D-tartaric acid. Helferich and Portz[4] discovered a novel method of resolution in which an ethereal solution is prepared from 1/20 mole of the *dl*-amine and 3/40 mole of 2,3,4,6-tetra-O-acetyl-D-glucose. A crystalline addition complex which separates on being split with hydrochloric acid affords the salt of the *d*-amine in 48% yield. The ethereal mother liquor serves for isolation of the *l*-amine.

Uses. These synthetic amines are preferred to the more usual alkaloids for resolution of racemic acids because the salts can be completely cleaved by passage through a column of Amberlite IR-120.[5]

[1]A. W. Ingersoll, *Org. Syn., Coll. Vol.*, **2**, 503 (1943)
[2]J. C. Robinson, Jr., and H. R. Snyder, *ibid.*, **3**, 717 (1955)
[3]A. W. Ingersoll, *ibid.*, **2**, 506 (1943)
[4]B. Helferich and W. Portz, *Ber.*, **86**, 1034 (1953); J. F. Tocanne and C. Asselineau, *Bull. soc.*, 3348 (1965)
[5]*Idem. ibid.*, 3346 (1965)

Phenylhydrazine, $C_6H_5NHNH_2$. Mol. wt. 108.14, b.p. 243°, dec.; pKb 8.80; sp. gr. 1.097. Suppliers: Dow, A, B, E, F, MCB. Preparation in 80–84% yield by reduction of benzenediazonium chloride with sodium sulfite.[1]

Stock solution. Measure 1 ml. of phenylhydrazine by pipet into a 10-ml. volumetric flask, add 3 ml. of acetic acid and swirl under the tap to cool to room temperature, and dilute with water to 10 ml. The solution contains 1 mmole of phenylhydrazine acetate per milliliter. If a substance to be tested for the presence of a carbonyl function is soluble in water to the extent of 8–10 microdrops in 1 ml. of water, dissolve 1 mmole of sample in 1 ml. of water and add 1 ml. of the stock solution. Separation of an oil or solid indicates the presence of a carbonyl function. If the sample is soluble to the extent of only about 4 microdrops in 1 ml. of water (diethyl ketone), treat 1 mmole of material with 1 ml. of water and add a few drops of methanol until solution is complete; then add 1 mole of stock solution. A sample insoluble in water can be dissolved in methanol, ethanol, or dioxane and treated with 1 ml. of stock solution.

Reaction with reducing sugars. For evidence on the mechanism of phenylosazone formation, see Chapman *et al.*[2]

Emil Fischer's "Anleitung zur Darstellung organischer Präparate"[3] describes the preparation of D-mannose by acid hydrolysis of 200 g. of scrap from the cutting of buttons from vegetable ivory, the seed of the tagua palm. The directions call for isolation of the sugar as the sparingly soluble D-mannose phenylhydrazone and for splitting the derivative by hydrazone exchange with benzaldehyde.

Porter-Silber reaction. Porter and Silber[4] found that steroids having the dihydroxyacetone side chain at C_{17} (1) react with phenylhydrazine in water–alcohol–sulfuric acid to give a yellow product with an absorption maximum at 410 mμ. The reaction is specific to the structure (1) and is useful for colorimetric determination of cortisone and other steroids having the same side chain. Both Barton *et al.*[5] and Lewbart and Mattox[6] established that the pigment is the 20-keto-21-phenylhydrazone (5). The latter investigators studied the reaction in the absence of phenylhydrazine and concluded that the essential steps are as formulated. The key step is a Mattox

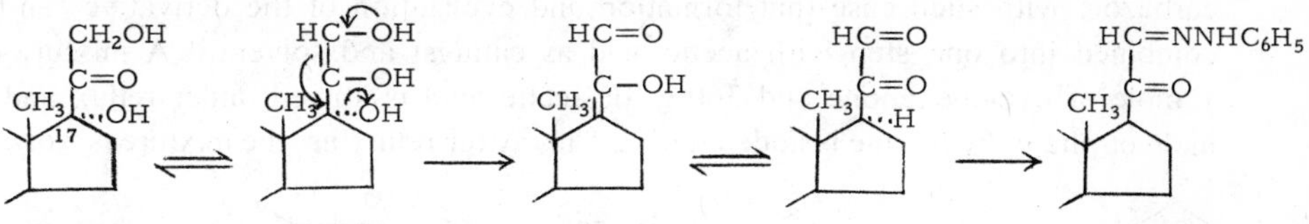

(1) Cortisone (2) (3) (4) (5)

rearrangement[7] of the enediol (2) to the enol (3). Ketonization of (3) gives the 17β-glyoxal (4), which combines with phenylhydrazine to give the pigment (5). Even in the presence of excess phenylhydrazine, the phenylhydrazone (5) was the only product isolated.

Cleavage of phthaloylpeptides. The conventional method of removing the N-protective phthaloyl group of a peptide involves reaction with hydrazine in alcoholic solution and heating the intermediate formed with dilute acid to effect hydrolysis. Boissonnas[8] found that cleavage can be effected more easily and in one step by heating in alcoholic solution with phenylhydrazine and a tertiary amine. For example, a solution of 535 mg. (2.05 mmoles) of phthaloyl-L-leucine, 0.5 ml. (2.1 mmoles) of tri-*n*-butylamine, and 0.405 ml. (4.1 mmoles) of phenylhydrazine in 3 ml. of ethanol

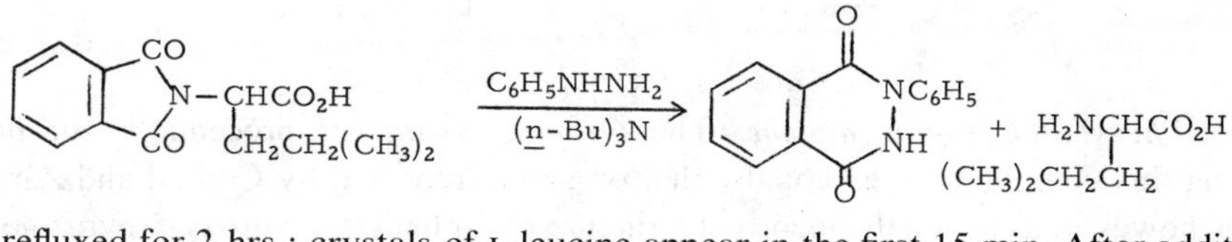

is refluxed for 2 hrs.; crystals of L-leucine appear in the first 15 min. After addition of 10 ml. of methyl ethyl ketone the mixture is refluxed for 15 min. more, cooled, and treated with 0.2 ml. of acetic acid. The precipitate is collected and washed with methyl ethyl ketone. The yield of L-leucine is 325 mg. (82.7%); the rotatory power is the same as that of the material from which the phthaloyl derivative was prepared.

Fischer indole synthesis. Emil Fischer discovered that pyruvic acid phenylhydrazone on being heated with zinc chloride loses ammonia with cyclization to

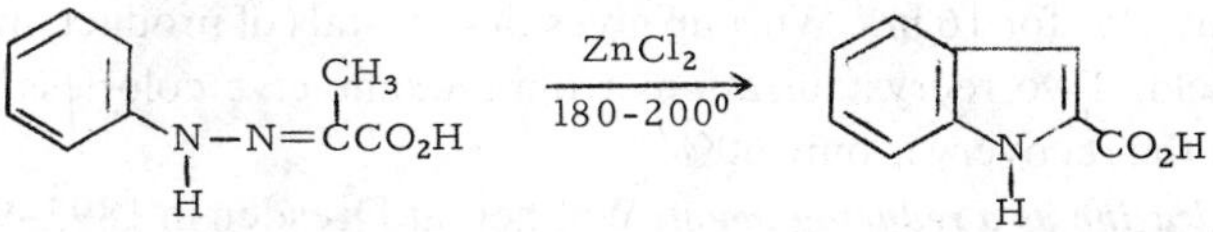

indole-2-carboxylic acid.[9] Acetone phenylhydrazone similarly affords 2-methylindole; the reaction is interpreted as follows:

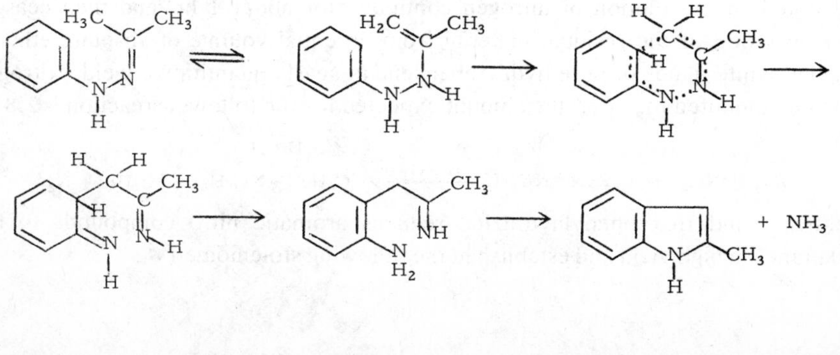

Acetophenone phenylhydrazone, heated with zinc chloride at 170°, affords 2-phenylindole in yield of 72–80%.[10]

Cyclohexanone phenylhydrazone undergoes cyclization to 1,2,3,4-tetrahydrocarbazole with such ease that formation and cyclization of the derivative can be combined into one step with acetic acid as catalyst and solvent.[11] A mixture of 1 mole of cyclohexanone and 360 g. of acetic acid is stirred under reflux and 1 mole of phenylhydrazine is added during 1 hr. After refluxing, the mixture is worked

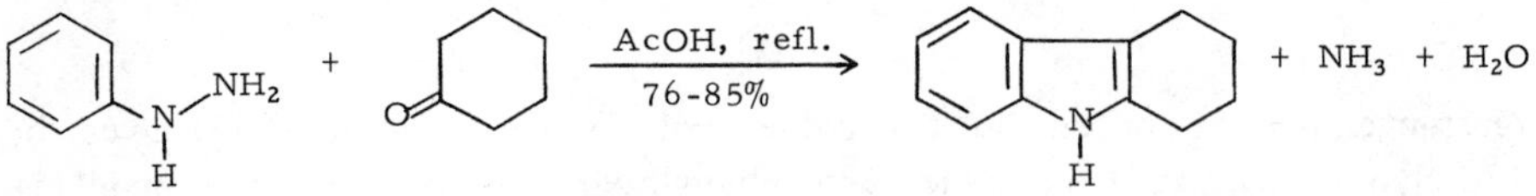
AcOH, refl.
76-85%
+ NH3 + H2O

up and the product (m.p. 115–116°) crystallized from methanol. 1,2-Benzo-3,4-dihydrocarbazole is prepared similarly from 1 mole of phenylhydrazine and 2 moles of concd. hydrochloric acid in 500 ml. of water, with addition of 1 mole of α-tetralone at the reflux temperature; yield, 82–87%.

The synthesis of 3-methyloxindole from β-propionylphenylhydrazine demonstrates application of the Fischer method to the construction of a dihydropyrrole ring.[12]

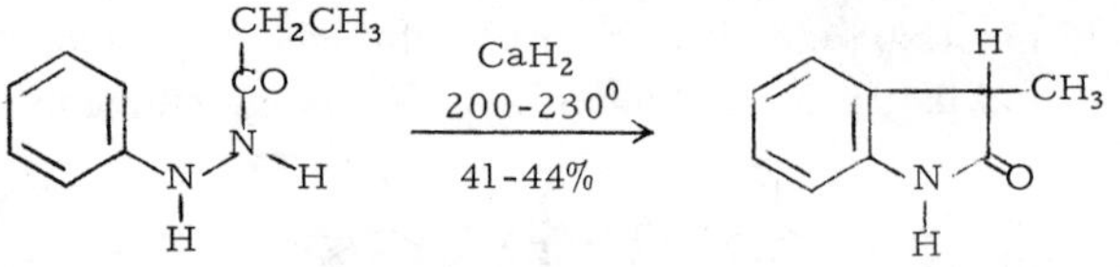
CaH2
200-230°
41-44%

1-Phenyl-3-amino-5-pyrazolone. The *Organic Syntheses* procedure[13] for preparing this compound is essentially that originally reported by Conrad and Zart,[14] who, however, assigned the incorrect structure of 1-phenyl-3-hydroxy-5-pyrazolone imide. A mixture of 1 mole each of ethyl cyanoacetate and phenylhydrazine and a

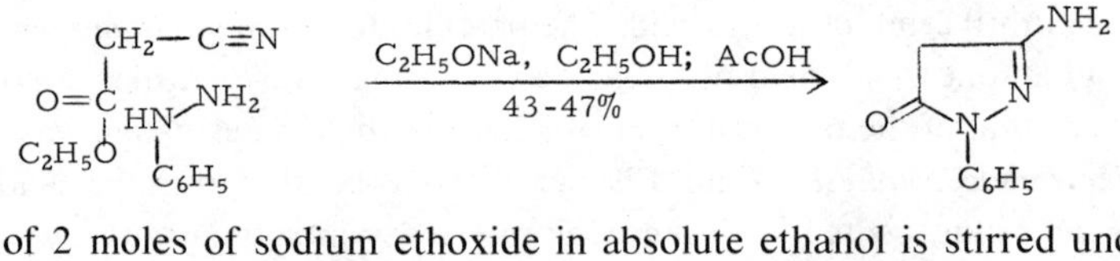
C2H5ONa, C2H5OH; AcOH
43-47%

solution of 2 moles of sodium ethoxide in absolute ethanol is stirred under reflux in an oil bath at 120° for 16 hrs. Workup gives tan crystals of product, m.p. 216–218°, in 43–47% yield. Two recrystallizations from dioxane give colorless product, m.p. 218–220°, but the recovery is only 60%.

Phenylhydrazine as a reducing agent. Walther, at Dresden in 1895–96,[15] observed by chance that phenylhydrazine reacts energetically with azobenzene at an elevated temperature. He heated a mixture of 0.1 mole of each reactant in a bath at 125–130° and noted that evolution of nitrogen continued for about 1 hr. and then ceased. Crystallization of the residual material from an equal volume of absolute ethanol afforded white plates of pure hydrazobenzene in nearly quantitative yield. Nitrogen evolved amounted to 2.7 g.; the amount expected for the following reaction is 2.8 g.:

$$C_6H_5NHNH_2 + C_6H_5N{=}NC_6H_5 \xrightarrow{125\text{-}130^\circ} C_6H_5\overset{H}{N}{-}\overset{H}{N}C_6H_5 + C_6H_6 + N_2$$

Walther found that phenylhydrazine reduces aromatic nitro compounds to the arylamines in high yield and established the following stoichiometry:

$$ArNO_2 + 3\ C_6H_5NHNH_2 \longrightarrow ArNH_2 + 3\ C_6H_6 + 2\ H_2O + 3\ N_2$$

Use of less than 3 equivalents of phenylhydrazine did not lead to products of reduction intermediate between nitrobenzene and aniline; such intermediates, if formed, are reduced more readily than nitrobenzene. Walther also showed that phenylhydrazine undergoes quantitative disproportionation on pyrolysis as follows:

$$2\ C_6H_5NHNH_2 \xrightarrow[3\ hrs.]{300^0} C_6H_6 + N_2 + C_6H_5NH_2 + NH_3$$

Since simpler methods were available for the reduction of azo and nitro compounds, over half a century elapsed before practical use was made of these early findings. Then Bredereck and von Schuh[16] encountered difficulty in the reduction of the nitro groups of a series of polynitroamido esters of formula (1). The compounds are insoluble in water and in aqueous acid. Hydrogenation in acetic acid

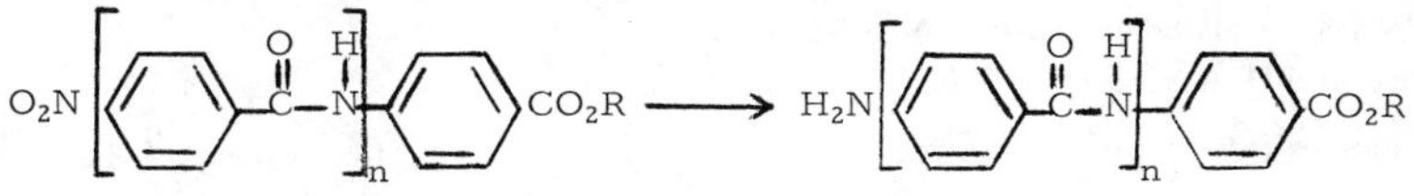

(1) (2)

n = 1, 2, 3; R = CH_3 or C_2H_5

solution was successful when $n = 1$ but not when $n = 2$. Reduction with phenylhydrazine according to Walther solved the problem. Phenylhydrazine proved to be a good solvent for the peptidelike compounds and was without effect on the peptide and ester functions. For reduction of the nitro amide methyl ester where $n = 2$, a mixture of 20 ml. of phenylhydrazine and 30 ml. of anisole was heated to a gentle boil and 7.5 g. of nitro compound was added; when gas evolution started, the flame was removed. When foaming decreased, a second 7.5 g. of compound was added, and heat was applied as required to maintain gas evolution. The amino product (2) began to separate in about 1 hr., and the reaction was done in 2 hrs. After cooling, the product was collected, washed with anisole and then with ether; yield 12.5 g. (93%). One crystallization from anisole gave analytically pure material.

Ochiai[17] used phenylhydrazine in ethereal solution to reduce 4-nitropyridine-N-oxide to 4-hydroxylaminopyridine N-oxide in high yield.

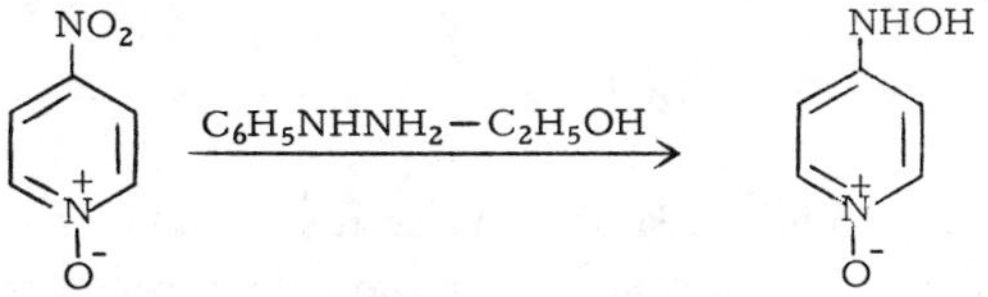
NO2
C6H5NHNH2—C2H5OH
NHOH

[1]G. H. Coleman, *Org. Syn., Coll. Vol.*, **1**, 442 (1941)
[2]O. L. Chapman, W. J. Welstead, Jr., T. J. Murphy, and R. W. King, *Am. Soc.*, **86**, 732 (1964)
[3]Eighth edition, p. 79, Braunschweig (1908)
[4]C. C. Porter and R. H. Silber, *J. Biol. Chem.*, **185**, 201 (1950)
[5]D. H. R. Barton, T. C. McMorris, and R. Segovia, *J. Chem. Soc.*, 2027 (1961)
[6]M. L. Lewbart, and V. R. Mattox, *J. Org.*, **29**, 513, 521 (1964)
[7]V. R. Mattox, *Am. Soc.*, **74**, 4340 (1952)
[8]I. Schumann and R. A. Boissonnas, *Helv.*, **35**, 2235, 2237 (1952)
[9]E. Fischer and F. Jourdan, *Ber.*, **16**, 2241 (1883); E. Fischer and O. Hess, *Ber.*, **17**, 559 (1884)
[10]R L. Shriner, W. C. Ashley, and E. Welch, *Org. Syn., Coll. Vol.*, **3**, 725 (1955)
[11]C. U. Rogers and B. B. Corson, *Am. Soc.*, **69**, 2910 (1947); *Org. Syn., Coll. Vol.*, **4**, 884 (1963)
[12]A. S. Endler and E. I. Becker, *ibid.*, **4**, 657 (1963)

[13]H. D. Porter and A. Weissberger, *ibid.*, **3**, 708 (1955)
[14]M. Conrad and A. Zart, *Ber.*, **39**, 2282 (1906)
[15]R. Walther, *J. prakt. Chem.*, **52**, 141 (1895); **53**, 433 (1896)
[16]H. Bredereck and H. von Schuh, *Ber.*, **81**, 215 (1948)
[17]E. Ochiai and H. Mitarashi, *Chem. Pharm. Bull.*, **11**, 1084 (1963)

Phenylhydrazine-*p*-sulfonic acid, $H_2NNHC_6H_4SO_3H$-*p*. Mol. wt. 188.21. Suppliers: Eastman, Fisher. Prepared in high yield by the sulfonation of phenylhydrazine.[1]

The reagent is suggested by W. Triebs[2] for the isolation of volatile ketones from natural sources. Condensation with a ketone is conducted by refluxing with the reagent in aqueous alcohol buffered with sodium acetate. The reaction mixture is extracted with ether for removal of nonketonic material, the aqueous solution is treated with hydrochloric acid, and the ketone liberated is removed by steam distillation.

[1]L. Claisen and P. Roosen, *Ann.*, **278**, 296 (1894)
[2]W. Treibs and H. Röhnert, *Ber.*, **84**, 433 (1951)

Phenyl isocyanate, $C_6H_5N{=}C{=}O$. Mol. wt. 119.11, b.p. 166°, sp. gr. 1.10, a liquid with an acrid odor and irritating to the eyes. Suppliers: Ott Chem. Co., A, B, E, MCB. Prepared by passing phosgene into a hot solution of aniline in toluene saturated with hydrogen chloride:[1]

$$C_6H_5NH_2 + COCl_2 \longrightarrow C_6H_5NHCOCl \xrightarrow[83\%]{} C_6H_5N{=}C{=}O$$

Oroshnik[2] found commercial material to contain a trace of acidic impurity which interfered with use of the substance as a dehydrating agent but which was removable by adding a few milliliters of a Grignard reagent and distilling rapidly at 20 mm.

Phenyl isocyanate reacts with ammonia to form phenylurea (1) and with aniline to give diphenylurea (2). When used as a dehydrating agent it apparently adds water

$$1.\quad C_6H_5N{=}C{=}O + NH_3 \longrightarrow C_6H_5NH{-}C({=}O){-}NH_2$$

$$2.\quad C_6H_5N{=}C{=}O + H_2NC_6H_5 \longrightarrow C_6H_5NH{-}C({=}O){-}NHC_6H_5$$

$$3.\quad C_6H_5N{=}C{=}O + H_2O \longrightarrow \left[C_6H_5NH{-}C({=}O)OH\right] \longrightarrow C_6H_5NH_2$$

with decarboxylation to aniline (3); the aniline then combines with fresh reagent as in (2) more rapidly than it is formed, since the sole nitrogen-containing product is diphenylurea.

Condensation of phenyl isocyanate with an aqueous solution of the sodium salt of cyanamide and acidification affords 1-cyano-3-phenylurea (4).[3]

$$4.\quad C_6H_5N{=}C{=}O \xrightarrow{NaNHCN} C_6H_5NH{-}C({=}O){-}\overset{-}{N}CN(\overset{+}{Na}) \xrightarrow[62-67\%]{} C_6H_5NH{-}C({=}O){-}NHCN$$

For catalyzed dimerization of phenyl isocyanate to diphenylcarbodiimide, *see* 3-Methyl-1-ethyl-3-phospholene-1-oxide.

Use as a dehydrating agent. In the course of the synthesis of compounds related to vitamin A, Oroshnik[2] explored the dehydration of the unsaturated carbinol-ether

(1) with acetic acid (12 hrs. at room temperature) and with phosphoryl chloride and pyridine in toluene (12 hrs. at room temperature) with unsatisfactory results.

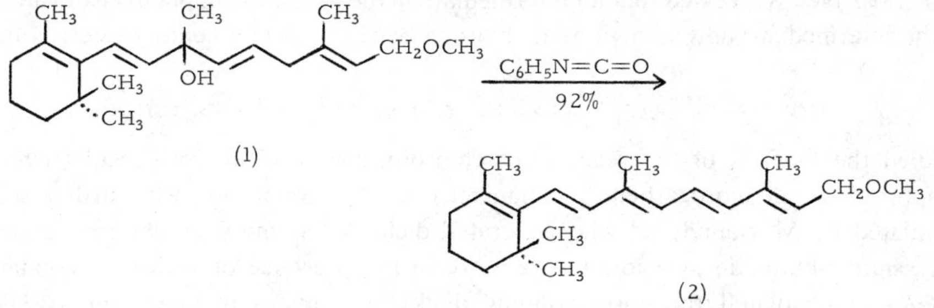
CH₂OCH₃
OH
C₆H₅N=C=O
92%
(1)
(2)

Dehydration with phenyl isocyanate afforded a more homogeneous product (still a mixture of isomers) of higher vitamin A activity. The reaction was carried out by treating a solution of 33 g. of (1) in 150 ml. of purified phenyl isocyanate with 2 ml. of 1.5 *M* ethylmagnesium bromide as catalyst and stirring the mixture at 95° for 3 hrs., during which time diphenylurea continuously precipitated (in the absence of catalyst no significant amount of diphenylurea was observed).

Dehydration of oximes to nitriles with the reagent also has been reported.[4] A solution of 0.05 mole of heptaldoxime and 5 drops of triethylamine in 30 ml. of benzene was treated with 0.1 mole of phenyl isocyanate in 10 ml. of benzene and refluxed for 2 hrs. The carbon dioxide evolved was collected as $BaCO_3$ and cor-

$$2\,CH_3(CH_2)_5CH{=}NOH + 2\,C_6H_5N{=}C{=}O \xrightarrow[-CO_2]{} 2\,CH_3(CH_2)_5C{\equiv}N + C_6H_5NH{-}\overset{O}{\overset{\|}{C}}{-}NHC_6H_5 \quad 93\%$$

responded to 82% of the theory. The diphenylurea which separated was obtained in 93% yield, and distillation of the filtrate afforded *n*-hexyl cyanide in 89% yield.

[1]D. V. N. Hardy, *J. Chem. Soc.*, 2011 (1934)
[2]W. Oroshnik, G. Karmas, and A. D. Mebane, *Am. Soc.*, **74**, 295 (1952)
[3]F. Kurzer and J. R. Powell, *Org. Syn., Coll. Vol.*, **4**, 213 (1963)
[4]T. Mukaiyama and H. Nohira, *J. Org.*, **26**, 782 (1961)

Phenyl isocyanide (isonitrile), $C_6H_5N{=}C$ or $C_6H_5\overset{+}{N}{\equiv}\overset{-}{C}$. Mol. wt. 103.12, b.p. 166°.

This interesting compound was discovered independently by A. W. Hofmann (Germany) and by A. Gautier (France) in 1867–68 as a product of the reaction of aniline with chloroform in the presence of potassium hydroxide.[1]

$$C_6H_5NH_2 + Cl_3CH + 3\,KOH \longrightarrow C_6H_5N{=}C + 3\,KCl + 3\,H_2O$$

The compound, a toxic oil, has a penetrating, characteristic, HCN-like odor and a bitter taste. A bluish color persists even after steam distillation. It reacts with sulfur to give phenylisothiocyanate (1), with chlorine to give N-(dichloromethylene)-

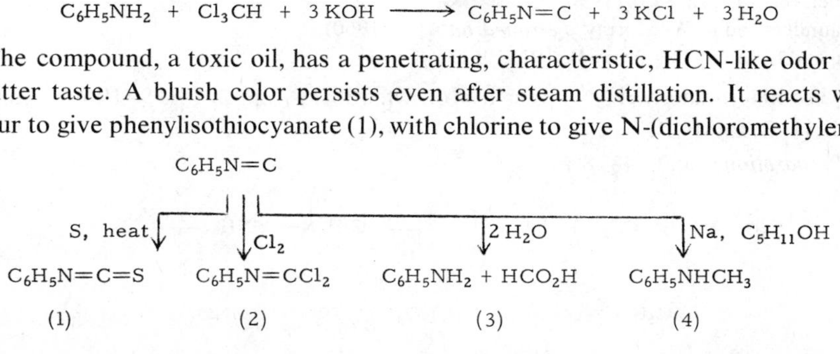
C₆H₅N=C
S, heat
Cl₂
2 H₂O
Na, C₅H₁₁OH
C₆H₅N=C=S
C₆H₅N=CCl₂
C₆H₅NH₂ + HCO₂H
C₆H₅NHCH₃
(1)
(2)
(3)
(4)

aniline (2), with water to give aniline and formic acid, and it is reduced by sodium and amyl alcohol to N-methylaniline (4).

In 1897 Nef[2] suggested that an intermediate in the formation of phenyl isocyanide is the intermediate now known as dichlorocarbene (1). A half-century later, Hine[3]

$$1.\quad CHCl_3 \xrightarrow{OH^-} :CCl_2 \xrightarrow{C_6H_5NH_2} C_6H_5\overset{H}{\overset{|}{N}}-\overset{H}{\overset{|}{C}}Cl_2 \xrightarrow{-2\,HCl} C_6H_5N{=}C$$

studied the kinetics of the reaction of chloroform with alkali with results which supported the dichlorocarbene mechanism (1). This conclusion was further substantiated by M. Saunders,[4] who generated dichlorocarbene from chloroform and potassium *t*-butoxide at a low temperature in the presence of various secondary amines and isolated the corresponding dialkylformamides in yields up to 31% (piperidine):

$$R_2NH \xrightarrow{:CCl_2} R_2\underset{H}{\underset{|}{\overset{+}{N}}}-\overset{-}{C}Cl_2 \longrightarrow R_2\underset{H}{\underset{|}{N}}-CCl_2 \xrightarrow{H_2O} R_2NCH{=}O$$

Ugi and Meyr[5] worked out an improved general procedure for the preparation of aryl isocyanides, illustrated by the case of *o*-tolyl isocyanide. A suspension of potassium *t*-butoxide is prepared under nitrogen from 2.6 g. atoms of potassium and

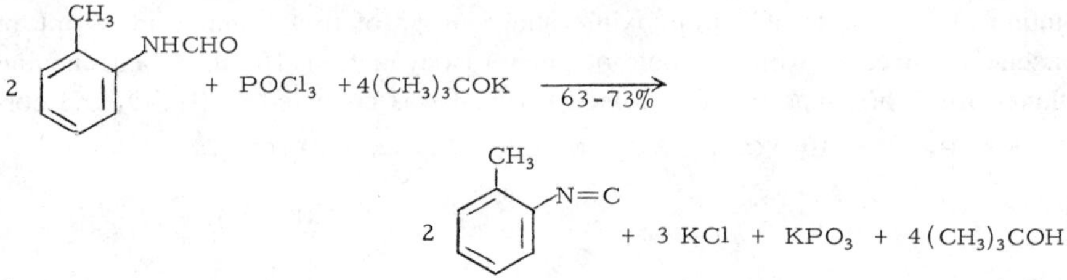
CH3
NHCHO
2
+ POCl3 +4(CH3)3COK
63-73%
CH3
N=C
2
+ 3 KCl + KPO3 + 4(CH3)3COH

1250 ml. of *t*-butanol and 1 mole of N-*o*-tolylformamide is added to the hot, stirred suspension, which becomes a clear solution in a few minutes. The solution is kept at 10–20° by cooling during addition, with stirring, of 0.60 mole of phosphoryl chloride in 30–40 min. Stirring is continued for 1 hr. and the reaction mixture is poured onto an ice-cold mixture of 50 g. of sodium bicarbonate in 5 l. of water. The isocyanide is collected by extraction with petroleum ether, washing, drying, and distillation. Phenyl isocyanide is obtained by this procedure in 56% yield; yields of other aryl isocyanides are in the range 41–80%.

[1]A. W. Hofmann, *Ann.*, **146**, 107 (1868); A. Gautier, *ibid.*, **146**, 119 (1868); each investigator had published earlier a preliminary account of the work.
[2]J. U. Nef, *Ann.*, **298**, 367 (1897)
[3]J. Hine, *Am. Soc.*, **72**, 2438 (1950)
[4]M. Saunders and R. W. Murray, *Tetrahedron*, **11**, 1 (1960)
[5]I. Ugi and R. Meyr, *Org. Syn.*, **41**, 101 (1961)

Phenylisothiocyanate, $C_6H_5N{=}C{=}S$. Mol. wt. 135.19, b.p. 221° (99–100°/15 mm.) Suppliers: B, E, MCB.

Preparation (yield 74–78%):[1]

$$C_6H_5NH_2 + CS_2 + NH_4OH \xrightarrow[-H_2O]{} C_6H_5\overset{H}{\overset{|}{N}}-\overset{S}{\overset{\|}{C}}SNH_4 \xrightarrow{Pb(NO_3)_2}$$

$$C_6H_5N{=}C{=}S + NH_4NO_3 + PbS + HNO_3$$

End-group analysis of proteins. A method introduced by Edman[2] involves selective elimination of the amino-terminal group. Reaction of this group with phenylisothiocyanate (a) gives the phenylthiocarbamyl derivative (b), which is cleaved

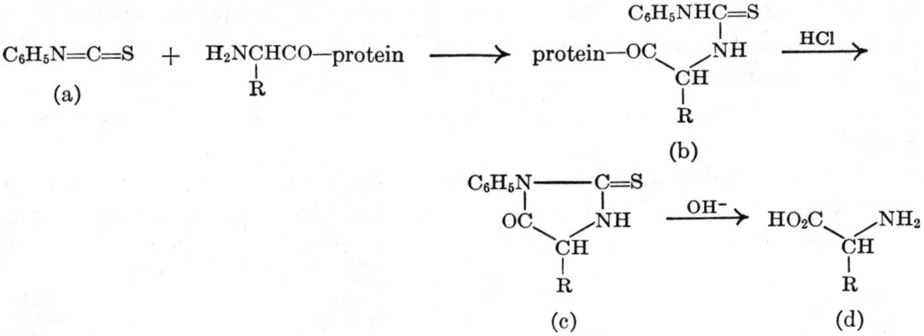
HCl
OH⁻
(a)
(b)
(c)
(d)

by hydrogen chloride to a phenylthiohydantoin (c). Finally, alkaline hydrolysis of (c) affords the free amino acid (d). An improved procedure is described by Sjöquist.[3]

[1]F. B. Dains, R. Q. Brewster, and C. P. Olander, *Org. Syn., Coll. Vol.*, **1**, 447 (1941)
[2]P. Edman, *Chem. Scand.*, **4**, 277 (1950); **10**, 761 (1956)
[3]J. Sjöquist, *Arkiv. Kemi*, **14**, 291 (1959)

Phenyllithium, C_6H_5Li. Mol. wt. 84.04. Supplier: Foote Mineral Co. offers a 20% solution in ether–benzene (1:3 by volume), 1 mole per bottle. Preparation and use in halogen–lithium exchange reactions.[1,2]

Modified Hofmann degradation. A suspension of trimethylcyclohexylammonium bromide in an ethereal solution of phenyllithium on being shaken for 72 hrs. at room temperature affords cyclohexene in high yield.[3] A similar degradation of

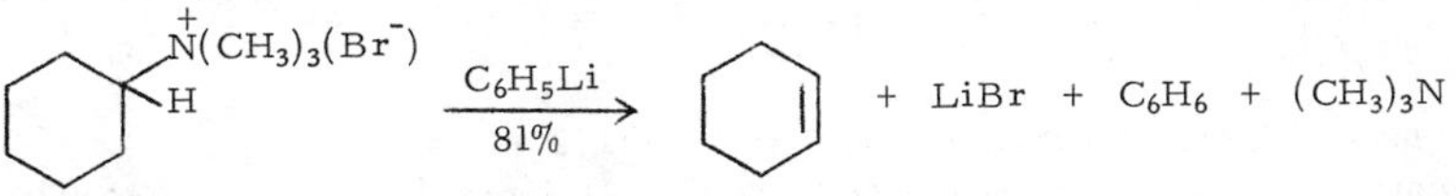
C6H5Li
81%
+ LiBr + C6H6 + (CH3)3N

trimethylcyclooctylammonium bromide gives a cyclooctene mixture containing 80% of the *cis* and 20% of the *trans* olefin. Regular Hofmann degradation gives 40% *cis* and 20% *trans*; elimination with potassium amide in liquid ammonia gives chiefly (85%) *trans*-cyclooctene.

Benzynes. Review of the formation of benzynes, for example by the reaction of phenyllithium on fluorobenzene, 1-fluoro- and 2-fluoronaphthalene.[4] For related observations, *see* Phenylsodium.

[1]R. G. Jones and H. Gilman, *Org. Reactions*, **6**, 339 (1951)
[2]G. Wittig, *Newer Methods of Preparative Organic Chemistry*, 576 (1948)
[3]J. Rabiant and G. Wittig, *Bull. soc.*, 798 (1957)
[4]G. Wittig, *Angew. Chem.*, **69**, 245 (1957)

N-Phenylmaleimide, (3). Mol. wt. 173.17, m.p. 89–90°. Suppliers: Aero, A, B.

Preparation.[1] A solution of 2 moles of maleic anhydride in ether is stirred under

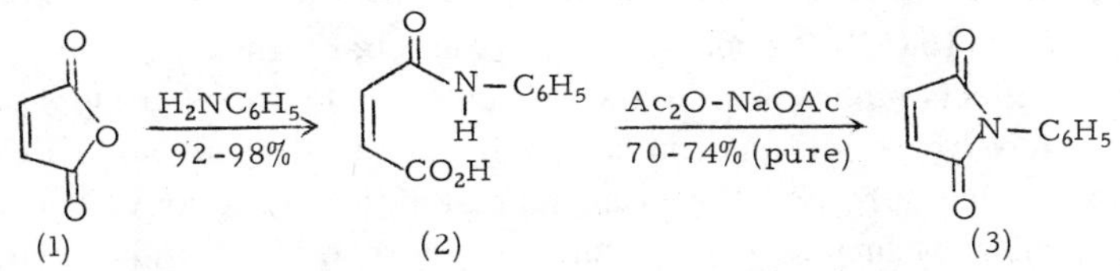
H2NC6H5
92-98%
Ac2O-NaOAc
70-74% (pure)
(1)
(2)
(3)

reflux during addition of a solution of 2 moles of aniline in 200 ml. of ether as rapidly as possible without flooding the condenser. The cream-colored amide–acid (2) is collected (m.p. 201–202°) and treated with 670 ml. of acetic anhydride and 65 g. of anhydrous sodium acetate. The mixture is heated on the steam bath with swirling until the amide–acid has dissolved (30 min.); it is then cooled somewhat and poured into 1.3 l. of ice water. The precipitated product, collected, washed, and dried, melts at 88–89°. Crystallization from cyclohexane gives canary yellow needles, m.p. 89–90°.

Dienophile. N-Phenylmaleimide is an active Diels-Alder dienophile which gives Diels-Alder adducts of melting point higher than the adducts from maleic anhydride.

[1]M. P. Cava, A. A. Deana, K. Muth, and M. J. Mitchell, *Org. Syn.*, **41**, 93 (1961)

N-Phenylmorpholine, 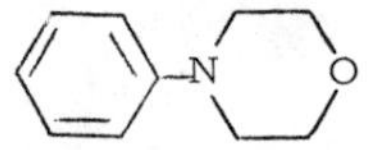
N
O

Mol. wt. 176.23, m.p. 58°, b.p. 268°. Preparation.[1] Suppliers: A, B, E, F, MCB.

This amine, structurally related to N,N-diethylaniline but more easily purified, has been used for dehydrobromination of the product of allylic bromination of vitamin A_1 ester (1) with N-bromosuccinimide to produce vitamin A_2 ester.[2]

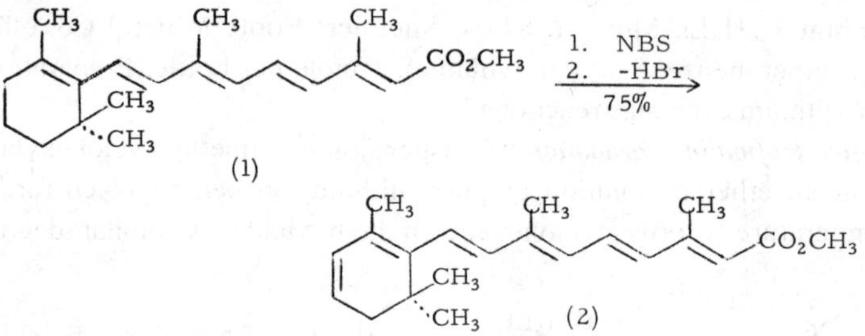
CH3
CH3
CH3
CO2CH3
1. NBS
2. -HBr
75%
CH3
CH3
(1)
CH3
CH3
CH3
CO2CH3
CH3
CH3
(2)

[1]H. Adkins and R. M. Simington, *Am. Soc.*, **47**, 1687 (1925)
[2]K. R. Farrar, J. C. Hamlet, H. B. Henbest, and E. R. H. Jones, *J. Chem. Soc.*, 2657 (1952)

Phenyl-β-naphthylamine, $C_{10}H_7NHC_6H_5$. Mol. wt. 219.27, m.p. 109°, b.p. 395°. Suppliers: du Pont, A, B, E, MCB.

The amine effectively inhibits polymerization of conjugated dienes.[1] Thus in a procedure[2] for the preparation of *trans*-1-phenyl-1,3-butadiene from cinnamaldehyde and methylmagnesium bromide, acidification, and extraction with ether, small

$$C_6H_5CH{=}CHCHO \xrightarrow{CH_3MgBr} C_6H_5CH{=}CH\underset{\displaystyle OMgBr}{CH}CH_3 \xrightarrow[72\text{-}75\%]{H_2SO_4} C_6H_5CH{=}CHCH{=}CH_2$$

amounts of phenyl-β-naphthylamine are added to the ethereal extract and to the crude hydrocarbon prior to distillation.

[1]O. Grummitt and E. I. Becker, *Am. Soc.*, **70**, 149 (1948)
[2]*Idem, Org. Syn., Coll. Vol.*, **4**, 771 (1963)

Phenylphosphonic dichloride (Dichlorophenylphosphine oxide), $C_6H_5POCl_2$. Mol. wt. 194.99, b.p. 258° (116–120°/6 mm.). Suppliers: Aldrich; Eastman.

The reagent effects replacement of the elements of a hydroxyl group by chlorine in nitrogen heterocycles, replacement being effected most readily when the oxygen is α or γ to the ring nitrogen.[1] For example, barbituric acid is converted smoothly into 2,4,6-trichloropyrimidine. A mixture of phosphoryl chloride ($POCl_3$) and

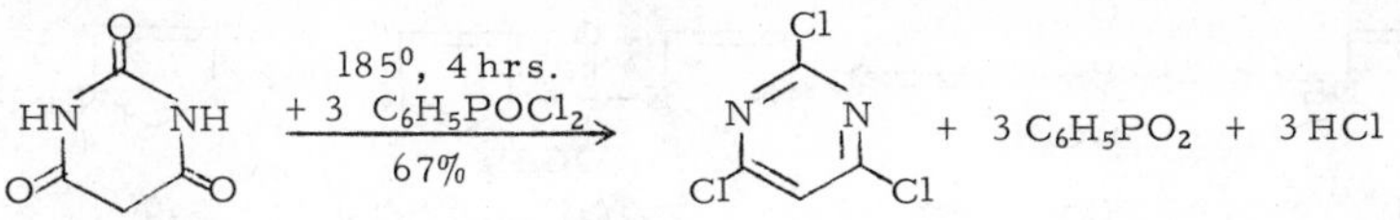

phosphorus pentachloride had previously been used, but the low boiling point of the former reagent (107°) usually required operation in a sealed tube. Both the reactivity and the high boiling point of phenylphosphonic dichloride make it an attractive alternative to phosphoryl chloride. The hydrolysis product, phenylphosphonic acid, is water soluble, as are its salts.

[1]M. M. Robison, *Am. Soc.*, **80**, 5481 (1958)

Phenyl phosphorodichloridate (Monophenylphosphoryl dichloride), $C_6H_5OPOCl_2$. Mol. wt 210.97, b.p. 103–106°/9 mm. Supplier: Aldrich. Preparation.[1,2]

The reagent is useful for the synthesis of α-glycerylphosphorylethanolamine (4), a moiety of cephalin and plasmalogens.[2] D-Acetone glycerol (1) is phosphorylated

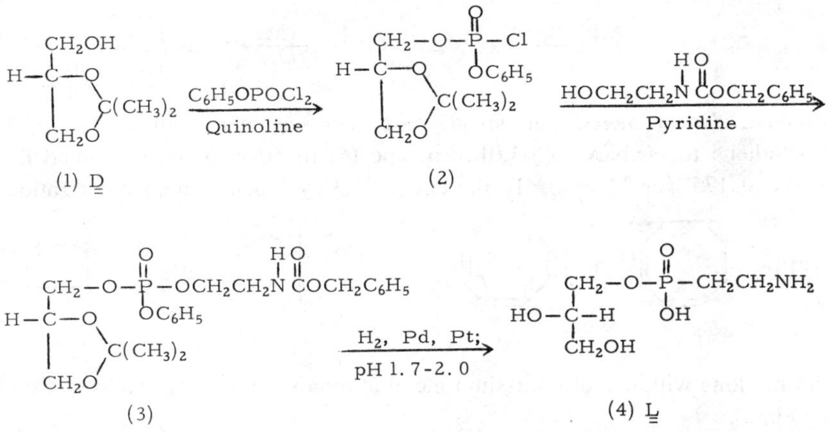

in the presence of quinoline to give (2), which is treated with N-carbobenzoxyethanolamine in pyridine to yield (3). The carbobenzoxy and phenyl groups are eliminated simultaneously by hydrogenolysis in the presence of both palladium and platinum, and acid hydrolysis to eliminate acetone is attended with inversion to the L-series (4). Baer has used this scheme in the synthesis of α-lecithins[3] and α-cephalins.[4]

Phenyl phosphorodichloridate is regarded as the reagent most widely used for the preparation of symmetrical dialkyl phosphates.[5]

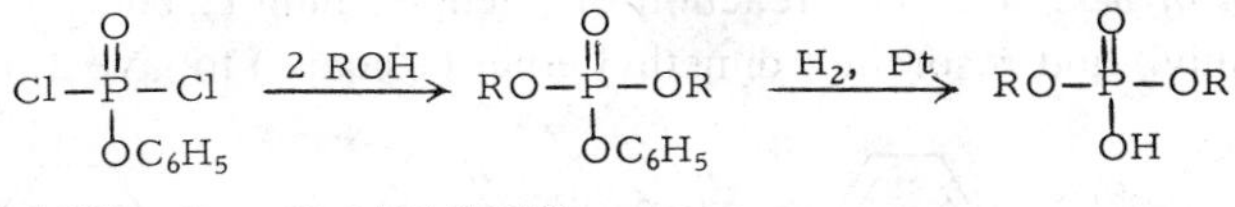

[1]P. Brigl and H. Müller, *Ber.*, **72**, 2121 (1939)
[2]E. Baer and H. C. Stancer, *Am. Soc.*, **75**, 4510 (1953)
[3]E. Baer and M. Kates, *Am. Soc.*, **72**, 942 (1950)
[4]E. Baer, J. Maurukas, and M. Russell, *Am. Soc.*, **74**, 152 (1952)
[5]D. M. Brown, *Advances in Organic Chemistry*, **3**, 91 (1963)

Phenylphosphorodi-(1-imidazolidate), (3). Mol. wt. 274.22, m.p. 90–92°. The reagent is prepared from imidazole and phenyl phosphorodichloridate in benzene[1] and used in peptide synthesis.[2]

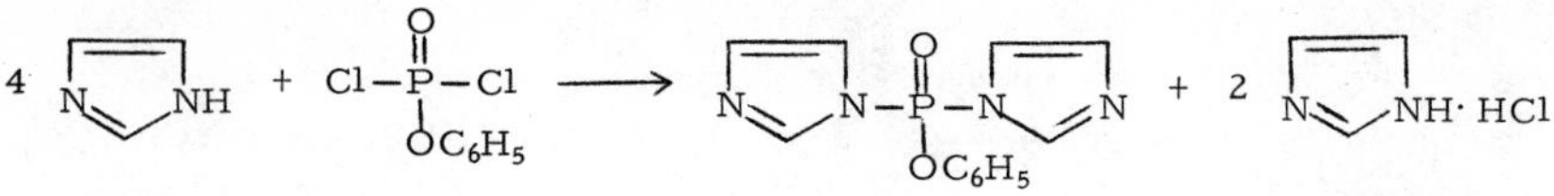

(1) (2) (3) (4)

[1]F. Cramer, H. Schaller, and H. A. Staab, *Ber.*, **94**, 1612 (1961)
[2]F. Cramer and H. Schaller, *Ber.*, **94**, 1634 (1961)

Phenylpotassium, C_6H_5K. Mol. wt. 116.20.

Preparation either from *n*-amyl chloride and benzene or from anisole:[1]

$$\underline{n}\text{-}C_5H_{11}Cl + 2\,K + C_6H_6 \longrightarrow C_5H_{12} + KCl + C_6H_5K$$

$$C_6H_5OCH_3 + 2\,K \longrightarrow CH_3OK + C_6H_5K$$

Metallation. This reagent metallates cumene predominantly at the α-position, as shown by carbonation to phenylisobutyric acid; in contrast, amylsodium metallates cumene almost exclusively in the nucleus.

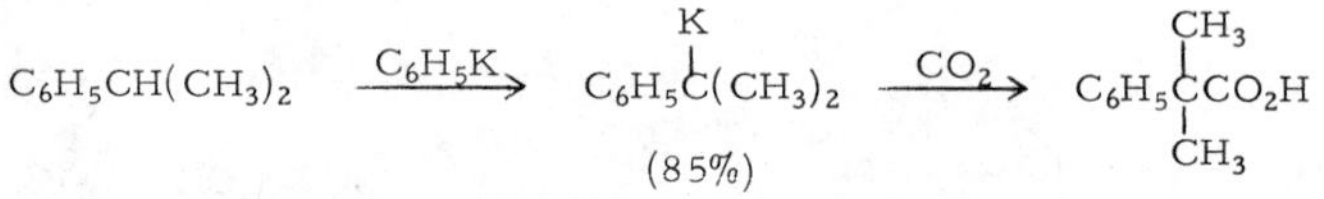

Isomerization of dienes. This strong base effects isomerization of 1,5- or 1,3-cyclooctadiene to *cis*-bicyclo[3.3.0]-Δ^2-octene (5) in 50–65% yield (stirred in an autoclave at 175° for 22 hrs.).[2] In the case of 1,3-cyclooctadiene, isomerization to

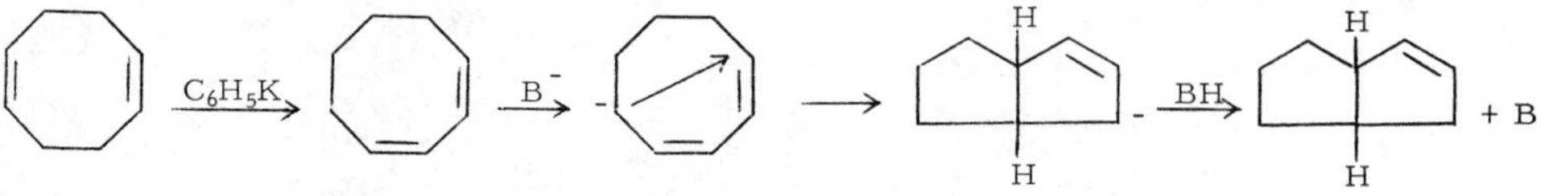

(1) (2) (3) (4) (5)

(5) can be done with use of potassium metal at atmospheric pressure (reflux under N_2 for 7 hrs.).

[1]A. A. Norton and E. J. Lanpher, *J. Org.*, **23**, 1636 (1958)
[2]R. Stapp and R. F. Kleinschmidt, *J. Org.*, **30**, 3006 (1965)

Phenylsodium, C_6H_5Na. Mol. wt. 100.10.

Preparation.[1] A suspension of 23 g. of finely divided sodium in 150 ml. of benzene is treated with 9 g. of chlorobenzene. When the initial exothermal reaction subsides, 50 g. more chlorobenzene is added at a suitable rate to maintain a temperature of 30–40°. Phenylsodium is obtained as a suspension in benzene in yield of 60–70%.

Production of benzynes. The reaction of phenylsodium (2 equiv.) with chlorobenzene (2 equiv.) and resorcinol dimethyl ether (1 equiv.) to give 2,6-dimethoxy-

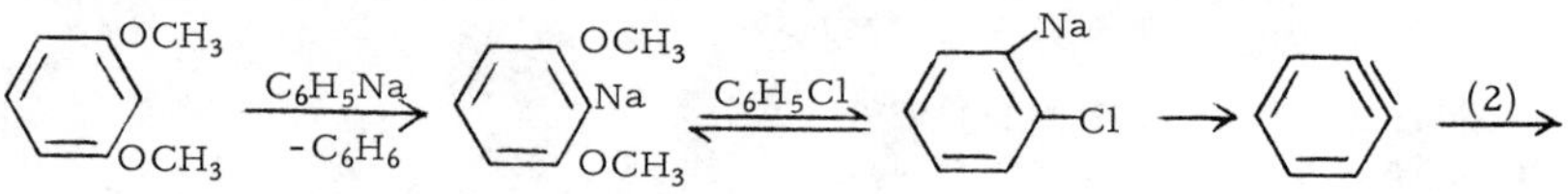

(1) (2) (3) (4)

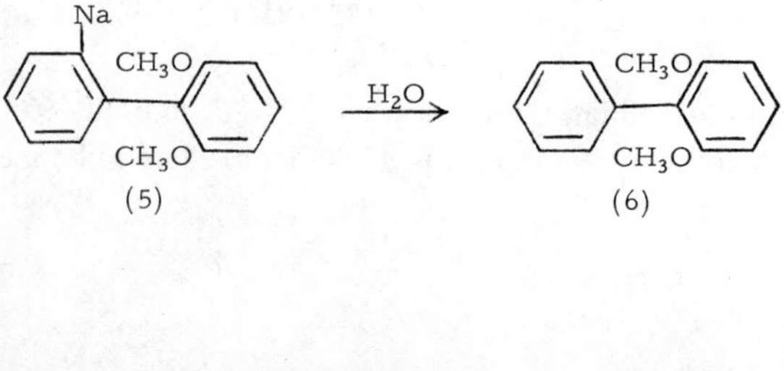

(5) (6)

diphenyl in 80% yield is interpreted as involving benzyne (4) as reactive species.[1] Evidence of the transient formation of benzyne was obtained by carrying out the reaction in the presence of furane as trapping agent and isolation of α-naphthol after suitable workup.

In the presence of a secondary or tertiary amine, an N- or C-arylated amine is obtained.[2]

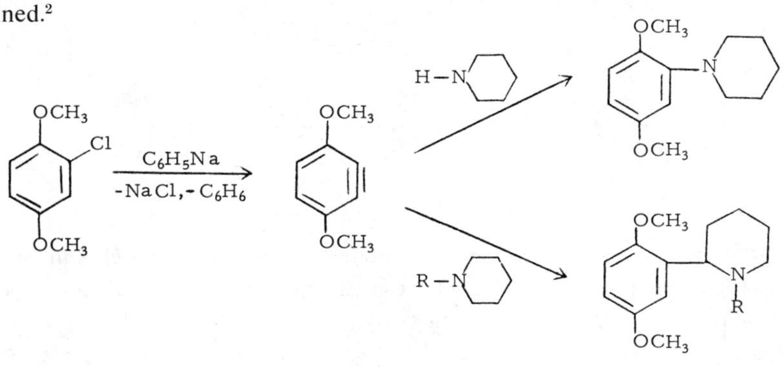

[1] G. Ehrhart, *Ber.*, **96**, 2042 (1963); for related reactions with *p*-dichlorobenzene and with 2-chloro-1,4-dimethoxybenzene see G. Ehrhart, *Ber.*, **97**, 74 (1964)
[2] G. Ehrhart and G. Seidl, *Ber.*, **97**, 1994 (1964)

***p*-Phenylsulfonylbenzoyl chloride**, *p*-$C_6H_5SO_2C_6H_4COCl$. Mol. wt. 137.70. Preparation.[1]

The reagent reacts with vitamin A_1, a sensitive allylic primary alcohol, in methylene chloride–pyridine to give a crystalline ester useful for isolation of the vitamin from biological material.[2]

[1] L. C. Newell, *Am. Chem. J.*, **20**, 302 (1898)
[2] C. K. Payne, F. J. Lotspeich, and R. F. Krause, *J. Org.*, **26**, 3535 (1961)

Phenylsulfur trifluoride, $C_6H_5SF_3$. Mol. wt. 166.17, m.p. *ca.* 0°, b.p. 60°/5 mm. (48°/2.6 mm.). The reagent is prepared[1] by reaction of diphenyl disulfide with silver difluoride in Freon-113, b.p. 47° (1, 1,2-trichloro-1,2,2-trifluoroethane). The reagent

$$C_6H_5SSC_6H_5 + 6\,AgF_2 \xrightarrow[56-61\%]{} 2\,C_6H_5SF_3 + 6\,AgF$$

slowly attacks glass but can be stored indefinitely in bottles of aluminum or of polytetrafluoroethylene plastic (Teflon). Like sulfur tetrafluoride, it reacts with carbonyl compounds to give difluoromethylene compounds ($R_2C{=}O \rightarrow R_2CF_2$) and with carboxylic acids to give trifluoromethyl derivatives ($RCO_2H \rightarrow RCF_3$). It is not so toxic as sulfur tetrafluoride and its use does not require pressure equipment constructed of fluorine-resistant alloy. Thus the reaction with benzaldehyde is conducted in a glass flask connected to a dry distillation column and heated to 100°; the pressure

$$C_6H_5CHO + C_6H_5SF_3 \xrightarrow[71-80\%]{} C_6H_5CHF_2 + C_6H_5SOF$$

is then reduced until the benzal fluoride distills (b.p. 45°/15 mm.). The second product, benzenesulfinyl fluoride, b.p. 60°/2.5 mm., is obtained in 82–89% yield.

[1] W. A. Sheppard, *Am. Soc.*, **84**, 3058 (1962); *Org. Syn.*, **44**, 39, 82 (1964)

4-Phenyl-1,2,4-triazoline-3,5-dione (3). Mol. wt. 149.13, carmine-red needles, dec. 160–180°. The reagent is prepared by oxidizing 4-phenylurazole (2) with *t*-butyl hypochloride in dry acetone at −50 to −78°.[1] The starting material was prepared by Thiele and Stange[2] from semicarbazide hydrochloride. Solutions are unstable at

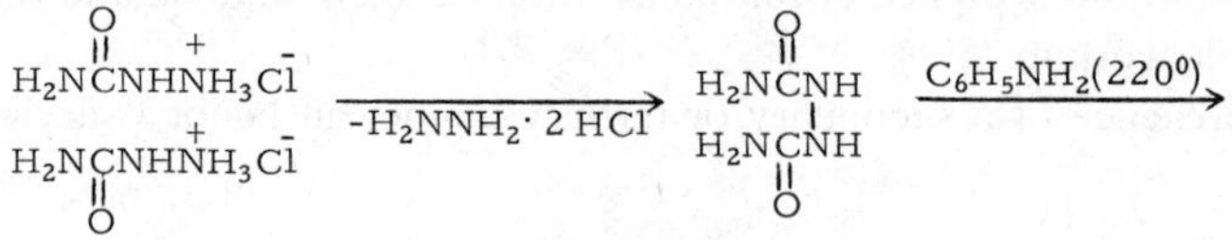
-H2NNH2 · 2 HCl
C6H5NH2(220°)

(1)

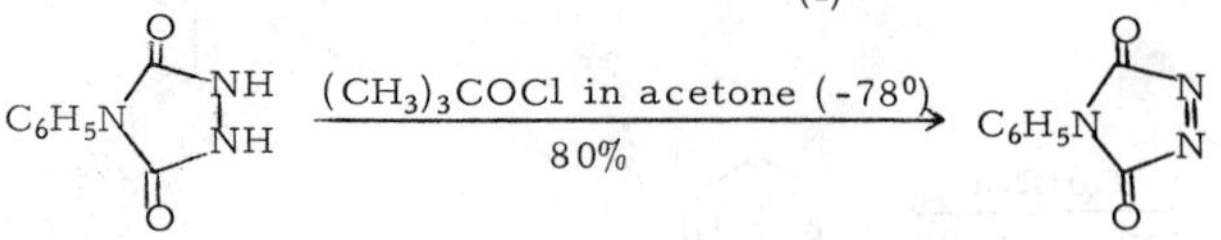
(CH3)3COCl in acetone (-78°)
80%

(2) (3)

room temperature but the substance can be used as a dienophile without isolation. Dimethyl azodicarboxylate is in some respects more reactive as a dienophile than maleic anhydride, but it fails to react, or reacts by allylic addition, with some dienes with which maleic anhydride reacts, a result ascribable to the *trans* configuration of the N═N bond. Constraining this bond to the *cis* configuration greatly enhances the dienophilic reactivity. Thus in acetone solution (3) reacts instantly with cyclopentadiene in acetone at −78° with discharge of the red color to form (4) and it reacts with butadiene at −50° to form (5). It is thus a dienophile of superior reactivity comparable to diazaquinones such as (6).[3]

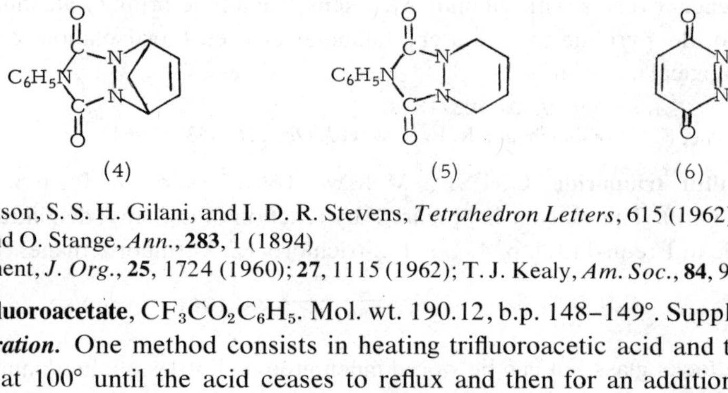

(4) (5) (6)

[1]R. C. Cookson, S. S. H. Gilani, and I. D. R. Stevens, *Tetrahedron Letters*, 615 (1962)
[2]J. Thiele and O. Stange, *Ann.*, **283**, 1 (1894)
[3]R. A. Clement, *J. Org.*, **25**, 1724 (1960); **27**, 1115 (1962); T. J. Kealy, *Am. Soc.*, **84**, 966 (1962)

Phenyl trifluoroacetate, $CF_3CO_2C_6H_5$. Mol. wt. 190.12, b.p. 148–149°. Supplier: KK.

Preparation. One method consists in heating trifluoroacetic acid and triphenyl phosphite at 100° until the acid ceases to reflux and then for an additional hour,

$$\underset{0.2\ m.}{CF_3CO_2H} + \underset{0.22\ m.}{(C_6H_5O)_3P} \xrightarrow[56\%\ \text{redistilled}]{100^0\ 1.5\ \text{hrs.}} CF_3CO_2C_6H_5 + (C_6H_5O)_2POH$$

followed by distillation.[1] A second method[2] involves reaction of sodium phenoxide with phosgene and treatment of the resulting chloroformate (1) with 1 equivalent each of trifluoroacetic acid and triethylamine in tetrahydrofurane.

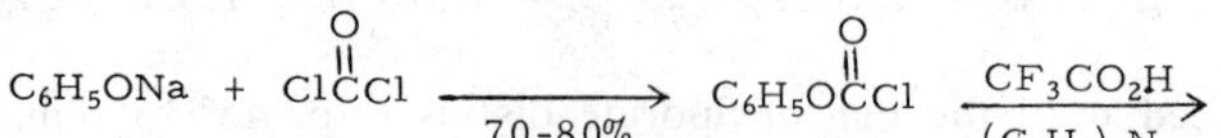
70-80%
CF3CO2H
(C2H5)3N

(1)

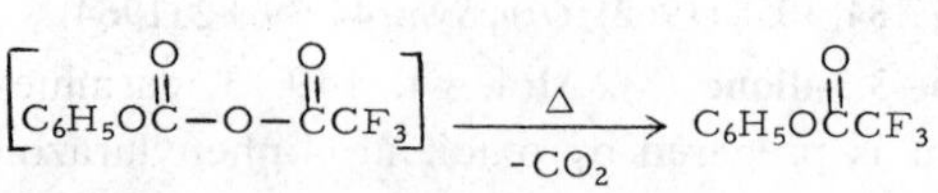
Δ
-CO2

Peptide synthesis. The reagent is used to prepare N-trifluoroacetyl derivatives of amino acids and peptides.[3] The substrate is warmed with 1.2–2 equivalents of the reagent in molten phenol at 120–150° for 2–20 min. Excess reagent and phenol are removed under vacuum or by solution in petroleum ether or carbon tetrachloride, from which the amino acid derivative crystallizes. The main advantage of the N-trifluoroacetyl protective group is that it can be removed under very mild conditions.

[1]L. Benoiton, H. N. Rydon, and J. E. Willett, *Chem. Ind.*, 1060 (1960)
[2]M. Green, *Chem. Ind.*, 435 (1961)
[3]F. Weygand and A. Röpsch, *Ber.*, **92**, 2095 (1959), and earlier papers

Phenyl(trihalomethyl)mercury:

I. $C_6H_5HgCBr_3$, mol. wt. 529.47, m.p. 119–120°
II. $C_6H_5HgCCl_3$, mol. wt., 396.09, m.p. 117–118°
III. $C_6H_5HgCBr_2Cl$, mol. wt. 485.01, m.p. 110–112°
IV. $C_6H_5HgCBrCl_2$, mol. wt. 440.55, m.p. 110–111°

Preparation. Reutov and Lovtsova[1] prepared phenyl(trichloromethyl)mercury (II) by reaction of phenylmercuric chloride with potassium *t*-butoxide and chloroform in benzene. Use of bromoform affords I. Seyferth and Burlitch,[2] who carried

$$C_6H_5HgCl + CHCl_3 + KOC(CH_3)_3 \longrightarrow \underset{\text{II}}{C_6H_5HgCCl_3} + KCl + (CH_3)_3COH$$

out reactions of this type at 0° with high-speed stirring, showed that II is the only product formed on reaction of phenylmercuric bromide with chloroform and $KOC(CH_3)_3$ and established that the reaction involves a simple displacement of halide ion by the trichloromethyl anion:

$$CHCl_3 \xrightarrow{KOC(CH_3)_3} \bar{C}Cl_3 \xrightarrow[51\%]{C_6H_5HgBr} \underset{\text{II}}{C_6H_5HgCCl_3}$$

The two mixed halides are prepared as follows:

$$C_6H_5HgCl + CHBr_2Cl \xrightarrow{KOC(CH_3)_3} \underset{\text{III}}{C_6H_5HgCBr_2Cl}$$

$$C_6H_5HgCl + CHBrCl_2 \xrightarrow{KOC(CH_3)_3} \underset{\text{IV}}{C_6H_5HgCBrCl_2}$$

Phenylmercuric chloride (m.p. 256–258°) is supplied by Metal Salts Corp., phenylmercuric bromide (m.p. 284–286°) by Wood Ridge Chem. Corp. Bromodichloromethane, b.p. 89–90°, is prepared in good yield by the bromination of chloroform with hydrogen bromide and aluminum bromide.

Schweizer and O'Neill,[3] unsuccessful in an attempt to follow the Reutov-Lovtsova procedure using commercial potassium *t*-butoxide, developed a convenient synthesis of phenyl(trichloromethyl)mercury in which phenylmercuric bromide is treated with trichloroacetate anion generated in large excess from ethyl trichloroacetate and commercial sodium methoxide. A mixture of 200 ml. of benzene, 0.18 mole of ethyl trichloroacetate, and 0.37 mole of pulverized phenylmercuric bromide is stirred for 15 min. in an ice bath and 0.154 mole of sodium methoxide is added all

$$C_6H_5HgBr + CCl_3CO_2C_2H_5 + CH_3ONa \xrightarrow[62\%]{\text{Benzene}} C_6H_5HgCCl_3 + CO_2 + C_2H_5OH$$

at once by the technique shown in Fig. A–1, an Erlenmeyer connected to one opening of the flask by means of a piece of Gooch tubing. The mixture is stirred with cooling for 1.5 hrs. and then quenched with water. Extraction with benzene and washing with cold ethanol gives satisfactory product, m.p. 114–115°, in 62% yield.

Generation of dihalocarbenes Seyferth's group[4] found that dibromocarbene can be generated in high yield under nonbasic conditions by the decomposition of phenyl (tribromomethyl)mercury (I) in refluxing benzene. A suspension of 0.105 mole of I in 50 ml. of benzene and 0.315 mole of cyclohexene was stirred under reflux for 2 hrs., during which time the starting mercurial dissolved and phenylmercuric bromide

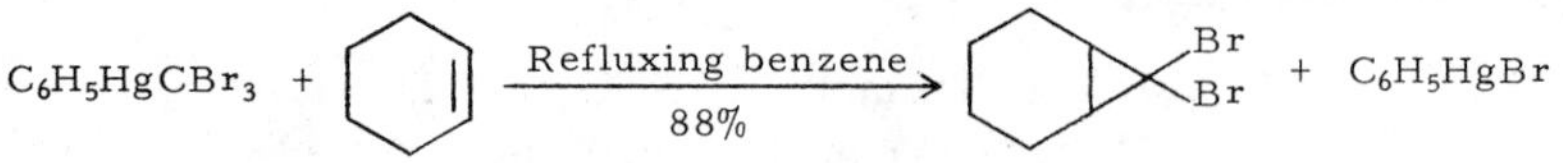
$C_6H_5HgCBr_3$ +
Refluxing benzene
88%
Br
Br
+ C_6H_5HgBr

precipitated. The latter compound, isolated in quantitative yield, was of satisfactory purity for reuse. Fractionation of the filtrate afforded 7,7-dibromonorcarane in 88% yield. Phenyl(trichloromethyl)mercury is much more stable and refluxing in benzene had to be extended for 48 hrs. to produce a comparable yield of 7,7-dichloronorcarane. The presence of a bromine atom in the trihalomethyl group greatly facilitates the reaction of the mercurial with an olefin. Thus $C_6H_5HgCBrCl_2$ reacts with

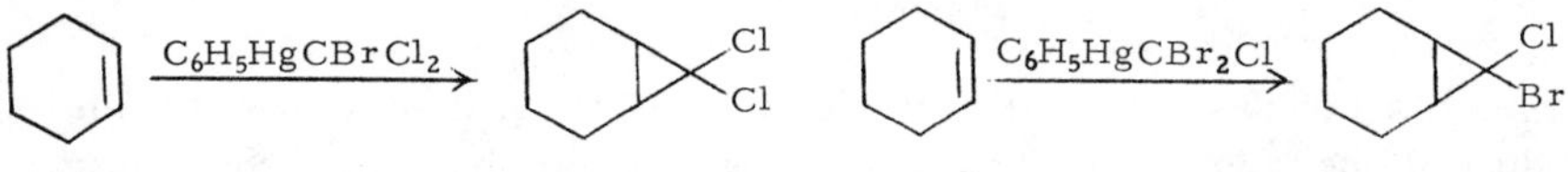
$C_6H_5HgCBrCl_2$
Cl
Cl
$C_6H_5HgCBr_2Cl$
Cl
Br

cyclohexene to give exclusively 7,7-dichloronorcarane in a period of only a few hours.[2] The reagent $C_6H_5HgCBr_2Cl$ serves excellently for the synthesis of the 7-bromo-7-chloro derivative.

A comparison of olefin reactivities toward $C_6H_5HgCBrCl_2$ in benzene at 80° with the reactivities of the same olefins toward sodium trichloroacetate in 1,2-dimethoxyethane at 80° established near identity of the relative reactivities toward both reagents, a result which favors the interpretation that both reactions involve free dichlorocarbene as an intermediate.[5] Of practical significance is the fact that yields are consistently higher by the mercurial route. Thus the latter route proved effective as applied to olefins of low reactivity toward dihalocarbenes generated by other procedures.[6] Examples are formulated:

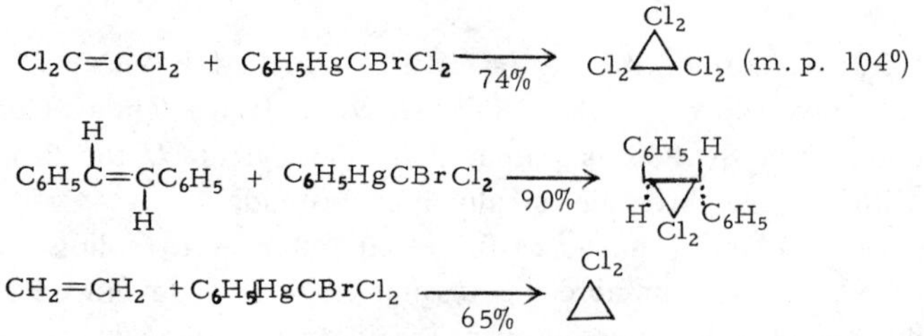
$Cl_2C{=}CCl_2$ + $C_6H_5HgCBrCl_2$
74%
Cl_2
Cl_2
Cl_2
(m. p. 104°)
H
$C_6H_5C{=}CC_6H_5$ + $C_6H_5HgCBrCl_2$
H
90%
C_6H_5 H
H
C_6H_5
Cl_2
$CH_2{=}CH_2$ + $C_6H_5HgCBrCl_2$
65%
Cl_2

The high yield of hexachlorocyclopropane is noteworthy for with dichlorocarbene generated from either ethyl trichloroacetate or chloroform the yield is only 0.2–1%.

In a useful variation of the mercurial route to dihalocarbenes, Seyferth *et al.*[7] found that dihalocarbenes can be generated under milder conditions than before by reaction, for example, of $C_6H_5HgCCl_3$ with an equivalent amount of sodium iodide in benzene–1,2-dimethoxyethane in the presence of cyclohexene. Dichloronorcarane

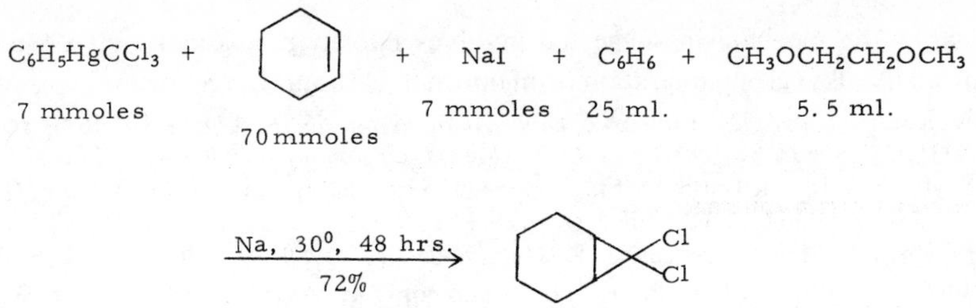

was obtained in good yield and both chloroform and phenylmercuric iodide were found present. The evidence points to the intermediacy of the trichloromethide ion:

$$C_6H_5HgCCl_3 + I^- \longrightarrow C_6H_5HgI + CCl_3^-$$

$$CCl_3^- \longrightarrow :ClCl_2 + Cl^-$$

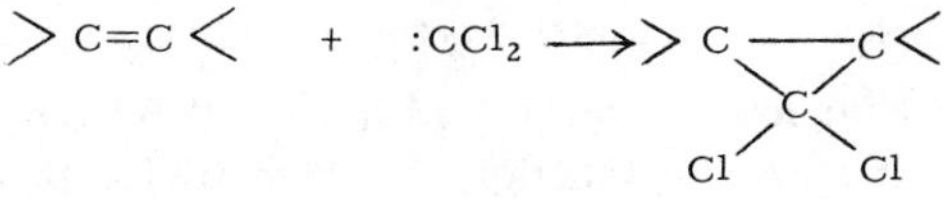

Reaction with carboxylic acids. A surprising finding is that phenyl(bromodichloromethyl)mercury reacts with carboxylic acids in benzene at 60–80° in about 45 min. to give dichloromethyl esters in high yield; phenylmercuric bromide is produced in virtually quantitative yield:[8]

$$(CH_3)_3CCO_2H + C_6H_5HgCBrCl_2 \xrightarrow[81\%]{} (CH_3)_3CCOOCHCl_2 + C_6H_5HgBr$$

A carbene mechanism is favored and the reaction is regarded as the first example in which dichlorocarbene functions as a nucleophile:

$$:CCl_2 + HO_2CR \longrightarrow HCl_2\overset{+}{C}\overset{-}{O_2}CR \longrightarrow CHCl_2O_2CR$$

Deoxygenation of pyridine N-oxide. Schweizer and O'Neill[9] showed that dichlorocarbene generated from a large excess of phenyl(trichloromethyl)mercury deoxygenates pyridine N-oxide to pyridine. The reaction was slow but the yield better than obtained with other dichlorocarbene precursors.

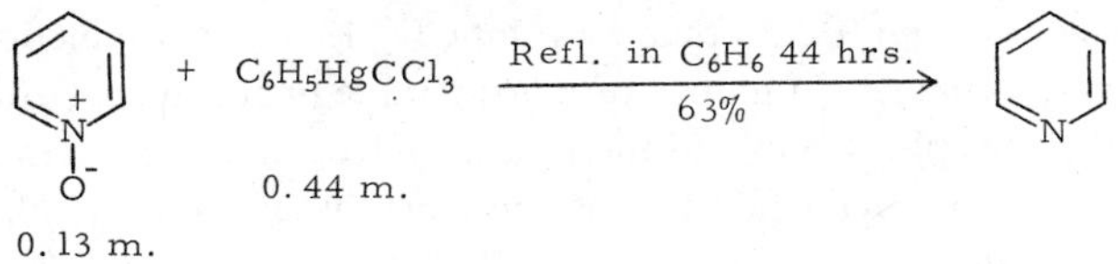

Cleavage of carbodiimides. Phenyl(bromodichloromethyl)mercury cleaves a carbodiimide to an isonitrile and an N-alkyldichloroazomethine:[10]

$$C_6H_5HgCCl_2Br + RN{=}C{=}NR \longrightarrow C_6H_5HgBr + RN{\equiv}C + RN{=}CCl_2$$

Olefin synthesis. Seyferth and Prokai[11] showed that a C_{2n+1}-terminal olefin (1) to (2) and reaction of the triorganoborane with phenyl(bromodichloromethyl)

$$\underset{(1)}{RCH{=}CH_2} \xrightarrow[\text{Argon, } C_6H_6]{B_2H_6} \underset{(2)}{(RCH_2CH_2)_3B} \xrightarrow{C_6H_5HgCCl_2Br} \underset{(3)}{RCH_2CH_2CH{=}CHCH_2R}$$

mercury. The mechanism suggested involves nucleophilic attack by CCl_2 at boron followed by alkyl group migration from boron to carbon.

[1]O. A. Reutov and A. N. Lovtsova, *Dokl. Akad. Nauk SSSR*, **139**, 622 (1961) [*C. A.*, **56**, 1469 (1962)]; see D. Seyferth and J. M. Burlitch, *J. Organometal. Chem.*, **4**, 127 (1965)
[2]D. Seyferth and J. M. Burlitch, *Am. Soc.*, **84**, 1757 (1962); procedure submitted to *Org. Syn.*
[3]E. E. Schweizer and G. J. O'Neill, *J. Org.*, **28**, 851 (1963)
[4]D. Seyferth, J. M. Burlitch, and J. K. Heeren, *J. Org.*, **27**, 1491 (1962); D. Seyferth, J. M. Burlitch, R. J. Minasz, J. Y.-P. Mui, H. D. Simmons, Jr., A. J. H. Treiber, and S. R. Dowd, *Am. Soc.*, **87**, 4259 (1965)
[5]D. Seyferth and J. M. Burlitch, *Am. Soc.*, **86**, 2730 (1964)
[6]D. Seyferth, R. J. Minasz, A. J.-H. Treiber, J. M. Burlitch, and S. R. Dowd, *J. Org.*, **28**, 1163 (1963)
[7]D. Seyferth, J. Y.-P. Mui, M. E. Gordon, and J. M. Burlitch, *Am. Soc.*, **87**, 681 (1965)
[8]D. Seyferth, J. Y.-P. Mui, and L. J. Todd, *Am. Soc.*, **86**, 2961 (1964)
[9]E. E. Schweizer and G. J. O'Neill, *J. Org.*, **28**, 2460 (1963)
[10]D. Seyferth and R. Damrauer, *Tetrahedron Letters*, 189 (1966)
[11]D. Seyferth and B. Prokai, *Am. Soc.*, **88**, 1834 (1966)

N-Phenyltrimethylacetimidoyl chloride (2). Mol. wt. 195.69, b.p. 112°/13 mm.

The reagent (2) is prepared by reaction of C-trimethylacetanilide (1) with PCl_5.[1] It converts carboxylic acids into acid chlorides in good yield. Used in peptide synthesis.

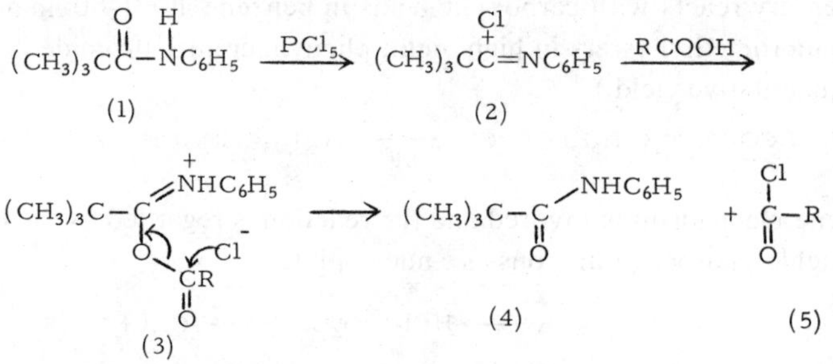

[1]F. Cramer and K. Baer, *Ber.*, **93**, 1231 (1960)

Phenyltrimethylammonium ethoxide, $[C_6H_5\overset{+}{N}(CH_3)_3]\overset{-}{O}C_2H_5$. Mol. wt. 217.30.

Codeine, the phenolic methyl ether of morphine, has considerably less addiction liability than morphine and is used widely as a local analgesic. Since it is present in opium in smaller amount (0.5%) than morphine (7–15%), its production by selective methylation of morphine is a matter of considerable importance. Ordinary methods attack both the phenolic group and the tertiary nitrogen, but Rodionow[1] found that morphine can be converted smoothly into codeine by reaction with phenyltrimethylammonium ethoxide.

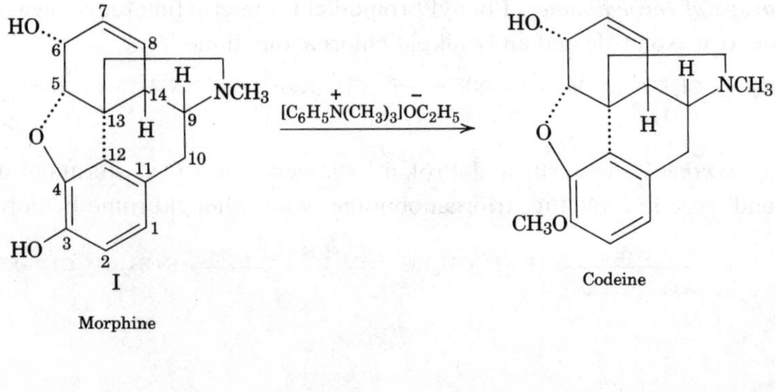
Morphine
Codeine

Snyder *et al.*[2] prepared the reagent in ethanol solution by dissolving 100 mg. of sodium in 2 ml. of ethanol and adding a solution of 1,037 mg. of phenyltrimethylammonium benzenesulfonate (Eastman practical trimethylphenylammonium

$$[C_6H_5\overset{+}{N}(CH_3)_3]\overset{-}{SO_3}C_6H_5 + NaOC_2H_5 \longrightarrow [C_6H_5\overset{+}{N}(CH_3)_3][\overset{-}{O}C_2H_5]$$

benzenesulfonate). Sodium benzenesulfonate precipitated and was removed by filtration. The solution was used for methylation of the phenolic alkaloid haplophytine, in which a phenolic hydroxyl is strongly hydrogen bonded to a carbonyl group. This method proved more satisfactory than use of diazomethane, which reacts extremely slowly. Attempted methylation with dimethyl sulfate and sodium hydroxide in a nitrogen atmosphere failed, as did attempted reaction with methyl iodide and potassium carbonate in boiling acetone.

[1]W. Rodionow, *Bull. Soc.*, **39**, 305 (1926)

[2]H. R. Snyder, H. F. Strohmayer, and R. A. Mooney, *Am. Soc.*, **80**, 3708 (1958)

Phenyltrimethylammonium perbromide (PTAB), $C_6H_5\overset{+}{N}(CH_3)_3Br_3^-$. Mol. wt. 375.96, m.p. 115.5–116.5°. Supplier: British Drug Houses. Preparation.[1,2]

The compound is a brominating agent similar to pyridinium bromide perbromide but more stable. It is very soluble in tetrahydrofurane and is employed in this solvent for the α-bromination of ketones and cyclic ketals, for example:[2]

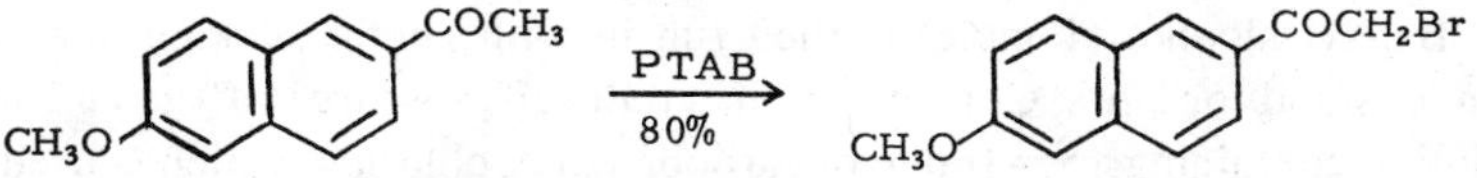

COCH₃
CH₃O
PTAB
80%
COCH₂Br
CH₃O

Of particular interest is the bromination of ethyleneketals of ketones, for example, the androstane derivative (1).[3] This affords the 16α-bromoketal (2), in which both

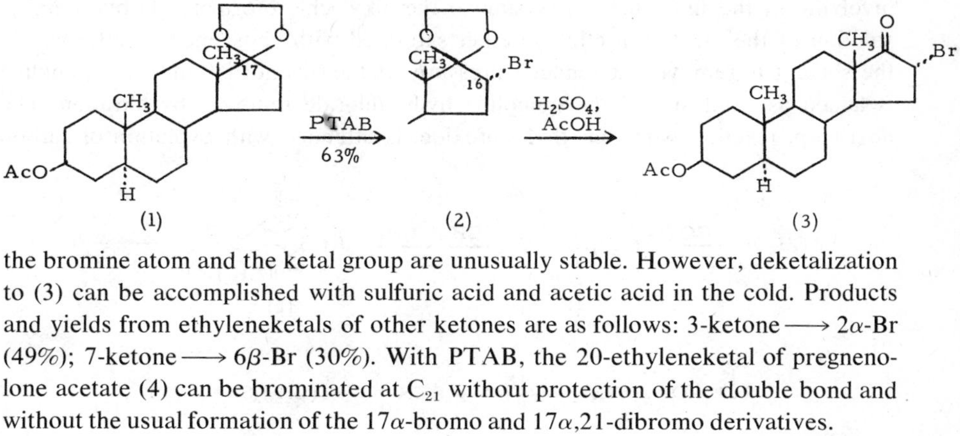

CH₃
17
CH₃
AcO
H
PTAB
63%
CH₃
Br
16
H₂SO₄,
AcOH
CH₃
Br
CH₃
AcO
H

(1) (2) (3)

the bromine atom and the ketal group are unusually stable. However, deketalization to (3) can be accomplished with sulfuric acid and acetic acid in the cold. Products and yields from ethyleneketals of other ketones are as follows: 3-ketone $\longrightarrow$ 2α-Br (49%); 7-ketone $\longrightarrow$ 6β-Br (30%). With PTAB, the 20-ethyleneketal of pregnenolone acetate (4) can be brominated at C_{21} without protection of the double bond and without the usual formation of the 17α-bromo and 17α,21-dibromo derivatives.

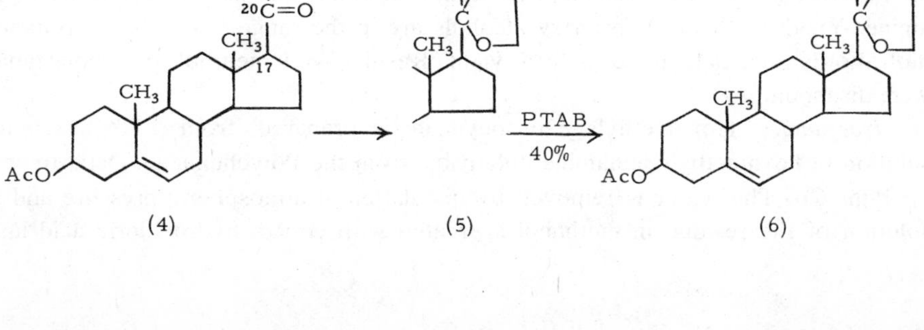

21
CH₃
20C=O
CH₃
17
CH₃
AcO
CH₃
C
CH₃
PTAB
40%
CH₂Br
C
CH₃
CH₃
AcO

(4) (5) (6)

[1]D. Vorländer and E. Siebert, *Ber.*, **52**, 283 (1919)
[2]W. S. Johnson, J. D. Bass, and K. L. Williamson, *Tetrahedron*, **19**, 861 (1963)
[3]A. Marquet and J. Jacques, *Tetrahedron Letters*, 24 (1959); A. Marquet, M. Dvolaitsky, H. B. Kagan, L. Mamlok, C. Ouannes, and J. Jacques, *Bull. Soc.*, 1822 (1961); A. Marquet and J. Jacques, *ibid.*, 90 (1962)

Phosgene, $COCl_2$. Mol. wt. 98.92, b.p. 7.6°, m.p. −127.8°. Phosgene is a colorless gas with a penetrating odor. It is highly toxic, with a dangerous delayed effect. Suppliers: Matheson; Niagara Chlorine Prod.; MCB (12.5% in benzene).

Purification.[1] Chlorine in phosgene can be detected by bubbling the gas rapidly through clean mercury; chlorine reacts with and discolors the metal whereas pure phosgene leaves it unchanged. Chlorine if present can be removed by bubbling the gas through two wash bottles containing cottonseed oil.

ROH ⟶ ROCOCl (See Ethyl chloroformate). Applied to benzyl alcohol, the reaction gives a product commonly known as carbobenzoxy chloride, which is widely used in peptide synthesis. A modification[2] of the original procedure of Bergmann and Zervas[3] is as follows. A flask containing 500 g. of toluene is weighed,

$$C_6H_5CH_2OH + ClCOCl \xrightarrow[91-94\%]{\text{Toluene}} C_6H_5CH_2OCOCl + HCl$$

cooled in ice, and gaseous phosgene is bubbled in to a gain in weight of 1.1 moles (1 hr.). Benzyl alcohol (1 mole) is then run in with gentle shaking at 0° and the solution let stand for 2 hrs. Concentration at reduced pressure (60°) gives 200–220 g. of a solution containing 155–160 g. of carbobenzoxy chloride, which is used as such for the preparation of N-carbobenzoxy amino acids.

Barton[4] devised a new method for the oxidation of primary alcohols to aldehydes involving in the first step conversion to the alkyl chloroformate (1) by adding a solution of the alcohol in ether to ether saturated with phosgene (15–20% w./v.); the solvent is removed at reduced pressure. If the alcohol is hindered, quinoline is added as catalyst and the quinoline hydrochloride removed by filtration. The next step, reaction with dimethyl sulfoxide, is attended with evolution of carbon

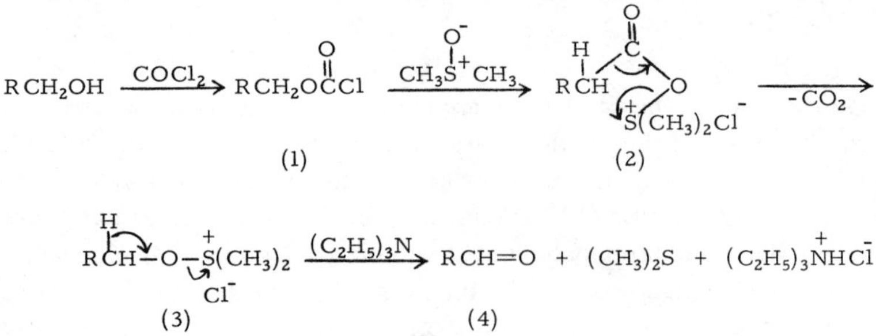
RCH2OH
COCl2
RCH2OCCl
CH3S+CH3
RCH
S(CH3)2Cl
-CO2
(1)
(2)
RCH–O–S(CH3)2
Cl
(C2H5)3N
RCH=O + (CH3)2S + (C2H5)3NHCl
(3)
(4)

dioxide and gives an intermediate (3) convertible into the aldehyde with triethylamine. Yields with lower primary alcohols are in the range 57–78%. 1,4-Butanediol affords succinaldehyde in 80% yield. Results with menthol and cholestanol were disappointing.

Isocyanates. Hexamethylene diisocyanate is prepared[5] from a 70% aqueous solution of hexamethylenediamine obtainable from the Polychemicals Department, du Pont Co. The water is removed by distillation at atmospheric pressure and a solution of the residue in methanol is treated with concd. hydrochloric acid and poured into acetone to precipitate the diammonium chloride (90–99%). A suspension of 0.5 mole of this salt in amylbenzene is stirred under reflux at 180–185°, and phosgene is passed in for 8–15 hrs. Fractionation of the filtered solution gives the diisocyanate with recovery of the amylbenzene.

$$\overset{-}{Cl}H_3\overset{+}{N}(CH_2)_6\overset{+}{N}H_3\overset{-}{Cl} + 2\,COCl_2 \xrightarrow[92\text{-}96\%]{} O{=}C{=}N(CH_2)_6N{=}C{=}O + 6\,HCl$$

In an earlier procedure[6] for the preparation of *p*-nitrophenyl isocyanate a solution of the free base in ethyl acetate is added gradually to ethyl acetate saturated with phosgene while fresh phosgene is passed into the reaction mixture; yield 85–95%.

An efficient continuous process for forming the isocyanates from amino ester hydrochlorides is described by Humphlett and Wilson.[7]

Isocyanides (Isonitriles). Phosgene in combination with triethylamine dehydrates monosubstituted alkylformamides to isocyanides (yields 60–80%).[8] The mechanism

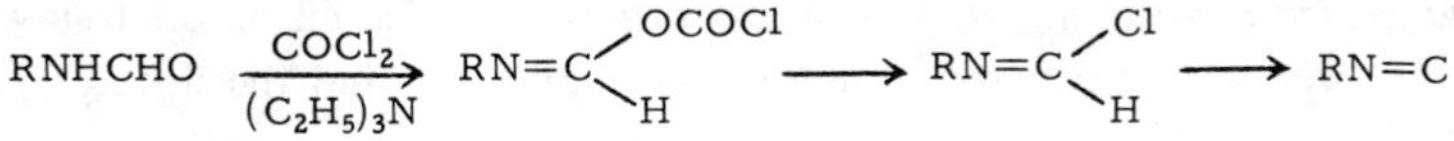

is not known. The starting material for the preparation of cyclohexyl isocyanide, N-cyclohexylformamide, is prepared[9] by slowly adding ethyl formate (3.52 moles) with stirring to cyclohexylamine (4 moles) with ice cooling. After 2 hrs. of refluxing

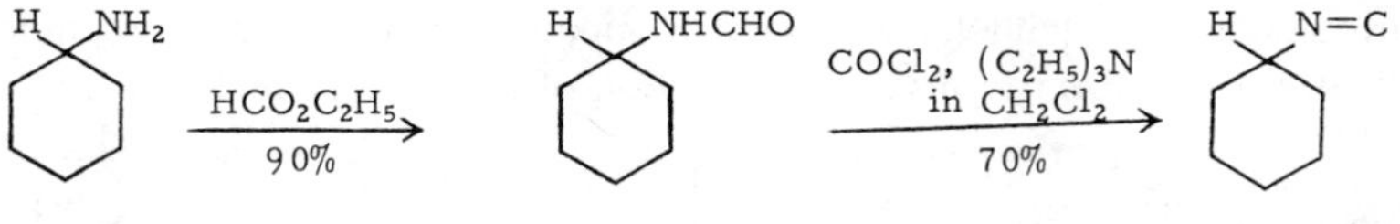

(1) (2) (3)

the product is distilled. A solution of 0.2 mole of (2) in 65 ml. of triethylamine and 200 ml. of methylene chloride is stirred with ice cooling, and 0.2 mole of phosgene is passed in. After the addition of 50 ml. of concd. sodium carbonate solution is added, the organic phase is separated, dried, and distilled.

Dehydration of amides. For efficient dehydration of the diamide (1), Linstead and co-workers[10] bubbled phosgene (dried by passage through H_2SO_4) into a suspension of (1) in pyridine, with cooling to control the temperature to 60–65°. The procedure also proved satisfactory for the conversion of (3) into (4).[11]

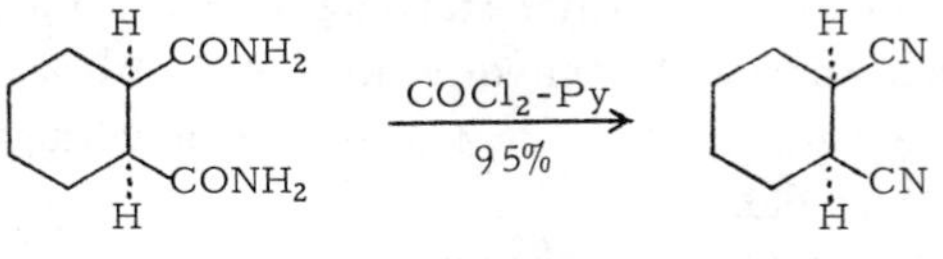

(1) (2)

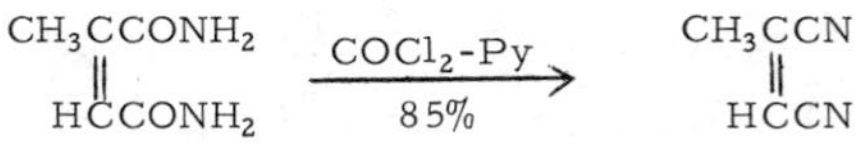

(3)

Cyclic carbonates. For conversion of catechol into phenylenecarbonate,[12] 1 mole of the phenol is dissolved under nitrogen in a deaerated solution of 2.2 moles of

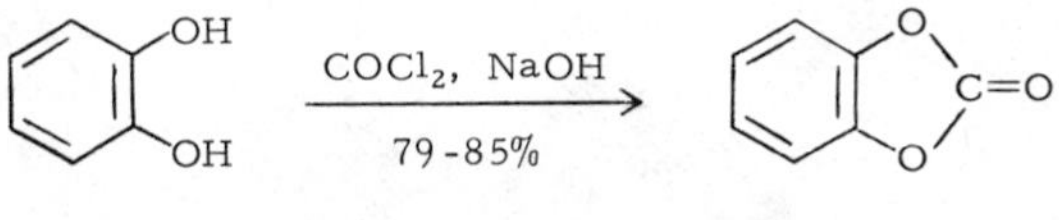

sodium hydroxide in 250 ml. of water, the solution is stirred in ice under nitrogen during the addition at 0–5° of a solution prepared by bubbling 2.23 moles of phosgene into 750 ml. of toluene at 0°. The solid that separates is collected and dissolved in the toluene layer, and the solution is dried and concentrated for crystallization of the product.

Anhydrides. Isatoic anhydride is prepared[13] by stirring an aqueous solution of

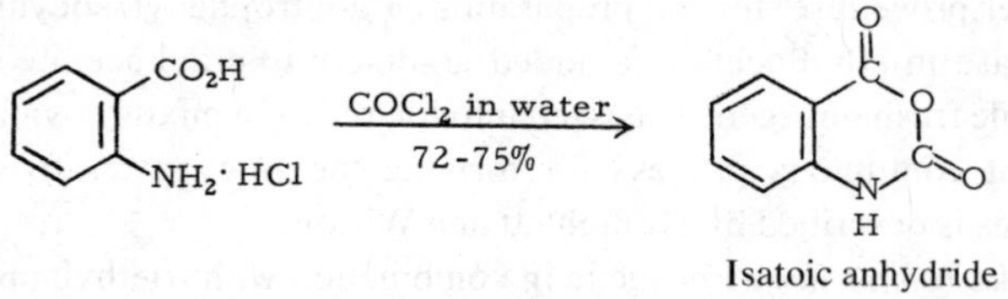
CO2H
NH2·HCl
COCl2 in water
72-75%
H
Isatoic anhydride

anthranilic acid hydrochloride and passing in phosgene with cooling to keep the temperature from rising above 50°. When separation of solid product slows down the gas absorption, a first crop of material is collected, and the filtrate processed as before.

Rinderknecht and Ma,[14] requiring a quantity of nicotinic acid anhydride, noted that the known methods suffer from the extreme sensitivity of nicotinic acid chloride, and employed with success a method analogous to the mixed anhydride method of peptide synthesis. Equimolar amounts of nicotinic acid and triethylamine were dissolved in chloroform to produce a solution of the salt (1), and the solution was

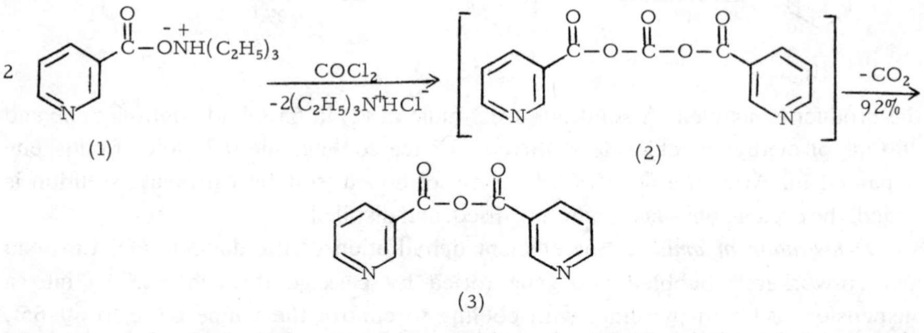
COCl2
-2(C2H5)3NHCl
-CO2
92%
(1)
(2)
(3)

stirred in an ice bath during dropwise addition of a 15.5% solution of phosgene in toluene (vigorous gas evolution). After standing overnight the mixture was evaporated to dryness in vacuum, and the anhydride (3) isolated by extraction with benzene and crystallization. Several aliphatic and aromatic anhydrides were obtained by this method in excellent yield.

Amino acid benzyl esters. Use of benzyl esters in the carbobenzoxy method of peptide synthesis has the advantage that both protective groups can be removed

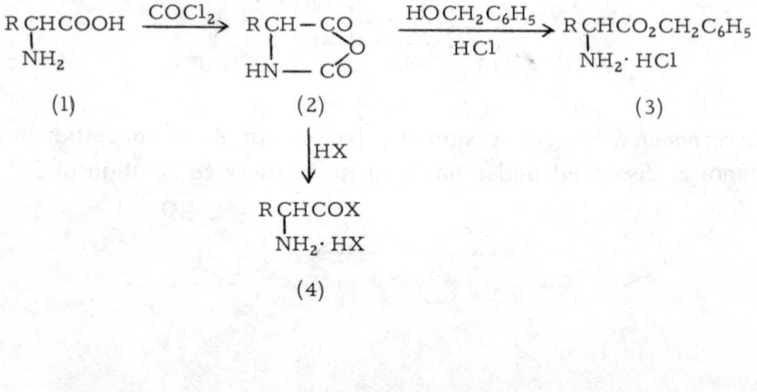
RCHCOOH
NH2
COCl2
RCH—CO
HN—CO
HOCH2C6H5
HCl
RCHCO2CH2C6H5
NH2·HCl
(1)
(2)
(3)
HX
RCHCOX
NH2·HX
(4)

simultaneously by catalytic hydrogenation. However, yields are poor in direct esterification with benzyl alcohol and a mineral acid. In a procedure introduced by Erlanger and Brand[15] (but credited to Max Bergmann) the amino acid is converted into the N-carboxyanhydride (2), which reacts with benzyl alcohol in the presence of hydrogen chloride to give the benzyl ester (3). In the original procedure the N-carboxy anhydride (2,5-oxazolidine) was obtained by an indirect method, but subsequent workers have used the reaction of the amino acid with phosgene.[16–18] Treatment of the anhydride (2) in dry dioxane with HCl or HBr gives the acid halide hydrohalide in analytically pure form.[18]

[1]M. W. Farlow, *Org. Syn., Coll. Vol.*, **4**, 524, Note 8 (1963)
[2]H. E. Carter, R. L. Frank, and H. W. Johnston, *ibid.*, **3**, 167 (1955)
[3]M. Bergmann and L. Zervas, *Ber.*, **65**, 1192 (1932)
[4]D. H. R. Barton, B. J. Garner, and R. H. Wightman, *J. Chem. Soc.*, 1855 (1964)
[5]M. W. Farlow, *Org. Syn., Coll. Vol.*, **4**, 521 (1963)
[6]R. L. Shriner, W. H. Horne, and R. F. B. Cox, *ibid.*, **2**, 453 (1943)
[7]W. J. Humphlett and C. V. Wilson, *J. Org.*, **26**, 2507 (1961)
[8]I. Ugi, W. Betz, V. Fetzer, and K. Offermann, *Ber.*, **94**, 2814 (1961); I. Ugi *et al.*, *Angew. Chem., Internat. Ed.*, **4**, 472 (1965)
[9]B. C. McKusick and M. E. Hermes, *Org. Syn.*, **41**, 14 (1961)
[10]G. E. Ficken, H. France, and R. P. Linstead, *J. Chem. Soc.*, 3730 (1954)
[11]P. M. Brown, D. B. Spiers, and M. Whalley, *J. Chem. Soc.*, 2882 (1957)
[12]R. S. Hanslick, W. F. Bruce, and A. Mascitti, *Org. Syn., Coll. Vol.*, **4**, 788 (1963)
[13]E. C. Wagner and M. F. Fegley, *ibid.*, **3**, 488 (1955)
[14]H. Rinderknecht and V. Ma, *Helv.*, **47**, 162 (1964); H. Rinderknecht and M. Guterstein, procedure submitted to *Org. Syn.*
[15]B. F. Erlanger and E. Brand, *Am. Soc.*, **73**, 3508 (1951)
[16]J. Rudinger and Z. Pravda, *Coll. Czech.*, **23**, 1947 (1958)
[17]M. Wilchek and A. Patchornik, *J. Org.*, **28**, 1874 (1963)
[18]M. Brenner and I. Photaki, *Helv.*, **39**, 1525 (1956)

Phosphine, PH_3. Mol. wt. 34.00. A highly toxic gas available in cylinders or prepared with a generator.[1]

In an alkaline medium, and in a system first purged with nitrogen, phosphine reduces aromatic nitro compounds to azoxy derivatives in high yield:[2]

$$ArNO_2 \xrightarrow[80\text{-}95\%]{PH_3} ArN{=}\overset{O^-}{\overset{|}{N^+}}Ar$$

Note, however, that under the same conditions 6-nitroquinoline affords both 6,6′-azoxyquinoline (1) and the oxygen-free 6,6′-azo-5,5′-diquinolyl (2).[3]

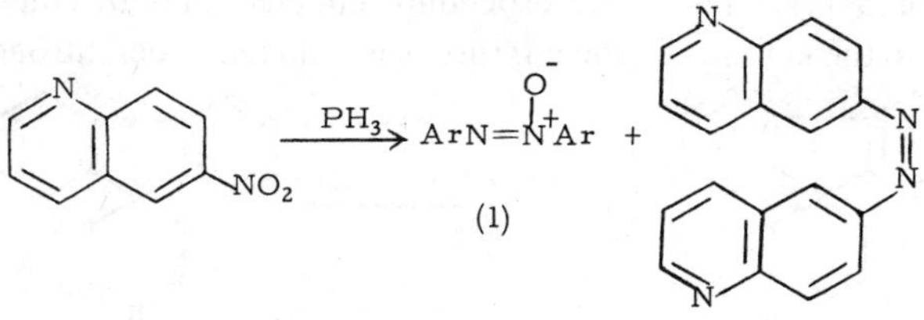

(2)

[1]W. A. Reeves, F. F. Flynn, and J. D. Guthrie, *Am. Soc.*, **77**, 3923 (1955)
[2]S. A. Buckler, L. Doll, F. K. Lind, and M. Epstein, *J. Org.*, **27**, 794 (1962)
[3]A. C. Bellaart, *Rec. trav.*, **83**, 718 (1964); *see also idem*, *Tetrahedron*, **21**, 3285 (1965)

Phosphonium iodide, PH_4I. Mol. wt. 161.91. A white solid which on heating dissociates to $PH_3 + HI$. Preparation:[1]

$$P_4 + 4\,I_2 \longrightarrow 2\,P_2I_4$$

$$10\,P_2I_4 + 13\,P_4 + 128\,H_2O \longrightarrow 40\,PH_4I + 32\,H_3PO_4 \text{ (yield 50-60\%)}$$

Robertson and Witkop[2] used phosphonium iodide in fuming hydriodic acid to reduce pyrrole-2-carboxamide (1) to an easily separated mixture of 3,4-dehydroprolinamide (2) and 3,4-dehydroproline (3).

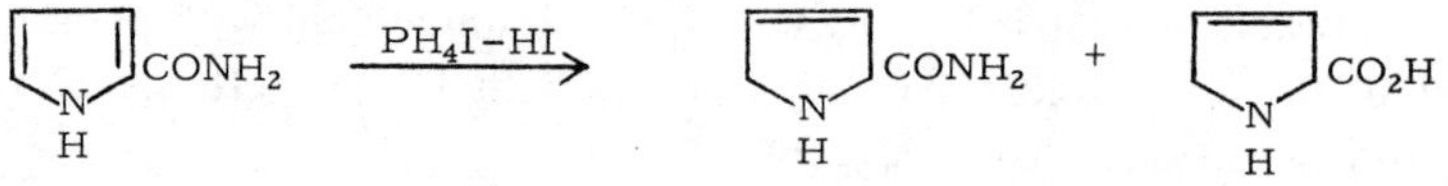
CONH₂
H
PH₄I−HI
CONH₂
H
+
CO₂H
H

(1) (2) 70% (3) 10%

[1]J. B. Work, *Inorg. Syn.*, **2**, 141 (1946)
[2]A. V. Robertson and B. Witkop, *Am. Soc.*, **84**, 1697 (1962)

Phosphoric acid, anhydrous, O═P(OH)$_3$. Mol. wt. 98.00. Preparation: phosphorus pentoxide (25 g.) is dissolved in 33 g. of 85% phosphoric acid.[1]

Wilson and Harris[2] (Merck) used the reagent to effect the synthesis of cocarboxylase (3) from pyridoxamine dihydrochloride (1). This route proved superior to one employing phosphoryl chloride.

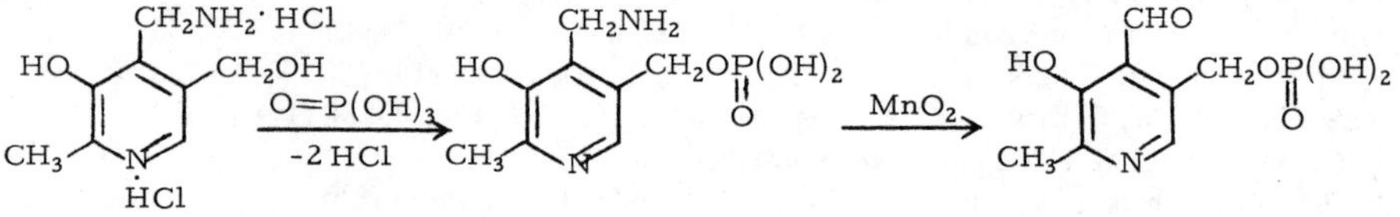
CH₂NH₂·HCl
HO
CH₂OH
CH₃
N
HCl
O=P(OH)₃
-2 HCl
CH₂NH₂
HO
CH₂OP(OH)₂
O
CH₃
N
MnO₂
CHO
HO
CH₂OP(OH)₂
O
CH₃
N

(1) (2) (3)

[1]R. E. Ferrel, H. S. Olcott, and H. Fraenkel-Conrat, *Am. Soc.*, **70**, 2101 (1948)
[2]A. N. Wilson and S. A. Harris, *Am. Soc.*, **73**, 4693 (1951)

Phosphoric acid (85%)–Acetic acid (1:1). This combination proved useful for the hydrolysis and decarboxylation of the iminonitrile (1) to the ketone (2); the mixture

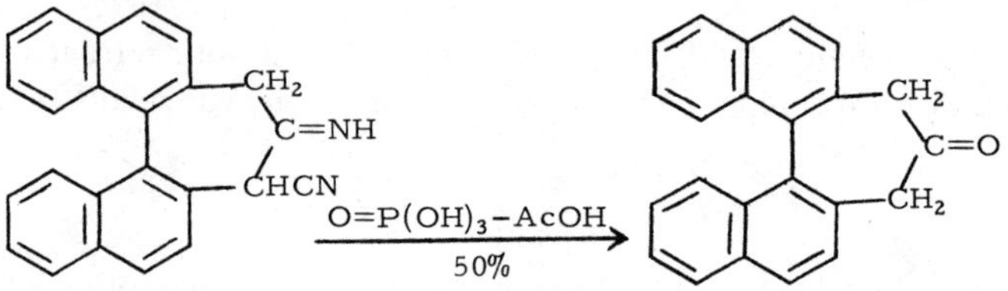
CH₂
C=NH
CHCN
O=P(OH)₃−AcOH
50%
CH₂
C=O
CH₂

(1) (2)

was refluxed for 5 hrs.[1] The same procedure effected smooth conversion of the aminonitrile (3) to the ketone (4); the mixture was refluxed under nitrogen for 16 hrs.[2]

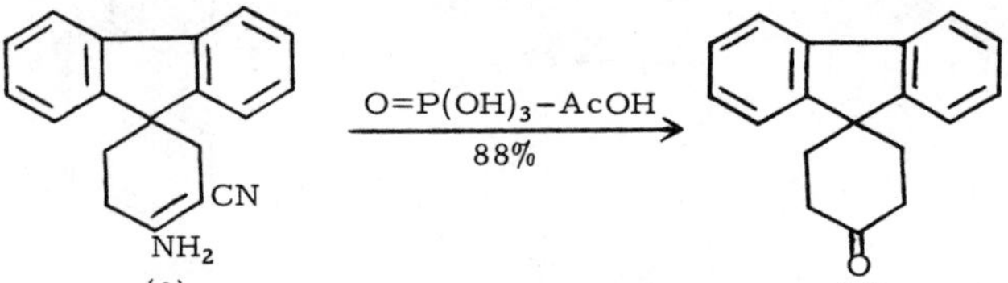
CN
NH₂
O=P(OH)₃−AcOH
88%
O

(3) (4)

[1]K. Mislow and F. A. McGinn, *Am. Soc.*, **80**, 6036 (1958)
[2]S. Baldwin, *J. Org.*, **26**, 3280 (1961)

Phosphoric acid–Formic acid. Braude and Coles[1] cyclized 5 g. of cyclohexenyl vinyl ketone (1) to 4,5,6,7-tetrahydroindane-1-one (2) by heating it with a mixture of 2 g. of 85% phosphoric acid and 6 g. of 98% formic acid at 80–90° under nitrogen for 4 hrs. The cyclization method proved useful in the synthesis of azulenes.[2,3]

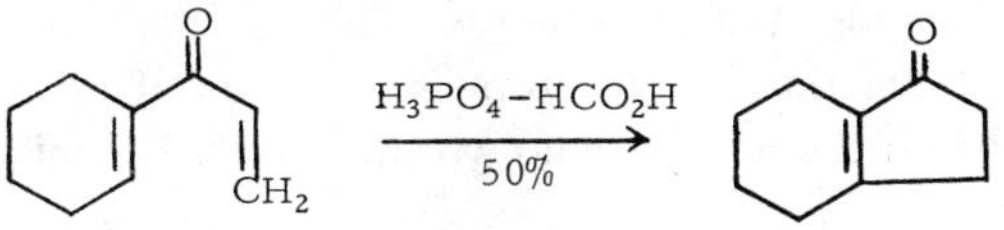

(1) (2)

[1]E. A. Braude and J. A. Coles, *J. Chem. Soc.*, 1430 (1952)
[2]E. A. Braude and W. F. Forbes, *J. Chem. Soc.*, 2208 (1953)
[3]A. M. Islam and R. A. Raphael, *J. Chem. Soc.*, 2247 (1953)

Phosphorodimorpholidic chloride, 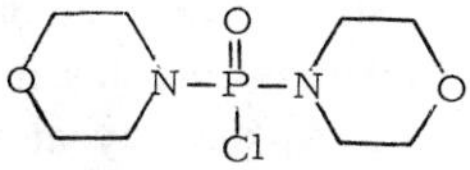 (1)

Mol. wt. 206.67, m.p. 81°, b.p. 137–140°/0.02 mm.

Preparation.[1] Morpholine (71 g.) was added slowly to 19 ml. of phosphoryl chloride in 200 ml. of benzene at 10–20°, the mixture was stirred for 3 hrs., filtered, evaporated, and the residue distilled. The bromide (liquid) was prepared similarly in chloroform (120 ml.) and the solution evaporated to 50 ml. and used directly.

Phosphorylation of alcohols.[1] A mixture of ethanol, the chloride (1), and 2,6-lutidine (2) was refluxed, the solution was evaporated, and the ethyl phosphorodimorpholidate (3) extracted with ether and distilled. Hydrolysis to the dihydrogen

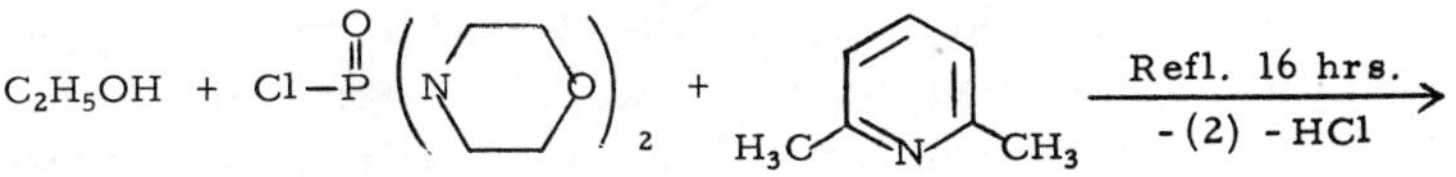

7 ml. (1) 8.5 g. (2) 4.5 ml.

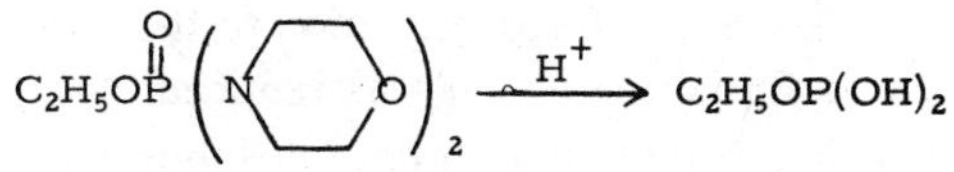

(3) (4)

phosphate, such as (4), can usually be done with a protonated ion-exchange resin or with hydrochloric acid. When the more reactive phosphorodimorpholidic bromide is used the reaction proceeds at room temperature. Cholesteryl phosphorodimorpholidate could not be hydrolyzed with the ion-exchange resin, and use of hydrochloric acid led to formation of cholesteryl chloride.[1] Riess[2] explored the reaction with and androstenolone but concluded that in this case phosphoryl chloride is superior.

[1]H. A. C. Montgomery and J. H. Turnbull, *J. Chem. Soc.*, 1963 (1958)
[2]J. Riess, *Bull. soc.*, 18 (1965)

Phosphorus, red, P_4. Mol. wt. $= 4 \times 30.975 = 123.90$.

ROH⟶RBr. For the preparation of ethyl bromide[1] a suspension of equal parts of red and yellow phosphorus in 95% ethanol is heated to gentle boiling, bromine is

$$\underset{7.5\text{ m.}}{C_2H_5OH} + \underbrace{P(\text{red}) + P(\text{yel.})}_{1.29\text{ g. atoms}} + \underset{2.36\text{ m.}}{Br_2} \xrightarrow[82\text{-}90\%]{100\text{-}140^\circ} C_2H_5Br$$

added during 3 hrs., and the product is distilled into ice and water. *n*-Butyl bromide and isobutyl bromide are prepared similarly but at higher temperatures. For the preparation of *t*-butyl bromide, red phosphorus alone is used and the conversion to PBr_3 is carried to completion before dropwise addition of *t*-butanol. For the conversion of glycerol into the α,γ-dibromohydrin,[2] glycerol is mixed intimately with

$$\underset{17.4\ \text{m.}}{3\ HOCH_2CHOHCH_2OH} + \underset{6.5\ \text{g. atoms}}{2\ \text{P(red)}} + \underset{17.5\ \text{m.}}{3\ Br_2} \xrightarrow[52\text{-}54\%]{80\text{-}100^\circ} 3\ BrCH_2CHOHCH_2Br$$

red phosphorus, an efficient stirrer is started, and bromine is added in 8 hrs. with control of the temperature at 80–100°.

ROH⟶RI. In the conversion of cetyl alcohol to the iodide,[3] red phosphorus alone is used and the amount taken is only half that required for the formation of PI_3 equivalent to the alcohol used. Apparently at the elevated reaction temperature

$$\underset{1\ \text{m.}}{3\ \underline{n}\text{-}C_{16}H_{33}OH} + \underset{0.32\ \text{g. atom}}{2\ \text{P(red)}} + \underset{1.06\ \text{g. atom}}{3\ I_2} \xrightarrow[85\%]{145\text{-}150^\circ\ 5\ \text{hrs.}} 3\ \underline{n}\text{-}C_{16}H_{33}I$$

phosphorus functions in part catalytically. At the very low reaction temperature involved in the preparation of methyl iodide,[4] the activity of the phosphorus is stepped up by use of a mixture of equal parts of red and yellow phosphorus and the amount

$$\underset{50\ \text{m.}}{3\ CH_3OH} + \underbrace{\text{P(red)} + \text{P(yel.)}}_{12.8\ \text{g. atoms}} + \underset{15.8\ \text{g. atom}}{3\ I_2} \xrightarrow[93\text{-}95\%]{} 3\ CH_3I$$

taken is in excess of the theory. Equal parts of red and yellow phosphorus are covered with methanol in a boiling flask fitted with a separatory funnel with a large chamber charged with iodine and with a tube carrying vapor (CH_3OH, CH_3I) to a reflux condenser which delivers into the separatory funnel and leaches iodine into the mixture of methanol and phosphorus.

In a structural study of fungichromin (I), an antifungal macrolide antibiotic, Cope *et al.*[5] used the following method for degradation to the parent hydrocarbon (III). Hydrogenation gave a decahydride which was reduced by lithium aluminum hydride to the polyol II, which was treated with red phosphorus in refluxing hydriodic acid. Reduction of the resulting iodine-containing product with lithium aluminum

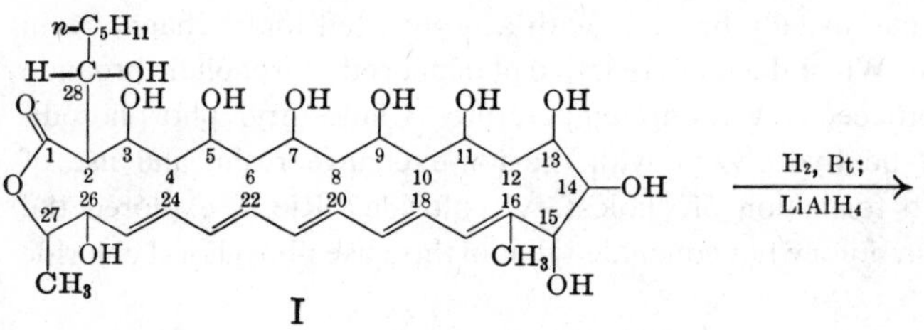
n-C5H11
H—C—OH
OH OH OH OH OH OH
OH
OH
CH3
CH3
OH
H2, Pt;
LiAlH4

I

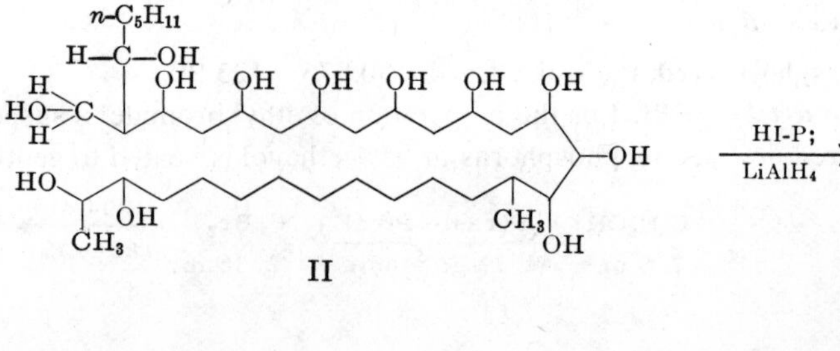
n-C5H11
H—C—OH
OH OH OH OH OH OH
HO—C
HO
OH
CH3
CH3
OH
HI-P;
LiAlH4

II

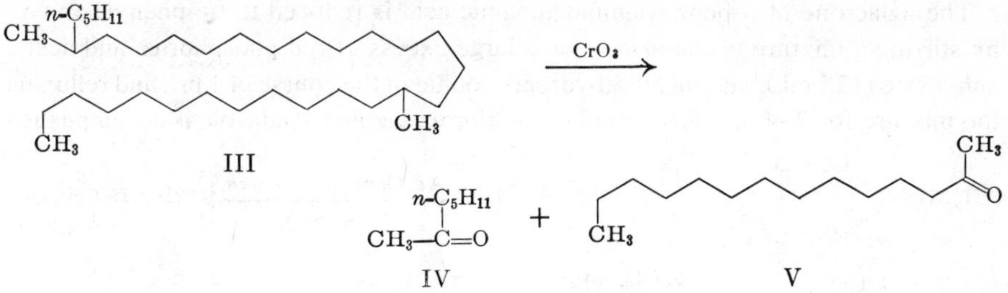
n-C5H11
CH3
CrO3
CH3
CH3
III
n-C5H11
CH3—C=O
+
CH3
CH3
O
IV
V

hydride and chromatography afforded a homogeneous hydrocarbon (III) in 13% overall yield. Analysis, molecular weight determination, and oxidation to IV and V established the structure. The method later was applied with success to the antibiotic rimocidin.[6]

Hell-Volhard-Zelinsky reaction. ε-Benzoylaminocaproic acid is converted into the α-bromo derivative by stirring a mixture of the acid with red phosphorus in a flask cooled in an ice bath, adding bromine dropwise, gradually increasing the temperature, and stirring on the steam bath until practical disappearance of bromine

$$\underset{0.64\ \text{m.}}{C_6H_5CONH(CH_2)_5CO_2H} + \underset{0.85\ \text{g. atom}}{P(\text{red})} + \underset{2.55\ \text{m.}}{Br_2} \longrightarrow C_6H_5CONH(CH_2)_4\underset{Br}{\underset{|}{C}}HCOBr$$

$$\xrightarrow[64\text{-}89\%\ \text{overall}]{H_2O} C_6H_5CONH(CH_2)_4\underset{Br}{\underset{|}{C}}HCO_2H$$

vapor.[7] The oily α-bromo acid bromide is stirred into water to effect hydrolysis to the solid α-bromo acid, which is collected, crushed, treated with SO_2 to remove excess bromine, and crystallized from ethanol.

In another example[8] bromine is added dropwise to a mixture of isobutyric acid and red phosphorus, and the mixture is warmed to 100° in the course of 6 hrs. Excess bromine and hydrogen bromide are removed by distillation at 30 mm., the α-bromo acid bromide is decanted from H_3PO_3 and distilled, and treated with zinc turnings

$$\underset{2.85\ \text{m.}}{(CH_3)_2CHCO_2H} + \underset{0.28\ \text{g. atom}}{P(\text{red})} + \underset{5.5\ \text{m.}}{Br_2} \xrightarrow[75\text{-}83\%]{} (CH_3)_2\underset{Br}{\underset{|}{C}}COBr$$

$$\xrightarrow[46\text{-}54\%]{Zn\text{-}EtOAc} (CH_3)_2C{=}C{=}O$$

and ethyl acetate. Distillation gives a 9–10% solution of dimethyl ketene in ethyl acetate.

Reduction. Efficient reduction of benzylic acid to diphenylacetic acid[9] is accomplished by allowing a mixture of the acid, red phosphorus, and iodine to stand until the iodine has reacted (15–20 min.), adding benzylic acid, refluxing for 2½ hrs.,

$$\underset{0.44\ \text{m.}}{(C_6H_5)_2C(OH)CO_2H} + \underset{0.48\ \text{g. atom}}{P(\text{red})} + \underset{0.0025\ \text{m.}}{I_2} \xrightarrow[94\text{-}97\%]{} (C_6H_5)_2CHCO_2H$$

filtering from phosphorus, and pouring the solution into a well-stirred solution of 20–25 g. of sodium bisulfite in 1 l. of water to remove excess iodine and precipitate the crude product, m.p. 141–144°.

The azlactone of α-benzoylaminocinnamic acid is reduced to DL-phenylalanine[10] by stirring a mixture of the azlactone, a large excess of red phosphorus, and acetic anhydride (125 ml.), adding 50% hydrogen iodide in the course of 1 hr., and refluxing the mixture for 3–4 hrs. Reduction of 3-chloroindazole to indazole is accomplished

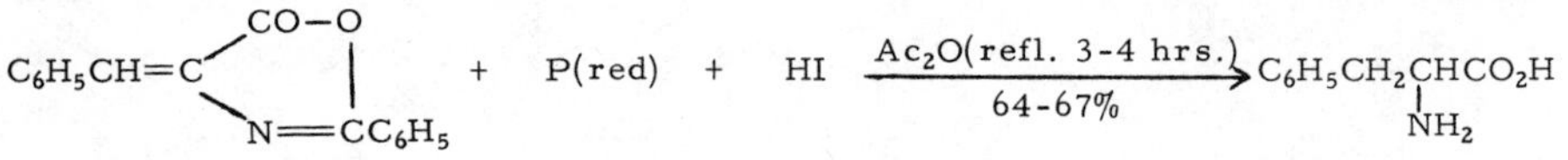

CO—O
$C_6H_5CH=C$
$N=CC_6H_5$
+ P(red) + HI
Ac_2O(refl. 3-4 hrs.)
64-67%
$C_6H_5CH_2CHCO_2H$
NH_2

0.1 m. 0.64 g. atom 0.76 m.

by refluxing a mixture of the compound with red phosphorus and 100 ml. of constant-boiling hydriodic acid for 24 hrs.; the solution is filtered through a sintered glass funnel to remove phosphorus.[11] The high yield obtained with 1.5 equivalents of

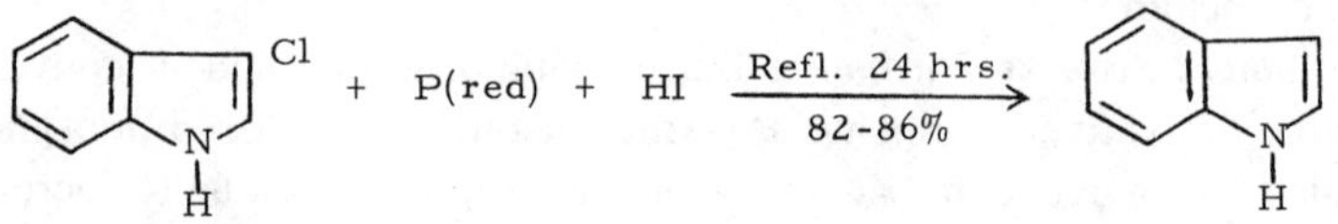

Cl
N
H
+ P(red) + HI
Refl. 24 hrs.
82-86%
N
H

0.1 m. 0.15 m. 0.88 m.

phosphorus and the fact that phosphorus was still present after the long reflux period suggests that the very large excess of this reagent specified in the preceding reduction may have been deleterious.

Demethylation. The last step in a synthesis of N-methyl-3,4-dihydroxyphenylalanine from vanillin and creatinine is demethylation of the phenolic monomethyl ether formulated.[12] A mixture of the ether, red phosphorus, 60 ml. of constant-boiling hydriodic acid, and 60 ml. of acetic anhydride is refluxed for 3 hrs.

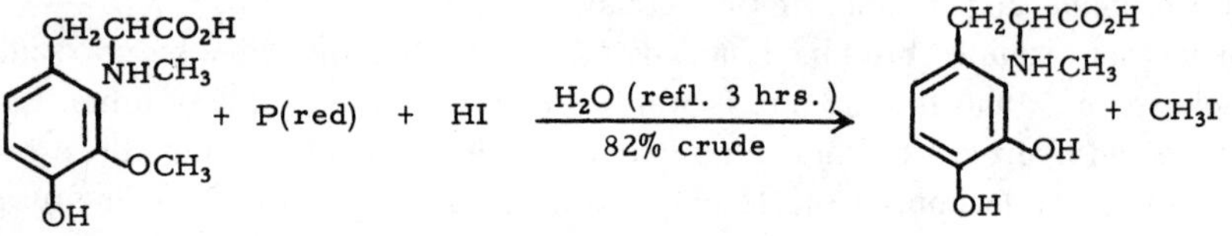

CH_2CHCO_2H
$NHCH_3$
OCH_3
OH
+ P(red) + HI
H_2O (refl. 3 hrs.)
82% crude
CH_2CHCO_2H
$NHCH_3$
OH
OH
+ CH_3I

0.5 m. 0.78 m. 0.52 m.

[1]R. H. Goshorn, T. Boyd, and E. F. Degering, *Org. Syn., Coll. Vol.*, **1**, 36 (1941)
[2]G. Braun, *ibid.*, **2**, 308 (1943)
[3]W. W. Hartman, J. R. Byers, and J. B. Dickey, *ibid.*, **2**, 322 (1943)
[4]H. S. King, *ibid.*, **2**, 399 (1943)
[5]A. C. Cope *et al.*, *Am. Soc.*, **84**, 2170 (1962)
[6]*Idem, ibid.*, **87**, 5452 (1965)
[7]J. C. Eck and C. S. Marvel, *Org. Syn., Coll. Vol.*, **2**, 74 (1943)
[8]C. W. Smith and D. G. Norton, *ibid.*, **4**, 348 (1963)
[9]C. S. Marvel, F. D. Hager, and E. C. Caudle, *ibid.*, **1**, 224 (1941)
[10]H. B. Gillespie and H. R. Snyder, *ibid.*, **2**, 489 (1943)
[11]E. F. M. Stephenson, *ibid.*, **3**, 475 (1955)
[12]V. Deulofeu and T. J. Guerrero, *ibid.*, **3**, 586 (1955)

Phosphorus diiodide, *see* Diphosphorus tetraiodide.

Phosphorus heptasulfide, P_4S_7. Mol. wt. 348.36, m.p. 305–310°, b.p. 523°. Suppliers: Oldbury Electrochemical Co., Eastman, Fisher.

The reagent is used for the synthesis of 3-methylthiophene from disodium methylsuccinate.[1] Rather elaborate provision is required for heating and insulating the equipment.

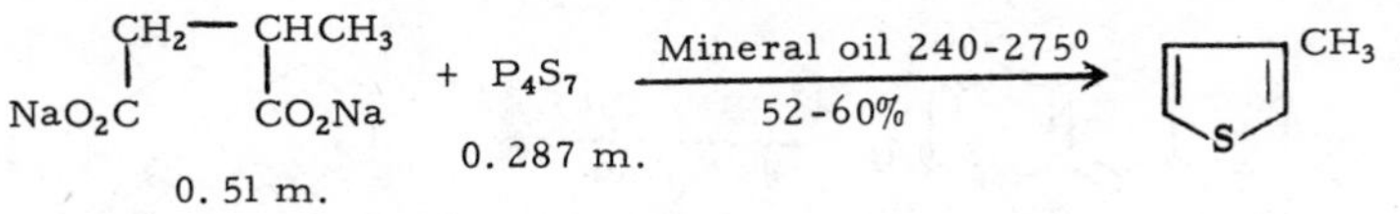

$$\underset{\underset{0.51 \text{ m.}}{\overset{NaO_2C \qquad CO_2Na}{}}}{\overset{CH_2 \!\!-\!\! CHCH_3}{|\qquad\quad|}} + P_4S_7 \xrightarrow[52-60\%]{\text{Mineral oil } 240-275^0} \underset{S}{\overset{CH_3}{\boxed{}}}$$

$$0.287 \text{ m.}$$

The "phosphorus trisulfide" used by Phillips[2] for similar reaction with disodium succinate for the preparation of thiophene in 25–30% yield was later shown to be somewhat impure phosphorus heptasulfide.[3] The reagent was prepared by placing an intimate mixture of finely powdered sulfur and red phosphorus in an earthenware flower pot embedded in sand, placed out-of-doors, dropping in a lighted match, and putting a heavily weighted cover in place. "The extremely vigorous reaction often gives an excellent display of fireworks."

[1]R. F. Feldkamp and B. F. Tullar, *Org. Syn., Coll. Vol.*, **4**, 671 (1963)
[2]R. Phillips, *ibid.*, **2**, 578 (1943)
[3]J. C. Pernert and J. H. Brown, *Chem. Eng. News*, **27**, 2143 (1949)

Phosphorus–Hydriodic acid; Zn–HCl–CH₃OH. An efficient method for the replacement of primary and secondary hydroxyl groups by hydrogen has proved useful in the investigation of hydroxy acid constituents of fats.[1] Thus an acid eventually characterized as 9,10,18-trihydroxyoctadecanoic acid (1) was refluxed with red phosphorus and hydriodic acid for 16 hrs. and the reduced product was recovered by ether extraction and refluxed with methanolic hydrochloric acid and zinc. The product was identified as methyl stearate (2).

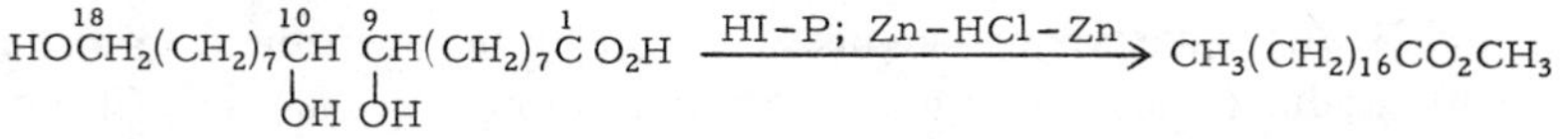

$$\underset{\underset{(1)}{\overset{\text{OH}\;\;\text{OH}}{|\quad|}}}{\overset{18}{\text{HOCH}_2}(\text{CH}_2)_7 \overset{10}{\text{CH}}\;\overset{9}{\text{CH}}(\text{CH}_2)_7 \overset{1}{\text{C}}\,\text{O}_2\text{H}} \xrightarrow{\text{HI–P; Zn–HCl–Zn}} \underset{(2)}{\text{CH}_3(\text{CH}_2)_{16}\text{CO}_2\text{CH}_3}$$

In another example[2] a new acid from *Sesquerella* seed oil eventually found to be (+)-14-hydroxy-*cis*-Δ¹¹-eicosenoic acid (3) was converted by hydrogenation and then reduction to a product identified as eicosanic acid (5).

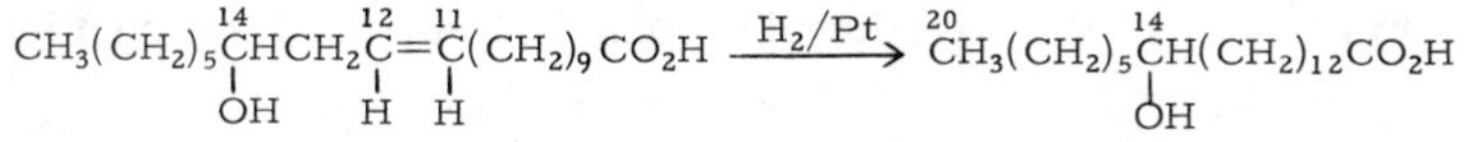

$$\underset{\underset{(3)}{\overset{\text{OH}\quad\;\text{H}\;\;\;\text{H}}{|\qquad\;|\quad\;|}}}{\text{CH}_3(\text{CH}_2)_5 \overset{14}{\text{CH}}\text{CH}_2\overset{12}{\text{C}}\!\!=\!\!\overset{11}{\text{C}}(\text{CH}_2)_9\,\text{CO}_2\text{H}} \xrightarrow{\text{H}_2/\text{Pt}} \underset{\underset{(4)}{\overset{\text{OH}}{|}}}{\overset{20}{\text{CH}}_3(\text{CH}_2)_5 \overset{14}{\text{CH}}(\text{CH}_2)_{12}\text{CO}_2\text{H}}$$

$$\xrightarrow{\text{HI–P; Zn–HCl–CH}_3\text{OH}} \underset{(5)}{\text{CH}_3(\text{CH}_2)_{18}\text{CO}_2\text{H}}$$

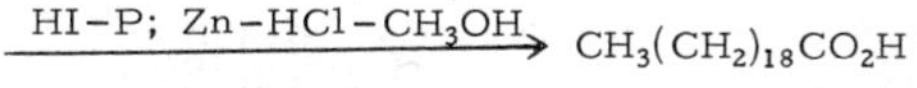

[1]G. D. Meakins and R. Swindells, *J. Chem. Soc.*, 1044 (1959)
[2]C. R. Smith, Jr., T. L. Wilson, T. K. Miwa, H. Zobel, R. L. Lohmar, and I. A. Wolff, *J. Org.*, **26**, 2903 (1961)

Phosphorus pentabromide, PBr₅. Mol. wt. 430.56, yellow.

Finely divided reagent can be prepared[1] by slowly adding the theoretical amount of PBr₃ to a cold, vigorously stirred solution of 240 g. of bromine in 600–1000 ml. of petroleum ether (b.p. 30–60°). The solvent was decanted and the phosphorus pentabromide washed several times with fresh solvent. Removal of the hydrocarbon in a vacuum desiccator gave 628 g. of reagent.

ArOH ⟶ *ArBr*. Kaslow and Marsh[1] mixed 5.5 g. of 2-hydroxy-6-methoxy-lepidine thoroughly with 18 g. of PBr₅ and heated the mixture at 70–80° and then at 120°. Hydrolysis with ice and water and workup afforded crude 2-bromo-6-methoxy-

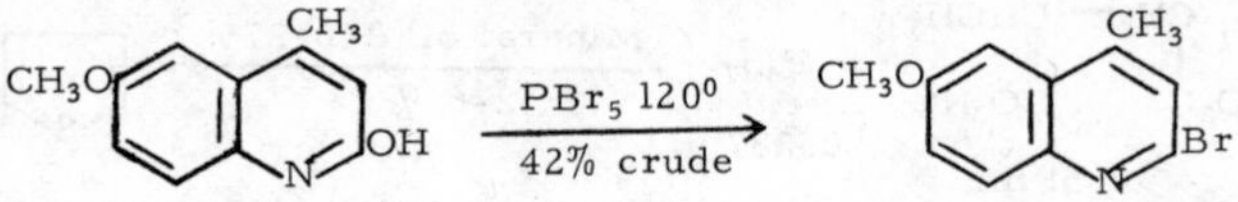
CH3
CH3O
OH
N
PBr5 120°
42% crude
CH3
CH3O
Br
N

lepidine, m.p. 137–142° (recrystallized 144–145°). 4-Hydroxyquinaldine similarly afforded 4-bromoquinaldine.

ROH⟶RBr. In the cyclohexane series reaction of diols with PBr_3 is attended with rearrangements. Thus treatment of the 1,3- and 1,4-cyclohexanediols with PBr_3 affords mixtures of *cis*- and *trans*-1,3- and 1,4-dibromides.[2] Eliel and Haber[3] found that *cis*-4-*t*-butylcyclohexanol reacts with PBr_5 to give *trans*-4-*t*-butylcyclohexyl bromide, together with a small amount of olefin and a mixture of dibromides. Thus PBr_5 is evidently superior to PBr_3 for conversion of secondary alcohols into bromides. However, the Hunsdiecker reaction seems to be the method of choice for preparation of cyclohexyl bromides.[3,4]

[1] C. E. Kaslow and M. M. Marsh, *J. Org.*, **12**, 456 (1947)
[2] B. Franzus and B. E. Hudson, Jr., *J. Org.*, **28**, 2238 (1963)
[3] E. L. Eliel and R. G. Haber, *J. Org.*, **24**, 143 (1959)
[4] E. N. Marvell and H. Sexton, *J. Org.*, **29**, 2919 (1964)

Phosphorus pentachloride, PCl_5. Mol. wt. 208.26, sublimes at 160°; slightly soluble in carbon disulfide.

The reagent can be prepared by passing chlorine into phosphorus trichloride to the proper gain in weight and distillation.[1]

Acid chlorides. Adams and Jenkins[1] converted *p*-nitrobenzoic acid to the acid chloride by gradually heating an equimolar mixture of the acid and PCl_5 until the

$$\underline{p}\text{-}NO_2C_6H_4CO_2H + PCl_5 \xrightarrow[90\text{-}96\%]{} \underline{p}\text{-}NO_2C_6H_4COCl + POCl_3 + HCl$$

3 m. 3 m.

evolution of hydrogen chloride abated, removing the phosphoryl chloride by distillation, and distilling the product at 155°/20mm. Feuer and Pier[2] used essentially the same procedure for the conversion of itaconic acid into the diacid chloride.

$$\begin{array}{c}CH_2{=}CCO_2H \\ \vert \\ CH_2CO_2H\end{array} + 2\ PCl_5 \xrightarrow[60\text{-}65\%]{} \begin{array}{c}CH_2{=}CCOCl \\ \vert \\ CH_2COCl\end{array}$$

0.5 m. 1.1m

Ether is used as solvent in a procedure by Braun and Cook[3] for the conversion of 2,3,4,5,6-penta-O-acetyl-D-gluconic acid into the acid chloride. The acid (25 g.) was shaken with 185 ml. of ether until most of it was dissolved, and phosphorus

$$CH_2(OAc)(CHOAc)_4CO_2H + PCl_5 \xrightarrow[80\text{-}92\%]{} CH_2(OAc)(CHOAc)_4COCl + POCl_3 + HCl$$

0.062 m. 0.072 m.

pentachloride was then added with shaking. The flask was closed with a calcium chloride tube and let stand overnight. Any solid was removed by filtration through a fritted glass funnel and the filtrate was concentrated to half-volume at reduced pressure, cooled at 0°, and a first crop collected. Concentration of the mother liquor afforded a second crop, bringing the total of acid chloride melting at 68–71° to 21–24 g. (80–92%).

Phthaloyl chloride is prepared by mixing phthalic anhydride and phosphorus pentachloride in a Claisen flask, heating at 150°, distilling the $POCl_3$ and then the product (which contains a little phthalic anhydride).[4]

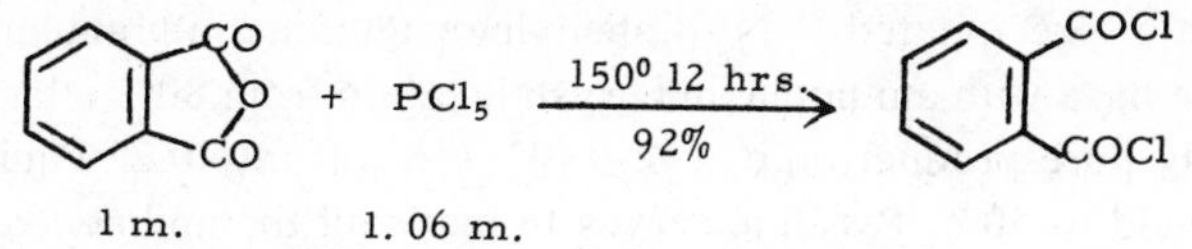
CO
O
CO
+ PCl5
150° 12 hrs.
92%
COCl
COCl

1 m. 1.06 m.

Sulfonyl chlorides. Adams and Marvel[5] prepared benzenesulfonyl chloride by heating sodium benzenesulfonate with PCl_5 in an oil bath at 170–180° for 15 hrs.; every 4 hrs. the flask was removed from the oil bath, stoppered, and shaken until the mass became pasty. The mixture was cooled, treated with ice and water, and

$$3\ C_6H_5SO_2ONa + PCl_5 \xrightarrow[75-80\%]{170-180°,\ 15\ hrs.} 3\ C_6H_5SO_2Cl + 2\ NaCl + NaPO_3$$

2.5 m. 1.2 m.

the sulfonyl chloride was separated and distilled at 145–150°/4 mm. In this case PCl_5 was taken in 44% excess of the theory; a note states that the yield was not improved by use of a still larger excess.

$RCONH_2 \longrightarrow RCN$. The reaction of cyanoacetamide (1.8 m.) with PCl_5 (0.7 m.)[6] appears to involve formation of $POCl_3$ in a first step (1) and its utilization in a second step (2). The solids were mixed in a mortar, and the mixture transferred

1. $CNCH_2CONH_2 + PCl_5 \longrightarrow CH_2(CN)_2 + POCl_3 + 2\ HCl$

2. $2\ CNCH_2CONH_2 + POCl_3 \longrightarrow 2\ CH_2(CN)_2 + HPO_3 + 3\ HCl$

quickly to a Claisen flask, which was evacuated at the water pump and heated cautiously. Distillation afforded $POCl_3$ and then malononitrile. Use of a larger excess of PCl_5 lowered the yield.

Taylor and Crovetti[7] found that nicotinamide-1-oxide reacts with a mixture of PCl_5 and $POCl_3$ to give 2-chloronicotinonitrile in moderate yield. The solids are mixed thoroughly, phosphoryl chloride is added slowly with stirring, and the

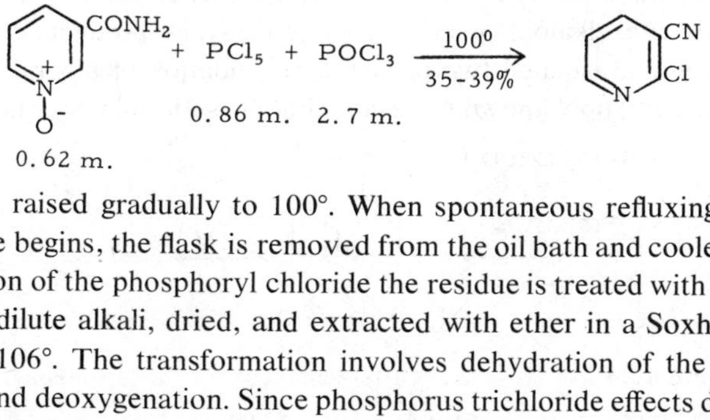
CONH2
N+
O-
+ PCl5 + POCl3
100°
35-39%
CN
Cl
N
0.86 m. 2.7 m.

0.62 m.

temperature is raised gradually to 100°. When spontaneous refluxing of the phosphoryl chloride begins, the flask is removed from the oil bath and cooled as required. After distillation of the phosphoryl chloride the residue is treated with ice and water, digested with dilute alkali, dried, and extracted with ether in a Soxhlet; the nitrile melts at 105–106°. The transformation involves dehydration of the amide group, chlorination, and deoxygenation. Since phosphorus trichloride effects deoxygenation of pyridine-1-oxides even at 0° (by formation of $POCl_3$), it is probably formed and utilized in the course of the reaction.

Imino chlorides. Hontz and Wagner[8] converted benzanilide into the imino chloride by heating a mixture with PCl_5 at 110° until evolution of hydrogen chloride had ceased (90 min.), slowly adding pyridine and then aniline, and heating at 160° until a red color was discharged (20 min.). The mixture was cooled to 90°, and

$$C_6H_5CONHC_6H_5 + PCl_5 \xrightarrow{110°} C_6H_5\overset{Cl}{\overset{|}{C}}{=}NC_6H_5 \xrightarrow[60-65\%\ overall]{0.46\ m.\ Py;\ 0.46\ m.\ C_6H_5NH_2} C_6H_5\overset{NHC_6H_5}{\overset{|}{C}}{=}NC_6H_5$$

0.46 m. 0.46 m.

addition of water precipitated N,N'-diphenylbenzamidine in granular form. Liberation of the free base with ammonia and crystallization from 80% ethanol (by weight) afforded nearly pure product, m.p. 142–144°. Use of only 0.25 equivalent of PCl_5 lowered the yield to 50%. Pyridine serves to make all the aniline available. It does not prevent combination of hydrogen chloride with the product, even when the amount is doubled. With pyridine present the mixture is an easily handled suspension; in the absence of pyridine the mixture solidifies to a cake.

For the Sonn-Müller synthesis of *o*-tolualdehyde, Williams *et al.*[9] prepared the imino chloride with benzene as solvent, distilled off the benzene and $POCl_3$, and dissolved the viscous residue in ether. A mixture of 0.26 mole of anhydrous stannous

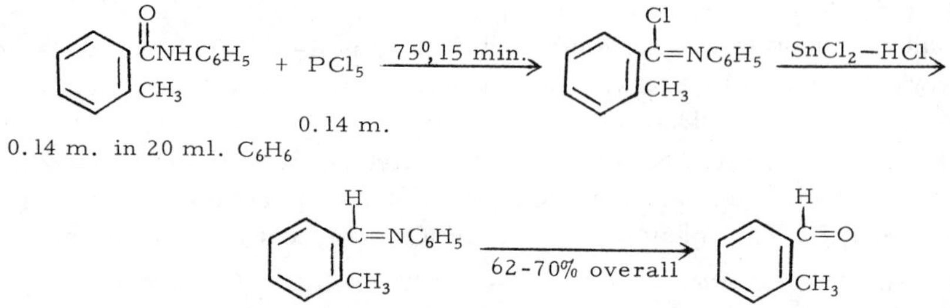
CNHC6H5
CH3
+ PCl5
0.14 m.
0.14 m. in 20 ml. C6H6
75°, 15 min.
Cl
C=NC6H5
CH3
SnCl2—HCl
H
C=NC6H5
CH3
62-70% overall
H
C=O
CH3

chloride and 225 ml. of ether was stirred under reflux while passing in hydrogen chloride until the solid was converted into a clear viscous lower layer. The source of hydrogen chloride was disconnected, and the ethereal solution of imino chloride was added. After stirring for ½ hr. the mixture was let stand for 12 hrs., treated with ice and water, and the aldehyde collected by steam distillation and ether extraction.

Protocatechualdehyde (3,4-dihydroxybenzaldehyde). The preparation of this aldehyde from piperonal was accomplished for the first time in 1871 at Tübingen by Rudolf Fittig, age 36, and his American co-worker Ira Remsen, age 25.[10] In earlier work on the alkaloid piperine and its cleavage products piperic acid and piperidine, Fittig had isolated two products of oxidation of piperic acid, piperonal and piperonylic acid, now known to be constituted as formulated. The aldehyde bore

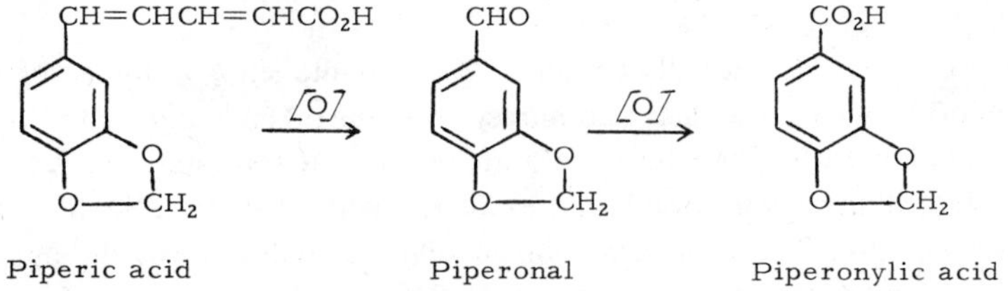
CH=CHCH=CHCO2H
[O]
CHO
[O]
CO2H
O
O—CH2
Piperic acid
Piperonal
Piperonylic acid

the usual relationship to the acid, but both substances, like piperic acid, contained two inert oxygen atoms which needed to be accounted for. For further characterization of piperonal, Fittig and Remsen treated the aldehyde with 1 equivalent of phosphorus pentachloride under mild conditions, distilled off the $POCl_3$ formed, and obtained a reaction product which distilled at 230–240° with some decomposition. Since the Claisen flask for vacuum distillation was not introduced until 1893, the product could not be obtained in analytically pure form, but it was nevertheless identified as piperonal chloride (2), for example by hydrolysis to piperonal (1). In an attempt to bring the two inert oxygen atoms into reaction, Fittig and Remsen explored the reaction of piperonal with 3 equivalents of PCl_5. The reaction seemed to stop with considerable PCl_5 still remaining, but gentle heating promoted further reaction and gave a liquid reaction mixture. Distillation removed PCl_5

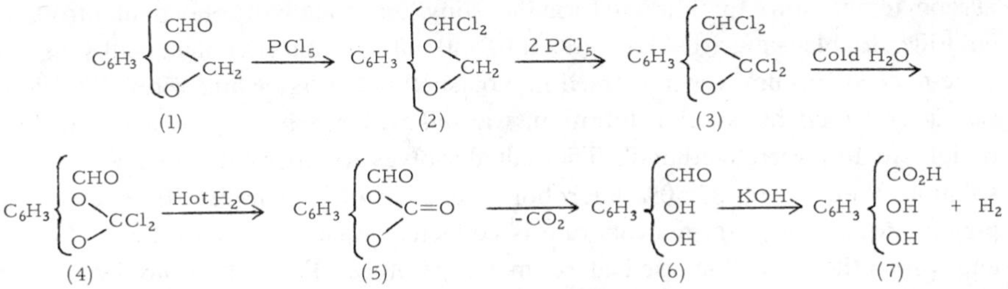

and then $POCl_3$ and gave a residue which boiled at about 280° but with considerable decomposition. However, the product was recognized as dichloropiperonal chloride (3) by its behavior on hydrolysis. On standing in contact with moist air, the product liberated hydrogen chloride and afforded a crystalline product, m.p. 90°, characterized as dichloropiperonal (4) hemihydrate. Hydrolysis with hot water gave a very unstable product recognized as the carbonate (5), which readily lost carbon dioxide to give a new aldehyde (6). Color tests and other properties suggested a relationship to protocatechuic acid, which had been obtained from various resins and other plant products by fusion with alkali. Attempts to oxidize the aldehydic function without attack of the two phenolic groups (*ortho*) at first failed, but smooth conversion of the aldehyde to the corresponding amine was accomplished by adding the substance to molten potassium hydroxide; a transient color soon disappeared, with evolution of hydrogen, and acidification of the cooled melt and extraction with ether afforded protocatechuic acid (7), as the hydrate, m.p. 199°.

An *Organic Syntheses* procedure[11] for the preparation of protocatechualdehyde is conducted in two steps: chlorination and hydrolysis. Phosphorus pentachloride is

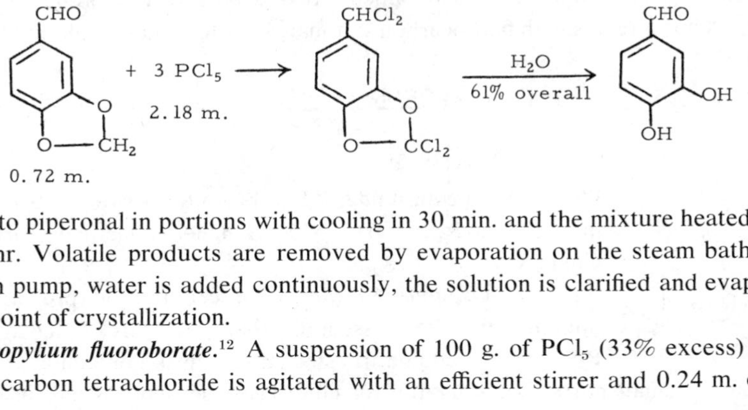

added to piperonal in portions with cooling in 30 min. and the mixture heated gently for 1 hr. Volatile products are removed by evaporation on the steam bath at the suction pump, water is added continuously, the solution is clarified and evaporated to he point of crystallization.

Tropylium fluoroborate.[12] A suspension of 100 g. of PCl_5 (33% excess) in 800 ml. of carbon tetrachloride is agitated with an efficient stirrer and 0.24 m. of 91%

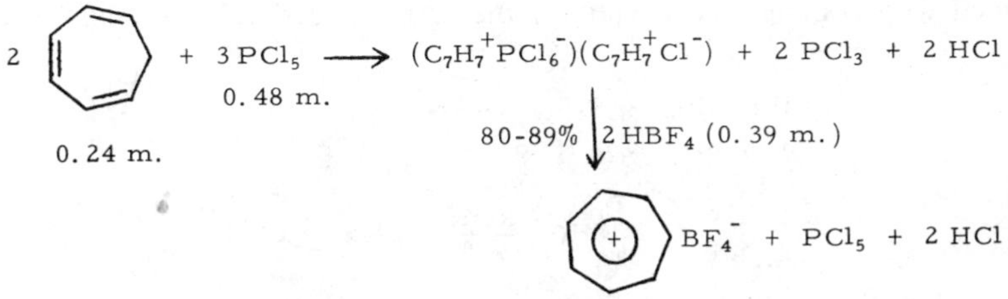

tropilidene (Shell Chemical Co., cycloheptatriene containing 9% of toluene) is added all at once. The mixture, which at first thickens and then thins again, is stirred at room temperature for 3 hrs. to form the tropylium hexachlorophosphate–tropylium chloride double salt. Absolute ethanol (400 ml.) is stirred magnetically in a 1-l. wide-necked Erlenmeyer immersed in a plastic bucket of ice and water. The double salt is collected by suction filtration, washed with fresh solvent, and transferred rapidly to the stirred ethanol. The salt dissolves exothermally to give a reddish solution; then 50 ml. of 50% fluoroboric acid is added. The resulting dense white precipitate of tropylium fluoroborate is collected, washed with a little cold ethanol and with ether, and air-dried at room temperature. The salt is nonhygroscopic, nonexplosive, and indefinitely stable.

Chlorination catalyst. Phosphorus pentachloride is employed as catalyst for the photochemical chlorination of *p*-chlorotoluene to *p*-chlorobenzal chloride.[13]

[1]R. Adams and R. L. Jenkins, *Org. Syn., Coll. Vol.*, **1**, 394 (1941)
[2]H. Feuer and S. M. Pier, *ibid.*, **4**, 554 (1963)
[3]C. E. Braun and C. D. Cook, *ibid.*, **41**, 79 (1961)
[4]E. Ott, *ibid.*, **2**, 528 (1943)
[5]R. Adams and C. S. Marvel, *ibid.*, **1**, 84 (1941)
[6]B. B. Corson, R. W. Scott, and C. E. Vose, *ibid.*, **2**, 379 (1943)
[7]E. C. Taylor, Jr., and A. J. Crovetti, *ibid.*, **4**, 166 (1963)
[8]A. C. Hontz and E. C. Wagner, *ibid.*, **4**, 383 (1963)
[9]J. W. Williams, C. H. Witten, and J. A. Krynitsky, *ibid.*, **3**, 818 (1955)
[10]R. Fittig and I. Remsen, *Ann.*, **159**, 144 (1871)
[11]J. S. Buck and F. J. Zimmermann, *Org. Syn., Coll. Vol.*, **2**, 549 (1943)
[12]K. Conrow, *Org. Syn.*, **43**, 101 (1963)
[13]W. L. McEwen, *Org. Syn., Coll. Vol.*, **2**, 133 (1943)

Phosphorus pentachloride–Zinc chloride, $PCl_5 \cdot 2\,ZnCl_2$, m.p. 190°.

The complex is prepared by adding PCl_5 to a well-stirred melt of zinc chloride until the desired composition is obtained.[1] It is a powerful chlorodeoxygenation reagent. Thus it reacts with fluorocarbon sulfonate salts to produce sulfonyl chlorides in high yield.

$$(CF_3SO_2O)_2Zn \xrightarrow[94\%]{PCl_5\cdot 2\,ZnCl_2} 2\,CF_3SO_2Cl$$

[1]G. Van Dyke Tiers, *J. Org.*, **28**, 1244 (1963)

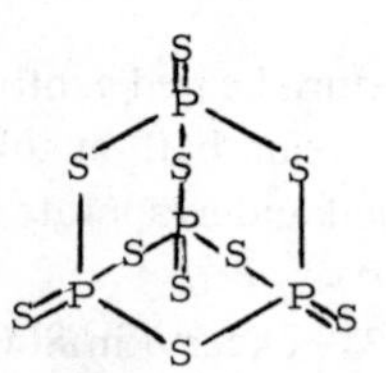

Phosphorus pentasulfide, P_4S_{10} (but often written P_2S_5). Yellow, m.p. 286–292°, b.p. 513–515°. Supplier: American Agricultural Chem. Co.

The *Organic Syntheses* procedure for preparing 2,4-dimethylthiazole[1] is essentially that of E. Merck.[2] Acetamide and P_4S_{10} are mixed and transferred to a flask containing 200 ml. of benzene. A small portion of a mixture of chloroacetone and 150 ml. of benzene is added with cooling, and eventually the rest is added and the mixture refluxed.

$$\underset{5.08\text{ m.}}{2\,CH_3CONH_2} + \underset{0.45\text{ m.}}{P_4S_{10}} \longrightarrow 2\,CH_3\overset{S}{\overset{\|}{C}}NH_2$$

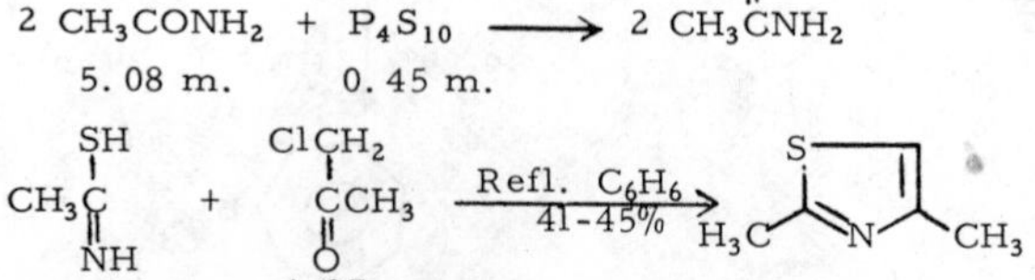
SH
CH3C
NH
ClCH2
CCH3
O
4.97 m.
Refl. C6H6
41-45%
S
H3C
N
CH3

[1]G. Schwarz, *Org. Syn., Coll. Vol.*, **3**, 332 (1955)
[2]E. Merck, Ger. patent 670,131 [*C.A.*, **33**, 2909 (1939)]

Phosphorus pentoxide, P_2O_5. Mol. wt. 141.95, subl. 250°.

RCONH₂ ⟶ RC≡N. In theory the dehydration of an amide should require only 0.33 mole of phosphorus pentoxide, but procedures for the dehydration of isobutyramide,[1] nicotinamide,[2] and fumaramide[3] all call for use of a large excess of reagent.

$$\underset{2\text{ m.}}{(CH_3)_2CHCONH_2} + \underset{2.1\text{ m.}}{P_2O_5} \xrightarrow[69-86\%]{200-220^\circ\ \ 8-10\text{ hrs.}} (CH_3)_2CHCN$$

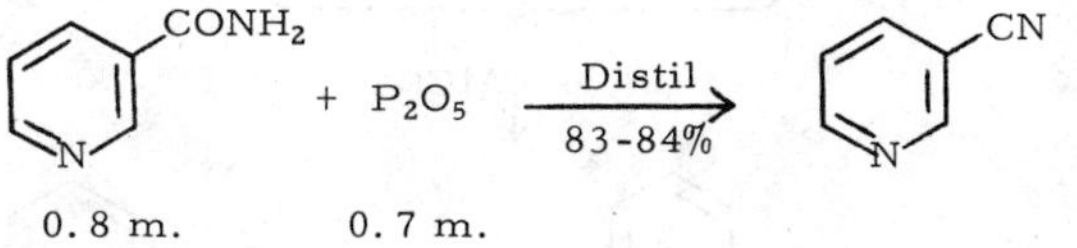

$$\underset{2\text{ m.}}{H_2NCOCH{=}CHCONH_2} + \underset{4.3\text{ m.}}{P_2O_5} \xrightarrow[75-80\%]{\text{Heat, distil}} NCCH{=}CHCN$$

The reactants are mixed by shaking, the mixture heated, and the product distilled at atmospheric pressure or at 20 mm. Dehydration of methyl sebacamate is done in tetrachloroethane,[4] and dehydration of chloroacetamide is carried out at the boiling

$$\underset{0.88\text{ m.}}{CH_3O_2C(CH_2)_8CONH_2} + \underset{1.34\text{ m.}}{P_2O_5} \xrightarrow[69-71\%]{CHCl_2CHCl_2\ 120-145^\circ} CH_3O_2C(CH_2)_8CN$$

$$\underset{2\text{ m.}}{ClCH_2CONH_2} + \underset{1.2\text{ m.}}{P_2O_5} \xrightarrow[62-70\%]{C_6H_3(CH_3)_3\ 166-174^\circ} 3\ ClCH_2CN$$

point of technical trimethylbenzene;[5] in each case the product is then distilled initially with part or all of the solvent.

Rearrangement catalyst. One method of preparing 1-acetylcyclohexene[6] involves refluxing a solution of 1-ethynylcyclohexanol in benzene with P_2O_5 for 2.5 hrs.

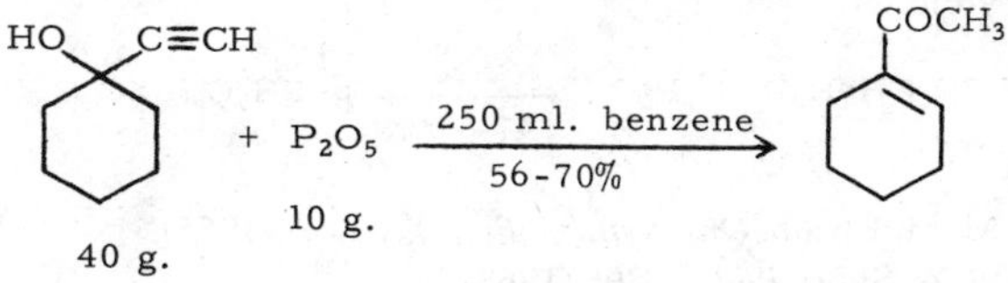

The solution is decanted, washed with sodium bicarbonate solution, dried, and evaporated. The product distills at 85–88°/22 mm.

Cyclodehydration. Attempts to cyclize the imide (1) with phosphoryl chloride, the usual reagent for Bischler-Napieralski ring closure, gave back only starting material, but the lactam (2) was obtained by refluxing (1) with P_2O_5 in xylene.[7]

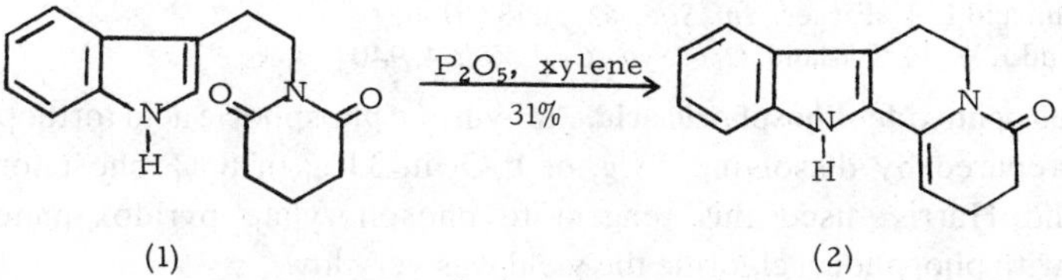

(1) (2)

J. W. Cook[8] introduced the method of cyclizing an *o*-aroylbenzoic acid to an anthraquinone by heating the keto acid with phosphorus pentoxide in nitrobenzene at 150–165°. For example, the isomeric acids (3) and (4) resulting from the Friedel-Crafts reaction of 1,2-naphthalic anhydride with thiophene are cyclized to the same quinone (5).[9]

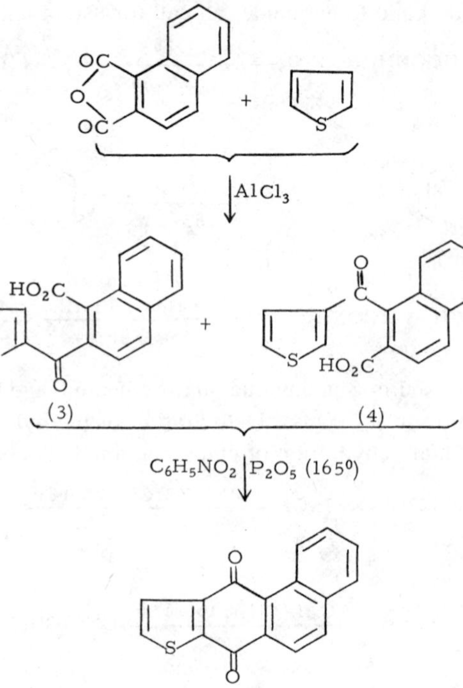
AlCl3
(3)
(4)
C6H5NO2 P2O5 (165°)

(5)

p-Toluenesulfonic anhydride. A procedure[10] for preparing this anhydride involves introduction of kieselguhr, asbestos, and glass, and gives material of wide melting point in variable yield.

$$\underset{1\ m.}{2\ \underline{p}\text{-}CH_3C_6H_4SO_2OH\cdot H_2O} + \underset{1.5\ m.}{P_2O_5} \xrightarrow[47\text{-}70\%]{} \underline{p}\text{-}CH_3C_6H_4SO_2OSO_2C_6H_4CH_3\text{-}\underline{p}$$

[1]R. E. Kent and S. M. McElvain, *Org. Syn., Coll. Vol.*, **3**, 493 (1955)
[2]P. C. Teague and W. A. Short, *ibid.*, **4**, 706 (1963)
[3]R. T. Bertz, *ibid.*, **4**, 489 (1963)
[4]W. S. Bishop, *ibid.*, **3**, 584 (1955)
[5]D. B. Reisner and E. C. Horning, *ibid.*, **4**, 144 (1963)
[6]J. H. Saunders, *ibid.*, **3**, 22 (1955)
[7]G. C. Morrison, W. Cetenko, and J. Shavel, Jr., *J. Org.* **29**, 2771 (1964)
[8]J. W. Cook, *J. Chem. Soc.*, 1472 (1932)
[9]R. B. Sandin and L. F. Fieser, *Am. Soc.*, **62**, 3098 (1940)
[10]L. Field and J. W. McFarland, *Org. Syn., Coll. Vol.*, **4**, 940 (1963)

Phosphorus pentoxide–Phosphoric acid. Anhydrous phosphoric acid (orthophosphoric acid) is prepared by dissolving 25 g. of P_2O_5 in 33 g. of 85% phosphoric acid.[1,2] Wilson and Harris[2] used this reagent to phosphorylate pyridoxamine dihydrochloride; with phosphoryl chloride the yield was very low.

Stone and Shechter have developed high-yield procedures for the preparation of alkyl iodides from alcohols, ethers, and olefins by reaction with a reagent described as 95% orthophosphoric acid in combination with potassium iodide. For the conversion of 1,6-hexanediol into 1,6-diiodohexane,[3] a mixture of 65 g. of P_2O_5 and 231 g. (135 ml.) of 85% phosphoric acid is stirred mechanically and let cool to room temperature, potassium iodide and 1,6-hexanediol are added, and the mixture is stirred and heated as indicated. The initially homogeneous solution separates into

$$\underset{0.5\text{ m.}}{HO(CH_2)_6OH} + \underset{2\text{ m.}}{2\,KI} + \underset{2\text{ m.}}{85\%\ H_3PO_4} + \underset{0.46\text{ m.}}{P_2O_5} \xrightarrow[83\text{-}85\%]{100\text{-}120^\circ\ 3\text{-}5\text{ hrs.}} I(CH_2)_6I$$

two phases, and finally a dense oil settles through the acid layer. Workup and distillation affords product of m.p. 8–10°.

By similar procedures tetrahydrofurane affords 1,4-diiodobutane,[4] and cyclohexene is converted into cyclohexyl iodide.[5]

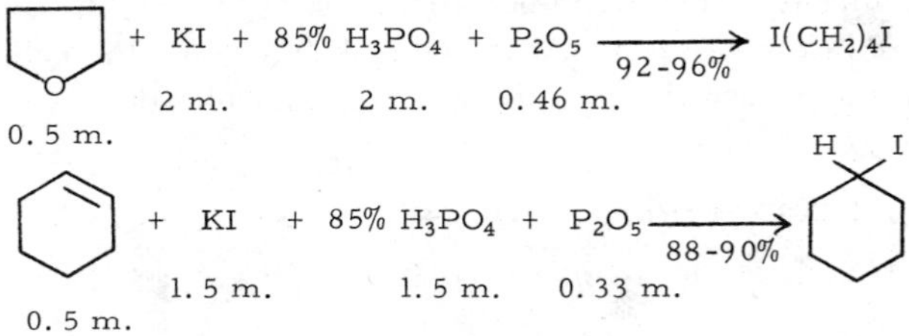

[1]R. E. Ferrel, H. S. Olcott, and H. Fraenkel-Conrat, *Am. Soc.*, **70**, 2101 (1948)
[2]A. N. Wilson and S. A. Harris, *Am. Soc.*, **73**, 4693 (1951)
[3]H. Stone and H. Shechter, *Org. Syn., Coll. Vol.*, **4**, 323 (1963)
[4]*Idem, ibid.*, **4**, 321 (1963)
[5]*Idem, ibid.*, **4**, 543 (1963)

Phosphorus triamide, hexamethyl-, *see* Hexamethylphosphorus triamide.

Phosphorus tribromide, PBr_3. Mol. wt. 270.72, b.p. 175° (168–170°/725 mm.), sp. gr. 2.852. Prepared in 90–95% yield by adding bromine to a stirred suspension of red phosphorus in carbon tetrachloride and fractionating.[1] Supplier: Dow Chem. Co.

ROH⟶RBr. Procedures[1-4] for the conversion of primary alcohols into the bromides vary considerably in reaction conditions, but the yields are all in the

$$\underset{7\text{ m.}}{(CH_3)_2CHCH_2OH} + \underset{2.56\text{ m.}}{PBr_3} \xrightarrow[55\text{-}60\%]{-10^\circ\text{ to }0^\circ,\ 4\text{ hrs.}} (CH_3)_2CHCH_2Br + P(OH)_3$$

$$\underset{0.92\text{ m.}}{3\,C(CH_2OH)_4} + \underset{1.85\text{ m.}}{4\,PBr_3} \xrightarrow[59\text{-}65\%]{170\text{-}180^\circ,\ 20\text{ hrs.}} 3\,C(CH_2Br)_4 + 4\,P(OH)_3$$

$$\underset{7\text{ m.}}{3\,C_2H_5OCH_2CH_2OH} + \underset{2.2\text{ m.}}{PBr_3} \xrightarrow[65\text{-}66\%]{\text{Refl.}} 3\,C_2H_5OCH_2CH_2Br + P(OH)_3$$

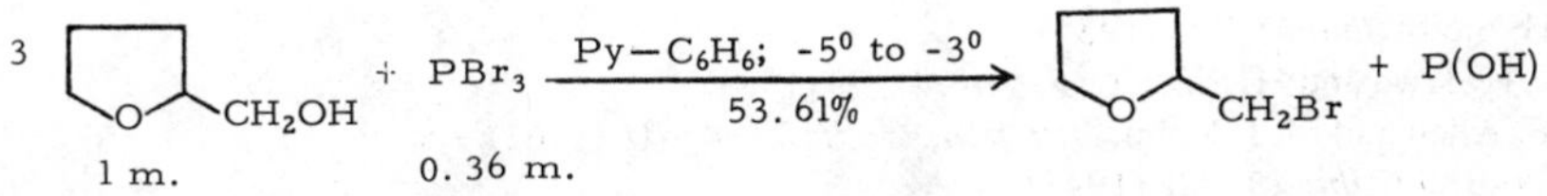

range 53–66%. In the first three procedures formulated, phosphorus tribromide was added gradually to the alcohol without solvent; in the fourth the reaction mixture contained pyridine and benzene.

The allylic secondary alcohol 1,5-hexadiene-3-ol reacts with phosphorus tribromide plus 2 drops of 48% hydrobromic acid to give in high yield a mixture of

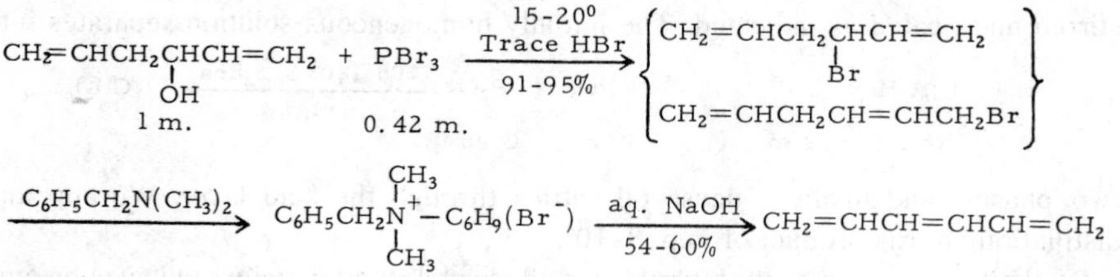

3-bromo-1,5-hexadiene and 1-bromo-2,5-hexadiene.[5] Quaternization of the mixture and treatment with base affords 1,3,5-hexatriene.

Hell-Volhard-Zelinsky reaction. Allen and Kalm[6] converted 2-methyldodecanoic acid into the α-bromo acid bromide by reaction with phosphorus tribromide and

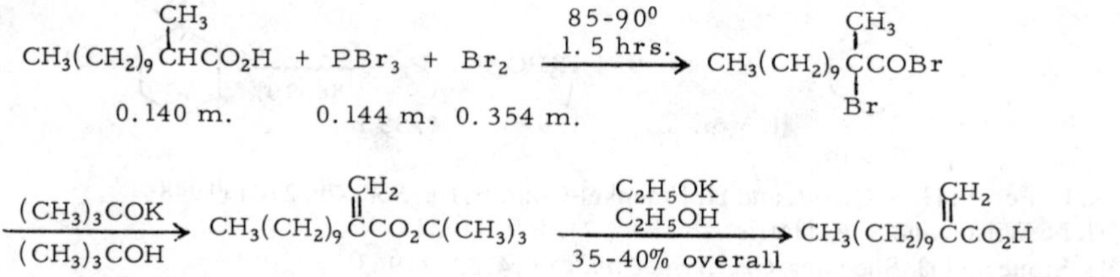

bromine. The bromine was added in portions, since the first mole reacts with PBr_3 to form PBr_5, which is rapidly consumed. The bromo acid bromide was converted into 2-methylenedodecanoic acid by the method formulated.

von Braun reaction.[7] von Braun's procedure for the preparation of pentamethylene dibromide also involves *in situ* formation of phosphorus pentabromide.

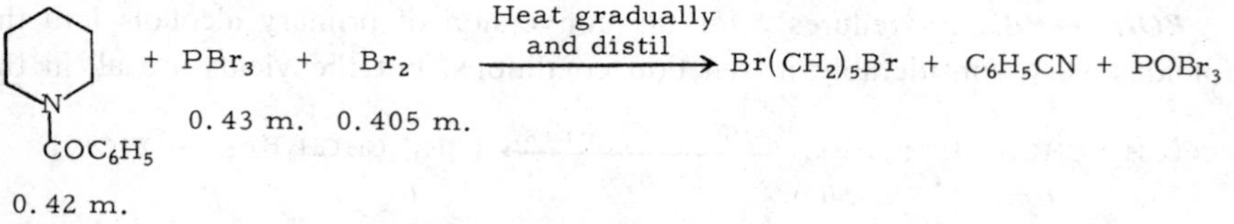

Bromine is added with cooling to a mixture of N-benzoylpiperidine and phosphorus tribromide, the temperature is gradually raised, at reduced pressure, and a mixture of pentamethylene dibromide and $POBr_3$ and benzonitrile is distilled at 20 mm. The organic layer is refluxed with hydrobromic acid to hydrolyze the benzonitrile, and the dibromide is separated by steam distillation, collected by ether extraction, and distilled.

[1]C. R. Noller and R. Dinsmore, *Org. Syn., Coll. Vol.*, **2**, 358 (1943)
[2]H. B. Schurink, *ibid.*, **2**, 476 (1943)
[3]G. C. Harrison and H. Diehl, *ibid.*, **3**, 370 (1955)
[4]L. H. Smith, *ibid.*, **3**, 793 (1955)
[5]J. C. H. Hwa and H. Sims, *Org. Syn.*, **41**, 49 (1961)
[6]C. F. Allen and M. J. Kalm, *Org. Syn., Coll. Vol.*, **4**, 616 (1963)
[7]J. von Braun, *ibid.*, **1**, 428 (1941)

Phosphorus trichloride, Mol. wt. 137.35, b.p. 76°, sp. gr. 1.57.

α-Bromination of acids. In a typical procedure[1] a mixture of caproic acid and bromine is treated with a catalytic amount of phosphorus trichloride and heated at 65–70° for 5–6 hrs.

$$\underset{1.72\text{ m.}}{CH_3(CH_2)_4CO_2H} + \underset{1.88\text{ m.}}{Br_2} \xrightarrow[83\text{-}89\%]{3\text{ ml. }PCl_3\ (65\text{-}70^0)} CH_3(CH_2)_3\underset{Br}{CH}CO_2H + HBr$$

Preparation of acid chlorides. A procedure for the preparation of desoxybenzoin[2] calls for heating a mixture of phenylacetic acid and PCl_3 on the steam bath for 1 hr., decanting the acid chloride and dissolving it in benzene, adding aluminum chloride, and refluxing the mixture for 1 hr. on the steam bath.

$$\underset{0.5\text{ m.}}{C_6H_5CH_2CO_2H} + \underset{0.25\text{ m.}}{PCl_3} \xrightarrow[-POCl_3]{90^0\ 1\text{ hr.}} C_6H_5CH_2COCl \xrightarrow[82\text{-}83\%\text{ overall}]{400\text{ ml. }C_6H_6;\ 0.56\text{ m. }AlCl_3;\ H_2O} C_6H_5CH_2COC_6H_5$$

Phosphorus compounds. Triethyl phosphite is prepared[3] by stirring a solution of N,N-diethylaniline and absolute ethanol in 1 l. of petroleum ether under reflux and adding a solution of PCl_3 in 400 ml. of petroleum ether in 30 min. After refluxing for 1 hr., the mixture is cooled, the N,N-diethylaniline hydrochloride is removed by filtration, the solvent is removed, and the product is distilled at 43–44°/10 mm.

$$\underset{1\text{ m.}}{PCl_3} + \underset{3\text{ m.}}{3\,C_2H_5OH} + \underset{3\text{ m.}}{3\,C_6H_5N(C_2H_5)_2} \xrightarrow[83\%]{\text{Refl. in PE}} (C_2H_5O)_3P + 3\,C_6H_5N(C_2H_5)_2\cdot HC$$

Diethylaniline is preferred to dimethylaniline or pyridine because the salt is nonhygroscopic and easily filtered.

Phenyldichlorophosphine (2) is prepared[4] by gradually heating a mixture of benzene, phosphorus trichloride, and aluminum chloride until a yellow solution of the complex (1) results (2 hrs.) and then refluxing for 1 hr. The heat source is

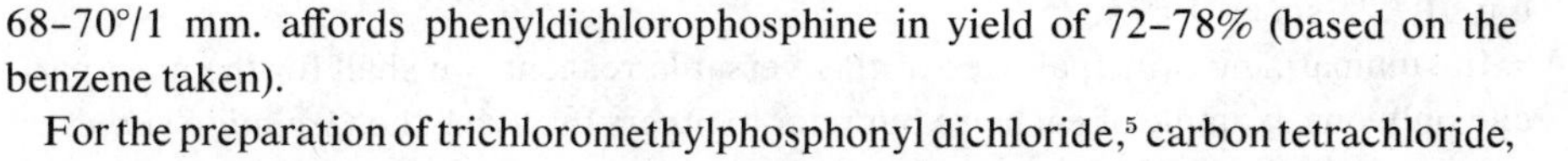

+ PCl3 + AlCl3
To refl.
PCl2 AlCl3
POCl3 (0.4 m.)
PCl2
+ AlCl3·POCl3

0.3 m. 1.2 m. 0.4 m. (1) (2) (3)

removed, and phosphoryl chloride is added to form the complex (3) and liberate the product (2). The complex (3) separates as a granular solid and is left as a residue on extraction with petroleum ether. Evaporation of the extract and distillation at 68–70°/1 mm. affords phenyldichlorophosphine in yield of 72–78% (based on the benzene taken).

For the preparation of trichloromethylphosphonyl dichloride,[5] carbon tetrachloride, aluminum chloride, and phosphorus trichloride are stirred slowly until thoroughly mixed, and heat is applied carefully until the reaction begins. The liquid boils vigorously and becomes thicker and finally the mixture sets to a solid. After cooling, the product is extracted with methylene chloride and obtained as a white crystalline solid, m.p. 155–156°, in 81–84% yield.

$$\underset{1.2\text{ m.}}{CCl_4} + \underset{1\text{ m.}}{AlCl_3} + \underset{1\text{ m.}}{PCl_3} \longrightarrow [Cl_3CPCl_3]^+[AlCl_4]^-$$

$$\xrightarrow{7\,H_2O} Cl_3C-\overset{O^-}{\underset{Cl}{P^+}}-Cl + AlCl_3\cdot 6\,H_2O + 2\,HCl$$

Peptide synthesis. Süs[6] has suggested use of the reagent in peptide synthesis, but the examples cited are not impressive.

Conjugated enones ⟶ chlorodienes. Ross and Martz[7] found that addition of 1.5 ml. of phosphorus trichloride to a solution of 3.53 g. of Δ^4-cholestene-3-one in 20–25 ml. of acetic acid produced a bright yellow solution from which 3-chloro-$\Delta^{3,5}$-cholestadiene began to separate in about 1 hr. The yield reported is for material

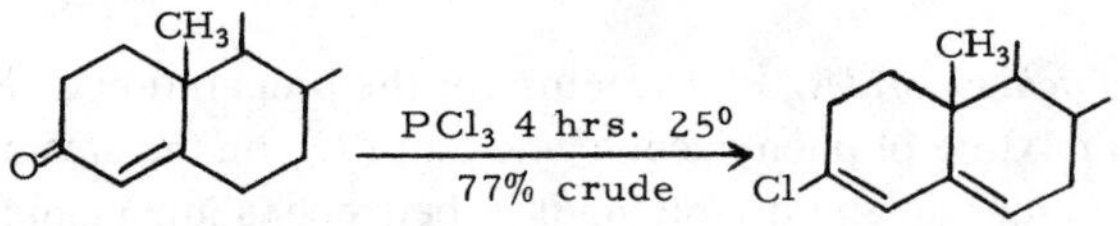

CH3
PCl3 4 hrs. 25°
77% crude
CH3
Cl

melting at 63–65°; crystallization from ether–95% ethanol gave pure material, m.p. 65–66°.

Deoxygenation of amine oxides. Since transformations sometimes can be accomplished more satisfactorily with a pyridine-1-oxide than with the pyridine itself, an efficient method of deoxygenation with PCl_3 introduced by Ochiai[8] has found wide usage. For example, 1 g. of 4-nitropyridine-1-oxide was suspended in 15 ml. of cold

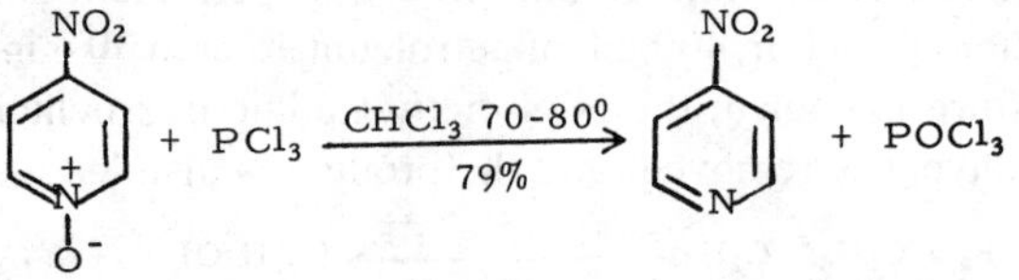

NO2
+ PCl3
CHCl3 70-80°
79%
NO2
+ POCl3

chloroform, 1.9 ml. of PCl_3 was added, and the mixture was heated for 1 hr. at 70–80°. The mixture was cooled, diluted with water, made alkaline with sodium hydroxide, and the 4-nitropyridine extracted with chloroform. The deoxygenation of 4-nitro-3-picoline-1-oxide was effected in 65% yield at a temperature below 10°.[9]

[1]H. T. Clarke and E. R. Taylor, *Org. Syn., Coll. Vol.*, **1**, 115 (1941)
[2]C. F. H. Allen and W. E. Barker, *ibid.*, **2**, 156 (1943)
[3]A. H. Ford-Moore and B. J. Perry, *ibid.*, **4**, 955 (1963)
[4]B. Buchner and L. B. Lockhart, Jr., *ibid.*, **4**, 784 (1963)
[5]K. C. Kennard and C. S. Hamilton, *ibid.*, **4**, 950 (1963)
[6]O. Süs, *Ann.*, **572**, 96 (1951)
[7]J. A. Ross and M. D. Martz, *J. Org.*, **29**, 2784 (1964)
[8]E. Ochiai, *J. Org.*, **18**, 534 (1953)
[9]W. Herz and L. Tsai, *Am. Soc.*, **76**, 4184 (1954)

Phosphoryl chloride (Phosphorus oxychloride), $POCl_3$. Mol. wt. 153.35, m.p. 1.2°, b.p. 107.2°, sp. gr. 1.675.

In summarizing principal uses of this versatile reagent, we shall for the most part cite only one example of each use but refer to others by giving the *Organic Syntheses* volume or *Collective Volume* number and page.

Phosphate esters. Phosphoryl chloride is the reagent of choice for the preparation of symmetrical trialkyl phosphates, for example tri-*n*-butyl phosphate.[1] After refluxing for 2 hrs. and cooling, water is added to dissolve the pyridine hydrochloride.

$$3\ \underline{n}\text{-}C_4H_9OH + POCl_3 + 3\ Py \xrightarrow[71-75\%]{\text{Benzene (refl.)}} O{=}P(OC_4H_9\text{-}\underline{n})_3 + 3\ Py^+HCl^-$$

Forrest and Todd[2] developed a procedure for treatment of riboflavin with phosphoryl chloride in pyridine containing a small amount of water to form the cyclic riboflavin-4′,5′-phosphate (2), which on acid hydrolysis yields riboflavin-5-phosphate (3), identical with the natural coenzyme.

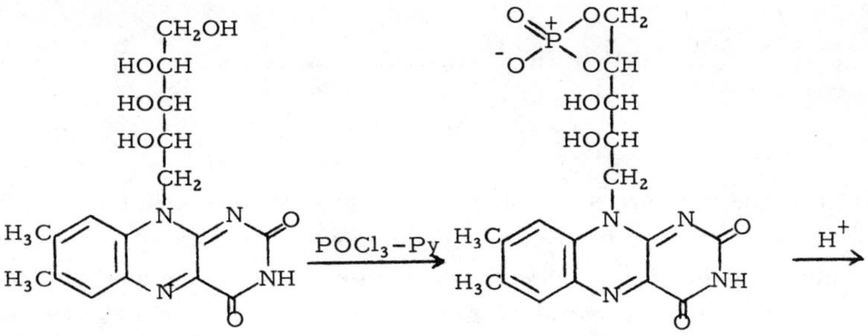

(1) (2)

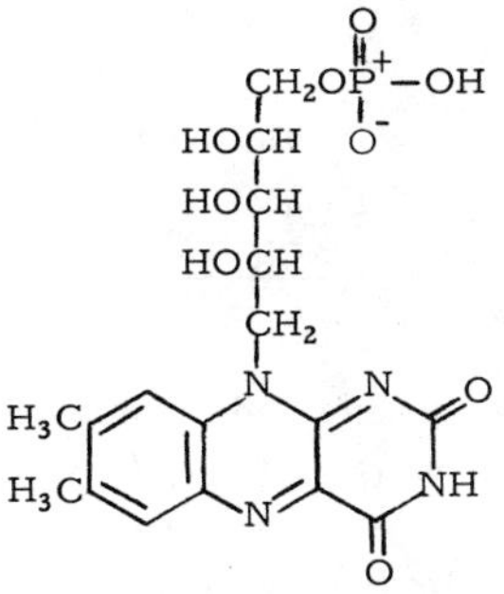

(3)

The same method was employed by Baddiley and Thain[3] for the synthesis of pantetheine-2′,4′-hydrogen phosphate (4).

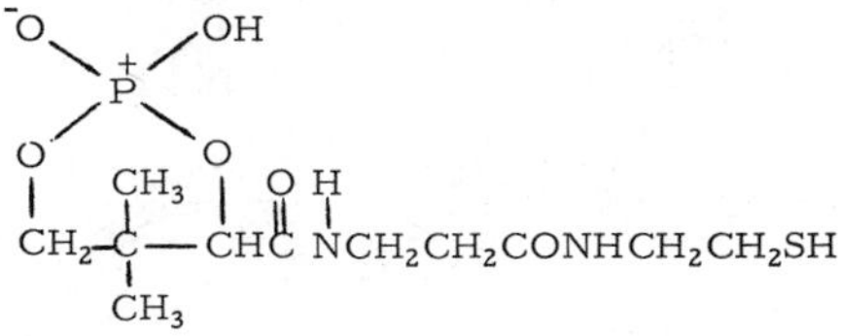

(4)

3β-Hydroxysteroids and 17β-hydroxysteroids can be directly phosphorylated to monoesters with phosphoryl chloride in good yield.[4] The intermediate dichloride from androstenolone has been characterized.

Phenyl esters. Preparation of diphenyl succinate.[5] The use of more than 0.9 mole of $POCl_3$ did not improve the yield and gave an inferior product. The mixture

$$\underset{1\text{ m.}}{\begin{array}{l}CH_2CO_2H\\ |\\ CH_2CO_2H\end{array}} + \underset{1\text{ m.}}{2\,C_6H_5OH} \xrightarrow[62\text{-}67\%]{POCl_3(0.9\text{ m.})} \begin{array}{l}CH_2CO_2C_6H_5\\ |\\ CH_2CO_2C_6H_5\end{array} + HPO_3 + 3HCl$$

is heated on the steam bath for 1.25 hrs., benzene is added, refluxing continued for 1 hr., and the hot benzene solution is decanted from the red sirupy residue and filtered by gravity. After extraction of the residue and concentration, diphenyl succinate separates as colorless crystals.

Another example: *O. S.*, **4**, 178.

Condensation with an N-formyl derivative to give an aldehyde. Preparation of 9-anthraldehyde.[6] On heating a mixture of the reactants on the steam bath with stirring, the anthracene dissolves to give a deep red solution and HCl is evolved.

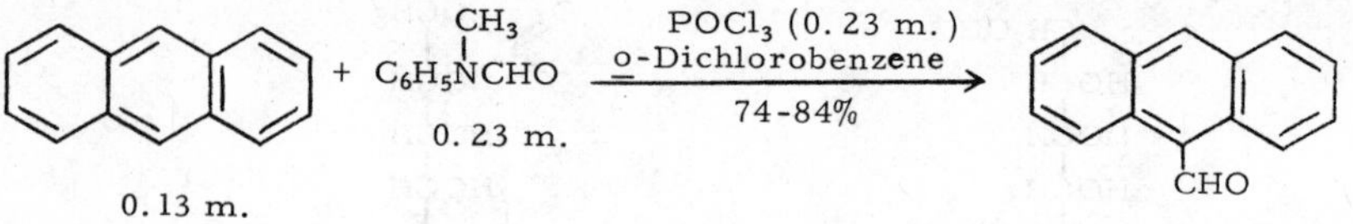

After $1\frac{1}{2}$ hrs. the mixture is cooled and treated with a solution of 140 g. of $NaOAc \cdot 3H_2O$ in 250 ml. of water to decompose a product of condensation of N-methylaniline with $POCl_3$. Steam distillation removes N-methylaniline and *o*-dichlorobenzene, and the residual reddish oil solidifies and on crystallization affords bright yellow aldehyde, m.p. 104.5–105°.

Other examples: pyrrole, *O. S.*, **4**, 831; thiophene, **4**, 915; indole **4**, 539; N,N-dimethylaniline, **4**, 331.

Amide ⟶ nitrile. Preparation of malononitrile.[7] The sodium chloride gives a

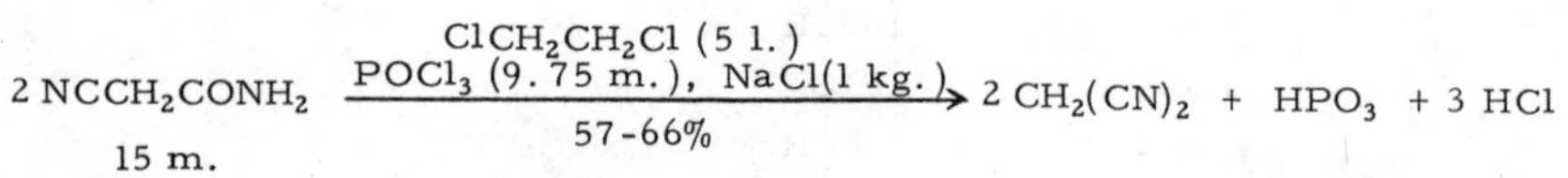

lighter-colored, granular solid easily removed by filtration and washed. The mixture is refluxed for 8 hrs., the cooled mixture filtered and washed, and the solution is fractionated.

In another example a mixture of PCl_5 and $POCl_3$ is used (*O. S.*, **4**, 166).

Preparation of isocyanides. Ugi, in an *Organic Syntheses* procedure,[8] used $POCl_3$ in pyridine to dehydrate N-cyclohexylformamide to cyclohexyl isocyanide, but in a later paper he expressed preference for phosgene (*see* Phosgene). The procedure preferred for aryl isocyanides is illustrated for *o*-tolyl isocyanide.[9] A suspension of potassium *t*-butoxide in *t*-butanol is prepared under nitrogen, N-*o*-

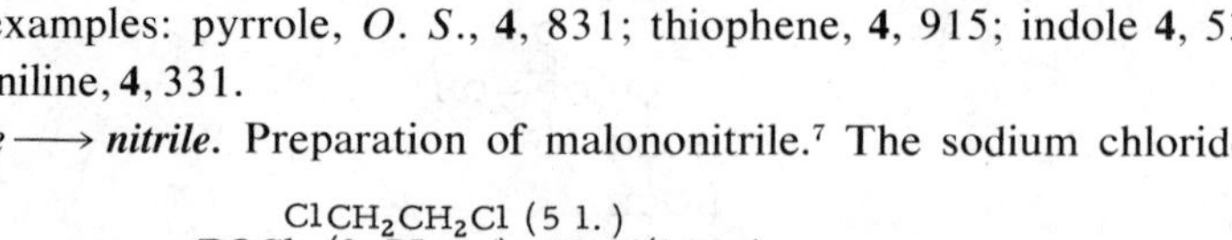

tolylformamide is added to the hot suspension, which soon becomes clear, and the solution is kept at 10–20° during addition of $POCl_3$. After 1 hr. at 30–35° the mixture is poured into a solution of 50 g. of sodium bicarbonate in 5 l. of water.

Dehydration of alcohols with $POCl_3$–Py. Butenandt and Schmidt-Thomé[10] tried various methods of dehydration for conversion of the epimeric mixture of 17-

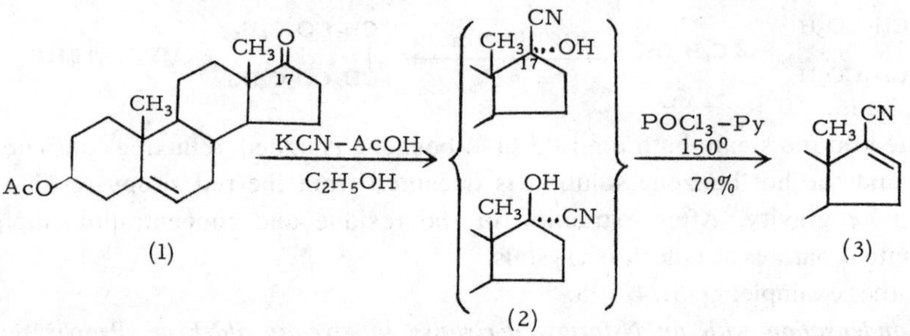

cyanohydrins from androstenolone acetate (1) but experienced difficulty from loss of hydrogen cyanide with reversion to the ketone. They then found that the reaction can be effected in high yield by heating 400 mg. of cyanohydrin mixture with 10 ml. of pyridine and 0.15 ml. of $POCl_3$ in a sealed tube for $1\frac{1}{2}$ hrs. at 150°.

This method of dehydration has become widely used by steroid chemists and proved of service in Sarett's synthesis of cortisone.[11] Thus dehydration of the 20-cyanohydrin (4) was accomplished with $POCl_3$ in pyridine at room temperature, and saponification gave (5) in very high yield.

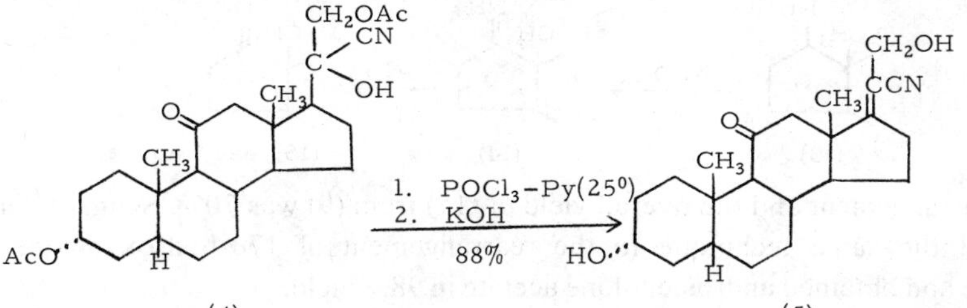
CH2OAc
CN
C
OH
CH3
CH3
AcO
H
1. POCl3–Py(25°)
2. KOH
88%
CH2OH
CCN
CH3
CH3
HO
H

(4) (5)

An observation by Bernstein's Lederle group[12] in the course of their synthesis of triamcinolone provides an interesting comparison of $POCl_3$–Py with $SOCl_2$–Py. The 3,20-bisketal (6) on reaction with former reagent at −5° suffered elimination of

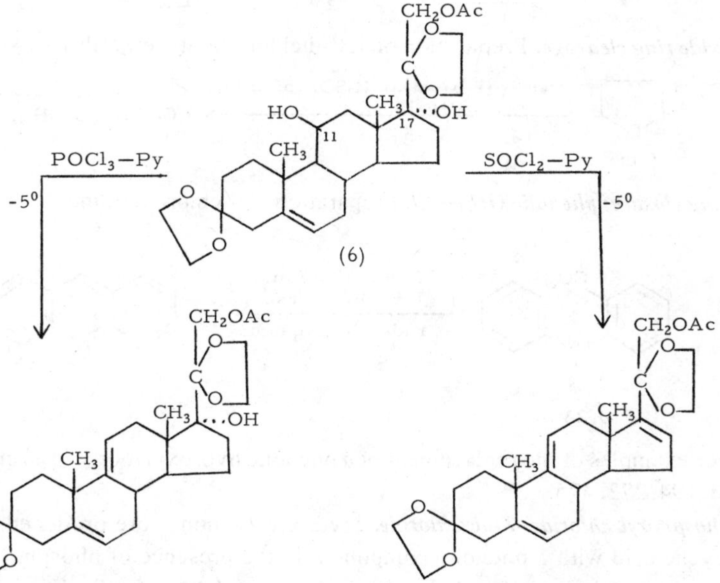
CH2OAc
HO
CH3
OH
17
11
CH3
POCl3–Py
-5°
SOCl2–Py
-5°
(6)
CH2OAc
CH3
OH
CH3
CH2OAc
CH3
CH3

(7) (8)

the 11β-hydroxyl group to produce a 9,11-double bond (7), whereas on reaction with thionyl chloride–pyridine both the 11β- and the 17α-hydroxyl groups were eliminated, giving (8). Thus $SOCl_2$–Py is the more powerful reagent, but $POCl_3$–Py has an advantage where selectivity is desired.

For a nonsteroid example, see *O.S.*, **4**, 444.

Beckmann rearrangement. Schmidt-Thomé[13] (at Hoechst AG) devised an ingenious degradation of pregnenolone acetate (9) to androstenolone acetate (1, above) involving conversion to the oxime and Beckmann rearrangement to the 17-acetylamine (11). Phosphoryl chloride in pyridine proved an effective agent for

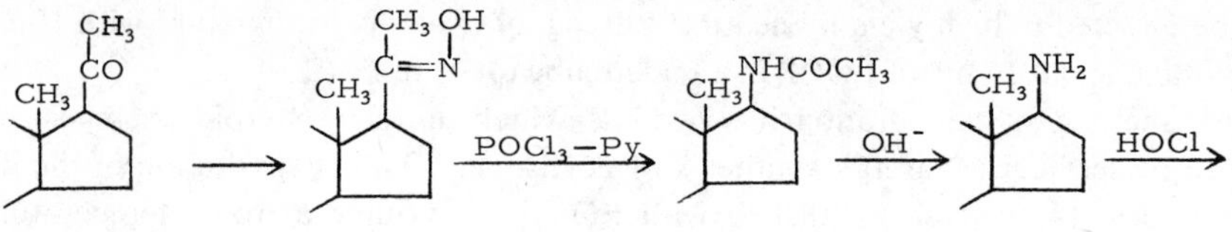
CH3
CO
CH3
CH3 OH
C=N
CH3
POCl3–Py
NHCOCH3
CH3
OH⁻
NH2
CH3
HOCl

(9) (10) (11) (12)

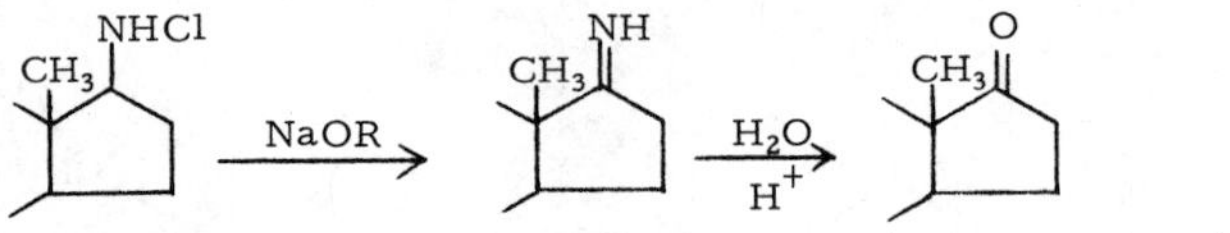
NHCl
CH3
NaOR
NH
CH3
H2O
H+
O
CH3

(13) (14) (15, see 1 above)

the rearrangement and the overall yield of (15) from (9) was 70%. Schmidt-Thomé[14] applied the same technique to the rearrangement of 17α-hydroxypregnenolone acetate and obtained androstenolone acetate in 98% yield.

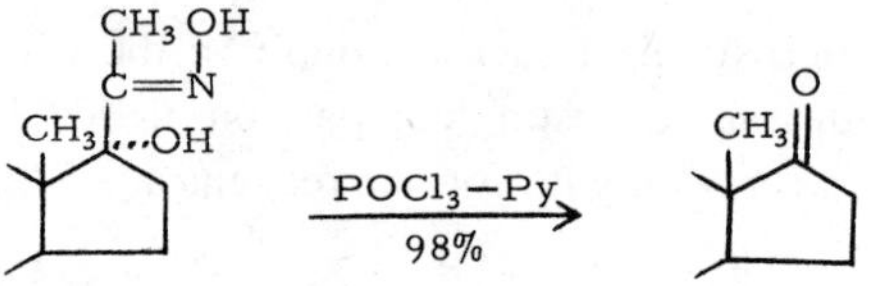
CH3 OH
C=N
CH3 ...OH
POCl3–Py
98%
O
CH3

Oxide ring cleavage. Preparation of 4,4′-dichlorodibutyl ether.[15]

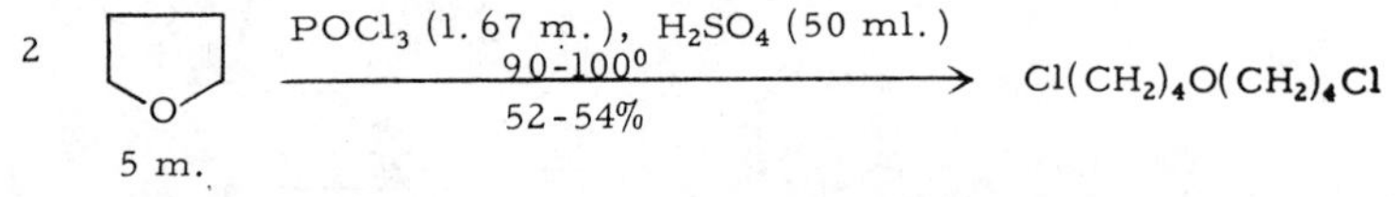
2
O
5 m.
POCl3 (1.67 m.), H2SO4 (50 ml.)
90-100°
52-54%
Cl(CH2)4O(CH2)4Cl

Ring closure; phenolic OH → Cl. Preparation of 9-chloroacridine.[16]

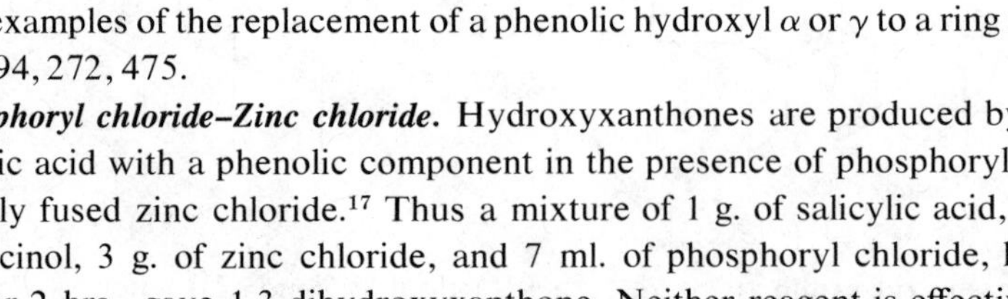
CO2H
N
H
0.23 m.
POCl3 (1.76 m.)
135-140° 2 hrs.
Crude yield quant.
Cl
N

Other examples of the replacement of a phenolic hydroxyl α or γ to a ring nitrogen: *O.S.*, **3**, 194, 272, 475.

Phosphoryl chloride–Zinc chloride. Hydroxyxanthones are produced by heating a salicyclic acid with a phenolic component in the presence of phosphoryl chloride and freshly fused zinc chloride.[17] Thus a mixture of 1 g. of salicylic acid, 1.4 g. of phloroglucinol, 3 g. of zinc chloride, and 7 ml. of phosphoryl chloride, heated at 60–70° for 2 hrs., gave 1,3-dihydroxyxanthone. Neither reagent is effective alone;

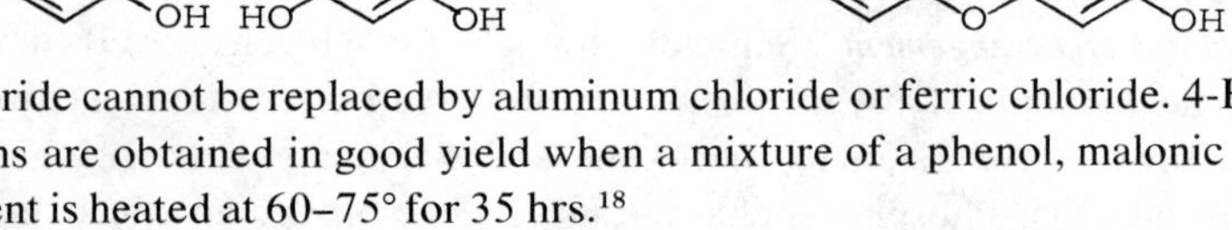
CO2H
OH
+
OH
HO
OH
POCl3–ZnCl2
O
OH
O
OH

zinc chloride cannot be replaced by aluminum chloride or ferric chloride. 4-Hydroxycoumarins are obtained in good yield when a mixture of a phenol, malonic acid, and the reagent is heated at 60–75° for 35 hrs.[18]

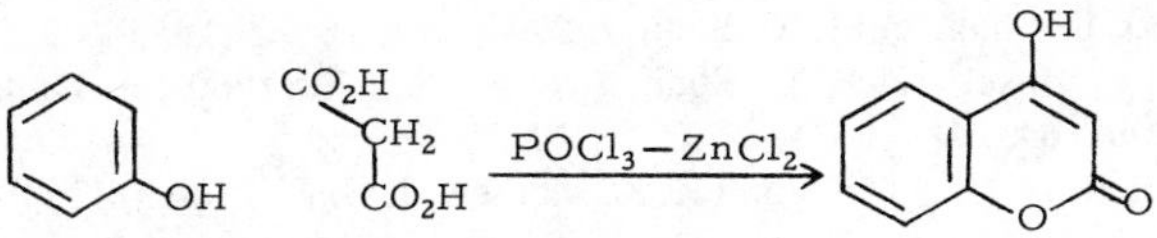
OH
CO_2H
CH_2
CO_2H
$POCl_3-ZnCl_2$
OH
O
O

Phosphoryl chloride–Stannous chloride–Pyridine. The reagent is used in one step of the Cornforth steroselective synthesis of olefins as illustrated by the example formulated (Me = methyl, Et = ethyl).[19] The steric course of the addition of Grignard

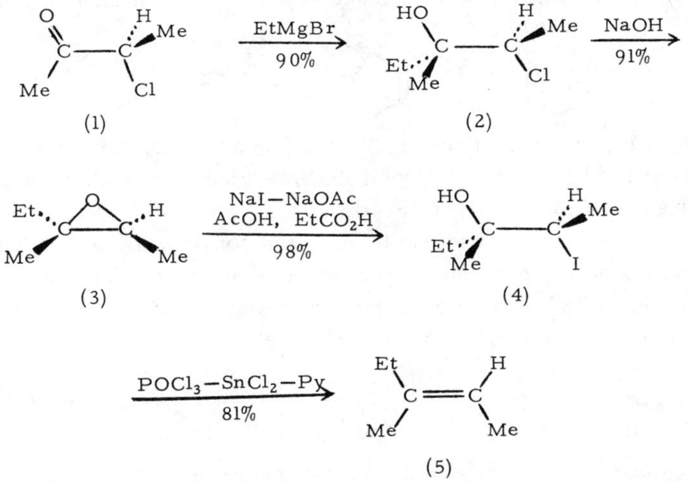
EtMgBr
90%
NaOH
91%
(1)
(2)
NaI–NaOAc
AcOH, $EtCO_2H$
98%
(3)
(4)
$POCl_3-SnCl_2-Py$
81%
(5)

reagent to the chloroketone (1) is predictable from the Cram-Prelog rule. The chlorohydrin (2) is not reducible directly to an olefin, but is converted via the epoxide (3) into the iodohydrin (4), of the same stereochemistry. Efficient cleavage of the epoxide is done with sodium iodide–sodium acetate in a mixture of acetic and propionic acids at −20°, adding the iodohydrin (4), and then phosphoryl chloride and stannous chloride in pyridine. The product is 80–85% sterically pure *trans*-3-methyl-2-pentene (5). The *cis*-isomer is obtained by a similar sequence starting with the reaction of EtCOCHClMe with MeMgBr.

[1]D. M. Brown, *Advances in Organic Chemistry*, **3**, 87 (1963); G. R. Dutton and C. R. Noller, *Org. Syn., Coll. Vol.*, **2**, 109 (1943)
[2]H. S. Forrest and A. R. Todd, *J. Chem. Soc.*, 3295 (1950)
[3]J. Baddiley and E. M. Thain, *J. Chem. Soc.*, 903 (1953)
[4]J. Riess, *Bull. soc.*, 18 (1965)
[5]G. H. Daub and W. S. Johnson, *Org. Syn., Coll. Vol.*, **4**, 390 (1963)
[6]L. F. Fieser, J. L. Hartwell, and J. E. Jones, *ibid.*, **3**, 98 (1955)
[7]A. R. Surrey, *ibid.*, **3**, 535 (1955)
[8]I. Ugi, R. Meyr, M. Lipinski, F. Bodesheim, and F. Rosendahl, *Org. Syn.*, **41**, 13 (1961)
[9]I. Ugi and R. Meyr, *ibid.*, **41**, 102 (1961)
[10]A. Butenandt and J. Schmidt-Thomé, *Ber.*, **71**, 1487 (1938); **72**, 182 (1939)
[11]L. H. Sarett, *Am. Soc.*, **70**, 1454 (1948); **71**, 2443 (1949)
[12]W. S. Allen and S. Bernstein, *Am. Soc.*, **77**, 1028 (1955); S. Bernstein, R. H. Lenhard, W. S. Allen, M. Heller, R. Littell, S. M. Stolar, L. I. Feldman, and R. H. Blank, *ibid.*, **78**, 5693 (1956)
[13]J. Schmidt-Thomé, *Ber.*, **88**, 895 (1955)
[14]*Idem*, *Ann.*, **603**, 43 (1957)
[15]K. Alexander and H. V. Towles, *Org. Syn., Coll. Vol.*, **4**, 266 (1963)
[16]A. Albert and B. Ritchie, *ibid.*, **3**, 53 (1955)

[17]P. K. Grover, G. D. Shah, and R. C. Shah, *J. Chem. Soc.*, 3982 (1955)
[18]V. R. Shah, J. L. Bose, and R. C. Shah, *J. Org.*, **25**, 677 (1960); F. Dallacker, P. Kratzer, and M. Lipp, *Ann.*, **643**, 97 (1961)
[19]J. W. Cornforth, R. H. Cornforth, and K. K. Mathew, *J. Chem. Soc.*, 112 (1959)

Phthalic anhydride. Mol. wt. 148.11, m.p. 131°.

Dehydration of alcohols.[1] A mixture of 0.5 mole of cyclohexanol, 0.5 mole of phthalic anhydride, and 2.5 g. of benzenesulfonic acid is heated gently at the boiling point for 30 min., and the distillate is saturated with sodium chloride. The

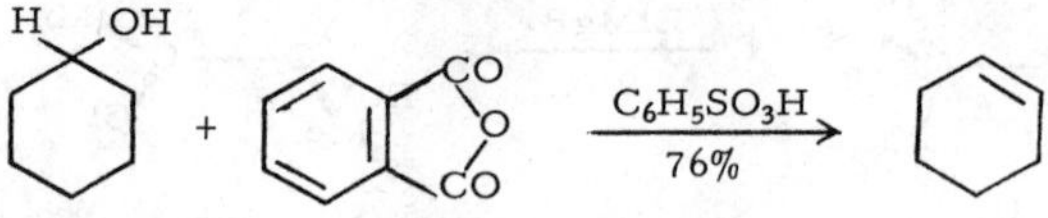
H
OH
CO
O
CO
+
$C_6H_5SO_3H$
76%

organic layer is separated, dried, and distilled and affords cyclohexene in 76% yield. Dehydration can be effected with phthalic anhydride alone without addition of the acid catalyst, but the reaction period is then 6 hrs.

Small amounts of butadiene can be prepared conveniently as follows from 1,3-butanediol, a starting material supplied by Aldrich and by Eastman: a mixture of 45 g. of the diol, 200 g. of phthalic anhydride, and 5 g. of benzenesulfonic acid on distillation for 2 hrs. affords butadiene in 47% yield.

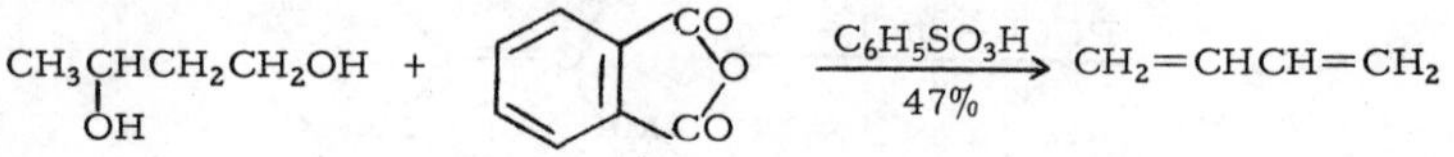
$CH_3CHCH_2CH_2OH$ +
OH
CO
O
CO
$C_6H_5SO_3H$
47%
$CH_2{=}CHCH{=}CH_2$

Phthaloylamino acids. In one procedure[2] a mixture of 0.1 mole of L-phenylalanine, 0.1 mole of finely ground phthalic anhydride, 150 ml. of toluene, and 1.3 ml. of triethylamine is refluxed under a water separator for 2 hrs., the solvent is removed in a rotary vacuum evaporator (Rinco Instrument Co.), and the solid residue is triturated with 200 ml. of water and 2 ml. of concd. hydrochloric acid and collected.

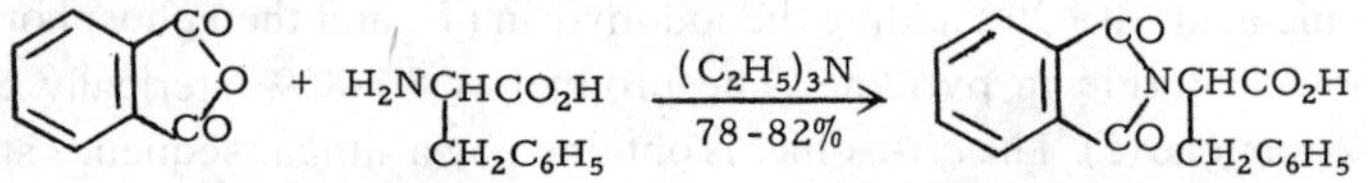
CO
O
CO
+ H_2NCHCO_2H
$CH_2C_6H_5$
$(C_2H_5)_3N$
78-82%
CO
$NCHCO_2H$
CO
$CH_2C_6H_5$

Crystallization from aqueous ethanol gives pure material. The reaction proceeds without racemization and is more convenient than the fusion process.

Another procedure[3] recommended for avoidance of racemization involves reaction of the amino acid with phthalic anhydride in aqueous dioxane in the presence of triethylamine to produce the N-substituted phthalamic acid and cyclization by addition of more triethylamine and dioxane and distillation until the boiling point of dioxane is reached.

Probably the most general method for introducing the phthaloyl group without racemization uses N-carboethoxyphthalimide, *which see*.

[1]H. Waldmann and F. Petru, *Ber.*, 287 (1950)
[2]A. K. Bose, *Org. Syn.*, **40**, 82 (1960)
[3]E. Hoffmann and H. Schiff-Shenhav, *J. Org.*, **27**, 4686 (1962)

Phthaloyl chloride, $C_6H_4(COCl)_2$. Mol. wt. 203.03, b.p. 281°, sp. gr. 1.41. Suppliers: Hooker, Monsanto, Aldrich, Eastman (the commercial material is about 94% pure; the impurity is phthalic anhydride).

Preparation.[1] A mixture of 1 mole of phthalic anhydride and 1.06 moles of phosphorus pentachloride is heated at 150° for 12 hrs., the temperature is raised

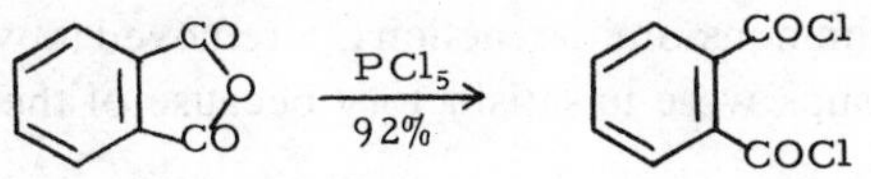

gradually to 250° to distil off most of the phosphoryl chloride, and phthaloyl chloride is collected at 131–133°/9–10 mm. (m.p. 11–12°).

Conversion into phthalyl chloride.[1] On heating a mixture of 105 g. of phthaloyl chloride and 75 g. of aluminum chloride on the steam bath for 8–10 hrs., a clear solution is obtained and on cooling the complex (2) sets to a hard solid. Small

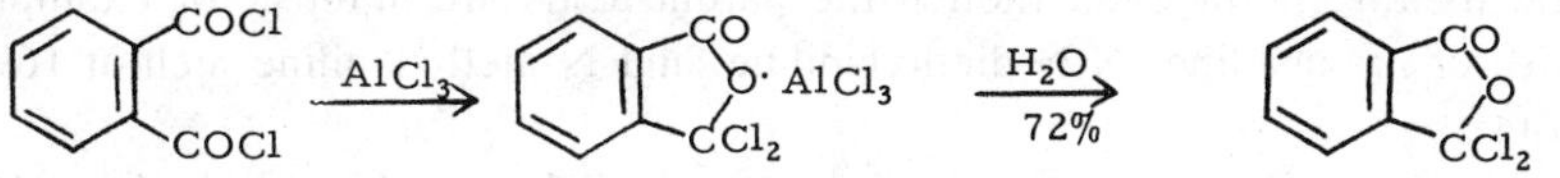

(1) (2) (3)

amounts are triturated with ice in a mortar, and the residual solid is dissolved in benzene. A water layer is separated, the solution is dried and evaporated, and the solid residue is extracted with petroleum ether in a Soxhlet apparatus and crystallized from petroleum ether. The yield of material melting at 87–89° is 72%; the melting point of phthalyl chloride is not sharp because the substance begins to revert to phthaloyl chloride.

Preparation of acid chlorides. Phthaloyl chloride is an excellent reagent for the conversion of acids and anhydrides into the acid chlorides, provided the boiling points are suitable for separation of products by distillation.[2] In the case of maleic and succinic anhydride a catalytic amount of zinc chloride is required. Maleic anhydride is isomerized in the process. Thus a mixture of 1 mole of maleic anhydride, 230 g. of commercial phthaloyl chloride, and 2 g. of anhydrous zinc chloride is

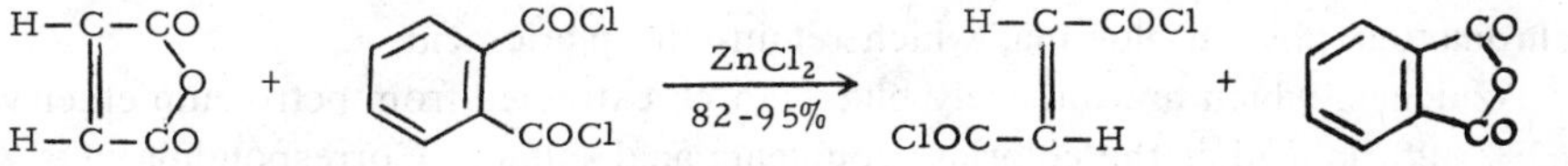

heated at 130–135° for 2 hrs., cooled slightly, and the fumaryl chloride is distilled and redistilled, b.p. 62–64°/13 mm. With monobasic acids no catalyst is required.

[1] E. Ott, *Org. Syn., Coll. Vol.*, **2**, 528 (1943)
[2] L. P. Kyrides, *ibid.*, **3**, 422 (1955)

Phthalyl alcohol, $HOCH_2C_6H_4CH_2OH$-*o*. Mol. wt. 138.16, m.p. 64°. Prepared by reduction of phthalic anhydride with lithium aluminum hydride in ether (87% yield).[1] In a novel and efficient synthesis of cyclopropanone hydrate, Grewe and Struve[2] used the *o*-xylylene acetal protective group (1) and, after elimination of

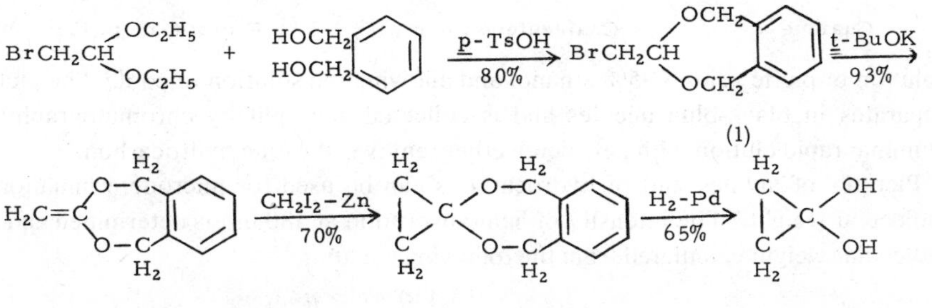

(2) (3) (4)

hydrogen bromide (2) and Simmons-Smith reaction (3), removed it by hydrogenolysis (4). In this case benzyl groups were unsatisfactory because of the ease of Claisen rearrangement.

[1]R. F. Nystrom and W. G. Brown, *Am. Soc.*, **69**, 1197 (1947)
[2]R. Grewe and A. Struve, *Ber.*, **96**, 2819 (1963)

Picric acid, 2,4,6-$(NO_2)_3C_6H_2OH$. Mol. wt. 229.10, m.p. 122.5°, pale yellow, pKa 0.80.

Picric acid forms ionic salts with amines, picrates, which are solids of characteristic melting point even though the parent acids are liquids; for example, the picrates of *m*-anisidine, N,N-diethylaniline, and N-methylaniline melt at 169, 142, and 144°.

Picric acid forms 1:1 π-complexes with naphthalene and higher hydrocarbons of higher melting point, lower solubility, and more intense color than either component, for example:

π-Base	π-Complex: color	M.p.
Naphthalene	Yellow	150°
Acenaphthene	Orange	162°
Anthracene	Red	142°
Phenanthrene	Yellow	145°
3,4-Benzpyrene	Purple-brown	198°
Methylcholanthrene	Purplish-black	182.5°

Picric acid complexes are useful for isolation and characterization. A complex can be split by extraction of a benzene solution with ammonium hydroxide or by chromatography on alumina, which retains the picric acid.

Azulenes, which are intensely blue, can be extracted from petroleum ether with 62% sulfuric acid as the colorless conjugate acid sulfates. Corresponding salts with picric acid, which are intensely colored, are useful for isolation of the hydrocarbons. Thus the guaiazulene present in an oily mixture obtained by dehydrogenation of technical guaiene with sulfur can be isolated by treating the oil with a hot saturated

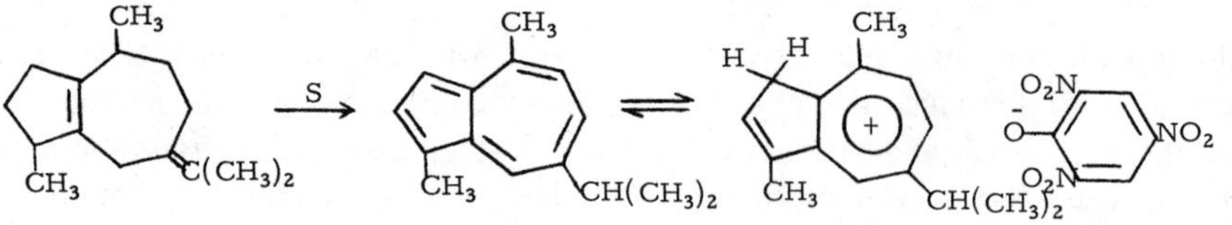
CH3
S
CH3
H
H
CH3
O2N
NO2
O2N
C(CH3)2
CH3
CH(CH3)2
CH(CH3)2

Guaiene Guaiazulene (m. p. 31°) Picrate, m. p. 123°

solution of picric acid in 95% ethanol and allowing the solution to cool.[1] The picrate separates in black-blue needles and is collected and split by chromatography on alumina; rapid elution with petroleum ether removes the blue hydrocarbon.

Picrates of amines and of hydrocarbons can be used for microdetermination of molecular weight.[2] The intensity of light absorption at 380 mμ is determined and the molecular weight calculated from the following equation:

$$\text{Mol. wt.} = \frac{13{,}440 \times C \times n}{\log(I_0/I)}$$

where C is the concentration in g./l and n is the molar ratio of picric acid to amine or hydrocarbon. The accuracy is of the order of ±2%.

[1]*Org. Expts.*, 298–301

[2]K. G. Cunningham, W. Dawson, and F. S. Spring, *J. Chem. Soc.*, 2305 (1951)

Picryl azide, 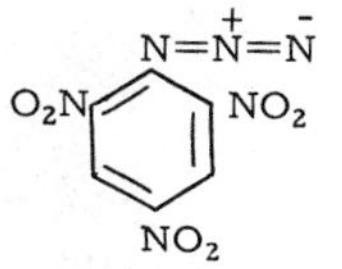

Mol. wt. 254.13, m.p. 93°.

Preparation. By reaction of picryl chloride with sodium azide[1] or by nitration of *o*- or *p*-nitroazidobenzene.[2] Like cyanogen azide (*which see*), the reagent (1) reacts with an olefin either to form a Schiff base (2), convertible into the corresponding ketone (3), or to form an aziridine (4). Simple olefins give Schiff bases. The reagent is much safer to handle than cyanogen azide.

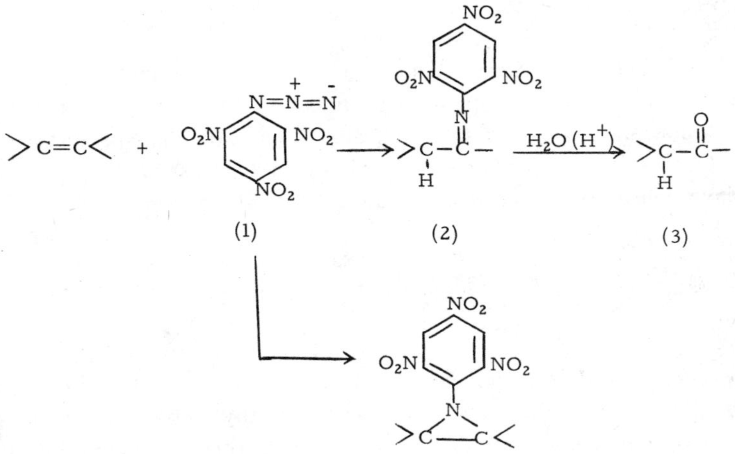
NO2
O2N
NO2
+ -
N=N=N
>C=C< +
O2N
NO2
NO2
N
>C–C–
H
H2O (H+)
O
>C–C–
H
(1)
(2)
(3)
NO2
O2N
NO2
N
>C–C<
(4)

[1]E. Schrader, *Ber.*, **50**, 777 (1917)

[2]A. S. Bailey, J. J. Merer, and J. E. White, *Chem. Comm.*, **1**, 4 (1965)

Picryl chloride, 2,4,6-$(NO_2)_3C_6H_2Cl$. Mol. wt. 247.50, m.p. 81°. Supplier: Eastman.

Picryl chloride reacts with azulene in boiling ethanol to give 1-picrylazulene in high yield.[1] 2,4-Dinitrochlorobenzene does not react with the hydrocarbon under similar conditions in ethanol or under Friedel-Crafts conditions.

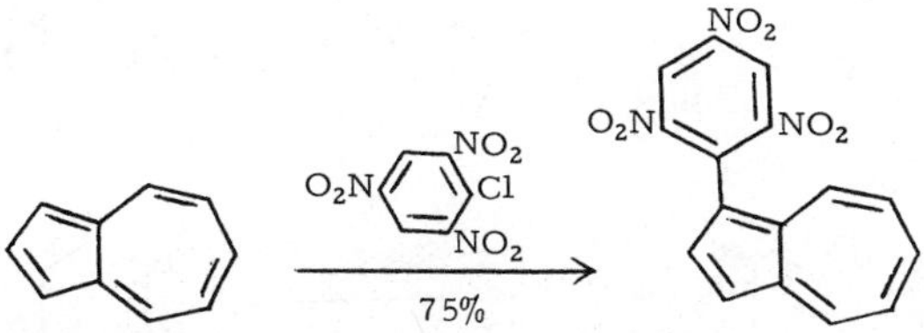
NO2
O2N
NO2
NO2
O2N
Cl
NO2
75%

Picryl chloride appears to be promising for formation of internucleotide bonds.[2]

[1]W. Treibs, K.-H. Jost, C. Kurpjun, and G. Grundke-Schroth, *Ber.*, **94**, 1728 (1961)

[2]F. Cramer *et al.*, *Angew. Chem., Internat. Ed.*, **2**, 43 (1963)

Piperidine, 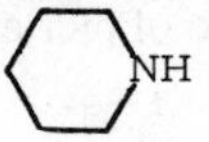

Mol. wt. 85.15, b.p. 105°, sp. gr. 0.86, pKb 2.9, Water-miscible. Suppliers (practical grade): Abbott, du Pont Dyes and Chem. Div., Hooker Chem. Corp., Reilly Tar and Chem., A, B, E, F, MCB.

von Braun degradation. Preparation of pentamethylene dibromide.[1] N-Benzoylpiperidine is treated with cooling with phosphorus tribromide, then bromine is added, and the mixture is heated until gas evolution ceases and then distilled. Probable intermediates are indicated.

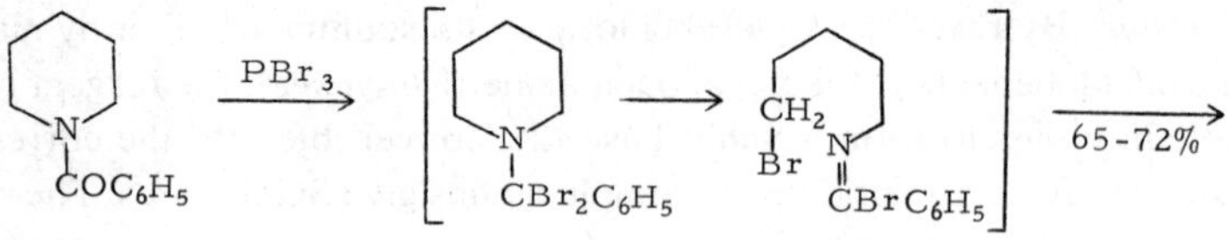
PBr3
CH2
Br
COC6H5
CBr2C6H5
CBrC6H5
65-72%

$$Br(CH_2)_5Br \quad + \quad C_6H_5CN \quad + \quad POBr_3$$

Ethylation. The following procedure illustrates a general method for ethylation of amines.[2] A flask flushed with nitrogen is charged with 4 moles of piperidine, 4.4 g. of sodium, and 5 g. of pyridine and heated briefly with high-speed stirring under nitrogen. The resulting dispersion (no H_2 is evolved) is transferred under nitrogen to an autoclave. The autoclave is pressurized with ethylene and stirred at 100°.

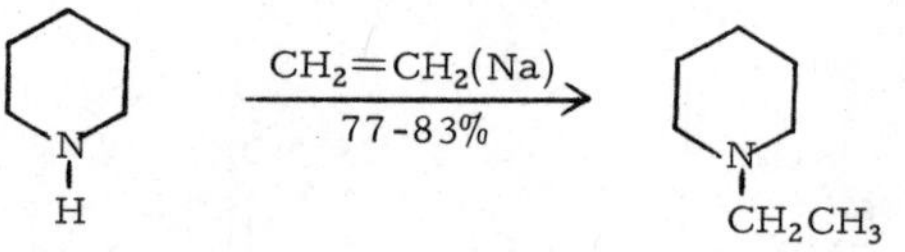
CH2=CH2(Na)
77-83%
H
CH2CH3

Preparation of 6-alkyl-Δ²-cyclohexenones. A procedure developed by Stork and White,[3] as illustrated by the preparation of the 6-methyl derivative, involves Birch reduction of *o*-toluidine to a mixture of dihydrides converted on mild acid hydrolysis to what proved to be a mixture of the unsaturated ketones, (3) and (4), and some

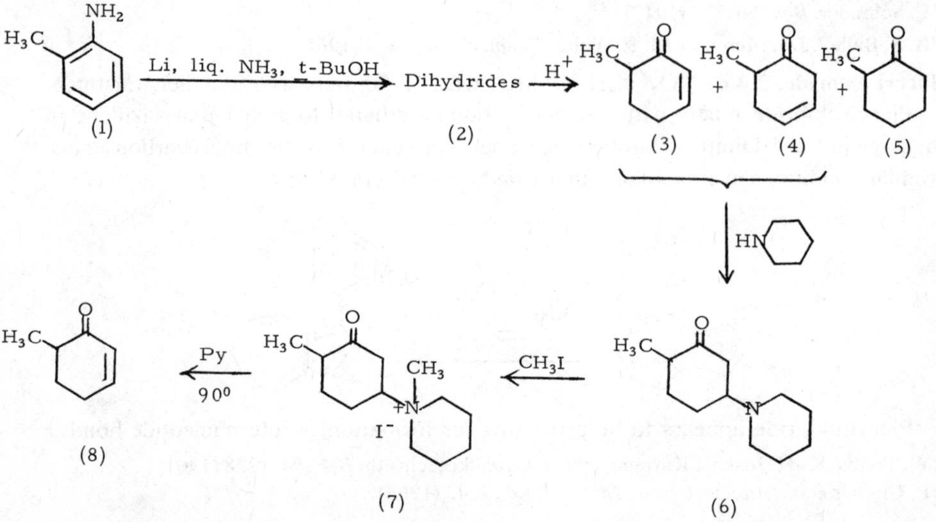
NH2
H3C
Li, liq. NH3, t-BuOH
Dihydrides
H+
(1)
(2)
(3)
(4)
(5)
HN
Py
90°
CH3
CH3I
I-
(8)
(7)
(6)

10–30% of the saturated ketone (5). When the mixture is refluxed briefly with piperidine, the amine adds smoothly to both (3) and (4) to give the piperidine ketone (6), while the saturated ketone (5) remains unchanged. A clean separation is effected by extraction of the piperidine derivative with dilute acid, and regeneration is accomplished by converting the base into the crystalline methiodide (7) and warming this with pyridine. The product is pure 6-methyl-Δ^2-cyclohexenone (8).

Claisen-Schmidt catalyst, R. Robinson's elegant method for the construction of ring A of steroids is illustrated by the synthesis of the bicyclic diketone (2).[4] A Michael addition of 0.5 mole of 2-methylcyclohexane-1,3-dione to methyl vinyl ketone under catalysis by methanolic potassium hydroxide gives (1), which is

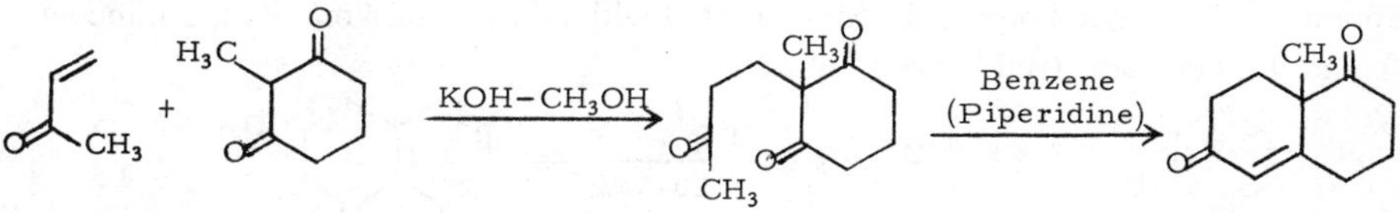
H3C
CH3
KOH–CH3OH
CH3
CH3
Benzene
(Piperidine)
CH3

(1) (2)

cyclized to (2) by refluxing a solution in benzene containing 3 ml. of piperidine. The overall yield is 63–65%.

Knoevenagel catalyst. The Knoevenagel reaction (or the Knoevenagel-Doebner modification) is usually understood as a condensation between an aldehyde or a ketone and a compound with a doubly activated methylene group such as malonic ester, ethyl acetoacetate, or ethyl cyanoacetate, catalyzed by piperidine or by piperidine acetate. However, in reporting an interesting finding applicable to the reactions just cited, Kuhn[5] described the self-condensation of crotonaldehyde under the influence of a catalyst as a Knoevenagel reaction. Kuhn and M. Hoffer had obtained

$$CH_3CH{=}CHCHO + CH_3CH{=}CHCHO \longrightarrow CH_3(CH{=}CH)_3CHO + H_2O$$

polyenealdehydes by the action of piperidine on acetaldehyde or on mixture of acetaldehyde and crotonaldehyde but had reported that no reaction occurs on treatment of pure crotonaldehyde with piperidine. K. Bernhauer and E. Waldan disputed this statement and asserted that reaction does occur and affords octatrienal. In reinvestigation Kuhn confirmed the report that no reaction occurs on treatment of pure crotonaldehyde with piperidine but found that if the aldehyde is first irradiated briefly with illuminated light, addition of piperidine promotes prompt self-condensation. The inferencc that light catalyzes autoxidation to crotonic acid was fully confirmed. Extraction of the "active crotonaldehyde" with base destroyed its activity; addition of crotonic acid or acetic acid to the pure aldehyde caused it to react on addition of piperidine. Thus the true catalyst was identified as a piperidine salt.

In the condensation of an aromatic aldehyde with pure malonic acid no other acid is needed and the reaction is conducted conveniently in pyridine.[6] The mixture is

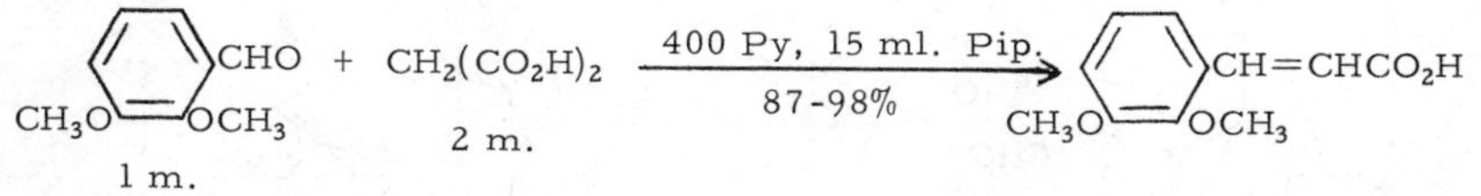
CHO
CH3O
OCH3
1 m.
+ CH2(CO2H)2
2 m.
400 Py, 15 ml. Pip.
87-98%
CH=CHCO2H
CH3O
OCH3

heated at 80–85° for 1 hr. and then refluxed. Malonic acid was taken in 100% excess not as an insurance measure but because in a trial run with no excess the yield was only 50%; decarboxylation of malonic acid evidently competes with Knoevenagel condensation.

A procedure for the condensation of benzaldehyde with malonic ester[7] was worked out at a time when the benzaldehyde available containing 2–8% of benzoic acid

$$\underset{0.66\ m.}{C_6H_5CHO} + \underset{0.63\ m.}{CH_2(CO_2C_2H_5)_2} \xrightarrow[89\text{-}91\%]{\text{Piperidine benzoate}} C_6H_5CH{=}C(CO_2C_2H_5)_2$$

was eminently satisfactory. With benzaldehyde containing 0.2% of benzoic acid the yield was only 71%; in this case addition of 2% benzoic acid is recommended.

The condensation of salicylaldehyde with malonic ester to give 2-carboethoxy-coumarin was carried out in absolute ethanol with addition of 4 ml. of piperidine and 0.5 ml. of acetic acid (refl. 3 hrs.).[8]

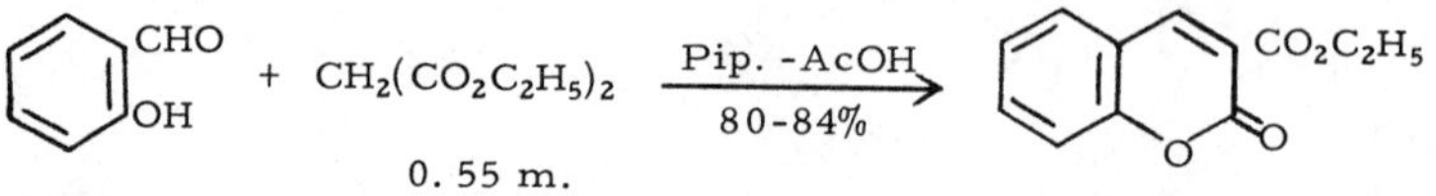
CHO
OH
+ $CH_2(CO_2C_2H_5)_2$
Pip. -AcOH
80-84%
$CO_2C_2H_5$
O
O

0. 55 m.

0. 5 m.

A procedure[9] for the condensation of acetaldehyde with 2 equivalents of ethyl acetoacetate calls for cooling 1.61 moles of ester to 0°, adding 0.78 mole of acetal-

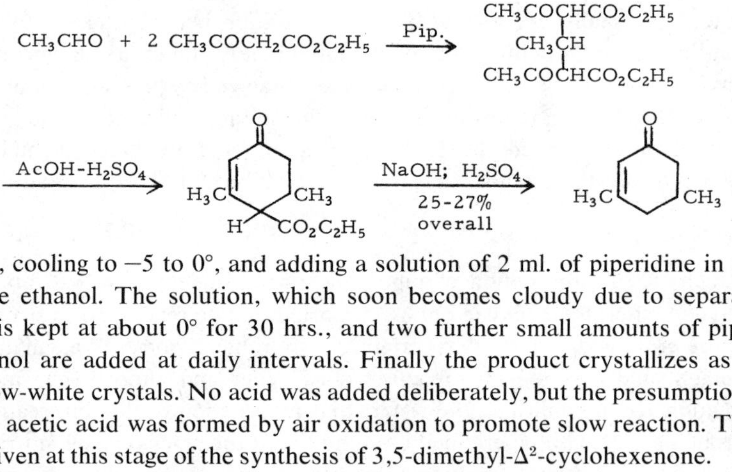
CH_3CHO + 2 $CH_3COCH_2CO_2C_2H_5$
Pip.
$CH_3COCHCO_2C_2H_5$
CH_3CH
$CH_3COCHCO_2C_2H_5$
$AcOH-H_2SO_4$
O
H_3C
CH_3
H
$CO_2C_2H_5$
NaOH; H_2SO_4
25-27%
overall
O
H_3C
CH_3

dehyde, cooling to −5 to 0°, and adding a solution of 2 ml. of piperidine in 5 ml. of absolute ethanol. The solution, which soon becomes cloudy due to separation of water, is kept at about 0° for 30 hrs., and two further small amounts of piperidine in ethanol are added at daily intervals. Finally the product crystallizes as a mass of yellow-white crystals. No acid was added deliberately, but the presumption is that enough acetic acid was formed by air oxidation to promote slow reaction. The yield is not given at this stage of the synthesis of 3,5-dimethyl-Δ^2-cyclohexenone.

Piperidine acetate also catalyzes the intramolecular cyclization of dialdehydes. Examples: cyclization of (1) to (2) by Woodward *et al.*[10]; conversion of (3) into

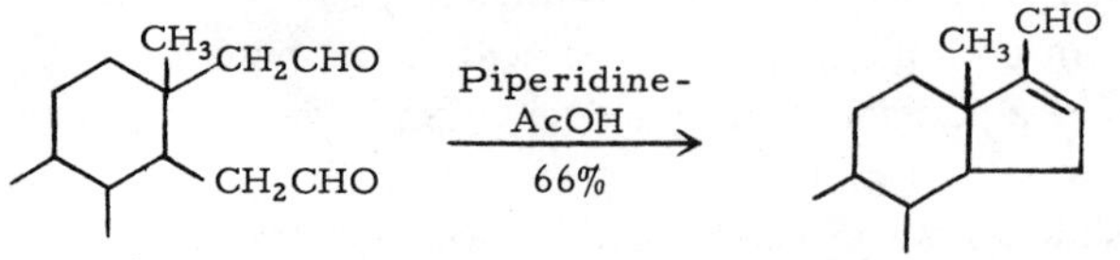
CH_3
CH_2CHO
CH_2CHO
Piperidine-
AcOH
66%
CH_3
CHO

(1) (2)

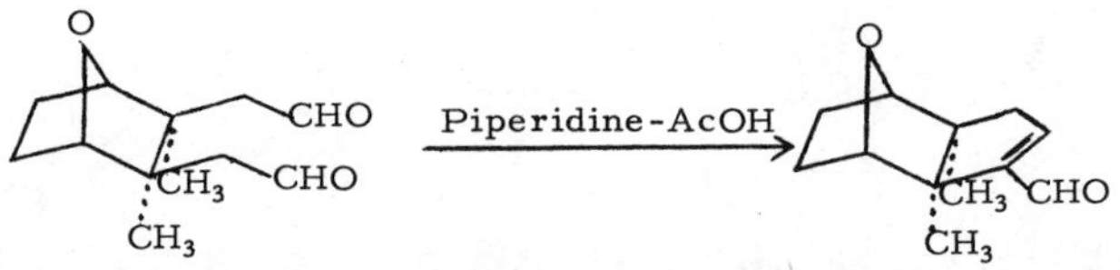
O
CHO
CH_3
CHO
CH_3
Piperidine-AcOH
O
CH_3
CHO
CH_3

(3) (4)

(4) by Stork *et al.*,[11] The reactions are carried out in benzene or in acetic acid at 60–70° under nitrogen.

Synthesis of a heterocycle. An efficient synthesis[12] of 3-cyano-6-methyl-2(lH)-pyridone (3) starts with 3-ketobutyraldehyde 1-dimethylacetal (1), available from Henley and Co. A mixture of 5 moles of (1) and 5.5 moles of cyanoacetamide (2)

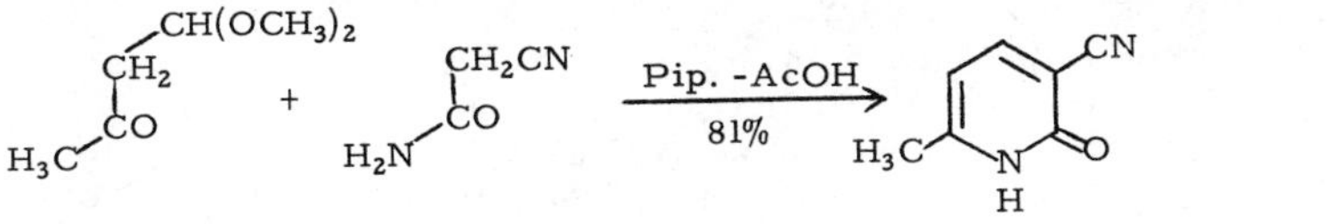
CH(OCH3)2
CH2
CO
H3C
+
CH2CN
CO
H2N
Pip. -AcOH
81%
CN
H3C
N
H
O

(1) (2) (3)

in 1 l. of water is treated with a solution prepared by adding about 25 ml. of piperidine to 100 ml. of 20% acetic acid to a pH of 9–10. Heating to about 80% produces a clear solution, which is then refluxed for 24 hrs.; the product starts to separate after about 1 hr.

Enamines. Herr and Heyl[13] of the Upjohn Co. introduced the method of protecting a steroid carbonyl function by acid-catalyzed reaction with piperidine or morpholine to form the enamine derivative; when desired, the carbonyl function can be restored by acid hydrolysis. A piperidyl enamine was employed in a different way in a high-yield method for the degradation of stigmasterol, a $\Delta^{5,22}$-diene-3β-ol of plant origin, to progesterone (4). Oppenauer oxidation gives stigmastadienone (1), which on selective ozonization in methylene chloride containing 1% of pyridine affords the aldehyde (2),[14] which is converted into the enamine (3) by refluxing with

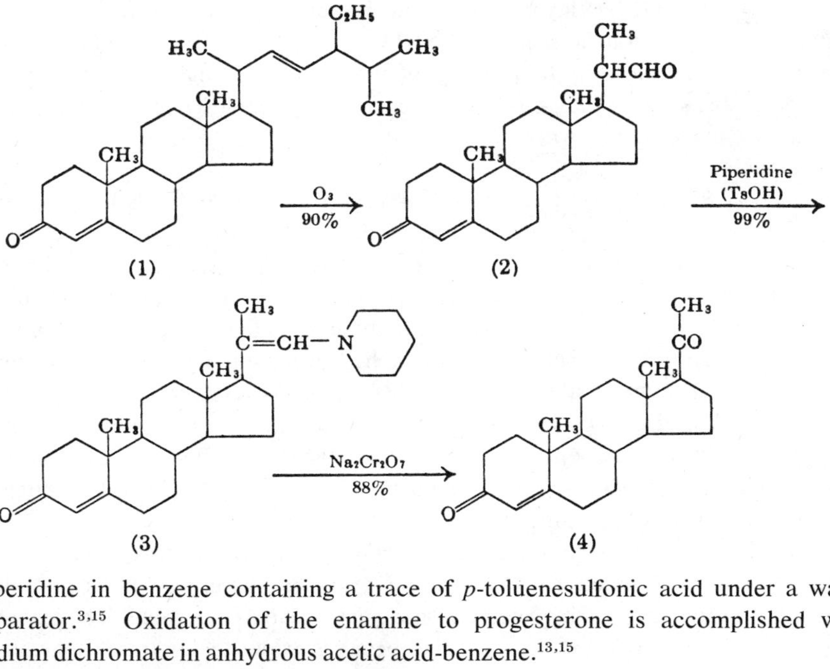
C2H5
H3C
CH3
CH3
CH3
CH3
O
O3
90%
CH3
CHCHO
CH3
CH3
O
Piperidine
(TsOH)
99%
CH3
C=CH— N
CH3
CH3
O
Na2Cr2O7
88%
CH3
CO
CH3
CH3
O

(1) (2) (3) (4)

piperidine in benzene containing a trace of *p*-toluenesulfonic acid under a water separator.[3,15] Oxidation of the enamine to progesterone is accomplished with sodium dichromate in anhydrous acetic acid-benzene.[13,15]

[1]J. von Braun, *Org. Syn., Coll. Vol.*, **1**, 428 (1941)
[2]J. Wollensak and R. D. Closson, *Org. Syn.*, **43**, 45 (1963)
[3]G. Stork and W. N. White, *Am. Soc.*, **78**, 4604 (1956)

[4]S. Ramachandran and M. S. Newman, *Org. Syn.*, **41**, 38 (1961)
[5]R. Kuhn, W. Badstübner, and C. Grundmann, *Ber.*, **69**, 98 (1936)
[6]J. Koo, M. S. Fish, G. N. Walker, and J. Blake, *Org. Syn., Coll. Vol.*, **4**, 327 (1963)
[7]C. F. H. Allen and F. W. Spangler, *ibid.*, **3**, 377 (1955)
[8]E. C. Horning, M. G. Horning, and D. A. Dimmig, *ibid.*, **3**, 165 (1955)
[9]E. C. Horning, M. O. Denekas, and R. E. Field, *ibid.*, **3**, 317 (1955)
[10]R. B. Woodward *et al.*, *Am. Soc.*, **74**, 4223 (1952)
[11]G. Stork *et al.*, *Am. Soc.*, **75**, 384 (1953)
[12]L. J. Binovi and H. G. Arlt, Jr., *J. Org.*, **26**, 1656 (1961)
[13]M. E. Herr and F. W. Heyl, *Am. Soc.*, **74**, 3627 (1952)
[14]G. Slomp, Jr., and J. L. Johnson, *Am. Soc.*, **80**, 915 (1958)
[15]D. A. Shepherd *et al.*, *Am. Soc.*, **77**, 1212 (1955)

Piperylene, $CH_3CH{=}CHCH{=}CH_2$. Mol. wt. 68.11, b.p. 44°. Suppliers: Enjay, K and K, Phillips Petroleum, Aldrich. The diene is prepared easily by reaction of crotonaldehyde with methylmagnesium chloride to give 3-pentene-2-ol[1] and dehydration over alumina at 450°.

$$CH_3CH{=}CHCHO \xrightarrow[81\text{-}86\%]{CH_3MgCl} CH_3CH{=}CH\underset{\displaystyle OH}{C}HCH_3 \xrightarrow[80\%]{Al_2O_3} CH_3CH{=}CHCH{=}CH_2$$

[1]E. R. Coburn, *Org. Syn., Coll. Vol.*, **3**, 696 (1955)

Pivalic acid, *see* Trimethylacetic acid.

Platinum catalysts. Mol. wts.: Pt, 195.23; chloroplatinic acid (platinic chloride), $PtCl_4 \cdot 2HCl \cdot 6H_2O$, 518.08; ammonium chloroplatinate, $(NH_4)_2PtCl_6$, 443.91. Suppliers: Baker and Co., J. Bishop and Co., Engelhard Industries. Used catalyst can be sent for recovery to Engelhard Industries. R. Adams *et al.*[1] give a procedure for recovery involving heating the residues with aqua regia, filtering, and evaporating to dryness. Purification is accomplished by precipitation of ammonium chloroplatinate, igniting the salt to produce platinum sponge, and dissolving this in aqua regia.[2] If Adams catalyst is to be prepared, the simplest procedure is to fuse ammonium chloroplatinate with sodium nitrate.[3]

See also: Heyn's catalyst.

Adams catalyst. $PtO_2 \cdot H_2O$, mol. wt. 245.25. Fusion of chloroplatinic acid with sodium nitrate at 500–550° and leaching of the cooled melt with water gives the brown oxide.[1] This is stable in storage and is activated by shaking with hydrogen and a solvent, either before or after addition of the substrate. Adams and Vorhees[4] describe a hydrogenation apparatus; a commercial model is available from Parr Instrument Co. Procedures are given for the hydrogenation of benzalacetophenone to benzylacetophenone[5] and of ethyl *p*-nitrobenzoate to ethyl *p*-aminobenzoate.[6]

Henze *et al.*[7] noted considerable variation in the activity of different batches of Adams catalyst and developed a procedure for the preparation of oxide of more reproducible activity which depends upon the essentially instantaneous heating of chloroplatinic acid to 520° in the presence of sodium nitrate.

Platinum black. An improvement over previous procedures described by Feulgen[8] involves reduction of a solution of chloroplatinic acid with alkaline formaldehyde, with vigorous shaking to coagulate colloidal material. This catalyst is useful for the destruction of excess hydrogen peroxide; the mixture (in alcohol–water) is stirred until evolution of oxygen ceases.[9]

Brown and Brown catalyst. Addition of sodium borohydride to a 10% solution of chloroplatinic acid produces a finely divided black precipitate of essentially pure platinum which is an active catalyst for the hydrogenation of olefins and acetylenes.[10] Reduction of the platinum salt in the presence of decolorizing carbon produces a supported catalyst of even greater activity.[11] The hydrogen for hydrogenation is generated internally and reacts more rapidly than external hydrogen. Procedure: a solution of sodium borohydride in ethanol stabilized with sodium hydroxide is added to a suspension of decolorizing carbon in a solution of chloroplatinic acid in ethanol to produce the supported catalyst. Excess hydrochloric acid is added, followed by the substance to be hydrogenated; a solid is added in a solution, a liquid olefin can be injected through a rubber port. Then stabilized borohydride solution is added from a dropping funnel at such a rate as to maintain atmospheric pressure. An automatic apparatus suitable for slow reactions or for large-scale hydrogenations[12] is available from Delmar Scientific Laboratories.

In an adaptation of the method of Brown and Brown as a student experiment[13] an Erlenmeyer flask with a side arm carrying a wired-on rubber bulb (for a medicine dropper) is charged with water, platinic chloride solution, and Norit, stabilized borohydride solution is added, followed by a solution of substrate (*endo*-norbornene-*cis*-5,6-dicarboxylic acid) and concd. hydrochloric acid. The flask is capped with a large serum stopper, which is wired on. Borohydride solution is injected through the rubber port with a syringe until the rubber bulb inflates to a sizable balloon. The syringe is withdrawn, and the flask swirled until deflation of the balloon indicates that further borohydride is required. The course of the reaction is easily followed, and a balloon pressure of about 10 p.s.i. promotes rapid reaction.

Active platinum catalyst. A catalyst regarded as more active than the unsupported catalyst of Brown and Brown is made by reduction of chloroplatinic acid with tribenzylsilane or other silicone hydride.[14]

Platinum–tin chloride catalyst.[15] A catalyst solution is prepared, with exclusion of oxygen, by dissolving 10 mmoles of $SnCl_2 \cdot 2\,H_2O$ and 1 mmole of chloroplatinic acid in 120 ml. of methanol. When a 1 : 1 mixture of ethylene and hydrogen at atmospheric pressure was admitted to the stirred catalyst solution, rapid and essentially quantitative hydrogenation occurred. A 1 : 1 mixture of acetylene and hydrogen gave ethane and ethylene in a 3 : 1 molar ratio. Since higher olefins are more difficult to hydrogenate with the complex catalyst, it appears that ease of hydrogenation parallels ability of the olefin to complex with platinum. A function of the stannous chloride is to stabilize platinum against reduction to the metal. The used catalyst solution is devoid of colloidal particles.

Selective hydrogenation. Although reduction of an unsaturated aldehyde to the unsaturated alcohol is usually a formidable problem, Tuley and Adams[16] reported reduction of cinnamaldehyde to cinnamyl alcohol in high yield using a mixture

$$C_6H_5CH{=}CHCHO + H_2 + Pt{-}PtO_2 \xrightarrow{FeCl_2-Zn(OAc)_2} C_6H_5CH{=}CHCH_2OH$$

of platinum oxide and platinum black in ethanol with added ferrous chloride and zinc acetate. The experiment was later confirmed by Rylander *et al.*,[17] who found that platinum oxide alone could be used, but that platinum is a highly specific metal for the reaction. Reduction of crotonaldehyde to crotyl alcohol was more difficult and required reactivation of the catalyst.

Platinum oxide on silicic acid. This modified Adams catalyst is very active at room temperature and is useful for determination of the degree of unsaturation because of short reaction time and sharp end point.[18] It can be used for hydrogenation of a methyl ester in ethanol without transesterification.[19]

[1]R. Adams, V. Voorhees, and R. L. Shriner, *Org. Syn., Coll. Vol.*, **1**, 463 (1941)
[2]E. Wichers, *Am. Soc.*, **43**, 1268 (1921)
[3]W. F. Bruce, *Am. Soc.*, **58**, 687 (1936); *Org. Syn., Coll. Vol.*, **1**, 466, Note 3 (1941)
[4]R. Adams and V. Voorhees, *ibid.*, **1**, 61 (1941)
[5]R. Adams, J. W. Kern, and R. L. Shriner, *ibid.*, **1**, 101 (1941)
[6]R. Adams and F. L. Cohen, *ibid.*, **1**, 240 (1941)
[7]V. L. Frampton, J. D. Edwards, Jr., and H. R. Henze, *Am. Soc.*, **73**, 4432 (1951)
[8]R. Feulgen, *Ber.*, **54**, 360 (1921)
[9]A. C. Cope and E. Ciganek, *Org. Syn., Coll. Vol.*, **4**, 612 (1963)
[10]H. C. Brown and C. A. Brown, *Am. Soc.*, **84**, 1493, 1494 (1962)
[11]*Idem, ibid.*, **84**, 2827, 2829 (1962)
[12]H. C. Brown, K. Sivasankaran, and C. A. Brown, *J. Org.*, **28**, 214 (1963)
[13]*Org. Expts.*, 86
[14]R. W. Bott, C. Eaborn, E. R. A. Peeling, and D. E. Webster, *Proc. Chem. Soc.*, 337 (1962)
[15]R. D. Cramer, E. L. Jenner, R. V. Lindsey, Jr., and U. G. Stolberg, *Am. Soc.*, **85**, 1691 (1963)
[16]W. F. Tuley and R. Adams, *Am. Soc.*, **47**, 3061 (1925)
[17]P. N. Rylander, N. Himelstein, and M. Kilroy, *Engelhard Ind. Techn. Bull.*, **4**, 49 (1963)
[18]F. A. Vandenheuvel, *Anal. Chem.*, **28**, 362 (1956)
[19]R. G. Ackman and R. D. Burgher, *J. Lipid Res.*, **5**, 130 (1964)

Platinum sulfide-on-carbon. This heterogeneous hydrogenation catalyst available from Engelhard Industries, although less active than palladium catalyst for the reduction of aromatic nitro compounds, has the advantage of being insensitive to poisons and of not reducing halogen substituents.[1] Thus 2,5-dichloronitrobenzene was reduced quantitatively to 2,5-dichloroaniline without dehalogenation.

[1]F. S. Dovell and H. Greenfield, *Am. Soc.*, **87**, 2767 (1965)

Polyoxyethylated lauryl alcohol (trade name Brij 30, Atlas Co.). A surfactant used in an efficient process for the ozonization of cycloolefins to dibasic acids in aqueous alkaline hydrogen peroxide.[1]

[1]M. I. Fremery and E. K. Fields, *J. Org.*, **28**, 2537 (1963)

Polyphosphate ester (PPE), $C_8H_{20}O_{12}P_4$. Mol. wt. 432.14. The reagent is prepared[1] by refluxing a mixture of 150 g. of phosphorus pentoxide (P_4O_{10}), 150 ml. of diethyl ether, and 300 ml. of chloroform until the solution is clear (15–30 hrs.), filtering through glass wool, and taking off the solvent in a rotary evaporator.

Material initially prepared by Langheld[2] and usually called ethyl metaphosphate was regarded by Rätz and Thilo[3] as a mixture of the cyclic esters (2) and (3). A later study by Calvin *et al.*[4] using NMR data indicated the probable presence of an additional cyclic ester and an open-chain ester.

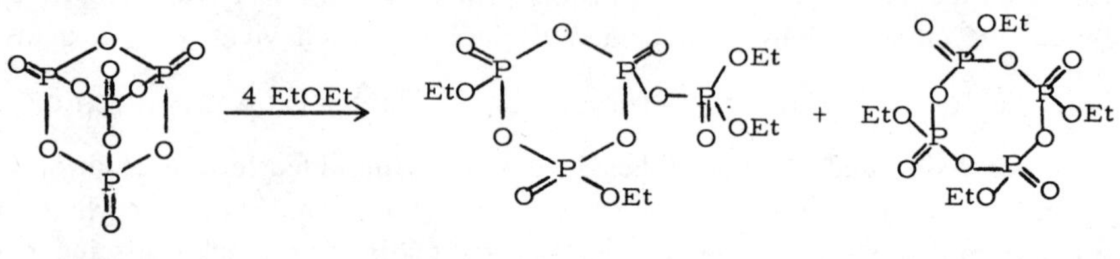
4 EtOEt
EtO
OEt
OEt
OEt
+
EtO
OEt
EtO
OEt
OEt

P_4O_{10} (1) (2) (3)

Reactions of particular biochemical interest are summarized in a review.[5] The reagent converts amino groups into reactive phosphamido groups capable of condensing with a carboxyl group. Amino acids and peptides can be converted into peptides of high molecular weight under mild conditions and without racemization. PPE also activates hydroxyl groups. Thus when a solution of D-glucose and methyl β-D-glucopyranoside (2) in dimethylformamide is treated with PPE, the glucose is

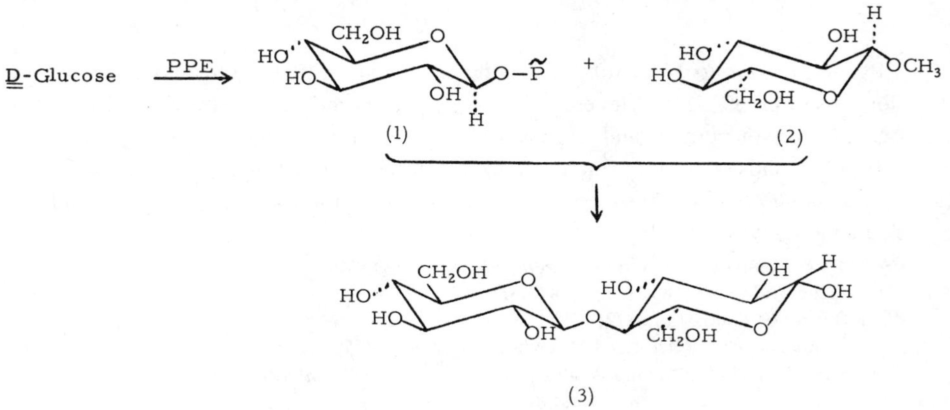

converted into "activated glucose" (1) which condenses with (2) with elimination of methanol to form in good yield a disaccharide identified by chromatography as pure cellobiose (3). "Activated glucose" has not been isolated; it is known not to be glucose-1-phosphate, for this ester does not react to form glucosides under the conditions described. Glucose is polymerized by the reagent into a long-chain, unbranched polyglucoside of a mean molecular weight of 50,000. Adenosine (6) can be prepared in good yield by the condensation of D-ribose (4) with adenine (5). Polynucleotides of high molecular weight can be prepared in good yield from various nucleotides.

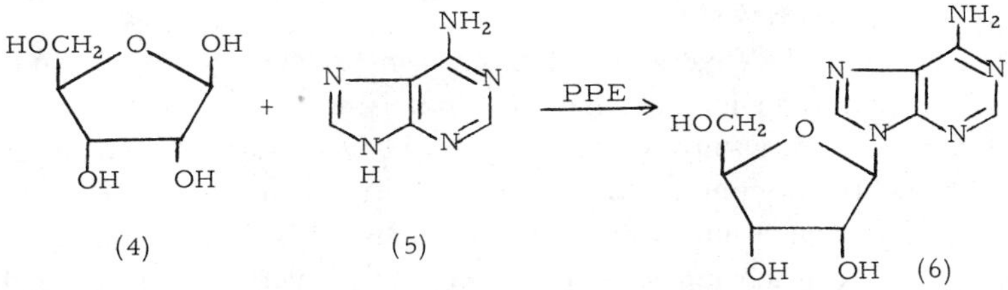

However, Hayes and Hansbury[6] found that although reaction of PPE with 5′-thymidylic acid led to polymerization, no high polymer with genuine phosphodiester linkages (polynucleotides) was formed.

Reagent for cyclodehydration. A Japanese group[7] prepared PPE as above and used the chloroform solution as such and found it to be an excellent reagent for the Bischler-Napieralski reaction. Thus phenylethylamides (1) are cyclized smoothly to dihydroisoquinolines (2) by refluxing with PPE in chloroform. Phenylpropylamides (3) similarly yield 3,4-dihydro-5H-2-benzopines (4). The same workers[8] report that

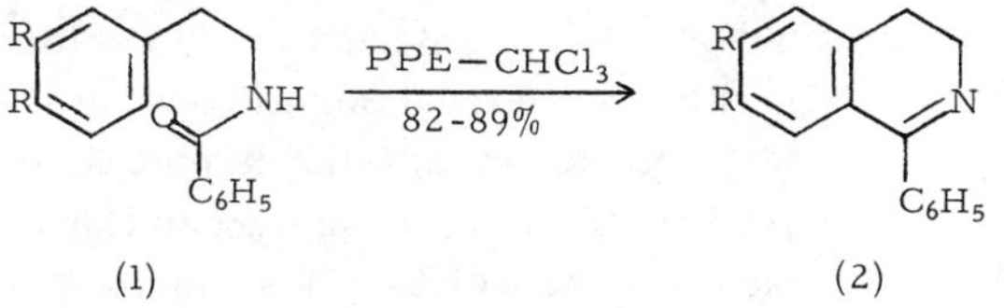

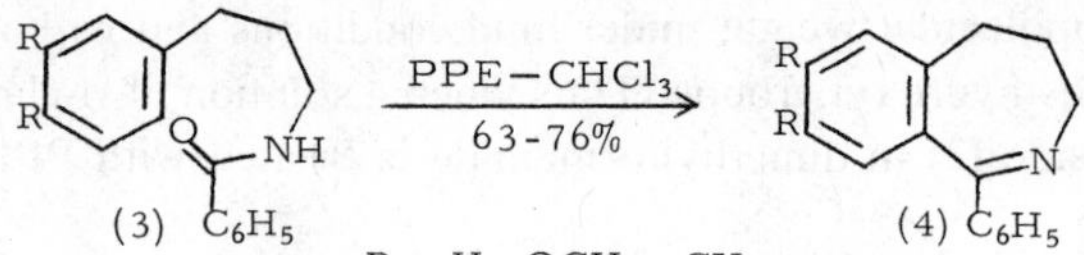

R = H, OCH_3, CH_3

PPE is superior to PPA for the synthesis of 2-substituted benzimidazoles by the condensation of *o*-phenylenediamine with carboxylic acids. An aliphatic acid is heated with the amine and excess PPE at 100° for 10 min. (yields about 70%). Aromatic acids require a temperature of 120° and a reaction period of 20–40 min. (yields 50–60%). PPE has been used successfully in the Fischer synthesis of indoles from phenylhydrazones.[9]

[1]W. Pollmann and G. Schramm, *Biochem. Biophys. Acta*, **80**, 1 (1964)
[2]K. Langheld, *Ber.*, **43**, 1857 (1910); **44**, 2076 (1911)
[3]R. Rätz and E. Thilo, *Ann.*, **572**, 173 (1951)
[4]G. Burckhardt, M. P. Klein, and M. Calvin, *Am. Soc.*, **87**, 591 (1965)
[5]G. Schramm, H. Grötsch, and W. Pollmann, *Angew. Chem., Internat. Ed.*, **1**, 1 (1962)
[6]F. N. Hayes and E. Hansbury, *Am. Soc.*, **86**, 4172 (1964)
[7]Y. Kanaoka, E. Sato, O. Yonemitsu, and Y. Ban, *Tetrahedron Letters*, No. 35, 2419 (1964)
[8]*Idem, Chem. Pharm. Bull.*, **12**, 793 (1964)
[9]Y. Kanaoka, Y. Ban, O. Yonemitsu, K. Irie, and K. Miyashita, *Chem. Ind.*, 473 (1965)

Polyphosphoric acid, 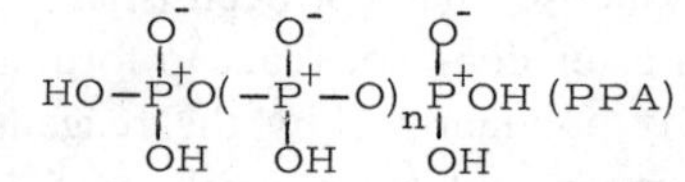

Supplier: Victor Chemical Works; phosphoric anhydride equivalent 82–84%. The reagent is a viscous liquid, almost a glass, but warming the container on the steam bath produces a mobile liquid which is suitable for easy pouring and which will remain so for days. Reviews.[1,2]

As a reagent for cyclodehydration, PPA is sometimes effective where sulfuric acid, hydrogen fluoride, or sodium aluminum chloride is not. It has good solvent power and contains anhydride groups which combine with water formed, with preservation of effective acidity. In contrast to sulfuric acid, PPA is not an oxidizing agent, it has no tendency to enter into aromatic substitutions, and it is less prone to promote rearrangements. A disadvantage is that, except at temperatures of 90° or higher, manual stirring of the viscous material is laborious. Hydrolysis of the reagent at the end of a reaction often is also laborious.

Mixtures of phosphoric acid and phosphorus pentoxide used by early workers probably varied in composition and degree of polymerization depending upon the time and temperature after mixing, for equilibrium at room temperature is very slow. Modern workers use the standard Victor material or else age a H_3PO_4–P_2O_5 mixture by heating.

Cyclodehydration. The first recorded use of the reagent provides a dramatic comparison in potency of PPA with three other reagents. In the synthesis of dibenzcoronene and coronene, Scholl and Meyer[3] effected the first of four ring closures of the hexacarboxylic acid (1) with concentrated sulfuric acid at room temperature. The red-violet product (2) was obtained in analytically pure form by precipitation from a solution in ammonia. Cyclization to (3) was accomplished with 20% oleum at 100°; the product was isolated as the violet-black sodium salt by adding sodium

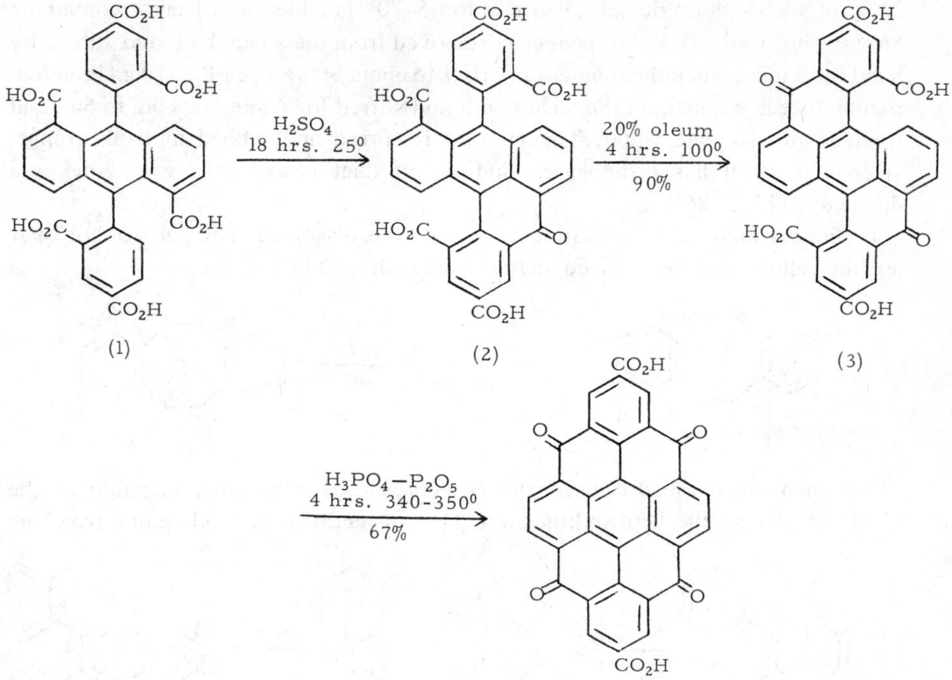

CO₂H
HO₂C
H_2SO_4
18 hrs. 25°
20% oleum
4 hrs. 100°
90%
$H_3PO_4-P_2O_5$
4 hrs. 340-350°
67%

(1) (2) (3) (4)

chloride to a solution in ammonium hydroxide. Finally, for double cyclization to (4), 1.5 g. of (3) was stirred with 10 g. of molten anhydrous phosphoric acid, 25 g. of phosphorus pentoxide was stirred in, and the mixture was heated at 340–350°. The product was precipitated by addition of hot water, dried, and stirred for 8 hrs. with concd. sulfuric acid at 100°, when most of the material dissolved to give a dark brown solution. Filtration through a sintered glass funnel and dilution with water gave analytically pure (4). Attempts to effect this cyclization with molten sodium aluminum chloride gave halogenated products.

In reporting a second instance of cyclization with a form of PPA, R. Robinson and co-workers,[4, 5] apparently unaware of the work of Scholl and Meyer, stirred 3 g. of the ketoacid (5) into a mixture of 10 g. of P_2O_5 and 10 ml. of sirupy phosphoric

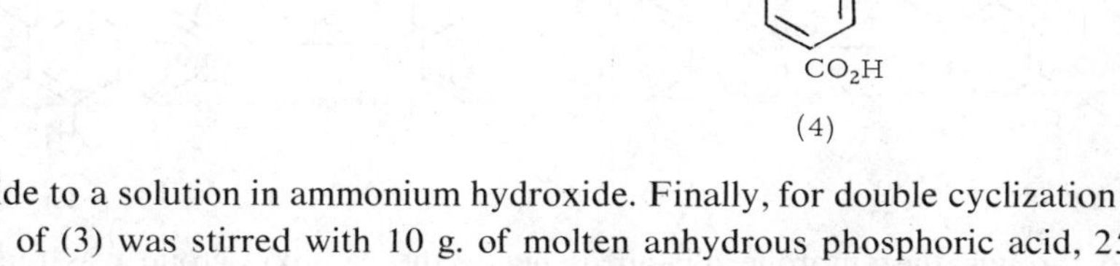

HO₂C
CH₂
CH₃O
$H_3PO_4-P_2O_5$
100°
25%

(5) (6)

acid (d. 1.75) and heated the mixture at 100°. The yield of (6) was low, but other reagents tried, including liquid hydrogen fluoride, gave no cyclized product. Birch, Jaeger, and Robinson[5] noted incidentally that by the same procedure γ-phenylbutyric acid affords α-tetralone in 86% yield and β-*m*-methoxyphenylpropionic acid affords 5-methoxy-1-indanone in 61% yield. The yield in the first case matches the upper yield of α-tetralone in an *Organic Syntheses* procedure using commercial PPA:[6]

22 g. of γ-phenylbutyric acid, warmed to 65–70°, is added to a beaker containing 80 g. of hot (90°) PPA, the beaker is removed from the steam bath and stirred by hand for 3 min., when the temperature rises to about 90°. More PPA (70 g.) is added, the mixture is warmed on the steam bath and stirred for 4 min., let cool to 60°, and treated with ice and water. The mixture is stirred and rubbed until all orange, viscous material has disappeared and the product is extracted with ether and distilled; yield 75–86%.

In another high-yield cyclization,[7] the acid (7) is dissolved in 50 parts of hot PPA and the yellow solution is stirred on the steam bath for 2 hrs.

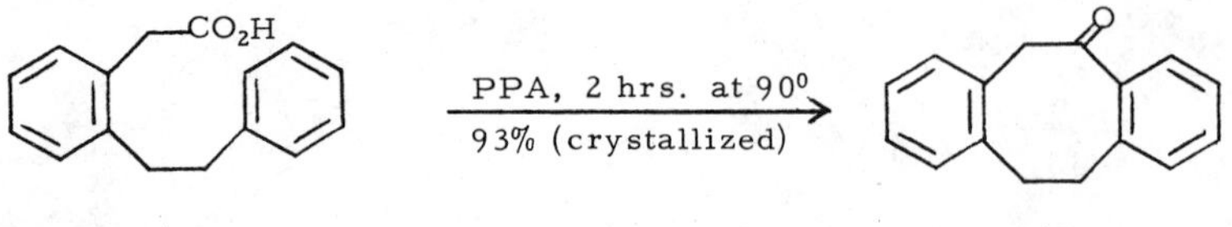
CO_2H
PPA, 2 hrs. at 90°
93% (crystallized)

(7) (8)

Newman and Seshadri[8] used the reagent successfully for cyclization of the aldehyde (10) to the benzanthracene (11). Conventional methods going from the

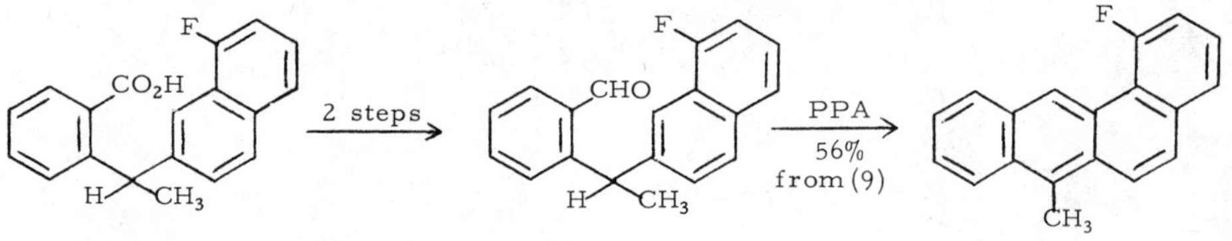
F
CO_2H
H
CH_3
2 steps
F
CHO
H
CH_3
PPA
56%
from (9)
F
CH_3

(9) (10) (11)

acid (9) through the anthrone all failed; hence the carboxyl group was reduced ($LiAlH_4$), the alcohol oxidized to the aldehyde (CrO_3–Py), and this was heated with PPA on the steam bath for 15 min. The intermediates were not purified, but the overall yield given is for pure product.

An example of several PPA cyclizations studied by Koo[9,10] is the preparation of ethyl 6,7-dimethoxy-3-methylindene-2-carboxylate (13). The keto ester (12) is added to PPA precooled to 5° in a beaker, and the mixture is stirred thoroughly

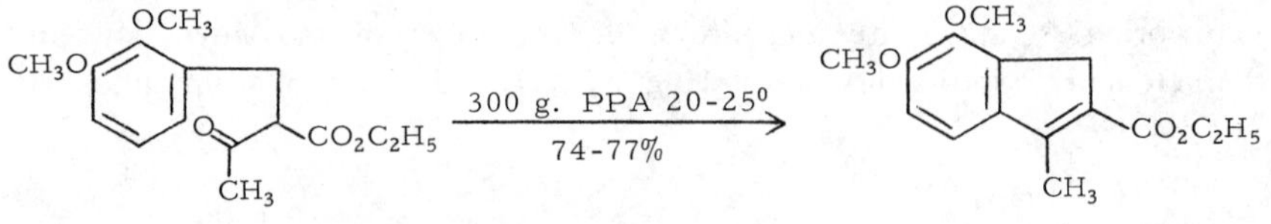
OCH_3
CH_3O
$CO_2C_2H_5$
CH_3
300 g. PPA 20-25°
74-77%
OCH_3
CH_3O
$CO_2C_2H_5$
CH_3

(12) 0.1 m. (13)

for 15 min. with a strong spatula while keeping the temperature at 20–25° by occasional ice cooling. The deep yellow paste which results is poured immediately into water, and the mixture stirred and triturated until all particles of yellow paste have been converted into colorless product. Extraction with chloroform gives a pale yellow oil which solidifies. Crystallization from ethanol gives colorless needles of pure (13).

Although the yield in the cyclization of (14) to (15) is very low, this reaction is better than the two other known routes to 9-keto-9H-pyrrolo [1,2a] indole (15), which forms bright yellow plates, m.p. 121–122° (Josey[11,12]). A novel feature of

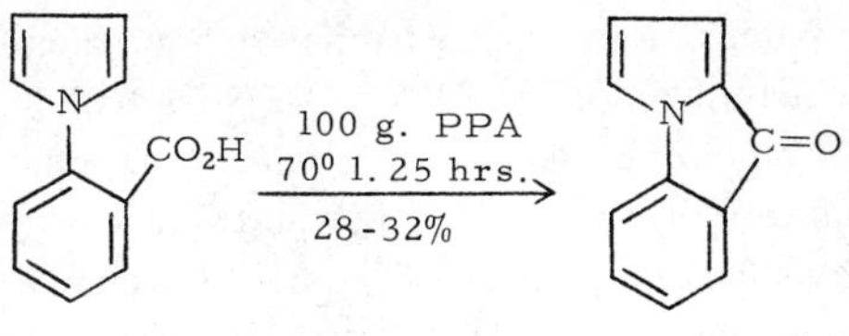

(14) 10 g. (15)

the workup is that the reaction mixture is poured into a rapidly stirred mixture of 200 ml. of ethyl acetate and 300 ml. of water. After 1 hr. the organic layer is separated and the water layer is extracted further with ethyl acetate. Chromatography separated (15) from other products.

Other cyclizations. Letsinger and co-workers[13] found that in hot PPA the ketone

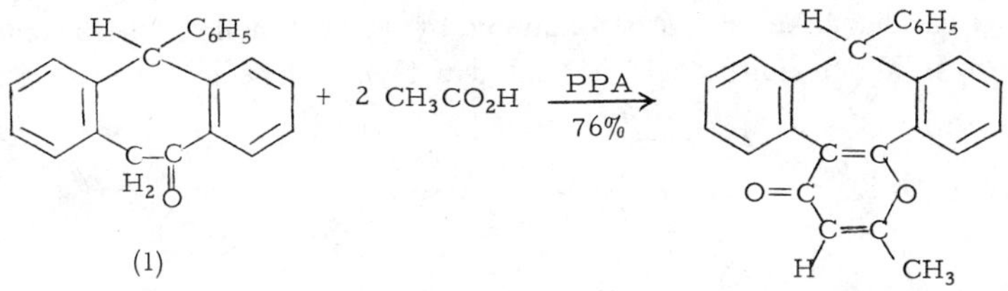

(1) (2)

(1) condenses with two moles of acetic acid to give the 4H-pyrane-4-one (2). In similar syntheses, dibenzyl ketone and acetic acid afford 2,6-dimethyl-3,5-diphenyl-

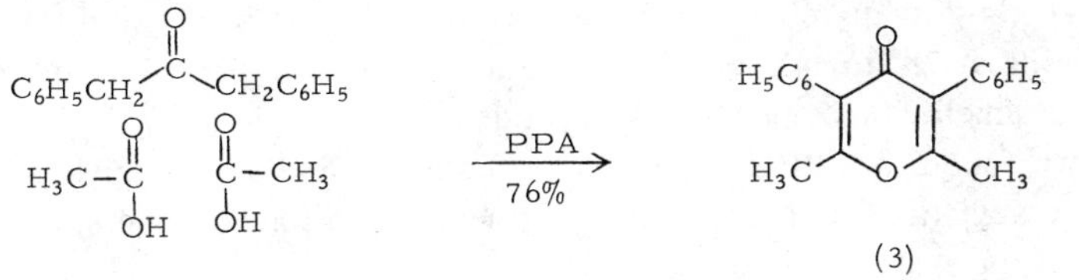

(3)

(4)

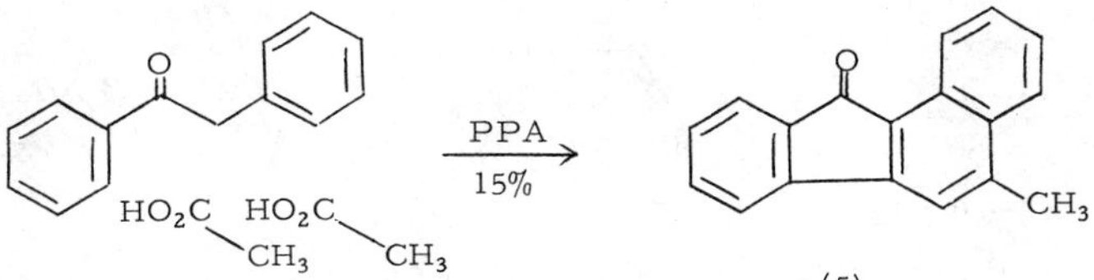

(5)

4H-pyrane-4-one (3),[14,15] and desoxybenzoin condenses with two moles of acetic acid to form the benzofluorenone (5), if in low yield. Heated alone in PPA, desoxy-

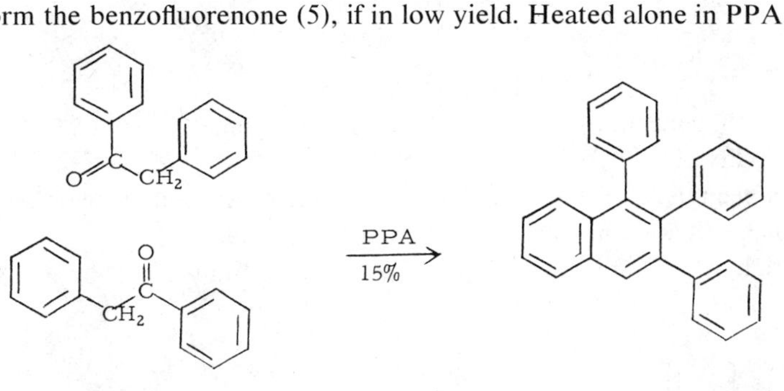

(4) (6)

benzoin underwent condensation to 1,2,3-triphenylnaphthalene.[14] A related reaction is the condensation of acetophenone in PPA to 1,3,5-triphenylbenzene.[15]

Acylation and alkylation of aromatic compounds. Benzophenones are obtained in good yield when a mixture of a phenolic ether and an alkoxybenzoic acid is treated with PPA prepared by stirring a mixture of 8 parts by weight of P_2O_5 and 5 parts by volume of 90% orthophosphoric acid (d1.75) at 85° for 30 min. (Ayers and Denney[16]). Thus veratrole and vanillic acid afforded 4-hydroxy-3,3′,4′-trimethoxybenzophenone.

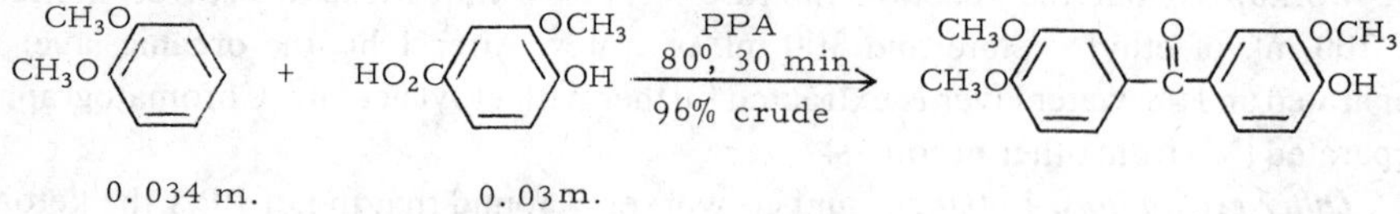
CH3O
CH3O
HO2C
OCH3
OH
PPA
80°, 30 min.
96% crude
CH3O
CH3O
O
C
OCH3
OH
0.034 m.
0.03 m.

Symmetrical diaryl ketones are obtainable by a novel method of transcarbonylation (Fuson[17]). When duroic acid is heated with *m*-xylene in PPA at 78° for 4.5

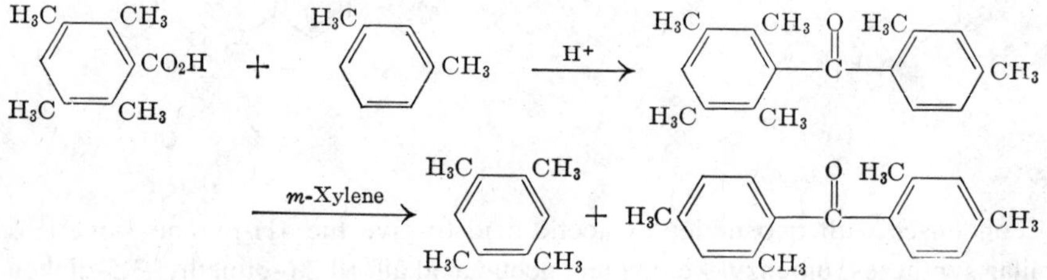
H3C
CH3
CO2H
H3C
CH3
+
H3C
CH3
H+
H3C
CH3
O
H3C
C
CH3
H3C
CH3
m-Xylene
H3C
CH3
H3C
CH3
+ H3C
O
H3C
C
CH3
CH3

hrs., duryl 2,4-dimethylphenyl ketone was obtained in 62% yield. When, after initial heating at 78° for 4.5 hrs., the mixture was heated for 5 hrs. more at 150°, 2,2′,4,4′-tetramethylbenzophenone was produced in 27% yield.

Gardner[18] found that PPA serves as both solvent and acid catalyst for Friedel-Crafts type acylations and alkylations of phenols and phenol ethers, as illustrated by the following examples:

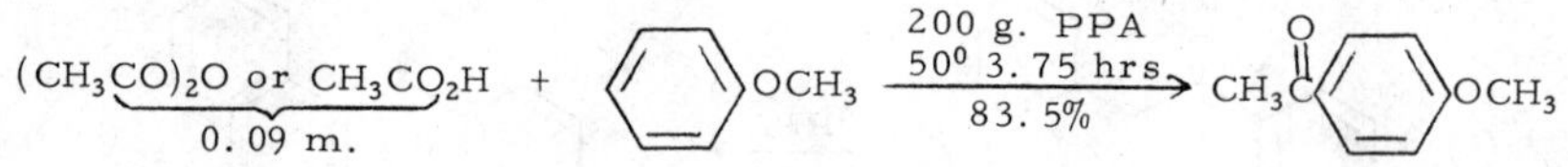
(CH3CO)2O or CH3CO2H
0.09 m.
+
OCH3
200 g. PPA
50° 3.75 hrs.
83.5%
CH3C
O
OCH3

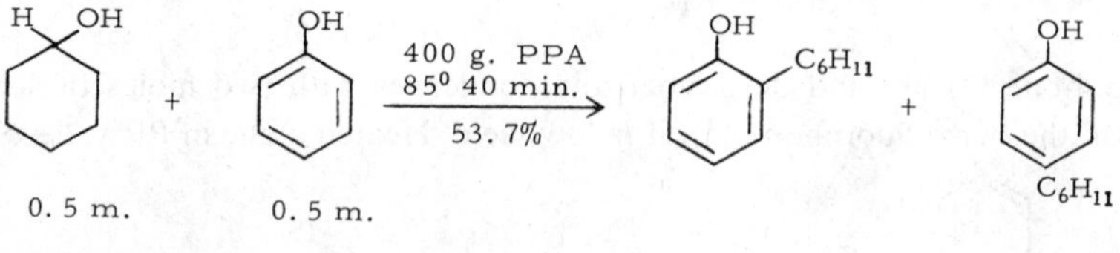
H
OH
+
OH
400 g. PPA
85° 40 min.
53.7%
OH
C6H11
+
OH
C6H11
0.5 m.
0.5 m.

Attempts with benzene, toluene, and cyclohexene failed completely because of lack of solubility of the hydrocarbons in PPA.

PPA has been used for the formylation of aromatic compounds with hexamethylenetetramine; yields are not high, but the method is simple.[19]

Acetylation of 2,3-dimethoxytoluene with acetic anhydride and PPA affords

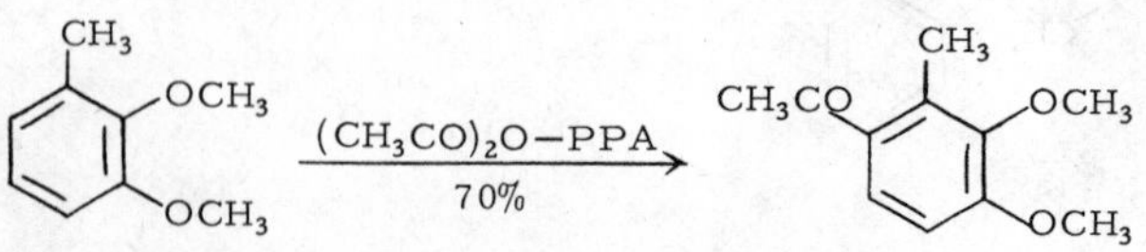
CH3
OCH3
OCH3
(CH3CO)2O—PPA
70%
CH3
CH3CO
OCH3
OCH3

the 6-acetyl derivative as the only isolated product; acetylation with acetyl chloride and aluminum chloride in CS_2, C_6H_6, or $C_6H_5NO_2$ gave the 5-(52–59%) and the 6-(41–45%) acetyl derivatives.[20]

Phenyl esters are obtained by heating an acid and a phenol in PPA (Bader[21]). In the example formulated the reaction mixture was cooled, treated with toluene,

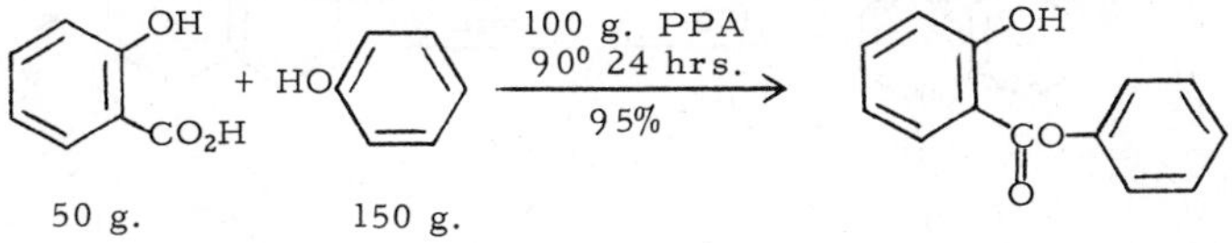
OH
+ HO
CO_2H
100 g. PPA
90° 24 hrs.
95%
OH
CO
O

50 g. 150 g.

and the toluene layer extracted with bicarbonate solution for recovery of 14 g. of salicylic acid.

Synthesis of heterocycles. An American Cyanamid group[22] showed that (2) can be prepared in high yield by mixing *o*-phenylenediamine and benzoic acid with sufficient PPA to give a stirrable paste, heating the mixture slowly to 250°, and

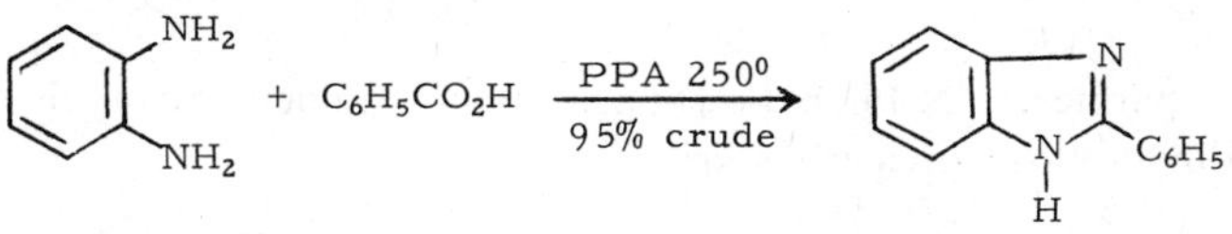
NH_2
NH_2
+ $C_6H_5CO_2H$
PPA 250°
95% crude
N
N
H
C_6H_5

(1) (2)

stirring at this temperature for 4 hrs. The method is very general. 2-Alkylbenzimidazoles can be obtained by using an aliphatic acid. Replacement of *o*-phenylenediamine by *o*-aminophenol or by *o*-aminothiophenol gives benzoxazoles and benzothiazoles. The method has been used also for the synthesis of bisbenzimidazoles[23] and bisbenzothiazoles.[24] See, however, claims for PPE on p. 894.

PPA is an efficient reagent for the Fischer indole synthesis.[25] Thus 2-phenylindole

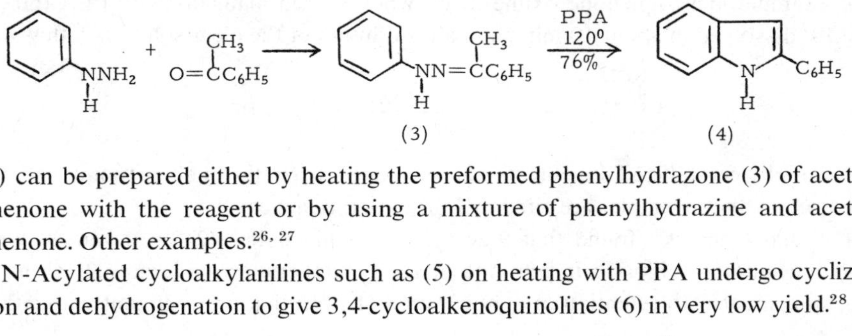
NNH_2
H
+
CH_3
$O{=}CC_6H_5$
CH_3
$NN{=}CC_6H_5$
H
PPA
120°
76%
N
H
C_6H_5

(3) (4)

(4) can be prepared either by heating the preformed phenylhydrazone (3) of acetophenone with the reagent or by using a mixture of phenylhydrazine and acetophenone. Other examples.[26, 27]

N-Acylated cycloalkylanilines such as (5) on heating with PPA undergo cyclization and dehydrogenation to give 3,4-cycloalkenoquinolines (6) in very low yield.[28]

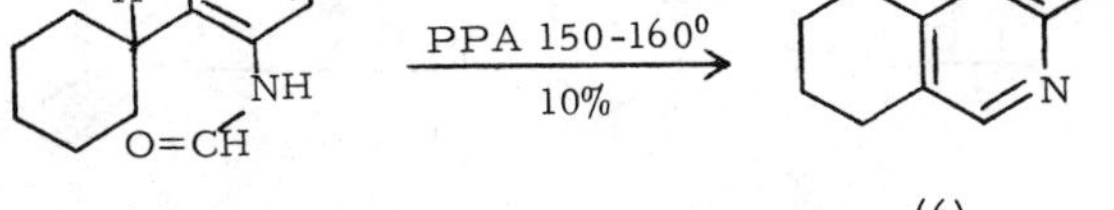
H
NH
O=CH
PPA 150-160°
10%
N

(5) (6)

Hurd and Hayao[29] found PPA very effective for the cyclodehydration of 3-aryloxypropionic acids to 4-chromanones, of 3-arylmercaptopropionic acids to 4-thiochromanones, and of N,N-diaryl-β-alanines to 2,3-dihydro-1-aryl-4(11)-quinolines.

Whereas other methods of cyclizing acids such as (1) to 4-ketotetrahydroquinolines (2) had required protection of the secondary amino hydrogen by tosylation, Koo[30] found that with PPA cyclization can be effected directly in one step.

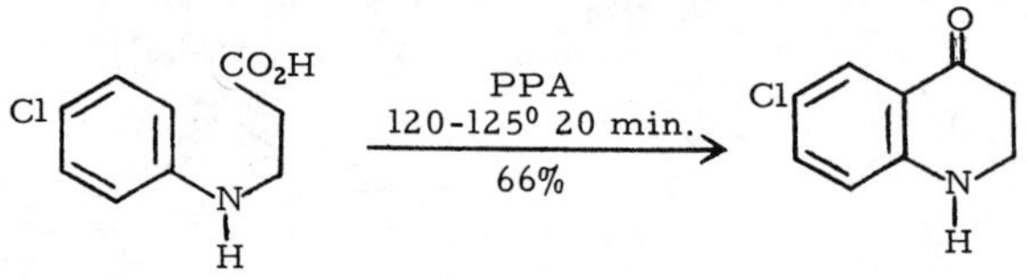
CO2H
Cl
N
H
PPA
120-125° 20 min.
66%
O
Cl
N
H

(1) (2)

The reagent proved useful for the cyclodehydration of 4-amino-5-arylamidopy-

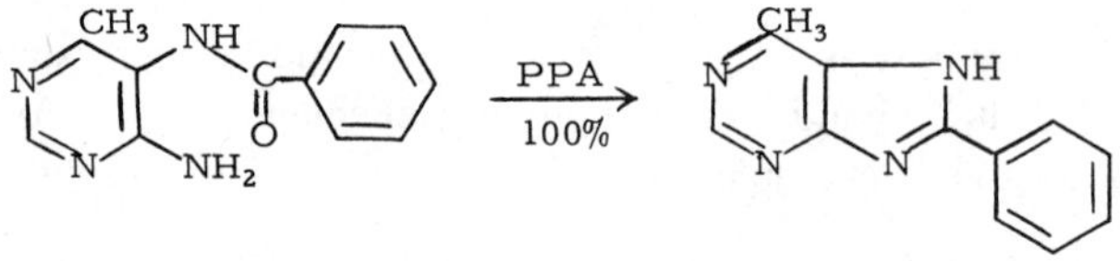
CH3
NH
N
C
O
N
NH2
PPA
100%
CH3
NH
N
N
N

rimidines to purines.[31] N-Phenyl-α,β-unsaturated amides are cyclized in high yield to 3,4-dihydro-4-phenylcarbostyrils.[32]

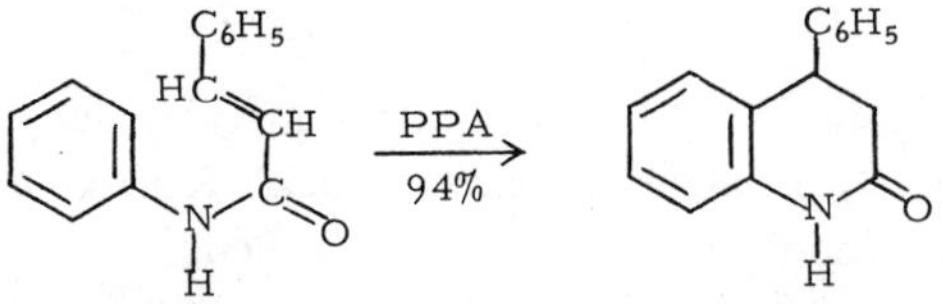
C6H5
HC
CH
N
C
O
H
PPA
94%
C6H5
N
O
H

Rearrangements. Horning and Stromberg found PPA an excellent solvent-reagent for effecting Beckmann rearrangement of ketoximes[33] and aldoximes.[34] For example, benzophenone oxime (2 g.) when stirred manually with PPA (60 g.) at 130° dissolved in about 10 min., and after dilution of the clear solution with water

$$C_6H_5\overset{\overset{\displaystyle NOH}{\|}}{C}C_6H_5 \xrightarrow[99\%]{\text{PPA } 130^0,\ 10 \text{ min.}} C_6H_5NHCOC_6H_5$$

benzanilide crystallized in pure form, as judged from the melting point and IR spectrum. Other examples are cited by Barnes and Beachem.[35]

Hill and Chortyk[36] found that 9-acetyl-*cis*-decalin oxime (2) on treatment with *p*-toluenesulfonyl chloride in pyridine at room temperature undergoes Beckmann rearrangement with retention of configuration (1), but that on treatment with either

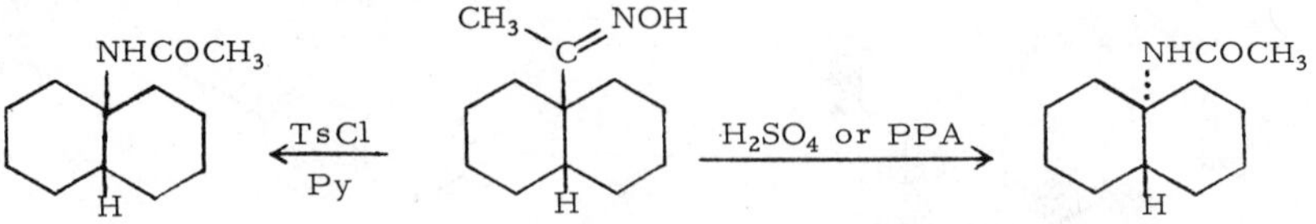
NHCOCH3
H
TsCl
Py
CH3
NOH
C
H
H2SO4 or PPA
NHCOCH3
H

(1) (2) (3)

concd. sulfuric acid or PPA at room temperature overnight it affords the *trans*-acetamide (3). Experimental support was found for the postulate that the inversion under acidic conditions involves fragmentation of the protonated oxime (4) into the 9-decalyl carbonium ion and acetonitrile (5) and recombination of these fragments.

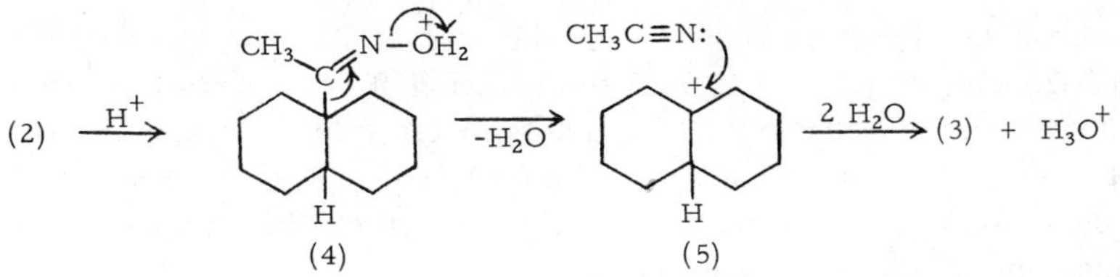

Fluorenone oxime was recovered unchanged after being heated with PPA at 120°, but at 180° it was rearranged in almost quantitative yield (Conley[37]). Conley found PPA to be a superior solvent-catalyst for the Schmidt rearrangement. He added sodium azide in portions to a mixture of a diaryl, aryl-alkyl, symmetrical or unsymmetrical ketones with 15–20 parts of PPA at a temperature in the range 25–75° until evolution of nitrogen ceased, and obtained amides in yields mainly of 80–90%. Doorenbos and Wu[38] studied the conversion of cholestanone and

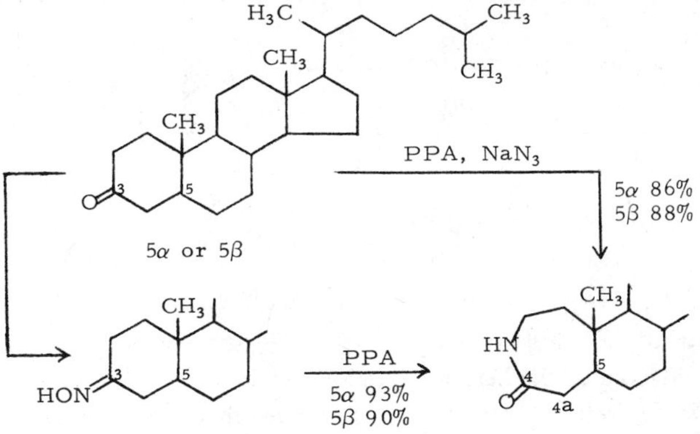

coprostanone into the 3-aza-A-homo-4-ones by Beckmann rearrangement of the oximes with PPA at 120–130° (10–20 min.) and by the Schmidt reaction in PPA at 50–60° for 10 hrs. Yields in all cases were close to 90%. For comparison they carried out the Schmidt reaction in benzene with a solution of hydrazoic and concd. sulfuric acid at room temperature and obtained the 5α- and 5β-azasteroids in yields of 81 and 71%.

In a kinetic study of the Beckmann rearrangement of acetophenone oximes, Pearson and Stone[39] found that rearrangement is 12–35 times as fast in PPA as it is in sulfuric acid and that the rates of rearrangement of substituted acetophenone oximes in PPA are nearly all the same. They conclude that the PPA reaction follows a mechanism different from that of the H_2SO_4 reaction.

The usual procedures for use of PPA in the Schmidt reaction call for a temperature in the range 60–100° or even higher. Stockel and Hall[40] report that improved yields can be obtained at room temperature (excess sodium azide is used). However, the yield depends upon the substituents present. Thus *p*-nitrobenzoic acid gives *p*-nitroaniline in only 50% yield, whereas *p*-methoxyaniline is obtained from *p*-methoxybenzoic acid in 80% yield.

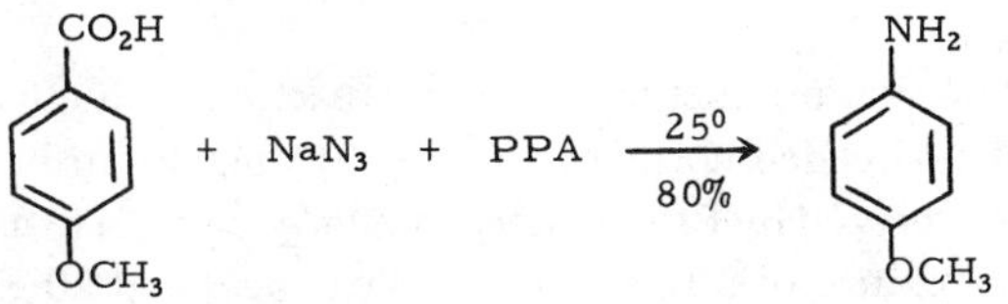

Universal Oil Products supplies a solid form of PPA on kieselguhr ("No. 2 Polymerization catalyst"). In a study of the action of PPA on *cis,trans,trans*-1,5,9-cyclododecatriene, Wellman *et al.*[41] found the UOP solid catalyst more convenient than the sirup. The solid catalyst can be removed by filtration, and extraction of the product from aqueous PPA is avoided. The reaction affords a mixture containing acenaphthene and decahydroacenaphthene.

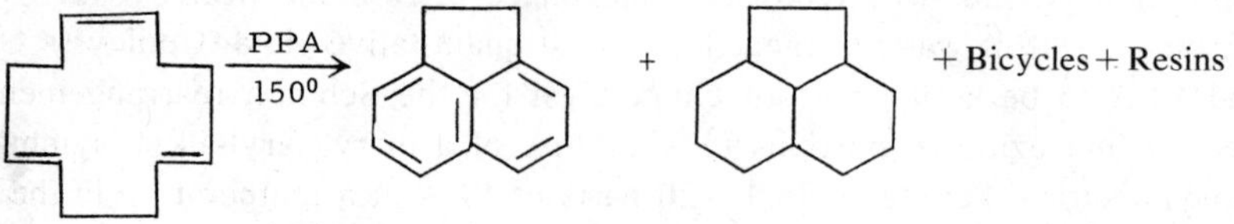
PPA
150°
+
+ Bicycles + Resins

Isomerization of γ-unsaturated alcohols. Colonge and Brunie[42] heated a mixture of 60 ml. of 85% phosphoric acid and 51 g. of P_2O_5 for 1½ hrs. on the steam bath,

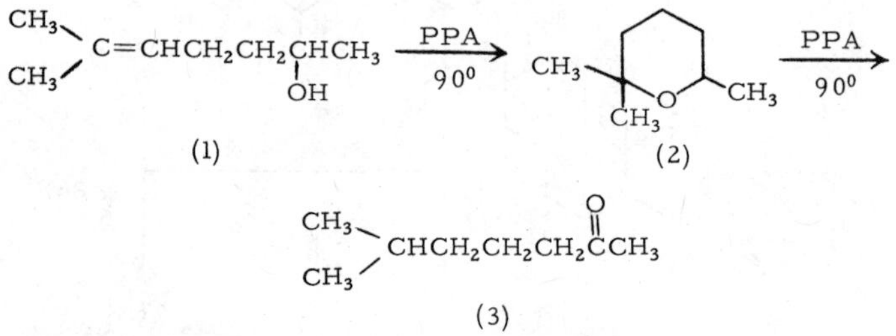
CH3
C=CHCH2CH2CHCH3
CH3
OH
PPA
90°
CH3
CH3
CH3
PPA
90°
(1)
(2)
CH3
CHCH2CH2CH2CCH3
CH3
O
(3)

added 20 g. of the alcohol (1), stirred the red⟶brown mixture for 1½ hrs. on the steam bath, and isolated the ketone (3) in 80% yield. The reaction evidently proceeds via 2,2,6-trimethyltetrahydropyrane (2), since this substance under the same conditions afforded (3) in 90% yield.

Epoxide fission. Tomoeda *et al.*[43] report interesting results on the fission of 4β,5β-epoxycoprostane-3-one (2) at room temperature. Acetolysis, catalyzed by

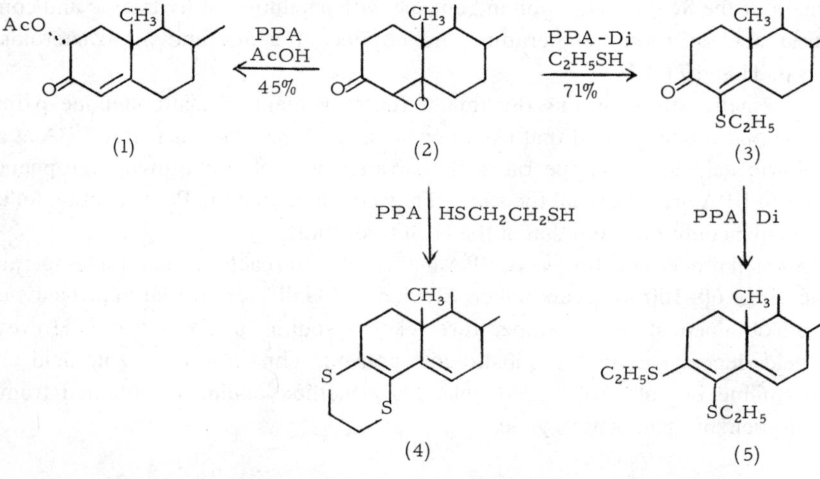
AcO
CH3
PPA
AcOH
45%
CH3
PPA-Di
C2H5SH
71%
CH3
SC2H5
(1)
(2)
(3)
PPA
HSCH2CH2SH
PPA
Di
CH3
S
S
CH3
C2H5S
SC2H5
(4)
(5)

PPA, gave the abnormal product (1), whereas reaction with ethanethiol in dioxane catalyzed by PPA proceeded normally to give (3); this on further reaction afforded (5). Ethanedithiol catalyzed by PPA reacted similarly with (2) to give (4).

Nitration.[44] A mixture of 50 g. of 100% nitric acid and 80 g. of PPA is stirred

at 60° until homogeneous, and a diethyl alkylmalonate is added dropwise in 15–30 min. with cooling to maintain a temperature of 60°. After 1 hr. more at 60° the

$$\underline{n}\text{-}C_4H_9CH(CO_2C_2H_5)_2 \xrightarrow[75\%]{HNO_3\text{-}PPA} \underline{n}\text{-}C_4H_9C(NO_2)(CO_2C_2H_5)_2$$

mixture is cooled and poured onto ice. Previously fuming nitric acid–acetic anhydride had been used, but this combination is potentially hazardous.

Hydrolysis. The enaminonitrile (1) is converted into the ketoamide (2) by hot PPA in high yield (1 g. ⟶ 1.09 g., nearly pure).[45] Sulfuric acid and sulfuric–acetic

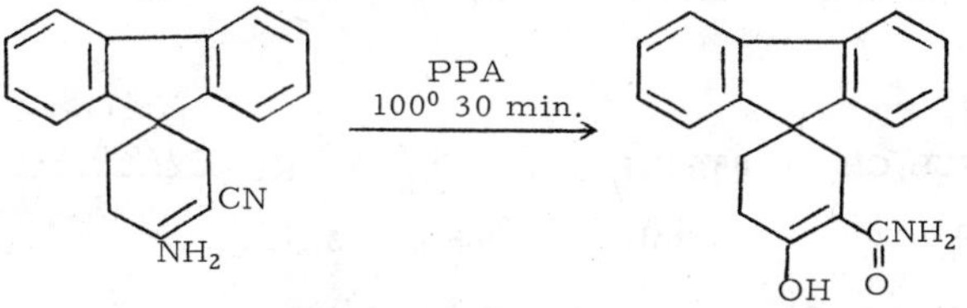

(1) (2)

acid mixtures gave water-soluble material; hydrochloric–acetic acid gave mixtures; basic hydrolysis gave none of the desired product.

Acids ⟶ amines or amides. Snyder and co-workers[46] found that certain aromatic acids are convertible into the corresponding amines by mechanically stirring a mixture of the acid, hydroxylamine hydrochloride, and PPA, gradually raising the

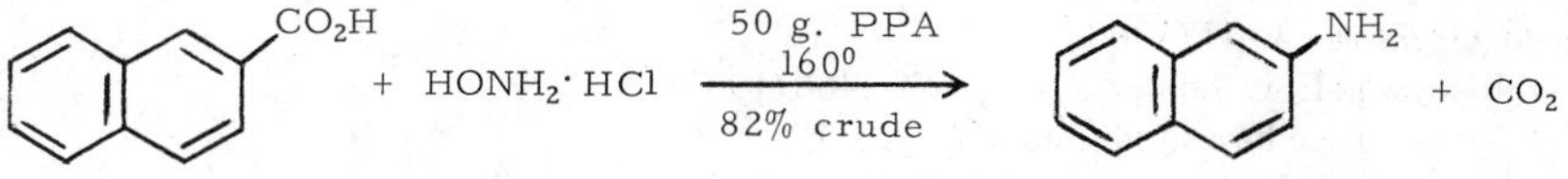

0.23 m.

temperature to 160°, and maintaining this temperature until evolution of carbon dioxide ceases. The reaction failed completely with *o*-, *m*-, and *p*-nitrobenzoic acids and with caprylic acid. Acetic acid reacts with very weakly basic aromatic amines to give amides. The reaction failed with benzoic acid.

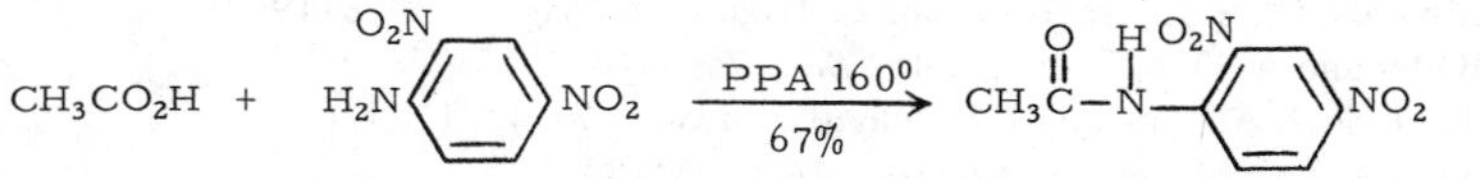

Nitromethane can serve as a source of hydroxylamine:[47]

$$CH_3NO_2 \xrightarrow{PPA} NH_2OH + CO$$

Thus a mixture of *p*-chlorobenzoic acid, nitromethane, and PPA when heated at 115° for 10 min. afforded *p*-chloroaniline in 60% yield.

Nitriles ⟶ amides. PPA is an excellent reagent for the hydrolysis of unhindered

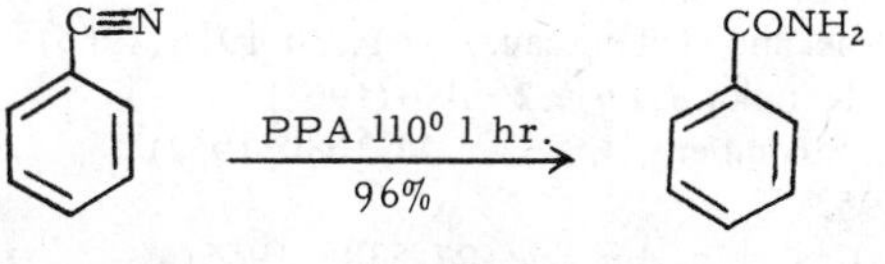

aromatic nitriles to amides (Snyder and Elston[48]). Cyanomesitylene did not react at 120° and at 155° it was converted into mesitylene.

Phosphorylation. Cherbuliez and Rubinowitz[49] used polyphosphoric acid for the preparation of monophosphates of amino alcohols and hydroxy esters. Hall and Khorana[50] used a mixture of 85% phosphoric acid and P_2O_5 for the synthesis of uridine-5′-phosphate.

Ether cleavage. Stone and Shechter[51] found that ethers are cleaved readily to iodides by treatment with 95% phosphoric acid and potassium iodide:

$$\text{ROR} + 2\,\text{KI} + 2\,H_3PO_4 \longrightarrow 2\,\text{RI} + 2\,KH_2PO_4 + H_2O$$

In applying the method to the ether (1), Cope *et al.*[52] used polyphosphoric acid prepared from phosphoric acid and P_2O_5; the yield of iodide (2) was practically quantitative.

$$\underset{(1)\ 0.90\ \text{g.}}{CH_3O(CH_2)_{10}\overset{\overset{CH_3}{|}}{C}HCH_2CH_3} + \underset{2\ \text{ml.}}{85\%\ H_3PO_4} + \underset{0.8\ \text{g.}}{P_2O_5} + \underset{3.32\ \text{g.}}{\text{KI}} \xrightarrow[99\%]{135\text{-}140^\circ,\ 5\,1/4\ \text{hrs.}} \underset{(2)}{I(CH_2)_{10}\overset{\overset{CH_3}{|}}{C}HCH_2CH_3}$$

[1]F. D. Popp and W. E. McEwen, *Chem. Rev.*, **58**, 321 (1958)
[2]F. Uhlig and H. R. Snyder, *Advances in Organic Chemistry*, **1**, 35 (1960)
[3]R. Scholl and K. Meyer, *Ber.*, **65**, 902 (1932)
[4]A. Koebner and R. Robinson, *J. Chem. Soc.*, 1994 (1938)
[5]A. J. Birch, R. Jaeger, and R. Robinson, *J. Chem. Soc.*, 582 (1945)
[6]H. R. Snyder and F. X. Werber, *Am. Soc.*, **72**, 2965 (1950); *Org. Syn., Coll. Vol.*, **3**, 798 (1955)
[7]A. C. Cope and R. D. Smith, *Am. Soc.*, **77**, 4596 (1955)
[8]M. S. Newman and S. Seshadri, *J. Org.*, **27**, 76 (1962)
[9]J. Koo, *Am. Soc.*, **75**, 1891 (1953)
[10]J. Koo, *Org. Syn.*, **40** 43 (1960)
[11]A. D. Josey and E. L. Jenner, *J. Org.*, **27**, 2466 (1962)
[12]A. D. Josey, procedure submitted to *Org. Syn.*
[13]R. L. Letsinger, J. D. Jamison, and A. S. Hussey, *J. Org.*, **26**, 97 (1960)
[14]R. L. Letsinger and J. D. Jamison, *Am. Soc.*, **83**, 193 (1961)
[15]T. L. Emmick and R. L. Letsinger, procedure submitted to *Org. Syn.*
[16]D. C. Ayres and R. C. Denney, *J. Chem. Soc.*, 4506 (1961)
[17]R. C. Fuson, G. R. Bakker, and B. Vittimberga, *Am. Soc.*, **81**, 4858 (1959)
[18]P. D. Gardner, *Am. Soc.*, **76**, 4550 (1951)
[19]D. A. Denton and H. Suschitzky, *J. Chem. Soc.*, 4741 (1963)
[20]J. D. Edwards, Jr., S. E. McGuire, and C. Hignite, *J. Org.*, **29**, 3028 (1964)
[21]A. R. Bader and A. D. Kontowicz, *Am. Soc.*, **75**, 5416 (1953)
[22]D. W. Hein, R. J. Alheim, and J. J. Leavitt, *Am. Soc.*, **79**, 427 (1957)
[23]L. L. Wang and M. M. Joullié, *Am. Soc.*, **79**, 5706 (1957)
[24]C. Rai and J. B. Braunworth, *J. Org.*, **26**, 3474 (1961)
[25]H. M. Kissman, D. W. Farnsworth, and B. Witkop, *Am. Soc.*, **74**, 3948 (1952)
[26]A. P. Gray and W. L. Archer, *Am. Soc.*, **79**, 3554 (1957)
[27]H. B. MacPhillamy, R. L. Dziemian, R. A. Lucas, and M. E. Kuehne, *Am. Soc.*, **80**, 2172 (1958)
[28]D. A. Denton, R. K. Smalley, and H. Suschitzky, *J. Chem. Soc.*, 2421 (1964)
[29]C. D. Hurd and S. Hayao, *Am. Soc.*, **76**, 5065 (1954)
[30]J. Koo, *J. Org.*, **26**, 2440 (1961); **28**, 1134 (1963)
[31]S.-C. J. Fu, E. Chinoporos, and H. Terzeau, *J. Org.*, **30**, 1916 (1965)
[32]R T. Conley and W. N. Knopka, *J. Org.*, **29**, 496 (1964)
[33]E. C. Horning and V. L. Stromberg, *Am. Soc.*, **74**, 2680 (1952)
[34]*Idem, ibid.*, **74**, 5151 (1952)
[35]R. A. Barnes and M. T. Beachem, *Am. Soc.*, **77**, 5388 (1955)
[36]R. K. Hill and O. T. Chortyk, *Am. Soc.*, **84**, 1064 (1962)
[37]R. T. Conley, *J. Org.*, **23**, 1330 (1958)
[38]N. J. Doorenbos and M. T. Wu, *J. Org.*, **26**, 2548 (1961)

[39]D. E. Pearson and R. M. Stone, *Am. Soc.*, **83**, 1715 (1961)
[40]R. F. Stockel and D. M. Hall, *Nature*, **197**, 787 (1963)
[41]W. F. Wellman, M. C. Brennerman, and M. P. Konecky, *J. Org.*, **30**, 1482 (1965)
[42]J. Colonge and J.-C. Brunie, *Bull. Soc.*, 1799 (1963)
[43]M. Tomoeda *et al.*, *Chem. Pharm. Bull.*, **12**, 383 (1964)
[44]J. P. Kispersky and K. Klager, *Am. Soc.*, **77**, 5433 (1955)
[45]S. Baldwin, *J. Org.*, **26**, 3280 (1961)
[46]H. R. Snyder, C. T. Elston, and D. B. Kellom, *Am. Soc.*, **75**, 2014 (1953)
[47]G. B. Bachman and J. E. Goldmacher, *J. Org.*, **29**, 2576 (1964)
[48]H. R. Snyder and C. T. Elston, *Am. Soc.*, **76**, 3039 (1954)
[49]E. Cherbuliez and J. Rabinowitz, *Helv.*, **39**, 1455, 1461 (1956); **41**, 1168 (1958)
[50]R. H. Hall and H. G. Khorana, *Am. Soc.*, **77**, 1871 (1955)
[51]H. Stone and H. Shechter, *J. Org.*, **15**, 491 (1950)
[52]A. C. Cope *et al.*, *Am. Soc.*, **87**, 5452 (1965)

Potassium. Atomic wt. 39.10, m.p. 62.3°. Suppliers: F, MCB.

Procedures for safe handling. For the preparation of a potassium *t*-alkoxide it is necessary to remove the outer oxide-coated layer from the 20-g. lumps of commercial potassium, weigh the clean metal, and transfer it under nitrogen to a flask containing the alcohol. Two alternative techniques are described below.

W. S. Johnson's procedure.[1,2] (The following is a brief outline; the original should be consulted for details and precautions). The oxidized surface of a lump of metal is cut off with a knife under xylene in a mortar and each scrap is transferred immediately with tweezers to a second mortar containing xylene. A clean piece of metal is removed with tweezers, blotted rapidly with a piece of filter paper, and introduced into a tared beaker containing xylene. For the preparation of potassium *t*-butoxide the weighed potassium is introduced to a flask flushed with nitrogen and containing *t*-butanol. All metal scraps and residue are decomposed immediately by placing the mortar at the rear of the hood, making ready a square of asbestos with which to cover the vessel if the liquid catches fire, and adding *t*-butanol in small portions from a medicine dropper at a rate such that the reaction does not become too vigorous.

D. E. Pearson's procedure.[3,4] The following procedure, which includes the preparation of potassium 2-methyl-2-butoxide, is considerably less wasteful of potassium than that described above, and it is regarded as simpler and less dangerous.

Two liters of *t*-amyl alcohol is dried by adding 10 g. of sodium, refluxing for several hours, and then distilling: b.p. 101–103.5°, $n_D^{25.6}$ 1.4018.

The reaction vessel is a 2-l. round-bottomed, three-necked flask equipped with reflux condenser and drying tube mounted in the side openings. The center neck is a 45/50 standard-taper joint to permit introduction of the rather large eggs of purified potassium described below. A nitrogen atmosphere is not necessary (hydrogen gas and solvent vapors protect the potassium) but if desired can be provided by the technique described by Johnson and Schneider[2] for alternately evacuating and introducing nitrogen into the reaction vessel. The flask is charged with 1 l. of dry *t*-amyl alcohol.

Purification of potassium is done in a 400-ml. beaker, with a watchglass at hand for covering the beaker, and with an ice-water bath for cooling rigid enough to prevent tipping of the beaker with introduction of water. The beaker is charged with 150 ml. of dry *t*-amyl alcohol* and the solvent is warmed to 70° by heating on a hot

*If the alcohol is replaced by a mixture of 120 ml. of dry toluene and 30 ml. of *t*-amyl alcohol, control of the hydrogen evolution is somewhat better.

plate. A lump of commercial potassium (usually about 20 g. or less) is added to the warm alcohol, and the beaker is covered immediately (otherwise small fragments of potassium tend to ignite). Evolution of hydrogen starts at once, and soon molten potassium flows out of the oxide shell and collects at the bottom of the beaker. At exactly this point the beaker with cover is transferred to the ice bath and held upright so that the potassium freezes to an egg. The evolution of hydrogen nearly ceases. The weight of the beaker plus contents is recorded, the potassium egg is lifted out of the alcohol with a pair of long tweezers (it glows with a blue fluorescence) and transferred to the reaction flask through the central opening; the weight of beaker and cover plus solvent, subtracted from the original weight, gives the amount of potassium added. Although exposure to air does not seem to have a deleterious effect, the egg during transfer can be held inside a filter funnel through which dry nitrogen is passed. The above process is repeated with further lumps of potassium with the same lot of *t*-amyl alcohol originally used for purification until about 1.5 g. atoms of metal has been added to the 1 l. of *t*-amyl alcohol. The "egg shell" residues contain very little potassium; decantation of the solvent and treatment with *t*-butanol destroys any active metal.

Refluxing the mixture of purified potassium in *t*-amyl alcohol for 3–4 hrs. afforded a solution calculated to be 0.19 *M* in potassium 2-methyl-2-butoxide and found by titration with standard acid to be 0.16 *M*. The appearance of the solution seems to be dependent upon the condition of the potassium used. Potassium that melts to a bright shiny pool of metal gives a clear light-colored solution of potassium 2-methyl-2-butoxide. If it melts to a crumbly, semi-solid liquid, the solution of alkoxide is darker colored.

t-Butanol cannot be used similarly to purify potassium because it reacts too rapidly with the metal. However, potassium purified by the *t*-amyl alcohol procedure described contains so little of this alcohol that it can be used to prepare potassium *t*-butoxide.

DePuy[5] has described a comparable method for purification of potassium by melting under heptane in a nitrogen atmosphere.

[1]W. S. Johnson and G. H. Daub, *Org. Reactions*, **6**, 42 (1951)
[2]W. S. Johnson and W. P. Schneider, *Org. Syn., Coll. Vol.*, **4**, 132 (1963)
[3]Contributed by D. E. Pearson, Vanderbilt University
[4]See also D. E. Pearson and O. D. Keaton, *J. Org.*, **28**, 1557 (1963)
[5]C. H. DePuy, G. F. Morris, J. S. Smith, and R. J. Smat, *Am. Soc.*, **87**, 2421 (1965)

Potassium acetate, CH_3CO_2K. Mol. wt. 98.15, m.p. 292°, solubility in 100 g. water: 217 g. at 0°, 396 g. at 90°. Suppliers: Baker, Fisher, MCB.

In the section on *Sodium acetate* we cite 10 procedures utilizing this reagent in ways other than for mere neutralization of a mineral acid amine salt; most of them are taken from *Organic Syntheses*. In contrast, this publication records only two procedures utilizing KOAc. One, by Bedoukian,[1] is for enolacetylation of an aldehyde, as part of a synthesis of an α-bromoaldehyde. Unfortunately no com-

$$\underset{0.25\text{ m.}}{CH_3(CH_2)_4CH_2CHO} + \underset{6\text{ m.}}{Ac_2O} + \underset{0.5\text{ m.}}{KOAc} \xrightarrow[45\text{-}50\%]{1\text{ hr. at }155\text{-}160°}$$

$$CH_3(CH_2)_4CH{=}CHOAc \xrightarrow[80\text{-}85\%]{Br_2-CCl_4;\ CH_3OH} CH_3(CH_2)_4\underset{Br}{\underset{|}{C}}HCH(OCH_3)_2 \xrightarrow[90\text{-}95\%]{HCl} CH_3(CH_2)_4\underset{Br}{\underset{|}{C}}HCHO$$

parison was made to determine the relative effectiveness of sodium acetate and potassium acetate. However, J. R. Johnson[2] made such a comparison in studying the Perkin reaction of furfural with acetic anhydride and a basic catalyst and found potassium acetate to be definitely superior: with KOAc a reaction period of 4 hrs. sufficed, whereas when NaOAc was used heating had to be continued for

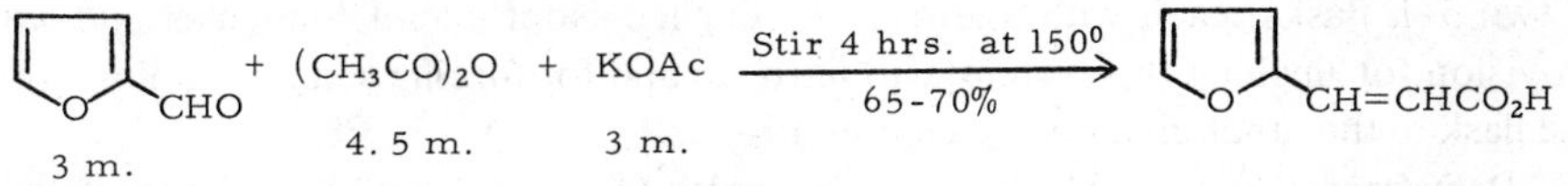
CHO
3 m.
+ (CH₃CO)₂O + KOAc
4. 5 m.
3 m.
Stir 4 hrs. at 150°
65-70%
CH=CHCO₂H

6–8 hrs. This result suggests the possibility that procedures utilizing sodium acetate could be improved by substitution of potassium acetate. This salt is lower melting than NaOAc and much more soluble in water; it forms no hydrate. The cost per mole is nearly the same: 42 cents (KOAc), and 40 cents (NaOAc).

[1]P. Z. Bedoukian, *Org. Syn., Coll. Vol.*, **3**, 127 (1955)
[2]J. R. Johnson, *ibid.*, **3**, 426 (1955)

Potassium acid acetylenedicarboxylate, $KO_2CC{\equiv}CCO_2H$ Suppliers: Allied Chem. Corp., National Aniline Div.; Baker.

Potassium amide, KNH_2. Mol. wt. 55.13.

Caution: Potassium amide is flammable and ignites on contact with moisture. Residues are destroyed by cautious treatment with ethanol or isopropanol. Reactions with liquid ammonia must be done in a hood.

***Preparation*; *benzylation of a nitrile*.** A procedure for preparing potassium amide,[1] as further clarified by Professor Hauser in a letter, is as follows. The reaction of potassium with liquid ammonia is carried out in an ordinary 1-l. three-necked flask equipped with an air condenser (without drying tube), a ball-sealed mechanical stirrer, and a dropping funnel. External cooling is not required because evaporation of the liquid ammonia furnishes ample cooling for a reaction which can be completed in 1–3 hrs., and in this case an air condenser is adequate. In the course of 3 hrs. the volume appears to decrease by no more than one third. For longer reaction periods a dry ice–acetone condenser can be used. A nitrogen atmosphere is not needed except for reactions conducted at −78° and in this case its function is to exclude atmospheric moisture.

Thus 500 ml. of commercial anhydrous liquid ammonia is introduced into the flask from a cylinder through an inlet tube. The liquid is stirred, and a 0.5-g. piece of potassium is removed from a container of kerosene, blotted with filter paper, and added. After the appearance of a blue color about 0.25 g. of ferric nitrate hydrate is added, followed by 0.5-g. pieces of potassium until 9 g. (0.23 mole) has been added. Discharge of the deep blue color after about 20 min. indicates complete conversion to potassium amide. In contrast to sodamide and lithium amide, which form suspensions in liquid ammonia, potassium amide appears to be mostly in solution, although the solutions are opaque and have been regarded by some workers as suspensions.

In the next step 0.23 mole of diphenylacetonitrile is added, and the resulting greenish-brown solution is stirred for 5 min. Then a solution of 0.24 mole of benzyl chloride in 100 ml. of anhydrous ether is added in the course of 10 min. The solution, which becomes orange, is stirred for 1 hr. and the ammonia is then evaporated on

$$(C_6H_5)_2CHCN \xrightarrow[NH_3]{KNH_2} (C_6H_5)_2\bar{C}CN(K^+) \xrightarrow[95\text{-}99\%]{C_6H_5CH_2Cl} (C_6H_5)_2C(CN)CH_2C_6H_5$$

the steam bath as 300 ml. of ether is being added. Addition of 300 ml. of water precipitates α,α,β-triphenylpropionitrile. The ether is removed by distillation, and crude, light tan product is collected and crystallized from ethanol; two crops of material, m.p. 126.5–127.5°, bring the yield to 95–97%.

Bunnett and co-workers[2] prepared potassium amide in an elaborate assembly of two 5-l. flasks, each with a stirrer and dry ice–isopropanol condenser and with provision for flushing the system with nitrogen and for forcing solution to flow from one flask to the other under nitrogen pressure.

Dehydrobromination. Alcohols and bromides of type (1) tend to undergo skeletal rearrangement on conversion to the olefins, particularly in the presence of an acidic reagent, to form internal olefins of type (3), rather than the normal terminal olefin (2). Hauser and co-workers[3] found that (1) on treatment with less than 1

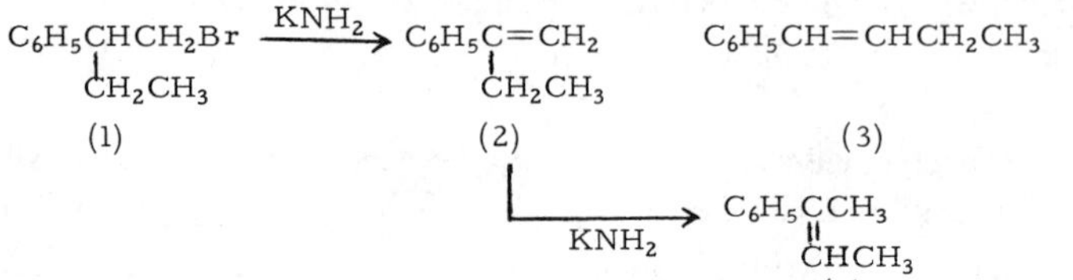
C6H5CHCH2Br
CH2CH3
KNH2
C6H5C=CH2
CH2CH3
C6H5CH=CHCH2CH3
KNH2
C6H5CCH3
CHCH3

(1) (2) (3) (4)

equivalent of potassium amide in liquid ammonia forms the terminal olefin (2) but that if excess base is used, the product is isomerized to the internal olefin (4). The product of rearrangement (3) was formed only in traces.

In an investigation of a modified Hofmann degradation, Wittig and Polster[4] found that HBr-elimination from trimethylcyclooctylammonium bromide with phenyllithium gave an olefin mixture containing 80% of *cis*-cyclooctene but that with potassium amide in liquid ammonia *trans*-cyclooctene predominated.

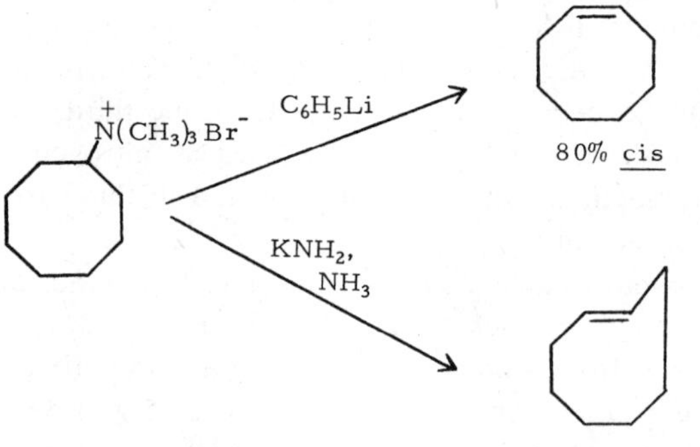
N(CH3)3 Br
C6H5Li
KNH2,
NH3

80% cis

85% trans

Wittig benzyl ether⟶carbinol rearrangement. Wittig had found that "diphenane" (1) is rearranged to 9-hydroxy-9,10-dihydrophenanthrene (4) by phenyllithium in

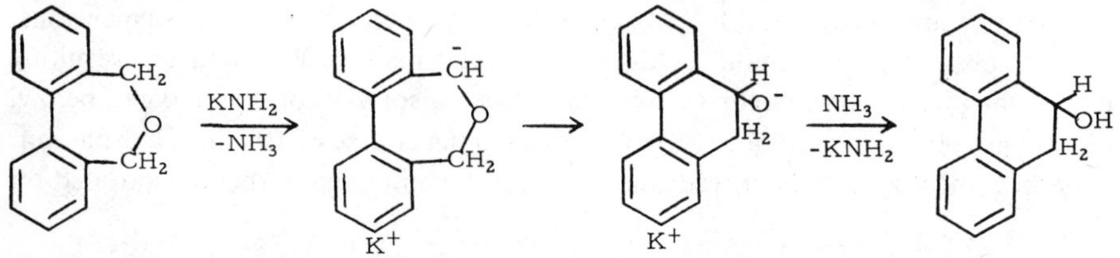
CH2
O
CH2
KNH2
-NH3
CH
O
CH2
K+
H
O
H2
K+
NH3
-KNH2
H
OH
H2

(1) (2) (3) (4)

one week. Hauser and co-workers[5] found that the reaction can be effected with potassium amide in liquid ammonia in 1 hr. in 90% yield.

Generation of a benzyne intermediate. A general principle of synthesis devised by Bunnett[2] involves creation of an intermediate benzyne having a nucleophilic center so located that it can add intramolecularly to the triple bond. An example is the synthesis of 3-acetyloxindole from *o*-chloroacetoacetanilide (Union Carbide Chem. Co.) with potassium amide in liquid ammonia. In another synthesis *o*- and *m*-halo isomers gave the same product.

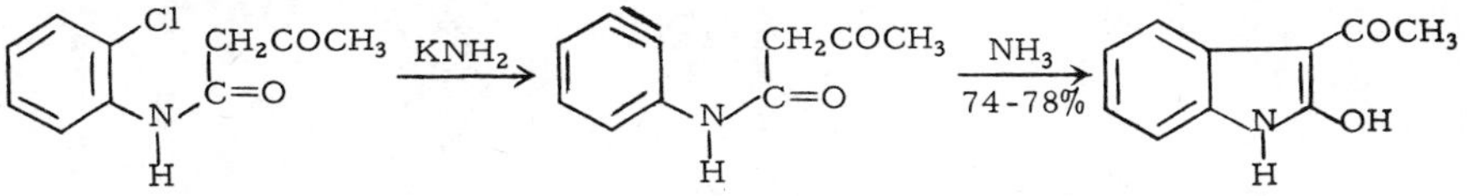

[1]C. R. Hauser and W. R. Dunnavant, *Org. Syn., Coll. Vol.*, **4**, 962 (1963)
[2]B. F. Hrutford and J. F. Bunnett, *Am. Soc.*, **80**, 2021 (1958); J. F. Bunnett, B. F. Hrutford, and S. M. Williamson, *Org. Syn.*, **40**, 1 (1960)
[3]C. R. Hauser, P. S. Skell, R. D. Bright, and W. B. Renfrow, *Am. Soc.*, **69**, 589 (1947)
[4]G. Wittig and R. Polster, *Ann.*, **612**, 102 (1958)
[5]A. J. Weinheimer, S. W. Kantor, and C. R. Hauser, *J. Org.*, **18**, 801 (1953); C. R. Hauser and S. W. Kantor, *Am. Soc.*, **73**, 1437 (1951); B. F. Hrutford and J. F. Bunnett, *Am. Soc.*, **80**, 2021 (1958)

Potassium azodicarboxylate, $KO_2CN{=}NCO_2K$. Mol. wt. 168.24. The salt can be prepared[1] by hydrolysis of azodicarboxamide (1), supplied by Aldrich. Acidification of the salt in the presence of azobenzene liberates the labile intermediate diimide

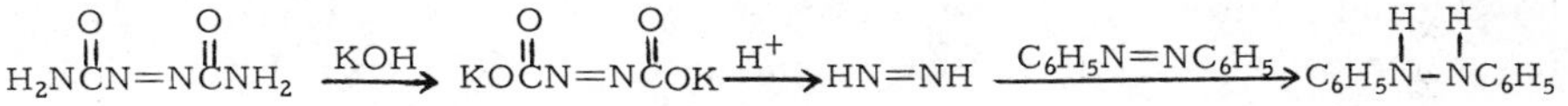

(1) (2) (3) (4)

(3), which effects smooth reduction to hydrazobenzene.[2] With some excess of reagent, oleic acid is reduced to stearic acid in 51% yield.

[1]J. Thiele, *Ann.*, **271**, 127 (1892)
[2]E. E. van Tamelen, R. S. Dewey, and R. J. Timmons, *Am. Soc.*, **83**, 3725 (1961)

Potassium bisulfate, $KHSO_4$. Mol. wt. 136.17, m.p. 197°. As an insurance measure, the salt can be freshly fused at 400° and ground before use. Suppliers: B, F, MCB.

Dehydration. Pyruvic acid was prepared by Berzelius in 1835 by destructive distillation of tartaric acid or glyceric acid, but the process is improved by use of potassium bisulfate as dehydration catalyst.[1] Thus a mixture of 2.7 moles of tartaric

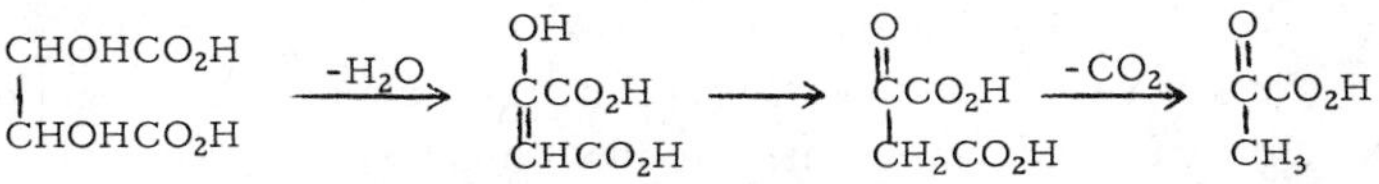

acid and 4.5 moles of $KHSO_4$ is ground in a mortar, placed in a 3-l. flask fitted with a short fractionating column, and heated in an oil bath at 200–220° until no further liquid distills. Fractionation gives 50–55% of pyruvic acid.

Selective dehydration of the diol (1) to (2) was accomplished by mixing 12 g. of material with 3 g. of powdered $KHSO_4$, distilling at 245–260°, washing the product

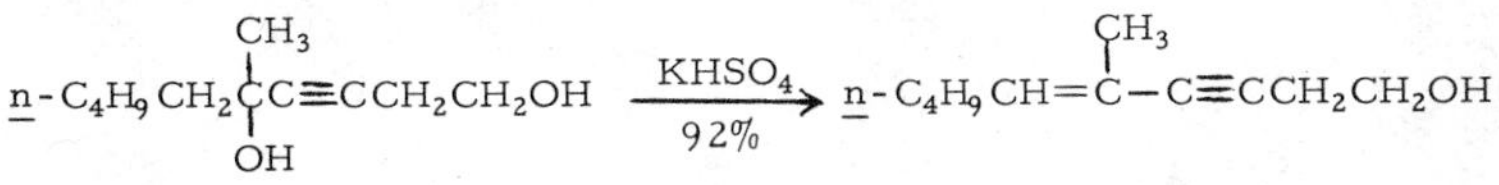

(1) (2)

in ether with bicarbonate solution, drying, and distilling.[2] The propargylic tertiary alcoholic group was eliminated without disturbance of the primary alcoholic function.

Inhoffen *et al.*[3] employed potassium bisulfate as a dehydrating agent in carotenoid synthesis as follows. The unsaturated triol monomethyl ether (3, 3 g.) heated with 3 g. of $KHSO_4$ at a bath temperature of 150° for 10 min. afforded 2.6 g. of the doubly unsaturated aldehyde (4).

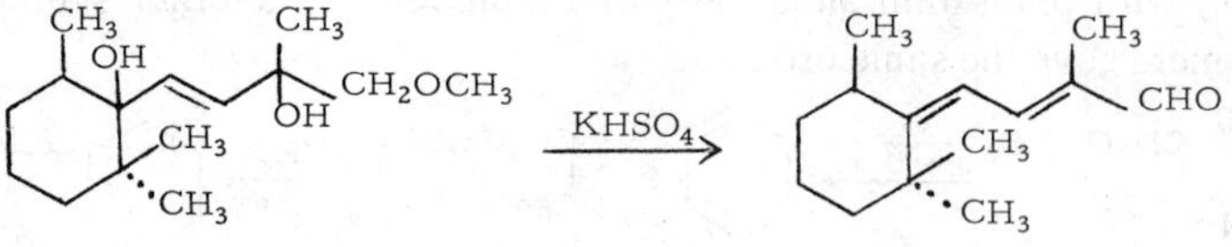
CH3
OH
CH3
CH2OCH3
OH
CH3
CH3
KHSO4
CH3
CH3
CHO
CH3
CH3

(3) (4)

Overberger and Saunders[4] prepared *m*-chlorostyrene (6) by charging a flask equipped with a dropping funnel and a short fractionating column with 12.5 g. of fused, powdered $KHSO_4$ and 0.05 g. of *p-t*-butylphenol (as antioxidant), heating the mixture in a bath at 220–230° at a pressure of 125 mm., and adding 145 g. of

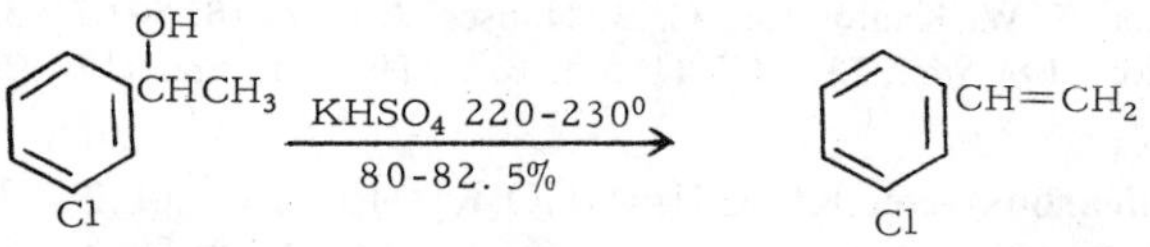
OH
CHCH3
Cl
KHSO4 220-230°
80-82.5%
CH=CH2
Cl

(5) (6)

the carbinol (5) dropwise at a rate sufficient to maintain a vapor temperature at the top of the column of 110–120°. The addition took 5.5–5.8 hrs. The dehydration product and water distilled together; separation of the organic phase and distillation at 20 mm. afforded *m*-chlorostyrene.

Hauser and co-workers[5] effected cyclodehydration of the diol (7) to the ether "diphenane" (8) by heating an intimate mixture with potassium bisulfate, cooling,

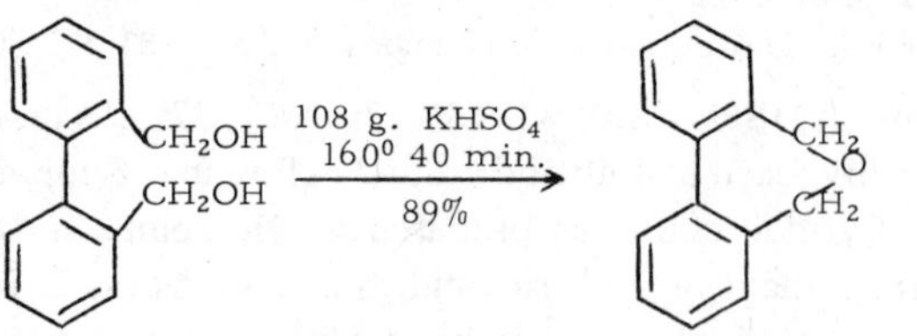
CH2OH
CH2OH
108 g. KHSO4
160° 40 min.
89%
CH2
O
CH2

(7) 37.6 g. (8)

adding water and benzene, passing the benzene through a short column of alumina to remove pigments, recovering the product, and crystallizing it from ligroin. "1,8-Naphthalane" was obtained in the same way in 81% yield.

Dehydrative condensation. Hirschmann, Miller, and Wendler[6] developed an improved synthesis of vitamin K_1 involving condensation of phytol with the 1-mono-acetate of 2-methyl-1,4-naphthohydroquinone in dioxane at 76° in the presence of

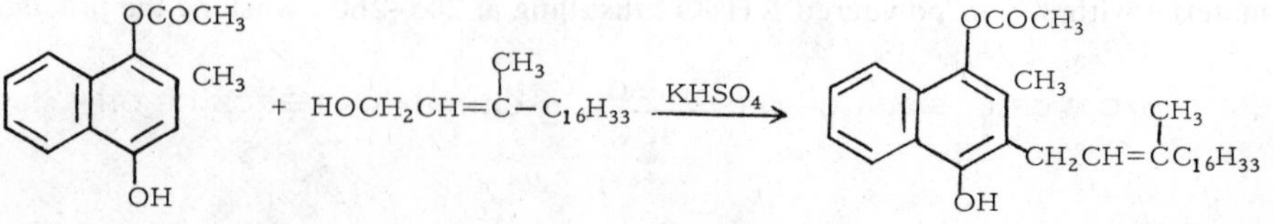
OCOCH3
CH3
OH
+ HOCH2CH=C—C16H33
CH3
KHSO4
OCOCH3
CH3
CH3
CH2CH=CC16H33
OH

potassium bisulfate. This acid catalyst is far more effective than oxalic acid. Hydrolysis and oxidation affords the vitamin.

[1]J. W. Howard and W. A. Fraser, *Org. Syn., Coll. Vol.*, **1**, 475 (1941)
[2]C. L. Leese and R. A. Raphael, *J. Chem. Soc.*, 2725 (1950)
[3]H. H. Inhoffen, H. Siemer, and K.-D. Möhle, *Ann.*, **585**, 126 (1954)
[4]C. G. Overberger and J. H. Saunders, *Org. Syn., Coll. Vol.*, **3**, 204 (1955)
[5]A. J. Weinheimer, S. W. Kantor, and C. R. Hauser, *J. Org.*, **18**, 801 (1953)
[6]R. Hirschmann, R. Miller, and N. L. Wendler, *Am. Soc.*, **76**, 4592 (1954)

Potassium *t*-butoxide, $(CH_3)_3COK$. Mol. wt. 112.21. White, hygroscopic powder. Supplier: MSA Research Corp. (bulletin available). Small lots are available in packages, for example, 6 × 25 g. (0.223 mole); 4 × ¼ lb. (1.001 moles).

Solvent	Solubility (25–26°), g./100 g.
Hexane	0.27
Toluene	2.27
Ether	4.34
t-Butanol	17.80
Tetrahydrofurane	25.00

The production material may contain traces of potassium hydroxide or potassium carbonate, but is satisfactory for most purposes. Purification can be effected by sublimation at 220°/1 mm., although at this temperature some material is lost by decomposition.

Precautions. The reagent is highly hygroscopic and is handled most satisfactorily in a dry box. Reactions are advisedly carried out in an inert atmosphere. It is caustic to skin and membranes. The reagent reacts exothermally with oxygen, and excessive exposure to oxygen or air at an elevated temperature presents a fire hazard. The flash point is 160–165°.

Preparation. Johnson and co-workers[1] give a procedure for the preparation of a solution of potassium *t*-butoxide in *t*-butanol under nitrogen. Skattebøl and Solomon[2] describe preparation of the solution by the same method and for evaporating it to solvent-free solid (for a summary, *see* Methyllithium, allene synthesis). The reagent can be prepared also from potassium purified with *t*-amyl alcohol (*see* Potassium, Pearson's procedure).

Potassium *t*-butoxide is probably the most powerful alkoxide known. It is a stronger base than sodium ethoxide and is free from oxidation-reduction complications sometimes encountered with sodium ethoxide. It is the reagent of choice for some reactions and a unique reagent for others.

Condensations (*see also* Potassium fluoride). The Stobbe condensation of benzophenone with diethyl succinate is carried out by adding these reactants to a

$$\underset{0.05\ m.}{(C_6H_5)_2C{=}O} + \underset{0.075\ m.}{\begin{array}{l}CH_2CO_2C_2H_5\\ |\\ CH_2CO_2C_2H_5\end{array}} \xrightarrow[92\text{-}94\%]{1.\ KOC(CH_3)_3\ (0.055\ m.)\quad 2.\ HCl} \begin{array}{l}\qquad\qquad\ CH_2CO_2H\\ \qquad\qquad\ |\\ (C_6H_5)_2C{=}CCO_2C_2H_5\end{array}$$

freshly prepared solution of potassium *t*-butoxide in *t*-butanol.[1] The Darzens condensation of cyclohexanone with ethyl chloroacetate is carried out similarly at 10–15°.[3]

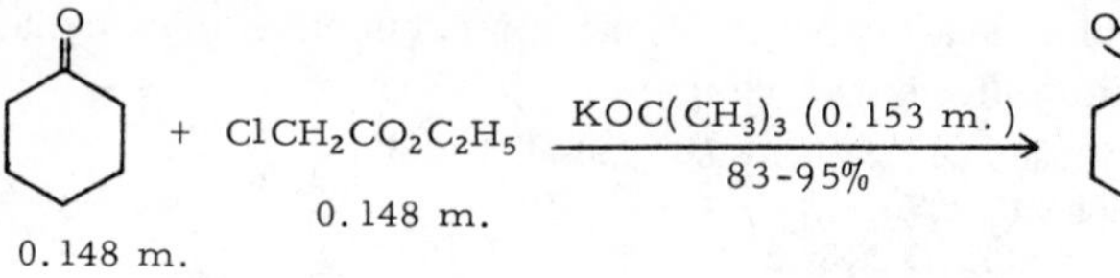

Potassium *t*-butoxide is used also as basic catalyst in a modified Darzens condensation in which a ketone is condensed with an α-halo nitrile rather than an α-halo ester to give a glycidonitrile.[4] Sodium ethoxide is unsatisfactory since it leads to a mixture of the epoxynitrile (I) and the imido ester (II).

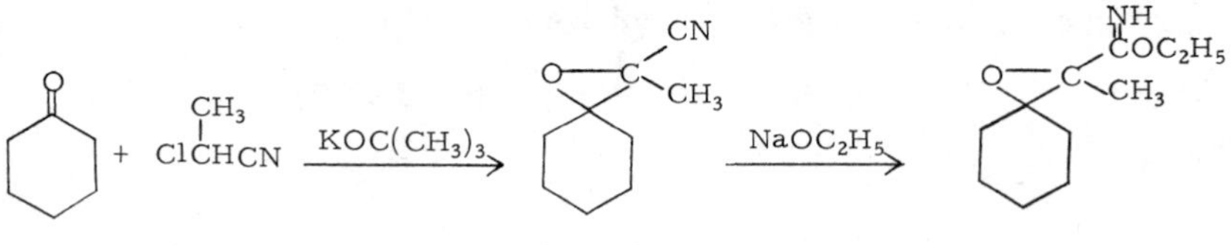

(I) 79% (II)

Thorpe condensation of the dinitrile (1) was effected successfully with the reagent in *t*-butanol;[5] attempted cyclisation with sodium ethoxide or sodium in dioxane

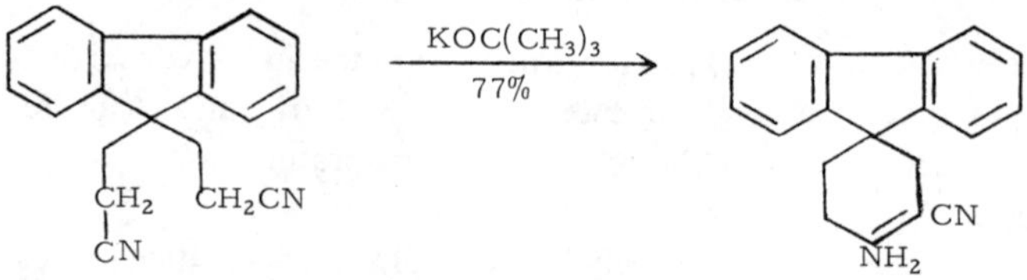

(1) (2)

failed. Leonard and Schimelpfenig[6] effected Dieckmann cyclization of α,ω-diesters with potassium *t*-butoxide in refluxing xylene under nitrogen with high-speed stirring and refluxed the products with alcoholic hydrochloric acid to effect hydrolysis and

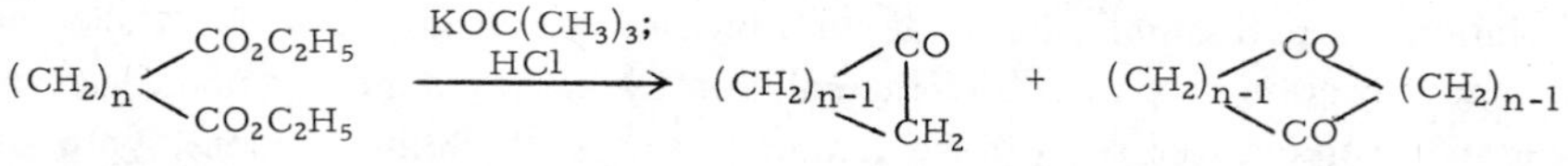

decarboxylation. C_{14}–C_{16}-Monoketones were obtained in yields of 24–48%; C_{18}–C_{20}-, C-$_{22}$, and C_{24}-diketones in somewhat lower yield. The method failed as applied to the production of C_8–C_{12} ring compounds. Leonard and co-workers used potassium *t*-butoxide to cyclize α-amino diesters to cyclic diamino diketones with rings up to 20-membered.[7]

Johnson and co-workers[8] found potassium *t*-butoxide the best base for the Dieckmann cyclization of the diester (3) to epiandrosterone (4). In this case both sodium methoxide and sodium hydride were ineffective.

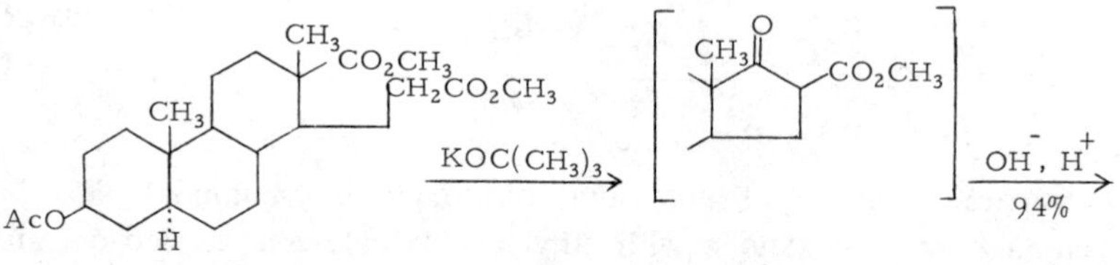

(3)

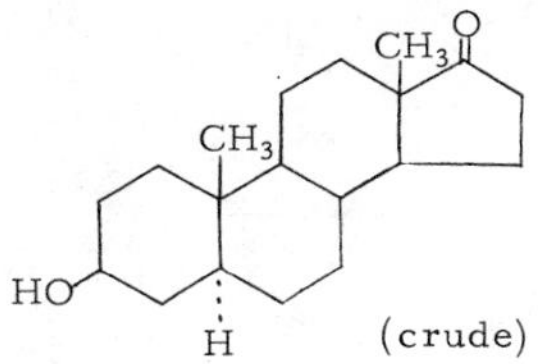

(4)

For effecting the aldol cyclization of the methyl ketone (5) to (6), a key intermediate in the total synthesis of cedrol, Stork and Clarke[9] used potassium *t*-butoxide in

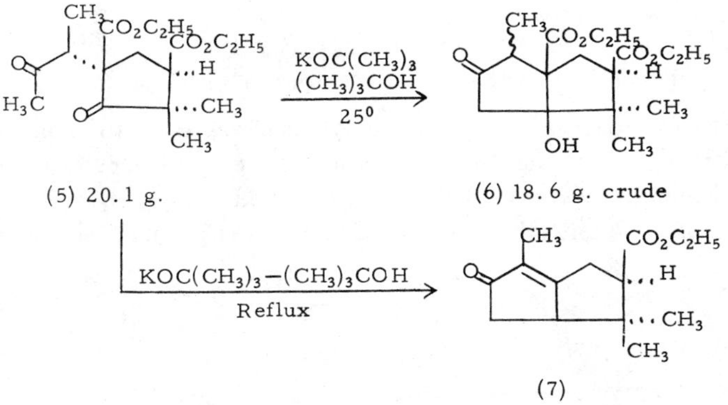

t-butanol at room temperature. Reaction at the reflux temperature gave the undesired product (7). The condensation of (5) to (6) could not be effected with aluminum *t*-butoxide or with *p*-toluenesulfonic acid.

The key step in a new synthesis of cyclic ketones by House and Babad[10] is cyclization with potassium *t*-butoxide in *t*-butanol of a phosphonium salt of type (8) to a cyclophosphorane (9). The reaction is considered to be related to a Dieckmann condensation. When $n = 2$–4 yields are 52–84%.

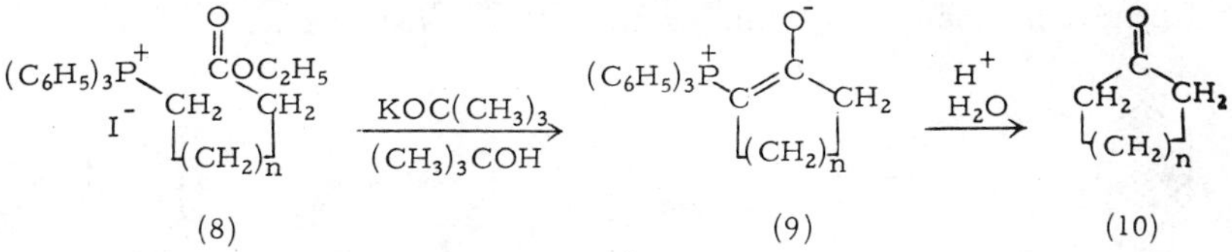

Isomerization of unsaturated compounds. Raphael[11] reported the finding that a 10% solution of the reagent in refluxing diethylene glycol dimethyl ether (b.p. 161°) rearranges diacetylenes to aromatic hydrocarbons in yields of about 65%. Thus 1,6-heptadiyne affords toluene.

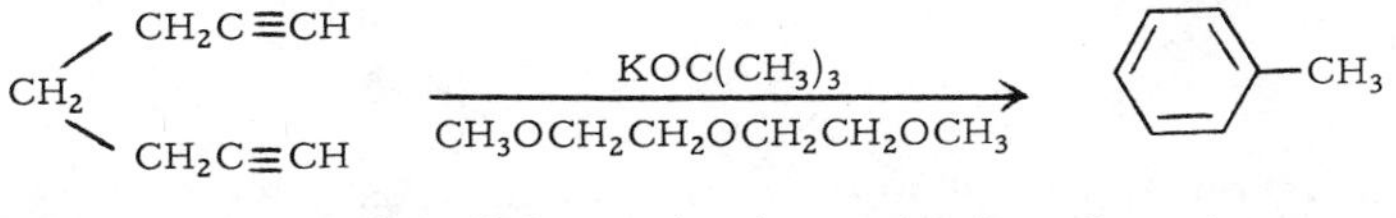

Ben-Efraim and Sondheimer[12] found that isomerization of *cis*-4-octene-1,7-diyne (1) by treatment with potassium *t*-butoxide in *t*-butanol at 60–65° for 15 min. gives, in addition to linear monomeric products, the aromatic dimers (4) and (5). Since Errede[13] has shown that hydrocarbons (4) and (5) are formed when *o*-xylylene (3)

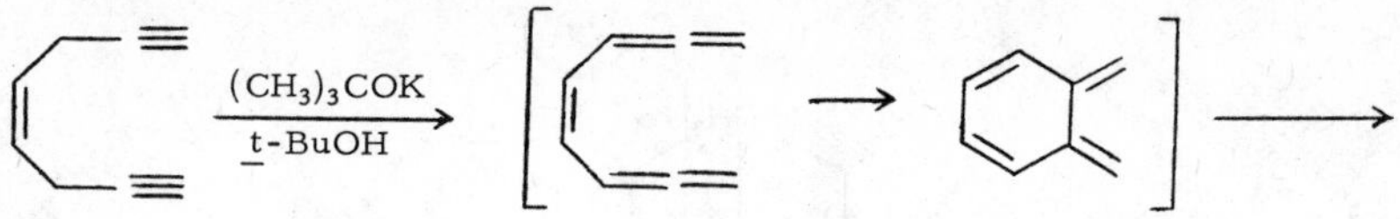
(CH3)3COK
t-BuOH

(1) (2) (3)

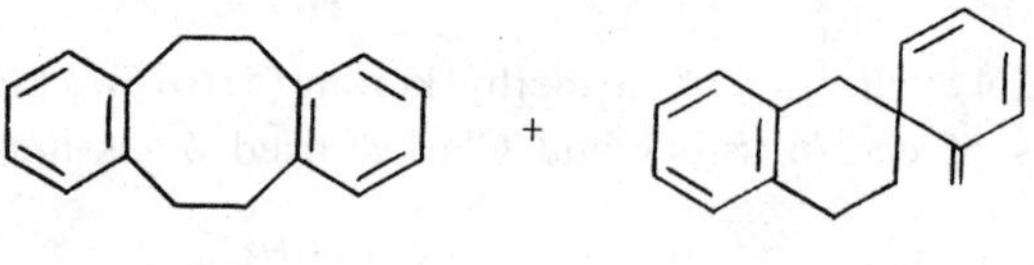

(4) (5)

Cyclo-di-*o*-xylylene Spiro-di-*o*-xylylene

is allowed to warm to room temperature, *o*-xylylene is assumed to be an intermediate. On isomerization under the same conditions, the *trans* isomer of (1) yields only linear conjugated products. 1,7-Diphenyl-1,6-heptadiyne (6) on treatment with the reagent in *t*-butanol affords the naphthalene derivative (7) in modest yield.[14]

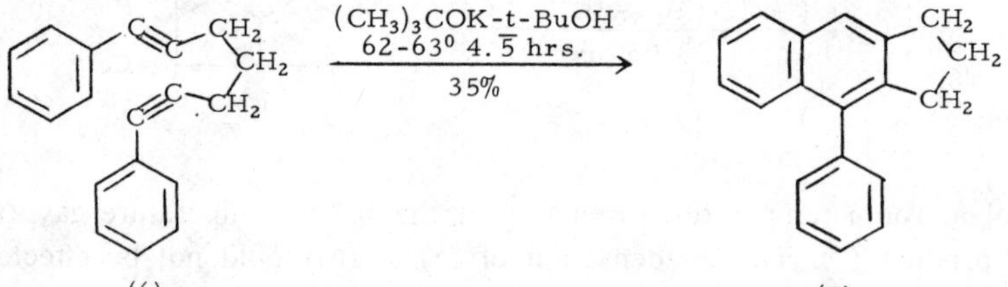
CH2
CH2
CH2
(CH3)3COK-t-BuOH
62-63° 4.5 hrs.
35%
CH2
CH2
CH2

(6) (7)

Ringold and Malhotra[15] found that when a Δ^4-3-ketosteroid is treated with 10 equivalents of potassium *t*-butoxide in *t*-butanol for 1.5 hrs. under nitrogen at room temperature and the resulting anion is protonated by rapid addition of acetic acid, the Δ^5-3-ketone is obtained in yield up to 95%. Deconjugation can be effected also in diethylene glycol dimethyl ether but is extremely slow in benzene.

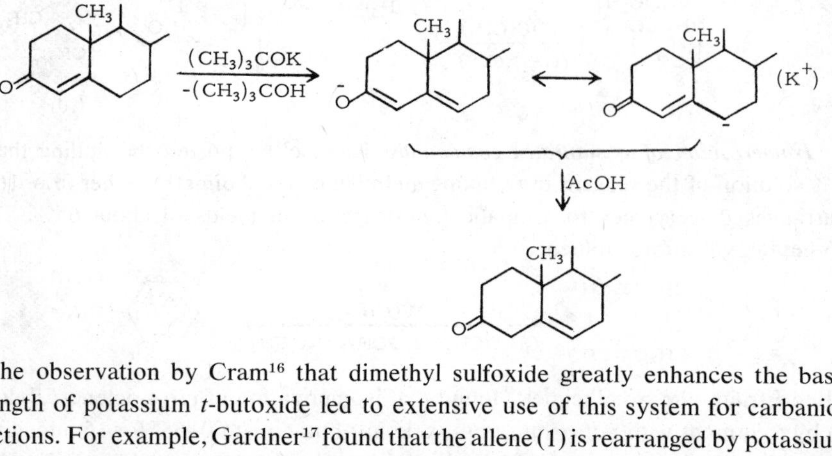
CH3
(CH3)3COK
-(CH3)3COH
CH3
CH3
(K+)
AcOH
CH3

The observation by Cram[16] that dimethyl sulfoxide greatly enhances the basic strength of potassium *t*-butoxide led to extensive use of this system for carbanion reactions. For example, Gardner[17] found that the allene (1) is rearranged by potassium *t*-butoxide in dimethyl sulfoxide at 70° in 1 hr. to *cis,cis*-cyclononadiene (2). Extension of the reaction period to 3 hrs. led to partial conversion of the 1,3-diene (2) into the

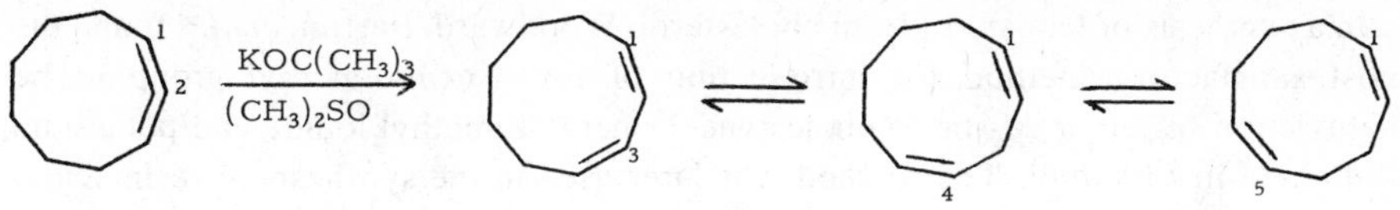

(1) (2) (3) (4)

1,4-diene (3) and the 1,5-diene (4). The conjugated isomer (2) lacks usual resonance stabilization because of the larger interplanar angle imposed by ring strain. The combination $(CH_3)_3COK$–DMSO rearranges *cis,cis,cis*-1,4,7-cyclononatriene (1) first to *cis,cis,cis*-1,3,6-cyclononatriene (2, 10 min. at 18°), then to *cis,cis,cis*-1,3,5-cyclononatriene (3, 1 hr. at 18°).[18] Finally, valence isomerization affords the bicyclic diene (4).

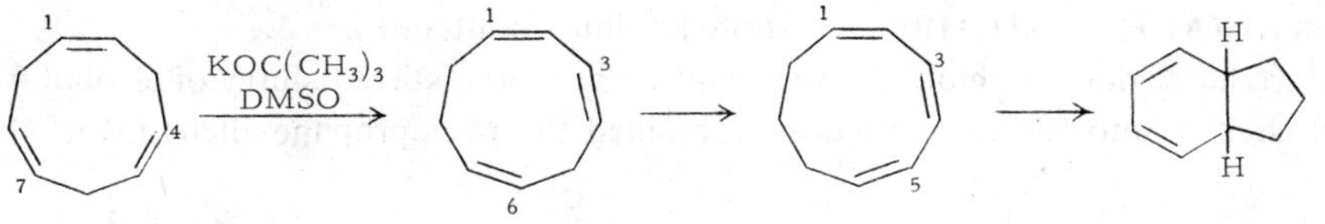

(1) (2) (3) (4)

Schriesheim[19] found a solution of potassium *t*-butoxide in dimethyl sulfoxide useful as a homogeneous basic medium for study of the rate constants, activation energies, and entropies for the isomerization of olefins. Birch *et al.*[20] found the reagent useful for the isomerization of (2), the primary product of Birch reduction of an estrogen methyl ether (1), to the conjugated diene (3).

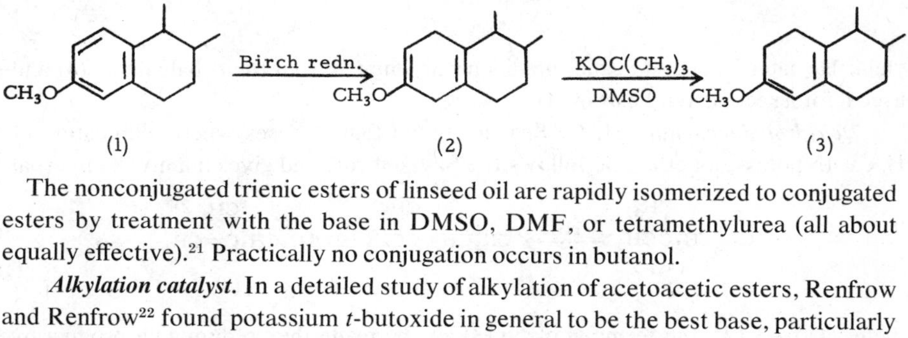

(1) (2) (3)

The nonconjugated trienic esters of linseed oil are rapidly isomerized to conjugated esters by treatment with the base in DMSO, DMF, or tetramethylurea (all about equally effective).[21] Practically no conjugation occurs in butanol.

Alkylation catalyst. In a detailed study of alkylation of acetoacetic esters, Renfrow and Renfrow[22] found potassium *t*-butoxide in general to be the best base, particularly for alkylation of α-substituted esters. In the Sarett synthesis of cortisone, potassium *t*-butoxide served well as base for effecting two successive alkylations of the 14-ketone (1). Methylation (CH_3I, *t*-BuOK)[23] gave a single stereoisomer, regarded as

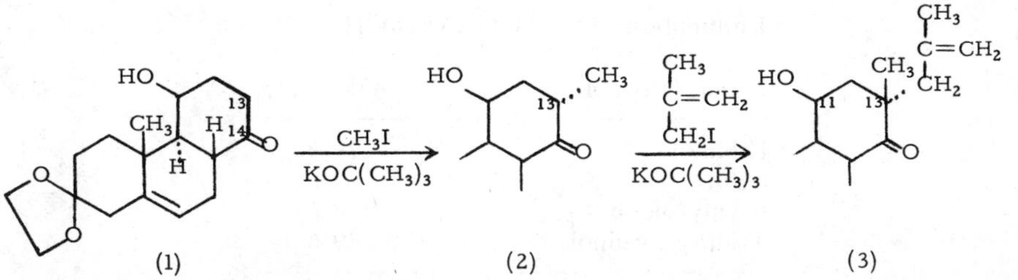

(1) (2) (3)

the equatorial 13α-methyl derivative (2). Alkylation of (2) with methallyl iodide (*t*-BuOK) gave a product the most stable form of which (3) is that with the smaller of the two groups at C_{13} axial (β) to the 11β-hydroxyl group.

In a synthesis of lanosterol from cholesterol, Woodward, Barton, *et al.*[24] found the most satisfactory method for introduction of the 4-*gem*-dimethyl group to be methylation of either Δ^4- or Δ^5-cholestene-3-one with methyl iodide and potassium *t*-butoxide in *t*-butanol. The method was later used in the synthesis of resin acids,

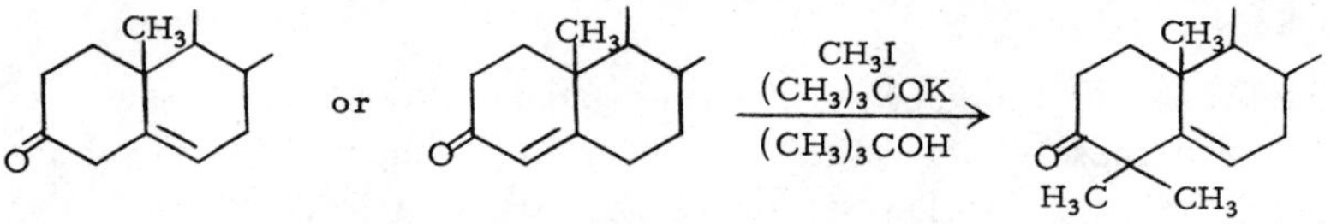

for example, by Ireland and Schiess,[25] who noted that a large excess of base is sometimes required to ensure complete enolization of the ketone.

Generation of carbenes. For an example of the generation of dibromocarbene from $KOC(CH_3)_3$ and $CHBr_3$, *see* Methyllithium (Skattebøl[2]).

Hartzler[26] added 3-chloro-3-methylbutyne-1 (1) to a stirred slurry of alcohol-free potassium *t*-butoxide in styrene and obtained the cyclopropane-allene (3) in 48%

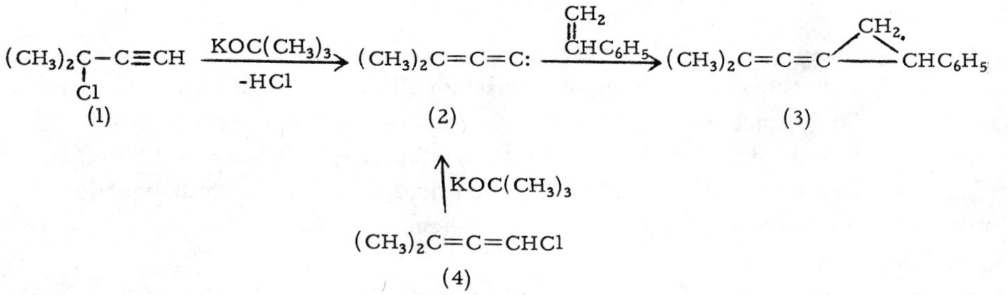

yield. He later[27] generated the vinylidenecarbene (2) from the α-haloallene (4) and used it for a second synthesis of (3).

Dehydrohalogenation. H. C. Brown[28] found that in cases where elimination of HX with potassium ethoxide follows the Saytzeff rule and gives mainly the internal

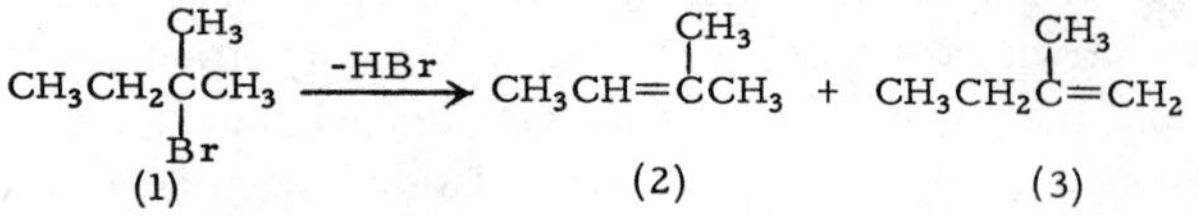

olefin (1) ⟶ (2), the terminal olefin (3) can be made the predominant product by use of potassium *t*-butoxide or bases of even greater steric requirement, as evident from the following results:

Eliminations of $CH_3CH_2C(Br)(CH_3)_2$ (1)

K-derivative of	1-Olefin (3)
Ethanol	29%
t-Butanol	72%
t-Amyl alcohol	78%
Triethylcarbinol	89%

Attempted direct dehydrobromination of 16-bromo-17-ketosteroids such as (6) gives very poor results owing apparently to extensive polymerization of the highly reactive cyclopentenone system. An expedient is to protect the keto group as the ethylene ketal (7) and to reflux the bromoketal (7) with potassium *t*-butoxide in xylene for 16 hrs.[29] Mild acid hydrolysis afforded 15,16-dehydroestrone methyl ether. Another example of the reaction sequences is reported by Sondheimer.[30]

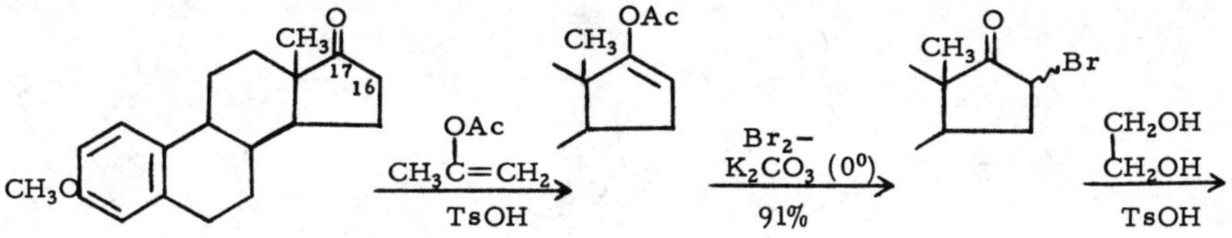
CH3O
OAc
CH3C=CH2
TsOH
Br2–K2CO3 (0°)
91%
CH2OH
CH2OH
TsOH

(4) (5) (6)

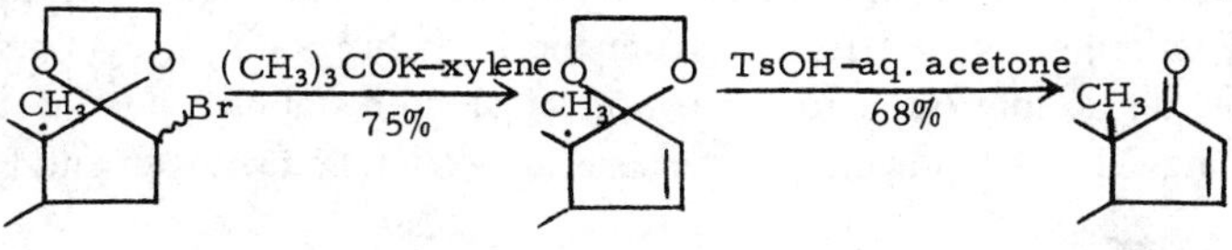
(CH3)3COK–xylene
75%
TsOH–aq. acetone
68%

(7) (8) (9)

McElvain and Kundiger[31] used potassium *t*-butoxide in *t*-butanol for the conversion of bromoacetal into ketene diethylacetal.

$$BrCH_2CH(OC_2H_5)_2 \xrightarrow[67\text{-}75\%]{160^\circ \quad (CH_3)_3COK-(CH_3)_3COH} CH_2{=}C(OC_2H_5)_2$$

Cason found [32] potassium *t*-butoxide the most satisfactory reagent for dehydrobromination of α-bromo acids to Δ^2-alkenoic acids. The bromo acid is added to a

$$CH_3(CH_2)_8CH_2\underset{\displaystyle Br}{C}HCO_2H \xrightarrow[74\%]{(CH_3)_3COK\ (3\ \text{equiv.})} CH_3(CH_2)_8CH{=}CHCO_2H$$

solution of 3 equivalents of alkoxide in *t*-butanol and the mixture refluxed for 4 hrs. Potassium hydroxide in methanol, ethanol, or propanol gave mixtures of Δ^2- and Δ^3-isomers. The method has been used for the preparation of *trans*-2-dodecenoic acid[33] and of 2-methylenedodecanoic acid.[34]

Dehydrohalogenation with rearrangement. Erickson and Wolinsky[35] have shown that the reaction of a bromomethylenecycloalkane with potassium *t*-butoxide can be used to generate highly strained cycloalkynes. Thus when ω-bromocamphene

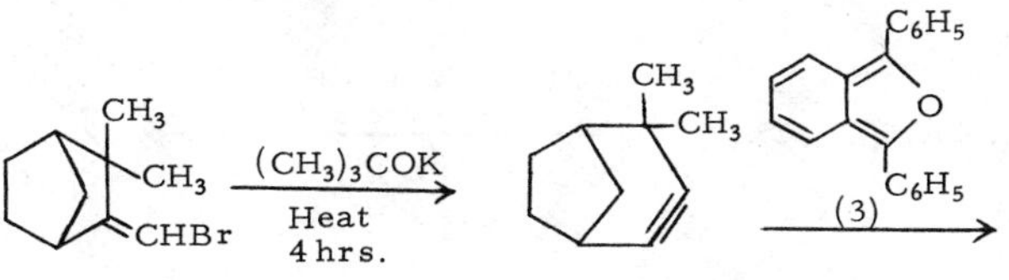
CH3
CH3
CHBr
(CH3)3COK
Heat
4 hrs.
CH3
CH3
C6H5
O
C6H5
(3)

(1) (2)

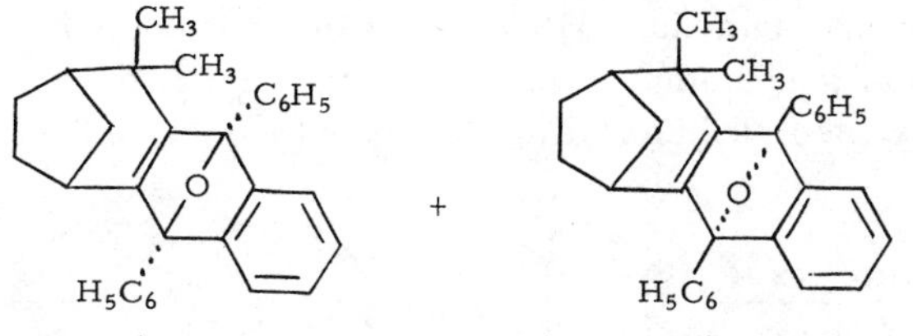
CH3
CH3
C6H5
O
H5C6
+
CH3
CH3
C6H5
O
H5C6

(4) (5)

(1) was heated for 4 hrs. with potassium *t*-butoxide and 1,3-diphenylisobenzofurane (3) as trapping agent, the formation of *endo* camphyne (2) was indicated by the isolation in 94% yield of a mixture of the *endo*- and *exo*-oxides (4) and (5). The reaction with bromomethylenecyclooctane gave a complex mixture of products (total yield 65%).

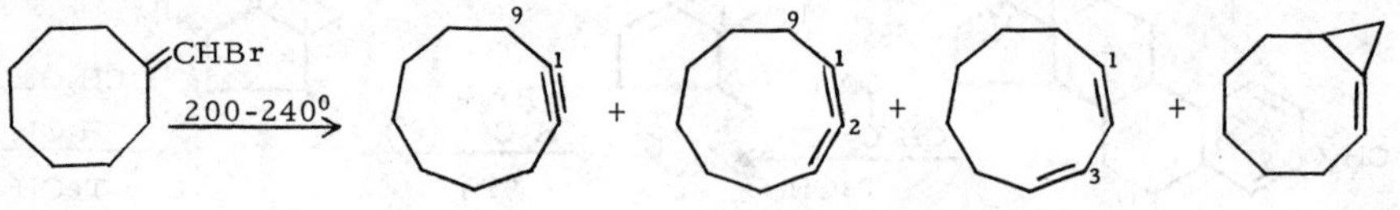
CHBr
200-240°

Gardner and co-workers[36] discovered that halocyclopropanes are dehydrohalogenated with ease by potassium *t*-butoxide in dimethyl sulfoxide. With this system the product is generally that formed from the initially produced cyclopropene by migration of the double bond to a position of greater stability outside the three-membered ring. This is followed in some cases by skeletal rearrangement. Examples:

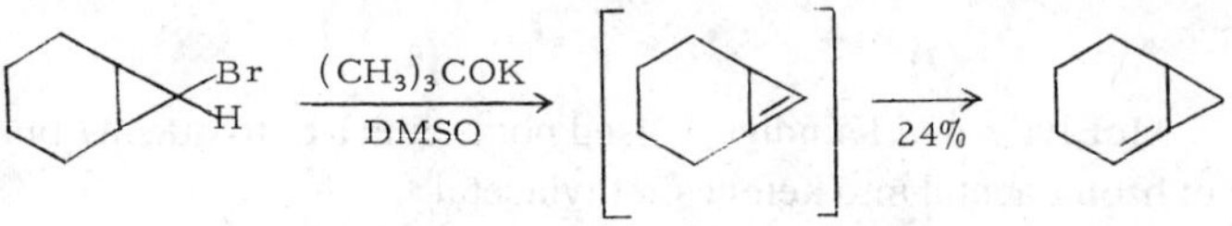
Br
H
(CH3)3COK
DMSO
24%

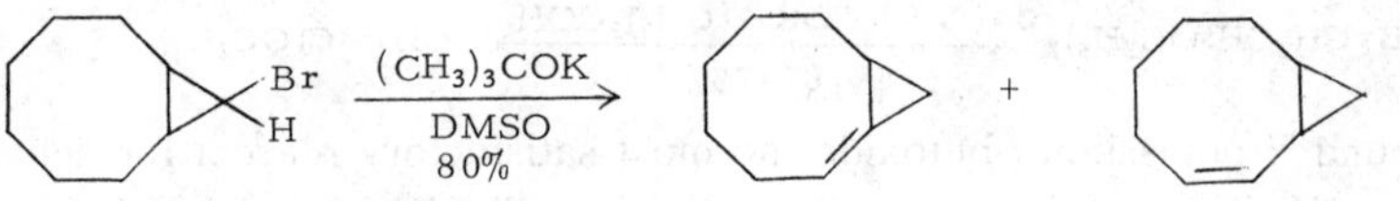
Br
H
(CH3)3COK
DMSO
80%

Main product   Minor product

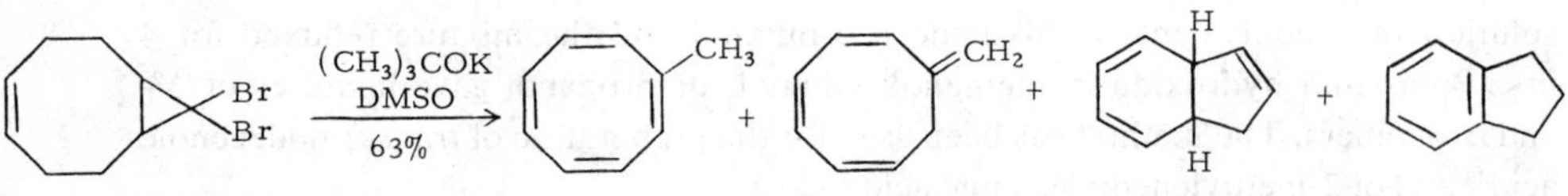
Br
Br
(CH3)3COK
DMSO
63%
CH3
CH2
H
H

53%   16%   23%   8%

Cleavage of tosylates and mesylates. Potassium *t*-butoxide was the most satisfactory base found for the cyclization of N,O-ditosyl-3-amino alcohols to azetidines (Vaughan[37]).

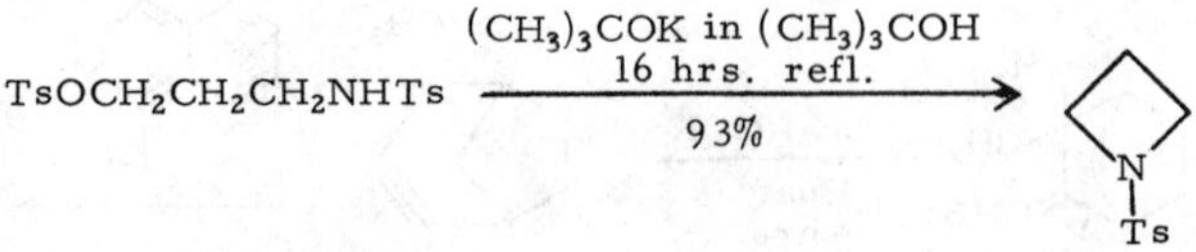
TsOCH2CH2CH2NHTs
(CH3)3COK in (CH3)3COH
16 hrs. refl.
93%
N
Ts

Chang[38] found that the mesylates of steroid 3-alcohols are cleaved stereospecifically by potassium *t*-butoxide (MSA Res. Corp. material) in benzene-dimethyl sulfoxide at room temperature to the corresponding 3-alcohols without inversion. In the examples cited olefin-formation was noted in yields of only 4 and 6%. The

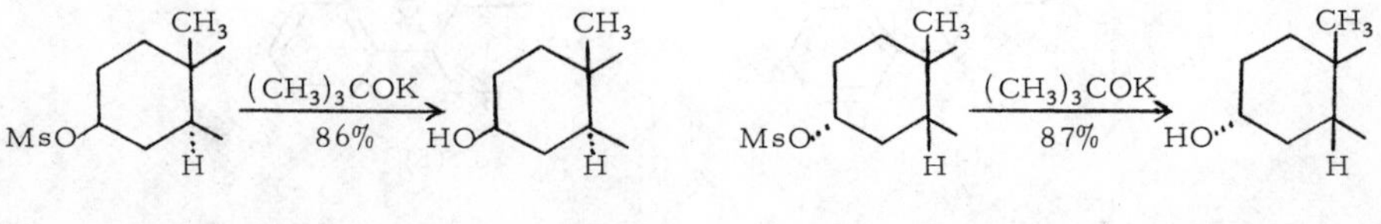
CH3
MsO
H
(CH3)3COK
86%
HO
(CH3)3COK
87%

corresponding tosylates under the same conditions afforded mainly olefins. Later, Chang and Wood[39] found that treatment of 12α-mesyloxycholane (1) with potassium *t*-butoxide in dimethyl sulfoxide at room temperature affords Δ^{11}-cholene (2) as the major product and cholene-12α-ol (3) as a minor product. No reaction occurred when the mesylate (1) was refluxed with potassium *t*-butoxide in *t*-butanol. A useful application of the finding is the reaction of cholesteryl tosylate in benzene

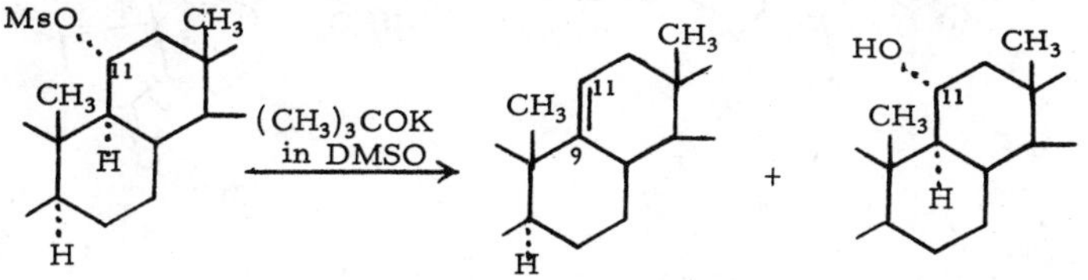
MsO
CH3
11
CH3
H
H
(CH3)3COK
in DMSO
CH3
11
CH3
9
H
+
HO
CH3
11
CH3
H

(1) (2) 65% (3) 29%

with potassium *t*-butoxide and dimethyl sulfoxide at room temperature to produce $\Delta^{3,5}$-cholestadiene in 92% yield. Snyder and Soto[40] studied the action of potassium *t*-butoxide in dimethyl sulfoxide on sulfonate esters of primary and secondary aliphatic alcohols and of conformationally flexible alicyclic alcohols. Reaction occurred under much milder conditions than in classical methods. Sulfonate esters of primary aliphatic alcohols afforded predominantly *t*-butyl esters. Esters of cyclic and secondary acyclic alcohols give alkenes in about 80% yield and only traces of ethers.

Arnold[41] reported a surprising difference between typical primary alkyl bromides and tosylates on being refluxed for extended periods with potassium *t*-butoxide in *t*-butanol: tosylates undergo displacement to give *t*-butyl ethers whereas bromides undergo elimination to the olefin (minor products of other reactions may be formed). Wood and Chang[42] found that, with DMSO as solvent, these reactions proceed in a matter of minutes, and with steroid tosylates and bromides noted the same difference in the nature of the reactions.

Another β-elimination to give an olefin observed by Schriesheim *et al.*[43] is the base-catalyzed decomposition of aliphatic sulfoxides and sulfones. With potassium *t*-butoxide in DMSO the reaction proceeds easily at 55°:

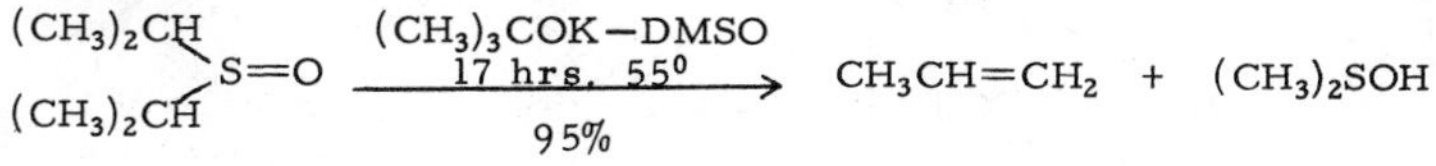
(CH3)2CH
S=O
(CH3)2CH
(CH3)3COK–DMSO
17 hrs. 55°
95%
CH3CH=CH2 + (CH3)2SOH

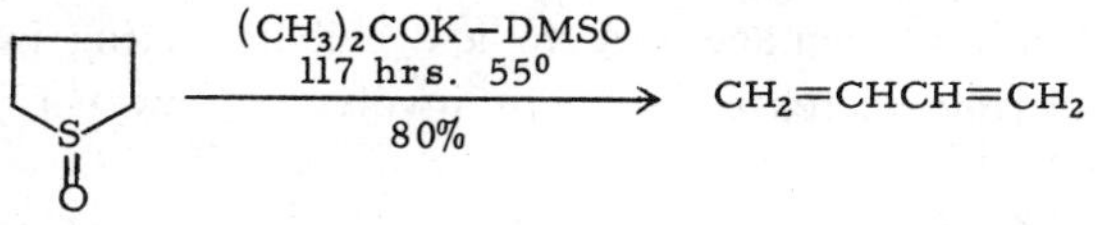
S
O
(CH3)2COK–DMSO
117 hrs. 55°
80%
CH2=CHCH=CH2

Rearrangements. Doering and Urban[44] refluxed a solution of dry potassium *t*-butoxide and benzil in benzene for 2 hrs. and isolated crude *t*-butyl benzilate in high yield. With sodium methoxide under the same conditions the yield of methyl benzilate, the product of benzilic acid rearrangement, was only 18%.

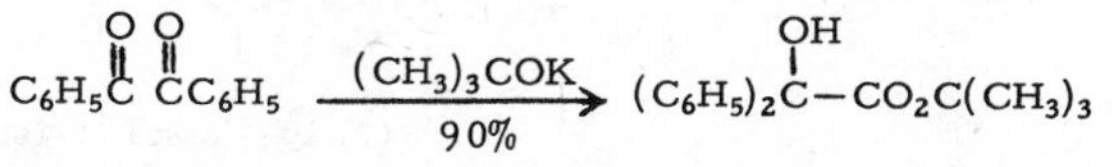
O O
C6H5C CC6H5
(CH3)3COK
90%
OH
(C6H5)2C–CO2C(CH3)3

In one phase of Eschenmoser's synthesis of colchicine,[45] a Diels-Alder addition of the pyrone derivative (1) with chloromethylmaleic anhydride gave the adduct

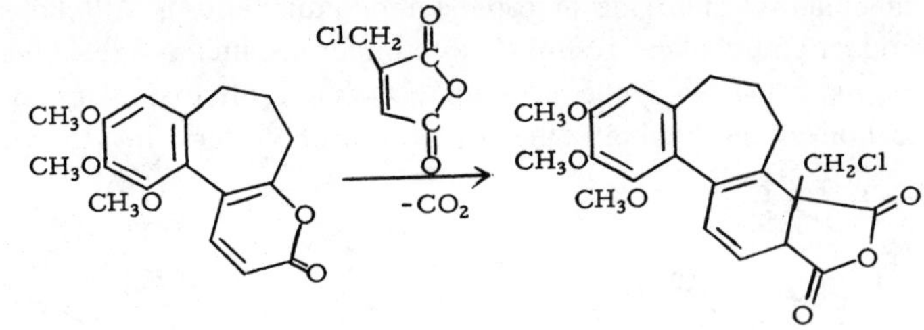

(1) (2)

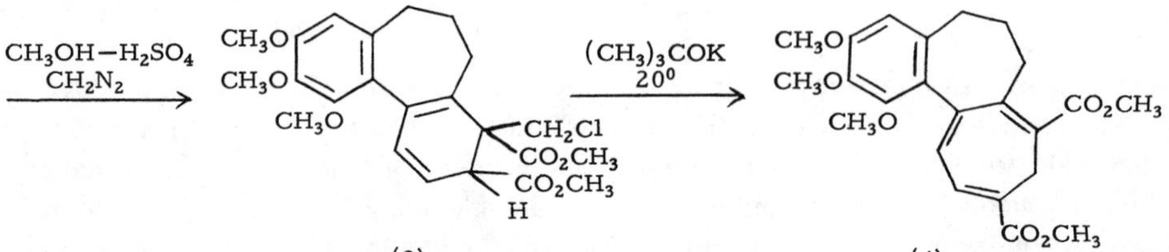

(3) (4)

(2), and this was transformed by acidolysis and diazomethane esterification into the chloromethyl diester (3). Potassium *t*-butoxide then effected ring expansion to (4), which contains two seven-membered rings.

Yates and Anderson[46] found that potassium *t*-butoxide rearranges 4-benzoyloxy-cyclohexanone (5) to 2-benzoylcyclopropanepropionic acid (10). The suggested reaction sequence is formulated.

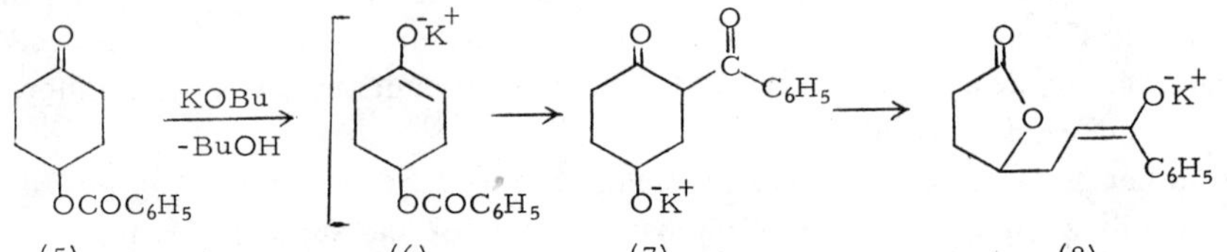

(5) (6) (7) (8)

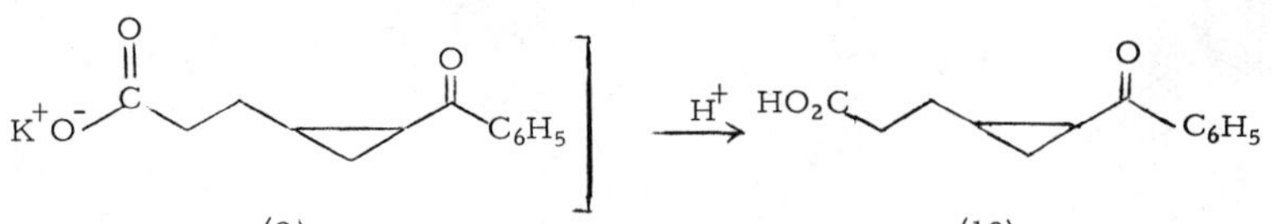

(9) (10)

Contraction of a cyclopentane ring to a cyclobutane ring is observed only rarely because of the increase in strain involved. However, Ghera[47] found

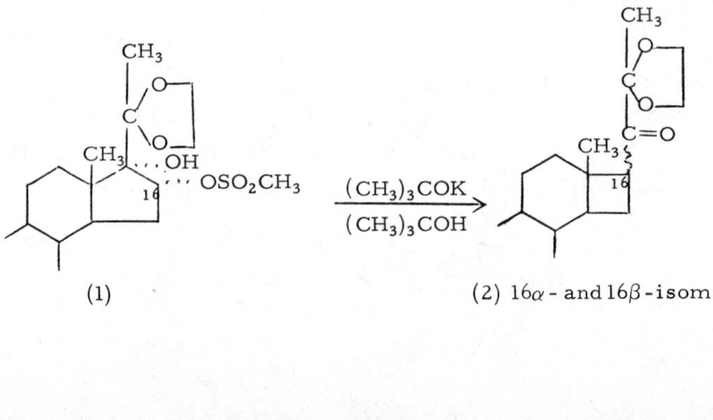

(1) (2) 16α- and 16β-isomers

that D-norsteroids (2) can be obtained by a pinacol-type rearrangement of a 16α-mesylate of type (1) induced by potassium *t*-butoxide in *t*-butanol.

Oppenauer oxidation. Since controlled oxidation of quinine (1) with chromic acid had afforded quininone (2) in only 3% yield, Woodward, Wendler, and Brutschy[48]

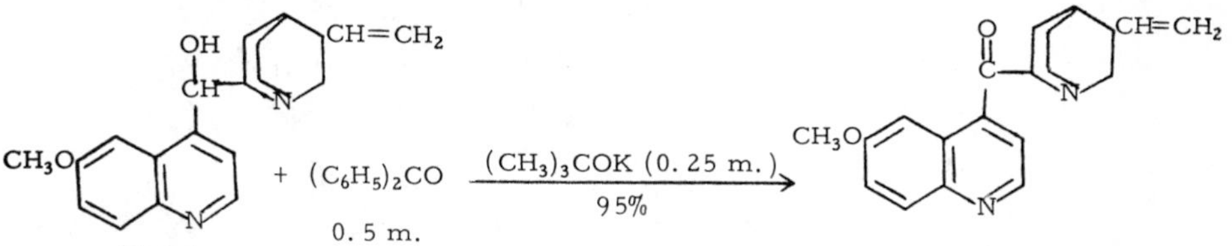

(1) 0.1 m. (2)

investigated the Oppenauer method. Trials with various aluminum alkoxides in combination with a variety of ketones as hydrogen acceptors were completely unsuccessful, but a highly satisfactory procedure was developed employing solvent-free potassium *t*-butoxide as base, benzene as solvent, and benzophenone as oxidant. A mixture of the reactants in the proportions indicated in 500 ml. of benzene was refluxed under nitrogen for 18 hrs. This method of oxidation has been used with excellent results by Rapoport[49] and by Gates[50] for oxidation of unactivated alcohols of the morphine series. In a newer procedure fluorenone is used as the hydride acceptor (*see* Fluorenone).

Autoxidation. Potassium *t*-butoxide is an excellent catalyst for the autoxidation of ketones and esters to α-hydroperoxides.[51] Thus when air is bubbled into a suspension of the solid alkoxide in a solution of methyl isopropyl ketone in ethylene glycol dimethyl ether and *t*-butanol at 8°, the hydroperoxide is formed in yield of 80% by titration or 40% by isolation.

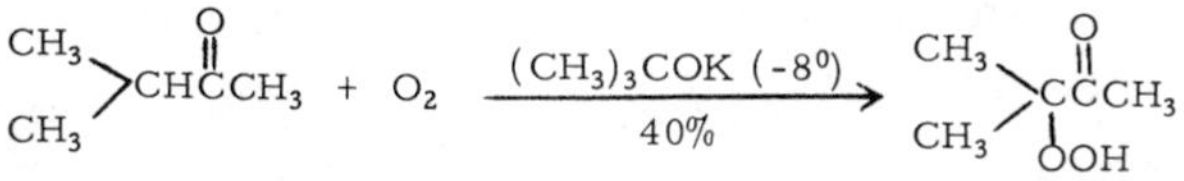

A British group[52] found that when a 20-ketosteroid is shaken with oxygen in the presence of potassium *t*-butoxide in *t*-butanol it is oxidized to the 17α-hydroperoxide; yields are in the range 30–60%. The hydroperoxides can be reduced to

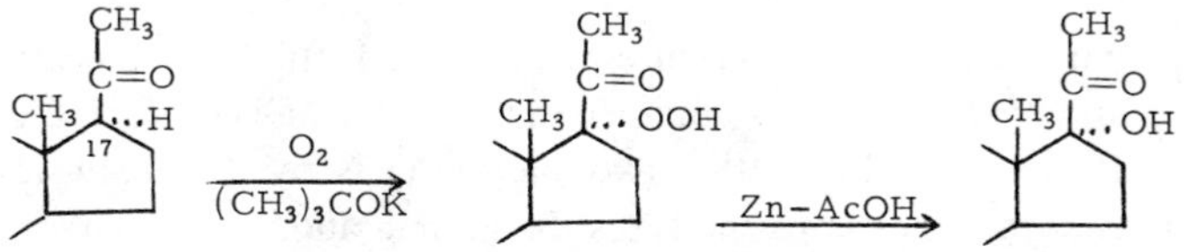

the 17α-alcohols with zinc and acetic acid in high yield. Under the same conditions (Camerino[53]), 3-keto-5β-steroids are oxidized to 4-hydroxy-Δ^4-3-ketones (18–50 hrs. at 25°), the 5α-isomers are oxidized to the 2-ketones (enolic forms), and cholestenone is oxidized in low yield to diosterol-I. Since in each case attack is at the site of enolization, oxygen evidently attacks the enolate anion.

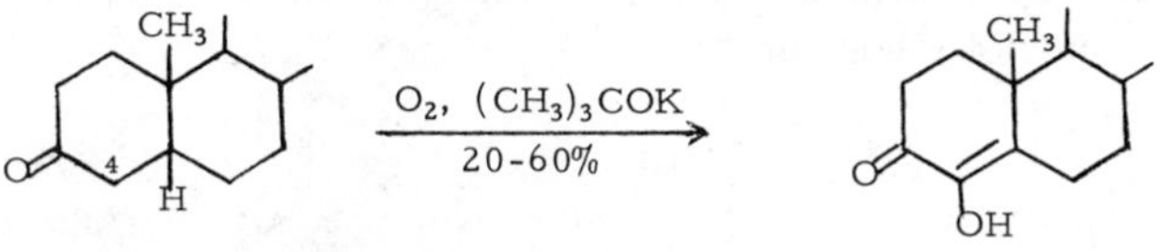

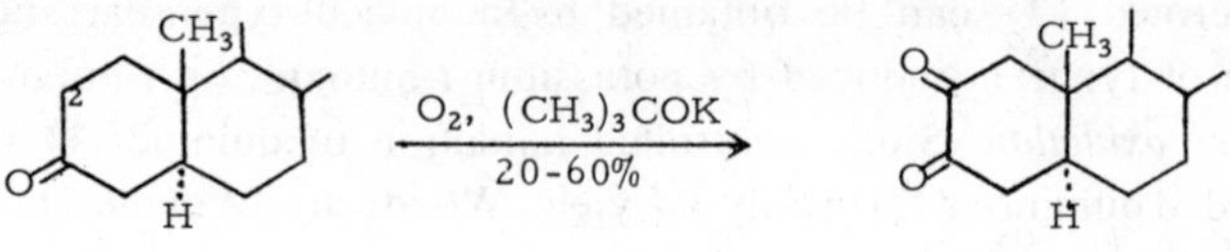

as enol

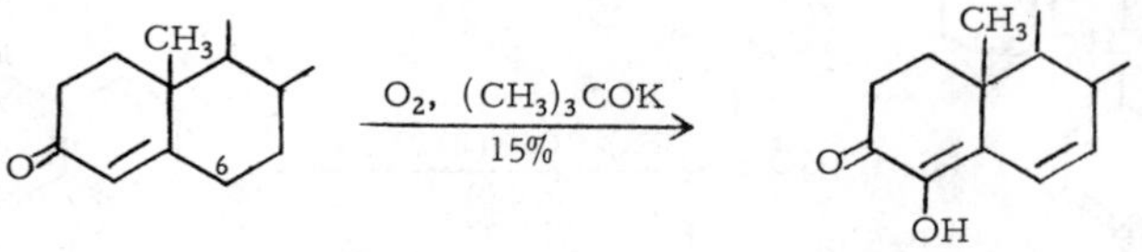

Diosterol-I

Similar autoxidation of limonin (1) in dry *t*-butanol containing potassium *t*-butoxide afforded the diosphenol (2).[54] This was the most useful method found in the lanosterol series for the preparation of diosphenols.

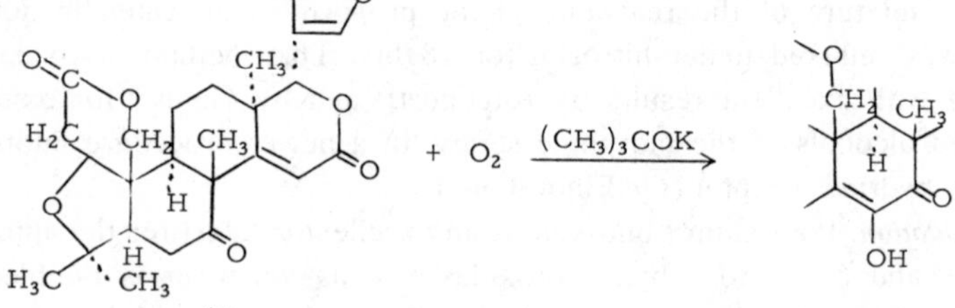

(1) (2)

Pines[55] found that potassium *t*-butoxide when heated to decomposition temperatures (250–300°) catalyzes the dehydrogenation of hydroaromatic hydrocarbons. For example, *d*-limonene and the alkoxide were sealed in an autoclave, the air was

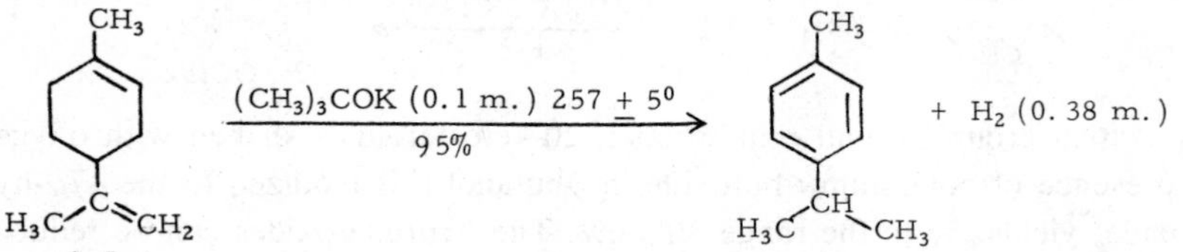

0. 5 m.

displaced by nitrogen, and the mixture was heated until the pressure remained constant (8.5 hrs.). *p*-Cymene was obtained in high yield, and considerable hydrogen was formed. It is assumed that the alkoxide anion is pyrolyzed to acetone and methide ion and that this ion catalyzes bond migration and dehydrogenation.

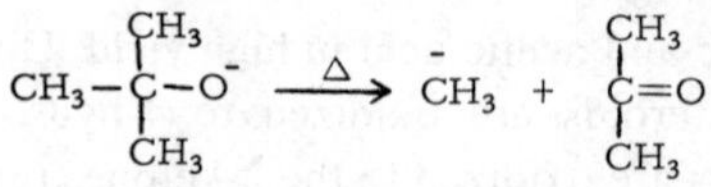

Potassium *t*-pentoxide gave similar results, whereas the yield dropped to 3–4% with potassium isopropoxide and to 1.5% with potassium methoxide.

The picolines are oxidized by O_2 in the system $(CH_3)_3COK$–DMF; no reaction occurred with *t*-butanol as solvent.[56]

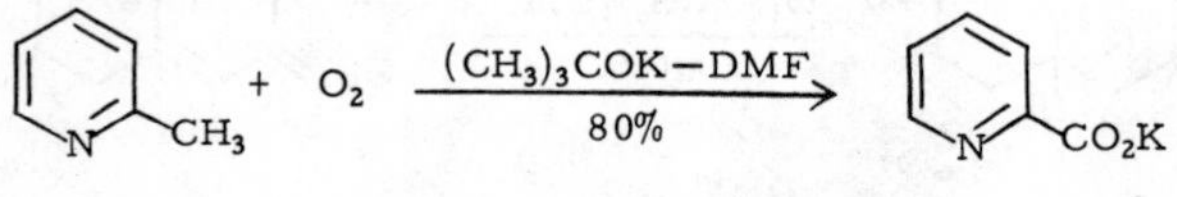

Autoxidations with DMSO as solvent are as follows:[57]

$$(C_6H_5)_2C{=}CHCH_2C_6H_5 + O_2 \xrightarrow{(CH_3)_3COK-DMSO} (C_6H_5)_2CO + C_6H_5CO_2H$$

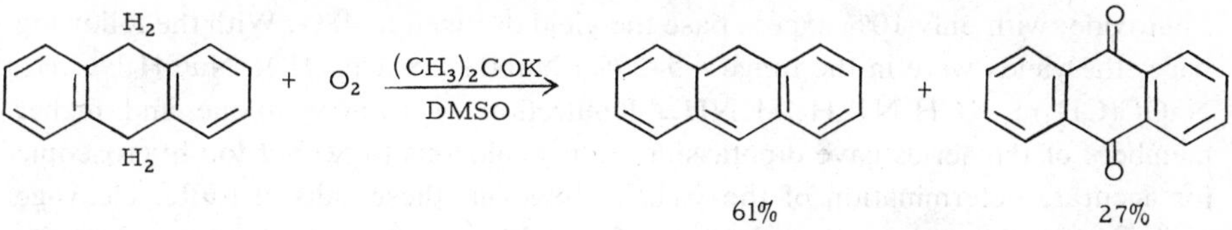

Cleavage of ketones. Swan[58] found that an unenolizable ketone such as benzophenone is cleaved smoothly by adding 3 equivalents of solid potassium *t*-butoxide to an ethereal solution of the ketone containing 1 mole of water and refluxing the mixture. A small neutral fraction consisted of pure triphenylcarbinol. Schrecker

$$C_6H_5\overset{O}{\overset{\|}{C}}C_6H_5 + H_2O \xrightarrow[90\%]{(CH_3)_3COK} C_6H_5\overset{O}{\overset{\|}{C}}OH + C_6H_6$$

and Hartwell[59] found the reaction useful for characterizing a hexasubstituted benzophenone derived from podophyllin; as solvent they used dioxane containing 1 equivalent of water.

Gassman and Zalar[60] found that ketone cleavage can be accomplished very efficiently by reaction with potassium *t*-butoxide in dimethyl sulfoxide at room temperature, as illustrated by the cleavage of benzopinacolone and dehydronorcamphene. In the second case use of potassium *t*-butoxide in refluxing *t*-butanol affords Δ^3-cyclopentenylacetic acid in yield of only 19%.

$$(C_6H_5)_3CCOC_6H_5 + KOC(CH_3)_3 \xrightarrow{DMSO\ 25^0} \underset{97\%}{(C_6H_5)_3CH} + \underset{100\%}{C_6H_5CO_2H}$$

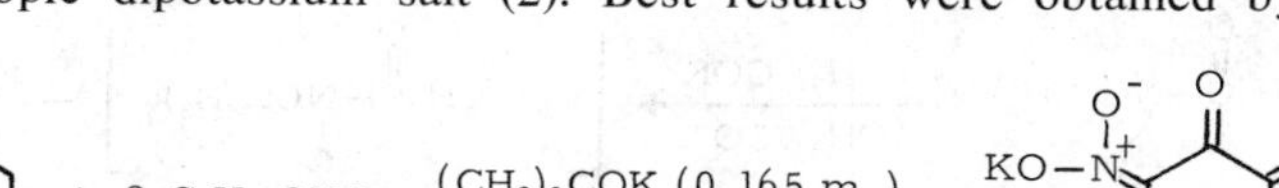

Nitration of ketones. Feuer and co-workers[61] made a careful study of the reagents and reaction conditions required for efficient reaction of cyclopentanone with amyl nitrate and a base to form 2,5-dinitrocyclopentanone, conveniently isolated as the nonhygroscopic dipotassium salt (2). Best results were obtained by stirring a

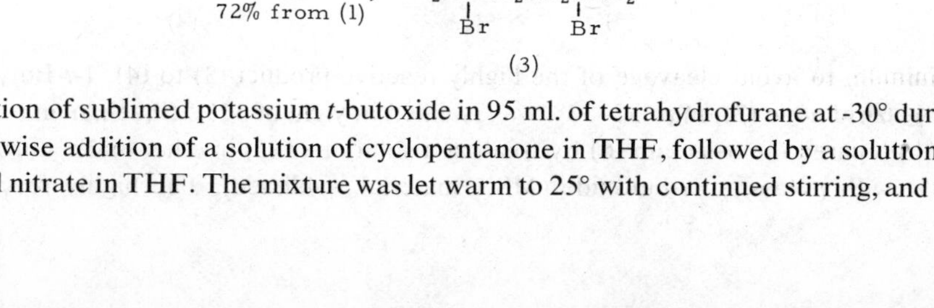

solution of sublimed potassium *t*-butoxide in 95 ml. of tetrahydrofurane at -30° during dropwise addition of a solution of cyclopentanone in THF, followed by a solution of amyl nitrate in THF. The mixture was let warm to 25° with continued stirring, and the green precipitate of dipotassium 2,5-dinitrocyclopentanone (2) was collected, washed, and dried. For purification, the salt was dissolved in water and precipitated with methanol. The best yield, 72%, was obtained with a 65% excess of potassium *t*-butoxide; with only 10% excess base the yield dropped to 48%. With the following bases the yields were in the range 4.5–37%: $NaNH_2$, $NaC(C_6H_5)_3$, $NaCH_3NC_6H_5$, $NaOC(CH_3)_3$, $KCH_3NC_6H_5$, KNH_2. Dinitration of cyclohexanone and higher members of the series gave dipotassium salts analogous to (2) but too hygroscopic for accurate determination of the yields. However, these salts all suffer cleavage analogous to the quantitative cleavage of (2) with aqueous potassium hypobromite at 0° to 1,1,4,4,-tetrabromo-1,4-dinitrobutane (3), isolated as a precipitate and purified by crystallization from hexane. Yields based on cleavage products analogous to (3) are as follows: cyclohexanone, 53%; cycloheptanone, 54%; cyclooctanone, 35%.

Preparation of isocyanides. A general method for the preparation of aryl isocyanides (Ugi[62]) is exemplified by the preparation of *o*-tolyl isocyanide. A suspension of potassium *t*-butoxide in *t*-butanol is prepared under nitrogen by stirring a mixture

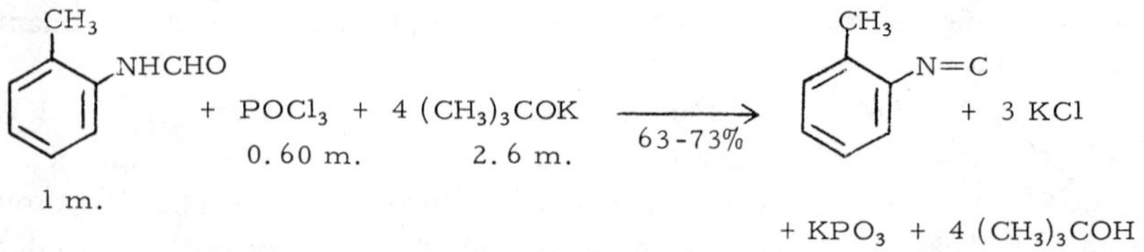

of potassium and the alcohol and letting the temperature rise until the metal melts and then to the point of refluxing. Potassium *t*-butoxide gradually precipitates and eventually forms a thick suspension. On addition of N-*o*-tolylformamide to the hot suspension a clear solution results in a few minutes. The solution is cooled to 10–20° and maintained at this temperature during addition of phosphoryl chloride in the course of 30–40 min. After 1 hr. at 30–35° the mixture is worked up and the product distilled.

Three-membered ring compounds. Synthesis of the first authentic α-lactam was achieved by Baumgarten[63] by slowly adding a solution of potassium *t*-butoxide in toluene to a solution of the amide (1) and *t*-butyl hypochlorite in toluene at 5°. Success of the procedure depended upon keeping the concentration of *t*-butanol at

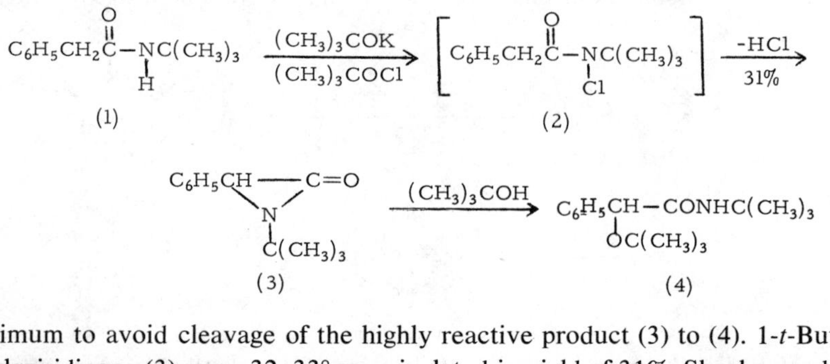

a minimum to avoid cleavage of the highly reactive product (3) to (4). 1-*t*-Butyl-3-phenylaziridinone (3), m.p. 32–33°, was isolated in yield of 31%. Sheehan and Lengyel[64] prepared the α-lactam (6) by treating a solution of (5) in ether under nitrogen at 25° with potassium *t*-butoxide (MSA material, sublimed). The highly reactive

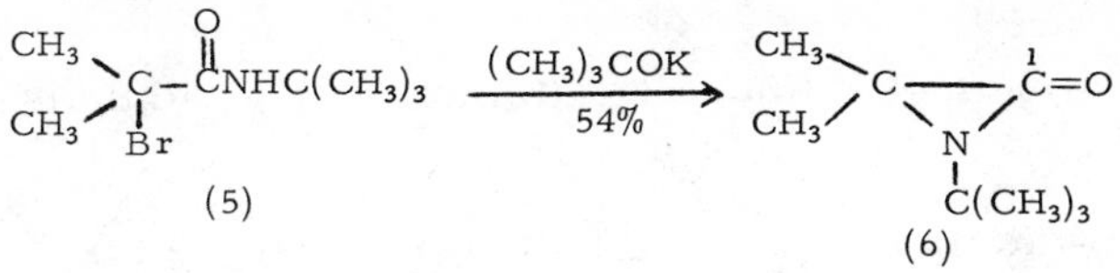

aziridinone (6) melts at 22–24°. Greene and Stowell[65] obtained di-*t*-butyldiaziridinone (8) by treatment of N,N′-di-*t*-butyl-N-chlorourea (7) either with potassium in

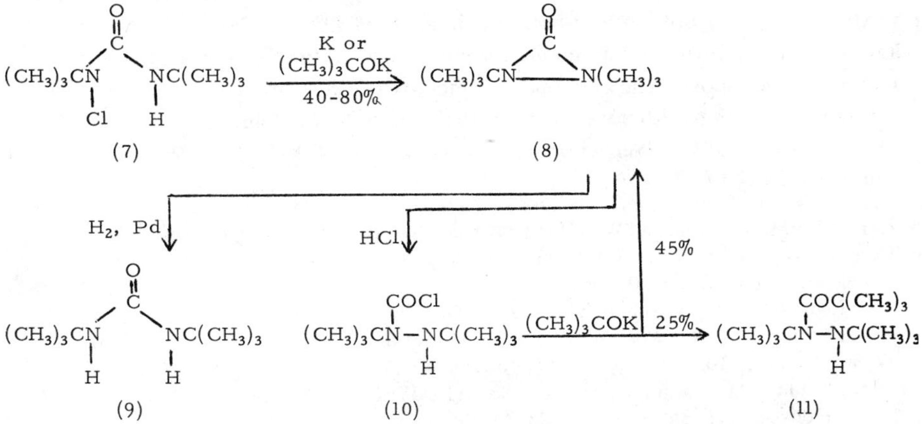

pentane or with potassium *t*-butoxide in *t*-butanol. The diaziridinone (m.p. 0–1°) possesses unusual thermal stability and stability towards nucleophiles. Hydrogenation cleaves the N–N bond to give (9); hydrogen chloride cleaves a C–N bond to give the acid chloride (10), which on reaction with potassium *t*-butoxide affords a mixture of (8) and (11).

Hydrolysis of hindered esters. Chang and Wood[66] found that potassium *t*-butoxide in dimethyl sulfoxide effectively cleaves hindered esters under conditions adjusted according to the degree of hindrance. Thus methyl dehydroabietate (1), in which the

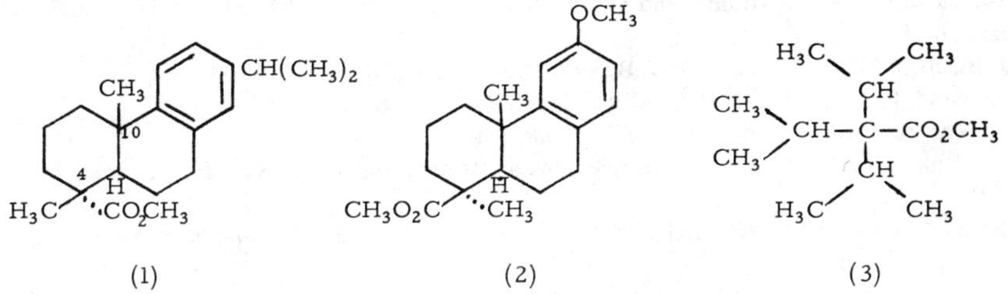

tertiary ester group is equatorial and not hindered by the 10-axial methyl group, is hydrolyzed in less than an hour at room temperature; methyl O-methylpodocarpate (2), with a strongly hindered axial ester group, requires 2 hrs. at 56°; and methyl triisopropyl acetate requires 4 hrs. at 100°.

Wittig reaction. Potassium *t*-butoxide in DMSO was found to be superior to butyllithium or sodium ethoxide in ethanol for generation of the ylide from acetone-1,3-bis(triphenylphosphonium) chloride.[67] Potassium *t*-butoxide is reported to markedly accelerate the Wittig reaction.[68] The 1:1 complex $(CH_3)_3COK$–$(CH_3)_3COH$ is more effective than solvent-free alkoxide.

$$(C_6H_5)_3\overset{+}{P}-CH_2\overset{O}{\overset{\|}{C}}CH_2-\overset{+}{P}(C_6H_5)_3 \;\; + \;\; 2\,(CH_3)_3COK \;\; + \;\; 2\,\underline{p}\text{-}ClC_6H_4CHO \xrightarrow[76\%]{DMSO}$$
$$\quad Cl^- \qquad\qquad\qquad Cl^-$$

$$\underline{p}\text{-}ClC_6H_4CH=CH\overset{O}{\overset{\|}{C}}CH=CHC_6H_4Cl\text{-}\underline{p} \;\; + \;\; 2\,(C_6H_5)_3PO$$

Wolff-Kishner reaction. Cram *et al.*[69] found that the reduction can be conducted at room temperature by use of dimethyl sulfoxide as solvent. For example, 1.96 g. of benzophenone hydrazone was added in portions over an 8-hr. period to a stirred mixture of 2 g. of potassium *t*-butoxide in 5 ml. of DMSO. Nitrogen was evolved, and the solution turned deep red. The yield of diphenylmethane was 90%. Grundon, Henbest, and Scott[70] showed that ketone hydrazones are converted into hydrocarbons by reaction with potassium *t*-butoxide in refluxing toluene.

[1]W. S. Johnson and W. P. Schneider, *Org. Syn., Coll. Vol.*, **4**, 132 (1963); W. S. Johnson and G. H. Daub, *Organic Reactions*, **6**, 1 (1951)
[2]L. Skattebøl and S. Solomon, procedure submitted to *Org. Synth.*
[3]R. H. Hunt, L. J. Chinn, and W. S. Johnson, *Org. Syn., Coll. Vol.*, **4**, 459 (1963)
[4]G. Stork, W. S. Worrall, and J. J. Pappas, *Am. Soc.*, **82**, 4315 (1960)
[5]S. Baldwin, *J. Org.*, **26**, 3280 (1961)
[6]N. J. Leonard and C. W. Schimelpfenig, Jr., *J. Org.*, **23**, 1708 (1958)
[7]N. J. Leonard *et al.*, *Am. Soc.*, **74**, 1704, 6251 (1952); **76**, 3193 (1954); **77**, 6234 (1955)
[8]W. S. Johnson, B. Bannister, and R. Pappo, *Am. Soc.*, **78**, 6331 (1956)
[9]G. Stork and F. H. Clarke, Jr., *Am. Soc.*, **83**, 3114 (1961)
[10]H. O. House and H. Babad, *J. Org.*, **28**, 90 (1963)
[11]G. Eglinton, R. A. Raphael, R. G. Willis, and J. A. Zabkiewicz, *J. Chem. Soc.*, 2597 (1964)
[12]D. A. Ben-Efraim and F. Sondheimer, *Tetrahedron Letters*, 313 (1963)
[13]L. A. Errede, *Am. Soc.*, **83**, 949 (1961)
[14]I. Iwai and J. Ide, *Chem. Pharm. Bull.*, **12**, 1094 (1964)
[15]H. J. Ringold and S. K. Malhotra, *Tetrahedron Letters*, 669 (1962); see also E. L. Shapiro *et al.*, *Steroids*, **3**, 183 (1964)
[16]D. J. Cram, B. Rickborn, and G. R. Knox, *Am. Soc.*, **82**, 6412 (1960)
[17]D. Devaprabhakara, C. G. Cardenas, and P. D. Gardner, *Am. Soc.*, **85**, 1553 (1963)
[18]J. W. H. Watthey and S. Winstein, *Am. Soc.*, **85**, 3715 (1963); D. S. Glass, J. W. H. Watthey, and S. Winstein, *Tetrahedron Letters*, 377 (1965)
[19]A. Schriesheim, R. J. Muller, and C. A. Rowe, Jr., *Am. Soc.*, **84**, 3164 (1962), and earlier papers cited
[20]A. J. Birch, J. M. H. Graves, and J. B. Siddall, *J. Chem. Soc.*, 4234 (1963)
[21]J. Ugelstad, B. Jenssen, and P. C. Mörk, *Chem. Scand.*, **16**, 323 (1962)
[22]W. B. Renfrow and A. Renfrow, *Am. Soc.*, **68**, 1801 (1946)
[23]R. M. Lukes, G. I. Poos, R. E. Beyler, W. F. Johns, and L. H. Sarett, *Am. Soc.*, **75**, 1707 (1953)
[24]R. B. Woodward and A. A. Patchett; D. H. R. Barton, D. A. J. Ives, and R. B. Kelly, *J. Chem. Soc.*, 1131 (1957)
[25]R. E. Ireland and P. W. Schiess, *J. Org.*, **28**, 6 (1963)
[26]H. D. Hartzler, *Am. Soc.*, **83**, 4990 (1961)
[27]H. D. Hartzler, *J. Org.*, **29**, 1311 (1964)
[28]J. C. Brown and I. Moritani, *Am. Soc.*, **75**, 4112 (1953)
[29]W. S. Johnson and W. F. Johns, *Am. Soc.*, **79**, 2005 (1957)
[30]F. Sondheimer, S. Burstein, and R. Mechoulam, *Am. Soc.*, **82**, 3209 (1960)
[31]S. M. McElvain and D. Kundiger, *Org. Syn., Coll. Vol.*, **3**, 506 (1955); for an application see R. Grewe and A. Struve, *Ber.*, **96**, 2819 (1963)
[32]J. Cason, N. L. Allinger, and G. Sumrell, *J. Org.*, **18**, 850 (1953)
[33]C. F. Allen and M. J. Kalm, *Org. Syn., Coll. Vol.*, **4**, 398 (1963)
[34]*Idem, ibid.*, **4**, 616 (1963)
[35]K. L. Erickson and J. Wolinsky, *Am. Soc.*, **87**, 1142 (1965)

[36]C. L. Osborn, T. C. Shields, B. A. Shoulders, J. F. Krause, H. V. Cortez, and P. D. Gardner, *Am. Soc.*, **87**, 3158 (1965)
[37]W. R. Vaughan, R. S. Klonowski, R. S. McElhinney, and B. B. Millward, *J. Org.*, **26**, 138 (1961)
[38]F. C. Chang, *Tetrahedron Letters*, 305 (1964)
[39]F. C. Chang and N. F. Wood, *Steroids*, **4**, 55 (1964); see also K. R. Bharucha and H. M. Schrenk, *Experientia*, **21**, 278 (1965)
[40]C. H. Snyder and A. R. Soto, *J. Org.*, **29**, 742 (1964)
[41]P. Veeravagu, R. T. Arnold, and E. W. Eigenmann, *Am. Soc.*, **86**, 3072 (1964)
[42]N. F. Wood and F. C. Chang, *J. Org.*, **30**, 2054 (1965)
[43]J. E. Hofmann, T. J. Wallace, P. A. Argabright, and A. Schriesheim, *Chem. Ind.*, 1243 (1963)
[44]W. von E. Doering and R. S. Urban, *Am. Soc.*, **78**, 5938 (1956)
[45]J. Schreiber, W. Leimgruber, M. Pesaro, P. Schudel, T. Threlfall, and A. Eschenmoser, *Helv.*, **44**, 540 (1961)
[46]P. Yates and C. D. Anderson, *Am. Soc.*, **85**, 2937 (1963)
[47]E. Ghera, *Tetrahedron Letters*, 4181 (1965)
[48]R. B. Woodward, N. L. Wendler, and F. J. Brutschy, *Am. Soc.*, **67**, 1425 (1945)
[49]H. Rapoport, R. Naumann, E. R. Bissell, and R. M. Bonner, *J. Org.*, **15**, 1103 (1950)
[50]M. Gates and G. Tschudi, *Am. Soc.*, **78**, 1380 (1956)
[51]H. R. Gersmann, H. J. W. Nieuwenhuis, and A. F. Bickel, *Proc. Chem. Soc.*, 279 (1962)
[52]E. J. Bailey, J. Elks, and D. H. R. Barton, *Proc. Chem. Soc.*, 214 (1960); E. J. Bailey, D. H. R. Barton, J. Elks, and J. F. Templeton, *J. Chem. Soc.*, 1578 (1962)
[53]B. Camerino, B. Patelli, and R. Sciaky, *Tetrahedron Letters*, 554 (1961)
[54]D. H. R. Barton, S. K. Pradhan, S. Sternhell, and J. F. Templeton, *J. Chem. Soc.*, 255 (1961)
[55]H. Pines and L. Schaap, *Am. Soc.*, **79**, 2956 (1957)
[56]W. Bartok, D. D. Rosenfeld, and H. Schriesheim, *J. Org.*, **28**, 410 (1963)
[57]D. H. R. Barton and D. W. Jones, *J. Chem. Soc.*, 3563 (1965)
[58]G. A. Swan, *J. Chem. Soc.*, 1408 (1948)
[59]A. W. Schrecker and J. L. Hartwell, *Am. Soc.*, **75**, 5924 (1953)
[60]P. G. Gassman and F. V. Zalar, *Tetrahedron Letters*, 3031, 3251 (1964)
[61]H. Feuer, J. W. Shepherd, and C. Savides, *Am. Soc.*, **78**, 4364 (1956)
[62]I. Ugi and R. Meyr, *Org. Syn.*, **41**, 101 (1961)
[63]H. E. Baumgarten, *Am. Soc.*, **84**, 4975 (1962)
[64]J. C. Sheehan and I. Lengyel, *Am. Soc.*, 1356 (1964)
[65]F. D. Greene and J. C. Stowell, *Am. Soc.*, **86**, 3569 (1964)
[66]F. C. Chang and N. F. Wood, *Tetrahedron Letters*, 2969 (1964)
[67]D. B. Denney and J. Song, *J. Org.*, **29**, 495 (1964)
[68]M. Schlosser and K. F. Christmann, *Angew. Chem., Internat. Ed.*, **3**, 636 (1964)
[69]D. J. Cram, M. R. V. Sahyun, and G. R. Knox, *Am. Soc.*, **84**, 1734 (1962)
[70]M. F. Grundon, H. B. Henbest, and M. D. Scott, *J. Chem. Soc.*, 1855 (1963); M. F. Grundon and M. D. Scott, *ibid.*, 5674 (1964)

Potassium chlorate, $KClO_3$. Mol. wt. 122.56. Suppliers: B, F.

Bromination. Use of potassium chlorate for removal of HBr in the bromination of simple ketones in aqueous solution was introduced in a German manufacturing

$$6\ CH_3COCH_3 + 3\ Br_2 + KClO_3 \longrightarrow 6\ CH_3COCH_2Br + KCl + 3\ H_2O$$

process.[1] Jones *et al.*[2] found this method suitable for the preparation of pure bromoacetone from acetone and of methyl 1-bromoethyl ketone from methyl ethyl ketone. Similarly methyl *n*-propyl ketone gave the two α-bromo derivatives.

$$CH_3COCH_2CH_2CH_3 + Br_2 + KClO_3 \xrightarrow[40\text{-}45\%]{\text{Water, } h\nu} CH_3COCH(Br)CH_2CH_3 + BrCH_2COCH_2CH_2CH_3$$

[1]J. F. Norris, *J. Ind. Eng. Chem.*, **11**, 828 (1919)
[2]J. R. Catch, D. F. Elliot, D. H. Hey, and E. R. H. Jones, *J. Chem. Soc.*, 272 (1948); J. R. Catch, D. H. Hey, E. R. H. Jones, and W. Wilson, *ibid.*, 276 (1948)

Potassium diazomethanedisulfonate (secondary). $(KO_3S)_2C=\overset{+}{N}=\overset{-}{N}$. Mol. wt. 278.36. The reagent was prepared by von Pechmann and Manck[1] by reaction of potassium cyanide in aqueous solution with potassium bisulfite to form the sparingly soluble primary (or acid) aminomethane disulfonate (1), which on reaction with

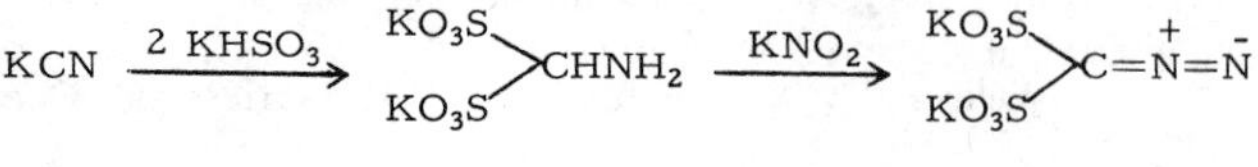
KCN
2 KHSO3
KO3S
KO3S
CHNH2
KNO2
KO3S
KO3S
C=N=N
+
-

(1) (2)

potassium nitrite affords secondary (or neutral) potassium diazomethanedisulfonate (2). The salt crystallizes from water in orange-yellow needles and prisms and is an unusually stable diazo compound. An improved procedure for preparing (1) has been reported.[2]

Kottenhahn[3] found that the reagent reacts readily with activated olefins such as acrylonitrile, methyl vinyl ketone, or ethyl acrylate to form Δ^1-pyrazolines (3), which readily rearrange to the Δ^2-pyrazolines (4). Since the reagent is not soluble

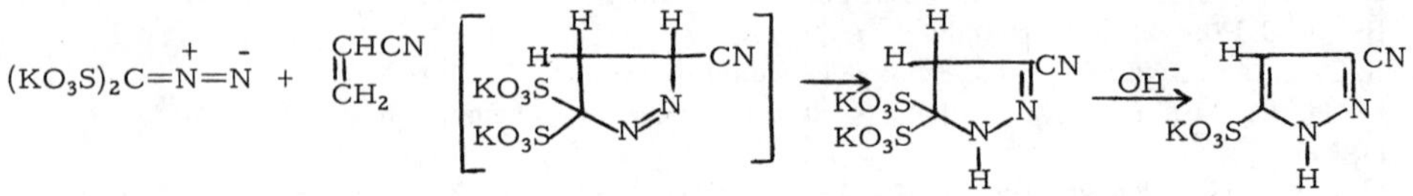
(KO3S)2C=N=N
+
-
CHCN
CH2
H
H
CN
KO3S
KO3S
N
N
H
H
CN
KO3S
KO3S
N
N
H
OH-
H
CN
KO3S
N
N
H

(2) (3) (4) (5)

in organic solvents, the reaction is carried out in water, and the olefin must have some solubility in water or in a water-alcohol mixture. Treatment of the Δ^2-pyrazoline disulfonate (4) with base effects elimination of the elements of $KHSO_3$ to give the pyrazole monosulfonate (5); when acid is used, the nitrile group is hydrolyzed to carboxyl.

[1]H. von Pechmann and Ph. Manck, *Ber.*, **28**, 2374 (1895)
[2]R. A. B. Bannard and J. H. Ross, *Can. J. Chem.*, **32**, 49 (1954)
[3]A. P. Kottenhahn, *J. Org.*, **28**, 3433 (1963)

Potassium ethoxide, C_2H_5OK. Mol. wt. 68.16.

Preparation. A suspension of the alkoxide in ether was prepared by Hershberg from potassium powdered as follows.[1] Potassium was cleaned by melting it under toluene, and 0.27 mole of the metal was covered with 150 ml. of toluene and heated to boiling on a hot plate. The flask was removed and closed with a ground glass stopper with a sealed-on stopcock. After one shake with the stopcock open to relieve superheating, the stopcock was closed and the flask given a few quick shakes to powder the metal. The mixture was allowed to cool undisturbed, nitrogen was admitted, the stopper was replaced by a distilling head carrying a flask into which the toluene could be decanted. The powdered metal was washed repeatedly with ether and converted into the ethoxide by reaction with 12.6 g. of ethanol diluted with 150 ml. of ether. Traces of potassium in the wash liquors were destroyed by treatment under reflux with ethanol diluted with ether.

Condensation catalyst. In preparing quinuclidone, Daeniker and Grob[2] effected the Dieckmann cyclization step with a suspension of potassium ethoxide in toluene. A mixture of clean potassium (2 g. atoms) and toluene was heated under nitrogen

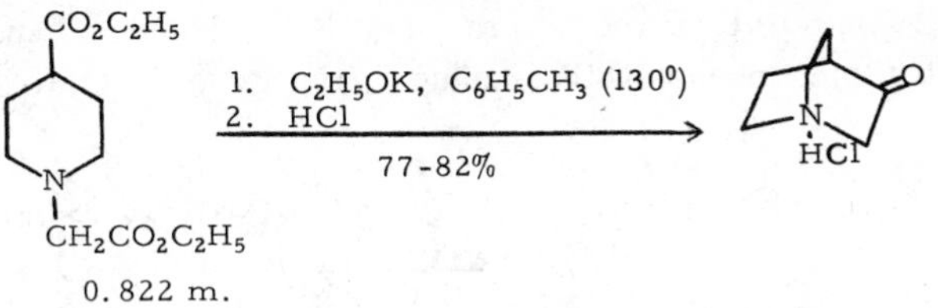

to melt the metal, which was then powdered by vigorous stirring (Hershberg stirrer). One equivalent of anhydrous ethanol was added from a pressure-equilizing addition funnel in 30 min. with continued stirring and heating. A solution of 1-carboethoxymethyl-4-carboethoxypiperidine in 500 ml. of toluene was added within 2 hrs., and the mixture heated and stirred for 3 hrs. more. The cyclized base was then extracted from the toluene with dilute hydrochloric acid and the extract refluxed for 15 hrs. to effect hydrolysis and decarboxylation.

The ester condensation of ethyl γ-phenylbutyrate with diethyl oxalate was effected by refluxing with potassium ethoxide in ether for 12 hrs., whereas with sodium ethoxide as condensing agent a reflux period of 24 hrs. was required.[1]

[1]E. B. Hershberg and L. F. Fieser, *Org. Syn., Coll. Vol.*, **2**, 195 (1943)
[2]H. V. Daeniker and C. A. Grob, *Org. Syn.*, **44**, 86 (1964)

Potassium ferricyanide [Potassium hexacyanoferrate (III)], $K_3Fe(CN)_6$. Mol. wt. 329.26. Suppliers: B, F, MCB.

Oxidation of phenols. Potassium ferricyanide in alkaline solution functions as a complex electron-abstracting ion, and many oxidations of phenols recorded in the literature[1] appear to involve oxidation of a phenolate anion to an oxygen radical, isomerization to a carbon radical, and dimerization of the carbon radical or combination with the oxygen radical. An interesting example is Barton's two-step synthesis of (±)-usnic acid.[2] Methylphloracetophenone (5 g.) was dissolved in a deaerated solution of 12.5 g. of sodium carbonate in water, the solution was cooled to 0°, and 1 equivalent of potassium ferricyanide in water was added with stirring under nitrogen. Evidently coupling of two carbon radicals gives the intermediate

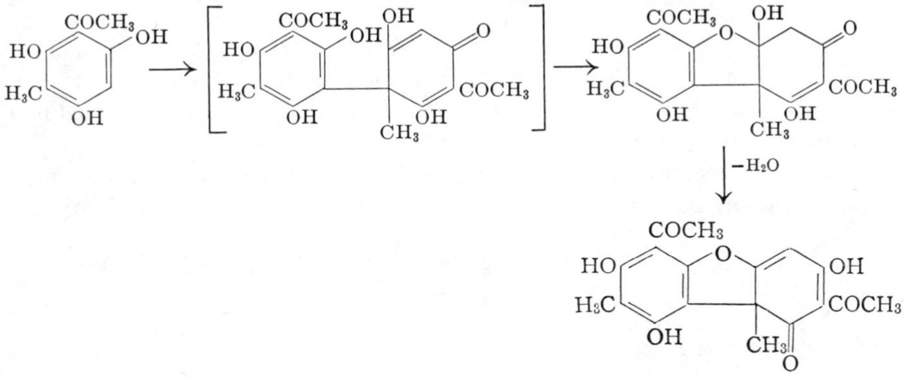

Usnic acid

formulated in brackets and the phenolic hydroxyl adds to the enone system to give the dimer, isolated by chromatography in 15% yield. Dehydration with sulfuric acid at 0° gave usnic acid in yield of about 2%.

That the first step in the oxidation of a phenol is formation of the oxygen radical has been proved conclusively by isolation of stable oxygen radicals. E. Müller and co-workers (1950–60)[3] prepared several radicals by shaking a benzene solution of a 2,4,6-trisubstituted phenol with alkaline ferricyanide. The aroxyl radical (2) is

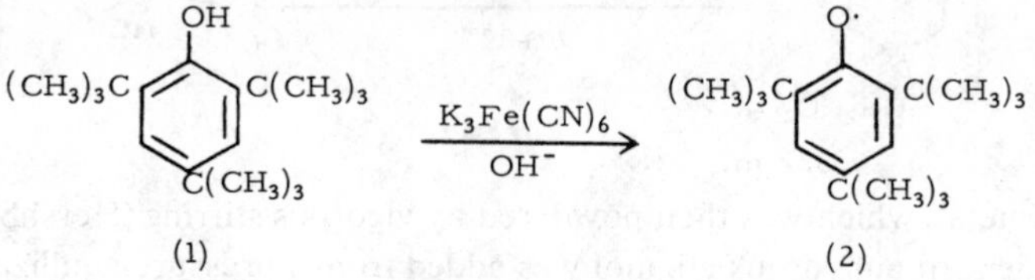
OH
(CH3)3C
C(CH3)3
C(CH3)3
(1)
K3Fe(CN)6
OH⁻
O·
(CH3)3C
C(CH3)3
C(CH3)3
(2)

deep blue, is a powerful oxidizing agent, and is completely monomeric in the solid state as well as in 0.1 *N* benzene solution. It is very sensitive to oxygen. The stable radical (4) was prepared by Galvin M. Coppinger[4] by oxidation of (3) with lead oxide in ether and by Kharasch and Joshi[5] by oxidation with alkaline ferricyanide.

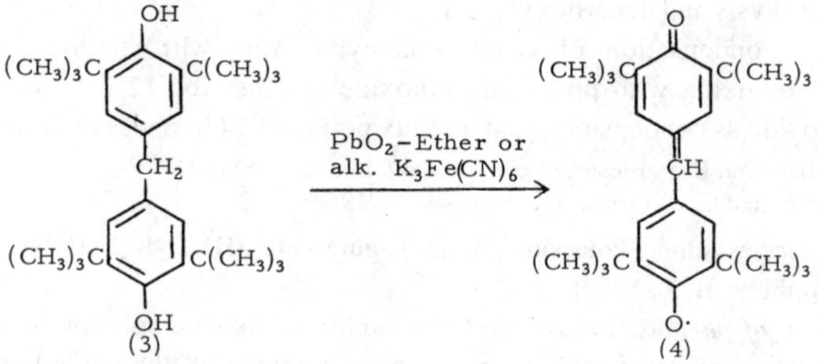
OH
(CH3)3C
C(CH3)3
CH2
(CH3)3C
C(CH3)3
OH
(3)
PbO2–Ether or
alk. K3Fe(CN)6
O
(CH3)3C
C(CH3)3
CH
(CH3)3C
C(CH3)3
O·
(4)

The substance forms deep blue needles, m.p. 157°. Bartlett[6] introduced use of the substance as a radical scavenger and named it galvinoxyl.

A 2-amino-1-phenol with a free position *para* to the hydroxyl group affords a dimeric oxidation product in high yield. Thus Butenandt and co-workers[7] oxidized

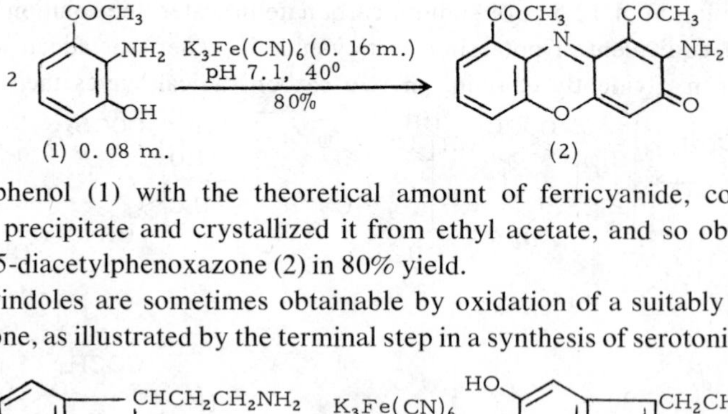
COCH3
NH2
2
OH
(1) 0. 08 m.
K3Fe(CN)6 (0. 16 m.)
pH 7. 1, 40°
80%
COCH3
COCH3
N
NH2
O
O
(2)

the aminophenol (1) with the theoretical amount of ferricyanide, collected the orange-red precipitate and crystallized it from ethyl acetate, and so obtained pure 3-amino-4,5-diacetylphenoxazone (2) in 80% yield.

Hydroxyindoles are sometimes obtainable by oxidation of a suitably substituted hydroquinone, as illustrated by the terminal step in a synthesis of serotonine (4).[8]

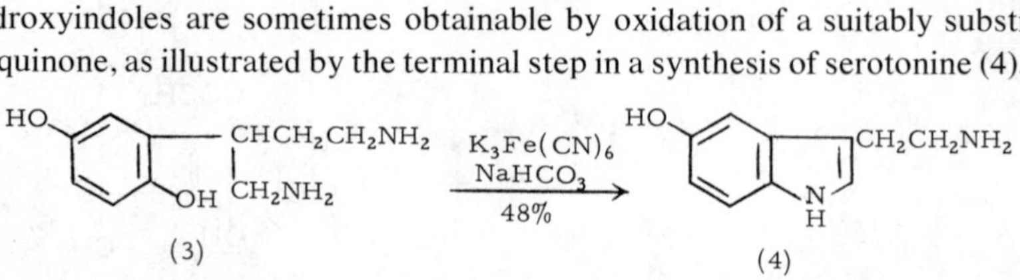
HO
CHCH2CH2NH2
OH
CH2NH2
(3)
K3Fe(CN)6
NaHCO3
48%
HO
CH2CH2NH2
N
H
(4)

In the oxidative ring closure of the benzophenone (5) to dehydrogriseofulvin (6), Taub *et al.*[9] found that under usual conditions (addition of the oxidant to the phenol) yields of (6) were in the order of 50–60%, with large amounts of the benzophenone recovered. However, an almost quantitative yield was obtained by the simple expedient of adding the substrate to a solution of the oxidant. A solution of 17.5 g. of potassium carbonate in 125 ml. of water was added to a solution of

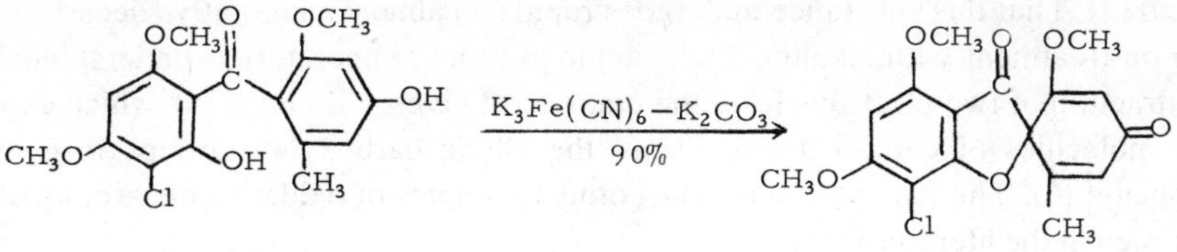
OCH3
OCH3
OH
CH3O
OH
Cl
CH3
K3Fe(CN)6−K2CO3
90%
OCH3
OCH3
CH3O
Cl
CH3

(5) (6)

1 g. of the benzophenone in 10–15 ml. of *t*-butanol, the *t*-butanol was removed by evaporation in vacuum, and the alkaline solution was added dropwise to a stirred solution of 4 g. of potassium ferricyanide in 50 ml. of water. The product precipitated immediately and was collected, washed, and dried; it was essentially single-spot on paper chromatography.

The *para* substituted phenol (7) gives the *o*-diphenoquinone (8) in high yield.[10]

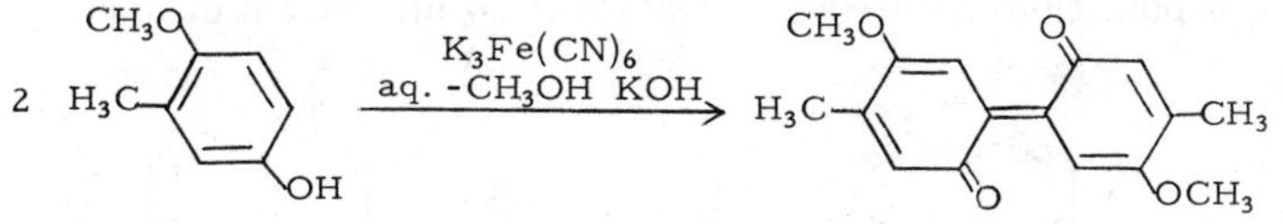
CH3O
2 H3C
OH
K3Fe(CN)6
aq. -CH3OH KOH
CH3O
H3C
CH3
OCH3

(7) (8)

Oxidative decarboxylation. Lohaus[11] briefly reported a synthesis of *p*-terphenyl involving in the terminal step a remarkably facile oxidative elimination of two carbonyl groups by reaction with alkaline ferricyanide. McDonald and Campbell[12] repeated the synthesis without isolation of intermediates and extended the scheme to the synthesis of a number of polyphenyls, including *p*-quinquephenyl.[13] However, Fieser and Haddadin[14] reinvestigated the synthesis of *p*-terphenyl and found that the diacid undergoing decarboxylation has a structure other than that which had been assumed. Their results are formulated. The Diels-Alder adduct (1) on brief treatment with methanolic potassium hydroxide is isomerized to the higher-melting *trans* ester (2), and prolonged treatment with base effects hydrolysis to the *trans*

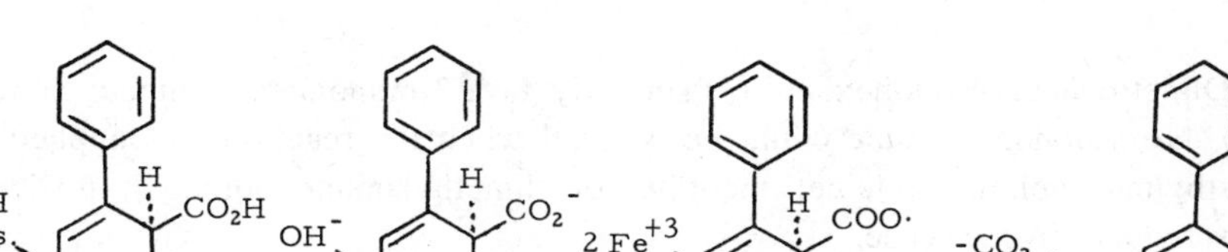
C6H5
C6H5
CO2CH3
CO2CH3
Refl. phenol
1 hr.
87%
C6H5 H
CO2CH3
CO2CH3
C6H5 H
KOH
15 min.
87%
C6H5
H
CO2CH3
H
CO2CH3
H5C6

(1) (2)

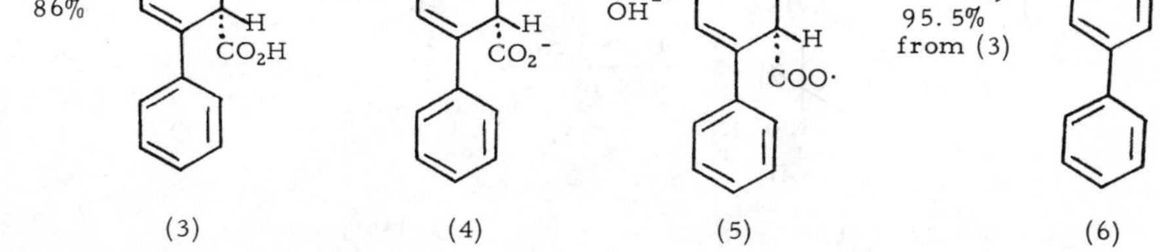
KOH
4 hrs.
86%
H
CO2H
H
CO2H
OH−
H
CO2−
H
CO2−
2 Fe+3
OH−
H
COO·
H
COO·
-CO2
95.5%
from (3)

(3) (4) (5) (6)

diacid (3). That this substance undergoes rapid and almost quantitative decarboxylation on treatment with alkaline ferricyanide at room temperature is understandable. Abstraction of two electrons from the dianion (4) gives a diradical (5), which expells two molecules of carbon dioxide from the allylic carboxylate groups to give *p*-terphenyl (6). The paper cited clarifies other examples of oxidative decarboxylation recorded in the literature.

Oxidation of methylaryl derivatives. W. A. Noyes established in early studies[15] that the methyl group of *o*- and *p*-nitrotoluene and of *o*-, *m*-, and *p*-toluenesulfonamide can be oxidized to carboxyl by boiling the material with a large excess of aqueous potassium ferricyanide. Although yields in other cases are very low, the reaction has been of some use in elucidation of structure. Thus Weissgerber and Kruber[16] on isolating 1,6-dimethylnaphthalene from coal tar characterized it by stirring 4 g. of the hydrocarbon with a total of 330 g. of potassium ferricyanide and 57 g. of potassium hydroxide in water for 24 hrs. Steam distillation removed

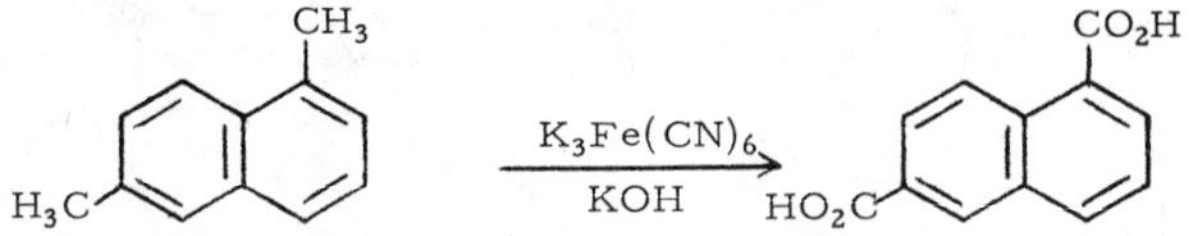
CH3
H3C
K3Fe(CN)6
KOH
CO2H
HO2C

unchanged hydrocarbon, and acidification and ether extraction afforded 0.5 g. of crude naphthalene-1,6-dicarboxylic acid. Similar experiences are reported by Ruzicka.[17]

Preparation of 1-methyl-2-pyridone (2). In a procedure by Prill and McElvain[18] pyridine is treated with an equivalent quantity of dimethyl sulfate to form 1-methylpyridinium methyl sulfate (1), and a solution of the salt in water is oxidized at 0° by simultaneous addition of solutions of ferricyanide and alkali.

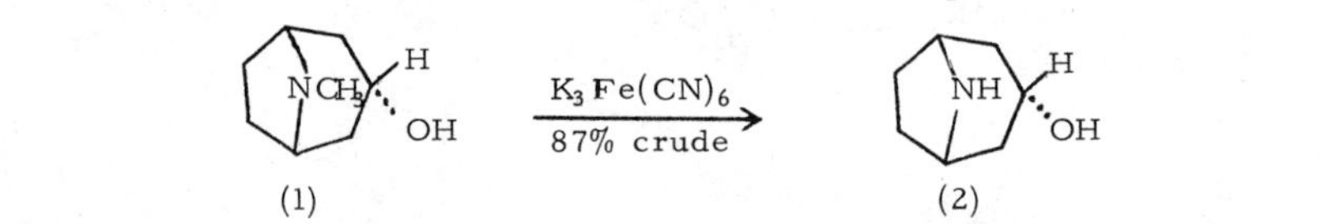
N
(CH3)2SO4 (1.83m.)
N
CH3
CH3SO4
K3Fe(CN)6 (3.65 m.)
NaOH (7. 5m.) 10°
65-70%
overall
N
CH3
O

1.83 m. (1) (2)

Demethylation of t-amines. Perrine,[19] on attempted oxidation of tropine (1) with alkaline ferricyanide to tropinone, obtained instead nortropine (2).

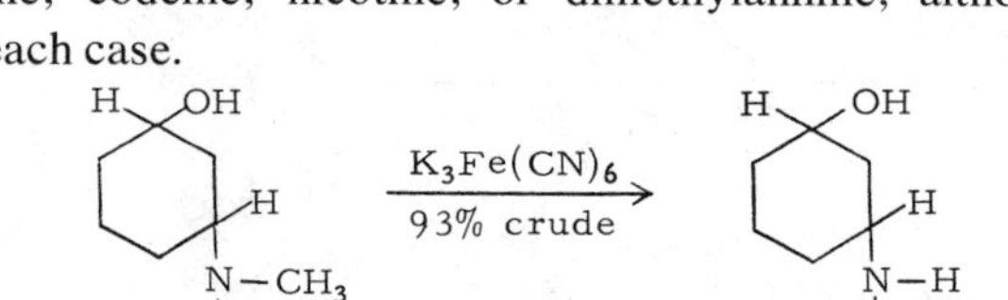
NCH3
H
OH
K3 Fe(CN)6
87% crude
NH
H
OH

(1) (2)

3-Dimethylaminocyclohexanol (3) similarly gave 3-monomethylaminocyclohexanol (4). No secondary amine could be isolated when the reaction was applied to N-methylmorpholine, codeine, nicotine, or dimethylaniline, although ferricyanide was reduced in each case.

H
OH
H
N−CH3
CH3
K3Fe(CN)6
93% crude
H
OH
H
N−H
CH3

(3) (4)

[1]B. S. Thyagarajan, *Chem. Rev.*, **58**, 439 (1958). For recent examples in the terpene series, *see* C. P. Falshaw, A. W. Johnson, and T. J. King, *J. Chem. Soc.*, 2422 (1963); A. C. Day, *ibid.*, 3001 (1964)
[2]D. H. R. Barton, A. M. Deflorin, and O. E. Edwards, *J. Chem. Soc.*, 530 (1956)
[3]E. Müller, A. Schick, R. Mayer, and K. Scheffler, *Ber.*, **93**, 2649 (1960). Paper XIV on oxygen radicals
[4]G. M. Coppinger, *Am. Soc.*, **79**, 501 (1957)
[5]M. S. Kharasch and B. S. Joshi, *J. Org.*, **22**, 1435 (1957)
[6]P. D. Bartlett *et al.*, *Am. Soc.*, **82**, 1756 (1960); **84**, 2596 (1962)
[7]A. Butenandt, U. Schiedt, and E. Biekert, *Ann.*, **588**, 106 (1954)
[8]J. Harley-Mason and A. H. Jackson, *J. Chem. Soc.*, 3651 (1954)
[9]D. Taub, C. H. Kuo, H. L. Slates, and N. L. Wendler, *Tetrahedron*, **19**, 1 (1963); *see also* D. Taub, C. H. Kuo, and N. L. Wendler, *J. Org.*, **28**, 2752 (1963)
[10]D. Schulte-Frohlinde and F. Erhardt, *Ann.*, **671**, 92 (1964)
[11]H. Lohaus, *Ann.*, **516**, 295 (1935)
[12]R. N. McDonald and T. W. Campbell, *J. Org.*, **24**, 1969 (1959)
[13]T. W. Campbell and R. N. McDonald, *Org. Syn.*, **40**, 85 (1960)
[14]L. F. Fieser and M. J. Haddadin, *Am. Soc.*, **86**, 2392 (1964)
[15]W. A. Noyes *et al.*, *Am. Chem. J.*, **5**, 97 (1883); **7**, 145, 149 (1885); **8**, 167, 176, 185, (1886)
[16]R. Weissgerber and O. Kruber, *Ber.*, **52**, 352 (1919)
[17]L. Ruzicka *et al.*, *Helv.*, **9**, 976 (1926); **14**, 238 (1931)
[18]E. A. Prill and S. M. McElvain, *Org. Syn.*, *Coll. Vol.*, **2**, 419 (1943)
[19]T. D. Perrine, *J. Org.*, **16**, 1303 (1951)

Potassium fluoride, KF. Mol. wt. 58.10, m.p. 860°, solubility in 100 g. H_2O: 92.3 at 18°, 150 at 80°. Suppliers: B, F, MCB.

Decarboxylation catalyst. The salt functions as a basic catalyst for the cyclization of adipic acid to cyclopentanone.[1] The best yield (81%) was obtained when the

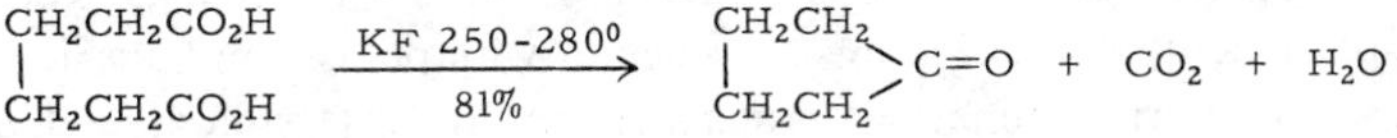
CH₂CH₂CO₂H
CH₂CH₂CO₂H
KF 250-280°
81%
CH₂CH₂
CH₂CH₂
C=O + CO₂ + H₂O

molar ratio of acid to catalyst was high (20:1) and was comparable to the yield of 75–80% in the classical procedure[2] using barium hydroxide in similar ratio (15:1).

Condensation catalyst. Papers by both Rand[1,3] and LeGoff[4] cite prior Japanese work on the ability of KF to catalyze the Knoevenagel condensation. Rand *et al.*[3] treated equimolecular amounts of a ketone and an active-methylene compound with 0.5 mole of a metal fluoride in ethanol or DMF and concluded that cesium and rubidium fluoride are somewhat superior to KF, whereas sodium and lithium fluoride are much inferior.

LeGoff[4] treated a mixture of 1,2,3-triphenylcyclopentadiene and 1,2,3-triphenylpropenone with potassium fluoride in dimethyl sulfoxide and so effected Michael-

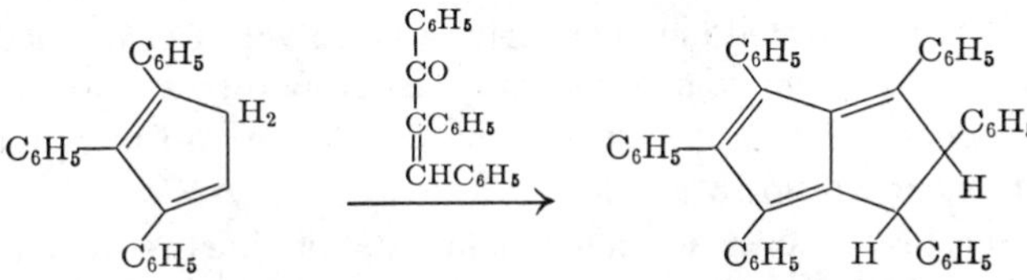
C₆H₅
C₆H₅
H₂
C₆H₅
C₆H₅
CO
CC₆H₅
CHC₆H₅
C₆H₅
C₆H₅
C₆H₅
C₆H₅
H
C₆H₅
H
C₆H₅

Knoevenagel dehydrative condensation to a bright yellow dihydrohexaphenylpentalene (53% yield; converted into hexaphenylpentalene with NBS in 77% yield). He suggests that the ability of F^- to function as a Lewis base is because HF_2^- forms a very strong hydrogen bond:

$$>CH_2 + 2F^- \longrightarrow >\bar{C}H + FHF^-$$

Rand and Dolinski[5] found that KF functions as a base in the Hofmann reaction of N-chlorobenzamide (1). Phenyl isocyanate (2) is formed but reacts further with starting material as shown to give (3).

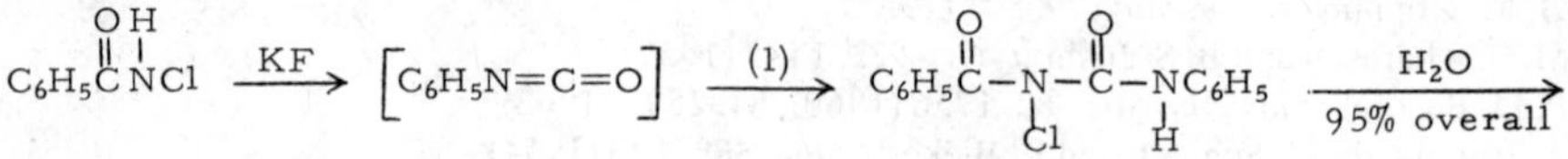
KF
(1)
H_2O
95% overall

(1) (2) (3)

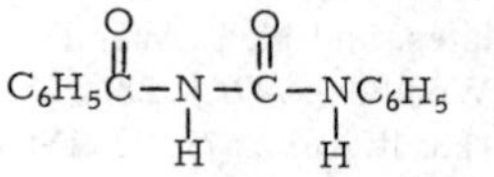

(4)

Alkyl fluorides. For the preparation of *n*-hexyl fluoride,[6] a round-bottomed flask equipped with a stirrer, a dropping tube, and a short fractionating column is charged with 2 moles of anhydrous, finely ground potassium fluoride and 200 g. of ethylene glycol, heated in an oil bath at 160–170°, and 1 mole of *n*-hexyl bromide is added

$$CH_3(CH_2)_4CH_2Br + KF \xrightarrow[40\text{-}45\%]{HOCH_2CH_2OH} CH_3(CH_2)_4CH_2F$$

dropwise in the course of 5 hrs. The product which distills is freed of 1-hexene by reaction with bromine and distilled at 91–92°.

Fluorinated aromatic compounds.[7] Finger and Kruse[8] found that an aromatic chlorine or bromine atom activated by an *o*- or *p*-nitro group is subject to nucleophilic displacement by fluoride ion on reaction with potassium fluoride in solvents such as succinonitrile, dimethyl formamide, or dimethyl sulfoxide. Unpublished work at Olin Mathieson Chemical Corporation[9,10] indicates that best results are obtained with dimethyl sulfoxide as solvent and with potassium fluoride that has been finely ground (<100 mesh) and dried in a vacuum oven at 100° for at least 4 hrs. prior to use. In a typical case[9] a mixture of *p*-nitrochlorobenzene, potassium fluoride, and dimethyl sulfoxide is stirred under reflux at 180° for 8–9 hrs., the cooled mixture

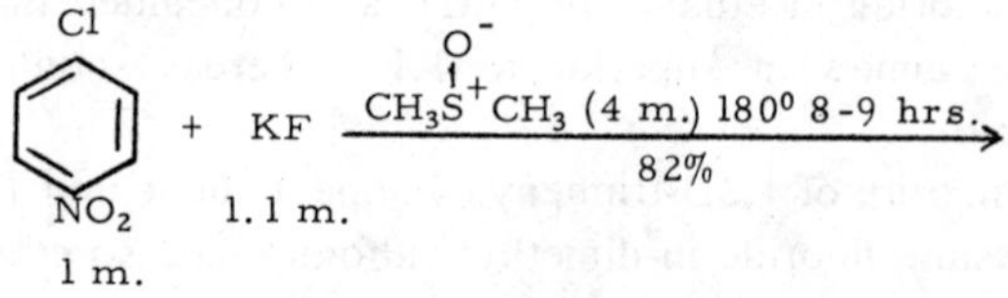
Cl
NO_2
1 m.
+ KF
1.1 m.
$CH_3\overset{+}{S}CH_3$ (4 m.) 180° 8-9 hrs.
82%
F
NO_2

is filtered, and the separated inorganic salts washed with fresh dimethyl sulfoxide (the extent of reaction can be determined by assaying the salt mixture for chloride ion). The product is separated by steam distillation. *o*-Fluoronitrobenzene is obtained in the same way in 58% yield[10].

Highest yields are obtained when the molar ratio of dimethyl sulfoxide to organic halide is 4:1. As the ratio is decreased, conversion and yield are both reduced. Kinetic studies under optimal conditions indicate that maximal conversion and yield are obtained in 8–9 hrs. When the reaction is continued beyond this time, the aryl halide reacts with the solvent to give by-products. The optimum temperature for the dimethyl sulfoxide system is 175–180°. At the reflux temperature (189°) the solvent slowly decomposes to give fragments which react with the aryl halide; for example, *p*-nitrophenylmethyl sulfide is formed when the above fluorination is conducted at 189°.

[1]L. Rand, W. Wagner, P. O. Warner, and L. R. Kovac, *J. Org.*, **27**, 1034 (1962)
[2]J. F. Thorpe and G. A. R. Kon, *Org. Syn., Coll. Vol.*, **1**, 192 (1941)
[3]L. Rand, J. V. Swisher, and C. J. Cronin, *J. Org.*, **27**, 3505 (1962)
[4]E. LeGoff, *Am. Soc.*, **84**, 3975 (1962)
[5]L. Rand and R. J. Dolinski, *J. Org.*, **30**, 48 (1965)
[6]A. I. Vogel, J. Leicester, and W. A. T. Macey, *Org. Syn., Coll. Vol.*, **4**, 525 (1963)
[7]Contributed by Eugene R. Shipkowski, Olin Mathieson Chemical Corporation
[8]G. C. Finger and C. W. Kruse, *Am. Soc.*, **78**, 6034 (1956)
[9]M. M. Boudakian and R. J. Polak, unpublished work at Olin Mathieson Chemical Corporation
[10]M. Lapkin, M. Ford, and E. R. Shipkowski, unpublished work at Olin Mathieson Chemical Corporation

Potassium hydride in oil, KH. Mol. wt. 40.11. Supplier: Ventron Corp.

The commercial material is a 40–50% dispersion of microcrystalline KH in an industrial white mineral oil (Bayol 85) which is freely soluble in hexane, heptane, diethyl ether, dibutyl ether, toluene, etc. Since potassium hydride is insoluble in these solvents (as well as in liquid ammonia and amines), the dispersion can be diluted with one of the inert organic solvents without impairing the activity.

Potassium hydride, being more basic and more active than sodium hydride, is suggested for use as a condensing agent in acetoacetic ester, Claisen, Stobbe, and related condensations which proceed with difficulty.

Handling procedures are the same as for sodium hydride in oil.

Potassium hydrosulfide, KSH. Zinner[1] describes a detailed preparation of an ethanolic solution and a procedure for its use for reaction with alkyl bromides to form alkylmercaptans, for example: CH_3CH_2SH (89%); $(CH_3)_2CHSH$ (79%); $C_6H_5CH_2SH$ (85%); $HSCH_2CH_2SH$ (69%).

[1]H, Zinner, *Ber.*, **86**, 825 (1953)

Potassium hydroxide, Mol. wt. 56.10. Commercially available pellets contain 85% KOH.

Ethanolic potassium hydroxide.[1] The preparation of alcoholic potassium hydroxide by refluxing potassium hydroxide pellets with 95% ethanol is a tedious process. If an efficient aspirator is available, the preparation is greatly facilitated by charging the pellets and solvent into a thick-walled filter flask of about 3–5 times the volume of ethanol to be used, stoppering the flask, marking the volume with a wax pencil mark, and connecting the side-arm to the suction pump. Initial vigorous boiling is accompanied by rapid solution of a large part of the alkali, and it may be necessary to swirl the flask to prevent loss of the alkaline solution. If boiling diminishes and the mixture begins to cool before all the solid has dissolved, the flask should be disconnected and warmed to complete solution. Evaporation of solvent during the process is compensated for by adding fresh ethanol to the original marked volume.

Agitation by stirring is also effective. A 1-l. three-necked flask was mounted over an ice bath, fitted with a stirrer with a curved Teflon blade and a thermometer, and charged with 75 g. of potassium hydroxide pellets and 300 ml. of 95% ethanol. With brisk stirring solution was complete in $2\frac{1}{2}$ min., with a temperature rise from 26 to 56°. When the ice bath was raised, the temperature was brought to 26° in $2\frac{1}{2}$ min.[2]

30% Methanolic potassium hydroxide is prepared by dissolving 680 g. of pellets in 2 l. of methanol.[3]

Cleavage of olefins and oxidation of aldehydes. In an investigation of 1840, Varrentrap[4] effected separation of oleic acid from saturated companions by extracting it as the ether-soluble lead salt and apparently obtained very pure acid. He carried out numerous analyses on the acid and its salts but despaired of the possibility of distinguishing between alternative formulas by analysis alone. Seeking chemical evidence, he treated oleic acid with nitric acid and discovered elaidic acid. In another experiment Varrentrap stirred a mixture of oleic acid, potassium hydroxide, and a few drops of water and gradually raised the temperature until the potassium hydroxide just melted. A vigorous reaction ensued with evolution of a gas identified as hydrogen. He poured the hot yellowish mixture onto a limited amount of water so that potassium soap separated as a solid which could be collected by filtration. Acidification liberated a solid acid which on two crystallizations melted constantly at 62° and was identified as palmitic acid. Elaidic acid gave the same cleavage product. A second cleavage product not identified at the time is acetic acid.

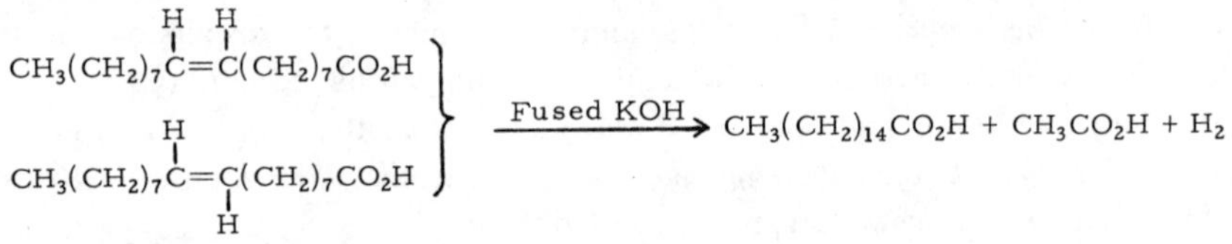

$$
\left.
\begin{array}{c}
\overset{\text{H}}{\underset{}{\text{CH}_3(\text{CH}_2)_7\overset{|}{\text{C}}}}=\overset{\text{H}}{\underset{}{\overset{|}{\text{C}}(\text{CH}_2)_7\text{CO}_2\text{H}}} \\[2ex]
\overset{\text{H}}{\underset{}{\text{CH}_3(\text{CH}_2)_7\overset{|}{\text{C}}}}=\underset{\underset{\text{H}}{|}}{\text{C}}(\text{CH}_2)_7\text{CO}_2\text{H}
\end{array}
\right\}
\xrightarrow{\text{Fused KOH}}
\text{CH}_3(\text{CH}_2)_{14}\text{CO}_2\text{H} + \text{CH}_3\text{CO}_2\text{H} + \text{H}_2
$$

More than 100 years later, Cornforth, Hunter, and Popják[5] made use of this reaction in determining the pattern of isotope distribution in cholesterol produced by biosynthesis from labeled acetic acid. One degradation liberated ring A in the form of 2-methylcyclohexanone (1), which was converted by the Schmidt reaction into the lactam (2), which in turn was converted by hydrolysis and methylation into the

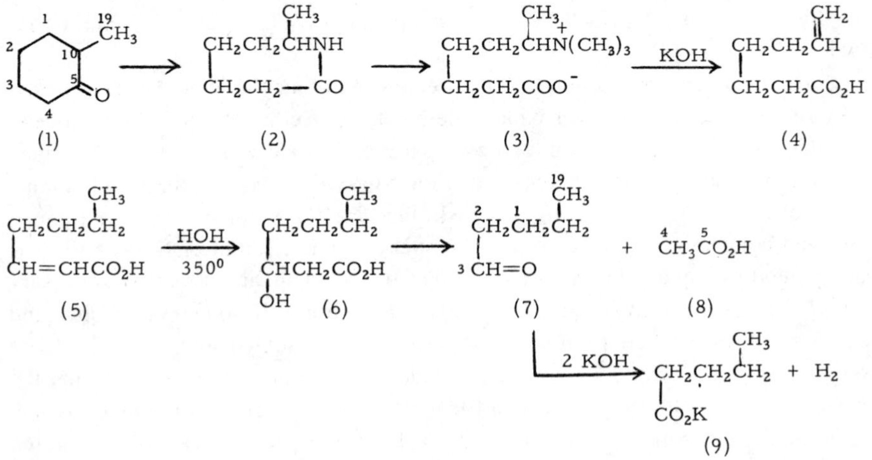

betaine (3). When this betaine was heated with potassium hydroxide, trimethylamine was eliminated smoothly, but the product was not the acid (4) with a terminal double bond. Instead, the double bond progressed down the chain by a series of allylic proton abstractions and equilibrations to form the α,β-unsaturated acid (5), but at this point cleavage occurred, probably by addition of water, to give (6) and reverse aldolization to the aldehyde (7) and acetic acid (8). Finally, at the elevated temperature (350°), the aldehyde reacted with potassium hydroxide to form the potassium carboxylate (9) and hydrogen.

In degrading ring D and the side chain of biosynthetic cholesterol, Cornforth, Gore, and Popják[6] demonstrated that potassium hydroxide fusion can be used to effect oxidation of an aldehyde to an acid. The aldehyde is converted into the oxime and this is fused with potassium hydroxide; the reaction may proceed through the nitrile as an intermediate:

$$RCHO \longrightarrow RCH{=}NOH \xrightarrow[150^0]{KOH} [RC{\equiv}N] \longrightarrow RCO_2K$$

Reaction of a free aldehyde with fused alkali to form the acid salt and hydrogen ($7 \longrightarrow 9$, above) is illustrated by a procedure for the preparation of vanillic acid.[7] A stainless steel beaker equipped with a Nichrome or Monel stirrer and heated on an

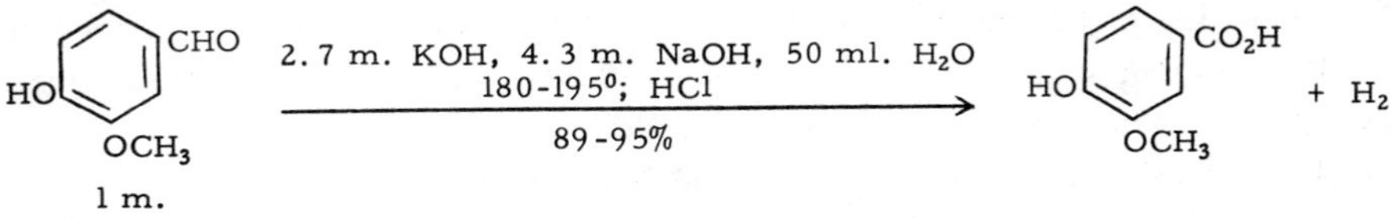
CHO
HO
OCH3
1 m.
2.7 m. KOH, 4.3 m. NaOH, 50 ml. H2O
180-195⁰; HCl
89-95%
CO2H
HO
OCH3
+ H2

electric hot plate is charged with pellets of KOH and of NaOH and a little water. The temperature of the melt is brought to 160°, and a portion of the vanillin is added. A vigorous reaction ensues and raises the temperature to 180–195°. This temperature is maintained by further addition of vanillin until 1 mole has been added. The workup includes brief treatment of a solution of the reaction mixture in water with sulfur dioxide to prevent the product from becoming brown in color.

Reduction. An observation made by us[8] over 30 years ago still awaits clarification. The lactone-diacid (1) on fusion with potassium hydroxide afforded, in low yield, the diphenylmethane triacid (2). Possibly (1) undergoes disproportionation to (2) and the corresponding benzophenonetricarboxylic acid.

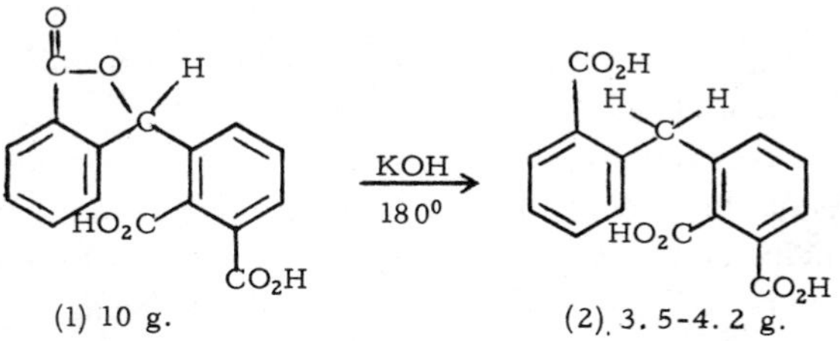
O
C—O
H
HO2C
CO2H
KOH
180⁰
CO2H
H
H
HO2C
CO2H

(1) 10 g. (2), 3.5–4.2 g.

Other reactions.[9]

[1]Contributed by Carl H. Snyder, University of Miami
[2]Experiment by L. F. Fieser
[3]H. O. House, *Org. Syn., Coll. Vol.*, **4**, 367 (1963)
[4]F. Varrentrapp, *Ann.*, **35**, 196 (1840)
[5]J. W. Cornforth, G. D. Hunter, and G. Popják, *Biochem. J.*, **54**, 590, 597 (1953)
[6]J. W. Cornforth, I. Y. Gore, and G. Popják, *Biochem. J.*, **65**, 94 (1957)
[7]I. A. Pearl, *Org. Syn., Coll. Vol.*, **4**, 974 (1963)
[8]L. F. and M. Fieser, *Am. Soc.*, **55**, 3010 (1933)
[9]See review of alkali fusion, B. C. L. Weedon, *Technique of Organic Chemistry*, **11**, 655 (1963)

Potassium hydroxide–Acetone. Roedig *et al.*[1] prepared perchlorohexadiene-1,5 (4) by known methods, namely addition of chloroform to perchloroethylene (1), dehydrochlorination to perchloropropylene (3), and coupling of two molecules of this with

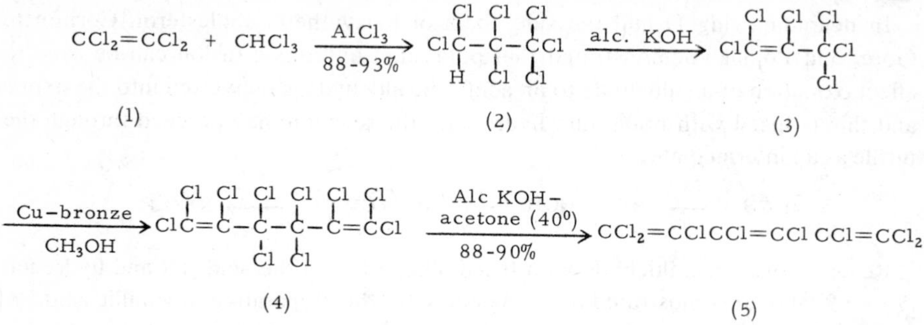

copper bronze. Perchlorohexadiene-1,5 (4) is incapable of undergoing dehydrohalogenation and, like other polyhalogen compounds of this special type, it undergoes facile dechlorination on reaction with alcoholic potassium hydroxide in acetone to give perchlorohexatriene-1,3,5 (5) in high yield.

[1]A. Roedig, G. Voss, and E. Kuchinke, *Ann.*, **580**, 24 (1953)

Potassium hypochlorite, KOCl. The reagent can be prepared from calcium hypochlorite and potassium carbonate.[1]

In preparing 5-formyl-4-phenanthroic acid by ozonization of pyrene in dimethylformamide, Dessy and Newman[2] stirred the solution of ozonide into 1% acetic acid, collected the solid which separated, and extracted it repeatedly with hot aqueous potassium hydroxide solution. When the dark brown filtrate was treated with potassium hypochlorite solution, let stand overnight, and then heated on the steam

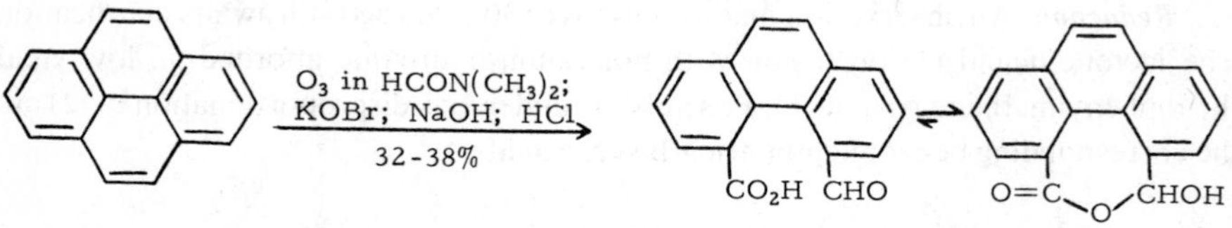

bath for 4 hrs., the color brightened to orange. Addition of strong sodium hydroxide then precipitated the product as the sodium salt.

Meyers[3] added potassium carbonate to a solution of calcium hypochlorite to pH 9–11 and found the filtered solution effective for oxidation of benzyl alcohols to the corresponding aldehydes. For example a mixture of the aqueous solution of KOCl with a solution of benzyl alcohol in methanol on being shaken overnight at room temperature, followed by extraction with benzene, afforded benzaldehyde in 77% yield.

[1]M. S. Newman and H. L. Holmes, *Org. Syn., Coll. Vol.*, **2**, 428 (1943)
[2]R. E. Dessy and M. S. Newman, *ibid.*, **4**, 484 (1963)
[3]C. Y. Meyers, *J. Org.*, **26**, 1046 (1961)

Potassium manganate, K_2MnO_4. Mol. wt. 197.13. When aqueous potassium permanganate is heated with potassium hydroxide, disproportionation occurs to give potassium manganate with evolution of oxygen:

$$4\ KMnO_4 + 4\ KOH \longrightarrow 4\ K_2MnO_4 + 2\ H_2O + O_2$$

Rigby[1] prepared potassium manganate by heating a solution of 1 kg. of 85% potassium hydroxide in 500 ml. of water at 120–140° during addition of 700 g. of $KMnO_4$ in 40–50 g. portions, allowing the effervescence (O_2) to settle after each addition. The resulting suspension of crystalline potassium manganate was heated at the boiling point for 1 hr. while being stirred with a steel spatula, with addition of water to compensate for evaporation. The mixture was then cooled, and the product was collected and drained well. It was an almost dry mass of fine purple crystals containing about 88% K_2MnO_4; yield about 1 kg. Although the solid is stable, even moderately alkaline solutions undergo slow disproportionation to MnO_2 and $KMnO_4$ at room temperature.

Rigby explored oxidation of a few unsaturated acids (cinnamic, oleic, cyclopentadiene-maleic acid) with potassium manganate in aqueous alkali at 0° and isolated the corresponding *cis*-diols in yields of about 20–40% (crude).

[1]W. Rigby, *J. Chem. Soc.*, 2452 (1956)

Potassium 2-methyl-2-butoxide, 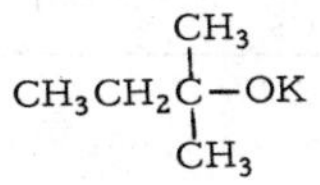

For preparation, *see* Potassium, Pearson procedure.

Lethargic oximation. Several failures to convert di-*o*-substituted acetophenones into oximes are recorded by Kadesch.[1] However, Greer and Pearson[2] discovered that 2,4,6-trimethylacetophenone oxime can be obtained in 40% yield by the simple expedient of allowing a piperidine solution of the ketone and hydroxylamine hydrochloride stand at room temperature for one month. The hindered oxime undergoes Beckmann rearrangement with extreme rapidity: in concd. sulfuric acid at 0° the reaction is about 94% complete in 75 min.; the reactivity is attributed to tilting of the α-oximinoethyl group at right angles to the ring with consequent loss of resonance interaction between the side chain and the ring. Later Pearson and Keaton[3] improved the procedure considerably by use of the powerful base potassium 2-methyl-2-butoxide in *t*-amyl alcohol. The reagents in the amounts indicated were added to 125 ml. of a solution of the alkoxide, and the flask was stoppered and let

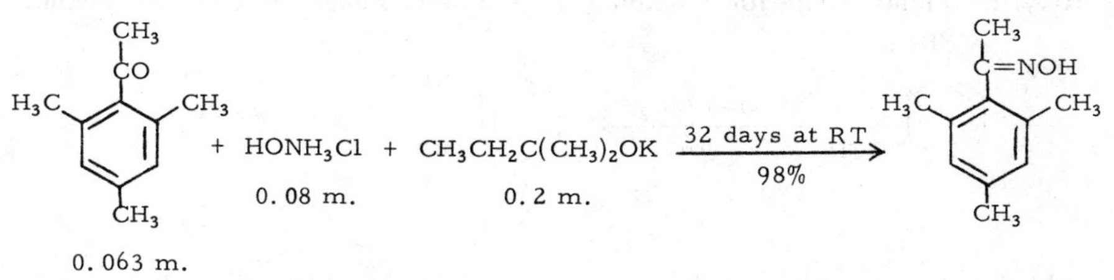
CH3
CO
H3C
CH3
CH3
0. 063 m.
+ HONH3Cl
0. 08 m.
+ CH3CH2C(CH3)2OK
0. 2 m.
32 days at RT
98%
CH3
C=NOH
H3C
CH3
CH3

stand at room temperature for 32 days. No change was noted other than the separation of potassium chloride. Workup afforded crude oxime, m.p. 98–101°, in 98% yield. Sublimation gave pure material, m.p. 101.5–102.5°. Refluxing the reaction mixture gave at most about 50% of the oxime, but the yield could not be raised above this figure, probably because of decomposition of hydroxylamine to ammonia, water, and other products. Other results are given in the table on p. 940.

Pearson coined the phrase "lethargic reaction" to describe one that proceeds slowly but cannot be forced because of destructive side reactions at elevated

Acetophenone (or ketone)	Time, days	Crude yield of oxime, %
2,4,6-Trimethyl-	32	98
	10	53
2,3,4,6-Tetramethyl-	180	90
	10	30
Pentamethyl-	420	81
2,6-Dimethyl-4-*t*-butyl-	180	95
Benzoylmesitylene	450	16

temperatures. The name draws attention to the potentialities for application of the technique to other reactions.

Potassium 2-methyl-2-butoxide is regarded as superior to potassium *t*-butoxide for the alkylation of 1,10-dimethyl-$\Delta^{1(9)}$-2-octalone.[4]

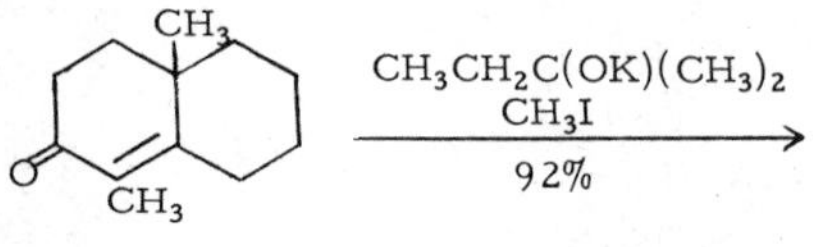
CH3
CH3CH2C(OK)(CH3)2
CH3I
92%
CH3
CH3
H3C
CH3

[1]R. G. Kadesch, *Am. Soc.*, **66**, 1207 (1944)
[2]F. Greer and D. E. Pearson, *Am. Soc.*, **77**, 6649 (1955)
[3]D. E. Pearson and O. D. Keaton, *J. Org.*, **28**, 1557 (1963)
[4]S. L. Mukherjee and P. C. Dutta, *J. Chem. Soc.*, 67 (1960)

Potassium nitrosodisulfonate (Fremy's salt), $\cdot O{-}N(SO_3K)_2$. Mol. wt. 268.33. A stable, red, water-soluble radical. Supplier: Aldrich.

The reagent is prepared[1] by oxidizing sodium hydroxylamine disulfonate with permanganate, removing the MnO_2 by filtration, and salting out with potassium chloride. Teuber and co-workers have studied extensively the oxidation of sub-

$$HO{-}N(SO_3Na)_2 \xrightarrow{[O]/KCl} \cdot O{-}N(SO_3K)_2$$

(colorless) (violet solution)

stituted phenols and derivatives of aniline to quinones.[2] Best results are obtained with at least two alkyl or alkoxyl substituents and a free position *o*- or *p*- to the hydroxyl or amino group, for example (1). Dann and Zeller[3] oxidized the phenol

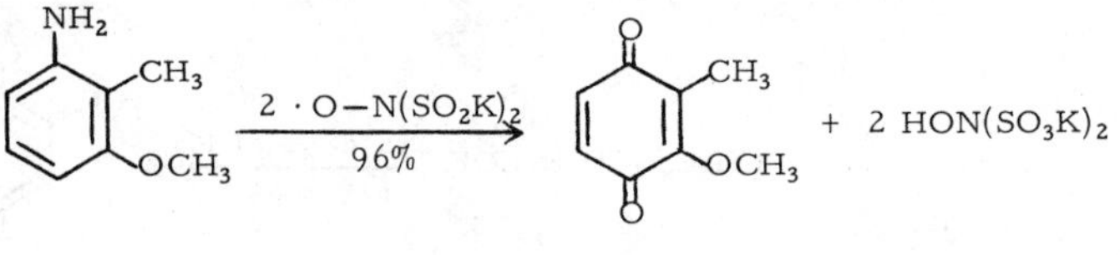
NH2
CH3
OCH3
2 · O−N(SO2K)2
96%
CH3
OCH3
+ 2 HON(SO3K)2

(1) (2)

(3) to khellinquinone (4) by dissolving 1 g. of material in 75 ml. of dimethylformamide in a round-bottomed flask with a glass stopper, adding all at once a solution of 3.1 g. of Fremy's salt and 1.2 g. of KH_2PO_4 in 230 ml. of water, stoppering the flask, and

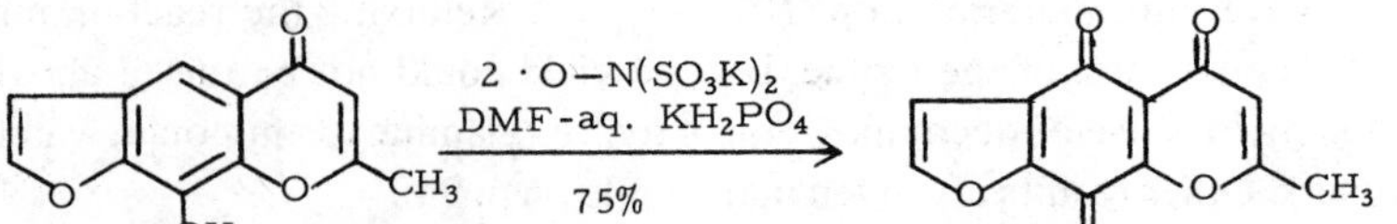
2 · O−N(SO3K)2
DMF-aq. KH2PO4
75%
CH3
OH
CH3

(3) (4)

shaking vigorously for a few minutes, until the violet color changed to red-brown. In another 10 min. yellow needles began to separate; yield 0.8 g.

Teuber and Staiger[4] found that the reagent not only dehydrogenates 2,3-dihydroindole but effects hydroxylation to produce 5-hydroxyindole. A Sandoz group[5]

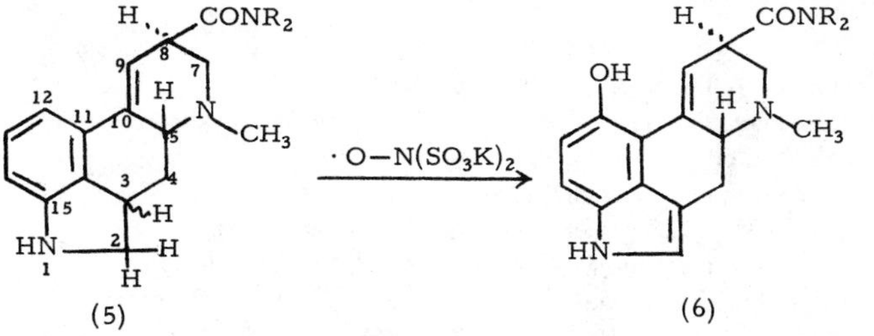

applied the reaction to effect the dehydrogenation-hydroxylation of 2,3-dihydrolysergic acid amides (5) to 12-hydroxylysergic acid amides (6).

A Lederle group,[6] interested in obtaining possible ultraviolet models for degradation products of the mitomycin group of antibiotics, explored the oxidation of various aminonaphthol derivatives with Fremy's salt with variable results. A solution

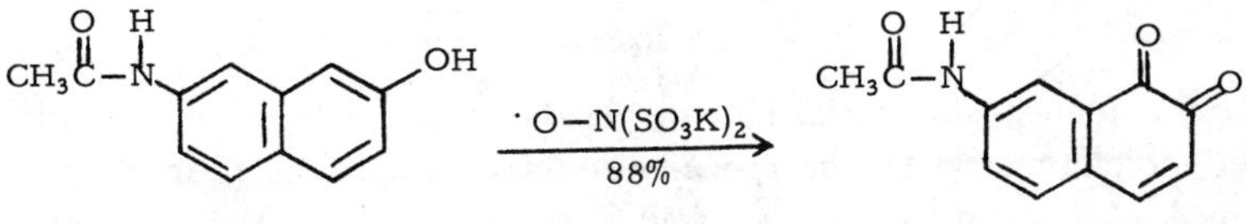

of 0.045 mole of 7-acetamido-2-naphthol in 200 ml. of methanol was added to an ice-cold solution of 0.1 mole of Fremy's salt in 2 l. of water and 400 ml. of 0.167 *M* KH_2PO_4. Brick red 7-acetamido-1,2-naphthoquinone separated at once. 8-Acetamido-2-naphthol by the same method afforded the *ortho* quinone in 73% yield, but the reaction failed completely with 5-acetamido-2-naphthol. 5-Amino-1-naphthol

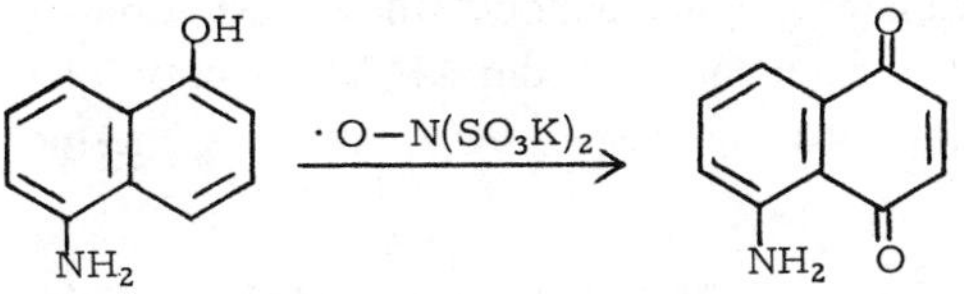

undergoes facile selective oxidation of the phenolic ring to give 5-amino-1,4-naphthoquinone "in high yield," but 8-amino-2-naphthol did not react.

Teuber[7] oxidized equilenin with potassium nitrosodisulfonate in aqueous acetone and isolated the bright red *o*-quinone in high yield.

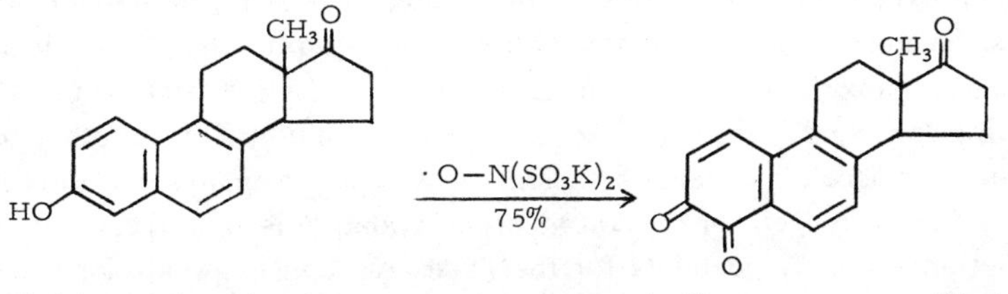

Pyrogallol is oxidized by the reagent in aqueous buffer to purpurogallin,[8] but the yield (39%) is lower than in oxidation with sodium iodate. Tropolone does not react and hence behaves more like a carboxylic acid than a phenol.

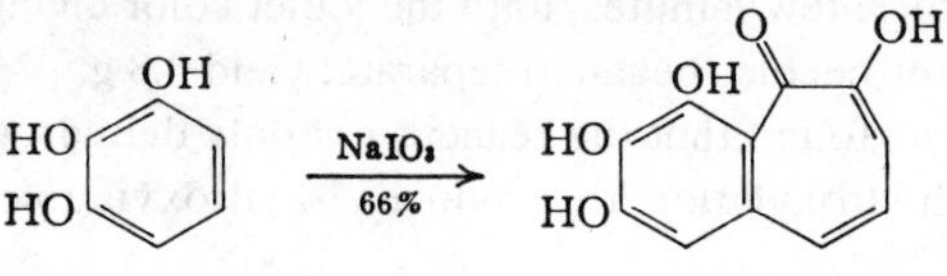

Purpurogallin

[1] H.-J. Teuber and G. Jellinek, *Ber.*, **85**, 95 (1952)
[2] Principal papers: H.-J. Teuber *et al.*, *Ber.*, **86**, 1036 (1953); **87**, 1841 (1954); **92**, 674 (1959)
[3] O. Dann and H.-G. Zeller, *Ber.*, **93**, 2829 (1960)
[4] H.-J. Teuber and G. Staiger, *Ber.*, **87**, 1251 (1954); **89**, 489 (1956)
[5] P. A. Stadler, A. J. Frey, F. Troxler, and A. Hofmann, *Helv.*, **47**, 756 (1964)
[6] W. A. Remers, P. N. James, and M. J. Weiss, *J. Org.*, **28**, 1169 (1963)
[7] H.-J. Teuber, *Ber.*, **86**, 1495 (1953)
[8] H.-J. Teuber and O. Glosauer, *Ber.*, **98**, 2643 (1965)

Potassium permanganate. $KMnO_4$. Mol. wt. 158.05, solubility in 100 parts of water: 2.83 g. at 0°, 32.5 g. at 75°.

Stoichiometry. It is usually assumed that the oxygen equivalent of potassium permanganate is 1.5, as represented for example in the following equation for the oxidation of a secondary alcohol:

$$3\ R_2CHOH + 2\ KMnO_4 \longrightarrow 3\ R_2CO + 2\ MnO_2 + 2\ KOH + 2\ H_2O$$

However, evidence summarized below leads us to question the validity of the accepted stoichiometry. In one procedure, permanganate taken in 103% excess of theory was all consumed, and the yield of product was 78–84%. In another, permanganate taken in 90% excess was all consumed, and the yield of product was 82%; when permanganate was taken in only 50% excess the yield dropped to 73%. Similar results are on record for the oxidation of an inorganic substrate, manganous sulfate. Ball, Goodwin, and Morton[1] prepared active manganese dioxide for oxidation of vitamin A to retinene by mixing aqueous solutions of equivalent amounts of potassium permanganate and manganous sulfate. Attenburrow *et al.*[2] prepared a considerably more active manganese dioxide by reaction of manganous sulfate in alkaline solution with permanganate in 85% excess of the theory according to the following equation:

$$2\ KMnO_4 + 3\ MnSO_4 + 4\ NaOH \longrightarrow 5\ MnO_2 + K_2SO_4 + 2\ Na_2SO_4$$

As in the case of organic substrates, the large excess of permanganate was all consumed. The brown precipitate was assumed to be MnO_2; the weight was recorded, but the substance was not analyzed.

A partial explanation of the observations cited is available from facts known to analytical chemists: under certain conditions potassium permanganate decomposes to manganese dioxide with evolution of oxygen.[3] A solution 0.04 *N* in sulfuric acid decomposes nearly 20 times as fast as a neutral solution. On the other hand, decomposition is accelerated also by alkali as well as by manganese dioxide. Thus decomposition of reagent in the course of an oxidation is an autocatalytic process. This reaction probably accounts for the excessive consumption of permanganate in the Attenburrow experiment.

An exploratory experiment by one of us demonstrated dramatically that oxygen is evolved in a typical permanganate oxidation of an organic substrate. A solu-

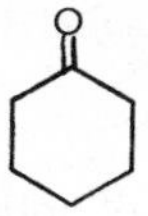 + $KMnO_4$ $\xrightarrow{50^0}$ $HO_2C(CH_2)_4CO_2H$ + MnO_2 + O_2

20 ml. (0.19 m.) | 91.5 g. (0.58 m.) (1.5 x theory) | 16.5 g. (58%) (0.11 m.) | 64.8 g. (75% pure) (0.56 m.) | 660 ml. (0.03 m.)

tion of permanganate in 50% excess of theory in 500 ml. of water was stirred mechanically and heated to 52° in a flask mounted above an ice-water bath and allowed to cool to 49°. Cyclohexanone (20 ml.) was then added and connection was made to a gas-collecting system (an Ace Glass stirring rod 8256-E fitted with a Teflon blade and mounted in an 8444 bearing lubricated with oil adequately retains the evolved oxygen). No alkali was required for catalysis, and the temperature was controlled to 50° by vigorous cooling. Oxygen evolution started at once and continued vigorously. After 47 min. intermittent heating was required to maintain a temperature of 50° until the supernatant liquor was colorless. The brown precipitate was collected, washed liberally, stirred with 500 ml. of water at 70°, and again collected and washed. Dried to constant weight, the material was found to be 75% pure MnO_2. The second wash liquor was evaporated to dryness, and the residue (8.9 g.) dissolved in a concentrate (130 ml.) of the first filtrate. Acidification and cooling gave a white precipitate (16.8 g.), which on extraction with ether in a Soxhlet afforded 16.5 g. of adipic acid, m.p. 152–154°. The oxygen evolved amounted to 660 ml. (N.P.T), or 0.03 mole, as compared to 0.11 mole of adipic acid formed. A balance is not possible for this experiment in the absence of information on the amount of permanganate consumed in overoxidation of adipic acid. The experiment suggests that routine determination of oxygen evolution might be revealing and that a quantitative study of high-yield permanganate oxidations might promote a better basis for the interpretation of mechanisms.[4] Such information might also suggest expedients for improvement of yields.

Isolation techniques. Since excess permanganate suffers autocatalytic destruction, an aqueous oxidation is almost invariably continued to an end point easily recognized by spotting a drop of brown suspension on a filter paper; a trace of permanganate will be evident from a pink color in the moist ring surrounding the brown spot.

In many procedures the voluminous brown precipitate is removed by filtration, washed liberally with water as in the experiment cited above, and even extracted in a Soxhlet for full recovery of product, which is often adsorbed very firmly on the precipitate. Vigorous boiling coagulates the precipitate and speeds up filtration, but the operation itself takes time. In any case one is left with a large volume of aqueous solution from which to recover the product, after acidification, by evaporation and solvent extraction. Where applicable, a much simpler process is to acidify the reaction mixture and pass in sulfur dioxide (or to add $NaHSO_3 + HCl$) to reduce MnO_2 to the soluble sulfate. A lengthy process of filtration and washing is avoided and, instead of increasing the volume of water, an inorganic solute is produced which descreases the solubility of the organic product. In reviewing procedures specifying MnO_2-filtration, we have wondered if many of them could not be improved by SO_2-reduction.

Aldehyde ⟶ acid. *n*-Heptanoic acid is prepared[5] by stirring a solution of 350 ml. of concd. sulfuric acid in 2.7 l. of water in an ice bath and, when the temperature

$$3\ \underline{n}\text{-}C_6H_{13}CHO + 2\ KMnO_4 + H_2SO_4 \xrightarrow[76\text{-}78\%]{} 3\ \underline{n}\text{-}C_6H_{13}CO_2H + K_2SO_4 + 2\ MnO_2 + H_2O$$

3 m. 2.15 m. (Theory = 2. 0)

falls to 15°, adding 3 moles of *n*-heptaldehyde, followed by a slight excess of potassium permanganate, added at such a rate that the temperature does not rise above 20°. Sulfur dioxide is passed into the mixture of suspended manganese dioxide until the solid is dissolved. The oily layer is separated, washed with water, dried, and distilled.

A procedure by Shriner and Kleiderer[6] for the oxidation of piperonal to piperonylic acid is as follows. A mixture of piperonal with 1.5 l. of water is heated on the steam bath with vigorous stirring to produce an emulsion, and a solution of permanganate

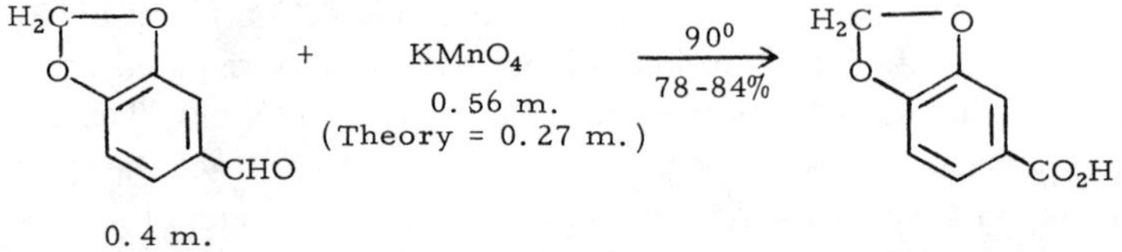

(in large excess) in 1.8 l. of water is added in the course of 45 min. After heating for 1 hr. more the solution is made alkaline, the hot mixture is filtered, and the manganese dioxide is washed with three 200-ml. portions of hot water. Acidification of the filtrate gives a precipitate which on crystallization from 95% ethanol affords pure piperonylic acid.

Note that the aldehyde oxidation in an acidic solution follows conventional stoichiometry whereas that in which the mixture was allowed to become alkaline does not.

$ArCH_3 \longrightarrow ArCO_2H$. The following two procedures both employ the MnO_2-filtration technique. In the first[7] the amount of permanganate taken is less than the

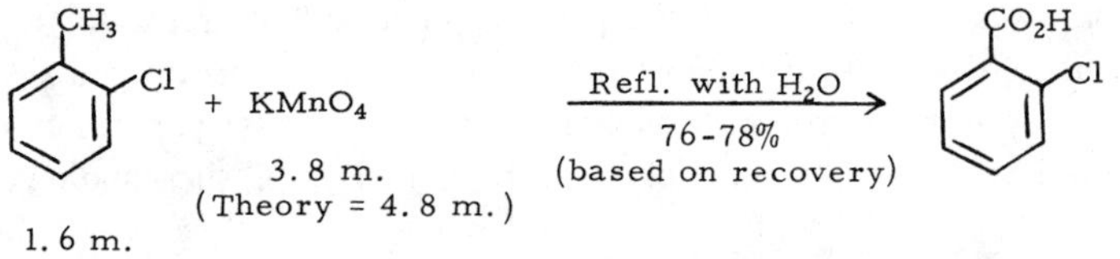

theory and starting material is recovered by steam distillation. In the second[8] α-picoline is heated with a slight excess of permanganate in 3 l. of water on the steam bath; the acidic product is isolated as the hydrochloride.

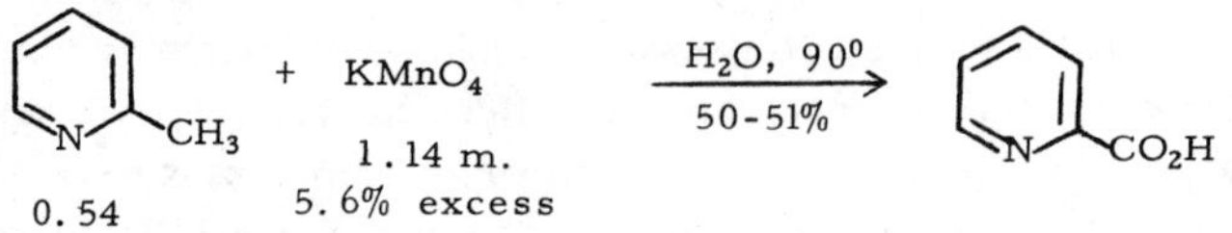

Cleavage of an aromatic ring. An *Organic Syntheses* procedure[9] for the oxidation of naphthalene to phthalonic acid (2) and for the decarboxylation of this substance, without isolation, via the bisulfite addition compound to phthalaldehydic acid in our opinion is not as satisfactory as a two-step procedure by Graebe and Trümpy[10] upon which it seems to have been based. The *O. S.* procedure is on one third the scale of the original and specifies addition of 1.5 l. of boiling permanganate solution in small portions over a 1½-hr. period, a laborious process. Graebe and Trümpy

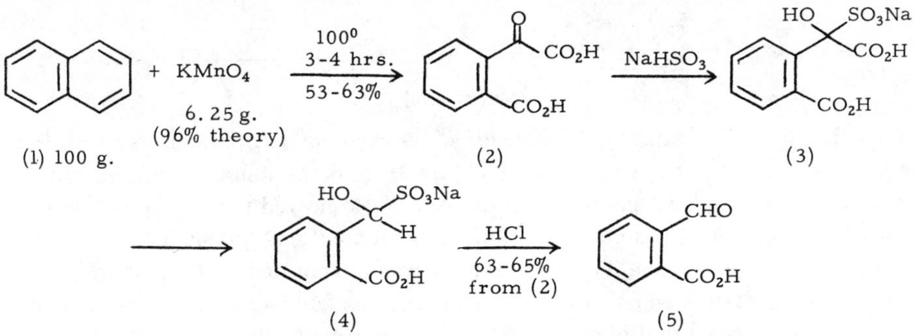

boiled a mixture of 100 g. of naphthalene, 625 g. of permanganate (96% of the theory), and $6\frac{1}{4}$ l. of water vigorously under reflux for 3–4 hrs. until a spot test showed the supernatant liquor to be colorless (mechanical stirring speeds up the reaction, prevents bumping, and permits rapid cooling). The mixture was cooled to solidify residual naphthalene and filtered to remove MnO_2 and hydrocarbon. In this case filtration is necessary for removal of starting material and it is easy and fast because manganese dioxide is converted by vigorous boiling into hard, granular particles. The solution was concentrated, neutralized with 220–240 g. of concd. sulfuric acid, and evaporated to dryness. The organic material was separated from potassium sulfate by ether extraction, and crystallization from a small volume of water effected separation of 8–10 g. (6–7%) of phthalic acid from the much more soluble phthalonic acid (80–95 g.). For conversion into phthalaldehydic acid a solution of 10 g. of phthalonic acid and 5.5 g. of Na_2CO_3 in water was evaporated to dryness and the salt added to 45 ml. of 40% sodium bisulfite solution at 60°. The solution was evaporated to dryness, the residue stirred with concd. hydrochloric acid to decompose the bisulfite compound (4), and evaporated to dryness again. Extraction with ether gave pure phthalaldehydic acid, m.p. 96°, in 63–65% yield.

For oxidation of quinoxaline to pyrazine-2,3-dicarboxylic acid, Jones and McLaughlin[11] took permanganate in 20% excess of the theory, and evidently it was

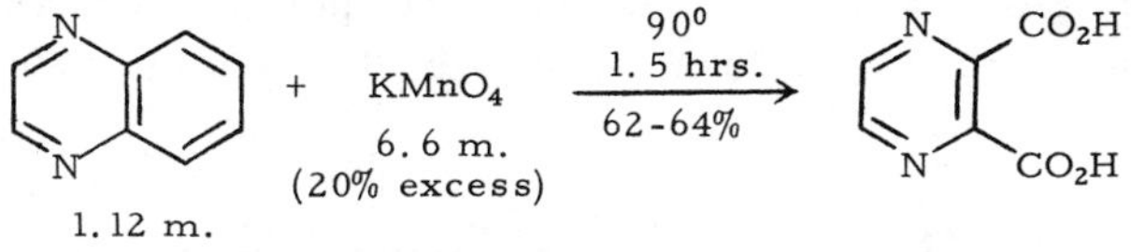

all consumed and there was no unchanged quinoxaline. A mixture of quinoxaline and 4 l. of water at 90° was stirred mechanically under reflux in a 12-l. flask, and a solution at 90–100° of 1050 g. of potassium permanganate in about 4 l. of water was added from a dropping funnel in a thin stream at a rate adjusted so that the reaction mixture boiled gently (1.5 hrs.). The mixture was cooled somewhat for easier handling, filtered, and the MnO_2 cake was removed and stirred to a smooth paste with 1 l. of water. The slurry was filtered and the washing repeated. (Evidently gentle boiling does not produce the compact, easily filterable MnO_2 that is formed on vigorous boiling). The recovery of the dicarboxylic acid from about 10 l. of filtrate is specific to the case at hand and presents no special points of interest. The yield given is for material crystallized from water.

Secondary alcohols ⟶ ketones. Cornforth[12] prepared ethyl pyruvate by stirring a mixture of ethyl lactate, 130 ml. of saturated magnesium sulfate solution, and 0.13

$$\underset{0.42\ m.}{CH_3CHOHCO_2C_2H_5} + \underset{0.35\ m.\ (25\%\ excess)}{KMnO_4} \xrightarrow[51\text{-}54\%]{15^0\ MgSO_4\text{-}NaH_2PO_4} CH_3COCO_2C_2H_5$$

mole of sodium dihydrogen phosphate dihydrate, and 500 ml. of olefin-free petroleum ether in a bath at 15° and adding 55 g. of powdered potassium permanganate during 25–30 min. Stirring was continued until a spot test showed the permanganate to be consumed, the petroleum ether solution was decanted, and the sludge washed with three portions of solvent. The total extract was evaporated under a short column and the residual oil shaken with two portions of saturated aqueous calcium chloride solution to remove ethyl lactate, and the ethyl pyruvate was distilled at 56–57°/20 mm. Further purification can be effected through the bisulfite compound.

Oxidation of the diol (1) to *o*-diacetylbenzene is conducted in a solution buffered with magnesium nitrate, apparently to precipitate hydroxide ion and so inhibit aldolization of the product (Goldschmidt[13]). A mixture of the diol and buffer in

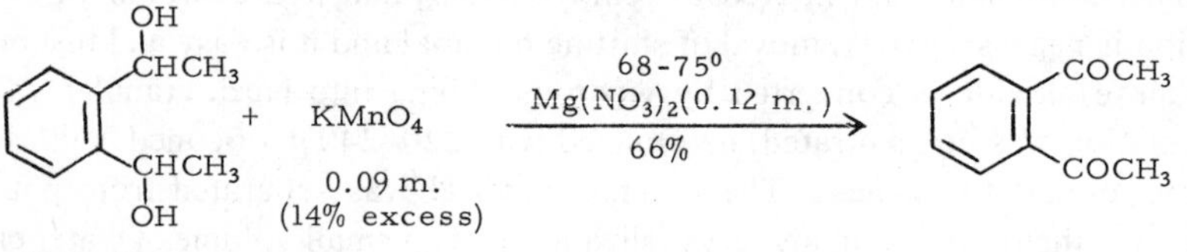
OH
CHCH3
CHCH3
OH
+ KMnO4
0.09 m.
(14% excess)
68-75⁰
Mg(NO3)2(0.12 m.)
66%
COCH3
COCH3

(1) 0.059 m.

250 ml. of water was stirred at 68–75° and finely powdered permanganate was added portionwise in the course of 3 hrs. The manganese dioxide was filtered, washed with ether, and extracted with ether in a Soxhlet, and the filtrate was extracted with ether for 12 hrs. in a percolator. Evaporation of the combined extracts and distillation at 110–116°/0.1 mm. gave *o*-diacetylbenzene in 66% yield. The oil solidified on scratching at −20°, and crystallization from ether–petroleum ether gave long needles, m.p. 39–40°.

This procedure is one which might be simplified by SO_2-reduction.

$ArCH_2CH_3 \longrightarrow ArCOCH_3$. Holsten and Pitts[14] prepared *p*-diacetylbenzene by dropwise addition of *p*-ethylacetophenone to a mixture, stirred at 60°, of potassium

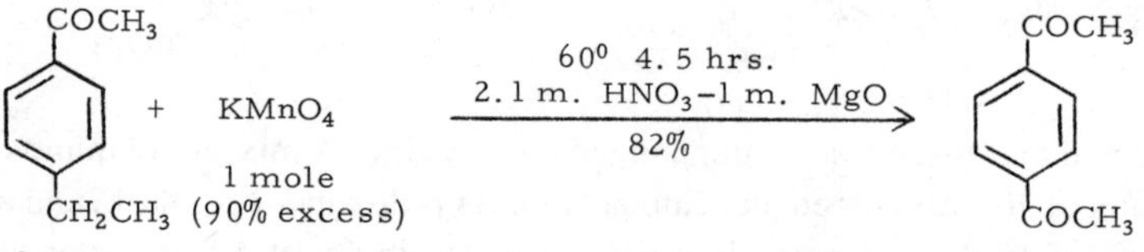
COCH3
CH2CH3
+ KMnO4
1 mole
(90% excess)
60⁰ 4.5 hrs.
2.1 m. HNO3-1 m. MgO
82%
COCH3
COCH3

0.4 m.

permanganate in 90% excess, nitric acid, and magnesium oxide in 1034 ml. of water with cooling to maintain a temperature of 60°. Stirring at 60° was continued for 4.5 hrs. After cooling, filtration gave a mixture of MnO_2 and the reaction product from which the latter was isolated in 82% yield by extraction with benzene. When the permanganate used was only 50% in excess of theory the yield dropped to 70–76%.

Olefin ⟶ acid(s). For the preparation of azelaic acid, Hill and McEwen[15] stirred permanganate with water at 35° in a 12-l. flask until solution was complete and added all at once an alkaline solution of crude ricinoleic acid prepared by saponification of castor oil. The temperature rose to 75° and after about 30 min. a

$$\underbrace{\underset{0.8\text{ m. (crude)}}{CH_3(CH_2)_5\overset{OH}{\overset{|}{C}}HCH_2CH{=}CH(CH_2)_7CO_2H} + \underset{1\text{ m.}}{KOH}}_{1.6\text{ l. } H_2O} + \underset{\substack{3.5\text{ m.}\\ 7.5\text{ l. } H_2O}}{KMnO_4} \xrightarrow[32\text{-}36\%]{35\text{-}75^\circ} HOOC(CH_2)_7CO_2H$$

spot test showed no permanganate color. For workup the mixture was divided into two portions. Each was acidified (H_2SO_4), cautiously because of foaming from liberated CO_2, and heated on the steam bath to coagulate (partially) the MnO_2, which was filtered, returned to the beaker, and boiled with 2 l. of water to dissolve adsorbed azelaic acid. Evaporation of the combined filtrate to about 4 l. and ice cooling gave crude azelaic acid, which on crystallization from water afforded satisfactory material, m.p. 104–106°.

In this instance SO_2-reduction might have eliminated hours of filtration, washing, and evaporation. It is possible that the yield could be improved by use of more permanganate. The presence of CO_2 in the reaction mixture indicates that some cleavage took place as follows:

$$CH_3(CH_2)_5\overset{OH}{\overset{|}{C}}HCH_2CH{=}CH(CH_2)_7CO_2H \xrightarrow{\text{11 Oxygen equivalents}}$$

$$CH_3(CH_2)_5CO_2H + CO_2 + CO_2 + HO_2C(CH_2)_7CO_2H$$

This process would require 5.85 moles of permanganate, whereas that used was only 3.5 moles. However, ricinoleic acid would appear to be a less promising starting material than oleic acid.

—C≡C— ⟶ —COCO—. A procedure by Khan and Newman[16] for oxidation of stearolic acid to 9,10-diketostearic acid is noteworthy for the high yield obtained

$$\underset{\substack{0.02\text{ m.}\\ (0.023\text{ m. KOH, 3 l. } H_2O)}}{CH_3(CH_2)_7C{\equiv}C(CH_2)_7CO_2H} + \underset{\substack{0.4\text{ m.}\\ (\text{Theory } 0.026)}}{KMnO_4} \xrightarrow[92\text{-}96\%]{CO_2\ 25^\circ} CH_3(CH_2)_7\overset{O}{\overset{\|}{C}}-\overset{O}{\overset{\|}{C}}(CH_2)_7CO_2H$$

by buffering the solution with carbon dioxide and for the fact that the permanganate was not all consumed. The acid was dissolved in dilute alkali by stirring, carbon dioxide was bubbled through the solution to adjust the pH to 7.5, a solution of excess permanganate in 300 ml. of water was added all at once, the temperature was controlled to 25°, and the solution was kept at pH 7–7.5 by bubbling in carbon dioxide. After 1 hr., sodium bisulfite and hydrochloric acid were added to destroy excess permanganate, reduce the MnO_2, and precipitate the diketo acid. Crystallization from absolute ethanol afforded, in two crops, material melting at 84.5–85°.

Bruson *et al.*[17] oxidized the hindered tetramethyltetralone (1) with a large excess of permanganate in a solution of 0.5 g. of sodium hydroxide in 255 ml. of water, added sulfuric acid and sodium bisulfite to dissolve the MnO_2, and characterized the white product which precipitated as the α-hydroxy acid (3). Ring-contraction must have resulted in oxidation to the α-diketone (2) and benzilic acid rearrangement in the increasingly alkaline medium. When the oxidation was carried out in the presence of magnesium sulfate as buffer to precipitate hydroxide ion, the reaction took considerably longer but afforded the diacid (4). The permanganate taken represented a 72% excess, and apparently it was all consumed since in the workup "the MnO_2 was removed by adding sodium bisulfite and the solution acidified with sulfuric acid and repeatedly extracted with ether."

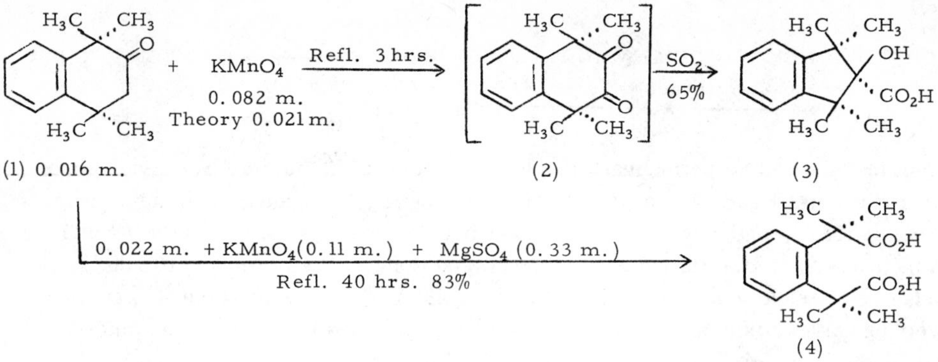

Olefin ⟶ vic-diol in aqueous medium. Unsaturated fatty acids are convertible by permanganate oxidation in alkaline solution into *vic*-diols in high yield. G. M. Robinson and R. Robinson[18] briefly report oxidation of oleic acid in 300 parts of water and one third part of potassium hydroxide at 0° by gradual addition of 0.5 *N*

$$CH_3(CH_2)_7\overset{H}{C}{=}\overset{H}{C}(CH_2)_7CO_2H \xrightarrow{KMnO_4-KOH} CH_3(CH_2)_7\underset{OH}{\overset{H}{C}}-\underset{OH}{\overset{H}{C}}(CH_2)_7CO_2H$$

potassium permanganate. Reduction of the MnO_2 with sulfur dioxide and crystallization of the product from alcohol–benzene gave *erythro*-9,10-dihydroxystearic acid (m.p. 132°) in "almost theoretical yield." Lapworth and Mottram[19] defined conditions for nearly perfect conversion into the diol and developed an analytical procedure applicable to a 5-g. sample of unsaturated fatty acid. The method is of little preparative value since it requires 1 l. of solution per gram of oleic acid. Traynard[20] obtained the *erythro*-diol in 70–75% yield with only 40 ml. of solution per gram of olefinic acid. Swern's group[21] found the permanganate oxidation of oleic acid to be pH-dependent. At a terminal pH of 11.8 the product was a mixture of the diol (60%) and the 9,10- and 10,9-hydroxyketostearic acids; at pH 9–9.5 the products were the diol (4%) and the α-ketols (60%). The oxidation of elaidic acid proved to be only slightly pH-dependent.

Under conditions similar to those under which cyclohexene affords *cis*-cyclohexane-1,2-diol in yield of 30–33%,[22, 23] Evans *et al.*[24] obtained DL-glyceraldehyde diethyl acetal from acrolein diethyl acetal in 67% yield. A suspension of the acetal

$$\underset{0.5\text{ m., }600\text{ ml. }H_2O}{CH_2{=}CHCH(OC_2H_5)_2} + \underset{\substack{0.5\text{ m. in }600\,H_2O\\(\text{Theory }0.33\text{ m.})}}{KMnO_4} \xrightarrow[67\%]{5^0} \underset{OH\ \ OH}{CH_2CHCH(OC_2H_5)_2}$$

in water was stirred at 5° and an aqueous solution of permanganate, taken in 52% excess, was added at the rate of about 25 ml. per minute. Soon after the addition was complete the mixture set to a gel. After standing for 2 hrs. the mixture was heated for 2 hrs. on the steam bath and the MnO_2 was collected and washed with water (apparently the permanganate had all been consumed). The filtrate was treated with 1.2 kg. of freshly dehydrated potassium carbonate. The layers were separated, the water layer was extracted with ether, and the product collected and distilled. A possible explanation for the relatively

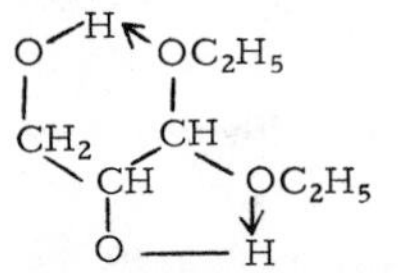

high yield in an oxidation conducted with excess permanganate is that the diol is stabilized by chelation.

Olefin ⟶ diol in anhydrous acetone. Tishler and co-workers[25] developed an efficient procedure for introducing a 17α-hydroxyl group into a 20-ketosteroid using permanganate for hydroxylation in place of osmium tetroxide. The 20-ketone is converted via the 20-cyanohydrin into the 17,20-unsaturated 20-nitrile (1). A solution of 90 g. of (1) in dry acetone was stirred at −5°, treated with piperidine

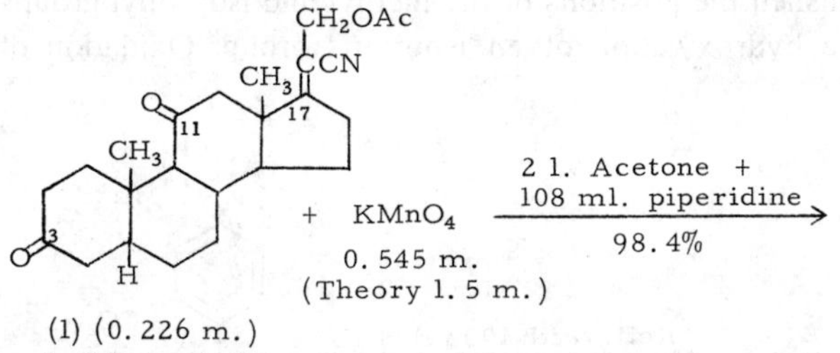
CH2OAc
CN
CH3
17
11
CH3
3
H
+ KMnO4
0. 545 m.
(Theory 1. 5 m.)
2 l. Acetone +
108 ml. piperidine
98. 4%
CH2OAc
C=O
CH3
OH
17
CH3
H

(1) (0. 226 m.)
In 2 l. acetone

and excess permanganate, stirred for 30 min., and treated below 2° with 18 ml. of acetic acid in 225 ml. of acetone. After 4 hrs. at 0–2°, 1.8 l. of chloroform was added, followed by 270 ml. of concd. hydrochloric acid in 1350 ml. of water and 126 g. of sodium bisulfite in 900 ml. of water. Extraction with chloroform and crystallization gave pure pregnane-17α,21-diol-3,11-20-trione 21-acetate.

Oxidation with aqueous permanganate–methylcyclohexane. In an early investigation aimed at elucidating the structure of ergosterol, Reindel and co-workers[26] oxidized ergosterol by an unusual procedure and obtained an oxygenated product which was identified a quarter of a century later in an investigation by M. Fieser, A. Quilico, *et al.*[27] The product proved to be a mixture of about 7 parts of an O_3-compound and 3 parts of an O_5-compound, and the main product was characterized as the triply unsaturated triol (2). It probably arises by *cis*-hydroxylation of the 5,6-double

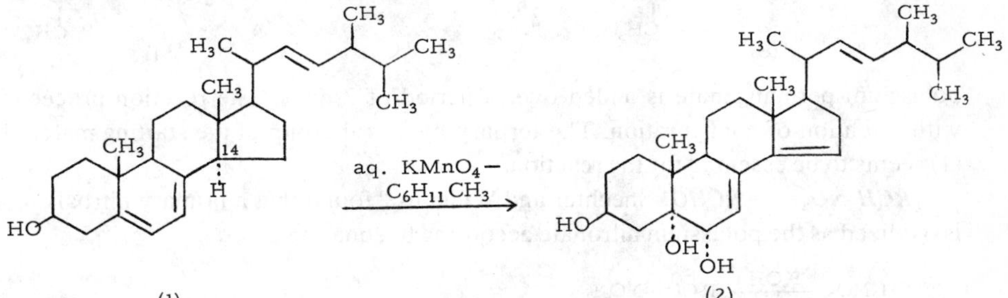
CH3
H3C
CH3
CH3
CH3
CH3
14
H
HO
aq. KMnO4–
C6H11CH3
H3C
CH3
CH3
CH3
CH3
CH3
HO
OH
OH

(1) (2)

bond, allylic hydroxylation at C_{14}, and elimination of water. Quilico developed a reproducible procedure for preparing ergosterol-free oxidation product in yield of 80% by weight using three oxygen equivalents of permanganate. A hot solution of 4 g. of ergosterol in 100 ml. of methylcyclohexane (a much better solvent for the sterol than *n*-alkanes) was treated with 75–80 ml. of 4% aqueous potassium permanganate preheated to 90–100°, and the flask was stoppered and shaken vigorously with frequent removal of the stopper for release of pressure. Manganese dioxide soon appeared, and the initially thick mass soon changed to a thin suspension. The permanganate was all consumed in 8–10 min. Sulfur dioxide was bubbled through the mixture, and the resulting suspension of white solid was transferred to a separatory funnel and washed with successive portions of water to extract inorganic salts. Steam distillation removed the hydrocarbon solvent, and evaporation and digestion with methanol afforded crystalline product.

—CH(CH₃)₂ ⟶ —C(OH)(CH₃)₂. An early observation by Bucher[28] which, coupled with earlier evidence, established the positions of the methyl and isopropyl groups in retene, demonstrated the hydroxylation of an isopropyl group. Oxidation of

(1)

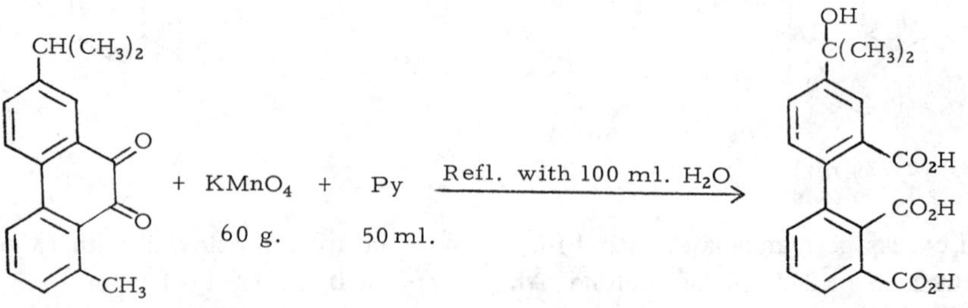
CH(CH3)2
CH3
+ KMnO4 + Py
60 g. 50 ml.
Refl. with 100 ml. H2O
OH
C(CH3)2
CO2H
CO2H
CO2H

(1) 10 g. (2) 2.2 g.

retenequinone (1) under the conditions indicated in the formulation afforded the product (2).

More recently Eastman and Quinn[29] found that the tertiary alcohol (3), which contains a tertiary hydrogen, is oxidized by permanganate to (4) in significant yield. The alcohol is suspended in water, and slightly more than 1 mole of powdered

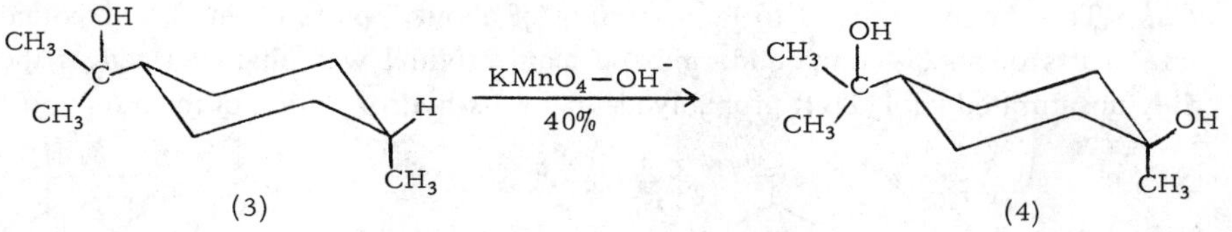
OH
CH3
C
CH3
H
CH3
KMnO4–OH⁻
40%
OH
CH3
C
CH3
OH
CH3

(3) (4)

potassium permanganate is added over a period of 2 days. The reaction proceeds with retention of configuration. The tertiary hydroxyl group of the starting material (1) seems to be essential for the reaction.

RCH_2NO_2 ⟶ RCHO. Shechter and Williams[30] found that a primary nitroalkane is oxidized as the potassium nitronate according to equation 2:

1. $RCH_2NO_2 \xrightarrow{KOH} RCH{=}NO_2K$

2. $3\,RCH{=}NO_2K + 2\,KMnO_4 + H_2O \longrightarrow 3\,RCHO + 2\,MnO_2 + 3\,KNO_2 + 2\,KOH$

Neutrality is maintained by addition of magnesium sulfate to precipitate hydroxide ion as magnesium hydroxide. Although the aldehyde produced is oxidized less rapidly than the nitronate, oxidation to the carboxylic acid becomes a competing reaction toward the end of the reaction unless the amount of permanganate used is 70–90% of the theory. A solution of the nitro compound in potassium hydroxide is

$$CH_3CH_2CH_2CH_2NO_2 + KMnO_4 + 2\,\underline{M}\,MgSO_4 \xrightarrow[89\%]{0\text{-}5^{\circ}} CH_3CH_2CH_2CHO$$

Moles x 10^3 = 6.88   6.22   20 ml.   (as DNP)

treated with magnesium sulfate solution, diluted to 500 ml., and permanganate solution is added dropwise with stirring at 0–5°. The aldehyde is steam distilled into aqueous hydrochloric acid saturated with 2,4-dinitrophenylhydrazine. The method is particularly useful for sensitive secondary nitro compounds, for example nitrocyclobutane.

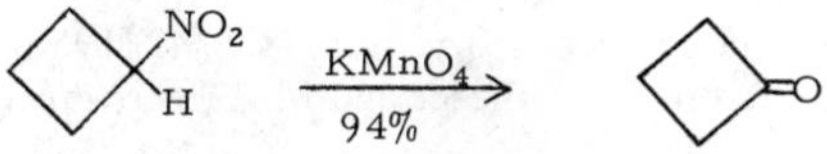
NO2
H
KMnO4
94%

$RNH_2 \longrightarrow RNO_2$. Kornblum and Jones[31] stirred a mixture of *t*-octylamine in 500 ml. of acetone with a solution of magnesium sulfate in 125 ml. of water and kept the temperature at 25–30° during the addition of a large excess of powdered permanganate over a period of about 30 min. Stirring at 25–30° was continued for 48

$$\underset{0.20\text{ m.}}{\underset{NH_2}{(CH_3)_3CCH_2C(CH_3)_2}} + \underset{\substack{1.20\text{ m.}\\(\text{Th.} = 0.40\text{ m.})}}{2\ KMnO_4} + \underset{(0.14\text{ m.})}{\text{aq. }MgSO_4} \xrightarrow[69-82\%]{25-30^0,\ 48\text{ hrs.}}$$

$$\underset{NO_2}{(CH_3)_3CCH_2C(CH_3)_2} + Mg(OH)_2 + K_2SO_4 + 2\ MnO_2$$

hrs. The acetone was removed by stirring at 30° and evacuation at the water pump. Steam distillation of the resulting viscous mixture gave a pale blue organic layer which was dried and distilled at 53–54°/3 mm. to give colorless 4-nitro-2,2,4-trimethylpentane, m.p. 23.5–23.7°. The apparent function of magnesium sulfate is to precipitate hydroxide ion and so prevent formation of the potassium isonitrosonate. At a temperature much above 30° potassium permanganate is rapidly consumed, "presumably by reaction with acetone." The method was used successfully for the synthesis of seven other trialkylnitromethanes.

$ArCOCH_3 \longrightarrow ArCHO$. An aryl methyl ketone obtained by a Friedel-Crafts or Fries reaction is oxidized with alkaline permanganate in pyridine at 10–15° to the arylglyoxylic acid, and this is decarboxylated by treatment at 140–160° with N,N-dimethyl-*p*-toluidine.[32]

$$ArCOCH_3 \xrightarrow{KMnO_4} ArCOCO_2H \xrightarrow[140-160^0]{CH_3C_6H_4N(CH_3)_2} ArCHO + CO_2$$

[1]S. Ball, T. W. Goodwin, and R. A. Morton, *Biochem. J.*, **42**, 516 (1948)

[2]J. Attenburrow, A. F. B. Cameron, J. H. Chapman, R. M. Evans, B. A. Hems, A. B. A. Jansen, and T. Walker, *J. Chem. Soc.*, 1094 (1952)

[3]I. M. Kolthoff and R. Belcher, *Volumetric Analysis*, **3**, 37–38 (1957)

[4]J. W. Ladbury and C. F. Cullis, *Chem. Rev.*, **58**, 403 (1958)

[5]J. R. Ruhoff, *Org. Syn., Coll. Vol.*, **2**, 315 (1943)

[6]R. L. Shriner and E. C. Kleiderer, *ibid.*, **2**, 538 (1943)

[7]H. T. Clarke and E. R. Taylor, *ibid.*, **2**, 135 (1943)

[8]A. W. Singer and S. M. McElvain, *ibid.*, **3**, 740 (1955)

[9]J. H. Gardner and C. A. Naylor, Jr., *ibid.*, **2**, 523 (1943)

[10]C. Graebe and F. Trümpy, *Ber.*, **31**, 369 (1898)

[11]R. G. Jones and K. C. McLaughlin, *Org. Syn., Coll. Vol.*, **4**, 824 (1963)

[12]J. W. Cornforth, *ibid.*, **4**, 467 (1963)

[13]S. Goldschmidt and A. Zoebelein, *Ber.*, **94**, 169 (1961)

[14]J. R. Holsten and E. H. Pitts, Jr., *J. Org.*, **26**, 4151 (1961)

[15]J. W. Hill and W. L. McEwen, *Org. Syn., Coll. Vol.*, **2**, 53 (1943)
[16]N. A. Khan and M. S. Newman, *J. Org.*, **17**, 1063 (1952)
[17]H. A. Bruson, F. W. Grant, and E. Bobko, *Am. Soc.*, **80**, 3633 (1958)
[18]G. M. Robinson and R. Robinson, *J. Chem. Soc.*, **127**, 175 (1925)
[19]A. Lapworth and E. N. Mottram, *J. Chem. Soc.*, **127**, 1628 (1925)
[20]J.-C. Traynard, *Bull. soc.*, **19**, 323 (1952)
[21]J. E. Coleman, C. Ricciuti, and D. Swern, *Am. Soc.*, **78**, 5342 (1956)
[22]Experiment by L. F. F. cited in L. F. Fieser and M. Fieser, *Advanced Organic Chemistry*, 191 (1961)
[23]M. F. Clarke and L. N. Owen, *J. Chem. Soc.*, 315 (1949), carried out the oxidation in ethanol–water with $MgSO_4$ as buffer at $-15°$ and obtained the diol in 33% yield.
[24]E. J. Witzemann, W. L. Evans, H. Hass, and E. F. Schroeder, *Org. Syn., Coll. Vol.*, **2**, 307 (1943)
[25]R. Tull, R. E. Jones, S. A. Robinson, and M. Tishler, *Am. Soc.*, **77**, 196 (1955)
[26]F. Reindel *et al.*, *Ann.*, **452**, 34 (1927); **460**, 212 (1928)
[27]M. Fieser, A. Quilico, A. Nickon, W. E. Rosen, E. J. Tarlton, and L. F. Fieser, *Am. Soc.*, **75**, 4066 (1953)
[28]J. E. Bucher, *Am. Soc.*, **32**, 374 (1910)
[29]R. H. Eastman and R. A. Quinn, *Am. Soc.*, **82**, 4249 (1960)
[30]H. Shechter and F. T. Williams, Jr., *J. Org.*, **27**, 3699 (1962)
[31]N. Kornblum and W. J. Jones, *Org. Syn.*, **43**, 87 (1963)
[32]J. Cymerman-Craig, J. W. Loder, and B. Moore, *Australian J. Chem.*, **9**, 222 (1956); *see also* J. L. Ferrari, I. M. Hunsberger, and H. S. Gutowsky, *Am. Soc.*, **87**, 1247 (1965)

Potassium peroxymonosulfate, 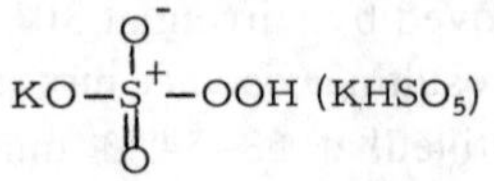

Mol. wt. 152.17. Preparation.[1]

A mixture of 1 mole of this stable salt of peroxymonosulfuric acid with 0.5 mole of potassium bisulfate and 0.5 mole of potassium sulfate is a powerful oxidizing agent comparable to Caro's acid but more stable.[2] This combination is not soluble in organic solvents, but according to Kennedy and Stock[2] can be used in mixtures of ethanol–water, acetic acid–water, and ethanol–acetic acid–water. Without giving fully definitive details and yields, these authors indicate that the $KHSO_5$–$KHSO_4$–K_2SO_4 reagent has wide application: cyclohexene ⟶ *trans*-cyclohexane-1,2-diol; toluene ⟶ benzoic acid; cyclic ketones ⟶ lactones (moderate yield); mercaptans ⟶ sulfonic acids (quantitative); primary arylamines and cycloalkylamines ⟶ nitroso compounds.

[1]S. E. Stephanon, U.S. patent 2,802,722 (1957)
[2]R. J. Kennedy and A. M. Stock, *J. Org.*, **25**, 1901 (1960)

Potassium persulfate and Ammonium persulfate, $K_2S_2O_8$, $(NH_4)_2S_2O_8$. Mol. wts. 270.33, 228.21. *See also* Caro's acid.

Phenols ⟶ diprotic phenols*[1] *(Elbs reaction*[2]*). An example is the oxidation of salicylaldehyde to gentisaldehyde:[3]

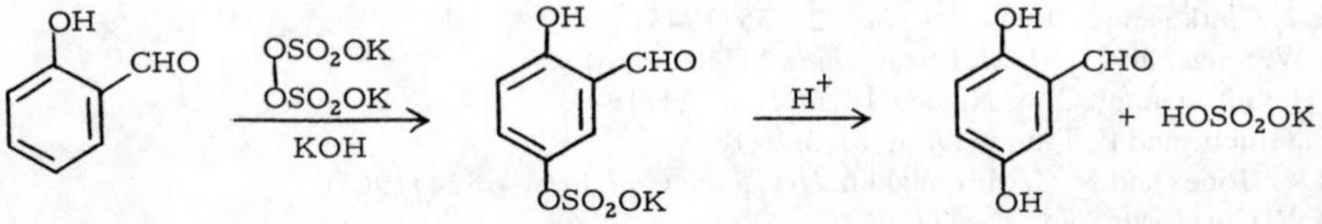

ArI ⟶ iodonium salt. An example is the preparation of diphenyliodonium-2-carboxylate[4] (IV) for use as a benzyne precursor.[5] In a simplified procedure[6] a

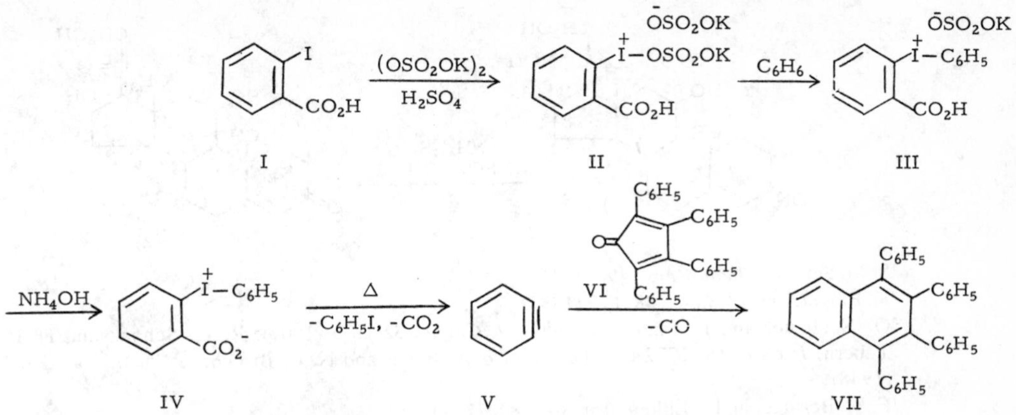

mixture of 2 g. of *o*-iodobenzoic acid and 2.6 g. of potassium persulfate is treated with 8 ml. of ice-cold sulfuric acid, and the mixture is swirled for 4–5 min. at 0° and let stand at 25° for 20 min. The resulting solution of II is cooled to 0°, 2 ml. of benzene is added, and the mixture is swirled and allowed to come to room temperature for conversion into III. With careful cooling, the mixture is diluted and then neutralized with ammonia in the presence of methylene chloride for extraction of IV, which crystallizes from water in the form of large prisms of the monohydrate; yield 2.1 g.

Arylamine⟶o-hydroxy sulfate ester.[7] In an alkaline medium, arylamines are oxidized by the reagent to *o*-hydroxy sulfate ester salts. Thus a solution of 5 g. of N,N-dimethylaniline in 250 ml. of water, 400 ml. of acetone, and 30 ml. of 2N potassium hydroxide was stirred at room temperature during addition in 8 hrs. of

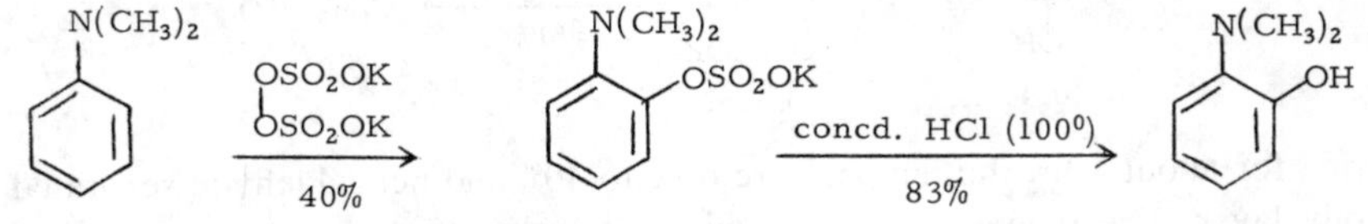

11.2 g. of potassium persulfate in saturated aqueous solution. The solution was evaporated to 250 ml., washed with ether, evaporated to dryness under reduced pressure, and the residue extracted with hot 95% ethanol. Dilution with ether (1.5 l.) precipitated *o*-dimethylaminophenyl potassium sulfate. Acidification of an aqueous solution of the salt afforded the corresponding hydrogen sulfate, whereas heating with concd. hydrochloric for 1 hr., followed by cooling and partial neutralization with alkali, gave *o*-dimethylaminophenol.

Hydrolysis of oximes. Brooks *et al.*[8] (Glaxo Labs.) found difficulty in regenerating steroid 3,20-diketones from their 3,20-dioximes by usual methods. Thus hydrolysis with hydrochloric acid generally resulted in hydrolysis of only the C_{20}-function. A successful method was found in oxidative hydrolysis with persulfuric acid. Thus 0.25 g. of cortisol 3,20-dioxime (1) was dissolved in a warm solution of 0.16 g. of ammonium persulfate in 50 ml. of warm *N* sulfuric acid, 50 ml. of methylene chloride was added, and the mixture shaken for 8 days at room temperature. Isolation of the steroid yielded 0.191 g. of cortisol (2). The reagent is successful because it oxidizes the liberated hydroxylamine; however, ferric salts could not be used in place of the persulfate.

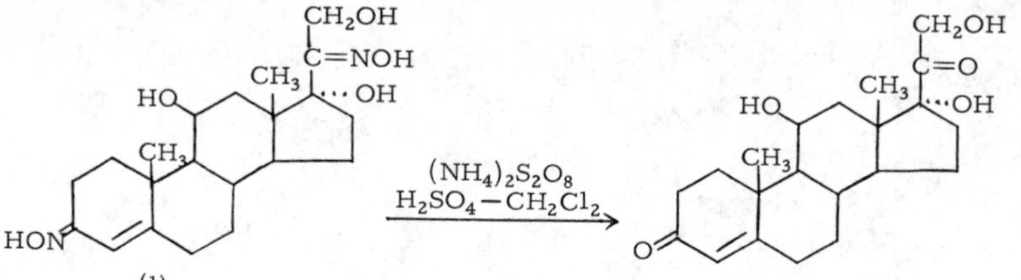

(1) (2)

[1]S. M. Sethna, *Chem. Revs.*, **49**, 91 (1951)
[2]K. Elbs, *J. prakt. Chem.*, **48**, 179 (1893)
[3]O. Neubauer and L. Flatow, *Z. physiol. Chem.*, **52**, 380 (1907): R. U. Schock and D. L. Tabern, *J. Org.*, **16**, 1772 (1951); *see also* W. Baker and N. C. Brown, *J. Chem. Soc.*, 2303 (1948)
[4]F. M. Beringer and I. Lillien, *Am. Soc.*, **82**, 725 (1960)
[5]E. Le Goff, *Am. Soc.*, **84**, 3786 (1962)
[6]*Org. Expts.*, 311–314
[7]E. Boyland, D. Mason, and P. Sims, *J. Chem. Soc.*, 3623 (1953)
[8]S. G. Brooks, R. M. Evans, G. F. H. Green, J. S. Hunt, A. G. Long, B. Mooney, and L. J. Wyman, *J. Chem. Soc.*, 4614 (1958)

Potassium persulfate, silver-catalyzed.[1] The reagent cleaves water-soluble *vic*-glycols to aldehydes, ketones, or both, in yields in the range 40–100%. For example,

$$\begin{matrix} R_1R_2C{-}OH \\ | \\ R_3R_4C{-}OH \end{matrix} + S_2O_8^{-2} \xrightarrow{Ag^+} R_1R_2C{=}O + R_3R_4C{=}O + 2\,SO_4^{-2} + 2\,H^+$$

potassium persulfate was added to a stirred solution of phenylethylene glycol in 300 ml. of water at 30° and 1.3 ml. of 10% silver nitrate solution was added. After

$$\underset{0.1\ m.}{C_6H_5\underset{OH}{C}HCH_2OH} + \underset{0.11\ m.}{\begin{matrix} OSO_2OK \\ | \\ OSO_2OK \end{matrix}} \xrightarrow[61\%]{Ag^+\ H_2O\ 30^0} C_6H_5CHO$$

stirring for about 1 hr. the temperature rose to 40°, and benzaldehyde separated in an oily layer. The reagent is not specific to *vic*-glycols; monoprotic alcohols and aldehydes are oxidized by silver-catalyzed persulfate. Kinetic study of the cleavage of *cis*- and *trans*-cyclohexane-1,2-diol showed strikingly identical rates of reaction for both isomers.

[1]F. P. Greenspan and H. M. Woodburn, *Am. Soc.*, **76**, 6345 (1954)

Potassium thiocyanate, KSCN. Mol. wt. 97.18.

The reagent reacts with alkyl halides, dialkyl sulfates[1] and tosylates[2] to form the thiocyano derivatives, which can be desulfurized with Raney nickel.[3] Potassium

$$RX(OTs) + KSCN \longrightarrow RSCN \xrightarrow{Ni\text{-}H_2} RH$$

thiocyanate in aqueous solution reacts readily with 3-methoxy-1,2-epoxypropane (1) to produce 2-methoxymethylthiirane (2).[4] Desulfurization of (2) is accomplished efficiently with triethyl phosphite to give (3).

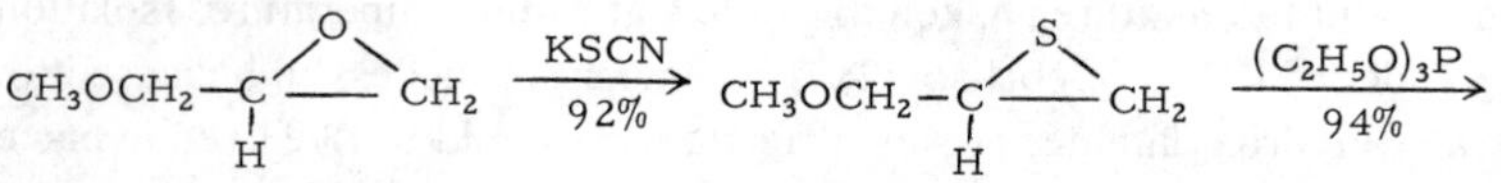

(1) (2)

$$CH_3OCH_2CH{=}CH_2 \quad + \quad (C_2H_5O)_3PS$$
$$(3) \qquad\qquad (4)$$

[1]P. Walden, *Ber.*, **40**, 3214 (1907)
[2]R. M. Hann, N. K. Richtmyer, H. W. Diehl, and C. S. Hudson, *Am. Soc.*, **72**, 561 (1950)
[3]H. R. Snyder, J. M. Stewart, and J. B. Ziegler, *Am. Soc.*, **69**, 2672 (1947)
[4]R. D. Schuetz and R. L. Jacobs, *J. Org.*, **26**, 3467 (1961)

Potassium thiolacetate, CH_3COSK, Mol. wt. 114.21. Preparation.[1]

Chapman and Owen[2] recommend reaction of primary tosylates with potassium thiolacetate as a useful method for the preparation of thiolacetates and thence, by hydrolysis, of thiols. Tetrahydrofurfuryl tosylate on reaction with the reagent

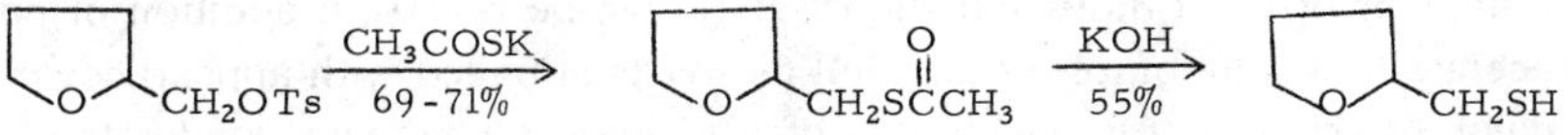

CH₂OTs
CH₃COSK
69-71%
CH₂SCCH₃
O
KOH
55%
CH₂SH

for 1 hr. in refluxing ethanol or acetone afforded the thiolacetate in good yield. The corresponding mesylate on reaction in refluxing acetone for 1 hr. afforded the ester in yield of only 20%; in a 6-hr. reaction period the yield rose to 60%. Secondary tosylates showed wide differences in behavior.

[1]C. Ulrich, *Ann.*, **109**, 272 (1859)
[2]J. H. Chapman and L. N. Owen, *J. Chem. Soc.*, 579 (1950)

Potassium triacetylosmate, $(CH_3CO_2)_3Os(=O)$—OK, and **Dipotassium tetramethylosmate**, $(CH_3O)_4Os(OK)_2$.

These reagents were developed by Criegee[1] for the characterization of *vic*-diols. Dipotassium tetramethylosmate (1) is a crystalline green solid readily made from osmium tetroxide and methanolic potassium hydroxide. It reacts in methanol with *cis*-diols and also with sufficiently flexible *trans*-diols. *trans*-Cyclohexane-1,2-diol,

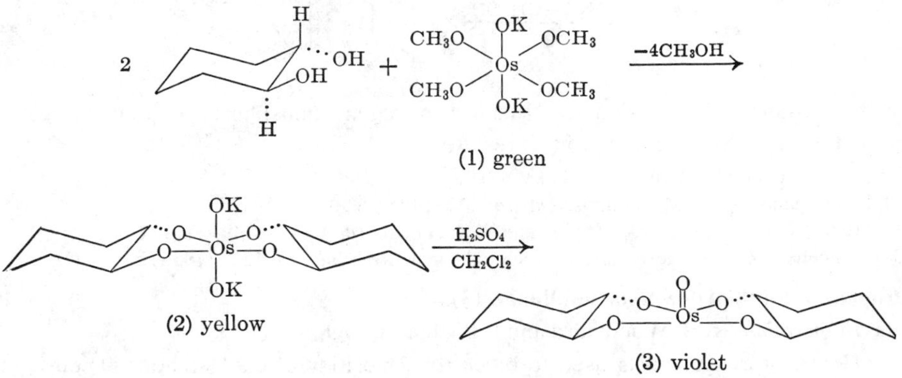

2
H
OH
OH
H
CH₃O
OK
OCH₃
Os
CH₃O
OCH₃
OK
−4CH₃OH
OK
O
Os
O
OK
H₂SO₄
CH₂Cl₂
O
Os

(1) green

(2) yellow

(3) violet

for example, dissolves in a solution of the reagent in methanol (KOH) with a color change, and the sparingly soluble salt (2) separates in high yield. The substance is very sensitive to water, but on acidification of a suspension of the salt in methylene chloride the diester (3) passes into the organic phase and is recoverable as a crystalline solid. Potassium triacetylosmate (6), is made by dissolving the tetramethylosmate (1) in acetic acid; at a suitable concentration, the cobalt-blue solution deposits crystals of (6). This substance reacts in acetic acid with *cis*-diols irreversibly,

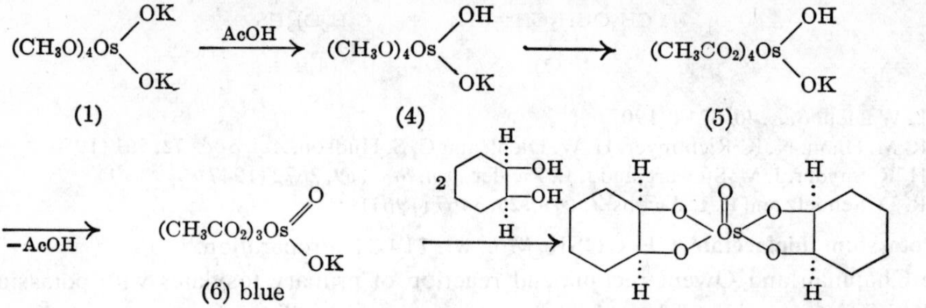

(7) + 2 AcOH + AcOK + H_2O

with discharge of the blue color and formation of the diester (7). It also reacts with a number of *trans*-diols, but the reaction can be reversed: addition of potassium acetate to a solution of a *trans*-diol diester is attended with appearance of the blue color of the salt (6); the diester of a *cis*-diol is unaffected under the same conditions.

[1]R. Criegee *et al.*, *Ann.*, **550**, 99 (1942); **599**, 81 (1956); *Ber.*, **90**, 417 (1957)

Potassium triphenylmethide, *see* Triphenylmethyl potassium.

Prévost's reagent, *see* Silver iodobenzoate.

Propane-1,3-dithiol, $HSCH_2CH_2CH_2SH$. Mol. wt. 108.23, b.p. 152°. Supplier: Aldrich. Preparation.[1]

Although Hauptmann[2] showed in parallel experiments that better yields of thioketals are obtained with this reagent than with ethanedithiol, propanedithiol has been used to only a limited extent,[3, 4] for example:[3]

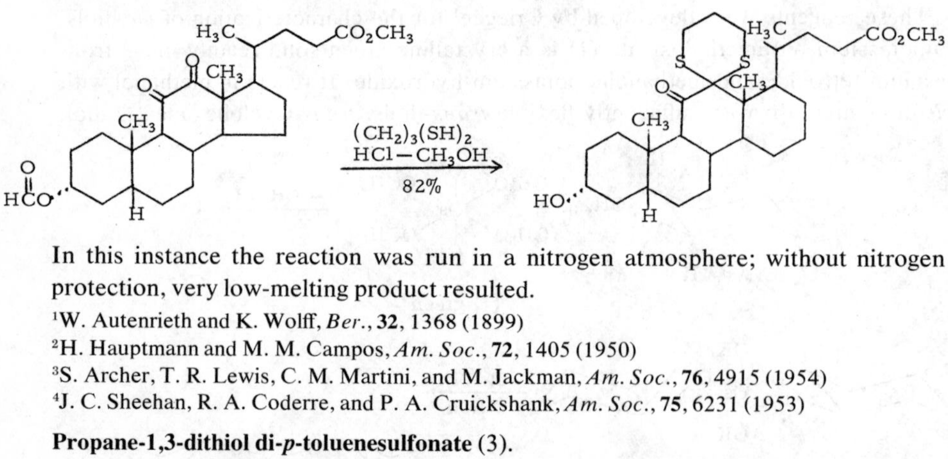

$(CH_2)_3(SH)_2$
$HCl-CH_3OH$
82%

In this instance the reaction was run in a nitrogen atmosphere; without nitrogen protection, very low-melting product resulted.

[1]W. Autenrieth and K. Wolff, *Ber.*, **32**, 1368 (1899)
[2]H. Hauptmann and M. M. Campos, *Am. Soc.*, **72**, 1405 (1950)
[3]S. Archer, T. R. Lewis, C. M. Martini, and M. Jackman, *Am. Soc.*, **76**, 4915 (1954)
[4]J. C. Sheehan, R. A. Coderre, and P. A. Cruickshank, *Am. Soc.*, **75**, 6231 (1953)

Propane-1,3-dithiol di-*p*-toluenesulfonate (3).

Preparation. R. B. Woodward and I. Pachter, unpublished.

Use.[1] The reagent was used to block the 2-position of cholestenone (1) during alkylation at C_4 in a synthesis of lanosterol. After alkylation the blocking group was removed by hydrogenolysis over Raney nickel.

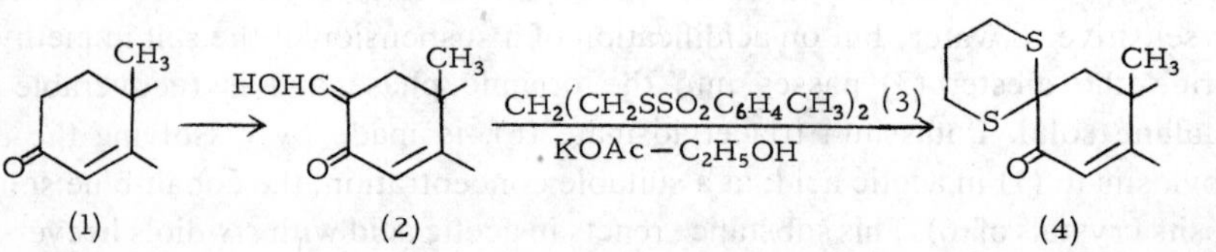

$CH_2(CH_2SSO_2C_6H_4CH_3)_2$ (3)
$KOAc-C_2H_5OH$

(1) (2) (4)

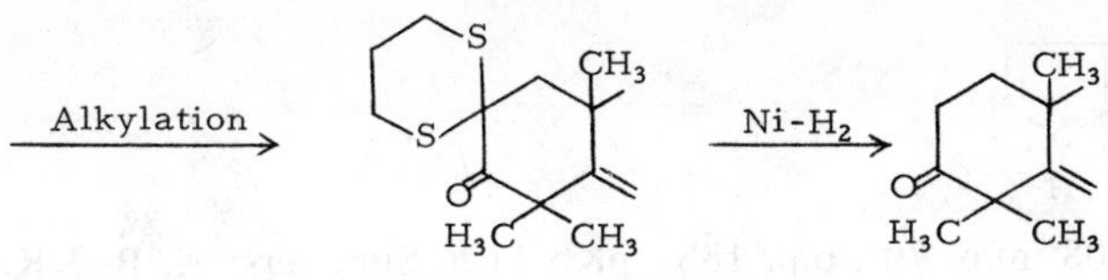

[1]R. B. Woodward, A. A. Patchett; D. H. R. Barton, D. A. J. Ives, and R. B. Kelly, *J. Chem. Soc.*, 1131 (1957)

β-Propiolactone, CH_2CH_2CO (the ends joined through O). Mol. wt. 72.06, b.p. 51°/10 mm., sp. gr. 1.15.

This highly reactive compound is made by B. F. Goodrich Chemical Co. by reaction of ketene with formaldehyde. Many reactions and uses are described by Gresham, Jansen, *et al.*[1] Barkley *et al.*[2] employed the reagent in one phase of a steroid total synthesis, namely for a Michael addition to the unsaturated system of

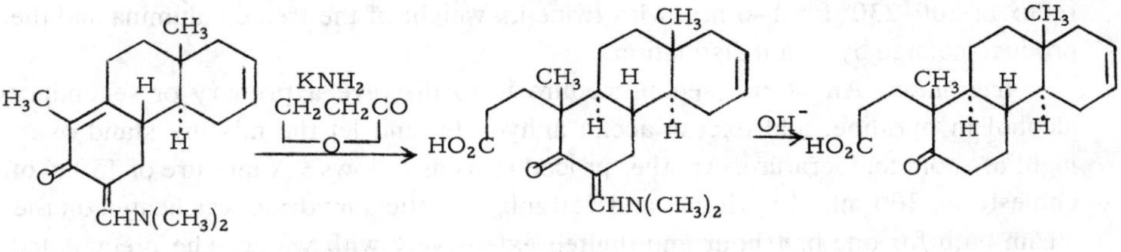

(1) (2) (3)

(1). After hydrolysis of (2), the yield of (3) from (1) was only 17%. The transformation was effected also with acrylonitrile in only slightly higher yield (22%).

In a new method of carboxyethylation[3] a ketone (3) is converted into the morpholine enamine (4), and this is heated with β-propiolactone to effect condensation

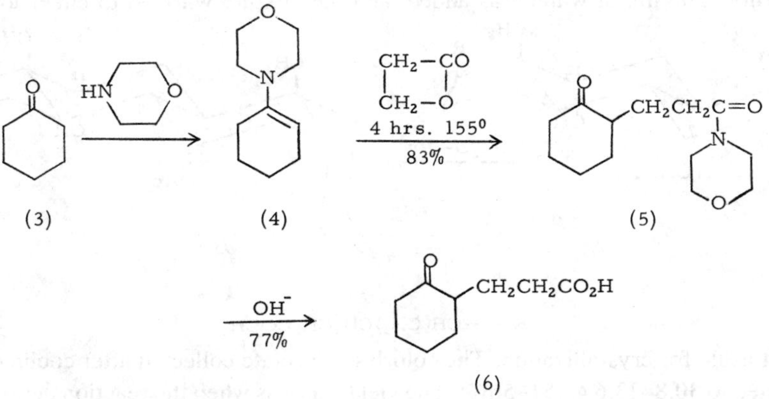

(3) (4) (5)

(6)

and rearrangement to the amide (5). Basic hydrolysis of the morpholide gives 2-ketocyclohexanepropionic acid (6).

[1]T. L. Gresham, J. E. Jansen, *et al.* (Papers I and XIII), *Am. Soc.*, **70**, 999 (1948); **74**, 1323 (1952)

[2]L. B. Barkley, M. W. Farrar, W. S. Knowles, H. Raffelson, and Q. E. Thompson, *Am. Soc.*, **76**, 5014 (1954)

[3]G. Schroll, P. Klemmensen, and S.-O. Lawesson, *Chem. Scand.*, **18**, 2201 (1964)

Pyrazole, 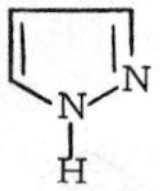
N
N
H

Mol. wt. 68.08, m.p. 70°, b.p. 188°, pKb 11.5. Suppliers: A, B, KK. For use as a bifunctional catalyst in peptide synthesis *see* 1,2,4-Triazole.

Pyridine, C_5H_5N. Mol. wt. 79.10, b.p. 115°, sp. gr. 0.98, pKb 8.8. Pure pyridine is available commercially (p. 1110) and remains absolutely dry when stored over molecular sieves (4 A), barium oxide, or calcium hydride. Potassium hydroxide pellets (80%) remove water only from very wet pyridine.

Dehydration (*see also* Phosphoryl chloride–pyridine; Thionyl chloride–pyridine). The dehydration of terpene alcohols with acidic reagents usually leads to several isomeric olefins. von Rudloff[1] found that dehydration of terpene and sesquiterpene alcohols with neutral alumina (Woelm 1) to which 1–2% of pyridine or quinoline has been added gives a single or only a few products. The alcohol is heated under reflux at 200–230° for 1–6 hrs. with twice its weight of the treated alumina and the product isolated by steam distillation.

Acetylation. An often used procedure is to dissolve a primary or secondary alcohol in pyridine, add excess acetic anhydride, and let the mixture stand overnight at room temperature. Another procedure is as follows.[2] A mixture of 150 g. of cholesterol, 300 ml. of pyridine, and 150 ml. of acetic anhydride was heated on the steam bath for one-half hour and diluted extensively with water. The precipitated granular product was collected and washed well; the yield of fully dry cholesteryl acetate was 166 g. (100%).

Selective conversion of methyl cholate into the 3,7-diacetate was accomplished[3] by dissolving 50 g. of ester in 100 ml. of dioxane and 100 ml. of pyridine, cooling to room temperature, and adding 150 ml. of acetic anhydride. After standing at 26–28° for 20 hrs., 200 ml. of water was added, and the mixture warmed to effect solution

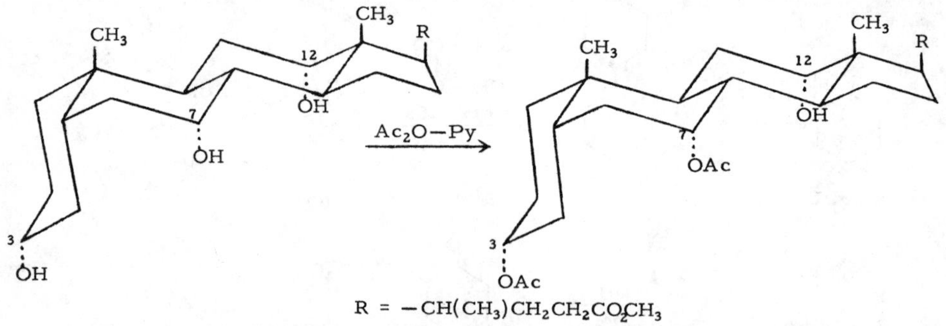
CH3
R
12
7
OH
OH
3
OH
Ac2O–Py
CH3
R
12
OH
7
OAc
3
OAc

$R = -CH(CH_3)CH_2CH_2CO_2CH_3$

and set aside for crystallization. The colorless diacetate collected after cooling to 3° amounted to 30.8–33.6 g. (51–56%). The yield was less when the reaction period was either shortened or extended. Suitable processing of the mother liquor afforded 20.3 g. of pure cholic acid, and the yield of methyl cholate 3,7-diacetate based on cholic acid consumed was 88.6%.

Other esters. For conversion of dodecanol into the tosylate, Marvel and Sekera[4] stirred and cooled a mixture of dodecanol and pyridine during the addition of

$$\underset{0.5\ m.}{CH_3(CH_2)_{10}CH_2OH} + \underset{2\ m.}{C_5H_5N} + \underset{0.55\ m.}{p\text{-}CH_3C_6H_4SO_2Cl} \xrightarrow[79-81\%]{} CH_3(CH_2)_{10}CH_2OSO_2C_6H_4CH_3\text{-}p$$

p-toluenesulfonyl chloride. After stirring at 20° for 3 hrs., the mixture was treated with hydrochloric acid and ice, and the precipitated product collected, washed, and drained well. It was transferred to a beaker, heated with 300 ml. of methanol, and the mixture warmed until the ester melted and then stirred continuously in a freezing bath to produce a fine suspension of the ester. This was air dried below 20° (m.p. 20–25°) and crystallized from petroleum ether to give material melting at 28–30°.

Wheeler[5] benzoylated *o*-hydroxyacetophenone by letting a mixture of the phenol, benzoyl chloride, and pyridine stand until the mildly exothermic reaction was over (15 min.) and pouring it into ice and hydrochloric acid. The product was crystallized from methanol.

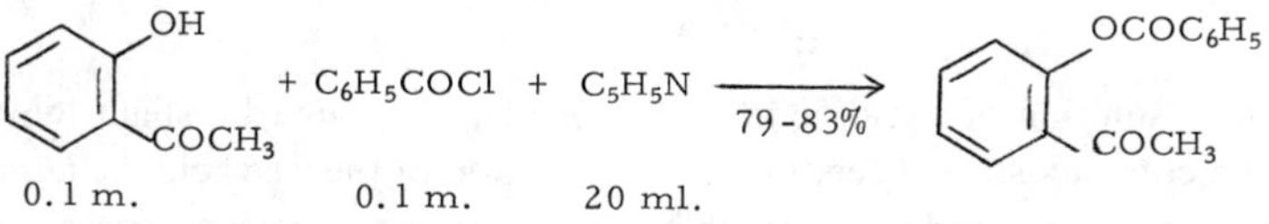
OH
COCH3
+ C6H5COCl + C5H5N
79-83%
OCOC6H5
COCH3
0.1 m.
0.1 m.
20 ml.

Condensation catalyst. Procedures for the condensation of malonic acid with furfural[6] and with *m*-nitrobenzaldehyde[7] are outlined in the formulations. The first uses more pyridine than the second and no other solvent, and the reaction period is

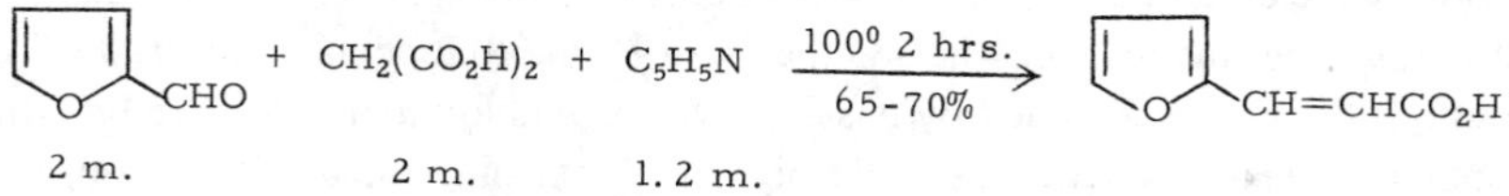
CHO
+ CH2(CO2H)2 + C5H5N
100° 2 hrs.
65-70%
CH=CHCO2H
2 m.
2 m.
1.2 m.

$$\underline{m}\text{-}NO_2C_6H_5CHO + CH_2(CO_2H)_2 + C_5H_5N \xrightarrow[\text{6-8 hrs.}]{\text{Refl. in EtOH}} \underline{m}\text{-}NO_2C_6H_4CH{=}CHCO_2H$$

1 m. 1.1 m. 0.3 m.

shorter than when ethanol is used as solvent. For the condensation of furfural with cyanoacetic acid, Patterson[8] added a small amount of ammonium acetate (no reason

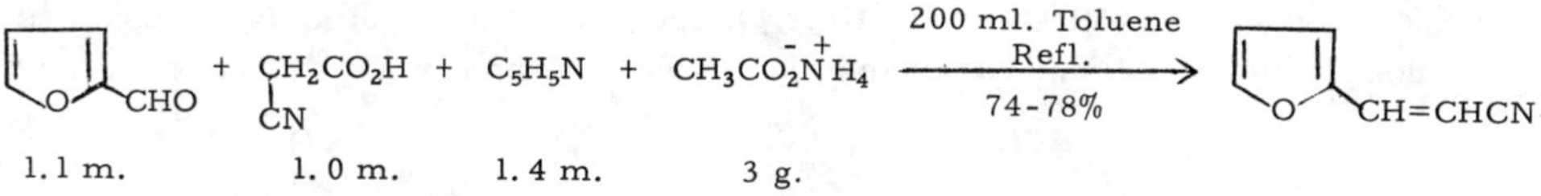
CHO
+ CH2CO2H + C5H5N + CH3CO2NH4
CN
200 ml. Toluene
Refl.
74-78%
CH=CHCN
1.1 m.
1.0 m.
1.4 m.
3 g.

given), used toluene as solvent, and refluxed the mixture under a Dean-Stark water separator for 20 hrs., although the theoretical amount of water was collected within 1 hr.

α-Amino acids from their hydrochlorides. For conversion of lysine dihydrochloride into the monohydrochloride, Eck and Marvel[9] added a solution of pyridine in 25 ml. of hot 95% ethanol to a hot solution of the dihydrochloride in 1 l. of 95% ethanol.

$$\underset{\overset{+}{N}H_3Cl^-}{CH_2}CH_2CH_2CH_2\underset{\overset{+}{N}H_3Cl^-}{CH}CO_2H + Py \xrightarrow[\text{91-94\%}]{\text{95\% EtOH}} \underset{\overset{+}{N}H_3Cl^-}{CH_2}CH_2CH_2CH_2\underset{NH_2}{CH}CO_2H + Py^{+}HCl^{-}$$

0.25 m. 0.32 m.

The crystalline monohydrochloride separated immediately and was collected after thorough cooling. The method was used also for the preparation of free α-aminoisobutyric acid[10] and α-amino-α-phenylpropionic acid[11] from their acid salts.

Preparation of carboxylic acid anhydrides. Adkins and Thompson[12] found that addition of furoyl chloride to a solution of pyridine in petroleum ether at −20° results in separation of an analytically pure 1 : 1 addition complex, m.p. about 60° dec. The complex reacts with 0.5 mole of water in benzene or dioxane to give the anhydride. Benzoic anhydride was obtained in this way in 97% yield. The anhydride can be obtained also by reaction of the complex with 1 mole of the free acid. This method was used in the preparation of heptoic anhydride.[13] Addition of the acid

$$\underset{0.1\ m.}{\underline{n}\text{-}C_6H_{13}COCl} + \underset{0.2\ m.}{Py} \longrightarrow \underline{n}\text{-}C_6H_{13}COCl\cdot Py \xrightarrow[78\text{-}83\%]{0.1\ m.\ \underline{n}\text{-}C_6H_{13}CO_2H}$$

$$(\underline{n}\text{-}C_6H_{13}CO)_2O + Py^+HCl^-$$

chloride to a solution of pyridine in 25 ml. of benzene caused a slight temperature rise and the complex started separating. On addition of the free acid the temperature rose to 60–65°, and pyridine hydrochloride separated. This hygroscopic salt was collected and washed with benzene; removal of the solvent and distillation afforded heptoic anhydride, b.p. 155–162°/2 mm.

An attempt to apply this method to alkanesulfonic acids was unsuccessful; the anhydrides were obtained by reaction of the acids with phosphorous pentoxide.[14]

Dehydrohalogenation. For the synthesis of 2-vinylthiophene, the crude 2-(α-chloroethyl)-thiophene obtained by passing hydrogen chloride into a stirred mixture of thiophene and paraldehyde was quaternized with pyridine and the salt pyrolyzed.[15]

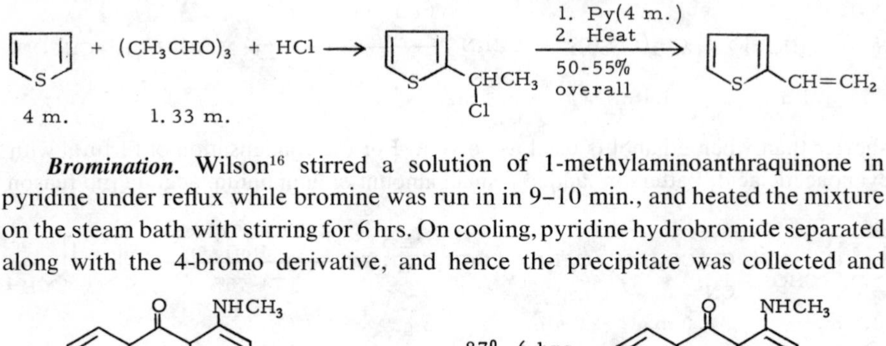
+ (CH₃CHO)₃ + HCl
1. Py(4 m.)
2. Heat
50-55%
overall
CHCH₃
Cl
CH=CH₂
S
4 m.
1. 33 m.

Bromination. Wilson[16] stirred a solution of 1-methylaminoanthraquinone in pyridine under reflux while bromine was run in in 9–10 min., and heated the mixture on the steam bath with stirring for 6 hrs. On cooling, pyridine hydrobromide separated along with the 4-bromo derivative, and hence the precipitate was collected and

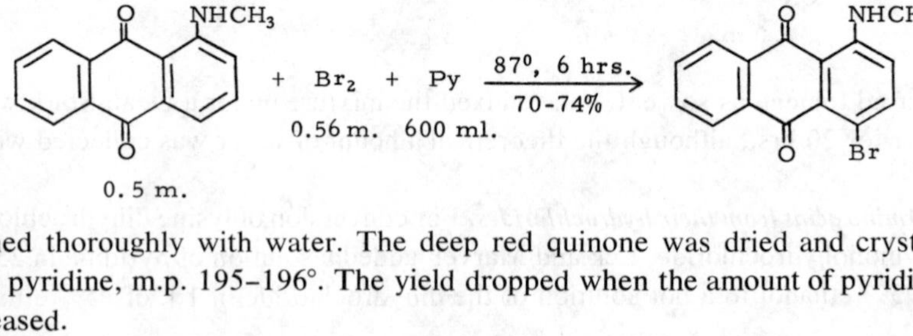
NHCH₃
O
+ Br₂ + Py
0.56 m. 600 ml.
87⁰, 6 hrs.
70-74%
Br
0. 5 m.

washed thoroughly with water. The deep red quinone was dried and crystallized from pyridine, m.p. 195–196°. The yield dropped when the amount of pyridine was increased.

Aromatic nitriles. Newman[17] added pyridine to a mixture of α-bromonaphthalene and cuprous cyanide and noticed an exothermic effect (due to formation of

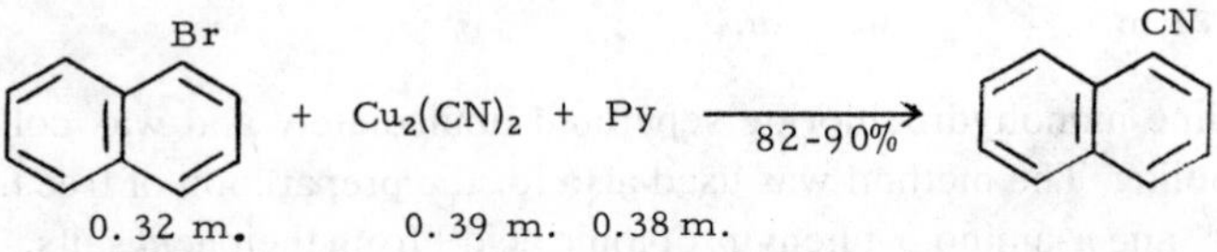
Br
+ Cu₂(CN)₂ + Py
82-90%
CN
0. 32 m.
0. 39 m. 0.38 m.

a pyridine–cuprous cyanide complex). The mixture was then heated under reflux for an appropriate period.

Rearrangements. *o*-Benzoyloxyacetophenone undergoes benzoyl migration to give *o*-hydroxydibenzoylmethane when a solution in pyridine is treated with 7 g.

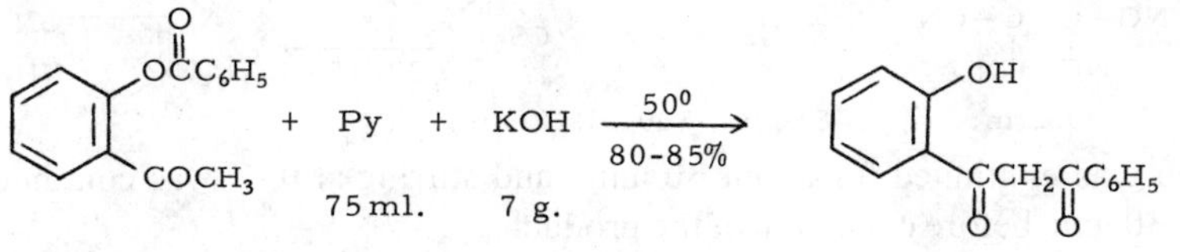

of hot powdered potassium hydroxide, prepared by pulverizing KOH pellets in a mortar previously "heated at 100°" (probably the mortar was heated at the steam bath temperature of about 87° and wiped dry).[18] On stirring for 15 min. a copious yellow precipitate of the potassium salt separated. After cooling, acidification with 10% acetic acid afforded the phenolic diketone as a light yellow precipitate.

1,4-Diphenyl-5-amino-1,2,3-triazole on being refluxed in pyridine for 24 hrs. rearranges to the acidic isomer 4-phenyl-5-anilino-1,2,3-triazole.[19] The solution is

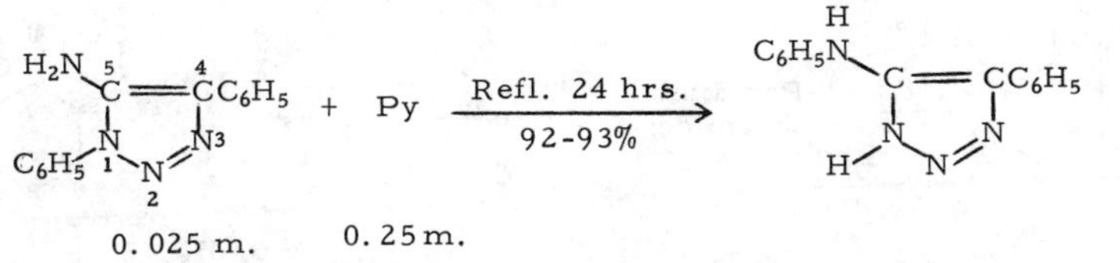

poured into ice-water and the product collected and crystallized from ethanol (m.p. 167–169°).

C-Acylation. A general reaction discovered by Dakin and West[20] is illustrated by the reaction of alanine with acetic anhydride and pyridine to afford 2-acetamido-2-butanone and carbon dioxide.[21] The mixture is stirred on the steam bath until

$$CH_3\underset{NH_2}{\underset{|}{CH}}CO_2H \ + \ (CH_3CO)_2O \ + \ Py \ \longrightarrow \ \left[CH_3\underset{NHCOCH_3}{\underset{|}{CH}}CO_2H \ \longrightarrow \right.$$

0.39 m.   2.35 m.   1.98 m.

$$\left. CH_3\underset{NHCOCH_3}{\overset{COCH_3}{\underset{|}{\overset{|}{C}}}}CO_2H \right] \ \xrightarrow[81-88\%]{-CO_2} \ CH_3\underset{NHCOCH_3}{\underset{|}{CH}}COCH_3$$

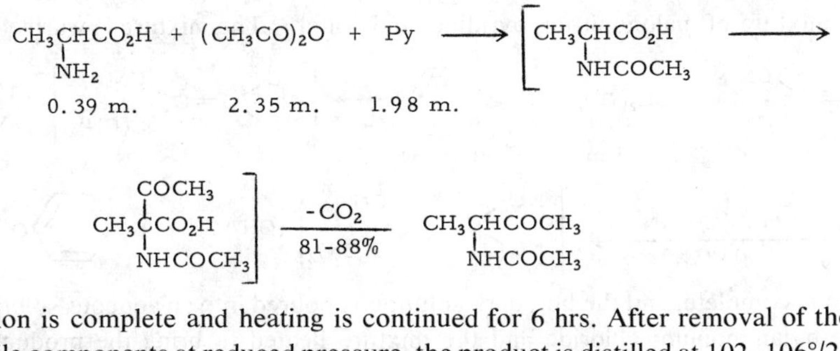

solution is complete and heating is continued for 6 hrs. After removal of the more volatile components at reduced pressure, the product is distilled at 102–106°/2 mm.

ROH ⟶ RX, *see* Phosphorous tribromide, Thionyl chloride.

Oxidation, *see* Copper sulfate.

Aniline extender. In the reaction of an imido chloride with aniline, pyridine is added to make all the aniline available for reaction.[22]

$$C_6H_5\overset{Cl}{\underset{}{\overset{|}{C}}}=NC_6H_5 \ + \ C_6H_5NH_2 \ + \ Py \ \xrightarrow[73-80\%]{} \ C_6H_5\overset{NHC_6H_5}{\underset{}{\overset{|}{C}}}=NC_6H_5 \ + \ Py^+HCl^-$$

0.46 m.   0.46 m.   0.46 m.

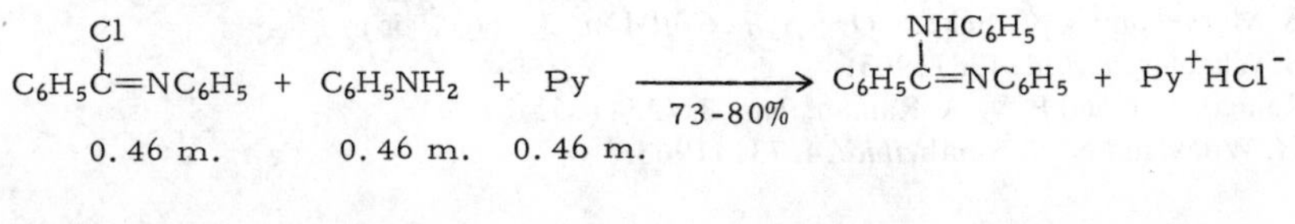

Special syntheses. Middleton[23] prepared 2,5-diamino-3,4-dicyanothiophene by passing hydrogen sulfide into a stirred solution of tetracyanoethylene in acetone and carbon disulfide at 0–5° and added pyridine rapidly. The solution became clear and then the yellow thiophene derivative began to separate. Hydrogen sulfide

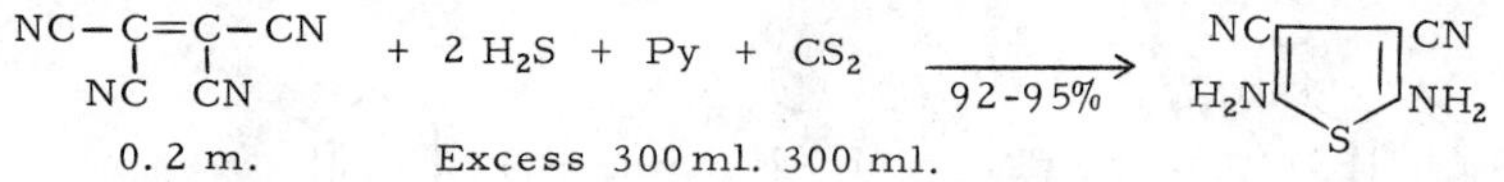
NC−C=C−CN
NC CN
+ 2 H₂S + Py + CS₂
0.2 m.
Excess 300 ml. 300 ml.
92-95%
NC
CN
H₂N
NH₂
S

addition was continued for about 30 min., and stirring at 0–5° was continued for a further 30 min. before collection of the product.

In the Kröhnke reaction a benzyl halide or α-haloketone is converted into the pyridinium salt, which reacts with *p*-nitrosodimethylaniline to give a nitrone; this,

$$C_6H_5CH_2Cl \xrightarrow{C_5H_5N} C_6H_5CH_2\overset{+}{N}C_5H_5(\overset{-}{Cl}) \xrightarrow{p\text{-}ONC_6H_4N(CH_3)_2}$$

$$C_6H_5CH{=}\overset{+}{N}(\overset{-}{O})C_6H_4N(CH_3)_2 \xrightarrow{H_2O\,(H^+)} C_6H_5CH{=}O$$

in turn, gives an aldehyde or ketone on acid hydrolysis. The reaction was employed by Ruzicka *et al.*[24] for the conversion of cholestanone via the 2α-bromo derivative

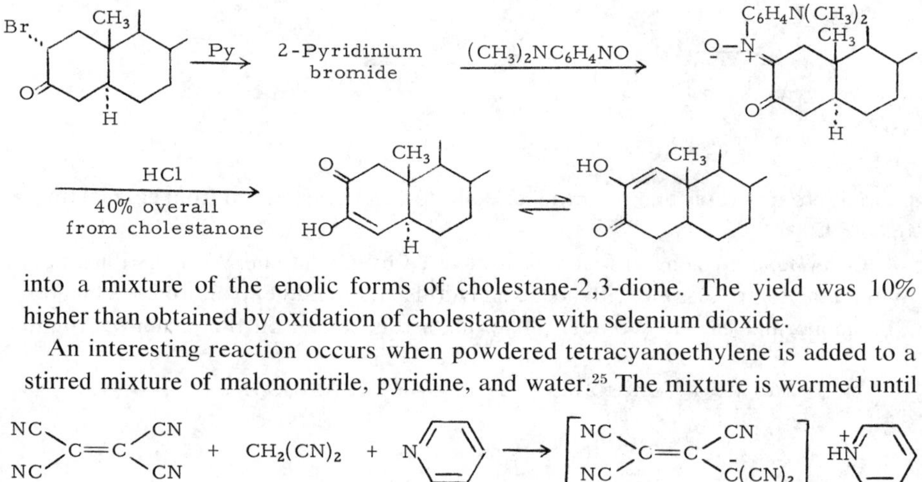
Br
CH₃
O
H
Py
2-Pyridinium
bromide
(CH₃)₂NC₆H₄NO
C₆H₄N(CH₃)₂
O−N
CH₃
O
H
HCl
40% overall
from cholestanone
CH₃
O
HO
H
HO
CH₃
O

into a mixture of the enolic forms of cholestane-2,3-dione. The yield was 10% higher than obtained by oxidation of cholestanone with selenium dioxide.

An interesting reaction occurs when powdered tetracyanoethylene is added to a stirred mixture of malononitrile, pyridine, and water.[25] The mixture is warmed until

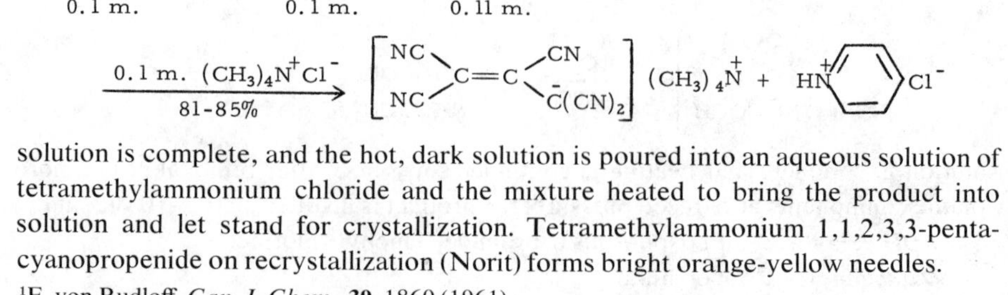
NC
CN
C=C
NC
CN
+ CH₂(CN)₂ +
N
0.1 m.
0.1 m.
0.11 m.
NC
CN
C=C
NC
C(CN)₂
HN
0.1 m. (CH₃)₄N⁺Cl⁻
81-85%
NC
CN
C=C
NC
C(CN)₂
(CH₃)₄N
+
HN
Cl⁻

solution is complete, and the hot, dark solution is poured into an aqueous solution of tetramethylammonium chloride and the mixture heated to bring the product into solution and let stand for crystallization. Tetramethylammonium 1,1,2,3,3-pentacyanopropenide on recrystallization (Norit) forms bright orange-yellow needles.

[1]E. von Rudloff, *Can. J. Chem.*, **39**, 1860 (1961)
[2]L. F. Fieser, *Am. Soc.*, **75**, 4400 (1953)
[3]L. F. Fieser, S. Rajagopalan, E. Wilson, and M. Tishler, *Am. Soc.*, **73**, 4133 (1951)
[4]C. S Marvel and V. C. Sekera, *Org. Syn., Coll. Vol.*, **3**, 366 (1955)
[5]T. S. Wheeler *ibid.*, **4**, 478 (1963)
[6]S. Rajagopalan and P. V. A. Raman, *ibid.*, **3**, 425 (1955)
[7]R. H. Wiley and N. R. Smith, *ibid.*, **4**, 731 (1963)

[8]J. M. Patterson, *Org. Syn.*, **40**, 46 (1960)
[9]J. C. Eck and C. S. Marvel, *Org. Syn.*, *Coll. Vol.*, **2**, 374 (1943)
[10]H. T. Clarke and H. J. Bean, *ibid.*, **2**, 29 (1943)
[11]R. E. Steiger, *ibid.*, **3**, 88 (1955)
[12]H. Adkins and Q. E. Thompson, *Am. Soc.*, **71**, 2242 (1949)
[13]C. F. H. Allen, C. J. Kibler, D. M. McLachlin, and C. V. Wilson, *Org. Syn.*, *Coll. Vol.*, **3**, 28 (1955)
[14]L. Field and P. H. Settlage, *Am. Soc.*, **76**, 1222 (1954)
[15]W. S. Emerson and T. M. Patrick, *Org. Syn.*, *Coll. Vol.*, **4**, 980 (1963)
[16]C. V. Wilson, *ibid.*, **3**, 575 (1955)
[17]M. S. Newman, *ibid.*, **3**, 631 (1955)
[18]T. S. Wheeler, *ibid.*, **4**, 479 (1963)
[19]E. Lieber, T. S. Chao, and C. N. R. Rao, *ibid.*, **4**, 380 (1963)
[20]H. D. Dakin and R. West, *J. Biol. Chem.*, **78**, 91 (1928)
[21]R. H. Wiley and O. H. Borum, *Org. Syn.*, *Coll. Vol.*, **4**, 5 (1963)
[22]A. C. Hontz and E. C. Wagner, *ibid.*, **4**, 383 (1963)
[23]W. J. Middleton, *ibid.*, **4**, 243 (1963)
[24]L. Ruzicka, Pl. A. Plattner, and M. Furrer, *Helv.*, **27**, 524 (1944)
[25]W. J. Middleton and D. W. Wiley, *Org. Syn.*, **41**, 99 (1961)

Pyridine borane, $C_5H_5NBH_3$. Mol. wt. 92.94. A pale yellow liquid which freezes to a white solid, m.p. 10–11°, pyridine borane is stable in dry air, nearly insoluble in water, and only slightly hydrolyzed by water. It is very soluble in alcohol or ether. Supplier: Callery Chem. Co. A convenient method of preparation from sodium borohydride and pyridine hydrochloride in pyridine solution is described by Taylor *et al.*[1]

Barnes, Graham, and Taylor[2] found that the reagent reduces aldehydes and ketones in refluxing ether, benzene, or toluene in yields in the range 24–94%. Acids and acid chlorides are reduced by this method in yields on only 21–40%, and esters do not react.

W. S. Johnson and co-workers[3] reduced estrone methyl ether (1) as follows. A solution of 0.38 mg. of (1) in 0.030 ml. of acetic acid was treated with a solution

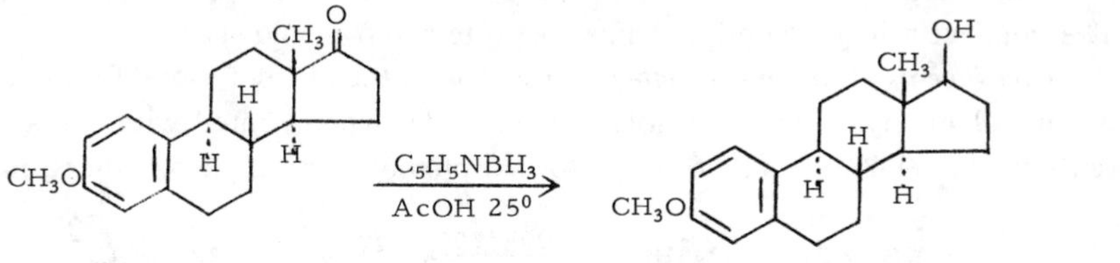
CH3
H
H
H
CH3O
C5H5NBH3
AcOH 25°
OH
CH3
H
H
H
CH3O

(1) (2)

of 0.23 mg. of pyridine borane in 0.005 ml. of acetic acid and let stand under nitrogen at room temperature for 3 hrs. After addition of 1 drop of concd. hydrochloric acid, the mixture was extracted with ether, and the extract washed, dried, and evaporated. TLC showed just one spot of Rf corresponding to that of estradiol methyl ether (2).

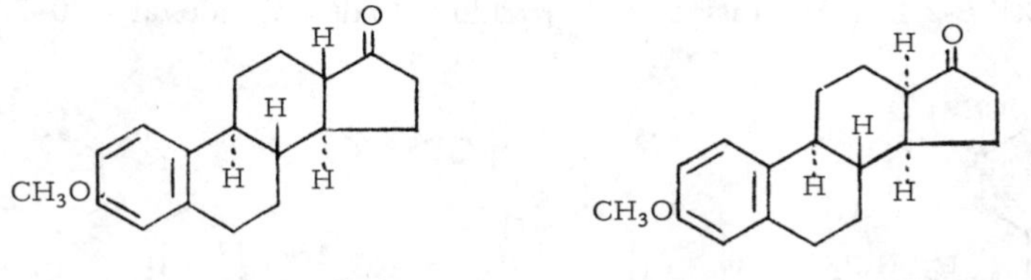
H
H
H
H
CH3O
H
H
H
H
CH3O

(3) (4)

The Johnson group found that the 17-keto-18-norsteroid (3), of C/D *trans* ring junction like (1), is similarly reduced by pyridine borane, whereas the C/D-*cis* isomer (4) is not.

Hawthorne[4] developed a new route to trialkylboranes involving reation of an olefin with pyridine borane in diglyme in a pressure bottle heated on the steam bath. The product from a terminal olefin on oxidation with hydrogen peroxide affords a primary alcohol in high overall yield.

$$3\ CH_3(CH_2)_5CH{=}CH_2 \xrightarrow[82\%]{C_5H_5NBH_3 \text{ in diglyme } (87^0)} [CH_3(CH_2)_5CH_2CH_2]_3BNC_5H_5$$

$$\xrightarrow[90\%]{H_2O_2} 3\ CH_3(CH_2)_5CH_2CH_2OH$$

[1]M. D Taylor, L. R. Grant, and C. A. Sands, *Am. Soc.*, **77**, 1506 (1955)
[2]R. P. Barnes, J. H. Graham, and M. D. Taylor, *J. Org.*, **23**, 1561 (1958)
[3]W. S. Johnson and K. V. Yorka, *Tetrahedron Letters*, **8**, 11 (1960); W. L. Meyer, D. D. Cameron, and W. S. Johnson, *J. Org.*, **27**, 1130 (1962); K. V. Yorka, M. L. Truett, and W. S. Johnson, *ibid.*, **27**, 4581 (1962)
[4]M. F. Hawthorne, *J. Org.*, **23**, 1788 (1958)

Pyridine hydrochloride, C_5H_5NHCl. Mol. wt. 115.56, m.p. 144°, b.p. 218°. Supplier: Eastman.

Preparation. Prey:[1] Hydrogen chloride is passed into a weighed flask containing pyridine through a tube extending above the surface of the liquid until 1 equivalent is absorbed. The solid is crystallized from chloroform–ethyl acetate and washed with ether.

Taylor and Grant:[2] Hydrogen chloride generated by dropping 24 ml. of concd. sulfuric acid onto 75 ml. of concd. hydrochloric acid and 50 g. of sodium chloride is passed into a solution of 69 g. of pyridine in 200 ml. of anhydrous ether until precipitation of the salt is complete. Excess ether is largely removed by distillation and finally removed by evacuation in a dessicator. Yield 98% of 99.8% pure salt. The salt is very hygroscopic and must be protected from moist air.

Cleavage of phenol methyl ethers. Prey[3] found the reagent useful for the cleavage of phenol methyl ethers. Anisole, for example, when heated with 3 parts of the reagent at 200–220° for 5–6 hrs. afforded phenol in 83% yield. Ethers of di- and

$$C_6H_5OCH_3 + Py^+HCl^- \xrightarrow{200\text{-}220^0} C_6H_5OH + Py^+CH_3Cl^-$$

triprotonic phenols are cleaved efficiently by this method, but diphenyl ether remains unattacked. Prey[4] suggested use of the reaction for methoxyl analysis by titration to determine the amount of reagent consumed.

Sheehan and co-workers[5] found that the 3-methyl ethers of both 16α,17β-estriol and 16β,17β-estriol on fusion with pyridine hydrochloride at 200–220° afford

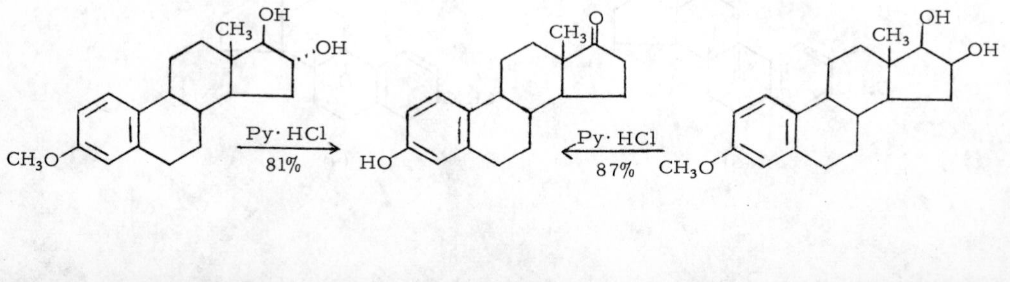
OH
CH3
OH
CH3O
Py·HCl
81%
CH3
O
HO
Py·HCl
87%
CH3O
OH
CH3
OH

estrone in good yield. The reagent not only cleaves the methyl ether group but also dehydrates the 16,17-glycol group, *cis* or *trans*.

Filler *et al.*[6] attempted to cleave the methyl ether (2) with strong acids or bases but found that the CF_3 group is also attacked, with formation of 5-nitrosalicylic acid (1). Cleavage to the desired phenol (3) without disturbance of the trifluoro-

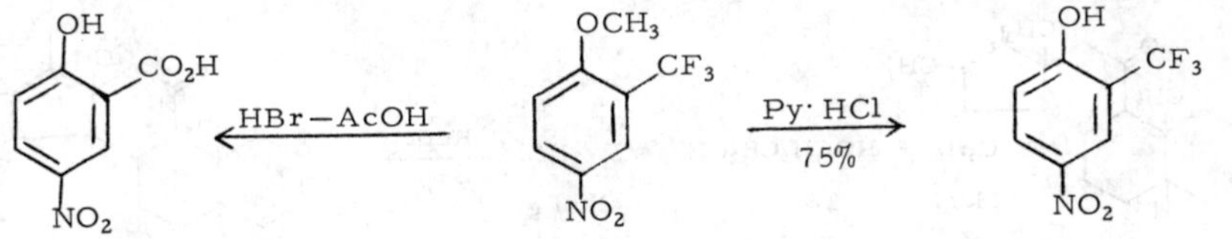

(1) (2) (3)

methyl group was accomplished by reaction with pyridine hydrochloride at 210° for 20 min. When the temperature was raised or the reaction period extended, the by-product (1) was formed in increasing amounts.

Erdtman and co-workers[7] attempted to demethylate the diphenyl derivative (4) with pyridine hydrochloride, but the reaction was attended with ring closure to

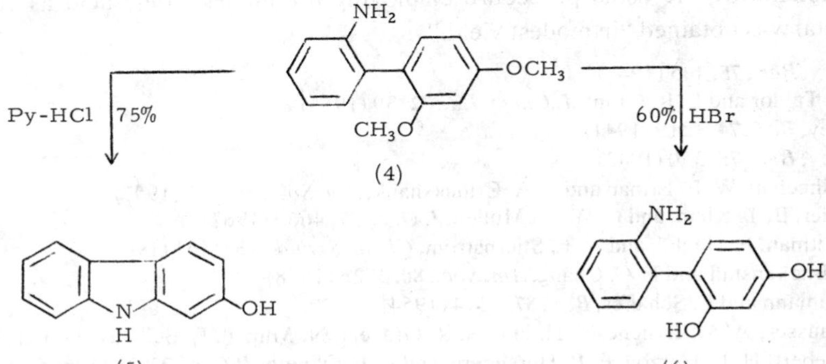

(4) (5) (6)

2-hydroxycarbazole (5). Demethylation to (6) was accomplished by refluxing the hydrobromide of (4) with hydrobromic acid for 24 hrs.

Other reactions. Blickenstaff and Chang[8] found that the tosylates of primary alcohols are readily converted into the corresponding chlorides by reaction with pyridine hydrochloride in solution in dimethylformamide at room temperature.

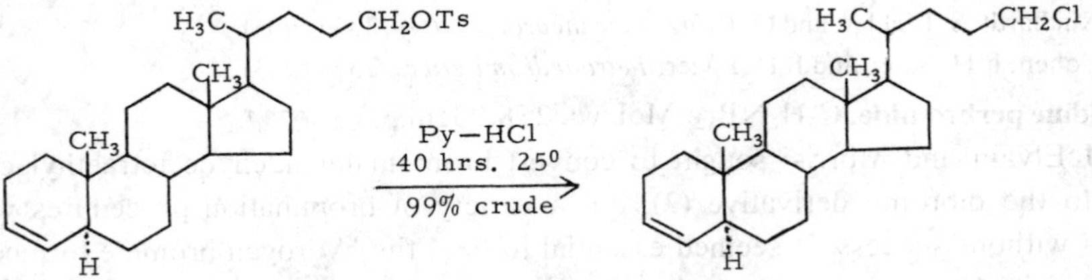

N-Alkylcarboxylic acid amides and anilides are dealkylated in good yield when heated with an excess of the reagent at 190–200°.[9]

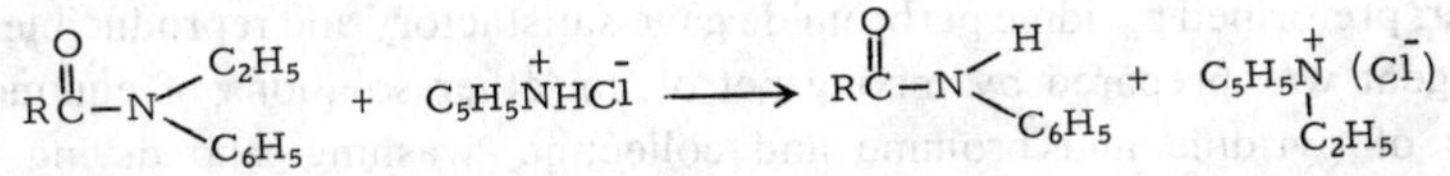

Ketalization. A Schering group[10] found pyridine hydrochloride to be a superior catalyst for the ketalization of 3α,17α-dihydroxy-16β-methylpregnane-11,20-dione. A mixture of the reactants, catalyst, and benzene was refluxed with stirring for 24

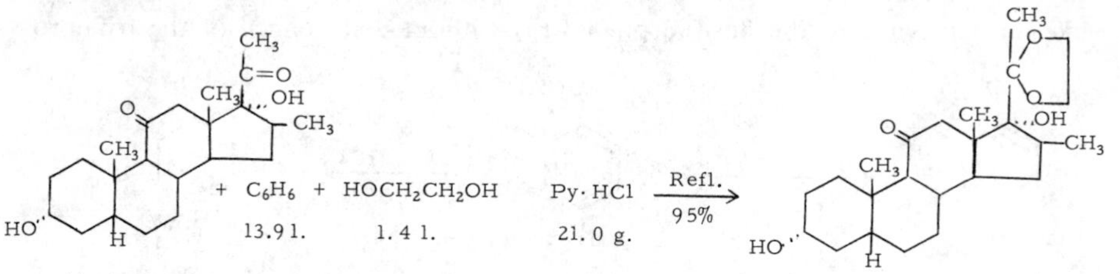
CH3
C=O
CH3
OH
CH3
O
CH3
HO
H
+ C6H6 + HOCH2CH2OH
13.9 l.
1.4 l.
Py·HCl
21.0 g.
Refl.
95%
CH3
O
C
O
CH3
OH
CH3
O
CH3
HO
H

(1) 139.8 g.

hrs. under a Dean-Stark trap packed with about 20 g. of anhydrous sodium sulfate to remove water formed. The mixture was concentrated to 3 l., 14 ml. of 50% sodium hydroxide solution was added, and the mixture was further concentrated until crystals appeared and poured onto 10 l. of ice and water. The solid product, collected, washed, and dried, was suitable for the next step; yield 95%. In earlier experiments by the usual procedure employing *p*-toluenesulfonic acid as catalyst the ketal was obtained "in modest yield."

[1]V. Prey, *Ber.*, **75**, 445 (1942)
[2]M. D. Taylor and L. R. Grant, *J. Chem. Ed.*, **32**, 39 (1955)
[3]V. Prey, *Ber.*, **74**, 1219 (1941)
[4]V. Prey, *Ber.*, **75**, 350 (1942)
[5]J. C. Sheehan, W. F. Erman and P. A. Cruickshank, *Am. Soc.*, **79**, 147 (1957)
[6]R. Filler, B. T. Khan, and C. W. McMullen, *J. Org.*, **27**, 4660 (1962)
[7]H. Erdtman, F. Haglid, and N. E. Stjernström, *Chem. Scand.*, **15**, 1761 (1961)
[8]R. T. Blickenstaff and F. C. Chang, *Am. Soc.*, **80**, 2726 (1958)
[9]D. Klamann and E. Schaffer, *Ber.*, **87**, 1294 (1954)
[10]R. Rausser, A. M. Lyncheski, H. Harris, R. Gracela, N. Murrill, E. Bellamy, D. Ferchinger, W. Gebert, H. L. Herzog, E. B. Hershberg, and E. P. Oliveto, *J. Org.*, **31**, 26 (1966)

Pyridine-N-oxide, $C_5H_5N^+—O^-$. Mol. wt. 95.10, m.p. 67°. Suppliers: A, B, E, F, MCB.

Reports from two laboratories[1,2] indicate that in refluxing benzene, toluene, or xylene this reagent oxidizes certain carboxylic acids or their anhydrides to aldehydes.

$$C_6H_5CH_2CO_2H + 2\ C_5H_5\overset{+}{N}—O^- + Ac_2O \longrightarrow C_6H_5CHO + 2\ C_5H_5N + 2\ AcOH + CO_2$$

[1]C Rüchardt, S. Eichler, and O. Krätz, *Tetrahedron Letters*, 233 (1965)
[2]T. Cohen, I. H. Song, and J. H. Fager, *Tetrahedron Letters*, 237 (1965)

Pyridine perbromide, $C_5H_5NBr_2$. Mol. wt. 238.93, m.p. 62–63°.[1]

McElvain and Morris[2] sought to convert bromomalonaldehyde tetraethylacetal (1) to the dibromo derivative (2), but a variety of bromination procedures were tried without success. It seemed essential to bind the hydrogen bromide formed to prevent its destructive action on the acetal groups, but this could not be accomplished by treating (1) with bromine in the presence of pyridine because the temperature of bromination is higher than that required for an oxidation of the amine by bromine. However, preformed pyridine perbromide gave satisfactory and reproducible results. The reagent was prepared by mixing petroleum ether solutions of equimolecular amounts of pyridine and bromine and collecting, washing, and drying the red

$$\underset{(1)\ 0.36\ \text{m.}}{(C_2H_5O)_2CHCHBrCH(OC_2H_5)_2} + \underset{0.38\ \text{m.}}{C_5H_5NBr_2} \xrightarrow[66\%]{90^\circ\ 2.5\ \text{hrs.}} \underset{(2)}{(C_2H_5O)_2CHCBr_2CH(OC_2H_5)_2}$$

precipitate. A mixture of the reagent with (1) was stirred and heated at 90°, when the red reagent gradually changed to the brown hydrobromide.

The only further mention of the reagent we have discovered is unfavorable. Djerassi and Scholz,[3] in introducing pyridinium hydrobromide perbromide as a reagent for the bromination of ketones, state that "attempts to employ pyridine perbromide, a reagent which should bind any hydrogen bromide liberated, proved disappointing."

[1]D. M. Williams, *J. Chem. Soc.*, 2783 (1931)

[2]S. M. McElvain and L. R. Morris, *Am Soc.*, **73**, 206 (1951)

[3]C. Djerassi and C. R. Scholz, *Am. Soc.*, **70**, 417 (1948)

Pyridinium hydrobromide perbromide, $C_5H_5\overset{+}{N}H\overset{-}{Br_3}$. Mol. wt. 319.86, m.p. 132–134° dec. Supplier: Arapahoe. This stable, crystalline solid is more convenient and agreeable to handle than bromine, and the small amounts required on a micro or semimicro scale can be measured accurately by weight. The reagent invariably is as satisfactory as free bromine, and instances are cited below in which it is distinctly superior.

Preparation (Fieser[1]). Mix 15 ml. of pyridine with 30 ml. of 48% hydrobromic acid and cool; add 25 g. of bromine gradually with swirling, cool, collect the product with use of acetic acid for rinsing and washing. Without drying the solid, crystallize it from 100 ml. of acetic acid. Yield of red needles, 33 g. (69%).

Bromination of ketones. Djerassi and Scholz[2] demonstrated use of the reagent for efficient α-bromination of steroid ketones. For example, a solution of 38 mg. of cholestanone (1) in 1 ml. of acetic acid was treated at 40–60° with 31 mg. (1 equiv.)

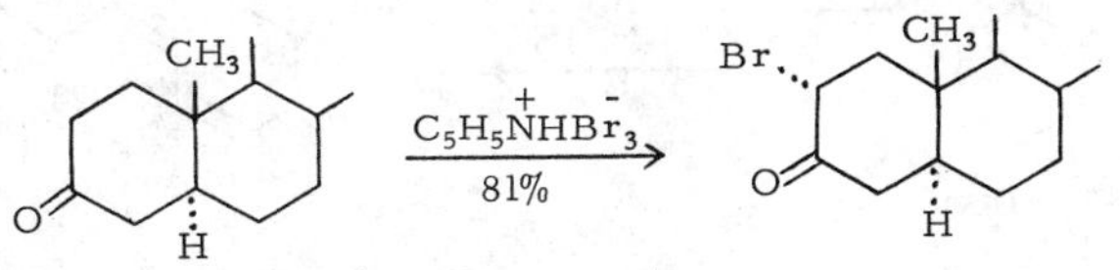
CH3
C5H5NHBr3
81%
H
Br
CH3
H

(1) (2)

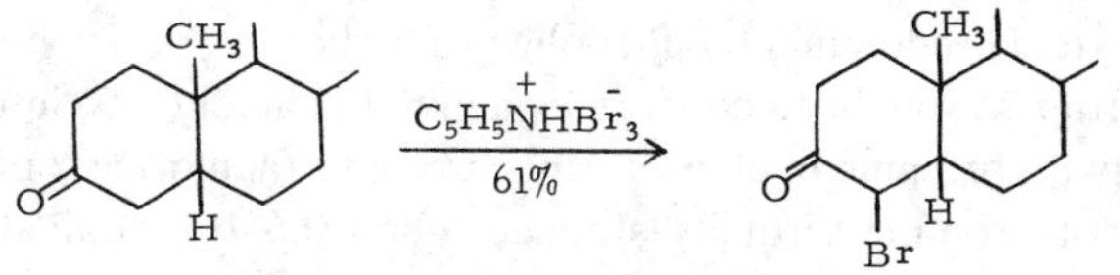
CH3
C5H5NHBr3
61%
H
CH3
H
Br

(3) (4)

of pyridinium hydrobromide perbromide. Hydrogen bromide was evolved, the solution became colorless, and within 1 minute crystals of 2α-bromocholestanone (2) appeared. After cooling, pure (2) was obtained in 81% yield. A 3-ketone of the 5β-series (3) afforded the 4β-bromo-3-ketone (4) in somewhat lower yield.

Olefin dibromides. Study of the bromination of *cis*- and *trans*-stilbene with bromine and with pyridinium hydrobromide perbromide has shown that the latter reagent possesses far greater stereoselectivity than bromine. From the reaction of *cis*-stilbene with free bromine in carbon disulfide, Straus[3] obtained two parts of *dl*-stilbene dibromide (*trans* addition) and one part of the *meso*-dibromide (*cis*-addition). Fieser[4] found that bromination of the hydrocarbon with pyridinium hydrobromide perbromide in acetic acid gave exclusively the *dl*-dibromide, and no trace of the sparingly soluble *meso*-dibromide was found. Breuer and Zincke[5] reported that reaction of *trans*-stilbene with bromine in chloroform or carbon disulfide with careful cooling to avoid substitution in the phenyl groups afforded a maximum of 79% of the *meso*-dibromide. Wislicenus and Seeler[6] used the same procedure and found that the much more soluble *dl*-dibromide can be isolated easily from the mother liquor in 13% yield. In sharp contrast with these results is the following procedure.[4] A suspension of 20 g. of *trans*-stilbene and 40 g. of pyridinium hydrobromide perbromide in 400 ml. of acetic acid was stirred mechanically

$$C_6H_5-\underset{H-\underset{|}{\overset{||}{C}}-C_6H_5}{\overset{|}{C}-H} \xrightarrow[93.5\%]{C_5H_5\overset{+}{N}HB\overset{-}{r}_3} \quad \underset{H-\underset{|}{\overset{|}{C}}-Br}{\overset{C_6H_5}{\underset{C_6H_5}{H-\overset{|}{C}-Br}}}$$

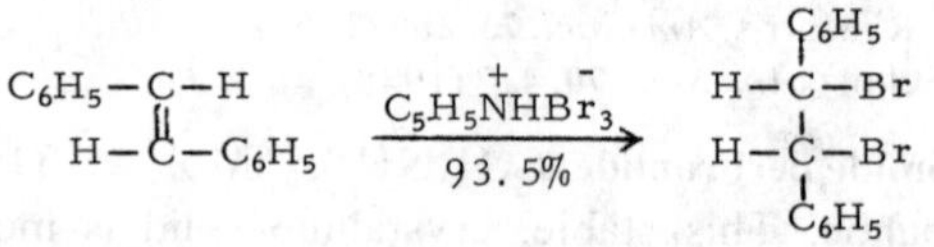

at room temperature. After one-half hour the reagent was consumed and a precipitate of colorless, crystalline, pure *meso*-dibromide was collected, dried, and found to weigh 35.3 g. (93.5%). Processing of the mother liquor afforded 0.3 g. more of the *meso*-dibromide and no *dl*-dibromide could be found in the small amount of residual oil.

Perelman *et al.*[7] of Eli Lilly and Co. found that the 19-nor-$\Delta^{5(10)}$-steroid (1, 19-nor-$\Delta^{5(10)}$-androstene-17β-ol-3-one) can be dehydrogenated smoothly to the 19-nor-$\Delta^{4,9}$-3-ketosteroid (2) by reaction with bromine in pyridine or, more conveniently,

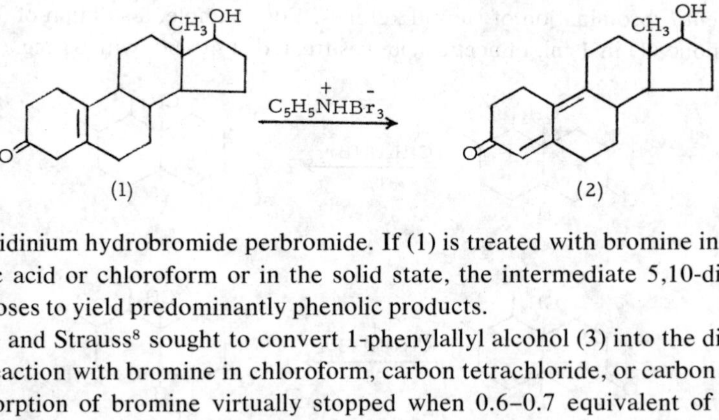

with pyridinium hydrobromide perbromide. If (1) is treated with bromine in solution in acetic acid or chloroform or in the solid state, the intermediate 5,10-dibromide decomposes to yield predominantly phenolic products.

Arcus and Strauss[8] sought to convert 1-phenylallyl alcohol (3) into the dibromide (5) by reaction with bromine in chloroform, carbon tetrachloride, or carbon disulfide but absorption of bromine virtually stopped when 0.6–0.7 equivalent of bromine had been added, and the only product isolated was 1,2,3-tribromo-1-phenylpropane,

$$CH_2=CH\underset{\underset{OH}{|}}{C}HC_6H_5 + Br_2 \xrightarrow{5-10\%} BrCH_2CH\underset{\underset{Br}{|}}{C}H\underset{\underset{Br}{|}}{C}_6H_5$$

(3)   (4)

$$\xrightarrow{C_5H_5\overset{+}{N}HB\overset{-}{r}_3} BrCH_2CH\underset{\underset{Br}{|}}{C}H\underset{\underset{OH}{|}}{C}_6H_5 + BrCH_2CH\underset{\underset{Br}{|}}{C}H\underset{\underset{Br}{|}}{C}_6H_5$$

(5) 47%   (4) 6%

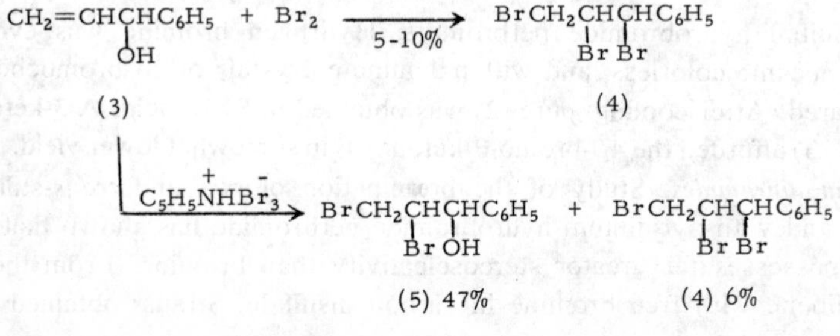

obtained in 5–10% yield. With pyridinium hydrobromide perbromide, however, the desired dibromide (5) was the major product and (4) was a minor by-product.

Freedman and Doorakian[9] report a striking case of the use of $Py^+HBr_3^-$ as a mild, selective brominating agent particularly suited for compounds prone to decomposition. When the tetraphenylcyclobutadiene–nickel bromide complex (6) was treated in methylene chloride with molecular bromine, extensive decomposition occurred, and only a trace of the desired dibromocyclobutene (7) was obtained. Bromination

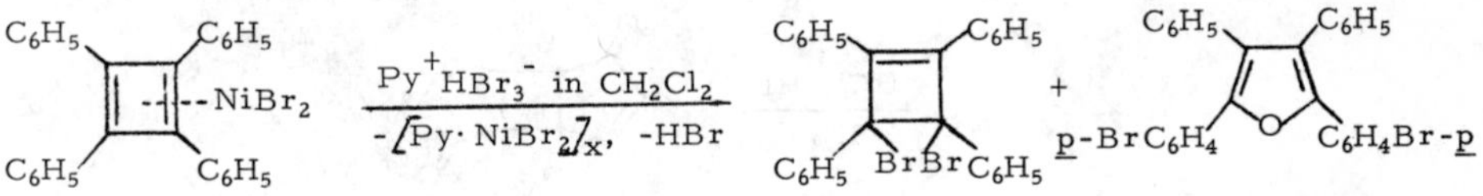
C6H5
C6H5
NiBr2
C6H5
C6H5
Py+HBr3- in CH2Cl2
-[Py·NiBr2]x, -HBr
C6H5
C6H5
C6H5 BrBr C6H5
+
C6H5
C6H5
p-BrC6H4
O
C6H4Br-p

(6) (7) (8)

with Py^+HBr_3 proved to be much better, but in the early stages of the study the yield of (7) varied erratically from 40 to 80%. Investigation showed that the production of (7) varied inversely with the formation of a by-product identified as 2,5-di-*p*-bromotetraphenylfurane (8). The formation of (8) requires water, and the source of the water was traced to the brominating agent. With reagent freshly prepared by the method of Fieser (above) the yield of (7) was 60% and that of (8) was 20%. When the material was allowed to age without protection from atmospheric moisture, the yield of (7) dropped to 10% and that of (8) increased to 65%. When anhydrous reagent was prepared *in situ* by dissolving 4.3 g. (7.5 mmol.) of (6) in 125 ml. of methylene chloride and adding 2.4 g. (15 mmol.) of anhydrous pyridinium bromide and then 2.4 g. (15 mmol.) of bromine and stirring overnight, evaporation of the solvent and trituration with acetone afforded pure (7) in yield of 3.3 g. (85%).

Glycidic acid⟶α,β-unsaturated aldehyde. In the course of a total synthesis of lysergic acid, Kornfeld *et al.*[10] treated the glycidic acid sodium salt (1) in acetonitrile solution successively with $Py^+HBr_3^-$ and then semicarbazide. The free aldehyde (4) was then obtained by exchange of the semicarbazide residue to pyruvic acid.

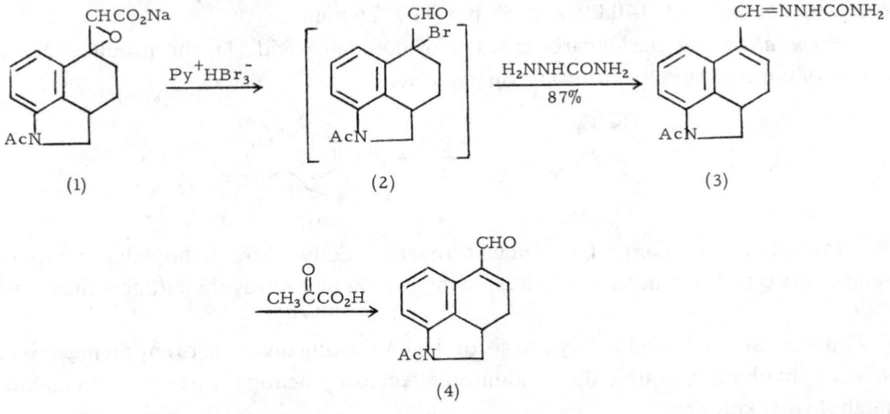
CHCO2Na
O
AcN
Py+HBr3-
CHO
Br
AcN
H2NNHCONH2
87%
CH=NNHCONH2
AcN
(1)
(2)
(3)
O
CH3CCO2H
CHO
AcN
(4)

Bromination of ethylene ketals. Eaton[11] found that the most satisfactory route to the doubly unsaturated diketone from the corresponding saturated diketone was reaction of the bis-ethyleneketal (1) with pyridinium hydrobromide perbromide to give the dibromide (2) and dehydrohalogenation with potassium *t*-butoxide in dimethyl sulfoxide.

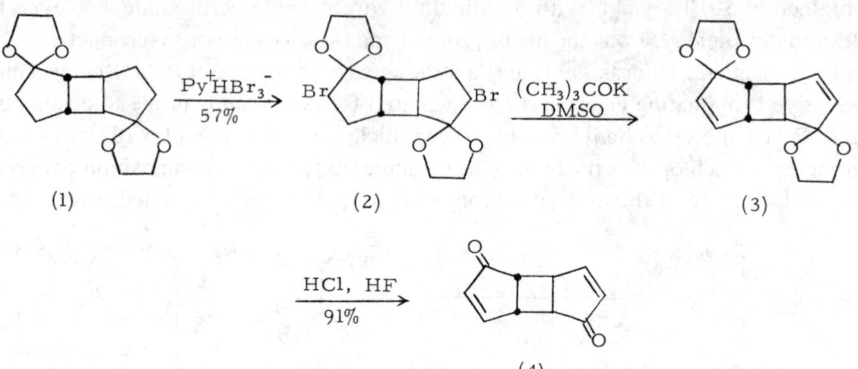
Py⁺HBr₃⁻
57%
Br
Br
(CH₃)₃COK
DMSO
HCl, HF
91%

(1) (2) (3)

(4)

Other reactions. Pyridinium hydrobromide perbromide is reported to be the most satisfactory reagent for the conversion of indole into 3-bromoindole.[12] The reagent can be used for the selective bromination of cyclic ketals in tetrahydrofurane solution,[13] but phenyltrimethylammonium perbromide is now the preferred reagent.

[1] *Org. Expts.*, 68
[2] C. Djerassi and C. R. Scholz, *Am. Soc.*, **70**, 417 (1948)
[3] F. Straus, *Ann.*, **342**, 262 (1905)
[4] L. F. Fieser, *J. Chem. Ed.*, **31**, 291 (1954)
[5] A Breuer and Th. Zincke, *Ann.*, **198**, 154 (1879)
[6] J. Wislicenus and F. Seeler, *Ber.*, **28**, 2694 (1895)
[7] M. Perelman, E. Farkas, E. J. Fornefeld, R. J. Kraay, and R. T. Rapala, *Am. Soc.*, **82**, 2402 (1960)
[8] C. L. Arcus and H. E. Strauss, *J. Chem. Soc.*, 2669 (1952)
[9] H. H. Freedman and G. A. Doorakian, *Tetrahedron*, **20**, 2181 (1964)
[10] E. C. Kornfeld, E. J. Fornefeld, G. B. Kline, M. J. Mann, D. E. Morrison, R. G. Jones, and R. B. Woodward, *Am. Soc.*, **78**, 3087 (1956)
[11] P. E. Eaton, *Am. Soc.*, **84**, 2344 (1962)
[12] K. Piers, C. Meimarogl, R. V. Jardine, and R. K. Brown, *Can. J. Chem.*, **41**, 2399 (1963)
[13] A. Marquet *et al.*, *Compt. rend.*, **248**, 984 (1959)

α-Pyrone (2). Mol. wt. 140.09, m.p. 5°, b.p. 110°/26 mm.

Preparation[1] by the decarboxylation of coumalic acid (1), the product of the action of concentrated sulfuric acid on malic acid.

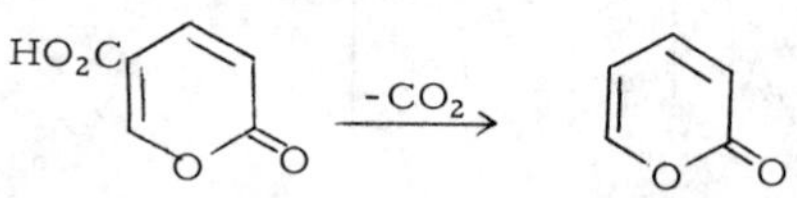
HO₂C
-CO₂

(1) (2)

Diels-Alder reactions. The reagent reacts readily with dienophiles at room temperature to form normal adducts; at higher temperatures the adducts may lose CO_2.[2]

Zimmerman[3] achieved the synthesis of the interesting diketone (3) by an ingenious scheme involving a double diene addition effected by heating α-pyrone with excess methyl vinyl ketone.

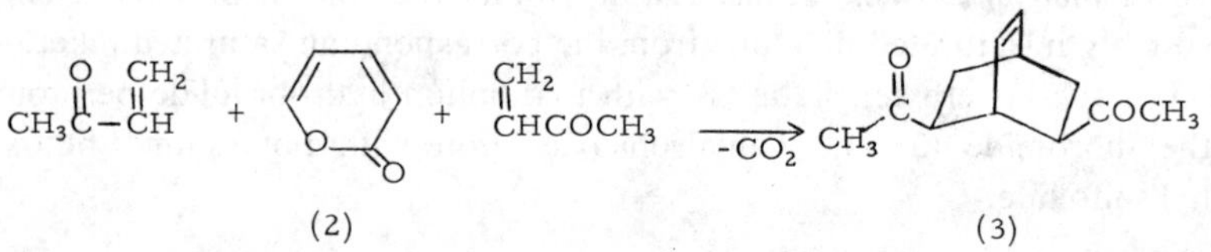
CH₃C–CH
CH₂
+
+
CH₂
CHCOCH₃
-CO₂
CH₃
COCH₃

(2) (3)

[1]H. E. Zimmerman, G. L. Grunewald, and R. M. Paufler, *Org. Syn.*, **46**, 101 (1966)
[2]L. F. Fieser and M. J. Haddadin, *Can. J. Chem.*, **43**, 1599 (1965)
[3]H. E. Zimmerman and R. M. Paufler, *Am. Soc.*, **82**, 1514 (1960)

Pyrophosphoryl tetrachloride, Cl_2P—O—PCl_2 (PPTC). Mol. wt. 251.78, b.p. 66–68°/0.01 mm., 90°/12 mm.

The reagent has been prepared in about 30% yield by heating phosphoryl chloride with phosphorus pentoxide in a sealed tube and fractionating the resulting mixture of products.[1] A second method involves reaction of phosphoryl chloride with a controlled amount of water: $2\,POCl_3 + H_2O \longrightarrow P_2O_3Cl_4 + 2\,HCl$.[2]

At a low temperature and without solvent, the reagent reacts with the primary alcoholic group of 2′,3′-isopropylidene nucleosides to give the 5′-dichlorophosphoric

$$RCH_2OH \;+\; Cl_2\overset{O}{\underset{}{P}}-O-\overset{O}{\underset{}{P}}Cl_2 \longrightarrow RCH_2O\overset{O}{\underset{Cl}{P}}-Cl \;+\; Cl-\overset{O}{\underset{Cl}{P}}-OH$$

$$\xrightarrow{H_2O}\; RCH_2O\overset{O}{\underset{OH}{P}}-OH$$

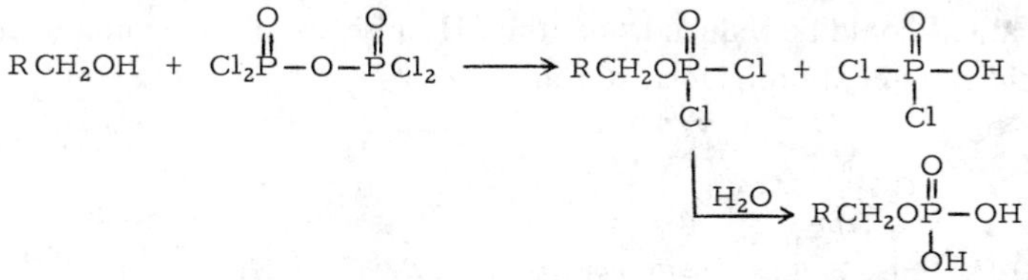

acid; this on treatment with water gives the 5′-monophosphate.[3] Yields are in the order of 80%. The method avoids basic catalysis and removal of organic residues.

Gruber and Lynen[4] found that in the presence of triethylamine the reagent reacts with the free alcoholic group of 2′,3′-isopropylidene-adenosine (1) to give the pyrophosphate ester (2). Hydrolysis, and removal of the isopropylidine group with formic acid gave material containing 22% of adenosine diphosphate (ADP). They

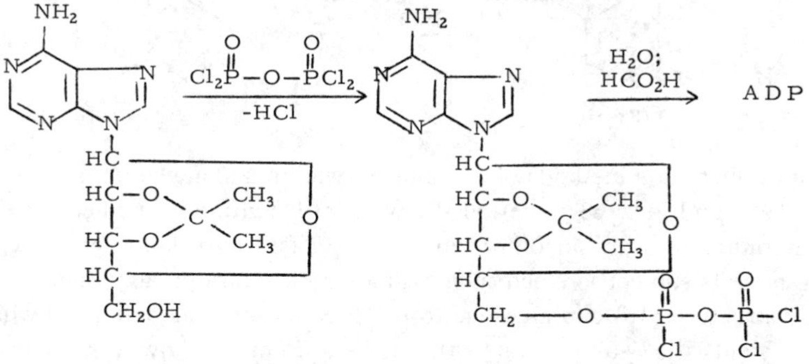

showed that the reagent can be used also for the preparation of unsymmetrical diesters of pyrophosphoric acid and succeeded in a synthesis of coenzyme A, if in low yield.

[1]H. Grunze, *Z. anorg. allg. Chem.*, **296**, 63 (1958)
[2]A. Besson, *Compt. rend.*, **124**, 1099 (1897); M. Becke-Goehring and J. Sambeth, *Angew. Chem.*, **69**, 640 (1957)
[3]W. Koransky, H. Grunze, and G. Münch, *Z. Naturforsch.*, **17b**, 191 (1962)
[4]W. Gruber and F. Lynen, *Ann.*, **659**, 139 (1962)

Pyrrolidine, 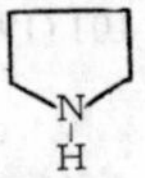

Mol. wt. 71.12, b.p. 88°, sp. gr. 0.87, pKb 2.9, water-miscible. Suppliers: Ansul; du Pont, Electrochem. Div., Henley, Columbia, A, B, E, F, MCB.

Enamines. Review of the formation and uses of enamines.[1] Of the three amines most commonly used for formation of enamines (morpholine, piperidine, pyrrolidine), pyrrolidine is the most reactive.[1]

Heyl and Herr[2] developed an efficient method for the conversion of Δ^4-androstene-3,17-dione into testosterone involving in the first step selective enamine formation at C_3. The diketone was refluxed with 4 equivalents of pyrrolidine and a trace of *p*-toluenesulfonic acid under a water separator, and the reaction was stopped when 1 mole of water had been collected. With the C_3 carbonyl function protected,[2] the keto group at C_{17} could be reduced smoothly. Hydrolysis of the enamine group with a weakly acidic buffer afforded testosterone.

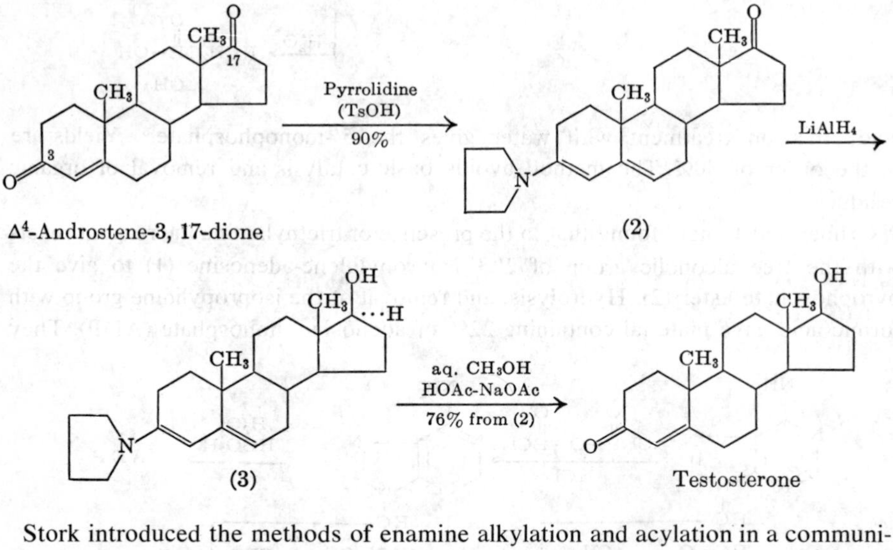
Pyrrolidine
(TsOH)
90%
LiAlH4
aq. CH3OH
HOAc-NaOAc
76% from (2)
Δ4-Androstene-3, 17-dione
(2)
(3)
Testosterone

Stork introduced the methods of enamine alkylation and acylation in a communication of 1954 and in 1963 reviewed the work of his group in the field and an extensive literature of work in other laboratories.[3] The pyrrolidine enamine (1) of cyclohexanone is subject to electrophilic attack by a saturated alkyl halide and the resonance structure (1b) can give rise to the product of C-alkylation (2) which on hydrolysis yields the α-alkylketone (5). Yields are limited, however, by the competing reaction of N-alkylation of (1a) to (3). With a halide of high electrophilic potency the interfering reaction is suppressed and yields of 50–70% are generally obtained with allyl halides, benzyl halides, propargyl halides, and α-halo ketones, esters, and nitriles. An unusual feature is the direction of enolization: the stable form of the enamine of an unsymmetrical ketone such as 2-methylcyclohexanone (5) is that in which the double bond carries the fewer substituents (7). Thus alkylation of (7) gives the 2,6-dialkyl derivative (6).

Alkylation of the enamines of aldehydes and ketones with electrophilic olefins such as α,β-unsaturated nitriles, esters, and ketones is a useful general method

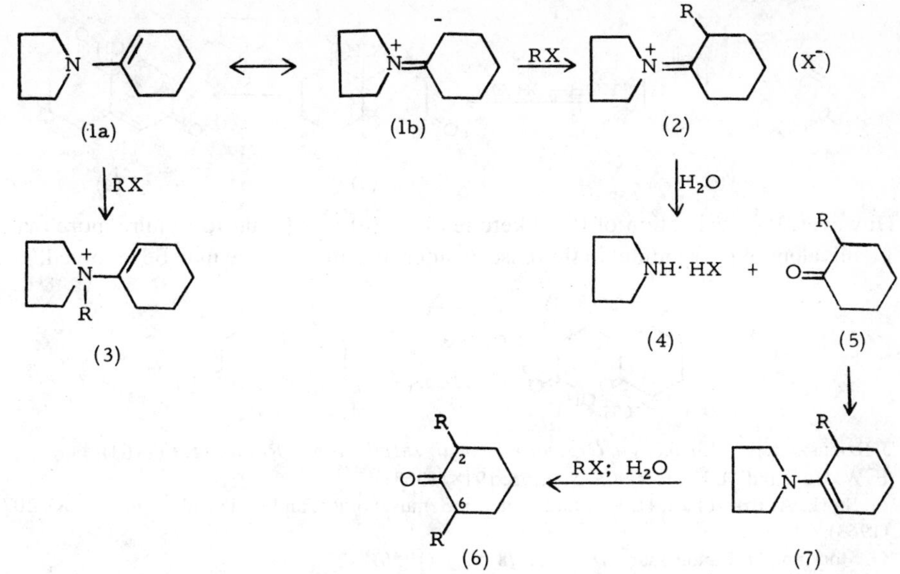
(1a)
(1b)
(2)
(3)
(4)
(5)
(6)
(7)
RX
H_2O
RX; H_2O
NH· HX
(X^-)

(Stork[4]); in this case N-alkylation is reversible and the yields of C-alkylation products often are in the range 60–85%. In the example formulated cyclopentanone

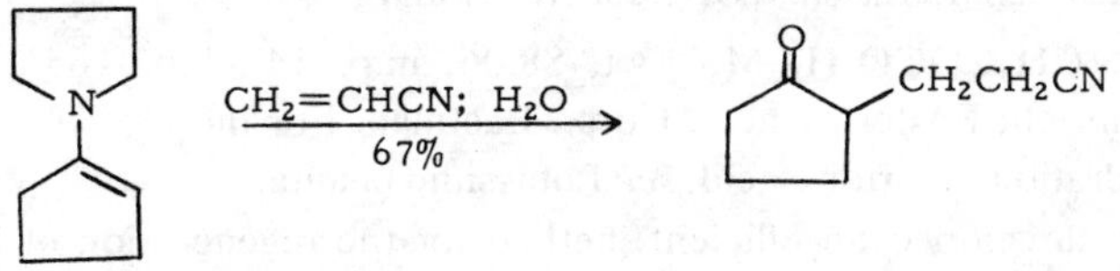
$CH_2{=}CHCN$; H_2O
67%
CH_2CH_2CN

enamine was refluxed with 1.3 equivalents of acrylonitrile in dioxane for 12 hrs. and the product hydrolyzed with water. Other solvents commonly employed are benzene, acetonitrile, and absolute ethanol.

Condensation with ninhydrin. Proline condenses with one equivalent of ninhydrin to give proline yellow (3) and with two equivalents to give proline red (4), which forms a deep blue anion.[5] In the first case, the reaction mixture smells strongly of pyrrolidine. An identical red pigment results from condensation of ninhydrin with pyrrolidine.

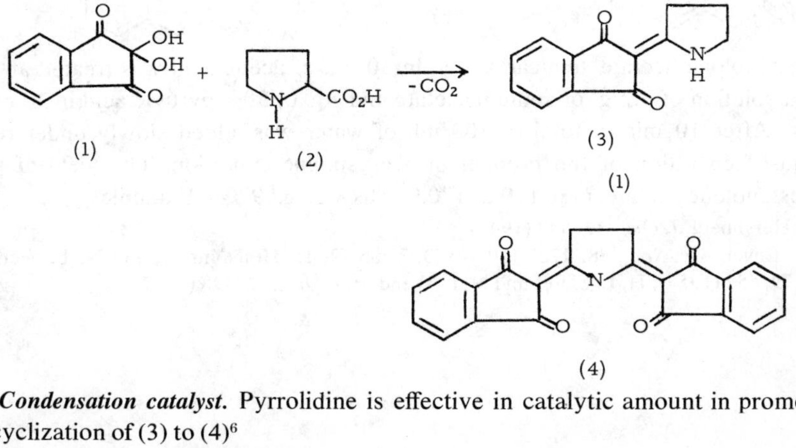
(1)
(2)
(3)
(1)
(4)
$-CO_2$
CO_2H

Condensation catalyst. Pyrrolidine is effective in catalytic amount in promoting the cyclization of (3) to (4)[6]

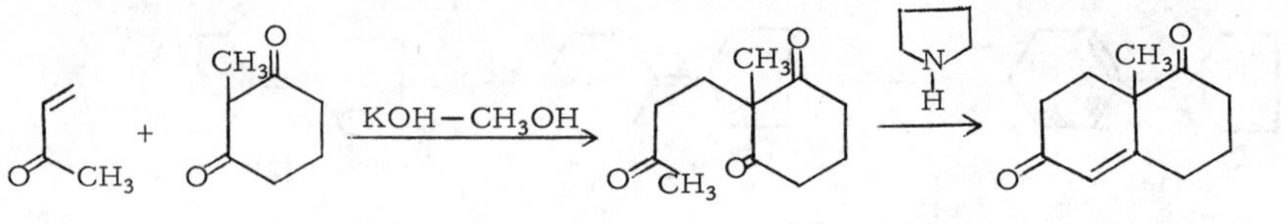

(3) (4)

However, the cyclization of the diketone (5) to (6) was found to require more than 1 equivalent of pyrrolidine; in this case an intermediate enamine may be involved.[7]

(5)

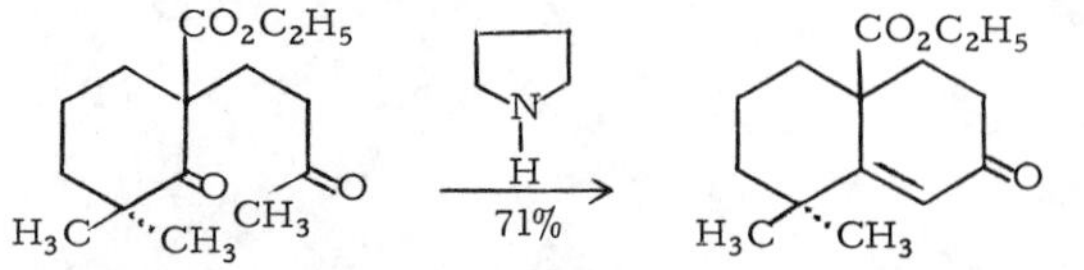

(6)

[1]J. Szmuszkovicz, *Advances in Organic Chemistry, Methods and Results*, **4**, 1 (1963)
[2]F. W. Heyl and M. E. Herr, *Am. Soc.*, **75**, 1918 (1953)
[3]G. Stork, A. Brizzolara, H. K. Landesman, J. Szmuszkovicz, and R. Terrell, *Am. Soc.*, **85**, 207 (1963)
[4]G. Stork and H. Landesman, *Am. Soc.*, **78**, 5128 (1956)
[5]W. Grassmann and K. v. Arnim, *Ann.*, **509**, 288 (1934)
[6]S. Ramachandran and M. S. Newman, *Org. Syn.*, **41**, 38 (1961)
[7]W. L. Meyer and A. S. Levinson, *J. Org.*, **28**, 2184 (1963)

Pyruvic acid, CH_3COCO_2H. Mol. wt. 88.06, m.p. 14°, b.p. 165°, sp. gr. 1.27. Suppliers: Aldrich; Eastern Chem. Corp., Eastman. For the preparation of pyruvic acid by dehydration of tartaric acid, *see* Potassium bisulfate.

Hershberg[1] developed an efficient method for the regeneration of ketones from their semicarbazones, phenylhydrazones, or oximes by exchange reaction. As applied to androstenolone acetate semicarbazone, the reaction is conducted in the presence of sodium acetate both to avoid deacetylation and to keep the exchanged semicarbazone in solution when the mixture is diluted with water. Thus 10 g. of pure

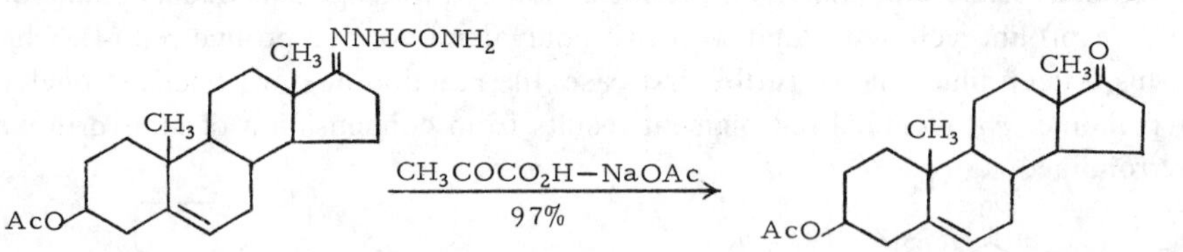

androstenolone acetate semicarbazone in 30 ml. of acetic acid was treated at 110° with a solution of 3.2 g. of sodium acetate and 7 g. of 50% pyruvic acid in 20 ml. of water. After 10 min. a total of 100 ml. of water was added slowly under reflux to cause separation of the product in a crystalline condition. The yield of pure androstenolone acetate, m.p. 170.2–170.9° was 8.28 g. (97%). Examples.[2]

[1]E. B. Hershberg, *J. Org.*, **13**, 542 (1948)
[2]C. R. Engel, *Am. Soc.*, **78**, 4727 (1956); D. Taub, R. D. Hoffsommer, and N. L. Wendler, *ibid.*, **79**, 452 (1957); H. L. Slates and N. L. Wendler, *J. Org.*, **22**, 498 (1957)

Q

Quinoline, C_9H_7N. Mol. wt., 129.16, b.p. 238°. Preparation by Skraup synthesis.[1] Suppliers: A, B, E, F, MCB.

Rosenmund reaction. The quinoline–sulfur poison of Rosenmund and Zetzsche[2] is prepared by refluxing 1 g. of sulfur with 6 g. of quinoline for 5 hrs. and diluting the resultant dark brown solution with purified xylene to 70 ml. In reducing β-naphthoyl chloride to β-naphthaldehyde, Hershberg and Cason[3] passed hydrogen

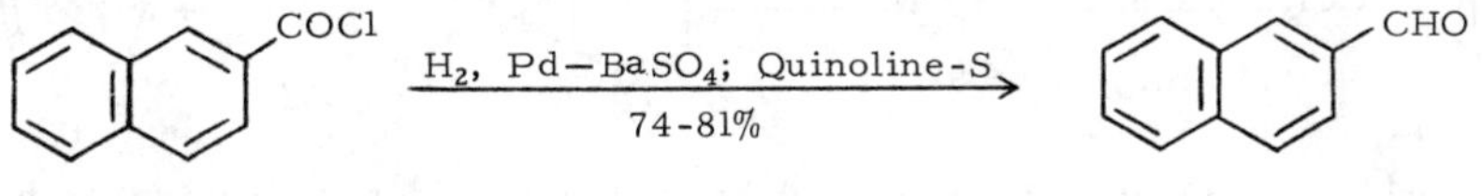
COCl
H_2, Pd−BaSO$_4$; Quinoline-S
74-81%
CHO

0.30 m.

gas into a stirred solution of β-naphthoyl chloride in 200 ml. of xylene after addition of 6 g. of palladium–barium sulfate catalyst and 0.6 ml. of the stock quinoline–sulfur poison solution, with arrangement for sweeping hydrogen chloride liberated into a solution of standard alkali. After the air had been displaced, the flask was heated in an oil bath at 140–150°, and the reaction was continued until the theoretical amount of hydrogen chloride had been collected.

Decarboxylation. Quinoline is useful as solvent for the decarboxylation of unsaturated acids because it is basic enough to form the required carboxylate anion and also because it boils at a temperature favorable for decarboxylation. Thus Burness[4] decarboxylated 3-methylfurane-2-carboxylic acid by refluxing a mixture of 25 g. of the acid, 50 g. of quinoline, and 4.5 g. of copper powder for 2–3 hrs. In

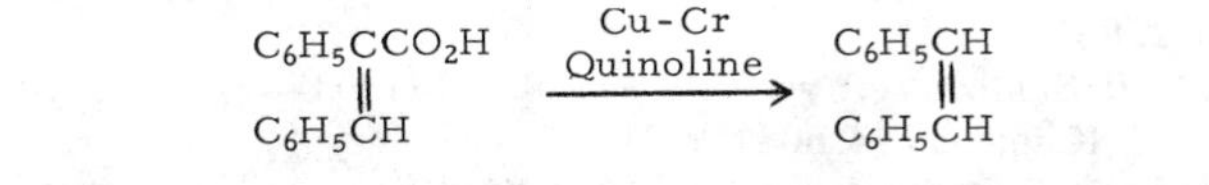
CH_3
CO_2H
Quinoline, Cu powder
83-89%
CH_3

an experiment for students,[5] *cis*-stilbene is prepared by refluxing for 10 min. a mixture of 2.5 g. of *cis*-α-phenylcinnamic acid, 0.2 g. of copper chromite catalyst, and 3 ml. of quinoline. In working out a procedure for the preparation of *m*-nitro-

$$\begin{matrix} C_6H_5CCO_2H \\ \| \\ C_6H_5CH \end{matrix} \xrightarrow[\text{Quinoline}]{\text{Cu-Cr}} \begin{matrix} C_6H_5CH \\ \| \\ C_6H_5CH \end{matrix}$$

styrene from the corresponding cinnamic acid, Wiley and Smith[6] did not take advantage of the convenient boiling point of quinoline, which permits unattended operation, but kept the temperature in the range 185–195° over a 2–3 hr. period.

$$\underline{m}\text{-}NO_2C_6H_4CH{=}CHCO_2H \xrightarrow[\text{56-60\%}]{\text{2 g. Cu powder, 60 ml. Quinoline; 185-195° for 2-3 hrs.}} \underline{m}\text{-}NO_2C_6H_4CH{=}CH_2$$

0.155 m.

Dehydrohalogenation. Allen and Kalm[7] converted 0.14 mole of 2-methyldodecanoic acid (1) into the α-bromo ester (2), and eliminated hydrogen bromide by heating (2) with 0.7 mole of quinoline for 3 hrs. at 160–170°.

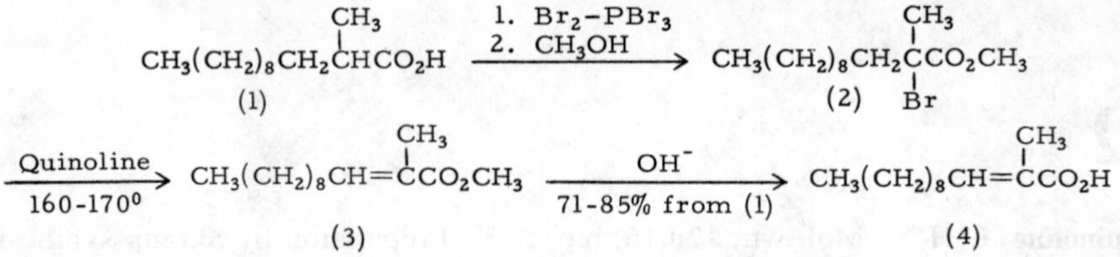

$$CH_3(CH_2)_8CH_2\overset{\overset{\displaystyle CH_3}{|}}{C}HCO_2H \xrightarrow[\text{2. }CH_3OH]{\text{1. }Br_2-PBr_3} CH_3(CH_2)_8CH_2\overset{\overset{\displaystyle CH_3}{|}}{\underset{\underset{\displaystyle Br}{|}}{C}}CO_2CH_3$$

(1)   (2)

$$\xrightarrow[160-170°]{\text{Quinoline}} CH_3(CH_2)_8CH=\overset{\overset{\displaystyle CH_3}{|}}{C}CO_2CH_3 \xrightarrow[\text{71-85\% from (1)}]{OH^-} CH_3(CH_2)_8CH=\overset{\overset{\displaystyle CH_3}{|}}{C}CO_2H$$

(3)   (4)

Schweizer and Parham[8] prepared 2-oxa-7,7-dichloronorcarane (6) by generating dichlorocarbene in the presence of dihydropyrane and converted the bicyclic compound (6) into 2,3-dihydro-6-chlorooxepine (7) in good yield by heating it with quinoline at 140–150°; (7) distilled as formed.

0.8 m. (5)   (6)   (7)

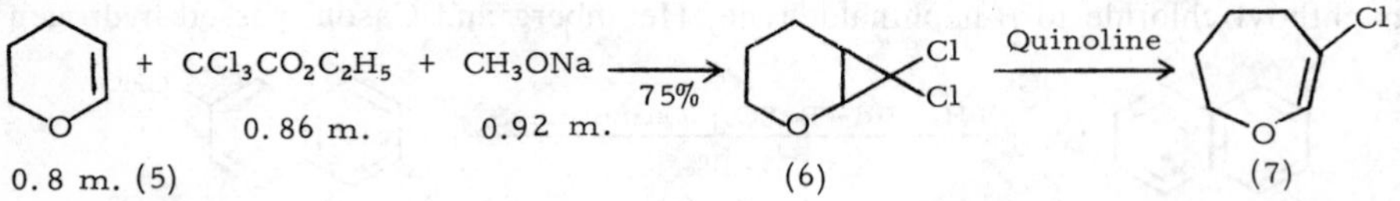

Similarly, the adduct of 1-ethoxycyclohexene and dichlorocarbene when heated with quinoline is converted into 1-ethoxy-1,3,5-cycloheptatriene (3).[9] This enol ether is hydrolyzed to 3,5-cycloheptadienone (4). When 7,7-dibromonorcarane

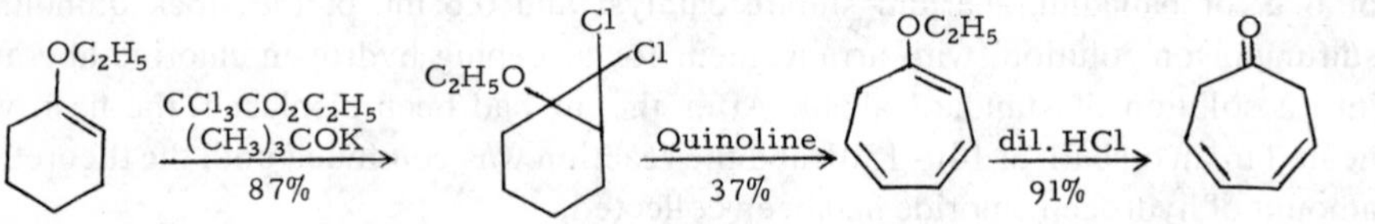

(1)   (2)   (3)   (4)

(5) is heated alone at 200° it gives a mixture of toluene (60%) and cycloheptatriene (8).[10] However, when it is heated with quinoline at 200° cycloheptatriene un-

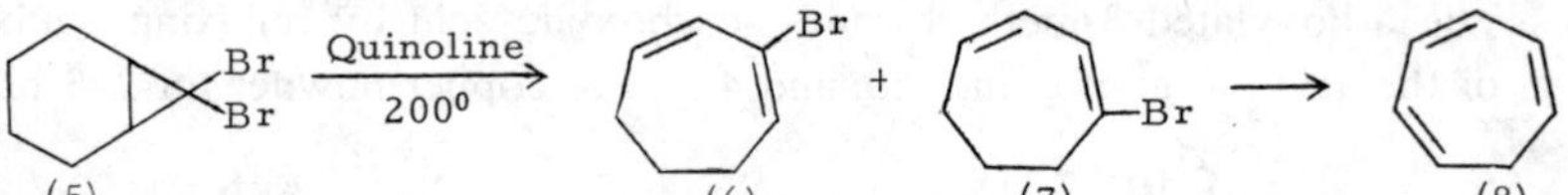

(5)   (6)   (7)   (8)

contaminated with toluene is obtained in 66% yield. Under carefully controlled conditions two intermediate bromocyclohepta-1,3-dienes can be isolated.

[1]H. T. Clarke and A. W. Davis, *Org. Syn., Coll. Vol.*, **1**, 478 (1941)
[2]K. W. Rosenmund and F. Zetzsche, *Ber.*, **54**, 436 (1921)
[3]E. B. Hershberg and J. Cason, *Org. Syn., Coll. Vol.*, **3**, 627 (1955)
[4]D. M. Burness, *ibid.*, **4**, 628 (1963)
[5]*Org. Expts.*, 226–228
[6]R. H. Wiley and N. R. Smith, *Org. Syn., Coll. Vol.*, **4**, 731 (1963)
[7]C. F. Allen and M. J. Kalm, *ibid.*, **4**, 608 (1963)
[8]E. E. Schweizer and W. E. Parham, *Am. Soc.*, **82**, 4085 (1960)
[9]W. E. Parham, R. W. Soeder, and R. M. Dodson, *Am. Soc.*, **84**, 1755 (1962)
[10]D. G. Lindsay and C. B. Reese, *Tetrahedron*, **21**, 1673 (1965)

Quinuclidine, 

Mol. wt. 111.18, m.p. 158–159°, sublimes, pKb 10.95, very soluble in water and in organic solvents. Preparation.[1] For complexes with organolithium compounds *see* Triethylenediamine.

[1]G. R. Clemo and T. P. Metcalfe, *J. Chem. Soc.*, 1989 (1937); V. Prelog *et al.*, *Ann.*, **532**, 69 (1937); **535**, 37 (1938)

R

Raney alloy, *see* Nickel–aluminum alloy.

Raney catalyst, *see* Nickel catalysts (a), Raney type.

Raney cobalt catalyst. The catalyst is prepared by the action of sodium hydroxide on a 40:60 cobalt–aluminum alloy available from Raney Catalyst Co. The procedure is that used by Billica and Adkins[1] for the preparation of W-7 Raney nickel.

For the hydrogenation of simple oximes, Raney cobalt is less effective than Raney nickel.[2] The 2-quinolylpyruvic acid ester oxime (1) absorbs two moles of hydrogen in the presence of Raney nickel and gives the quinolylalanine ester (2),

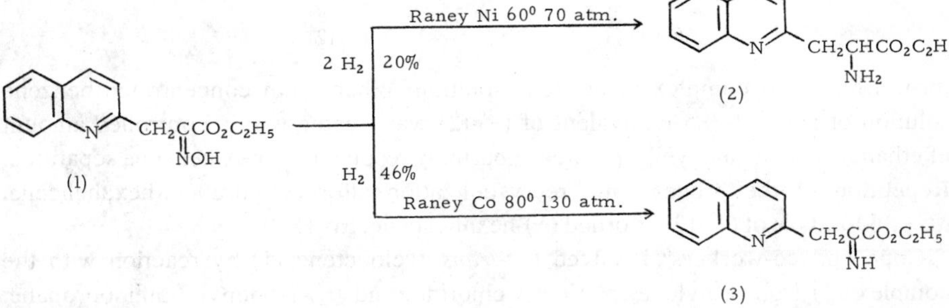
Raney Ni 60° 70 atm.
2 H2 20%
H2 46%
Raney Co 80° 130 atm.
CH2CCO2C2H5
NOH
(1)
CH2CHCO2C2H5
NH2
(2)
CH2CCO2C2H5
NH
(3)

but in the presence of Raney cobalt the reaction stops at the stage of the α-iminopropionic ester (3).[3] Raney cobalt is also less reactive than Raney nickel in desulfurizations[4] and in the hydrogenation of nitriles.[5]

[1]H. R. Billica and H. Adkins, *Org. Syn., Coll. Vol.*, **3**, 176 (1955)
[2]W. Reeve and J. Christian, *Am. Soc.*, **78**, 860 (1956)
[3]W. Ried and H. Schiller, *Ber.*, **86**, 730 (1953)
[4]G. M. Badger, N. Kowanko, and W. H. F. Sasse, *J. Chem. Soc.*, 440 (1959)
[5]F. E. Gould, G. S. Johnson, and A. F. Ferris, *J. Org.*, **25**, 1658 (1960)

Reinecke salt, $NH_4[Cr(NH_3)_2(SCN)_4]\cdot H_2O$. Mol. wt. 354.47. Preparation.[1] Suppliers: B, E, F, MCB.

The reagent is of value as a precipitant for primary and secondary amines, proline, hydroxyproline, and certain other amino acids.

[1]H. D. Dakin, *Org. Syn., Coll. Vol.*, **2**, 555 (1943)

Resolving agents, *see also* 3β-Acetoxy-Δ^5-etianic acid; L-(+)-2,3-Butanedithiol; Dehydroabietylamine. Velluz[1] and Eliel[2] survey methods and list agents and procedures for their preparation. Additional methods and agents are listed below.

In a novel synthesis of L-lysine starting with caprolactam, Brenner and Rickenbacher[3] resolved the intermediate DL-α-aminocaprolactam with L-pyrrolidone-

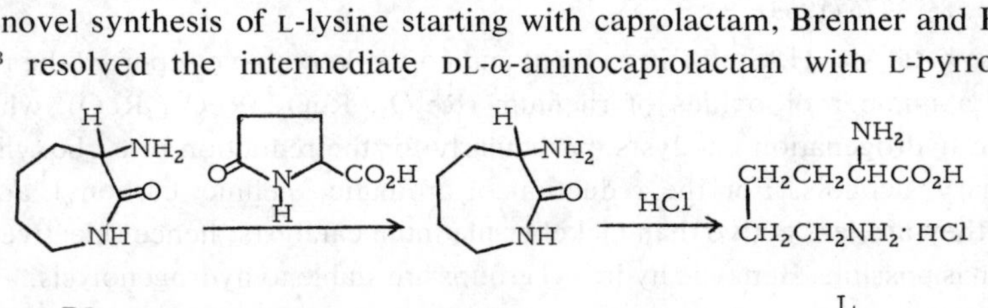
H
NH2
NH
CO2H
H
HCl
CH2CH2CHCO2H
CH2CH2NH2·HCl
DL
L
L

carboxylic acid, prepared from L-glutamic acid. By an ingenious scheme of crystallization, splitting with hydrochloric acid, racemization, and recycling, the DL-α-aminocaprolactam was converted into about equal parts of L-lysine hydrochloride and DL-α-aminocaprolactam, with recovery of the resolving agent.

Newman and Lednicer[4] resolved hexahelicene (1) with an optically active π-acid complexing agent (2) prepared from 2,4,5,7-tetranitrofluorenone and an optically active hydroxylamine. *dl*-Hexihelicene (1) and the π-acid (2), both pale yellow,

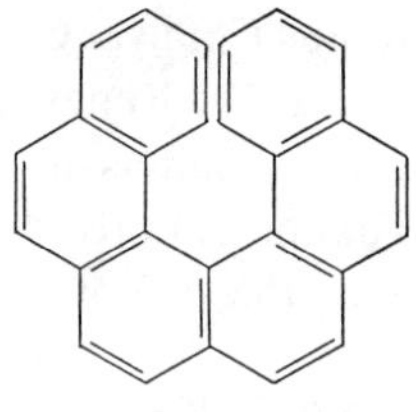

(1)

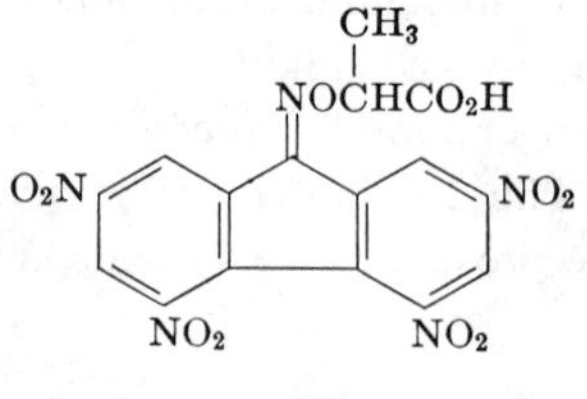

(2)

form a deep red complex in benzene solution. When a hot concentrated benzene solution of (1) with 0.5 equivalent of (−)-(2) was treated with a controlled amount of ethanol and washed, yellow plates of partially resolved (−)-hexihelicene separated. Repetition of the process and recrystallization afforded pure (−)-hexahelicene, αD −3640°. Use of (+)-(2) afforded (+)-hexahelicene, αD +3707°.

Cope and co-workers[5] resolved (±)-*trans*-cyclooctene (1) by reaction with the complex (2) from ethylene, platinous chloride, and (+)-1-phenyl-2-aminopropane.

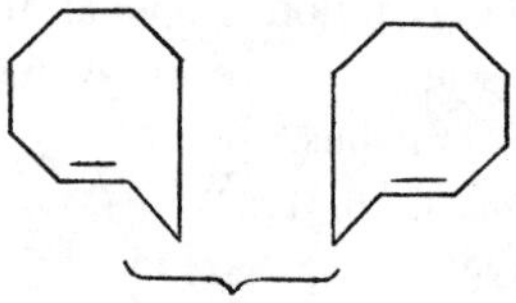

(1)

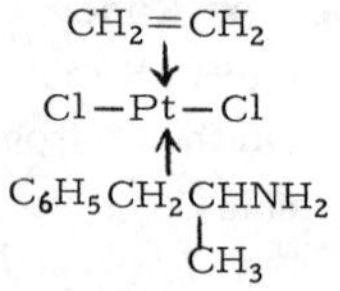

(2)

Displacement of ethylene in the complex by the cycloalkene gave a mixture of diastereoisomeric complexes separable by crystallization at −20°. Liberation of the (−)-hydrocarbon from the complex with potassium cyanide gave a product of αD −411°.

[1]L. Velluz, *Substances Naturelle de Synthèse*, **9**, 119–174 (1954)
[2]E. L. Eliel, "Stereochemistry of Carbon Compounds," Chapter 4, McGraw-Hill Book Co., New York (1962)
[3]M. Brenner and H. R. Rickenbacher, *Helv.*, **41**, 181 (1958)
[4]M. S. Newman and D. Lednicer, *Am. Soc.*, **78**, 4765 (1956)
[5]A. C. Cope, C. R. Ganellin, H. W. Johnson, Jr., T. V. Van Auken, and H. J. S. Winkler, *Am. Soc.*, **85**, 3276 (1963)

Rhenium catalysts. H. Smith Broadbent and co-workers[1] have reported the preparation of a number of oxides of rhenium (Re_2O_7, ReO_2, ReO_3, ReO), which are effective hydrogenation catalysts, particularly for the reduction of carboxylic acids to primary alcohols. For the reduction of aromatic, olefinic, carbonyl, and nitro groups they are less active than nickel or platinum catalysts; hence selective hydrogenation is possible. Benzylic hydroxyl groups are stable to hydrogenolysis.

[1]H. S. Broadbent *et al.*, *J. Org.*, **24**, 1847 (1959); **27**, 4400 (1962); **28** 2343, 2345, 2347 (1963)

Rhenium heptaselenide, Re_2Se_7. Broadbent and Whittle[1] describe the preparation of this hydrogenation catalyst and report that it is more reactive in the reduction of C=O than of C=C (unless conjugated). It does not effect hydrogenolysis of the C—S link.

[1]H. S. Broadbent and C. W. Whittle, *Am. Soc.*, **81**, 3587 (1959)

Rhodium (5%) on alumina, $Rh–Al_2O_3$. Supplier: Engelhard Industries.

Hydrogenation of phenols. The catalyst has the advantage of promoting hydrogenation of phenols with minimal hydrogenolysis. Burgstahler and Bithos[1] hydrogenated gallic acid under the conditions indicated and isolated all-*cis*-hexahydrogallic

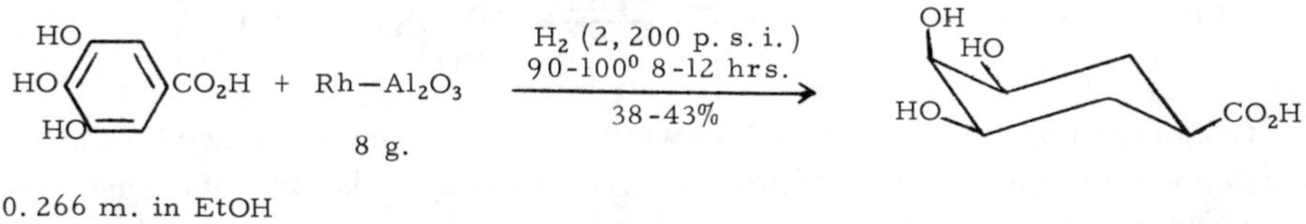
HO
HO
HO
CO2H + Rh–Al2O3
8 g.
H2 (2, 200 p. s. i.)
90-100° 8-12 hrs.
38-43%
OH
HO
HO
CO2H
0. 266 m. in EtOH

acid as the only crystalline product. The yield was over twice the overall yield (13–19%) of a previous two-step process in which gallic acid was hydrogenated with Raney nickel in basic solution to a dihydro intermediate which was then hydrogenated with a platinum catalyst. Smith and Stump[2] and Kaye and Matthews[3] also obtained better results in the hydrogenation of mono and polyprotic phenols with $Rh–Al_2O_3$ than with other catalysts. Meyers *et al.*[4] found the catalyst distinctly

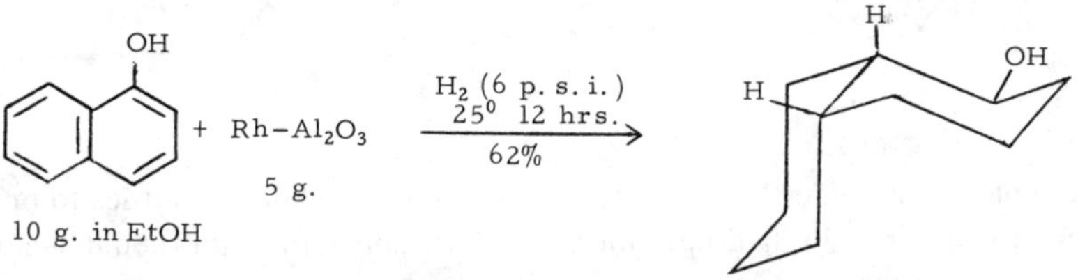
OH
+ Rh–Al2O3
5 g.
10 g. in EtOH
H2 (6 p. s. i.)
25° 12 hrs.
62%
H
H
OH

superior to Raney nickel or platinum for conversion of α-naphthol to *cis,cis*-1-decalol. Hydrogenation in ethanol under very mild conditions and two crystallizations afforded the pure *cis,cis*-alcohol. Gas chromatography indicated that hydrogenolysis amounted to only 3%.

Low-temperature hydrogenation of resorcinol over rhodium–alumina in alkaline solution affords a convenient procedure for the preparation of 1,3-cyclohexanedione.[5]

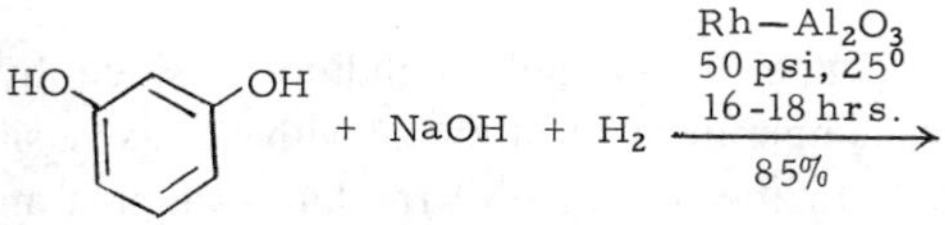
HO
OH
+ NaOH + H2
Rh–Al2O3
50 psi, 25°
16-18 hrs.
85%
O
O

Hydrogenation of benzyl alcohols. Stocker[6] found that hydrogenolysis is completely avoided by use of rhodium on alumina. Under very mild conditions,

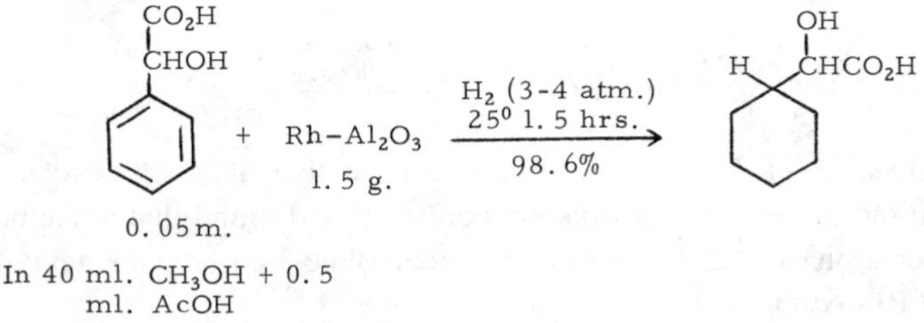
CO2H
CHOH
+ Rh–Al2O3
1. 5 g.
0. 05 m.
In 40 ml. CH3OH + 0. 5 ml. AcOH
H2 (3-4 atm.)
25° 1. 5 hrs.
98. 6%
OH
H CHCO2H

(+)-mandelic acid afforded (±)-hexahydromandelic acid and D-mandelic acid gave pure D-hexahydromandelic acid. Note that the procedure calls for use of a small quantity of acetic acid. In repeating Stocker's work with acetic acid present, Rylander[7] found the reduction to proceed exactly as reported; when the acetic acid was left out there was no reduction at all!

Hydrogenation of C=C, C=NOH, C≡N. Ham and Coker[8] found rhodium–alumina the catalyst of choice for hydrogenation of vinylic and allylic halides with minimal hydrogenolysis, for example:

$$\underset{0.334\text{ m.}}{ClCH{=}CHCH_2Cl} + \underset{2.4\text{ g.}}{Rh{-}Al_2O_3} \xrightarrow[100^0\ 15\text{ min.}]{H_2\,(500\text{ p. s. i.})} \underset{48\%}{Cl(CH_2)_3Cl} + \underset{34\%}{CH_3CH_2CH_2Cl}$$

Cycloheptanone oxime can be converted into the amine in good yield by high-pressure hydrogenation of the oxime with Raney nickel and ammonia,[9] but Freifelder[10] found that the reaction is so strongly exothermal that on a large scale it may get out of hand. He then found that the hydrogenation can be accomplished smoothly by low-pressure hydrogenation in the presence of rhodium–alumina. The temperature was allowed to rise spontaneously to 60° and kept there until reduction

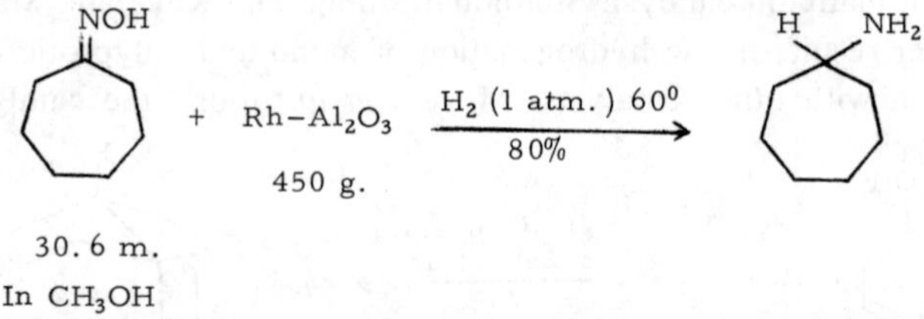
NOH
+ Rh–Al2O3
450 g.
H2 (1 atm.) 60°
80%
H NH2
30.6 m.
In CH3OH

was complete. Freifelder[11] effected the reduction of 16 aliphatic nitriles to primary amines in good yield by hydrogenation in a Parr apparatus with rhodium–alumina at room temperature and low pressure. Of particular interest in the reduction of

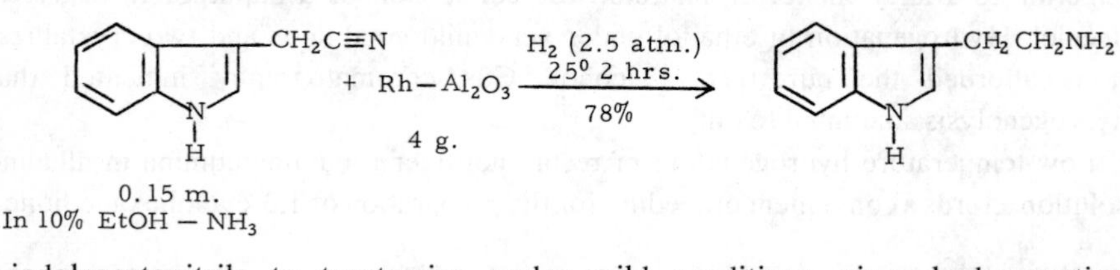
CH2C≡N
N
H
+ Rh–Al2O3
4 g.
H2 (2.5 atm.)
25° 2 hrs.
78%
CH2 CH2NH2
0.15 m.
In 10% EtOH – NH3

3-indoleacetonitrile to tryptamine under mild conditions, since hydrogenation with other catalysts at higher temperatures is attended with extensive side reactions.

Hydrogenolysis of ketals.[12] In the presence of rhodium–alumina and a trace of hydrochloric acid, a ketal can be reductively cleaved to an ether and an alcohol. Platinum and ruthenium were almost inactive, and palladium was about half as active as rhodium.

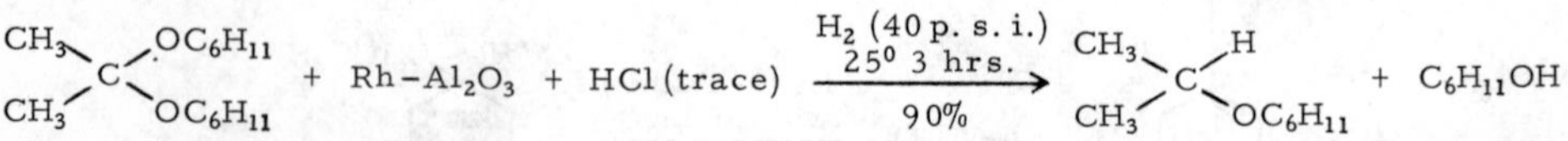
CH3
CH3
OC6H11
OC6H11
+ Rh–Al2O3 + HCl (trace)
H2 (40 p. s. i.)
25° 3 hrs.
90%
H
+ C6H11OH

Hydrogenation of heterocycles. For a prospective alkaloid synthesis, Rapala *et al.*[13] required the *cis*-decahydroisoquinoline (2) and found that it can be obtained by hydrogenation of 1,2,3,4-tetrahydroisoquinoline-3-carboxylic acid (1) in the presence of $Rh{-}Al_2O_3$.

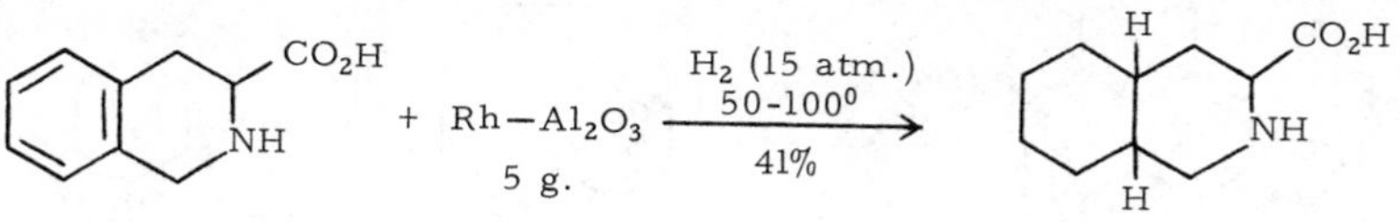
CO2H
NH
+ Rh–Al2O3
5 g.
H2 (15 atm.)
50-100°
41%
H
CO2H
NH
H

20 g. in 5% CH_3OH

(1) (2)

Freifelder[14] found that pyridinecarboxylic acids can be hydrogenated smoothly over Rh–Al_2O_3 in ammonia solution. A particularly significant example is the hydrogenation of nicotinic acid to nipecotic acid, since this had not previously been accomplished without decarboxylation.

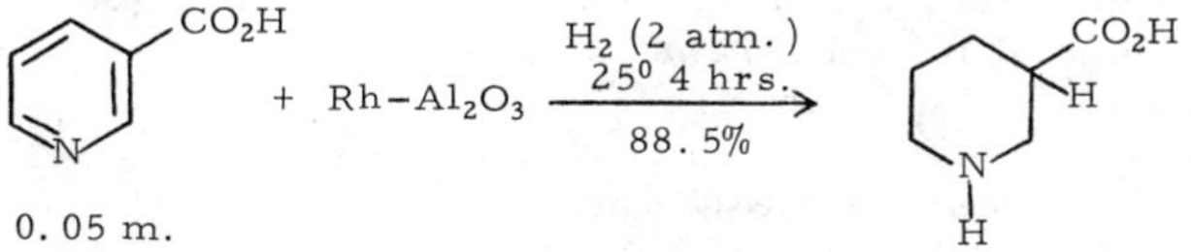
CO2H
N
+ Rh–Al2O3
H2 (2 atm.)
25° 4 hrs.
88.5%
CO2H
H
N
H

0.05 m.

In 50 ml. H_2O + 5 ml. aq. NH_3

In synthetic work in the Senecio alkaloid series, Adams and co-workers[15] prepared a 1-hydroxypyrrolizidine of the desired stereochemistry (2) by Rh–Al_2O_3 hydrogenation of (1), a reaction involving complete saturation of the dienone system.

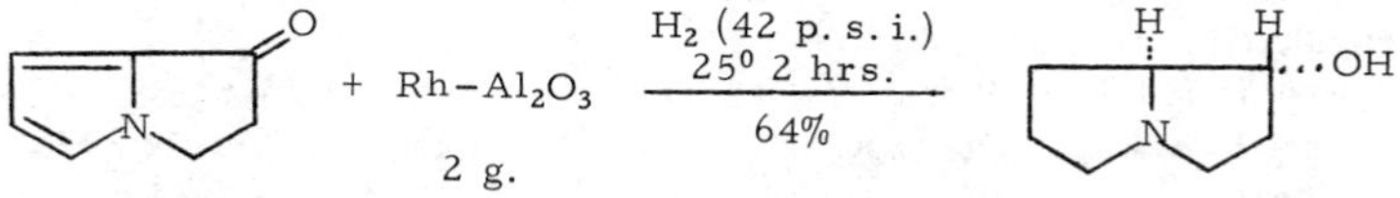
O
N
+ Rh–Al2O3
2 g.
H2 (42 p.s.i.)
25° 2 hrs.
64%
H
H
...OH
N

(1) 14 g. in AcOH (2)

Only one stereoisomer was detected. Other investigators had attempted without success to reduce (1) by hydrogenation with platinum oxide or with palladized charcoal.

The catalyst has been used successfully for the complete hydrogenation of quinoxaline[16] and for reduction of the 4,5-double bond of pyrimidine nucleosides and nucleotides.[17]

Dehydrogenation. Newman and Lednicer[18] found rhodium–alumina to be an effective catalyst for the transfer of hydrogen from hexahydrohexahelicene (1) to benzene to produce hexahelicene (2). Similar exchange over palladium catalyst

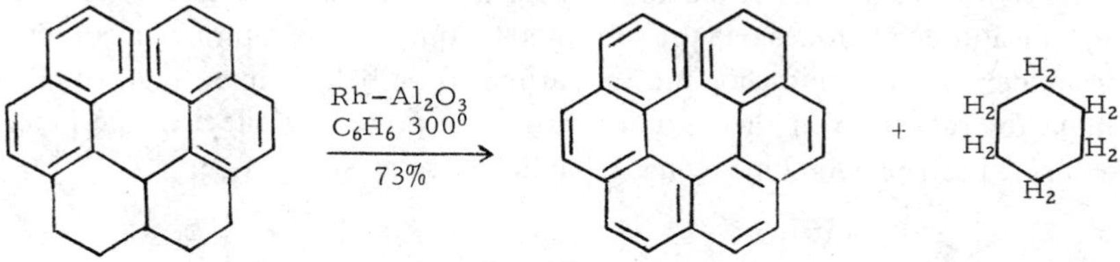
Rh–Al2O3
C6H6 300°
73%
+
H2
H2
H2
H2
H2
H2

(1) (2)

proceeded in poor yield. Refluxing of hexahydrohexahelicene with 2,3-dichloro-5,6-dicyano-1,4-benzoquinone in xylene for 7 hrs. gave a dihydro derivative, which could not be dehydrogenated further. Hydrogen exchange with benzene over Rh–Al_2O_3 is also the method of choice for the dehydrogenation of (3) to pyrocene (4). Anderson and Anderson[19] placed a sealed tube containing 0.24 g. of (3), 0.1 g. of catalyst, and 15 ml. of benzene in a rocker-type autoclave packed with glass wool and charged with benzene and agitated the mixture at 290°.

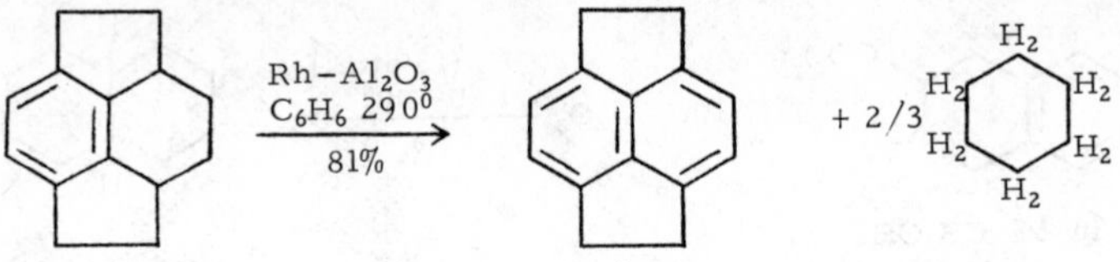
Rh−Al₂O₃
C₆H₆ 290°
81%
+ 2/3
(3)
(4)

Hydrogenation of aryl amines. Rhodium–alumina appears to be the catalyst of choice for the low-temperature hydrogenation of anilines, particularly alkoxy derivatives, to cyclohexylamines.[20] Hydrogenolysis is not extensive, and usually secondary amines are formed in only small amounts.

[1]A. W. Burgstahler and Z. J. Bithos, *Am. Soc.*, **82**, 5466 (1960); *Org. Syn.*, **42**, 62 (1962)
[2]H. A. Smith and B. L. Stump, *Am. Soc.*, **83**, 2739 (1961)
[3]I. A. Kaye and R. S. Matthews, *J. Org.*, **28**, 325 (1963)
[4]A. I. Meyers, W. N. Beverung, and G. Garcia-Munoz, *J. Org.*, **29**, 3427 (1964); A. I. Meyers and W. N. Beverung, procedure submitted to *Org. Syn.*
[5]J. C. Sircar and A. I. Meyers, *J. Org.*, **30**, 3206 (1965)
[6]J. H. Stocker, *J. Org.*, **27**, 2288 (1962)
[7]P. N. Rylander (Engelhard Industries), private communication
[8]G. E. Ham and W. P. Coker, *J. Org.*, **29**, 194 (1964)
[9]A. C. Cope, R. A. Pike, and C. F. Spencer, *Am. Soc.*, **75**, 3212 (1953)
[10]M. Freifelder, W. D. Smart, and G. R. Stone, *J. Org.*, **27**, 2209 (1962)
[11]M. Freifelder, *Am. Soc.*, **82**, 2386 (1960)
[12]W. L. Howard and J. H. Brown, Jr., *J. Org.*, **26**, 1026 (1961)
[13]R. T. Rapala, E. R. Lavagnino, E. R. Shepard, and E. Farkas, *Am. Soc.*, **79**, 3770 (1957)
[14]M. Freifelder, *J. Org.*, **27**, 4046 (1962); **28**, 602, 1135 (1963)
[15]R. Adams, S. Miyano, and D. Fleš, *Am. Soc.*, **82**, 1466 (1960)
[16]H. S. Broadbent *et al.*, *Am. Soc.*, **82**, 189 (1960)
[17]W. E. Cohn and D. G. Doherty, *Am. Soc.*, **78**, 2863 (1956)
[18]M. S. Newman and D. Lednicer, *Am. Soc.*, **78**, 4765 (1956)
[19]A. G. Anderson, Jr., and R. G. Anderson, *J. Org.*, **22**, 1197 (1957)
[20]M. Freifelder, Y. H. Ng, and P. F. Helgren, *J. Org.*, **30**, 2485 (1965)

Rhodium (5%) on carbon (Norit), Rh–C. Supplier: Engelhard Industries.

Breitner *et al.*[1] of Engelhard Industries report comparisons of the activities of commercial preparations of Rh–C, Ru–C, Pt–C, and Pd–C (all on Norit). Rhodium–C and ruthenium–C appear to be the catalysts of choice for the hydrogenation of ketones in a neutral or basic medium. These catalysts can be used for the reduction of α,β-unsaturated ketones, via the saturated ketones, to the saturated alcohols.

Freifelder *et al.*[2] found rhodium on carbon to be better suited than rhodium on alumina for reduction of the pyridine ring. The poisoning effect of the piperidine base formed can be overcome by use of sufficient catalyst.

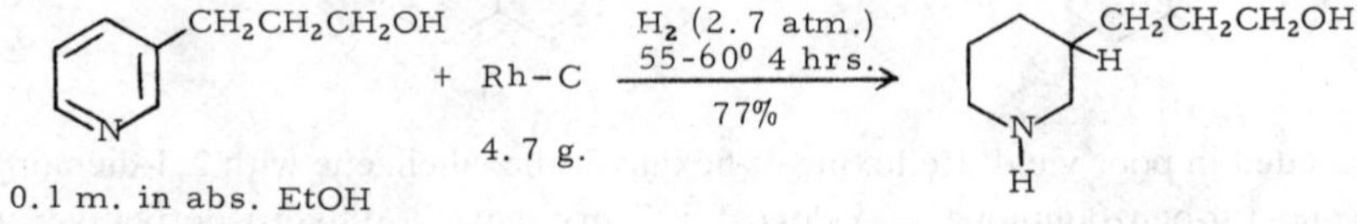
CH₂CH₂CH₂OH
+ Rh−C
4.7 g.
0.1 m. in abs. EtOH
H₂ (2.7 atm.)
55-60° 4 hrs.
77%
CH₂CH₂CH₂OH
H

Schweizer and Light[3] found 5% rhodium on carbon satisfactory for the stepwise hydrogenation of 3 H-pyrrolizine (1).

Refluxing norbornadiene with 5% rhodium on carbon converts the hydrocarbon in nearly quantitative yield into a product of which 70–80% is a mixture of dimers and 20–30% is a trimer.[4]

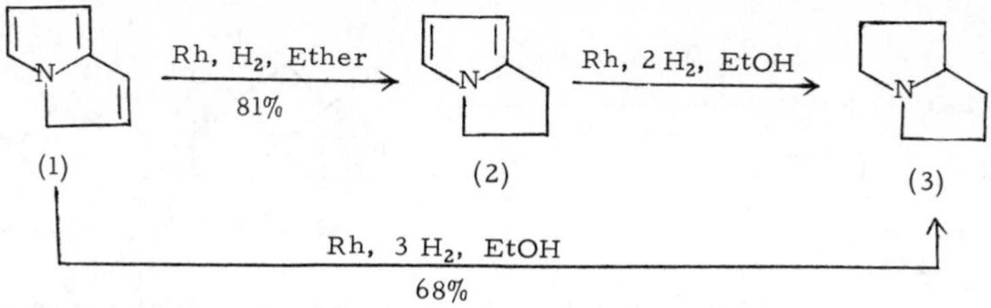

[1]E. Breitner, E. Roginski, and P. N. Rylander, *J. Org.*, **24**, 1855 (1959)
[2]M. Freifelder, R. M. Robinson, and G. R. Stone, *J. Org.*, **27**, 284 (1962)
[3]E. E. Schweizer and K. K. Light, *Am. Soc.*, **86**, 2963 (1964)
[4]J. J. Mrowca and T. J. Katz, *Am. Soc.*, **88**, 4012 (1966)

Rochelle salt (Sodium potassium tartrate tetrahydrate).
$NaO_2CCH(OH)CH(OH)CO_2K \cdot 4H_2O$. Mol. wt. 282.23. Solubility in 100 g. H_2O:26 g. at 0°, 66 g. at 26°.

The tartrate ion forms a complex with aluminum ion and keeps it in solution. After Oppenauer oxidation of cholesterol in refluxing toluene and removal of most

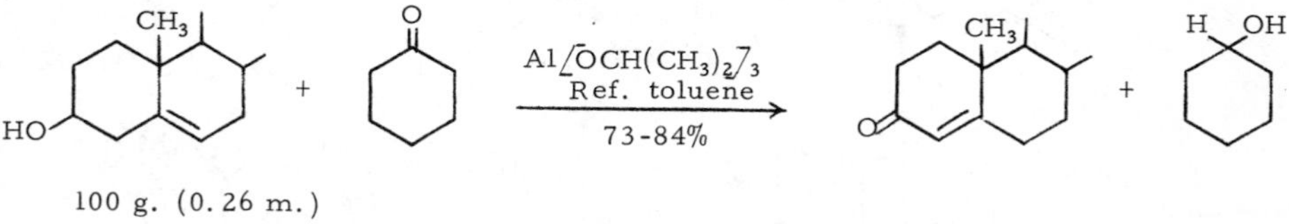

of the volatile material, Eastham and Teranishi[1] added 400 ml. of a saturated aqueous solution of Rochelle salt to the murky reaction mixture with stirring until the organic layer became clear and orange. The mixture was then steam distilled, and the residue cooled and extracted with chloroform.

After reduction of cholestenone with lithium tri-*t*-butoxyaluminum hydride, Burgstahler and Nordin[2] hydrolyzed the organometallic complex by treatment with ice, sodium hydroxide, and Rochelle salt.

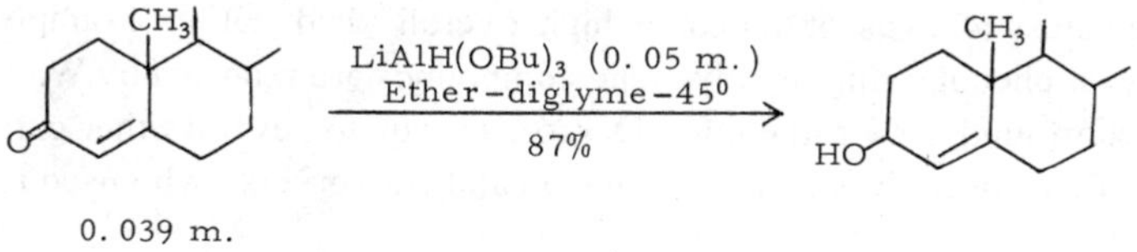

[1]J. F. Eastham and R. Teranishi, *Org. Syn.*, *Coll. Vol.*, **4**, 192 (1963)
[2]A. W. Burgstahler and I. C. Nordin, *Am. Soc.*, **83**, 198 (see p. 205) (1961)

Rubidium fluoride, RbF. Supplier: Amer. Potash and Chem. Corp. Catalyst for the Knoevenagel reaction, *see* Cesium fluoride.

Ruthenium hydrogenation catalyst. A 5% ruthenium–carbon (Norit) catalyst and a 5% Ru–Al_2O_3 catalyst are supplied by Engelhard Industries. Some investigators employ ruthenium dioxide, which is reduced to the metal *in situ*.

Carbonyl groups. Breitner *et al.*[1] found Engelhard Ru-C and Rh-C distinctly superior to Pt-C and Pd-C for the hydrogenation of ketones in neutral or basic medium. Hasek *et al.*[2] attempted reduction of tetramethyl-1,3-cyclobutanedione with platinum, palladium, and rhodium catalysts but the results were very poor. With copper–chromium oxide and supported nickel catalysts yields of diols were moderate,

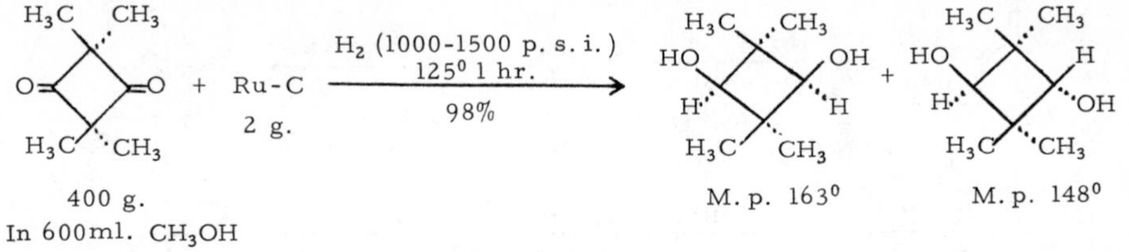
H₂ (1000-1500 p.s.i.)
125° 1 hr.
98%
Ru-C
2 g.
400 g.
In 600ml. CH₃OH
M.p. 163°
M.p. 148°

and the product contaminated with low-melting by-products. With commercial 5% ruthenium–C results were outstanding. Hydrogenation in methanol proceeded rapidly to give in nearly quantitative yield a mixture of about equal parts of the *cis*- and *trans*-diol.

Aromatic rings. Ireland and Schiess[3] effected efficient reduction of the phenolic ring of (1, from podocarpic acid) by hydrogenation with added ruthenium dioxide

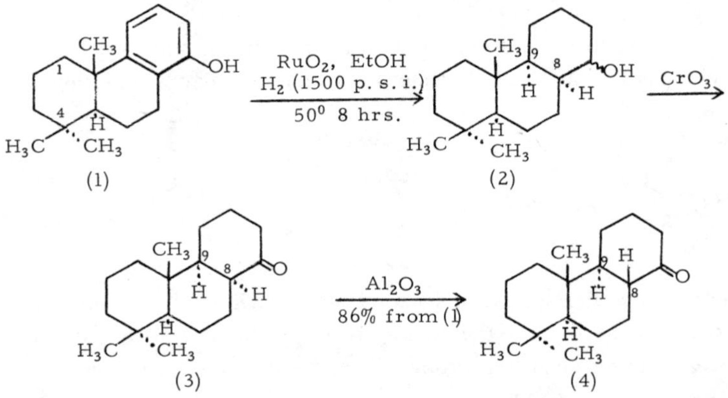
RuO₂, EtOH
H₂ (1500 p.s.i.)
50° 8 hrs.
CrO₃
Al₂O₃
86% from (1)
(1)
(2)
(3)
(4)

in ethanol under mild conditions. The product was a mixture of stereoisomers which appeared to be derived largely from *cis*-rear attack of the 8,9-double bond, for Jones CrO_3-oxidation gave a labile ketone (3) which was epimerized at C_8 on alumina to give the stable ketone (4). By carrying out the reaction sequence without isolation of intermediates, (4) was obtained in high overall yield. Other examples of the reduction of a phenolic ring in polycyclic compounds are reported by Walton *et al.*[4] (Merck), who found 10% ruthenium–Darco superior to several other catalysts for avoidance of hydrogenolysis, and by Johnson and co-workers,[5] who used ruthenium dioxide.

Rapala and Farkas[6] found that ruthenium hydrogenation of estrone, estradiol, and related steroids (1) gives rise to 3β-hydroxy-5α,10α-estranes (2) as the major

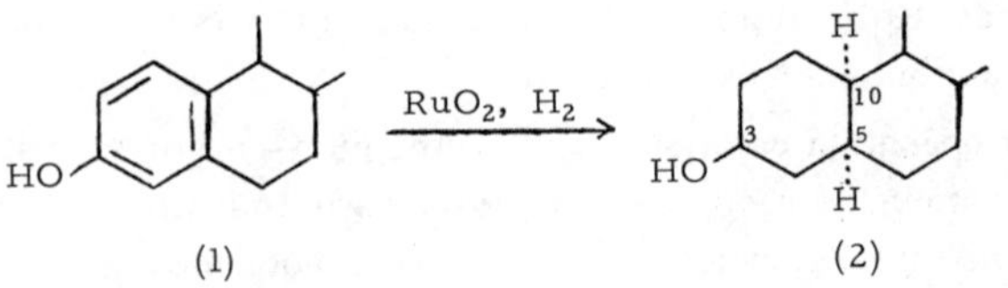
RuO₂, H₂
HO
(1)
(2)

products. Since the reaction probably proceeds through a 3-keto intermediate, the stereochemistry of the product at all three new centers of asymmetry follows the pattern of rear attack. In confirmatory work, Counsell[7] later hydrogenated estradiol in alkaline solution in the presence of RuO_2 at 65° and at a hydrogen pressure of 1475 p.s.i. and isolated the 19-norsteroid of type (2) in yield of 78%.

Rapala and Farkas[8] found an entry into another series of 19-norsteroids in ruthenium hydrogenation of 19-nortestosterone (3), followed by oxidation with N-bromoacetamide to give 5β,10β-estrane-3,17-dione (4) as the major product.

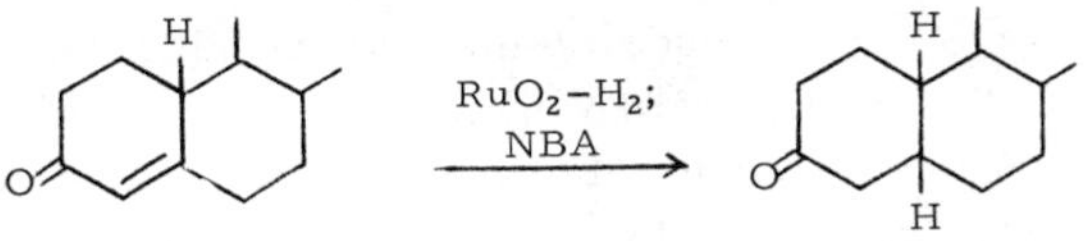
H
RuO_2–H_2;
NBA
H
H

(3) (4)

Kolobielski[9] investigated the hydrogenation of the *endo*-adduct (5) of maleic anhydride and anthracene at high pressure and moderate temperature and found that the reduction stopped sharply after reduction of one of the two benzene rings.

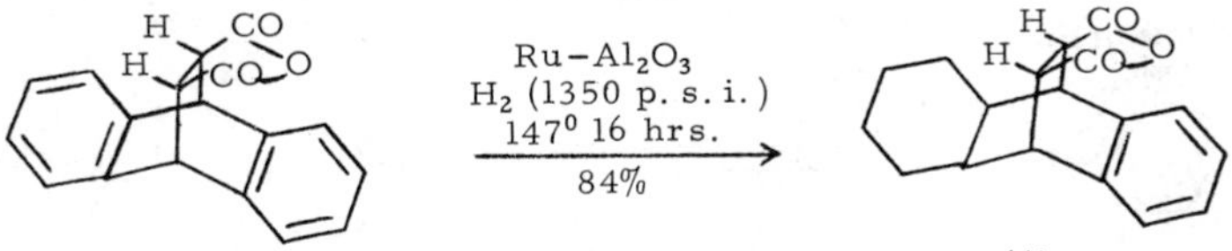
H CO
H CO O
Ru–Al_2O_3
H_2 (1350 p. s. i.)
147° 16 hrs.
84%
H CO
H CO O

(5) (6)

That the second ring is not hydrogenated probably is because the adduct molecule is bent and only one ring can lie flush on the catalyst surface.

Johnson *et al.*[10] prepared *trans*-decalin-1,4-dione by hydrogenation of 1,4-naphthoquinone over RuO_2, followed by chromic acid oxidation of the resulting diol and isomerization of the *cis*- to the *trans*-dione.

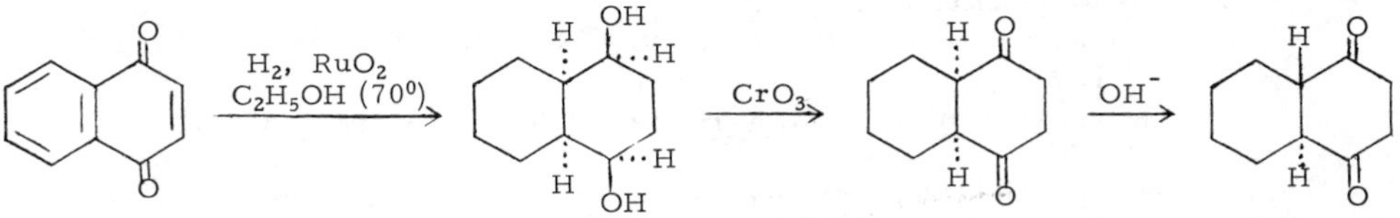
H_2, RuO_2
C_2H_5OH (70°)
OH
H
H
H
H
OH
CrO_3
H
H
OH^-
H
H

Freifelder and Stone[11] found that phenylalkylamines can be reduced to cyclohexylalkylamines in excellent yield by ruthenium hydrogenation. Under similar conditions, pyridines are reduced efficiently to the corresponding piperidines.[12]

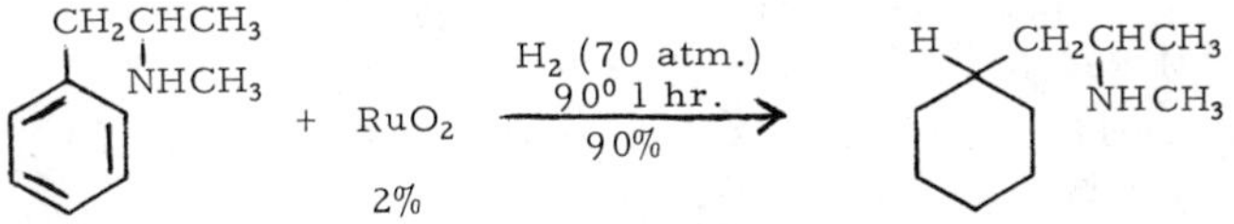
CH_2CHCH_3
$NHCH_3$
+ RuO_2
2%
H_2 (70 atm.)
90° 1 hr.
90%
H
CH_2CHCH_3
$NHCH_3$

Reduction of C=C and C≡C. Berkowitz and Rylander[13] report that hydrogenation in the presence of 5% ruthenium on Norit selectively reduces monosubstituted olefins in the presence of di- and trisubstituted olefins. Thus, in mixtures, (1) is

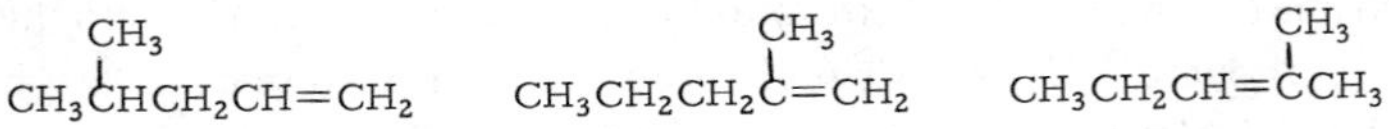
CH_3
$CH_3CHCH_2CH{=}CH_2$
CH_3
$CH_3CH_2CH_2C{=}CH_2$
CH_3
$CH_3CH_2CH{=}CCH_3$

(1) (2) (3)

reduced in preference to either (2) or (3). Selectivity unique to ruthenium is evident also from the reduction of octene-1 in the presence of octene-2. The hydrogenations were conducted with a dispersion of the olefin in water. Acetylenes were found to be hydrogenated readily to alkanes but not convertible by selective semihydrogenation into olefins.

In contrast to the behavior of usual catalysts, hydrogenations over ruthenium are very sensitive to the solvent, and often no reduction occurs unless water is present.[14] This is true for both low- and high-pressure hydrogenations and even if the substrate is water-insoluble.

Although ruthenium is the most active known catalyst for hydrogenation of the carbonyl group, it is possible to reduce α,β-unsaturated aldehydes to saturated aldehydes quantitatively.[15] The Engelhard workers, however, regard palladium as the catalyst of choice for this conversion. For hydrogenation to the saturated alcohol, they prefer a two-step process: Pd-reduction of the double bond, followed by Ru-reduction of the aldehyde group.

$ArNO_2 \longrightarrow ArNHNHAr$, Aromatic nitro compounds are conveniently reduced to hydrazo compounds by hydrazine and 5% Ru–C in 5% alcoholic potassium hydroxide.[16] A 5% Pt–C catalyst can be used, but precautions are required to prevent further reduction to amines.

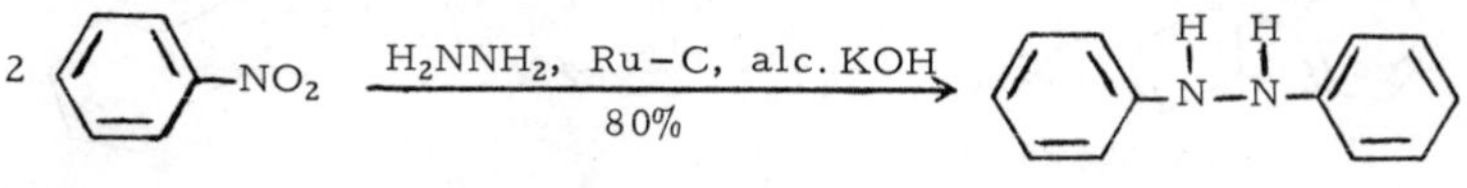
2
NO2
H2NNH2, Ru–C, alc. KOH
80%
H H
N–N

[1]E. Breitner, E. Roginski, and P. N. Rylander, *J. Org.*, **24**, 1855 (1959)
[2]R. H. Hasek, E. U. Elam, J. C. Martin, and R. G. Nations, *J. Org.*, **26**, 700 (1961)
[3]R. E. Ireland and P. W. Schiess, *J. Org.*, **28**, 6 (1963)
[4]E. Walton *et al.*, *Am. Soc.*, **78**, 4760 (1956)
[5]W. S. Johnson, E. R. Rogier, and J. Ackerman, *Am. Soc.*, **78**, 6322 (1956)
[6]R. T. Rapala and E. Farkas, *J. Org.*, **23**, 1404 (1958)
[7]R. E. Counsell, *Tetrahedron*, **15**, 202 (1961)
[8]R. T. Rapala and E. Farkas, *Am. Soc.*, **80**, 1008 (1958)
[9]M. Kolobielski, *J. Org.*, **28**, 1883 (1963)
[10]W. S. Johnson, D. S. Allen, Jr., R. R. Hindersinn, G. N. Sausen, and R. Pappo, *Am. Soc.*, **84**, 2181 (1962)
[11]M. Freifelder and G. R. Stone, *Am. Soc.*, **80**, 5270 (1958); *J. Org.*, **27**, 3568 (1962)
[12]*Idem*, *J. Org.*, **26**, 3805 (1961)
[13]L. M. Berkowitz and P. N. Rylander, *J. Org.*, **24**, 708 (1959)
[14]P. N. Rylander, N. Rakoneza, D. Steele, and M. Bollinger, *Engelhard Ind. Techn. Bull.*, **4**, 95 (1963)
[15]P. N. Rylander, N. Himelstein, and M. Kilroy, *ibid.*, **4**, 49 (1963)
[16]S. Pietra and M. Res, *Ann. chim. (Rome)*, **48**, 299 (1958)

Ruthenium tetroxide, RuO_4. Mol. wt. 165.10, yellow needles, m.p. 25.5°; the yellow-red solution in carbon tetrachloride is stable.

Preparation. Berkowitz and Rylander[1] prepared the reagent by oxidation of ruthenium trichloride in dilute hydrochloric acid with aqueous sodium bromate solution in a rather elaborate apparatus. Later workers[2] found it desirable to remove bromine from this material by extraction of a solution in carbon tetrachloride with aqueous bicarbonate solution. A simple preparative procedure described by Nakata[3] starts with black ruthenium dioxide, supplied by Engelhard Industries. A suspension of 0.4 g. of RuO_2 in 50 ml. of carbon tetrachloride was stirred at 0° and treated with a solution of 3.2 g. of sodium metaperiodate in 50 ml. of water. The black oxide dissolves in about 1 hr. The clear carbon tetrachloride layer is separated and filtered through glass wool. A rough analysis can be made by treating an aliquot with ethanol to reduce the tetroxide to the black dioxide, which is collected and weighed.[4]

Solvents. Djerassi and Engle,[4] in an exploratory comparison of the reagent with osmium tetroxide, found that ether, benzene, and pyridine, usual solvents for OsO_4, are in this case completely useless since they react instantly and violently with

RuO_4. They found carbon tetrachloride and chloroform to be the only satisfactory common solvents, and most workers have used carbon tetrachloride. Corey *et al.*[5] used Freon 11 (CCl_3F, b.p. 22°) for oxidation of norborneol-*d* with RuO_4 on a microscale to norcamphor-*d*.

Olefins, *see also* Catalytic method (p. 989). Djerassi and Engle[4] report one experiment on the oxidation of phenanthrene with ruthenium tetroxide in carbon tetrachloride; the reaction mixture contained considerable starting material and a small amount of phenanthrenequinone. Berkowitz and Rylander[1] oxidized cyclohexene in the same way and obtained adipaldehyde in low yield as the only isolated product; they regard the method as unsatisfactory for the production of aldehydes and acids because these substances are strongly adsorbed on the ruthenium dioxide formed.

Dean and Knight[2] found the reaction useful for oxidation of 3-alkylidene-2′-grisenes (1) to grisene-3-ones (2). Yields were only 5–30%, but ozonolysis and chromic acid oxidation failed completely.

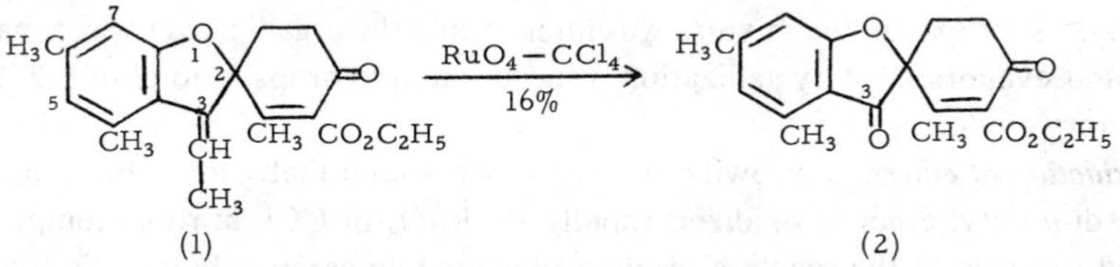
H₃C
7
1
2
3
5
O
CH₃
CH
CH₃
CO₂C₂H₅
RuO₄—CCl₄
16%
(1)
(2)

Castells and co-workers[6] developed a method for determining the position of an isolated double bond in a steroid consisting in refluxing with osmium tetroxide in ether, reducing the osmate ester with lithium aluminum hydride in THF, cleaving the diol with lead tetraacetate, and characterizing the dicarbonyl compound by infrared analysis. A disubstituted double bond (Δ^1-cholestene) generates two aldehyde groups, with bands at 2704 cm^{-1} (weak) and 1730 cm^{-1} (strong). A trisubstituted bond, as in (1) or (2), gives rise to an aldehyde group and to a keto group in a six-membered ring, with a band of 1705 cm^{-1}. The method differentiates

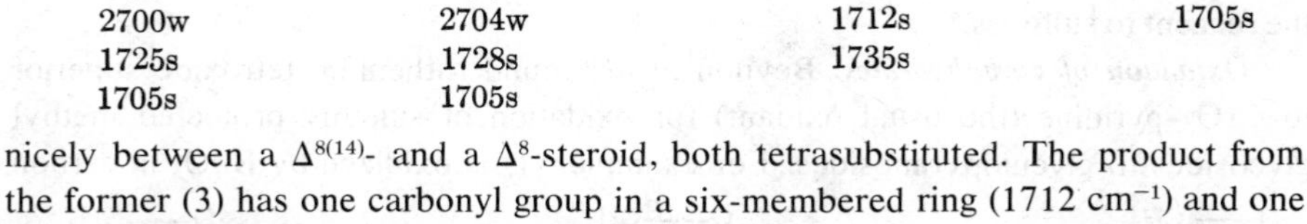
CH₃
CH₃
R

(1)	(2)	(3)	(4)
2700w	2704w	1712s	1705s
1725s	1728s	1735s	
1705s	1705s		

nicely between a $\Delta^{8(14)}$- and a Δ^8-steroid, both tetrasubstituted. The product from the former (3) has one carbonyl group in a six-membered ring (1712 cm^{-1}) and one in a five-membered ring which absorbs at the much higher frequency of 1735 cm^{-1}; the product from a Δ^8-ene (4) has two keto groups in a ten-membered ring and absorbs only at 1705 cm^{-1}. Snatzke and Fehlhaber[7] simplified the method by using ruthenium tetroxide to form the dicarbonyl compound in one step. A 5–10 mg. sample of steroid was oxidized with RuO_4 in CCl_4 for 1 hr., excess reagent was destroyed with methanol, and the solution was diluted and filtered from RuO_2.

$R_2CHOH \longrightarrow R_2C{=}O$. In a preliminary experiment Berkowitz and Rylander[1] oxidized *trans*-cyclohexane-1,2-diol in water as solvent and isolated in 15% yield a product characterized as probably cyclohexane-1,2-dione. Nakata[3] oxidized several steroid secondary alcohols to ketones in very high yield. For example a

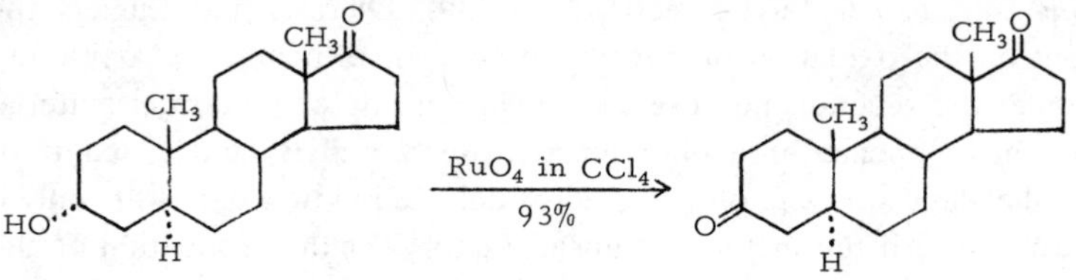

solution of 0.295 g. of 5α-androstane-3α-ol-17-one in 30 ml. of carbon tetrachloride was covered with 2 ml. of water, stirred, and a solution of RuO_4 in CCl_4 was added until a yellow color persisted (most of the black dioxide was retained in the aqueous phase and the function of the water apparently was to make the yellow color of excess reagent more easily discernible). Excess reagent was destroyed by addition of 1 ml. of isopropanol, the mixture was filtered, and the organic layer was separated, dried, and evaporated. Crystallization afforded, in two crops, a total of 0.272 g. of satisfactory dione.

Oxidation of ethers. Berkowitz and Rylander[1] found that, after a brief induction period, di-*n*-butyl ether is oxidized rapidly by RuO_4 in CCl_4 at room temperature. Infrared analysis of the reaction product indicated an essentially quantitative yield of *n*-butyl *n*-butyrate.

A synthesis of aldosterone from an alkaloid precursor by Wolff *et al.*[8] required, in a crucial step, the oxidation of the ether (1) to the lactone (2). Chromic acid

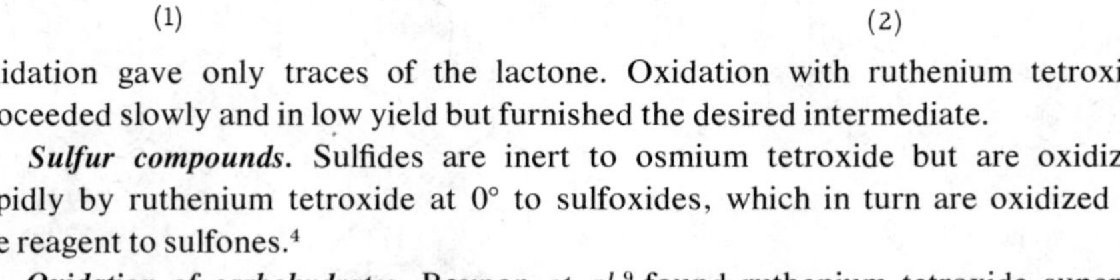

(1) (2)

oxidation gave only traces of the lactone. Oxidation with ruthenium tetroxide proceeded slowly and in low yield but furnished the desired intermediate.

Sulfur compounds. Sulfides are inert to osmium tetroxide but are oxidized rapidly by ruthenium tetroxide at 0° to sulfoxides, which in turn are oxidized by the reagent to sulfones.[4]

Oxidation of carbohydrates. Beynon *et al.*[9] found ruthenium tetroxide superior to CrO_3–pyridine (the usual oxidant) for oxidation of suitably protected methyl glycosides to glyculopyranosides. For example, (1) is oxidized by RuO_4 in carbon

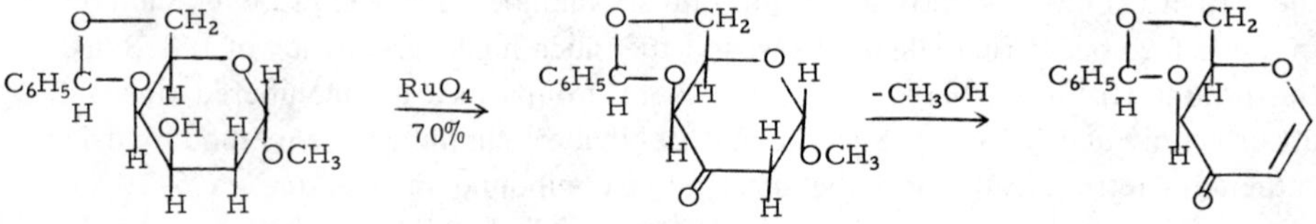

(1) (2) (3)

tetrachloride at 20° to give (2) in good yield. When the oxidation was done with CrO_3 in pyridine at 80°, methanol was eliminated to give (3) as the major product. The di-O-isopropylidene-D-glucofuranose (4) was oxidized smoothly to the corresponding ketone (5).

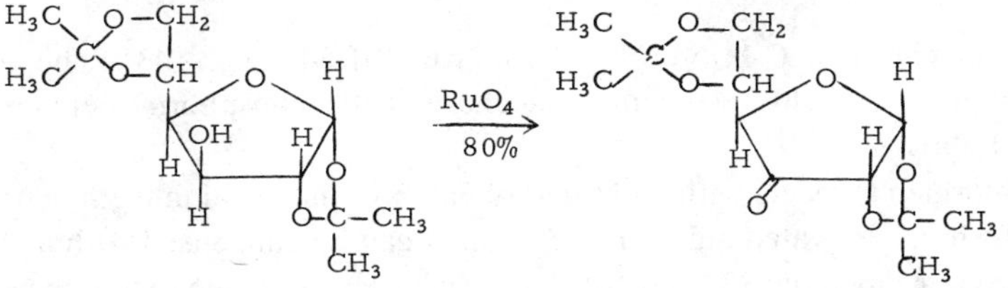

RuO4
80%

(4) (5)

Beynon's procedure requires preparation of the oxidant from ruthenium dioxide and oxidation of the alcohol with at least one equivalent of the tetroxide. Parikh and Jones[10] improved the procedure by combining the two steps, that is, treating the carbohydrate in chloroform or carbon tetrachloride with a trace amount of ruthenium tetroxide (20 mg. per gram of sugar), and adding a 5% aqueous solution of sodium metaperiodate (1.3 equiv.) with vigorous stirring and control to pH 6–7 (bicarbonate). The reaction is complete when the color changes from black to yellow. This procedure gives high yields of keto sugars, it uses only small amounts of costly ruthenium dioxide, and it saves time.

[1]L. M. Berkowitz and P. N. Rylander, *Am. Soc.*, **80**, 6682 (1958)
[2]F. M. Dean and J. C. Knight, *J. Chem. Soc.*, 4745 (1962)
[3]H. Nakata, *Tetrahedron*, **19**, 1959 (1963)
[4]C. Djerassi and R. R. Engle, *Am. Soc.*, **75**, 3838 (1953)
[5]E. J. Corey, J. Casanova, Jr., P. A. Vatakencherry, and R. Winter, *Am. Soc.*, **85**, 169 (1963)
[6]J. Castells, G. D. Meakins, and R. Swindells, *J. Chem. Soc.*, 2917 (1962)
[7]G. Snatzke and H.-W. Fehlhaber, *Ann.*, **663**, 123 (1963)
[8]M. E. Wolff, J. F. Kerwin, F. F. Owings, B. B. Lewis, and B. Blank, *J. Org.*, **28**, 2729 (1963)
[9]P. J. Beynon, P. M. Collins, and W. G. Overend, *Proc. Chem. Soc.*, 342 (1964); P. J. Beynon, P. M. Collins, P. T. Doganges, and W. G. Overend, *J. Chem. Soc.*, (C) 1131 (1966)
[10]V. M. Parikh and J. K. N. Jones, *Can. J. Chem.*, **43**, 3452 (1965)

S

γ-Saccharin chloride, $C_7H_4NO_2SCl$. Mol. wt. 201.64, m.p. 143°. The reagent is prepared in 35% yield by heating saccharin with phosphorus pentachloride at 170° for 1½ hrs.[1]

The chloride (1) reacts with an N-carbobenzoxyamino acid in methylene chloride at 0° to form an activated anhydride (2) which can be condensed with a free amino acid to give a dipeptide (3) and saccharin, which can be separated because of its greater acidity.[2]

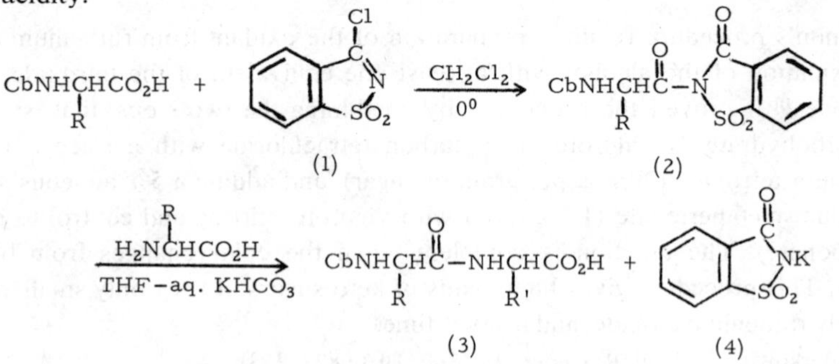

(1) (2) (3) (4)

[1]E. Stephen and H. Stephen, *J. Chem. Soc.*, 490 (1957)
[2]F. Micheel and M. Lorenz, *Tetrahedron Letters*, 2119 (1963)

Selenium, Se. At. wt. 78.96. Suppliers: Alfa, F, MCB.

Dehydrogenation. Early structural studies of bicyclic sesquiterpenes and tricyclic diterpenes by dehydrogenation with sulfur, with elimination of angular methyl or carboxyl groups if present, with formation of alkyl substituted naphthalenes or phenanthrenes, provided valuable information about the structures of the natural products. The method has the disadvantage that sulfur is reactive enough to combine with unsaturated intermediates and yields usually are very low. In 1925, when the evidence regarding the structure of cholesterol suggested the accompanying partial formula (reproduced from the *Berichte*), Diels and Gädke[1] sought to clarify the problem by dehydrogenating cholesterol to an aromatic hydrocarbon. Trials with sulfur and with platinum black were unpromising, but when 30 g. of cholesterol was heated with palladium charcoal to an unspecified temperature, water was eliminated, a volatile fragment from the side chain distilled, and extraction of the residue afforded 1.2–1.5 g. of material melting at 180–200° and later[2] found to contain chrysene.

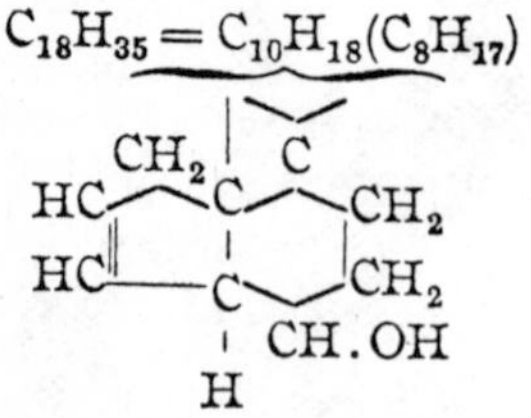

Hoping to obtain a still larger fragment by operating under milder conditions, Diels and co-workers in 1927 investigated dehydrogenation with selenium.[3,4] A higher temperature is required than with sulfur, but even so selenium is much less destructive than sulfur. Results with simpler compounds were encouraging.[4] Dehydrogenation of 30 g. of abietic acid afforded 22 g. of retene, whereas A. Vesterberg had obtained only 3.2 g. of the hydrocarbon from the reaction of 50 g.

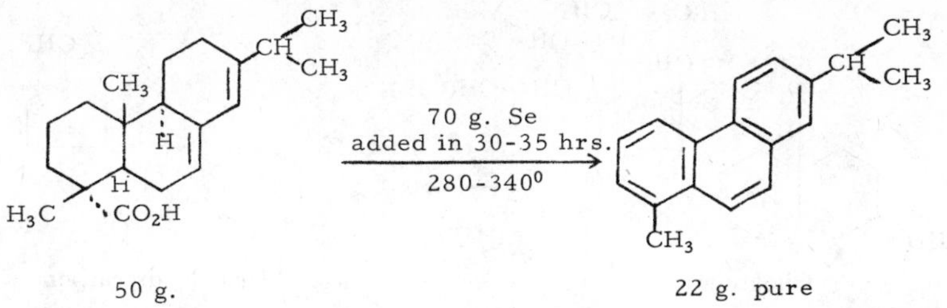
CH3
CH
CH3
CH3
H
H
H3C
CO2H
70 g. Se
added in 30-35 hrs.
280-340°
CH3
CH
CH3
CH3
50 g.
22 g. pure

of the resin acid with sulfur. Triphenylene was obtained from the dodecahydride in 32% yield with selenium.

Dehydrogenation of cholesterol[4] presented a much more difficult problem, for a very complex mixture resulted and isolation of crystalline products required use of a large batch of starting material and extensive processing. Each of five 40-g. batches of cholesterol was heated in a bath eventually maintained at 330° and 50 g. of selenium was added in 3–4 portions in the course of 30 hrs. The effluent gas was led into the flame of a Bunsen burner to destroy H_2Se, which burns with a blue flame which serves as a guide to the progress of the reaction. The cooled batches were combined, extracted with ether, and separated into 5 fractions boiling from 200 to over 285° at 12 mm. and totaling about 100 g. Processing in various ways afforded a few grams of a hydrocarbon melting at 127–128° and a much smaller amount of a second hydrocarbon melting at 219–220° ($C_{25}H_{24}$). Diels assigned to the more abundant degradation product the correct formula $C_{18}H_{16}$ and showed that it can be identified most securely in the form of a yellow nitroso derivative $C_{18}H_{13}O_2N$, m.p. 239°, but in the next five years published nothing further on the structure. Since by 1933 the discovery of vitamin D and the C_{18} and C_{19} sex hormones had stimulated several groups to take up steroid research, publications began to appear in this year from a number of laboratories on the formation and structure of a substance that became known as the Diels hydrocarbon.[5] The hydrocarbon was obtained by dehydrogenation of ergosterol and of cholic acid. The absorption spectrum suggested the structure of 1,2-cyclopentenophenanthrene (C_{17}) or the 1′,2′, or 3′-methyl derivative (C_{18}), and for a time the formula $C_{17}H_{16}$ was under consideration. Synthesis of the four hydrocarbons mentioned still left a decision open between 1,2-cyclopentenophenanthrene and its 3′-methyl derivative, for mixtures of the synthetic hydrocarbons showed no melting-point depression and the Diels method of identification was ignored. Finally, in 1934, Diels and Klare convincingly confirmed the growing agreement that the substance is 3′-methyl-1,2-cyclopentenophenanthrene by showing that 3 g. of synthetic 1,2-cyclopentenophenanthrene obtained from J. W. Cook on treatment in ether with oxides of nitrogen gave no trace of the nitroso derivative, m.p. 239°, formed by the Diels hydrocarbon, which alone has a tertiary hydrogen atom.

The problem had been undertaken for clarification of the structure of cholesterol, but in the 9 years required for its solution the structure of cholesterol was deduced from other evidence (in 1932). The interpretation of the dehydrogenation is that the angular methyl group at C_{13} migrates to C_{17} with expulsion of the side chain. However, selenium dehydrogenation proved to be a valuable tool in the investigation of estrogens[6] and steroidal alkaloids[7] and in establishing the steroidal character of the sapogenins.[8]

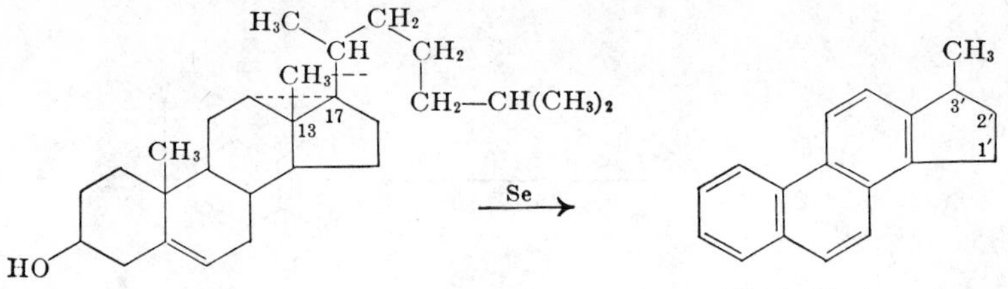

H_3C
CH_2
CH
CH_2
CH_3
$CH_2—CH(CH_3)_2$
17
13
CH_3
HO
Se
CH_3
3′
2′
1′

Cholesterol   Diels Hydrocarbon

A quantitative study of the selenium dehydrogenation of guaiene to guaiazulene and of three hydroaromatic compounds is presented by Silverwood and Orchin,[9] who suggest a mechanism. Of practical interest is their finding that in the dehydrogenation of 2.5 mmoles of guaiene with 7.5 mmoles of selenium at 290° for 1 hr. the yield is improved from 14 to 29.6% by addition of 15 mmoles of oleic acid as hydrogen acceptor.

Selenophenols. Foster[10] describes a procedure for the reaction of phenylmagnesium bromide with selenium in refluxing ether in an atmosphere of hydrogen to produce, after hydrolysis, selenophenol.

$$\underset{0.5\text{ m.}}{C_6H_5MgBr} + \underset{0.5\text{ m.}}{Se} \xrightarrow{\text{Refl. ether}} C_6H_5SeMgBr \xrightarrow[43\text{-}54\%]{HCl} C_6H_5SeH$$

Isomerization catalyst. Oleic acid is isomerized to elaidic acid by a variety of catalysts, of which selenium is the most convenient. The reaction is conducted at

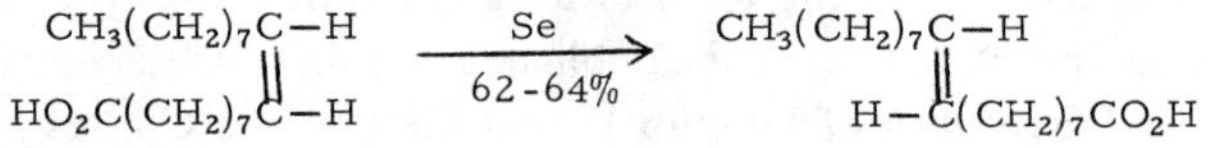

$CH_3(CH_2)_7C—H$
$HO_2C(CH_2)_7C—H$
Se
62-64%
$CH_3(CH_2)_7C—H$
$H—C(CH_2)_7CO_2H$

Oleic acid   Elaidic acid

220–225° for 1 hr. in an atmosphere of carbon dioxide or nitrogen.[11] Under these conditions ω-fluorooleic acid is elaidized in 75% yield.[12]

[1]O. Diels and W. Gädke, *Ber.*, **58**, 1231 (1925)
[2]*Idem, ibid.*, **60**, 140 (1927)
[3]O. Diels and A. Karstens, *Ber.*, **60**, 2323 (1927)
[4]O. Diels, W. Gädke, and P. Körding, *Ann.*, **459**, 1 (1927)
[5]L. F. Fieser and M. Fieser, "Steroids," 83–87, Reinhold (1959)
[6]Ref. 5, pp. 455–461
[7]Ref. 5, pp. 870–871, 883–884
[8]Ref. 5, pp. 812–813
[9]H. A. Silverwood and M. Orchin, *J. Org.*, **27**, 3401 (1962)
[10]D. G. Foster, *Org. Syn., Coll. Vol.*, **3**, 771 (1955)
[11]D. Swern and J. T. Scanlan, *Biochem. Prep.*, **3**, 118 (1953)
[12]R. E. A. Dear and F. L. M. Pattison, *Am. Soc.*, **85**, 622 (1963)

Selenium dioxide, Selenious acid, SeO_2, H_2SeO_3. Mol. wts. 110.96, 128.98. Suppliers: Fairmont Chemical Co.; K and K Laboratories; Matheson, Coleman and Bell (SeO_2); Fisher (H_2SeO_3). Review and procedures for conversion of recovered Se into SeO_2.[1]

In oxidations and dehydrogenations with either preformed selenious acid or selenium dioxide and water, usually in acetic acid, ethanol, or dioxane, the solution turns yellow and then red, and later some of the selenium separates in a red form difficult to filter, some is retained in colloidal solution, and some is bound to an organic substrate. Elimination of all selenium from a reaction mixture often presents difficulty. Sometimes refluxing, if feasible, may convert colloidal red material into black, metallic selenium. Sometimes chromatography on a column of alumina provided with a layer of commercial precipitated silver is helpful.[2] Other expedients are suggested in procedures which follow, but a generally applicable method (other than distillation) is still to be developed.

Oxidation. Use of selenium dioxide for the oxidation of reactive methylene groups to carbonyl groups was introduced by H. L. Riley, Morley, and Friend.[3] A procedure developed by H. A. Riley and Gray[4] for the preparation of phenylglyoxal calls for heating a solution of acetophenone in dioxane with either selenious acid, or equivalent amounts of selenium dioxide and water, and stirring the mixture

$$\underset{\text{1 m. in 600 dioxane}}{C_6H_5COCH_3} + \underset{\text{1 m.}}{H_2SeO_3} \xrightarrow[69-72\%]{\text{Stir under refl. 4 hrs.}} C_6H_5COCHO$$

under reflux. The product is isolated by distillation at 25 mm. The preparation of glyoxal bisulfite[5] is similar, except that paraldehyde is taken in large excess and the solvent is a mixture of 540 ml. of dioxane to which 40 ml. of acetic acid is added

$$\underset{\text{270 ml.}}{(CH_3CHO)_3} + \underset{\text{1.72 m.}}{3\ H_2SeO_3} \xrightarrow{\text{Refl. 6 hrs.}} H\overset{O}{\overset{\|}{C}}\overset{O}{\overset{\|}{C}}H \xrightarrow[72-74\%]{NaHSO_3} OHCCHO\cdot 2\ NaHSO_3\cdot H_2O$$

both to serve as accelerator and inhibit rearrangement to glycolic acid. Oxidation of cyclohexanone to cyclohexane-1,2-dione[6] is done by adding a solution of selenious acid in 500 ml. of dioxane and 100 ml. of water over a period of 3 hrs. with stirring and cooling to a large excess of cyclohexanone and then heating for 5 hrs. on the steam bath. The product is distilled twice at 16 mm.

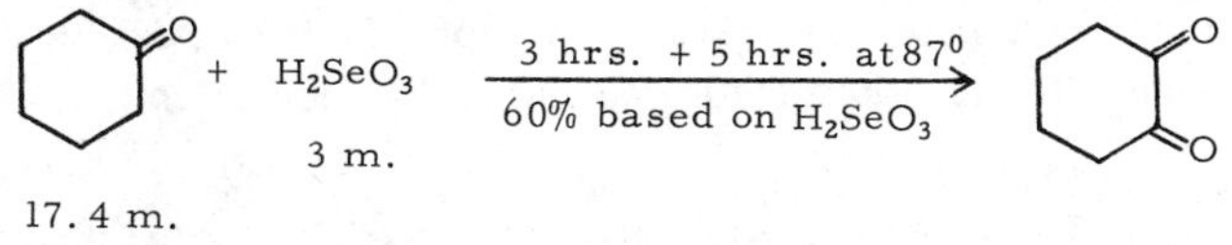
+ H_2SeO_3
3 hrs. + 5 hrs. at 87°
60% based on H_2SeO_3
3 m.
17.4 m.

Corey and Schaefer[7] studied the kinetics of the oxidation of desoxybenzoin to benzil and suggested a mechanism for the reaction. They also developed an efficient synthesis of α-keto esters involving oxidation of an α-bromoketone in an anhydrous solvent, as illustrated by a procedure for the preparation of ethyl benzoyl formate.[8]

$$\underset{\text{15 g.}}{C_6H_5\overset{O}{\overset{\|}{C}}CH_2Br} + \underset{\text{9 g.}}{SeO_2} \longrightarrow \left[C_6H_5\overset{O}{\overset{\|}{C}}\overset{O}{\overset{\|}{C}}Br\right] \xrightarrow[70\%]{C_2H_5OH} C_6H_5\overset{O}{\overset{\|}{C}}\overset{O}{\overset{\|}{C}}OC_2H_5$$

A solution of the reactants in 75 ml. of hot ethanol was refluxed for 12 hrs. and diluted with water. The product was extracted with ether and distilled at 97–98°/2 mm.

Radlick[9] describes a short, if low-yield, preparation of tropone from 1,3,5-cycloheptatriene (Shell Chem. Corp.) by oxidation with selenium dioxide in buffered aqueous dioxane; the function of the buffer is not known, but it raises the yield by a factor of about two.

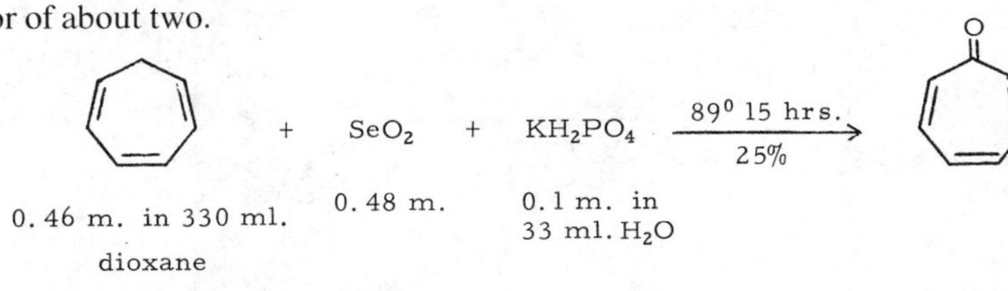
+ SeO_2 + KH_2PO_4
89° 15 hrs.
25%
0.46 m. in 330 ml.
dioxane
0.48 m.
0.1 m. in
33 ml. H_2O

Meinwald *et al.*[10] found that in the oxidation of the chloroketone (1) a purer product could be obtained by carrying out the reaction in bromobenzene (rather than toluene or xylene), followed by an aqueous workup.

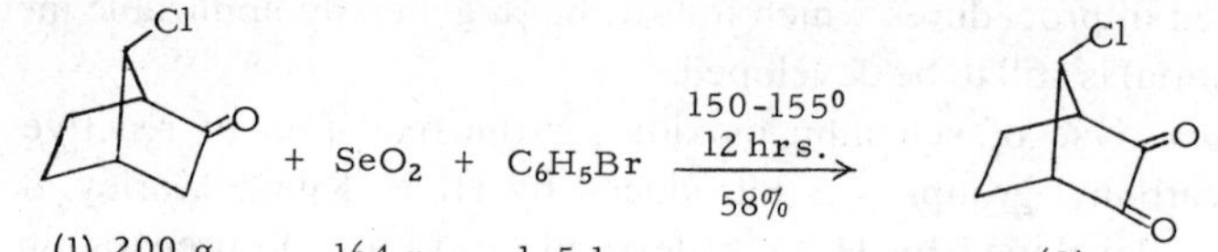
Cl
+ SeO2 + C6H5Br
150-155°
12 hrs.
58%
Cl

(1) 200 g. 164 g. 1.5 l. (2)

Allylic hydroxylation. From investigations of his own and analysis of the literature, Guillemonat[11] proposed the following rules regarding the allylic hydroxylation of olefins by selenium dioxide:

(1) Hydroxylation occurs α to the more highly substituted end of the double bond:

$$CH_3-\overset{CH_3}{\overset{|}{C}}=CHCH_3 \xrightarrow{SeO_2} CH_3-\overset{CH_2OH}{\overset{|}{C}}=CHCH_3$$

(2) The order of facility of oxidation is $CH_2 > CH_3 > CH$:

$$CH_3CH_2-\overset{CH_3}{\overset{|}{C}}=CHCH_3 \xrightarrow{SeO_2} \underset{34\%}{CH_3\underset{OH}{\underset{|}{CH}}-\overset{CH_3}{\overset{|}{C}}=CHCH_3} + \underset{1\%}{CH_3CH_2\overset{CH_2OH}{\overset{|}{C}}=CHCH_3}$$

The following example shows that the first rule takes precedence over the second.

$$CH_3-\overset{CH_3}{\overset{|}{C}}=CHCH_2CH_3 \xrightarrow{SeO_2} CH_3-\overset{CH_2OH}{\overset{|}{C}}=CHCH_2CH_3$$

(3) When the double bond is in a ring, oxidation whenever possible occurs within the ring and again α to the more substituted end of the double bond:

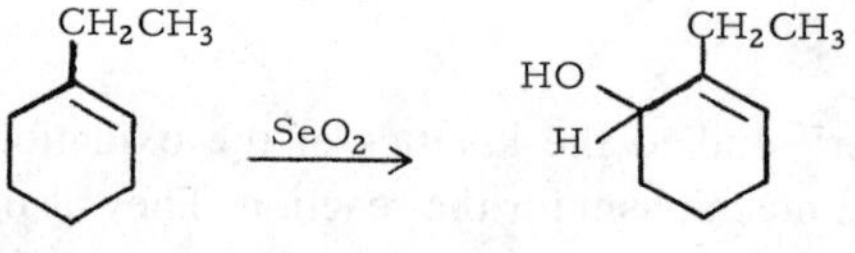
CH2CH3
SeO2
CH2CH3
HO
H

(4) Oxidation of a terminal double bond affords a primary alcohol with allylic migration of the double bond:

$$CH_3(CH_2)_2CH_2CH=CH_2 \xrightarrow{SeO_2} CH_3(CH_2)_2CH=CHCH_2OH$$

Guillemonat noted also that oxidation of CH to a tertiary alcohol often is accompanied by dehydrogenation to a conjugated diene. In considering experiments carried out over 20 years after appearance of Guillemonat's paper it is interesting to see how well results conform to predictions.

For a discussion of possible mechanisms, see Wiberg and Nielsen.[12]

Sondheimer *et al.*[13] oxidized the α,β-unsatured ester (1) in acetic acid solution and obtained the lactone (2) in good yield. Nakazaki and Naemura[14] obtained the

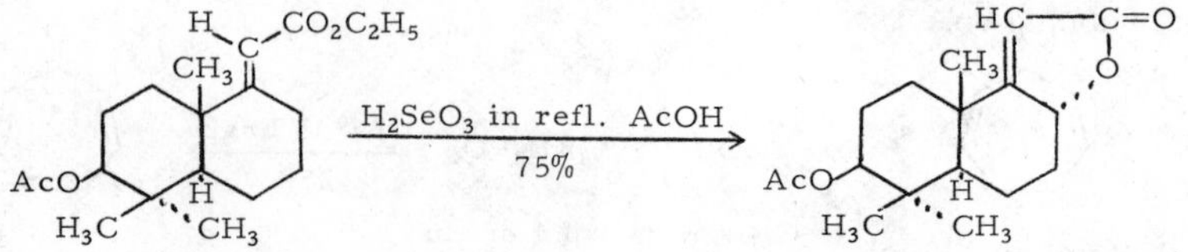
H
CO2C2H5
C
CH3
AcO
H
H3C
CH3
H2SeO3 in refl. AcOH
75%
HC
C=O
CH3
O
AcO
H
H3C
CH3

(1) (2)

acid (4) by reductive cleavage of the 6β-lactone (3) with zinc dust in acetic acid and oxidized it by the stereospecific procedure of Abe *et al.*[15] to the (*trans*) 6α-

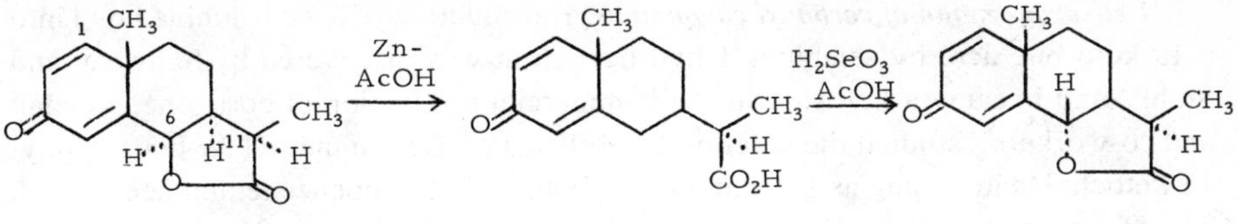

(3) (4) (5)

lactone (5), (−)-β-santonin. The oxidation of both (1) and (4) is possible in only one direction, but it is worth noting that both reactions follow the rule of rear attack.

Sondheimer *et al.*[16] oxidized the α,β-unsaturated ester (6) with selenium dioxide as the key step in the synthesis of digitoxigenin acetate (7). In accordance with the Guillemonat rules, hydroxylation occurs α to the more highly substituted

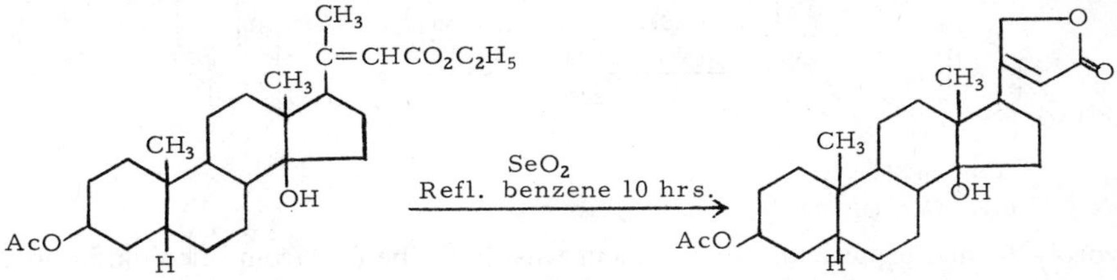

(6) (7)

carbon of the double bond and CH_3 is attacked in preference to CH. Another paper from the Sondheimer group[17] reports 17α-hydroxylation of cardiac aglycones

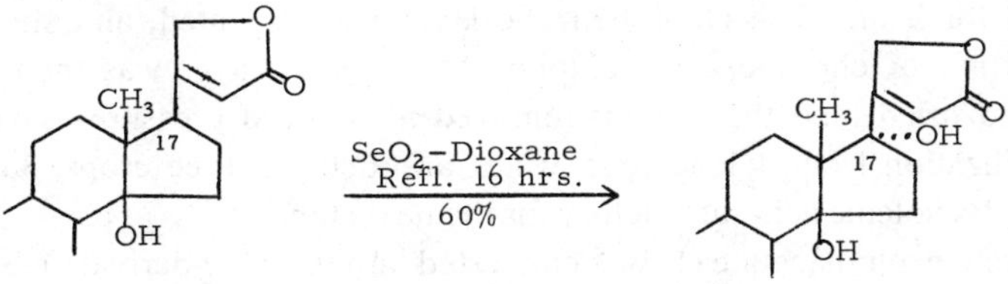

(8) (9)

(8). In this case the Guillemonat rules are not applicable because one carbon α to the more highly substituted olefinic carbon carries an oxygen substituent.

Rosenheim and Starling's[18] oxidation of cholesterol (10) to Δ⁵-cholestene-3β,4β-diol (11) involves attack α to the more highly substituted unsaturated carbon. The workup specified is to add 100 g. of sodium acetate and heat briefly to convert

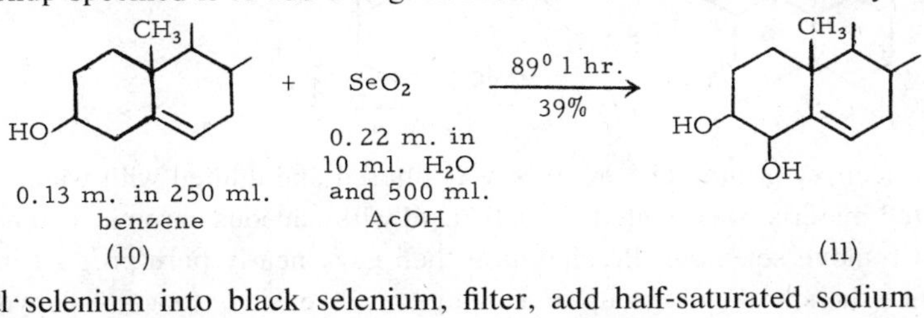

(10) (11)

colloidal·selenium into black selenium, filter, add half-saturated sodium chloride solution, collect the product, digest it with petroleum ether, and crystallize twice to obtain inch-long colorless needles, m.p. 176–177°. One of us has repeated this experiment and found elimination of selenium difficult and the results highly variable.

On two occasions the yield was in the range 60–70% with no evident change in the procedure, but the results could not be reproduced later.

Dehydrogenation of carbonyl compounds. Introduction of a 9,11-double bond into a 12-keto bile acid by selenium dehydrogenation was discovered by Schwenk and Stahl[19] and became a key step in the commercial production of cortisone. Kendall and co-workers[20] studied the reaction carefully as applied to methyl 3α-benzoyloxy-12-ketocholanate, using as solvent a 4:1 mixture of chlorobenzene and acetic acid. They found that a trace of hydrochloric acid slowed down the reaction but increased the yield. Thus with no added mineral acid the yield reached a maximum of 67% in a reflux period of 24 hrs. but with the mixture 0.0006 normal in HCl refluxing for 72 hrs. afforded the product in 84% yield. The product was to be isolated after

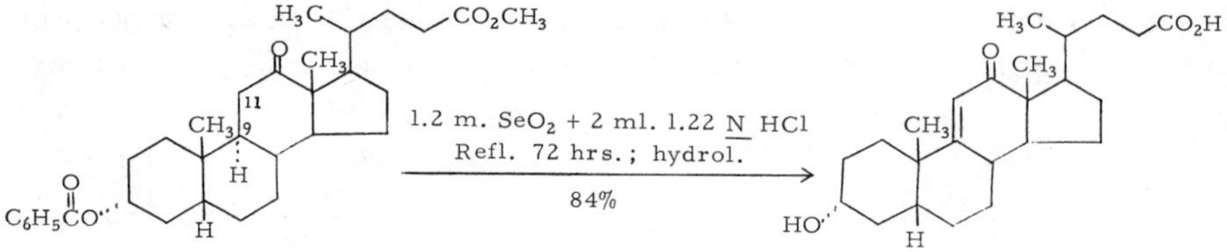

H₃C
CO₂CH₃
CH₃
11
CH₃ 9
H
C₆H₅CO
H
1.2 m. SeO₂ + 2 ml. 1.22 N HCl
Refl. 72 hrs.; hydrol.
84%
CO₂H
HO

1 m. in 3560 ml.
C_6H_5Cl—AcOH (4:1)

hydrolysis, and a particular requirement was that it be free from selenium in order to permit reduction of the carbonyl group by platinum-catalyzed hydrogenation. Of many expedients tried, the only one found effective was treatment with chromic acid. The reaction mixtures from three oxidations of 508 g. of keto ester benzoate were combined, filtered, treated with 150 g. of chromic anhydride in 150 ml. of water, and stirred for 2 hrs. The chlorobenzene layer was decanted, and stirred with a second portion of chromic acid solution. The organic layer was then separated, washed, and dried, and the solvent removed at reduced pressure. Saponification and crystallization from 4:1 acetone–water afforded, in three crops, 3α-hydroxy-12-keto-$\Delta^{9(11)}$-cholenic acid satisfactory for the next step.

The 12-keto ester diacetate (1) was converted into the $\Delta^{9(11)}$-derivative by refluxing 82 g. in 800 ml. of acetic acid with 48 g. of SeO_2 for 18 hrs.[21] Hydrochloric acid was not added because the reaction product is sensitive to acid-catalyzed elimination

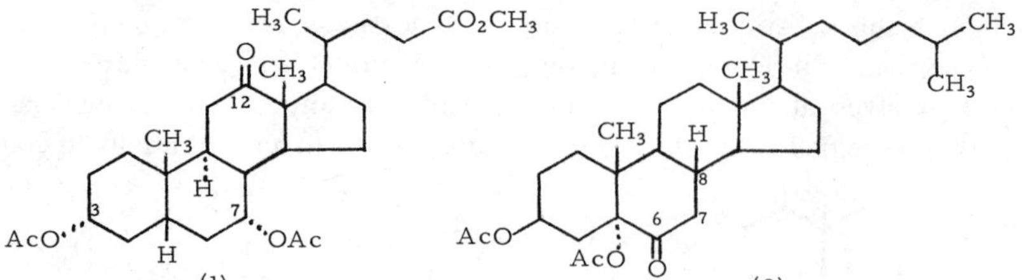

H₃C
CO₂CH₃
CH₃
12
CH₃
H
3
7
AcO
OAc
H
H₃C
CH₃
CH₃
CH₃
CH₃
H
8
6
7
AcO
AcO

(1) (2)

of the 7α-acetoxy group. The mixture was filtered and diluted with water, and the precipitated material was treated in acetic acid with aqueous chromic acid at 20° for 4 hrs., to remove selenium. Precipitation then gave nearly pure product in 75.5% yield. Since the spiroketal group of a sapogenin is sensitive to acids, dehydrogenation of 12-ketones in this series is carried out in *t*-butanol containing some pyridine.[22]

That the reaction is not applicable to all steroid ketones, even to those having a tertiary hydrogen β to the carbonyl group, is evident from the fact that

6-keto 3β,5α-diacetoxycholestane (2) resisted all attempts to introduce a 7,8-double bond by refluxing with selenium dioxide in acetic acid or nitrobenzene.[23]

In view of the negative result just cited, it is surprising that 3-ketosteroids of either the 5α or 5β series, as well as Δ^1- and Δ^4-3-ketosteroids, are dehydrogenated by selenium dioxide in *t*-butanol to $\Delta^{1,4}$-dienones (4) in good yield. The discovery was made simultaneously and independently in 1956 by a Ciba-Basel group[24] and

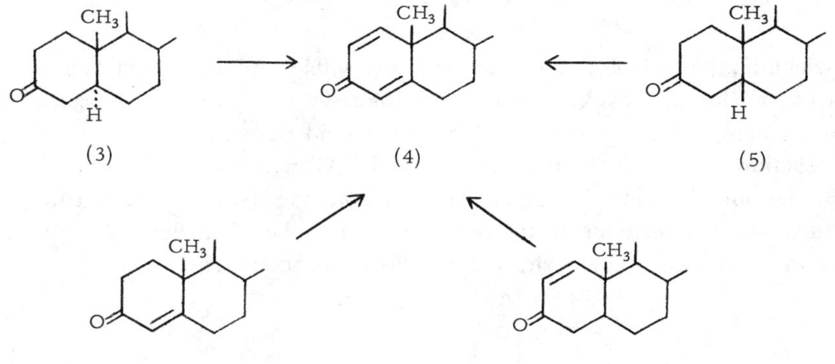
CH3
O
H

(3) (4) (5) (6) (7)

by an Organon group.[25] Hershberg's group at Schering[26] had discovered the year before that microbiological dehydrogenation of Δ^4-3-ketosteroids cortisone and cortisol (6) to the $\Delta^{1,4}$-dienones prednisone and prednisolone (4) markedly enhances biological potency. Dehydrogenation via the bromoketone or with lead tetraacetate had been effected in only very low yield. Hence the new one-step process, applicable to a variety of starting ketones, was of enormous value for the production of new hormone analogs. The Schering group missed discovery of the reaction by a slight margin. In 1954, seeking a new method for introducing a 4,5-double bond into the saturated 3-ketone (8) and so shortening the last steps of the bile acid synthesis of

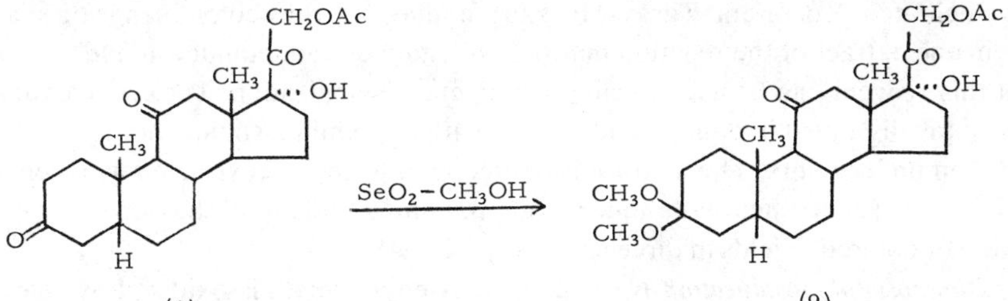
CH2OAc
CO
CH3
OH
O
H
SeO2–CH3OH
CH3O
CH3O

(8) (9)

cortisone, they investigated the action of selenium dioxide on the ketone but chose methanol as the solvent.[27] No selenium was deposited and the crystalline product resulting in good yield proved to be the dimethylketal (9). The reaction of 3-keto-steroids with ethylene glycol and selenium dioxide in methylene chloride was later used for the preparation of 3-ethyleneketals.[28] Evidently selenium dioxide functions as a dehydrating agent, and the selenious acid formed catalyzes ketal formation. Keto groups at C_{11} and C_{20} are inert to methanol–SeO_2, as are Δ^4-3-ketosteroids.[29]

An example of use of the dehydrogenation reaction reported by Bernstein and Littell[30] is as follows. A mixture of 2 g. of (10), 16α-acetoxycortisol 21-acetate, 2.2 g. of SeO_2, and 200 ml. of *t*-butanol was refluxed under nitrogen for 24 hrs. and filtered. The filtrate was evaporated to dryness at reduced pressure, and, for removal

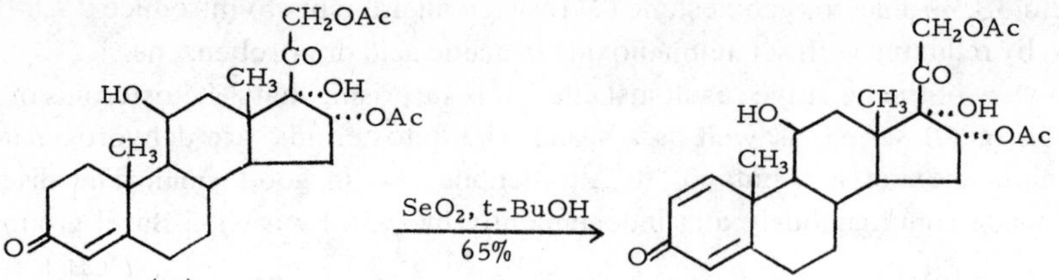
SeO2, t-BuOH
65%

(10) (11)

of selenium, the residue was stirred for 2 hrs. with 50 ml. of methanol and 3 g. of a deactivated Raney nickel catalyst (see footnote 22 of the paper). The same procedure was employed in another case by Heller and Bernstein.[31] Other examples of the reaction are reported by Fried *et al.*[32] and by Allen and Austin.[33]

Barnes and Barton[34] found that selenium dioxide dehydrogenates triterpenoid 1,4-diketones to enediones if the two hydrogens to be eliminated are *cis* but not if they are *trans*. Reaction was effected by refluxing in acetic acid.

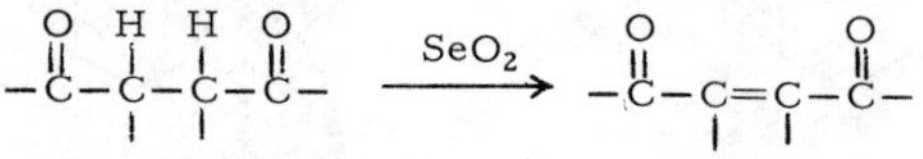
SeO2

Hill[35] showed that phenylmaleic anhydride, a useful dienophile, can be prepared easily by reaction of phenylsuccinic acid with selenium dioxide in acetic anhydride.

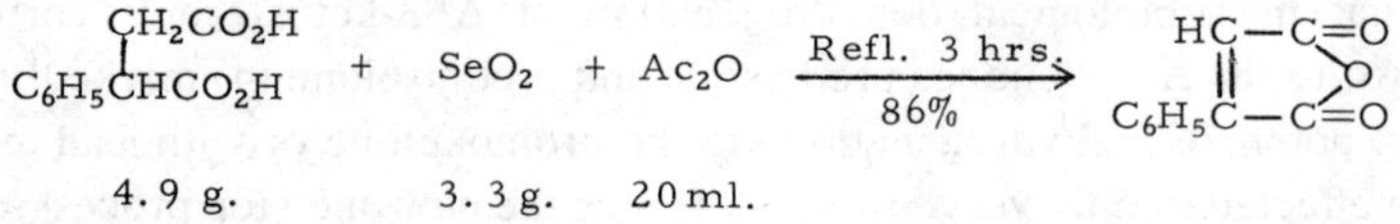
Refl. 3 hrs.
86%
4.9 g.
3.3 g.
20 ml.

The phenyl group appears to be essential, since succinic acid and ethylsuccinic acid are not attacked by the reagent.

In the dehydrogenation of Δ^4-3-ketosteroids to $\Delta^{1,4}$-3-ketosteroids, yields are generally less than 50% because of the formation of a selenosteroid of structure not yet elucidated. Kocór and Tuszy-Maczka[36] noticed a strong color change on shaking a benzene extract of the reaction mixture with aqueous ammonium sulfide and found that this reagent can be used to effect complete deselenization. The crude oxidation product is dissolved in ethanol and treated with ammonium sulfide, at first overnight and then under reflux. The mixture is diluted with water and extracted with benzene, and the extract washed with aqueous ammonium sulfide until the aqueous layer is no longer colored. Yields in three cases were 75–90%.

Diagnosis of unsaturation types. Δ^7-Cholestenyl acetate is oxidized by selenium dioxide in acetic acid–benzene to give 7α-acetoxy-$\Delta^{8(14)}$-cholestene (2), probably by allylic hydroxylation at C_{14}, allylic rearrangement, and acetylation.[2] Since oxidation occurs readily at room temperature whereas cholesterol is attacked only at 55–60°, the reaction can be used to detect small amounts of Δ^7-cholestenol

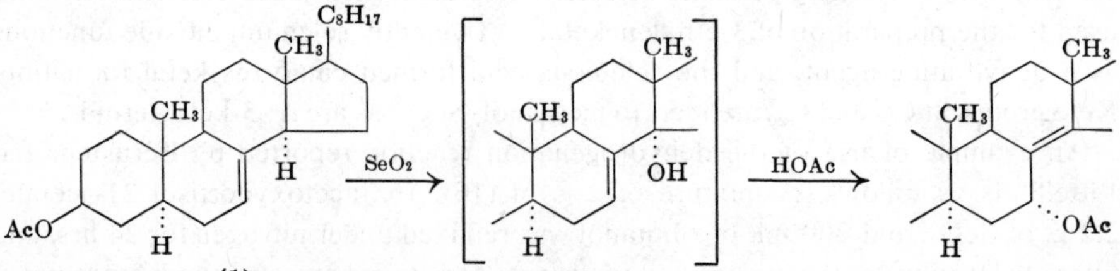
SeO2
HOAc

(1) (2)

occurring as a companion of cholesterol in most tissues.[37] By tests conducted with 1-mg. samples in melting-point capillaries,[38] gall stone cholesterol containing about 3% of the Δ^7-isomer is easily differentiated from pure cholesterol. The selenium dioxide test is specific to steroids of the types (1) and (3)–(6), that is to Δ^7- and Δ^8-stenols of the 5α-series, to $\Delta^{7,9(11)}$- and $\Delta^{6,8}$-dienes of this series and to $\Delta^{5,7}$-dienes.

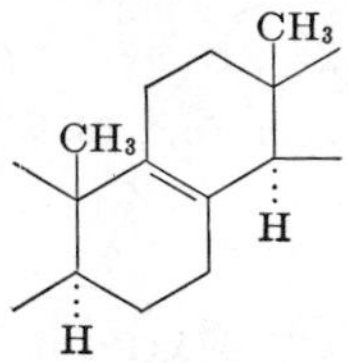
(3)

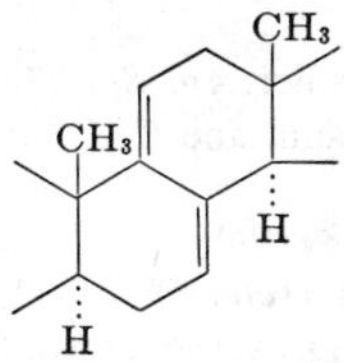
(4)

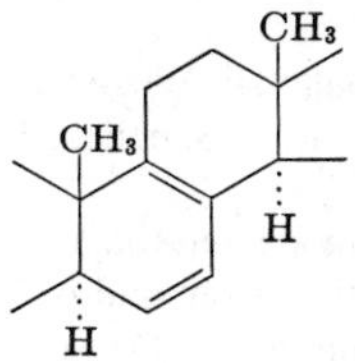
(5)

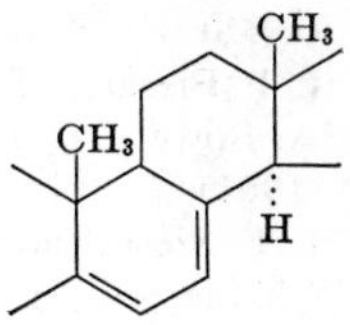
(6)

Thus a positive result is given only by 5α- or Δ^5-steroids having a double bond adjacent to the 14α-hydrogen atom. The 5β-isomers of compounds of types (1), (3), and (4) give negative tests, perhaps because the folding back of ring A to form a cage renders the 14α-hydrogen inaccessible to rear attack. Other structural types

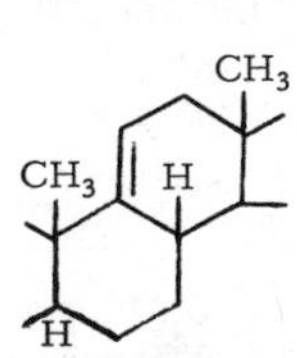
5α or β
(7)

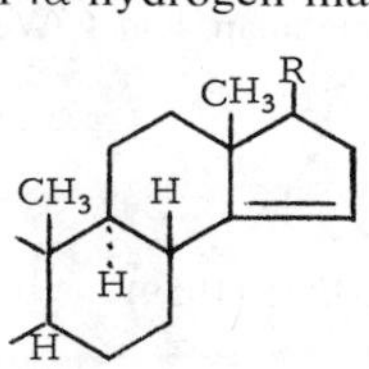
5α or β
(8)

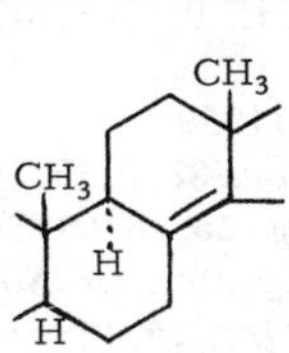
5α or β
(9)

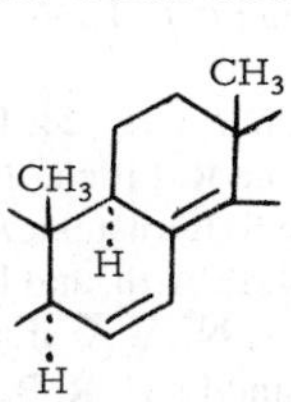
(10)

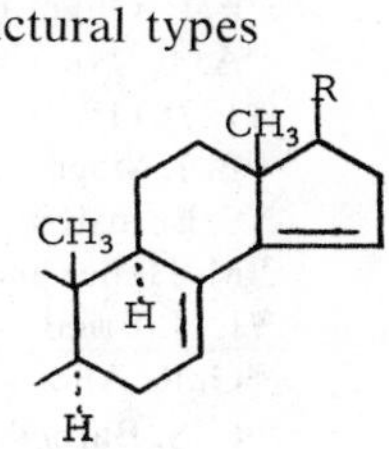
(11)

with which the test is negative are listed in formulas (7)–(11). The behavior of (7) and (8) shows that an activated 8β-hydrogen is not vulnerable, and the behavior of (9), (10), and (11) shows that an activated α-hydrogen at C_9 does not show the sensitivity of one at C_{14}.

The nature of the product resulting from selenium dioxide oxidation of an ene or diene depends upon the specific structure and the reaction conditions. Oxidation of Δ^7-cholestenyl acetate (1) in acetic acid gives the $\Delta^{8(14)}$-7α-acetoxy compound (2); oxidation in ethanol benzene gives the $\Delta^{8(14)}$-7α-ethyl ether.[2] In other instances an ene gives a diene, a diene gives a triene.

[1]N. Rabjohn, *Org. Reactions*, **5**, 331 (1949)
[2]L. F. Fieser and G. Ourisson, *Am. Soc.*, **75**, 4404 (1953)
[3]H. L. Riley, J. F. Morley, and N. A. C. Friend, *J. Chem. Soc.*, 1875 (1932)
[4]H. A. Riley and A. R. Gray, *Org. Syn., Coll. Vol.*, **2**, 509 (1943)
[5]A. R. Ronzio and T. D. Waugh, *ibid.*, **3**, 438 (1955)
[6]C. C. Hach, C. V. Banks, and H. Diehl, *ibid.*, **4**, 229 (1963)
[7]E. J. Corey and J. P. Schaefer, *Am. Soc*, **82**, 918 (1960); *see also* J. P. Schaefer and B. Horvath, *Tetrahedron Letters*, 2023 (1964)
[8]J. P. Schaefer and E. J. Corey, *J. Org.*, **24**, 1827 (1959)
[9]P. Radlick, *J. Org.*, **29**, 960 (1964)
[10]J. Meinwald, C. B. Jensen, A. Lewis, and C. Swithenbank, *J. Org.*, **29**, 3469 (1964)
[11]A. Guillemonat, *Ann. chim.*, **11**, 143 (1939)
[12]K. B. Wiberg and S. D. Nielsen, *J. Org.*, **29**, 3353 (1964)
[13]N. Danieli, Y. Mazur, and F. Sondheimer, *Tetrahedron Letters*, 310 (1961)
[14]M. Nakazaki and K. Naemura, *Chem. Ind.*, 1708 (1964)

[15]Y. Abe *et al.*, *Am. Soc.*, **78**, 1422 (1956)
[16]N. Danieli, Y. Mazur, and F. Sondheimer, *Am. Soc.*, **84**, 875 (1962)
[17]*Idem*, *Tetrahedron Letters*, 1281 (1962)
[18]O. Rosenheim and W. W. Starling, *J. Chem. Soc.*, 377 (1937)
[19]E. Schwenk and E. Stahl, *Arch. Biochem.*, **14**, 125 (1947)
[20]B. F. McKenzie, V. R. Mattox, L. L. Engel, and E. C. Kendall, *J. Biol. Chem.*, **173**, 271 (1948)
[21]L. F. Fieser, S. Rajagopalan, E. Wilson, and M. Tishler, *Am. Soc.*, **73**, 4133 (1951)
[22]A. Bowers, E. Denot, M. B. Sanchez, F. Neumann, and C. Djerassi, *J. Chem. Soc.*, 1859 (1961)
[23]L. F. Fieser and S. Rajagopalan, *Am. Soc.*, **71**, 3938 (1949)
[24]Ch. Meystre, H. Frey, W. Voser, and A. Wettstein, *Helv.*, **39**, 734 (1956)
[25]S. Szpilfogel, T. Posthumus, M. De Winter, and D. A. van Dorp, *Rec. trav.*, **75**, 475 (1956). At about the same time H. J. Ringold, G. Rosenkranz, and F. Sondheimer, *J. Org.*, **21**, 239 (1956), reported that refluxing 1 g. of testosterone with selenium dioxide in benzene containing a little water for 64 hrs. afforded 0.35 g. of $\Delta^{1,4}$-androstadiene-17β-ol-3-one.
[26]H. L. Herzog, A. Nobile, S. Tolksdorf, W. Charney, E. B. Hershberg, and P. L. Perlman, *Science*, **121**, 176 (1955)
[27]E. P. Oliveto, C. Gerold, and E. B. Hershberg, *Am. Soc.*, **76**, 6113 (1954)
[28]A. L. Nussbaum, T. L. Popper, E. P. Oliveto, S. Friedman, and I. Wender, *Am. Soc.*, **81**, 1228 (1959)
[29]B. J. Magerlein, *J. Org.*, **24**, 1564 (1959)
[30]S. Bernstein and R. Littell, *Am. Soc.*, **82**, 1235 (1960)
[31]M. Heller and S. Bernstein, *J. Org.*, **26**, 3876 (1961)
[32]J. H. Fried, G. E. Arth, and L. H. Sarett, *Am. Soc.*, **81**, 1235 (1959)
[33]G. R. Allen and N. A. Austin, *J. Org.*, **26**, 4574 (1961)
[34]C. S. Barnes and D. H. R. Barton, *J. Chem. Soc.*, 1419 (1953)
[35]R. K. Hill, *J. Org.*, **26**, 4745 (1961)
[36]M. Kocór and M. Tuszy-Maczka, *Bull. acad. Polon., Sci. Chem.*, **9**, 405 (1961) (in English)
[37]L. F. Fieser, *Am. Soc.*, **75**, 4395 (1953)
[38]*Org. Expts.*, 74–75

Selenium oxychloride, $SeOCl_2$. Mol. wt. 165.87, b.p. 177.6°, sp. gr. 2.44. Suppliers: Fisher, K and K Labs.

Schaefer and Sonnenberg[1] added the reagent with cooling to a solution of acetophenone in benzene and let the solution stand at room temperature. Separation of the

$$\underset{0.2\text{ m.}}{2\ C_6H_5COCH_3} + \underset{0.1\text{ m.}}{SeOCl_2} \longrightarrow \underset{(1)}{[C_6H_5COCH_2]_2SeCl_2} \xrightarrow[54\%]{\Delta} \underset{(2)}{2\ C_6H_5\overset{O}{\overset{\|}{C}}CH_2Cl} + Se$$

crystalline reaction product (1) was complete in 1 hr., and pyrolysis afforded α-chloroacetophenone (2) in moderate yield.

[1]J. P. Schaefer and F. Sonnenburg, *J. Org.*, **28**, 1128 (1963)

Semicarbazide hydrochloride, $HCl\cdot H_2NNHCONH_2$. Mol. wt. 111.54, m.p. 173° (free base m.p. 96°, unstable). Suppliers: A, B, E, F. MCB.

Preparation of a semicarbazone.[1] Prepared a stock solution of 1.11 g. of the hydrochloride in 5 ml. of water; 0.5 ml. of the solution contains 1 mmole of reagent. To 0.5 ml. of the solution add 1 mmole of acetophenone and enough methanol (1 ml.) to produce a clear solution; then 10 micro drops of pyridine, and warm gently on the steam bath until crystals begin to separate (m.p. 198°).

[1]*Org. Expts.*, 97

Sephadex. Suppliers: Pharmacia Fine Chemicals, Uppsala, Sweden; Pharmacia Fine Chemicals Inc., 800 Centennial Ave., Piscataway, N. J.

Sephadex is a chemically modified dextran obtained by the fermentation of cane sugar. Due to its cross-linked structure, it can act as a molecular sieve and separate materials of low molecular weight from companion substances of higher molecular weight by a process known as gel filtration. Sephadex is available in five types: G-25, G-50, G-75, G-100, G-200; G-25 being the most tightly cross-linked and G-200 being the least. Thus Sephadex G-100 separates substances of molecular weight smaller than about 100,000 from larger molecules; G-200 separates substances of molecular weights below 200,000; G-50 releases those of molecular weight 7,000–10,000.

These materials were introduced in 1959 by workers in Uppsala[1,2] for the fractionation of water-soluble substances. Small grains of Sephadex swell in an aqueous medium and a column of the swollen grains forms a stationary gel phase consisting of cross-linked dextrose chains. Large molecules are prevented from entering the gel phase and move in the aqueous phase surrounding the grains. Smaller molecules penetrate the pores of the gel and pass through the column in the water phase inside as well as outside the grains and so travel further and are eluted later than the large molecules.

The first publication[1] described use of Sephadex for desalting serum proteins; the process is similar to dialysis. Probably the most important use is for the separation of proteins, peptides, and amino acids, investigated in detail by Porath.[3] Sephadex was allowed to swell in a phosphate buffer (pH 6.8), transferred to a chromatography tube, and a solution of the mixture to be separated was slowly filtered through the column. Group separations are readily achieved; fractionation within a group may require use of a dextran gel of lower degree of cross linking. An advantage is that a gel can be regenerated easily in the column and can be reused over a period of months.

Gross and Witkop[4] found gel filtration over Sephadex G-25 invaluable for the separation of fragments resulting from the cleavage of methionine–peptide bonds in ribonuclease by reaction with cyanogen bromide. In this case 0.2*N* acetic acid was used as solvent.

The A. B. Pharmacia brochure on Sephadex in gel filtration listed about 100 references in the period 1959–62 on the use of Sephadex in the isolation, separation, and purification of proteins, peptides, and amino acids. Brochure No. 1, 1965, lists 638 references and gives cross references to abstract cards.

Sephadex has been used to a lesser extent in the nucleic acid field.[5,6] The G-50 type will separate RNA of high molecular weight from nucleosides or bases. However, Sephadex does not separate the complex mixture of mono- to hexanucleotides obtained by enzymic hydrolysis of RNA. Sephadex-25 has been used as the supporting medium for partition chromatography of yeast-soluble RNA.[7]

In the carbohydrate field it has been found possible to separate mixtures of oligosaccharides differing in size by only one glucose unit.[8] Dextrans of low molecular weight have been separated by gel filtration.[9] Partial resolution of a number of starch dextrins on Sephadex columns has been reported.[10] Thus amylodextrin can be separated completely from amylose. The method is particularly useful for polysaccharides because they show only small differences in their properties and usually differ only in molecular size.

[1]J. Porath and P. Flodin, *Nature*, **183**, 1657 (1959)

[2]B. Gelotte and A.-B. Krantz, *Chem. Scand.*, **13**, 2127 (1959)
[3]J. Porath, *Biochem. Biophys. Acta*, **39**, 193 (1960)
[4]E. Gross and B. Witkop, *J. Biol. Chem.*, **237**, 1856 (1962)
[5]B. Gelotte, *Naturwissenschaften*, **48**, 554 (1961)
[6]S. Zadražil, Z. Šormová, and F. Šorm, *Coll. Czech.*, **26**, 2643 (1961)
[7]K. Tanaka, H. H. Richards, and G. L. Cantoni, *Biochem. Biophys. Acta*, **61**, 846 (1962)
[8]P. Flodin and K. Aspburg, IUB/IUBS Symposium on "Biological Structure and Function," Stockholm, 1960
[9]K. A. Granath and P. Flodin, *Makromolekular Chem.*, **48**, 160 (1961)
[10]P. Nordin, *Arch. Biochem. Biophys.*, **99**, 101 (1962)

Silicon tetraisocyanate, $Si(N{=}C{=}O)_4$. Mol. wt. 196.16, m.p. 26°, b.p. 185.6°. Supplier: K and K Laboratories. The reagent can be prepared in 85% yield by the reaction of silver cyanate with silicon tetrachloride.[1]

Silicon isocyanate reacts exothermally in benzene solution with primary and secondary amines to give the corresponding ureas in yields of 97–100%.[1] Silicon

$$4\ RR'NH + Si(NCO)_4 \longrightarrow 4\ RR'N\overset{\overset{O}{\|}}{C}NH_2 + SiO_2 + 2\ H_2O$$

tetraisothiocyanate similarly gives thioureas. The method is said to be much superior to an earlier procedure using aqueous cyanic or thiocyanic acid.

[1]R. G. Neville and J. J. McGee, *Can. J. Chem.*, **41**, 2123 (1963)

Silver acetate, CH_3CO_2Ag. Mol. wt. 163.90. Suppliers: B, F, MCB. In studying the reaction of *vic*-dihalides with silver acetate in acetic acid, Winstein and Buckles[1] discovered that addition of a small amount of water alters the course of the reaction. The observation provided the basis for a simple procedure[2] for the preparation of the otherwise difficultly accessible *dl*-hydrobenzoin, of interest because it can be resolved by crystallization. When *meso*-stilbene dibromide (I) is heated with silver

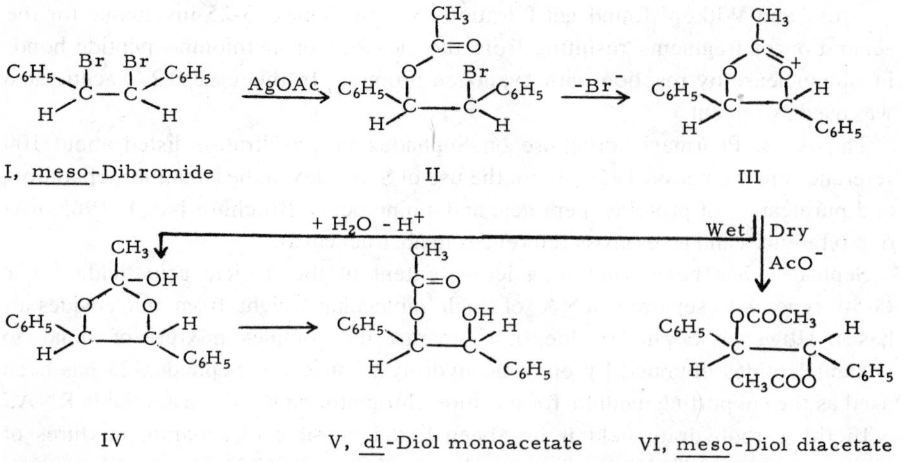
AgOAc
-Br⁻
I, meso-Dibromide
II
III
+ H₂O - H⁺
Wet
Dry
AcO⁻
IV
V, dl-Diol monoacetate
VI, meso-Diol diacetate

acetate in anydrous acetic acid the product is the *meso*-diol diacetate (VI), presumably formed via the bromohydrin acetate II and the oxonium ion III with inversion in reactions II⟶III and III⟶VI. When the *meso*-dibromide I (2 g.) is heated for 10 min. on the steam bath with silver acetate (2 g.), acetic acid (25 ml.), and water (1 ml.), a substantial amount of the *dl*-diol monoacetate V is found. The interpretation[1] is that water reacts with the oxonium ion III to produce IV, the conjugate acid of the orthomonoacetate, which then affords V. Since V is the sole monoacetate present, it is easily separated from the mixture containing both diols and their diacetates by chromatography. Saponification and crystallization affords pure *dl*-hydrobenzoin in overall yield of 23%.

Woodward and Brutcher[3] developed a procedure for the *cis*-hydroxylation of olefins which combines halogenation with acetolysis in wet acetic acid. Thus the synthetic intermediate (1) on treatment in acetic acid with iodine (1.05 equiv.),

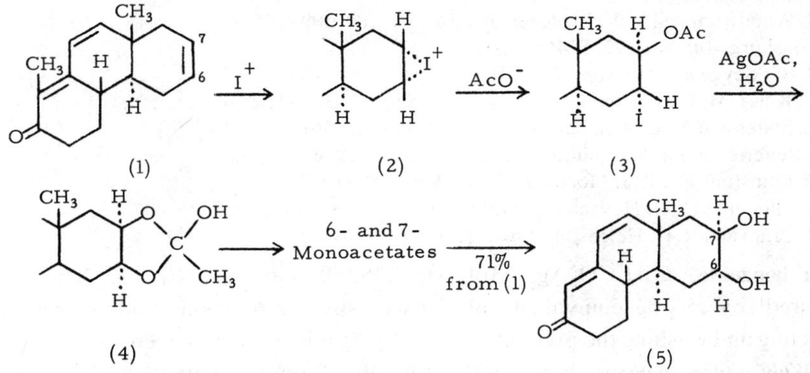

I^+
AcO⁻
AgOAc, H_2O
6- and 7- Monoacetates
71% from (1)
(1)
(2)
(3)
(4)
(5)

water (1 equiv.), and silver acetate at 90–95° for 3 hrs., gave a mixture of monoacetates, which on saponification afforded the 6β,7β-diol (5) in 71% yield. Scrutiny of the mother liquor yielded 2.5% of the 6α,7α-diol, the major product of osmium tetroxide hydroxylation of (1). The reagent thus provides a means for producing in quantity the opposite *cis*-glycol of that available through the osmate ester.

Other examples of this novel method of *cis*-hydroxylation are reported by Ginsburg,[4] Berkley *et al.*,[5] Klass *et al.*,[6] Slates and Wendler,[7] Jefferies and Milligan,[8] and Gunstone and Morris.[9] In the *cis*-hydroxylation of Δ^2-cholestene the yield of the 2β-3β-diol is improved from 50–65% to 81% by carrying out the reaction at 20° under nitrogen for 12 hrs.[10] (standard procedure: 45–90° for 1–20 hrs.)

Hydrolysis of gem-dihalides. Cram and Helgeson[11] brominated [2.2]paracyclophane (II) with N-bromosuccinimide in refluxing carbon tetrachloride in the presence

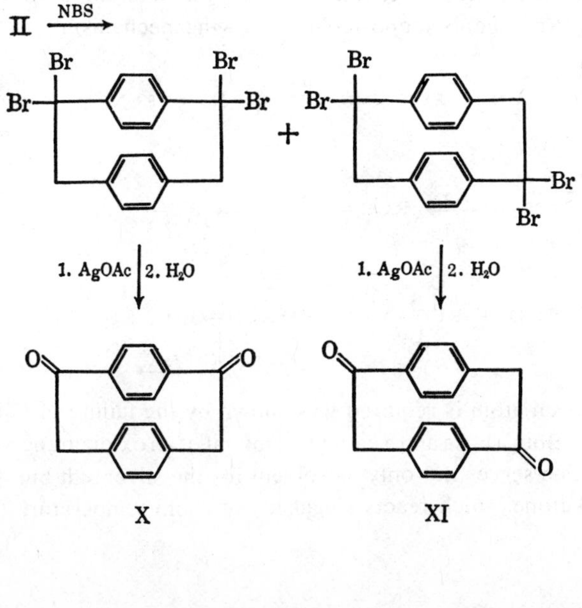

II
NBS
1. AgOAc
2. H_2O
1. AgOAc
2. H_2O

X XI

of a catalytic amount of dibenzoyl peroxide and with UV irradiation and, by crystallization and chromatography, isolated the two tetrabromides formulated in total yield of about 46%. Each tetrabromide on hydrolysis with silver acetate in refluxing acetic acid afforded the corresponding diketone, 1,9-diketo[2.2]paracyclophane (X) or the 1,10-isomer (XI) in yield of about 65%.

[1]S. Winstein and R. E. Buckles, *Am. Soc.*, **64**, 2787 (1942)
[2]*Org. Expts.*, Chapt. 43
[3]R. B. Woodward and F. V. Brutcher, Jr., *Am. Soc.*, **80**, 209 (1958)
[4]D. Ginsburg, *Am. Soc.*, **75**, 5746 (1953)
[5]L. B. Barkley *et al.*, *Am. Soc.*, **76**, 5014 (1954)
[6]D. L. Klass, M. Fieser, and L. F. Fieser, *Am. Soc.*, **77**, 3829 (1955)
[7]H. L. Slates and N. L. Wendler, *Am. Soc.*, **78**, 3749 (1956)
[8]P. R. Jefferies and B. Milligan, *J. Chem. Soc.*, 2363 (1956)
[9]F. D. Gunstone and L. J. Morris, *J. Chem. Soc.*, 487 (1957)
[10]P. S. Ellington, D. G. Hey, and G. D. Meakins, *J. Chem. Soc.*, 1327 (1966)
[11]D. J. Cram and R. C. Helgeson, *Am. Soc.*, **88**, 3515 (1966)

Silver benzoate, $C_6H_5CO_2Ag$. Mol. wt. 228.99. Suppliers: E, KK. The salt is prepared[1] by mixing equivalent solutions of silver nitrate and sodium benzoate, collecting and washing the precipitate, and drying it in a vacuum oven.

Wolff rearrangement. Finding that the Wolff rearrangement of diazoketones with silver oxide and methanol gave erratic, highly variable results, Newman and Beal[1] developed a procedure for carrying out the reaction in homogeneous solution under mild conditions. A solution of diazoacetophenone in methanol was treated at room temperature with a few drops of a filtered solution of 0.004 mole of silver

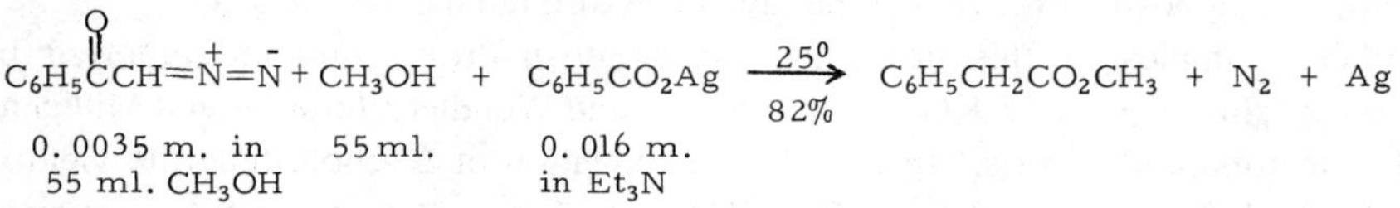

$$C_6H_5\overset{O}{\overset{\|}{C}}CH=\overset{+}{N}=\overset{-}{N} + CH_3OH + C_6H_5CO_2Ag \xrightarrow[82\%]{25^0} C_6H_5CH_2CO_2CH_3 + N_2 + Ag$$

0.0035 m. in   55 ml.   0.016 m.
55 ml. CH_3OH in Et_3N

benzoate in 9.1 ml. of triethylamine; the mixture turned black and nitrogen evolution commenced. When gas evolution slackened, more silver benzoate was added and the process was repeated as required until the reaction stopped and the nitrogen collected was 92% of the theoretical amount. The catalyst solution used was about 0.5 equivalent. Test experiments supported the following mechanism:

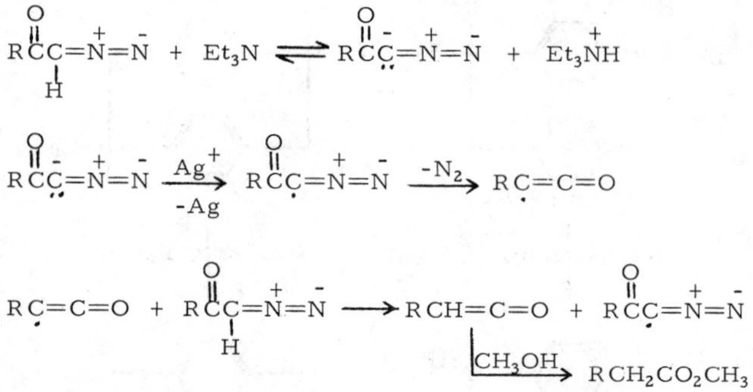

$$R\overset{O}{\overset{\|}{C}}\underset{H}{\overset{|}{C}}=\overset{+}{N}=\overset{-}{N} + Et_3N \rightleftharpoons R\overset{O}{\overset{\|}{C}}\overset{-}{\underset{\cdot\cdot}{C}}=\overset{+}{N}=\overset{-}{N} + Et_3\overset{+}{N}H$$

$$R\overset{O}{\overset{\|}{C}}\overset{-}{\underset{\cdot\cdot}{C}}=\overset{+}{N}=\overset{-}{N} \xrightarrow[-Ag]{Ag^+} R\overset{O}{\overset{\|}{C}}\underset{\cdot}{C}=\overset{+}{N}=\overset{-}{N} \xrightarrow{-N_2} R\underset{\cdot}{C}=C=O$$

$$R\underset{\cdot}{C}=C=O + R\overset{O}{\overset{\|}{C}}\underset{H}{\overset{|}{C}}=\overset{+}{N}=\overset{-}{N} \longrightarrow RCH=C=O + R\overset{O}{\overset{\|}{C}}\underset{\cdot}{C}=\overset{+}{N}=\overset{-}{N}$$
$$\xrightarrow{CH_3OH} RCH_2CO_2CH_3$$

That an α-hydrogen atom is required was shown by the failure of $C_2H_5COC(CH_3)$ $=\overset{+}{N}=\overset{-}{N}$ to react. Both a base and a catalytic amount of an oxidizing agent are required, and trimethylamine serves not only as solvent for the silver salt but as base. In the case of a diazoketone which reacts sluggishly at room temperature the reaction is carried out in *t*-butanol as solvent, for then the temperature can be raised even to the boiling point; when a solution of silver benzoate and triethylamine is heated the salt is reduced to silver and its catalytic effectiveness destroyed.

The reaction is almost completely stereospecific with retention of configuration; with other catalysts 20–30% racemization occurs.[2] Other examples.[3]

[1]M. S. Newman and P. F. Beal III, *Am. Soc.*, **72**, 5163 (1950)
[2]K. B. Wiberg and T. W. Hutton, *Am. Soc.*, **78**, 1640 (1956)
[3]J. Klein and E. D. Bergmann, *J. Org.*, **22**, 1019 (1957)

Silver carbonate, Ag_2CO_3. Mol. wt. 275.77. Sensitive to light. Preparation.[1] Suppliers: F, MCB.

For the preparation of 2,3,4,6-tetra-O-acetyl-β-D-glucopyranose,[1] a solution of 2,3,4,6-tetra-O-acetyl-α-D-glucopyranosyl bromide in 125 ml. of acetone is cooled in an ice bath and treated with a small amount of water, and silver carbonate,

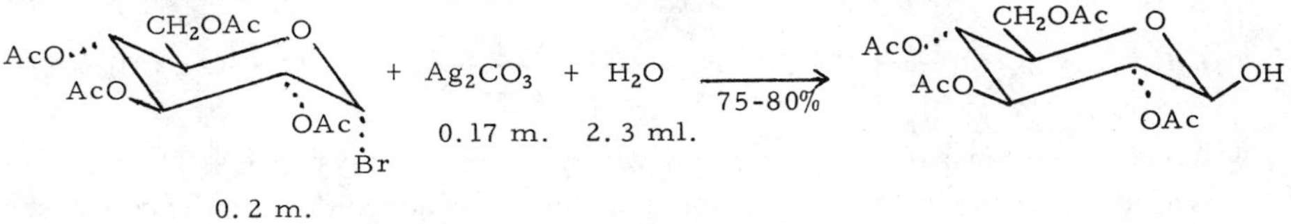
CH₂OAc
O
AcO
AcO
OAc
Br
0. 2 m.
+ Ag₂CO₃ + H₂O
0. 17 m. 2. 3 ml.
75-80%
CH₂OAc
O
AcO
AcO
OH
OAc

freshly prepared in dim light, is added in the course of 15 min. After shaking for 30 min. longer the solution is filtered, and the filtrate is evaporated at reduced pressure and low temperature. One crystallization from ether–ligroin gives nearly pure product, m.p. 132–134°.

[1]C. M. McCloskey and G. H. Coleman, *Org. Syn., Coll. Vol.*, **3**, 434 (1955)

Silver chlorate, $AgClO_3$. Mol. wt. 191.34, m.p. 230°, 10 g. dissolves in 100 g. H_2O at 15°. Supplier: KK.

The reagent has been used to a limited extent in combination with a catalytic amount of osmium tetroxide to hydroxylate water-soluble olefinic acids[1] and alcohols[2] to *cis-vic*-diols:

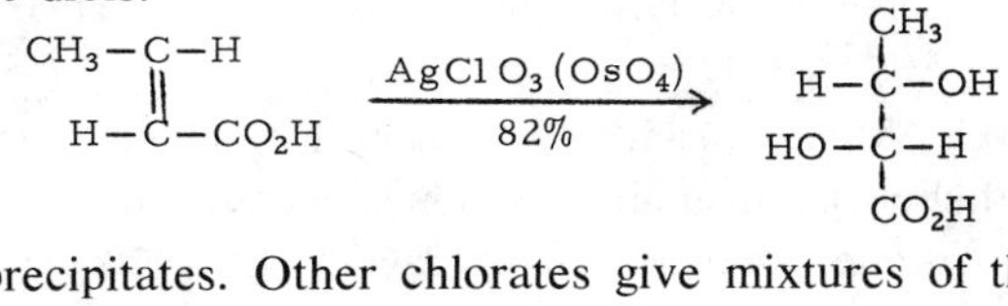
CH₃–C–H
H–C–CO₂H
AgCl O₃ (OsO₄)
82%
CH₃
H–C–OH
HO–C–H
CO₂H

Silver chloride precipitates. Other chlorates give mixtures of the glycol and the chlorohydrin.

[1]Géza Braun, *Am. Soc.*, **51**, 228 (1929)
[2]Th. Posternak and H. Friedli, *Helv.*, **36**, 251 (1953)

Silver chromate, Ag_2CrO_4 Mol. wt. 331.77. Suppliers: F, KK.

A key step in the bile acid synthesis of cortisone discovered by Kendall and

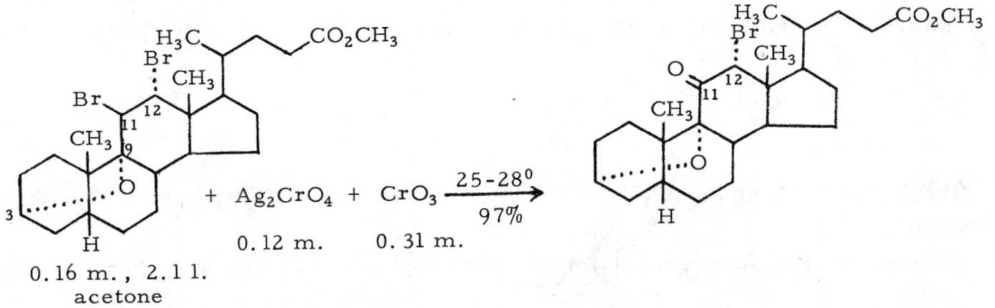
H₃C
CO₂CH₃
Br
CH₃
Br
12
11
CH₃
9
O
3
H
0. 16 m., 2. 1 l.
acetone
+ Ag₂CrO₄ + CrO₃
0. 12 m. 0. 31 m.
25-28°
97%
H₃C
CO₂CH₃
Br
CH₃
O
12
11
CH₃
O
H

co-workers[1] is an oxidative hydrolysis of the $11\beta,12\alpha$-dibromide formulated to the 12α-bromo-11-ketone. The reaction is dependent upon activation of the 11β-bromine substituent by the oxidic oxygen at C_9, since the corresponding steroid without the $3\alpha,9\alpha$-oxide bridge resists attack. The reaction was carried out by stirring a suspension of the dibromide in acetone, adding silver chromate and a solution of chromic anhydride in 200 ml. of water, and keeping the temperature at 25–28° by cooling for 2 hrs., when the solid had all dissolved. Sulfuric acid (77 ml. of 5 *N* acid) was added, and the solution filtered and concentrated at reduced pressure.

[1]R. B. Turner, V. R. Mattox, L. L. Engel, B. F. McKenzie, and E. C. Kendall, *J. Biol. Chem.*, **166**, 345 (1946)

Silver cyanide, AgCN. Mol. wt. 133.90. Suppliers: B, E. F.

For the preparation of ethyl isocyanide,[1] silver cyanide is added to ethyl iodide with stirring and heating on the steam bath until the mixture turns to a brown,

$$\underset{3.4\text{ m.}}{C_2H_5I} + \underset{3.4\text{ m.}}{AgCN} \longrightarrow C_2H_5N{=}C\cdot AgI \xrightarrow[47-55\%]{2\,KCN(H_2O)} C_2H_5N{=}C + KAg(CN)_2 + KI$$

viscous liquid consisting of a complex which sometimes crystallizes. The stirrer is raised to a position just above the liquid, and water is added through the condenser, followed by a solution of potassium cyanide. Ethyl isocyanide separates as a brown layer.

[1]H. L. Jackson and B. C. McKusick, *Org. Syn., Coll. Vol.*, **4**, 438 (1963)

Silver dibenzyl phosphate, $AgO\overset{O}{\overset{\|}{P}}(OCH_2C_6H_5)_2$. Mol. wt. 385.11, m.p. 230° dec.

Sarett *et al.*[1] worked out an improved method of preparation and used the reagent for the synthesis of cortisone 21-phosphate. Zervas[2] describes other examples of use as a phosphorylating agent.

[1]F. A. Cutler, Jr., J. P. Conbere, R. M. Lukes, J. F. Fisher, H. E. Mertel, R. Hirschmann, J. M. Chemerda, L. H. Sarett, and K. Pfister, 3rd, *Am. Soc.*, **80**, 6300 (1958)

[2]L. Zervas, *Naturwiss.*, **27**, 317 (1939)

Silver difluoride, AgF_2. Mol. wt. 145.85. Suppliers: KK, Harshaw (85%).

For reaction with diphenyl disulfide to produce phenylsulfur trifluoride, Sheppard[1] added the contents of one can of Harshaw reagent (435–470 g. of a black powder)

$$\underset{0.458\text{ m.}}{(C_6H_5S)_2} + \underset{3.1\text{m.}}{6\ AgF_2} \xrightarrow[55-60\%]{Cl_2FCCClF_2} 2\ C_6H_5SF_3 + 6\ AgF$$

to a flask flushed with dry nitrogen, added 500 ml. of Freon 113 (b.p. 47°, available from du Pont), and stirred the mixture under nitrogen during gradual addition of diphenyl disulfide at a temperature of 35–40° (45–60 min.). The reaction mixture is filtered under nitrogen to remove yellow silver fluoride and the product collected by distillation.

[1]W. A. Sheppard, *Org. Syn.*, **44**, 82 (1964)

Silver diphenylphosphate, $(C_6H_5O)_2\overset{O}{\overset{\|}{P}}OAg$. Mol. wt. 357.05. Preparation.[1] Supplier: Eastman.

Posternak[2] used the reagent for the synthesis of α-D-glucose 1-phosphate (4) as follows:

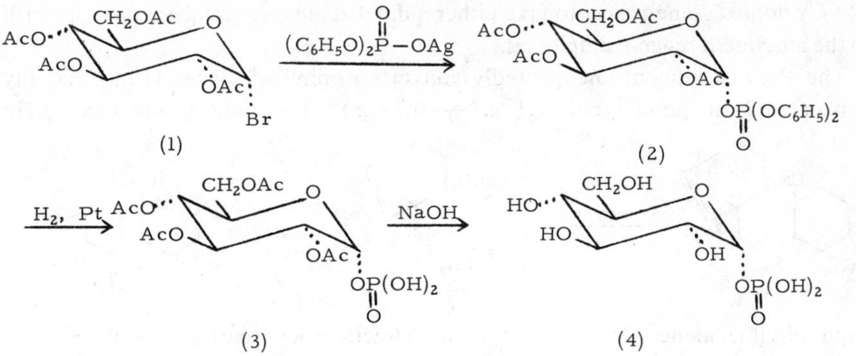

(1)

(2)

(3)

(4)

Tetra-O-acetyl-α-D-glucopyranosyl bromide (1) usually reacts with inversion at C_1 but it reacts with silver diphenylphosphate with retention of configuration. The phosphate ester (4) was isolated as the crystalline potassium salt.

[1]T. Posternak, *J. Biol. Chem.*, **180**, 1269 (1949)
[2]T. Posternak, *Am. Soc.*, **72**, 4824 (1950)

Silver iododibenzoate (Prévost reagent[1]). Reviews.[2]

Preparation. The reagent can be prepared by refluxing a mixture of silver

$$2 \ C_6H_5CO_2Ag \ + \ I_2 \longrightarrow Ag(C_6H_5CO_2)_2I \ + \ AgI$$

$$2 \times 228.99 = 457.98 \quad 253.82 \quad\quad 477.10$$

benzoate and iodine in benzene.[3] The crystalline complex which separates can be purified by extraction with benzene in a Soxhlet.

Conversion of olefins into vic-glycol dibenzoates. This characteristic reaction is carried out in benzene and proceeds as follows:

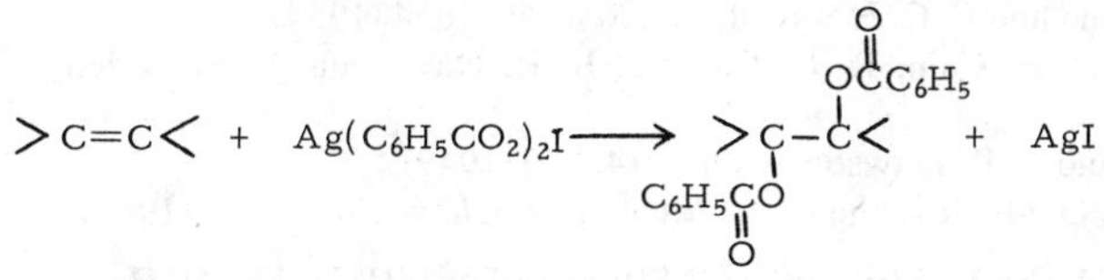

The precipitated silver iodide is removed by filtration and the dibenzoate isolated by evaporation. The reaction can be carried out with preformed reagent or with reagent generated *in situ*. As part of the aldehyde synthesis formulated, Hershberg[3] refluxed a mixture of 0.1 mole of allylbenzene, 45.8 g. of silver benzoate, 25.4 g.

$$C_6H_5MgBr \xrightarrow[82\%]{CH_2=CHCH_2Br} C_6H_5CH_2CH=CH_2 \xrightarrow[85\%]{2\ C_6H_5CO_2Ag\ +\ I_2}$$

$$\underset{\underset{OCOC_6H_5}{|}}{C_6H_5CH_2CHCH_2OCOC_6H_5} \xrightarrow[84\%]{OH^-} \underset{\underset{OH}{|}}{C_6H_5CH_2CHCH_2OH} \xrightarrow[72\%]{Pb(OAc)_4} C_6H_5CH_2CHO$$

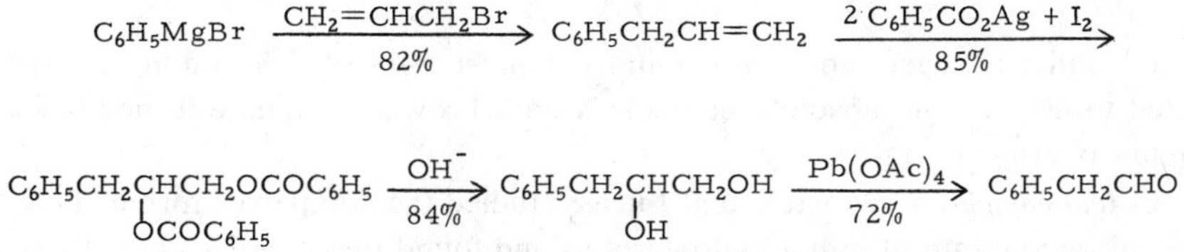

of iodine, and 300 ml. of benzene for 15 hrs. with exclusion of moisture. Filtration from silver iodide and workup afforded 3-phenylpropane-1,2-diol, and saponification and glycol cleavage gave phenylacetaldehyde.

Test cases show that the Prévost reaction proceeds by *trans* addition. Thus it adds to methyl oleate to give, after hydrolysis, the low-melting *threo*-9,10-dihydroxy-stearic acid in 75% yield and with methyl elaidate to give the *erythro*-diol (69%).[4]

$\Delta^{1,4}$-Cyclohexadiene reacts to give either a dibenzoate or a tetrabenzoate, according to the amount of reagent employed.[5]

The Prévost reagent unexpectedly converts isophyllocladene (1) into the allylic benzoate (2) in good yield (1.1 g. ⟶ 0.79 g.).[6] The same product is obtained

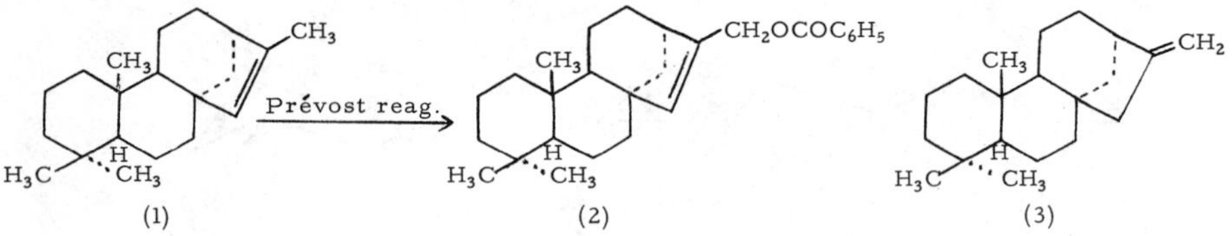

from phyllocladene (3) because iodine effects prior rearrangement of this to isophyllocladene (1).

Silver salts of other acids have been employed in the Prévost reaction, as summarized by Wilson.[2] As a means of preparing *trans*-glycols, the method seems to us less convenient and more costly than reaction with performic acid and hydrolysis. The presence of free iodine may lead to complications. A methoxybenzene with an olefinic side chain gave a glycol having an iodine substituent *para* to the methoxyl group.[7]

The Prévost reagent converts Δ^2-cholestene into both possible *trans*-diol dibenzoates.[8] The diaxial $2\beta,3\alpha$-isomer predominates over the diequatorial $2\alpha,3\beta$-isomer in the ratio of 2:1, but difficulty in the separation limits the total yield to 31%.

[1]C. Prévost, *Compt. rend.*, **196**, 1129 (1933); **197**, 1661 (1933)
[2]C. V. Wilson, *Org. Reactions*, **9**, 332 (1957); F. D. Gunstone, "Advances in Organic Chemistry, Methods and Results," **1**, 122 (1960)
[3]E. B. Hershberg, *Helv.*, **17**, 351 (1934)
[4]H. Wittcoff and S. E. Miller, *Am. Soc.*, **69**, 3138 (1947)
[5]G. E. McCasland and E. C. Horswill, *Am. Soc.*, **76**, 1654 (1954)
[6]L. H. Briggs, B. F. Cain, R. C. Cambie, B. R. Davis, and P. S. Rutledge, *J. Chem. Soc.*, 1850 (1962)
[7]M. Sletzinger and C. R. Dawson, *J. Org.*, **14**, 670 (1949)
[8]P. S. Ellington, D. G. Hey, and G. D. Meakins, *J. Chem. Soc.*, 1327 (1966)

Silver nitrate, $AgNO_3$. Mol. wt. 169.89. Suppliers: B, E, F, MCB.

Oxidation, *see also* Silver oxide. In one step of a synthesis of linoleic acid, Walborsky and co-workers[1] oxidized $\Delta^{9,12}$-stearadiyneal to the corresponding acid by stirring a solution of 2.34 g. of aldehyde and 2.32 g. of silver nitrate in absolute

$$CH_3(CH_2)_4C{\equiv}CCH_2C{\equiv}C(CH_2)_7CHO \xrightarrow[78\%]{AgNO_3-NaOH} CH_3(CH_2)_4C{\equiv}CCH_2C{\equiv}C(CH_2)_7CO_2H$$

ethanol under nitrogen and adding in 10 min. 4.5 ml. of 5 *N* sodium hydroxide diluted to 40 ml. with absolute ethanol. A dark brown precipitate turned black on standing overnight.

π-Acid complexes. Winstein and Lucas[2] studied the complexes formed between silver nitrate and unsaturated hydrocarbons and found that in some cases these are solids useful for isolation and purification. Stille and Frey[3] investigated the Diels-Alder addition of norbornadiene (1) and cyclopentadiene (2) and obtained the 1:1 adduct (3) as the chief product, along with small amounts of the 1:2 adduct and of cyclopentadiene dimer, trimer, and tetramer. Redistilled adduct (3) of about 97% purity was obtained in completely pure form through the silver nitrate complex (4), prepared by adding 10 ml. of a saturated solution of silver nitrate in water

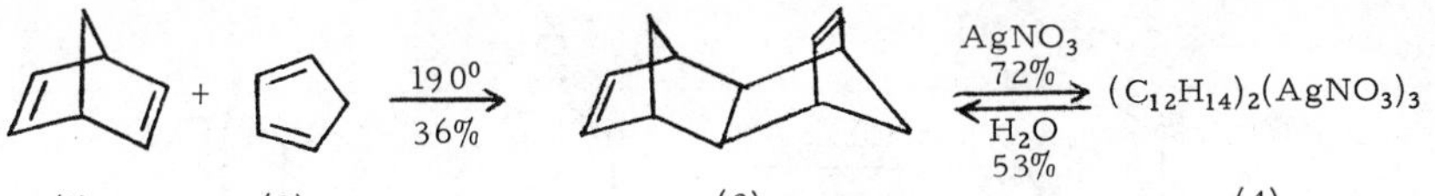
+
190°
36%
AgNO3
72%
H2O
53%
(C12H14)2(AgNO3)3

(1) (2) (3) (4)

slowly with stirring to 7.9 g. (0.05 mole) hydrocarbon. The mixture set up to a hard gel, which was removed by filtration and crystallized from hot absolute ethanol. Thus obtained, the complex melted at 157–158° dec. and was analytically pure. For recovery of the hydrocarbon the complex (14.8 g.) was dissolved in 150 ml. of warm water and the solution subjected to continuous ether extraction. Hydrocarbon (5), which is a dihydro derivative of (3) but which was prepared by addition of

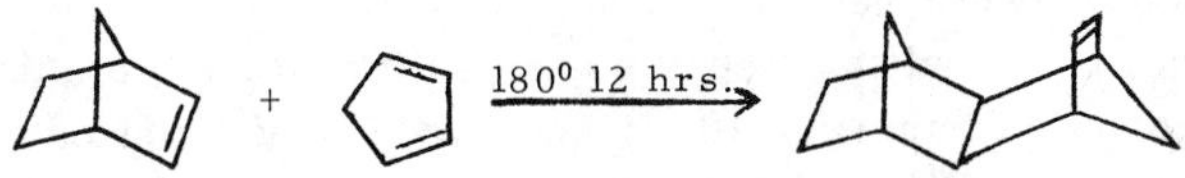
+
180° 12 hrs.

(4) (5)

cyclopentadiene to norbornene (4),[4] was purified by Stille and Witherell[5] for a rate study through its 1:1 silver nitrate complex. Pure complex was obtained in 77% yield and split by refluxing 100 g. of complex with 400 ml. of ammonium hydroxide for 3 hrs., cooling, and extracting with ether (yield 71%).

Complex formation provides a simple method for purifying straight-chain α,ω-dienes containing 12 or more carbon atoms.[6] $\Delta^{1,11}$-Dodecadiene prepared according to Drahowzal[7] presents particular difficulty because it co-distills with 1,6-dichlorohexane. On addition of 1 mole of crude oily diene to a saturated solution of 2.37

$$Cl(CH_2)_6Cl \longrightarrow ClMg(CH_2)_6MgCl \xrightarrow{2\ CH_2{=}CHCH_2Cl} CH_2{=}CH(CH_2)_8CH{=}CH_2$$

moles of silver nitrate in 243 ml. of distilled water with rapid stirring a solid complex separated and was collected, washed with cold absolute ethanol, air dried, and crystallized from hot ethanol.[8] The purified complex was decomposed with water and the hydrocarbon collected by ether extraction and distilled. A gas chromatogram showed only one peak. The yield was 18.9% based on the 1,6-dichlorohexane used.

Silica gel impregnated with aqueous silver nitrate and dried at 120° displays highly selective adsorption properties with respect to the geometry, degree of substitution, and number of double bonds of the unsaturated lipid.[9] Thus, by elution with mixtures of benzene or ether with petroleum ether, sharp separations were made of a mixture of methyl stearate, elaidate, and oleate and of a mixture of methyl oleate, linoleate, and linolenate. With the use of this π-acid complexing adsorbent, de Vries[9] effected quantitative separation of cholesterol and cholestanol; this separation has been achieved by chromatography on alumina, but with considerably greater difficulty.[10] Silver nitrate-impregnated silica gel also has been shown to effect separation of 1-oleodistearin and 2-oleodistearin.[11]

Pesnelle and Ourisson[12] used $AgNO_3$-impregnated silica gel to advantage in separation of the mixture of the hydrocarbons (2) and (3) resulting on dehydration of the sesquiterpene alcohol (1). The major product (2, *trans* elimination) has a trisubstituted double bond; in the minor product (3, α-gurjunene, *cis* elimination) the double bond is tetrasubstituted and this isomer was retained more strongly

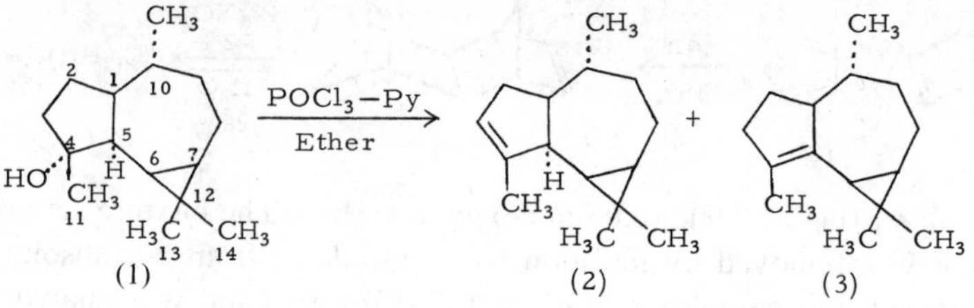

on the selective adsorbent, in accordance with a rule for monoethylenic sesquiterpenes that Rf values are between 0.90 and 0.95 for the tetrasubstituted types, between 0.60 and 0.75 for the trisubstituted type, and between 0.25 and 0.40 for hydrocarbons having a methylene group.

Gunstone *et al.*[13] found that separation of saturated and unsaturated fatty acid glycerides by crystallization at -10° is improved by addition of silver nitrate in methanol to complex with the unsaturated components. Morris[14] reports separation of *cis* and *trans* fatty acids and oxygenated fatty acids by thin-layer chromatography on silica gel impregnated with silver nitrate. With untreated silica there was no separation.

Wolovsky[15] describes the oxidative coupling of 1,5-hexadiyne, followed by prototropic rearrangement by heating with potassium *t*-butoxide in *t*-butanol–benzene at 100°. The resulting mixture of unsaturated hydrocarbons was easily separated by elution chromatography on a column of alumina impregnated with 20% silver nitrate into the tetradehydro[18]annulene (III) and the two isomeric tridehydro[18]annulenes (I) and (II).

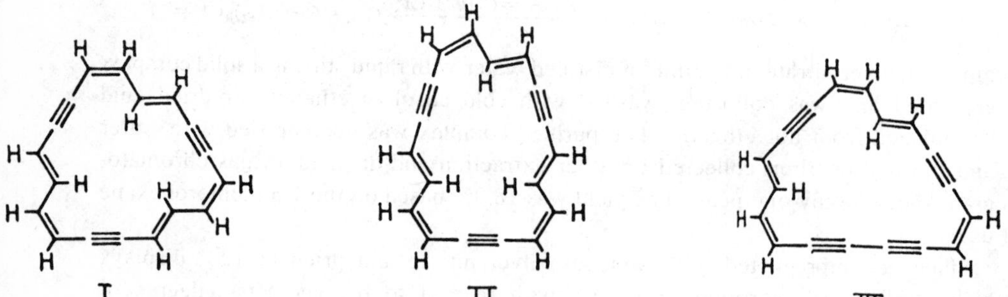

I II III

Silver nitrate complexes are very useful for separation of the mixture of 1,3-, 1,4-, and 1,5-cyclooctadienes obtained by sodium–alcohol reduction of cyclooctatetraene.[16] The three adducts differ in stability in aqueous silver nitrate. All are stable at 0–5°; that from the 1,3-diene dissociates at 30°; that from the 1,4-diene dissociates above 60°; that from the 1,5-isomer is stable at 90–100°. The pure dienes are recovered by steam distillation.

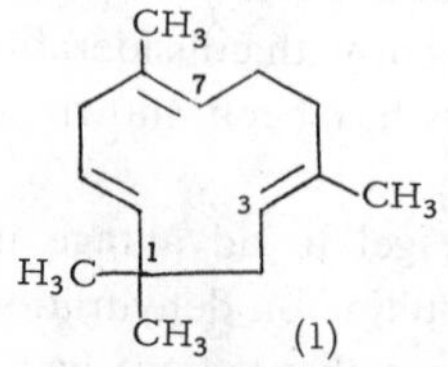

(1)

Humulene (1) is readily purified as the silver nitrate adduct, $C_{15}H_{24}\cdot 2\,AgNO_3$. It is regenerated by steam distillation or by treatment with aqueous ammonia.[17]

Dehydrobromination. Silver nitrate in ethanol proved to be a very effective agent for dehydrobromination of the ketone (1).[18]

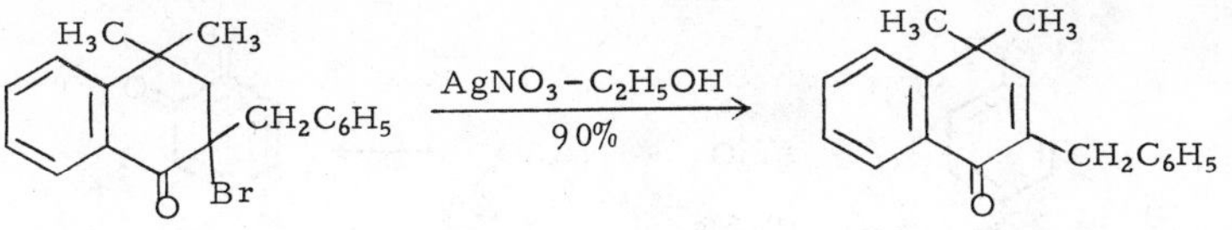

(1) (2)

[1]H. M. Walborsky, R. H. Davis, and D. R. Howton, *Am. Soc.*, **73**, 2593 (1951)
[2]S. Winstein and H. J. Lucas, *Am. Soc.*, **60**, 836 (1938)
[3]J. K. Stille and D. A. Frey, *Am. Soc.*, **81**, 4273 (1959)
[4]S. B. Soloway, *Am. Soc.*, **74**, 1027 (1952)
[5]J. K. Stille and D. R. Witherell, *Am. Soc.*, **86**, 2188 (1964)
[6]Contributed by Paul D. Klimstra, G. D. Searle and Co.
[7]F. Drahowzal, *Monatsh.*, **82**, 793 (1951)
[8]S. Wawzonek, P. D. Klimstra, R. E. Kallio, and J. E. Stewart, *Am. Soc.*, **82**, 1421 (1960)
[9]B. de Vries, *Chem. Ind.*, 1049 (1962)
[10]L. F. Fieser, *Am. Soc.*, **75**, 4395 (1953)
[11]C. B. Barrett, M. S. J. Dallas, and F. B. Padley, *Chem. Ind.*, 1050 (1962)
[12]P. Pesnelle and G. Ourisson, *J. Org.*, **30**, 1744 (1965)
[13]F. D. Gunstone, R. J. Hamilton, and M. I. Qureshi, *J. Chem. Soc.*, 319 (1965)
[14]L. J. Morris, *Chem. Ind.*, 1238 (1962)
[15]R. Wolovsky, *Am. Soc.*, **87**, 3638 (1965)
[16]W. J. Jones, *J. Chem. Soc.*, 312 (1954)
[17]R. P. Hildebrand and M. D. Sutherland, *Australian J. Chem.*, **14**, 272 (1961); *see also* N. P. Damodaran and S. Dev, *Tetrahedron Letters*, 1977 (1965)
[18]N. H. Cromwell, R. P. Ayer, and P. W. Foster, *Am. Soc.*, **82**, 130 (1960)

Silver nitrite, $AgNO_2$. Mol. wt. 153.89, dec. 140°, solubility in 100 g. H_2O 1.06 g. at 60°, light-sensitive. Suppliers: Mallinckrodt, Fisher. Preparation.[1-3]

Victor Meyer[4] refluxed amyl iodide with silver nitrite and obtained a mixture of the alkyl nitrite and nitroalkane. By improved procedures[1,2] it is possible to obtain pure nitroparaffins from primary straight-chain halides in high yield. An example is the preparation of 1-nitrooctane.[2] 1-Bromooctane is added in 2 hrs. with stirring

$$\underset{0.5\text{ m.}}{CH_3(CH_2)_7Br} + \underset{0.75\text{ m.}}{AgNO_2} \xrightarrow[75\text{-}80\%]{\text{150 ml. Ether, 66 hrs.}} CH_3(CH_2)_7NO_2 + AgBr$$

under reflux and cooling to a suspension of silver nitrite in ether. The mixture is stirred for 24 hrs. in the ice bath, and then for 40 hrs. at room temperature. After filtration and removal of solvent, fractionation affords 14% of 1-octyl nitrite (b.p. 37°/3 mm.) and then 1-nitrooctane (b.p. 66°/2 mm.).

[1]N. Kornblum, B. Taub, and H. E. Ungnade, *Am. Soc.*, **76**, 3209 (1954); N. Kornblum, *Org. Reactions*, **12**, 101 (1962)
[2]N. Kornblum and H. E. Ungnade, *Org. Syn., Coll. Vol.*, **4**, 724 (1963)
[3]C. W. Plummer and N. L. Drake, *Am. Soc.*, **76**, 2720 (1954)
[4]V. Meyer and O. Stüber, *Ber.*, 203 (1872)

Silver oxide, Ag_2O. Mol. wt. 231.76. Suppliers: Fisher, Mallinckrodt.

Oxidation of hydroquinones. *o*-Benzoquinone, a highly reactive red compound sensitive even to water, eluded all attempts to obtain it by oxidation of catechol until Willstätter and Pfannenstiel[1] found that the reaction can be effected with silver oxide prepared by precipitation from aqueous solution and washed extensively with several portions of distilled water, then with acetone, and finally with absolute ether. A solution of catechol in absolute ether was treated with anhydrous sodium

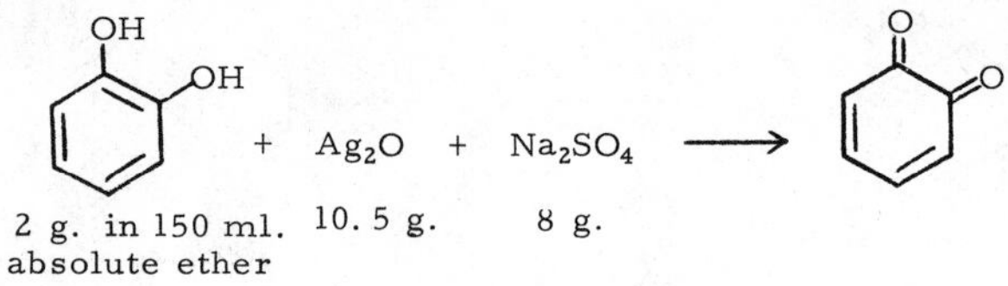
OH
OH
+ Ag2O + Na2SO4
O
O
2 g. in 150 ml. absolute ether
10. 5 g.
8 g.

sulfate, to absorb the water formed, and the dry silver oxide added with shaking until there was no further deepening in color. On concentration of the filtered solution *o*-benzoquinone separated in shiny, bright red plates. The yield is low. The method

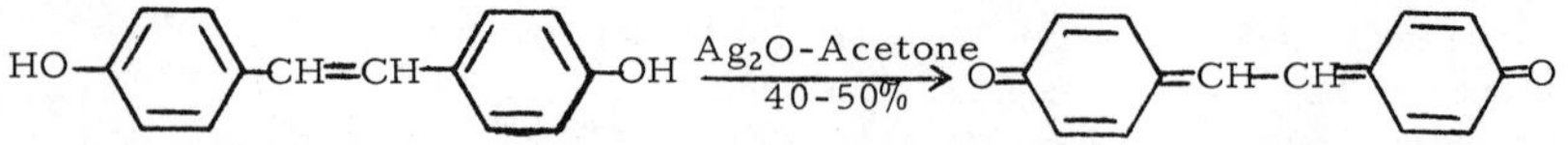
HO
CH=CH
OH
Ag2O-Acetone
40-50%
O
CH-CH
O

is useful for the preparation of other sensitive quinones of high oxidation potential such as stilbenequinone.[2] The hydroquinone is refluxed in acetone with a large excess of reagent.

Silver oxide is the reagent of choice for oxidation of vitamin K_1 hydroquinone,[3] because this substance is readily soluble in ether and because the conversion is quantitative; evaporation of the filtered solution gives analytically pure K_1, a yellow oil. In this case commercial reagent is satisfactory.

***ArCHO* ⟶ *ArCO$_2$H*.** For oxidation of 3-thenaldehyde to 3-thenoic acid, Campaigne and LeSuer[4] mixed solutions of silver nitrate and excess alkali to

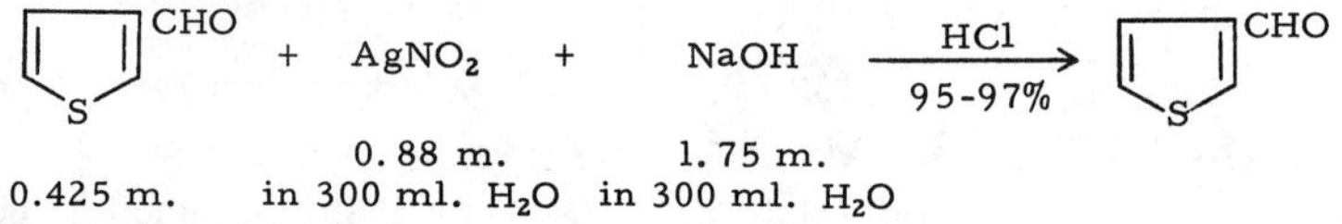
CHO
S
+ AgNO2 + NaOH
HCl
95-97%
CHO
S
0.425 m.
0. 88 m. in 300 ml. H2O
1. 75 m. in 300 ml. H2O

precipitate brown silver oxide, added the aldehyde with stirring, removed the black precipitate of silver by filtration, and acidified the filtrate. The black silver can be dissolved in concd. nitric acid for reuse.

For oxidation of vanillin to vanillic acid, Pearl[5] used an economical procedure which is unique to *p*-hydroxybenzaldehydes. Whereas aldehydes of other types, like 3-thenaldehyde, require two moles of silver nitrate for alkaline oxidation, a *p*-hydroxyaldehyde requires only one mole of reagent.[6] Pearl mixed aqueous

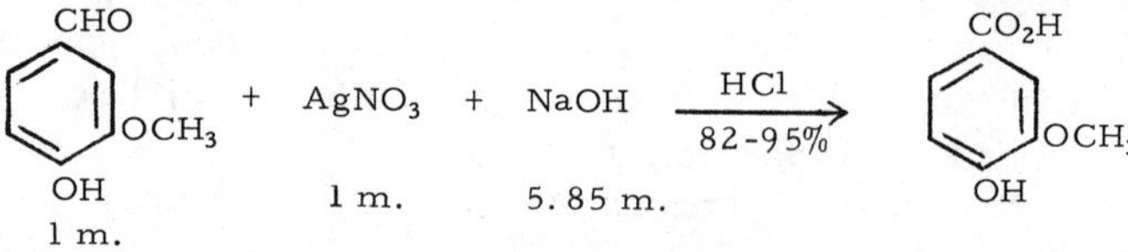
CHO
OCH3
OH
1 m.
+ AgNO3 + NaOH
1 m.
5. 85 m.
HCl
82-95%
CO2H
OCH3
OH

solutions of 1 mole each of silver nitrate and sodium hydroxide, stirred the mixture for 5 min., and collected the silver oxide and washed it free of nitrates with water. The wet oxide was covered with 2 l. of water and treated with 4.85 moles of sodium hydroxide pellets with vigorous stirring, the temperature was adjusted to 55–60°, and 1 mole of vanillin was added. The reaction, which begins in a few minutes and is strongly exothermal, involves two stages. The first stage is oxidation by Ag_2O in the presence of alkali to give sodium vanillate and spongy silver metal. The second stage involves conversion of more vanillin into the acid by reaction with spongy silver and alkali (note the hugh excess of NaOH). After stirring for 10 min. the mixture was filtered, and the filtrate and washings treated with sulfur dioxide to prevent the product from becoming tan. Acidification afforded white needles melting at 209–210° (pure material 210–211°). Pearl[7] discusses the possibility of a special Cannizzaro reaction catalyzed by silver, but if this were true the maximum yield would be 0.5 mole by Ag_2O oxidation and 0.25 mole by the Cannizzaro reaction; the actual yield is well above 75%. Pearl[8] obtained vanillic acid in 89–95% yield by fusion with a mixture of sodium and potassium hydroxide at 180–195°, and in this case the reaction involves evolution of hydrogen:

$$ArCHO + 2\,KOH \longrightarrow ArCO_2K + H_2 + H_2O$$

It would be interesting to know if hydrogen is evolved in the reaction with spongy silver and aqueous alkali at 55–60°.

Oxidation of sugars. Evans and co-workers[9] studied the oxidation of glucose, fructose, mannose, and galactose in 0.025 *M* solution with freshly precipitated silver oxide alone and with added alkali. The oxidation products, carbon dioxide, oxalic acid, formic acid, and glycolic acid, can all be determined quantitatively, and by removing the mixture of silver and silver oxide by filtration and extracting the silver oxide with ammonium hydroxide the oxygen consumed can be calculated from the weight of residual silver.

Oxidation of hydrazones. Schroeder and Katz[10] found that oxidation of the hydrazone (1) by the usual mercuric oxide procedure gave erratic results and found silver oxide to be a much better oxidant. When (1) was shaken with silver oxide and magnesium sulfate in ether, the diaryldiazomethane was obtained in 99% yield. Mercuric oxide failed also to oxidize the hydrazone (2), but the diazo compound was

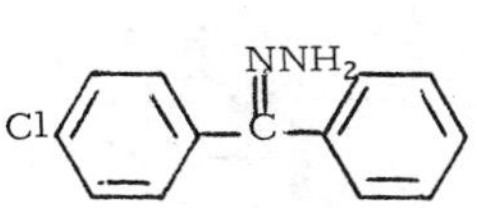
NNH2
Cl
C

(1)

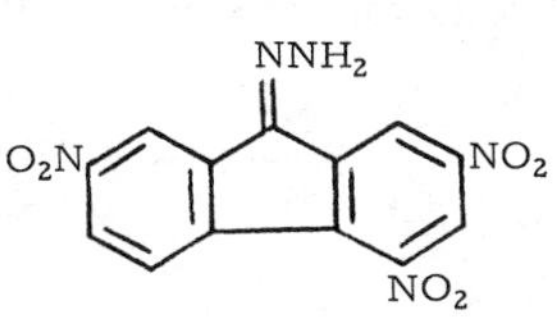
NNH2
O2N
NO2
NO2

(2)

obtained in 89% yield by refluxing (2) with silver oxide in tetrahydrofurane for 5 hrs.[11] Hüttel *et al.*[12] also found silver oxide superior to mercuric oxide.

Dehydrohalogenation. For dehydrohalogenation of the allylic bromide (1), Steiner and Schinz[13] precipitated silver oxide from 5 g. of $AgNO_3$, washed it with

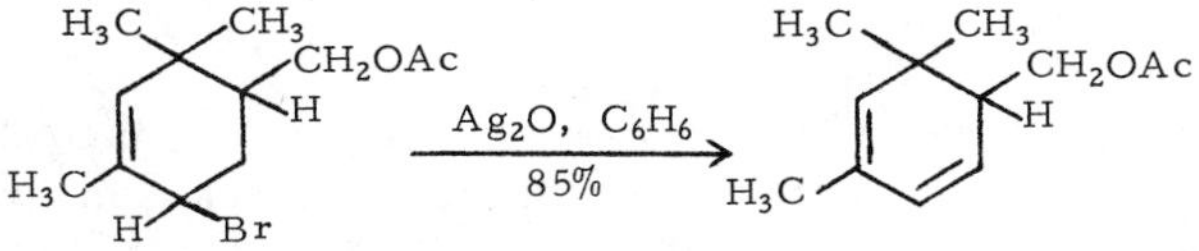
H3C
CH3
CH2OAc
H
H3C
H
Br
Ag2O, C6H6
85%
H3C
CH3
CH2OAc
H
H3C

(1) (2)

water and then ethanol, and dried it in high vacuum. The oxide was added with cooling to a solution of 4 g. of the halide (1) in benzene and the mixture shaken for 15 hrs.

Schmidlin and Wettstein[14] converted the *trans*-diequatorial bromohydrin (3) to the epoxide (4) by treating a solution of 1.132 g. of material in 21.5 ml. of pyridine with 4.143 g. of freshly precipitated and dried silver oxide in the dark for 48 hrs., adding 2.072 g. more silver oxide, and shaking for 72 hrs.

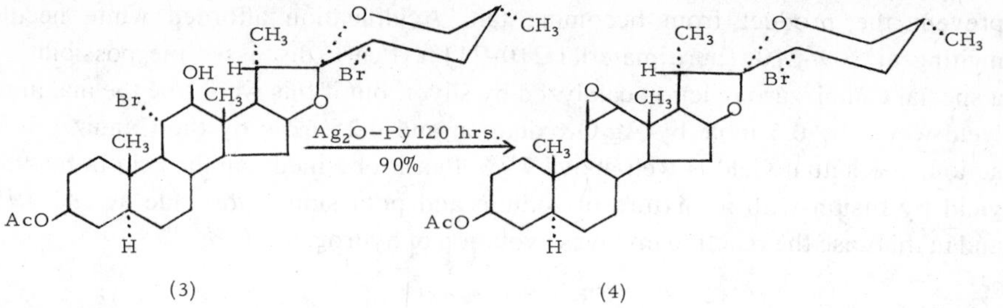

Silver oxide converts homoallylic halides into cyclopropane derivatives, for example:[15]

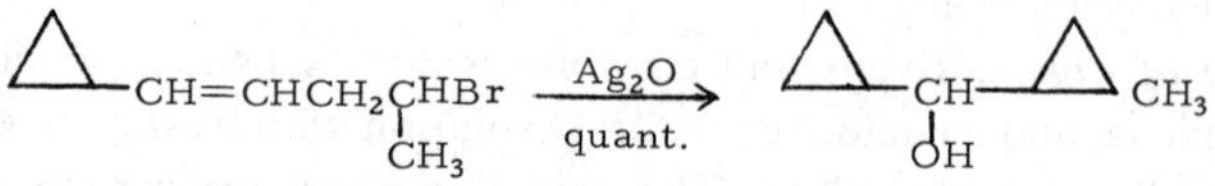

Potassium carbonate effects the same reaction, but yields are lower.

Amino acids from their hydrochlorides. β-Alanine can be isolated from an aqueous solution of the crude hydrochloride by stirring with moist, precipitated silver oxide, filtering, and treating the filtrate with hydrogen sulfide.[16] For recovery of ϵ-aminocaproic acid from the hydrochloride, Eck[17] used two portions of powdered litharge and 1 portion each of precipitated lead hydroxide and silver oxide, followed by treatment with hydrogen sulfide.

Synthesis of polyhydroxy-L-proline (6). In an improved procedure described by Katchalski *et al.*,[18] O-acetylhydroxy-L-proline (2) was converted into the carbonyl chloride (3), and this was cyclized to the Leuchs anhydride (4) by treatment with

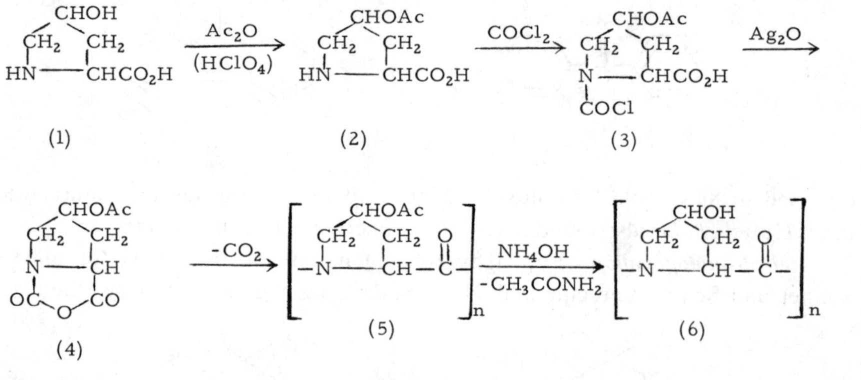

silver oxide in dry acetone. The crystalline anhydride on polymerization in pyridine yielded poly-O-acetylhydroxyproline (5) and this on deacetylation with ammonium hydroxide afforded the polymer (6).

Decomposition of diazoketones. Diazoketones of the general formula (1) when treated with freshly prepared silver oxide in dioxane decompose at room temperature

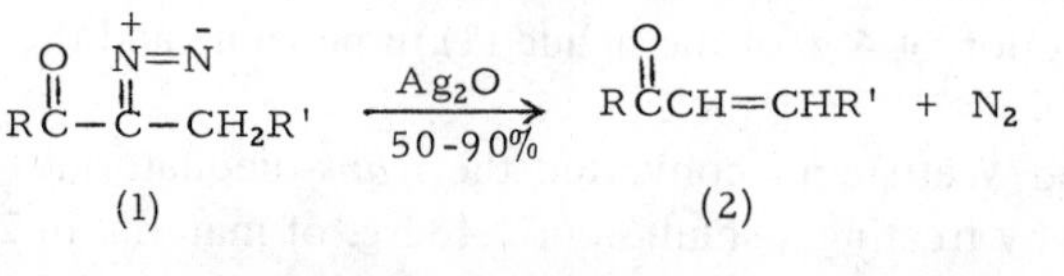

to α,β-unsaturated ketones.[19] The reaction can also be realized by photolysis, but in lower yield. The reaction proceeds through an intermediate ketocarbene.

[1]R. Willstätter and A. Pfannenstiel, *Ber.*, **37**, 4744 (1904)
[2]K.-H. König, W. Schulze, and G. Möller, *Ber.*, **93**, 555 (1960)
[3]L. F. Fieser, *Am. Soc.*, **61**, 3467 (1939)
[4]E. Campaigne and W. M. LeSuer, *Org. Syn., Coll. Vol.*, **4**, 919 (1963)
[5]I. A. Pearl, *ibid.*, **4**, 972 (1963)
[6]We are indebted to R. L. Shriner, senior checker of the procedure, for explaining the nature of this special reaction.
[7]I. A. Pearl, *Am. Soc.*, **68**, 429 (1946); *J. Org.*, **12**, 79 (1947)
[8]I. A. Pearl, *Org. Syn., Coll. Vol.*, **4**, 974 (1963)
[9]W. L. Evans *et al.*, *J. Org.*, **1**, 1 (1936)
[10]W. Schroeder and L. Katz, *J. Org.*, **19**, 718 (1954)
[11]A. K. Colter and S. S. Wang, *J. Org.*, **27**, 1517 (1962)
[12]R. Hüttel, J. Riedl, H. Martin, and K. Franke, *Ber.*, **93**, 1425 (1960)
[13]U. Steiner and H. Schinz, *Helv.*, **34**, 1176 (1951)
[14]J. Schmidlin and A. Wettstein, *Helv.*, **36**, 1241 (1953)
[15]M. Hanack and H. Eggensperger, *Ber.*, **96**, 1259 (1963); M. Hanack and K. Görler, *ibid.*, **96**, 2121 (1963)
[16]H. T. Clarke and L. D. Behr, *Org. Syn., Coll. Vol.*, **2**, 19 (1945)
[17]J. C. Eck, *Org. Syn., Coll. Vol.*, **2**, 28 (1945)
[18]J. Kurtz, G. D. Fasman, A. Berger, and E. Katchalski, *Am. Soc.*, **80**, 393 (1958)
[19]V. Franzen, *Ann.*, **602**, 199 (1957)

Silver sulfate, Ag_2SO_4. Mol. wt. 311.83. Suppliers: B, E, F, MCB. Derbyshire and Waters[1] found that addition of silver sulfate to a solution of bromine in sulfuric acid produces a highly reactive brominating electrophile which probably is $AgBr_2^+$. Aromatic compounds deactivated by *m*-directing substituents, for example nitrobenzene and 2,4-dinitrotoluene, can be brominated in good yield. Silver bromide is formed and is filtered off.

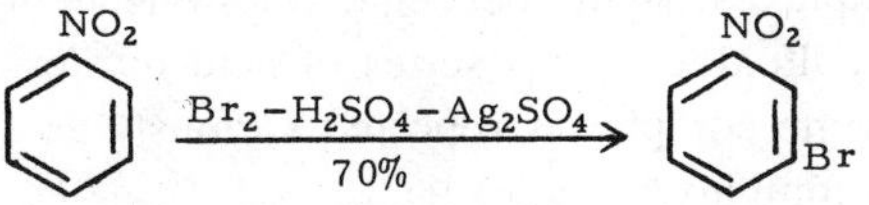
NO2
Br2–H2SO4–Ag2SO4
70%
NO2
Br

1-Phenylpyrazole (1) on bromination under ordinary conditions is brominated in the 4-position of the pyrazole ring, but Lynch and co-workers[2] found that under the

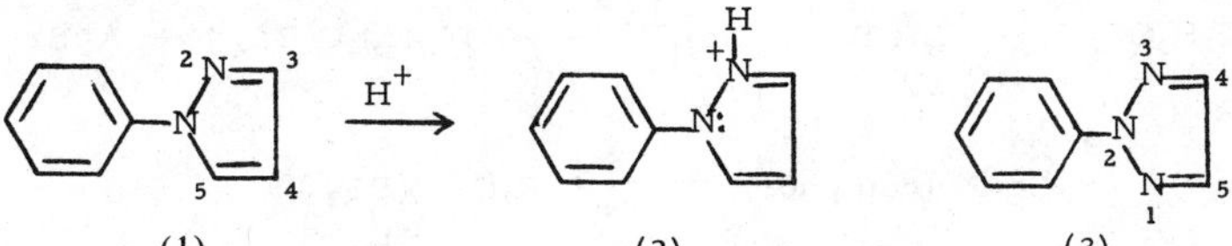
2
N
3
N
5
4
H+
H
+
N
N:
3
N
4
N
2
N
5
1

(1) (2) (3)

Waters conditions the product is 1-*p*-bromophenylpyrazole (62% yield). In the strongly acidic medium, protonation gives the conjugate acid (2) in which the pyrazole ring is deactivated toward electrophiles, whereas the +T-effect of the 1-nitrogen atom facilitates *p*-substitution. The reagent brominates unreactive 2-phenyl-1,2,3,2H-triazole (3) to give the *p*-bromophenyl derivative in high yield.[3]

[1]D. H. Derbyshire and W. A. Waters, *J. Chem. Soc.*, 573 (1950)
[2]M. A. Khan, B. M. Lynch, and Y.-Y. Hung, *Can. J. Chem.*, **41**, 1540 (1963)
[3]B. M. Lynch, *ibid.*, **41**, 2380 (1963)

Silver tetrafluoroborate, $AgBF_4$. Mol. wt. 194.70, dec. 200°. Suppliers: Allied Chem. Corp., Gen. Chem. Div.; Ozark-Mahoning Co.

Preparation. A procedure described by Meerwein *et al.*[1] involving reaction of boron trifluoride etherate with a suspension of silver oxide in nitromethane is reported to have resulted in "an explosion of great violence."[2] A simple and safe method of preparing pure silver tetrafluoroborate involves reaction of boron trifluoride with silver monofluoride in nitromethane[3] or benzene.[4]

$$AgF + BF_3 \xrightarrow[90\%]{CH_3NO_2} AgBF_4$$

Procedure of Olah and Quinn. Anhydrous silver tetrafluoroborate is prepared most conveniently by reaction of silver fluoride (Harshaw Chem. Co.) with boron trifluoride in nitromethane.[3] A suspension of 1 mole of silver fluoride in 170 ml. of nitromethane is stirred magnetically while passing in boron trifluoride gas with protection from air moisture. The temperature rises to 60° and is checked there by cooling. Addition of boron trifluoride is stopped when the silver fluoride has almost completely dissolved to give a colorless but murky solution. Dry nitrogen is passed through the solution for 1 hr. to remove excess boron trifluoride, the solution is filtered under nitrogen through a sintered glass disk, and the nitromethane is removed by distillation at reduced pressure at 70°. The off-white solid residue is placed in a mortar in a dry box, covered with *n*-pentane, and ground to a fine particle size. The silver fluoroborate is pumped dry at 70° and obtained in yield of 86–94%. It is stored under *n*-pentane.

Complexes. Silver tetrafluoroborate is insoluble in cyclohexane but very soluble in water, ether, toluene, and nitromethane, and moderately soluble in benzene and cyclohexene.[1,5] Solubility in the unsaturated hydrocarbons is attributed to π-bond formation between the unsaturated compound and silver ion. Complexes have been observed with both π-electron donors and lone-pair donors; thus Meerwein[1] prepared stable complexes with benzene, mesitylene, cycloheptatriene, diethyl ether, dimethyl sulfoxide, and a series of acid nitriles. However, the silver tetrafluoroborate–benzene complex is reported variously as a 1:1 complex,[1] a 1:2 complex,[4] and as a 2:3 complex.[6]

Alkylation. Meerwein[1] reported efficient ethylation of ethers and ketones with ethyl bromide and silver tetrafluoroborate in ethylene dichloride at room temperature:

$$C_2H_5OC_2H_5 + C_2H_5Br + AgBF_4 \xrightarrow[89\%]{} (C_2H_5)_3\overset{+}{O}(BF_4^-) + AgBr$$

$$R_2C{=}O \text{ (camphor)} \xrightarrow[59\%]{} R_2C{-}\overset{+}{\underset{C_2H_5}{O}}(BF_4^-)$$

$$(C_2H_5O)_2C{=}O \xrightarrow[72\%]{} (C_2H_5O)_2C{=}\overset{+}{O}C_2H_5(BF_4^-)$$

In the case of a γ-bromo ester, intramolecular alkylation can be realized:

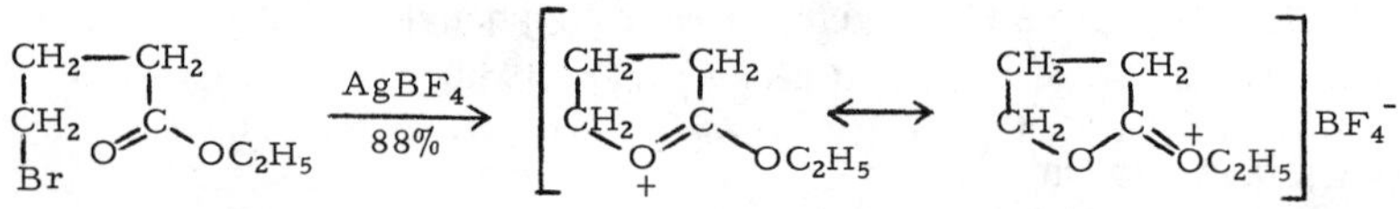

Eschenmoser *et al.*[7] used the reagent in a novel synthesis of N-methylamino acids. L-Tryptophane methyl ester hydrochloride (1) was condensed with γ-chlorobutyryl chloride in the presence of pyridine, and the derivative (2) was cyclized to the iminolactone (3) with a slight excess of silver tetrafluoroborate in methylene chloride–

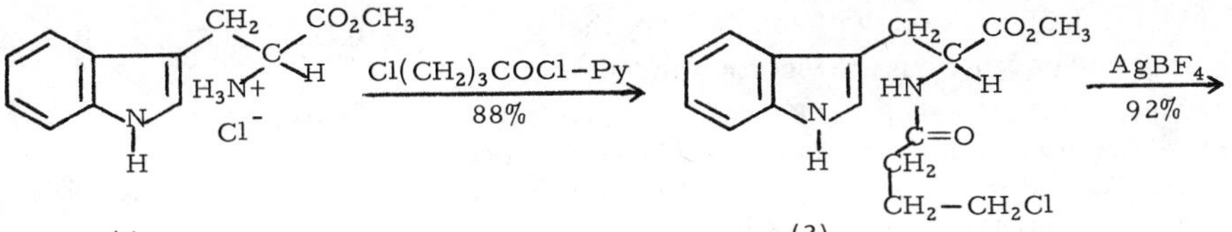

(1) (2)

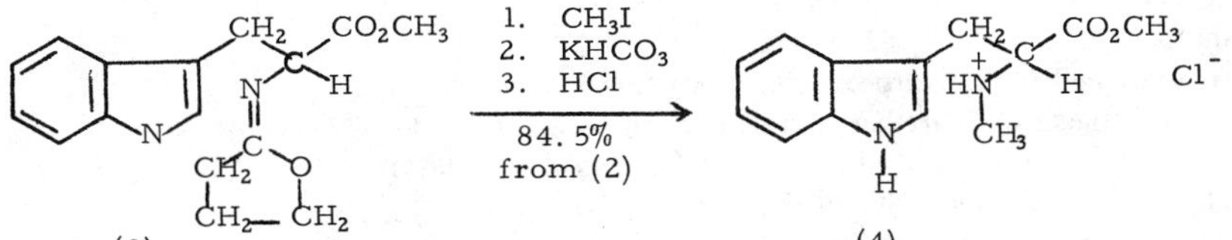

(3) (4)

benzene at −20 to +20°. The iminolactone was then alkylated with methyl iodide in acetonitrile (41 hrs. at 30°) to the quaternary monomethyl derivative, from which the protective group was removed by treatment with potassium bicarbonate (2 hrs. at 20°). The product was isolated as the hydrochloride (4).

$R_2CHBr \longrightarrow R_2C{=}O$. Lemal and Fry[2] used the reagent in a modification of Kornblum's method of oxidation with dimethyl sulfoxide. 3-Nortricyclyl bromide (1) was added to a solution of an equivalent amount of silver tetrafluoroborate in dimethyl sulfoxide, whereupon silver bromide began to precipitate. After 1 hr.,

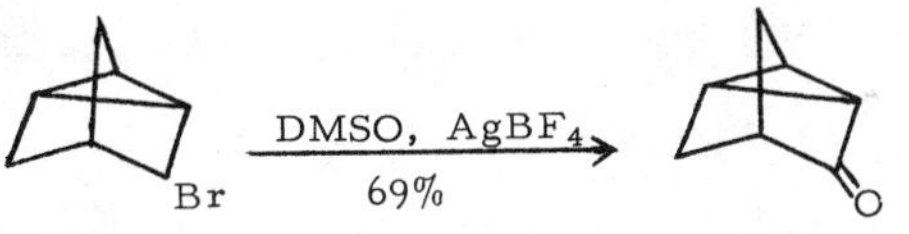

(1) (2)

when formation of the dimethyl-3-nortricyclysulfonium salt appeared to be complete, excess triethylamine was added, and the mixture (dark brown) let stand at room temperature and then heated briefly on the steam bath. The mixture was then filtered and the ketone (2) recovered by ether extraction.

$Ar_2CHCl \longrightarrow Ar_2CHNCHO_2C_2H_5$. Zimmerman and Paskovich[8] used the reagent for conversion of dimesityl chloromethane (1) into the ethyl carbamate (2). A solution of 0.00385 mole of silver tetrafluoroborate in 5 g. of ethyl carbamate was treated at

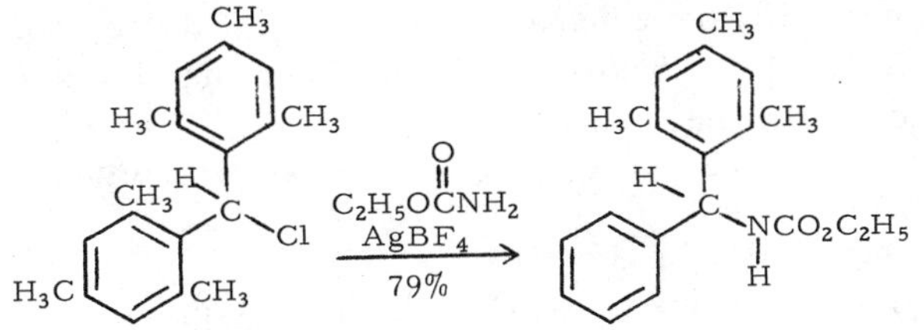

(1) (2)

60° with a solution of 0.0035 mole of the chloride (1) in 20 ml. of dioxane. Silver chloride precipitated at once, and after 5 min. heating on the steam bath water was added and the product extracted with chloroform.

Catalyst for electrophilic aromatic substitution. In a preliminary note describing work completed just before the Hungarian revolution, Oláh and co-workers[9] reported that silver tetrafluoroborate is effective as a cation-forming agent in electrophilic aromatic substitutions. It is a suitable catalyst for nitration with

NO_2Cl,[10] halogenation with elementary halogen,[11] and formylation with formyl fluoride.[12] The reactions are vigorous and almost quantitative.

[1]H. Meerwein, V. Hederich, and K. Wunderlich, *Arch. Pharm.*, **291**, 541 (1958)
[2]D. M. Lemal and A. J. Fry, *Tetrahedron Letters*, 775 (1961)
[3]G. A. Olah and H. W. Quinn, *J. Inorg. Nucl. Chem.*, **14**, 295 (1960); private communication from Drs. Olah and Quinn of Dow Chemical
[4]K. Heyns and H. Paulsen, *Angew. chem.*, **72**, 349 (1960)
[5]A. G. Sharpe, *J. Chem. Soc.*, 4538 (1952)
[6]D. W. A. Sharp and A. G. Sharpe, *J. Chem. Soc.*, 1855 (1956)
[7]H. Peter, M. Brugger, J. Schreiber, and A. Eschenmoser, *Helv.*, **46**, 577 (1963)
[8]H. E. Zimmerman and D. H. Paskovich, *Am. Soc.*, **86**, 2149 (1964)
[9]G. A. Olah, A. Pavláth, and S. Kuhn, *Chem. Ind.*, 50 (1957)
[10]S. J. Kuhn and G. A. Olah, *Am. Soc.*, **83**, 4564 (1961)
[11]G. A. Olah, S. J. Kuhn, S. H. Flood, and B. A. Hardie, *Am. Soc.*, **86**, 1039 (1964); G. A. Olah, S. J. Kuhn, and B. A. Hardie, *ibid.*, **86**, 1055 (1964)
[12]G. A. Olah and S. J. Kuhn, *ibid.*, **82**, 2380 (1960)

Silver tosylate, *p*-$CH_3C_6H_4SO_3Ag$. Mol. wt. 279.08. Supplier: KK.

Kornblum and co-workers[1] prepared the reagent quantitatively by mixing equivalent amounts of silver oxide and *p*-toluenesulfonic acid monohydrate in acetonitrile (with protection from light). After one-half hour silver chloride was removed by filtration and the filtrate was evaporated to dryness.

The reagent reacts with alkyl halides to give tosylates and these on oxidation with a mixture of dimethyl sulfoxide and sodium bicarbonate give aldehydes.[1] Thus

$$\underset{7\text{ g.}}{CH_3(CH_2)_5CH_2I} + \underset{11\text{ g. in }100\text{ ml. }CH_3CN}{\underline{p}\text{-}CH_3C_6H_4SO_2OAg} \xrightarrow{0^{\circ},\ \text{then } 25^{\circ}} CH_3(CH_2)_5CH_2OSO_2C_6H_4CH_3\text{-}\underline{p}$$

$$\xrightarrow[70\%\text{ overall as DNP}]{150\text{ ml. DMSO} + 20\text{ g. }NaHCO_3\ (150^{\circ})} CH_3(CH_2)_5CHO$$

the oily tosylate from 1-iodoheptane is extracted with ether and added under nitrogen to a mixture of DMSO and $NaHCO_3$ preheated to 150°. A reaction period of 3 min. suffices for the oxidation. When a benzyl halide is used the oxidation is conducted at 100° for 5 min.

Ordinarily the reaction of silver tosylate with a secondary or tertiary halide yields only products resulting from elimination.[2] However, if the reaction is carried out at a temperature of −25° or below in a minimum amount of acetonitrile, esters of even tertiary alcohols can be isolated.[3]

[1]N. Kornblum, W. J. Jones, and G. J. Anderson, *Am. Soc.*, **81**, 4113 (1959)
[2]W. D. Emmons and A. F. Ferris, *Am. Soc.*, **75**, 2257 (1953)
[3]H. M. R. Hoffmann, *J. Chem. Soc.*, 6748 (1965)

Silver trifluoroacetate, CF_3CO_2Ag. Mol. wt. 220.90. Supplier: KK.

Preparation.[1] Trifluoroacetic acid is added to freshly precipitated silver oxide and water, and the solution is filtered and evaporated to dryness under reduced pressure. The white solid is purified by extraction with ether in a Soxhlet; yield 88%.

Iodination. Janssen and Wilson[1] used the reagent in combination with iodine for iodination of veratrole. A mixture of veratrole and silver trifluoroacetate was stirred, and a solution of iodine in 1.6 l. of chloroform was added in the course of 2 hrs. After 1 hr. more the silver iodide was removed by filtration, and the filtrate washed and evaporated and the product distilled.

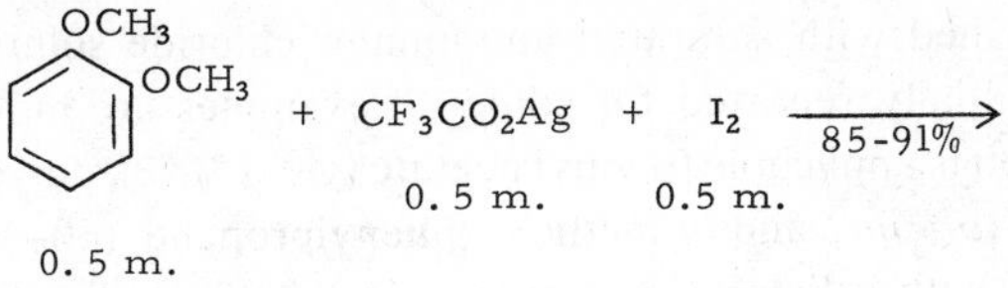

OCH3
OCH3
+ CF3CO2Ag + I2
85-91%
0. 5 m. 0.5 m.
0. 5 m.
OCH3
OCH3
I

Oxidation. Newman and Reid[2] found the reagent superior to mercuric oxide for Curtius oxidation of dihydrazones of benzils to give diarylacetylenes. A mixture of benzil dihydrazone and silver trifluoroacetate in 250 ml. of acetonitrile is stirred at room temperature and triethylamine is added in the course of 150 min. The course

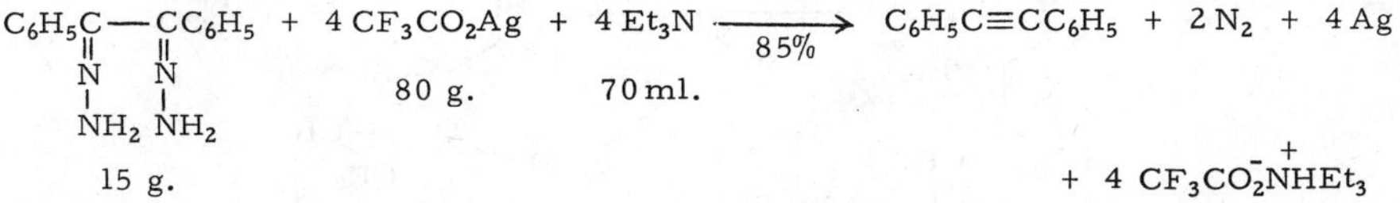

C6H5C—CC6H5 + 4 CF3CO2Ag + 4 Et3N
85%
C6H5C≡CC6H5 + 2 N2 + 4 Ag
N N
NH2 NH2
80 g. 70 ml.
15 g.
+ 4 CF3CO2NHEt3

of the reaction is followed by measuring the evolution of nitrogen, which continues for about 6 hrs. The mixture is then poured into 200 ml. of concd. ammonia solution.

$> CBr_2 \longrightarrow > C{=}O$. Cava[3] found silver trifluoroacetate to be an excellent reagent for the hydrolysis of 1,1,2,2-tetrabromobenzocyclobutene (1) to the α-diketone (2).

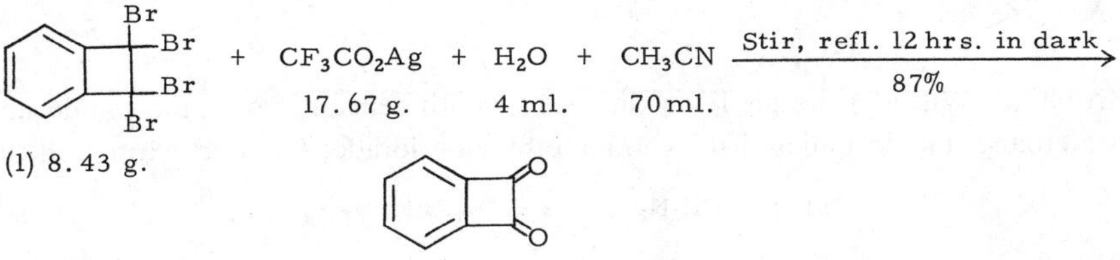

Br
Br
Br
Br
+ CF3CO2Ag + H2O + CH3CN
Stir, refl. 12 hrs. in dark
87%
17. 67 g. 4 ml. 70 ml.

(1) 8. 43 g.

(2)

Olefin addition. Addition of iodine and Δ^2-cholestene to a solution of silver trifluoroacetate in methylene chloride generates iodine trifluoroacetate which adds to the olefin to give 3α-iodo-2β-trifluoroacetatoxycholestane in 72% yield.[4] Lithium aluminum hydride effects reduction and deacetylation to cholestane-2β-ol.

[1] D. E. Janssen and C. V. Wilson, *Org. Syn., Coll. Vol.*, **4**, 547 (1963)
[2] M. S. Newman and D. E. Reid, *J. Org.*, **23**, 665 (1958)
[3] M. P. Cava, D. R. Napier, and R. J. Pohl, *Am. Soc.*, **85**, 2076 (1963)
[4] D. G. Hey, *J. Chem. Soc.* (C), 1331 (1966)

Simmons-Smith reagent, ICH_2ZnI or dimer.

Nature of the reagent. The du Pont chemists[1] who discovered the reagent later defined precise directions for its generation and use in reaction with cyclohexene to form norcarane.[2] The reagent is generated by reaction of methylene iodide in

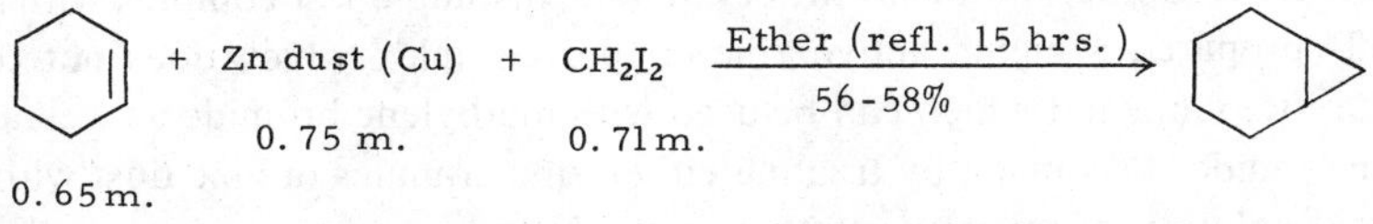

+ Zn dust (Cu) + CH2I2
Ether (refl. 15 hrs.)
56-58%
0. 75 m. 0. 71 m.
0. 65 m.

ether with an active zinc-copper couple prepared essentially according to Shank and Shechter[3] from acid-washed zinc dust and copper sulfate solution and washed with water, absolute ethanol, and absolute ether. A suspension of the couple in absolute ether is treated with a crystal of iodine followed by the olefin and methylene iodide. After a suitable reflux period, the ethereal solution is decanted from the finely divided copper, washed with saturated ammonium chloride solution, dried, and evaporated. Yields initially reported for some 30 examples are mostly in the range 30–60%. The reaction is applicable to vinyl acetate (yield 30%), dihydropyrane (65%), styrene (32%), 1-(*o*-, *m*-, and *p*-methoxy)phenylpropene (60–70%). The reaction of excess reagent with D-limonene to give a mono adduct affords evidence of a considerable steric requirement of the reagent.

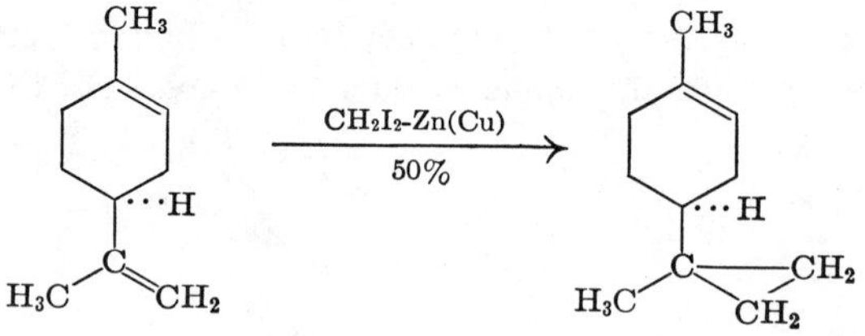
CH3
CH2I2-Zn(Cu)
50%
H
H3C
CH2
CH3
H
CH2
H3C
CH2

D-Limonene (+ 106°) + 51°

The reagent was initially regarded[1] as iodomethylzinc iodide, ICH_2ZnI, a relatively strongly bonded complex of carbene and zinc iodide (a) in which the carbon atom is

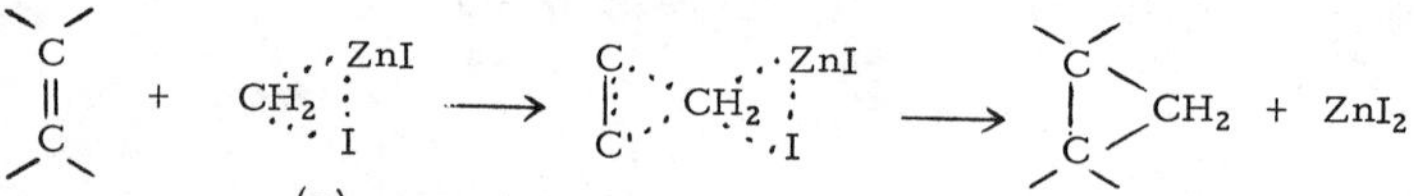
C
C
+
ZnI
CH2
I
C
C
ZnI
CH2
I
C
CH2 + ZnI2
C

(a)

electrophilic. Wittig[4] generated the Simmons-Smith reagent from zinc iodide and diazomethane and formulated it as iodomethylzinc iodide. Hoberg[5] expressed the

$$ZnI_2 + CH_2N_2 \longrightarrow ICH_2ZnI + N_2$$

view that the Simmons-Smith reaction does not involve a carbene intermediate but proceeds through a simple addition of the highly reactive iodomethylzinc iodide:

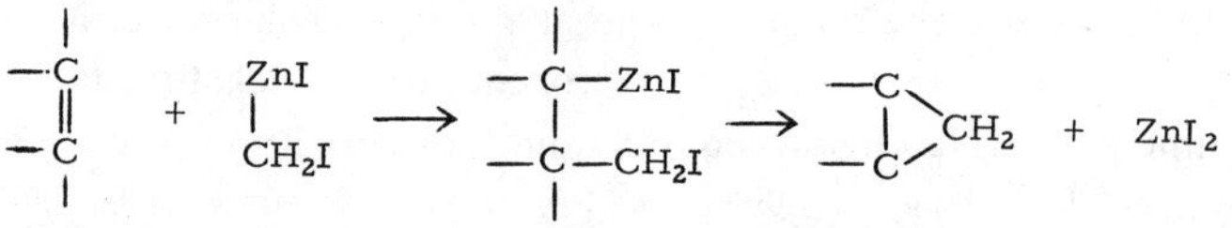
C
C
+
ZnI
CH2I
C—ZnI
C—CH2I
C
CH2 + ZnI2
C

However, du Pont chemists[6,7] later expressed the view that the reagent is probably not the simple monomeric species but is bis(iodomethyl)zinc · zinc iodide, $(ICH_2)_2Zn \cdot ZnI_2$.

If either the olefin or the cyclopropane produced is sensitive to zinc iodide, a Lewis acid, the usual procedure can be modified by using as solvent diethyl ether containing 1 equivalent of glyme (1,2-dimethoxyethane).[7] Zinc iodide is removed quantitatively from the solution as the crystalline, insoluble 1 : 1 complex with glyme.

LeGoff[8] prepared a highly active zinc-copper complex which does not require activation by iodine and which can be used with methylene bromide as well as with methylene iodide. It is made by treating either zinc granules or zinc dust with a hot acetic acid solution of cupric acetate monohydrate and then washing with acetic acid and ether.

In the course of synthesizing spirocyclic hydrocarbons such as that formulated, Wineman *et al.*[9] improved the procedure by removal of methylene iodide prior to workup by adding 2-methylbutene-2 and more zinc-copper couple and refluxing for another 24 hrs.

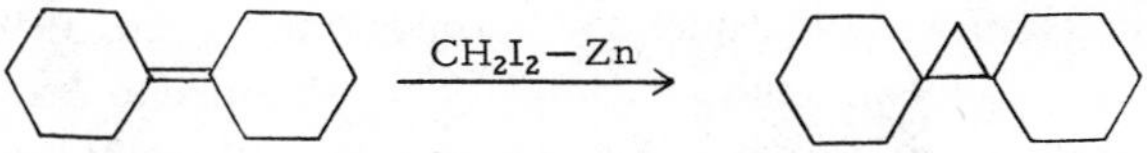

Steric course. Cyclopropane formation with the Simmons-Smith reagent is a stereospecific *cis*- addition; for example *cis*- and *trans*-hexene-3 give pure *cis*- and *trans*-1,2-diethylcyclopropane.

Corey *et al.*[10] found that the Simmons-Smith reaction is facilitated by a hydroxyl substituent. Thus Δ^3-cyclopentenol reacts more rapidly and in higher yield than the corresponding acetate or than cyclopentadiene. Since the cyclopropane ring is *cis* to the hydroxyl group, Corey suggests that the oxygen atom of the starting material is coordinated with the zinc. Dauben and Berezin[11] found that the hydroxyl

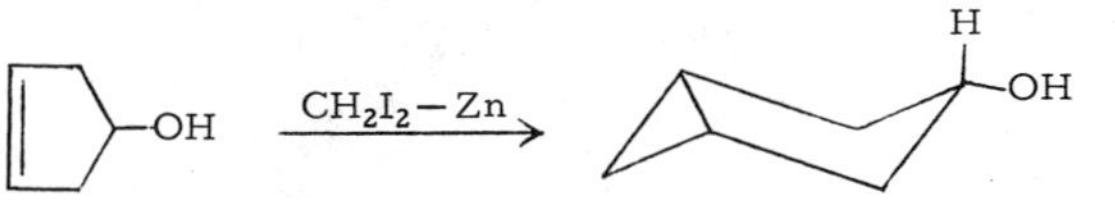

(1) (2)

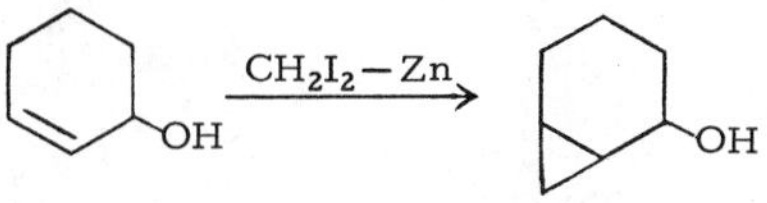

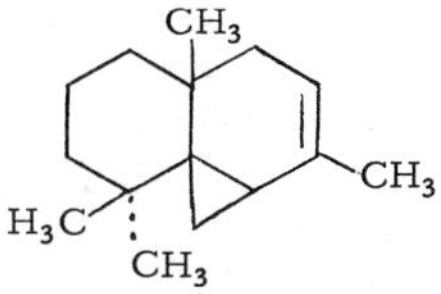

(3) (4) (5)

group of the allylic alcohol (3) similarly accelerates the Simmons-Smith reaction and that the cyclopropane ring and hydroxyl group are in the *cis* relationship (4). The latter finding set the stage for a stereospecific synthesis of the hydrocarbon (5),[12] which proved to be identical with the sesquiterpene (±)-thujopsene; the stereochemistry was established in this way.

It is striking that usually only one isomer is formed. Thus the du Pont workers[7] found that norbornene (6) gives the *exo*-adduct (7); none of the *endo*-isomer was detected. Winstein *et al.*[13] found that *cis,cis,cis*-1,4,7-cyclononatriene (8) reacts with the reagent to give in high yield the all-*cis* adduct (9).

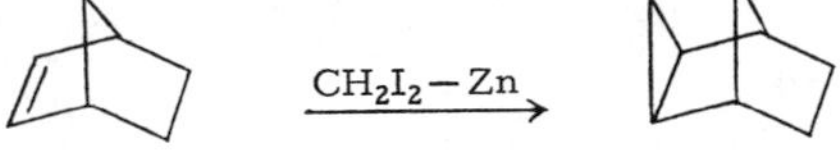

(6) (7)

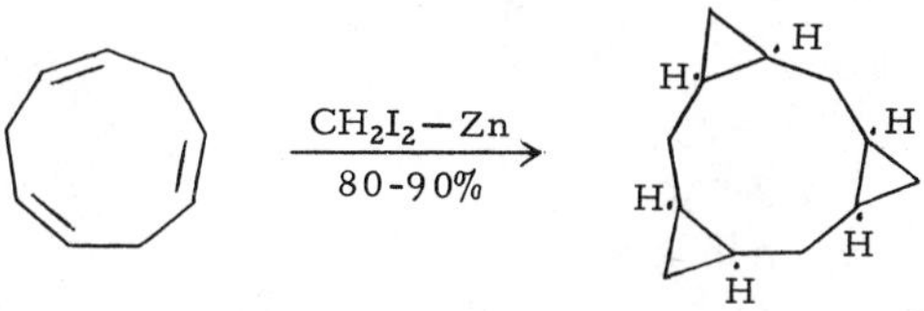

(8) (9)

Reaction with acetylenes. Sterculic acid was synthesized by refluxing stearolic acid in ether with methylene iodide and zinc-copper couple.[14] Fractional crystallization

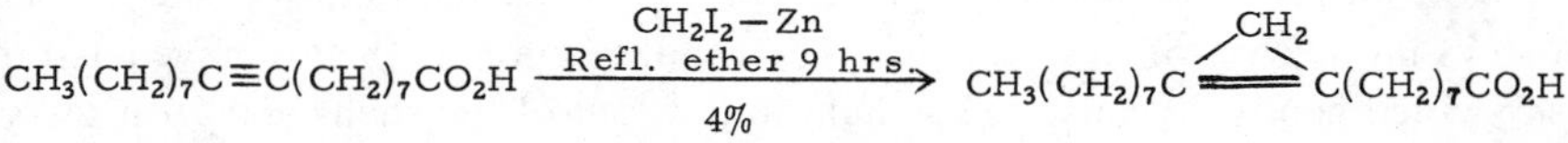

of the urea adducts afforded the pure cyclopropene acid (m.p. 19°) in yield of 4% after a reaction period of 9 hrs.; the yield after 12 hrs. dropped to 2.5%.

The reagent reacts with a terminal acetylene to give a methyl derivative as the

$$C_6H_5C{\equiv}CH \xrightarrow{CH_2I_2-Zn} \underset{37.5\%}{C_6H_5C{\equiv}CCH_3} + \underset{6\%}{C_6H_5CH{=}C{=}CH_2}$$

main product.[15] A mixture of 1,2-dimethoxyethane and ether is recommended as solvent; in the former solvent alone the reaction is fast and may be uncontrollable.

Wasserman and Clagett[16] effected the synthesis of 1-substituted cyclopropanols starting with a 1-alkoxyvinyl ester (2) derivable from an alkoxyacetylene (1). In the

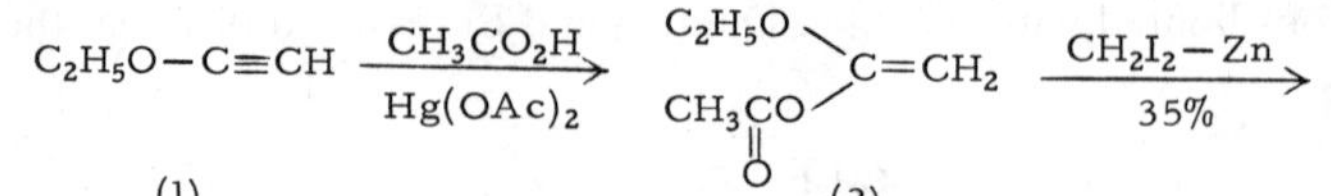
C2H5O−C≡CH
CH3CO2H
Hg(OAc)2
C2H5O
C=CH2
CH3CO
O
CH2I2−Zn
35%

(1) (2)

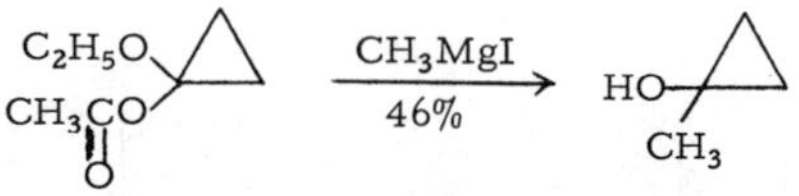
C2H5O
CH3CO
O
CH3MgI
46%
HO
CH3

(3) (4)

preparation of 1-ethoxycyclopropyl acetate (3), for example, glyme was added to precipitate zinc iodide and protect the sensitive product (3), which was converted into 1-methylcyclopropanol (4) by a Grignard reaction.

[1]H. E. Simmons and R. D. Smith, *Am. Soc.*, **81**, 4256 (1959)
[2]R. D. Smith and H. E. Simmons, *Org. Syn.*, **41**, 72 (1961)
[3]R. S. Shank and H. Shechter, *J. Org.*, **24**, 1825 (1959)
[4]G. Wittig and K. Schwarzenbach, *Ann.*, **650**, 1 (1961); G. Wittig and F. Wingler, *Ann.*, **656**, 18 (1962)
[5]H. Hoberg, *Ann.*, **656**, 1, 15 (1962)
[6]E. P. Blanchard and H. E. Simmons, *Am. Soc.*, **86**, 1337 (1964)
[7]H. E. Simmons, E. P. Blanchard, and R. D. Smith, *Am. Soc.*, **86**, 1347 (1964)
[8]E. LeGoff, *J. Org.*, **29**, 2048 (1964)
[9]S. D. Koch, R. M. Kliss, D. V. Lopiekes, and R. J. Wineman, *J. Org.*, **26**, 3122 (1961)
[10]E. J. Corey and R. L. Dawson, *Am. Soc.*, **85**, 1782 (1963); E. J. Corey and H. Uda, *Am. Soc.*, **85**, 1788 (1963)
[11]W. G. Dauben and G. H. Berezin, *Am. Soc.*, **85**, 468 (1963)
[12]W. G. Dauben and A. C. Ashcraft, *Am. Soc.*, **85**, 3673 (1963)
[13]R. S. Boikess and S. Winstein, *Am. Soc.*, **85**, 343 (1963); P. Radlick and S. Winstein, *Am. Soc.*, **85**, 344 (1963)
[14]N. T. Castellucci and C. E. Griffin, *Am. Soc.*, **82**, 4107 (1960)
[15]L. V. Quang and P. Cadiot, *Bull. soc.*, 1525 (1965)
[16]H. H. Wasserman and D. C. Clagett, *Tetrahedron Letters*, 341 (1964)

Sodium, Na. Atomic wt. 23.00, m.p. 97.5°.

For precautions for safe handling, *see* Potassium.

Preparation of clean sodium.[1] The cleaning of sodium by cutting off the oxidized surfaces with a knife under xylene is tedious and wasteful. The following technique facilitates the operation, saves metal, and minimizes scrap disposal. Considerably more sodium than will be needed is placed in a wide-mouthed Erlenmeyer flask or tall beaker and covered with dry xylene. The flask is heated until the sodium melts but stays in the cage of surface oxide. The heating source is then removed and the flask swirled gently to cause the sodium to flow out of the shells and form several globules. The flask is then cooled without much agitation in order that the sodium globules do not unite. As a precaution, dry nitrogen can be passed through the solution during cooling. When the metal has solidified, the clear globules are removed with a pointed iron rod or other device, dried rapidly with filter paper or a *dry* towel, and added to the weighed container used to measure the amount of sodium needed. The oxide shells contain very little sodium; the solvent is decanted, and treatment of the residue with *t*-butanol destroys traces of active metal.

Sodium suspension (sodium shot or sand). Hershberg[2] prepared sodium suspension by melting the metal under xylene and shaking, by the technique used for powdering potassium (except for the solvent) described *under* Potassium ethoxide. An adaptation of this method safe enough for use by beginning students has been described.[3] Marvel and King[4] placed sodium and xylene in a flask equipped with a mechanical stirrer and condenser, heated the mixture until the sodium melted, stirred vigorously to break up the sodium to fine particles, and removed the heat source until the sodium solidified. Hafner and Kaiser[5] prepared a suspension of sodium (23 g.) in toluene (150 ml.) in a flask fitted with a reflux condenser, ground-glass stopper, and a Vibromischer (available from A. G. für Chemie-Apparatebau). The toluene was heated to boiling, the melted metal was dispersed by the Vibromischer, and the mixture was quickly cooled. In a nitrogen atmosphere the toluene was removed by decantation and replaced by tetrahydrofurane. Elsinger[6] used a similar technique.

Sodium dispersions. Dispersions containing particles in the order of 10 microns in diameter and having one hundred times the surface area of sodium sand are best prepared prior to use according to directions available from U. S. Industrial Chemicals Co.; the technique and equipment are not elaborate. U. S. I. does not sell dispersions but will supply pint or quart size samples for trial. Solvents include hydrocarbons and ethers for the most part in the boiling range 100–200°. A dispersing agent such as oleic acid or aluminum laurate is often used to the extent of 0.2–1.0%. Use of dispersions often leads to higher yields, faster reactions, lower temperatures, and improved control.

Sodium on alumina. This high-surface sodium catalyst, prepared by adding sodium to dry alumina with stirring at 150° under nitrogen,[7] is effective for isomerization of butenes and pentene-1. In contrast to acid-catalyzed isomerization, no skeletal rearrangement occurs. The results indicate some stereoselectivity. Thus butene-1 is isomerized initially to about equal amounts of *cis*- and *trans*-butene-2; eventually the thermodynamic equilibrium mixture rich in the *trans*-isomer is obtained.[8] The catalyst was used to effect almost quantitative conversion of methylenecyclobutane into 1-methylcyclobutene.[9]

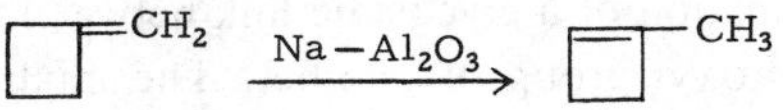

CH2
Na−Al2O3
CH3

[1]Contributed by Melvin S. Newman, Ohio State University
[2]E. B. Hershberg and L. F. Fieser, *Org. Syn., Coll. Vol.*, **2**, 195 (1943)
[3]*Org. Expts.*, 142–143
[4]C. S. Marvel and W. B. King, *Org. Syn., Coll. Vol.*, **1**, 252 (1941)
[5]K. Hafner and H. Kaiser, *Org. Syn.*, **44**, 94 (1964)
[6]F. Elsinger, *ibid.*, **45**, 7 (1965)
[7]H. Pines and W. O. Haag, *J. Org.*, **23**, 328 (1958)
[8]W. O. Haag and H. Pines, *Am. Soc.*, **82**, 387 (1960)
[9]J. Shabtai and E. Gil-Av, *J. Org.*, **28**, 2893 (1963)

Sodium acetate, CH_3CO_2Na. Mol. wt. 82.04, m.p. 324°, solubility in 100 g. water: 46.5 g. at 20°, 170 g. at 100°.

Trihydrate. Mol. wt. 136.09. Use of the reagent for neutralization of a reaction mixture or liberation of a reagent from a salt (hydrazine sulfate[1]) requires no comment. Where anhydrous reagent is required, that now available commercially is entirely satisfactory (early procedures often called for use of material that had been freshly fused and powdered).

Acetylation. A comparative student experiment with salicylic acid and acetic anhydride demonstrates that common acetylation catalysts fall into the following order of relative effectiveness: concd. H_2SO_4 > boron trifluoride etherate > pyridine > sodium acetate.[2] Although it is a relatively weak catalyst, sodium acetate is completely nondestructive and can be employed in much larger than truly catalytic amounts. An example is the acetylation of furylcarbinol.[3] A mixture of the reactants and solvent benzene was heated on the steam bath with stirring to prevent caking

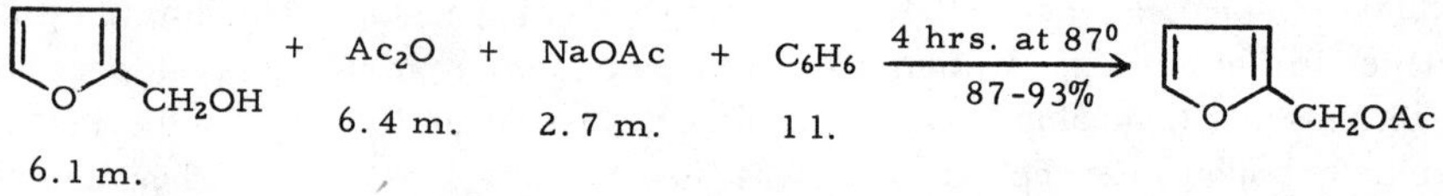
CH₂OH
6.1 m.
+ Ac₂O
6.4 m.
+ NaOAc
2.7 m.
+ C₆H₆
1 l.
4 hrs. at 87°
87-93%
CH₂OAc

of the sodium acetate and after 4 hrs. poured into water. After stirring to decompose excess anhydride, the benzene layer was separated, washed, dried, and evaporated, and the product distilled.

In the preparation of gentiobiose octaacetate (Helferich and Leete[4]), 15 g. of emulsin and 20 ml. of toluene was added to a solution of glucose in distilled water, the mixture was let stand in a glass-stoppered flask for 5 weeks with occasional

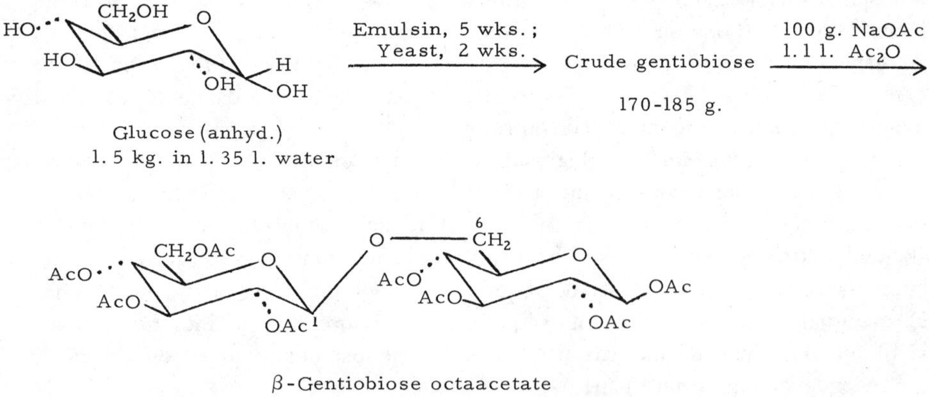
CH₂OH
HO
OH
H
Glucose (anhyd.)
1.5 kg. in 1.35 l. water
Emulsin, 5 wks.;
Yeast, 2 wks.
Crude gentiobiose
170-185 g.
100 g. NaOAc
1.1 l. Ac₂O
CH₂OAc
AcO
OAc
6
CH₂
β-Gentiobiose octaacetate

shaking for enzymic formation of a glycosidic link between the 1β-hydroxyl of one molecule and the 6-hydroxyl group of another. The mixture was boiled, diluted with 8.5 l. of water, and filtered. The filtrate was treated with a solution of 56 g. of baker's yeast in 650 ml. of water and let stand at 28–32° for 12–14 days in a flask protected with a trap for the escape of carbon dioxide. The yeast, by fermentation, removes glucose not transformed into gentiobiose. The mixture was then boiled with an excess of powdered calcium carbonate and filtered. Evaporation at 20–30 mm. then gave a thick sirup of crude gentiobiose. Sodium acetate and acetic anhydride were added, and the mixture heated cautiously to the boiling point (20 min.). The solution was poured into water, and the dark brown product dried and extracted with ether in a Soxhlet. Two crystallizations from methanol gave 69–78 g. of pure gentiobiose octaacetate.

A procedure[5] for the ***reductive acetylation*** of a quinone, a benzanthrone, or an α-diketone is as follows. A suspension of 1 g. of substance in 5–6 ml. of acetic anhydride is treated with 1 g. of zinc dust and 0.2 g. of anhydrous sodium acetate until the colored material has disappeared and the supernatant liquor is colorless or pale yellow, depending upon the purity of the starting material. After short boiling to complete the reaction, acetic acid is added as required to dissolve the product and part of the zinc acetate that has separated, the solution is filtered at the boiling point from zinc and zinc acetate, the residue is washed with hot solvent, and the total filtrate is boiled under a reflux condenser and treated cautiously with sufficient water to hydrolyze the excess acetic anhydride and then to produce a saturated solution.

Condensations. For the condensation of veratraldehyde with hippuric acid by reaction with acetic anhydride and sodium acetate to produce the azlactone, Buck and Ide[6] used a full equivalent of sodium acetate. Heated on a hot plate with constant

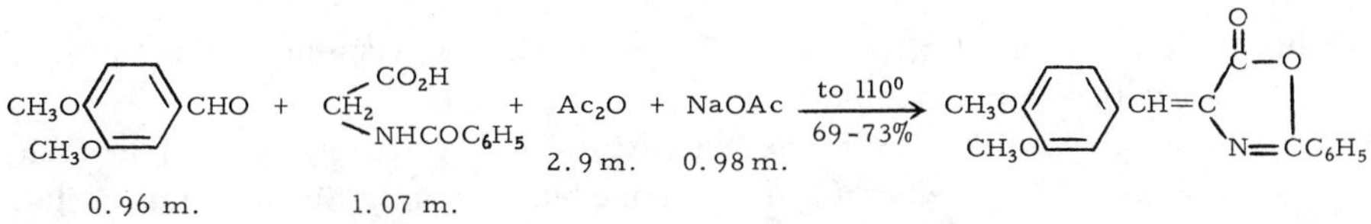

CH3O
CH3O
CHO
CO2H
CH2
NHCOC6H5
Ac2O
NaOAc
to 110°
69-73%
2.9 m.
0.98 m.
0.96 m.
1.07 m.
CH=C
C—O
N=CC6H5

shaking, the mixture became almost solid and then (at about 110°) gradually liquefied and turned deep yellow. Heating was then continued on the steam bath for 2 hrs., during which time the product partially separates as deep yellow crystals. Ethanol (400 ml.) is added with cooling as required, and the product let crystallize.

Perkin-type condensations of *m*-nitrobenzaldehyde[7] and of phthalic anhydride[8] are formulated. In the second case sodium acetate is taken in catalytic amount, and the water produced in the formation of benzalphthalide distills as formed.

$$\underset{0.33\text{ m.}}{\underline{m}\text{-}NO_2C_6H_4CHO} + \underset{0.68\text{ m.}}{(CH_3CO)_2O} + \underset{0.48\text{ m.}}{CH_3CO_2Na} \xrightarrow[74\text{-}77\%]{13\text{ hrs. at }180^\circ} \underline{m}\text{-}NO_2C_6H_5CH{=}CHCO_2H$$

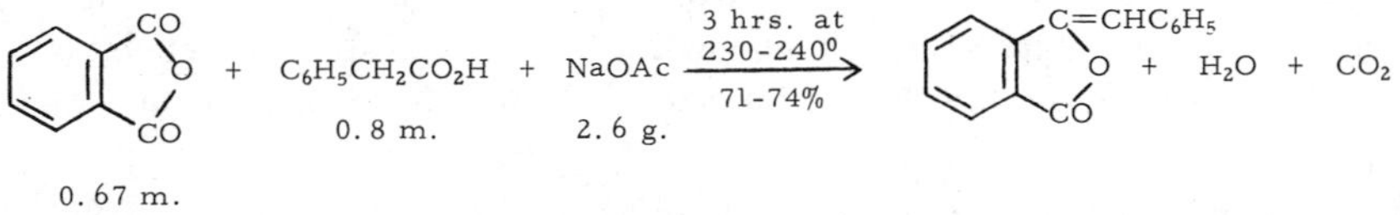

CO
O
CO
C6H5CH2CO2H
NaOAc
3 hrs. at
230-240°
71-74%
0.8 m.
2.6 g.
0.67 m.
C=CHC6H5
H2O
CO2

In bromination. In the addition of bromine to commercial cholesterol (0.39 m.) in ether–acetic acid at 25° the yield of $5\alpha,6\beta$-dibromide rose from 73 to 84% on addition of 0.06 m. of sodium acetate.[9] The function of the additive is not known.

Dehydrohalogenation. α-Bromobenzalacetone has been prepared from benzalacetone by addition of bromine[10] and elimination of HBr by reaction with sodium

$$\underset{0.68\text{ m.}}{C_6H_5CH{=}CHCOCH_3} + \underset{0.68\text{ m.}}{Br_2} + \underset{360\text{ ml.}}{CCl_4} \xrightarrow[52\text{-}57\%]{10\text{-}20^\circ} \underset{0.33\text{ m.}}{C_6H_5\underset{Br}{CH}\,\underset{Br}{CH}COCH_3}$$

$$\xrightarrow[64\text{-}73\%]{0.37\text{ m. NaOAc; 250 ml. 95\% EtOH, refl. 4 hrs.}} C_6H_5CH{=}\underset{Br}{C}COCH_3$$

acetate in refluxing ethanol.[11] Since the second reaction requires conditions much more drastic than the first and since the yield is much lower than in the bromination of cholesterol, it would be interesting to explore the possibility of improving the yield by use of sodium acetate as additive, either in CCl_4 or in ether–AcOH.

Dehalogenation. A procedure for the preparation of lepidine from 2-chlorolepidine calls for hydrogenation in acetic acid containing sodium acetate.[12] The

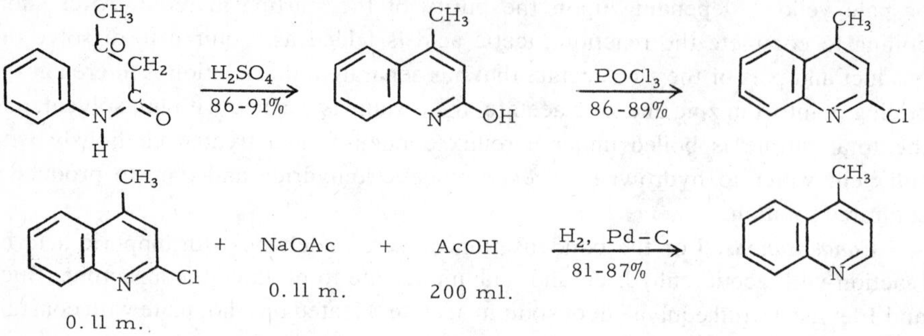

CH_3
CO
CH_2
N
H
O
H_2SO_4
86-91%
OH
$POCl_3$
86-89%
Cl
+ NaOAc + AcOH
0.11 m.
200 ml.
0.11 m.
H_2, Pd-C
81-87%

starting material is readily available from acetoacetanilide via 4-methylcarbostyril[13] and reaction with phosphoryl chloride.[14]

Salt-free sodium sulfonates.[15] The blue dye formulated is required in pure, salt-free form for use in a procedure for determining blood volume. Since sodium acetate is moderately soluble in hot alcohol, elimination of sodium chloride can be accom-

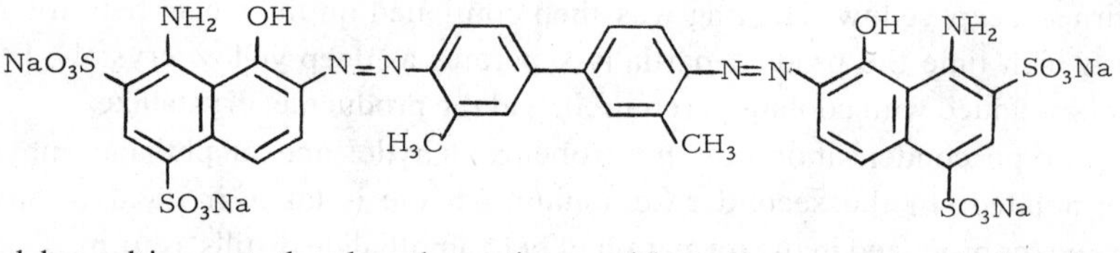

NH_2
OH
NaO_3S
N=N
H_3C
CH_3
SO_3Na

plished by salting out the dye three times with sodium acetate, drying the solid, grinding it, and passing it through a 40-mesh sieve. Four extractions with hot 95% ethanol should remove all the sodium acetate, as determined by adding 3 drops of concd. sulfuric acid to 10 ml. of the alcoholic filtrate and cooling in ice; a white precipitate will form if the solution contains 2 mg. or more of sodium acetate.

Other reactions. *p*-Nitrobenzyl acetate is formed by displacement from the halide.[16]

$$\underset{1.46\ m.}{\underline{p}\text{-}NO_2C_6H_4CH_2Cl} + \underset{2.74\ m.}{NaOAc} + \underset{375\ ml.}{AcOH} \xrightarrow[78-82\%]{Refl.\ 8-10\ hrs.} \underline{p}\text{-}NO_2C_6H_4CH_2OCOCH_3$$

In an alcoholic solution of sodium acetate trihydrate, hydrogen sulfide adds to methyl acrylate to give dimethyl β-thiodipropionate.[17]

$$\underset{1.74\ m.}{2\ CH_2{=}CHCO_2CH_3} + \underset{0.77\ m.}{AcONa \cdot 3\,H_2O} + \underset{800\ ml.}{95\%\ EtOH} \xrightarrow[71-81\%]{H_2S} S\begin{matrix} \diagup CH_2CH_2CO_2CH_3 \\ \diagdown CH_2CH_2CO_2CH_3 \end{matrix}$$

α-Dicarbonyl compounds. Kornblum and Frazier[18] report an efficient synthesis of glyoxals, glyoxalate esters, and α-diketones involving reaction of an α-keto nitrate ester with sodium acetate trihydrate (or the anhydrous acetate); dimethyl sulfoxide is the preferred solvent.

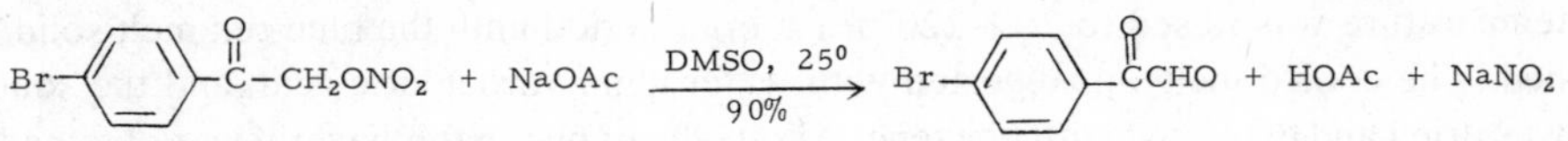

[1]C. D. Nenitzescu and E. Solomonica, *Org. Syn., Coll. Vol.*, **2**, 496 (1943)
[2]*Org. Expts.*, 246
[3]The Miner Laboratories, *Org. Syn., Coll. Vol.*, **1**, 285 (1941)
[4]B. Helferich and J. F. Leete, *ibid.*, **3**, 428 (1955)
[5]L. F. Fieser, *Experiments in Organic Chemistry*, 2nd Ed., 399
[6]J. S. Buck and W. S. Ide, *Org. Syn., Coll. Vol.*, **2**, 55 (1943)
[7]F. K. Thayer, *ibid.*, **1**, 398 (1941)
[8]R. Weiss, *ibid.*, **2**, 61 (1943)
[9]L. F. Fieser, *ibid.*, **4**, 196 (1963)
[10]N. H. Cromwell and R. Benson, *ibid.*, **3**, 105 (1955)
[11]N. H. Cromwell, D. J. Cram, and C. E. Harris, *ibid.*, **3**, 125 (1955)
[12]F. W. Neumann, N. B. Sommer, C. E. Kaslow, and R. L. Shriner, *ibid.*, **3**, 519 (1955)
[13]W. M. Lauer and C. E. Kaslow, *ibid.*, **3**, 580 (1955)
[14]C. E. Kaslow and W. M. Lauer, *ibid.*, **3**, 194 (1955)
[15]J. L. Hartwell and L. F. Fieser, *ibid.*, **2**, 145 (1943)
[16]W. W. Hartman and E. J. Rahrs, *ibid.*, **3**, 650 (1955)
[17]E. A. Fehnel and M. Carmack, *ibid.*, **4**, 669 (1963)
[18]N. Kornblum and H. W. Frazier, *Am. Soc.*, **88**, 865 (1966)

Sodium acetylide CH≡CNa. Mol. wt. 48.03.

Reagent for use in the synthesis of 1-ethynylcyclohexanol is prepared by passing acetylene into 1 l. of liquid ammonia with stirring and slowly adding 23 g. of sodium.[1]

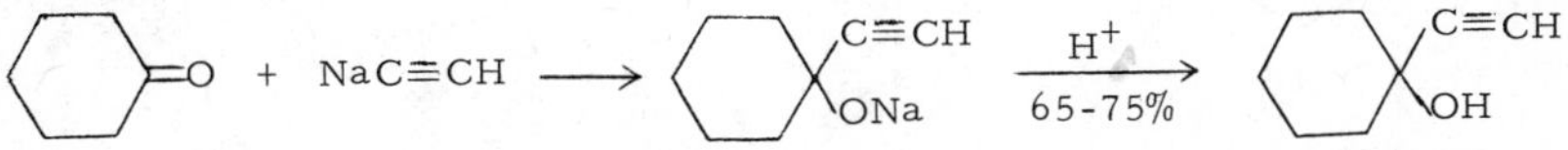

Cyclohexanone (1 mole) is added over a 1-hr. period and the ammonia is allowed to evaporate. The solid residue is decomposed with ice and water, acidified, and the product extracted with ether.

[1]J. H. Saunders, *Org. Syn., Coll. Vol.*, **3**, 416 (1955)

Sodium aluminum chloride, $NaAlCl_4$. Mol. wt. 191.80 (NaCl=58.45; $AlCl_3$= 133.35).

In all but one of the investigations cited below the reagent was prepared by mixing 1 equivalent of NaCl thoroughly with 2 equivalents of $AlCl_3$ and heating the mixture over a free flame until a clear melt results (180–200°). If this is done in a test tube and the tube is then swirled gently as the temperature slowly falls, the melt begins to solidify at 87° and can be kept semifluid by heating on the steam bath. If a 1:1 mixture is treated in the same way some solid remains undissolved at 180–250°; on cooling the fluid part starts to crystallize at 150° and becomes solid at 130°. The 1:2 reagent provides a powerful means of effecting dehydrations, condensations, and cyclizations at temperatures ranging from about 90 to 220°.

Zahn and Ochwat[1] stirred a melt of 20 g. of NaCl and 100 g. of $AlCl_3$ at 180°, added a mixture of 10 g. of maleic anhydride and 11 g. of hydroquinone. The

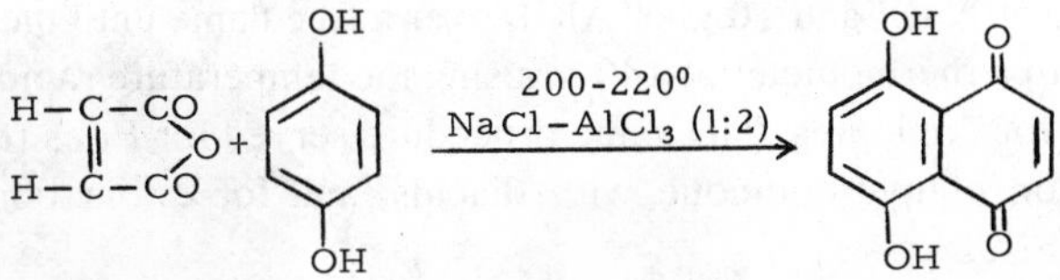

temperature was raised to 200–220° for a brief period until the blue-red melt solidified. The cooled melt was digested with water and hydrochloric acid, and the solid was dried and extracted with benzene. About 4 g. of pure naphthazarin was obtained. No yield is reported for a similar condensation of phthalic anhydride with the diketo form of dihydronaphthazarin to give 1,4,5,8-tetrahydroxynaphthacenequinone.

Fieser and Peters,[2] who in the following year became Fieser and Fieser,[3] used the 1:2 reagent for cyclization of the keto ester (1) to *peri*-succinoylacenaphthene (2) and for cyclization of the keto acid (3) to the dimethyl-1,2-benzanthraquinone (4);

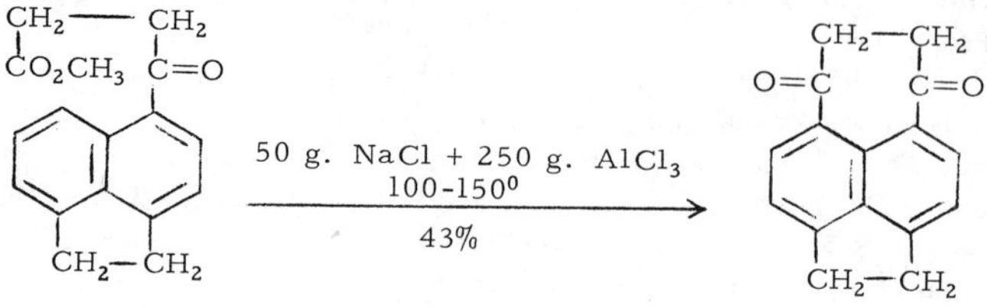
50 g. NaCl + 250 g. AlCl3
100-150°
43%

(1) 40 g.

(2)

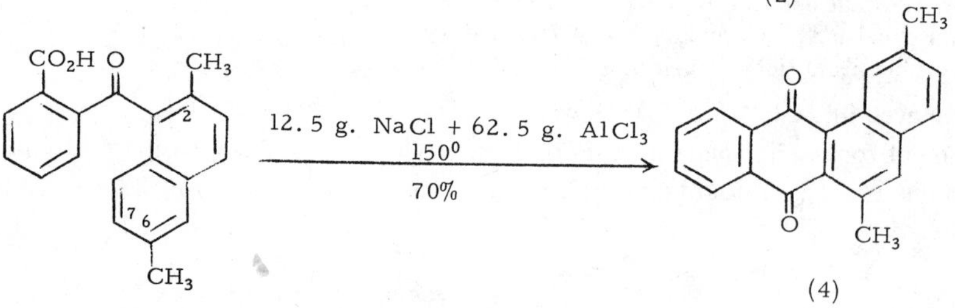
12.5 g. NaCl + 62.5 g. AlCl3
150°
70%

(4)

(3) 10 g.

the latter reaction probably involves migration of the aroyl group from C_1 (α-position) to the β-position C_7.[4] Similar procedures were employed for effecting the Scholl ring closure of 1-benzoyl-2-naphthol to 4-hydroxybenzanthrone[5] and of 3-benzoylperinaphthane (5) to 2,1′-trimethylenebenzanthrone-10 (6)[6]. Although the yield of

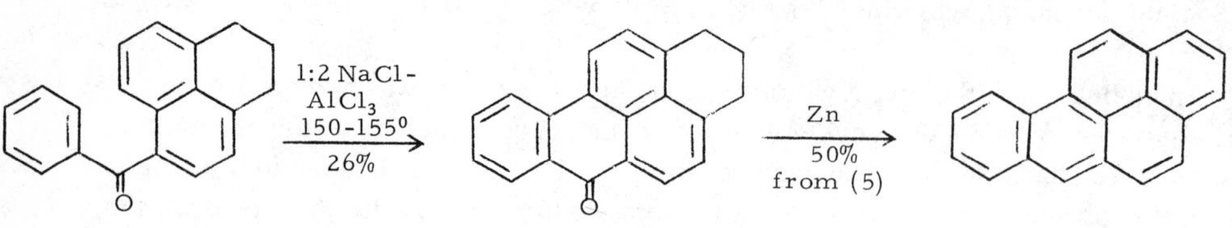
1:2 NaCl-AlCl3
150-155°
26%
Zn
50%
from (5)

(5) (6) (7)

isolated (6) was poor, utilization of the total crude reaction mixture in the final step of zinc dust distillation raised the overall yield of the end product, 3,4-benzpyrene (7), to 50%.

Thomson and co-workers[7] effected cyclizations such as β-phenylpropionic acid → 1-indanone (85%), α-phenylbutyric acid → α-tetralone (73%) by stirring a mixture of 2 g. of NaCl and 10 g. of $AlCl_3$ over a free flame until molten, stirring in 2 g. of acid with a thermometer at 140°, raising the temperature rapidly and keeping it at 180–200° for 2 minutes. The same procedure served for Fries rearrangements, for condensation of hydroquinone with diacids, and for cyclization of aryl vinyl ketones.

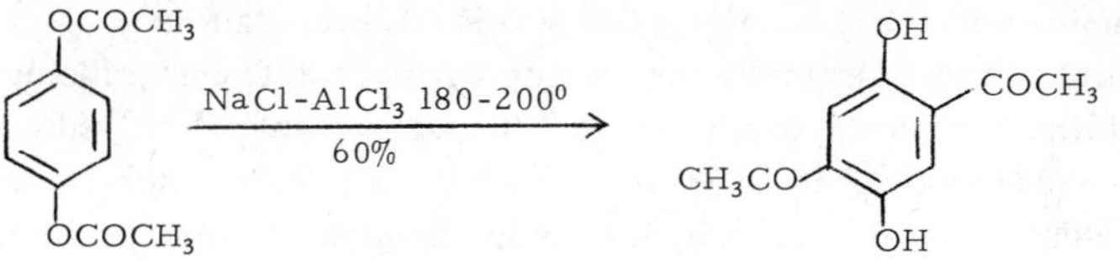

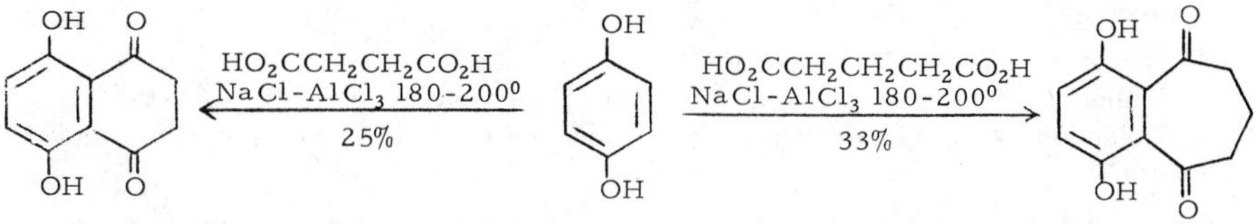

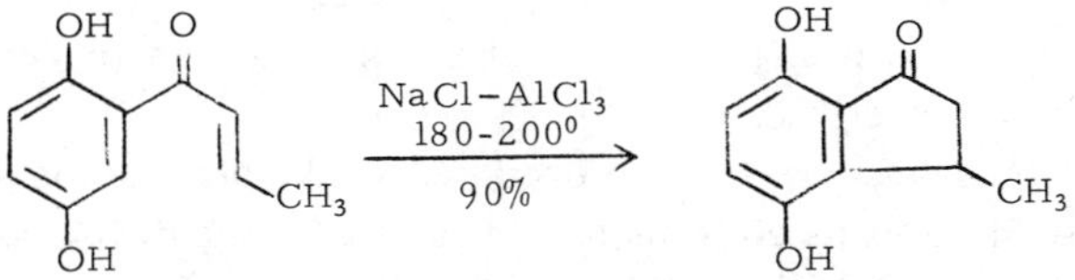

Norris and Klemka[8] employed the 1:1 double salt prepared by heating 30 g. of NaCl and 68 g. of $AlCl_3$ at 230–250° for 1 hr., pouring the melt into a beaker, and stirring to a meal as it solidified. The reagent was kept in a well-stoppered bottle and powdered before use. An aliphatic or aromatic amide in a distilling flask was heated

$$C_6H_5CONH_2 \xrightarrow[-HCl]{AlCl_3} C_6H_5CONHAlCl_2 \longrightarrow C_6H_5CN + HCl + AlOCl$$

continuously with the reagent over a free flame until evolution of hydrogen chloride nearly ceased, and the nitrile was then distilled from the charred residue; yields 60–90%.

Partridge *et al.*[9] effected cyclodehydrogenation of the triazabenzanthracene (1) to tricycloquinazoline (2) with sodium aluminum chloride in much higher yield

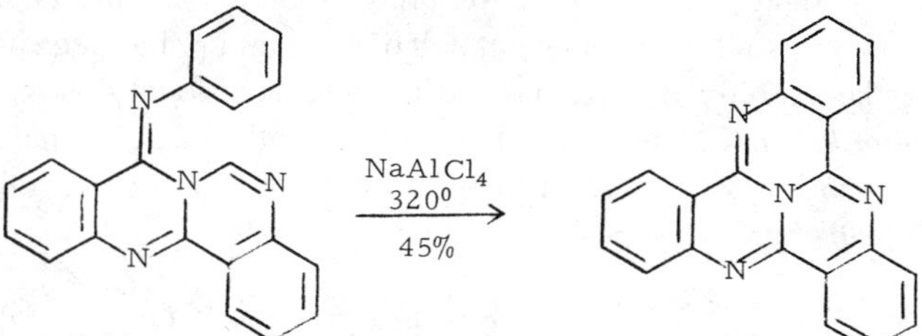

(1) (2)

(45%) than with palladium charcoal (3%). With sulfur at 280° or in refluxing dimethylformamide, (1) did not evolve hydrogen sulfide.

[1]K. Zahn and P. Ochwat, *Ann.*, **462**, 72 (1928)
[2]L. F. Fieser and Mary A. Peters, *Am. Soc.*, **54**, 4347 (1932)
[3]For wedding announcement see Louis F. and Mary Fieser, *Am. Soc.*, **55**, 3012, footnote 8 (1933)
[4]L. F. Fieser and M. A. Peters, *Am. Soc.*, **54**, 3742 (1932); L. F. Fieser and M. Fieser, *Am. Soc.*, **55**, 3347 (1933)
[5]L. F. Fieser, *Am. Soc.*, **53**, 3546 (1931)
[6]L. F. Fieser and E. B. Hershberg, *Am. Soc.*, **60**, 1658 (1938)
[7]D. B. Bruce, A. J. S. Sorrie, and R. H. Thomson, *J. Chem. Soc.*, 2403 (1953)
[8]J. F. Norris and A. J. Klemka, *Am. Soc.*, **62**, 1432 (1940)
[9]M. W. Partridge, S. A. Slorach, and H. J. Vipond, *J. Chem. Soc.*, 3670 (1964)

Sodium aluminum hydride, $NaAlH_4$. Mol. wt. 54.01. Preparation.[1]

The hydride is similar to lithium aluminum hydride as a reducing agent.[2] In tetrahydrofurane or tetrahydrofurane–pyridine at −65 to −45° it reduces esters of carboxylic acids to aldehydes in yields of 50–85%. Aliphatic aldehydes are obtained in higher yields than aromatic aldehydes. Under the same conditions lithium aluminum hydride gives lower yields.[3]

[1] H. Clasen, *Angew. chem.*, **73**, 322 (1961)
[2] A. E. Finholt, E. C. Jacobson, A. E. Ogard, and P. Thompson, *Am. Soc.*, **77**, 4163 (1955)
[3] L. I. Zakharkin, V. V. Gavrilenko, D. N. Maslin, and I. M. Khorlina, *Tetrahedron Letters*, 2087 (1963)

Sodium amalgam (dilute). A 1.2% amalgam is semisolid at room temperature and melts completely at 50°; amalgams of higher concentration (except 38–41%) are solids which can be pulverized.

Preparation. *Organic Syntheses* procedures for the preparation of sodium amalgam[1-3] do not seem to us as convenient and as safe as one developed at the Merck Laboratories for the preparation of 2% amalgam.[4]

Sodium (6.9 g., 0.3 mole) is placed in a 250-ml. round-bottomed three-necked flask fitted with nitrogen inlet and outlet tubes in the side openings and a dropping funnel in the center opening. The flask is flushed out with nitrogen and the funnel charged with 340 g. of mercury. About 10 ml. of mercury is added and the flask is warmed slightly with a free flame until the reaction starts. Once this point is reached little further heating is required, for the reaction can be kept in progress by slow addition of mercury until the total amount has been added. At the end, the still hot, molten amalgam is poured onto a piece of Transite board. While it is still warm and soft, the silvery amalgam is crushed in a mortar and transferred to a tightly stoppered bottle. Amalgam can be analyzed for sodium before use by titration of a 10-g. sample with 0.1 *N* sulfuric or hydrochloric acid.

The method is equally applicable to the preparation of 3% amalgam. In this case the mixture begins to solidify when about half of the mercury has been added, and the mixture is then kept molten by external heating, with occasional shaking.

Reduction of ketones. Xanthone is reduced to xanthhydrol by shaking a suspension of the ketone in 95% ethanol at 60–70° with the liquid amalgam from 9 g. of sodium and 55 ml. of mercury.[1]

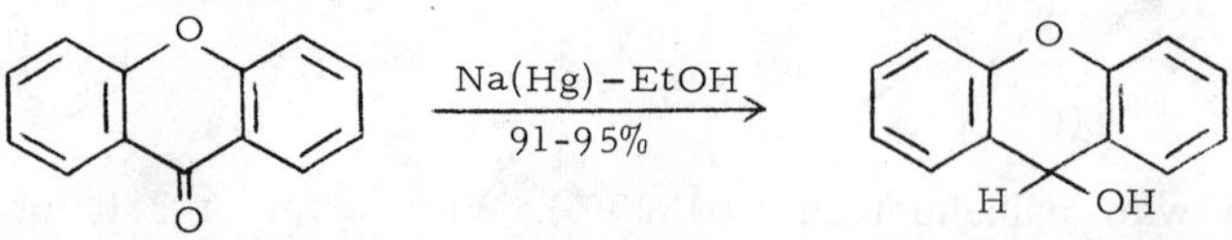

O
O
Na(Hg) – EtOH
91-95%
O
H
OH

Triphenylmethylsodium. A 1% sodium amalgam is added to a solution of triphenylchloromethane in 1.5 l. of anhydrous ether in a 2-l. bottle with a ground-glass

$$\underset{0.236\ \text{m.}}{(C_6H_5)_3CCl} + \underset{0.5\ \text{m.}}{2\ NaHg} \xrightarrow[\text{quant.}]{} Na^+\,\bar{C}(C_6H_5)_3 + NaCl + 2\ Hg$$

stopper, the stopper is greased with Lubriseal, and the stoppered bottle is shaken mechanically and cooled with wet towels.[3] The red color of triphenylmethylcarbanion becomes apparent in a few minutes and the reaction is done in 3 hrs. The mixture is then cooled to room temperature, and the sodium chloride is allowed to settle. Separation of the ethereal solution of triphenylmethylsodium from sodium chloride and mercury is then accomplished by a technique of filtration under nitrogen pressure.

Reduction of α,β-unsaturated carbonyl compounds. The following example is taken from Emil Fischer's classical "Anleitung,"[5] the first laboratory manual for beginning students.

Powdered cinnamic acid (10 g.) is dissolved in 50 g. of water and the calculated amount of sodium hydroxide in a 200-ml. stoppered flask. Then 2.5% sodium amalgam is added gradually with shaking; with amalgam of good quality 200–250 g.

$$C_6H_5CH{=}CHCO_2H \xrightarrow{\text{NaOH, NaHg; HCl}} C_6H_5CH_2CH_2CO_2H$$

should suffice. The alkaline solution is decanted and acidified, and the precipitate is crystallized "aus nicht zu wenig Wasser . . . Ausbeute sehr gut."

In a synthesis of N-methyl-3,4-dihydroxyphenylalanine (6) from vanillin (1) and creatinine (2), one step is a sodium amalgam reduction of (3) ⟶ (4), carried out by adding 3% amalgam in portions to a stirred suspension of (3) in water.[6] The solid dissolves, and the initial orange-red color of the solution slowly fades as the reduction proceeds. With good agitation, decolorization is complete in 45–60 min.

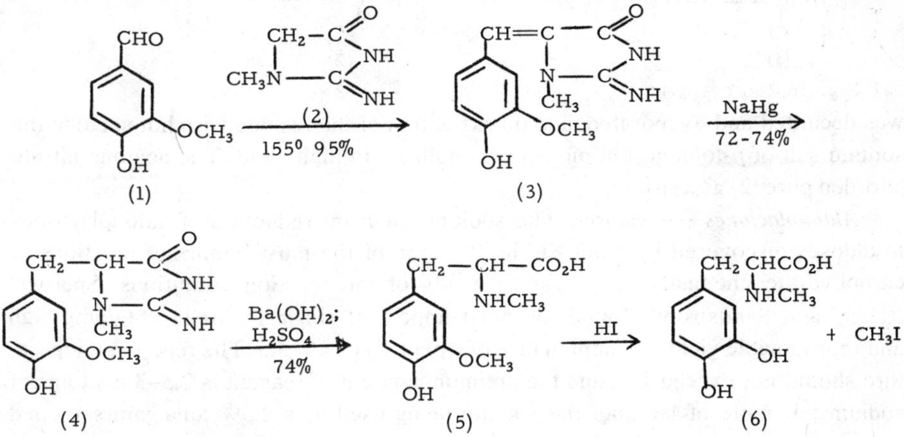

Emde degradation.[7] The Emde degradation consists in treating a quaternary halide in alcoholic or aqueous solution with sodium amalgam. The quaternary chloride (1a) in the Emde degradation affords the same product (2) as is obtained by the Hofmann degradation of the quaternary base (1b). Further degradation of (3) by

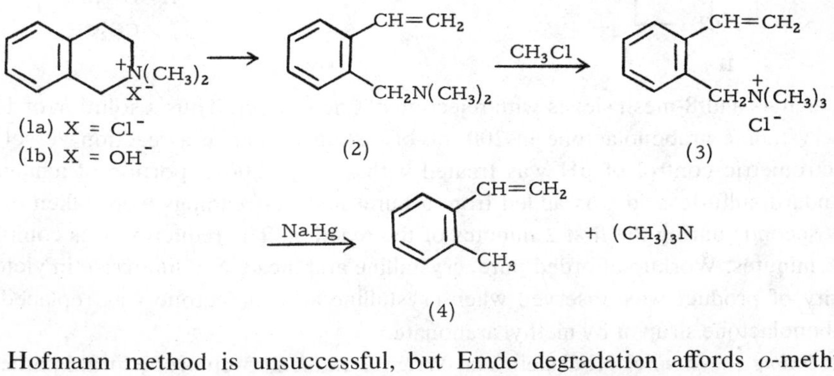

the Hofmann method is unsuccessful, but Emde degradation affords *o*-methylstyrene (4).[8] In a procedure worked out by Brasen and Hauser[9] for the preparation

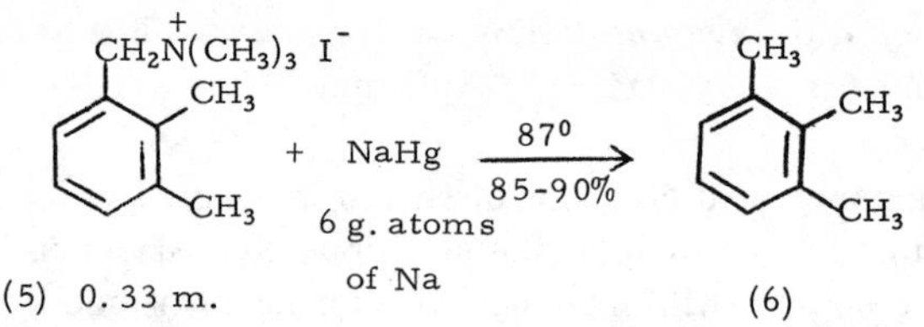
NaHg
87°
85-90%
6 g. atoms
of Na

(5) 0.33 m. (6)

Suspended in 2 l. H_2O

of hemimellitene (6), a suspension of the quaternary iodide (5) is stirred on the steam bath during addition of a very large excess of sodium amalgam.

Reductive detosylation. Grob and Prins[10] stirred a methanol solution of the ditosylate (1) and added sodium amalgam in 10 portions over 7 hrs. The solution

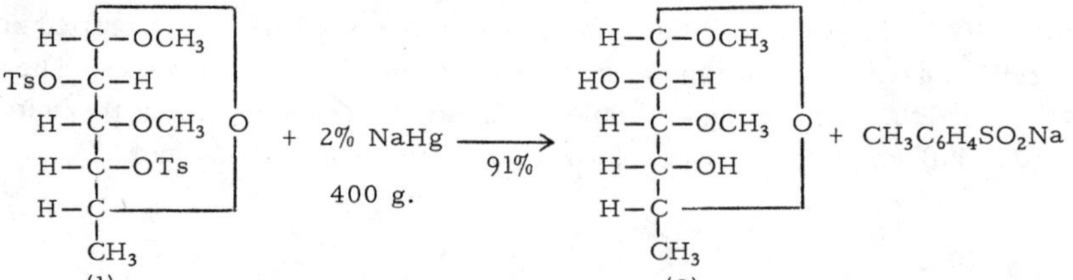
2% NaHg
400 g.
91%
CH3C6H4SO2Na

(1) (2)

7.2 g. in 200 CH_3OH

was decanted and evaporated, and on extraction of the residue with hot acetone the sodium salt of *p*-toluenesulfonic acid crystallized in high yield. The acetone filtrate afforded pure (2) as a sirup.

Aldonolactones ⟶ aldoses. The sodium amalgam reduction of aldonolactones to aldoses discovered by Emil Fischer[11] is one of the most important reactions of carbohydrate chemistry. In a careful study of the reaction conditions, Sperber, Zaugg, and Sandstrom[12] found the most important single factor in obtaining high and reproducible yields is maintenance of a pH range of 3–3.5. The reaction temperature should not exceed 15°, and the optimum amount of reagent is 2.5–3 g. atoms of sodium per mole of lactone, the sodium being used as a 2.5% amalgam screened

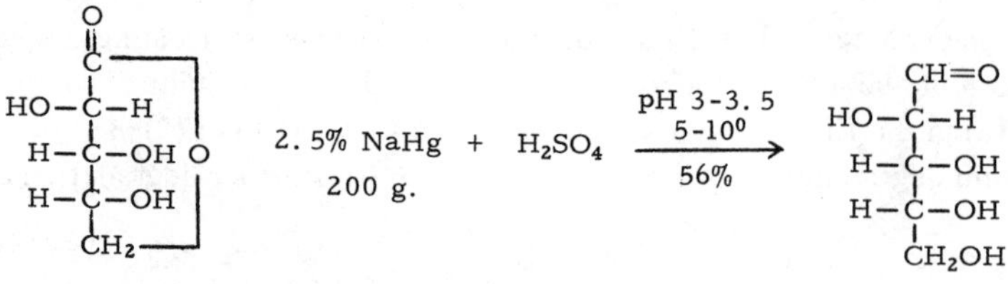
2.5% NaHg + H2SO4
200 g.
pH 3-3.5
5-10°
56%

11 g.

through 4- and 8-mesh sieves with rejection of fine powder. Thus a solution of 11 g. of crystalline arabonolactone in 200 ml. of distilled water in a reaction vessel for electrometric control of pH was treated with a single 200-g. portion of amalgam. Standard sulfuric acid was added from a buret and pH readings were taken every few seconds during the first 2 minutes of the reaction. The reduction was complete in 8 minutes. Workup afforded pure, crystalline arabinose. No difference in yield or purity of product was observed when crystalline arabonolactone was replaced by arabonolactone sirup or by methyl arabonate.

Oxime ⟶ amine. Hochstein and Wright[13] reduced 5-chlorofurfuraldoxime by stirring an alcoholic solution for 1 hr. while sodium amalgam and acetic acid were

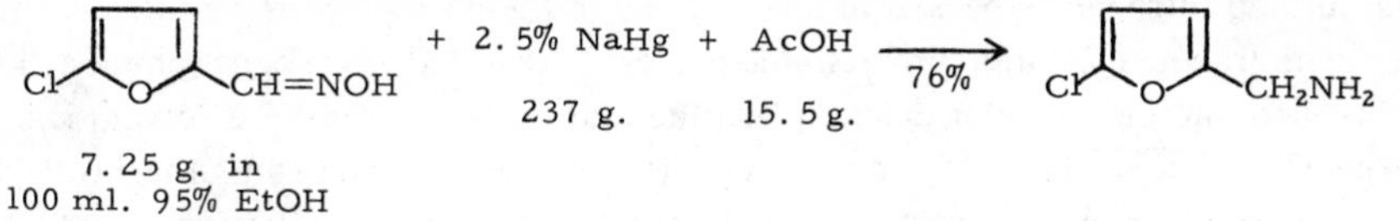
+ 2.5% NaHg + AcOH
237 g.
15.5 g.
76%
Cl
O
CH=NOH
CH₂NH₂
7.25 g. in
100 ml. 95% EtOH

added proportionately. The solution was diluted and extracted with ether, the amine was then liberated with alkali and extracted with ether.

[1]A. F. Holleman, *Org. Syn., Coll. Vol.*, **1**, 554 (1941)
[2]W. R. Brasen and C. R. Hauser, *Org. Syn., Coll. Vol.*, **4**, 508 (1963)
[3]W. B. Renfrow, Jr. and C. R. Hauser, *ibid.*, **2**, 607 (1943)
[4]Contributed by Max Tishler
[5]E. Fischer, "Anleitung zur Darstellung organischen Präparate," 8th edition, p. 39, Braunschweig (1908)
[6]V. Deulofeu and T. J. Guerrero, *Org. Syn., Coll. Vol.*, **3**, 586 (1955)
[7]H. Emde, *Ber.*, **42**, 2590 (1909); *Ann.*, **391**, 88 (1912)
[8]H. Emde and H. Kull, *Arch. Pharm.*, **274**, 173 (1936)
[9]W. R. Brasen and C. R. Hauser, *Org. Syn., Coll. Vol.*, **4**, 508 (1963)
[10]C. A. Grob and D. A. Prins, *Helv.*, **28**, 840 (1945)
[11]E. Fischer, *Ber.*, **22**, 2204 (1889); **23**, 930 (1890)
[12]N. Sperber, H. E. Zaugg, and W. M. Sandstrom, *Am. Soc.*, **69**, 915 (1947)
[13]F. A. Hochstein and G. F. Wright, *Am. Soc.*, **71**, 2257 (1949)

40% Sodium amalgam (liquid).

Preparation.[1] A 600-ml. beaker is charged with enough paraffin (about 80 g.) to form a layer about one half inch thick when molten, weighed, and heated on a hot plate to a temperature above the melting point of sodium (97.5°). About 100 g. (97 ml.) of sodium is cut into small pieces, cleaned by brief immersion in alcohol, and quickly transferred to the molten paraffin. The weight of sodium is determined by weighing the beaker while still hot. While the sodium is still molten a piece of 4-mm. glass tubing with one end drawn down to 1 mm. is supported so that the tip extends into the sodium layer; because of possible spattering, the tube should not be held in the hand. Mercury (1.50 g. per gram of sodium) is introduced slowly through the addition tube by means of a dropper, and after each addition the mixture is stirred gently by rotating the beaker. The heat of amalgamation is sufficient to keep the amalgam molten until all the mercury has been added (30–50 min.). While the paraffin is still semisolid, glass rods are thrust through the paraffin on opposite sides of the beaker. When the paraffin has been allowed to cool and solidify, withdrawal of the rods provides a hole through which the amalgam can be poured, and a vent hole. For storage the amalgam is transferred at 20–30° to a bottle containing a layer of ligroin (b.p. 60–90°). It is dispensed by syringe.

The phase diagram[2] indicates that amalgams containing 38–41% sodium are liquid in the temperature range 20–30°. Such an amalgam is conveniently manipulated after slight warming or cooling, as required.

Carboxylation of aromatic hydrocarbons.[1] An 8-oz. narrow-mouthed bottle is

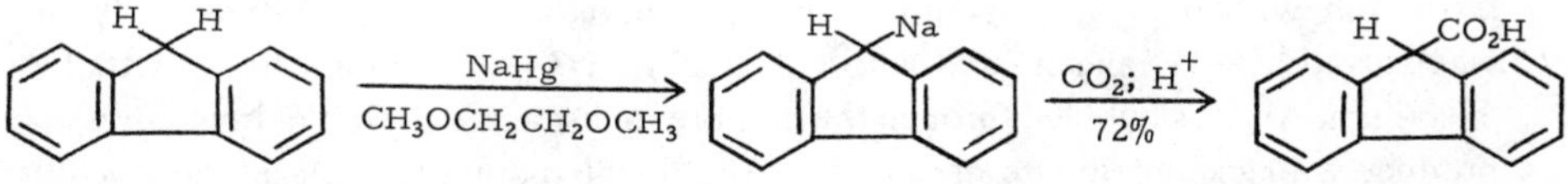
H H
NaHg
CH₃OCH₂CH₂OCH₃
H Na
CO₂; H⁺
72%
H CO₂H

charged with 16.6 g. (0.1 mole) of fluorene (m.p. 113–114°) and capped with a rubber serum stopper. Two 2-inch 20-gauge syringe needles are inserted through the stopper and purified nitrogen is passed in through one of them for 15 min. and this needle is then withdrawn. 1,2-Dimethoxyethane is freshly distilled from benzophenone ketyl or distilled and stored over calcium hydride and 32 ml. of the pure solvent is introduced into the bottle by syringe. The needles are removed, and the bottle and contents weighed. After inserting a needle for exit of displaced nitrogen, about 80 ml. of 40% sodium amalgam (a ten-fold excess) is introduced by syringe.

The sodium content of the amalgam can be determined both before and after the reaction by placing one drop of amalgam in excess standard 0.1 *N* hydrochloric acid and back titrating with 0.1 *N* sodium hydroxide. The globule of mercury is dried and weighed.

The reaction mixture is shaken for 1 hr., or until the temperature has dropped to the ambient temperature, whichever is longer. The organosodium compound forms a brown-black solution in the ether layer, which can be decanted into a slurry of dry ice in dimethoxyethane after the amalgam has been frozen by inversion of the bottle in a dry ice–acetone bath. The amalgam is washed by adding dimethoxyethane, melting the amalgam, and refreezing it.

The carbonated mixture is allowed to come to room temperature, water is added, and the two-phase system is extracted three times with benzene to remove neutral material. Acidification of the water phase with concd. hydrochloric acid precipitates fluorenone-9-carboxylic acid as a white product, m.p. 224–225° (15 g., 72%). Crystallization from acetic acid raises the melting point to 230–231°.

[1]Contributed by J. K. Blatchford, Saul I. Shupack, and Milton Orchin, University of Cincinnati

[2]M. Sittig, "Sodium, its Manufacture, Properties, and Uses," 54, Reinhold Publishing Corp., (1956)

Sodium amide, $NaNH_2$. Mol. wt. 39.02, m.p. 210°. Suppliers: Roberts Chem. Co., Fisher. Commercial sodium amide stored in a once opened bottle absorbs moisture, decomposes, and may turn yellow and become potentially explosive; such material should be destroyed by reaction with solid ammonium chloride:

$$NaNH_2 + NH_4Cl \longrightarrow NaCl + 2\ NH_3$$

Because of this instability, most procedures for use of the reagent include directions for its preparation. However, Fisher supplies material in two small sizes which come in amber bottles with wax-sealed screwcaps encased in plastic-wrapped steel cans (Fig. S-1): a 6-pack of 4-g. bottles (0.1 mole), and a 6-pack of 20-g. bottles (0.5 mole).

Preparation. Most laboratory procedures are based upon the discovery by Nieuwland and co-workers[1] that the reaction between sodium and liquid ammonia is accelerated markedly by a black catalyst prepared originally by reaction of a

$$2\ Na + 2\ NH_3 \longrightarrow 2\ NaNH_2 + H_2$$

ferric salt with oxides of sodium in liquid ammonia. Liquid ammonia (500 ml.) was stirred at −33° and treated with 0.3 g. of $Fe(NO_3)_3 \cdot 6\,H_2O$ and then with 1 g. of sodium. Air was bubbled through the mixture to discharge the initial blue color and produce a black precipitate of catalyst. Additional sodium (25 g.), added portionwise, was then consumed in about 20 min. with discharge of the blue color and production of a grey suspension of sodium amide.

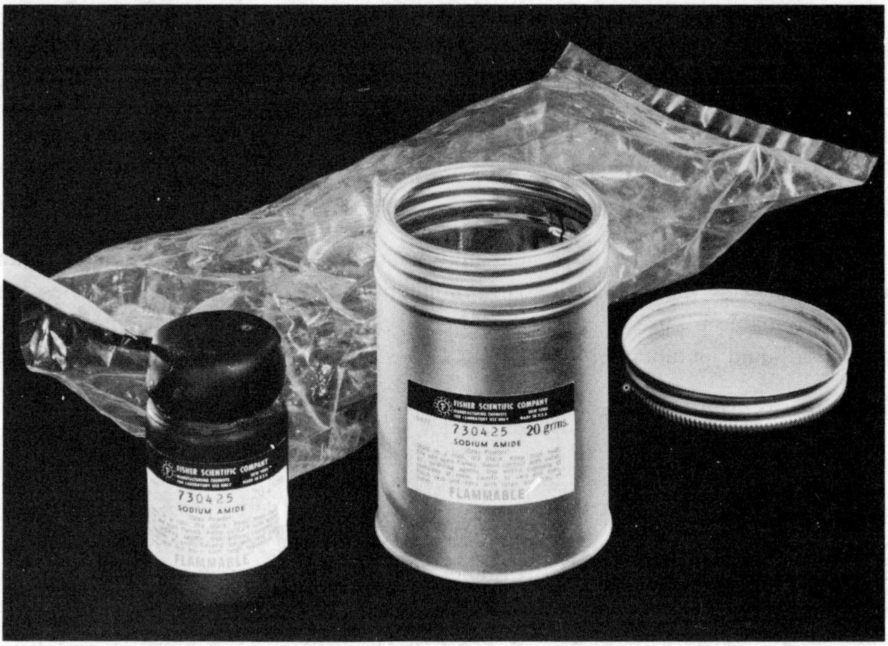
FISHER SCIENTIFIC COMPANY
730425 20 grms
SODIUM AMIDE
FLAMMABLE
FISHER SCIENTIFIC COMPANY
730425
SODIUM AMIDE
FLAMMABLE

Fig. S-1 No-waste sealed bottle of 0.5 mole of sodium amide

The solubility of sodium amide prepared in this way is 1 mole per liter of liquid ammonia at −33°.

Subsequent workers found air oxidation of the initial small portion of sodium unnecessary and merely wait until evolution of hydrogen has ceased and the black catalyst has separated before adding the main batch of sodium. Satisfactory procedures which differ in some details are described by Hauser,[2,3] by E R. H. Jones,[4] and by Khan.[5] Hauser and several other investigators employ neither a cooling bath nor a condenser and compensate for evaporation of ammonia by adding fresh ammonia as required; evaporating ammonia provides protection from oxygen. Both Jones and Khan use a dry ice condenser, a nitrogen atmosphere, a gas absorption trap, and a bubbler. Khan uses ferric chloride for production of the catalyst. A Hershberg stirrer is usually preferred; a stirrer with a Teflon blade should not be used.

Dehydrohalogenation (and debromination). Nieuwland and co-workers[1] stirred a suspension of 2 moles of sodium amide in 1.5 l. of liquid ammonia at −33° and added dropwise 1 mole of β-bromostyrene. When the vigorous reaction was over, the

$$\underset{1\text{ m.}}{C_6H_5CH{=}CHBr} + \underset{2\text{ m.}}{2\ NaNH_2} \longrightarrow C_6H_5C{\equiv}CNa \xrightarrow[75\%]{H_2O} C_6H_5C{\equiv}CH$$

mixture was stirred for 10 minutes more and treated with 600 ml. of water. Phenylacetylene was obtained in good yield, and some starting material was recovered. By the same procedure styrene dibromide suffered dehydrobromination to phenylacetylene in 64% yield. Although under some conditions internal acetylenes can be rearranged to allenes and to terminal acetylenes, the Notre Dame workers found that 3-nonyne is not isomerized by sodium amide in 16 hrs. at −34°.

For the preparation of cyclohexylpropyne (2), Lespieau and Bourguel[6] ground excess commercial sodium amide in a mortar with 200 ml. of mineral oil, stirred the suspension in a bath at 160–165°, and added the vinyl bromide (1) dropwise in 1½ hrs.

$$\underset{(1)\ 1\ m.}{C_6H_{11}CH_2\underset{\displaystyle Br}{C}{=}CH_2} + \underset{3.1\ m.}{2\ NaNH_2} \xrightarrow{160\text{-}165°} C_6H_{11}CH_2C{\equiv}CNa \xrightarrow[66\%]{HCl} \underset{(2)}{C_6H_{11}CH_2C{\equiv}CH}$$

For the conversion of oleic acid into stearolic acid via the dibromide, Khan and co-workers[5] found that dehydrobromination with sodium amide doubles the yield obtainable by use of ethanolic potassium hydroxide. A suspension of sodium amide in liquid ammonia prepared from 1.87 g. atoms of sodium was stirred under

$$\underset{0.35\ m.}{CH_3(CH_2)_7CH{=}CH(CH_2)_7CO_2H} \xrightarrow[Ether]{Br_2} CH_3(CH_2)_7\underset{\displaystyle Br}{CH}\,\underset{\displaystyle Br}{CH}(CH_2)_7CO_2H \xrightarrow[52\text{-}62\%\ overall]{3\ NaNH_2;\ HCl} CH_3(CH_2)_7C{\equiv}C(CH_2)_7CO_2H$$

a dry ice-acetone condenser, and an ethereal solution of oleic acid dibromide was introduced slowly from a dropping funnel. After 6 hrs. of continuous stirring, 60 g. of solid ammonium chloride was added in portions to destroy excess sodium amide and the ammonia was allowed to evaporate. The solid salt remaining was warmed with water under nitrogen and acidified, and stearolic acid was obtained in pure form by two crystallizations from petroleum ether (m.p. 46–46.5°).

A similar process was used for the conversion of Δ^{10}-undecylenic acid into Δ^{10}-undecynoic acid.[7]

Two dehydrohalogenations involving ether cleavage are described by the Jones group. One is the preparation of ethoxyacetylene from diethylchloroacetal.[4] A suspension of sodium amide in liquid ammonia was agitated manually in a flask

$$\underset{0.50\ m.}{ClCH_2CH(OC_2H_5)_2} + \underset{1.65\ g.\ at.Na}{3\ NaNH_2} \xrightarrow{Liq.\ NH_3} NaC{\equiv}COC_2H_5 \xrightarrow[57\text{-}60\%]{NaCl\text{-}H_2O} HC{\equiv}COC_2H_5$$

equipped with a cold-finger condenser cooled with dry ice, and diethylchloroacetal was added in 15–20 min. After 15 min. more the ammonia was evaporated in a stream of nitrogen, and the flask containing the highly pyrophoric sodio derivative was cooled to −70° in a dry ice–trichloroethylene bath. A saturated solution of sodium chloride precooled to −20° was added all at once and volatile material was distilled into a receiver cooled to −70°. The distillate was neutralized with aqueous sodium dihydrogen phosphate and the aqueous layer was frozen by cooling with dry ice and the supernatant liquor dried and distilled. Fractionation afforded pure ethoxyacetylene, b.p. 49–51°/749 mm. The second reaction involving ether cleavage affords 4-pentyne-1-ol starting with tetrahydrofurfuryl chloride.[8]

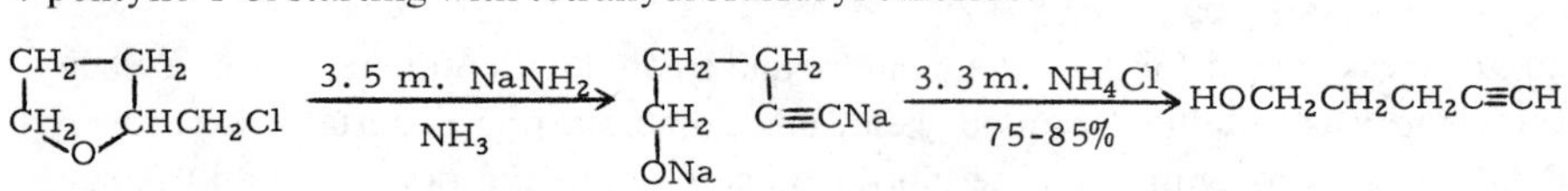
3.5 m. NaNH2
NH3
3.3 m. NH4Cl
75-85%
HOCH2CH2CH2C≡CH

1 m.

For the preparation of 2-butyne-1-ol (4), Whiting *et al.*[9] prepared a solution of sodium amide in a 4-l. Dewar flask equipped with a plastic paddle and mechanical stirrer (3 l. of NH_3, 1.5 g. of hydrated ferric nitrate, 2.8 g. atoms of sodium) and added 1.26 moles of 3-chloro-2-butene-1-ol (2) over a period of 30 min. The mixture was stirred overnight, then 2.8 moles of solid ammonium chloride was added in

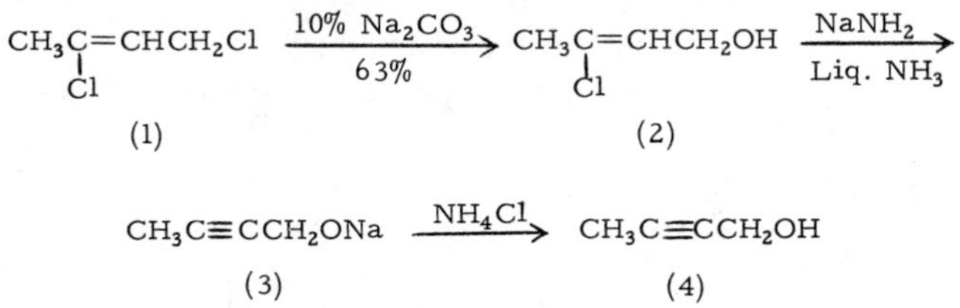

portions at a rate permitting control of the exothermal reaction. The mixture was transferred to a stainless steel bucket and let stand overnight in a hood for evaporation of the ammonia. The product was collected by ether extraction and distilled at 55°/8 mm.

Dehydrohalogenation with sodium amide also provides a useful method for closing an aziridine ring, as illustrated by the preparation of N-ethylallenimine (1-ethyl-2-methylenaziridine).[10]

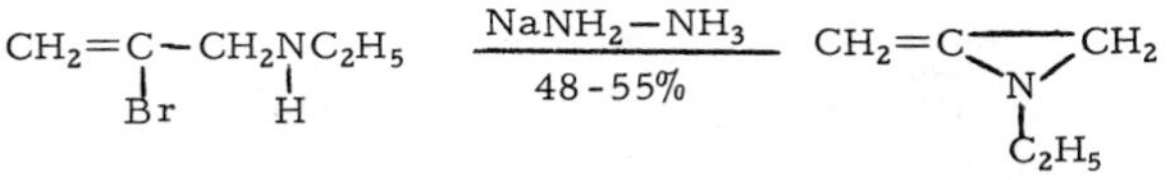

RC≡CNa and derived products. Tetrolic acid is prepared by conversion of propyne into the sodio derivative, carbonation, and acidification.[11] For the preparation of 1-hexyne, Campbell and Campbell[12] passed acetylene into 3 l. of liquid

$$CH_3C\equiv CH \xrightarrow{NaNH_2} CH_3C\equiv CNa \xrightarrow{CO_2} CH_3C\equiv CCO_2Na \xrightarrow[50-59\%]{HCl} CH_3C\equiv CCO_2H$$

ammonia and gradually added 4 g. atoms of sodium at such a rate that the entire solution did not turn blue (each piece of sodium was attached to an iron fish hook mounted on a flexible iron wire with which the sodium could be lowered into the

$$HC\equiv CH \ + \ Na \xrightarrow[-H_2]{liq.\ NH_3} HC\equiv CNa \xrightarrow[70-77\%]{n\text{-}C_4H_9Br} n\text{-}C_4H_9C\equiv CH$$

ammonia or withdrawn). The acetylene was shut off and *n*-butyl bromide (4 moles) was added dropwise. Finally 500 ml. of ammonium hydroxide was added dropwise, followed by distilled water. The workup afforded pure 1-hexyne, b.p. 71–72°.

Generation of reactive anions. In a typical alkylation of a ketone, Vanderwerf and Lemmerman[13] prepared a grey suspension of sodium amide under nitrogen (1.5 l. of ammonia, a trace of ferric nitrate, and 2.05 g. atoms of sodium), added 1.2 l. of ether as rapidly as vaporization of ammonia permitted, warmed the mixture

1.
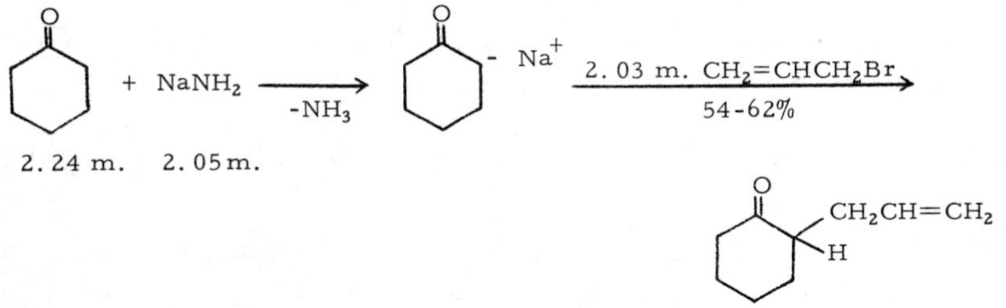
+ NaNH2
-NH3
Na+
2. 03 m. CH2=CHCH2Br
54-62%
2. 24 m.
2. 05m.
CH2CH=CH2
H

to remove ammonia, added cyclohexanone, and stirred and heated the mixture under reflux for 3 hrs. After cooling, an ethereal solution of allyl bromide was added continuously (exothermal reaction), and the reaction was completed by refluxing for 3 hrs.

Other examples are indicated briefly in formulations 2–5. The sodium amide used in the preparation of 2-methyl-3-butyne-3-ol (2) was moistened with heptane, ground

2.[14] $CH_3\underset{CH_3}{\overset{CH_3}{C}}=O \xrightarrow[-NH_3]{NaNH_2} CH_3\underset{CH_2(Na^+)}{\overset{}{C}}=O \xrightarrow{HC\equiv CH} CH_3\underset{ONa}{\overset{CH_3}{C}}-C\equiv CH \xrightarrow[40-46\%]{H^+} CH_3\underset{OH}{\overset{CH_3}{C}}-C\equiv CH$ 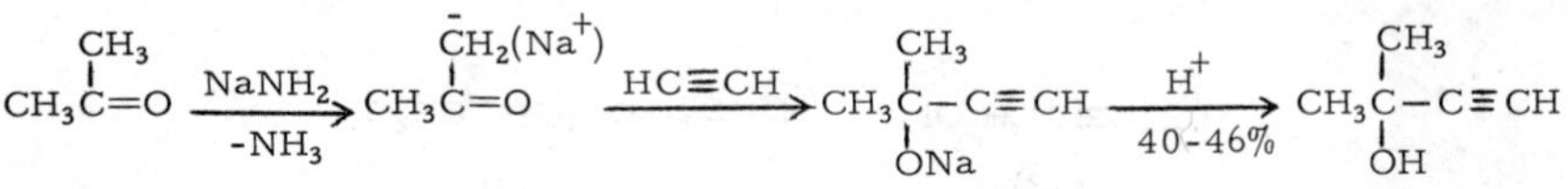

3.[15] $CH_3COCH_2CH(CH_3)_2 \xrightarrow[-NH_3]{NaNH_2} {}^-CH_2COCH_2CH(CH_3)_2(Na^+) \xrightarrow[-C_2H_5OH]{(CH_3)_2CHCH_2CO_2C_2H_5}$ 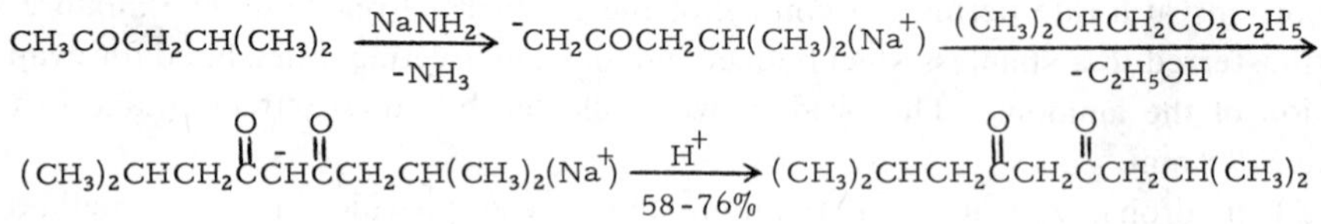

$(CH_3)_2CHCH_2\overset{O}{\overset{\|}{C}}\overset{-}{C}H\overset{O}{\overset{\|}{C}}CH_2CH(CH_3)_2(Na^+) \xrightarrow[58-76\%]{H^+} (CH_3)_2CHCH_2\overset{O}{\overset{\|}{C}}CH_2\overset{O}{\overset{\|}{C}}CH_2CH(CH_3)_2$

4.[3] $C_6H_5CH_2CO_2H \xrightarrow{2\ NaNH_2} C_6H_5\overset{-}{C}HCO_2^-(Na^+)_2 \xrightarrow[Et_2O]{C_6H_5CH_2Cl}$ 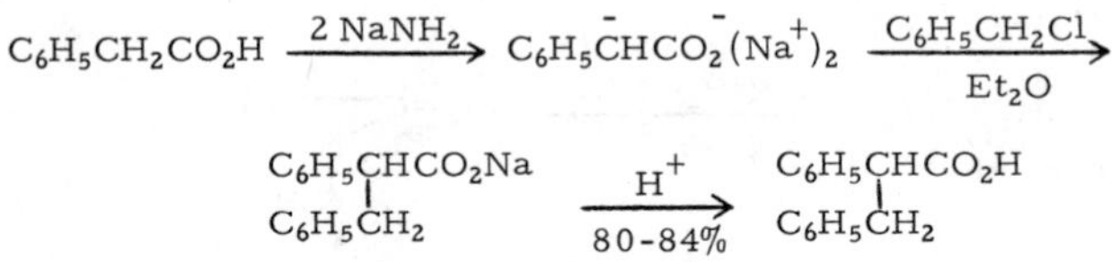

$\underset{C_6H_5CH_2}{\overset{C_6H_5CHCO_2Na}{}} \xrightarrow[80-84\%]{H^+} \underset{C_6H_5CH_2}{\overset{C_6H_5CHCO_2H}{}}$

5.[16] 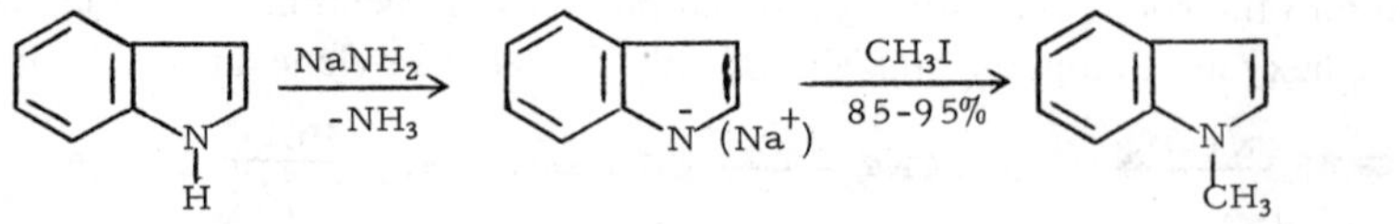

to a fine powder, and the solvent allowed to evaporate. A suspension of the amide in ether was stirred in a bath at −10° while acetylene was passed in to form the adduct and sweep out the ammonia. In example 3 (Hauser), sodium amide was prepared from liquid ammonia, sodium, and iron catalyst, and the ammonia was displaced by ether before addition of an ethereal solution of methyl isobutyl ketone. After formation of the sodio derivative, ethyl isovalerate was added in ether to offset ester condensation. Diisovalerylmethane was isolated in pure form through the copper salt. In example 4 (Hauser), phenylacetic acid was added directly to a suspension of sodium amide in liquid ammonia and benzyl chloride was added to the resulting dark green suspension. On completion of the alkylation the ammonia was displaced by ether. Several other alkylations of phenylacetic acid and its esters are reported by Hauser and co-workers.[17] Example 5, N-methylation of indole, was accomplished substantially in the same way as 4.

Use of sodium amide for generation of an ion for reaction with a ketone is illustrated by the synthesis of α,β-diphenylcinnamonitrile.[18]

$C_6H_5CH_2CN \xrightarrow{NaNH_2} C_6H_5\overset{-}{C}HCN(Na^+) \xrightarrow[50-66\%]{(C_6H_5)_2C=O} (C_6H_5)_2C=\underset{CN}{\overset{}{C}}C_6H_5$

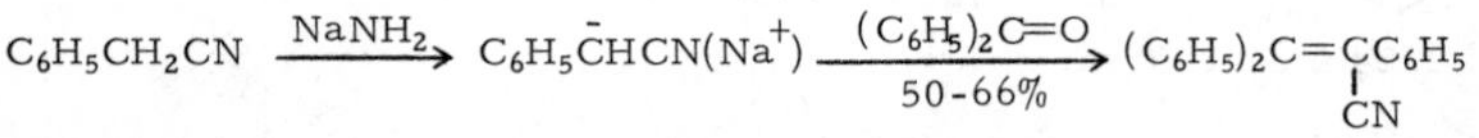

Verley's synthesis of 2-methylindole[19] involves cyclization of an intermediate anion. The yield is that reported by Allen and VanAllan.[20]

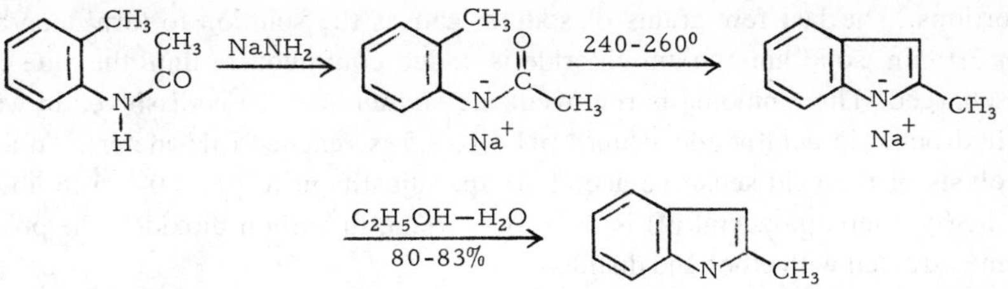

In a detailed study of the generation of dicarbanions from β-dicarbonyl compounds, Harris and Harris[21] found that sodium amide is superior to potassium amide, sodium methylsulfinylmethylide, or sodium hydride.

$$RCOCH_2COCH_3 \xrightarrow[NH_3]{NH_2^-} R\overset{-}{C}OCHCO\overset{-}{C}H_2 \xrightarrow{CO_2} RCOCH_2COCH_2CO_2H$$

Other reactions. The Darzens condensation,[22] for example of acetophenone with ethyl chloroacetate to produce phenylmethylglycidic ester, was carried out

$$C_6H_5\overset{CH_3}{\underset{}{C}}{=}O + ClCH_2CO_2C_2H_5 \xrightarrow[62-64\%]{NaNH_2} C_6H_5\overset{CH_3}{\underset{O}{C}}{-}CHCO_2C_2H_5$$

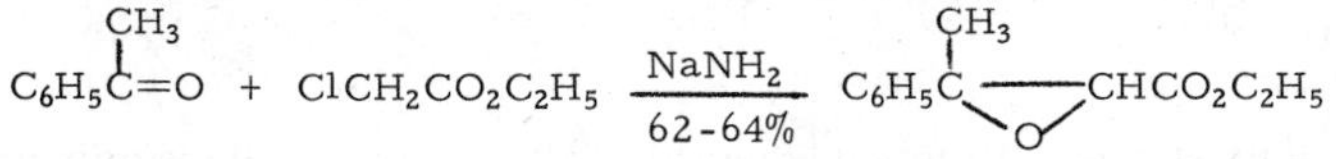

by Allen and VanAllan[23] by adding finely powdered sodium amide to a solution of the reactants in benzene.

Brasen and Hauser[24] describe a procedure for effecting reductive rearrangement of benzyltrimethylammonium iodide (1) to the tertiary amine (2).

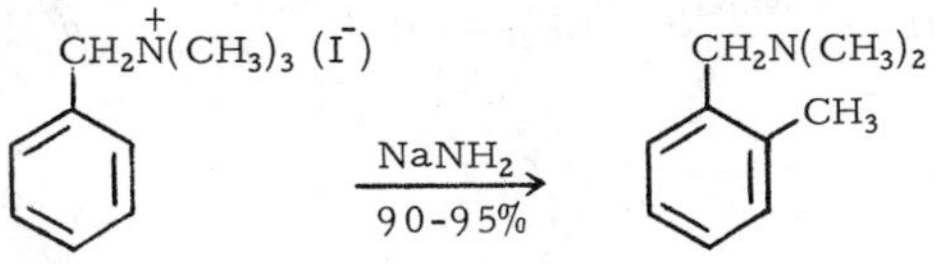

Bunnett and co-workers[25] used sodium amide to effect condensation of sodium naphthalene-β-sulfonate with piperidine to produce N-β-naphthylpiperidine.

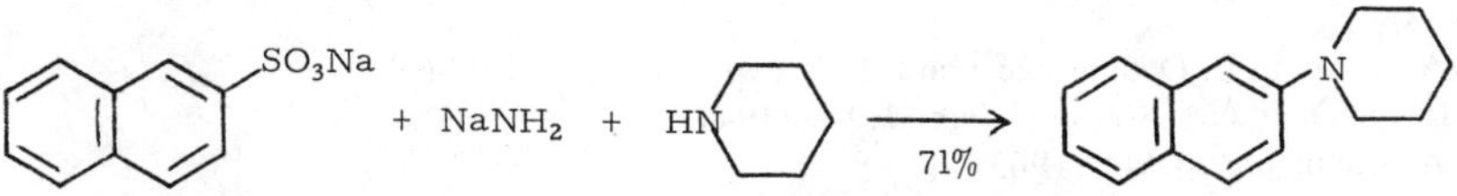

Parham and Wynberg[26] describe a procedure for the reaction of diethyl bromo-acetal (1) with sodium sulfide and sulfur to give a product of the average composition of the tetrasulfide (2). A mixture of this substance (1.38 m.) and ether is stirred at -35 to $-45°$, liquid ammonia is added, followed by 8.7 g. atoms of sodium, added

$$BrCH_2CH(OC_2H_5)_2 + Na_2S \cdot 9 H_2O + 2 S \xrightarrow[83-92\%]{EtOH}$$

(1) 3 m.   2.25 m.   4.5 g.atoms

$$(C_2H_5O)_2CHCH_2SSSSCH_2CH(OC_2H_5)_2 \xrightarrow{NaNH_2-NH_3} 2 (C_2H_5O)_2CHCH_2SNa$$

(2)   (3)

$$\xrightarrow[71-83\% \text{ from (2)}]{HCl, CO_2} 2 (C_2H_5O)_2CHCH_2SH$$

(4)

in portions. The last few grams of sodium causes the solution to turn deep blue. After 30 min. solid ammonium chloride is added continuously until the blue color is discharged. The ammonia is removed, the sodium salt (3) is dissolved in water, and hydrochloric acid is added until pH 8.0–8.5 is reached (pH meter). To avoid hydrolysis of the acid-sensitive acetal group, adjustment to pH 8.0–8.5 to liberate free diethyl mercaptoacetal (4) is done by passing in carbon dioxide. The product is then extracted with ether and distilled.

Cyclopropene synthesis. Fisher and Applequist[27] treated methallyl chloride (1) with sodium amide in refluxing tetrahydrofurane and obtained 1-methylcyclopropene (2) in reasonable yield; the reaction probably involves formation and cyclization of the vinylcarbene. Closs and Krantz[28] found that cyclopropene (3) itself can be

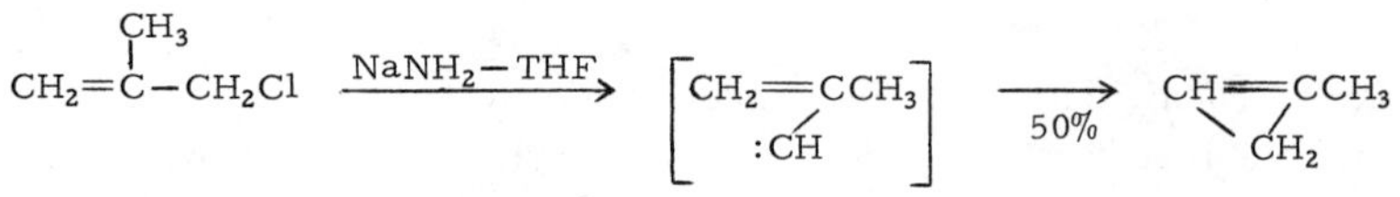
CH_3
$CH_2=C-CH_2Cl$
$NaNH_2-THF$
$CH_2=CCH_3$
:CH
50%
$CH=CCH_3$
CH_2

(1) (2)

obtained in the same way from allyl chloride. Although the yield is only about 10%, the method is far simpler than earlier methods. Cyclopropene produced in this way reacts with cyclopentadiene at 0° to give the Diels-Alder adduct (4) in 10% yield.

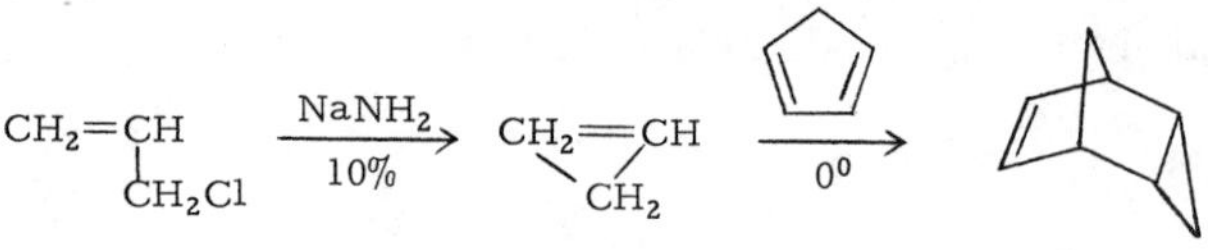
$CH_2=CH$
CH_2Cl
$NaNH_2$
10%
$CH_2=CH$
CH_2
0°

(3) (4)

[1]T. H. Vaughn, R. R. Vogt, and J. A. Nieuwland, *Am. Soc.*, **56**, 2120 (1934)
[2]C. R. Hauser, J. T. Adams, and R. Levine, *Org. Syn., Coll. Vol.*, **3**, 291 (1955)
[3]C. R. Hauser and W. R. Dunnavant, *Org. Syn.*, **40**, 38 (1960)
[4]E. R. H. Jones, G. Eglinton, M. C. Whiting, and B. L. Shaw, *Org. Syn., Coll. Vol.*, **4**, 404 (1963)
[5]N. A. Khan, F. E. Deatherage, and J. B. Brown, *ibid.*, **4**, 851 (1963)
[6]R. Lespieau and M. Bourguel, *ibid.*, **1**, 191 (1941)
[7]N. A. Khan, *ibid.*, **4**, 969 (1963)
[8]E. R. H. Jones, G. Eglinton, and M. C. Whiting, *ibid.*, **4**, 755 (1963)
[9]P. J. Ashworth, G. H. Mansfield, and M. C. Whiting, *ibid.*, **4**, 128 (1963)
[10]A. T. Bottini and R. E. Olsen, *Org. Syn.*, **44**, 53 (1964)
[11]J. C. Kauer, and M. Brown, *ibid.*, **42**, 97 (1962)
[12]K. N. Campbell and B. K. Campbell, *Org. Syn., Coll. Vol.*, **4**, 117 (1963)
[13]C. A. Vanderwerf and L. V. Lemmerman, *ibid.*, **3**, 44 (1955)
[14]D. D. Coffman, *ibid.*, **3**, 320 (1955)
[15]C. R. Hauser, J. T. Adams, and R. Levine, *bid.*, **3**, 291 (1955)
[16]K. T. Potts and J. E. Saxton, *Org. Syn.*, **40**, 68 (1960)
[17]C. R. Hauser and W. J. Chambers, *Am. Soc.*, **78**, 4942 (1956); R. B. Meyer and C. R. Hauser, *J. Org.*, **26**, 3696 (1961); W. G. Kenyon, R. B. Meyer, and C. R. Hauser, *J. Org.*, **28**, 3108 (1963); W. M. Kauser, W. G. Kenyon, and C. R. Hauser, procedure submitted to *Org. Syn.*
[18]S. Wawzonek and E. M. Smolin, *Org. Syn., Coll. Vol.*, **4**, 387 (1963)
[19]M. A. Verley, *Bull. soc.*, **35**, 1039 (1924)
[20]C. F. H. Allen and J. VanAllan, *Org. Syn., Coll. Vol.*, **3**, 597 (1955)
[21]T. M. Harris and C. M. Harris, *J. Org.*, **31**, 1032 (1966)

[22]G. Darzens, *Compt. rend.*, **139**, 1215 (1904)
[23]C. F. H. Allen and J. VanAllan, *Org. Syn., Coll. Vol.*, **3**, 727 (1955)
[24]W. R. Brasen and C. R. Hauser, *ibid.*, **4**, 585 (1963)
[25]J. F. Bunnett, T. K. Brotherton, and S. M. Williamson, *Org. Syn.*, **40**, 74 (1960)
[26]W. E. Parham and H. Wynberg, *Org. Syn., Coll. Vol.*, **4**, 295 (1963)
[27]F. Fisher and D. E. Applequist, *J. Org.*, **30**, 2089 (1965)
[28]G. L. Closs and K. D. Krantz, *J. Org.*, **31**, 638 (1966)

Sodium–Ammonia. For the reduction of a phenol to an aromatic hydrocarbon a solution of the phenol and diethyl hydrogen phosphite in carbon tetrachloride is treated with triethylamine and allowed to stand for 24 hrs. for complete separation of triethylamine hydrochloride (1). The phenol diethyl phosphate ester is collected,

$$1.\quad ArOH + HO{-}P(OC_2H_5)_2 + (C_2H_5)_3N + CCl_4 \longrightarrow ArO\overset{O^-}{\overset{|}{P^+}}(OC_2H_5)_2 + CHCl_3 + (C_2H_5)_3NHCl$$

$$2.\quad ArO{-}\overset{O^-}{\overset{|}{P^+}}(OC_2H_5)_2 + 2\,Na + NH_3 \longrightarrow ArH + NaNH_2 + NaO\overset{O^-}{\overset{|}{P^+}}(OC_2H_5)_2$$

dissolved in tetrahydrofurane, and reduced with sodium and liquid ammonia.[1] The reaction can be run on a 50–200 mg. sample and yields are in the range 18–52%.[2] Other examples.[3]

[1]G. W. Kenner and N. R. Williams, *J. Chem. Soc.*, 522 (1955)
[2]S. W. Pelletier and D. M. Locke, *J. Org.*, **23**, 131 (1958)
[3]J. Fishman and M. Tomasz, *J. Org.*, **27**, 365 (1962); E. Caspi, E. Cullen, and P. K. Grover, *J. Chem. Soc.*, 212 (1963)

Sodium–Ammonia–Ethanol. The combination of sodium with liquid ammonia and ethanol is commonly employed in the Birch reduction (of phenol ethers), *which see.*

Paquette and Nelson[1] found this combination superior to the Bouveault-Blanc method for the selective reduction of the ester acid (1). Small portions of sodium

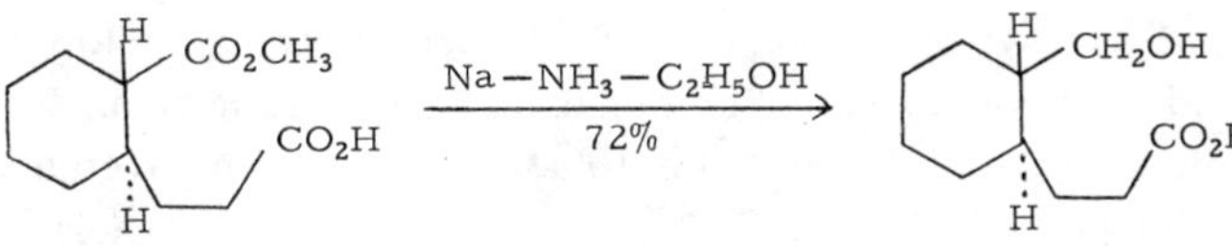
H
CO_2CH_3
CO_2H
H
$Na-NH_3-C_2H_5OH$
72%
H
CH_2OH
CO_2H
H

(1) (2)

were added to a stirred solution of (1) in ammonia and ethanol; the solution was stirred until most of the ammonia had evaporated, water and hydrochloric acid were added, and the product was extracted with ether.

[1]L. A. Paquette and N. A. Nelson, *J. Org.*, **27**, 2272 (1962)

Sodium *t*-amylate, *see* Sodium 2-methyl-2-butoxide.

Sodium azide, $NaN{=}\overset{+}{N}{=}\bar{N}$. Mol. wt. 65.02, solubility in 100 g. water: 39 g. at 0°, 55 g. at 100°. Suppliers: du Pont; Alfa, E, F, MCB. *See also* Hydrazoic acid.

Synthesis of isocyanates.[1] In a typical procedure[2] a mixture of lauroyl chloride and acetone is added from a dropping funnel with cooling and stirring to a solution of sodium azide in water at a rate such as to maintain a temperature of 10–15°. After the reaction is complete (*ca.* 1 hr.), the lower aqueous layer is removed carefully by suction through a glass capillary tube. The upper layer of crude acyl azide is added slowly to 500 ml. of benzene that has been warmed to 60°. Thermal elimination

$$\underset{\substack{0.5\text{ m.}\\ \text{In 150 ml. acetone}}}{n\text{-}C_{11}H_{23}COCl} + \underset{0.7\text{ m.}}{NaN_3} \xrightarrow[\text{water}]{10\text{-}15^0} n\text{-}C_{11}H_{23}\overset{\overset{O}{\|}}{C}-N=\overset{+}{N}=\overset{-}{N}$$

$$\xrightarrow[\text{81-86\% overall}]{\text{In } C_6H_6 \text{ at } 60\text{-}70^0} n\text{-}C_{11}H_{23}N=C=O + N_2$$

of nitrogen gives a radical which undergoes rearrangement to the isocyanate. After about 1 hr. the benzene is removed, and undecyl isocyanate is distilled at 3 mm.

Curtius reaction. The overall process of converting an acid through the azide and isocyanate into an amine is known as the Curtius reaction.[3] For application of the Curtius reaction to *cis*-2-phenylcyclopropanecarboxylic acid, J. Weinstock[4] found the route to the azide through the acid chloride unsatisfactory because of ready isomerization to the *trans* acid chloride. An efficient alternative method utilizes a mixed anhydride intermediate of a type found serviceable in the synthesis of peptides. A solution of the acid in aqueous acetone was treated with triethylamine to

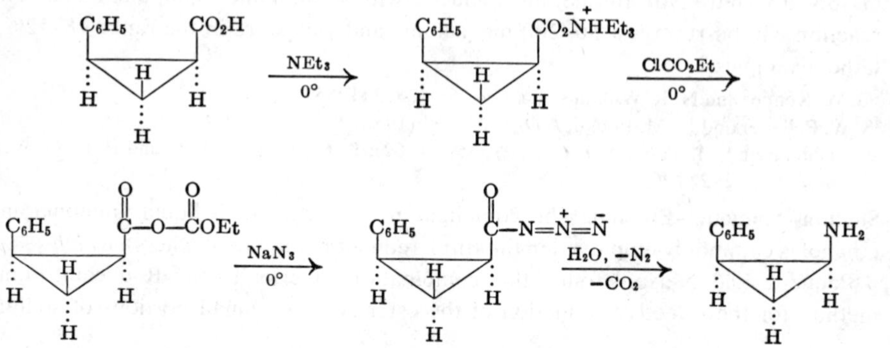

form the amine salt, which on addition of ethyl chloroformate afforded the mixed anhydride. Addition of aqueous sodium azide gave the acyl azide, which was extracted with ether and decomposed in toluene solution. The overall yield of pure *cis*-2-phenylcyclopropylamine was 77%.

Munch-Petersen[5] prepared *m*-nitrobenzazide by adding *m*-nitrobenzoyl chloride in acetone dropwise to a stirred solution of sodium azide in water. The azide separates

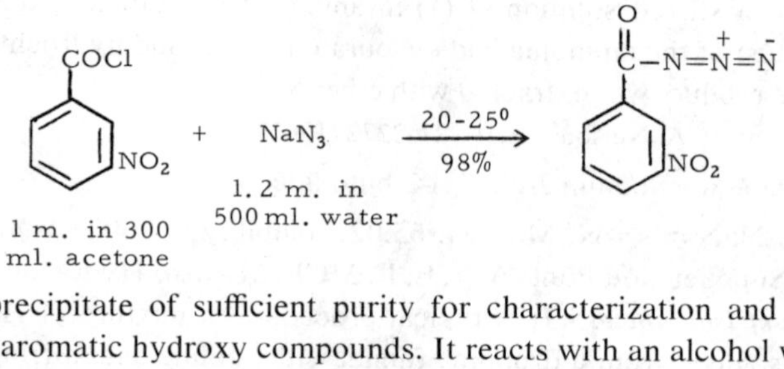

as a white precipitate of sufficient purity for characterization and estimation of aliphatic and aromatic hydroxy compounds. It reacts with an alcohol with liberation of nitrogen with rearrangement to the aryl isocyanate, to which the alcohol adds to give an aryl urethane:[6]

$$Ar\overset{\overset{O}{\|}}{C}-N=\overset{+}{N}=\overset{-}{N} \xrightarrow{-N_2} ArN=C=O \xrightarrow{ROH} ArNHCOOR$$

These substances are nicely crystalline derivatives in which the nitro group can be titrated with titanous chloride. With amines it forms *m*-nitrophenylureas:[6]

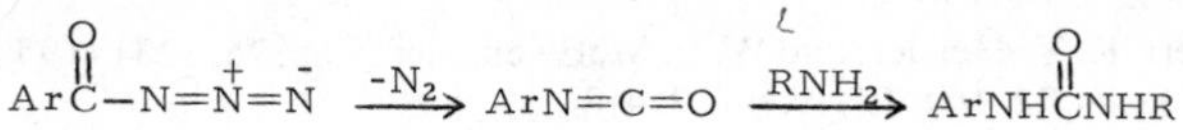

Aryl azides. In one synthesis[7] of benzofurazane oxide (3) an aqueous solution of sodium azide is added to a solution of diazotized *o*-nitroaniline at 0–5°. Nitrogen is evolved, and *o*-nitrophenylazide separates as a light cream to colorless solid. Crystallization from 95% ethanol gives material (m.p. 52–54°) satisfactory for the

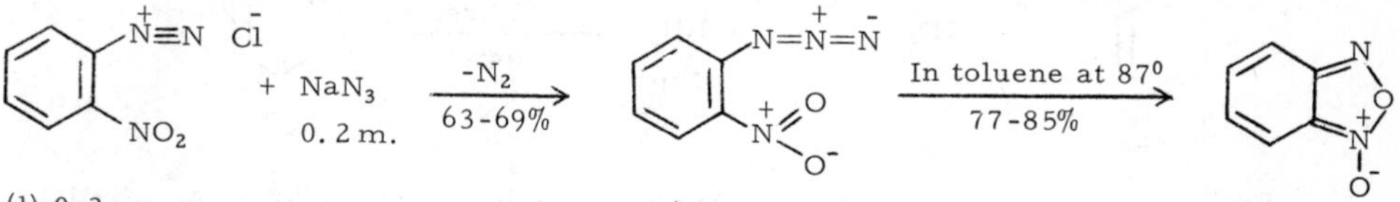

(1) 0. 2 m. (2) (3)

next step of thermal decomposition in toluene. Benzofurazane oxide can be prepared more conveniently by hypochlorite oxidation of *o*-nitroaniline.[8]

Reaction with epoxides. Vanderwerf and co-workers[9] added a saturated aqueous solution of sodium azide (in slight excess) to a refluxing solution of propylene oxide in dioxane and established that the main product is 1-azido-2-propanol. Cyclopentene oxide reacts to give *trans*-2-azidocyclopentanol.

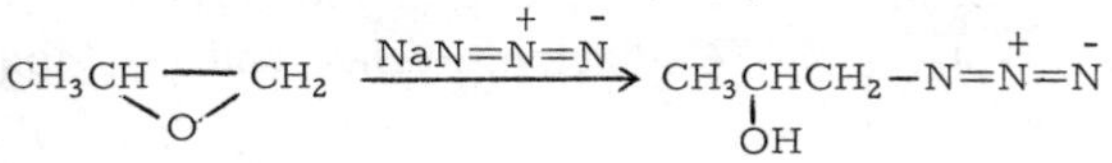

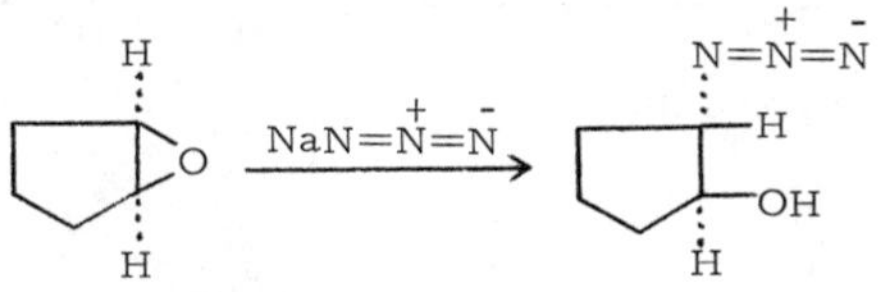

Spot test.[10] Mercaptans and thioketones markedly catalyze the reaction of sodium azide with iodine with liberation of nitrogen, and this catalyst is used for detection of the sulfur containing compounds.

$$2\ NaN_3 + I_2 \longrightarrow 2\ NaI + 3\ N_2$$

Benzofurazane synthesis.[11] Benzofurazanes can be prepared by the spontaneous decomposition (100°) of *o*-nitrosophenylazides. Thus 1,3-dichloro-2-nitrosobenzene (2) reacts with sodium azide to give 4-chlorobenzofurazane (3), m.p. 84°.

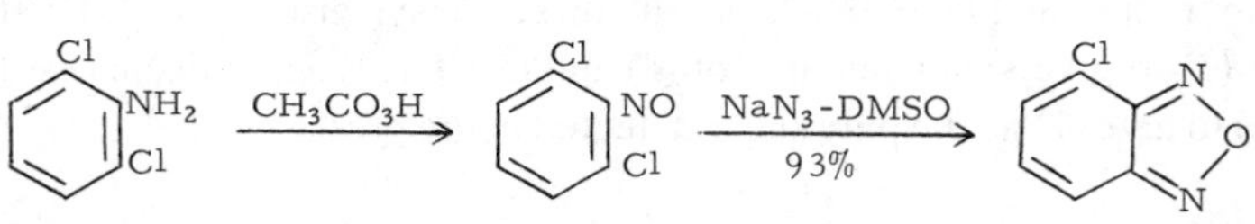

(1) (2) (3)

[1]G. Schroeter, *Ber.*, **42**, 3356 (1909)
[2]C. F. H. Allen and A. Bell, *Org. Syn., Coll. Vol.*, **3**, 846 (1955)
[3]Th. Curtius, *J. prakt. Chem.*, [2], **50**, 275 (1894)
[4]J. Weinstock, *J. Org.*, **26**, 3511 (1961)
[5]J. Munch-Petersen, *Org. Syn., Coll. Vol.*,**4**, 715 (1963)

[6]See references in preceding paper.
[7]P. A. S. Smith and J. H. Boyer, *Org. Syn., Coll. Vol.*, **4**, 75 (1963)
[8]F. B. Mallory, *ibid.*, **4**, 74 (1963)
[9]C. A. Vanderwerf, R. Y. Heisler, and W. E. McEwen, *Am. Soc.*, **76**, 1231 (1954)
[10]F. Feigl, "Spot Tests," **2**, 164, Elsevier (1954)
[11]A. J. Boulton, P. B. Ghosh, and A. R. Katritzky, *Tetrahedron Letters*, 2887 (1966)

Sodium benzenesulfonate, $C_6H_5SO_3Na$. Mol. wt. 180.17. Suppliers: B, KK, MCB.

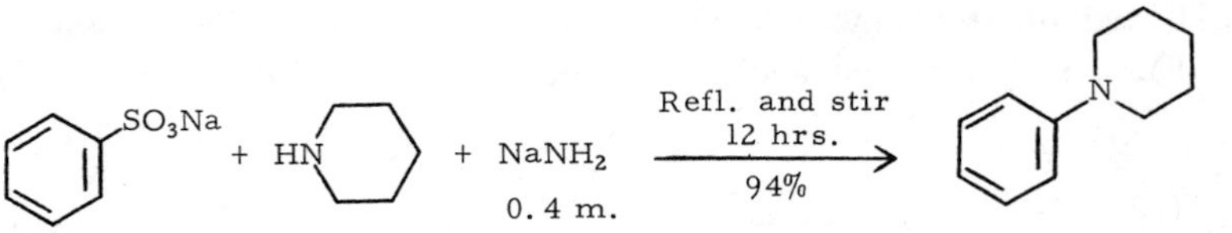
SO3Na
+ HN
+ NaNH2
0.4 m.
Refl. and stir
12 hrs.
94%
N
0.2 m.
75 ml.

Bunnett and co-workers[1] obtained N-phenylpiperidine in high yield from sodium benzenesulfonate, excess piperidine, and sodium amide after extended refluxing.
[1]J. F. Bunnett, T. K. Brotherton, and S. M. Williamson, *Org. Syn.*, **40**, 74 (1960)

Sodium benzoate, $C_6H_5CO_2Na$. Mol. wt. 144.11. Solubility in 100 g. H_2O: 62.5 g. at 25°, 76.9 g. at 100°. Suppliers: A, F, MCB.

Baker and co-workers[1] found this salt, in refluxing dimethylformamide, to be one of the most potent nucleophilic reagents. Thus, unaided by a neighboring group effect, it displaces the C_4 secondary tosylate group of the sugar (1) with inversion of configuration (S_N2 displacement). Codington, Fecher, and Fox[2] found that sodium

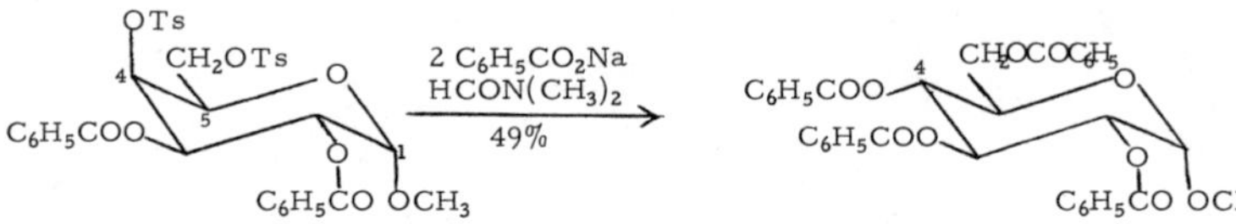
OTs
CH2OTs
C6H5COO
C6H5CO OCH3
2 C6H5CO2Na
HCON(CH3)2
49%
CH2OCOC6H5
C6H5COO
C6H5COO
C6H5CO OCH3

(1) (2)

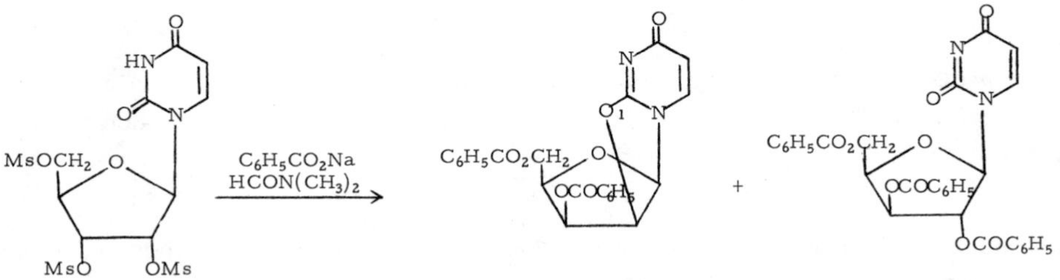
HN
MsOCH2
MsO
OMs
C6H5CO2Na
HCON(CH3)2
C6H5CO2CH2
OCOC6H5
+
C6H5CO2CH2
OCOC6H5
OCOC6H5

(3) (4) 10-15% (5) 7-9%

benzoate in refluxing DMF displaces all three mesyl groups of 2′3′5′-trimesyloxyuridine (3) to give small amounts of (4) and (5). The reagent displaces the mesyl group of (6) to give (7) as the only sugar detected in the product.[3]

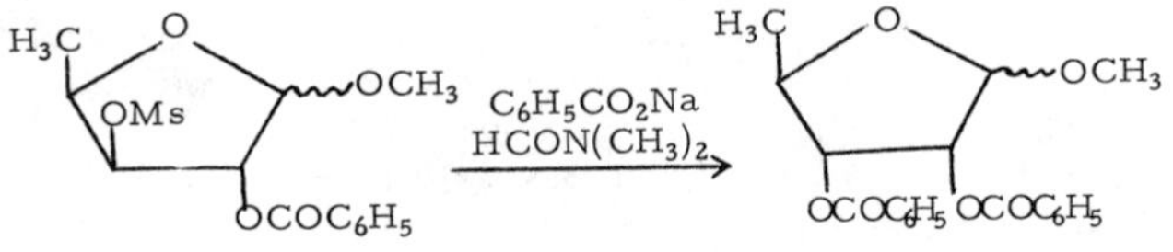
H3C
OMs
OCH3
OCOC6H5
C6H5CO2Na
HCON(CH3)2
H3C
OCH3
OCOC6H5 OCOC6H5

(6) (7)

[1]E. J. Reist, L. Goodman, and B. R. Baker, *Am. Soc.*, **80**, 5775 (1958); E. J. Reist, R. R. Spencer, and B. R. Baker, *J. Org.*, **24**, 1618 (1959)
[2]J. F. Codington, R. Fecher, and J. J. Fox, *Am. Soc.*, **82**, 2794 (1960)
[3]K. J. Ryan, H. Arzoumanian, E. M. Acton, and L. Goodman, *Am. Soc.*, **86**, 2497 (1964)

Sodium benzylate, $C_6H_5CH_2ONa$. Mol. wt. 130.13.

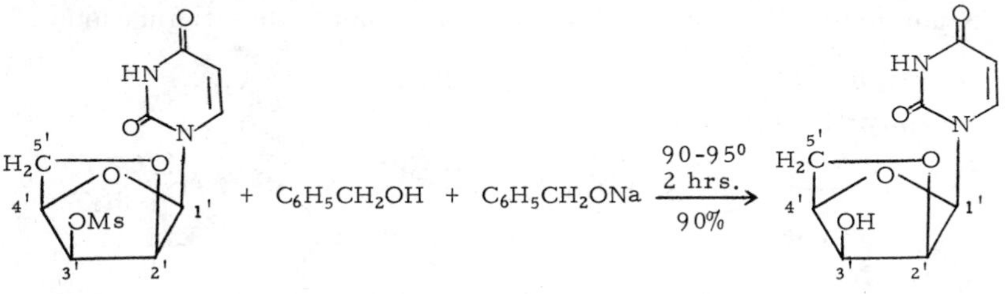
HN
O
N
5'
H2C
O
O
4'
OMs
1'
3'
2'
+ C6H5CH2OH + C6H5CH2ONa
90-95°
2 hrs.
90%
OH

(1) 0.15 g. (2)

Sodium benzylate in benzyl alcohol removes the mesyl group of the 3′-mesylate (1) with retention of the oxide linkage and without inversion to give (2), 1-(2′,5′-oxy-β-D-lyxofuranosyl)uracil.[1,2] A reaction of this type without inversion has been designated as an S_N2S reaction. The solution of sodium benzylate is prepared from benzyl alcohol and a solution of sodium methoxide in methanol.

[1]J. F. Codington, I. L. Doerr, and J. J. Fox, *J. Org.*, **30**, 476 (1965)
[2]The name used in the reference cited, "1-(2′,5′-epoxy-β-D-lyxofuranosyl," in our opinion is ill advised. The best practice, we think, is to reserve the term "epoxide" to describe a 1,2-oxide resulting from the addition to the two carbon atoms of a double bond (from *epi*, on, upon).

Sodium bismuthate, $NaBiO_3$. Mol. wt. 280.00. Suppliers: Mallinckrodt, B, F, MCB.

Rigby[1] found this reagent to be an oxidizing agent somewhat similar in scope to lead tetraacetate and periodic acid. It is used in combination with phosphoric acid, usually in aqueous alcohol, aqueous dioxane, or aqueous acetic acid and it is reduced to insoluble bismuth phosphate. Exhaustion of the reagent is evident from discharge of the orange-yellow color. *vic*-Glycols are cleaved to the carbonyl compounds; if an aldehyde is formed it is not oxidized further by the reagent. In a qualitative comparison of the oxidation of *cis*- and *trans*-cyclohexane-1,2-diol to adipic dialdehyde, Rigby noticed no marked difference in reaction rate. Other oxidations are as follows:

$$RCH(OH)CO_2H \longrightarrow RCHO$$

$$R_2C(OH)CO_2H \longrightarrow R_2C{=}O$$

$$RCOCH(OH)R \longrightarrow RCO_2H + O{=}CHR$$

$$HCO_2H \longrightarrow CO_2$$

Stoll and co-workers[2] developed routes to glutardialdehyde from cyclopentadiene and from cyclopentanone, both of which involve oxidation of *trans*-cyclopentane-1,2-diol with sodium bismuthate. A solution of 3 g. of the diol in 10 ml. of acetic acid and 8 ml. of water was stirred, and 12 g. of $NaBiO_3$ and 14 ml. of 33% phosphoric

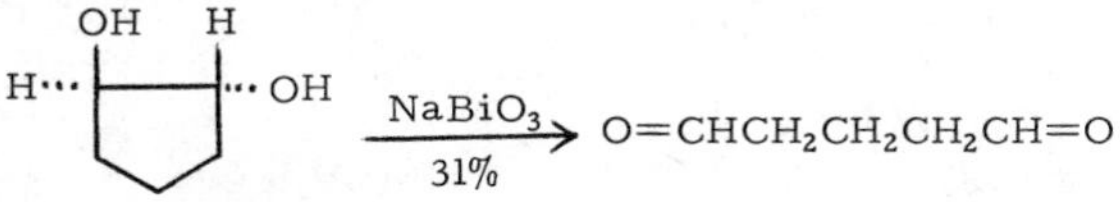
OH H
H···
··· OH
NaBiO3
31%
O=CHCH2CH2CH2CH=O

acid were added in portions with control of the temperature to 30°. The mixture was filtered for removal of bismuth phosphate and unchanged reagent, and the dialdehyde was isolated as the bis-*p*-nitrophenylhydrazone.

Brooks and Norymberski[3] used sodium bismuthate for the analytical determination of adrenocortical hormone side chains of types (1)–(5). A 21-desoxyketol (6) alone is not attacked. Oxidations (1)–(5) can be followed by determination of the

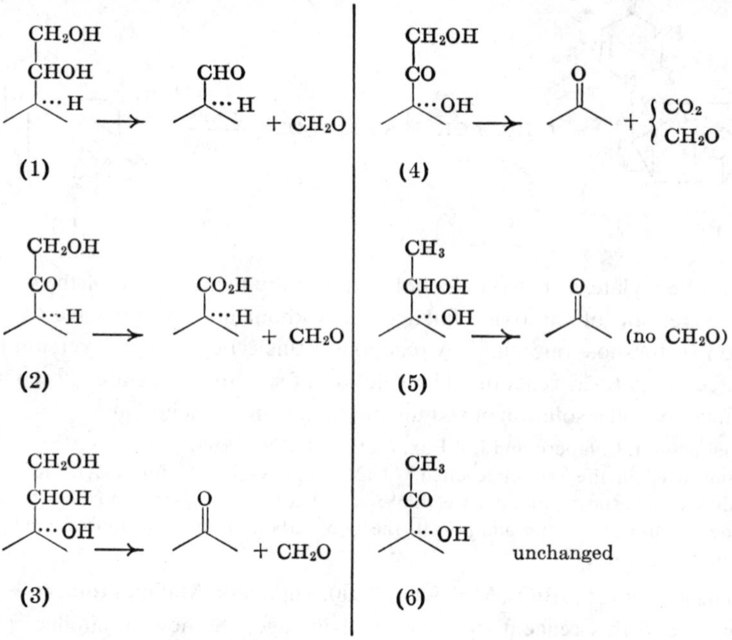
CH2OH
CHOH
H
CHO
H
+ CH2O
(1)
CH2OH
CO
OH
O
+ CO2 CH2O
(4)
CH2OH
CO
H
CO2H
H
+ CH2O
(2)
CH3
CHOH
OH
O
(no CH2O)
(5)
CH2OH
CHOH
OH
O
+ CH2O
(3)
CH3
CO
OH
unchanged
(6)

formaldehyde formed, if any, and by assay for 17-ketosteroids. To assay for total 17α-hydroxypregnanes, for example in a urinary sample, the mixture can be reduced with sodium borohydride prior to oxidation in order to convert unreactive ketals (6) into α-glycols (5) oxidizable to 17-ketones.

[1]W. Rigby, *J. Chem. Soc.*, 1907 (1950)
[2]A. Stoll, A. Lindenmann, and E. Jucker, *Helv.*, **36**, 268 (1953)
[3]C. J. W. Brooks and J. K. Norymberski, *Biochem. J.*, **55**, 371 (1953)

Sodium bistrimethylsilylamide (2). Mol. wt. 183.39, m.p. 165–167°. Preparation from hexamethyldisilazane and a suspension of sodium amide in benzene:[1]

$$\underset{(1)}{(CH_3)_3Si\overset{H}{N}Si(CH_3)_3} + NaNH_2 \longrightarrow \underset{(2)}{(CH_3)_3Si\overset{Na}{N}Si(CH_3)_3} + NH_3$$

The solution so obtained can be used for synthetic purposes, or the base can be isolated as a white crystalline product by removal of the solvent.

The base converts enolizable ketones almost quantitatively into the corresponding enolates. Non-enolizable ketones react in the manner formulated for benzophenone:[2]

$$(C_6H_5)_2C{=}O + NaN\begin{matrix} Si(CH_3)_3 \\ Si(CH_3)_3 \end{matrix} \longrightarrow (C_6H_5)_2C\begin{matrix} ONa \\ N\begin{matrix} SiN(CH_3)_3 \\ SiN(CH_3)_3 \end{matrix} \end{matrix} \longrightarrow$$

$$(C_6H_5)_2C{=}NSiN(CH_3)_3 + NaOSiN(CH_3)_3$$

Both carbonyl groups of 1,4-benzoquinone condense with the reagent.

The base is useful for promoting sterically hindered condensations, for example ethyl isobutyrate ⟶ ethyl isobutyrylisobutyrate (30%).[3] It is useful for effecting

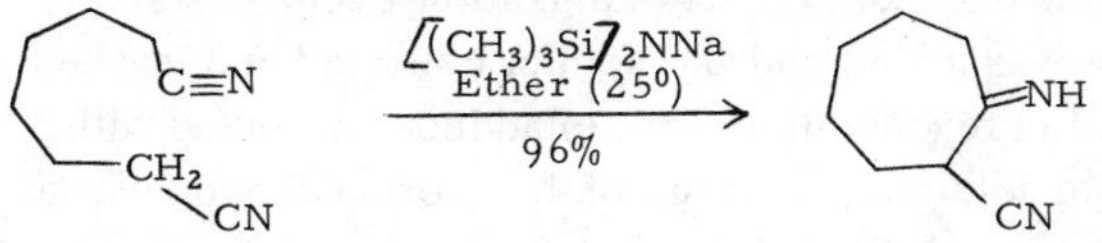
C≡N
CH2
CN
[(CH3)3Si]2NNa
Ether (25°)
96%
NH
CN

cyclization of suberodinitrile to α-cyanocycloheptanone imide and as base for the Wittig reaction (benzophenone ⟶ 1,1-diphenylethylene, 92% yield).

[1]U. Wannagat and H. Niederprüm, *Ber.*, **94**, 1540 (1961)
[2]C. Krüger, E. G. Rochow, and U. Wannagat, *Ber.*, **96**, 2131 (1963)
[3]C. R. Krüger and E. G. Rochow, *Angew. Chem., Internat. Ed.*, **2**, 617 (1963)

Sodium bisulfite, $NaHSO_3$. Mol. wt. 104.07 Suppliers: B, E, F, MCB. *See also* Sodium metabisulfite.

Aldehydes. Conversion to solid sodium bisulfite addition product with excess reagent, removal of nonaldehydic material by washing with alcohol or ether, and regeneration usually with acid, base, or sodium carbonate, provides a convenient method of purification. Examples: syringic aldehyde,[1] *n*-hexaldehyde.[2]

The conversion of formaldehyde to diethylaminoacetonitrile[3] or of benzaldehyde[4] to mandelonitrile is conveniently done by first converting the aldehyde to its bisulfite addition compound.

$$C_6H_5\underset{\displaystyle OH}{\underset{|}{C}}HSO_3Na + KCN \longrightarrow C_6H_5\underset{\displaystyle OH}{\underset{|}{C}}HCN + KNaSO_3$$

$$(C_2H_5)_2NH + HOCH_2SO_3Na + KCN \longrightarrow (C_2H_5)_2NCH_2CN + KNaSO_3 + H_2O$$

Glyoxal is isolated as the addition product $O{=}CHCH{=}O\cdot 2NaHSO_3\cdot 2H_2O$.[5]

Although some methyl ketones form adducts on treatment with excess reagent, *o*-methoxyphenylacetone is sufficiently inert to the reagent that the crude material can be freed of aldehydic starting material by washing a toluene solution with aqueous bisulfite solution.[6]

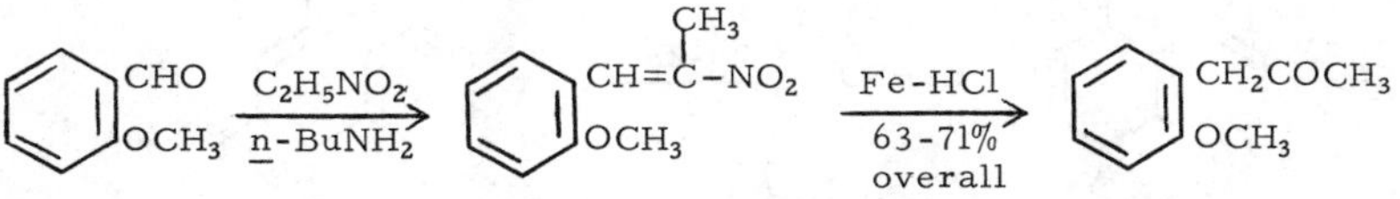
CHO
OCH3
C2H5NO2
n-BuNH2
CH3
CH=C–NO2
OCH3
Fe-HCl
63-71%
overall
CH2COCH3
OCH3

Cyclic ketones. Diazomethane ring expansion of cyclohexanone gives cycloheptanone in yields as high as 63% and smaller amounts of cyclooctanone and methylenecyclohexane oxide.[7] The cycloheptanone is easily separated from the

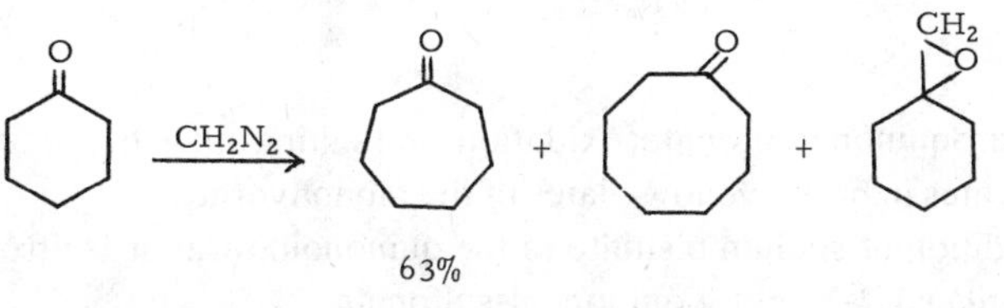
O
CH2N2
O
63%
+
O
+
CH2
O

mixture by shaking an ethereal solution with aqueous sodium bisulfite to form the adduct of cycloheptanone; cyclooctanone does not form a bisulfite adduct.[8]

α-Keto esters. Cornforth[9] purified ethyl pyruvate by conversion to the bisulfite compound in small batches. The ester (2.2 ml.) in a long test tube was underlayered with 3.6 ml. of saturated bisulfite solution, the tube was chilled in a freezing mixture, and the layers shaken together. Crystallization occurred rapidly and after 3 min. 10 ml. of ethanol was added and the adduct collected and washed with ethanol and ether; yield 3.0 g. For regeneration, 16 g. of adduct was mixed with 32 ml. of saturated magnesium sulfate solution, and 5 ml. of 40% formaldehyde was added. After shaking, the oil formed was collected with ether, dried, and distilled; yield 5.5 g. of ethyl pyruvate.

Ethyl benzoylformate has been purified by extraction from benzene with bisulfite solution and regeneration with sulfuric acid.[10]

Substituted orthoquinones. Crude phenanthrenequinone prepared by oxidation of technical (90%) phenanthrene can be freed from anthraquinone and other contaminants by triturating it with 4–6 portions of hot 40% bisulfite solution, filtering, cooling, collecting the adduct, and adding hydrochloric acid to a suspension in water.[11]

Fieser[12] found that alkylation of the silver salt of 2-hydroxy-1,4-naphthoquinone with methyl iodide in ether or benzene gives a mixture of about equal parts of the *para* and *ortho* quinone methyl ethers and that the mixture is easily separated by treatment with sodium bisulfite solution. The *para* quinone dissolves to a negligible

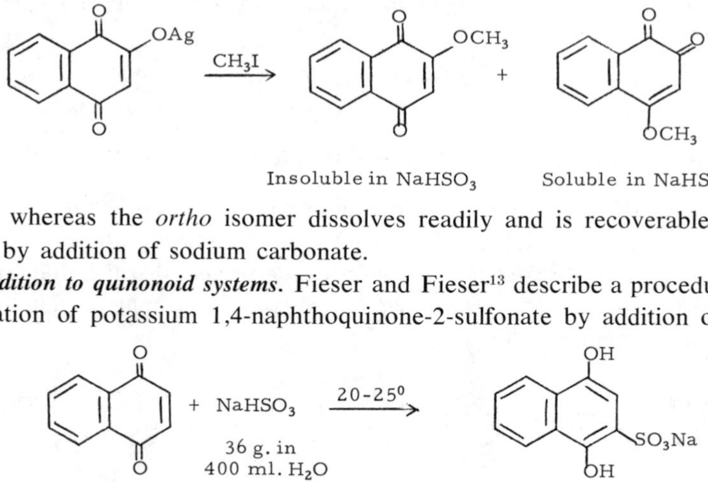
OAg
CH_3I
OCH_3
+
OCH_3

Insoluble in $NaHSO_3$ Soluble in $NaHSO_3$

extent, whereas the *ortho* isomer dissolves readily and is recoverable from the filtrate by addition of sodium carbonate.

Addition to quinonoid systems. Fieser and Fieser[13] describe a procedure for the preparation of potassium 1,4-naphthoquinone-2-sulfonate by addition of bisulfite

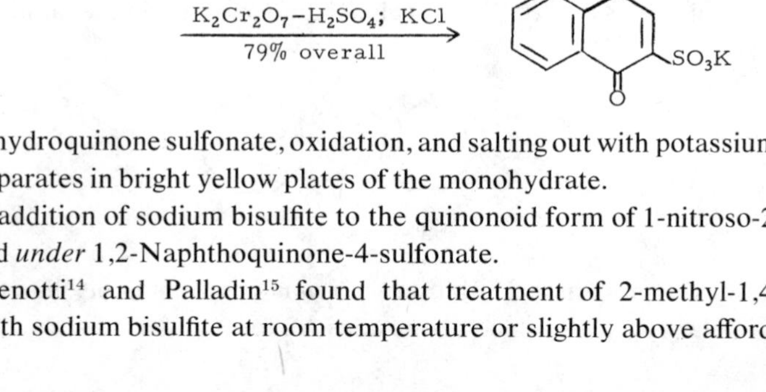
+ $NaHSO_3$
20-25°
36 g. in
400 ml. H_2O
32 g.
OH
SO_3Na
OH
$K_2Cr_2O_7$–H_2SO_4; KCl
79% overall
SO_3K

to give the hydroquinone sulfonate, oxidation, and salting out with potassium chloride. The salt separates in bright yellow plates of the monohydrate.

The 1,4-addition of sodium bisulfite to the quinonoid form of 1-nitroso-2-naphthol is described *under* 1,2-Naphthoquinone-4-sulfonate.

Both Menotti[14] and Palladin[15] found that treatment of 2-methyl-1,4-naphthoquinone with sodium bisulfite at room temperature or slightly above affords a mono-addition product which is reconvertible to methylnaphthoquinone on treatment with acid or base. If the reaction is conducted at 70–100°, the labile addition product is converted irreversibly into sodium 2-methylnaphthohydroquinone-3-sulfonate.

Addition to an α,β-unsaturated ester. Cope and Hancock[16] purified the ester (1) by stirring the crude material with aqueous bisulfite, extracting the turbid aqueous solution with benzene, and regenerating (1) by neutralizing the clarified aqueous

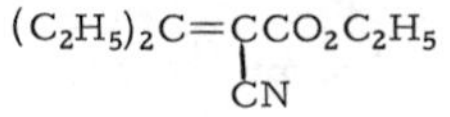

(1)

solution with sodium hydroxide. Presumably the water-soluble product is the 1,4-adduct which, unlike an adduct to a quinone, is incapable of aromatization.

[1]C. F. H. Allen and G. W. Leubner, *Org. Syn., Coll. Vol.*, **4**, 866 (1963)
[2]G. B. Bachman, *ibid.*, **2**, 323 (1943)
[3]C. F. H. Allen and J. A. VanAllan, *ibid.*, **3**, 275 (1955)
[4]*Org. Expts.*, 109
[5]A. R. Ronzio and T. D. Waugh, *Org. Syn., Coll. Vol.*, **3**, 438 (1955)
[6]R. V. Heinzelman, *ibid.*, **4**, 573 (1963)
[7]E. P. Kohler, M. Tishler, H. Potter, and H. T. Thompson, *Am. Soc.*, **61**, 1059 (1939)
[8]E. Mosettig and A. Burger, *Am. Soc.*, **52**, 3456 (1930)
[9]J. W. Cornforth, *Org. Syn., Coll. Vol.*, **4**, 468 (1963)
[10]B. B. Corson, R. A. Dodge, S. A. Harris, and R. K. Hazen, *ibid.*, **1**, 241 (1941)
[11]R. Wendland and J. LaLonde, *ibid.*, **4**, 757 (1963)
[12]L. F. Fieser, *Am. Soc.*, **48**, 2922 (1926)
[13]L. F. Fieser and Mary Fieser, *Am. Soc.*, **57**, 491 (1935)
[14]A. R. Menotti, *Am. Soc.*, **65**, 1209 (1943)
[15]A. V. Palladin, *Doklady Akad. Nauk S.S.S.R.*, **41**, 258 (1943) [*C.A.*, **38**, 4014 (1944)]
[16]A. C. Cope and E. M. Hancock, *Org. Syn., Coll. Vol.*, **3**, 399 (1955)

Sodium borohydride, $NaBH_4$. Mol. wt. 37.85.

History. Sodium borohydride was discovered by H. I. Schlesinger and H. C. Brown and co-workers at the University of Chicago in 1943.[1] The preparative method involves reaction of methyl borate with sodium hydride at elevated temperatures:

$$4\ NaH + B(OCH_3)_3 \longrightarrow NaBH_4 + 3\ NaOCH_3$$

In 1949 Chaikin and W. G. Brown[2] reported that the hydride in aqueous or alcoholic solution effects remarkably efficient and specific reduction of aldehydes, ketones, and acid chlorides containing other reducible groups, and in 1953 a series of papers by Schlesinger, H. C. Brown, and co-workers[3] gave details of the preparation and chemistry of the alkali metal hydrides and of diborane, and another paper[4] described uses of sodium borohydride as a reducing agent and for generation of hydrogen. Metal Hydrides Incorporated (MHI, now Ventron Corp.) obtained the first license under the patents and applications in 1950, and the first commercial sales occurred in that year. During the next few years the price of borohydrides dropped as improvements were made and new markets developed, for example in the textile industry, in the pulp and paper field, and in the petrochemical area. A large-scale manufacturing plant built by MHI for the U. S. Navy in 1957 operated at 100% capacity for two years and was then purchased by MHI.

Available forms. Ventron Corp. supplies sodium borohydride as powder and as pellets of about 99% purity. The only active impurities are sodium methoxide

(0.3%) and sodium hydroxide (0.2%). At a time when commercial samples showed analyses of from 79 to 94% purity, purification was accomplished by crystallization from diglyme[5] or from isopropylamine.[6]

Ventron also offers "Sodium borohydride-SWS," a stabilized water solution of average composition 40% sodium hydroxide and 12% sodium borohydride, sp. gr. 1.4. It is stable indefinitely and can be used as such for reduction of carbonyl compounds or it can be diluted with water, methanol, or ethanol. Most aldehydes are reduced so rapidly that condensations promoted by the alkali do not interfere. For reduction of an α,β-unsaturated aldehyde, however, the strongly basic reagent is unsatisfactory and should be diluted to the desired concentration and neutralized with carbon dioxide. The neutralized reagent should be used immediately, since the borohydride is no longer stable and decomposes at 25° at the rate of 4.5% per hr.

Solvents. In contrast to lithium aluminum hydride, sodium borohydride is insoluble in ether and soluble in methanol and ethanol. It is more soluble in methanol than in ethanol, but since it reacts with methanol at an appreciable rate and only

Solubility, g. per 100 g. of solvent[7]

Solvent	t	g.	Solvent	t	g.
Water	0°	25	Tetrahydrofurane	20°	0.1
Water	60°	88.5	1,2-Dimethoxyethane	20°	0.8
Methanol (reacts)	20°	16.4	Diglyme	25°	5.5
Ethanol (reacts slowly)	20°	4.0	Dimethylformamide	20°	18.0

slowly with ethanol, ethanol is preferred as a reaction solvent.[8] The reagent is still more stable in isopropanol, and this solvent has been used for kinetic studies of the reduction of aldehydes and ketones.[9]

$NaBH_4$ is a much milder reducing agent than $LiAlH_4$. In hydroxylic solvents it reduces aldehydes and ketones rapidly at 25° but is essentially inert to other functional groups: epoxides, esters, lactones, carboxylic acids and salts, nitrile and nitro groups.[8] Acid chlorides are reduced rapidly in diglyme or dioxane.

$$4\ R_2C{=}O + Na^+BH_4 \longrightarrow Na^+[B(OCHR_2)_4]^- \xrightarrow{H_2O} 4\ R_2CHOH$$

Reduction of ketones. Bernstein and co-workers[10] found the reduction of the 11-ketosteroid (1) to cortisol 3,20-bisethyleneketal (2) with lithium aluminum hydride unsatisfactory with respect to yield, safety hazard, etc., but found that the reduction can be accomplished in high yield with an exceedingly large excess of sodium

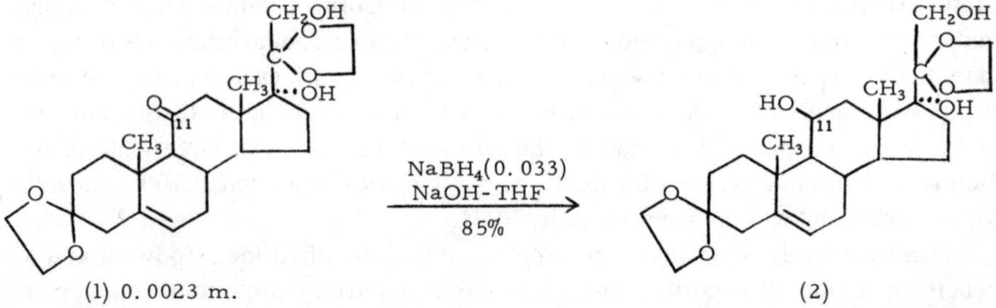
CH2OH
CH3
OH
11
CH3
NaBH4(0. 033)
NaOH-THF
85%
CH2OH
HO
CH3
OH
11
CH3

(1) 0. 0023 m. (2)

borohydride in a mixture of tetrahydrofurane and aqueous sodium hydroxide after refluxing for 20 hrs.

In the reduction of 4 moles of a ketone with 1 mole of sodium borohydride, a slow first step is followed by three rapid reactions. Jones and Wise[11] suggest a possible interpretation and present experimental evidence in support of their postulate.

Esters. Under usual conditions of experimentation, carboxylic esters are not reduced by sodium borohydride. Thus 3-keto bile acid esters can be reduced at C_3 without disturbance of the ester function.[12] Methyl 4β-bromo-3-ketocholanate

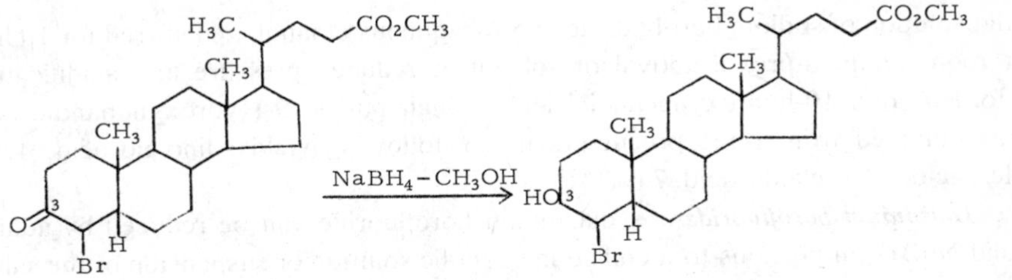
H3C
CO2CH3
CH3
CH3
3
O
H
Br
NaBH4−CH3OH
H3C
CO2CH3
CH3
CH3
HO
3
H
Br

(1) (2) 3α + 3β

(1) is reduced by $NaBH_4$ in methanol to a mixture of the 3α- and 3β-bromohydrins (2).[13] In such reactions it is necessary to use as solvent for the reduction the alcohol corresponding to the ester group of the keto acid, since sodium borohydride is an effective catalyst for ester interchange.

Although esters are sufficiently unreactive to permit selective reductions such as those described, esters of aliphatic, aromatic, and heterocyclic acids can be reduced in varying degree with a large excess of sodium borohydride in methanol.[14]

Halides. Sodium borohydride reduces secondary and tertiary alkyl halides capable of forming relatively stable carbonium ions, for example benzhydryl

$$(C_6H_5)_2CHCl \xrightarrow{OH^-} (C_6H_5)_2\overset{+}{C}H \xrightarrow[94\%]{NaBH_4} (C_6H_5)_2CH_2$$

chloride.[15] The reaction is accelerated by water. Thus 0.5 mole of halide was treated at 50° with 65 volume % of a solution of diglyme 4 *M* in $NaBH_4$ and 1 *M* in NaOH (to minimize hydrolysis of the borohydride).

Tosylhydrazones. The condensation of a C_3-, C_7-, C_{17}-, or C_{20}-ketosteroid with *p*-toluenesulfonylhydrazine and reduction of the resulting tosylhydrazone with excess $NaBH_4$ in dioxane provides a means for elimination of the carbonyl oxygen

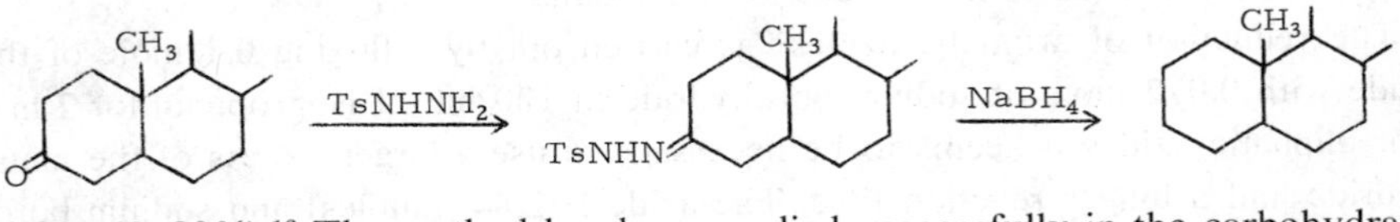
CH3
O
TsNHNH2
CH3
TsNHN
NaBH4
CH3

(yields 60–80%).[16] The method has been applied successfully in the carbohydrate field (KBH_4 used).[17] Thus D-glucose tosylhydrazone (5 g.) on reduction gives 1-desoxy-D-glucitol.

Ozonides. The reagent reduces an ozonide directly to the alcoholic products or product, and the ozonide need not be isolated. Sousa and Bluhm[18] ozonized cyclohexene in chloroform at 0° to disappearance of the IR band at 6.2 μ, added ethanol to dissolve the resinous ozonide, and stirred the solution vigorously during addition of a solution of $NaBH_4$ in 50% aqueous ethanol. Lithium aluminum hydride can be

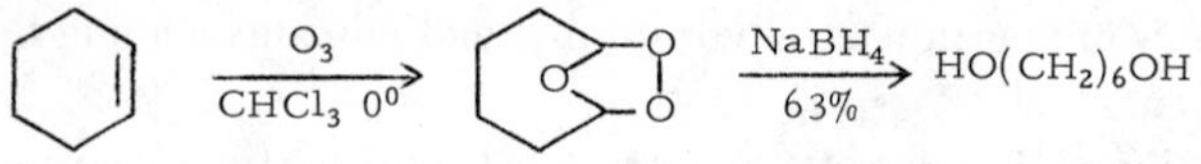

used, but only in a compatible solvent (an ether or a hydrocarbon) and not if the desired product is a hydroxy acid. Diaper and Mitchell[19] ozonized Δ^{10}-undecylenic acid in methanol at 0°, added the ozonide solution dropwise with stirring to an ice

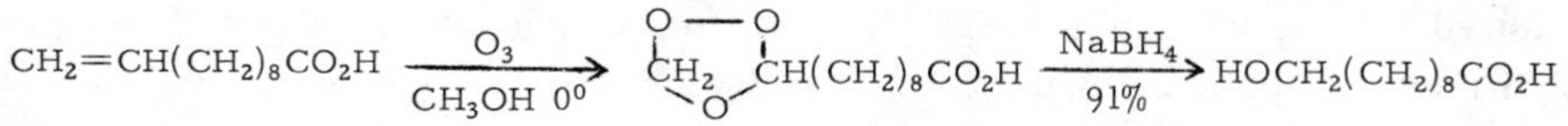

cold solution of sodium borohydride in 50% aqueous ethanol, and stirred for 10 hrs. at room temperature. Removal of solvent at reduced pressure and acidification afforded solid 10-hydroxydecanoic acid of high purity. 9-Hydroxynonanoic acid was obtained from three precursors in the following yields: linoleic acid, 42%; oleic acid, 50%; elaidic acid, 74%.[20]

Diazonium borofluorides. A diazonium borofluoride can be reduced by adding solid $NaBH_4$ in portions to a chilled methanolic solution or suspension of the salt.[21]

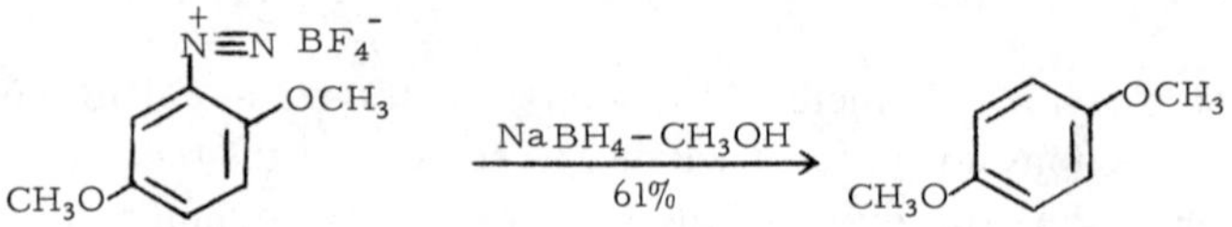

Alternatively, a chilled solution of $NaBH_4$ in dimethylformamide is added to a chilled solution of the diazonium salt in the same solvent.

König's salt. König's salt (red), the unsaturated system of which is derived from a pyridine ring, is reduced by $NaBH_4$ in aqueous ethanol at 0° to an enamine which on acid hydrolysis affords pentadienal (R. Grewe[22]).

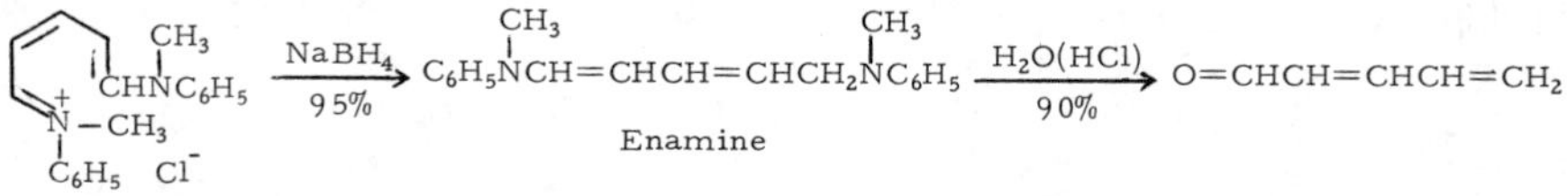

König's salt

Reduction of azides to amines.[23] Reduction of an azide to the corresponding amine by the conventional method of hydrogenation is impractical if the molecule contains sulfur. In this case an alternative procedure is reduction with sodium borohydride in isopropanol, provided no other easily reduced group is present. This procedure can be used with acid-sensitive compounds.

The reduction of aromatic azides[24] is carried out by refluxing 0.1 mole of the azide with 0.072 mole of sodium borohydride in 100 ml. of isopropanol for 2 hrs. For aliphatic azides it seems to be necessary to use a larger excess of the metal hydride and a longer reaction time. The azide I (2.84 mmoles) and sodium borohydride (6.61 mmoles) are heated together in 10 ml. of refluxing isopropanol for

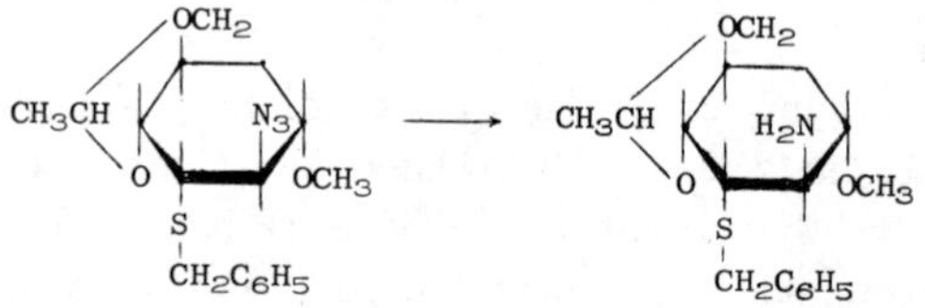

I II

16 hrs. The solvent is evaporated and the residue is partioned between methylene chloride and water. After washing the methylene chloride layer with water and drying, the solvent is evaporated to give the amine II in better than 90% yield.

This method avoids the hazards involved with the use of lithium aluminum hydride and is considerably less complicated.

Hydrogenation. For use of $NaBH_4$ for the preparation of active catalysts and *in situ* generation of hydrogen, *see* Platinum catalysts, Brown and Brown catalyst.

Applications in cellulose and sugar chemistry.[25]

Sodium borohydride–aluminum chloride.[26] A mixture of freshly prepared solutions of $NaBH_4$ and $AlCl_3$ in diglyme in the ratio $3\,NaBH_4 : 1\,AlCl_3$ gives a clear solution which has powerful reducing properties.[27] The order of mixing is not important, but usually the $AlCl_3$ solution is added to a solution of $NaBH_4$ and the compound to be reduced. Like $NaBH_4$ the mixed hydride reduces aldehydes, ketones, and acid chlorides to alcohols, but in addition it reduces esters, lactones, and acids (but not sodium salts of acids). Since in the reduction of an acid one extra hydride equivalent is required to react with the acidic hydrogen, it is preferable to reduce an ester rather than the acid. Epoxides are reduced readily at 25°, but anhydrides react sluggishly at 25° and require a temperature of 75°. Nitriles are reduced to primary amines. Unsubstituted amides form sodium salts and are not reduced; disubstituted amides are reduced readily:

$$C_6H_5CON(C_2H_5)_2 \xrightarrow{NaBH_4-AlCl_3} C_6H_5CH_2N(C_2H_5)_2$$

Pyridine-N-oxide is reduced cleanly to pyridine at 25° but further reduction occurs at higher temperatures. Nitro compounds and aromatic halides are not reduced.

The reagent is not suitable for reduction of a compound having a reactive double bond for it reacts to form an alkylborane.[28] The reagent thus can be used to effect

$$9 \, >C=C< \; + \; 3\,NaBH_4 \; + \; AlCl_3 \longrightarrow 3\,(>CH-\underset{|}{C}<)_3B \; + \; 3\,NaCl \; + \; AlH_3$$

hydroboration, but it is inferior to diborane since only three of the four hydrogens of $NaBH_4$ are utilized; the fourth is lost in the form of aluminum hydride.

$NaBH_4$–BF_3. Pettit and Piatak[29] found this combination effective for the reduction of lactones to cyclic ethers and of esters to ethers. For reduction of

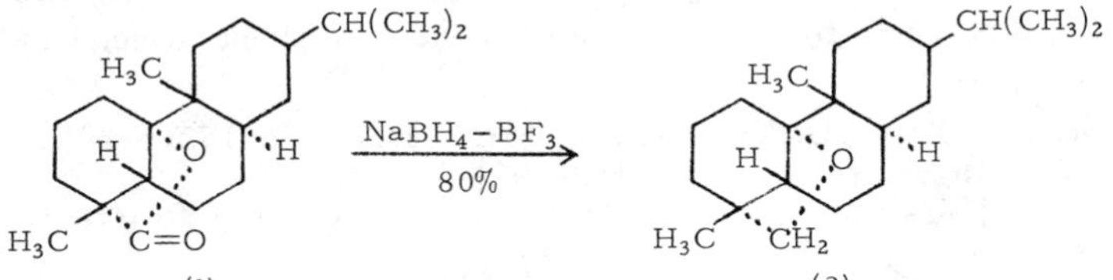
CH(CH3)2
H3C
H
O
H
NaBH4–BF3
80%
H3C
C=O
CH(CH3)2
H3C
H
O
H
H3C
CH2

(1) (2)

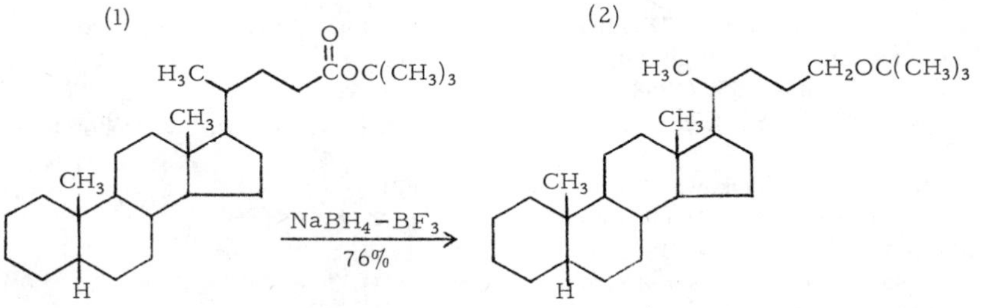
H3C
COC(CH3)3
CH3
CH3
H
NaBH4–BF3
76%
H3C
CH2OC(CH3)3
CH3
CH3
H

(3) (4)

dihydroabietic γ-lactone (1) to the oxide (2), best results were obtained with a reagent corresponding to 30 moles of boron trifluoride etherate and 2 moles of sodium borohydride per mole of lactone in diglyme–tetrahydrofurane. Since both $LiAlH_4$ and $LiBH_4$ seemed somewhat less satisfactory than $NaBH_4$, the combination $NaBH_4$–BF_3 was used for comparing the behavior of three isomeric butyl cholanates. Reduced by the above procedure, the *t*-butyl ester (3) afforded the *t*-butyl ether (4) in 76% yield. The yield from the *sec*-butyl ester was 41% and from the *n*-butyl ester it was 7%. This reduction of a *t*-butyl ester provides an efficient route to a *t*-butyl ether.

$NaBH_4$–Pd–C. Aromatic nitro compounds are not normally reduced by $NaBH_4$ but can be reduced smoothly in solution in water or methanol with sodium borohydride catalyzed by 10% palladium charcoal.[30] Azobenzene is reduced by this method to hydrazobenzene and not to aniline.

ROH ⟶ RX. Sodium borohydride, in combination with a halogen, converts an alcohol or an ether into an alkyl halide.[31]

$$3\ ROH + 2\ X_2 + NaBH_4 \longrightarrow 3\ RX + H_3BO_3 + NaX + 2\ H_2$$

$$3\ ROR + 2\ X_2 + NaBH_4 \longrightarrow 3\ RX + B(OR)_3 + NaX + 2\ H_2$$

Reduction of nitroolefins. Hassner and Heathcock[32] found that 6-nitrocholesteryl acetate can be reduced with sodium borohydride in ethanol to the saturated 6α-nitro-5α-steroid.

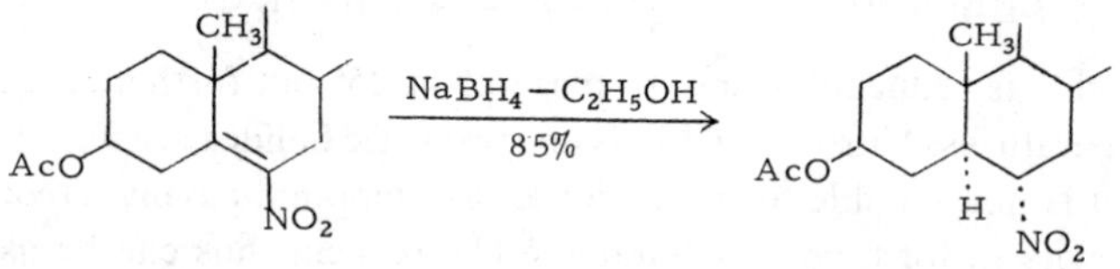
CH3
AcO
NO2
NaBH4–C2H5OH
85%
CH3
AcO
H
NO2

Reduction of alkyl halides. $LiAlH_4$ is satisfactory for reduction of primary and secondary halides or tosylates to the hydrocarbons, but in the case of a tertiary halide the product is predominantly the olefin. Alkyl halides and tosylates, even tertiary, are reduced in good to high yield by sodium borohydride in 65% aqueous diglyme.[33] When a relatively stable carbonium ion incapable of elimination is formed, yields are high, but yields are still satisfactory when elimination is possible. The reaction is very slow in the absence of water. A homogeneous solution required for kinetic studies is prepared from 80% (volume) aqueous diglyme, which can be made 1.80 *M* in the reagent.

[1]H. I. Schlesinger and H. C. Brown, U. S. Patents 2,461,661, 2,534,533, 2,683,721
[2]S. W. Chaikin and W. G. Brown, *Am. Soc.*, **71**, 122 (1949)
[3]H. I. Schlesinger, H. C. Brown, *et al.*, *Am. Soc.*, **75**, 186, 191, 192, 195, 199, 205 (1953)
[4]*Idem, ibid.*, **75**, 215 (1953)
[5]H. C. Brown *et al.*, *Am. Soc.*, **77**, 6209 (1955); *Tetrahedron*, **1**, 214 (1957)
[6]J. D. Cocker and T. G. Halsall, *J. Chem. Soc.*, 3441 (1957)
[7]Technical Bulletin No. 550 (1958)
[8]Reviews: E. Schenker, *Angew. Chem.*, **73**, 81 (1961); H. C. Brown, "Hydroboration," 242 (1963)
[9]H. C. Brown and K. Ichikawa, *Am. Soc.*, **84**, 373 (1962)
[10]W. S. Allen, S. Bernstein, and R. Littell, *Am. Soc.*, **76**, 6116 (1954)
[11]W. M. Jones and H. E. Wise, Jr., *Am. Soc.*, **84**, 997 (1962)
[12]H. Heymann and L. F. Fieser, *Am. Soc.*, **73**, 5252 (1951)
[13]L. F. Fieser and R. Ettorre, *Am. Soc.*, **75**, 1700 (1953)

[14]M. S. Brown and H. Rapoport, *J. Org.*, **28**, 3261 (1963)
[15]H. C. Brown and H. M. Bell, *J. Org.*, **27**, 1928 (1962)
[16]L. Caglioti and P. Grasselli, *Chem. Ind.*, 153 (1964)
[17]A. N. de Belder and H. Weigel, *Chem. Ind.*, 1689 (1964)
[18]J. A. Sousa and A. L. Bluhm, *J. Org.*, **25**, 108 (1960)
[19]D. G. M. Diaper and D. L. Mitchell, *Can. J. Chem.*, **38**, 1976 (1960)
[20]F. L. Benton and A. A. Kiess, *J. Org.*, **25**, 470 (1960)
[21]J. B. Hendrickson, *Am. Soc.*, **83**, 1251 (1961)
[22]R. Grewe and W. von Bonin, *Ber.*, **94**, 234 (1961)
[23]Contributed by James E. Christensen, Stanford Research Inst.
[24]P. A. S. Smith, J. H. Hall, and R. O. Kan, *Am. Soc.*, **84**, 485 (1962)
[25]"The Borohydrides in Cellulose and Sugar Chemistry," Metal Hydrides Inc., 1959
[26]M. N. Rerick, "The Metal Hydrides," Metal Hydrides Inc. (Ventron)
[27]H. C. Brown and B. C. Subba Rao, *Am. Soc.*, **78**, 2582 (1956)
[28]*Idem, ibid.*, **78**, 5694 (1956)
[29]G. R. Pettit and D. M. Piatak, *J. Org.*, **27**, 2127 (1962)
[30]T. Neilson, H. C. S. Wood, and A. G. Wylie, *J. Chem. Soc.*, 371 (1962)
[31]N. H. Long and G. F. Freeguard, *Chem. Ind.*, 223 (1965)
[32]A. Hassner and C. Heathcock, *J. Org.*, **29**, 1350 (1964)
[33]H. M. Bell and H. C. Brown, *Am. Soc.*, **88**, 1473 (1966)

Sodium bromate (or **iodate**), $NaBrO_3$ ($NaIO_3$). Mol. wt. 150.91 (197.91). Solubility in 100 g. water: $NaBrO_3$: 27.5 g. at 0°, 90.9 g. at 100°; $NaIO_3$: 2.5 g. at 0°, 24 g. at 100°. Suppliers: F, MCB.

Dehn[1] found sodium bromate (or iodate) a superior reagent for effecting a few specific oxidations. One is for oxidation of benzoin to benzil under conditions such

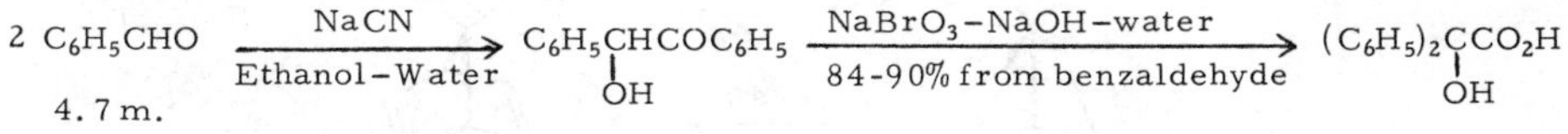
2 C6H5CHO
4.7 m.
NaCN
Ethanol–Water
C6H5CHCOC6H5
OH
NaBrO3–NaOH–water
84–90% from benzaldehyde
(C6H5)2CCO2H
OH

that the diketone rearranges to benzilic acid. In a procedure[2] for the preparation of benzilic acid on a large scale the benzoin obtained from benzaldehyde[3] is washed but not dried and added in portions with stirring to a solution of 0.76 mole of $NaBrO_3$ (1.08 × theory) and 12.5 moles of sodium hydroxide in 880 ml. of water. The mixture is heated on the steam bath with stirring, with addition of more water (800 ml.) as required for thinning until a test portion is completely soluble in water (5–6 hrs.). The mixture is diluted with water, filtered from a little benzhydrol, and acidified. The benzilic acid precipitated is directly pure, m.p. 149–150°.

On comparing various reagents which had been used for the oxidation of pyrogallol to purpurogallin, A. G. Perkin and Steven[4] concluded that the best one tried was sodium nitrite in acetic acid, but their yield was only 30–40%. Evans and Dehn[1] were able to double this yield by use of sodium iodate. They stirred a solution of 10 g. of pyrogallol in a little water and gradually added a cold solution of 8 g. of sodium iodate in 100 ml. of water. A smooth reaction ensued and the pigment

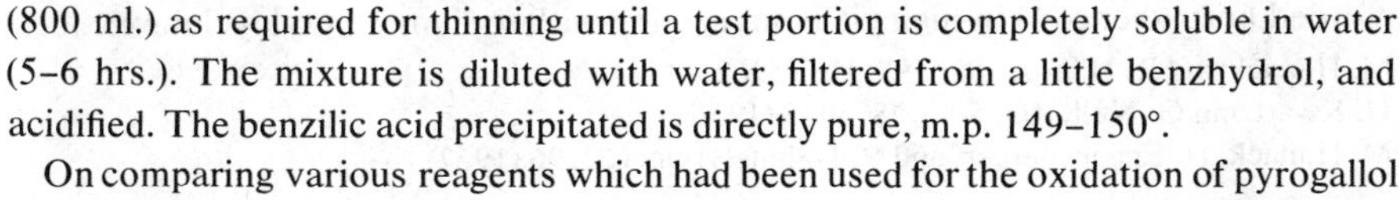
2
OH
HO
HO
NaIO3
2
O
O
HO
71% overall
OH
O
OH
HO
HO
+ CO2

separated in good yield. The structure of purpurogallin was not known until 1945. Later Horner,[5] by a trapping technique, established that the initial oxidation product is 3-hydroxy-1,2-benzoquinone and suggested a reasonable mechanism for its conversion to purpurogallin.[6, 7]

[1]T. W. Evans and W. M. Dehn, *Am. Soc.*, **52**, 3649 (1930)
[2]D. A. Ballard and W. M. Dehn, *Org. Syn.*, *Coll. Vol.*, **1**, 89 (1941)
[3]R. Adams and C. S. Marvel, *ibid.*, **1**, 94 (1941)
[4]A. G. Perkin and A. B. Steven, *J. Chem. Soc.*, **83**, 194 (1903)
[5]L. Horner and S. Göwecke, *Ber.*, **94**, 1267 (1961)
[6]L. Horner, K. H. Weber, and W. Dürckheimer, *Ber.*, **94**, 2881 (1961)
[7]For a formulation, *see* L. F. Fieser and M. Fieser, "Topics in Organic Chemistry," 567–568, Reinhold Publishing Corp. (1963)

Sodium–*t*-Butanol–THF, *see* Lithium–*t*-butanol–THF.

Sodium 2-*n*-butylcyclohexoxide,

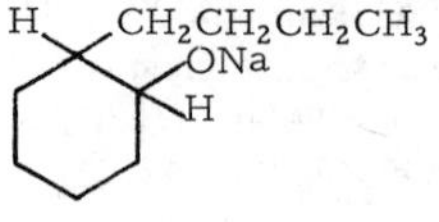

2-*n*-Butylcyclohexanol is supplied by Aldrich and by Fluka.

The reagent is prepared by dissolving 24 g. of sodium in 400 g. of 2-butylcyclohexanol.[1] It is an effective reagent for dehydrohalogenation of even very unreactive

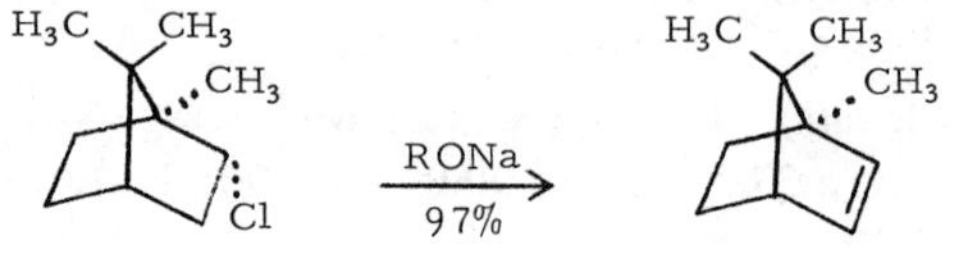

(1) (2)

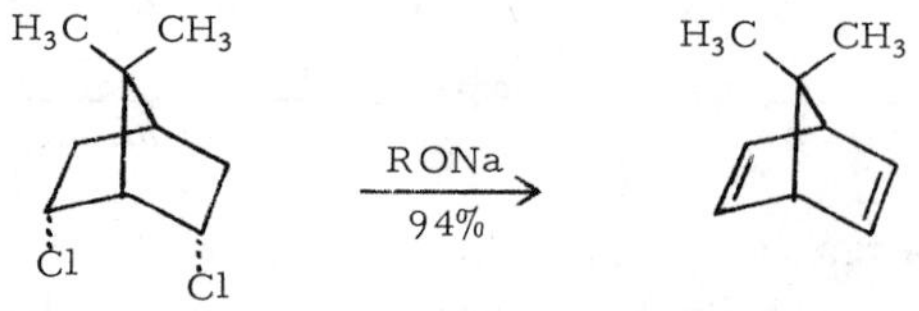

(3) (4)

compounds such as bornyl chloride (1). 2,6-Dichlorocamphane (3) resisted attempted dehydrohalogenation by lithium diethylamide and by potassium *t*-butoxide[2] but afforded bornadiene (4) in high yield on reaction with sodium 2-butylcyclohexoxide.[3]

[1]M. Hanack and R. Hähnle, *Ber.*, **95**, 191 (1962)
[2]H. Kwart and G. Null, *Am. Soc.*, **78**, 5943 (1956)
[3]M. Hanack, H. Eggensperger, and R. Hähnle, *Ann.*, **652**, 96 (1962)

Sodium chlorate, $NaClO_3$. Mol. wt. 106.46. Suppliers: B, F, MCB.

Chlorination. α-Chloroanthraquinone is obtainable in almost quantitative yield by a procedure originated by Ullmann and Ochsner[1] and perfected as an *Organic Syntheses* preparation.[2] A solution of potassium anthraquinone-α-sulfonate and

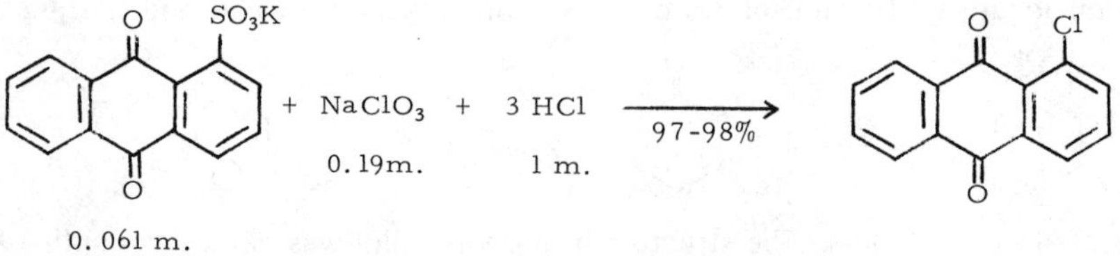

hydrochloric acid in 500 ml. of water is stirred with a Hershberg tantalum wire stirrer under reflux, and a solution of sodium chlorate in 100 ml. of water is added from a Hershberg dropping funnel (Fig. P-1) over a period of 3 hrs. After an additional hour, the precipitated bright yellow α-chloroanthraquinone is collected, washed, and dried; it is of high purity.

Oxidation. Milas[3] developed a procedure for the preparation of fumaric acid by oxidation of furfural with sodium chlorate catalyzed by vanadium pentoxide. A mixture of sodium chlorate, 1 l. of water, and 2 g. of vanadium pentoxide is

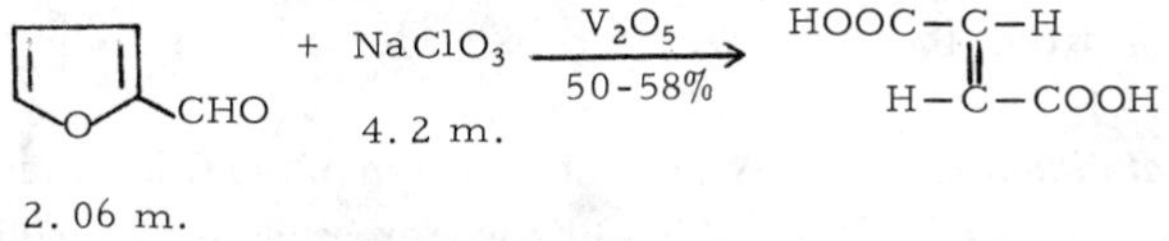

CHO
2.06 m.
+ NaClO3
4.2 m.
V2O5
50-58%
HOOC−C−H
H−C−COOH

stirred at 70–75°, and furfural is added at such a rate as to maintain a vigorous reaction for 70–80 min. The mixture is stirred at 70–75° for 10–11 hrs., let stand at room temperature overnight, and a first crop of crude fumaric acid is collected. Evaporation and treatment with hydrochloric acid gives a second crop; pure material is obtained by crystallization from dilute hydrochloric acid.

With vanadium pentoxide as catalyst in an acidic medium, sodium chlorate oxidizes hydroquinone to quinone. A procedure by Underwood and Walsh[4] calls for stirring a mixture of the reactants in 1 l. of 2% sulfuric acid until the green

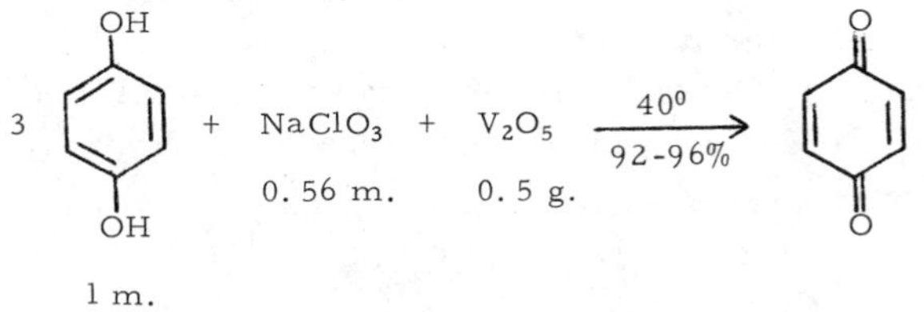

OH
3
OH
1 m.
+ NaClO3 + V2O5
0.56 m.
0.5 g.
40°
92-96%
O
O

quinhydrone initially formed is converted into yellow quinone (about 3 hrs.), with rise in temperature to 40°. After cooling, the product was collected; extraction of filtrate and washings with benzene afforded 12–14 g. more material.

Anthracene was oxidized to anthraquinone by warming a mixture of 90 g. of finely powdered hydrocarbon, 0.5 g. of vanadium pentoxide, 76 g. of sodium chlorate, 1 l. of acetic acid, and 200 ml. of 2% sulfuric acid under reflux until a vigorous reaction set in.[4] After eventual brief refluxing, anthraquinone was obtained in 88–91% yield. The method is not suitable for the oxidation of hydrocarbons of the naphthalene or phenanthrene series to the quinones or for oxidation of acenaphthene or fluorene.

In an efficient procedure for the preparation of 2,5-bis[2-carboxyanilino]-1,4-benzoquinone, Braun and Mecke[5] used sodium chlorate as oxidant, catalyzed by ammonium vanadate.

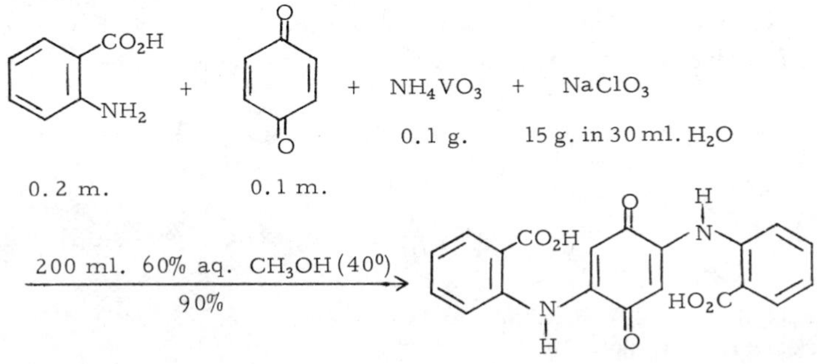

CO2H
NH2
0.2 m.
+
O
O
0.1 m.
+ NH4VO3 + NaClO3
0.1 g.
15 g. in 30 ml. H2O
200 ml. 60% aq. CH3OH (40°)
90%
CO2H
N
H
O
O
H
N
HO2C

[1]F. Ullmann and P. Ochsner, *Ann.*, **381**, 1(1911); German Patents 205,195, 228,875
[2]W. J. Scott and C. F. H. Allen, *Org. Syn., Coll. Vol.*, **2**, 128 (1943)
[3]N. A. Milas, *Org. Syn., Coll. Vol.*, **2**, 302 (1943)
[4]H. W. Underwood, Jr., and W. L. Walsh, *ibid.*, **2**, 553 (1943)
[5]W. Braun and R. Mecke, *Ber.*, **99**, 1991 (1966)

Sodium chlorodifluoroacetate, $CClF_2CO_2Na$. Mol. wt. 152.48

Preparation from the acid (Aldrich, KK Labs.) and sodium hydroxide in methanol.[1]

Precursor of difluorocarbene. When a suspension of the salt in diglyme containing cyclohexene is refluxed for 30 hrs., 7,7-difluoronorcarane is obtained in 11% yield.[2] When a mixture of the salt, an aldehyde, and triphenyl phosphine in 50 ml. of 1,2-dimethoxyethane is refluxed in a nitrogen atmosphere until 0.11 mole of carbon

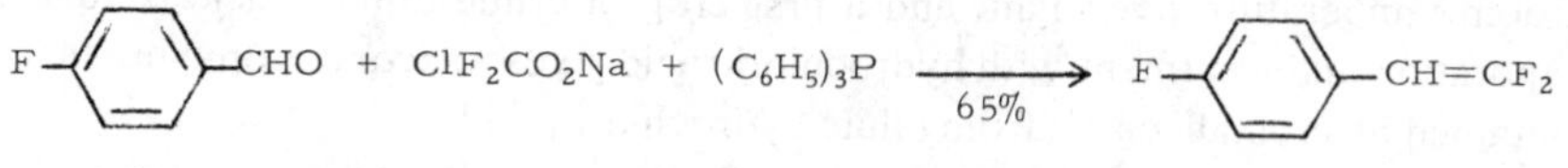

0.1 m. 0.11 m. 0.11 m.

dioxide has been evolved (72 hrs.), the difluorocarbene combines with triphenylphosphine to form an ylide, and this reacts with the aldehyde to give the 1,1-difluoroolefin.[3] Extension of the reaction to ketones was, on the whole, unpromising.[4] However, Burton and Herkes[5] found that the reagent reacts with α,α,α-trifluoroacetophenone to give 2-phenylpentafluoropropene in high yield.

$$C_6H_5COCF_3 + 2\,(C_6H_5)_3P + CClF_2CO_2Na \xrightarrow[80\%]{\text{Diglyme } 105^\circ,\ 20\text{ hrs. }(N_2)} C_6H_5C(=CF_2)-CF_3$$

50 mml. 100 mml. 100 mml.

[1]S. A. Fuqua, W. G. Duncan, and R. M. Silverstein, *J. Org.*, **30**, 1027 (1965); procedure submitted to *Org. Syn.*
[2]J. M. Birchall, G. W. Cross, and R. N. Haszeldine, *Proc. Chem. Soc.*, 81 (1960)
[3]S. A. Fuqua, W. G. Duncan, and R. M. Silverstein, *Tetrahedron Letters*, 1461 (1964)
[4]*Idem., ibid.*, 521 (1965); *J. Org.*, **30**, 2543 (1965)
[5]D. J. Burton and F. E. Herkes, *Tetrahedron Letters*, 1883 (1965)

Sodium cobalt carbonylate, $Na^+Co(CO)_4^-$. Colorless powder insoluble in water. It is prepared[1] from dicobalt octacarbonyl and sodium amalgam in ether solution under nitrogen. If a solution in another solvent is required the ether is evaporated in vacuum and the new solvent is added with exclusion of air.

Heck and Breslow[2] found that alkyl halides, sulfates, and sulfonates undergo carboalkoxylation in the presence of carbon monoxide, an alcohol, a base, and a catalytic amount of sodium cobalt carbonylate, as illustrated for the reaction of 1-iodooctane in methanol with the strongly basic hindered amine dicyclohexylethylamine as base. Use of an unhindered amine leads to formation of the amide.

$$\underline{n}\text{-}C_8H_{17}I + CO + CH_3OH + (C_6H_{11})_2NCH_2CH_3 \xrightarrow[56\%]{Na^+Co(CO)_4^-,\ \text{toluene }(50^\circ)}$$

$$\underline{n}\text{-}C_8H_{17}CO_2CH_3 + (C_6H_{11})_2\overset{+}{N}HCH_2CH_3\ (I^-)$$

[1]W. Hieber, G. Braun, and W. Beck, *Ber.*, **93**, 901 (1960)
[2]R. F. Heck and D. S. Breslow, *Am. Soc.*, **85**, 2779 (1963)

Sodium chromate, anhydrous, Na_2CrO_4. Mol. wt. 162.00. Supplier: Fisher.

A patent[1] described use of this reagent in a mixture of acetic acid and acetic anhydride for the oxidation of Δ^5-steroids to Δ^5-7-ketosteroids. Searle chemists have oxidized Δ^5-steroids by this procedure and also with *t*-butyl chromate[2] and seem to

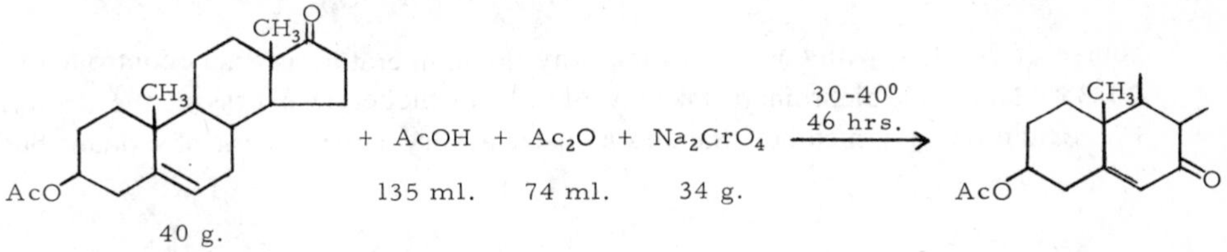
CH3
CH3
AcO
40 g.
+ AcOH + Ac2O + Na2CrO4
135 ml. 74 ml. 34 g.
30-40°
46 hrs.
CH3
AcO

prefer sodium chromate.[3] Their procedure, as applied to pregnenolone acetate, is indicated in the formulation. The yield of crude product, ε 10,300, was 79%.

[1]W. C. Meuley, U. S. Patent 2,505,646 (1950) [*C. A.*, **44**, 6894 (1950)]
[2]C. W. Marshall, R. E. Ray, I. Laos, and B. Riegel, *Am. Soc.*, **79**, 6308 (1957)
[3]N. W. Atwater, *Am. Soc.*, **83**, 3071 (1961)

Sodium cyanate, NaOC≡N. Mol. wt. 65.02. Suppliers: Aceto Chem. Co.; Alfa; Fairmont Chem. Co.

Loev and Kormendy[1] found that hitherto difficultly obtainable carbamates of tertiary alcohols can be prepared by simply stirring the substrate, sodium cyanate, and trifluoroacetic acid in an inert solvent at room temperature. For the preparation of *t*-butyl carbamate[2] a suspension of sodium cyanate in a solution of *t*-butanol in

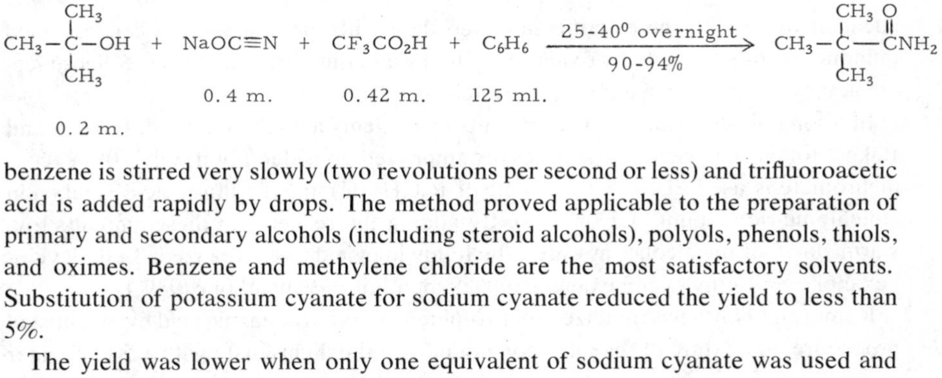
CH3
CH3—C—OH
CH3
0.2 m.
+ NaOC≡N + CF3CO2H + C6H6
0.4 m. 0.42 m. 125 ml.
25-40° overnight
90-94%
CH3 O
CH3—C—CNH2
CH3

benzene is stirred very slowly (two revolutions per second or less) and trifluoroacetic acid is added rapidly by drops. The method proved applicable to the preparation of primary and secondary alcohols (including steroid alcohols), polyols, phenols, thiols, and oximes. Benzene and methylene chloride are the most satisfactory solvents. Substitution of potassium cyanate for sodium cyanate reduced the yield to less than 5%.

The yield was lower when only one equivalent of sodium cyanate was used and increased to a maximum with two equivalents.[3] The lower yield at the lower mole ratio is assumed to be due to loss of some HOCN. The lower yield on rapid stirring is rationalized as due to the same factor.

[1]B. Loev and M. F. Kormendy, *J. Org.*, **28**, 3421 (1963)
[2]B. Loev, M. F. Kormendy, and M. M. Goodman, procedure submitted to *Org. Syn.*
[3]B. Loev, private communication

Sodium dichromate dihydrate, $Na_2Cr_2O_7 \cdot 2H_2O$. Mol. wt. 298.05. Oxygen equivalents per mole = 3.

Aqueous H_2SO_4. Sulfuric acid accelerates oxidation by hexavalent chromium and keeps the reduced reagent in solution as chromous sulfate, as illustrated by the equation for the oxidation of α,γ-dichlorohydrin. Conant and Quayle's procedure[1] calls for stirring a mixture of the alcohol and dichromate and 225 ml. of water and adding a cooled mixture of 245 ml. of concd. sulfuric acid and 115 ml. of water in the

$$3\ ClCH_2CH(OH)CH_2Cl + Na_2Cr_2O_7 \cdot 2\,H_2O + 4\ H_2SO_4 \xrightarrow[68\text{-}75\%]{20\text{-}35^0} 3\ ClCH_2C(=O)CH_2Cl + Na_2SO_4 + Cr_2(SO_4)_3 + 9\ H_2O$$

2.3 m. 1.3 m. (1.3 x theory) 4.3 m.

course of 7–8 hrs. with cooling. In this way the temperature is easily controlled to 20–35°. Here the dichromate taken was only in moderate excess of the theory. For oxidation of L-menthol, Sandborn[2] used a very large excess of oxidant but

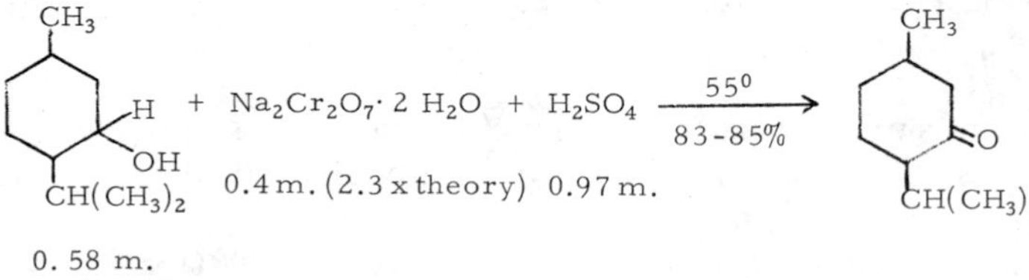

without experimental justification. In this case the reaction was controlled to 55° by gradual addition of the alcohol to the dichromate-acid solution. For half-oxidation of *n*-butanol and conversion into *n*-butyl butyrate in an aqueous medium, Robertson[3] limited the oxidant to the theoretical amount, but the low yield may well be as-

$$\underline{n}\text{-}C_4H_9OH + Na_2Cr_2O_7 \cdot 2\ H_2O + H_2SO_4 \xrightarrow[41\text{-}47\%]{20\text{-}35^0} \underline{n}\text{-}C_3H_7CO_2C_4H_9\text{-}\underline{n}$$

3.24 m. 1.07 m. (theory) 4.3 m.

sociated more with the esterification step than with the oxidation. Vliet[4] obtained quinone in 76–81% yield by oxidation of hydroquinone with 1.6 times the theoretical amount of dichromate in dilute sulfuric acid at 30°.

In a long needed study of the amounts of reagents actually required, Hussey and Baker[5] found that yields of ketones are improved considerably if only 20% excess dichromate is used (0.4 mole per mole of R_2CHOH) and if sulfuric acid is taken in stoichiometric amount (1.33 moles). By adding the reagents in these amounts to a warm slurry of the alcohol in water, the following yields were realized: 4-ethylcyclohexanol ⟶ 4-ethylcyclohexanone (90%); menthol ⟶ menthone (94%).

Kamm and Matthews[6] oxidized *p*-nitrotoluene to *p*-nitrobenzoic acid by an unusual procedure. A mixture of the nitro compound and the dichromate with 1.5 l. of water

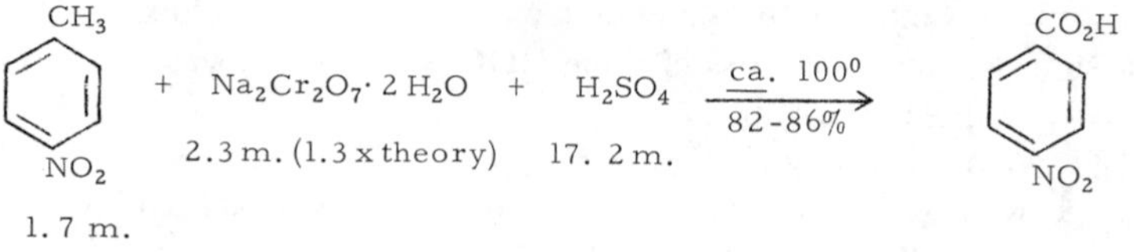

was stirred, and 1,700 g. of sulfuric acid was allowed to flow in during about 30 min. The mixture was then heated to gentle boiling for about one-half hour. Use of the large excess of acid shortened the reaction time from 40 hrs. to about 1 hr. Trinitrotoluene was oxidized similarly but with slow addition of the dichromate to control the temperature to 45–55°.[7]

Two-phase oxidation. For the preparation of cholestanone, Bruce[8] stirred a solution of dichromate, sulfuric acid, and acetic acid in water and added, with cooling,

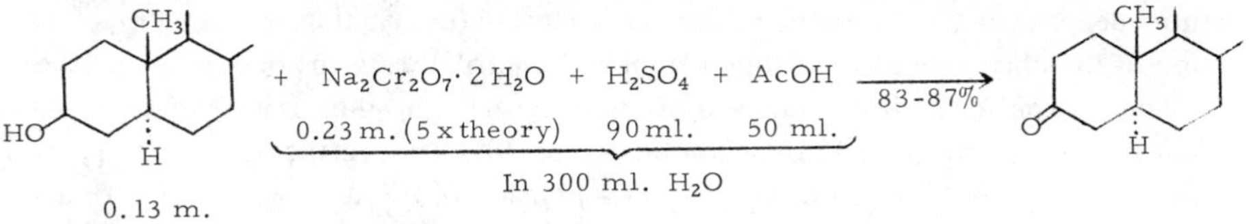
CH3
HO
H
0.13 m.
+ Na2Cr2O7· 2 H2O + H2SO4 + AcOH
0.23 m. (5 x theory) 90 ml. 50 ml.
In 300 ml. H2O
83-87%
CH3
O
H

a solution of cholestanol in 500 ml. of benzene. After stirring at 25–30° for 6 hrs., the layers were separated and the aqueous layer extracted with benzene.

Johnson and co-workers[9] oxidized 2-methylcyclohexanol by a similar process but without use of a huge excess of oxidant. An aqueous solution of the reagents in-

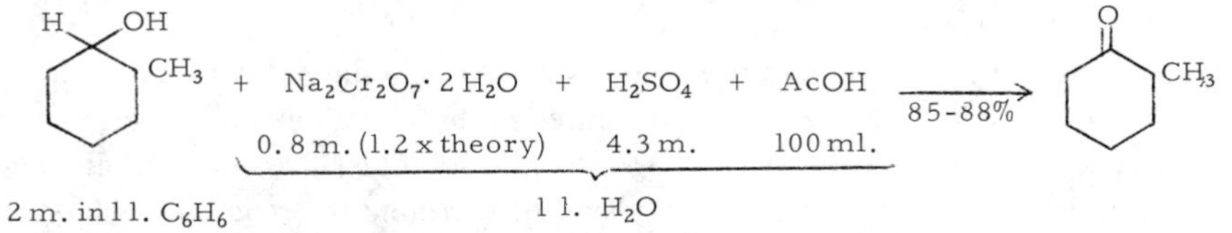
H OH
CH3
2 m. in 1 l. C6H6
+ Na2Cr2O7· 2 H2O + H2SO4 + AcOH
0. 8 m. (1.2 x theory) 4.3 m. 100 ml.
1 l. H2O
85-88%
O
CH3

dicated was stirred and kept at 10° during addition of a benzene solution of the alcohol in 2.5 hrs. Stirring was continued for 3 hrs. more.

In water at 250°. A highly efficient oxidation procedure first described in the patent literature and developed further by Friedman[10] is illustrated by Friedman's procedure for the oxidation of 2,3-dimethylnaphthalene to naphthalene-2,3-dicarboxylic acid.[11] An autoclave charged with the reagents indicated is shaken con-

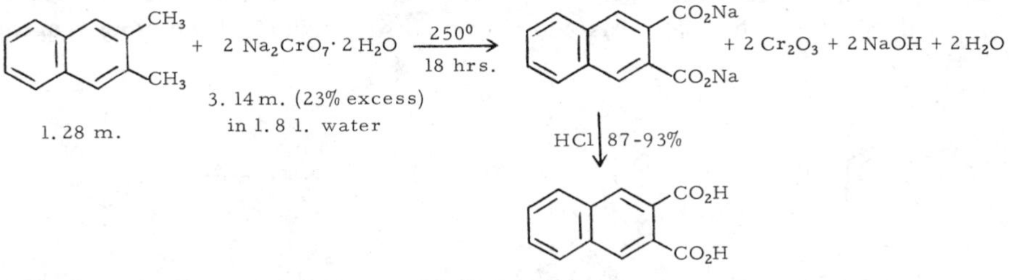
CH3
CH3
1. 28 m.
+ 2 Na2CrO7· 2 H2O
3. 14 m. (23% excess)
in 1. 8 l. water
250°
18 hrs.
CO2Na
CO2Na
+ 2 Cr2O3 + 2 NaOH + 2 H2O
HCl 87-93%
CO2H
CO2H

tinuously at 250° for 18 hrs. The cooled mixture is filtered to remove green hydrated chromium oxide, and the filtrate is acidified. Newman and Boden[12] oxidized four methylbenzo[c]phenanthrenes by this method to the carboxylic acids in yields of 85–95%. Reitsema and Allphin[13] reported oxidation of ethylbenzene to phenylacetic acid by this method, but the claim has not been confirmed.

In anhydrous acetic acid. Sodium dichromate dihydrate is soluble in acetic acid without addition of water; a 25% solution in hot acetic acid can be cooled to 10° with no separation of solid.[14] In a procedure for the preparation of cyclohexanone,[15]

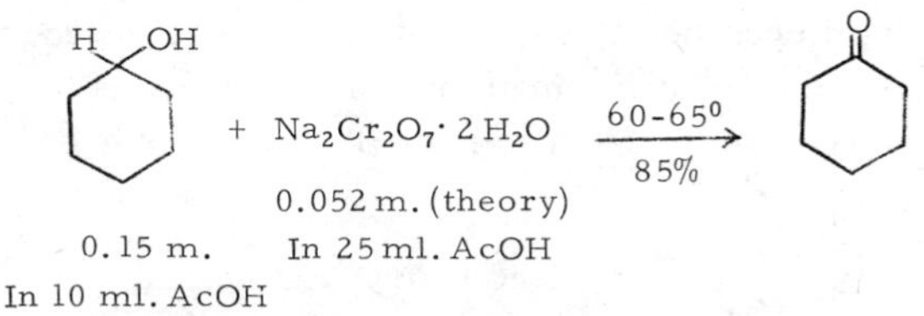
H OH
0. 15 m.
In 10 ml. AcOH
+ Na2Cr2O7· 2 H2O
0.052 m. (theory)
In 25 ml. AcOH
60-65°
85%
O

solutions of the alcohol and of the dichromate in acetic acid are adjusted to 15° and mixed. The temperature rises to 60° and is kept there for 15 min.; after a further spontaneous rise to 65° (25–30 min.) the temperature begins to fall, and the solution turns deep green. A low temperature at the start of the reaction permits initial formation of the chromate ester of the alcohol and contributes to the high yield of ketone.

The ketone formed on oxidation of cholesterol-5α,6β-dibromide rapidly darkens on exposure to strong light or when heated to 70° but nevertheless was prepared in high yield by one of us as follows.[16] A suspension of the dibromide in 2 l. of acetic

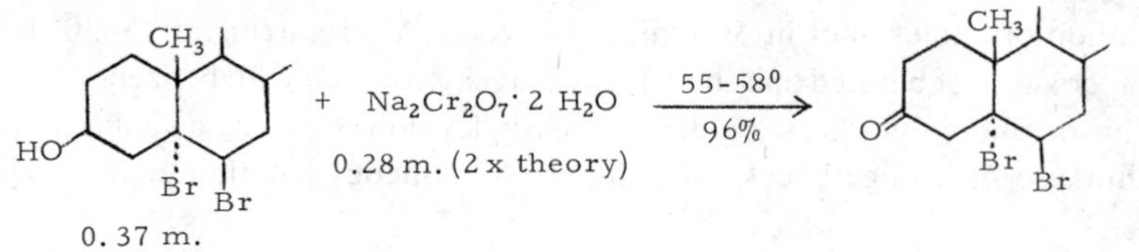
CH3
HO
Br
Br
0.37 m.
+ Na2Cr2O7· 2 H2O
0.28 m. (2 x theory)
55-58°
96%
CH3
O
Br
Br

acid is stirred at room temperature, and a solution of sodium dichromate dihydrate in 2 l. of acetic acid preheated to 90° is poured in through a funnel. The temperature of the reaction mixture soon rises to 55–58°, the solid dissolves in 3–5 min., and in 2 min. more an ice bath is raised on a jack to surround the reaction flask, stirring is stopped, and the product let crystallize. The mixture is then cooled with stirring to 25°, 400 ml. of water is added and the temperature brought to 15°; the crystalline, pure product is collected and washed thoroughly with methanol. Should another occasion arise to run this reaction, I (L.F.F.) would use the calculated amount of dichromate.

A procedure[17] for the preparation of acenaphthenequinone based upon a postwar disclosure of German work[18] used anhydrous acetic acid as solvent and ceric acetate as catalyst. The dichromate was added to a stirred mixture of the other reagents

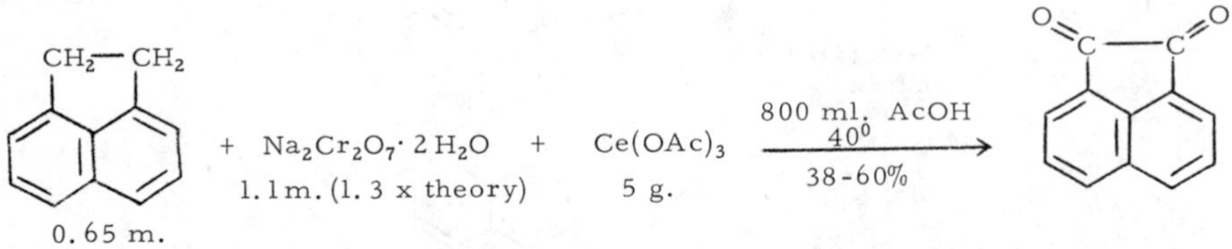
CH2—CH2
0.65 m.
+ Na2Cr2O7· 2 H2O +
1.1 m. (1.3 x theory)
Ce(OAc)3
5 g.
800 ml. AcOH
40°
38-60%
O
C
C
O

at 40° and stirring was continued at room temperature for 8 hrs. more. The α-diketone was purified through the bisulfite addition product. The high variation in yield suggests that the procedure is in need of further standardization.

In anhydrous acetic acid–benzene. This solvent mixture, introduced by one of us in a study of the oxidation of cholesterol, has these advantages: benzene is a better solvent for the sterol than acetic acid; less acetylation occurs; oxidation can be conducted in homogeneous solution at temperatures ranging from 0°[19] to 121°.[20] Experiments conducted in this way led to isolation of no less than six neutral products and two diacids derived from cholesterol (and not from its companions). A procedure[21] for the preparation of Δ^4-cholestene-3,6-dione includes a novel procedure for workup. A mixture of dichromate and acetic acid was heated and swirled to effect solution and cooled to 15°. A solution of cholesterol in benzene was cooled to

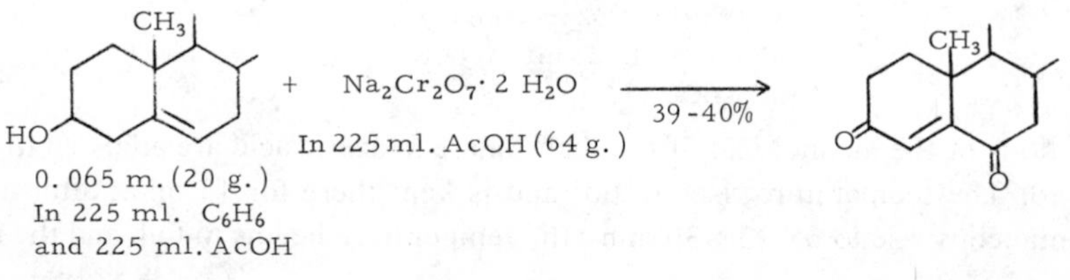
CH3
HO
0.065 m. (20 g.)
In 225 ml. C6H6
and 225 ml. AcOH
+ Na2Cr2O7· 2 H2O
In 225 ml. AcOH (64 g.)
39-40%
CH3
O
O

20°, an equal volume of acetic acid was added, and the solution cooled to 15°. On pouring in the dichromate solution a thick orange paste of cholesteryl chromate separated. The flask was immersed in an ice-water bath that was let stand unattended in a refrigerator for 40 hrs. The temperature soon dropped to 0° and the chromate ester dissolved in a few hours. The brown solution was poured into a separatory funnel and shaken with 225 ml. of 30–60° petroleum ether. After brief standing the mixture separated to a reddish upper hydrocarbon layer and a smaller, very dark lower layer containing chromium compounds and acetic acid. The lower layer was discarded, and the upper layer shaken with 50 ml. of water and let settle. A fairly large lower layer was drawn off and discarded and the process was repeated. The now light-colored hydrocarbon layer was then extracted repeatedly with Claisen's alkali for selective removal of enedione as the yellow enolate. Each yellow extract was run out into a separatory funnel containing hydrochloric acid, ice, and ether for collection of the free enedione, which crystallized from methanol–ether in yellow plates.

Four other products can be obtained easily by oxidation of cholesterol in the same way but at high temperatures. Oxidation at 60°, separation of the acidic

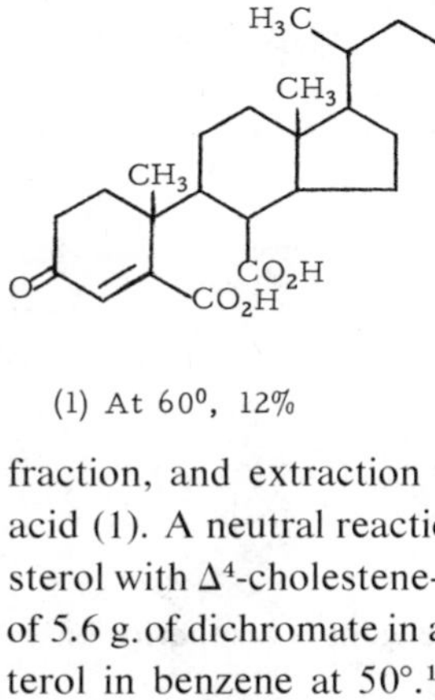

(1) At 60°, 12%

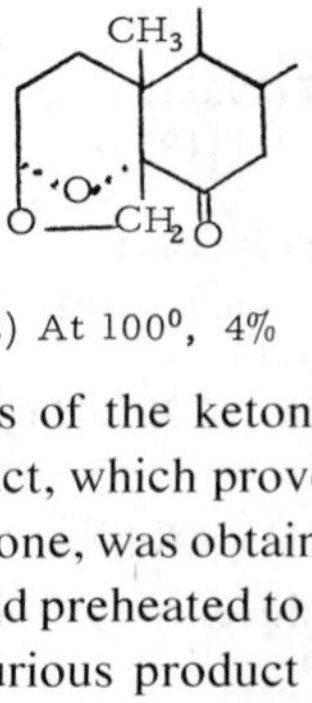

(2) At 100°, 4%

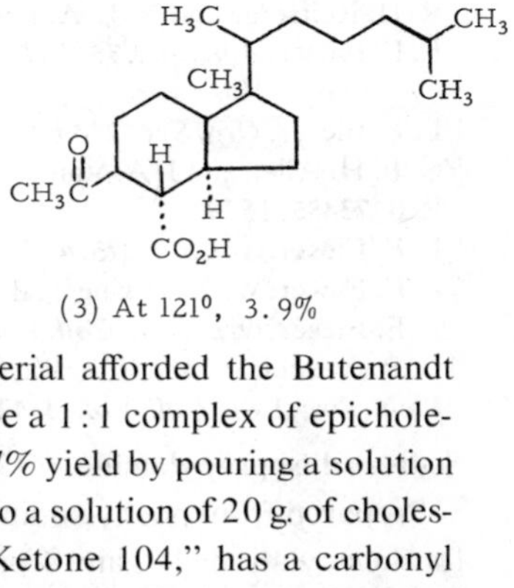

(3) At 121°, 3.9%

fraction, and extraction from this of the ketonic material afforded the Butenandt acid (1). A neutral reaction product, which proved to be a 1 : 1 complex of epicholesterol with Δ^4-cholestene-6β-ol-3-one, was obtained in 7% yield by pouring a solution of 5.6 g. of dichromate in acetic acid preheated to 90° into a solution of 20 g. of cholesterol in benzene at 50°.[19] The curious product (2), "Ketone 104," has a carbonyl group inert to Girard's reagent and is very stable to oxidation. The best preparative procedure involves oxidation under such drastic conditions that all other neutral products are destroyed.[20] A solution of 150 g. of dichromate in acetic acid was adjusted to 70° and poured into a solution of cholesterol in benzene at the boiling point and the solution was refluxed for 1 hr. at a liquid temperature of 100°. Ketone 104 was then easily isolated from the small neutral fraction. The further oxidation product duoannelic acid (3) was obtained by proceeding as in the preparation of Ketone 104 until a first vigorous reaction subsided, distilling off solvent to a liquid temperature of 121°, and refluxing for 4 hrs. The neutral fraction afforded 1.1% of Ketone 104, and treatment of the acidic fraction with Girard's reagent gave a ketonic fraction which afforded pure duoannelic acid in 3.9% yield.

In acetic acid–acetic anhydride. Use of this combination in dichromate oxidations is described by Rieveschl and Ray in a patent[22] and in a procedure for the preparation of fluorenone-2-carboxylic acid.[23] A solution of 2-acetylfluorenone in acetic acid was heated on the steam bath with stirring, and powdered dichromate was added in portions in about 45 min. The solution was then brought to a gentle reflux over a

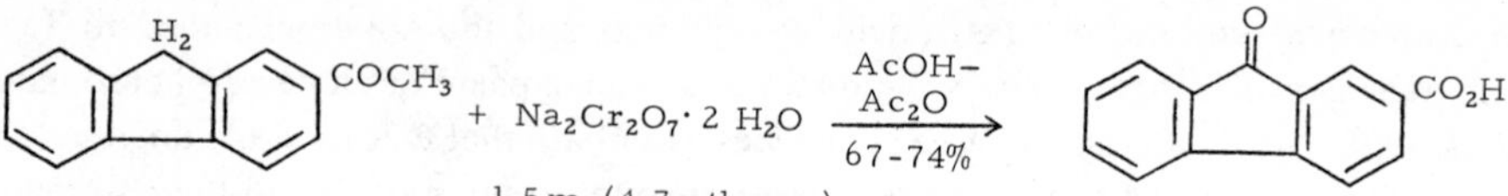
H2
COCH3
+ Na2Cr2O7· 2 H2O
AcOH–
Ac2O
67-74%
O
CO2H

0.24 m. 1.5 m. (4.7 x theory)

In 650 ml. AcOH

flame and 200 ml. of acetic anhydride was added in the course of 1.5 hrs. Refluxing was continued for 8 hrs. more. The workup included conversion to the potassium salt and a small neutral fraction afforded 1–5 g. of pure 2-acetylfluorenone.

[1]J. B. Conant and O. R. Quayle, *Org. Syn., Coll. Vol.*, **1**, 211 (1941)
[2]L. T. Sandborn, *ibid.*, **1**, 340 (1941)
[3]G. R. Robertson, *ibid.*, **1**, 138 (1941)
[4]E. B. Vliet, *ibid.*, **1**, 482 (1941)
[5]A. S. Hussey and R. H. Baker, *J. Org.*, **25**, 1434 (1960)
[6]O. Kamm and A. O. Matthews, *Org. Syn., Coll. Vol.*, **1**, 392 (1941)
[7]H. T. Clarke and W. W. Hartman, *ibid.*, **1**, 540 (1941)
[8]W. F. Bruce, *ibid.*, **2**, 139 (1943)
[9]E. W. Warnhoff, D. G. Martin, and W. S. Johnson, *ibid.*, **4**, 164 (1963)
[10]L. Friedman, D. L. Fishel, and H. Shechter, *J. Org.*, **30**, 1453 (1965)
[11]L. Friedman, *Org. Syn.*, **43**, 80 (1963)
[12]M. S. Newman and H. Boden, *J. Org.*, **26**, 1759 (1961)
[13]R. H. Reitsema and N. L. Allphin, *J. Org.*, **27**, 27 (1962)
[14]L. F. Fieser, *Am. Soc.*, **75**, 4377 (1953)
[15]*Org. Expts.*, 106
[16]L. F. Fieser, *Org. Syn., Coll. Vol.*, **4**, 197 (1963)
[17]C. F. H. Allen and J. A. VanAllan, *ibid.*, **3**, 1 (1955)
[18]P. B. **73485**, 1579
[19]L. F. Fieser, *Am. Soc.*, **75**, 4377, 4386, 4395 (1953)
[20]L. F. Fieser, W.-Y. Huang, and T. Goto, *Am. Soc.*, **82**, 1688 (1960)
[21]L. F. Fieser, *Org. Syn., Coll. Vol.*, **4**, 189 (1963)
[22]G. Rieveschl, Jr., and F. E. Ray, U.S. patent 2,377,040 [*C.A.*, **39**, 3305 (1945)]
[23]*Idem*, *Org. Syn., Coll. Vol.*, **3**, 420 (1955)

Sodium diisopropylamide, $(i\text{-}C_3H_7)_2NNa$. Mol. wt. 123.18.

Prepared from phenylsodium and diisopropylamine, the reagent is a superior base for the acylation of 3- and 4-picoline.[1]

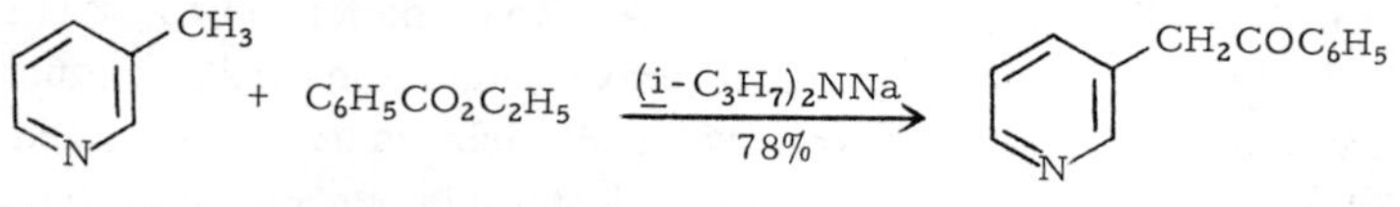
CH3
N
+ C6H5CO2C2H5
(i-C3H7)2NNa
78%
CH2COC6H5
N

[1]S. Raynolds and R. Levine, *Am. Soc.*, **82**, 472 (1960)

Sodium disulfide (or polysulfide) solution, Na_2S_2. For use in the preparation of thiosalicylic acid,[1] the reagent is prepared by heating and stirring a mixture of 1.1 moles of $Na_2S\cdot 9H_2O$, 290 ml. of water, and 1.1 moles of powdered sulfur until a brownish red solution results. A solution of 40 g. of sodium hydroxide in 100 ml. of water is added, and the solution is cooled in a salt-ice bath and stirred at 0–5° during addition of a solution of diazotized anthranilic acid. This Sandmeyer-type displacement gives a disulfide reducible to thiosalicylic acid.

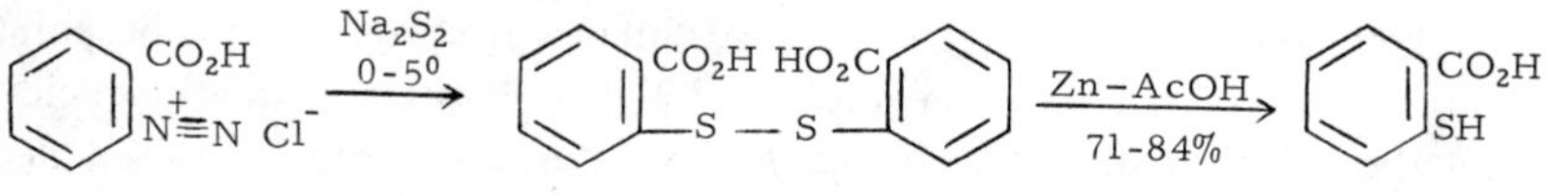
CO2H
N≡N Cl
Na2S2
0-5°
CO2H HO2C
S — S
Zn–AcOH
71-84%
CO2H
SH

1 m.

In the preparation of 2,4,5-triaminonitrobenzene starting with *m*-dichlorobenzene,[2] reduction of 0.114 mole of the diaminodinitrobenzene to the triaminonitrobenzene (red needles) is effected by heating a mixture of 0.125 mole of $Na_2S \cdot 9H_2O$, 0.20

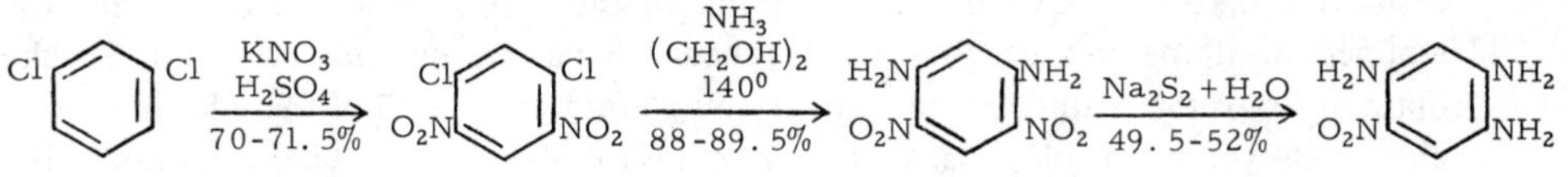
Cl
Cl
KNO3
H2SO4
70-71.5%
Cl
Cl
O2N
NO2
NH3
(CH2OH)2
140°
88-89.5%
H2N
NH2
O2N
NO2
Na2S2 + H2O
49.5-52%
H2N
NH2
O2N
NH2

mole of sulfur, and 125 ml. of water and adding the clear orange-red solution dropwise during 1.5 hrs. to a well-stirred slurry of the diaminodinitrobenzene in 150 ml. of water under reflux.

The reaction of *p*-nitrotoluene with alkaline sodium polysulfide involves an interesting disproportionation: the methyl group is oxidized to —CHO and the nitro group is reduced to NH_2.[3] After refluxing for 3 hrs. the resulting deep red

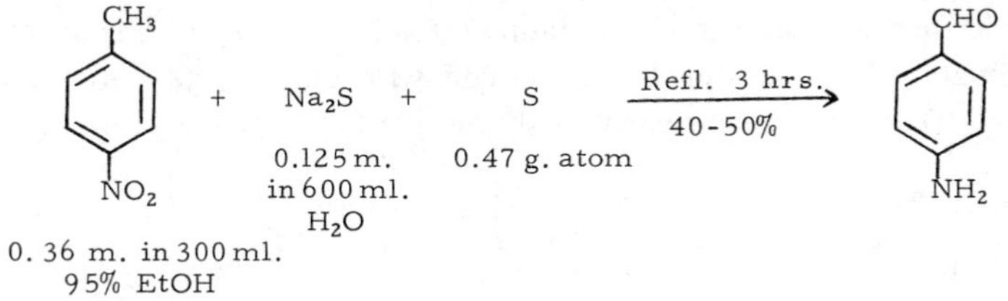
CH3
NO2
0.36 m. in 300 ml.
95% EtOH
+
Na2S
0.125 m.
in 600 ml.
H2O
+
S
0.47 g. atom
Refl. 3 hrs.
40-50%
CHO
NH2

solution is steam distilled to remove ethanol, *p*-toluidine, and starting material, and then cooled for crystallization of *p*-aminobenzaldehyde.

[1]C. F. H. Allen and D. D. MacKay, *Org. Syn., Coll. Vol.*, **2**, 580 (1943)
[2]J. H. Boyer and R. S. Buriks, *ibid.*, **40**, 96 (1960)
[3]E. Campaigne, W. M. Budde, and G. F. Schaefer, *ibid.*, **4**, 31 (1963)

Sodium ethoxide, C_2H_5ONa. Mol. wt. 68.06. Supplier: K and K Labs.

Solution in ethanol. For preparation of 6–10% solutions of the ethoxide in absolute ethanol, investigators conventionally add sodium to the alcohol in lumps or slices, with or without mechanical stirring, with or without a nitrogen atmosphere, and cool or heat as required until the metal is dissolved. Commercial anhydrous ethanol now available is satisfactory. A glycerol-sealed stirrer[1] has been used,[2] also a magnetic stirrer.[3] The hydrogen liberated should be vented into a hood. Two inverse procedures are described below; the second is shorter.

Tishler procedure.[4] A 2-l. flask equipped with a dropping funnel and a condenser protected with a calcium chloride tube is flushed with nitrogen and charged with 23 g. (1 g. atom) of clean sodium. Anhydrous ethanol (300 ml.) is run in intermittently, the rate of addition being such as to maintain rapid refluxing without reaching the melting point of sodium (97.5°). When the addition is complete the mixture is heated on the steam bath (90°) until all the sodium has reacted (1 mole/300 ml.). The process takes 2–3 hrs.

Zaugg procedure.[5] A 1-l. flask fitted with a dropping funnel, a stirrer, a reflux condenser protected with a drying tube, and a heating mantle is charged with 23 g. of clean sodium, and 300 ml. of absolute ethanol is placed in the dropping funnel. About 50 ml. of ethanol is added in a rapid stream and, after the initial vigorous refluxing has subsided, the mixture is heated until the sodium melts in the thick slurry of sodium ethoxide. Stirring is then started, and the remainder of the ethanol is added rapidly, dropwise or in a thin stream, at such a rate that the heat of reaction maintains the sodium in the molten state (clearly detectable by the silvery appearance of the agitated globules). Excessively rapid addition of ethanol cools the mixture below the melting point of the metal; in that event addition is stopped until the mixture has been reheated and the sodium remelted. At the optimum rate the sodium is consumed completely at the end of the addition and a clear, light yellow to colorless, solution results (1 mole/300 ml.). The entire operation requires about 30 minutes.

Alcohol-free reagent can be made by melting sodium under xylene and stirring[6] or shaking[7] to produce powdered sodium. The solvent is decanted, the metal is washed repeatedly with ether and covered with ether, and one equivalent of anhydrous ethanol is added. The mixture is allowed to stand overnight to complete the reaction and give a suspension of sodium ethoxide.

$ArCHO + CH_3CO_2C_2H_5$, $C_6H_5CH_2CN$. In 1890 Ludwig Claisen,[8] who three years earlier had proposed use of sodium ethoxide as a condensing agent, explored the possibility of condensing benzaldehyde with ethyl acetate to produce ethyl cinnamate (3), in possible rivalry with the Perkin reaction. First trials with a

$$\underset{(1)}{C_6H_5CHO} + \underset{(2)}{CH_3CO_2C_2H_5} \xrightarrow{EtONa} \underset{(3)}{C_6H_5CH{=}CHCO_2C_2H_5} + H_2O$$

$$C_6H_5CHO \xrightarrow{NaOEt} \underset{(4)}{C_6H_5CO_2CH_2C_6H_5}$$

solution of sodium ethoxide in ethanol were unpromising: some of the desired ester (3) was formed, but a characteristic odor led to the finding as a by-product of considerable benzyl benzoate (4), a known product of the action of alcoholic sodium ethoxide on benzaldehyde alone. Treatment of a mixture of benzaldehyde and ethyl acetate with alcohol-free sodium ethoxide afforded the desired ester (3) in 30–40% yield by weight, but considerable benzyl benzoate (4) was also formed. Claisen then found that the condensation proceeds with great ease and without by-products when 1 g. atom of sodium is added to excess ethyl acetate, followed by 1 mole of benzaldehyde. After disappearance of the sodium, addition of the calculated amount of acetic acid and workup afforded recovered ethyl acetate and gave ethyl cinnamate in yield of 57–62%. Later,[6] a standardized procedure used 1.26 g. atom of sodium (powdered), 47 m. of ethyl acetate, and 3.5 ml. of ethanol per mole of benzaldehyde and raised the yield to 68–74%. When only 1 equivalent of sodium was used the yield (60%) corresponded to that obtained by Claisen. Benzyl cyanide is so much more reactive than ethyl acetate that it condenses satisfactorily with benzaldehyde in 95% ethanol in the presence of catalytic amount (7 g.) of sodium ethoxide.[9]

$$\underbrace{\underset{1\ m.}{C_6H_5CHO} + \underset{1\ m.}{C_6H_5CH_2CN}}_{\text{In 650 ml. 95\% EtOH}} \xrightarrow[87\text{-}97\%]{EtONa} C_6H_5CH{=}C(CN)C_6H_5$$

Condensation of esters with diethyl oxalate. Ester condensation under catalysis by sodium ethoxide is a reversible reaction, and McElvain and co-workers[10] showed

$$RCH_2COOC_2H_5 + \begin{matrix} CO_2C_2H_5 \\ | \\ CO_2C_2H_5 \end{matrix} \underset{}{\overset{EtONa}{\rightleftharpoons}} R\underset{}{\overset{\displaystyle COCO_2C_2H_5 \atop |}{C}}HCO_2C_2H_5 + C_2H_5OH$$

that yields can be improved by removing the ethanol from the equilibrium mixture by distillation. Floyd and Miller[11] employed this technique in the condensation of ethyl stearate with a large excess of diethyl oxalate. A mixture of the esters was

$$\underset{1\text{ m.}}{\underline{n}\text{-}C_{16}H_{33}CH_2CO_2C_2H_5} + \underset{4\text{ m.}}{\begin{matrix} CO_2C_2H_5 \\ | \\ CO_2C_2H_5 \end{matrix}} + \underset{1\text{ m.}}{C_2H_5ONa} \longrightarrow \underline{n}\text{-}C_{16}H_{33}\overset{\displaystyle COCO_2C_2H_5 \atop |}{C}HCO_2C_2H_5$$

$$\xrightarrow[68.5\text{-}71\%\text{ overall}]{160\text{-}170^0} \underline{n}\text{-}C_{16}H_{33}CH(CO_2C_2H_5)_2$$

added to a solution of 1 mole of sodium ethoxide in 300 ml. of ethanol, and ethanol was removed by distillation at 50–60°/100 mm. in the course of 2–3 hrs. Distillation at 15 mm. pressure removed excess oxalic ester, and neutralization with acetic acid and extraction with ether afforded the α-keto ester, which was decarbonylated to the malonic ester. When diethyl oxalate was taken in theoretical amount the yield was 10–15% lower.

In condensing diethyl succinate with diethyl oxalate to form triethyl oxalylsuccinate, Bottorff and Moore[12] achieved a high yield with all three reactants taken in theoretical amount by the expedient of removing the ethanol prior to addition of the two esters. A solution of 23 g. of sodium in 356 ml. of ethanol was distilled at

$$\underset{1\text{ m.}}{\begin{matrix} CH_2CO_2C_2H_5 \\ | \\ CH_2CO_2C_2H_5 \end{matrix}} + \underset{1\text{ m.}}{\begin{matrix} CO_2C_2H_5 \\ | \\ CO_2C_2H_5 \end{matrix}} + \underset{1\text{ m.}}{C_2H_5ONa} \xrightarrow[86\text{-}91\%]{\text{Ether}\ 25^0} \begin{matrix} COCO_2C_2H_5 \\ | \\ CHCO_2C_2H_5 \\ | \\ CH_2CO_2C_2H_5 \end{matrix}$$

atmospheric pressure until the mixture became pasty, toluene was added in portions in amounts sufficient to permit stirring and prevent splattering, until all the ethanol was removed and the contents of the flask reached a temperature of 105° (when the toluene distillation was omitted the yield was lower by 5–10%). The slurry of sodium ethoxide was cooled, 650 ml. of ether was added, followed by diethyl oxalate (yellow solution) and diethyl succinate. After 12 hrs. at room temperature the sodio derivative was hydrolyzed by addition of water. Separation of the layers and ether extraction gave the keto ester as a yellow oil. This ester (not distillable) was converted by acid hydrolysis–decarboxylation into α-ketoglutaric acid (yield 73–83%). Trimethyl oxalylsuccinate had been obtained from 1 mole each of the same component esters in 82–83% yield by reaction with the potassium ethoxide prepared from 1 g. atom of potassium and 2.6 moles of ethanol.[13]

For the condensation of ethyl γ-phenylbutyrate with diethyl oxalate to produce the oxalyl ester (3), Hershberg and Fieser[7] prepared ethanol-free sodium ethoxide by adding 0.27 mole of ethanol to a suspension of 0.27 g. atom of powdered sodium in 150 ml. of ether, added to the resulting suspension of alkoxide an ethereal solution of diethyl oxalate in portions to produce a pale yellow solution, added an ethereal solution of the second ester, and refluxed for 24 hrs. (when potassium ethoxide was

$$\underset{(1)\ 0.26\text{ m.}}{C_6H_5CH_2CH_2CH_2CO_2C_2H_5} + \underset{(2)\ 0.39\text{ m.}}{\begin{matrix} CO_2C_2H_5 \\ | \\ CO_2C_2H_5 \end{matrix}} + \underset{0.27\text{ m.}}{C_2H_5ONa} \longrightarrow$$

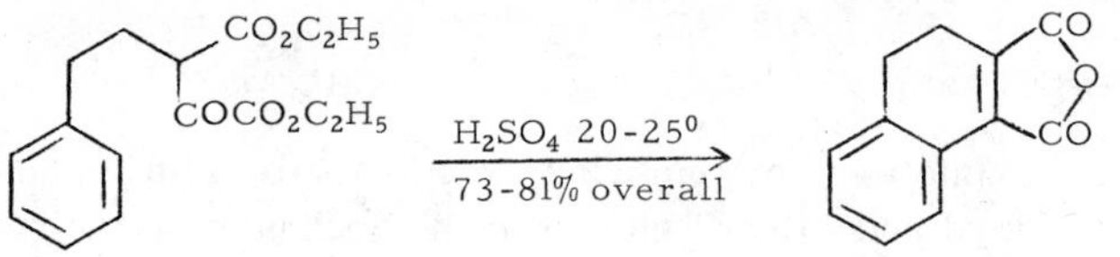
CO2C2H5
COCO2C2H5
H2SO4 20-25°
73-81% overall
CO
O
CO

(3) (4)

used as the condensing agent the reaction was done in 12 hrs.). The resulting deep red solution was cooled and neutralized with sulfuric acid, and the oxalyl ester (3) obtained by ether extraction as a pale yellow oil was cyclized with concd. sulfuric acid to 3,4-dihydronaphthalic anhydride (4), a useful dienophile.

Formylation. For α-formylation of the ester (1), Holmes and Trevoy[14] prepared ethanol-free sodium ethoxide by adding 0.41 mole of ethanol in ether to a suspension of 0.41 g. atom of sodium in ether and refluxing for 10–11 hrs. The suspension of ethoxide was cooled to −10 to −15° and an ethereal solution of the two esters was

CH3O
CH3O
CH2CH2CH2CO2C2H5
+ HCO2C2H5 + C2H5ONa
Ether
51-55%

0.4 m.

(1) 0.19 m.

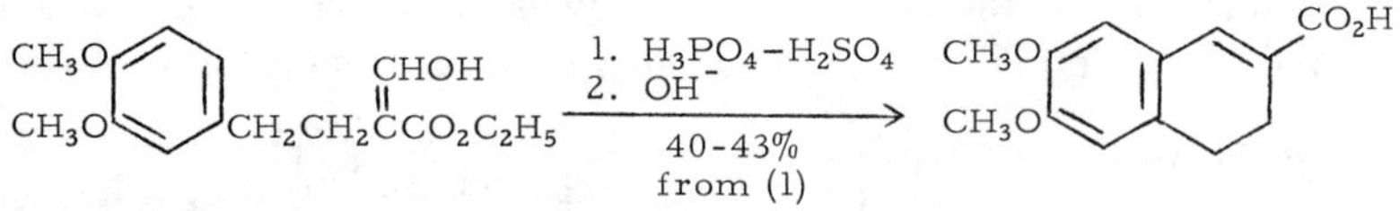
CH3O
CH3O
CHOH
CH2CH2CCO2C2H5
1. H3PO4−H2SO4
2. OH−
40-43%
from (1)
CH3O
CH3O
CO2H

(2) (3)

added with stirring at −10°. After 4 hrs. at −10° and 72 hrs. at 25°, workup afforded the formyl derivative (2) as a nondistillable oil which on cyclization and saponification afforded 6,7-dimethoxy-3,4-dihydro-2-naphthoic acid (3). Although in the ester condensation ethyl formate and sodium ethoxide were both taken in twice the theoretical amounts, separation of the formyl derivative (2) as the water-soluble enolate left a residue of 38–42% of unchanged ethyl γ-veratrylbutyrate.

Ainsworth[15] obtained a much better yield in the formylation of cyclohexanone in a process in which sodium ethoxide was generated *in situ* from 1 equivalent of sodium and the ethanol eliminated in the condensation. A mixture of cyclohexanone,

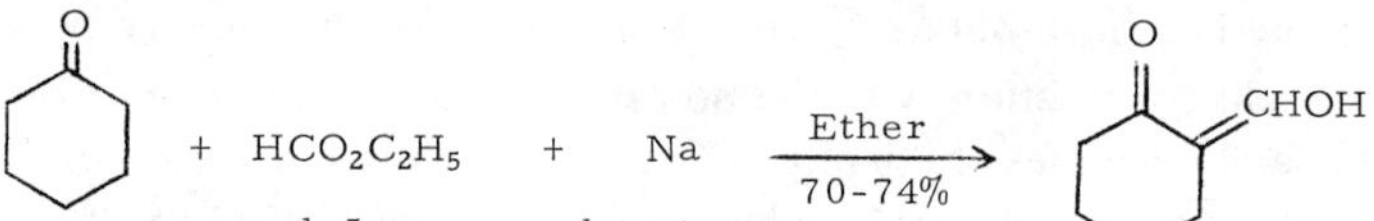
O
+ HCO2C2H5 + Na
Ether
70-74%
O
CHOH

1.5 m. 1 g. atom

1 m.

ethyl formate, sodium, and 2 l. of ether was stirred mechanically, the reaction was initiated by addition of 5 ml. of anhydrous ethanol.

Other ester condensations. Magnani and McElvain[16] prepared dibenzoylmethane by adding dry sodium ethoxide in portions with heating and stirring to a mixture of

$$C_6H_5CO_2C_2H_5 + CH_3COC_6H_5 + NaOEt \xrightarrow{150\text{-}160^\circ} C_6H_5COCH_2COC_6H_5$$

4 m. 0.5 m. 0.65 m.

ethyl benzoate and acetophenone and distilling the ethanol as formed to displace the equilibrium.

p-Chlorobenzyl cyanide is sufficiently reactive to condense with ethyl phenylacetate in high yield in a solution of sodium ethoxide in anhydrous ethanol.[17] Benzyl

$$\underline{p}\text{-}ClC_6H_4CH_2CN + C_6H_5CH_2CO_2C_2H_5 + \underset{\substack{0.5\ m.\ in\\150\ ml.\ ethanol}}{C_2H_5ONa} \xrightarrow[86\text{-}92\%]{Refl.} \underline{p}\text{-}ClC_6H_4CH(CN)COCH_2C_6H_5$$

cyanide undergoes carboethoxylation on condensation with diethyl carbonate under similar conditions.[18]

$$\underset{0.5\ m.}{C_6H_5CH_2CN} + \underset{2.5\ m.}{C_2H_5OCOC_2H_5} + \underset{\substack{0.5\ m.\ in\\300\ ml.\ ethanol}}{C_2H_5ONa} \xrightarrow[70\text{-}78\%]{Refl.} C_6H_5CH(CN)CO_2C_2H_5$$

Alkylation of malonic ester. This alkylation can be done with a solution of sodium ethoxide in ethanol and requires one mole of reagent per alkyl group introduced. An example is Perkin's synthesis of diethyl cyclobutane-1,1-dicarboxylate by alkylation with trimethylene chlorobromide.[19] When trimethylene dibromide was

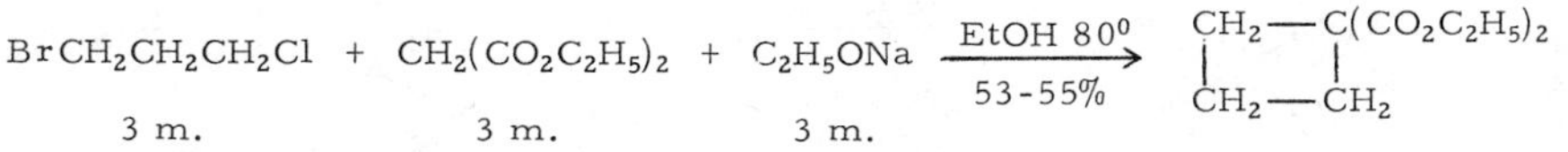

used for the alkylation the yield of product, isolated as the diacid, was only 21–23%.[20]

For monomethylation of malonic ester, Weiner[21] used methyl bromide, which gives none of the dimethyl derivative. Methyl bromide was bubbled into a solution

$$\underset{2.1\ m.}{CH_3Br} + \underset{2\ m.}{CH_2(CO_2C_2H_5)_2} + \underset{\substack{2\ m.\ in\\1\ l.\ C_2H_5OH}}{C_2H_5ONa} \xrightarrow[79\text{-}83\%]{} CH_3CH(CO_2C_2H_5)_2$$

of the ester in a solution of the ethoxide in ethanol for about 4 hrs.; the reaction was mildly exothermic and the mixture occasionally reached the boiling point. Before the final distillation, Weiner introduced a purification step learned from A. Michael. The crude ester (about 300 g.) was shaken for exactly one minute with a cold solution of 10 g. of sodium hydroxide in 30 ml. of water; the alkaline layer was then drawn off, and the ester washed with dilute acid, dried, and distilled. Michael had shown that this extraction completely removes unchanged malonic ester while hardly attacking diethyl methylmalonate.

Other alkylations.[2, 22, 23]

Michael condensation. Cason[24] prepared β-methylglutaric anhydride starting with the condensation of methyl crotonate with diethyl malonate, carried out by

$$\underset{0.6\ m.}{CH_3CH{=}CHCO_2CH_3} + \underset{0.72\ m.}{CH_2(CO_2C_2H_5)_2} + \underset{0.61\ m.}{C_2H_5ONa} \xrightarrow{C_2H_5OH}$$

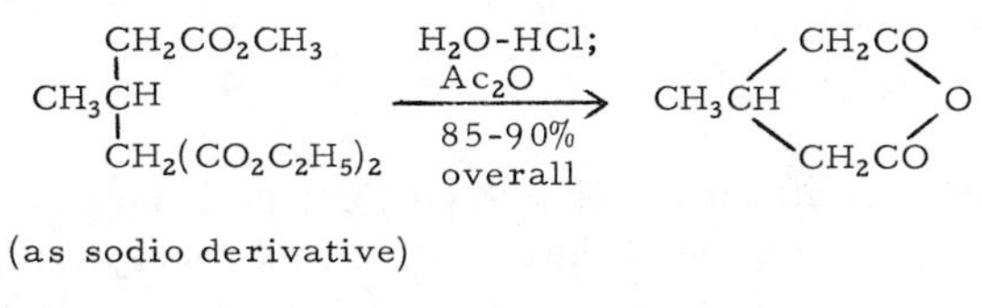
CH2CO2CH3
CH3CH
CH2(CO2C2H5)2
(as sodio derivative)
H2O-HCl; Ac2O
85-90% overall
CH3CH
CH2CO
O
CH2CO

adding a mixture of the two esters to a solution of the ethoxide in ethanol with stirring under reflux. After refluxing for 1 hr. longer the alcohol was largely removed by distillation, a solution of the sodio derivative in water was acidified, and the mixture refluxed to effect hydrolysis and decarboxylation; water and ethanol were removed by distillation, and the diacid was cyclized with acetic anhydride. Another high-yield condensation conducted in much the same way affords tetraethyl propane-1,1,2,3-tetracarboxylate (Clarke and Murray[25]).

$$C_2H_5O_2C\underset{H\overset{|}{C}CO_2C_2H_5}{\overset{\parallel}{C}CH} \; + \; CH_2(CO_2C_2H_5)_2 \; + \; C_2H_5ONa \; \xrightarrow[\substack{93-94\%}]{\text{Refl. }C_2H_5OH} \; C_2H_5O_2C\underset{CH_2CO_2C_2H_5}{\overset{|}{C}}CHCH(CO_2C_2H_5)_2$$

4.1 m.   5 m.   4 m.

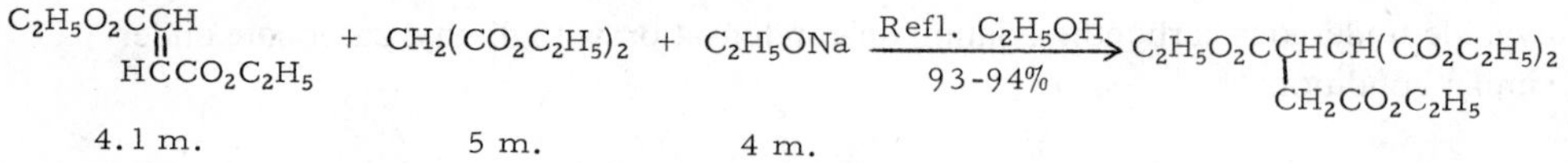

Dimedon, 5,5-dimethylcyclohexane-1,3-dione, is prepared by an interesting process discovered by Vorländer.[26] The reaction of 1 mole each of diethyl malonate, mesityl oxide, and sodium ethoxide in absolute ethanol probably involves a Michael addition followed by an intramolecular ester condensation to close the ring to give the diketo ester formulated (procedure of Shriner and Todd[27]). Aqueous potassium

(as enolate)

(as enolate)   Dimedone

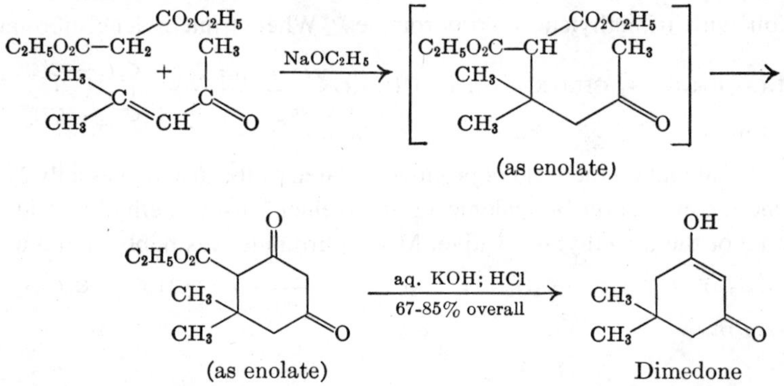

hydroxide is added to effect saponification, and the free acid is then liberated with hydrochloric acid and decarboxylated.

Synthesis of heterocycles. The synthesis of chelidonic acid from acetone, diethyl oxalate, and ethanolic sodium ethoxide, discovered by Claisen[28] and perfected by Riegel and Zwilgmeyer,[29] involves initially a double ester condensation to give acetonedioxalic ester as the sodio derivative. The free diester is liberated by neutralization and heated with hydrochloric acid to effect cyclization through the dienol and hydrolysis.

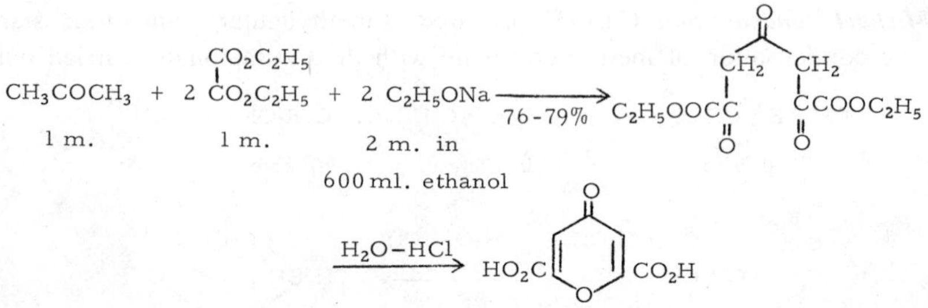

The condensation of ethanedithiol with ethylene dibromide to give *p*-dithiane is effected with a solution of sodium ethoxide at the reflux temperature.[30] The product,

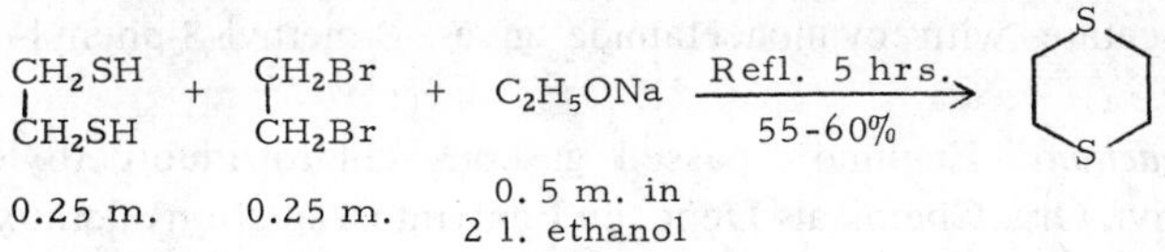

a solid, is steam distilled with one of the receiving systems shown in Fig. S-2; solid that threatens to plug the system can be dislodged by brief interruption of the cooling.

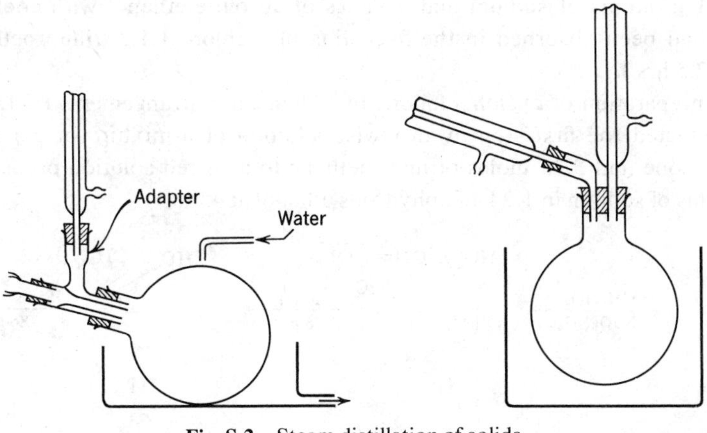
Adapter
Water

Fig. S-2 Steam distillation of solids

Ethanolic sodium ethoxide is employed also to effect condensation of guanidine (taken as the hydrochloride) with ethyl cyanoacetate to produce 2,4-diamino-6-hydroxypyrimidine.[31] With urea as reactant in place of guanidine, the product is

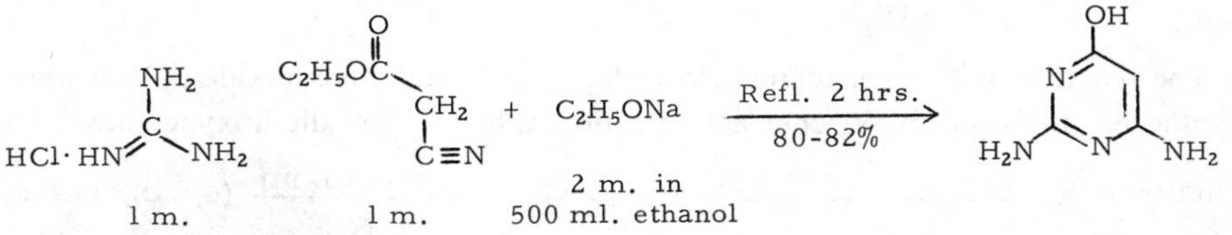

4-amino-2,6-dihydroxypyrimidine.[32] The condensation of thiourea with ethyl ethoxymethylene cyanoacetate involves ring closure in both possible directions to

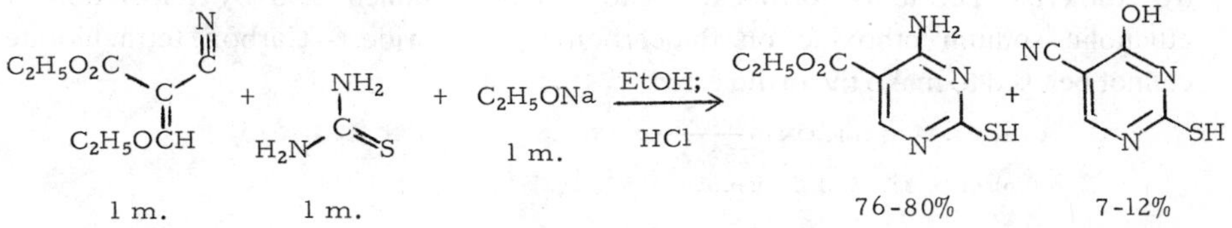

give (after acidification, etc.) 2-mercapto-4-amino-5-carboethoxypyrimidine and 2-mercapto-4-hydroxy-5-cyanopyrimidine.[33] Condensation of ethyl (1-phenylethyl-

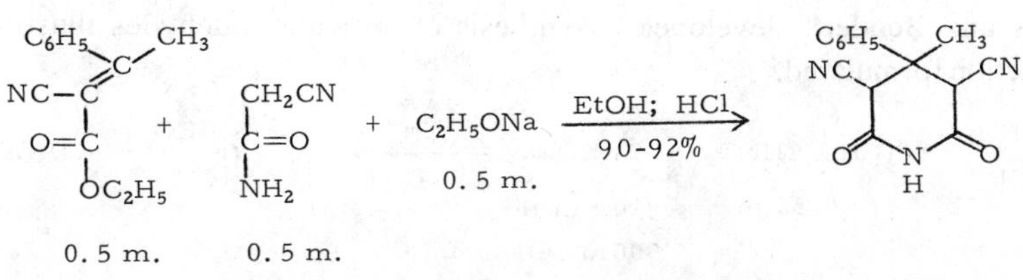

idene)cyanoacetate with cyanoacetamide gives β-methyl-β-phenyl-α,α′-dicyano-glutarimide.[34]

Other reactions. Englund[35] passed gaseous chlorotrifluoroethylene (Kinetic Chemicals Div., Org. Chemicals Dept., du Pont) into a solution of ethoxide prepared

$$CFCl{=}CF_2 \;+\; C_2H_5OH \xrightarrow[92\text{-}97\%]{C_2H_5ONa} HCFClCF_2OC_2H_5$$

from 0.11 g. atoms of sodium and 5 moles of absolute ethanol with cooling until 2 moles had been absorbed in the formation of 2-chloro-1,1,2-trifluoroethyl ethyl ether (2–2.5 hrs.).

In the preparation of cycloheptanone by Tiffeneau rearrangement, H. J. Dauben *et al.*[36] effected the first step by dropwise addition of a mixture of 2.5 moles of cyclohexanone and 3.25 moles of nitromethane to a stirred solution prepared from 2.5 g. atoms of sodium in 1.2 l. of anhydrous ethanol at 45°.

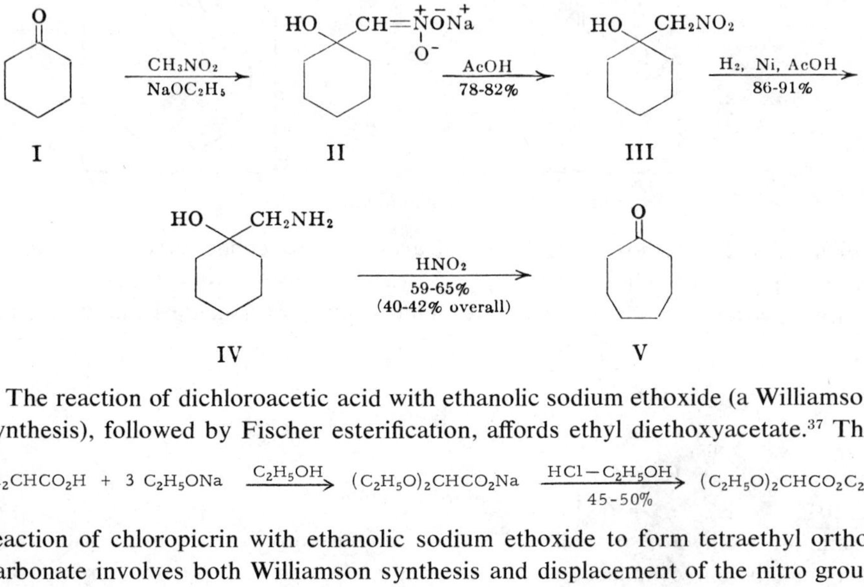
CH3NO2
NaOC2H5
HO
CH=NONa
O-
AcOH
78-82%
HO
CH2NO2
H2, Ni, AcOH
86-91%
I
II
III
HO
CH2NH2
HNO2
59-65%
(40-42% overall)
IV
V

The reaction of dichloroacetic acid with ethanolic sodium ethoxide (a Williamson synthesis), followed by Fischer esterification, affords ethyl diethoxyacetate.[37] The

$$Cl_2CHCO_2H \;+\; 3\;C_2H_5ONa \xrightarrow{C_2H_5OH} (C_2H_5O)_2CHCO_2Na \xrightarrow[45\text{-}50\%]{HCl-C_2H_5OH} (C_2H_5O)_2CHCO_2C_2H_5$$

reaction of chloropicrin with ethanolic sodium ethoxide to form tetraethyl orthocarbonate involves both Williamson synthesis and displacement of the nitro group by ethoxyl.[38] Tetraethyl orthocarbonate can be obtained also by the action of ethanolic sodium ethoxide on thiocarbonyl perchloride.[39] Carbon tetrachloride cannot be used to make the ortho ester.

$$\underset{0.61\text{ m.}}{CCl_3NO_2} \;+\; \underset{3\text{ m. in 1 l. ethanol}}{4\;C_2H_5ONa} \xrightarrow[46\text{-}49\%]{} C(OC_2H_5)_4 \;+\; 3\;NaCl \;+\; NaNO_2$$

$$Cl_3CSCl \xrightarrow{C_2H_5ONa} Cl_3CSOC_2H_5 \xrightarrow{3\;C_2H_5ONa} (C_2H_5O)_3CSOC_2H_5 \xrightarrow[71\%\text{ overall}]{} (C_2H_5O)_4C \;+\; S$$

Hass and Bender[40] developed a synthesis of aromatic aldehydes illustrated by the reaction formulated.

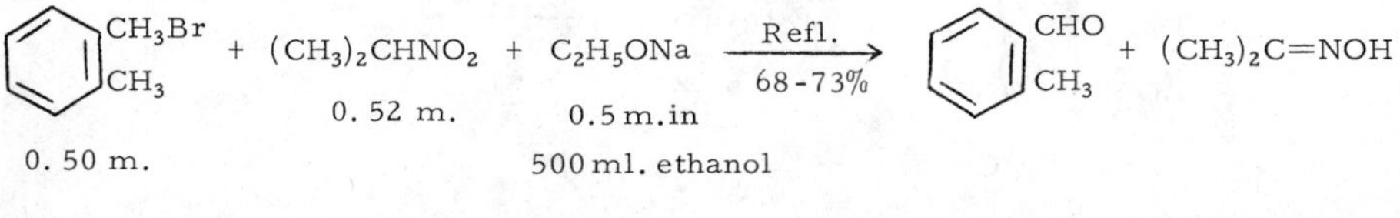
CH3Br
CH3
+ (CH3)2CHNO2 + C2H5ONa
Refl.
68-73%
CHO
CH3
+ (CH3)2C=NOH
0. 52 m.
0.5 m.in
0. 50 m.
500 ml. ethanol

It will be evident from this survey that the reagent introduced by Claisen has found a wide variety of uses.

[1]L. P. Kyrides, *Org. Syn., Coll. Vol.*, **3**, 368 (1955)
[2]R. B. Moffett, *ibid.*, **4**, 291 (1963)
[3]S. M. McElvain and D. H. Clemens, *ibid.*, **4**, 662 (1963)
[4]Contributed by Max Tishler, Merck Sharp and Dohme
[5]Contributed by Howard E. Zaugg, Abbott Laboratories
[6]C. S. Marvel and W. B. King, *Org. Syn., Coll. Vol.*, **1**, 252 (1941)
[7]E. B. Hershberg and L. F. Fieser, *ibid.*, **2**, 194 (1943)
[8]L. Claisen, *Ber.*, **23**, 977 (1890)
[9]S. Wawzonek and E. M. Smolin, *Org. Syn., Coll. Vol.*, **3**, 715 (1955)
[10]S. M. McElvain *et al.*, *Am. Soc.*, **51**, 3124 (1929); **55**, 1697 (1933)
[11]D. E. Floyd and S. E. Miller, *Org. Syn., Coll. Vol.*, **4**, 141 (1963)
[12]E. M. Bottorff and L. L. Moore, *Org. Syn.*, **44**, 67 (1964)
[13]L. Friedman and E. Kosower, *Org. Syn., Coll. Vol.*, **3**, 510 (1955)
[14]H. L. Holmes and L. W. Trevoy, *ibid.*, **3**, 300 (1955)
[15]C. Ainsworth, *ibid.*, **4**, 536 (1963)
[16]A. Magnani and S. M. McElvain, *ibid.*, **3**, 251 (1955)
[17]S. B. Coan and E. I. Becker, *ibid.*, **4**, 174 (1963)
[18]E. C. Horning and A. F. Finelli, *ibid.*, **4**, 461 (1963)
[19]R. P. Mariella and R. Raube, *ibid.*, **4**, 288 (1963)
[20]G. B. Heisig and F. H. Stodola, *ibid.*, **3**, 213 (1955)
[21]N. Weiner, *ibid.*, **2**, 279 (1943)
[22]R. Adams and R. M. Kamm, *ibid.*, **1**, 250 (1941)
[23]C. F. Allen and M. J. Kalm, *ibid.*, **4**, 616 (1963)
[24]J. Cason, *ibid.*, **4**, 630 (1963)
[25]H. T. Clarke and T. F. Murray, *ibid.*, **1**, 272 (1941)
[26]D. Vorländer and J. Erig, *Ann.*, **294**, 314 (1897)
[27]R. L. Shriner and H. R. Todd, *Org. Syn., Coll. Vol.*, **2**, 200 (1943)
[28]L. Claisen, *Ber.*, **24**, 111 (1891)
[29]E. R. Riegel and F. Zwilgmeyer, *Org. Syn., Coll. Vol.*, **2**, 126 (1943)
[30]R. G. Gillis and A. B. Lacey, *ibid.*, **4**, 396 (1963)
[31]J. A. VanAllan, *ibid.*, **4**, 245 (1963)
[32]W. R. Sherman and E. C. Taylor, Jr., *ibid.*, **4**, 247 (1963)
[33]T. L. V. Ulbricht, T. Okuda, and C. C. Price, *ibid.*, **4**, 566 (1963)
[34]S. M. McElvain and D. H. Clemens, *ibid.*, **4**, 662 (1963)
[35]B. Englund, *ibid.*, **4**, 184 (1963)
[36]H. J. Dauben, Jr., H. J. Ringold, R. H. Wade, D. L. Pearson, and A. G. Anderson, Jr., *ibid.*, **4**, 221 (1963)
[37]R. B. Moffett, *ibid.*, **4**, 427 (1963)
[38]J. D. Roberts and R. E. McMahon, *ibid.*, **4**, 457 (1963)
[39]H. Tieckelmann and H. W. Post, *J. Org.*, **13**, 265 (1948)
[40]H. B. Hass and M. L. Bender, *Org. Syn., Coll. Vol.*, **4**, 932 (1963)

Sodium fluoride, NaF. Mol. wt. 42.00. Suppliers: B, F, MCB.

In the preparation of L-tryptophane from 600 g. of casein suspended in 3.2 l. of water at 37° by digestion with commercial pancreatin (two 20-g. batches), a solution of 60 g. of Na_2CO_3 and 6 g. of NaF in 100 ml. of water is poured in prior to addition of the enzyme as a thin paste in water.[1] The mixture is covered with a layer of toluene, diluted to 6 l., shaken thoroughly, and let stand at 37° for 5 days; a second batch of enzyme is added and the mixture let incubate for 12 days more. "The sodium fluoride probably inhibits the action of the oxidases." Workup affords 4.0–4.1 g. of pure L-tryptophane and 17.0–18.2 g. of pure L-tyrosine; L-tyrosine is obtained more readily and as the primary product by hydrolysis of silk with hydrochloric acid, neutralization of the acid with sodium hydroxide, and acidification with acetic acid.

[1]G. J. Cox and H. King, *Org. Syn., Coll. Vol.*, **2**, 612 (1943)

Sodium hydrazide, H_2NNHNa.[1] Mol. wt. 54.04. The reagent is a crystalline yellow solid which explodes at temperatures above 70° or in the presence of oxygen. It is prepared by reaction of sodium amide with anhydrous hydrazine under nitrogen at 0°, for most purposes as a suspension in ether, diisopropyl ether, benzene, or a mixture of these solvents.[2]

$$NaNH_2 + H_2NNH_2 \longrightarrow H_2NNHNa + NH_3$$

Reduction.[3] In the presence of hydrazine the reagent reduces many unsaturated and aromatic hydrocarbons. Examples:

trans-Stilbene ⟶ Dibenzyl (92% yield)
$\Delta^{1,3,5,7}$-Cyclooctatetraene ⟶ $\Delta^{1,3,5}$-Cyclooctatriene (85%)
Acenaphthylene ⟶ Acenaphthene (81%)
Anthracene ⟶ 9,10-Dihydroanthracene (90%)

Reductive cleavage of olefins.[4] In refluxing ether under nitrogen, sodium hydrazide reductively cleaves olefins of the type (1). Yields of the nitrogen-free products (3) are as a rule over 80%. An isolated double bond is not attacked.

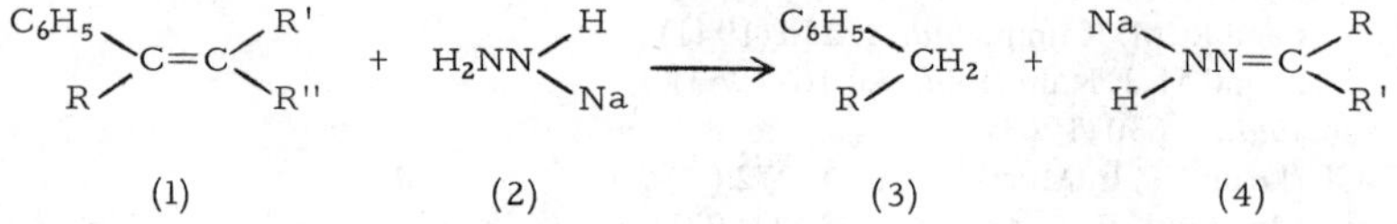

(1) (2) (3) (4)

Reaction with conjugated dienes.[5] Sodium hydrazide in the presence of hydrazine reacts with isoprene to form the azine (1) and the pyrazole (2).

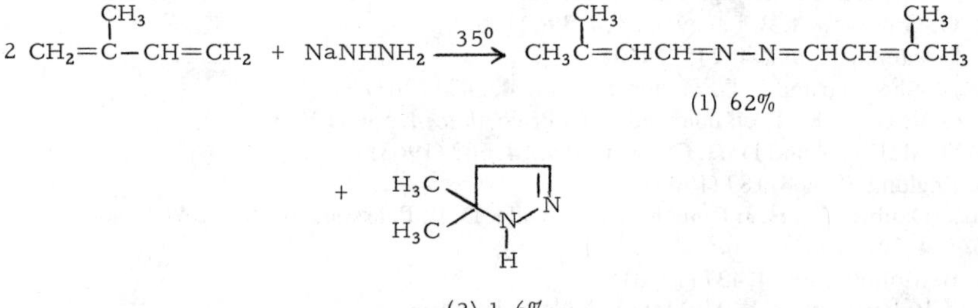

(1) 62%

(2) 1.6%

Nucleophilic attack of the pyridine ring.[6] Sodium hydrazide in hydrazine reacts with pyridine, isoquinoline, and 4-methylquinoline to give α-hydrazine derivatives in yields of 25–76%.

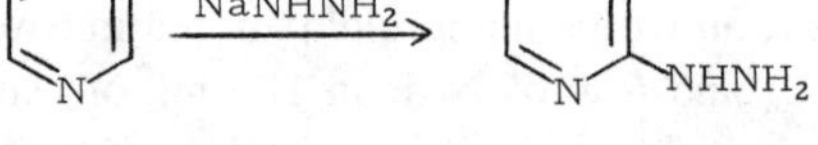

Reaction with nitriles.[7] The reagent reacts with a nitrile at 0° in diethyl or diisopropyl ether to give an adduct which on hydrolysis gives an amidrazone, a compound of a type comparatively inaccessible by other methods.

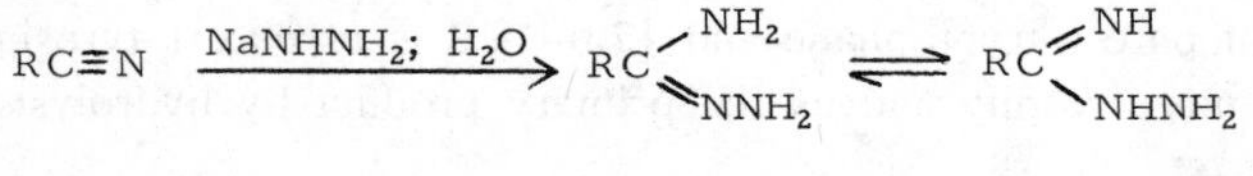

[1]Review: T. Kauffmann, *Angew. Chem., Internat. Ed.*, **3**, 342 (1964)
[2]T. Kauffmann, C. Kosel, and D. Wolf, *Ber.*, **95**, 1540 (1962)
[3]T. Kauffmann, C. Kosel, and W. Schoeneck, *Ber.*, **96**, 999 (1963)
[4]T. Kauffmann, H. Henkler, C. Kosel, J. Schulz, and R. Weber, *Angew. Chem., Internat. Ed.*, **1**, 456 (1962)
[5]T. Kauffmann and H. Müller, *Ber.*, **96**, 2206 (1963)
[6]T. Kauffmann, J. Hansen, C. Kosel, and W. Schoeneck, *Ann.*, **656**, 103 (1962)
[7]T. Kauffmann, S. Spaude, and D. Wolf, *Ber.*, **97**, 3436 (1964)

Sodium hydride, NaH. Mol. wt. 24.01. Supplier: Ventron Corp. The reagent is available as a granular solid and, since about 1958, as a dispersion of gray powder (av. particle size 25 microns) in an industrial white oil.[1] The dispersion contains 50–55% NaH by weight; the exact concentration is specified on each shipment. The oil is freely miscible with hydrocarbons and ethers and may be diluted with hexane, ether, THF, toluene, etc., without impairing the activity. Sodium hydride is insoluble in all inert organic solvents, as well as in liquid ammonia. It reacts violently with water. The reaction of the oil dispersion with water is easily controlled; it is vigorous but not hazardous as with the dry material, and ignition of hydrogen is infrequent.

Ester condensation. The self-condensation of ethyl acetate requires one equivalent of sodium hydride to form sodio ethyl acetoacetate and a second to react with

$$\underset{1\text{ m.}}{2\ CH_3CO_2C_2H_5} + \underset{\substack{1\text{ m.}\\ \text{Dispersion}}}{2\ NaH} \xrightarrow[88\%]{\text{Toluene, 1 hr. }78^0} CH_3\overset{\overset{\displaystyle O^-Na^+}{|}}{C}{=}CHCO_2C_2H_5 + C_2H_5ONa + H_2$$

the ethanol eliminated. Under the mild conditions possible with use of the oil dispersion and indicated in the formulation,[1] the sodium ethoxide formed does not appear to participate in the ester condensation. The yield is high, and the reaction period is short, for there is no necessity for removal of ethanol, as when sodium ethoxide is used as condensing agent. Furthermore, the course of the reaction can be followed by measuring the hydrogen evolved.

For a student preparation of ninhydrin starting with the condensation of dimethyl phthalate with ethyl acetate, one of us developed a technique for powdering sodium which seemed safe for use by beginners.[2] Whatever hazard remained was eliminated by Gruen and Norcross[3] by using sodium hydride dispersion in place of powdered sodium. Their yield is better than that obtained with powdered sodium (36–72%). Both bases promote ester interchange to give the methyl ester enolate.

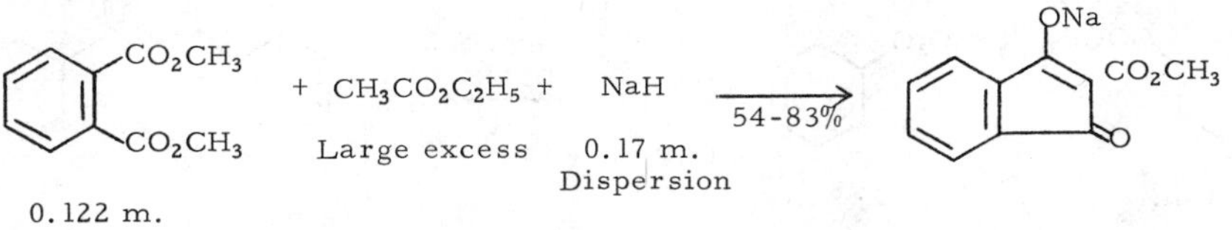

CO2CH3
CO2CH3
0.122 m.
+ CH3CO2C2H5 +
Large excess
NaH
0.17 m.
Dispersion
54-83%
ONa
CO2CH3
O

Roberts and McElvain[4] had failed to effect self-condensation of ethyl isovalerate with sodium ethoxide, even with provision for forcing the reaction by distillation

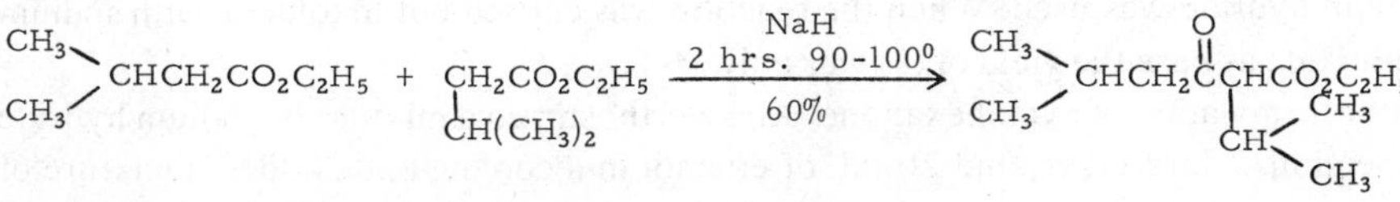

CH3
CH3
CHCH2CO2C2H5 + CH2CO2C2H5
CH(CH3)2
NaH
2 hrs. 90-100⁰
60%
CH3
CH3
CHCH2CCHCO2C2H5
O
CH
CH3
CH3

of ethanol, but Swamer and Hauser[5] obtained the hindered β-keto ester with use of sodium hydride.

Stobbe condensations. W. S. Johnson and Daub,[6] having found potassium *t*-butoxide superior to sodium ethoxide as a base for carrying out the Stobbe condensation, found an even better base in sodium hydride. When a mixture of benzophenone, diethyl succinate, and sodium hydride was stirred at room temperature,

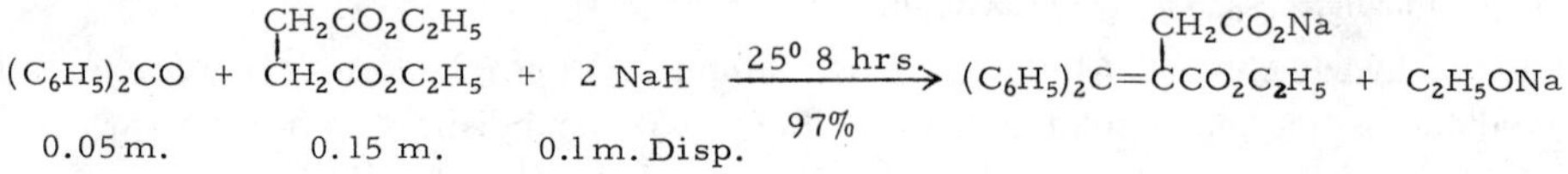

no appreciable reaction occurred, even at 87°, until a little ethanol (10 drops) was added. The reaction then proceeded readily at room temperature. After 5 hrs. ether was added to facilitate stirring, and hydrogen evolution ceased after about 8 hrs. Hinckley and co-workers[1] used the sodium hydride dispersion in hexane at 40°, found addition of ethanol unnecessary for initiation of the reaction, and obtained the pure condensation product in 95.7% yield in a reaction period of only 1 hr.

Dieckmann condensation. At the time of submission of an *Organic Syntheses* procedure for the cyclization of diethyl adipate to 2-carboethoxycyclopentanone, the reaction had been effected with sodium, sodium amide, and sodium ethoxide. Pinkney[7] selected sodium, conducted the reaction at 110–115° for 7 hrs., and

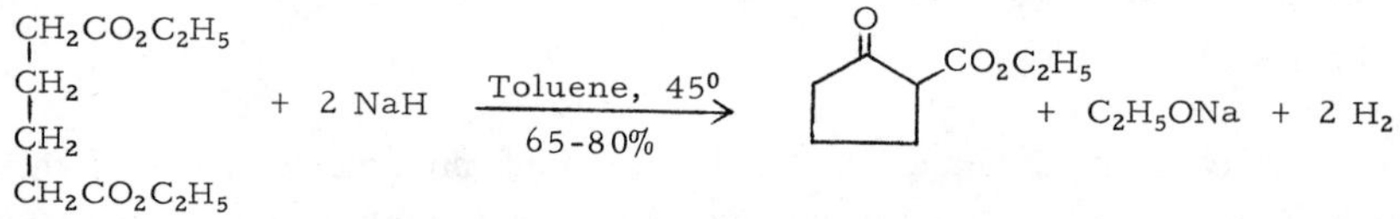

obtained the keto ester in yield of 74–81%. Later Hinckley and co-workers[1] added the ester to a slurry of sodium hydride dispersion in toluene at 45°, and the product was obtained in 65–80% yield after a reaction period of 1.5 hrs.

Acylation of ketones. Noting that on alkylation of malononitrile in ethanol or benzene with sodium ethoxide as base the yield is only 70% and formation of imide esters is an important side reaction, Bloomfield[8] was led to try the non-nucleophilic base sodium hydride in dimethyl sulfoxide, a relatively non-nucleophilic solvent capable of dissolving intermediate salts. With this combination he obtained dimethylmalononitrile in 60% yield. He then studied the acylation of the ketone (1) with the

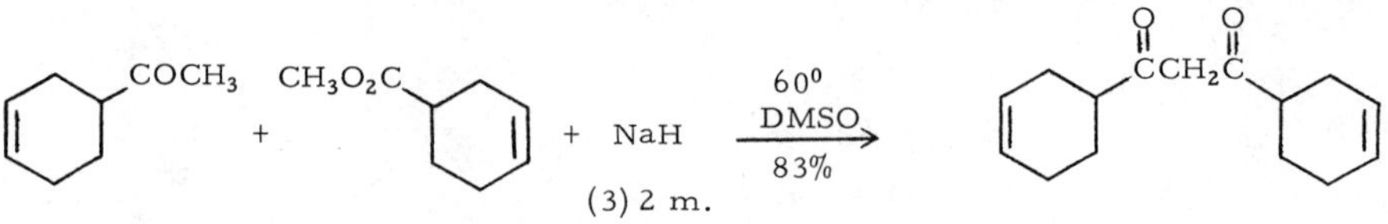

(1) 1 m. (2) 2 m. (4)

ester (2) by the same method and with the reagents taken in the ratio indicated; the diketone (4) was obtained in high yield. The yield was lower when only 1 mole of sodium hydride was used. When the reaction was carried out in toluene with sodium methoxide as base the yield of (4) was only 36%.

For formylation of cyclohexanone, Ainsworth[9] stirred a mixture of sodium hydride dispersion, 2 l. of ether, and 20 ml. of ethanol in a cooling bath, added a mixture of

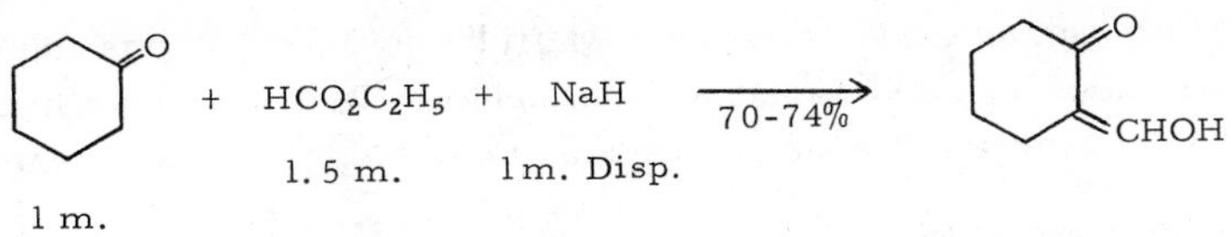
O
+ $HCO_2C_2H_5$ + NaH
70-74%
CHOH
1.5 m.
1 m. Disp.
1 m.

cyclohexanone and ethyl formate in 1 hr., stirred for 6 hrs., and let the mixture stand overnight. The yield was the same as when sodium was used as base (cut in 1-cm. cubes). For the formylation of cholestenone, Weisenborn, Remy, and Jacobs[10]

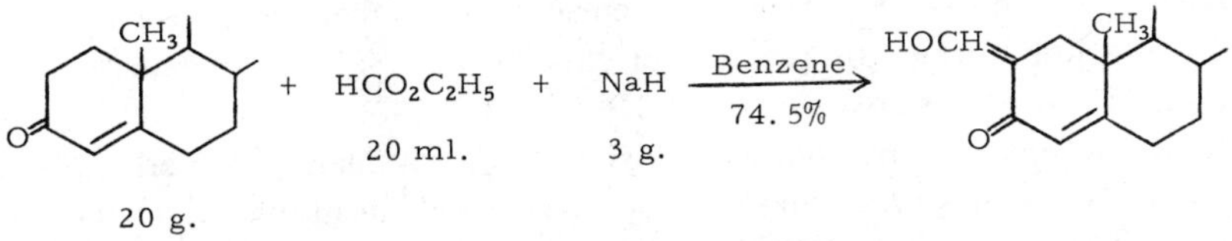
CH_3
+ $HCO_2C_2H_5$ + NaH
Benzene
74.5%
HOCH
CH_3
20 ml.
3 g.
20 g.

added solid sodium hydride to a solution of the ketone and ethyl formate in 250 ml. of benzene. Reaction commenced at once with separation of the enolate salt, and after two days the remaining NaH was decomposed by addition of 20 ml. of methanol in 100 ml. of ether and the salt was collected. The free enol was liberated by shaking a suspension of the salt with hydrochloric acid and eventually crystallized. Testosterone was formylated in the same way (83%).

Alkylation and nitration of malonic esters. Zaugg,[11] using sodium hydride for conversion of *n*-butylmalonic ester into its sodio derivative, found that the rate of alkylation of this substance with *n*-butyl bromide in dimethylformamide is 1,000 times the rate in benzene. Emmons and Freeman[12] developed a procedure for the nitration of active methylene compounds with acetone cyanohydrin nitrate by conversion to the sodio derivative with sodium hydride. The reaction product (3)

$$\underset{(1)}{CH_2(CO_2C_2H_5)_2} \xrightarrow{NaH} \underset{(2)}{Na^+\bar{C}H(CO_2C_2H_5)_2} \xrightarrow[45\%]{(CH_3)_2C(CN)ONO_2} \underset{(3)}{O_2NCH(CO_2C_2H_5)_2} + (CH_3)_2CO + NaCN$$

tends to abstract sodium ion from (2), but by taking sodio malonic ester in three-fold excess it was possible to realize a 45% yield.

Darzens condensation. Hinckley[1] found that the condensation of α-halo esters with ketones when effected with sodium hydride dispersion proceeds smoothly and without usual side reactions.

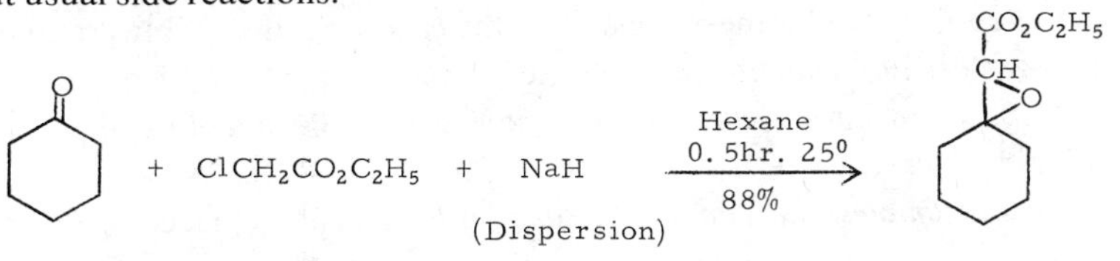
O
+ $ClCH_2CO_2C_2H_5$ + NaH
(Dispersion)
Hexane
0.5hr. 25°
88%
$CO_2C_2H_5$
CH
O

Intramolecular alkylation of a ketone. The key step in Whitlock's synthesis of twistane[13] was cyclization of the keto-tosylate with solid NaH in dimethylformamide.

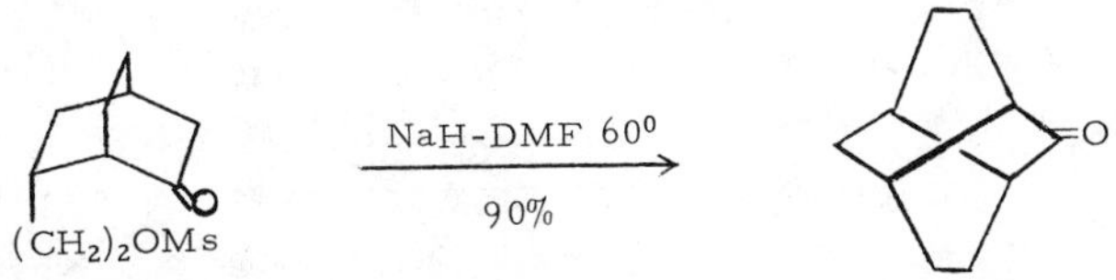
$(CH_2)_2OMs$
NaH-DMF 60°
90%
O

(1) d- or l- (2) d- or l-

Dehydrohalogenation. Newman and Merrill[14] developed a procedure for dehydrohalogenation of ethyl cinnamate dibromide by adding sodium hydride to a solution of the ester in benzene and heating under reflux with occasional addition

$$\underset{0.65\text{ m.}\ \ 300\text{ ml. benzene}}{C_6H_5CHBrCHBrCO_2CO_2C_2H_5} + \underset{1.33\text{ m.}}{2\ NaH} + C_2H_5OH \xrightarrow[51\%]{\text{Refl. 3 hrs.; saponify; crystallize twice}} C_6H_5C{\equiv}CCO_2H$$

of a 2-ml. portion of absolute ethanol. The fact that sodium hydride alone does not react with the dibromide in refluxing benzene shows that the effective reagent is sodium ethoxide, formed by reaction of ethanol with NaH and regenerated by the reaction: $C_2H_5ONa + HBr \longrightarrow C_2H_5OH + NaBr$. Note that ethyl cinnamate dibromide is converted by sodium ethoxide in ethanol mainly into ethyl β-ethoxycinnamate. Newman's procedure was used for the dehydrobromination of (1), which had proved inert to boiling 2,4,6-collidine.[15] Sodium hydride in boiling benzene in a

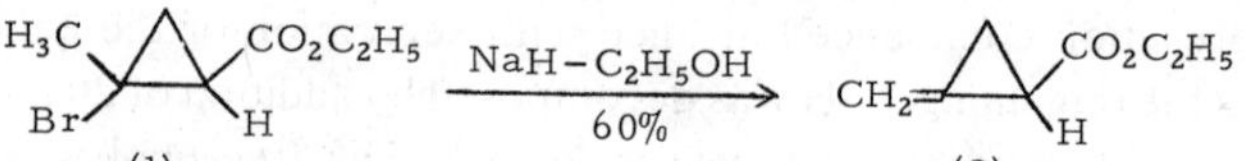

(1) (2)

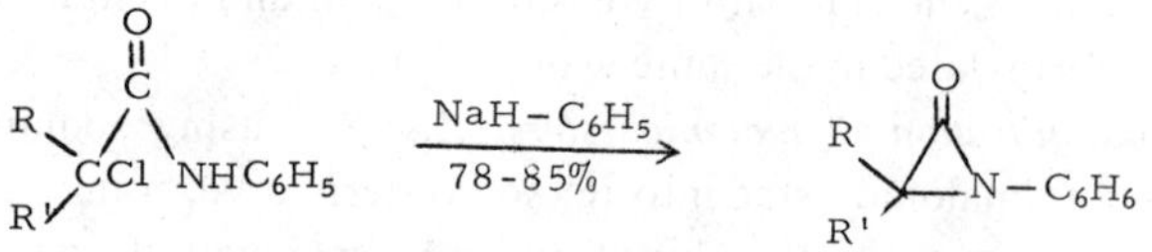

(3) (4)

nitrogen atmosphere converts α-halogenocarboxylic amides (3) into aziridinones (4).[16]

Synthesis of cyclopropanes. McCoy[17] found that an α-chloro (or bromo) ester condenses with an α,β-unsaturated ester in benzene or an ether under the influence of sodium hydride dispersed in oil to give a cyclopropanedicarboxylic ester (solid NaH was ineffective). In the general procedure a mixture of 1 mole of each component and 1 mole of sodium hydride dispersion is stirred with benzene or toluene

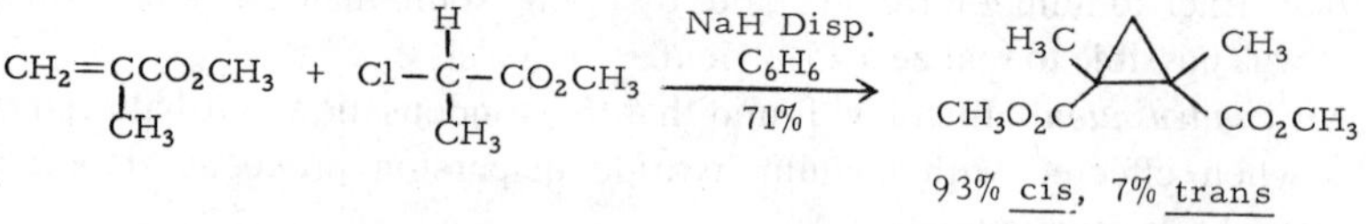

or an ether at 20–30°. Under these conditions the *cis* isomer invariably predominates. The more stable *trans* isomers become the predominant products when a solvent of high dielectric constant is used, for example dimethylformamide or hexamethylphosphoramide–benzene.

Alkylation of amines and amides. Major and Peterson[18] replaced the two amino

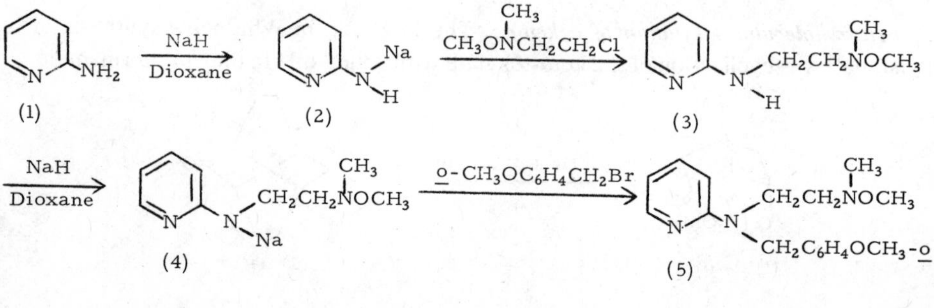

hydrogens of 2-aminopyridine with two different groups, as shown in the formulation. The amine, (1) or (3), is converted into the sodio derivative, which is treated with a substituted alkyl halide.

Fones[19] alkylated a number of N-substituted amides with sodium hydride as base and found it more convenient than sodium. Marvel and Moyer[20] used this method

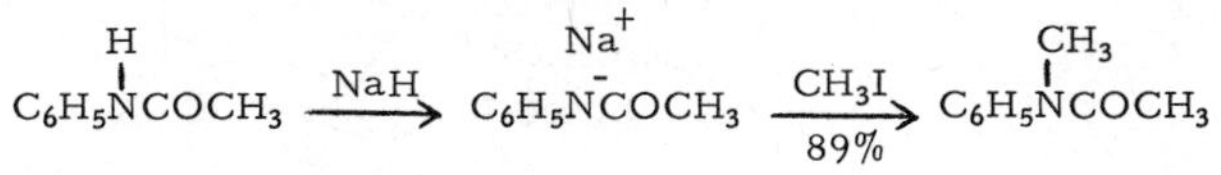

for alkylation of ϵ-caprolactam. A mixture of the lactam and sodium hydride in toluene was stirred and refluxed under nitrogen for 10 hrs., a solution of *n*-butyl bromide in xylene was added, and the mixture refluxed for 4 hrs.

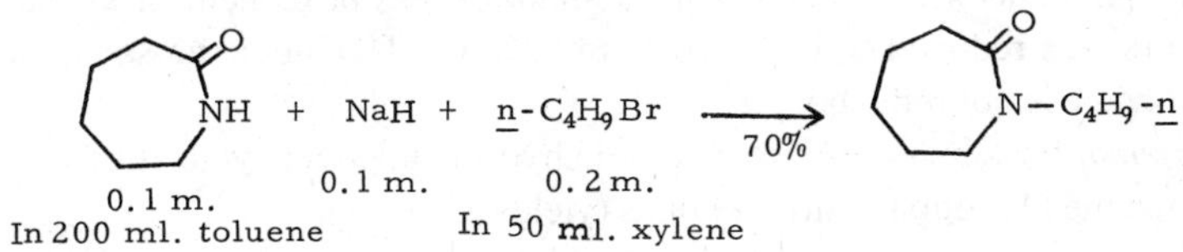

Benzyl tosylates. Usual preparative methods are unsatisfactory as applied to benzyl tosylates because these esters are highly reactive and unstable. A method introduced by Kochi and Hammond[21] consists in first forming a suspension of the

$$\underset{0.047\text{ m.}}{C_6H_5CH_2OH} + \underset{0.10\text{ m.}}{NaH} \xrightarrow[\text{15 hrs.}]{\text{Refl. in ether}} C_6H_5CH_2ONa \xrightarrow[80\%]{0.097\text{ m. TsCl},\ -10^0} C_6H_5CH_2OTs$$

sodium benzylate by stirring an ethereal solution of the alcohol with sodium hydride with exclusion of moisture and air; use of glass beads in conjunction with a paddle-type stirrer was found helpful in crushing the sodium hydride *in situ*. The suspension was cooled to $-10°$, an ethereal solution of tosyl chloride added, and the mixture stirred at $-10°$ for 2 hrs. and then at 25° for 1 hr. When the solution was filtered and cooled in a dry ice bath the tosylate separated in fine white needles.

The hydroxyl group of the benzyl alcohol (1) is strongly hydrogen bonded to the nitro group and Tarbell *et al.*[22] attempted tosylation in pyridine without success,

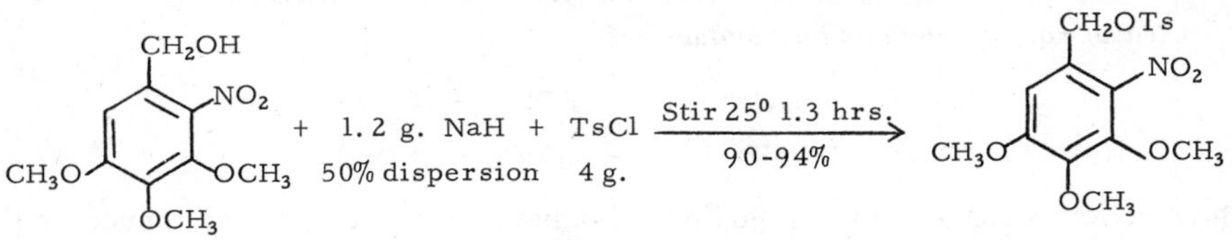

(1) 6 g. in 140 ml. C_6H_6 (2)

but the sodium hydride method proved highly satisfactory. By using sodium hydride dispersion in benzene and adding tosyl chloride at the start they were able to effect rapid and efficient conversion at room temperature.

Benzylation. Tate and Bishop[23] found sodium hydride an excellent base for the benzylation of carbohydrates with benzyl chloride. For example, methyl α-D-glucopyranoside was converted into the tetrabenzyl ether in high yield and without by-products whereas in the usual procedure, with powdered KOH as base, the yield was only 33% and dibenzyl ether was a major product.

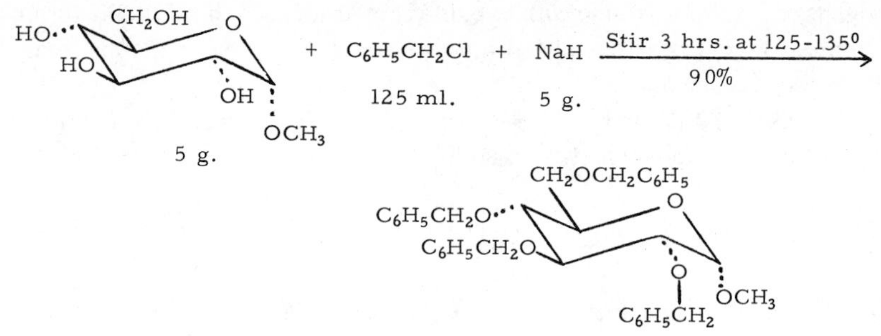
CH₂OH
HO
HO
O
OH
OCH₃
5 g.
+ C₆H₅CH₂Cl + NaH
125 ml.
5 g.
Stir 3 hrs. at 125-135°
90%
CH₂OCH₂C₆H₅
C₆H₅CH₂O
C₆H₅CH₂O
C₆H₅CH₂
OCH₃

Reduction. Ketones which have no α-hydrogen are reduced by sodium hydride. Thus benzophenone when heated with a slight excess of reagent in xylene at 145° for six hours was reduced to benzhydrol in 83% yield.[5] Under the same conditions methyl benzoate is not reduced.

Cyclization. Folkers *et al.*[24] used sodium hydride to effect cyclization of quinones of the coenzyme Q group to chromenones (yields 45–90%).

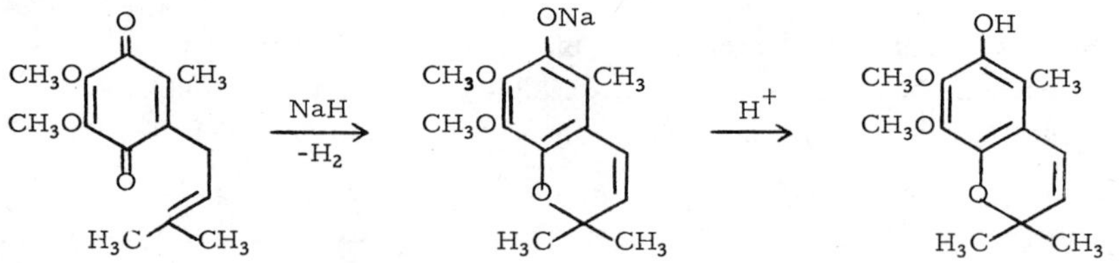
CH₃O
CH₃O
CH₃
H₃C
CH₃
NaH
-H₂
ONa
H⁺
OH

Sodium phenolates. A suspension of sodium hydride in toluene is used to advantage in preparation of sodium salts of phenols suspended in toluene.[25]

Aroylation of β-diketones with esters. Sodium hydride in refluxing 1,2-dimethoxyethane is superior to potassium amide in liquid ammonia for the terminal aroylation

$$C_6H_5COCH_2COCH_3 + C_6H_5CO_2C_2H_5 \xrightarrow[87\%]{NaH} C_6H_5COCH_2COCH_2COC_6H_5$$

of β-diketones with esters to produce triketones.[26] The most convenient procedure involves addition of the β-diketone (1 equiv.) and ester (1.2 equiv.) to the reagent (4 equiv.). In the example cited the yield obtained with KNH_2 was 62%.

Cleavage of formates and formamides:[27]

$$ROCHO + NaH \longrightarrow RONa + CO + H_2$$

$$R_2NCHO + NaH \longrightarrow R_2NNa + CO + H_2$$

The reaction is carried out in refluxing 1,2-dimethoxyethane and is continued until gas evolution stops (20 min.). Yields are 30–80%.

[1] A. A. Hinckley, "Sodium Hydride Dispersions," Metal Hydrides, Inc., July, 1964
[2] *Org. Expts.*, 141
[3] H. Gruen and B. E. Norcross, *J. Chem. Ed.*, **42**, 268 (1965)
[4] D. C. Roberts and S. M. McElvain, *Am. Soc.*, **59**, 2007 (1937)
[5] F. W. Swamer and C. R. Hauser, *Am. Soc.*, **68**, 2647 (1946)
[6] G. H. Daub and W. S. Johnson, *Am. Soc.*, **70**, 418 (1948); **72**, 501 (1950)
[7] P. S. Pinkney, *Org. Syn.*, *Coll. Vol.*, **2**, 116 (1943)
[8] J. J. Bloomfield, *J. Org.*, **26**, 4112 (1961); **27**, 2742 (1962)
[9] C. Ainsworth, *Org. Syn.*, *Coll. Vol.*, **4**, 536 (1963)
[10] F. L. Weisenborn, D. C. Remy, and T. L. Jacobs, *Am. Soc.*, **76**, 552 (1954)

[11]H. E. Zaugg *et al.*, *Am. Soc.*, **82**, 2895, 2903 (1960); *J. Org.*, **26**, 644 (1960)
[12]W. D. Emmons and J. P. Freeman, *Am. Soc.*, **77**, 4391 (1955)
[13]H. W. Whitlock, Jr., *Am. Soc.*, **84**, 3412 (1962)
[14]M. S. Newman and S. H. Merrill, *Am. Soc.*, **77**, 5549 (1955)
[15]J. H. Carbon, W. B. Martin, and L. R. Swett, *Am. Soc.*, **80**, 1002 (1958)
[16]S. Sarel and H. Leader, *Am. Soc.*, **82**, 4752 (1960)
[17]L. L. McCoy, *Am. Soc.*, **80**, 6568 (1958); **84**, 2246 (1962); *J. Org.*, **25**, 2078 (1960)
[18]R. T. Major and L. H. Peterson, *J. Org.*, **22**, 579 (1957)
[19]W. S. Fones, *J. Org.*, **14**, 1099 (1949)
[20]C. S. Marvel and W. W. Moyer, Jr., *J. Org.*, **22**, 1065 (1957)
[21]J. K. Kochi and G. S. Hammond, *Am. Soc.*, **75**, 3443 (1953)
[22]K. I. H. Williams, S. E. Cremer, F. W. Kent, E. J. Sehm, and D. S. Tarbell, *Am. Soc.*, **82**, 3982 (1960)
[23]M. E. Tate and C. T. Bishop, *Can. J. Chem.*, **41**, 1801 (1963)
[24]B. O. Linn, C. H. Shunk, E. L. Wong, and K. Folkers, *Am. Soc.*, **85**, 239 (1963); A. F. Wagner, P. E. Wittreich, B. Arison, N. R. Trenner, and K. Folkers, *ibid.*, **85**, 1178 (1965)
[25]F. M. Elkobaisi and W. J. Hickinbottom, *J. Chem. Soc.*, 2431 (1958)
[26]M. L. Miles, T. M. Harris, and C. R. Hauser, *J. Org.*, **30**, 1007 (1965)
[27]J. C. Powers, R. Seidner, and T. G. Parsons, *Tetrahedron Letters*, 1713 (1965)

Sodium hydrosulfite (Sodium dithionite), $Na_2S_2O_4$; *see also* Fieser's solution. Mol. wt. 174.11, solubility in 100 g. water: 22 g. at 20°. Suppliers: Mallinckrodt; MCB.

This is a powerful reagent for reduction in alkaline solution. At a pH lower than 7, hydrosulfite decomposes with liberation of sulfur. If used in an unbuffered aqueous solution initially neutral, say for the reduction of an azo compound, the solution becomes progressively acidic because of the formation of sodium bisulfite.

$$\text{ArN{=}NAr} + 2\,Na_2S_2O_4 + 2\,H_2O \longrightarrow 2\,\text{ArNH}_2 + 4\,NaHSO_3$$

Quinones. Sodium hydrosulfite reduces most quinones to their hydroquinones; indeed a quinone of the benzene or naphthalene series can be identified by discharge of color with hydrosulfite to a colorless product and restoration of the color by oxidation. Anthraquinones can be recognized by reduction with alkaline hydrosulfite to a blood red vat containing the anthrahydroquinone anion; on shaking with air the red vat color is discharged. Vatting sometimes affords a useful means of purification or separation. Naphthacenequinone and higher linear benzologs are exceptional in that they are not reduced by alkaline hydrosulfite.[1] *lin*-Naphthindazole-4,9-quinone is exceptional in that it is colorless, but it gives a characteristic red vat with alkaline hydrosulfite.[2]

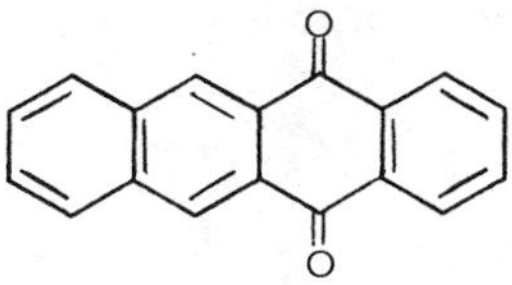

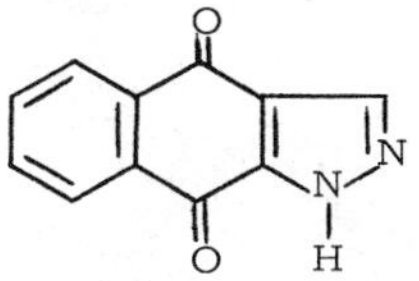
O
N
N
H
O

2-Methyl-1,4-naphthohydroquinone is easily prepared[3] by dissolving 2 g. of the quinone in 20 ml. of ether by warming and shaking the solution with a fresh solution of 4 g. of sodium hydrosulfite in 30 ml. of water. The solution passes through a brown phase (quinhydrone) and becomes colorless to pale yellow in a few minutes. After drawing off the aqueous layer the ether is shaken with saturated salt solution, filtered through sodium sulfate, and evaporated to a small volume. On addition of petroleum ether the hydroquinone separates as a white or grayish powder (1.9 g.).

***Anthraquinone* ⟶ *anthrone*.**[4] A mixture of 5 g. of anthraquinone, 6 g. of sodium hydroxide, and 15 g. of sodium hydrosulfite and 130 ml. of water is heated and swirled to form the red vat and then refluxed for 45 min. The red color gives way to the yellow color of anthranol anion, and, as the alkali is neutralized by the $NaHSO_3$ formed, anthranol gives way to the more stable anthrone, which is collected and washed.

***—N=N— ⟶ —NH₂ + H₂N—*.** A convenient route from a phenol to a quinone involves coupling with diazotized sulfanilic acid, reduction of the azo compound with aqueous sodium hydrosulfite, purification through the aminophenol hydrochloride, and oxidation with ferric chloride, as in the preparation of 1,2-naphthoquinone.[5]

***—NO₂ ⟶ —NH₂*.** An example is the reduction of Martius Yellow as the ammonium salt to 2,4-diamino-1-naphthol.[6]

***C-Nitroso compounds*.** As an alternative to diazo coupling and reduction, a phenol is often converted into the *o*- or *p*-nitroso derivative and this is reduced, as exemplified by Sherman and Taylor's synthesis of diaminouracil.[7] The rose red

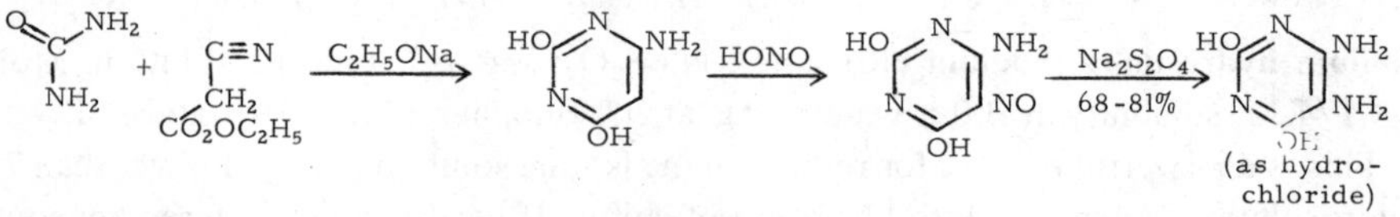
NH2
NH2
C≡N
CH2
CO2C2H5
C2H5ONa
HO
N
NH2
OH
HONO
NO
Na2S2O4
68-81%
(as hydro-
chloride)

nitroso compound was stirred with water and heated, and hydrosulfite was added to discharge the red color and give a tan suspension of diaminouracil bisulfite salt. This was then transformed into the hydrochloride.

***N-Nitroso compounds*.** Overberger and co-workers[8] found that reduction of N-nitrosodibenzylamine with sodium hydrosulfite is attended with liberation of nitrogen

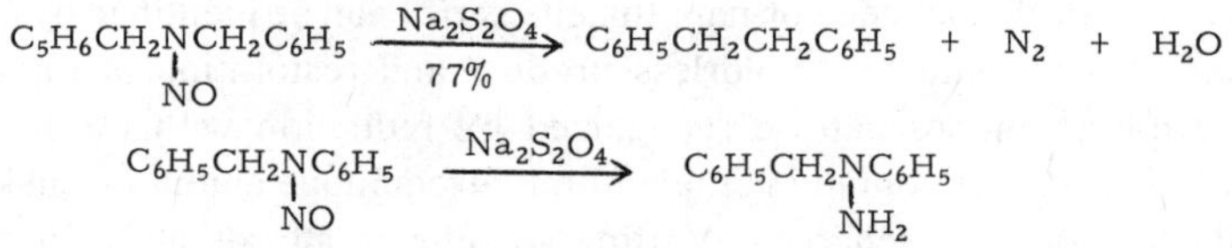
C5H6CH2NCH2C6H5
NO
Na2S2O4
77%
C6H5CH2CH2C6H5 + N2 + H2O
C6H5CH2NC6H5
NO
Na2S2O4
C6H5CH2NC6H5
NH2

and rearrangement to dibenzyl. Mixed benzylaryl or diaryl-N-nitrosoamines are reduced to hydrazines.

***Dihydropyridines*.** Sodium hydrosulfite has been used extensively for the reduction of pyridinium salts to 1,4-dihydropyridines.[9] The reaction is important because the products are models of reduced dihydrophosphorpyridine nucleotide (DPNH).

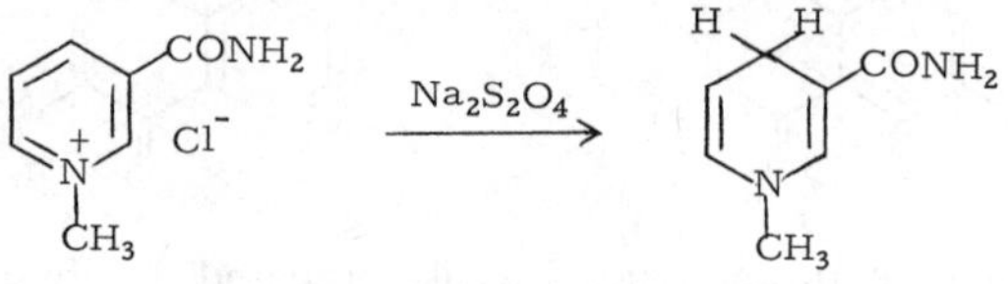
CONH2
Cl⁻
N
CH3
Na2S2O4
H
H
CONH2
N
CH3

[1] L. F. Fieser, *Am. Soc.*, **53**, 2329 (1931)
[2] L. F. Fieser and Mary A. Peters, *Am. Soc.*, **53**, 4080 (1931)
[3] *Org. Expts.*, 250–251
[4] *Org. Expts.*, 199
[5] *Ibid.*, 249–250; L. F. Fieser, *Org. Syn., Coll. Vol.*, **2**, 36, 430 (1943)
[6] L. F. Fieser and M. Fieser, *Am. Soc.*, **56**, 1565 (1934); *Org. Expts.*, 280
[7] W. R. Sherman and E. C. Taylor, Jr., *Org. Syn., Coll. Vol.*, **4**, 247 (1963)

[8]C. G. Overberger, J. G. Lombardino, and R. G. Hiskey, *Am. Soc.*, **80**, 3009 (1958)
[9]P. Karrer *et al.*, *Helv.*, **19**, 1028 (1936); **20**, 55, 72, 418, 622 (1937); **21**, 223, 1174 (1938); **29**, 1152 (1946); **32**, 960 (1949); D. Mauzerall and F. H. Westheimer, *Am. Soc.*, **77**, 2261 (1955); R. F. Hulton and F. H. Westheimer, *Tetrahedron*, **3**, 73 (1958)

Sodium hydroxide solutions. Fisher offers standardized solutions of various concentrations, for example 5 *N*, 2.5 *N* (10%), 1 *N*, 0.1 *N* and also concentrates.

Sodium hypobromite, NaOBr.

Preparation of aqueous solution: from bromine and sodium hydroxide;[1] by adding sodium bromide to a solution of sodium hypochlorite.[2]

Hofmann reaction. In the preparation of 3-aminopyridine from nicotinamide, bromine is added to an aqueous solution of sodium hydroxide with stirring in a

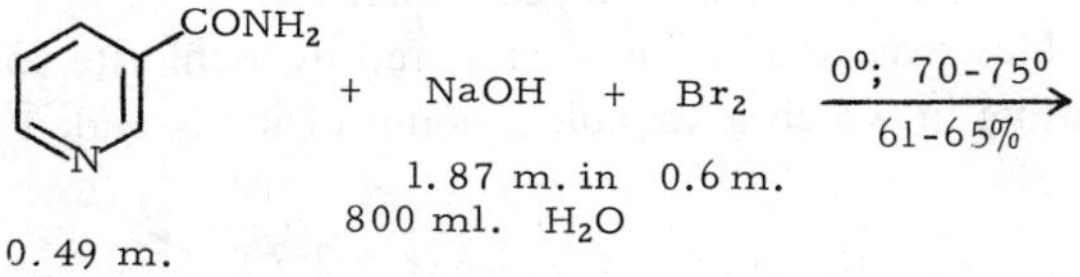
CONH2
N
0.49 m.
+ NaOH + Br2
1.87 m. in 0.6 m.
800 ml. H2O
0°; 70-75°
61-65%
NH2
N

salt-ice bath.[1] When the temperature reaches 0° the amide is added all at once. When the amide is all dissolved (15 min.) the temperature is raised to 70–75° for 45 min. The solution is cooled, saturated with sodium chloride, and extracted with ether in a continuous extractor (15–20 hrs.). Workup then affords a dark red product from which the color is discharged by heating a solution in ligroin with Norit and sodium hydrosulfite and further crystallization. Pure material is white, m.p. 63–64°.

Oxidative cleavage of ketones. Levine and Stephens[3] showed that hypohalite oxidation of ketones to acids is not limited to methyl ketones. Thus sodium hypobromite oxidizes propiophenone to benzoic acid in 96% yield, probably as follows:

$$C_6H_5COCH_2CH_3 \xrightarrow{2\,NaOBr} C_6H_5COCBr_2CH_3 \xrightarrow{NaOH} C_6H_5COCOCH_3 \xrightarrow[96\%]{NaOBr\ +\ NaOH}$$

$$C_6H_5CO_2Na + NaBr + H_2O + CH_3CO_2Na$$

The reagent oxidizes cyclohexanone to adipic acid in 82% yield and cyclopentanone to glutaric acid in fair yield.[4]

Other reactions. Kergomard[2] used reagent prepared from sodium hyprochlorite for generation of HOBr for additions, for example to cyclohexene. A solution of 110 g. of sodium bromide in 250 ml. of water was treated at 15° with 500 ml. of

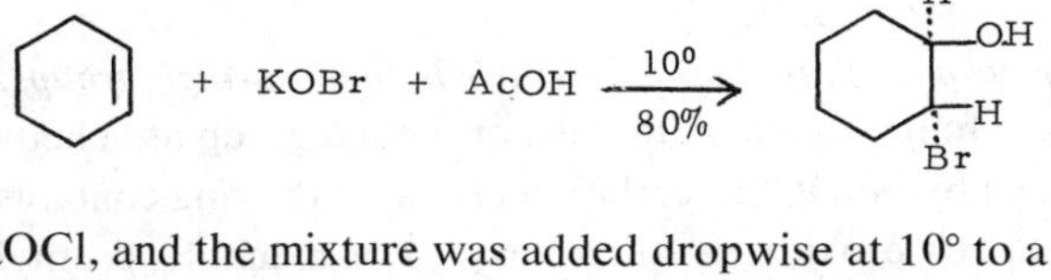
+ KOBr + AcOH
10°
80%
H
OH
H
Br

commercial NaOCl, and the mixture was added dropwise at 10° to a stirred mixture of 100 ml. of cyclohexene, 100 ml. of acetic acid, and 750 ml. of water. In a similar manner phthalimide afforded N-bromophthalimide in 70% yield.

Sodium hypobromite reacts with a terminal acetylene to give the 1-bromoalkyne.[5] A solution of the reagent was prepared from aqueous alkali and bromine, phenylacetylene (a liquid) was added, and the mixture was shaken mechanically and vigorously for 60 hrs. at room temperature.

$$C_6H_5C{\equiv}CH \xrightarrow[73\text{-}83\%]{NaOBr} C_6H_5C{\equiv}CBr$$

[1]C. F. H. Allen and C. N. Wolf, *Org. Syn., Coll. Vol.*, **4**, 45 (1963)
[2]A. Kergomard, *Bull. Soc.*, 2360 (1961)
[3]R. Levine and J. R. Stephens, *Am. Soc.*, **72**, 1642 (1950)
[4]M. W. Farrar, *J. Org.*, **22**, 1708 (1957)
[5]S. I. Miller, G. R. Ziegler, and R. Wieleseck, *Org. Syn.*, **45**, 86 (1965)

Sodium hypochlorite solution, NaOCl. Supplier: MCB (5.0% available chlorine). Chlorox, sold as a household bleaching agent, is satisfactory; contains 5.25% of NaOCl.

Preparation. From chlorine and sodium hydroxide[1] (see also references 5 and 6); from calcium hypochlorite (MCB) and sodium carbonate.[1]

Haloform oxidation.[2] Newman and Holmes[1] prepared hypochlorite solution by the method of E. C. Sterling in which a suitable amount of ice is added to a cold

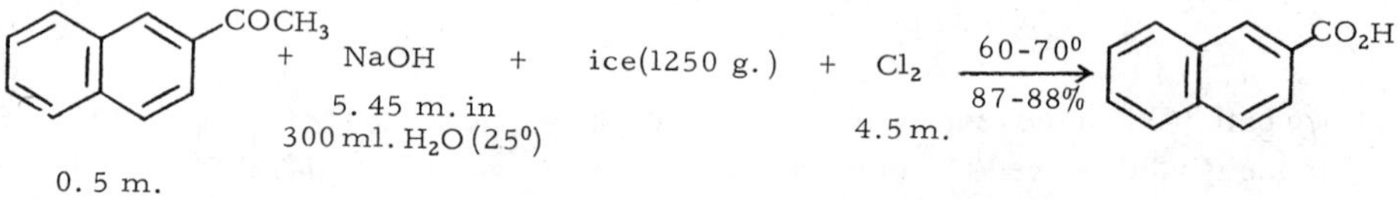
COCH3
0. 5 m.
NaOH
5. 45 m. in
300 ml. H2O (25°)
ice(1250 g.)
Cl2
4.5 m.
60-70°
87-88%
CO2H

solution of alkali, and chlorine is passed in without external cooling to the proper gain in weight, at which point a solution at 0° is obtained. Newman and Holmes fitted the flask with a thermometer and an efficient stirrer, warmed the solution to 55°, added methyl β-naphthyl ketone, and maintained a temperature of 60–70° by cooling. About 30 min. after the exothermal reaction was over, bisulfite solution was added to destroy excess hypochlorite and the mixture was cooled and acidified. Crystallization from 95% ethanol gave pure β-naphthoic acid, m.p. 184–185°.

The amount of hypochlorite used in this reaction is three times the theory, but no evidence is presented that this huge excess is needed. For oxidation of dimedone to β,β-dimethylglutaric acid, Smith and McLeod[3] used only 1.5 times the theoretical

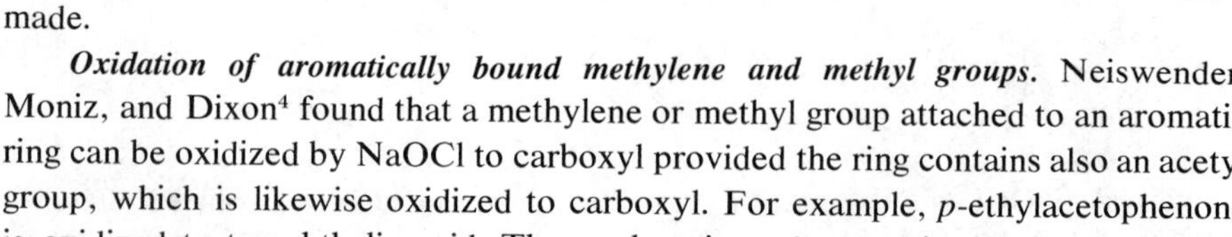
OH
H3C
CH3
0. 5 m.
in aq. KOH
NaOH
5. 45 m. in
300 ml. water
ice(1250 g.)
Cl2
2. 27 m.
35-40°
81-91%
(CH3)2C
CH2CO2H
CH2CO2H
+ CHCl3

amount of hypochlorite, and it seems possible that a further reduction could be made.

Oxidation of aromatically bound methylene and methyl groups. Neiswender, Moniz, and Dixon[4] found that a methylene or methyl group attached to an aromatic ring can be oxidized by NaOCl to carboxyl provided the ring contains also an acetyl group, which is likewise oxidized to carboxyl. For example, *p*-ethylacetophenone is oxidized to terephthalic acid. The explanation advanced is that in the alkaline medium the ketone (1) forms a resonance-stabilized anion (2), which reacts further as shown in the formulation. *p*-Methylacetophenone afforded terephthalic acid in 47% yield. In another example 2-acetyl-9,10-dihydrophenanthrene on oxidation with sodium hypochlorite at pH 12–13 afforded diphenyl-2,2′,4-tricarboxylic acid. Phenanthrenequinone-2-carboxylic acid is postulated as an intermediate and,

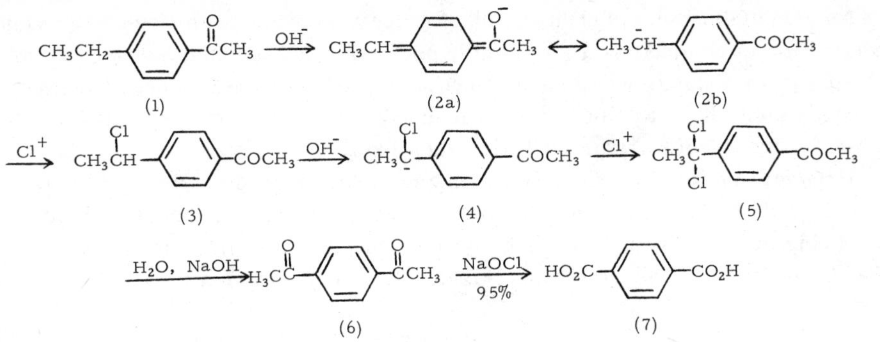

indeed, oxidation of phenanthrenequinone with sodium hypochlorite afforded diphenic acid in 94% yield.

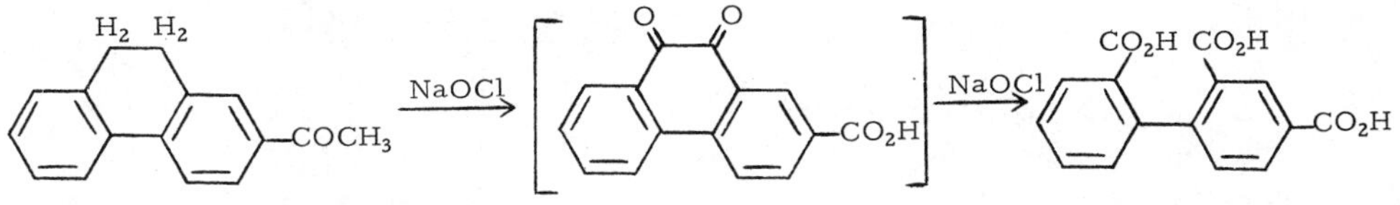

Other oxidations. The reagent oxidizes ammonia to hydrazine,[5] *o*-nitroaniline to benzofurazane oxide,[6] and iodobenzene dichloride to iodoxybenzene.[7] In an

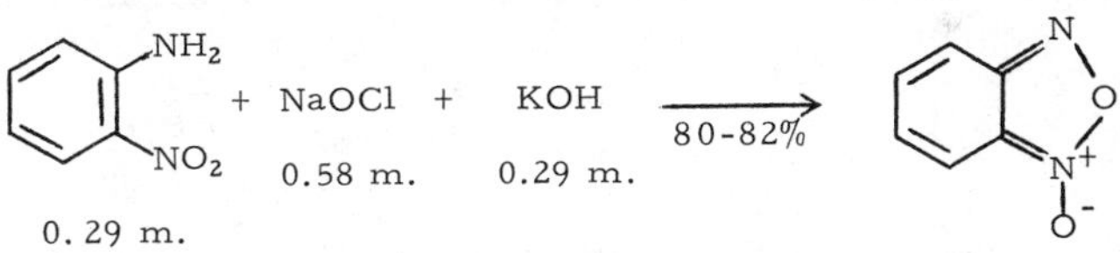

alkaline medium sodium hypochlorite oxidizes methylenediamine sulfate to diaziridine in yield of 26%.[8]

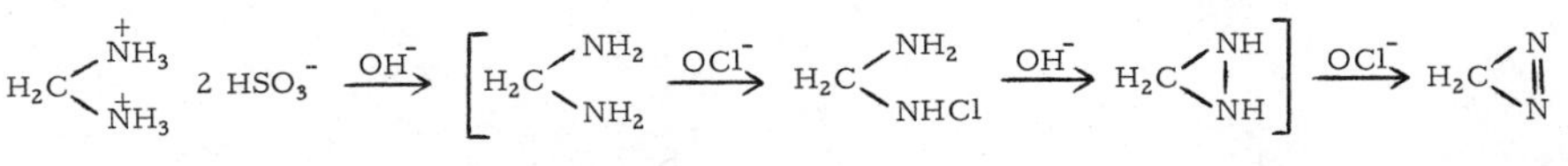

Epoxidation of α,β-unsaturated compounds. Marmor[9] found that on addition of Chlorox to a solution of benzalacetophenone in pyridine the yellow color faded almost immediately and addition of water precipitated the epoxide in high yield.

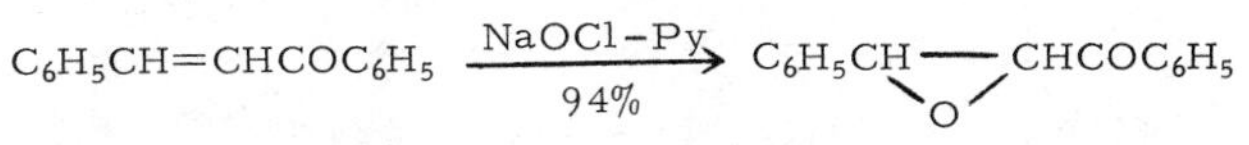

1,4-Naphthoquinone on reaction with sodium hypochlorite in dioxane affords the 2,3-epoxide in 71.5% yield.

Oxidative decarboxylation of amino acids. A recent report by a Merck group[10] illustrates a reaction which had previously been applied to natural amino acids.[11]

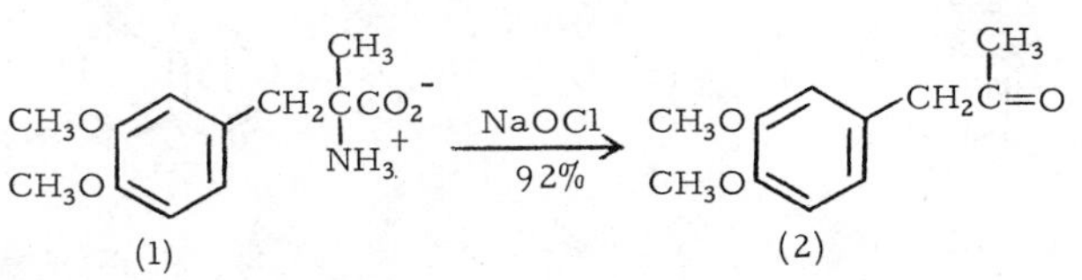

A solution of 4 mmoles of (1) in 25 ml. of water was covered with 10 ml. of benzene, and 14 ml. of commercial 5% sodium hypochlorite was added dropwise in 20 min. A spot test on starch-iodide paper taken after each addition was negative after 30 seconds until the end point had been reached. The layers were separated and the basic aqueous layer was extracted with ether–benzene.

Halooxy substitutions. On reinvestigating a curious product obtained by Green and Rowe[12] by treatment of 2,4-dinitroaniline in alkaline methanol with sodium hypochlorite, Mallory and Varimbi[13] established the true structure to be that of 5-chloro-4-methoxybenzofurazane-1-oxide (4, in equilibrium with the less stable

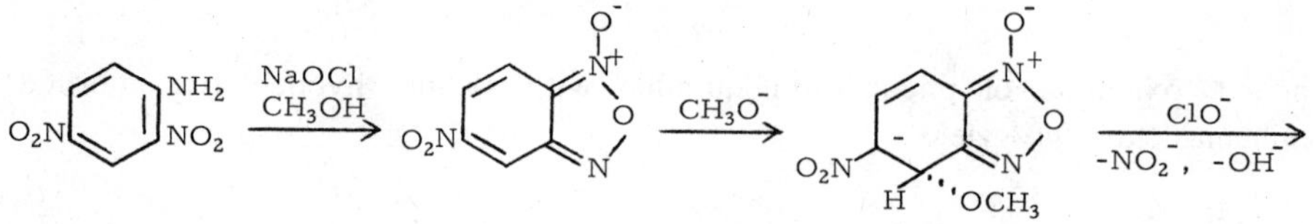
NaOCl
CH₃OH
CH₃O⁻
ClO⁻
-NO₂⁻, -OH⁻

(1) (2) (3)

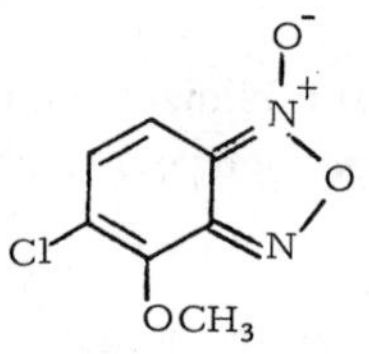

(4)

3-oxide). The mechanism postulated for this unusual reaction is supported by various pieces of evidence. 2,3-Dinitroaniline (5) enters into a similar reaction,[13] as do the benzotriazole[14] (6) and the diphenylquinoxaline (7).[14]

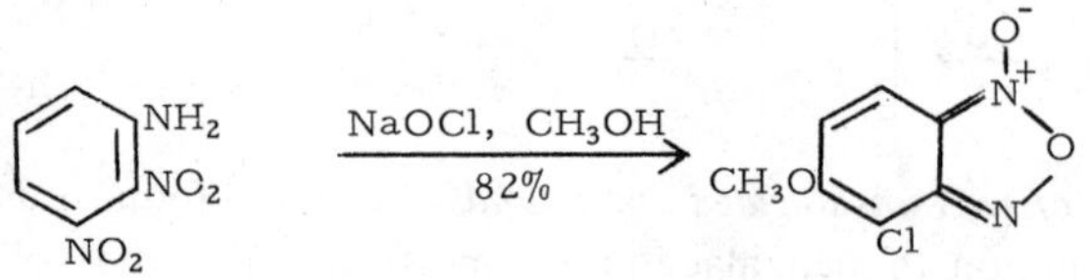
NaOCl, CH₃OH
82%

(5)

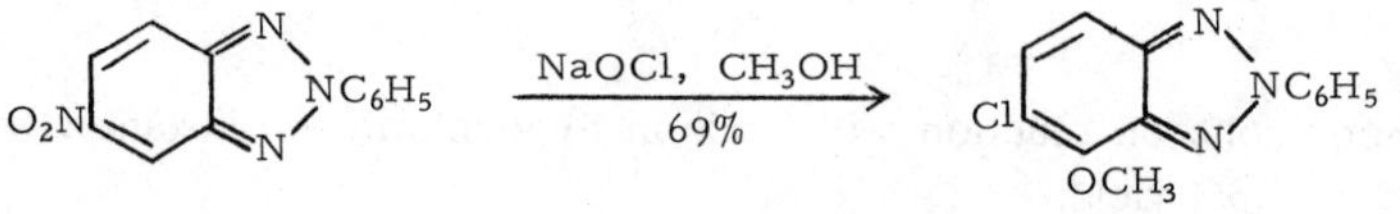
NaOCl, CH₃OH
69%

(6)

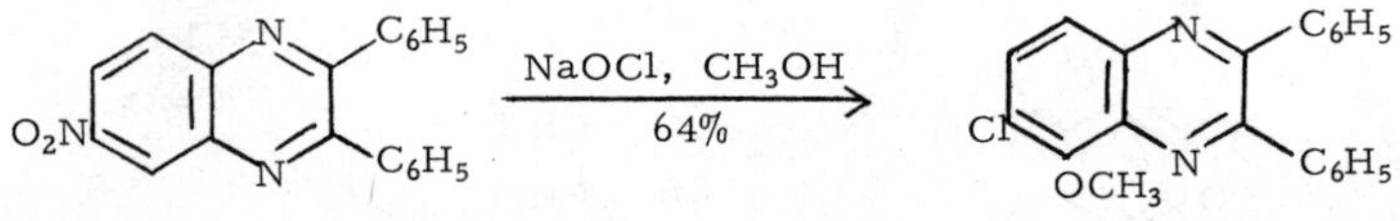
NaOCl, CH₃OH
64%

(7)

[1]M. S. Newman and H. L. Holmes, *Org. Syn., Coll. Vol.*, **2**, 428 (1943)
[2]R. C. Fuson and B. A. Bull, *Chem. Rev.*, **15**, 275 (1934)
[3]W. T. Smith and G. L. McLeod, *Org. Syn., Coll. Vol.*, **4**, 345 (1963)
[4]D. D. Neiswender, Jr., W. B. Moniz, and J. A. Dixon, *Am. Soc.*, **82**, 2876 (1960)
[5]R. Adams and B. K. Brown, *Org. Syn., Coll. Vol.*, **1**, 309 (1941)
[6]F. B. Mallory, *ibid.*, **4**, 74 (1963)
[7]M. W. Formo and J. R. Johnson, *ibid.*, **3**, 486 (1955)
[8]R. Ohme and E. Schmitz, *Ber.*, **97**, 297 (1964)
[9]S. Marmor, *J. Org.*, **28**, 250 (1963)
[10]H. L. Slates, D. Taub, C. H. Kuo, and N. L. Wendler, *J. Org.*, **29**, 1424 (1964)
[11]See preceding reference, note 4.
[12]A. G. Green and F. M. Rowe, *J. Chem. Soc.*, **101**, 2452 (1912)
[13]F. B. Mallory and S. P. Varimbi, *J. Org.*, **28**, 1656 (1963)
[14]F. B. Mallory, C. S. Wood, and B. M. Hurwitz, *J. Org.*, **29**, 2605 (1964)

Sodium hypophosphite, $NaH_2PO_2 \cdot H_2O$. Mol. wt. 106.01, solubility in 100 g. H_2O: 100 g. at 25°. Suppliers: F; MCB.

Raney nickel liberates hydrogen from water in the presence of sodium hypophosphite, which becomes oxidized to sodium phosphite. This reducing system in aqueous acetic acid or aqueous acetic acid–pyridine reduces nitriles to aldehydes at room temperature and pressure; yields 50–90%.[1]

[1]O. G. Backeberg and B. Staskun, *J. Chem. Soc.*, 3961 (1962)

Sodium iodide, NaI. Mol. wt. 149.91. Soluble in water, ethanol, acetone, acetic acid, dimethylformamide, methyl ethyl ketone.

Displacement of Cl or Br. Bunnett and Conner[1] improved a previous procedure for the preparation of 2,4-dinitroiodobenzene from the chloro compound by using dimethylformamide as solvent in place of ethylene glycol. Sodium iodide was taken in fivefold molar excess, presumably on the ground that the reaction is reversible.

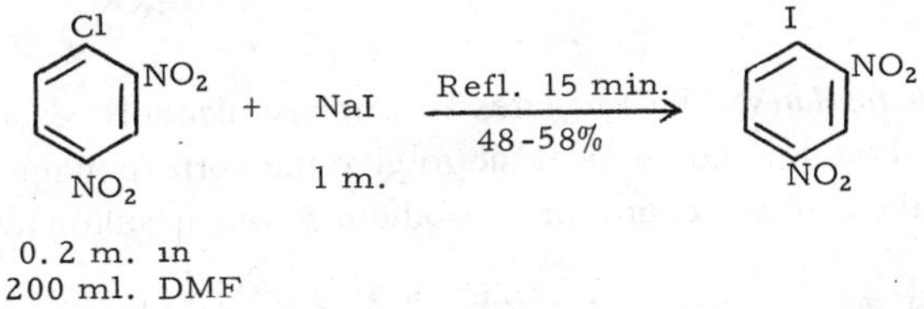
Cl
NO2
NO2
+ NaI
Refl. 15 min.
48-58%
1 m.
I
NO2
NO2
0. 2 m. in
200 ml. DMF

However, Ford-Moore[2] effected a displacement of chlorine by iodine in much higher yield by heating the chloride (1) with a modest excess of sodium iodide in methyl

$$\underset{(1)\ 0.7\ m.}{C_6H_5CO_2CH_2CH_2Cl} + \underset{1.1\ m.}{NaI} \xrightarrow[78-81\%]{\text{1. 2 l. Methyl ethyl ketone, 24 hrs. at } 87^0} \underset{(2)}{C_6H_5CO_2CH_2CH_2I}$$

ethyl ketone for 24 hrs. Schurink[3] converted pentaerythrityl tetrabromide into the tetraiodide in very high yield by reaction in methyl ethyl ketone with 0.67 mole of sodium iodide, whereas the theoretical amount is 0.52 mole.

$$\underbrace{\underset{0.13\ m.}{C(CH_2Br)_4} + \underset{0.67\ m.}{NaI}}_{\text{In 240 ml. of methyl ethyl ketone}} \xrightarrow[89-98\%]{87^0\ 48\ hrs.} C(CH_2I)_4$$

Displacement of bromine by iodine is a key step in a procedure developed by a Syntex group for the conversion of a 3-keto-5α-steroid (1) into the Δ^4-3-ketosteroid.[4]

Bromination gives the 2α,4α-dibromoketone (2), which when refluxed with sodium iodide in acetone for 20 hrs. affords the Δ^4-2α-iodo-3-ketone (3). This is deiodinated

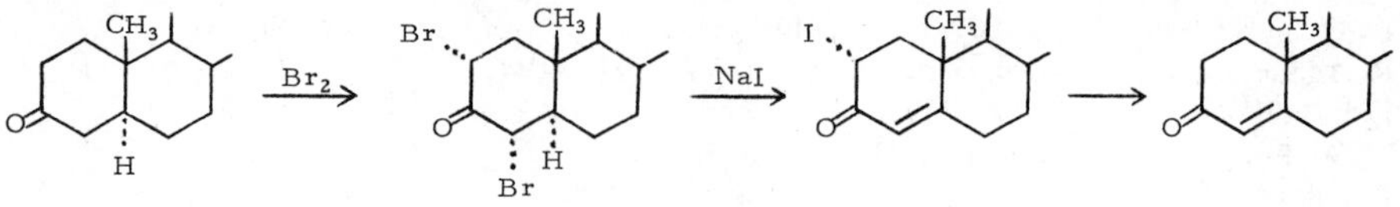

(1) (2) (3) (4)

by reduction with chromous chloride or zinc dust to the Δ^4-3-ketosteroid (4). The intermediate iodoketone (3) need not be isolated and the overall yield is 50–60%. In applying this procedure for the conversion of 4,5α-dihydrocortisone 21-acetate into cortisone acetate, a Glaxo Laboratories group[5] initially obtained the desired product in only 10% yield. Isolation from the reaction mixture of a saturated ketone suggested that the intermediate iodoketone (3) was being reduced by hydrogen iodide. Indeed the yield was improved by addition of a second iodoketone (iodoacetone) to react competitively with the hydrogen iodide. Schreiber[6] used the Glaxo modification to advantage in the synthesis of a solanum alkaloid and by increasing the proportion of iodoacetone achieved a yield of 73%.

Displacement catalyst. In the preparation of *p*-methoxybenzyl cyanide from the chloride, a small amount of sodium iodide apparently functions as catalyst, although no evidence is presented that this is the case.[7]

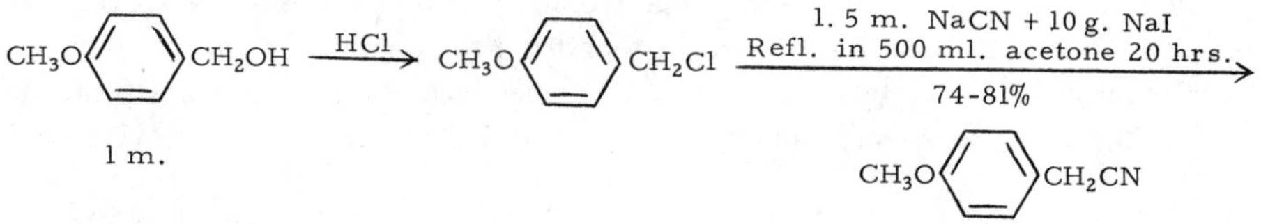

Reaction with tosylates. The tosylates of primary alcohols react readily with a 10% solution of sodium iodide in acetone to give the corresponding iodides. Thus, as measured by the rate of separation of sodium *p*-toluenesulfonate, the reactions

$$C_6H_5CH_2OSO_2C_6H_4CH_3 + \underset{\text{2 equiv.}}{NaI} \xrightarrow[97\%]{\text{Acetone 2 hrs. } 25^0} C_6H_5CH_2I + CH_3C_6H_4SO_3Na$$

of the tosylates of benzyl alcohol and methanol are nearly complete at room temperature in 2 hrs.[8] Under the same conditions cyclohexanyl tosylate reacts to the extent of only 3%.

Under forcing conditions the group $-CH_2OTs$ can be reduced to methyl. Thus Bowers and Ringold[9] prepared 21-desoxycortisone (II) by treating cortisone (I)

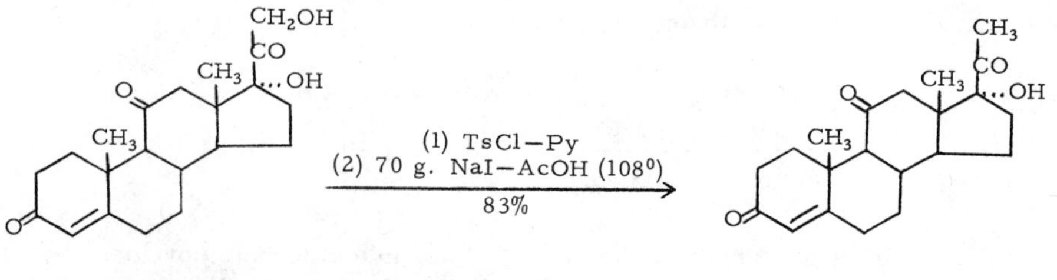

I 20 g. II

with *p*-toluenesulfonyl chloride in pyridine at 0° and refluxing the crude product with a huge excess of sodium iodide in 1 l. of acetic acid for 1 hr. (in Mexico City the boiling point is about 10° below that at sea level). This method of reduction permits preservation of the carbonyl groups.

Dehalogenation of vic-dihalides. Schoenheimer[10] used sodium iodide in acetone for recovery of cholesterol from the 5α,6β-dibromide. Barton and Miller[11] found the isomeric 5β,6α-dibromide very resistant to debromination and so established that the *trans*-diaxial configuration is optimal for facile elimination.

Stevens and French[12] synthesized diphenylketene-*p*-tolylimine (4) by dechlorination of the precursor (3) by refluxing it with one equivalent amount of sodium iodide in acetone for 90 min. The product (4) is a bright yellow solid which reacts with water, ethanol, and chlorine in the same way that ketenes react.

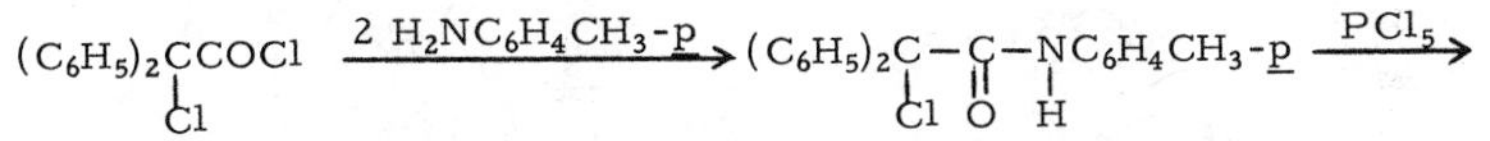
(C6H5)2CCOCl
Cl
2 H2NC6H4CH3-p
(C6H5)2C—C—NC6H4CH3-p
Cl O H
PCl5

(1) (2)

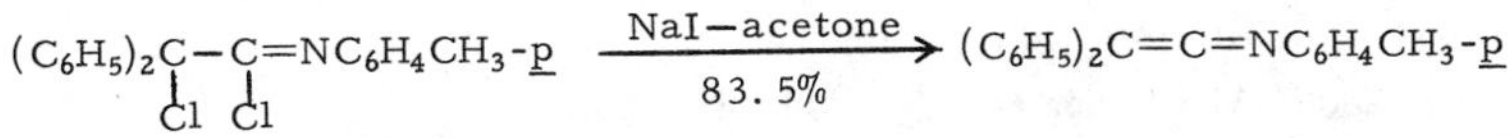
(C6H5)2C—C=NC6H4CH3-p
Cl Cl
NaI—acetone
83.5%
(C6H5)2C=C=NC6H4CH3-p

(3) (4)

vic-Ditosylates ⟶ olefins. Like *vic*-dihalides, *vic*-ditosylates (or dimesylates) of the type $RCH(OTs)CH_2OTs$ are converted by sodium iodide in acetone into the unsaturated products: $RCH{=}CH_2$.[13] The elimination is slow if both groups are secondary.[14] Slates and Wendler[15] examined the dimesylates of the four 2,3-diols derived from 5α-spirostane-12-one and found that (1) and (2) react with sodium iodide in acetone at 100° to give the Δ²-olefin in high yield whereas under the same conditions (3) and (4) are essentially unchanged. The result is a surprise; from

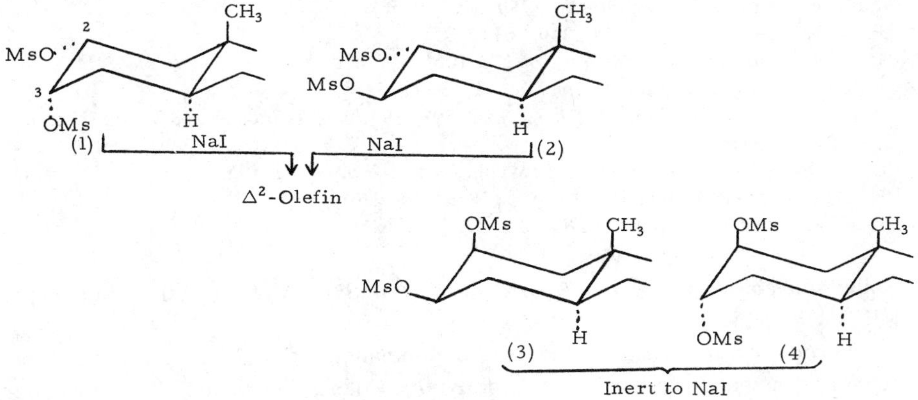
CH3
2
MsO
3
OMs
H
(1)
NaI
CH3
MsO
MsO
H
NaI
(2)
Δ2-Olefin
OMs
CH3
MsO
H
(3)
OMs
CH3
OMs
H
(4)
Inert to NaI

analogy to the debromination of the 5,6-dibromides, the *trans*-diaxial isomer (4) would be expected to be the most prone to elimination with iodide ion. Possibly the failure of (3) and (4) to react is associated with hindrance of the 2β-axial mesylate group by the axial angular methyl group.

Reduction. In preparing 2,2′-diphenaldehyde, Bailey and Erickson[16] ozonized phenanthrene in methanol and reduced the resulting peroxidic product to the dialdehyde by treating it with 1.5 times the theory with sodium iodide in acetic acid.

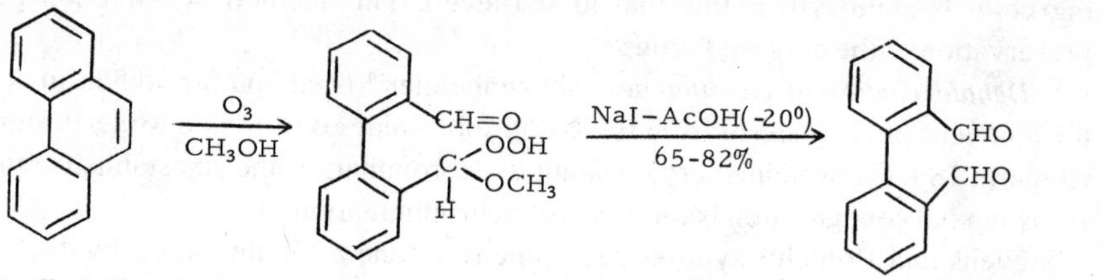

O_3
CH_3OH
CH=O
OOH
C
OCH_3
H
NaI–AcOH(-20°)
65-82%
CHO
CHO

Dealkylation of ethers. Sodium iodide in acetic acid is a mild reagent for effective dealkylation of alkoxypyrimidines.[17] For example (1) is converted into (2) in 82%

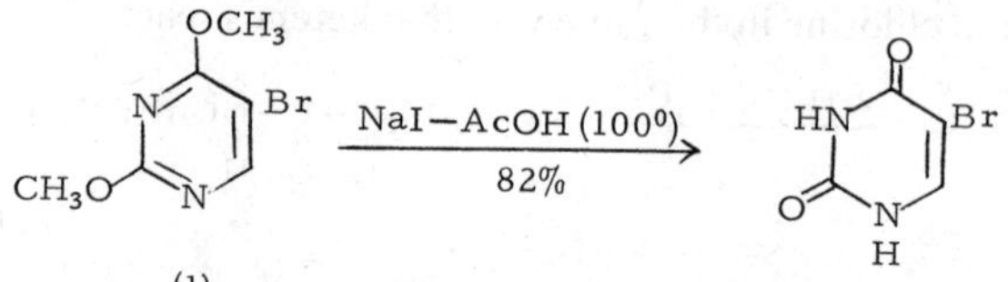

OCH_3
N
Br
CH_3O
N
NaI–AcOH (100°)
82%
O
HN
Br
O
N
H

(1) (2)

yield in 4 hrs. at 100°. The same reaction is effected by hydrogen bromide in acetic acid in nearly quantitative yield in 2 hrs. at 70°.

[1]J. F. Bunnett and R. M. Conner, *Org. Syn.*, **40**, 34 (1964)

[2]A. H. Ford-Moore, *Org. Syn., Coll. Vol.*, **4**, 84 (1963)

[3]H. B. Schurink, *ibid.*, **2**, 476 (1943)

[4]G. Rosenkranz, O. Mancera, J. Gatica, and C. Djerassi, *Am. Soc.*, **72**, 4077 (1950); G. Rosenkranz, J. Pataki, St. Kaufmann, J. Berlin, and C. Djerassi, *ibid.*, **72**, 4081 (1950); F. Sondheimer, G. Rosenkranz, O. Mancera, and C. Djerassi, *ibid.*, **75**, 2601 (1953)

[5]R. M. Evans, J. C. Hamlet, J. C. Hunt, P. G. Jones, A. G. Long, J. F. Oughton, L. Stephenson, T. Walker, and B. M. Wilson, *J. Chem. Soc.*, 4356 (1956)

[6]K. Schreiber and H. Rönsch, *Ann.*, **681**, 196 (1965)

[7]K. Rorig, J. D. Johnston, R. W. Hamilton, and T. J. Telinski, *Org. Syn., Coll. Vol.*, **4**, 576 (1963)

[8]R. S. Tipson, M. A. Clapp, and L. H. Cretcher, *J. Org.*, **12**, 133 (1947)

[9]A. Bowers and H. J. Ringold, *Am. Soc.*, **80**, 3091 (1958)

[10]R. Schoenheimer, *J. Biol. Chem.*, **110**, 461 (1935)

[11]D. H. R. Barton and E. Miller, *Am. Soc.*, **72**, 1066 (1950)

[12]C. L. Stevens and J. C. French, *Am. Soc.*, **75**, 657 (1953)

[13]P. Bladon and L. N. Owen, *J. Chem. Soc.*, 598 (1950); A. B. Foster and W. G. Overend, *ibid.*, 3452 (1951).

[14]R. P. Linstead, L. N. Owen, and R. F. Webb, *J. Chem. Soc.*, 1211 (1953)

[15]H. L. Slates and N. L. Wendler, *Am. Soc.*, **78**, 3749 (1956)

[16]P. S. Bailey and R. E. Erickson, *Org. Syn.*, **41**, 41 (1961)

[17]T. L. V. Ulbricht, *J. Chem. Soc.*, 3345 (1961)

Sodium metabisulfite, $Na_2S_2O_5$ (plus $H_2O \rightarrow 2\,NaHSO_3$). Mol. wt. 190.11. Suppliers: KK, MCB.

Purification of 3-keto-Δ¹-steroids.[1,2] Bromination of a 3-keto-5α-steroid and dehydrobromination to the 3-keto-Δ¹-derivative (3) is accompanied by the following

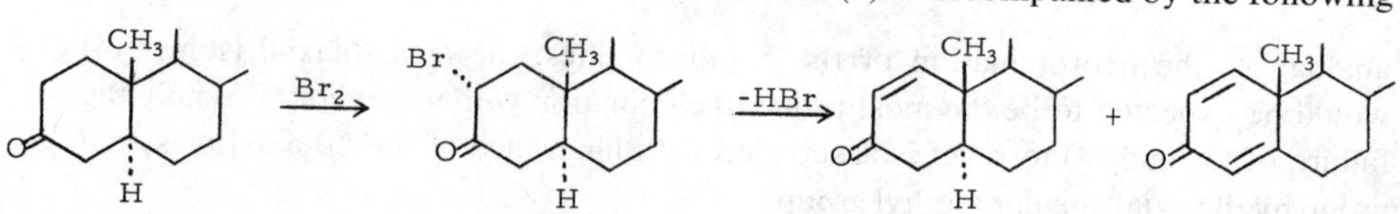

CH_3
Br_2
Br
CH_3
-HBr
CH_3
+
CH_3
H
H
H

(1) (2) (3) (4)

by-products removable with difficulty by chromatography or crystallization: starting material (1), the 2α-bromoketone (2), and possibly some 3-keto-$\Delta^{1,4}$-steroid. A convenient and commercially applicable process for the isolation of pure (3) is as follows. A solution of the crude reaction mixture [0.05 mole, estimated to contain 50% of (3)] in 250 ml. of methanol is mixed with a solution of 40 g. of sodium metabisulfite in 210 ml. of water for 2–3 min. and immediately extracted 3 times with methylene chloride; starting material (1) is left in the aqueous phase as the bisulfite adduct. The methylene chloride layer is washed with water, dried, and evaporated, and a solution of the residue in 150 ml. of methanol is refluxed for 1–1.5 hrs. with a solution of 28 g. of sodium metabisulfite in 110 ml. of water for conversion of (3) into the bisulfite addition compound. Water (100 ml.) is added, and the solution cooled and extracted three times with methylene chloride. For regeneration of (3) the aqueous layer is refluxed for 15 min. with a solution of 5 g. of sodium hydroxide in 150 ml. of water. After cooling, extraction with methylene chloride affords the pure 3-keto-Δ^1-steroid, λ_{max}228.5 mμ, log ε 4.004.

[1]Contributed by Paul D. Klimstra, G. D. Searle and Co.
[2]P. D. Klimstra, U.S. patent 3,018,296

Sodium methoxide, CH_3ONa. Mol. wt. 54.03, soluble in pentane. Suppliers: Olin Mathieson Chem. Corp., Fisher, MCB.

Solution in methanol. A solution of 1 mole of the reagent in 300 ml. of methanol can be prepared by the inverse Tishler procedure for the preparation of sodium ethoxide. An inverse procedure similar to that of Zaugg's procedure for preparing sodium ethoxide differs only in that the sodium is cut in small pieces and not melted.[1]

Use as a base. For the conversion of dibenzoyl peroxide into sodium perbenzoate, sodium methoxide is preferred to sodium ethoxide because it is more soluble in methanol and does not precipitate provided that the temperature does not fall below

$$\underset{0.21\text{ m.}}{(C_6H_5COO)_2} + \underset{\substack{0.22\text{ m. in}\\ 100\text{ ml. }CH_3OH}}{2\,CH_3ONa} \xrightarrow{HCCl_3\ 0^\circ} 2\,C_6H_5CO_3Na + C_6H_5CO_2CH_3$$

$$2\,C_6H_5CO_3Na \xrightarrow[82.5\text{-}86\%]{H_2SO_4} C_6H_5CO_3H(\text{in } CHCl_3)$$

−5°.[2] It is used as base in Kolbe electrolytic coupling in methanol solution; a catalytic amount suffices since it is regenerated at the cathode.[3] Commercial sodium methoxide

$$\underset{1\text{ m.}}{CH_3O_2C(CH_2)_8CO_2H} + \underset{\substack{0.05\text{ m. in}\\ 500\text{ ml. }CH_3OH}}{CH_3ONa} \xrightarrow[68\text{-}74\%]{\text{Electrol.}} CH_3O_2C(CH_2)_{16}CO_2CH_3$$

is the preferred base for generation of dichlorocarbene from ethyl trichloroacetate, for example for reaction with dihydropyrane.[4] A flask flushed with nitrogen is

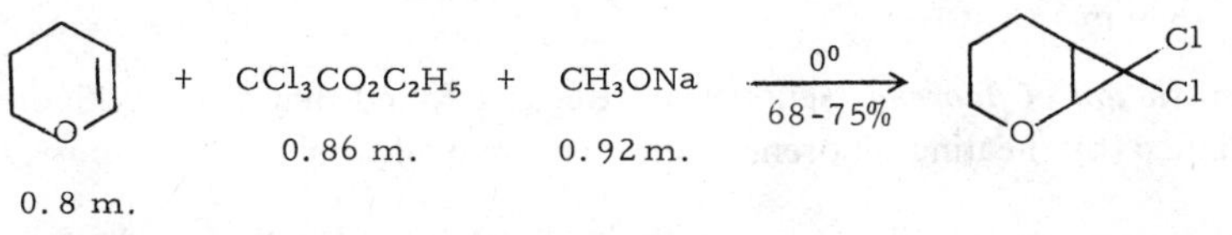
+ CCl₃CO₂C₂H₅ + CH₃ONa
0° 68-75%
Cl
Cl
O
0.86 m.
0.92 m.
0.8 m.

charged with sodium methoxide powder, dihydropyrane, and 600 ml. of pentane. The light yellow solution is stirred for 15 min. in an ice–water bath, the ester is run in in 3–4 min., and the mixture is stirred for 6 hrs. at 0° and overnight at 25°.

Dehydrohalogenation. The first step in the preparation of dibenzoylmethane from benzalacetophenone dibromide is elimination of one bromine atom as HBr and displacement of the other one by methoxyl.[5] The terminal step is acid hydrolysis of the enol ether.

$$\underset{\underset{0.5\ m.}{\text{Br Br}}}{C_6H_5CH\,CHCOC_6H_5} + \underset{\text{1 m. in 230 ml. } CH_3OH}{2\ CH_3ONa} \longrightarrow \underset{OCH_3}{C_6H_5C{=}CHCOC_6H_5} \xrightarrow[\text{74-80\% (crude)}]{H_2O{-}HCl} C_6H_5COCH_2COC_6H_5$$

A postulated intermediate in the Favorsky rearrangement is the product of dehydrochlorination. For rearrangement of 2-chlorocyclohexanone, Goheen and Vaughan[6] stirred a suspension of commercial sodium methoxide in 330 ml. of ether, added the

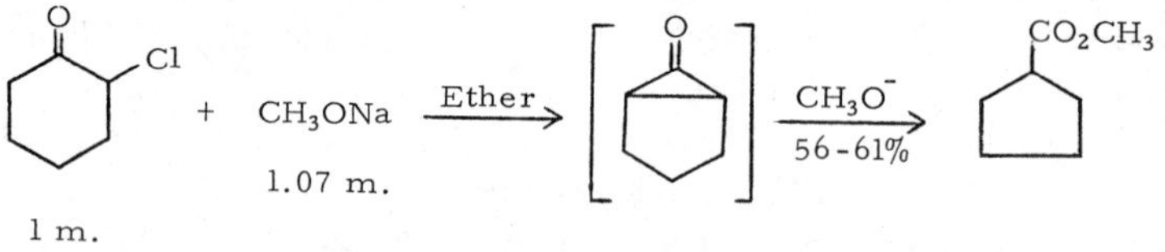

chloro ketone in 40 min., and stirred under reflux for 2 hrs. Sodium methoxide was taken in slight excess since use of just one equivalent lowered the yield.

Baumgarten and Petersen developed a general method for the synthesis of α-amino ketones illustrated by the conversion of α-phenylethylamine into phenacylamine.[7] The conversion of the N,N-dichloroamine (1) into the cyclic intermediate

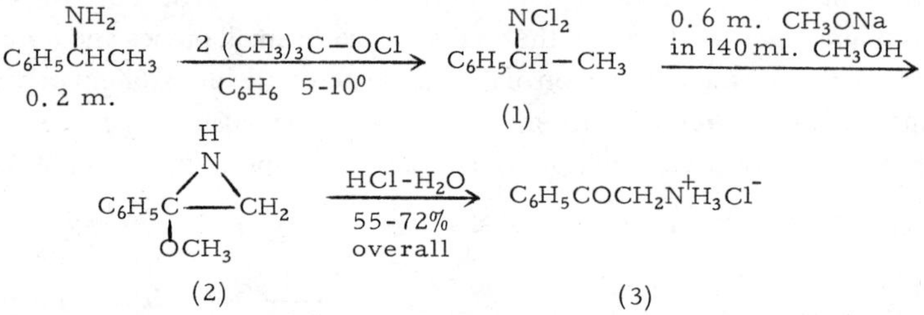

(2) was accomplished by adding a solution of sodium methoxide in methanol to a solution of (1) in benzene.

Nucleophilic displacement. One nitro group of 1,3,5-trinitrobenzene is displaced by methoxyl on brief refluxing with sodium methoxide in methanol.[8]

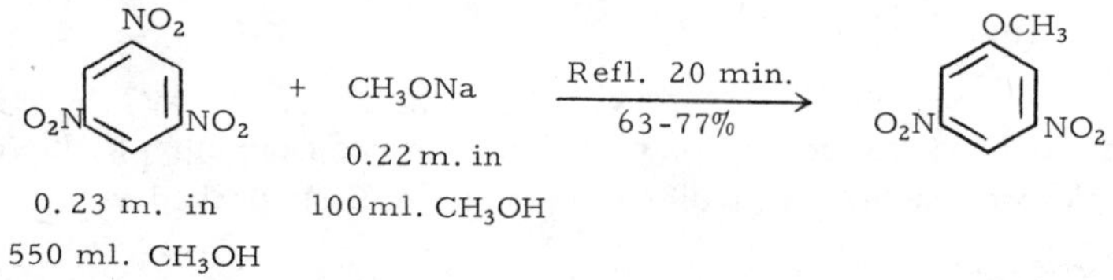

Methylation of fluorene. Schoen and Becker[9] found that 9-*n*-alkylfluorenes can be obtained by heating fluorene with a normal alcohol and the corresponding

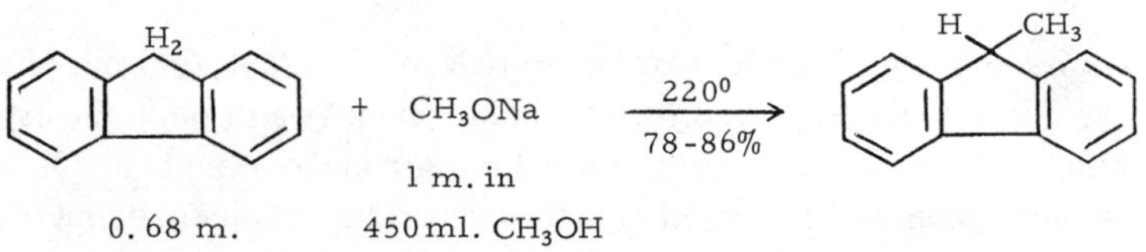

sodium alkoxide in an autoclave at 210–220°, as illustrated for the preparation of 9-methylfluorene.[10]

The mechanism suggested is as follows:

$$Ar_2CH_2 \xrightarrow{RCHO} Ar_2C{=}CHR \xrightarrow{RCH_2ONa} Ar_2C(Na)(CH_2R) + RCHO$$

$$\downarrow RCH_2OH$$

$$Ar_2C(H)(CH_2R) + RCH_2ONa$$

Condensation catalyst. In the synthesis of *trans*-stilbene by the phosphonate modification of the Wittig reaction formulated, Seus and Wilson[11] effected the terminal condensation with sodium methoxide in dimethylformamide. The reaction

$$C_6H_5CH_2Br + P(OC_2H_5)_3 \rightarrow C_6H_5CH_2\overset{OC_2H_5}{\overset{|}{P^+}}(OC_2H_5)_2Br^- \xrightarrow[85\%]{-C_2H_5Br} C_6H_5CH_2\overset{O^-}{\overset{|}{P^+}}(OC_2H_5)_2$$

(5) (6)

$$C_6H_5CH_2\overset{O^-}{\overset{|}{P^+}}(OC_2H_5)_2 + O{=}CHC_6H_5 \xrightarrow[85\%]{NaOCH_3} C_6H_5\overset{H}{C}{=}\underset{H}{C}C_6H_5$$

$$+ (C_2H_5O)_2POONa + CH_3OH$$

can be done also in 1,2-dimethoxyethane with sodium hydride but the yield is lower.[12] A synthesis of *trans,trans*-1,4-diphenyl-$\Delta^{1,3}$-butadiene by the method of Seus and Wilson has been developed as a student experiment.[13]

A synthesis of dicyclopropyl ketone developed by Hart and co-workers[14] involves in the first step a sodium methoxide-catalyzed self-condensation of γ-butyrolactone (1) to the dimeric product (2). Solvent methanol is removed by distillation, eventually

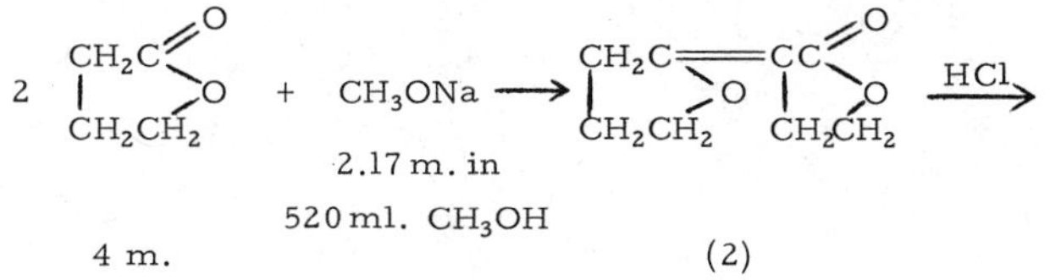

4 m. (1) (2)

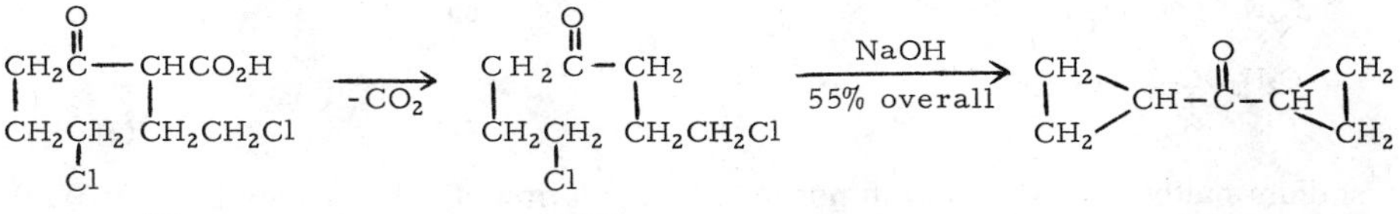

(3) (4) (5)

at reduced pressure. Heating with concentrated hydrochloric acid effects conversion into the keto dichloroacid (3) and decarboxylation to (4), which is cyclized by aqueous alkali to dicyclopropyl ketone (5). Intermediates (2) and (4) were both isolated.

Other condensations effected under the influence of sodium methoxide are described in the next section.

Heterocycles. Dropwise addition of a mixture of 0.8 mole each of acetone and ethyl formate to a suspension of 0.86 mole of sodium methoxide in ether at 0° effected condensation to sodio formylacetone (1).[15] Removal of the ether and treatment of the residual sodium enolate with 0.3 mole of cyanoacetamide in 400 ml. of

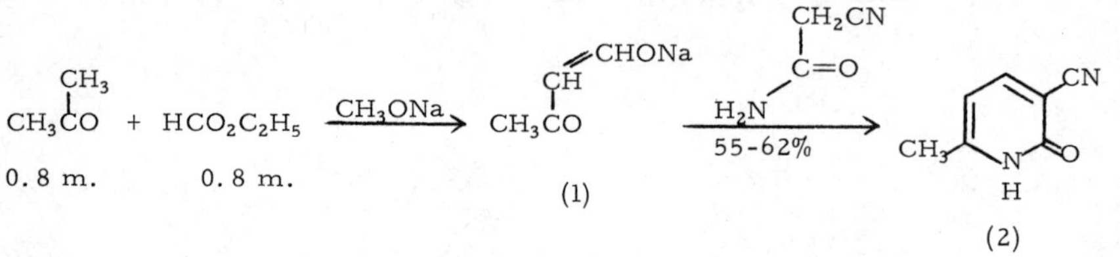

water and a solution of 8 ml. of acetic acid in 20 ml. of water made just basic by the addition of piperidine, followed by refluxing for 2 hrs. and acidification with acetic acid, gave a yellow precipitate of 3-cyano-6-methyl-2(1)-pyridone (2).

In the synthesis of 2-amino-4-anilino-6-chloromethyl-*s*-triazine from phenylbiguanide hydrochloride (0.3 mole) and ethyl chloroacetate (0.3 mole), the hydrochloride is added to a solution of 0.3 mole of sodium methoxide in methanol, the solution is filtered from sodium chloride, ethyl chloroacetate is added, and the mixture is stirred at room temperature for 14 hrs.[16] The product precipitates and is crystallized from dioxane.

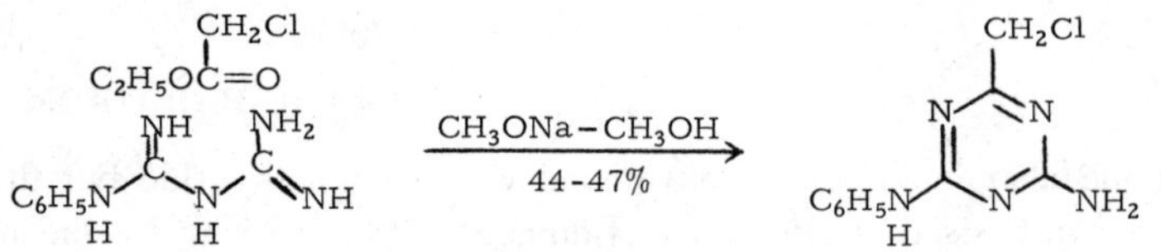

3-*n*-Heptyl-5-cyanocytosine (3) is obtained by a synthesis[17] starting with the conversion of *n*-heptylamine into *n*-heptylurea (1). This is then refluxed with malononitrile and triethyl orthoformate to produce crystalline 3-*n*-heptylureidomethylenemalononitrile (2). Cyclization to (3) is effected by adding commercial

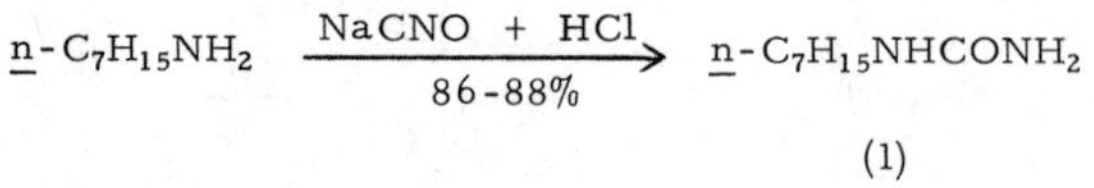

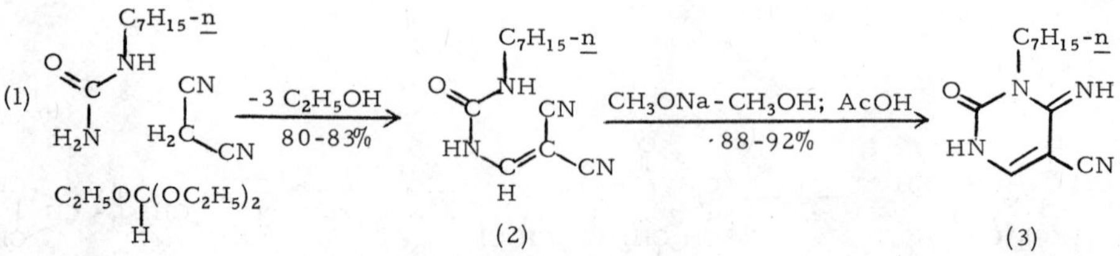

sodium methoxide (0.16 m.) in portions to a solution of 0.145 mole of (2) in 70 ml. of methanol and letting the mixture stand at 25° for 3 days. The salt which separates is dissolved in water, and the product precipitated with acetic acid and crystallized from alcohol.

2-Thio-6-methyluracil is prepared[18] by heating a mixture of 1 mole of ethyl acetoacetate, 1 mole of thiourea, and 3.3 moles of commercial sodium methoxide on the steam bath and evaporating to dryness. A solution of the salt in water is

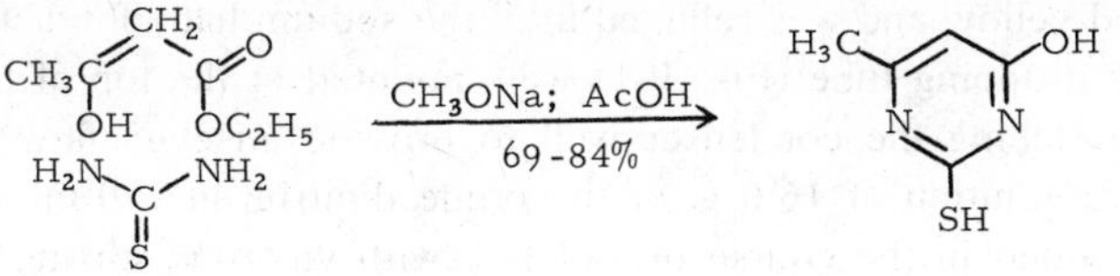

acidified with acetic acid. Parabanic acid is prepared by similar condensation of urea with diethyl oxalate.[19] Sodium ethoxide in ethanol can be used, but the yield is lower. For the sodium methoxide-catalyzed condensation of phenyl azide with phenylacetonitrile, *see* Phenyl azide, Heterocycles.

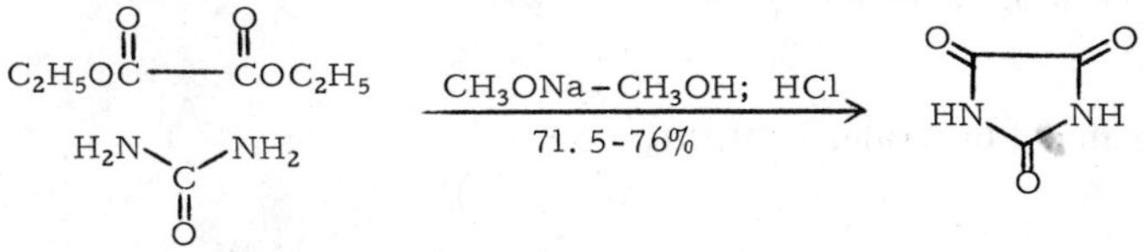

[1]H. E. Baumgarten and J. M. Petersen, *Org. Syn.*, **41**, 82 (1961)
[2]G. Braun, *Org. Syn.*, *Coll. Vol.*, **1**, 431 (1941)
[3]S. Swann, Jr., and W. E. Garrison, Jr., *Org. Syn.*, **41**, 33 (1961)
[4]W. E. Parham, E. E. Schweizer, and S. A. Mierzwa, Jr., *ibid.*, **41**, 76 (1961)
[5]C. F. H. Allen, R. D. Abell, and J. B. Normington, *Org. Syn.*, *Coll. Vol.*, **1**, 205 (1941)
[6]D. W. Goheen and W. R. Vaughan, *ibid.*, **4**, 594 (1963)
[7]H. E. Baumgarten and J. M. Petersen, *Org. Syn.*, **41**, 82 (1961)
[8]F. Reverdin, *Org. Syn.*, *Coll. Vol.*, **1**, 219 (1941)
[9]K. L. Schoen and E. I. Becker, *Am. Soc.*, **77**, 6030 (1955)
[10]*Idem*, *Org. Syn.*. *Coll. Vol.*, **4**, 623 (1963)
[11]E. J. Seus and C. V. Wilson, *J. Org.*, **26**, 5243 (1961)
[12]W. S. Wadsworth, Jr., and W. D. Emmons, *Am. Soc.*, **83**, 1733 (1961)
[13]*Org. Expts.*, 119
[14]O. E. Curtis, Jr., J. M. Sandri, R. E. Crocker, and H. Hart, *Org. Syn.*, *Coll. Vol.*, **4**, 278 (1963)
[15]R. P. Mariella, *ibid.*, **4**, 210 (1963)
[16]C. G. Overberger and F. W. Michelotti, *ibid.*, **4**, 29 (1963)
[17]B. B. Kehm and C. W. Whitehead, *ibid.*, **4**, 515 (1963)
[18]H. M. Foster and H. R. Snyder, *ibid.*, **4**, 638 (1963)
[19]J. I. Murray, *ibid.*, **4**, 744 (1963)

Sodium N-methylanilide, $C_6H_5N(CH_3)Na$. Mol. wt. 129.14.

The reagent is used in the Ziegler cyclization of dinitriles.[1] The greatly simplified technique of Kohler and Schroeder[2] was used by Fry and Fieser[3] and by Cope and

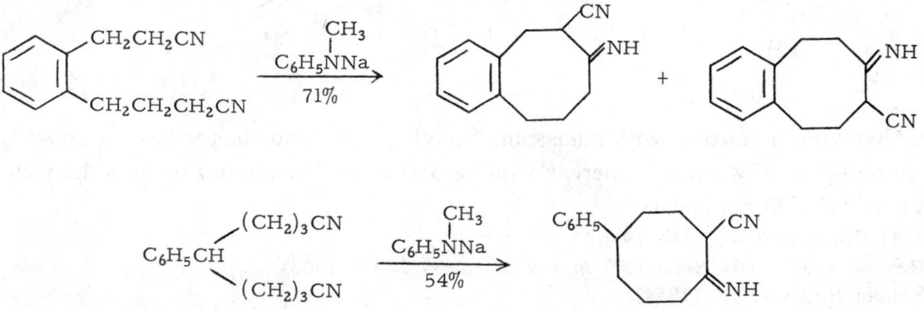

Cotter[4] in the examples formulated. In the first case a solution of 18.3 g. of naphthalene (as assistant) in 415 ml. of ether was treated with 5.5 g. of cleaned sodium wire and stirred under nitrogen, and 31.6 ml. of N-methylaniline was added. The

mixture turned yellow and was refluxed until the sodium had all reacted. A Hershberg capillary dropping tube (Fig. P-1) was mounted at the top of the condenser with the tip touching the condenser wall to provide an even flow of liquid and charged with a solution of 16.6 g. of the crude dinitrile in 250 ml. of ether. The solution was added in the course of 144 hrs. with vigorous stirring and refluxing (reasonably high dilution).

[1]K. Ziegler, H. Eberle and H. Ohlinger, *Ann.*, **504**, 94 (1933); K. Ziegler and R. Aurnhammer, *Ann.*, **513**, 43 (1934)
[2]H. E. Schroeder, Ph.D. Thesis, Harvard University, 1938
[3]E. M. Fry and L. F. Fieser, *Am. Soc.*, **62**, 3489 (1940)
[4]A. C. Cope and R. J. Cotter, *J. Org.*, **29**, 3467 (1964)

Sodium 2-methyl-2-butoxide, $CH_3CH_2\overset{CH_3}{\underset{CH_3}{C}}ONa$.

Since the reagent is soluble in aromatic hydrocarbon solvents, a solution (1–3 molar) can be prepared by refluxing a solution of *t*-amyl alcohol in benzene, toluene, or xylene with excess sodium wire or ribbon under nitrogen until evolution of hydrogen ceases.[1] After cooling under nitrogen the solution remains clear, and a sample can be titrated with standard acid using phenolphthalein as indicator. Alkylation can then be carried out in a homogeneous solution. This alkoxide is an effective base for use in the alkylation of ketones, particularly those that tend to undergo self-condensation.[2] Thus isophorone (1) on alkylation with 1,3-dichloro-2-butene and sodium 2-methyl-2-butoxide in benzene gives the ketone (2) in 50% yield,[3] whereas on alkylation with sodium *t*-butoxide the yield is only 20%.[2] The octalone (3) on

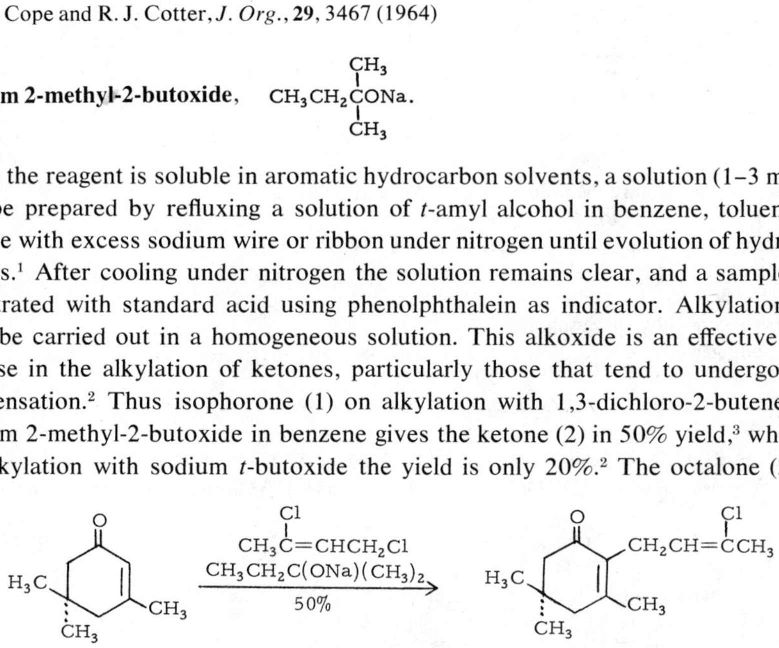
Cl
CH3C=CHCH2Cl
CH3CH2C(ONa)(CH3)2
50%
H3C
CH3
CH3
H3C
CH3
Cl
CH2CH=CCH3
CH3
CH3

(1) (2)

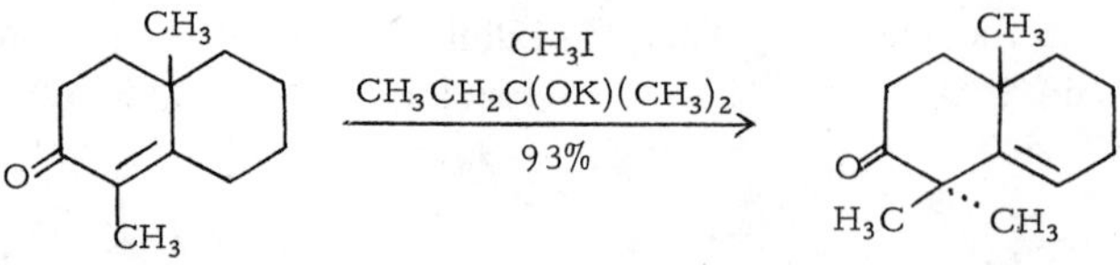
CH3I
CH3CH2C(OK)(CH3)2
93%
CH3
CH3
CH3
H3C
CH3

(3) (4)

methylation in benzene with potassium 2-methyl-2-butoxide as the base affords the ketone (4) in 93% yield,[4] whereas with potassium *t*-butoxide in *t*-butanol the yield is only 24%.[5] Other papers.[6]

[1]J.-M. Conia, *Bull. soc.*, 533 (1950)
[2]Review: J.-M. Conia, *Record of Chemical Progress*, **24**, 43 (1963)
[3]S. Julia, *Bull. soc.*, 780 (1954)
[4]S. L. Mukherjee and P. C. Dutta, *J. Chem. Soc.*, 67 (1960)
[5]M. Yanagita, M. Hirakura, and F. Seki, *J. Org.*, **23**, 841 (1958)
[6]J.-M. Conia and co-workers, *Bull. soc.*, 690, 943 (1954); 1040, 1392 (1956); 1064 (1957); 1929 (1960); 836 (1961); *Tetrahedron*, **16**, 45 (1961); *Tetrahedron Letters*, 505 (1962)

Sodium naphthalene, *see* Naphthalene-sodium.

Sodium nitrite, $NaNO_2$. Mol. wt. 69.01.

Uses of the reagent listed below do not include diazotization, nitrosation of phenols and tertiary aromatic amines, and simple N-nitrosation of amines.

Isonitroso derivatives. Procedures for the reaction of ethyl acetoacetate[1] and diethyl malonate[2] with sodium nitrite and aqueous acetic acid to form the isonitroso derivatives describe merely the first step in a synthetic sequence.

$$HON{=}C(CO_2C_2H_5)_2 \qquad CH_3\overset{O}{\overset{\|}{C}}-\overset{NOH}{\overset{\|}{C}}CO_2C_2H_5$$

Hydrazide ⟶ azide. In the synthesis of putrescine dihydrochloride,[3] diethyl adipate is converted into the dihydrazide, which on reaction with nitrous acid affords

$$(CH_2)_4(CO_2C_2H_5)_2 \xrightarrow{H_2NNH_2} (CH_2)_4(CONHNH_2)_2 \xrightarrow[5-10^0]{aq.\ NaNO_2-HCl}$$

$$(CH_2)_4(\overset{O}{\overset{\|}{C}}N{=}\overset{+}{N}{=}\overset{-}{N})_2 \xrightarrow[-N_2]{} (CH_2)_4(N{=}C{=}O)_2 \xrightarrow[73-77\%\ overall]{H_2O-HCl} (CH_2)_4(NH_2\cdot HCl)_2$$

the diazide. This reactive substance is extracted with ether, the solution added to benzene, and the mixture is heated gently on the steam bath. As the ether distils, rearrangement to the diisocyanate begins with liberation of nitrogen.

Ring expansion. An example of the Tiffeneau rearrangement is the conversion of 1-aminomethylcyclohexanol into cycloheptanone.[4] Gutsche[5] found that 2-phenylcycloheptanone can be prepared by conversion of ethyl N-benzylcarbamate

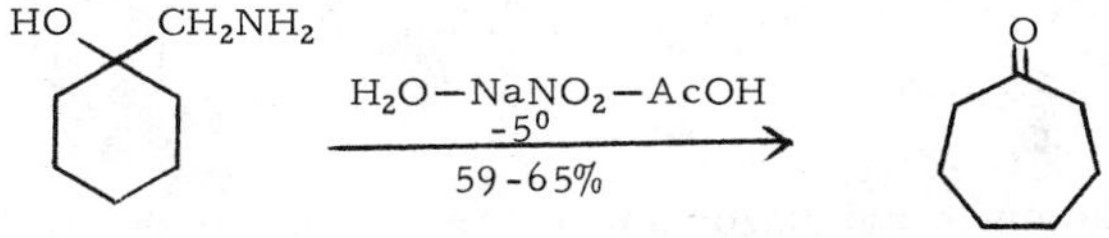
HO
CH₂NH₂
H₂O−NaNO₂−AcOH
-5⁰
59-65%
O

into the N-nitroso derivative and reaction of this with cyclohexanone in methanol in the presence of powdered potassium carbonate. The postulated mechanism is formulated.

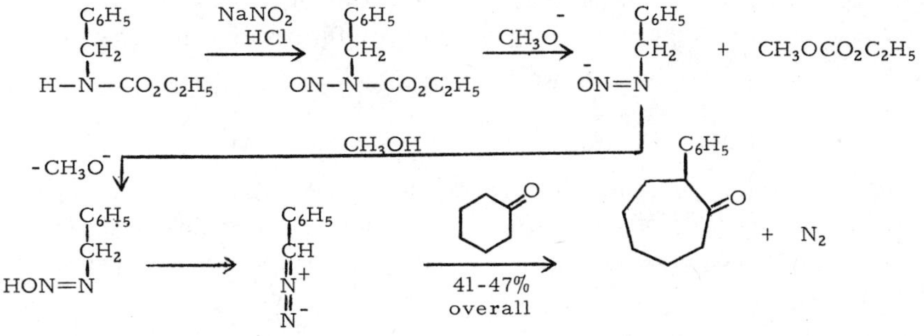
C₆H₅
CH₂
H−N−CO₂C₂H₅
NaNO₂
HCl
ON−N−CO₂C₂H₅
CH₃O⁻
ON=N
+ CH₃OCO₂C₂H₅
CH₃OH
-CH₃O⁻
HON=N
CH
N⁺
N⁻
41-47%
overall
+ N₂

Hydrazine ⟶ azide. Phenyl azide can be prepared by stirring a suspension of phenylhydrazine hydrochloride in dilute hydrochloric acid overlaid with ether,

$$C_6H_5NHNH_2\cdot HCl \xrightarrow[65-68\%]{NaNO_2(0-5^0)} C_6H_5N{=}\overset{+}{N}{=}\overset{-}{N}$$

cooling well, and dropping in aqueous sodium nitrite.[6] The workup involves ether extraction and cautious vacuum distillation behind a protective screen.

Ethyl diazoacetate is prepared by the action of sodium nitrite on glycine ethyl ester hydrochloride.[7]

$$Cl^- H_3\overset{+}{N}CH_2CO_2C_2H_5 + NaNO_2 \xrightarrow[85\%]{0^\circ - 2^\circ} \overset{-}{N}{=}\overset{+}{N}{=}CHCO_2C_2H_5$$

Nitromethane by the Kolbe method (1872). In this method the chlorine atom of sodium chloroacetate is displaced by the nitro group, and the solution heated gently to effect decarboxylation.[8]

$$ClCH_2CO_2Na \xrightarrow{NaNO_2} NO_2CH_2CO_2Na \xrightarrow[35-38\%]{80-85^\circ} CH_3NO_2 + NaHCO_3$$

Sandmeyer reaction. In the preparation of 1,4-dinitronaphthalene from 4-nitro-1-naphthylamine Hodgson and co-workers[9] used a special procedure to diazotize the weakly basic amine. Ten grams of powdered sodium nitrite was dissolved in 50 ml. of concd. sulfuric acid with ice cooling. A solution of 10 g. of the amine in 100 ml. of acetic acid was cooled to 20°, and the resulting thin slurry was dropped slowly into the cold nitrosylsulfuric acid with mechanical stirring and cooling. Ether was stirred in at 0° to precipitate the crystalline if somewhat sticky diazonium sulfate, which was collected and dissolved in 100 ml. of ice water. A copper reagent suitable for effecting Sandmeyer decomposition was prepared by mixing aqueous solutions of 50 g. each of copper sulfate and sodium sulfite and collecting the greenish brown precipitate. This was stirred with an aqueous solution of 100 g. of sodium nitrite,

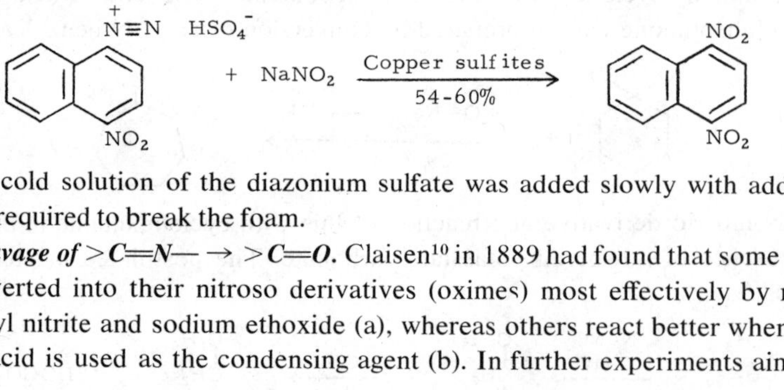

N≡N HSO4
NO2
+ NaNO2
Copper sulfites
54-60%
NO2
NO2

and the cold solution of the diazonium sulfate was added slowly with addition of ether as required to break the foam.

Cleavage of >C=N— → >C=O. Claisen[10] in 1889 had found that some ketones are converted into their nitroso derivatives (oximes) most effectively by reaction with amyl nitrite and sodium ethoxide (a), whereas others react better when hydrochloric acid is used as the condensing agent (b). In further experiments aimed at a

$$\text{(a)}\quad C_6H_5COCH_3 + C_5H_{11}NO_2 + C_2H_5ONa \longrightarrow C_6H_5COCH{=}NOH$$

$$\text{(b)}\quad CH_3COCH_3 + C_5H_{11}NO_2 + HCl \longrightarrow CH_3COCH{=}NOH$$

decision as to which procedure is, in general, the better he found that it is important to avoid use of an excess of amyl nitrite, for this converts the initially formed nitroso compound into a dicarbonyl product. Thus nitrosopropiophenone is converted into benzoylacetyl:

$$C_6H_5CO\underset{\underset{NOH}{\|}}{C}CH_3 + C_5H_{11}NO_2 \longrightarrow C_6H_5COCOCH_3 + C_5H_{11}OH + N_2O$$

In investigating nitrosocamphor, Claisen then found that by simply adding sodium nitrite to a solution of the nitroso compound in acetic acid the substance is converted smoothly into camphorquinone. Claisen's method for the cleavage of isonitroso

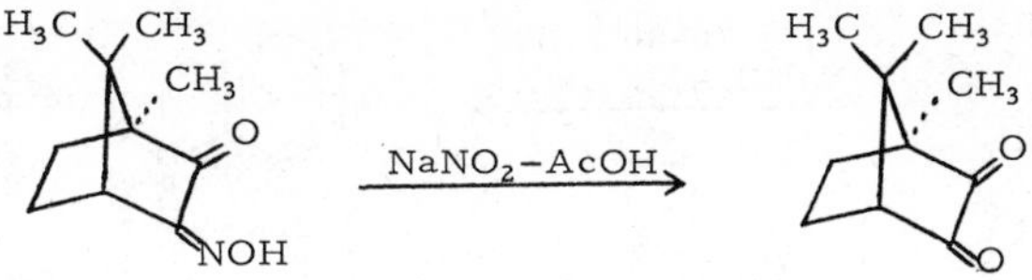
H_3C
CH_3
CH_3
O
NOH
$NaNO_2$–AcOH
H_3C
CH_3
CH_3
O
O

compounds or of oximes has proved to be generally applicable to compounds containing the grouping >C=N—. Wieland and Grimm[11] studied the reaction with isotopic labeling and found that the carbonyl oxygen is derived from the oxime, which supports the postulate of a three-membered ring intermediate:

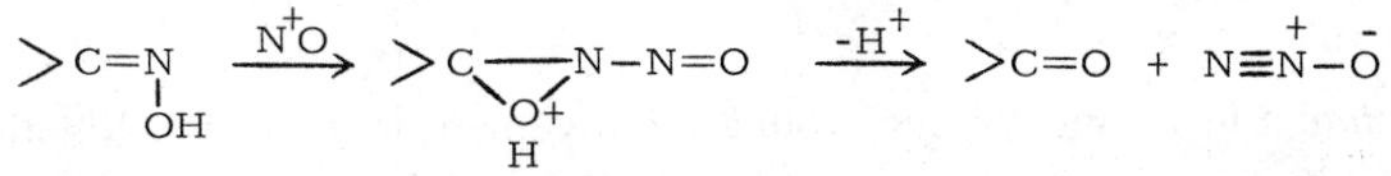

Nitrous acid cleaves the pentaacetyl oximes and semicarbazones of glucose and galactose to give the aldehyde pentaacetates.[12] The reagent cleaves benzaldehyde semicarbazone,[13] phenylosazones,[14] and 4-cinnolinecarboxaldehyde hydrazone.[15]

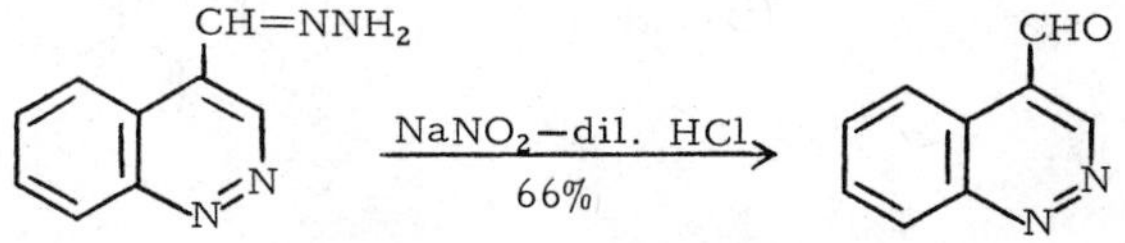
CH=NNH_2
$NaNO_2$–dil. HCl
66%
CHO
N
N
N
N

An Organon group[16] found steroid semicarbazones easy to split in acetic acid solution by adding sodium nitrite in the cold or with gentle heating; on addition of water the ketone separates, sometimes in crystalline form. A Schering group[17] cleaved hydrocortisone-3,20-bissemicarbazone by addition of excess sodium nitrite to a solution of the derivative in dilute hydrochloric acid at 5° and obtained hydrocortisone in 69.5% yield. Cleavage by exchange with pyruvic acid gave poor results.

Bestmann and co-workers[18] developed a synthesis of α-ketoaldehydes which involves in the first step reaction of a diazo ketone with triphenylphosphine to form a phosphorazane (3). This can be converted into the α-ketoaldehyde (6) by hydrolysis

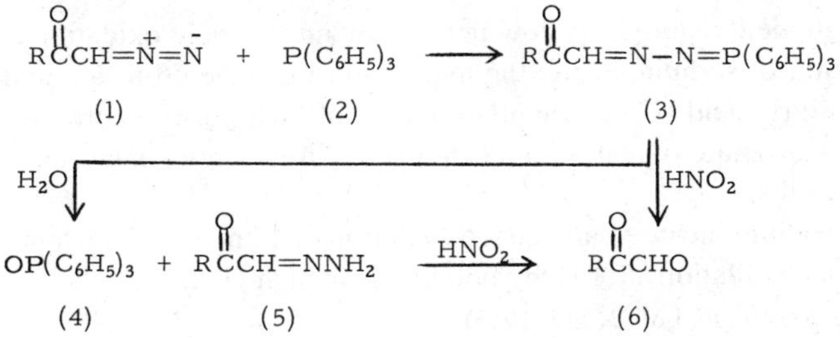
RCCH=N=N
$P(C_6H_5)_3$
RCCH=N–N=$P(C_6H_5)_3$
(1)
(2)
(3)
H_2O
HNO_2
$OP(C_6H_5)_3$
RCCH=NNH_2
HNO_2
RCCHO
(4)
(5)
(6)

and treatment of the hydrazone with nitrous acid or by direct treatment with nitrous acid. Phenylglyoxal was obtained in this way in overall yield of 70% based on benzoyl chloride.

Heterocycles. 1,2,3-Benzotriazole is prepared[19] by a method discovered by Ladenburg in 1876.[20] A solution of *o*-phenylenediamine in acetic acid and water is cooled to 5°, and an aqueous solution of sodium nitrite is added all at once with vigorous stirring in an ice bath. The temperature rises to 70–80° and the solution turns dark green and then clear orange-red. The product is obtained as an oil which

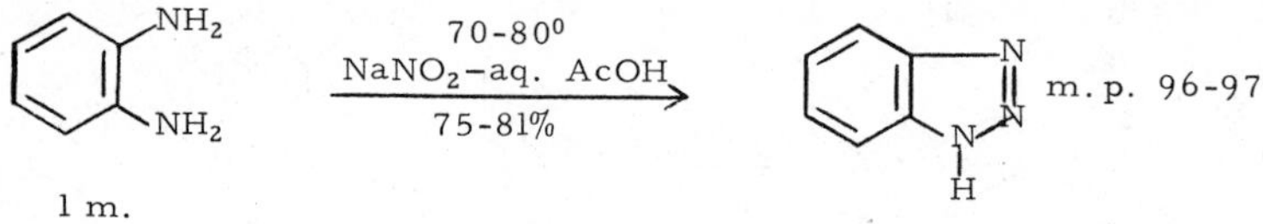

NH2
NH2
70-80°
NaNO2–aq. AcOH
75-81%
N
N
N
H
m.p. 96-97°

1 m.

soon solidifies, and it is purified by distillation and then crystallization. The higher-melting 5-nitroindazole is prepared similarly and crystallized from methanol.[21]

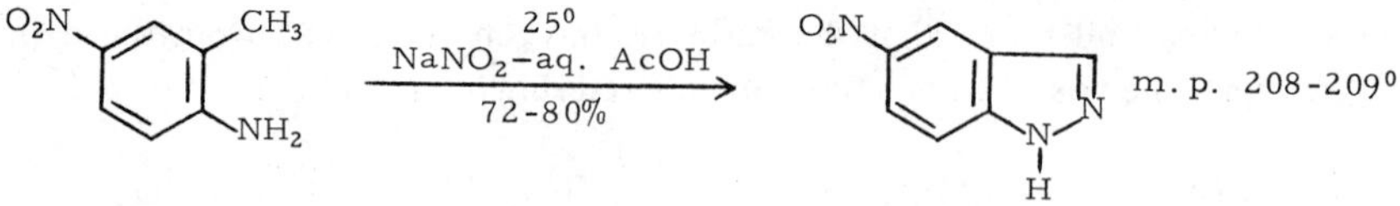

O2N
CH3
NH2
25°
NaNO2–aq. AcOH
72-80%
O2N
N
N
H
m.p. 208-209°

Oxidation. Fieser and Peters[22] found a convenient method for oxidation of the heterocyclic hydroquinone (1) to the quinone (2) consisting in adding excess aqueous

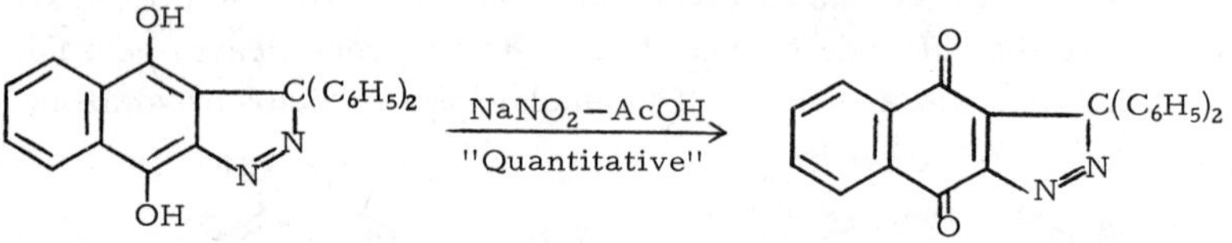

OH
C(C6H5)2
N
N
OH
NaNO2–AcOH
"Quantitative"
O
C(C6H5)2
N
N
O

(1) (2)

sodium nitrite to a solution of (1) in acetic acid; on dilution with water the quinone separates as yellow microcrystals. One of us[23] later found this method useful in perfecting a diene synthesis of 1,4-naphthoquinone from 1,4-benzoquinone. Although 5,8-dihydro-1,4-naphthoquinone (4) is light sensitive and highly reactive, it can be

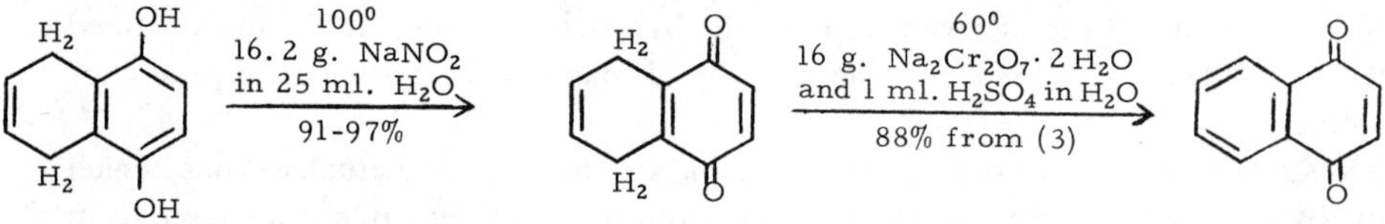

H2
OH
H2
OH
100°
16.2 g. NaNO2
in 25 ml. H2O
91-97%
H2
O
H2
O
60°
16 g. Na2Cr2O7·2H2O
and 1 ml. H2SO4 in H2O
88% from (3)
O
O

(3) 0.1 m. in 150 ml. AcOH (4) (5)

obtained in high yield as bright yellow needles by nitrous acid oxidation at 100°. If, after the addition of sodium nitrite, the temperature is lowered to 60° and aqueous dichromate–sulfuric acid added, naphthoquinone of high purity is produced in high yield. Direct dichromate oxidation of (3) to (5) requires a more elaborate procedure for satisfying results.

Addition of sodium nitrite to an aqueous solution of 4-aminothymol hydrochloride at 60° and steam distillation affords thymoquinone in high yield.[24]

[1]H. Fischer, *Org. Syn., Coll. Vol.*, **3**, 513 (1955)

[2]A. J. Zambito and E. E. Howe, *Org. Syn.*, **40**, 21 (1960)

[3]P. A. S. Smith, *Org. Syn., Coll. Vol.*, **4**, 819 (1963)

[4]H. J. Dauben, Jr., H. J. Ringold, R. H. Wade, D. L. Pearson, and A. G. Anderson, Jr., *ibid.*, **4**, 221 (1963)

[5]C. D. Gutsche, *Am. Soc.*, **71**, 3513 (1949); C. D. Gutsche and H. E. Johnson, *ibid.*, **77**, 109 (1955); *Org. Syn., Coll. Vol.*, **4**, 780 (1963)

[6]R. O. Lindsay and C. F. H. Allen, *ibid.*, **3**, 710 (1955)

[7]E. B. Womack and A. B. Nelson, *ibid.*, **3**, 392 (1955)

[8]F. C. Whitmore and M. G. Whitmore, *ibid.*, **1**, 401 (1941)

[9]H. H. Hodgson, A. P. Mahadevan, and E. R. Ward, *ibid.*, **3**, 341 (1955)

[10]L. Claisen and O. Manasse, *Ber.*, **22**, 526, 530 (1889)
[11]T. Wieland and D. Grimm, *Ber.*, **96**, 275 (1963)
[12]M. L. Wolfrom, L. W. Georges, and S. Soltzberg, *Am. Soc.*, **56**, 1794 (1934)
[13]K. von Auwers and B. Ottens, *Ber.*, **58**, 2067 (1925)
[14]H. Ohle, G. Henseke, and A. Czyzewski, *Ber.*, **86**, 316 (1953)
[15]R. N. Castle and M. Onda, *J. Org.*, **26**, 4465 (1961)
[16]St. Goldschmidt and W. L. C. Veer, *Rec. trav.*, **66**, 238 (1947)
[17]E. P. Oliveto, R. Rausser, L. Weber, E. Shapiro, D. Gould, and E. B. Hershberg, *Am. Soc.*, **78**, 1736 (1956)
[18]H.-J. Bestmann, O. Klein, L. Göthlich, and H. Buckschewski, *Ber.*, **96**, 2259 (1963)
[19]R. E. Damschroder and W. D. Peterson, *Org. Syn., Coll. Vol.*, **3**, 106 (1955)
[20]K. Ladenburg, *Ber.*, **9**, 219 (1876)
[21]H. D. Porter and W. D. Peterson, *Org. Syn., Coll. Vol.*, **3**, 660 (1955)
[22]L. F. Fieser and M. Peters, *Am. Soc.*, **53**, 4080 (1931)
[23]L. F. Fieser, *Am. Soc.*, **70**, 3165 (1948)
[24]E. Kremers, N. Wakeman, and R. M. Hixon, *Org. Syn., Coll. Vol.*, **1**, 511 (1941)

Sodium 2-nitropropanenitronate, $(CH_3)_2C{=}\overset{\overset{\displaystyle O^-}{|}}{N^+}{-}O^-Na^+$. Mol. wt. 111.09.

A solution of the reagent is prepared by adding 2-nitropropane to a solution of sodium ethoxide in absolute ethanol and adding more ethanol to dissolve any nitronate that precipitates. Hass and Bender[1] used the reagent for conversion of benzyl halides into aldehydes. The components were stirred at room temperature for about

$$\underline{o}\text{-}CH_3C_6H_4CH_2Cl + (CH_3)_2C{=}\overset{\overset{O^-}{|}}{N^+}{-}O^-Na^+ \xrightarrow[68-73\%]{C_2H_5OH} \underline{o}\text{-}CH_3C_6H_4CHO + (CH_3)_2C{=}NOH + NaCl$$

15 hrs., the solution filtered from sodium chloride, and the filtrate diluted with water and extracted with ether. Washing with aqueous sodium hydroxide removes acetoxime and excess 2-nitropropane. The method is not suitable for the preparation of nitrobenzaldehydes and fails with some polysubstituted benzyl halides.

Blomquist and co-workers[2] used the method successfully for the preparation of [10]paracyclophane-12-carboxyldehyde (2). An *Organic Syntheses* procedure[3] for

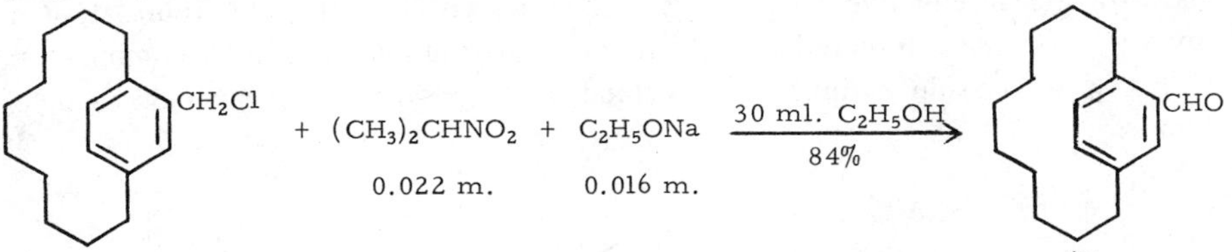
CH₂Cl
+ (CH₃)₂CHNO₂ + C₂H₅ONa
0.022 m.
0.016 m.
30 ml. C₂H₅OH
84%
CHO

(1) 0.015 m. (2)

the preparation of α-naphthaldehyde by the Sommelet reaction in 75–82% yield afforded the aldehyde (2) at best in 20% yield.

The reaction of the reagent with 9,10-bis(chloromethyl)anthracene provides the simplest synthesis of anthracene-9,10-dicarboxaldehyde.[4] Dimethyl sulfoxide is preferred to dimethylformamide as solvent.

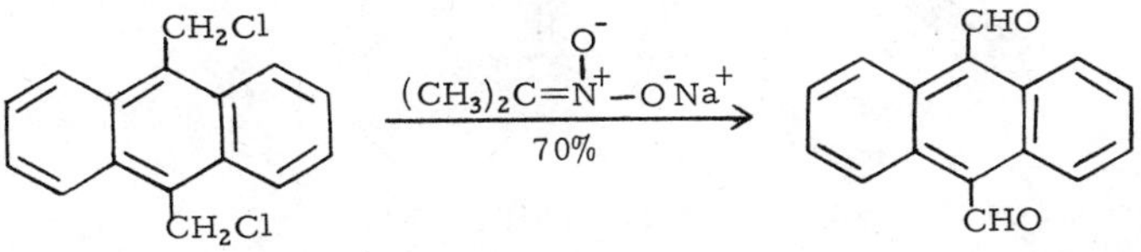
CH₂Cl
O⁻
(CH₃)₂C=N⁺–O⁻Na⁺
70%
CH₂Cl
CHO
CHO

[1]H. B. Hass and M. L. Bender, *Am. Soc.*, **71**, 1767 (1949)
[2]A. T. Blomquist, R. E. Stahl, Y. C. Meinwald, and B. H. Smith, *J. Org.*, **26**, 1687 (1961)

[3]S. J. Angyal, J. R. Tetaz, and J. G. Wilson, *Org. Syn., Coll. Vol.*, **4**, 690 (1963)
[4]B. H. Klanderman, *J. Org.*, **31**, 2618 (1966)

Sodium perborate, $NaBO_3 \cdot 4H_2O$. Mol. wt. 153.88. Suppliers: Mallinckrodt, F, MCB.

With acetic acid as solvent, sodium perborate oxidizes *p*-substituted derivatives of aniline to azobenzenes at temperatures of 35–60°.[1] An example [2] is formulated:

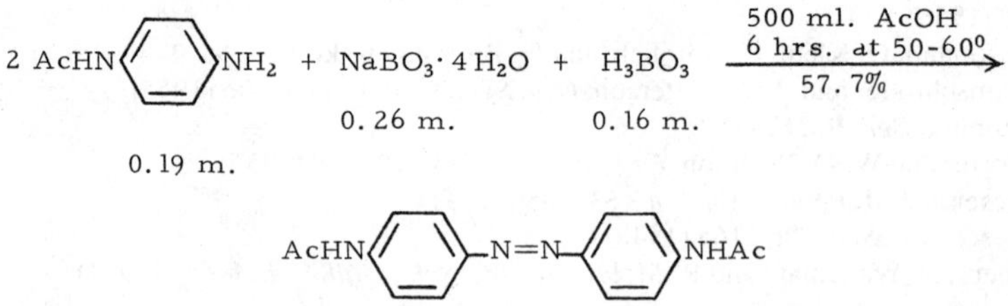
2 AcHN
NH₂
+ NaBO₃·4H₂O +
H₃BO₃
500 ml. AcOH
6 hrs. at 50-60°
57.7%
0.26 m.
0.16 m.
0.19 m.
AcHN
N=N
NHAc

Addition of boric acid improves the yield. The reactants initially dissolve and after about 40 min. the product begins to separate.

See also Perphthalic acid for the reaction of sodium perborate with phthalic anhydride.

[1]S. M. Mehta and M. V. Vakilwala, *Am. Soc.*, **74**, 563 (1952)
[2]P. Santurri, F. Robbins, and R. Stubbings, *Org. Syn.*, **40**, 18 (1960)

Sodium persulfate, $Na_2S_2O_8$. Mol. wt. 238.11. Suppliers: Alfa, British Drug Houses.

Oxidations with sodium persulfate are markedly catalyzed by silver ions; the actual reagent is presumed to be $S_2O_8^{2-}Ag^+$. Toluene is oxidized to benzaldehyde (50% yield) together with coupled products, and benzyl alcohol is oxidized to benzaldehyde (75% yield).[1] Phenols are oxidized to resinous polyphenols.[2]

[1]R. G. R. Bacon and J. R. Doggart, *J. Chem. Soc.*, 1332 (1960)
[2]R. G. R. Bacon, R. Grime, and D. J. Munro, *ibid.*, 2275 (1954)

Sodium phenolate, C_6H_5ONa. Magnesium methyl carbonate (*which see*) has been shown to be an effective agent for carboxylating active methylene groups. Italian investigators[1] have found that sodium (or potassium) phenolate in combination with carbon dioxide carboxylates cyclohexanone in dimethylformamide to give

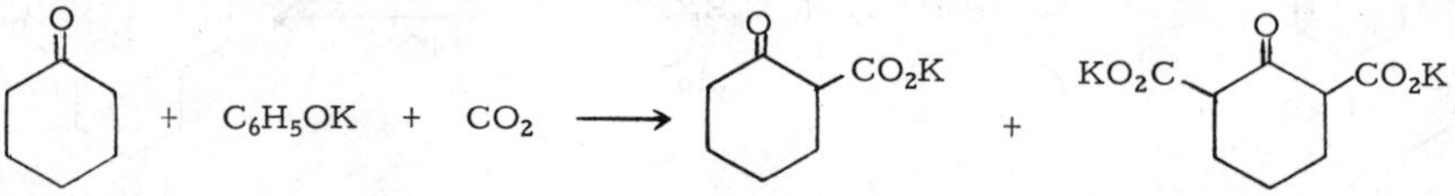
O
+ C₆H₅OK + CO₂
CO₂K
+
KO₂C
CO₂K

the cyclohexanonemonocarboxylate and cyclohexanone-2,6-dicarboxylate. Nuclear reaction as in the Kolbe synthesis is prevented, when necessary, by substituting or deactivating the reactive positions or by using a substance lacking hydrogen in such positions, such as the dimethylhydroxythiazole I.

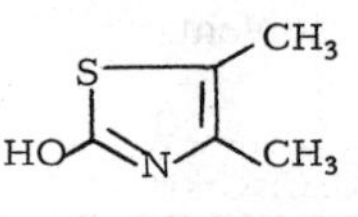
CH₃
S
HO
N
CH₃

I

[1]G. Bottaccio and G. P. Chiusoli, *Chem. Comm.*, 618 (1966)

Sodium–Potassium alloy (1:5, liquid). The alloy is prepared by heating 1 part of sodium and 5 parts of potassium under xylene until melted and carefully mixing the molten metals with a stirring rod and keeping the alloy in one large globule. On cooling, the alloy remains liquid and can be transferred by pipet.[1] The reagent reacts to give potassium, not sodium derivatives, for example:

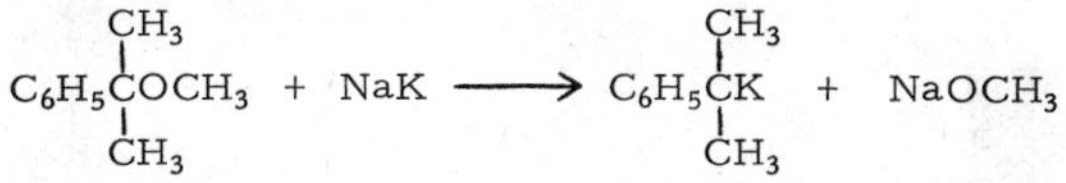

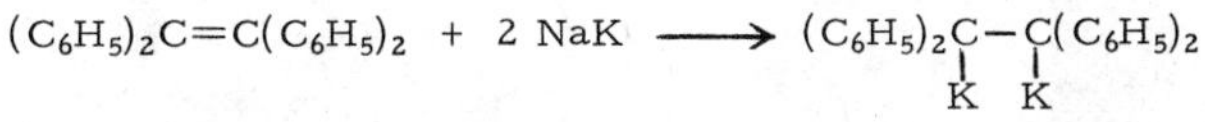

$$(C_6H_5)_3CLi + NaK \longrightarrow (C_6H_5)_3CK$$

Metalation of benzene is effected by reaction with an alkyllithium and sodium–potassium alloy in ether.[2]

$$RLi + \xrightarrow{NaK} RK \xrightarrow{C_6H_6} RH + C_6H_5K$$

This alloy was used by Ziegler[3] to cleave methyl ethers and by Müller and Bunge[4] to cleave diphenyl ether:

$$C_6H_5OC_6H_5 + 2\ NaK \longrightarrow C_6H_5OK + C_6H_5K \xrightarrow{CO_2;\ H^+} C_6H_5OH + C_6H_5CO_2H$$

Markiewitz and Dawson[5] separated olefinic components of the poison ivy principle "urushiol" in active phenolic form by benzylation, chromatography, and debenzylation with a dispersion of sodium–potassium alloy in ligroin. With this reagent debenzylation was effected without alteration in the position or geometrical configuration of mutliple olefinic bonds separated from one another by only one methylene group (sodium isomerizes such compounds to conjugated systems, which are then reduced by the metal).

Sodium itself is not satisfactory for effecting acyloin condensation in the thiophene series, but sodium–potassium alloy can be used. Thus with this reagent (2) was obtained readily from (1).[6]

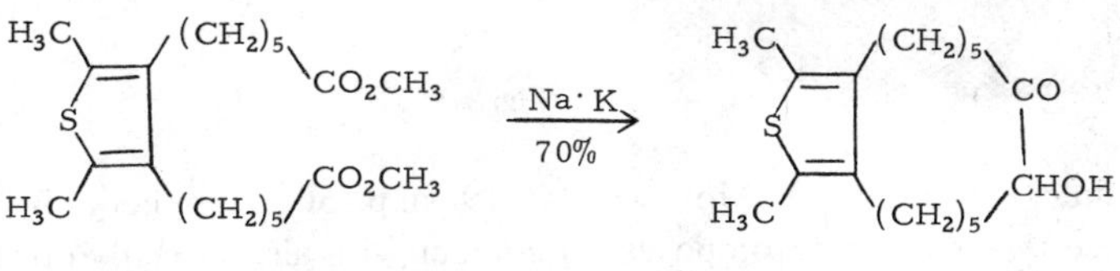

(1) (2)

[1]H. Gilman and R. V. Young, *J. Org.*, **1**, 315 (1936)
[2]D. Bryce-Smith and E. E. Turner, *J. Chem. Soc.*, 861 (1953)
[3]K. Ziegler and B. Schnell, *Ann.*, **437**, 227 (1924)
[4]E. Müller and W. Bunge, *Ber.*, **69**, 2164 (1936)
[5]K. H. Markiewitz and C. R. Dawson, *J. Org.*, **30**, 1610 (1965)
[6]Ya. L. Gol'dfarb, S. Z. Taits, and L. I. Beten'kii, *Tetrahedron*, **19**, 1851 (1963)

Sodium sulfate, anhydrous, Na_2SO_4. Mol. wt. 142.05. Suppliers: Mallinckrodt, Merck, MCB.

The reagent has a high capacity of absorbing water because at temperatures below 33° it forms the hydrate $Na_2SO_4 \cdot 10H_2O$. A useful technique for drying an ethereal extract involves shaking with saturated salt solution to remove most of the water and running the solution from the separatory funnel through a cone of drying agent on a filter paper. Sodium sulfate, being granular, is easier to pour into a filter than powdery magnesium sulfate, conveniently from the handy dispenser shown in Fig. S-3 (Mallinckrodt's rectangular bottle saves shelf space).

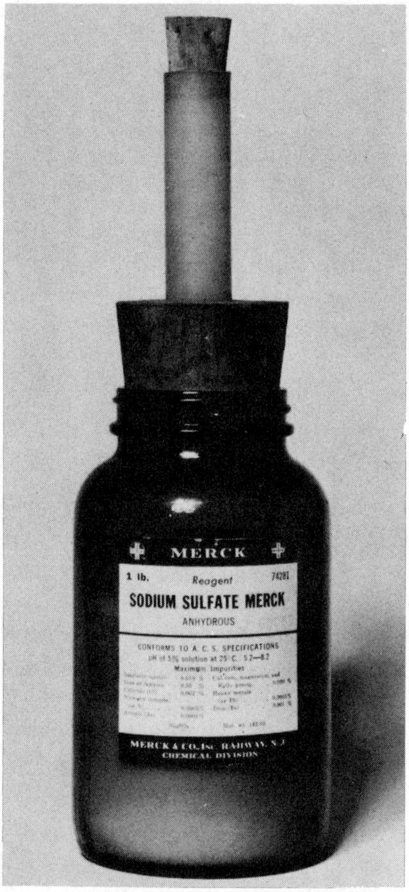
MERCK
1 lb.
Reagent
74281
SODIUM SULFATE MERCK
ANHYDROUS
CONFORMS TO A. C. S. SPECIFICATIONS
MERCK & CO., INC.
CHEMICAL DIVISION

Fig. S-3

Sodium sulfide, $Na_2S \cdot 9H_2O$. Mol. wt. 240.19, m.p. 50°. Suppliers: B, F, MCB, Merck. Since the reagent decomposes on contact with air, a freshly opened bottle should be used.

The reagent is used in 10% excess for selective reduction of one nitro group of 2,4-dinitrophenol with fair efficiency.[1] Note that the group reduced is that *ortho*

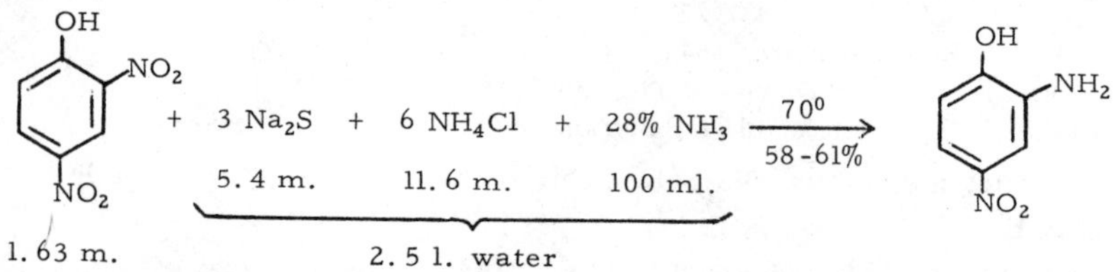
OH
NO2
NO2
1.63 m.
+ 3 Na2S + 6 NH4Cl + 28% NH3
5.4 m.
11.6 m.
100 ml.
2.5 l. water
70°
58-61%
OH
NH2
NO2

to the hydroxyl group. In the nearly quantitative selective reduction of 2,4-dinitronaphthol with stannous chloride the group reduced is that *para* to the hydroxyl group (*see* Stannous chloride).

In the preparation of di(*p*-aminophenyl) disulfide,[2] sodium sulfide first displaces the activated halogen atom and then reduces the nitro group. The terminal step is

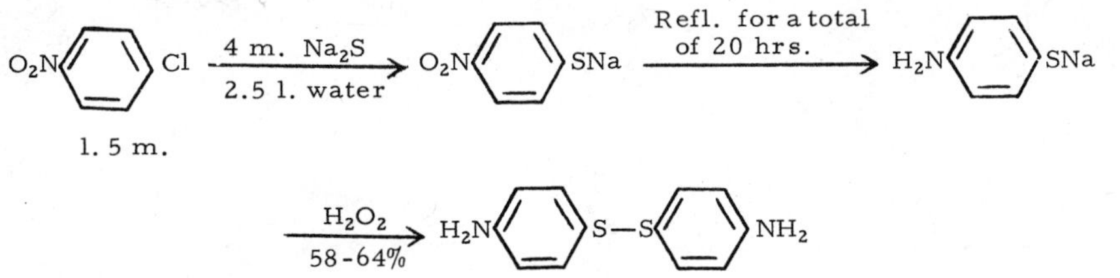

an oxidative coupling with hydrogen peroxide. The displacement of chlorine involved in the preparation of β-thiodiglycol proceeds rapidly.[3]

$$\underset{\substack{0.37\ m.\\ 20\%\ \text{aqueous solution}}}{HOCH_2CH_2Cl} + \underset{2.05\ m.}{Na_2S} \xrightarrow[79-86\%]{45\ \text{min. at}\ 87^0} HOCH_2CH_2SCH_2CH_2OH$$

In the first step of a preparation of diethyl mercaptoacetal[4] a mixture of sodium sulfide and sulfur in ethanol is refluxed until a deep red solution results, the heat is

$$\underset{3\ m.}{BrCH_2CH(OC_2H_5)_2} + \underset{2.5\ m.}{Na_2S\cdot 9\ H_2O} + \underset{4.5\ g.\ atoms}{S} + \underset{2.7\ l.}{95\%\ EtOH} \xrightarrow[83-92\%]{\text{Refl. 4 hrs.}}$$

$$(C_2H_5O)_2CHCH_2-(S)_4-CH_2CH(OC_2H_5)_2 \xrightarrow{\text{Na-liq. } NH_3;\ HCl} HSCH_2CH(OC_2H_5)_2$$

removed, and diethyl bromoacetal is added in 30 min. Sodium bicarbonate is added as buffer, most of the ethanol is distilled, water is added, and the polysulfide is extracted with ether. This material is a mixture of average composition of the tetrasulfide, and the percentage yield is calculated on this basis. If sodium sulfide is used considerable monosulfide is formed, and this is not subsequently reduced by sodium in liquid ammonia. Polysulfides are easily reduced.

[1]W. W. Hartman and H. L. Silloway, *Org. Syn., Coll. Vol.*, **3**, 83 (1955)
[2]C. C. Price and G. W. Stacy, *ibid.*, **3**, 86 (1955)
[3]E. M. Faber and G. E. Miller, *ibid.*, **2**, 576 (1943)
[4]W. E. Parham and H. Wynberg, *ibid.*, **4**, 295 (1963)

Sodium thiocyanate, NaSCN. Mol. wt. 81.09. Suppliers: B, E, F, MCB.

For the preparation of isopropyl thiocyanate[1] a suspension of sodium thiocyanate

$$\underset{5\ m.}{(CH_3)_2CHBr} + \underset{5.5\ m.}{NaSCN} \xrightarrow[76-79\%]{\substack{\text{Refl. 6 hrs.}\\ 1250\ \text{ml. } 90\%\ EtOH}} (CH_3)_2CHSCN$$

in 90% ethanol is stirred and refluxed during addition in 1 hr. of isopropyl bromide, and the mixture is refluxed with stirring for 6 hrs. Sodium chloride is removed by filtration, most of the alcohol is removed by distillation, water is added, and the product extracted with ether.

Use of the reagent for construction of a heterocyclic ring is illustrated by the preparation of 2-amino-6-methylbenzothiazole.[2] Concentrated sulfuric acid is added dropwise with stirring to a solution of *p*-toluidine in chlorobenzene, sodium

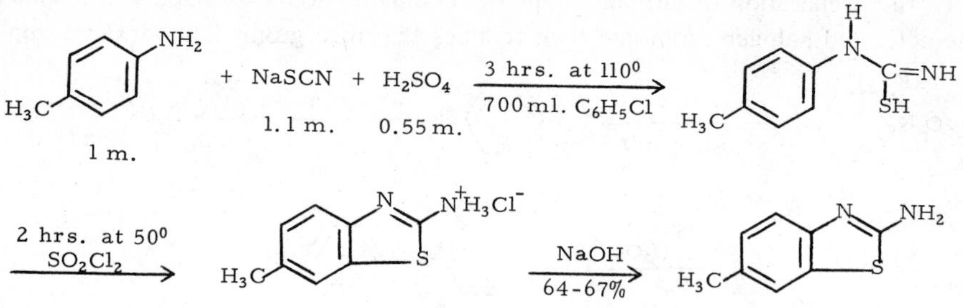
NH_2
H_3C
1 m.
+ NaSCN + H_2SO_4
1.1 m. 0.55 m.
3 hrs. at 110°
700 ml. C_6H_5Cl
H
N
C=NH
SH
H_3C
2 hrs. at 50°
SO_2Cl_2
N
$\overset{+}{N}H_3Cl^-$
S
H_3C
NaOH
64-67%
N
NH_2
S
H_3C

thiocyanate is added to the resulting fine suspension of *p*-toluidine sulfate, and the mixture is heated for 3 hrs. at 110° to produce *p*-tolylthiourea and sodium sulfate. The mixture is cooled to 30°, sulfuryl chloride is added at 30–50°, and heating at 50° is continued for 2 hours. The product separates as the hydrochloride, which is collected and dissolved in water. After steam distillation to remove remaining chlorobenzene the product is precipitated with ammonium hydroxide and crystallized from ethanol.

[1] R. L. Shriner, *Org. Syn., Coll. Vol.*, **2**, 366 (1943)
[2] C. F. H. Allen and J. VanAllan, *ibid.*, **3**, 76 (1955)

Sodium thiophenoxide, $C_6H_5S^-Na^+$. Mol. wt. 142.17.

Although this substance is a weaker base than either sodium ethoxide or sodium phenoxide, it has much greater nucleophilic power.[1] Sheehan and Daves[2] used it in dimethylformamide at or below room temperature to cleave sensitive ester functions (principally phenacyl) at the C—O bond to generate sodium salts of the corresponding carboxylic acids in good to excellent yield, as illustrated for phenacyl benzoate. The reagent is prepared by adding thiophenol to a molar proportion of

$$\underset{1\text{ m.}}{C_6H_5\overset{O}{\overset{\|}{C}}CH_2O-\overset{O}{\overset{\|}{C}}C_6H_5} + \underset{2\text{ m.}}{C_6H_5SNa} \xrightarrow[3\text{ min.}]{\text{DMF } 25^0} C_6H_5\overset{O}{\overset{\|}{C}}CH_2SC_6H_5 + \underset{87\%}{C_6H_5CO_2Na}$$

finely dispersed sodium in ether, stirring for 72 hrs., and collecting the salt by filtration and storing it in a vacuum dessicator. It is particularly useful for removing a masking ester function in a synthetic series involving sensitive molecules such as penicillins and peptides. Thus sodium thiophenoxide selectively removed phenacyl esters from molecules containing the highly labile phthalimido system.

Sodium thiophenoxide appears to be the best reagent for the selective demethylation of quaternary ammonium salts.[3] 2-Butanone is used as solvent. The reaction involves S_N2 attack by the phenoxide anion on the N-methyl group. An example is the demethylation of (±)-laudanosine methochloride to laudanosine. A limitation

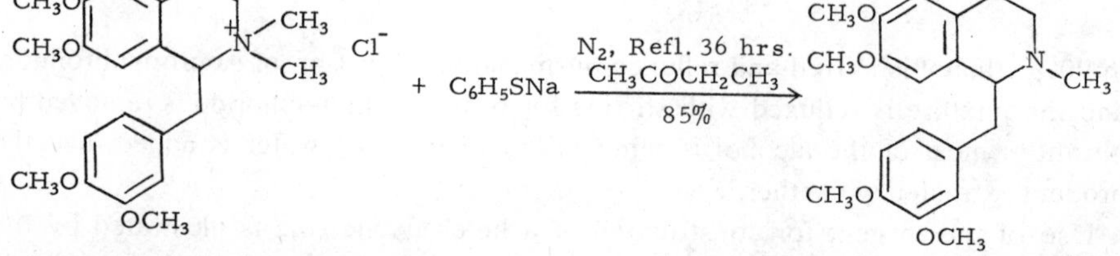
CH_3O
CH_3O
CH_3
CH_3
Cl^-
CH_3O
OCH_3
+ C_6H_5SNa
N_2, Refl. 36 hrs.
$CH_3COCH_2CH_3$
85%
CH_3O
CH_3O
N
CH_3
CH_3O
OCH_3

is that an ester group if present is converted into the thiophenyl derivative of the corresponding carboxylic acid.[2]

[1]P. B. D. de la Mare and C. A. Vernon, *J. Chem. Soc.*, 41 (1956)
[2]J. C. Sheehan and G. D. Daves, Jr., *J. Org.*, **29**, 2006 (1964)
[3]M. Shamma, N. C. Deno, and J. F. Remar, *Tetrahedron Letters*, 1375 (1966)

Sodium thiosulfate, $Na_2S_2O_3 \cdot 5H_2O$. Mol. wt. 248.21. Suppliers: B, F, MCB.

Titration with sodium thiosulfate solution is useful for determining the amount or purity of iodine, monochlorourea, N-bromosuccinimide,[1] etc.

[1]E. P. Oliveto and C. Gerold, *Org. Syn., Coll. Vol.*, **4**, 104 (1963)

Sodium trichloroacetate, CCl_3CO_2Na. Mol. wt. 185.39. Supplier: Robeco Chemicals Inc.

Preparation. (a)[1] Place 12.8 g. of trichloroacetic acid in a 125-ml. Erlenmeyer with a side arm and, with cooling in an ice bath, pour in about nine tenths of an ice cold solution of 3.2 g. of sodium hydroxide pellets in 12 ml. of water. Add a drop of 0.04% Bromocresol Green solution to produce a faint yellow color, and with a capillary dropping tube titrate the solution to an end point where a single drop produces a change from yellow to blue. Close the flask with a rubber stopper, mount it within the rings of a steam bath and wrap a towel around it for maximal heat, connect it to the suction pump, and turn on the water at full speed. The evaporation needs no attention and can be completed in 15–20 minutes. The weight of dry powder is close to the theory (14.5 g.).

(b)[2] A solution of trichloroacetic acid in absolute methanol was neutralized to the phenolphthalein end point with a freshly prepared solution of sodium methoxide while keeping the temperature below 20°. The excess methanol was removed by a rotary evaporator at 20° leaving the salt as a white solid, which was dried at 50° over P_2O_5 for 20 hrs.

Generation of dichlorocarbene. Thermal decomposition of the salt in an aprotic solvent containing a reactive olefin generates dichlorocarbene for addition to the olefin.[3-5] In a student experiment[1] in which *cis,cis*-1,5-cyclooctadiene affords a mixture of two crystalline bis-adducts, the main solvent is tetrachloroethylene, selected because of its appropriate boiling point (121°) and low degree of reactivity. However, since sodium trichloroacetate has little if any solubility in this solvent and does not react at the boiling point, a little diglyme is used as co-solvent. Diglyme

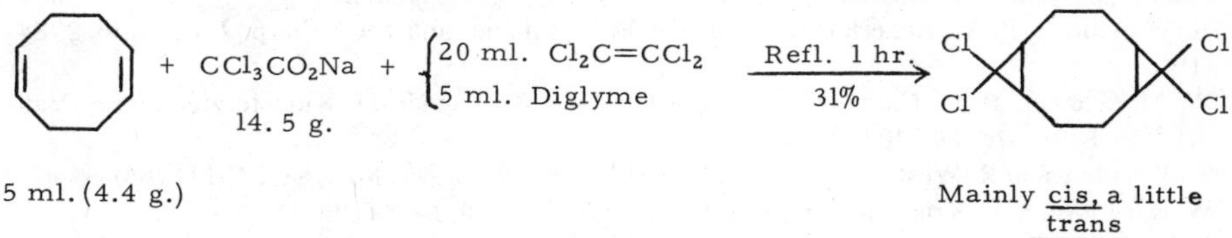

+ CCl₃CO₂Na +
14. 5 g.
20 ml. Cl₂C=CCl₂
5 ml. Diglyme
Refl. 1 hr.
31%
Cl
Cl
Cl
Cl
5 ml. (4.4 g.)
Mainly cis, a little trans

is so powerful a solvent for the salt that if used alone it has to be added in portions with control of the temperature to 120° and the solution becomes very black. With the solvent mixture recommended the salt can be added all at once and the mixture let reflux unattended, for then the salt dissolves slowly as the reaction proceeds. What little color develops is eliminated, after steam distillation, by washing the crude product with methanol.

Fields[6] found this precursor superior to others for the generation of CCl_2 for insertion into the C—H bond of cumene. Thus when a mixture of cumene and sodium trichloroacetate was refluxed in glyme until CO_2 was no longer evolved,

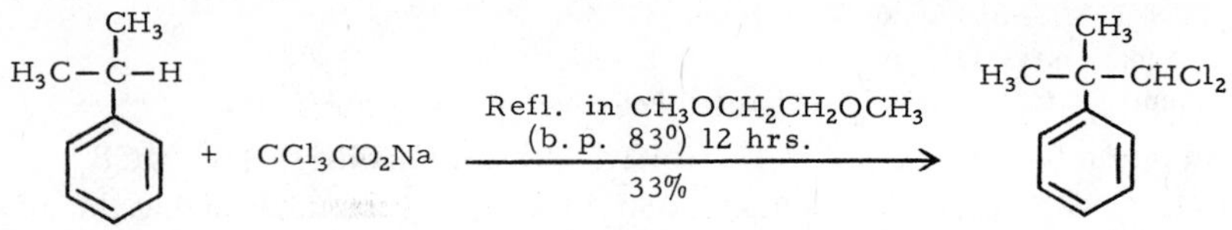

CH3
H3C–C–H
+ CCl3CO2Na
Refl. in CH3OCH2CH2OCH3
(b.p. 83°) 12 hrs.
33%
CH3
H3C–C–CHCl2

β,β-dichloro-*t*-butylbenzene was obtained in 33% yield. Generation of dichlorocarbene from chloroform and potassium *t*-butoxide or from ethyl trichloroacetate and sodium methoxide gave the same product in yields of only 0.5 and 5%.

To effect addition of dichlorocarbene to trichloroethylene to produce pentachlorocyclopropane, Tobey and West[4] heated and stirred a mixture of 1,600 g. of commercial sodium trichloroacetate containing 6% of water with 2.5 l. of trichloroethylene under a water take off trap for removal of all water (2 hrs.). Evolution of

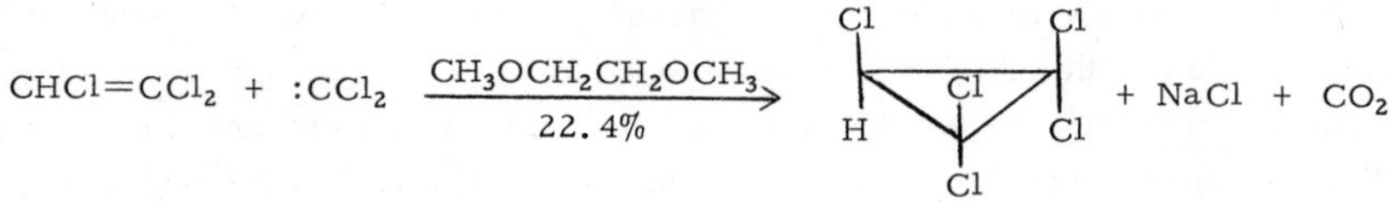

CHCl=CCl2 + :CCl2
CH3OCH2CH2OCH3
22.4%
Cl
Cl
Cl
H
Cl
Cl
+ NaCl + CO2

(b.p. 56°/7 mm.)

carbon dioxide began only after addition of 750 ml. of dimethoxyethane (glyme, b.p. 83°) and further heating and continued for 2 days, with gradual darkening of the pot contents. Drying the sodium salt under vacuum did not improve the yield of product.

Trichloromethylation of anhydrides.[2] When the salt is decomposed in 1,2-dimethoxyethane in the presence of an olefinic anhydride, reaction occurs exclusively at a carbonyl group of the anhydride, apparently by the trichloromethyl anion precursor of the carbene, to give the trichloromethyl addition product. Yields are erratic, ranging from 80 to 8%.

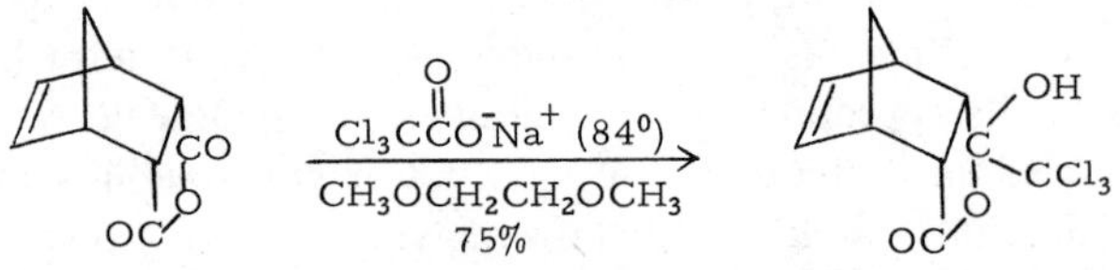

CO
O
OC
Cl3CCO⁻Na⁺ (84°)
CH3OCH2CH2OCH3
75%
OH
C
CCl3
O
OC

[1]L. F. Fieser and D. H. Sachs, *J. Org.*, **29**, 1113 (1964); *Org. Expts.*, 205

[2]A. Winston, J. P. M. Bederka, W. G. Isner. P. C. Juliano, and J. C. Sharp, *J. Org.*, **30**, 2784 (1965)

[3]W. M. Wagner, *Proc. Chem. Soc.*, 229 (1959); W. M. Wagner, H. Kloosterziel, and S. Van der Ven, *Rec. trav.*, **80** 740 (1961)

[4]S. W. Tobey and R. West, *Tetrahedron Letters*, 1179 (1963); *Am. Soc.*, **88**, 2481 (1966)

[5]W. R. Moore, S. E. Krikorian, and J. E. LaPrade, *J. Org.*, **28**, 1404 (1963)

[6]E. K. Fields, *Am. Soc.*, **84,** 1744 (1962)

Sodium triethoxyaluminum hydride, $NaAlH(OC_2H_5)_3$. Mol. wt. 37.95.

Prepared by the reaction of aluminum ethoxide with sodium hydride, it is a colorless, microcrystalline powder stable for several weeks in the absence of air and is used in ether or THF.[1] A somewhat milder reducing agent than lithium aluminum hydride, it is recommended particularly for the reduction of aromatic nitriles to aldehydes: $C_6H_5CN \longrightarrow C_6H_5CHO$ (95%).

[1]G. Hesse and R. Schrödel, *Ann.*, **607**, 24 (1957)

Sodium trimethoxyborohydride, $NaBH(OCH_3)_3$. Mol. wt. 127.93, m.p. 230°. Suppliers: MSA Research Corp., KK (10-g. lots).

Preparation from sodium hydride and methyl borate.[1]

$$NaH + B(OCH_3)_3 \xrightarrow{THF} NaBH(OCH_3)_3$$

The reagent is used for the reduction of aldehydes, ketones, acid anhydrides, and acid chlorides. Esters and nitriles are reduced only slowly at elevated temperatures.[2]

[1]H. C. Brown, H. I. Schlesinger, I. Sheft, and D. M. Ritter, *Am. Soc.*, **75**, 192 (1953)
[2]H. C. Brown and E. J. Mead, *ibid.*, **75**, 6263 (1953)

Solvents (*see also* Diels-Alder solvents). Table 1 lists 51 pure solvents available commercially in the United States. They are arranged in four solubility groups, and the members of each group are listed in order of increasing boiling point.

One supplier, Burdick and Jackson Laboratories, makes the point that their solvents are all distilled in glass through efficient columns after physical and/or chemical pretreatment as required. Boiling range is at most 1–2°. Dimethylformamide, dimethyl sulfoxide, and N-methylpyrrolidone are distilled in vacuum, but vacuum boiling points are not recorded in the table. Sensitive solvents are packed under nitrogen, in brown bottles if required. Most products are essentially pure and dry and contain no residue. Exceptions are dioxane and 2-methoxyethanol, which contain 10 ppm. of *p*-benzylaminophenol as antioxidant; chloroform and diethyl ether are stablized with ethanol. Shipping containers are 1-gallon glass jugs with finger-grip handles and Teflon-lined caps. Ultraviolet, infrared, and gas chromatographic data are available on request for most of the solvents offered.

Table 1 Pure solvents available commercially

Solvent	B.p.	Suppliers[a]
Aprotic solvents insoluble in water		
Isopentane, $(CH_3)_2CHCH_2CH_3$	28°	MCB
Pentane	35–36°	BJ, MCB, USTC
Methylene chloride	40–41°	BJ, Mall., MCB
Petroleum ether	30–60°	BJ
Carbon disulfide	46.3°	Mall., MCB, USTC
Cyclopentane	48.5–49.5°	MCB
Neohexane, $(CH_3)_3CCH_2CH_3$	49–50°	BJ
Isohexane, $(CH_3)_2CHCH_2CH_2CH_3$	59–63°	BJ
Chloroform	60–61°	BJ, MCB, USTC, E
n-Hexane	68–69°	BJ, Mall., MCB, USTC
Carbon tetrachloride	76–77°	BJ, Mall., MCB, USTC, E
Benzene	80.1°	BJ, Mall., MCB, USTC, E
Cyclohexane	80.8°	BJ, Mall., MCB, E
Ethylene dichloride, $ClCH_2CH_2Cl$	83.8°	MCB, USTC, E
1-Chloro-2,2-dimethylpropane	84–85°	MCB
n-Heptane	95–99°	BJ, Mall., MCB, USTC
Isooctane, $(CH_3)_3CCH_2CH(CH_3)_2$	99–100°	BJ, Mall., MCB, E
Methylcyclohexane	100.5–101.5°	MCB, E
Toluene	109–111°	BJ, MCB, USTC
Tetrachloroethylene	120.8°	MCB
m-Xylene	138–139°	MCB
Bromoform	149.6°	MCB, E

Table 1 (continued) Pure solvents available commercially

Solvent	B.p.	Suppliers[a]
Aprotic solvents partially soluble in water		
Methyl formate, 30 g./100 g. $H_2O^{20°}$	32°	MCB, E
Diethyl ether, 7.5 g./100 g. $H_2O^{20°}$	34.6°	BJ, MCB, E
Ethyl formate, 11 g./100 g. $H_2O^{18°}$	54°	MCB
Ethyl acetate, 8.5 g./100 g. $H_2O^{15°}$	77–78°	BJ, Mall., MCB
Methyl ethyl ketone, 37 g./100 g. $H_2O^{25°}$	79–80°	BJ
Ethyl propionate, 2.4 g./100 g. $H_2O^{20°}$	99.1°	MCB
Nitromethane, 9.5 g./100 g. $H_2O^{20°}$	100–102°	MCB, E
Isobutanol, 10 g./100 g. $H_2O^{15°}$	108–109°	BJ
Isobutyl acetate, 0.6 g./100 g. $H_2O^{25°}$	115–117°	MCB
n-Butyl acetate, 0.7 g./100 g. $H_2O^{25°}$	126.5°	McB
Aprotic solvents, water-miscible		
Acetone	56–57°	BJ, Mall., MCB, USTC, E
Tetrahydrofurane	65.4°	MCB, BJ
Acetonitrile	81–83°	BJ, MCB, E
Glyme, $CH_3OCH_2CH_2OCH_3$	83°	BJ
Dioxane	100–102°	BJ, MCB
Pyridine	115.2°	Mall., MCB, USTC, E
Dimethylformamide	153°	BJ, MCB, E
Diglyme	161°	BJ
Dimethyl sulfoxide	189°	BJ
Triglyme	222°	BJ
1-Methyl-2-pyrrolidone	222°	BJ
Protonic solvents, partially soluble in water		
Butanol-2, 12.5 g./100 g. $H_2O^{20°}$	99–100°	BJ, MCB
Butanol-1, 9 g./100 g. $H_2O^{15°}$	116–118°	BJ, MCB
Isomyl alcohol, 2.8 g./100 g. $H_2O^{15°}$	131.5°	MCB
Protonic solvents, water-miscible		
Methanol	64.7°	BJ, Mall., MCB, USTC, E
Isopropanol	82.3°	BJ, Mall., MCB, USTC, E
2-Methoxyethanol	123–124°	BJ
Glycerol	290°	MCB

[a]MCB = Matheson Coleman and Bell; USTC = United States Testing Co.; Mall. = Mallinckrodt Chemical Works; BJ = Burdick and Jackson Laboratories; E = Eastman

The Mallinckrodt SpectrAr and Eastman Spectro Grade lines of spectrophotometric solvents were developed specifically to satisfy the need for solvents of extremely high purity required by modern analytical techniques. These solvents provide a high degree of freedom from impurities absorbing throughout the UV, visible, and IR regions of the spectrum. High purity and low limits for residue-on-evaporation make them particularly useful when small amounts of products are to be isolated by procedures requiring large volumes of solvent. A Mallinckrodt brochure giving specifications includes typical absorption curves for each of the

SpectrAR solvents. The premium solvents offered by Matheson Coleman and Bell and by United States Testing Co. are described as Spectro quality and Spectroline Reagents, respectively; in each case spectrographic control data are available.

Solvent wash bottles. The Nalgene[R] 500-ml. polyethylene wash bottles are recommended as containers for common solvents (see Fig. S-4). The name, symbol, or formula of a solvent can be written on a bottle with a Magic Marker or wax pencil. A volume up to 25 ml. per squeeze can be delivered from the spout into a graduate; a larger volume can be poured out of the central opening. For crystallizations and for the quick cleaning of apparatus it is convenient to have available one bottle for each frequently used solvent.

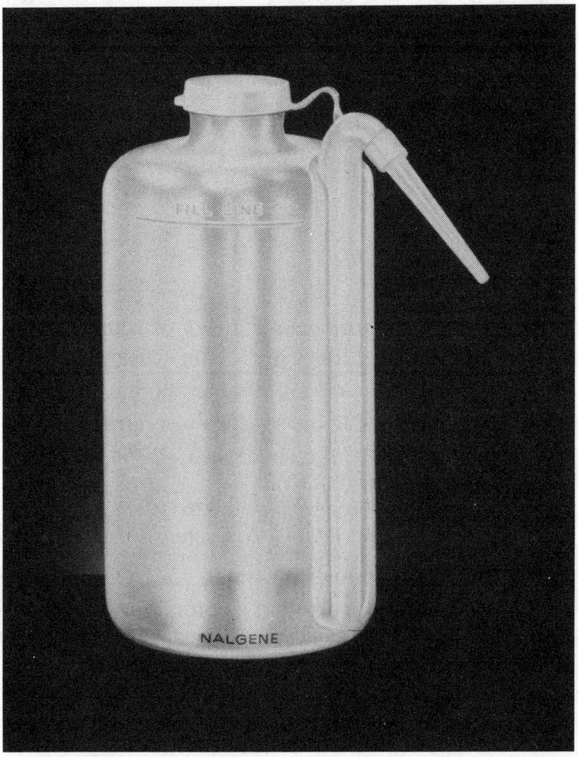
NALGENE

Fig. S-4 Polyethylene wash bottle[a]

Stannic chloride, $SnCl_4$. Mol. wt. 260.53, m.p. -30°, b.p. 114°, sp. gr. 2.226. Suppliers: B, F, MCB.

As a catalyst for the Friedel-Crafts reaction, stannic chloride is milder than aluminum chloride. Thiophene is so reactive that it can be acylated with stannic

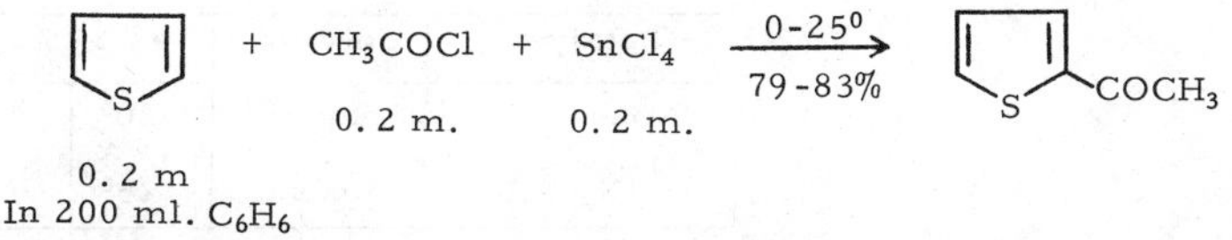

+ CH₃COCl + SnCl₄
0-25°
79-83%
COCH₃
S
0. 2 m.
0. 2 m.
0. 2 m
In 200 ml. C₆H₆

chloride as catalyst and benzene as solvent.[1] Aluminum chloride causes polymerization of thiophene. Benzene serves as solvent also in the intramolecular cyclization of γ-phenylbutyryl chloride.[2]

[a]Photograph courtesy of The Nalge Co., Inc.

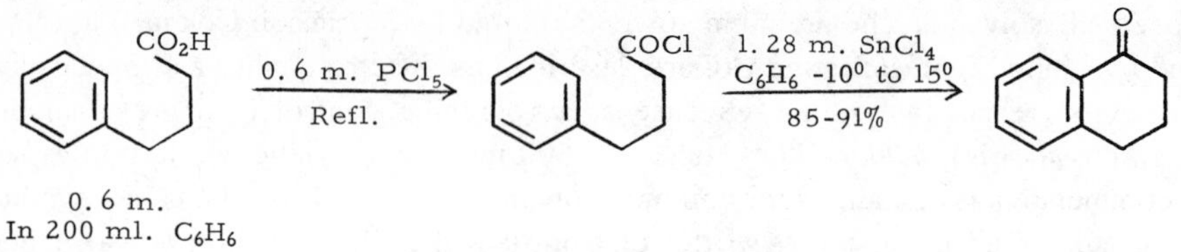

Stannic chloride is a moderately effective catalyst for the condensation of ethyl ketomalonate with *p*-xylene.[3] A mixture of the ester and hydrocarbon is cooled in

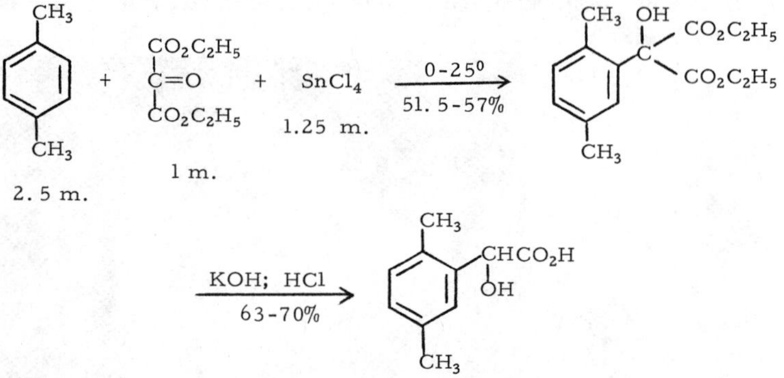

an ice bath during dropwise addition of stannic chloride, and the mixture is stirred at room temperature for 3 hrs. After workup, saponification, and decarboxylation, 2,5-dimethylmandelic acid is obtained pure by crystallization from benzene.

The reagent is used in the synthesis of tetraethyltin.[4]

[1]J. R. Johnson and G. E. May, *Org. Syn., Coll. Vol.*, **2**, 8 (1943)
[2]G. D. Johnson, *ibid*, **4**, 900 (1963)
[3]J. L. Riebsomer and J. Irvine, *ibid.*, **3**, 326 (1955)
[4]G. J. M. Van Der Kerk and J. G. A. Luijten, *ibid.*, **4**, 881 (1963)

Stannic chloride pentahydrate, $SnCl_4 \cdot 5\,H_2O$. Mol. wt. 350.61. Suppliers: B, F, MCB.

Lutz and Bailey[1] reported in a communication that this reagent not only catalyzes the reaction of methyl vinyl ketone or acrolein with isoprene but markedly effects the isomer distribution, as illustrated for the former dienophile.

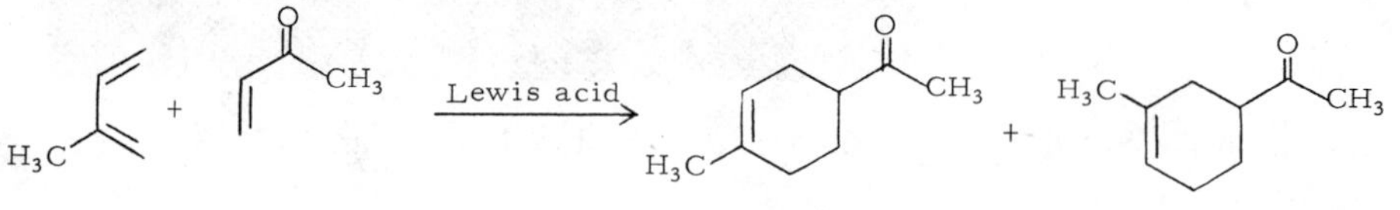

I II III IV

Catalyst	Solvent	Temp.	% III	% IV
None	Toluene	120°	71	29
$SnCl_4 \cdot 5\,H_2O$	Benzene	25°	93	7

[1]E. F. Lutz and G. M. Bailey, *Am. Soc.*, **86**, 3899 (1964)

Stannous bromide, $SnBr_2$. Mol. wt. 278.53. Suppliers: Alfa, KK.

m-Chlorobenzaldehyde can be prepared from *m*-nitrobenzaldehyde by reduction with stannous chloride and hydrochloric acid, diazotization by addition of sodium nitrite to the cooled acidic mixture, and addition of cuprous chloride and hydrochloric acid.[1] *m*-Bromobenzaldehyde prepared in the same way with use of CuBr–HBr contains about 20% of *m*-chlorobenzaldehyde, but this defect is corrected by use of stannous bromide and HBr for the initial reduction; a solution of the reagent can be prepared from mossy tin and 46% hydrobromic acid.[2]

[1]J. S. Buck and W. S. Ide, *Org. Syn., Coll. Vol.*, **2**, 130 (1943)
[2]F. T. Tyson, *ibid*, **2**, 132 (1943)

Stannous chloride, $SnCl_2 \cdot 2H_2O$. Mol. wt. 225.65. Suppliers: F, MCB.

$ArNO_2 \longrightarrow ArNH_2$. Efficient reduction of an aromatic nitro compound to the corresponding amine often can be accomplished satisfactorily by adding the nitro compound to a solution of stannous chloride dihydrate in concd. hydrochloric acid (1 ml. per g.); examples.[1-3] Oelschläger[4] suggests that in the reduction of a nitro compound on a large scale it may be advantageous to replace a part of the hydrochloric acid by sodium chloride to moderate the reaction. Sparingly soluble dinitrodurene is dissolved in acetic acid and reduced by addition of a solution of stannous chloride in concd. hydrochloric acid.[5] 1,5-Dinitro-2,6-dihydroxynaphthalene is reduced in 80% yield with stannous chloride in acetic acid to the air-sensitive diamine dihydrochloride.[6]

Hodgson and Smith[7] found that with use of just 3 equivalents of stannous chloride 2,4-dinitro-1-naphthol can be reduced selectively to 2-nitro-4-amino-1-napthol. A suspension of the dinitro compound in concd. hydrochloric acid and ethanol was

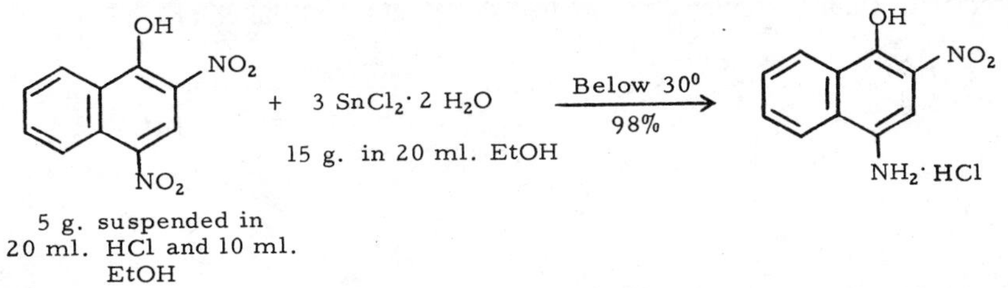
OH
NO2
NO2
+ 3 SnCl2· 2 H2O
15 g. in 20 ml. EtOH
Below 30°
98%
OH
NO2
NH2· HCl
5 g. suspended in
20 ml. HCl and 10 ml.
EtOH

stirred during addition in 1 hr. of a solution of stannous chloride in ethanol. The reduction product separated as pale yellow needles of the hydrochloride.

Antioxidant. In the conversion of 1-amino-2-naphthol or 4-amino-1-naphthol to the hydrochloride, or on recrystallization of such an air-sensitive hydrochloride, addition of a small amount of stannous chloride (2 g. per 0.5 mole) serves as an efficient antioxidant.[8]

$-CH_2X \longrightarrow -CH_3$. Rinkes[9] devised a procedure for the preparation of 5-methylfurfural (4) involving acid hydrolysis of sucrose (1), conversion of the fructose

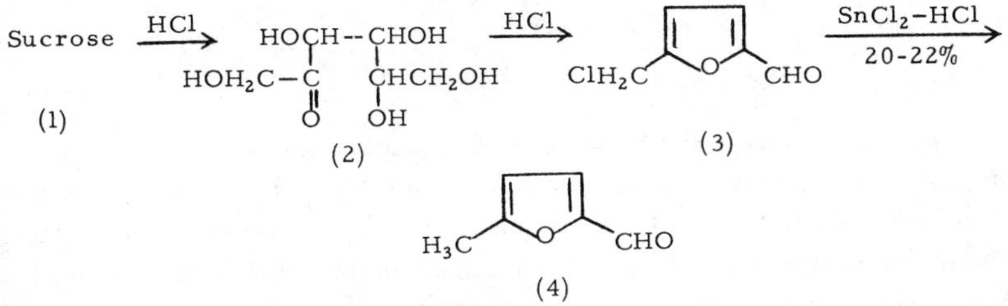
Sucrose
(1)
HCl
HOCH--CHOH
HOH2C—C
CHCH2OH
O
OH
(2)
HCl
ClH2C
O
CHO
(3)
SnCl2–HCl
20-22%
H3C
O
CHO
(4)

component (2) into the chloromethylfurane derivative (3), and reduction with stannous chloride to (4).

Sandin and Fieser[10] developed an efficient synthesis of the potently carcinogenic 9,10-dimethyl-1,2-benzanthracene from 1,2-benzanthraquinone (1) involving addition of methyl Grignard reagent, reaction of the adduct (2) with hydriodic acid to give the 9-methyl-10-iodomethyl derivative (3) and reduction with stannous chloride in dioxane–hydrochloric acid.

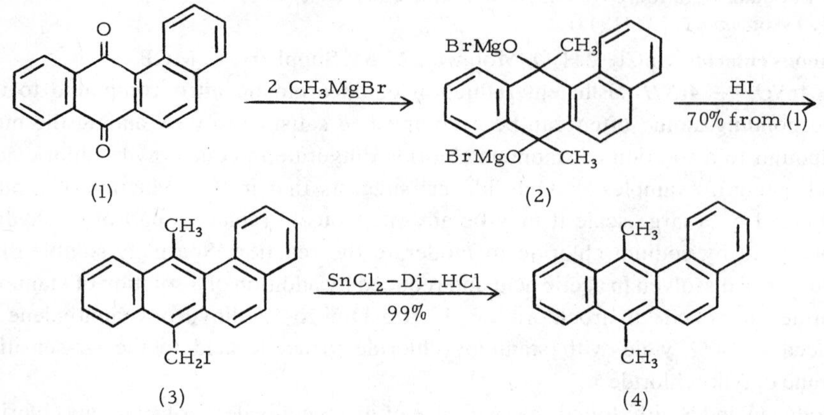

Aziridine synthesis. Closs and Brois[11] devised a new synthesis in which a tetraalkylethylene is converted into the blue monomeric chloronitroso derivative; this is reduced with stannous chloride in hydrochloric acid, and the chloroamine without

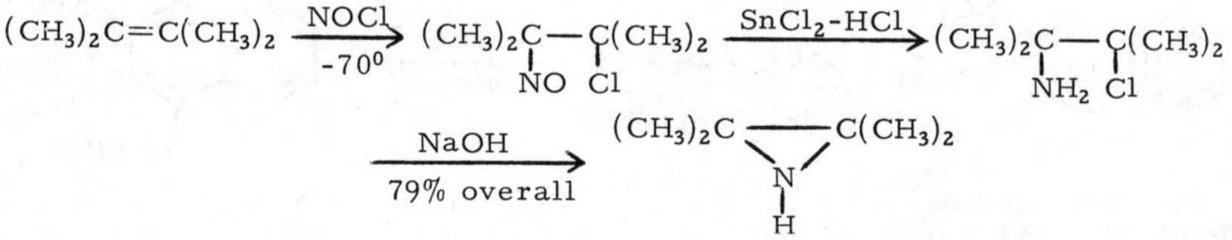

being isolated is treated with base to close the ring and produce, for example, 2,2,3,3-tetramethylaziridine. Other reducing agents tried and found ineffective included $LiAlH_4$, $NaBH_4$, Al(Hg), Zn–AcOH. The synthesis is applicable only to tetrasubstituted ethylenes.

Cleavage of N-nitrosoamines. Buck and Ferry[12] purified N-ethyl-*m*-toluidine by conversion into the N-nitroso derivative and cleavage of this by reduction with stannous chloride.

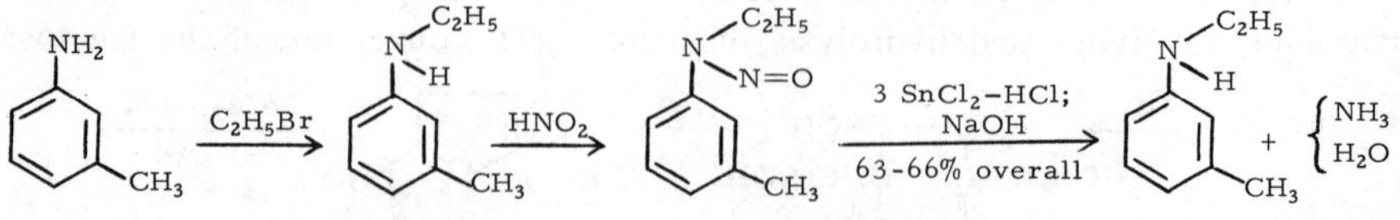

Cleavage of 2,4-dinitrophenylhydrazones. Demaecker and Martin[13] found that the 2,4-dinitrophenylhydrazones of saturated and unsaturated 3-ketosteroids can be cleaved in high yield (84–98%) by refluxing with a large excess of acetone to effect partial hydrazone interchange and then adding stannous chloride–hydrochloric acid to reduce one or both nitro groups. Cullinane and Edwards[14] found the method useful for characterization and purification of the oily *o*-hydroxyketones obtained by the Fries reaction. 2-Hydroxy-3-methylacetophenone 2,4-dinitrophenylhydrazone was refluxed with acetone until a clear solution resulted, a solution of stannous chloride and hydrochloric acid was added, and refluxing was continued for 30 min. If, in

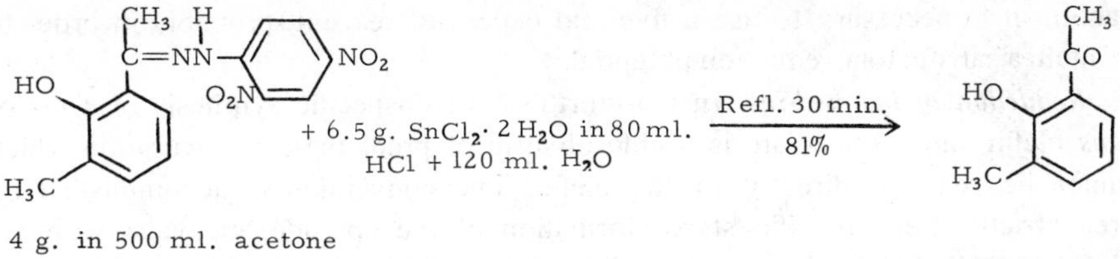
+ 6.5 g. SnCl2· 2 H2O in 80 ml. HCl + 120 ml. H2O
Refl. 30 min.
81%
4 g. in 500 ml. acetone

place of 6.5 g. of stannous chloride dihydrate the amount used was either 6.0 g. or 7.0 g., the yield dropped from 81 to 70%.

Quinones. Sager[15] developed a procedure for the preparation of tetrahydroxy-1,4-benzoquinone in which 30% glyoxal solution is added to a bicarbonate-buffered solution of sodium sulfite and a brisk stream of air is drawn through the solution.

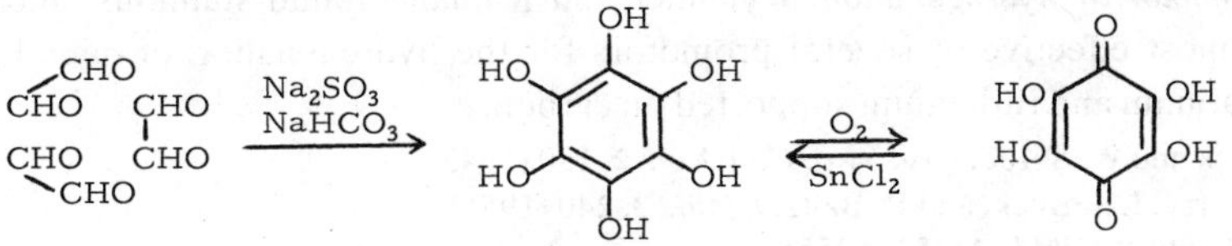
Na2SO3
NaHCO3
O2
SnCl2

Greenish black crystals of the disodium salt of tetrahydroxyquinone separate, and acidification affords glistening black crystals of tetrahydroxyquinone in 8% yield. Reduction of this with stannous chloride in hydrochloric acid affords hexahydroxybenzene in 70–77% yield.

Reduction of a quinone beyond the hydroquinone stage is noted only in the anthracene series. In the preferred laboratory procedure for the preparation of anthrone[16] a mixture of anthraquinone, stannous chloride in concd. hydrochloric acid,

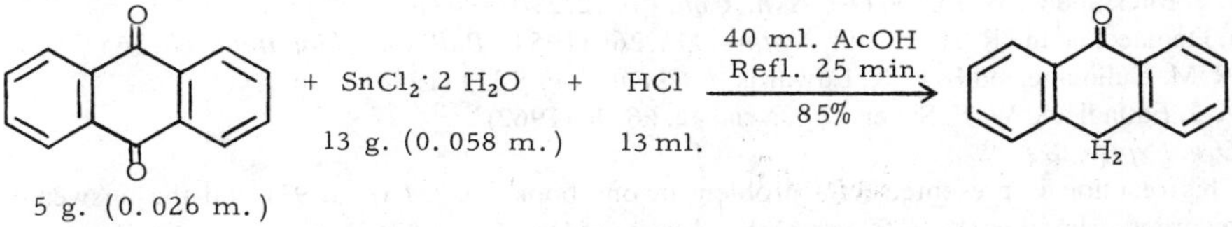
+ SnCl2· 2 H2O + HCl
13 g. (0.058 m.)
13 ml.
40 ml. AcOH
Refl. 25 min.
85%
5 g. (0.026 m.)

and acetic acid is refluxed gently until crystals of anthraquinone have disappeared completely (8–10 min.) and then for 15 min. longer. The solution is heated on the steam bath and water is added in 1-ml. portions to the point of saturation (about 12 ml.). Anthrone is obtained as pale yellow needles of high purity.

Reduction of $(C_6H_5)_3COH$.[17] Reduction of triphenylcarbinol with stannous chloride in acetic acid at the steam-bath temperature gives the triphenylmethyl

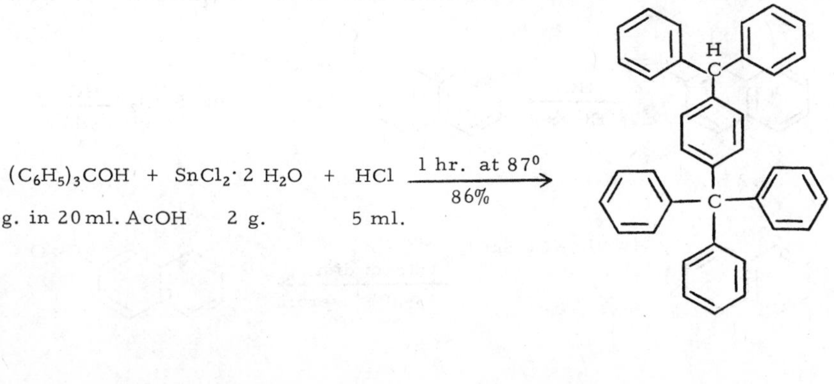
(C6H5)3COH + SnCl2· 2 H2O + HCl
1 g. in 20 ml. AcOH
2 g.
5 ml.
1 hr. at 87°
86%

radical which dimerizes to give *p*-benzhydryltetraphenylmethane (m.p. 224°). Theoretically the reduction requires 0.5 mole of stannous chloride, but for efficient reaction it is necessary to use a five-fold excess of reagent, probably in order to maintain a satisfactory reduction potential.

Reduction of iodohydrins. In Cornforth's[18] stereospecific synthesis of a *cis* or *trans* olefin, an intermediate is a chlorohydrin of predictable configuration which cannot be reduced directly to the olefin. The conversion is accomplished by three strictly stereospecific steps: formation of the epoxide, cleavage with HI (NaI—AcOH—$EtCO_2H$), and reduction of the resulting iodohydrin with stannous chloride, phosphoryl chloride, and pyridine.

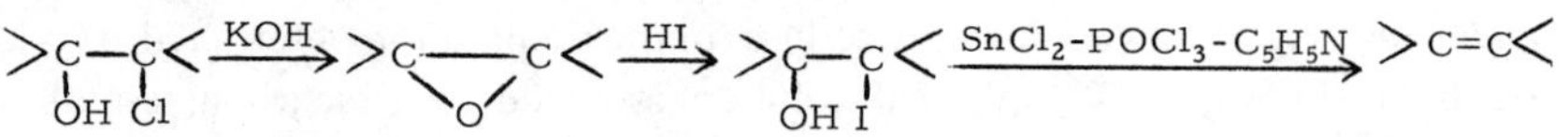

KOH
HI
$SnCl_2$-$POCl_3$-C_5H_5N
OH Cl
OH I

Promotor of hydrogenation. Rylander and Kaplan[19] found stannous chloride to be the most effective of several promotors for the hydrogenation of heptaldehyde with platinum and ruthenium supported on carbon.

[1]J. S. Buck and W. S. Ide, *Org. Syn., Coll. Vol.*, **2**, 130 (1943)
[2]C. W. Ferry, J. S. Buck, and R. Baltzly, *ibid.*, **3**, 240 (1955)
[3]R. B. Woodward, *ibid.*, **3**, 453 (1955)
[4]H. Oelschläger, *Ann.*, **641**, 81 (1961)
[5]L. I. Smith, *Org. Syn., Coll. Vol.*, **2**, 254 (1943)
[6]H. Paul and G. Zimmer, *J. prakt. Chem.*, **18**, 219 (1962)
[7]H. H. Hodgson and E. W. Smith, *J. Chem. Soc.*, 671 (1935)
[8]L. F. Fieser, *Org. Syn., Coll. Vol.*, **2**, 36, 40 (1943)
[9]I. J. Rinkes, *ibid.*, **2**, 393 (1943)
[10]R. B. Sandin and L. F. Fieser, *Am. Soc.*, **62**, 3098 (1940)
[11]G. L. Closs and S. J. Brois, *Am. Soc.*, **82**, 6068 (1960)
[12]J. S. Buck and C. W. Ferry, *Org. Syn., Coll. Vol.*, **2**, 290 (1943)
[13]J. Demaecker and R. H. Martin, *Nature*, **173**, 266 (1954); *Bull. soc. chim. Belg.*, **68**, 365 (1959)
[14]N. M. Cullinane and B. F. R. Edwards, *J. Chem. Soc.*, 1311 (1958)
[15]A. J. Fatiadi and W. F. Sager, *Org. Syn.*, **42**, 66, 90 (1962)
[16]*Org. Expts.*, 199
[17]This reaction is presented as a problem in one book (*Org. Expts.*, 93) and the answer is recorded in another (L. F. F. and M. F., *Advanced Org. Chem.*, 353)
[18]J. W. Cornforth, R. H. Cornforth, and K. K. Mathew, *J. Chem. Soc.*, 112 (1959)
[19]P. N. Rylander and J. Kaplan, *Engelhard Ind. Tech. Bull.*, **2**, 48 (1961)

Stannous chloride, anhydrous, $SnCl_2$. Mol. wt. 189.61, m.p. 247°. Suppliers: MCB (Reagent), E and F (Pract.). Prepared from the dihydrate and acetic anhydride.[1]

The ether-soluble anhydrous reagent is used in the preparation of β-naphthaldehyde by the Stephen reaction[1] and of *o*-tolualdehyde by the Sonn-Müller reaction.[2] A

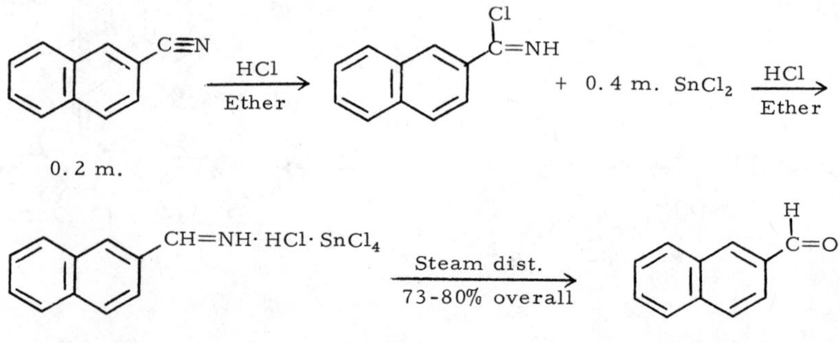

C≡N
HCl
Ether
0.2 m.
Cl
C=NH
+ 0.4 m. $SnCl_2$
HCl
Ether
CH=NH· HCl· $SnCl_4$
Steam dist.
73-80% overall
H
C=O

solution of the reagent in ether is saturated with hydrogen chloride and β-naphthonitrile is added in ether. The mixture is again saturated with hydrogen chloride, stirred for 1 hour, and let stand overnight for complete separation of the yellow aldimine-stannic chloride. The ethereal solution is decanted and the solid submitted to distillation with superheated steam (8–10 hr., 8–10 l. of distillate). In the preparation of *o*-tolualdehyde the reduction of the imidyl chloride is carried out similarly. Steam distillation of the product from an acidic medium in this case takes only 1 hour.

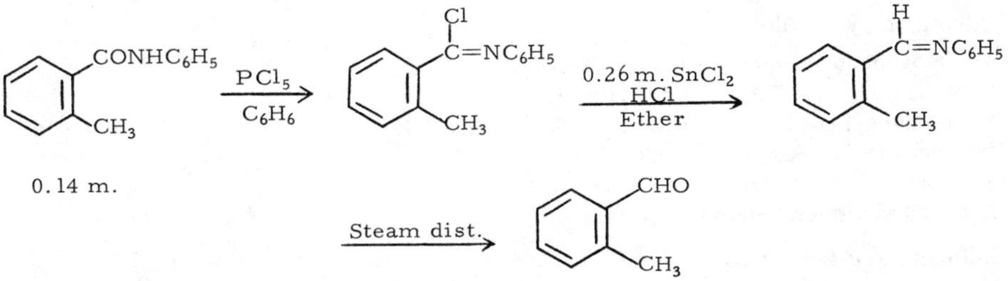
CONHC6H5
CH3
0.14 m.
PCl5
C6H6
Cl
C=NC6H5
CH3
0.26 m. SnCl2
HCl
Ether
H
C=NC6H5
CH3
Steam dist.
CHO
CH3

[1]J. W. Williams, *Org. Syn., Coll. Vol.*, **3**, 626 (1955)
[2]J. W. Williams, C. H. Witten, and J. A. Krynitsky, *ibid.*, **3**, 818 (1955)

Stearic anhydride, $(C_{17}H_{35}CO)_2O$. Mol. wt. 550.92, m.p. 71°. Suppliers: F, KK, BDH.

Prepared by refluxing stearic acid with acetic anhydride, the reagent is recommended for use in place of acetic anhydride for the analytical determination of primary and secondary alcohols and phenols.[1] The substrate is refluxed with a known weight of anhydride in xylene, the excess reagent is hydrolyzed with water, and a stearic acid is titrated.

[1]B. D. Sully, *Analyst*, **87**, 940 (1962)

Succinoyl peroxide, $HO_2CCH_2CH_2\overset{O}{\overset{\|}{C}}OO\overset{O}{\overset{\|}{C}}CH_2CH_2CO_2H$. Mol. wt. 243.16. Suppliers of reagent of 90–95% purity: KK, Establishment Electrochemical Work, Munich.

The reagent is prepared from succinic anhydride and aqueous hydrogen peroxide and is used for the epoxidation of olefins.[1] A satisfactory solvent is dimethylformamide containing a little water; water is required for hydrolysis of the peroxide.

[1]J. Blum, *Compt. rend.*, **248**, 2883 (1959)

Sulfamic acid, H_2NSO_2OH. Mol. wt. 97.10.

The reagent is used to destroy excess nitrous acid in diazotizations.

Sulfoacetic acid, CH_3COOSO_2OH. A solution of the reagent is prepared by dropwise addition with ice cooling of concd. sulfuric acid to acetic anhydride.[1] The crystalline product was isolated by Stillich.[2]

Hauser and co-workers[3] converted aliphatic ketones into their enol acetates by reaction with ketene in the presence of 0.5% sulfoacetic acid.

Schneider and Sack[4] prepared 2,4,6-trimethylpyrylium perchlorate by adding 8 ml. of concd. sulfuric acid with cooling to 33 ml. of acetic anhydride, adding 8 ml. of mesityl oxide and, after 24 hrs., heating to 40–45°. Addition of ice and then perchloric acid afforded 3.5 g. of the pyrylium perchlorate. In an improved procedure[5] perchloric acid is employed as the acid catalyst.

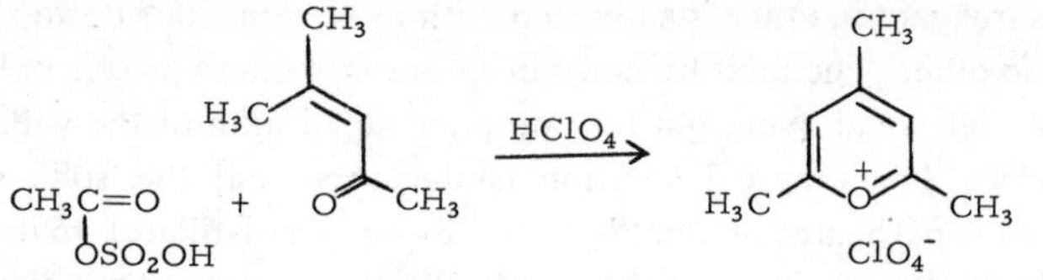

The Liebermann-Burchard test for unsaturated sterols involves addition of sulfoacetic acid to a solution of the sterol in chloroform and is characterized by a brilliant array of colors.[6]

[1]M. Franchimont, *Compt. rend.*, **92**, 1054 (1881)
[2]O. Stillich, *J. prakt. Chem.*, **73**, 541 (1906)
[3]F. G. Young, F. C. Frostick, Jr., J. J. Sanderson, and C. R. Hauser, *Am. Soc.*, **72**, 3635 (1950)
[4]W. Schneider and A. Sack, *Ber.*, **56**, 1786 (1923)
[5]K. Hafner and H. Kaiser, *Org. Syn.*, **44**, 101 (1964)
[6]L. F. and M. Fieser, "Steroids," p. 342

Sulfolane, *see* Tetramethylene sulfone.

Sulfosalicylic acid,

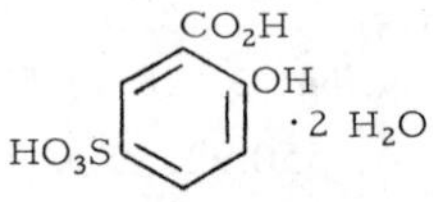

Mol. wt. 254.22; soluble in water, alcohol, ether. Suppliers (fine crystals): E, F, MCB.

Kendall and McKenzie[1] esterified crude β-bromopropionic acid (about 4 m.) by heating an ethanol–carbon tetrachloride solution containing 10 g. of sulfosalicylic acid under a water separator and obtained ethyl β-bromopropionate in overall yield of 85–87%.

An Upjohn group[2] developed a technique for determining the relative rates of enol acetylation of ketosteroids with acetic anhydride and an acid catalyst and established that sulfosalicylic acid has increased catalytic power over *p*-toluenesulfonic acid.

The reagent has been used as catalyst for the preparation of 1,2-cyclic glycerol acetals by transacetalation.[3] Thus a mixture of 30 g. of hexadecanal dimethyl acetal,

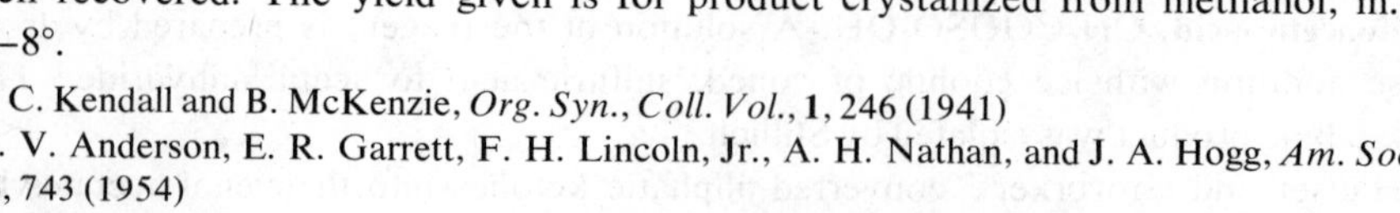

18 g. of glycerol, and 50 mg. of sulfosalicylic acid was heated with stirring at 130–170° under a distilling head for collection of methanol until the theoretical amount had been recovered. The yield given is for product crystallized from methanol, m.p. 47–8°.

[1]E. C. Kendall and B. McKenzie, *Org. Syn., Coll. Vol.*, **1**, 246 (1941)
[2]H. V. Anderson, E. R. Garrett, F. H. Lincoln, Jr., A. H. Nathan, and J. A. Hogg, *Am. Soc.*, **76**, 743 (1954)
[3]C. Piantadosi, C. E. Anderson, E. A. Brecht, and C. L. Yarbro, *Am. Soc.*, **80**, 6613 (1958); C. Piantadosi, A. F. Hirsch, C. L. Yarbro, and C. E. Anderson, *J. Org.*, **28**, 2425 (1963)

Sulfur, S. At. wt. 32.07, pale yellow, m.p. 120°, b.p. 444.6°, slightly soluble in CS_2.

Dehydrogenation.[1] For the dehydrogenation of hydroaromatic compounds, sulfur is simpler than selenium and sometimes preferable to either selenium or catalytic dehydrogenation. 3,4-Dihydro-1,2-naphthalic anhydride (for synthesis *see* Sodium ethoxide) was dehydrogenated by placing a mixture with 1 equivalent of sulfur in a 50-ml. Claisen flask with a sealed-on distilling flask as receiver.[2] The flask was

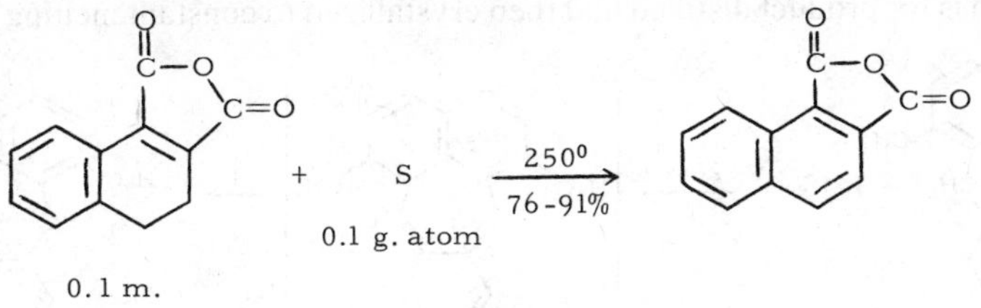

immersed in a Wood's metal bath or a mixture of 10 parts of KNO_3 with 7.5 parts of $NaNO_2$ (m.p. about 150°) preheated to 230–235°. The flask was shaken until the globule of sulfur had dissolved and the temperature was raised to 250° for 30 min. 1,2-Naphthalic anhydride was distilled at 12–13 mm. and crystallized from benzene–ligroin to give light yellow needles, m.p. 166–167°. 1-Phenyl-Δ^1-dialin has been dehydrogenated by a similar procedure.[3]

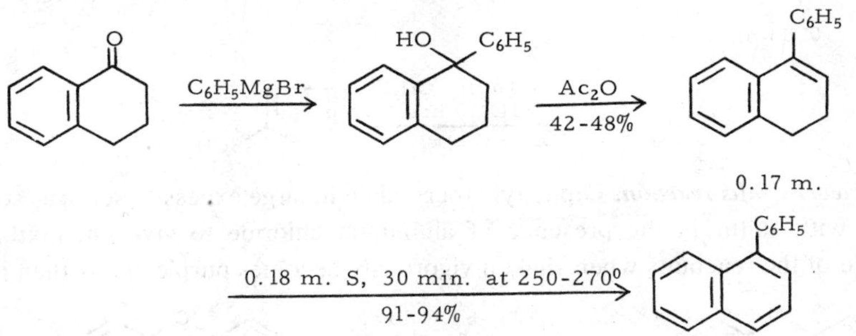

Cocker *et al.*[4] found the tetralin derivative (1) to be dehydrogenated normally by sulfur at 230–240° but to lose the *n*-butyl group to give (3) on reaction with selenium at 330–350°. Dehydrogenation with palladium charcoal at 260–280° gave a mixture of (2) and (3).

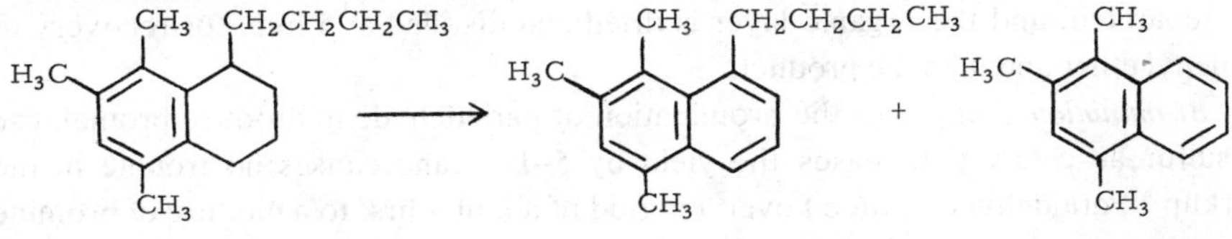

(1) (2) (3)

On the other hand, for dehydrogenations requiring elimination of angular methyl groups selenium usually is superior to sulfur, as is evident from results obtained with cholesterol and with retene (*see* Selenium). Selenium probably is superior to sulfur also for the dehydrogenation of hydroazulenes, although the procedure most widely used has been heating with sulfur in a bath maintained at 220°. A convenient student experiment consists in refluxing guaiene with sulfur in triglyme (b.p. 222°) for 1 hr. and recovering the guaiazulene by steam distillation and conversion to the crystalline picrate.[5]

A probable factor limiting the yield in the sulfur-dehydrogenation of an azulene precursor is reaction of sulfur with an unsaturated intermediate. One of us[6] encountered a sulfur-containing reaction product in an attempt to dehydrogenate

1,6-dimethyl-7,12-dihydropleiadene (1) to the anthracene-like pleiadene (2). The product has the composition of the Diels-Alder adduct (3) of (2) with sulfur. The yield given is for product distilled and then crystallized to constant melting point.

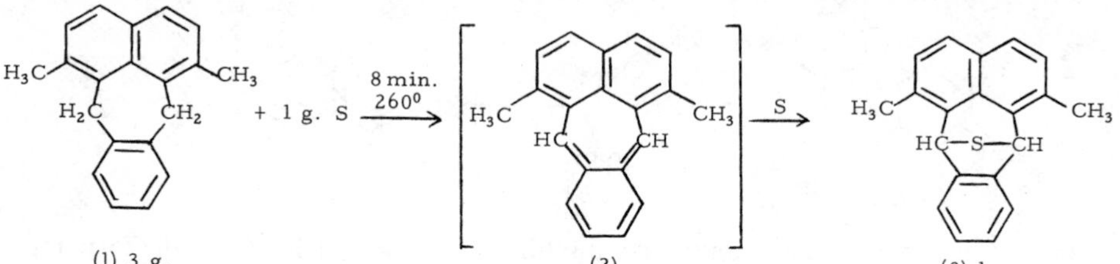

H₃C
CH₃
H₂C
CH₂
+ 1 g. S
8 min.
260°
HC
CH
S
HC—S—CH

(1) 3 g. (2) (3) 1 g.

Grignard reaction. The Grignard reagent from 2-iodothiophene reacts with sulfur to form a derivative suitable for methylation.[7]

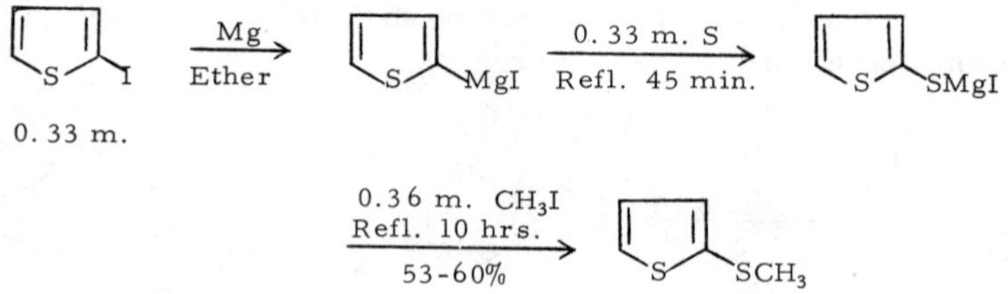

Mg
Ether
MgI
0. 33 m. S
Refl. 45 min.
SMgI
0. 33 m.
0.36 m. CH₃I
Refl. 10 hrs.
53-60%
SCH₃

Friedel-Crafts reaction. Diphenyl ether, taken in large excess to serve as solvent, reacts with sulfur in the presence of aluminum chloride to give phenoxthin.[8] A mixture of the reactants when shaken vigorously becomes purple and is then heated

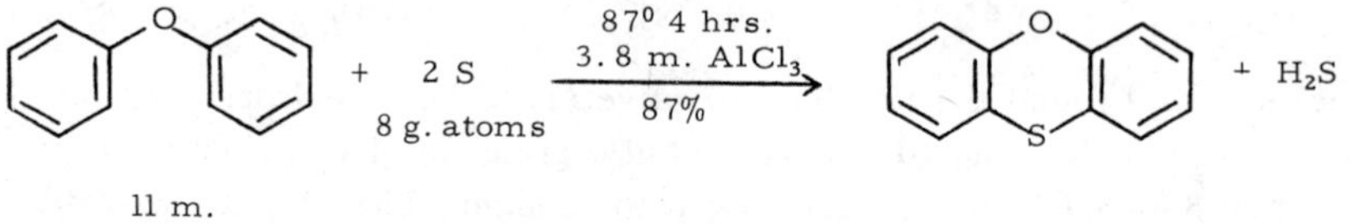

+ 2 S
8 g. atoms
87° 4 hrs.
3. 8 m. AlCl₃
87%
+ H₂S
11 m.

on the steam bath for 4 hrs. and poured onto ice and hydrochloric acid. The layers are separated, and the organic layer is dried and distilled at 5 mm. for recovery of diphenyl ether and then the product.

Bromination catalyst. In the bromination of paraldehyde to produce bromal, use of sulfur as catalyst increases the yield by 5–10% and causes no trouble in the workup.[9] Paraldehyde is added over a period of about 4 hrs. to a mixture of bromine

$$\underset{0.52\ \text{m.}}{(CH_3CHO)_3} + \underset{4.5\ \text{m.}}{9\ Br_2} + \underset{1.5\ \text{g.}}{S} \xrightarrow[52\text{-}57\%]{} 3\ CBr_3CHO + 9\ HBr$$

and sulfur, and external heating is required to maintain a temperature of 60–80° for 2 hrs. more.

Willgerodt reaction.[10] In its original form this reaction involved heating an aryl alkyl ketone in a sealed tube at 210–230° with an aqueous solution of yellow ammonium polysulfide, prepared by dissolving sulfur in ammonium sulfide solution. The product is an aryl substituted aliphatic acid amide, together with some of the corresponding carboxylic acid and often the hydrocarbon. An example is the

$$C_6H_5COCH_3 + (NH_4)_2S_x \longrightarrow \underset{50\%}{C_6H_5CH_2CONH_2} + \underset{13.5\%}{C_6H_5CH_2CO_2NH_4}$$

reaction of acetophenone to give phenylacetamide and ammonium phenylacetate.[11] The reaction initially found only limited application because of the modest yields and the necessity of using a sealed tube. Fieser and Kilmer[12] improved the procedure by use of dioxane as solvent, which permits operations at a temperature of 160°. By this procedure 1-acetoacenaphthene was converted into 1-acenaphthylacetic acid in 57% yield. By comparison, the alternative route of hypohalite oxidation,

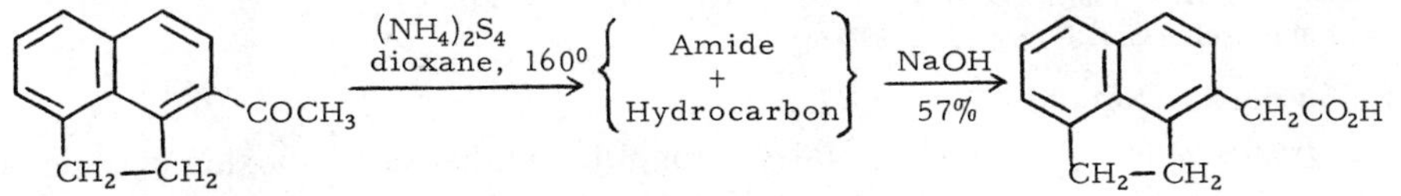
COCH3
CH2—CH2
(NH4)2S4
dioxane, 160°
Amide
+
Hydrocarbon
NaOH
57%
CH2CO2H
CH2—CH2

conversion through the acid chloride to the diazoketone, and Arndt-Eistert reaction afforded 1-acenaphthylacetic acid in overall yield of only 50%.

An important modification of the reaction introduced by Kindler[13] consists in heating the ketone with an equivalent amount of sulfur and an amine which is resistant to oxidation; the product is a thioamide which on hydrolysis yields the carboxylic acid. Morpholine is a particularly favorable amine component; for this variation *see* Morpholine.

[1]Pl. A. Plattner and E. C. Armstrong, "Dehydrogenation with Sulfur, Selenium, and Platinum Metals," *Newer Methods of Preparative Organic Chemistry*, p. 21, Interscience (1948)
[2]E. B. Hershberg and L. F. Fieser, *Org. Syn., Coll. Vol.*, **2**, 423 (1943)
[3]R. Weiss, *ibid.*, **3**, 729 (1955)
[4]W. Cocker, B. E. Cross, J. T. Edward, D. S. Jenkinson, and J. McCormick, *J. Chem. Soc.*, 2355 (1953)
[5]*Org. Expts.*, 298
[6]L. F. Fieser, *Am. Soc.*, **55**, 4977 (1933)
[7]J. Cymerman-Craig and J. W. Loder, *Org. Syn., Coll. Vol.*, **4**, 667 (1963)
[8]C. M. Suter and C. E. Maxwell, *ibid.*, **2**, 485 (1943)
[9]F. A. Long and J. W. Howard, *ibid.*, **2**, 87 (1943)
[10]Reviews: M. Carmack and M. A. Spielman, *Org. Reactions*, **3**, 83 (1946); R. Wegler, E. Kühle, and W. Schäfer, *Newer Methods of Preparative Organic Chemistry*, **3**, 1 (1964)
[11]C. Willgerodt and F. H. Merk, *J. prakt. Chem.*, [2], **80**, 192 (1909)
[12]L. F. Fieser and G. W. Kilmer, *Am. Soc.*, **62**, 1354 (1940)
[13]K. Kindler, *Ann.*, **431**, 187 (1923)

Sulfur dichloride, SCl_2. Mol. wt. 102.98. Dark red liquid, b.p. 59°, sp. gr. 1.621. Suppliers: Hooker Chem. Co., MCB. Preparation.[1]

Three groups independently investigated the addition of this reagent to cyclic dienes.[2-4] Under proper conditions of dilution, sulfur dichloride adds to nonconjugated cyclic dienes to give sulfur-bridged systems. Thus Corey and Block[2] stirred a solution of 1,5-cyclooctadiene in methylene chloride at −50° under nitrogen,

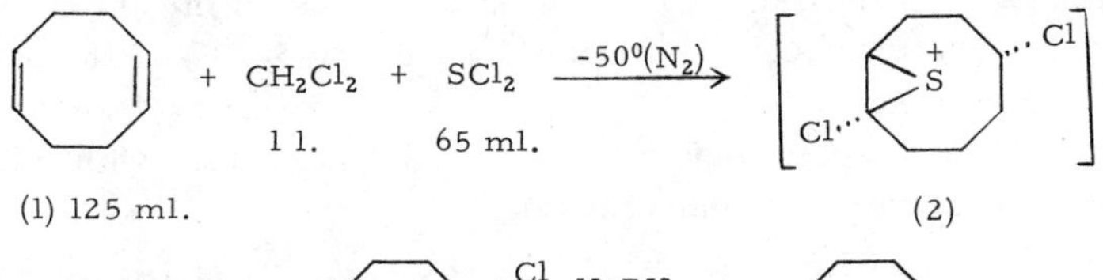
+ CH2Cl2 + SCl2
1 l.
65 ml.
-50°(N2)
Cl
S
+
Cl

(1) 125 ml. (2)

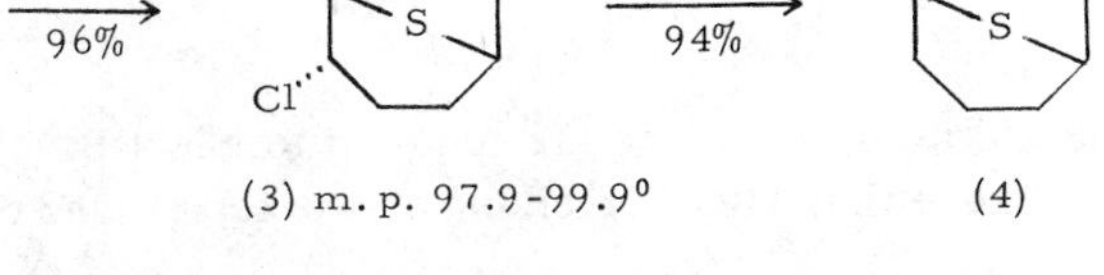
96%
Cl
S
Cl
NaBH4
94%
S

(3) m. p. 97.9-99.9° (4)

added sulfur dichloride with cooling, and eventually obtained in high yield a crystalline adduct characterized as 2,6-dichloro-9-thiabicyclo[3.3.1]nonane (3). The 1,2-episulfonium ion (2) is suggested as an intermediate. The chlorine atoms can be removed reductively in high yield to give 9-thiabicyclo[3.3.1]nonane (4).

[1]G. Brauer, "Handbuch der Präparativen Anorganische Chemie," **1**, 336 (1960)
[2]E. J. Corey and E. Block, *J. Org.*, **31**, 1663 (1966)
[3]E. D. Weil, K. J. Smith, and R. J. Gruber, *J. Org.*, **31**, 1669 (1966)
[4]F. Lautenschlaeger, *J. Org.*, **31**, 1679 (1966)

Sulfur dioxide, SO_2. Mol. wt. 64.07, b.p. −10°.

Isomerization of polyenes. A Pfizer group[1] found that sulfur dioxide in pyridine isomerized ergosteryl acetate (1) into the $\Delta^{6,8(14)}$- or B_2-isomer (4) probably via the

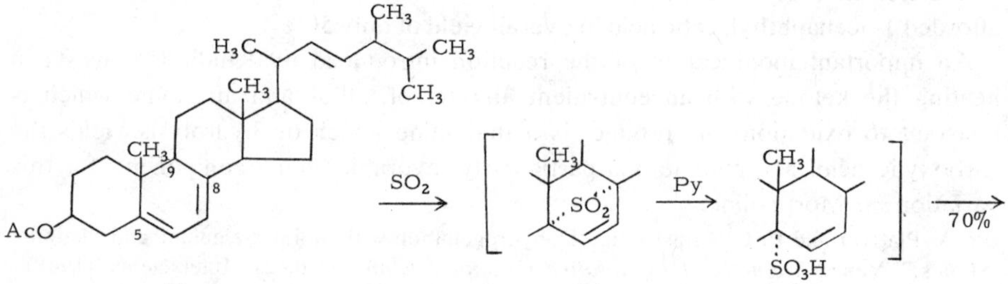
H3C
CH3
CH3
CH3
CH3
9
8
5
AcO
SO2
CH3
SO2
Py
CH3
SO3H
70%

(1) (2) (3)

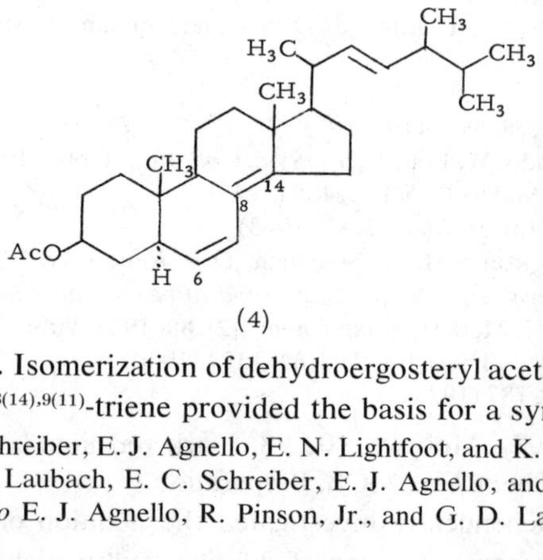
CH3
H3C
CH3
CH3
CH3
CH3
14
8
AcO
H 6

(4)

Diels-Alder adduct (2). Isomerization of dehydroergosteryl acetate, a $\Delta^{5,7,9(11)}$-triene, by SO_2-Py to the $\Delta^{6,8(14),9(11)}$-triene provided the basis for a synthesis of cortisone.

[1]G. D. Laubach, E. C. Schreiber, E. J. Agnello, E. N. Lightfoot, and K. J. Brunings, *Am. Soc.*, **75**, 1514 (1953); G. D. Laubach, E. C. Schreiber, E. J. Agnello, and K. J. Brunings, *ibid.*, **78**, 4743 (1956); *see also* E. J. Agnello, R. Pinson, Jr., and G. D. Laubach, *ibid.*, **78**, 4756 (1956)

Sulfur monochloride, S_2Cl_2. Mol. wt. 135.03. Red-yellow liquid, b.p. 137°, sp. gr. 1.687. Suppliers: F, MCB.

Purification. J. Szmuszkovicz[1] recommends distillation of 500 g. from 20 g. of sulfur and 5 g. of charcoal; the fraction boiling at 135–137° is stored in a dark bottle in the refrigerator.

Diphenyl sulfide can best be prepared by reaction of benzene with sulfur monochloride in the presence of aluminum chloride.[2]

$$\underset{16\text{ m.}}{2\,C_6H_6} + \underset{3\text{ m.}}{S_2Cl_2} + \underset{3.48\text{ m.}}{AlCl_3} \xrightarrow[81-83\%]{10-30^0} C_6H_5SC_6H_5 + S + 2\,HCl$$

Although extremely drastic conditions are required to effect usual electrophilic substitutions in the pyridine ring, Hunsberger and co-workers[3] found that by heating pyridine with bromine and sulfur monochloride at a moderate temperature, 3,5-dibromopyridine can be prepared in modest yield. Characteristic of the reaction is the entry of bromine β to the ring nitrogen regardless of the substituents already present. Hunsberger postulates that the reaction involves intermediates of types (2)–(4).

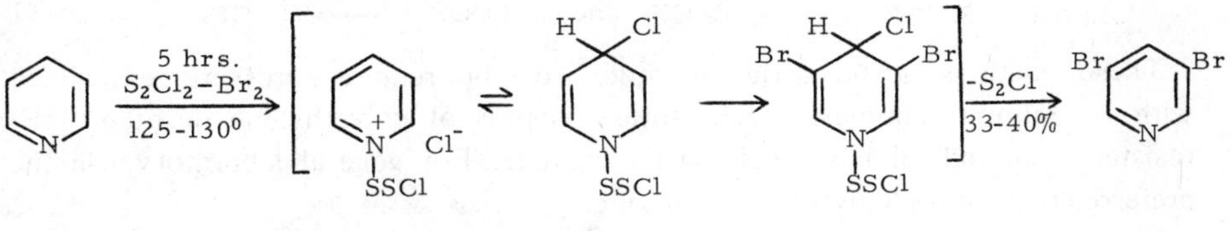
5 hrs.
$S_2Cl_2-Br_2$
125-130°
Cl^-
SSCl
H
Cl
Br
$-S_2Cl$
33-40%

(1) (2) (3) (4) (5)

[1]Private communication
[2]W. W. Hartman, L. A. Smith, and J. B. Dickey, *Org. Syn., Coll. Vol.*, **2**, 242 (1943)
[3]E. E. Garcia, C. V. Greco, and I. M. Hunsberger, *Am. Soc.*, **82**, 4430 (1960)

Sulfurous acid monomethyl ester N,N-diethylamide, $CH_3O\overset{O}{\overset{\|}{S}}N(C_2H_5)_2$. Mol. wt. 151.23, b.p. 73–74°/10 mm.

The reagent is prepared by reaction of diethylamine with chlorosulfonic acid methyl ester.[1] In pyridine solution at room temperature it reacts with primary and secondary carboxylic acids after several days to give the carboxylic acid N,N-diethylamide. Tertiary acids and α-amino acids do not react.[2]

$$RCOOH + CH_3O\overset{O}{\overset{\|}{S}}N(C_2H_5)_2 \longrightarrow R\overset{O}{\overset{\|}{C}}N(C_2H_5)_2$$

[1]G. Zinner, *Ber.*, **91**, 966 (1958)
[2]H. G. O. Becker and K. F. Funk, *J. prakt. Chem.*, **14**, 55 (1961)

Sulfur sesquioxide, S_2O_6. The preparation of this reagent and its use for the conversion of 1,5-dinitronaphthalene into naphthazarin is described in the patent literature[1] and in papers by Charrier[2] and by one of us.[3] In our procedure 25 g. of sulfur is dissolved in 375 ml. of 18% fuming sulfuric acid and the solution is added to a suspension of 50 g. of 1,5-dinitronaphthalene in 230 ml. of concd. sulfuric acid and

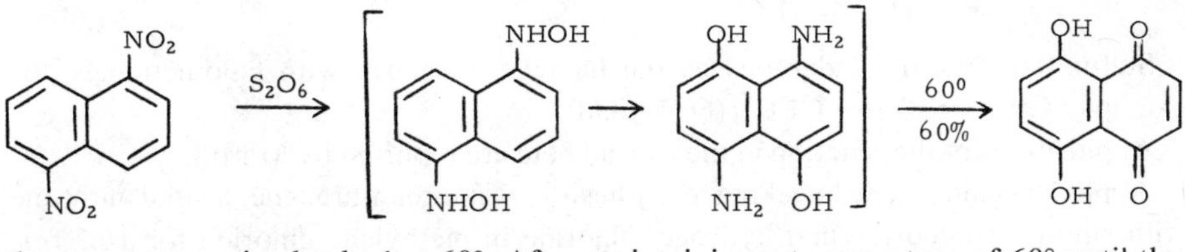
NO_2
S_2O_6
NHOH
OH
NH_2
60°
60%
O

the temperature is checked at 60°. After maintaining a temperature of 60° until the reaction appeared to subside, the mixture was cooled and poured onto ice.

[1]German patent 71,386
[2]G. Charrier and G. Tocco, *Gazz.*, **53**, 431 (1923)
[3]L. F. Fieser, *Am. Soc.*, **50**, 439 (1928)

Sulfur tetrafluoride, SF_4. Mol. wt. 108.07. Suppliers of 95% pure reagent: Matheson, du Pont Dyes and Chem. Div. (1-lb. cylinders).[1]

Caution. The gaseous reagent is about twice as toxic as phosgene. It is readily hydrolyzed to hydrogen fluoride, hence equipment for its use must be thoroughly dry.

Preparation. It is prepared most conveniently by refluxing sulfur dichloride and sodium fluoride in acetonitrile.[2]

$$3\ SCl_2 + 4\ NaF \longrightarrow SF_4 + S_2Cl_2 + 4\ NaCl$$

Reactions.[3] The reactions most characteristic of sulfur tetrafluoride are:

$$>C{=}O \longrightarrow\ >CF_2 \quad \text{and} \quad -COOH \longrightarrow -CF_3$$

These reactions can be carried out at low cost, but require a pressure vessel lined with Hastelloy-C stainless steel, storage vessels of polyethylene or other HF-resistant material, and a specialized technique. For general laboratory use the preferred reagent is phenylsulfur trifluoride.

The reaction of a carboxylic acid takes place in 2 steps:

$$RCOOH + SF_4 \longrightarrow RCOF + HF + SOF_2$$

$$RCOF + SF_4 \longrightarrow RCF_3 + SOF_2$$

The first step occurs at room temperature, but the second requires a temperature of 100–200°. In the conversion of heptanoic acid into 1,1,1-trifluoroheptane[4] the reactants are heated in a pressure vessel with agitation for 4 hrs. at 100° and for 6 hrs.

$$\underset{0.2\ m.}{CH_3(CH_2)_5CO_2H} + \underset{0.57\ m.}{2\ SF_4} \xrightarrow[70-80\%]{100-130^0} CH_3(CH_2)_5CF_3$$

at 130°. Carboxylic anhydrides and acid chlorides give the same products as the corresponding carboxylic acids but temperatures of 300–350° are required. Esters react at a lower temperature in the presence of boron trifluoride as catalyst.

The reaction can be applied to amino acids if carried out in hydrogen fluoride, which serves to protect the amino group.[5] If the amino acid is optically active, optical activity is retained. Thus L-leucine gives levorotatory 3-methyl-1-(trifluoromethyl)-butylamine.

$$CH_3CH(CH_3)CH_2CH(NH_2)CO_2H \xrightarrow[22\%]{8\ hrs.\ at\ 120^0,\ 0.93\ m.\ SF_4\ +\ 2.5\ m.\ HF} CH_3CH(CH_3)CH_2CH(NH_2)CF_3$$

L-Leucine (0.4 m.)

Sulfur tetrafluoride undergoes partial halogen exchange with halomethanes, for example: $CCl_4 + SF_4 \longrightarrow CFCl_3$ (66% yield).[6]

Applications of the reaction in the steroid field are reported by Martin.[7]

Applequist and Searle,[8] seeking a synthesis of 9-fluoroanthracene, heated anthrone with sulfur tetrafluoride and hydrogen fluoride in methylene chloride for 16.3 hrs. at 69° but, surprisingly, obtained 10,10-difluoranthrone in high yield.

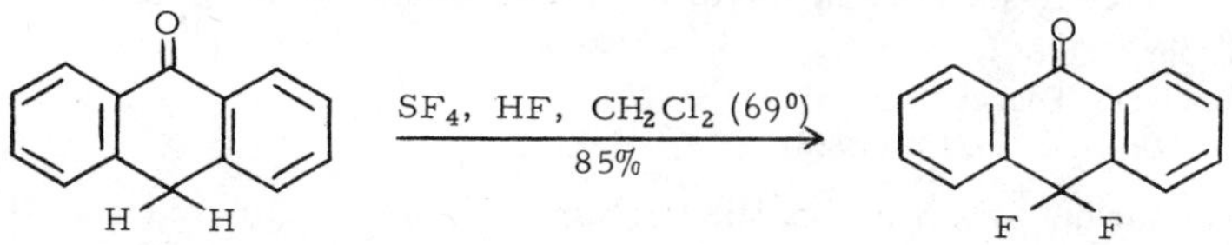
SF₄, HF, CH₂Cl₂ (69⁰)
85%
H
H
F
F

[1]Technical Information Bulletin, 1959

[2]C. W. Tullock, F. S. Fawcett, W. C. Smith, and D. D. Coffman, *Am. Soc.*, **82**, 539 (1960)

[3]For a review of du Pont work, see W. C. Smith, *Angew. Chem., Internat. Ed.*, **1**, 467 (1962)

[4]W. R. Hasek, W. C. Smith, and V. A. Engelhardt, *Am. Soc.*, **82**, 543 (1960)
[5]M. S. Raasch, *J. Org.*, **27**, 1406 (1962)
[6]C. W. Tullock, R. A. Carboni, R. J. Harder, W. C. Smith, and D. D. Coffman, *Am. Soc.*, **82**, 5107 (1960)
[7]D. G. Martin and F. Kagan, *J. Org.*, **27**, 3164 (1962); D. G. Martin and J. E. Pike, *ibid.*, **27**, 4086 (1962)
[8]D. E. Applequist and R. Searle, *J. Org.*, **29**, 987 (1964)

Sulfur trioxide, SO_3. Mol. wt. 80.07, m.p. 17°, b.p. about 45°. This reagent is available from the General Chemical Division, Allied Chemical and Dye Corp., under the name Sulfan B, a γ-form stabilized with an inhibitor to prevent polymerization. It is used in organic work chiefly in the form of complexes with dioxane, dimethylformamide, and pyridine.

A 3–5*M* solution of the reagent in nitromethane reacts with an aryl halide at 0° to give a sulfonic acid anhydride, which precipitates from the solution and which affords a sulfonic acid on hydrolysis.[1]

$$3\ SO_3 + 2\ C_6H_5I \xrightarrow[70\%]{CH_3NO_2} (IC_6H_4)_2S_2O_5 + H_2SO_4$$

[1]N. H. Christensen, *Chem. Scand.*, **15**, 1507 (1961)

Sulfur trioxide–Dimethylformamide, $SO_3 \cdot HCON(CH_3)_2$. Mol. wt. 153.17.

Garbrecht[1] prepared the reagent by stirring 10–11 l. of dimethylformamide under a condenser with protection against moisture, cooling to 0–5°, and dropping in 2 lbs. of crude sulfur trioxide in the form of Sulfan B over a period of 4–5 hrs. Some crystalline complex separated, but stirring was continued until it had all dissolved.

Synthesis of amides. Garbrecht[1] used the reagent in an improved method for the conversion of (+)-lysergic acid into amide derivatives. The acid, as monohydrate,

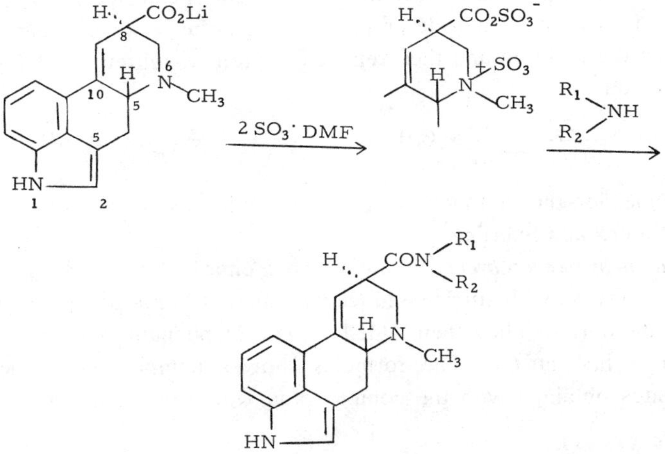
H
CO_2Li
8
H
6
N
CH_3
10
5
5
HN
1
2
$2\ SO_3 \cdot DMF$
H
$CO_2SO_3^-$
SO_3
H
N
CH_3
R_1
NH
R_2
H
CON
R_1
R_2
H
N
CH_3
HN

is dissolved with lithium hydroxide monohydrate in methanol, dimethylformamide is added, and distillation, eventually in vacuum, removes methanol and then water. Addition of 2 equivalents of the reagent complex gives a mixed anhydride and this reacts with 1 equivalent of a secondary amine to afford the acid amide.

Cyclodehydration. The reagent has been used to advantage for cyclization in the synthesis of azlactones,[2] and butenolides.[3–5] N,N′-Diacylhydrazines are cyclized by the reagent to 1,3,4-oxadiazoles in yield of about 90%.[5]

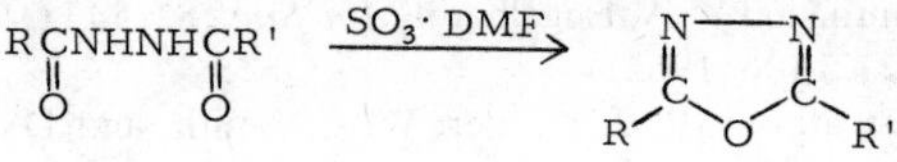
RCNHNHCR'
SO₃·DMF

Peptide synthesis:[6]

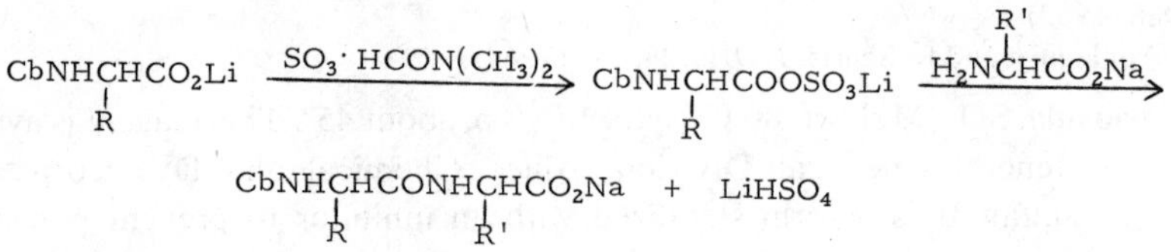
CbNHCHCO₂Li
SO₃·HCON(CH₃)₂
CbNHCHCOOSO₃Li
H₂NCHCO₂Na
CbNHCHCONHCHCO₂Na + LiHSO₄

[1]W. I. Garbrecht, *J. Org.*, **24**, 368 (1959)
[2]E. Baltazzi and E. A. Davis, *Chem. Ind.*, 929 (1962)
[3]*Idem, ibid.*, 1653 (1962)
[4]Y. S. Rao and R. Filler, *Chem. Ind.*, 280 (1964)
[5]E. Baltazzi and A. J. Wysocki, *Chem. Ind.*, 1080 (1963)
[6]G. W. Kenner and R. J. Stedman, *J. Chem. Soc.*, 2069 (1952); D. W. Clayton, J. A. Farrington, G. W. Kenner, J. M. Turner, *ibid.*, 1398 (1957); N. F. Albertson, *Org. Reactions*, **12**, 255–260 (1962)

Sulfur trioxide–Dioxane, $SO_3 \cdot OC_4H_8O$. Mol. wt. 168.17.

Preparation. Rondestvedt and Bordwell[1] cooled 800 g. of ethylene dichloride and distilled in sulfur trioxide from 60% fuming sulfuric acid or Sulfan B to a gain in weight of 300 g. The sulfur trioxide solution was cooled below −5°, and dioxane equivalent to the sulfur trioxide was added dropwise with stirring, when the complex separated as fine granules. This complex is much more reactive than sulfur trioxide–pyridine; it is hydrolyzed by cold water.

Sulfonation of olefins. Rondestvedt and Bordwell sulfonated styrene by adding 1 mole of the hydrocarbon in 2 volumes of ethylene dichloride to the suspension of complex in ethylene dichloride prepared as described above. The mixture was cooled during the addition and then refluxed; styrene-β-sulfonic acid was isolated as the sodium salt.

$$C_6H_5CH{=}CH_2 \xrightarrow[ClCH_2CH_2Cl]{OC_4H_8O\cdot SO_3} C_6H_5CH{=}CHSO_3H \xrightarrow[58\text{-}65\%]{NaOH} C_6H_5CH{=}CHSO_3Na$$

Other studies of the sulfonation of olefins with this reagent are reported by Bordwell,[2] Suter,[3] and Sperling.[4]

Sulfonation of phenanthrene. Solomon and Hennessy[5] added 88 g. (1.1 moles) of Sulfan B dropwise with stirring and ice cooling to 100 ml. of dioxane in 350 ml. of ethylene dichloride. They then added 1 mole of phenanthrene and heated the mixture for 5 hrs. at 60°. The formulas show a comparison of the yields of monosulfonates obtained with the complex and with concd. sulfuric acid.[6]

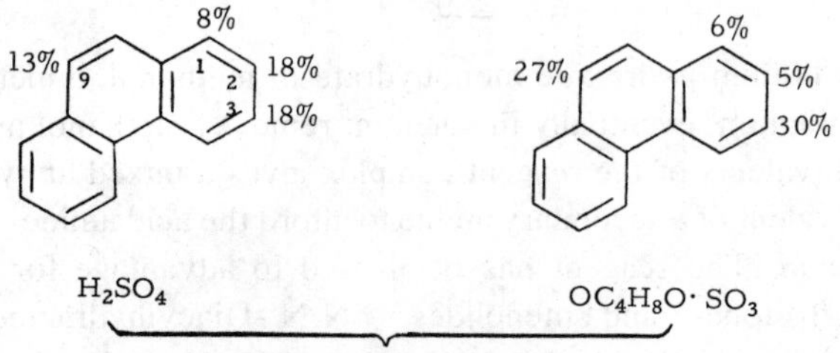
8%
13%
18%
18%
6%
27%
5%
30%
H₂SO₄
OC₄H₈O·SO₃

Sulfonation at 60°

Sulfonation of aldehydes and ketones. Truce and Alfieri[7] used the reagent for conversion of simple aldehydes and ketones into the α-sulfonates in yields of about 70%.

[1]C. S. Rondestvedt, Jr., and F. G. Bordwell, *Org. Syn., Coll. Vol.*, **4**, 846 (1963)
[2]F. G. Bordwell *et al., Am. Soc.*, **81**, 2002 (1959) and preceding papers of the series
[3]C. M. Suter and W. E. Truce, *Am. Soc.*, **66**, 1105 (1944)
[4]R. Sperling, *J. Chem. Soc.*, 1925 (1949)
[5]M. G. Solomon and D. J. Hennessy, *J. Org.*, **22**, 1649 (1957)
[6]L. F. Fieser, *Org. Syn., Coll. Vol.*, **2**, 482 (1943)
[7]W. E. Truce and C. C. Alfieri, *Am. Soc.*, **72**, 2740 (1950)

Sulfur trioxide–Pyridine. $SO_3 \cdot C_5H_5N$. Mol. wt. 159.16, m.p. 175°.

Preparation.[1] (a) Pyridine (1 mole) is slowly dropped with cooling and stirring into a mixture of crushed sulfur trioxide (1 mole) and carbon tetrachloride (4 parts), and the crystalline product is collected and washed with a little ice water to remove a trace of pyridine–H_2SO_4.

(b) Chlorosulfonic acid (0.5 mole) is dropped with cooling and stirring into a solution of pyridine (1 mole) in carbon tetrachloride; the product is washed with ice water to remove pyridine hydrochloride:

$$2\ C_2H_5N + ClSO_3H \longrightarrow C_5H_5N \cdot SO_3 + C_5H_5\overset{+}{N}H\overset{-}{Cl}$$

Sulfate esters. Baumgarten[2] showed that the reagent sulfates substances such as hydrazine, diethylamine, phenol, and naphthalene. It has been used in preparation of carbohydrate sulfate esters,[3] sterol sulfate esters,[4] phenol sulfate esters.[5]

2-Hydroxy-3-alkyl-1,4-naphthoquinone antimalarials can be separated from metabolites hydroxylated in the side chain by conversion of the hydroxylated metabolites into water-soluble sulfate esters and extraction with water.[6] Hydroxyhydrolapachol (1), which was identified as a metabolite of hydrolapachol and which

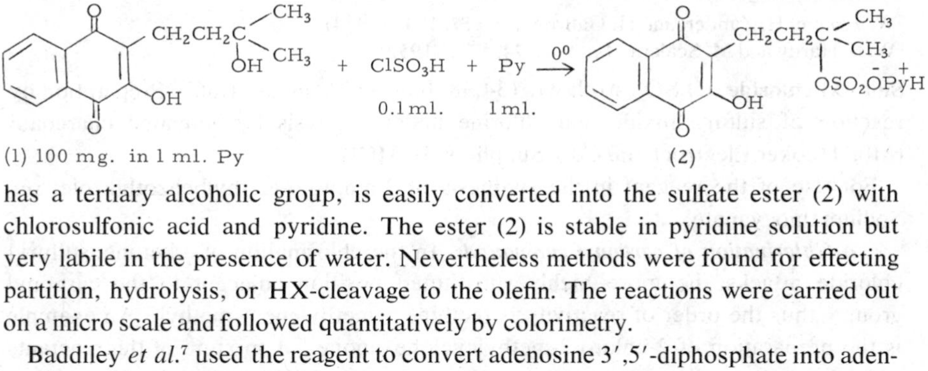

(1) 100 mg. in 1 ml. Py (2)

has a tertiary alcoholic group, is easily converted into the sulfate ester (2) with chlorosulfonic acid and pyridine. The ester (2) is stable in pyridine solution but very labile in the presence of water. Nevertheless methods were found for effecting partition, hydrolysis, or HX-cleavage to the olefin. The reactions were carried out on a micro scale and followed quantitatively by colorimetry.

Baddiley *et al.*[7] used the reagent to convert adenosine 3′,5′-diphosphate into adenosine 3′-phosphate-5′-sulfatophosphate. In the presence of an appropriate enzyme, this "active sulfate" transfers the sulfate group to a variety of substrates.

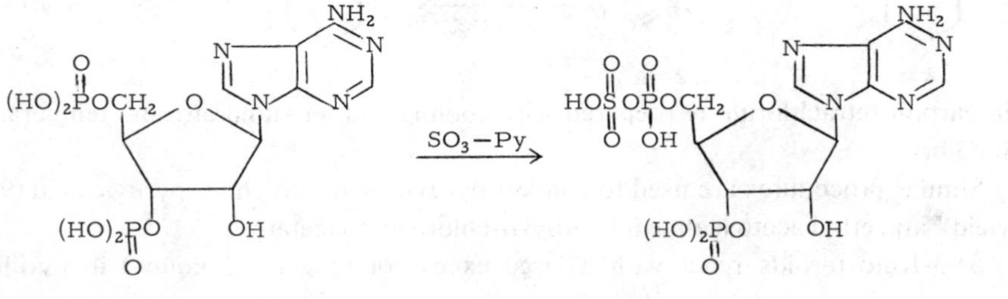

Klass[8] found that aliphatic aldehydes react with the complex in refluxing ethylene dichloride to give betaine salts in good yield. Acetone and benzaldehyde did not react under the same conditions.

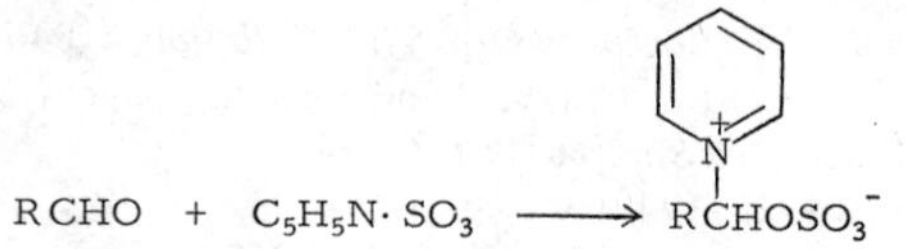
RCHO + $C_5H_5N \cdot SO_3$ → $R\overset{|}{C}HOSO_3^-$

[1]P. Baumgarten, *Ber.*, **59**, 1166 (1926)
[2]*Idem, ibid.*, **59**, 1976 (1926)
[3]R. B. Duff, *J. Chem. Soc.*, 1597 (1949)
[4]A. E. Sobel, I. J. Drekter, and S. Natelson, *J. Biol. Chem.*, **115**, 381 (1936); A. E. Sobel and P. E. Spoerri, *Am. Soc.*, **63**, 1259 (1941); **64**, 361 (1942)
[5]G. N. Burkhardt and A. Lapworth, *J. Chem. Soc.*, 684 (1926)
[6]L. F. Fieser, *Am. Soc.*, **70**, 3232 (1948)
[7]J. Baddiley, J. G. Buchanan, R. Letters, and A. R. Sanderson, *J. Chem. Soc.*, 1731 (1959); J. Baddiley and A. R. Sanderson, *Biochem. Prepns.*, **10**, 3 (1963)
[8]D. L. Klass, *J. Org.*, **29**, 2666 (1964)

Sulfur trioxide–Trimethylamine, $(CH_3)_3N^+—SO_3^-$. Mol. wt. 139.18 m.p. 236–240° dec., slowly hydrolyzed by water but can be crystallized from water. Supplier: Borden Chemical.

The reagent is a mild sulfonating agent. It reacts with alcohols and phenols to form amine salts of the sulfuric acid ester.[1] A mixture of 0.05 mole each of

$$\underline{p}\text{-}C_6H_5C_6H_4OH + (CH_3)_3\overset{+}{N}-SO_3^- \longrightarrow \underline{p}\text{-}C_6H_5C_6H_4OSO_3^-\,(CH_3)_3\overset{+}{N}$$

o- and *p*-phenylphenol was separated by treatment with 0.05 mole of the reagent in weakly alkaline solution and extraction with ether; *p*-phenyphenol reacted selectively to give the pure sulfate ester.[2]

[1]W. Traube, H. Zander, and H. Gaffron, *Ber.*, **57**, 1045 (1924)
[2]W. B. Hardy and M. Scalera, *Am. Soc.*, **74**, 5212 (1952)

Sulfuryl chloride, Cl_2SO_2. Mol. wt. 134.98, b.p. 69.2°, sp. gr. 1.667. Preparation by reaction of sulfur dioxide and chlorine under catalysis by activated charcoal.[1] (Mfg. Hooker Electrochem. Co.). Suppliers: E, MCB.

For use of the reagent in the synthesis of 2-amino-6-methylbenzothiazole, *see* Sodium thiocyanate.

α-Chlorination of carbonyl compounds. In the chlorination of a ketone, sulfuryl chloride attacks the more highly substituted position adjacent to the carbonyl group;[2] thus the order of reactivity is methine > methylene > methyl.[3] An example is the preparation of 2-chloro-2-methylcyclohexanone.[4] A mixture of the reactants

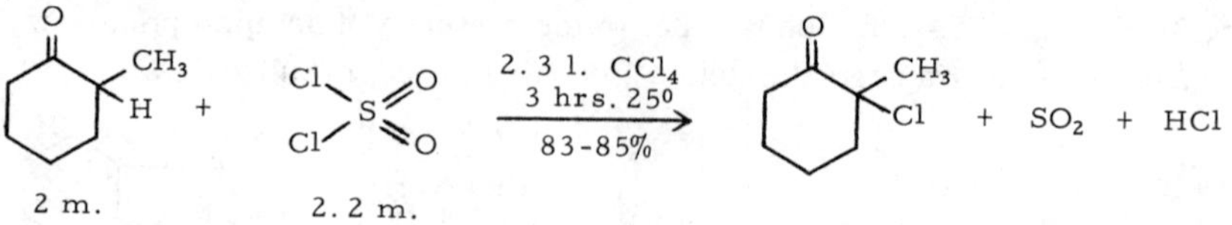
O
CH3
H
+
Cl
S
O
Cl
O
2.3 l. CCl4
3 hrs. 25°
83-85%
O
CH3
Cl
+ SO2 + HCl
2 m.
2.2 m.

in carbon tetrachloride is prepared with cooling and let stand at room temperature for 3 hrs.

Similar procedures are used to convert pyruvic acid into chloropyruvic acid (98% yield)[5] and ethyl acetoacetate into ethyl α-chloroacetoacetate.[6]

Δ^4-3-Ketosteroids react with a large excess of reagent (2 equiv.) in pyridine solution to give 4-chloro-Δ^4-3-ketosteroids in yields of 70–90%.[7] According to a Danish patent[8] if benzene is used as solvent, 2-chloro-Δ^4-3-ketosteroids are ob-

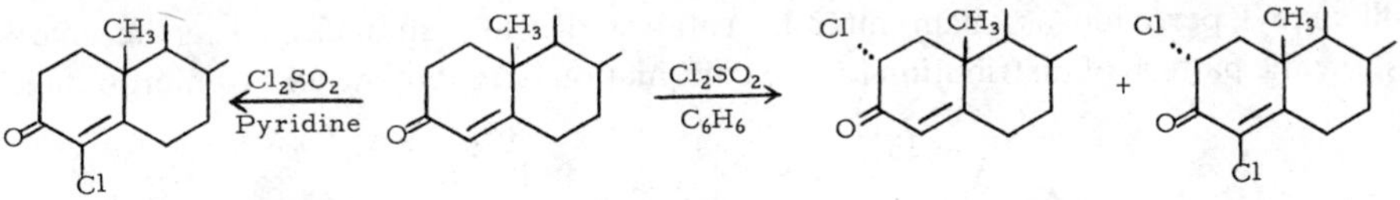

CH3
Cl2SO2
Pyridine
Cl
CH3
Cl2SO2
C6H6
Cl
CH3
+
Cl
CH3
Cl

tained. Yasuda[9] chlorinated testosterone acetate in benzene and characterized the product as a mixture of 2α-chlorotestosterone acetate and 2α,4-dichlorotestosterone acetate.

Wyman and Kaufman[3] report that chlorination of acetone and methyl ethyl ketone affords mainly the α,α-dichloro derivatives even though the ketone and sulfuryl chloride are taken in equivalent amounts. α-Chlorination of aldehydes has been reported,[10] but the reaction appears to yield chiefly cyclic ether trimers and linear polyethers.[11]

Sulfuryl chloride is superior to chlorine for the chlorination of N-benzoyl-ϵ-caprolactam.[12] The product on ring opening, ammonolysis, and hydrolysis is converted into DL-lysine in 73% overall yield from ϵ-caprolactam.

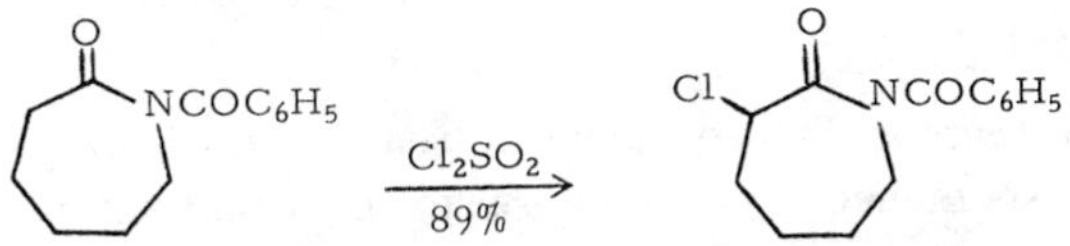

O
NCOC6H5
Cl2SO2
89%
Cl
O
NCOC6H5

Chlorination of aromatics. Although benzene is sufficiently inert to sulfuryl chloride to be used as a solvent (see above), the reagent dichlorinates ethyl 4-hydroxybenzoate under mild conditions.[13] The reaction is the first step in a synthesis of 2,6-dichlorophenol.

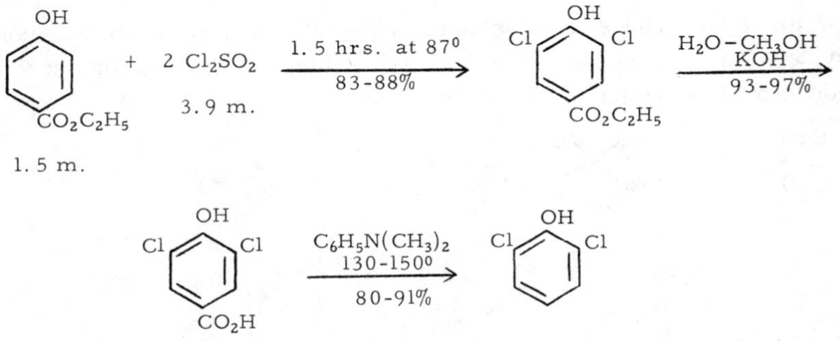

OH
CO2C2H5
1. 5 m.
+ 2 Cl2SO2
3. 9 m.
1. 5 hrs. at 87⁰
83-88%
OH
Cl
Cl
CO2C2H5
H2O-CH3OH
KOH
93-97%
OH
Cl
Cl
CO2H
C6H5N(CH3)2
130-150⁰
80-91%
OH
Cl
Cl

Thiophene and alkylthiophenes are readily chlorinated in the α-position by the reagent.[14] Thus thiophene affords 2-chlorothiophene (43%) and 2,5-dichlorothiophene; 3-methylthiophene gives 2-chloro-3-methylthiophene in 79% yield. Sulfuryl chloride is an excellent reagent for the selective chlorination of 2-thenylamines in the 5-position.[15]

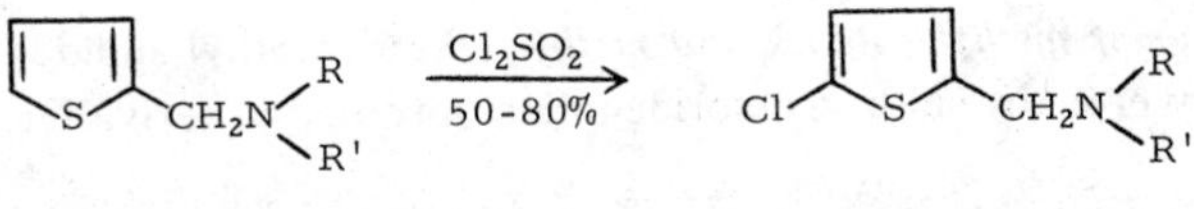

S
CH2N
R
R'
Cl2SO2
50-80%
Cl
S
CH2N
R
R'

Aliphatic hydrocarbons. M. S. Kharasch and H. C. Brown[16] developed a method for the low-temperature chlorination of alkanes utilizing sulfuryl chloride and

$$RH + Cl_2SO_2 \xrightarrow[40\text{-}80\%]{(C_6H_5COO)_2} RCl + HCl + SO_2$$

dibenzoyl peroxide as chain-initiating catalyst. Results summarized in the charts show the pattern of distribution for the chlorination of *n*-heptane and 1-chlorobutane.

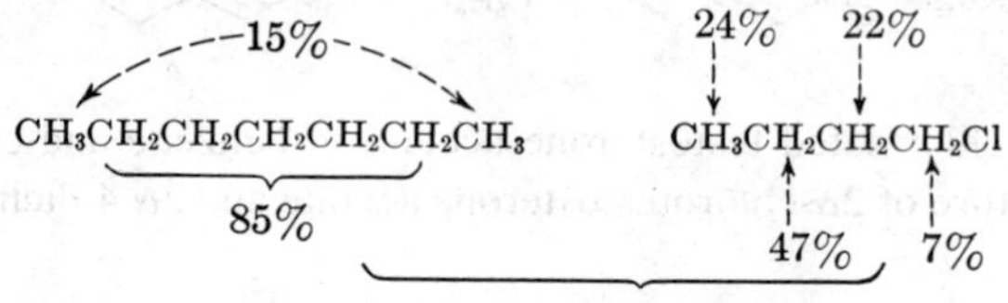
15%
$CH_3CH_2CH_2CH_2CH_2CH_2CH_3$
85%
24%
22%
$CH_3CH_2CH_2CH_2Cl$
47%
7%

Kharasch chlorination

Cleavage of the S-benzyl link. Such a cleavage is involved in a procedure by N. Kharasch and Langford[17] for the preparation of 2,4-dinitrobenzenesulfenyl chloride.

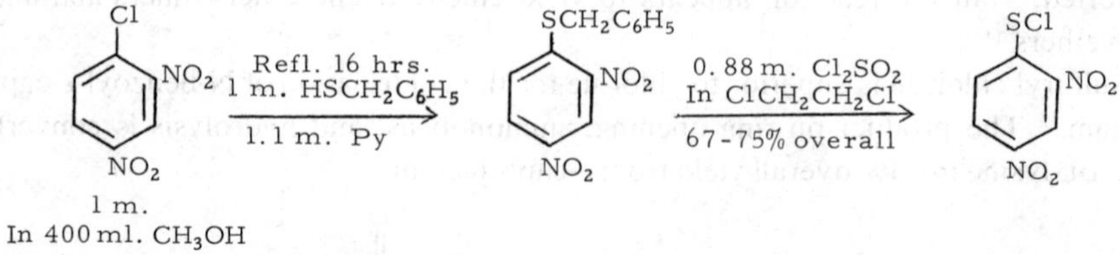
Cl
NO_2
NO_2
1 m.
In 400 ml. CH_3OH
Refl. 16 hrs.
1 m. $HSCH_2C_6H_5$
1.1 m. Py
$SCH_2C_6H_5$
NO_2
NO_2
0.88 m. Cl_2SO_2
In $ClCH_2CH_2Cl$
67-75% overall
SCl
NO_2
NO_2

Reaction with benzoin. Sulfuryl chloride reacts with benzoin to give a cyclic dienol ester which spontaneously decomposes to benzil and sulfur dioxide.[18]

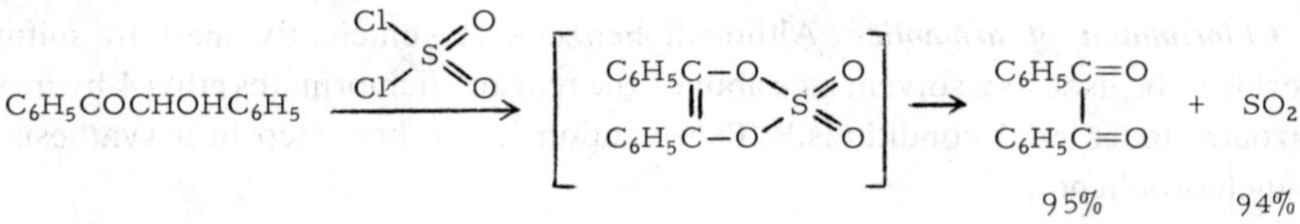
$C_6H_5COCHOHC_6H_5$
95%
94%
+ SO_2

n-Butyl sulfate. This ester is obtained in 74–83% yield by heating 1.6 moles of sulfuryl chloride with 3.2 moles of *n*-butyl sulfite and distilling off the *n*-butyl chloride formed.[19] The steps appear to be as follows:

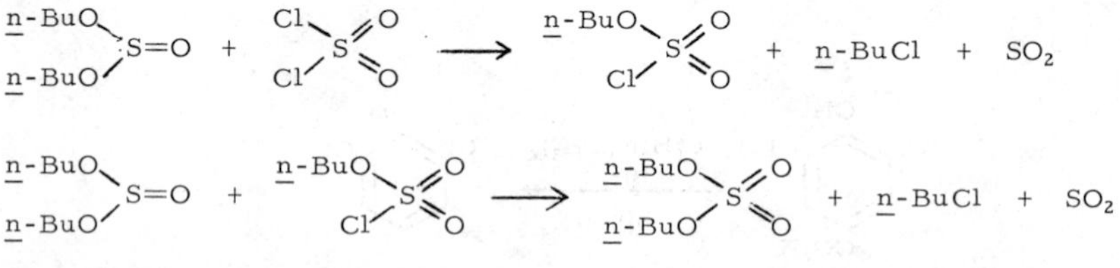
+ n-BuCl + SO_2
+ n-BuCl + SO_2

Esterification catalyst. Sulfuryl chloride is an effective esterification catalyst, particularly useful for peptides since no racemization occurs.[20] The N-acylated amino acid or peptide is dissolved in methanol containing a trace of the catalyst, and the mixture is allowed to stand for 48 hrs. at 20°. Yields are generally in the range 85–95%. Benzyl esters are prepared similarly except that *sym*-tetrachloroethane is used as solvent.

Chlorination of methyl sulfides and methyl ethers. Methyl sulfides[21] and methyl ethers[22] are converted by sulfuryl chloride into chloromethyl derivatives.

$$RSCH_3 + SO_2Cl_2 \longrightarrow RSCH_2Cl + SO_2 + HCl$$

$$C_6H_5OCH_3 + SO_2Cl_2 \xrightarrow[93\text{-}95\%]{} C_6H_5OCH_2Cl + SO_2 + HCl$$

Sulfanilides.[23] A convenient route to sulfanilides involves reaction of an aniline with sulfuryl chloride in pyridine at 0° (in the absence of pyridine products of oxidation and chlorination are produced).

$$2\ \underline{p}\text{-ClC}_6\text{H}_4\text{NH}_2 + \text{ClSO}_2\text{Cl} \xrightarrow[68\%]{\text{Py } 0^\circ} \underline{p}\text{-ClC}_6\text{H}_4\overset{\text{H}}{\text{N}}\text{SO}_2\overset{\text{H}}{\text{N}}\text{C}_6\text{H}_4\text{Cl-}\underline{p}$$

Reagent BMC. This powerful perchlorinating reagent is named from the initials of the Spanish discoverers.[24] As required for the perchlorination of an organic substrate, the reagent is prepared from sulfur monochloride (S_2Cl_2, 5 g.), sulfuryl chloride (850 ml.), and aluminum chloride (2.5 g.) in sulfuryl chloride (750 ml.). Ballester[25] describes the preparation of some highly strained aromatic chlorohydrocarbons by this method, for example, perchlorophenanthrene and perchlorotoluene.

[1]H. R. Allen and R. N. Maxson, *Inorg. Syn.*, **1**, 114 (1939)

[2]B. Tchoubar and O. Sackur, *Compt. rend.*, **208**, 1020 (1939); P. Delbaere, *Bull. soc. chim. Belg.*, **51**, 1 (1942)

[3]D. P. Wyman and P. R. Kaufman, *J. Org.*, **29**, 1956 (1964); D. P. Wyman, P. R. Kaufman, and W. R. Freeman, *ibid.*, **29**, 2706 (1964)

[4]E. W. Warnhoff, D. G. Martin, and W. S. Johnson, *Org. Syn., Coll. Vol.*, **4**, 162 (1963)

[5]E. J. Cragoe, Jr., and C. M. Robb, *Org. Syn.*, **40**, 54 (1960)

[6]W. R. Boehme, *Org. Syn., Coll. Vol.*, **4**, 590 (1963)

[7]H. Mori, *Chem. Pharm. Bull.*, **10**, 429 (1962)

[8]Danish patent 83,631 (1957) [*C.A.*, **53**, 11452 (1959)]

[9]K. Yasuda, *Chem. Pharm. Bull.*, **12**, 1217 (1964)

[10]C. L. Stevens, E. Farkas, and B. Gillis, *Am. Soc.*, **76**, 2695 (1954)

[11]D. P. Wyman, P. R. Kaufman, and W. R. Freeman, *J. Org.*, **29**, 2706 (1964)

[12]R. Tull, R. C. O'Neill, E. P. McCarthy, J. J. Pappas, and J. M. Chemerda, *J. Org.*, **29**, 2425 (1964)

[13]D. S. Tarbell, J. W. Wilson, and P. E. Fanta, *Org. Syn., Coll. Vol.*, **3**, 267 (1955)

[14]E. Campaigne and W. M. LeSuer, *Am. Soc.*, **70**, 415 (1948)

[15]H. C. Godt, Jr., and R. E. Wann, *J. Org.*, **27**, 1459 (1962)

[16]M. S. Karasch and H. C. Brown, *Am. Soc.*, **61**, 2142 (1939); H. C. Brown and A. B. Ash, *ibid.*, **77**, 4019 (1955); G. A. Russell and H. C. Brown, *ibid.*, **77**, 4031 (1955); M. C. Ford and W. A. Waters, *J. Chem. Soc.*, 1851 (1951); 2240 (1952); A. Mooradian and J. B. Cloke, *Am. Soc.*, **68**, 785 (1946)

[17]N. Kharasch and R. B. Langford, *Org. Syn.*, **44**, 47 (1964)

[18]L. F. Fieser and Y. Okumura, *J. Org.*, **27**, 2247 (1962)

[19]C. M. Suter and H. L. Gerhart, *Org. Syn., Coll. Vol.*, **2**, 111 (1943)

[20]E. Taschner and C. Wasielewski, *Ann.*, **640**, 136,139 (1961)

[21]W. E. Truce, G. H. Birum, and E. T. McBee, *Am. Soc.*, **74**, 3594 (1952); F. G. Bordwell and B. M. Pitt, *ibid.*, **77**, 572 (1955)

[22]C. S. Davis and G. S. Longheed, procedure submitted to *Org. Syn.*

[23]E. W. Parnell, *J. Chem. Soc.*, 4367 (1960)

[24]M. Ballester, C. Molinet, and J. Castañer, *Am. Soc.*, **82**, 4254 (1960)

[25]M. Ballester, *Bull. soc.*, 7 (1966)

T

Tetra-n-butylammonium iodotetrachloride, $n\text{-}Bu_4NICl_4$. Mol. wt. 511.20, m.p. 137–139° dec. Preparation.[1]

The reagent reacts with alkenes and alkynes with stereospecific *trans*-addition of chlorine. Thus *cis*- and *trans*-stilbene react to give, respectively, the *meso*- and *dl*-dichlorides.[2] A solution of 0.005 to 0.06 mole of olefin in 25 ml. of ethylene dichloride is treated with one equivalent of the reagent and let stand 3 days at room temperature; yield of each stilbene dichloride, 50%. Acetophenone afforded phenacyl chloride in 54% yield.

[1]A. I. Popov and R. E. Buckles, *Inorg. Syn.*, **5**, 176 (1957)
[2]R. E. Buckles and D. F. Knaack, *J. Org.*, **25**, 20 (1960)

N,2,4,6-Tetrachloroacetanilide, $Cl_3C_6H_2NClCOCH_3$. Mol. wt. 272.96, m.p. 75°.

Preparation from 2,4,6-trichloroacetanilide and sodium hypochlorite.[1]

Chlorination. In the presence of dibenzoyl peroxide as initiator, the reagent chlorinates alkanes, cycloalkanes, and alkylbenzenes (side chain).[2]

[1]K. N. Ayad, C. Beard, R. F. Garwood, and W. J. Hickinbottom, *J. Chem. Soc.*, 2981 (1957)
[2]J. R. B. Boocock and W. J. Hickinbottom, *J. Chem. Soc.*, 1234 (1963)

Tetrachlorocyclopentadienone (3). This reactive Diels-Alder diene is unknown as the monomer but can be generated starting with the readily available hexachlorocyclopentadiene (1, Hooker Chem. Corp., Aldrich). Newcomer and McBee[1] found

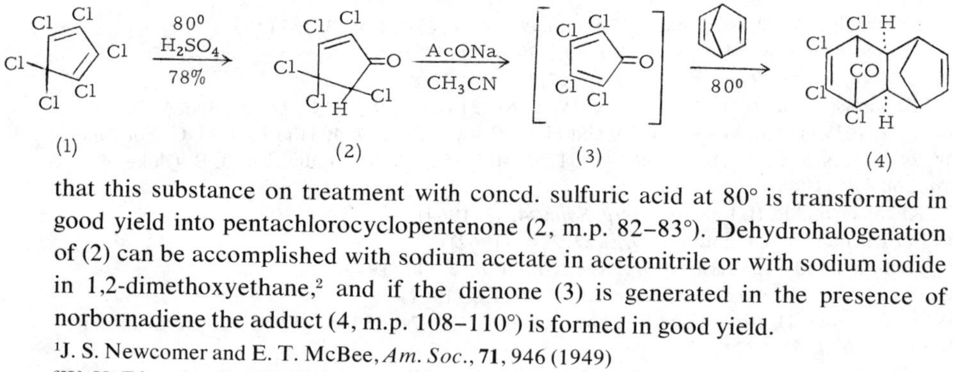
80°
H₂SO₄
78%
AcONa
CH₃CN
80°

(1) (2) (3) (4)

that this substance on treatment with concd. sulfuric acid at 80° is transformed in good yield into pentachlorocyclopentenone (2, m.p. 82–83°). Dehydrohalogenation of (2) can be accomplished with sodium acetate in acetonitrile or with sodium iodide in 1,2-dimethoxyethane,[2] and if the dienone (3) is generated in the presence of norbornadiene the adduct (4, m.p. 108–110°) is formed in good yield.

[1]J. S. Newcomer and E. T. McBee, *Am. Soc.*, **71**, 946 (1949)
[2]W. H. Dietsche, *Tetrahedron Letters*, 201 (1966)

Tetrachlorofurane, Mol. wt. 205.87, b.p. 61°/10 mm.

The reagent (3) is prepared from tetrahydrofurane by chlorination to hexachlorodihydrofurane (2) and dechlorination with magnesium.[1] It has been used as a Diels-

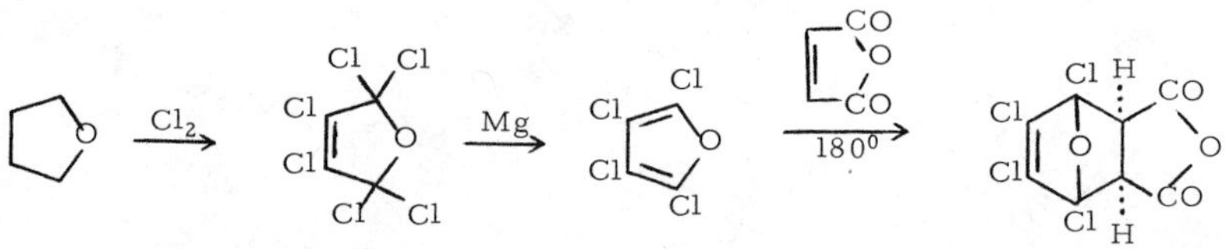
Cl₂
Mg
180°

(1) (2) (3) (4)

Alder diene; for example it reacts with maleic anhydride in an autoclave at 180° to give the adduct (4). With the less reactive dienophiles acrylic acid and cyclohexene, a small amount of hydroquinone or of trichloroacetic acid is added as catalyst.

[1]H. Krzikalla and H. Linge, *Ber.*, **96**, 1751 (1963)

Tetracyanoethylene (TCNE), $(N{\equiv}C)_2C{=}C(C{\equiv}N)_2$. Mol. wt. 128.10, m.p. 200° (sealed capillary), light beige crystals, can be crystallized from 9 times its weight of chlorobenzene. Suppliers: A, E.

Preparation.[1] Bromine is added in the course of 2.5 hrs. to a mixture of malonylnitrile and potassium bromide in 900 ml. of water to produce the solid dibromomalononitrile–potassium bromide complex. A mixture of this complex with benzene is stirred, precipitated copper powder is added, and the mixture is stirred and

$$\underset{1.5\text{ m.}}{4\,CH_2(CN)_2} + \underset{3.05\text{ m.}}{8\,Br_2} + \underset{0.63\text{ m.}}{KBr} \xrightarrow[85\text{-}89\%]{5\text{-}10^0} KBr[Br_2C(CN)_2]_4 + 8\,HBr$$

$$\underset{0.25\text{ m.}}{KBr[Br_2C(CN)_2]_4} + \underset{1.57\text{ g. at.}}{4\,Cu} + \underset{1\text{ l.}}{C_6H_6} \xrightarrow{\text{Stir and refl. 10-16 hrs.}} (NC)_2C{=}C(CN)_2$$

refluxed to complete the reaction. The product is extracted with hot benzene and recrystallized from chlorobenzene.

π-Complexes. TCNE is a powerful π-acid and forms colored π-complexes with aromatic compounds, as shown in the table.[2] Hexamethylbenzene is an enormously

π-Complexes of TCNE in CH_2Cl_2

	Assoc. const. (K)	λ_{max}, mμ	ε
Benzene	2.00	384	3,570
Toluene	3.70	406	3,330
o-Xylene	6.97	430	3,860
Hexamethylbenzene	263	545	4,390
Hexaethylbenzene	5.11	550	56
Naphthalene	11.7	550	1,240
Pyrene	29.5	724	1,137

more powerful π-base than benzene; one of the two partners lies over the other with maximal overlap between the π-molecular orbitals of the two. Since all twelve carbon atoms of hexamethylbenzene lie in a plane, the methyl groups do not interfere with close approach of the second component to about 5 Å. Tetraethylbenzene, by contrast, is not coplanar. TCNE forms a π-complex with cyclohexene, but K is only about one tenth that of the benzene complex. The complex of benzene is yellow, that of naphthalene is red, and that of anthracene is brilliant green.

Diels-Alder reactions.[3] The green color of the TCNE-anthracene π-complex is only transient, for it soon gives way to the colorless Diels-Alder adduct. When butadiene is added in slight excess to an ice-cold solution of TCNE in tetrahydrofurane, the solution at first takes on the bright yellow color of a π-complex, and within a few minutes the colorless adduct begins to separate. Under the same conditions the reaction of maleic anhydride with butadiene required many hours for completion.

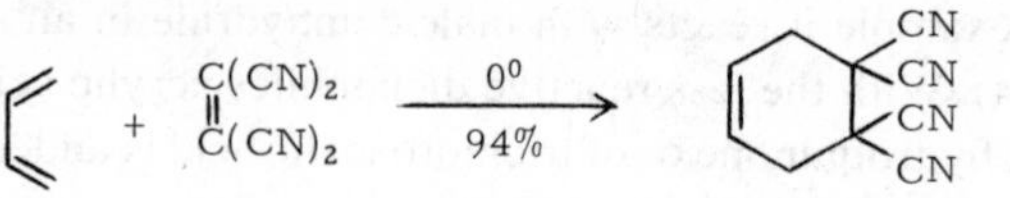

In a kinetic study of the Diels-Alder addition of 24 dienophiles with cyclopentadiene and 9,10-dimethylanthracene, TCNE was found to be the most reactive.[4] Paquette[5] treated the azepinone (1) with maleic anhydride or N-phenylmaleimide in the temperature range 35–120° but failed to obtain a crystalline adduct. Reaction occurred when (1) was heated with dimethyl acetylenedicarboxylate without solvent at 130°, but the yield of adduct was only 20%. In contrast, a solution of equivalent amounts of (1) and TCNE in tetrahydrofurane at room temperature had an initial violet-brown color which faded completely in 30 min. to pale yellow and the adduct (2) was obtained in quantitative yield.

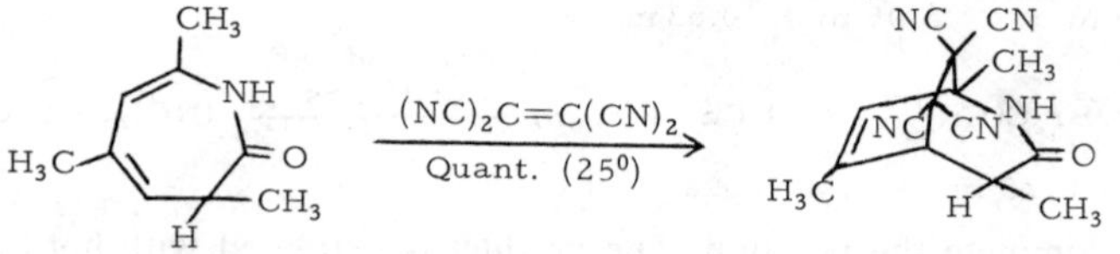

(1) (2)

Additions giving cyclobutanes. Blomquist and Meinwald[6] found that a diene incapable of forming a Diels-Alder product may react with tetracyanoethylene to form a cyclobutane. Thus Williams[7] investigated the reaction of the diene (1) with TCNE at room temperature and found that the principal product (2) was that

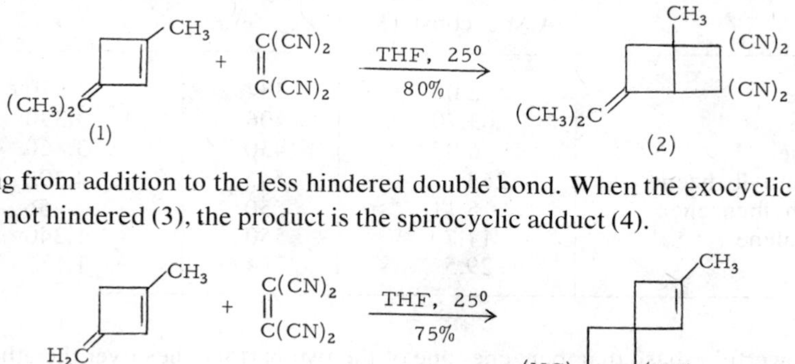

(1) (2)

resulting from addition to the less hindered double bond. When the exocyclic double bond is not hindered (3), the product is the spirocyclic adduct (4).

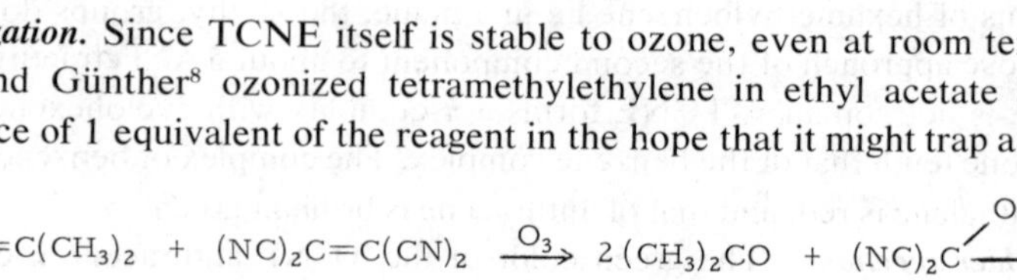

(3) (4)

Ozonization. Since TCNE itself is stable to ozone, even at room temperature, Criegee and Günther[8] ozonized tetramethylethylene in ethyl acetate at −70° in the presence of 1 equivalent of the reagent in the hope that it might trap a postulated

$$(CH_3)_2C{=}C(CH_3)_2 + (NC)_2C{=}C(CN)_2 \xrightarrow{O_3} \underset{80\%}{2\,(CH_3)_2CO} + \underset{98\%}{(NC)_2C\overset{O}{\frown}C(CN)_2}$$

intermediate "carbonyloxide," $>\overset{+}{C}{-}O{-}\overset{-}{O} \leftrightarrow > C{=}\overset{+}{O}{-}\overset{-}{O}$ by 1:3-dipolar addition. Surprisingly, the products were acetone and crystalline TCNE oxide. The mechanism is not clear, but the observation provides a new method for the cleavage of ozonides to produce aldehydes and ketones directly in good yield. Thus Munavalli and Ourisson[9] found that ozonization of longifolene (1) under standard conditions gives the epoxide, an abnormal product, but that in the presence of TCNE the

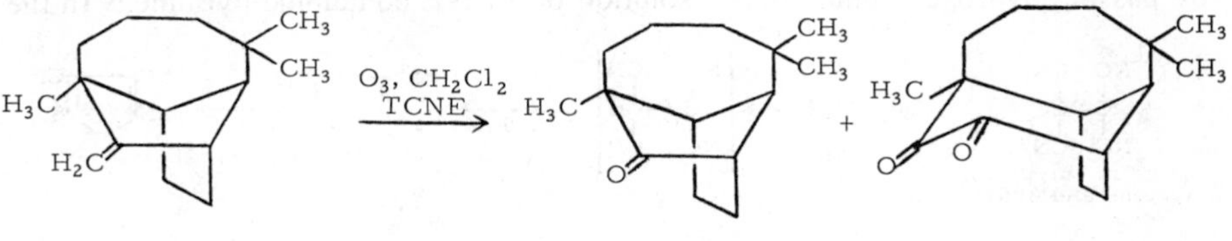
CH3
CH3
H3C
H2C
O3, CH2Cl2
TCNE
CH3
CH3
H3C
O
+
CH3
CH3
H3C
O
O

(1) (2) 30% (3) 16%

expected ketone (2) is the major product and is accompanied by a small amount of the dione (3) resulting from ring expansion.

Reaction with alcohols. In the presence of urea, TCNE reacts with alcohols, for example ethylene glycol, with elimination of hydrogen cyanide.[10] The product,

$$(NC)_2C{=}C(CN)_2 + \begin{matrix}HOCH_2\\ |\\ HOCH_2\end{matrix} + H_2NCONH_2 \xrightarrow[77\text{-}85\%]{70\text{-}75^0} (NC)_2C{=}C\begin{matrix}O{-}CH_2\\ |\\ O{-}CH_2\end{matrix}$$

0.2 m. in 300 ml. acetone and 300 ml. CS_2 — 50 ml. — 0.067 m.

dicyanoketene ethyleneketal, can be converted in one-step reactions to a pyrimidine, a pyrazole, or an isoxazole.

Aromatic substitution. TCNE reacts with N,N-dimethylaniline under mild conditions in dimethylformamide to give the *p*-tricyanovinyl derivative.[11] This substance forms dark blue needles and is a dye. Under similar conditions 2,6-

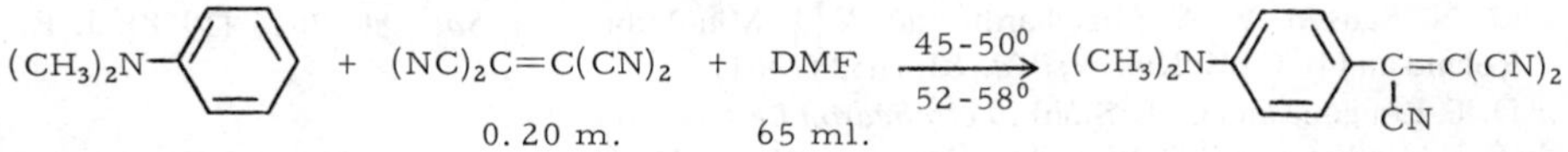
(CH3)2N
+ (NC)2C=C(CN)2 + DMF
45-50°
52-58°
(CH3)2N
C=C(CN)2
CN

0.22 m. — 0.20 m. — 65 ml.

dimethylphenol gives the 4-tricyanovinyl derivative, pyrrole gives 2-tricyanovinylpyrrole, and phenanthrene gives 9-tricyanovinylphenanthrene.[12]

Dehydrogenation. TCNE is an effective reagent for aromatization of 1,4-dihydrobenzenoids:[13]

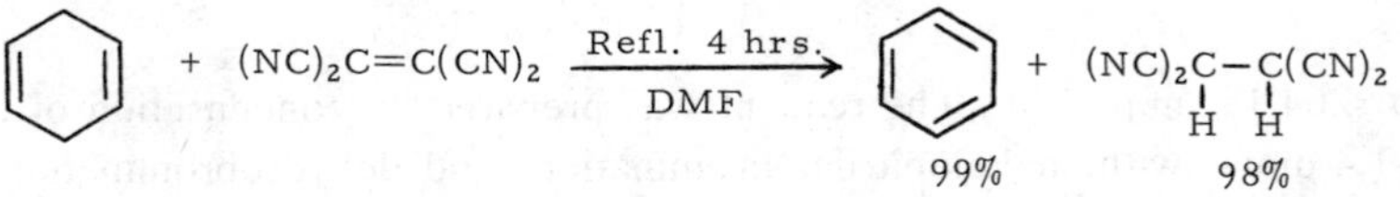
+ (NC)2C=C(CN)2
Refl. 4 hrs.
DMF
+ (NC)2C−C(CN)2
H H

99% 98%

No reaction is observed with cyclohexene, acenaphthene, 9,10-dihydrophenanthrene, tetralin, etc.

Other reactions. Like phenols and *t*-arylamines, malononitrile is substituted readily by TCNE.[14] The reaction is conducted in the presence of pyridine and the

$$(NC)_2C{=}C(CN)_2 + CH_2(CN)_2 + Py \xrightarrow{25\text{ ml. }H_2O} (NC)_2C{=}\underset{\displaystyle CN}{\underset{|}{C}}{-}\bar{C}(CN)_2[\overset{+}{Py}H]$$

0.1 m. — 0.1 m. — 0.11 m.

$$\xrightarrow[81\text{-}85\%\text{ overall}]{(CH_3)_3\overset{+}{N}Cl\text{ in 500 ml. }H_2O} (NC)_2C{=}\underset{\displaystyle CN}{\underset{|}{C}}{-}\bar{C}(CN)_2[(CH_3)_3\overset{+}{N}] + Py\cdot HCl$$

product isolated as bright yellow-orange needles of tetramethylammonium 1,1,2,3,3-pentacyanopropenide.

An interesting reaction affording 2,5-diamino-3,4-dicyanothiophene is carried out by passing hydrogen sulfide into a solution of TCNE containing pyridine.[15] In the

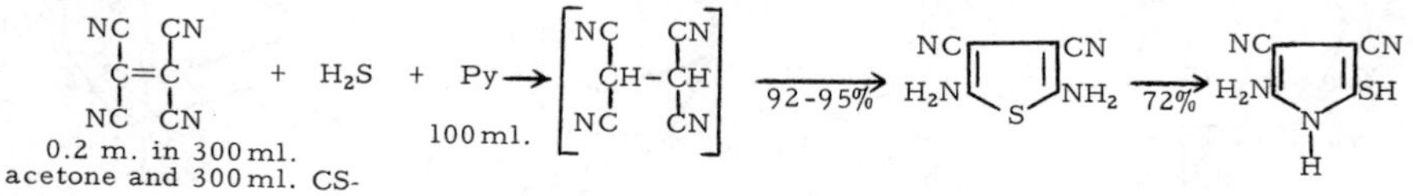
+ H₂S + Py
0.2 m. in 300 ml.
acetone and 300 ml. CS-
100 ml.
92-95%
72%

absence of pyridine or other base the reaction stops with the formation of tetracyanoethane and sulfur. The diaminodicyanothiophene is isomerized to 2-amino-3,4-dicyano-5-mercaptopyrrole by heating it with aqueous alkali until dissolved and acidifying the solution. Thiophenes and pyrroles containing amino or thiol groups are usually very unstable but those containing the two electron-withdrawing cyano groups are stable.

[1]R. A. Carboni, *Org. Syn., Coll. Vol.*, **4**, 877 (1963)
[2]R. E. Merrifield and W. D. Phillips, *Am. Soc.*, **80**, 2778 (1958)
[3]W. J. Middleton, R. E. Heckert, E. L. Little, and C. G. Krespan, *Am. Soc.*, **80**, 2783 (1958); C. A. Stewart, Jr., *J. Org.*, **28**, 3320 (1963)
[4]J. Sauer, H. Wiest, and A. Mielert, *Ber.*, **97**, 3183 (1964)
[5]L. A. Paquette, *J. Org.*, **29**, 3447 (1964)
[6]A. T. Blomquist and Y. C. Meinwald, *Am. Soc.*, **79**, 5316 (1957)
[7]J. K. Williams, *Am. Soc.*, **81**, 4013 (1959)
[8]R. Criegee and P. Günther, *Ber.*, **96**, 1564 (1963)
[9]S. Munavalli and G. Ourisson, *Bull. Soc.*, 729 (1964)
[10]C. L. Dickinson and L. R. Melby, *Org. Syn., Coll. Vol.*, **4**, 276 (1963)
[11]B. C. McKusick and L. R. Melby, *ibid.*, **4**, 953 (1963)
[12]G. N. Sausen, V. A. Engelhardt, and W. J. Middleton, *Am. Soc.*, **80**, 2815 (1958); J. R. Roland and B. C. McKusick, *ibid.*, **83**, 1652 (1961)
[13]D. T. Longone and G. L. Smith, *Tetrahedron Letters*, 205 (1962)
[14]W. J. Middleton and D. W. Wiley, *Org. Syn.*, **41**, 99 (1961)
[15]W. J. Middleton, V. A. Engelhardt, and B. S. Fisher, *Am. Soc.*, **80**, 2822 (1958)

7,7,8,8-Tetracyanoquinonedimethane,

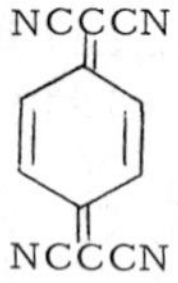
NCCCN
NCCCN

Mol. wt. 204.18, m.p. 296°. The reagent was prepared by condensation of cyclohexane-1,4-dione with malononitrile, bromination, and dehydrobromination with pyridine.[1] The compound has properties similar to those of tetracyanoethylene and is a π-acid of comparable strength. The equilibrium constant for π-complex formation with pyrene is 78.4 as compared to 29.5 for the tetracyanoethylene–pyrene complex.

[1]D. S. Acker and W. R. Hertler, *Am. Soc.*, **84**, 3370 (1962)

Tetracyclone, *see* Tetraphenylcyclopentadienone.

Tetraethylammonium acetate $(C_2H_5)_4\overset{+}{N}(\overset{-}{O}COCH_3)$. Mol. wt. 189.29, viscous oil. Preparation.[1,2]

The reagent reacts with (−)α-phenylethyl chloride in acetone solution to give

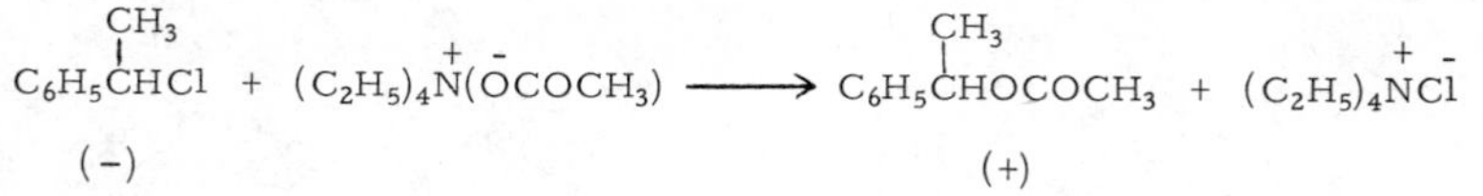

$$\underset{(-)}{C_6H_5\overset{CH_3}{\overset{|}{C}}HCl} + (C_2H_5)_4\overset{+}{N}(\overset{-}{O}COCH_3) \longrightarrow \underset{(+)}{C_6H_5\overset{CH_3}{\overset{|}{C}}HOCOCH_3} + (C_2H_5)_4\overset{+}{N}\overset{-}{Cl}$$

(+)α-phenylethyl acetate, racemization being less than in the solvolysis with acetic acid.[1] It reacts with allylic halides without anionotropic rearrangement.[2,3]

On experiencing difficulty in the preparation of the 2′,5′-diacetate of uridine-3′-phosphate (1) by usual methods, Khorana et al.[4] found that the desired derivative could be obtained by treatment of (1) in pyridine with a tenfold excess of acetic anhydride and one equivalent of tetraethylammonium acetate as a source of acetate ions. The reagent was prepared by neutralizing aqueous tetraethylammonium hydroxide (Eastman, MCB) with acetic acid and rendered anhydrous by repeated evaporation with pyridine.

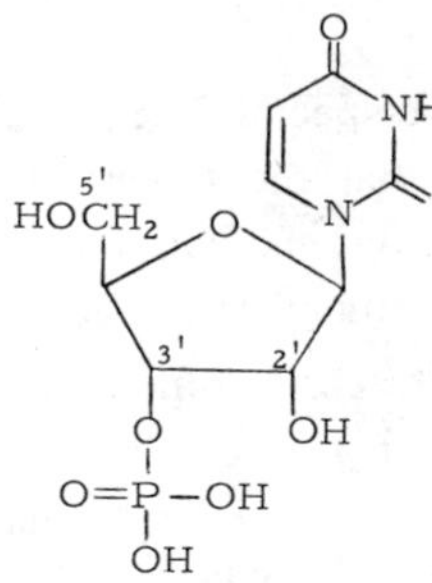

[1] J. Steigman and L. P. Hammett, Am. Soc., **59**, 2536 (1937)

[2] R. Riemschneider and R. Nehring, Ann., **660**, 42 (1962)

[3] J. D. Roberts, W. G. Young, and S. Winstein, Am. Soc., **64**, 2157 (1942); L. N. Owen and P. N. Smith, J. Chem. Soc., 4035 (1952); J. R. B. Campbell, A. M. Islam, and R. A. Raphael, ibid., 4096 (1956)

[4] D. H. Rammler, Y. Lapidot, and H. G. Khorana, Am. Soc., **85**, 1989 (1963)

Tetraethylammonium chloride, $(C_2H_5)_4\overset{+}{N}\overset{-}{Cl}$. Mol. wt. 165.71. Suppliers: A, B, E, F. Preparation.[1]

In solution in acetonitrile, the reagent functions as a dehydrohalogenation catalyst.[1,2]

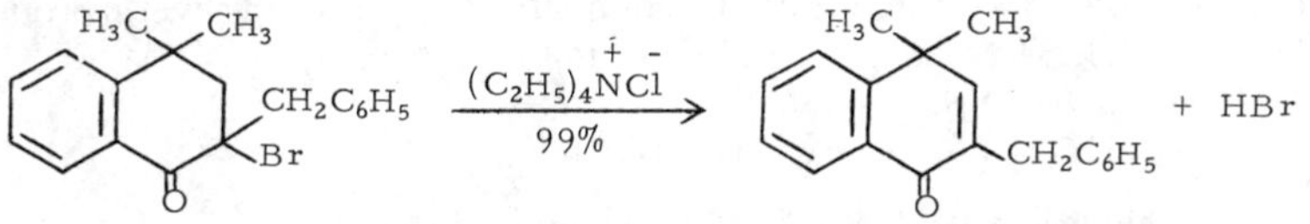

[1] G. Coppens, D. N. Kevill, and N. H. Cromwell, J. Org., **27**, 3299 (1962)

[2] D. N. Kevill, G. A. Coppens, and N. H. Cromwell, J. Org., **28**, 567 (1963)

Tetraethylammonium formate, $(C_2H_5)_4\overset{+}{N}\overset{-}{O}\overset{\overset{\displaystyle O}{\|}}{C}H$.[1] This reagent, a hygroscopic crystalline solid, is useful for the epimerization of secondary alcohols. It is prepared by mixing equivalent amounts of 90% formic acid and 25% aqueous tetraethylammonium hydroxide (MCB), removing most of the water by distillation in vacuum, and eliminating traces of water by azeotropic distillation with benzene. The crystalline product is protected carefully from moisture. Example of use:

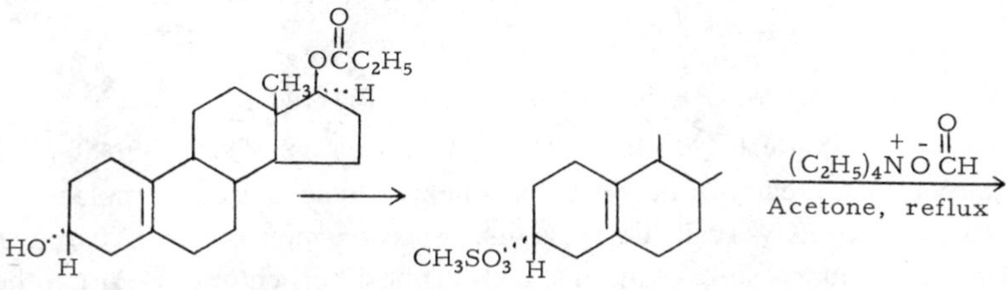

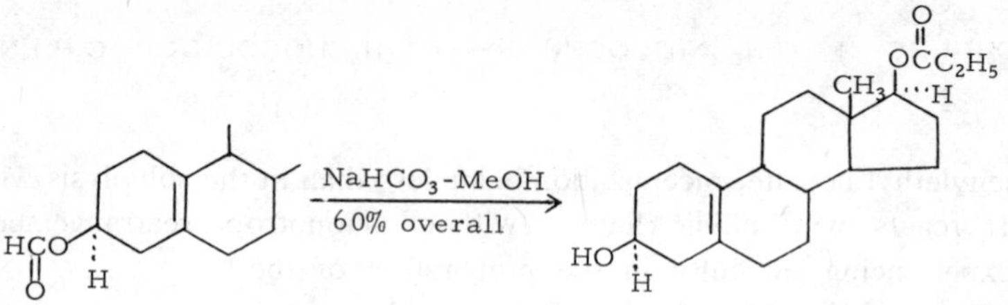
NaHCO3-MeOH
60% overall
HCO
H
HO
H
CH3
OCC2H5
H

The special advantage of this epimerizing agent, for example over potassium acetate, is that the intermediate formate ester can be hydrolyzed under such very mild conditions that other ester groups will remain intact.

[1]Contributed by S. G. Levine, North Carolina State Univ.

Tetraethylammonium hydroxide, $(C_2H_5)_4N^+\bar{O}H$. Preparation and use as described for tri-*n*-hexylethylammonium hydroxide.[1]

[1]G. Weimann, H. Schaller, and H. G. Khorana, *Am. Soc.*, **85**, 3835 (1963); R. Lohrmann and H. G. Khorana, *Am. Soc.*, **86**, 4188 (1964)

Tetraethyl orthocarbonate. $C(OC_2H_5)_4$. Mol. wt. 192.25, b.p. 158°, $n^{25}D$ 1.3907, sp. gr. 0.92.

Chloropicrin is added during 2 hrs. to a stirred solution of sodium ethoxide in

$$CCl_3NO_2 + 4\ C_2H_5ONa \xrightarrow[46\text{-}49\%]{} C(OC_2H_5)_4 + 3\ NaCl + NaNO_2$$

absolute ethanol at 58–60°; the solvent is largely removed by distillation, and the mixture is treated with water and extracted with ether.[1]

Hofmann[2] found that the ortho ester reacts with aqueous ammonia to form guanidine. The reaction with aniline can be conducted to give any of several products.[3]

[1]J. D. Roberts and R. E. McMahon, *Org. Syn., Coll. Vol.*, **4**, 457 (1963)
[2]A. W. von Hofmann, *Ann.*, **139**, 114 (1866)
[3]H. Tieckelmann and H. W. Post, *J. Org.*, **13**, 268 (1948)

Tetraethyl phosphonosuccinate (see 3). Mol. wt. 309.28.

The reagent is prepared by treating the half-ester of maleic acid with triethyl phosphite and converting the product into the phosphonate carbanion by stirring with a slurry of 50% sodium hydride in 1,2-dimethoxyethane.[1] The carbanion reacts with carbonyl compounds like a Wittig ylide but is more reactive.

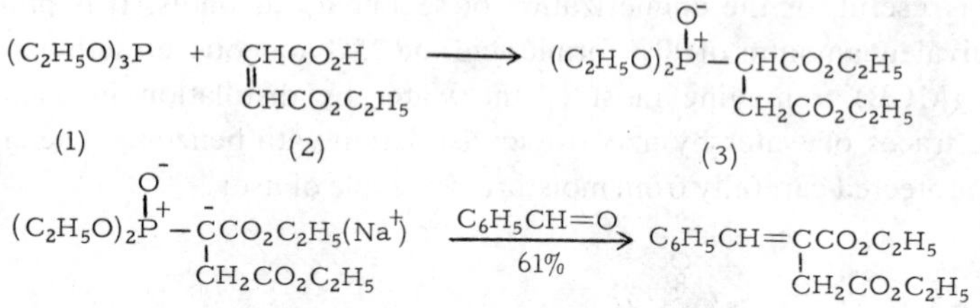
(C2H5O)3P + CHCO2H / CHCO2C2H5 → (C2H5O)2P(O−)(+)—CHCO2C2H5 / CH2CO2C2H5
(1)
(2)
(3)
(C2H5O)2P(O−)(+)—C̄CO2C2H5(Na+) / CH2CO2C2H5
C6H5CH=O
61%
C6H5CH=CCO2C2H5 / CH2CO2C2H5

[1]W. S. Wadsworth, Jr., and W. D. Emmons, *Am. Soc.*, **83**, 1733 (1961)

Tetraethyl pyrophosphite, $(C_2H_5O)_2POP(OC_2H_5)_2$. Mol. wt. 258.19, b.p. 88°/1 mm.

Preparation. An early procedure[1] has been replaced by a simpler method[2] using two commercially available reagents. A suspension of silver carbonate in benzene is stirred during slow addition of diethyl phosphorochloridite. An exothermal reaction ensues with evolution of gas, and the product is distilled from the reaction mixture without separating the solid:

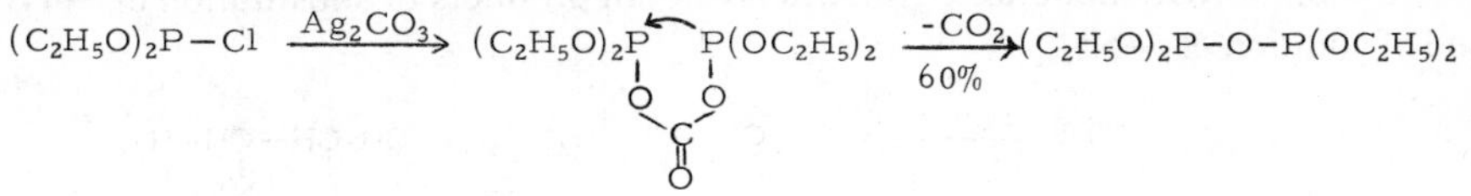

Use in peptide synthesis.[1,3]

$$\underset{\displaystyle R}{CbNHCHCO_2H} + \underset{\displaystyle R'}{H_2NCHCO_2C_2H_5} + (C_2H_5O)_2POP(OC_2H_5)_2 \longrightarrow$$

$$\underset{\displaystyle R\qquad\quad R'}{CbNHCHCONHCHCO_2C_2H_5} + 2\,(C_2H_5O)_2POH$$

[1]G. W. Anderson, J. Blodinger, and A. D. Welcher, *Am. Soc.*, **74**, 5309 (1952)
[2]D. Samuel and B. L. Silver, *J. Org.*, **28**, 1155 (1963)
[3]V. du Vigneaud, C. Ressler, J. M. Swan, C. W. Roberts, and P. G. Katsoyannis, *Am. Soc.*, **76**, 3115 (1954); D. T. Gish and V. du Vigneaud, *ibid.*, **79**, 3579 (1957); C. Ressler and V. du Vigneaud, *ibid.*, **79**, 4511 (1957); P. G. Katsoyannis, D. T. Gish, and V. du Vigneaud, *ibid.*, **79**, 4516 (1957)

Tetrafluoroboric acid, HBF_4. Whiting and co-workers[1-3] prepare 20–30% solutions of the reagent by absorbing gaseous hydrogen fluoride and gaseous boron trifluoride in sulfolane (tetramethylene sulfone) and store it in polytetrafluoroethylene containers. It is diluted as required in glass apparatus. Of eleven acids or Lewis acids tested for their ability to isomerize $\Delta^{1(9)}$-octalin and *trans*-Δ^1-octalin to an equilibrium mixture of Δ^9-octalin, $\Delta^{1(9)}$-octalin, and the other four isomers in the ratio of 90:9:1, tetrafluoroboric acid in benzene–sulfolane proved to be the most efficient.[1]

A solution of phenylhydroxylamine in the reagent is converted into the cation, $C_6H_5NH^+$, which reacts with aromatic compounds to give mixtures of products

$$C_6H_6 \xrightarrow[\text{22\% (total yield)}]{C_6H_5NH^+} \underbrace{\underset{10}{C_6H_5NHC_6H_5} + \underset{67}{4\text{-}H_2NC_6H_4\cdot C_6H_5} + \underset{28}{2\text{-}H_2NC_6H_4\cdot C_6H_5}}_{\text{\% of distribution}}$$

of aminophenylation in low yield.[2] A solution of 85% hydrogen peroxide in the reagent contains the ion $H_3O_2^+$, a powerful but nonselective oxidant.[3]

[1]J. W. Powell and M. C. Whiting, *Proc. Chem. Soc.*, 412 (1960)
[2]J. H. Parish and M. C. Whiting, *J. Chem. Soc.*, 4713 (1964)
[3]R. W. Alder and M. C. Whiting, *J. Chem. Soc.*, 4707 (1964)

Tetrafluorohydrazine, F_2NNF_2. Mol. wt. 104.02, b.p. -73°.

The reagent is prepared by reaction of nitrogen trifluoride with various metals but must be handled cautiously because explosions may result from mixture with organic compounds unless oxygen is rigorously excluded.[1] It is a versatile reagent for introduction of the difluoroamino group. Thus photolysis of an α-diketone in the presence of N_2F_4 gives N,N-difluoroamides. It reacts thermally with aldehydes to

$$RCOCOR + F_2NNF_2 \xrightarrow{h\nu} 2\ RCONF_2$$

give N,N-difluoroamides.[2] Irradiation of an alkane and tetrafluorohydrazine at 2337 Å promotes a radical reaction leading to alkyldifluoramines.[3] *n*-Butane gives

$$RCHO + H_2NNF_2 \xrightarrow{150^\circ} RCONF_2 + HNF_2$$

1- and 2-difluoroaminobutane. Alkenes give both products of substitution of $—NF_2$ and of addition of $—NF_3$.

$$CH_3CH{=}CHCH_3 \xrightarrow{H_2NNH_2} CH_3CH{=}CHCH_2NF_3 + \underset{F}{CH_3CH}-\underset{NF_2}{CHCH_3}$$

[1]C. B. Colburn and A. Kennedy, *Am. Soc.*, **80**, 5004 (1958)
[2]R. C. Petry and J. P. Freeman, *Am. Soc.*, **83**, 3912 (1960)
[3]C. L. Bumgardner, *Tetrahedron Letters*, 3683 (1964)

Tetrahydrofurane (THF). Mol. wt. 72.10, b.p. 65.4°, sp. gr. 0.88, water-miscible. Suppliers of premium grade: BJ, MCB.

Purification.[1] A 1-l. portion of tetrahydrofurane is stirred magnetically in a flask fitted with a reflux condenser protected by a calcium chloride drying tube, and 2–4 g. of lithium aluminum hydride is added in small portions. A sample of the clear liquid is withdrawn and added to a few milliliters of water; if vigorous evolution of hydrogen does not occur, an additional 2-3 g. of $LiAlH_4$ is added. The tetrahydrofurane is then distilled at atmospheric pressure. Since $LiAlH_4$ decomposes at temperatures above 150°, the distillation should be stopped well short of dryness. The purified THF should be stored under nitrogen.

A Ventron bulletin recommends addition of the metal hydride from an Erlenmeyer attached to the flask by Gooch tubing (Fig. A-1). Diethyl ether and dioxane may be dried effectively with lithium aluminum hydride without hazard provided that the solvent is not distilled to dryness.

Solvent effects, *see also under* Grignard reagents, vinyl halides, co-solvents.

THF is useful in the malonic ester synthesis since it dissolves many sodio derivatives.[2] Use of aqueous THF facilitates diazotization of salts of aminopolyphenyls which are sparingly soluble in water[3] and is recommended also for the Schiemann reaction.[4] It is superior to ether as solvent for coupling of an acetylenic Grignard reagent with a propargylic halide.[5] Cremlyn and Chisholm[6] found THF superior to dioxane or Cellosolve as solvent for the hydrogenation of cholesterol at ordinary temperature and pressure with perchloric acid as catalyst.

Grignard reagents can be prepared readily in a hydrocarbon medium (toluene, xylene) if 1 mole of THF is present per mole of halide.[7] The process is economical, and there is no fire hazard since the THF is bound to the Grignard reagent.

THF exerts a remarkable stabilizing influence on α-chloroalkyllithium compounds making it possible to prepare them and to store them indefinitely at −100°.[8] Thus α,α-dichlorobenzyllithium (2) is prepared by reaction of benzotrichloride with

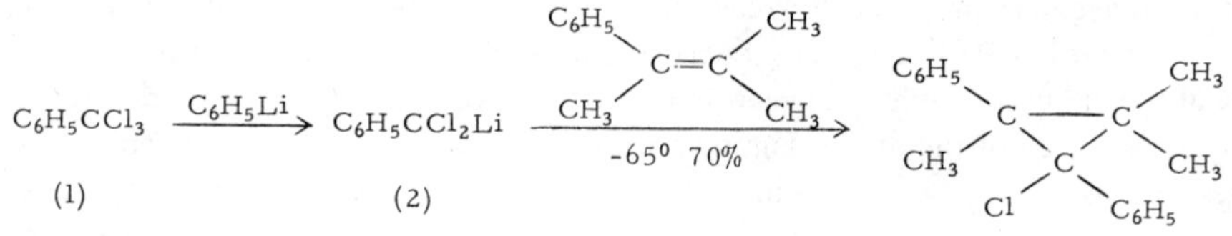
C6H5CCl3
C6H5Li
C6H5CCl2Li
C6H5
CH3
C=C
CH3
CH3
-65° 70%
C6H5
CH3
CH3
CH3
Cl
C6H5

(1) (2) (3)

phenyllithium, and on decomposition at −65° in the presence of an olefin it yields a cyclopropane derivative (3).

Methyllithium shows greatly enhanced reactivity in THF over ether. Thus methyllithium and tetraphenylphosphonium bromide react very slowly in ether (4 months, N_2) to give a mixture of several phosphorus-containing compounds,[9] but when the reaction is carried out in THF–ether (4:1) the following reaction can be realized:[10]

$$(C_6H_5)_4\overset{+}{P}\overset{-}{Br} + CH_3Li \xrightarrow{THF} (C_6H_5)_3\overset{+}{P}-\overset{-}{C}H_2 + C_6H_6 + LiBr$$

Benzene is formed in almost quantitative yield and the ylide in at least 70% yield, as demonstrated by reaction with cyclohexanone.

6-Methyl-2-bromonaphthalene does not form a Grignard reagent satisfactorily in ether or ether–benzene, but with use of THF the reagent is obtained readily in high yield.[11]

Reactions. THF is the starting material for preparation of the four compounds listed below:

Tetramethylene chlorohydrin:[12]

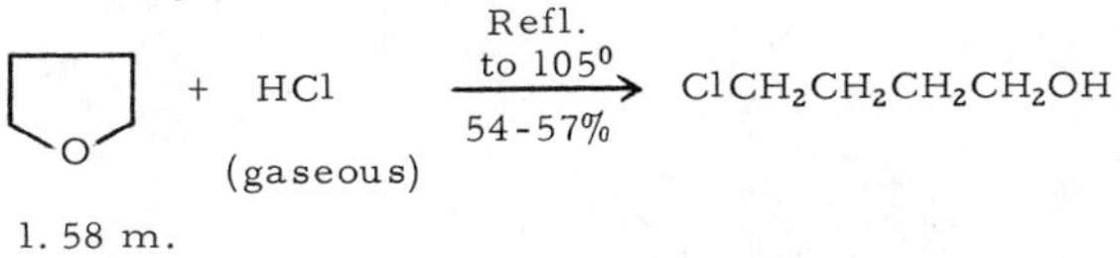

4-Chlorobutyl benzoate:[13]

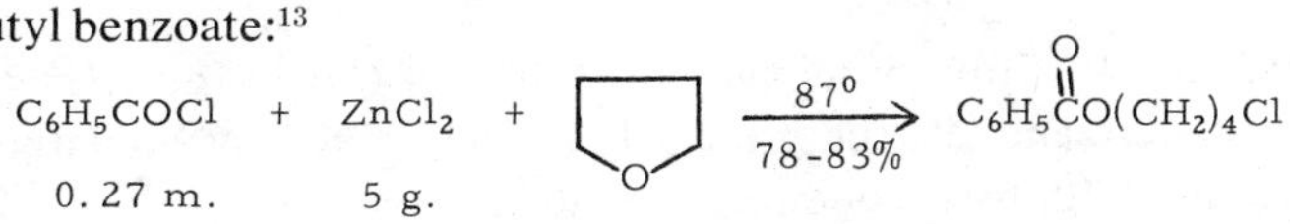

1,4-Diiodobutane:[14]

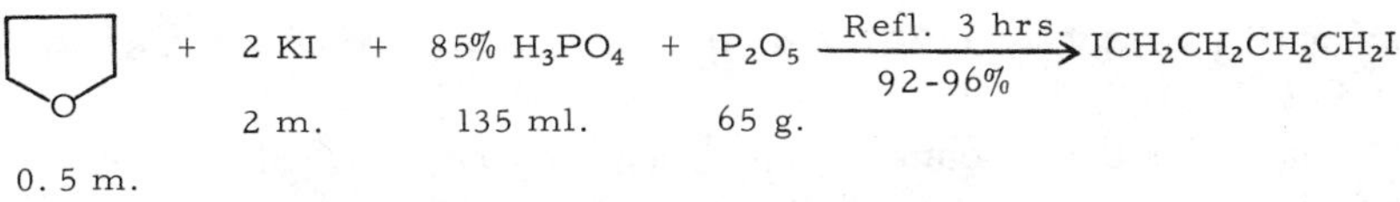

4,4′-Dichlorodibutyl ether:[15]

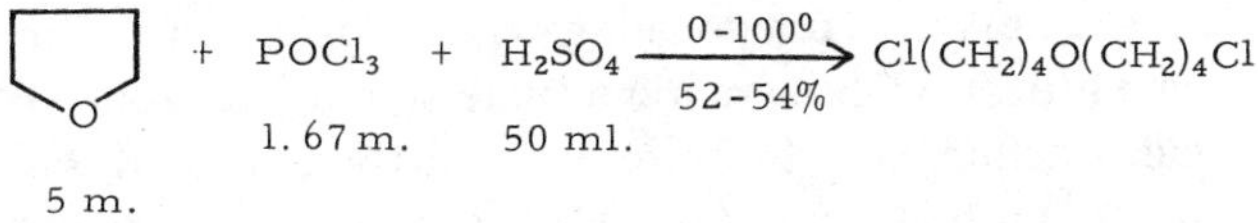

[1]Contributed by George Zweifel, University of California, Davis, California
[2]S.-O. Lawesson and T. Busch, *Chem. Scand.*, **13**, 1717 (1959)
[3]J. A. Cade and A. Pilbeam, *Chem. Ind.*, 1578 (1959)
[4]T. L. Fletcher and M. J. Namkung, *Chem. Ind.*, 179 (1961)
[5]S. N. Eğe, R. Wolovsky, and W. J. Gensler, *Am. Soc.*, **83**, 3080 (1961)
[6]R. J. W. Cremlyn and M. Chisholm, *J. Chem. Soc.*, 5117 (1965)
[7]T. Leigh, *Chem. Ind.*, 426 (1965)
[8]D. F. Hoeg, D. I. Lusk, and A. L. Crumbliss, *Am. Soc.*, **87**, 4147 (1965)
[9]G. Wittig and G. Geissler, *Ann.*, **580**, 44 (1953)
[10]D. Seyferth, W. B. Hughes, and J. K. Heeren, *Am. Soc.*, **87**, 3467 (1965)
[11]H. Lapin, *Chimia*, **18**, 141 (1964)
[12]D. Starr and R. M. Hixon, *Org. Syn.*, *Coll. Vol.*, **2**, 571 (1943)
[13]M. E. Synerholm, *ibid*, **3**, 187 (1955)
[14]H. Stone and H. Shechter, *ibid.*, **4**, 321 (1963)
[15]K. Alexander and H. V. Towles, *ibid*, **4**, 266 (1963)

Tetrakis-(N-dihydropyridyl)-aluminate, *see* Lithium aluminum hydride–Pyridine.

Tetrakis-phosphorus trichloride-nickel (0), $Ni(PCl_3)_4$. Preparation.[1]

Cyclotetramerization of acetylenes.[2] Although several nickel-carbonyl-phosphine catalysts trimerize acetylenes to benzene derivatives, this CO-free catalyst induces tetramerization to cyclooctatetraenes. The reaction, however, is extremely limited in scope. Of a number of acetylenes examined, only the highly reactive methyl and ethyl esters of propiolic acid reacted and these gave both tetramers and trimers. The reaction, carried out at room temperature in cyclohexane and benzene under nitrogen is attended with a rise in temperature to 50–60° and is complete in about 20 min. Methyl propiolate was converted into 1,2,4,6-tetracarbomethoxycyclo-octatetraene (1) in 83% yield, the remaining product (17%) being 1,2,4-tricarbo-

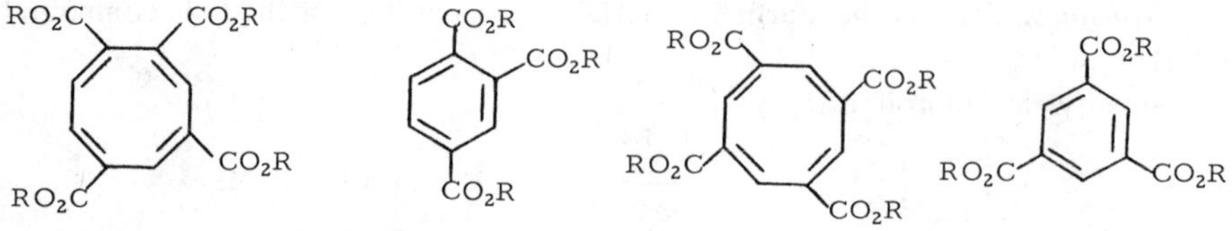

RO_2C
CO_2R

(1) (2) (3) (4)

R = CH_3 or C_2H_5

methoxybenzene (2). Ethyl propiolate gave 28% of (1), 1% of 1,3,5,7-tetracarbo-ethoxycyclooctatetraene (3); the principal product (71%) was a mixture of (2) and (4), the 1,2,4- and 1,3,5-tricarboethoxy derivatives of benzene.

[1] J. W. Irvine, Jr., and G. Wilkinson, *Science*, **113**, 742 (1951)
[2] J. R. Leto and M. F. Leto, *Am. Soc.*, **83**, 2944 (1961)

Tetralin, $C_{10}H_{12}$. Mol. wt. 132.20, b.p. 207–208°, sp. gr. 0.973, n^{20}D 1.5461. Suppliers: A, B, E, F, MCB.

Commercial tetralin contains naphthalene as the principal impurity and this interferes with the preparation of tetralin-1-hydroperoxide[1] or with use of the hydrocarbon as hydrogen donor in hydrogen-transfer reactions.[2] An early purification procedure[1] is uninviting: fractionation; extraction in turn with mercury (to remove sulfur impurities), with mercuric acetate solution (to remove olefins), and with sulfuric acid; fractionation. More recently Bass[2] sulfonated the crude hydrocarbon with concd. sulfuric acid and added ammonium chloride to precipitate ammonium tetralin-6-sulfonate. The salt was crystallized until pure and hydrolyzed by steam distillation from sulfuric acid solution. Distillation from sodium gave material showing no ultraviolet bands characteristic of naphthalene.

[1] H. B. Knight and D. Swern, *Org. Syn., Coll. Vol.*, **4**, 895 (1963)
[2] K. C. Bass, *J. Chem. Soc.*, 3498 (1964)

Tetramethylammonium acetate, $(CH_3)_4\overset{+}{N}\overset{-}{O}_2CCH_3$. Mol. wt. 133.19.

Preparation.[1] An aqueous solution of tetramethylammonium hydroxide is neutralized with acetic acid and evaporated to dryness under reduced pressure. The solid residue is crystallized from acetone and dried in a vacuum dessicator.

Halogen displacement. Steigman and Hammett[2] found that the reaction of α-phenylethyl chloride with the reagent is second order with predominant inversion of configuration. By contrast, solvolysis with acetic acid is essentially first order and considerable racemization takes place. Allylic chlorides react with the reagent in dry acetone to give the corresponding acetates with no rearrangement and with predominant inversion of configuration.[1,3]

Fieser and Romero[4] found that treatment of 2α-bromocholestane-3-one with potassium acetate in refluxing acetic acid affords a complex of 2α- and 4α-acetoxy-cholestane-3-one (2 and 3), inseparable by chromatography. Williamson and

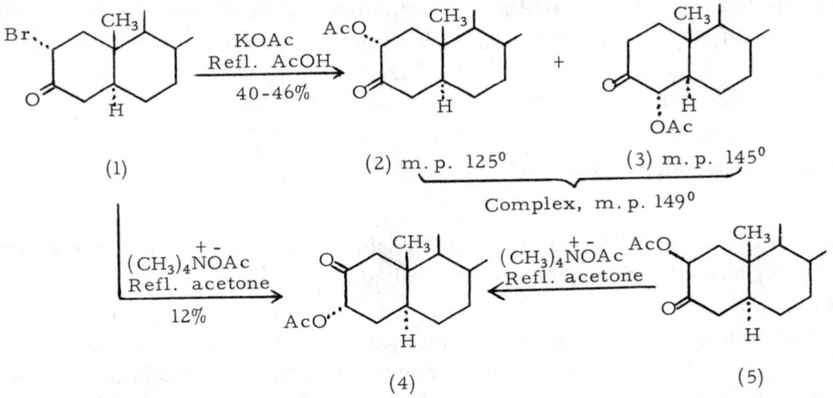
KOAc
Refl. AcOH
40-46%
(1)
(2) m.p. 125°
(3) m.p. 145°
Complex, m.p. 149°
$(CH_3)_4\overset{+}{N}\overset{-}{O}Ac$
Refl. acetone
12%
$(CH_3)_4\overset{+}{N}\overset{-}{O}Ac$
Refl. acetone
(4)
(5)

Johnson[5] confirmed this finding and also prepared an identical complex from an equimolecular mixture of the components. They then treated the α-bromo-3-ketone (1) with tetramethylammonium acetate in refluxing acetone and obtained a mixture of acetoxyketones from which only 3α-acetoxycholestane-2-one (4) could be isolated in pure form. 2β-Acetoxycholestane-3-one (5), the expected product, on treatment with the reagent is isomerized to (4). Treatment of 4α-bromocholestane-3-one with tetramethylammonium acetate in acetone (25°) gave Δ^4-cholestene-3-one (20%) and 4% of the complex of (3) and (4).

[1]H. L. Goering, T. D. Nevitt, and E. F. Silversmith, *Am. Soc.*, **77**, 4042 (1955)

[2]J. Steigman and L. P. Hammett, *Am. Soc.*, **59**, 2536 (1937); see also A. Streitwieser, Jr., and J. R. Wolfe, Jr., *ibid.*, **79**, 903 (1957)

[3]J. D. Roberts, W. G. Young, and S. Winstein, *Am. Soc.*, **64**, 2157 (1942)

[4]L. F. Fieser and M. A. Romero, *Am. Soc.*, **75**, 4716 (1953)

[5]K. L. Williamson and W. S. Johnson, *J. Org.*, **26**, 4563 (1961)

Tetramethylammonium borohydride, $(CH_3)_4NBH_4$. Mol. wt. 89.00. Supplier: Ventron Corp.

This is a stable white solid which has the same reducing properties as sodium borohydride but has the advantage that it is more soluble in nonpolar solvents, even benzene.[1] The reagent is somewhat less sensitive to hydroxylic solvents than sodium borohydride; thus it is stable in ethanol and ethylene glycol, both of which react with sodium borohydride. Similar available reagents are:

Tetraethylammonium borohydride $(C_2H_5)_4NBH_4$, 145.10
Cetyltrimethylammonium borohydride, n-$C_{16}H_{33}(CH_3)_3NBH_4$, 299.39
Tricaprylmethylammonium borohydride, $(n$-$C_8H_{17})_3CH_3NBH_4$, 383.54

[1]E. A. Sullivan and A. A. Hinckley, *J. Org.*, **27**, 3731 (1962)

Tetramethylammonium bromide, $(CH_3)_4N^+Br^-$. Mol. wt. 153.67. Suppliers: E, F.

The reagent is more active than sodium acetate as catalyst for the reductive acetylation of quinones with zinc dust and acetic anhydride.[1]

[1]L. F. Fieser, "Experiments in Organic Chemistry," 2nd ed., 399

Tetramethylammonium tribromide, $(CH_3)_4NBr_3$. Mol. wt. 313.90, m.p. 118.5°. Preparation from tetramethylammonium bromide and bromine.[1]

This is a mild brominating agent.[2] In benzene solution in the presence of dibenzoyl peroxide, hydrocarbons such as toluene undergo benzylic bromination. In acetic acid solution aromatic substitution is observed. In the reaction with cyclohexene the only product obtained was that of addition (1,2-dibromocyclohexane, 80%) regardless of the solvent and in the presence or absence of dibenzoyl peroxide.

[1]F. D. Chattaway and G. Hoyle, *J. Chem. Soc.*, **123**, 654 (1923)
[2]M. Avramoff, J. Weiss, and O. Schächter, *J. Org.*, **28**, 3256 (1963)

Tetramethylene sulfone (sulfolane), 
SO2

Mol. wt. 120.17, m.p. 28°, b.p. 130/6.5 mm., dipole moment 4.69, dielectric constant 42, water-miscible. Suppliers of practical grade: A, Shell Chem. Co., Phillips Petroleum Co.

Purification. The commercial solvent, made by addition of SO_2 to butadiene and hydrogenation, contains various contaminants. One method of purification[1] is by passage through a column of Linde molecular sieves and fractionation. Another[2] is by repeated vacuum distillation from sodium hydroxide pellets until a mixture of 1 ml. with 1 ml. of 100% sulfuric acid developes no color in 5 min.; the solvent is then distilled again from calcium hydride. Another[3] involves treatment with permanganate and sulfuric acid to remove oxidizable impurities, addition of enough sodium pyrosulfite to give a clear solution, extraction with methylene chloride, treatment with P_2O_5, decantation, and distillation.

Uses. Tetramethylene sulfone has high solvent power for aromatics and has been used extensively by Olah and co-workers for Friedel-Crafts type nitrations[4] and for studies of the mechanism of nitronium tetrafluoroborate nitration of alkylbenzenes[5] and halobenzenes[6] in homogeneous solution. It is a superior solvent for quaternization of tertiary amines with alkyl halides, since it has a high dielectric constant and does not enter into side reactions observed with nitrobenzene and dimethylformamide.[7] For example in the synthesis of the acridizinium salt (3), Bradsher and Parham[8] effected quaternization of (1) with benzyl bromide in tetramethylene sulfone at room temperature in excellent yield. Several other salts analagous to (2) were obtained in good yield and in crystalline form with use of tetramethylene sulfone, whereas with other solvents the products were colored

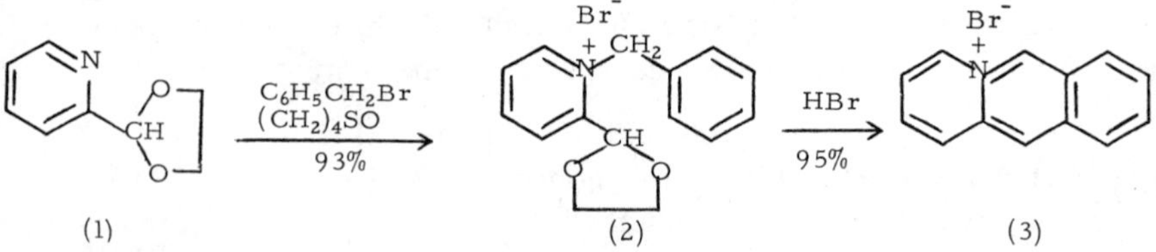
C6H5CH2Br
(CH2)4SO
93%
Br-
+ CH2
HBr
95%
Br-
+
(1) (2) (3)

oils. Arnett and Douty[2] characterize the solvent as very weakly basic and with feeble tendency to solvate cations; it is similar to nitromethane in this respect but easier to handle.

Fuller[9] found sulfolane a particularly effective reaction medium for halogen exchange in the reaction of perchloro and perbromo aromatic compounds with potassium fluoride. Its superiority to other aprotic solvents is attributed to its high boiling point (reflux temperature 230–240°), good thermal and chemical stability, and low anion solvation. It was found useful for kinetic studies of the quaternization of peptide derivatives and of poly-4-vinylpyridine.[7,10]

[1]D. J. Cram, B. Rickborn, C. A. Kingsbury, and P. Haberfield, *Am. Soc.*, **83**, 3678 (1961)
[2]E. M. Arnett and C. F. Douty, *Am. Soc.*, **86**, 409 (1964)
[3]R. W. Alder and M. C. Whiting, *J. Chem. Soc.*, 4707 (1964)
[4]S. J. Kuhn and G. A. Olah, *Am. Soc.*, **83**, 4564 (1961)
[5]G. A. Olah, S. J. Kuhn, and S. H. Flood, *Am. Soc.*, **83**, 4571 (1961)
[6]G. A. Olah, S. J. Kuhn, and S. H. Flood, *Am. Soc.*, **83**, 4581 (1961)
[7]B. D. Coleman and R. M. Fuoss, *Am. Soc.*, **77**, 5472 (1955)
[8]C. K. Bradsher and J. C. Parham, *J. Org.*, **28**, 83 (1963)
[9]G. Fuller, *J. Chem. Soc.*, 6264 (1965)
[10]C. L. Arcus and W. A. Hall, *J. Chem. Soc.*, 5995 (1964)

Tetramethylene sulfoxide, 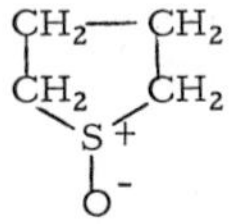

Mol. wt. 104.17, b.p. 45°/3 mm. Supplier: Aldrich. The reagent can be purified by distillation at reduced pressure over Linde 13 X molecular sieves. It should be stored at reduced temperature since the shelf life at room temperature is about 3 months.[1]

Like dimethyl sulfoxide, the cyclic sulfoxide is a superior reagent for the oxidation of thiols to disulfides.[2]

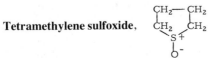

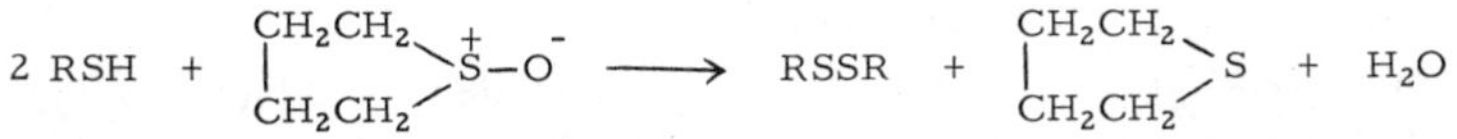

[1]Private communication from T. J. Wallace, Esso Research and Engineering Co.
[2]T. J. Wallace, *Am. Soc.*, **86**, 2018 (1964); T. J. Wallace and J. J. Mahon, *Am. Soc.*, **86**, 4099 (1964); *J. Org.*, **30**, 1502 (1965)

Tetramethylguanidine, $(CH_3)_2NC(=NH)N(CH_3)_2$. Mol. wt. 115.18, b.p. 165°. Suppliers of practical grade: E, F.

The reagent was used as catalyst for Michael condensation of 3β-hydroxy-17β-nitro-5β-androstane with methyl acrylate.[1] When either triton B or sodium ethoxide was used the yield was only 10%.

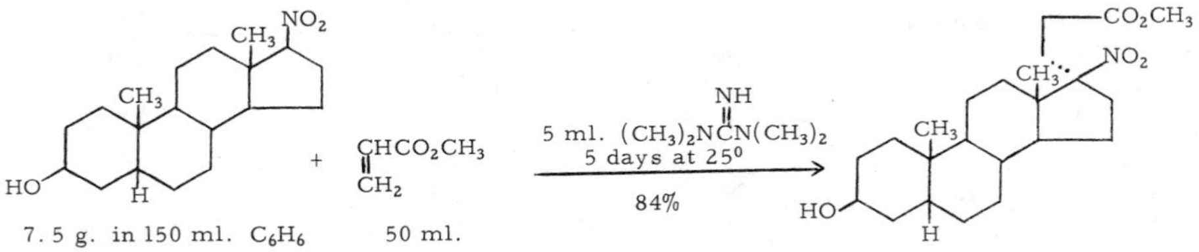

[1]L. N. Nysted and R. R. Burtner, *J. Org.*, **27**, 3175 (1962)

Tetramethylthiourea, $(CH_3)_2NC(=S)N(CH_3)_2$. Mol. wt. 132.24, m.p. 78°.

Of several sulfur-containing compounds studied as poisons in the Rosenmund reduction of benzoyl chloride, tetramethylthiourea was the most effective in preventing hydrogenation beyond the benzaldehyde stage.[1]

[1]S. Affrossman and S. J. Thomson, *J. Chem. Soc.*, 2024 (1962)

Tetramethylurea, $(CH_3)_2NC(=O)N(CH_3)_2$. Mol. wt. 116.17, b.p. 176.5°, water-miscible. Suppliers: J. Deere Chem. Co. (bulletin available), du Pont Ind. and Biochem. Dept., Frinton Labs.

Preparation. From dimethylamine and phosgene:[1]

$$2\ (CH_3)_2NH + ClCOCl \xrightarrow[-(CH_3)_2\overset{+}{N}H_2\overset{-}{Cl}]{} (CH_3)_2NCOCl \xrightarrow{2\ (CH_3)_2NH} (CH_3)_2NCON(CH_3)_2$$

From trimethylamine and phosgene:[2]

$$(CH_3)_3N + ClCOCl \longrightarrow [(CH_3)_3\overset{+}{N}COCl]\overset{-}{Cl} \longrightarrow CH_3Cl + (CH_3)_2NCON(CH_3)_2$$

As solvent. Tetramethylurea is similar as solvent to pyridine (b.p. 115°) but higher boiling. It is miscible in all proportions not only with water but with all common organic solvents, even petroleum ether. It is an excellent solvent for aromatic and heterocyclic compounds.

Deamination. A new deamination procedure involves precipitation of the aryldiazonium hexafluoroborate, for example as described by Rutherford and Redmond,[3] and spontaneous and exothermal decomposition of this salt in tetramethylurea, which functioned satisfactorily where other nonaqueous solvents failed.[4] Yields are best (75–85%) when the aromatic ring carries an electron-withdrawing substituent (NO_2, CO_2H, Br). In the case of aniline, *o*- and *p*- toluidine, and *o*-anisidine yields are in the range 25–30%. The dry salt is added to tetramethylurea in portions at room temperature, and the temperature is controlled to 65°; when evolution of nitrogen has ceased, the solution is poured into water.

Ullmann reaction. The solvent is recommended as diluent in the Ullmann reaction in place of dimethylformamide (b.p. 153°) when a higher reaction temperature is required.[2]

Alkylation of enolates. Determination of the rate constants for alkylation, for example of sodio-*n*-butylmalonic ester with *n*-butyl bromide in benzene solution in the presence of various additives, showed that at a concentration of less than 5% tetramethylurea increases the reaction rate many fold. Dimethylformamide was slightly more effective, but higher yields are reported[2] when tetramethylurea was used.

[1]W. Michler and C. Escherich, *Ber.*, **12**, 1162 (1879)
[2]A. Lüttringhaus and H. W. Dirksen, *Angew. Chem., Internat. Ed.*, **3**, 260 (1964)
[3]K. G. Rutherford and W. Redmond, *Org. Syn.*, **43**, 12 (1963)
[4]*Idem., J. Org.*, **28**, 568 (1963)

2,4,5,7-Tetranitrofluorenone,

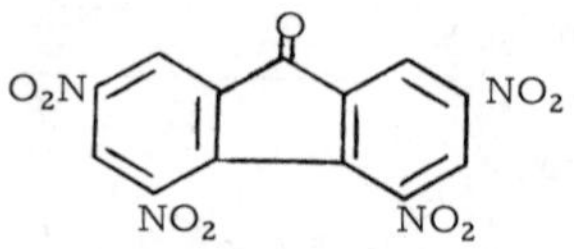
O₂N
NO₂
NO₂
NO₂

Mol. wt. 360.19, m.p. 254.5°. Supplier: Frinton Labs. Prepared in 77% yield by nitration of fluorenone.[1]

The reagent forms π-complexes with effective π-base hydrocarbons (naphthalene, anthracene, hexamethylbenzene), which melt about 50° higher than the corresponding complexes of 2,4,7-trinitrofluorenone.[1] The following order of π-acid strength has been established:[2]

2,4,5,7-tetranitrofluorenone > 2,4,7-trinitrofluorenone >
1,3,5-trinitrobenzene > picryl chloride

[1]M. S. Newman and W. B. Lutz, *Am. Soc.*, **78**, 2469 (1956)
[2]A. K. Colter and L. M. Clemens, *Am. Soc.*, **87**, 847 (1965)

(+)- and (−)-α-(2,4,5,7-Tetranitro-9-fluorenylideneaminooxy)-propionic acid. Mol. wt. 447.27, m.p. 200°.

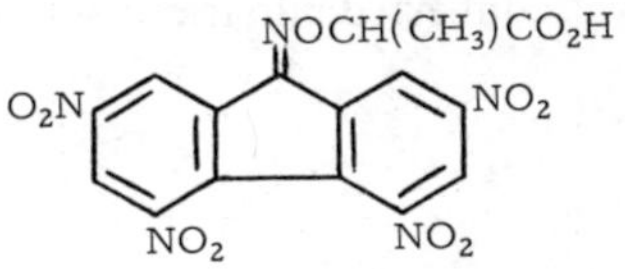
NOCH(CH3)CO2H
O2N
NO2
NO2
NO2

Newman and co-workers[1] prepared these optically active π-acid complexing agents and used them to effect resolution of overcrowded polycyclic aromatic hydrocarbons such as hexihelicene.[2]

[1]M. S. Newman and W. B. Lutz, *Am. Soc.*, **78**, 2469 (1956); P. Block, Jr., and M. S. Newman, procedure submitted to *Org. Syn.*
[2]M. S. Newman and D. Lednicer, *Am. Soc.*, **78**, 4765 (1956)

Tetranitromethane, $C(NO_2)_4$. Mol. wt. 196.04, m.p. 13.8°, b.p. 126°, sp. gr. 1.639. Suppliers: A, C, KK.

Preparation. The *Organic Syntheses* procedure[1] involves reaction of acetic

$$\underset{0.5\text{ m.}}{4\ (CH_3CO)_2O} + \underset{\substack{0.5\text{ m.}\\(\text{anhydrous})}}{HNO_3} \xrightarrow[57\text{-}65\%]{\substack{\text{Below }10^{\circ}\text{, then}\\ 7\text{ days at }25^{\circ}}} C(NO_2)_4 + 7\ CH_3CO_2H + CO_2$$

anhydride with anhydrous nitric acid. At the time this method was selected, the more efficient method of Darzens and Lévy[2] using ketene was not available.

$$4\ CH_2{=}C{=}O + 4\ HNO_3 \xrightarrow[90\%]{} C(NO_2)_4 + CO_2 + 3\ CH_3CO_2H$$

Test for unsaturation. Tetranitromethane forms colored π-complexes with many olefins and other unsaturated compounds and is hence useful in detection of unsaturation. Since the reagent is costly, a test is advisedly made in a melting-point capillary tube;[3] since it is very volatile but freezes in a refrigerator, the reagent is mixed with an equal volume of chloroform and stored in the cold. A test is made, for example, by introducing an 0.5-mm. layer of cholesterol in a melting-point capillary and drawing down a second capillary to a fine tip to form a pipet. The pipet is dipped into the $CHCl_3$–$C(NO_2)$ solution to introduce a 5-mm. column, the tip is inserted in the mouth of the first tube, and the solution is shaken down by a whipping motion. The yellow color of the complex is particularly apparent when the tube is viewed against a white paper.

The color increases in intensity from yellow to dark red with increasing unsaturation and is positive for compounds having a double bond inert to hydrogenation. The π-complex with benzene is yellow, that with naphthalene orange. It is positive with cyclopropane, negative with cyclobutane. It is feeble with α,β-unsaturated carbonyl compounds, negative with alkynes and with allylic alcohols.

Heilbronner[4] developed a technique for differentiating double bond types involving measurement of the wave length of absorption of an olefin complex with tetranitromethane at a standard extinction coefficient. Each alkyl substituent produces a shift of about 56 mμ to higher wave length, as shown by the following data for steroid olefins: disubstituted double bond, Δ^2, λ 478mμ; trisubstituted, Δ^4, λ 530 mμ;

Δ^5 (2 examples), λ 530, 531 mμ; tetrasubstituted, $\Delta^{8(14)}$ (3 examples), λ 570, 559, 580 mμ.

Nitration. In improvement of an earlier procedure, Anderson and co-workers[5] treated azulene in pyridine with a slight excess of tetranitromethane in ethanol. The color changed to red in 18 min., and on working up the mixture by extraction and chromatography 1-nitroazulene was obtained in high yield. Presumably the reagent was converted into trinitromethane.

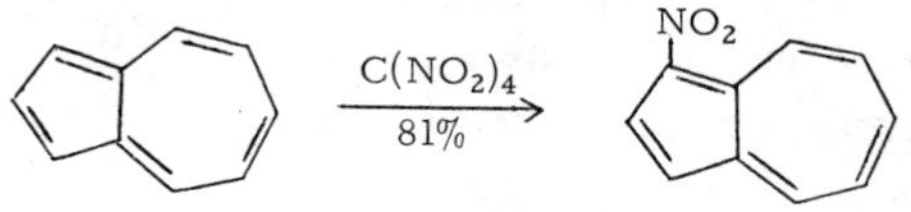
NO2
C(NO2)4
81%

Riordan *et al.*[6] found the reagent suitable for nitration of tyrosine and of tyrosyl residues of proteins. At pH 8 nitration displaces two protons with formation of 3-nitrotyrosine.

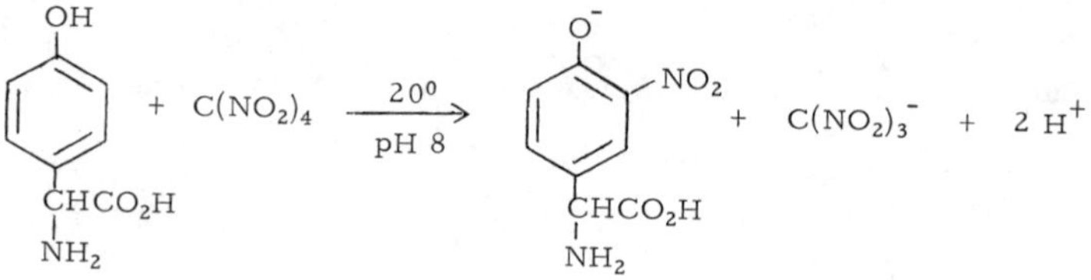
OH
+ C(NO2)4
20°
pH 8
CHCO2H
NH2
O-
NO2
+ C(NO2)3-
+ 2 H+
CHCO2H
NH2

[1]P. Liang, *Org. Syn., Coll. Vol.*, **3**, 803 (1955)
[2]G. Darzens and G. Lévy, *Compt. rend.*, **229**, 1081 (1949)
[3]*Org. Expts.*, 75
[4]E. Heilbronner, *Helv.*, **36**, 1121 (1953)
[5]A. G. Anderson, Jr., R. Scotoni, Jr., E. J. Cowles, and C. G. Fritz, *J. Org.*, **22**, 1193 (1957)
[6]J. F. Riordan, W. E. C. Wacker, and B. L. Vallee, *Am. Soc.*, **88**, 4104 (1966)

Tetra-*p*-nitrophenyl pyrophosphate. The reagent (2) is prepared *in situ* by the reaction of di-*p*-nitrophenyl hydrogen phosphate (1) with di-*p*-tolylcarbodiimide in anhydrous

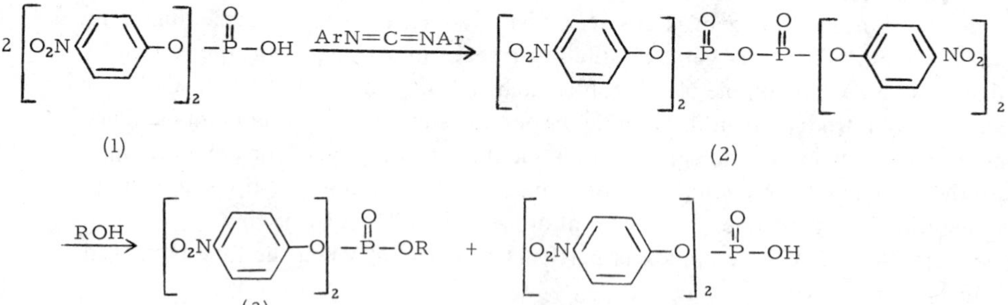
2
O2N
O
P
OH
ArN=C=NAr
O2N
O
P
O
P
O
NO2
ROH
O2N
O
P
OR
+
O2N
O
P
OH

(1) (2) (3)

dioxane.[1] It reacts with an alcohol without basic catalysis to give the neutral di-*p*-nitrophenyl phosphate ester (3). One of the *p*-nitrophenyl groups is removed under mild alkaline conditions (30 min. at 25° with 1 *N* alkali), the second is removed by 1 *N* alkali after several hours at 100°. The groups are also removable by hydrogenolysis; the first in a neutral medium and the second in the presence of an acid catalyst. Guanosine-5′-phosphate was prepared in yield of 70% using this reagent.[2] Phosphoryl chloride in this case gave at best a yield of 20%.

[1]J. G. Moffatt and H. G. Khorana, *Am. Soc.*, **79**, 3741 (1957)
[2]R. W. Chambers, J. G. Moffatt, and H. G. Khorana, *Am. Soc.*, **79**, 3747 (1957)

Tetraphenylcyclopentadienone, Mol. wt. 384.45, m.p. 220°, purple plates. Suppliers: A, KK.

Preparation. The reagent is readily prepared by the condensation of benzil with dibenzyl ketone in the presence of a basic catalyst. Dilthey's original procedure,[1] adopted by *Organic Syntheses*,[2] specifies ethanol as solvent and a solution of potassium hydroxide in ethanol as the base. This procedure suffers from the low boiling point of the alcohol and the limited solubility of both potassium hydroxide and the reaction product in ethanol. In an improved procedure[3, 4] use of the better solvent triethylene glycol permits operation at a higher temperature, and use of the readily soluble benzyltrimethylammonium hydroxide as base eliminates the step of dissolving potassium hydroxide in ethanol. A mixture of 0.2 mole each of benzil and

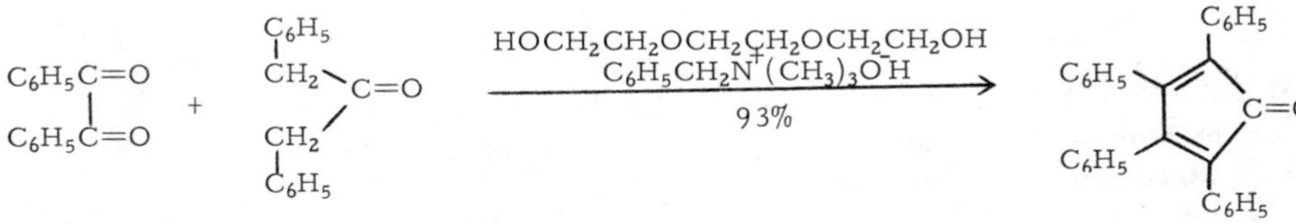

$C_6H_5C=O$
$C_6H_5C=O$
+
C_6H_5
CH_2
C=O
CH_2
C_6H_5
$HOCH_2CH_2OCH_2CH_2OCH_2CH_2OH$
$C_6H_5CH_2\overset{+}{N}(CH_3)_3\overset{-}{O}H$
93%
C_6H_5
C_6H_5
C=O
C_6H_5
C_6H_5

dibenzyl ketone in 200 ml. of triethylene glycol is adjusted to 100°, 20 ml. of a 40% solution of benzyltrimethylammonium hydroxide in methanol is added, and the solution is swirled once for mixing and let stand for crystallization, which starts in less than a minute. When cold, the crystallizate is thinned with ethanol, and the deep purple plates washed with ethanol until the wash liquor is pure pink. For recrystallization the material is dissolved in triethylene glycol (10 ml./g.) at 220°, and the solution let stand (large plates) or stirred magnetically while cooling (small plates); the recovery is 92%.

Diels-Alder reaction. Because of its reactivity and color, which visualizes the reaction, the reagent has been brought into reaction with a large number of dienophiles.[5] It reacts with maleic anhydride in refluxing benzene (b.p. 80°) with retention of the carbonyl bridge to give (3),[6] whereas in refluxing chlorobenzene (b.p. 132°)[6] or bromobenzene[7] (b.p. 155.5°) carbon monoxide is evolved and the product is tetraphenyldihydrophthalic anhydride (4). The reaction of the reagent with dimethyl

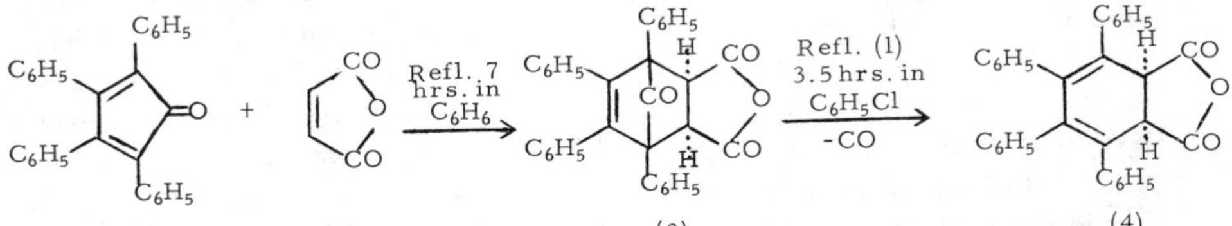

C_6H_5
C_6H_5
C_6H_5
C_6H_5
+
CO
O
CO
Refl. 7 hrs. in
C_6H_6
C_6H_5
H
CO
CO
O
CO
C_6H_5
H
C_6H_5
C_6H_5
Refl. (1) 3.5 hrs. in
C_6H_5Cl
-CO
C_6H_5
H
CO
O
CO
H
C_6H_5
C_6H_5
C_6H_5

(1) (2) (3) (4)

acetylenedicarboxylate to give dimethyl tetraphenylphthalate, first reported by Dilthey,[6] was studied by one of us in a wide range of solvents with results which are summarized in the section on Diels-Alder solvents. For examples of the reaction of the reagent with diphenylacetylene and with benzyne, *see* Diphenylacetylene and Diphenyliodonium-*o*-carboxylate.

Allen and co-workers[7] succeeded in obtaining the adduct of the dienone with ethylene by operating at high pressure at 100° and adding aluminum chloride as catalyst. Without the catalyst no appreciable reaction occurred at 100°, and at 180° the reaction was accompanied by decarbonylation to give (6) and (7).

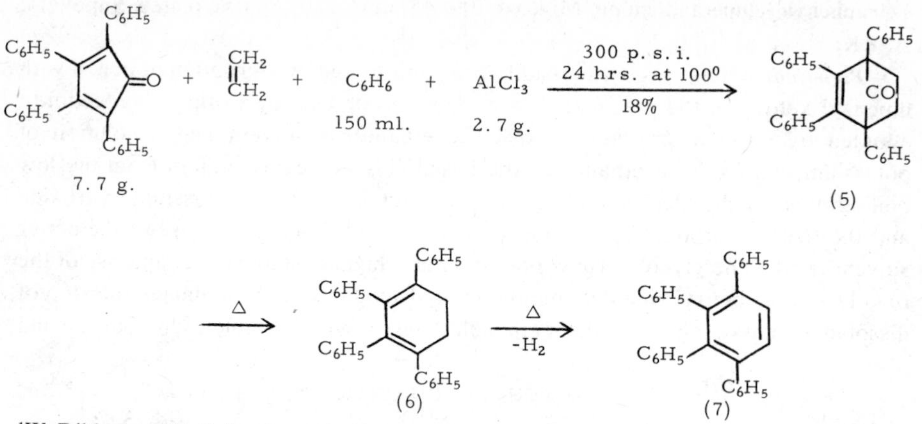
7.7 g.
150 ml.
2.7 g.
300 p.s.i.
24 hrs. at 100°
18%
(5)
(6)
(7)

[1]W. Dilthey and F. Quint, *J. prakt. Chem.*, (2), **128**, 146 (1930)
[2]J. R. Johnson and O. Grummitt, *Org. Syn., Coll. Vol.*, **3**, 806 (1955)
[3]*Org. Expts.*, 301; L. F. Fieser, *Org. Syn.*, **46**, 44 (1966)
[4]L. F. Fieser and M. J. Haddadin, *Can J. Chem.*, **43**, 1599 (1965)
[5]C. F. H. Allen, *Chem. Revs.*, **37**, 209 (1945): **62**, 653 (1962)
[6]W. Dilthey, I. Thewalt, and O. Trösken, *Ber.*, **67**, 1959 (1934)
[7]C. F. H. Allen, R. W. Ryan, Jr., and J. A. VanAllan, *J. Org.*, **27**, 778 (1962)

Thallium triacetate, $Tl(OCOCH_3)_3$. Mol. wt. 381.53, m.p. 195°.

Preparation according to Winstein and C. B. Anderson.[1] A mixture of thallic oxide (Fairmont Chem. Co.) and acetic acid was stirred at 65° until all the brown-

$$\underset{50\text{ g.}}{Tl_2O_3} + \underset{300\text{ ml.}}{CH_3CO_2H} \xrightarrow[99.5\%]{65^0} 2\ Tl(OCOCH_3)_3 + 3\ H_2O$$

black mass had gone into solution (*ca.* 1 day). The solution was filtered and on cooling deposited 73 g. (87.5%) of white crystals. Concentration of the mother liquor afforded a second crop of 10 g. (12%). The solid is collected by decantation and dried *in vacuo*, since on standing in the air it turns brown. Dried *in vacuo* over P_2O_5 and stored in a stoppered flask, the reagent is stable for at least 1 year. It is poisonous.

Oxidation of olefins. Kabbe[2] explored briefly the oxidation of cyclohexene, styrene, and *o*-allylphenol and expressed the view that thallium triacetate is intermediate between lead tetraacetate and mercuric acetate. Anderson and Winstein[3] followed the oxidation of cyclohexene in acetic acid at room temperature (several days) by vapor-phase chromatography and accounted quantitatively for all the five products formed: the *cis*-and *trans*-diacetates (1 and 2), the ring-contracted diacetate (3) and aldehyde (4), and the product of allylic oxidation (5). In dry solvent the *trans*-diacetate (2) predominated, and in moist solvent the *cis*-diacetate (1) predominated.

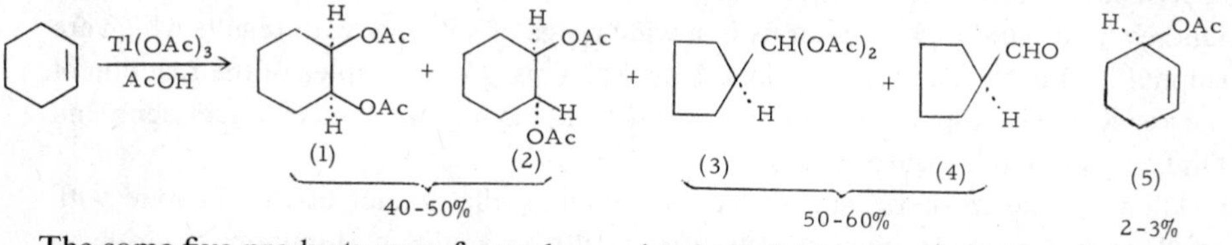
Tl(OAc)3
AcOH
(1)
(2)
(3)
(4)
(5)
40-50%
50-60%
2-3%

The same five products were formed on oxidation with lead tetraacetate but here the amount of the allylic acetate (5) increased to 37%. Oxidation with mercuric acetate gave the allylic acetate (5) exclusively.

Cleavage of cyclopropanes. Ouellette[4] found that thallium triacetate cleaves phenylcyclopropane to give essentially a single product, 1-phenyl-1,3-diacetoxypropane (6). Analysis (v.p.c.) showed the presence of only a trace of *trans*-cinnamyl

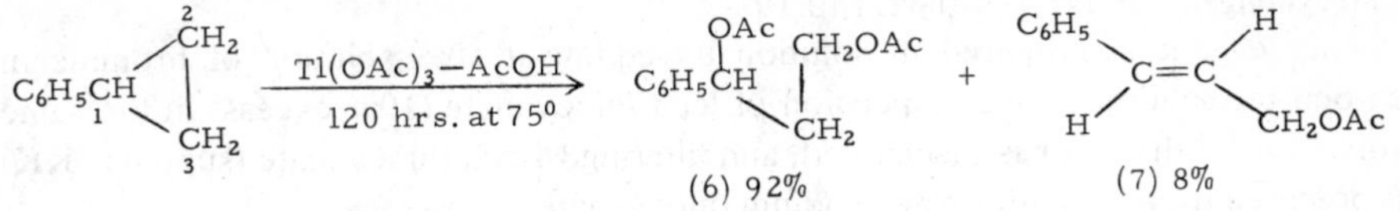

acetate (7). Use of the moisture-sensitive thallium triacetate can be avoided; thallium oxide can be used as the reagent since under the reaction conditions it is converted into thallium triacetate.

In a parallel oxidation of phenylcyclopropane with lead tetraacetate[5] in acetic acid at 75° the reagent gradually dissolved, and the reaction was complete in 10 hrs. (compare 120 hrs.). Products (6) and (7) resulting from cleavage of the C_1—C_2

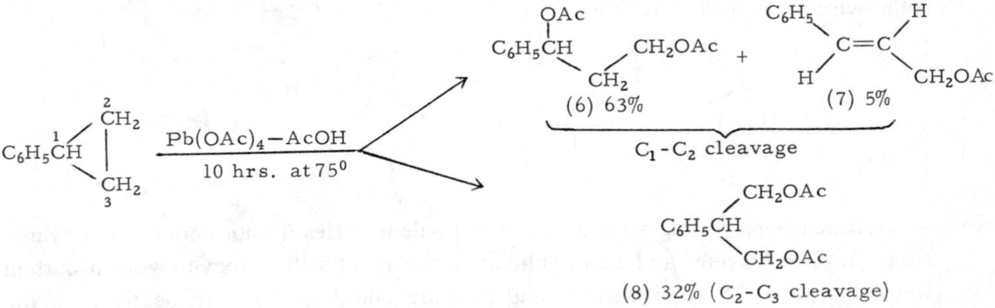

bond were formed, as in the oxidation with thallium triacetate, but the diacetate (8) resulting from cleavage of the C_2—C_3 bond accounted for one third of the reaction mixture.

[1]S. Winstein, private communication
[2]H.-J. Kabbe, *Ann.*, **656**, 204 (1962)
[3]C. B. Anderson and S. Winstein, *J. Org.*, **28**, 605 (1963)
[4]R. J. Ouellette, D. L. Shaw, and A. South, Jr., *Am. Soc.*, **86**, 2744 (1964)
[5]R. J. Ouellette and D. L. Shaw, *Am. Soc.*, **86**, 1651 (1964)

N,N′-Thiocarbonyldiimidazole, 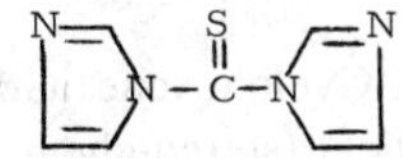

Mol. wt. 178.22, m.p. 105–106°.

The reagent is prepared by reaction of thiophosgene with two equivalents of imidazole.[1] It reacts with a primary amine at room temperature to form an imidazole-1-thiocarboxamide (2), which decomposes as formed to the isothiocyanate (3) and imidazole.[2]

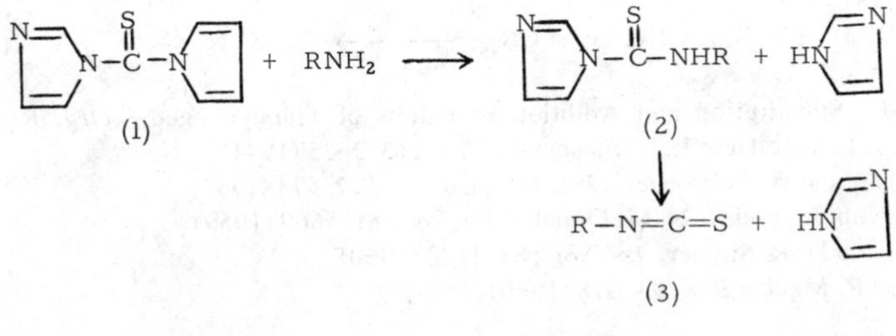

For use of the reagent in a stereospecific olefin synthesis, *see* Trimethyl phosphite.

[1]H. A. Staab and G. Walther, *Ann.*, **657**, 98 (1962)
[2]*Idem, ibid.*, **657**, 104 (1962)

Thiocyanogen, $(SCN)_2$. Mol. wt. 140.19

The reagent is prepared in solution by adding a 10% solution of bromine in carbon tetrachloride to a suspension of lead thiocyanate (10% excess) in the same solvent until the color is discharged, and filtering.[1] Lead thiocyanate (supplier: KK) is prepared from lead nitrate and sodium thiocyanate.

In its reactions thiocyanogen is somewhat like bromine without a catalyst. Direct replacement of hydrogen by the thiocyano group, thiocyanation, is limited to aromatic amines and phenols, and a few very reactive hydrocarbons. Aniline is converted into 4-thiocyanoaniline in 97% yield, *o*-toluidine into 4-thiocyano-2-methylaniline in 80% yield, anthranilic acid into 5-thiocyanoanthranilic acid in 80% yield. Phenols likewise are thiocyanated in the *para* position if available, otherwise in the *ortho* position.

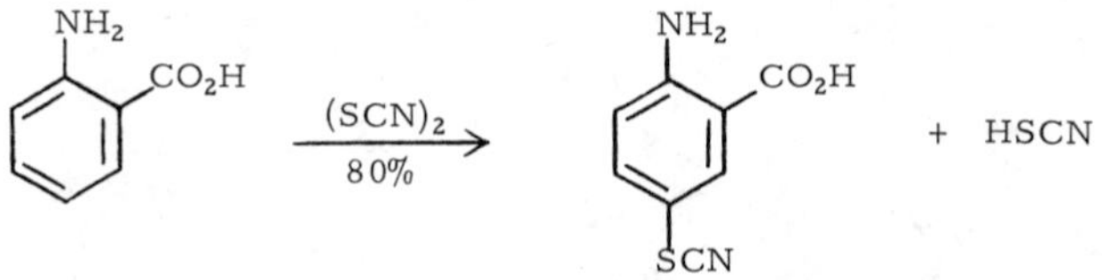
NH2
CO2H
(SCN)2
80%
NH2
CO2H
SCN
+ HSCN

Hydrocarbons of the benzene and naphthalene series do not undergo thiocyanation. 3,4-Benzpyrene and methylcholanthrene react with thiocyanogen in carbon tetrachloride at 25° to give the 5- and 15-thiocyano derivatives, respectively, in the yields indicated.[1] Anthracene affords 9,10-dithiocyanoanthracene in lower yield.

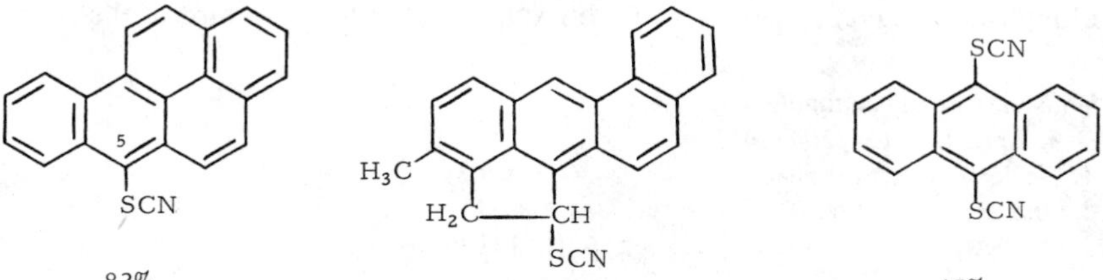
5
SCN
H3C
H2C
CH
SCN
SCN
SCN

82% 87% 45%

A second reaction is addition. Cyclohexene adds to the reagent to give 1,2-dithiocyanocyclohexane in 73% yield (stereo-chemistry unknown). Butadiene and isoprene give the 1,4-adducts in yields of 80 and 19%.

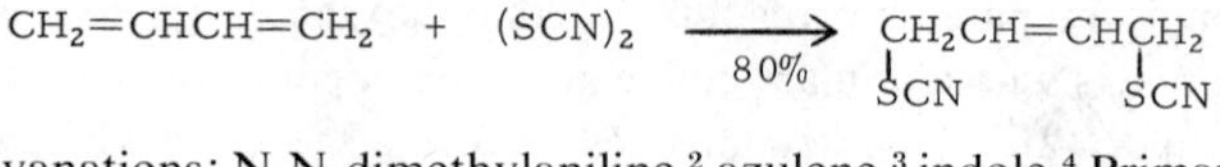
CH2=CHCH=CH2 + (SCN)2
80%
CH2CH=CHCH2
SCN SCN

Other thiocyanations: N,N-dimethylaniline,[2] azulene,[3] indole.[4] Primary and secondary amines give thiocyanamides.[5]

$$R_2NH + (SCN)_2 \xrightarrow[50\text{-}70\%]{} R_2NSCN$$

[1]J. L. Wood, "Substitution and Addition Reactions of Thiocyanogen," *Org. Reactions*, **3**, 240 (1946); J. L. Wood and L. F. Fieser, *Am. Soc.*, **63**, 2323 (1941)
[2]R. Q. Brewster and W. Schroeder, *Org. Syn., Coll. Vol.*, **2**, 574 (1943)
[3]A. G. Anderson, Jr., and R. N. McDonald, *Am. Soc.*, **81**, 5669 (1959)
[4]M. S. Grant and H. R. Snyder, *Am. Soc.*, **82**, 2742 (1960)
[5]M. Kuhn and R. Mecke, *Ber.*, **93**, 618 (1960)

Thiocyanogen chloride, ClSC≡N. A solution of the reagent is prepared by adding 1 mole of lead thiocyanate, $Pb(SCN)_2$, mol. wt. 323.38 (Fisher supplies practical grade) to a solution of 2 moles of chlorine in acetic acid and filtering off the lead chloride.

In solution in chloroform or toluene, the reagent adds rapidly to ethylene or to cyclohexene at −60° to give 2-chloroethyl thiocyanate or 2-chlorocyclohexyl thiocyanate, both in yield of about 80%.[1]

Aryl ethers and anilides, which are unreactive to thiocyanogen, react readily with thiocyanogen chloride in acetic acid to give thiocyano derivatives in high yield.[2]

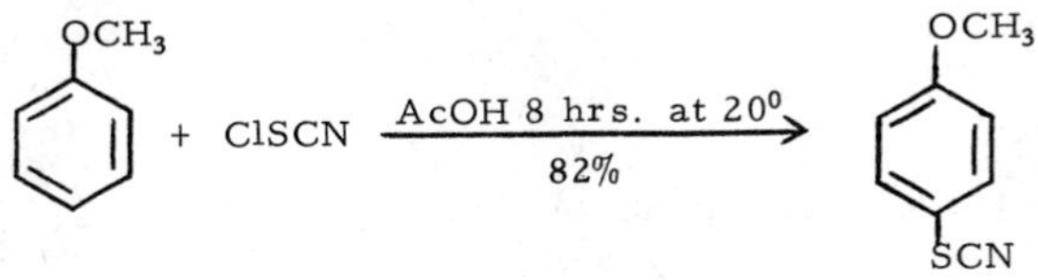

A synthesis of N-benzylthieno[2,3-b]pyrrole (7) demonstrates ready substitution of a pyrrole ring.[3]

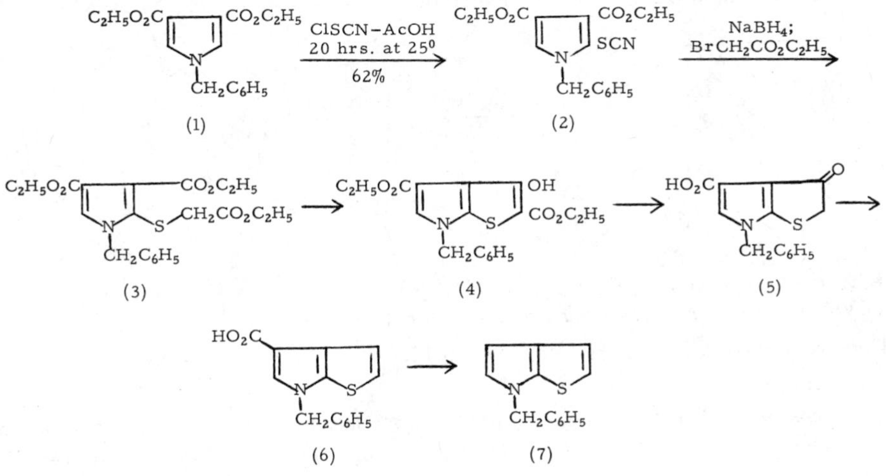

[1]A. B. Angus and R. G. R. Bacon, *J. Chem. Soc.*, 774 (1958)
[2]R. G. R. Bacon and R. G. Guy, *J. Chem. Soc.*, 318 (1960)
[3]R. K. Olsen and H. R. Snyder, *J. Org.*, **30**, 184 (1965)

Thioglycolic acid, $HSCH_2CO_2H$. Mol. wt. 92.12, m.p. −16.5°, b.p. 96°/3 mm., pKa 3.83, sp. gr. 1.325. Suppliers: A, B, E, F, MCB. The reagent has a strong, unpleasant odor and is readily oxidized by air. The sodium and ammonium salts are used for cold waving of hair and for modifying wool since they effect reductive cleavage of —S—S— cross links in keratin.

Peptide synthesis.[1] The unwanted side reactions often encountered in the synthesis of peptides containing methionine can be eliminated by the temporary conversion of the thioether function of methionine into the sulfoxide at any stage of a peptide synthesis. The sulfoxide oxygen is introduced without formation of the sulfone when a small excess of hydrogen peroxide is used and its elimination is easily achieved by reduction with thioglycolic acid. In contrast to the carbobenzoxy (Cb) derivative of methionine, elimination of the carbobenzoxy group of the corresponding sulfoxide proceeds smoothly on mild treatment with concd. hydrochloric acid. Thus all attempts

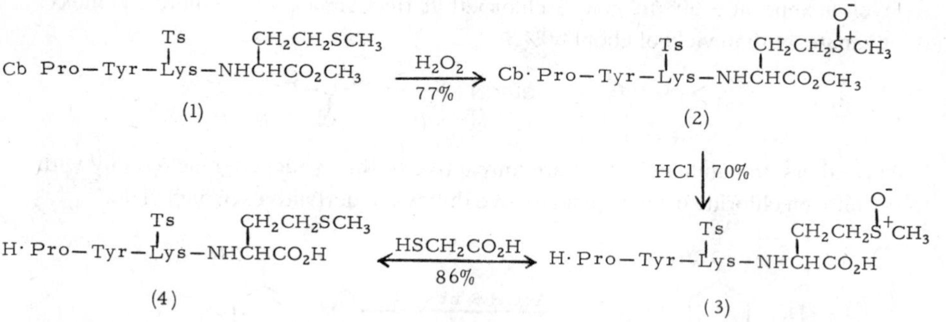

to effect direct decarbobenzoxylation of (1) failed but the desired product (4) was obtained readily via the sulfoxides (2) and (3).

Isomerization of olefins. Thioglycolic acid isomerizes *trans,trans,cis*-1,5,9-cyclododecatriene (1) to the *trans,trans,trans*-isomer (2) at room temperature (exothermal reaction).[2] Conventional methods of isomerization are unsuccessful.[3]

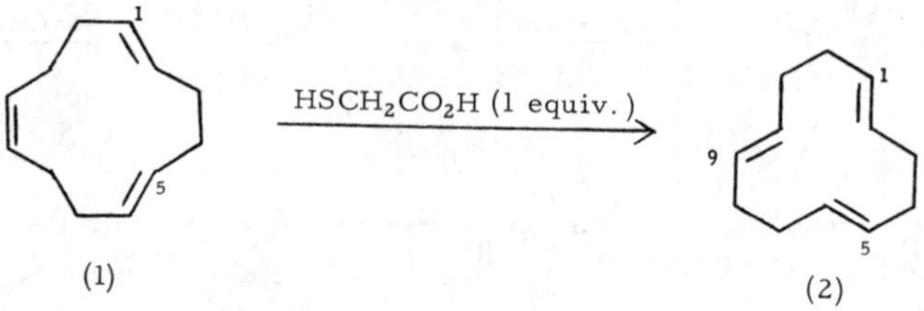

[1]B. Iselin, *Helv.*, **44**, 61 (1961)
[2]E. W. Duck and J. M. Locke, *Chem. Ind.*, 507 (1965)
[3]G. Wilke, *Angew. Chem., Internat. Ed.*, **2**, 105 (1963)

Thiolacetic acid, CH_3COSH. Mol. wt. 76.12, b.p. 87° (34°/100 mm.), sp. gr. 1.07. Suppliers: A, B, E, F, MCB.

Preparation. The reagent can be prepared by passing hydrogen sulfide (available in a cylinder from Matheson Co.) into acetic anhydride in the presence of a small amount of powdered sodium hydroxide as catalyst until the proper gain in weight has been realized.[1]

$$\underset{1\text{ m.}}{(CH_3CO)_2O} + H_2S + \underset{1\text{ g.}}{NaOH} \xrightarrow{6\text{ hrs. at }50\text{-}55^0} CH_3\overset{O}{\overset{\|}{C}}SH + CH_3\overset{O}{\overset{\|}{C}}OH$$

Radical addition to olefins.[2] Holberg[3] and Ipatieff[4] studied this reaction and found that thiolacetic acid adds almost exclusively in the anti-Markownikoff direction:

$$CH_2{=}C(CH_3)_2 + CH_3COSH \xrightarrow[60\%]{3\text{ hrs. at }100^0} CH_3COSCH_2CH(CH_3)_2$$

A British group[5] found that yields are improved by addition of a few drops of ascaridole or a small amount of dibenzoyl peroxide. Bordwell and Hewett[6] showed that the addition is predominantly *trans*. Thus addition of thiolacetic acid to 1-methylcyclohexene and to 1-methylcyclopentene, followed by alkaline hydrolysis to the thiol, in each case gave chiefly the products with *cis*-oriented substituents.

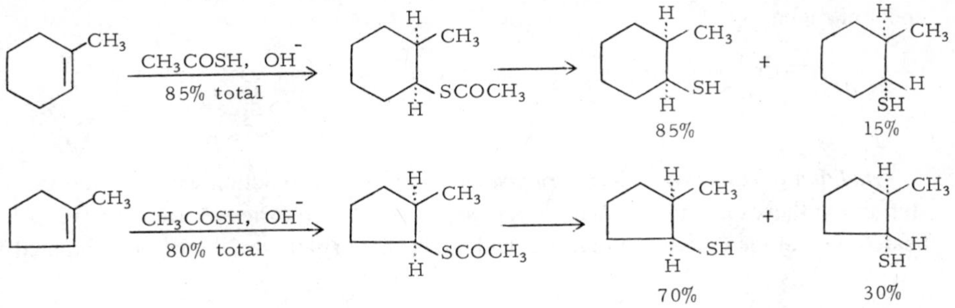
CH3
CH3COSH, OH-
85% total
H
CH3
SCOCH3
H
H
CH3
SH
H
85%
+
H
CH3
H
SH
15%
CH3
CH3COSH, OH-
80% total
H
CH3
SCOCH3
H
H
CH3
SH
H
70%
+
H
CH3
H
SH
30%

Radical addition of thiolacetic acid to terminal olefins is the key step in a new method for the preparation of alkylsulfonic acids.[7] The thiolacetate is obtained in

$$RCH{=}CH_2 + CH_3COSH \xrightarrow[55\text{-}91\%]{h\nu,\ N_2} RCH_2CH_2SCOCH_3 \xrightarrow[90\%]{90\%\ H_2O_2-CH_3CO_2H,\ 60^0} RCH_2CH_2SO_3H$$

high purity and good yield and is oxidized to the sulfonic acid with peracetic acid generated *in situ* from hydrogen peroxide and acetic acid.

Radical addition to alkynes.[2] Thiolacetic acid reacts vigorously with alkynes to give mono and di adducts. In some cases the yield is improved by irradiation or addition of a peroxide, but the main value of initiation lies in improved reproducibility of the experiments. The reaction provides a means of converting a terminal acetylene into an aldehyde.[8]

$$\underset{0.2\text{ m.}}{\underline{n}\text{-}C_4H_9C{\equiv}CH} \xrightarrow{0.2\text{ m. }CH_3COSH} \underset{53\%}{\underline{n}\text{-}C_4H_9CH{=}\underset{H}{C}{-}\overset{O}{\overset{\|}{S}}CCH_3} + \underset{19\%}{\underline{n}\text{-}C_4H_9\underset{SCOCH_3}{CH}{-}CH_2SCOCH_3}$$

$$\xrightarrow[57\%]{H_2NNHCONH_2} \underline{n}\text{-}C_4H_9CH_2\underset{H}{C}{=}NNHCONH_2$$

This reaction is the key step in a total synthesis of linoleic acid by Walborsky.[9]

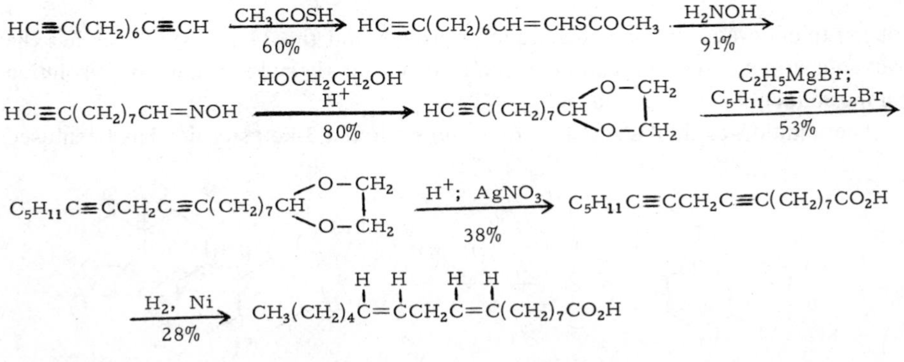
HC≡C(CH2)6C≡CH
CH3COSH
60%
HC≡C(CH2)6CH=CHSCOCH3
H2NOH
91%
HC≡C(CH2)7CH=NOH
HOCH2CH2OH
H+
80%
HC≡C(CH2)7CH
O-CH2
O-CH2
C2H5MgBr;
C5H11C≡CCH2Br
53%
C5H11C≡CCH2C≡C(CH2)7CH
O-CH2
O-CH2
H+; AgNO3
38%
C5H11C≡CCH2C≡C(CH2)7CO2H
H2, Ni
28%
H H H H
CH3(CH2)4C=CCH2C=C(CH2)7CO2H

Blomquist and Wolinsky[10] sealed a mixture of 20 g. of propargyl alcohol, 1 equivalent of thiolacetic acid, and 0.1 g. of dibenzoyl peroxide in a Pyrex tube and irradiated it for 1 month. Distillation gave 30 g. of a mixture of dithiol acetates and possibly trithiol acetates and this on acid hydrolysis gave BAL, b.p. 86–90°/1 mm. under nitrogen.

$$CH\equiv CCH_2OH \ + \ CH_3COSH \longrightarrow Mixture \xrightarrow{HCl} CH_2CHCH_2OH$$
$$\underset{SH \ SH}{\qquad\qquad\qquad\qquad\qquad\qquad\qquad\qquad}$$

BAL

Addition to α,β-unsaturated carbonyl compounds. This reaction was an important step in syntheses of thioctic acid developed in two laboratories. A Lederle group[11] heated the ketone (2) with thiolacetic acid on the steam bath for 20 min. and obtained

$$C_2H_5O_2C(CH_2)_4COCl \ + \ CH_2{=}CH_2 \xrightarrow{AlCl_3} C_2H_5O_2C(CH_2)_4\overset{O}{\overset{\|}{C}}CH{=}CH_2 \xrightarrow[91\%]{CH_3COSH}$$

(1)   (2)

$$C_2H_5O_2C(CH_2)_4\overset{O}{\overset{\|}{C}}CH_2CH_2\overset{O}{\overset{\|}{S}}CCH_3 \xrightarrow{NaBH_4} C_2H_5O_2C(CH_2)_4\underset{OH}{CHCH_2CH_2}\overset{O}{\overset{\|}{S}}CCH_3 \xrightarrow{OH^-}$$

(3)   (4)

$$HO_2C(CH_2)_4\underset{OH \quad SH}{CHCH_2CH_2} \xrightarrow[\substack{2. \ NaOH}]{1. \ HI + S{=}C(NH_2)_2} HO_2C(CH_2)_4\underset{SH \quad SH}{CHCH_2CH_2} \xrightarrow[Fe^{+3}]{O_2} \underset{S}{\overset{S}{\diagdown}}(CH_2)_4CO_2H$$

(5)   (6)   (7)

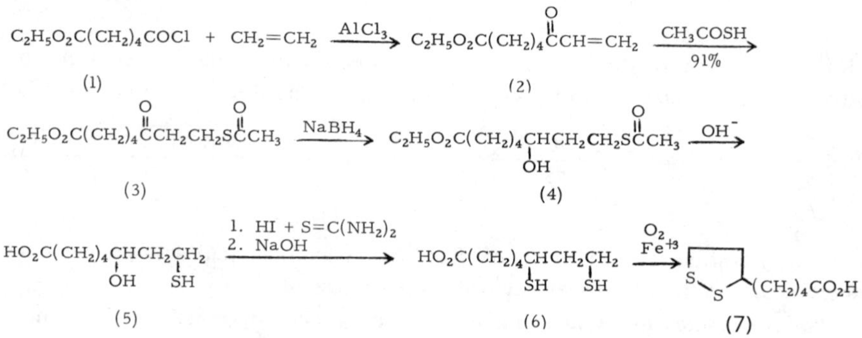

the product of 1,4-addition (3) in high yield. In a Merck synthesis[12] the reagent was added to the α,β-unsaturated acid (8) at room temperature (17 days). The resolution

$$C_2H_5O_2C(CH_2)_4CH{=}CHCO_2H \xrightarrow[91\%]{CH_3COSH} C_2H_5O_2C(CH_2)_4\underset{SCOCH_3}{CHCH_2CO_2H} \xrightarrow{SOCl_2}$$

(8)   (9)

$$C_2H_5O_2C(CH_2)_4\underset{SCOCH_3}{CHCH_2COCl} \xrightarrow[\substack{2. \ NaOH}]{1. \ NaBH_4} HO_2C(CH_2)_4\underset{SH}{CHCH_2CH_2OH} \xrightarrow[NaOH]{S{=}C(NH_2)_2, \ HBr}$$

(10)   (11)

$$\underset{SH \quad SH}{CH_2CH_2CH(CH_2)_4CO_2H} \xrightarrow{I_2-KI} \underset{S}{\overset{S}{\diagdown}}(CH_2)_4CO_2H$$

(12)   (13)

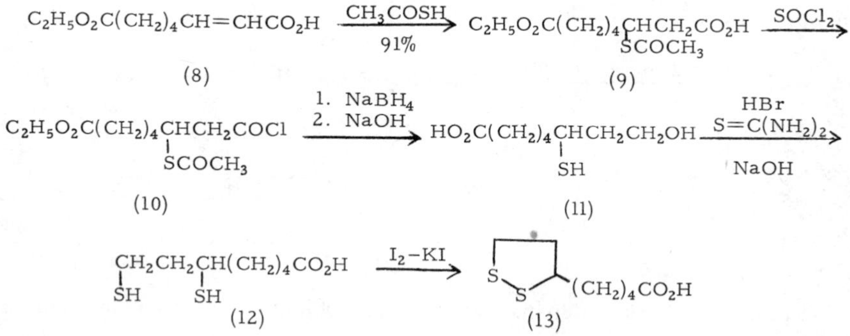

of (±)-thioctic acid has not been accomplished, and the Merck synthesis has the advantage that resolution can be introduced at an early stage, namely by resolution of the adduct (9) with *l*-ephedrine.

The addition of thiolacetic acid to α,β-unsaturated 3-ketosteroids has been used

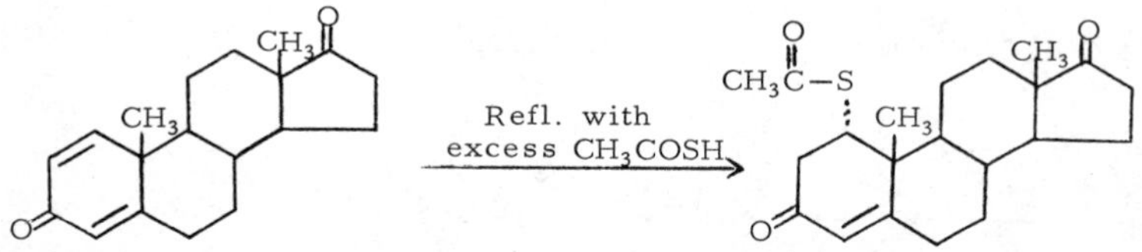

to obtain compounds of enhanced biological activity (irradiation is not necessary).[13] Thus $\Delta^{1,4}$-3-ketosteroids react to give 1α-acetylthio-Δ^{4}-3-ketosteroids; $\Delta^{4,6}$-3-ketosteroids give 7α-acetylthio-Δ^{4}-3-ketones.

Displacement of bromine. Bonner[14] treated tetracetyl-α-D-glucosyl bromide in chloroform with thiolacetic acid and KOH (both in 10% excess) and obtained pentaacetyl-1-thio-β-D-glucose in good yield.

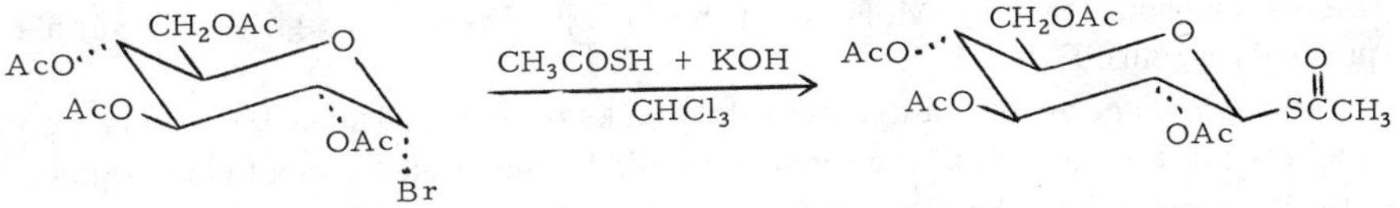

S-Acetylation. Thiolacetic acid, or an aqueous solution of the sodium salt, has been used for S-acetylation of coenzyme A[15] and of pantetheine.[16] A Merck group[16]

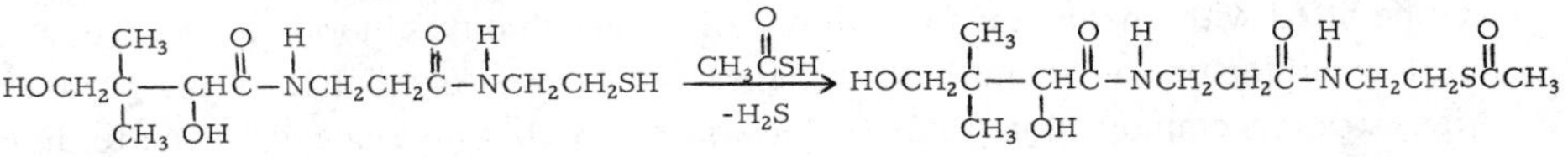

Pantetheine S-Acetylpantetheine

dissolved 10 g. of pantetheine in 60 ml. of water, added 20 ml. of thiolacetic acid (a large excess), and let the mixture stand at room temperature overnight, when it had become homogeneous. Concentration under reduced pressure left S-acetylpantetheine as a viscous pale yellow oil of analytical purity. The yield was quantitative.

[1]E. K. Ellingboe, *Org. Syn., Coll. Vol.*, **4**, 928 (1963)
[2]C. Walling, "Free Radicals in Solution," pp. 316–317, John Wiley and Sons (1957)
[3]B. Holberg, *Arkiv Kemi Min. Geol.*, **12B**, No. 47, 3 (1938)
[4]V. N. Ipatieff and B. S. Friedman, *Am. Soc.*, **61**, 71 (1939)
[5]R. Brown, W. E. Jones, and A. R. Pinder, *J. Chem. Soc.*, 2123 (1951)
[6]F. G. Bordwell and W. A. Hewett, *Am. Soc.*, **79**, 3493 (1957)
[7]J. S. Showell, J. R. Russell and D. Swern, *J. Org.*, **27**, 2853 (1962)
[8]H. Bader, L. C. Cross, I. Heilbron, and E. R. H. Jones, *J. Chem. Soc.*, 619 (1949); H. Behringer, *Ann.*, **564**, 219 (1949)
[9]H. M. Walborsky, R. H. Davis, and D. R. Howton, *Am. Soc.*, **73**, 2590 (1951)
[10]A. T. Blomquist and J. Wolinsky, *J. Org.*, **23**, 551 (1958)
[11]M. W. Bullock, J. A. Brockman, Jr., E. L. Patterson, J. V. Pierce, M. H. von Saltza, F. Sanders, and E. L. R. Stokstad, *Am. Soc.*, **76**, 1828 (1954)
[12]E. Walton, A. F. Wagner, F. W. Bachelor, L. H. Peterson, F. W. Holly, and K. Folkers, *Am. Soc.*, **77**, 5144 (1955)
[13]R. M. Dodson and R. C. Tweit, *Am. Soc.*, **81**, 1224 (1959); J. A. Cella and R. C. Tweit, *J. Org.*, **24**, 1109 (1959)
[14]W. A. Bonner, *Am. Soc.*, **73**, 2659 (1951); see also D. Horton and M. L. Wolfrom, *J. Org.*, **27**, 1794 (1962)
[15]I. B. Wilson, *Am. Soc.*, **74**, 3205 (1952)
[16]E. Walton, A. N. Wilson, F. W. Holly, and K. Folkers, *Am. Soc.*, **76**, 1146 (1954)

Thionyl bromide, $SOBr_2$. Mol. wt. 207.89, b.p. 68°/40 mm., sp. gr. 2.68.

The reagent is prepared in good yield from thionyl chloride and potassium bromide.[1] For economical conversion of an alcohol to the bromide[2] the alcohol is treated in ether at −10° with 0.5 mole of thionyl chloride and 1 mole of pyridine; the solution is filtered from precipitated pyridine hydrochloride, evaporated, and the residual sulfite $(RO)_2SO$, is heated with 0.5 mole of thionyl bromide.

In an earlier procedure[3] the alcohol was dissolved in benzene below 100° and thionyl bromide was added dropwise.

$$\underset{\displaystyle OH}{CH_3CH}(CH_2)_4N(C_2H_5)_2 \xrightarrow[88\%]{SOBr_2} \underset{\displaystyle Br}{CH_3CH}(CH_2)_4N(C_2H_5)_2$$

[1]M. J. Frazer and W. Gerrard, *Chem. Ind.*, 280 (1954); see also *Inorg. Syn.*, **1**, 113 (1939)
[2]M. J. Frazer, W. Gerrard, G. Machell, and B. D. Shepherd, *Chem. Ind.*, 931 (1954)
[3]R. C. Elderfield, C. B. Kremer, S. M. Kupchan, O. Birstein, and G. Cortes, *Am. Soc.*, **69**, 1258 (1947)

Thionyl chloride, $SOCl_2$. Mol. wt. 118.98, b.p. 75–76°, sp. gr. 1.67. Suppliers (purified reagent): E, F, MCB.

Manipulation. Since thionyl chloride attacks rubber, and since the usual reaction products are HCl and SO_2, the reagent should be manipulated in all-glass apparatus with silicon grease on the glass joints, and provision should be made for entrainment of the gases. The best practice is to carry out the reaction in a round-bottomed flask fitted with a reflux condenser and drying tube and mounted in a hood; the flask later can be fitted with an adapter for removal of excess thionyl chloride at the pressure of an aspirator on the steam bath and then for vacuum distillation of the product. Many workers employ thionyl chloride in large excess, but the evidence seems to show that no excess is needed in case the reaction is run at room temperature or if the mixture is refluxed under an efficient condenser cooled with water at 20° or below.

Purification.[1] Triphenyl phosphite (160 ml.) was added to 1 liter of technical thionyl chloride, with stirring, over a period of 0.5 hr. (stirring minimizes local reaction of the phosphite with thionyl chloride). The mixture was fractionated through a 12-in. column packed with glass helices and connected to a reflux distilling head equipped with calcium chloride drying tube. After a small fore-run, thionyl chloride, b.p. 75.5–76°, was collected. The initial pot temperature was 79° and when it rose above 85–86° the distillate (b.p. 76°) was decidedly yellow. A yield of 60% of partially purified thionyl chloride was obtained (80% if the yellow fractions are included).

Distillation of partially purified material with 15–20 ml. per liter of triphenyl phosphite gave pure thionyl chloride, b.p. 76°, in greater than 90% yield. Distillation was terminated when the pot temperature reached 84°. In small quantities the thionyl chloride appeared colorless; in larger quantities a very faint yellow. Pure thionyl chloride is reported to be faintly yellow. The visible spectrum of partially purified material was almost comparable to that of pure thionyl chloride.

RCOOH ⟶ RCOCl. Most of the *Organic Syntheses* procedures call for use of the reagent in large excess. One submitted by Cason[2] ($CH_3O_2CCH_2CH_2COCl$) specifies a 100% excess, but checkers C. F. H. Allen and C. V. Wilson obtained the same yield (90–93%) with only 20% excess. In preparing the related $CH_3O_2C(CH_2)COCl$, Bishop[3] used 27% excess thionyl chloride, refluxed the mixture for 5 hrs., and the yield was 83–86%. In the preparation of another ester acid chloride, $CH_3O_2C(CH_2)_4COCl$, the reaction mixture was let stand overnight at room temperature, and use of the reagent in 100% excess seems unwarranted.[4] Womack and McWhirter[5] refluxed a mixture of 1.0 mole each of cinnamic acid and thionyl chloride for 1 hr., treated the crude acid chloride with 1 mole of phenol, and obtained phenyl cinnamate in yield of 83–89%. A procedure for the conversion of

$$\underset{(1)\ 0.22\ m.}{C_6H_5N{=}NC_6H_4CO_2H} + \underset{3.5\ m.}{SOCl_2} + \underset{0.47\ m.}{Na_2CO_3} \xrightarrow[89\%]{\text{Refl. 1.5 hrs.}} \underset{(2)}{C_6H_5N{=}NC_6H_4COCl}$$

the acid (1) into the orange-red acid chloride (2), m.p. 94.5–95.5°, specifies addition to the reaction mixture of solid sodium carbonate, which is claimed to prevent decomposition and tar formation.[6]

A procedure from our laboratory[7] is mentioned in order to call attention to the highly efficient apparatus (Fig. A-3) used for steam distillation of the product, α-tetralone. The round-bottomed flask on the left functions as a liquid seal or vapor

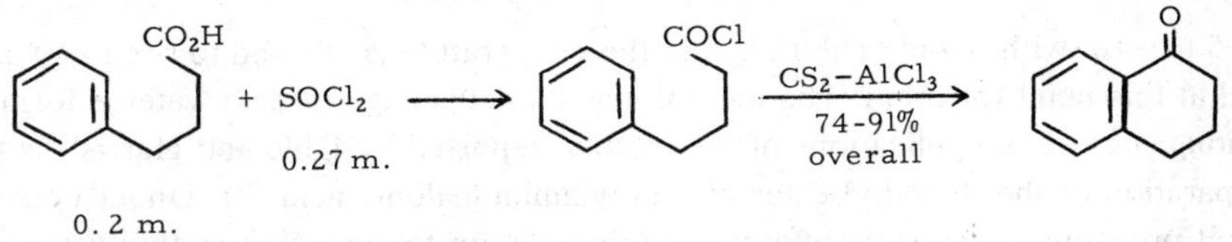
CO2H
+ SOCl2
0.27 m.
0.2 m.
COCl
CS2–AlCl3
74-91%
overall

trap; it always retains some condensate except when, for inspection or at the end, it is emptied by diverting the stream of cooling water. Nitrobenzene can be distilled at a rate of 400 g. per hr.

Allen and co-workers[8] at the Eastman laboratories developed a procedure for the preparation of oleoyl chloride and other acid chlorides sensitive to heat in a continuous reaction with countercurrent distillation such that the heat-sensitive acid chloride is heated for only a few minutes. Oleoyl chloride of satisfactory purity was obtained in yield of 97–98%.

The preparation of pure, colorless phenylacetyl chloride and other acid chlorides with a reactive α-hydrogen presents some difficulties because of condensations giving colored products. Buckles and Cooper[9] developed a procedure which consistently gives colorless phenylacetyl chloride. A mixture of the acid with 1 equivalent of thionyl chloride is provided with a boiling stone to promote spontaneous

$$\underset{4\text{ m.}}{C_6H_5CH_2CO_2H} + \underset{4\text{ m.}}{SOCl_2} \xrightarrow[80\text{-}85\%]{25\text{ hrs. }30\text{-}40^0} C_6H_5CH_2COCl$$

evolution of gases, swirled occasionally at room temperature as long as gases are evolved, and then warmed to 35–40° to complete the reaction. One-fourth volume of benzene is added, and all volatile material is carefully removed by low-temperature distillation with two dry ice–acetone traps, at first at the pressure of an aspirator and then at 1 mm. After removal of all volatile material, which would give rise to a red color, the acid chloride is distilled at 55–57°/1 mm.

As used in the preparation of α-bromocarboxylic esters, the thionyl chloride serves not only as a reagent for the formation of the acid chloride but as solvent for the subsequent bromination.[10] The acid is refluxed with thionyl chloride, then bromine is added and the mixture heated further. The method is suitable for both mono- and dicarboxylic acids.[11]

For the preparation of chlorides of carboxylic and sulfonic acids which do not react with thionyl chloride alone, *see* Dimethylformamide–Thionyl chloride.

Sulfonyl chlorides. The examples formulated show that thionyl chloride reacts readily with either a free sulfonic acid[12] or a hydrated sodium salt.[13]

$$\underset{1.5\text{ m.}}{CH_3SO_3H} + \underset{2\text{ m.}}{SOCl_2} \xrightarrow[71\text{-}83\%]{3.5\text{ hrs. at }90^0} CH_3SO_2Cl$$

$$\underset{0.2\text{ m.}}{\underline{p}\text{-}CH_3C_6H_4SO_3Na\cdot 2H_2O} + \underset{1.5\text{ m.}}{SOCl_2} \xrightarrow[86\text{-}92\%]{2\text{ hrs. at }25^0} \underline{p}\text{-}CH_3C_6H_4SO_2Cl$$

Amino acid esters. The combination of methanol and thionyl chloride is recommended by Brenner[14] for the esterification of amino acids (the Fischer procedure usually is troublesome). A mixture of the amino acid and methanol is treated

$$\underset{\substack{|\\NH_2\\1\ m.}}{RCHCO_2H} + \underset{1.1\ m.}{SOCl_2} + \underset{8\ m.}{CH_3OH} \xrightarrow[70-80\%]{40^0} \underset{\substack{|\\NH_2\cdot HCl}}{RCHCO_2CH_3} + SO_2 + HCl$$

at −5 to −10° with thionyl chloride, and the temperature is allowed to rise to 40° and held at this point for 2 hrs. The method has the advantage that no water is formed. Among successful applications of the method reported by Uhle and Harris[15] is the preparation of the dimethyl ester of methylaminomalonic acid (2). Under conventional procedures the acid suffered decarboxylation to sarcosine methyl ester (3).

$$\underset{(1)}{CH_3NHCH(CO_2H)_2} \longrightarrow \underset{(2)}{CH_3NHCH(CO_2CH_3)_2} \qquad \underset{(3)}{CH_3NHCH_2CO_2CH_3}$$

Patel and Price[16] used thionyl chloride as acid catalyst and dehydrating agent in the preparation of benzyl esters of amino acids, as illustrated for L-phenylalanine.

$$\underset{\substack{|\\NH_2\\3.3\ g.}}{C_6H_5CH_2CHCO_2H} + \underset{125\ ml.}{C_6H_5CH_2OH} + \underset{20\ ml.}{SOCl_2} \xrightarrow[90\%]{5^0} \underset{\substack{|\\NH_3Cl}}{C_6H_5CH_2CHCO_2CH_2C_6H_5}$$

ROH⟶RCl. Reaction conditions similar to those used for the preparation of acid chlorides are often adequate, as illustrated for the preparation of β-chloroethyl methyl sulfide.[17] A solution of thionyl chloride was added dropwise to a stirred

$$\underset{\substack{1.63\ m.\\ \text{In 200 ml. } CHCl_3}}{CH_3SCH_2CH_2OH} + \underset{\substack{1.7\ m.\\ \text{In 135 ml. } CHCl_3}}{SOCl_2} \xrightarrow[75-85\%]{6\ hrs.} CH_3SCH_2CH_2Cl$$

solution of the alcohol in the same solvent in the course of 2 hrs. A similar procedure was used for the preparation of ethyl α-chlorophenylacetate.[18]

$$\underset{\substack{|\\OH\\0.75\ m.}}{C_6H_5CHCO_2C_2H_5} + \underset{0.82\ m.}{SOCl_2} \xrightarrow[81-85\%]{\substack{16\ hrs.\ at\ 25^0,\\ refl.\ 30\ min.}} \underset{\substack{|\\Cl}}{C_6H_5CHCO_2C_2H_5}$$

Darzens[19] carried out the reaction in the presence of one mole of pyridine or other tertiary amine. In the example formulated, the intermediate chlorosulfinate separated

$$\underset{\substack{|\\OH\\1\ m.}}{CH_3CHCO_2C_2H_5} + \underset{1.05\ m.}{SOCl_2} + \underset{1\ m.}{Py} \xrightarrow[95\%]{} \underset{\substack{|\\Cl}}{CH_3CHCO_2C_2H_5} + Py\cdot HCl$$

at first as a mass of crystals which changed to a liquid layer without gas evolution. Heating at 110° then completed the reaction with loss of SO_2 and HCl. Darzens noted that the tertiary carbinol formulated affords the α,β-unsaturated ester,

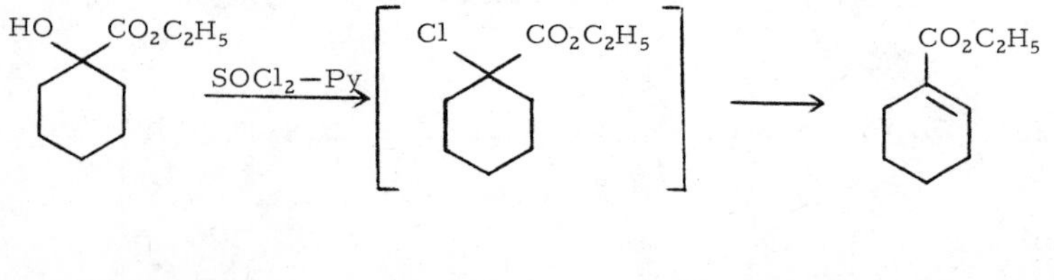
HO
$CO_2C_2H_5$
$SOCl_2-Py$
Cl
$CO_2C_2H_5$
$CO_2C_2H_5$

probably via the chloride. Gerrard and co-workers[20] used a modified procedure in which 1 mole of the alcohol is treated in ether at $-10°$ with 0.5 mole of thionyl

$$\underset{1\text{ m.}}{\underline{n}\text{-BuOH}} + \underset{0.5\text{ m.}}{SOCl_2} + \underset{1\text{ m.}}{Py} \longrightarrow \underset{0.5\text{ m.}}{(\underline{n}\text{-BuO})_2S{=}O} + Py \cdot HCl$$

$$(\underline{n}\text{-BuO})_2S{=}O + \underset{0.5\text{ m.}}{SOCl_2} \longrightarrow 2\ \underline{n}\text{-BuOS(=O)Cl} + SO_2$$

$$2\ \underline{n}\text{-BuOS(=O)Cl} \longrightarrow 2\ \underline{n}\text{-BuCl} + SO_2$$

chloride, and the resulting solution of *n*-butyl sulfite is filtered from precipitated pyridine hydrochloride. The solvent is removed, and the residue is heated with 0.5 mole of thionyl chloride, when the rapidly formed chlorosulfinite slowly decomposes to *n*-butyl chloride and SO_2.

Brooks and Snyder[21] used the Darzens procedure in the conversion of tetrahydrofurfuryl alcohol into the chloride.

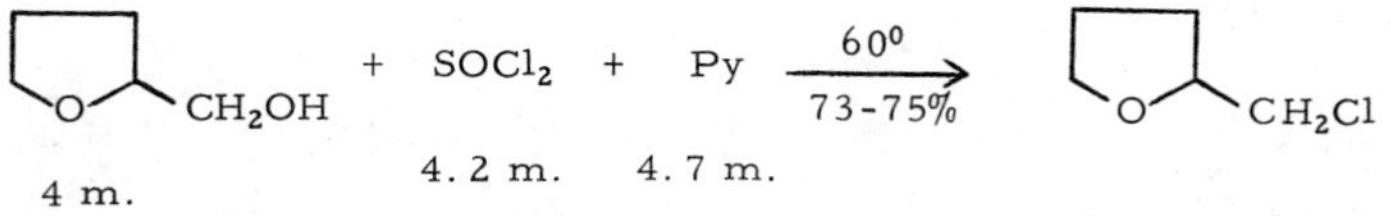
CH₂OH
+ SOCl₂ + Py
60⁰
73-75%
CH₂Cl
4 m.
4.2 m.
4.7 m.

In a student preparation of *trans*-stilbene,[22] benzoin is warmed on the steam bath with thionyl chloride, excess reagent is evaporated at the suction pump, the desyl chloride (2) is reduced with sodium borohydride in ethanol, and the solution of stereoisomeric chlorohydrins (3) is treated with zinc dust and acetic acid and refluxed for 1 hr. Crystallization affords diamond-shaped iridescent plates of pure *trans*-stilbene,

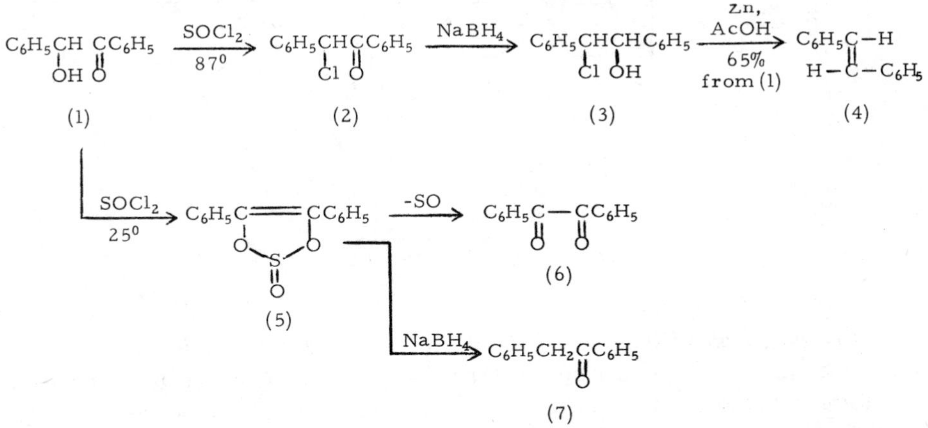
SOCl₂
87⁰
NaBH₄
Zn, AcOH
65% from (1)
(1)
(2)
(3)
(4)
SOCl₂
25⁰
-SO
(5)
(6)
NaBH₄
(7)

m.p. 124–125°. Most of the students performing this experiment at the Massachusetts Institute of Technology obtained *trans*-stilbene, but a few obtained a low-melting product identified as desoxybenzoin. Investigation[23,24] showed that if the mixture of benzoin and thionyl chloride is not heated at once but let stand for an appreciable time at room temperature (or below) another reaction sets in affording the sulfate ester of benzoin enediol (5). If let stand, this ester decomposes to benzil; if treated with sodium borohydride it is reduced to desoxybenzoin.

Dehydration. For a comparison showing that thionyl chloride in pyridine is a more powerful reagent for the dehydration of steroid alcohols than phosphoryl chloride in pyridine *see* Phosphoryl chloride.

A procedure for the preparation of 2-ethylhexanonitrile[25] shows that the conditions required for the dehydration of an amide are much the same as for the preparation of an acid chloride.

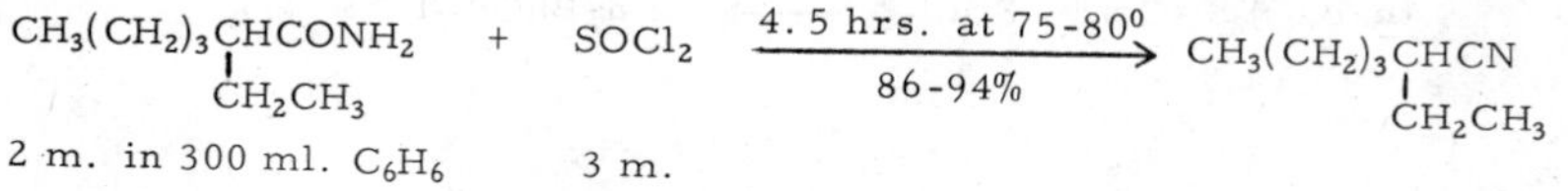

von Braun reaction. Vaughan and Carlson[26] found that, for cleavage of an amide to an alkyl chloride and a nitrile, thionyl chloride offers some advantages over

$$RNHCOR' + SOCl_2 \longrightarrow RCl + R'CN + SO_2 + HCl$$

phosphorus pentachloride, the reagent originally used by von Braun. The reagent is used in excess (3–4 equiv.), and nitromethane is useful as solvent when it does not interfere with the isolation of products.

n-Butyl sulfite. This ester is prepared by adding thionyl chloride to *n*-butanol over a period of 2 hrs.[27]

$$\underset{9\text{ m.}}{2\ \underline{n}\text{-}C_4H_9OH} + \underset{4.2\text{ m.}}{SOCl_2} \xrightarrow[77-84\%]{35-45^0} (\underline{n}\text{-}C_4H_9O)_2SO + 2\ HCl$$

Pyridine-4-sulfonic acid. The formation of an arylsulfonic acid by reaction with thionyl chloride is a special reaction of pyridine derivatives; a possible sequence of events is indicated in the formulation.[28]

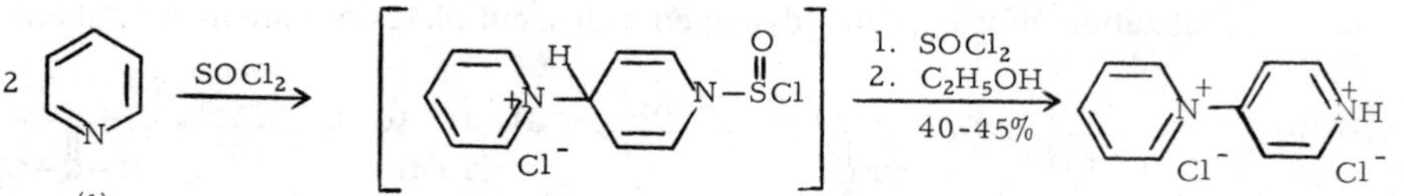

(1) (2) (3)

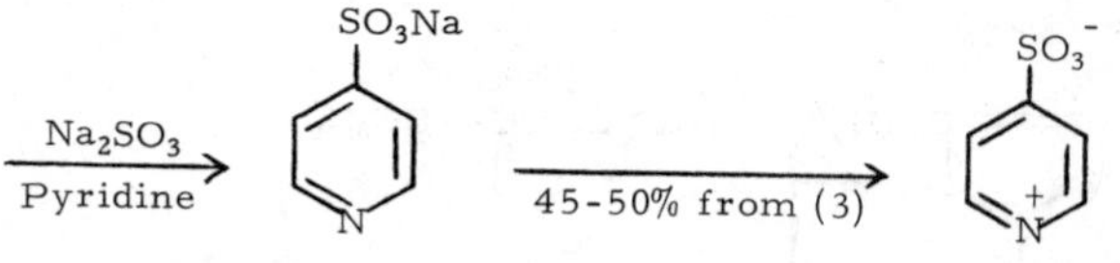

(4) (5)

Dehydrogenation. Büchi and Lukas[29] report a novel reaction in which thionyl chloride effects dehydrogenation of (1) to (4). Treatment of the pyrrolinone (1) with thionyl chloride at room temperature affords the yellow product (4) in high

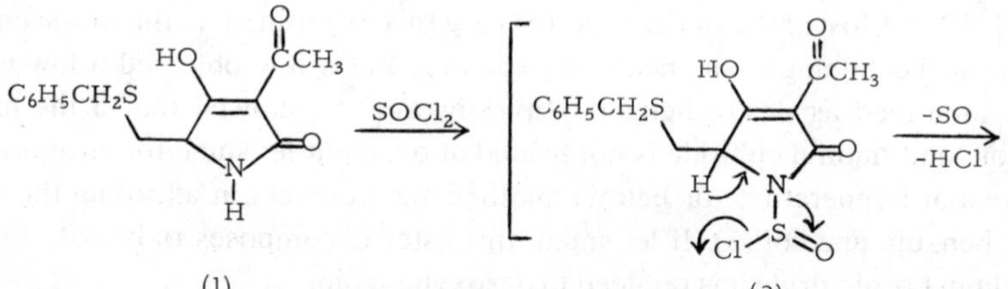

(1) (2)

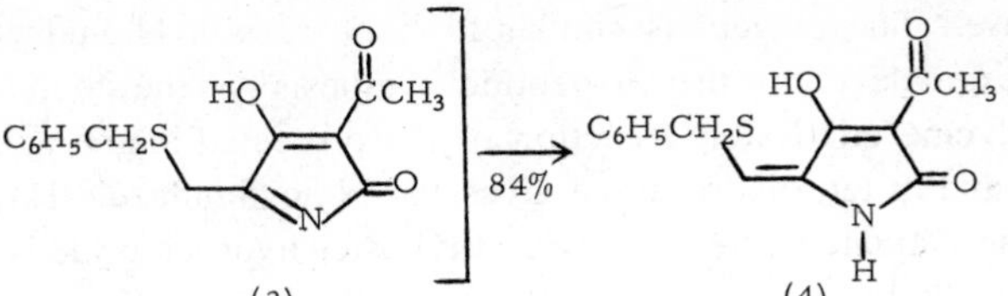

(3) (4)

yield. The chlorosulfite (2) and its decomposition product (3) are suggested intermediates. The N-methyl derivative of (1) is stable to thionyl chloride.

[1]Contributed by L. Friedman and W. P. Wetter, Case Institute of Technology.
[2]J. Cason, *Org. Syn., Coll. Vol.*, **3**, 169 (1955)
[3]W. S. Bishop, *ibid.*, **3**, 613 (1955)
[4]L. J. Durham, D. J. McLeod, and J. Cason, *ibid.*, **4**, 556 (1963)
[5]E. B. Womack and J. McWhirter, *ibid.*, **3**, 714 (1955)
[6]G. H. Coleman, G. Nichols, C. M. McCloskey, and H. D. Anspon, *ibid.*, **3**, 712 (1955)
[7]E. L. Martin and L. F. Fieser, *ibid.*, **2**, 569 (1943)
[8]C. F. H. Allen, J. R. Byers, and W. J. Humphlett, *ibid.*, **4**, 739 (1963)
[9]R. E. Buckles and J. A. Cooper, procedure submitted to *Org. Syn.*
[10]E. Schwenk and D. Papa, *Am. Soc.*, **70**, 3626 (1948)
[11]N. S. Radin, private communication
[12]P. J. Hearst and C. R. Noller, *Org. Syn., Coll. Vol.*, **4**, 571 (1963)
[13]F. Kurzer, *ibid.*, **4**, 937 (1963)
[14]M. Brenner and W. Huber, *Helv.*, **36**, 1109 (1953)
[15]F. C. Uhle and L. S. Harris, *Am. Soc.*, **78**, 381 (1956); F. C. Uhle, *J. Org.*, **27**, 4081 (1962)
[16]R. P. Patel and S. Price, *J. Org.*, **30**, 3575 (1965)
[17]W. R. Kirner and W. Windus, *Org. Syn., Coll. Vol.*, **2**, 136 (1943)
[18]E. L. Eliel, M. T. Fisk, and T. Prosser, *ibid.*, **4**, 169 (1963)
[19]G. Darzens, *Compt. rend.*, **152**, 1601 (1911)
[20]M. J. Frazer, W. Gerrard, G. Machell, and B. D. Shepherd, *Chem. Ind.*, 931 (1954)
[21]L. A. Brooks and H. R. Snyder, *Org. Syn., Coll. Vol.*, **3**, 698 (1955)
[22]*Org. Expts.*, 219
[23]L. F. Fieser and Y. Okumura, *J. Org.*, **27**, 2247 (1962)
[24]Y. Okumura, *J. Org.*, **28**, 1075 (1963)
[25]J. A. Krynitsky and H. W. Carhart, *Org. Syn., Coll. Vol.*, **4**, 436 (1963)
[26]W. R. Vaughan and R. D. Carlson, *Am. Soc.*, **84**, 769 (1962)
[27]C. M. Suter and H. L. Gerhart, *Org. Syn., Coll. Vol.*, **2**, 112 (1943)
[28]R. F. Evans, H. C. Brown, H. C. van der Plas, *ibid.*, **43**, 97 (1963)
[29]G. Büchi and G. Lukas, *Am. Soc.*, **86**, 5654 (1964)

N,N′-Thionyldiimidazole (II). Mol. wt. 182.21, m.p. 78–79°.

The reagent is prepared by the reaction of thionyl chloride in tetrahydrofurane with 4 equivalents of imidazole; the solution, filtered from imidazole hydrochloride,

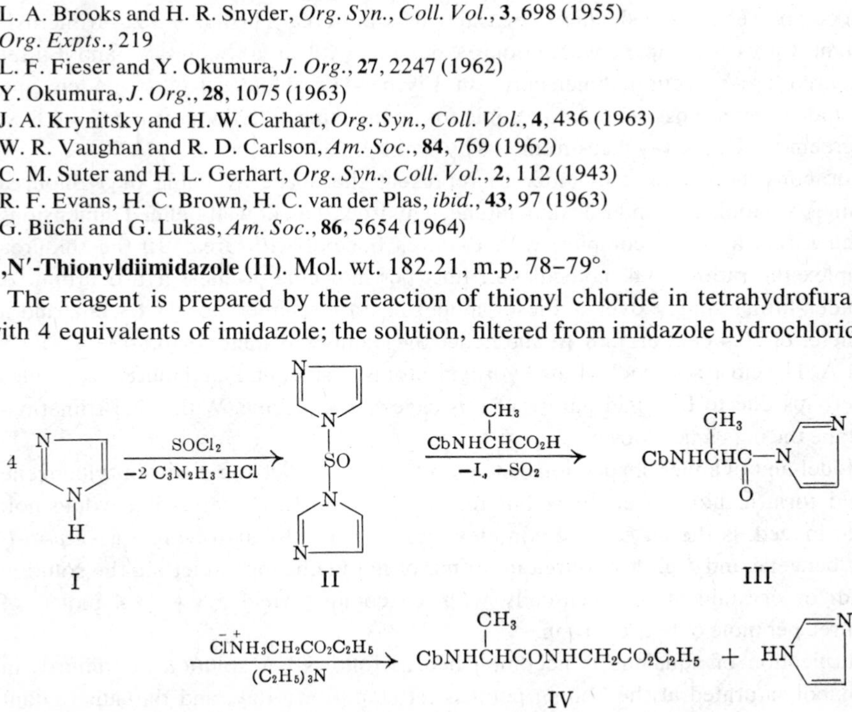

IV

is ready for use.[1] The reagent is similar to N,N′-carbonyldiimidazole but more reactive. An example of its use in peptide synthesis[2] is the coupling of Cb-DL-alanine with glycine ethylester. Reaction of the protected amino acid with N,N′-thionyldiimidazole in tetrahydrofurane gives the N-acylimidazole III, which is the equivalent of an activated ester. Glycine ethyl ester hydrochloride is present from the start, and addition of triethylamine liberates the free amino ester for coupling with intermediate III to give the dipeptide derivative IV.

[1]H. A. Staab and K. Wendel, *Angew. Chem.*, **73**, 26 (1961); H. A. Staab and G. Walther, *Ann.*, **657**, 98 (1962)
[2]T. Wieland and K. Vogeler, *Angew. Chem.*, **73**, 435 (1961)

Thiosemicarbazide, $H_2N\overset{\displaystyle H}{\overset{|}{N}}-\overset{\displaystyle S}{\overset{\|}{C}}NH_2$. Mol. wt. 91.14, m.p. 180–181° dec. Suppliers: Olin Mathieson, A, B, E, F, MCB.

Preparation of heterocycles, *see* 1,2,4-Triazole.

Thiourea, $S{=}C(NH_2)_2$. Mol. wt. 76.12, m.p. 178°, solubility: 9.2 g./100 g. water at 13°. Suppliers: Eastern, A, B, E, F.

Thiourea inclusion complexes (*see* Urea, inclusion complexes). Apparently unaware of the urea complexes, Angla[1] found that thiourea combines with many substrates to form what he regarded as unstable stoichiometric compounds containing 3 moles of thiourea. It is now clear that they are nonstoichiometric complexes analogous to the urea complexes but with a capacity for trapping somewhat larger molecules, since sulfur is a larger atom than oxygen. Schiessler and Flitter[2] tested a number of hydrocarbons for formation or nonformation of adducts with both urea and thiourea but did not analyze the complexes. Straight-chain alkanes which form urea complexes are too slender to be trapped by thiourea. The bulky 2,2,4-trimethylpentane forms a complex with thiourea but not with urea. Schiessler and Flitter measured cross section dimensions on Fischer-Hirschfelder-Taylor models and concluded that a cross section favorable for complexing with thiourea is 5.8 × 6.8 Å, in agreement with X-ray measurements by Smith.[3]

For construction of a cylinder to represent the space available in a thiourea channel, we took as standard adamantane, a hydrocarbon of well-defined dimensions which forms a stable complex with thiourea but not with urea.[4] In the thiourea complex the ratio is 3.4 molecules of host per molecule of guest hydrocarbon. A cylinder fitting snugly over a Fieser model has a circumference of 83 cm. and a diameter of 26.4 cm., or 5.28 Å, and hence the stand-off distance is (6.80 − 5.28)/2 = 0.64 Å. The close approach of the hydrogen atoms to surrounding thiourea molecules is perhaps due to the rigid nature of this cage-ring molecule. With 2,2,4-trimethylpentane the fit is much looser.

Model-and-cylinder inspection led one of us to predict the *p*-di-*t*-butylbenzene would form a thiourea complex but that its 2,5-dimethoxy derivative would not. Such, indeed, is the case.[5] The complex was prepared by dissolving 3 g. of *p*-di-*t*-butylbenzene and 5 g. of thiourea in 50 ml. of hot methanol and letting the solution stand for crystallization, eventually with ice cooling; yield 5.8 g., 4.4 moles, of thiourea per mole of hydrocarbon.

Purification of squalene is accomplished as follows.[6] A solution of thiourea in methanol saturated at the boiling point is let stand for a day, and the supernatant solution filtered. Squalene (100 mg.) is added to 10 ml. of the saturated solution, and the vessel is closed with a wired-on stopper and placed in a rotary shaker for 6–8 hrs. Globules of hydrocarbon gradually disappear, and crystals of the complex separate. The crystalline product is collected, and squalene is regenerated by addition of water and extraction of the hydrocarbon into petroleum ether.

In squalene each of the four middle bonds must have the two chain-methylene groups in either the *cis* or the *trans* configuration; formula I shows them all *trans*, as deduced by Nicolaides and Laves[7] by a novel method. From X-ray diffraction

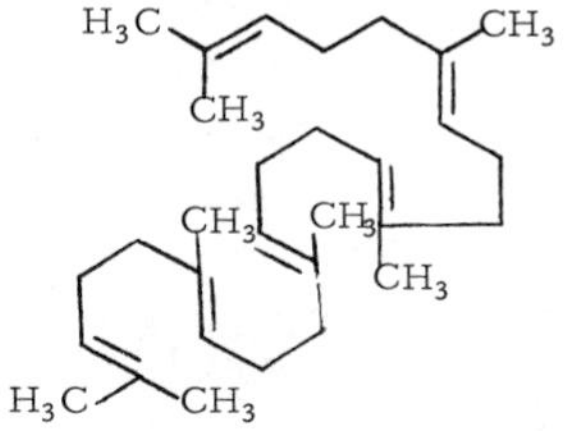

I

patterns of single crystals of urea and thiourea inclusion complexes, it was possible to measure the length of a saturated compound and compare it with that of olefinic derivatives. Comparison of the lengths of the urea complexes of oleic acid (*cis*) and of elaidic acid (*trans*) with that of the stearic acid urea complex showed that one *trans* double bond shortens a molecule by 0.19 Å, whereas a *cis* double bond shortens it by 0.88 Å. Comparison of the thiourea complexes of perhydrosqualene and squalene showed that the four double bonds in question shorten the molecule by only 0.73 Å or 4×0.182 Å. On the assumption that the effect of isolated double bonds is the same for thiourea complexes as for urea complexes, the evidence shows that the four double bonds concerned are all *trans*, as in I. The molar ratio of thiourea to squalene in the complex is calculated from X-ray data to be 14.65.[8]

A Delft group[9] separated the two isomeric 4-isopropylcyclohexanecarboxylic acids by making use of the observation that the high-melting form (94–94.5°), shown to be *trans*, forms a thiourea complex, whereas the low-melting *cis* form (40–41°) does not. They noted that the methyl esters of both isomers form complexes. Without fully explaining the difference, the Netherlands chemists suggest that it is associated with the finding of Nicolaides and Laves[7] that fatty acids are included in urea in the form of their dimers.

Preparation of mercaptans (thiols). An alkyl halide, for example lauryl bromide, reacts with thiourea to give the isothiourea hydrobromide, which on alkaline hydrolysis affords the mercaptan. In the preparation of lauryl mercaptan[10] the

$$\underline{n}\text{-}C_{12}H_{25}Br + H_2N\overset{\overset{S}{\|}}{C}NH_2 + 95\% \ C_2H_5OH \xrightarrow[3 \ hrs.]{Refl.} \left[\underline{n}\text{-}C_{12}H_{25}SC\overset{NH}{\underset{NH_2}{\diagdown}} \cdot HBr \right]$$

0.5 m.   0.5 m.   250 ml.

$$\xrightarrow[79-83\%]{Refl. \ 2 \ hrs. \ with \ aq. \ NaOH} \underline{n}\text{-}C_{12}H_{25}SH$$

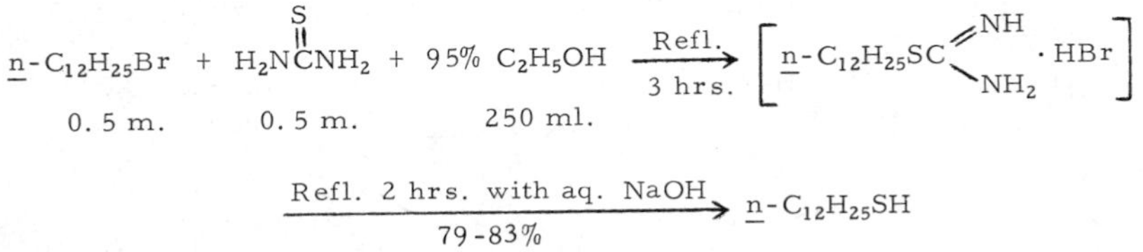

intermediate is not isolated, but aqueous alkali is added and the mixture refluxed. In the preparation of ethanedithiol from ethylene bromide the intermediate ethylene-dithiuronium bromide separates from 95% ethanol in yield of 90% and is hydrolyzed in a separate step in 55–62% yield.[11] That a benzyl-type alcohol can be used as starting material is illustrated by the preparation of 2-furfuryl mercaptan.[12]

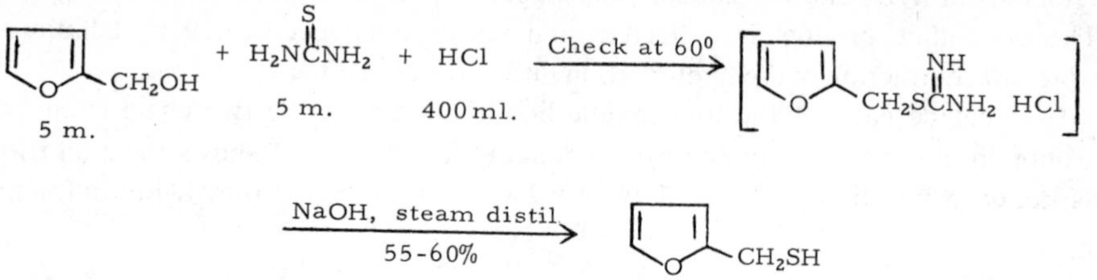

Derivatives. Procedures are formulated for the preparation of S-methylisothiourea sulfate[13] and guanidinoacetic acid.[14]

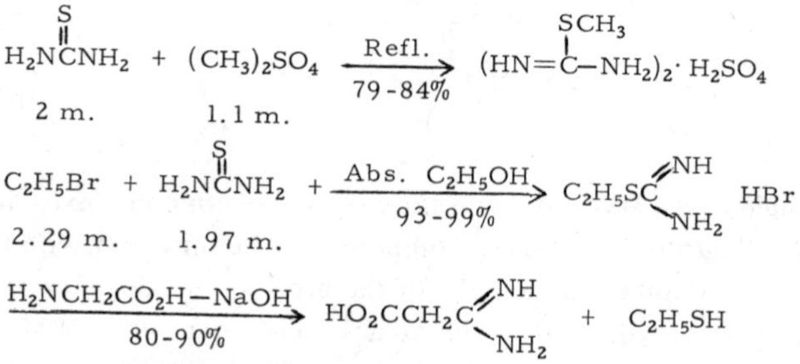

Epoxides ⟶ thiiranes. Epoxides, at least those obtained in good yield by the method formulated, are converted into thiiranes in yields mainly in the range 90–95% by reaction with thiourea in the presence of dilute sulfuric acid at 5–10°.[15]

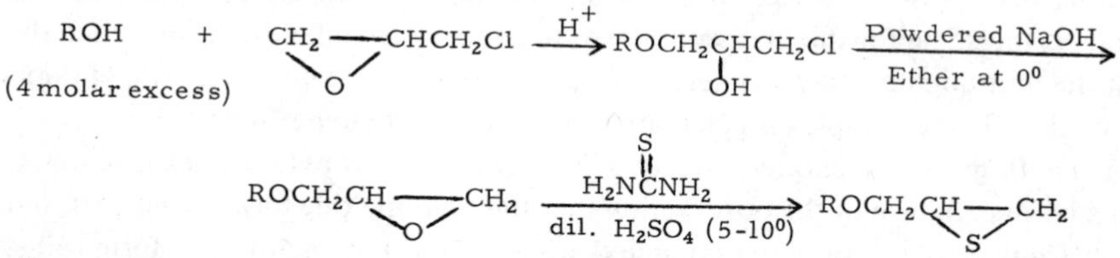

Heterocycles, *see also* Sodium ethoxide, heterocycles. 2-Amino-4-methylthiazole (1) is obtained by refluxing a mixture of chloroacetone and thiourea in 200 ml. of water, cooling, and adding 200 g. of solid sodium hydroxide.[16] The formulation shows participation of thiourea as the enol in this reaction as well as in reaction (2)

1.

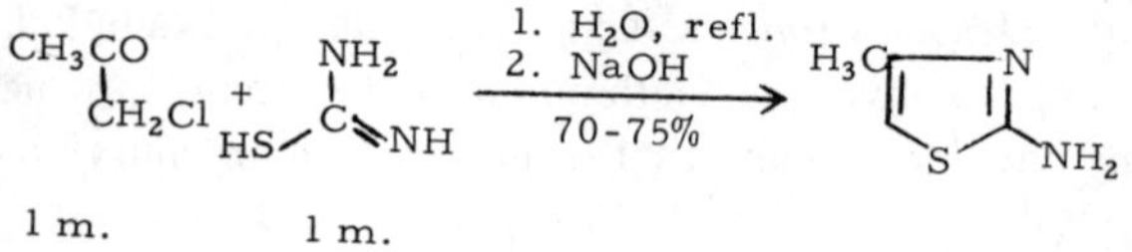

2.

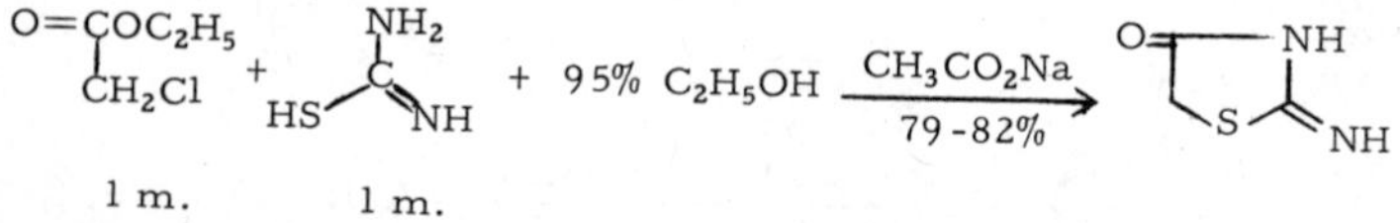

3.

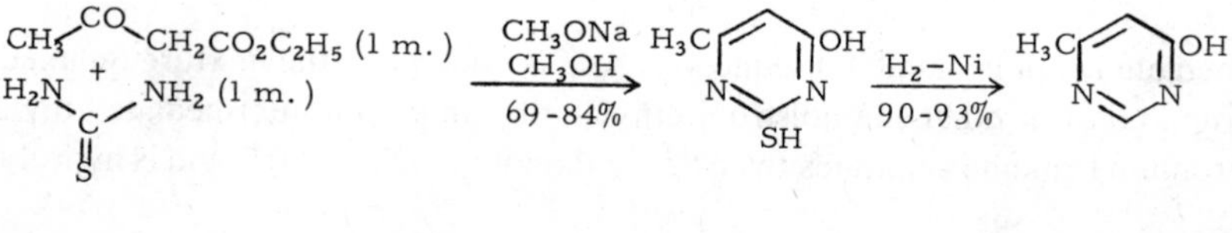

4. 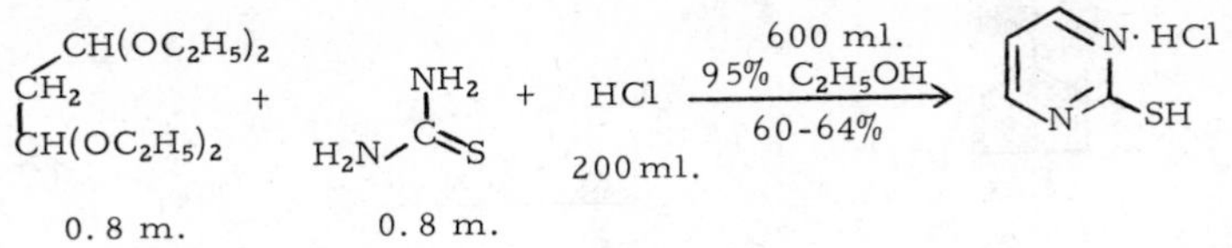

leading to pseudothiohydantoin.[17] The product of reaction sequence (3) is 4-methyl-6-hydroxypyrimidine.[18] Sequence (4), using 1,1,3,3-tetraethoxypropane (supplied by Eastman), affords 2-mercaptopyrimidine hydrochloride.[19]

Debromination. *vic*-Dibromides are debrominated by thiourea in good yield. The reaction is conducted in ethanol at 50° for 8 hrs.[20]

[1]B. Angla, *Bull. soc.*, **16**, 12 (1949); *Ann. chim.*, [12], **4**, 639 (1949)
[2]R. W. Schiessler and D. Flitter, *Am. Soc.*, **74**, 1720 (1952)
[3]A. E. Smith, *J. Chem. Phys.*, **18**, 150 (1950)
[4]S. Landa and V. Macháček, *Coll. Czech.*, **5**, 1 (1933); S. Landa and S. Hála, *ibid.*, **24**, 93 (1959)
[5]*Org. Expts.*, 184, 187
[6]K. Bloch, private communication
[7]N. Nicolaides and F. Laves, *Am. Soc.*, **76**, 2596 (1954); **80**, 5752 (1958)
[8]N. Nicolaides, private communication
[9]H. van Bekkum, A. A. B. Kleis, D. Medema, P. E. Verkade, and B. M. Wepster, *Rec. Trav.*, **81**, 833 (1962)
[10]G. G. Urquhart, J. W. Gates, Jr., and R. Connor, *Org. Syn., Coll. Vol.*, **3**, 363 (1955)
[11]A. J. Speziale, *ibid.*, **4**, 401 (1963)
[12]H. Kofod, *ibid.*, **4**, 491 (1963)
[13]P. R. Shildneck and W. Windus, *ibid.*, **2**, 411 (1943)
[14]E. Brand and F. C. Brand, *ibid.*, **3**, 440 (1955)
[15]R. D. Schuetz and R. L. Jacobs, *J. Org.*, **26**, 3467 (1961)
[16]J. R. Byers and J. B. Dickey, *Org. Syn., Coll. Vol.*, **2**, 31 (1943)
[17]C. F. H. Allen and J. A. VanAllan, *ibid.*, **3**, 751 (1955)
[18]H. M. Foster and H. R. Snyder, *ibid.*, **4**, 638 (1963)
[19]D. G. Crosby, R. V. Berthold, and H. E. Johnson, *Org. Syn.*, **43**, 68 (1963)
[20]K. M. Ibne-Rasa, N. Muhammad, and Hasibullah, *Chem. Ind.*, 1418 (1966)

Thoria (Thorium dioxide), ThO_2. Mol. wt. 264.05. Suppliers: Alfa, Fisher.

Vapor-phase dehydration of 2-alcohols with thoria as catalyst affords almost exclusively the 1-olefin:[1]

$$CH_3(CH_2)_4CH_2CH(OH)CH_3 \xrightarrow[350-400^0]{ThO_2} \underset{95-97\%}{CH_3(CH_2)_4CH_2CH{=}CH_2} + \underset{3-5\%}{CH_3(CH_2)_4CH{=}CHCH_3}$$

In contrast, dehydration of 2-alcohols over alumina yields mixtures rich in the 2-olefins. The most active catalysts were prepared by calcining thorium oxalate at 350–450° for a few hours.

[1]A. J. Lundeen and R. Van Hoozer, *Am. Soc.*, **85**, 2180 (1963)

Tigloyl chloride, 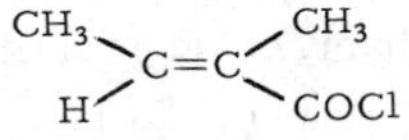

Mol. wt. 118.57, b.p. 69°/35 mm.

Preparation by the action of phosphorus trichloride on tiglic acid[1] (supplied by Aldrich).

Protection of OH-groups.[2] Conversion of an alcohol (1) into the tiglic acid ester (2) provides protection during synthetic operations, and the protective group is removed by cleavage with OsO_4–HIO_4 to the pyruvate ester (3), which is cleaved very readily by weak base.

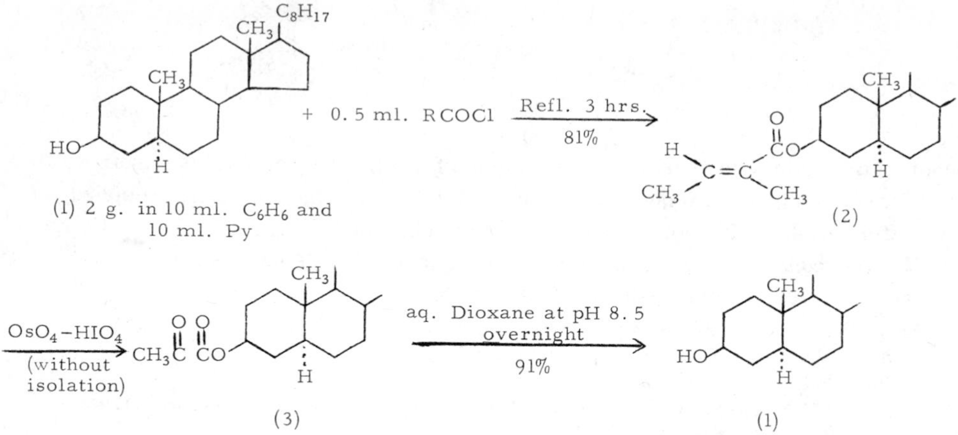
C8H17
CH3
CH3
HO
H
+ 0.5 ml. RCOCl
Refl. 3 hrs.
81%
(1) 2 g. in 10 ml. C6H6 and
10 ml. Py
CH3
O
CO
H
C=C
CH3
CH3
H
(2)
OsO4–HIO4
(without isolation)
CH3C CO
CH3
H
(3)
aq. Dioxane at pH 8.5
overnight
91%
CH3
HO
H
(1)

[1]G. Barger, W. F. Martin, and W. Mitchell, *J. Chem. Soc.*, 1822 (1937)
[2]S. M. Kupchan, A. D. J. Balon, and E. Fujita, *J. Org.*, **27**, 3103 (1962)

Tin, Sn. At. wt. 118.70. Suppliers: F (granular, mossy, 20-mesh, 30-mesh); MCB (granular, mossy, 200-mesh powder).

In the classical student preparation of aniline[1,2] granular tin and hydrochloric acid is just as satisfactory as the more expensive stannous chloride because the product is isolated, after alkalinization, by steam distillation. However, for reduction of anthraquinone to anthrone the $SnCl_2$–HCl–AcOH method (*see* Stannous chloride) seems preferable to the Sn–HCl–AcOH method[3] because it eliminates a troublesome filtration of the hot acid solution and because the yield is 10% better. In those cases where the metal alone has been used it is not clear whether the chloride was tried and found unsatisfactory or merely not tried. A procedure[4] for the reduction of anisoin to desoxyanisoin specifies use of 200-mesh tin in 40% excess of the theory, and a note states that reduction in the amount of metal reduces the yield.

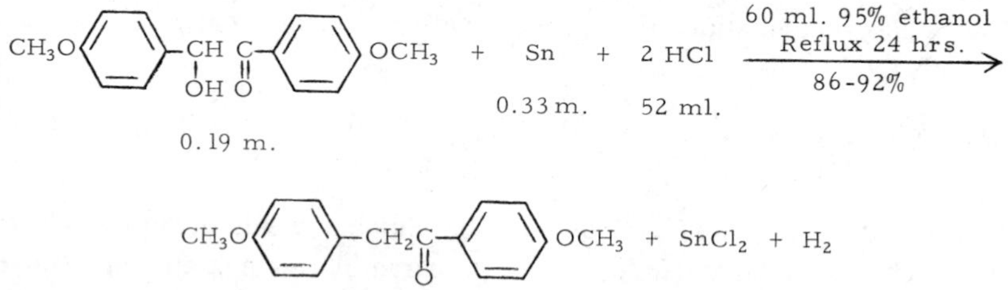
CH3O
CH C
OH O
OCH3
+ Sn + 2 HCl
60 ml. 95% ethanol
Reflux 24 hrs.
86-92%
0.33 m.
52 ml.
0.19 m.
CH3O
CH2C
O
OCH3 + SnCl2 + H2

Procedures for the reduction of 2,6-dibromo-4-nitrophenol to the aminophenol,[5] of nitrobarbituric acid to the amine (uramil),[6] and for the preparation of phloroglucinol[7] call for the use of the metal but probably could be carried out with stannous chloride.

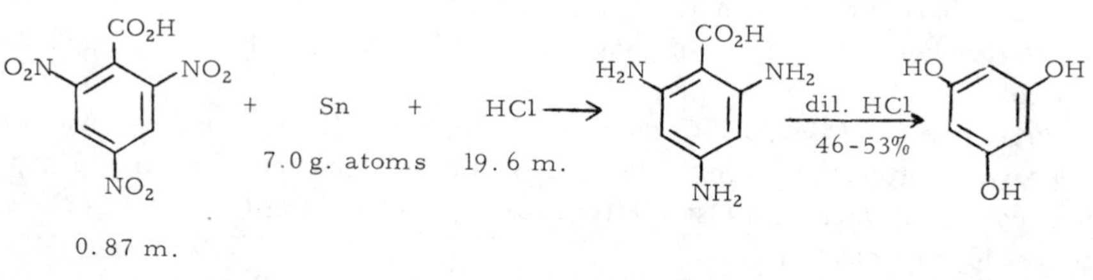
CO2H
O2N
NO2
NO2
0.87 m.
+ Sn + HCl
7.0 g. atoms
19.6 m.
CO2H
H2N
NH2
NH2
dil. HCl
46-53%
HO
OH
OH

In 1886 Guido Goldschmiedt reduced papaverine with tin and hydrochloric acid and isolated 1,2,3,4-tetrahydropapaverine and a crystalline base now known as pavine, the structure of which was established by Battersby and Binks.[8]

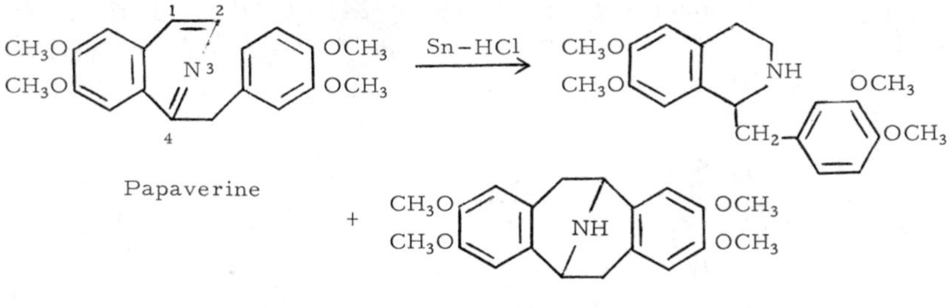

Papaverine

Pavine

Amalgamated tin. Schaefer[9] prepared this reagent by shaking a mixture of 100 g. of 30 mesh tin, 15 g. of mercuric chloride, and 100 ml. of water in a stoppered flask for a few minutes until all the tin appeared to have a shiny coating of mercury. Amalgamated tin and hydrochloric acid reduces conjugated enediones smoothly and without side reactions to the saturated diketones. In contrast, reduction with zinc and acetic acid is often attended with serious side reactions.

$$\underset{5\text{ g.}}{C_6H_5\overset{O}{\overset{\|}{C}}CH{=}CH\overset{O}{\overset{\|}{C}}C_6H_5} + \underset{10\text{ g.}}{Sn(Hg)} + \underset{150\text{ ml.}}{95\%\ C_2H_5OH} + \underset{20\text{ ml.}}{\text{concd. HCl}}$$

$$\xrightarrow[90\%]{} C_6H_5\overset{O}{\overset{\|}{C}}CH_2CH_2\overset{O}{\overset{\|}{C}}C_6H_5$$

[1]E. Fischer, "Anleitung zur Darstellung organischer Präparate, 8th Ed., Friedr. Vieweg and Sohn, 1908
[2]*Org. Expts.*, 177
[3]K. H. Meyer, *Org. Syn.*, *Coll. Vol.*, **1**, 60 (1941)
[4]P. H. Carter, J. C. Craig, R. E. Lack, and M. Moyle, *Org. Syn.*, **40**, 16 (1960)
[5]W. W. Hartman, J. B. Dickey, and J. G. Stampfli, *Org. Syn.*, *Coll. Vol.*, **2**, 175 (1943)
[6]W. W. Hartman and O. E. Sheppard, *ibid.*, **2**, 617 (1943)
[7]H. T. Clarke and W. W. Hartman, *ibid.*, **1**, 455 (1941)
[8]A. R. Battersby and R. Binks, *J. Chem. Soc.*, 2888 (1955)
[9]J. P. Schaefer, *J. Org.*, **25**, 2027 (1960)

Titanium tetrachloride, $TiCl_4$. Mol. wt. 189.73, b.p. 136°, sp. gr. 1.726. Suppliers: B, F, MCB.

Friedel-Crafts reaction. The reagent is not so active as aluminum chloride, for example alkylation of benzene with primary alkyl halides proceeds poorly.[1] However, titanium tetrachloride effectively catalyzes alkylation with the more reactive *t*-alkyl halides and has the advantage of being soluble in organic solvents. Benzene

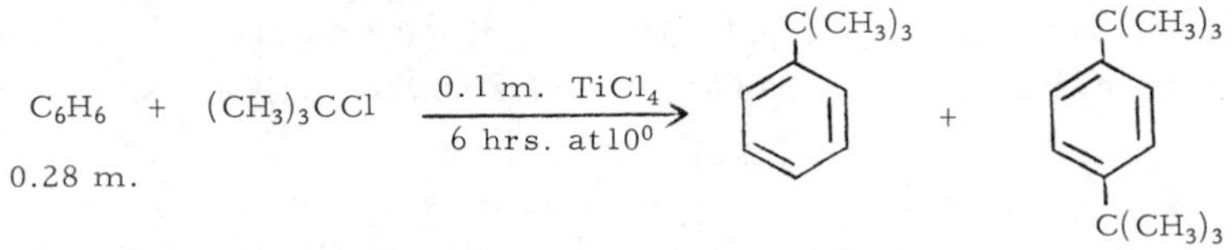
C6H6 + (CH3)3CCl
0.28 m.
0.1 m. TiCl4
6 hrs. at 10°
C(CH3)3
C(CH3)3
C(CH3)3

4% 84%

reacts with *t*-butyl chloride in 6 hrs. at 10° to give chiefly *p*-di-*t*-butylbenzene and a little *t*-butylbenzene. Under the same conditions toluene reacts in 1–3 hrs. to give *p*-*t*-butyltoluene in 77% yield.

Fries reaction. Titanium tetrachloride is useful as catalyst for the Fries rearrangement of phenol esters to *o*- or *p*-hydroxyketones.[2,3] For example, *p*-cresyl acetate is rearranged in high yield in nitrobenzene solution.[3]

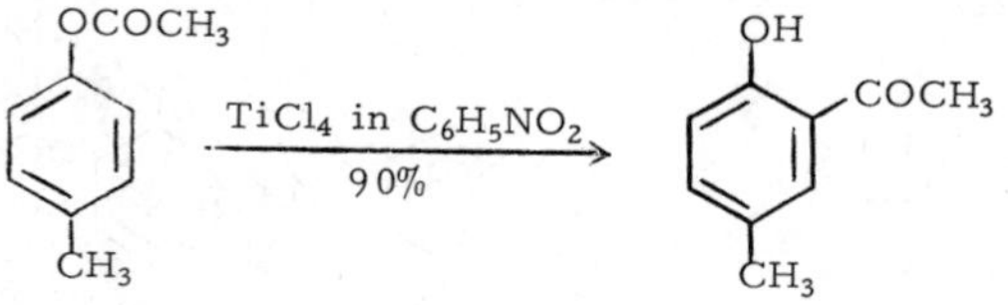

Taub *et al.*[4] at Merck used titanium tetrachloride to effect rearrangement of the phenol ester (1) to the benzophenone (2). In this case use of aluminum chloride led mainly to fragmentation into the components and the best yield of (2) was 5%.

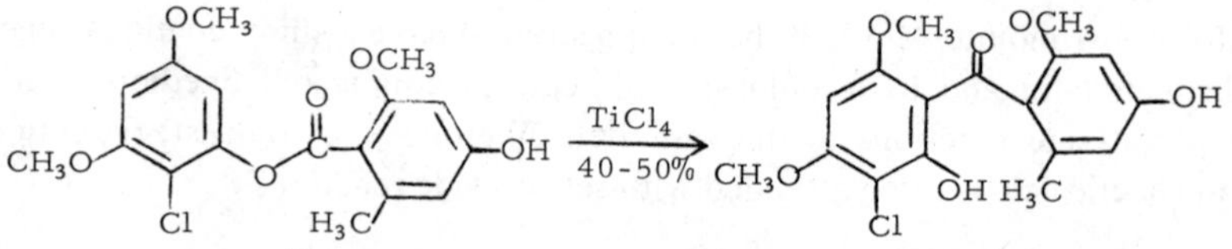

(1) (2)

Grignard Reagents. Finkbeiner and Cooper[5] found that a small amount of titanium tetrachloride effects isomerization of isopropylmagnesium bromide in ether solution to *n*-propylmagnesium bromide and postulated that the reaction involves an elimination-addition sequence involving formation of propylene and its conversion into *n*-propylmagnesium bromide by exchange with the Grignard reagent. This suggested the novel method of forming a Grignard reagent by catalyzed exchange of a terminal olefin with *n*-propylmagnesium chloride, which is not rearranged by $TiCl_4$. Indeed when an ethereal solution of 4-methyl-1-pentene and an equivalent amount of *n*-propylmagnesium chloride and 0.03 equivalent of $TiCl_4$ was refluxed for 18 hrs. and then treated with acetaldehyde, 6-methyl-2-heptanol was obtained in 37% yield. In 10 cases, yields are in the range 22–60%. An internal double bond does not react.

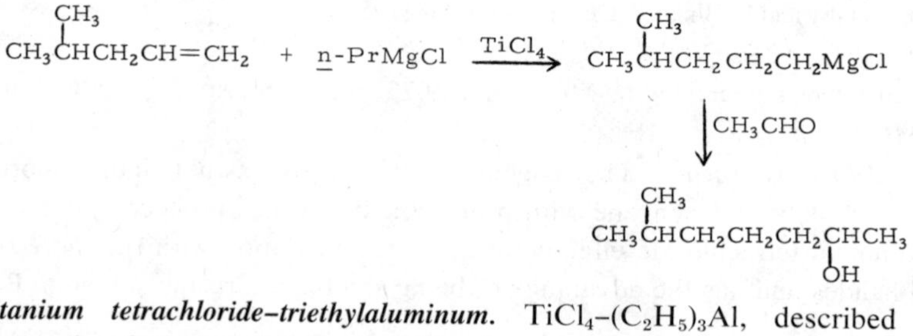

Titanium tetrachloride–triethylaluminum. $TiCl_4$–$(C_2H_5)_3Al$, described as a Ziegler catalyst,[6] catalyzes the polymerization of α-olefins. This combination also catalyzes the cyclic trimerization of acetylenes to benzene derivatives.[7]

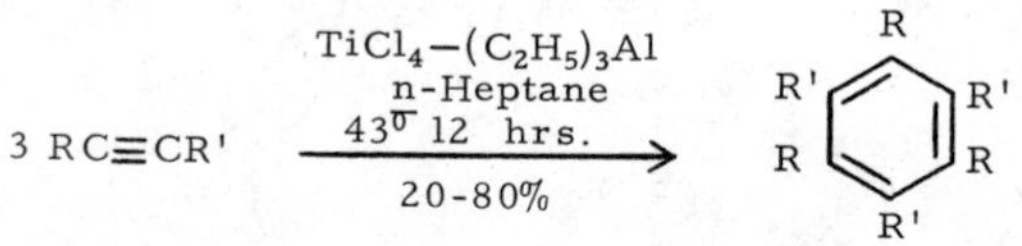

Reductive coupling of alcohols. van Tamelen and Schwartz[8] used the reagent in the reductive coupling of an alcohol to form a hydrocarbon: 2 ROH $\longrightarrow$ RR. The alcohol is converted by means of sodium hydride into the alkoxide (1), and this is

$$\underset{}{ROH} \xrightarrow{NaH} \underset{(1)}{RONa} \xrightarrow{TiCl_4} \underset{(2)}{(RO)_2TiCl_2} \xrightarrow{K} \underset{(3)}{(RO)_2Ti} \xrightarrow{100\text{-}140^0} \underset{(4)}{RR + TiO_2}$$

treated with titanium tetrachloride to form the intermediate (2), which is reduced with potassium metal to the titanium dialkoxide (3). When heated at 100–140° this is converted into the hydrocarbon and titanium dioxide. The yield of dibenzyl from benzyl alcohol was 51%.

[1]N. M. Cullinane and D. M. Leyshon, *J. Chem. Soc.*, 2944 (1954)
[2]N. M. Cullinane, E. T. Lloyd, and J. K. Tudball, *J. Chem. Soc.*, 3894 (1954)
[3]N. M. Cullinane and B. F. R. Edwards, *J. Chem. Soc.*, 3016 (1957); N. M. Cullinane, R. A. Woolhouse, and B. F. R. Edwards, *ibid.*, 3842 (1961)
[4]D. Taub, C. H. Juo, H. L. Slates, and N. L. Wendler, *Tetrahedron*, **19**, 1 (1963)
[5]H. L. Finkbeiner and G. D. Cooper, *J. Org.*, **26**, 4779 (1961)
[6]G. Natta and I. Pasquori, "Advances in Catalysis and Related Subjects," **11**, 1 (1959); N. G. Gaylord and H. F. Neark, "Linear and Stereoregular Addition Polymers," Interscience, N. Y. (1959)
[7]E. F. Lutz, *Am. Soc.*, **83**, 2551 (1961); H. Hopff and A. Gati, *Helv.*, **48**, 509 (1965)
[8]E. E. van Tamelen and M. A. Schwartz, *Am. Soc.*, **87**, 3277 (1965)

Tollens reagent.[1]

Test for reducing power.[2] Measure 2 ml. of 5% silver nitrate solution into a test tube, add 1 ml. of 10% sodium hydroxide to precipitate silver oxide, and make a dilute solution of ammonia by mixing 1 ml. of concd. ammonia solution with 10 ml. of water. Add 0.5 ml. of this solution to the precipitated silver oxide, stopper the tube, and shake. Repeat until the precipitate just dissolves (3 ml., avoid an excess) and dilute the solution to a volume of 10 ml.

Clean five test tubes by adding a few milliliters of 10% sodium hydroxide to each and heating them in a water bath and then empty and rinse with distilled water. In each tube place one micro drop of a 0.1 *M* solution of a sugar or of *n*-butyraldehyde and 1 ml. of the test solution. Let the reaction proceed first at room temperature and note the order of reactivity, as judged both from the color and by the time of first appearance of silver. After a few minutes put the tubes in the heating bath. The test shows the order of reactivity to be: fructose > glucose > lactose > maltose > *n*-butyraldehyde.

Tollens reagent is convenient for oxidation of retinene to vitamin A acid (yield 50%).[3]

[1]B. Tollens, *Ber.*, **15**, 1635 (1882)
[2]*Org. Expts.*, 132
[3]R. K. Barua and H. B. Barua, *Biochem. J.*, **92**, 21C (1964)

Toluene diisocyanate, $CH_3C_6H_3(N{=}C{=}O)_2$.[1] Mol. wt. 174.16.

The material used was a commercial 80:20 mixture of the 2,4- and 2,6-isomers available from du Pont ("Hylene TM") or National Aniline ("Mondur TD-80," "Mobay," or "Nacconate 80"). This mixture boils at about 240°.

The reagent is useful for the preparation of an aliphatic isocyanate by an exchange reaction with an aliphatic isothiocyanate. It is particularly well suited to the preparation of allyl isocyanate and methallyl isocyanate since the corresponding isothiocyanates are readily available. Allyl-type isocyanates are available by other known methods but these involve inconveniences not experienced in the present method.

Preparation of allyl isocyanate. One equivalent of allyl isothiocyanate (supplier: Baker) is dissolved in 3 equivalents of toluene diisocyanate in a flask fitted with a column packed with glass helices and heated in the range 190–210°. The reaction is

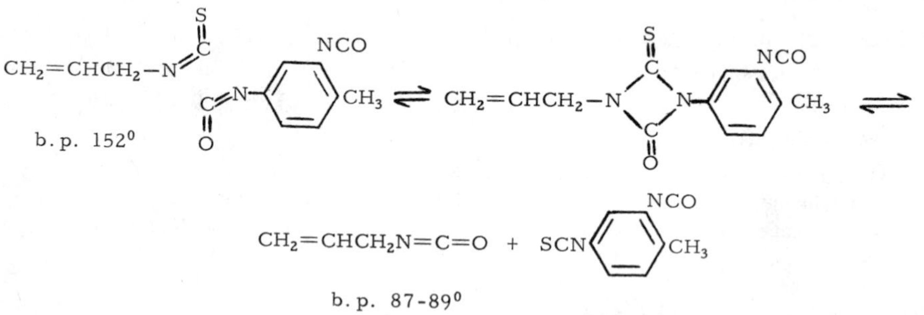
CH₂=CHCH₂–N=C=S
NCO
CH₃
b.p. 152°
CH₂=CHCH₂–N
NCO
CH₃
CH₂=CHCH₂N=C=O + SCN
NCO
CH₃
b.p. 87-89°

believed to proceed through the intermediate formation of the uretidine thione ketone; removal of the volatile allyl isocyanate shifts the equilibrium in the desired direction. The column is operated at total reflux until adequate allyl isocyanate has formed in the column head and this is then distilled. The reaction tends to be exothermal above 210°, particularly in advanced stages of the exchange, and is terminated at the first indication of an exotherm. Sparingly soluble polymers are formed in the reaction vessel. The yield of allyl isocyanate is 23%. Methallyl isocyanate, b.p. 107–109°, is obtained in yield of 31%.

[1]Contributed by William E. Erner, Air Products and Chemicals, Inc.

p-Toluenesulfonic acid (Tosic acid), p-$CH_3C_6H_4SO_3H \cdot H_2O$ ($TsOH \cdot H_2O$). Mol. wt. 190.22, m.p. 105°. Suppliers: A, B, E, F, MCB.

p-Toluenesulfonic acid is as effective an acid catalyst as sulfuric acid and it is generally preferred to sulfuric acid because it is less damaging to reactants and is a solid. Isolation is accomplished by diluting the sulfonation mixture with a moderate amount of water and adding concd. hydrochloric acid to decrease the solubility; the acid crystallizes on cooling as the monohydrate.

Esterification. The preparation of γ-chloropropyl acetate was accomplished[1] by heating the components in benzene with a little tosic acid under a Vigreaux column

$$\underset{3\text{ m.}}{CH_3CO_2H} + \underset{2\text{ m.}}{HOCH_2CH_2CH_2Cl} + \underset{2\text{ g.}}{TsOH\cdot H_2O} + \underset{300\text{ ml.}}{C_6H_6} \xrightarrow[93-95\%]{} CH_3CO_2CH_2CH_2CH_2Cl$$

connected to a condenser leading to an automatic separator arranged for return of the lighter liquid to the reaction flask[2] (a Clarke-Rahrs methyl ester column[3] is also recommended). Azeotropic distillation required 7–9 hrs. The Clarke-Rahrs column was used in the preparation of methyl pyruvate (distillation for about 1.5 days).[4] Conversion of *cis*-Δ^4-tetrahydrophthalic anhydride into the corresponding

$$\underset{1\text{ m.}}{CH_3COCO_2H} + \underset{1\text{ m.}}{CH_3OH} + \underset{0.2\text{ g.}}{TsOH\cdot H_2O} + \underset{350\text{ ml.}}{C_6H_6} \xrightarrow[65-71\%]{150-155^0} CH_3COCO_2CH_3$$

diethyl ester was done by refluxing the reaction mixture in a bath at 95–105° for 12–16 hrs., adding 270 ml. of toluene, azeotropic distillation, addition of more ethanol and refluxing, addition of toluene, and azeotropic distillation.[5]

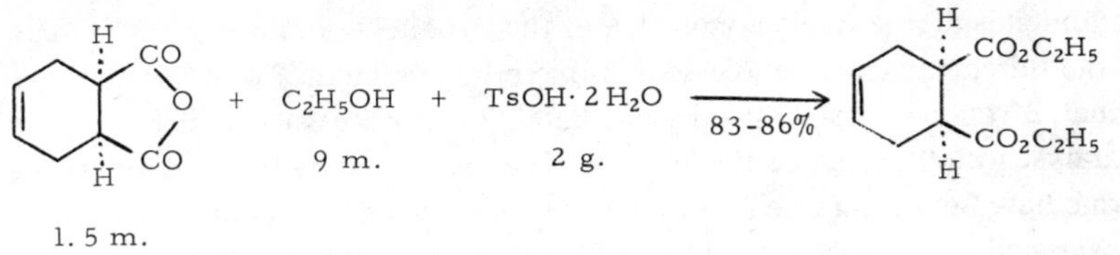

Transesterification. In the conversion of methyl acrylate to *n*-butyl acrylate,[6] hydroquinone was added as polymerization inhibitor. The mixture was heated at

$$\underset{10\ m.}{CH_2{=}CHCO_2CH_3} + \underset{5\ m.}{\underline{n}\text{-}BuOH} + \underset{10\ g.}{TsOH \cdot 2\,H_2O} \xrightarrow[78\text{-}94\%]{} CH_2{=}CHCO_2Bu\text{-}\underline{n}$$

first under total reflux, a methanol–methyl acrylate azeotrope was then distilled as formed, followed (in 6–10 hrs.) by methyl acrylate and then the butyl ester.

Enol etherification. 3-Ethoxy-Δ^2-cyclohexenone, the enol ether of dihydroresorcinol, was obtained similarly by refluxing followed by azeotropic distillation in 6–8 hrs.[7]

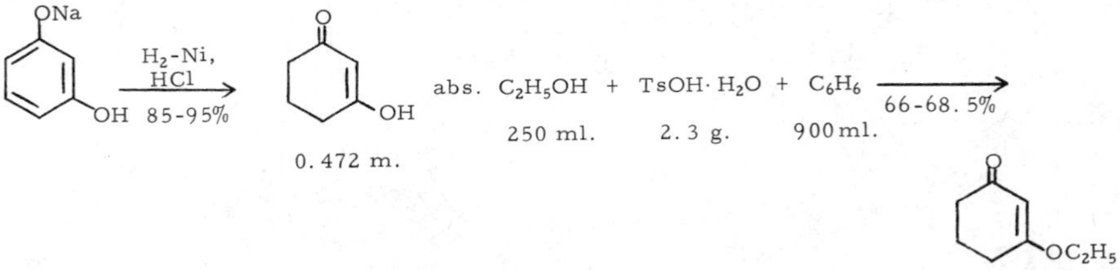

Ketalization. Conversion of glycerol into the acetonide was accomplished by refluxing and stirring a mixture of the components and catalyst in 300 ml. of petroleum ether (b.p. 35–55°) for 21–36 hrs. under a water separator.[8] The mixture was neutralized with sodium acetate, filtered, evaporated, and the product distilled.

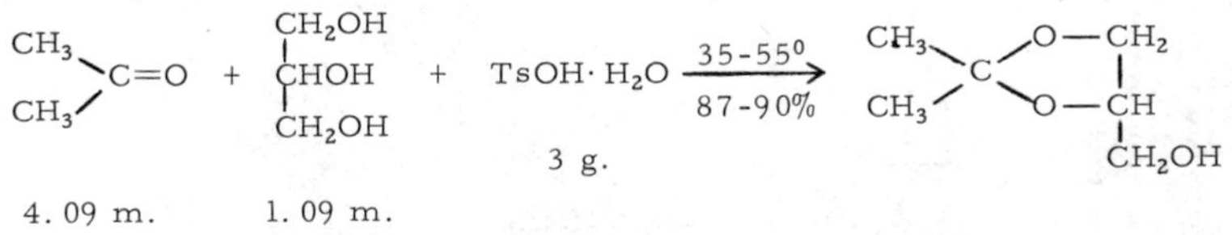

The conversion of a Δ^4-3-ketosteroid into the ethyleneketal is attended with migration of the double bond from the 4,5- to the 5,6-position.[9] In applying the reaction to cholestenone, a Lederle group[10] took the reagents in the amounts shown and

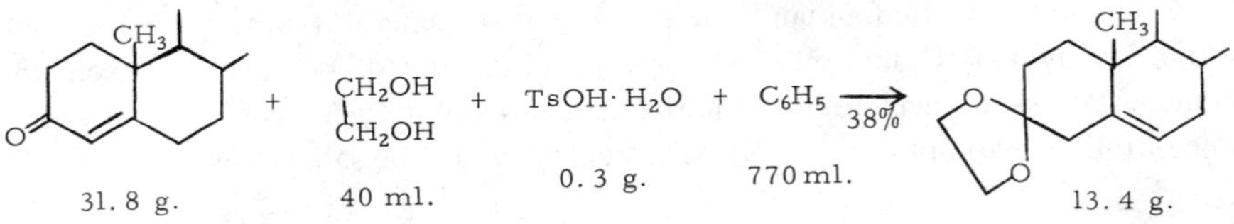

stirred the mixture under reflux for 4.5 hrs. (continuous water-removal adapter). The cooled solution was treated with saturated sodium bicarbonate solution, and the product recovered from the benzene layer and crystallized. The yield given is for material melting about 2° low and is very low. The probable explanation emerged from later work by Petersen and Sowers,[11] who studied the effect of varying the proportion of catalyst in experiments with 1.02 g. of cholestenone. With only 0.0052 g.

of *p*-toluenesulfonic acid monohydrate, the product, obtained pure in 77% yield, was the unrearranged Δ^4-cholestene-3-one ethylene ketal. To obtain the rearranged product, Δ^5-cholestene-3-one ethylene ketal, it was necessary to increase the amount of catalyst to 0.05 g. Since the amount of catalyst used by the Lederle workers was intermediate between these two values, it evidently led to a mixture of the two ketals.

Tosic acid has been used similarly for the preparation of ethylenethioketals. It is a milder catalyst than boron fluoride etherate for condensation of a ketone with ethanedithiol and has been used, in acetic acid solution, for selective conversions, for example of Δ^4-androstene-3,11,17-trione (0.60 g.) into the 3-ethyleneketal (0.40 g.).[12]

Acetylation. Cholestane-3β,5α,6β-triol on acetylation with acetic anhydride in pyridine affords the 3,6-diacetate; the triacetate can be prepared by brief heating

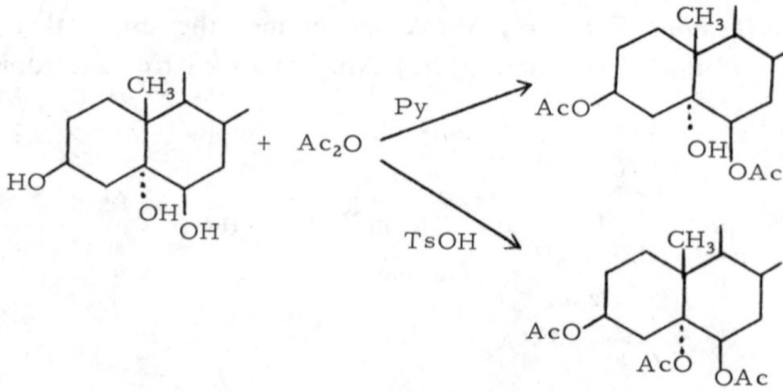

CH3
HO
OH
OH
+ Ac2O
Py
TsOH
AcO
OH
OAc
AcO
AcO
OAc

of the triol with acetic anhydride and tosic acid at 100°.[13] A number of steroids having a 5β-hydroxyl group have resisted attempted acetylation by all known methods. Turner[14] and a Merck group[15] independently found that the tertiary 17α-hydroxyl group of a steroid can be acetylated readily at room temperature with acetic anhydride

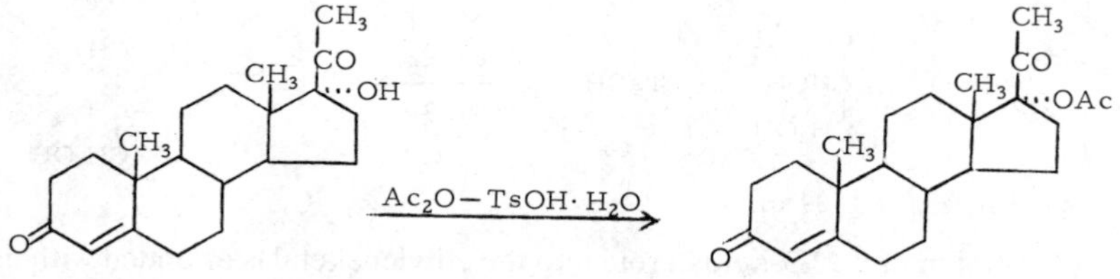

CH3
CO
OH
OAc
Ac2O—TsOH·H2O

and tosic acid as catalyst. 17α-Acetoxyprogesterone became readily available by this reaction and is a useful oral progestogen.

Enolacetylation. Bedoukian,[16] in developing a general synthesis of α-bromoketones, sought a reliable catalyst for the conversion of ketones into their enol acetates. Potassium acetate, phosphoric acid, and sulfuric acid were tried in various concentrations without success, but tosic acid proved to be satisfactory.

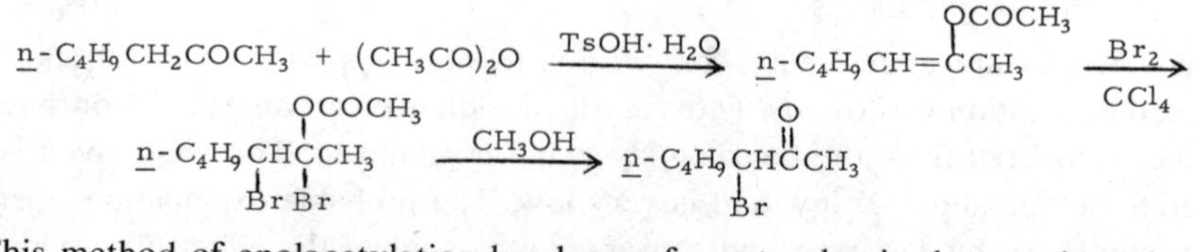

n-C4H9CH2COCH3 + (CH3CO)2O
TsOH·H2O
n-C4H9CH=CCH3
OCOCH3
Br2
CCl4
n-C4H9CHCCH3
Br Br
CH3OH
n-C4H9CHCCH3
Br

This method of enolacetylation has been of great value in the steroid series, for example in Gallagher's synthesis of 17α-hydroxycorticoids,[17] including cortisone,[18] through the sequence (1) ⟶ (4). This procedure for enolacetylation, described in detail only for monoketones, is as follows.[19] A solution of 2 mmoles of the ketone

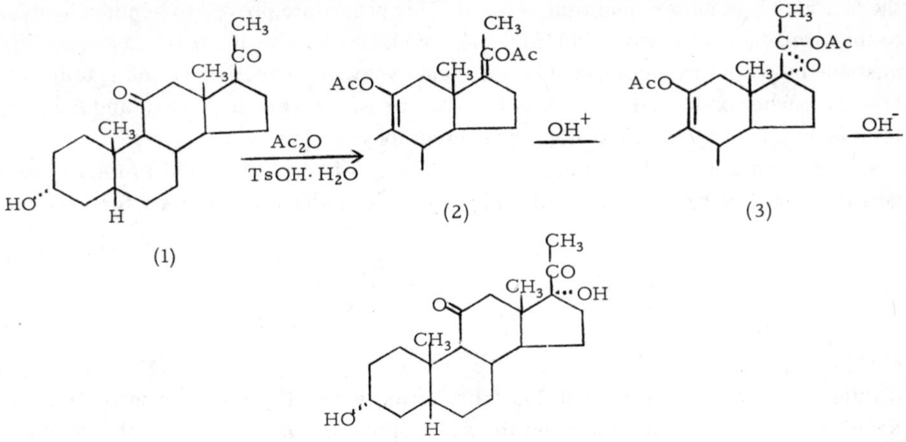

(4)

and 2 mmoles of $TsOH \cdot H_2O$ in 75 ml. of acetic anhydride was distilled slowly through an unpacked column until most of the acetic anhydride had been removed (4–5 hrs.). After cooling and addition of water, the product was extracted with ether and chromatographed; yields 60–70%.

Enamines. In preparing the morpholine enamine from cyclohexanone, Hünig and co-workers[20] carried out the azeotropic distillation with a water separator

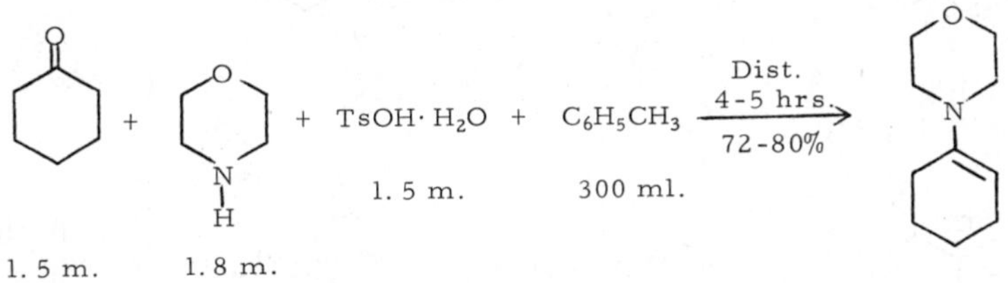

described by Natelson and Gottfried.[21] They used excess morpholine because the water that separates contains a substantial amount of this base.

Dehydration. Johnson and co-workers[22] in a study of the energy difference between the chair and boat forms of cyclohexane required the isomeric lactones (2) and (4) and were able to synthesize the required γ-hydroxy acids (1) and (3). Lactonization of (1) was accomplished by heating a solution of 1 g. of (1) and 0.4 g. of

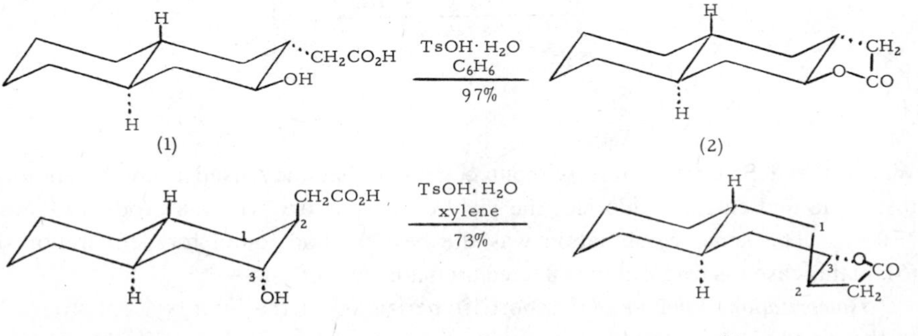

(3) (4)

$TsOH \cdot H_2O$ in 600 ml. of benzene at reflux with stirring under nitrogen for 10 min. and collecting the distillate in a water separator. Workup and crystallization afforded the lactone (2) in nearly quantitative yield. This procedure proved to be inapplicable to the *trans*-diaxial hydroxy acid (3), which can lactonize only if one ring assumes the unstable boat conformation (4). The same catalyst was ineffective in boiling toluene; however, when xylene was used as the solvent water was eliminated, and lactone (4) was obtained in 73% yield after chromatography.

Cyclodehydration of butane-1,2,4-triol to 3-hydroxytetrahydrofurane was accomplished by heating the triol with tosic acid under a column set for vacuum

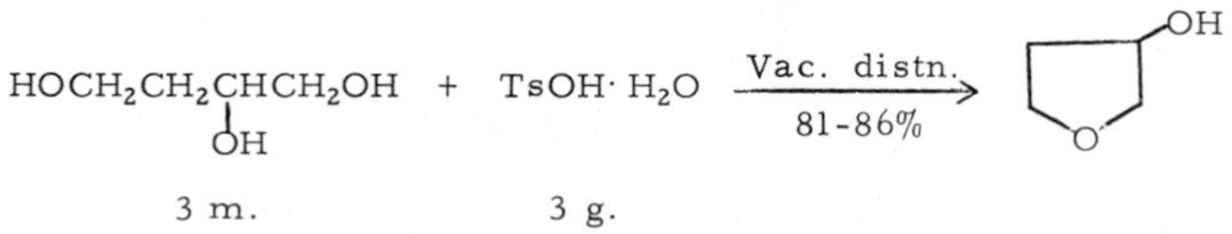
HOCH₂CH₂CHCH₂OH + TsOH· H₂O
OH
Vac. distn.
81-86%
OH
O
3 m.
3 g.

distillation.[23] Distillation over a 2–2.5 hr. period gave 300–306 g. of material, b.p. 85–87°/22 mm. Redistillation from the same apparatus gave 50–60 g. of a fraction boiling at 42–44°/24 mm. and regarded as mainly water, a small intermediate fraction, and 215–231 g. of the pure product, b.p. 93–95°/26 mm.

Tosic acid in boiling benzene was the most satisfactory reagent found for dehydration of the ketol ester (1).[24] In reporting the dehydration of the ketol (3) to (4),

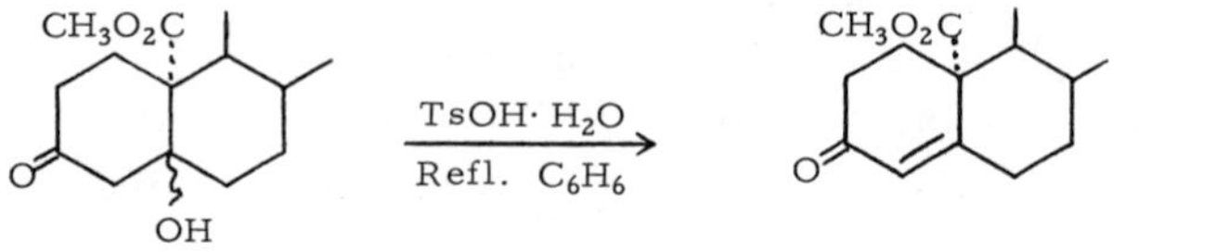
CH₃O₂C
O
OH
TsOH· H₂O
Refl. C₆H₆
CH₃O₂C
O

(1) (2)

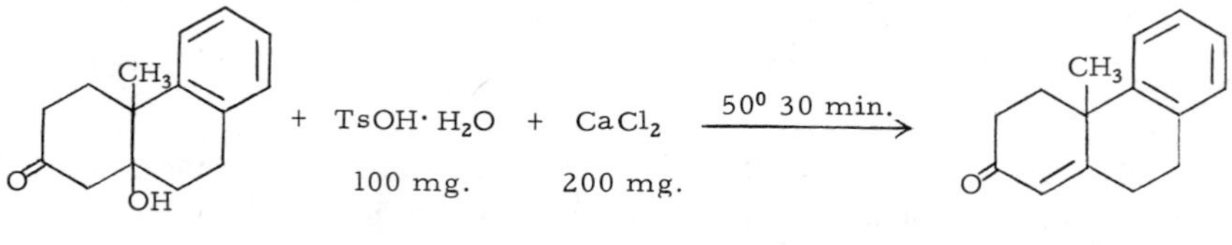
CH₃
O
OH
+ TsOH· H₂O + CaCl₂
100 mg. 200 mg.
50⁰ 30 min.
CH₃
O

(3) 200 mg.
in 25 ml. C_6H_6

(4)

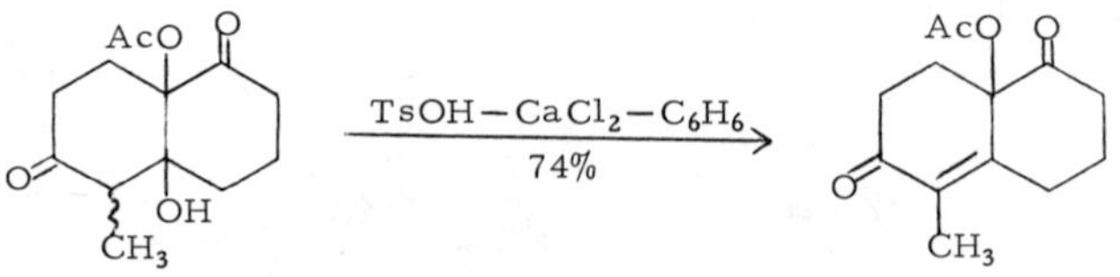
AcO
O
O
OH
CH₃
TsOH—CaCl₂—C₆H₆
74%
AcO
O
O
CH₃

(5) (6)

Wenkert and Stevens[25] state without comment that they used a combination of tosic acid and calcium chloride; the yield of satisfactory reaction product (4) was 170 mg. The same combination was the best method found for dehydration of (5); in this case tosic acid alone effected aromatization of ring A.[26]

Isomerization. Isler *et al.*[27] report isomerization of the hydroxyisophorone (2) to the dione (3) by refluxing a solution of 100 g. of (2) and 2 g. of $TsOH \cdot H_2O$ in

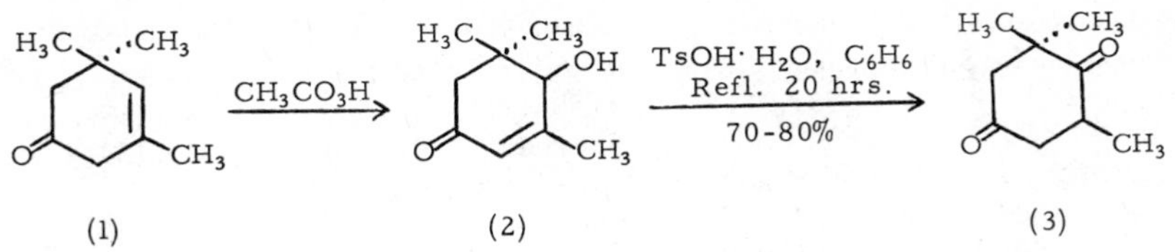
H₃C
CH₃
CH₃CO₃H
OH
TsOH·H₂O, C₆H₆
Refl. 20 hrs.
70-80%
(1)
(2)
(3)

200 ml. of benzene. After shaking with solid sodium bicarbonate the solution was filtered through a column of alumina, concentrated and diluted with petroleum ether; the dione (3) separated as white needles.

Aromatic phenylation. Kaslow and Summers[28] prepared 3-nitrodiphenyl from *m*-nitroaniline by diazotization, conversion to *m*-nitrophenyl-N,N-dimethyltriazine, and reaction of this with benzene in the presence of tosic acid as catalyst. A solution

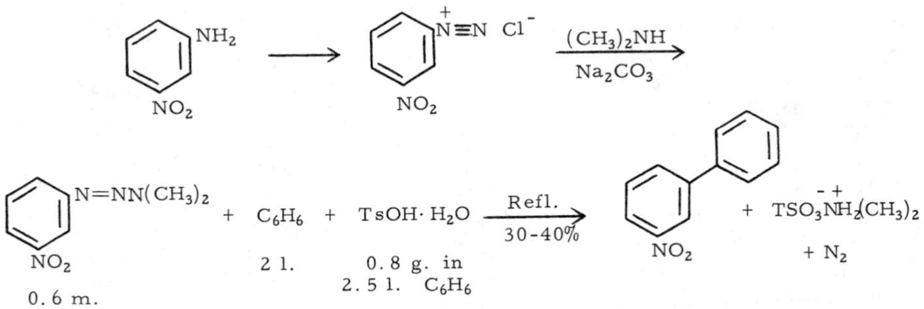
NH₂
NO₂
N≡N Cl⁻
(CH₃)₂NH
Na₂CO₃
N=NN(CH₃)₂
+ C₆H₆ + TsOH·H₂O
Refl.
30-40%
+ TSO₃NH₂(CH₃)₂
+ N₂
0.6 m.
2 l.
0.8 g. in
2.5 l. C₆H₆

of the triazine in benzene was refluxed with stirring during slow addition of a benzene solution of the catalyst. At the end of the reaction water was added and the product recovered from the benzene layer, distilled at 0.1 mm., and crystallized.

p-Toluenesulfonic anhydride. A procedure[29] for preparing this derivative calls for use of inert additive to facilitate mixing of the solid reactants and extraction of

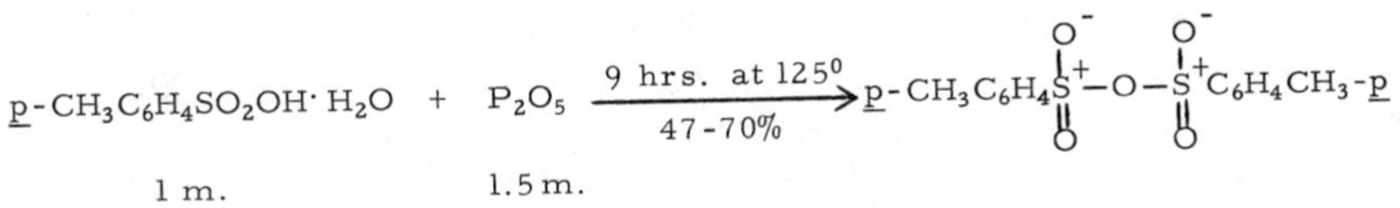
p-CH₃C₆H₄SO₂OH·H₂O + P₂O₅
9 hrs. at 125°
47-70%
p-CH₃C₆H₄S-O-SC₆H₄CH₃-p
1 m.
1.5 m.

the product: kieselguhr is mixed with P_2O_5 and asbestos is mixed with tosic acid. The total mixture gets hot and after a time is heated in an oil bath. The anhydride is recovered by repeated extraction with ethylene dichloride and crystallized from benzene–ether.

Alcoholysis of nitriles. Tosic acid is regarded as the most satisfactory reagent for this reaction.[30] A mixture of equimolecular amounts of the nitrile, the alcohol, and

$$RCN + R'OH + TsOH \cdot H_2O \xrightarrow[25\text{-}85\%]{\text{Refl.}} RCO_2R' + TsO^-NH_4^+$$

tosic acid is refluxed for several hours, then water is added to dissolve the ammonium salt, and the ester is isolated in the usual way.

[1]C. F. H. Allen and F. W. Spangler, *Org. Syn., Coll. Vol.*, **3**, 203 (1955)
[2]For drawing of apparatus, see *ibid.*, **1**, 422 (1941)
[3]*Synth. Org. Chem.*, **9**, No. 3 (May 1936), Eastman Kodak Co.
[4]A. Weissberger and C. J. Kibler, *Org. Syn., Coll. Vol.*, **3**, 610 (1943)
[5]A. C. Cope and E. C. Herrick, *ibid.*, **4**, 304 (1963)
[6]C. E. Rehberg, *ibid.*, **3**, 146 (1955)

[7]W. F. Gannon and H. O. House, *Org. Syn.*, **40**, 41 (1960)
[8]M. Renoll and M. S. Newman, *Org. Syn., Coll. Vol.*, **3**, 502 (1955)
[9]E. Fernholtz, U. S. Patent 2,378,918 [*C.A.*, **39**, 5051 (1945)]
[10]R. Antonucci, S. Bernstein, R. Littell, K. J. Sax and J. H. Williams, *J. Org.*, **17**, 1341 (1952)
[11]Q. R. Petersen and E. E. Sowers, *J. Org.*, **29**, 1627 (1964)
[12]J. W. Ralls and B. Riegel, *Am. Soc.*, **76**, 4479 (1952)
[13]M. Davis and V. Petrow, *J. Chem. Soc.*, 2536 (1949)
[14]R. B. Turner, *Am. Soc.*, **74**, 4220 (1952)
[15]Huang-Minlon, E. Wilson, N. L. Wendler, and M. Tishler, *Am. Soc.*, **74**, 5394 (1952)
[16]P. Z. Bedoukian, *Am. Soc.*, **67**, 1430 (1945)
[17]T. H. Kritchevsky and T. F. Gallagher, *Am. Soc.*, **73**, 184 (1951); B. A. Koechlin, T. H. Kritchevsky, and T. F. Gallagher, *ibid.*, **73**, 189 (1951)
[18]T. H. Kritchevsky, D. L. Garmaise, and T. F. Gallagher, *Am. Soc.*, **74**, 483 (1952)
[19]C. W. Marshall, T. H. Kritchevsky, S. Lieberman, and T. F. Gallagher, *Am. Soc.*, **70**, 1837 (1948)
[20]S. Hünig, E. Lücke, and W. Brenninger, *Org. Syn.*, **41**, 65 (1961)
[21]S. Natelson and S. Gottfried, *Org. Syn., Coll. Vol.*, **3**, 381 (1955)
[22]W. S. Johnson, V. J. Bauer, J. L. Margrave, M. A. Frisch, L. H. Dreger, and W. N. Hubbard, *Am. Soc.*, **83**, 606 (1961)
[23]H. Wynberg and A. Bantjes, *Org. Syn., Coll. Vol.*, **4**, 534 (1963)
[24]F. Sondheimer, R. Mechoulam, and M. Sprecher, *Tetrahedron*, **20**, 2473 (1964)
[25]E. Wenkert and T. E. Stevens, *Am. Soc.*, **78**, 2318 (1956)
[26]T. A. Spencer, K. K. Schmiegel, and W. W. Schmiegel, *J. Org.*, **30**, 1626 (1965)
[27]O. Isler, H. Lindlar, M. Montavon, R. Rüegg, G. Saucy, and P. Zeller, *Helv.*, **39**, 2041 (1956)
[28]C. E. Kaslow and R. M. Summers, *Org. Syn., Coll. Vol.*, **4**, 718 (1963)
[29]L. Field and J. W. McFarland, *ibid.*, **4**, 940 (1963)
[30]F. L. James and W. H. Bregan, *J. Org.*, **23**, 1225 (1958)

p-Toluenesulfonyl azide (Tosyl azide),

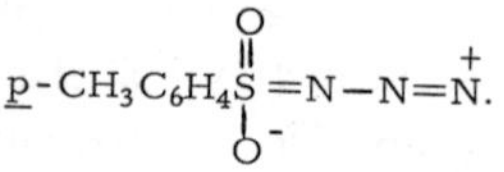

Mol. wt. 197.22, m.p. 22°.

Preparation from tosyl chloride and sodium azide in 83% yield.[1,2]

Diazo compounds. Doering and De Puy[2] treated cyclopentadienyllithium with tosyl azide and obtained diazocyclopentadiene, a remarkably stable dark red compound (b.p. 53°/50 mm.) in 35% yield. The initially formed triazine decomposes

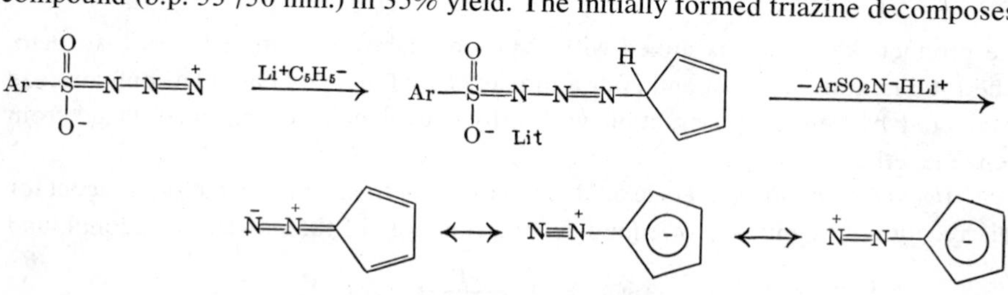

spontaneously with elimination of the lithium salt of *p*-toluenesulfonamide. The reagent reacts with β-diketones under basic catalysis (potassium ethoxide, aqueous-ethanolic KOH) to give α-diazo-β-diketones.[3] Thus dimedone yields 2-diazodimedone

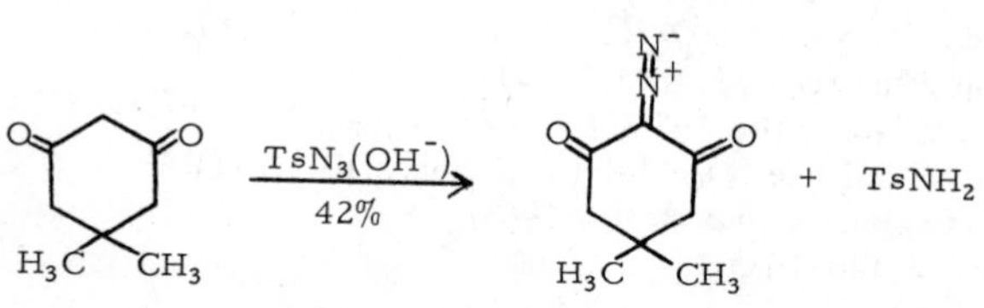

as bright yellow needles, m.p. 108°. Anthrone is converted into 9-diazoanthrone (I) in high yield by magnetically stirring a mixture with tosyl azide in ethanol at room temperature and adding piperidine.[4] If 3 g. of 9-diazoanthrone (I, brown) is

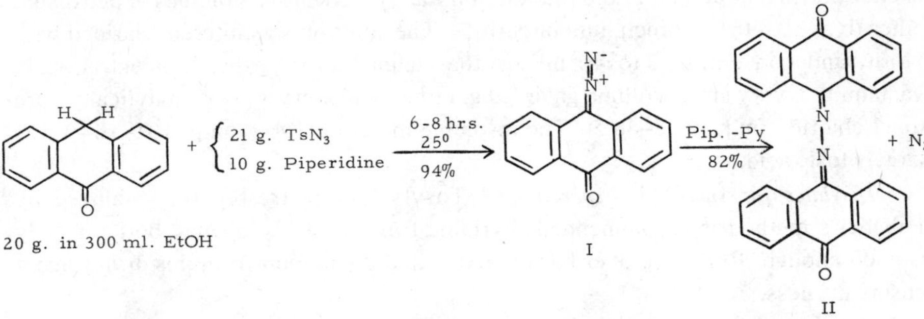
21 g. TsN3
10 g. Piperidine
6-8 hrs.
25°
94%
Pip. -Py
82%
+ N2
20 g. in 300 ml. EtOH
I
II

stirred with a mixture of 20 ml. each of piperidine and pyridine with gentle warming, nitrogen is evolved and the product is the orange-red anthraquinoneazine (II).

Tedder and Webster[5] noted a reaction analogous to that of Doering and De Puy.[2] The dry sodium salt of β-naphthol reacts with tosyl azide in ether to give the triazine

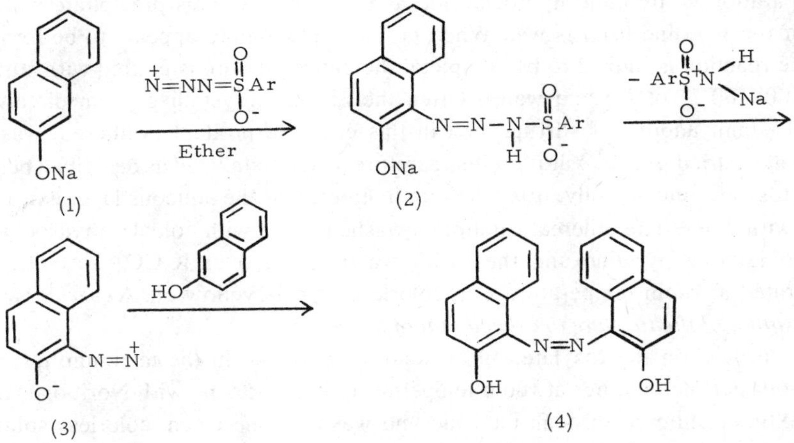
Ether
ONa
(1)
(2)
(3)
(4)
HO
OH
N=N

(2), which decomposes spontaneously with elimination of sodium *p*-toluenesulfonamide and formation of the diazo-oxide (3); in the alkaline medium this couples with β-naphthol to give 2,2′-dihydroxy-1,1′-azonaphthalene (4). The reaction is restricted to the more reactive phenols: the naphthols, resorcinol, phloroglucinol, and 3-methoxyphenol. The salt of a less reactive phenol yields the *p*-toluenesulfonate and sodium azide.

[1]T. Curtius and G. Kraemer, *J. prakt. Chem.*, **125**, 323 (1930)
[2]W. von E. Doering and C. H. De Puy, *Am. Soc.*, **75**, 5955 (1953)
[3]M. Regitz, *Ann.*, **676**, 101 (1964)
[4]M. Regitz, *Ber.*, **97**, 2742 (1964)
[5]J. M. Tedder and B. Webster, *J. Chem. Soc.*, 4417 (1960)

***p*-Toluenesulfonyl chloride (Tosyl chloride).** Mol. wt. 190.65, m.p. 67.5–68.5°, b.p. 146°/15 mm. Suppliers: A, E, F, MCB.

Purification. Commercial material of premium grade is satisfactory if taken from a freshly opened bottle but on long storage without protection may contain substantial amounts of tosic acid. Pelletier,[1] having tried out two published procedures for purification and found them unsatisfactory, devised the following method. Ten grams of very impure chloride, m.p. 66–86°, was dissolved in the minimum amount of chloroform (about 25 ml.) and the solution was diluted with 5 volumes of petroleum ether (b.p. 30–60°) to precipitate impurities. The solution was filtered, clarified with Norit, and concentrated to 40 ml. on the steam bath. Further concentration in vacuum to a very small volume gave 7.0 g. of fine white crystals of analytically pure tosyl chloride, m.p. 67.5–68.5°. The insoluble material (1.4 g., m.p. 101–104°) was largely tosic acid.

Tosylates for solvolysis experiments.[2] Tosyl chloride freshly recrystallized by Pelletier's method is recommended. Pyridine from a freshly opened bottle usually is good enough. Refluxing over barium oxide and distillation from this drying agent ensure dryness.

A solution of 3–5 g. of the alcohol in 50–75 ml. of pyridine in a 125-ml. glass-stoppered Erlenmeyer is cooled to 0° and treated with 1 molar excess of tosyl chloride.[3] After solution is complete the flask is placed in a refrigerator for 12–24 hrs. (a hindered alcohol may require a longer period). The reaction can be followed by the development of color (yellow, brown, and often pinkish purple), followed by separation of pyridine hydrochloride as long needles. This precipitate will not form if the pyridine used is wet. When no more precipitate appears to be forming and the reaction is judged to be complete, the entire mixture is poured with stirring into 300–400 g. of ice and water. Often the tosylate crystallizes immediately or after 15 min. additional stirring, and in this case the product is filtered, washed with water, dried *in vacuo* at room temperature, and crystallized as described below. If the tosylate remains oily, it is taken up in ether, and the aqueous layer extracted twice with ether. The ethereal solution is washed twice with cold 1:1 hydrochloric acid to remove pyridine and then with water, dried over K_2CO_3–Na_2SO_4, and evaporated at room temperature to a colorless or pale yellow oil. *Never allow the temperature of the tosylate to exceed that of the room.*

For purification the tosylate, oil or solid, is dissolved in the minimum quantity of 30–60° petroleum ether at room temperature. After stirring with Norit or Darco, the mixture is filtered through filter aid and washed. The clean, colorless solution is cooled slowly to −75° in a dry ice–acetone bath with scratching to induce crystallization and to avoid oiling out. Once crystallization starts it proceeds very well, but cooling to −75° is completed before filtering. The precipitate is not sucked completely dry but rather transferred to a vial for drying *in vacuo* at or below room temperature.

If all attempts to induce crystallization fail, the oiling out process is completed by cooling to −75° to produce a white glass which sometimes slowly solidifies to crystals on standing at −75°; in this case the solid is broken up and filtered as above. If the compound remains a viscous glass the supernatant liquid is poured off; tosylates are so sparingly soluble in petroleum ether that little except contaminants is lost by this process. Indeed an analytically pure oil often can be obtained by repeating the process one or two more times and drying the oil *in vacuo* at room temperature.

Although many tosylates are stable at room temperature, they all should be stored in the cold. Some tosylates, particularly unsaturated ones, decompose in a few hours at 25°. Such unstable esters should be prepared just prior to use, and samples sent out for analysis should be marked "Unstable, analyze promptly."

Tosylates show a sharp and intense doublet of IR bands at 8.4 and 8.5 μ.

Very few tosylates melt without some decomposition. They give yellow or dark colored melts. Sometimes the melting point depends upon the temperature of the bath before insertion of the sample or upon the rate of heating.

In case a tosylate cannot be obtained analytically pure or cannot be analyzed, solvolysis experiments may still be valid because the linearity of the first order plot and a comparison of the experimental infinity titer with that calculated are checks on purity. Also tosylates prepared from a mixture of isomers may yield valuable information; if the rates of the components differ by a factor of 10 or more the mixture can be analyzed kinetically.[4]

Selective tosylation. One step of Johnson's synthesis of aldosterone[5] required selective tosylation of the primary alcoholic group of the ketal diol (1) in order that the secondary group could be oxidized to produce the keto tosylate (2) required for

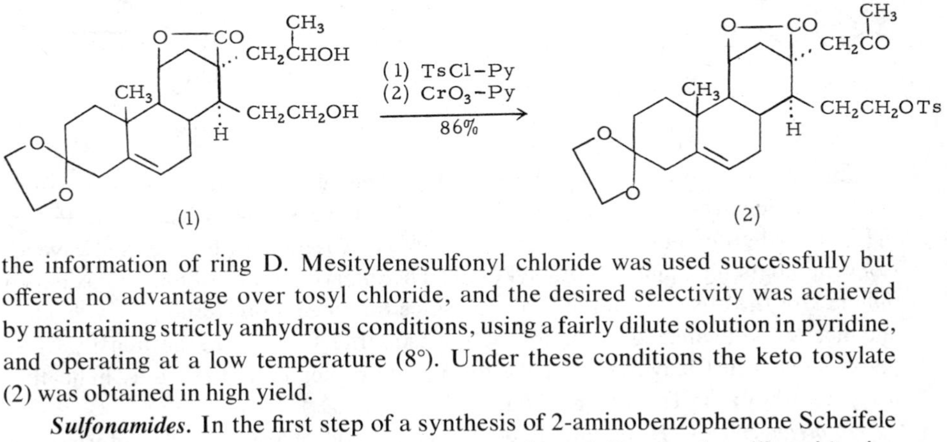
CH3
O
CO
CH2CHOH
CH3
CH2CH2OH
H
O
O
(1) TsCl-Py
(2) CrO3-Py
86%
CH3
O
CO
CH2CO
CH3
CH2CH2OTs
H
O
O

(1) (2)

the information of ring D. Mesitylenesulfonyl chloride was used successfully but offered no advantage over tosyl chloride, and the desired selectivity was achieved by maintaining strictly anhydrous conditions, using a fairly dilute solution in pyridine, and operating at a low temperature (8°). Under these conditions the keto tosylate (2) was obtained in high yield.

Sulfonamides. In the first step of a synthesis of 2-aminobenzophenone Scheifele and De Tar[6] converted anthranilic acid into *p*-toluenesulfonylanthranilic acid using sodium carbonate as base; if sodium hydroxide is used the main product is the tosic

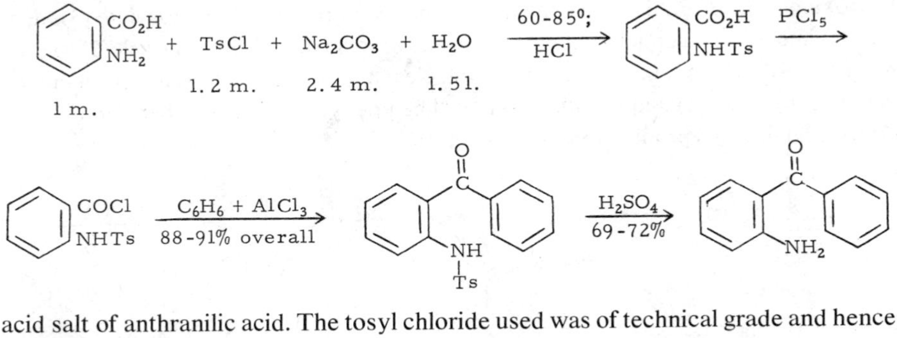
CO2H
NH2
+ TsCl + Na2CO3 + H2O
1 m.
1. 2 m.
2. 4 m.
1. 5 l.
60-85°;
HCl
CO2H
NHTs
PCl5
COCl
NHTs
C6H6 + AlCl3
88-91% overall
O
C
NH
Ts
H2SO4
69-72%
O
C
NH2

acid salt of anthranilic acid. The tosyl chloride used was of technical grade and hence was taken in 20% excess.

In a procedure for the preparation of their diazomethane precursor, *p*-tolylsulfonylmethylnitrosamide, de Boer and Backer[7] carried out the reaction of tosyl chloride with methylamine by the Schotten-Baumann procedure, that is with the alternating addition of acid chloride and sodium hydroxide solution.

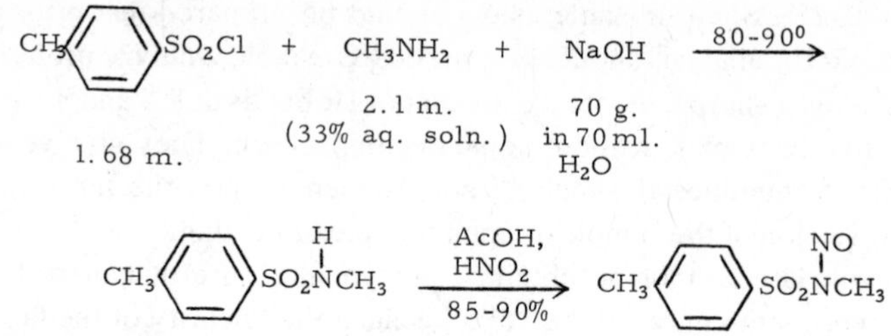
CH3 SO2Cl + CH3NH2 + NaOH 80-90°
2.1 m. (33% aq. soln.)
70 g. in 70 ml. H2O
1.68 m.
CH3 SO2NCH3 H
AcOH, HNO2
85-90%
CH3 SO2NCH3 NO

Conversion of formamides into isonitriles. In the course of a synthesis of dihydroconessine from 3β-acetoxy-Δ^5-bisnorcholenic acid, Corey and Hertler[8] treated the formamide (1) with tosyl chloride in pyridine in the expectation of merely tosylating the 3β-hydroxyl group. The product, however, was not (3) but the tosylate isonitrile

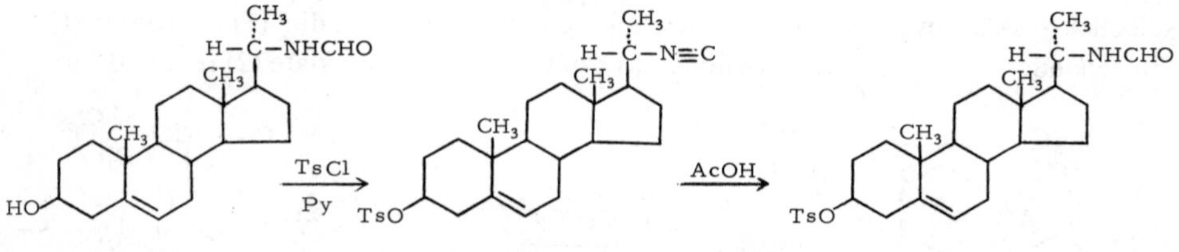
H–C–NHCHO
TsCl
Py
H–C–N≡C
AcOH
H–C–NHCHO
HO
TsO
TsO

(1) (2) (3)

(2); in the first step the TsCl–Py reagent had removed the elements of water from the formamide group to produce an isonitrile and excess reagent had then esterfied the hydroxyl group at C_3. Treatment of (1) with just 1 equivalent of reagent gave an isonitrile containing almost no tosyl ester. The intervention of isonitrile was easily nullified for the objective of the synthesis since treatment of the crude (2) with ether and acetic acid effected hydration of the isonitrile to give the desired (3). The unexpected facile dehydration of the formamide prompted a brief study indicating the generality of the TsCl–Py method.[9]

Rearrangements. Hurd and Bauer[10] added tosyl chloride in chloroform to a suspension of the sodium salt of benzohydroxamic acid (1) in the same solvent, noted a vigorous reaction, and characterized the product as benzo-(phenylcarbonylhydroxamic) acid (6). They postulate that the initial product is the tosylate (3), that this is a much stronger acid than (1) and so abstracts sodium from the salt (2) to give the salt (4), which undergoes concerted cleavage and Lossen rearrangement to phenylisocyanate (5) and sodium tosylate. The phenylisocyanate is then captured by (1) to yield the final product (6).

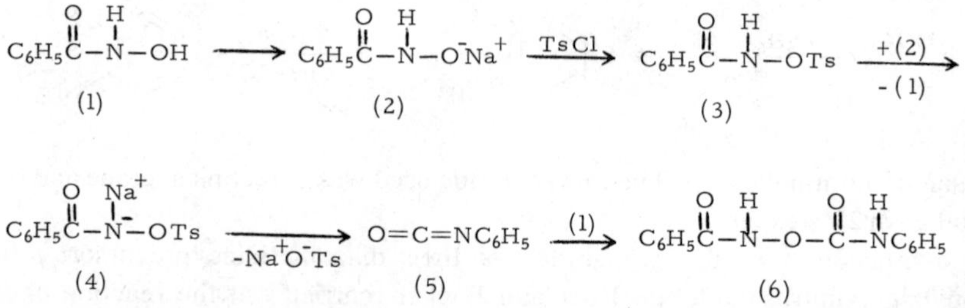
C6H5C–N–OH
(1)
C6H5C–N–O⁻Na⁺
(2)
TsCl
C6H5C–N–OTs
(3)
+(2)
-(1)
C6H5C–N–OTs
(4)
-Na⁺O⁻Ts
O=C=NC6H5
(5)
(1)
C6H5C–N–O–C–NC6H5
(6)

Tosyl chloride is also useful as a reagent for effecting Beckmann rearrangements,[11,12] for example of 9-acetyl-*cis*-decalin oxime.[12]

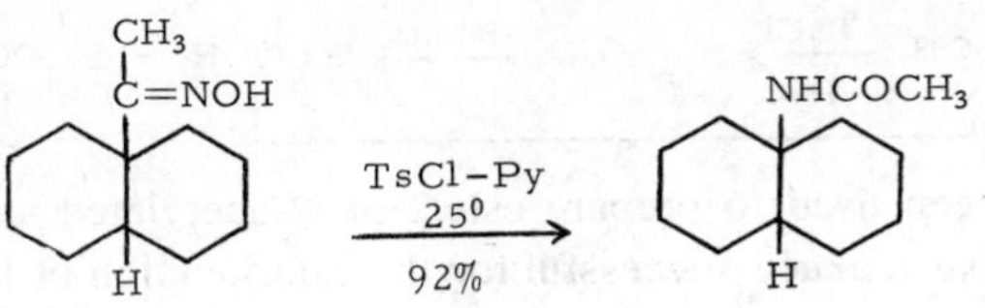

Methyl p-tolyl sulfone. In the method of Field and Clark[13] tosyl chloride is reduced with sodium sulfite in a solution buffered with sodium bicarbonate to sodium

$$\underline{p}\text{-}CH_3C_6H_4\overset{O}{\overset{\|}{S^+}}(O^-)-Cl \;+\; Na_2SO_3 \;+\; NaHCO_3 \;+\; H_2O \xrightarrow{\text{Stir at 70-80}^0\text{ for 3 hrs.}}$$
$$\text{2.54 m.} \qquad \text{4.76 m.} \qquad \text{5 m.} \qquad \text{2.4 l.}$$

$$\underline{p}\text{-}CH_3C_6H_4\overset{O}{\overset{\|}{S}}ONa \xrightarrow[78\text{-}82\%]{(CH_3O)_2SO_2 \;+\; NaHCO_3} \underline{p}\text{-}CH_3C_6H_4SO_2CH_3$$

p-toluenesulfinate, which separates as a solid. The salt is collected and stirred with dimethyl sulfate, sodium bicarbonate, and water for conversion to the sulfone, which is extracted with benzene.

Dehydration. Wintersteiner and Moore[14] effected dehydration of 3β-acetoxy-cholestane-7α-ol by reaction with tosyl chloride in refluxing pyridine and obtained a mixture of stenyl acetates rich in the Δ^7-isomer and convertible into pure $\Delta^{8(14)}$-cholestenyl acetate on catalytic rearrangement.

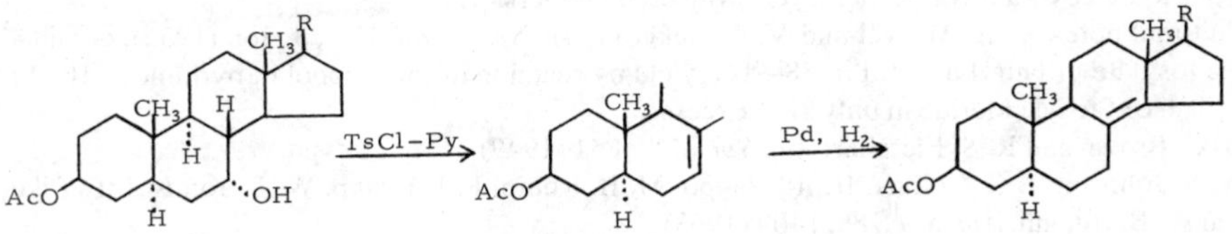

(and isomers)

In structural studies on oxytetracycline (Terramycin) and chlorotetracycline (Aureomycin) a Pfizer group[15] found that tosyl chloride in cold pyridine converts the antibiotics into nitrile derivatives through dehydration of the primary carbox-amido group. They then found that simpler amides are dehydrated to nitriles in good yield by adding 1 equivalent of tosyl chloride slowly to a mixture of the amide with 2.25 equivalents of pyridine at a rate such as to maintain a temperature of about 70°. The mechanism proposed involves O-tosylation.

$$R\overset{O}{\overset{\|}{C}}-NH_2 \xrightarrow[-Py\cdot HCl]{ArSO_2Cl,\ Py} R\overset{OSO_2Ar}{\overset{|}{C}}=NH \xrightarrow{Py} RC\equiv N \;+\; Py^+H(ArSO_3^-)$$

Sheehan and co-workers[16] used tosyl chloride and triethylamine to effect dehydration of ureas to carbodiimides.

$$R\overset{H}{\overset{|}{N}}-\overset{O}{\overset{\|}{C}}-\overset{H}{\overset{|}{N}}R' \xrightarrow[(C_2H_5)_3N]{TsCl} RN=C=NR'$$

Esterification. Esters are formed in high yield when a solution of an acid (1 equiv.) and an alcohol (1 equiv.) in pyridine is treated with tosyl chloride (2 equivs.).[17] The reaction is considered to involve intermediate formation of the acid anhydride.

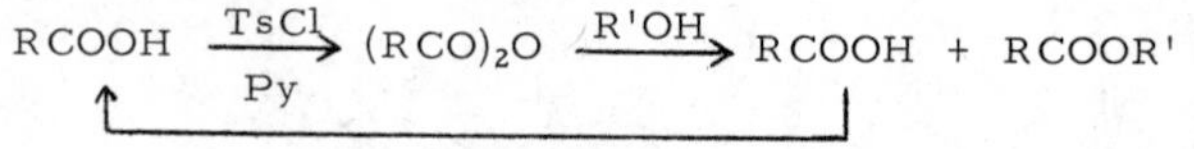
RCOOH
TsCl
Py
(RCO)₂O
R'OH
RCOOH + RCOOR'

The method has been used to prepare esters of N-acetylated amino acids (yields 51–97%)[18] and is particularly successful for the esterification of tertiary acetylenic alcohols.[19]

N-Protective group. Tosyl chloride reacts with amino acids in alkaline solution to give N-tosylamino acids.[20] The protective group is removed by reductive cleavage with sodium in liquid ammonia. The tosyl group is particularly useful for protection of the ω-amino groups of lysine and ornithin peptides.

Decarbonylation. Sheehan and Frankenfeld[21] treated α-anilino-α,α-diphenylacetic acid (1) with tosyl chloride in pyridine with the expectation of obtaining the N-tosyl derivative but instead characterized the products as benzophenone anil (3), carbon monoxide, and pyridinium tosylate. They postulate the intermediate formation of the mixed anhydride (2).

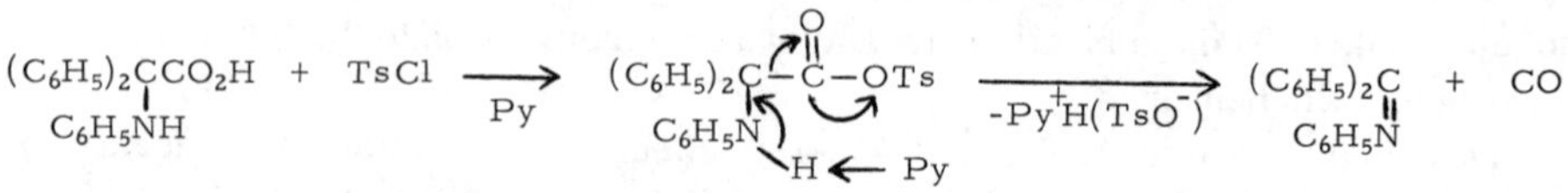
(C₆H₅)₂CCO₂H
C₆H₅NH
+ TsCl
Py
(C₆H₅)₂C
C–OTs
O
C₆H₅N
H ← Py
-PyH(TsO)
(C₆H₅)₂C
C₆H₅N
+ CO

(1) (2) (3)

[1]S. W. Pelletier, *Chem. Ind.*, 1034 (1953)
[2]Contributed by Paul von R. Schleyer, Princeton University
[3]Editorial note – C. S. Marvel and V. C. Sekera, *Org. Syn., Coll. Vol.*, **3**, 366 (1955), obtained the tosylate of lauryl alcohol in 88–90% yield by reaction of the alcohol in pyridine at 10–20° with use of tosyl chloride in only 10% excess.
[4]H. C. Brown and R. S. Fletcher, *Am. Soc.*, **71**, 1851 (1949)
[5]W. S. Johnson, J. C. Collins, Jr., R. Pappo, M. B. Rubin, P. J. Kropp, W. F. Johns, J. E. Pike, and W. Bartmann, *Am. Soc.*, **85**, 1409 (1963)
[6]H. J. Scheifele and D. F. De Tar, *Org. Syn., Coll. Vol.*, **4**, 34 (1963)
[7]Th. J. de Boer and H. J. Backer, *ibid.*, **4**, 943 (1963)
[8]E. J. Corey and W. R. Hertler, *Am. Soc.*, **81**, 5209 (1959)
[9]W. R. Hertler and E. J. Corey, *J. Org.*, **23**, 1221 (1958)
[10]C. D. Hurd and L. Bauer, *Am. Soc.*, **76**, 2791 (1954)
[11]P. Oxley and W. F. Short, *J. Chem. Soc.*, 382 (1947)
[12]R. K. Hill and O. T. Chortyk, *Am. Soc.*, **84**, 1065 (1962)
[13]L. Field and R. D. Clark, *Org. Syn., Coll. Vol.*, **4**, 674 (1963)
[14]O. Wintersteiner and M. Moore, *Am. Soc.*, **65**, 1503, 1507 (1943)
[15]C. R. Stephens, E. J. Bianco, and F. J. Pilgrim, *Am. Soc.*, **77**, 1701 (1955)
[16]J. C. Sheehan, P. A. Cruickshank, and G. L. Boshart, *J. Org.*, **26**, 2525 (1961)
[17]J. H. Brewster and C. J. Ciotti, Jr., *Am. Soc.*, **77**, 6214 (1955)
[18]G. Blotny, J. F. Biernat, and E. Taschner, *Ann.*, **663**, 194 (1963)
[19]G. F. Hennion and S. O. Barrett, *Am. Soc.*, **79**, 2146 (1957); J. Klosa, *Angew. Chem.*, **69**, 135 (1957)
[20]R. A. Boissonnas, *Advances in Org. Chem.*, **3**, 175 (1963)
[21]J. C. Sheehan and J. W. Frankenfeld, *J. Org.*, **27**, 628 (1962)

_p_-Toluenesulfonyl chloride–Dimethylformamide. Tosyl chloride combines with DMF to give a complex formulated as follows:[1]

$$(CH_3)_2\overset{+}{N}{=}CHOSO_2C_6H_4CH_3\text{-}\underline{p}\ (\overset{-}{Cl})$$

(1)

It is an excellent reagent for the formylation of alcohols. Thus testosterone and tosyl chloride in DMF give testosterone formate in 79% yield.[2] The reagent reacts with arylamines to produce sulfonanilides (2) and/or formamidines (3).

$$(1) \xrightarrow{ArNH_2} \underline{p}\text{-}CH_3C_6H_4SO_2NHAr + (CH_3)_2\overset{+}{N}{=}CHNHAr\ (\underline{p}\text{-}CH_2C_6H_4SO_2\bar{O})$$

(2) (3)

[1]H. K. Hall, Jr., *Am. Soc.*, **78**, 2717 (1956)
[2]J. D. Albright, E. Benz, A. E. Lanzilotti, and L. Goldman, *Chem. Comm.*, 413 (1965)

2-*p*-Toluenesulfonylethyl chloroformate, *p*-$CH_3C_6H_4SO_2CH_2CH_2OCOCl$.[1] Mol. wt. 262.72, m.p. m.p. 48°.

Preparation. Sodium *p*-toluenesulfinate is heated with excess 2-chloroethanol in dimethylformamide, and the resulting sulfone in dry benzene is stirred with excess phosgene.

$$\underset{1\text{ m.}}{\underline{p}\text{-}CH_3C_6H_4SO_2Na} + \underset{3\text{ m.}}{ClCH_2CH_2OH} \xrightarrow[70\%]{HCON(CH_3)_2,\ 3\text{ hrs.}} \underline{p}\text{-}CH_3C_6H_4SO_2CH_2CH_2OH$$

$$\xrightarrow[82\%]{COCl_2 \text{ in } C_6H_6\ 0\text{-}25^0} \underline{p}\text{-}CH_3C_6H_4SO_2CH_2CH_2OCOCl$$

N-Protective group. The chloroformate reacts with an amino acid in aqueous dioxane in the presence of magnesium oxide to give the N-protected amino acid in yield of 85–92%. Unlike the usual N-protective groups, the *p*-toluenesulfonylethoxycarbonyl group is stable to acid and to catalytic hydrogenation. The group is removed when desired by treatment with excess aqueous ethanolic *N* sodium hydroxide at room temperature or with a strongly basic ion-exchange resin. It is used in peptide synthesis by both the azide and the *p*-nitrophenyl ester coupling procedures.

[1]A. T. Kader and C. J. M. Stirling, *J. Chem. Soc.*, 258 (1964)

***p*-Toluenesulfonylhydrazide**, *p*-$CH_3C_6H_4SO_2NHNH_2$. (Tosylhydrazide, $TsNHNH_2$). Mol. wt. 186.23, m.p. 104–107°. Suppliers: A, E, F.

Preparation.[1] A solution of tosyl chloride in tetrahydrofurane is stirred at 10°,

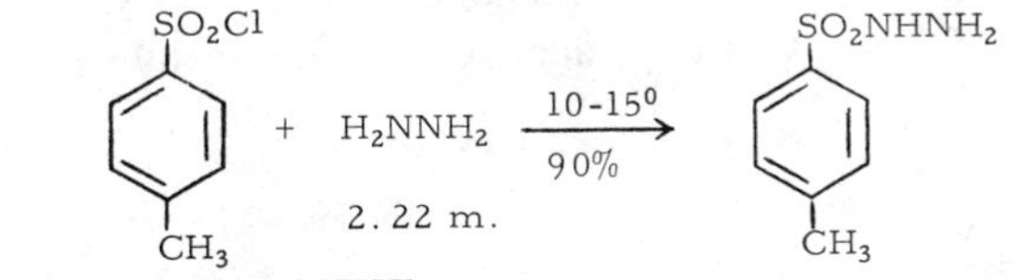
SO2Cl
+ H2NNH2
10-15⁰
90%
2.22 m.
CH3
SO2NHNH2
CH3

1.05 m. in 340 ml. THF

and a solution of hydrazine is added at such a rate as to maintain a temperature of 10–15°. After 15 min. more the lower layer is drawn off, and the upper layer is washed with saturated salt solution, filtered through a drying agent, evaporated, and treated with petroleum ether to precipitate the product.

Elimination reactions. The tosylhydrazone obtained on condensation of the reagent with benzyl methyl ketone reacts with a solution of sodium in ethylene glycol to give a diazo compound which decomposes with elimination of nitrogen to form an olefin.[2,3] 2-Methylpropanal tosylhydrazone similarly reacts with sodium methoxide in diethylene glycol to give chiefly 2-methylpropene, but when the

$$C_6H_5CH_2C(CH_3){=}NNHTs \xrightarrow[-\,TsNa]{NaOCH_2CH_2OH} C_6H_5CH_2C(CH_3){=}\overset{+}{N}{=}\overset{-}{N} \longrightarrow C_6H_5CH{=}CHCH_3 + N_2$$

reaction is carried out in an aprotic solvent such as diethylene glycol diethyl ether, 2-methylpropene and methylcyclopropane are formed in about equal amounts.[4] The tosylhydrazones of cycloalkanones decompose on treatment with sodium methoxide in diethylene glycol diethyl ether to give the diazohydrocarbons in 60–70% yield.[5]

The reaction is useful for conversion of an α-diketone into the corresponding α-diazoketone.[6] 1,2-Indanedione, prepared via the 2-oximido derivative, reacts with tosylhydrazide to form the monotosylhydrazone. Treatment with base effects elimina-

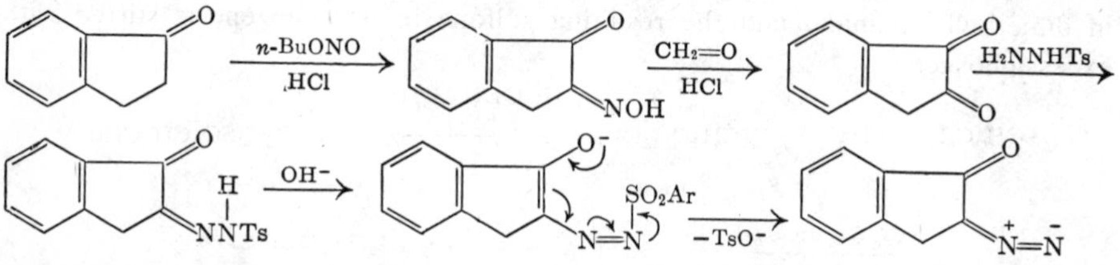

tion of the tosyloxy anion with formation of 2-diazo-1-indanone in good yield. 2,6-Dimethyl-1,4-benzoquinone yields 2,6-dimethyl-1,4-benzoquinone-4-diazide, and 4,5-dimethyl-1,2-benzoquinone yields 4,5-dimethyl-1,2-benzoquinone-1-diazide.[7]

Lithium salts of tosylhydrazones can be decomposed by vacuum pyrolysis (0.2–0.3 mm) at 70–140° to give diazo compounds.[8] The tosylhydrazone can be prepared

$$(CH_3)_2CHCH{=}O \xrightarrow{TsNHNH_2} (CH_3)_2CHCH{=}NNHTs \xrightarrow{BuLi}$$

$$(CH_3)_2CHCH{=}N\overset{+}{N}Ts\ \overset{-}{Li} \xrightarrow[65\text{-}75\%]{} (CH_3)_2CHCH{=}\overset{+}{N}{=}\overset{-}{N} + TsLi$$

in situ from equivalent amounts of the aldehyde or ketone and tosylhydrazide in tetrahydrofurane and converted into the salt by treatment with butyllithium. The method is of particular value for preparing diazo compounds when the carbonyl precursor is more readily available than the amine.

Tosylhydrazones of α,β-unsaturated aldehydes on base-catalyzed decomposition afford cyclopropenes.[9]

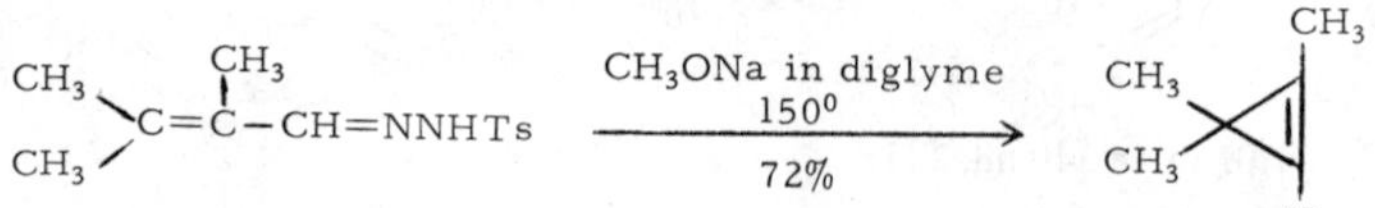

Carbon-skeleton rearrangements occur on decomposition of the tosylhydrazones of pinacolone and camphor to give 2,3-dimethyl-2-butene and camphene.[2] Treat-

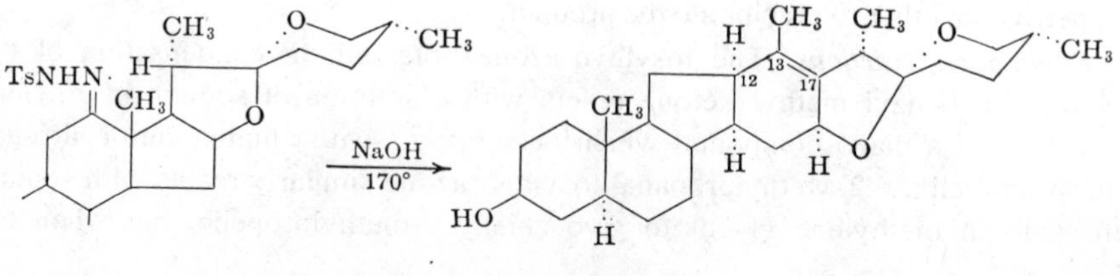

ment of the tosylhydrazone of hecongenin with hot alkali effects elimination and rearrangement contraction of ring C and enlargement of ring D.[10]

Characterization of sugars. Glucose reacts with the reagent to form "ein schönes Hydrazon" from which the sugar can be easily regenerated with benzaldehyde.[11] D-Ribose reacts in methanol to give the highly crystalline tosylhydrazone (m.p. 164°) in quantitative yield.[12] Similar derivatives are obtained from L-arabinose, D-xylose, D-lyxose, and L-fucose.

Dechlorination. 5-Chloracridine reacts with tosylhydrazide in chloroform solution to form a salt, which when heated with alkali decomposes to give acridine.[13]

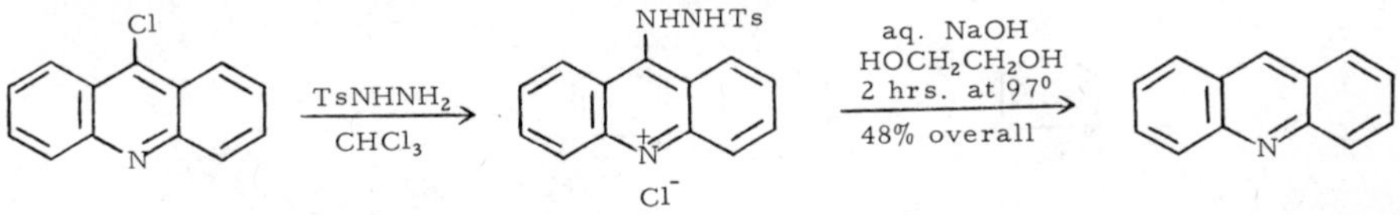
Cl
N
TsNHNH₂
CHCl₃
NHNHTs
N
Cl⁻
aq. NaOH
HOCH₂CH₂OH
2 hrs. at 97°
48% overall
N

The method makes possible the dechlorination of 5-chloroacridines containing reducible groups (NO_2, CN).

Generation of diimide, *see* Diimide.

Abnormal reaction. Chang[14] treated 12-ketocholane with the reagent in refluxing ethanol and obtained, in addition to the expected tosylhydrazone (2), the azine (3),

$$\underset{(1)}{>C=O} \longrightarrow \underset{(2)}{>C=NNHSO_2Ar} \quad + \quad \underset{(3)}{>C=NN=C<}$$

evidently derived by slow decomposition of the reagent to hydrazine. Addition of hydrochloric acid as catalyst made the formation of (2) sufficiently fast that no azine was formed.

[1]L. Friedman, R. L. Litle, and W. R. Reichle, *Org. Syn.*, **40**, 93 (1960)
[2]W. R. Bamford and T. S. Stevens, *J. Chem. Soc.*, 4735 (1952)
[3]Other examples: J. W. Powell and M. C. Whiting, *Tetrahedron*, **7**, 305 (1959); C. H. DePuy and D. H. Froemsdorf, *Am. Soc.*, **82**, 634 (1960)
[4]L. Friedman and H. Shechter, *Am. Soc.*, **81**, 5512 (1959)
[5]L. Friedman and H. Shechter, *Am. Soc.*, **83**, 3159 (1961)
[6]M. P. Cava, R. L. Litle, and D. R. Napier, *Am. Soc.*, **80**, 2257 (1958); A. T. Blomquist and F. W. Schlaefer, *Am. Soc.*, **83**, 4547 (1961)
[7]W. Ried and R. Dietrich, *Ber.*, **94**, 387 (1961)
[8]G. M. Kaufman, J. A. Smith, G. G. Vander Stouw, and H. Shechter, *Am. Soc.*, **87**, 935 (1965)
[9]G. L. Closs and L. E. Closs, *Am. Soc.*, **83**, 2015 (1961)
[10]R. Hirschmann, C. S. Snoddy, Jr., C. F. Hiskey, and N. L. Wendler, *Am. Soc.*, **76**, 4013 (1954)
[11]K. Freudenberg and F. Blümmel, *Ann.*, **440**, 45 (1924)
[12]D. G. Easterby, L. Hough, and J. K. N. Jones, *J. Chem. Soc.*, 3416 (1951)
[13]A. Albert and R. Royer, *J. Chem. Soc.*, 1148 (1949)
[14]F. C. Chang, *J. Org.*, **30**, 2053 (1965)

***p*-Toluenesulfonylmethylnitrosoamide (Diazald)**, *see* Diazomethane.

Tosyl chloride, *see p*-Toluenesulfonyl chloride.

Tosyl perchlorate, $p\text{-}CH_3C_6H_4SO_2^+ClO_4^-$. Mol. wt. 254.60. The reagent is prepared in nitromethane solution by reaction of tosyl bromide with silver perchlorate at 0°. It is a powerful sulfonation reagent which reacts with even the relatively inert halobenzenes to give sulfones.[1]

$$[p\text{-}CH_3C_6H_4SO_2]^+ClO_4^- + ArH \longrightarrow p\text{-}CH_3C_6H_4SO_2Ar + HClO_4$$

[1]F. Klages and F. E. Malecki, *Ann.*, **691**, 15 (1966)

1,2,4-Triazole, see (3). Mol. wt. 69.06, m.p. 121°, b.p. 260°. Supplier: Aldrich.

Preparation.[1] Thiosemicarbazide is condensed with formic acid to produce 1-formyl-3-thiosemicarbazide (1); this is cyclized with alkali to produce 1,2,4-

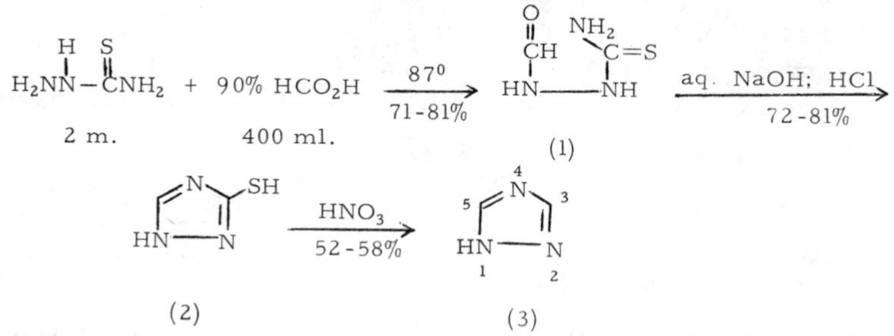

triazole-3(5)-thiol (2), and the thiol group is removed by oxidation (*see* Nitric acid, oxidation).

Catalyst for peptide bond formation. Beyerman[2] found that 1,2,4-triazole, which possesses both a weakly acidic group (N_1) and a weakly basic group (N_2), markedly catalyzes peptide bond formation. In the condensation of N-benzoylglycine cyanomethyl ester with benzylamine in acetonitrile the yield of N-benzoylglycylbenzylamine was increased from 20 to 80% in the presence of this catalyst. Addition of 1,2,4-trizaole to a solution in acetonitrile of the *p*-nitrophenyl ester of glycine, L-leucine, or L-phenylalanine caused precipitation of a polypeptide in a few minutes. The catalytic effectiveness of 1,2,4-triazole is attributed to the fact that the acidic and basic groups occupy adjacent positions and so permit formation of a cyclic transition state favorable for a concerted displacement. Pyrazole and 2-hydroxypyridine also meet this condition and are catalytically effective. Imidazole is trifunctional but the functions are not adjacent and the compound is only very weakly effective. Beyerman's group at Delft and Weygand's group at Munich in a joint study[3] found that 1,2,4-triazole effects no noticeable racemization. Slight racemization was noted in one case with 1,2,3-triazole. On the other hand, use of imidazole led to racemization in several cases.

[1]C. Ainsworth, *Org. Syn.*, **40**, 99 (1960)
[2]H. C. Beyerman and W. M. van der Brink, *Proc. Chem. Soc.*, 266 (1963)
[3]H. C. Beyerman *et al.*, F. Weygand, *et al.*, *Rec. trav.*, **84**, 213 (1965)

ω-Tribromoacetophenone, $C_6H_5COCBr_3$. Mol. wt. 356.87, m.p. 66°, b.p. 176°/16 mm.

Kröhnke[1] described the preparation of the reagent and its use for the selective bromination of methyl ketones and for side-chain bromination of aromatic hydrocarbons.

[1]F. Kröhnke, *Ber.*, **69**, 921 (1936); F. Kröhnke and K. Ellegast, *ibid.*, **86**, 1556 (1953)

Tri-i-butylaluminum $[(CH_3)_2CHCH_2]_3Al$. Mol. wt. 198.32, m.p. 4.3°, b.p. 86°/10 mm. (in order to avoid decomposition to di-*i*-butylaluminum hydride and isobutene, it is advisable to distil the material at 0.5 mm.), sp. gr. 0.7859. Supplier: Texas Alkyls. Preparation.[1]

Reduction of ketones. The reduction of the carbonyl group with trialkylaluminum compounds was first reported by Meerwein (triethylaluminum).[2] Ziegler[3] introduced tri-*i*-butylaluminum for this purpose and showed that the reaction proceeds as follows with formation of isobutene:

$$R_2C{=}O + Al(C_4H_9\text{-}\underline{i})_3 \longrightarrow R_2CHO{-}Al(C_4H_9\text{-}\underline{i})_2 + (CH_3)_2C{=}CH_2$$

$$\downarrow H_2O$$

$$R_2CHOH$$

In the reduction of ketones only one isobutyl group is utilized.

Haubenstock and Davidson[4] found that reduction of 3,3,5-trimethylcyclohexanone (dihydroisophorone) with the reagent is highly stereospecific but dependent upon the reaction conditions. When a solution of the ketone in benzene is added to excess reagent the product consists of 96% of the axial *trans*-alcohol; in this kinetically controlled reduction the reagent approaches on the less hindered α-face to give

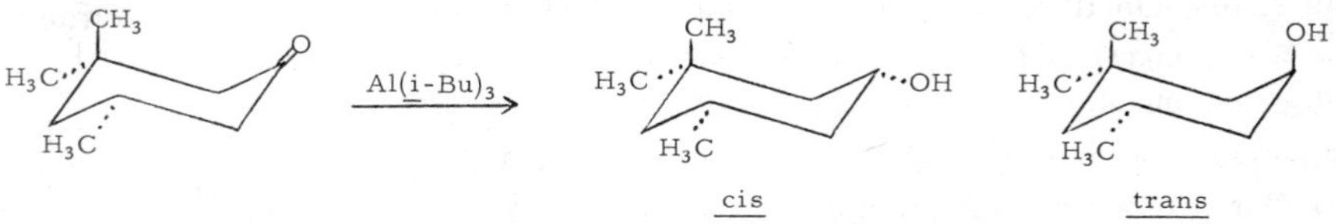
CH3
H3C
H3C
Al(i-Bu)3
CH3
H3C
OH
H3C
CH3
OH
H3C
H3C
cis
trans

the β-alcohol. However, if excess ketone is present, equilibration is possible and the exclusive product is the thermodynamically more stable *cis*-alcohol. Reduction is regarded as proceeding through a cyclic transition state, as in a Meerwein-Ponndorf reduction.

Reduction of unsaturated lactones. Minato and Nagasaki[5] found that α,β-unsaturated γ-lactones are reduced by either tri-*i*-butylaluminum or di-*i*-butylaluminum hydride in tetrahydrofurane to give the corresponding furane derivative, for example (2).

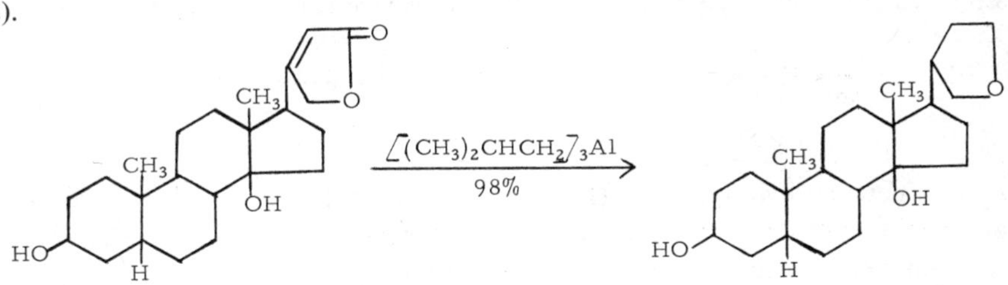
O
CH3
O
CH3
OH
HO
H
[(CH3)2CHCH2]3Al
98%
CH3
O
CH3
OH
HO
H

(1) (2)

[1]K. Ziegler, H.-G. Gellert, H. Lehmkuhl, W. Pfohl, and K. Zosel, *Ann.*, **629**, 1 (1960)
[2]H. Meerwein, G. Hinz, H. Majert, and H. Sönke, *J. prakt. Chem.*, (2), **147**, 226 (1936)
[3]K. Ziegler, K. Schneider, and J. Schneider, *Ann.*, **623**, 9 (1959); K. Ziegler in "Organometallic Chemistry," 236–238, Reinhold (1960)
[4]H. Haubenstock and E. B. Davidson, *J. Org.*, **28**, 2772 (1963)
[5]H. Minato and T. Nagasaki, *Chem. Ind.*, 899 (1965); *Chem. Comm.*, 377 (1965)

Tri-*n*-butylamine, $(CH_3CH_2CH_2CH_2)_3N$. Mol. wt. 185.35, b.p. 216.5°, sp. gr. 0.78. Suppliers: A, B, E, F, MCB.

Dehydrohalogenation. Lyle and Lyle[1] established that a readily prepared product of the action of sulfuric acid on the diol (2) is not an epoxide, as previously supposed, but 2,2-diphenylcycloheptanone (3). This hindered ketone could be brominated, but attempted dehydrohalogenation with pyridine or picoline, even at the boiling point of tetralin, gave poor results. Reaction with N,N-dimethylaniline led to abstraction

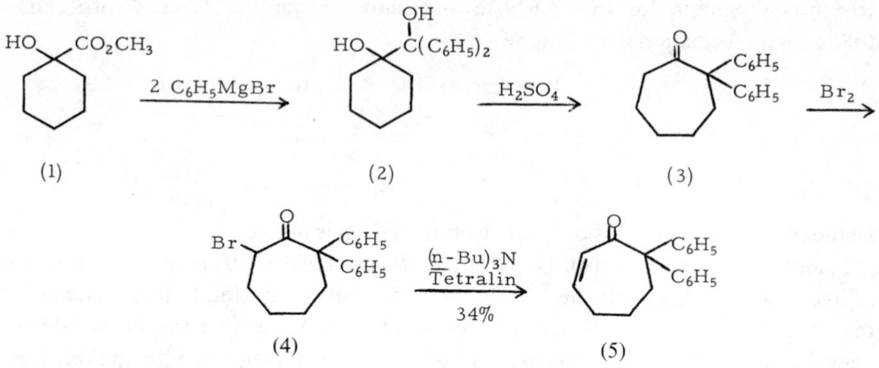

of bromine with reformation of (3).[2] Finally the objective was realized by refluxing for 12 hrs. a mixture of 5 g. of (4), 8.6 ml. of tri-*n*-butylamine, and 25 ml. of tetralin.

[1]R. E. Lyle and G. G. Lyle, *Am. Soc.*, **74**, 4059 (1952)

[2]R. E. Lyle and R. A. Covey, *Am. Soc.*, **75**, 4973 (1953)

Tri-*n*-butylborane and **Tri-*i*-butylborane,** $(C_4H_9)_3B$. Mol. wt. 182.16, b.p. 109°/20 mm., 91°/9 mm., and 68°/7 mm. Supplier: MSA Res. Corp.

Homogeneous hydrogenation. The facile addition of boron hydrides to olefins, coupled with the fact that a trialkylborane can be hydrogenolized to a dialkylborane and an alkane, suggested that trialkylboranes would serve as homogeneous catalysts for hydrogenation.[1] Although any simple alkylborane should be effective, the two primary tributylboranes offer convenience in handling. Thus cyclohexene or caprylene containing 3.8% of tri-*n*-butylborane is hydrogenated quantitatively in 3 hrs. at 200° and a hydrogen pressure of 1000 p.s.i. Hydrogenolysis of the metal-to-carbon bond of the reagent gives dibutylborohydride (2), which adds to the olefin

$$\underset{(1)}{BBu_3} \xrightarrow[-BuH]{H_2} \underset{(2)}{HBBu_2} \xrightarrow{RCH=CH_2} \underset{(3)}{RCH_2CH_2BBu_2} \xrightarrow{3\ H_2} \underset{(4)}{RCHCH_3} + \underset{(2)}{HBBu_2}$$

to give a new trialkylborane (3). Hydrogenolysis gives the alkane (4) with regeneration of dibutylborohydride (2) for recycling. When all the olefin has been consumed, the boron is converted into some form of metal hydride.

[1]E. J. DeWitt, F. L. Ramp, and L. E. Trapasso, *Am. Soc.*, **83**, 4672 (1961); *J. Org.*, **27**, 4368 (1962)

Tri-*n*-butylboroxine, $(n\text{-}C_4H_9BO)_3$. Mol. wt. 198.16, b.p. 133.5°/16 mm.

Preparation.[1] A mixture of tri-*n*-butylborane and anhydrous boric oxide is refluxed under nitrogen for 40 hrs. and the product distilled. Under the same conditions tri-*sec*-butylborane reacted with rearrangement to give tri-*n*-butylboroxine.

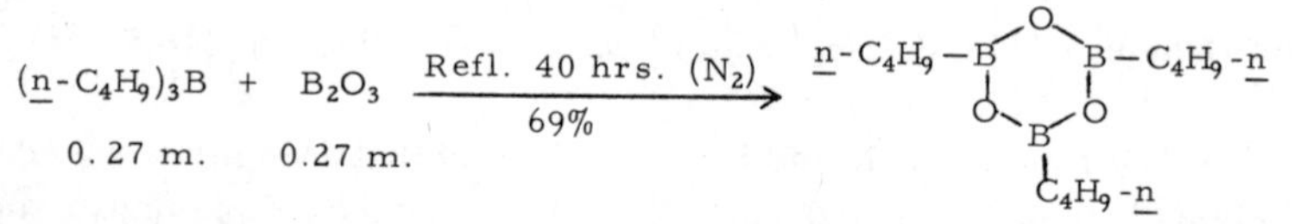

Separation of cis- and trans-diols. Brown and Zweifel[2] found that *cis*-cyclohexane-1,4-diol reacts with excess tri-*n*-butylboroxine to form a cyclic boronate of reasonable volatility (b.p. 200°/0.05 mm.), whereas the *trans*-isomer forms a nonvolatile linear

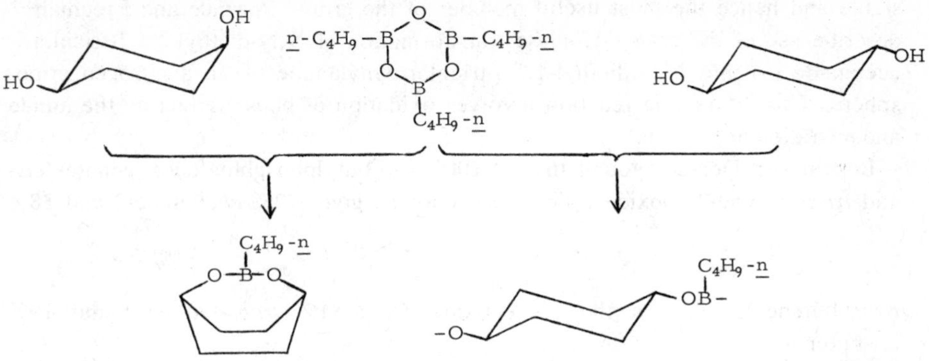

polymeric boronate. Thus 5 g. of a mixture of the isomers was treated with excess reagent, and the water of esterification removed by azeotropic distillation with benzene. Vacuum distillation gave a viscous product from which the *cis*-diol (1.42 g.) was recovered by transesterification with ethylene glycol, removal of ethylene glycol boronate by vacuum distillation, and crystallization of the residue. Treatment of the polymeric boronate in the same way afforded 2.25 g. of the *trans*-isomer. The *cis*- and *trans*-cyclohexane-1,3-diols were separated in the same way, but the boronates from *cis*- and *trans*-cyclohexane-1,2-diol did not differ enough in boiling point for easy separation.

In applying this method to carbohydrate 1,3-diols, a British group[3] preferred to liberate carbohydrate 1,3-diols from their boronates by transesterification with propane-1,3-diol rather than with ethylene glycol. The boronate was treated in acetone with propane-1,3-diol and the solution evaporated to dryness. Extraction with petroluem ether removed the exchanged boronate and crystallization of the residue afforded the carbohydrate component.

[1]G. F. Hennion, P. A. McCusker, E. C. Ashby, and A. J. Rutkowski, *Am. Soc.*, **79**, 5194 (1957)
[2]H. C. Brown and G. Zweifel, *J. Org.*, **27**, 4708 (1962)
[3]R. J. Ferrier, D. Prasad, A. Rudowski, and I. Sangster, *J. Chem. Soc.*, 3330 (1964)

Tri-*n*-butylphosphine, $(n\text{-}C_4H_9)_3P$. Mol. wt. 202.32, b.p. 240° (130°/20 mm.), pKa 8.4, colorless liquid with a garliclike odor, immiscible with water but miscible with most organic solvents. Prepared by reaction of *n*-butylmagnesium bromide with

$$3\ \underline{n}\text{-}C_4H_9MgBr + PCl_3 \xrightarrow[57\%]{\text{Ether; } NH_4Cl} (\underline{n}\text{-}C_4H_9)_3P + 3\ MgBrCl$$

phosphorus trichloride.[1] Suppliers: Carlisle Chemical Works (1 gal. minimum), Metal and Thermite Chem. Co., Westvaco Mineral Prod., KK.

The trialkylphosphines are all subject to air oxidation, but, as the size of the alkyl group increases, the sensitivity falls and then rises. Tri-*n*-butylphosphine is the most

$$\underset{(1)}{CCl_3COCl} + \underset{(2)}{HN(C_2H_5)_2} \xrightarrow[89-92\%]{\text{Ether-aq. NaOH}} \underset{(3)}{CCl_3CON(C_2H_5)_2} \xrightarrow[69-74\%]{(\underline{n}\text{-}C_4H_9)_3P}$$

$$\underset{(4)}{Cl_2C{=}C(Cl)N(C_2H_5)_2} + \underset{(5)}{(\underline{n}\text{-}C_4H_9)_3\overset{+}{P}{-}\overset{-}{O}}$$

stable and hence the most useful member of the group. Speziale and Freeman[2, 3] describe use of the reagent for the transformation of N,N-diethyl-2,2,2-trichloro-acetamide (3) into N,N-diethyl-1,2,2-trichlorovinylamine (4) in a nitrogen atmosphere. This interesting reaction involves oxidation of phosphorous by the amide and migration of halogen.

Boskin and Denney[4] found that at 100° tri-*n*-butylphosphine deoxygenates *cis*- and *trans*-butene-2-epoxides. The *trans* epoxide gives 72% *cis*-butene-2 and 28%

$$>C\overset{}{\underset{O}{----}}C< \quad + \quad (\underline{n}\text{-}C_4H_9)_3P \quad \longrightarrow \quad >C=C< \quad + \quad (\underline{n}\text{-}C_4H_9)_3PO$$

trans-butene-2, whereas the *cis* epoxide gives 81% *trans*-butene-2 and 19% *cis*-butene-2.

Haloketones are reductively dehalogenated by reaction with a trialkylphosphine.[5]

$$\underset{\displaystyle |}{\overset{\displaystyle |}{-}}\overset{X}{\underset{\displaystyle |}{C}}-\overset{O}{\overset{\|}{C}}- \; + \; R_3P \; \longrightarrow \; -C=\overset{\overset{+}{O}PR_3\;(X^-)}{\underset{|}{C}}- \; \xrightarrow{H_2O} \; -\overset{H}{\underset{|}{C}}-\overset{O}{\overset{\|}{C}}- \; + \; HX \; + \; R_3PO$$

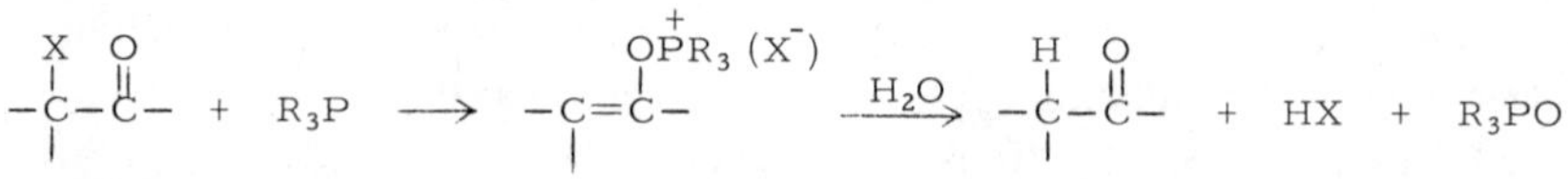

[1]G. B. Kauffman and L. A. Teter, *Inorg. Syn.*, **6**, 87 (1960)
[2]A. J. Speziale and R. C. Freeman, *Am. Soc.*, **82**, 903 (1960)
[3]*Idem, Org. Syn.*, **41**, 21 (1961)
[4]M. J. Boskin and D. B. Denney, *Chem. Ind.*, 330 (1959)
[5]H. Hoffmann and H. J. Diehr, *Angew. Chem., Internat. Ed.*, **3**, 145 (1964)

Tri-*n*-butylphosphine oxide, $(n\text{-}C_4H_9)_3P^+\text{—}O^-$. Mol. wt. 218.32, very hygroscopic white solid, m.p. 51–52° from CCl_4, b.p. 300°. Supplier: Metal and Thermit Chem. Co.

The reagent catalyzes the isomerization of epoxides to carbonyl compounds.[1] Thus *trans*-stilbene epoxide reacts to give desoxybenzoin and diphenylacetaldehyde

$$\overset{C_6H_5}{\underset{H}{}}\!\!\diagdown\!\underset{}{\overset{O}{\underset{C----C}{\bigtriangleup}}}\!\!\diagup\!\overset{H}{\underset{C_6H_5}{}} \xrightarrow{(n\text{-}C_4H_9)_3PO} C_6H_5COCH_2C_6H_5 \quad + \quad (C_6H_5)_2CHCHO$$

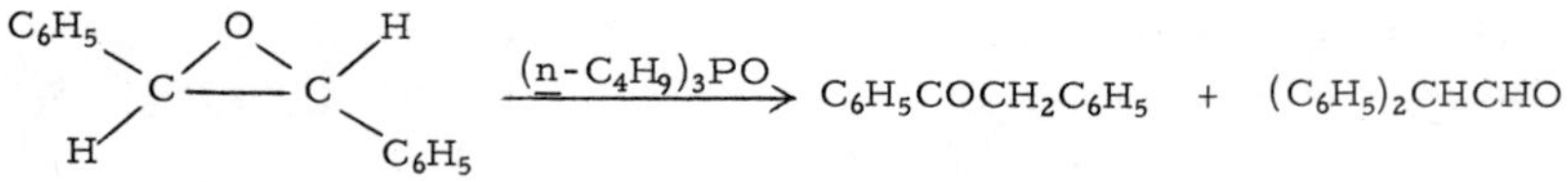

in the ratio of 9:1. *cis*-Stilbene epoxide gives the same products. *trans*-4-Octene epoxide is converted into 4-octanone only if *m*-chloroperbenzoic acid is present in the reaction mixture.

[1]D. E. Bissing and A. J. Speziale, *Am. Soc.*, **87**, 1405 (1965)

Tri-*n*-butyltin hydride, $(n\text{-}C_4H_9)_3SnH$. Mol. wt. 291.04, b.p. 76°/0.7 mm. Supplier: Alfa.

The reagent is prepared according to Van Der Kerk *et al.*[1] by lithium aluminum hydride reduction of tri-*n*-butyltin (supplier: MCB). It is a water-white liquid, which can be kept for some time if dry. Moisture converts it into tri-*n*-butyltin hydroxide.

Reduction of organic halides. This reaction was discovered by van der Kerk *et al.*,

$$R_3SnH \; + \; R'X \; \longrightarrow \; R_3SnX \; + \; R'H$$

and several examples have been reported since then.[2, 3] Kuivila[3] presented convincing evidence that the reductions are radical reactions and effected the stepwise reduction of benzotrichloride to toluene:[4] $C_6H_5CCl_3 \longrightarrow C_6H_5CHCl_2 \longrightarrow C_6H_5CH_2Cl \longrightarrow C_6H_5CH_3$.

Seyferth *et al.*[5] found that the *gem*-dibromocyclopropanes readily available by addition of dibromocarbene to olefins can be reduced to monobromocyclopropanes with tri-*n*-butyltin hydride in good yield if the temperature is kept from rising above 40°. For example, 1,1-dibromo-2,2,3-trimethylcyclopropane is reduced by one equivalent of the reagent under nitrogen to a mixture of isomers, assigned the configurations shown on the basis of the NMR spectra. Complete reduction to the

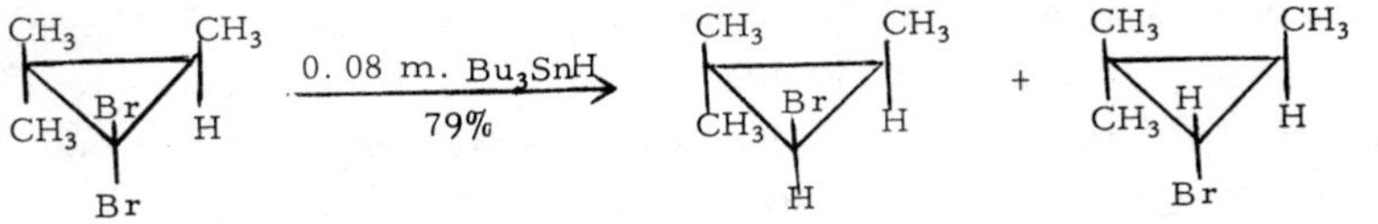
0. 08 m. Bu_3SnH
79%
+

0. 08 m. — 1 part, J_{AX} = 7. 2 cps — 4 parts, J_{AX} = 3. 8 cps

cyclopropane also was easily accomplished, as illustrated for the adduct of tetramethylethylene and dibromocarbene.[6] Reduction of 7,7-dichloronorcarane requires

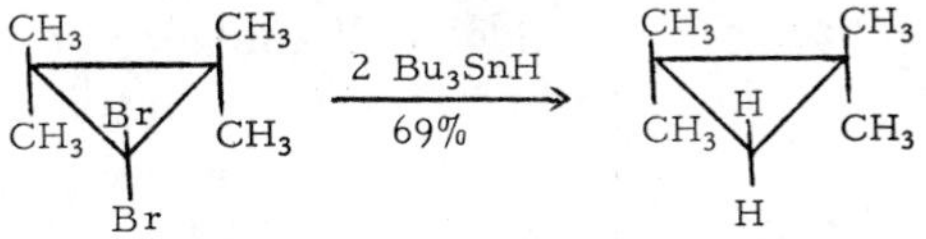
2 Bu_3SnH
69%

a temperature of about 140°, and preferential reduction of the bromine atom of 7-bromo-7-chloronorcarane is easily accomplished. Selective reduction of 1-chloro-1-fluorocyclopropanes to monofluorocyclopropanes is also reported.[6]

Kuivila[3] found that the reagent reduces acid chlorides to aldehydes at room temperature.

Other reductions. A limitation to the reductions cited is that the reagent can react with other organic functions such as the carbonyl group of aldehydes and ketones,[7] olefinic and acetylenic unsaturation.[1, 8]

House *et al.*[9] found the reagent superior to Raney nickel or platinum for hydrogenolysis of the carbon–iodine bond of the iodolactone (1).

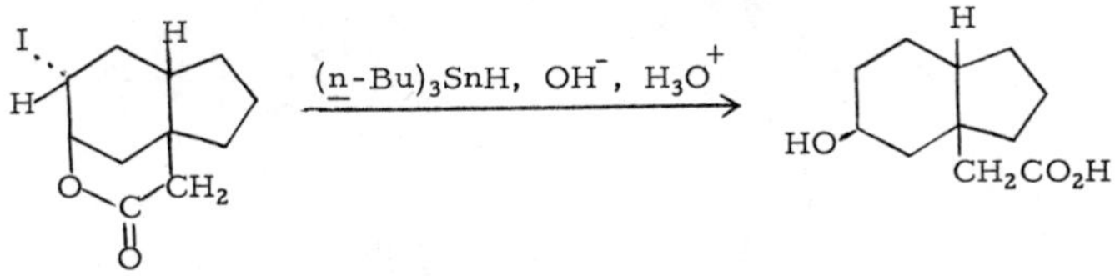
($\underline{n}$-Bu)$_3$SnH, OH^-, H_3O^+

(1) (2)

[1]G. J. M. Van Der Kerk, J. G. Noltes, and J. G. H. Suijten, *J. Appl. Chem.*, **7**, 366 (1937)

[2]L. A. Rothman and E. I. Becker, *J. Org.*, **25**, 2203 (1960); E. J. Kupchik and R. E. Connolly, *ibid.*, **26**, 4747 (1961); E. J. Kupchik and R. J. Kiesel, *Chem. Ind.*, 1654 (1962)

[3]H. G. Kuivila, *J. Org.*, **25**, 284 (1961); H. G. Kuivila and E. J. Walsh, Jr., *Am. Soc.*, **88**, 571 (1966)

[4]H. G. Kuivila, L. W. Menapace, and C. R. Warner, *Am. Soc.*, **84**, 3584 (1962)

[5]D. Seyferth, H. Yamazaki, and D. L. Alleston, *J. Org.*, **28**, 703 (1963)

[6]J. P. Oliver, U. V. Rao, and M. T. Emerson, *Tetrahedron Letters*, 3419 (1964)

[7]H. G. Kuivila and O. F. Beumel, Jr., *Am. Soc.*, **80**, 3798 (1958); **83**, 1246 (1961)

[8]G. J. van der Kerk and J. G. Noltes, *J. Appl. Chem.*, **9**, 106 (1959)

[9]H. O. House, S. G. Boots, and V. K. Jones, *J. Org.*, **30**, 2519 (1965)

Trichloramine, NCl_3. *Toxic* and potentially *explosive*, particularly in an aqueous medium, the reagent should be prepared in a nonaqueous solvent behind a safety shield.

Preparation.[1] A suspension of 1.2 m. of calcium hypochlorite in 600 ml. of water and 900 ml. of *o*-dichlorobenzene is cooled to 0° in a dry ice–acetone bath, and a solution of 1.2 m. of ammonium chloride, 150 ml. of concd. hydrochloric acid, and 450 ml. of water is added from a dropping funnel with good stirring below 10°. After an additional 15 min., the bright yellow organic phase is separated, washed with water, dried with sodium sulfate, titrated iodometrically, and stored at 0°.

Friedel-Crafts reaction.[2] The aluminum chloride-catalyzed reaction of the reagent with cumene follows roughly the stoichiometry indicated in the equation.

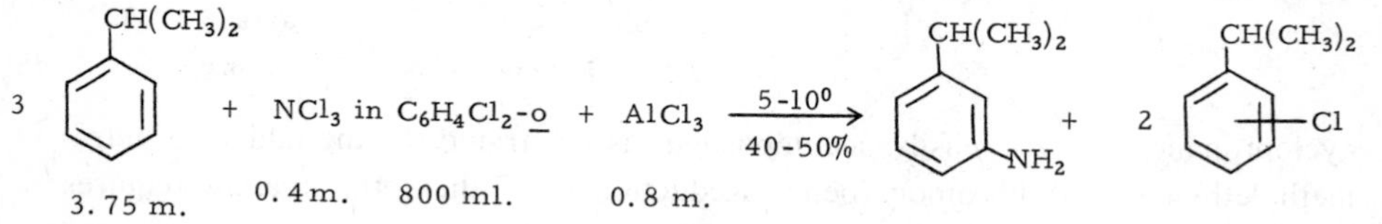
CH(CH3)2
3
+ NCl3 in C6H4Cl2-o + AlCl3
5-10°
40-50%
CH(CH3)2
NH2
+ 2
CH(CH3)2
Cl
3.75 m.
0.4 m.
800 ml.
0.8 m.

Cumene and aluminum chloride are added to the solution of reagent in *o*-dichlorobenzene prepared as described. By similar reaction with trichloroamine and aluminum chloride in methylene chloride at 0° under nitrogen, methylcyclohexane is converted into 1-methyl-1-aminocyclohexane in yield of 86–87%.

[1] P. Kovacic and J. A. Levinsky, Case Inst. Techn., procedure submitted to *Org. Syn.*; P. Kovacic *et al.*, *Am. Soc.*, **86**, 1650 (1964)

[2] P. Kovacic and S. S. Chaudhary, procedure submitted to *Org. Syn.*; P. Kovacic, R. J. Hopper, S. S. Chaudhary, J. A. Levisky, and V. A. Liepkalns, *Chem. Comm.*, 232 (1966); P. Kovacic and J. F. Gormish, *Am. Soc.*, **88**, 3819 (1966)

Trichloroacetic acid, CCl_3COOH. Mol. wt. 163.40, m.p. 58°, b.p. 196°, sp. gr. 1.62, pKa 0.08. Suppliers, B, E, MCB. Preparation by oxidation of chloral with nitric acid containing oxides of nitrogen.[1]

Trichloroacetic acid is recommended as catalyst in the diazotization of anthranilic acid for the preparation of benzenediazonium-2-carboxylate, a benzyne precursor.[2]

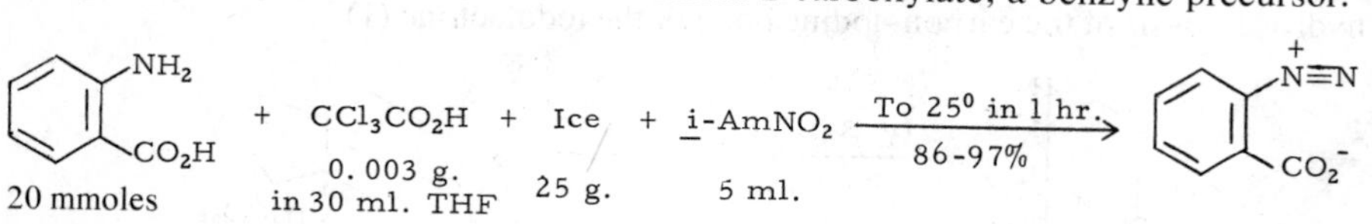
NH2
CO2H
20 mmoles
+ CCl3CO2H + Ice + i-AmNO2
0.003 g.
in 30 ml. THF
25 g.
5 ml.
To 25° in 1 hr.
86-97%
+
N≡N
CO2-

The initially cold reaction mixture is stirred mechanically and allowed to come to room temperature in about one hour. The diazonium salt that separates, being explosive, is collected on a plastic Büchner funnel with a plastic spatula, washed with cold THF, washed with the solvent to be used in the next reaction, and used as a slurry in this solvent. Acetic acid and strong mineral acids are wholly unsatisfactory as catalysts.

[1] G. D. Parkes and R. G. W. Hollingshead, *Chem. Ind.*, 222 (1954)

[2] F. M. Logullo and L. Friedman, Case Inst. Techn., procedure submitted to *Org. Syn.*

Trichloroacetonitrile, CCl_3CN. Mol. wt. 144.40, m.p. 44°, b.p. 84.6/741 mm., sp. gr. 1.44. Suppliers: Aldrich, Dr. Theodor Schuchardt.

Preparation. F. Cramer[1] states that the method of Steinkopf[2] has been used in his laboratory with little variation: 200 g. of trichloroacetamide and 250 g. of phosphorus pentoxide are well mixed and heated to about 200° in an oil bath for several hours. During this time trichloroacetonitrile distills and it is redistilled over P_2O_5; yield 70–80%.

Uses. In replacement of a hydroxyl group by chlorine, the alcohol is treated with the reagent in ether solution at −15° and dry hydrogen chloride is passed in.[2] The reaction proceeds with complete inversion of configuration. The reagent is useful

$$ROH + CCl_3CN + HCl \longrightarrow RCl + CCl_3CONH_2$$

for the selective esterification of phosphoric acid (1).[3] It converts a phosphate ester into the symmetrical pyrophosphate (2) or, in the presence of an alcohol, into the diester. Trichloroacetonitrile is somewhat similar to the carbodiimides, but

$$(1)\quad C_6H_5CH_2OH + H_3PO_4 \xrightarrow[77\%]{CCl_3CN + (C_2H_5)_3N} C_6H_5CH_2OP(=O)(OH)-OH + CCl_3CONH_2$$

$$(2)\quad 2\ \underline{p}\text{-}ClC_6H_4OP(=O)(OH)-OH \xrightarrow[91\%]{CCl_3CN + Py} \underline{p}\text{-}ClC_6H_4OP(=O)(OH)O-P(=O)(OH)OC_6H_4Cl\text{-}\underline{p} + CCl_3CONH_2$$

it does react with phosphoric acid diesters or with carboxylic acids. It has been used for the preparation of pyrophosphate esters of terpene alcohols.[4]

[1]F. Cramer, private communication
[2]W. Steinkopf, *Ber.*, **41**, 2541 (1908)
[3]F. Cramer and H.-J. Baldauf, *Ber.*, **92**, 370 (1959)
[4]F. Cramer, W. Rittersdorf, and W. Böhm, *Ann.*, **654** 180 (1962)

Trichloromethanesulfonyl bromide, CCl_3SO_2Br. Mol. wt. 262.37, m.p. 139°.

The reagent is prepared from the corresponding chloride as follows:[1]

$$CCl_3SO_2Cl + KCN \xrightarrow{\text{liq. } SO_2} CCl_3SO_2K \xrightarrow{Br_2} CCl_3CSO_2Br$$

In the presence of light or a peroxide initiator it reacts with cyclohexane, cyclopentane, or toluene to give cyclohexyl bromide, cyclopentyl bromide, or benzyl bromide, respectively.

[1]R. P. Pinnell, E. S. Huyser, and J. Kleinberg, *J. Org.*, **30**, 38 (1965)

Trichloromethanesulfonyl chloride, Cl_3CSO_2Cl. Mol. wt. 217.89, m.p. 140–141°. Supplier of practical grade, m.p. 137–140°: Eastman.

Preparation. The reagent is prepared in 78% yield by oxidation of trichloromethanesulfenyl chloride (Stauffer Chem. Co.) with hydrogen peroxide in acetic acid.[1]

Radical chlorination. Under radical conditions the reagent functions as a chlorinating agent:[2]

$$RH + Cl_3CSO_2Cl \xrightarrow[(C_6H_5COO)_2]{h\nu} RCl + Cl_3CH + SO_2$$

Typical high-yield reactions are: $C_6H_5CH_3 \longrightarrow C_6H_5CH_2Cl$, $p\text{-}BrC_6H_4CH_3 \longrightarrow p\text{-}BrC_6H_4CH_2Cl$, cyclohexane⟶cyclohexyl chloride. Considerable selectivity is noted in this method of chlorination. Thus *n*-hexane and ethylbenzene give only 2-chlorohexane and α-chloroethylbenzene. In the chlorination of *n*-heptane, 2-chloroheptane is obtained in 50% yield but statistical substitution occurs at the carbon atoms further along in the chain.

The reaction sequence is regarded as the following:

$$R\cdot + Cl_3CSO_2Cl \longrightarrow RCl + Cl_3C\dot{S}O_2$$

$$Cl_3C\dot{S}O_2 + RH \longrightarrow R\cdot + Cl_3CSO_2H$$

$$Cl_3CSO_2H \longrightarrow Cl_3CH + SO_2$$

[1]G. Sosnovsky, *J. Org.*, **26**, 3506 (1961)
[2]E. S. Huyser, *Am. Soc.*, **82**, 5246 (1960); E. S. Huyser and B. Giddings, *J. Org.*, **27**, 3391 (1962); E. S. Huyser, H. Schimke, and R. L. Burham, *ibid.*, **28**, 2141 (1963)

Trichloromethyllithium, Cl_3CLi (see also pp. 223–224).

The reagent is prepared as required by adding bromotrichloromethane (supplied in premium grade by MCB) to a slurry of methyllithium in ether at −115°.[1] At −80° the reagent decomposes exothermally to give a mixture of tetrahaloethylenes. It reacts with cyclohexene at −100° to give dichloronorcarane in high yield, and the available evidence indicates that the reaction probably does not involve dichlorocarbene.

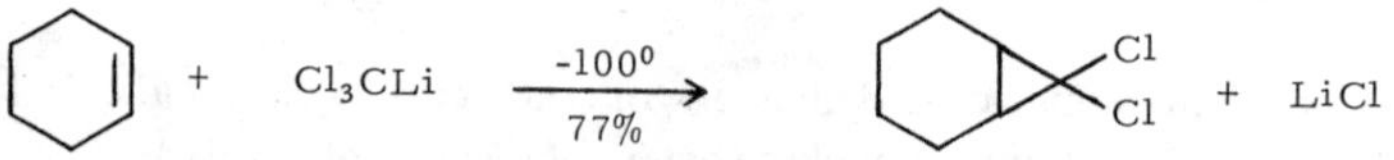
+
Cl3CLi
-100°
77%
Cl
Cl
+
LiCl

Köbrich *et al.*[2] prefer THF to ether as solvent because the reagent is more stable in this solvent.

[1]W. T. Miller, Jr. and D. M. Whalen, *Am. Soc.*, **86**, 2089 (1964)
[2]G. Köbrich, K. Flory, and H. R. Merkle, *Tetrahedron Letters*, 973 (1965)

2,4,5-Trichlorophenol, $Cl_3C_6H_2OH$. Mol. wt. 197.46, m.p. 62–63; pKa 9.45. Suppliers: A, B, E, F, MCB.

A comparison of the rates of reaction of many substituted phenyl esters with N-protected α-amino acids showed that the 2,4,5-trichlorophenyl esters are more reactive than *p*-nitrophenyl esters and are promising new active derivatives for the synthesis of peptides.[1] One advantage is that the ester group does not interfere with catalytic hydrogenation for removal of the carbobenzoxy group. The esters are prepared by condensation of the N-protected amino acid and the phenol with dicyclohexylcarbodiimide. In the synthesis of a particular hexapeptide, where other methods were unsatisfactory because of poor yields or impure products, Bentley *et al.*[2] obtained the hexapeptide in 97% yield through the trichlorophenyl ester.

[1]J. Pless and R. A. Boissonnas, *Helv.*, **46**, 1609 (1963)
[2]P. H. Bentley, H. Gregory, A. H. Laird, and J. S. Morley, *J. Chem. Soc.*, 6130 (1964)

Triethanolamine, $(HOCH_2CH_2)_3N$. Mol. wt. 149.19, m.p. 21°, b.p. 279°/150 mm., sp. gr. 1.12. Suppliers: Baker, E, F, MCB.

Linstead and co-workers[1] found this base the most satisfactory reagent for the condensation of aliphatic aldehydes with malonic acid to produce pure β,γ-unsaturated acids, for example Δ^3-hexenoic acid. A mixture of 1 mole each of aldehyde,

$$\underset{1\ m.}{CH_3CH_2CH_2CHO} + \underset{1\ m.}{CH_2(CO_2H)_2} + \underset{1\ m.}{(HOCH_2CH_2)_3N} \xrightarrow[40\text{-}42\%]{-CO_2} CH_3CH_2CH{=}CHCH_2CO_2H$$

acid, and catalyst is shaken until homogeneous, let stand for 2 days, and then heated on the steam bath overnight, when evolution of carbon dioxide has ceased. Losses arise from autocondensation of the aldehyde and from its reaction with 2 moles of malonic acid.

Gardner and co-workers[2] developed an efficient synthesis of pimelic acid (4) from furfural (1) involving conversion to furylacrylic acid (2) and esterification, which proceeds with cleavage of the ring and disproportionation to give diethyl γ-ketopimelate (3). Huang-Minlon reduction by the usual procedure in diethylene

(1)   (2)

$$C_2H_5O_2C(CH_2)_2\overset{\overset{O}{\|}}{C}(CH_2)_2CO_2C_2H_5 \xrightarrow[99.5\%]{H.-M.} HO_2C(CH_2)_5CO_2H$$

(3)   (4)

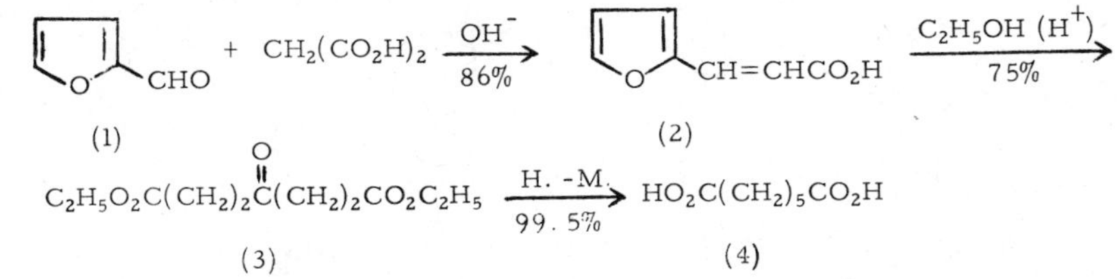

glycol or ethylene glycol presented the difficulty that on isolation of the reduction product by continuous ether extraction a substantial amount of the diol solvent was extracted. The problem was resolved by using triethanolamine as solvent in the Huang-Minlon reduction and acidifying the reaction mixture prior to extraction with ether in an automatic apparatus.

[1]R. P. Linstead, E. G. Noble, and E. J. Boorman, *J. Chem. Soc.*, 557 (1933)
[2]P. D. Gardner, L. Rand, and G. R. Haynes, *Am. Soc.*, **78**, 3425 (1956)

Triethylaluminum, $Al(C_2H_5)_3$. Mol. wt. 114.16. Suppliers: Ethyl Corp., KK, Alfa.

Preparation. The best procedure is by a two-step process.[1]

$$Al + 1\ 1/2\ H_2 + 2\ (C_2H_5)_3Al \longrightarrow 3\ Al(C_2H_5)_2H$$

$$3\ Al(C_2H_5)_2H + 3\ CH_2{=}CH_2 \longrightarrow 3\ Al(C_2H_5)_3$$

Reaction with nitriles and isocyanates.[2] Triethylaluminum reacts with nitriles in the molar ratio 2:1 to form ethyl ketones in high yield. With a 1:1 ratio no reaction occurs. The reaction of triethylaluminum with benzonitrile to form propiophenone in 77% yield is interpreted as involving a cyclic transition state:

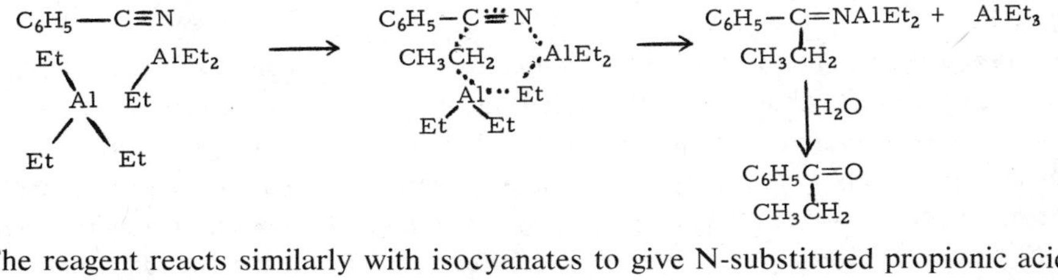

The reagent reacts similarly with isocyanates to give N-substituted propionic acid amides in yields of 90–95%.

$$RN{=}C{=}O \xrightarrow{Al(C_2H_5)_3} CH_3CH_2CONHR$$

Addition of hydrogen cyanide. The 1,4-addition of hydrogen cyanide to cholestenone under usual conditions affords the 5α- and 5β-cyano ketones in the ratio 1:1. However, when the reaction is run in ether, tetrahydrofurane, or benzene with hydrogen cyanide in the presence of triethylaluminum with a ratio of ketone:

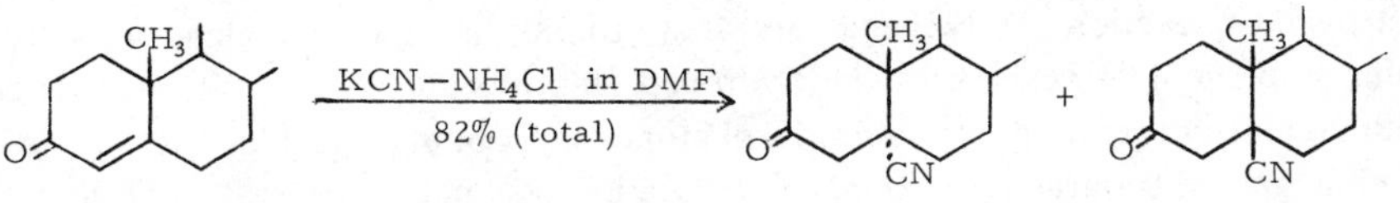

Ratio 1:1

HCN:$(C_2H_5)_3Al$ of 1:2:3 the yield is about the same (85%) but the ratio of 5α- to 5β-nitrile is 2:1.[3] With some ketones the stereoselectivity obtainable with the additive is even higher. In the case of the model ketone (1) the ratio of *trans* to *cis* product is 24:1.

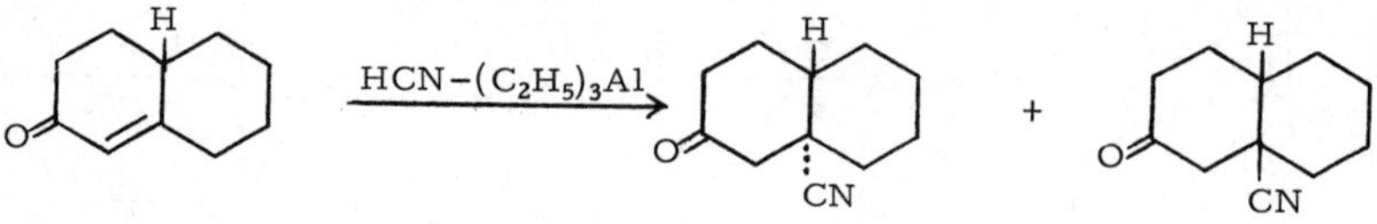
H
HCN–(C₂H₅)₃Al
H
CN
H
CN

(1)

The combination triethylaluminum–HCN in the molar ratio 6:10 cleaves steroidal epoxides to give the diaxial products in high yield:[4]

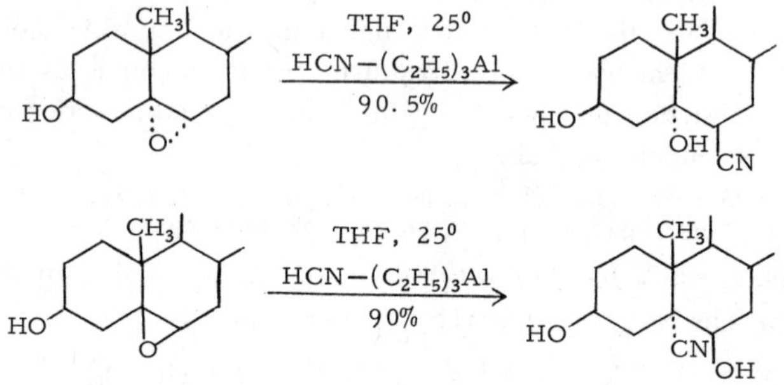
CH₃
HO
THF, 25°
HCN–(C₂H₅)₃Al
90.5%
CH₃
HO
OH
CN
CH₃
HO
THF, 25°
HCN–(C₂H₅)₃Al
90%
CH₃
HO
CN
OH

[1]K. Ziegler, H.-G. Gellert, H. Lehmkuhl, W. Pfohl, and K. Zosel, *Ann.*, **629**, 1 (1960)
[2]H. Reinheckel and D. Jahnke, *Ber.*, **97**, 2661 (1964)
[3]W. Nagata, M. Yoshioka, and S. Hirai, *Tetrahedron Letters*, 461 (1962); W. Nagata, T. Terasawa, and T. Aoki, *ibid.*, 865 (1963); W. Nagata, I. Kikkawa, and M. Fujimoto, *Bull. Chem. Pharm.*, **11**, 226 (1963)
[4]W. Nagata, M. Yoshioka, and T. Okumura, *Tetrahedron Letters*, 847 (1966)

Triethylamine, $(C_2H_5)_3N$. Mol. wt. 101.19, b.p. 89.5°, sp. gr. 0.73, pKb 3.36. Suppliers: A, B, E, F.

Purification. The reagent is let stand over sodium hydroxide and distilled from a mixture with about 2% of phenyl isocyanate[1] or naphthyl isocyanate.[2]

Preparation of diazoketones. The usual procedure for preparing a diazoketone is to add the acid chloride slowly to an ethereal solution of 2 moles of diazomethane; the hydrogen chloride is then destroyed by reaction (2). If the order is reversed and only 1 mole of diazomethane is used, the diazoketone formed in the reaction (1) is

$$1.\quad R\overset{\overset{\displaystyle O}{\|}}{C}{-}Cl + CH_2{=}\overset{+}{N}{=}\overset{-}{N} \longrightarrow R\overset{\overset{\displaystyle O}{\|}}{C}{-}CH{=}\overset{+}{N}{=}\overset{-}{N} + HCl$$

$$2.\quad HCl + CH_2{=}\overset{+}{N}{=}\overset{-}{N} \longrightarrow CH_3Cl + N_2$$

$$3.\quad R\overset{\overset{\displaystyle O}{\|}}{C}CH{=}\overset{+}{N}{=}\overset{-}{N} + HCl \longrightarrow R\overset{\overset{\displaystyle O}{\|}}{C}CH_2Cl + N_2$$

destroyed in reaction (3). Newman and Beal[3] found that if 1 equivalent of triethylamine is present to react with the hydrogen chloride only 1 equivalent of diazomethane is required and aromatic diazoketones are obtained in high yield. Use of pyridine gave discouraging results. This variation was not successful with aliphatic acid chlorides; thus caproyl chloride gave a mixture of products.

Ketenes and derived products. Triethylamine dehydrohalogenates an acid chloride having an α-hydrogen atom to give a ketene isolable as the ketene dimer, which can be converted into a β-ketoacid or a symmetrical ketone (Sauer[1,4]). An example is the preparation of laurone from lauroyl chloride. An ethereal solution of the acid

$$2\ C_{10}H_{21}CH_2\overset{\underset{\textstyle |}{Cl}}{C}{=}O \xrightarrow[\substack{-(C_2H_5)_3NHCl}]{\overset{(C_2H_5)_3N}{0.7\ m.}} C_{10}H_{21}CH{=}C{=}O \longrightarrow$$

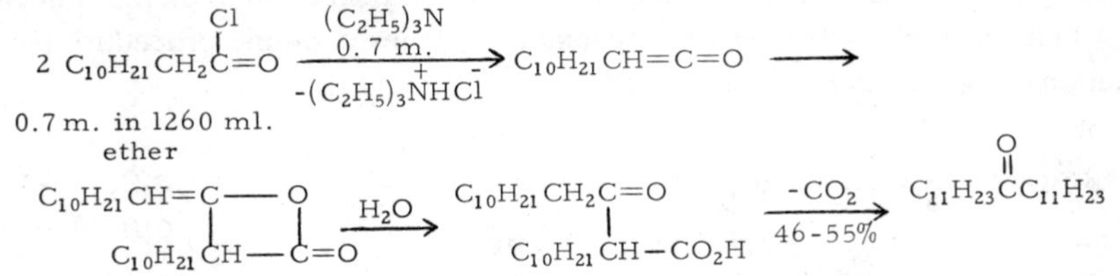

0.7 m. in 1260 ml.
ether

$$\underset{\underset{\textstyle C_{10}H_{21}}{\overset{\textstyle |}{}}}{C_{10}H_{21}CH{=}C}\!\!-\!\!\underset{\underset{\textstyle CH}{\overset{\textstyle |}{}}}{O}\!\!-\!\!\overset{\textstyle O}{\underset{\textstyle C{=}O}{}} \xrightarrow{H_2O} \underset{\underset{\textstyle C_{10}H_{21}CH{-}CO_2H}{}}{C_{10}H_{21}CH_2C{=}O} \xrightarrow[46-55\%]{-CO_2} C_{11}H_{23}\overset{O}{\overset{||}{C}}C_{11}H_{23}$$

chloride is stirred and cooled during rapid addition of an equivalent amount of triethylamine, and the mixture is let stand at room temperature. The triethylamine hydrochloride is removed by extraction with 2% sulfuric acid, and the wet solution of decylketene dimer is heated with 2% sulfuric acid to drive off the ether, hydrolyze the dimer, and decarboxylate the β-ketoacid. The laurone formed is distilled at 3 mm. and crystallized from acetone. Another example is the synthesis of 6-ketohexadecanedioic acid from δ-carbomethoxyvaleryl chloride.[2]

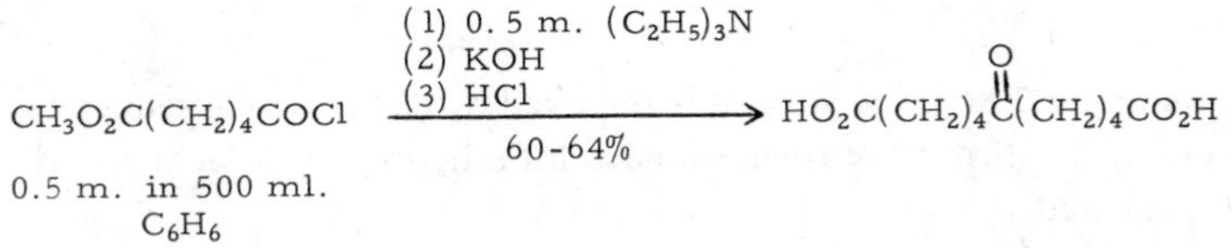

$$CH_3O_2C(CH_2)_4COCl \xrightarrow[60-64\%]{\substack{(1)\ 0.5\ m.\ (C_2H_5)_3N \\ (2)\ KOH \\ (3)\ HCl}} HO_2C(CH_2)_4\overset{O}{\overset{||}{C}}(CH_2)_4CO_2H$$

0.5 m. in 500 ml.
C_6H_6

Mixed anhydride synthesis. A method widely used for the synthesis of peptides[5] provides a general method of acylation, illustrated by the benzoylation of diethyl malonate.[6] Benzoic acid is converted into the triethylamine salt in toluene, and the solution is treated at 0° with ethyl chloroformate to form the mixed benzoic-carbonic anhydride, with precipitation of triethylamine hydrochloride. The second component, ethoxymagnesium malonic ester, is prepared from diethyl malonate in ethanol–ether

$$C_6H_5CO_2H\ +\ N(C_2H_5)_3 \qquad\qquad CH_2(CO_2C_2H_5)_2\ +\ Mg\ +\ CCl_4$$

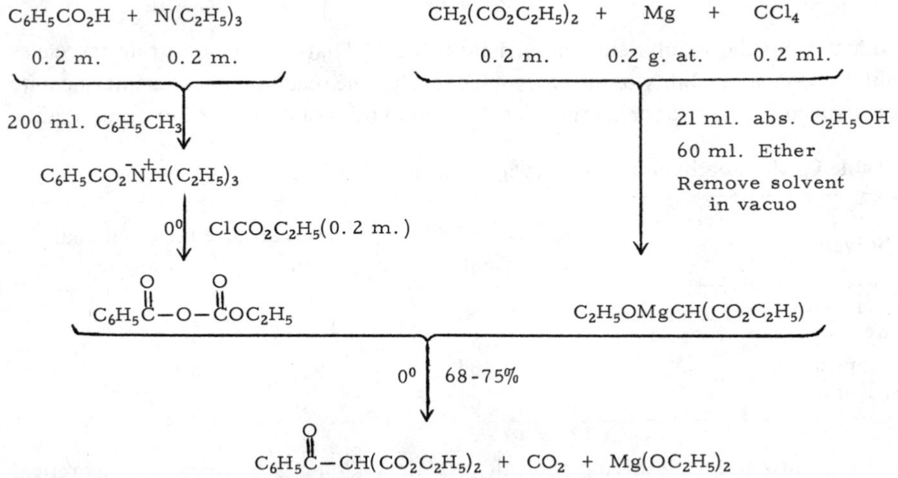

0.2 m.   0.2 m.   0.2 m.   0.2 g. at.   0.2 ml.

200 ml. $C_6H_5CH_3$ 21 ml. abs. C_2H_5OH
                                                       60 ml. Ether
$C_6H_5CO_2\overset{-}{N}\overset{+}{H}(C_2H_5)_3$ Remove solvent
                                                       in vacuo
      $0°$ | $ClCO_2C_2H_5(0.2\ m.)$

$$C_6H_5\overset{O}{\overset{||}{C}}{-}O{-}\overset{O}{\overset{||}{C}}OC_2H_5 \qquad\qquad C_2H_5OMgCH(CO_2C_2H_5)$$

            $0°$ | $68-75\%$

$$C_6H_5\overset{O}{\overset{||}{C}}{-}CH(CO_2C_2H_5)_2\ +\ CO_2\ +\ Mg(OC_2H_5)_2$$

with a catalytic amount of carbon tetrachloride. After removal of the solvents a solution of the residue in ether is added at 0° to the mixed anhydride; a rapid reaction occurs with evolution of carbon dioxide. Diethyl benzoylmalonate is isolated by distillation and refractionation.

Condensation catalyst. Triethylamine serves as basic catalyst for the Perkin condensation of *o*-nitrobenzaldehyde with phenylacetic acid and acetic anhydride to give α-phenyl-*trans-o*-nitrocinnamic acid.[7] A student experiment on the condensation of benzaldehyde with the same reactants includes a simple procedure for the separation of the *cis-* and *trans*-isomers.[8]

$$\underline{o}\text{-NO}_2\text{C}_6\text{H}_4\text{CHO} + \overset{\text{CO}_2\text{H}}{\underset{\text{CH}_2\text{C}_6\text{H}_5}{|}} + (\text{C}_2\text{H}_5)_3\text{N} + \text{Ac}_2\text{O} \xrightarrow[71-72\%]{\text{Refl. 15 min.}} \underline{o}\text{-NO}_2\text{C}_6\text{H}_4-\overset{}{\underset{}{\text{C}}}-\text{H}$$

0. 2 m.	0.29 m.	0. 2 m.	100 ml.

$$\text{C}_6\text{H}_5-\text{C}-\text{CO}_2\text{H}$$

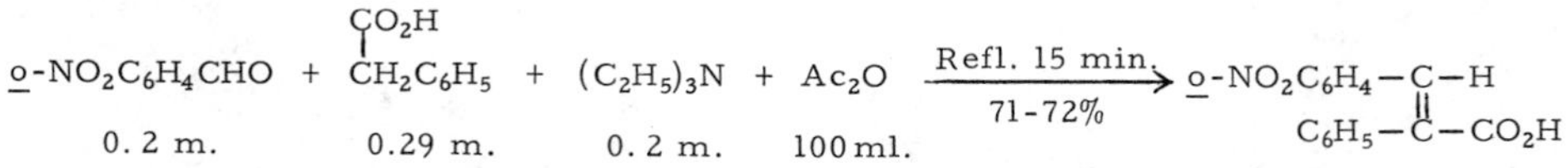

The reagent catalyzes the formation of hydrazones[9] and the reaction of amino acids with phthalic anhydride to form protected phthaloylamino acids, as illustrated for L-phenylalanine.[10] A mixture of the reactants indicated is refluxed under a water

$$\text{C}_6\text{H}_5\text{CH}_2\text{CHCO}_2\text{H}$$

$$+ \ \underline{\text{L}}\text{-C}_6\text{H}_5\text{CH}_2\overset{}{\underset{\text{NH}_2}{\text{CHCO}_2\text{H}}} + \text{C}_6\text{H}_5\text{CH}_3 + (\text{C}_2\text{H}_5)_3\text{N} \xrightarrow[91.5-95\%]{\text{Refl.}}$$

0. 1 m.	0. 1 m.	150 ml.	1. 3 ml.

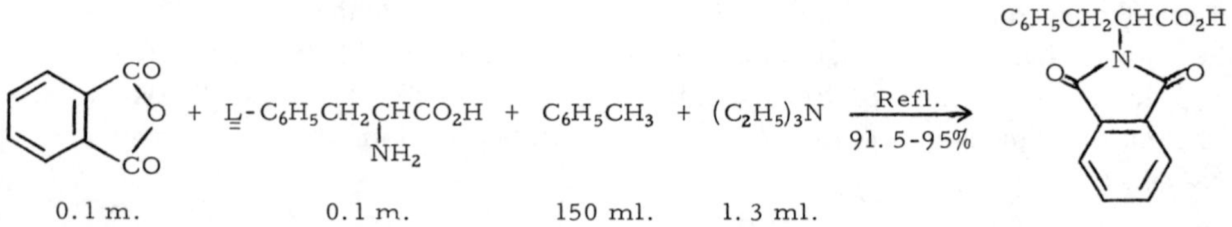

separator for about 2 hrs., the solvent is removed by distillation at reduced pressure, and the residue is stirred with water and a little hydrochloric acid to disintegrate lumps and collected.

Triethylamine is an effective catalyst for the cyanoethylation reaction.[11] Thus when a solution of acetylacetone, acrylonitrile, and triethylamine in a mixture of *t*-butanol (45 ml.) and water (15 ml.) was allowed to stand at 25° for 2 days, γ,γ-diacetylpimelonitrile separated in high yield and purity. A striking solvent effect is

$$\text{CH}_3\text{COCH}_2\text{COCH}_3 + 2 \ \text{CH}_2{=}\text{CHCN} \xrightarrow[77\%]{(\text{C}_2\text{H}_5)_3\text{N}} \text{NCCH}_2\text{CH}_2\overset{\text{COCH}_3}{\underset{\text{COCH}_3}{\text{CCH}_2\text{CH}_2\text{CN}}}$$

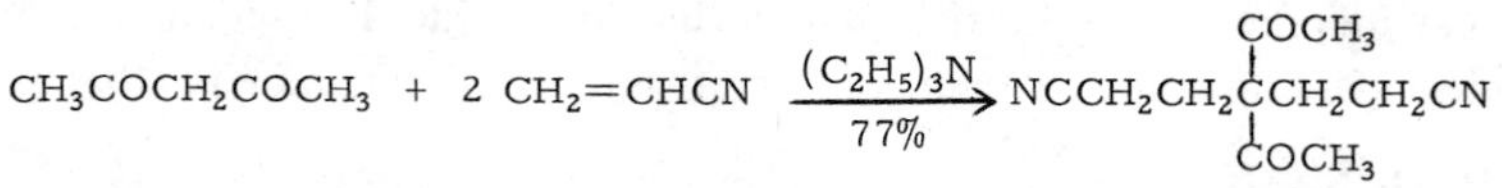

evident from the results summarized in Table 1. That the reaction rate increases with the solvating ability of the solvent means that the reaction involves intermediate ionic species. The cyanoethylation of the methiodides of 4- and of 2-methylpyridine

Table 1 Cyanoethylation of acetylacetone at 25°

Solvent	Ultimate yield, %	Time required to obtain half the ultimate yield, hrs.
1:1 (vol.) 95% Ethanol–Water	50	2
95% Ethanol	57	23
Isopropanol	67	89
t-Butanol	54	106

was carried out by dissolving the salt in water, adding 2–3 times the theoretical amount of acrylonitrile and 0.3 mole of triethylamine per mole of nitrile, and refluxing the mixture. The 4-methyl derivative gave the tris-(2-cyanoethyl) compound, and

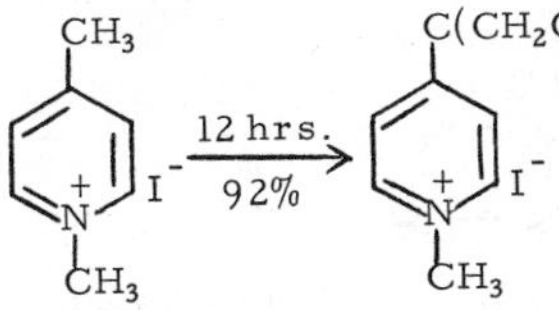

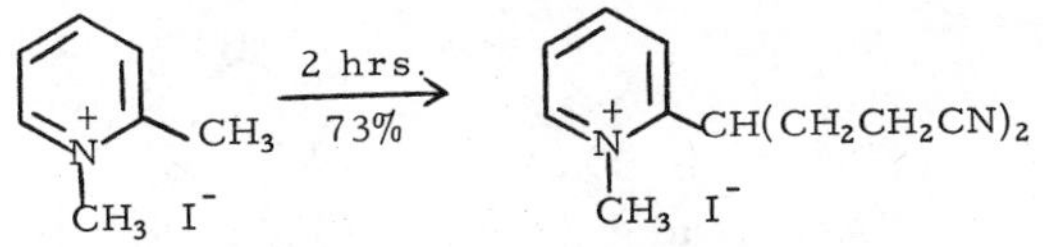

the 2-methyl derivative give only the bis-2-cyanoethyl derivative. The latter compound had been obtained in 34% yield with sodium methoxide as catalyst.

Triethylamine was used by a Ciba-Basel group in different ways in two successive steps of a steroid total synthesis.[12] Alkylation of the 1,3-diketone (2) with the halide (1) was accomplished by reaction in *t*-butanol under catalysis by triethylamine. Intramolecular aldolization of (3) with elimination of water to close ring C required

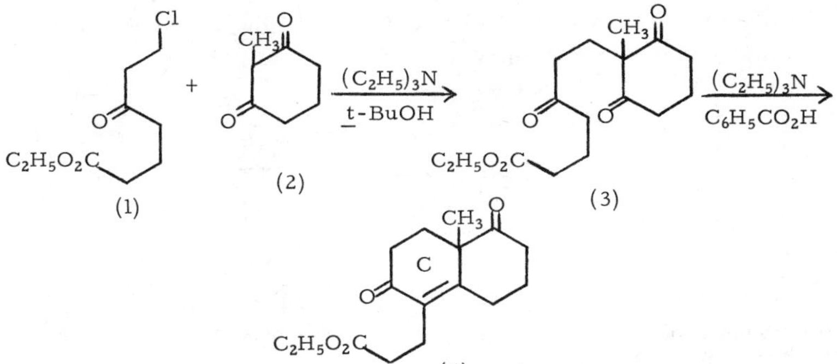

(1) (2) (3) (4)

more drastic conditions, and (3) is extremely sensitive, especially to bases. The cyclization was accomplished in good yield by treating a solution of (3) in xylene with equivalent amounts of benzoic acid and triethylamine to produce triethylamine benzoate and refluxing the solution under a water separator.

$RCOCH_2X \longrightarrow 2RCOCH_2OCOR'$. Moreland[13] developed a superior method for conversion of α-haloketones and other active halogen compounds into esters using a solution of the triethylamine salt of the acid in an organic solvent for reaction with the halide. For example equivalent amounts of benzoic acid and triethylamine were

$$C_6H_5CO_2H + N(C_2H_5)_3 \xrightarrow{\text{Acetone}} C_6H_5CO_2^- \overset{+}{N}H(C_2H_5)_3 \xrightarrow[85\%]{C_6H_5COCH_2Br} C_6H_5COCH_2OCOC_6H_5$$

dissolved in acetone, one equivalent of phenacyl bromide was added, and the homogeneous solution let stand at room temperature for 2 hrs. With the less reactive *p*-nitrobenzyl chloride, it was necessary to add a catalytic amount of sodium iodide and reflux for 2 hrs. (50–70% yield). The reaction has proved useful for the conversion of 21-iodo-ketosteroids into the 21-acetoxy compounds.[14]

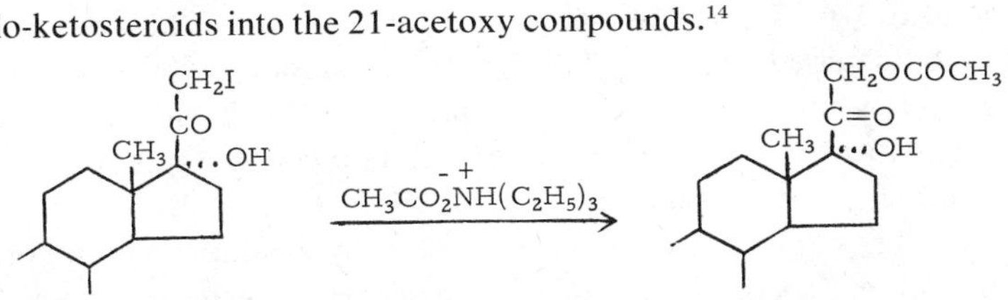

Dehydrohalogenation. Triethylamine effects cyclization of N-chloroformyl amino acids to the anhydrides in acetone or dioxane.[15] The reagent has the advantage over other bases that the hydrochloride separates readily from the solution.

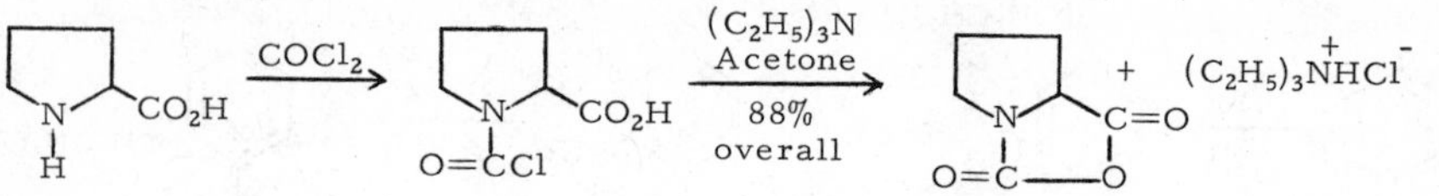

Dicyanocarbene has been generated by dehydrobromination of bromomalononitrile with triethylamine in the presence of tetramethylethylene as trapping agent and solvent.[16]

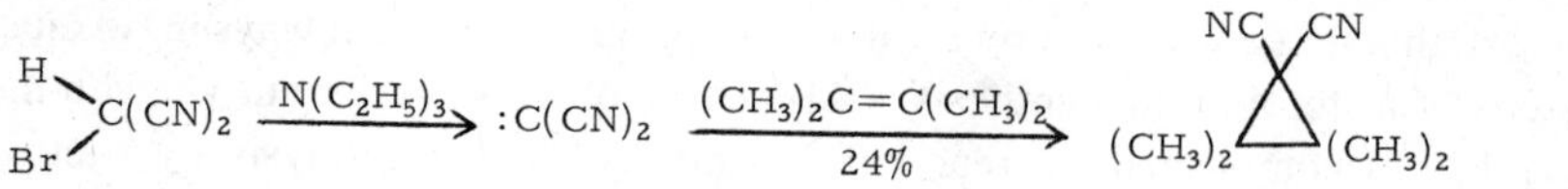

Breslow and co-workers[17] used triethylamine to effect Favorski rearrangement of α,α'-dibromodibenzyl ketone (1) to (2) and dehydrohalogenation of this intermediate to produce diphenylcyclopropenone (3).

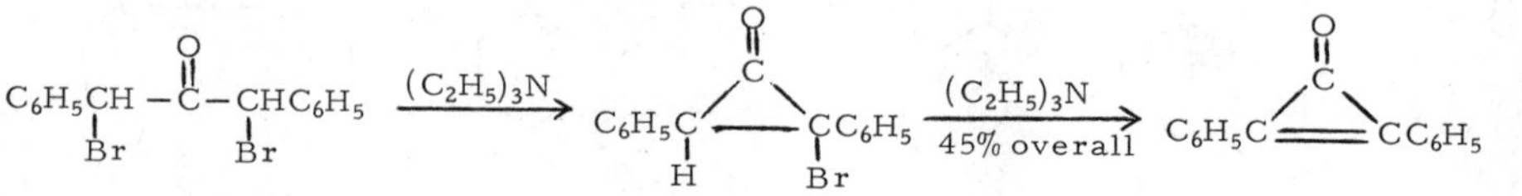

(1) (2) (3)

Price and Judge[18] effected dehydrobromination of α-bromo-γ-butyrolactone (2) to γ-crotonolactone (3) by reaction with triethylamine in ether.

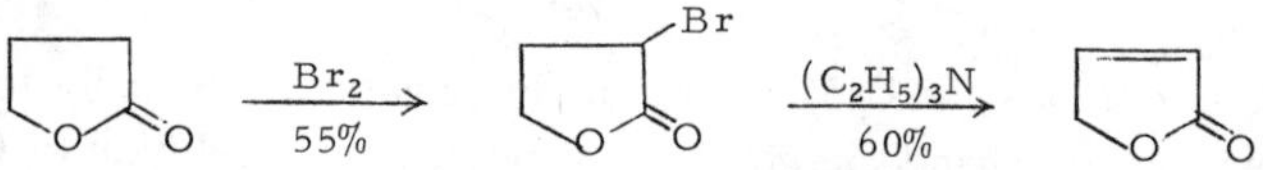

(1) (2) (3)

Reductive acetylation.[19] A suspension of 0.5 g. each of 2-methyl-1,4-naphthoquinone and zinc dust in 3 ml. of acetic anhydride is treated at 25° with one drop of triethylamine and swirled; a transient red color appears, the temperature rises, and in about 2 minutes the color is discharged. The mixture is boiled for a minute or two and extracted with a total of 10 ml. of hot acetic acid in portions. A carborundum boiling stone is added to the filtered solution, which is kept at the boiling point while water is added slowly to the saturation point (20–25 ml.); the colorless solution on cooling deposits 0.6 g. of pure 2-methyl-1,4-naphthohydroquinone diacetate, m.p. 112–113°. A variation of convenience in the preparation of a low-melting product such as vitamin K_1 hydroquinone diacetate[20] is to extract the reaction mixture with ether, wash the filtered extract with dilute hydrochloric acid, and then shake it vigorously with 3 or 4 portions of dilute alkali to destroy the acetic anhydride; evaporation of the dried ethereal solution affords the product. An earlier procedure[21] specified use of pyridine as basic catalyst, but this has the distinct disadvantage that the reaction mixture tends to acquire a yellow color owing to a side reaction between acetic anhydride, pyridine, and zinc. Use of triethylamine obviates the difficulty.

Esters can be prepared, usually in good yield, by heating a mixture of the acid, an alkyl halide, and triethylamine for a few hours.[22] The ester is washed free from the amine hydrohalide and purified by a standard procedure.

Grignard reagents. A Grignard reagent can be prepared in benzene or toluene solution if a stoichiometric amount of triethylamine is used as complexing agent.[23]

Starting the reaction is more difficult than in diethyl ether, but the method avoids use of the more expensive and hazardous solvent.

[1]J. C. Sauer, *Org. Syn., Coll. Vol.*, **4**. 560 (1963)
[2]L. J. Durham, D. J. McLeod, and J. Cason, *ibid.*, **4**, 555 (1963)
[3]M. S. Newman and P. Beal, III, *Am. Soc.*, **71**, 1506 (1949); see also V. Franzen, *Ann.*, **602**, 199 (1957)
[4]J. C. Sauer, *Am. Soc.*, **69**, 2444 (1947)
[5]J. R. Vaughan, Jr., *Am. Soc.*, **73**, 3547 (1951); R. A. Boissonnas, *Helv.*, **34**, 874 (1951); T. Wieland and H. Bernhard, *Ann.*, **572**, 190 (1951)
[6]J. A. Price and D. S. Tarbell, *Org. Syn., Coll. Vol.*, **4**, 285 (1963)
[7]D. F. De Tar, *ibid.*, **4**, 730 (1963)
[8]*Org. Expts.*, 224
[9]D. H. R. Barton, R. E. O'Brien, and S. Sternhell, *J. Chem. Soc.*, 470 (1962)
[10]A. K. Bose, *Org. Syn.*, **40**, 82 (1960)
[11]J. A. Adamcik and E. J. Miklasiewicz, *J. Org.*, **28**, 336 (1963), J. A. Adamcik and R. J. Flores, *ibid*, **29**, 572 (1964)
[12]P. Wieland, H. Ueberwasser, G. Anner, and K. Miescher, *Helv.*, **36**, 376 (1953)
[13]W. T. Moreland, Jr., *J. Org.*, **21**, 820 (1956)
[14]E. S. Rothman, T. Perlstein, and M. E. Wall, *J. Org.*, **25**, 1966 (1960)
[15]A. A. Randall, *J. Chem. Soc.*, 374 (1962); W. Dvonch and H. E. Alburn, *J. Org.*, **29**, 3719 (1964)
[16]J. S. Swenson and D. J. Renaud, *Am. Soc.*, **87**, 1394 (1965)
[17]R. Breslow, T. Eicher, A. Krebs, R. A. Peterson, and J. Posner, *Am. Soc.*, **87**, 1323 (1965)
[18]C. C. Price and J. M. Judge, *Org. Syn.*, **45**, 22 (1965)
[19]L. F. Fieser, "Expts. in Org. Chem.," 2nd ed., 399 (1941)
[20]L. F. Fieser, *Am. Soc.*, **61**, 3467 (1939)
[21]L. F. Fieser, W. P. Campbell, E. M. Fry, and M. Gates, *Am. Soc.*, **61**, 3216 (1939)
[22]R. H. Mills, M. W. Farrar, and O. J. Weinkauff, *Chem. Ind.*, 2144 (1962)
[23]E. C. Ashby and R. Reed, *J. Org.*, **31**, 971 (1966)

Triethylenediamine (TED), 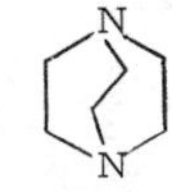

Mol. wt. 112.18, m.p. 159–160°, b.p. 174°, pKa_1 2.95, pKa_2 8.60. Solubility 46 g./100. H_2O at 26°; about 50 g./100 g. benzene at 26°. Suppliers: Houdry Chem. Co., Columbia, Howe and French.

Screttas and Eastham[1] found that this base forms crystalline complexes with organometallic compounds of magnesium, lithium, and zinc. Thus on addition of a benzene solution of TED to a solution of butyllithium a crystalline complex separates; the composition is $Bu_2Li_2-TED-Bu_2Li_2$. Triethylamine (1) and quinuclidine (2) form similar complexes but they are considerably more soluble than the corresponding TED complex.

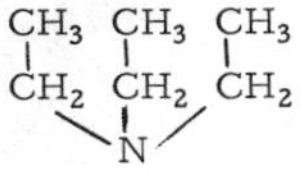

(1)

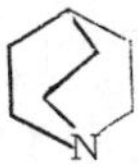

(2)

The complexes are useful for characterization; for lithium reagents, most of which are liquids, the amine complexes provide the first nondestructive technique for obtaining crystalline derivatives. The complexes are also useful because they can be used in conventional reactions of the organometallic compounds. The three bases have also been found to catalyze reactions of alkyllithiums, particularly metalation. Thus reaction of toluene with an alkyllithium to form benzyllithium is easily promoted by all three complexing bases.

[1]C. G. Screttas and J. F. Eastham, *Am. Soc.*, **87**, 3276 (1965)

Triethylenetetramine, $H_2NCH_2CH_2NHCH_2CH_2NHCH_2CH_2NH_2$. Mol. wt. 146.24, b.p. 155–161°/18 mm. Suppliers: B, E, F, MCB.

Metal-bearing enzymes such as leucine amino peptidase catalyze the hydrolysis of N-terminal peptide bonds through a process involving chelation between the enzyme, the substrate, and the metal ion. Collman[1] reported the selective N-terminal hydrolysis of simple peptides by *cis*-hydroxyaquotriethylenetetramine cobalt (III)

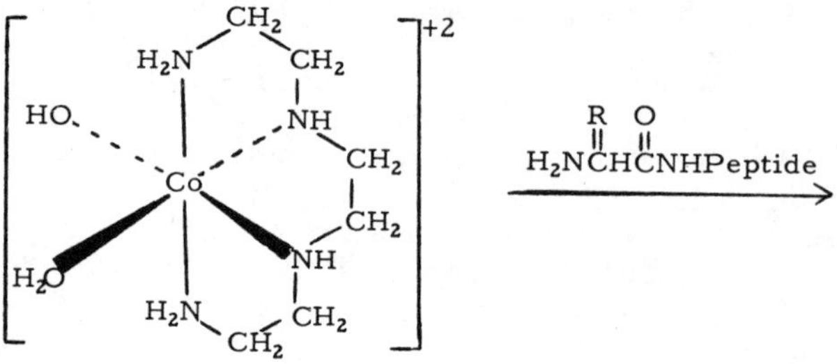
HO
H2O
Co
H2N
NH
CH2
+2
R O
H2NCHCNHPeptide

(1)

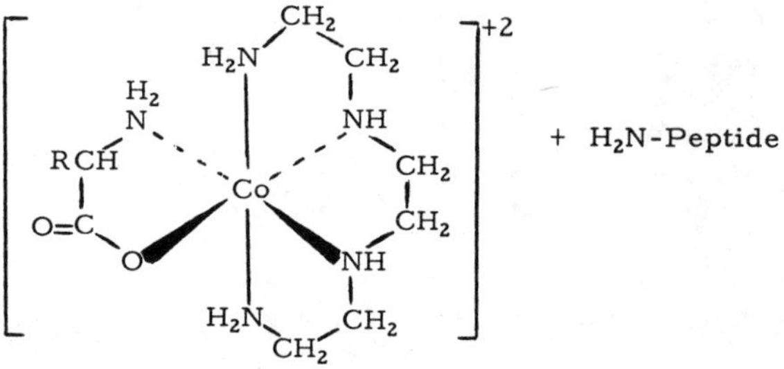
RCH
O=C
H2
N
O
Co
H2N
NH
CH2
+2
+ H2N-Peptide

(2)

ions. Formula (1) shows the *cis*-α-form, one of two possible stereochemical forms of the complex. In aqueous solution of pH 7–8 at 65° the complex rapidly effects hydrolysis of the N-terminal amino acid residue and converts it into the inert metal complex (2). Although stoichiometric rather than catalytic, the process is perhaps the best model to date for the *in vitro* action of *exo*-metal peptidases.

[1]J. P. Collman and D. A. Buckingham, *Am. Soc.*, **85**, 3039 (1963)

Triethyl orthoformate, $HC(OC_2H_5)_3$. Mol. wt. 148.20, b.p. 146°, sp. gr. 0.94. Suppliers: Eastern, Kay-Fries, A, B, E, F, MCB.

The ester is prepared[1] by adding sodium over a period of 2 hrs. to a solution of chloroform in absolute ethanol:

$$2\ CHCl_3 + 6\ C_2H_5OH + 6\ Na \xrightarrow[45\%]{} 2\ HC(OC_2H_5)_3 + 6\ NaCl + 3\ H_2$$

Preparation of aldehydes. A Grignard reagent reacts with triethyl orthoformate with displacement of one ethoxy group and formation of an acetal, which can be hydrolyzed with dilute acid to the aldehyde. In the synthesis of *n*-hexaldehyde,[2]

$$\underline{n}\text{-}C_5H_{11}MgBr \xrightarrow{HC(OC_2H_5)_3} \underline{n}\text{-}C_5H_{11}CH(OC_2H_5)_2 \xrightarrow[45\text{-}50\%\ \text{overall}]{HCl} \underline{n}\text{-}C_5H_{11}CHO$$

the crude acetal is hydrolyzed by heating it with dilute sulfuric acid and distilling the aldehyde into sodium bisulfite solution. The solution of bisulfite addition compound is steam distilled to remove *n*-amyl alcohol, treated with sodium bicarbonate, and the free aldehyde is removed by steam distillation.

The preparation of phenanthrene-9-aldehyde[3] starts with crude 9-bromophenanthrene obtained by bromination of technical phenanthrene (90%) and purified only by distillation (m.p. 54–56°; crystallization[4] gives pure material, m.p. 65–66°). The magnesium (50.3 g.) is placed in a 5-l. flask provided with a stirrer, a 500-ml. separatory funnel, a nitrogen-inlet tube, and a large reflux condenser fitted at the top with a 1-l. separatory funnel. The crude bromophenanthrene (514 g.) is

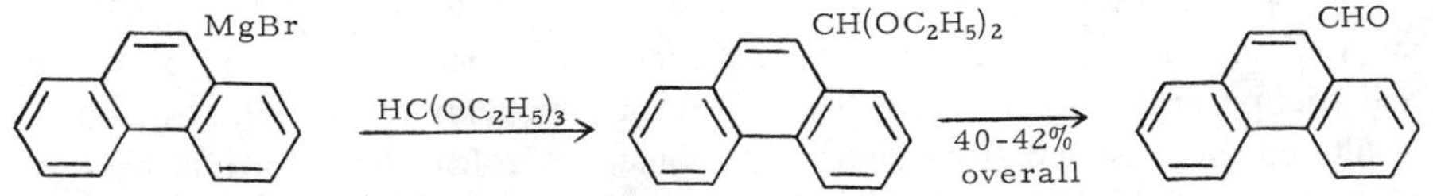
MgBr
HC(OC2H5)3
CH(OC2H5)2
40-42%
overall
CHO

melted and placed in the smaller separatory funnel and kept at about 70° by careful heating with a microburner. Ether (1 l.) is placed in the large separatory funnel. After the air has been displaced with nitrogen, ether and melted bromophenanthrene are run into the flask at rates sufficient to maintain gentle refluxing and to discharge both funnels at the same time. After a time, the Grignard reagent begins to precipitate on the sides of the flask, and 1 l. of benzene is added to dissolve it. When the reaction is complete, triethyl orthoformate (296.4 g.) is added gradually from the lower funnel, and the mixture is refluxed for 6 hrs. The mixture is then cooled in an ice bath, 1 l. of 10% hydrochloric acid is run in, and the ether–benzene layer is separated and evaporated at reduced pressure. One liter of 25% sulfuric acid is added to the residual acetal, and the mixture is refluxed for 12 hrs. to effect hydrolysis. The crude aldehyde is purified through the bisulfite addition compound, since anthracene-9-aldehyde from anthracene present in the starting material forms no addition compound and is thus eliminated; crystallization gives pure aldehyde, m.p. 101°.

A route to α,β-acetylenic aldehydes is based on the fact that triethyl orthoformate in the presence of zinc iodide as catalyst reacts with a terminal acetylene with elimination of ethanol and formation of an acetal, as illustrated for the preparation of phenylpropargyl aldehyde.[5] The reactants are heated neat with the catalyst to about 135°, and ethanol is removed by distillation (about 1 hr.).

$$C_6H_5C\equiv CH + H\dot{C}(OEt)_3 \xrightarrow[72-78\%]{ZnI_2} C_6H_5C\equiv CCH(OC_2H_5)_2 + C_2H_5OH$$

$$\downarrow H_2O\ (H^+)$$

$$C_6H_5C\equiv CCH{=}O$$

Triethyl orthoformate has been used for formylation of phenols in methylene chloride in the presence of aluminum chloride.[6]

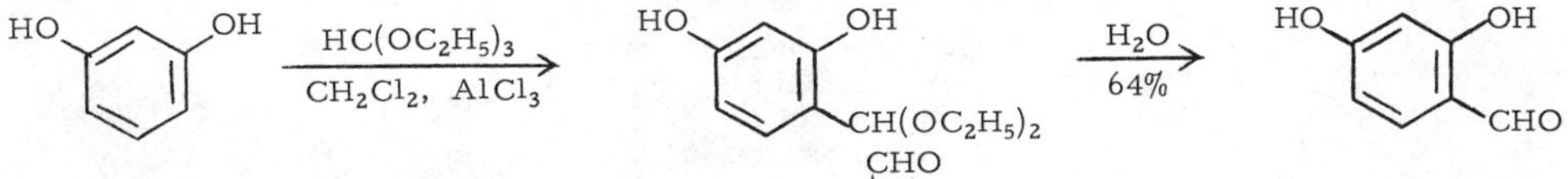
HO
OH
HC(OC2H5)3
CH2Cl2, AlCl3
HO
OH
CH(OC2H5)2
H2O
64%
HO
OH
CHO

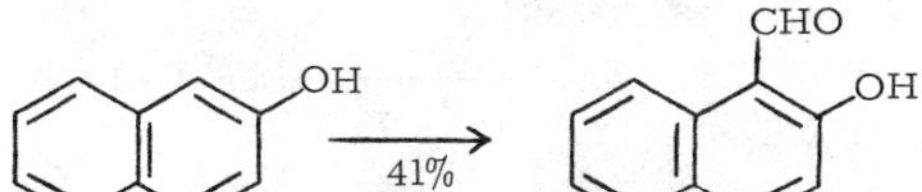
OH
41%
CHO
OH

Diethyl acetals and ketals. Claisen[7] showed that triethyl orthoformate reacts with aldehydes and ketones to form diethyl acetals and diethyl ketals. The reaction

$$R_2C{=}O + HC(OC_2H_5)_3 \longrightarrow R_2C(OC_2H_5)_2 + HCO_2C_2H_5$$

usually is carried out in absolute ethanol in the presence of a catalytic amount of ferric chloride, ammonium chloride, or ammonium nitrate either at room temperature (6.8 hrs.) or at the boiling point (30 min.). An example is the conversion of the sensitive acrolein into its more stable diethyl acetal.[8] A warm solution of 3 g. of

$$\underset{0.79\ m.}{CH_2{=}CHCHO} + \underset{0.97\ m.}{HC(OC_2H_5)_3} \xrightarrow[72\text{-}80\%]{NH_4NO_3} CH_2{=}CHCH(OC_2H_5)_2 + HCO_2C_2H_5$$

ammonium nitrate in 50 ml. of absolute ethanol is added to a mixture of acrolein and triethyl orthoformate; after 7 hrs. at room temperature the light red solution is filtered and distilled from sodium carbonate. Acrolein diethyl acetal has been obtained in 62% yield from the aldehyde, ethanol, and *p*-toluenesulfonic acid as catalyst.[9]

A synthesis of phenylpropargyl aldehyde less efficient than that described above is of interest in demonstrating acetal formation as a means of protecting an aldehyde function during dehydrohalogenation with a strong base.[10]

$$C_6H_5CH{=}CHCHO \xrightarrow{Br_2} [C_6H_5CHBrCHBrCHO] \xrightarrow[75\text{-}85\%]{K_2CO_3} C_6H_5CH{=}CBrCHO$$

$$\xrightarrow[82\text{-}86\%]{HC(OC_2H_5)_3} C_6H_5CH{=}CBrCH(OC_2H_5)_2 \xrightarrow[80\text{-}86\%]{KOH} C_6H_5C{\equiv}CCH(OC_2H_5)_2$$

$$\xrightarrow[70\text{-}81\%]{dil.\ H_2SO_4} C_6H_5C{\equiv}CCHO$$

For a useful synthetic sequence involving reaction of an aldehyde diethylacetal with an alkyl vinyl ether, *see* Ethyl vinyl ketone.

Reaction with reactive-methylene compounds. Claisen[11] developed a method for the preparation of diethyl ethoxymethylenemalonate by heating a mixture of triethyl orthoformate, malonic ester, and acetic anhydride with a catalytic amount of zinc chloride. The yield cited is that obtained in a standardized procedure.[12]

$$C_2H_5OCH(OC_2H_5)_2 + CH_2(CO_2C_2H_5)_2 + 2\,(CH_3CO)_2O \xrightarrow[50\text{-}60\%]{ZnCl_2}$$

$$C_2H_5OCH{=}CH(CO_2C_2H_5)_2 + 2\ CH_3CO_2C_2H_5 + 2\ CH_3CO_2H$$

The reaction of N-*n*-heptylurea (1) with triethyl orthoformate (2) and malononitrile (3) to form 3-*n*-heptylureidomethylenemalononitrile (4) probably involves initial condensation of (2) and (3) to form ethoxymethylenemalononitrile (5).[13]

$$\underset{(1)}{n\text{-}C_7H_{15}NHCONH_2} + \underset{(2)}{HC(OC_2H_5)_3} + \underset{(3)}{CH_2(CN)_2} \xrightarrow[80\text{-}83\%]{}$$

$$\underset{(4)}{n\text{-}C_7H_{15}NHCONHCH{=}C(CN)_2}$$

$$(2) + (3) \longrightarrow \underset{(5)}{\underset{OC_2H_5}{CH}{=}C(CN)_2} \xrightarrow{(1)} (4) + C_2H_5OH$$

Preparation of enol ethers. Condensation of a 3-keto-5α-steroid (1) with triethyl orthoformate in ethanol containing a trace of hydrogen chloride affords the diethyl ketal (2) in good yield; when refluxed in xylene the ketal loses a molecule of ethanol and affords the Δ^2-enol ether (3).[14] A Δ^4-3-ketosteroid (4) when condensed with

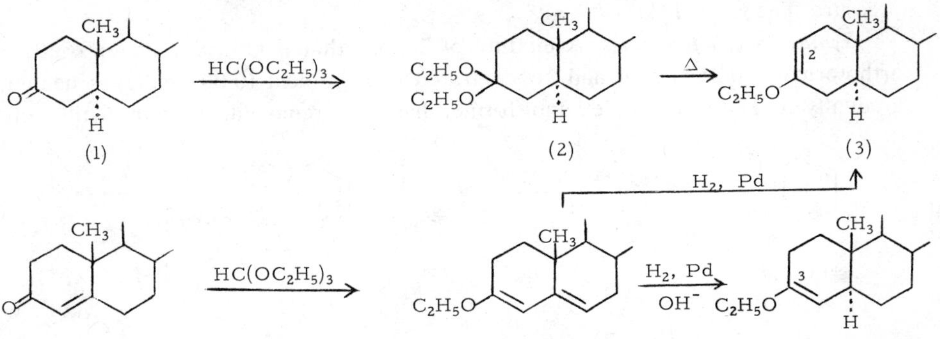

CH3
HC(OC2H5)3
C2H5O
H
Δ
2
H2, Pd
OH⁻
3

(1) (2) (3) (4) (5) (6)

triethyl orthoformate in the same way affords the dienol ethyl ether (5) directly, rather than the diethyl ketal. Under ordinary conditions selective hydrogenation of the dienol ether (5) is attended with bond migration, for the product is the Δ^2-enol ether (3). However, hydrogenation of (5) in the presence of base (NaOH, pyridine) saturates the 5,6-double bond to give the Δ^3-enol ethyl ether (6).[15] Conversion of cyclohexane-1,3-diones into enol ethers has been accomplished with use of triethyl orthoformate.[16]

Reaction with primary arylamines. Investigations of Wichelhaus[17] and of Claisen[18] of the reaction of aniline with triethyl orthoformate were later clarified by Roberts and Vogt,[19] who showed that by adjustment of the catalyst and the experimental conditions it is possible to produce in good yield any one of three products from the two reactants. The initial product in each use is ethyl N-phenylformimidate (1).

$$C_6H_5NH_2 + (C_2H_5O)_2\underset{H}{C}-OC_2H_5 \longrightarrow C_6H_5N{=}\underset{H}{C}-OC_2H_5 + C_2H_5OH$$

(1)

$$(1) \xrightarrow{C_6H_5NH_2} C_6H_5N{=}\underset{H}{C}-\underset{H}{N}C_6H_5 + C_2H_5OH \qquad (2)$$

$$(1) \xrightarrow{(H_2SO_4)\ 140^0} C_6H_5\underset{C_2H_5}{N}-CH{=}O \qquad (3)$$

In the absence of an acid catalyst, (1) reacts with aniline to form N,N′-diphenylformamidine (2), which also results in the presence of hydrogen chloride as catalyst if 2 moles of aniline are present per mole of triethyl orthoformate. Sulfuric acid is a specific catalyst for the allylic rearrangement of (1) to N-ethylformanilide (3). Thus (3) can be prepared from 0.50 mole of aniline, 0.75 mole of triethyl orthoformate, and 0.02 mole of concentrated sulfuric acid by distilling off the ethanol produced in the formation of (1) and then raising the temperature to 140° to effect rearrangement. By this procedure *p*-chloroaniline has been converted into N-ethyl-*p*-chloroformanilide in yields of 87–92%.[20] Hydrogen chloride catalyzes the formation of (2) but not the rearrangement to (3). A procedure for the preparation of ethyl

N-phenylformimidate[21] calls for heating 1.01 moles of aniline with 1 ml. of concentrated hydrochloric acid to remove the water present in the acid, adding 1.50 moles of triethyl orthoformate, distilling ethanol until the amount collected is close to the theoretical amount (2 hrs.), and then fractionating the mixture at reduced pressure. The yield of (1) is 78–84%.

Synthesis of heterocycles. A mixture of 2-aminothiophenol (0.17 mole), triethyl orthoformate (0.25 mole), and concentrated sulfuric acid (0.007 mole) is heated gradually to 180° until the ethanol formed has been removed; vacuum distillation

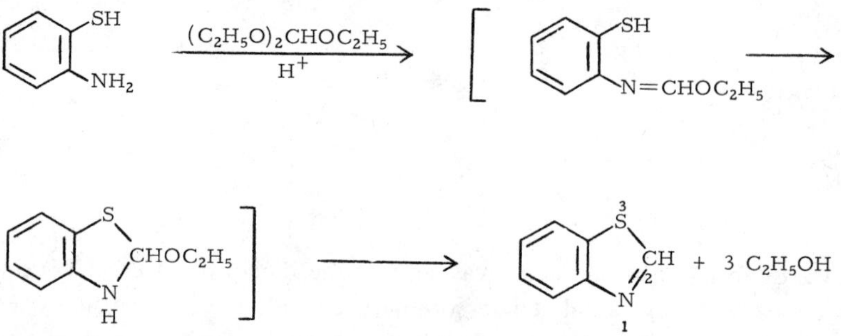
SH
NH2
(C2H5O)2CHOC2H5
H+
SH
N=CHOC2H5
S
CHOC2H5
N
H
3
S
CH + 3 C2H5OH
2
N
1

then affords benzothiazole.[22] Use of ortho esters of the type $RC(OC_2H_5)_3$ gives 2-alkylbenzothiazoles. Applied to *o*-aminophenol, the reaction affords benzoxazoles. Yields of 75–85% are reported.

Steroid ortho esters. Under acid catalysis, triethyl orthoformate reacts with a steroid containing the dihydroxyacetone side chain (1) to form a protected derivative (2).[23] Under the conditions used a Δ^4-3-keto group is not converted into the

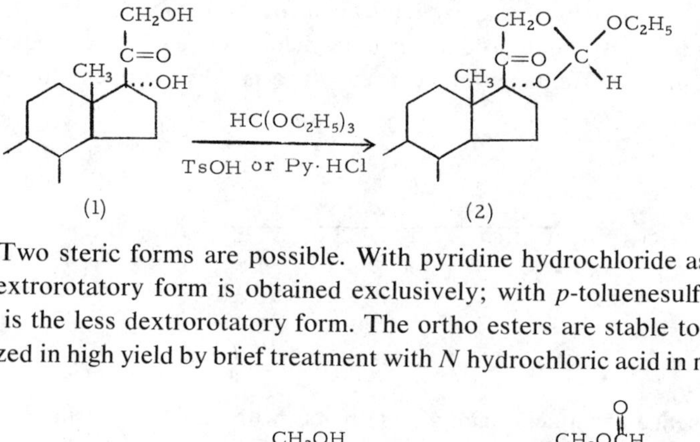
CH2OH
C=O
CH3
OH
HC(OC2H5)3
TsOH or Py·HCl
CH2O
OC2H5
C=O
C
CH3
O
H

(1) (2)

enol ether. Two steric forms are possible. With pyridine hydrochloride as catalyst the more dextrorotatory form is obtained exclusively; with *p*-toluenesulfonic acid the product is the less dextrorotatory form. The ortho esters are stable to base but are hydrolyzed in high yield by brief treatment with *N* hydrochloric acid in methanol.

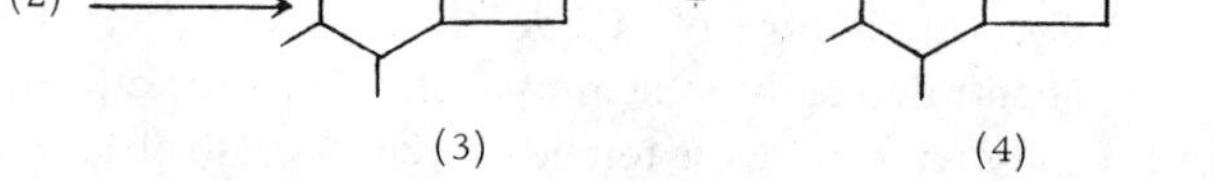
(2)
(COOH)2
CH2OH
C=O
O
CH3
OCH
+
O
CH2OCH
C=O
CH3
OH

(3) (4)

Hydrolysis with oxalic acid gives a mixture of the 17α-monoformate (3) and the 21-monoformate (4), separable by chromatography. This route provides the only route to 21-hydroxy-17-monoesters.

Esterification. Carboxylic acids are converted into their ethyl esters when heated with an excess of the reagent. Even hindered acids such as 2,4,6-trimethylbenzoic acid are esterified.[24] An acid catalyst usually is not necessary and may be deleterious; however, esterification of nicotinic acid and of hippuric acid was accomplished with *p*-toluenesulfonic acid in dimethylformamide.

Cyclic amidines.[25] Cyclization of (1) to (2) is effected by refluxing the base with triethyl orthoformate. The reagent supplies carbon atom 6 of (2).

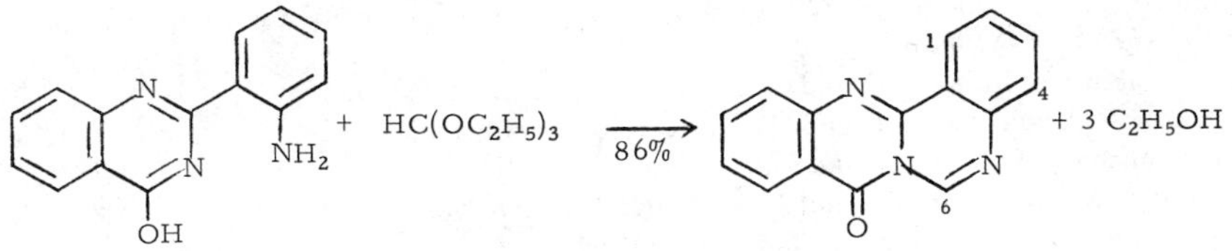

(1) (2)

Oxaadamantanes. Vogl, Anderson, and Simons[26] found that scyllitol (1) reacts with triethyl orthoformate in dimethyl sulfoxide at 200° to form hexaoxadiamantane (2, m.p. 303–305°). The product is of interest because it is related to the hydrocarbon

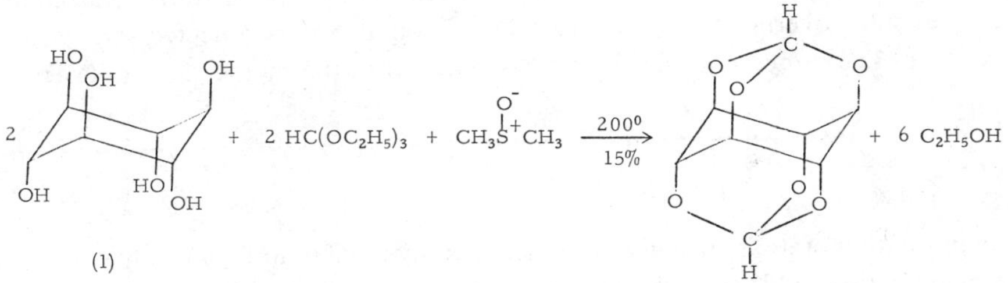

(1) (2)

diamantane (formerly congressane), the second member of the adamantane series. Stetter and Steinacker[27] had prepared trioxaadamantane (4) by reaction of *cis*-phloroglucitol (3) with triethyl orthoformate in methanol containing hydrogen chloride.

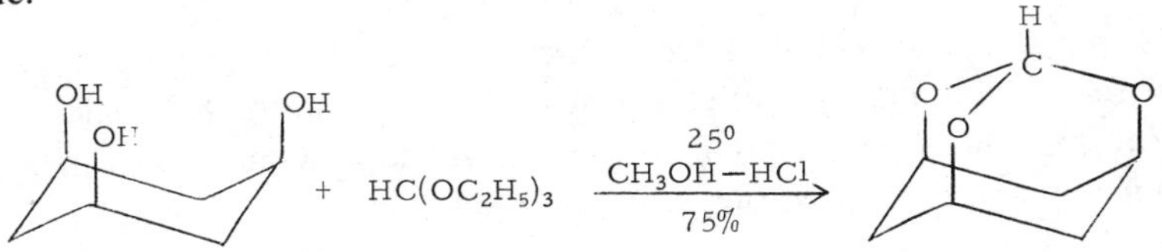

(3) (4)

[1]W. E. Kaufmann and E. E. Dreger, *Org. Syn., Coll. Vol.*, **1**, 258 (1941)
[2]G. B. Bachman, *ibid.*, **2**, 323 (1943)
[3]C. A. Dornfeld and G. H. Coleman, *ibid.*, **3**, 701 (1955)
[4]C. A. Dornfeld, J. E. Callen, and G. H. Coleman, *ibid.*, **3**, 134 (1955)
[5]B. W. Howk and J. C. Sauer, *Am. Soc.*, **80**, 4607 (1958); *Org. Syn., Coll. Vol.*, **4**, 801 (1963)
[6]H. Gross, A. Rieche, and G. Matthey, *Ber.*, **96**, 308 (1963)
[7]L. Claisen, *Ber.*, **26**, 2729 (1893); *Ann.*, **297**, 76 (1897)
[8]J. A. VanAllan, *Org. Syn., Coll. Vol.*, **4**, 21 (1963)
[9]D. I. Weisblat *et al.*, *Am. Soc.*, **75**, 5893 (1953)
[10]C. F. H. Allen and C. O. Edens, Jr., *Org. Syn., Coll. Vol.*, **3**, 731 (1955)
[11]L. Claisen, *Ber.*, **29**, 1005 (1896); **31**, 1019 (1898); **40**, 3903 (1907)

[12]W. E. Parham and L. J. Reed, *Org. Syn., Coll. Vol.*, **3**, 395 (1955)

[13]B. B. Kehm and C. W. Whitehead, *ibid.*, **4**, 515 (1963)

[14]A. Serini and H. Köster, *Ber.*, **71**, 1766 (1938); H. H. Inhoffen *et al., ibid.*, **84**, 361 (1951)

[15]R. Gardi, P. P. Castelli, and A. Ercoli, *Tetrahedron Letters*, 497 (1962)

[16]E. G. Meek, J. H. Turnbull, and W. Wilson, *J. Chem. Soc.*, 811 (1953)

[17]H. Wichelhaus, *Ber.*, **2**, 115 (1869)

[18]L. Claisen, *Ann.*, **287**, 360 (1895)

[19]R. M. Roberts and P. J. Vogt, *Am. Soc.*, **78**, 4778 (1956)

[20]*Idem, Org. Syn., Coll. Vol.*, **4**, 420 (1963)

[21]*Idem, ibid.*, **4**, 464 (1963)

[22]G. L. Jenkins, A. M. Knevel, and C. S. Davis, *J. Org.*, **26**, 274 (1961)

[23]R. Gardi, R. Vitali, and A. Ercoli, *Tetrahedron Letters*, 448 (1961)

[24]H. Cohen and J. D. Mier, *Chem. Ind.*, 349 (1965)

[25]K. Butler, M. W. Partridge, and J. A. Waite, *J. Chem. Soc.*, 4970 (1960); M. W. Partridge, S. A. Slorach, and H. J. Vipond, *ibid.*, 3670 (1964)

[26]O. Vogl, B. C. Anderson, and D. M. Simons, *Tetrahedron Letters*, 415 (1966)

[27]H. Stetter and K. H. Steinacker, *Ber.*, **86**, 790 (1953)

Triethyloxonium fluoroborate, $(C_2H_5)_3O^+BF_4^-$. Mol. wt. 189.90, m.p. 92°.

Preparation.[1] An ethereal solution of epichlorohydrin is added dropwise to a solution of boron trifluoride etherate in ether under reflux. The mixture is stirred for 2 hrs., let stand overnight, and the crystalline oxonium salt is collected and washed with ether. The boric acid ester formed is retained in the mother liquor. Triethyl-

$$3 \quad \overset{ClCH_2CH}{\underset{CH_2}{|}}O + 4(C_2H_5)_2\overset{+}{O}BF_3 + 2(C_2H_5)_2O \xrightarrow{90\%} 3(C_2H_5)_3O^+BF_4^- + \left[\overset{ClCH_2CH-O}{\underset{C_2H_5O\dot{C}H_2}{|}}\right]_3 B$$

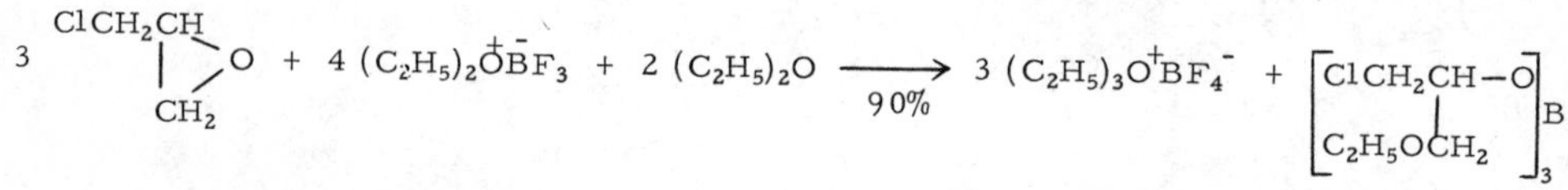

oxonium fluoroborate is insoluble in ether, readily soluble in liquid sulfur dioxide, and moderately soluble in methylene chloride. For one use, *see* Dimethylformamide diethyl acetal.

Reactions observed by Meerwein et al.[2] The reagent ethylates alcohols, phenols, and carboxylic acids. A phenol or acid will react with the reagent in the form of the

$$(C_2H_5)_3O^+BF_4^- + ROH \longrightarrow (C_2H_5)_2O + ROC_2H_5 + BF_3 + HF$$

sodium salt in aqueous solution, since it reacts more readily with the salt than with water. It reacts with ammonia to give a mixture of di- and triethylamine and with pyridine to give an N-ethylpyridinium salt. Urea and acetamide are converted at room temperature into salts. Nitriles afford nitrilium salts:

$$CH_3C\equiv N + (C_2H_5)_3O^+BF_4^- \xrightarrow{89\%} (CH_3C\equiv \overset{+}{N}C_2H_5)BF_4^- + (C_2H_5)_2O$$

Ketones and esters react to give salts, for example:

$$R_2C=O + (C_2H_5)_3O^+BF_4^- \longrightarrow [R_2C=\overset{+}{O}C_2H_5]BF_4^-$$

Nitronic esters (2). These previously little known compounds are obtained in quantitative yield by reaction with sodium salts of nitroparaffins at 0°.[3] Methylene

$$RR'C=\overset{-}{N}O_2\overset{+}{Na} + (C_2H_5)_3\overset{+}{O}BF_4^- \longrightarrow RR'C=\overset{+}{N}\overset{O^-}{\underset{OC_2H_5}{\diagdown}} + NaBF_4 + (C_2H_5)_2O$$

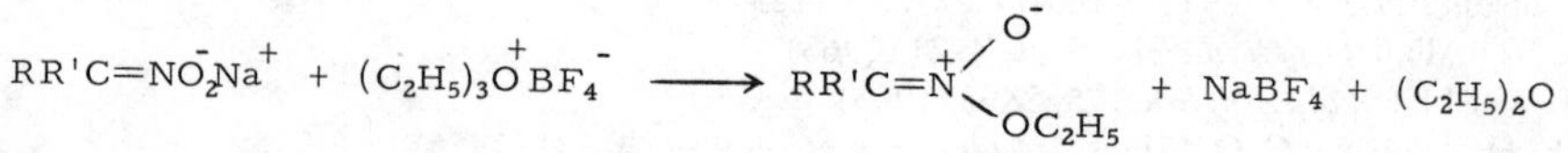

(1)   (2)

chloride is the solvent of choice because sodium fluoroborate and the reagent (taken in excess) are insoluble and can be removed by filtration.

Debenzoylation. In the course of the synthesis of a tricycline, Muxfeldt and Rogalski[4] employed Meerwein's reagent for the removal of an N-benzoyl group.

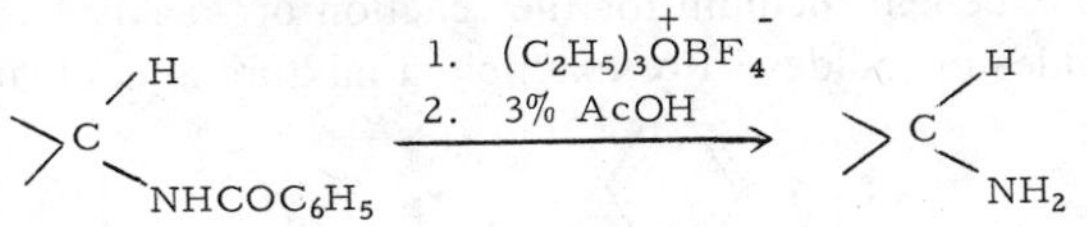
H
C
NHCOC6H5
1. (C2H5)3OBF4
2. 3% AcOH
H
C
NH2

Treatment with the reagent was followed by hydrolysis in dioxane with 3% aqueous acetic acid.

Imino ethers. Harley-Mason and Leeney[5] found that triethyloxonium fluoroborate converts oxindole (1) into the enol ether (2, m.p. 110°) and that (2) on being heated in vacuum is isomerized to the imino ether (3, m.p. 63°). If (3) is heated above its

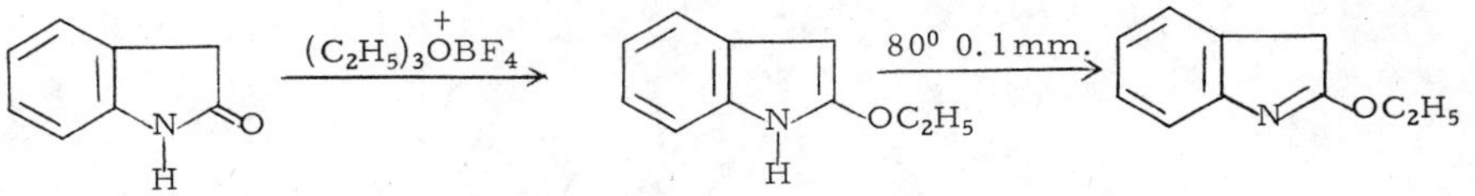
(C2H5)3OBF4
80° 0.1mm.
OC2H5
OC2H5

(1) (2) (3)

melting point, it solidified and then remelts at 110°, reconversion to (2) having taken place. Solutions of (2) and (3) in carbon tetrachloride have identical spectra and evidently contain both (2) and (3). In a synthesis of the corrin system, which characterizes the vitamin B_{12} structure, Eschenmoser *et al.*[6] made repeated use of Meerwein's reagent. An example:

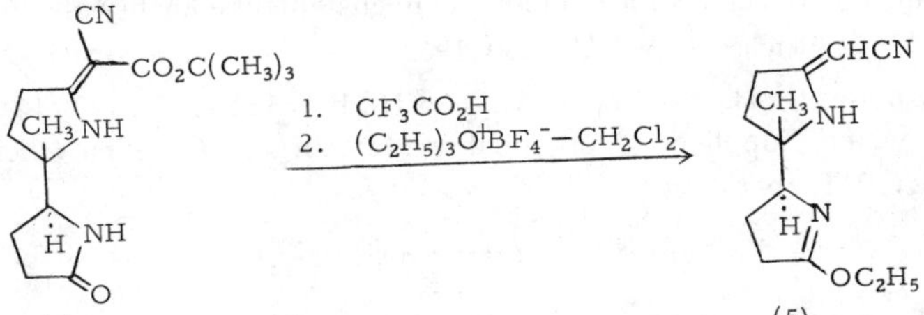
CN
CO2C(CH3)3
CH3 NH
H NH
1. CF3CO2H
2. (C2H5)3O+BF4−—CH2Cl2
CHCN
CH3 NH
H N
OC2H5

(4) (5)

Paquette[7] found the usual reagents (dimethyl sulfate, benzenesulfonyl chloride–pyridine) ineffective for conversion of the amide function of 1,3-dihydro-3,5,7-trimethyl-2H-azepinone-2 (6) into an imino derivative but that this conversion was effected readily with triethyloxonium fluoroborate to give the 3H-azepine (7).

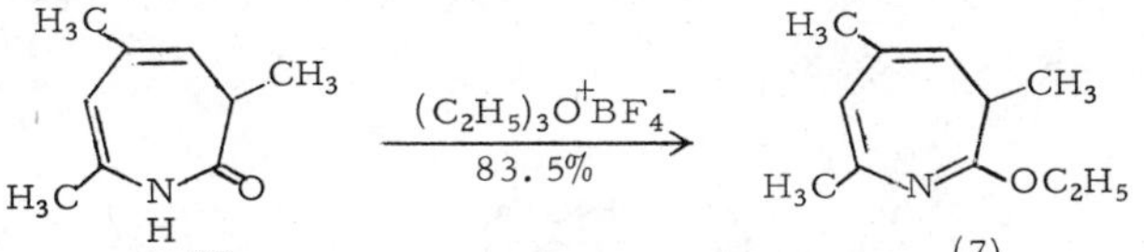
H3C
CH3
H3C
(C2H5)3O+BF4−
83.5%
H3C
CH3
H3C
OC2H5

(6) (7)

[1]H. Meerwein, E. Battenberg, H. Gold, E. Pfeil, and G. Willfang, *J. prakt.*, [2], **154**, 83 (1940); H. Meerwein, *Org. Syn.*, **46**, 113 (1966)

[2]H. Meerwein *et al.*, *J. prakt.*, **147**, 257 (1937); *Ber.*, **89**, 209, 2060 (1956)

[3]N. Kornblum and R. A. Brown, *Am. Soc.*, **86**, 2681 (1964)

[4]H. Muxfeldt and W. Rogalski, *Am. Soc.*, **87**, 933 (1965)

[5]J. Harley-Mason and T. J. Leeney, *Proc. Chem. Soc.*, 368 (1964)

[6]A. Eschenmoser *et al.*, *Angew. Chem., Internat. Ed.*, **3**, 490 (1964)
[7]L. A. Paquette, *Am. Soc.*, **86**, 4096 (1964)

Triethyl phosphate, $(C_2H_5O)_3P{=}O$. Mol. wt. 182.16, b.p. 215–216°, sp. gr. 1.068, solubility in 100 g. water, 100 g. at 25°. Suppliers of practical grade: A, E, F, MCB.

This ester is an excellent medium for the reaction of weakly basic amines with alkyl or aryl bromides or iodides.[1] For example a mixture of equivalent amounts of

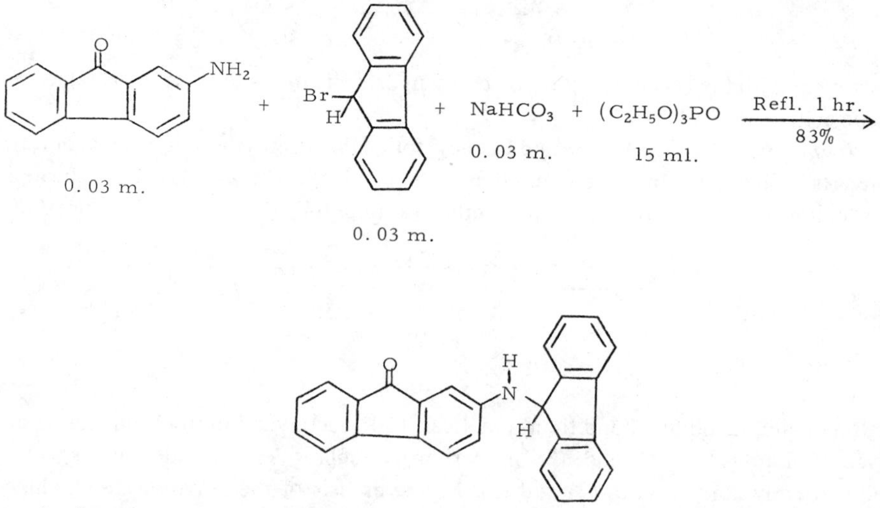
NH2
Br
H
+
+ NaHCO3 + (C2H5O)3PO
Refl. 1 hr.
83%
0.03 m.
15 ml.
0.03 m.
0.03 m.
H
N
H

2-aminofluorenone, 9-bromofluorene, and sodium bicarbonate is refluxed in triethyl phosphate for one hour, cooled, and diluted with water to precipitate the product, which crystallizes from chloroform–methanol in glistening reddish-violet needles.

[1]H.-L. Pan and T. L. Fletcher, *J. Org.*, **27**, 3639 (1962)

Triethyl phosphite, $(C_2H_5O)_3P$. Mol. wt. 166.16, b.p. 155–157°, (43–44°/10 mm.), insoluble in water. Suppliers of practical grade: Virginia-Carolina Chem. Corp., A, B, E, F, MCB. Preparation:[1]

$$PCl_3 + 3\ C_2H_5OH + 3\ C_6H_5N(C_2H_5)_2 \xrightarrow[83\%]{} (C_2H_5O)_3P + 3\ C_6H_5N(C_2H_5)_2\ HCl$$

Purification by distillation from metallic sodium.[2]

Preparation of phosphonates for modified Wittig reaction.[3] In a typical synthesis[4] benzyl chloride is added to triethyl phosphite to form benzyltriethylphosphonium chloride, which on brief refluxing loses ethyl chloride to form diethyl benzylphos-

$$\underset{5\ ml.}{C_6H_5CH_2Cl} + \underset{7.7\ ml.}{(C_2H_5O)_3P} \longrightarrow C_6H_5CH_2\overset{OC_2H_5}{\overset{|}{P^+}}(OC_2H_5)_2(Cl^-) \xrightarrow[-C_2H_5Cl]{Refl.\ 1\ hr.}$$

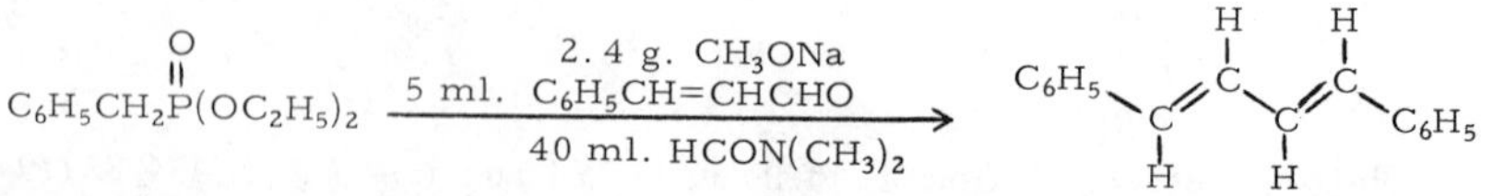
O
C6H5CH2P(OC2H5)2
2.4 g. CH3ONa
5 ml. C6H5CH=CHCHO
40 ml. HCON(CH3)2
C6H5
C6H5
H
H
H
H

phonate. After cooling, this is added to a suspension of sodium methoxide in dimethylformamide and the mixture is swirled in an ice bath while cinnamaldehyde is added. Crystallization of the product affords *trans,trans*-diphenylbutadiene in yield of 64% based on benzyl chloride.

Bicyclic phosphites. Trimethylolethane when heated with triethyl phosphite and a trace of triethylamine affords a crystalline phosphite, m.p. 94°.[5]

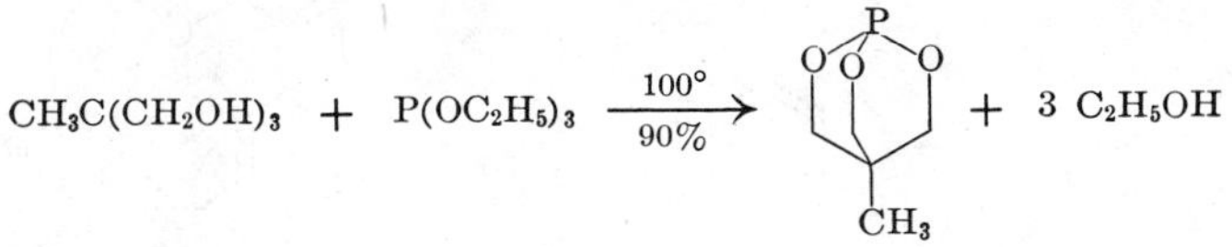

Alkylation. An example is the preparation of 1,2,3,4,5-pentachloro-5-ethylcyclopentadiene by ethylation of hexachlorocyclopentadiene (Mark[6]). A solution of

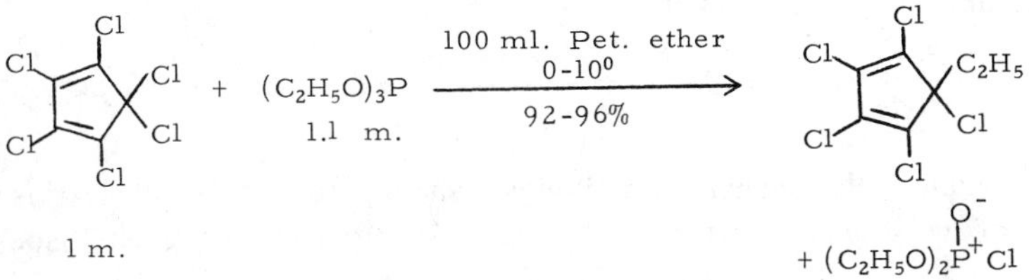

triethyl phosphite in petroleum ether is stirred in a salt-ice bath and a solution of hexachlorocyclopentadiene in the same solvent is added in the course of 4–6 hrs. The product of alkylation can be separated from diethyl phosphorochloridate [chlorodiethoxyphosphorus (V) oxide] by hydrolysis of the latter compound or by distillation.

Dechlorination. Mark[7] dechlorinated decachlorobiscyclopentadienyl (1) with triethyl phosphite and obtained perchlorofulvalene (2) as deep violet crystals.

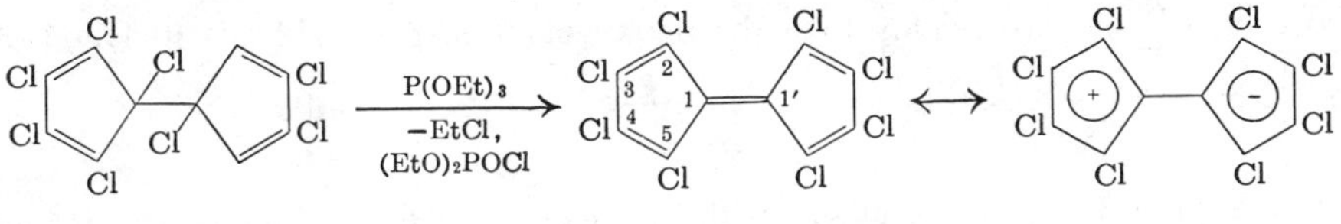

(1) (2) (3)

Reduction of hydroperoxides. Triethyl phosphite reduces hydroperoxides vigorously but smoothly to the corresponding alcohols.[8]

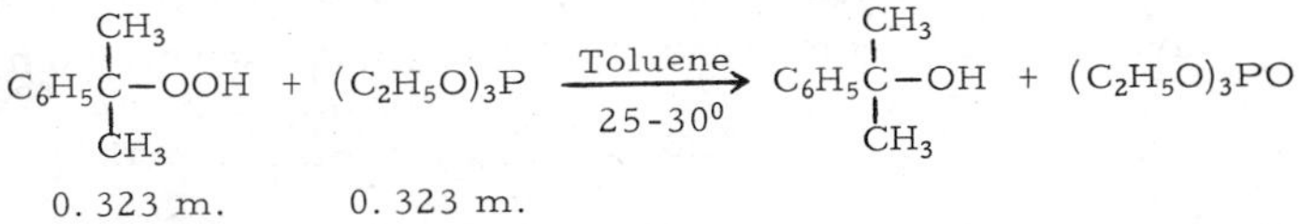

Deoxygenation. The reagent abstracts oxygen from a diaroyl peroxide to give the diaroyl anhydride and triethyl phosphate.[9]

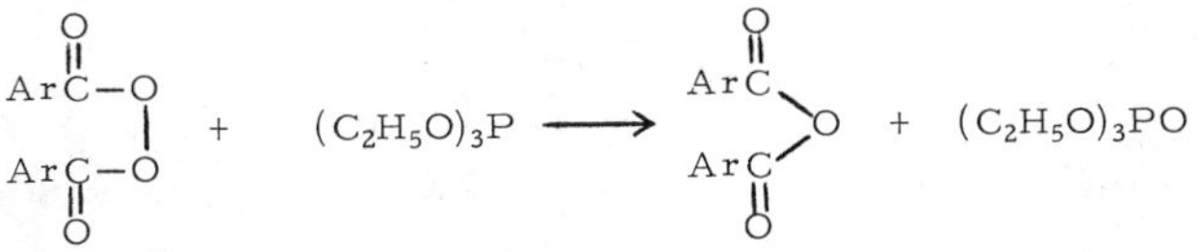

When phthalic anhydride is heated with excess triethyl phosphite as solvent, diphthalyl separates in nearly pure condition (Ramirez[10]). As a working hypothesis it is suggested that the initial product may be the carbene. Trimethyl phosphite is a less satisfactory reagent because of the lower reaction temperature.

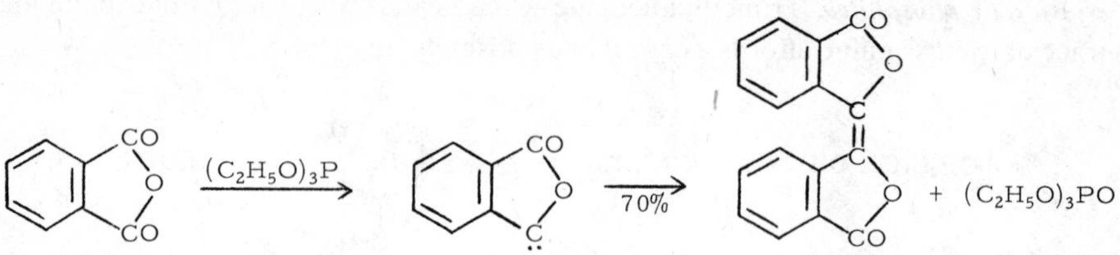
CO
O
CO
$(C_2H_5O)_3P$
CO
O
C
70%
CO
O
C
C
O
CO
+ $(C_2H_5O)_3PO$

Ethylene oxide and propylene oxide release oxygen to triethyl phosphite at 150–175°,[11] but the reaction is not general. Thus the 2-butene epoxides give only small amounts of the olefins at 200°.[12]

$$\underset{\diagdown\,O\,\diagup}{CH_3CH{-}CH_2} + (C_2H_5O)_3P \xrightarrow{150\text{-}175^0} CH_3CH{=}CH_2 + (C_2H_5O)_3PO$$

The reaction of the reagent with alkyl isocyanates to form isonitriles[13] is another example of deoxygenation (the reaction did not proceed with phenyl isocyanate).

$$RN{=}C{=}O + (C_2H_5O)_3P \longrightarrow RN \overset{\rightarrow}{\equiv} C + (C_2H_5O)_3PO$$

Diphenylketene when heated with the reagent at 215° under nitrogen is converted into diphenylacetylene.[13] This reaction involves deoxygenation and migration of a phenyl group.

$$(C_6H_5)_2C{=}C{=}O + (C_2H_5O)_3P \xrightarrow[65\%]{215^0} C_6H_5C{\equiv}CC_6H_5 + (C_2H_5O)_3PO$$

Desulfurization. Desulfurization of olefin episulfides (thiiranes) to olefins proceeds much more readily than the deoxygenation of epoxides.[14] Reaction of

$$\underset{\diagdown\,S\,\diagup}{>C{-}C<} + (C_2H_5O)_3P \longrightarrow >C{=}C< + (C_2H_5O)_3PS$$

cis- and *trans*-3-butene episulfides with the reagent leads to removal of sulfur to give *cis*- and *trans*-2-butene with a stereospecificity of over 97%.[12]

Hoffmann *et al.*[15] found that triethyl phosphite when heated with an aliphatic alcohol other than ethyl undergoes uncatalyzed transesterification in three distinct steps which proceed at about equal rates. A mercaptan, on the other hand, is desulfurized to give the hydrocarbon and triethyl thionophosphate, both in good

$$\underset{0.5\ m.}{\underline{n}\text{-}C_8H_{17}SH} + \underset{0.5\ m.}{(C_2H_5O)_3P} \xrightarrow[\text{on refl. several hours}]{h\nu\ 6.25\ hrs.} \underset{88\%}{\underline{n}\text{-}C_8H_{18}} + \underset{92\%}{(C_2H_5O)_3PS}$$

yield.[16] Walling and Rabinowitz[17] established that the reaction proceeds by a radical chain mechanism. A related reaction, brought about by slow distillation of a mixture of diethyl disulfide and triethyl phosphite, gives diethyl sulfide and triethyl mono-

$$C_2H_5SSC_2H_5 + (C_2H_5O)_3P \longrightarrow C_2H_5SC_2H_5 + (C_2H_5O)_2\overset{O}{\overset{\|}{P}}SC_2H_5$$

thiophosphate.[18] In another variation, a disulfide reacts with triethyl phosphite in the presence of carbon monoxide to give a thioester via a radical reaction.[19]

$$\underline{n}\text{-}C_4H_9SSC_4H_9\text{-}\underline{n} + CO + (C_2H_5O)_3P \xrightarrow[99\%]{} \underline{n}\text{-}C_4H_9\overset{O}{\overset{\|}{C}}{-}SC_4H_9\text{-}\underline{n} + (C_2H_5O)_3PS$$

Reduction of N-functions. Triethyl phosphite converts nitrosobenzene into azoxybenzene, probably via an intermediate azene.[20] With more reagent azoxybenzene is reduced to azobenzene.[13]

$$C_6H_5NO \xrightarrow{(C_2H_5O)_3P} [C_6H_5N\cdot] \xrightarrow{C_6H_5NO} C_6H_5N{=}\overset{O^-}{\overset{|}{N^+}}C_6H_5$$

$$C_6H_5N{=}\overset{O^-}{\overset{|}{N^+}}C_6H_5 + (C_2H_5O)_3P \xrightarrow[91\%]{} C_6H_5N{=}NC_6H_5 + (C_2H_5O)_3PO$$

The reaction of 2-nitrodiphenyl with triethyl phosphite at 160° under nitrogen to give carbazole involves initial reduction of the nitrogen function.[21] 2-Nitrosodiphenyl is converted similarly by triethyl phosphite into carbazole.[22]

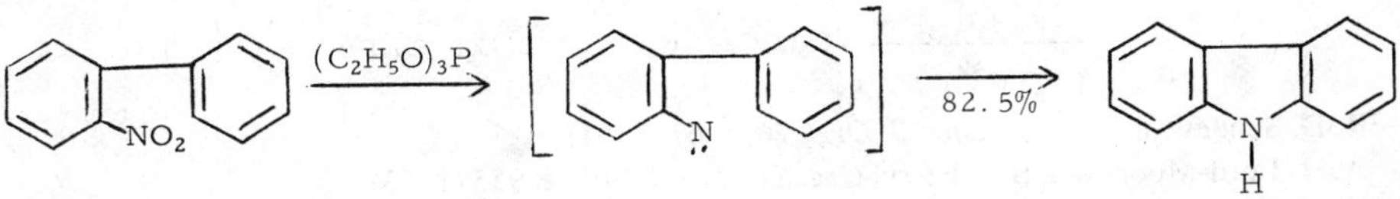

NO2
(C2H5O)3P
N
82.5%
N
H

Triethyl phosphite also effects reductive cyclization of the anil (1) to the imidazole (2).[23]

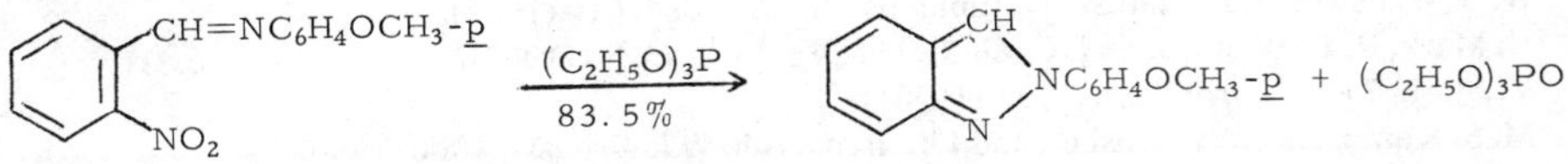

CH=NC6H4OCH3-p
NO2
(C2H5O)3P
83.5%
CH
NC6H4OCH3-p + (C2H5O)3PO
N

(1) (2)

A related reaction is the reduction of a β-alkyl-*o*-nitrostyrene (3) to the alkylindole (4).[24]

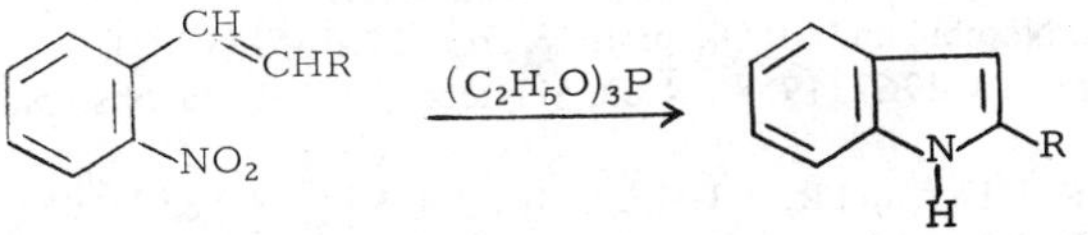

CH
CHR
NO2
(C2H5O)3P
N
R
H

(3) (4)

Similarly, 2-nitrophenyl phenyl sulfide (1) is reductively cyclized to phenothiazine (2) and 2′-nitrochalcone (3) affords 3-styrylanthranil (4).[25]

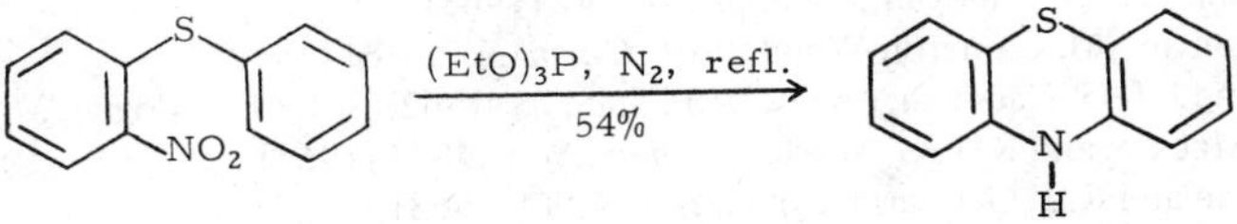

S
NO2
(EtO)3P, N2, refl.
54%
S
N
H

(1) (2)

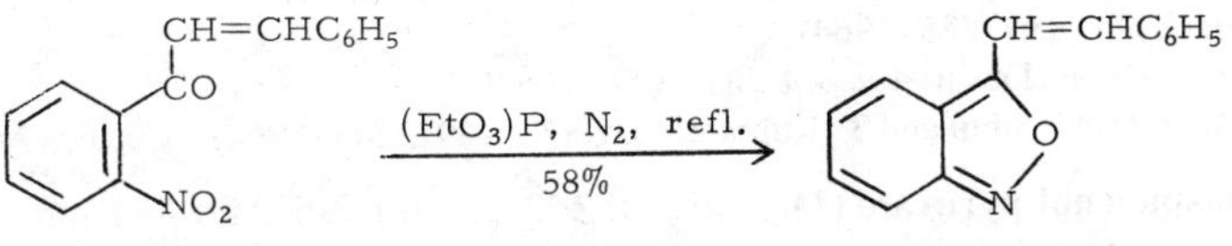

CH=CHC6H5
CO
NO2
(EtO3)P, N2, refl.
58%
CH=CHC6H5
O
N

(3) (4)

Grundmann[26] used triethyl phosphite for the reduction of furoxanes (5) to furazanes (6). Triphenyl phosphite also can be used for this reduction but triethyl phosphite is preferred for preparative purposes since the triethyl phosphate formed is water-soluble and easily eliminated.

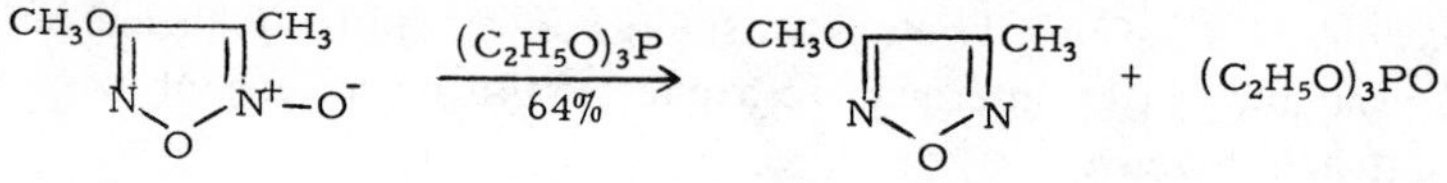

(5) (6)

Adducts with α-diketones. Triethyl phosphite combines with α-diketones to form 1:1 adducts.[27] When the adduct from benzil is heated at 215–225° under nitrogen with additional reagent, diphenylacetylene and triethyl phosphate are obtained.[28]

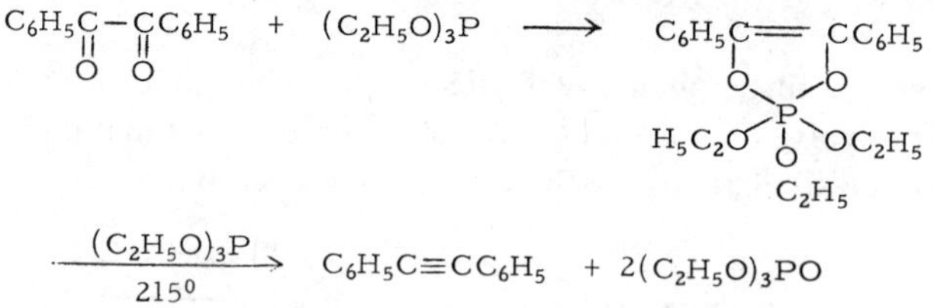

[1]R. D. Schuetz and R. L. Jacobs, *J. Org.*, **26**, 3467 (1961)
[2]A. H. Ford-Moore and B. J. Perry, *Org. Syn., Coll. Vol.*, **4**, 955 (1963)
[3]W. S. Wadsworth, Jr., and W. D. Emmons, *Am. Soc.*, **83**, 1733 (1961); E. J. Seus and C. V. Wilson, *J. Org.*, **26**, 5243 (1961)
[4]*Org. Expts.*, 119
[5]W. S. Wadsworth, Jr., and W. D. Emmons, *Am. Soc.*, **84**, 610 (1962)
[6]V. Mark, R. E. Wann, and H. C. Godt, Jr., *Org. Syn.*, **43**, 90 (1963)
[7]V. Mark, *Tetrahedron Letters*, 333 (1961)
[8]M. S. Kharasch, R. A. Mosher, and I. S. Bengelsdorf, *J. Org.*, **25**, 1000 (1960)
[9]A. J. Burn, J. I. G. Cadogan, and P. J. Bunyan, *J. Chem. Soc.*, 1527 (1963)
[10]F. Ramirez, H. Yamanaka, and O. H. Basedow, *Am. Soc.*, **83**, 173 (1961)
[11]C. B. Scott, *J. Org.*, **22**, 1118 (1957)
[12]N. P. Neureiter and F. G. Bordwell, *Am. Soc.*, **81**, 578 (1959)
[13]T. Mukaiyama, H. Nambu, and M. Okamoto, *J. Org.*, **27**, 3651 (1962).
[14]R. E. Davis, *J. Org.*, **23**, 1767 (1958); R. D. Schuetz and R. L. Jacobs, *ibid.*, **23**, 1799 (1958); see also ref. 1
[15]F. W. Hoffmann, R. J. Ess, and R. P. Usinger, Jr., *Am. Soc.*, **78**, 5817 (1956)
[16]F. W. Hoffmann, R. J. Ess, T. C. Simmons, and R. S. Hanzel, *Am. Soc.*, **78**, 6414 (1956)
[17]C. Walling and R. Rabinowitz, *Am. Soc.*, **81**, 1243 (1959)
[18]H. I. Jacobson, R. G. Harvey, and E. V. Jensen, *Am. Soc.*, **77**, 6064 (1955)
[19]C. Walling, O. H. Basedow, and E. S. Savas, Am. Soc., **82**, 2181 (1960)
[20]P. J. Bunyan and J. I. G. Cadogan, *J. Chem. Soc.*, 42 (1963)
[21]J. I. G. Cadogan and M. Cameron-Wood, *Proc. Chem. Soc.*, 361 (1962)
[22]P. J. Bunyan and J. I. G. Cadogan, *Proc. Chem. Soc.*, 78 (1962); J. I. G. Cadogan, M. Cameron-Wood, R. K. Mackie, and R. J. G. Searle, *J. Chem. Soc.*, 4831 (1965)
[23]J. I. G. Cadogan and R. J. G. Searle, *Chem. Ind.*, 1282 (1963)
[24]R. J. Sundberg, *J. Org.*, **30**, 3604 (1965)
[25]J. I. G. Cadogan, R. K. Mackie, and M. J. Todd, *Chem. Comm.*, 491 (1966)
[26]C. Grundmann, *Ber.*, **97**, 575 (1964)
[27]F. Ramirez and N. B. Desai, *Am. Soc.*, **85**, 3056 (1963)
[28]T. Makaiyama, H. Nambu, and T. Kumamoto, *J. Org.*, **29**, 2243 (1964)

Triethyl phosphoenol pyruvate (1). Mol. wt. 252.21., b.p. 95–100°/0.1 mm. Prepared by reaction of ethyl bromopyruvate and triethyl phosphate, the reagent reacts with

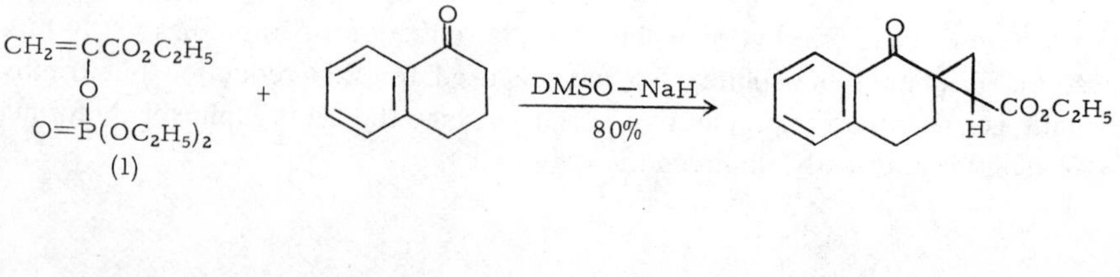

(1)

an active-methylene compound in the presence of a strong base to give a cyclopropane derivative.[1]

[1]U. Schmidt, *Angew. Chem., Internat. Ed.*, **4**, 238 (1965)

Triethyl phosphonoacetate, $(C_2H_5O)_2\overset{O^-}{\underset{+}{P}}CH_2CO_2C_2H_5$. Mol. wt. 224.20, b.p. 109°/0.80 mm., 140°/10 mm. Supplier: A.

The reagent is prepared by heating triethyl phosphite with ethyl bromoacetate and is used in the modified Wittig reaction.[1] Sodium hydride abstracts an activated α-hydrogen atom to give an anion salt (2) which functions like an ylide in reacting with an aldehyde or ketone to give an α,β-unsaturated ester (3). 1,2-Dimethoxyethane is used as solvent. The anion salt (2) reacts with diphenylketene to form the allene (5) and with styrene epoxide to form the cyclopropane (6).

$$(C_2H_5O)_2\overset{O^-}{\underset{+}{P}}CH_2CO_2C_2H_5 \xrightarrow[-H_2]{NaH} (C_2H_5O)_2\overset{O^-}{\underset{+}{P}}\overset{-}{C}HCO_2C_2H_5(Na^+)$$

$$(1) \qquad\qquad (2)$$

$$\xrightarrow[70\%]{(CH_3)_2C=O} (CH_3)_2C=CHCO_2C_2H_5 \ + \ (C_2H_5O)_2\overset{O^-}{\underset{+}{P}}ONa$$

$$(3) \qquad\qquad (4)$$

$$(C_2H_5O)_2\overset{O^-}{\underset{+}{P}}\overset{-}{C}HCO_2C_2H_5(Na^+) \ + \ (C_6H_5)_2C=C=O \xrightarrow{32\%} (C_6H_5)_2C=C=CHCO_2C_2H_5 \ + \ (4)$$

$$(2) \qquad\qquad\qquad\qquad (5)$$

$$(C_2H_5O)_2\overset{O^-}{\underset{+}{P}}\overset{-}{C}HCO_2C_2H_5(Na^+) \ + \ C_6H_5\overset{O}{\overset{\diagup\diagdown}{CH-\!\!-CH_2}} \xrightarrow{42\%} C_6H_5\overset{CHCO_2C_2H_5}{\overset{\diagup\diagdown}{CH-\!\!-CH_2}}$$

$$(2) \qquad\qquad\qquad\qquad (6)$$

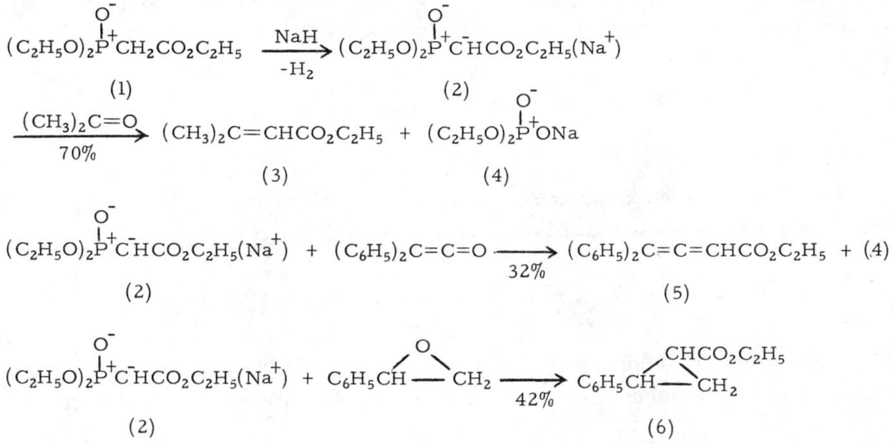

Wadsworth and Emmons[2] stirred a mixture of 50% sodium hydride dispersion in mineral oil with benzene and added triethyl phosphonoacetate slowly at 30–35° under nitrogen.

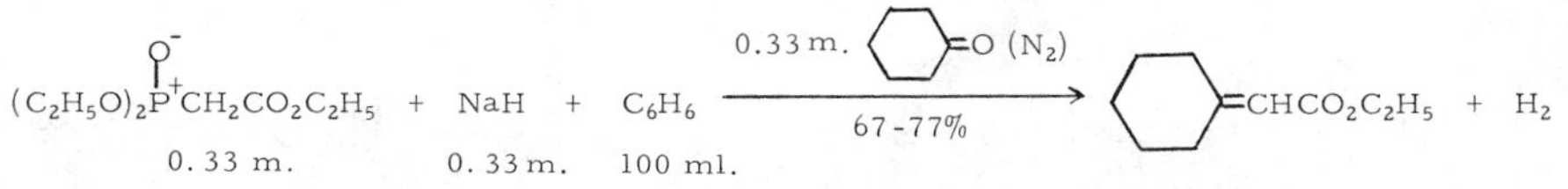

$$(C_2H_5O)_2\overset{O^-}{\underset{+}{P}}CH_2CO_2C_2H_5 \ + \ NaH \ + \ C_6H_6 \xrightarrow[67-77\%]{0.33\,m. \ \bigcirc\!=\!O\ (N_2)} \bigcirc\!=\!CHCO_2C_2H_5 \ + \ H_2$$

$$0.33\,m. \qquad\quad 0.33\,m. \qquad 100\ ml.$$

Hydrogen was evolved vigorously, and a nearly colorless solution resulted. Cyclohexanone was then added slowly and allowed to react at 20–30° and then at 60–65°. The reaction product, ethyl cyclohexylideneacetate, was collected at 48–49°/0.02 mm.

The reagent is selective toward ketosteroids.[3] 3-Ketosteroids and Δ⁴-3-keto-steroids react readily, but keto groups at C_6, C_7, and C_{17} are unreactive to the reagent. Japanese workers have used sodium amide as base and tetrahydrofurane as solvent.[4]

[1]W. S. Wadsworth, Jr., and W. D. Emmons, *Am. Soc.*, **83**, 1733 (1961); J. Wolinsky and K. L. Erickson, *J. Org.*, **30**, 2208 (1965)
[2]W. S. Wadsworth, Jr., and W. D. Emmons, *Org. Syn.*, **45**, 44 (1965)
[3]A. K. Bose and R. T. Dahill, Jr., *Tetrahedron Letters*, 959 (1963); *J. Org.*, **30**, 505 (1965)
[4]H. Takahashi, K. Fujiwara, and M. Ohta, *Bull. Chem. Soc., Japan*, **35**, 1498 (1962)

Triethyl phosphonoiodoacetate, $(C_2H_5O)_2\overset{O^-}{\underset{}{P^+}}CHICO_2C_2H_5$.

The reagent is generated *in situ* by treatment of triethyl phosphonoacetate with sodium hydride in 1,2-dimethoxyethane, followed by addition of iodine.[1] It reacts with benzaldehyde in the presence of two equivalents of sodium hydride to give ethyl propiolate.

$$(C_2H_5O)_2\overset{O^-}{P^+}CH_2CO_2C_2H_5 \xrightarrow[CH_3OCH_2CH_2OCH_3]{\begin{array}{l}1.\ \ NaH\\2.\ \ I_2\end{array}} (C_2H_5O)_2\overset{O^-}{\underset{I}{P^+}}CHCO_2C_2H_5$$

$$\xrightarrow[57\%]{C_6H_5CHO + 2\ NaH} C_6H_5C{\equiv}CCO_2C_2H_5 + (C_2H_5O)_2\overset{O^-}{P^+}ONa + NaI + H_2$$

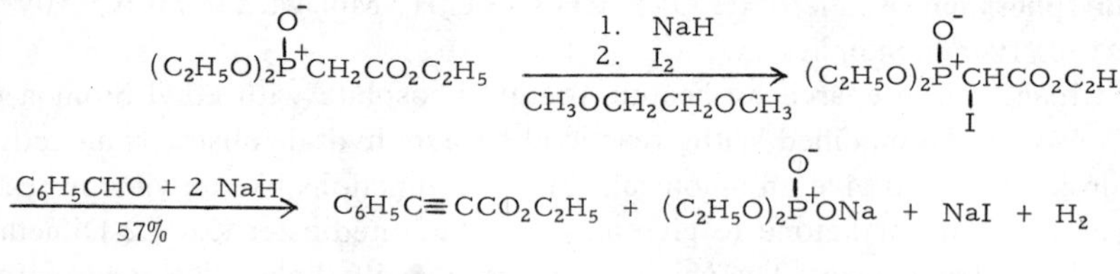

[1]W. S. Wadsworth, Jr. and W. D. Emmons, *Am. Soc.*, **83**, 1733 (1961)

Triethylsilane, $(C_2H_5)_3SiH$. Mol. wt. 116.28, b.p. 107°/733 mm. Supplier: Pierce Chem. Co. The reagent is prepared by the reaction of trichlorosilane with ethylmagnesium bromide.[1]

$$3\ C_2H_5MgBr + Cl_3SiH \xrightarrow[70-80\%]{} (C_2H_5)_3SiH$$
$$12.6\ m. \qquad 3\ m.$$

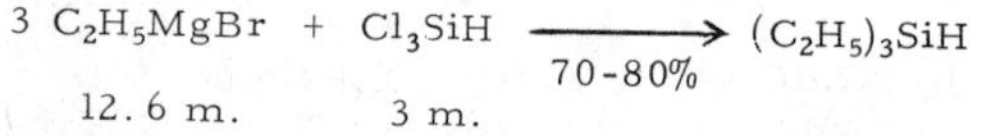

Reaction with acetals and ketals. Noncyclic acetals and ketals on reaction with triethylsilane in the presence of 5–10% of zinc chloride are converted into ethers.[2]

$$C_6H_5CH{\overset{OC_2H_5}{\underset{OC_2H_5}{\diagup\!\diagdown}}} + (C_2H_5)_3SiH \xrightarrow{ZnCl_2} C_6H_5CH_2OC_2H_5 + (C_2H_5)_3SiOC_2H_5$$

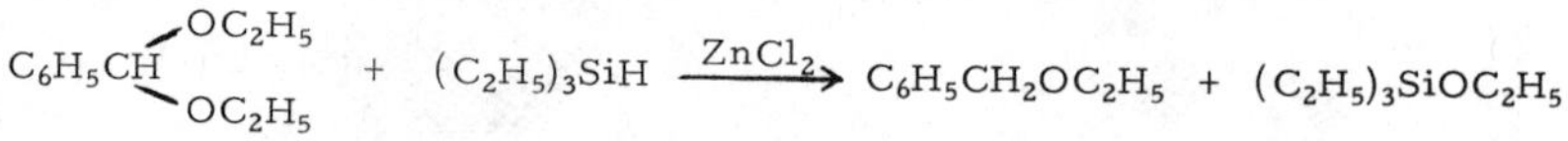

Cleavage of N-carbobenzoxy derivatives. In a general procedure for selective cleavage, the N-carbobenzoxy amino acid or peptide is treated with excess reagent

$$C_6H_5CH_2O\overset{O}{\overset{\|}{C}}-\overset{H}{\underset{}{N}}-\overset{R}{\underset{}{C}}HCO_2H + (C_2H_5)_3SiH + PdCl_2 + (C_2H_5)_3N \xrightarrow[-H_2]{Refl.}$$
$$0.01\ m. \qquad\qquad 0.04\ m. \qquad 50\ mg. \qquad 1\ drop$$

$$C_6H_5CH_2O\overset{O}{\overset{\|}{C}}-\overset{H}{\underset{}{N}}-\overset{R}{\underset{}{C}}HCO_2Si(C_2H_5)_3 \xrightarrow[-C_6H_5CH_3,\ -CO_2]{(C_2H_5)_3SiH} (C_2H_5)_3Si-\overset{H}{\underset{}{N}}-\overset{R}{\underset{}{C}}HCO_2Si(C_2H_5)_3$$

$$\xrightarrow{CH_3OH} H_2N\overset{R}{\underset{}{C}}HCO_2H + 2\ (C_2H_5)_3SiOCH_3$$

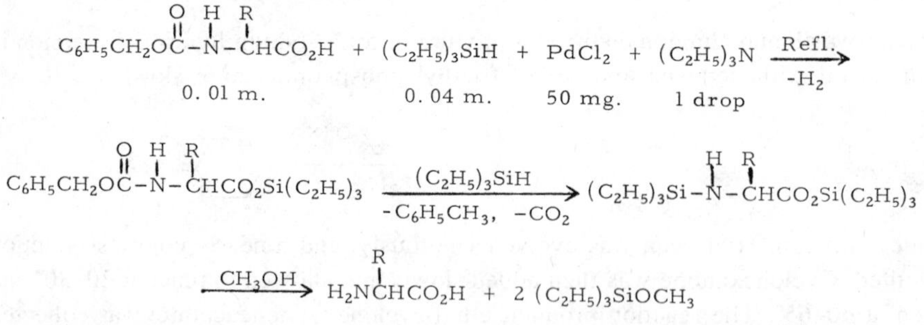

(to serve as solvent) and with catalytic amounts of palladium chloride and triethylamine.[3] The mixture is refluxed for 3 hrs., filtered, and the free amino acid or peptide is precipitated by the addition of methanol. The method affords the free amino acid; cleavage with HBr in acetic acid gives the hydrobromide. Benzyl esters are also cleaved, but S-benzyl groups are not affected.

[1]F. C. Whitmore, E. W. Pietrusza, and L. H. Sommer, *Am. Soc.*, **69**, 2108 (1947)
[2]E. Frainnet and C. Esclamadon, *Compt. rend.*, **254**, 1814 (1962)
[3]L. Birkofer, E. Bierwirth, and A. Ritter, *Ber.*, **94**, 821 (1961)

Triethyl N-tricarboxylate, $N(CO_2C_2H_5)_3$. Mol. wt. 233.22, b.p. 146–147°/12 mm. Suppliers: E, F.

The reagent is prepared[1] by warming a suspension of sodium wire with a solution of urethane in ether under reflux until the greater part of the metal has been converted into a gelatinous white precipitate of sodiourethane. The stirrer is started and heating under reflux continued. Then ethyl chloroformate is added with cooling and the mixture is stirred overnight (reaction 1). The mixture is filtered and the

$$1.\quad H_2NCO_2C_2H_5 + 2\ ClCO_2C_2H_5 + 2\ Na \xrightarrow[51-57\%]{} N(CO_2C_2H_5)_3 + 2\ NaCl + H_2$$

$$2.\quad 2\ N(CO_2C_2H_5)_3 + 5\ H_2NNH_2 + H_2O \longrightarrow HN(CONHNH_2)_2$$

$$+\ 3\ H_2NNHCO_2C_2H_5 + CO_2 + 3\ C_2H_5OH + NH_3$$

product recovered by distillation. This triester reacts with hydrazine as in (2) to form two products.[2] The first, diaminobiuret, is obtained as an ethanol-insoluble solid, m.p. 205° dec., in 69–75% yield. The second, ethyl hydrazinecarboxylate, is recovered from the filtrate by distillation and crystallized; m.p. 52°, yield 90–95%.

[1]O. Diels, *Ber.*, **36**, 740 (1903); C. F. H. Allen and A. Bell, *Org. Syn., Coll. Vol.*, **3**, 415 (1955)
[2]C. F. H. Allen and A. Bell, *Org. Syn., Coll. Vol.*, **3**, 404 (1955)

Trifluoroacetic acid, CF_3CO_2H. Mol. wt. 114.03, b.p. 72–72.5°, pKa 0.3 (a stronger acid than trichloroacetic acid). Suppliers: General Chemical Div., Allied Chemical, Eastern, A, B, E, F, MCB, PRC.

Preparation of carbamates. Carbamates of alcohols, even tertiary (which are easily dehydrated), can be obtained in good yield by stirring the alcohol with 2 equivalents each of sodium cyanate and trifluoroacetic acid in benzene or methylene chloride at room temperature for a few hours (Loev and Kormendy[1]). A mildly

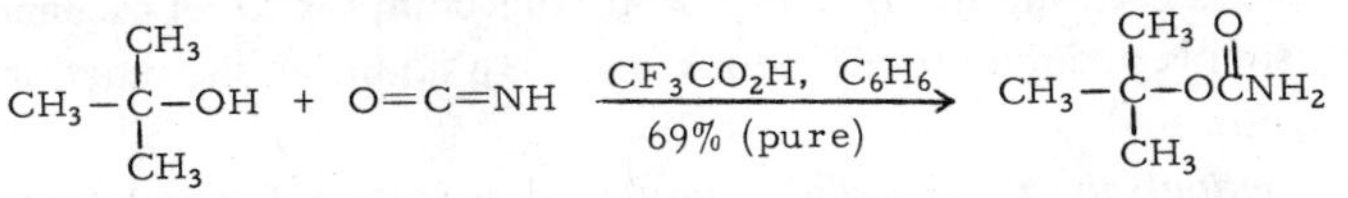
CH3
CH3—C—OH + O=C=NH
CH3
CF3CO2H, C6H6
69% (pure)
CH3 O
CH3—C—OCNH2
CH3

m.p. 107-108°

exothermal reaction ensues, and some gaseous cyanic acid bubbles out of the system. *t*-Butyl carbamate was isolated by adding water, separating the organic layer, evaporating the solvent, and crystallizing the crude product (92% yield, m.p. 98–101°) from water. With trichloroacetic acid, the yield was only about 50%. The fact that other very strong acids (hydrochloric acid, methanesulfonic acid) gave only traces of product led the authors[1] to suggest participation of trifluoroacetic acid in a cyclic transition state or a mixed anhydride. The method is generally applicable. Even the relatively unstable propargyl alcohol gives a carbamate.

Condensation-cyclization. Woods[2] developed a simple synthesis of coumarins by condensation of a phenol with a β-keto ester in the presence of trifluoroacetic acid. However, in the example formulated the position of ring closure is activated by

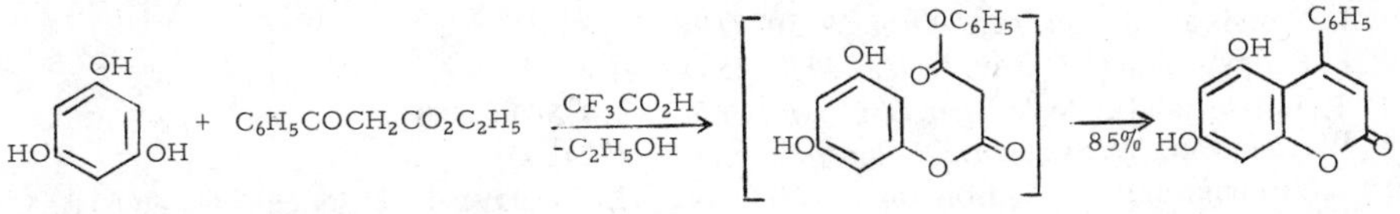
OH
HO
OH
+ C6H5COCH2CO2C2H5
CF3CO2H
-C2H5OH
OC6H5
OH
HO
O
O
85%
C6H5
OH
HO
O
O

ortho and *para* hydroxyl groups. Phenol, catechol, 4,6-dichlororesorcinol, hydroquinone, and the cresols all failed to react. The reaction was applied successfully to 1-naphthol, but 2-naphthol failed to react.[3]

Hydrogenation solvent. Trifluoroacetic acid appears to be the best solvent for the platinum-catalyzed hydrogenation of ketones.[4] The rate in this solvent is approximately three times the rate in acetic acid. A relatively high concentration of the ketone should be used, since the reactions are slow in dilute solution. The alcohol is frequently obtained as the ether.

Silver trifluoroacetate. For use in combination with iodine for the 4-iodination of veratrole,[5] the salt is prepared by adding trifluoroacetic acid to a suspension of precipitated silver oxide in water, filtering, and evaporating the filtrate to dryness. Extraction of the dry salt with ether in a Soxhlet and evaporation of the solvent gives colorless, crystalline material in 88% yield.

Peptide synthesis. Protective N-carbobenzoxy groups of amino acids and peptides can be cleaved in good yield by refluxing in trifluoroacetic acid.[6] The acid is recommended as solvent for the HBr-cleavage of protective groups in place of acetic acid, which permits partial acetylation of the hydroxyl group of serine.[7] Addition of diethyl phosphate removes any trace of bromine which may be formed. Trifluoroacetic acid removes protective *t*-butoxy groups of hydroxyamino acids at or below room temperature without racemization and without cleavage of peptide bonds.[8] *t*-Butyl and *t*-butyloxycarbonyl protective groups are cleaved by 90% trifluoroacetic acid in "outstanding" yield.[9]

Cleavage of benzyl ethers. The acid cleaves aromatic benzyl ethers at room temperature.[10]

Solvent for kinetic studies. Brown and Wirkala[11] found trifluoroacetic acid a superior medium for kinetic studies of aromatic electrophilic substitutions (bromination, nitration, mercuration). In contrast to the complex kinetics encountered in acetic acid, simple second-order kinetics apply. In addition, the reactions are faster in trifluoroacetic acid.

Hydrogenation solvent. The acid is preferred as solvent for hydrogenation (Pt or Rh) of pterines, since complete and selective hydrogenation of the pyrazine ring of pterine, its N-methylated derivative, and of folic acid is effected in short time.[12]

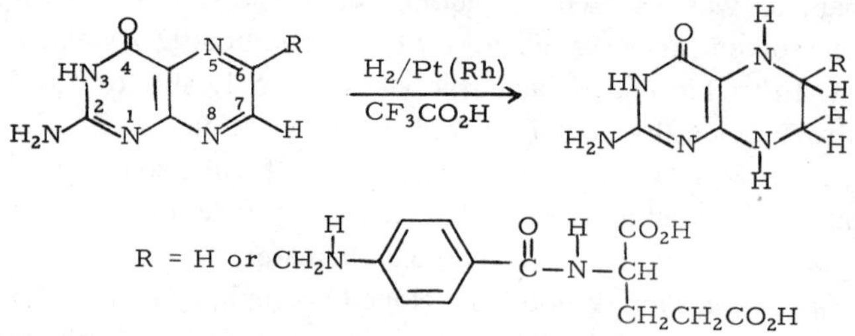
HN
H₂N
N
R
H
H₂/Pt (Rh)
CF₃CO₂H
R = H or CH₂N
CO₂H
C—N—CH
CH₂CH₂CO₂H

[1]B. Loev and M. F. Kormendy, *J. Org.*, **28**, 3421 (1963)
[2]L. L. Woods and J. Sapp, *J. Org.*, **27**, 3703 (1962)
[3]L. L. Woods and J. Sterling, *J. Org.*, **29**, 502 (1964)
[4]P. E. Peterson and C. Casey, *J. Org.*, **29**, 2325 (1964)
[5]D. E. Janssen and C. V. Wilson, *Org. Syn., Coll. Vol.*, **4**, 547 (1963)
[6]F. Weygand and W. Steglich, *Z. Naturforsch.*, **14B**, 472 (1959)
[7]St. Guttmann and R. A. Boissonnas, *Helv.*, **42**, 1257 (1959); E. D. Nicolaides and H. A. DeWald, *J. Org.*, **28**, 1926 (1963)
[8]H. C. Beyerman and J. S. Bontekoe, *Proc. Chem. Soc.*, 249 (1961); *Rec. trav.*, **81**, 691 (1962)
[9]R. Schwyzer, A. Costopanagiotis, and P. Sieber, *Helv.*, **46**, 870 (1963)
[10]J. P. Marsh, Jr., and L. Goodman, *J. Org.*, **30**, 2491 (1965)

[11]H. C. Brown and R. A. Wirkkala, *Am. Soc.*, **88**, 1447, 1453, 1456 (1966)
[12]A. Bobst and M. Viscontini, *Helv.*, **49**, 875 (1966)

Trifluoroacetic anhydride, $(CF_3CO)_2O$. Mol. wt. 210.04, b.p. 39.5°, sp. gr. 1.48. Suppliers: B, E, MCB, PRC. Review.[1] Preparation: $C_6H_5CF_3 \longrightarrow m\text{-}NO_2C_6H_4CF_3 \longrightarrow m\text{-}NH_2C_6H_4CF_3 \longrightarrow HO_2CCF_3 \longrightarrow$ anhydride (P_2O_5).[2]

Trifluoroacetylation. Trifluoroacetic anhydride acylates primary and secondary alcohols with great ease, but the esters are so extremely sensitive to hydrolysis that it is sometimes necessary to avoid contact with water in the workup. Thus Tatlow and co-workers[3] refluxed briefly a mixture of *p*-nitrobenzyl alcohol, trifluoroacetic anhydride, and sodium trifluoroacetate and distilled the mixture with four portions of carbon tetrachloride at reduced pressure to remove excess anhydride and trifluoroacetic acid. The ester was then extracted from the sodium trifluoroacetate

$$\underset{0.407\text{ g.}}{p\text{-}NO_2C_6H_4CH_2OH} + \underset{2\text{ ml.}}{(CF_3CO)_2O} + \underset{0.161\text{ g.}}{CF_3CO_2Na} \xrightarrow[73\%]{\text{Refl.}} \underset{\text{m.p. }47^0}{p\text{-}NO_2C_6H_4CH_2OCOCF_3}$$

with petroleum ether, and the solution concentrated for crystallization. The ester is hydrolyzed rapidly by water. Applications in syntheses in the sugar series are reported in another paper.[4]

Lardon and Reichstein[5] found trifluoroacetylation useful for protecting steroid 11α- and 11β-hydroxyl groups during operations required for transforming a 17β-carboxyl group into $—COCH_2OH$. With these large, water-insoluble substances unwanted hydrolysis during workup presented no problem, and the protective group could be removed easily when desired with a weak base. Acylation was done without solvent or in dioxane or pyridine. An axial 11β-hydroxyl group resists acetylation in pyridine but reacts readily with trifluoroacetic anhydride at room temperature. Thus a solution of the 3β,11β-diol (1) in the anhydride was let stand

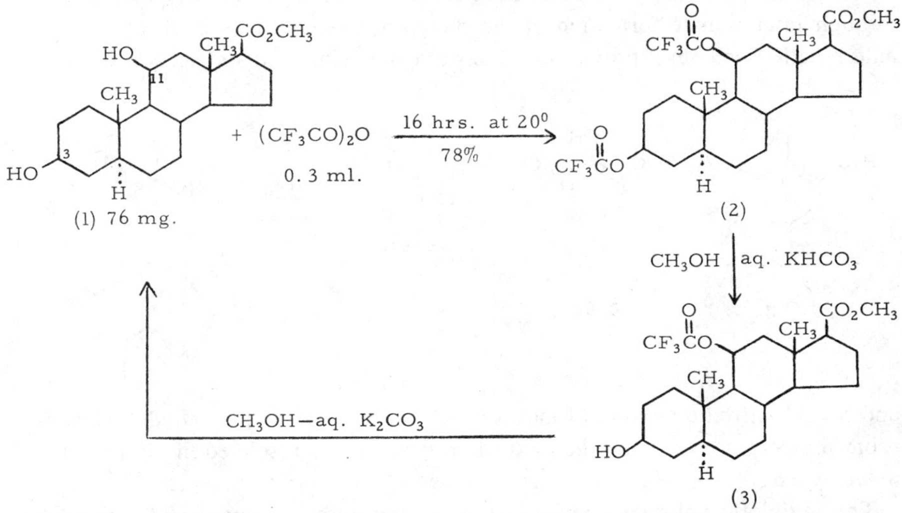
CO2CH3
CH3
HO
11
CH3
3
HO
H
(1) 76 mg.
+ (CF3CO)2O
0. 3 ml.
16 hrs. at 20°
78%
CF3CO
(2)
CH3OH
aq. KHCO3
(3)
CH3OH–aq. K2CO3

for 16 hrs. and extracted with ether. The solution was washed neutral with soda solution and dried, and evaporation afforded the crude ditrifluoroacetyl derivative (2) in good yield. One crystallization gave pure material. For selective removal of the group at C_3, a solution of 51 mg. of (2) in 8 ml. of methanol was treated with a solution of 200 mg. of potassium bicarbonate in 6 ml. of water and let stand for 2 days at 20°. Evaporation, extraction with ether, and crystallization afforded 21 mg. of the pure 11-monotrifluoroacetate (3). Treatment of this substance with aqueous-methanolic potassium carbonate removed the highly hindered group at C_{11} and gave (1).

Mixed anhydrides. Emmons *et al.*[6] found that trifluoroacetic anhydride reacts readily with carboxylic acids to form mixed anhydrides, isolable in yields of 50–65%.

$$\underset{2.5\text{ m.}}{C_6H_5CO_2H} + \underset{2.0\text{ m.}}{(CF_3CO)_2O} \xrightarrow[55\%]{\text{Refl. 30 min.}} C_6H_5\overset{O}{\overset{\|}{C}}O\overset{O}{\overset{\|}{C}}CF_3 + CF_3CO_2H$$

Thus a mixture of benzoic acid and the anhydride was stirred under reflux for 30 min., the trifluoroacetic anhydride and acid were distilled in vacuum, and the benzoyl trifluoroacetate was distilled at 100°/0.5 mm. Infrared study of this reaction in *n*-butyl ether and in acetonitrile showed that equilibrium is established immediately and that benzoyl trifluoroacetate is formed quantitatively.

Duckworth[7] used the reagent to prepare anhydrides of malonic acids, which are of interest because on pyrolysis they afford ketenes.

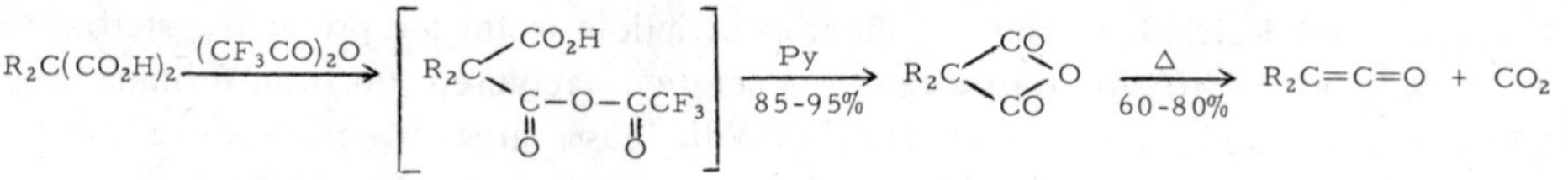

Esterification. In the presence of trifluoroacetic anhydride an alcohol reacts with an acid according to the following scheme:[1]

$$ROH + R'CO_2H + (CF_3CO)_2O \longrightarrow R'CO_2R + 2\ CF_3CO_2H$$

From the work of Emmons cited above it is evident that the esterification proceeds through intermediate formation of the mixed anhydride. This is an exceptionally mild reaction and yet it provides a means of esterifying hindered acids. Thus Parish

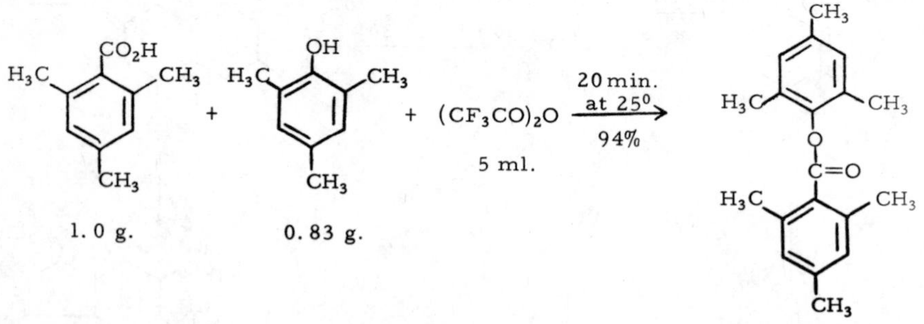

and Stock[8] stirred a mixture of mesitoic acid and mesitol in the anhydride briefly at room temperature, took up the mixture in benzene, and isolated mesityl mesitoate in high yield.

The antibiotic chloroamphenicol (1) contains both a primary and a secondary alcoholic function, and partial acylation by usual techniques affords primary monoesters of value for parenteral administration and as tasteless derivatives for pediatric use. Almirante and Tosolini[9] devised a method for the preparation of monoesters of the secondary alcoholic function based upon one previously used by Schmidt

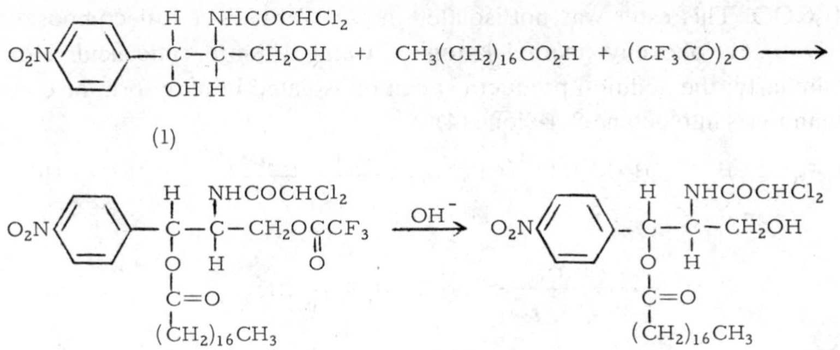

(1) (2) (3)

and Staab[2] in the sugar series. Reaction of chloramphenicol (1) with stearic acid and trifluoroacetic anhydride produces the derivative (2) with the stearoyl group on the secondary alcoholic function and the trifluoroacetyl group on the primary function. A weak base eliminates the trifluoroacetyl group to give (3), and the net result is selective stearoylation of the secondary alcoholic group in the presence of a primary alcoholic function.

Acylation. The reagent catalyzes the acylation of activated aromatic compounds by reaction with carboxylic acids.[10] γ-Phenylbutyric acid is converted into α-

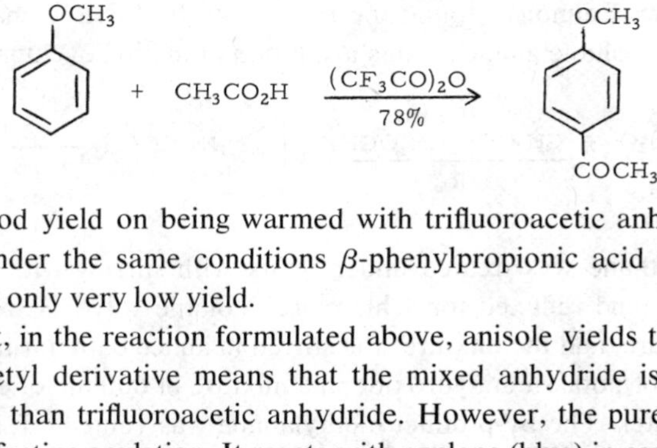

tetralone in good yield on being warmed with trifluoroacetic anhydride at 60–70° for 3 hrs.[11] Under the same conditions β-phenylpropionic acid is converted into indane-1-one in only very low yield.

The fact that, in the reaction formulated above, anisole yields the acetyl and not the trifluoroacetyl derivative means that the mixed anhydride is a more reactive acylating agent than trifluoroacetic anhydride. However, the pure anhydride alone is capable of effecting acylation. It reacts with azulene (blue) in carbon tetrachloride without catalyst at room temperature to give l-trifluoroacetylazulene (red) in high yield.[12]

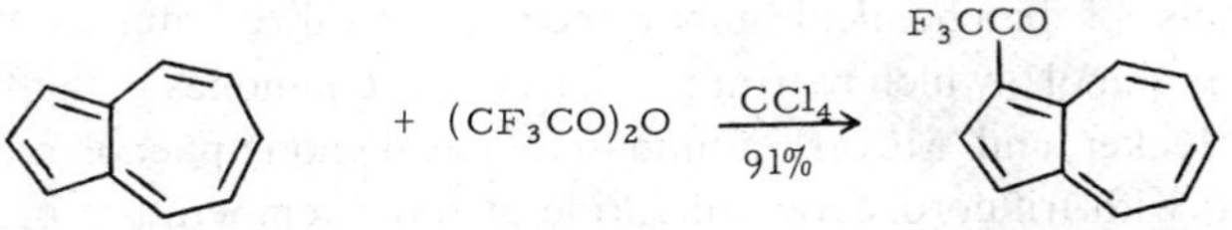

Henne and Tedder[13] report interesting, if low-yield, acylations of olefins and acetylenes. A mixture of cyclohexene, acetic acid, and trifluoroacetic anhydride warms up to about 37° with formation of the ester (1), formally the product of addition of the mixed anhydride and regarded as resulting from attack by $CH_3\overset{+}{C}{=}O$

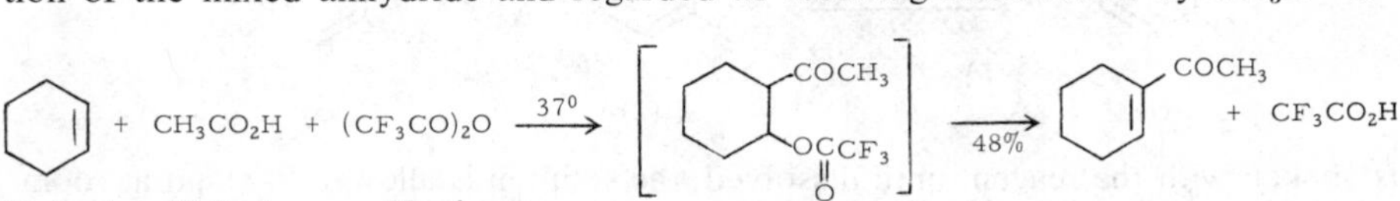

18.5 ml. 10.5 ml. 27 ml. (1) (2)

and CF_3COO. This ester was not isolated in pure form since it decomposes spontaneously to cyclohexenyl methyl ketone (2) and trifluoroacetic acid. Hexyne-1 reacts similarly; the addition product (3) can be isolated in pure form or converted by methanolysis into octane-2,4-dione (4).

$$\underset{11.4\ \text{ml.}}{\underline{n}\text{-}C_4H_9C\equiv CH} + \underset{5.7\ \text{ml.}}{CH_3CO_2H} + \underset{14.5\ \text{ml.}}{(CF_3CO)_2O} \xrightarrow{18\ \text{hrs. at}\ 27^0} \underset{(3)}{\underline{n}\text{-}C_4H_9\underset{\substack{|\\ OCCF_3\\ \|\\ O}}{C}=CHCOCH_3}$$

$$\xrightarrow[20\%\ \text{overall}]{CH_3OH} \underset{(4)}{\underline{n}\text{-}C_4H_9\underset{\substack{\|\\ O}}{C}CH_2\underset{\substack{\|\\ O}}{C}CH_3}$$

Harfenist[14] found the reagent far superior to phosphorus pentoxide for cyclization of the phenothiazine derivative (5). The acid was refluxed briefly in benzene containing an equivalent amount of the anhydride.

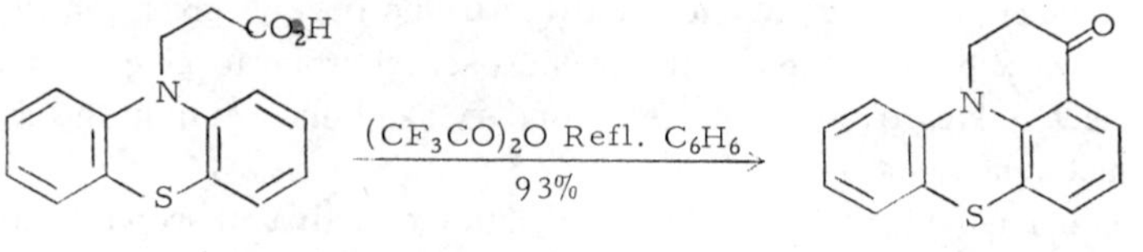
CO2H
N
S
(CF3CO)2O Refl. C6H6
93%
O
N
S

(5) (6)

Rearrangements. Emmons[15] found the reagent useful for Beckmann rearrangements giving water-soluble amides. Thus a solution of methyl cyclopropyl ketoxime

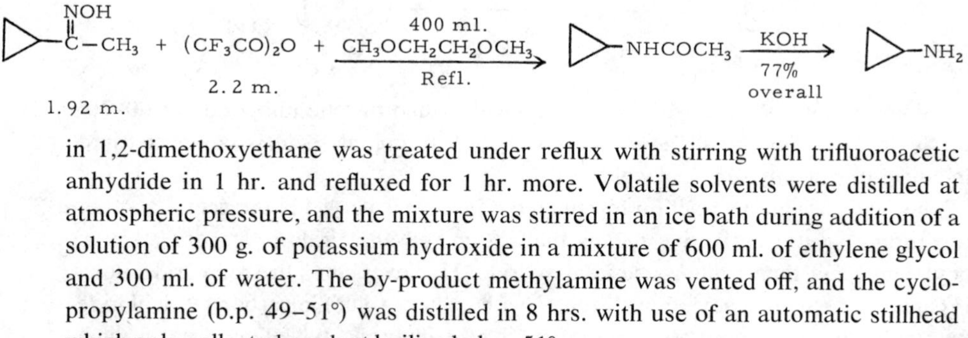
NOH
C–CH3 + (CF3CO)2O + CH3OCH2CH2OCH3
400 ml.
Refl.
2.2 m.
1.92 m.
NHCOCH3
KOH
77%
overall
NH2

in 1,2-dimethoxyethane was treated under reflux with stirring with trifluoroacetic anhydride in 1 hr. and refluxed for 1 hr. more. Volatile solvents were distilled at atmospheric pressure, and the mixture was stirred in an ice bath during addition of a solution of 300 g. of potassium hydroxide in a mixture of 600 ml. of ethylene glycol and 300 ml. of water. The by-product methylamine was vented off, and the cyclopropylamine (b.p. 49–51°) was distilled in 8 hrs. with use of an automatic stillhead which only collected product boiling below 51°.

p-Alkylphenols of biochemical importance are oxidized during intermediary metabolism to *p*-quinols, which rearrange to alkylhydroquinones under the influence of enzymes. Hecker and Meyer[16] found that the dienone-phenol rearrangement can be effected with trifluoroacetic anhydride at room temperature in quantitative yield. Toluquinol (1) yields toluhydroquinone (2) and cresorcinol (3). The dienone

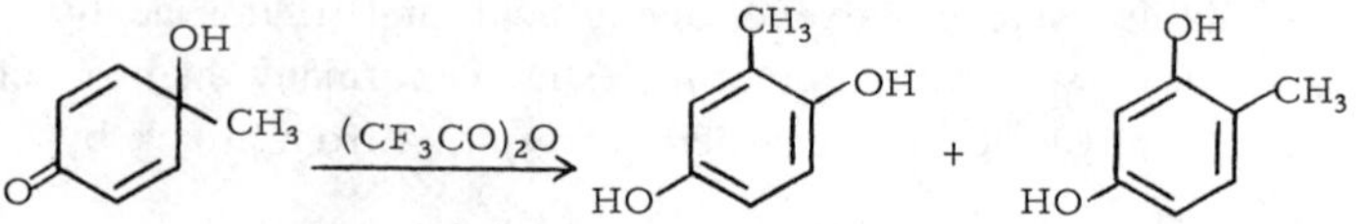
OH
CH3
(CF3CO)2O
CH3
OH
HO
+
OH
CH3
HO

(1) (2) (3)

is shaken with the reagent until dissolved, the solution is allowed to stand at room temperature for 72 hrs., and the trifluoro residue is removed by hydrolysis with aqueous dioxane. The methyl and tetrahydropyranyl ethers of (1) yield only the hydroquinone and no resorcinol derivative. Tetralin-*p*-quinol (4) and its ethers rearrange exclusively to 5,8-dihydroxytetralin (5) or its monoethers. The steroid

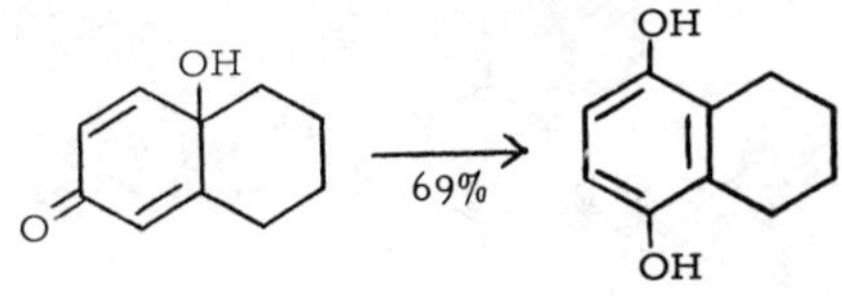
OH
69%
OH
OH

(4) (5)

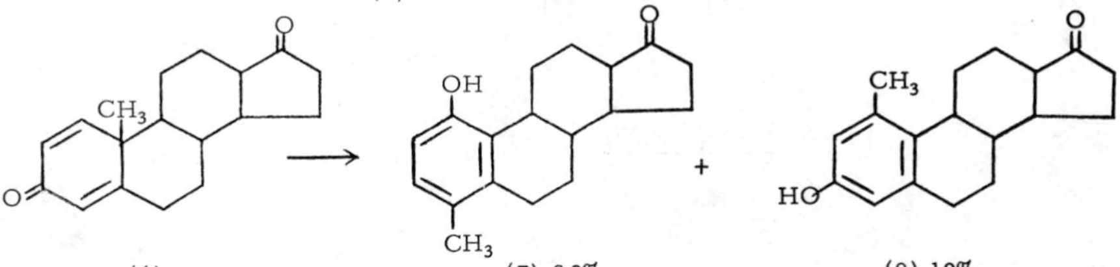
CH3
OH
CH3
CH3
HO
+

(6) (7) 80% (8) 10%

dienone (6) gives the products (7) and (8). Hecker[17] also showed that the dienone ester (9) is rearranged by the reagent to the aromatic product (10).

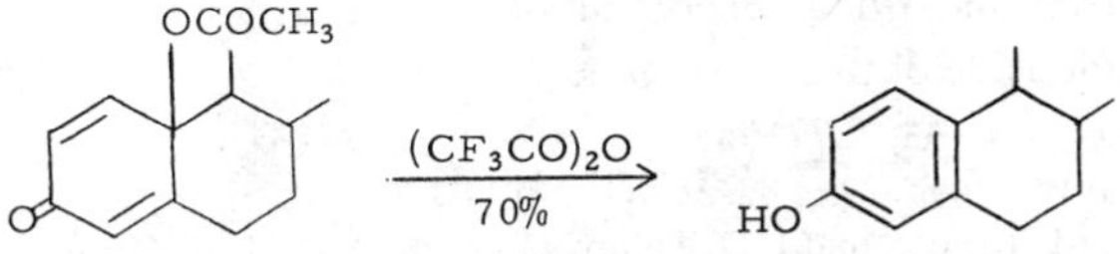
OCOCH3
(CF3CO)2O
70%
HO

(9) (10)

Rutherford and Newman[18] effected the Schmidt reaction on phenanthrene-4-carboxylic acid by dissolving 5 g. of the acid in 100 ml. of a solution containing equal

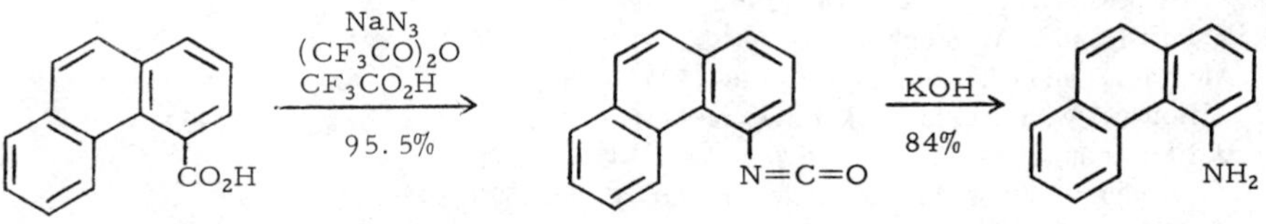
NaN3
(CF3CO)2O
CF3CO2H
95.5%
CO2H
N=C=O
KOH
84%
NH2

volumes of trifluoroacetic anhydride and trifluoroacetic acid, cooling to 0–5°, and adding excess sodium azide portionwise with stirring. Nitrogen was evolved rapidly, and the isocyanate crystallized from the solution in nearly quantitative yield. It is noteworthy that no mineral acid was required and that the reaction did not proceed when either the anhydride or the acid was used alone.

Preparation of secondary nitramines. The reagent is used for the nitrolysis of *sec*-acylamines to produce *sec*-nitramines.[19]

$$2\ HNO_3 + (CF_3CO)_2O \longrightarrow NO_2^+ + NO_3^- + 2\ CF_3CO_2H$$

$$NO_2^+ + R_2\ddot{N}{-}Ac \longrightarrow O_2N{-}NR_2 + Ac^+$$

Synthesis of glycerides. Trifluoroacetic anhydride has been used for the direct synthesis of a glyceride from glycerol and a fatty acid; monoglycerides (1- and 2-) can be obtained by using appropriately blocked glycerol.[20]

***O*-Protection and *N*-protection.** Newman[21] used the trifluoroacetyl group to block the hydroxyl and amino groups in the 1-halosugar (1); after condensation of (2) with

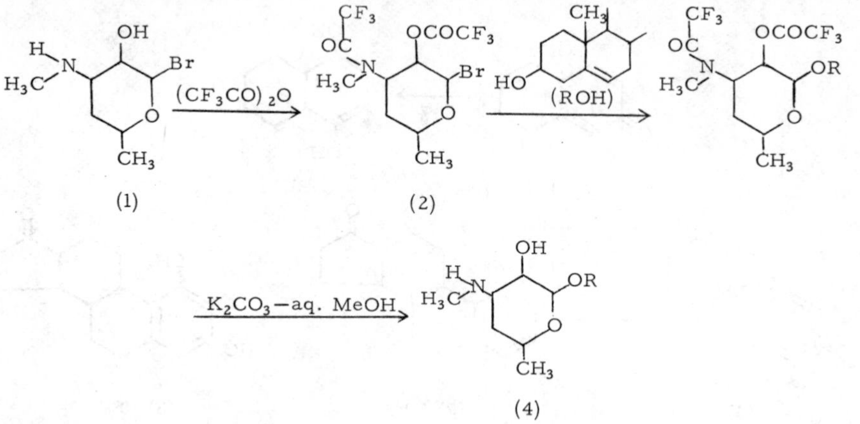

an alcoholic component, for example, cholesterol, the protective groups were removed by treatment at room temperature with aqueous-methanolic potassium carbonate. Wolfrom and Bhat[22] used trifluoroacetyl as the N-blocking group in the synthesis of a nucleoside of D-glucosamine.

[1]J. M. Tedder, *Chem. Rev.*, **55**, 787 (1955)
[2]O. Th. Schmidt and W. Staab, *Ber.*, **87**, 388 (1954)
[3]E. J. Bourne, C. E. M. Tatlow, and J. C. Tatlow, *J. Chem. Soc.*, 1367 (1950)
[4]E. J. Bourne, M. Stacey, C. E. M. Tatlow, and J. C. Tatlow, *J. Chem. Soc.*, 826 (1951)
[5]A. Lardon and T. Reichstein, *Helv.*, **37**, 388, 443 (1954)
[6]W. D. Emmons, K. S. McCallum, and A. F. Ferris, *Am. Soc.*, **75**, 6047 (1953). See also E. J. Bourne, M. Stacey, J. C. Tatlow and R. Worrall, *J. Chem. Soc.*, 2006 (1954)
[7]A. C. Duckworth, *J. Org.*, **27**, 3146 (1962)
[8]R. C. Parish and L. M. Stock, *J. Org.*, **30**, 927 (1965)
[9]L. Almirante and G. Tosolini, *J. Org.*, **26**, 177 (1961)
[10]E. J. Bourne, M. Stacey, J. C. Tatlow, and J. M. Tedder, *J. Chem. Soc.*, 718 (1951)
[11]R. J. Ferrier and J. M. Tedder, *J. Chem. Soc.*, 1435 (1957)
[12]A. G. Anderson, Jr., and R. G. Anderson, *J. Org.*, **27**, 3578 (1962)
[13]A. L. Henne and J. M. Tedder, *J. Chem. Soc.*, 3628 (1953)
[14]M. Harfenist, *J. Org.*, **28**, 1834 (1963)
[15]W. D. Emmons, *Am. Soc.*, **79**, 6522 (1957)
[16]E. Hecker and E. Meyer, *Angew. Chem., Internat. Ed.*, **3**, 229 (1964); *Ber.*, **97**, 1926 (1964)
[17]E. Hecker, *Ber.*, **97**, 1940 (1964)
[18]K. G. Rutherford and M. S. Newman, *Am. Soc.*, **79**, 213 (1957)
[19]J. H. Robson and J. Reinhart, *Am. Soc.*, **77**, 2453 (1955); M. B. Frankel, C. H. Tieman, C. R. Vanneman, and M. H. Gold, *J. Org.*, **25**, 744 (1960)
[20]E. J. Bourne, M. Stacey, J. C. Tatlow, and J. M. Tedder, *J. Chem. Soc.*, 2976 (1949); P. F. E. Cook and A. J. Showler, *ibid.*, 4594 (1965)
[21]H. Newman, *J. Org.*, **30**, 1287 (1965)
[22]M. L. Wolfrom and H. B. Bhat, *Chem. Comm.*, 146 (1966)

α,α,α-Trifluoroacetophenone, $C_6H_5COCF_3$. Mol. wt. 174.12, b.p. 151°/740 mm. Preparation.[1] Supplier: Pierce Chem. Co.

Use of the reagent for the synthesis of β-substituted perfluoroolefins is illustrated by a procedure for the preparation of 2-phenylperfluoropropene.[2] A mixture of equimolecular amounts of α,α,α-trifluoroacetophenone, triphenylphosphine, and

$$\underset{0.25\ \text{m.}}{C_6H_5COCF_3} + \underset{0.25\ \text{m.}}{(C_6H_5)_3P} + \underset{0.5\ \text{m.}}{CClF_2CO_2Na} \xrightarrow[50\text{-}60\%]{\text{Diglyme, } 140^0 \text{ under } N_2} C_6H_5\underset{CF_3}{C}{=}CF_2 + (C_6H_5)_3PO$$

sodium chlorodifluoroacetate (taken in 100% excess for generation of CF_2) is heated in diglyme in a nitrogen atmosphere.

[1]K. T. Dishart and R. Levine, *Am. Soc.*, **78**, 2268 (1956)
[2]F. E. Herkes and D. J. Burton, procedure submitted to *Org. Syn.*

Trifluoroacetyl hypohalites, CF_3CO_2X. Prepared by interaction of bromine or iodine with silver trifluoroacetate.[1]

The reagents have been used to effect halogenation of aromatic compounds; examples:[2]

$$C_6H_5I + CF_3CO_2I \xrightarrow[77\%]{} p\text{-}IC_6H_4I$$

$$C_6H_5CH_3 + CF_3CO_2Br \xrightarrow[90\%]{} p\text{-}BrC_6H_4CH_3$$

$$C_6H_5NH_2 + CF_3CO_2I \xrightarrow[51\%]{} p\text{-}IC_6H_4NH_2$$

[1]A. L. Henne and W. F. Zimmer, *Am. Soc.*, **73**, 1362 (1951)
[2]R. N. Haszeldine and A. G. Sharpe, *J. Chem. Soc.*, 993 (1952)

N-Trifluoroacetyl imidazole (1). Mol. wt. 176.10, m.p. 136–137°.

Preparation.[1] The reagent is prepared from imidazole and trifluoroacetic anhydride in tetrahydrofurane or from N,N′-carbonyldiimidazole and trifluoroacetic acid in THF.

Carboxylic acid anhydrides.[2] When 1 mole of the reagent is treated with 2 moles of a carboxylic acid in tetrahydrofurane, the mixed anhydride (2) and then the pure anhydride (3) are formed in equilibrium reactions, but the sparingly soluble salt (4) of trifluoroacetic acid and imidazole separates and drives the reactions to completion. Thus on standing at room temperature for about 5 hrs. the pure anhydride is obtained in high yield.

(1) (2)

2. $$RCO{-}O{-}COCF_3 + RCO_2H \rightleftharpoons RCO{-}O{-}COR + CF_3CO_2H$$

(3)

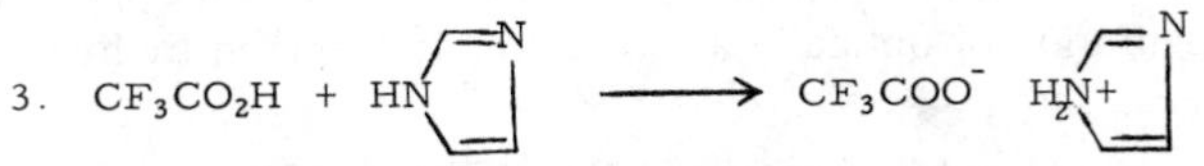

(4)

The reagent is regarded as superior to N,N′-carbonyldiimidazole for the preparation of *p*-nitrophenyl esters of amino acids; however, the method is less satisfactory than the standard DCC procedure.[3]

[1]H. A. Staab and G. Walther, *Ber.*, **95**, 2070 (1962)
[2]H. A. Staab, G. Walther, and W. Rohr, *Ber.*, **95**, 2073 (1962)
[3]H. D. Law, *J. Chem. Soc.*, 3897 (1965)

Trifluoroiodomethane (Trifluoromethyl iodide), CF_3I. Mol. wt. 195.92, a gas. Supplier: Pierce Chem. Co.

Trifluoromethylation. Under irradiation in pyridine, the reagent reacts with a steroidal 3-ethyl enol ether (1) to give the 6-trifluoromethyl derivative (2).[1] Acid-catalyzed hydrolysis gives the equatorial 6α-trifluoromethyl-Δ^4-3-ketosteroid (3) in good yield.

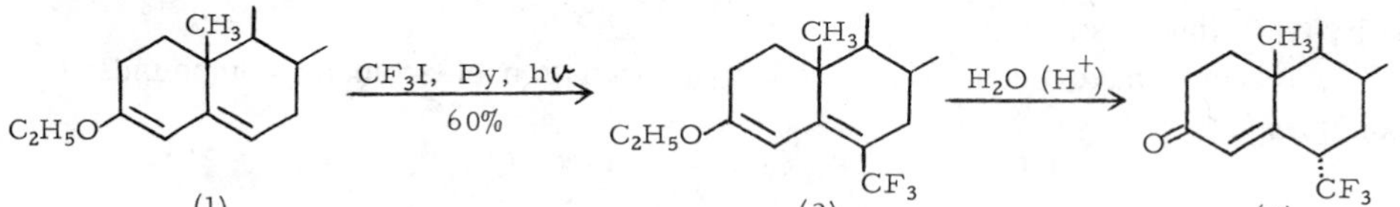
CH3
C2H5O
CF3I, Py, hν
60%
CH3
C2H5O
CF3
H2O (H+)
CH3
O
CF3

(1) (2) (3)

Radical addition to olefins has been studied extensively by R. N. Haszeldine; for a summary with references, see Walling.[2]

[1]W. O. Gotfredsen and S. Vangedal, *Chem. Scand.*, **15**, 1786 (1961)
[2]C. Walling, "Free Radicals in Solution," p. 251, Wiley (1957)

Triglyme, $CH_3OCH_2CH_2OCH_2CH_2OCH_2CH_2OCH_3$. Mol. wt. 178.22, b.p. 222°. Water-miscible. Suppliers: Ansul Chem. Co., (pure) BJ.

Tri-*n*-hexylethyl ammonium hydroxide, $(n\text{-}C_6H_{13})_3NC_2H_5(OH)$. Mol. wt. 315.57.

Preparation.[1] The reagent is prepared by reaction of tri-*n*-hexylamine with ethyl iodide in ethyl acetate and conversion of the resulting insoluble quaternary iodide into the hydroxide form by passage through Dowex-1 ion exchange resin.

Acetylation assistant. Khorana[1] used the reagent in the acetylation of a trinucleotide that was practically insoluble in all solvents. The strongly basic reagent formed a salt with the nucleotide and effected solubilization in pyridine; acetylation with acetic anhydride then proceeded smoothly.

[1]H. Schaller and H. G. Khorana, *Am. Soc.*, **85**, 3841 (1963)

Triisopinocampheyldiborane. This hydroboration reagent is prepared *in situ* by addition of 3 moles of α-pinene to 1 mole of diborane in tetrahydrofuran at 0°.[1] In

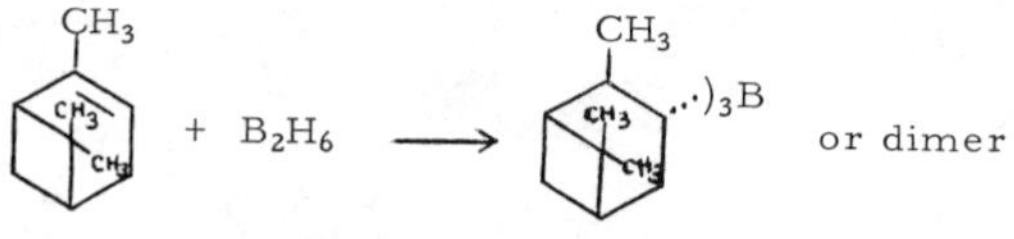
CH3
+ B2H6
CH3
)3B
or dimer

contrast to diisopinocampheylborane, the reagent reacts at a reasonable rate with *trans* and hindered olefins. Since both the D- and L-forms of α-pinene are available from natural sources, a reagent prepared from one or the other can be used to determine the configuration of optically active alcohols prepared by hydroboration and oxidation.

[1]H. C. Brown, N. R. Ayyangar, and G. Zweifel, *Am. Soc.*, **86**, 1071 (1964)

2,4,6-Triisopropylbenzenesulfonyl chloride (TPS), 2,4,6-$[(CH_3)_2CH]_3C_6H_2SO_2Cl$. Mol. wt. 302.86, m.p. 96–97°. Prepared by chlorosulfonation of 1,3,5-triisopropylbenzene.[1,2] Supplier: Aldrich.

The reagent is preferred to mesitylenesulfonyl chloride (*which see*) for formation of the $C_{3'}$-$C_{2'}$-internucleotide bond.[2] The main advantage is that the reagent shows a much reduced rate of sulfonation of the 5′-hydroxyl group of nucleosides.

[1]A. Newton, *Am. Soc.*, **65**, 2439 (1943)

[2]R. Lohrmann and H. G. Khorana, *Am. Soc.*, **88**, 829 (1966)

Triisopropyl phosphite, $[(CH_3)_2CHO]_3P$. Mol. wt. 208.03, b.p. 60–61°/10 mm., 43.5°/1 mm. Suppliers of practical or technical grades: Virginia-Carolina Chem. Corp., Baker, E. Preparation.[1]

Dechlorination. This mild dechlorinating agent is used as triethyl phosphite is used for the preparation of perchlorofulvalene.[2]

Diisopropyl methylphosphonate. This ester is prepared by adding triisopropyl phosphite gradually to methyl iodide under reflux and applying heat after the initial exothermal reaction has subsided.[3]

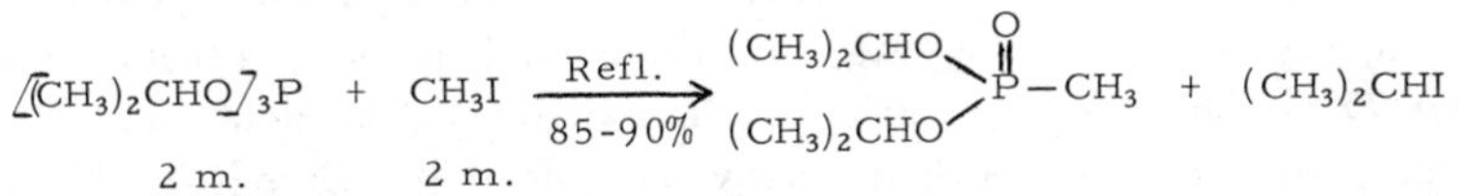
[(CH3)2CHO]3P + CH3I
Refl.
85-90%
(CH3)2CHO
(CH3)2CHO
P—CH3 + (CH3)2CHI
2 m.
2 m.

[1]A. H. Ford-Moore and B. J. Perry, *Org. Syn., Coll. Vol.*, **4**, 955 (1963)

[2]V. Mark, *Tetrahedron Letters*, 333 (1961); *Org. Syn.*, **46**, 93 (1966)

[3]A. H. Ford-Moore and B. J. Perry, *Org. Syn., Coll. Vol.*, **4**, 325 (1963)

Trimethylacetic acid (Pivalic acid), $(CH_3)_3CCO_2H$. Mol. wt. 102.13, m.p. 35°, b.p. 164°. Suppliers: B, E, F, MCB.

Preparation: carbonation of *t*-butylmagnesium chloride (61–70%);[1] hypobromite oxidation of pinacolone (71–74%).[2]

[1]S. V. Puntambeker and E. A. Zoellner, *Org. Syn., Coll. Vol.*, **1**, 524 (1941)

[2]L. T. Sandborn and E. W. Bousquet, *ibid.*, **1**, 526 (1941)

Trimethylacetyl chloride (Pivaloyl chloride), $(CH_3)_3CCOCl$. Mol. wt. 120.58, b.p. 105–106°. Supplier: Aldrich.

Peptide synthesis. In an early study of the mixed anhydride peptide synthesis, Vaughan and Osato[1] briefly mentioned pivaloyl chloride as a reagent but regarded it as slightly inferior to isovaleroyl chloride. Zaoral[2] reintroduced the reagent, partly with the view of supressing carbonyl activity of the auxiliary acid component, and obtained yields better than obtained with *s*-butyl chloroformate. Kenner *et al.*[3] found the method particularly useful for the synthesis of peptides of the sterically hindered α-methylalanine. When the triethylammonium salt of Cb-α-methylalanine is treated with pivaloyl chloride the crystalline mixed anhydride is formed quantitatively:

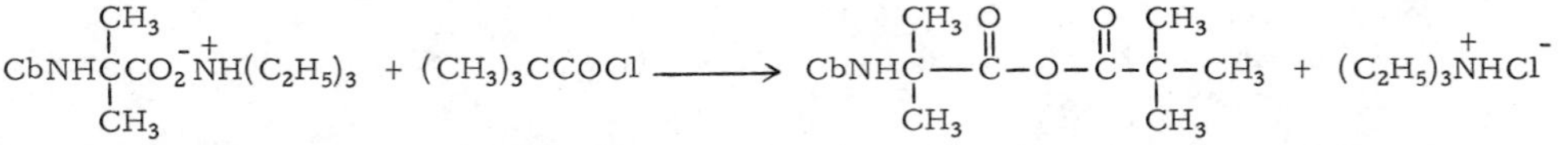
CH3
CbNHCCO2 NH(C2H5)3 + (CH3)3CCOCl
CH3
CH3 O O CH3
CbNHC—C—O—C—C—CH3 + (C2H5)3NHCl
CH3 CH3

Schwyzer and Lieber[4] applied this method for the coupling of peptides at the point of a proline residue.

[1]J. R. Vaughan, Jr., and R. L. Osato, *Am. Soc.*, **73**, 5553 (1951)

[2]M. Zaoral, *Coll. Czech.*, **27**, 1273 (1962)

[3]M. T. Leplawy, D. S. Jones, G. W. Kenner, and R. C. Sheppard, *Tetrahedron*, **11**, 39 (1960)

[4]R. Schwyzer and P. Sieber, *Nature*, **199**, 172 (1963)

Trimethylamine borane, $(CH_3)_3N^+B^-H_3$. Mol. wt. 72.96, m.p. 94°, b.p. 171°, white crystals, stable for hours at 125°. Solubility (g./l. at 30°): water, 14.8; methanol,

10; ether, very soluble; hexane, 7.4; benzene, very soluble; acetic acid, soluble. Suppliers: Callery Chem. Co., Peninsular ChemResearch, MCB.

Preparation. The reagent is prepared by reaction of trimethylamine with diborane or with borine carbonyl (BH_3CO).[1] In a new procedure developed by Ethyl Corp.[2] an activated aluminum powder and aluminum chloride are added to a solution of triphenyl borate in trimethylamine and the mixture is agitated at 180° under 2,000 p.s.i. of hydrogen.

$$2\ (CH_3)_3N + 2\ B(OC_6H_5)_3 + 2\ Al + 3\ H_2 \xrightarrow[99\%]{1\ hr.\ 180^0} 2\ (CH_3)_3NBH_3 + 2\ Al(OC_6H_5)_3$$

Reduction of ketones. In usual solvents trimethylamine borane reacts with ketones only very slowly even at steam bath temperature, but in the presence of boron trifluoride etherate it is an effective reducing agent; 4-*t*-butylcyclohexanone is reduced rapidly to a mixture of 46% of the *cis* alcohol and 54% of the *trans* isomer.[3]

Reduction of Schiff bases. Schiff bases, for example benzylideneaniline from benzaldehyde and aniline, are reduced rapidly and efficiently by the reagent in solution or suspension in acetic acid.[4] Unlike other borohydrides, trimethylamine borane

$$3\ C_6H_5CH{=}NC_6H_5 + (CH_3)_3NBH_3 + 3\ CH_3CO_2H \xrightarrow[84\%]{} 3\ C_6H_5CH_2N(COCH_3)C_6H_5 + (CH_3)_3N + B(OH)_3$$

can be used in boiling acetic acid with little loss in active hydrogen (25% excess reagent is recommended). Borane appears to be the effective reagent and is liberated from the complex as it is utilized; yields are often high, and the following groups are not affected: NO_2, Cl, OH, CO_2H, CO_2CH_3, SO_2NH_2. The reaction is unique in that the secondary amine formed is acetylated by acetic acid.

Reaction with olefins. Olefins are converted into trialkylboranes when heated with the reagent in an autoclave at 100–200° in the absence of solvent.[5] The reaction

$$(CH_3)_3N{:}BH_3 \rightleftharpoons (CH_3)_3N + 1/2\ (BH_3)_2 \xrightarrow[80\text{-}95\%]{3\ RCH=CH_2} (RCH_2CH)_3B$$

is believed to involve dissociation into diborane. An interior olefin such as 2-hexene produces the isomerized product, for example, tri-*n*-hexylborane.

$$CH_3CH_2CH_2CH{=}CHCH_3 \xrightarrow{(CH_3)_3NBH_3} (CH_3CH_2CH_2CH_2CH_2CH_2)_3B$$

[1] A. B. Burg and H. I. Schlesinger, *Am. Soc.*, **59**, 780 (1939)
[2] E. C. Ashby and W. E. Foster, *Am. Soc.*, **84**, 3407 (1962)
[3] W. M. Jones, *Am. Soc.*, **82**, 2528 (1960)
[4] J. H. Billman and J. W. McDowell, *J. Org.*, **27**, 2640 (1962)
[5] E. C. Ashby, *Am. Soc.*, **81**, 4791 (1959)

Trimethylamine oxide, $(CH_3)_3N{-}O$. Mol. wt. 75.11, m.p. 213–214°. Preparation.[1] Drying by azeotropic distillation with benzene, heating at 105°/200 mm., and sublimation did not remove all the water.[2]

Aldehyde synthesis. The reagent can be used to convert a halide or tosylate into the corresponding aldehyde:[3]

$$RCH_2Br(Ts) + (CH_3)_3\overset{+}{N}-\overset{-}{O} \longrightarrow RCH_2O\overset{+}{N}(CH_3)_3(Br^-) \longrightarrow RCH{=}O + (CH_3)_3N\cdot HBr$$

The bromide or tosylate and the amine oxide are refluxed and stirred in chloroform for 30 min.; yields are in the range 30–60%. Aldehydes prepared in this way were found satisfactory for use in the Wittig reaction, whereas aldehydes prepared by the Rosenmund reaction contained undesirable impurities which react with Wittig reagents.[4]

[1]W. J. Hickinbottom, "Reactions of Organic Compounds," p. 277, Longmans Green and Co., N.Y. (1936)
[2]J. J. Monagle, *J. Org.*, **27**, 3851 (1962)
[3]V. Franzen and S. Otto, *Ber.*, **94**, 1360 (1961)
[4]L. D. Bergelson and M. M. Shemyakin, *Angew. Chem., Internat. Ed.*, **3**, 250 (1964)

Trimethylammonium formate, $5\,HCO_2H \cdot 2\,N(CH_3)_3$. Mol. wt. 348.35, b.p. 91–93°/18 mm. This liquid salt of constant high boiling point and very weak acidity is prepared by introducing trimethylamine gas into ice-cooled 80% formic acid until the solution becomes basic.[1]

Reducing action. The reagent reductively cleaves N-acylaminomethyl and N-sulfonamido methyl compounds.[1] Formic acid also can be used, but yields are lower.

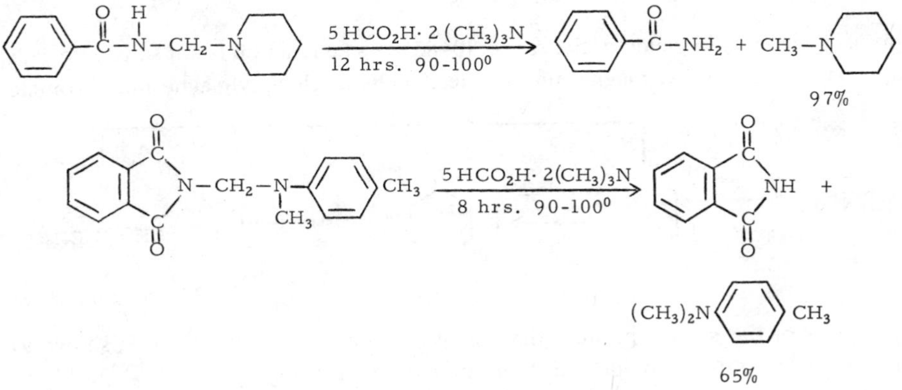
5 HCO₂H· 2 (CH₃)₃N
12 hrs. 90-100⁰
97%
5 HCO₂H· 2(CH₃)₃N
8 hrs. 90-100⁰
65%

[1]M. Sekiya and K. Ito, *Chem. Pharm. Bull. (Japan)*, **12**, 677 (1964)

Trimethylboroxine, $(CH_3BO)_3$. Mol. wt. 124.75, b.p. 79–80°.

Preparation. This compound was prepared initially in 12% yield by dehydration of methylboronic acid synthesized from trimethyl borate and methylmagnesium halide.[1] Then Rathke and Brown[2] found a practical preparative method in the reaction of

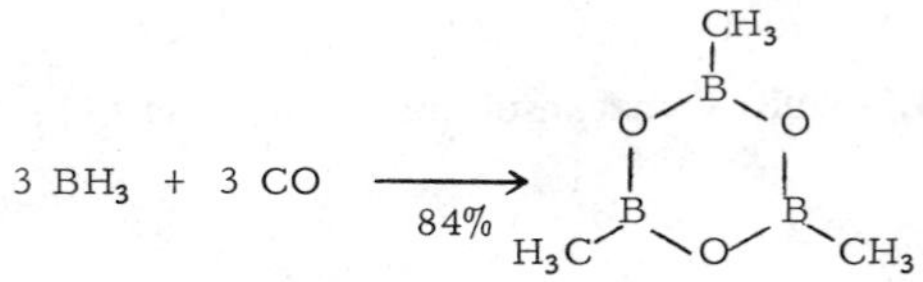
3 BH₃ + 3 CO
84%

carbon monoxide with diborane in tetrahydrofurane solution. Distillation of the reaction mixture provided trimethylboroxine in excellent yield.

Separation of cis- and trans-diols. Tri-*n*-butylboroxine (*which see*) is useful for the separation and identification of isomeric *cis-trans* diols. Trimethylboroxine possesses the obvious advantages for this application of greater volatility of the cyclic esters and greater stability to atmospheric oxygen.

[1]M. E. D. Hillman, *Am. Soc.*, **84**, 4715 (1962); **85**, 982, 1636 (1963)
[2]M. W. Rathke and H. C. Brown, *Am. Soc.*, **88**, 2606 (1966)

Trimethylchlorosilane, $(CH_3)_3SiCl$. Mol. wt. 108.65, b.p. 57.3°. Supplier: General Electric Co., Silicone Products Dept.

Hydroxyl, amino, and carboxyl groups can be silylated by reaction with this reagent in the presence of triethylamine or piperidine, but N-trimethylsilylacetamide is a more useful reagent for this purpose.[1]

$$ROH + (CH_3)_3SiCl + (C_2H_5)_3N \longrightarrow ROSi(CH_3)_3 + (C_2H_5)_3\overset{+}{N}H\overset{-}{Cl}$$

[1]L. Birkofer and A. Ritter, *Angew. Chem., Internat. Ed.*, **4**, 417 (1965)

Trimethylene oxide, $(CH_2)_3O$. Mol. wt. 58.08, b.p. 48°. Supplier: Aldrich. Preparation from 3-chloropropyl acetate.[1]

Grignard reaction. The reagent reacts with benzylmagnesium chloride, phenylmagnesium chloride, phenyllithium, and 1-naphthylmagnesium bromide to give

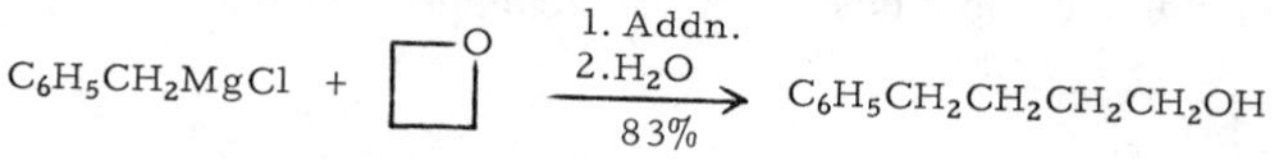
C6H5CH2MgCl +
1. Addn.
2.H2O
83%
C6H5CH2CH2CH2CH2OH

primary alcohols in yields in the range 80–85%.[2] The reaction is not satisfactory as applied to secondary magnesium halides. Thus cyclohexylmagnesium bromide

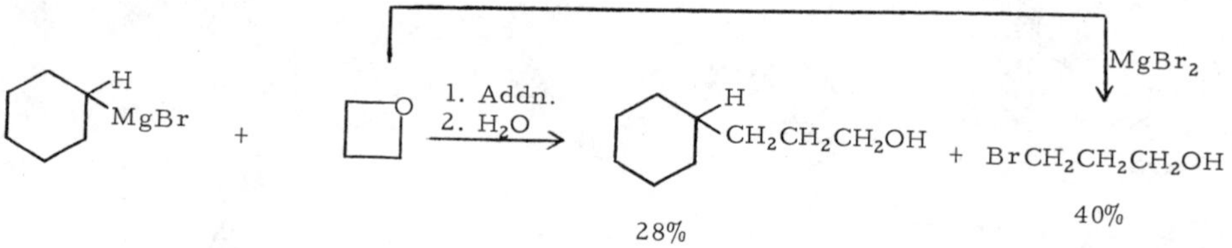
H
MgBr
+
1. Addn.
2. H2O
H
CH2CH2CH2OH
28%
MgBr2
+ BrCH2CH2CH2OH
40%

gives less of the normal product than of by-product 3-bromopropanol-1, shown to arise from trimethylene oxide and magnesium bromide.

[1]C. R. Noller, *Org. Syn., Coll. Vol.*, **3**, 835 (1955)
[2]S. Searles, *Am. Soc.*, **73**, 124 (1951)

Trimethyloxonium fluoroborate, $(CH_3)_3O^+BF_4^-$. Mol. wt. 147.92, m.p. 141–143°. Preparation.[1]

Like triethyloxonium fluoroborate, the salt is a potent alkylating agent. It is easier to make than trimethyloxonium 2,4,6-trinitrobenzenesulfonate but does not keep as well in storage.

[1]H. Meerwein, *Org. Syn.*, **46**, 120 (1966)

Trimethyloxonium 2,4,6-trinitrobenzenesulfonate. Mol. wt. 311.25, m.p. 181–183°. Preparation.[1]

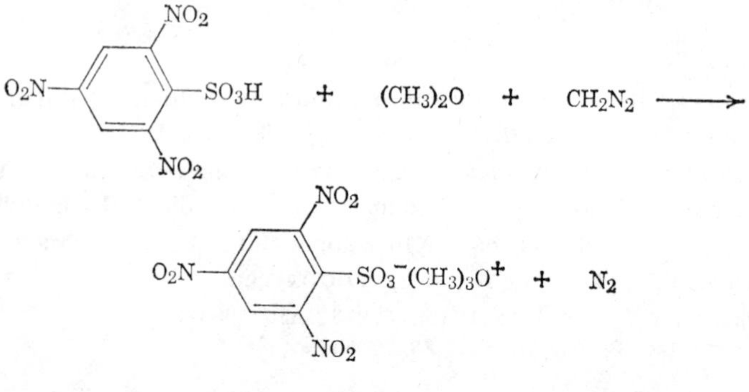
NO2
O2N
SO3H
NO2
+ (CH3)2O + CH2N2
NO2
O2N
SO3⁻(CH3)3O⁺ + N2
NO2

This potent alkylating agent is more laborious to make than trimethyloxonium fluoroborate, but it is nonhygroscopic and keeps better in storage.

[1]G. K. Helmkamp and D. J. Pettitt, *Org. Syn.*, **46**, 122 (1966)

Trimethylphenylammonium tribromide, *see* Phenyltrimethylammonium tribromide.

Trimethyl phosphite, $(CH_3O)_3P$. Mol. wt. 124.08, b.p. 111°. Suppliers: A, B, E, F, MCB.

Adducts with quinones, α-diketones, and α-keto esters. Ramirez and co-workers[1] found that chloranil and other 1,4-benzoquinones react readily with trimethyl phosphite to form the methyl ethers of the hydroquinone monophosphates, for example (3), and presented evidence that the primary product is the dipolar ion (2).

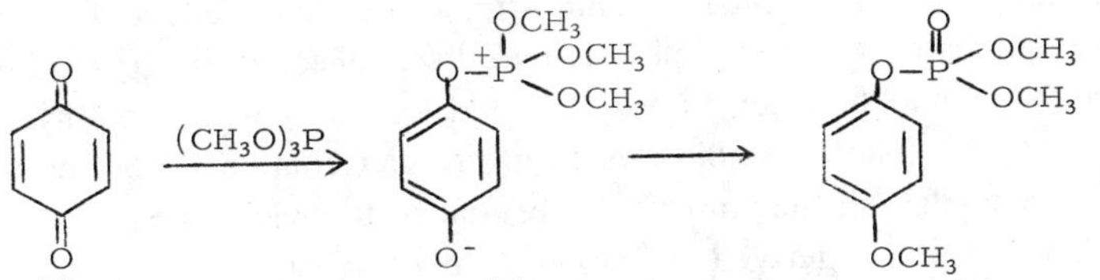
(CH₃O)₃P

(1) (2) (3)

Phenanthrenequinone reacts with trimethyl phosphite in benzene at 20° to give the colorless crystalline cyclic unsaturated oxyphosphorane (5) in quantitative yield.

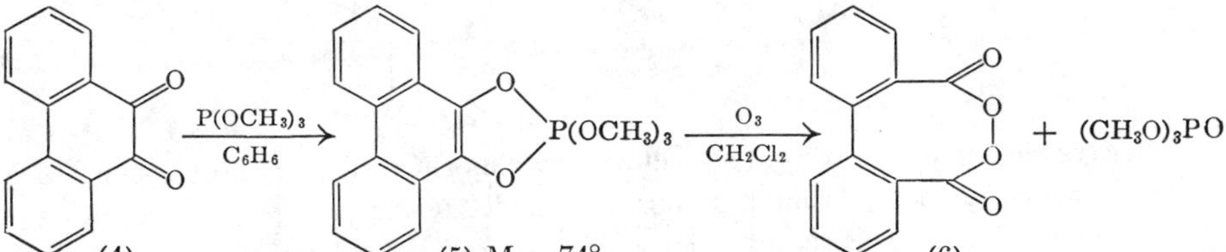
P(OCH₃)₃
C₆H₆
O₃
CH₂Cl₂
+ (CH₃O)₃PO

(4) (5) M.p. 74° (6)

Ozonolysis of the adduct affords diphenoyl peroxide (6). Benzil and diacetyl give adducts melting at 50 and at 46° respectively.

Trimethyl phosphite reacts with methyl pyruvate to give a liquid mixture of the two possible diastereoisomeric cyclic orthophosphates; on treatment with one equivalent of 2*N* sodium hydroxide these yield the diastereoisomeric dimethyl C,C-dimethyltartrates.[2] The reaction represents a novel method for the formation of a C–C bond.

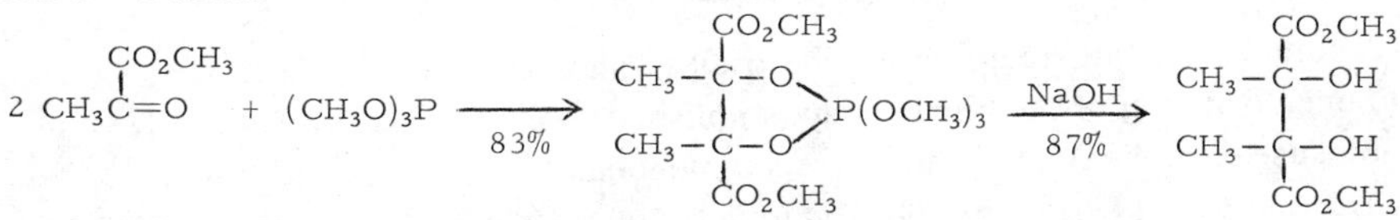
2 CH₃C=O + (CH₃O)₃P
83%
NaOH
87%

Olefin synthesis. Corey and Winter[3] devised a novel olefin synthesis from 1,2-diols utilizing in the first step the reaction with thiocarbonyldiimidazole (2), a reagent prepared from 2 moles of imidazole and 1 mole of thiophosgene. Corey postulated that treatment of the cyclic thionocarbonate (3) with a reagent which would abstract sulfur to form the carbene (4) would effect *cis* elimination to the olefin and found that trimethyl phosphite effects this reaction cleanly and in high yield. *dl*-Hydrobenzoin treated in the same way afforded pure *trans*-stilbene in yield of 87%. The method of Corey and Winter has been used for the synthesis of unsaturated sugars.[4]

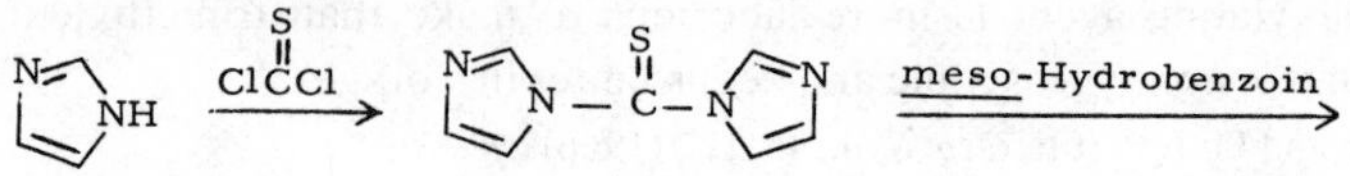

(1) (2)

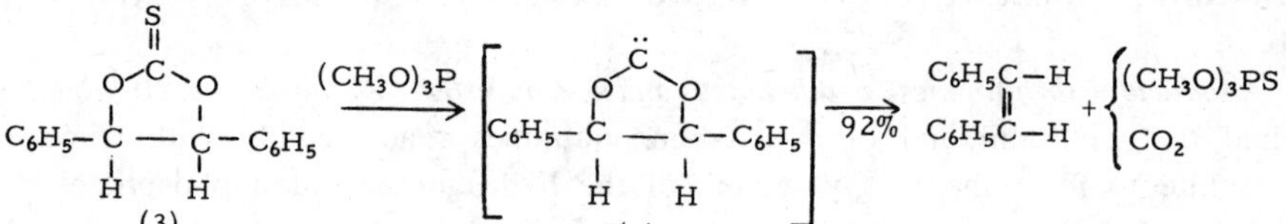

(3) (4) (5)

Dehydrohalogenation. Hunziker and Müllner[5] effected allylic bromination of cholesteryl benzoate in refluxing petroleum ether, collected the crystalline mixture of 7α- and 7β-bromocholesteryl benzoate, and treated it with trimethyl phosphite in refluxing xylene. The crystalline reaction product contained 56% of 7-dehydrocholesteryl benzoate and the amount corresponded to a yield of 52%. It is assumed that the dienic ester is derived from the 7α-cholesteryl benzoate present in the bromination product (*trans* elimination).

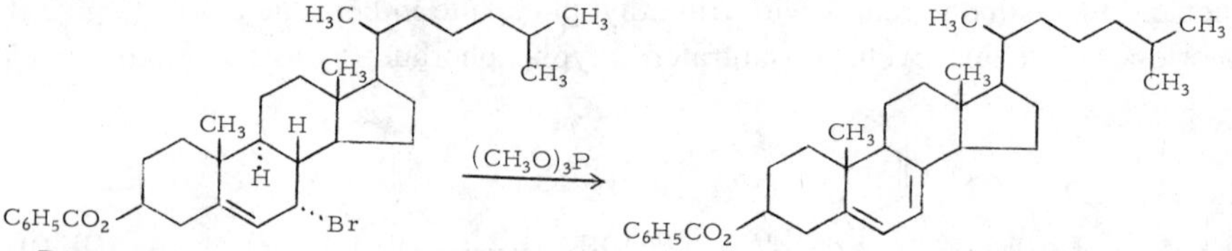

Deoxygenation. Ozonolysis of an olefin in methanol at −40° gives a hydroperoxide and a carbonyl fragment. Trimethyl phosphite, added to the mixture at −40°, reduces the hydroperoxide to the carbonyl compound.[6]

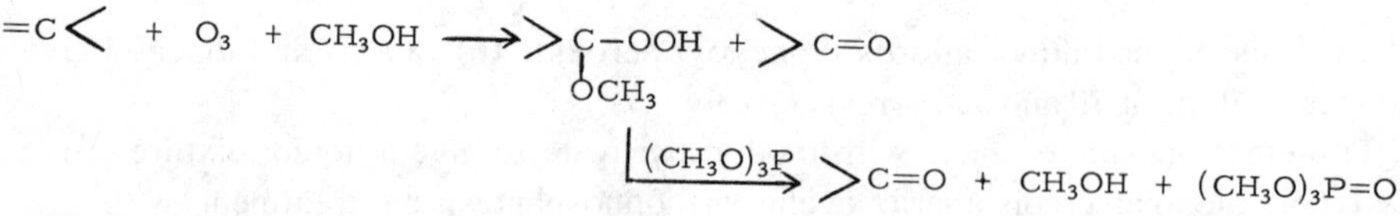

Nitrile oxides are readily deoxygenated in high yield by trimethyl or triethyl phosphite.[7] The reaction is carried out by heating the reactants in benzene for

$$RC{\equiv}\overset{+}{N}{-}\overset{-}{O} + (CH_3O)_3P \longrightarrow RC{\equiv}N + (CH_3O)_3PO$$

5–10 minutes on the steam bath. The trialkyl phosphate formed is water soluble, and hence isolation of the nitrile is accomplished easily. In case the nitrile is water soluble, use of triphenylphosphine is recommended.

Debromination. Trimethyl phosphite debrominates *vic*-dibromides in which one or both halogens is adjacent to a carbonyl group.[8] Use of sodium iodide leads to tars; zinc dust reduces the double bond.

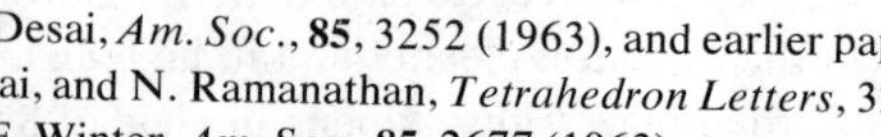

[1] F. Ramirez and N. B. Desai, *Am. Soc.*, **85**, 3252 (1963), and earlier papers cited
[2] F. Ramirez, N. B. Desai, and N. Ramanathan, *Tetrahedron Letters*, 323 (1963)
[3] E. J. Corey and R. A. E. Winter, *Am. Soc.*, **85**, 2677 (1963)
[4] D. Horton and W. N. Turner, *Tetrahedron Letters*, 2531 (1964)

[5]F. Hunziker and F. X. Müllner, *Helv.*, **41**, 70 (1958)
[6]W. S. Knowles and Q. E. Thompson, *J. Org.*, **25**, 1031 (1960)
[7]C. Grundmann and H.-D. Frommeld, *J. Org.*, **30**, 2077 (1965)
[8]S. Dershowitz and S. Proskauer, *J. Org.*, **26**, 3595 (1961)

Trimethylsilane, $(CH_3)_3SiH$. Mol. wt. 74.20, b.p. 6.7°, m.p. −136°. Preparation.[1] Suppliers: Baker, Columbia, Peninsular ChemResearch. According to N. C. Cook,[2] the Peninsular material contains about 0.2% $(CH_3)_2SiH_2$, which can be stripped out by a regular fractionating column with acetone at −30 to −78° circulating in the cold finger.

1,4-Dihydropyridine. Although several N-ethyl and N-aryldihydropyridines were known, the first of five possible isomeric unsubstituted dihydropyridines was prepared by Cook and Lyons.[3] Palladium-catalyzed addition of trimethylsilane to pyridine gives 1-trimethylsilyl-1,4-dihydropyridine (b.p. 57°), and this on treatment with methanol containing 0.1% potassium hydroxide gave 1,4-dihydropyridine and

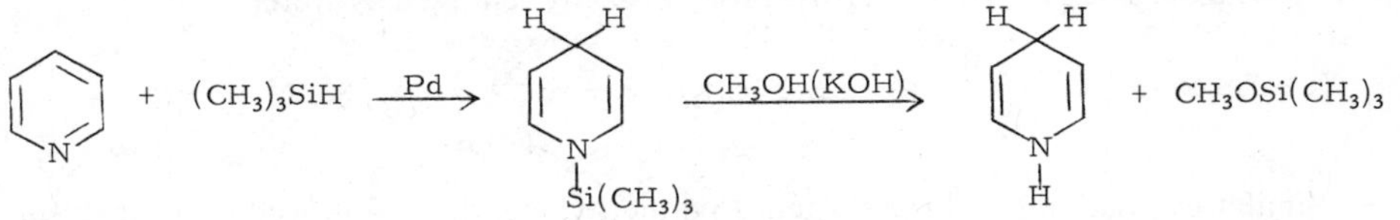
+ (CH3)3SiH
Pd
H H
N
Si(CH3)3
CH3OH(KOH)
H H
N
H
+ CH3OSi(CH3)3

I II

methoxytrimethylsilane. Compounds I and II have to be rigorously protected from air since oxygen reacts readily with both materials.

[1]S. Tannenbaum, S. Kaye, and G. F. Lewenz, *Am. Soc.*, **75**, 3753 (1953)
[2]Private communication
[3]N. C. Cook and J. E. Lyons, *Am. Soc.*, **87**, 3283 (1965)

N-Trimethylsilylacetamide, $CH_3CONHSi(CH_3)_3$. Mol. wt. 131.25 b.p. 84°/13 mm.

Preparation: by the condensation of trimethylchlorosilane with acetamide in the presence of triethylamine.[1]

The reagent silylates alcohols under very mild conditions; the only by-product is neutral acetamide. A carbohydrate on brief fusion with the reagent is converted

$$ROH + CH_3CONHSi(CH_3)_3 \longrightarrow ROSi(CH_3)_3 + CH_3CONH_2$$

into a persilylated derivative. By the fusion method, glucose affords 1,2,3,4,6-pentakis-O-trimethylsilylglucose, but the reaction is moderated by conducting it in pyridine solution and in this case, surprisingly, the primary alcoholic hydroxyl remains unattacked and the product is 1,2,3,4-tetrakis-O-trimethylsilylglucose (2).[2] The ready availability of this substance opened the way to a synthesis of the gentiobiose derivative (4), obtained by conversion of (2) into the sodio derivative (3), condensation with acetobromoglucose, and desilylation with aqueous or alcoholic hydrochloric acid.

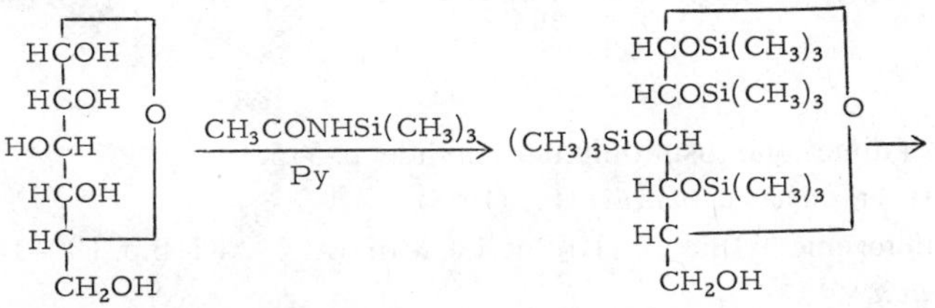
HCOH
HCOH
HOCH
HCOH
HC
CH2OH
O
CH3CONHSi(CH3)3
Py
HCOSi(CH3)3
HCOSi(CH3)3
(CH3)3SiOCH
HCOSi(CH3)3
HC
CH2OH
O

(1) (2)

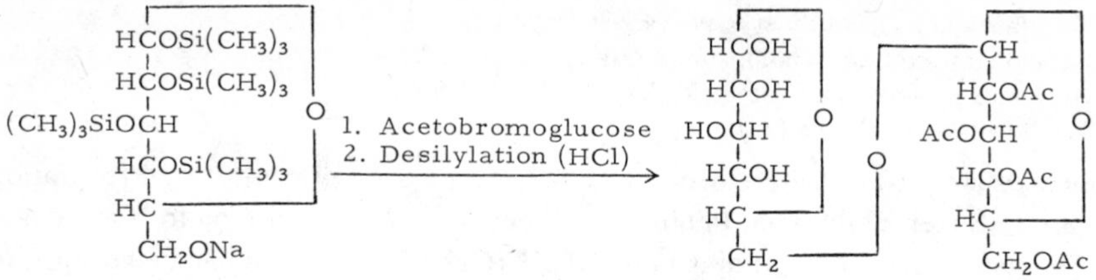

(3) (4)

[1]L. Birkofer, A. Ritter, and H. Dickopp, *Ber.*, **96**, 1473 (1963)
[2]L. Birkofer, A. Ritter, and F. Bentz, *Ber.*, **97**, 2196 (1964)

Trimethylsilyl azide, $(CH_3)_3SiN{=}N{=}N$.[1] Mol. wt. 115.21, m.p. 95°. This stable azide can be prepared in 87% yield by reaction of trimethylchlorosilane with sodium azide in THF, filtration from sodium chloride, and distillation. The reagent can be used in place of the explosive hydrazoic acid for the synthesis of 4-phenyl-1,2,3-triazole since the trimethylsilyl group is easily removed by water.

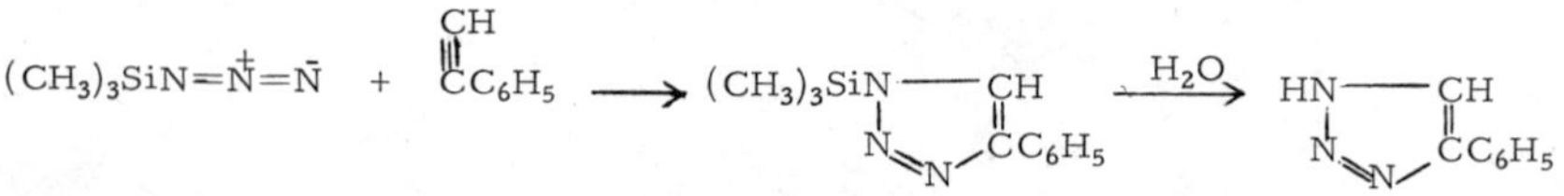

Similar cycloaddition of the reagent to dimethylacetylene, followed by hydrolysis, affords 4,5-dimethyl-1,2,3-triazole.[2]

[1]L. Birkofer and A. Ritter, *Angew. Chem., Internat. Ed.*, **4**, 417 (1965)
[2]L. Birkofer and P. Wegner, procedure submitted to *Org. Syn.*

Trimethylsulfonium iodide, $(CH_3)_3S^+I^-$. Mol. wt. 204.08. Supplier: Aldrich.

This precursor of dimethylsulfonium methylide is prepared in almost quantitative yield by mixing dimethyl sulfide (6 g.) and methyl iodide (14 g.).[1] The sulfonium iodide separates as a solid cake and after 24 hrs. it is collected, crystallized from ethanol, and washed with ether.

[1]H. J. Emeléus and H. G. Heal, *J. Chem. Soc.*, 1126 (1946)

Trimethylsulfoxonium iodide $(CH_3)_3\overset{+}{S}{=}O\ (I^-)$. Mol. wt. 220.08, dec. 200°. Supplier: Aldrich.

The reagent is prepared by refluxing dimethyl sulfoxide with methyl iodide for many days.[1] The tendency to revert to dimethyl sulfoxide makes the substance an effective methylating agent.

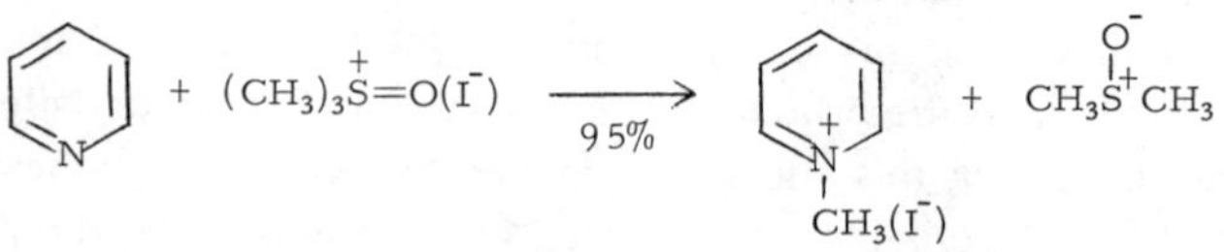

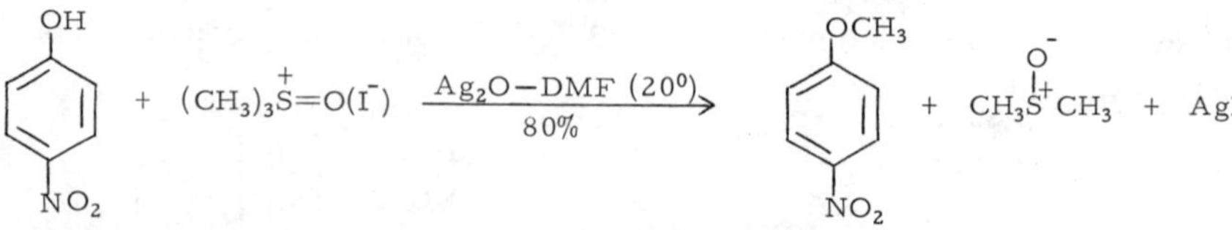

Precursor of dimethyloxosulfonium methylide, p. 315.

[1]R. Kuhn and H. Trischmann, *Ann.*, **611**, 117 (1958)

Trimethyl(trifluoromethyl)tin, $(CH_3)_3SnCF_3$. Mol. wt. 232.81, b.p. 100–101°.

Preparation:[1]

$$(CH_3)_3Sn\cdot Sn(CH_3)_3 + CF_3I \longrightarrow (CH_3)_3SnI + (CH_3)_3SnCF_3$$

Generation of difluorocarbene. When heated in a sealed tube in the absence of air for 20 hrs., the reagent decomposes smoothly to trimethyltin fluoride and perfluoro-

$$3\ (CH_3)_3SnCF_3 \xrightarrow{150^0} 3\ (CH_3)_3SnF + 3\ :CF_2 \longrightarrow F_2C{\diagdown\!\!\!\diagup}^{CF_2}_{CF_2}$$

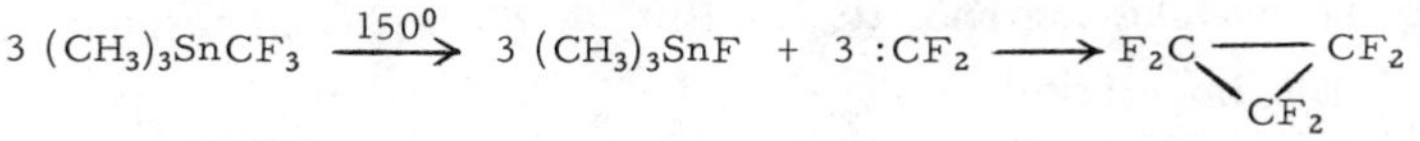

cyclopropane.[1] Perfluorocyclopropane was obtained also on decomposition of the reagent in the presence of excess tetrafluoroethylene.

Seyferth and co-workers[2] found that difluorocarbene can be generated at 80° by reaction of the reagent with sodium iodide in 1,2-dimethoxyethane. The filtered

$$\underset{100\ mmoles}{\bighexagon} + \underset{11.8\ mmoles}{(CH_3)_3SnCF_3} + \underset{15\ mmoles}{NaI} \xrightarrow[73\%]{\overset{12\ hrs.\ at\ 80^0}{CH_3OCH_2CH_2OCH_3}} \overset{F}{\underset{F}{\diagup}}$$

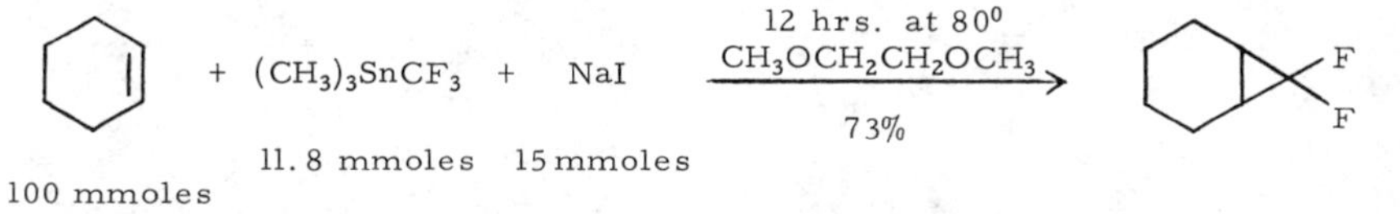

reaction mixture from cyclohexene was shown by gas chromatographic analysis to contain trimethyltin iodide (90%) and 7,7-difluoronorcarane (73%).

[1] H. C. Clark and C. J. Willis, *Am. Soc.*, **82**, 1888 (1960)
[2] D. Seyferth, J. Y.-P. Mui, M. E. Gordon, and J. M. Burlitch, *Am. Soc.*, **87**, 681 (1965)

1,3,5-Trinitrobenzene, $C_6H_3(NO_2)_3$. Mol. wt. 213.10, m.p. 122°. The reagent is prepared by oxidizing trinitrotoluene with sodium dichromate in dilute sulfuric acid at 45–55°,[1] extracting the 2,4,6-trinitrobenzoic acid from unchanged TNT with dilute alkali, and heating the filtrate to the boiling point to effect decarboxylation; overall yield 43–46%.[2]

π-Complexes with reactive hydrocarbons. Useful for characterization and isolation, the π-acid forms complexes of melting point intermediate between that of the complexes with picric acid and with trinitrofluorenone. Thus the melting point and color of the methylcholanthrene complexes are: PA, 182.5°, purplish black; TNB, 204.5°, dark red; TNF, 254°, green.

[1] H. T. Clarke and W. W. Hartman, *Org. Syn.*, *Coll. Vol.*, **1**, 543 (1941)
[2] *Idem, ibid.*, **1**, 541 (1941)

2,4,7-Trinitrofluorenone,

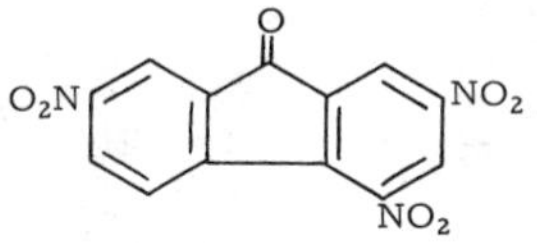

Mol. wt. 315.19, m.p. 175–176°. Suppliers: A, B, E, F, KK, MCB. Preparation from fluorenone in yield of 75–78%.[1]

The reagent forms π-complexes useful for the isolation and characterization of suitably reactive aromatic hydrocarbons.[2]

[1] E. O. Woolfolk and M. Orchin, *Org. Syn.*, *Coll. Vol.*, **3**, 837 (1955)
[2] M. Orchin and E. O. Woolfolk, *Am. Soc.*, **68**, 1727 (1946); M. Orchin, L. Reggel, and E. O. Woolfolk, *ibid.*, **69**, 1225 (1947)

Tri-*p*-nitrophenyl phosphorotrithioite, $P(SC_6H_4NO_2-p)_3$. Mol. wt. 397.28, m.p. 166–167°, yellow.

Peptide synthesis. Preparation[1] and use in the synthesis of cyclic peptides where the C-terminal amino acid is glycine (with an optically active acid, racemization is extensive).[2]

[1] J. A. Farrington, P. J. Hextall, G. W. Kenner, and J. M. Turner, *J. Chem. Soc.*, 1407 (1957)

[2]G. W. Kenner, P. J. Thomson, and J. M. Turner, *J. Chem. Soc.*, 4148 (1958); G. W. Kenner and A. H. Laird, *Chem. Comm.*, 305 (1965)

Triphenyldibromophosphorane, $(C_6H_5)_3PBr_2$, *see* Triphenylphosphine dibromide.

Triphenylmethyl, *see* Trityl.

Triphenylphosphine, $(C_6H_5)_3P$. Mol. wt. 262.28, m.p. 79–80°. Suppliers, Techn.: Metal and Thermit Corp.; Premium: A, Alfa, E, F, MCB.

Wittig reaction. In a procedure by Wittig and Schoellkopf[1] for the synthesis of methylenecyclohexane a solution of triphenylphosphine in benzene in a pressure

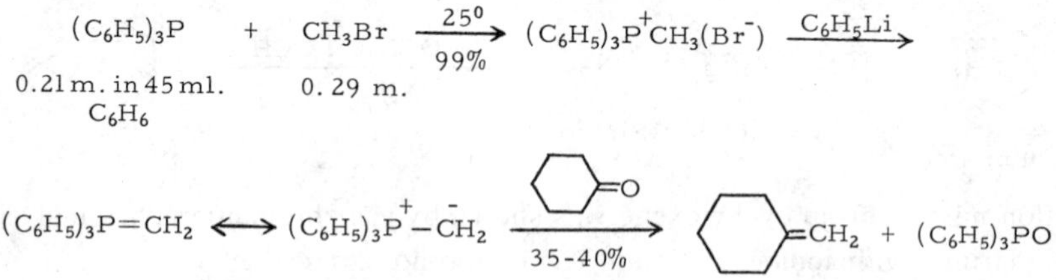

(C6H5)3P
+
CH3Br
25°
99%
(C6H5)3P+CH3(Br−)
C6H5Li
0.21 m. in 45 ml. C6H6
0. 29 m.
(C6H5)3P=CH2
(C6H5)3P+−CH2−
35-40%
=CH2 + (C6H5)3PO

bottle is cooled in a salt–ice bath, previously condensed methyl bromide is added, and the bottle is sealed and let stand at room temperature for 2 days. Triphenylmethylphosphonium bromide separates as a white solid, which is collected with the aid of about 500 ml. of hot benzene. A 0.1 m. portion of this salt is added in portions to an ethereal solution of 0.1 m. of phenyllithium with stirring under nitrogen to produce the Wittig reagent, or ylide, and cyclohexanone is added to complete the reaction.

In a synthesis of *trans,trans*-1,4-diphenyl-1,3-butadiene[2] the required phosphonium chloride is prepared by heating a mixture of triphenylphosphine and cinnamyl chloride with stirring under reflux for 12 hrs. The salt, which separates as

$$\underset{0.26\ m.}{C_6H_5CH{=}CHCH_2Cl} + \underset{0.35\ m.}{(C_6H_5)_3P} + \underset{500\ ml.}{Xylene} \xrightarrow[91\text{-}93\%]{Refl.\ 12\ hrs.}$$

$$C_6H_5CH{=}CHCH_2\overset{+}{P}(C_6H_5)_3(\overset{-}{Cl}) \xrightarrow{LiOC_2H_5} C_6H_5CH{=}CHCH{=}P(C_6H_5)_3 \xrightarrow[60\text{-}67\%]{C_6H_5CH{=}O}$$

$$C_6H_5CH{=}CHCH{=}CHC_6H_5$$

a colorless solid, is collected and dried, and a 0.145 m. portion is dissolved, along with 0.155 m. of benzaldehyde, in 200 ml. of anhydrous ethanol. The ylide is then generated in the presence of the carbonyl component by addition of a solution of lithium ethoxide in anhydrous ethanol.

The first step in a synthesis of *p*-quinquephenyl[3] is carried out by stirring and heating under reflux for 3 hrs. a mixture of *p*-xylylene dichloride, triphenylphosphine, and 1 l. of dimethylformamide. The Wittig reaction of cinnamaldehyde with lithium ethoxide as base gives a tetraene, which then adds two moles of diethyl acetylenedicarboxylate to give the bis-adduct. Hydrolysis of this adduct is formulated as involving bond migration to give the *trans,trans*-tetraacid in accordance with evidence reported later in a similar synthesis of *p*-terphenyl.[4]

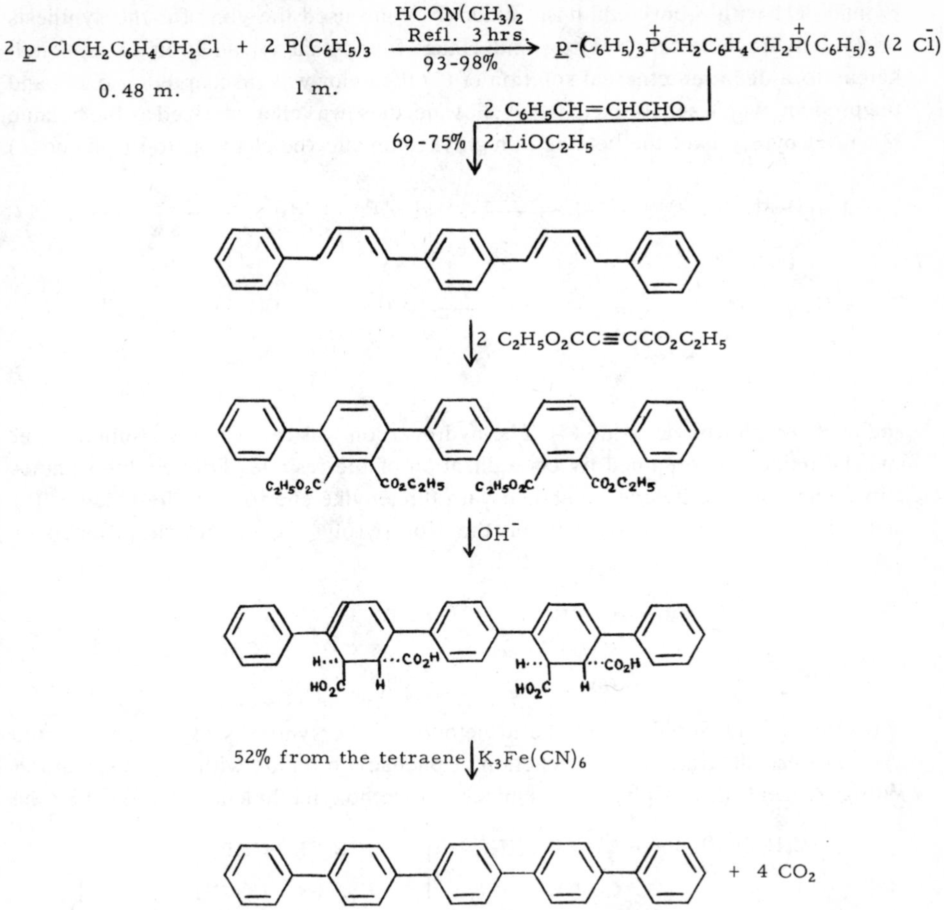
HCON(CH3)2
Refl. 3 hrs.
93-98%
2 p-ClCH2C6H4CH2Cl + 2 P(C6H5)3
p-(C6H5)3P+CH2C6H4CH2P+(C6H5)3 (2 Cl−)
0.48 m.
1 m.
2 C6H5CH=CHCHO
69-75%
2 LiOC2H5
2 C2H5O2CC≡CCO2C2H5
C2H5O2C
CO2C2H5
C2H5O2C
CO2C2H5
OH−
H
CO2H
HO2C
H
H
CO2H
HO2C
H
52% from the tetraene
K3Fe(CN)6
+ 4 CO2

The nucleophilic character of triphenylphosphine and the electrophilic character of carbenes suggested to two groups of investigators that the two reagents should combine to form ylides of a new type to use in expansion of the Wittig reaction. Speziale *et al.*[5] found that triphenylphosphine on treatment in pentane with chloroform and potassium *t*-butoxide at 0° gives yellow triphenylphosphine dichloromethylene, which reacts with benzophenone to yield 1,1-diphenyl-2,2-dichloroethylene (1).

1. $(C_6H_5)_3P \xrightarrow{:CCl_2} (C_6H_5)_3P{=}CCl_2 \xrightarrow[46\%]{(C_6H_5)_2CO} (C_6H_5)_2C{=}CCl_2$

2. $(C_6H_5)_3P + n\text{-BuLi} + CH_2Cl_2 \longrightarrow (C_6H_5)_3P{=}CHCl \xrightarrow[46\%]{C_6H_5COCH_3} C_6H_5C(CH_3){=}CHCl$ *cis* and *trans*

Seyferth *et al.*[6] generated chlorocarbene from *n*-butyllithium and methylene chloride in the presence of triphenylphosphine to produce triphenylphosphine chloromethylene, which reacts with acetophenone to give a mixture of *cis*- and *trans*-1-chloro-2-phenyl-1-propene (2).

Fagerlund and Idler[7] prepared triphenylphosphine isopropylidene (2) by heating triphenylphosphine with isopropyl bromide at 150° and treating the phosphonium bromide (1) with *n*-butyllithium in ether (N_2) and used the ylide for the synthesis of 24-dehydrocholesterol. Wittig and Haag[8] found that on addition of diphenylketene to a deep red ethereal solution of (2) the color was discharged at once, and the product, which separated in pale yellow needles, was characterized as the betaine (3). Brief pyrolysis of the betaine in high vacuum effected cleavage to the allene (4)

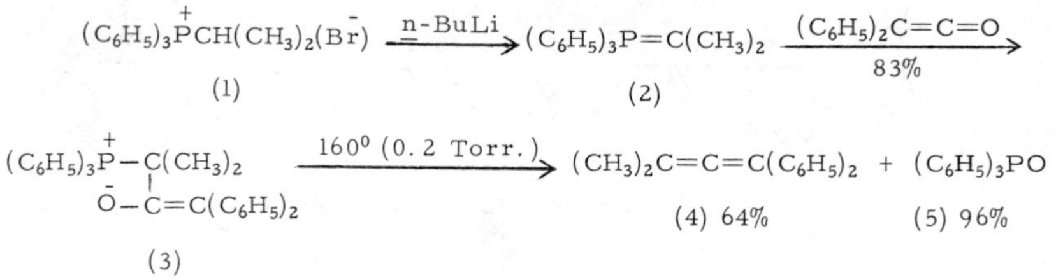

and triphenylphosphine oxide (5). The hydrocarbon was recovered by sublimation, and the oxide was obtained by crystallization of the residue. Triphenylphosphine-ethylidene and -methylene combined with diphenylketene to give the betaines (6) and (7) in yields of 45 and 12%, but these on pyrolysis gave dimeric products of unestablished structure.

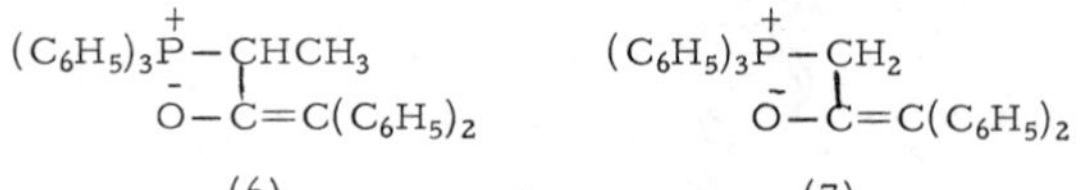

Bestmann and Schulz[9] describe a method for the synthesis of α,β-unsaturated γ-keto esters illustrated by the reaction of phenacyl bromide with two moles of the Wittig reagent (2), triphenylphosphinecarbomethoxymethylene. Presumably the

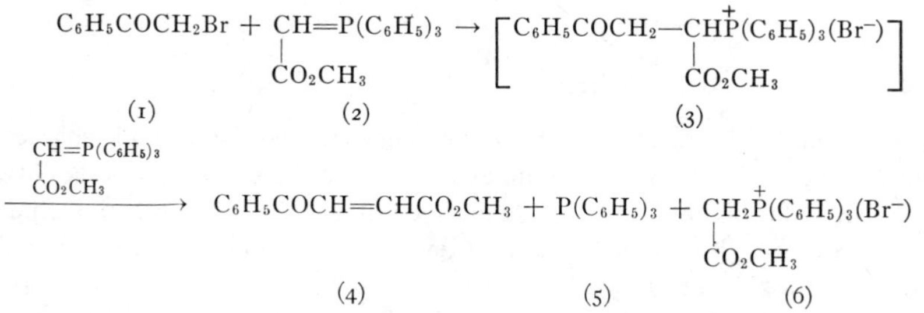

intermediate phosphonium salt (3) reacts with a second mole of reagent (2) with elimination of triphenylphosphine and production of the keto ester (4) and the phosphonium salt (6), from which the reagent (2) can be regenerated. A synthesis of aldehydes reported by Wittig[10] is illustrated in the formulation:

$$\underset{(1)}{C_6H_5OCH_2SO_2Na} \xrightarrow{PCl_5} \underset{(2)}{C_6H_5OCH_2Cl} \xrightarrow{(C_6H_5)_3P}$$

$$\underset{(3)}{C_6H_5OCH_2P^+(C_6H_5)_3(Cl^-)} \xrightarrow{C_6H_5Li} \underset{(4)}{C_6H_5OCH{=}P(C_6H_5)_3} \xrightarrow{(C_6H_5)_2CO}$$

$$(C_6H_5)_2C{=}CHOC_6H_5 \xrightarrow[\text{Ether}]{HClO_4} (C_6H_5)_2CHCHO$$

Bergelson and Shemyakin[11] discovered the possibility of controlling the steric course of the Wittig reaction by appropriate selection of the environmental conditions

$$C_6H_5CH_2\overset{+}{P}(C_6H_5)_3Cl^- \xrightarrow{\text{Base}} C_6H_5CH=P(C_6H_5)_3 \xrightarrow{CH_3CH_3CHO}$$

(1)   (2)

$$C_6H_5CH=CHCH_2CH_3$$

(3)

and the structure of the reactants. In the reaction formulated leading to β-ethylstyrene (3) the *cis/trans* ratio varied widely with the conditions as shown in the following examples:

Solvent	Additive	Base	*cis/trans*
Benzene		BuLi	26:74
Benzene	Aniline	BuLi	40:60
Benzene	LiBr	BuLi	91:9
Benzene	LiI	BuLi	93:7
Ether		EtONa	31:69
Dimethylformamide (DMF)	LiI	EtONa	96:4

Lithium bromide and iodide were added to the benzene solution in the form of highly disperse suspensions prepared by neutralizing a benzene solution of butyllithium with dry hydrogen bromide or iodide; lithium chloride had no effect.

When the ylide molecule carries an electron acceptor group at the ylide carbon (4), the reaction yields practically only the *trans* product (5), regardless of the presence of halide ions or of the solvent used:

$$(C_6H_5)_3P=CHCO_2Et \xrightarrow{C_6H_5CHO} C_6H_5\overset{\overset{H}{|}}{C}=\underset{\underset{H}{|}}{C}CO_2Et + (C_6H_5)_3PO$$

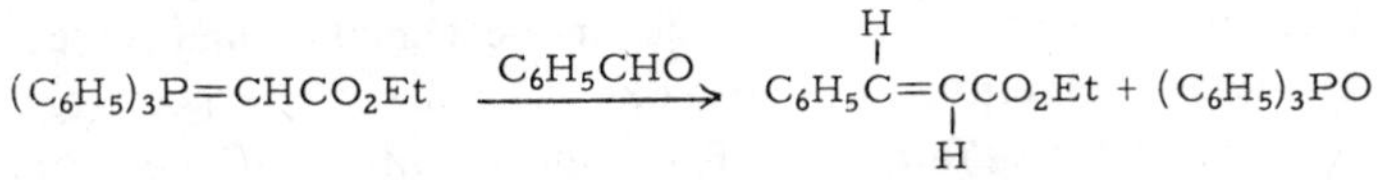

(4)   (5)

The Russian investigators found that a carboethoxy group at a distance from the ylide carbon exerts no effect on the steric course of the reaction, and on this basis

$$R\overset{\overset{H}{|}}{C}=\overset{\overset{H}{|}}{C}CH_2CH=P(C_6H_5)_3 \xrightarrow[\text{DMF, I}^-]{OCH(CH_2)_nCO_2CH_3} R\overset{\overset{H}{|}}{C}=\overset{\overset{H}{|}}{C}CH_2\overset{\overset{H}{|}}{C}=\overset{\overset{H}{|}}{C}(CH_2)_nCO_2CH_3$$

$$R\overset{\overset{H}{|}}{C}=\overset{\overset{H}{|}}{C}CH_2CHO \xrightarrow{(C_6H_5)_3P=CH(CH_2)_nCO_2CH_3} \Big\uparrow \text{DMF, I}^-$$

$$R\overset{\overset{H}{|}}{C}=\overset{\overset{H}{|}}{C}CH_2CH_2CH=P(C_6H_5)_3 \xrightarrow[\text{DMF, I}^-]{OCH(CH_2)_nCO_2CH_3}$$

$$R\overset{\overset{H}{|}}{C}=\overset{\overset{H}{|}}{C}CH_2CH_2\overset{\overset{H}{|}}{C}=\overset{\overset{H}{|}}{C}(CH_2)_nCO_2H$$

$$CH_3(CH_2)_3\overset{\overset{H'}{|}}{C}=\overset{\overset{H}{|}}{C}-\overset{\overset{H}{|}}{C}=\overset{\overset{H}{|}}{C}CHO \xrightarrow[\text{DMF, I}^-]{(C_6H_5)_3P=CH(CH_2)_7CO_2Et}$$

$$CH_3(CH_2)_3\overset{\overset{H}{|}}{C}=\overset{\underset{H}{|}}{C}-\overset{\overset{H}{|}}{C}=\overset{\underset{H}{|}}{C}-\overset{\overset{H}{|}}{C}=\overset{\overset{H}{|}}{C}(CH_2)_7CO_2Et$$

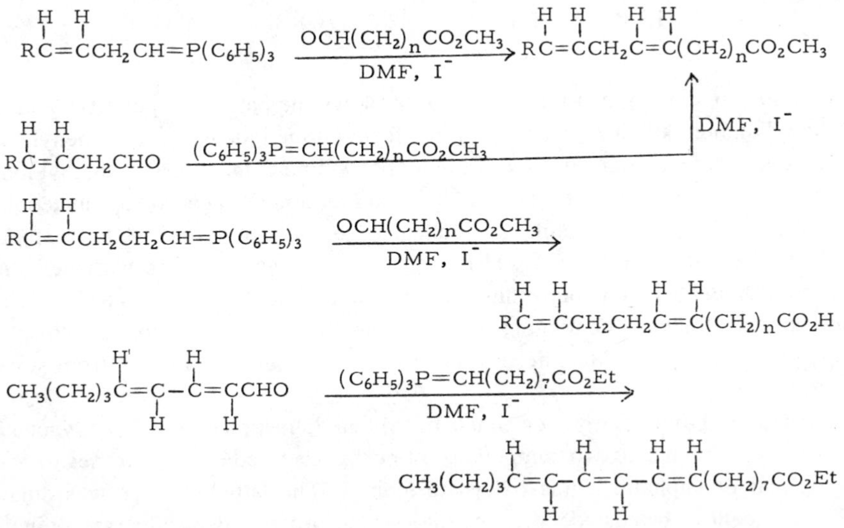

developed several highly stereospecific syntheses of natural *cis*-ethylenic fatty acids, some of which are formulated.

The Wittig method of olefin synthesis has been extended to include aliphatic aldehydes, for example gaseous formaldehyde and paraformaldehyde, acetaldehyde, propionaldehyde, *n*-hexaldehyde, acrolein.[12]

Corey and co-workers[13] found that when powdered sodium hydride is stirred into excess dimethyl sulfoxide under nitrogen at 65–70° hydrogen is evolved with formation of a solution of the highly reactive sodium methylsulfinyl methylide (1). This salt reacts with ethyltriphenylphosphonium bromide at room temperature to form the Wittig reagent (2), as shown by its rapid reaction with benzophenone to form 1,1-diphenylpropene-1 (3). The strongly basic character of methylsulfinyl

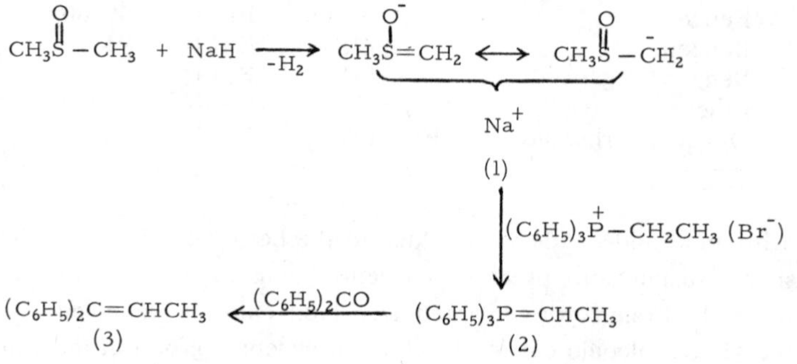

$CH_3\overset{O}{\overset{\|}{S}}-CH_3$ + NaH
$-H_2$
$CH_3\overset{O^-}{\overset{|}{S}}=CH_2$ ⟷ $CH_3\overset{O}{\overset{\|}{S}}-\bar{C}H_2$
Na^+
(1)
$(C_6H_5)_3\overset{+}{P}-CH_2CH_3\ (Br^-)$
$(C_6H_5)_2C=CHCH_3$
(3)
$(C_6H_5)_2CO$
$(C_6H_5)_3P=CHCH_3$
(2)

carbanion is shown by its rapid reaction with triphenylmethane to produce a deep red solution containing triphenylmethyl carbanion. The methylene homolog of (2) prepared in the same way reacts with cyclohexanone, camphor, and cholestane-3-one to give the methylene derivatives in yields of 86, 73, and 69%, respectively.

Deoxygenation and desulfurization. The high polarity of triphenylphosphine is evident from comparison of the dipole moments of the triphenyl derivatives of group 5 elements:

$(C_6H_5)_3N$	0.26 D	$(C_6H_5)_3Sb$	0.57 D
$(C_6H_5)_3P$	1.45 D	$(C_6H_5)_3Bi$	0 D
$(C_6H_5)_3As$	1.07 D		

The ability to combine with methyl iodide follows the pattern of polarity. Whereas triphenyl-amine, -stibene, and -bismuthine form no methiodides and triphenylarsine reacts only at a high temperature, triphenylphosphine combines with methyl iodide at room temperature in a strongly exothermal reaction.[14] The strongly nucleophilic character of triphenylphosphine is evident also from its reaction with fluoroacetylene to form the salt $(C_6H_5)_3P^+C{\equiv}CH(F^-)$[15] and by its reaction with carbenes, cited above. Although triphenylphosphine is substantially stable to air, it is attacked by a wide variety of oxygen-containing compounds with formation of triphenylphosphine oxide. Thus the reagent has been utilized for the removal of oxygen from several substrate types.

Horner and Hoffmann[16] reported use of triphenylphosphine for the deoxygenation of azoxy compounds to azo compounds, of peroxides to ethers, of nitrones to Schiff bases, and of aliphatic amine oxides to amines. The latter reaction was done in refluxing acetic acid and under these conditions the N-oxides of pyridine and quinoline were stable. Since these N-oxides provide useful routes to 2- and 4-substitution products, a method of deoxygenation more general than with use of phosphorus trichloride would be of value. Howard and Olszewski[17] found that at a high enough temperature triphenylphosphine serves the purpose admirably. The optimum procedure is to heat the reactants in the absence of a solvent to a temperature at which the amine distills from the reaction mixture. Triphenylphosphine oxide can be isolated by crystallization of the residue from methanol–water. 2-Picoline, 4-picoline,

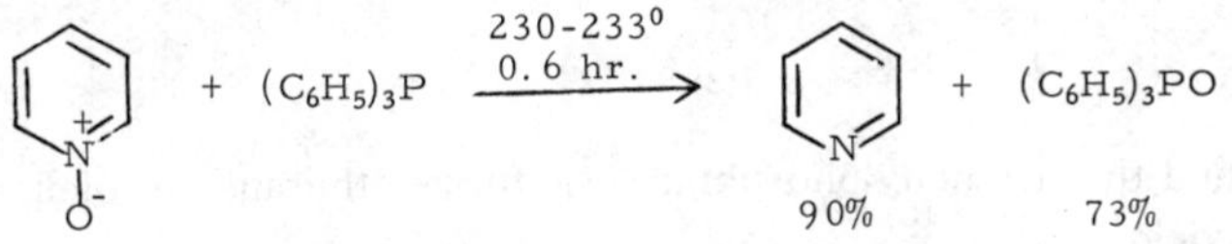
+ $(C_6H_5)_3P$
230-233°
0.6 hr.
+ $(C_6H_5)_3PO$
90%
73%

4-methoxypyridine, and quinoline were obtained satisfactorily from the N-oxides by the same procedure, but the method failed in the case of 4-nitropyridine. The only solvent found suitable for the reaction was triethylene glycol, but there is no apparent advantage in its use.

Wittig and Haag[18] found that styrene oxide on being heated with triphenylphosphine to 165° is deoxygenated to styrene in yield of about 50%. The reaction acquires preparative interest as applied to epoxides not available from olefins such as the β-phenylglycidic ester (1) obtained by the Darzens condensation. Heated with an

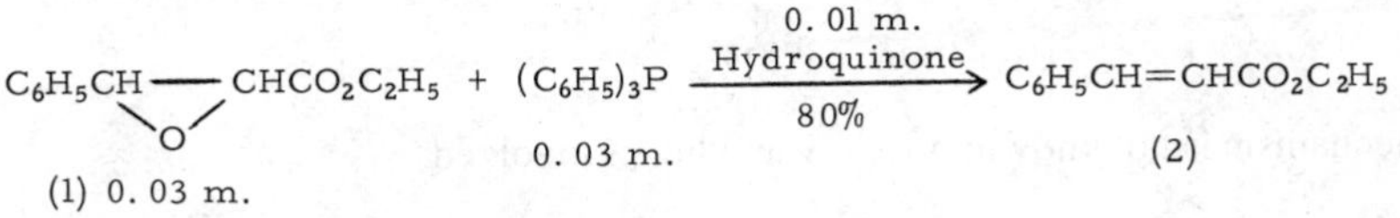
C_6H_5CH—$CHCO_2C_2H_5$ + $(C_6H_5)_3P$
0.01 m.
Hydroquinone
80%
$C_6H_5CH{=}CHCO_2C_2H_5$
0.03 m.
(2)
(1) 0.03 m.

equivalent amount of triphenylphosphine at 210°, this ester afforded ethyl cinnamate (2) in 61% yield. However on addition of hydroquinone as acid catalyst and to inhibit polymerization, the reaction started at 125° and was completed at 170° with production of ethyl cinnamate in 80% yield.

Boyer and Ellzey[19] obtained benzofurazane by refluxing *o*-dinitrosobenzene with triphenylphosphine in 95% ethanol for 30 min. (red solution) and separating the

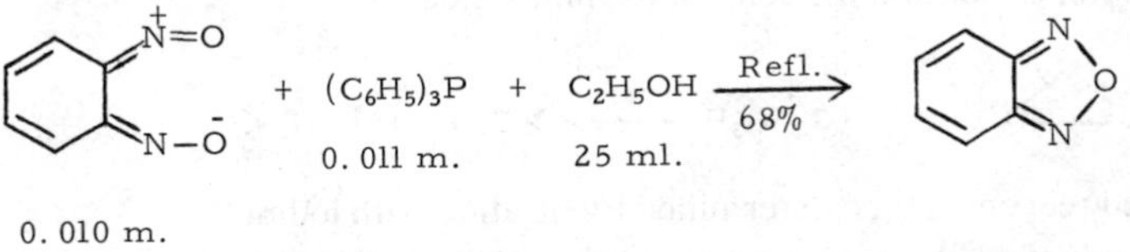
+ $(C_6H_5)_3P$ + C_2H_5OH
Refl.
68%
0.011 m.
25 ml.
0.010 m.

product by steam distillation. Substitution of triethylphosphine or tri-*n*-butylphosphine for triphenylphosphine afforded benzofurazane in comparable yield.

Weis[20] used triphenylphosphine successfully for deoxygenation of the Diels-Alder adduct (1) of 3,4-dicyanofurane with dicyanoacetylene. A suspension of the adduct (1) in acetonitrile reacts with triphenylphosphine even at room temperature to form the dark blue, sparingly soluble betaine (3), probably formed by initial addition to the double bond to give the precursor betaine (2). Pyrolysis of (3) affords 1,2,4,5-tetracyanobenzene in good yield along with triphenylphosphine oxide. One other adduct similar to (1) was cleaved similarly, but three others failed to react with triphenylphosphine.

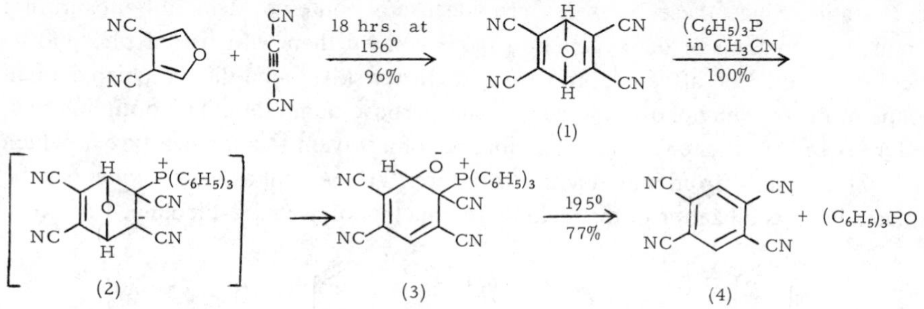

Davis[21] found that triphenylphosphine desulfurizes thiiranes to olefins in 3 days at 25° in high yield.

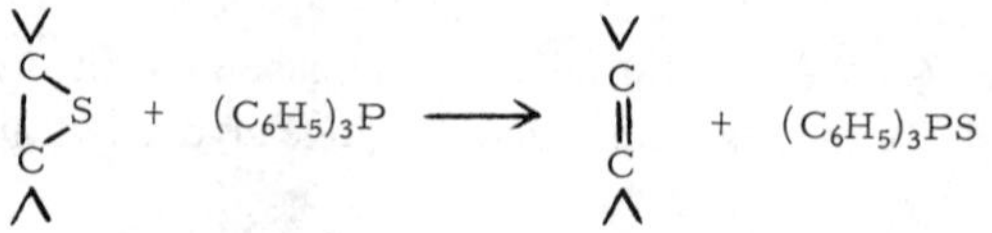

Sulfoxides are deoxygenated to sulfides by triphenylphosphine in carbon tetrachloride.[22]

$$\underline{p}\text{-}CH_3OC_6H_4\overset{O^-}{\overset{|}{S^+}}C_6H_4OCH_3\text{-}\underline{p} + (C_6H_5)_3P + CCl_4 \xrightarrow[90\%]{\text{Refl. 2 hrs.}} \underline{p}\text{-}CH_3OC_6H_4SC_6H_4OCH_3\text{-}\underline{p}$$

0.01 m.   0.02 m.   100 ml.

The mechanism is not known; possibly an ylide is involved:

$$2\,(C_6H_5)_3P + CCl_4 \xrightarrow[-(C_6H_5)_3PCl_2]{} (C_6H_5)_3P{=}CCl_2 \xrightarrow[-(C_6H_5)_3PO]{R_2SO}$$

$$R_2S{=}CCl_2 \xrightarrow{(C_6H_5)_3PCl_2} R_2S + CCl_4 + (C_6H_5)_3P$$

Determination of ozonides.[23] Treatment of an ozonide in alcoholic solution with an excess of triphenylphosphine for 3 days at room temperature in the absence of oxygen (which could oxidize the reagent directly) leads to the corresponding carbonyl compounds and triphenylphosphine oxide:

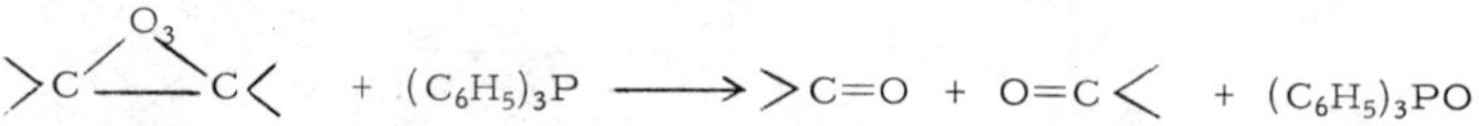

The unused reagent is then determined by titration with iodine.

Debromination. Horner and co-workers[24] used the reagent for reductive displacement of bromine or iodine atoms in *ortho* or *para* positions of phenols. An *ortho* halogen can be replaced preferentially. A mixture of the phenol, one equivalent

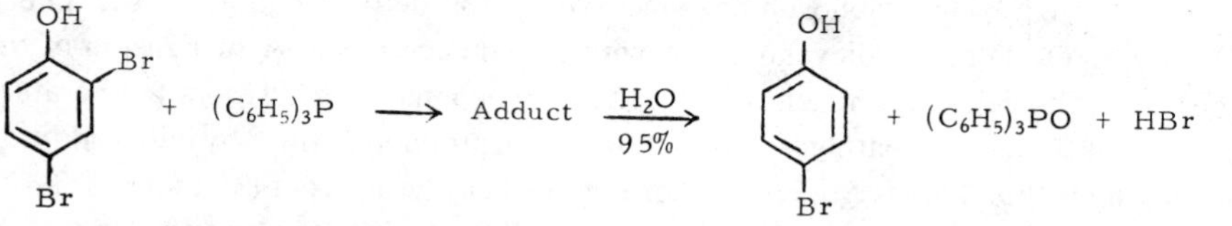

of triphenylphosphine, and benzene in a sealed tube is heated at 100–150° for 1–5 hrs., and the resulting adduct is treated with 2 *N* sodium hydroxide.

Triphenylphosphine reacts with phenacyl bromide by nucleophilic displacement of bromine to give phenacyltriphenylphosphonium bromide.[25] In contrast, secondary and tertiary α-bromoketones react with triphenylphosphine in refluxing benzene–methanol to give the debrominated ketone.[26] In 5 examples yields are in the range 60–70%.

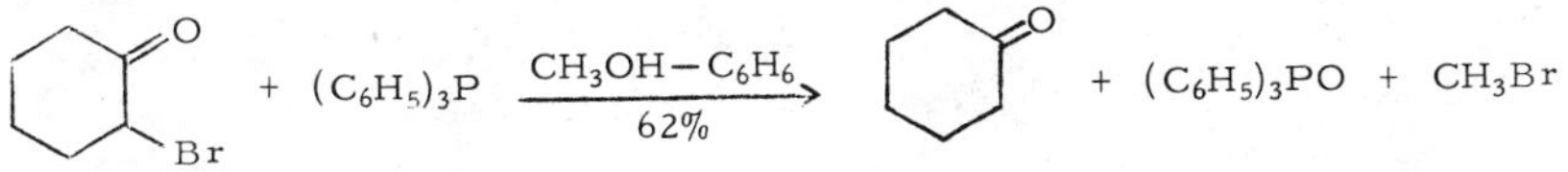
O
Br
+ (C6H5)3P
CH3OH−C6H6
62%
O
+ (C6H5)3PO + CH3Br

The reagent debrominates methyl and ethyl methacrylate dibromide to the unsaturated esters in yields of 50–65%.[27]

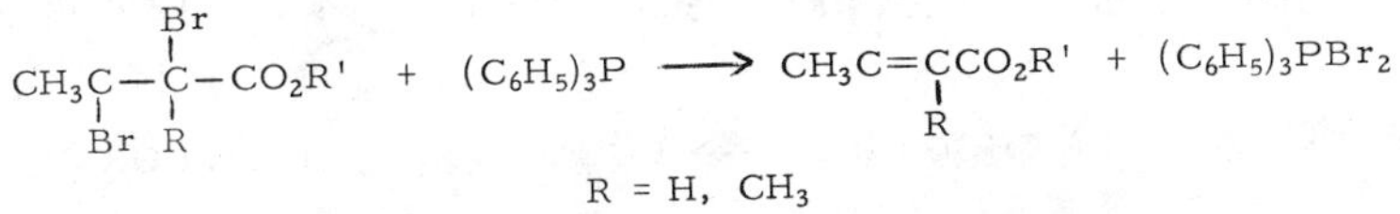
Br
CH3C−C−CO2R'
Br R
+ (C6H5)3P
CH3C=CCO2R' + (C6H5)3PBr2
R
R = H, CH3

The reaction of triphenylphosphine with bromomethyltriphenylphosphonium bromide in refluxing ethanol led, unexpectedly, to methyltriphenylphosphonium bromide in high yield.[28]

$$(C_6H_5)_3\overset{+}{P}CH_2Br(\overset{-}{Br}) + (C_6H_5)_3P \xrightarrow{C_2H_5OH} \underset{83\%}{(C_6H_5)_3\overset{+}{P}CH_3(\overset{-}{Br})} + \underset{87\%}{(C_6H_5)_3PO} + C_2H_5Br$$

Formation of nitriles and acetylenes. An unusual reaction described by Trippett and Walker[29] is the reaction of 1-bromo-1-nitrooctane with two equivalents of triphenylphosphine to give octanonitrile and triphenylphosphine oxide as its hydrobromide. The reaction is postulated to involve attack on an oxygen atom of the

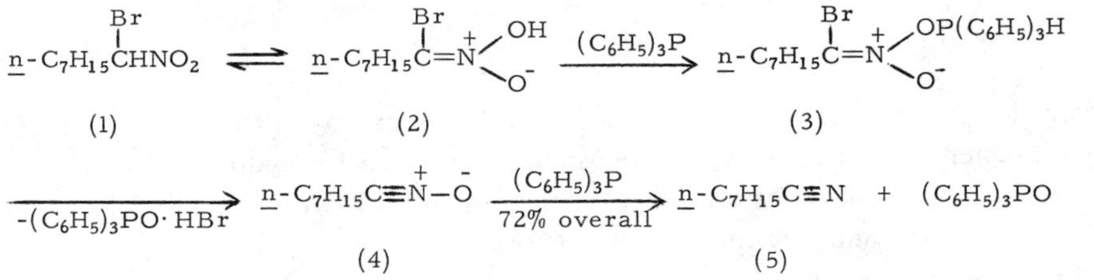
Br
n-C7H15CHNO2
(1)
Br
n-C7H15C=N
OH
O
(2)
(C6H5)3P
Br
n-C7H15C=N
OP(C6H5)3H
O
(3)
-(C6H5)3PO· HBr
n-C7H15C≡N−O
(4)
(C6H5)3P
72% overall
n-C7H15C≡N + (C6H5)3PO
(5)

aci-form (2), elimination of triphenylphosphine oxide hydrobromide to give the nitrile oxide (4, not isolated), and deoxygenation to the nitrile (5). N-Bromophenylacetamide ($C_6H_5CH_2CONHBr$) similarly gave phenylacetonitrile ($C_6H_5CH_2CN$, 60%), and α-phenylphenacyl chloride (6) gave in boiling benzene diphenylacetylene in 90% yield.

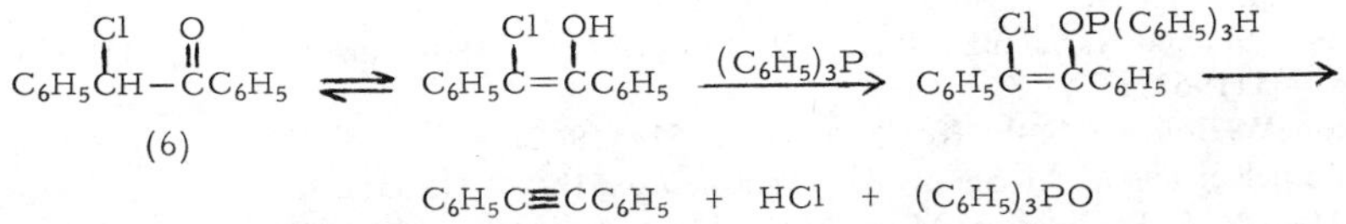
Cl O
C6H5CH−CC6H5
(6)
Cl OH
C6H5C=CC6H5
(C6H5)3P
Cl OP(C6H5)3H
C6H5C=CC6H5
C6H5C≡CC6H5 + HCl + (C6H5)3PO

Synthesis of ketoaldehydes and keto esters. Staudinger and Meyer[30] noted that triphenylphosphine combines with aliphatic diazo compounds to form compounds of a new type named phosphazines. Later Bestmann and co-workers developed the

$$R_2C=\overset{+}{N}=\overset{-}{N} \ + \ (C_6H_5)_3P \ \longrightarrow \ R_2C=N-N=P(C_6H_5)_3 \ \longleftrightarrow \ R_2\overset{-}{C}-N=N-\overset{+}{P}(C_6H_5)_3$$

following synthesis of α-keto aldehydes.[31] A diazoketone reacts with triphenylphosphine in ether with prompt separation of the crystalline α-ketotriphenylphosphazine in high yield. The phosphazine is cleaved by nitrous acid to the α-keto aldehyde.

$$\underset{O}{\overset{O}{\underset{\|}{R\overset{}{C}CH=\overset{+}{N}=\overset{-}{N}}}} \ + \ P(C_6H_5)_3 \ \longrightarrow \ \underset{O}{\overset{O}{\underset{\|}{R\overset{}{C}CH=N-N=P(C_6H_5)_3}}}$$

$$\xrightarrow{HNO_2} \ \underset{}{\overset{O}{\underset{\|}{R\overset{}{C}CH=O}}} \ + \ N_2 \ + \ (C_6H_5)_3PO$$

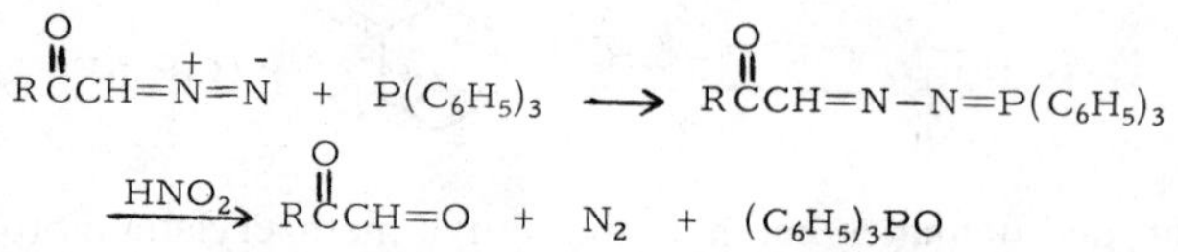

Another synthesis[32] starts with an α-diazo-β-keto ester (3) readily prepared from an acid chloride and diazoacetic ester. The ester (3) reacts with triphenylphosphine

$$\underset{(1)}{\overset{O}{\underset{\|}{R\overset{}{C}Cl}}} \ + \ 2 \ \underset{(2)}{\overset{-}{N}=\overset{+}{N}=CHCO_2C_2H_5} \ \longrightarrow \ \underset{(3)}{\overset{O}{\underset{\|}{R\overset{}{C}\underset{\overset{\|}{\underset{\|}{\overset{N^+}{N^-}}}}{C}CO_2C_2H_5}}} \ + \ \underset{(4)}{ClCH_2CO_2C_2H_5}$$

$$\underset{(3)}{\overset{O}{\underset{\|}{R\overset{}{C}\underset{\overset{\|}{\underset{\|}{\overset{N^+}{N^-}}}}{C}CO_2C_2H_5}}} \ \xrightarrow{P(C_6H_5)_3} \ \underset{(5)}{\overset{O}{\underset{\|}{R\overset{}{C}\underset{N-N=P(C_6H_5)_3}{C}CO_2C_2H_5}}} \ \xrightarrow{HNO_2} \ \underset{(6)}{\overset{O\ \ \ O}{\underset{\|\ \ \ \|}{R\overset{}{C}\ \overset{}{C}CO_2C_2H_5}}}$$

$$\Big\downarrow H_2O-ZnCl_2$$

$$\underset{(7)}{\overset{O}{\underset{\|}{R\overset{}{C}\underset{NNH_2}{C}CO_2C_2H_5}}} \ \xrightarrow[-N_2]{base} \ \underset{(8)}{\overset{O}{\underset{\|}{R\overset{}{C}CH_2CO_2C_2H_5}}}$$

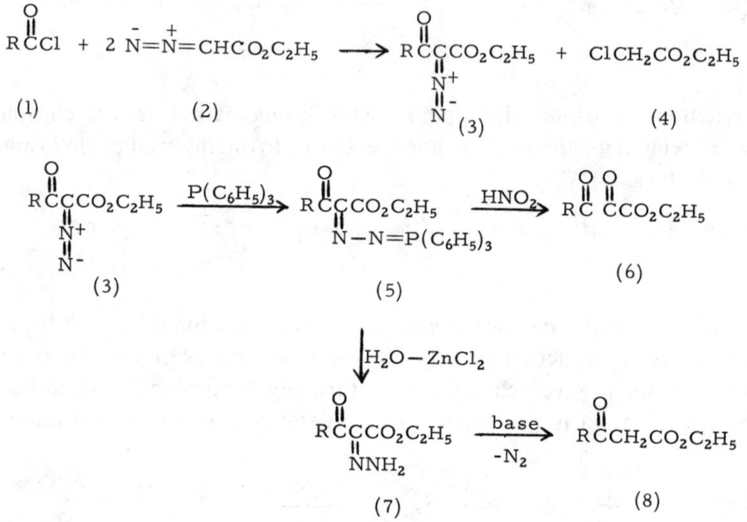

in acetic acid to form the phosphazine (5), which is split by nitrous acid to the α,β-diketo ester (6). Alternately, the phosphazine (5) can be hydrolyzed to (7), the hydrazone of a β-keto ester. When warmed at 60–70° with N-methylpiperidine, (7) loses nitrogen and gives the β-keto ester (8).

[1] G. Wittig and U. Schoellkopf, *Org. Syn.*, **40**, 66 (1960)

[2] R. N. McDonald and T. W. Campbell, *ibid.*, **40**, 36 (1960)

[3] T. W. Campbell and R. N. McDonald, *ibid.*, **40**, 85 (1960)

[4] L. F. Fieser and M. J. Haddadin, *Am. Soc.*, **86**, 2392 (1964)

[5] A. J. Speziale, G. J. Marco, and K. W. Ratts, *Am. Soc.*, **82**, 1260 (1960)

[6] D. Seyferth, S. O. Grim, and T. O. Read, *Am. Soc.*, **82**, 1510 (1960)

[7] U. H. M. Fagerlund and D. R. Idler, *Am. Soc.*, **79**, 6473 (1957)

[8] G. Wittig and A. Haag, *Ber.*, **96**, 1535 (1963)

[9] H. J. Bestmann and H. Schulz, *Ber.*, **95**, 2921 (1962); H. J. Bestmann, F. Seng, and H. Schulz, *Ber.*, **96**, 465 (1963)

[10] G. Wittig, W. Böll, and K.-H. Krück, *Ber.*, **95**, 2514 (1962)

[11] L. D. Bergelson and M. M. Shemyakin, *Tetrahedron*, **19**, 149 (1963)

[12] C. F. Hauser, T. W. Brooks, M. L. Miles, M. A. Raymond, and G. B. Butler, *J. Org.*, **28**, 372 (1963)

[13] R. Greenwald, M. Chaykovsky, and E. J. Corey, *J. Org.*, **28**, 1128 (1963)

[14] W. C. Davies and W. P. G. Lewis, *J. Chem. Soc.*, 1599 (1934)

[15] W. J. Middleton and W. H. Sharkey, *Am. Soc.*, **81**, 803 (1959)

[16]L. Horner and H. Hoffmann, *Angew. Chem.*, **68**, 480 (1956)
[17]E. Howard, Jr., and W. F. Olszewski, *Am. Soc.*, **81**, 1483 (1959)
[18]G. Wittig and W. Haag, *Ber.*, **88**, 1654 (1955)
[19]J. H. Boyer and S. E. Ellzey, Jr., *J. Org.*, **26**, 4684 (1961)
[20]C. D. Weis, *J. Org.*, **27**, 3520 (1962)
[21]R. E. Davis, *J. Org.*, **23**, 1767 (1958)
[22]J. P. A. Castrillón and H. H. Szmant, *J. Org.*, **30**, 1338 (1965)
[23]O. Lorenz, *Anal. Chem.*, **37**, 101 (1965); O. Lorenz and C. R. Parks, *J. Org.*, **30**, 1976 (1965)
[24]H. Hoffmann, L. Horner, H. G. Wippel, and D. Michael, *Ber.*, **95**, 523 (1962)
[25]F. Ramirez and S. Dershowitz, *J. Org.*, **22**, 41 (1957)
[26]I. J. Borowitz and L. I. Grossman, *Tetrahedron Letters*, 471 (1962)
[27]C. C. Tung and A. J. Speziale, *J. Org.*, **28**, 1521 (1963)
[28]D. W. Grisley, Jr., *Tetrahedron Letters*, 435 (1963)
[29]S. Trippett and D. M. Walker, *J. Chem. Soc.*, 2976 (1960)
[30]H. Staudinger and J. Meyer, *Helv.*, **2**, 619 (1919)
[31]H. J. Bestmann, H. Buckschewski, and H. Leube, *Ber.*, **92**, 1345 (1959); H. J. Bestmann, O. Klein, L. Göthlich, and H. Buckschewski, *Ber.*, **96**, 2259 (1963)
[32]H. J. Bestmann and H. Kolm, *Ber.*, **96**, 1948 (1963); H. J. Bestmann and O. Klein, *Ann.*, **676**, 97 (1964)

Triphenylphosphine–Carbon tetrachloride. This combination converts primary and secondary alcohols into the corresponding chlorides under mild, essentially neutral conditions.[1]

A carboxylic acid is converted into the acid chloride, possibly as follows:[2]

$$(C_6H_5)_3P + CCl_4 \longrightarrow (C_6H_5)_3\overset{+}{P}CCl_3(\overset{-}{Cl}) \xrightarrow[-CHCl_3]{RCOOH}$$

$$RCOO\overset{+}{P}(C_6H_5)_3\overset{-}{Cl} \longrightarrow RCOCl + (C_6H_5)_3PO$$

[1]I. M. Downie, J. B. Holmes, and J. B. Lee, *Chem. Ind.*, 900 (1966)
[2]J. B. Lee, *Am. Soc.*, **88**, 3440 (1966)

Triphenylphosphine dibromide and dichloride, $(C_6H_5)_3PX_2$. Mol. wt.: dibromide, 422.11; dichloride, 333.19. Eastman supplies the dibromide as Dibromotriphenylphosphorane, 9155.

Reaction with alcohols and phenols. Horner and co-workers[1] introduced use of the reagents for the preparation of alkyl and aryl halides from alcohols or phenols, and Wiley and co-workers[2] established that the reagents have considerable advantage over phosphorus pentahalides in the C—OH ⟶ C—X conversion[2] and discussed the mechanism.[3] For the preparation of *n*-butyl bromide the Wiley group added bromine under nitrogen to a solution of *n*-butanol and triphenylphosphine in dimethylformamide until 2 drops persisted in giving the solution an orange tint. Volatile

$$\underset{0.1\text{ m.}}{\underline{n}\text{-}C_4H_{11}OH} + \underset{0.107\text{ m.}}{(C_6H_5)_3P} + \underset{100\text{ ml.}}{HCON(CH_3)_2} + Br_2 \xrightarrow[91\%]{55^0\ (N_2)} \underline{n}\text{-}C_4H_{11}Br + (C_6H_5)_3PO + HBr$$

material was removed by distillation, water was added, and the layers separated. For the preparation of neopentyl chloride, triphenylphosphine was first chlorinated in carbon tetrachloride followed by exhaustive evaporation of the solvent. The reagent was taken up in DMF, neopentyl alcohol was added, and the mixture refluxed for 1 hr. Workup gave, in 92% yield, neopentyl chloride, which gave no precipitate with alcoholic silver nitrate. The conversion of secondary pentyl and neopentyl alcohols to the pure corresponding halides shows that substitution occurs without rearrangement. The conversion of cyclopentyl and cyclohexyl alcohols into the bromides shows the absence of a tendency to effect elimination. Phenols are converted smoothly into bromides at elevated temperatures without position isomers being formed.

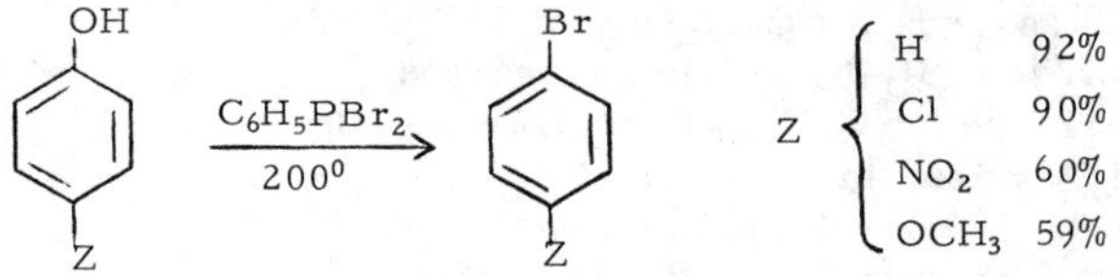
OH
$C_6H_5PBr_2$
200°
Br
Z
Z
Z
H 92%
Cl 90%
NO_2 60%
OCH_3 59%

Levy and Stevenson[4] heated cholesterol and cholestanol with a 10:1 molar ratio of $(C_6H_5)_3PBr_2$ in dimethylformamide on the steam bath for 20 hrs. and obtained 3β-bromo-Δ^5-cholestene and 2α-bromocholestane in yields of about 80%.

The 4,4-dimethyl-3β-ol system of triterpenoids and trimethylsteroids, which undergoes rearrangement on treatment with phosphorus pentahalides, reacts with $(C_6H_5)_3PBr_2$ to give chiefly the Δ^2-ene, together with some 2α,3β-dibromide. A 4,4-dimethyl-3-ketone gave the Δ^1-3-one. The reactivity of the three hydroxyl groups of methyl cholate to the reagent is the usual one: $3\alpha > 7\alpha > 12\alpha$.

Benzaldehyde reacts with triphenylphosphine dichloride to give benzal chloride in 59% yield, but cyclohexanone gives 1-chlorocyclohexene-1 in 45% yield.[1] Benzoic acid affords benzoyl chloride in 63% yield.

Triphenylphosphine dibromide reacts with carboxylic acids and anhydrides to give acid bromides.[4a]

$$RCO_2H + (C_6H_5)_3PBr_2 \xrightarrow[50-80\%]{} RCOBr + (C_6H_5)_3PO + HBr$$

Schaefer and Weinberg[5] found the Wiley procedure useful for the conversion of (+)-*endo*-norbornanol (1) into (+)-*exo*-norbornyl bromide (3). Under mild reaction

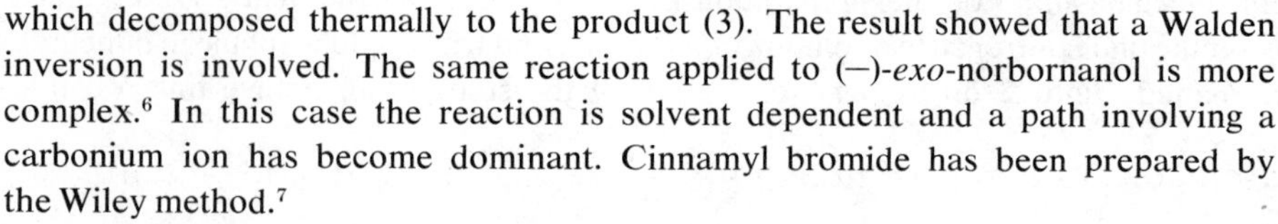
H
OH
$(C_6H_5)_3PBr_2$
-HBr
H
$\overset{+}{O}P(C_6H_5)_3(Br^-)$
Δ
Br
H
+ $(C_6H_5)_3PO$

(1) (2) (3)

conditions it was possible to isolate an intermediate phosphonium bromide salt (2), which decomposed thermally to the product (3). The result showed that a Walden inversion is involved. The same reaction applied to (−)-*exo*-norbornanol is more complex.[6] In this case the reaction is solvent dependent and a path involving a carbonium ion has become dominant. Cinnamyl bromide has been prepared by the Wiley method.[7]

Cleavage of ethers.[8] The reagent, prepared in solution as required, cleaves aliphatic ethers to ethyl bromides at moderate temperatures in benzonitrile, acetonitrile (slower), or chlorobenzene. For example, bromine was added slowly to a

$$\underset{0.052\ m.}{(C_6H_5)_3P} + \underset{0.052\ m.}{Br_2} \xrightarrow{50\ ml.\ C_6H_5CN} (C_6H_5)_3PBr_2 \xrightarrow[78\%]{125^0 \quad 0.5\ m.\ (\underline{n}\text{-}C_5H_{11})_2O}$$

$$2\ \underline{n}\text{-}C_5H_{11}Br + (C_6H_5)_3PO$$

cooled solution of triphenylphosphine in benzonitrile, the mixture was heated to 125°, and di-*n*-pentyl ether was added. Gas chromatographic analysis showed that the ether was no longer present after about 4 hrs. Tetrahydrofurane afforded 1,4-dibromobutane in 75% yield. The unsaturation in allylic ethers was not affected by the phosphorane reagent.

[1]L. Horner, H. Oediger, and H. Hoffmann, *Ann.*, **626**, 26 (1959)
[2]G. A. Wiley, R. L. Hershkowitz, B. M. Rein, and B. C. Chung, *Am. Soc.*, **86**, 964 (1964)
[3]G. A. Wiley, B. M. Rein, and R. L. Hershkowitz, *Tetrahedron Letters*, 2509 (1964)
[4]D. Levy and R. Stevenson, *Tetrahedron Letters*, 341 (1965); *idem, J. Org.*, **30**, 3469 (1965)
[4a]H.-J. Bestmann and L. Mott, *Ann.*, **693**, 132 (1966)
[5]J. P. Schaefer and D. S. Weinberg, *J. Org.*, **30**, 2635 (1965)
[6]*Idem, ibid.*, **30**, 2639 (1965)
[7]A procedure for the preparation of cinnamyl bromide by the Wiley method has been submitted to *Org. Syn.* by J. P. Schaefer, J. G. Higgins, and P. K. Shenoy.
[8]A. G. Anderson, Jr., and F. J. Freenor, *Am. Soc.*, **86**, 5037 (1964)

Triphenyl phosphite, $(C_6H_5O)_3P$. Mol. wt. 310.27, m.p. 25°, b.p. 210°/1 mm. Suppliers: B, E, F, MCB. Preparation.[1]

Preparation of alkyl halides. The reagent reacts with an alcohol and a hydrogen halide to form the halide (1).[2] The hydrogen halide can be replaced by an alkyl halide

$$1.\quad (C_6H_5O)_3P + ROH + HX \longrightarrow RX + H\overset{O^-}{\overset{|}{P^+}}(OC_6H_5)_2 + C_6H_5OH$$

$$2.\quad (C_6H_5O)_3P + ROH + R'X \longrightarrow RX + R'\overset{O^-}{\overset{|}{P^+}}(OC_6H_5)_2 + C_6H_5OH$$

$$3.\quad (C_6H_5O)_3P + ROH + X_2 \longrightarrow RX + X\overset{O^-}{\overset{|}{P^+}}(OC_6H_5)_2 + C_6H_5OH$$

(2)[2] or by a halogen (3).[3] If the alcohol is unsaturated the reaction can be carried out in two steps (4).

$$4.\quad (C_6H_5O)_3P \xrightarrow{X_2} (C_6H_5O)_3PX_2 \xrightarrow{CH_2{=}CHCH_2OH} CH_2{=}CHCH_2X + XPO(OC_6H_5)_2 + C_6H_5OH$$

[1]H. B. Gottlieb, *Am. Soc.*, **54**, 748 (1932)
[2]S. R. Landauer and H. N. Rydon, *J. Chem. Soc.*, 2224 (1953)
[3]D. G. Coe, S. R. Landauer, and H. N. Rydon, *J. Chem. Soc.*, 2281 (1954)

Triphenyl phosphite dibromide, $(C_6H_5O)_3PBr(Br)$. Mol. wt. 470.11.

This is the preferred reagent for the conversion of primary or secondary acetylenic or allenic alcohols into the bromides.[1] Pyridine is used to neutralize the hydrogen bromide, which otherwise would add to a double or triple bond.

[1]D. K. Black, S. R. Landor, A. N. Patel, and P. F. Whiter, *Tetrahedron Letters*, 483 (1963)

Triphenyl phosphite methiodide, $(C_6H_5O)_3PCH_3(I)$. Mol. wt. 452.22, m.p. 146°.

Landauer and Rydon[1] describe the preparation of the reagent and its use for the conversion of alcohols to alkyl iodides. Kornblum and Iffland[2] found neopentyl

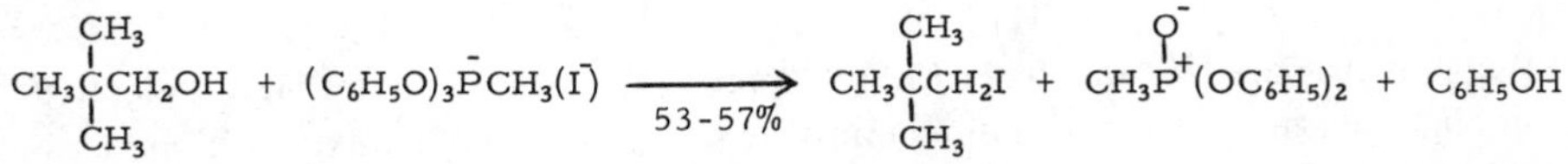

iodide prepared in this way to contain 6% of *t*-amyl iodide but were able to obtain pure material by modifying the procedure of workup.

The reagent has been used for the synthesis of iodoallenes and iodoacetylenes[3] and also to effect direct iodination of the sugar moiety in nucleosides.[4]

[1]S. R. Landauer and H. N. Rydon, *J. Chem. Soc.*, 2224 (1953)
[2]N. Kornblum and D. C. Iffland, *Am. Soc.*, **77**, 6653 (1955)
[3]C. S. L. Baker, P. D. Landor, S. R. Landor, and A. N. Patel, *J. Chem. Soc.*, 4348 (1965)
[4]J. P. H. Verheyden and J. G. Moffatt, *Am. Soc.*, **86**, 2093 (1964)

Triphenyltin chloride, $(C_6H_5)_3SnCl$. Mol. wt. 385.46, m.p. 104–106°, soluble in benzene and in methanol, insoluble in water. Suppliers: Metal and Thermit Corp., B, E, F, KK, MCB.

Preparation of allyllithium.[1] Allyltriphenyltin is prepared as follows: a suspension of magnesium in ether is stirred under reflux, a solution of allyl bromide and triphenyltin chloride in tetrahydrofurane is run in during 7 hrs., more benzene is added, and the mixture is refluxed overnight. After hydrolysis with saturated ammonium chloride solution, the allyltriphenyltin is recovered from the organic layer and crystallized from ligroin (m.p. 73–74%). In the next step a solution of 0.127 m. of allyltriphenyltin is stirred under nitrogen, and an ethereal solution of 0.127 m. of phenyllithium is added. Tetraphenyltin precipitates rapidly and after 30 min.

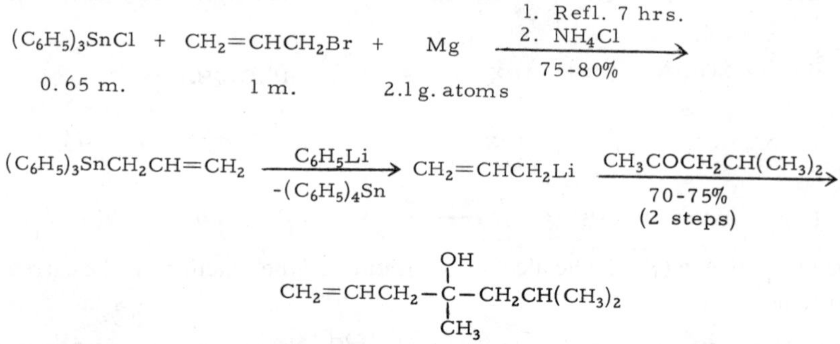
$(C_6H_5)_3SnCl$ + $CH_2{=}CHCH_2Br$ + Mg
0.65 m.
1 m.
2.1 g. atoms
1. Refl. 7 hrs.
2. NH_4Cl
75-80%
$(C_6H_5)_3SnCH_2CH{=}CH_2$
C_6H_5Li
$-(C_6H_5)_4Sn$
$CH_2{=}CHCH_2Li$
$CH_3COCH_2CH(CH_3)_2$
70-75%
(2 steps)
OH
$CH_2{=}CHCH_2{-}C{-}CH_2CH(CH_3)_2$
CH_3

the solution of allyllithium is treated with an ethereal solution of 0.12 m. of 4-methyl-2-pentanone, added at a rate such that moderate reflux is maintained. After hydrolysis with water, the mixture is filtered to remove solid tetraphenyltin, and the product, 4,6-dimethyl-Δ^1-heptene-4-ol, is recovered from the organic phase.

This is the only method known for preparation of pure allyllithium. It can be used also for preparation of vinyllithium.

[1]D. Seyferth and M. A. Weiner, *Org. Syn.*, **41**, 30 (1961)

Triphenyltin hydride, $(C_6H_5)_3SnH$. Mol. wt. 351.01, m.p. 26–28°, b.p. 174°/6 mm.

Preparation[1] by $LiAlH_4$ reduction of triphenyltin chloride. Suppliers: B, E, F, KK, MCB.

Reducing action. The reagent reduces organic halides and amines to the corresponding hydrocarbons.[2,3] The ease of replacement of halogen is I > Br > Cl[4]. 1-Bromo-4-chlorobenzene is reduced to chlorobenzene in 75% yield.[5] Other papers.[6–8]

Reagent prepared *in situ* from $LiAlH_4$ and $(C_6H_5)_3SnCl$ is more stereoselective than zinc dust and ethanolic sodium ethoxide for reduction of a mixture of epimeric 7-chloro-7-phenylnorcaranes (1) prepared as indicated. Thus the mixture affords 80% of *endo*-7-phenylnorcarane, 1% of the *exo*-isomer, and 19% of an unidentified olefin.[9]

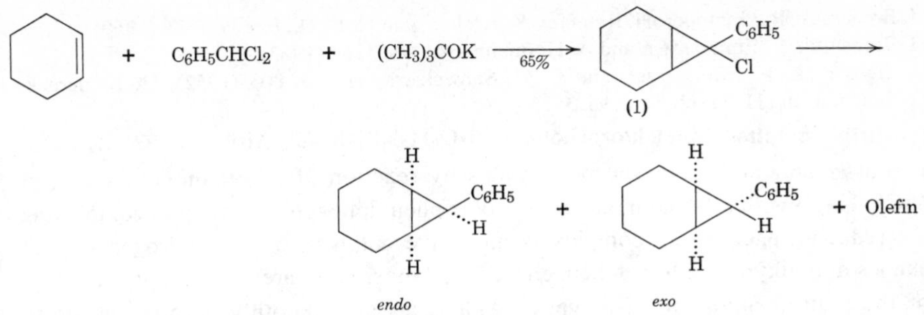
C6H5CHCl2
(CH3)3COK
65%
C6H5
Cl
(1)
C6H5
H
H
H
C6H5
H
H
H
+ Olefin
endo
exo

[1]A. E. Finholt, A. C. Bond, Jr., K. E. Wilzbach, and H. I. Schlesinger, *Am. Soc.*, **69**, 2692 (1947); G. J. M. Van Der Kerk, J. G. Noltes, and J. G. A. Luijten, *J. Appl. Chem. (London)*, **7**, 366 (1957); H. G. Kuivila and O. F. Beumel, Jr., *Am. Soc.*, **83**, 1246 (1961)
[2]R. K. Ingham, S. D. Rosenberg, and H. Gilman, *Chem. Rev.*, **60**, 459 (1960)
[3]E. J. Kupchik and R. E. Connolly, *J. Org.*, **26**, 4747 (1961); E. J. Kupchik and R. J. Kiesel, *Chem. Ind.*, 1654 (1962); *idem, J. Org.*, **29**, 764 (1964)
[4]H. G. Kuivila, L. W. Menapace, and C. R. Warner, *Am. Soc.*, **84**, 3584 (1962)
[5]L. A. Rothman and E. I. Becker, *J. Org.*, **25**, 2203 (1960)
[6]D. H. Lorenz, P. Shapiro, A. Stern, and E. I. Becker, *J. Org.*, **28**, 2332 (1963)
[7]E. J. Kupchik and R. J. Kiesel, *J. Org.*, **29**, 3690 (1964)
[8]H. G. Kuivila, "Advances in Organometallic Chemistry," **1**, 81 (1964)
[9]F. R. Jensen and D. B. Patterson, *Tetrahedron Letters*, 3837 (1966)

Tris-dimethylaminophosphine, *see* Hexamethylphosphorous triamide.

Tris-formaminomethane, $HC(NHCHO)_3$. Mol. wt. 145.12, m.p. 164.5°.

Preparation. The procedure originally described by Bredereck *et al.*[1] is advantageously replaced by the following, more fully perfected procedure kindly supplied by Professor Bredereck. A mixture of 15.4 g. (0.1 mole) of diethyl sulfate and 45 g. (1 mole) of formamide is heated for 2 hrs. at 70–80° and a pressure of 12 mm. After standing for 3 hrs. in a cooling closet the crystals that separate are collected and washed with absolute methanol. The yield of crude product, m.p. 158–159° is 7.2 g. Recrystallization from water gives pure material, m.p. 164.5°, yield 5.7 g. (39%).

Synthesis of heterocycles. The substance when heated to 160° gives *s*-triazine, formed from formylformamidine as intermediate.[2]

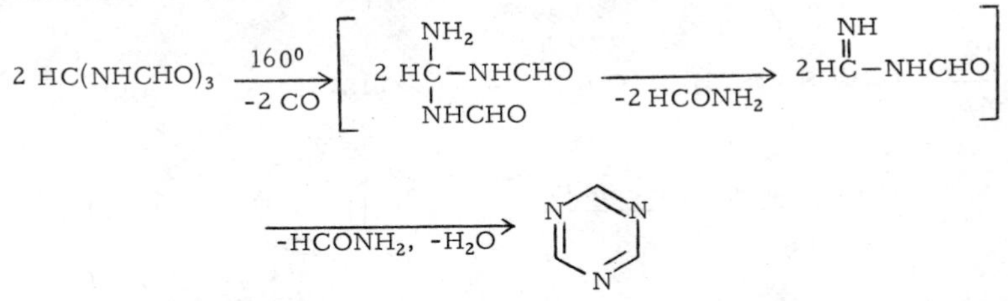
2 HC(NHCHO)3
160°
-2 CO
NH2
2 HC—NHCHO
NHCHO
-2 HCONH2
NH
2 HC—NHCHO
-HCONH2, -H2O
N
N
N

The reagent reacts with active-methylene compounds to give pyrimidines:[3]

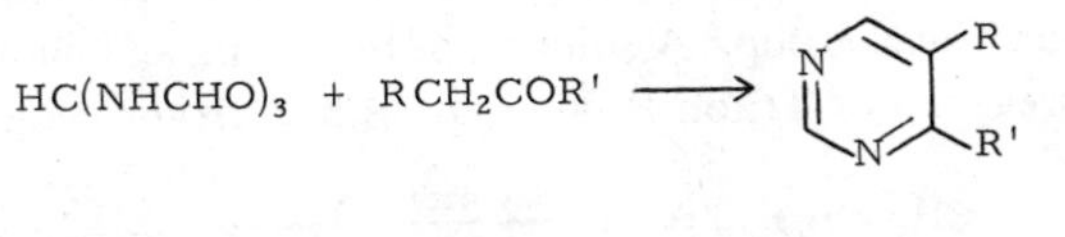
HC(NHCHO)3 + RCH2COR'
R
N
N
R'

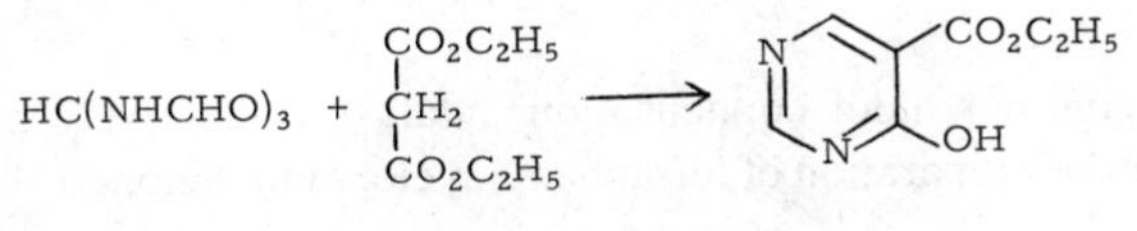
CO2C2H5
HC(NHCHO)3 + CH2
CO2C2H5
N
CO2C2H5
N
OH

[1]H. Bredereck, R. Gompper, H. Rempfer, K. Klemm, and H. Keck, *Ber.*, **92**, 329 (1959)
[2]H. Bredereck, F. Effenberger, and A. Hofmann, *Ber.*, **96**, 3260 (1963)
[3]H. Bredereck, F. Effenberger, and E. H. Schweizer, *Ber.*, **95**, 803 (1962); H. Bredereck, G. Simchen, and H. Traub, *ibid.*, **98**, 3883 (1965)

Tris-(triphenylphosphine)chlororhodium, $[(C_6H_5)_3P]_3RhCl$. Mol. wt. 925.21. This reagent is obtained as red-purple crystals by reaction of an ethanolic solution of $RhCl_3 \cdot 3H_2O$ with a 6-fold molar excess of triphenylphosphine, acting as complexing and reducing agent. The complex is an active catalyst for the hydrogenation of alkenes and alkynes (ethanol–benzene, 250°, atmospheric pressure).[1] It can be used for the hydroformylation of alkynes.[1] Aldehydes are smoothly decarbonylated to paraffins:[2]

$$RCHO + [(C_6H_5)_3P]_3RhCl \longrightarrow RH + RhCl(CO)[(C_6H_5)_3P]_2 + (C_6H_5)_3P$$

[1]J. F. Young, J. A. Osborn, F. H. Jardine, and G. Wilkinson, *Chem. Comm.*, 131 (1965); F. H. Jardine, J. A. Osborn, G. Wilkinson, and J. F. Young, *Chem. Ind.*, 560 (1965)
[2]J. Tsuji and K. Ohno, *Tetrahedron Letters*, 3969 (1965)

Triton B (Benzyltrimethylammonium hydroxide), $C_6H_5CH_2(CH_3)_3NOH$. Mol. wt. 167.25; soluble in both hydroxylic and nonhydroxylic solvents. Suppliers of a 40% solution in methanol: A, B, E, F, KK.

Condensation catalyst. Because it is a strong base soluble in a variety of solvents, Triton B is a preferred catalyst for many condensations. An example of a Michael reaction is the condensation of 2-nitropropane with methyl acrylate to produce methyl γ-methyl-γ-nitrovalerate.[1]

$$\underbrace{\underset{1\ m.}{(CH_3)_2CHNO_2} + \underset{1\ m.}{CH_2{=}CHCO_2CH_3}}_{\text{In 50 ml. dioxane}} + \text{Trit. B} \xrightarrow[80-86\%]{85-100^\circ} (CH_3)_2\underset{NO_2}{\underset{|}{C}}CH_2CH_2CO_2CH_3$$

Gardner and co-workers[2] found Triton B a superior catalyst for the aldol-type condensation of aromatic aldehydes with ethylidenemalonic esters, one step in a new synthesis of benzosuberones.

$$C_6H_5CHO + CH_3CH{=}C(CO_2C_2H_5)_2 \xrightarrow[98\%]{\text{Triton B; hydrol.}} C_6H_5CH{=}CHCH{=}C(CO_2H)_2$$

$$\xrightarrow[84\%]{H_2-Pd} C_6H_5CH_2CH_2CH_2CH(CO_2H)_2 \xrightarrow[84\%]{-CO_2} C_6H_5CH_2CH_2CH_2CH_2CO_2H$$

$$\xrightarrow[67\%]{\text{Polyphosphoric acid } (95^\circ)}$$

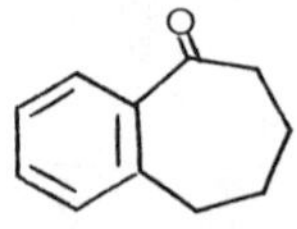

(as the ethyleneketal)

The reagent effectively catalyzes the hydroxymethylation of acetonitrile derivatives containing at least one α-aryl group.[3] A solution of the nitrile and paraformaldehyde in pyridine in the presence of Triton B gives the β-hydroxypropionitrile, often in

$$(C_6H_5)_2CHCN + CH_2{=}O + \text{Trit. B} \xrightarrow[97\%]{22\text{ hrs. at }25^\circ} (C_6H_5)_2\underset{CN}{\underset{|}{C}}CH_2OH$$

excellent yield. Another aldol condensation catalyzed effectively by Triton B is that described for the preparation of tetraphenylcyclopentadienone.

Another reaction for which Triton B is a particularly satisfactory catalyst is cyanoethylation.[4] In the steroid total synthesis by the Woodward group,[5] Triton B-catalyzed cyanoethylation initiated construction of ring A.

Carbanion formation. Sprinzak[6] found that Triton B in pyridine solution converts compounds of the cyclopentadiene series into carbanions highly sensitive to air oxidation. He prepared a solution of the base in pyridine by adding the solvent to the commercial 40% solution in methanol and evaporated the methanol at reduced pressure. Addition of this solution to a solution of fluorene in pyridine produced an orange-yellow solution which absorbed oxygen rapidly; after stirring for 40 min.

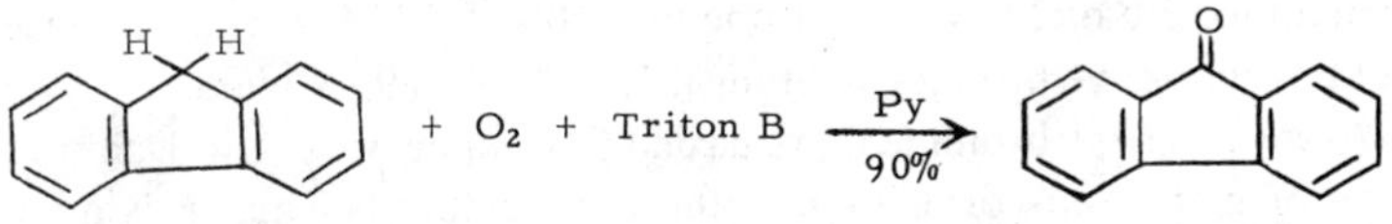
H H
+ O2 + Triton B
Py
90%
O

crystallized fluorenone was isolated in high yield. At low temperature (−15 to −40°) 9-alkylfluorenes are converted into 9-hydroperoxy-9-alkylfluorenes; at a temperature

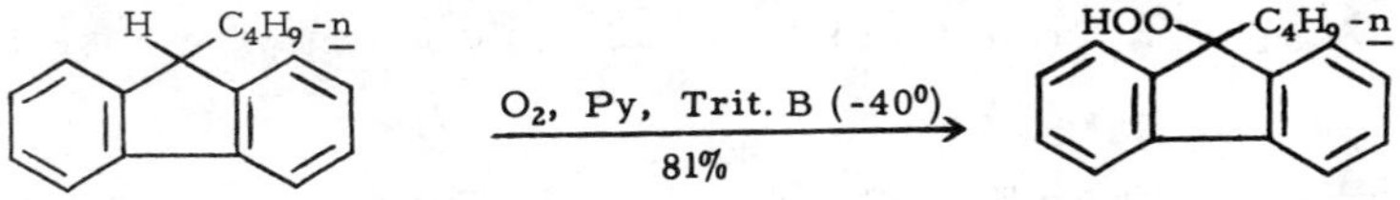
H C4H9-n
O2, Py, Trit. B (-40°)
81%
HOO C4H9-n

of 40° the product is the corresponding 9-alkyl-9-fluorenol. Indene gives an amorphous product of complex structure, but oxidation of 2,3-diphenylindene at −40° gives about equal amounts of the ketone and the hydroperoxide. Esters of diaryl-

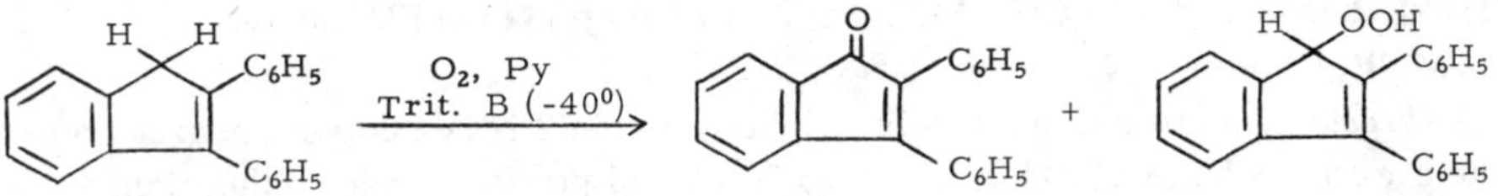
H H
C6H5
C6H5
O2, Py
Trit. B (-40°)
O
C6H5
C6H5
+
H OOH
C6H5
C6H5

acetic acids and 2,3-diarylpropionic acids also undergo autoxidation to hydroperoxides in pyridine–Triton B in yields of 30–45%; α-hydroxy esters and ketones are also formed.[7] The scope of the autoxidation is expanded by use of potassium *t*-butoxide in an aprotic solvent, as illustrated by the reaction of isopropyl methyl ketone.[8]

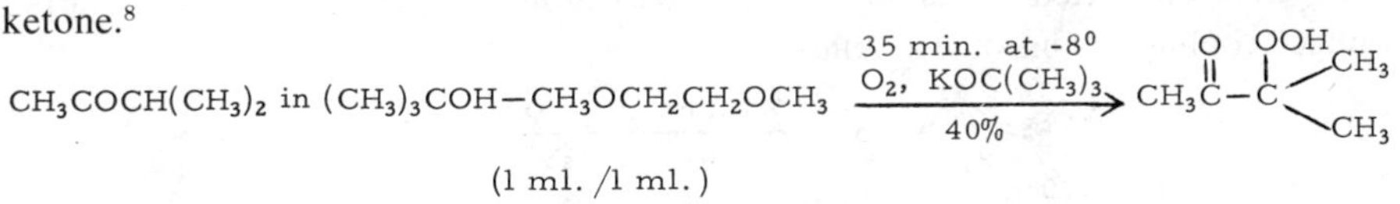
CH3COCH(CH3)2 in (CH3)3COH−CH3OCH2CH2OCH3
(1 ml. /1 ml.)
35 min. at -8°
O2, KOC(CH3)3
40%
O OOH
CH3C−C
CH3
CH3

Aldol-type condensation of aldehydes with indene and fluorene can be achieved by reaction in a nitrogen atmosphere in pyridine containing Triton B.[9]

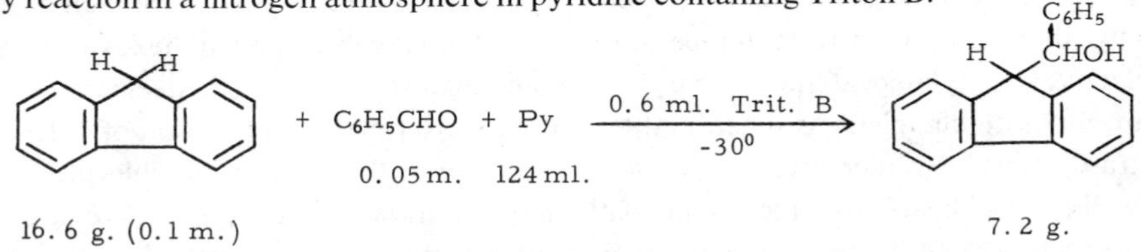
H H
16.6 g. (0.1 m.)
+ C6H5CHO + Py
0.05 m.
124 ml.
0.6 ml. Trit. B
-30°
C6H5
H CHOH
7.2 g.

[1]R. B. Moffett, *Org. Syn., Coll. Vol.*, **4**, 652 (1963); other examples: G. N. Walker, *Am. Soc.*, **77**, 3664 (1955); S. Julia and Y. Bonnet, *Compt. rend.*, **243**, 2079 (1956)

[2]P. D. Gardner, W. J. Horton, G. Thompson, and R. R. Twelves, *Am. Soc.*, **74**, 5527 (1952); W. J. Horton and L. L. Pitchforth, *J. Org.*, **25**, 131 (1960)

[3]M. Avramoff and Y. Sprinzak, *J. Org.*, **26**, 1284 (1961)

[4]H. A. Bruson, *Org. Reactions*, **5**, 79 (1949); J. C. Sheehan and C. E. Mumaw, *Am. Soc.*, **72**, 2127 (1950)
[5]R. B. Woodward, F. Sondheimer, D. Taub, K. Heusler, and W. M. McLamore, *Am. Soc.*, 4223 (1952)
[6]Y. Sprinzak, *Am. Soc.*, **80**, 5449 (1958)
[7]M. Avramoff and Y. Sprinzak, *Am. Soc.*, **85**, 1655 (1963)
[8]H. R. Gersmann, H. J. W. Nieuwenhuis, and A. F. Bickel, *Proc. Chem. Soc.*, 279 (1962)
[9]E. Ghera and Y. Sprinzak, *Am. Soc.*, **82**, 4945 (1960)

Trityl bromide, $(C_6H_5)_3CBr$. Mol. wt. 323.23, m.p. 153–154°. Supplier: Aldrich.

Preparation.[1] The procedure is similar to that given for trityl chloride.[2] In a typical run a mixture of 220.8 g. of triphenylcarbinol, 122.9 g. of acetyl bromide, and 100 ml. of benzene gave bromide melting at 153–154° in 90% yield.

Trityl esters. Trityl bromide is regarded[1] as superior to trityl chloride for the conversion of metal salts of acids into the trityl esters because it is more reactive and less hygroscopic and can be used with the dry silver, sodium, or potassium salt of the acid. Benzene usually serves satisfactorily as solvent.

$$RCO_2Ag\ (K,\ Na) + (C_6H_5)_3CBr \xrightarrow[85\text{-}95\%]{C_6H_6} RCO_2C(C_6H_5)_3 + AgBr$$

$$R = CH_3,\ C_2H_5,\ (CH_3)_3C,\ C_6H_5,\ C_6H_5CH{=}CH$$

[1]K. D. Berlin, L. H. Gower, J. W. White, D. E. Gibbs, and G. P. Sturm, *J. Org.*, **27**, 3595 (1962)
[2]W. E. Bachmann, *Org. Syn., Coll. Vol.*, **3**, 841 (1955)

Trityl chloride, $(C_6H_5)_3CCl$. Mol. wt. 278.78, m.p. 111–112°. Suppliers: A, B, E, F, MCB, Frinton, KK.

Preparation. In one procedure[1] triphenylcarbinol is heated with benzene and a large excess of acetyl chloride, added in eleven portions under reflux. Petroleum

$$(C_6H_5)_3COH + CH_3COCl + C_6H_6 \xrightarrow[79\text{-}83\%\ (1st\ crop)]{With\ cooling,\ then\ refl.} (C_6H_5)_3CCl$$

ether is added with cooling, and the crystalline product collected and washed (m.p. 111–112°). The Friedel-Crafts reaction employed in a second procedure[2] is started with ice cooling and finished at reflux.

$$\underset{5.2\ m.}{C_6H_6} + \underset{470\ ml.}{CS_2} + \underset{4.51\ m.}{AlCl_3} \xrightarrow[69\text{-}75\%\ (1st\ crop)]{With\ cooling,\ then\ refl.} (C_6H_5)_3CCl$$

O-Protective group. Helferich *et al.*[3] showed that primary and secondary alcohols and phenols can be converted into their trityl ethers by reaction with trityl chloride in pyridine solution at room temperature and established that an aldohexose can be selectively tritylated at the primary alcoholic function.[4] Use of this selectivity in carbohydrate chemistry is illustrated by a procedure for the preparation of 1,2,3,4-tetra-O-acetyl-β-D-glucopyranoside (4).[5] A mixture of the sugar, trityl chloride, and pyridine was heated on the steam bath until solution was complete, and acetic anhydride was added to the hot solution of the 6-trityl derivative (2) to acetylate the four secondary alcoholic functions. Pouring the solution in a thin stream into a mechanically stirred mixture of ice-water and acetic acid gave a granular solid, a mixture of the 1β-isomer (3) and the α-isomer. Digestion of the dried solid with ether removed most of the α-isomer, and two crystallizations from 95% ethanol gave

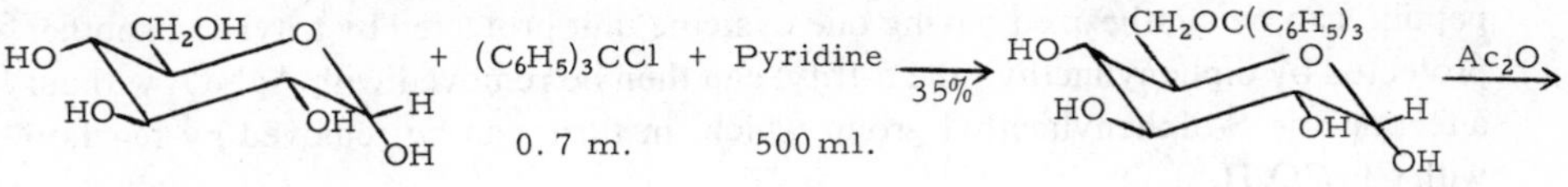
+ $(C_6H_5)_3CCl$ + Pyridine
0.7 m.
500 ml.
35%
Ac_2O

(1) 0.67 m. (2)

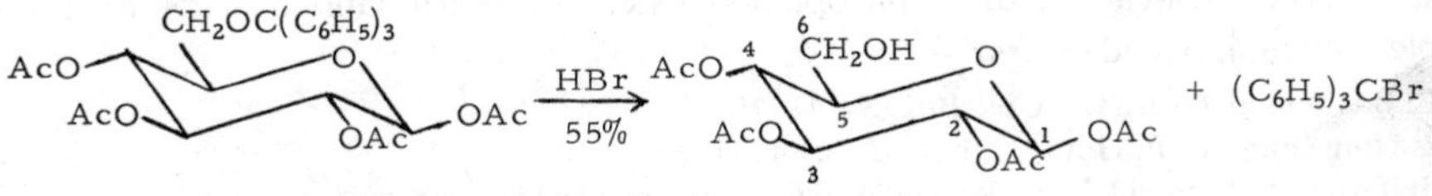
HBr
55%
+ $(C_6H_5)_3CBr$

(3) 35% from (1) (4)

pure (3). For removal of the trityl group a solution of the trityltetraacetate (3) in acetic acid was adjusted to 10°, a solution of dry hydrogen bromide in acetic acid was added, the solution was shaken for 45 seconds, the trityl bromide which separated was removed at once by filtration, and the filtrate poured immediately into water. Extraction with chloroform and evaporation gave a sirup, which crystallized at once when rubbed with ether and was recrystallized from chloroform-ether.

In a synthesis of coenzyme uridine-diphosphate-glucose (UDPG), Todd and co-workers[6] protected the primary 5-hydroxyl group of uridine (5) by tritylation, benzylated the 2- and 3-hydroxyl groups (6), and removed the trityl group with 80% acetic acid.

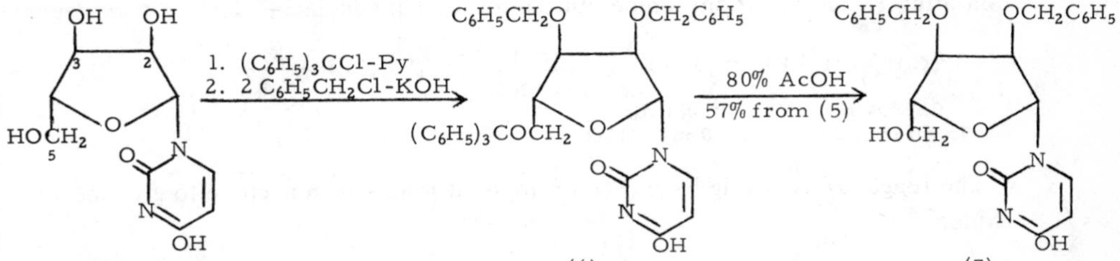
1. $(C_6H_5)_3CCl$-Py
2. 2 $C_6H_5CH_2Cl$-KOH
80% AcOH
57% from (5)

(5) (6) (7)

***S*-Protective group.**[7] Cysteine is easily converted into the S-protected trityl derivative by interaction of the hydrochloride with trityl chloride in dimethylformamide; under these conditions the amino group is not affected. The blocking group can be cleaved by Na–NH_3, by HBr in AcOH, and by CF_3CO_2H. However, the first method has the disadvantage that existing S–S links in a peptide would be cleaved, and the two methods of acid cleavage also cleave N-carbobenzoxy groups. The S-trityl group is split initially at 0° by treatment with a methanolic solution of silver nitrate and pyridine. A silver mercaptide is formed and this is converted into the SH compound by treatment with concd. hydrochloric acid in dimethylformamide:

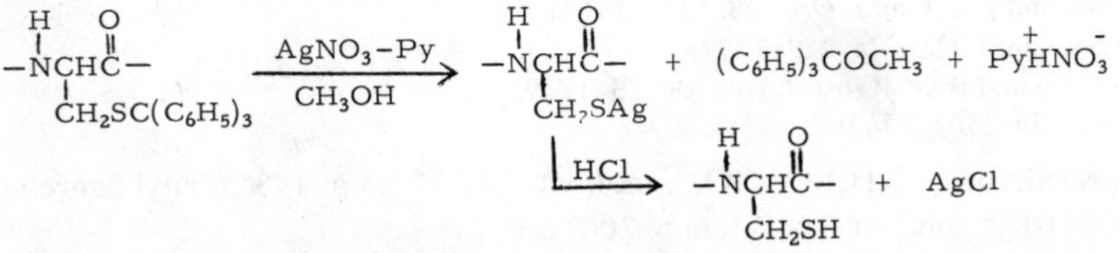
$AgNO_3$–Py
CH_3OH
+ $(C_6H_5)_3COCH_3$ + $Py\overset{+}{H}NO_3^-$
HCl
+ AgCl

A diphenylmethyl group is also useful for S-protection; it is stable to metal salts but can be removed by the other methods: HBr–AcOH or CF_3CO_2H. Thus a peptide can be synthesized having one cysteine unit protected by trityl and another protected by diphenylmethyl; the S-trityl can then be removed with $AgNO_3$ without affecting the S-diphenylmethyl group which, in turn, can be removed by reaction with CF_3CO_2H.

N-Protective group. Although the trityl group has received some consideration for the N-protection of amino acids,[8] a serious limitation is that, with the exception of the glycine derivative, the *p*-nitrophenyl esters of N-tritylamino acids do not couple with amino acid esters.[9]

[1]W. E. Bachmann, *Org. Syn., Coll. Vol.*, **3**, 841 (1955)
[2]C. R. Hauser and B. E. Hudson, Jr., *ibid.*, **3**, 842 (1955)
[3]B. Helferich, P. E. Speidel, and W. Toeldte, *Ber.*, **56**, 766 (1923)
[4]B. Helferich, L. Moog, and A. Jünger, *Ber.*, **58**, 872 (1925)
[5]D. D. Reynolds and W. L. Evans, *Org. Syn., Coll. Vol.*, **3**, 432 (1955)
[6]G. W. Kenner, A. R. Todd, and R. F. Webb, *J. Chem. Soc.*, 2843 (1954); A. M. Michelson and A. Todd, *ibid.*, 3459 (1956)
[7]L. Zervas and I. Photaki, *Am. Soc.*, **84**, 3887 (1962)
[8]L. Zervas and D. M. Theodoropoulos, *Am. Soc.*, **78**, 1359 (1956); R. A. Boissonnas, *Advances in Org. Chem.*, **3**, 172 (1963); G. Amiard, R. Heymes, and L. Velluz, *Bull. Soc.*, **22**, 191 (1955)
[9]L. Zervas, D. Borovas, and E. Gazis, *Am. Soc.*, **85**, 3660 (1963)

Trityllithium, $(C_6H_5)_3CLi$. Mol. wt. 250.25, red in solution.

Gilman and Gaj[1] prepared this reagent conveniently by adding an ethereal solution of *n*-butyllithium to a solution of triphenylmethane in tetrahydrofurane; carbonation after 15 min. at room temperature afforded triphenylacetic acid in good yield.

$$\underset{\text{in 50 ml. THF}}{\underset{0.0204\text{ m.}}{(C_6H_5)_3CH}} + \underset{\text{in 20 ml. }Et_2O}{\underset{0.021\text{ m.}}{BuLi}} \xrightarrow[-BuH]{25^0\ (N_2)} (C_6H_5)_3CLi \xrightarrow[78\%]{CO_2} (C_6H_5)_3CCO_2H$$

The reagent is a strong base. For example, it reacts with acetone to give mesityl oxide:[2]

$$CH_3COCH_3 \xrightarrow[-(C_6H_5)_3CH]{(C_6H_5)_3CLi} \overset{-}{C}H_2COCH_3(\overset{+}{Li}) \xrightarrow[-LiOH]{(CH_3)_2CO} (CH_3)_2C{=}CHCOCH_3$$

House and Trost[3] prepared an approximately 1 *M* solution of this reagent in 1,2-dimethoxyethane for use in an extensive study of compositions of enolate anions from ketones[3] and enol acetates[4] as follows. The solvent was removed from 15 ml. of a 1 *M* ethereal solution of methyllithium under reduced pressure, and the residual solid was dissolved in 15 ml. of 1,2-dimethoxyethane. Triphenylmethane (18.3 mmoles) was added, and the mixture stirred under nitrogen until a test for methyllithium was negative. Reaction of an enol acetate with trityllithium can then be carried out as a titration and stopped when the solution is pale pink.

[1]H. Gilman and B. J. Gaj, *J. Org.*, **28**, 1725 (1963)
[2]P. Tomboulian, *J. Org.*, **24**, 229 (1959)
[3]H. O. House and B. M. Trost, *J. Org.*, **30**, 1341 (1965)
[4]*Idem, ibid.*, **30**, 2502 (1965)

Trityl perchlorate, $(C_6H_5)_3C^+ClO_4^-$, mol. wt. 342.77, m.p. 143°. **Trityl fluoroborate,** $(C_6H_5)_3C^+BF_4^-$, mol. wt. 330.13, m.p. 200° dec.

Preparation.[1] The reagents are prepared by reaction of pure triphenylcarbinol with 71% perchloric acid or 48% fluoroboric acid in the presence of acetic anhydride for removal of water (yields 76 and 92%). In general the fluoroborate is the preferred reagent because it is more stable than the perchlorate.

Hydride abstraction. Dauben *et al.*[2] prepared tropylium salts by reaction of cycloheptatriene with trityl perchlorate or fluoroborate in solution in acetonitrile or liquid sulfur dioxide at −20°. The reaction was of service also for the synthesis of heptalene (6).[3]

H
H
$+ (C_6H_5)_3\overset{+}{C}BF_4^-$
+
BF_4^- + $(C_6H_5)_3CH$

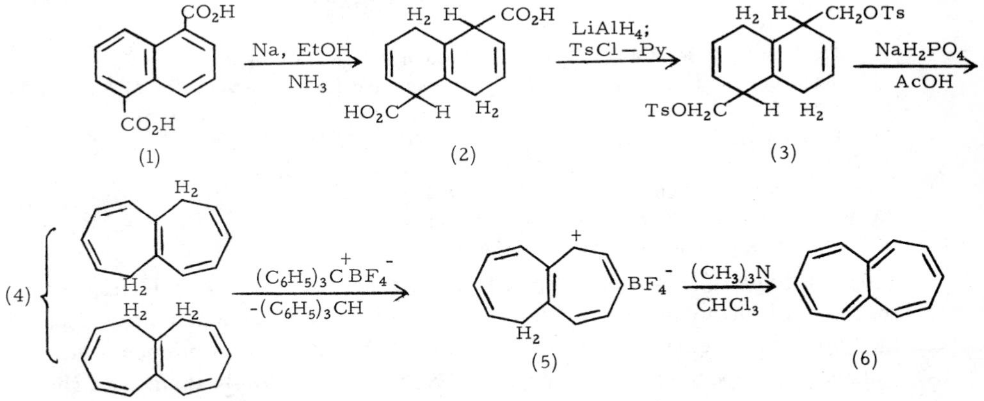
CO_2H
CO_2H
(1)
Na, EtOH
NH_3
H_2 H CO_2H
HO_2C H H_2
(2)
$LiAlH_4$;
TsCl−Py
H_2 H CH_2OTs
$TsOH_2C$ H H_2
(3)
NaH_2PO_4
AcOH
H_2
H_2
(4)
H_2 H_2
$(C_6H_5)_3\overset{+}{C}BF_4^-$
$-(C_6H_5)_3CH$
+
BF_4^-
H_2
(5)
$(CH_3)_3N$
$CHCl_3$
(6)

Dehydrogenation. The perchlorate is an effective reagent for the aromatization of hydroaromatic compounds. It converts 9,10-dihydroanthracene into anthracene in quantitative yield.[4] It provides the best means available for the conversion of perinaphthanones into perinaphthenones[4] and for the dehydrogenation of chromanones to chromones.[5]

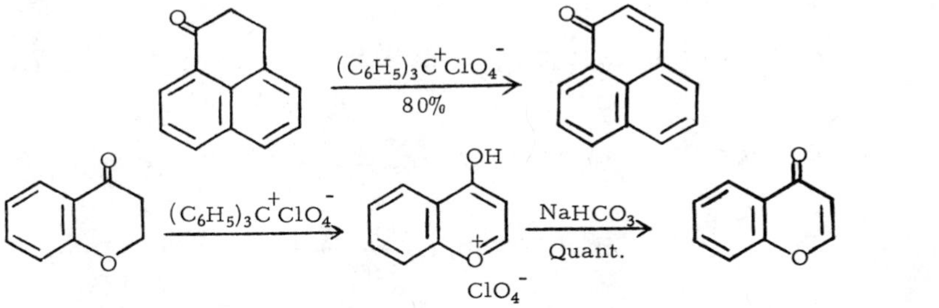
$(C_6H_5)_3\overset{+}{C}ClO_4^-$
80%
$(C_6H_5)_3\overset{+}{C}ClO_4^-$
OH
+
ClO_4^-
$NaHCO_3$
Quant.

Synthesis of heterocycles. The 1,5-diketone (1) on reaction with trityl perchlorate undergoes dehydrogenative cyclization to the pyrylium salt (2), which is transformed by reaction with ammonium acetate into the pyridine derivative (3).[6]

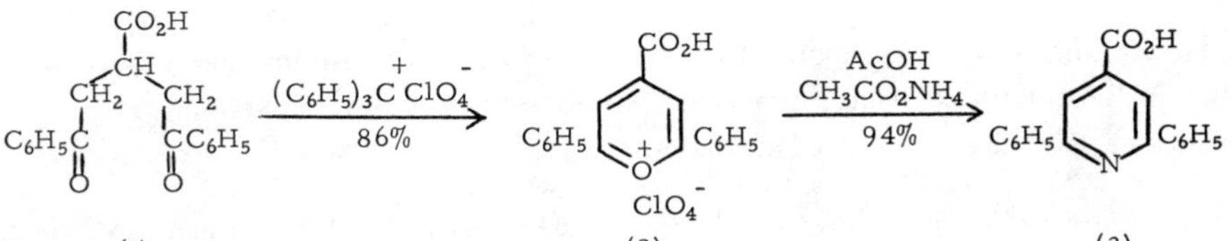
CO_2H
CH
CH_2 CH_2
C_6H_5C CC_6H_5
O O
$(C_6H_5)_3\overset{+}{C}ClO_4^-$
86%
CO_2H
C_6H_5 C_6H_5
+
O
ClO_4^-
AcOH
$CH_3CO_2NH_4$
94%
CO_2H
C_6H_5 C_6H_5
N

(1) (2) (3)

Reaction with diazo compounds.[7] Trityl perchlorate is a very efficient catalyst for the decomposition of diphenyldiazomethane with dimerization to tetraphenylethylene. The reagent reacts with ethyl diazoacetate in acetonitrile to give ethyl

$$2\ (C_6H_5)_2C{=}\overset{+}{N}{=}\overset{-}{N} \xrightarrow[97\%]{(C_6H_5)_3\overset{+}{C}\overset{-}{ClO_4}} (C_6H_5)_2C{=}C(C_6H_5)_2 + 2\ N_2$$

triphenylacrylate. The reaction is considered to involve addition, rearrangement, and elimination.

$$(C_6H_5)_3\overset{+}{C}\overset{-}{ClO_4} + \overset{-}{N}{=}\overset{+}{N}{=}CHCO_2C_2H_5 \xrightarrow[72\%]{0^0} (C_6H_5)_2C{=}\underset{}{\overset{C_6H_5}{\overset{|}{C}}}CO_2C_2H_5 + N_2 + HClO_4$$

[1]H. J. Dauben, Jr., L. R. Honnen, and K. M. Harmon, *J. Org.*, **25**, 1442 (1960)
[2]H. J. Dauben, Jr., F. A. Gadecki, K. M. Harmon, and D. L. Pearson, *Am. Soc.*, **79**, 4557 (1957)
[3]H. J. Dauben, Jr., and D. J. Bertelli, *Am. Soc.*, **83**, 4657, 4659 (1961)
[4]W. Bonthrone and D. H. Reid, *J. Chem. Soc.*, 2773 (1959)
[5]A. Schönberg and G. Schütz, *Ber.*, **93**, 1466 (1960)
[6]M. Siemiatycki and R. Fugnitto, *Bull. Soc.*, 538 (1961)
[7]H. W. Whitlock, Jr., *Am. Soc.*, **84**, 2807 (1962)

Tritylpotassium. $(C_6H_5)_3CK$ (deep red). In one method[1] the reagent is prepared by reaction of triphenylmethane with potassium in liquid ammonia and replacement of the ammonia by ether to give a suspension of the reagent and provide a medium suitable for acylations, alkylations, and carbonations. Tritylsodium also can be prepared from triphenylmethane in the same way, but it is destroyed when the ammonia is replaced by ether; the preparation of an ethereal solution from trityl chloride and sodium takes considerably longer than the preparation of tritylpotassium from triphenylmethane. Moreover, triphenylmethane is recoverable in good yield from a condensation in a form suitable for reuse. Typical reactions utilizing tritylpotassium are: an ester condensation (1[1]), an acylation (2[1]), an alkylation (3[1]), and a carbonylation (4[2]).

1. $$2\ (CH_3)_2CHCH_2CH_2CO_2C_2H_5 \xrightarrow[64\%]{(C_6H_5)_3CK} (CH_3)_2CHCH_2\underset{\underset{(CH_3)_2CHCH_2CH_2CO}{|}}{C}HCO_2C_2H_5$$

2. $$C_6H_5COCl + (CH_3)_2CHCH_2CO_2C_2H_5 \xrightarrow[55\%]{(C_6H_5)_3CK} (CH_3)_2CH\underset{\underset{C_6H_5CO}{|}}{C}HCO_2C_2H_5$$

3. $$C_2H_5I + (CH_3)_2CHCH_2CO_2C_2H_5 \xrightarrow[53\%]{(C_6H_5)_3CK} (CH_3)_2CH\underset{\underset{C_2H_5}{|}}{C}HCO_2C_2H_5$$

4. 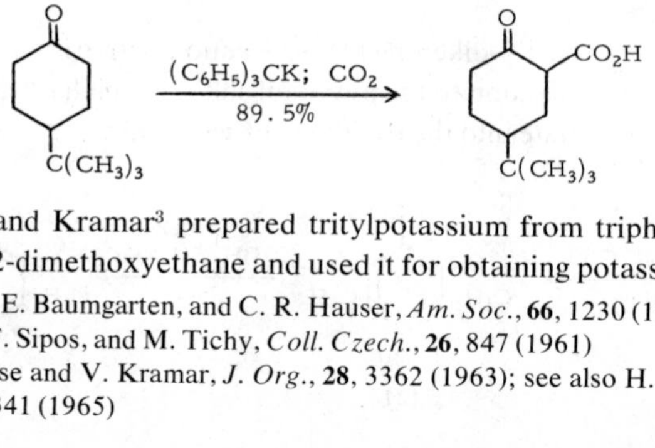

House and Kramar[3] prepared tritylpotassium from triphenylmethane and potassium in 1,2-dimethoxyethane and used it for obtaining potassium enolates.

[1]R. Levine, E. Baumgarten, and C. R. Hauser, *Am. Soc.*, **66**, 1230 (1944)
[2]J. Sicher, F. Sipos, and M. Tichy, *Coll. Czech.*, **26**, 847 (1961)
[3]H. O. House and V. Kramar, *J. Org.*, **28**, 3362 (1963); see also H. O. House and B. M. Trost, *ibid.*, **30**, 1341 (1965)

Tritylsodium, $(C_6H_5)_3CNa$. Red solution in ether.

Preparation.[1] A solution of trityl chloride in ether is shaken mechanically with

$$\underset{0.226\ m.}{(C_6H_5)_3CCl} + \underset{1.5\ l.}{\text{Ether}} + \underset{1.50\ g.}{1\%\ NaHg} \longrightarrow (C_6H_5)_3CNa + NaCl + Hg$$

1% sodium amalgam with some cooling for about 3 hrs., let stand for the sodium chloride to settle, and the red supernatant solution is transferred by siphon under nitrogen pressure. Use of solid 3% sodium amalgam permits preparation of a solution about five times as concentrated as that described.

Acylation and alkylation. An example is the benzoylation of ethyl isobutyrate.[2] The ester is added to an ethereal solution of tritylsodium from 1% amalgam and after

$$\underset{0.187\ m.}{(CH_3)_2CHCO_2C_2H_5} \xrightarrow[-(C_6H_5)_3CH]{0.187\ m.\ (C_6H_5)_3CNa \text{ in ether}} (CH_3)_2\bar{C}CO_2C_2H_5\ Na^+ \xrightarrow[50\text{-}55\%]{0.187\ m.\ C_6H_5COCl} C_6H_5COC(CH_3)_2CO_2C_2H_5$$

10 min. at room temperature an ethereal solution of benzoyl chloride is added. The solution becomes warm, and sodium chloride begins to separate. After several hours most of the ether is distilled, dilute acetic acid is added, and the mixture is worked up.

Similar procedures apply to the alkylation of esters[3] and for the carbonation and carboethoxylation of esters[4] and ketones.[5] In a synthesis of α-amyrin from glycyrrhetic acid, Corey and Cantrall[6] effected clean methylation of the 20-ketone-3-benzoate (1) to the 19β-methyl derivative (2) by treating a solution of (1) in dioxane

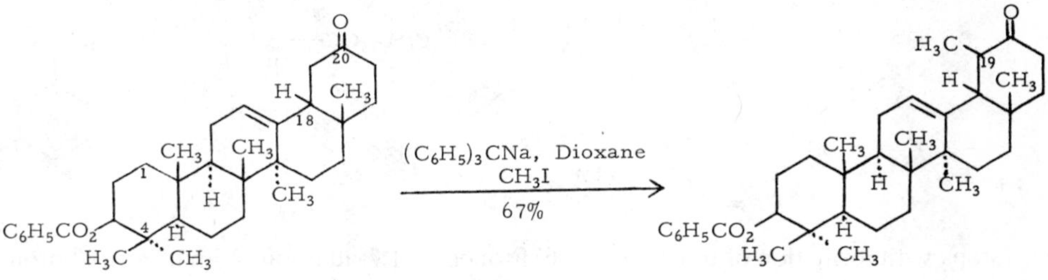
(C6H5)3CNa, Dioxane
CH3I
67%

(1) (2)

with an ethereal solution of tritylsodium until a red color persisted, and methylating with a large excess of methyl iodide to minimize dimethylation. The excess tritylsodium is destroyed instantaneously by methyl iodide, presumably with formation of triphenylethane.

Polymerization of dienes. Ziegler[7] noted that tritylsodium converts butadiene

$$(C_6H_5)_3CNa \xrightarrow{CH_2=CHCH=CH_2} \underset{(1)}{(C_6H_5)_3CCH_2\bar{C}HCH=CH_2\ Na^+} \xrightarrow{(C_6H_5)_3B}$$

$$\underset{(2)}{(C_6H_5)_3CCH_2CH(\bar{B}(C_6H_5)_3)CH=CH_2\ (Na^+)} \xrightarrow{\text{Hydrol.}} \underset{(3)}{(C_6H_5)_3CCH_2CH_2CH=CH_2}$$

into a polymer mixture, probably by 1,2,- and 1,4-addition, followed by addition of the new sodio derivatives to butadiene. Then Wittig and Schloeder[8] found that triphenylboron stops the reaction at the first step by combining with the 1,2-adduct (1) to form the complex (2). The complex is soluble in water without decomposition, but when boiled with dilute mineral acid it is hydrolyzed to 5,5,5-triphenylpentene-1 (3). Under the same conditions isoprene is converted into hydrocarbon (4).

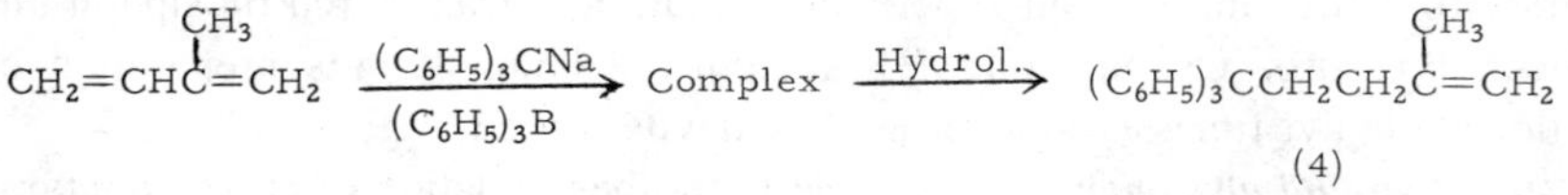
CH3
CH2=CHC=CH2
(C6H5)3CNa
(C6H5)3B
Complex
Hydrol.
CH3
(C6H5)3CCH2CH2C=CH2
(4)

In the absence of an additive, tritylsodium polymerizes acenaphthylene (5).[9] When the reaction is carried out in the presence of triphenylaluminum the reaction

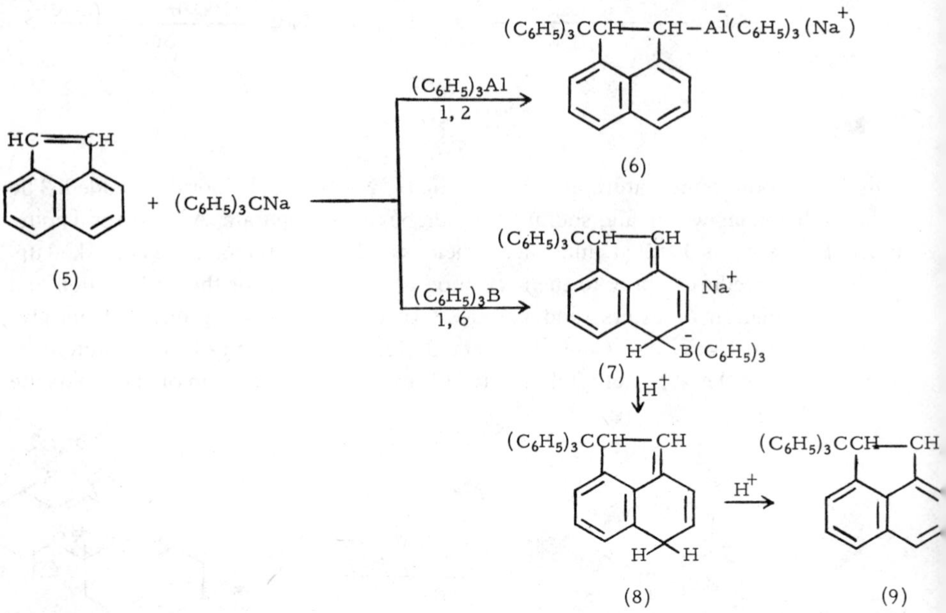
(C6H5)3CCH—CH—Al(C6H5)3 (Na+)
(C6H5)3Al
1, 2
(6)
HC=CH
+ (C6H5)3CNa
(5)
(C6H5)3CCH—CH
Na+
(C6H5)3B
1, 6
H
B(C6H5)3
(7)
H+
(C6H5)3CCH—CH
H
H
(8)
H+
(C6H5)3CCH—CH
(9)

stops with formation of the complex (6) formed by 1,2-addition. With triphenylboron as the coupling agent, 1,6-addition takes place to give the complex (7). This can be converted by careful acid hydrolysis to (8), which is readily isomerized to 7-tritylacenaphthene (9).

Tritylsodium has been used in the preparation of acid phthalate derivatives of tertiary alcohols by the following procedure.[10] An ethereal solution of tritylsodium

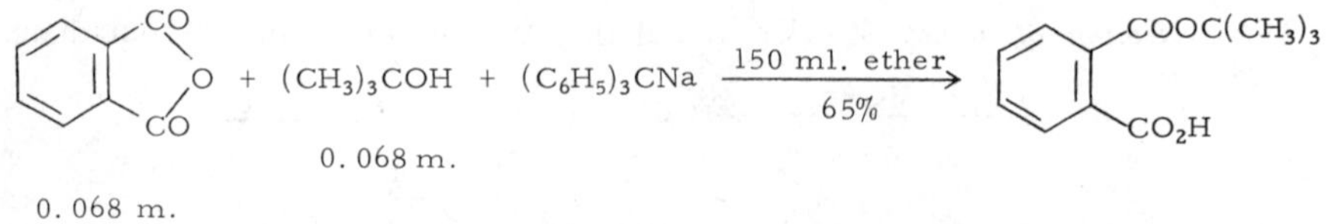
CO
O
CO
+ (CH3)3COH + (C6H5)3CNa
150 ml. ether
65%
COOC(CH3)3
CO2H
0. 068 m.
0. 068 m.

was added to a stirred solution of *t*-butanol in ether until a persistent red coloration indicated a slight excess of the base, phthalic anhydride was added, and stirring was continued for 2 hrs. Workup and crystallization afforded pure half-ester, m.p. 87–88°. This ester decomposes with liberation of isobutene at 151–155°.

[1]W. B. Renfrow, Jr., and C. R. Hauser, *Org. Syn., Coll. Vol.*, **2**, 607 (1943)
[2]C. R. Hauser and W. B. Renfrow, Jr., *ibid.*, **2**, 268 (1943)
[3]B. E. Hudson, Jr., and C. R. Hauser, *Am. Soc.*, **62**, 2457 (1940); **63**, 3156 (1941)
[4]E. Baumgarten and C. R. Hauser, *Am. Soc.*, **66**, 1037 (1944)
[5]E. Baumgarten, R. Levine, and C. R. Hauser, *Am. Soc.*, **66**, 862 (1944)
[6]E. J. Corey and E. W. Cantrall, *Am. Soc.*, **81**, 1745 (1959)
[7]K. Ziegler, F. Dersch, and H. Wollthan, *Ann.*, **511**, 13 (1934)
[8]G. Wittig and H. Schloeder, *Ann.*, **592**, 38 (1955); G. Wittig and D. Wittenberg, *ibid.*, **606**, 1 (1957)
[9]G. Wittig, H. G. Reppe, and T. Eicher, *Ann.*, **643**, 47 (1961)
[10]K. G. Rutherford, J. M. Prokipcak, and D. P. C. Fung, *J. Org.*, **28**, 582 (1963)

Tritylsulfenyl chloride, $(C_6H_5)_3CSCl$. Mol. wt. 310.84, m.p. 137°, yellow. Preparation.[1]

N-Protective group. Zervas[2] introduced use of the reagent for the N-protection of amino acid esters, including the activated *p*-nitrophenyl esters of use in peptide coupling. When required, the protective group is split by treatment with the theoretical amount of hydrogen chloride in alcohol.

$$(C_6H_5)_3CSCl + H_2NCHRCO_2C_6H_4NO_2\text{-}\underline{p} \longrightarrow$$

$$\underset{(1)}{(C_6H_5)_3CS{-}NHCHRCO_2C_6H_4NO_2\text{-}\underline{p}} \xrightarrow{H_2NCHR'CO_2R''}$$

$$\underset{(2)}{(C_6H_5)_3CS{-}NHCHRCO{-}NHCHR'CO_2R''} \xrightarrow{HCl}$$

$$\underset{(3)}{NH_2CHRCO{-}NHCHR'CO_2R''}$$

[1]D. Vorländer and E. Mittag, *Ber.*, **52**, 413 (1919)
[2]L. Zervas, D. Borovas, and E. Gazis, *Am. Soc.*, **85**, 3660 (1963)

Tropylium fluoroborate, $C_7H_7^+BF_4^-$. Mol. wt. 177.95, dec. about 200°. For preparation, *see* Cycloheptatriene.

Use of the reagent for the preparation of substituted tropilidenes is described in papers by Conrow and in literature cited.[1]

[1]K. Conrow, *Am. Soc.*, **81**, 5461 (1959); **83**, 2343 (1961); K. Conrow and D. N. Naik, *J. Med. Chem.*, **6**, 69 (1963)

U

Urea, H_2NCONH_2. Mol. wt. 60.06, m.p. 133°, 1 g. dissolves in 1 ml. water, 10 ml. 95% ethanol, 1 ml. boiling 95% ethanol, 20 ml. abs. alcohol, 6 ml. methanol; almost insoluble in ether or chloroform, soluble in concd. HCl.

Urea inclusion complexes. Normal alkanes having seven or more carbon atoms form inclusion complexes in which hydrogen-bonded urea molecules are oriented in a helical lattice in which a straight-chain hydrocarbon fits. The guest molecule is not bonded to the host but merely trapped in the channel. Discovered in 1940 by Bengen,[1] the hydrocarbon-urea complexes have been studied particularly by Schlenk,[2] Schiessler and Flitter,[3] and Smith (X-ray analysis).[4] The principal findings are as follows.

As the hydrocarbon chain is lengthened, more urea molecules are required to extend the channel but the ratios given in Table 1 show that the host-guest relationship is not stoichiometric. For preparation of a complex, a liquid hydrocarbon is added to a solution saturated with urea at room temperature; an adduct if formed separates as a voluminous crystallizate, usually directly pure. Recovery of the hydrocarbon is accomplished by treating a complex with water, to dissolve the urea, and ether to extract the hydrocarbon. The channel into which a normal alkane fits will not accommodate a thick, branched-chain hydrocarbon such as 2,2,4-trimethylpentane (isooctane), which forms no complex. A methyl group centered on a C_{13}-chain inhibits complex formation.

Table 1 Molecules of urea per molecule of *n*-alkane

Alkane	Ratio	Alkane	Ratio
C_6	No complex	C_{11}	8.7
C_7	6.1	C_{12}	9.7
C_8	7.0	C_{16}	12.3
C_9	7.3	C_{24}	18.0
C_{10}	8.3	C_{28}	21.2

The *n*-decane-urea complex can be demonstrated as shown in the photograph with the models developed by one of us[5] and a sheet of cellulose acetate to represent the inside of the complex (Fig. U-1). The cylinder fits snugly over the model and defines the total space, not that actually occupied, since van der Waals forces prevent actual contact and keep the atoms at a certain stand-off distance. The diameter represented is $14.3 \times 0.2 = 2.86$ Å, but X-ray measurements indicate the true diameter to be 5.30 Å. Half the difference, or 1.22 Å, then represents the stand-off distance all the way around the circumference. This estimate applies only to the complexes of normal alkanes and may be a maximal value. Thus 3-nonyne forms a urea complex but the model cannot be fitted into the 14.3-cm. cylinder. It does

$CH_3CH_2C{\equiv}CCH_2CH_2CH_2CH_2CH_3$ — 3-Nonyne

$CH_3C{\equiv}C-C{\equiv}C-C{\equiv}CCH_3$ — 2,4,6-Octatriyne

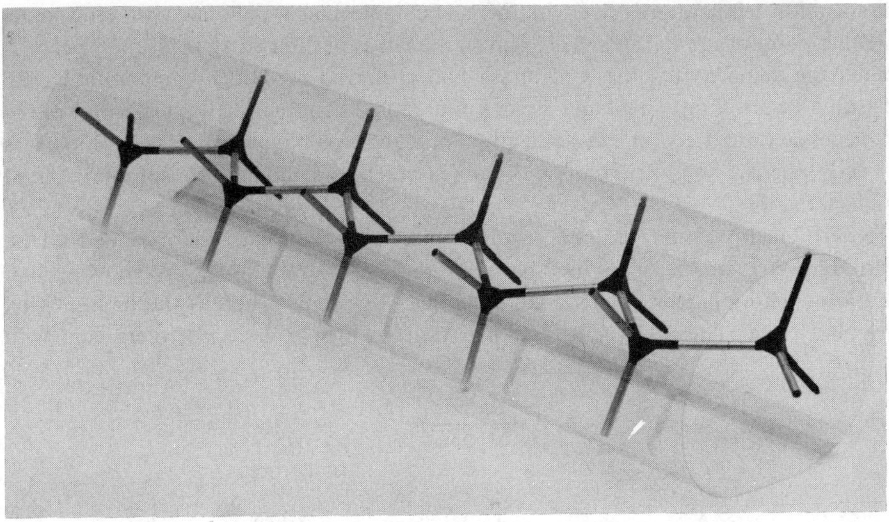

Fig. U-1 *n*-Decane in a 14.3-cm. cylinder (circumference 45 cm.)

fit into a cylinder of diameter 16.2 cm., or 3.24 Å, and here the stand-off distance is only 1.03 Å. Thus *n*-decane finds more room in the channel than is actually needed. On the other hand, the guest molecule must fill the bulk of the available space in order to produce a stable complex. Thus the model of 2,4,6-octatriyne is very slender because of the three linear acetylenic groups, and the hydrocarbon forms no urea complex. A model of 7-methyltridecane does not fit into even the 16.2-cm. cylinder, in accord with the finding that the hydrocarbon forms no urea complex.

$$\underset{1}{CH_3}CH_2CH_2CH_2CH_2CH_2\underset{7}{\overset{\overset{\displaystyle CH_3}{|}}{C}}HCH_2CH_2CH_2CH_2CH_2\underset{13}{CH_3}$$

Swern and co-workers[6] studied the urea complexes of higher fatty acids and showed that each has a characteristic dissociation temperature at which transparent crystals of the complex turn opaque. It will be evident from the values given in Table 2 that these constants are useful for identification. In a given series the

Table 2 Fatty acid-urea complexes

No. C	Acid	Ratio urea: acid		Dissoc. temp.
		Molar	Weight	
8	Caprylic	6.7	2.8	73°
9	Pelargonic	7.4	2.8	80.5°
10	Capric	8.0	2.8	85°
12	Lauric	9.7	2.9	92.5°
13	Tridecylic	11.8	3.3	96°
14	Myristic	10.6	2.8	103°
16	Palmitic	12.0	2.8	114°
18	Stearic	14.7	3.1	126°
18	Oleic (*cis*)	13.6	2.9	110°
18	Elaidic (*trans*)	13.6	2.9	116°
18	*threo*-9,10-Dihydroxystearic	14.7	2.8	107°

dissociation temperatures rise with increasing molecular weight and converge on the melting point of urea. Table 2 also gives the ratios of urea: fatty acid; the pattern is much the same as for the *n*-alkanes. The preferred method of preparation is to dissolve 1 part of fatty acid and 5 parts of urea in 20 parts of methanol, with heating if necessary, and to let crystallization proceed spontaneously. The complex is directly pure. A solid acid can be dissolved in isopropanol and the solution diluted with methanol.

Swern[7] neatly confirmed the *erythro-* and *threo*-configurations assigned to the 9,10-dihydroxystearic acids melting, respectively, at 132 and 95° by showing that the low-melting isomer forms a urea complex (Table 2) whereas the high-melting acid does not. The 132° acid resulting from *cis*-hydroxylation of oleic acid with

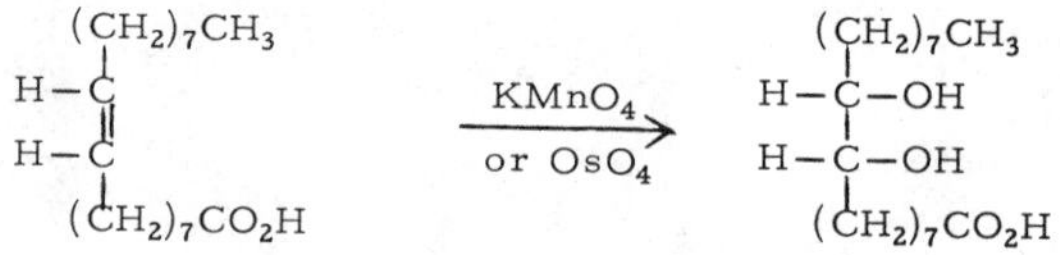
$(CH_2)_7CH_3$
H−C
H−C
$(CH_2)_7CO_2H$
$KMnO_4$
or OsO_4
$(CH_2)_7CH_3$
H−C−OH
H−C−OH
$(CH_2)_7CO_2H$

(1) (2) Erythro

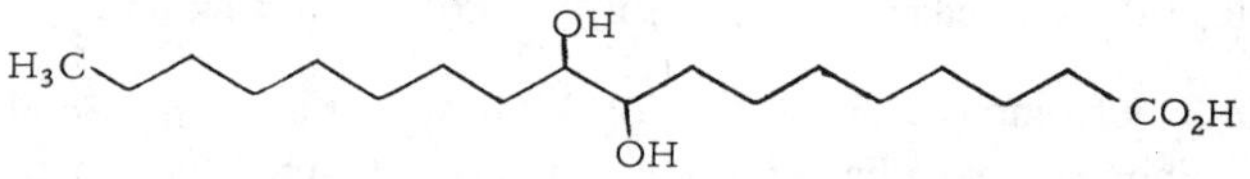
OH
H_3C
CO_2H
OH

(2a)

permanganate is written conventionally as in (2), but when the chain is extended as in (2a) it is evident that the hydroxyl groups are on opposite sides of the chain. Inspection of a model shows that these protruding groups prevent entrance of the chain into the 16.2-cm. urea cylinder. In the *threo*-diol the hydroxyl groups are in the skew conformation and the molecule is capable of insertion.

Schlenk and Holman[8] report isolation of 97–98% pure methyl oleate from olive oil methyl esters via the urea complex. In a student experiment[9] olive oil (10 g.) is saponified (KOH in triethylene glycol, 5 min. at 160°), and the hydrolyzate (9.5 g.) is crystallized from acetone (70 ml.) at −15° to remove saturated acids. Evaporation of the filtrate gives 5–7 g. of a mixture of oleic acid and linoleic acid, which is dissolved with 11 g. of urea in 50 ml. of hot methanol. On cooling, the oleic acid-urea complex separates as long, heavy spars (10–12 g.). The explanation of the separation previously advanced[9] was that oleic acid has two straight hydrocarbon chains long enough for complexing, whereas linoleic has only one. This argument is invalidated by Swern's finding that *erythro*-9,10-dihydroxystearic acid, which has the same potentiality for two-channel complexing as oleic acid, forms no complex. Furthermore, linoleic acid forms a complex, if with difficulty. Schlenk and Holman[8] applied to several of the acids of Table 2 a procedure (1 acid:5 urea:30 methanol) essentially like that of Swern and obtained complexes of similar composition. Treated in the

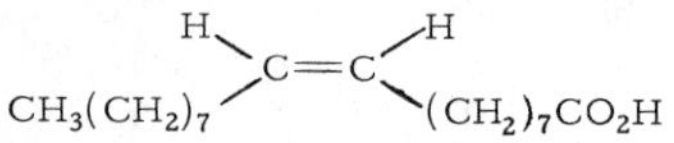
H
H
C=C
$CH_3(CH_2)_7$
$(CH_2)_7CO_2H$

Oleic acid

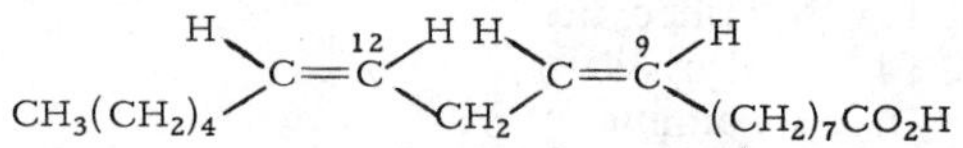
H
12
H H
9
H
C=C
C=C
$CH_3(CH_2)_4$
CH_2
$(CH_2)_7CO_2H$

Linoleic acid

same way, linoleic acid afforded no crystallizate. However, a complex containing 13.1 moles of urea per mole of acid was obtained under forcing conditions, namely with a ratio of 1:3:3.2 and by heating the mixture on the steam bath with stirring. Although no evidence was presented to exclude the possibility that the *cis,cis*-$\Delta^{9,12}$-acid was isomerized in the process, X-ray measurements by Nicolaides and Laves[10] show that shortening of the molecule for the two *cis* double bonds of linoleic acid-urea is very close to twice that of oleic acid-urea. Re-inspection of models and cylinders in the light of evidence not previously known to us leads to the following conclusions. Oleic acid does not fit into the 14.3-cm. cylinder for *n*-alkanes but is accommodated by the 16.2-cm. cylinder required for 3-nonyne. Linoleic acid cannot be squeezed into this larger cylinder but requires a slight expansion of the channel, a slight compression of the molecule, or both, corresponding to the forcing condition required for formation of the complex.

Linstead and Whalley[11] showed that urea complexes can be used for effective separation of straight- and branched-chain carboxylic esters. Swern's group[12] used the method to advantage in working up autoxidized methyl oleate for isolation of long-chain hydroperoxides. These peroxides behave like branched-chain compounds because of the bulky peroxide group and remain in solution when nonperoxidic components of the mixture are precipitated as urea complexes.

Transamination. An acid, on being heated with urea, is converted into the corresponding amide.[13,14]

$$RCOOH + H_2NCONH_2 \xrightarrow{\Delta} RCONH_2 + CO_2 + NH_3$$

An example is the preparation of sebaconitrile and ω-cyanopelargonic acid.[15] A mixture of sebacic acid and urea is stirred and heated at 160° for 4 hrs., at which

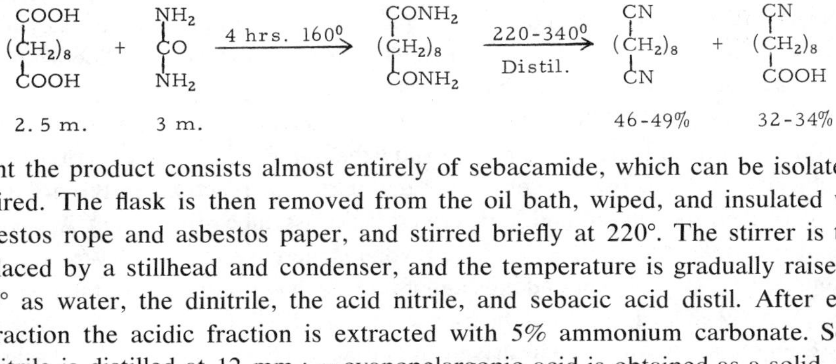
COOH
(CH2)8
COOH
NH2
CO
NH2
4 hrs. 160°
CONH2
(CH2)8
CONH2
220-340°
Distil.
CN
(CH2)8
CN
CN
(CH2)8
COOH
2.5 m.
3 m.
46-49%
32-34%

point the product consists almost entirely of sebacamide, which can be isolated if desired. The flask is then removed from the oil bath, wiped, and insulated with asbestos rope and asbestos paper, and stirred briefly at 220°. The stirrer is then replaced by a stillhead and condenser, and the temperature is gradually raised to 340° as water, the dinitrile, the acid nitrile, and sebacic acid distil. After ether extraction the acidic fraction is extracted with 5% ammonium carbonate. Sebaconitrile is distilled at 12 mm.; ω-cyanopelargonic acid is obtained as a solid, m.p. 48–49°. The amount of urea specified is that found by experiment to be required. When the amount taken was 2 moles per mole of sebacic acid, the yield of sebaconitrile was only 27%.

Preparation of t-butylamine.[16] *t*-Butylurea, obtained from Eastman or Fisher or prepared from *t*-butanol, urea, and concd. sulfuric acid, is heated with phthalic anhydride to produce *t*-butylphthalimide, which is cleaved by refluxing with aqueous-alcoholic hydrazine to phthalhydrazide and *t*-butylamine. After cooling, the mixture is made acid (HCl) and the heterocyclic product removed by filtration. Suitable processing of the filtrate affords pure *t*-butylamine hydrochloride.

$$\underset{2\text{ m.}}{(CH_3)_3COH} + \underset{1\text{ m.}}{H_2NCONH_2} + \underset{1.98\text{ m.}}{\text{concd. } H_2SO_4} \xrightarrow[31\text{-}33\%]{20\text{-}25^\circ} (CH_3)_3CNH\overset{O}{\overset{\|}{C}}NH_2$$

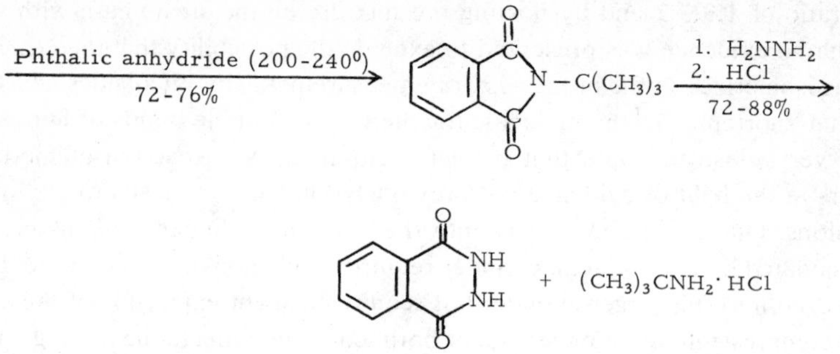

Phthalic anhydride (200-240°)
72-76%
N − C(CH3)3
1. H2NNH2
2. HCl
72-88%
NH
NH
+ (CH3)3CNH2 · HCl

Catalyst. Tetracyanoethylene can be crystallized unchanged from methyl or ethyl alcohol. However, if urea is present as catalyst a clean reaction occurs to give a dicyanoketene ketal.[17] An example is the preparation of dicyanoketene ethyleneketal.[18] A suspension of finely divided tetracyanoethylene in a solution of

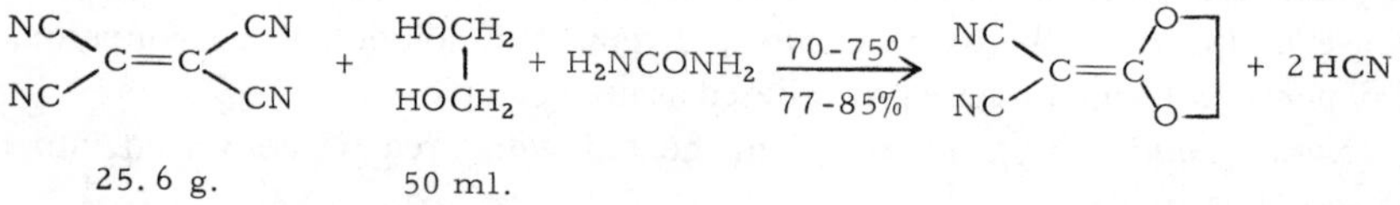

NC
CN
C=C
NC
CN
HOCH2
HOCH2
+ H2NCONH2
70-75°
77-85%
+ 2 HCN
25.6 g.
50 ml.

urea in ethylene glycol is swirled at 70–75° until solution is complete (15 min.), and the brownish solution is cooled for crystallization of the product of satisfactory purity.

The function of the urea is not known. Addition of urea to a solution of tetracyanoethylene in ethanol produces a brilliant purple color; as hydrogen cyanide is evolved, the color fades, and the solution becomes nearly colorless. Pyridine and zinc acetate also catalyze the reaction, but urea is the best catalyst found.

Chlorohydrination.[19] Urea is chlorinated in an apparatus designed for rapid absorption of a gas. Figures U-2 and U-3 show the principles of two ways of realizing this objective but not the assemblies for this particular reaction. In the first (Fig. U-2), described by Russell and Vanderwerf,[20] a glass tube stirrer shaft AA terminates at the top as the inner tube of the revolving mercury seal C. Gas-inlet tube B is clamped so that the lower end dips about 8 cm. below the surface of the mercury. When properly aligned in the ball bearing mountings D, E, E, the movable assembly turns freely and smoothly when driven by the pulley F and motor G. A discarded automobile generator makes an excellent aligning bearing D. To prevent clogging, the direction of stirring should be away from the orifice.

Figure U-3 shows a liquid-sealed stirrer and gas-delivery tube (*b*) provided with a glass rod slipped through a section of suction tubing; should the delivery tube become plugged, it can be freed with the glass rod.[21] The stirrer sleeve (*a*) is extended well into the flask so that the liquid itself seals the stirrer. A simple rubber slip joint can also be used; a stirrer sleeve similar to (*a*) but extending only a short distance into the flask is provided with a 2-cm. section of rubber tubing which projects above the end of the tube and fits snugly around the stirrer shaft. The point of contact between rubber and shaft is lubricated with glycerol.

For the chlorohydrination, a mixture of urea, calcium carbonate, and water is stirred at 0–15°, and a rapid stream of chlorine is passed in until the weight has increased by 95 g. (30–40 min.). Water is added to the suspension, the mixture is filtered, and the filtrate containing monochlorourea is kept cold. The filter cake is extracted with further portions of water, the filtrates are combined, treated with ice, then with acetic acid to liberate hypochlorous acid, and with cyclopentene,

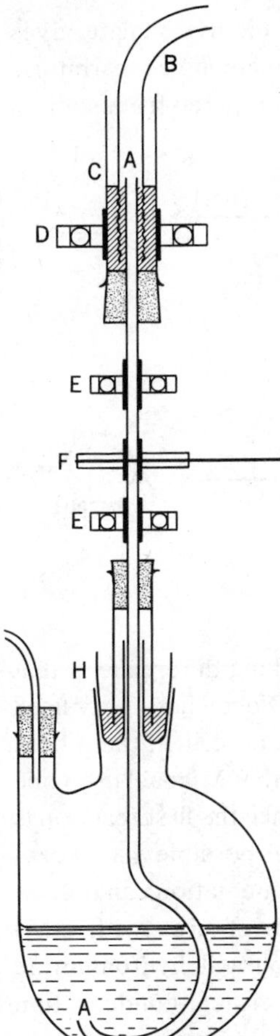
B
C
A
D
E
F
E
G
H
A

Fig. U-2

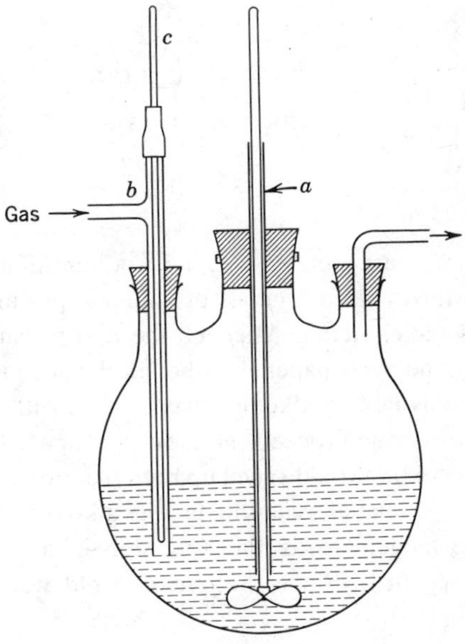
c
b
a
Gas

Fig. U-3

taken in excess. The flask is stirred and kept packed in ice until the top layer disappears and a heavy oil separates at the bottom (12–15 hrs.). The aqueous liquor

$$\underset{2.5\ m.}{H_2NCONH_2} + \underset{1.25\ m.}{CaCO_3} + \underset{150\ ml.}{H_2O} \xrightarrow[0\text{-}15^0]{Cl_2} H_2NCONHCl$$

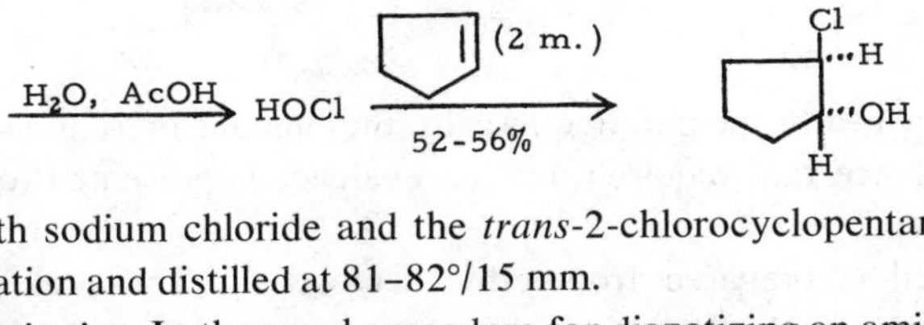
H₂O, AcOH
HOCl
(2 m.)
52-56%
Cl
H
OH
H

is saturated with sodium chloride and the *trans*-2-chlorocyclopentanol is removed by steam distillation and distilled at 81–82°/15 mm.

Use in diazotization. In the usual procedure for diazotizing an amine salt, sodium nitrite is added until a test with starch-KI paper shows that an excess is present.

Before use of the solution in the next step, the excess reagent can be destroyed by addition of urea.[22,23]

Stabilizer. Wilson[24] recommends that freshly distilled 2-furfuryl alcohol be treated with 0.5–1% of urea as stabilizer.

Heterocycles. Base-catalyzed condensation of urea with diethyl oxalate gives parabanic acid (Murray[25]); condensation with diethyl malonate gives barbituric acid (Dickey and Gray[26]). The first procedure uses a solution prepared from sodium

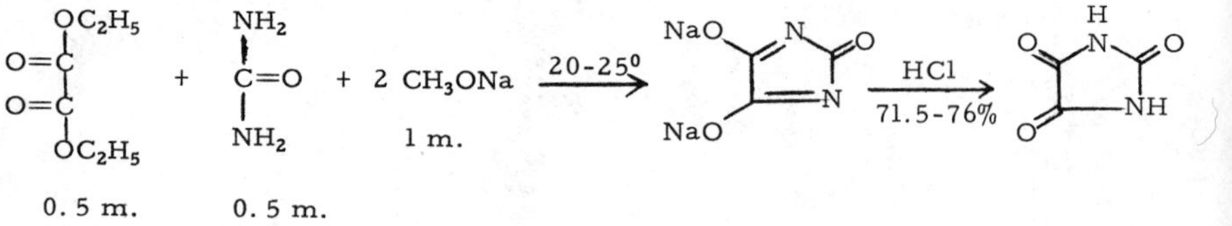

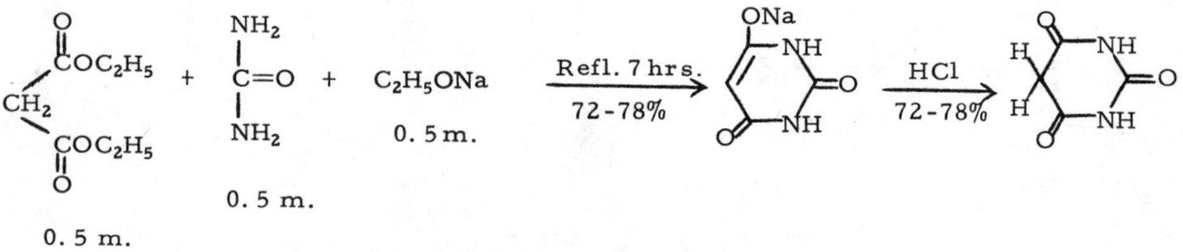

and methanol, the second, sodium in absolute ethanol. A striking difference is that Murray used 2 moles of alkoxide per mole of ester whereas Dickey and Gray used 1 mole. Arthur Michael, the first to carry out either reaction, described them both in the same paper.[27] Although the accounts are brief, it is clear that Michael used one equivalent of alkoxide in each case with satisfactory results. Since the first reaction is faster and proceeds at a lower temperature than the second, it is possible that excess alkoxide would speed up both reactions, but only further experimentation can tell.

A related reaction is the alkoxide-catalyzed condensation of urea with ethyl cyanoacetate as the first step in a synthesis of 5,6-diaminouracil hydrochloride (Taylor[28]). Since the overall yield was high, the yield in the critical condensation

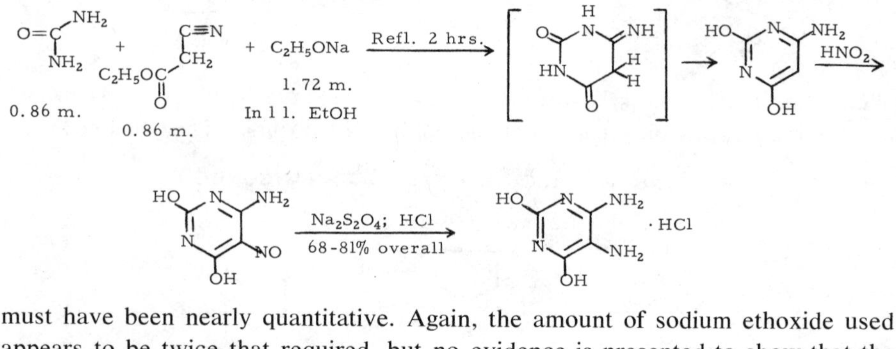

must have been nearly quantitative. Again, the amount of sodium ethoxide used appears to be twice that required, but no evidence is presented to show that the excess is beneficial.

6-Methyluracil is prepared from ethyl acetoacetate and urea by a two-step process.[29] Finely powdered urea is stirred into a mixture of small amounts of absolute ethanol and hydrochloric acid in a crystallizing dish which is placed in a

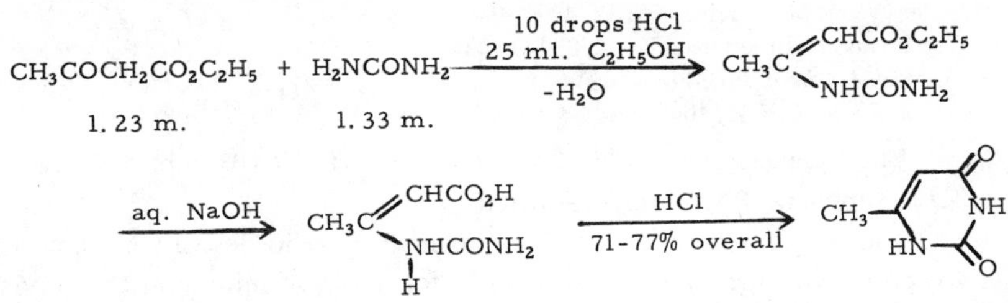

vacuum dessicator over concd. sulfuric acid for elimination of water formed in the condensation to β-uraminic crotonic ester. The dessicator is evacuated continuously at the water pump with daily replacement of the sulfuric acid. When the crude product is thoroughly dry (5–7 days), it is stirred into dilute sodium hydroxide solution at 95°. The clear solution is cooled to 65° and stirred during slow addition of acid. The pyrimidine precipitates almost immediately and is obtained as a colorless powder of high purity.

The condensation of benzoin with urea to produce 4,5-diphenylglyoxalone is carried out by refluxing the reactants in acetic acid.[30]

Displacement of aromatic halogen. Ashton[31] reported an "improved" procedure for conversion of 2,4-dinitrochlorobenzene into 2,4-dinitroaniline by refluxing with aqueous ammonia in ethanol for 3 hrs. Ferry[32] states that brief refluxing with urea in ethanol "is much shorter and gives excellent yields." The fact is that the yields were 53 and 57% and that in each case the crystallized product melted about 10° below the accepted value.

[1] M. F. Bengen, German patent application, March 18 (1940)
[2] W. Schlenk, Jr., *Ann.*, **565**, 204 (1949); **573**, 142 (1951); *Fortschr. chem. Forsch.*, **2**, 92 (1951)
[3] R. W. Schiessler and D. Flitter, *Am. Soc.*, **74**, 1720 (1952)
[4] A. E. Smith, *J. Chem. Phys.*, **18**, 150 (1950)
[5] L. F. Fieser, *J. Chem. Ed.*, **40**, 457 (1963); **42**, 408 (1965)
[6] H. B. Knight, L. P. Witnauer, J. E. Coleman, W. R. Noble, Jr., and D. Swern, *Anal. Chem.*, **24**, 1331, (1952)
[7] D. Swern, L. P. Witnauer, and H. B. Knight, *Am. Soc.*, **74**, 1655 (1952)
[8] H. Schlenk and R. T. Holman, *Am. Soc.*, **72**, 5001 (1950)
[9] *Org. Expts.*, 162
[10] N. Nicolaides and F. Laves, *Am. Soc.*, **76**, 2596 (1954); **80**, 5752 (1958)
[11] R. P. Linstead and M. Whalley, *J. Chem. Soc.*, 2987 (1950)
[12] J. E. Coleman, H. B. Knight, and D. Swern, *Am. Soc.*, **74**, 4886 (1952)
[13] B. S. Biggs and W. S. Bishop, *Am. Soc.*, **63**, 944 (1941)
[14] E. Cherbuliez and F. Landolt, *Helv.*, **29**, 1438 (1946)
[15] B. S. Biggs and W. S. Bishop, *Org. Syn., Coll. Vol.*, **3**, 768 (1955)
[16] L. I. Smith and O. H. Emerson, *ibid.*, **3**, 151 (1955)
[17] W. J. Middleton and V. A. Engelhardt, *Am. Soc.*, **80**, 2788 (1958)
[18] C. L. Dickinson and L. R. Melby, *Org. Syn., Coll. Vol.*, **4**, 276 (1963)
[19] H. B. Donahoe and C. A. Vanderwerf, *ibid.*, **4**, 157 (1963)
[20] R. R. Russell and C. A. Vanderwerf, *Ind. Eng. Chem., Anal. Ed.*, **17**, 269 (1945)
[21] L. F. Fieser, "Experiments in Organic Chemistry," 2nd Ed., 310, D. C. Heath (1941)
[22] H. E. Ungnade and E. F. Orwoll, *Org. Syn., Coll. Vol.*, **3**, 130 (1955)
[23] C. E. Kaslow and R. M. Summers, *ibid.*, **4**, 718 (1963)
[24] W. C. Wilson, *ibid.*, **1**, 279 (1941)
[25] J. I. Murray, *ibid.*, **4**, 744 (1963)
[26] J. B. Dickey and A. R. Gray, *ibid.*, **2**, 60 (1943)
[27] A. Michael, *J. prakt. Chem.*, [2], **35**, 456, 458 (1887)
[28] W. R. Sherman and E. C. Taylor, Jr., *Org. Syn., Coll. Vol.*, **4**, 247 (1963)

[29] J. J. Donleavy and M. A. Kise, *ibid.*, **2**, 422 (1943)
[30] B. B. Corson and E. Freeborn, *ibid.*, **2**, 231 (1943)
[31] A. A. Ashton, *J. Chem. Ed.*, **40**, 545 (1963)
[32] N. Ferry, *J. Chem. Ed.*, **41**, 404 (1964)

Urethane (Ethyl carbamate), $H_2NCO_2C_2H_5$. Mol. wt. 89.10, m.p. 50°, b.p. 183°, sp. gr. 1.11. Suppliers: FMC Corp., Aldrich.

Treated with sodium in ether, urethane forms a disodio derivative which reacts with two equivalents of ethyl chloroformate to form triethyl nitrogentricarboxylate.[1]

$$H_2NCO_2C_2H_5 \xrightarrow{2\ Na} Na_2NCO_2C_2H_5 \xrightarrow[51\text{-}57\%]{2\ ClCO_2C_2H_5} N(CO_2C_2H_5)_3$$

[1] O. Diels, *Ber.*, **36**, 736 (1903); C. F. H. Allen and A. Bell, *Org. Syn.*, *Coll. Vol.*, **3**, 415 (1955)

V

Vilsmeier reagent, *see* Dimethylformamide–Phosphoryl chloride.

Vinyl acetate, $CH_2{=}CHOCOCH_3$. Mol. wt. 86.09, b.p. 72–73°, stabilized with hydroquinone. Suppliers: B, E, F, MCB.

Vinyl interchange.[1] A mixture, prepared under nitrogen, of freshly distilled vinyl acetate and lauric acid is warmed to dissolve the acid, mercuric acetate is

$$\underset{0.4\ m.}{CH_3(CH_2)_{10}CO_2H} + \underset{2.4\ m.}{CH_3CO_2CH{=}CH_2} + \underset{1.6\ g.}{Hg(OAc)_2} + \underset{0.15\ ml.}{100\%\ H_2SO_4}$$

$$\xrightarrow[53\text{-}59\%]{\text{Refl. 3 hrs.}} CH_3(CH_2)_{10}CO_2CH{=}CH_2$$

added followed by sulfuric acid, and the mixture is refluxed for 3 hrs. Sodium acetate trihydrate (0.83 g.) is added to neutralize the acid, excess vinyl acetate is removed by distillation, and vinyl laurate is distilled and redistilled, b.p. 120–120.5°/2 mm.

Dienes via cyclobutanes.[2] In the synthetic sequence formulated, the addition step is carried out by charging a high-pressure autoclave with redistilled vinyl acetate, 1 g. of hydroquinone, and 3 drops of dipentene or terpinolene (to inhibit polymerization of the fluoroolefin). The vessel is closed, cooled in dry ice-acetone, evacuated, and charged with chlorotrifluoroethylene (available in 1-lb. and 5-lb. cylinders from

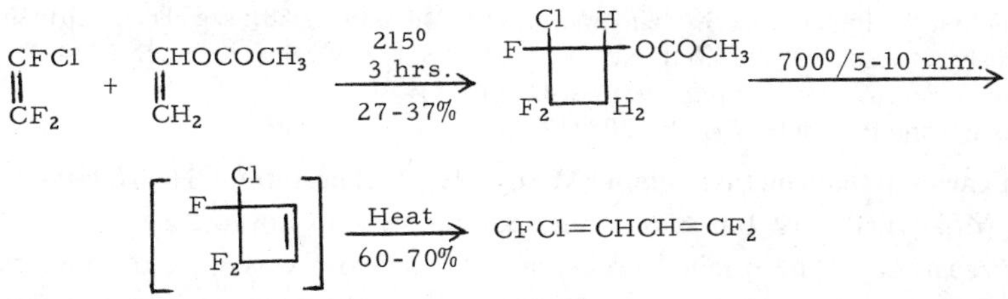
CFCl
CF2
CHOCOCH3
CH2
215°
3 hrs.
27-37%
Cl H
F
OCOCH3
F2
H2
700°/5-10 mm.
Cl
F
F2
Heat
60-70%
CFCl=CHCH=CF2

Matheson Co.). The mixture is heated with agitation to 215° and kept at this temperature for 3 hrs. The reaction mixture is black and viscous, and the yield of 2-chloro-2,3,3-trifluorocyclobutyl acetate is low. The next step, involving elimination of acetic acid followed by rupturing of the carbon chain, is done in a Vycor glass reaction tube heated in an electric furnace.

Peptide synthesis. Vinyl esters of amino acids, prepared by transesterification with vinyl acetate, have been used as activated esters in peptide synthesis.[3] The coupling reaction is best carried out in ethyl cyanoacetate, for this solvent suppresses the formation of colored products derived from liberated acetaldehyde. Racemization appears to be slight.

[1] D. Swern and E. F. Jordan, Jr., *Org. Syn., Coll. Vol.*, **4**, 977 (1963)
[2] R. E. Putnam, B. T. Anderson, and W. H. Sharkey, *Org. Syn.*, **43**, 17 (1963)
[3] F. Weygand and W. Steglich, *Angew. Chem.*, **73**, 757 (1961)

Vinyl β-carbomethoxyethyl ketone, $CH_2{=}CHCOCH_2CH_2CO_2CH_3$. Mol. wt., 142.15, b.p. 103°/16 mm. Polymerizes readily, must be stabilized with hydroquinone.

Preparation:[1]

$$CH_2{=}CH_2 + ClC(=O)CH_2CH_2CO_2CH_3 \xrightarrow{AlCl_3} ClCH_2CH_2C(=O)CH_2CH_2CO_2CH_3$$

$$\xrightarrow{(C_2H_5)_3N} CH_2{=}CHC(=O)CH_2CH_2CO_2CH_3$$

Synthesis of aromatics. Buchta[2] introduced use of the reagent for the construction of polycyclic compounds capable of aromatization. Thus Michael addition of the reagent (2) to 2-carboethoxy-1-tetralone (1) gives (3), which undergoes intramolecular Stobbe condensation to (4). Ring closure with liquid hydrogen fluoride gives (5), and zinc dust distillation gives pyrene (6). The same scheme served for the synthesis of the heptacyclic peropyrene.[3]

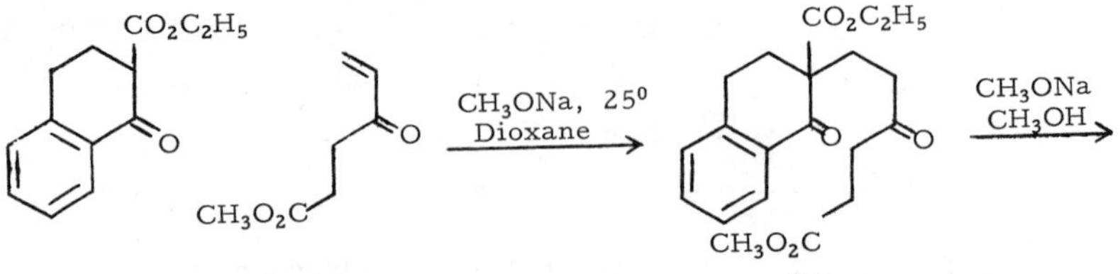
CO2C2H5
CH3O2C
CH3ONa, 25°
Dioxane
CO2C2H5
CH3O2C
CH3ONa
CH3OH

(1) (2) (3)

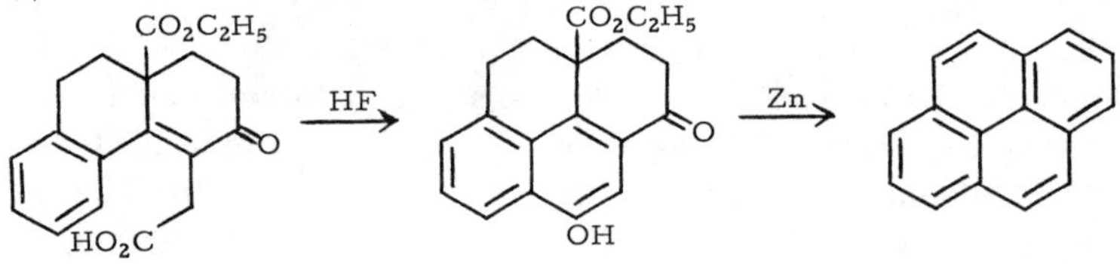
CO2C2H5
HO2C
HF
CO2C2H5
OH
Zn

(4) (5) (6)

[1]E. Buchta, W. Bayer, and G. Heinz, *Naturwiss.*, **45**, 439 (1958); see also C. Grundmann and W. Ruske, *Ber.*, **86**, 939 (1953); I. N. Nazarov and S. I. Zav'yalov, *C. A.*, **47**, 5364 (1953)
[2]E. Buchta and W. Bayer, *Naturwiss.*, **45**, 440 (1958)
[3]E. Buchta and P. Vincke, *Ber.*, **98**, 208 (1965)

Vinyl carbomethoxymethyl ketone (Methyl acryloyl acetate), $CH_2{=}CHCOCH_2CO_2$-CH_3. Mol. wt. 128.12, b.p. 65°/19 mm; polymerizes on storage.

Preparation. The original procedure of Nazarov[1] was applied on a larger scale by Wenkert *et al.*[2] and by Hohenlohe-Oehringen[3] (the Austrian investigator prepared the carboethoxymethyl derivative; yield in the first step 29%). Wenkert's group slowly added to a suspension of 1.5 moles of methyl sodioacetoacetate in ether an ethanol solution of 1.6 moles of β-ethoxypropionyl chloride (the acid is supplied by Aldrich). When acylation was completed, sodium chloride was removed by filtration, and the filtrate was saturated with ammonia to effect ketonic cleavage. Vacuum distillation from *p*-toluenesulfonic acid eliminated ethanol and gave the reagent.

$$\underset{1.6\ m.}{C_2H_5OCH_2CH_2COCl} + \underset{1.5\ m.}{CH_3CO\bar{C}HCO_2CH_3(\overset{+}{Na})} \longrightarrow CH_3COCH(COCH_2CH_2OC_2H_5)CO_2CH_3 \xrightarrow[29\%]{NH_3}$$

$$C_2H_5OCH_2CH_2C(=O)CH_2CO_2CH_3 \xrightarrow[79\%]{TsOH\ vac.\ dist.\ 120\text{-}130^\circ} CH_2{=}CHC(=O)CH_2CO_2CH_3 + C_2H_5OH$$

Synthesis of carboalkylated ring systems. Nazarov[1] demonstrated the usefulness of the reagent by synthesis of the diketo ester (4) from 2-methylcyclohexane-1,3-

dione (1) and the carboethoxymethyl reagent (2). Base-catalyzed Michael addition gave (3) and vacuum distillation effected aldolization to (4).

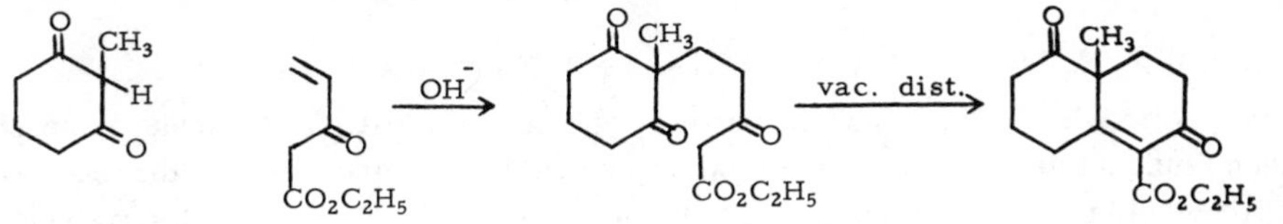
OH⁻
vac. dist.
CH3
H
CO2C2H5

(1) (2) (3) (4)

Hohenlohe-Oehringen[3] used the carboethoxymethyl reagent for a Mannich condensation with methylamine and phenylacetaldehyde to form the 4-piperidone (5) in moderate yield.

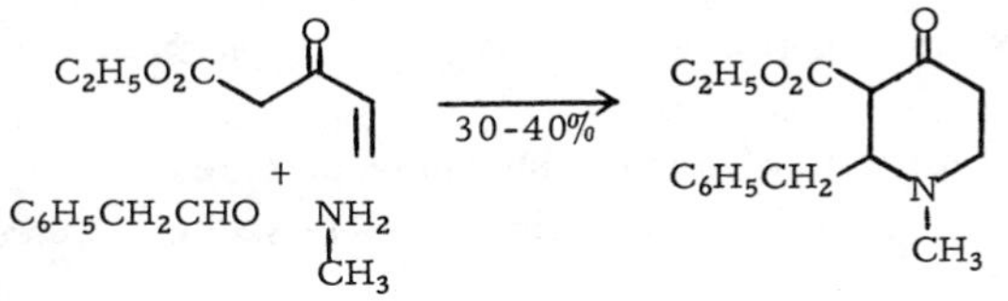
C2H5O2C
C6H5CH2CHO
NH2
CH3
30-40%
C6H5CH2
N

(5)

Wenkert et *al.*[2] report two examples of use of the reagent for accomplishing Robinson annellation to products containing a carbomethoxyl group. One, the reaction with (6) to give (7), is a key step in a synthesis of resin acids. The second, conversion of (8) to (9), was a step in a problem of synthesis in the gibberellic acid field.

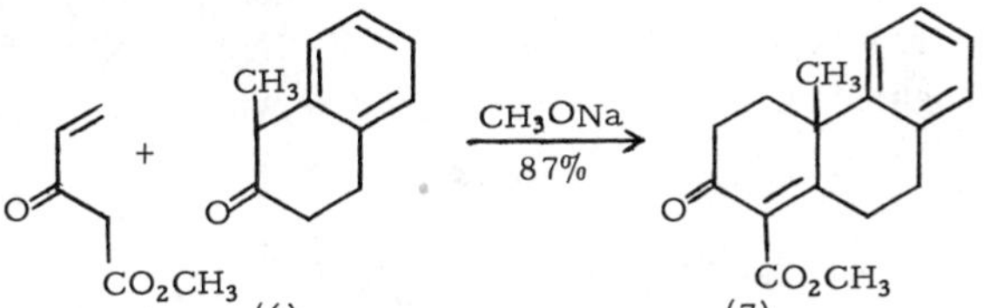
CH3
CH3ONa
87%
CO2CH3

(6) (7)

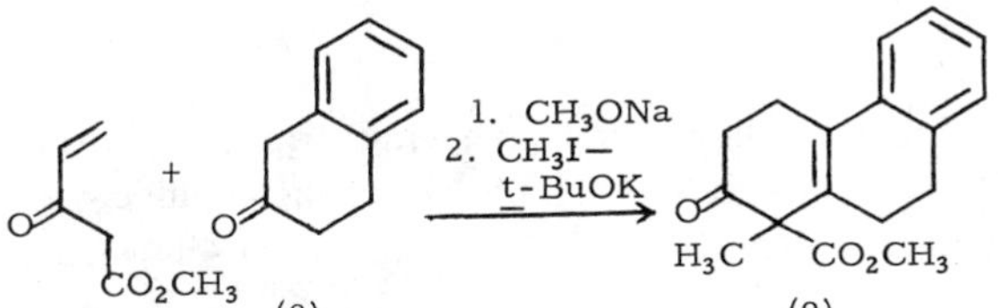
1. CH3ONa
2. CH3I–
t-BuOK
H3C
CO2CH3

(8) (9)

[1]I. N. Nazarov and S. I. Zav'yalov, *Zh. Obshch. Khim.*, **23**, 1703 (1953); English translation, *ibid.*, **23**, 1793 (1953); *C. A.*, **48**, 13667 (1954)

[2]E. Wenkert, A. Afonso, J. B-son Bredenberg, C. Kaneko, and A. Tahara, *Am. Soc.*, **86**, 2038 (1964)

[3]K. Hohenlohe-Oehringen, *Monatsh.*, **93**, 576 (1962)

Vinyllithium, $CH_2{=}CHLi$.

Juenge and Seyferth[1] prepared vinyllithium in ethereal solution by two methods. Phenyllithium in ether reacts with tetravinyllead with precipitation of tetraphenyllead (m.p. 230°) and the filtered solution contains vinyllithium. A second method involves metal exchange between lithium and tetravinyltin (Metal and Thermit Corp., Peninsular ChemResearch); tin separates as a black precipitate and is the only by-product. Both methods were used for isolation of vinyllithium as a violently

1. $4\ C_6H_5Li + Pb(CH{=}CH_2)_4 \longrightarrow 4\ CH_2{=}CHLi + Pb(C_6H_5)_4$

2. $4\ Li + Sn(CH{=}CH_2)_4 \longrightarrow 4\ CH_2{=}CHLi + Sn$

3. $4\ C_6H_5Li + Sn(CH{=}CH_2)_4 \longrightarrow 2\ CH_2{=}CHLi + Sn(C_6H_5)_4$

pyrophoric white solid, which affords ethylene on hydrolysis. Seyferth and Weiner[2] then found a third method, which produces vinyllithium in good yield, in the reaction of phenyllithium with tetravinyltin and demonstrated several reactions, for example:

$$CH_2{=}CHLi + CO_2 \xrightarrow[37\%]{H_2O} CH_2{=}CHCO_2H$$

$$CH_2{=}CHLi + (CH_3)_2C{=}O \xrightarrow[74\%]{H_2O} CH_2{=}CHC(OH)(CH_3){-}CH_3$$

$$(\underline{n}\text{-}C_4H_9)_3SnCl + CH_2{=}CHLi \xrightarrow[74\%]{} (\underline{n}\text{-}C_4H_9)_3SnCH{=}CH_2 + LiCl$$

West and Glaze[3] found that vinyllithium can be prepared directly from vinyl chloride and a lithium dispersion containing about 2% of sodium (preferred temperature 0–10°).

[1]E. C. Juenge and D. Seyferth, *J. Org.*, **26**, 563 (1961)
[2]D. Seyferth and M. A. Weiner, *Am. Soc.*, **83**, 3583 (1961)
[3]R. West and W. H. Glaze, *J. Org.*, **26**, 2096 (1961)

Vinyl triphenylphosphonium bromide, $CH_2{=}CH\overset{+}{P}(C_6H_5)_3Br^-$. Mol. wt. 269.24, m.p. 188–190°, the salt produces a sneezing allergic reaction.

Preparation.[1,2] A mixture of β-bromophenetole, triphenylphosphine, and phenol as solvent is stirred at 90° for 48 hrs., and the crystalline white phenoxyethyltriphenylphosphonium bromide is precipitated with ether and refluxed with

$$\underset{0.5\ m.}{C_6H_5OCH_2CH_2Br} + \underset{0.5\ m.}{(C_6H_5)_3P} + \underset{1\ lb.}{\text{Phenol}} \xrightarrow{90^0\ 48\ hrs.} C_6H_5OCH_2CH_2\overset{+}{P}(C_6H_5)_3Br^-$$

$$\xrightarrow[75\text{-}85\%\ overall]{CH_3CO_2C_2H_5} CH_2{=}CH\overset{+}{P}(C_6H_5)_3Br^- + C_6H_5OH$$

ethyl acetate to eliminate phenol and give vinyl triphenylphosphonium bromide.

Michael additions. The reagent undergoes nucleophilic conjugate addition reactions with R(Ar)OH, R(Ar)SH, R_2NH, for example (Schweizer and Bach[1]):

$$C_6H_5SH + CH_2{=}CH\overset{+}{P}(C_6H_5)_3(Br^-) \xrightarrow[74\%]{} C_6H_5SCH_2CH_2\overset{+}{P}(C_6H_5)_3(Br^-)$$

Ring synthesis. The vinyl group of the reagent supplies a two-carbon unit required for the construction of heterocyclic and carbocyclic rings. Thus reaction with the sodium salt of salicylaldehyde in salicyaldehyde at 110°, followed by

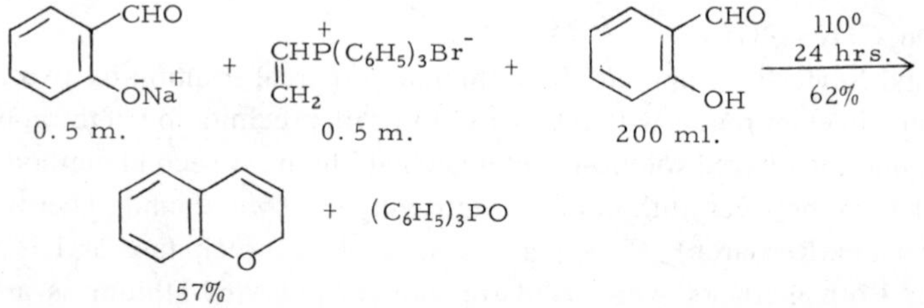
CHO
ONa
0. 5 m.
CHP(C6H5)3Br
CH2
0. 5 m.
CHO
OH
200 ml.
110⁰
24 hrs.
62%
(C6H5)3PO
57%

removal of the phenolic material with alkali-ether and distillation afforded 3,4-chromene.[3] Acetoin similarly afforded 2,5-dihydro-2,3-dimethylfurane.[4]

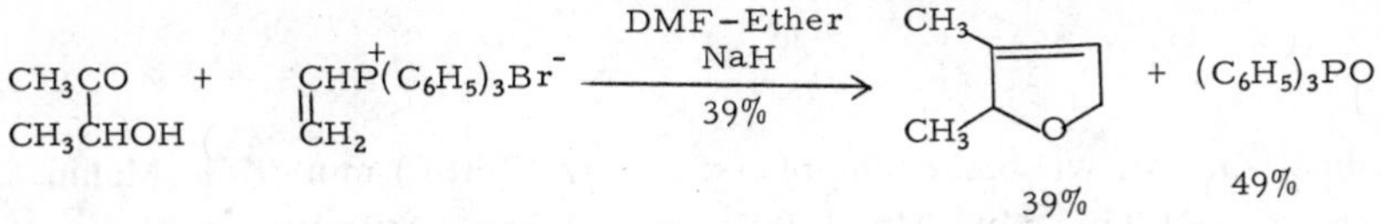
DMF–Ether
NaH
39%
CH3CO
CH3CHOH
CHP(C6H5)3Br
CH2
CH3
CH3
+ (C6H5)3PO
39%
49%

A further example is the synthesis of 3H-pyrrolizine from the reagent and pyrrole-2-aldehyde (Eastman) in ether with sodium hydride as condensing agent.

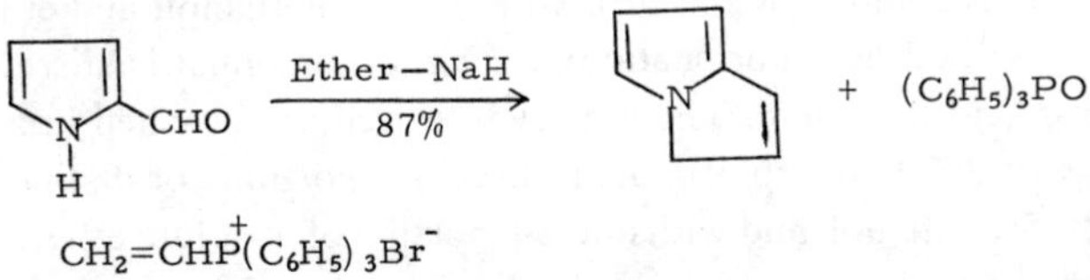
Ether–NaH
87%
CHO
N
H
N
+ (C6H5)3PO
CH2=CHP(C6H5)3Br

[1]E. E. Schweizer and R. D. Bach, *J. Org.*, **29**, 1746 (1964)
[2]*Idem*, procedure submitted to *Org. Syn.*
[3]E. E. Schweizer, *Am. Soc.*, **86**, 2744 (1964)
[4]E. E. Schweizer and K. K. Light, *Am. Soc.*, **86**, 2963 (1964)

Z

Zinc, dust, Zn. At. wt. 65.38. Suppliers: Fisher (Certif.) min. 97%; Mallinckrodt (best grade, AR) 95% min.; Merck 90% min. Once a bottle has been opened, the reagent may deteriorate owing to oxidation. Procedures of activation (other than alloy formation) may involve chiefly removal of inert oxide. (1) Wash several times with 5% hydrochloric acid, wash in turn with water, methanol, and ether, and dry.[1] (2) Another,[2] described as a purification: 1.2 kg. of zinc dust is stirred with 3 l. of 2% hydrochloric acid for 1 min. The zinc is collected and washed in a beaker with one 3-l. portion of 2% hydrochloric acid, three 3-l. portions of distilled water, two 2-l. portions of 95% ethanol, and with one 2-l. portion of absolute ether. The material is thoroughly dried and any lumps are broken up in a mortar. (3) Stir with 10% hydrochloric acid 2 min., collect, wash with water, then acetone.[3] (4) *See* Zinc, for Reformatsky.

Gattermann's method for the analysis of zinc dust is to add dilute HCl containing a few drops of $PtCl_2$ solution to a weighed sample (100 mg.) in a flask flushed with carbon dioxide and to sweep the hydrogen into a nitrometer charged with potassium hydroxide solution.[4]

Zinc dust–acetic acid. Reduction. N-Nitroso-N-methylaniline is reduced to α-

$$\underset{0.73\ m.}{C_6H_5\overset{CH_3}{\overset{|}{N}}NO} + \underset{200\ g.}{Zn} + \underset{200\ ml.}{AcOH} + \underset{300\ ml.}{H_2O} \xrightarrow[52-56\%]{10-20^0,\ then\ 80^0} C_6H_5\overset{CH_3}{\overset{|}{N}}NH_2$$

methyl-α-phenylhydrazine by zinc in aqueous acetic acid;[5] the same conditions serve for reduction of nitroguanidine to aminoguanidine in 63–64% yield.[2]

16-Oximino-17-ketosteroids (1) are reduced by zinc dust in refluxing acetic acid to 16-keto-17β-hydroxysteroids.[6] The reaction proceeds through the 16ξ-amino-17-ketone (2).[7] A related reduction reported by Mauthner and by Windaus in 1903[8] is the reduction of 6-nitrocholesteryl acetate (5) to 6-ketocholestanyl acetate (6).

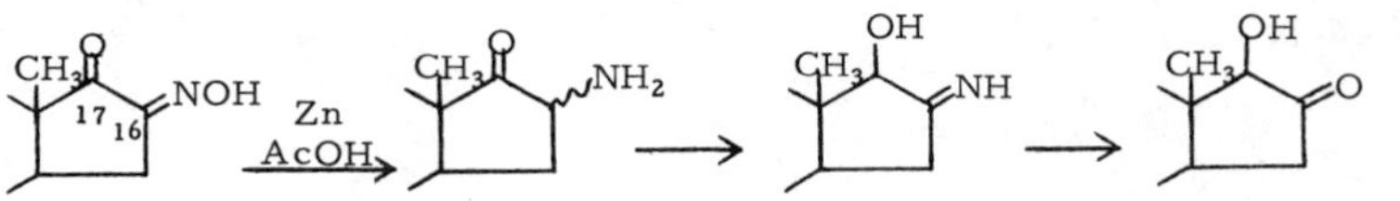
CH3
O
NOH
17
16
Zn
AcOH
NH2
OH
NH
O

(1) (2) (3) (4)

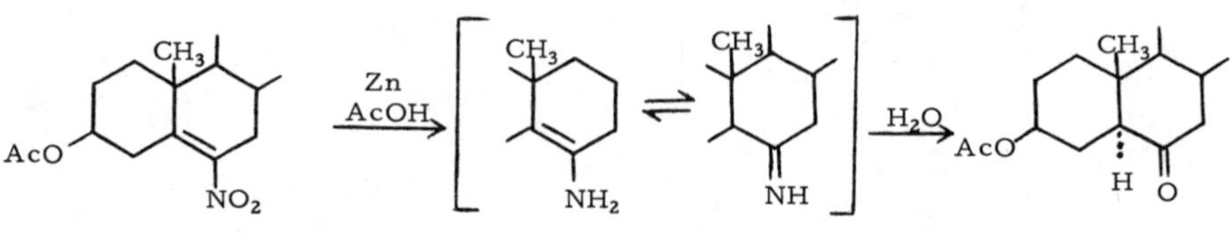
CH3
AcO
NO2
Zn
AcOH
NH2
NH
H2O
H
O

(5) (6)

The finding of Windaus[9] that a yellow product of oxidation of cholesterol later characterized as Δ^4-cholestene-3,6-dione (7) is reduced quantitatively by zinc and acetic acid to the colorless dione (8) provided useful evidence of structure (7).

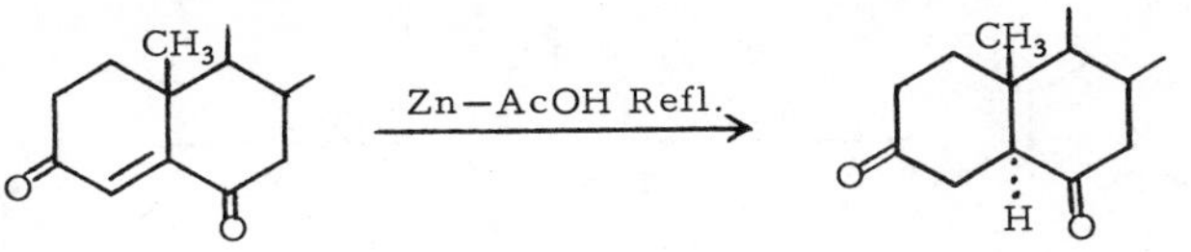

(7) (8)

7,11-Diketo-A[8]-steroids are similarly reduced to the saturated 7,11-diketones.[10] The ergosterol-derived yellow diene-7,11-dione (9) is reduced selectively at the double bond flanked by two carbonyl groups to give (11).[11] Actually the reaction involves *cis* addition (10) and isomerization through the enol to the more stable *trans* product (11).[12]

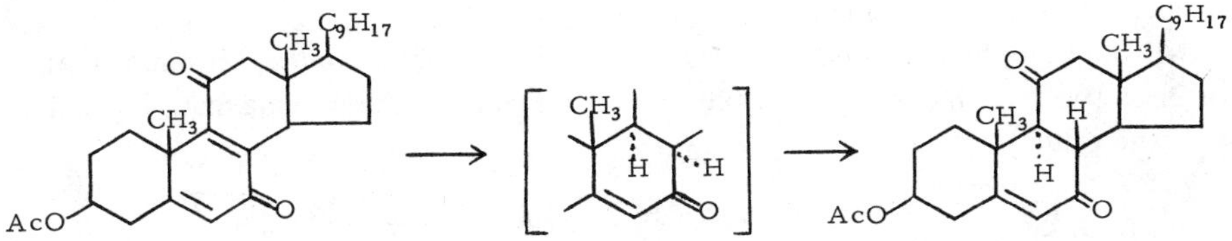

(9) (10) (11)

The conjugated bile acid dienone (12) is reduced by zinc dust and acetic acid to the nonconjugated enone (13).[13] Note, however, that similar reduction of β-

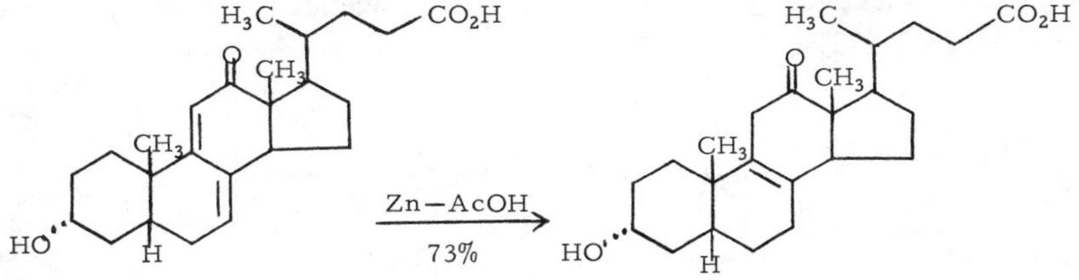

(12) (13)

cyperone (14) gives the conjugated enone (15).[14] In this case the same product was obtained on hydrogenation in the presence of Lindlar catalyst.

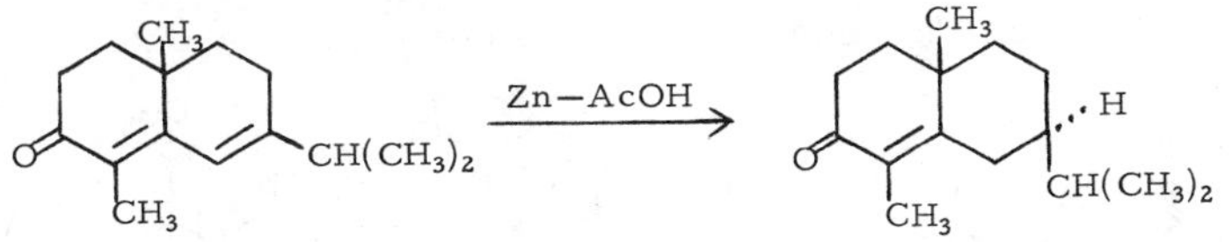

(14) (15)

Acetates of α-ketols or their vinylogs are subject to reductive elimination of the acetoxy group by treatment with zinc dust and acetic acid. Thus both 6α- and 6β-acetoxy-Δ^4-cholestene-3-one (16) are reduced smoothly to Δ^4-cholestene-3-one (17); the free alcohols are reduced somewhat less smoothly.[15] In the case of steroidal

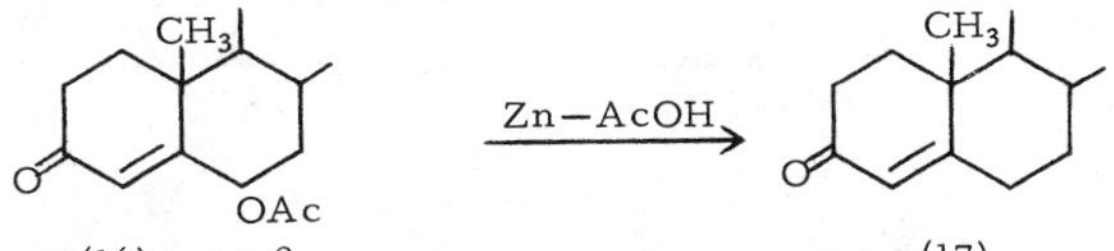

(16) α or β (17)

ring C ketol acetates, the configuration of the acetoxyl group is important, and elimination occurs in good yield only if the group is axial.[16] Deacetoxylation of a

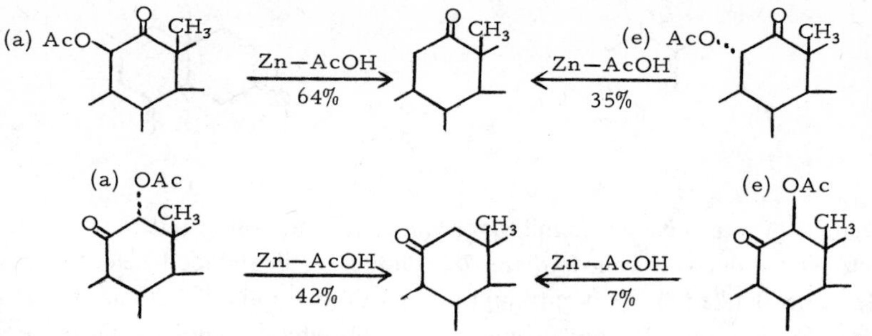
(a) AcO
CH3
Zn—AcOH
64%
CH3
(e) AcO
Zn—AcOH
35%
CH3
(a) OAc
CH3
Zn—AcOH
42%
CH3
Zn—AcOH
7%
(e) OAc
CH3

17α-acetoxy-20-ketosteroid was effected in 89% yield, whereas the yield from the epimeric 17β-acetoxy-20-ketosteroid was only 46%.[17]

3-Bromothiophene can be prepared by selective reduction of the α-bromine atoms of 2,3,5-tribromothiophene.[18] At the end of the reaction the organic material distills with the water.

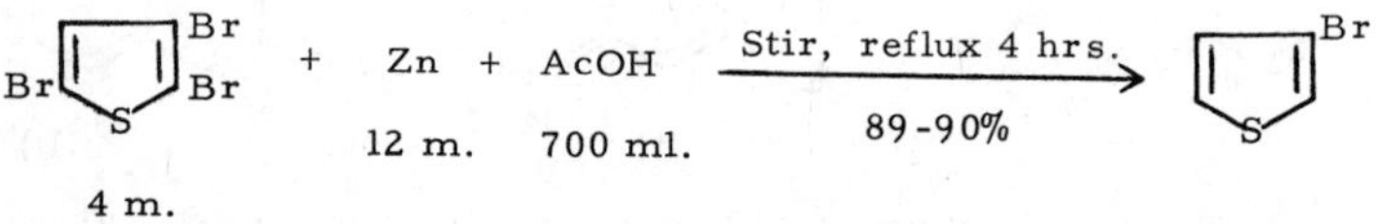
Br
Br
Br
S
+ Zn + AcOH
12 m. 700 ml.
4 m.
Stir, reflux 4 hrs.
89-90%
Br
S

Steroid ketoxides of types (1) and (3), both of which result on dichromate oxidation of Δ^7-cholestenyl acetate, are reduced by zinc dust and acetic acid to the corresponding conjugated ketones, (2) and (4).[19]

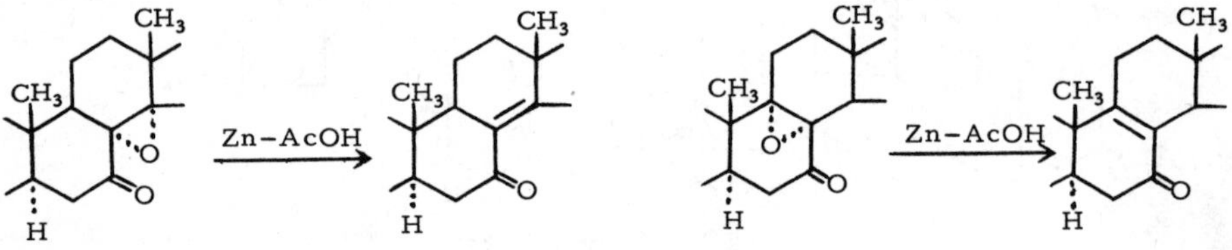
CH3
CH3
O
H
Zn—AcOH
CH3
CH3
H
CH3
CH3
O
H
Zn—AcOH
CH3
CH3
H

(1) (2) (3) (4)

Reduction of Δ^4-3-ketosteroids with 4,000-fold quantity of zinc dust in acetic acid affords Δ^3-5α-steroids and/or Δ^3-5β-steroids in small amounts.[19a]

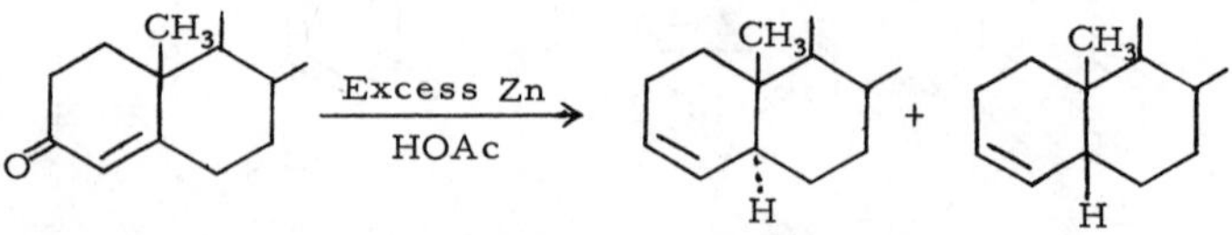
CH3
Excess Zn
HOAc
CH3
H
+
CH3
H

Dehalogenation, dehalohydrination. *See also* β-Chloroethyl ethyl ether. The debromination of cholesterol dibromide with zinc dust requires only a catalytic amount of acetic acid and once the reaction has started it proceeds with vigor, like

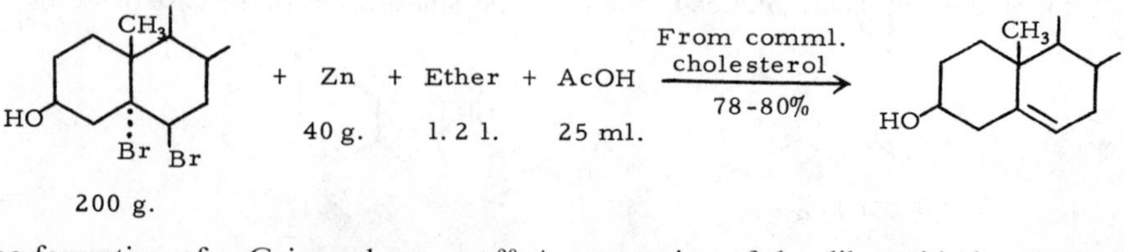
CH3
HO
Br Br
200 g.
+ Zn + Ether + AcOH
40 g. 1. 2 l. 25 ml.
From comml.
cholesterol
78-80%
CH3
HO

the formation of a Grignard reagent.[20] A suspension of the dibromide in ether and acetic acid is stirred vigorously and two or three 5-g. portions of zinc are added until boiling begins. An ice bath is raised into place during addition of the remaining metal. Workup involves addition of water, extraction with ether, and crystallization from ether–methanol. By a similar procedure 5α,6β-dibromocholestane-3-one affords Δ^5-cholestene-3-one in overall yield of 71–72% from cholesterol.[20]

The *erythro*-chlorohydrin formed as the major product of the reduction of desyl chloride is converted by zinc dust–acetic acid into *trans*-stilbene.[21] Some, if not all,

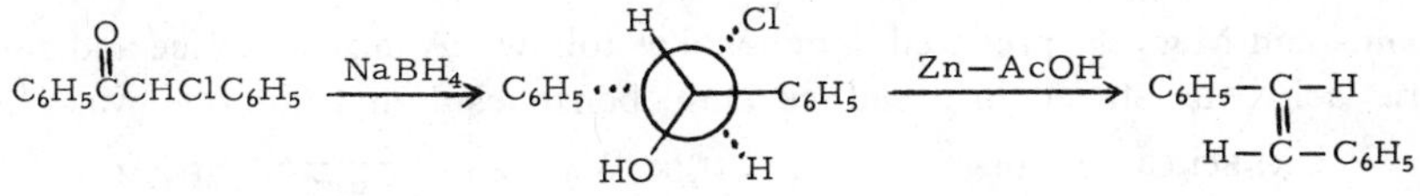
C₆H₅CCHClC₆H₅
O
NaBH₄
C₆H₅
H
Cl
C₆H₅
HO
H
Zn–AcOH
C₆H₅–C–H
H–C–C₆H₅

(+) or (-)

steroid *trans*-halohydrins are similarly converted into the corresponding olefinic products.[22] Cornforth[23] employed the combination: zinc dust–AcOH–NaI–NaOAc for conversion of epoxides to olefins via the intermediate iodohydrins.

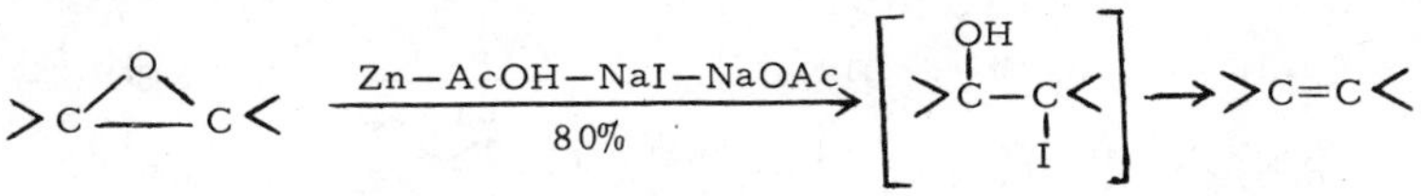
Zn–AcOH–NaI–NaOAc
80%
OH
I

Debromination of 1,2-bisbromomethylcyclooctatetraene to give 7,8-dimethylene-$\Delta^{1,3,5}$-cyclooctatriene, the exocyclic double bonds of which constitute a system highly reactive to dienophiles, was effected in practically quantitative yield with zinc dust in DMF at room temperature.[23a]

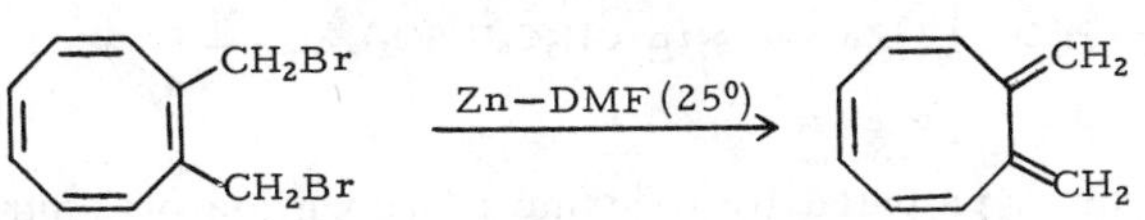
CH₂Br
CH₂Br
Zn–DMF (25°)
CH₂
CH₂

Zinc dust–mineral acid. Reduction. For the reduction of sebacoin to cyclodecanone, Cope and co-workers[24] added hydrochloric acid slowly to a stirred

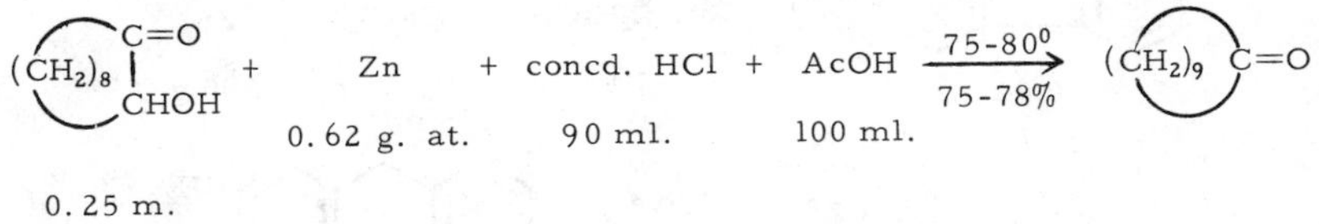
(CH₂)₈
C=O
CHOH
+ Zn + concd. HCl + AcOH
75-80°
75-78%
(CH₂)₉ C=O
0.62 g. at.
90 ml.
100 ml.
0.25 m.

mixture of sebacoin, Mallinckrodt technical grade zinc dust, and acetic acid, and kept the temperature at 75–80° by cooling. When Mallinckrodt AR grade zinc dust was used, the lower temperature of 50–55° had to be maintained.

Levene[25] reduced cetyl iodide to *n*-hexadecane by heating a stirred mixture of

$$\underset{1\ \text{m.}}{\underline{n}\text{-}C_{16}H_{33}I} + \underset{5\ \text{g. at.}}{Zn} + \underset{915\ \text{ml.}}{AcOH} + HCl \xrightarrow[85\%]{25\ \text{hrs.}} \underline{n}\text{-}C_{16}H_{34} + HI$$

the alkyl halide, zinc dust, and acetic acid on the steam bath and saturating the mixture with hydrogen chloride every 5 hrs.

The combination zinc dust–1 *N* sulfuric acid reduces fluorocurarine chloride (1) to the dihydro derivative (2).[26] The zinc dust is activated by etching with sulfuric acid, washing with water and ethanol, and drying. Further examples.[27]

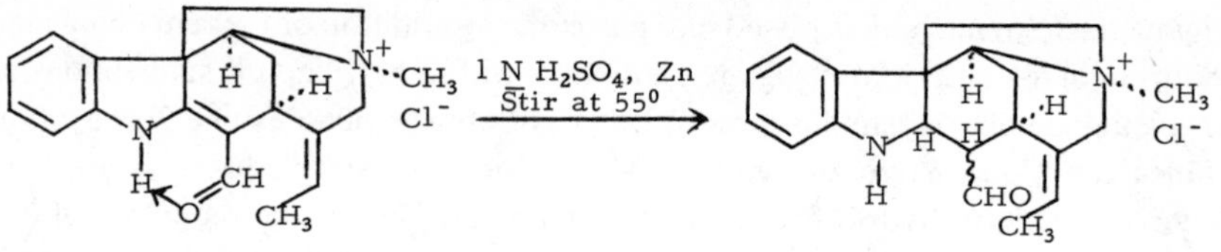
1 N H₂SO₄, Zn
Stir at 55°
CH₃
Cl⁻
CHO

(1) (2)

Adams and Marvel[28] prepared thiophenol as follows. A mixture of ice and concd. sulfuric acid was stirred in a salt-ice bath, benzenesulfonyl chloride was added

$$\underset{3.4\ \text{m.}}{C_6H_5SO_2Cl} + \underset{7.2\ \text{kg.}}{\text{Ice}} + \underset{2.4\ \text{kg.}}{\text{concd. } H_2SO_4} + \underset{1.2\ \text{kg.}}{Zn} \xrightarrow[91\%]{} C_6H_5SH$$

slowly, followed by zinc dust. When the initial exothermal reaction subsided, the mixture was heated with continued stirring for about 7 hrs. Steam distillation followed by distillation (71°/15 mm.) gave 340 g. of pure product.

Zinc dust, neutral medium. Reduction. On gradual addition of zinc dust to a

$$\underset{4.1\ \text{m.}}{C_6H_5NO_2} + \underset{4.7\ \text{m.}}{NH_4Cl} + \underset{8\ \text{l.}}{H_2O} + \underset{8.1\ \text{g. at.}}{Zn} \xrightarrow[62-68\%]{60-65^\circ} C_6H_5NHOH$$

stirred mixture of nitrobenzene and aqueous ammonium chloride the temperature rises, and reduction to phenylhydroxylamine proceeds.[29]

Reduction of *p*-toluenesulfonyl chloride with zinc dust in aqueous suspension and treatment of the zinc salt with sodium carbonate affords sodium *p*-toluenesulfinate.[30]

$$\underset{2.6\ \text{m.}}{2\ \underline{p}\text{-}CH_3C_6H_4SO_2Cl} + \underset{3\ \text{l.}}{H_2O} + \underset{6\ \text{g. at.}}{Zn} \longrightarrow (\underline{p}\text{-}CH_3C_6H_4SO_2)_2Zn \xrightarrow[64\%]{Na_2CO_3} \underline{p}\text{-}CH_3C_6H_4SO_2Na$$

2-Aminofluorene is prepared by refluxing a suspension of 2-nitrofluorene in a mixture of ethanol and aqueous calcium chloride solution.[31] The procedure is

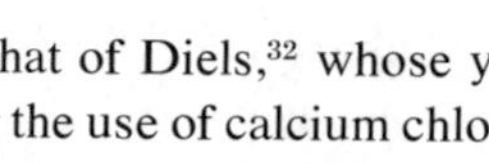
NO₂

0.14 m.

$$+ \underset{1\ \text{l.}}{78\%\ C_2H_5OH} + \underset{10\ \text{g. in 15 ml. } H_2O}{CaCl_2} + \underset{300\ \text{g.}}{Zn}$$

$$\xrightarrow[78-82\%]{\text{Refl. 2 hrs.}}$$

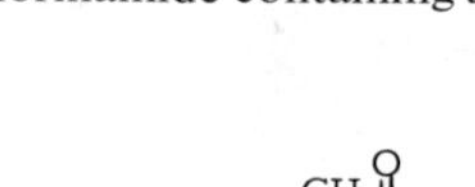
NH₂

essentially that of Diels,[32] whose yield was even better (85%); his paper records no rationale for the use of calcium chloride.

Tsuda[1] used zinc activated by method 1 above for reductive rearrangement of the steroid (1) to (2). Pyridine, piperidine, or dimethylformamide containing a little water can be used as solvent.

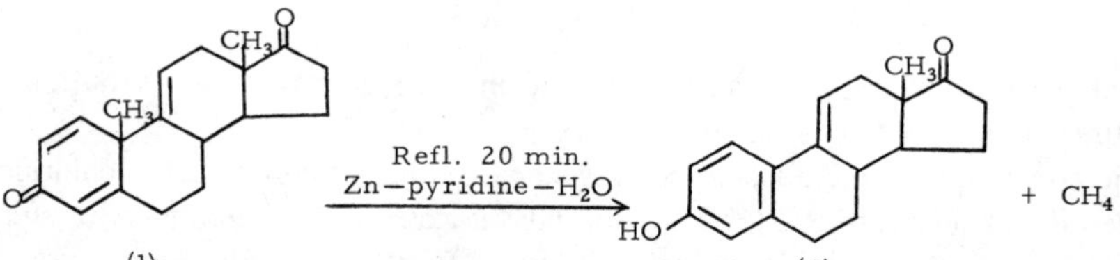
CH₃
CH₃
Refl. 20 min.
Zn–pyridine–H₂O
HO
CH₃
+ CH₄

(1) (2)

Liquid aromatic amines, for example aniline or isoquinoline,[33] often become badly discolored on storage as the result of air oxidation. Distillation from a pinch of zinc dust affords almost water-clear liquid. Kuhn and Winterstein[34] found that the violet carotinoid ester-acid bixin can be reduced very smoothly to its yellow dihydride with zinc dust in pyridine containing a small amount of acetic acid. Thus on addition of

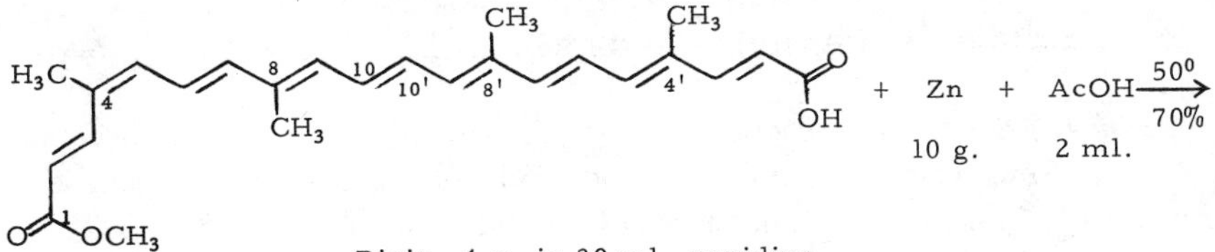

Bixin, 4 g. in 30 ml. pyridine

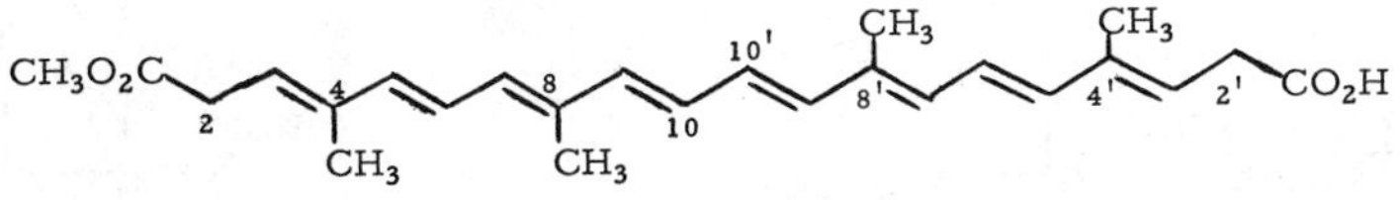

2, 2'-Dihydrobixin

zinc dust and acetic acid to a solution of bixin in pyridine at 50° and shaking, the violet color is discharged within a few seconds, and on filtration and dilution of the light brown filtrate with methanol the bright yellow zinc salt of 2,2′-dihydrobixin separates. Crystallization from benzene–acetic acid affords yellow plates of 2,2′-dihydrobixin. Bixin is not reduced by zinc dust–pyridine or by zinc dust–acetic acid, even on prolonged heating. The Zn–Py–AcOH method affords a route to dihydro derivatives of cyanine dyes[35] and of cyanidines[36] not obtainable by other methods.

Dehalogenation, dehydrohalogenation. An early procedure for debromination of 5α,6β-dibromocholestane-3-one without isomerization of the nonconjugated ketone initially formed involved reaction with zinc dust in hot ethanol,[37] but the reaction proceeds better in ether containing a catalytic amount of acetic acid (see above[20]).

The combination zinc dust–methanol serves for dechlorination of tetrachlorodifluoroethane (Genetron-131, m.p. 38–40°, Gen. Chem. Div., Allied Chem. and Dye Corp.) to 1,1-dichloro-2,2-difluoroethylene; a small amount of zinc chloride apparently functions as catalyst.[38]

$$CCl_3CClF_2 + Zn + CH_3OH + ZnCl_2 \xrightarrow[89-95\%]{} CCl_2{=}CF_2$$

0.6 m. in 50 ml. CH_3OH 42.2 g. 150 ml. 0.2 g.

Ordinary zinc dust is not suitable for debromination of the vinylic dibromide (1) to the cumulene (2), but can be activated by etching with sulfuric acid, drying, and treatment with a small amount of iodine.[39]

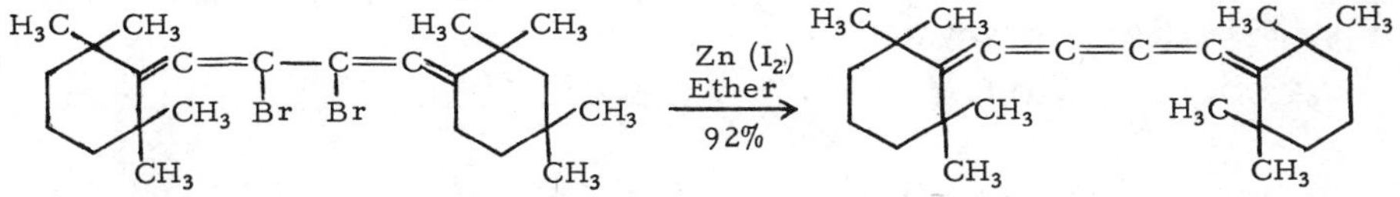

(1) (2)

In a synthesis of 1,4-pentadiene by the method formulated[40] the last step involves an elimination of a bromohydrin ether. This reaction is effected with zinc in *n*-butanol containing a small amount of anhydrous zinc chloride at a temperature at which 1,4-pentadiene distills (b.p. 26–27°/740 mm.).

$$CH_3CHO + C_2H_5OH \xrightarrow{HCl} CH_3CH(Cl)OC_2H_5 \xrightarrow{Br_2} BrCH_2CH(Br)OC_2H_5$$

$$\xrightarrow{CH_2{=}CHCH_2MgBr} CH_2{=}CHCH_2CH(OC_2H_5)CH_2Br \xrightarrow[72\text{-}76\%]{Zn(ZnCl_2),\ n\text{-}BuOH} CH_2{=}CHCH_2CH{=}CH_2$$

Serini reaction.[41] This reaction consists in refluxing a *vic*-diol monoacetate of type (1) or (3) with zinc dust in toluene and results in elimination of acetic acid with formation of a 20-ketone (2, 4) inverted at C_{17}. Discussion of mechanism.[42]

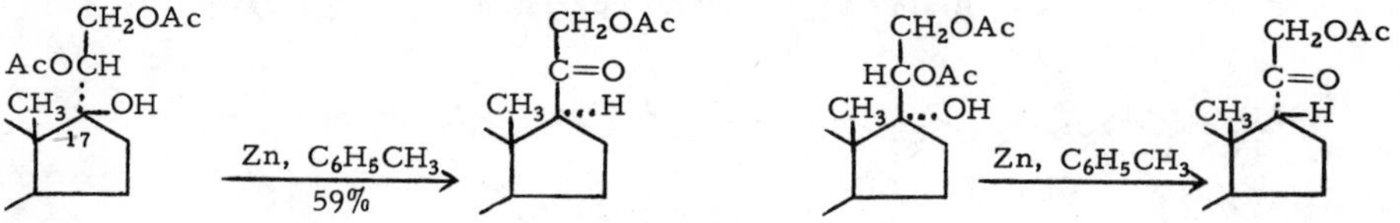

(1) (2) (3) (4)

Zinc dust distillation. See review.[43]

Zinc dust and alkali. Reduction. Beautifully fluorescent anthracene of high purity can be prepared[44] by refluxing a mixture of anthrone, zinc dust, dilute alkali, and a

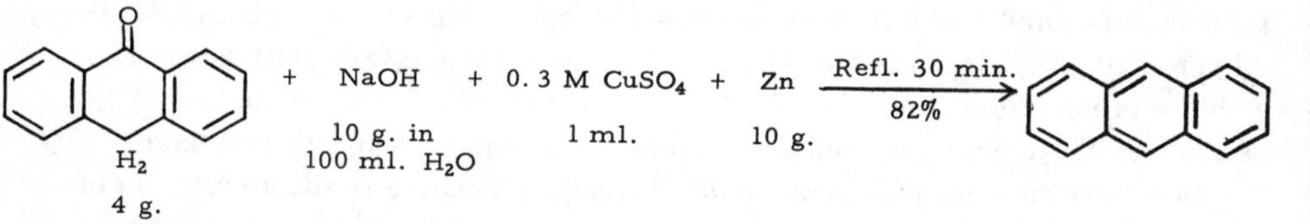

trace of copper sulfate briefly, collecting a grey cake of hydrocarbon, zinc, and zinc oxide and washing it with methanol, and dissolving the organic material in hot benzene for crystallization. Aroylbenzoic acids resulting from Friedel-Crafts acylations with phthalic anhydride, for example (1), are reduced in the same way in high yield.[45]

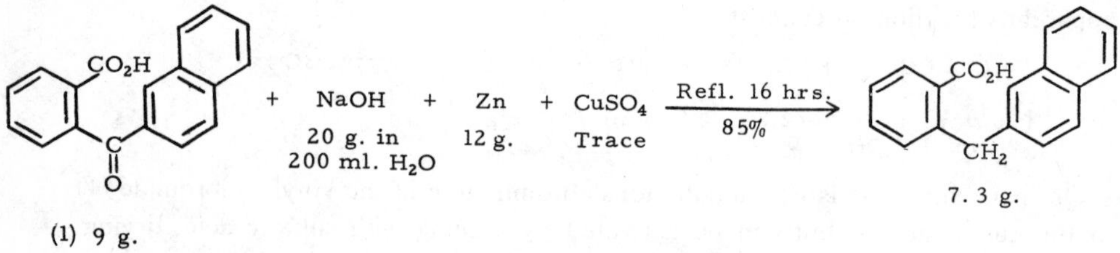

Reduction of benzophenone by a similar procedure but in an alcoholic medium stops at the stage of benzhydrol.[46]

$$\underset{200\ g.}{C_6H_5COC_6H_5} + \underset{200\ g.}{NaOH} + \underset{200\ g.}{Zn} + \underset{2\ l.}{95\%\ EtOH} \xrightarrow[96\text{-}97\%\ crude]{Exotherm.\ to\ 70^0} (C_6H_5)_2CHOH$$

The *Organic Syntheses* procedure[47] for reduction of nitrobenzene to azobenzene calls for refluxing for 20 hrs. a mixture of the nitro compound (208 ml.), zinc dust (265 g.), methanol (2.5 l.), and a solution of 325 g. of sodium hydroxide in 750 ml. of water. In developing a procedure suitable for use as a student experiment,[48] one of us used triethylene glycol to permit operation at a higher temperature, omitted the water,

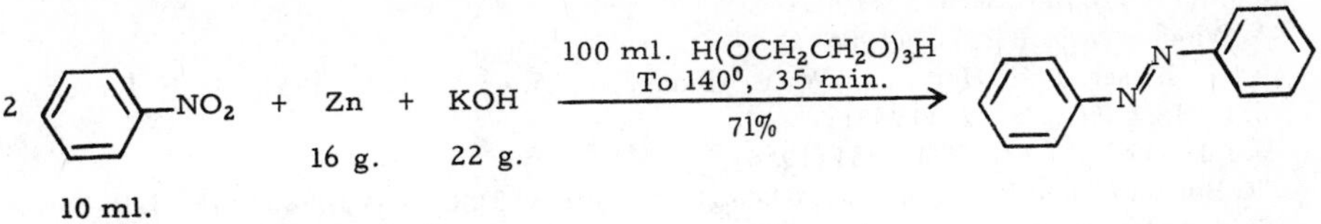

and replaced NaOH by KOH, since the latter is often a more efficient alkali, and so reduced the reaction period to 35 minutes. Gentle heating of the mixture initiates a strongly exothermal reaction which, if checked first at 80–85°, can be controlled and completed in 15 minutes at 135–140°. Surprisingly, very little zinc is consumed. Azobenzene can be obtained in the same yield from nitrobenzene, potassium hydroxide, and triethylene glycol without any zinc, but the period of heating must be extended from 15 minutes to 1 hour. The reducing agent is treithylene glycol, which is transformed into aldehydic material (formation of 2,4-dinitrophenylhydrazone). Zinc is beneficial in accelerating the reaction if used in adequate quantity, but it is not effective in truly catalytic amount.

Martin[49] prepared *o*-phenylenediamine by stirring a hot mixture of *o*-nitroaniline, aqueous alkali, and ethanol and adding zinc dust in portions at a rate sufficient to

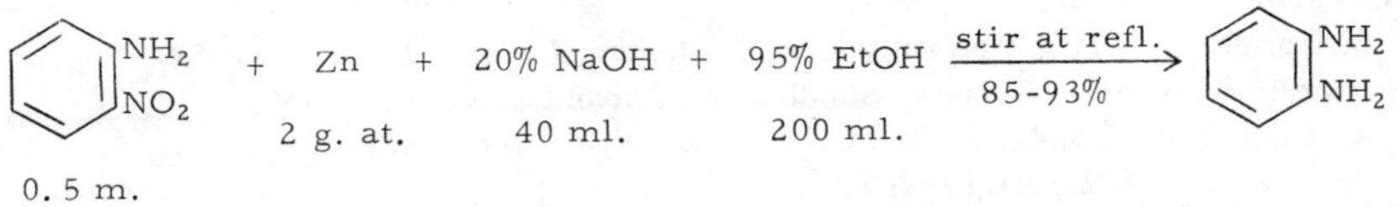

maintain gentle refluxing. After continued stirring for about 1 hr. the color changed from deep red to nearly colorless, and workup afforded material melting at 97–100° (pure, m.p. 99–101°). Use of excess zinc did not alter the yield.

The reduction of ketoximes to primary amines by zinc dust often proceeds better with concd. ammonia solution than with acid.[50] The reaction proceeds particularly well with diaryl ketoximes (80–90% yield); with aryl alkyl ketoximes yields are lower (50–60%). The oxime, zinc, ammonium acetate, and concd. ammonia solution are refluxed in ethanol.

Windaus[51] found that ergosterol peroxide is reduced smoothly to "Triol-I" by zinc dust in refluxing alcoholic alkali.

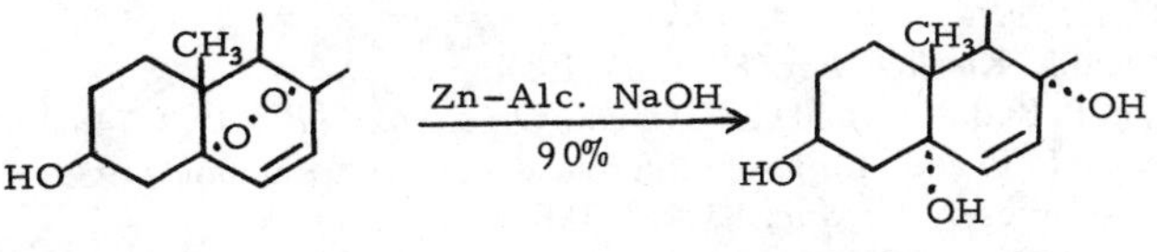

See also Zinc amalgam; Zinc copper couple; Zinc, for Reformatsky.

[1]K. Tsuda, E. Ohki, and S. Nozoe, *J. Org.*, **28**, 783 (1963)
[2]R. L. Shriner and F. W. Neumann, *Org. Syn., Coll. Vol.*, **3**, 73 (1955)
[3]R. L. Frank and P. V. Smith, *ibid.*, **3**, 410 (1955)
[4]L. Gattermann, "Practical Organic Chemistry," translation of 3rd ed., 390.
[5]W. W. Hartman and L. J. Roll, *Org. Syn., Coll. Vol.*, **2**, 418 (1943)
[6]F. H. Stodola, E. C. Kendall, and B. F. McKenzie, *J. Org.*, **6**, 841 (1941); F. H. Stodola and E. C. Kendall, *ibid.*, **7**, 336 (1942); N. S. Leeds, D. K. Fukushima, and T. F. Gallagher, *Am. Soc.*, **76**, 2943 (1954)

[7]A. Hassner and A. W. Coulter, *Steroids*, **4**, 281 (1964)
[8]J. Mauthner and W. Suida, *Monatsh.*, **15**, 85 (1894); **24**, 648 (1903); A. Windaus, Habilitationsschrift (1903); *Ber.*, **36**, 3752 (1903)
[9]A. Windaus, *Ber.*, **39**, 2249 (1906)
[10]L. F. Fieser, J. E. Herz, and W.-Y. Huang, *Am. Soc.*, **73**, 2397 (1951); L. F. Fieser and J. E. Herz, *ibid.*, **75**, 121 (1953)
[11]J. Elks *et al.*, *J. Chem. Soc.*, 451 (1954)
[12]R. Budziarek and F. S. Spring, *J. Chem. Soc.*, 956 (1953); C. S. Barnes and D. H. R. Barton, *ibid.*, 1419 (1953)
[13]L. F. Fieser and S. Rajagopalan; E. Wilson and M. Tishler, *Am. Soc.*, **73**, 4133 (1951)
[14]R. Howe and F. J. McQuillin, *J. Chem. Soc.*, 2670 (1956)
[15]L. F. Fieser, *Am. Soc.*, **75**, 4377 (1953)
[16]R. S. Rosenfeld and T. F. Gallagher, *Am. Soc.*, **77**, 4367 (1955)
[17]R. S. Rosenfeld, *Am. Soc.*, **79**, 5540 (1957)
[18]S. Granowitz and T. Raznikiewicz, *Org. Syn.*, **44**, 9 (1964)
[19]L. F. Fieser, *Am. Soc.*, **75**, 4395 (1953); H. Heusser, G. Saucy, R. Anliker, and O. Jeger, *Helv.*, **35**, 2090 (1952)
[19a]J. McKenna, J. K. Norymberski, and R. D. Stubbs, *J. Chem. Soc.*, 2502 (1959); A. Crastes de Paulet, and J. Bascoul, *Bull. soc.*, 939 (1966)
[20]L. F. Fieser, *Org. Syn.*, *Coll. Vol.*, **4**, 195 (1963)
[21]*Org. Expts.*, 219
[22]S. Mori, *J. Chem. Soc. Japan*, **70**, 303 (1949); L. F. Fieser and R. Ettorre, *Am. Soc.*, **75**, 1700 (1953); L. F. Fieser and X. A. Dominguez, *ibid.*, **75**, 1704 (1953); L. F. Fieser and W.-Y. Huang, *ibid.*, **75**, 4837 (1953)
[23]J. W. Cornforth, R. H. Cornforth, and K. K. Mathew, *J. Chem. Soc.*, 112 (1959)
[23a]J. A. Elix, M. V. Sargent, and F. Sondheimer, *Chem. Comm.*, 508 (1966)
[24]A. C. Cope, J. W. Barthel, and R. D. Smith, *Org. Syn.*, *Coll. Vol.*, **4**, 218 (1963)
[25]P. A. Levene, *ibid.*, **2**, 320 (1943)
[26]W. von Philipsborn, K. Bernauer, H. Schmid, and P. Karrer, *Helv.*, **42**, 461 (1959)
[27]P. N. Edwards and G. F. Smith, *J. Chem. Soc.*, 152 (1961); C. Djerassi *et al.*, *Tetrahedron Letters*, 235 (1962); J. P. Kutney, R. T. Brown, and E. Piers, *Am. Soc.*, **86**, 2286 (1964)
[28]R. Adams and C. S. Marvel, *Org. Syn.*, *Coll. Vol.*, **1**, 504 (1941)
[29]O. Kamm, *ibid.*, **1**, 445 (1941)
[30]F. C. Whitmore and F. H. Hamilton, *ibid.*, **1**, 492 (1941)
[31]W. E. Kuhn, *ibid.*, **2**, 447 (1943)
[32]O. Diels, *Ber.*, **34**, 1758 (1901)
[33]J. Weinstock and V. Boekelheide, *Org. Syn.*, *Coll. Vol.*, **4**, 642 (1963)
[34]R. Kuhn and A. Winterstein, *Ber.*, **65**, 646 (1932)
[35]Idem, *ibid.*, **65**, 1737 (1932)
[36]Idem, *ibid.*, **65**, 1742 (1932)
[37]A. Butenandt and J. Schmidt-Thomé, *Ber.*, **69**, 882 (1936)
[38]J. C. Sauer, *Org. Syn.*, *Coll. Vol.*, **4**, 268 (1963)
[39]F. Bohlmann and K. Kieslich, *Ber.*, **87**, 1363 (1954)
[40]O. Grummit, E. P. Budewitz, and C. C. Chudd, *Org. Syn.*, *Coll. Vol.*, **4**, 748 (1963)
[41]For a review, see L. F. and M. Fieser, "Steroids," 628–631, Reinhold (1959)
[42]T. Goto and L. F. Fieser, *Am. Soc.*, **83**, 251 (1961)
[43]Z. Valenta, *Techniques of Organic Chemistry*, **XI**, Part II, 643 (1963)
[44]*Org. Expts.*, 200
[45]L. F. Fieser and E. B. Hershberg, *Am. Soc.*, **62**, 49 (1940)
[46]F. Y. Wiselogle and H. Sonneborn, III, *Org. Syn.*, *Coll. Vol.*, **1**, 90 (1941)
[47]H. E. Bigelow and D. B. Robinson, *Org. Syn.*, *Coll. Vol.*, **3**, 103 (1955)
[48]*Org. Expts.*, 190
[49]E. L. Martin, *Org. Syn.*, **2**, 501 (1943)
[50]J. C. Jochims, *Monatsh.*, **94**, 677 (1963)
[51]A. Windaus and O. Linsert, *Ann.*, **465**, 148 (1928); A. Windaus, W. Bergmann, and A. Lüttringhaus, *ibid.*, **472**, 195 (1929)

Zinc, for Reformatsky. The form of zinc used and the procedure for its activation vary considerably. Fieser and Johnson[1] obtained highly active metal by immersing ***30-mesh zinc*** in hot (100°) concd. sulfuric acid containing a few drops of concd. nitric acid. After about 10 min. the zinc surface became very bright and the acid was largely decanted and water was added. A vigorous reaction with the dilute acid was allowed to proceed for a few minutes and the metal was then collected, washed thoroughly with water and then with acetone, and dried. A suspension of this metal in a solution of the ketone I and methyl bromoacetate was heated nearly

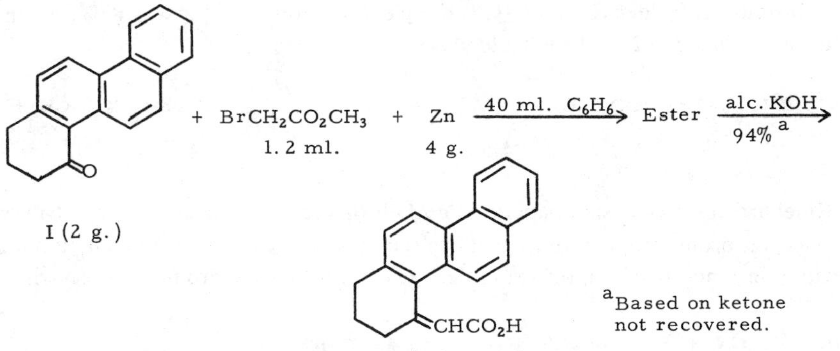
+ $BrCH_2CO_2CH_3$ + Zn
1.2 ml.
4 g.
40 ml. C_6H_6
Ester
alc. KOH
94% [a]
$CHCO_2H$

I (2 g.)

II (+ isomer)

[a]Based on ketone not recovered.

to boiling and treated with a crystal of iodine, when the reaction started and soon became vigorous. After about an hour, and before the reaction was complete, the zinc became completely coated over with an oily gum. At this point the mixture was worked up, the ester was hydrolyzed, and unchanged starting ketone was separated from acid II, a mixture of double-bond isomers which was to be hydrogenated.

For effecting reaction of propargyl bromide with methoxybutanone to give propargylmethyl-β-methoxylethylcarbinol, Oroshnik *et al.*[2] heated 200 g. of acid-washed and dried ***zinc dust*** gently with 0.5 g. of iodine in a loosely stoppered Erlenmeyer until iodine vapors were no longer perceptible, as suggested in a patent,[3] which states also that a trace of copper ethyl acetoacetate has a beneficial effect. However, Dr. Oroshnik later informed us[4] that subsequent experiences showed that

$$\underset{0.42\text{ m.}}{HC{\equiv}CCH_2Br} + \underset{0.82\text{ m.}}{CH_3\overset{\text{O}}{\overset{\|}{C}}CH_2CH_2OCH_3} + \underset{0.92\text{ m.}}{Zn(I_2)} + \underset{15\text{ ml.}-100\text{ ml.}}{C_6H_6\text{-Ether}} + \underset{0.2\text{ g.}}{\text{Cu-enolate}}$$

$$\xrightarrow[34\%]{\text{Refl.; } H_2O(H^+)} HC{\equiv}CCH_2\underset{\text{OH}}{\overset{CH_3}{C}}CH_2CH_2OCH_3$$

neither the iodine-activated zinc dust nor the copper derivative was necessary and that smooth reactions were obtained with a variety of ketones with reagent grade granulated zinc. In the example formulated the ketone was taken in twice the theoretical amount to make up for loss of ketone by polymerization under the reaction conditions.

For the Reformatsky reaction of benzaldehyde with ethyl bromoacetate to form ethyl β-phenyl-β-hydroxypropionate, Hauser and Breslow[5] purified ***zinc dust*** by washing it rapidly with dilute sodium hydroxide solution, water, dilute acetic acid, water, ethanol, acetone, and ether. It was then dried in a vacuum over at 100°.

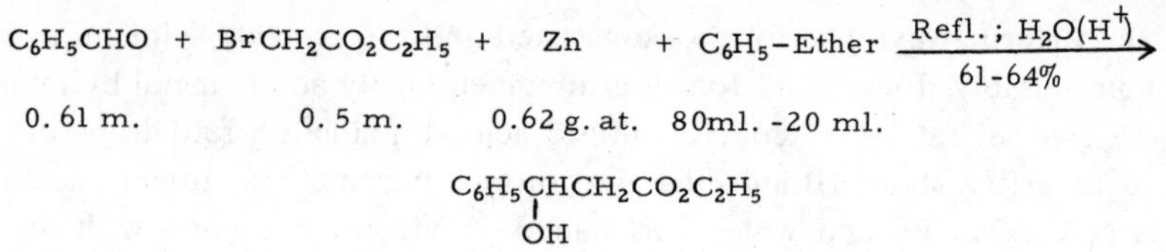

$$C_6H_5CHO \ + \ BrCH_2CO_2C_2H_5 \ + \ Zn \ + \ C_6H_5-Ether \xrightarrow[61-64\%]{Refl.\ ;\ H_2O(H^+)}$$

0. 61 m.   0. 5 m.   0.62 g. at.   80ml. -20 ml.

$$C_6H_5CHCH_2CO_2C_2H_5$$
$$\underset{OH}{|}$$

For the synthesis of ethyl 2,4-dimethyl-3-hydroxyoctanoate from 2-methylhexanal and ethyl α-bromopropionate, Rinehart and Perkins[6] used freshly sandpapered *zinc foil* which had been cut into narrow strips and rolled loosely. With a ratio of zinc:bromoester:aldehyde of 3:3:1, the yield of hydroxy ester was 87%; when the ratio was reduced to 2:3:1 the yield was lowered to 68%.

$$n\text{-}C_4H_9\underset{\underset{CH_3}{|}}{C}HCHO \ + \ Br\underset{\underset{CH_3}{|}}{C}HCO_2C_2H_5 \ + \ Zn \ + \ C_6H_6 \xrightarrow[87\%]{Refl.\ ;\ H_2SO_4} n\text{-}C_4H_9\underset{\underset{CH_3}{|}}{C}H\overset{\overset{OH}{|}}{C}H\underset{\underset{CH_3}{|}}{C}HCO_2C_2H_5$$

0. 5 m.   1.5 m.   1.5 g. at.   1 l.

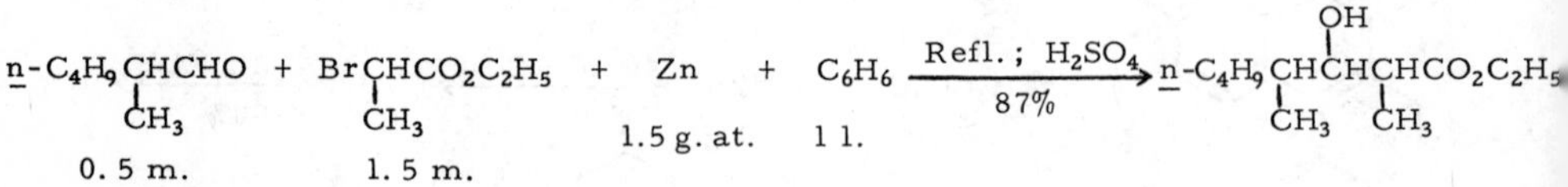

Rinehart[7] also used sandpapered *zinc foil* for the reaction of *n*-capronitrile with *sec*-butyl α-bromopropionate to produce *sec*-butyl α-*n*-caproylpropionate, but added a catalytic amount of cupric bromide.[8] The ratio of zinc:bromo ester:nitrile was 1.5:1.5:1.

$$CH_3(CH_2)_4CN \ + \ Br\underset{\underset{CH_3}{|}}{C}HCO_2\underset{\underset{C_2H_5}{|}}{C}HCH_3 \ + \ Zn \ + \ CuBr_2 \ + \ C_6H_6 \xrightarrow[50-58\%]{Refl.\ ;\ H_2O(H^+)}$$

0. 4 m.   0. 6 m.   0. 60   0. 4 g.   9 00 ml.

$$CH_3(CH_2)_4\overset{\overset{O}{\parallel}}{C}\underset{\underset{CH_3}{|}}{C}HCO_2\underset{\underset{C_2H_5}{|}}{C}HCH_3$$

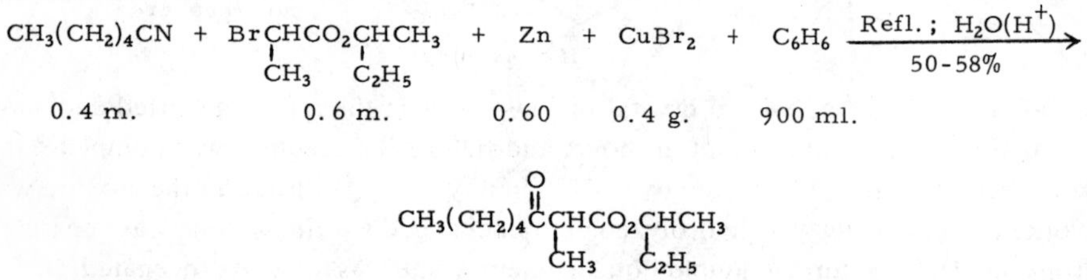

[1]L. F. Fieser and W. S. Johnson, *Am. Soc.*, **62**, 575 (1940); see also W. R. Vaughan, S. C. Bernstein, and M. E. Larber, *J. Org.*, **30**, 1790 (1965)

[2]W. Oroshnik, G. Karmas, and A. D. Mebane, *Am. Soc.*, **74**, 3807 (1952)

[3]Roche Products, Ltd., British patent 617, 482 (1949) [*C.A.*, **43**, 5798 (1949)]

[4]W. Oroshnik, private communication

[5]C. R. Hauser and D. S. Breslow, *Org. Syn., Coll. Vol.*, **3**, 408 (1955)

[6]K. L. Rinehart, Jr., and E. G. Perkins, *ibid.*, **4**, 444 (1963)

[7]K. L. Rinehart, Jr., *ibid.*, **4**, 120 (1963)

[8]Dr. Rinehart replied to our query as follows: "I doubt that the cupric bromide had a great deal to do with the yield, and could be dispensed with; sandpapering the zinc, however, it is a useful procedure for obtaining active zinc foil."

Zinc, other forms.

Debromination. In a procedure for the isolation of linoleic acid from the split acids of sunflower-seed oil,[1] bromination of the acid mixture affords, after recrystallization, 72–76% of pure tetrabromostearic acid. A solution of this material in

$$CH_3(CH_2)_4\overset{\overset{H}{|}}{\underset{\underset{Br}{|}}{C}}-\overset{\overset{Br}{|}}{\underset{\underset{H}{|}}{C}}CH_2\overset{\overset{H}{|}}{\underset{\underset{Br}{|}}{C}}-\overset{\overset{Br}{|}}{\underset{\underset{H}{|}}{C}}(CH_2)_7CO_2H \ + \ Zn \ + \ abs.\ C_2H_5OH \xrightarrow{Refl.\ ;\ HCl\ and\ refl.}$$

26-30 g.   30 g.   85 ml.

$$CH_3(CH_2)_4\overset{\overset{H}{|}}{C}=\overset{\overset{H}{|}}{C}CH_2\overset{\overset{H}{|}}{C}=\overset{\overset{H}{|}}{C}(CH_2)_7CO_2C_2H_5$$

12-15 g.

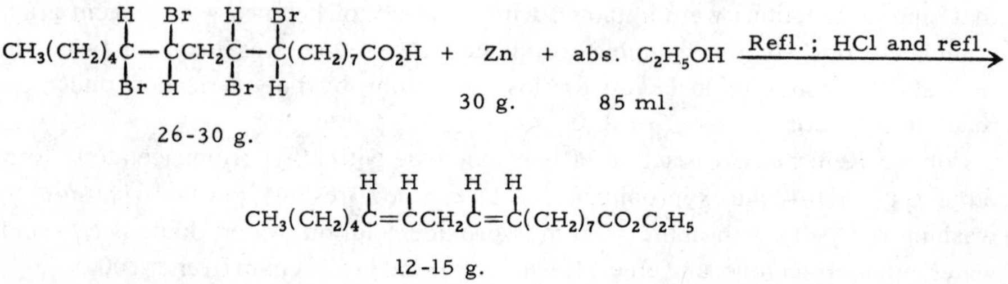

absolute ethanol is treated with 30 g. of 20-mesh granulated zinc, the mixture is warmed gently until an exothermal reaction sets in, and eventually refluxed. A solution of hydrogen chloride in ethanol is added and the mixture refluxed to promote esterification, and ethyl linoleate is isolated and distilled at 175°/2.5 mm.

α-Bromo-isobutyryl bromide is converted into dimethylketene by reaction with ***zinc turnings*** in ethyl acetate.[2] The reaction is carried out at 300 mm. pressure in

$$\underset{0.48\ \text{m.}}{(CH_3)_2C(Br)-C(Br)=O} + \underset{0.61\ \text{g. at.}}{Zn} + \underset{300\ \text{ml.}}{CH_3CO_2C_2H_5} \xrightarrow[46\text{-}54\%]{} \underset{9\text{-}10\%\ \text{solution in}\ CH_3CO_2C_2H_5}{(CH_3)_2C=C=O}$$

a nitrogen atmosphere. Dimethylketene distills along with ethyl acetate and is obtained in 9–10% solution in ethyl acetate.

Zinc activated with copper was found effective for the dehalogenation of trichloroacetyl bromide to dichloroketene:[3]

$$CCl_3-\overset{O}{\overset{\|}{C}}-Br + Zn \xrightarrow[60\text{-}70\%]{} Cl_2C=C=O$$

A solution of 4 g. of hydrated copper sulfate in water was stirred with 60 g. of zinc dust for 2 hrs., and the couple was collected and dried by washing with acetone.

With tetrahydrofurane as solvent, ***zinc turnings*** made from electrolytic zinc can be used to form reactive organometallic reagents from allyl bromide and propargyl bromide similar to Grignard reagents.[4] For example, reaction of allyl bromide with an equivalent amount of sinc in THF, followed by addition of acetone, affords dimethylallylcarbinol in the good yield.

$$CH_2=CHCH_2Br + Zn \xrightarrow[68\%]{(CH_3)_2CO} CH_2=CHCH_2C(CH_3)(OH)-CH_3$$

[1]J. W. McCutcheon, *Org. Syn., Coll. Vol.*, **3**, 526 (1955)
[2]C. W. Smith and D. G. Norton, *ibid.*, **4**, 348 (1963)
[3]W. T. Brady, H. G. Liddell, and W. L. Vaughn, *J. Org.*, **31**, 626 (1966)
[4]M. Gaudemar, *Bull. Soc.*, 974 (1962)

Zinc amalgam, Zn(Hg).

Clemmensen reduction. This method for the reduction of aryl alkyl ketones to the corresponding hydrocarbons, for example of acetophenone to ethylbenzene,

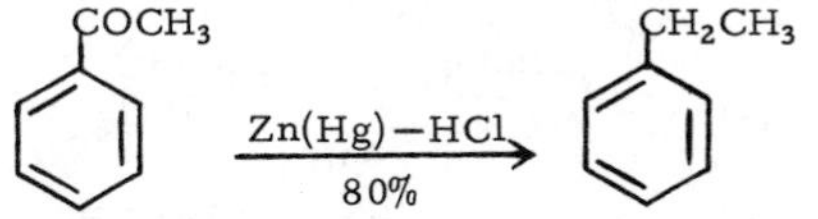
COCH3
CH2CH3
Zn(Hg)−HCl
80%

was introduced by Clemmensen,[1] who was born and trained in Denmark and made a career as an industrial chemist in New Jersey. As originally applied, the method consisted in refluxing the ketone with amalgamated zinc and concd. hydrochloric acid. The scope and limitations of the method are discussed in reviews.[2]

Martin[3] introduced a generally useful improvement consisting in adding a layer of toluene, which probably is beneficial partly because it keeps undissolved material out of contact with the metal and partly because in the two-phase system reduction occurs at such a high dilution that polymolecular reactions are inhibited. For the reduction of 50 g. of β-benzoylpropionic acid, Martin amalgamated 120 g. of mossy zinc by stirring a mixture with 12 g. of mercuric chloride, 6 ml. of concd. hydrochloric acid, and 150 ml. of water for 5 minutes and decanting the aqueous solution. After addition of 75 ml. of water, 175 ml. of concd. hydrochloric acid, 100 ml. of toluene, and 50 g. of the keto acid the mixture was refluxed briskly for 24–30 hrs., with addition of a 50 ml. portion of concd. hydrochloric acid every 6 hrs. The mixture

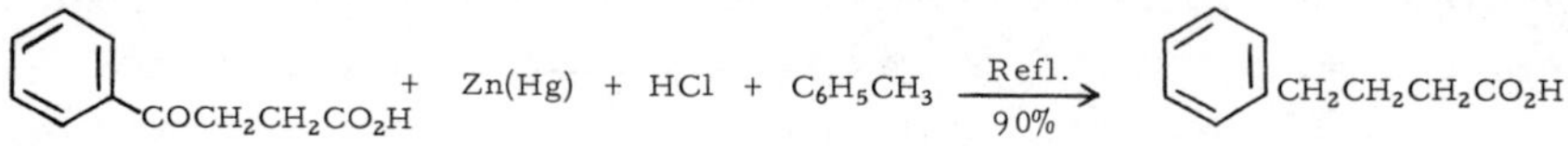
COCH₂CH₂CO₂H
+ Zn(Hg) + HCl + C₆H₅CH₃
Refl.
90%
CH₂CH₂CH₂CO₂H

was cooled, and the aqueous solution diluted and extracted with ether. The combined ether-toluene mixture was washed, dried, and evaporated, and the γ-phenylbutyric acid distilled; yield 90% of pure acid, m.p. 46–48°. In a parallel run by Martin without addition of toluene the yield was 72–78%.

The Martin modification is generally applicable but may require slight adjustment. Thus if the ketone is very sparingly soluble in water, addition of a small amount of acetic acid may be required (3–5 ml. per 50 g. of ketone). A substance containing phenolic methoxyl groups may suffer some demethylation, and remethylation may be required.

Sherman[4] introduced two improvements which were found very useful for the reduction of keto acids obtained by succinoylation of aromatics.[5] The zinc, prior to amalgamation, is melted in a casserole and poured into a large volume of water. The reduction is carried out according to Martin except that the reaction flask, heated conveniently with a Glass-Col mantle, is provided with a Hershberg stirrer to effect vigorous agitation. A typical charge is: 93 g, of β-naphthoylpropionic acid, 185 ml. of sulfur-free toluene, 41 ml. of acetic acid, 140 ml. of water, 185 g. of freshly poured zinc amalgamated with 16.3 g. of mercuric chloride, 322 ml. of concd. hydrochloric acid (added cautiously). Fresh portions of acid (70 ml.) are added at the end of the first, second, and third hour (yield 87%). The Sherman improvements reduced the reaction time from 36–40 hrs. to 4–6 hrs. without material changes in yield.

In the case of phenolic ketones the preferred procedure utilizes dilute ethanol as solvent. Read and Wood[6] stirred the reactants formulated under reflux until a

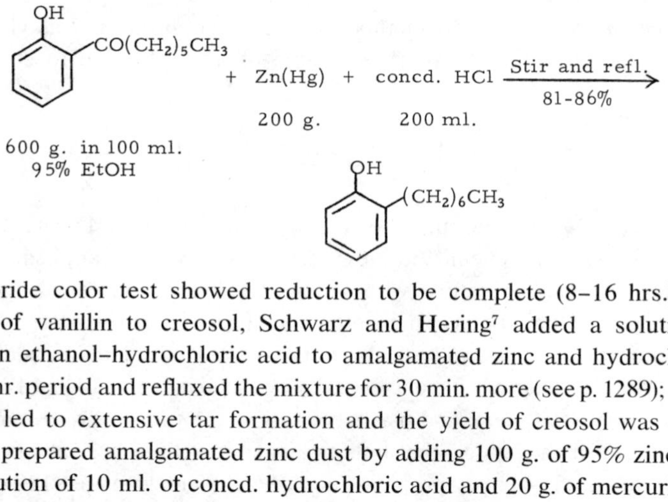
OH
CO(CH₂)₅CH₃
+ Zn(Hg) + concd. HCl
Stir and refl.
81-86%
200 g.
200 ml.
600 g. in 100 ml.
95% EtOH
OH
(CH₂)₆CH₃

ferric chloride color test showed reduction to be complete (8–16 hrs.). For the reduction of vanillin to creosol, Schwarz and Hering[7] added a solution of the aldehyde in ethanol–hydrochloric acid to amalgamated zinc and hydrochloric acid over an 8-hr. period and refluxed the mixture for 30 min. more (see p. 1289); the Martin procedure led to extensive tar formation and the yield of creosol was only 49%.

Caesar[8] prepared amalgamated zinc dust by adding 100 g. of 95% zinc dust to a stirred solution of 10 ml. of concd. hydrochloric acid and 20 g. of mercuric chloride

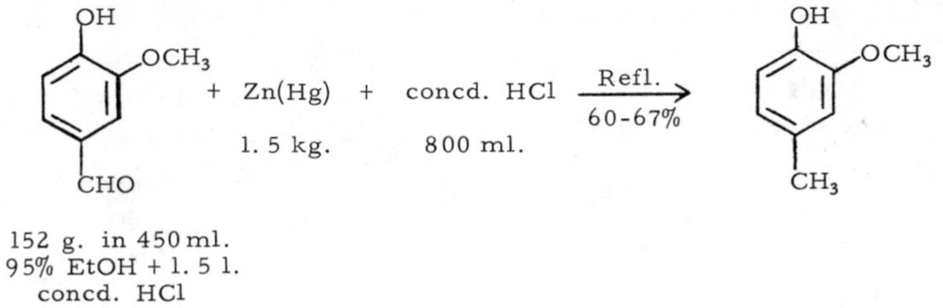

in 300 ml. of distilled water. After stirring for 10–15 min., lumps are crushed, and the zinc is collected and washed with a total of 500 ml. of distilled water containing a trace of hydrochloric acid. It is then washed with ethanol and with ether and dried. Naphthalene-1,5-disulfonyl chloride is reduced to naphthalene-1,5-dithiol

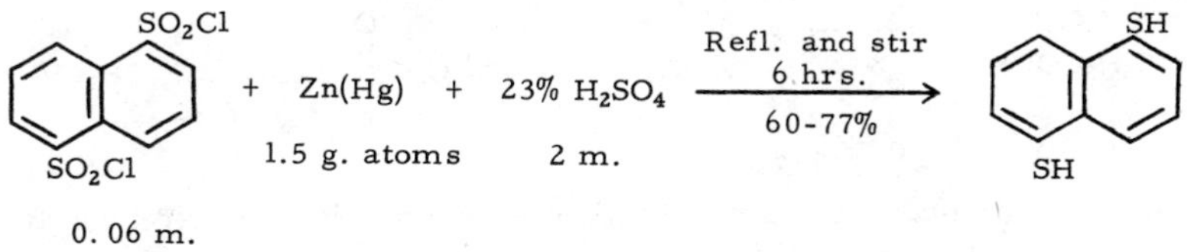

by heating and stirring a mixture of the finely powdered sulfonyl chloride, amalgamated zinc dust, and 33% sulfuric acid. After standing overnight, the solid is collected and extracted with ether for recovery of the dithiol. When unamalgamated zinc dust was used considerable insoluble yellow polymeric material was noted at this point.

[1]E. Clemmensen, *Ber.*, **46**, 1837 (1913); **47**, 51, 681 (1914)
[2]E. L. Martin, *Org. Reactions*, **1**, 155 (1942); D. Staschewski, *Angew. Chem.*, **71**, 726 (1959)
[3]E. L. Martin, *Am. Soc.*, **58**, 1438 (1936)
[4]C. S. Sherman, private communication reported in the next reference
[5]L. F. Fieser *et al.*, *Am. Soc.*, **70**, 3197 (1948)
[6]R. R. Read and J. Wood, Jr., *Org. Syn.*, *Coll. Vol.*, **3**, 444 (1955)
[7]R. Schwarz and H. Hering, *ibid.*, **4**, 203 (1963)
[8]P. D. Caesar, *ibid.*, **4**, 695 (1963)

Zinc chloride, $ZnCl_2$. Mol. wt. 136.29, m.p. 283°, deliquescent.

***n-BuOH* ⟶ *n-BuCl*.** The displacement is effected with two moles each of HCl and $ZnCl_2$.[1]

$$\underline{n}\text{-}C_4H_9OH + \text{concd. } HCl + ZnCl_2 \xrightarrow[76\text{-}78\%]{150\text{-}155^0} \underline{n}\text{-}C_4H_9Cl$$

5 m. 10 m. 10 m.

Acetylation. The reagent is an effective catalyst for the acetylation of *t*-butanol.[2] The method is simpler than either a procedure using acetyl chloride and N,N-dimethylaniline (63–68%)[3] or one using acetyl chloride and magnesium (45–55%).[4]

$$(CH_3)_3COH + (CH_3CO)_2O + ZnCl_2 \xrightarrow[53\text{-}60\%]{\text{Refl. 2 hrs.}} (CH_3)_3COCOCH_3$$

2.1 m. 2.1 m. 0.5 g.

For the acetylation of D-glucose-δ-lactone (Pfizer and Co.),[5] crushed fused zinc chloride is shaken with acetic anhydride until most of the solid has dissolved. A mechanical stirrer is started, the flask is cooled in an ice bath, and the lactone is added slowly with cooling to keep the temperature below 65°. After an hour in the ice bath the solution is kept at room temperature for 24 hrs. and poured into 1 l. of water and stirred until the excess anhydride is destroyed. The lactone ring is cleaved at

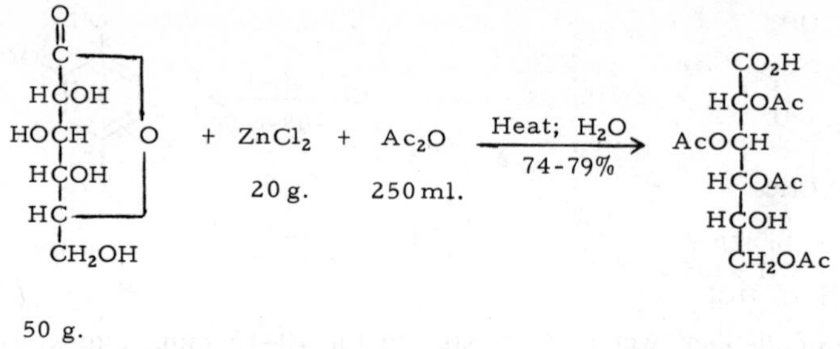
+ ZnCl2 + Ac2O
20 g.
250 ml.
Heat; H2O
74-79%
50 g.

this point, and the product is 2,3,4,6-tetra-O-acetyl-D-gluconic acid monohydrate.

Oxide fission. Benzoyl chloride reacts with tetrahydrofurane in the presence of a catalytic amount of zinc chloride to give 4-chlorobutyl benzoate.[6] The reaction is

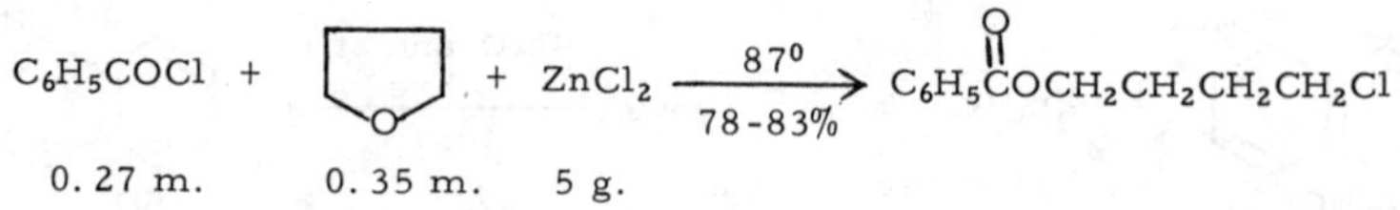
C6H5COCl +
+ ZnCl2
87°
78-83%
C6H5COCH2CH2CH2CH2Cl
0. 27 m.
0. 35 m.
5 g.

exothermal at the outset. Acetic anhydride reacts similarly with tetrahydrofurfuryl alcohol to give 1,2,5-triacetoxypentane.[7]

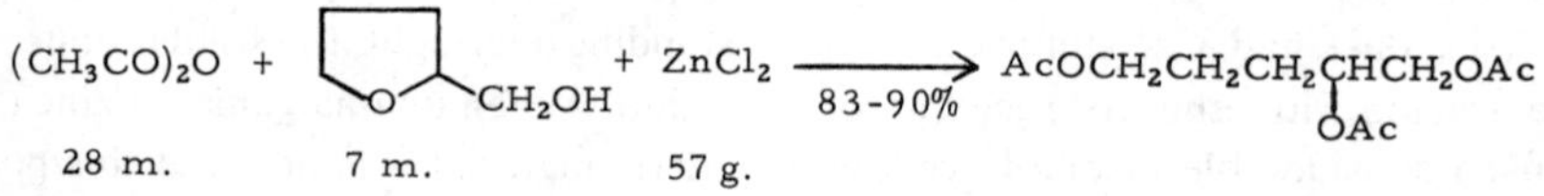
(CH3CO)2O +
CH2OH
+ ZnCl2
83-90%
AcOCH2CH2CH2CHCH2OAc
OAc
28 m.
7 m.
57 g.

Acid chloride exchange. In the presence of a small amount of zinc chloride, phthaloyl chloride reacts with maleic anhydride with both chloride exchange and isomerization to fumaryl chloride.[8]

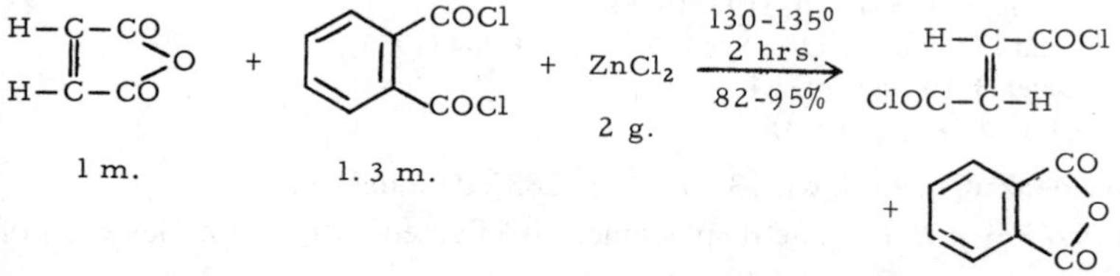
+ ZnCl2
2 g.
130-135°
2 hrs.
82-95%
1 m.
1. 3 m.

Amination. For the conversion of 2-hydroxy-3-naphthoic acid into 3-amino-2-naphthoic acid, an autoclave is charged successively with aqueous ammonia, zinc chloride, and 2-hydroxy-3-naphthoic acid, and closed.[9] With constant agitation

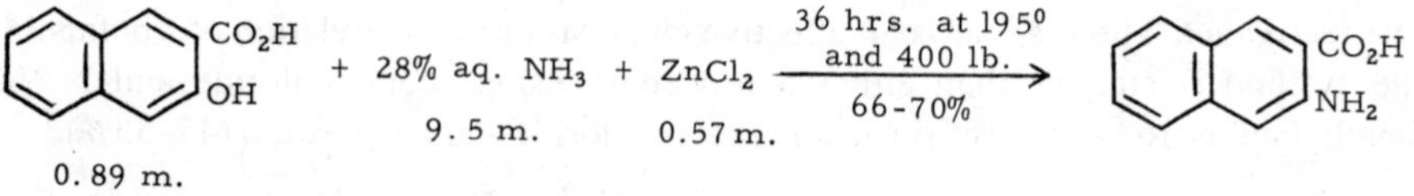
+ 28% aq. NH3 + ZnCl2
9. 5 m.
0.57 m.
36 hrs. at 195°
and 400 lb.
66-70%
0. 89 m.

the temperature is raised during 3 hrs. to 195°, and this temperature is maintained for 36 hrs. The workup is rather lengthy.

Fischer indole synthesis. The synthesis of a 2-substituted indole by the method of Fischer[10] can be interpreted as shown in the formulation. An *Organic Syntheses* preparation[11] of 2-phenylindole specified heating acetone phenylhydrazone with five times its weight of zinc chloride at 170° and adding 4 parts of sand to prevent solidification on cooling (yield of 2-phenylindole 72–80%). Chapman *et al.*[12] studied various cyclizing agents (concd. hydrochloric acid, dry HCl, BF_3, polyphosphoric

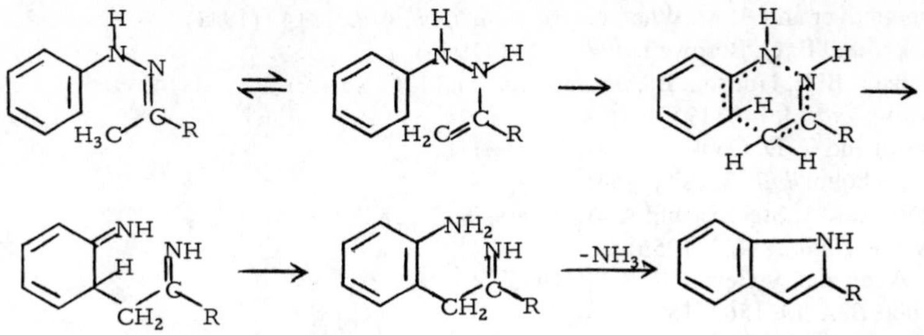

acid, Cu_2Cl_2, and various proportions of $ZnCl_2$) and found zinc chloride to be the most effective catalyst. 2-Methylindole was obtained in 85% yield by heating acetone phenylhydrazone with an equal weight of zinc chloride in cumene, as a high-boiling solvent.

Nencki reaction.[13] The general reaction, zinc chloride-catalyzed acylation of a phenol, is illustrated by the conversion of resorcinol into resacetophenone[14] and of pyrogallol into gallacetophenone.[15] The procedures are similar except that acetic acid is used in the first and acetic anhydride in the second.

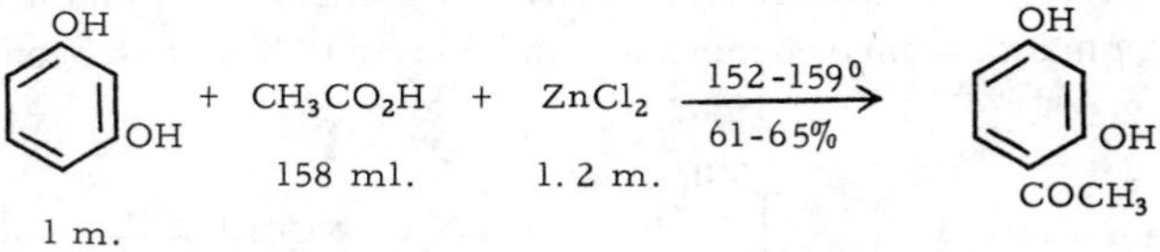

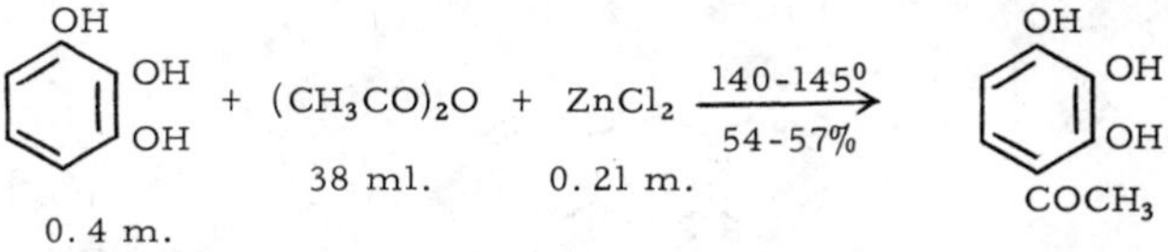

Hoesch reaction.[16] Phloroacetophenone is prepared as follows.[17] A mixture of phloroglucinol, acetonitrile, finely powdered zinc chloride, and anhydrous ether is

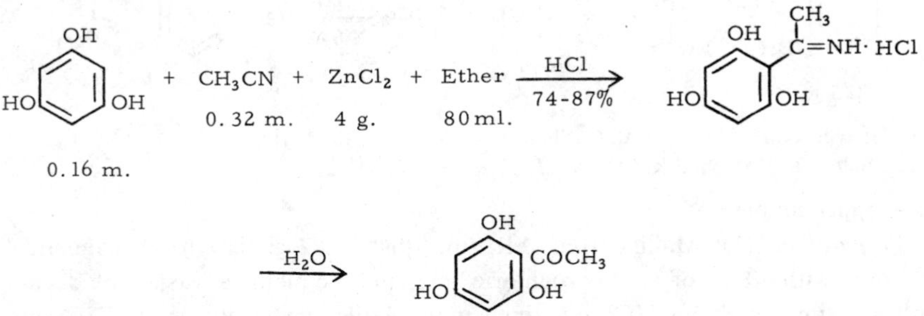

cooled in ice-salt mixture, and a rapid stream of hydrogen chloride is passed in for 2 hrs. with occasional swirling. The flask is let stand in an ice chest for 24 hrs., and the mixture, now pale orange, is treated again with hydrogen chloride for 2 hrs. and stoppered and let stand in the ice chest for 3 days, when the ketimine hydrochloride separates as a bulky orange-yellow precipitate. The ether is decanted and the ketimine hydrochloride is washed with ether and hydrolyzed by refluxing it with 1 l. of water for 2 hrs. The hot solution is clarified with Norit, filtered, and let stand for crystallization.

[1]J. E. Copenhaver and A. M. Whaley, *Org. Syn., Coll. Vol.*, **1**, 142 (1941)
[2]R. H. Baker and F. G. Bordwell, *ibid.*, **3**, 141 (1955)
[3]C. R. Hauser, B. E. Hudson, B. Abramovitch, and J. C. Shivers, *ibid.*, **3**, 142 (1955)
[4]A. Spassow, *ibid.*, **3**, 144 (1955)
[5]C. E. Braun and C. D. Cook, *ibid.*, **41**, 79 (1961)
[6]M. E. Synerholm, *ibid.*, **3**, 187 (1955)
[7]O. Grummitt, J. A. Stearns, and A. A. Arters, *ibid.*, **3**, 833 (1955)
[8]L. P. Kyrides, *ibid.*, **3**, 422 (1955)
[9]C. F. H. Allen and A. Bell, *ibid.*, **3**, 78 (1955)
[10]E. Fischer, *Ber.*, **19**, 1563 (1886)
[11]R. L. Shriner, W. C. Ashley, and E. Welch, *Org. Syn., Coll. Vol.*, **3**, 725 (1955)
[12]N. B. Chapman, K. Clarke, and H. Hughes, *J. Chem. Soc.*, 1424 (1965)
[13]M. Nencki and N. Sieber, *J. prakt. Chem.*, [2], **23**, 147 (1881); M. Nencki, *Ber.*, **32**, 2414 (1899)
[14]S. R. Cooper, *Org. Syn., Coll. Vol.*, **3**, 761 (1955)
[15]I. S. Badhwar and K. Venkataraman, *ibid.*, **2**, 304 (1943)
[16]K. Hoesch, *Ber.*, **48**, 1122 (1915)
[17]K. C. Gulati, S. R. Seth, and K. Venkataraman, *Org. Syn., Coll. Vol.*, **2**, 522 (1943)

Zinc chloride–Phosphoryl chloride, $ZnCl_2$–$POCl_3$. This combination is used for the synthesis of hydroxyxanthones from an *o*-hydroxybenzoic acid and a phenol[1] and of 4-hydroxycoumarins from a phenol and a malonic acid.[2] Neither reagent is effective alone.

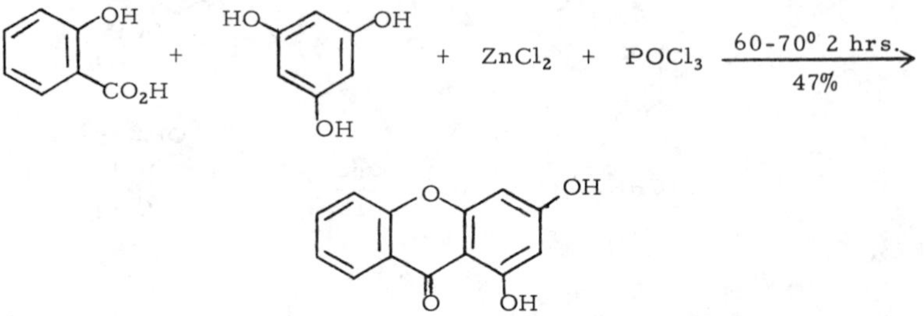
OH
CO₂H
HO
OH
OH
+ ZnCl₂ + POCl₃
60-70⁰ 2 hrs.
47%
OH
OH

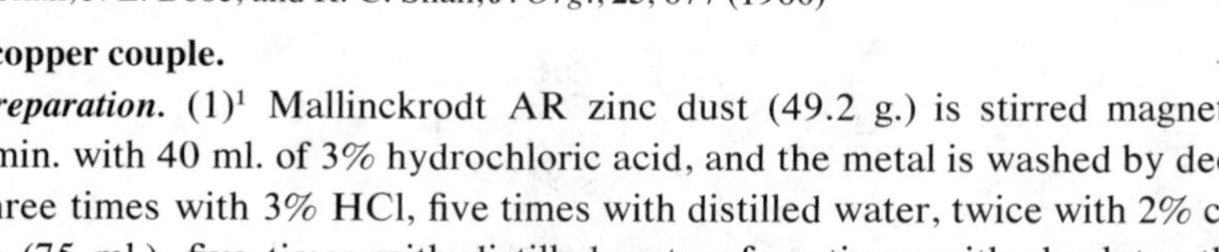
OH
225 g.
HOC
CH₂
HOC=O
247 g.
+ ZnCl₂ + POCl₃
979 g.
657 g.
60-65⁰
35 hrs.
64%
OH

[1]P. K. Grover, G. D. Shah, and R. C. Shah, *J. Chem. Soc.*, 3982 (1955)
[2]V. R. Shah, J. L. Bose, and R. C. Shah, *J. Org.*, **25**, 677 (1960)

Zinc–copper couple.

Preparation. (1)[1] Mallinckrodt AR zinc dust (49.2 g.) is stirred magnetically for 1 min. with 40 ml. of 3% hydrochloric acid, and the metal is washed by decantation three times with 3% HCl, five times with distilled water, twice with 2% copper sulfate (75 ml.), five times with distilled water, four times with absolute ethanol, and five times with absolute ether. The couple is sucked dry on a Büchner under a rubber dam, stored overnight in a dessicator over P_2O_5, and is then ready for use. It deteriorates rapidly in moist air. Similar procedure: Corbin *et al.*[2]

(2)[1,3] An equally active couple can be made by heating a mixture of cupric oxide (30 g.) and zinc dust (240 g.) in a stream of hydrogen–nitrogen at 500°.

(3)[3] A 5–8% zinc–copper alloy is made by melting zinc with clean brass turnings and casting into bars, which are turned into five shavings.

Reduction. The couple required for the preparation of phthalide[4] is prepared by stirring 2.75 g. atoms of zinc dust to a paste with a solution of 1 g. of copper sulfate in 35 ml. of water. Aqueous alkali is added, the mixture is stirred mechanically and cooled to 5°, and phthalimide is added at 5–8° in about 30 min. The workup, after

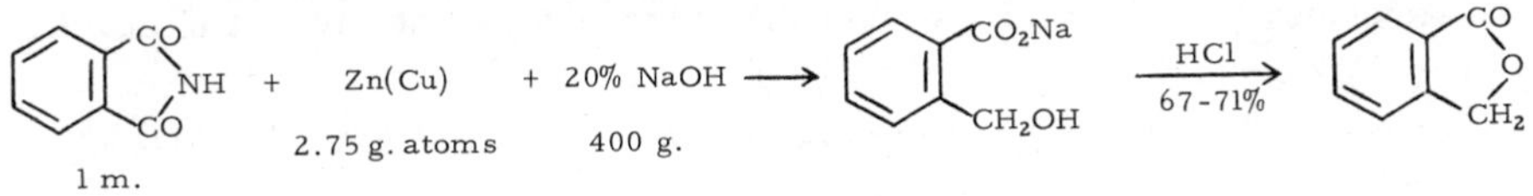

acidification, includes crystallization from water. H. T. Clarke and D. Blumenthal state that "In checking these directions complete failure was repeatedly encountered with good commercial grades of zinc dust, and only when the metal was activated with copper sulfate did reduction proceed at all."

Debromination. Benzylic bromination of 3-bromo-1-phenylpropane gives the 1,3-dibromo derivative in nearly 100% yield.[2] A mixture of 500 ml. of dimethylformamide and 2 g. atoms of zinc copper couple (method 2) is stirred at 7° and the

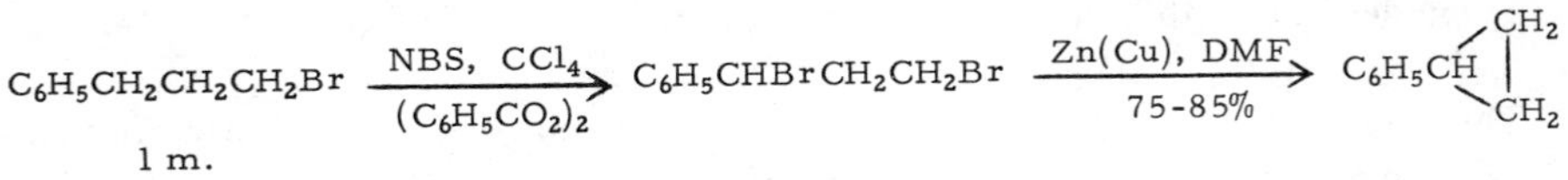

1,3-dibromo-1-phenylpropane is added at a rate such as to maintain a temperature of 7–9°. Workup affords cyclopropylbenzene, b.p. 170–175° (98.5% pure).

Preparation of diethylzinc.[3] Zinc–copper couple prepared by either method 2 or 3 is placed in a dry flask equipped with a heavy stirrer and a reflux condenser and

$$\underset{0.5\text{ m.}}{2\,C_2H_5I} + \underset{0.5\text{ m.}}{2\,C_2H_5Br} + \underset{2\text{ g. atoms}}{4\,Zn(Cu)} \xrightarrow[86\text{-}89\%]{} Zn(C_2H_5)_2$$

covered with ethyl iodide and ethyl bromide. On stirring and heating to reflux there is usually an induction period of 0.5–1.5 hrs. before the reaction starts; it is then controlled by mild cooling or heating. The spontaneously flammable compound is obtained in satisfactory purity by distillation at 8 mm. from the reaction mixture; dry carbon dioxide or nitrogen is then admitted to the apparatus.

Reaction with methylene iodide and an olefin: *see* Simmons-Smith reaction.

[1]R. D. Smith and H. E. Simmons, *Org. Syn.*, **41**, 72 (1961)
[2]T. F. Corbin, R. C. Hahn, and H. Shechter, *ibid.*, **44**, 30 (1964)
[3]C. R. Noller, *Org. Syn.*, *Coll. Vol.*, **2**, 184 (1943)
[4]J. H. Gardner and C. A. Naylor, Jr., *ibid.*, **2**, 526 (1943)

Zinc cyanide, $Zn(CN)_2$. Mol. wt. 117.42, sp. sol. H_2O. Suppliers: Fisher, MCB.

Mesitaldehyde can be prepared[1] from mesitylene by the Gattermann synthesis[2] as modified by Adams.[3] A mixture of the hydrocarbon, zinc cyanide, and tetrachloroethane is stirred at room temperature while a rapid stream of hydrogen

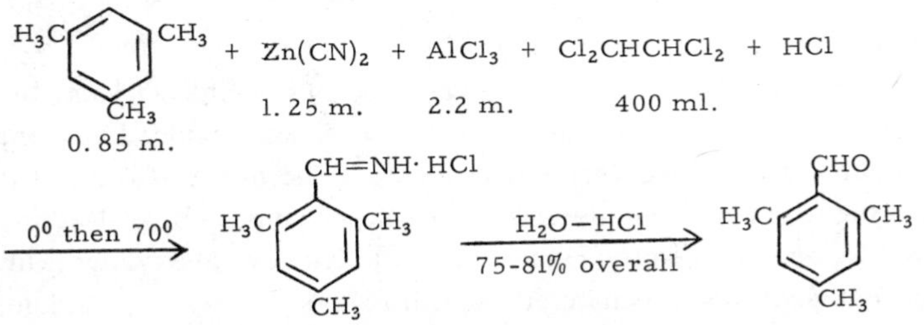

chloride is passed in until the zinc cyanide is decomposed (about 3 hrs.). The flask is immersed in an ice bath, and aluminum chloride is added with vigorous stirring. The ice bath is removed, passage of HCl is resumed and the temperature allowed to rise to 70° in about an hour and kept there for another hour. After suitable workup, which includes hydrolysis of the aldimine hydrochloride, the product is distilled at 118–121°/16 mm.

[1]R. C. Fuson, E. C. Horning, S. P. Rowland, and M. L. Ward, *Org. Syn., Coll. Vol.*, **3**, 549 (1955)
[2]L. Gattermann, *Ber.*, **31**, 1149 (1898); *Ann.*, **357**, 313 (1907)
[3]R. Adams and I. Levine, *Am. Soc.*, **45**, 2373 (1923)

Zinc hydrosulfite, ZnS_2O_4. Mol. wt. 193.50, readily soluble in water. Supplier: Virginia Chemicals.

The commercial material is 86–88% pure. It has a powerful reducing action at pH 3–7.5, and is maximally effective at pH 5. The reagent is thus effective at pH 4, where sodium hydrosulfite is rapidly decomposed.

Zinc iodide, ZnI_2. Mol. wt. 319.18, m.p. 446°, solubility in 100 parts water: 430 g. at 0°, 510 g. at 100°.

The reagent catalyzes the reaction of phenylacetylene with triethyl orthoformate to give phenylpropargylaldehyde diethyl acetal.[1] A mixture of the reactants is

$$\underset{0.5\text{ m.}}{C_6H_5C{\equiv}CH} + \underset{0.5\text{ m.}}{HC(OC_2H_5)_3} + \underset{3\text{ g.}}{ZnI_2} \xrightarrow[72-78\%]{135-210^0} C_6H_5C{\equiv}CCH(OC_2H_5)_2 + C_2H_5OH$$

heated in a flask equipped with a nitrogen inlet, a thermometer, and a short fractionating column, and ethanol is slowly distilled in the course of about 1 hr. After suitable processing the product is distilled at 99–100°/2 mm. Zinc nitrate appears to be equivalent to zinc iodide as a catalyst. Zinc chloride is effective but the heating time is 2–3 times as long and the yield is 64–70%.

[1]B. W. Howk and J. C. Sauer, *Org. Syn., Coll. Vol.*, **4**, 801 (1963)

Zinc oxide, ZnO. Mol. wt. 81.38. Suppliers: Mallinckrodt, MCB.

In a Glaxo synthesis of vitamin A from cyclohexanone,[1] one step involved dehydration of the alcohol (2), or elimination of acetic acid from the corresponding acetate (4), to produce the eneyne (5). This transformation proved to be surprisingly

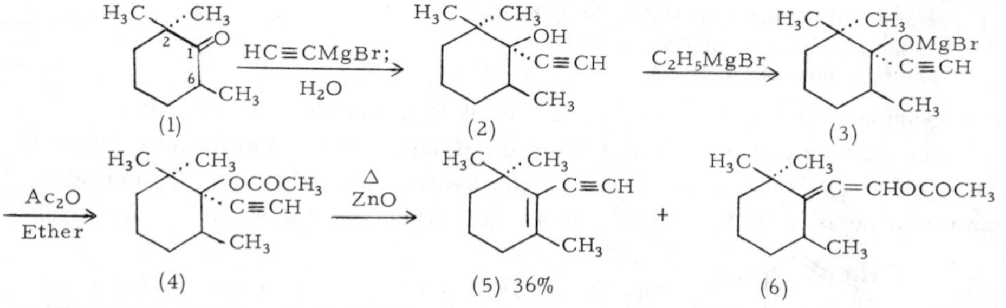
HC≡CMgBr; H2O
C2H5MgBr
Ac2O Ether
Δ ZnO
(1)
(2)
(3)
(4)
(5) 36%
(6)

difficult with the free alcohol (2) but was achieved, at least in 36% yield, by pyrolysis of the acetate (4) in silicon oil in the presence of zinc oxide. Later Landor and Landor[2] identified a second reaction product as the allene (6) and, as a further example of the rearrangement, described the smooth conversion of 1-ethinyl-2,2,6,6-tetramethylcyclohexyl acetate (7) into the tetramethylcyclohexylidenevinyl acetate (9) in excellent yield. A six-membered cyclic transition state (8) is postulated.

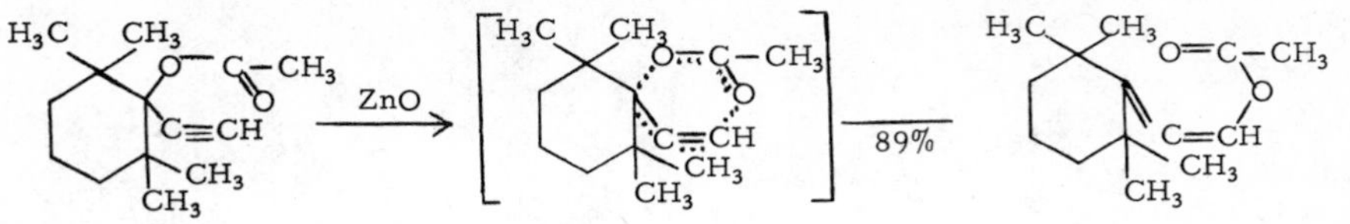

H3C CH3
O C-CH3
O
C≡CH
CH3
CH3
ZnO
89%
O=C-CH3
C=CH

(7) (8) (9)

Raphael *et al.*[3] made use of this facile transformation for the synthesis of the allene acetate (12), hydrolyzed by mild acid treatment to the α,β-unsaturated aldehyde.

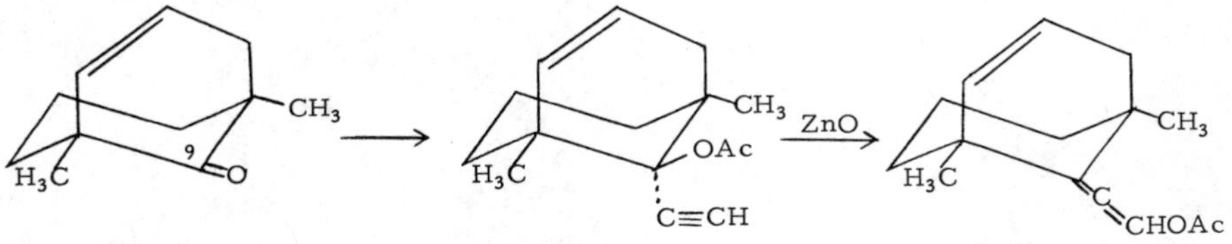

CH3
H3C
9
O
OAc
C≡CH
ZnO
CHOAc

(10) (11) (12)

[1]J. Attenburrow *et al.*, *J. Chem. Soc.*, 1094 (1952)
[2]P. D. Landor and S. R. Landor, *J. Chem. Soc.*, 1015 (1956)
[3]J. Martin, W. Parker, and R. A. Raphael, *Chem. Comm.*, 633 (1965)

Zirconium tetrachloride, $ZrCl_4$. Mol. wt. 233.05, sp. gr. 2.8, subl. above 300°. Supplier: Titanium Alloy Manufacturing Co.

In a brief preliminary note Heine, Cottle, and Van Mater[1] reported comparative results showing that zirconium tetrachloride is generally superior to aluminum chloride for the Friedel-Crafts *p*-acetylation of toluene with acetyl chloride. However, the results were less consistent with $ZrCl_4$ than with $AlCl_3$. Gore and Hoskins[2] studied the Friedel-Crafts reaction of benzoyl chloride with anthracene in ethylene dichloride at 0° in the presence of 14 different catalysts. The best yields of very pure 9-benzoylanthracene were with zirconium tetrachloride (97% yield after $1\frac{1}{2}$ hrs.) and with aluminum chloride (86% yield after $1\frac{1}{4}$ hrs., 99% yield after 99 hrs.).

[1]H. W. Heine, D. L. Cottle, and H. L. Van Mater, *Am. Soc.*, **68**, 524 (1946)
[2]P. H. Gore and J. A. Hoskins, *J. Chem. Soc.*, 5666 (1964)

SUPPLIERS

The suppliers referred to most frequently, the abbreviations which we have used for them, and their addresses are as follows:

(A) Aldrich Chemical Co., 2371 North 30th St., Milwaukee, Wis. 53210
(B) J. T. Baker Chemical Co., Phillipsburg, N. J. 08865
(E) Eastman Organic Chemicals Dept., Rochester 3, N. Y.
(F) Fisher Scientific Co., 717 Forbes Ave., Pittsburgh, Pa. 15219
(MCB) Matheson, Coleman and Bell, 2909 Highland Ave., Norwood (Cincinnati) 12, Ohio

Other suppliers of chemicals, solvents, or apparatus cited are listed below.

Abbott Laboratories, Chemical Marketing Division, North Chicago, Ill.
Ace Glass Inc., Vineland, N. J.
Aceto Chemical Co., 40–40 Lawrence St., Flushing, N. Y.
Aero Chemical Corp., 338 Wilson Ave., Newark, N. J.
Air Reduction Chemical and Carbide Co., 150 East 42nd St., New York 17, N. Y.
Alfa Inorganics, Inc., 8 Congress St., Beverly, Mass.
Allied Chemical Co., Marcus Hook, Pa.
Allied Chemical Corp., Baker and Adamson Products, General Chemical Division, 40 Rector St., New York 6, N. Y.
Allied Chemical Corp., Solvay Process Division, 61 Broadway, New York 6, N. Y.
American Agricultural Chemical Co., Dept. E-10, 100 Church St., New York 7, N. Y.
American Platinum Works, Newark, N. J.
American Potash and Chemical Corp., 3000 W. Sixth St., Los Angeles 54, Calif.
Ansul Chemical Co., 1 Stanton St., Marinette, Wisc.
Antara Chemicals, Division of General Aniline and Film Corp., 435 Hudson St., New York 14, N. Y.
Arapahoe Chemicals, Division of Syntex Corp., 2855 Walnut St., Boulder, Colo. 80301
Baker and Co., Inc., Newark, N. J.
Beacon Chemical Industries, Inc., 33 Richdale Ave., Cambridge, Mass. 02140
Bishop, J., and Co. Platinum Works, Malvern, Pa.
Borden Chemical Co., Division of the Borden Co., 350 Madison Ave., New York 17, N. Y.
Bower, Henry, Chemical Mfg. Co., Gray's Ferry Rd. and 29th St., Philadelphia, 46, Pa.
British Drug Houses Ltd., Laboratory Chemicals Division, Poole, Dorset, England. U. S. distributor: Ealing Corp., Cambridge, Mass. 02140
Burdick and Jackson Laboratories, Muskegon, Mich.
Calbiochem, 3625 Medford St., Los Angeles 63, Calif.
Callery Chemical Co., Callery, Pa.
Carbide and Carbon Chemicals Co., 30 East 42nd St., New York, N. Y.
Carborundum Co., Niagara Falls, N. Y.
Carlisle Chemical Works, Inc., Reading, Ohio 45215
Chattern Chemicals, Fine Chemicals Division of the Chattanooga Medical Co., Chattanooga 9, Tenn.
Chemical Intermediates and Research Labs., Box 146, Cuyahoga Falls, Ohio
Chemical Procurement Laboratories, Inc., 18-17 130th St., College Point 56, N. Y.
Chemical Samples Co., 4692 Kenney Rd., Columbus, Ohio 43221
Ciba Chemical and Dye Co., Division of Ciba Corporation, Fairlawn, N. J.
Coaltar Chemicals Corp., 430 Lexington Ave., New York, N. Y.
Columbia Organic Chemicals, Inc., 1012 Drake St., Columbia, S. C.
Commercial Solvents Corp., 260 Madison Ave., New York 16, N. Y.
Crown Zellerbach Corp., Chemical Products Division, Camas 97, Wash.
Davidson Chemical Co., Division of W. R. Grace and Co., 101 N. Charles St., Baltimore 3, Md.
Deere, John, Chemical Co., Tulsa and Pryor, Okla.
Delmar Scientific Laboratories, Inc., 317 Madison St., Maywood, Ill.
Dow Chemical Company, Midland, Mich.
du Pont de Nemours, E. I., and Co., Inc.,

Electrochemicals Department, Wilmington 98, Del.

du Pont de Nemours, E. I., and Co., Inc., Industrial and Biochemicals Department, Wilmington 98, Del.

Eastern Chemical Corp., 34 Spring St., Newark 4, N. J.

Eastman Chemical Products, Inc., Subsidiary of Eastman Kodak Company, Kingsport, Tenn.

Engelhard Industries, Inc., 113 Astor St., Newark, N. J.

Enjay Chemical Company, A Division of Humble Oil and Refining Co., 60 West 49th St., New York 20, N. Y.

Ethyl Corp., 100 Park Ave., New York 17, N. Y.

Fairmont Chemical Co., Inc., 136 Liberty St., New York 6, N. Y.

Farchan Research Laboratories, Anderson Rd., Wickliffe, Ohio

Fluka AG Chemische Fabrik, Buchs SG., Switzerland

FMC Corporation, Inorganic Chemicals Division, 633 Third Ave., New York 17, N. Y.

Foote Mineral Co., Route 100, Exton, Pa.

Frinton Laboratories, P. O. Box 301, South Vineland, N. J.

Fritzsche Brothers, Inc., 76 Ninth Ave., New York 11, N. Y.

General Aniline and Film Corp., 435 Hudson St., New York 14, N. Y.

Gilman Paint and Varnish Co., Chattanooga, Tenn.

Goodrich, B. F., Chemical Co., 3135-T Euclid Ave., Cleveland, Ohio

Grace, W. R., and Co., Raney Catalyst Division, Chattanooga, Tenn.

Hammons, W. A., Drierite Co., Xenia, Ohio

Harshaw Chemical Co., 1945 E. 97th St., Cleveland 6, Ohio

Hengar Co., 1711 Spruce St., Philadelphia 3, Pa.

Henley and Co., Inc., 202 E. 44th St., New York 17, N. Y.

Hercules, Inc., 910 Market St., Wilmington 99, Del.

Hooker Chemical Corp., 1300 47th St., Niagara Falls, N. Y.

Houdry Chem. Co., 1339 Chestnut St., Philadelphia, Pa.

Howe and French, Inc., 99 Broad St., Boston, Mass.

International Minerals and Chemicals Corp., 5401 Old Orchard Rd., Skokie, Ill.

Jefferson Chemical Co., Inc., P. O. Box 303, Houston 1, Texas

Johnson, Matthey, Ltd., 78 Hatton Garden, London, E. C.1, England

K and K Laboratories, Inc., 177–10 93rd Ave., Jamaica 22, N. Y.

Kaplop Laboratories, 6710 West Jefferson, Detroit 17, Mich.

Kay-Fries Chemicals, Inc., 360 Lexington Ave., New York 17, N. Y.

Koch-Light Laboratories (formerly L. Light Co.), Colburn, Bucks, England

Koppers Company, Inc., Chemicals and Dyestuffs Division, 1450 Koppers Bldg., Pittsburgh 19, Pa.

Lithium Corporation of America, Inc., 500 5th Ave., New York 36, N. Y.

Lucidol Division, Wallace and Tiernan, Inc., 1740 Military Rd., Buffalo 5, N. Y.

Mackay, A. D., Inc., 198 Broadway, New York 38, N. Y.

McKesson and Robbins, Inc., Chemical Department, 155 E. 44th St., New York 17, N. Y.

Mallinckrodt Chemical Works, St. Louis 47, Mo.

Mann Research Labs., Inc., 136 Liberty St., New York 16, N. Y.

Matheson Co., Inc., P. O. Box 85, East Rutherford, N. J.

Matheson, Coleman and Bell Div. of the Matheson Co., Inc., P. O. Box 85, East Rutherford, N. J.

Merck Chemical Division, Rahway, N. J.

Metal and Thermit Corp., Rahway, N. J.

Metalloy Corp., Rand Tower, Minneapolis, Minn.

Metalsalts Corp., 200 Wagaraw Road, Hawthorne, N. J.

Monsanto Chemical Co., 800 N. Lindbergh Blvd., St. Louis 66, Mo.

MSA (Mine Safety Appliance) Research Corp., Callery, Pa.

Niacet Chemicals Division, Niagara Falls, N. Y.

Niagara Chlorine Products Co., Inc., Lockport, N. Y.

Oldbury Electrochemical Co., Niagara Falls, N. Y.

Olin Mathieson Chemical Corp., Chemicals Division, 745 Fifth Ave., New York 22, N. Y.

Orgmet, Hamstead, N. H.

Ott Chemical Co., 500 Agard Road, Muskegon, Mich.

Ozark-Mahoning Co., 310 West 6th St., Tulsa 19, Okla.

Parr Instrument Co., Moline, Ill.

Peninsular ChemResearch, Inc., P. O. Box 14318, Gainesville, Fla.

Pennsalt Chemicals Corp., Industrial Chemicals Division, 3 Penn Center, Philadelphia 2, Pa.
Pfister Chemical Works, Inc., Ridgefield, N. J.
Pfizer, Chas., and Co., Inc., Chemical Sales Division, 235 East 42nd St., New York 17, N. Y.
Pharmacia Fine Chemicals, Inc., 800 Centennial Ave., Piscataway, N. J. 08854
Phillips Petroleum Co., Market Development Division, Bartlesville, Okla.
Pierce Chemical Co., P. O. Box 117, Rockford, Ill.
Pilot Chemicals, Inc., 36 Pleasant St., Watertown 72, Mass.
Quaker Oats Co., Chemicals Division, 341 Merchandise Mart Plaza, Chicago 54, Ill.
Raney Catalyst Co., Inc., 1322 Hamilton National Bank Bldg., Chattanooga 2, Tenn.
Robeco Chemicals Inc., 25 E. 26th St., New York 10, N. Y.
Roberts Chemical Co., Nitro, W. Va.
Rohm and Haas Co., Washington Sq., Philadelphia 5, Pa.
Schuchardt, Dr. Theodor, Ainmillerstrasse 25, München 13, Germany
Sharples Chemicals Co., Philadelphia, Pa.
Shell Chemical Co., 110 West 51st St., New York 20, N. Y.
Smith, G. Frederick, Chemical Co., P. O. Box 5906 Station D, 867 McKinley Ave., Columbus 22, Ohio
Standard Oil of Indiana, 910 So. Michigan Ave., Chicago, Ill.
Stauffer Chemical Co., 380 Madison Ave., New York 17, N. Y.
Stepan Chemical Co., Maywood Chemical Works Division, 100 West Hunter Ave., Maywood, N. J.
Syntex S.A., Apartado Postal 2679, Mexico, D. F., Mexico
Texaco Inc., 135 East 42nd St., New York 17, N. Y.
Texas Alkyls, Inc., P. O. Box 988, Pasadena, Texas
Thomas, Arthur H. Co., Vine St. and Third, P. O. Box 779, Philadelphia, Pa.
Titanium Alloy Manufacturing Division, National Lead Co., 111 Broadway, New York 6, N. Y.
Troemner, Henry, Inc., 6824 Greenway Ave., Philadelphia, Pa.
Trylon Chemicals, Inc., Lock Haven, Pa.
Union Carbide Chemicals Co., Div. of Union Carbide Corp., 270 Park Ave., New York 17, N. Y.
Universal Oil Products Co., 30 Algonquin Rd., Des Plaines, Ill.
U. S. Industrial Chemicals Co., Division of National Distillers and Chemical Corp., 99 Park Ave., New York 16, N. Y.
Ventron Corp., 12–24 Congress St., Beverly, Mass.
Victor Chemical Works, Div. of Stauffer Chemical Co., 155 North Wacker Drive, Chicago 6, Ill.
Virginia-Carolina Chemical Corp., 401 E. Main St., Richmond 8, Va.
Virginia Chemicals Inc., West Norfolk, Va.
Welsback Corp., Ozone Process Division, Westmoreland and Stokley Sts., Philadelphia 29, Pa.
Westvaco Mineral Products, 161 East 42nd St., New York, N. Y.
Westville Chemical Corp., Route 110, Monroe, Conn.
Wilkens-Anderson Co., 4525 W. Division St., Chicago 51, Ill.
Wilson Laboratories, 4221 South Western Blvd., Chicago 9, Ill.
Wood Ridge Chemical Corp., Park Place East, Wood Ridge, N. J.
Yeda Research and Development Co. Ltd., P. O. Box 26, Rehovoth, Israel

INDEX OF APPARATUS

INDEX OF REAGENTS ACCORDING TO TYPES

AUTHOR INDEX

SUBJECT INDEX

Page numbers referring to reagents are indicated in **boldface**.